Gwasanaethau Llyfrgell a Gwybodaeth
Library and Information Services

Prif Swyddfa
Headquarters

Hafod Elfyn
Penrhos Road
Bangor
Gwynedd
LL57 2BQ

Tel: (01248) 385522

Fax: (01248) 385510

Cyngor Cefn Gwlad Cymru
Countryside Council for Wales

BirdLife Conservation Series No. 12

BIRDS IN EUROPE

Population estimates, trends and conservation status

Compiled by
Ian Burfield and Frans van Bommel

On behalf of
The BirdLife International European Partnership

With contributions from
Umberto Gallo-Orsi, Szabolcs Nagy, Canan Orhun,
Rob Pople (species accounts) and Roland van Zoest (maps)

Project managers
Ian Burfield (2003–2004) and Des Callaghan (2002)

With sponsorship from

BirdLife Suomi
The BirdLife Partner
in Finland

Ligue pour la Protection des Oiseaux
The BirdLife Partner
in France and its supporters

Vogelbescherming Nederland
The BirdLife Partner
in the Netherlands

Sveriges Ornitologiska Förening
The BirdLife Partner
in Sweden

SVS/BirdLife Schweiz
The BirdLife Partner
in Switzerland

European Commission

Netherlands Ministry of Agriculture, Nature and Food Quality through the PIN/MATRA Funds of the Ministry of Foreign Affairs

Dutch Postcode Lottery

Swedish Environmental Protection Agency

Birds, our common heritage
Celebrating 25 years of the European Birds Directive

BirdLife INTERNATIONAL
Together for birds and people

COUNTDOWN 2010
Halt the loss of biodiversity

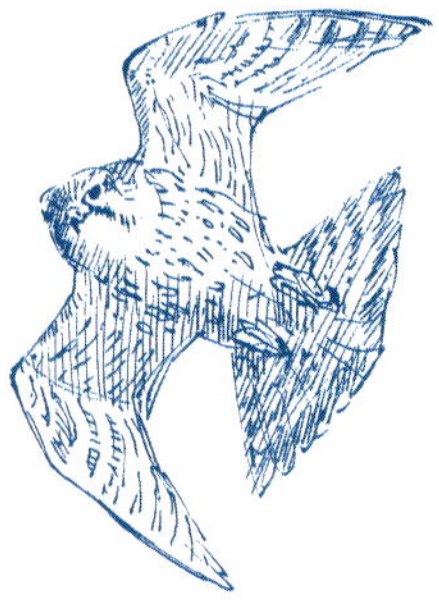

Dedication *Birds in Europe* is dedicated to Colin Bibby in memory of his outstanding contributions to ornithology, bird conservation and BirdLife International, with special gratitude for his pioneering work in monitoring bird populations.

Recommended citation
BirdLife International (2004) *Birds in Europe: population estimates, trends and conservation status.* Cambridge, UK: BirdLife International. (BirdLife Conservation Series No. 12).

Wellbrook Court, Girton Road, Cambridge CB3 0NA, United Kingdom
Tel: +44 1223 277318 Fax: +44 1223 277200 Email: birdlife@birdlife.org
Internet: www.birdlife.org

For further information about BirdLife International's work in Europe, please contact: BirdLife International European Division Office, Droevendaalsesteeg 3a, PO Box 127, 6700 AC, Wageningen, The Netherlands
Tel: +31 317 478831 Fax: +31 317 478844 Email: birdlife@birdlife-europe.nl

BirdLife International is a UK-registered charity no. 1042125

ISBN 0-946888-52-3 (Softback)
ISBN 0-946888-53-1 (Hardback)

British Library-in-Publication Data
A catalogue record for this book is available from the British Library

First published 2004 by BirdLife International

Designed and produced by the **Nature**Bureau Limited, 36 Kingfisher Court, Hambridge Road, Newbury, Berkshire RG14 5SJ, United Kingdom

Printed by Information Press, Oxford, United Kingdom

Available from the Natural History Book Service Ltd, 2–3 Wills Road, Totnes, Devon TQ9 5XN, United Kingdom
Tel: +44 1803 865913 Fax: +44 1803 865280 Email: nhbs@nhbs.co.uk
Internet: www.nhbs.com/services/birdlife.html

BIRDS IN EUROPE
Population estimates, trends and conservation status

NATIONAL COORDINATORS AND CONTRIBUTORS

ALBANIA

T. Bino, G. Jorgo and B. Hallmann
With contributions from: R. Bashford, F. Bego, A. Crivelli, P. Defos du Rau, M. Fasola, W. Fremuth, M. Gauthier-Clerc, Y. Kayser, R. Landsdown, N. Peja, D. Tempelman, C. Tourenq, D. Vangeluwe, A. Wilson, M. Zekhuist, H. Zoto, General Directorate of Forestry and Pastures, and Museum of Natural Sciences, Tirana

ANDORRA

J. Palau
With contributions from: J. Crozier, J. Dalmau, B. Dore, R. Mariné and A. Matschke

ARMENIA

L. V. Balyan, M. G. Ghasabyan and V. Y. Ananian
With contributions from: K. Aghababyan, N. Margaryan, V. Hakobyan, M. Maregasparyan, L. Janoian, and the Birds of Armenia project (BoA)

AUSTRIA

M. Dvorak and J. Frühauf
With contributions from: G. Amann, H.-M. Berg, G. Bierbaumer, M. Brader, K. Buschenreiter, J. Feldner, C. Gassan, A. Grüll, W. Jiresch, A. Landmann, G. Kilzer, R. Kilzer, C. Medicus-Arnold, R. Probst, N. Pühringer, A. Schuster, L. Slotta-Bachmayr, P. Rass, P. Sackl, O. Samwald, S. Stadler, H. Steiner, S. Wagner, W. Weißmair, S. Werner and T. Zuna-Kratky

AZERBAIJAN

E. H. Sultanov

BELARUS

L. Vergeichik
With contributions from: V. Dombrovski, A. Kozulin, E. Mongin, T. Pavluschik, P. Pinchuk, I. Samusenko, N. Yakovets and B. Yaminski

BELGIUM

A. Anselin, W. van den Bossche and J.-P. Jacob

BOSNIA & HERZEGOVINA

D. Kotrošan and S. Obratil

BULGARIA

D. Georgiev and P. Iankov
With contributions from: A. Antonov, V. Delov, D. Demerdjiev, S. Dereliev, M. Dimitrov, A. Kovachev, M. Kurtev, N. Petkov, D. Plachiiski, K. Ruskov, S. Spassov, S. Stoychev, P. Shurulinkov, Z. Spiridonov, G. Stoyanov and V. Vassilev

CROATIA

A. Radović
With contributions from: A. Delic, V. Dumbovic-Ruzic, D. Kovacic, G. Lukač, J. Mikuska, T. Mikuska, J. Mužinić, G. Piasevolli, B. Rubinić, M. Schneider-Jacoby, M. Setina, B. Štumberger, Z. Tadić and Z. Vasilik

CYPRUS

M. Charalambides
With contributions from: C. Charalambides, J. Dawes, P. Flint, J. Gordon, S. Iezekiel, N. Kasinis, D. Pomeroy and D. Whaley

CZECH REPUBLIC

K. Šťastný and V. Bejček
With contributions from: T. Bělka, S. Bureš, J. Cepák, J. Čejka, J. Flousek, M. Frencl, K. Hudec, J. Hora, D. Horal, J. Chytil, B. Kloubec, I. Kunstmüller, V. Lemberk, P. Lumpe, P. Málková, P. Musil, J. Pavelka, K. Pavelka, F. Pojer, Z. Polášek, K. Poprach, L. Schröpfer, M. Šálek, J. Ševčík, V. Tejrovský, R. Vacík, M. Vavřík and P. Voříšek

DENMARK

T. Nyegaard, K. N. Flensted and H. Heldbjerg
With contributions from: H. E. Jørgensen, P. Lange, P. Lyngs, L. Novrup, E. Østergaard and P. Sunde

FAROE ISLANDS

B. Olsen

GREENLAND

D. Boertmann

ESTONIA

J. Elts

FINLAND

R. A. Väisänen and T. Lehtiniemi
With contributions from: M. Ellermaa, M. Hario, H. Lindén and J. Valkama

FRANCE

B. Deceuninck
With contributions from: E. Leroy, J. Sériot and M. Weltz

GEORGIA

A. Gavashelishvili, Z. Javakhishvili and R. Gokhelashvili
With contributions from: G. Darchiashvili (NACRES) and L. Rusishvili (GCCW)

GERMANY

H. Hötker, H. G. Bauer, C. Sudfeldt and J. Wahl
With contributions from: R. K. Berndt, P. Berthold, J. Blew, M. Boschert, P. Boye, S. Garthe, K. Günther, M. Jöbges, W. Knief, J. Mooij, T. Ryslavy, P. Südbeck, K. Witt, Rote-Liste-Gremium Deutschland, Dachverband Deutscher Avifaunisten and Deutscher Rat für Vogelschutz

GREECE

G. Handrinos and S. Papoulia
With contributions from: H. Alivizatos, T. Akriotis, S. Bourdakis, D. Bousbouras, A. J. Crivelli, G. Catsadorakis, B. Hallmann, S. Kazantzidis, T. Nazirides, M. Panagiotopoulou, C. Papaconstantinou, C. Papazoglou, D. Portolou, Th. Skartsi, S. Xirouchakis and S. Zogaris

HUNGARY

K. Nagy
With contributions from: J. Bagyura, G. Halmos, Cs. Moskát, B. Szabó, T. Szép and Z. Waliczky

ICELAND

Ó. Einarsson and G. Pétursson

REPUBLIC OF IRELAND

S. Newton
With contributions from: O. Crowe, D. Coombes, P. Hillis, O. Merne, BirdWatch Ireland, National Parks and Wildlife Service, and the Heritage Council and the Wildfowl and Wetlands Trust

ITALY

M. Gustin and P. Brichetti

LATVIA

E. Račinskis and A. Stīpniece
With contributions from: A. Auniņš, A. Avotiņš, J. Baumanis, U. Bergmanis, A. Celmiņš, M. Janaus, O. Keišs, V. Ķerus, J. Lipsbergs, R. Matrozis, O. Opermanis, A. Petriņš, J. Priednieks, M. Strazds and J. Vīksne

LIECHTENSTEIN

G. Willi

LITHUANIA

L. Raudonikis
With contributions from: G.Brazaitis, K. Castren, K. Čepėnas, M. Dagys, D. Dementavičius, G. Graţulevičius, M. Mačiulis, M. Mečionis, M. Kirstukas, P. Kurlavičius, Ţ. Preikđa, S. Skuja, V. Stanevičius, J. Stratford, B. Šablevičius, R. Treinys, R. Žydelis, and all those that contributed to the Lithuanian IBA Database and to the wintering and breeding waterfowl monitoring schemes

LUXEMBOURG

P. Lorgé
With contributions from: T. Conzemius, M. Jans, E. Melchior, G. Mirgain, N. Paler, R. Peltzer, J. Schmitz and J. Weiss

THE FORMER YUGOSLAV REPUBLIC OF MACEDONIA

E. Stoynov
With contributions from: B. Hallman and M. Velevski

MALTA

J. J. Borg

MOLDOVA

N. Zubkov and J. Serghei

NETHERLANDS

C. A. Borggreve
With contributions from: C. van Turnhout (SOVON – Dutch Centre for Field Ornithology)

NORWAY

I. J. Øien and T. Berge

SVALBARD

I. J. Øien

POLAND

P. Chylarecki and A. Sikora

PORTUGAL, MADEIRA AND AZORES

H. Costa, G. Elias, R. Tomé, M. Dias, T. Catry and I. Catry
With contributions from: J. Almeida, J. Bried (Azores), F. Canário, A. C. Cardoso, M. Conde, L. Costa, C. Cruz, P. Faria (Azores), H. Feith, P. Geraldes (Madeira), M. J. Groz (Azores), M. Lecoq, D. Leitão, P. Marques, A. Meirinho (Azores), F. Moreira, J. Nunes (Madeira), M. Nunes (Madeira and Azores), C. Pacheco, L. Reino, S. Rosa, V. Encarnação, F. Zino (Madeira), and Associação A Rocha

ROMANIA

A. D. Sándor
With contributions from: G. Ardelean, G. Bănică, L. Béres, S. Daróczi, C. Gache, D. Hulea, D. Ionescu, M. A. Kelemen, A. Kis, J. B. Kiss, F. Kósa, I. Kovács, W. J. Muller, D. Munteanu, A. Nagy, N. Onea, P. Pap, T. Papp, T. Petre, E. Petrescu, T. Sike, D. Z. Szabó, J. Szabó, J. Thökölyi, I. Urák, P. Weber, R. Zeitz, and the Danube Delta Biosphere Reserve Monitoring Group

RUSSIA

A. Mischenko, V. Belik, O. Borodin, S. Bakka, V. Sarychev, V. Galushin, E. Ravkin, J. Krasnov, E. Lebedeva, O. Sukhanova, S. Volkov and A. Mezhnev
With contributions from: A. Antonchikov, D. Bogomolov, V. Frolov, V. Garanin, A. Ivanov, I. Karyakin, S. Kossenko, A. Kostin, V. Krivenko, A. Leonov, V. Morozov, I. Muraviev, V. Piskunov, I. Rakhimov, A. Sharikov, P. Tomkovich and V. Vinogradov

SERBIA & MONTENEGRO

S. Puzović and D. V. Simić
With contributions from: J. Gergelj, I. Ham, I. Hulo, T. B. Jovanović, M. Ružić, D. Saveljić, M. Šćiban, N. Stojnić, M. Tucakov, O. Vizi and M. Vučanović

SLOVAKIA

R. Rybanic
With contributions from: P. Kanuch and all those that contributed to the *Distribution of Birds in Slovakia*

SLOVENIA

L. Bozic
With contributions from: F. Bracko, D. Denac, M. Kercek, T. Mihelic and S. Polak

SPAIN & CANARY ISLANDS

Coordinator: R. Martí
Compilers: B. Molina, J. A. Lorenzo (Canary Islands) and R. Martí
With contributions from: J. Alonso (Canary Islands), J. C. Atienza, R. Barone (Canary Islands), E. De Juana, J. C. Del Moral, I. Fernández Aransay, C. González (Canary Islands), A. Madroño, A. Ruiz, and all those that contributed to the *Red Data Book of the Birds of Spain* and to the *Atlas of Breeding Birds of Spain*

SWEDEN

B. Welander

SWITZERLAND

H. Schmid

TURKEY

G. Eken and B. Kurt
With contributions from: A. Aykurt, Ö. Balkız, S. Barış, M. Biricik, B. Demirci, E. Gem, S. İsfendiyaroğlu, S. Karauz, C. O. Kıraç, G. M. Kirwan, G. Magnin, O. Onmuş, N. Özbağdatlı, U. Özesmi, J. Tavares, E. Vaassen, G. Welch and H. Welch

UKRAINE

G. Gavris and O. Dudkin
With contributions from: T. Ardamatskaya, M. Beskaravajnyj, A. Bokotey, S. Domashevskiy, G. Fesenko, I. Gorban, V. Grischenko, N. Knysh, I. Krivitskij, Y. Kuzmenko, S. Loparev, Y. Milobog, V. Popenko, L. Potish, K. Redinov, A. Shevtsov, I. Skilskiy, N. Slyusar, L. Taranenko, V. Tsitsyura, A. Tsvelykh and V. Vetrov

UNITED KINGDOM

D. Noble and R. Gregory
With contributions from: N. Aebischer, A. Banks, M. Eaton, A. Joys, J. Marchant, I. Mitchell, A. Musgrove, M. O'Brien, M. Parsons, M. Raven, R. Thewlis and S. Wotton

GIBRALTAR

J. E. Cortes, C. Perez and K. Bensusan
With contributions from: P. Acolina, A. Fortuna, H. van Gils, P. Rocca, R. Rutherford and A. Yome

CONTENTS

FOREWORDS

IN 1994, BirdLife International published *Birds in Europe*—the first ever review of the conservation status of all wild bird species in Europe. National population and trend data were collected in almost every European country, producing an impressive database and creating a foundation for bird conservation work in Europe. Now, ten years on, BirdLife publishes this comprehensive update, which covers every European country and thus represents a truly pan-European inventory. Coinciding with the enlargement of the European Union, this analysis provides us with important baseline data for our future work, and allows us to assess European efforts towards biodiversity conservation and sustainable development. Europe's wild birds have not fared well over the last ten years. Although we have succeeded in stabilising the populations of some threatened species—not least with BirdLife's help—the proportion of species with an unfavourable conservation status has increased from 38% to 43%. This is worrying: as BirdLife demonstrates, birds are indicators of the state of wider biodiversity. The conclusion we must draw is that we need to do more to save our birds and meet our 2010 deadline to halt biodiversity decline. In this endeavour, this fully revised edition of *Birds in Europe* will be a unique source of information to help us effectively target and prioritise our conservation efforts. BirdLife International has reinforced its credibility as the authority on the status of birds and their habitats across Europe. Once again, I would like to congratulate BirdLife International for the clarity and quality of its work.

Margot Wallström,
European Commissioner for the Environment

THE Dutch Government has been a long-time supporter of the activities of BirdLife International, especially since the move of the European Division Office to Wageningen in the Netherlands. In 2000, we welcomed *Important Bird Areas in Europe: priority sites for conservation*, and now—four years later—*Birds in Europe*, a reassessment of bird population trends in Europe. Our Ministry of Agriculture, Nature and Food Quality has helped with the financing of this study for the past three years, and we proudly welcome its publication. At a pan-European level, it will be helpful in developing conservation policies and in implementing international conservation frameworks, such as the Bern Convention and the Bonn Convention. Within the European Union, as we work towards fulfilling our commitment to halt biodiversity declines by 2010, this book is an important tool for monitoring the effects of our policy decisions. Birds have proven to be useful indicators because they are widely distributed and well monitored by the extensive network of ornithologists across Europe. Our congratulations go to BirdLife's European Division for successfully compiling this information. The BirdLife Partnership and its many volunteers have again demonstrated their capability to produce a valuable standard reference.

Giuseppe B. Raaphorst,
Director, Department of Nature,
The Netherlands Ministry of Agriculture,
Nature and Food Quality

ON behalf of the UNEP Convention on the Conservation of Migratory Species of Wild Animals (CMS), I warmly welcome the arrival on the conservation scene of this fully revised edition of *Birds in Europe*. For the past 25 years, CMS has provided a global platform for the conservation and sustainable management of migratory animals and the places they inhabit. To do this, the Convention has performed its fundamental duty of bringing together the Range States of migratory species, but at the same time we have always included and valued our non-governmental organisation partners. Among the latter, BirdLife International has played a key and growing role, and earlier in 2004 our two organisations signed a Memorandum of Cooperation. The Memorandum recognises the key importance of understanding the status of species. This is vital, for instance, in the selection of species for the Appendices of CMS, which include numerous European birds. Several CMS Agreements also cover these species, including those for Slender-billed Curlew, Great Bustard and Aquatic Warbler. The largest CMS Agreement of all, the African Eurasian Waterbird Agreement (AEWA), aims to provide conservation benefits to some 140 European species. This new BirdLife publication provides us with state-of-the-art knowledge of all these species. It is a first-class publication of which all the contributors should be truly proud. Although there has been much encouraging progress—for instance, in the production and delivery of Species Action Plans—we need constantly to challenge a growing legion of threats to our migratory birds. I can say with some certainty that this book will be an invaluable tool in helping us to do so. As I am also a new arrival (as the chief officer at UNEP-CMS), I expect both to use it myself, and to develop further our productive relations with BirdLife and other NGOs in conserving migratory species everywhere.

Robert Hepworth,
Executive Secretary, UNEP-CMS

TWENTY-FIVE years ago, the Birds Directive was established for the protection of wild birds and their habitats within the European Community. This was the first piece of Community legislation addressing Europe's biological diversity. Many people and organisations have been involved directly and indirectly in implementing the objectives and requirements of this directive—from Community officials, through national and local public servants, non-governmental organisations and site managers, to members of the general public. There has been some progress through the directive but, as shown by this book, birds remain under threat, and in many areas the threats and pressures are growing. This book provides a timely review of the state of birds in Europe, highlighting changes since the original version was published in 1994. It is also an excellent example of the role of a multi-national network of data collectors, many of whom are volunteers who collect the data in their spare time. It provides reliable, quality-assured information on biodiversity trends across Europe, and it makes a vital contribution to the monitoring and achievement of the target to halt biodiversity loss by 2010. The European Environment Agency will be working with BirdLife International to build on its experience, to extend its approach to other international networks monitoring other taxa, and to develop a set of indicators on the state of Europe's biodiversity, the services it provides and the threats it faces. Everyone acknowledges the wide appeal of birds, and most can recognise birds as indicators of the state of the environment, sustainable development and progress towards the 2010 target. This book reinforces the case for concern and increased action. Let us make sure—by using *Birds in Europe* and working together with all interested organisations, networks and the general public—that the loss of biodiversity in general, and birds in particular, is halted by 2010. We do not have another 25 years to meet the objectives of the Birds Directive.

Prof. Jacqueline McGlade,
Executive Director, European Environment Agency

AT the European Council summit held in June 2001, heads of state and government of the European Union committed themselves to halt the loss of biodiversity by 2010, and to make this a goal in the European Union Strategy for Sustainable Development. However, given the current rapid decline in biodiversity globally, and the ever-increasing extent and intensity of many human activities, the objective of halting the decline in biodiversity by 2010 will require unprecedented efforts to adapt our activities to the needs of natural systems. IUCN spearheads the Countdown 2010 initiative to join key decision-makers and civil society in halting the loss of biodiversity by 2010. We are joined in this effort by our Members, which include institutions and organisations such as BirdLife International. BirdLife has been an IUCN member since 1949, and is a vital resource in our collective efforts to achieve conservation results around the globe. This fully revised edition of *Birds in Europe* provides us with an important analysis of population trends of European birds, enabling us to make scientifically sound decisions as we channel scarce resources towards conservation action that will halt biodiversity loss. This publication also paves the way for other regions around the world, by providing the first continental-level application of the IUCN Red Listing process. IUCN applauds the efforts that have gone into this volume, and encourages others to undertake similar studies for other taxa as we try to reach our 2010 goals. With six more years to go, *Birds in Europe* will be a key tool and a model for the conservation community at large.

Achim Steiner,
Director General, IUCN – The World Conservation Union

HAVING recently taken the helm as Chairman of BirdLife International, I am proud to welcome you, the reader, to this important publication. BirdLife International's strength comes from its diverse network of organisations, which covers more than 100 countries around the globe. Through this vast network, we are able to make the link from local and national levels to regional and global scales. Our jointly agreed strategy, which is based on firm scientific foundations and coupled with our broad grassroots representation, enables us to reach our targets successfully. This book is yet another testimony to the power of the BirdLife network. During the past three years, we have liaised closely with national coordinators across Europe to update the information provided in the original version of *Birds in Europe* published in 1994. This comprehensively revised edition includes all countries (including those for which information was not previously available), and provides data on populations and trends of all 526 regularly occurring European species. We publish this update to ensure that bird conservation efforts in Europe remain well informed and based on sound science. It is a unique and powerful reference tool and, like the original version, it will be the benchmark for any future evaluation of the status of birds in Europe. However, documenting population trends is just the beginning. Even more importantly, this update provides a solid basis for future actions to improve the status of threatened bird species, particularly those concentrated in the continent and for which Europe therefore has a special responsibility. Concerted effort is now required to persuade governments and stakeholders at all levels to change land management and agricultural policies. Essential ecological networks must be maintained or restored, and threats to bird populations need to be removed or substantially reduced. We must also be ready to respond to newly emerging threats, such as climate change, whose pace and scope demand the development and implementation of entirely new strategies for bird conservation and land management. I am confident that the BirdLife Partnership will rise to this challenge with strength and professionalism.

Peter Johan Schei,
Chairman, BirdLife International

ACKNOWLEDGEMENTS

This review is based on the huge amount of work carried out by thousands of ornithologists, both volunteers and professionals, throughout Europe over the last few decades. Without their expertise and commitment, this study would not have been possible. Although we cannot thank them individually, we owe them a great debt and wish to recognise the important contribution that each has made to bird conservation in Europe by participating in bird censusing or monitoring.

In particular, we wish to thank the national coordinators and contributors (listed at the front of this book), and the numerous organisations that they represent or with which they are associated. All of them showed great enthusiasm for this project from the moment it was launched in 2002, and diligently collated the information required to provide the national data sets that form the core of this publication.

The basis for this review was of course the original edition of *Birds in Europe*, compiled by Graham Tucker and Melanie Heath and published in 1994. This new assessment owes a huge amount to its predecessor, and in the process of compiling it we have gained even more appreciation and respect for the massive efforts that went into producing the first version, long before the days of e-mail and attachable spreadsheets.

The data used in the original edition were collected in close collaboration with the European Ornithological Atlas project of the European Bird Census Council (EBCC). The Executive Committee of the EBCC were also extremely supportive of the current project, for which we thank them: David Gibbons, Richard Gregory, Ward Hagemeijer, Anny Anselin, Przemek Chylarecki, Martin Flade, Ruud Foppen, Lorenzo Fornasari, Elena Lebedeva, Uygar Özesmi, David Noble, Tibor Szep, Juha Tiainen and Petr Voøíšek.

We are very grateful to the past and present members of BirdLife's European Committee for supporting our efforts and motivating us to get the job done: Mika Asikainen, Alison Duncan, Jan Eljsted, József Fidlóczky, Alistair Gammell, Lars Lindell, Costas Papaconstantinou, Alejandro Sánchez, Thomas Tennhardt, Alexandre Vintchevski and Johanna Winkelman.

Des Callaghan started the ball rolling on this project in early 2002, by establishing the network of national coordinators and initiating the consultation on revising the criteria used to identify Species of European Conservation Concern (SPECs). He also set up the Internet portal (with Crispen Wilson), calculated the generation lengths of all European species, and collated and checked a considerable amount of the data before leaving BirdLife at the end of 2002. Umberto Gallo-Orsi then took over the running of the project for several months, before passing on the baton to Ian Burfield (IB) in May 2003. Checking, correcting and finalising the data took many more months, but was completed in early 2004, when Frans van Bommel (FvB) was employed to help with the analysis, mapping and compilation of the book. Rob Pople joined the team later to write and edit the texts justifying each species's status, and to proofread the rest of the book. Most chapters were written by the compilers, but many include contributions from Szabolcs Nagy, Canan Orhun and other team members.

Among our other colleagues at the European Division Office in Wageningen, we thank Güven Eken for his early efforts in fundraising for the project, which were continued by Umberto Gallo-Orsi. Marianne de Rijk provided invaluable administrative support throughout the project, and did her best to keep us all sane by dragging us away from our computers for a short lunch break every day. Sara Sharpe gave us her expert opinion on draft designs and suggested how we could best communicate the results of the project, and Jane Kuenen helped us find library resources. We thank Szabolcs Nagy for fruitful discussions throughout the project on practically every major issue, and also acknowledge the very important contributions of several volunteers: Esra Basak, Javier Carrera, Çiğdem Coşkun Hepcan, Nezihe Hewson, Peter Mateev and Aderonke Ojei. Canan Orhun kept us all in line and averted numerous crises along the way.

At the BirdLife Secretariat in Cambridge, Alison Stattersfield, Stuart Butchart and Leon Bennun provided scientific advice and encouragement at all stages. Martin Sneary dealt with all our data management requests, and made it possible for national coordinators to submit data over the Internet. He also developed the searchable database for the web site that accompanies this book, allowing users to view species factsheets online. The BirdLife Taxonomic Working Group (particularly Nigel Collar, Lincoln Fishpool and Sue Shutes) kept us abreast of their decisions and helped us by fast-tracking several European species through their review process. Our thanks also go to many other colleagues who helped in different ways, including Rosina Abudulai, Mark Balman, Gilly Banks, Lisa Canessa, Beverley Childs, Janet Chow, Mike Evans, Marco Lambertini, Chris Mills, Caroline Pridham, Mike Rands, Mary Saunders and Chris Spreadbury.

The communications team in Cambridge played a particularly important role in the production of this book. We thank Adrian Long for sharing his experiences of earlier projects and suggesting how best to compile a publication of this size. Richard Thomas helped oversee the book's production and made sure it adhered to the standards of the BirdLife Conservation Series. He also helped Umberto Gallo-Orsi to gather the illustrations of each species (see 'Artists' acknowledgements'), and—with Sue Shutes and Sally Fisher—painstakingly checked and corrected all the data references compiled by FvB. Mich Herrick showed great creativity in turning IB's half-baked idea for the front cover into an eye-catching design, and Ed Parnell developed the web site that allows most of the information in this book to be accessed via the Internet.

Vital support and encouragement were given in other ways by several colleagues within the Alterra Research Institute. The technical assistance of Roland van Zoest was fundamental in helping FvB to automate the process of creating maps for all species. Andre Jellema also provided valuable help with the ArcView GIS software, which was donated to BirdLife through the ESRI Conservation Programme. Sharing our corridor with Wetlands International enabled us to discuss many ideas and problems with Ward Hagemeijer, Simon Delany and Lieuwe Haanstra, all of whom gave their help willingly and shared unpublished data with us.

Beyond the countless queries dealt with in good humour by the national coordinators and contributors, the following experts provided helpful comments or additional information on particular species or issues: Bart Ebbinge, Martin Flade, Geoff Hilton, Kees Koffijberg, Alberto Madroño, Dilia Menezes, Ingar Jostein Øien, Paulo Oliveira, John O'Sullivan, Stefan Pihl, Iván Ramírez, Martin Schneider-Jacoby and Māris Strazds. We thank Ole Thorup, David Stroud and Helen Baker for sharing the results of two major International Wader Study Group projects with us before publication, and Petr Voříšek for doing the same with data from the Pan-European Common Bird Monitoring Scheme. Fiona Sanderson and Paul Donald provided information on species' habitat associations and migration strategies, and Jörn Scharlemann solved several database queries.

Many people provided valuable contributions during the lengthy process of revising the SPEC criteria, and the consultation exercise involved all national coordinators and many others in the BirdLife network. Particularly useful comments were received from John Cortes, Mark Eaton, Peter Flint, Richard Gregory, Michael Grell, Julian Hughes, Verena Keller, Alexander Mischenko, Szabolcs Nagy and Dave Pritchard. The members of BirdLife's Birds and Habitats Directive Task Force also made a major contribution to this discussion, and a working group was established to consider all comments and finalise the criteria to be used in both *Birds in Europe* and *Birds in the European Union*. This group, which took its lead from a thorough scoping paper compiled by Eduard Osieck (then Task Force coordinator), also comprised IB, Clairie Papazoglou, Paul Donald, Ariel Brunner, Nicola Crockford and Zoltán Waliczky.

We are highly indebted to the team responsible for design, layout and print management at NatureBureau, who turned a complicated manuscript into an attractive, user-friendly book. Peter Creed in particular was an indefatigable source of optimism, radiating calm under pressure and frequently sacrificing evenings and weekends to ensure deadlines were met. Substantial contributions were also made by Barbara Creed, Simon Green and Helen Dobie. The whole team's professionalism, enthusiasm and flexibility made the tight schedules involved in the production process tolerable, and guaranteed the high quality of the end product.

The translation of the book's summary from English into nine other European languages was undertaken with great care by the staff of BirdLife Partners. In the order in which the translations appear in this book, we thank: Konstantin Kreiser and Claus Mayr; Ramón Martí and Blas Molina; Bernard Deceuninck; Claudio Celada and Marco Gustin; Johanna Winkelman; Aleksandra Priwieziencew and Piotr Kamont; Andrey Nedossekin and Alexander Mischenko; Lars Lindell; Burcu Arık and Dicle Tuba Kılıç.

We also acknowledge the vital support given by many friends and relatives over the three years of work on this book. In particular, IB thanks Katherine Bongaerts for her patience and support, and for enduring the seemingly unending late nights and weekends of work required to finish the project. Tobias Heffelaar put up with a good deal of grumbling during the daily commute between Utrecht and Wageningen. Back at work, Szabolcs Nagy, Zoltán Waliczky and others took on several of IB's designated tasks for 2004 to enable the analyses for this book and *Birds in the European Union* to be completed in time, while FvB coped admirably with everything that IB threw at him.

Finally, we extend our deepest thanks to the sponsoring organisations whose financial support enabled this project to be completed. Like many other aspects of BirdLife's European Programme, the Netherlands Ministry of Agriculture, Nature and Food Quality supported it from the very beginning, through the PIN/MATRA Funds of the Ministry of Foreign Affairs. As in 1994, Naturvårdsverket, the Swedish Environmental Protection Agency, made a valuable contribution to help us finalise the publication. The European Commission also made additional funds available in 2004, to help ensure that the book would be ready to launch in November at a conference celebrating the 25th anniversary of the EU Wild Birds Directive. We are particularly grateful to several BirdLife Partners for their significant contributions, and to the staff who arranged them: BirdLife Finland (Mika Asikainen), the Ligue pour la Protection des Oiseaux and its supporters (Alison Duncan), Vogelbescherming Nederland (Johanna Winkelman, who also secured funds from the Dutch Postcode Lottery), the Swedish Ornithological Society (Lars Lindell) and SVS-BirdLife Switzerland (Werner Müller).

ARTISTS' ACKNOWLEDGEMENTS

The great majority of the line drawings come from the *EBCC Atlas of European Breeding Birds: their distribution and abundance* produced by the European Bird Census Council. From the original set, some 40 images have been changed, and four artists have contributed new drawings especially for this publication. We thank Christine Isherwood, Richard Johnson, Hans Larsson and Māris Strazds for these contributions. All artists and the species drawings they contributed are credited below.

Richard Allen
Phoenicurus ochruros, Phoenicurus phoenicurus, Saxicola rubetra, Saxicola torquata, Oenanthe isabellina, Oenanthe oenanthe, Oenanthe pleschanka, Oenanthe hispanica, Oenanthe leucura

S. Mark Andrews
Passer moabiticus

Norman Arlott
Pelecanus crispus, Saxicola dacotiae, Sylvia melanothorax

Christof Bobzin
Falco biarmicus, Falco pelegrinoides, Bonasa bonasia, Lagopus lagopus, Lagopus mutus, Alectoris chukar, Alectoris graeca, Alectoris rufa, Alectoris barbara, Perdix perdix, Phasianus colchicus, Rallus aquaticus, Porzana porzana, Porzana parva, Porzana pusilla, Crex crex, Porphyrio porphyrio, Grus grus, Grus virgo, Cursorius cursor, Calidris ferruginea, Otus brucei, Ketupa zeylonensis, Ceryle rudis

George Brown
Turdus viscivorus

John Busby / RSPB
Falco columbarius (Dedication), *Tyto alba, Hirundo rustica, Turdus torquatus, Turdus merula, Turdus philomelos, Turdus iliacus, Sylvia undata* (Dedication), *Ficedula hypoleuca* (Dedication), *Panurus biarmicus* (Dedication), *Pyrrhula pyrrhula*

Jens Overgaard Christensen
Phalacrocorax carbo, Phalacrocorax aristotelis, Pernis apivorus, Circus aeruginosus, Circus cyaneus, Circus macrourus, Circus pygargus, Aquila chrysaetos, Tetrao tetrix, Tetrao urogallus, Coturnix coturnix, Nyctea scandiaca, Surnia ulula, Glaucidium passerinum, Strix uralensis, Strix nebulosa, Asio flammeus, Caprimulgus europaeus, Caprimulgus ruficollis

Przemek Chylarecki
Philomachus pugnax, Larus minutus

Tomasz Cofta
Cinclus cinclus, Troglodytes troglodytes, Luscinia luscinia, Luscinia megarhynchos, Locustella lanceolata, Locustella naevia, Locustella fluviatilis, Acrocephalus melanopogon, Acrocephalus schoenobaenus, Acrocephalus agricola, Acrocephalus dumetorum, Acrocephalus palustris, Acrocephalus scirpaceus, Hippolais caligata, Hippolais languida, Hippolais icterina, Sylvia mystacea, Sylvia nana, Sylvia nisoria, Sylvia curruca, Sylvia communis, Sylvia borin, Sylvia atricapilla, Phylloscopus trochiloides, Phylloscopus borealis, Phylloscopus inornatus, Phylloscopus bonelli, Phylloscopus sibilatrix, Phylloscopus sindianus, Phylloscopus collybita, Phylloscopus trochilus, Regulus ignicapilla, Emberiza schoeniclus, Emberiza pallasi

Koen Devos
Milvus migrans, Milvus milvus, Haliaeetus albicilla, Gypaetus barbatus, Accipiter gentilis, Accipiter nisus, Accipiter brevipes, Falco columbarius, Falco subbuteo, Falco eleonorae, Charadrius dubius, Charadrius hiaticula, Charadrius alexandrinus, Charadrius leschenaultii, Actitis hypoleucos, Arenaria interpres, Motacilla flava, Motacilla citreola, Motacilla cinerea, Motacilla alba

Jon Fjeldså
Panurus biarmicus, Aegithalos caudatus, Parus palustris, Parus lugubris, Parus montanus, Parus cinctus, Parus ater, Parus caeruleus, Parus cyanus, Parus major, Remiz pendulinus

Sylvia Gandini
Larus ichthyaetus, Larus melanocephalus, Coracias garrulus, Upupa epops, Bombycilla garrulus, Sitta krueperi, Sitta whiteheadi, Sitta neumayer, Emberiza cia, Emberiza cineracea, Emberiza hortulana, Emberiza buchanani, Emberiza caesia, Emberiza rustica, Emberiza pusilla, Miliaria calandra

Simon Gillings
Gallinago media, Gallinago stenura, Larus armenicus, Picus canus, Dendrocopos medius, Dendrocopos minor, Picoides tridactylus, Nucifraga caryocatactes, Corvus monedula

Susan Gough
Puffinus gravis, Ammoperdix griseogularis, Vanellus indicus, Lymnocryptes minimus, Apus affinis, Halcyon smyrnensis, Jynx torquilla, Picus viridis, Dendrocopos syriacus, Anthus gustavi, Oenanthe cypriaca, Regulus teneriffae, Perisoreus infaustus, Pica pica, Pyrrhocorax graculus, Pyrrhocorax pyrrhocorax, Corvus corone, Corvus corax, Sturnus unicolor, Petronia xanthocollis

Alan Harris
Puffinus griseus, Pycnonotus xanthopygos, Prinia gracilis

Ren Hathaway
Aquila heliaca, Pandion haliaetus, Larus genei, Columba livia, Streptopelia turtur, Merops apiaster

Paul Hirst
Gavia immer, Gavia adamsii, Podiceps grisegena, Bulweria bulwerii, Calonectris diomedea, Puffinus yelkouan, Hydrobates pelagicus, Oceanodroma leucorhoa, Oceanodroma castro, Morus bassanus, Polysticta stelleri, Stercorarius pomarinus, Stercorarius parasiticus, Stercorarius longicaudus, Catharacta skua, Larus ridibundus, Larus canus, Larus fuscus, Larus argentatus, Larus cachinnans, Larus glaucoides, Larus hyperboreus, Larus marinus, Xema sabini, Rissa tridactyla, Pagophila eburnea, Uria lomvia, Alle alle

Alan Hitchings
Calidris minuta, Calidris temminckii

Mark Hulme
Plegadis falcinellus, Platalea leucorodia, Phoenicopterus roseus, Cygnus olor, Cygnus columbianus, Cygnus cygnus, Tadorna ferruginea, Tadorna tadorna, Somateria mollissima, Somateria spectabilis, Histrionicus histrionicus, Clangula hyemalis, Melanitta nigra, Melanitta fusca, Bucephala islandica, Gallinula chloropus, Fulica atra, Fulica cristata, Himantopus himantopus, Recurvirostra avosetta

Christine Isherwood
Puffinus puffinus, Chen caerulescens

Richard Johnson
Puffinus mauretanicus, Geronticus eremita, Accipiter badius, Calidris bairdii, Merops persicus, Ammomanes deserti, Calandrella cheleensis

Mike Langman / RSPB
Erithacus rubecula

Hans Larsson
Serinus corsicana

Ernest Leahy
Fringilla coelebs, Fringilla montifringilla, Serinus serinus, Serinus citrinella, Carduelis chloris, Carduelis carduelis, Carduelis spinus, Carduelis cannabina, Carduelis flavirostris, Carduelis flammea, Carduelis hornemanni, Loxia leucoptera, Loxia curvirostra, Loxia scotica, Loxia pytyopsittacus, Carpodacus erythrinus, Pinicola enucleator, Coccothraustes coccothraustes

Joaquín López-Rojas
Otus scops, Bubo bubo, Athene noctua, Strix aluco, Asio otus, Hirundo rupestris, Hirundo daurica, Ficedula semitorquata, Ficedula albicollis, Serinus canaria

Dan Powell
Haematopus ostralegus, Haematopus meadewaldoi, Charadrius asiaticus, Eudromias morinellus, Pluvialis apricaria, Pluvialis squatarola, Vanellus spinosus, Vanellus gregarius, Phalaropus lobatus, Phalaropus fulicarius, Sterna nilotica, Sterna caspia, Sterna bengalensis, Sterna sandvicensis, Sterna dougallii, Sterna hirundo, Sterna paradisaea, Chlidonias hybrida, Chlidonias niger, Chlidonias leucopterus, Pterocles orientalis, Pterocles alchata, Clamator glandarius, Cuculus canorus, Cuculus saturatus, Tichodroma muraria, Certhia familiaris, Certhia brachydactyla

Dan Powell / RSPB
Garrulus glandarius

Pavel Procházka
Anser fabalis, Anser brachyrhynchus, Anser albifrons, Anser erythropus, Anser anser, Branta canadensis, Branta leucopsis, Branta bernicla, Anas penelope, Anas strepera, Anas crecca, Anas platyrhynchos, Anas acuta, Anas querquedula, Anas clypeata, Netta rufina, Aythya ferina, Aythya fuligula, Aythya marila, Lanius collurio, Lanius minor, Lanius senator, Lanius nubicus

José Projecto
Columba trocaz, Apus unicolor, Monticola saxatilis, Monticola solitarius, Cettia cetti, Cisticola juncidis, Hippolais pallida, Hippolais polyglotta, Sylvia undata, Sylvia conspicillata, Sylvia cantillans, Sylvia melanocephala, Sylvia hortensis, Muscicapa striata, Ficedula hypoleuca, Emberiza cirlus

Petr Rob
Galerida cristata, Lullula arborea, Alauda arvensis, Riparia riparia, Delichon urbica

Dave Showler
Streptopelia senegalensis

Michal Skakuj
Puffinus assimilis, Pelagodroma marina, Falco tinnunculus, Falco vespertinus, Falco cherrug, Falco rusticolus, Falco peregrinus, Tetrao mlokosiewiczi, Tetraogallus caucasicus, Tetraogallus caspius, Francolinus francolinus, Vanellus leucurus, Melanocorypha bimaculata, Melanocorypha leucoptera, Melanocorypha yeltoniensis, Eremophila alpestris, Anthus berthelotii, Phoenicurus erythrogastrus, Oenanthe deserti, Zoothera dauma, Turdus ruficollis, Sylvia sarda, Sylvia rueppelli, Sitta tephronota, Petronia brachydactyla, Serinus pusillus, Rhodopechys sanguinea, Rhodopechys mongolica, Carpodacus rubicilla, Calcarius lapponicus, Plectrophenax nivalis, Emberiza leucocephalos, Emberiza citrinella, Emberiza aureola, Emberiza bruniceps, Emberiza melanocephala

Māris Strazds
Botaurus stellaris, Ixobrychus minutus, Nycticorax nycticorax, Ardeola ralloides, Bubulcus ibis, Egretta garzetta, Casmerodius albus, Ardea cinerea, Ardea purpurea, Ciconia nigra, Ciconia ciconia, Vanellus vanellus, Calidris canutus, Calidris alba, Calidris maritima, Calidris alpina, Limicola falcinellus, Gallinago gallinago, Scolopax rusticola, Limosa limosa, Limosa lapponica, Numenius phaeopus, Numenius tenuirostris, Numenius arquata, Tringa erythropus, Tringa totanus, Tringa stagnatilis, Tringa nebularia, Tringa ochropus, Tringa glareola, Xenus cinereus, Sterna albifrons, Columba oenas, Columba palumbus, Streptopelia decaocto, Aegolius funereus, Alcedo atthis, Dryocopus martius, Dendrocopos major, Dendrocopos leucotos, Anthus campestris, Turdus pilaris, Locustella luscinioides, Acrocephalus arundinaceus, Hippolais olivetorum, Phylloscopus brehmii, Phylloscopus canariensis, Regulus regulus, Ficedula parva, Parus cristatus, Sitta europaea, Lanius excubitor, Corvus frugilegus

Thelma K. Sykes
Gavia stellata, Gavia arctica, Tachybaptus ruficollis, Podiceps cristatus, Podiceps auritus, Podiceps nigricollis, Fulmarus glacialis, Bucephala clangula, Mergellus albellus, Mergus serrator, Mergus merganser, Uria aalge, Alca torda, Cepphus grylle, Fratercula arctica

Carl Christian Tofte
Phalacrocorax pygmeus, Circaetus gallicus, Buteo buteo, Buteo rufinus, Buteo lagopus, Aquila pomarina, Aquila clanga, Aquila nipalensis, Hieraaetus pennatus, Hieraaetus fasciatus, Tachymarptis melba, Apus apus, Apus pallidus, Apus caffer

Juan Varela
Pterodroma madeira, Pterodroma feae, Branta ruficollis, Marmaronetta angustirostris, Aythya nyroca, Oxyura leucocephala, Elanus caeruleus, Neophron percnopterus, Gyps fulvus, Aegypius monachus, Aquila adalberti, Falco naumanni, Turnix sylvatica, Tetrax tetrax, Chlamydotis undulata, Otis tarda, Larus audouinii, Rhodostethia rosea, Columba bollii, Columba junoniae, Chersophilus duponti, Melanocorypha calandra, Calandrella brachydactyla, Calandrella rufescens, Galerida theklae, Erythropygia galactotes, Acrocephalus paludicola, Cyanopica cyanus, Fringilla teydea, Rhodopechys obsoleta, Bucanetes githagineus, Pyrrhula murina

Jan Wilczur
Pelecanus onocrotalus, Prunella modularis, Prunella montanella, Prunella ocularis, Prunella atrogularis, Prunella collaris, Luscinia calliope, Luscinia svecica, Tarsiger cyanurus, Irania gutturalis, Oenanthe finschii, Oenanthe xanthoprymna

Jos Zwarts
Burhinus oedicnemus, Glareola pratincola, Glareola nordmanni, Anthus hodgsoni, Anthus trivialis, Anthus pratensis, Anthus cervinus, Anthus spinoletta, Anthus petrosus, Oriolus oriolus, Sturnus vulgari, Sturnus roseus, Passer domesticus, Passer hispaniolensis, Passer montanus, Petronia petronia, Montifringilla nivalis

SUMMARY

AIM

Birds in Europe (2004) or *BiE2* is the second review of the conservation status of all wild birds in Europe. Like its 1994 predecessor, *Birds in Europe* (*BiE1*), it identifies priority species (Species of European Conservation Concern, or SPECs) in order that conservation action can be taken to improve their status.

SCOPE

The geographical scope is continent-wide, extending from Greenland in the west to the Urals in the east, and from Svalbard in the north to the Canary Islands in the south. Increased political stability in the Balkans and the Caucasus has allowed data to be collected from all European countries for the first time.

DATA COLLECTION

Data were collected through a network of national coordinators, who sought input from relevant experts, monitoring organisations and regional contributors. The data derive from fieldwork carried out by thousands of ornithologists, including countless volunteers.

For each species, national data were gathered on breeding population size (in or around the year 2000) and trend (over the period 1990–2000). Where available, equivalent winter population data were also collected, mainly for waterbirds. In total, some 14,000 population/trend records were received, including many of higher quality than in *BiE1*. Together with the existing trend and range data from 1970–1990, these new population data were used to reassess each species's conservation status in Europe.

STATUS ASSESSMENT

For *BiE1*, a set of quantitative criteria was developed to identify SPECs according to their global and European status, and to classify them by the proportion of their global population or range in Europe. For *BiE2*, an extensive consultation process concluded that these criteria could be strengthened by incorporating the IUCN Red List Criteria, which represent the accepted system for assessing species' relative extinction risk. The recent publication of guidelines for applying the IUCN Red List Criteria at a regional level made integrating them into the existing SPEC criteria relatively simple.

Following the system used in *BiE1*, each species was assigned to one of five categories:

Category	European species of global conservation concern	Conservation status in Europe	Global population or range concentrated in Europe
SPEC 1	Yes	–	–
SPEC 2	No	Unfavourable	Yes
SPEC 3	No	Unfavourable	No
Non-SPEC^{E}	No	Favourable	Yes
Non-SPEC	No	Favourable	No

A species is of global conservation concern if its status is classified as Threatened, Near Threatened or Data Deficient under the IUCN Red List Criteria at global level. It has an Unfavourable Conservation Status in Europe if its European population is classified as Threatened under a regional application of the IUCN Red List Criteria, or if its population is any of the following (as in *BiE1*): small and non-marginal, declining moderately, depleted following earlier declines, or highly localised. A species is concentrated in Europe if more than 50% of its global breeding or wintering population or range occurs in Europe.

RESULTS

Of the 524 species assessed, 226—or 43% of the European avifauna—have an Unfavourable Conservation Status in Europe (Figure 1). Of these, 40 species (7.6%) are classified as SPEC 1, 45 (8.6%) as SPEC 2 and 141 (26.9%) as SPEC 3. All these percentages exceed those in *BiE1*, when 195 species (38% of the 511 assessed) were classified as SPECs.

Figure 1. Percentage of European bird species in each category in *BiE1* (1994) and *BiE2* (2004), rounded to the nearest 1%.

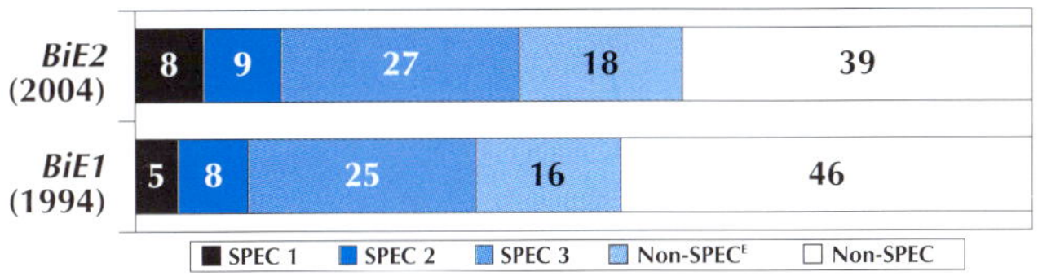

DISCUSSION

The increased number of SPEC 1 species reflects the reclassification (under the revised criteria) of globally Near Threatened species, which were previously listed as SPEC 2 or 3, but are clearly of global conservation concern. However, the increased number of SPEC 2 and 3 species is truly alarming, because it means that the European conservation status of many more birds (45 species) has changed from Favourable to Unfavourable than vice versa (14 species).

Given the magnitude of declines during 1970–1990, particularly those affecting farmland birds, the populations of many SPECs remain heavily depleted. However, a few species have recovered and are now classified as having a Favourable Conservation Status in Europe. The recovery of Peregrine Falcon *Falco peregrinus* is a good example, illustrating the benefits of targeted conservation action.

Many SPEC 1 species are also increasing in Europe, due to the effective implementation of species action plans (SAPs) over the past decade. It will take time before these species can be reclassified, but progress to date indicates that co-operative actions that are planned well and funded adequately can indeed reverse declines and restore species to a more favourable status.

Of the 129 species that declined significantly during 1970–1990, 79 (61%) continued to decline during the 1990s, including many farmland birds, waders and raptors. Their plight is particularly worrying—and they have now been joined by 35 declining species formerly considered to have a Favourable Conservation Status in Europe. These include many long-distance migrants, several waterbirds, and some of Europe's most familiar species, such as House Sparrow *Passer domesticus* and Common Starling *Sturnus vulgaris*.

These are sobering facts when most governments have pledged to reduce the rate of biodiversity loss by 2010, and the European Union is committed to halting this loss completely. For most taxa, assessing whether these targets are met will be very difficult, but birds are an exception. The main requirement is modest long-term support for monitoring, both to sustain existing schemes and to develop and implement strategies for other species. This would allow governments to meet their reporting obligations, and facilitate status reviews like this every decade (with *BiE3* scheduled for 2014).

The time left to meet these targets is short, so it is vital that biodiversity concerns are integrated fully into all sectoral policies that affect the environment. Europe already benefits from some of the finest biodiversity conservation legislation in the world. The Birds Directive, the Bern Convention and the Convention on Migratory Species were all landmarks when they were adopted 25 years ago, and have already achieved a huge amount. Yet, as *BiE2* demonstrates, many challenges remain, and the need to apply these tools to maximum effect for biodiversity will only increase over the next 25 years.

CONCLUSIONS

The overall message from *BiE2* is as clear as that from *BiE1*. Birds in Europe continue to be threatened by widespread environmental change, and many populations are now in deeper trouble than a decade ago. As birds are good environmental indicators, the ongoing decline of so many species sends clear signals about the state of European biodiversity and the health of the wider environment.

Given the scale of the problem, the massive and urgent response called for in *BiE1* is now even more pressing. Action must be taken immediately—not only to stop the continuing loss of Europe's once rich and abundant avifauna, but also to show serious commitment to halting biodiversity loss by 2010.

ZUSAMMENFASSUNG

ZIEL

Birds in Europe (2004) oder *BiE2* ist der zweite Überblick über den Erhaltungsstatus aller wildlebenden Vogelarten Europas. Wie sein Vorgänger *Birds in Europe* (1994) oder *BiE1*, identifiziert *BiE2* prioritäre Arten (Species of European Conservation Concern, im folg. SPECs) mit dem Ziel, Maßnahmen für eine Verbesserung ihres Status zu ermöglichen.

UNTERSUCHTER RAUM

Der behandelte Raum umfasst den gesamten Kontinent, von Grönland im Westen bis zum Ural im Osten, von Spitzbergen im Norden bis zu den Kanarischen Inseln im Süden. Die Stabilisierung der politischen Lage auf dem Balkan und im Kaukasus ermöglichte es, nun erstmals Daten aus allen europäischen Ländern zu erfassen.

DATENSAMMLUNG

Die Datensammlung erfolgte über ein Netzwerk nationaler Koordinatoren, unterstützt durch Experten, Monitoring-Organisationen und regional Mitwirkende. Die Daten selbst beruhen auf der Feldarbeit tausender Ornithologen, einschließlich ungezählter Ehrenamtlicher.

Für jede Art wurden nationale Daten erfasst, sowohl zur Größe der Brutpopulation (um das Jahr 2000) als auch zum Trend (über die Periode 1990–2000). Wo erhältlich, wurden zusätzlich die Winterpopulationen erfasst (v.a. für Wasservögel). Insgesamt wurden etwa 14.000 Populations- und Trenddatensätze aufgenommen, darunter viele von höherer Qualität als in *BiE1*. Zusammen mit den vorliegenden Trend- und Verbreitungsdaten aus den Jahren 1970–1990 wurden die neuen Populationsdaten dazu benutzt, den Erhaltungsstatus jeder europäischen Art neu zu bestimmen.

BESTIMMUNG DES ERHALTUNGSSTATUS

Für *BiE1* wurde eine Reihe quantitativer Kriterien entwickelt, um SPECs gemäß ihres globalen und europäischen Status festzulegen, und um sie nach ihrem Anteil an der Weltpopulation bzw. ihrer Verbreitung in Europa zu klassifizieren. Ausführliche Konsultationen ergaben, dass man für *BiE2* die Kriterien stärken könne, indem man die Rote Liste-Kriterien der IUCN mit einbezieht, da sie als System für die Einschätzung des relativen Aussterberisikos von Arten weltweit anerkannt sind. Die kürzlich publizierten Richtlinien zur Anwendung dieser IUCN-Kriterien auf regionaler Ebene machten es relativ einfach, diese in die existierenden SPEC-Kriterien zu integrieren. Gemäß des System von *BiE1*, wurde jede Art einer von fünf Kategorien zugeordnet:

Kategorie	Europäische Art von globalem Natur-schutzbelang	Erhaltungs-status in Europa	Globale Population oder Verbreitungsgebiet konzentriert in Europa
SPEC 1	Ja	–	–
SPEC 2	Nein	Ungünstig	Ja
SPEC 3	Nein	Ungünstig	Nein
Non-SPEC[E]	Nein	Günstig	Ja
Non-SPEC	Nein	Günstig	Nein

Eine Art ist von globalem Naturschutzbelang, wenn sie gemäß den Rote Liste-Kriterien der IUCN auf globaler Ebene als „Threatened", „Near Threatened" oder „Data Deficient" eingestuft ist. Sie hat einen ungünstigen Erhaltungsstatus in Europa, wenn ihre europäische Population nach den IUCN-Kriterien auf regionaler Ebene als „Threatened" eingestuft ist, oder wenn ihre Population eines der folgenden Merkmale aufweist (analog zu *BiE1*): klein und nicht-marginal, mäßiger Rückgang, dezimiert nach früherem Rückgang oder stark lokalisiert. Eine Art wird als „konzentriert in Europa" eingestuft, wenn mehr als 50% der globalen Brut- oder Winterpopulation oder des Verbreitungsgebiets in Europa zu verzeichnen ist.

Abb.1. Prozentwerte der europäischen Vogelarten nach den Kategorien von *BiE1* (1994) und *BiE2* (2004).

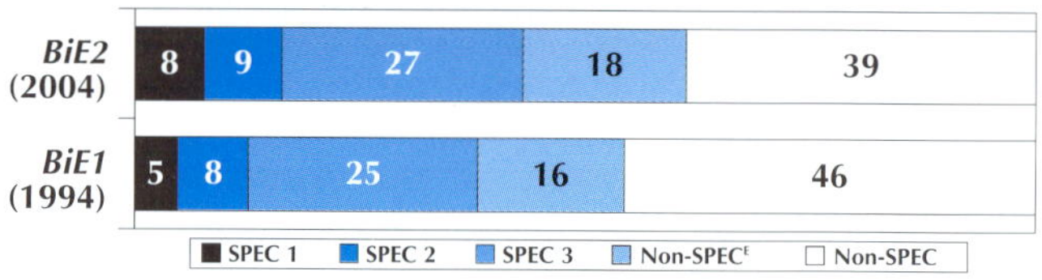

ERGEBNISSE

Von der 524 untersuchten Arten haben 226—oder 43% der europäischen Auvifauna—einen ungünstigen Erhaltungsstatus in Europa (Abb.1). Von diesen fallen 40 Arten (7.6%) unter die Kategorie SPEC 1, 45 (8.6%) unter SPEC 2 und 141 (26.9%) unter SPEC 3. All diese Prozentwerte übertreffen diejenigen von *BiE1*, wonach 195 Arten (38% der 511 untersuchten) als SPEC klassifiziert waren.

DISKUSSION

Die gestiegene Anzahl von SPEC 1-Arten geht auf die Neuklassifizierung (durch die angepassten Kriterien) von Arten der Kategorie „Near Threatened" auf globaler Ebene zurück, die zuvor als SPEC 2 oder SPEC 3 gelistet waren, aber doch klar von globalem Naturschutzbelang sind. Dagegen ist die angestiegene Zahl von SPEC 2- und SPEC 3-Arten wirklich alarmierend, da sie zeigen, dass sich der Status von viel mehr Arten (45) von „günstig" zu „ungünstig" verschlechtert hat, als umgekehrt (14).

Angesichts gewaltiger Bestandsrückgänge zwischen 1970 und 1990 sind die Populationen vieler SPECs bis heute stark gefährdet; dies gilt v.a. für Vögel der Agrarlandschaft. Jedoch haben sich auch einige wenige Arten erholt und sind nun in Europa mit einem günstigen Status gelistet. Die Erholung des Wanderfalken (*Falco peregrinus*) ist ein gutes Beispiel für den positiven Effekt gezielter Naturschutzmaßnahmen.

Auch viele SPEC 1-Arten nehmen in Europa aufgrund der effektiven Umsetzung von „Arten-Aktionsplänen" im letzten Jahrzehnt wieder zu. Es wird Zeit brauchen, bevor diese Arten neu klassifiziert werden können. Aber der bisherige Fortschritt zeigt, dass gemeinsames, gut geplantes und ausreichend finanziertes Handeln tatsächlich Bestandsrückgänge umkehren und Arten zu einem günstigeren Status zurückführen kann.

Von 129 Arten, die in *BiE1* aufgrund von Rückgängen in der Zeit 1970–1990 als SPECs gelistet waren, hat sich für 79 (61%) der negative Trend in den 90er Jahren fortgesetzt. Ihr Schicksal ist besonders besorgniserregend—zumal nun weitere 35 Arten hinzu gekommen sind, die zuvor einen günstigen Status aufgewiesen hatten. Zur letzteren Gruppe gehören viele ziehende Watvögel, Sperlingsvögel, einige Enten und Seevögel, sowie einige der häufigsten Arten Europas, wie Haussperling (*Passer domesticus*) und Star (*Sturnus vulgaris*).

Dies sind die ernüchternden Fakten—wogegen die meisten Regierungen versprochen haben, den Rückgang der Biodiversität bis 2010 zu bremsen, und die Europäischen Union diesen Rückgang völlig aufhalten will. Bei den meisten Tier- und Pflanzengruppen wird es sehr schwierig sein festzustellen, ob diese Ziele erreicht werden, aber Vögel sind eine Ausnahme. Am dringendsten wird eine maßvolle, langfristige Unterstützung für Monitoring benötigt, sowohl um existierende Programme fortzuführen, als auch um Strategien für andere Arten zu entwickeln und umzusetzen. Dies würde es den Regierungen ermöglichen, ihre Berichtspflichten zu erfüllen, und es erleichtern, jedes Jahrzehnt Statusberichte wie den vorliegenden herauszugeben (*BiE3* ist für 2014 geplant).

Die Zeit, die bleibt, um diese Ziele zu erreichen, ist kurz. Daher ist es unerlässlich, die Belange der Biodiversität voll in alle umweltrelevanten Sektoren der Politik zu integrieren. Europa profitiert schon heute von einer der besten Naturschutzgesetzgebungen der Welt. Die EG-Vogelschutzrichtlinie, die Berner Konvention und die Konvention zur Erhaltung der wandernden Tierarten (Bonner Konvention) waren zum Zeitpunkt ihrer Ratifizierung vor 25 Jahren bedeutende Meilensteine und haben schon viel erreicht. Dennoch zeigt *BiE2*, dass noch viele Herausforderungen bleiben und auch in den nächsten 25 Jahren die Notwendigkeit weiter wachsen wird, diese Instrumente zum maximalen Nutzen der Biodiversität anzuwenden.

SCHLUSSFOLGERUNGEN

Die zusammenfassende Botschaft von *BiE2* ist so eindeutig wie die von *BiE1*: Die europäischen Vögel bleiben bedroht durch weitreichende Umweltveränderungen, und viele Populationen sind sogar noch größeren Problemen ausgesetzt als vor einem Jahrzehnt. Da Vögel gute Umweltindikatoren sind, spricht der andauernde Rückgang vieler Arten eine klare Sprache über den Zustand der europäischen Biodiversität und der Gesundheit unserer Umwelt allgemein.

Angesichts der Größenordnung dieses Problems ist die schon in *BiE1* gestellte Forderung nach massivem und entschlossenem Handeln nun noch drängender. Es müssen sofort Taten folgen—nicht nur, um den stetigen Verlust an der einst reichen und vielfältigen europäischen Vogelwelt zu stoppen, sondern auch um ein ernsthaftes Engagement für den Stopp des Artenschwunds bis 2010 unter Beweis zu stellen!

RESUMEN

OBJETIVO

Birds in Europe (2004) *BiE2* es la segunda revisión del estado de conservación de todas las especies de aves silvestres de Europa. Como la de 1994, *Birds in Europe* (*BiE1*), identifica las especies prioritarias (Especies que requieren medidas de conservación en Europa; SPEC en sus siglas en inglés) de modo que puedan adoptarse las medidas necesarias para mejorar su estado de conservación.

ÁMBITO

Su ámbito geográfico es continental y abarca desde Groenlandia, al oeste, hasta los Urales, al este, y desde las Islas Svalbard, al norte, hasta las Islas Canarias, al sur. La creciente estabilidad política en los Balcanes y el Cáucaso permitió, por primera vez, obtener información de todos los países europeos.

COMPILACIÓN DE DATOS

Una red de coordinadores nacionales recopiló las aportaciones de expertos, organizaciones con programas de seguimiento y colaboradores regionales. Los datos proceden del trabajo de campo de miles de ornitólogos, muchos de ellos voluntarios.

Para cada especie se obtuvieron datos nacionales sobre el tamaño de su población reproductora (en 2000 o próxima a ese año) y su tendencia (en el periodo 1990–2000). Se obtuvieron datos similares sobre poblaciones invernantes, principalmente de aves acuáticas. En conjunto se compilaron unos 14.000 registros, muchos de mayor calidad que en *BiE1*. Estos nuevos datos de población se utilizaron, junto con los de tendencia y área de distribución en 1970–1990, para reevaluar el estado de conservación de cada especie en Europa.

EVALUACIÓN DEL ESTADO DE CONSERVACIÓN

Para *BiE1* se desarrollaron criterios cuantitativos para identificar las SPEC según su estado a escala global y europea, y clasificarlas según la proporción en Europa de su población y área de distribución globales. Para *BiE2*, y tras un amplio proceso de consulta, los criterios existentes se fortalecieron al incorporar los de la Lista Roja de la UICN, aceptados para evaluar el riesgo de extinción relativo de las especies. Las normas de aplicación de dichos criterios de la UICN a escala regional, hicieron relativamente sencilla su integración en los ya existentes para identificar las SPEC.

Como en *BiE1*, cada especie fue asignada a una de estas cinco categorías:

Categoría	Especies europeas que requieren medidas de conservación a nivel global	Estado de conservación en Europa	Población o distribución global concentradas en Europa
SPEC 1	Sí	–	–
SPEC 2	No	Desfavorable	Sí
SPEC 3	No	Desfavorable	No
No SPEC^{E}	No	Favorable	Sí
No SPEC	No	Favorable	No

Una especie se considera que requiere medidas de conservación a nivel global si se clasifica como Amenazada, Casi Amenazada o con Datos Insuficientes, según los criterios de la UICN a escala mundial. Tiene un Estado de Conservación Desfavorable en Europa si su población europea se clasifica como Amenazada, según la aplicación regional de dichos criterios de la UICN, o si su población corresponde a alguna de las siguientes categorías (como en *BiE1*): pequeña y no marginal, en moderado declive, mermada tras declives previos, o muy localizada. Una especie está concentrada en Europa si se incluye aquí más del 50% de su población global, reproductora o invernante, o más del 50% de su área de distribución.

RESULTADOS

De las 524 especies evaluadas, 226 (43% de la avifauna europea) tienen un Estado de Conservación Desfavorable en Europa (Figura 1). De ellas, 40 (7,6%) se clasifican como SPEC 1, 45 (8,6%) como SPEC 2 y 141 (26,9%) como SPEC 3. Todos estos porcentajes superan los obtenidos para *BiE1*, cuando 195 especies (38% de las 511 evaluadas) fueron clasificadas como SPEC.

Figura 1. Porcentaje de especies de aves europeas en cada categoría en *BiE1* (1994) y en *BiE2* (2004).

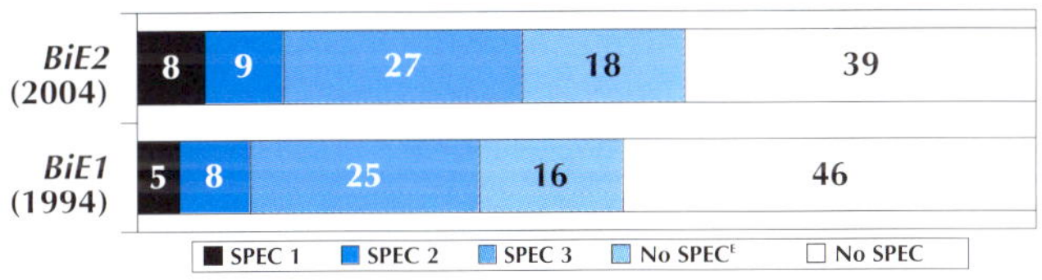

DISCUSIÓN

El mayor número de SPEC 1 refleja la reclasificación (según los criterios revisados) de las especies globalmente Casi Amenazadas, que fueron clasificadas previamente como SPEC 2 ó 3, pero que son de clara preocupación conservacionista global. Sin embargo, el mayor número de SPEC 2 y 3 resulta realmente alarmante porque significa que el estado de conservación europeo de muchas aves ha cambiado más de Favorable a Desfavorable (45 especies) que viceversa (sólo 14 especies).

Dada la magnitud de los declives registrados durante 1970–1990, especialmente en aves de medios agrícolas, muchas SPEC mantienen poblaciones muy reducidas. Sin embargo, unas pocas especies se han recuperado y han sido reclasificadas con un Estado de Conservación Favorable en Europa. El caso del Halcón Peregrino *Falco peregrinus* ilustra los beneficios de acciones de conservación bien dirigidas.

Muchas especies SPEC 1 también están en aumento en Europa, debido a la aplicación efectiva de sus planes de acción durante la década pasada. Pasará tiempo antes de que puedan ser reclasificadas, pero los progresos obtenidos indican que una cooperación bien planificada y financiada puede invertir la disminución de esas especies y devolverlas a estados de conservación más favorables.

De las 129 especies clasificadas como SPEC en *BiE1* por su disminución durante 1970–1990, 79 (61%) continuaron en regresión durante la década de 1990. Su situación es especialmente preocupante y a ellas se han añadido otras 35 especies consideradas antes con un Estado de Conservación Favorable en Europa, que incluyen muchos limícolas y paseriformes migradores, algunos patos y aves marinas, y algunas especies europeas muy abundantes, como el Gorrión Común *Passer domesticus* y el Estornino Pinto *Sturnus vulgaris*.

Estas situaciones resultan determinantes ahora que muchos gobiernos se han comprometido a reducir la tasa de pérdida de biodiversidad para 2010, y la Unión Europea incluso a detener completamente esa pérdida. Para muchos taxones puede ser muy difícil evaluar si estos objetivos se alcanzan, pero las aves constituyen una excepción. El principal requisito es un modesto apoyo a largo plazo para su seguimiento, tanto para los programas ya existentes como para desarrollar nuevas estrategias para otras especies. Esto permitiría a los gobiernos elaborar sus obligados informes, y facilitará revisiones como ésta cada década (*BiE3* está prevista para 2014).

El plazo para alcanzar estos objetivos es corto, y por ello es vital que la preocupación por nuestra biodiversidad se integre plenamente en todas las políticas sectoriales que afecten al medio ambiente. Europa ya se beneficia de alguna de la mejor legislación conservacionista del mundo. La Directiva Aves, el Convenio de Berna y el Convenio sobre Especies Migradoras constituyeron hitos importantes cuando fueron aprobados hace 25 años, y ya han alcanzado logros muy destacados. Como demuestra *BiE2* quedan todavía muchos desafíos, y en los próximos 25 años deberá incrementarse la aplicación más efectiva de esas herramientas para conseguir el máximo beneficio para la biodiversidad.

CONCLUSIONES

El mensaje final de *BiE2* es tan claro como lo fue el de *BiE1*. En Europa las aves siguen amenazadas por cambios ambientales generalizados, y muchas poblaciones afrontan ahora problemas más serios que hace una década. Dado que las aves son buenos indicadores ambientales, la regresión actual de tantas especies lanza claras señales de alarma sobre la situación de la biodiversidad europea y la salud del medio ambiente.

Dada la magnitud del problema, la urgente respuesta solicitada en *BiE1* resulta ahora aún más apremiante. Deben adoptarse medidas inmediatas no sólo para frenar esa regresión continua de la, una vez, rica y abundante avifauna europea, sino también como muestra del serio compromiso para detener la pérdida de biodiversidad en 2010.

RÉSUMÉ

OBJECTIF

Birds in Europe (2004) *BiE2*—est la seconde révision du statut de conservation de l'ensemble des espèces d'oiseaux indigènes du continent européen. De même que dans la première édition de 1994, *Birds in Europe* (*BiE1*), on y identifie les espèces prioritaires, celles qui méritent une attention spéciale en Europe—*Species of European Conservation Concern*—SPECs, afin qu'elles puissent bénéficier d'actions de conservation pour améliorer leur statut.

CONTEXTE GÉOGRAPHIQUE

La portée géographique est l'ensemble du continent européen, s'étendant depuis le Groenland à l'ouest, jusqu'à l'Oural à l'est, et du Spitzberg au nord, jusqu'au îles Canaries au sud. L'amélioration de la stabilité politique dans les Balkans et le Caucase a permis la collecte de données dans tous les pays européens pour la première fois.

COLLECTE DES DONNÉES

Les données ont été collectées par le réseau des coordinateurs nationaux, appuyé par la contribution d'experts, des organisations impliquées dans les suivis ornithologiques et des contributeurs spécialisés dans le statut des oiseaux au niveau continental. Les données proviennent de travaux de terrain menés par des milliers d'ornithologues dont bon nombre sont bénévoles.

Pour chaque espèce, les données nationales ont été réunies sur la taille des populations nicheuses (autour de l'année 2000), ainsi que sur les tendances (période s'étendant de 1990 à 2000). Quand disponibles, les données équivalentes sur les populations présentes en hiver ont également été rassemblées, essentiellement pour ce qui concerne les oiseaux d'eau. Au total, près de 14.000 données sur les populations et tendances ont été reçues. Beaucoup des informations collectées sont plus précises que celles présentées dans l'édition précédente (*BiE1*). Associées aux données déjà existantes de la période comprise entre 1970 et 1990, ces nouvelles informations ont été utilisées pour ré-évaluer le statut de conservation de chaque espèce présente en Europe.

EVALUATION DU STATUT

Dans la première édition, un ensemble de critères quantitatifs a été développé pour identifier les espèces prioritaires en Europe—SPECs selon leur statut au niveau mondial et européen. Cela a permis de les classer selon la proportion de leur population ou leur aire de distribution comprise en Europe. Pour la seconde édition, à la suite d'une large consultation, il a été décidé de renforcer les critères existants en y introduisant ceux des listes rouges UICN. Ces derniers constituent la référence pour évaluer les risques d'extinction des espèces. La publication récente de la méthode de détermination des critères des listes rouges UICN au niveau continental a permis leur intégration relativement simple dans l'évaluation des espèces prioritaires en Europe.

Par l'application de la méthode de classement suivie dans la première version—*BiE1*; chaque espèce se retrouve dans une des cinq catégories suivantes :

Categorie	Espèce européenne menacée au niveau mondial	Statut de Conservation en Europe	Population mondiale ou aire de distribution concentrée en Europe
SPEC 1	Oui	–	–
SPEC 2	Non	Défavorable	Oui
SPEC 3	Non	Défavorable	Non
Non-SPEC^E	Non	Favorable	Oui
Non-SPEC	Non	Favorable	Non

Une espèce est considérée comme mondialement menacée si elle est classée comme Menacée, Quasi Menacée ou Insuffisamment Connue selon les critères mondiaux de la Liste Rouge de l'UICN. Elle a un statut de conservation défavorable en Europe si sa population européenne est classée comme Menacée selon l'application continentale des critères UICN, ou si sa population est comprise dans l'une des catégories suivantes (comme dans *BiE1*) : de petite taille et non-marginale, en déclin modéré, de taille très modeste suite à des déclins antérieurs ou si cette population est très confinée. Une espèce est concentrée en Europe si plus de 50 % des ses populations nicheuses ou hivernantes, ou 50 % des son aire de distribution sont incluses dans le continent européen.

RÉSULTATS

Des 524 espèces évaluées, 226—ou 43 % de l'avifaune européenne—présentent un statut de conservation défavorable en Europe (Figure 1). Parmi ces dernières, 40 espèces (7,6%) sont classées en SPEC 1, 45 (8,6%) en SPEC 2 et 141 (26,9%) en SPEC 3. Tous ces pourcentages dépassent ceux de *BiE1*, il y avait alors 195 espèces (38% des 511 évaluées) qui étaient classées en SPECs.

Figure 1. Pourcentages des espèces d'oiseaux d'Europe dans chacune des catégories de *BiE1* (1994) et de *BiE2* (2004).

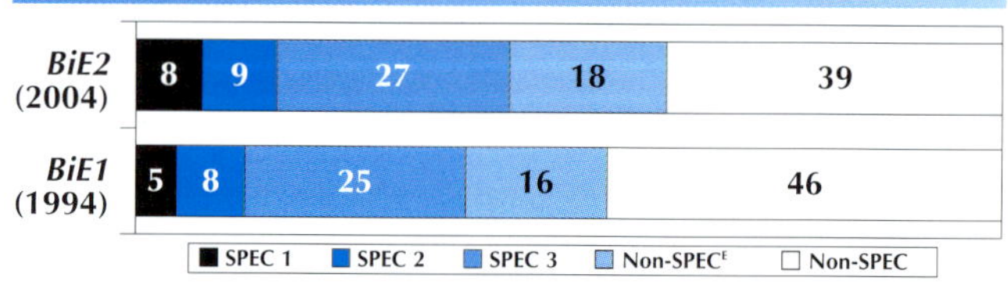

DISCUSSION

L'augmentation du nombre d'espèces classées en SPEC 1 est surtout le reflet du nouveau classement (selon les critères révisés) des espèces Quasi Menacées au niveau mondial, classées précédemment comme SPEC 2 ou 3, bien qu'en situation mondialement préoccupante.

L'augmentation du nombre d'espèces classées en SPEC 2 et 3 est réellement préoccupante. Cela signifie que le statut de conservation européen des oiseaux qui sont passés de favorable à défavorable (45 espèces) est beaucoup plus important que l'inverse (14 espèces).

Compte tenu de l'ampleur des déclins observés durant la période allant de 1970 à 1990, particulièrement ceux qui affectent les espèces présentes dans les espaces agricoles, les populations de beaucoup d'espèces menacées sont de petite taille. Il y en a cependant quelques-unes dont la situation s'est améliorée et qui ont été reconsidérées comme ayant un statut de conservation favorable en Europe. L'amélioration de la situation du Faucon pèlerin *Falco peregrinus* constitue un bon exemple, illustrant les bénéfices engendrés par les actions de conservation spécifiques.

Beaucoup d'espèces classées en SPEC 1 augmentent également en Europe, en raison de la mise en application de plans d'action « espèces » durant la dernière décennie. Cela prendra encore du temps avant qu'elles puissent être reclassifiées, mais une amélioration constatée actuellement indique que des actions conjointes bien planifiées et bénéficiant de financements adéquats permettent d'inverser les déclins et de restaurer les effectifs des espèces, les ramenant vers un statut de conservation plus favorable.

Des 129 espèces qui étaient classées comme SPEC dans *BiE1* en raison de déclins observés entre 1970 et 1990, 79 (61%) ont poursuivi leur déclin pendant les années 1990. Leur situation est particulièrement préoccupante—d'autant qu'elles ont été rejointes par 35 espèces qui étaient considérées comme bénéficiant d'un statut de conservation favorable en Europe. On compte parmi celles-ci de nombreux limicoles migrateurs et des passereaux, quelques canards et des oiseaux marins, des espèces parmi les plus communes d'Europe, telles que le Moineau domestique *Passer domesticus* et l'Etourneau sansonnet *Sturnus vulgaris*.

Ces exemples incitent à la réflexion à l'heure où la plupart des gouvernements se sont engagés à réduire la dégradation de la biodiversité d'ici à 2010. L'Union Européenne s'est, elle, engagée à arrêter complètement cette perte de biodiversité. Pour la plupart des taxons, il sera difficile de vérifier dans quelle mesure ces objectifs seront atteints, sauf pour les oiseaux qui constituent une exception. Il s'en dégage qu'un soutien modeste aux suivis à long terme est nécessaire, permettant la poursuite des programmes existants et le développement des stratégies pour les espèces qui n'en bénéficient pas encore. Ceci donnerait la possibilité aux gouvernements de tenir leurs obligations et de faciliter la révision des statuts tous les dix ans, telle que celle qui est réalisée dans l'ouvrage présent, la parution de *BiE3* étant prévue pour 2014.

Le temps qu'il reste pour atteindre ces objectifs est court, il est donc vital d'intégrer la préservation de la biodiversité dans les politiques sectorielles concernant l'environnement. L'Europe bénéficie déjà d'une des meilleures législations au monde pour ce qui concerne le maintien de la biodiversité. On peut considérer l'adoption il y a 25 ans de la Directive Oiseaux, de la Convention de Berne et de la Convention sur les Espèces Migratrices comme des faits saillants, qui ont déjà débouché sur de nombreuses applications pratiques. Mais, comme *BiE2* le démontre, il reste à surmonter de nombreux défis. L'application de ces outils au bénéfice de la biodiversité ne s'améliorera que dans les 25 années à venir.

■ CONCLUSIONS

Le message général qui ressort de *BiE2* va dans le sens de celui de *BiE1*. Les oiseaux d'Europe subissent plus que jamais des menaces dues aux changements environnementaux de grande échelle. De nombreuses populations se retrouvent maintenant dans une situation moins confortable qu'il y a une décennie. Dans la mesure où les oiseaux sont de bons indicateurs environnementaux, le déclin actuel d'autant d'espèces en dit long sur l'état de la biodiversité européenne et la santé de l'environnement en général.

Compte tenu de l'étendue du problème, le message délivré par *BiE1*, appelant à une réponse directe et urgente, est à considérer maintenant comme impérieux. Une action doit être entreprise immédiatement—pas seulement pour arrêter la perte continue de l'avifaune d'Europe qui est réputée riche et abondante, mais aussi afin de s'engager à enrayer la perte de biodiversité d'ici à 2010.

■ RIASSUNTO

■ SCOPO DEL LAVORO

Birds in Europe (2004) *BiE2* è la seconda analisi dello stato di conservazione di tutte le specie di uccelli selvatici in Europa. Come il suo predecessore del 1994 *Birds in Europe* (*BiE1*), individua le specie prioritarie (specie d'interesse conservazionistico in Europa *Species of European Conservation Concern* o SPECs) in modo da poter attuare azioni di conservazione volte a migliorarne lo status.

■ COPERTURA GEOGRAFICA

Il lavoro copre l'intero continente europeo dalla Groenlandia ad ovest fino agli Urali ad est; dalle Svalbard a nord fino alle Isole Canarie a sud. L'attuale stabilità politica nei Balcani e nel Caucaso ha permesso per la prima volta la raccolta dei dati da tutti i paesi Europei.

■ RACCOLTA DEI DATI

I dati sono stati raccolti attraverso una rete di coordinatori nazionali che hanno ottenuto informazioni da esperti, organizzazioni che curano il monitoraggio delle specie, e collaboratori regionali. I dati provengono dal lavoro sul campo di migliaia di ornitologi compresi innumerevoli volontari.

Per ciascuna specie sono stati raccolti i dati nazionali sulle dimensioni delle popolazioni nidificanti (dati in generale per l'anno 2000) e sulle tendenze (per il periodo 1990—2000). Quando disponibili, e ciò è accaduto principalmente per gli uccelli acquatici, sono stati raccolti anche gli analoghi dati per le popolazioni svernanti. In totale sono stati raccolti 14.000 dati di popolazione/tendenza, molti dei quali di qualità superiore a quelli del 1994 (*BiE1*). Assieme ai dati esistenti per il periodo 1970–1990 questi nuovi dati sono stati utilizzati per definire nuovamente lo stato di conservazione di ciascuna specie in Europa.

■ VALUTAZIONE DELLO STATUS

In *BiE1* vennero definiti una serie di criteri quantitativi per identificare le SPEC sulla base del loro status globale ed europeo e per classificarle secondo la proporzione della loro popolazione globale o del loro areale in Europa. In *BiE2* un ampio processo consultivo ha evidenziato la necessità di consolidare i criteri incorporando quelli utilizzati per la Lista Rossa IUCN, che rappresenta un sistema universalmente accettato per definire il rischio di estinzione relativo di ciascuna specie. La recente pubblicazione delle linee guida per l'applicazione dei criteri IUCN a livello regionale ha reso il lavoro di integrazione relativamente semplice.

In base al sistema utilizzato in *BiE1*, ciascuna specie è stata assegnata ad una delle cinque categorie:

Categoria	Specie europea di interesse conservazionistico a livello globale	Status di conservazione in Europa	Popolazione o areale concentrati in Europa
SPEC 1	Si	–	–
SPEC 2	No	Sfavorevole	Si
SPEC 3	No	Sfavorevole	No
Non-SPEC^E	No	Favorevole	Si
Non-SPEC	No	Favorevole	No

Una specie è di interesse conservazionistico a livello globale se è classificata come "Minacciata", "Quasi Minacciata" o "Insufficientemente Conosciuta" sulla base dei Criteri IUCN per la Lista Rossa mondiale. Il suo stato di conservazione è considerato sfavorevole in Europa se la popolazione europea è classificata come "Minacciata" sulla base dell'applicazione regionale dei criteri IUCN o se la sua popolazione è (come in *BiE1*) piccola e non marginale, in declino numerico moderato, ridotta a seguito di decremento numerico occorso in passato, o fortemente localizzata. Una specie è considerata concentrata in Europa se più del 50% della sua popolazione nidificante o svernante o del suo areale è in Europa.

■ RISULTATI

Delle 524 specie prese in considerazione, 226—43% dell'avifauna europea—hanno uno status di conservazione sfavorevole in Europa (Figura 1). 40 specie (7.6%) sono classificate SPEC 1, 45 (8.6%) sono SPEC 2, e 141 (26.9%) sono SPEC 3. Tutte queste percentuali sono superiori a quelle in *BiE1* dove 195 (38% delle 511 specie analizzate) erano classificate come SPEC 1–3.

Figura 1. Percentuali delle specie europee in ciascna categoria in *BiE1* (1994) e in *BiE2* (2004).

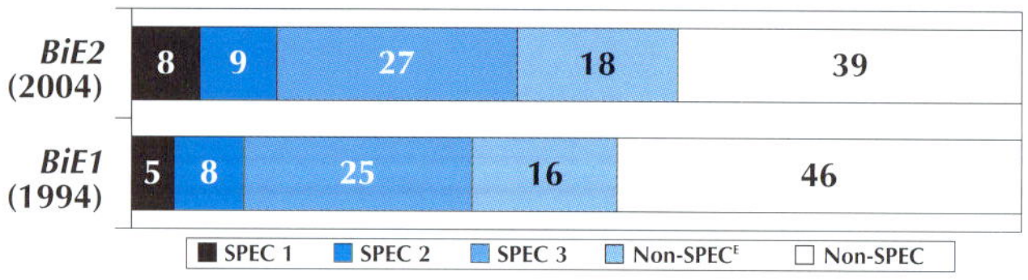

■ DISCUSSIONE

L'aumento delle specie SPEC 1 rispecchia la ri-classificazione (sulla base dei nuovi criteri) delle specie "Quasi Minacciate"a livello globale che erano precedentemente classificate SPEC 2 o 3, ma che sono di chiaro interesse conservazionistico globale. Comunque l'aumentato numero di SPEC 2 e 3 è veramente allarmante in quanto per un maggior numero di specie di uccelli (45) lo stato di conservazione in Europa è passato da "Favorevole" a "Sfavorevole" mentre per sole 14 specie si è verificato un cambiamento in direzione opposta.

Le popolazioni di molte specie, in particolare quelle legate agli ambienti agricoli, a seguito del declino occorso nel periodo 1970–1990, sono ancora ampiamente al di sotto delle dimensioni che avevano nel passato. Tuttavia, alcune specie hanno recuperato ed il loro status è oggi considerato "Favorevole" in Europa. Il recupero del Falco pellegrino *Falco peregrinus* è un buon esempio dei risultati di azioni mirate di conservazione.

Anche le popolazioni di numerose SPEC 1 stanno aumentando in Europa a seguito dell'efficace messa in pratica dei piani d'azione nel corso dell'ultimo decennio. Ci vorrà tempo prima che queste specie possano essere classificate in una categoria di rischio inferiore, ma i progressi finora registrati indicano che azioni di conservazione svolte in cooperazione, ben pianificate e sufficientemente finanziate, possono davvero arrestare il declino di molte specie facendole ritornare ad uno stato di conservazione favorevole.

Delle 129 specie classificate SPEC in *BiE1* a causa del loro declino numerico durante il periodo 1970–1990, 79 (61%) hanno continuato a diminuire durante gli anni novanta. La loro situazione è davvero preoccupante—e sono ora in compagnia di altre 35 specie che erano considerate precedentemente con stato di conservazione "Favorevole" in Europa. Tra queste vi sono molte specie di limicoli migratori e passeriformi, numerosi anatidi e uccelli marini e alcune delle specie europee piú note come la Passera oltremontana *Passer domesticus* e lo Storno *Sturnus vulgaris*.

Si tratta di segnali molto preoccupanti soprattutto in considerazione dell'impegno di molti governi a ridurre il tasso di perdita di biodiversità entro il 2010 e dell'impegno dell'Unione Europea ad arrestarne completamente la perdita. Valutare se questi obiettivi saranno raggiunti è molto difficile per molti taxa, ma gli uccelli sono un'eccezione. C'è bisogno di un modesto ma continuo sostegno economico per il loro monitoraggio, sia per mantenere i programmi di monitoraggio esistenti, sia per sviluppare e mettere in pratica nuovi programmi di monitoraggio per le specie attualmente scoperte. Ciò permetterebbe ai governi di rispettare i propri obblighi di fornire periodicamente rapporti sulla situazione della biodiversità (all'Unione Europa e/o ad altre convenzioni internazionali) e renderebbe piú facile preparare future revisioni come questa (*BiE3* è previsto per il 2014).

Il tempo rimasto per raggiungere questi obiettivi è poco, perciò è fondamentale che la conservazione della biodiversità sia completamente integrata in tutte le politiche settoriali che hanno un impatto sull'ambiente. In Europa esistono alcune delle migliori leggi al mondo per la conservazione della biodiversità. La Direttiva Uccelli, la Convenzione di Berna e la Convenzione di Bonn rappresentarono pietre miliari quando furono adottate 25 anni fa ed hanno già ottenuto enormi risultati. Ciononostante, come BiE2 dimostra, molte sfide rimangono aperte e la necessità di utilizzare questi strumenti per il massimo effetto sulla biodiversità potrà solo aumentare nei prossimi 25 anni.

■ CONCLUSIONI

Il messaggio generale di *BiE2* è chiaro come quello di *BiE1*. Gli uccelli in Europa continuano ad essere minacciati da diffuse alterazioni ambientali e molte popolazioni sono oggi in condizioni peggiori rispetto a dieci anni fa. Dal momento che gli uccelli sono buoni indicatori ambientali, il continuo declino di un numero così elevato di specie fotografa in modo esplicito lo stato della biodiversità europea e della salute dell'ambiente in generale.

Data la dimensione del fenomeno, la necessità di una reazione urgente e in grande scala già evidenziata in *BiE1* è oggi ancora piú pressante. Debbono essere intraprese immediatamente azioni volte non solo a fermare la continua perdita della avifauna europea—un tempo ricca ed abbondante—, ma anche a dimostrare un serio impegno ad arrestare la perdita di biodiversità entro il 2010.

■ SAMENVATTING

■ DOEL

Birds in Europe (2004) *BiE2* is het tweede overzicht waarin voor alle in het wild levende vogels in Europa de beschermingsstatus is bepaald. Net als in het in 1994 verschenen eerste overzicht (*BiE1*) zijn die vogelsoorten opgenomen waarvoor speciale aandacht nodig is teneinde de beschermingsstatus van deze vogelsoorten te verbeteren (Species of European Conservation Concern, of SPECs; "Europese aandachtssoorten").

■ GEBIED

Het bestreken gebied betreft geheel Europa: van Groenland in het westen tot de Oeral in het oosten, en van Spitsbergen in het noorden tot de Canarische Eilanden in het zuiden. Ook zijn gegevens van de in begin jaren negentig voor gegevensverzameling politiek nog te instabiele Balkan en Kaukasus opgenomen.

■ GEGEVENSVERZAMELING

De gegevens werden verzameld via een netwerk van nationale coördinatoren, die op hun beurt deskundigen, organisaties gericht op gegevensverzameling en regionale medewerkers raadpleegden. Op deze manier droegen duizenden professionele en amateur-ornithologen aan de gegevensverzameling bij.

Voor iedere soort werden gegevens verzameld over omvang van de landelijke broedpopulatie (in of rond het jaar 2000) en landelijke trend in de aantallen (1990–2000). Waar mogelijk werden ook gegevens over de winterpopulaties bijeengebracht, met name voor watervogels. In totaal werden 14,000 verschillende opgaven over populatieomvang en aantalstrends ontvangen, die vaak van betere kwaliteit zijn dan die beschikbaar waren voor *BiE1*. Door vergelijking van de al voor handen zijnde gegevens uit 1970–1990 en de nieuwe gegevens voor *BiE2* kon voor iedere soort de beschermingsstatus in Europa opnieuw worden beoordeeld.

■ BEPALING BESCHERMINGSSTATUS

Voor *BiE1* werden kwantitatieve criteria ontwikkeld om de beschermingsstatus van soorten te beoordelen aan de hand van hun wereldwijde en Europese voorkomen en trends. Hierbij is een tweedeling aangebracht tussen soorten die vooral in Europa voorkomen (meer dan helft van de broed- of winterpopulatie) en soorten waarvan het merendeel buiten Europa voorkomt. Voor *BiE2* is tevens rekening gehouden met de IUCN Rode Lijst Criteria, welke algemeen aanvaard zijn voor het inschatten van het relatieve uitsterfrisico van soorten.

Iedere soort is ingedeeld in een van de vijf categorieën uit *BiE1*:

Categorie	In Europa voorkomende vogelsoort die op wereldschaal bedreigd is	Beschermings-status in Europa	Wereldpopulatie of verspreidings-gebied geconcentreerd in Europa
SPEC 1	Ja	–	–
SPEC 2	Nee	Ongunstig	Ja
SPEC 3	Nee	Ongunstig	Nee
Non-SPEC^E	Nee	Gunstig	Ja
Non-SPEC	Nee	Gunstig	Nee

Een soort wordt als een wereldwijde aandachtssoort (Species of Global Conservation Concern) beschouwd als deze volgens de op wereldschaal geldende IUCN Rode Lijst Criteria in een van de volgende drie categorieën valt: bedreigd (Threatened), gevoelig (Near Threatened), of onvoldoende gegevens voorhanden (Data Deficient). Een soort heeft een ongunstige beschermingsstatus (Unfavourable Conservation Status) in Europa als de Europese populatie volgens de regionale IUCN Rode Lijst Criteria moet worden aangemerkt als bedreigd (Threatened), of wanneer de populatie voldoet aan een van de volgende criteria van *BiE1*: klein en niet-marginaal voorkomend, matig afnemend, populatie stabiel na eerdere sterke afname, sterk lokaal voorkomend.

■ RESULTATEN

Van de 524 in *BiE2* in behandelde Europese vogelsoorten hebben 226 (43%) een ongunstige beschermingsstatus in Europa (Figuur 1). Hiervan konden er 40 (7.6%) worden aangemerkt als SPEC 1, 45 (8.6%) als SPEC 2 en 141 (26.9%) als SPEC 3. Alle percentages liggen hoger dan in *BiE1*, waarin 195 soorten (38% van de 511 behandelde soorten) werden aangemerkt als SPEC.

Figuur 1. Procentuele verdeling van de Europese vogelsoorten over de vijf onderscheiden categorieën (SPEC 1–3 ongunstig, non-SPEC gunstige beschermingsstatus).

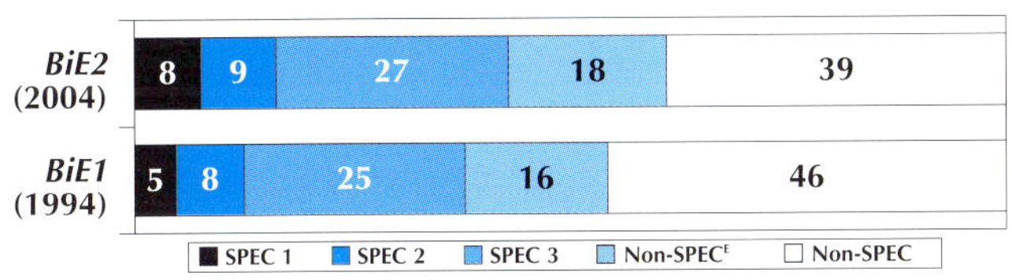

■ DISCUSSIE

De toename in de categorie SPEC 1 reflecteert de herindeling (volgens de herziene criteria) van de op wereldschaal gevoelige (Near Threatened) soorten die voorheen als SPEC 2 of 3 waren opgenomen, maar die nu als aandachtssoorten op wereldschaal (SPEC 1) aangemerkt worden. Echter, het toegenomen aantal SPEC 2 en SPEC 3 is alarmerend: van maar liefst 45 vogelsoorten is de beschermingsstatus veranderd van gunstig naar ongunstig, terwijl voor slechts 14 soorten het omgekeerde het geval is (status gewijzigd van ongunstig naar gunstig).

Wanneer ook de omvang van de afnamen in de periode 1970–1990 in ogenschouw wordt genomen, moet worden geconstateerd dat van veel SPECs de huidige populaties maar een fractie zijn van weleer; dit geldt vooral voor vogels van het agrarische landschap. Slechts enkele soorten konden zich herstellen en worden nu ingedeeld bij de soorten met een gunstige beschermingsstatus in Europa. Het herstel van de Slechtvalk (*Falco peregrinus*) is hiervan een goed voorbeeld, en geeft het succes dat gerichte beschermingsacties kunnen hebben goed weer.

Veel SPEC 1-soorten nemen in Europa toe dankzij effectieve uitvoering van soortbeschermingsplannen (Species Action Plans; SAPs) in de afgelopen 10 jaar. Het duurt nog wel even voordat de beschermingsstatus van deze soorten opgewaardeerd zal kunnen worden, maar de eerste resultaten tonen aan dat goed geplande gezamenlijke acties met voldoende financiële middelen wel degelijk de achteruitgang in aantallen een halt kunnen toeroepen of populaties weer in omvang kunnen laten toenemen.

Van de 129 soorten die in *BiE1* vanwege aantalsachteruitgang in de periode 1970–1990 werden aangemerkt als SPECs, bleken er 79 (61%) in de periode 1991–2000 verder in aantallen achteruit gegaan te zijn. Dit is verontrustend—vooral ook omdat nog eens 35 soorten die voorheen een gunstige beschermingsstatus in Europa hadden, zich nu bij deze groep hebben gevoegd. Het gaat daarbij om een flink aantal trekkende steltlopers en zangvogels, om verschillende soorten eenden en zeevogels, en om voorheen in Europa algemene soorten als Huismus en Spreeuw.

Dit zijn ontnuchterende feiten. En dat in een tijd waarin veel overheden hebben toegezegd het verlies aan biodiversiteit uiterlijk in 2010 te zullen hebben gereduceerd, terwijl de Europese Unie zelfs op zich heeft genomen deze afname dan geheel gestopt te hebben. Voor de meeste planten- en diergroepen zal het bijzonder moeilijk zijn vast te stellen of deze doelen zijn gehaald, maar vogels vormen hierop een uitzondering. De belangrijkste voorwaarde hierbij is een relatief bescheiden langetermijn-bijdrage aan monitoringsinspanningen, zowel aan bestaande programma's als aan het opzetten en uitvoeren van strategieën voor andere soorten. Een dergelijke bijdrage zal overheden eveneens helpen hun rapportageverplichtingen na te komen, en zou tienjaarlijkse statusherzieningen zoals de onderhavige mede mogelijk maken (met een *BiE3* beschikbaar in 2014).

De tijd die nog rest om deze biodiversiteitsdoelen te halen is kort. Het is dus van cruciaal belang dat behoud van biodiversiteit volledig wordt geïntegreerd in alle beleidssectoren die een negatieve invloed op natuur en milieu hebben. Europa profiteert al van de beste wetgeving voor het behoud van biodiversiteit die er wereldwijd beschikbaar is. De Vogelrichtlijn, de Bern Conventie en de Conventie voor het behoud van trekkende diersoorten waren mijlpalen toen zij 25 jaar terug werden aangenomen, en met deze wetgeving in de hand is al veel bereikt. Maar *BiE2* toont aan dat er desondanks nog veel uitdagingen overblijven, en dat de noodzaak om deze instrumenten maximaal voor het behoud van biodiversiteit in te zetten de komende 25 jaar alleen maar zal toenemen.

■ CONCLUSIES

De boodschap in *BiE2* is net zo duidelijk als die in *BiE1*. Vogels in Europa blijven bedreigd door grootschalige veranderingen in landschap en milieu, en veel populaties zijn er slechter aan toe dan tien jaar geleden. Omdat vogels goede milieu-indicatoren zijn, is de voortgaande afname bij zo veel vogelsoorten dan ook een duidelijk signaal over de staat waarin de Europese biodiversiteit en het Europese milieu verkeren.

De omvang van het probleem in ogenschouw nemende kan worden gesteld dat de grootschalige respons waartoe *BiE1* al opriep alleen nog maar urgenter is geworden. Er moet onmiddellijk actie ondernomen worden—niet alleen om het voortgaande verlies in de eens zo rijke Europese vogelwereld te stoppen, maar ook om te tonen dat de gedane toezeggingen om in 2010 het biodiversiteitsverlies gestopt te hebben, serieus zijn.

STRESZCZENIE

CEL
"*Birds in Europe*" (2004) *BiE2*. Druga edycja książki zawiera uaktualnione dane dotyczące statusu ochronnego wszystkich dziko żyjących ptaków w europie. Podobnie jak w poprzednim wydaniu z 1994 (*BiE1*) roku wskazuje gatunki priorytetowe (Species of European Conservation Concern, or SPECs), w przypadku których podjęcie zabiegów ochronnych przyczyni się do poprawy ich statusu.

ZASIĘG
Zasięg geograficzny obejmuje teren od Grenlandii na wschodzie, gór Ural na zachodzie, wyspę Svalbard na północy po Wyspy Kanaryjskie na południu. Wzrost stabilności politycznej na Bałkanach i na Kaukazie pozwolił na zebranie po raz pierwszy danych we wszystkich europejskich krajach tych regionów.

ZBIERANIE DANYCH
Dane były zbierane poprzez sieć koordynatorów krajowych, którzy angażowali we współpracę i pozyskiwali dane od ekspertów, organizacji zajmujących się monitoringiem oraz regionalnych współpracowników. Dane pochodzą od tysięcy ornitologów zawodowych i amatorów zbierających dane w terenie.

Dla każdego gatunku w poszczególnych krajach zebrano dane o wielkości populacji lęgowej (około 2000 roku) i trendów populacyjnych (na podstawie danych z lat 1990–2000). Tam gdzie było to możliwe zbierano dane dotyczące populacji ptaków zimujących, głównie ptaków wodnych. W sumie zebrano około 14000 rekordów, zawierających wiele lepszej jakości danych niż te, którymi posłużono się w poprzednim wydaniu. Nowe dane, razem z aktualnymi trendami i danymi zebranymi w latach 1970–1990 zostały użyte do opracowania statusu ochronnego wszystkich europejskich ptaków.

KRYTERIA
Do pierwszego wydania "*Birds in Europe*" zestaw kryteriów ilościowych został wprowadzony do identyfikacji SPECs w porównaniu do globalnego i europejskiego statusu i do klasyfikacji ich w oparciu o udział populacji europejskiej w całej populacji światowej.

Do drugiej edycji, po licznych konsultacjach stwierdzono, że obowiązujące wcześniej kryteria mogą być wzmocnione przez włączenie kryteriów obowiązujących przy zbieraniu przez IUCN danych do Czerwonej Księgi, które są powszechnie akceptowane do określania stopnia zagrożenia gatunku wyginięciem. Ostatnie publikacje z wytycznymi dotyczącymi kryteriów do Czerwonej Księgi IUCN na poziomie regionalnym mogą stosunkowo łatwo połączone z istniejącymi kryteriami SPEC

Klasyfikacaja zastosowana w książce "*Birds in Europe*":

Kategoria	Gatunek zagrożony globalnie	Status ochronny w Europie	Gatunek koncentruje się w Europie
SPEC 1	Tak	–	–
SPEC 2	Nie	Niekorzystny	Tak
SPEC 3	Nie	Niekorzystny	Nie
Non-SPEC[E]	Nie	Korzystny	Tak
Non-SPEC	Nie	Korzystny	Nie

Gatunek zagrożony globalnie to taki, który według kryteriów Czerwonej Listy IUCN został zakwalifikowany jako: zagrożony, bliski zagrożenia lub o którym brak wystarczających danych na poziomie globalnym. Ma on niekorzystny status ochronny w Europie jeśli jego europejska populacja została uznana jako zagrożona na poziomie regionu zgodnie z kryteriami Czerwonej Listy IUCN lub gdy jego populacja mieści się w następujących kryteriach (na podstawie BiE1): mała i niemarginalna, powoli zmniejszająca liczebność, wyginęła jako wynik wcześniejszego zmniejszania się lub jest izolowana. Gatunek koncentruje się w Europie jeśli ponad 50% jego globalnej lęgowej lub zimującej populacji lub zasięgu znajduje się w Europie.

Rysunek 1. Procentowy udział europejskiej populacji ptaków w każdej z kategorii w *BiE1* (1994) i w *BiE2* (2004).

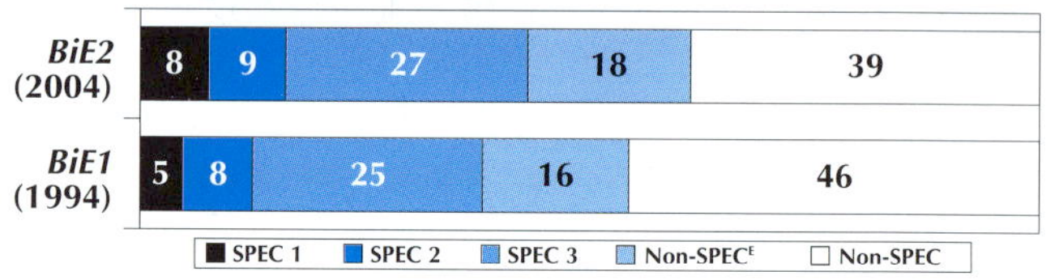

WYNIKI
Z 524 analizowanych gatunków 226 (43%) ma niekorzystny status ochronny w Europie (Rysunek 1). 40 z nich (7.6%) zostało sklasyfikowanych jako SPEC 1, 45 gatunków (8.6%) jako SPEC 2, a 141 gatunki (26.9%) jako SPEC 3. Udziały procentowe poszczególnych kategorii są wyższe od tych z pierwszego wydania książki gdzie 195 gatunków (38%) z 511 analizowanych zostało zakwalifikowanych do którejś z kategorii SPEC.

DYSKUSJA
Wzrost liczby gatunków zakwalifikowanych jako SPEC 1 wynika ze zmiany kryteriów klasyfikacji gatunków bliskich zagroenia w skali globalnej, które poprzednio znalazły się w kategoriach SPEC 2 i 3. Jednak wzrost liczby gatunków w kategoriach SPEC 2 i 3 jest alarmujący ponieważ oznacza, że status ochronny większej liczby gatunków (45) zmienił się z korzystnego na niekorzystny niż na odwrót (zmiana statusu z niekorzystnego na korzystny dokonała się w przypadku 14 gatunków).

Po ogromnym spadku w latach 1970–1990, zwłaszcza dotyczącym ptaków krajobrazu rolnego, wielkości populacji gatunków z kategorii SPEC uległy znacznemu uszczupleniu. Jednak kilka gatunków zostało przeklasyfikowanych i ich status ochronny jest korzystny. Wzrost liczebności sokoła wędrownego *Falco peregrinus* jest dobrym przykładem pokazującym korzyści płynące z działań ochronnych.

Wiele gatunków z kategorii SPEC 1 również wykazuje wzrost liczebności, dzięki efektywnemu wprowadzeniu planów ochrony w ciągu ostatniej dekady. Przeklasyfikowanie tych gatunków zajmie dużo czasu, jednak pokazuje, że odpowiednio przygotowana, przeprowadzona i finansowana akcja może odwrócić trend spadkowy i przywrócić gatunki do bardziej korzystnego statusu.

Ze 129 gatunków, które znalazły się w kategoriach SPEC w pierwszym wydaniu książki, wykazujących spadek liczebności w latach 1970–1990, 79 gatunków (61%) nadal wykazuje tendencję spadkową w latach dziewięćdziesiątych. Ich sytuacja jest niepokojąca. Do tych gatunków dołączyły 35 gatunki, które dawniej posiadały korzystny status ochronny. Do gatunków tych należy wiele migrujących siewek, ptaków wróblowatych kilka gatunków blaszkodziobych, ptaków morskich oraz jedne z najpospolitszych gatunków takich jak: wróbel *Paser doemsticus* i Szpak *Sturnus vulgaris*

Większość rządów zobowiązała się do zmniejszenia tempa ubożenia bioróżnorodności do roku 2010, a Unia Europejska planuje całkowite powstrzymanie tego niekorzystnego trendu. Dla większości taksonów ocena czy udało się osiągnąć te plany będzie bardzo trudna. Jednak ptaki są w tym przypadku wyjątkiem. Głównym wymogiem jest tu wspieranie długoterminowego monitoringu, zarówno w celu podtrzymania już prowadzonych działań, jak i wprowadzenie strategii dla innych gatunków. Pozwoliłoby to rządom na wywiązanie się z przyjętych zobowiązań i przygotowanie oceny statusu bioróżnorodności co 10 lat (podobnej do planowanej w BiE3 na 2014 rok)

Pozostało bardzo niewiele czasu, aby spełnić te wymagania, dlatego też bardzo istotne jest, aby problemy ochrony bioróżnorodności włączyć we wszystkie sektory, których polityka ma wpływ na środowisku naturalne. Europa korzysta już z jednego z najlepszych na świecie prawodawstwa w dziedzinie ochrony bioróżnorodności. Dyrektywa Ptasia, Konwencja Berneńska i

Konwencja o Ochronie Gatunków Migrujących były kamieniami milowymi już w chwili ich przyjęcia 25 lat temu i dzięki nim udało się wiele osiągnąć. Wciąż jednak pozostało wiele wyzwań w tej dziedzinie, jak to przedstawia BiE2, a potrzeba wykorzystania tego prawodawstwa w celu osiągnięcia jak najlepszych warunków ochrony bioróżnorodności będzie się jeszcze zwiększała w ciągu nadchodzących 25 lat.

■ PODSUMOWANIE

Ogólne przesłanie książki jest tak samo jasne jak poprzedniego wydania. Ptaki w Europie są ciągle zagrożone zmianami w środowisku, a wiele gatunków jest w gorszej sytuacji niż w poprzedniej dekadzie. Ptaki jako indykator zmian w środowisko ze stałą tendencją spadkową obserwowaną u wielu gatunków jasno pokazują stan bioróżnorodności oraz ogólny stan środowiska w Europie.

Skala problemu oraz potrzeba szybkiego i zakrojonego na szeroką skalę działania wydaje się być jeszcze bardziej pilna. Zabiegi ochronne muszą być podjęte natychmiast, nie tylko po to by zatrzymać pogarszanie się stanu przyrody w Europie bogatą niegdyś awifauną ale także by pokazać zdecydowane zaangażowanie w zatrzymanie spadku bioróżnorodości do 2010 roku.

■ РЕЗЮМЕ

■ ЦЕЛЬ

Birds in Europe (2004) *BiE2*—это второй обзор природоохранного статуса всех диких птиц в Европе. Как и в предыдущем издании 1994 года *Birds in Europe* (*BiE1*), здесь определены приоритетные виды птиц (Европейские Виды птиц, статус которых вызывает наибольшее беспокойство—Species of European Conservation Concern, или SPECs), на которых в первую очередь должны быть направлены природоохранные действия для улучшения их статуса.

■ МАСШТБ

Географический масштаб проекта охватывает всю Европу и прилегающие территории от Гренландии на западе до Уральских гор на Востоке и от Шпицбергена на севере до Канарских островов на юге. Увеличившаяся политическая стабильность на Балканах и Кавказе позволила впервые собрать данные из всех европейских стран.

■ СБОР ДАННЫХ

Сбор данных проводился через сеть национальных координаторов, которые получали информацию от местных экспертов, организаций, осуществляющих мониторинг авифауны, и иных региональных представителей, предоставивших свои данные. Первичные данные были получены путем полевых исследований, выполненных тысячами орнитологов, включая многочисленных добровольцев.

Для каждого вида птиц собирали национальную информацию о величине размножающейся популяции (в 2000 году или ближайшие годы) и популяционных трендах (за период 1990–2000). По возможности, такие же данные собирали о зимующих популяциях птиц (преимущественно водоплавающих и околоводных). В целом было получено около 14000 записей о состоянии и трендах популяций, причем существенная их часть превосходила по точности данные *BiE1*. В совокупности с данными 1970-1990 гг., новые сведения о популяциях птиц и их трендах были использованы для переоценки природоохранного статуса всех птиц Европы.

■ ОЦЕНКА СТАТУСА

Для выявления Европейских Видов птиц, статус которых вызывает наибольшее беспокойство (SPECs) в *BiE1* был разработан набор количественных критериев, учитывающих мировой и европейский статус каждого вида, и позволивший классифицировать их по доле от глобальной популяции или общего ареала на территории Европы. В *BiE2*, после интенсивных консультаций, было принято решение, что критерии могут быть улучшены, если объединить их с Красным Списком МСОП, который отражает общепринятую систему оценки риска возможного исчезновения того или иного вида. Недавняя публикация руководства к Критериям Красного Списка МСОП для регионов сделала процесс их интеграции в уже существующие критерии SPEC относительно простым.

Следуя системе, использованной в *BiE1*, каждый вид птиц был отнесен к одной из пяти категорий.

Категория	Европейский вид, находящийся под глобальной угрозой исчезновения	Природоохранный статус в Европе	Мировая популяция или ареал приурочены к Европе
SPEC 1	Да	–	–
SPEC 2	Нер	Неблагоприятный	Да
SPEC 3	Нер	Неблагоприятный	Нер
Non-SPEC[E]	Нер	Благоприятный	Да
Non-SPEC	Нер	Благоприятный	Нер

Вид птиц отнесен к категории «Европейский вид, находящийся под глобальной угрозой исчезновения» если его статус в Красном Списке МСОП отмечен как Угрожаемый, Близкий к Угрожаемому, Недостаточно изученный. Вид имеет Неблагоприятный природоохранный статус, если его Европейская популяция классифицируется как Угрожаемая в рамках регионального Красного Списка МСОП, или если его популяция относится к одной из следующих категорий принятых в *BiE1*: маленькая не краевая, слабо снижающая численность, уменьшившаяся вследствие предшествовавшего снижения численности или сильно локализованная. Вид считается приуроченным к Европе, если более 50% его ареала или мировой популяции (размножающейся или зимующей) находятся в Европе.

■ РЕЗУЛЬТАТЫ

Из 524 оцененых видов 226, или 43% Европейской авифауны, имеют Неблагоприятный природоохранный статус в Европе (Рисунок 1). Из них 40 видов (7.6%) классифицированы как SPEC 1, 45 (8.6%), как SPEC 2 и 141 (26.9%), как SPEC 3. Все эти представленные проценты превосходят таковые в *BiE1*, когда 195 видов (38% из 511 оцененных) были классифицированы как SPECs.

Рисунок 1. Процент европейских видов птиц каждой категории в книге Птицы Европы BiE1(1994) и Птицы Европы 2 BiE2 (2004).

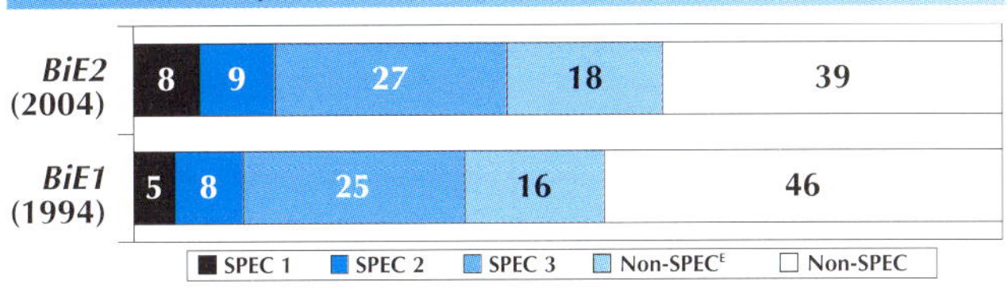

■ ОБСУЖДЕНИЕ

Увеличившееся число птиц, отнесенных к категории SPEC1, отражает переклассификацию (по обновленным критериям) Близких к Угрожаемым видов, которые ранее были перечислены в категории SPEC 2 или 3, но однозначно вызывают общемировую обеспокоенность их статусом. Увеличившееся число видов в категориях SPEC 2 и 3—по-настоящему тревожный сигнал, потому, что это значит, что природоохранный статус значительно большего числа европейских видов птиц (63 вида) ухудшился от Благоприятного к Неблагоприятному, в то время, как только 10 видов улучшили свой статус.

Учитывая величину сокращения численности в течение 1970–1990, особенно птиц агроландшафтов, популяции многих видов SPECs остаются сильно истощенными. Тем не менее, популяции нескольких видов восстановились и были переведены в раздел видов с Благоприятным природоохранным статусом в Европе. Хорошим примером, иллюстрирующим преимущества целевых природоохранных акций, может служить восстановление численности сапсана (*Falco peregrinus*)

Большое количество видов категории SPEC1 также увеличивает свою численность в Европе, благодаря эффективному выполнению планов действий по конкретным видам (SAPs) за прошедшее десятилетие. Пройдет немало времени до того, как природоохранный статус этих видов можно будет пересмотреть, но прогресс, достигнутый на сегодняшний день, показывает, что совместные, хорошо спланированные и в достаточной мере профинансированные усилия, действительно могут остановить падение численности и восстановить популяцию до более благоприятного уровня.

Из 129 видов, классифицированных как SPECs в *BiE1* из-за спада их численности в 1970–1990 гг., у 79 видов (61%) спад численности продолжился и в 90-х годах. Их состояние вызывает особенное беспокойство. Более того, теперь их ряды пополнились еще 35 видами, ранее относящихся к группе с Благоприятным природоохранным статусом. Они включают большое количество мигрирующих через Европу куликов и воробьинообразных, некоторых утиных и морских птиц, несколько самых обычных Европейских видов, таких как домовый воробей *Passer domesticus* и скворец *Sturnus vulgaris*.

Это факты, которые заставляют трезво взглянуть на обещание правительств уменьшить объемы потерь биоразнообразия к 2010 году и обязательства Европейского Союза полностью остановить процесс деградации биоразнообразия. Для большинства таксонов затруднительна оценка того, насколько поставленные задачи были достигнуты, но птицы – исключение из общего правила. Главное требование—это скромная финансовая поддержка мониторинга, поддержка существующих схем, разработки и выполнения стратегий сохранения для других видов. Это позволит правительствам достичь заявленных обязательств и содействовать таким обзорам состояния видов каждое десятилетие (очередной выпуск *BiE3* запланирован на 2014 год).

Осталось мало времени, чтобы достичь поставленных целей, поэтому жизненно необходимо интегрировать проблемы биоразнообразия во все сферы политических решений, затрагивающих окружающую среду. Европа уже находится в боле выгодном положении, обладая одним из лучших в мире законодательств об охране природы. Принятые 25 лет назад Директива Европейского Союза по диким птицам, Боннская и Бернская конвенции стали вехами в деле охраны природы и уже достигли многого. Действительно, как показывает *BiE2*, еще остается много проблем и нам необходимо максимально использовать все эти возможности для того, чтобы в следующие 25 лет биоразнообразие только увеличивалось.

■ ЗАКЛЮЧЕНИЯ

Послание, которое несет нам *BiE2*, настолько же ясно, как и таковое в *BiE1*. Птицам Европы все еще угрожает широкомасштабное преобразование окружающей среды, и популяции многих видов птиц сейчас находятся в еще большей беде, чем 10 лет назад. Птицы—великолепный индикатор состояния окружающей среды, поэтому продолжающийся спад численности такого большого числа видов птиц однозначно сигнализирует нам о состоянии европейского биоразнообразия и состояния окружающей среды в целом.

Учитывая масштаб проблемы, теперь еще более актуален широкомасштабный и срочный ответ, к которому призывали в *BiE1*. Действовать надо прямо сейчас и не только для того, чтобы остановить продолжающуюся потерю некогда многочисленной и разнообразной авифауны Европы, но и для того, чтобы показать серьезность наших намерений прекращения потерь биоразнообразия к 2010 году.

■ SAMMANFATTNING

■ MÅL

Birds in Europe (2004) *(BiE2)* är den andra översikten av fågelskyddsstatusen för alla vilda fågelarter i Europa. Liksom föregångaren från 1994, *Birds in Europe (BiE1)*, redovisar den prioriterade arter (Species of European Conservation Concern, s.k. SPEC:s) med avsikt att fågelskyddsinsatser skall kunna genomföras för att förbättra arternas status.

■ OMFATTNING

Den geografiska spännvidden omfattar hela kontinenten, från Grönland i väster till Ural i öster och från Svalbard (Spetsbergen) i norr till Kanarieöarna i söder. Ökad politisk stabilitet på Balkan och i Kaukasus har medfört att uppgifter för första gången har kunnat samlas in från samtliga europeiska länder.

■ INSAMLING AV FAKTA

Datainsamlingen gjordes med hjälp av ett nätverk av nationella sammanställare, vilka tog hjälp av experter, ansvariga för räkningar och monitoring liksom regionala uppgiftslämnare. Faktauppgifterna bygger på fältarbete genomfört av tusentals ornitologer, framför allt ideella krafter.

För varje art rapporterades antalet häckande i varje land (omkring år 2000) och utvecklingen (för perioden 1990–2000). När uppgifter fanns insamlades också data om övervintrande bestånd, framför allt för vattenfåglar. Sammanlagt rapporterades omkring 14 000 uppgifter om populationer och/eller trender, många av dessa utförligare än i *BiE1*. De nya populationsuppgifterna användes—tillsammans med de utvecklings- och utbredningsuppgifter som fanns från 1970–1990—för att ompröva fågelskyddsstatusen för varje europeisk art.

■ BEDÖMNING AV STATUS

Vid framtagandet av *BiE1* utarbetades ett antal kvantitativa kriterier för att identifiera s.k. SPEC:s. Dessa kriterier utgick från fågelarternas globala och europeiska status, där arterna klassificerades utifrån den globala andelen resp. utbredningen i Europa. I samband med *BiE2* visade ett större samråd att befintliga kriterier kunde förstärkas genom att arbeta in IUCN:s kriterier för rödlistning, vilka utgör ett accepterat system för att bedöma arters risk för utrotning. Den senaste sammanställningen av regler för hur IUCN:s kriterier för rödlistning ska användas på regional nivå gjorde att de ganska enkelt kunde integreras med gällande SPEC-kriterier.

Utifrån det system som användes i *BiE1*, så fördes varje art till en av fem kategorier.

Kategori	Europeisk art av globalt fågelskydds-intresse	Fågelskydds-status i Europa	Global population eller ut-bredningsområde i Europa
SPEC 1	Ja	–	–
SPEC 2	Nej	Ogynnsam	Ja
SPEC 3	Nej	Ogynnsam	Nej
Ej SPEC[E]	Nej	Gynnsam	Ja
Ej SPEC	Nej	Gynnsam	Nej

En art har globalt fågelskyddsintresse om dess status klassificeras som Hotad, Missgynnad eller Kunskapsbrist enligt IUCN:s globala kriterier för rödlistning. Den har en Ogynnsam bevarandestatus i Europa om dess europeiska population klassificeras som Hotad utifrån en regional tillämpning av IUCN:s kriterier för rödlistning, eller om för dess population något av följande gäller (såsom i *BIE1*):

liten och ej marginell, måttlig nedgång, påtagligt reducerad efter tidigare nedgång eller mycket lokal. En art anses koncentrerad till Europa om mer än 50% av dess globala häcknings- eller vinterpopulation resp. utbredning påträffas i Europa.

RESULTAT

Av de 524 arter som har bedömts, beräknas 226—dvs. 43% av den europeiska fågelfaunan—ha en Ogynnsam bevarandestatus i Europa (Figur 1). Av dessa klassificerades 40 arter (7.6%) som SPEC 1, 45 (8.6%) som SPEC 2 och 141 (26.9%) som SPEC 3. Alla dessa andelar är större än de i *BiE1*, då 195 arter (38% av 511 bedömda) klassificerades som SPEC-arter.

Figur 1. Procentandel av europeiska fågelarter i varje kategori i *BiE1* (1994) och i *BiE2* (2004).

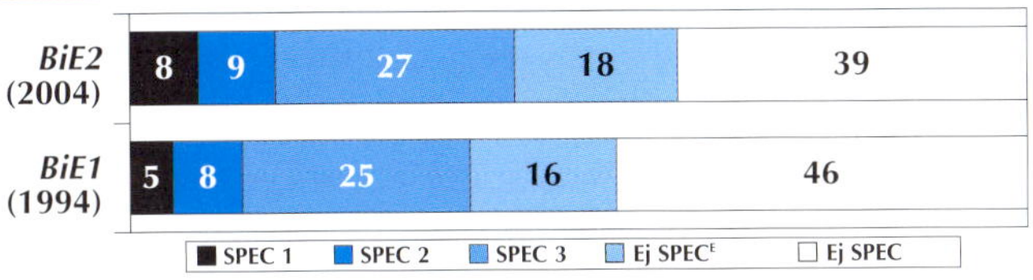

DISKUSSION

Det ökade antalet SPEC 1-arter återspeglar omklassificeringen (utifrån reviderade kriterier) av globalt Missgynnade arter, vilka tidigare listades som SPEC 2 och 3, men där de helt klart är av globalt fågelskyddsintresse. Dock är det ökade antalet SPEC 2 och 3 verkligt alarmerande, därför att det betyder att den europeiska hotbilden för många fler fåglar (45 arter) har förändrats från Gynnsam till Missgynnad än tvärtom (14 arter).

Med kännedom om de omfattande minskningarna i antal 1970–1990, särskilt vad gäller jordbrukslandskapets fåglar, har för populationerna av många SPEC-arter skett en fortsatt kraftig åderlåtning. Dock har några arter återhämtat sig och har omklassificerats till Gynnsam bevarandestatus i Europa. Återhämtningen hos pilgrimsfalk *Falco peregrinus* är ett bra exempel, som illustrerar vinsterna i ett målinriktat fågelskyddsarbete.

Många SPEC 1-arter ökar också i Europa beroende på ett effektivt arbete med handlingsplaner under det senaste årtiondet. Det kommer att dröja innan dessa arter kan omklassificeras, men hittillsvarande framsteg visar att samordnade aktioner som är välplanerade och tillfredsställande finansierade verkligen kan vända nedgångar och återföra arter till en mer gynnsam status.

Av de 129 arter som i *BiE1*, utifrån nedgångar 1970–1990, listades som SPEC-arter, fortsatte 79 (61%) att minska under 1990-talet. Deras belägenhet är särskilt oroande—och de har nu fått efterföljare i 35 arter som tidigare ansågs ha en Gynnsam bevarandestatus i Europa. Bland dessa arter finns flera flyttande vadare och tättingar, flera änder och havsfåglar liksom några av Europas mest välkända arter, såsom gråsparv *Passer domesticus* och stare *Sturnus vulgaris*.

Detta är den nyktra verkligheten när de flesta regeringar har utfäst sig att reducera förlusten i biologisk mångfald till år 2010, och EU har bestämt att helt sätta stopp för dessa förluster. För de flesta taxa blir det svårt att avgöra om dessa mål uppnås, men fåglar utgör ett undantag. Det huvudsakliga kravet är ett någorlunda långsiktig stöd för monitoring, både för att fortsätta med pågående övervakning och för att utveckla och påbörja program för andra arter. Detta skulle kunna möjliggöra för regeringar att uppfylla rapporteringsskyldigheter liksom att underlätta en översyn av status—på samma sätt som den här—varje decennium (med *BiE3* planerad för 2014).

Den tid som återstår för att möta de utsatta målen är kort, så det är väsentligt att frågor om biologisk mångfald till fullo implementeras inom allt sektorsansvar som påverkar miljön. Europa har redan fördelen av kanske den bästa lagstiftningen i världen med avseende på biologisk mångfald. Fågeldirektivet, Bern- och Bonnkonventionerna var alla milstolpar när de antogs för 25 år sen, och de har redan uppnått väldigt mycket. Men, som *BiE2* visar, återstår många utmaningar, och kravet på att använda dessa verktyg för att optimera den biologiska mångfalden kommer bara att öka under de nästkommande 25 åren.

SLUTSATSER

Det helt överskuggande budskapet från *BiE2* är lika klart som det från *BiE1*. Europas fåglar fortsätter att vara hotade på grund av omfattande miljöförändringar, och många populationer har det nu värre än för ett decennium sen. Eftersom fåglar är utmärkta miljöindikatorer utgör den pågående nedgången för många arter en tydlig signal om den biologiska mångfalden i Europa och tillståndet för miljön i ett större sammanhang.

Utifrån problemets storlek är de massiva och angelägna krav på reaktion som framfördes i *BiE1* nu än mer påkallade. Insatser måste göras omgående—inte bara för att förhindra den pågående utarmningen av Europas en gång så rika och flödande fågelfauna, utan också för att på ett trovärdigt sätt göra halt i förlusterna i biologisk mångfald till 2010.

ÖZET

AMAÇ

Birds in Europe (2004), *BiE2* Avrupa'da doğal olarak yaşayan bütün yabani kuş türlerinin koruma durumları ile ilgili ikinci değerlendirme çalışmasıdır. 1994 yılında basılan (*Birds in Europe*, *BiE1*) adlı yayın gibi bu değerlendirme de kuşlara yönelik koruma çalışmalarının gerçekleştirilmesi için öncelikli türleri (Avrupa'da Koruma Öncelikli Türler—SPEC) belirlemektedir.

ÇALIŢMA ALANI

Batıda Grönland'dan doğuda Ural'lara, kuzeyde Svalbard'dan güneyde Kanarya adalarına kadar olan alanı kapsamaktadır. Balkan ve Kafkas bölgelerindeki politik istikrar sayesinde bu çalışma kapsamında ilk defa tüm Avrupa ülkelerinden veri toplanabilmiştir.

VERİ TOPLAMA

Veriler, konuyla ilgili tüm ulusal uzmanlar ve kurumlarla iletişim içinde olan ulusal koordinatörler aracılığıyla toplanmıştır. Bu veriler, binlerce ornitolog ve çok sayıda gönüllü tarafından yapılan arazi çalışmaları sonucunda elde edilmiştir.

Her bir tür için ulusal ölçekte üreme popülasyonunun büyüklüğü (2000 yılı süresince ve civarında) ve 1990–2000 yılları arasındaki değişimi hakkında veri toplanmıştır. Mümkün olan durumlarda, çoğunlukla su kuşları olmak üzere, kış popülasyonlarının verileri de toplanmıştır. Sonuç olarak, çoğu *BiE1*'deki kayıtlardan daha kaliteli olan, yaklaşık 14,000 popülasyon/değişim kaydı derlenmiştir. 1970–1990 yılları aralığına ait veriler ve elde edilen yeni popülasyon verileri, her türün Avrupa'daki koruma durumunun yeniden değerlendirmesi için kullanılmıştır.

KORUMA ÖNCELİKLERİNİN BELİRLENMESİ

BiE1'de, kuş türlerinin koruma önceliklerini değerlendirmek ve Avrupa'daki dağılım oranlarına göre sınıflandırmak için bir dizi nicel kriter geliştirilmiştir. *BiE2*'de ise türlerin yok olma riskini değerlendirmek için IUCN Kırmızı Liste kriterlerini kullanmanın daha uygun olacağı düşünülmüştür. IUCN Kırmızı Liste kriterlerini bölgesel ölçekte uygulamak için hazırlanan rehber, bu kriterlerin SPEC kategorilerini belirlemek üzere kullanılmasını kolaylaştırmıştır.

Bu envanterde, *BiE1*'de olduğu gibi, her bir tür aşağıdaki beş kategoriden birine yerleştirilmiştir. Tablodaki "non-SPEC" kısaltması, Avrupa ölçeğinde koruma önceliği olmayan türler için kullanılmıştır.

Kategori	Küresel ölçekte koruma önceliği olan türler	Avrupa ölçeğindeki koruma durumu	Küresel popülasyonu ya da dağılımı Avrupa'da yoğunlaşmış türler
SPEC 1	Evet	–	–
SPEC 2	Hayır	Durumu kötü	Evet
SPEC 3	Hayır	Durumu kötü	Hayır
Non-SPEC^E	Hayır	Durumu iyi	Evet
Non-SPEC	Hayır	Durumu iyi	Hayır

Bir tür IUCN Kırmızı Liste Kriterlerine göre küresel ölçekte Tehlike Altında (CR, EN, VU), Tehlike Altına Girmeye Yakın (NT) ya da Yetersiz Bilgi (DD) olarak sınıflandırılmışsa küresel ölçekte koruma önceliği olan bir tür olarak kabul edilmiştir (soldan ikinci sütun). Eğer bir tür IUCN kırmızı liste kriterlerininin bölgesel olarak uygulanması sonucunda tehlike altında olarak sınıflandırılıyorsa, bu türün Avrupa ölçeğindeki koruma durumu kötü olarak kabul edilmiştir. Ayrıca, aşağıdaki kriterlerden herhangi birini sağlayan türlerin de Avrupa ölçeğindeki koruma durumu kötü olarak kabul edilmiştir: Küçük ama marjinal olmayan popülasyona sahip türler, orta derecede azalan türler, daha önce bulundukları bazı bölgelerden tümüyle kaybolmuş türler veya dar bir alanda yoğunlaşmış türler (soldan üçüncü sütun). Şayet bir türün tüm dünyadaki üreme veya kışlama popülasyonunun veya dağılış alanının yarıdan fazlası Avrupa'daysa, bu tür Avrupa'da yoğunlaşmış tür olarak sınıflandırılmıştır (en sağdaki sütün). (*) ile işaretli non-SPEC kategorisi, koruma önceliği olmamakla birlikte Avrupa'da yoğunlaşmış türleri tanımlamaktadır.

BULGULAR

Değerlendirme sonucunda, 524 Avrupa kuş türünden 226'ü (%43'u) SPEC kategorilerinden birine yerleştirilmiştir (Şekil 1). Bunlardan 40'ı (%7.6) SPEC 1, 45'i (%8.6) SPEC 2 ve 141'si (%26.9) SPEC 3 olarak sınıflandırılmıştır. *BiE1*'de SPEC olarak sınıflandırılan 195 (değerlendirilen 511 türün %38'i) tür için geçerli olan tüm oranlar artmıştır.

Şekil 1. Avrupa'nın Kuşları (1994) ve Avrupa'nın Kuşları 2'de (2004) her bir kategori için türlerin oranları.

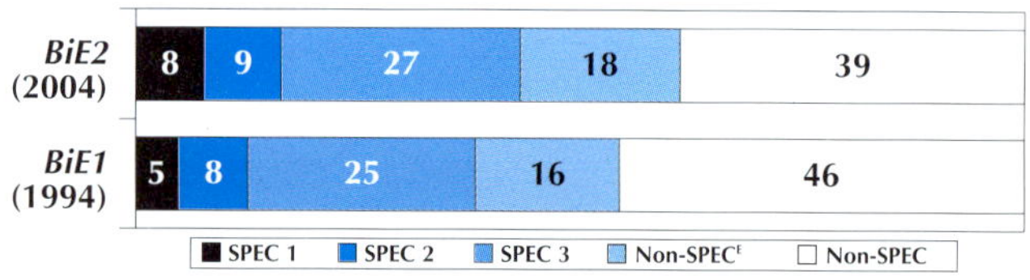

TARTIŞMA

SPEC 1 kategorisndeki artış, daha önce SPEC 2 ve SPEC 3 olarak sınıflandırılan ve küresel ölçekte Tehlike Altına Girmeye Yakın (NT) olan türlerin durumlarındaki değişimi yansıtmaktadır ve bu türlerin dünya ölçeğinde korunması gerekmektedir. Bununla birlikte SPEC 2 ve SPEC 3 türlerinin oranlarının artması ciddi bir tehlikenin işaretidir. Avrupa ölçeğinde bir çok kuş türünün (45 tür) durumu iyiden kötüye doğru değişmiştir. Öte yandan, sadece 14 türün durumu kötüden iyiye doğru değişmiştir.

1970–1990 yılları arasındaki verilere bakıldığında, özellikle tarım alanlarında görülen kuşlar olmak üzere SPEC kategorisindeki pek çok türün ciddi bir biçimde azalmaya devam ettiği görülmektedir. Buna rağmen koruma çalışmaları sayesinde birkaç tür kendini toparlayabilmiş ve "Avrupa Ölçeğinde Koruma Durumu İyi" olarak sınıflandırılmıştır. Gök doğan *Falco peregrinus,* bu durum için iyi bir örnektir.

Bazı SPEC 1 türleri de, tür eylem planlarının son on yıl süresince etkili bir şekilde uygulanması sonucunda Avrupa'da artmıştır. Bu türlerin yeniden sınıflandırılabilmesi için zaman gerekmektedir. Elde edilen sonuçlar, iyi planlanmış ve yeteri kadar bütçesi olan eylem planlarının türlerin azalmasını geriye çevirebileceğini göstermektedir.

BiE1'ye göre 1970–1990 yılları arasındaki azalma nedeniyle SPEC olarak listelenen 129 türün, 79'i (%61) 1990'lı yıllar boyunca azalmaya devam etmiştir. Bu 79 türün durumu özellikle endişe vericidir ve ne yazık ki bu türlere 1970–1990 yılları arasında sayısı azalmayan 35 tür daha eklenmiştir. Eklenen yeni türler, bir çok göçmen kıyı kuşu ve ötücü kuş türünü, birkaç ördek ve deniz kuşunu ve ayrıca Avrupa'daki en yaygın türlerden serçe *Passer domesticus* ve sığırcığı *Sturnus vulgaris* da içermektedir.

Bunlar, dünya devletlerinin 2010 yılına kadar biyolojik çeşitlilik kaybının azalması ve Avrupa Birliği'nin bu kaybın tamamen durması için onay verdiği bir dönemde üzerinde dikkatle durulması gereken bulgulardır. Bir çok tür grubu için bu hedeflere ulaşılması zor olabilir ama kuşlar istisnadır. Temel gereksinim, var olan koruma planların uygulanması ve diğer türler için de koruma planlarının geliştirilmesi ve uygulanması için uzun vadeli desteğin sağlanmasıdır. Bu destek, hükümetlerin yükümlülüklerini yerine getirmeleri ve on yılda bir değerlendirme çalışmalarının yapılmasını kolaylaştıracaktır (*BiE3* 2014 yılında yapılacaktır).

Bu hedeflere ulaşmak için geriye kalan zaman azdır ve bu nedenle biyolojik çeşitlilik ile ilgili sorunların çevreyi etkileyen tüm sektörel politikalara entegre edilmesi çok önemlidir. Avrupa kıtası, doğayı koruyan yasal düzenlemelerden halihazırda büyük yarar görmüştür. AB Kuş Direktifi, Bern Sözleşmesi ve Göçmen Türlerle ilgili sözleşme, 25 yıl önce bir dönüm noktası olmuş ve bu sözleşmeler sayesinde daha şimdiden oldukça önemli başarılar elde edilmiştir. Buna rağmen, bu kaynakta görüldüğü üzere bir çok sorun devam etmektedir ve biyolojik çeşitliliğin korunabilmesi için gerekli adımların gelecek 25 yıl boyunca artarak atılması gerekecektir.

DEĞERLENDİRME

BiE1'de olduğu gibi *BiE2*'nin de verdiği mesaj açıktır. Avrupa'nın kuşları büyük çevresel değişimler nedeniyle ciddi tehdit altındadır ve ne yazık ki bir çok kuş türü on yıl öncesine göre şimdi daha da kötü durumdadır. Sağlıklı bir çevrenin göstergesi olan kuşların azalmaya devam etmesi, Avrupa'daki yaşam kalitesinin bozulduğunun açık bir işaretidir.

BiE1'de yoğun olarak belirtilen eylem çağrısı şimdi daha da acildir. Bir an önce eyleme geçilmeli ve 2010 yılına kadar yalnızca kuş türlerinin değil, aynı zamanda tüm biyolojik çeşitliliğin kaybının durması için ciddi önlemler alınmalıdır.

INTRODUCTION

There is growing recognition that protecting biodiversity—the sum total of all biotic variation, from the level of genes to ecosystems (Purvis and Hector 2000)—is in our self-interest. Biological resources are the foundations upon which we build our civilisations. Biodiversity conservation directly affects issues such as health, water, sanitation and many other aspects of our quality of life.

At the United Nations Millennium Summit in 2000, the world's political leaders adopted the Millennium Development Goals, one of which was to ensure environmental sustainability. At the World Summit on Sustainable Development in 2002, governments pledged to reduce significantly the rate of biodiversity loss by 2010. In 2001, the European Council in Gothenburg in 2001 adopted the more ambitious target of halting the decline of biodiversity in the European Union by 2010. A similar target was agreed at a pan-European level in 2003 at the 5th Ministerial Conference on Environment for Europe in Kiev.

BIRDS AS BIODIVERSITY INDICATORS

There is little doubt that biodiversity is declining, but for most taxa trends are difficult to quantify due to lack of data. However, birds are better studied than any other group of organism, and thus are well placed to provide us with information on the overall health of our environment (e.g. Furness *et al.* 1993; Donald *et al.* 2001; Gregory *et al.* 2003). Birds are good bio-indicators because they occupy a high trophic level (and thus reflect environmental changes occurring at lower trophic levels), occur in a wide range of ecosystems, their taxonomy and identification is well known, their territorial behaviour (songs and displays) allows them to be censused readily during the breeding season, it is possible to collect large quantities of data in a highly efficient manner using skilled volunteers, and they have great public resonance across European cultures (Bibby 1999; Gregory *et al.* in press).

ASSESSING THE STATUS OF BIRDS IN EUROPE

This publication is the second assessment of the conservation status of all naturally occurring wild birds in Europe. Updated data on the population sizes and trends of all species were collected from 52 countries and territories in Europe, through the BirdLife Partnership and other selected organisations, and were then analysed at a continental level. A decade ago, *Birds in Europe* (Tucker and Heath 1994) assessed the conservation status of all European species for the first time. This was followed by the *EBCC Atlas of European breeding birds* (Hagemeijer and Blair 1997), which provided a continent-wide analysis of species' distributions and summary population data. *European bird populations* (BirdLife International/European Bird Census Council 2000) provided comprehensive data on population size, and population and range trends, for all species.

WHY A NEW ASSESSMENT?

The main aim of this book is to provide a new assessment of the conservation status of bird species at a pan-European scale. Repeated assessments of conservation status serve two major purposes: ensuring that bird conservation priorities reflect status changes; and providing indicators for measuring the progress towards conservation targets, such as those set for 2010 at the World Summit on Sustainable Development.

This assessment also provides the first application of the IUCN Red List Criteria (IUCN 2003) at a European scale. A regional approach to identifying threatened species complements global conservation status assessments, and provides a more comprehensive framework for international conservation treaties (such as the Bern Convention and the African-Eurasian Migratory Waterbird Agreement) and legislation (such as the EU Wild Birds Directive).

In addition, the information provided here helps to put national priorities into a continental context, thus maximising the effectiveness of local and national conservation measures.

PRIORITISING CONSERVATION ACTION

Quantitative data and analysis of bird populations and their distribution is the scientific basis for setting priorities for conservation actions at a European scale. A comprehensive conservation strategy for birds encompasses three components (BirdLife International 2004b):

- species conservation;
- site conservation;
- conservation of the wider environment.

Species conservation

Although the declines of many species are driven primarily by the loss or degradation of their habitats, threats that are unrelated to habitat change (e.g. persecution, overexploitation) may also be important contributing factors. Information provided in this review will help to identify those species that require improved legislation at a national or international level, as well as the legally hunted species whose sustainable use will necessitate the development of management plans. In addition, the results of this assessment will highlight those species that should become the focus of concerted conservation action at an international level.

Site conservation

Numerous species concentrate at a relatively small number of sites at some point in their annual life cycle. Therefore, they can often be conserved effectively by the protection of a coherent network of these sites. Several international conservation treaties (e.g. the Ramsar Convention, the Bern Convention's Emerald Network, the EU Wild Birds Directive, the Baltic Sea Special Protection Areas under the Helsinki Convention, and the Mediterranean Special Protection Areas under the Barcelona Convention) aim to protect networks of key sites in Europe. The Important Bird Area (IBA) Programme of BirdLife International aims to identify key sites for threatened and congregatory species, and to protect them through national and international designations (Grimmett and Jones 1989; Heath and Evans 2000). The results of this review will provide further guidance on the species that should be protected within the framework of these site networks.

Conservation of the wider environment

Many threats to European birds (and biodiversity in general) stem from the continuing erosion of the quality and extent of habitats across the wider landscape. Habitat loss and degradation is driven by the increasing intensity of human uses of the environment. Habitat conservation at the appropriate scale cannot be achieved solely by the protection of representative areas—a wider approach is needed.

BirdLife International has produced a conservation strategy for the wider environment, applying a species-led approach to set objective targets for habitat conservation (Tucker and Evans 1997). Since then, many of the recommendations have been reflected in policies developed within the European Union (e.g. the Cardiff process, 6th Environmental Action Plan) and at a pan-European level (e.g. in the Kiev Declaration).

National governments and regional or local authorities can encourage the conservation of the wider environment in the way that they formulate their plans and use subsidies and other broad measures. Ideally, these concepts should be integrated fully into land-use policies and regulations across all sectors of the economy, and at all scales. The results of this review should be used to revisit habitat conservation priorities, and to measure the success of attempts to integrate biodiversity considerations into sectoral policies at a continental level.

PLANS FOR THE FUTURE

The vast amount of data presented in this book will be extremely useful for further analyses, which should provide deeper insights into the conservation needs of wild birds and the impacts on their populations of land-use policies and natural resource use.

These analyses could include studies at different geographical or geopolitical scales, or focusing on different taxonomic or ecological groups. The first example of such an analysis is *Birds in the European Union* (BirdLife International 2004c), which is published simultaneously with this book and which applies the same criteria to identify species with an Unfavourable conservation status at the scale of the EU25. This illustrates the potential of these data for measuring the performance of national and supra-national initiatives, as well as for identifying changes since the previous assessment and helping to re-evaluate conservation priorities.

The dynamic nature of bird populations means their numbers can alter rapidly over relatively short periods of time. Regular updates of the conservation status of Europe's birds are therefore essential, both to assess the effectiveness of conservation efforts, and to ensure that species in most need of attention receive it promptly. The next update of national population and trend data is scheduled for 2006–2008, when BirdLife's European Partnership will publish estimates for the period 2000–2005. The next comprehensive reassessment of the conservation status of European birds is currently scheduled for 2012–2014.

REFERENCES

Bibby, C. J. (1999) Making the most of birds as environmental indicators. *Ostrich* 70: 81–88.

BirdLife International/European Bird Census Council (2000) *European bird populations: estimates and trends*. Cambridge, UK: BirdLife International (BirdLife Conservation Series no. 10).

BirdLife International/European Bird Census Council (2003) *Birds as biodiversity indicators for sustainability: a pan-European strategy*. Sandy, UK: Royal Society for the Protection of Birds.

BirdLife International (2004a) *Threatened birds of the world 2004*. CD-ROM. Cambridge, UK: BirdLife International.

BirdLife International (2004b) *A strategy for birds and people: responding to our changing world*. Cambridge, UK: BirdLife International.

BirdLife International (2004c) *Birds in the European Union: a status assessment*. Wageningen, The Netherlands: BirdLife International.

Donald, P. F., Green, R. E. and Heath, M. F. (2001) Agricultural intensification and the collapse of Europe's farmland bird populations. *Proceedings of the Royal Society of London B* 268: 25–29.

Furness, R. W., Greenwood, J. J. D. and Jarvis, P. J. (1993) Can birds be used to monitor the environment? Pp.1–41 in R. W. Furness and J. D. D. Greenwood, eds. *Birds as monitors of environmental change*. London: Chapman and Hall.

Gregory, R. D., Noble, D., Field, R., Marchant, J., Raven, M. and Gibbons, D. W. (2003) Using birds as indicators of biodiversity. *Ornis Hungaria* 12–13: 11–24.

Gregory, R. D., van Strien, A. J., Vorisek, P., Gmelig Meyling, A. W., Noble, D. G., Foppen, R. P. B. and Gibbons, D.W. (in press) Developing indicators for European birds. *Philosophical Transactions of the Royal Society of London B*.

Grimmett, R. F. A. and Jones, T. A., eds. (1989) *Important Bird Areas in Europe*. Cambridge, U.K.: International Council for Bird Preservation (ICBP Technical Publication no. 9).

Hagemeijer, E. J. M. and Blair, M. J., eds. (1997) *The EBCC Atlas of European breeding birds: their distribution and abundance*. London: T. and A. D. Poyser.

Heath, M. F. and Evans, M. I., eds. (2000) *Important Bird Areas in Europe: priority sites for conservation*. 2 vols. Cambridge, UK: BirdLife International (BirdLife Conservation Series no. 8).

IUCN (2003) *Guidelines for application of IUCN Red List Criteria at regional levels. Version 3.0*. IUCN Species Survival Commission. Gland, Switzerland and Cambridge, UK: IUCN.

Purvis, A. and Hector, H. (2000) Getting the measure of biodiversity. *Nature* 405: 212–219.

Tucker, G. M. and Evans, M. I. (1997) *Habitats for birds in Europe: a conservation strategy for the wider environment*. Cambridge, UK: BirdLife International (BirdLife Conservation Series no. 6).

Tucker, G. M. and Heath, M. F. (1994) *Birds in Europe: their conservation status*. Cambridge, UK: BirdLife International (BirdLife Conservation Series no. 3).

DATA COLLECTION

The geographical scope of this publication is defined in Figure 1, and is the same as that covered by the original edition of *Birds in Europe* (hereafter *BiE1*; Tucker and Heath 1994), and also by *Important Bird Areas in Europe* (Heath and Evans 2000). It includes the Atlantic archipelagos of the Azores, Madeira and the Canary Islands, as well as western Russia (east to the Ural mountains and Ural river), Greenland, Svalbard, Iceland, the Faroe Islands, Turkey, Cyprus and the Caucasus. Political instability in the Caucasus and parts of the former Yugoslavia meant that data were not available from Armenia, Azerbaijan, Georgia, Bosnia and Herzegovina, Serbia and Montenegro, or the Former Yugoslav Republic of Macedonia in *BiE1*. The current review includes data from all these countries, and hence its scope is truly pan-European.

PAN-EUROPEAN NETWORK

To reassess the conservation status of Europe's birds reliably and comprehensively, it was necessary to obtain updated population data for all species and from every European country. This was achieved via a continent-wide network of national coordinators—drawn from within and beyond BirdLife's European Partnership—who are listed at the front of this book. To ensure as broad a consensus as possible, national coordinators sought collaboration from relevant experts, monitoring organisations and regional contributors, as well as Partners in other countries. Consequently, the data gathered and presented here are based on a huge amount of fieldwork carried out over the last few decades by thousands of ornithologists, including countless volunteers. This review therefore represents a massive input of effort and dedication, the result of which is a more comprehensive assessment than would be possible for any other class of organism in Europe.

DATA COLLECTION

For all wild bird species occurring naturally and regularly in Europe (see 'Data analysis' on p. 10 for details of those considered), the following data were collected from each country:

- breeding population size (in or around the year 2000);
- breeding population trend (over the period 1990–2000).

Figure 1. The geopolitical units for which data are presented in this review.

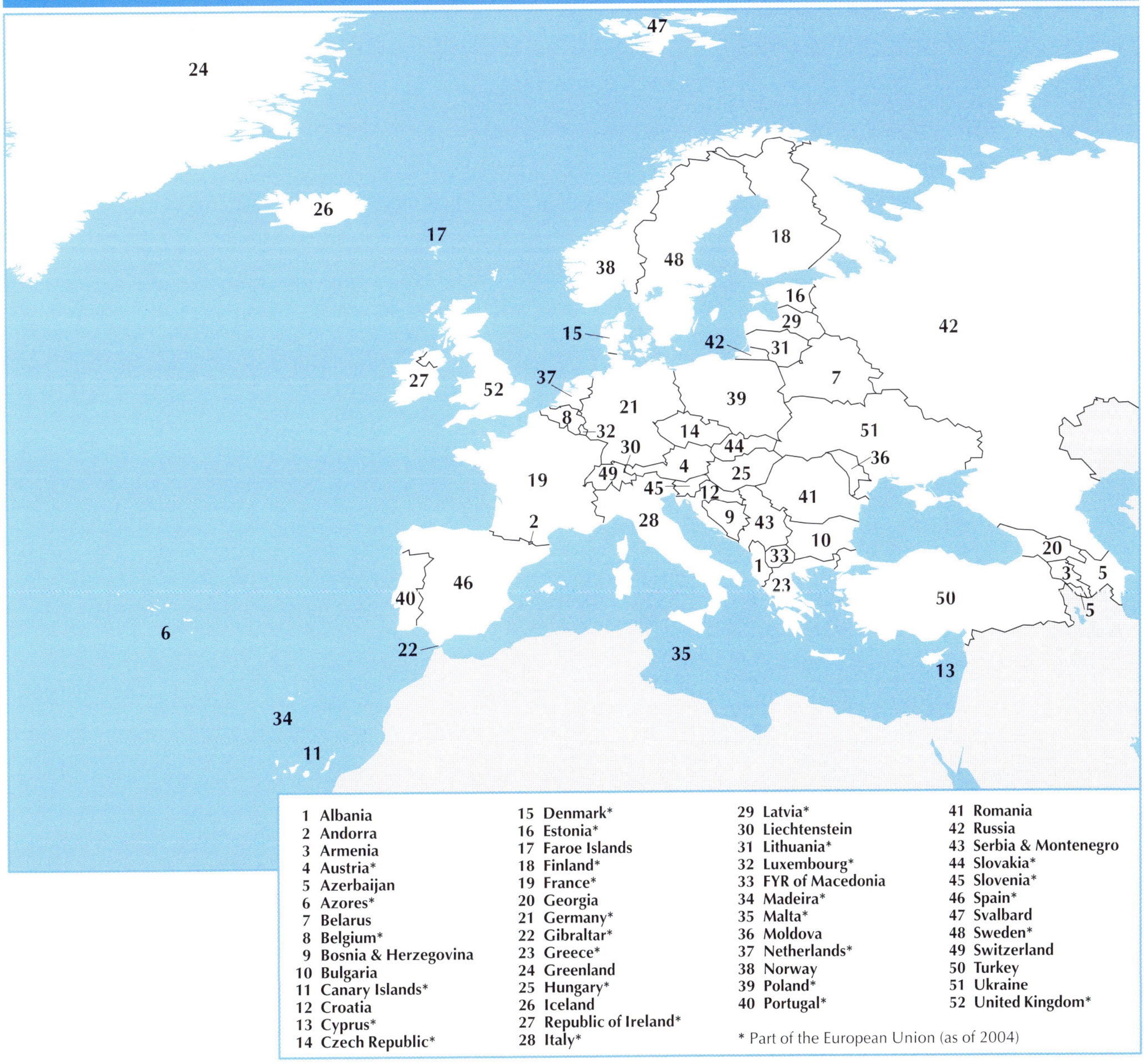

1 Albania	15 Denmark*	29 Latvia*	41 Romania
2 Andorra	16 Estonia*	30 Liechtenstein	42 Russia
3 Armenia	17 Faroe Islands	31 Lithuania*	43 Serbia & Montenegro
4 Austria*	18 Finland*	32 Luxembourg*	44 Slovakia*
5 Azerbaijan	19 France*	33 FYR of Macedonia	45 Slovenia*
6 Azores*	20 Georgia	34 Madeira*	46 Spain*
7 Belarus	21 Germany*	35 Malta*	47 Svalbard
8 Belgium*	22 Gibraltar*	36 Moldova	48 Sweden*
9 Bosnia & Herzegovina	23 Greece*	37 Netherlands*	49 Switzerland
10 Bulgaria	24 Greenland	38 Norway	50 Turkey
11 Canary Islands*	25 Hungary*	39 Poland*	51 Ukraine
12 Croatia	26 Iceland	40 Portugal*	52 United Kingdom*
13 Cyprus*	27 Republic of Ireland*		
14 Czech Republic*	28 Italy*		

* Part of the European Union (as of 2004)

Where available, equivalent midwinter population data were also collected, mainly for species covered by the International Waterbird Census run by Wetlands International (Box 1). For waders, many data were collected in parallel with those provided for two projects undertaken by the International Wader Study Group (Box 2).

Wherever possible, national coordinators supplied population trend data as precise percentage changes over the 1990–2000 period. For a number of widespread common species in certain countries, this information was identical to that used by the Pan-European Common Bird Monitoring Scheme (PECBMS; Box 3). Where such precise data were not available, national coordinators supplied data on trend direction and magnitude using a set of categories and codes (Table 1). Trend categories ranged from -5 to +5, with the sign indicating the direction of the change. Stable populations were represented by a value of zero. There were thus 11 population trend categories, as well as special codes for fluctuations, new breeders and national extinctions.

The reference sources used to evaluate population sizes and trends were also recorded. Estimates were often derived from a combination of published literature sources and unpublished survey data. Derivation of figures sometimes involved interpretation and

Box 1. Wetlands International and the International Waterbird Census.

Wetlands International is a leading global organisation dedicated solely to the work of wetland conservation and sustainable management. Its mission is to sustain and restore wetlands, their resources and biodiversity for future generations through research, information exchange and conservation activities worldwide. Well-established networks of experts, and close partnerships with key organisations, provide Wetlands International with the essential tools for catalysing conservation activities based on sound science in over 120 countries around the globe.

The International Waterbird Census (IWC) is undertaken by one of the most extensive and productive of these networks, and has been coordinated by Wetlands International (and its predecessor, IWRB) since 1967. It is nearly global in scope, but is most active in Europe. Every January, more than 10,000 counters (mostly volunteers) across Europe census over 15 million waterbirds. The census is site-based and uses rigorous standardised methods, so the data collected have high conservation value and a growing scientific impact. The IWC aims to contribute significantly to international efforts to conserve waterbirds and their wetland habitats. To achieve this, it uses information gathered over the long term to estimate population sizes of waterbird species, to monitor changes in numbers and distribution of their populations, and to assess the importance of individual sites, as part of a network, for waterbirds during the non-breeding season. A sophisticated data management system allows powerful analyses, including long-term waterbird population trend analyses at an international level using TRIM software (Pannekoek and van Strien 2001), while the calculation of multi-species population indices provides summaries of this information.

IWC is funded largely by national government agencies, which value its contribution to meeting their obligations under international environmental treaties and agreements. The publication *Waterbird Population Estimates*, which is updated every three years, covers 868 species worldwide. It forms the definitive basis of the waterbird criteria used by the Ramsar Convention on Wetlands to designate wetlands of international importance, and by the European Union to designate wetland Special Protection Areas (SPAs). The African-Eurasian Migratory Waterbird Agreement (AEWA) under the Convention on Migratory Species also owes much to the IWC, which provides the scientific basis for a high proportion of its activities.

For more information on Wetlands International and the IWC, visit www.wetlands.org

Box 2. International Wader Study Group.

The International Wader Study Group (IWSG) is a voluntary association of amateur and professional ornithologists interested in all aspects of wader biology. It currently has more than 650 members worldwide, giving it rapid access to an extensive network of experience and information. To maintain contact between its members and to facilitate information exchange, the IWSG holds annual conferences and publishes its journal *Wader Study Group Bulletin* three times per year. It also acts as the Specialist Group on waders for both Wetlands International (Box 1) and the IUCN Species Survival Commission.

Through its network, the IWSG organises cooperative studies on major topics relating to wader biology and conservation, publishing the results in special issues of its journal. Two such projects of great relevance to *Birds in Europe* were undertaken at the same time as this book. The first (Stroud *et al.* 2004) re-evaluated the population sizes and trends of all species of migratory waders in Africa and western Eurasia in the 1990s, while the second (Thorup in press) updated population estimates for all waders breeding in Europe in or around the year 2000. The networks supplying data for these projects (particularly the latter) overlapped extensively with the one used to compile this book. This overlap—along with the coordinators' willingness to share information—greatly improved efficiency, and allowed any discrepancies to be investigated and resolved.

For more information on the IWSG and the two projects mentioned above, visit www.waderstudygroup.org

Box 3. European Bird Census Council and the Pan-European Common Bird Monitoring Scheme.

The European Bird Census Council (EBCC) brings together ornithologists from all European countries, particularly those organisations, institutions and individuals interested in and responsible for monitoring bird populations and their distribution. It encourages bird monitoring studies that aim to improve the conservation and management of bird populations (especially atlases and monitoring of common birds), and promotes monitoring that is rigorously planned with clear objectives. The EBCC also runs major international projects, such as the European Breeding Bird Atlas (Hagemeijer & Blair 1997; see also Box 4), and the Pan-European Common Bird Monitoring Scheme (PECBMS; see below). It promotes the development of indicators of the changing ability of European landscapes to support wildlife, and works closely with international conservation organisations (including BirdLife) to encourage links between ornithologists and policy makers. It also facilitates communication and collaboration among its members via its journal *Bird Census News*, and through its programme of conferences and workshops. Since 1969, it has hosted 15 major international conferences across Europe (all with published proceedings) on a three-to-four year cycle.

The Pan-European Common Bird Monitoring Scheme (PECBMS) is a joint initiative of the EBCC and BirdLife International. It was developed during the 1990s, and formally launched in 2002 with seed funding from the RSPB (the BirdLife Partner in the United Kingdom). Its main goal is to collate national data on the breeding population trends of common, well-monitored birds in many European countries, and then use them to develop and promote indicators of the general state of biodiversity across Europe. The data are derived from large-scale sampling schemes based on volunteer fieldwork, all of which employ standardised methods. After collation, the data are analysed using TRIM software (Pannekoek and van Strien 2001) to generate national and multinational trend indices for individual species. These results are then combined to produce composite indices (or indicators) for groups of species associated with particular habitats, for example farmland birds. In its first year, the PECBMS collated data on 48 species from 18 European countries, which enabled it to produce the first ever multinational, multi-species indicators for farmland specialists, forest specialists and generalist species. If long-term financial support can be secured, these figures will continue to grow as data from other existing schemes are mobilised, new schemes are established, and more species are covered. The scheme's strength is its ability to produce updated indices on an annual basis, thereby offering a unique and powerful way of assessing the sustainability of land-use policies and the effectiveness of conservation measures.

For more information on the EBCC and the PECBMS, visit www.ebcc.info

Table 1. Categories and codes used for recording population trend direction and magnitude.

Trend direction (codes)	Trend magnitude categories (%) 0–19	20–29	30–49	50–79	>80
Increasing (+)	+1	+2	+3	+4	+5
Decreasing (-)	-1	-2	-3	-4	-5
Stable (0)	0	n/a	n/a	n/a	n/a
Fluctuating (F)[a]	n/a	F2	F3	F4	F5
New breeder (N)[b]	n/a	n/a	n/a	n/a	n/a
Extinct (X)[c]	n/a	n/a	n/a	n/a	n/a

a Species that underwent interannual changes exceeding 20% during 1990–2000, but whose numbers remained broadly stable over the decade as a whole.
b Species that began to breed regularly during 1990–2000, either for the first time or as part of a recolonisation.
c Species that became nationally extinct during 1990–2000, or which were recorded during 1970–1990 in Tucker and Heath (1994) but not since.

extrapolation by national coordinators, and personal consultation with other experts. Consequently, national coordinators were asked to indicate the reliability of each estimate of population size and trend using data quality codes, following the system used in *BiE1*:

1 Poorly known, with no quantitative data available.
2 Generally well known, but only poor or incomplete quantitative data available.
3 Reliable quantitative data available (e.g. atlas, survey or monitoring data) for the whole period and country.

Although data on breeding and wintering range trends during 1970–1990 were presented in *BiE1*, no equivalent data were collected for 1990–2000, for two reasons. First, for most species, 10 years is too short a period over which to detect changes in range size, unless their numbers are also changing very rapidly. Second, few European countries repeated the atlas surveys necessary to assess range changes between 1990–2000. Nevertheless, range data retain an important role in conservation status assessments, and the next generation of atlases has now either begun to appear or is in preparation. Consequently, range trends for the period 1990–2010 should be available for many countries by 2010, and will therefore be used in the next edition of *Birds in Europe* (currently scheduled for 2012–2014).

DATA MANAGEMENT

All data were checked by staff in BirdLife's European Division, and any queries were referred back to national coordinators for their comment and approval before amendment. Once verified, some 10,000 breeding records and 4,000 winter records were added to the BirdLife/EBCC European Bird Database (Box 4), which forms part of BirdLife's World Bird Database (Box 5). Together with the existing trend data from 1970–1990, these population and trend data for 1990–2000 were used in the quantitative reassessment of each species's European conservation status, as described in the next chapter. Further data checks for specific species were made during the assessment process, and a final comprehensive check was made by circulating the status maps for all species to national coordinators before publication.

DATA AVAILABILITY

One of BirdLife's key principles is to maximise the availability and use of its data for conservation purposes. Much of the information in this book is also available (as species factsheets and tables) over the Internet, via the URL at the bottom of each page. Users from outside the BirdLife International Partnership can also request additional data from the European Bird Database through the web site (providing that guidelines on the use and dissemination of data stipulated in the Data Access Policy are followed).

Box 4. European Bird Database.

In 1990, when BirdLife International launched the Dispersed Species Project that culminated in *BiE1*, the EBCC Atlas Project had already been underway for five years. During this time, a large amount of information had been collected on population sizes and trends and habitat preferences for most breeding species in Europe. Many of these data were also required for the Dispersed Species Project. To avoid duplication of effort and to ensure consistency between the data sets, it was therefore decided to combine work into a single cooperative project. BirdLife used its European network—in collaboration with the EBCC Atlas contributors—to update existing data, obtain data from other countries, and acquire midwinter population data and other information required for the Dispersed Species Project. All data were stored in the newly created BirdLife International/EBCC **European Bird Database**.

Data for the subset of species of European conservation concern were first published in *BiE1* (Tucker and Heath 1994). Partial distribution and population data were published for all species in the *EBCC Atlas of European breeding birds* (Hagemeijer and Blair 1997). *European bird populations* (BirdLife International/EBCC 2000) presented population data for all species, including updated population size estimates for 27% of species, but contained no other new information. Hence, to date, the data collection process has had four main phases:

- collecting population and trend data for *BiE1* (1990–1993);
- collecting distribution data for the *EBCC Atlas* (1985–1995);
- updating some population data for *European bird populations* (1996–1999);
- comprehensive updating of population and trend data for *BiE2* (2002–2004).

The next update is scheduled for 2006–2008, when BirdLife intends to publish population and trend estimates for the period 2000–2005.

Box 5. World Bird Database.

The European Bird Database forms part of the **World Bird Database**—a tool developed by BirdLife International to manage, analyse and report on the full spectrum of its data on species and sites, for the purpose of conserving birds and their habitats. The World Bird Database is a distribution network that enables the flow of information between the people who collect or update it, those who collate and verify it, and those who analyse it to set targets and influence policy and decision-making. It exemplifies BirdLife's philosophy of taking conservation action based on sound science.

Key benefits of the World Bird Database are its ability to:

- manage and validate a large volume of information, especially data on globally threatened species and Important Bird Areas (IBAs);
- analyse trends and monitor changes;
- link site (IBA) data with species data, e.g. for international conventions, directives and other initiatives or programmes;
- determine the conservation status of sites or species internationally, regionally and nationally;
- produce focused, targeted reports for specific purposes;
- improve the sharing of information and expertise between BirdLife Partners;
- improve electronic links to non-BirdLife data and information, such as socio-economic and non-bird data, for use in analyses;
- link to geographical information systems (GIS) for presentation and analytical purposes.

The World Bird Database runs in single-user and multi-user environments, and can also be made available over the Internet. The development tools (Microsoft Visual Basic and Microsoft Access) offer suitable upgrade paths when new technologies become available. The World Bird Database is simple to use and intuitive in operation, adopting the same 'look and feel' as other widely used Microsoft Windows products.

REFERENCES

BirdLife International/European Bird Census Council (2000) *European bird populations: estimates and trends.* Cambridge, UK: BirdLife International (BirdLife Conservation Series no. 10).

Hagemeijer, E. J. M. and Blair, M. J., eds. (1997) *The EBCC Atlas of European breeding birds: their distribution and abundance.* London: T. & A. D. Poyser.

Heath, M. F. and Evan, M. I., eds. (2000) *Important Bird Areas in Europe: priority sites for conservation.* 2 volumes. Cambridge, UK: BirdLife International (BirdLife Conservation Series no. 8).

Pannekoek, J. and van Strien, A. J. (2001) *TRIM 3 Manual. TRends and Indices for Monitoring Data.* Research paper no. 0102. Voorburg, The Netherlands: Statistics Netherlands. [Software freely available at www.ebcc.info]

Stroud, D. A., Davidson, N. C., West, R., Scott, D. A., Haanstra, L., Thorup, O., Ganter, B. and Delany, S., compilers (2004) Status of migratory wader populations in Africa and Western Eurasia in the 1990s. *International Wader Studies* 15: 1–259.

Thorup, O., compiler (in press) Breeding waders in Europe: a year 2000 assessment. *International Wader Studies* 14.

Tucker, G. M. and Heath, M. F. (1994) *Birds in Europe: their conservation status.* Cambridge, UK: BirdLife International (BirdLife Conservation Series no. 3).

STATUS ASSESSMENT

CRITERIA DEVELOPMENT

Historical background

The aim of this assessment is to identify species of conservation concern on a European scale. In the early 1990s, no objective criteria existed for assessing a species's conservation status at a regional level. When compiling the original *Birds in Europe* (Tucker and Heath 1994, hereafter '*BiE1*'), BirdLife therefore aligned its criteria with the relevant articles of the EU Birds Directive (Box 1) to develop a policy-relevant system by which species were allocated a European threat status (see Appendix 6 for details). Endangered corresponded with Article 4.1(a), Vulnerable with Article 4.1(b), and Rare and Localised with Article 4.1(c), whereas Declining referred to Articles 2 and 4.1(d). Species classified as Secure had a Favourable conservation status, but all others had an Unfavourable conservation status, and were therefore treated as Species of European Conservation Concern (SPECs).

Box 1. Selected provisions of the EU Wild Birds Directive (79/409/EEC).

Article 1 states that the Directive relates to the conservation of all species of wild birds occurring naturally in the European territory of the Member States, and that it applies to birds, their eggs, nests and habitats.

Article 2 requires Member States to take measures to maintain the population of the species referred to in Article 1 at a level that corresponds in particular to ecological, scientific and cultural requirements (while taking account of economic and recreational requirements), or to adapt the population of these species to that level.

Article 4.1 requires Member States to take special habitat conservation measures to ensure the survival and reproduction, in their area of distribution, of species listed on Annex I that are: (a) in danger of extinction; (b) vulnerable to specific changes in their habitats; (c) considered rare, because of small populations or restricted local distribution; or (d) in need of particular attention, owing to the specific nature of their habitat.

In particular, Member States are required to classify the most suitable territories in number and size as special protection areas for the conservation of these species, as well as regularly occurring migratory species (covered by Article 4.2), taking into account their protection requirements in the geographical sea and land area where the Directive applies.

The Directive also states that trends and variations in population levels should be taken into account as a background for evaluations. For details of the species listed on Annex I as of 2004, see Appendix 3.

Incorporating the IUCN Red List Criteria

Since *BiE1* was published, SPEC categories have been used widely in national and regional priority-setting exercises across Europe, and have become well known among conservationists and decision-makers. For the sake of comparison and consistency, it is important to retain as much stability in their structure as possible. However, the SPEC system should also have the flexibility to be adapted over time, particularly in the light of new and potentially beneficial developments. One such development was the publication in 2003 of guidelines for applying the IUCN Red List Criteria at a regional level (IUCN 2003a). At a global level, these criteria are firmly established as a valuable tool for assessing species' relative extinction risk (classifying those with a high risk as Critically Endangered, Endangered or Vulnerable; see Appendix 7), and thereby helping to set priorities for conservation action.

The new guidelines make it possible to assess species' relative extinction risk at a European level, using data from within the region. Initially, the IUCN Red List Criteria are applied to the regional population as specified by IUCN (2001). This preliminary classification may then be adjusted if there are populations outside the region that could affect the species's regional extinction risk (for example, by exerting a 'rescue effect', whereby immigration into the region may prevent local extinction; IUCN 2003a). In these cases, the preliminary threat category is downgraded to a level that more accurately reflects the species's regional extinction risk. All species with a relatively high risk of extinction—at either a global or European level—are clearly of conservation concern. Given the advantages of the IUCN system, and following an extensive consultation process (involving the *BiE2* national coordinators and many others in the BirdLife network), it was concluded that the *BiE1* criteria for Endangered and Vulnerable (see Appendix 6) should be replaced by the IUCN Red List Criteria in *BiE2*.

Reconciling extinction risk and conservation status

The IUCN Red List Criteria classify species solely on the basis of their relative extinction risk (IUCN 2001). However, as discussed above, Unfavourable conservation status has a much broader definition. This is spelt out clearly in Article 1 of the EU Habitats Directive (Box 2), which is currently applied by the European Commission as a working definition of Article 2 of the Birds Directive (CEC 2004).

Box 2. Selected provisions of the EU Habitats Directive (92/43/EEC).

Article 1(i) defines the conservation status of a species as 'the sum of the influences acting on the species concerned that may affect the long-term distribution and abundance of its populations in the European territory of the Member States'.

It states that a species's conservation status will be taken as Favourable when:

- population dynamics data on the species concerned indicate that it is maintaining itself on a long-term basis as a viable component of its natural habitats; and
- the natural range of the species is neither being reduced nor is likely to be reduced for the foreseeable future; and
- there is, and will probably continue to be, a sufficiently large habitat to maintain its populations on a long-term basis.

No species meeting the IUCN Red List Criteria at a regional level can be considered to have a Favourable conservation status in Europe. To be classified as Vulnerable (the lowest of the three IUCN threatened categories) a species must undergo a reduction in population size of at least 30% over 10 years or three generations (or have a small population or geographic range; see Appendix 7). It is difficult to claim that a species experiencing a decline of this magnitude is maintaining its population, that its range is stable, and that it remains a viable component of its habitat. Crucially, however, this does not mean that the opposite is true: species that are not threatened as defined by the IUCN Red List Criteria do not necessarily have a Favourable conservation status.

Many bird species remain widely distributed in Europe, although their populations and ranges have suffered significant long-term declines, owing mainly to habitat loss or degradation. Typically, these species have declined at a rate that does not exceed 30% over 10 years or three generations, and hence does not trigger IUCN Red List Criterion A. In many cases, these declines continue to the present day, although often at a reduced rate because of the heavy losses already suffered. If the IUCN Red List Criteria alone were used to assess conservation status, then species that are depleted or declining only moderately could move from Unfavourable to Favourable without any genuine improvement in their conservation status (provided that the size of their population or range does not trigger Criteria B, C or D). This is because Criterion A applies a 'moving time window' approach, which considers only the last 10 years or three generations.

Based on the definition of Favourable conservation status in Box 2, it was concluded that the SPEC criteria should continue to highlight species that are depleted or declining moderately as having an Unfavourable conservation status, even though they are not threatened by imminent extinction. IUCN (2001) acknowledges this distinction, stating clearly that although the Red List focuses attention on taxa at the highest risk, it is not the only means of setting conservation priorities. In the context of the EU directives and other international conservation agreements (such as the Bern and Bonn

Conventions; see Appendix 3), it would be misleading to assess conservation status based solely on the IUCN Red List Criteria. This falls short of the guidelines for determining Favourable conservation status, and also risks losing an important function of the term, i.e. steering the implementation of the relevant directives and conventions. Consequently, in *BiE2*, the threat status resulting from a regional application of the IUCN Red List Criteria forms only part of the evidence for assessing species' conservation status.

Interpreting the Near Threatened concept in Europe

According to IUCN (2003b), a species should be classified as Near Threatened if it does not currently qualify for Critically Endangered, Endangered or Vulnerable, but is close to qualifying, or is likely to qualify in the near future. Estimates of range and population size or decline should therefore be 'close' to the thresholds for Vulnerable (see Appendix 7), especially if there is a high degree of uncertainty or the species meets some of the sub-criteria. The crucial point is that Near Threatened is not triggered using quantitative criteria, but in the context of a species's proximity to the thresholds for another category (as well as, for instance, its ecological susceptibility, or the nature of the threats facing it). In other words, there is considerable latitude for interpretation.

Consequently, having decided to apply the IUCN Red List Criteria at a European level, it was agreed that the existing *BiE1* criteria of Rare, Localised and Declining could legitimately be interpreted as an expansion of the IUCN category of Near Threatened, and hence be applied as in *BiE1*. Only two minor amendments were necessary: an adjustment to the Declining criterion (because trend data were collected over 10, rather than 20, years for *BiE2*), and the introduction of the Depleted criterion, to highlight species that have not yet recovered from historical declines (see Box 3).

It was also agreed that all globally Near Threatened species occurring in Europe should be categorised as SPEC 1. This was not the case in *BiE1*, when such species were classified as SPEC 2 or 3. Nevertheless, these species are—by definition—also of global conservation concern, and thus deserve to be ranked alongside those meeting the IUCN Red List Criteria at a global level. This minor revision also ensured consistency with two of the criteria used to identify Important Bird Areas at a global (A1) and European Union (C1) level, which refer to 'sites that regularly hold significant numbers of a globally threatened species, or other species of global conservation concern' (Heath and Evans 2000).

Conclusions regarding the revised criteria

The revised SPEC list presented in this review includes species meeting the IUCN Red List Criteria at a European level, and those meeting the additional 'Near Threatened' criteria derived mainly from *BiE1*. Whilst not at imminent risk of regional extinction, the latter also have an Unfavourable conservation status in Europe, and are hence deserving of special conservation measures. This approach should be regarded as a legitimate interpretation of the Near Threatened concept at European level, taking into account the fact that a species's conservation status depends on more than just its relative extinction risk. Thus, it remains consistent with the definition and interpretation of Favourable conservation status in the EU directives (Boxes 1 and 2) and in other international conventions and agreements.

In summary, the only significant differences between this system and that used in *BiE1* are:

- the reallocation of globally Near Threatened species from SPEC 2 and 3 to SPEC 1, thereby placing all species of global conservation concern in the same category;

Box 3. Classification of European threat status.

Each species is initially assessed against the IUCN Red List Criteria (IUCN 2001) at a European level, and then against the additional criteria derived mainly from *BiE1* (Tucker and Heath 1994). All population size thresholds refer to minimum population estimates. In descending order of threat, a species is evaluated as:

- **Critically Endangered (CR)** if its European population meets any of the IUCN Red List Criteria for Critically Endangered (see Appendix 7). Such species have an Unfavourable conservation status in Europe because they are considered to be facing an extremely high risk of extinction in the wild (IUCN 2001).
- **Endangered (EN)** if its European population meets any of the IUCN Red List Criteria for Endangered (see Appendix 7). Such species have an Unfavourable conservation status in Europe because they are considered to be facing a very high risk of extinction in the wild (IUCN 2001).
- **Vulnerable (VU)** if its European population meets any of the IUCN Red List Criteria for Vulnerable (see Appendix 7). Such species have an Unfavourable conservation status in Europe because they are considered to be facing a high risk of extinction in the wild (IUCN 2001).
- **Declining (D)** if its European population does not meet any IUCN Red List Criteria, but declined by more than 10% over 10 years (i.e. 1990–2000) or three generations, whichever is longer. Such species have an Unfavourable conservation status in Europe because they are unable to maintain their populations and/or natural ranges in the long-term. [*BiE1* classified species as SPECs if the size of their population or range declined between 1970–1990 by **20%** or more in 33–65% of the population (or by 50% or more in 12–24% of the population). Given the shorter time period covered by *BiE2*, an overall decline exceeding **10%** is comparable with this approach.]
- **Rare (R)** if its European population does not meet any IUCN Red List Criteria and is not Declining, but numbers fewer than 10,000 breeding pairs (*or* 20,000 breeding individuals *or* 40,000 wintering individuals[1]), and is not marginal[2] to a larger non-European population. Such species have an Unfavourable conservation status in Europe because the small size of their population renders them more susceptible to accelerated declines as a result of:
  - break-up of social structure;
  - loss of genetic diversity;
  - large-scale population fluctuations and catastrophic chance events;
  - existing or potential exploitation, persecution or disturbance by humans.
- **Depleted (H)** if its European population does not meet any IUCN Red List Criteria and is not Rare or Declining, but has not yet recovered from a moderate or large decline suffered during 1970–1990 (see Appendix 6), which led to its classification as Endangered, Vulnerable or Declining in *BiE1*. Such species have an Unfavourable conservation status in Europe because they have already undergone a population decline of the type that various directives, conventions and agreements intend to prevent, and have not yet recovered.
- **Localised (L)** if its European population does not meet any IUCN Red List Criteria and is not Declining, Rare or Depleted, but is heavily concentrated, with more than 90% of the European population occurring at 10 or fewer sites (as listed in Heath and Evans 2000). Such species have an Unfavourable conservation status in Europe because their dependence on a small number of sites renders them more susceptible to accelerated declines as a result of:
  - large-scale population fluctuations and catastrophic chance events;
  - existing or potential exploitation, persecution or disturbance by humans.
- **Secure (S)** if its European population does not meet any of the criteria listed above. Such species have a Favourable conservation status in Europe.

In addition, a species is considered to be:

- **Data Deficient (DD)** if there is inadequate information to make a direct, or indirect, assessment of its risk of extinction based on its distribution and/or population status. A species in this category may be well studied, and its biology well known, but appropriate data on its abundance and/or distribution in Europe are lacking. Data Deficient is therefore not a category of threat (IUCN 2001).
- **Not Evaluated (NE)** if its European population has not yet been evaluated against the criteria.

[1] Only wintering populations of waterbirds of the families Anatidae, Haematopodidae, Charadriidae and Scolopacidae are considered, because these are typically the species with well-monitored winter populations.

[2] Marginal European populations are those that may experience significant immigration from neighbouring non-European populations (the combined total of which exceed 10,000 pairs), and are thus at a reduced risk of extinction resulting from small population size.

Box 4. Example calculation of overall European population trend.

This example illustrates the steps involved in the trend calculations made for each species, using data for Black-throated Loon *Gavia arctica* (see p. 28). Population and trend data for this species are as follows:

Gavia arctica	2000 breeding population (pairs)			1990–2000 population trend		Back-calculated 1990 breeding population (pairs)	
Country	Minimum	Maximum	Geomean	Direction	Magnitude (%)	'Best-case' (geomean)	'Worst-case' (geomean)
Belarus	15	30	21	Stable	+/- 0–19	21	21
Estonia	5	10	7	Stable	+/- 0–19	7	7
Finland	8,000	10,000	8,944	Increasing	+ 10	8,131	8,131
Latvia	0*	5	2	Declining	- 30–49	3	4
Lithuania	3	5	4	Fluctuating	+/- 30–49	4	4
Norway	2,000	5,000	3,162	Declining	- 0–19	3,162	3,904
Russia	35,000	70,000	49,497	Declining	- 20–29	61,872	69,715
Sweden	5,500	7,000	6,205	Stable	+/- 0–19	6,205	6,205
UK	155	190	172	Increasing	+ 17	147	147
Total (pairs)	50,678	92,240	68,015			79,552	88,137

* Substituted with 1 when calculating geometric mean

1. The upper and lower limits of the 1990–2000 trend estimate from each country were applied to the geomean national population estimate, to back-calculate the most likely minimum and maximum population sizes for 1990. [For national populations that remained stable or fluctuated during 1990–2000, the 2000 geomean population estimate was taken as a reasonable estimate of the 1990 population estimate, i.e. no back-calculation was necessary. National populations whose trend was unknown during 1990–2000 were excluded from calculations, but were taken into account when assessing whether a species's status was provisional or not; see also Box 6.]
   Example The Norwegian population of *G. arctica* declined by 0–19% during 1990–2000, leaving 2,000–5,000 pairs in the year 2000, with a geomean (hereafter just 'mean') of 3,162 pairs. Back-calculating from the mean gave a best-case–worst-case estimate of 3,162–3,904 pairs in 1990.

2. The back-calculated population estimates from each country were summed to give a European best-case–worst-case population estimate for 1990.
   Example *G. arctica* bred in nine European countries during 1990–2000. The sum of the nine national back-calculated population estimates for 1990 was 79,552–88,137 pairs.

3. The mean European population estimate for 2000 was compared to the values obtained in step 2, to calculate the best- and worst-case trend scenarios during 1990–2000.
   Example The mean European population estimate for *G. arctica* in 2000 was 68,015 pairs. Comparison with the values obtained in step 2 indicated that the European population declined by 15–23% during 1990–2000:

   Best-case overall trend scenario during 1990–2000 = (79,552 - 68,015) / 79,552 x 100 = -15%
   Worst-case overall trend scenario during 1990–2000 = (88,137 - 68,015) / 88,137 x 100 = -23%

4. For species with a generation length of 3.3 years or less (i.e. most passerines), the calculations ended here, because 10 years is the appropriate time period for assessing trends against IUCN Red List Criterion A (see Appendix 7). When the species was assessed against the criteria, the worst-case trend obtained in step 3 was compared with the relevant thresholds to determine the species's status. It was also used to allocate each species to one of the verbal trend categories in Box 6.
   Example If *G. arctica* were a short-lived species with a generation length of <3.3 years, the worst-case trend calculated in step 3 (-23%) would not exceed the IUCN Red List Criterion A decline threshold for Vulnerable (≥30%). However, it does exceed 10%, so the species would be evaluated as Declining, with a verbal trend of 'moderate decline'.

5. For species with a generation length exceeding 3.3 years (i.e. most non-passerines), further calculations were required to extrapolate the trend obtained in step 3 to the appropriate three-generation time period (see Box 5) for assessment against IUCN Red List Criterion A (see Appendix 7). This involved first calculating the annual rate of population change during 1990–2000.
   Example *G. arctica* has a generation length of 7 years, so its trend must be assessed over 21 years. Assuming that most species show exponential increases or decreases over time (following IUCN 2003b), the species's annual rate of population change during 1990–2000 (-1.4% to -2.1%) was calculated as follows:

   Annual best-case trend = $(((1 + 0.15)^{1/10}) - 1) \times 100\%$ = 1.4%
   Annual worst-case trend = $(((1 + 0.23)^{1/10}) - 1) \times 100\%$ = 2.1%

6. The annual rate of change was then extrapolated to the appropriate time period.
   Example If *G. arctica* continued to decline at the same annual rate for three generations, then it would have declined overall by between 25% and 36%:

   Overall best-case trend extrapolated to three generations = $(1 - ((1 - 0.014)^{21})) \times 100\%$ = -25%
   Overall worst-case trend extrapolated to three generations = $(1 - ((1 - 0.021)^{21})) \times 100\%$ = -36%

7. When a species with a generation length exceeding 3.3 years was assessed against the criteria, the worst-case trend obtained in step 6 was compared with the relevant thresholds to determine the species's status. It was also used to allocate each species to one of the verbal trend categories in Box 6.
   Example Taking into account the generation length of *G. arctica*, the worst-case trend obtained in step 6 met IUCN Red List Criterion A for Vulnerable, because the decline exceeded 30%. Consequently, the species was evaluated as Vulnerable, and was allocated a verbal trend category of 'large decline'.

Note: *G. arctica* underwent a large decline during 1970–1990, so it is very likely that an overall decline exceeding 30% has taken place over the last three generations. Consequently, the species meets IUCN Red List Criterion A2 (see Appendix 7). However, some other species have generation lengths exceeding 10 or even 20 years (especially certain seabirds; see Appendix 4). Even if such species declined during both 1970–1990 and 1990–2000, it is difficult (without further evidence) to justify extrapolating their recent trends back beyond 1970. In these cases, provided that there was no evidence to suggest that recent trends are likely to change, they were extrapolated into the future to invoke IUCN Red List Criterion A4, which takes into account both past and future trends (see Appendix 7).

- the replacement of the original Endangered and Vulnerable criteria with the IUCN Red List Criteria;
- the introduction of the Depleted criterion, to highlight species that declined significantly during 1970–1990 and have yet to recover (although their declines have slowed or ceased).

By retaining a system that closely mirrors the one applied in *BiE1*, the results of the two assessments can be compared with confidence. This is important, because any changes to a list of species of conservation concern should reflect genuine changes in status (or improved knowledge), rather than changes resulting solely from the application of a different set of criteria. Overall, these revised criteria draw attention to all species of European conservation concern, and ensure that a species is assigned Favourable conservation status only if it can be regarded as Secure in the long term.

DATA ANALYSIS

The starting point for this assessment was the list of 514 species assessed in *BiE1*, taking into account the relevant changes in taxonomy and nomenclature over the last decade (see Appendix 4 for details). Species that breed or winter in Europe only occasionally were not included, as Europe is not within their natural range. Four species that occur regularly in Europe, but on passage only, were not assessed: Sooty Shearwater *Puffinus griseus*, Great Shearwater *Puffinus gravis*, Curlew Sandpiper *Calidris ferruginea* and Slender-billed Curlew *Numenius tenuirostris*. However, the first and last of these are of global conservation concern (BirdLife International 2004; IUCN 2004) and are thus classified automatically as SPEC 1. Consequently, the total number of species whose SPEC status was evaluated was 524 (of a total of 526 occurring regularly in Europe).

For each species, the assessment of European threat status was based on four main parameters:

- minimum European population size (in or around the year 2000)
- minimum European range size
- European population trend during 1970–1990
- European population trend during 1990–2000

Population size

As described in the preceding chapter, all national population size estimates were supplied as ranges with minimum and maximum values. To calculate the minimum European population size, minimum national values were summed. For some analyses, the geometric mean (or 'geomean') European population size was required. This was obtained by calculating the geomean of each national population estimate (substituting minimum estimates of 0 for 1, where necessary), and summing all national geomeans. This method provides a better estimate than simply taking the geomean of the minimum and maximum European population sizes (Hagemeijer and Blair 1997).

Range size

For most species, the size of their European range was derived from the *EBCC Atlas* (Hagemeijer and Blair 1997), which maps the breeding ranges of almost every European species at a resolution of 50-km squares. Nevertheless, the *Atlas* does not cover Greenland, Turkey, Cyprus, the Canary Islands or most of Russia. For species whose ranges include these regions, the figures derived from the *Atlas* were adjusted accordingly, mainly with reference to the maps in del Hoyo *et al.* (1992–2003) and Snow and Perrins (1998). Given the relatively coarse scale of the *Atlas* and the nature of the adjustments made, estimates of range size probably conformed more closely to a species's Extent of Occurrence than to its Area of Occupancy (IUCN 2001). Consequently, range size was generally assessed in relation to IUCN Red List Criterion B1, rather than B2 (see Appendix 7).

1970–1990 population trend

BiE1 identified all species that experienced a moderate or large population decline during 1970–1990. For the purposes of the current review, it was also necessary to determine which species were stable and which increased during 1970–1990. This was done by applying criteria analogous to those used to identify declines in *BiE1* (see Appendix 6 for details). Thus, a species that increased during 1970–1990 by 20% or more in 33–65% of its population, or by 50% or more in 12–24% of its population (where the total size of increasing populations exceeded that of declining populations), was classified as having undergone a *moderate historical increase*. Similarly, a species that increased during 1970–1990 by 20% or more in over 66% of its population, or by 50% or more in over 25% of its population (where the total size of the increasing populations exceeded that of declining populations), was classified as having undergone a *large historical increase*. Species that met neither these criteria, nor those for historical declines, were classified as *historically stable* (or *unknown*, if no historical trend information was available).

1990–2000 population trend

Since *BiE1* was published, continuing improvements in survey and monitoring techniques in many European countries have increased the accuracy of the population trend data available. Although most trend estimates supplied for *BiE2* were still banded ranges (i.e. minimum to maximum), they were generally more precise than in *BiE1*. These improvements allowed the calculation of 'best-case' and 'worst-case' European trend scenarios for each species, using geomean population sizes. A worked example of the methods used to perform these calculations, annotated to explain each step, is provided in Box 4. For some long-lived species, the trends obtained for 1990–2000 were then extrapolated to three generation lengths (see Box 5 for details of the calculation of generation lengths). Taking the worst-case trend scenario (in accordance with the precautionary principle), each species was then allocated a verbal trend category as outlined in Box 6.

Box 5. Calculation of generation length.

IUCN (2001) defines generation length as 'the average age of the parents of the current cohort (i.e. newborn individuals in the population)'. Generation lengths (GL) were calculated using the equation:

$$GL = ((2-m)/2m) + b$$

where m = mean adult mortality in a stable population and b = mean age at first breeding in a stable population.

For many species, m was not available from the literature (primarily Cramp *et al.* 1977–1994). Thus m was estimated using two proxy life-history variables, mean clutch size (s) and mean age at first breeding (b), using the equation:

$$\ln(m) = -1.096 - 0.4215b + 0.1961s + 0.0229b^2 - 0.0097s^2$$

This equation described 88% of the variance in m for the 149 European bird species for which reliable estimates of m were available.

The generation lengths calculated with this method are given in Appendix 4.

Box 6. Allocation of 1990–2000 trends to verbal categories.

Worst-case trend scenario 1990–2000	1990–2000 trend category
≥30% decline	Large decline
10–29% decline[1]	Moderate decline
0–9% decline[2]	Small decline
<10% decline and <10% increase[2]	Stable
<0–9% increase[2]	Small increase
10–29% increase	Moderate increase
≥30% increase	Large increase
Unknown (insufficient data[3])	Unknown

1 An exception was made in cases where a species occurred (or was heavily concentrated) in a single country, and where that country reported a declining trend of 0–19% for 1990–2000. Allocating such species a 'moderate decline' (on the basis of a worst-case trend of -19%) would have inflated the SPEC list with species that probably declined only slightly overall. It would not have served the purpose of this book to list such species alongside those that underwent genuine moderate declines (≥10%) during 1990–2000. To avoid this, such species were assessed as having undergone a 'small decline'.

2 Species undergoing small declines and small increases were only distinguished from stable populations if both the worst-case and best-case trends were in the same direction. For example, a species with a worst-case trend of -8% and a best-case trend of +7% was classified as 'stable'. However, a species with a worst-case trend of -8% and a best-case trend of -5% was classified as having undergone a 'small decline'.

3 When trend data were available for less than 50% of a species's European population, it was not possible to calculate overall trends with confidence. Such species were allocated an overall trend of 'unknown', and are a clear priority for improved monitoring in the future.

Integrating trend information from 1970–1990 and 1990–2000

Differences in the precision and quality of the data from 1970–1990 and 1990–2000 meant that it would have been difficult to combine them to obtain a single overall trend for 1970–2000. Nevertheless, in the absence of an agreed historical baseline or specific targets for recovery, trends from 1990–2000 were assessed in the context of trends from 1970–1990. Under the Depleted criterion (see Box 3), a species was assigned Unfavourable conservation status if its population underwent a moderate or large decline during 1970–1990 and did not recover fully during 1990–2000. Following the same logic, a species was allocated Favourable conservation status if its population increased during 1970–1990 but then declined during 1990–2000, provided that:

- the extent of the recent decline did not exceed that of the earlier increase;
- the recent decline did not meet any IUCN Red List Criteria.

Allocation of provisional status

IUCN (2003b) stresses that Red List assessments should follow a precautionary approach, rather than an evidentiary one. Nevertheless, the limitations of the data meant that on occasions it was prudent to allocate European threat status categories only provisionally. This was done to highlight the conditional nature of their status, which could plausibly have been different if more complete and/or better quality data had been available. In *BiE1*, provisional status was assigned when more than 50% of a species's population size or trend data were of poor or unknown quality. In the current assessment, a more flexible process was used, which permitted the allocation of provisional status on a case-by-case basis. For most species, the quality of the population size data was not relevant in this process, as their European populations far exceeded the thresholds for Unfavourable conservation status. Although the quality of the population trend data was far more relevant, the most crucial factor was the likelihood of better-quality information subsequently revealing the species's current status to be incorrect. The approach is best illustrated using examples:

1. The Russian trend for Black-throated Loon *Gavia arctica* (p. 28) for 1990–2000 was of poor quality, but—because of the size of the Russian population—it had a large influence on the overall trend. If the extent of the Russian decline was underestimated, the species could have declined by more than 50% over three generations, and thus may qualify as Endangered. Conversely, if the extent of the Russian decline was overestimated, the species could have declined by less than 30% over three generations, and thus may instead qualify as Declining. Consequently, it was prudent to evaluate the species's Vulnerable status as provisional.
2. Common Raven *Corvus corax* underwent a large increase across Europe during 1970–1990, and increased slightly during 1990–2000. Nevertheless, the proportion of its European population with good or medium quality trend data was lower in 1990–2000 (31%) than in 1970–1990 (35%), when its status was only provisionally evaluated as Secure. Given the size of its population and range, and the continuing population increase, it is extremely unlikely that any better-quality information that becomes available will reveal the species's status to be anything other than Secure. Consequently, despite the quality of its population trend data, the species's Secure status is not considered to be provisional.

Assessing overall population trends using only quantitative data (i.e. restricting the analysis to countries with good or medium quality data for each species) would have introduced considerable regional bias to the assessment process. Similarly, it was not deemed appropriate to 'weight' national trends according to the data quality codes supplied by national coordinators. Bird populations in east and south-east Europe, for instance, are relatively poorly known, largely owing to the smaller number of ornithologists in these regions. Nevertheless, population trends of species in these areas may well differ from those in other (potentially better-studied) regions of Europe. Any weighting process would have obscured such trends.

It is important to note that data were not collected on range trends during 1990–2000, and that very little reliable information was available on future population or range trends. Consequently, the vast majority of assessments were based on current population and range sizes, and on recent (1970–2000) population trends. Had more information been available on recent range trends, and on projected population and range trends, it is likely that many more species would have been assessed as having an Unfavourable conservation status. Thus, the results of this assessment should be viewed as conservative.

STATUS ASSESSMENT

Having calculated all the parameters described above, each species was assessed against the criteria in Box 3, following the procedure outlined in Figure 1. As in *BiE1*, this resulted in species being classified into one of five categories, depending on their global conservation status, their European threat status and the proportion of their global population or range in Europe (Box 7). The first

Box 7. Categories of Species of European Conservation Concern (SPECs) and Non-SPECs.

- **SPEC 1** European species of global conservation concern, i.e. classified as Critically Endangered, Endangered, Vulnerable, Near Threatened or Data Deficient under the IUCN Red List Criteria at a global level (BirdLife International 2004; IUCN 2004).
- **SPEC 2** Species whose global populations are concentrated in Europe, and which have an Unfavourable conservation status in Europe.
- **SPEC 3** Species whose global populations are not concentrated in Europe, but which have an Unfavourable conservation status in Europe.
- **Non-SPEC^E** Species whose global populations are concentrated in Europe, but which have a Favourable conservation status in Europe. [Non-SPEC^E corresponds with the SPEC 4 category in *BiE1*. The name of the category has been changed because the species it contains are not SPECs.]
- **Non-SPEC** Species whose global populations are not concentrated in Europe, and which have a Favourable conservation status in Europe.

Figure 1. The procedure for classifying Species of European Conservation Concern (SPECs).

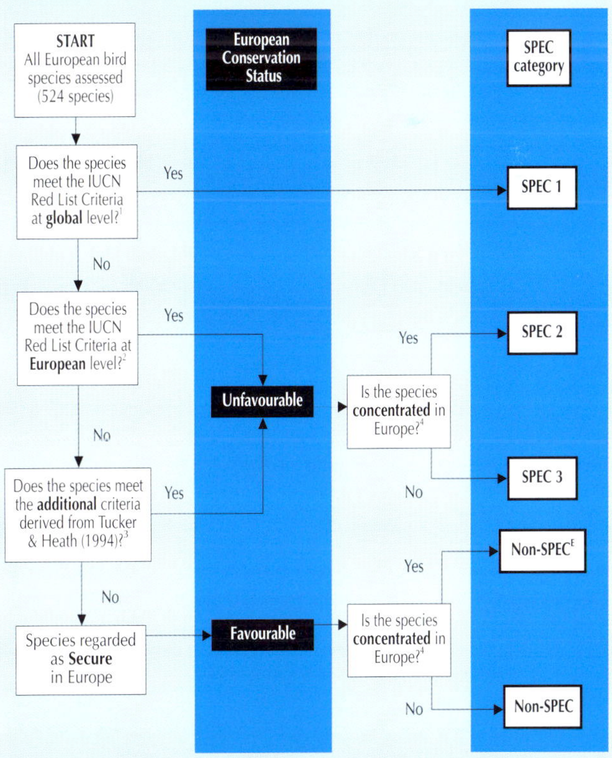

1 Species classified as Critically Endangered, Endangered, Vulnerable, Near Threatened or Data Deficient under the IUCN Red List Criteria at a global level (BirdLife International 2004; IUCN 2004).
2 Species classified as Critically Endangered, Endangered, Vulnerable under the IUCN Red List Criteria at European level, following the guidelines in IUCN (2003a).
3 Species classified as Declining, Rare, Depleted or Localised in Europe, based on the criteria developed by Tucker and Heath (1994).
4 Concentrated in Europe: species with more than 50% of their breeding or wintering population or range in Europe, according to range maps in Cramp *et al.* (1977–1994) or del Hoyo *et al.* (1992–2003), or to global population estimates where available (mostly for waterbirds, in Wetlands International 2002).

three categories together represent Species of European Conservation Concern (SPECs)—species that are either of global conservation concern (SPEC 1) or have an Unfavourable conservation status in Europe (SPEC 2 and 3). A species is considered to have an Unfavourable conservation status if its European population is classified as Critically Endangered, Endangered or Vulnerable under a regional application of the IUCN Red List Criteria, or as Declining, Rare, Depleted or Localised in Europe under the additional criteria derived from *BiE1* (Box 3). All assessments were based on breeding season data, unless a species qualified for a higher category on the basis of winter data.

Integration of breeding and wintering population data

For certain well-monitored waterbirds (i.e. species in the families Anatidae, Haematopodidae, Charadriidae and Scolopacidae), the assessment process was carried out independently on data for both the breeding and wintering populations. Through schemes such as the International Waterbird Census (see "Data collecting" p. 4), the winter populations of many species are monitored more closely than their breeding populations. In many cases, it is easier to census a species when it congregates in winter than when it is dispersed over an extensive (and often remote) breeding area. For some species, however, underlying population trends can be obscured by demographic factors, often related to interannual variation in weather conditions. In some years, for instance, birds that usually winter in Europe may be forced to move elsewhere by harsh winter conditions, whilst in others, birds that usually winter outside Europe may show marked influxes into the region.

Consequently, European threat status and SPEC categories were allocated principally on the basis of breeding data, provided that the resulting category was the same as or higher than that obtained using winter data. For species qualifying as SPECs on both breeding and wintering data, appropriate conservation measures should cover both seasons. However, for species qualifying as SPECs based solely on winter declines, conservation measures on the breeding grounds may be ineffectual if the causes of declines on the wintering grounds are not also addressed. Therefore, if a species qualifies as a SPEC based solely on winter data, this is always clearly indicated.

Use and interpretation of the criteria and categories

In general, conservation priority should be allocated according to SPEC category, with the highest importance given to species of global concern (SPEC 1), especially those threatened with global extinction. Europe has a special responsibility to ensure that the status of these species does not deteriorate within its territory, because any such deterioration would increase their (already relatively high) risk of extinction. However, when setting priorities for action, species' European threat status should also be taken into account. In the light of the limited resources available, this may sometimes involve trade-offs—for example, prioritising action for an Endangered SPEC 2 species above that of a Rare or Depleted SPEC 1 species.

Furthermore, a species's SPEC category and European threat status alone may not necessarily indicate the importance of, or urgency for, conservation actions at the national level. Assessment of this should also take into account: the proportion of a species's global and European population occurring in the country; the status of its national population; the potential for (and cost of) successful action; a species's potential as a 'flagship' to promote conservation; the effects of action on other species; and other logistical, political and strategic considerations.

Consequently, the absence of a species from the SPEC list, or its allocation to a low European threat status, does not in itself justify its exclusion from national conservation actions, or from regional European programmes (e.g. within the European Union). Indeed, the maintenance of regional European and national bird diversities and population levels is highly desirable. On the other hand, recognition as a Species of European Conservation Concern should be seen as additional justification for such conservation measures.

REFERENCES

BirdLife International (2004) *Threatened birds of the world 2004*. CD-ROM. Cambridge, UK: BirdLife International.

CEC [Commission of the European Communities] (2004) *Guidance document on hunting under Council Directive 79/409/EEC on the conservation of wild birds*. Brussels: European Commission.

Cramp, S. *et al.* (1977–1994) *The birds of the western Palearctic*. Volumes 1–9. Oxford, UK: Oxford University Press.

Hagemeijer, E. J. M. and Blair, M., eds. (1997) *The EBCC Atlas of European breeding birds: their distribution and abundance*. London: T. & A. D. Poyser.

Heath, M. F. and Evans, M. I., eds. (2000) *Important Bird Areas in Europe: priority sites for conservation*. 2 volumes. Cambridge, UK: BirdLife International (BirdLife Conservation Series no. 8).

del Hoyo, J. *et al.* (1992–2003) *Handbook of the birds of the world*. Volumes 1–8. Barcelona: Lynx Edicions.

IUCN (2001) *IUCN Red List Categories and Criteria: Version 3.1*. IUCN Species Survival Commission. Gland, Switzerland and Cambridge, UK: IUCN.

IUCN (2003a) *Guidelines for application of IUCN Red List Criteria at regional levels: Version 3.0*. IUCN Species Survival Commission. Gland, Switzerland and Cambridge, UK: IUCN.

IUCN (2003b) *Guidelines for using the IUCN Red List Categories and Criteria*. IUCN Species Survival Commission. Gland, Switzerland and Cambridge, UK: IUCN.

IUCN (2004) *The 2004 IUCN Red List of threatened species*. www.redlist.org

Snow, D. W. and Perrins, C. M. (1998) *The birds of the western Palearctic: concise edition*. 2 volumes. Oxford, UK: Oxford University Press.

Tucker, G. M. and Heath, M. F. (1994) *Birds in Europe: their conservation status*. Cambridge, UK: BirdLife International (BirdLife Conservation Series no. 3).

Wetlands International (2002) *Waterbird population estimates: third edition*. Wageningen, The Netherlands: Wetlands International (Wetlands International Global Series no. 12).

DATA PRESENTATION

The main section of this book (pages 27–290) comprises individual accounts for each of the 526 species considered in this review (see 'Data analysis', p. 10). Each account includes: an illustration of the species; a table and a map presenting national and European population size and trend data; a graph comparing the quality of these data with those collected for *BiE1* (Tucker and Heath 1994); a summary of the species's status at a European and global level; and a short text justifying each species's European conservation status. The following examples (taken from a variety of species) explain how to interpret each of these components. Each account is also available as a species factsheet over the Internet, via the URL at the bottom of each page in this book.

All scientific and common names follow standard taxonomies (see Box 1). The names of many European species have changed since *BiE1*, but because space is limited, the index to this book includes only those names in current use (i.e. those appearing at the top of each species account). However, to allow easy cross-referencing, Appendix 4 lists all the names used in this book and in *BiE1*, highlighting those that have changed in the interim. It also lists the page numbers on which each species is treated in a number of other key references.

Most of the **illustrations** originally appeared in the *EBCC Atlas* (Hagemeijer and Blair 1997), but a number derive from other sources, and some were drawn specifically for this publication. All artists (and the drawings they contributed) are credited in the artists' acknowledgements at the front of this book.

Gavia stellata
RED-THROATED LOON

SPEC 3 (1994: 3) Status (Depleted)
Criteria Large historical decline
European IUCN Red List Category —
Criteria —
Global IUCN Red List Category —
Criteria —

Country	Breeding pop. size (pairs)	Year(s)	Trend	Mag.%	References
Denmark					
Faroe Is.	15 – 15	95	(0)	(0–19)	
Greenland	(5,000 – 30,000)	90–00	?	–	
Finland	900 – 1,100	99–01	**0**	**0–19**	
Iceland	(1,000 – 2,000)	78–94	?	–	*1*
Rep. Ireland	**3 – 6**	97–02	**-**	**20–29**	
Norway	(2,000 – 5,000)	90–03	-	0–19	
Svalbard	(500 – 1,000)	90–03	(0)	(0–19)	
Russia	(20,000 – 50,000)	90–00	(0)	(0–19)	*71,100,128,138*
Sweden	**1,200 – 1,400**	99–00	0	0–19	
UK	**935 – 1,500**	94	0	0–19	*7*
Total (approx.)	32,000 – 92,000	**Overall trend** Stable			
Breeding range	>2,000,000 km²	**Gen. length** 7		**% Global pop.** 5–24	

Country	Winter pop. size (individuals)	Year(s)	Trend	Mag.%	References
Croatia	(300 – 500)	02	?	–	*16,26*
Denmark	(5,000 – 50,000)	99–00	(0)	(0–19)	*25*
Estonia	5,000 – 23,000	98	(0)	(0–19)	*1*
France	300 – 1,000	98–02	?	–	*9*
Germany	**14,500 – 14,500**	87–03	?	–	*4*
Greece	**50 – 200**	96–00	F	**>80**	
Iceland	100 – 1,000	78–94	?	–	*48*
Rep. Ireland	**860 – 1,100**	94–00	0	0–19	
Latvia	**10,000 – 10,000**	92–94	(F)	(>80)	*27*
Lithuania	2,000 – 3,000	92–02	F	20–29	
Netherlands	(6,800 – 6,800)	85–93	(+)	(–)	*3,4,5,8,9,11,12*
Poland	(1,000 – 2,500)	87–99	?	–	*113,114,129*
Romania	20 – 300	90–00	0	0–19	*60,38*
Ukraine	(50 – 250)	90–00	F	20–29	
UK	4,900 – 4,900	80–86	(0)	(0–19)	
Total (approx.)	>51,000	**Overall trend** Stable			
% in European IBAs	9 – 20	**Gen. length** 7		**% Global pop.** 5–24	

Gavia stellata is a widespread breeder across much of northern Europe, which accounts for less than a quarter of its global breeding range. Its European breeding population is relatively small (<92,000 pairs), and underwent a large decline between 1970–1990, notably in Russia and Fennoscandia. Although the species was stable overall during 1990–2000, with stable trends in most countries within its European range, its population has clearly not yet recovered to the level that preceded its decline. Consequently, it is provisionally evaluated as Depleted.

The justification is a brief text summarising the rationale for each species's European conservation status. It describes the highest criteria met using non-technical language but with some standard terminology.

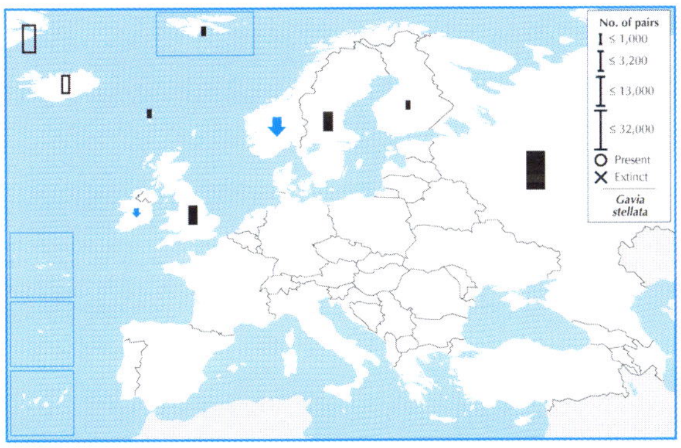

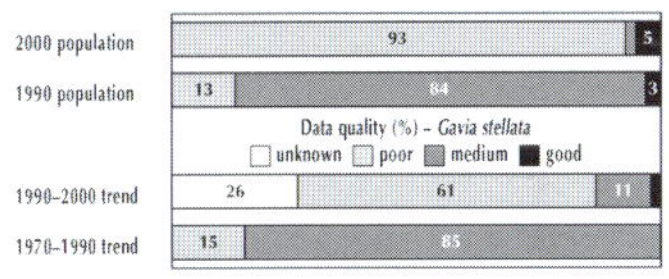

Box 1. Taxonomy of European birds.

BirdLife International maintains its own taxonomic list of all the world's bird species because there are so many different global, regional, national, site and family taxonomic checklists, and thus many differences of opinion and much confusion over the taxonomic rank of certain species. BirdLife's list of all the world's bird species is a work in progress, but is based on:

- well-recognised and established sources adopted by the BirdLife Taxonomic Working Group (BTWG; see below)—these sources are reviewed each year and, where possible and appropriate, updated or revised;
- peer-reviewed papers (published in the major ornithological journals), which have themselves been reviewed by the BTWG;
- original taxonomic research conducted by BirdLife researchers and published in BirdLife's Red Data Books and, more recently, by the BTWG, notably concerning particularly controversial and complex taxa (usually where there are important conservation considerations);
- some deviations from the adopted sources where treatment is judged to be mistaken and/or controversial.

Main taxonomic sources used by BirdLife for European birds

- **Precedence:** Sibley and Monroe (1990, 1993) is currently the default used by BirdLife for the recognition of species limits, unless the species is treated by Cramp *et al.* (1977–1994), which takes precedence for species endemic to Europe only.
- **Nomenclature:** BirdLife aims to follow David and Gosselin (2002) for consistent gender agreement of scientific names, which may be *contra* the taxonomic sources. Otherwise, names usually follow the sources above, although global consistency is taken into account (e.g. where generic names are changed regionally but not consistently for the whole group). Where species limits are recognised by more than one source, but different nomenclature is used, global consistency is again taken into account.
- **Sequence:** the names and sequence of orders and families follows Morony *et al.* (1975), whereas the species sequence within these families follows Sibley and Monroe (1990, 1993).
- **Common names:** for each species, the common name is taken from the source used to determine the scientific name, following the order of precedence outlined above.
- **Global versus regional:** globally, the sources above are used to ensure worldwide consistency in BirdLife's work. Nationally, BirdLife Partners may use other regional/national sources that are particularly relevant in this context.

BirdLife Taxonomic Working Group (BTWG)

The BTWG sets the standards by which BirdLife takes and implements decisions on bird taxonomy and nomenclature. It aims to ensure that BirdLife's list of all the world's bird species evolves in a structured, documented, transparent and defensible way. In terms of standards for species recognition, the BTWG takes a conservative but pragmatic view and advocates the Biological Species Concept. The BTWG is in the process of developing a set of criteria by which species rank can be consistently assessed where this is necessary (e.g. newly described or split species). As currently formulated, these criteria involve weighting morphological and acoustic differences from the nearest believed relative. They will be published in due course.

Box 2. Interpreting the status information.

The **status summary** indicates the European conservation status of each species, based on its global conservation status, its European threat status, and the proportion of its global population or range in Europe (for details, see the preceding chapter). If the European threat status has been downgraded following the IUCN guidelines (IUCN 2003), this is indicated with the ▼ symbol, with the number of symbols denoting the number of steps by which the status has been downgraded.

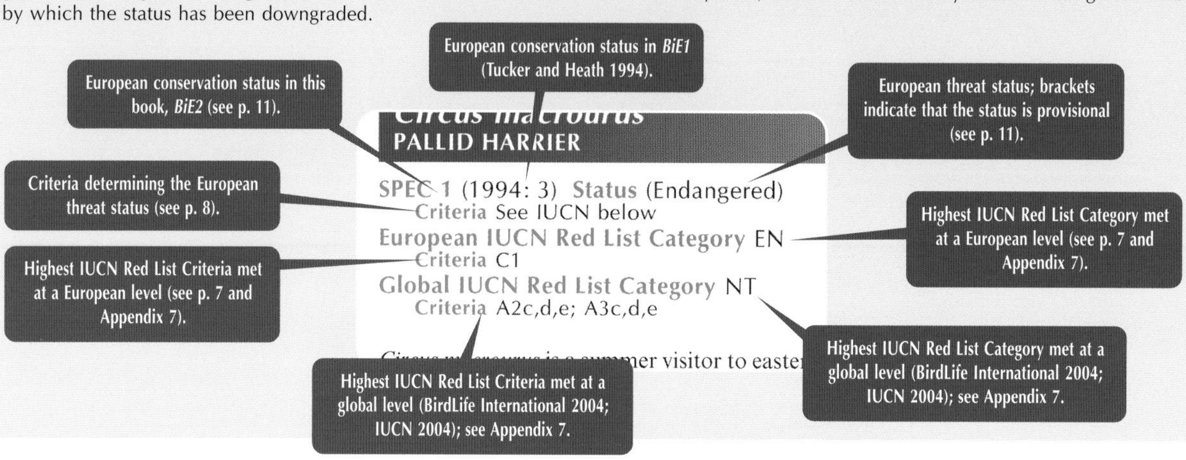

Box 3. Interpreting the data table.

The **data table** has been generated directly from the European Bird Database, presenting the data that were gathered during this review and used to reassess each species's conservation status (see 'Data collection', p. 3). The table also provides full details of the data presented in the map and summarised in the justification paragraph.

For most species, data were available on breeding populations only; these are shown in black type. If reliable winter data were available (as with many waterbirds), these are shown in blue type. The data on which a species qualifies appear at the top of the table. Due to space limitations, it has not always been possible to list all the data from the non-qualifying season. In some of these cases, partial data are shown (from countries with particularly large populations). In others, only the overall European population figures are shown.

Country	Breeding pop. size (pairs)	Year(s)	Trend	Mag.%	References
Armenia	(20 – 30)	97–02	+	0–19	
Georgia	Present	03	?	–	
Russia	**80 – 100**	99–02	**0**	**0–19**	*16,69,74*
Spain	**250 – 1,000**	02	+	**>80**	*10,16*
Turkey	**200 – 250**	01	-	**30–49**	
Total (approx.)	550 – 1,400	**Overall trend** Moderate increase			
Breeding range	>100,000 km²	**Gen. length.** <3.3		**% Global pop.** 5–24	
Country	**Winter pop. size (individuals)**	**Year(s)**	**Trend**	**Mag.%**	**References**
Albania	**0 – 2**	95–02	-	**>80**	
Azerbaijan	2,000 – 7,000	96–02	0	0–19	
Bulgaria	**12 – 2,260**	97–02	+	**>80**	
Cyprus	0 – 6	94–02	?	-	
France	**0 – 2**	98–02	**0**	**0–19**	*9*
Greece	**260 – 2,200**	95–99	**F**	**30–49**	
Romania	50 – 700	90–00	F	50–79	*60,49*
Spain	**420 – 2,400**	90–01	+	**>80**	*19*
Turkey	**3,000 – 12,000**	91–01	**F**	**>80**	
Ukraine	0 – 8	90–00	0	0–19	
Total (approx.)	>5,700	**Overall trend** Fluctuating			
% in European IBAs	>90	**Gen. length** <3.3		**% Global pop.** 75–94	

Country Nation states are listed alphabetically, with the territories of Greenland, Svalbard, the Faroe Islands, Gibraltar, the Azores, Madeira and the Canary Islands listed under their respective nation states. Some names are abbreviated because space is limited. [The names of states and territories are based on the names used by the International Standards Organisation. These geographical designations do not imply the expression of any opinion whatsoever on the part of BirdLife International concerning the legal status of any country, territory or area, or concerning the delimitation of its frontiers or boundaries.]

Breeding population size The minimum and maximum figures for breeding population size, normally in pairs, but occasionally in individuals, as indicated in the column header. Where both figures are identical, they should be interpreted as a maximum rather than a minimum estimate. All numbers are rounded to two significant figures. 'Present' indicates that a species occurs but that no other information is available.

Winter population size The minimum and maximum midwinter population size, in individuals.

Year(s) The year(s) to which the population size estimate refers.

Trend The overall **direction** of the population trend during 1990–2000 (for details, see 'Data collection', p. 5):

- \+ increasing
- 0 stable (overall change less than 20%)
- \- decreasing
- F fluctuating (interannual changes of at least 20%, but no clear trend overall)
- N new breeder within period
- X went extinct within period
- ? unknown (no trend data)

Mag. % The overall **magnitude** of the population trend during 1990–2000, as a percentage (for details, see 'Data collection', p. 5). If the magnitude of the trend is unknown, it is denoted with the symbol: '–'.

References All literature sources used by national coordinators for making population estimates are listed in the final column of the table, using a numbered code. These are cross-referenced to the list of 'Data references by country' (pp. 342–370), and also to the European Bird Database, which represents the most comprehensive source of reference information on bird populations in Europe. Where no source is specified, the data are either personal assessments by national coordinators and contributors, or are held in national or regional databases in an unpublished format.

Data quality The reliability of the data is indicated by their allocation to one of three categories (see 'Data collection', p. 5):

- **Bold** – reliable quantitative data (e.g. atlas, survey or monitoring data) available for the whole period and country (verification code 3);
- Normal – generally well known, but only poor or incomplete quantitative data available (verification code 2);
- (Bracketed) – poorly known, with no quantitative data available (verification code 1).

Data quality varies greatly between species and countries. Consequently, these codes should be examined closely when interpreting the table. Particular caution should be applied when using any individual figures based on poor, non-quantitative data. When used in combination with other European estimates, such data are extremely useful indicators of broad population size and change, but individually they may be subject to considerable error.

Total (approx.) The total European population estimate, i.e. the sum of all national population estimates, rounded to two significant figures. This figure should also be regarded only as a guideline estimate (except where comprehensively based on high-quality data), due to the considerable margin of error that is sometimes involved. Only minimum totals are presented for winter data to avoid double-counting of birds moving between countries.

Overall trend The overall trend of the European population during 1990–2000, calculated as described in 'Data analysis' (see Box 4 on p. 9) and then allocated to a verbal trend category (see Box 6 on p. 10).

Breeding range The estimated size of the European breeding range, derived from the *EBCC Atlas* (Hagemeijer and Blair 1997) as described in 'Data analysis' (p. 10).

Gen. length Generation length, i.e. the average age (in years) of the parents of the current cohort, calculated as described in 'Data analysis' (see Box 5 on p. 10).

% Global pop. The approximate percentage of the global breeding or wintering population or range lying within Europe, according to range maps in Cramp *et al.* (1977–1994) or del Hoyo *et al.* (1992–2003), or to global population estimates where available (mostly for waterbirds, in Wetlands International 2002). Seven categories are used: <5%, 5–24%, 25–49%, 50–74%, 75–94%, >95%, 100%.

% in European IBAs For most European species, the estimated size of the wintering range is unknown. Consequently, the percentage of the European wintering population occurring in Important Bird Areas is given instead (based on Heath and Evans 2000), to give a rough indication of the degree of concentration in winter.

Box 4. Interpreting the data quality graph.

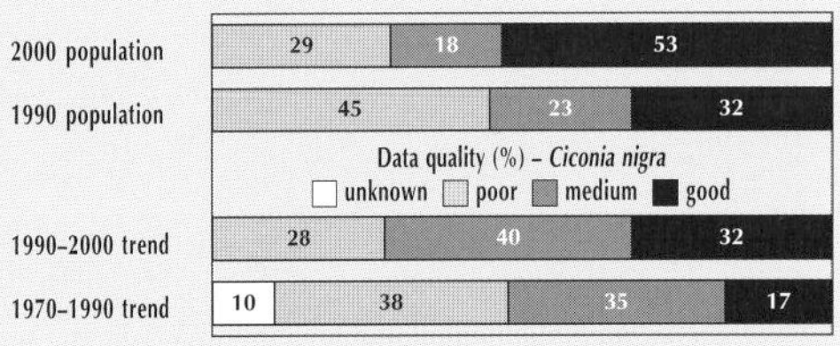

The **data quality graph** provides a visual comparison of the overall quality of the population and trend data gathered for this review with those collected for *BiE1*. All figures were calculated using the geometric means of the national population estimates for 1990 (*BiE1*) and 2000 (this book), each of which was associated with one of the data quality codes described in Box 3. The number in each section of each bar represents the percentage of the species's population comprising data of that quality (figures below 2% are not shown, due to lack of space). Thus, the example to the right clearly illustrates that the quality of the data on the population size and trend of Black Stork *Ciconia nigra* has improved since *BiE1*.

Only one data quality graph is shown for each species. In most cases it refers to breeding data and appears in black type, but if a species qualifies on winter data, the data quality graph reflects this and appears in blue type. For species that were not assessed during the 1970–1990 period, or for which no data on population sizes or trends were available, certain bars do not contain any data. In some cases, lack of space below long data tables means that the data quality graphs have been displaced to nearby pages. In these cases, page references just below the table and in the centre of the displaced graph allow easy navigation between them.

Box 5. Interpreting the map.

The **map** presents the population sizes and trends of the species in each country or territory. It gives a broad impression of the species's status in Europe, allowing rapid identification of countries in which it occurs, the approximate size and relative importance of national populations, and the population trend in every country during 1990–2000. This allows patterns in distribution and trend to be easily assessed and compared. The accompanying data table should be consulted for more precise data on numbers and trends, and for information on data quality.

All maps are based on data from the European Bird Database, analysed and mapped using a geographical information system software package (ArcView 3.2 GIS; ESRI 1999). As with the data tables, information is presented for breeding populations unless the species qualifies on winter data only. A standard base map of Europe in the Robinson projection has been used, including appropriate insets for Svalbard, the Azores, Madeira and the Canary Islands (due to space limitations).

A symbol appears in each country where the species occurs: the *shape* and *colour* of the symbol represent the population trend during 1990–2000, whereas the *size* of the symbol corresponds to the population size. [However, the *position* of the symbol bears no relation to the species's distribution with each country.]

Symbol shape and colour

Population trends for the 1990–2000 period, as presented in the data table for each species, are represented by six types of symbol:

Symbol size

Each symbol type—with the exception of present and extinct—may occur in up to four different population size classes. The overall height of the symbol corresponds to a range of population sizes. The population sizes ranges vary between species, but the upper threshold of each range is always shown in the legend accompanying each map. [However, symbol size is *not* directly proportional to population size.]

To assign each national population to one of the size ranges, the following calculations were made. First, for each country, the geometric mean of the minimum and maximum population sizes was calculated, to standardise values logarithmically. Then, using ArcView 3.2 GIS, the geometric means of all national populations were divided into four size ranges using an 'optimal ranging' method, which allowed for maximum distinction between groups but minimum distinction within groups. Depending on the amount of data for each species, these groups may have covered a range of values or just one.

A sample map:
Red-crested Pochard
Netta rufina

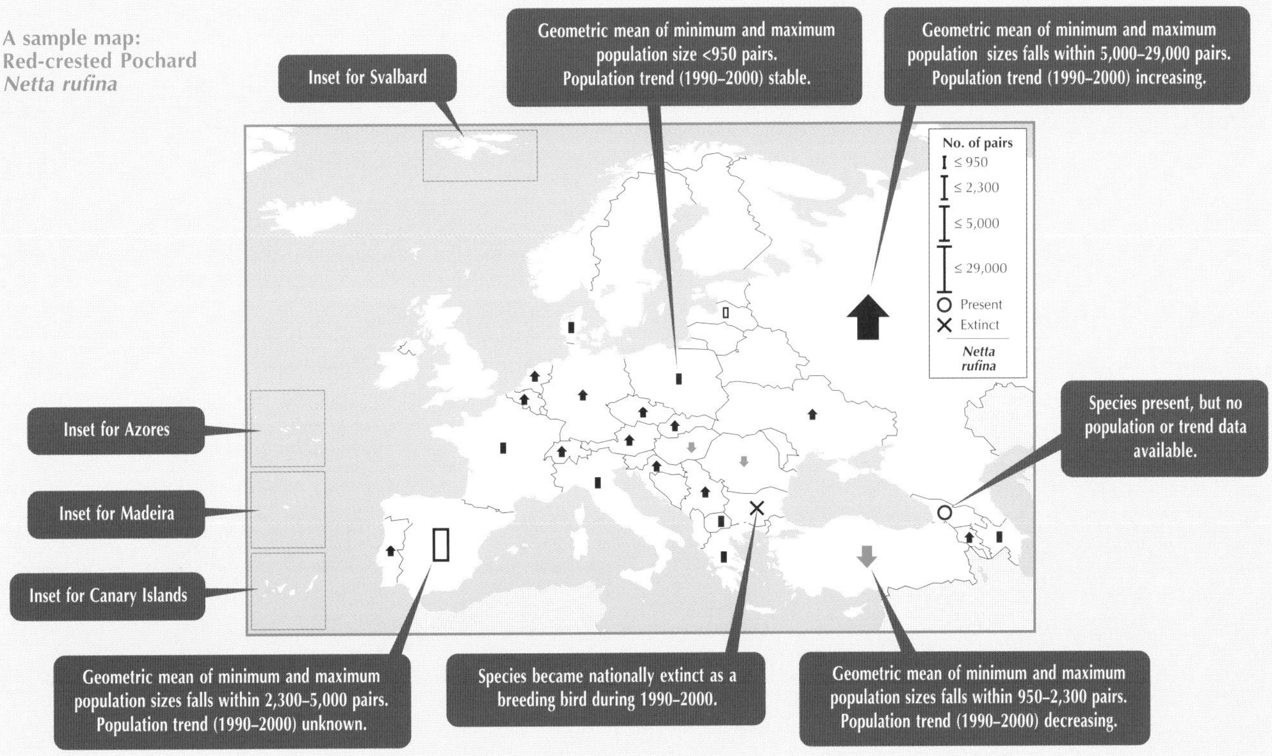

REFERENCES

Cramp, S. *et al.* (1977–1994) *Birds of the western Palearctic.* Volumes 1–9. Oxford, UK: Oxford University Press.

David, N. and Gosselin, M. (2002) Gender agreement of avian species names. *Bull. Brit. Orn. Club* 122: 14–68.

ESRI (1999) *ArcView 3.2 GIS software.* Redlands, California: Environmental Systems Research Institute.

Hagemeijer, E. J. M. and Blair, M. J., eds. (1997) *The EBCC Atlas of European breeding birds: their distribution and abundance.* London: T. & A. D. Poyser.

Heath, M. F. and Evans, M. I., eds. (2000) *Important Bird Areas in Europe: priority sites for conservation.* 2 volumes. Cambridge, UK: BirdLife International (BirdLife Conservation Series no. 8).

del Hoyo, J. *et al.* (1992–2003) *Handbook of the birds of the world.* Volumes 1–8. Barcelona: Lynx Edicions.

IUCN (2003) *Guidelines for application of IUCN Red List Criteria at regional levels: Version 3.0.* IUCN Species Survival Commission. Gland, Switzerland and Cambridge, UK: IUCN.

Morony, J. J., Bock, W. J. and Farrand, J. (1975) *Reference list of the birds of the world.* New York: Department of Ornithology, American Museum of Natural History.

Sibley, C. G. and Monroe, B. L. (1990) *Distribution and taxonomy of the birds of the world.* New Haven, USA: Yale University Press.

Sibley, C. G. and Monroe, B. L. (1993) *A supplement to "Distribution and taxonomy of the birds of the world".* New Haven, USA: Yale University Press.

Tucker, G. M. and Heath, M. F. (1994) *Birds in Europe: their conservation status.* Cambridge, UK: BirdLife International (BirdLife Conservation Series no. 3).

Wetlands International (2002) *Waterbird population estimates: third edition.* Wageningen, The Netherlands: Wetlands International (Wetlands International Global Series no. 12).

OVERVIEW OF RESULTS

This review (*BiE2*) reveals that 226 (43%) of the 526 bird species occurring regularly in Europe have an Unfavourable conservation status and therefore qualify as Species of European Conservation Concern (SPECs). Of the 524 species assessed, 40 (7.6%) are classified as SPEC 1, 45 (8.6%) as SPEC 2 and 141 (26.9%) as SPEC 3 (Figure 1; see Appendix 1 for a summary of the conservation status of all species). The percentage of species in each of these three categories is higher than in the 1994 edition of *Birds in Europe* (*BiE1*), when 38% of species (195 of the 511 assessed) were classified as SPEC 1, 2 or 3 (Figure 1).

The greatest proportional increase between the two assessments was in the number of SPEC 1 species, i.e. those of global conservation concern. However, this was primarily caused by the reclassification of globally Near Threatened species (previously listed as SPEC 2 or 3) under the revised criteria. Nevertheless, 24 species listed as SPEC 1 (5% of all species assessed) remain globally threatened or Data Deficient (Table 1), and are hence of the highest conservation importance. Of particular concern are the globally threatened species concentrated in or endemic to Europe, notably Zino's Petrel *Pterodroma madeira*, Balearic Shearwater *Puffinus mauretanicus*, Spanish Imperial Eagle *Aquila adalberti*, White-tailed Laurel Pigeon *Columba junoniae*, Fuerteventura Chat *Saxicola dacotiae* and Azores Bullfinch *Pyrrhula murina*. Although an encouraging number of SPEC 1 species increased during 1990–2000, a larger number declined, starkly highlighting the challenges ahead.

Figure 1. Percentage of European bird species in each SPEC category in *BiE1* and *BiE2*.

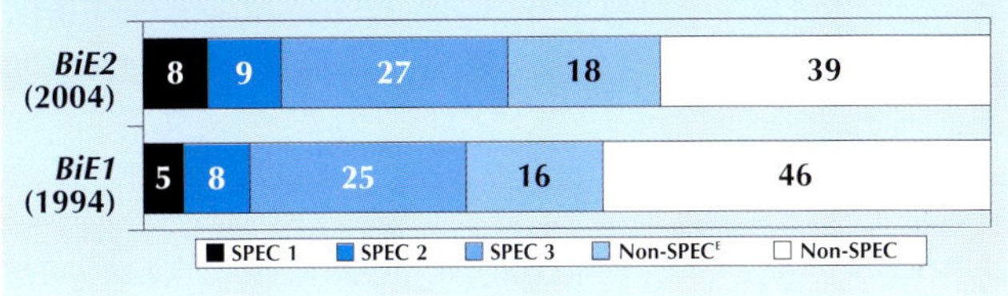

Table 1. The European status of species of global conservation concern.

Species	Global threat status [1]	European threat status [2]	Percentage in Europe [3]	1990–2000 European population trend
Pterodroma madeira Zino's Petrel	Critically Endangered	(Critically Endangered)	100	Stable
Pterodroma feae Fea's Petrel	Near Threatened	Vulnerable	25–49	Stable
Puffinus griseus Sooty Shearwater	Near Threatened	Not Evaluated [4]	—	—
Puffinus mauretanicus Balearic Shearwater	Critically Endangered	Critically Endangered	100	Large decline
Phalacrocorax pygmeus Pygmy Cormorant	Near Threatened	Secure	75–94	Moderate increase
Pelecanus crispus Dalmatian Pelican	Vulnerable	Rare	25–49	Moderate increase
Geronticus eremita Northern Bald Ibis	Critically Endangered	Critically Endangered	5–24	Stable
Anser erythropus Lesser White-fronted Goose	Vulnerable	Endangered	5–24	Moderate decline
Branta ruficollis Red-breasted Goose	Vulnerable	Vulnerable	100 (winter)	Fluctuating (winter)
Marmaronetta angustirostris Marbled Teal	Vulnerable	(Vulnerable)	5–24	Moderate decline
Aythya nyroca Ferruginous Duck	Near Threatened	(Vulnerable)	25–49	Large decline
Oxyura leucocephala White-headed Duck	Endangered	Vulnerable	5–24	Moderate increase
Haliaeetus albicilla White-tailed Eagle	Near Threatened	Rare	50–74	Large increase
Aegypius monachus Cinereous Vulture	Near Threatened	Rare	25–49	Large increase
Circus macrourus Pallid Harrier	Near Threatened	(Endangered)	25–49	Large decline
Aquila clanga Greater Spotted Eagle	Vulnerable	Endangered	25–49	Large decline
Aquila heliaca Imperial Eagle	Vulnerable	Rare	25–49	Stable
Aquila adalberti Spanish Imperial Eagle	Endangered	(Endangered)	100	Small increase
Falco naumanni Lesser Kestrel	Vulnerable	Depleted	25–49	Small decline
Falco cherrug Saker Falcon	Endangered	Endangered	<5	Large decline
Tetrao mlokosiewiczi Caucasian Grouse	Data Deficient	Data Deficient	>95	Unknown
Crex crex Corncrake	Near Threatened	Depleted	50–74	Fluctuating
Tetrax tetrax Little Bustard	Near Threatened	Vulnerable	75–94	Moderate decline
Chlamydotis undulata Houbara Bustard	Vulnerable	(Vulnerable)	<5	Unknown
Otis tarda Great Bustard	Vulnerable	Vulnerable	50–74	Stable
Glareola nordmanni Black-winged Pratincole	Data Deficient	Endangered	25–49	Large decline
Vanellus gregarius Sociable Lapwing	Critically Endangered	Critically Endangered	5–24	Large decline
Gallinago media Great Snipe	Near Threatened	Declining	50–74	Moderate decline
Numenius tenuirostris Slender-billed Curlew	Critically Endangered	Not Evaluated [4]	—	—
Larus audouinii Audouin's Gull	Near Threatened	Localised	75–94	Large increase
Columba trocaz Madeira Laurel Pigeon	Near Threatened	(Rare)	100	Small increase
Columba bollii Dark-tailed Laurel Pigeon	Near Threatened	(Rare)	100	Small increase
Columba junoniae White-tailed Laurel Pigeon	Endangered	Endangered	100	Unknown
Saxicola dacotiae Fuerteventura Chat	Endangered	Endangered	100	Small decline
Acrocephalus paludicola Aquatic Warbler	Vulnerable	(Vulnerable)	>95	Moderate decline
Fringilla teydea Blue Chaffinch	Near Threatened	Rare	100	Stable
Loxia scotica Scottish Crossbill	Data Deficient	Data Deficient	100	Unknown
Pyrrhula murina Azores Bullfinch	Endangered	(Endangered)	100	Decline
Emberiza cineracea Cinereous Bunting	Near Threatened	(Rare)	>95	Small decline
Emberiza aureola Yellow-breasted Bunting	Near Threatened	Declining	25–49	Moderate decline

[1] BirdLife International (2004); IUCN (2004). Two further European species listed by these authorities are not dealt with here because they are globally extinct: Canary Islands Oystercatcher *Haematopus meadewaldoi* and Great Auk *Pinguinis impennis*.
[2] Parentheses indicate that the status is provisional.
[3] Percentage of the global range (or population where known) within Europe.
[4] Occurs in Europe only as a passage migrant.

Beyond the 226 species with an Unfavourable conservation status in Europe, a further 94 species (18% of those assessed) are categorised as Non-SPEC[E] (which equates directly to the SPEC 4 category used in *BiE1*). Although these species currently have a Favourable conservation status, the majority of their global breeding or wintering populations are concentrated in Europe. If their conservation status were to become Unfavourable, such species would immediately qualify for SPEC 2, and in some cases they could quickly become globally threatened, especially where their populations or ranges are small. Europe therefore has a particular responsibility for these species, and should take special measures to safeguard them and to monitor their numbers.

Table 2 summarises the European threat status of all 226 species with an Unfavourable conservation status in Europe (i.e. SPEC 1, 2 and 3). Whereas almost 30% of species meet the IUCN Red List Criteria at European level (i.e. Critically Endangered, Endangered or Vulnerable), half of all SPECs qualify as either Declining (27%) or Depleted (23%), highlighting the ongoing effects of recent or historical population declines. Despite remaining stable during 1990–2000, populations of 15% of species are sufficiently small (but not marginal to a large non-European population) to qualify as Rare, and a further 4% qualify as Localised because of their heavy dependence on a small number of key sites. As in *BiE1*, insufficient data were available to assess the European Threat Status of Caucasian Grouse *Tetrao mlokosiewiczi* and Scottish Crossbill *Loxia scotica*, both of which are also considered Data Deficient at a global level. However, various studies of these species are currently in progress, the results of which should allow their status to be formally evaluated in the next assessment.

Table 2. The European threat status of all species with an Unfavourable Conservation Status in Europe (i.e. SPEC 1, 2 and 3).

European threat status	SPEC 1	SPEC 2	SPEC 3	Total
Critically Endangered	4	—	5	9
Endangered	9	—	11	20
Vulnerable	9	5	24	38
Declining	2	19	41	62
Rare	8	4	21	33
Depleted	2	15	34	51
Localised	1	2	5	8
Data Deficient	2	—	—	2
Secure	1	—	—	1
Not Evaluated	2	—	—	2
Total	40	45	141	226

EUROPEAN RED LIST

In total, 67 species meet the IUCN Red List Criteria for Critically Endangered, Endangered or Vulnerable at a European level (Table 3), and hence qualify for the European Red List. Two of the nine Critically Endangered species are concentrated within Europe (Zino's Petrel *Pterodroma madeira* and Balearic Shearwater *Puffinus mauretanicus*), but most are at the very limit of their global distribution in Europe (notably in Turkey and Spain). Although Zino's Petrel may be downlisted to Endangered in due course (owing to improved knowledge of its population size), many of the other Critically Endangered species will become extinct in Europe if current trends continue unabated. Of the 20 species qualifying as Endangered, 16 are declining in Europe, and two of these are endemic to single Macaronesian islands: Fuerteventura Chat *Saxicola dacotiae* and Azores Bullfinch *Pyrrhula murina*. Thirty-eight species are Vulnerable in Europe, 30 of which are declining within the region. A number of these species—notably Cory's Shearwater *Calonectris diomedea*, Levant Sparrowhawk *Accipiter brevipes*, Northern Lapwing *Vanellus vanellus*, Black-tailed Godwit *Limosa limosa* and European Roller *Coracias garrulus*—are of particular concern, as they are concentrated within Europe and have recently undergone substantial declines.

WINNERS AND LOSERS

A total of 32 species underwent a large decline in Europe during 1990–2000, whereas 25 species experienced a large increase during the same period (Table 4). The list of declining species includes a

Table 3. SPEC categories and population trends of all species meeting IUCN Red List Criteria at a European level.

Species[1]	SPEC category	Population trend[2]
Critically Endangered in Europe		
***Pterodroma madeira* Zino's Petrel**	**SPEC 1**	**Stable**
***Puffinus mauretanicus* Balearic Shearwater**	**SPEC 1**	**Large decline**
***Geronticus eremita* Northern Bald Ibis**	**SPEC 1**	**Stable**
Turnix sylvatica Small Buttonquail	SPEC 3	Unknown
Fulica cristata Red-knobbed Coot	SPEC 3	Fluctuating
***Vanellus gregarius* Sociable Lapwing**	**SPEC 1**	**Large decline**
Otus brucei Pallid Scops-owl	SPEC 3	Large decline
Ketupa zeylonensis Brown Fish-owl	SPEC 3	Large decline
Ceryle rudis Pied Kingfisher	SPEC 3	Large decline
Endangered in Europe		
***Anser erythropus* Lesser White-fronted Goose**	**SPEC 1**	**Moderate decline**
Aythya marila Greater Scaup	SPEC 3W	Large decline (winter)
Neophron percnopterus Egyptian Vulture	SPEC 3	Large decline
***Circus macrourus* Pallid Harrier**	**SPEC 1**	**Large decline**
***Aquila clanga* Greater Spotted Eagle**	**SPEC 1**	**Large decline**
Aquila nipalensis Steppe Eagle	SPEC 3	Large decline
***Aquila adalberti* Spanish Imperial Eagle**	**SPEC 1**	**Small increase**
Hieraaetus fasciatus Bonelli's Eagle	SPEC 3	Large decline
***Falco cherrug* Saker Falcon**	**SPEC 1**	**Large decline**
Cursorius cursor Cream-coloured Courser	SPEC 3	Unknown
***Glareola nordmanni* Black-winged Pratincole**	**SPEC 1**	**Large decline**
Charadrius leschenaultii Greater Sand Plover	SPEC 3	Moderate decline
Charadrius asiaticus Caspian Plover	SPEC 3	Unknown
***Columba junoniae* White-tailed Laurel Pigeon**	**SPEC 1**	**Unknown**
Apus affinis Little Swift	SPEC 3	Small decline
Halcyon smyrnensis White-throated Kingfisher	SPEC 3	Large decline
Ammomanes deserti Desert Lark	SPEC 3	Large decline
Melanocorypha yeltoniensis Black Lark	SPEC 3	Large decline
***Saxicola dacotiae* Fuerteventura Chat**	**SPEC 1**	**Small decline**
***Pyrrhula murina* Azores Bullfinch**	**SPEC 1**	**Decline**
Vulnerable in Europe		
Gavia arctica Arctic Loon	SPEC 3	Large decline
***Pterodroma feae* Fea's Petrel**	**SPEC 1**	**Stable**
Calonectris diomedea Cory's Shearwater	SPEC 2	Large decline
Pelagodroma marina White-faced Storm-petrel	SPEC 3	Stable
Cygnus columbianus Tundra Swan	SPEC 3W	Large decline (winter)
Branta bernicla Brent Goose	SPEC 3W	Large decline (winter)
***Branta ruficollis* Red-breasted Goose**	**SPEC 1W**	**Fluctuating (winter)**
Tadorna ferruginea Ruddy Shelduck	SPEC 3	Large decline
***Marmaronetta angustirostris* Marbled Teal**	**SPEC 1**	**Moderate decline**
***Aythya nyroca* Ferruginous Duck**	**SPEC 1**	**Large decline**
Bucephala islandica Barrow's Goldeneye	SPEC 3	Fluctuating
***Oxyura leucocephala* White-headed Duck**	**SPEC 1**	**Moderate increase**
Milvus migrans Black Kite	SPEC 3	Large decline
Gypaetus barbatus Lammergeier	SPEC 3	Moderate decline
Accipiter brevipes Levant Sparrowhawk	SPEC 2	Moderate decline
Buteo rufinus Long-legged Buzzard	SPEC 3	Large decline
Falco vespertinus Red-footed Falcon	SPEC 3	Large decline
Falco biarmicus Lanner Falcon	SPEC 3	Moderate decline
Tetraogallus caspius Caspian Snowcock	SPEC 3	Moderate decline
Alectoris chukar Chukar	SPEC 3	Large decline
Ammoperdix griseogularis See-see Partridge	SPEC 3	Moderate decline
Perdix perdix Grey Partridge	SPEC 3	Large decline
***Tetrax tetrax* Little Bustard**	**SPEC 1**	**Moderate decline**
***Chlamydotis undulata* Houbara Bustard**	**SPEC 1**	**Unknown**
***Otis tarda* Great Bustard**	**SPEC 1**	**Large decline**
Burhinus oedicnemus Eurasian Thick-knee	SPEC 3	Large decline
Vanellus spinosus Spur-winged Lapwing	SPEC 3	Moderate decline
Vanellus indicus Red-wattled Lapwing	SPEC 3	Stable
Vanellus vanellus Northern Lapwing	SPEC 2	Large decline
Limosa limosa Black-tailed Godwit	SPEC 2	Large decline
Sterna nilotica Gull-billed Tern	SPEC 3	Moderate decline
Uria lomvia Thick-billed Murre	SPEC 3	Large decline
Coracias garrulus European Roller	SPEC 2	Large decline
Calandrella cheleensis Asian Short-toed Lark	SPEC 3	Small increase
Erythropygia galactotes Rufous-tailed Scrub-robin	SPEC 3	Large decline
Oenanthe xanthoprymna Rufous-tailed Wheatear	SPEC 3	Moderate decline
Prinia gracilis Graceful Prinia	SPEC 3	Moderate decline
***Acrocephalus paludicola* Aquatic Warbler**	**SPEC 1**	**Moderate decline**

[1] Species of global conservation concern (i.e. SPEC 1) are highlighted in bold.
[2] Population trends refer to 10 years (1990–2000) for species with a generation length of <3.3 years, and three generations for all others.

Table 4. Species that underwent rapid population declines or increases in Europe during 1990–2000.

Species[1]	SPEC category
Rapid decline[2]	
Gavia arctica Arctic Loon	SPEC 3
***Puffinus mauretanicus* Balearic Shearwater**	**SPEC 1**
Cygnus columbianus Tundra Swan	SPEC 3W
Branta bernicla Brent Goose	SPEC 3W
Tadorna ferruginea Ruddy Shelduck	SPEC 3
***Aythya nyroca* Ferruginous Duck**	**SPEC 1**
Aythya marila Greater Scaup	SPEC 3W
Milvus migrans Black Kite	SPEC 3
Neophron percnopterus Egyptian Vulture	SPEC 3
***Circus macrourus* Pallid Harrier**	**SPEC 1**
Buteo rufinus Long-legged Buzzard	SPEC 3
***Aquila clanga* Greater Spotted Eagle**	**SPEC 1**
Aquila nipalensis Steppe Eagle	SPEC 3
Hieraaetus fasciatus Bonelli's Eagle	SPEC 3
Falco vespertinus Red-footed Falcon	SPEC 3
Falco cherrug Saker Falcon	SPEC 1
Alectoris chukar Chukar	SPEC 3
Perdix perdix Grey Partridge	SPEC 3
Burhinus oedicnemus Eurasian Thick-knee	SPEC 3
***Glareola nordmanni* Black-winged Pratincole**	**SPEC 1**
***Vanellus gregarius* Sociable Lapwing**	**SPEC 1**
Vanellus vanellus Northern Lapwing	SPEC 2
Limosa limosa Black-tailed Godwit	SPEC 2
Uria lomvia Thick-billed Murre	SPEC 3
Otus brucei Pallid Scops-owl	SPEC 3
Ketupa zeylonensis Brown Fish-owl	SPEC 3
Halcyon smyrnensis White-throated Kingfisher	SPEC 3
Ceryle rudis Pied Kingfisher	SPEC 3
Coracias garrulus European Roller	SPEC 2
Ammomanes deserti Desert Lark	SPEC 3
Melanocorypha yeltoniensis Black Lark	SPEC 3
Erythropygia galactotes Rufous-tailed Scrub-robin	SPEC 3
Rapid increase[2]	
Fulmarus glacialis Northern Fulmar	Non-SPEC
Morus bassanus Northern Gannet	Non-SPEC^{E}
Phalacrocorax carbo Great Cormorant	Non-SPEC
Casmerodius albus Great Egret	Non-SPEC
Phoenicopterus roseus Greater Flamingo	SPEC 3
Cygnus olor Mute Swan	Non-SPEC^{E}
Cygnus cygnus Whooper Swan	Non-SPEC^{E}W
Anser albifrons Greater White-fronted Goose	Non-SPEC
Anser anser Greylag Goose	Non-SPEC
Branta leucopsis Barnacle Goose	Non-SPEC^{E}
***Haliaeetus albicilla* White-tailed Eagle**	**SPEC 1**
Gyps fulvus Eurasian Griffon	Non-SPEC
***Aegypius monachus* Cinereous Vulture**	**SPEC 1**
Grus grus Common Crane	SPEC 2
Grus virgo Demoiselle Crane	Non-SPEC
Catharacta skua Great Skua	Non-SPEC^{E}
***Larus audouinii* Audouin's Gull**	**SPEC 1**
Larus fuscus Lesser Black-backed Gull	Non-SPEC^{E}
Larus cachinnans Yellow-legged Gull	Non-SPEC^{E}
Larus marinus Great Black-backed Gull	Non-SPEC^{E}
Sterna caspia Caspian Tern	SPEC 3
Uria aalge Common Murre	Non-SPEC
Streptopelia senegalensis Laughing Dove	Non-SPEC
Apus caffer White-rumped Swift	Non-SPEC
Petronia brachydactyla Pale Rock-finch	Non-SPEC

[1] Species of global conservation concern (i.e. SPEC 1) are highlighted in bold.
[2] Rapid trends equate to changes exceeding 30%, measured over 10 years (1990–2000) for species with a generation length of <3.3 years, and three generations for all others.

number of seabirds, waterfowl, raptors and waders, as well as a suite of species dependent on steppe habitats (e.g. Pallid Harrier *Circus macrourus*, Steppe Eagle *Aquila nipalensis*, Black-winged Pratincole *Glareola nordmanni*, Sociable Lapwing *Vanellus gregarius* and Black Lark *Melanocorypha yeltoniensis*). Meanwhile, the increasing species include a number of seabirds, large waterfowl and gulls (especially colonial breeders), as well as several raptors. Many of these have benefited from a combination of site protection measures, protection from persecution and even anthropogenic effects (e.g. discards from fishing vessels). All three SPEC 1 species on the list (White-tailed Eagle *Haliaeetus albicilla*, Cinereous Vulture *Aegypius monachus* and Audouin's Gull *Larus audouinii*) have responded to conservation efforts following the development of Species Action Plans (see Appendix 3 for a list of all European species covered by such plans).

Overall, the European conservation status of 45 species changed from Favourable to Unfavourable between the two assessments (Table 5). In practice, five of these were not formally assessed in 1994, and hence do not necessary represent true changes in status. However, the rest comprise a broad range of species, including several ducks and waders, a number of long-distance migrants (e.g. Northern House-martin *Delichon urbica*, Northern Wheatear *Oenanthe*

Table 5. Species whose European Conservation Status changed from Favourable to Unfavourable (or vice versa) between *BiE1* and *BiE2*.

Species[1]	SPEC category
Favourable to Unfavourable	
Podiceps auritus Horned Grebe	SPEC 3
***Puffinus griseus* Sooty Shearwater[2]**	**SPEC 1**
***Puffinus mauretanicus* Balearic Shearwater[2]**	**SPEC 1**
***Geronticus eremita* Northern Bald Ibis[2]**	**SPEC 1**
Anas clypeata Northern Shoveler	SPEC 3
Aythya ferina Common Pochard	SPEC 2
Aythya fuligula Tufted Duck	SPEC 3
Milvus milvus Red Kite	SPEC 2
Ammoperdix griseogularis See-see Partridge	SPEC 3
Vanellus indicus Red-wattled Lapwing	SPEC 3
Vanellus vanellus Northern Lapwing	SPEC 2
Philomachus pugnax Ruff	SPEC 2
Gallinago gallinago Common Snipe	SPEC 3
Tringa erythropus Spotted Redshank	SPEC 3
Actitis hypoleucos Common Sandpiper	SPEC 3
Larus genei Slender-billed Gull	SPEC 3
Larus armenicus Armenian Gull	SPEC 2
Uria lomvia Thick-billed Murre	SPEC 3
Otus brucei Pallid Scops-owl	SPEC 3
Ketupa zeylonensis Brown Fish-owl	SPEC 3
Apus unicolor Plain Swift	SPEC 2
Apus affinis Little Swift	SPEC 3
Halcyon smyrnensis White-throated Kingfisher	SPEC 3
Ceryle rudis Pied Kingfisher	SPEC 3
Upupa epops Eurasian Hoopoe	SPEC 3
Ammomanes deserti Desert Lark	SPEC 3
Calandrella cheleensis Asian Short-toed Lark[2]	SPEC 3
Delichon urbica Northern House-martin	SPEC 3
Erythropygia galactotes Rufous-tailed Scrub-robin	SPEC 3
Oenanthe oenanthe Northern Wheatear	SPEC 3
Oenanthe xanthoprymna Rufous-tailed Wheatear	SPEC 3
Prinia gracilis Graceful Prinia	SPEC 3
Phylloscopus bonelli Bonelli's Warbler	SPEC 2
Phylloscopus sibilatrix Wood Warbler	SPEC 2
Phylloscopus sindianus Mountain Chiffchaff	SPEC 3
Parus palustris Marsh Tit	SPEC 3
Parus cristatus Crested Tit	SPEC 2
Sitta krueperi Krüper's Nuthatch	SPEC 2
Sturnus vulgaris Common Starling	SPEC 3
Passer domesticus House Sparrow	SPEC 3
Passer montanus Eurasian Tree Sparrow	SPEC 3
Carduelis cannabina Eurasian Linnet	SPEC 2
***Pyrrhula murina* Azores Bullfinch[2]**	**SPEC 1**
***Emberiza aureola* Yellow-breasted Bunting**	**SPEC 1**
Miliaria calandra Corn Bunting	SPEC 2
Unfavourable to Favourable	
Hydrobates pelagicus European Storm-petrel	Non-SPEC^{E}
Morus bassanus Northern Gannet	Non-SPEC^{E}
Branta leucopsis Barnacle Goose	Non-SPEC^{E}
Netta rufina Red-crested Pochard	Non-SPEC
Gyps fulvus Eurasian Griffon	Non-SPEC
Falco peregrinus Peregrine Falcon	Non-SPEC
Recurvirostra avosetta Pied Avocet	Non-SPEC
Limosa lapponica Bar-tailed Godwit	Non-SPEC
Prunella ocularis Radde's Accentor	Non-SPEC^{E}
Saxicola torquata Common Stonechat	Non-SPEC
Oenanthe cypriaca Cyprus Wheatear	Non-SPEC^{E}
Hippolais olivetorum Olive-tree Warbler	Non-SPEC^{E}
Sylvia melanothorax Cyprus Warbler	Non-SPEC^{E}
Bucanetes githagineus Trumpeter Finch	Non-SPEC

[1] Species of global conservation concern (i.e. SPEC 1) are highlighted in bold.
[2] Species not assessed in *BiE1* (1994).

oenanthe and Wood Warbler *Phylloscopus sibilatrix*), and even certain very common and widespread species that are now declining across much of Europe (e.g. Common Starling *Sturnus vulgaris* and House Sparrow *Passer domesticus*). These are accompanied by several farmland birds, which join the long list of species associated with agricultural habitats that remain on the SPEC list.

Over the same period, the status of 14 species changed from Unfavourable to Favourable. These include a number of previously Localised species (e.g. Northern Gannet *Morus bassanus* and Pied Avocet *Recurvirostra avosetta*) that, owing to recent population increases and range expansion, are no longer concentrated at such a small number of sites. Others species whose status has improved include Eurasian Griffon *Gyps fulvus* and Peregrine Falcon *Falco peregrinus*, whose populations have recovered as a consequence of conservation efforts, and hence no longer qualify as Rare. The conservation status of certain other species (e.g. Radde's Accentor *Prunella ocularis* and the two Cyprus endemics, Cyprus Wheatear *Oenanthe cypriaca* and Cyprus Warbler *Sylvia melanothorax*) has been re-evaluated in the light of improved information on the size of their populations.

SPEC DISTRIBUTION AND TRENDS

The numbers of SPECs and Non-SPECs breeding in each country are listed in Table 6 and summarised in Figure 2 (see Appendix 2 for a comprehensive list of species by country). Although this simple analysis does not take into account the proportions of each species's European population in each country, it does indicate broad levels of responsibility for the conservation of SPECs. Most importantly,

Table 6. The numbers and population trends of SPECs and Non-SPECs breeding in each geopolitical unit covered by this review (see also Figures 2, 3 and 4).

	SPEC category								1990–2000 population trends of all SPECs				1990–2000 population trends of all species			
Country	SPEC 1	SPEC 2	SPEC 3	Subtotal SPEC 1–3	Non-SPEC^E	Non-SPEC	Total	Country specific[1]	Decline	Stable	Increase	Unknown	Decline	Stable	Increase	Unknown
Albania	6	26	64	**96**	45	83	**224**	–	47	23	–	26	89	76	3	56
Andorra	–	12	25	**37**	29	45	**111**	–	10	22	4	1	17	74	19	1
Armenia	10	23	62	**95**	32	98	**225**	–	21	36	7	31	47	86	22	70
Austria	6	24	49	**79**	48	90	**217**	–	6	53	20	–	16	157	44	–
Azerbaijan	12	21	74	**107**	44	102	**253**	1	9	92	3	3	11	223	14	5
Belarus	6	24	57	**87**	49	84	**220**	–	15	65	5	2	16	169	30	5
Belgium	1	20	42	**63**	48	72	**183**	–	22	19	22	–	45	78	59	1
Bosnia and Herzegovina	4	16	42	**62**	33	63	**158**	–	1	11	1	49	3	23	1	131
Bulgaria	11	28	71	**110**	54	96	**260**	–	14	61	25	10	27	165	50	18
Croatia	8	29	63	**100**	53	93	**246**	–	41	29	24	6	104	55	75	12
Cyprus	2	10	29	**41**	15	38	**94**	2	3	26	12	–	5	59	29	1
Czech Republic	6	24	47	**77**	49	83	**209**	–	31	22	20	4	47	90	67	5
Denmark	2	22	46	**70**	52	79	**201**	–	26	19	23	2	58	78	61	4
Faroe Islands	–	7	12	**19**	28	23	**70**	–	–	19	–	–	1	69	–	–
Greenland	1	2	12	**15**	15	33	**63**	6	1	7	1	6	5	24	11	23
Estonia	4	23	52	**79**	55	88	**222**	–	29	36	13	1	55	102	58	7
Finland	6	16	54	**76**	56	115	**247**	–	27	24	25	–	85	80	82	–
France	5	35	76	**116**	64	101	**281**	1	37	40	27	12	76	91	84	30
Georgia	11	25	70	**106**	41	99	**246**	–	8	12	1	85	12	18	1	215
Germany	6	25	61	**92**	63	98	**253**	–	44	29	19	–	80	97	76	–
Greece	9	31	68	**108**	51	93	**252**	–	38	64	6	–	66	169	14	3
Hungary	9	25	58	**92**	46	76	**214**	–	13	59	19	1	19	139	54	2
Iceland	1	8	21	**30**	26	28	**84**	1	1	8	7	14	7	17	19	41
Republic of Ireland	1	16	30	**47**	43	61	**151**	–	17	14	2	14	37	55	37	22
Italy	7	31	66	**104**	53	93	**250**	–	29	48	21	6	40	137	53	20
Latvia	6	22	52	**80**	51	87	**218**	–	29	37	8	6	43	128	33	14
Liechtenstein	1	8	25	**34**	33	61	**128**	–	14	15	4	1	32	71	24	1
Lithuania	5	23	51	**79**	45	81	**205**	–	28	40	11	–	51	124	27	3
Luxembourg	1	13	33	**47**	36	54	**137**	–	26	13	5	3	40	71	21	5
FYR of Macedonia	8	27	61	**96**	47	87	**230**	–	16	75	5	–	23	189	15	3
Malta	–	4	12	**16**	9	10	**35**	–	10	5	1	–	14	15	6	–
Moldova	9	19	48	**76**	38	56	**170**	–	30	34	11	1	52	94	23	1
Netherlands	2	22	46	**70**	51	77	**198**	–	32	16	14	8	50	66	61	21
Norway	4	18	50	**72**	55	107	**234**	–	24	36	12	–	48	142	44	–
Svalbard	–	3	8	**11**	13	21	**45**	–	2	9	–	–	5	34	6	–
Poland	7	25	57	**89**	55	90	**234**	–	37	31	20	1	51	118	62	3
Portugal	7	23	65	**95**	35	71	**201**	–	19	60	14	2	30	133	30	8
Azores	2	2	10	**14**	9	8	**31**	1	1	6	–	7	1	7	–	23
Madeira	3	5	12	**20**	9	12	**41**	2	–	10	–	10	–	11	2	28
Romania	12	29	65	**106**	49	97	**252**	–	21	55	30	–	32	151	68	1
Russia	23	34	106	**163**	69	171	**403**	22	54	46	21	42	72	166	49	116
Serbia and Montenegro	9	29	65	**103**	52	92	**247**	–	46	43	14	–	64	125	58	–
Slovakia	7	24	55	**86**	48	82	**216**	–	37	34	9	6	51	120	34	11
Slovenia	4	21	46	**71**	47	81	**199**	–	30	37	4	–	55	129	14	1
Spain	10	29	80	**119**	49	93	**261**	4	33	21	22	43	47	35	46	133
Canary Islands	6	6	27	**39**	15	23	**77**	6	8	2	11	18	14	4	23	36
Sweden	4	22	59	**85**	56	111	**252**	–	48	17	16	4	117	65	62	8
Switzerland	2	19	41	**62**	46	81	**189**	–	24	29	9	–	49	95	45	–
Turkey	19	32	97	**148**	53	118	**319**	10	104	21	12	11	171	95	31	22
Ukraine	17	29	72	**118**	53	101	**272**	–	54	47	17	–	78	136	58	–
United Kingdom	3	24	43	**70**	58	82	**210**	1	31	13	21	5	76	43	79	12
Gibraltar	1	–	9	**10**	12	9	**31**	–	1	9	–	–	1	26	4	–

[1] The number of species that are confined to that country within Europe.

Figure 2. Numbers of SPECs and Non-SPECs breeding in each geopolitical unit covered by this review.

all countries have SPECs and therefore all have some responsibility for the conservation of these species as a whole.

The highest absolute numbers of SPECs are found in Russia, Turkey, Spain, Ukraine and France, each of which (as in *BiE1*) holds more than 50% of all species with an Unfavourable conservation status in Europe. Russia alone has breeding populations of almost three-quarters of all SPECs, due partly to its great size and the associated high diversity of habitats, including tundra, boreal and temperate forests, mountains, steppe, deserts, numerous wetlands and marine habitats of the Arctic, Baltic, Black and Caspian Seas. Furthermore, being at the eastern edge of Europe, the region holds a number of SPECs that have predominantly Asian ranges but which occur marginally in European Russia, such as Sociable Lapwing *Vanellus gregarius*, Caspian Plover *Charadrius asiaticus*, Black Lark *Melanocorypha yeltoniensis* and Black-throated Accentor *Prunella atrogularis*.

However, as illustrated by the size of the circles in Figure 2, these countries are also amongst the largest and most species-rich in Europe. When considering the number of SPECs as a proportion of all species, it becomes evident that the number of SPECs in Russia, France and Ukraine is roughly proportional to the size of their avifaunas, and thus similar to the overall European figure (43%). On the other hand, Spain, Turkey, Portugal, Malta, the Azores, Madeira and the Canary Islands all have disproportionately high numbers of SPECs, highlighting the importance of these Mediterranean countries and Macaronesian islands for bird conservation in Europe. Along with Russia (22 species) and Greenland (6), these countries (or territories) also host important numbers of species of birds breeding nowhere else in Europe, especially Turkey (10), the Canary Islands (6) and Spain (4).

Mirroring the pattern described above, the highest absolute numbers of SPEC 1 species are also found in Russia, Turkey and Ukraine, all of which support breeding populations of around half of Europe's 40 species of global conservation concern. Taking the total size of national avifaunas into account, these three countries remain near the top of the list, with more than 5% of their breeding species classified as SPEC 1. However, topping this list are the Macaronesian islands, all of whose breeding avifaunas comprise more than 6% SPEC 1 species, with the Canary Islands nearing 8%. This highlights the susceptibility of many birds confined to small islands around the world, and shows that Europe is no exception to the rule.

Figures 3 and 4 summarise the trends of individual species in each country during 1990–2000. Considering first the trends of all species (Figure 3), it is clear that birds breeding in Turkey are in particular trouble, with more than 50% of species in decline. Croatia and Albania also have a high proportion of declining species, indicating that the problems affecting birds in south-east Europe extend well into the Balkans. It is therefore worrying that the trends of so many species remain unknown in countries like Georgia, Bosnia and Herzegovina and Armenia, as well as in south-west Russia. However, the situation is clearly far from secure in northern Europe, where large numbers of species are declining in both Sweden and Finland, and where the trends of many key populations of birds breeding in Greenland, Iceland and north-west Russia remain unknown.

Considering the trends of SPECs alone (Figure 4), Turkey once again stands out, with 70% of its breeding SPECs in decline. Elsewhere, more than half of all SPECs breeding in Malta, Sweden and Luxembourg are declining, and more than a third of all SPECs are declining in many other countries (including most of those in west-central Europe). As described above, the trends of many SPECs breeding in certain countries in the Balkans and the Caucasus are unknown. Given their importance for so many species, it is also disturbing that the trends of such high proportions of SPECs are unknown in the Macaronesian islands, Spain and Iceland.

TRENDS BETWEEN THE TWO ASSESSMENTS

Comparing the overall population trends of all species between 1970–1990 and 1990–2000 shows that the numbers of stable (c.50%) and declining species (c.25%) were broadly comparable across

Figure 3. Population trends during 1990–2000 of all species breeding in each geopolitical unit covered by this review.

Figure 4. Population trends during 1990–2000 of all SPECs breeding in each geopolitical unit covered by this review.

Table 7. A comparison of overall population trends between 1970–1990 and 1990–2000.

	Overall trend 1990–2000[1]				
Overall trend 1970–1990[1]	Decline	Stable	Increase	Unknown	Total
Decline	63	45	8	19	135
Stable	36	166	8	48	258
Increase	11	35	31	8	85
Unknown	11	20	3	14	48
Total	121	266	50	89	526

[1] Figures represent the number of species in each trend category. Where a species qualifies as a SPEC based on winter data, winter trends are used.

both periods (Table 7). However, the number of increasing species fell by over 40% during 1990–2000. Furthermore, many more of the species that were stable during 1970–1990 declined than increased during 1990–2000. Based on this, it seems likely that a higher proportion of species whose trends are currently unknown are actually declining than increasing. Nevertheless, the number of formerly declining species that are now stable did exceed the number of formerly stable species that are now declining. The relatively high proportion of species with unknown trends remains a cause of considerable concern.

Overall, the quality of the data on both the size and trend of species' breeding populations has improved since the publication of *BiE1* (Figure 5). Data of good or medium quality are now available on the population sizes of almost two-thirds of all European species, and population estimates were available for all species. Although good- and medium-quality trend estimates are now available for over 50% of species, there remains a large number of species for which no overall trend estimate is available. These species are amongst those whose European Threat Status could only be evaluated provisionally (see Appendix 1), and for which improved monitoring is a priority. Many of these species—including numerous SPECs—are restricted to the same parts of Europe, such as Greenland and Iceland, Spain and the Macaronesian Islands, and much of south-east and eastern Europe, particularly Russia and Turkey (see Figure 2).

Figure 5. The percentage of data (breeding) in each quality band in *BiE1* and *BiE2*.

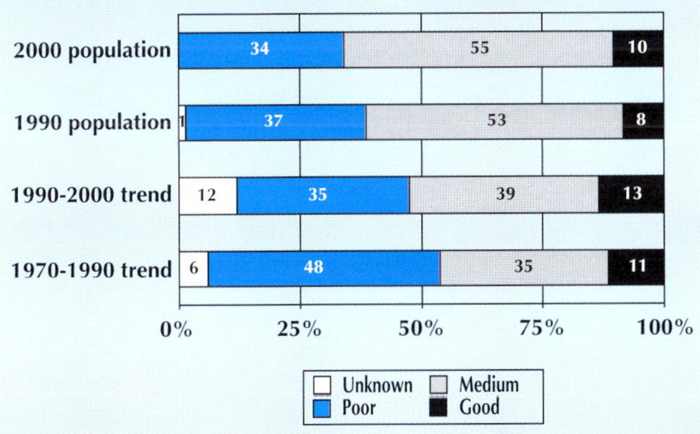

TRENDS BY TAXONOMIC ORDER

The high proportion, and absolute number, of waterfowl (Anseriformes), raptors (Acciptriformes and Falconiformes), gamebirds (Galliformes) and shorebirds (Charadriiformes) in decline is highlighted by Figure 6. The situation of two smaller orders, the sandgrouse (Pteroclidiformes) and the kingfishers, rollers and allies (Coraciiformes), is also cause for considerable concern. On the other hand, groups like the gannets, cormorants and pelicans (Pelecaniformes) and the flamingoes (Phoenicopteriformes) have fared well, and many species in other groups (e.g. the Procellariiformes, Ciconiiformes, Strigiformes and Piciformes) were stable overall. Nevertheless, the fact that trends were unknown for almost a quarter of the species in two large orders, the shorebirds (Charadriiformes) and the passerines (Passeriformes), is very worrying, and highlights the need for improved monitoring.

Figure 6. Population trends across different taxonomic orders (figures represent numbers of species).

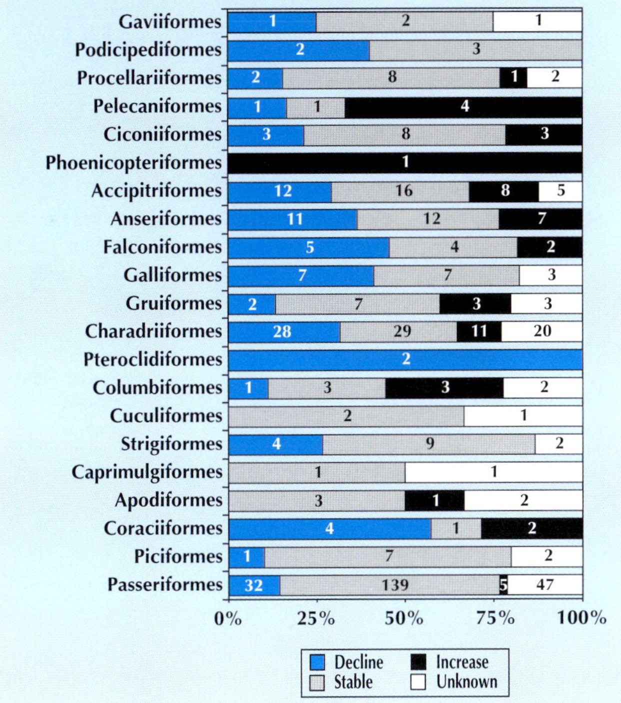

TRENDS BY MIGRATION STRATEGY

When comparing population trends by migration strategy (Figure 7), long-distant migrants stand out as a group in particular trouble, with a large number of species declining, only a handful increasing, and the trends of almost a quarter unknown. For these and other migrant species, there is a pressing need for research to identify the factors responsible for their decline. If (as seems likely) some of these factors are found to be operating outside Europe, then it will be necessary to implement agreements and initiatives to assist their conservation in the non-European parts of their range. The African-Eurasian Migratory Waterbird Agreement (AEWA) under the Bonn Convention is an excellent example of such an agreement, whose importance will only increase over time if the trends identified in this review continue.

Figure 7. Population trends by migration strategy (figures represent numbers of species).

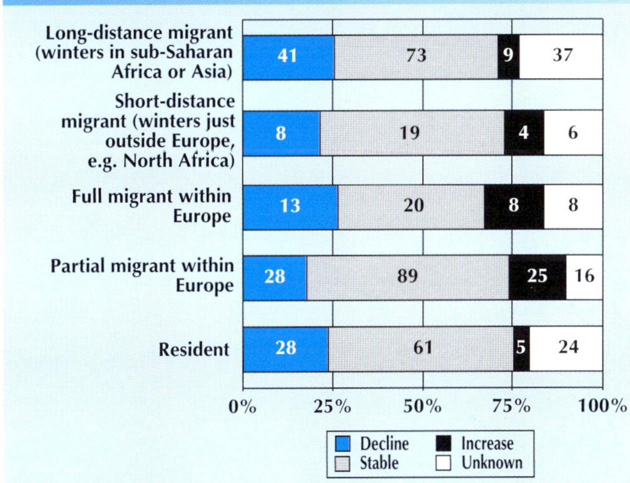

TRENDS BY HABITAT ASSOCIATION

Analysing the fortunes of species according to their habitat associations highlights the continuing decline of farmland and grassland birds: 58% of species (with known trends) associated primarily with this habitat declined during 1990–2000. An above-

Figure 8. Population trends by habitat association (figures represent numbers of species).

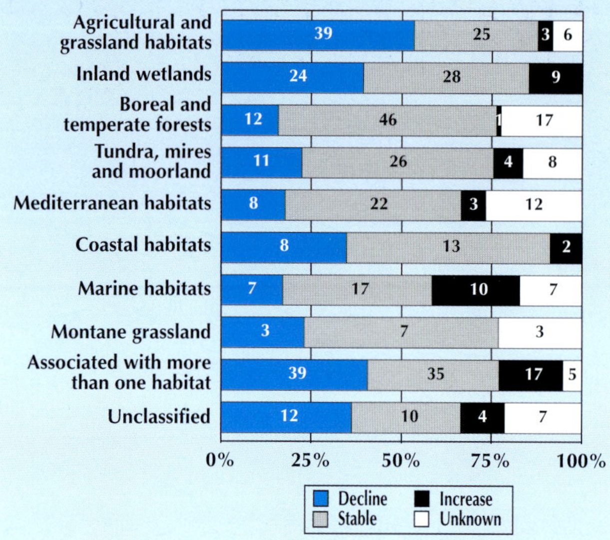

average percentage of species of inland wetlands (39%) also declined, as did those associated with two or more habitats (43%), many of which were long-distance migrants (see above). When considering the percentage of species that increased in each habitat category, birds associated with marine habitats fared best (of those with known trends, 29% increased). Similarly, the habitat category with the highest proportion of stable species was boreal and temperate forest, in which 78% of species with known trends were stable. Although species associated with Mediterranean habitats also fared relatively well, this habitat category held the highest percentage of species with an unknown trend. Hence, it is a clear priority for improved monitoring in the future.

FURTHER ANALYSES

This overview of results aspires only to describe some broad patterns emerging from the data collected for this review. Unlike the equivalent section in *BiE1*, it does not attempt to deal with such topics as the threats facing SPECs, regional differences in population trends, or the underlying causes of the trends observed. All of these are vital questions, and further research is already underway to investigate them within and beyond the BirdLife Partnership. The results of this research will be published in a series of papers in the scientific literature in the coming years.

CONCLUSIONS AND RECOMMENDATIONS

This review shows that the number of bird species with an Unfavourable conservation status in Europe has increased since the previous assessment (Tucker and Heath 1994). Although some species have improved in status over the past decade, many more have deteriorated, and overall the total number of Species of European Conservation Concern (SPECs) has increased from 38% to 43% of the European avifauna.

This implies that efforts to date to halt biodiversity loss have not been sufficient to stop the decline of many bird populations in Europe. Hence, there is a clear need to improve the effectiveness of current conservation efforts, and to ensure that there is adequate funding and political commitment to continue their implementation in the future.

Despite the stark messages conveyed in Tucker and Heath (1994), many European bird species – particularly those associated with agricultural habitats – are either still declining or are heavily depleted. Nevertheless, there are encouraging signs that in the European Union (EU) some nature conservation legislation is yielding results (see BirdLife International 2004 for a full review). To ensure wider success, these conservation measures should be enhanced and replicated elsewhere, both within and beyond the EU25.

The countries of eastern and south-eastern Europe hold the highest proportion of species with an Unfavourable conservation status. These countries are also home to many of the continent's endemic and restricted-range species, as well as significant populations of more widely distributed species that are of global conservation concern. Alarmingly, this region is also where information about population sizes and trends is most sparse. Considerable investment is needed to improve the infrastructure for conservation and build capacity for monitoring in these areas.

PRIORITIES FOR CONSERVATION

Although species are dealt with individually in this review, a species-by-species approach to their conservation is not recommended. Instead, BirdLife International advocates a conservation strategy encompassing three complementary approaches (see also 'Introduction', p. 1). In combination, the three elements—species conservation, site conservation, and conservation of the wider environment—form a framework for generic conservation actions that can be applied to species groups, communities or habitats. This review highlights a number of conservation priorities relevant to each of these approaches.

Species conservation

The protection of species through conservation action may take many forms, including legislation, monitoring, research, prioritisation, management of populations, and land acquisition and management. These actions may be taken for species that are threatened with extinction at a local, national or global scale.

- The results of this review should be incorporated into the existing legal instruments for conservation (see Appendix 3). In particular, the annexes and appendices of the relevant directives, conventions and agreements should take into account the status of all SPECs.
- National legislation and conservation strategies based on national 'Red Lists' should take into account SPEC categories and the proportion of the global and European population occurring in the country (see Appendix 2).
- Existing legislation should be strengthened and better enforced to reduce the illegal trade and persecution of wild birds. Conflict resolution measures such as compensation should be used in cases of economic damage.
- Sustainable exploitation of legally hunted species should be based on management plans that apply an adaptive approach. Levels of exploitation should be set on the basis of reliable population monitoring and bag data. For migratory species, collaboration at a flyway level is needed to take into account interannual changes in numbers.
- Species Action Plans (SAPs) are currently available for a number of species with an Unfavourable conservation status in Europe (see Appendix 3). The implementation of SAPs is an effective means of improving the status of some of Europe's most threatened birds (Nagy and Crockford 2004). Incentives should be made available to ensure the successful development and implementation of more SAPs.

Site conservation

The protection of key sites has a crucial role in any effective conservation strategy. Several international treaties call for the selection and protection of sites based on their importance for birds and other biodiversity. The identification of Important Bird Areas (e.g. Heath and Evans 2000) provides the information needed by national governments to establish a network of protected areas for birds, and to ensure their adequate management.

- The population data and updated list of SPECs contained in this book (see Appendix 1) should be taken into account when identifying sites at a continental or national level.
- Adequate funding should be made available to identify gaps in the protected area network arising from these new results, and to fill any gaps by designating new protected areas.
- Site protection should be implemented effectively and integrated fully in all land-use policies. The indirect impacts of other policies on protected areas should also be taken into consideration.
- Every protected area should have a management plan that addresses the ecological needs and conservation of the species for which it was designated. All such plans should be implemented fully.

Conservation of the wider environment

Many common and widespread species continued to decline during 1990–2000, and the populations of many others remained heavily depleted following the large declines suffered between 1970–1990. Traditional site-based measures are not adequate to ensure the protection of many of these dispersed species, underlining the importance of integrating conservation objectives into all land-use policies that impact on biodiversity.

- Regulations governing agriculture, forestry, water management and fisheries should incorporate safeguards for biodiversity, particularly measures to prevent damage to habitats.
- Environmental assessment procedures should ensure that negative impacts of new developments and policies are minimised.
- Financial support to agriculture, forestry and fisheries should avoid subsidising activities that lead to the loss of biodiversity. Instead, it should be used to promote management practices that enhance habitat availability and reduce mortality, and to resolve conflicts between conservation objectives and the economic interests of land-users.
- If migratory species are shown to be declining because of factors operating in passage and wintering areas outside Europe, overseas development aid should be directed towards environmental programmes that take into account these species' habitat requirements. This underlines the importance of collaboration under international conventions and agreements, such as the African-Eurasian Migratory Waterbird Agreement.

PRIORITIES FOR MONITORING

The overall quality of the information supplied for this review has improved since the last assessment. Nevertheless, the data available for certain species and countries were insufficient to assess population size and trends reliably, and consequently the status of numerous species remains provisional (see Appendix 1). To further improve the reliability and accuracy of future assessments of conservation status, adequate censusing and monitoring is essential in all countries. Ideally, all species should be monitored, but financial restrictions and limited capacity mean that priorities must be set. BirdLife International advocates a Pan-European Bird Monitoring Strategy with three complementary strands, focusing on common and widespread birds, threatened species and key sites.

- Long-term monitoring requires a long-term commitment. Although governments have international reporting obligations, many national bird monitoring schemes—including some of those forming the basis of the Pan-European Common Bird Monitoring Scheme (PECBMS; p. 4) and the International Waterbird Census (IWC; p. 4)—currently receive little or no state funding, and face uncertain futures. Sustaining them would require extremely modest long-term investment.
- Many European countries (especially in the east and south-east) still have no formal schemes for monitoring even common and widespread birds. The governments of these countries should address this problem as a matter of urgency, drawing on the expertise of organisations like BirdLife, the EBCC and Wetlands International where necessary. Novel methods should also be used to harness the potential of citizen science for monitoring via the Internet (as in Turkey, for example, through www.kusbank.org).
- At a continental level, Europe must also report on its progress towards halting biodiversity loss. National data can be combined through schemes like the PECBMS or the IWC to produce indicators suitable for this purpose (Gregory *et al.* in press). Modest but sustained funding would guarantee the future of these schemes, and allow the inclusion of data from more countries and species.
- National working groups should be established to carry out systematic monitoring of all species covered by European Species Action Plans. Without regular surveys of these threatened species, it is impossible to target conservation measures appropriately and evaluate their effectiveness.
- The process of identifying and protecting key sites for birds and other wildlife increasingly requires state management and monitoring of protected areas. Governments should take advantage of this opportunity to involve local communities in collecting the data needed to meet their reporting requirements. BirdLife Partners across Europe have been developing a network of 'caretakers' and 'site support groups' at Important Bird Areas for this purpose. Modest state funding for such schemes would ensure regular data for reporting purposes, as well as increasing public participation in biodiversity conservation.

PRIORITIES FOR RESEARCH

Well-designed monitoring schemes should act as early warning systems, drawing attention to problems as soon as they manifest a measurable impact. They should also detect improvements in status, thereby informing governments about the effectiveness of conservation measures put in place in response to these problems. Nevertheless, monitoring alone cannot usually identify the root causes of a decline, pinpoint the mechanisms involved, or develop suitable remedies. This requires detailed scientific research.

- Over the past decade, intensive research has uncovered the causes of decline of many farmland birds, and proposed measures to reverse this trend. Given the ongoing decline of farmland birds across much of Europe, this research should continue, building on the lessons learnt elsewhere. This applies particularly in central and eastern Europe, where agricultural changes may differ from those in the west. The same approach should also be extended to other habitats.
- The causes of declines of many migratory species, particularly long-distance migrants, are currently unknown. This area is an urgent priority for research and international collaboration, as it is likely that—in at least some species—the declines reflect factors operating outside Europe.
- There is increasing evidence that the current rapid rate of climate change will have a profound effect on European birds (e.g. Collingham *et al.* in prep.). Changes in the distribution of habitats will probably cause range shifts in many species, affecting the functionality of current protected area networks. Research should develop models to assess how species' predicted future distributions can be accommodated in existing protected area networks under different scenarios.
- Similar models should also be developed to support environmental assessments of the effects of proposed changes to far-reaching policies, such as the EU Common Agricultural Policy.
- The ongoing shift in protected area objectives—from establishing site networks to managing them—creates a need for research on how to set the right targets for conservation at population, site and network levels. Methods for establishing baselines, setting targets and assessing network coherence should all be priorities for research.
- Much research has already been carried out on individual species, but it is often difficult to access. This existing science should be inventoried and presented to decision-makers in non-technical and accessible ways. Mechanisms are also required to improve information exchange between researchers, such as user-driven databases of publications on the ecology, declines and recoveries of particular species or communities. This would reduce wasteful repetition and focus new research on policy-relevant issues.

TOWARDS A RENEWED BIODIVERSITY POLICY FOR EUROPE

The results of this review clearly indicate that – despite some noteworthy successes – conservation efforts to date have been insufficient to halt biodiversity loss in Europe. Nevertheless, it should be noted that the integration of biodiversity concerns into sectoral policies has only started in recent years, and its effects—in terms of the trends of European bird populations—may take some time to become evident.

With only a few years remaining, it is still possible to achieve the target of halting biodiversity loss by 2010—but only if all European governments take decisive action now, and make good on their pledges in the Declarations from Malahide and Kiev. Funding should be made available to all relevant stakeholders, from national ministries to local land managers, as well as non-governmental organisations, which often represent the most cost-effective way to implement and monitor conservation actions on the ground. Only by working in cooperation at international, national and local levels can we stem the continuing loss of Europe's rich and abundant biodiversity.

REFERENCES

BirdLife International (2004) *Birds in the European Union: a status assessment*. Wageningen, The Netherlands: BirdLife International.

Collingham, Y. C., Willis, S. G., Green, R. E. and Huntley, B. (in prep.) A climatic atlas of European breeding birds.

Gregory, R. D., van Strien, A. J., Vorisek, P., Gmelig Meyling, A. W., Noble, D. G., Foppen, R. P. B. and Gibbons, D. W. (in press) Developing indicators for European birds. *Phil. Trans. R. Soc. Lond. B.*

Heath, M. F. and Evans, M. I., eds. (2000) *Important Bird Areas in Europe: priority sites for conservation*. 2 vols. Cambridge, UK: BirdLife International (BirdLife Conservation Series no. 8).

Nagy, S. and Crockford, N. (2004) *Review of implementation of species action plans for threatened birds within the framework of the Birds Directive*. Wageningen, The Netherlands: BirdLife International (Research report to the European Commission, DG Environment). http://europa.eu.int/comm/environment/nature/nature_conservation/focus_wild_birds/species_birds_directive/pdf/action_plans_review_final.pdf

Tucker, G. M. and Heath, M. F. (1994) *Birds in Europe: their conservation status*. Cambridge, UK: BirdLife International (BirdLife Conservation Series no. 3).

SPECIES ACCOUNTS

This section comprises individual accounts for each of the 526 species considered in this review. Each account includes: an illustration of the species; a table and a map presenting national and European population size and trend data; a graph comparing the quality of these data with those collected in the previous assessment; a summary of the species's status at a European and global level; and a short text justifying each species's European conservation status.

For a detailed explanation of how to interpret each of these components, see 'Data presentation' (p. 13). Each account is also available as a species factsheet over the Internet, via the URL at the bottom of each page in this book. The appendices that follow this section summarise much of the key information from the species accounts in large tables, which contain data for all species. They also provide additional information that could not be included in the species accounts, due to space limitations.

Gavia stellata
RED-THROATED LOON

SPEC 3 (1994: 3) **Status** (Depleted)
Criteria Large historical decline
European IUCN Red List Category —
Criteria —
Global IUCN Red List Category —
Criteria —

Country	Breeding pop. size (pairs)	Year(s)	Trend	Mag.%	References
Denmark					
Faroe Is.	15 – 15	95	(0)	(0–19)	
Greenland	(5,000 – 30,000)	90–00	?	–	
Finland	900 – 1,100	99–01	**0**	**0–19**	
Iceland	(1,000 – 2,000)	78–94	?	–	*1*
Rep. Ireland	**3 – 6**	97–02	**–**	**20–29**	
Norway	(2,000 – 5,000)	90–03	–	0–19	
Svalbard	(500 – 1,000)	90–03	(0)	(0–19)	
Russia	(20,000 – 50,000)	90–00	(0)	(0–19)	*71,100,128,138*
Sweden	**1,200 – 1,400**	99–00	0	0–19	
UK	**935 – 1,500**	94	0	0–19	*7*
Total (approx.)	32,000 – 92,000	**Overall trend** Stable			
Breeding range	>2,000,000 km²	**Gen. length** 7		**% Global pop.** 5–24	

Country	Winter pop. size (individuals)	Year(s)	Trend	Mag.%	References
Croatia	(300 – 500)	02	?	–	*16,26*
Denmark	(5,000 – 50,000)	99–00	(0)	(0–19)	*25*
Estonia	5,000 – 23,000	98	(0)	(0–19)	*1*
France	300 – 1,000	98–02	?	–	*9*
Germany	**14,500 – 14,500**	87–03	?	–	*4*
Greece	**50 – 200**	96–00	**F**	**>80**	
Iceland	100 – 1,000	78–94	?	–	*48*
Rep. Ireland	**860 – 1,100**	94–00	0	0–19	
Latvia	**10,000 – 10,000**	92–94	(F)	(>80)	*27*
Lithuania	2,000 – 3,000	92–02	F	20–29	
Netherlands	(6,800 – 6,800)	85–93	(+)	(–)	*3,4,5,8,9,11,12*
Poland	(1,000 – 2,500)	87–99	?	–	*113,114,129*
Romania	20 – 300	90–00	0	0–19	*60,38*
Ukraine	(50 – 250)	90–00	F	20–29	
UK	4,900 – 4,900	80–86	(0)	(0–19)	
Total (approx.)	>51,000	**Overall trend** Stable			
% in European IBAs	9 – 20	**Gen. length** 7		**% Global pop.** 5–24	

Gavia stellata is a widespread breeder across much of northern Europe, which accounts for less than a quarter of its global breeding range. Its European breeding population is relatively small (<92,000 pairs), and underwent a large decline between 1970–1990, notably in Russia and Fennoscandia. Although the species was stable overall during 1990–2000, with stable trends in most countries within its European range, its population has clearly not yet recovered to the level that preceded its decline. Consequently, it is provisionally evaluated as Depleted.

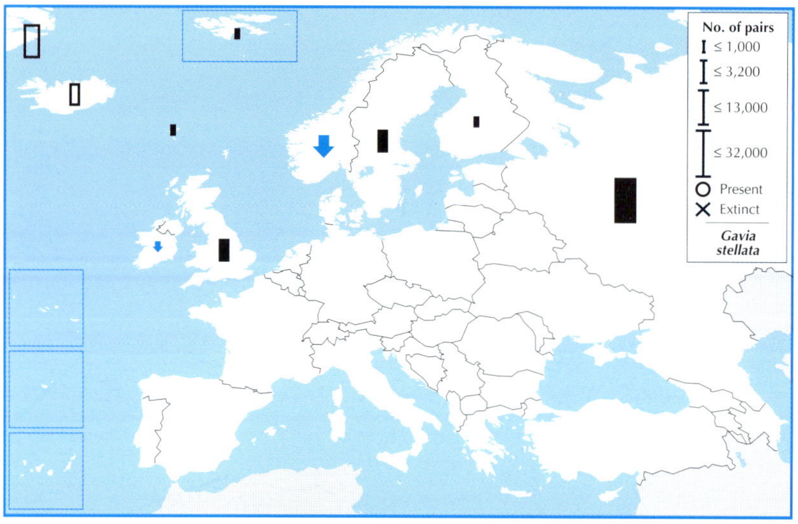

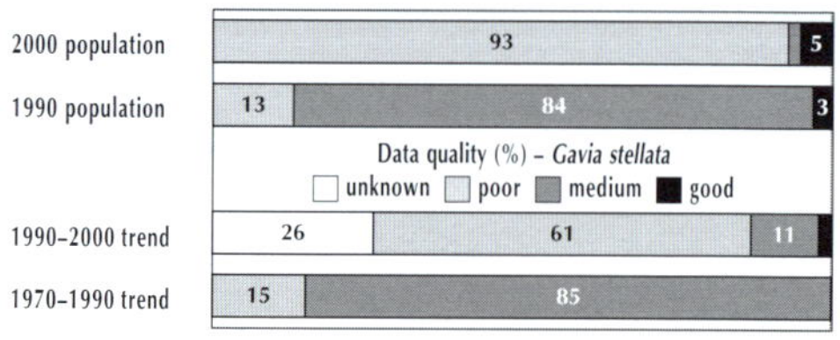

Gavia arctica
ARCTIC LOON

SPEC 3 (1994: 3) **Status** (Vulnerable)
Criteria See IUCN below
European IUCN Red List Category VU
Criteria A2b
Global IUCN Red List Category —
Criteria —

Country	Breeding pop. size (pairs)	Year(s)	Trend	Mag.%	References
Belarus	**15 – 30**	97–00	0	0–19	
Estonia	5 – 10	98	0	0–19	*1*
Finland	8,000 – 10,000	99–01	**+**	**10**	
Latvia	0 – 5	00–03	–	30–49	*22*
Lithuania	3 – 5	99–01	F	30–49	*20*
Norway	2,000 – 5,000	94	–	0–19	*34*
Russia	(35,000 – 70,000)	90–00	(–)	(20–29)	*74,100,128,138*
Sweden	**5,500 – 7,000**	99–00	0	0–19	
UK	**155 – 190**	94	**+**	**17**	
Total (approx.)	51,000 – 92,000	**Overall trend** Large decline			
Breeding range	>3,000,000 km²	**Gen. length** 7		**% Global pop.** 5–24	

Country	Winter pop. size (individuals)	Year(s)	Trend	Mag.%	References
Croatia	(5,000 – 6,000)	02	?	–	*16,70*
Denmark	(500 – 500)	92–93	?	–	*27*
Estonia	200 – 1,000	98	(0)	(0–19)	*1*
France	300 – 800	98–02	?	–	*9*
Germany	**3,250 – 3,250**	87–03	?	–	*4*
Greece	**50 – 400**	96–00	**F**	**>80**	
Italy	150 – 350	02	0	0–19	*22*
Latvia	100 – 500	92–94	(F)	(>80)	*32*
Lithuania	200 – 500	92–02	(F)	(30–49)	
Netherlands	(250 – 250)	85–93	(F)	(–)	*3,4,5,8,9,11,12*
Poland	2,500 – 10,000	87–99	?	–	*113,114*
Romania	50 – 500	90–00	0	0–19	*60,38*
Turkey	**3,000 – 4,000**	91–01	(0)	(0–19)	
Ukraine	(1,000 – 2,500)	90–00	F	30–49	
UK	700 – 700	80–86	(+)	(0–19)	
Total (approx.)	>17,000	**Overall trend** Unknown			
% in European IBAs	14 – 20	**Gen. length** 7		**% Global pop.** 5–24	

Gavia arctica is a widespread breeder across much of northern Europe, which accounts for less than a quarter of its global breeding range. Its European breeding population is relatively small (<92,000 pairs), and underwent a large decline between 1970–1990. Although most European populations—including sizeable ones in Sweden and Finland—were stable or increased during 1990–2000, the species continued to decline in Norway and its Russian stronghold, and underwent a large decline (>30%) overall. Consequently, it is provisionally evaluated as Vulnerable.

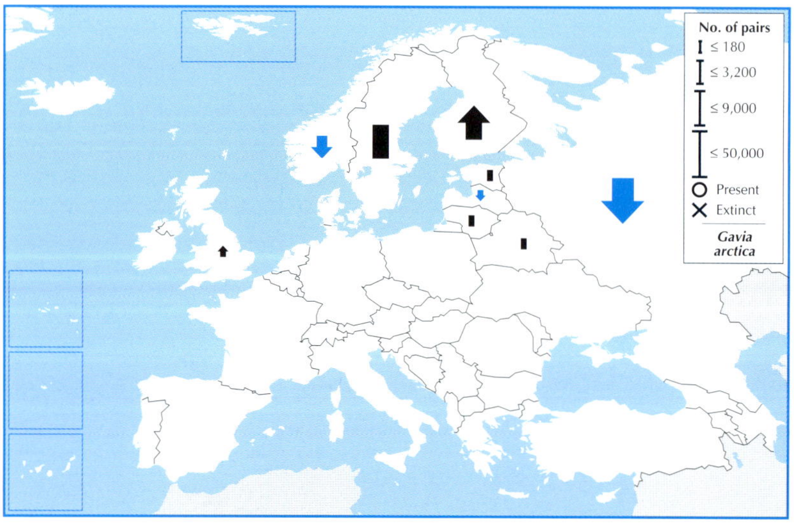

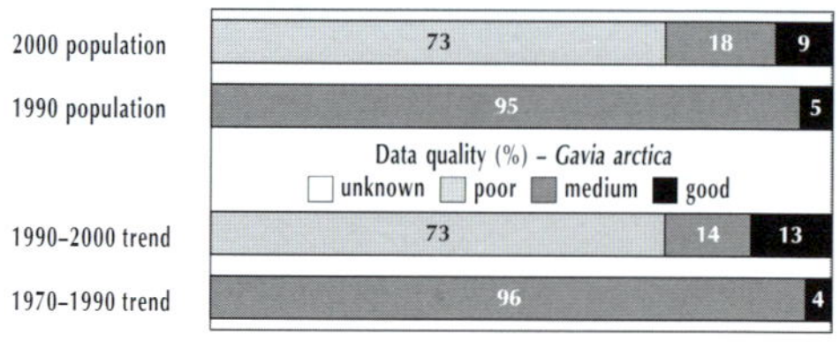

Gavia immer
COMMON LOON

Non-SPEC (1994: —) **Status** (Secure)
Criteria —
European IUCN Red List Category —▼
Criteria —
Global IUCN Red List Category —
Criteria —

Gavia immer has a predominantly North American breeding range, but also breeds in the European Arctic. Its European breeding population is very small (as few as 700 pairs), but was stable between 1970–1990. Trends were not available for Greenland and Iceland during 1990–2000, but there is no evidence to suggest that the species declined. Although the size of the European population could render it susceptible to the risks affecting small populations, it is marginal to a much larger non-European population. Consequently, the species is provisionally evaluated as Secure.

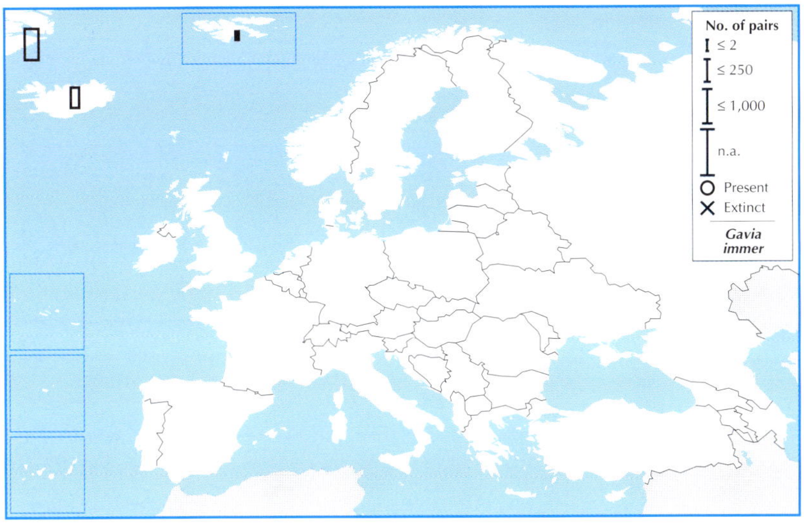

Country	Breeding pop. size (pairs)	Year(s)	Trend	Mag.%	References
Denmark					
Greenland	(500 – 2,000)	90–00	?	–	
Iceland	(200 – 300)	78–94	?	–	*1,21*
Norway					
Svalbard	0 – 3	95–03	(0)	(0–19)	
Total (approx.)	700 – 2,300	**Overall trend** Unknown			
Breeding range	>100,000 km²	**Gen. length.** 7		**% Global pop.** <5	

Country	Winter pop. size (individuals)	Year(s)	Trend	Mag.%	References
Belgium	**2 – 5**	95–00	**0**	**0–19**	*1*
Denmark					
Faroe Is.	10 – 50	92	?	–	*3*
Greenland	(0 – 100)	90–00	?	–	
France	250 – 500	98–02	?	–	*9*
Germany	**0 – 50**	95–00	?	–	*3*
Iceland	200 – 1,000	78–94	?	–	*48*
Rep. Ireland	**1,400 – 1,400**	94–00	0	0–19	
Italy	(0 – 1)	02	?	–	*22*
Netherlands	**6 – 6**	99–01	?	–	*3,4,8,9,11,12*
Norway	**1,000 – 1,100**	95–00	(0)	(0–19)	*67*
Portugal	Present	02	?	–	
Romania	**0 – 1**	90–00	?	–	
Slovakia	0 – 2	90–99	?	–	*4*
Switzerland	**0 – 3**	98–02	**0**	**0–19**	
UK	2,500 – 3,000	74–84	(+)	(0–19)	
Total (approx.)	>5,400	**Overall trend** Stable			
% in European IBAs	>10	**Gen. length** 7		**% Global pop.** <5	

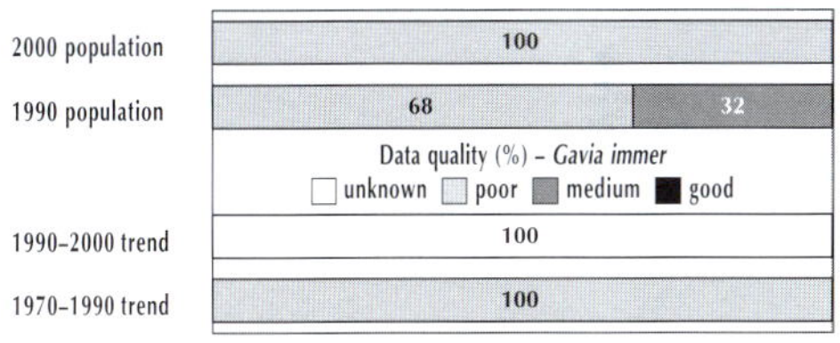

Gavia adamsii
YELLOW-BILLED LOON

Non-SPEC (1994: —) **Status** (Secure)
Criteria —
European IUCN Red List Category —▼▼
Criteria —
Global IUCN Red List Category —
Criteria —

Gavia adamsii breeds outside of Europe in the high Arctic, but a very small population (as few as 500 individuals) winters in Europe—mainly along the Norwegian coast. Trend data were not available for the European wintering population between 1970–1990, but the key population in Norway was stable during 1990–2000. Although the size of the European population could render it susceptible to the risks affecting small populations, it is marginal to a much larger non-European population. Consequently, the species is provisionally evaluated as Secure.

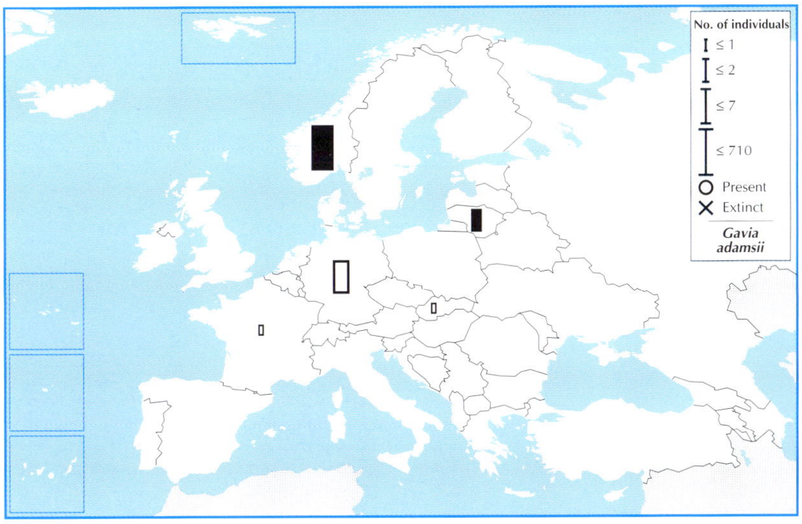

Country	Winter pop. size (individuals)	Year(s)	Trend	Mag.%	References
France	**0 – 2**	98–02	?	–	*9*
Germany	**0 – 50**	95–00	?	–	*3*
Lithuania	0 – 5	92–02	**0**	**0–19**	
Norway	500 – 1,000	93	(0)	(0–19)	*68*
Slovakia	0 – 1	90–99	?	–	*4*
Total (approx.)	>500	**Overall trend** Stable			
% in European IBAs	39–41	**Gen. length** 7		**% Global pop.** 5–24	

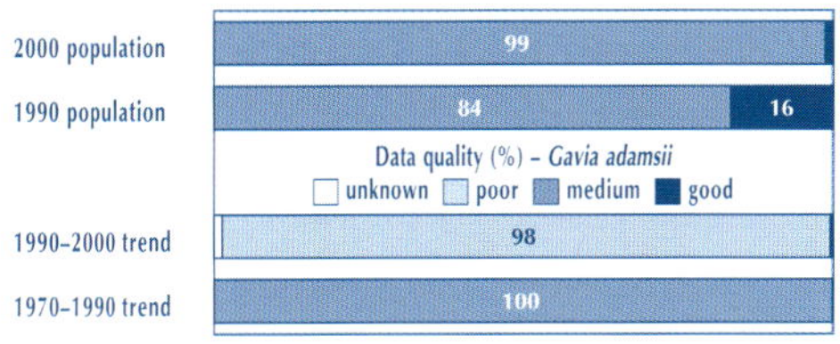

Tachybaptus ruficollis
LITTLE GREBE

Non-SPEC (1994: —) **Status** Secure
Criteria —
European IUCN Red List Category —
Criteria —
Global IUCN Red List Category —
Criteria —

Tachybaptus ruficollis is a widespread breeder across much of western and central Europe, which accounts for less than a quarter of its global breeding range. Its European breeding population is relatively large (>99,000 pairs), and was stable between 1970–1990. Although there were declines in a few countries—notably the sizeable Turkish population—during 1990–2000, populations in the vast majority of its European range were stable or increased, and the species remained stable overall. Consequently, it is evaluated as Secure.

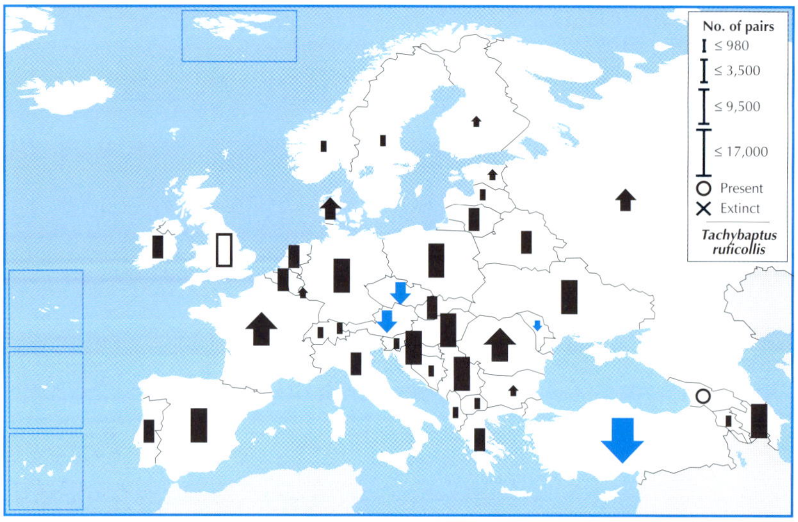

Country	Breeding pop. size (pairs)	Year(s)	Trend	Mag.%	References
Albania	120 – 540	96–02	0	0–19	*6*
Armenia	**800 – 1,200**	98–02	**0**	**0–19**	*11*
Austria	1,200 – 2,000	98–02	–	30–49	
Azerbaijan	4,000 – 8,000	96–02	(0)	(0–19)	
Belarus	**2,000 – 2,400**	97–00	0	0–19	
Belgium	1,000 – 1,600	95–02	0	0–19	*1*
Bosnia & HG	(10 – 15)	90–00	(0)	(0–19)	
Bulgaria	**500 – 900**	95–00	+	**>80**	
Croatia	(6,000 – 7,000)	02	0	0–19	*54*
Cyprus	**80 – 160**	94–02	**0**	**0–19**	
Czech Rep.	**2,000 – 4,000**	00	**–**	**30–49**	
Denmark	2,000 – 3,000	00	(+)	(0–19)	*3*
Estonia	5 – 15	98	+	20–29	*1*
Finland	10 – 30	99–01	+	100	
France	4,000 – 8,000	98–02	+	0–9	*4*
Georgia	Present	03	?	–	
Germany	6,000 – 9,800	95–99	(0)	(0–19)	
Greece	(1,500 – 2,000)	95–00	(0)	(0–19)	
Hungary	(9,000 – 10,000)	90–93	0	0–19	*14*
Rep. Ireland	1,000 – 2,500	88–91	0	0–19	
Italy	3,000 – 4,000	03	0	0–19	*6*
Latvia	(300 – 800)	90–00	(0)	(0–19)	*23,17*
Liechtenstein	3 – 5	98–00	0	0–19	
Lithuania	(1,000 – 2,000)	99–01	(0)	(0–19)	*20*
Luxembourg	**60 – 75**	01–02	+	**30–49**	
Macedonia	(200 – 500)	99–00	(0)	(0–19)	
Moldova	50 – 100	90–00	–	30–49	
Netherlands	**1,800 – 2,500**	98–00	**F**	**40**	*1*
Norway	**23 – 37**	00	**0**	**0–19**	*29*
Poland	7,500 – 10,000	90–00	(0)	(0–19)	*1*
Portugal	(1,000 – 5,000)	02	(0)	(0–19)	
Romania	(7,000 – 12,000)	98–02	(+)	(0–19)	
Russia	(2,000 – 5,500)	90–00	+	20–29	*8,66,178*
Serbia & MN	6,000 – 9,000	90–02	0	0–19	*203,29,155,156, 100,13,10,225,227*
Slovakia	1,000 – 3,000	90–99	0	0–19	
Slovenia	600 – 1,000	94	(0)	(0–19)	
Spain	(5,200 – 8,000)	98–02	(F)	(–)	*10*
Sweden	300 – 400	99–00	F	20–29	
Switzerland	**600 – 800**	93–96	0	10–19	
Turkey	13,000 – 20,000	01	(–)	(0–19)	
Ukraine	3,500 – 6,900	90–00	F	20–29	
UK	3,800 – 13,000	00	?	–	*5,31*

Total (approx.) 99,000 – 170,000 **Overall trend** Stable
Breeding range >4,000,000 km² **Gen. length.** 4 **% Global pop.** 5–24
(See p. 31, bottom, for data quality graph)

Winter pop. size (individuals)
Total (approx.) >72,000 **Overall trend** Small increase
% in European IBAs 12–25 **Gen. length** 4 **% Global pop.** 5–24

Podiceps cristatus
GREAT CRESTED GREBE

Non-SPEC (1994: —) **Status** Secure
Criteria —
European IUCN Red List Category —
Criteria —
Global IUCN Red List Category —
Criteria —

Podiceps cristatus is a widespread breeder across much of Europe, which accounts for less than half of its global breeding range. Its European breeding population is large (>300,000 pairs), and underwent a large increase between 1970–1990. Although the species was stable or increased across much of Europe during 1990–2000, certain populations—notably in Finland, Sweden and Poland—suffered declines, and the species underwent a moderate decline (>10%) overall. Nevertheless, this decline is still outweighed by earlier increases, and consequently it is evaluated as Secure.

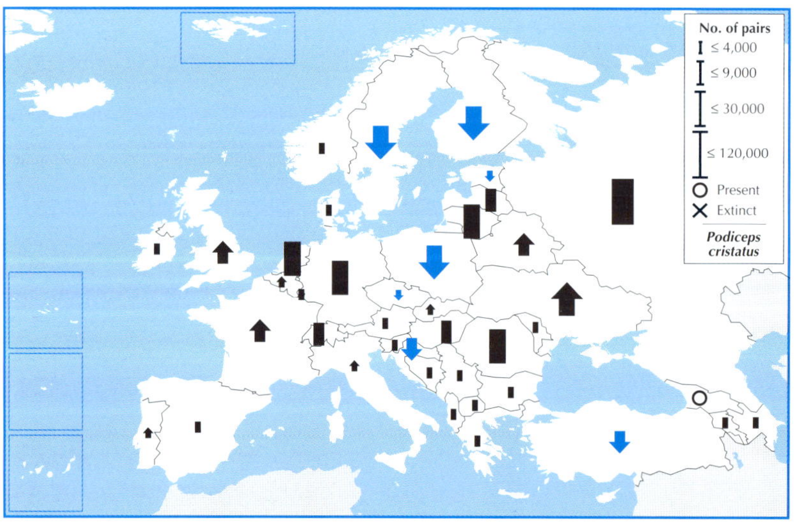

Country	Breeding pop. size (pairs)	Year(s)	Trend	Mag.%	References
Albania	110 – 230	96–02	0	0–19	*6*
Armenia	**800 – 1,500**	97–02	**0**	**0–19**	
Austria	1,000 – 1,200	98–02	0	0–19	
Azerbaijan	2,500 – 5,000	96–02	(0)	(0–19)	
Belarus	**8,000 – 10,000**	97–00	+	**0–9**	
Belgium	1,400 – 1,900	95–02	+	20–29	*1*
Bosnia & HG	(12 – 20)	90–00	(F)	(–)	
Bulgaria	**400 – 700**	95–02	**0**	**0–9**	
Croatia	(4,000 – 5,000)	02	(–)	(0–19)	*54,26*
Czech Rep.	**2,500 – 5,000**	00	**–**	**20–29**	
Denmark	3,500 – 4,500	00	(0)	(0–19)	*3*
Estonia	(2,000 – 3,000)	98	–	20–29	1
Finland	25,000 – 35,000	99–01	**–**	**40**	
France	6,000 – 10,000	98–02	+	**20–29**	*4*
Georgia	Present	03	?	–	
Germany	16,000 – 26,000	95–99	0	0–19	
Greece	(800 – 1,500)	95–00	(0)	(0–19)	
Hungary	(7,000 – 9,000)	90–93	0	0–19	*14*
Rep. Ireland	1,000 – 2,500	88–91	0	0–19	
Italy	3,000 – 3,500	03	+	**30–49**	*6*
Latvia	5,000 – 8,000	90–00	0	0–19	*23,17*
Lithuania	(15,000 – 20,000)	99–01	(0)	(0–19)	*20*
Luxembourg	**15 – 20**	01–02	**0**	**0–19**	
Macedonia	200 – 500	90–00	(0)	(0–9)	
Moldova	2,000 – 2,500	90–00	0	0–19	
Netherlands	**13,000 – 16,000**	98–00	**0**	**5**	*1*
Norway	**250 – 400**	90–02	(0)	(0–19)	*27*
Poland	15,000 – 25,000	90–00	–	0–19	*1*
Portugal	300 – 500	02	(+)	(–)	
Romania	(20,000 – 30,000)	98–02	(0)	(0–19)	
Russia	90,000 – 150,000	90–00	0	0–19	*3,8,51,52,104*
Serbia & MN	**2,800 – 4,000**	90–02	0	0–19	*138,1,105,29, 155,100,227*
Slovakia	500 – 1,000	80–99	+	20–29	
Slovenia	**100 – 200**	95–00	0	0–19	
Spain	(2,300 – 3,400)	92	(0)	(0–19)	*13,12,10*
Sweden	15,000 – 25,000	99–00	–	30–49	
Switzerland	**4,500 – 5,500**	93–96	0	0–19	
Turkey	**6,000 – 8,000**	01	–	0–19	
Ukraine	14,500 – 16,700	90–00	+	5–19	
UK	6,100 – 6,100	00	+	**30**	*22,31*

Total (approx.) 300,000 – 450,000 **Overall trend** Moderate decline
Breeding range >5,000,000 km² **Gen. length.** 5 **% Global pop.** 25–49
(See p. 31, bottom, for data quality graph)

Winter pop. size (individuals)
Total (approx.) >240,000 **Overall trend** Small increase
% in European IBAs 31–37 **Gen. length** 5 **% Global pop.** 25–49

Podiceps grisegena
RED-NECKED GREBE

Non-SPEC (1994: —) **Status** Secure
Criteria —
European IUCN Red List Category —
Criteria —
Global IUCN Red List Category —
Criteria —

Podiceps grisegena is a widespread breeder across much of central, eastern and northern Europe, which accounts for less than a quarter of its global breeding range. Its European breeding population is relatively small (<56,000 pairs), but was stable between 1970–1990. Although there were declines in countries such as Ukraine and Poland during 1990–2000, other sizeable populations in Russia and Finland were stable or increased, and the species declined only slightly overall. Consequently, it is evaluated as Secure.

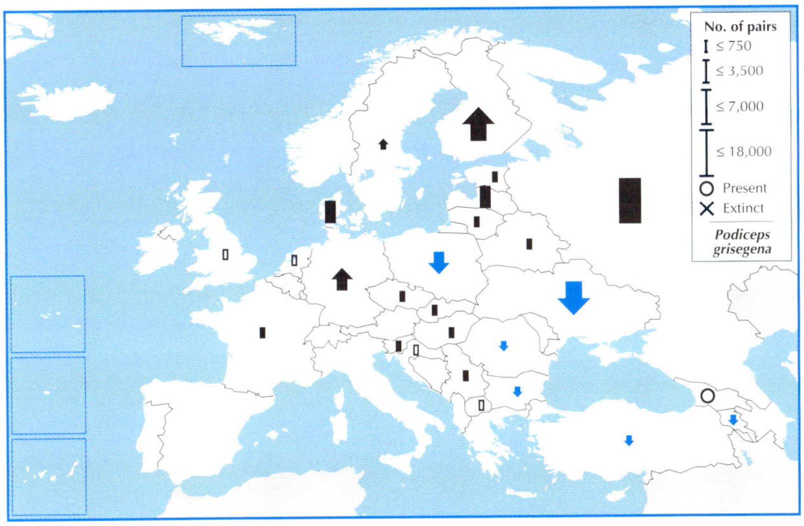

Country	Breeding pop. size (pairs)	Year(s)	Trend	Mag.%	References
Armenia	**2 – 8**	97–02	**–**	**30–49**	
Belarus	**50 – 100**	97–00	0	0–19	
Bulgaria	**30 – 70**	95–02	**–**	**0–19**	
Croatia	(1 – 10)	02	?	–	*57*
Czech Rep.	**1 – 5**	00	**0**	**0–19**	
Denmark	1,000 – 2,000	00	(0)	(0–19)	*3*
Estonia	200 – 300	98	0	0–19	*1*
Finland	6,000 – 8,000	99–01	+	**15**	
France	**0 – 1**	90–00	**0**	**0–19**	*6*
Georgia	Present	03	?	–	
Germany	1,500 – 2,600	95–99	+	0–19	
Hungary	(80 – 150)	90–93	0	0–19	*14*
Latvia	1,500 – 2,000	90–00	(0)	(0–19)	*23,17*
Lithuania	(150 – 200)	99–01	(F)	(30–49)	*20*
Macedonia	0 – 10	90–00	?	–	*7*
Netherlands	**5 – 10**	98–00	?	–	*1*
Poland	3,000 – 4,000	90–00	(–)	(0–19)	*2*
Romania	(650 – 850)	98–02	(–)	(0–19)	
Russia	12,000 – 25,000	90–00	0	0–19	*8,59*
Serbia & MN	**10 – 20**	90–02	**0**	**0–19**	*102,118,29,155, 227*
Slovakia	5 – 20	80–99	F	20–29	
Slovenia	**3 – 6**	90–00	**F**	**>80**	
Sweden	**700 – 800**	99–00	+	10–19	
Turkey	280 – 350	01	–	0–19	
Ukraine	5,000 – 9,000	90–00	–	20–29	
UK	**2 – 2**	96–00	?	–	
Total (approx.)	32,000 – 56,000	**Overall trend** Small decline			
Breeding range	>3,000,000 km²	**Gen. length.** 5		**% Global pop.** 5–24	

Winter pop. size (individuals)

Total (approx.)	>4,400	**Overall trend** Stable	
% in European IBAs	22–46	**Gen. length** 5	**% Global pop.** 5–24

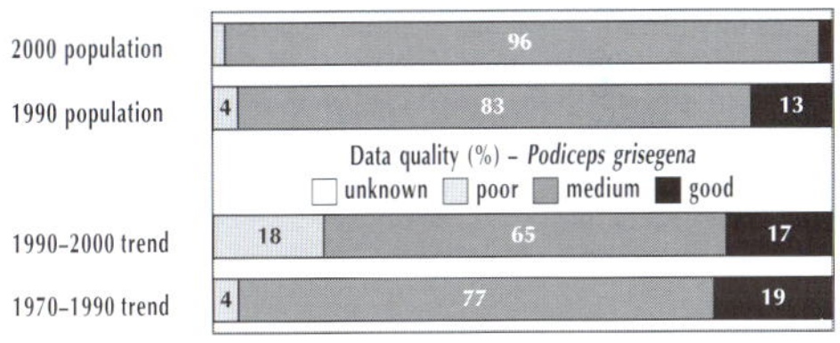

Podiceps auritus
HORNED GREBE

SPEC 3 (1994: —) **Status** (Declining)
Criteria Moderate recent decline
European IUCN Red List Category —
Criteria —
Global IUCN Red List Category —
Criteria —

Podiceps auritus breeds mainly in northern Europe, which accounts for less than a quarter of its global breeding range. Its European breeding population is relatively small (<11,000 pairs), but was stable between 1970–1990. Although it remained stable or increased in several countries during 1990–2000 (and no trend data were available for Russia), the sizeable populations in Finland and Sweden declined, and the species underwent a moderate decline (>10%) overall. Consequently, this previously Secure species is now provisionally evaluated as Declining.

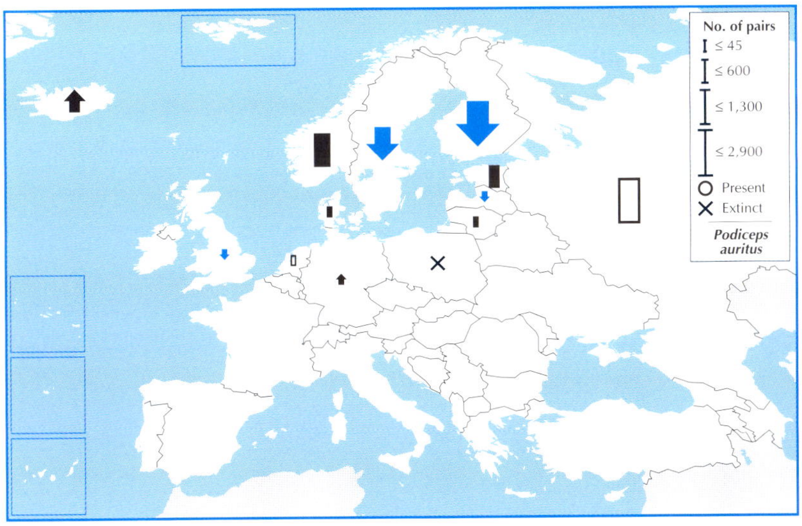

Country	Breeding pop. size (pairs)	Year(s)	Trend	Mag.%	References
Denmark	**0 – 2**	98–01	**0**	**0–19**	*4,5,6,7*
Estonia	300 – 400	98	0	0–19	*1*
Finland	2,000 – 4,000	99–01	**–**	**30**	
Germany	**2 – 2**	95–99	**+**	**50–79**	
Iceland	500 – 700	02	+	30–49	*4,28*
Latvia	(20 – 50)	90–00	(–)	(50–79)	*17*
Lithuania	1 – 10	99–01	0	0–19	*20*
Netherlands	**0 – 2**	98–00	?	–	*1*
Norway	(1,000 – 1,500)	90–03	0	0–19	
Poland	**0 – 0**	90–00	**–**	**X**	*2*
Russia	1,500 – 3,000	90–00	?	–	*3,68,98,134,141*
Sweden	900 – 1,200	99–00	–	20–29	
UK	**42 – 48**	96–00	**–**	**62**	
Total (approx.)	6,300 – 11,000	**Overall trend** Moderate decline			
Breeding range	>2,000,000 km²	**Gen. length.** 5		**% Global pop.** 5–24	

Winter pop. size (individuals)

Total (approx.)	>3,000	**Overall trend** Small increase	
% in European IBAs	30–39	**Gen. length** 5	**% Global pop.** 5–24

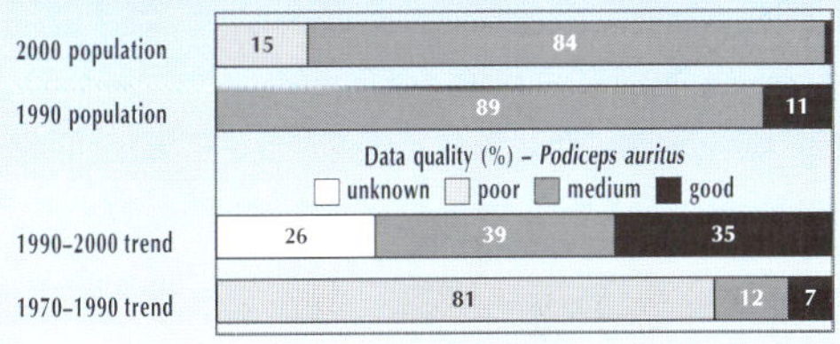

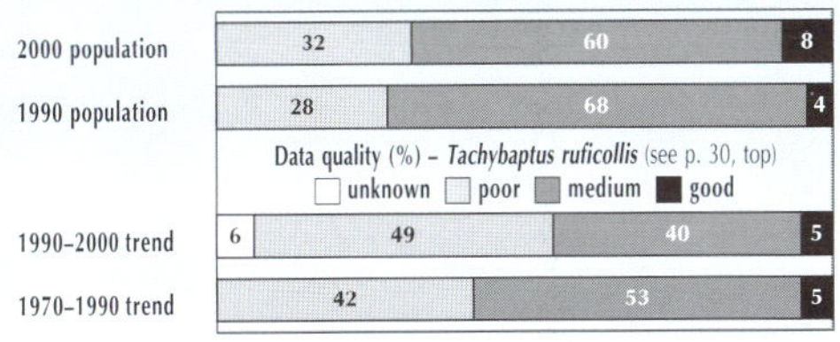

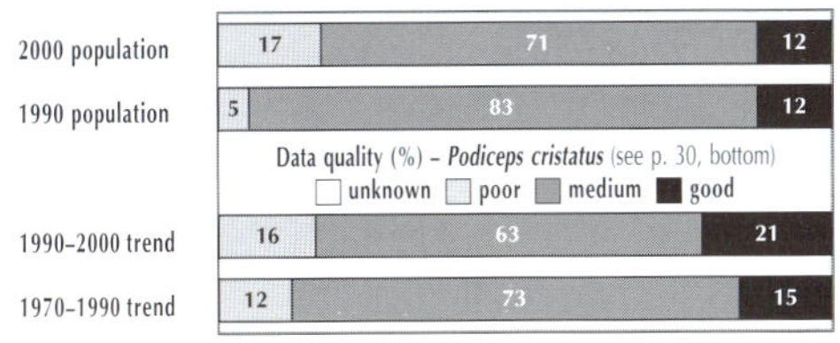

Podiceps nigricollis
BLACK-NECKED GREBE

Non-SPEC (1994: —) **Status** Secure
Criteria —
European IUCN Red List Category —
Criteria —
Global IUCN Red List Category —
Criteria —

Podiceps nigricollis is a widespread but patchily distributed breeder across much of Europe, which accounts for less than a quarter of its global breeding range. Its European breeding population is relatively small (<96,000 pairs), but underwent a large increase between 1970–1990. Although there were declines in a number of countries—most notably Romania—during 1990–2000, key populations in Russia and Ukraine were stable, and the species declined only slightly overall. Consequently, it is evaluated as Secure.

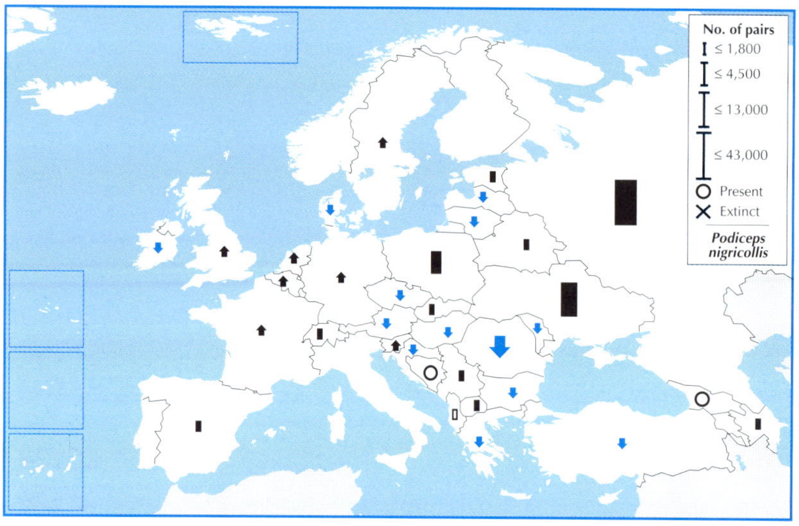

Country	Breeding pop. size (pairs)	Year(s)	Trend	Mag.%	References
Albania	5 – 25	96–02	?	–	*6*
Austria	**20 – 60**	98–02	–	**>80**	
Azerbaijan	(250 – 500)	96–02	(0)	(0–19)	
Belarus	**250 – 500**	97–00	0	0–19	
Belgium	**204 – 243**	95–02	+	**50–79**	*1*
Bosnia & HG	Present	85–89	?	–	
Bulgaria	**10 – 30**	95–02	(–)	(20–29)	
Croatia	20 – 100	02	–	50–79	*70*
Czech Rep.	**400 – 800**	00	–	**>80**	
Denmark	**200 – 300**	99–00	(–)	(0–19)	*16,3*
Estonia	(0 – 10)	98	0	0–19	*1*
France	**1,200 – 1,500**	95–00	+	0–9	*4*
Georgia	Present	03	?	–	
Germany	**1,500 – 2,000**	95–99	+	**50–79**	
Greece	0 – 20	95–00	(–)	(0–19)	
Hungary	(600 – 1,000)	90–93	–	20–49	*14*
Rep. Ireland	(0 – 50)	88–91	(–)	(>80)	
Latvia	(25 – 50)	90–00	(–)	(0–19)	*23*
Lithuania	**150 – 200**	99–01	–	0–19	*20*
Macedonia	(10 – 50)	99–00	(0)	(0–9)	
Moldova	30 – 60	90–00	–	20–29	
Netherlands	**300 – 500**	98–00	+	**116**	*1*
Poland	4,000 – 5,000	95–00	0	0–19	*2*
Romania	(2,500 – 3,500)	98–02	(–)	(0–19)	
Russia	30,000 – 60,000	90–00	0	0–19	*3,8,52*
Serbia & MN	**100 – 150**	90–02	0	0–19	*1,29,155,156, 100,225,10,227*
Slovakia	100 – 250	80–99	F	20–29	
Slovenia	0 – 3	96–00	+	N	
Spain	(300 – 750)	98–02	(F)	(–)	*10,16*
Sweden	**105 – 105**	99–00	+	**380–400**	
Switzerland	**2 – 10**	93–96	**F**	**20–29**	
Turkey	800 – 1,600	01	–	0–19	
Ukraine	10,000 – 16,500	90–00	F	30–49	
UK	**31 – 53**	96–00	+	**48**	
Total (approx.)	53,000 – 96,000	**Overall trend** Small decline			
Breeding range	>2,000,000 km²	**Gen. length.** 5		**% Global pop.** 5–24	

	Winter pop. size (individuals)		
Total (approx.)	>70,000	**Overall trend** Small increase	
% in European IBAs	42–47	**Gen. length** 5	**% Global pop.** 5–24

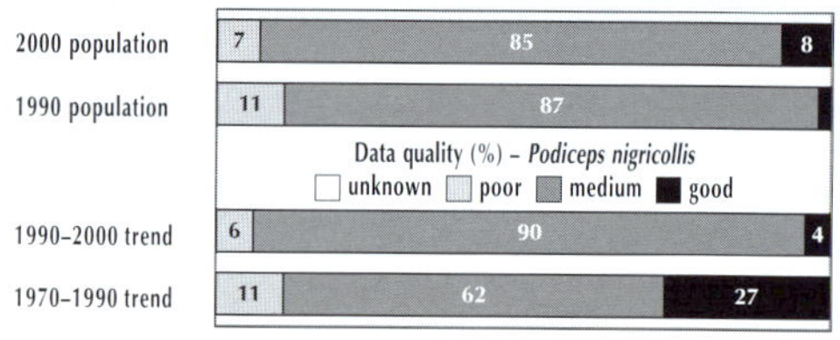

Fulmarus glacialis
NORTHERN FULMAR

Non-SPEC (1994: —) **Status** Secure
Criteria —
European IUCN Red List Category —
Criteria —
Global IUCN Red List Category —
Criteria —

Fulmarus glacialis is a widespread breeder in coastal areas of north-west Europe, which accounts for less than half of its global breeding range. Its European breeding population is very large (>2,800,000 pairs), and underwent a large increase between 1970–1990. Although the species declined slightly in the United Kingdom during 1990–2000, other European populations—including key ones in Iceland, Svalbard and the Faroes—increased or were stable, and the species underwent a large increase overall. Consequently, it is evaluated as Secure.

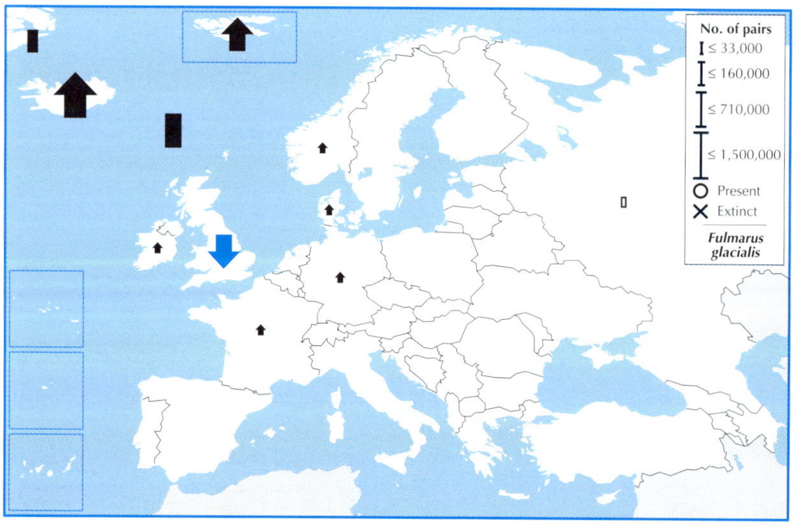

Country	Breeding pop. size (pairs)	Year(s)	Trend	Mag.%	References
Denmark	**0 – 2**	98–01	+	N	*4,5,6,7*
Faroe Is.	600,000 – 600,000	95	(0)	(0–19)	
Greenland	(120,000 – 200,000)	90–00	(0)	(0–19)	*5*
France	**1,300 – 1,350**	00	+	**20–29**	*1*
Germany	**102 – 102**	95–99	+	**30–49**	
Iceland	1,000,000 – 2,000,000	78–94	(+)	(20–29)	*1*
Rep. Ireland	**33,000 – 33,000**	99–02	+	**88**	
Norway	**7,000 – 8,000**	90–01	+	**0–19**	*35,4,31*
Svalbard	(500,000 – 1,000,000)	90–01	(+)	(20–29)	
Russia	(1,000 – 2,500)	90–00	?	–	*154,155*
UK	**506,000 – 506,000**	98–02	–	**3**	*20*
Total (approx.)	2,800,000 – 4,400,000	**Overall trend** Large increase			
Breeding range	>500,000 km²	**Gen. length.** 31		**% Global pop.** 25–49	

Country	Winter pop. size (individuals)	Year(s)	Trend	Mag.%	References
Denmark	(200 – 200)	92–93	?	–	*27*
Faroe Is.	500,000 – 3,000,000	92	?	–	*3*
Greenland	(10,000 – 100,000)	90–00	(F)	(–)	
France	(100 – 500)	98–02	?	–	*9*
Iceland	(1,000,000 – 5,000,000)	78–94	?	–	*48*
Total (approx.)	>1,500,000	**Overall trend** Unknown			
% in European IBAs	Unknown	**Gen. length** 31		**% Global pop.** 5–24	

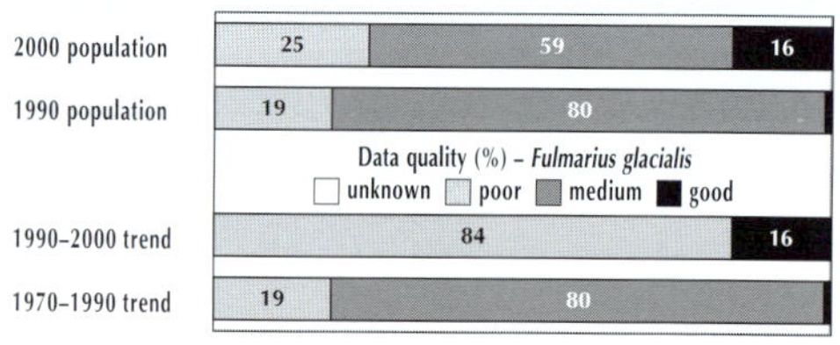

Pterodroma madeira
ZINO'S PETREL

SPEC 1 (1994: 1) **Status** Critically Endangered
Criteria See IUCN below
European IUCN Red List Category CR
Criteria D1
Global IUCN Red List Category CR
Criteria D1

Country	Breeding pop. size (pairs)	Year(s)	Trend	Mag.%	References
Portugal					
Madeira	**30 - 40**	00	**0**	**0–19**	*32*
Total (approx.)	30 - 40	**Overall trend** Stable			
Breeding range	<20 km²	**Gen. length.** 16		**% Global pop.** 100	

Pterodroma madeira breeds only in Madeira, where it has an extremely small breeding range (<20 km²) in the central mountain massif. Its breeding population was tiny (as few as 20 pairs), but stable between 1970–1990. Since 2000, the population has numbered more than 30 pairs, and if it remains at this level until 2005, the species will warrant downlisting to Endangered. However, in the meantime, this globally threatened species is provisionally evaluated as Critically Endangered in Europe.

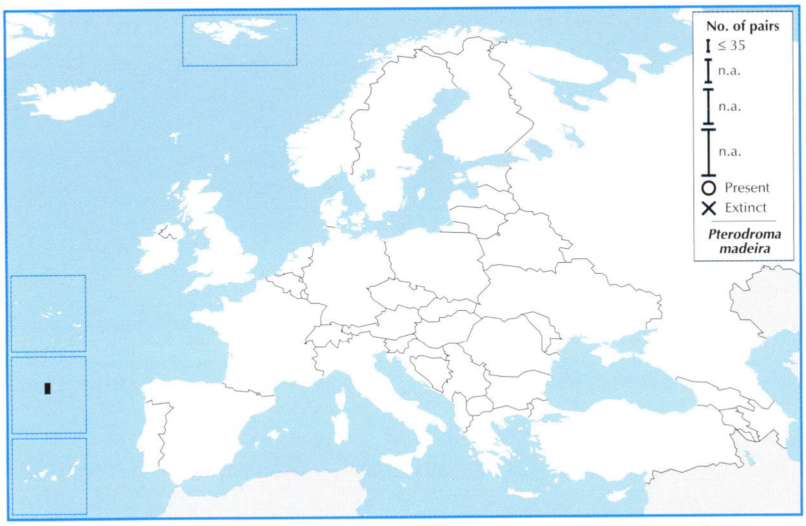

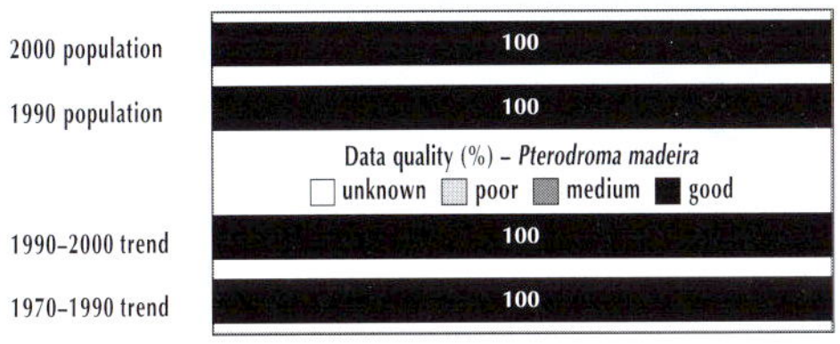

Pterodroma feae
FEA'S PETREL

SPEC 1 (1994: 1) **Status** Vulnerable
Criteria See IUCN below
European IUCN Red List Category VU
Criteria D1; D2
Global IUCN Red List Category NT
Criteria D1; D2

Country	Breeding pop. size (pairs)	Year(s)	Trend	Mag.%	References
Portugal					
Azores	**0 - 1**	02	?	–	*36*
Madeira	**170 - 260**	01	**0**	**0–19**	*29*
Total (approx.)	170 - 260	**Overall trend** Stable			
Breeding range	<20 km²	**Gen. length.** 16		**% Global pop.** 25–49	

Pterodroma feae has an extremely small breeding range (<20 km²) on Bugio, one of the Deserta Islands to the south-east of Madeira, with the majority of its global population breeding outside Europe in the Cape Verde Islands. Breeding may also occur in the Azores. Its European breeding population is very small (as few as 170 pairs), but was stable between 1970–1990. The species remained stable during 1990–2000, but as a consequence of its very small population and range, this globally Near Threatened species is evaluated as Vulnerable in Europe.

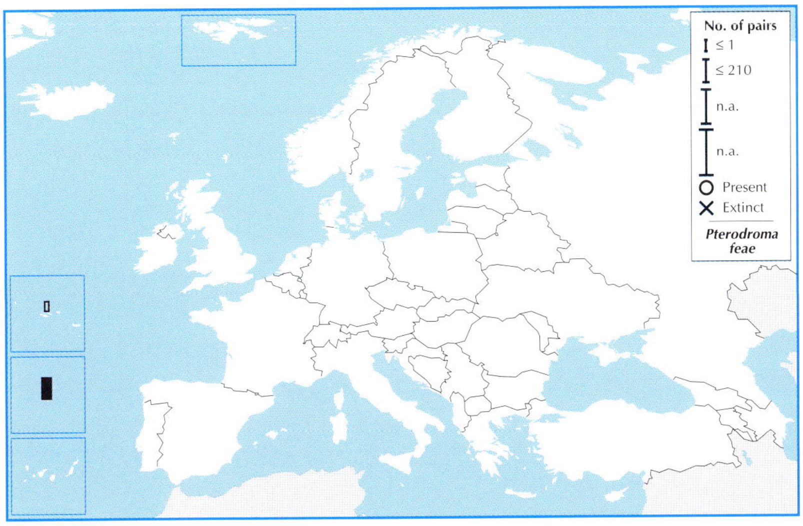

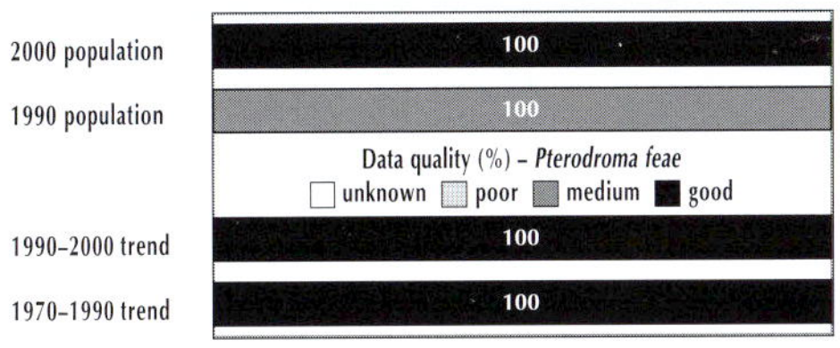

Bulweria bulwerii
BULWER'S PETREL

SPEC 3 (1994: 3) **Status** (Rare)
Criteria <10,000 pairs
European IUCN Red List Category —
Criteria —
Global IUCN Red List Category —
Criteria —

Country	Breeding pop. size (pairs)	Year(s)	Trend	Mag.%	References
Portugal					
Azores	**30 – 42**	00–03	0	0–19	
Madeira	(6,000 – 8,000)	94	(0)	(0–19)	
Spain					
Canary Is.	1,000 – 1,000	87	?	–	*23,28*
Total (approx.)	7,000 – 9,000	**Overall trend** Stable			
Breeding range	>20,000 km²	**Gen. length.** 24		**% Global pop.** 5–24	

Bulweria bulwerii breeds in the Azores, Madeira and the Canary Islands, which together account for less than a quarter of its global breeding range. Its European breeding population is small (as few as 7,000 pairs), and underwent a moderate decline between 1970–1990. Although the trend in the Canary Islands during 1990–2000 was unknown, the species remained stable in its stronghold in Madeira, and was stable overall. Nevertheless, its population size renders it susceptible to the risks affecting small populations, and consequently it is provisionally evaluated as Rare.

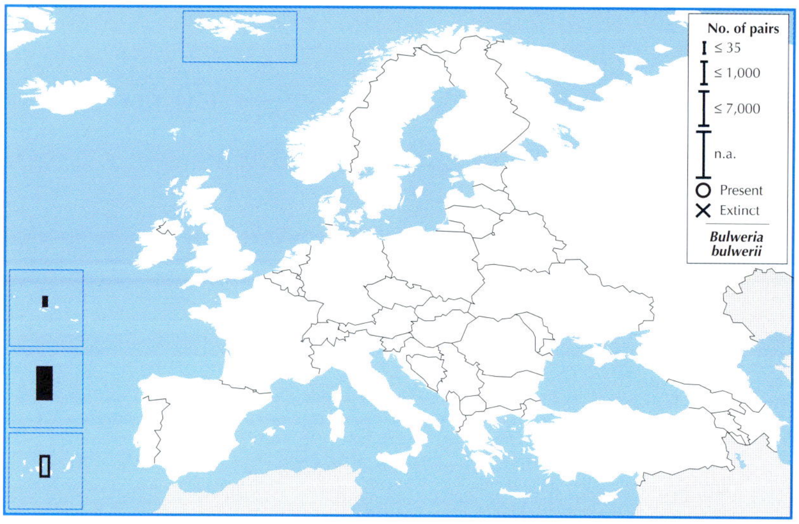

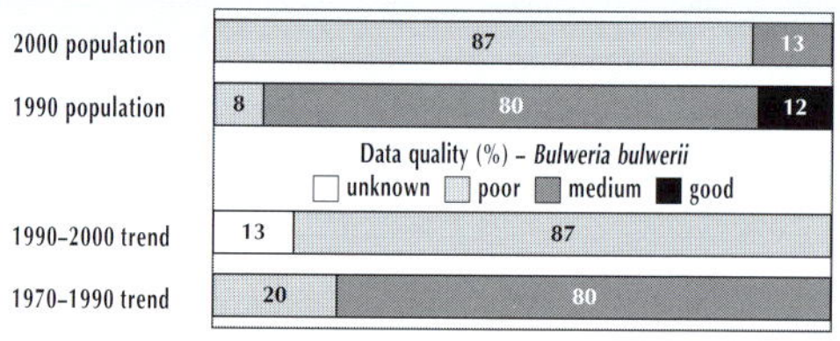

Calonectris diomedea
CORY'S SHEARWATER

SPEC 2 (1994: 2) **Status** (Vulnerable)
Criteria See IUCN below
European IUCN Red List Category VU
Criteria A4b
Global IUCN Red List Category —
Criteria —

Country	Breeding pop. size (pairs)	Year(s)	Trend	Mag.%	References
Croatia	(800 – 1,000)	02	(–)	(50–79)	*16*
France	**970 – 1,200**	95–00	**0**	**0–19**	*1*
Greece	5,000 – 5,000	95–00	(0)	(0–19)	
Italy	15,000 – 18,000	03	0	0–19	*6*
Malta	6,090 – 7,130	90–02	–	0–19	*2*
Portugal	**200 – 250**	99–03	(+)	(–)	*5*
Azores	188,000 – 188,000	96–01	0	0–19	*38*
Madeira	16,500 – 25,000	00	(0)	(0–19)	*30*
Spain	(2,500 – 10,000)	98–02	(–)	(20–29)	*10,16*
Canary Is.	30,000 – 30,000	87	?	–	*31,28*
Turkey	(0 – 200)	01	?	–	
Total (approx.)	270,000 – 290,000	**Overall trend** Large decline			
Breeding range	>100,000 km²	**Gen. length.** 18		**% Global pop.** 75–94	

Calonectris diomedea breeds on the rocky coasts and islands of the Mediterranean and Atlantic, with Europe constituting >75% of its global breeding range. Its European breeding population is large (>270,000 pairs), but underwent a large decline between 1970–1990. Although declines abated in many countries during 1990–2000, the species continued to decline overall at a rate that—if sustained—probably equates to a large decline (>30%) over three generations. As a consequence of this continuing decline, the species is provisionally evaluated as Vulnerable.

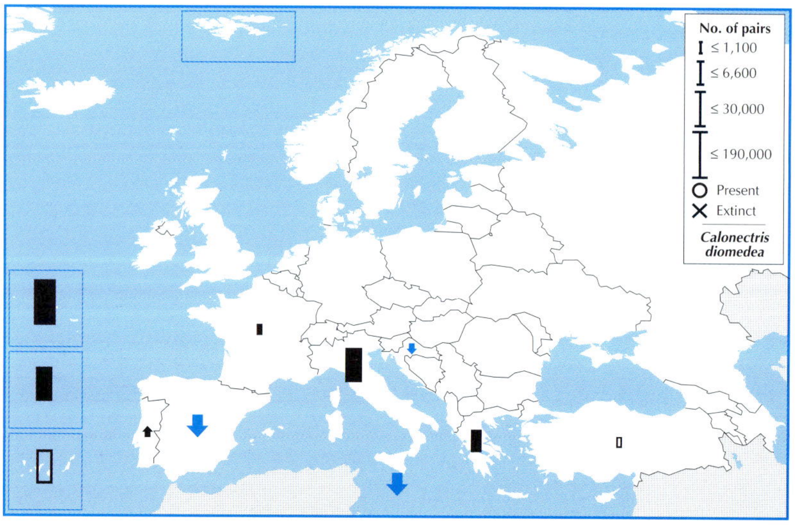

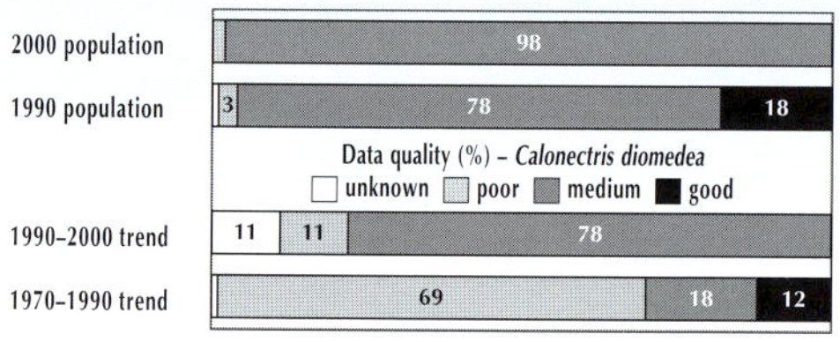

Puffinus gravis
GREAT SHEARWATER

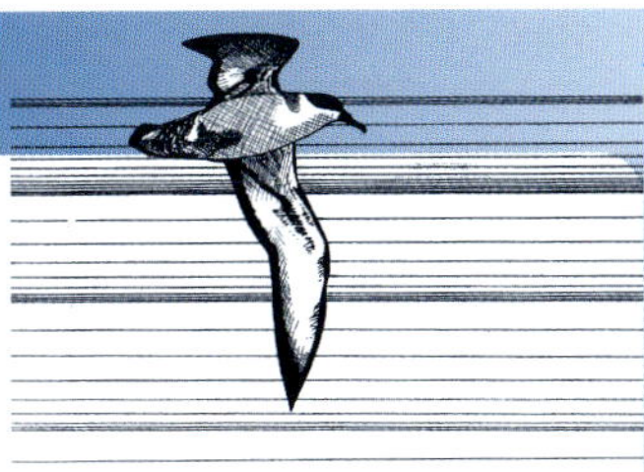

NE (1994: NE) **Status** Not Evaluated
Criteria —
European IUCN Red List Category —
Criteria —
Global IUCN Red List Category —
Criteria —

Puffinus gravis breeds on three island groups in the South Atlantic—Tristan da Cunha, Gough Island and the Falkland Islands—but spends most of its non-breeding season (the boreal summer) in the North Atlantic. Although substantial numbers (>100,000 individuals) occur off the coast of Greenland during the non-breeding season, the species is essentially a passage visitor to European waters (mainly in August–September). Consequently, its status in Europe is Not Evaluated.

Country	Winter pop. size (individuals)	Year(s)	Trend	Mag.%	References
Denmark					
Greenland	(100,000 – 500,000)	90–00	(-)	(0–19)	

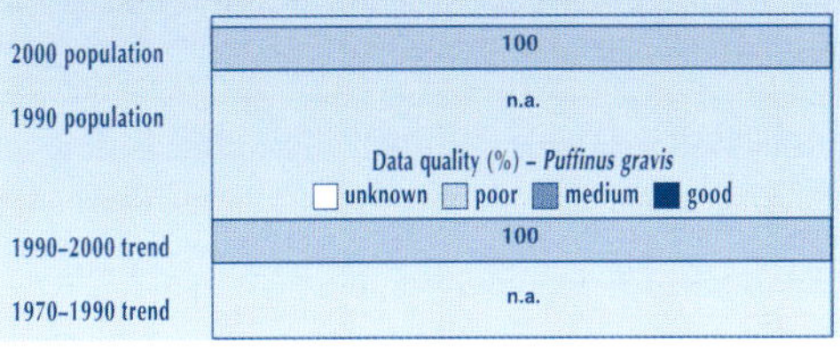

Puffinus griseus
SOOTY SHEARWATER

SPEC 1 (1994: NE) **Status** Not Evaluated
Criteria —
European IUCN Red List Category —
Criteria —
Global IUCN Red List Category NT
Criteria A2d,e; A3d,e

Puffinus griseus breeds on subantarctic islands in the Southern Ocean, migrating to the temperate zones of the North Atlantic and North Pacific during its non-breeding season (the boreal summer). It occurs mainly off the North American coast during the first half of the non-breeding season, and is only present in significant numbers in European waters during (and just prior to) its southern passage (mainly in August–October). Consequently, the European status of this globally Near Threatened species is Not Evaluated.

Puffinus puffinus
MANX SHEARWATER

SPEC 2 (1994: 2) **Status** (Localised)
Criteria ≥90% breed at ≤10 sites
European IUCN Red List Category —
Criteria —
Global IUCN Red List Category —
Criteria —

Puffinus puffinus breeds almost exclusively within Europe, nesting mostly on offshore islands in the north-east Atlantic. Its European breeding population is large (>350,000 pairs), and was stable between 1970–1990. Although trend data were not available for the stronghold population in the United Kingdom during 1990–2000, there was no evidence to suggest that the species declined overall. Nevertheless, more than 90% of the European breeding population occurs at 10 sites (see Appendix 5), and consequently the species is provisionally evaluated as Localised.

Country	Breeding pop. size (pairs)	Year(s)	Trend	Mag.%	References
Denmark					
Faroe Is.	25,000 – 25,000	95	(0)	(0–19)	
France	**140 – 200**	96–00	+	0–19	*4*
Iceland	7,000 – 10,000	91	?	-	*20*
Rep. Ireland	**33,000 – 33,000**	99–02	(0)	(0–19)	
Portugal					
Azores	115 – 240	99	0	0–19	*36*
Madeira	(1,500 – 2,500)	00	(0)	(0–19)	
Spain					
Canary Is.	(250 – 1,000)	87–01	(-)	(0–19)	*26,28*
UK	**281,000 – 320,000**	98–02	?	-	*20*
Total (approx.)	350,000 – 390,000	**Overall trend** Unknown			
Breeding range	>50,000 km²	**Gen. length.** 18		**% Global pop.** >95	

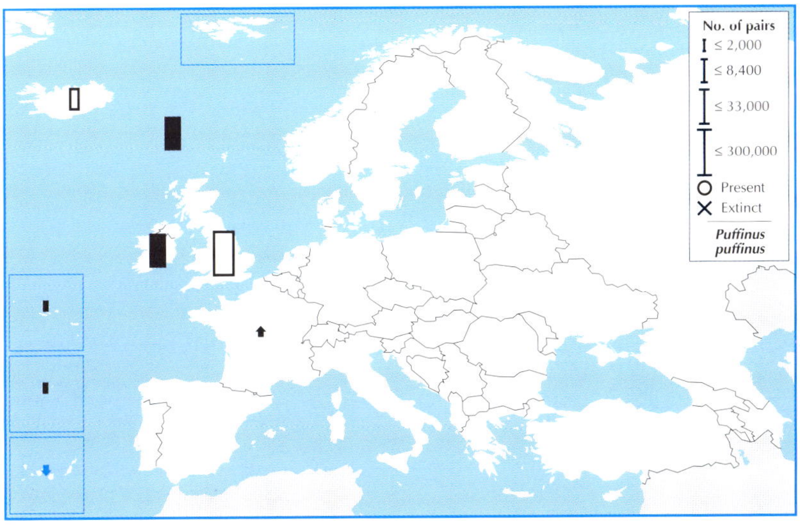

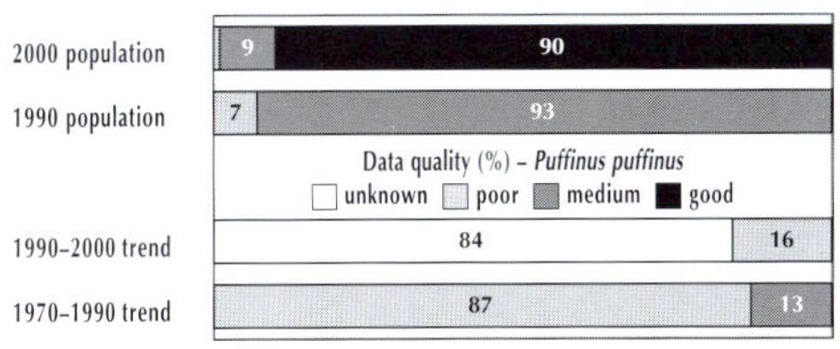

Puffinus mauretanicus
BALEARIC SHEARWATER

SPEC 1 (1994: NE) **Status** Critically Endangered
Criteria See IUCN below
European IUCN Red List Category CR
Criteria A4b,c,e; B2a+b(ii,iii,iv,v)
Global IUCN Red List Category CR
Criteria A4b,c,e; B2a+b(ii,iii,iv,v)

Country	Breeding pop. size (pairs)	Year(s)	Trend	Mag.%	References
Spain	**1,650 – 2,050**	98–02	–	30–49	*14,16,12*
Total (approx.)	1,650 – 2,050	**Overall trend** Large decline			
Breeding range	<100 km²	**Gen. length.** 18		**% Global pop.** 100	

Puffinus mauretanicus (formerly treated as a subspecies of *P. yelkouan*) has an extremely small global breeding range (<100 km²) confined to the Balearic Islands (Spain). Its breeding population is small (as few as 1,650 pairs), and underwent a moderate decline between 1970–1990. The species continued to decline during 1990–2000 at a rate that—if sustained—would equate to an extremely large decline (>80%) over three generations. As a consequence of this continuing decline and its extremely small range, this globally threatened species is evaluated as Critically Endangered.

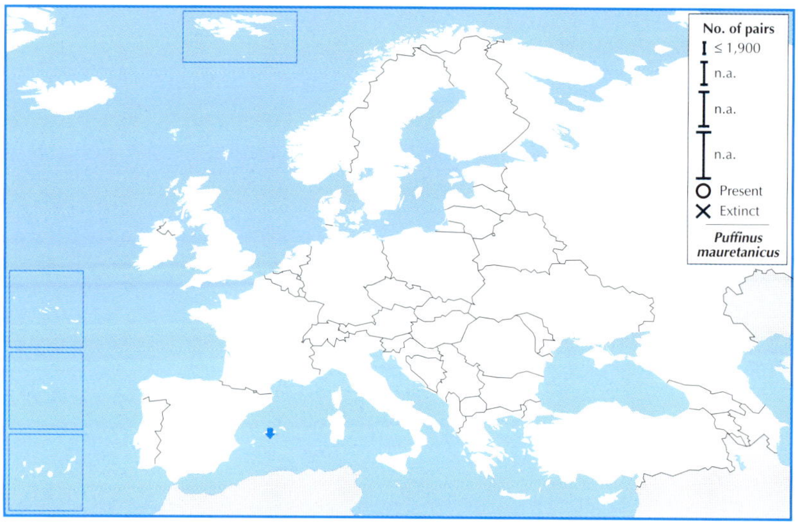

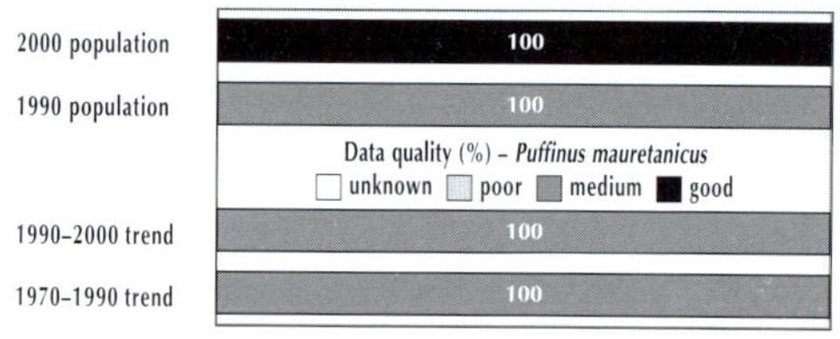

Puffinus yelkouan
YELKOUAN SHEARWATER

Non-SPEC^E (1994: 4) **Status** Secure
Criteria —
European IUCN Red List Category —
Criteria —
Global IUCN Red List Category —
Criteria —

Country	Breeding pop. size (pairs)	Year(s)	Trend	Mag.%	References
Albania	(1 – 10)	92–02	?	–	
Bulgaria	(0 – 10)	95–02	?	–	
Croatia	(50 – 100)	02	(–)	(80–89)	*70*
France	**240 – 360**	96–00	0	0–19	*1*
Greece	4,000 – 7,000	95–00	(0)	(0–19)	
Italy	7,000 – 14,000	03	0	0–19	*6*
Malta	1,400 – 1,560	90–02	0	0–19	*2*
Spain	(50 – 250)	98–02	?	–	*10*
Turkey	(0 – 10,000)	01	?	–	
Total (approx.)	13,000 – 33,000	**Overall trend** Stable			
Breeding range	>100,000 km²	**Gen. length.** 18		**% Global pop.** >95	

Puffinus yelkouan breeds in the central and eastern Mediterranean, with >95% of its population occurring in Europe. The European breeding population is relatively small (<33,000 pairs), but was stable between 1970–1990. Although trends during 1990–2000 were not available for certain countries—including the poorly known but potentially sizeable population in Turkey—the species was stable in its strongholds in Italy and Greece, and remained stable overall. Consequently, it is evaluated as Secure.

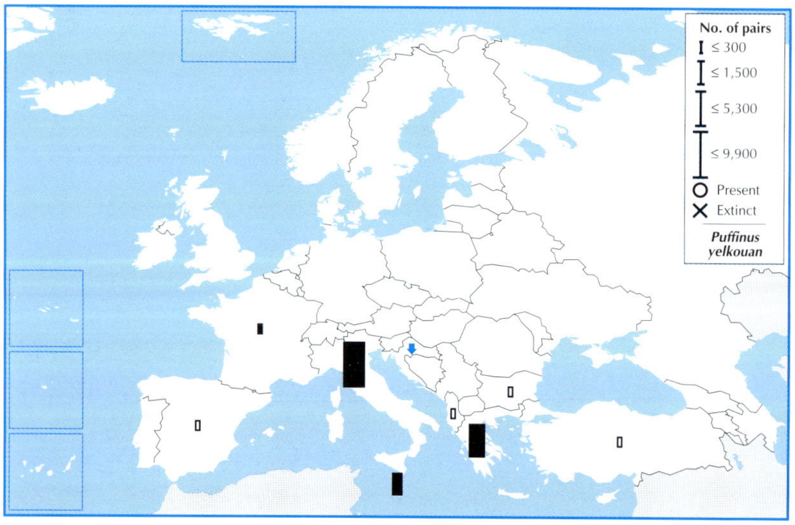

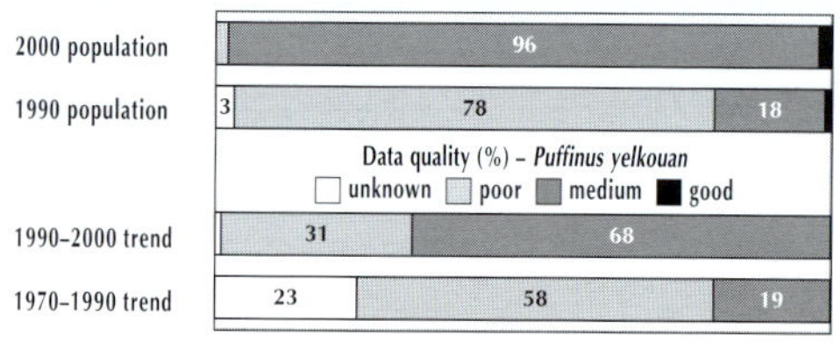

Puffinus assimilis
LITTLE SHEARWATER

SPEC 3 (1994: 3) **Status** (Rare)
Criteria <10,000 pairs
European IUCN Red List Category —
Criteria —
Global IUCN Red List Category —
Criteria —

Country	Breeding pop. size (pairs)	Year(s)	Trend	Mag.%	References
Portugal					
Azores	800 - 1,500	97	0	0–19	*36*
Madeira	(4,000 - 5,000)	94–95	(0)	(0–19)	*31*
Spain					
Canary Is.	(400 - 400)	87	?	-	*31,28*
Total (approx.)	5,200 - 6,900	**Overall trend** Stable			
Breeding range	>20,000 km²	**Gen. length** 18		**% Global pop.** 5–24	

Puffinus assimilis breeds in the Azores, Madeira and the Canary Islands, which together account for less than a quarter of its global breeding range. Its European breeding population is small (as few as 5,200 pairs), and underwent a moderate decline between 1970–1990. Although the trend in the Canary Islands during 1990–2000 was unknown, the species was stable in the Azores and its stronghold Madeira, and was stable overall. Nevertheless, its population size renders it susceptible to the risks affecting small populations, and consequently it is provisionally evaluated as Rare.

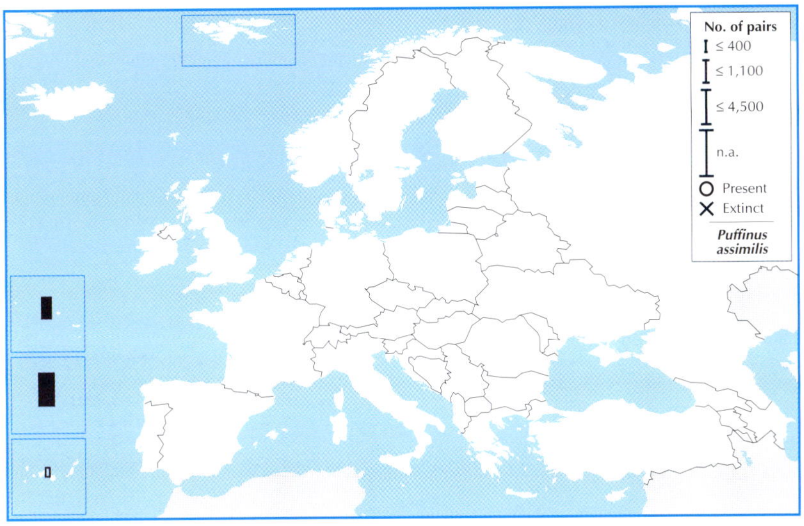

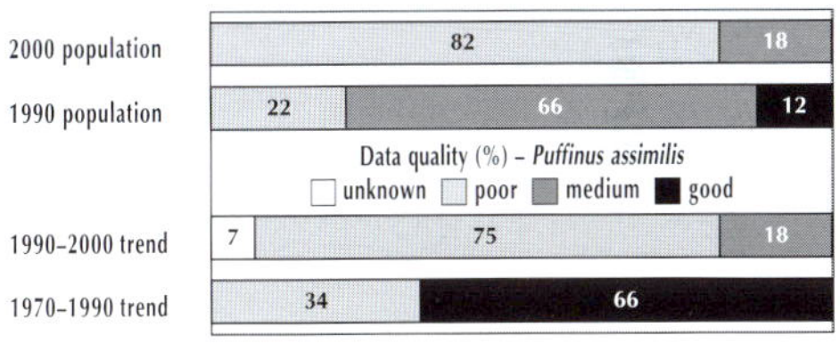

Pelagodroma marina
WHITE-FACED STORM-PETREL

SPEC 3 (1994: 3) **Status** Vulnerable
Criteria See IUCN below
European IUCN Red List Category VU
Criteria D2
Global IUCN Red List Category —
Criteria —

Country	Breeding pop. size (pairs)	Year(s)	Trend	Mag.%	References
Portugal					
Madeira	**61,000 - 61,000**	99	0	0–19	
Spain					
Canary Is.	**50 - 60**	00–01	F	20–29	*30*
Total (approx.)	61,000 - 61,000	**Overall trend** Stable			
Breeding range	<10 km²	**Gen. length** 16		**% Global pop.** 5–24	

Pelagodroma marina breeds on the Selvagens (in the Madeiran archipelago) and on islets off Lanzarote (in the Canary Islands), with Europe accounting for less than a quarter of its global breeding range. Its European breeding population is relatively small (c.61,000 pairs), but was stable between 1970–1990. Although the species remained stable overall during 1990–2000, the entire European breeding population is confined to fewer than six locations, with >99% of birds breeding in an area smaller than 3 km² on the Selvagens. Consequently, it is evaluated as Vulnerable.

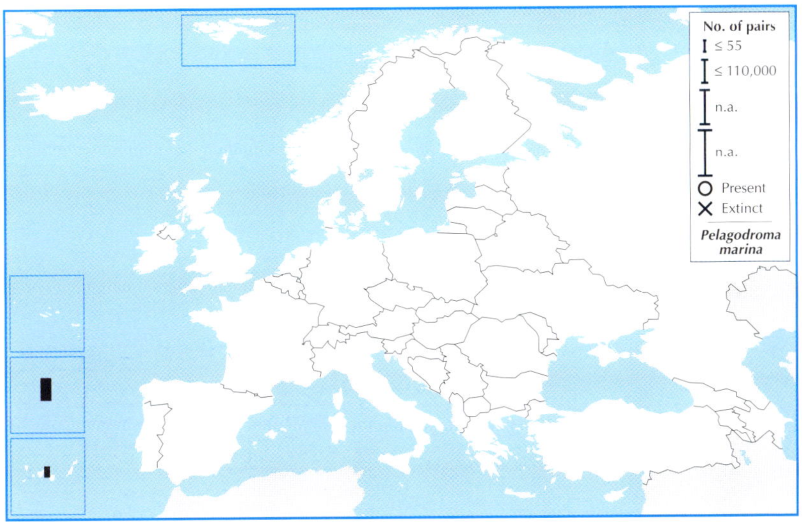

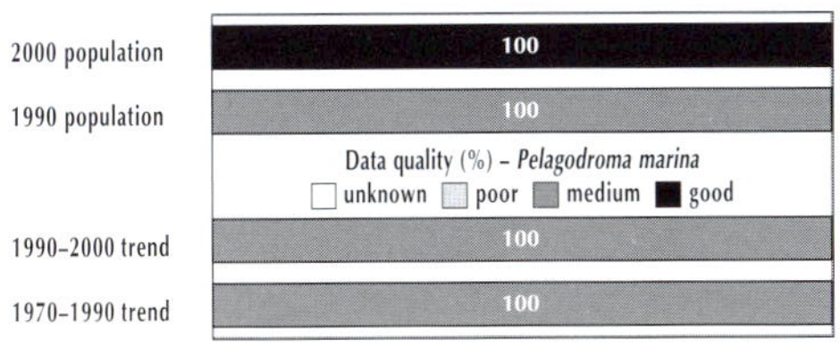

Hydrobates pelagicus
EUROPEAN STORM-PETREL

Non-SPEC[E] (1994: 2) **Status** (Secure)
Criteria —
European IUCN Red List Category —
Criteria —
Global IUCN Red List Category —
Criteria —

Country	Breeding pop. size (pairs)	Year(s)	Trend	Mag.%	References
Croatia	(1 – 10)	02	?	–	*16*
Denmark					
Faroe Is.	250,000 – 250,000	95	(0)	(0–19)	
France	**600 – 700**	97–00	**0**	**0–19**	*1*
Greece	(10 – 30)	95–00	(0)	(0–19)	
Iceland	50,000 – 100,000	91	?	–	*20*
Rep. Ireland	**99,950 – 99,950**	99–02	(0)	(0–19)	
Italy	1,700 – 2,500	03	0	0–19	*6*
Malta	5,025 – 8,025	90–02	0	0–19	*2*
Norway	(1,000 – 10,000)	90–02	(0)	(0–19)	*4*
Spain	(4,410 – 6,900)	98–02	(–)	(0–19)	*10,16*
Canary Is.	1,000 – 1,000	87	(–)	(0–19)	*31,28,30*
UK	**21,100 – 33,500**	95–02	?	–	*20*
Total (approx.)	430,000 – 510,000	**Overall trend** Stable			
Breeding range	>100,000 km²	**Gen. length.** 14		**% Global pop.** >95	

Hydrobates pelagicus breeds almost exclusively within Europe, nesting on offshore islands in the north-east Atlantic and western Mediterranean. Its European breeding population is large (>430,000 pairs), and was stable during 1970–1990. Although the species declined in Spain and the Canary Islands during 1990–2000, most European populations were stable, and the species remained stable overall. As a consequence of an improved knowledge of its distribution and numbers, this previously Localised species is now provisionally evaluated as Secure.

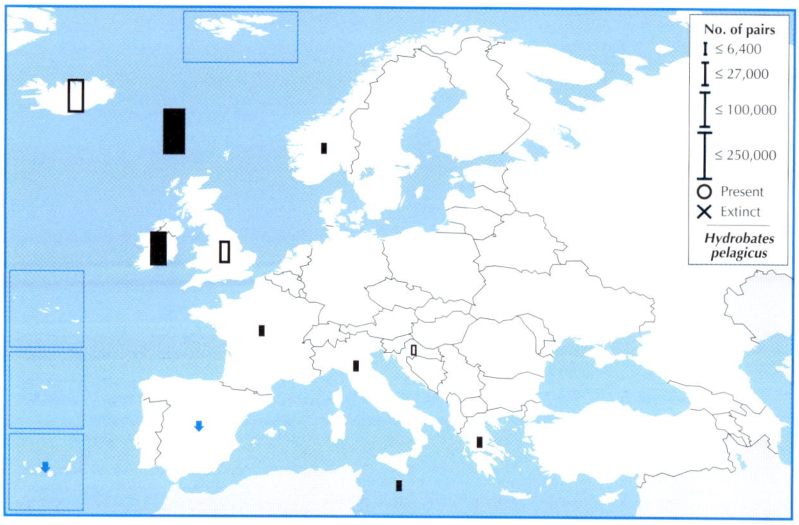

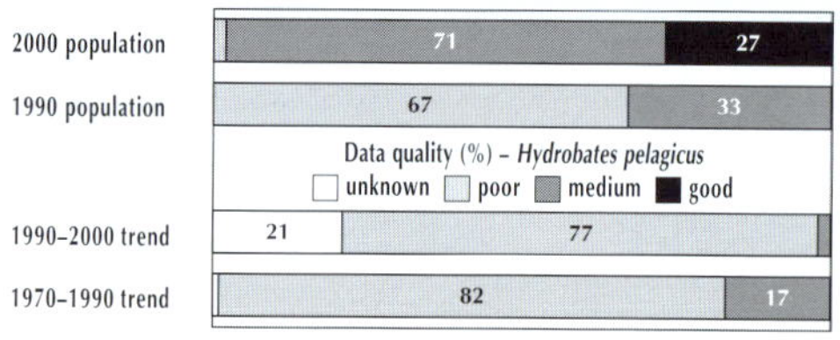

Oceanodroma leucorhoa
LEACH'S STORM-PETREL

SPEC 3 (1994: 3) **Status** (Localised)
Criteria ≥90% breed at ≤10 sites
European IUCN Red List Category —
Criteria —
Global IUCN Red List Category —
Criteria —

Country	Breeding pop. size (pairs)	Year(s)	Trend	Mag.%	References
Denmark					
Faroe Is.	1,000 – 1,000	95	(0)	(0–19)	
Iceland	80,000 – 150,000	91	?	–	*20*
Rep. Ireland	**310 – 310**	99–02	(0)	(0–19)	
Norway	(100 – 1,000)	90–02	(0)	(0–19)	*4*
UK	**36,400 – 64,900**	99–01	?	–	*20*
Total (approx.)	120,000 – 220,000	**Overall trend** Unknown			
Breeding range	>20,000 km²	**Gen. length.** 9		**% Global pop.** 5–24	

Oceanodroma leucorhoa breeds on remote islands in north-western Europe, which accounts for less than a quarter of its global breeding range. Its European breeding population is large (>120,000 pairs), and was stable between 1970–1990. Although trends were not available for key populations in Iceland and the United Kingdom during 1990–2000, there was no evidence to suggest that the species declined. Nevertheless, more than 90% of the European breeding population occurs at 10 sites (see Appendix 5), and consequently the species is provisionally evaluated as Localised.

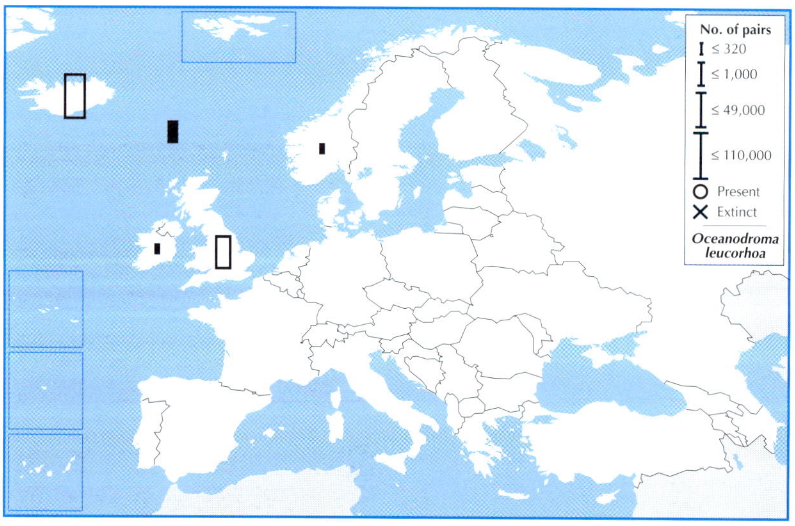

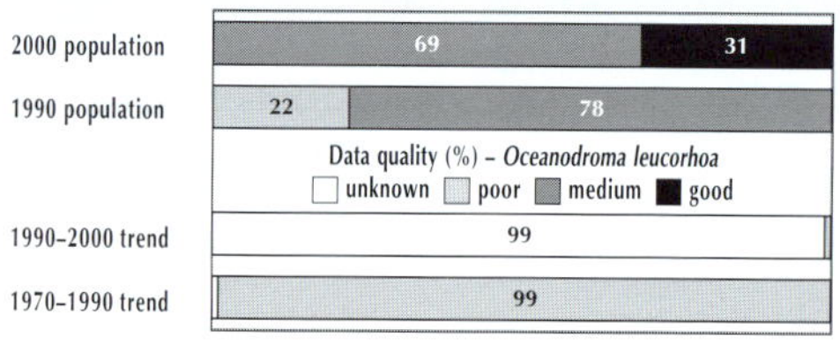

Oceanodroma castro
BAND-RUMPED STORM-PETREL

SPEC 3 (1994: 3) **Status** (Rare)
Criteria <10,000 pairs
European IUCN Red List Category —
Criteria —
Global IUCN Red List Category —
Criteria —

Country	Breeding pop. size (pairs)	Year(s)	Trend	Mag.%	References
Portugal	200 – 400	02	0	0–19	*4*
Azores	900 – 1,250	97	0	0–19	*36*
Madeira	(2,000 – 2,500)	94	(0)	(0–19)	
Spain					
Canary Is.	550 – 600	87-03	?	–	*25*
Total (approx.)	3,700 – 4,800	**Overall trend** Stable			
Breeding range	>20,000 km²	**Gen. length.** 16		**% Global pop.** 25–49	

Oceanodroma castro breeds in the Azores, Madeira, the Canary Islands and Portugal, with Europe accounting for less than half of its global breeding range. Its European breeding population is small (as few as 3,700 pairs), and underwent a moderate decline between 1970–1990. Although the trend in the Canary Islands during 1990–2000 was unknown, the species was stable elsewhere within its European range. Nevertheless, its population size renders it susceptible to the risks affecting small populations, and consequently it is provisionally evaluated as Rare.

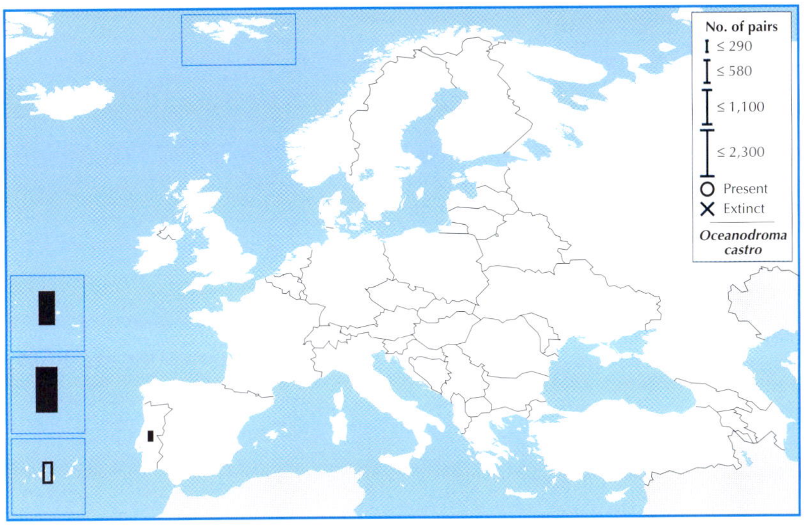

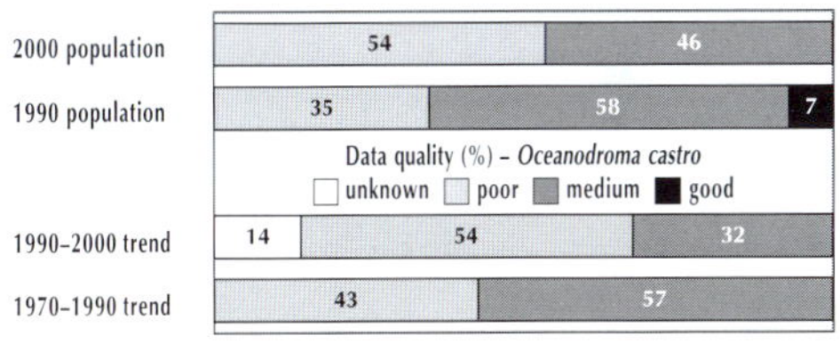

Morus bassanus
NORTHERN GANNET

Non-SPEC[E] (1994: 2) **Status** Secure
Criteria —
European IUCN Red List Category —
Criteria —
Global IUCN Red List Category —
Criteria —

Country	Breeding pop. size (pairs)	Year(s)	Trend	Mag.%	References
Denmark					
Faroe Is.	2,000 – 2,000	95	(0)	(0–19)	
France	**16,000 – 16,000**	01	+	**>80**	*6*
Germany	**70 – 70**	95–99	+	**50–79**	
Iceland	**23,000 – 25,000**	84–94	+	**0–19**	*7*
Rep. Ireland	**33,000 – 33,000**	99–02	+	**19**	
Norway	**4,200 – 4,200**	02	+	**20**	*35*
Russia	**50 – 60**	90–00	+	**N**	*80,154,155*
UK	**226,600 – 226,600**	98–00	+	**40**	*20*
Total (approx.)	300,000 – 310,000	**Overall trend** Large increase			
Breeding range	>50,000 km²	**Gen. length.** 21		**% Global pop.** 75–94	

Morus bassanus breeds at a relatively small number of sites in north-western Europe, which holds >75% of its global breeding population. Its European breeding population is large (>300,000 pairs), and increased substantially between 1970–1990. The species continued to increase across the majority of its European range during 1990–2000, and underwent a large increase overall. As a consequence of the range expansion accompanying its population growth, this previously Localised species is now evaluated as Secure.

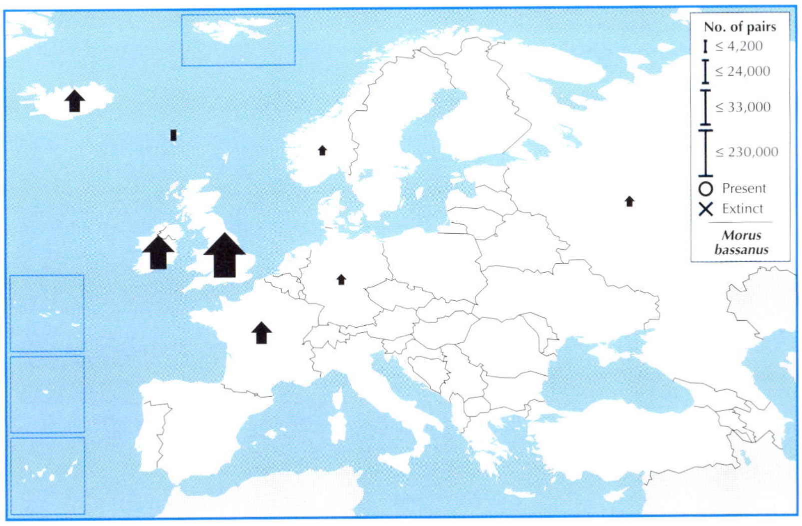

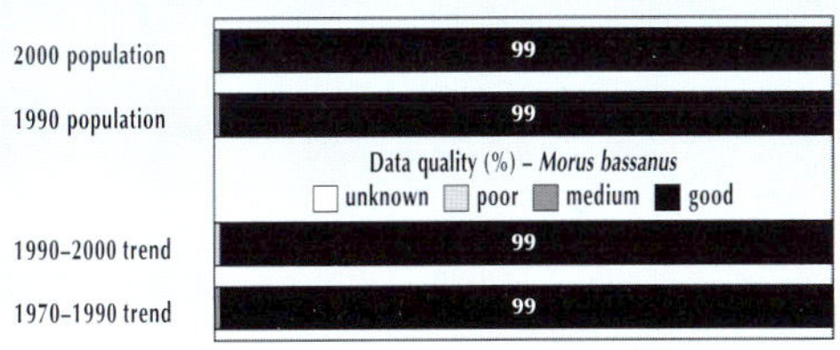

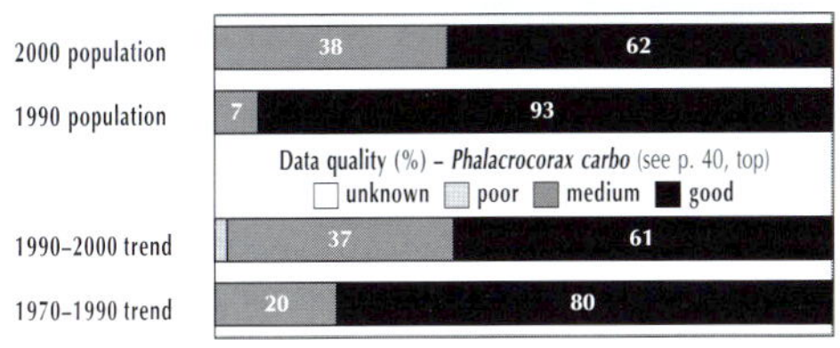

Phalacrocorax carbo
GREAT CORMORANT

Non-SPEC (1994: —) **Status** Secure
Criteria —
European IUCN Red List Category —
Criteria —
Global IUCN Red List Category —
Criteria —

Phalacrocorax carbo breeds patchily across much of Europe, which accounts for less than half of its global breeding range. Its European breeding population is large (>310,000 pairs), and underwent a large increase between 1970–1990. The species continued to increase during 1990–2000, with almost all national trends either stable or increasing, including those of key populations in Denmark, Ukraine and Russia. Consequently, it is evaluated as Secure.

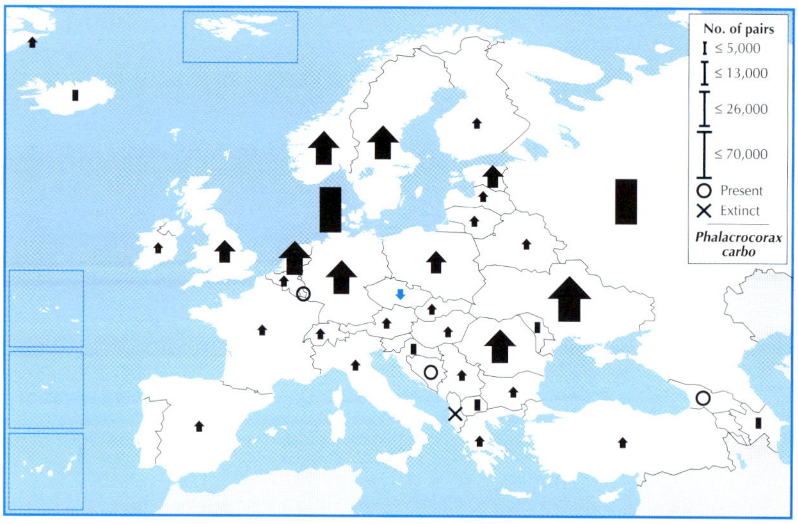

Country	Breeding pop. size (pairs)	Year(s)	Trend	Mag.%	References
Albania	0 – 0	96–02	-	X	
Austria	**0 – 33**	03	+	N	
Azerbaijan	2,000 – 4,000	96–00	(0)	(0–19)	
Belarus	**1,300 – 1,500**	97–00	+	50–79	*2*
Belgium	**980 – 1,000**	01–02	+	**>80**	*1*
Bosnia & HG	Present	85–89	?	–	*1*
Bulgaria	**2,000 – 2,800**	95–02	+	**>80**	
Croatia	**2,000 – 3,000**	02	**0**	**0–19**	*26*
Czech Rep.	**170 – 190**	00	-	**50–79**	
Denmark	**36,000 – 41,000**	97–00	**0**	**0–19**	*1*
Greenland	5,000 – 5,000	95–00	+	0–19	*4*
Estonia	**9,000 – 10,000**	98	+	**50–79**	*1*
Finland	**800 – 1,200**	02	+	N	
France	3,350 – 3,350	00	+	**50–79**	*1*
Georgia	Present	03	?	–	
Germany	**16,800 – 16,800**	95–99	+	**0–19**	
Greece	**4,300 – 4,300**	02	+	**20–29**	
Hungary	**1,800 – 3,000**	95–02	+	**>80**	
Iceland	**2,600 – 3,700**	75–94	(F)	(–)	*13*
Rep. Ireland	**4,550 – 4,550**	99–02	+	**0–19**	
Italy	**880 – 880**	00	+	**30–49**	*18*
Latvia	400 – 500	90–00	+	>80	*17*
Lithuania	**2,500 – 3,000**	99–01	+	**>80**	*20*
Luxembourg	Present	02	?	–	
Macedonia	(400 – 600)	98–00	(F)	(30–49)	
Moldova	300 – 500	90–00	0	0–19	
Netherlands	**18,400 – 19,500**	98–00	+	**10**	*1*
Norway	**20,000 – 25,000**	96–01	+	**0–19**	*35,20,63*
Poland	**12,500 – 12,500**	00	+	**200–230**	*2*
Romania	**18,000 – 20,000**	99–02	+	**0–19**	*47*
Russia	35,000 – 60,000	90–00	0	0–19	*8,11,120,59,60, 82,120,154,155*
Serbia & MN	**2,100 – 2,400**	00–02	+	**50–79**	*163,65,186,212, 230,138,231*
Slovakia	**50 – 250**	80–99	+	50–79	
Spain	**0 – 50**	98–02	+	**>80**	*10*
Sweden	**25,000 – 26,000**	99–00	+	**>80**	
Switzerland	**0 – 7**	00–02	+	**N**	
Turkey	**3,000 – 4,500**	01	+	**30–49**	
Ukraine	65,000 – 75,000	90–00	+	50–79	
UK	**9,100 – 9,100**	99–02	+	**27**	*20*

Total (approx.) 310,000 – 370,000 **Overall trend** Large increase
Breeding range >1,000,000 km² **Gen. length.** 11 **% Global pop.** 25–49
(See p. 39, bottom, for data quality graph)

Winter pop. size (individuals)

Total (approx.) >420,000 **Overall trend** Large increase
% in European IBAs 41–44 **Gen. length** 11 **% Global pop.** 25–49

Phalacrocorax aristotelis
EUROPEAN SHAG

Non-SPEC[E] (1994: 4) **Status** (Secure)
Criteria —
European IUCN Red List Category —
Criteria —
Global IUCN Red List Category —
Criteria —

Phalacrocorax aristotelis breeds in coastal areas of north-western and southern Europe, which constitutes >75% of its global breeding range. Its European breeding population is relatively small (<81,000 pairs), but increased substantially between 1970–1990. Although it was stable or increased in most countries during 1990–2000, there were declines in the United Kingdom, and the species underwent a moderate decline (>10%) overall. Nevertheless, this decline is probably outweighed by earlier increases, and consequently the species is provisionally evaluated as Secure.

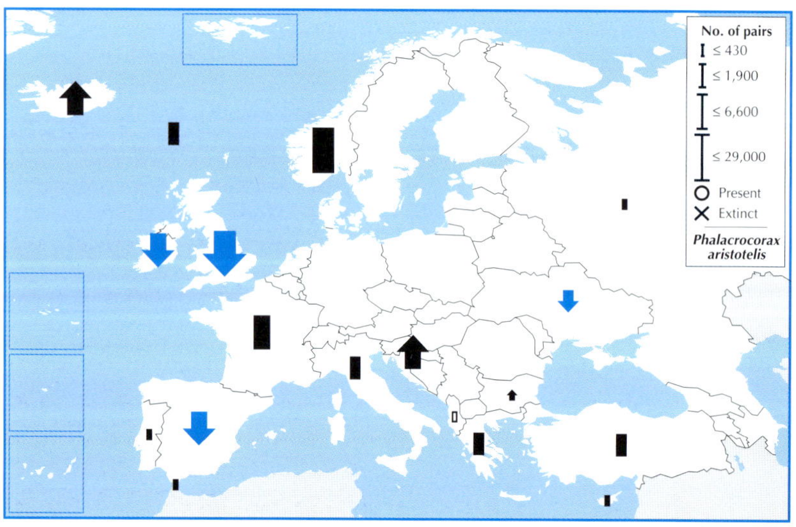

Country	Breeding pop. size (pairs)	Year(s)	Trend	Mag.%	References
Albania	(0 – 20)	92–02	?	–	
Bulgaria	**180 – 250**	01–02	+	**>80**	
Croatia	(2,500 – 5,000)	02	(+)	(20–29)	*16*
Cyprus	80 – 120	94–02	0	0–19	
Denmark					
Faroe Is.	1,500 – 1,500	95	(0)	(0–19)	
France	**6,300 – 6,400**	00	**0**	**0–19**	*1*
Greece	1,000 – 1,200	95–00	(0)	(0–19)	
Iceland	6,200 – 7,000	75	(+)	(–)	*11,1*
Rep. Ireland	**3,400 – 3,400**	99–02	-	**27**	
Italy	**1,600 – 2,200**	03	0	0–19	*6*
Norway	**15,500 – 16,500**	80–01	**0**	**0–19**	*35*
Portugal	100 – 150	02	0	0–19	*2,5*
Russia	400 – 450	90–00	0	0–19	*11,82,154,155*
Spain	4,390 – 4,390	98–02	(–)	(0–19)	*10,16*
Turkey	900 – 1,800	01	0	0–19	
Ukraine	1,500 – 1,700	90–98	-	10–19	
UK	**28,900 – 28,900**	98–02	-	**25**	*20*
Gibraltar	**5 – 10**	00	0	0–19	

Total (approx.) 75,000 – 81,000 **Overall trend** Moderate decline
Breeding range >750,000 km² **Gen. length.** 10 **% Global pop.** 75–94

Country	Winter pop. size (individuals)	Year(s)	Trend	Mag.%	References
Bulgaria	**139 – 468**	97–01	+	**>80**	
Croatia	(2,500 – 5,000)	02	(+)	(20–29)	*16*
Denmark					
Faroe Is.	3,000 – 6,000	92	?	–	*3*
France	1,000 – 2,500	98–02	0	0–19	*9*
Greece	(1,500 – 3,000)	95–00	(0)	(0–19)	
Iceland	30,000 – 40,000	78–94	?	–	*48*
Italy	(500 – 1,000)	02	(0)	(0–19)	*22*
Netherlands	8 – 8	99–01	?	–	*3,4,8,9,11,12*
Norway	**50,000 – 70,000**	93	(0)	(0–19)	*68*
Serbia & MN	(2 – 2)	90–02	(F)	(50–100)	
Slovakia	0 – 1	90–99	?	–	*4*
Slovenia	5 – 20	95–00	F	0–100	
Turkey	3,000 – 6,000	91–01	0	0–19	
Ukraine	250 – 1,000	90–00	(–)	(0–19)	

Total (approx.) >92,000 **Overall trend** Stable
% in European IBAs 6–8 **Gen. length** 10 **% Global pop.** 75–94

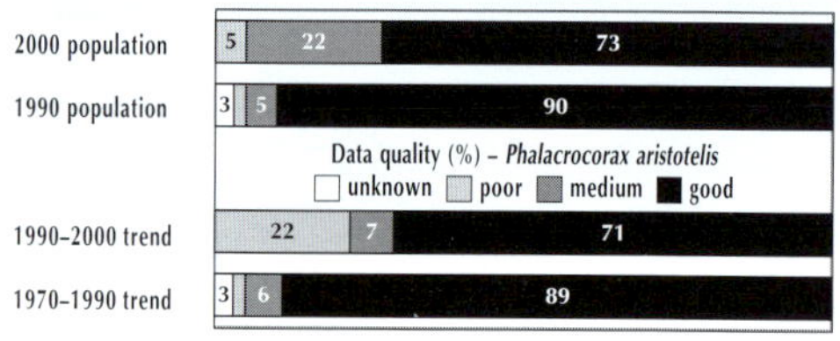

Phalacrocorax pygmeus
PYGMY CORMORANT

SPEC 1 (1994: 2) **Status** Secure
Criteria —
European IUCN Red List Category —
Criteria —
Global IUCN Red List Category NT
Criteria A2c; A3c

Phalacrocorax pygmeus breeds patchily in southern and south-eastern Europe, which constitutes >75% of its global breeding range. Its European breeding population is relatively small (<39,000 pairs), and underwent a moderate decline between 1970–1990. Although declines continued in a few countries during 1990–2000, key populations in Azerbaijan and Romania were stable or increased, and the species underwent a moderate increase overall. Consequently, this globally Near Threatened species—previously assessed as Vulnerable in Europe—is now evaluated as Secure.

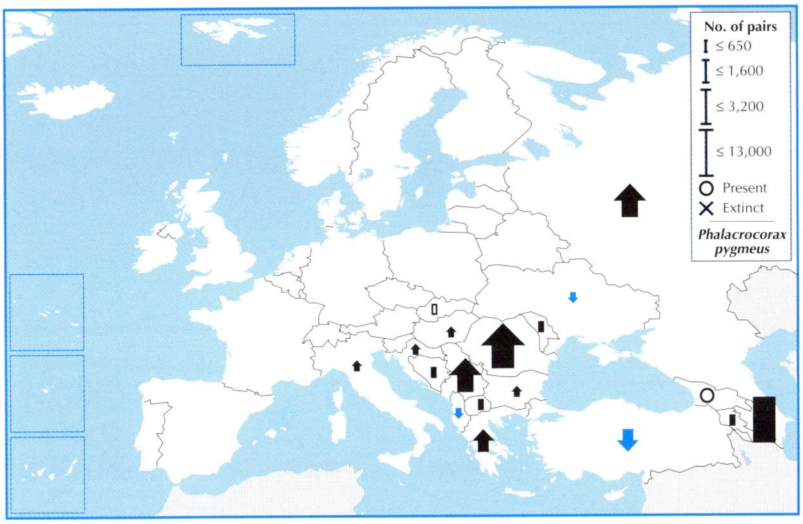

Country	Breeding pop. size (pairs)	Year(s)	Trend	Mag.%	References
Albania	0 – 25	96–02	–	>80	
Armenia	200 – 400	98–02	0	0–19	
Azerbaijan	8,000 – 12,000	96–00	0	0–19	
Bosnia & HG	(50 – 60)	90–00	(0)	(0–19)	
Bulgaria	350 – 400	97–01	+	>80	
Croatia	10 – 25	02	(+)	(N)	70
Georgia	Present	03	?	–	
Greece	1,250 – 1,310	97	+	0–19	
Hungary	80 – 190	98–01	+	>80	
Italy	120 – 130	01	+	30–49	6
Macedonia	(100 – 150)	99–00	(0)	(0–19)	
Moldova	8 – 12	90–00	F	20–29	
Romania	11,500 – 14,000	99–02	+	0–19	47
Russia	2,000 – 5,000	90–00	+	50–79	8,33,120,172
Serbia & MN	2,400 – 2,800	00–02	+	10–29	163,65,145,230, 138,216,3
Slovakia	0 – 1	80–99	?	–	
Turkey	1,300 – 1,800	01	–	30–49	
Ukraine	550 – 750	90–98	–	0–19	
Total (approx.)	28,000 – 39,000	Overall trend Moderate increase			
Breeding range	>100,000 km²	Gen. length. 5		% Global pop. 75–94	

(See p. 41, bottom, for data quality graph)

Country	Winter pop. size (individuals)	Year(s)	Trend	Mag.%	References
Albania	974 – 1,914	95–02	(F)	(–)	
Armenia	(1,500 – 3,000)	98–02	–	20–29	
Azerbaijan	10,000 – 15,000	96–02	0	0–19	
Bosnia & HG	(50 – 60)	90–00	(0)	(0–19)	
Bulgaria	3,500 – 12,000	97–01	+	>80	
Croatia	500 – 1,000	02	(F)	(–)	16,26
Georgia	360 – 360	02	?	–	1
Greece	34,500 – 34,500	99	F	30–49	
Hungary	1 – 20	97–01	F	>80	24
Italy	20 – 200	02	+	30–49	22
Macedonia	(3,520 – 4,000)	97–00	(0)	(0–19)	
Moldova	0 – 3	90–00	+	>80	
Romania	1,500 – 4,000	90–00	0	0–19	60
Serbia & MN	5,000 – 15,000	90–02	+	10–29	
Slovakia	0 – 60	90–99	?	–	4
Slovenia	0 – 23	90–00	F	0–100	
Turkey	1,500 – 22,000	91–01	F	>80	
Ukraine	100 – 1,800	90–00	+	>80	
Total (approx.)	>63,000	Overall trend Fluctuating			
% in European IBAs	>90	Gen. length 5		% Global pop. 50–74	

Pelecanus onocrotalus
GREAT WHITE PELICAN

SPEC 3 (1994: 3) **Status** Rare
Criteria <10,000 pairs
European IUCN Red List Category —
Criteria —
Global IUCN Red List Category —
Criteria —

Pelecanus onocrotalus is a patchily distributed summer visitor to south-east Europe, which accounts for less than a quarter of its global breeding range. Its European breeding population is small (as few as 4,100 pairs), but underwent a large increase between 1970–1990. The stronghold population in Romania was stable during 1990–2000, and although there were slight declines in Turkey, the species was stable overall. Nevertheless, its population size renders it susceptible to the risks affecting small populations, and consequently it is evaluated as Rare.

Country	Breeding pop. size (pairs)	Year(s)	Trend	Mag.%	References
Bulgaria	2 – 2	00	?	N	
Georgia	Present	03	?	–	
Greece	50 – 100	90–98	F	30–49	
Romania	3,500 – 4,000	90–02	0	0–19	47
Russia	400 – 450	90–00	+	20–29	6,8,33,60,74
Turkey	180 – 420	01	–	0–19	
Ukraine	14 – 150	96–99	+	N	
Total (approx.)	4,100 – 5,100	Overall trend Stable			
Breeding range	>50,000 km²	Gen. length. 11		% Global pop. 5–24	

Country	Winter pop. size (individuals)	Year(s)	Trend	Mag.%	References
Azerbaijan	150 – 850	96–02	(F)	(–)	
Bulgaria	1 – 22	97–01	F	50–100	
Cyprus	0 – 25	94–02	?	–	
Greece	1 – 20	95–99	0	0–19	
Macedonia	0 – 20	99–00	(F)	(50–79)	
Romania	12 – 40	90–00	F	50–79	60
Serbia & MN	(0 – 2)	90–02	F	50–100	
Turkey	60 – 80	91–01	+	30–49	
Total (approx.)	>200	Overall trend Small increase			
% in European IBAs	4–8	Gen. length 11		% Global pop. <5	

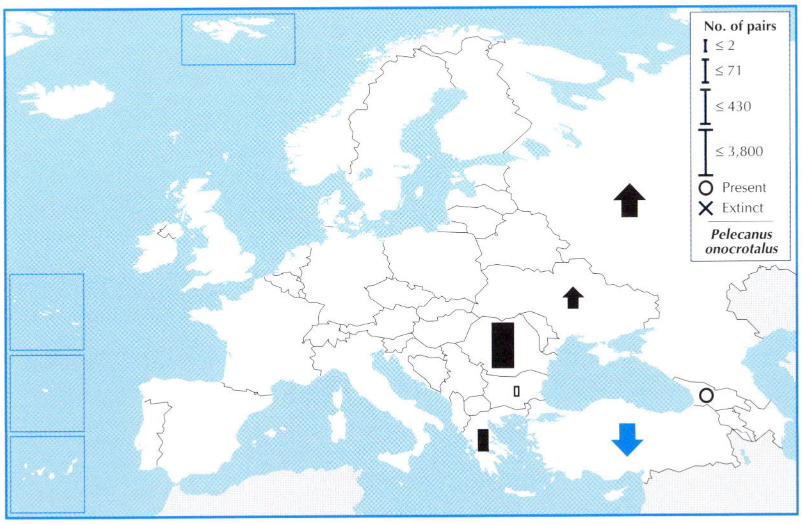

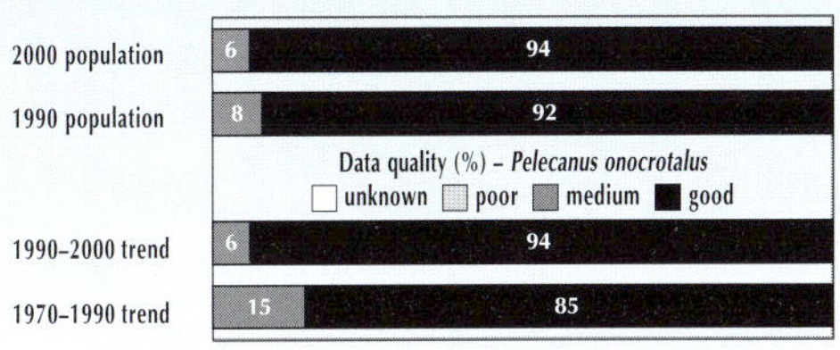

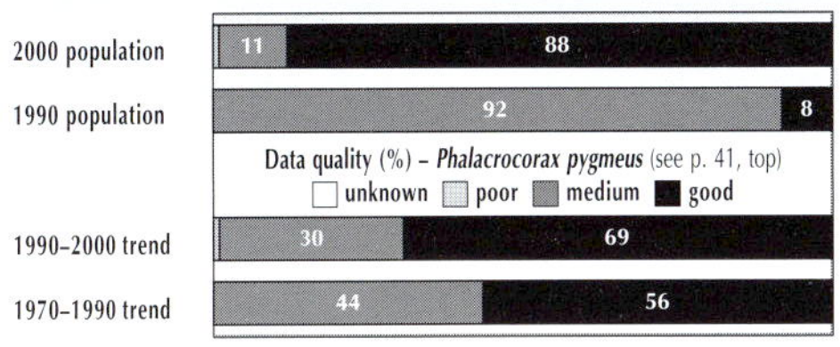

Pelecanus crispus
DALMATIAN PELICAN

SPEC 1 (1994: 1) **Status** Rare
Criteria <10,000 pairs
European IUCN Red List Category —
Criteria —
Global IUCN Red List Category VU
Criteria A2c; A3c

Pelecanus crispus breeds locally in south-eastern Europe, which accounts for less than half of its global breeding range. Its European breeding population is small (as few as 1,600 pairs), but increased substantially between 1970–1990. Despite marked declines elsewhere within its global range during 1990–2000, the species underwent a moderate increase overall in Europe. Nevertheless, its population size still renders it susceptible to the risks affecting small populations, and consequently this globally threatened species is evaluated as Rare in Europe.

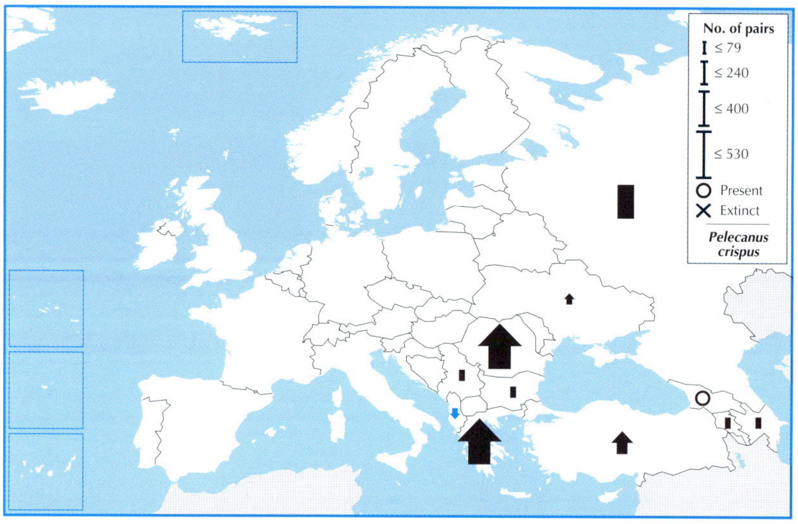

Country	Breeding pop. size (pairs)	Year(s)	Trend	Mag.%	References
Albania	**19 – 19**	92–02	–	**50–79**	*4*
Armenia	5 – 8	97–02	0	0–19	
Azerbaijan	3 – 10	96–00	(0)	(0–19)	
Bulgaria	**49 – 128**	96–01	**F**	**>80**	
Georgia	Present	03	?	–	
Greece	**500 – 550**	95–97	**+**	**20–29**	
Romania	**400 – 550**	90–02	**+**	**0–19**	*47*
Russia	**350 – 450**	90–00	0	0–19	*6,8,74,120,33,60*
Serbia & MN	**4 – 7**	00–02	F	50–79	*1,230,231,216, 156,220*
Turkey	**220 – 250**	01	+	20–29	
Ukraine	3 – 14	90–00	+	>80	
Total (approx.)	1,600 – 2,000	**Overall trend** Moderate increase			
Breeding range	>50,000 km²	**Gen. length.** 11		**% Global pop.** 25–49	

Country	Winter pop. size (individuals)	Year(s)	Trend	Mag.%	References
Albania	**91 – 186**	95–02	(F)	(–)	
Azerbaijan	**200 – 400**	96–02	–	50–79	
Bulgaria	**423 – 607**	97–01	**+**	**>80**	
Greece	**930 – 1,700**	95–99	**+**	**0–19**	
Macedonia	**10 – 36**	97–99	(F)	(50–79)	*8*
Romania	**20 – 50**	90–00	**F**	**50–79**	*60*
Serbia & MN	**5 – 50**	90–02	F	50–100	
Turkey	**1,300 – 1,600**	91–01	**+**	**50–79**	
Total (approx.)	>3,000	**Overall trend** Fluctuating			
% in European IBAs	>90	**Gen. length** 11		**% Global pop.** 25–49	

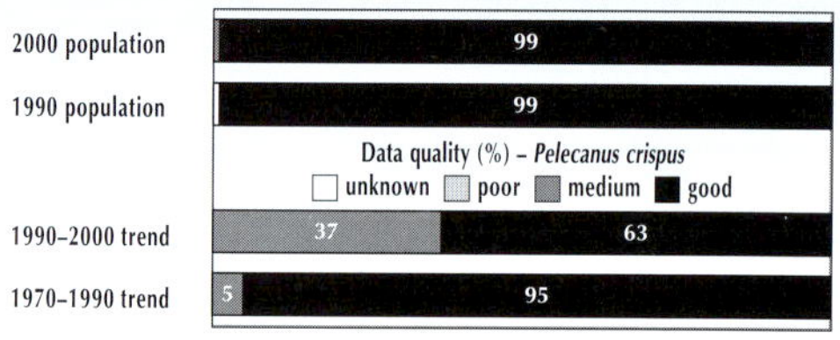

Botaurus stellaris
GREAT BITTERN

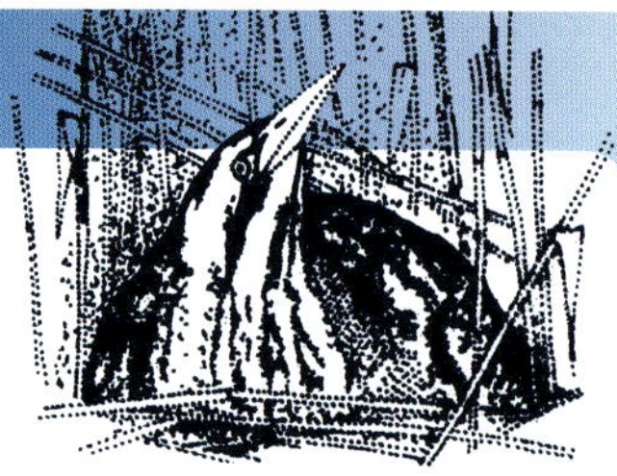

SPEC 3 (1994: 3) **Status** Depleted
Criteria Large historical decline
European IUCN Red List Category —
Criteria —
Global IUCN Red List Category —
Criteria —

Botaurus stellaris is a widespread but patchily distributed breeder across much of Europe, which accounts for less than half its global breeding range. Its European breeding population is relatively small (<54,000 pairs), and underwent a large decline between 1970–1990. Although the species was stable overall during 1990–2000, with stable or increasing trends across much of Europe (Ukraine being a notable exception), its population has clearly not yet recovered to the level that preceded its decline. Consequently, it is evaluated as Depleted.

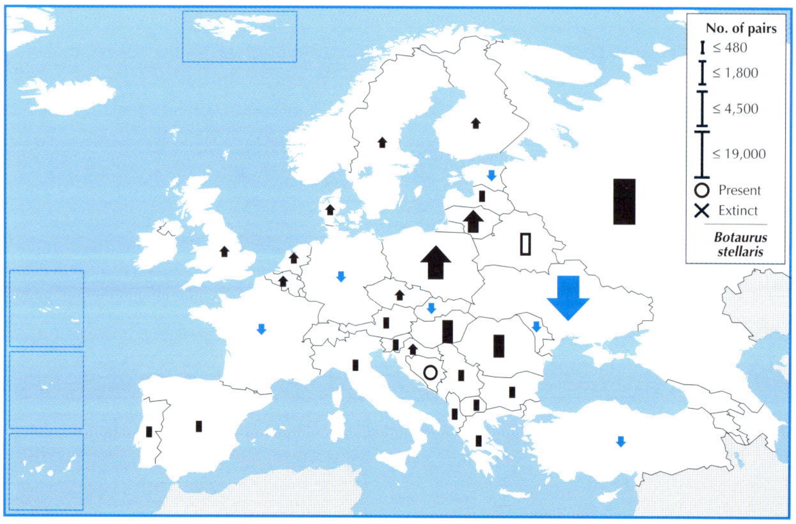

Country	Breeding pop. size (pairs)	Year(s)	Trend	Mag.%	References
Albania	5 – 15	96–02	0	0–19	
Austria	100 – 130	98–02	0	0–19	
Belarus	**950 – 1,200**	97–02	?	–	
Belgium	**12 – 20**	95–02	**+**	**0–19**	*1*
Bosnia & HG	Present	85–89	?	–	*1*
Bulgaria	20 – 70	95–02	(0)	(0–19)	
Croatia	60 – 100	02	(+)	(>80)	*54,16*
Czech Rep.	**30 – 40**	00	**+**	**30–49**	
Denmark	200 – 300	99–00	+	10–19	*16*
Estonia	100 – 150	98	–	20–29	*1*
Finland	200 – 300	99–01	**+**	**100**	
France	**270 – 317**	00	–	0–9	*6*
Germany	360 – 620	95–99	–	20–29	
Greece	**10 – 15**	02	0	0–19	
Hungary	**700 – 1,000**	98–01	0	0–19	
Italy	50 – 70	03	F	30–49	*6*
Latvia	300 – 500	00–03	(0)	(0–19)	*23*
Lithuania	**800 – 1,000**	99–01	+	0–19	*20*
Macedonia	(1 – 10)	99–00	(0)	(0–19)	
Moldova	30 – 40	90–00	–	20–29	
Netherlands	**200 – 250**	98–00	**+**	**34**	*1*
Poland	**4,100 – 4,800**	02–03	(+)	(0–19)	*10,11*
Portugal	0 – 3	02	(F)	(–)	
Romania	(1,500 – 2,000)	99–02	(0)	(0–19)	
Russia	(13,000 – 25,000)	90–00	0	0–19	*8,31,33,44,70, 73,104*
Serbia & MN	200 – 300	95–02	0	0–19	*1,29,78,101,100, 203,155*
Slovakia	50 – 100	80–99	–	30–49	
Slovenia	**0 – 2**	90–00	**F**	**>80**	
Spain	0 – 50	98–02	(F)	(–)	*10,16*
Sweden	**360 – 400**	99–00	**+**	**10–19**	
Turkey	(300 – 600)	01	(–)	(0–19)	
Ukraine	10,000 – 15,000	90–00	–	5–9	
UK	**21 – 21**	97–01	**+**	**16**	
Total (approx.)	34,000 – 54,000	**Overall trend** Stable			
Breeding range	>3,000,000 km²	**Gen. length.** 4		**% Global pop.** 25–49	

	Winter pop. size (individuals)				
Total (approx.)	>2,400	**Overall trend** Unknown			
% in European IBAs	4–7	**Gen. length** 4		**% Global pop.** 25–49	

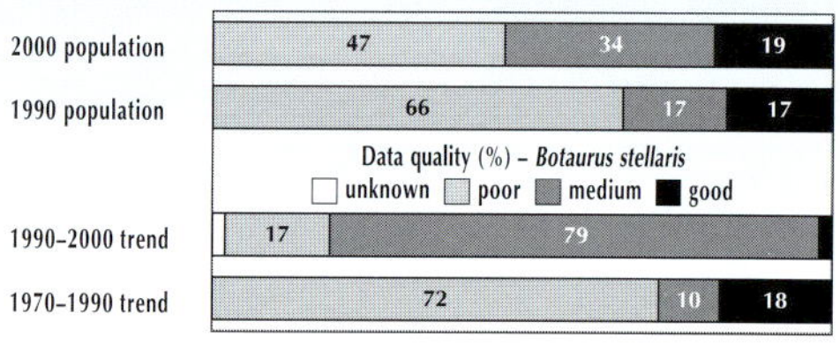

Ixobrychus minutus
LITTLE BITTERN

SPEC 3 (1994: 3) **Status** (Depleted)
Criteria Large historical decline
European IUCN Red List Category —
Criteria —
Global IUCN Red List Category —
Criteria —

Ixobrychus minutus is a widespread summer visitor to much of central and southern Europe, which accounts for less than a quarter of its global breeding range. Its European breeding population is relatively small (<120,000 pairs), and underwent a large decline between 1970–1990. Although the species was stable overall during 1990–2000, with stable or increasing trends across much of its European range (Turkey being a notable exception), its population has clearly not yet recovered to the level that preceded its decline. Consequently, it is provisionally evaluated as Depleted.

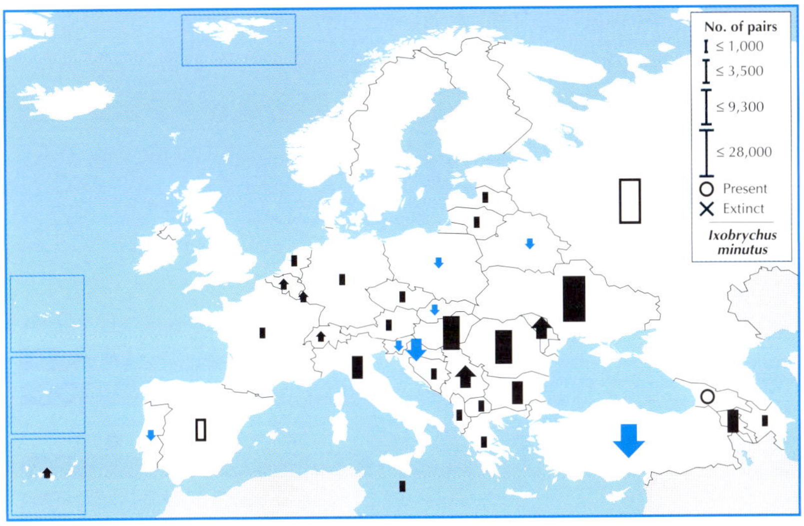

Country	Breeding pop. size (pairs)	Year(s)	Trend	Mag.%	References
Albania	40 – 370	96–02	0	0–19	*6*
Armenia	1,000 – 2,500	98–02	(0)	(0–19)	
Austria	150 – 300	98–02	(0)	(0–19)	
Azerbaijan	(250 – 2,500)	96–00	(0)	(0–19)	
Belarus	300 – 600	97–00	–	10–19	
Belgium	**22 – 31**	95–02	+	**0–19**	*1*
Bosnia & HG	(8 – 10)	90–00	(0)	(0–19)	
Bulgaria	**900 – 1,500**	95–02	0	0–19	
Croatia	(800 – 2,000)	02	(–)	(0–19)	*54,26*
Czech Rep.	**60 – 80**	00	**0**	**0–19**	
France	250 – 300	98–01	(0)	(0–19)	*6*
Georgia	Present	03	?	–	
Germany	**90 – 120**	95–99	**0**	**0–19**	
Greece	500 – 2,000	01	0	0–19	
Hungary	**4,000 – 6,000**	95–02	0	0–19	
Italy	1,300 – 2,300	03	0	0–19	*6*
Latvia	(5 – 20)	90–00	(0)	(0–19)	*22,17*
Lithuania	(1 – 10)	99–01	(F)	(>80)	*20*
Luxembourg	**2 – 4**	02	+	**100**	
Macedonia	200 – 300	99–00	0	0–19	
Malta	**1 – 2**	97–00	**0**	**0–19**	*3*
Moldova	1,800 – 2,300	90–00	+	20–29	
Netherlands	**10 – 30**	98–00	**0**	**0–19**	*1*
Poland	700 – 700	95–00	(–)	(20–29)	*9*
Portugal	(150 – 500)	02	(–)	(–)	
Romania	(8,500 – 10,000)	98–02	(0)	(0–19)	
Russia	(15,000 – 50,000)	90–00	?	–	*8,51,52,73,75*
Serbia & MN	3,000 – 4,000	00–02	+	10–19	*1,29,78,100,101, 155,144,227*
Slovakia	200 – 400	80–99	–	30–49	
Slovenia	30 – 60	96–00	–	20–29	
Spain	(1,900 – 2,300)	92	?	–	*13,12,10*
Canary Is.	1 – 2	97–03	+	N	*28,25*
Switzerland	**80 – 120**	93–96	+	10–29	
Turkey	6,000 – 9,000	01	(–)	(0–19)	
Ukraine	13,200 – 22,300	90–00	0	0–19	
Total (approx.)	60,000 – 120,000	**Overall trend** Stable			
Breeding range	$>3,000,000$ km^2	**Gen. length.** <3.3		**% Global pop.** 5–24	

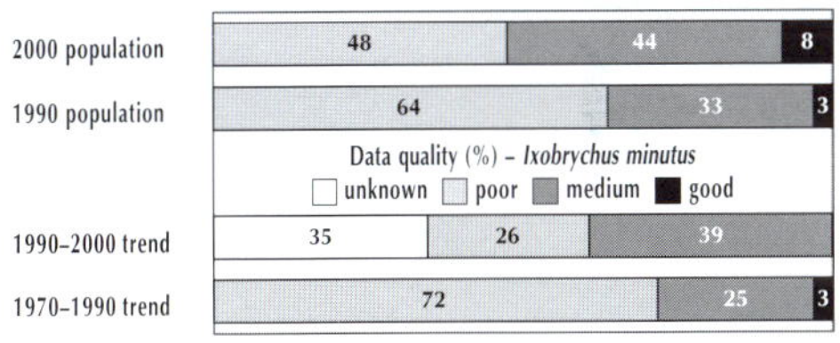

Nycticorax nycticorax
BLACK-CROWNED NIGHT-HERON

SPEC 3 (1994: 3) **Status** Depleted
Criteria Moderate historical decline
European IUCN Red List Category —
Criteria —
Global IUCN Red List Category —
Criteria —

Nycticorax nycticorax is a widespread summer visitor to much of the southern half of Europe, which accounts for less than a quarter of its global breeding range. Its European breeding population is relatively small (<87,000 pairs), and underwent a moderate decline between 1970–1990. Although the species was stable overall during 1990–2000—with stable, fluctuating or increasing trends across the vast majority of its European range—its population has not yet recovered to the level that preceded its decline. Consequently, it is evaluated as Depleted.

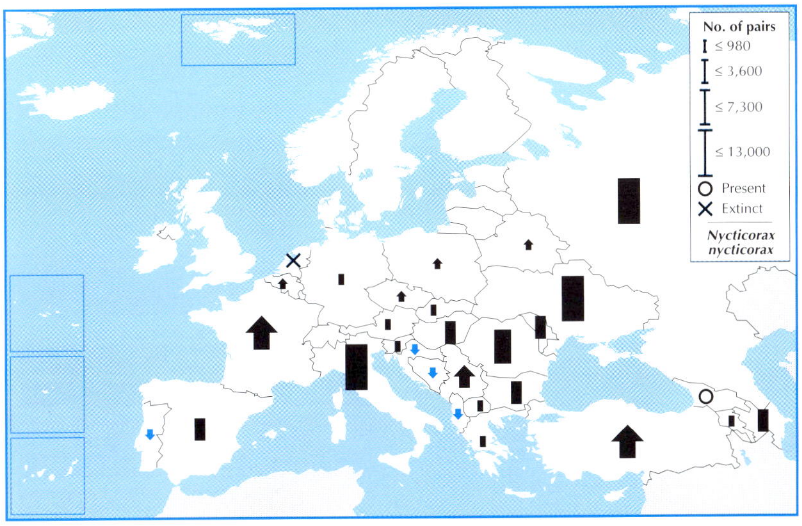

Country	Breeding pop. size (pairs)	Year(s)	Trend	Mag.%	References
Albania	0 – 10	92–02	–	**>80**	*4*
Armenia	50 – 250	98–02	(0)	(0–19)	
Austria	40 – 70	98–02	0	0–19	
Azerbaijan	2,500 – 5,000	96–00	0	0–19	
Belarus	5 – 27	99–00	+	N	
Belgium	**13 – 14**	00–02	+	**0–19**	*1*
Bosnia & HG	(4 – 5)	90–00	(–)	(50–79)	
Bulgaria	**1,800 – 2,500**	95–02	0	0–19	
Croatia	500 – 600	02	–	20–29	*26*
Czech Rep.	**480 – 500**	00	+	**30–49**	
France	4,500 – 5,500	98–02	+	0–9	*4*
Georgia	Present	03	?	–	
Germany	**35 – 35**	95–99	**0**	**0–19**	
Greece	800 – 1,200	95–00	0	0–19	
Hungary	**2,400 – 3,600**	95–02	0	0–19	
Italy	**12,000 – 14,000**	03	**F**	**20–29**	*6*
Macedonia	**180 – 200**	99	0	0–19	
Moldova	1,500 – 1,800	90–00	0	0–19	
Netherlands	**0 – 0**	98–00	–	**X**	*1*
Poland	**513 – 513**	02	+	**480–490**	*2,7,8*
Portugal	**30 – 120**	02	(–)	(–)	
Romania	6,500 – 8,000	96–02	0	0–19	*47,51*
Russia	10,000 – 15,000	90–00	0	0–19	8,60,72
Serbia & MN	**2,200 – 2,500**	00–02	+	10–19	*163,65,67,115, 145,230,138*
Slovakia	**100 – 400**	90–99	0	0–19	
Slovenia	(3 – 5)	00	(F)	(–)	
Spain	(2,170 – 4,430)	98–02	(F)	(–)	*10*
Turkey	4,000 – 8,000	01	+	30–49	
Ukraine	10,400 – 12,900	90–00	F	30–49	
Total (approx.)	63,000 – 87,000	**Overall trend** Stable			
Breeding range	$>1,000,000$ km^2	**Gen. length.** 5		**% Global pop.** 5–24	

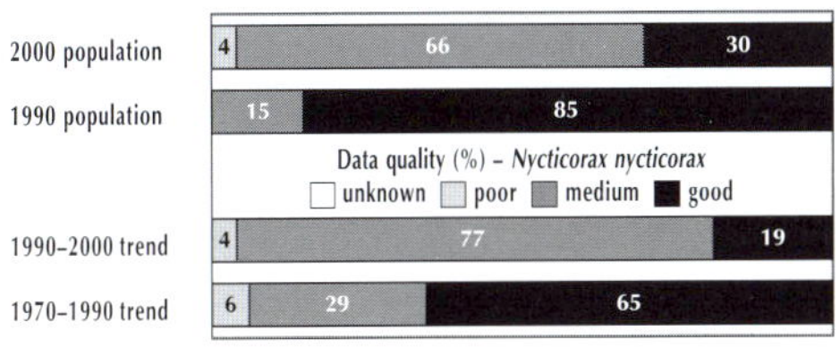

Ardeola ralloides
SQUACCO HERON

SPEC 3 (1994: 3) **Status** (Declining)
Criteria Moderate continuing decline
European IUCN Red List Category —
Criteria —
Global IUCN Red List Category —
Criteria —

Ardeola ralloides is a widespread but patchily distributed summer visitor to much of southern Europe, which accounts for less than a quarter of its global breeding range. Its European breeding population is relatively small (<27,000 pairs), and underwent a large decline between 1970–1990. Although most populations—notably sizeable ones in Romania and Azerbaijan—were stable or increased during 1990–2000, other significant populations in Turkey and Russia declined, and the species underwent a moderate decline (>10%) overall. Consequently, it is provisionally evaluated as Declining.

Country	Breeding pop. size (pairs)	Year(s)	Trend	Mag.%	References
Albania	0 – 48	92–02	–	**>80**	*4*
Armenia	330 – 830	97–02	0	0–19	
Azerbaijan	2,500 – 5,000	96–00	0	0–19	
Bosnia & HG	(10 – 12)	90–00	(0)	(0–19)	
Bulgaria	**300 – 650**	95–02	**0**	**0–19**	
Croatia	100 – 500	02	+	0–19	*16,26*
France	**120 – 130**	97–00	0	0–19	*5*
Georgia	Present	03	?	–	
Greece	400 – 700	95–00	0	0–19	
Hungary	300 – 410	95–02	0	0–19	
Italy	550 – 650	03	**0**	**0–19**	*6*
Macedonia	100 – 150	99–00	(F)	(30–49)	
Moldova	**15 – 20**	90–00	–	0–19	
Portugal	**0 – 10**	02	(F)	(–)	
Romania	5,500 – 6,500	00–02	0	0–19	*47*
Russia	1,500 – 2,000	90–00	–	**20–29**	*8,60,120,178*
Serbia & MN	**750 – 900**	00–02	0	0–19	*163,65,67,115, 145,230,138*
Slovakia	**0 – 5**	80–99	?	–	
Spain	850 – 1,100	98–02	+	0–19	*10,16*
Turkey	(4,000 – 6,000)	01	(–)	(0–19)	
Ukraine	**1,100 – 1,300**	90–00	+	**5–9**	
Total (approx.)	18,000 – 27,000	**Overall trend** Moderate decline			
Breeding range	>500,000 km²	**Gen. length.** 5		**% Global pop.** 5–24	

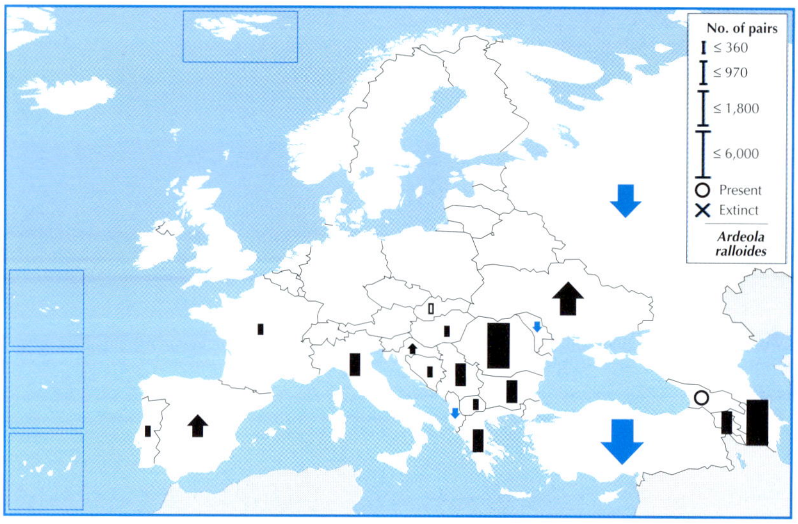

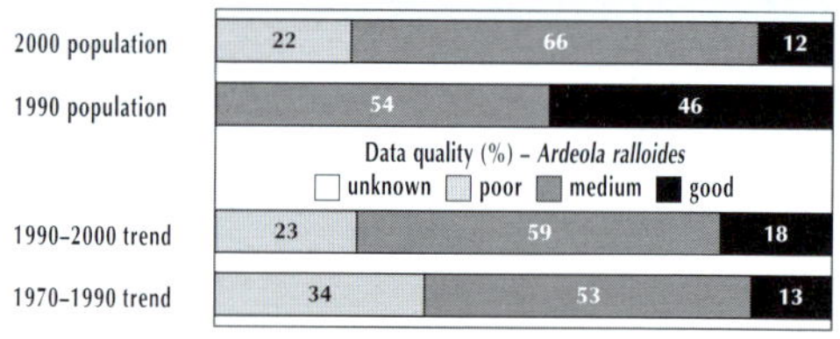

Bubulcus ibis
CATTLE EGRET

Non-SPEC (1994: —) **Status** Secure
Criteria —
European IUCN Red List Category —
Criteria —
Global IUCN Red List Category —
Criteria —

Bubulcus ibis breeds mainly in Iberia but also patchily elsewhere in southern Europe, which accounts for less than a quarter of its global breeding range. Its European breeding population is relatively small (<150,000 pairs), but underwent a large increase between 1970–1990. The stronghold populations in Spain and Portugal continued to increase during 1990–2000, and populations were stable or increased elsewhere in its European range. Consequently, it is evaluated as Secure.

Country	Breeding pop. size (pairs)	Year(s)	Trend	Mag.%	References
Armenia	25 – 30	98–02	?	–	
Azerbaijan	3,500 – 7,000	96–00	0	0–19	
France	4,500 – 5,500	98–02	0	0–19	*4*
Georgia	Present	03	?	–	
Greece	1 – 1	91–92	?	–	
Italy	**720 – 760**	00	+	**30–49**	*18*
Netherlands	**0 – 3**	98–00	?	–	1
Portugal	25,000 – 35,000	02	(+)	(–)	
Romania	8 – 20	93–02	+	0–19	*47*
Russia	30 – 50	90–00	?	–	*33,12*
Spain	20,000 – 100,000	98–02	+	0–19	*10*
Canary Is.	**58 – 135**	97–03	+	N	*28,25*
Turkey	150 – 200	01	+	0–19	
Total (approx.)	54,000 – 150,000	**Overall trend** Small increase			
Breeding range	>100,000 km²	**Gen. length.** 5		**% Global pop.** 5–24	

Country	Winter pop. size (individuals)	Year(s)	Trend	Mag.%	References
France	**5,000 – 10,000**	98–02	+	**30–49**	*9*
Italy	1,200 – 1,200	98	+	**>80**	*22*
Portugal	46,400 – 46,400	92–93	(+)	(–)	*28*
Spain	(7,100 – 43,000)	90–01	(+)	(>80)	
Turkey	**20 – 40**	91–01	F	**>80**	
Total (approx.)	>60,000	**Overall trend** Large increase			
% in European IBAs	23–27	**Gen. length** 5		**% Global pop.** <5	

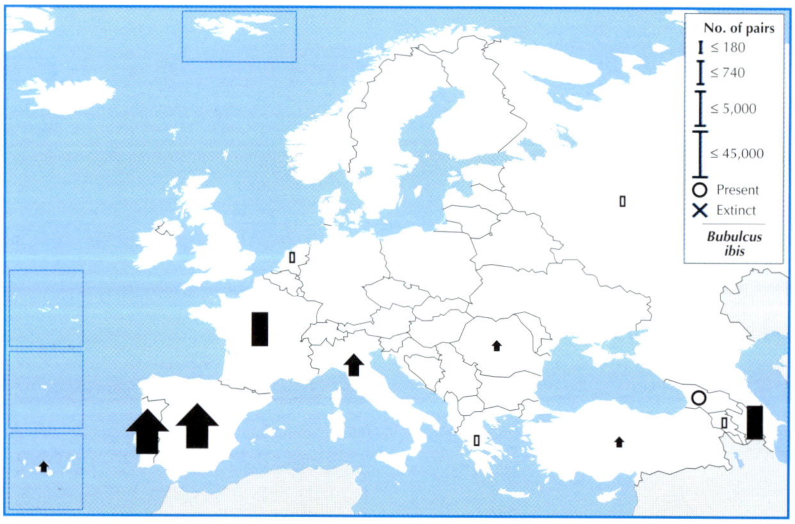

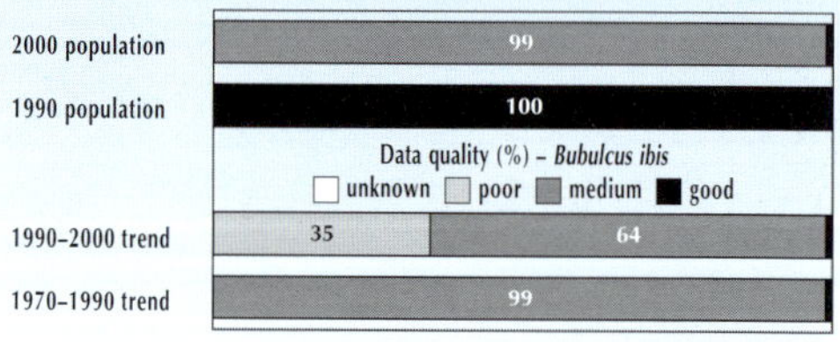

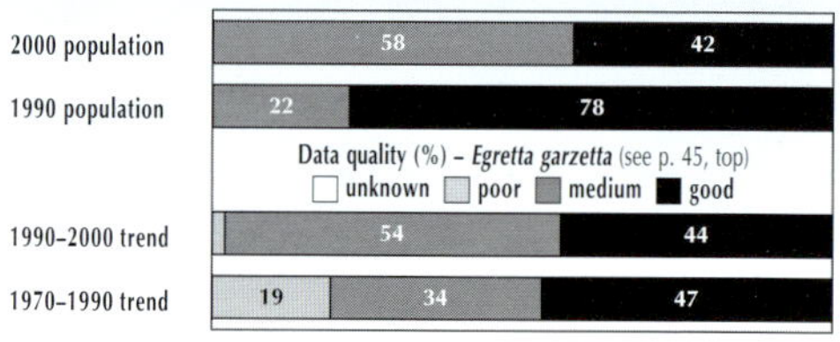

Egretta garzetta
LITTLE EGRET

Non-SPEC (1994: —) **Status** Secure
Criteria —
European IUCN Red List Category —
Criteria —
Global IUCN Red List Category —
Criteria —

Egretta garzetta is a widespread but patchily distributed breeder in southern Europe, which accounts for less than a quarter of its global breeding range. Its European breeding population is relatively small (<94,000 pairs), but increased between 1970–1990. Although there were declines in a few countries during 1990–2000, populations across most of its European range—including sizeable ones in Spain, France, Italy and Azerbaijan—increased or were stable, and consequently the species is evaluated as Secure.

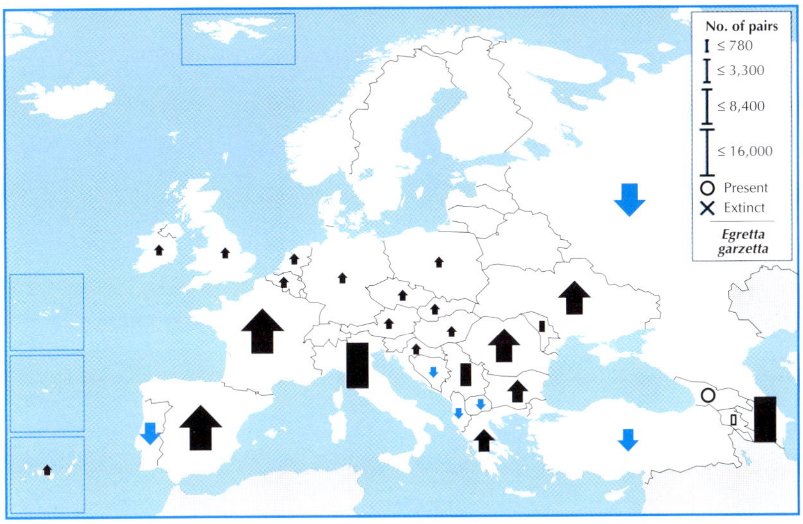

Country	Breeding pop. size (pairs)	Year(s)	Trend	Mag.%	References
Albania	2 – 5	92–02	–	**>80**	*4*
Armenia	50 – 250	97–02	?	–	
Austria	**3 – 5**	98–02	+	**N**	
Azerbaijan	8,000 – 12,000	96–00	0	0–19	
Belgium	**17 – 17**	01–03	+	**>80**	*1*
Bosnia & HG	(15 – 20)	90–00	(–)	(50–79)	
Bulgaria	**1,400 – 2,000**	95–02	+	**50–79**	
Croatia	**500 – 600**	02	+	>80	*70,26*
Czech Rep.	**1 – 4**	00	+	**50–79**	
France	**11,000 – 13,000**	98–02	+	**10–19**	*4*
Georgia	Present	03	?	–	
Germany	**0 – 2**	95–99	+	**N**	
Greece	**1,500 – 1,800**	96–98	+	**0–19**	
Hungary	**600 – 1,000**	98–01	+	20–29	
Rep. Ireland	**12 – 55**	97–01	+	**N**	
Italy	**15,000 – 16,000**	03	F	**20–29**	*6*
Macedonia	20 – 100	96–00	–	0–9	
Moldova	400 – 500	90–00	0	0–19	
Netherlands	**5 – 20**	98–00	+	N	*1*
Poland	**1 – 1**	03	+	**N**	*2,108*
Portugal	1,000 – 1,500	02	(–)	(–)	
Romania	4,000 – 5,000	98–02	+	0–19	47,26
Russia	7,000 – 10,000	90–00	–	0–19	*8,60,120*
Serbia & MN	**1,000 – 1,200**	00–02	F	20–29	*163,65,230,138, 216,231*
Slovakia	**0 – 30**	80–99	+	20–29	
Spain	10,000 – 20,000	98–02	+	20–29	*10*
Canary Is.	**1 – 2**	97–03	+	**N**	28,25
Turkey	2,800 – 3,800	01	–	20–29	
Ukraine	4,100 – 4,600	90–00	+	**0–19**	
UK	**68 – 77**	00	+	**N**	
Total (approx.)	68,000 – 94,000	**Overall trend** Small increase			
Breeding range	>1,000,000 km²	**Gen. length.** 5		**% Global pop.** 5–24	

(See p. 44, bottom, for data quality graph)

Country	Winter pop. size (individuals)	Year(s)	Trend	Mag.%	References
France	**12,000 – 18,000**	98–02	+	**30–49**	*9*
Greece	**1,000 – 2,000**	95–99	+	**0–19**	
Italy	5,000 – 9,000	02	0	0–19	*22*
Portugal	3,500 – 3,500	92–93	0	0–19	*28*
Spain	(3,200 – 8,500)	90–01	(+)	(30–49)	
Total (approx.)	>28,000	**Overall trend** Large increase			
% in European IBAs	33–47	**Gen. length** 5		**% Global pop.** <5	

Casmerodius albus
GREAT EGRET

Non-SPEC (1994: —) **Status** Secure
Criteria —
European IUCN Red List Category —
Criteria —
Global IUCN Red List Category —
Criteria —

Casmerodius albus is a widespread but patchily distributed breeder across much of Europe, which accounts for a tiny proportion of its global breeding range. Its European breeding population is relatively small (<24,000 pairs), but increased between 1970–1990. The species continued to undergo a large increase during 1990–2000, with positive trends across almost all of its European range—including key populations in Ukraine and Hungary—and the sizeable Russian population stable. Consequently, the species is evaluated as Secure.

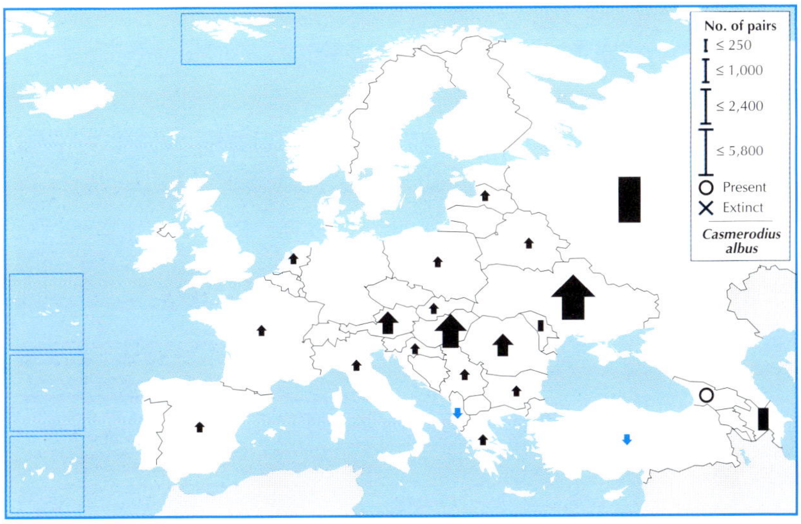

Country	Breeding pop. size (pairs)	Year(s)	Trend	Mag.%	References
Albania	(0 – 2)	96–02	–	>80	*4*
Austria	**580 – 720**	98–02	+	**>80**	
Azerbaijan	250 – 800	96–00	0	0–19	
Belarus	**10 – 30**	97–00	+	N	
Bulgaria	**4 – 20**	99–02	+	**50–79**	
Croatia	**10 – 50**	02	+	**50–79**	*54,26*
France	**15 – 20**	00	+	**>80**	1
Georgia	Present	03	?	–	
Greece	(10 – 30)	98–00	(+)	(0–19)	
Hungary	**1,800 – 3,000**	97–02	+	**50–79**	
Italy	**37 – 45**	00	+	**50–79**	*18*
Latvia	2 – 5	00–03	+	20–29	*22,17,26,13,14*
Moldova	**10 – 15**	90–00	0	0–19	
Netherlands	**1 – 11**	98–00	+	271	*1*
Poland	**20 – 25**	02	+	**N**	*2,4,5,6*
Romania	900 – 1,100	00–02	+	0–19	*47*
Russia	3,000 – 10,000	90–00	0	0–19	*8,24,73,114, 120,60*
Serbia & MN	200 – 300	00–02	+	20–29	*163,138,65,115, 155,156,7,8*
Slovakia	**0 – 80**	80–99	+	20–29	
Spain	**0 – 50**	98–02	+	**>80**	*10*
Turkey	50 – 150	01	–	**50–79**	
Ukraine	4,500 – 7,300	90–00	+	**30–49**	
Total (approx.)	11,000 – 24,000	**Overall trend** Large increase			
Breeding range	>250,000 km²	**Gen. length.** 5		**% Global pop.** <5	

Country	Winter pop. size (individuals)	Year(s)	Trend	Mag.%	References
Azerbaijan	1,000 – 2,000	96–02	–	30–49	
Bulgaria	**596 – 869**	97–01	+	**>80**	
Croatia	1,000 – 2,000	02	(+)	(–)	*16,26*
France	**900 – 2,500**	98–02	+	**30–49**	*9*
Greece	**1,250 – 1,950**	95–99	(0)	(0–19)	
Hungary	250 – 1,000	90–00	F	>80	*24*
Italy	300 – 5,000	02	+	>80	*22*
Romania	250 – 1,000	90–00	0	0–19	60
Serbia & MN	1,000 – 2,000	90–02	+	40–69	
Turkey	**1,500 – 2,500**	91–01	0	0–19	
Total (approx.)	>8,600	**Overall trend** Moderate increase			
% in European IBAs	45–74	**Gen. length** 5		**% Global pop.** <5	

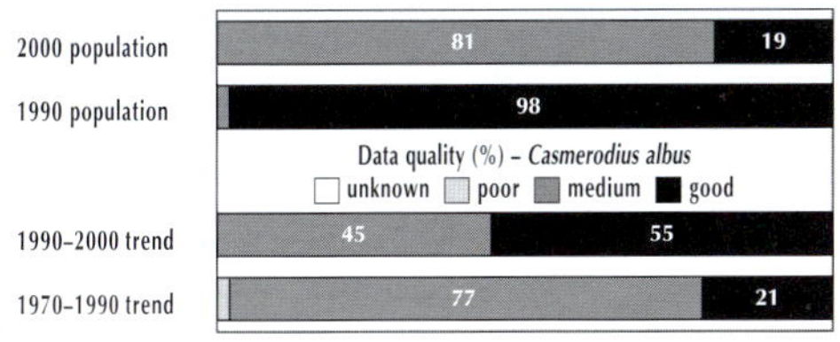

Ardea cinerea
GREY HERON

Non-SPEC (1994: —) **Status** Secure
Criteria —
European IUCN Red List Category —
Criteria —
Global IUCN Red List Category —
Criteria —

Ardea cinerea is a widespread breeder across much of Europe, which accounts for less than a quarter of its global breeding range. Its European breeding population is large (>210,000 pairs), and underwent a large increase between 1970–1990. Although there were declines in the Ukrainian population during 1990–2000, most other European populations—including sizeable ones in France, Germany and Russia—increased or were stable, and the species underwent a moderate increase overall. Consequently, it is evaluated as Secure.

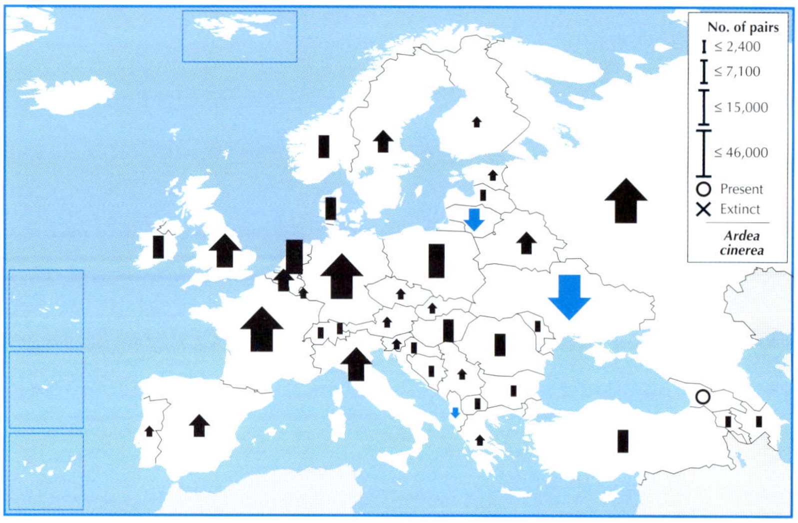

Country	Breeding pop. size (pairs)	Year(s)	Trend	Mag.%	References
Albania	1 – 3	92–02	**–**	**>80**	*4*
Armenia	50 – 250	98–02	0	0–19	
Austria	**1,300 – 1,500**	98–02	**+**	**20–29**	
Azerbaijan	250 – 1,000	96–00	0	0–19	
Belarus	**4,500 – 5,500**	97–00	+	0–9	
Belgium	**3,200 – 3,600**	00–02	**+**	**50–79**	*1*
Bosnia & HG	(7 – 10)	90–00	(0)	(0–19)	
Bulgaria	**1,000 – 1,400**	95–02	**0**	**0–9**	
Croatia	**1,500 – 3,000**	02	**0**	**0–19**	*16,26*
Czech Rep.	**1,900 – 2,300**	00	**+**	**>80**	
Denmark	(6,000 – 7,000)	00	0	0–19	
Estonia	1,200 – 1,500	98	+	0–19	*1*
Finland	200 – 400	99–01	+	500	
France	28,000 – 32,000	98–02	**+**	**10–19**	*4*
Georgia	Present	03	?	–	
Germany	**24,000 – 27,500**	95–99	**+**	**20–29**	
Greece	(600 – 800)	95–00	(+)	(0–19)	
Hungary	**2,500 – 3,500**	97–02	**0**	**0–19**	
Rep. Ireland	2,500 – 10,000	88–91	0	0–19	
Italy	**10,000 – 11,000**	03	+	30–49	*6*
Latvia	1,100 – 1,500	90–00	0	0–19	*23,17*
Liechtenstein	**10 – 20**	98–00	+	0–9	
Lithuania	3,000 – 6,000	99–01	–	0–19	*20*
Luxembourg	**50 – 70**	00–02	**+**	**50–79**	
Macedonia	400 – 500	99–00	(0)	(0–19)	
Moldova	1,200 – 1,400	90–00	0	0–19	
Netherlands	**10,000 – 12,750**	98–00	**0**	**10**	*1*
Norway	(5,000 – 10,000)	90–02	(0)	(0–19)	
Poland	**9,000 – 10,000**	90–00	(0)	(0–19)	*2*
Portugal	600 – 700	02	(+)	(–)	
Romania	(3,500 – 4,500)	98–02	(0)	(0–19)	*47*
Russia	35,000 – 60,000	90–00	+	20–29	*4,8,24,25,30,52,60, 61,90,93,104,112, 113,114,120,126*
Serbia & MN	**2,200 – 2,500**	00–02	+	20–29	*163,138,65,67, 186,230*
Slovakia	300 – 700	80–99	+	20–29	
Slovenia	**500 – 600**	90–00	+	10–19	
Spain	5,000 – 6,000	98–02	+	>80	*10*
Sweden	**5,000 – 6,500**	99–00	**+**	**160–180**	
Switzerland	**1,300 – 1,400**	93–96	0	0–19	
Turkey	2500 – 5,000	01	(0)	(0–19)	
Ukraine	23,800 – 32,900	90–00	–	0–19	
UK	**14,800 – 14,800**	00	+	**14**	*5*

Total (approx.) 210,000 – 290,000 **Overall trend** Moderate increase
Breeding range >5,000,000 km² **Gen. length.** 5 **% Global pop.** 5–24

(See p. 46, bottom, for data quality graph)

Winter pop. size (individuals)

Total (approx.) >86,000 **Overall trend** Moderate increase
% in European IBAs 7–9 **Gen. length** 5 **% Global pop.** 5–24

Ardea purpurea
PURPLE HERON

SPEC 3 (1994: 3) **Status** (Declining)
Criteria Moderate continuing decline
European IUCN Red List Category —
Criteria —
Global IUCN Red List Category —
Criteria —

Ardea purpurea is a widespread summer visitor to much of the southern half of Europe, which accounts for less than a quarter of its global breeding range. Its European breeding population is relatively small (<42,000 pairs), and underwent a large decline between 1970–1990. Although many populations—notably those in western and central Europe—were stable or increased during 1990–2000, the species continued to decline in several eastern European countries, and underwent a moderate decline (>10%) overall. Consequently, it is provisionally evaluated as Declining.

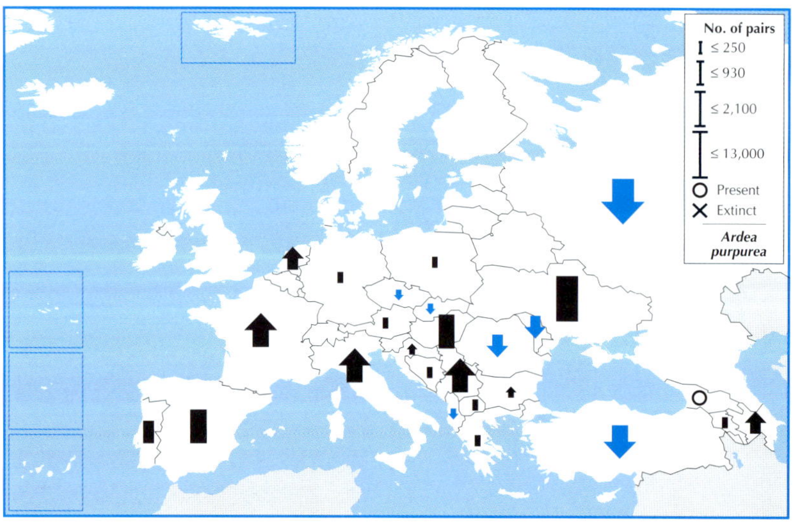

Country	Breeding pop. size (pairs)	Year(s)	Trend	Mag.%	References
Albania	0 – 10	96–02	**–**	**>80**	*4*
Armenia	80 – 330	98–02	0	0–19	
Austria	**200 – 300**	98–02	**0**	**0–19**	
Azerbaijan	250 – 500	96–00	+	0–19	
Bosnia & HG	(4 – 5)	90–00	(0)	(0–19)	
Bulgaria	**100 – 190**	95–02	**+**	**>80**	
Croatia	140 – 150	02	+	30–49	*70,25*
Czech Rep.	**1 – 5**	00	**–**	**50–79**	
France	**2,000 – 2,200**	97–00	**+**	**10–19**	*4*
Georgia	Present	03	?	–	
Germany	**21 – 34**	95–99	**0**	**0–19**	
Greece	(105 – 150)	95–00	(0)	(0–19)	
Hungary	**900 – 1,500**	98–01	0	0–19	
Italy	1,800 – 2,000	03	+	10–19	*6*
Macedonia	5 – 10	90–00	(F)	(30–49)	
Moldova	250 – 350	90–00	–	20–29	
Netherlands	**370 – 445**	98–00	**+**	**55**	*1*
Poland	**0 – 3**	90–00	**0**	**0–19**	*3*
Portugal	**350 – 500**	02	**0**	**0–19**	
Romania	(850 – 1,000)	00–02	(–)	(0–19)	
Russia	(10,000 – 15,000)	90–00	(–)	(20–29)	*8,60,112,114, 120*
Serbia & MN	1,100 – 1,300	00–02	+	10–19	*163,65,67,115, 145,230,138*
Slovakia	**20 – 60**	80–99	–	30–49	
Spain	2,000 – 2,000	98–02	(F)	(–)	*10*
Turkey	1,500 – 2,000	01	–	20–29	
Ukraine	6,700 – 11,900	90–00	0	0–19	

Total (approx.) 29,000 – 42,000 **Overall trend** Moderate decline
Breeding range >1,000,000 km² **Gen. length.** 5 **% Global pop.** 5–24

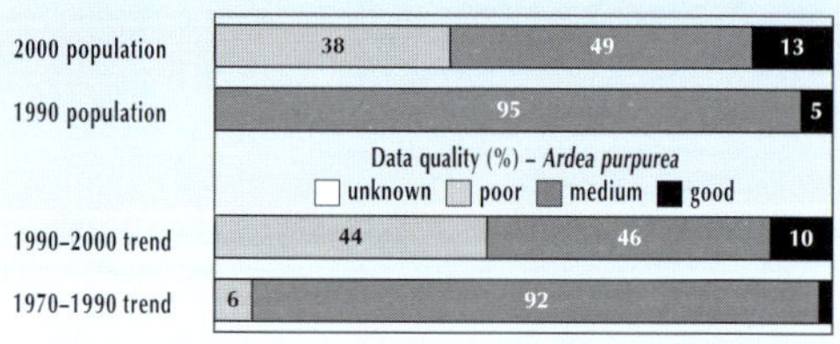

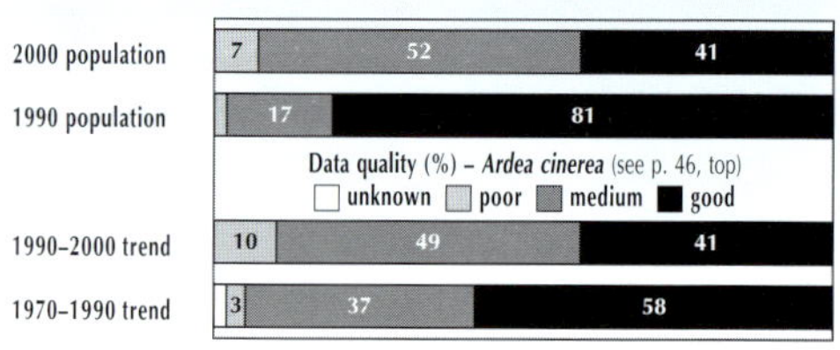

Ciconia nigra
BLACK STORK

SPEC 2 (1994: 3) **Status** Rare
Criteria <10,000 pairs
European IUCN Red List Category —
Criteria —
Global IUCN Red List Category —
Criteria —

Ciconia nigra is a widespread summer visitor to central and eastern Europe, with a patchier distribution in the west and a partly resident population in Iberia. Its European breeding population is small (as few as 7,800 pairs), but constitutes >50% of the global population, and was stable between 1970–1990. Although many populations increased during 1990–2000, declines in the Baltic states meant that the species remained stable overall. Nevertheless, its population size renders it susceptible to the risks affecting small populations. Consequently, it is evaluated as Rare.

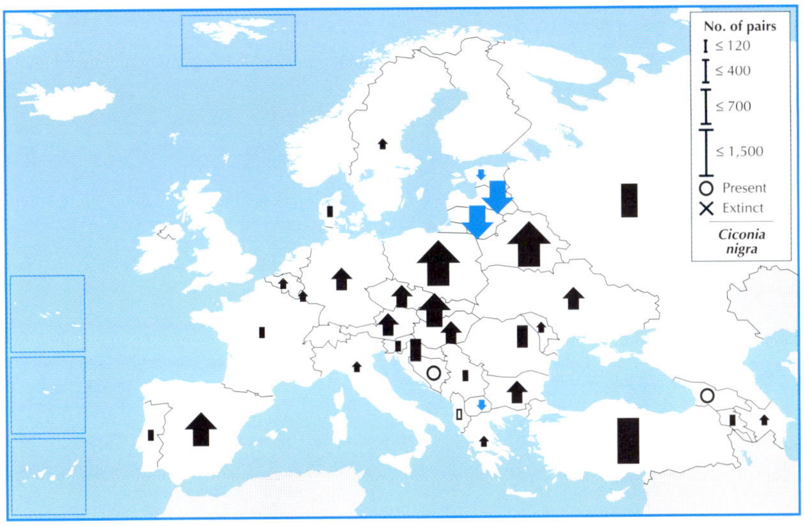

Country	Breeding pop. size (pairs)	Year(s)	Trend	Mag.%	References
Albania	10 – 20	02	?	–	
Armenia	**4 – 6**	97–02	**0**	**0–19**	*1*
Austria	**200 – 300**	98–02	**+**	**20–29**	
Azerbaijan	(50 – 200)	96–00	(+)	(20–29)	
Belarus	**950 – 1,300**	97–02	**+**	**0–19**	
Belgium	**31 – 41**	00–02	**+**	**50–79**	*1*
Bosnia & HG	Present	85–89	?	–	
Bulgaria	**180 – 220**	97–02	**+**	**0–19**	
Croatia	(350 – 400)	02	(0)	(0–19)	*70*
Czech Rep.	**300 – 400**	00	**+**	**30–49**	
Denmark	(0 – 2)	98–01	0	0–19	*4,5,6,7*
Estonia	**80 – 120**	98	–	20–29	*1*
France	**13 – 13**	02	0	0–19	*6*
Georgia	Present	03	?	–	
Germany	**330 – 390**	95–99	**+**	**20–29**	
Greece	30 – 50	95–00	+	0–19	
Hungary	**210 – 280**	95–02	**+**	**20–29**	*9*
Italy	4 – 6	03	+	0–19	*6*
Latvia	**500 – 700**	90–00	–	30–49	*18*
Lithuania	(600 – 800)	99–01	–	20–29	*20*
Luxembourg	**4 – 7**	02	**+**	**N**	
Macedonia	45 – 50	98–00	–	0–9	
Moldova	8 – 12	90–00	**+**	**30–49**	
Poland	**1,100 – 1,200**	00–01	+	5–14	*2*
Portugal	**85 – 96**	97	**0**	**0–19**	*16,11,12*
Romania	(160 – 250)	96–02	(0)	(0–19)	*35*
Russia	400 – 700	90–00	(0)	(0–19)	*4,6,23,31,33,60, 72,103,113,127*
Serbia & MN	**110 – 120**	90–02	0	0–19	*151,196,208,43, 155,16*
Slovakia	400 – 600	80–99	+	20–29	
Slovenia	25 – 35	90–00	0	0–19	
Spain	250 – 1,000	98–02	+	0–19	*10,16*
Sweden	**1 – 2**	99–00	**+**	**>80**	
Turkey	(1,000 – 2,000)	01	(0)	(0–19)	
Ukraine	**330 – 480**	90–00	**+**	**0–19**	

Total (approx.) 7,800 – 12,000 **Overall trend** Stable
Breeding range >3,000,000 km² **Gen. length.** 8 **% Global pop.** 50–74

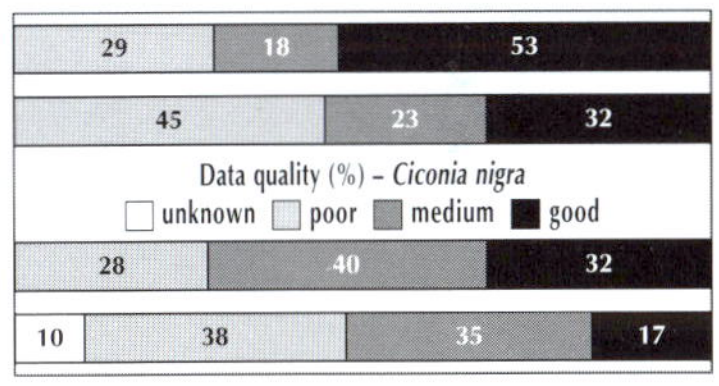

Ciconia ciconia
WHITE STORK

SPEC 2 (1994: 2) **Status** Depleted
Criteria Large historical decline
European IUCN Red List Category —
Criteria —
Global IUCN Red List Category —
Criteria —

Ciconia ciconia is a widespread summer visitor to much of Europe, which constitutes >75% of its global breeding range. Its European breeding population is large (>180,000 pairs), but underwent a large decline between 1970–1990. Although the species experienced a moderate increase overall during 1990–2000—with stable or increasing trends across most of Europe—its population has not yet recovered to the level that preceded its decline. Consequently, it is evaluated as Depleted.

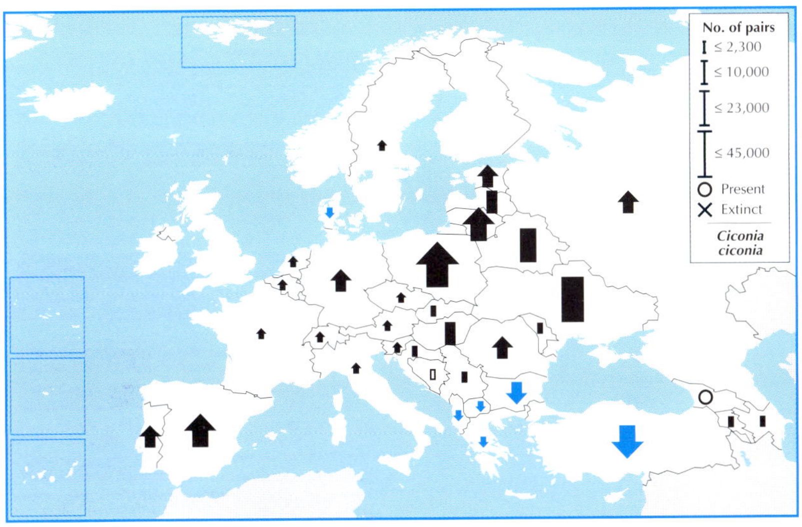

Country	Breeding pop. size (pairs)	Year(s)	Trend	Mag.%	References
Albania	10 – 20	96–02	–	0–19	
Armenia	1,000 – 1,500	97–02	0	0–19	
Austria	**365 – 415**	98–02	**+**	**30–49**	
Azerbaijan	(1,000 – 5,000)	96–00	(0)	(0–19)	
Belarus	**10,300 – 13,300**	97–99	**0**	**0–19**	
Belgium	**45 – 46**	95–02	**+**	**50–79**	*1*
Bosnia & HG	(50 – 50)	84	?	–	
Bulgaria	**4,200 – 4,200**	94	**–**	**20–24**	
Croatia	1,000 – 1,500	02	(0)	(0–19)	*29,54,26,25*
Czech Rep.	**931 – 954**	00	**+**	**30–49**	
Denmark	**1 – 3**	98–01	**–**	**50–79**	*4,5,6,7*
Estonia	**3,000 – 4,000**	98	**+**	**50–79**	*1*
France	**646 – 655**	00	**+**	**>80**	*6*
Georgia	Present	03	?	–	
Germany	**4,300 – 4,400**	95–99	**+**	**20–29**	
Greece	(2,000 – 2,500)	95–00	(–)	(0–19)	
Hungary	**4,800 – 5,600**	98–02	**0**	**0–19**	*12*
Italy	**50 – 60**	03	**+**	**0–19**	*6*
Latvia	**9,500 – 10,500**	94–03	**0**	**0–19**	*16,35*
Lithuania	**12,500 – 13,000**	99–01	+	0–19	*20*
Macedonia	800 – 1,200	00	–	30	
Moldova	**400 – 600**	90–00	**0**	**0–19**	
Netherlands	**330 – 396**	98–00	**+**	**271**	*1*
Poland	**44,000 – 46,000**	00–01	+	20–29	*13,23,12*
Portugal	4,000 – 6,000	02	(+)	(–)	
Romania	**4,000 – 5,000**	96–02	**+**	**0–19**	*43,58,35,36*
Russia	**5,500 – 7,500**	90–00	+	20–29	*10,102,103,114*
Serbia & MN	**1,100 – 1,250**	99–02	**0**	**0–19**	*28,155,127,192, 128a,24*
Slovakia	**1,000 – 1,350**	90–99	0	0–19	
Slovenia	**195 – 205**	99–00	+	0–9	
Spain	**16,600 – 16,600**	94	+	50–79	*10,9*
Sweden	**3 – 3**	99–00	**+**	**N**	
Switzerland	**167 – 182**	97–01	**+**	**0–9**	
Turkey	15,000 – 35,000	01	(–)	(0–19)	
Ukraine	**26,200 – 32,400**	90–00	**0**	**0–19**	

Total (approx.) 180,000 – 220,000 **Overall trend** Moderate increase
Breeding range >4,000,000 km² **Gen. length.** 8 **% Global pop.** 75–94

Winter pop. size (individuals)

Total (approx.) >9,000 **Overall trend** Moderate increase
% in European IBAs 7–8 **Gen. length** 8 **% Global pop.** <5

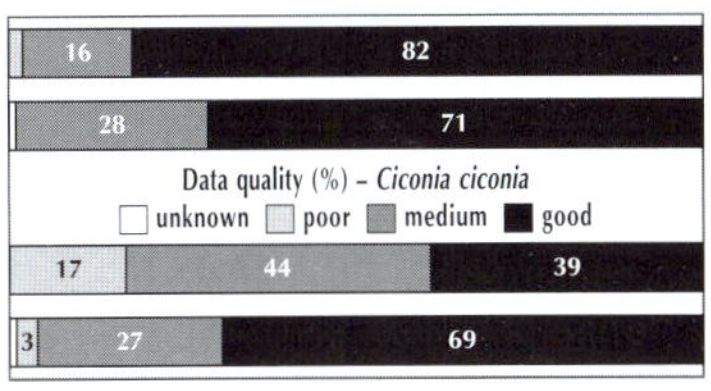

Plegadis falcinellus
GLOSSY IBIS

SPEC 3 (1994: 3) **Status** (Declining)
Criteria Moderate continuing decline
European IUCN Red List Category —
Criteria —
Global IUCN Red List Category —
Criteria —

Country	Breeding pop. size (pairs)	Year(s)	Trend	Mag.%	References
Albania	0 – 0	96–02	**-**	**X**	*4*
Armenia	20 – 80	97–02	+	0–19	
Azerbaijan	4,000 – 6,000	96–00	0	0–19	
Bosnia & HG	Present	85–89	?	–	
Bulgaria	**50 – 130**	95–02	**-**	**>80**	
Croatia	**1 – 10**	02	?	–	*70,16,54*
France	**0 – 4**	95–02	**0**	**0–19**	*6*
Greece	**150 – 200**	01	**F**	**>80**	
Hungary	**5 – 20**	98–01	**F**	**20–49**	
Italy	**11 – 15**	01	0	0–19	*6*
Macedonia	0 – 50	98–00	(F)	(-)	*7*
Moldova	**2 – 5**	90–00	**0**	**0–19**	
Romania	**2,500 – 2,800**	00–02	(-)	(0–19)	*47*
Russia	**7,000 – 8,000**	90–00	0	0–19	*6,8,33,60,120,178*
Serbia & MN	**11 – 14**	00–02	-	10–29	*1,163,65,29,155*
Spain	**393 – 423**	02	+	**>80**	*10,16*
Turkey	500 – 1,000	01	**-**	**30–49**	
Ukraine	**1,700 – 3,600**	90–98	**F**	**20–29**	
Total (approx.)	16,000 – 22,000	**Overall trend** Moderate decline			
Breeding range	>100,000 km²	**Gen. length.** 5		**% Global pop.** <5	

Plegadis falcinellus is a patchily distributed summer visitor to southern and south-eastern Europe, which accounts for a tiny proportion of its global breeding range. Its European breeding population is relatively small (<22,000 pairs), and underwent a moderate decline between 1970–1990. Although key populations in Russia and Azerbaijan were stable during 1990–2000, the species continued to decline in parts of south-eastern Europe, and underwent a moderate decline (>10%) overall. Consequently, it is provisionally evaluated as Declining.

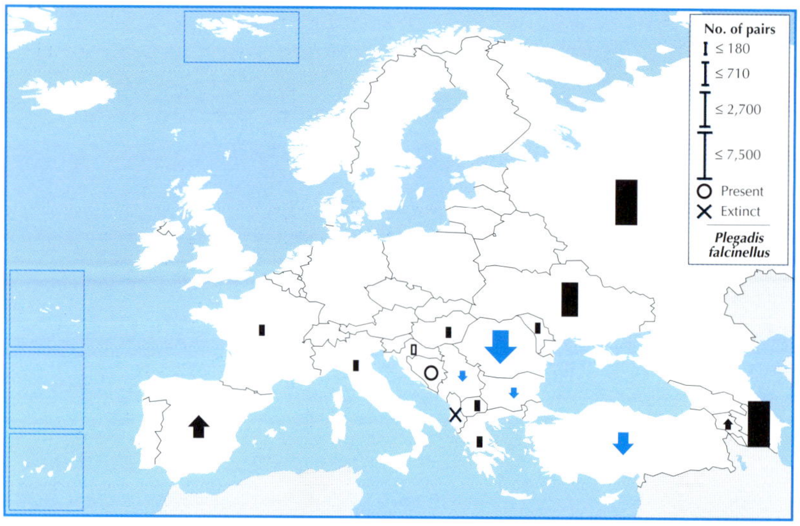

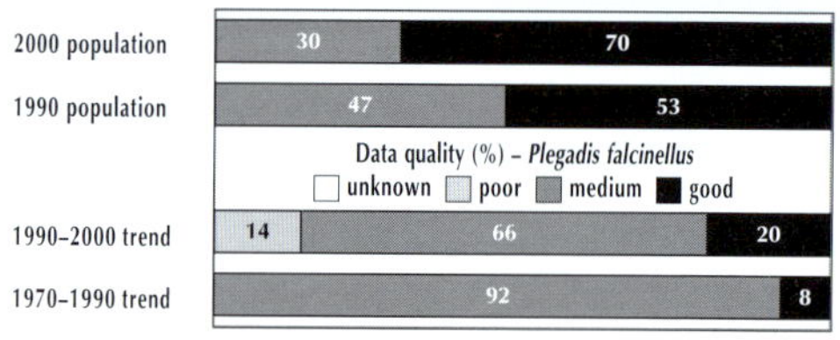

Geronticus eremita
NORTHERN BALD IBIS

SPEC 1 (1994: NE) **Status** Critically Endangered
Criteria See IUCN below
European IUCN Red List Category CR
Criteria D1
Global IUCN Red List Category CR
Criteria C2a(ii)

Country	Breeding pop. size (pairs)	Year(s)	Trend	Mag.%	References
Turkey	**15 – 15**	01	**0**	**0–19**	
Total (approx.)	15 – 15	**Overall trend** Stable			
Breeding range	<100 km²	**Gen. length.** 8		**% Global pop.** 5–24	

Geronticus eremita breeds mostly outside Europe in Morocco and (to a lesser extent) Syria, but a tiny breeding population of 15 pairs also persists at Bireçik in Turkey. Although this population underwent a large decline between 1970–1990, it was stable during 1990–2000. Nevertheless, as a consequence of the tiny size of its European population, this globally threatened species is evaluated as Critically Endangered in Europe.

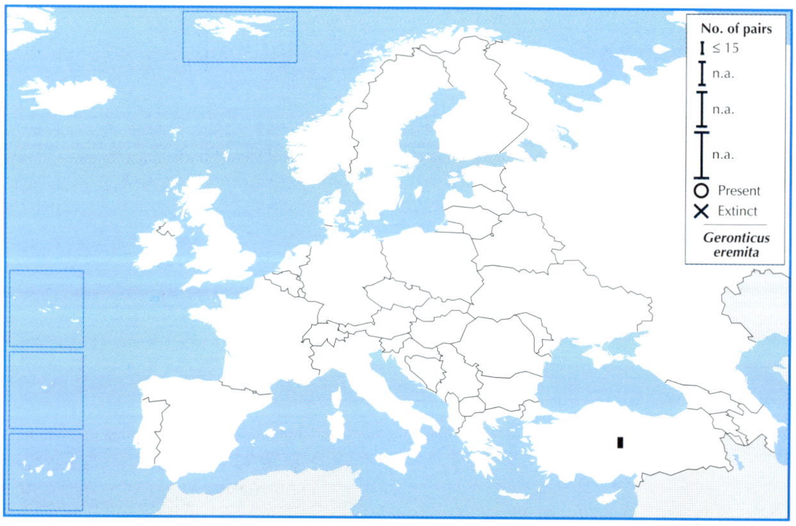

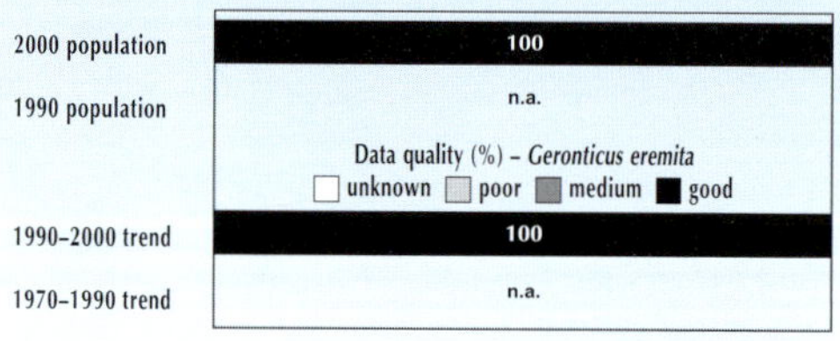

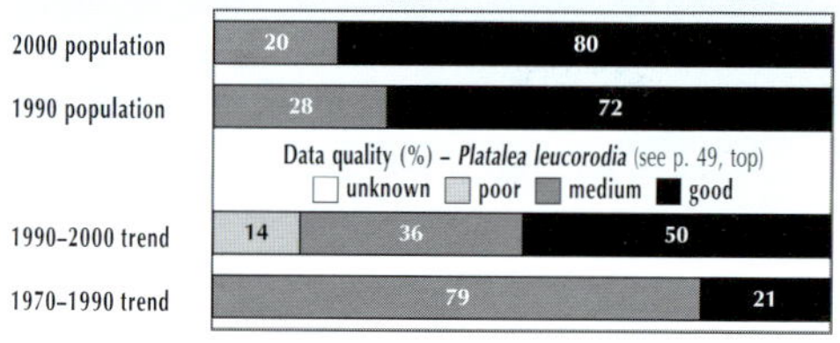

Platalea leucorodia
EURASIAN SPOONBILL

SPEC 2 (1994: 2) **Status** Rare
Criteria <10,000 pairs
European IUCN Red List Category —
Criteria —
Global IUCN Red List Category —
Criteria —

Platalea leucorodia is a widespread but patchily distributed breeder across much of southern Europe, which holds just over 50% of its global population. Its European breeding population is small (as few as 8,900 pairs), and underwent a large decline between 1970–1990. Although the sizeable Russian population continued to decline during 1990–2000, the species increased or was stable across most of the rest of Europe, and was stable overall. Nevertheless, its population size renders it susceptible to the risks affecting small populations, and consequently it is evaluated as Rare.

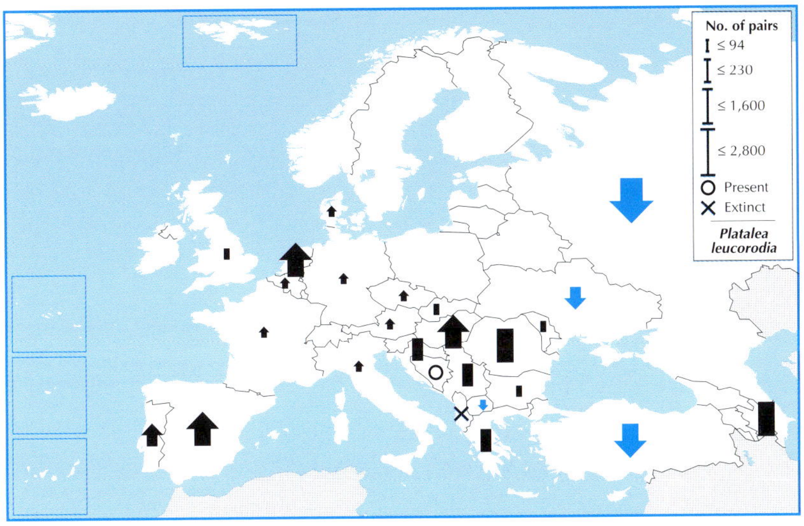

Country	Breeding pop. size (pairs)	Year(s)	Trend	Mag.%	References
Albania	0 – 0	96–02	–	X	4
Austria	**40 – 77**	98–02	+	**>80**	
Azerbaijan	800 – 2,500	96–00	(0)	(0–19)	
Belgium	**3 – 5**	00–02	+	N	1
Bosnia & HG	Present	85–89	?	–	
Bulgaria	**60 – 100**	00–02	F	**20–29**	
Croatia	**80 – 180**	02	0	**0–19**	16,44
Czech Rep.	**3 – 5**	00	+	**30–49**	
Denmark	**3 – 7**	98–01	+	N	4,5,6,7
France	**91 – 98**	01	+	**>80**	6
Germany	**42 – 42**	95–99	+	**>80**	
Greece	**150 – 200**	98–99	F	**30–49**	
Hungary	**850 – 1,200**	98–02	+	**50–79**	
Italy	**77 – 80**	01	+	**>80**	6
Macedonia	0 – 20	90–00	(–)	(0–19)	7
Moldova	0 – 20	90–00	0	**0–19**	
Netherlands	**1,000 – 1,270**	98–00	+	**97**	1
Portugal	**100 – 200**	02	(+)	(–)	
Romania	**1,100 – 1,500**	90–02	0	0–19	47
Russia	**2,500 – 3,000**	90–00	–	20–29	6,8,33,60,120,178
Serbia & MN	**190 – 220**	00–02	0	**0–19**	163,65,67,138, 155,156
Slovakia	**0 – 35**	80–99	F	30–49	
Spain	**1,000 – 2,500**	00–01	+	**0–19**	10,16
Turkey	600 – 1,200	01	–	**30–49**	
Ukraine	**200 – 250**	90–98	–	**10–19**	
UK	**1 – 7**	99–00	F	N	
Total (approx.)	8,900 – 15,000	**Overall trend** Stable			
Breeding range	>250,000 km²	**Gen. length.** 9		**% Global pop.** 50–74	

(See p. 48, bottom, for data quality graph)

Country	Winter pop. size (individuals)	Year(s)	Trend	Mag.%	References
Albania	**3 – 24**	95–02	(F)	(–)	
Bulgaria	**0 – 30**	97–01	F	**>80**	
Croatia	50 – 100	02	(F)	(–)	16,26
France	**180 – 360**	98–02	+	**30–49**	9
Greece	**140 – 330**	95–99	F	**30–49**	
Italy	**60 – 400**	02	+	**20–29**	22
Portugal	140 – 180	96–01	(F)	(–)	18
Serbia & MN	**30 – 150**	90–02	F	50–100	
Turkey	**50 – 200**	91–01	F	**>80**	
Total (approx.)	>700	**Overall trend** Small increase			
% in European IBAs	54–87	**Gen. length** 9		**% Global pop.** <5	

Phoenicopterus roseus
GREATER FLAMINGO

SPEC 3 (1994: 3) **Status** Localised
Criteria ≥90% breed at ≤10 sites
European IUCN Red List Category —
Criteria —
Global IUCN Red List Category —
Criteria —

Phoenicopterus roseus breeds irregularly at a small number of sites in southern Europe, which holds less than half of its global breeding population. The European breeding population is relatively small (<58,000 pairs), but increased substantially between 1970–1990. The species also underwent a large increase during 1990–2000, increasing in France and remaining broadly stable in Turkey and Spain. Nevertheless, more than 90% of the European breeding population occurs at 10 sites (see Appendix 5), and consequently the species is evaluated as Localised.

Country	Breeding pop. size (pairs)	Year(s)	Trend	Mag.%	References
Azerbaijan	100 – 500	96–00	(0)	(0–19)	
Cyprus	**4 – 40**	01	+	N	
France	**15,300 – 15,300**	01	+	**50–79**	6
Italy	**2,900 – 3,300**	00	+	**>80**	18
Spain	**23,000 – 23,000**	01	(F)	(–)	10
Turkey	**15,000 – 16,000**	01	0	**0–19**	
Total (approx.)	56,000 – 58,000	**Overall trend** Large increase			
Breeding range	>20,000 km²	**Gen. length.** 16		**% Global pop.** 25–49	

Country	Winter pop. size (individuals)	Year(s)	Trend	Mag.%	References
Albania	**0 – 270**	95–02	(F)	(–)	
Azerbaijan	**7,000 – 14,000**	96–02	+	**30–49**	
Cyprus	**5,000 – 15,000**	94–02	0	**0–9**	
France	**30,000 – 58,000**	98–02	+	**30–49**	9
Greece	**5,800 – 11,200**	95–99	+	**0–19**	
Italy	**8,000 – 19,000**	02	+	**0–19**	22
Portugal	2,600 – 7,000	96–01	(+)	(–)	18
Slovenia	Present	90–00	?	–	
Spain	**14,000 – 38,000**	90–01	+	**50–79**	19
Turkey	**35,000 – 55,000**	91–01	+	**0–19**	
Total (approx.)	>110,000	**Overall trend** Large increase			
% in European IBAs	75–84	**Gen. length** 16		**% Global pop.** 25–49	

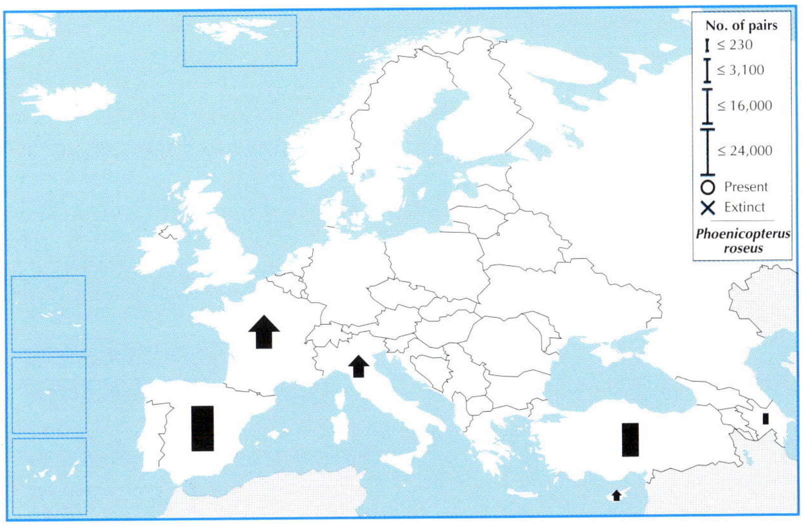

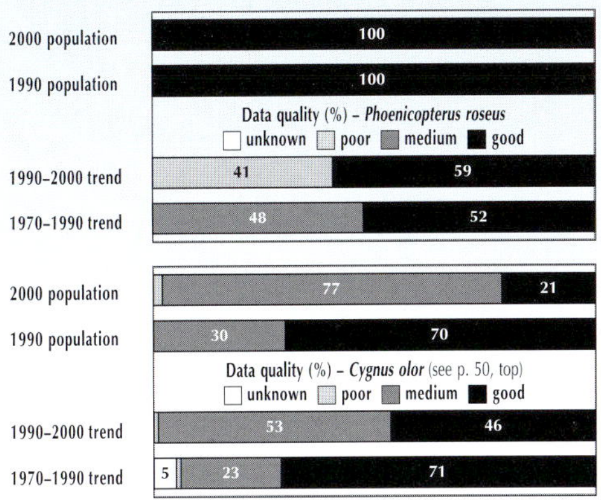

Cygnus olor
MUTE SWAN

Non-SPEC[E] (1994: —) **Status** Secure
Criteria —
European IUCN Red List Category —
Criteria —
Global IUCN Red List Category —
Criteria —

Cygnus olor is a widespread but patchily distributed breeder across much of Europe, which probably constitutes >50% of its global breeding range. Its European breeding population is relatively small (<120,000 pairs), but underwent a moderate increase between 1970–1990. Although there were declines in a few countries during 1990–2000, most European populations—including key ones in the United Kingdom and Russia—increased or were stable, and the species underwent a large increase overall. Consequently, it is evaluated as Secure.

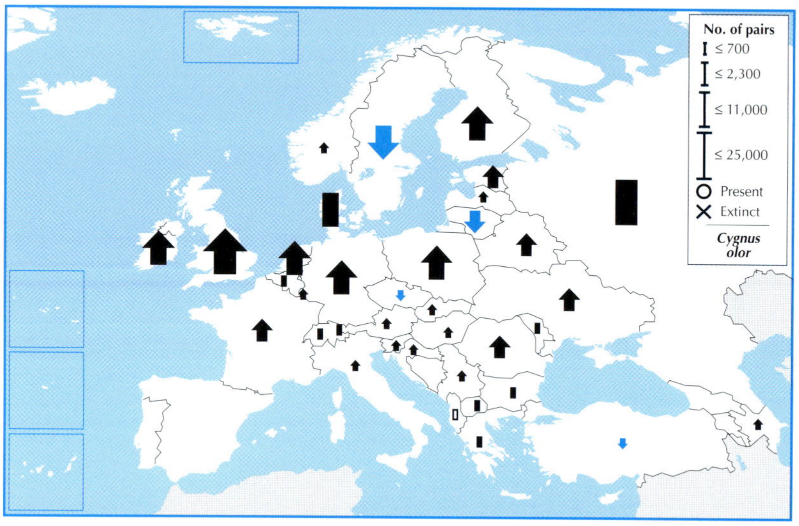

Country	Breeding pop. size (pairs)	Year(s)	Trend	Mag.%	References
Albania	0 – 5	96–02	?	–	
Austria	350 – 500	98–02	+	20–29	
Azerbaijan	(2 – 5)	96–00	(+)	(0–19)	
Belarus	750 – 900	97–02	+	0–9	
Belgium	341 – 469	01–02	0	0–19	*1*
Bulgaria	**12 – 30**	95–02	**0**	**0–9**	
Croatia	25 – 100	02	+	N	*16,54*
Czech Rep.	**440 – 500**	00	**-**	**20–29**	
Denmark	4,500 – 5,000	00	0	0–19	*3*
Estonia	2,000 – 2,500	98	+	20–29	*1*
Finland	6,000 – 8,000	99–02	**+**	**300–400**	
France	1,000 – 1,600	98–02	+	10–19	*4*
Germany	7,700 – 13,400	95–99	+	20–29	
Greece	**50 – 80**	98	(0)	(0–19)	
Hungary	**220 – 250**	98–02	**+**	**>80**	
Rep. Ireland	2,500 – 10,000	88–91	+	0–19	
Italy	**300 – 500**	03	**+**	**30–49**	*6*
Latvia	600 – 800	90–00	+	30–49	*17*
Liechtenstein	**1 – 2**	98–00	**0**	**0–19**	
Lithuania	(1,000 – 1,500)	99–01	–	20–29	*20*
Luxembourg	**35 – 45**	00–02	**+**	**0–19**	
Macedonia	10 – 25	99–00	(0)	(0–19)	
Moldova	**40 – 70**	90–00	**0**	**0–19**	
Netherlands	**5,500 – 6,500**	98–00	**+**	**60**	*1*
Norway	400 – 800	90–02	**+**	**0–19**	*27*
Poland	**6,500 – 7,000**	01	+	0–19	*13,14*
Romania	(750 – 1,000)	90–02	(+)	(0–19)	
Russia	15,000 – 20,000	90–00	0	0–19	*8,85,102,114,131,141*
Serbia & MN	**50 – 60**	00–02	**+**	**50–79**	*1,155,99,120,82,78*
Slovakia	100 – 400	80–99	+	50–79	
Slovenia	**50 – 70**	95–00	+	30–49	
Sweden	**4,500 – 7,500**	99–00	**-**	**0–9**	
Switzerland	**450 – 600**	93–96	0	0–19	
Turkey	**10 – 20**	01	**-**	**30–49**	
Ukraine	950 – 1,550	90–00	+	0–19	
UK	23,900 – 25,600	00	**+**	**71**	*5,31*

Total (approx.)	86,000 – 120,000	**Overall trend** Large increase	
Breeding range	>3,000,000 km²	**Gen. length.** 7	**% Global pop.** 50–74

(See p. 49, bottom, for data quality graph)

	Winter pop. size (individuals)		
Total (approx.)	>260,000	**Overall trend** Fluctuating	
% in European IBAs	25–31	**Gen. length** 7	**% Global pop.** 50–74

Cygnus columbianus
TUNDRA SWAN

SPEC 3W (1994: 3W) **Status** Vulnerable
Criteria See IUCN below
European IUCN Red List Category VU
Criteria A2b
Global IUCN Red List Category —
Criteria —

Cygnus columbianus breeds mainly in Russia, but winters locally in western and south-eastern Europe, which accounts for less than a quarter of its global wintering range. Its European wintering population is relatively large (>23,000 individuals), and was stable between 1970–1990. Although most wintering populations were stable or increased during 1990–2000, there were substantial declines in the Republic of Ireland and Netherlands, and the species underwent a large decline (>30%) overall. Consequently, this previously Localised species is now evaluated as Vulnerable.

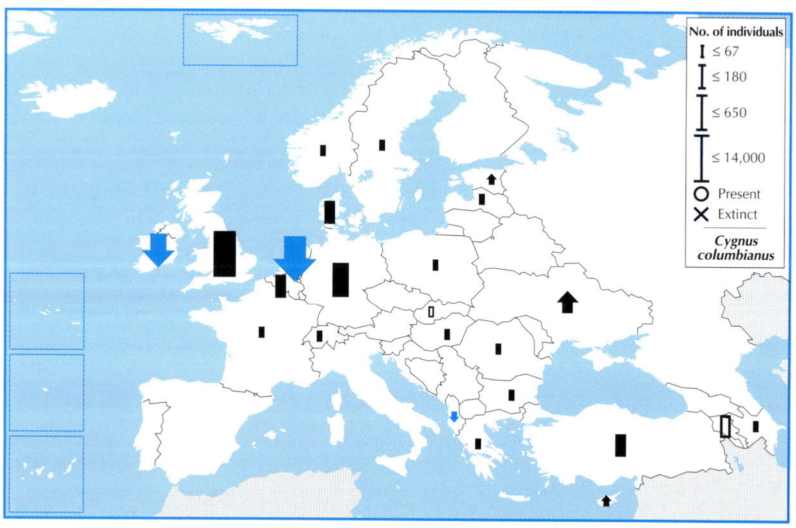

Country	Winter pop. size (individuals)	Year(s)	Trend	Mag.%	References
Albania	**0 – 2**	95–02	**-**	**>80**	
Armenia	(50 – 250)	90–00	?	–	
Azerbaijan	5 – 10	96–02	0	0–19	
Belgium	**40 – 470**	95–00	**0**	**0–19**	*1*
Bulgaria	**0 – 76**	97–01	**F**	**>80**	
Cyprus	**14 – 14**	02	(+)	(N)	
Denmark	**150 – 200**	99–00	0	0–19	*25,26*
Estonia	0 – 10	98	+	**N**	*1*
France	**45 – 100**	98–02	**0**	**0–19**	*9*
Germany	**130 – 3,200**	95–00	**F**	**>80**	
Greece	**46 – 46**	01	**F**	**>80**	
Hungary	0 – 50	90–00	F	>80	*23*
Rep. Ireland	**350 – 1,000**	94–00	**-**	**50–79**	
Latvia	**0 – 10**	90–99	**0**	**0–19**	*36*
Netherlands	**14,000 – 14,000**	99–01	**-**	**49**	*3,4,8,9,11,12*
Norway	**22 – 56**	91–99	**0**	**0–19**	
Poland	**1 – 30**	95–00	0	0–19	*117*
Romania	1 – 6	90–00	0	0–19	*44,60*
Slovakia	0 – 4	90–99	?	–	*4*
Sweden	**0 – 10**	98–01	**0**	**0–19**	
Switzerland	**0 – 10**	98–02	**0**	**0–19**	
Turkey	**80 – 160**	91–01	**0**	**0–19**	
Ukraine	(50 – 250)	94	+	20–29	
UK	**8,200 – 8,200**	94–99	**0**	**3**	*44,49*

Total (approx.)	>23,000	**Overall trend** Large decline	
% in European IBAs	>90	**Gen. length** 9	**% Global pop.** 5–24

Country	Breeding pop. size (pairs)	Year(s)	Trend	Mag.%	References
Denmark					
Greenland	0 – 1	95–00	?	–	*3*
Lithuania	**1 – 1**	97–03	**0**	**0–19**	*20*
Russia	9,000 – 11,000	90–00	+	**20–29**	*74,85,101*

Total (approx.)	9,000 – 11,000	**Overall trend** Moderate increase	
Breeding range	>100,000 km²	**Gen. length.** 9	**% Global pop.** 5–24

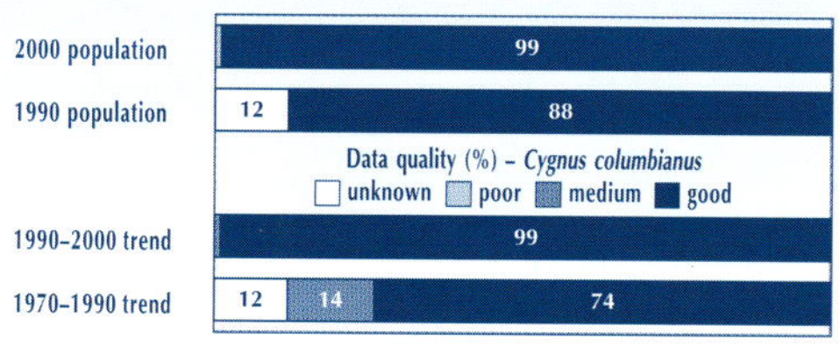

Cygnus cygnus
WHOOPER SWAN

Non-SPEC^E W (1994: 4W) **Status** Secure
Criteria —
European IUCN Red List Category —
Criteria —
Global IUCN Red List Category —
Criteria —

Cygnus cygnus breeds mainly in Iceland, Fennoscandia and northern Russia, but winters patchily across much of Europe, which constitutes >50% of its global wintering range. Its European wintering population is relatively large (>65,000 individuals), and was stable between 1970–1990. Although there were declines in a handful of countries during 1990–2000, most European wintering populations—including key ones in Denmark and Germany—were broadly stable or increased, and the species underwent a large increase overall. Consequently, it is evaluated as Secure.

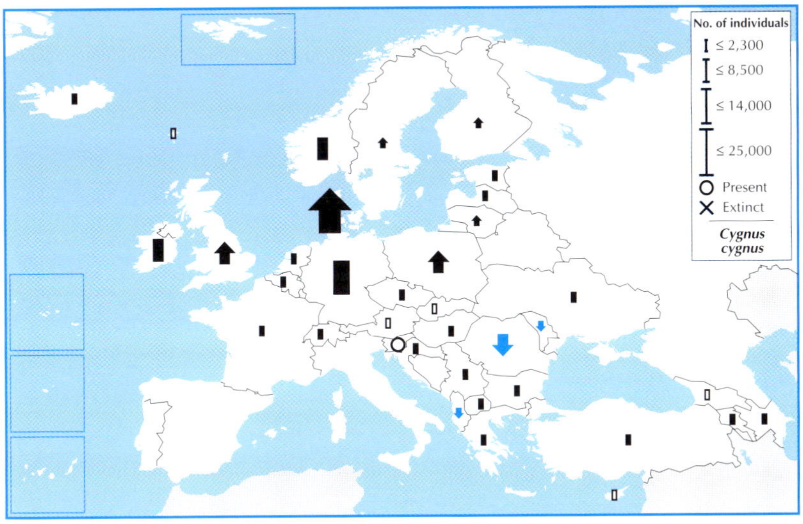

Country	Winter pop. size (individuals)	Year(s)	Trend	Mag.%	References
Azerbaijan	200 – 600	96–02	(F)	(–)	
Albania	0 – 8	95–02	–	>80	
Armenia	50 – 250	97–02	0	0–19	
Austria	90 – 140	97–99	?	–	
Azerbaijan	200 – 600	96–02	(F)	(–)	
Belgium	10 – 35	95–00	0	0–19	1
Bulgaria	80 – 1,280	97–01	F	50–500	
Croatia	11 – 100	02	(F)	(–)	16,26
Cyprus	0 – 6	94–02	?	–	
Czech Rep.	0 – 10	90–00	0	0–19	
Denmark	23,000 – 27,000	99–00	+	>80	25,26
Faroe Is.	0 – 20	92	?	–	3
Estonia	200 – 500	98	F	20–29	1
Finland	300 – 1,000	99–01	+	100–200	
France	30 – 75	98–02	0	0–19	9
Georgia	350 – 350	02	?	–	1
Germany	12,000 – 20,000	95–00	F	30–49	
Greece	500 – 500	03	F	>80	
Hungary	0 – 8	97–01	F	>80	24
Iceland	830 – 1,200	90–00	0	0–19	57,49,50
Rep. Ireland	7,900 – 9,100	94–00	0	0–19	
Latvia	20 – 100	90–99	0	0–19	29
Lithuania	0 – 50	92–02	+	>80	
Macedonia	0 – 10	90–00	(F)	(20–29)	
Moldova	0 – 5	90–00	–	20–29	
Netherlands	1,600 – 1,600	99–01	0	0–19	3,4,8,9,11,12
Norway	4,700 – 4,800	93–96	(0)	(0–19)	68,70
Poland	2,000 – 4,000	92–00	+	>80	115,117
Romania	2,000 – 4,500	90–00	–	0–19	60,14
Serbia & MN	(1 – 5)	90–02	(F)	(50–79)	
Slovakia	0 – 10	90–99	?	–	4
Slovenia	Present	90–00	?	–	
Sweden	1,000 – 4,000	98–01	+	0–19	
Switzerland	11 – 54	98–02	F	50–79	
Turkey	(500 – 1,500)	91–01	F	>80	
Ukraine	1,000 – 5,000	90–95	F	20–29	
UK	6,900 – 6,900	94–99	+	23	48,49

Total (approx.) >65,000 **Overall trend** Large increase
% in European IBAs >90 **Gen. length** 9 **% Global pop.** 50–74
(See p. 52, top, for data quality graph)

Breeding pop. size (pairs)
Total (approx.) 16,000 – 21,000 **Overall trend** Large increase
Breeding range >2,000,000 km² **Gen. length.** 9 **% Global pop.** 25–49

Anser fabalis
BEAN GOOSE

Non-SPEC^E W (1994: —) **Status** Secure
Criteria —
European IUCN Red List Category —
Criteria —
Global IUCN Red List Category —
Criteria —

Anser fabalis breeds in Fennoscandia and northern Russia, and winters patchily in western, central and south-east Europe, which probably holds >50% of its global wintering population. Its European wintering population is large (>390,000 individuals), and underwent a large increase between 1970–1990. Although there were declines in a number of countries—notably the Czech Republic—during 1990–2000, key wintering populations in Germany and the Netherlands were stable or increased, and the species was stable overall. Consequently, it is evaluated as Secure.

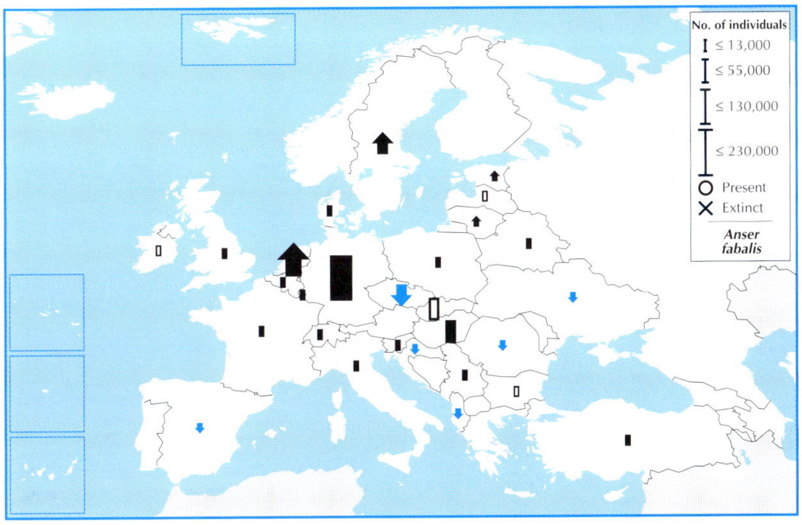

Country	Winter pop. size (individuals)	Year(s)	Trend	Mag.%	References
Albania	0 – 1	95–02	–	>80	
Belarus	0 – 50	91–00	0	0–19	3
Bulgaria	0 – 5	96–01	?	–	
Croatia	10,000 – 15,000	02	(–)	(–)	26,70
Czech Rep.	5,000 – 80,000	90–00	–	>80	
Denmark	10,000 – 12,000	99–00	0	0–19	25,26
Estonia	(0 – 5)	98	+	N	1
France	2,700 – 5,000	98–02	0	0–19	9
Germany	170,000 – 290,000	94–99	0	0–19	
Hungary	30,000 – 100,000	90–00	F	30–49	21,24
Rep. Ireland	0 – 6	94–00	?	–	
Italy	100 – 350	02	0	0–19	22
Latvia	0 – 10	90–00	?	–	
Lithuania	1 – 10	92–02	+	>80	
Luxembourg	50 – 150	00–02	0	0–19	
Netherlands	130,000 – 130,000	99–01	+	117	3,4,8,9,11,12
Poland	1,200 – 115,000	92–97	(0)	(0–19)	118
Romania	120 – 500	90–00	–	0–19	60,40,42
Serbia & MN	2,000 – 6,000	90–02	F	20–39	
Slovakia	10,000 – 35,000	90–99	?	–	4
Slovenia	150 – 4,000	90–00	F	0–100	
Spain	1 – 268	90–01	–	>80	19
Sweden	15,000 – 30,000	98–01	+	0–19	
Switzerland	19 – 52	98–02	F	50–79	
Turkey	(10 – 100)	91–01	(F)	(>80)	
Ukraine	(250 – 1,000)	90–00	(–)	(10–19)	
UK	950 – 950	89–99	(0)	(0–19)	49

Total (approx.) >390,000 **Overall trend** Stable
% in European IBAs 70–91 **Gen. length** 7 **% Global pop.** 50–74

Country	Breeding pop. size (pairs)	Year(s)	Trend	Mag.%	References
Finland	(1,500 – 2,000)	99–01	(0)	(0–19)	
Norway	50 – 200	00	–	0–19	19
Russia	135,000 – 140,000	90–00	0	0–19	85,164
Sweden	800 – 1,200	99–00	–	0–19	

Total (approx.) 140,000 – 140,000 **Overall trend** Stable
Breeding range >1,000,000 km² **Gen. length.** 7 **% Global pop.** 25–49

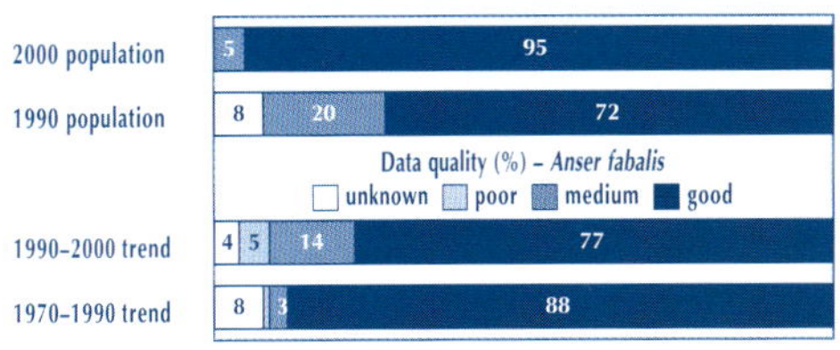

Anser brachyrhynchus
PINK-FOOTED GOOSE

Non-SPEC^{E} (1994: 4) **Status** Secure
Criteria —
European IUCN Red List Category —
Criteria —
Global IUCN Red List Category —
Criteria —

Anser brachyrhynchus breeds only in Svalbard, Iceland and east Greenland, with the entire global breeding range hence confined to Europe. The European breeding population is relatively small (<69,000 pairs), but increased substantially between 1970–1990. All three populations continued to increase during 1990–2000, and the species underwent a moderate increase overall. Consequently, it is evaluated as Secure.

Country	Breeding pop. size (pairs)	Year(s)	Trend	Mag.%	References
Denmark					
Greenland	(2,500 – 5,000)	95–00	(+)	(0–19)	*2*
Iceland	35,000 – 45,000	96–00	+	0–19	*5,18*
Norway					
Svalbard	**12,250 – 18,800**	02	+	**30–49**	
Total (approx.)	50,000 – 69,000	**Overall trend** Moderate increase			
Breeding range	>500,000 km²	**Gen. length.** 7		**% Global pop.** 100	

Country	Winter pop. size (individuals)	Year(s)	Trend	Mag.%	References
Belgium	**20,100 – 36,900**	95–00	+	**>80**	*1*
Denmark	**21,000 – 23,000**	99–00	0	0–19	*25,26*
France	**0 – 3**	98–02	**0**	**0–19**	*9*
Germany	(50 – 200)	94–99	?	–	
Hungary	Present	90–00	F	>80	*23*
Rep. Ireland	**32 – 47**	94–00	F	20–29	
Netherlands	**4,300 – 4,300**	99–01	**0**	**0–19**	*3,4,8,9,11,12*
Romania	(1 – 5)	90–00	(0)	(0–19)	*60,42*
Slovenia	Present	90–00	?	–	
Sweden	**30 – 80**	98–01	+	**0–19**	
UK	**241,000 – 241,000**	94–99	+	**24**	*48,49*
Total (approx.)	>290,000	**Overall trend** Large increase			
% in European IBAs	>90	**Gen. length** 7		**% Global pop.** 100	

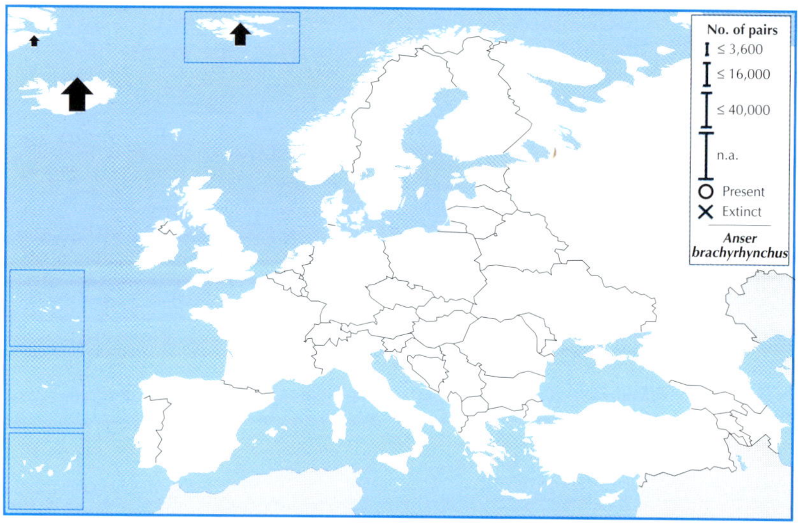

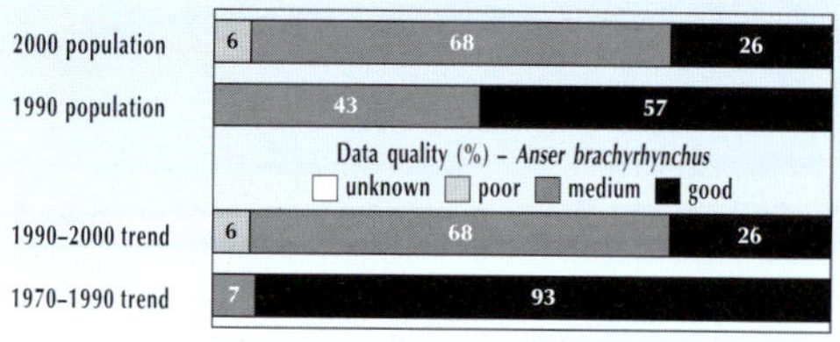

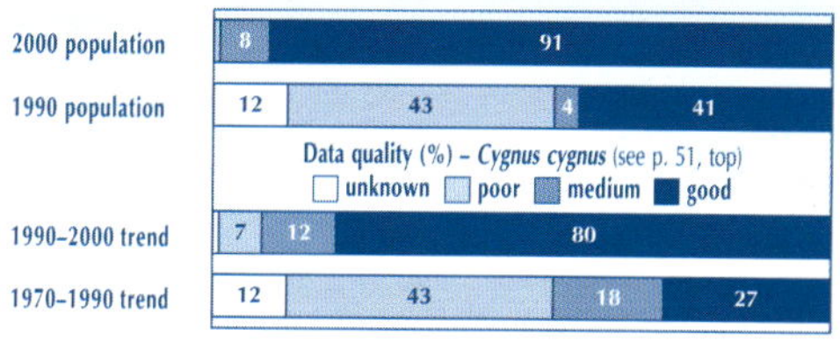

Anser albifrons
GREATER WHITE-FRONTED GOOSE

Non-SPEC (1994: —) **Status** Secure
Criteria —
European IUCN Red List Category —
Criteria —
Global IUCN Red List Category —
Criteria —

Anser albifrons breeds in Greenland and arctic Russia, with Europe accounting for less than a quarter of its global breeding range. Its European breeding population is relatively small (<72,000 pairs), but was stable between 1970–1990. Although the population in Greenland declined during 1990–2000, the species increased in its Russian stronghold, and underwent a large increase overall. Consequently, it is evaluated as Secure.

Country	Breeding pop. size (pairs)	Year(s)	Trend	Mag.%	References
Denmark					
Greenland	1,700 – 2,000	90–95	-	0–19	*2*
Russia	60,000 – 70,000	90–00	+	20–29	*85*
Total (approx.)	62,000 – 72,000	**Overall trend** Large increase			
Breeding range	>500,000 km²	**Gen. length.** 7		**% Global pop.** 5–24	

Country	Winter pop. size (individuals)	Year(s)	Trend	Mag.%	References
Azerbaijan	6,000 – 14,000	96–02	0	0–19	
Belgium	**23,100 – 34,100**	95–00	+	**>80**	*1*
Bulgaria	**33,000 – 400,000**	97–01	**F**	**>80**	
Croatia	**5,000 – 10,000**	02	(+)	(–)	*16,26,70*
Cyprus	500 – 1,200	94–02	0	0–19	
Czech Rep.	**1,000 – 30,000**	90–00	**-**	**>80**	
Denmark	**2,500 – 3,000**	99–00	+	**>80**	*25,26*
Germany	**210,000 – 450,000**	94–99	**0**	**0–19**	
Greece	**3,300 – 11,500**	95–99	**F**	**30–49**	
Hungary	**5,000 – 75,000**	90–00	F	>80	*21,24*
Rep. Ireland	**11,000 – 12,000**	94–00	**-**	**0–19**	
Italy	50 – 1,000	02	+	0–19	*22*
Macedonia	300 – 800	90–00	(F)	(20–29)	
Moldova	0 – 1,200	90–00	+	>80	
Netherlands	**640,000 – 640,000**	99–01	**0**	**0–19**	*3,4,8,9,11,12*
Poland	**150 – 15,000**	92–97	+	0–19	*118*
Romania	66,000 – 260,000	90–00	-	0–19	*60,14,42*
Serbia & MN	10,000 – 15,000	90–02	+	0–19	
Slovakia	3,000 – 5,000	90–99	?	–	*4*
Slovenia	100 – 1,800	90–00	F	0–100	
Sweden	**300 – 7,000**	98–01	+	**>80**	
Turkey	(30,000 – 60,000)	91–01	(–)	(0–19)	
Ukraine	(20,000 – 100,000)	90–00	+	20–29	
UK	26,800 – 26,800	91–99	+	30	*48,49*
Total (approx.)	>1,100,000	**Overall trend** Stable			
% in European IBAs	>90	**Gen. length** 7		**% Global pop.** 25–49	

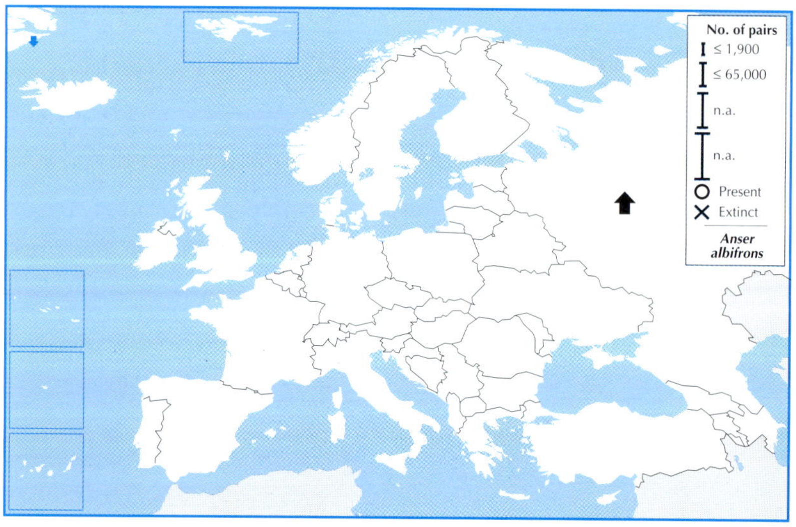

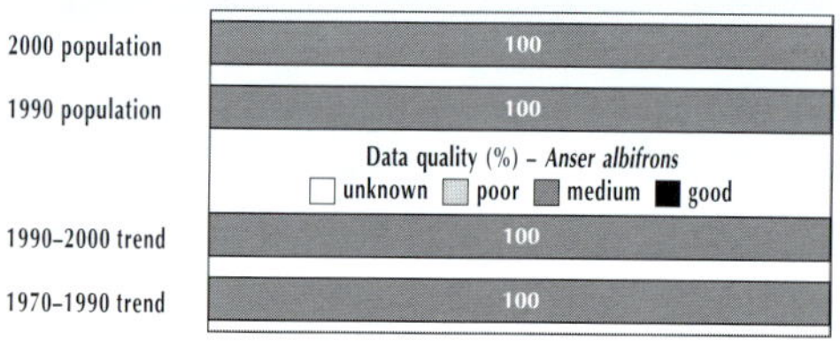

Anser erythropus
LESSER WHITE-FRONTED GOOSE

SPEC 1 (1994: 1) **Status** Endangered
Criteria See IUCN below
European IUCN Red List Category EN
Criteria C1
Global IUCN Red List Category VU
Criteria A2b,c,d; A3b,c,d

Anser erythropus breeds mainly in northern Siberia, but its global breeding range also extends into northern European Russia and Fennoscandia. Its European breeding population is very small (as few as 240 pairs), but was stable between 1970–1990. Although the small reintroduced Swedish population increased during 1990–2000, the species declined elsewhere in its European range, and underwent a moderate decline (>20%) overall. As a consequence of this continuing decline and its very small population, this globally threatened species is evaluated as Endangered in Europe.

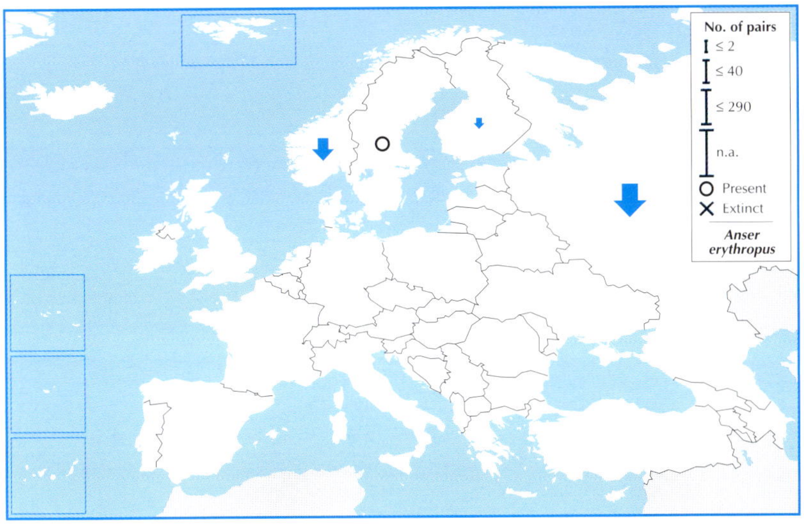

Country	Breeding pop. size (pairs)	Year(s)	Trend	Mag.%	References
Finland	0 – 5	99–01	**-**	**50**	
Norway	**35 – 45**	02	**-**	**5**	*44*
Russia	200 – 400	95–00	-	20–29	*74,85,107*
Sweden	**[5 – 5]**	[99–00]	**[+]**	**[0–9]**	*Reintroduction*
Total (approx.)	240 – 450	**Overall trend** Moderate decline			
Breeding range	>100,000 km²	**Gen. length**. 7		**% Global pop.** 5–24	

Country	Winter pop. size (individuals)	Year(s)	Trend	Mag.%	References
Azerbaijan	1,500 – 10,000	96–02	0	0–19	
Belgium	**[1 – 3]**	[95–00]	**[0]**	**[0–19]**	*Reintroduction*
Bulgaria	2 – 15	96–02	?	–	
Croatia	(1 – 20)	02	**F**	**>80**	*26*
Germany	[20 – 80]	[94–99]	[?]	[–]	*Reintroduction*
Greece	**71 – 71**	99	**-**	**0–19**	
Hungary	**0 – 20**	90–00	F	>80	*24*
Macedonia	0 – 2	90–00	(F)	(20–29)	
Netherlands	**[46 – 46]**	[99–01]	**[+]**	**[>80]**	*Reintroduction*
Romania	(31 – 50)	90–00	(-)	(50–79)	*60*
Slovakia	0 – 15	90–99	?	–	*4*
Sweden	**[0 – 1]**	[98–01]	**[0]**	**[0–19]**	*Reintroduction*
Turkey	(10 – 100)	91–01	?	–	
Ukraine	**300 – 1,200**	90–00	(-)	(-)	
Total (approx.)	>1,900	**Overall trend** Moderate decline			
% in European IBAs	>60	**Gen. length** 7		**% Global pop.** 25–49	

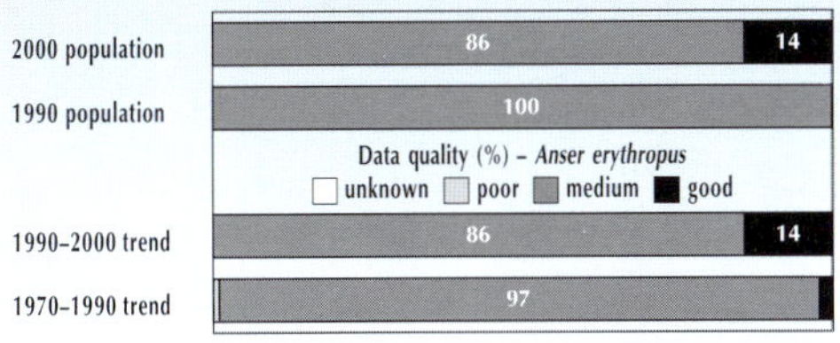

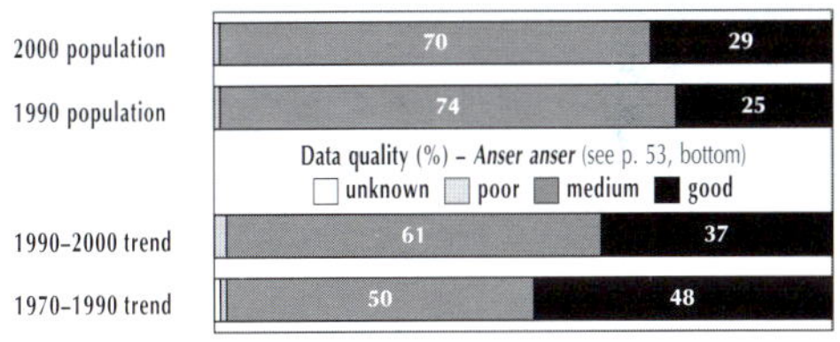

Anser anser
GREYLAG GOOSE

Non-SPEC (1994: —) **Status** Secure
Criteria —
European IUCN Red List Category —
Criteria —
Global IUCN Red List Category —
Criteria —

Anser anser is a widespread but patchily distributed breeder across much of Europe, which accounts for less than half of its global breeding range. Its European breeding population is large (>120,000 pairs), and increased substantially between 1970–1990. Although key populations in Iceland and Russia declined during 1990–2000, there were increases in many other European populations—notably those in Norway, Sweden, Germany and the United Kingdom—and the species underwent a large increase overall. Consequently, it is evaluated as Secure.

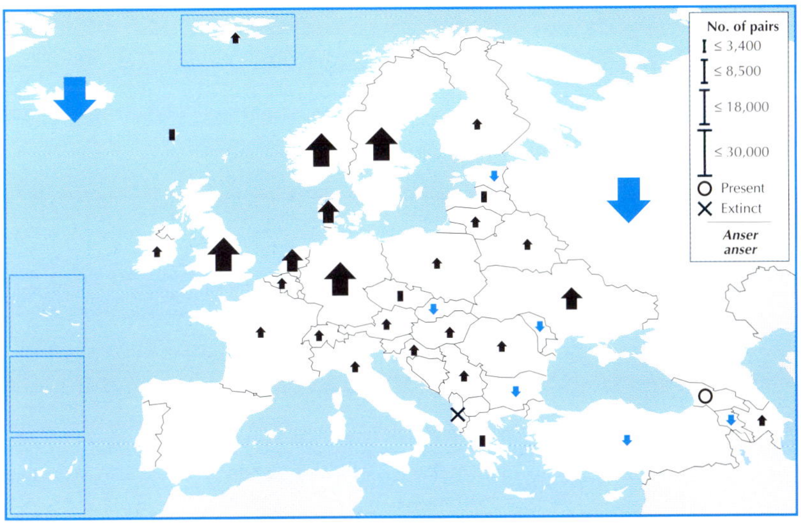

Country	Breeding pop. size (pairs)	Year(s)	Trend	Mag.%	References
Albania	0 – 0	96–02	-	X	
Armenia	50 – 250	97–02	-	30–49	
Austria	500 – 700	98–02	**+**	**50–79**	
Azerbaijan	(200 – 2,000)	96–00	(+)	(>80)	
Belarus	**50 – 250**	97–02	+	50–79	
Belgium	**1,200 – 1,300**	95–02	+	50–79	*1*
Bulgaria	**5 – 15**	96–02	**-**	**0–19**	
Croatia	**50 – 150**	02	**+**	**20–29**	*26,57*
Czech Rep.	**620 – 650**	00	**0**	**0–19**	
Denmark	6,000 – 10,000	00	+	50–79	
Faroe Is.	75 – 75	95	(0)	(0–19)	
Estonia	1,000 – 1,200	98	-	20–29	*1*
Finland	2,500 – 3,500	99–01	+	115	
France	**73 – 74**	01	**+**	**>80**	*6*
Georgia	Present	03	?	–	
Germany	10,000 – 18,000	95–99	+	50–79	
Greece	20 – 40	95–00	0	0–19	
Hungary	**1,100 – 1,400**	95–02	**+**	**0–19**	
Iceland	20,000 – 30,000	90–00	-	0–19	*1,18*
Rep. Ireland	250 – 1,000	88–91	(+)	(0–19)	
Italy	**150 – 200**	03	+	30–49	*6*
Latvia	30 – 80	90–00	0	0–19	*23,17*
Lithuania	100 – 150	99–01	(+)	(0–19)	*20*
Moldova	20 – 50	90–00	-	30–49	
Netherlands	**8,000 – 9,000**	98–00	**+**	**624**	*1*
Norway	**10,000 – 12,000**	02	**+**	**0–19**	*17*
Svalbard	0 – 5	95–03	+	0–19	
Poland	**3,200 – 3,600**	90–94	+	30–49	*18*
Romania	(1,000 – 1,300)	90–02	(+)	(0–19)	
Russia	20,000 – 45,000	90–00	-	20–29	*3,8,24,33,85, 102,114*
Serbia & MN	**100 – 130**	95–02	**+**	**50–69**	*1,126a,29,143a, 96,26,155*
Slovakia	15 – 80	80–99	-	30–49	
Sweden	**15,000 – 20,000**	99–00	**+**	**>80**	
Switzerland	**6 – 12**	93–96	**+**	**50–79**	
Turkey	400 – 800	01	-	30–49	
Ukraine	5,000 – 7,100	90–00	+	30–49	
UK	15,600 – 15,800	00	+	**118**	*12,31*
Total (approx.)	120,000 – 190,000	**Overall trend** Large increase			
Breeding range	>2,000,000 km²	**Gen. length**. 7		**% Global pop.** 25–49	

(See p. 53, top, for data quality graph)

	Winter pop. size (individuals)				
Total (approx.)	>390,000	**Overall trend** Large increase			
% in European IBAs	67–77	**Gen. length** 7		**% Global pop.** 25–49	

Chen caerulescens
SNOW GOOSE

Non-SPEC (1994: NE) **Status** (Secure)
Criteria —
European IUCN Red List Category —▼▼
Criteria —
Global IUCN Red List Category —
Criteria —

Country	Breeding pop. size (pairs)	Year(s)	Trend	Mag.%	References
Denmark					
Greenland	(200 – 2,000)	90–00	(+)	(0–19)	2
Total (approx.)	200 – 2,000	Overall trend Small increase			
Breeding range	>20,000 km^2	Gen. length 7		% Global pop. <5	

Chen caerulescens breeds predominantly in Siberia and arctic North America, but its breeding range just extends into Europe in western Greenland. Its European breeding population is very small (as few as 200 pairs), and its trend between 1970–1990 was unknown, but it underwent a small increase during 1990–2000. Although the size of the European population could render it susceptible to the risks affecting small populations, it is marginal to a much larger non-European population. Consequently, the species is provisionally evaluated as Secure.

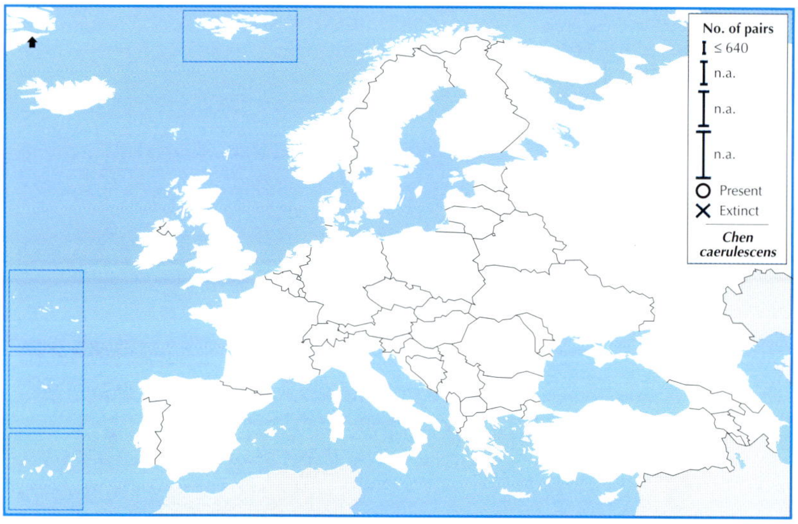

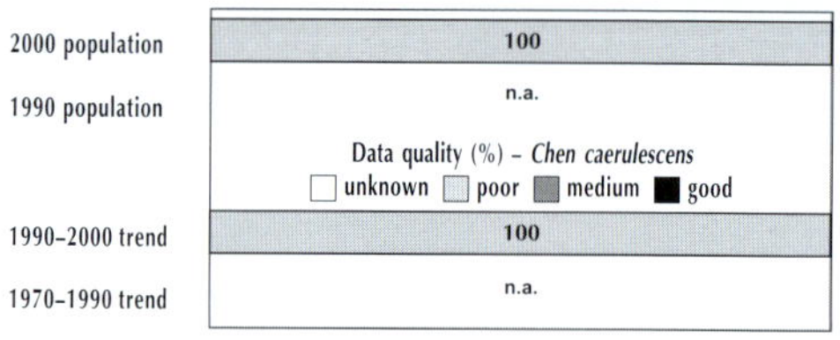

Branta canadensis
CANADA GOOSE

Non-SPEC (1994: NE) **Status** (Secure)
Criteria —
European IUCN Red List Category —▼
Criteria —
Global IUCN Red List Category —
Criteria —

Country	Breeding pop. size (pairs)	Year(s)	Trend	Mag.%	References
Denmark					
Greenland	(2,500 – 10,000)	95–00	(+)	(0–19)	9
Total (approx.)	2,500 – 10,000	Overall trend Small increase			
Breeding range	>20,000 km^2	Gen. length 7		% Global pop. <5	

Branta canadensis has a predominantly North American breeding distribution, which just extends into Europe in western Greenland. The native European breeding population is small (as few as 2,500 pairs), and its trend between 1970–1990 was unknown, but it underwent a small increase during 1990–2000. Although the size of the non-feral European population could render it susceptible to the risks affecting small populations, it is marginal to a much larger non-European population. Consequently, the species is provisionally evaluated as Secure.

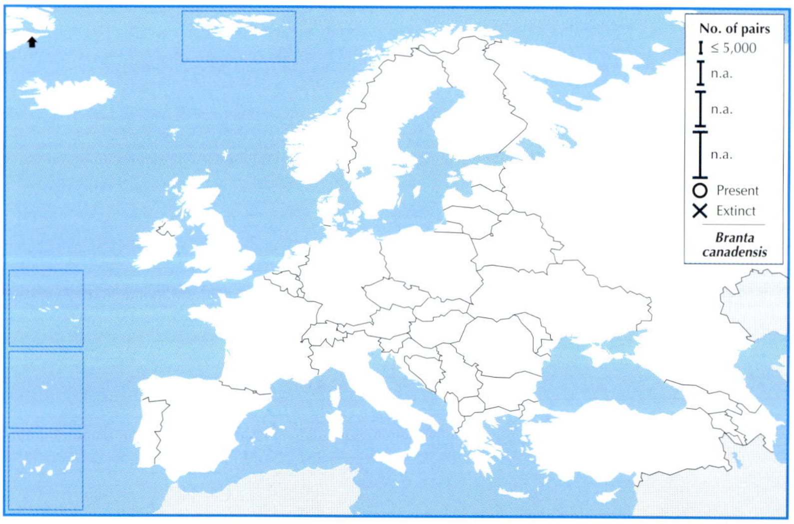

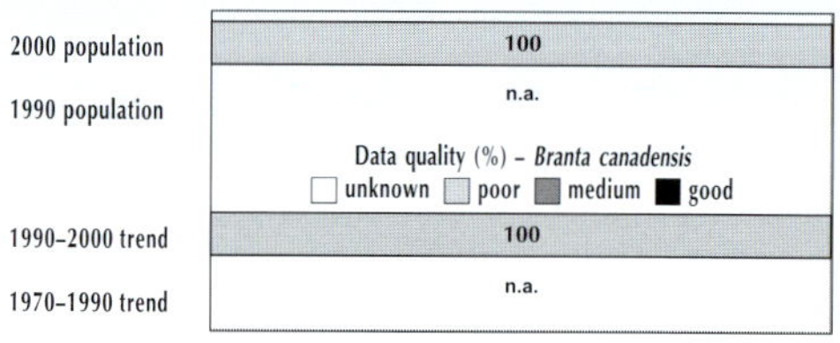

Branta leucopsis
BARNACLE GOOSE

Non-SPEC[E] (1994: 4/2W) **Status** Secure
Criteria —
European IUCN Red List Category —
Criteria —
Global IUCN Red List Category —
Criteria —

Country	Breeding pop. size (pairs)	Year(s)	Trend	Mag.%	References
Belgium	**145 – 198**	95–02	+	50–79	*1*
Denmark	**100 – 150**	01	+	**N**	*7*
Faroe Is.	**1 – 1**	91	(0)	(0–19)	
Greenland	(2,500 – 5,000)	95–00	+	0–19	
Estonia	**100 – 130**	98	+	50–79	*1*
Finland	**300 – 500**	01–02	+	**1,900**	
Germany	**32 – 32**	95–99	+	**50–79**	
Iceland	**0 – 50**	00	+	**>80**	*21*
Netherlands	**750 – 1,100**	98–00	+	**3,476**	*1*
Norway	**70 – 80**	02	+	**0–19**	*53*
Svalbard	**9,500 – 11,000**	02	+	**30–49**	
Russia	23,000 – 30,000	90–00	+	20–29	*1,49,85*
Sweden	**4,500 – 5,500**	99–00	+	**>80**	
Total (approx.)	41,000 – 54,000	**Overall trend** Large increase			
Breeding range	>50,000 km²	**Gen. length.** 7		**% Global pop.** 100	

Country	Winter pop. size (individuals)	Year(s)	Trend	Mag.%	References
Bulgaria	0 – 2	96–02	?	–	
Croatia	(1 – 5)	02	(F)	(–)	*26*
Denmark	**13,000 – 15,000**	99–00	+	**>80**	*25,26*
Estonia	0 – 200	98	F	**20–29**	*1*
France	**17 – 54**	98–02	0	**0–19**	*9*
Germany	**19,000 – 57,500**	94–99	F	**50–79**	
Hungary	0 – 5	97–01	F	>80	*24*
Rep. Ireland	**8,100 – 8,700**	94–00	0	**0–19**	
Latvia	**0 – 10**	90–01	?	–	*32*
Netherlands	**260,000 – 260,000**	99–01	+	**62**	*3,4,8,9,11,12*
Poland	(0 – 50)	90–00	(+)	(0–19)	*114*
Spain	0 – 50	90–01	0	**0–19**	
Sweden	**50 – 100**	98–01	+	**>80**	
UK	67,000 – 67,000	94–99	+	83	*48,49*
Total (approx.)	>370,000	**Overall trend** Large increase			
% in European IBAs	83–92	**Gen. length** 7		**% Global pop.** 100	

Branta leucopsis has an entirely European distribution, breeding mainly in the far north, and wintering in north-west Europe. Its breeding population is relatively small (<54,000 pairs), but increased substantially between 1970–1990. Furthermore, all significant national populations increased during 1990–2000, and the species continued to undergo a large increase overall. As a result of the expansion of its wintering range accompanying this population growth, the species no longer qualifies as Localised in winter. Consequently, it is evaluated as Secure.

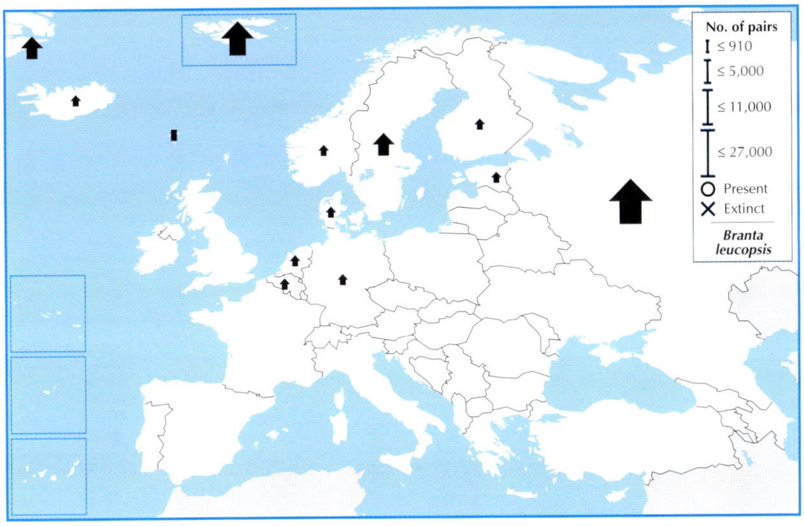

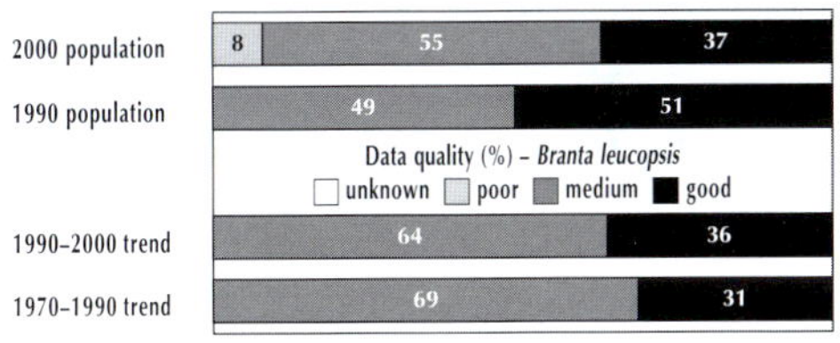

Branta bernicla
BRENT GOOSE

SPEC 3W (1994: 3) **Status** Vulnerable
Criteria See IUCN below
European IUCN Red List Category VU
Criteria A2b
Global IUCN Red List Category —
Criteria —

Country	Winter pop. size (individuals)	Year(s)	Trend	Mag.%	References
Bulgaria	0 – 2	96–02	?	–	
Denmark	**7,500 – 8,200**	99–00	+	20–29	*25,26*
France	**80,000 – 115,000**	98–02	-	**30–49**	*9*
Germany	**600 – 4,500**	94–99	-	**30–49**	
Rep. Ireland	**11,000 – 14,000**	94–00	F	**20–29**	
Netherlands	**39,000 – 39,000**	99–01	-	**32**	*3,4,8,9,11,12*
Poland	(0 – 10)	90–00	(0)	(0–19)	*114*
Romania	0 – 4	90–00	0	0–19	*60,40*
UK	101,000 – 101,000	94–99	-	12	*48,49*
Total (approx.)	>240,000	**Overall trend** Large decline			
% in European IBAs	>90	**Gen. length** 7		**% Global pop.** 25–49	

Country	Breeding pop. size (pairs)	Year(s)	Trend	Mag.%	References
Denmark					
Greenland	(100 – 150)	95–00	(F)	(–)	*2*
Norway					
Svalbard	500 – 1,500	95–03	**0**	**0–19**	
Russia	(400 – 600)	90–00	?	–	*74,85,99,137,139*
Total (approx.)	1,000 – 2,300	**Overall trend** Stable			
Breeding range	>100,000 km²	**Gen. length.** 7		**% Global pop.** <5	

Branta bernicla is a rare breeder in the European Arctic, but winters mainly in coastal areas of western Europe, which accounts for less than half of its global wintering range. Its European wintering population is large (>240,000 individuals), and increased between 1970–1990. Although a few populations increased or were broadly stable during 1990–2000, key wintering populations (of the subspecies *B. b. bernicla*) in the United Kingdom, France and the Netherlands declined, and the species underwent a large decline (>30%) overall. Consequently, it is evaluated as Vulnerable.

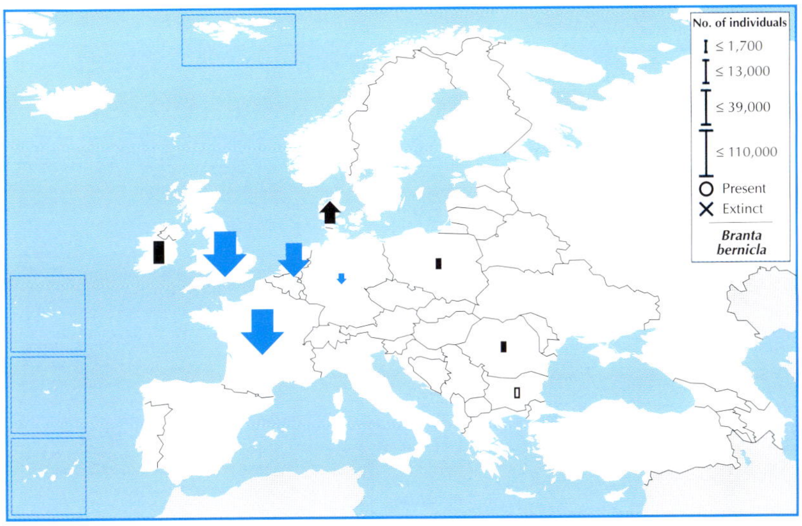

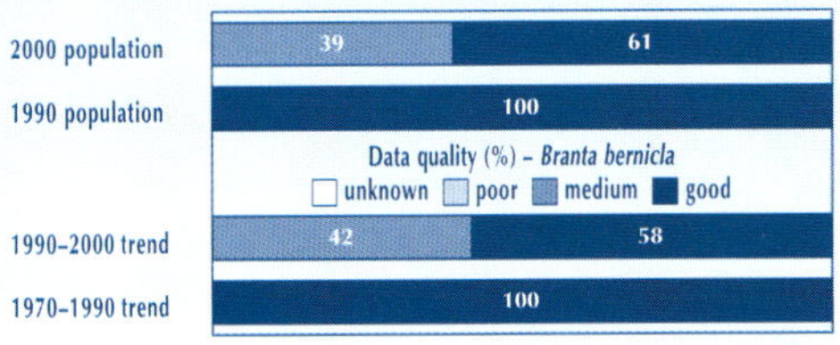

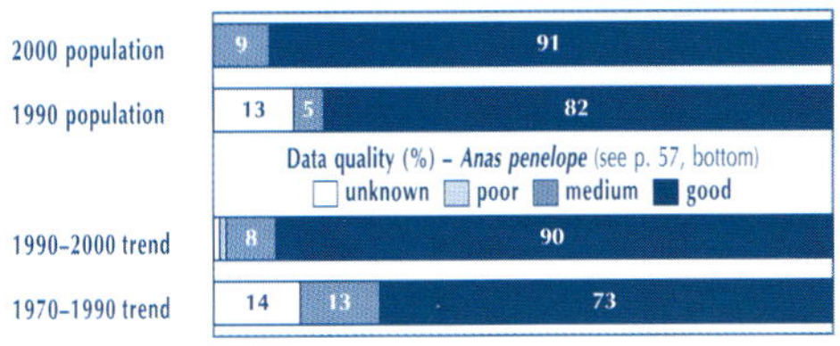

Branta ruficollis
RED-BREASTED GOOSE

SPEC 1W (1994: 1W) **Status** Vulnerable
Criteria See IUCN below
European IUCN Red List Category VU
Criteria B2a+b(iii)
Global IUCN Red List Category VU
Criteria B2a+b(iii)

Branta ruficollis breeds in northern Siberia, but >95% of its global population winters around the Black Sea in south-east Europe. The European wintering population is relatively large (>27,000 individuals), and increased substantially between 1970–1990. The species was broadly stable overall during 1990–2000, but its small wintering range (<20,000 km²) is threatened by changes in land-use. As a consequence of its small, declining range and the small number of sites used during the winter (see Appendix 5), this globally threatened species is evaluated as Vulnerable in Europe.

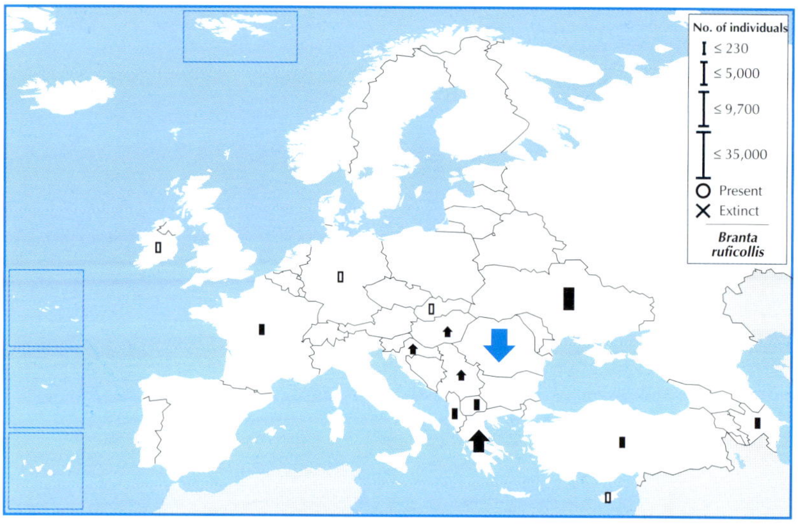

Country	Winter pop. size (individuals)	Year(s)	Trend	Mag.%	References
Albania	**0 – 2**	95–02	(F)	(–)	
Azerbaijan	25 – 300	96–02	0	0–19	
Bulgaria	**17,000 – 68,000**	96–03	**F**	**0–200**	
Croatia	**1 – 16**	02	(+)	(–)	*26*
Cyprus	**0 – 5**	94–02	?	–	
France	**0 – 1**	98–02	**0**	**0–19**	*9*
Germany	(10 – 35)	94–99	?	–	
Greece	**2,300 – 2,300**	03	**+**	**0–19**	
Hungary	1 – 40	97–01	+	50–79	*24*
Rep. Ireland	**0 – 1**	94–00	?	–	
Macedonia	**12 – 12**	03	(F)	(50–79)	
Moldova	0 – 450	90–00	+	>80	
Romania	4,300 – 21,500	90–00	–	50–79	*60,14,42*
Serbia & MN	10 – 50	90–02	+	10–29	
Slovakia	0 – 5	90–99	?	–	4
Turkey	100 – 500	91–01	**F**	**>80**	
Ukraine	2,500 – 10,000	98–99	**F**	**30–49**	
Total (approx.)	>27,000	**Overall trend** Fluctuating			
% in European IBAs	>90	**Gen. length** 7		**% Global pop.** >95	

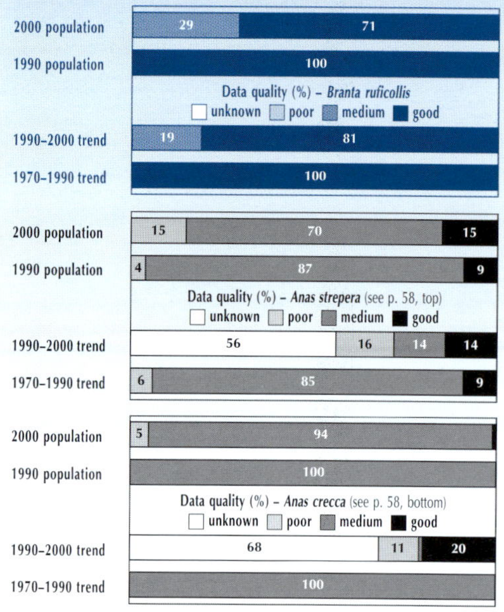

Tadorna ferruginea
RUDDY SHELDUCK

SPEC 3 (1994: 3) **Status** (Vulnerable)
Criteria See IUCN below
European IUCN Red List Category VU
Criteria A2b
Global IUCN Red List Category —
Criteria —

Tadorna ferruginea is a widespread breeder across much of south-eastern Europe, which accounts for less than a quarter of its global breeding range. Its European breeding population is relatively small (<33,000 pairs), and underwent a large decline between 1970–1990. Although the trend in Russia during 1990–2000 was unknown, the other key population in Turkey declined, and the species continued to decline at a rate that—on top of earlier declines—probably equates to a large decline (>30%) over three generations. Consequently, it is provisionally evaluated as Vulnerable.

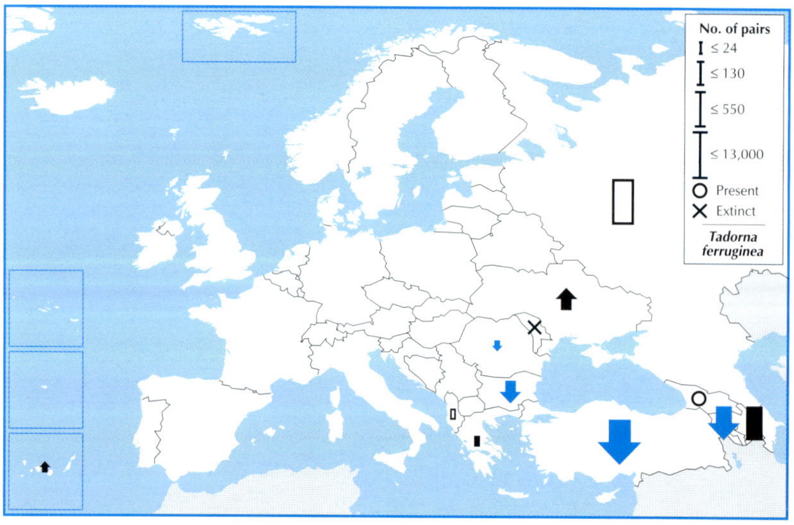

Country	Breeding pop. size (pairs)	Year(s)	Trend	Mag.%	References
Albania	(0 – 5)	02	?	–	
Armenia	250 – 1,000	98–02	–	20–29	*11*
Azerbaijan	(300 – 1,000)	96–00	(0)	(0–19)	
Bulgaria	**60 – 120**	96–02	–	0–19	
Georgia	Present	03	?	–	
Greece	(15 – 40)	95–00	**F**	**50–79**	
Moldova	**0 – 0**	00	–	**X**	
Romania	(20 – 25)	90–02	(–)	(0–19)	
Russia	8,500 – 16,000	90–00	?	–	*6,8,14,33,48, 85,114,119*
Spain					
Canary Is.	**3 – 7**	97–03	**+**	**N**	*28,25*
Turkey	(10,000 – 15,000)	01	(–)	(0–19)	
Ukraine	80 – 190	90–00	**+**	**0–19**	
Total (approx.)	19,000 – 33,000	**Overall trend** Large decline			
Breeding range	>250,000 km²	**Gen. length.** 6		**% Global pop.** 5–24	

Country	Winter pop. size (individuals)	Year(s)	Trend	Mag.%	References
Armenia	1,000 – 2,500	98–02	?	–	
Azerbaijan	500 – 3,000	96–02	+	>80	
Bulgaria	**0 – 24**	97–01	**F**	**>80**	
Cyprus	0 – 80	94–02	?	–	
Greece	**150 – 150**	03	**F**	**>80**	
Macedonia	(0 – 5)	90–00	(F)	(20–29)	
Turkey	9,000 – 12,000	91–01	+	**0–19**	
Ukraine	(50 – 250)	90–00	(0)	(0–19)	
Total (approx.)	>11,000	**Overall trend** Moderate increase			
% in European IBAs	>60	**Gen. length** 6		**% Global pop.** 5–24	

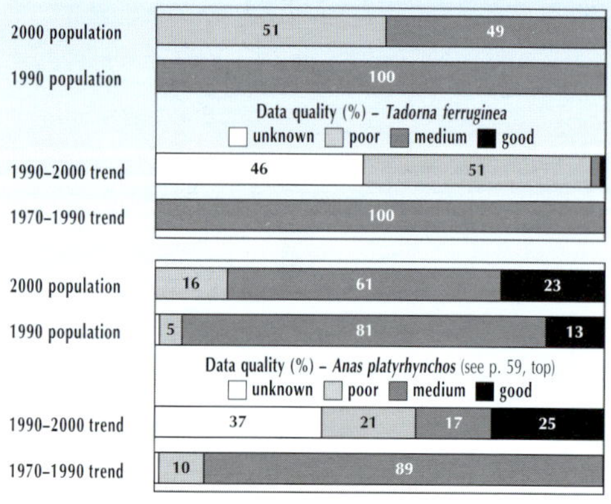

Tadorna tadorna
COMMON SHELDUCK

Non-SPEC (1994: —) **Status** Secure
Criteria —
European IUCN Red List Category —
Criteria —
Global IUCN Red List Category —
Criteria —

Tadorna tadorna is a widespread breeder in coastal areas of north-west and south-east Europe, which accounts for less than half of its global breeding range. Its European breeding population is relatively small (<65,000 pairs), but increased moderately between 1970–1990. Although there were declines in Sweden and the United Kingdom during 1990–2000, other sizeable populations—notably in Germany, Netherlands and Russia—were stable or increased, and the species was stable overall. Consequently, it is evaluated as Secure.

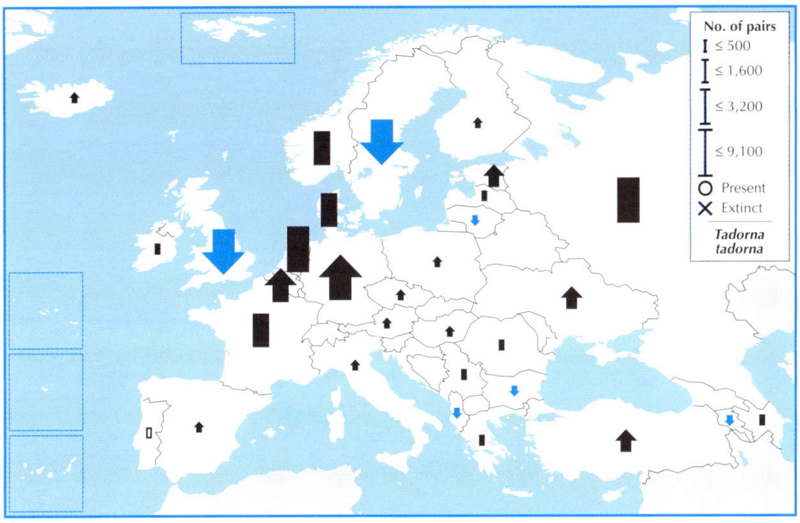

Country	Breeding pop. size (pairs)	Year(s)	Trend	Mag.%	References
Albania	(2 – 5)	96–02	(–)	(0–19)	
Armenia	0 – 50	98–02	(–)	(–)	
Austria	30 – 40	98–02	+	N	
Azerbaijan	(250 – 800)	96–00	(0)	(0–19)	
Belgium	2,100 – 2,800	95–02	+	50–79	1
Bulgaria	20 – 40	95–02	–	20–29	
Czech Rep.	1 – 5	00	+	N	
Denmark	2,000 – 3,000	00	0	0–19	3
Estonia	700 – 1,000	98	+	20–29	1
Finland	200 – 400	99–01	+	150	
France	2,500 – 3,500	00–03	0	0–19	1,11
Germany	5,400 – 6,300	95–99	+	20–29	
Greece	(60 – 80)	95–00	F	30–49	
Hungary	0 – 2	96–02	+	N	
Iceland	0 – 50	00	+	N	38,24,25
Rep. Ireland	250 – 1,000	88–91	(0)	(0–19)	
Italy	99 – 129	00	+	30–49	18
Latvia	50 – 60	90–94	(0)	(0–19)	20
Lithuania	30 – 50	99–01	–	0–19	20
Netherlands	5,000 – 8,000	98–00	0	6	1
Norway	(2,000 – 5,000)	90–03	(0)	(0–19)	
Poland	120 – 140	95–00	+	10–19	19,20,21
Portugal	0 – 2	02	?	N	
Romania	(120 – 160)	90–02	(0)	(0–19)	
Russia	7,500 – 11,000	90–00	0	0–19	6,8,33,60,76,85
Serbia & MN	2 – 3	90–02	0	0–19	1,135a,156
Spain	125 – 150	98–02	+	>80	10,16
Sweden	6,000 – 8,000	99–00	–	10–19	
Turkey	600 – 1,200	01	+	0–19	
Ukraine	1,400 – 1,700	90–00	+	0–9	
UK	5,800 – 10,800	00	(–)	(–)	22,31
Total (approx.)	42,000 – 65,000	Overall trend Stable			
Breeding range	>1,000,000 km²	Gen. length. 6		% Global pop. 25–49	

	Winter pop. size (individuals)		
Total (approx.)	>280,000	Overall trend Small decline	
% in European IBAs	60–63	Gen. length 6	% Global pop. 25–49

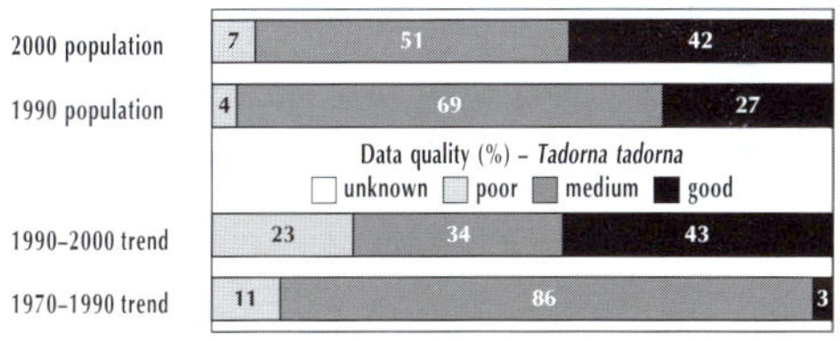

Anas penelope
EURASIAN WIGEON

Non-SPEC[E]W (1994: —) **Status** Secure
Criteria —
European IUCN Red List Category —
Criteria —
Global IUCN Red List Category —
Criteria —

Anas penelope breeds in northern Europe, and winters in coastal areas of the Mediterranean, the Black Sea and north-west Europe, which together probably hold >50% of its global population. Its European wintering population is very large (>1,700,000 individuals), and underwent a moderate increase between 1970–1990. Although there were substantial declines in Spain and Azerbaijan during 1990–2000, wintering populations were stable or increased across the rest of Europe, and the species was probably stable overall. Consequently, it is evaluated as Secure.

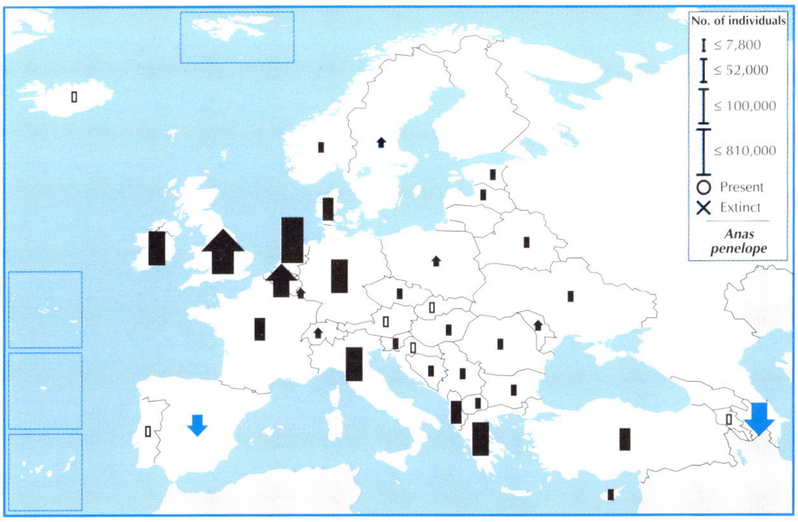

Country	Winter pop. size (individuals)	Year(s)	Trend	Mag.%	References
Albania	11,000 – 52,000	95–02	(F)	(–)	
Armenia	250 – 1,000	97–02	?	–	
Austria	100 – 600	97–99	?	–	
Azerbaijan	50,000 – 200,000	96–02	–	>80	
Belarus	0 – 50	91–00	0	0–19	3
Belgium	45,000 – 107,000	95–00	+	>80	1
Bosnia & HG	(65 – 100)	90–00	(0)	(0–19)	
Bulgaria	200 – 2,800	97–01	F	>80	
Croatia	(1,000 – 5,000)	02	?	–	16,26,70
Cyprus	200 – 800	94–02	0	0–9	
Czech Rep.	10 – 100	90–00	0	0–19	
Denmark	28,000 – 30,000	99–00	0	0–19	25,26
Estonia	0 – 5	98	0	0–19	1
France	37,000 – 60,000	98–02	F	30–49	9
Germany	40,000 – 210,000	95–00	F	50–79	
Greece	50,000 – 80,000	95–99	0	0–19	
Hungary	30 – 300	97–98	F	>80	24,21
Iceland	500 – 2,000	78–94	?	–	48
Rep. Ireland	65,000 – 80,000	94–00	0	0–19	
Italy	40,000 – 100,000	02	0	0–19	22
Latvia	0 – 10	90–99	0	0–19	29
Luxemhourg	10 – 20	00–02	0	0–19	
Macedonia	100 – 900	97–99	(0)	(0–19)	8
Moldova	20 – 200	90–00	+	20–29	
Netherlands	810,000 – 810,000	99–01	0	0–19	3,4,8,9,11,12
Norway	500 – 2,000	93–96	(0)	(0–19)	68,70
Poland	(50 – 250)	90–00	(+)	(0–19)	114
Portugal	3,150 – 19,100	96–01	?	–	18
Romania	500 – 1,500	90–00	0	0–19	60,42
Serbia & MN	1,500 – 5,000	90–02	F	10–29	
Slovakia	1 – 45	95–99	?	–	6,5
Slovenia	550 – 1200	95–00	F	20–29	
Spain	20,000 – 124,000	90–01	–	50–79	19
Sweden	5,000 – 8,000	98–01	+	>80	
Switzerland	900 – 1,550	98–02	+	60–114	
Turkey	45,000 – 60,000	91–01	0	0–19	
Ukraine	2,500 – 10,000	90–00	0	0–19	
UK	426,000 – 426,000	94–99	+	15	48,49
Total (approx.)	>1,700,000	Overall trend Stable			
% in European IBAs	77–93	Gen. length <3.3		% Global pop. 50–74	

(See p. 55, bottom, for data quality graph)

	Breeding pop. size (pairs)		
Total (approx.)	300,000 – 360,000	Overall trend Unknown	
Breeding range	>3,000,000 km²	Gen. length. <3.3	% Global pop. 25–49

Anas strepera
GADWALL

SPEC 3 (1994: 3) **Status** (Depleted)
Criteria Large historical decline
European IUCN Red List Category —
Criteria —
Global IUCN Red List Category —
Criteria —

Anas strepera is a widespread breeder across much of Europe, which accounts for less than a quarter of its global breeding range. Its European breeding population is relatively small (<96,000 pairs), and underwent a large decline between 1970–1990. Although declines continued in a few countries during 1990–2000 and the trend in Russia was unknown, the species was probably stable overall. Nevertheless, it is likely that the its total population size remains below the level that preceded its decline, and consequently it is provisionally evaluated as Depleted.

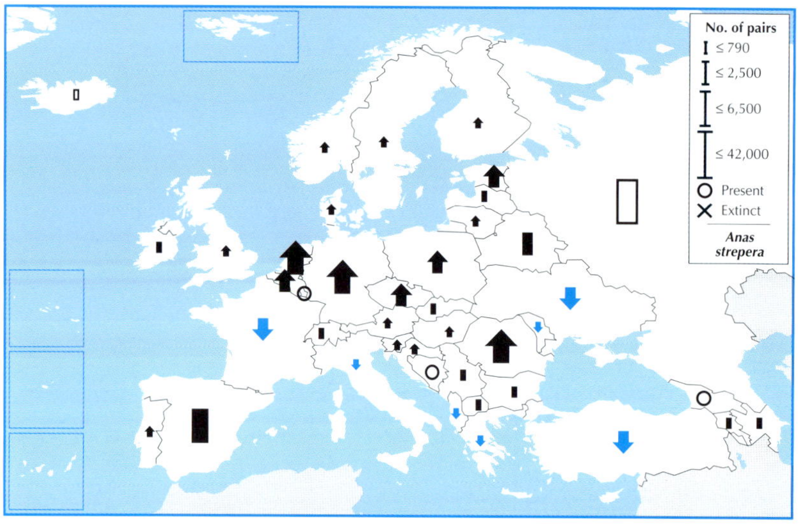

Country	Breeding pop. size (pairs)	Year(s)	Trend	Mag.%	References
Albania	(0 – 5)	96–02	(–)	(0–19)	6
Armenia	250 – 1,000	00–02	0	0–19	12
Austria	**200 – 300**	98–02	+	**50–79**	
Azerbaijan	(100 – 200)	96–00	(0)	(0–19)	
Belarus	**1,000 – 1,500**	97–00	0	0–19	
Belgium	**1,100 – 1,200**	01–02	+	**>80**	1
Bosnia & HG	Present	85–89	?	–	
Bulgaria	**40 – 80**	98–02	F	**20–29**	
Croatia	**40 – 50**	02	(+)	(>80)	70,16
Czech Rep.	**1,800 – 2,500**	00	+	**20–29**	
Denmark	(300 – 500)	00	(+)	(30–49)	3
Estonia	700 – 1,200	98	+	20–29	1
Finland	150 – 300	99–01	+	**100**	
France	900 – 1,000	00	(–)	(10–19)	1
Georgia	Present	03	?	–	
Germany	2,700 – 5,000	95–99	+	20–29	
Greece	(10 – 20)	95–00	–	0–19	
Hungary	(100 – 200)	90–93	(+)	(0–19)	14
Iceland	(200 – 300)	91	?	–	12
Rep. Ireland	30 – 100	90	(0)	(0–19)	
Italy	50 – 100	03	–	0–19	6
Latvia	(100 – 300)	90–00	(0)	(0–19)	23,17
Lithuania	(100 – 150)	99–01	(+)	(0–19)	20
Luxembourg	Present	02	?	–	
Macedonia	5 – 30	90–00	(0)	(0–19)	
Moldova	30 – 60	90–00	–	50–79	
Netherlands	**6,000 – 7,000**	98–00	+	**42**	1
Norway	(1 – 10)	02	+	**0–19**	63
Poland	2,000 – 2,200	95–00	+	30–49	2
Portugal	150 – 250	02	(+)	(–)	
Romania	(3,500 – 5,000)	90–02	(+)	(0–19)	
Russia	32,000 – 55,000	90–00	?	–	8,24,85
Serbia & MN	40 – 60	90–02	0	0–19	1,155,143a,100,13
Slovakia	50 – 80	90–99	0	0–19	
Slovenia	3 – 5	96–00	+	N	
Spain	(2,500 – 3,900)	98–02	(F)	(–)	10
Sweden	400 – 800	99–00	+	**0–19**	
Switzerland	**3 – 10**	93–96	F	**20–29**	
Turkey	(2,000 – 3,000)	01	(–)	(0–19)	
Ukraine	650 – 1,400	90–00	–	20–29	
UK	790 – 790	90	+	0–19	6
Total (approx.)	60,000 – 96,000	**Overall trend** Unknown			
Breeding range	>3,000,000 km²	**Gen. length** <3.3		**% Global pop.** 5–24	

(See p. 56, top, for data quality graph)

Winter pop. size (individuals)

Total (approx.)	>90,000	**Overall trend** Large increase	
% in European IBAs	50–51	**Gen. length** <3.3	**% Global pop.** 5–24

Anas crecca
COMMON TEAL

Non-SPEC (1994: —) **Status** (Secure)
Criteria —
European IUCN Red List Category —
Criteria —
Global IUCN Red List Category —
Criteria —

Anas crecca is a widespread breeder across much of Europe (especially in the north-east), which accounts for less than a quarter of its global breeding range. Its European breeding population is large (>920,000 pairs), and was stable between 1970–1990. Although the species declined in a number of countries—notably Finland—during 1990–2000, and the trend of the stronghold population in Russia was unknown, it remained stable in many other countries and probably declined only slightly overall. Consequently, the species is provisionally evaluated as Secure.

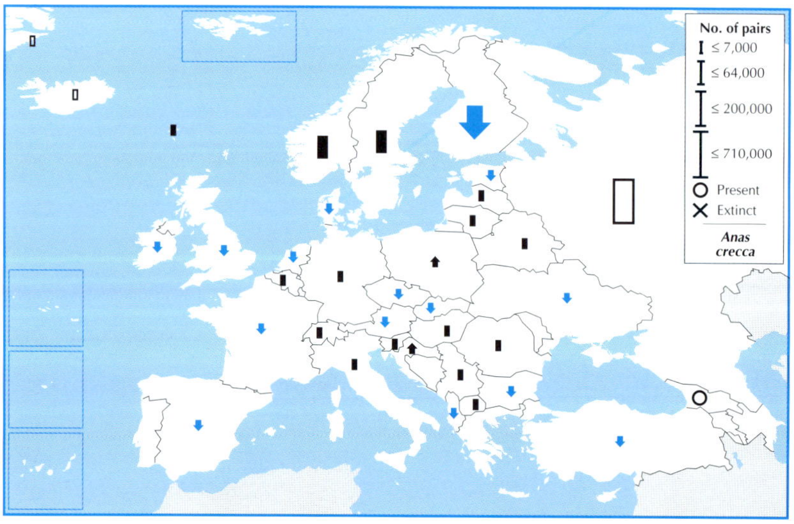

Country	Breeding pop. size (pairs)	Year(s)	Trend	Mag.%	References
Albania	2 – 7	96–02	(–)	(0–19)	6
Austria	(70 – 120)	98–02	–	50–79	
Belarus	**6,000 – 8,000**	97–00	**0**	**0–19**	
Belgium	**500 – 700**	95–02	0	0–19	1
Bulgaria	**2 – 15**	98–02	–	0–19	
Croatia	(50 – 100)	02	(+)	(>80)	70,16
Czech Rep.	**60 – 100**	00	–	**50–79**	
Denmark	(200 – 300)	93–96	(–)	(–)	3
Estonia	2,500 – 3,000	98	–	20–29	1
Denmark					
Faroe Is.	0 – 2	83	(0)	(0–19)	
Greenland	(0 – 10)	90–00	?	–	
Finland	150,000 – 250,000	99–01	–	**10**	
France	(200 – 500)	95–00	(–)	(10–19)	1
Georgia	Present	03	?	–	
Germany	3,700 – 5,800	95–99	0	0–19	
Hungary	5 – 15	95–02	(0)	(0–19)	
Iceland	(3,000 – 5,000)	78–94	?	–	1
Rep. Ireland	250 – 1,000	88–91	(–)	(30–49)	
Italy	20 – 50	03	0	0–19	6
Latvia	2,000 – 5,000	90–00	(0)	(0–19)	17
Lithuania	(2,000 – 3,000)	99–01	0	0–19	20
Macedonia	(1 – 10)	90–00	(0)	(0–19)	
Netherlands	**2,000 – 2,500**	98–00	–	**33**	1
Norway	(30,000 – 50,000)	90–03	(0)	(0–19)	
Svalbard	0 – 5	95–03	0	0–19	
Poland	1,300 – 1,700	95–00	(+)	(0–19)	1
Romania	(2 – 20)	90–02	(F)	(50–79)	
Russia	665,000 – 740,000	90–00	?	–	85
Serbia & MN	10 – 20	90–02	0	0–19	1,143a,155,101,100
Slovakia	10 – 40	80–99	–	30–49	
Slovenia	(0 – 10)	90–00	(F)	(>80)	
Spain	7 – 42	98–01	–	0–19	10,16
Sweden	50,000 – 80,000	99–00	(0)	(0–19)	
Switzerland	**0 – 5**	93–96	**0**	**0–19**	
Turkey	(600 – 900)	01	(–)	(0–19)	
Ukraine	(1,500 – 2,500)	90–00	(–)	(0–19)	
UK	1,600 – 2,800	88–91	(–)	(30–49)	8
Total (approx.)	920,000 – 1,200,000	**Overall trend** Unknown			
Breeding range	>6,000,000 km²	**Gen. length** <3.3		**% Global pop.** 5–24	

(See p. 56, top, for data quality graph)

Winter pop. size (individuals)

Total (approx.)	>730,000	**Overall trend** Small decline	
% in European IBAs	70–93	**Gen. length** <3.3	**% Global pop.** 5–24

Anas platyrhynchos
MALLARD

Non-SPEC (1994: —) **Status** (Secure)
Criteria —
European IUCN Red List Category —
Criteria —
Global IUCN Red List Category —
Criteria —

Anas platyrhynchos is a widespread breeder across most of Europe, which accounts for less than a quarter of its global breeding range. Its European breeding population is very large (>3,300,000 pairs), and was stable between 1970–1990. Although the species declined in several countries during 1990–2000 (and the Russian trend was unknown), key populations in the Netherlands, Germany and Poland were stable, and the species probably declined only slightly overall. Consequently, it is provisionally evaluated as Secure.

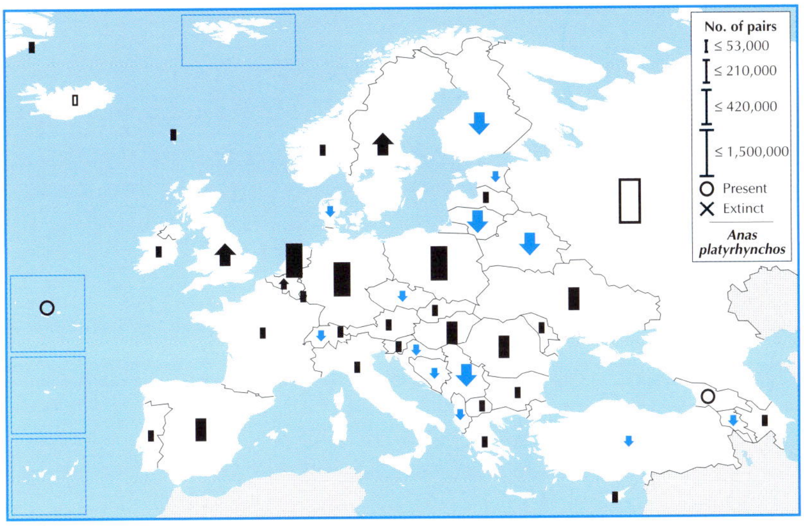

Country	Breeding pop. size (pairs)	Year(s)	Trend	Mag.%	References
Albania	50 – 200	96–02	(–)	(0–19)	*6*
Armenia	2,500 – 10,000	99–02	–	10–19	
Austria	(10,000 – 20,000)	98–02	(0)	(0–19)	
Azerbaijan	(1,000 – 5,000)	96–00	(0)	(0–19)	
Belarus	**60,000 – 100,000**	97–00	**–**	**20–29**	
Belgium	25,000 – 35,000	95–02	+	0–19	*1*
Bosnia & HG	(200 – 250)	90–00	(–)	(10–19)	
Bulgaria	1,500 – 4,000	96–02	0	0–9	
Croatia	(10,000 – 35,000)	02	(–)	(30–49)	*54*
Cyprus	20 – 60	98–00	(0)	(0–19)	
Czech Rep.	**25,000 – 45,000**	00	**–**	**20–29**	
Denmark	(20,000 – 50,000)	00	(–)	(30)	
Faroe Is.	300 – 300	95	(0)	(0–19)	
Greenland	(5,000 – 10,000)	90–00	(F)	(–)	
Estonia	25,000 – 30,000	98	–	20–29	*1*
Finland	150,000 – 250,000	99–01	**–**	**10**	
France	(30,000 – 60,000)	00	(0)	(0–19)	*1*
Georgia	Present	03	?	–	
Germany	210,000 – 470,000	95–99	0	0–19	
Greece	(100 – 1,000)	95–00	0	0–19	
Hungary	(100,000 – 150,000)	90–93	(0)	(0–19)	*14*
Iceland	(10,000 – 15,000)	78–94	?	–	*1*
Rep. Ireland	10,000 – 20,000	88–91	0	0–19	
Italy	(10,000 – 20,000)	03	(0)	(0–19)	*6*
Latvia	30,000 – 40,000	90–00	0	0–19	*23,17*
Liechtenstein	5 – 10	98–00	**0**	**0–19**	
Lithuania	(70,000 – 100,000)	99–01	(–)	(20–29)	*20*
Luxembourg	**1,000 – 1,500**	01–02	**0**	**0–19**	
Macedonia	650 – 1,000	90–00	(0)	(0–19)	
Moldova	3200 – 4,000	90–00	0	0–19	
Netherlands	**350,000 – 500,000**	98–00	**0**	**8**	*1*
Norway	(40,000 – 70,000)	90–03	0	0–19	
Poland	**200,000 – 400,000**	00	(0)	(0–19)	*23*
Portugal	(3,000 – 10,000)	02	(0)	(0–19)	
Azores	Present	02	?	–	
Romania	(120,000 – 200,000)	90–02	(0)	(0–19)	
Russia	1,300,000 – 1,700,000	90–00	?	–	*85*
Serbia & MN	75,000 – 95,000	90–02	–	10–29	*1,143a,172a, 67a,13,101*
Slovakia	12,000 – 20,000	90–99	0	0–19	
Slovenia	10,000 – 20,000	94	(0)	(0–19)	
Spain	(65,500 – 100,000)	92	(F)	(–)	*13,12,10*
Sweden	175,000 – 250,000	99–00	+	**50–79**	
Switzerland	**10,000 – 20,000**	93–96	–	0–9	
Turkey	(3,000 – 6,000)	01	(–)	(0–19)	
Ukraine	72,000 – 112,000	90–00	0	0–19	
UK	**63,000 – 158,900**	00	+	**25**	*5,31*

Total (approx.) 3,300,000 – 5,100,000 **Overall trend** Small decline
Breeding range >8,000,000 km² **Gen. length.** <3.3 **% Global pop.** 5–24
(See p. 56, bottom, for data quality graph)

Winter pop. size (individuals)

Total (approx.) >3,700,000 **Overall trend** Small decline
% in European IBAs 21–26 **Gen. length** <3.3 **% Global pop.** 5–24

Anas acuta
NORTHERN PINTAIL

SPEC 3 (1994: 3) **Status** (Declining)
Criteria Moderate continuing decline
European IUCN Red List Category —
Criteria —
Global IUCN Red List Category —
Criteria —

Anas acuta is a widespread breeder in much of northern and parts of central Europe, which accounts for less than a quarter of its global breeding range. Its European breeding population is large (>320,000 pairs), but underwent a large decline between 1970–1990. Although it was stable or increased across much of its European range during 1990–2000, the stronghold population in Russia continued to decline, and the species underwent a moderate decline (>10%) overall. Consequently, it is provisionally evaluated as Declining.

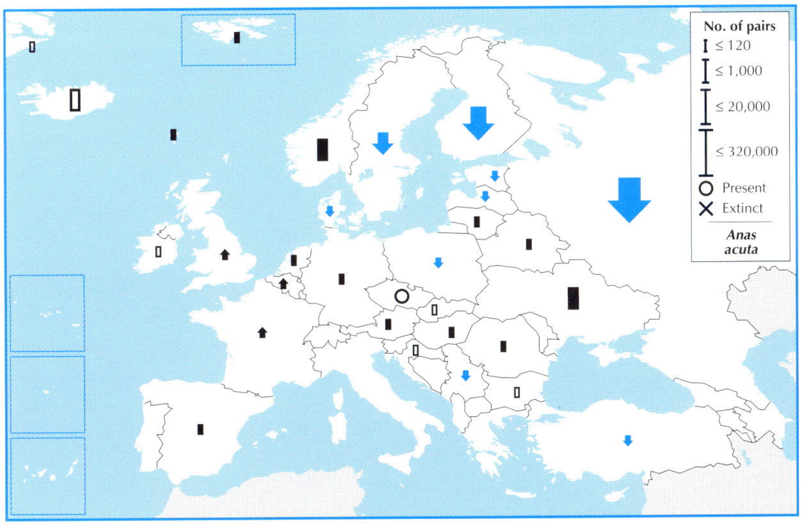

Country	Breeding pop. size (pairs)	Year(s)	Trend	Mag.%	References
Austria	**1 – 5**	98–02	**0**	**0–19**	
Belarus	**70 – 150**	97–00	0	0–19	
Belgium	**9 – 12**	00–02	+	0–19	*1*
Bulgaria	0 – 5	95–02	?	–	
Croatia	(1 – 5)	02	?	–	*70,16,44*
Czech Rep.	Present	00	?	–	
Denmark	(50 – 100)	00	–	20–29	*3*
Faroe Is.	(0 – 1)	81	(0)	(0–19)	
Greenland	(0 – 10)	90–00	?	–	
Estonia	50 – 100	98	–	20–29	*1*
Finland	15,000 – 25,000	99–01	**–**	**15**	
France	**0 – 7**	94–00	+	**>80**	*5*
Germany	**15 – 24**	95–99	**0**	**0–19**	
Hungary	(30 – 50)	90–93	(0)	(0–19)	*14*
Iceland	(400 – 600)	91	?	–	*12*
Rep. Ireland	2 – 2	88–91	?	–	
Latvia	(0 – 20)	90–00	(–)	(30–49)	*17*
Lithuania	(5 – 20)	99–01	(F)	(30–49)	*20*
Netherlands	**20 – 30**	98–00	F	22	*1*
Norway	(500 – 2,000)	90–03	(0)	(0–19)	
Svalbard	0 – 5	95–03	0	0–19	
Poland	**10 – 20**	95–00	**–**	**60–79**	*24*
Romania	(0 – 5)	90–02	(F)	(50–79)	*37*
Russia	300,000 – 325,000	90–00	(–)	(20–29)	*85*
Serbia & MN	8 – 15	90–02	–	20–29	*1,29,155,13, 26,101*
Slovakia	0 – 10	80–99	?	–	
Spain	0 – 50	98–02	(F)	(–)	*10,16*
Sweden	(600 – 1,100)	99–00	(–)	(20–29)	
Turkey	80 – 160	01	(–)	(0–19)	
Ukraine	300 – 900	90–00	F	20–29	
UK	**12 – 40**	96–00	+	**72**	

Total (approx.) 320,000 – 360,000 **Overall trend** Moderate decline
Breeding range >1,000,000 km² **Gen. length.** <3.3 **% Global pop.** 5–24

Winter pop. size (individuals)

Total (approx.) >120,000 **Overall trend** Moderate decline
% in European IBAs >80 **Gen. length** <3.3 **% Global pop.** 5–24

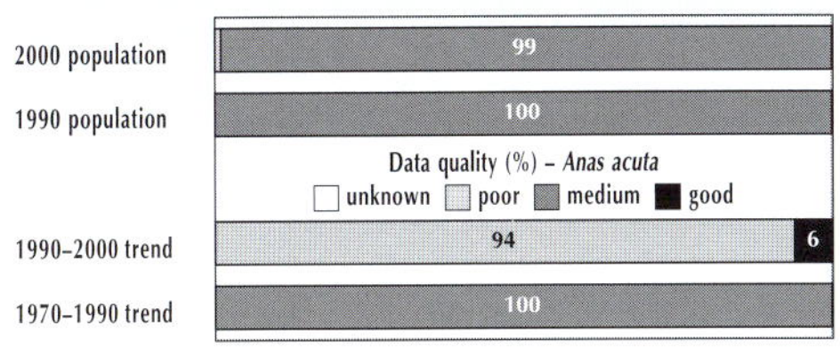

Anas querquedula
GARGANEY

SPEC 3 (1994: 3) **Status** (Declining)
Criteria Moderate continuing decline
European IUCN Red List Category —
Criteria —
Global IUCN Red List Category —
Criteria —

Anas querquedula is a widespread summer visitor to much of Europe, which accounts for less than half of its global breeding range. Its European breeding population is large (>390,000 pairs), but underwent a large decline between 1970–1990. Although the trend of the stronghold population in Russia during 1990–2000 was unknown, the species continued to decline across most of its European range—including sizeable populations in Belarus and Ukraine—and probably underwent a moderate decline (>10%) overall. Consequently, it is provisionally evaluated as Declining.

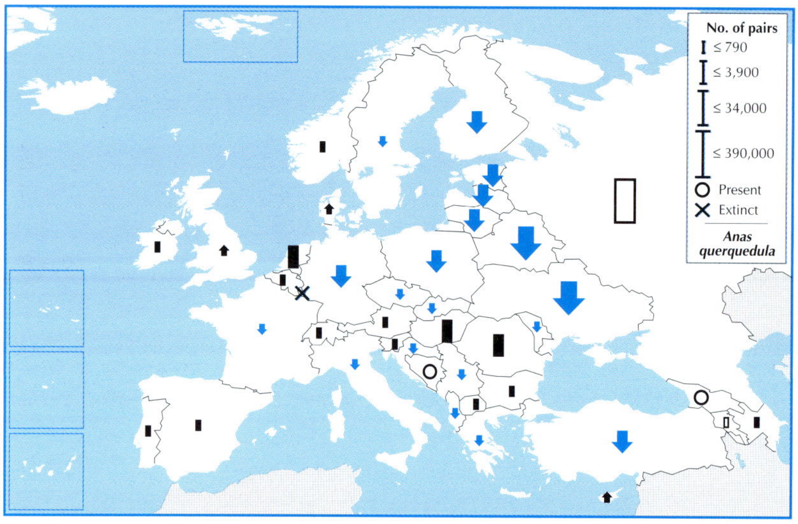

Country	Breeding pop. size (pairs)	Year(s)	Trend	Mag.%	References
Albania	20 – 60	96–02	(–)	(0–19)	6
Armenia	20 – 80	97–02	?	–	
Austria	**70 – 200**	98–02	**0**	**0–19**	
Azerbaijan	(0 – 200)	96–00	(0)	(0–19)	
Belarus	**25,000 – 45,000**	97–02	**–**	**10–19**	
Belgium	**200 – 200**	95–02	**0**	**0–19**	*1*
Bosnia & HG	Present	90–03	?	–	
Bulgaria	**80 – 250**	96–02	0	0–19	
Croatia	(50 – 500)	02	(–)	(50–79)	*70,26*
Cyprus	**3 – 3**	02	**+**	**N**	
Czech Rep.	**60 – 120**	00	**–**	**30–49**	
Denmark	300 – 500	02	(+)	(20–29)	*16,3*
Estonia	2,000 – 2,500	98	–	20–29	*1*
Finland	1,000 – 4,000	99–01	**–**	**60**	
France	(270 – 500)	98–02	(–)	(10–19)	*4*
Georgia	Present	03	?	–	
Germany	1,200 – 1,900	95–99	–	20–29	
Greece	(0 – 10)	95–00	–	0–19	
Hungary	(1,200 – 1,500)	90–93	(0)	(0–19)	*14*
Rep. Ireland	2 – 2	88–91	(F)	(–)	
Italy	350 – 500	03	–	10–19	*6*
Latvia	1,000 – 2,000	90–00	(–)	(0–19)	*23,17*
Lithuania	**2,000 – 3,000**	99–01	(–)	(30–49)	*20*
Luxembourg	**0 – 0**	02	**–**	**X**	
Macedonia	(1 – 10)	90–00	(0)	(0–19)	
Moldova	150 – 200	90–00	–	20–29	
Netherlands	**1,600 – 1,900**	98–00	**0**	**14**	*1*
Norway	5 – 20	02	**F**	**30–49**	*63*
Poland	2,000 – 3,500	95–00	(–)	(0–19)	1
Portugal	0 – 5	02	0	0–19	
Romania	(3,000 – 5,000)	90–02	(0)	(0–19)	
Russia	320,000 – 475,000	90–00	?	–	*85*
Serbia & MN	650 – 950	90–02	–	0–19	*1,143a,155,100, 13,101,100,96,78*
Slovakia	100 – 200	80–99	–	30–49	
Slovenia	20 – 30	90–00	(F)	(>80)	
Spain	50 – 250	98–02	(F)	(–)	*10,16*
Sweden	250 – 400	99–00	–	20–29	
Switzerland	**0 – 2**	93–96	**0**	**0–19**	
Turkey	(1,500 – 2,500)	01	(–)	(0–19)	
Ukraine	22,500 – 35,400	90–00	–	0–19	
UK	**23 – 114**	96–00	**+**	**99**	

Total (approx.) 390,000 – 590,000 — **Overall trend** Unknown
Breeding range >5,000,000 km² — **Gen. length**. <3.3 **% Global pop.** 25–49

(See p. 61, top, for data quality graph)

Anas clypeata
NORTHERN SHOVELER

SPEC 3 (1994: —) **Status** (Declining)
Criteria Moderate recent decline
European IUCN Red List Category —
Criteria —
Global IUCN Red List Category —
Criteria —

Anas clypeata is widespread breeder across much of Europe, which accounts for less than a quarter of its global breeding range. Its European breeding population is large (>170,000 pairs), and was stable between 1970–1990. Although no trend data were available for the stronghold in Russia during 1990–2000, several countries—notably the Netherlands—suffered marked declines, and the species probably underwent a moderate decline (>10%) overall. Consequently, this previously Secure species is now provisionally evaluated as Declining.

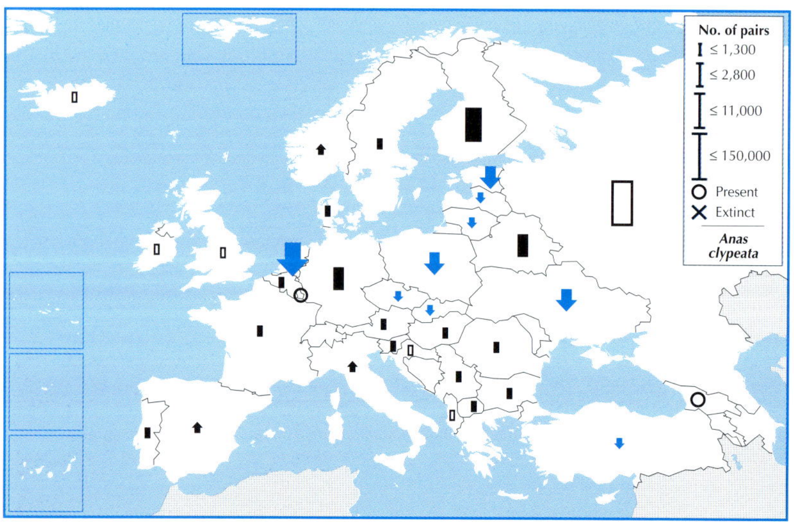

Country	Breeding pop. size (pairs)	Year(s)	Trend	Mag.%	References
Albania	(0 – 5)	96–02	?	–	*6*
Austria	**160 – 300**	98–02	**0**	**0–19**	
Belarus	**1,000 – 6,400**	97–00	0	0–19	
Belgium	**800 – 1,100**	95–02	**0**	**0–19**	*1*
Bulgaria	**12 – 25**	98–02	0	0–9	
Croatia	(50 – 100)	02	?	–	*70,16*
Czech Rep.	**70 – 120**	00	**–**	**50–79**	
Denmark	800 – 1,000	02	0	0–9	*21*
Estonia	2,500 – 3,000	98	–	20–29	*1*
Finland	10,000 – 12,000	99–01	**0**	**0–19**	
France	700 – 1,300	00	0	0–19	*1*
Georgia	Present	03	?	–	
Germany	2,100 – 3,300	95–99	**0**	**0–19**	
Hungary	(500 – 600)	90–93	(0)	(0–19)	*14*
Iceland	(0 – 50)	00	?	–	*21*
Rep. Ireland	20 – 100	88–91	?	–	
Italy	150 – 200	03	+	10–19	*6*
Latvia	200 – 400	90–00	–	30–49	*17*
Lithuania	**150 – 300**	99–01	(–)	(20–29)	*20*
Luxembourg	Present	02	?	–	
Macedonia	(1 – 5)	90–00	(0)	(0–19)	
Netherlands	**8,000 – 9,000**	98–00	**–**	**24**	*1*
Norway	(100 – 300)	90–03	+	0–19	
Poland	1,300 – 2,000	95–00	(–)	(0–19)	*1,2*
Portugal	0 – 5	02	(0)	(0–19)	
Romania	(250 – 350)	90–02	(0)	(0–19)	
Russia	(140,000 – 160,000)	90–00	?	–	*85*
Serbia & MN	50 – 70	90–02	0	0–19	*1,29,155,13,82, 26,100*
Slovakia	10 – 40	80–99	–	30–49	
Slovenia	(0 – 10)	90–00	(F)	(>80)	
Spain	(50 – 250)	98–02	(+)	(0–19)	*10,16*
Sweden	1,000 – 1,500	99–00	0	0–19	
Turkey	40 – 80	01	(–)	(30–49)	
Ukraine	1,900 – 3,100	90–00	–	0–19	
UK	1,000 – 1,500	88–91	?	–	*15*

Total (approx.) 170,000 – 210,000 — **Overall trend** Unknown
Breeding range >5,000,000 km² — **Gen. length**. <3.3 **% Global pop.** 5–24

(See p. 61, top, for data quality graph)

Winter pop. size (individuals)

Total (approx.) >200,000 — **Overall trend** Moderate decline
% in European IBAs 83–93 — **Gen. length** <3.3 **% Global pop.** 5–24

Marmaronetta angustirostris
MARBLED TEAL

SPEC 1 (1994: 1) **Status** (Vulnerable)
Criteria See IUCN below
European IUCN Red List Category VU
Criteria C1; C2a(i); D1
Global IUCN Red List Category VU
Criteria A2c,d; A3c,d

Marmaronetta angustirostris breeds locally in southern Europe, which accounts for less than a quarter of its global breeding range. Its European breeding population is very small (as few as 390 pairs), and underwent a large decline between 1970–1990. Although most populations were stable or increased during 1990–2000, the large Turkish population declined, and the species probably underwent a moderate decline (>10%) overall. As a consequence of this decline and its very small population, this globally threatened species is provisionally evaluated as Vulnerable in Europe.

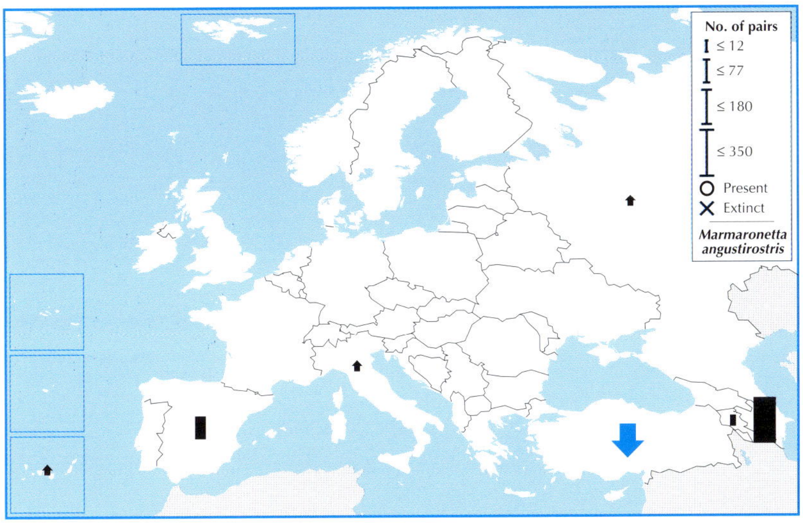

Country	Breeding pop. size (pairs)	Year(s)	Trend	Mag.%	References
Armenia	5 – 30	99–02	0	0–19	*1*
Azerbaijan	(200 – 600)	96–00	(0)	(0–19)	
Italy	**1 – 2**	00–01	+	N	*6*
Russia	1 – 10	95–00	(+)	(20–29)	*33,60,74*
Spain	**30 – 200**	94–01	(F)	(–)	*10,16*
Canary Is.	**1 – 4**	97–03	+	**N**	*28,25*
Turkey	150 – 200	01	–	20–29	
Total (approx.)	390 – 1,000	**Overall trend** Moderate decline			
Breeding range	>50,000 km²	**Gen. length.** <3.3 **% Global pop.** 5–24			

Country	Winter pop. size (individuals)	Year(s)	Trend	Mag.%	References
Azerbaijan	400 – 1,000	96–02	+	>80	
Spain	**2 – 221**	90–01	+	**>80**	*19*
Turkey	0 – 20	91–01	–	**50–79**	
Total (approx.)	>400	**Overall trend** Large increase			
% in European IBAs	>70	**Gen. length** <3.3 **% Global pop.** <5			

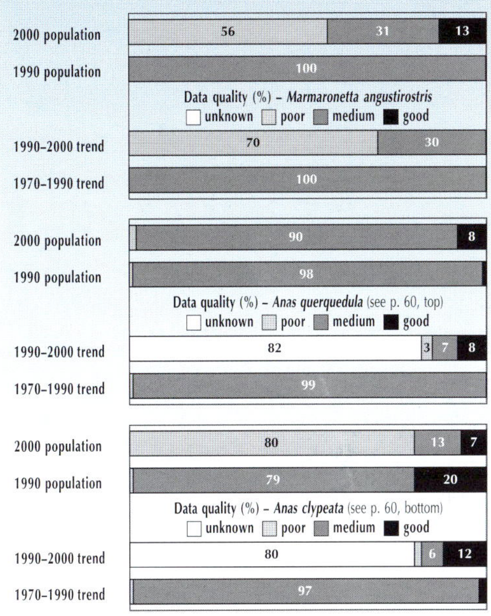

Netta rufina
RED-CRESTED POCHARD

Non-SPEC (1994: 3) **Status** (Secure)
Criteria —
European IUCN Red List Category —
Criteria —
Global IUCN Red List Category —
Criteria —

Netta rufina is a widespread but patchily distributed breeder in west-central and southern Europe, which accounts for less than half of its global breeding range. Its European breeding population is relatively small (<59,000 pairs), and declined markedly between 1970–1990. Although the species declined in a few countries during 1990–2000, many populations (including the Russian stronghold) increased or were stable, and it underwent a moderate increase overall. This increase probably outweighs the earlier decline, and consequently the species is provisionally evaluated as Secure.

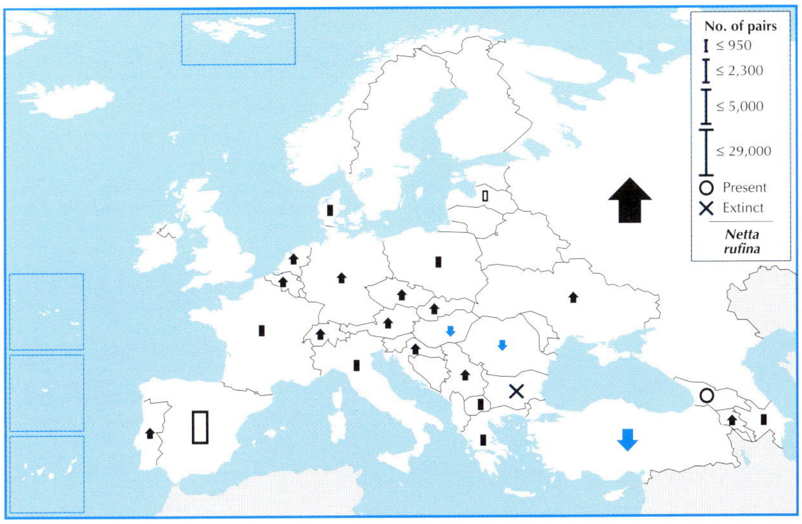

Country	Breeding pop. size (pairs)	Year(s)	Trend	Mag.%	References
Armenia	80 – 330	97–02	+	0–19	
Austria	**150 – 250**	98–02	+	**50–79**	
Azerbaijan	(300 – 3,000)	96–00	(0)	(0–19)	
Belgium	**1 – 1**	01–02	+	**N**	*1*
Bulgaria	**0 – 0**	96–02	–	**X**	
Croatia	10 – 20	02	+	>80	*70,44*
Czech Rep.	**200 – 250**	00	+	**20–29**	
Denmark	**0 – 1**	98–01	**0**	**0–19**	*4,5,6,7*
France	650 – 850	00–02	(0)	(0–19)	*4*
Georgia	Present	03	?	–	
Germany	**420 – 540**	95–99	+	**20–29**	
Greece	(0 – 5)	95–00	(F)	(>80)	
Hungary	20 – 50	95–02	(–)	(0–19)	
Italy	40 – 60	03	0	0–19	*6*
Latvia	(0 – 5)	02	?	–	*22,26*
Macedonia	3 – 9	90–00	(0)	(0–19)	
Netherlands	**120 – 170**	98–00	+	**1279**	*1*
Poland	**15 – 20**	95–00	0	0–19	*2*
Portugal	(40 – 100)	02	(+)	(–)	
Romania	(500 – 600)	00–02	(–)	(0–19)	
Russia	20,000 – 40,000	90–00	+	20–29	*8,85*
Serbia & MN	**0 – 2**	90–02	+	**N**	*1,182,29,155*
Slovakia	**10 – 30**	80–99	+	20–29	
Spain	(2,500 – 10,000)	98–02	?	–	*10,16*
Switzerland	**50 – 100**	98–02	+	**>80**	
Turkey	1,800 – 2,800	01	–	30–49	
Ukraine	**150 – 200**	90–00	+	**0–9**	
Total (approx.)	27,000 – 59,000	**Overall trend** Moderate increase			
Breeding range	>750,000 km²	**Gen. length.** <3.3 **% Global pop.** 25–49			

Country	Winter pop. size (individuals)	Year(s)	Trend	Mag.%	References
Armenia	2,500 – 10,000	97–02	0	10–19	
Azerbaijan	**50,000 – 250,000**	96–02	+	**30–49**	
France	**1,500 – 3,000**	98–02	**F**	**50–79**	*9*
Germany	**3,500 – 13,000**	98–02	+	**>80**	
Spain	**7,900 – 19,000**	90–01	**0**	**0–19**	*19*
Switzerland	**6,700 – 16,000**	99–01	+	**167–303**	
Turkey	6,000 – 12,000	91–01	0	0–19	
Ukraine	(2,500 – 10,000)	90–00	(0)	(0–19)	
Total (approx.)	>84,000	**Overall trend** Moderate increase			
% in European IBAs	>90	**Gen. length** <3.3 **% Global pop.** 25–49			

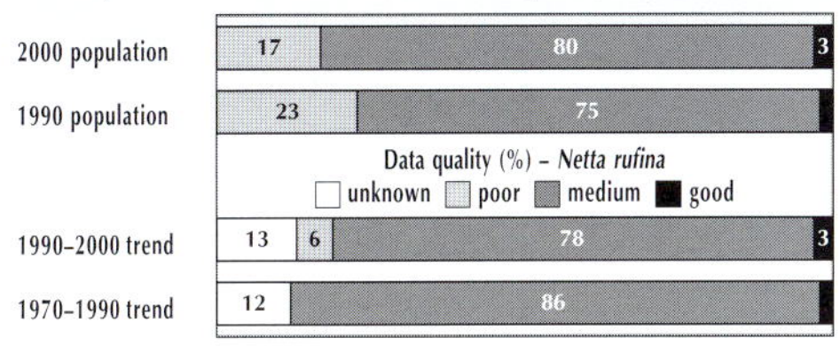

Aythya ferina
COMMON POCHARD

SPEC 2 (1994: 4) **Status** (Declining)
Criteria Moderate recent decline
European IUCN Red List Category —
Criteria —
Global IUCN Red List Category —
Criteria —

Aythya ferina is a widespread breeder across much of Europe, which probably constitutes >50% of its global breeding range. Its European breeding population is large (>210,000 pairs), and was stable between 1970–1990. Although populations were stable or increased across much of Europe during 1990–2000, there were declines in several countries—including the key Russian population—and the species underwent a moderate decline (>10%) overall. Consequently, this previously Secure species is now provisionally evaluated as Declining.

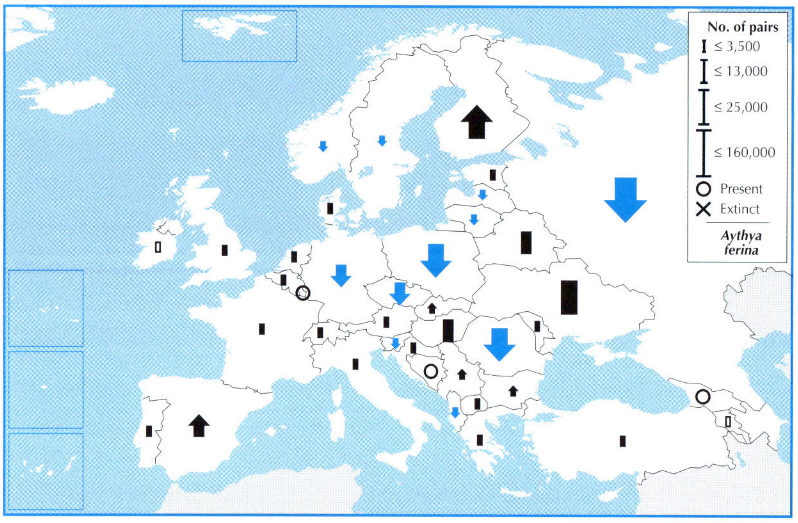

Country	Breeding pop. size (pairs)	Year(s)	Trend	Mag.%	References
Albania	0 – 10	96–00	(–)	(0–19)	*6*
Armenia	80 – 330	97–02	?	–	
Austria	**200 – 300**	98–02	**0**	**0–19**	
Belarus	**6,000 – 8,000**	97–02	**0**	**0–19**	
Belgium	653 – 916	95–02	0	0–19	*1*
Bosnia & HG	Present	85–89	?	–	
Bulgaria	**90 – 170**	98–02	**+**	**50–79**	
Croatia	(1,000 – 5,000)	02	(0)	(0–19)	*16*
Czech Rep.	**9,000 – 17,000**	00	**–**	**10–19**	
Denmark	(400 – 600)	00	0	0–19	*3*
Estonia	1,000 – 2,000	98	0	0–19	1
Finland	15,000 – 20,000	99–01	**+**	**25**	
France	2,000 – 3,500	98–02	0	0–19	*4*
Georgia	Present	03	?	–	
Germany	4,500 – 7,500	95–99	–	0–19	
Greece	(0 – 20)	95–00	(F)	(30–49)	
Hungary	(5,000 – 10,000)	90–93	(0)	(0–19)	*14*
Rep. Ireland	0 – 50	88–91	?	–	
Italy	300 – 400	03	0	0–19	*6*
Latvia	1,500 – 2,000	90–00	–	30–49	*17*
Lithuania	3,000 – 4,000	99–01	(–)	(20–29)	*20*
Luxembourg	Present	02	?	–	
Macedonia	15 – 60	90–00	(0)	(0–19)	
Moldova	900 – 1,300	90–00	**0**	**0–19**	
Netherlands	**1,700 – 2,100**	98–00	**0**	**5**	*1*
Norway	0 – 10	99	–	0–19	*63*
Poland	20,000 – 30,000	90–00	(–)	(0–19)	*2*
Portugal	20 – 50	02	0	0–19	
Romania	(15,000 – 20,000)	95–02	(–)	(0–19)	
Russia	95,000 – 265,000	90–00	(–)	(20–29)	*10*
Serbia & MN	800 – 1,000	90–02	+	10–19	*1,82,29,155,13, 143a,204*
Slovakia	500 – 1,000	80–99	+	20–29	
Slovenia	10 – 40	90–00	–	30–49	
Spain	2,500 – 10,000	98–02	(+)	(0–19)	*10*
Sweden	**1,000 – 1,700**	99–00	**–**	**50–79**	
Switzerland	**3 – 10**	93–96	**0**	**0–19**	
Turkey	2,500 – 3,500	01	(0)	(0–19)	
Ukraine	17,300 – 25,900	90–00	0	0–19	
UK	**324 – 426**	96–00	**F**	**69**	
Total (approx.)	210,000 – 440,000	**Overall trend** Moderate decline			
Breeding range	>5,000,000 km²	**Gen. length.** <3.3 **% Global pop.** 50–74			

(See p. 64, bottom, for data quality graph)

	Winter pop. size (individuals)	
Total (approx.)	>790,000	**Overall trend** Moderate decline
% in European IBAs	>80	**Gen. length** <3.3 **% Global pop.** 50–74

Aythya nyroca
FERRUGINOUS DUCK

SPEC 1 (1994: 1) **Status** (Vulnerable)
Criteria See IUCN below
European IUCN Red List Category VU
Criteria A2b
Global IUCN Red List Category NT
Criteria A2c,d; A3c,d

Aythya nyroca is a widespread but patchily distributed breeder across much of Europe, which accounts for less than half of its global breeding range. Its European breeding population is relatively small (<18,000 pairs), and declined substantially between 1970–1990. Although the species was stable or increased in many countries during 1990–2000, several populations—notably key ones in Romania and Croatia—declined, and the species underwent a large decline (>30%) overall. Consequently, this globally Near Threatened species is provisionally evaluated as Vulnerable in Europe.

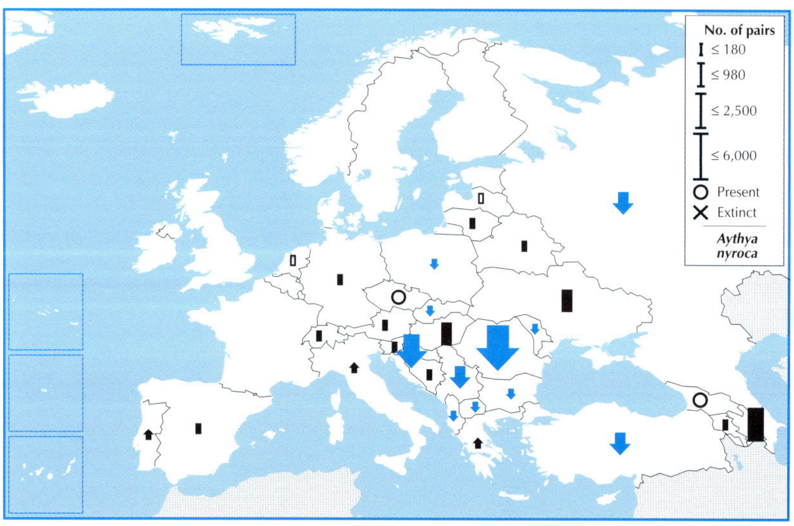

Country	Breeding pop. size (pairs)	Year(s)	Trend	Mag.%	References
Albania	10 – 30	96–02	(–)	(0–19)	
Armenia	**5 – 30**	97–02	**0**	**0–19**	
Austria	50 – 150	98–02	0	0–19	
Azerbaijan	1,000 – 3,000	96–00	0	0–19	
Belarus	50 – 200	97–01	0	0–19	
Bosnia & HG	(8 – 10)	90–00	(0)	(0–19)	
Bulgaria	**125 – 225**	98–02	**–**	**5–19**	
Croatia	2,000 – 3,000	02	–	50–79	*16,26,44*
Czech Rep.	Present	00	?	–	
Georgia	Present	03	?	–	
Germany	**0 – 3**	95–99	**0**	**0–19**	
Greece	**130 – 250**	02	**+**	**20–29**	
Hungary	**550 – 1,000**	97–02	**0**	**0–19**	
Italy	**70 – 100**	03	+	0–19	*6*
Latvia	(0 – 5)	90–00	?	–	*23,7,4,17,19*
Lithuania	10 – 30	99–01	F	20–29	*20*
Macedonia	20 – 50	90–00	(–)	(20–29)	
Moldova	70 – 150	90–00	–	50–79	
Netherlands	**0 – 3**	98–00	?	–	*1*
Poland	**30 – 40**	00–03	–	80–89	*25,26*
Portugal	**0 – 2**	02	**+**	**N**	*7*
Romania	(5,500 – 6,500)	96–02	(–)	(0–19)	*55*
Russia	500 – 1,150	90–00	(–)	(–)	*33,74,129,172, 175,178*
Serbia & MN	**420 – 570**	95–02	–	0–19	*164a,29,230, 155,13,143a*
Slovakia	**5 – 20**	80–99	–	30–49	
Slovenia	(0 – 10)	90–00	(F)	(>80)	
Spain	1 – 10	98–02	(F)	(–)	*10,16*
Switzerland	**0 – 1**	90–00	**0**	**0–19**	
Turkey	800 – 1,200	01	–	20–29	
Ukraine	300 – 600	90–00	0	10–19	
Total (approx.)	12,000 – 18,000	**Overall trend** Large decline			
Breeding range	>1,000,000 km²	**Gen. length.** <3.3 **% Global pop.** 25–49			

	Winter pop. size (individuals)	
Total (approx.)	>3,400	**Overall trend** Stable
% in European IBAs	>80	**Gen. length** <3.3 **% Global pop.** 5–24

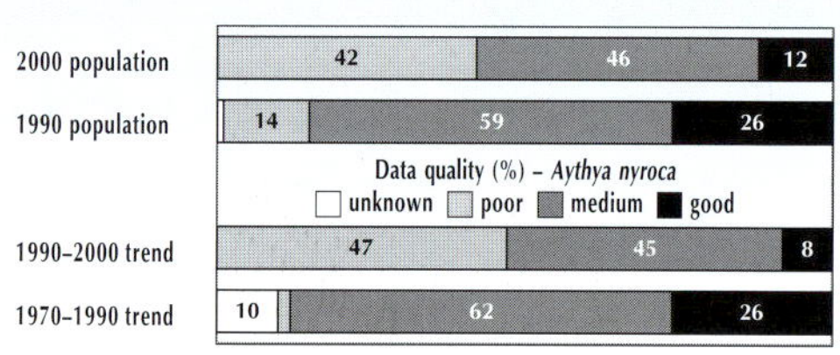

Aythya fuligula
TUFTED DUCK

SPEC 3 (1994: —) **Status** (Declining)
Criteria Moderate recent decline
European IUCN Red List Category —
Criteria —
Global IUCN Red List Category —
Criteria —

Aythya fuligula is a widespread breeder across much of Europe, which accounts for less than half its global breeding range. Its European breeding population is large (>730,000 pairs), and was stable between 1970–1990. Although the species was stable or increased across much of Europe during 1990–2000, there were declines in several north-eastern European countries—notably Finland and Russia—and the species underwent a moderate decline (>10%) overall. Consequently, this previously Secure species is now provisionally evaluated as Declining.

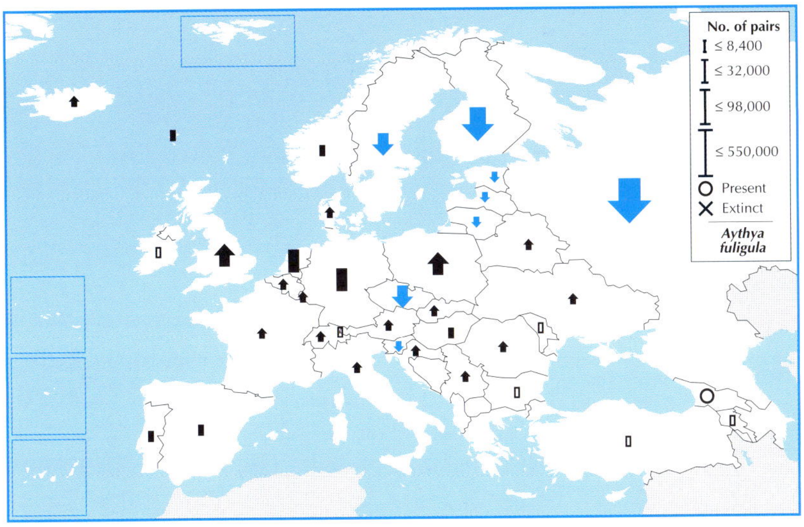

Country	Breeding pop. size (pairs)	Year(s)	Trend	Mag.%	References
Armenia	(0 – 50)	97–02	?	–	
Austria	700 – 1,000	98–02	+	20–29	
Belarus	4,000 – 6,000	97–02	**+**	**0–9**	
Belgium	1,900 – 2,600	95–02	+	0–19	*1*
Bulgaria	0 – 3	98–02	?	–	
Croatia	(25 – 50)	02	(+)	(>80)	*70,16*
Czech Rep.	**12,000 – 24,000**	00	**–**	**20–29**	
Denmark	(1,000 – 2,000)	00	(+)	(>80)	*3*
Faroe Is.	(0 – 1)	92–93	(0)	(0–19)	
Estonia	5,000 – 7,000	98	–	20–29	*1*
Finland	80,000 – 120,000	99–01	**–**	**25**	
France	500 – 800	98–02	+	0–9	*4*
Georgia	Present	03	?	–	
Germany	11,000 – 16,000	95–99	0	0–19	
Hungary	30 – 70	97–02	(0)	(0–19)	
Iceland	5,000 – 8,000	00	+	0–19	*21,9*
Rep. Ireland	1,000 – 2,500	88–91	?	–	
Italy	**40 – 50**	02	+	0–19	*6*
Latvia	800 – 1,200	90–00	–	0–19	*17*
Liechtenstein	**0 – 1**	98–00	?	–	
Lithuania	4,000 – 6,000	99–01	–	20–29	*20*
Luxembourg	**2 – 5**	00–02	**+**	**>80**	
Moldova	50 – 120	90–00	?	N	
Netherlands	**14,000 – 18,000**	98–00	**0**	**7**	*1,2*
Norway	(7,000 – 10,000)	90–03	(0)	(0–19)	
Poland	15,000 – 25,000	90–00	(+)	(0–19)	*27*
Portugal	0 – 5	02	**F**	**N**	
Romania	(60 – 120)	90–02	(+)	(0–19)	*2*
Russia	530,000 – 560,000	90–00	(–)	(20–29)	*85*
Serbia & MN	60 – 90	90–02	+	20–29	*1,82,29,155,143a, 227,229,225,10*
Slovakia	250 – 500	80–99	+	50–79	
Slovenia	30 – 60	90–00	–	10–19	
Spain	(0 – 50)	98–02	(F)	(–)	*10*
Sweden	(20,000 – 50,000)	99–00	–	10–19	
Switzerland	**100 – 200**	98–02	**+**	**30–49**	
Turkey	(100 – 200)	01	?	–	
Ukraine	**1,550 – 2,950**	90–00	**+**	**20–29**	
UK	**10,200 – 11,500**	00	+	25	*5,25*

Total (approx.) 730,000 – 880,000 **Overall trend** Moderate decline
Breeding range >5,000,000 km² **Gen. length**. <3.3 **% Global pop.** 25–49
(See p. 64, bottom, for data quality graph)

Winter pop. size (individuals)

Total (approx.) >1,200,000 **Overall trend** Moderate decline
% in European IBAs 58–61 **Gen. length** <3.3 **% Global pop.** 25–49

Aythya marila
GREATER SCAUP

SPEC 3W (1994: 3W) **Status** Endangered
Criteria See IUCN below
European IUCN Red List Category EN
Criteria A2b
Global IUCN Red List Category —
Criteria —

Aythya marila breeds in northern Europe, and winters mainly in north-western Europe and the Black Sea, which together account for less than half of its global wintering range. Its European wintering population is large (>120,000 individuals), and was stable between 1970–1990. Although most wintering populations were broadly stable during 1990–2000, there were substantial declines in north-western Europe, and the species probably underwent a very large decline (>50%) overall. Consequently, this previously Localised species (see Appendix 5) is now evaluated as Endangered.

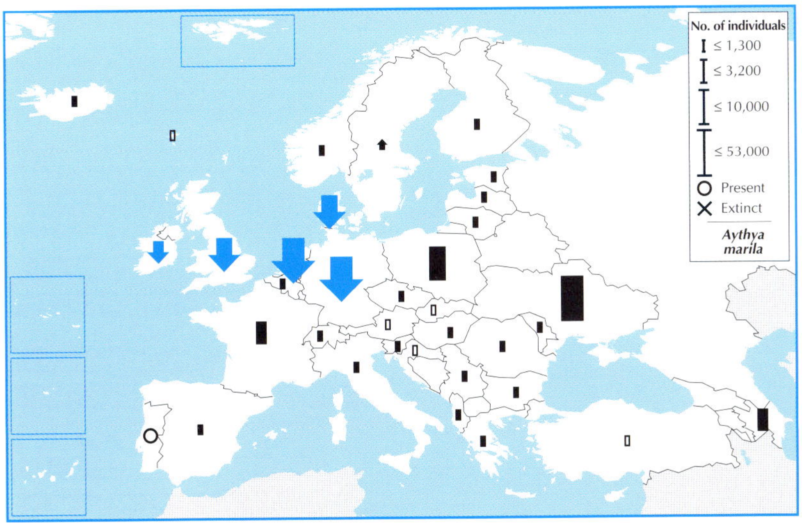

Country	Winter pop. size (individuals)	Year(s)	Trend	Mag.%	References
Albania	**1 – 6**	95–02	(F)	(–)	
Austria	**72 – 131**	97–99	?	–	
Azerbaijan	1,500 – 6,500	96–02	0	0–19	
Belgium	**0 – 30**	95–00	**0**	**0–19**	*1*
Bulgaria	**1 – 162**	97–00	**F**	**>80**	
Croatia	**11 – 100**	02	?	–	*26*
Czech Rep.	**1 – 50**	90–00	0	0–9	
Denmark	**9,000 – 11,000**	99–00	–	30–49	*25,26*
Faroe Is.	0 – 20	92	?	–	*3*
Estonia	100 – 500	98	0	0–19	*1*
Finland	10 – 50	99–01	(0)	(0–19)	
France	**2,000 – 3,000**	98–02	**0**	**0–19**	*9*
Germany	20,000 – 80,000	95–00	–	60–77	
Greece	**0 – 7**	95–99	**0**	**0–19**	
Hungary	5 – 30	97–01	F	>80	*24*
Iceland	**10 – 150**	78–94	**0**	**0–19**	*48*
Rep. Ireland	**1,500 – 3,000**	94–00	–	>80	
Italy	200 – 400	02	**0**	**0–19**	*22*
Latvia	0 – 200	90–01	F	50–79	*32*
Lithuania	0 – 10	92–02	(0)	(0–19)	
Moldova	20 – 235	90–00	0	0–19	
Netherlands	**53,000 – 53,000**	99–01	**–**	**75**	*3,4,8,9,11,12*
Norway	500 – 2,000	93–96	(0)	(0–19)	*60,70*
Poland	5,000 – 15,000	92–97	F	>80	*116*
Portugal	Present	02	?	–	
Romania	5 – 35	90–00	F	50–79	*60*
Serbia & MN	5 – 20	90–02	F	20–39	
Slovakia	0 – 70	95–99	?	–	*6,5*
Slovenia	0 – 20	90–00	F	0–100	
Spain	**4 – 27**	90–01	0	0–19	
Sweden	**1,000 – 1,500**	98–01	**+**	**60**	
Switzerland	**14 – 44**	98–02	**F**	**50–79**	
Turkey	20 – 100	91–01	?	–	
Ukraine	(20,000 – 100,000)	90–00	F	20–29	
UK	**9,200 – 9,200**	90–99	**–**	**31**	*49*

Total (approx.) >120,000 **Overall trend** Large decline
% in European IBAs >90 **Gen. length** <3.3 **% Global pop.** 25–49

Breeding pop. size (pairs)

Total (approx.) 180,000 – 190,000 **Overall trend** Stable
Breeding range >1,000,000 km² **Gen. length**. <3.3 **% Global pop.** 25–49

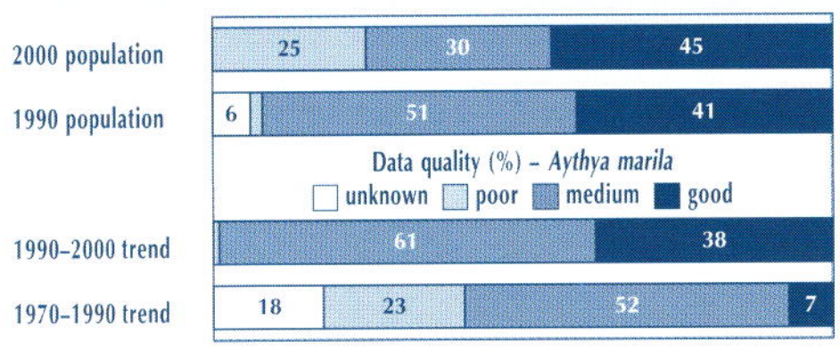

Somateria mollissima
COMMON EIDER

Non-SPEC[E] (1994: —) **Status** Secure
Criteria —
European IUCN Red List Category —
Criteria —
Global IUCN Red List Category —
Criteria —

Somateria mollissima breeds in coastal areas of north-west and northern Europe, which holds >50% of its global population. Its European breeding population is large (>840,000 pairs), and underwent a large increase between 1970–1990. Although there were small declines in Greenland and Svalbard during 1990–2000, the species was stable or increased across most of its European range, and underwent a small increase overall. Consequently, it is evaluated as Secure.

Country	Breeding pop. size (pairs)	Year(s)	Trend	Mag.%	References
Denmark	25,000 – 25,000	00–02	0	0–19	*17*
Faroe Is.	3,500 – 3,500	95	(0)	(0–19)	
Greenland	15,000 – 25,000	95–00	(–)	(0–19)	*12*
Estonia	15,000 – 20,000	98	+	20–29	*1*
Finland	140,000 – 160,000	01–02	F	30–49	
France	0 – 20	97–00	+	30–49	*5*
Germany	1,400 – 1,500	95–99	0	0–19	
Iceland	200,000 – 350,000	00	0	0–19	*16*
Rep. Ireland	250 – 1,000	88–91	(+)	(0–19)	
Italy	1 – 1	99–00	+	0–19	*18*
Netherlands	8,000 – 10,000	98–00	0	10	*1*
Norway	100,000 – 150,000	90–01	+	0–19	*35*
Svalbard	13,500 – 27,500	90–01	–	0–19	
Poland	0 – 1	97–00	?	–	*28*
Russia	(15,000 – 25,000)	90–00	?	–	*11,82,85,154*
Sweden	270,000 – 360,000	99–00	+	10–19	
Switzerland	1 – 3	93–96	0	0–19	
Ukraine	700 – 1,100	90–98	+	30–49	
UK	31,600 – 31,600	88–91	(+)	(50–79)	*8*
Total (approx.)	840,000 – 1,200,000	**Overall trend** Small increase			
Breeding range	>1,000,000 km²	**Gen. length.** 5		**% Global pop.** 50–74	

Country	Winter pop. size (individuals)	Year(s)	Trend	Mag.%	References
Denmark	320,000 – 370,000	99–00	–	30–49	*25,26*
Faroe Is.	5,000 – 15,000	92	?	–	*3*
Greenland	350,000 – 630,000	99–00	(–)	(0–19)	*19*
France	2,000 – 4,200	98–02	F	50–79	*9*
Germany	350,000 – 350,000	87–03	F	20–29	*4*
Iceland	(100,000 – 1,000,000)	78–94	?	–	*48*
Netherlands	120,000 – 120,000	99–01	0	0–19	*3,4,8,9,11,12*
Norway	400,000 – 500,000	93–96	(0)	(0–19)	
Poland	1,000 – 2,000	92–97	0	0–19	*116*
Russia	1,500 – 1,500	97	?	–	*179*
Sweden	7,000 – 30,000	98–01	+	>80	
UK	80,000 – 80,000	90–99	–	6	*49*
Total (approx.)	>1,700,000	**Overall trend** Small decline			
% in European IBAs	14–22	**Gen. length** 5		**% Global pop.** 50–74	

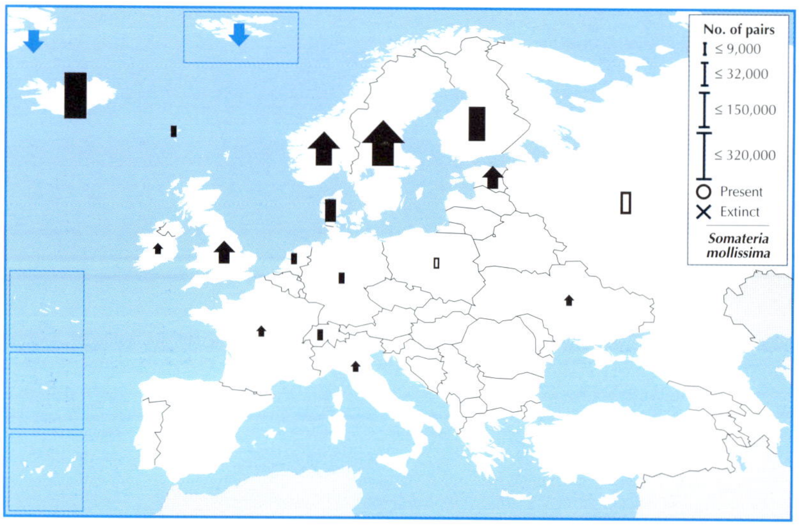

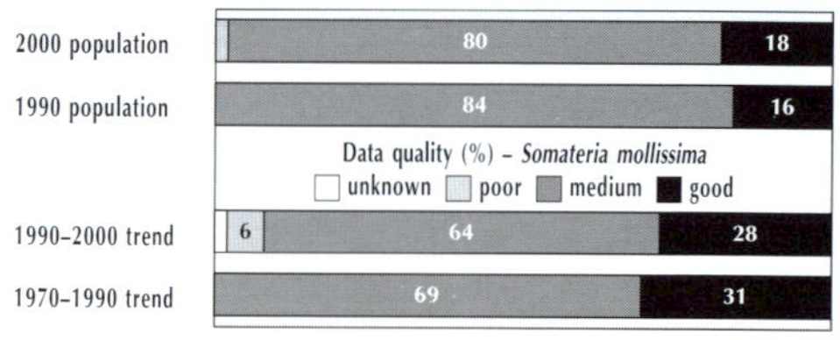

Somateria spectabilis
KING EIDER

Non-SPEC (1994: —) **Status** (Secure)
Criteria —
European IUCN Red List Category —
Criteria —
Global IUCN Red List Category —
Criteria —

Somateria spectabilis breeds in Russia, Svalbard and Greenland, which together account for a tiny proportion of its global breeding range. Its European breeding population is relatively small (<46,000 pairs), but was stable between 1970–1990. Although the trend of the Greenland population during 1990–2000 was unknown, the species was stable in Russia and Svalbard, and probably remained stable overall. Consequently, it is provisionally evaluated as Secure.

Country	Breeding pop. size (pairs)	Year(s)	Trend	Mag.%	References
Denmark					
Greenland	(2,000 – 5,000)	90–00	?	–	
Norway					
Svalbard	300 – 1,000	95	(0)	(0–19)	
Russia	(35,000 – 40,000)	90–00	(0)	(0–19)	*85*
Total (approx.)	37,000 – 46,000	**Overall trend** Stable			
Breeding range	>500,000 km²	**Gen. length.** 5		**% Global pop.** <5	

Country	Winter pop. size (individuals)	Year(s)	Trend	Mag.%	References
Denmark					
Greenland	300,000 – 500,000	90–00	(–)	(0–19)	*20*
Estonia	0 – 1	98	?	–	*1*
Iceland	10 – 50	90–99	(F)	(–)	*65,66,69,70,71, 72,73,74,58,59*
Norway	50,000 – 100,000	93–96	(0)	(0–19)	*68,70*
Russia	304 – 304	97	?	–	*179*
Total (approx.)	>350,000	**Overall trend** Small decline			
% in European IBAs	>40	**Gen. length** 5		**% Global pop.** 5–24	

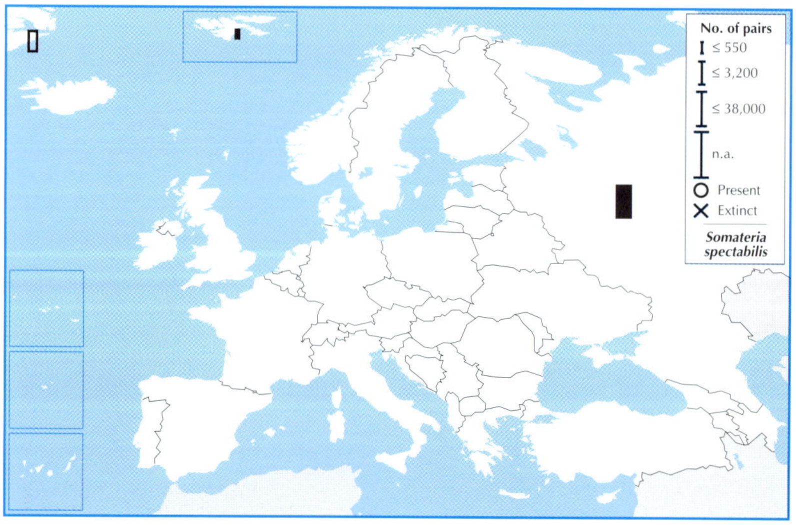

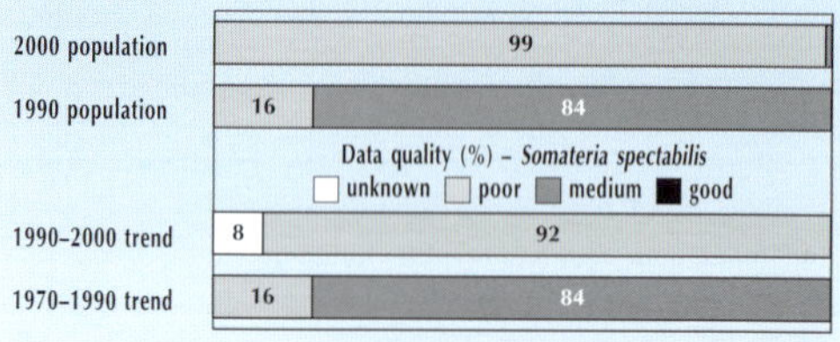

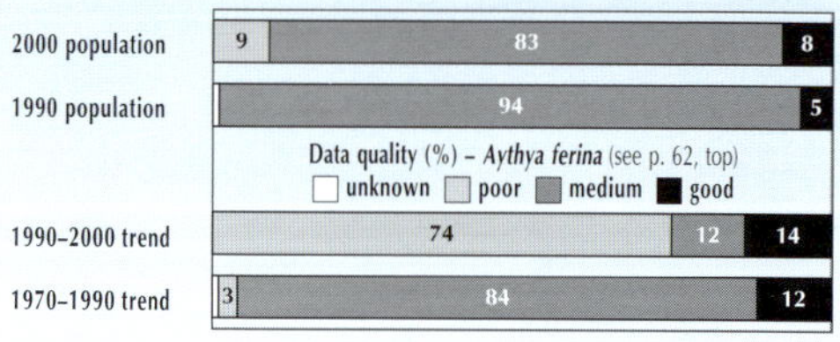

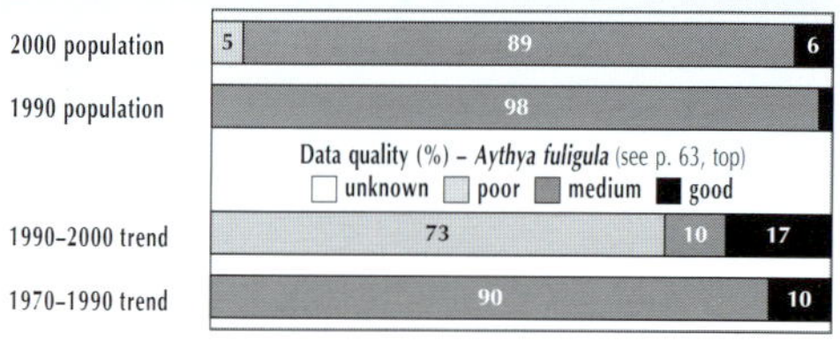

Polysticta stelleri
STELLER'S EIDER

SPEC 3W (1994: 1W) **Status** Localised
Criteria ≥90% winter at ≤10 sites
European IUCN Red List Category —
Criteria —
Global IUCN Red List Category —
Criteria —

Country	Winter pop. size (individuals)	Year(s)	Trend	Mag.%	References
Estonia	2,500 - 6,000	98	0	0-19	1
Finland	(150 - 500)	99–00	(0)	(0-19)	
Latvia	0 - 20	90–01	0	0-19	32
Lithuania	400 - 600	92–02	-	>80	
Norway	4,000 - 13,000	80–03	F	50-79	
Russia	640 - 640	97	?	-	179
Sweden	5 - 20	98–01	0	0-19	
Total (approx.)	>7,700	Overall trend Fluctuating			
% in European IBAs	>90	Gen. length 4		% Global pop. <5	

Country	Breeding pop. size (pairs)	Year(s)	Trend	Mag.%	References
Russia	(5 - 50)	90–00	?	-	85
Total (approx.)	5 - 50	Overall trend Unknown			
Breeding range	<20,000 km²	Gen. length. 4		% Global pop. <5	

Polysticta stelleri breeds (in trivial numbers) in European Russia, and winters in the Baltic and off the northern coast of Norway, which together account for a tiny proportion of its global wintering range. Its European wintering population is small (as few as 7,700 individuals), but was broadly stable during both 1970–1990 and 1990–2000. Nevertheless, more than 90% of the European wintering population is confined to just 10 sites (see Appendix 5). Consequently, the species is evaluated as Localised in winter.

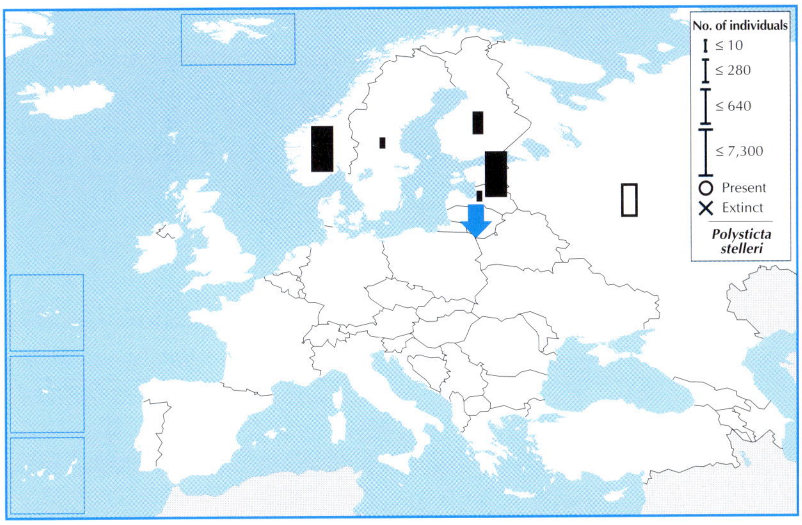

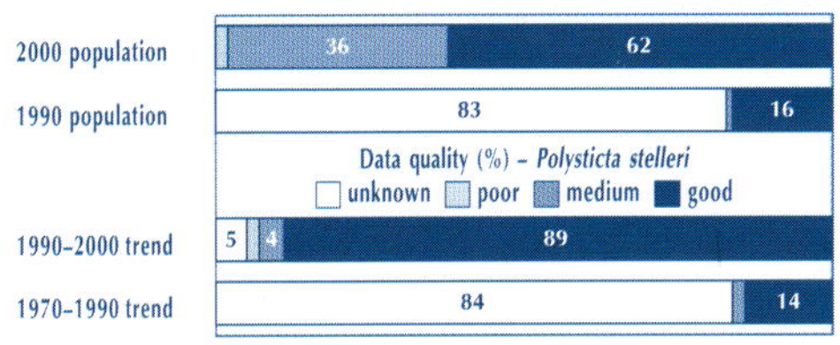

Histrionicus histrionicus
HARLEQUIN DUCK

SPEC 3 (1994: 3) **Status** (Rare)
Criteria <10,000 pairs
European IUCN Red List Category —
Criteria —
Global IUCN Red List Category —
Criteria —

Country	Breeding pop. size (pairs)	Year(s)	Trend	Mag.%	References
Denmark					
Greenland	(1,000 - 5,000)	90–00	?	-	13,14
Iceland	3,000 - 5,000	98–01	?	-	10
Total (approx.)	4,000 - 10,000	Overall trend Unknown			
Breeding range	>100,000 km²	Gen. length. 4		% Global pop. <5	

Country	Winter pop. size (individuals)	Year(s)	Trend	Mag.%	References
Denmark					
Greenland	(10,000 - 15,000)	98–00	?	-	
Iceland	12,000 - 16,000	98–01	?	-	52
Total (approx.)	>22,000	Overall trend Unknown			
% in European IBAs	Unknown	Gen. length 4		% Global pop. <5	

Histrionicus histrionicus breeds in Greenland and Iceland, with Europe accounting for a tiny proportion of its global breeding range. Its European breeding population is small (as few as 4,000 pairs), but increased substantially between 1970–1990. Although no trend data were available for either country during 1990–2000, there was no evidence to suggest that the species's status deteriorated since 1990. Nevertheless, its population size probably still renders it susceptible to the risks affecting small populations, and consequently it is provisionally evaluated as Rare.

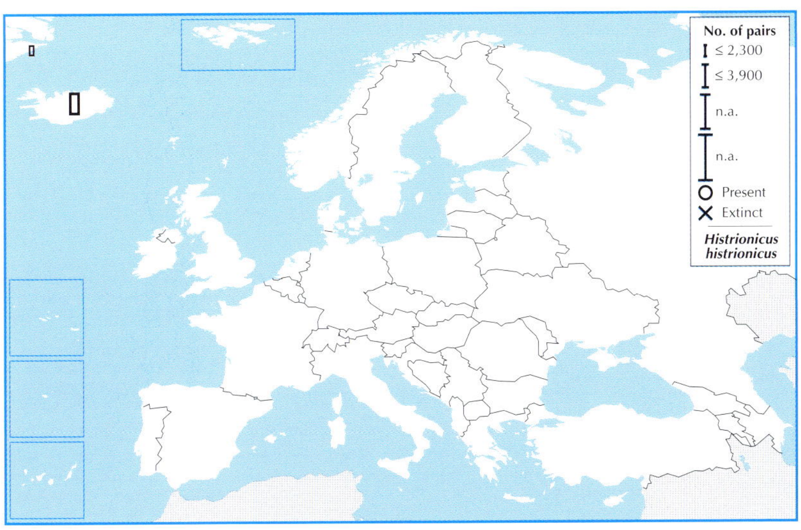

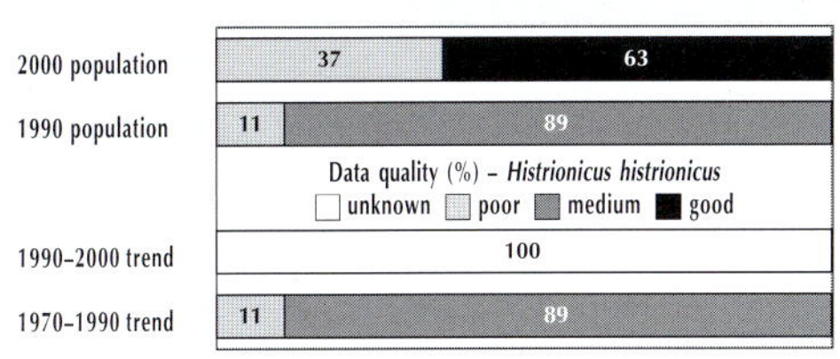

Clangula hyemalis
LONG-TAILED DUCK

Non-SPEC (1994: —) **Status** (Secure)
Criteria —
European IUCN Red List Category —
Criteria —
Global IUCN Red List Category —
Criteria —

Country	Breeding pop. size (pairs)	Year(s)	Trend	Mag.%	References
Denmark					
Greenland	(10,000 – 30,000)	90–00	?	–	
Finland	(1,500 – 2,000)	99–01	(0)	(0–19)	
Iceland	(2,000 – 3,000)	00	?	–	*21*
Norway	(5,000 – 10,000)	90–03	(0)	(0–19)	
Svalbard	500 – 1,000	90–03	(0)	(0–19)	
Russia	(670,000 – 700,000)	90–00	(0)	(0–19)	*85*
Sweden	(1,000 – 2,000)	99–00	?	–	
Total (approx.)	690,000 – 750,000	**Overall trend** Stable			
Breeding range	>1,000,000 km²	**Gen. length.** 4		**% Global pop.** 25–49	

Country	Winter pop. size (individuals)	Year(s)	Trend	Mag.%	References
Azerbaijan	10 – 150	96–02	(0)	(0–19)	
Denmark	2,000 – 3,000	99–00	(–)	(20–29)	*25,26*
Faroe Is.	100 – 1,000	92	?	–	*3*
Greenland	(67,000 – 133,000)	99–00	?	–	*19*
Estonia	100,000 – 700,000	98	(0)	(0–19)	*1*
Finland	(3,000 – 10,000)	99–01	(0)	(0–19)	
Germany	**596,000 – 596,000**	87–03	?	–	*4*
Iceland	10,000 – 100,000	78–94	?	–	*48*
Rep. Ireland	**168 – 260**	94–00	(F)	(20–29)	
Latvia	**350,000 – 350,000**	92–94	(0)	(0–19)	*27,32*
Lithuania	(10,000 – 20,000)	92–02	(F)	(>80)	
Netherlands	(200 – 800)	99–01	(–)	(–)	*3,4,8,9,11,12*
Norway	**80,000 – 120,000**	93–96	(0)	(0–19)	*68,70*
Poland	(100,000 – 250,000)	92–97	?	–	*116*
Russia	196 – 196	97	?	–	*179*
Sweden	800,000 – 1,300,000	98–01	0	0–19	
UK	**16,300 – 16,300**	90–00	**–**	**32**	*49*
Total (approx.)	>2,100,000	**Overall trend** Stable			
% in European IBAs	69–83	**Gen. length** 4		**% Global pop.** 25–49	

Clangula hyemalis is a widespread breeder in northernmost Europe, which accounts for less than half of its global breeding range. Its European breeding population is large (>690,000 pairs), and was stable between 1970–1990. Although trends in Greenland, Iceland and Sweden during 1990–2000 were unknown, populations were stable elsewhere—including in the Russian stronghold—and the species probably remained stable overall. Consequently, it is provisionally evaluated as Secure.

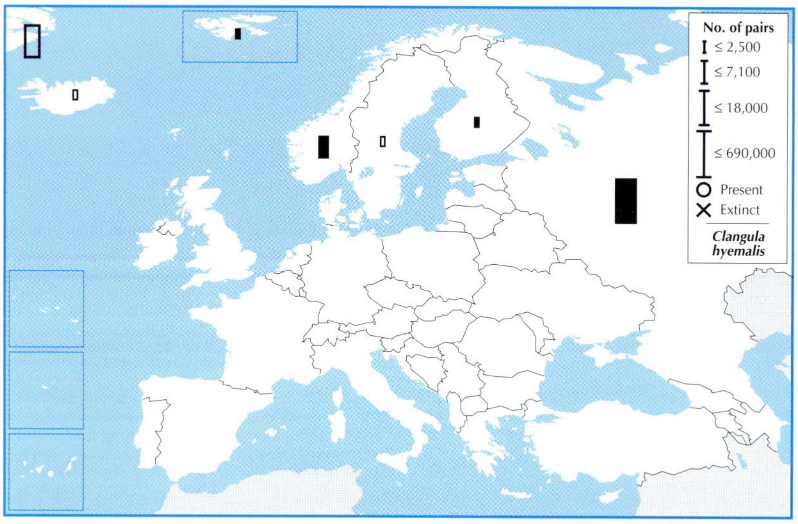

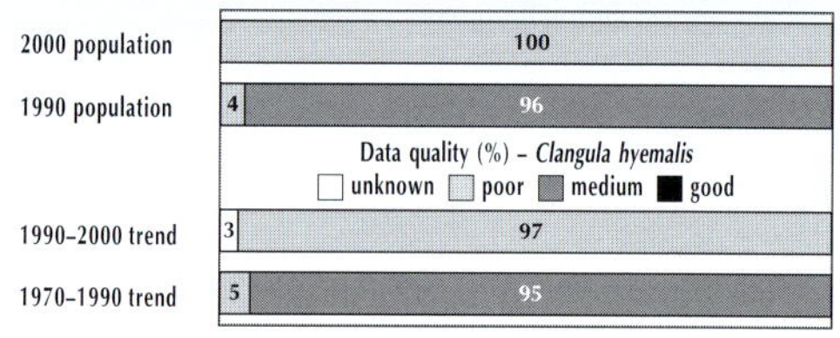

Melanitta nigra
BLACK SCOTER

Non-SPEC (1994: —) **Status** (Secure)
Criteria —
European IUCN Red List Category —
Criteria —
Global IUCN Red List Category —
Criteria —

Country	Breeding pop. size (pairs)	Year(s)	Trend	Mag.%	References
Finland	(1,000 – 2,000)	99–01	(0)	(0–19)	
Iceland	300 – 500	00	+	0–19	*21,9*
Rep. Ireland	**80 – 100**	95–99	**0**	**0–19**	
Norway	(1,000 – 5,000)	90–03	(0)	(0–19)	
Svalbard	0 – 10	90–03	(0)	(0–19)	
Russia	(100,000 – 120,000)	90–00	(0)	(0–19)	*85*
Sweden	(1,500 – 3,000)	99–00	(0)	(0–19)	
UK	77 – 77	96	**0**	**1**	*34*
Total (approx.)	100,000 – 130,000	**Overall trend** Stable			
Breeding range	>1,000,000 km²	**Gen. length.** 4		**% Global pop.** 5–24	

Country	Winter pop. size (individuals)	Year(s)	Trend	Mag.%	References
Belgium	**900 – 9,600**	95–00	**0**	**0–19**	*1*
Denmark	**240,000 – 240,000**	99–00	–	30–49	*25,26*
Estonia	50 – 600	98	**F**	**20–29**	1
Finland	(20 – 200)	99–01	(0)	(0–19)	
France	**23,000 – 45,000**	98–02	–	30–49	*9*
Germany	**179,000 – 179,000**	87–03	?	–	*4*
Rep. Ireland	**15,000 – 20,000**	94–00	F	20–29	
Italy	150 – 400	02	0	0–19	22
Latvia	**1,000 – 5,000**	92–94	–	50–79	*27,32*
Lithuania	250 – 1,000	92–02	F	>80	
Netherlands	**69,000 – 69,000**	99–01	**0**	**0–19**	*3,4,8,9,11,12*
Norway	2,000 – 6,000	93–96	(0)	(0–19)	
Poland	(20,000 – 100,000)	92–97	(F)	(–)	*116*
Portugal	(6,000 – 27,000)	90–02	(–)	(–)	*27*
Spain	**300 – 16,000**	90–01	?	–	*19*
Sweden	**1,000 – 5,000**	98–01	**0**	**0–19**	
UK	**50,000 – 50,000**	90–99	**+**	**83**	*49*
Total (approx.)	>610,000	**Overall trend** Small decline			
% in European IBAs	73–85	**Gen. length** 4		**% Global pop.** 25–49	

Melanitta nigra breeds in northern Europe, which accounts for less than a quarter of its global breeding range. Its European breeding population is relatively large (>100,000 pairs), and was stable between 1970–1990. Populations in most countries—including the Russian stronghold—were stable during 1990–2000, and thus the species remained stable overall. Consequently, it is provisionally evaluated as Secure.

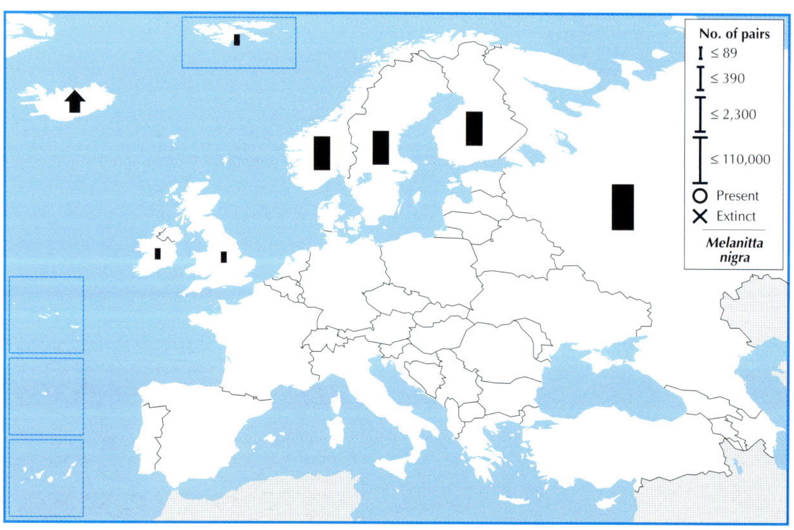

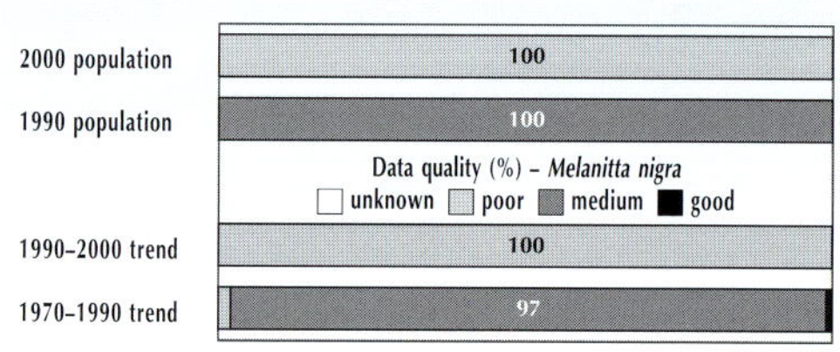

Melanitta fusca
WHITE-WINGED SCOTER

SPEC 3 (1994: 3W) **Status** (Declining)
Criteria Moderate recent decline
European IUCN Red List Category —
Criteria —
Global IUCN Red List Category —
Criteria —

Country	Breeding pop. size (pairs)	Year(s)	Trend	Mag.%	References
Estonia	500 – 900	98	+	20–29	*1*
Finland	14,000 – 16,000	99–01	+	10	
Georgia	**20 – 50**	94–02	0	0–19	
Norway	(500 – 1,500)	90–03	(0)	(0–19)	
Russia	(60,000 – 70,000)	90–00	(-)	(20–29)	*85*
Sweden	10,000 – 14,000	99–00	-	20–29	
Turkey	(60 – 90)	01	(-)	(0–19)	
Total (approx.)	85,000 – 100,000	**Overall trend** Moderate decline			
Breeding range	>1,000,000 km²	**Gen. length.** 4		**% Global pop.** 5–24	

Country	Winter pop. size (individuals)	Year(s)	Trend	Mag.%	References
Belgium	**5 – 100**	95–00	**0**	**0–19**	*1*
Croatia	(50 – 100)	02	(F)	(-)	*26*
Denmark	1,700 – 2,000	99–00	(-)	(30–49)	*25,26*
Estonia	20,000 – 200,000	98	(0)	(0–19)	*1*
Finland	20 – 200	99–01	0	0–19	
France	**2,000 – 3,440**	98–02	**-**	**30–49**	*9*
Germany	**51,000 – 51,000**	87–03	?	–	*4*
Italy	100 – 300	02	0	0–19	*22*
Latvia	**5,000 – 140,000**	92–94	-	50–79	*27,32*
Lithuania	20,000 – 50,000	92–02	**F**	**50–79**	
Netherlands	**350 – 350**	99–01	(F)	(-)	*3,4,8,9,11,12*
Norway	**25,000 – 30,000**	93–96	(0)	(0–19)	*68,70*
Poland	(10,000 – 20,000)	92–97	(F)	(-)	*116*
Russia	87 – 132	97–99	?	–	*179*
Serbia & MN	**20 – 380**	90–02	F	50–100	
Spain	**7 – 72**	90–01	?	–	*19*
Sweden	**1,000 – 2,500**	98–01	**0**	**0–19**	
Turkey	400 – 800	91–01	0	0–19	
UK	**3,000 – 3,000**	90–99	**0**	**0**	*49*
Total (approx.)	>140,000	**Overall trend** Moderate decline			
% in European IBAs	>90	**Gen. length** 4		**% Global pop.** 5–24	

Melanitta fusca breeds in Fennoscandia and northern Russia (with a disjunct population in the Caucasus), which together account for less than a quarter of its global breeding range. Its European breeding population is relatively small (<100,000 pairs), but was stable between 1970–1990. Although some populations were stable or increased during 1990–2000, the species declined in Russia and Norway, and underwent a moderate decline (>10%) overall. Consequently, this species (which is still Localised in winter; see Appendix 5) is now provisionally evaluated as Declining.

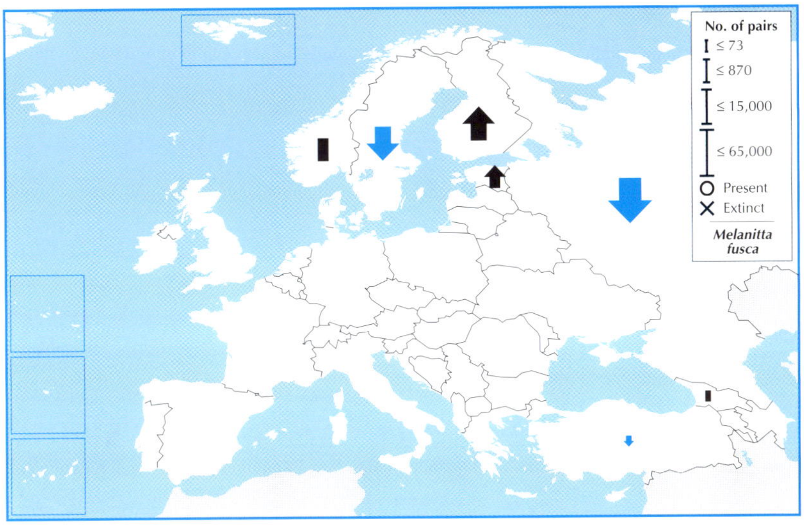

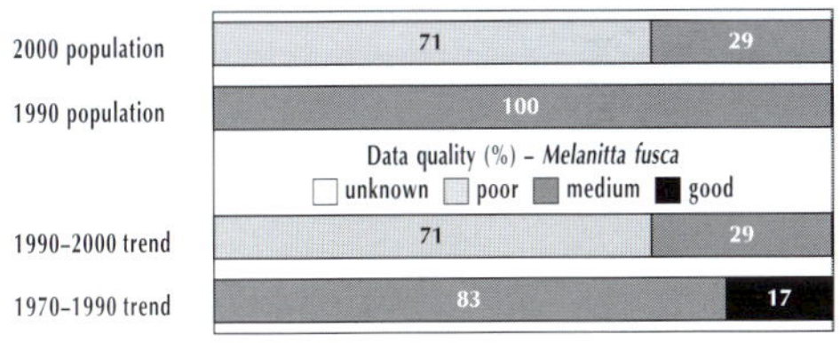

Bucephala islandica
BARROW'S GOLDENEYE

SPEC 3 (1994: 3) **Status** Vulnerable
Criteria See IUCN below
European IUCN Red List Category VU
Criteria D2
Global IUCN Red List Category —
Criteria —

Country	Breeding pop. size (pairs)	Year(s)	Trend	Mag.%	References
Iceland	**500 – 600**	90–00	**F**	**30–49**	*21,28,2*
Total (approx.)	500 – 600	**Overall trend** Fluctuating			
Breeding range	<20,000 km²	**Gen. length.** 4		**% Global pop.** <5	

Country	Winter pop. size (individuals)	Year(s)	Trend	Mag.%	References
Iceland	**1,000 – 1,650**	90–00	**F**	**30–49**	*56,60,51*
Total (approx.)	>1,000	**Overall trend** Fluctuating			
% in European IBAs	>90	**Gen. length** 4		**% Global pop.** <5	

Bucephala islandica breeds in Europe only in Iceland, which accounts for a tiny proportion of its global breeding range. Its European breeding population is very small (as few as 500 pairs), but was broadly stable between 1970–1990, and—despite substantial fluctuations—remained broadly stable overall during 1990–2000. Nevertheless, the entire European breeding population is confined to just a handful of locations in north-eastern Iceland, with the vast majority (85–90%) concentrated at just one site (Lake Mývatn and the River Laxá). Consequently, it is evaluated as Vulnerable.

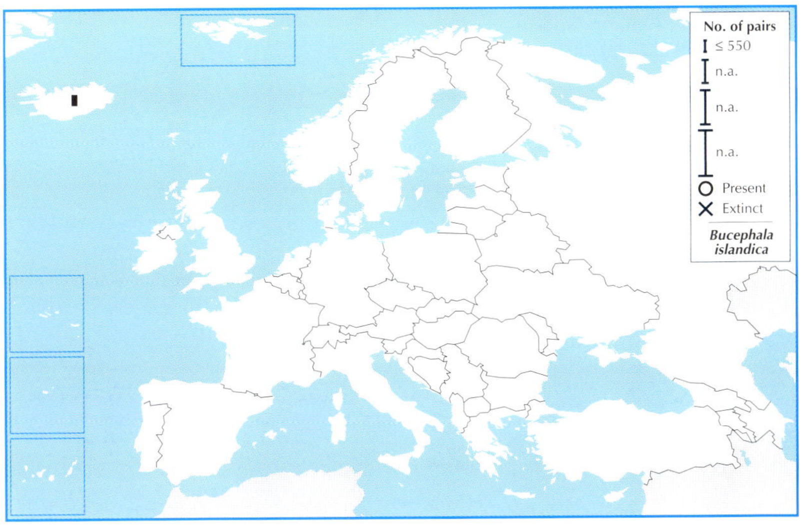

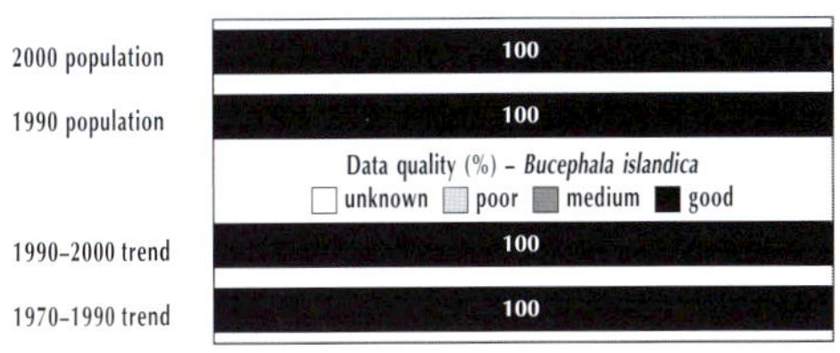

Bucephala clangula
COMMON GOLDENEYE

Non-SPEC (1994: —) **Status** (Secure)
Criteria —
European IUCN Red List Category —
Criteria —
Global IUCN Red List Category —
Criteria —

Bucephala clangula is a widespread breeder in northern and central Europe, which holds less than half of its global population. Its European breeding population is large (>490,000 pairs), and was stable between 1970–1990. Although the species declined slightly in Sweden during 1990–2000, other European populations—including the sizeable ones in Finland and Russia—increased or were stable, and the species increased slightly overall. Consequently, it is provisionally evaluated as Secure.

Country	Breeding pop. size (pairs)	Year(s)	Trend	Mag.%	References
Austria	**5 – 10**	98–02	+	N	
Belarus	**800 – 1,400**	97–02	+	**0–9**	
Czech Rep.	**70 – 90**	00	**0**	**0–19**	
Denmark	**63 – 76**	98–01	+	**>80**	*4,5,6,7*
Estonia	(1,000 – 1,500)	98	+	20–29	*1*
Finland	200,000 – 250,000	99–01	+	**20**	
France	**0 – 1**	99	**0**	**0–19**	
Germany	1,720 – 3,050	95–99	+	30–49	
Hungary	0 – 1	98–02	+	N	
Latvia	300 – 600	90–00	(0)	(0–19)	*23,17*
Lithuania	**1,500 – 2,000**	99–01	+	20–29	*20*
Netherlands	**15 – 20**	98–00	+	759	*1*
Norway	(10,000 – 20,000)	90–02	(0)	(0–19)	
Poland	1,200 – 1,500	90–00	(+)	(0–19)	*2*
Romania	**0 – 3**	96–02	+	**0–19**	
Russia	200,000 – 210,000	90–00	(0)	(0–19)	*85*
Sweden	**75,000 – 100,000**	99–00	–	**6**	
Ukraine	10 – 45	90–00	F	30–49	
UK	**200 – 200**	98	(0)	(0–19)	
Total (approx.)	490,000 – 590,000	**Overall trend** Small increase			
Breeding range	>3,000,000 km²	**Gen. length.** 4		**% Global pop.** 25–49	

Country	Winter pop. size (individuals)	Year(s)	Trend	Mag.%	References
Denmark	**60,000 – 70,000**	99–00	**0**	**0–19**	*25,26*
Estonia	10,000 – 20,000	98	+	50–79	*1*
Finland	(10,000 – 50,000)	99–01	**F**	**10–29**	
Germany	**56,000 – 72,000**	95–00	**0**	**4–19**	
Hungary	10,000 – 15,000	97–02	F	>80	*24,21*
Netherlands	**16,000 – 16,000**	99–01	**0**	**0–19**	*3,4,8,9,11,12*
Norway	**15,000 – 20,000**	93–96	(0)	(0–19)	
Poland	25,000 – 41,000	92–97	+	0–19	*115*
Serbia & MN	**20,000 – 35,000**	90–02	F	10–29	
Sweden	**25,000 – 40,000**	98–01	+	**>80**	
UK	**35,000 – 35,000**	94–99	–	**3**	*48,49*
Total (approx.)	>310,000	**Overall trend** Small increase			
% in European IBAs	36–41	**Gen. length** 4		**% Global pop.** 25–49	

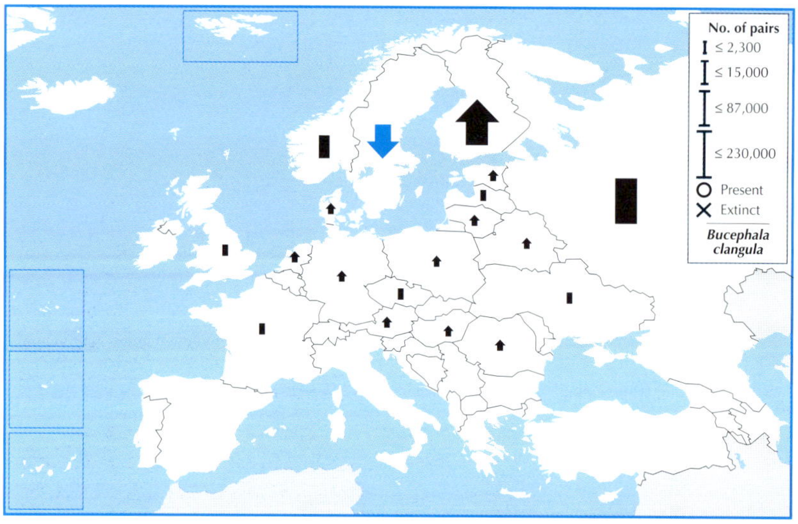

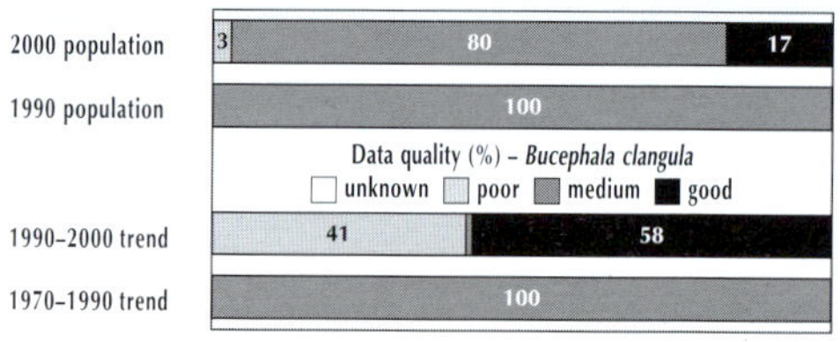

Mergellus albellus
SMEW

SPEC 3 (1994: 3) **Status** (Declining)
Criteria Moderate continuing decline
European IUCN Red List Category —
Criteria —
Global IUCN Red List Category —
Criteria —

Mergellus albellus breeds mainly in northern Russia and Fennoscandia, with Europe accounting for less than a quarter of its global breeding range. Its European breeding population is small (as few as 5,300 pairs), and its range contracted markedly between 1970–1990. Although most European populations—including the sizeable one in Finland—were stable or increased during 1990–2000, the stronghold population in Russia declined, and the species underwent a moderate decline (>10%) overall. Consequently, it is provisionally evaluated as Declining.

Country	Breeding pop. size (pairs)	Year(s)	Trend	Mag.%	References
Belarus	**15 – 30**	91–00	**0**	**0–19**	
Finland	1,000 – 2,000	99–01	(0)	(0–19)	
Norway	10 – 20	90–03	+	**0–19**	*24*
Romania	**0 – 5**	90–02	+	**0–19**	
Russia	(4,000 – 6,000)	90–00	(–)	(20–29)	*85*
Sweden	**250 – 350**	99–00	+	10–19	
Total (approx.)	5,300 – 8,400	**Overall trend** Moderate decline			
Breeding range	>1,000,000 km²	**Gen. length.** 4		**% Global pop.** 5–24	

Country	Winter pop. size (individuals)	Year(s)	Trend	Mag.%	References
Azerbaijan	100 – 700	96–02	+	>80	
Belgium	**100 – 970**	95–00	+	**>80**	*1*
Bulgaria	**23 – 742**	97–01	**F**	**>80**	
Croatia	1,000 – 1,000	02	?	–	*26*
Denmark	**600 – 800**	99–00	+	**>80**	*25,26*
Estonia	100 – 300	98	+	20–29	*1*
Finland	(50 – 500)	99–01	(0)	(0–19)	
France	**199 – 600**	98–02	**F**	**50–79**	*9*
Georgia	**681 – 681**	02	?	–	*1*
Germany	**2,500 – 9,000**	95–00	**F**	**30–49**	*4*
Greece	**69 – 297**	95–99	**F**	**30–49**	
Hungary	200 – 600	97–01	F	>80	*21,22,24*
Latvia	**20 – 200**	90–99	–	**50–79**	*29*
Lithuania	**50 – 200**	99–02	–	**>80**	
Netherlands	**3,300 – 3,300**	99–01	**0**	**0–19**	*3,4,8,9,11,12*
Norway	**300 – 315**	97–99	+	**0–19**	*71*
Poland	1,000 – 30,000	92–97	F	>80	*116*
Romania	1,400 – 2,600	90–00	0	0–19	*60,42*
Serbia & MN	**3,700 – 13,000**	90–02	+	10–29	
Slovakia	100 – 700	90–99	?	–	*4*
Sweden	**2,000 – 3,000**	98–01	+	**>80**	
Turkey	250 – 1,000	91–01	**F**	**>80**	
Ukraine	(400 – 600)	90–00	(–)	(20–29)	
UK	**390 – 390**	94–99	+	**48**	*49*
Total (approx.)	>19,000	**Overall trend** Small increase			
% in European IBAs	>80	**Gen. length** 4		**% Global pop.** 5–24	

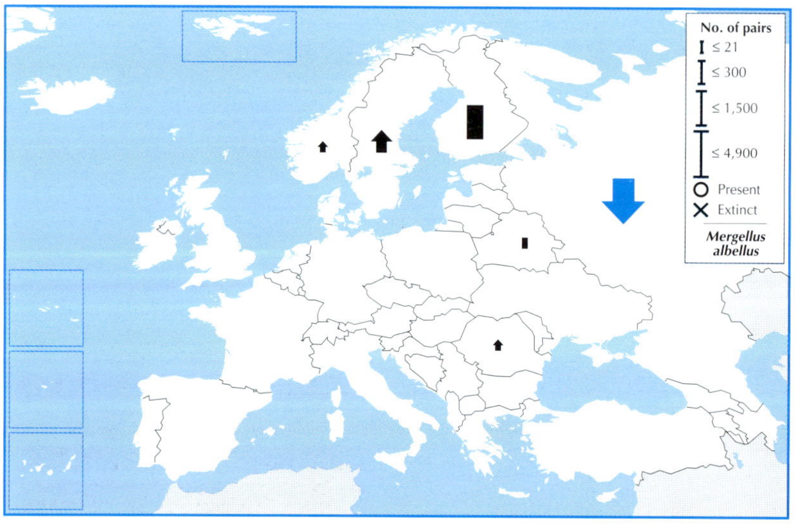

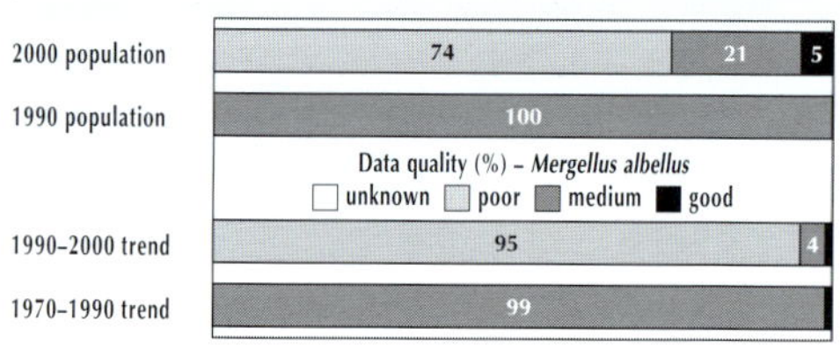

Mergus serrator
RED-BREASTED MERGANSER

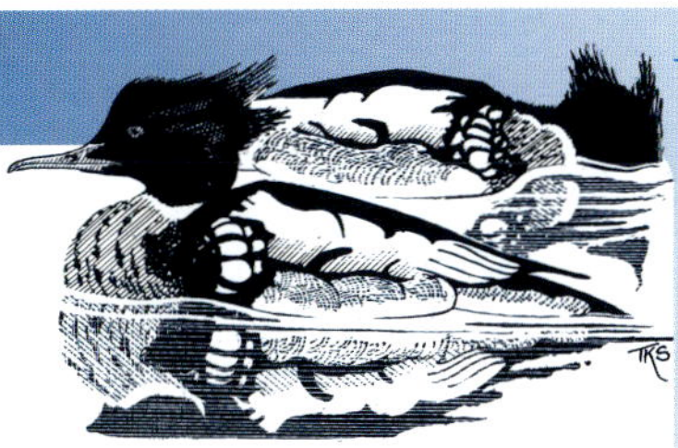

Non-SPEC (1994: —) **Status** (Secure)
Criteria —
European IUCN Red List Category —
Criteria —
Global IUCN Red List Category —
Criteria —

Mergus serrator is a widespread breeder across much of northern Europe, which accounts for less than a quarter of its global breeding range. Its European breeding population is relatively small (<120,000 pairs), but was stable between 1970–1990. Although the species declined in Norway, Sweden and Russia during 1990–2000, it was stable across much of its European range, and the large Finnish population increased markedly. The species probably underwent only a small decline overall, and consequently it is provisionally evaluated as Secure.

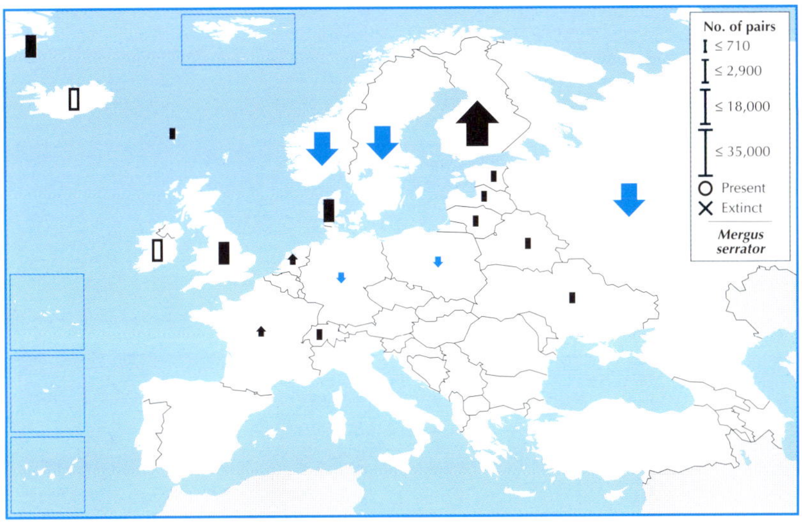

Country	Breeding pop. size (pairs)	Year(s)	Trend	Mag.%	References
Belarus	10 – 20	97–00	0	0–19	
Denmark	2,000 – 3,000	93–96	(0)	(0–19)	3
Faroe Is.	150 – 150	95	(0)	(0–19)	
Greenland	(1,000 – 5,000)	90–00	(0)	(0–19)	
Estonia	(500 – 1,000)	98	0	0–19	1
Finland	30,000 – 40,000	99–01	+	**15**	
France	**2 – 8**	96–00	+	**>80**	5
Germany	**340 – 410**	95–99	–	**0–19**	
Iceland	(2,000 – 4,000)	00	?	–	21
Rep. Ireland	1,000 – 2,500	88–91	?	–	
Latvia	(10 – 20)	90–00	(0)	(0–19)	23,17
Lithuania	(1 – 10)	99–01	(F)	(50–79)	20
Netherlands	**35 – 45**	98–00	+	**423**	1
Norway	(10,000 – 30,000)	90–03	–	0–19	4,41,27
Poland	**15 – 15**	99	–	**70–79**	29,30
Russia	(9,000 – 12,000)	90–00	(–)	(20–29)	85
Sweden	**14,000 – 18,000**	99–00	–	**10–19**	
Switzerland	**0 – 1**	93–96	**0**	**0–19**	
Ukraine	**384 – 384**	98	0	0–9	
UK	**2,400 – 2,400**	88–91	(0)	(0–19)	8
Total (approx.)	73,000 – 120,000	**Overall trend** Small decline			
Breeding range	>2,000,000 km²	**Gen. length.** 4		**% Global pop.** 5–24	

Country	Winter pop. size (individuals)	Year(s)	Trend	Mag.%	References
Denmark	**10,000 – 15,000**	99–00	0	0–19	25,26
Greenland	(5,000 – 20,000)	90–00	(0)	(0–19)	
France	**3,600 – 4,500**	98–02	**F**	**30–49**	9
Germany	**13,500 – 13,500**	95–00	–	41–56	4
Greece	**300 – 2,000**	96–00	**0**	**0–19**	
Iceland	(5,000 – 15,000)	78–94	?	–	48
Rep. Ireland	**2,600 – 2,800**	94–00	0	0–19	
Italy	1,000 – 2,000	02	F	20–29	22
Netherlands	**5,400 – 5,400**	99–01	**0**	**0–19**	3,4,8,9,11,12
Norway	**25,000 – 30,000**	93–96	(0)	(0–19)	68,70
Poland	(1,000 – 3,000)	92–97	?	–	116
Sweden	**3,000 – 6,000**	98–01	+	**50**	
Turkey	600 – 1,200	91–01	(0)	(0–19)	
UK	10,200 – 10,200	86–91	+	**23**	44,49
Total (approx.)	>89,000	**Overall trend** Small decline			
% in European IBAs	31–44	**Gen. length** 4		**% Global pop.** 5–24	

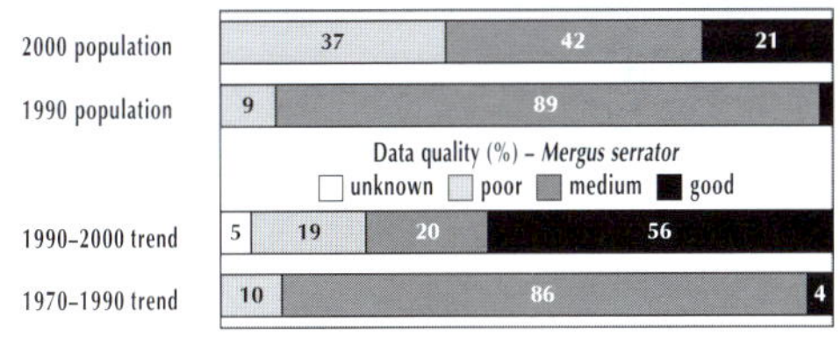

Mergus merganser
COMMON MERGANSER

Non-SPEC (1994: —) **Status** (Secure)
Criteria —
European IUCN Red List Category —
Criteria —
Global IUCN Red List Category —
Criteria —

Mergus merganser is a widespread breeder in northern Europe, but occurs more patchily farther south, with Europe accounting for less than a quarter of its global range. Its European breeding population is relatively small (<74,000 pairs), but increased between 1970–1990. Although populations increased or were stable across most of Europe during 1990–2000, key ones in Finland and Russia declined, and the species declined slightly overall. Nevertheless, this recent decline is still outweighed by earlier increases, and consequently it is provisionally evaluated as Secure.

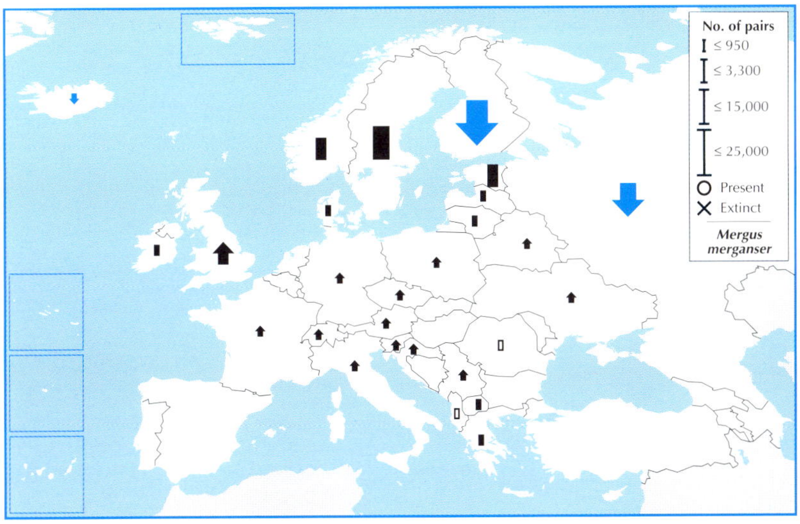

Country	Breeding pop. size (pairs)	Year(s)	Trend	Mag.%	References
Albania	2 – 5	99–00	?	–	
Austria	150 – 200	98–02	+	**>80**	
Belarus	40 – 50	92–00	+	0–9	
Croatia	**4 – 6**	02	+	N	70
Czech Rep.	**1 – 3**	00	+	**50–79**	
Denmark	**30 – 36**	98–00	**0**	**0–19**	4,5,6,7
Estonia	1,500 – 2,000	98	0	0–19	1
Finland	20,000 – 30,000	99–01	–	**20**	
France	**180 – 200**	96–00	+	**>80**	5
Germany	490 – 640	95–99	+	30–49	
Greece	**10 – 30**	97	0	0–19	
Iceland	(100 – 300)	00	(–)	(–)	1,21
Rep. Ireland	0 – 50	88–91	0	0–19	
Italy	**3 – 3**	00	+	0–19	18
Latvia	(100 – 150)	90–00	(0)	(0–19)	23,17
Lithuania	700 – 1,000	99–01	(0)	(0–19)	20
Macedonia	(3 – 5)	90–00	(0)	(0–19)	
Norway	(1,000 – 5,000)	90–03	(0)	(0–19)	
Poland	900 – 1,000	95–00	+	40–59	2,31
Romania	**1 – 1**	03	?	–	33
Russia	(8,000 – 9,000)	90–00	(–)	(20–29)	85
Serbia & MN	**20 – 25**	00–02	+	20–29	1,108,155
Slovenia	4 – 12	91–00	+	N	
Sweden	10,000 – 20,000	99–00	0	0–19	
Switzerland	**450 – 500**	93–96	+	20–29	
Ukraine	10 – 25	90–00	+	**30–49**	
UK	**2,900 – 3,600**	00	+	**24**	5,31
Total (approx.)	47,000 – 74,000	**Overall trend** Small decline			
Breeding range	>3,000,000 km²	**Gen. length.** 4		**% Global pop.** 5–24	

Country	Winter pop. size (individuals)	Year(s)	Trend	Mag.%	References
Denmark	**13,000 – 17,000**	99–00	–	10–19	25,26
Finland	10,000 – 100,000	99–01	+	**20–49**	
Germany	**32,000 – 45,000**	95–00	+	**8–30**	4
Latvia	**1,500 – 12,000**	90–99	**F**	**>80**	29
Lithuania	10,000 – 20,000	92–02	F	30–49	
Poland	40,000 – 80,000	92–97	F	30–49	115
Sweden	**8,000 – 14,000**	98–01	**0**	**0–19**	
UK	**16,100 – 16,100**	94–99	+	**6**	44,49
Total (approx.)	>150,000	**Overall trend** Small increase			
% in European IBAs	69–84	**Gen. length** 4		**% Global pop.** 5–24	

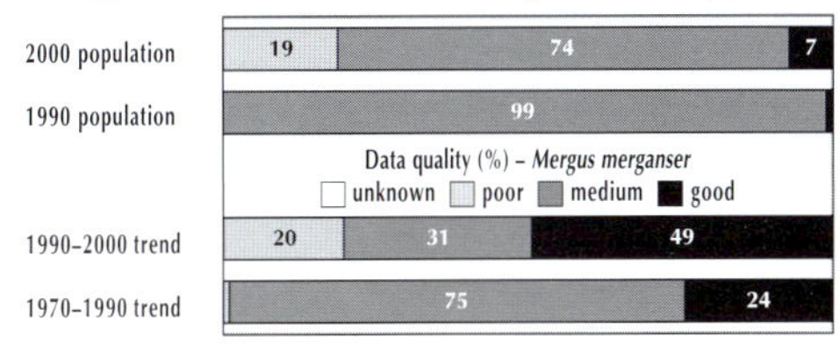

Oxyura leucocephala
WHITE-HEADED DUCK

SPEC 1 (1994: 1) **Status** Vulnerable
Criteria See IUCN below
European IUCN Red List Category VU
Criteria A3e
Global IUCN Red List Category EN
Criteria A2b,c,d,e

Oxyura leucocephala has a disjunct breeding distribution in Spain and south-eastern Europe, which accounts for less than a quarter of its global breeding range. Its European breeding population is small (as few as 550 pairs), but increased between 1970–1990. Although the species declined in Turkey during 1990–2000, the key Spanish population increased, and it increased markedly overall. However, it may undergo a large future decline (>30%) owing to the risk of hybridisation with *O. jamaicensis*. Consequently, this globally threatened species is evaluated as Vulnerable in Europe.

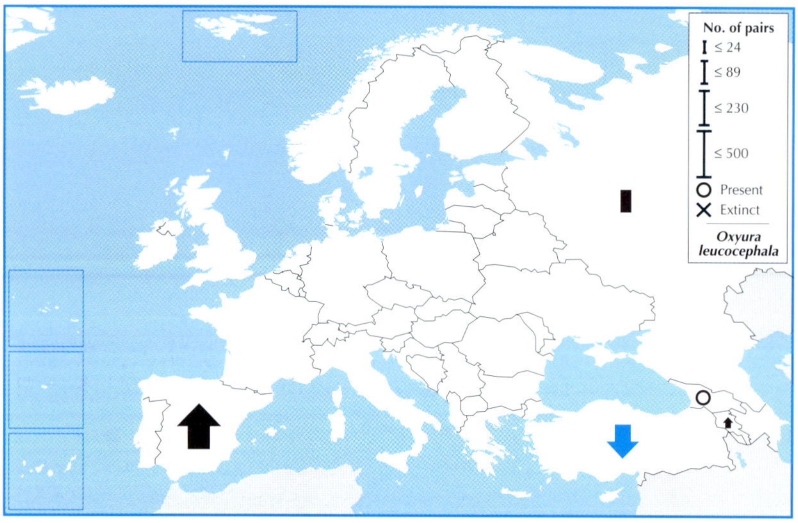

Country	Breeding pop. size (pairs)	Year(s)	Trend	Mag.%	References
Armenia	(20 – 30)	97–02	+	0–19	
Georgia	Present	03	?	–	
Russia	**80 – 100**	99–02	**0**	**0–19**	*16,69,74*
Spain	**250 – 1,000**	02	+	**>80**	*10,16*
Turkey	**200 – 250**	01	**-**	**30–49**	
Total (approx.)	550 – 1,400	**Overall trend** Moderate increase			
Breeding range	>100,000 km²	**Gen. length.** <3.3		**% Global pop.** 5–24	

Country	Winter pop. size (individuals)	Year(s)	Trend	Mag.%	References
Albania	**0 – 2**	95–02	**-**	**>80**	
Azerbaijan	2,000 – 7,000	96–02	0	0–19	
Bulgaria	**12 – 2,260**	97–02	+	**>80**	
Cyprus	0 – 6	94–02	?	–	
France	**0 – 2**	98–02	**0**	**0–19**	*9*
Greece	**260 – 2,200**	95–99	**F**	**30–49**	
Romania	50 – 700	90–00	F	50–79	*60,49*
Spain	**420 – 2,400**	90–01	+	**>80**	*19*
Turkey	**3,000 – 12,000**	91–01	**F**	**>80**	
Ukraine	0 – 8	90–00	0	0–19	
Total (approx.)	>5,700	**Overall trend** Fluctuating			
% in European IBAs	>90	**Gen. length** <3.3		**% Global pop.** 75–94	

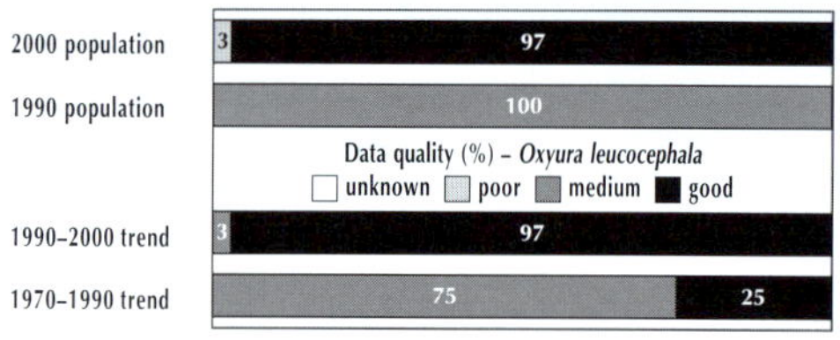

Pernis apivorus
EUROPEAN HONEY-BUZZARD

Non-SPEC[E] (1994: 4) **Status** (Secure)
Criteria —
European IUCN Red List Category —
Criteria —
Global IUCN Red List Category —
Criteria —

Pernis apivorus is a widespread summer visitor to Europe, which constitutes >75% of its global breeding range. Its European breeding population is large (>110,000 pairs), and was stable between 1970–1990. Although there were declines in countries such as Finland and Sweden during 1990–2000, key populations in Russia, Belarus and France were stable, and the species remained stable overall. Consequently, it is provisionally evaluated as Secure.

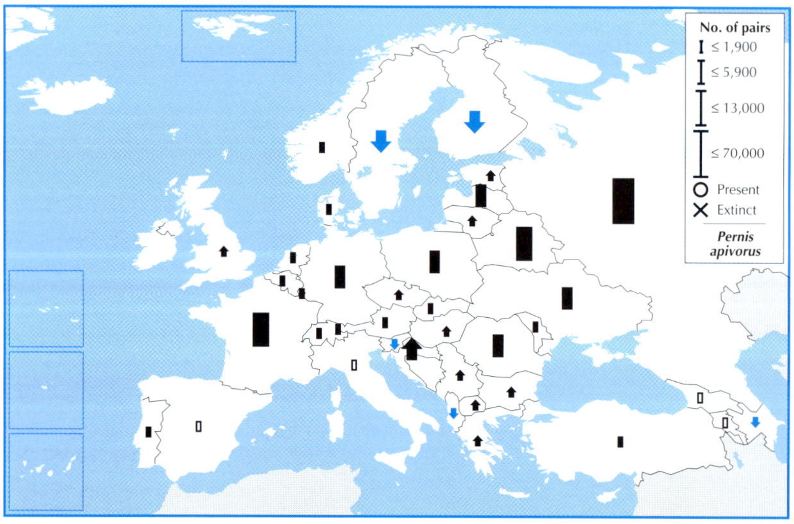

Country	Breeding pop. size (pairs)	Year(s)	Trend	Mag.%	References
Albania	(20 – 100)	96–02	(–)	(0–19)	
Armenia	50 – 250	98–02	?	–	
Austria	1,400 – 2,500	98–02	0	0–19	
Azerbaijan	(100 – 300)	96–00	(–)	(20–29)	
Belarus	**8,000 – 11,000**	97–02	0	0–19	
Belgium	464 – 702	01–02	0	0–19	*1*
Bulgaria	150 – 300	96–02	+	0–19	
Croatia	(2,500 – 5,000)	02	(+)	(>80)	*70,16,54*
Czech Rep.	**800 – 1,200**	00	+	**30–49**	
Denmark	650 – 650	00	**0**	**0–19**	*14*
Estonia	800 – 1,200	98	+	20–29	*1*
Finland	3,000 – 4,000	98–02	**-**	**20**	
France	10,600 – 15,000	00–02	(0)	(0–19)	*8*
Georgia	200 – 600	94–03	?	–	
Germany	3,800 – 5,200	95–99	0	0–19	
Greece	1,000 – 2,000	95–00	(+)	(0–19)	
Hungary	**500 – 650**	98–01	+	**>80**	*1*
Italy	(600 – 1,000)	03	?	–	*6*
Latvia	(2,000 – 3,000)	90–00	(0)	(0–19)	
Liechtenstein	(2 – 5)	98–00	(0)	(0–19)	
Lithuania	(300 – 400)	99–01	(+)	(20–29)	*20*
Luxembourg	**100 – 150**	02	**0**	**0–19**	
Macedonia	220 – 400	90–00	(+)	(20–29)	*3,6*
Moldova	**40 – 60**	90–00	**0**	**0–19**	
Netherlands	**500 – 650**	98–00	0	0–19	*1*
Norway	(500 – 1,500)	90–02	(0)	(0–19)	
Poland	2,000 – 4,000	90–00	(0)	(0–19)	*32*
Portugal	(50 – 200)	02	(0)	(0–19)	
Romania	2,000 – 2,600	90–02	(0)	(0–19)	*62*
Russia	(60,000 – 80,000)	94–02	(0)	(0–19)	
Serbia & MN	700 – 900	97–02	+	50–69	*1,158,195,239,155*
Slovakia	900 – 1,300	90–99	0	0–19	
Slovenia	600 – 800	94	(–)	(0–19)	
Spain	(900 – 1,300)	98–02	?	–	*10*
Sweden	**5,400 – 6,400**	99–00	**-**	**10–19**	
Switzerland	**400 – 600**	93–96	(0)	(0–19)	
Turkey	(500 – 800)	01	(0)	(0–19)	
Ukraine	2,000 – 2,500	90–00	0	0–19	
UK	**33 – 69**	00	+	**68**	*2*
Total (approx.)	110,000 – 160,000	**Overall trend** Stable			
Breeding range	>5,000,000 km²	**Gen. length.** 9		**% Global pop.** 75–94	

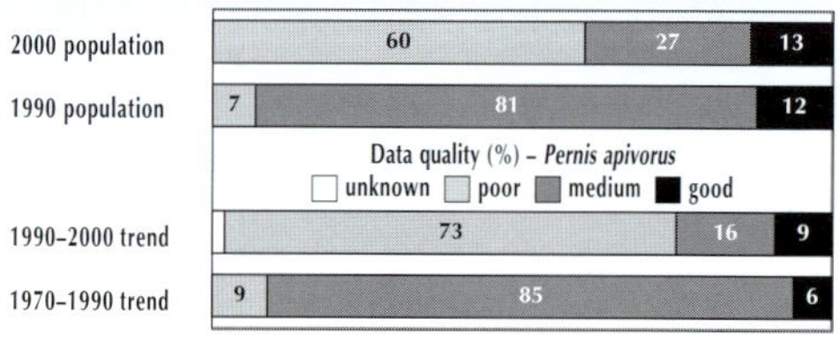

Elanus caeruleus
BLACK-WINGED KITE

SPEC 3 (1994: 3) **Status** Rare
Criteria <10,000 pairs
European IUCN Red List Category —
Criteria —
Global IUCN Red List Category —
Criteria —

Country	Breeding pop. size (pairs)	Year(s)	Trend	Mag.%	References
France	**7 – 7**	00–02	+	**>80**	*7*
Portugal	(300 – 1,000)	02	(+)	(-)	
Spain	(500 – 1,000)	98–02	(+)	(0–19)	*10,16*
Total (approx.)	810 – 2,000	**Overall trend** Small increase			
Breeding range	>100,000 km²	**Gen. length.** 5		**% Global pop.** <5	

Elanus caeruleus is resident in Iberia and south-west France, with Europe accounting for a tiny proportion of its global range. Its European breeding population is very small (as few as 810 pairs), but increased substantially between 1970–1990, and continued to increase—albeit at a slower rate—during 1990–2000. Nevertheless, its population size still renders it susceptible to the risks affecting small populations, and consequently it is evaluated as Rare.

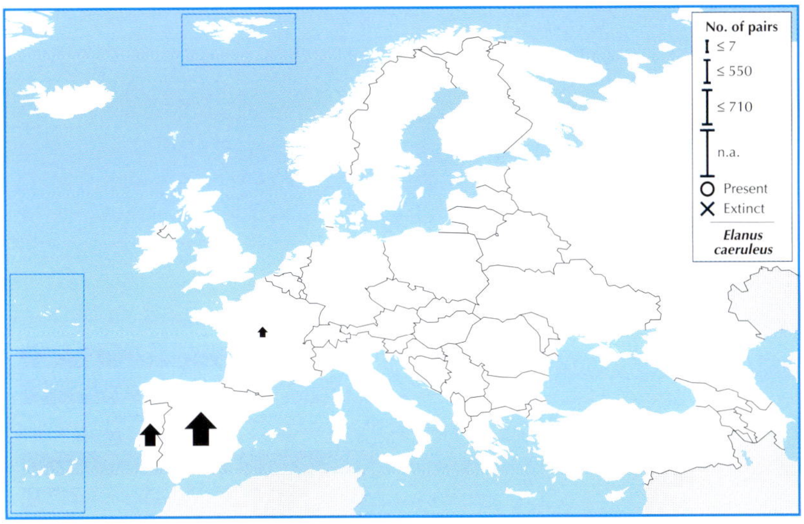

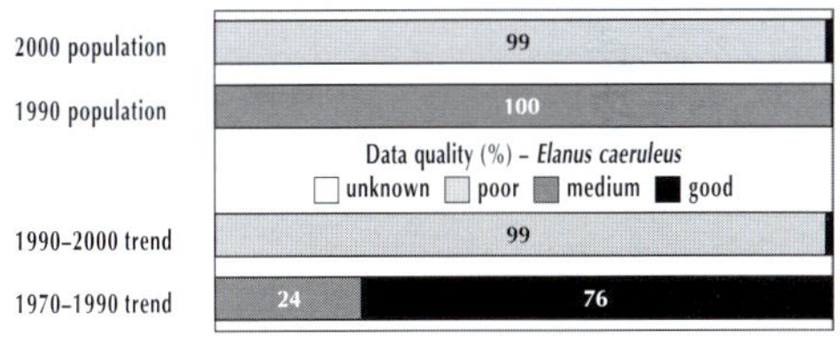

Milvus migrans
BLACK KITE

SPEC 3 (1994: 3) **Status** (Vulnerable)
Criteria See IUCN below
European IUCN Red List Category VU
Criteria A2b
Global IUCN Red List Category —
Criteria —

Country	Breeding pop. size (pairs)	Year(s)	Trend	Mag.%	References
Albania	(0 – 5)	96–02	?	–	
Armenia	20 – 80	97–02	–	30–49	
Austria	50 – 100	98–02	0	0–19	
Azerbaijan	(100 – 500)	96–00	(–)	(20–29)	
Belarus	**190 – 220**	97–02	–	0–9	
Belgium	18 – 28	95–02	+	50–79	*1*
Bosnia & HG	Present	85–89	?	–	
Bulgaria	80 – 140	96–02	0	0–19	
Croatia	(100 – 500)	02	(–)	(0–19)	*16*
Czech Rep.	**40 – 60**	00	**–**	**30–49**	
Estonia	1 – 5	98	0	0–19	*1*
Finland	(10 – 15)	98–02	(0)	(0–19)	
France	22,500 – 26,300	00–02	+	10–19	*8,7*
Georgia	200 – 400	94–03	(–)	(0–19)	
Germany	2,700 – 4,100	95–99	0	0–19	
Greece	5 – 20	95–00	(–)	(0–19)	
Hungary	**90 – 120**	97–02	**–**	**50–79**	*1*
Italy	700 – 1,200	03	0	0–19	*6*
Latvia	(10 – 30)	90–00	(–)	(20–29)	*18*
Liechtenstein	3 – 4	98–00	0	0–19	
Lithuania	**50 – 80**	99–01	–	30–49	*20*
Luxembourg	**35 – 45**	00–02	**+**	**10–19**	
Macedonia	**10 – 25**	99–00	(–)	(20–29)	*3*
Moldova	**35 – 60**	90–00	**0**	**0–19**	
Netherlands	**0 – 1**	98–00	?	–	*1*
Poland	**300 – 400**	98	(–)	(20–29)	*33*
Portugal	(800 – 1,600)	02	(0)	(0–19)	
Romania	120 – 160	00–02	(–)	(0–19)	*62*
Russia	30,000 – 50,000	94–02	–	20–29	*12,22,23,33,43, 104,114*
Serbia & MN	50 – 70	97–02	**–**	**0–19**	*1,170,155*
Slovakia	40 – 60	80–99	–	50–79	
Slovenia	1 – 3	90–00	0	0–19	
Spain	(2,500 – 10,000)	98–02	(–)	(0–19)	*10,16*
Sweden	**0 – 2**	99–00	**0**	**0–19**	
Switzerland	**1,200 – 1,500**	93–96	(–)	(0–19)	
Turkey	(1,000 – 1,500)	01	–	20–29	
Ukraine	1,500 – 1,800	90–00	–	30–49	
Total (approx.)	64,000 – 100,000	**Overall trend** Large decline			
Breeding range	>5,000,000 km²	**Gen. length.** 6		**% Global pop.** 5–24	

Milvus migrans is a widespread summer visitor to much of Europe, which accounts for less than a quarter of its global breeding range. Its European breeding population is relatively small (<100,000 pairs), and declined substantially between 1970–1990. Although some populations—notably the French one—were stable or increased during 1990–2000, the species declined across much of Europe, and declined overall at a rate that—on top of earlier declines—equates to a large decline (>30%) over three generations. Consequently, it is provisionally evaluated as Vulnerable.

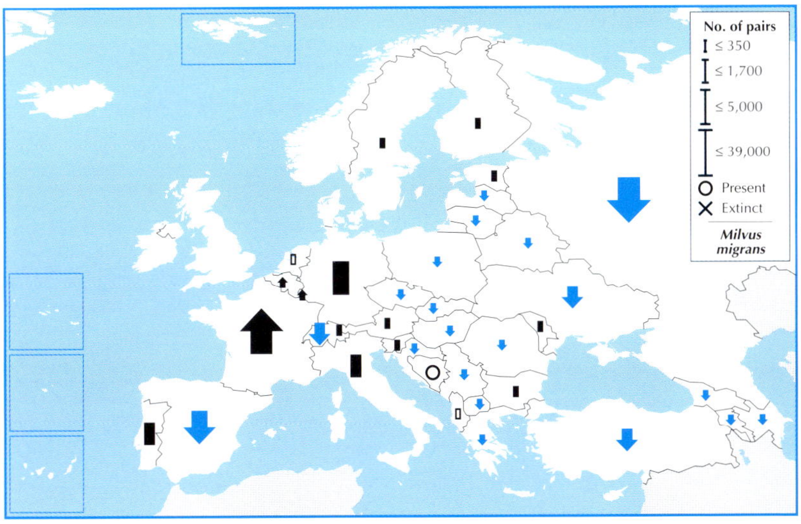

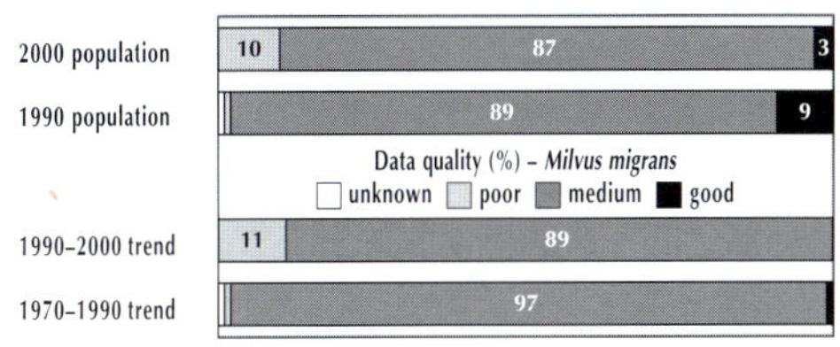

Milvus milvus
RED KITE

SPEC 2 (1994: 4) **Status** Declining
Criteria Moderate recent decline
European IUCN Red List Category —
Criteria —
Global IUCN Red List Category —
Criteria —

Milvus milvus breeds mainly in Iberia and west-central Europe, which constitutes >95% of its global breeding range. Its European breeding population is relatively small (<25,000 pairs), but was stable between 1970–1990. Although trends were stable or increased in several countries during 1990–2000, key populations in Germany, France and Spain declined, and the species underwent a moderate decline (>10%) overall. Consequently, this previously Secure species is now evaluated as Declining.

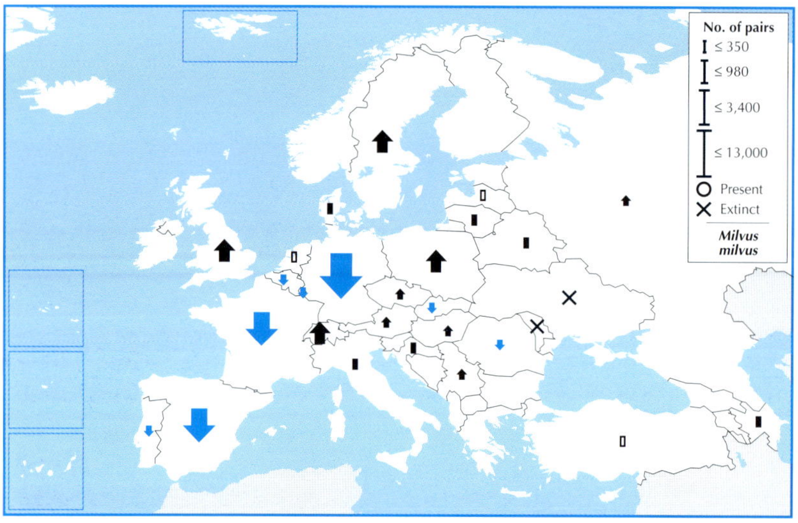

Country	Breeding pop. size (pairs)	Year(s)	Trend	Mag.%	References
Austria	5 – 10	98–02	+	>80	
Azerbaijan	(0 – 2)	96–00	(0)	(0–19)	
Belarus	3 – 10	94–97	0	0–19	
Belgium	51 – 81	95–02	–	0–19	1
Croatia	(3 – 5)	02	(0)	(0–19)	70,57
Czech Rep.	70 – 100	00	+	>80	
Denmark	17 – 22	01–02	0	0–19	8
France	3,000 – 3,800	00–02	–	10–19	8,7
Germany	10,500 – 14,000	95–99	–	0–19	
Hungary	4 – 10	97–02	+	>80	1
Italy	300 – 400	03	0	0–19	6
Latvia	(0 – 3)	90–00	?	–	23,18
Lithuania	(1 – 10)	99–01	(F)	(>80)	20
Luxembourg	35 – 45	02	–	10–19	
Moldova	0 – 0	00	–	X	
Netherlands	0 – 1	98–00	?	–	1
Poland	650 – 700	98	+	20–29	33
Portugal	50 – 100	02	(–)	(–)	8
Romania	(0 – 5)	95–02	(–)	(0–19)	62
Russia	5 – 10	99–02	+	20–29	74,94,171
Serbia & MN	3 – 5	98–02	+	50–79	1,171,146,155
Slovakia	15 – 20	80–99	–	30–49	
Spain	1,900 – 2,700	98–02	–	30–49	10,16
Sweden	800 – 850	99–00	+	50–79	
Switzerland	800 – 1,200	93–96	+	0–19	
Turkey	(0 – 10)	01	?	–	
Ukraine	0 – 0	00	–	X	
UK	372 – 490	00	+	284	42
Total (approx.)	19,000 – 25,000	Overall trend Moderate decline			
Breeding range	>1,000,000 km²	Gen. length. 6		% Global pop. >95	

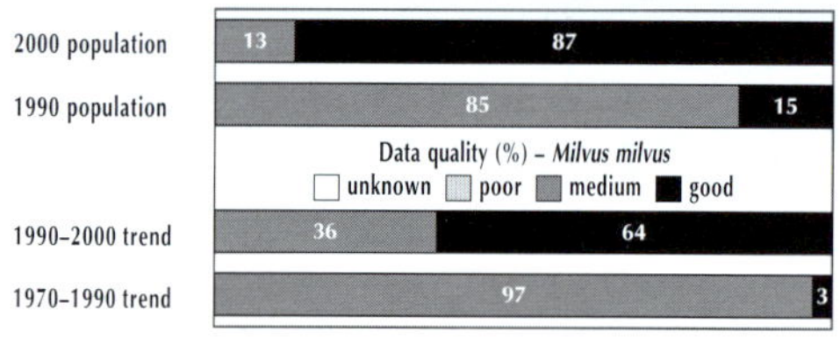

Haliaeetus albicilla
WHITE-TAILED EAGLE

SPEC 1 (1994: 3) **Status** Rare
Criteria <10,000 pairs
European IUCN Red List Category —
Criteria —
Global IUCN Red List Category NT
Criteria C2a(i)

Haliaeetus albicilla is a widespread but patchily distributed breeder in parts of northern, eastern and central Europe, which probably now holds >50% of the global population. Its European breeding population is small (as few as 5,000 pairs), but increased substantially between 1970–1990. With the exception of a few countries in south-eastern Europe, the species continued to increase across the continent during 1990–2000, and underwent a large increase overall. Consequently, this globally Near Threatened species is evaluated as Rare in Europe.

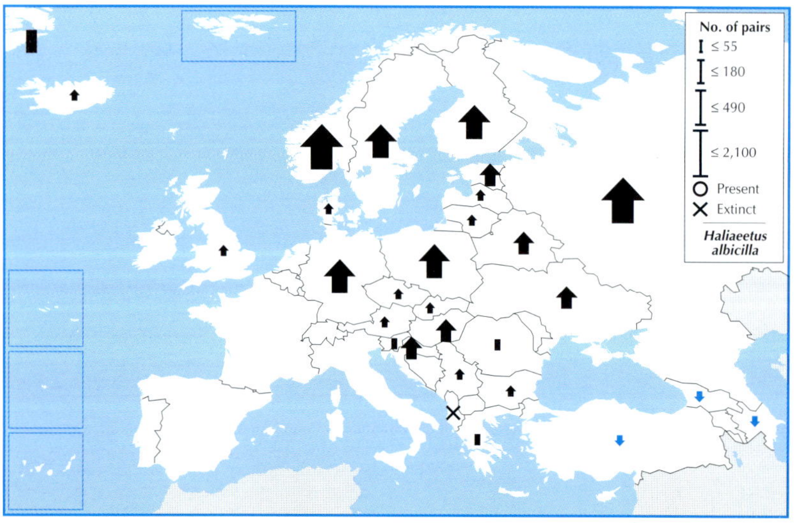

Country	Breeding pop. size (pairs)	Year(s)	Trend	Mag.%	References
Albania	0 – 0	00–02	–	X	
Austria	4 – 4	03	+	N	
Azerbaijan	5 – 10	96–00	(–)	(20–29)	
Belarus	85 – 105	97–02	+	0–9	
Bulgaria	7 – 10	01–02	+	>80	
Croatia	80 – 90	02	+	>80	70,16,44
Czech Rep.	21 – 28	00	+	>80	
Denmark	7 – 7	01	+	N	7
Greenland	150 – 200	80–90	0	0–19	10
Estonia	70 – 80	98	+	20–29	1
Finland	230 – 270	01–02	+	180	
Georgia	1 – 3	94–03	–	>80	
Germany	321 – 326	95–99	+	30–49	
Greece	6 – 6	02	0	0–19	
Hungary	60 – 100	97–02	+	>80	1
Iceland	50 – 50	02	+	20–29	21,34,35,36
Latvia	30 – 40	90–00	+	30–49	21
Lithuania	40 – 50	99–01	+	30–49	20
Norway	1,900 – 2,200	02	+	0–19	16
Poland	450 – 520	98	+	40–49	33,35
Romania	28 – 33	95–02	0	0–19	62
Russia	1,000 – 2,000	94–02	+	20–29	2,62,63,74,86, 94,103
Serbia & MN	52 – 58	00–02	+	50–69	1,68,183,155,16
Slovakia	0 – 5	80–99	+	20–29	
Slovenia	1 – 3	90–00	0	0–19	
Sweden	260 – 280	99–00	+	80–99	
Turkey	8 – 15	01	–	50–79	
Ukraine	80 – 100	90–00	+	30–49	
UK	16 – 16	96–00	+	144	
Total (approx.)	5,000 – 6,600	Overall trend Large increase			
Breeding range	>2,000,000 km²	Gen. length. 16		% Global pop. 50–74	

(See p. 73, top, for data quality graph)

Country	Winter pop. size (individuals)	Year(s)	Trend	Mag.%	References
Denmark					
Greenland	(300 – 500)	90–00	(0)	(0–19)	
Finland	1,000 – 2,000	01–02	+	50–150	
Hungary	200 – 400	00–02	?	–	24
Poland	(1,000 – 2,500)	95–02	+	>80	
Sweden	1,000 – 1,500	98–01	+	>80	
Ukraine	260 – 370	96–01	+	30–49	
Total (approx.)	>4,700	Overall trend Large increase			
% in European IBAs	20–23	Gen. length 16		% Global pop. 50–74	

Gypaetus barbatus
LAMMERGEIER

SPEC 3 (1994: 3) **Status** (Vulnerable)
Criteria See IUCN below
European IUCN Red List Category VU
Criteria C1; C2a(i)
Global IUCN Red List Category —
Criteria —

Country	Breeding pop. size (pairs)	Year(s)	Trend	Mag.%	References
Albania	(0 – 5)	02–03	?	–	
Andorra	**0 – 1**	01	**0**	**0–19**	*1,3*
Armenia	**4 – 10**	99–02	**-**	**0–9**	*3*
Azerbaijan	(10 – 30)	96–00	(0)	(0–19)	
France	**40 – 40**	03	**+**	**30–49**	*8,7*
Georgia	**19 – 25**	94–03	**0**	**0–19**	
Greece	**4 – 4**	02	**-**	**20–29**	
Italy	**1 – 3**	98–03	+	30–49	*6*
Macedonia	0 – 4	96–00	(–)	(30–49)	*2,3*
Russia	(50 – 100)	94–02	(0)	(0–19)	*26,72,74,94,153*
Spain	**81 – 81**	01	**+**	**>80**	*10,16*
Turkey	400 – 700	01	(–)	(0–19)	
Total (approx.)	610 – 1,000	**Overall trend** Moderate decline			
Breeding range	>250,000 km²	**Gen. length.** 15		**% Global pop.** 5–24	

Gypaetus barbatus is resident in Turkey and the Caucasus and locally elsewhere in southern Europe, which accounts for less than a quarter of its global range. Its European breeding population is very small (as few as 610 pairs), but was stable between 1970–1990. Although the species was stable or increased in most countries during 1990–2000, there were declines in the Turkish stronghold, and the species underwent a moderate decline (>10%) overall. As a consequence of this decline and its small population, it is provisionally evaluated as Vulnerable.

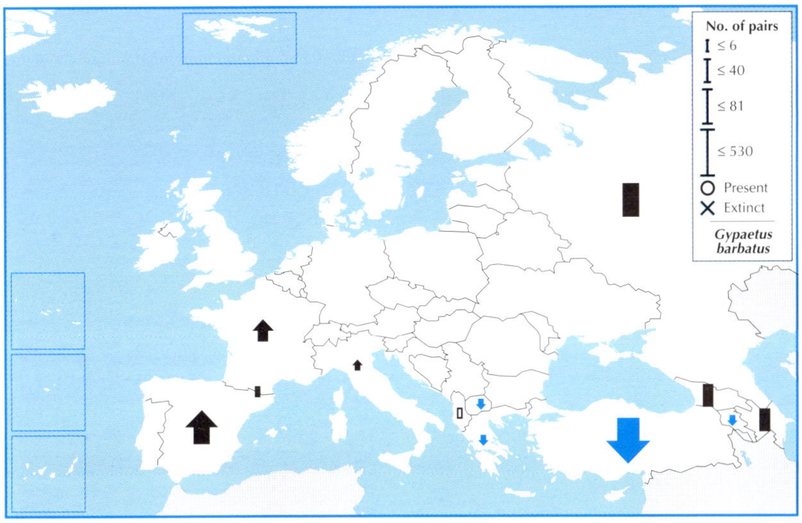

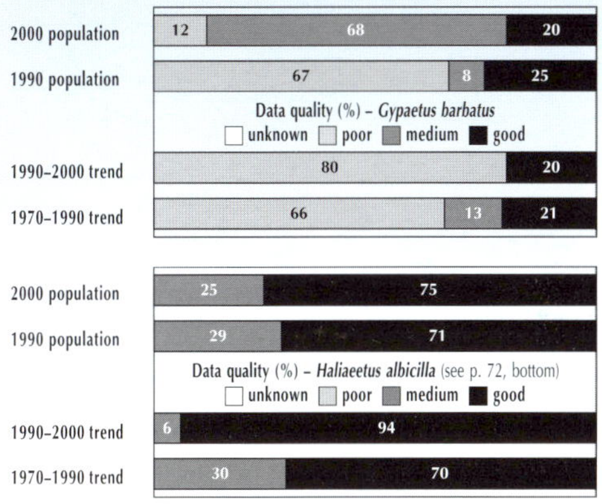

Neophron percnopterus
EGYPTIAN VULTURE

SPEC 3 (1994: 3) **Status** Endangered
Criteria See IUCN below
European IUCN Red List Category EN
Criteria A2b
Global IUCN Red List Category —
Criteria —

Country	Breeding pop. size (pairs)	Year(s)	Trend	Mag.%	References
Albania	30 – 60	02–03	(–)	(0–19)	
Armenia	50 – 70	99–02	–	0–9	
Azerbaijan	(100 – 300)	96–00	(0)	(0–19)	
Bosnia & HG	Present	85–89	?	–	
Bulgaria	**65 – 80**	00–02	**-**	**50–79**	
France	**69 – 75**	00–03	**+**	**0–19**	*8,7*
Georgia	100 – 140	94–03	**0**	**0–19**	
Greece	100 – 150	95–00	(–)	(10–19)	
Italy	15 – 20	03	–	20–29	*6*
Macedonia	45 – 60	00	F	20–29	*2,3*
Moldova	0 – 2	90–00	**-**	**>80**	
Portugal	**83 – 84**	00	(–)	(–)	*9,11,12*
Russia	(70 – 120)	94–02	(–)	(20–29)	*26,72,74,94,153*
Serbia & MN	**2 – 3**	00–02	**0**	**0–19**	*1,58,44,47,155, 43,97*
Spain	**1,300 – 1,500**	00	**-**	**20–29**	*10,3*
Canary Is.	**22 – 26**	97–03	**-**	**0–19**	*28,21*
Turkey	1,500 – 3,000	01	–	30–49	
Total (approx.)	3,500 – 5,600	**Overall trend** Large decline			
Breeding range	>1,000,000 km²	**Gen. length.** 14		**% Global pop.** 25–49	

Neophron percnopterus is a widespread but patchily distributed breeder in southern Europe, which accounts for less than half of its global breeding range. Its European breeding population is small (as few as 3,500 pairs), and declined substantially between 1970–1990. The species continued to decline in most countries—including its key populations in Spain and Turkey—during 1990–2000, and underwent a large decline overall. On top of earlier declines, this equates to a very large decline (>50%) over three generations. Consequently, it is evaluated as Endangered.

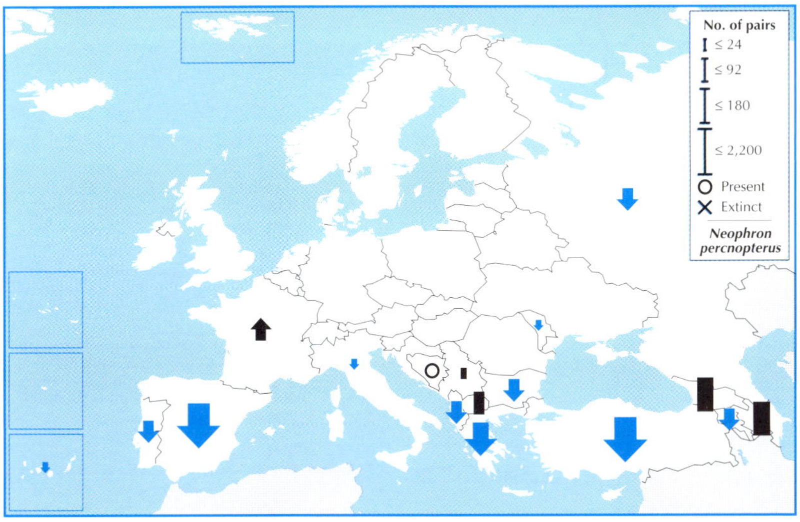

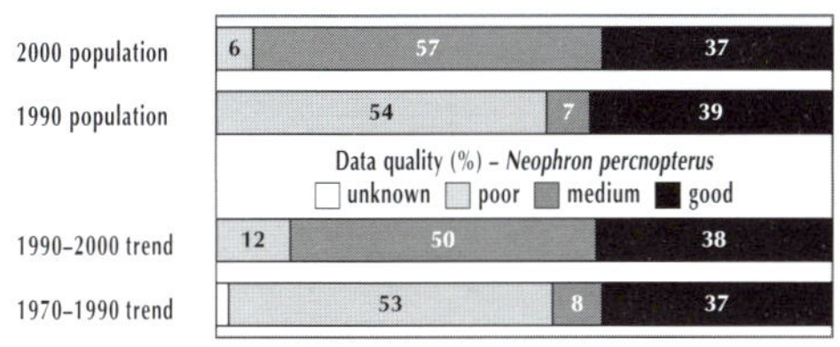

Gyps fulvus
EURASIAN GRIFFON

Non-SPEC (1994: 3) **Status** Secure
Criteria —
European IUCN Red List Category —
Criteria —
Global IUCN Red List Category —
Criteria —

Country	Breeding pop. size (pairs)	Year(s)	Trend	Mag.%	References
Albania	10 – 50	02–03	(–)	(50–79)	
Armenia	15 – 60	99–02	–	30–39	*6*
Azerbaijan	(10 – 50)	96–00	(0)	(0–19)	
Bulgaria	14 – 30	98–01	+	>80	
Croatia	90 – 100	02	0	0–19	*70*
Cyprus	8 – 10	98–02	0	0–9	
France	589 – 639	03	+	50–79	*8,7*
Georgia	60 – 70	94–03	–	0–19	
Greece	173 – 194	00–02	0	0–19	
Italy	37 – 42	03	+	0–19	*6*
Macedonia	30 – 50	00	0	10–19	*2,3*
Portugal	267 – 272	99	(+)	(–)	*1*
Russia	(200 – 400)	94–02	(–)	(20–29)	*26,72,74,94,153*
Serbia & MN	55 – 60	01–02	+	50–79	*109,44,155*
Spain	17,300 – 18,100	99	+	>80	*10,2*
Turkey	300 – 600	01	–	50–79	
Ukraine	8 – 10	90–00	–	20–29	
Total (approx.)	19,000 – 21,000	**Overall trend** Large increase			
Breeding range	>1,000,000 km²	**Gen. length.** 16		**% Global pop.** 25–49	

Gyps fulvus is a widespread but patchily distributed resident in southern Europe, which accounts for less than half of its global range. Its European breeding population is relatively small (<21,000 pairs), but increased substantially between 1970–1990. Although the species continued to decline in Turkey and the Caucasus during 1990–2000, most other populations—including the Spanish stronghold—increased or were stable, and the species underwent a large increase overall. Consequently, this previously Rare species is now evaluated as Secure.

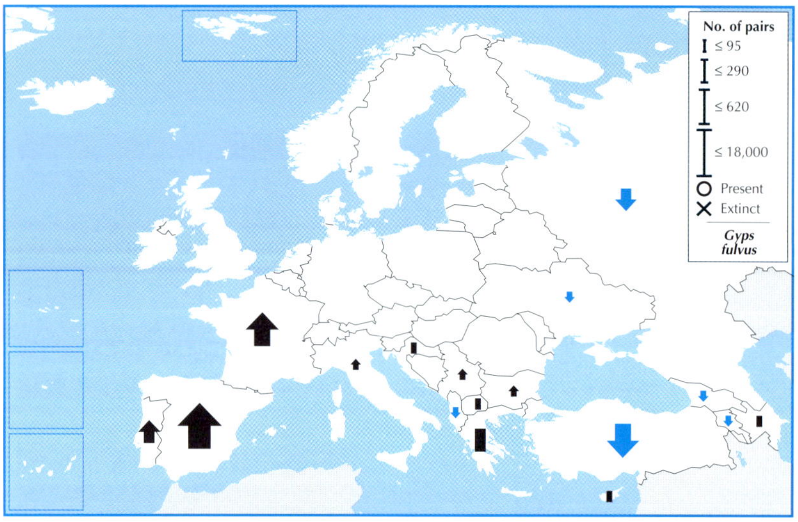

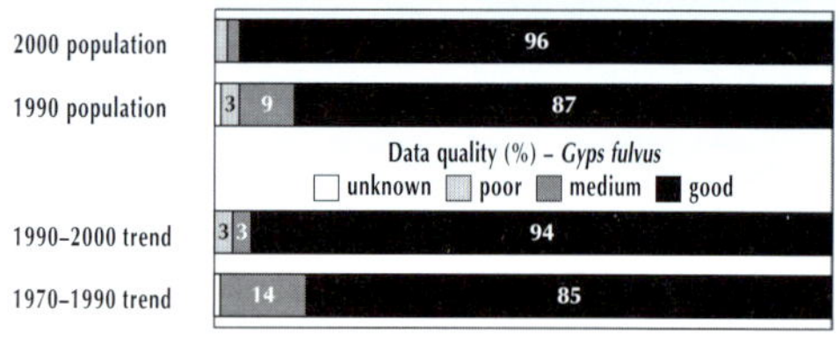

Aegypius monachus
CINEREOUS VULTURE

SPEC 1 (1994: 3) **Status** Rare
Criteria <10,000 pairs
European IUCN Red List Category —
Criteria —
Global IUCN Red List Category NT
Criteria C1

Country	Breeding pop. size (pairs)	Year(s)	Trend	Mag.%	References
Armenia	8 – 15	01–02	–	20–29	
Azerbaijan	(10 – 30)	96–00	(–)	(30–49)	
Bulgaria	1 – 1	93	?	–	
France	8 – 10	02	+	>80	*8,7*
Georgia	20 – 30	94–03	–	20–29	
Greece	21 – 21	02	0	0–19	
Macedonia	0 – 4	00	0	0–19	*2,3*
Portugal	0 – 5	02	+	N	
Russia	(30 – 70)	94–02	(–)	(20–29)	*26,74,94,153*
Spain	1,400 – 1,400	01	+	20–29	*10,16*
Turkey	300 – 400	01	–	0–19	
Ukraine	2 – 3	90–00	–	20–29	
Total (approx.)	1,800 – 1,900	**Overall trend** Large increase			
Breeding range	>500,000 km²	**Gen. length.** 16		**% Global pop.** 25–49	

Aegypius monachus is a patchily distributed resident in southern Europe, which accounts for less than half of its global breeding range. Its European breeding population is small (as few as 1,800 pairs), but increased markedly between 1970–1990. Although the species declined across much of south-eastern Europe during 1990–2000, numbers in the Spanish stronghold increased, and the species underwent a large increase (>30%) overall. Consequently, this globally Near Threatened species is evaluated as Rare in Europe.

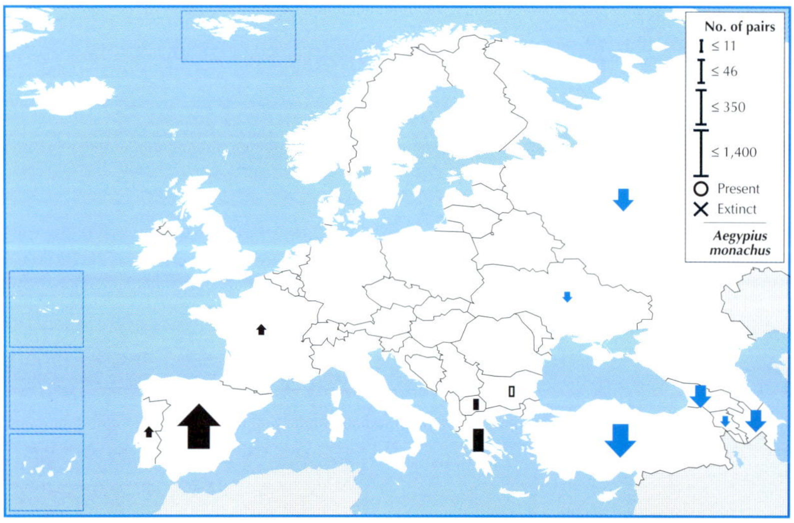

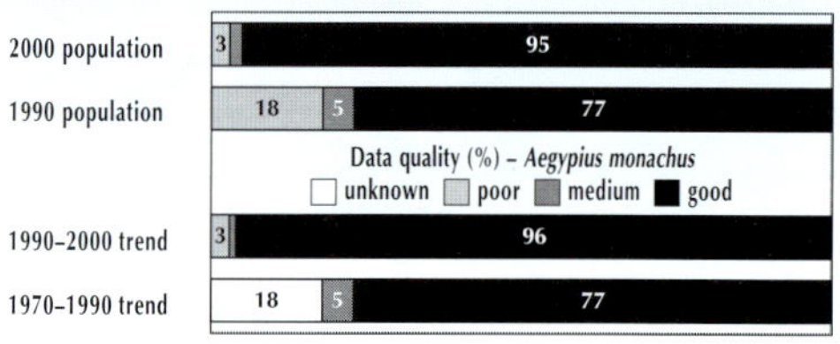

Circaetus gallicus
SHORT-TOED SNAKE-EAGLE

SPEC 3 (1994: 3) **Status** (Rare)
Criteria <10,000 pairs
European IUCN Red List Category —
Criteria —
Global IUCN Red List Category —
Criteria —

Circaetus gallicus is a widespread summer visitor to much of Europe, which accounts for less than half of its global breeding range. Its European breeding population is small (as few as 8,400 pairs), but was stable between 1970–1990. Although there were declines in Turkey and several smaller populations during 1990–2000, the species was stable or increased across most of Europe—including the sizeable population in France—and the species probably only declined slightly overall. Consequently, it is provisionally evaluated as Rare.

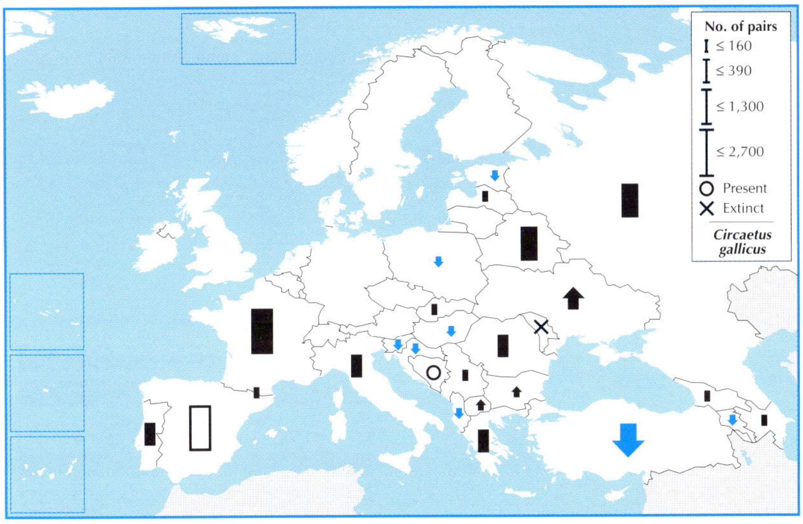

Country	Breeding pop. size (pairs)	Year(s)	Trend	Mag.%	References
Albania	50 – 100	96–02	(–)	(0–19)	
Andorra	3 – 5	98	0	0–19	*1,3*
Armenia	20 – 80	98–02	–	0–9	
Azerbaijan	(50 – 300)	96–00	(0)	(0–19)	
Belarus	**600 – 700**	97–02	**0**	**0–19**	
Bosnia & HG	Present	90–03	?	–	
Bulgaria	80 – 200	96–02	+	50–79	
Croatia	(100 – 250)	02	(–)	(50–79)	*70,54,26*
Estonia	5 – 8	98	–	20–29	*1*
France	2,400 – 2,900	00–03	0	0–19	*8,7*
Georgia	15 – 25	94–03	0	0–19	
Greece	300 – 500	95–00	0	0–19	
Hungary	**30 – 50**	97–02	**–**	**20–29**	*1*
Italy	(350 – 400)	03	0	0–19	*6*
Latvia	(5 – 20)	90–00	(F)	(–)	*23*
Macedonia	**50 – 150**	99–00	(+)	(20–29)	*3*
Moldova	**0 – 0**	00	**–**	**X**	
Poland	**10 – 15**	90–00	(–)	(30–49)	*34*
Portugal	(250 – 600)	02	(0)	(0–19)	
Romania	(220 – 300)	95–02	(0)	(0–19)	*62*
Russia	(500 – 1,000)	94–02	(0)	(0–19)	*2,6,26,33,60,67,72,74,76*
Serbia & MN	130 – 170	97–02	0	0–19	*1,159,168,155*
Slovakia	(20 – 25)	80–99	0	0–19	
Slovenia	10 – 15	99–00	–	0–19	
Spain	(2,000 – 3,000)	98–02	?	–	*10*
Turkey	(1,000 – 1,500)	01	–	0–19	
Ukraine	160 – 300	90–00	+	0–19	
Total (approx.)	8,400 – 13,000	**Overall trend** Small decline			
Breeding range	>3,000,000 km²	**Gen. length.** 13		**% Global pop.** 25–49	

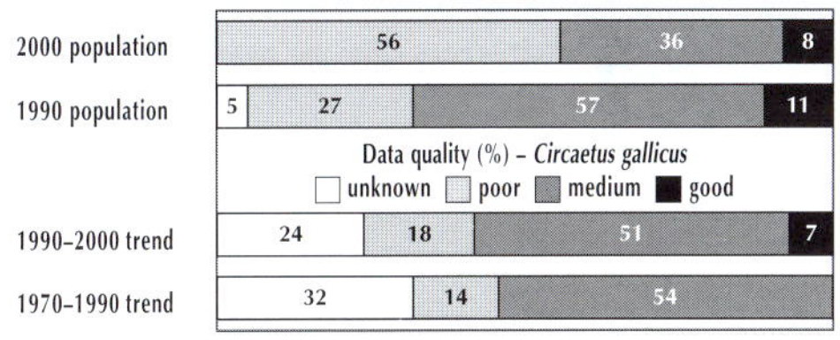

Circus aeruginosus
WESTERN MARSH-HARRIER

Non-SPEC (1994: —) **Status** Secure
Criteria —
European IUCN Red List Category —
Criteria —
Global IUCN Red List Category —
Criteria —

Circus aeruginosus is a widespread breeder across much of Europe, which accounts for less than half of its global breeding range. Its European breeding population is relatively small (<140,000 pairs), but increased between 1970–1990. Although there were some declines in south-eastern Europe during 1990–2000, populations elsewhere increased—notably in Ukraine and Russia—or were stable, and the species underwent a moderate increase (>10%) overall. Consequently, it is evaluated as Secure

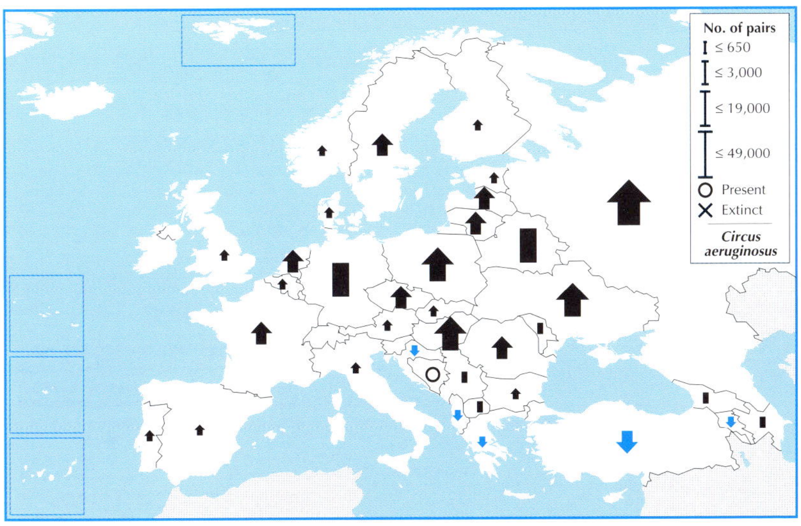

Country	Breeding pop. size (pairs)	Year(s)	Trend	Mag.%	References
Albania	5 – 15	95–02	(–)	(0–19)	
Armenia	50 – 80	97–02	–	0–9	
Austria	300 – 400	98–02	+	**20–29**	
Azerbaijan	(100 – 500)	96–00	(0)	(0–19)	
Belarus	**6,000 – 9,000**	97–02	**0**	**0–19**	
Belgium	**169 – 208**	95–02	+	**0–19**	*1*
Bosnia & HG	Present	90–03	?	–	
Bulgaria	**80 – 150**	95–02	+	**50–79**	
Croatia	40 – 100	02	(–)	(20–29)	*70,26,16*
Czech Rep.	**1,300 – 1,700**	00	+	**30–49**	
Denmark	650 – 650	00	+	**>80**	*14*
Estonia	350 – 600	98	+	20–29	*1*
Finland	**350 – 500**	98–02	+	**25**	
France	**1,600 – 2,200**	00–02	+	10–19	*8,7*
Georgia	80 – 100	94–03	0	0–19	
Germany	5,500 – 8,400	95–99	0	0–19	
Greece	50 – 80	95–00	–	0–19	
Hungary	5,200 – 6,700	99–02	+	20–29	*19,1*
Italy	170 – 220	03	+	0–19	*6*
Latvia	1,000 – 1,500	90–00	+	50–79	*25*
Lithuania	2,500 – 3,500	99–01	(+)	(0–19)	*20*
Macedonia	**30 – 80**	99–00	(0)	(0–19)	*3*
Moldova	**50 – 70**	90–00	**0**	**0–19**	
Netherlands	**1,300 – 1,450**	98–00	+	**10**	*1*
Norway	0 – 5	90–03	+	0–19	*63*
Poland	**6,500 – 8,000**	02–03	(+)	(0–19)	*10*
Portugal	**110 – 116**	98	(+)	(–)	*14*
Romania	(1,700 – 2,500)	98–02	(+)	(0–19)	*62*
Russia	40,000 – 60,000	94–02	+	20–29	*2,104,132,134*
Serbia & MN	280 – 350	97–02	0	0–19	*1,69,155,100,101,13,78*
Slovakia	400 – 500	80–99	+	20–29	
Spain	250 – 1,000	98–02	+	30–49	*10*
Sweden	**1,400 – 1,500**	99–00	+	**30–49**	
Turkey	1,500 – 2,000	01	–	0–19	
Ukraine	13,800 – 23,600	90–00	+	5–19	
UK	**151 – 151**	96–00	+	**87**	*24*
Total (approx.)	93,000 – 140,000	**Overall trend** Moderate increase			
Breeding range	>5,000,000 km²	**Gen. length.** 6		**% Global pop.** 25–49	

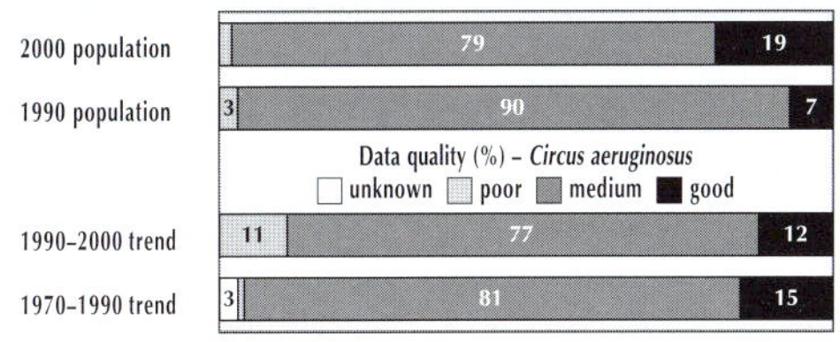

Circus cyaneus
NORTHERN HARRIER

SPEC 3 (1994: 3) **Status** Depleted
Criteria Large historical decline
European IUCN Red List Category —
Criteria —
Global IUCN Red List Category —
Criteria —

Circus cyaneus is a widespread but patchily distributed breeder across much of northern and central Europe, which accounts for less than a quarter of its global breeding range. Its European breeding population is relatively small (<59,000 pairs), and underwent a large decline between 1970–1990. These declines abated to some extent during 1990–2000, with many populations stabilising, and the species declined only slightly overall. Nevertheless, its population remains far below the level that preceded its decline, and consequently it is evaluated as Depleted.

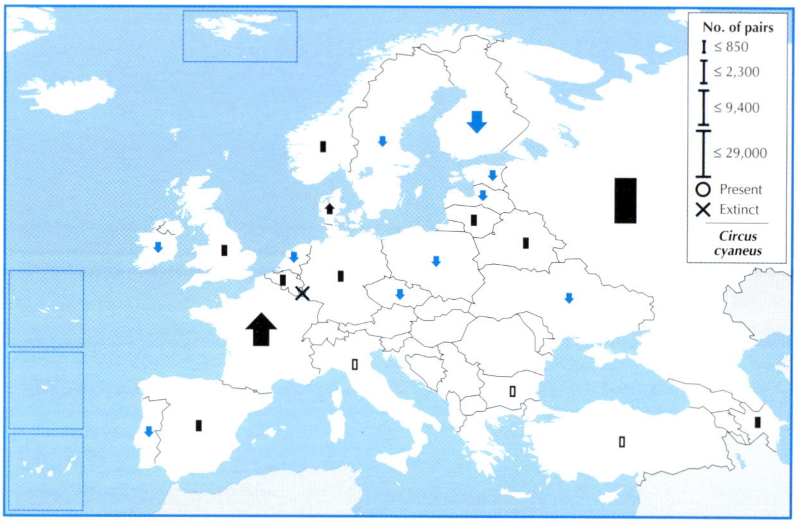

Country	Breeding pop. size (pairs)	Year(s)	Trend	Mag.%	References
Azerbaijan	(50 – 200)	96–00	(0)	(0–19)	
Belarus	**520 – 700**	97–02	(0)	(0–19)	
Belgium	**0 – 1**	95–02	**0**	**0–19**	*1*
Bulgaria	(0 – 2)	95–02	?	–	
Czech Rep.	**30 – 50**	00	**-**	**30–49**	
Denmark	**0 – 7**	98–01	+	N	*4,5,6,7*
Estonia	150 – 200	98	–	20–29	*1*
Finland	1,500 – 3,500	98–02	–	30	
France	7,800 – 11,200	00–02	+	0–9	*8,7*
Germany	**48 – 76**	95–99	**0**	**0–19**	
Rep. Ireland	**102 – 131**	98–00	–	20–29	
Italy	0 – 1	98–03	?	–	*6*
Latvia	(5 – 20)	90–00	(–)	(0–19)	*23*
Lithuania	(0 – 5)	99–01	(F)	(>80)	*20*
Luxembourg	**0 – 0**	02	**-**	**X**	
Netherlands	**85 – 105**	98–00	**-**	**40**	*1*
Norway	(5 – 50)	90–03	(0)	(0–19)	*63*
Poland	10 – 30	95–00	(–)	(40–69)	*2,35,36*
Portugal	10 – 30	02	(–)	(–)	
Russia	20,000 – 40,000	94–02	0	0–19	*2,68,73,103*
Spain	(250 – 1,000)	98–02	(0)	(0–19)	*10*
Sweden	800 – 900	99–00	–	10–19	
Turkey	(0 – 10)	01	?	–	
Ukraine	10 – 25	90–00	**-**	**0–19**	
UK	**500 – 640**	98	**0**	**3**	*30*
Total (approx.)	32,000 – 59,000	**Overall trend** Small decline			
Breeding range	>4,000,000 km²	**Gen. length.** 6		**% Global pop.** 5–24	

Country	Winter pop. size (individuals)	Year(s)	Trend	Mag.%	References
Armenia	250 – 1,000	97–02	?	–	
Austria	(300 – 600)	98–02	(F)	(>80)	
Croatia	(500 – 1,000)	02	?	–	*16*
Hungary	1,500 – 3,000	00–02	?	–	*24*
Netherlands	400 – 600	00	–	30–49	*4,5,6,7,9,10*
Poland	(1,000 – 2,500)	90–00	(–)	(20–29)	*121*
Serbia & MN	(300 – 800)	90–02	(F)	(10–29)	
Slovakia	3,000 – 5,500	90–99	?	–	*4*
Ukraine	(250 – 1,000)	90–00	(F)	(20–29)	
Total (approx.)	>8,500	**Overall trend** Unknown			
% in European IBAs	18–23	**Gen. length** 6		**% Global pop.** 5–24	

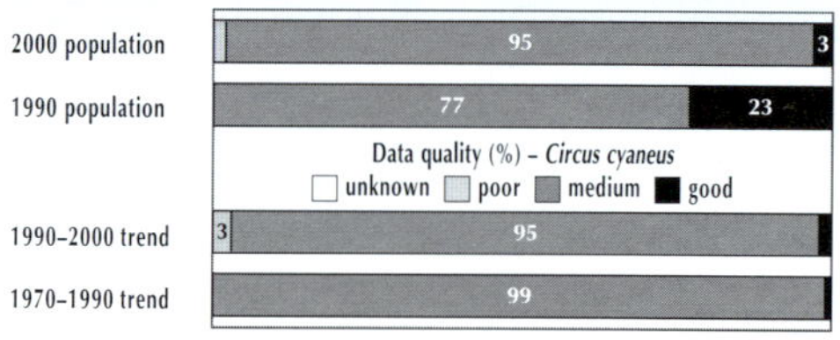

Circus macrourus
PALLID HARRIER

SPEC 1 (1994: 3) **Status** (Endangered)
Criteria See IUCN below
European IUCN Red List Category EN
Criteria C1
Global IUCN Red List Category NT
Criteria A2c,d,e; A3c,d,e

Circus macrourus is a summer visitor to eastern Europe, which accounts for less than half of its global breeding range. Its European breeding population is very small (as few as 310 pairs), and declined substantially between 1970–1990. It continued to decline in its Russian stronghold during 1990–2000, and declined overall at a rate that—on top of earlier declines—equates to a large decline (>20%) over two generations. As a consequence of this decline and its very small population, this globally Near Threatened species is provisionally evaluated as Endangered in Europe.

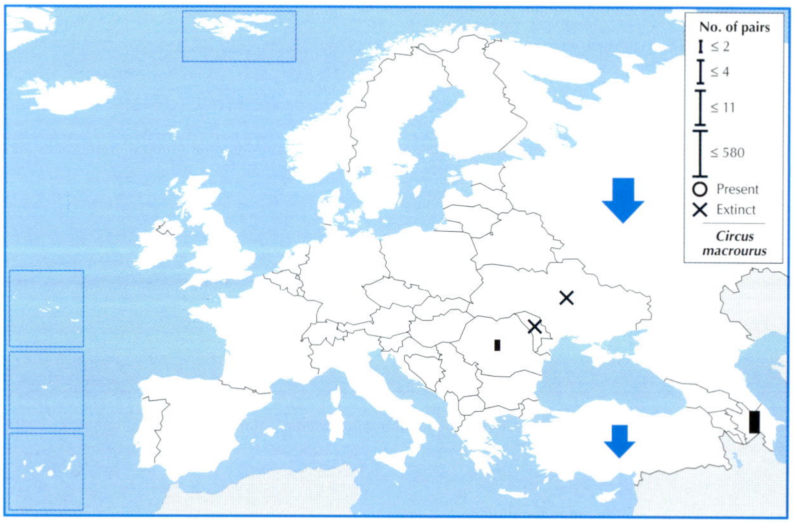

Country	Breeding pop. size (pairs)	Year(s)	Trend	Mag.%	References
Azerbaijan	(0 – 20)	96–00	(0)	(0–19)	
Moldova	0 – 0	96–00	–	X	
Romania	0 – 6	90–02	(F)	(50–79)	*62*
Russia	(300 – 1,100)	94–01	(–)	(20–29)	*15,157*
Turkey	(5 – 25)	01	**-**	**30–49**	
Ukraine	**0 – 0**	00	**-**	**X**	
Total (approx.)	310 – 1,200	**Overall trend** Large decline			
Breeding range	>1,000,000 km²	**Gen. length.** 6		**% Global pop.** 25–49	

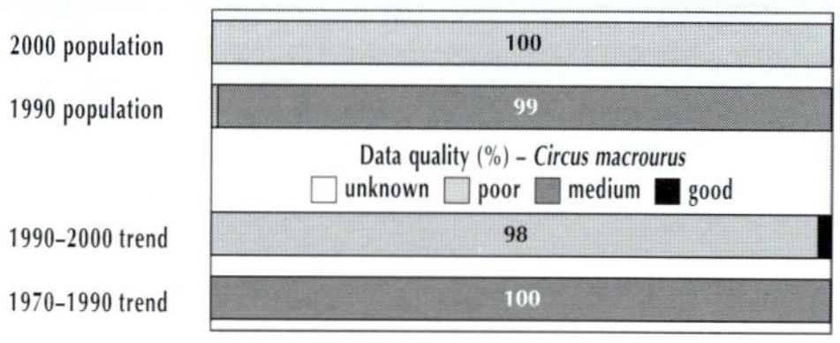

Circus pygargus
MONTAGU'S HARRIER

Non-SPEC[E] (1994: 4) **Status** Secure
Criteria —
European IUCN Red List Category —
Criteria —
Global IUCN Red List Category —
Criteria —

Circus pygargus has a widespread but patchy breeding distribution in Europe, which constitutes >50% of its global breeding range. Its European breeding population is relatively small (<65,000 pairs), but underwent a large increase between 1970–1990. This trend continued during 1990–2000, with declines in a small number of countries more than compensated for by stable or increasing trends across the vast majority of its European range. Consequently, it is evaluated as Secure.

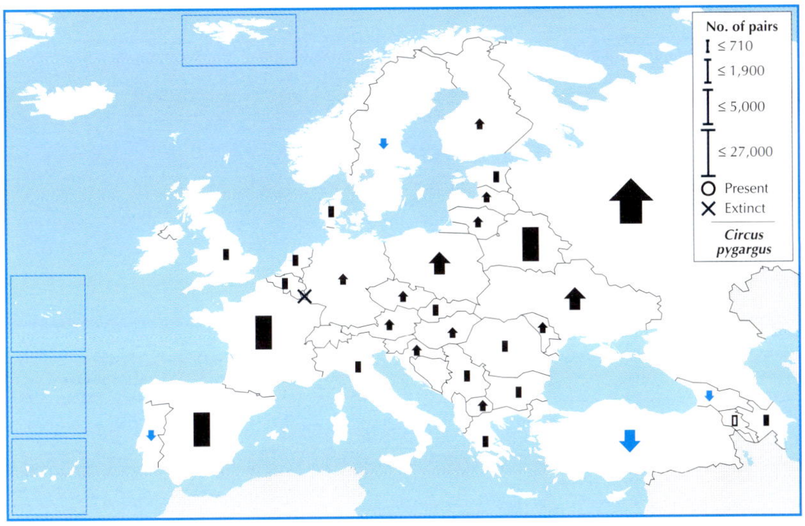

Country	Breeding pop. size (pairs)	Year(s)	Trend	Mag.%	References
Armenia	50 – 250	98–02	?	–	
Austria	15 – 25	98–02	+	**20–29**	
Azerbaijan	(200 – 500)	96–00	(0)	(0–19)	
Belarus	**3,000 – 5,000**	97–02	**0**	**0–19**	
Belgium	**6 – 12**	95–02	**0**	**0–19**	*1*
Bulgaria	80 – 180	95–02	(F)	(30–49)	
Croatia	30 – 50	02	(+)	(>80)	*70,16*
Czech Rep.	**80 – 120**	00	+	**300**	
Denmark	**36 – 45**	01–02	**0**	**0–19**	*7,19*
Estonia	100 – 200	98	0	0–19	*1*
Finland	2 – 10	98–02	+	100	
France	**3,800 – 5,100**	00–02	0	0–19	*8,7*
Georgia	15 – 30	94–03	–	30–49	
Germany	**234 – 283**	95–99	+	**0–19**	
Greece	(10 – 30)	95–00	(0)	(0–19)	
Hungary	**100 – 200**	95–02	+	**50–79**	*1*
Italy	260 – 380	03	0	0–19	*6*
Latvia	100 – 200	90–00	(+)	(30–49)	*23*
Lithuania	200 – 300	99–01	(+)	(0–19)	*20*
Luxembourg	**0 – 0**	02	**–**	**X**	
Macedonia	**100 – 140**	99–00	(+)	(20–29)	*3*
Moldova	4 – 6	90–00	+	0–19	
Netherlands	**29 – 45**	98–00	**0**	**10**	*1*
Poland	1,300 – 2,500	00–02	(+)	(>80)	*2,23*
Portugal	500 – 1,000	02	(–)	(–)	
Romania	0 – 12	90–02	(0)	(0–19)	*62*
Russia	20,000 – 35,000	94–01	+	**20–29**	*2,12,22,73,104, 132,134*
Serbia & MN	4 – 6	00–02	0	0–19	*1,70,200,10a,155*
Slovakia	30 – 80	80–99	F	20–29	
Spain	(2,500 – 10,000)	98–02	(0)	(0–19)	*10,16*
Sweden	**55 – 75**	99–00	–	0–9	
Turkey	800 – 1,200	01	–	20–29	
Ukraine	1,500 – 2,400	90–00	+	0–19	
UK	**6 – 6**	96–00	**0**	**5**	
Total (approx.)	35,000 – 65,000	**Overall trend** Moderate increase			
Breeding range	>3,000,000 km²	**Gen. length.** 6		**% Global pop.** 50–74	

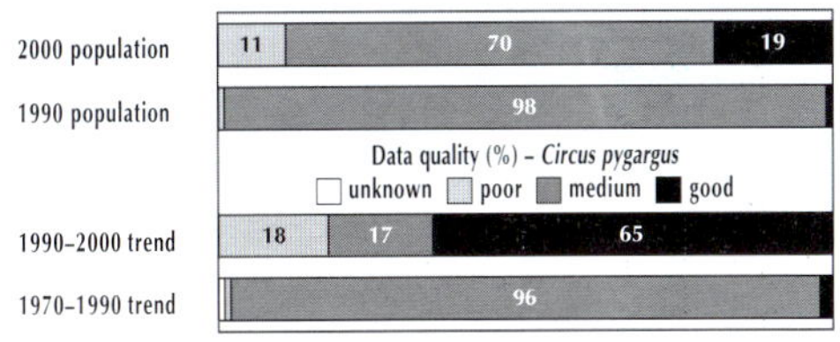

Accipiter gentilis
NORTHERN GOSHAWK

Non-SPEC (1994: —) **Status** Secure
Criteria —
European IUCN Red List Category —
Criteria —
Global IUCN Red List Category —
Criteria —

Accipiter gentilis is a widespread resident across most of Europe, which accounts for less than half of its global range. Its European breeding population is large (>160,000 pairs), and increased substantially between 1970–1990. Although there were declines in several countries during 1990–2000, these were more than compensated for by positive trends elsewhere—notably in the Russian stronghold—and the species underwent a moderate increase overall. Consequently, it is evaluated as Secure.

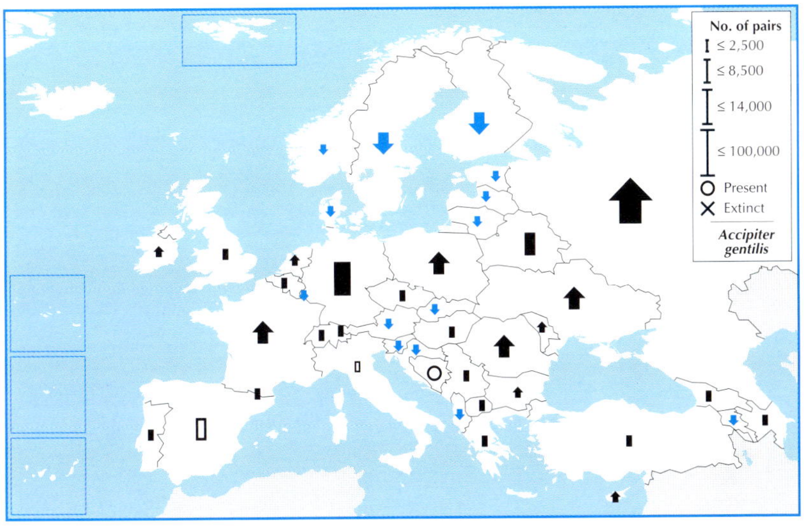

Country	Breeding pop. size (pairs)	Year(s)	Trend	Mag.%	References
Albania	(100 – 300)	96–02	(–)	(20–29)	
Andorra	6 – 8	99–02	0	0–19	*1,3*
Armenia	0 – 50	97–02	–	0–19	
Austria	1,400 – 2,200	98–02	(–)	(20–29)	
Azerbaijan	(50 – 200)	96–00	(0)	(0–19)	
Belarus	**3,500 – 5,000**	97–02	0	0–19	
Belgium	476 – 820	01–02	0	0–19	*1*
Bosnia & HG	Present	90–03	?	–	
Bulgaria	500 – 1,500	96–02	+	0–19	
Croatia	(500 – 1,000)	02	(–)	(>80)	*16*
Cyprus	30 – 60	94–02	+	0–19	
Czech Rep.	**1,800 – 2,500**	00	**0**	**0–9**	
Denmark	600 – 600	00	–	0–19	*14*
Estonia	700 – 1,000	98	–	20–29	*1*
Finland	4,500 – 5,500	98–02	**–**	**10**	
France	4,600 – 6,600	00–02	+	0–9	*8,7*
Georgia	230 – 350	94–03	0	0–19	
Germany	11,500 – 15,000	95–99	0	0–19	
Greece	(1,000 – 1,200)	95–00	(0)	(0–19)	
Hungary	1,200 – 3,100	99–02	0	0–19	*19,1*
Rep. Ireland	0 – 2	88–91	(+)	(N)	
Italy	(500 – 800)	03	?	–	*6*
Latvia	2,000 – 3,000	90–00	(–)	(0–19)	*18,23*
Liechtenstein	2 – 4	98–00	0	0–19	
Lithuania	500 – 800	99–01	(–)	(20–29)	*20*
Luxembourg	**50 – 60**	00–02	**–**	**10–19**	
Macedonia	300 – 700	99–00	(0)	(0–19)	*3*
Moldova	**40 – 70**	90–00	+	0–19	
Netherlands	**1,800 – 2,000**	98–00	+	**89**	*1*
Norway	**1,400 – 2,000**	00	**–**	**0–19**	*22*
Poland	5,000 – 10,000	90–00	(+)	(0–19)	*23*
Portugal	(50 – 200)	02	(0)	(0–19)	
Romania	**5,000 – 7,000**	98–02	+	0–19	*61*
Russia	90,000 – 110,000	98–02	+	**30–49**	*8,12,22,61,23, 43,104,114,134*
Serbia & MN	1,600 – 2,100	97–02	0	0–19	*1,71,202,155, 32,172a,67a*
Slovakia	1,600 – 1,800	80–99	–	30–49	
Slovenia	500 – 600	00	(–)	(0–19)	
Spain	(3,500 – 6,500)	98–02	?	–	*10,16*
Sweden	2,500 – 5,000	99–00	–	10–19	
Switzerland	**1,400 – 1,600**	93–96	0	0–19	
Turkey	(500 – 800)	01	(0)	(0–19)	
Ukraine	6,900 – 10,300	90–00	+	**20–29**	
UK	**294 – 299**	96–00	**F**	**87**	
Total (approx.)	160,000 – 210,000	**Overall trend** Moderate increase			
Breeding range	>7,000,000 km²	**Gen. length.** 6		**% Global pop.** 25–49	

(See p. 78, bottom, for data quality graph)

Accipiter nisus
EURASIAN SPARROWHAWK

Non-SPEC (1994: —) **Status** Secure
Criteria —
European IUCN Red List Category —
Criteria —
Global IUCN Red List Category —
Criteria —

Accipiter nisus is a widespread breeder across most of Europe, which accounts for less than half of its global breeding range. Its European breeding population is large (>340,000 pairs), and underwent a large increase between 1970–1990. Although there were declines in a few countries during 1990–2000, populations were stable or increasing across most of Europe—including the key one in Russia—and the species increased slightly overall. Consequently, it is evaluated as Secure.

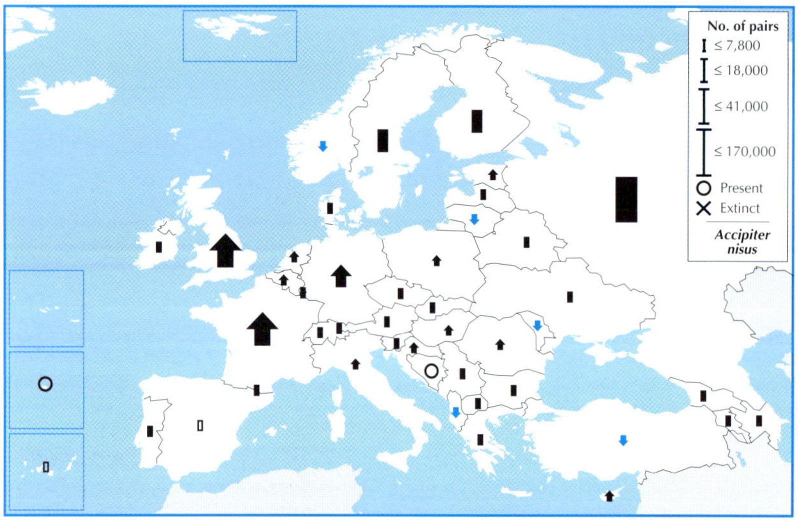

Country	Breeding pop. size (pairs)	Year(s)	Trend	Mag.%	References
Albania	100 – 300	96–02	(–)	(20–29)	
Andorra	15 – 20	99–02	0	0–19	*1,3*
Armenia	100 – 150	97–02	0	0–19	
Austria	6,000 – 8,000	98–02	(0)	(0–19)	
Azerbaijan	(200 – 500)	96–00	(0)	(0–19)	
Belarus	6,500 – 8,500	97–02	0	0–19	
Belgium	3,000 – 4,600	01–02	+	20–29	*1*
Bosnia & HG	Present	90–03	?	–	
Bulgaria	1,000 – 3,000	96–02	0	0–9	
Croatia	(5,000 – 7,000)	02	(+)	(>80)	*16*
Cyprus	**1 – 1**	96	+	**N**	
Czech Rep.	**3,500 – 4,500**	00	**0**	**0–19**	
Denmark	3,500 – 3,500	00	0	0–19	*14*
Estonia	1,700 – 2,700	98	+	20–29	*1*
Finland	12,000 – 15,000	98–02	**0**	**0–19**	
France	26,600 – 42,600	00–02	+	0–19	*8,7*
Georgia	950 – 2,150	94–03	0	0–19	
Germany	14,400 – 21,000	95–99	+	20–29	
Greece	(1,000 – 3,000)	95–00	(0)	(0–19)	
Hungary	1,000 – 2,000	99–02	+	>80	*19,1*
Rep. Ireland	2,500 – 10,000	88–91	0	0–19	
Italy	(2,000 – 4,000)	03	+	0–19	*6*
Latvia	2,000 – 3,000	90–00	(0)	(0–19)	*18,23*
Liechtenstein	10 – 15	98–00	0	0–19	
Lithuania	(2,000 – 4,000)	99–01	(–)	(20–29)	*20*
Luxembourg	200 – 300	00–02	**0**	**0–19**	
Macedonia	500 – 1,200	99–00	(0)	(0–19)	*3*
Moldova	**80 – 120**	90–00	–	20–29	
Netherlands	**4,000 – 5,000**	98–00	+	**205**	*1*
Norway	(3,000 – 6,000)	94–03	–	0–19	*51*
Poland	(2,500 – 10,000)	00–02	(+)	(0–19)	*2,23*
Portugal	(300 – 1,500)	02	(0)	(0–19)	
Madeira	Present	02	?	–	
Romania	(1,200 – 1,400)	98–02	+	20–29	*62*
Russia	160,000 – 180,000	94–02	0	0–19	*3,12,22,104, 114,134*
Serbia & MN	1,000 – 1300	97–02	0	0–19	*1,157,202,155*
Slovakia	1,500 – 2,000	90–99	0	0–19	
Slovenia	1,000 – 1,500	94	(0)	(0–19)	
Spain	(6,000 – 10,000)	98–02	?	–	*10,16*
Canary Is.	250 – 1,000	97–03	?	–	*28,25*
Sweden	15,000 – 20,000	99–00	0	0–19	
Switzerland	**3,000 – 4,000**	93–96	0	0–19	
Turkey	(3,000 – 7,000)	01	(–)	(0–19)	
Ukraine	4,500 – 7,600	90–00	0	0–9	
UK	**40,100 – 40,100**	00	+	**18**	*5,31*

Total (approx.) 340,000 – 450,000 **Overall trend** Small increase
Breeding range >7,000,000 km² **Gen. length.** 4 **% Global pop.** 25–49
(See p. 78, bottom, for data quality graph)

Accipiter badius
SHIKRA

Non-SPEC (1994: NE) **Status** (Secure)
Criteria —
European IUCN Red List Category —▼▼▼▼
Criteria —
Global IUCN Red List Category —
Criteria —

Accipiter badius has a predominantly Asian and African distribution, which just extends into south-easternmost Europe in Azerbaijan. Its European breeding population is tiny (as few as 10 pairs), and though its trend between 1970–1990 was unknown, the species probably remained stable during 1990–2000. Although the size of the European population could render it susceptible to the risks affecting small populations, it is marginal to a much larger non-European population. Consequently, the species is provisionally evaluated as Secure.

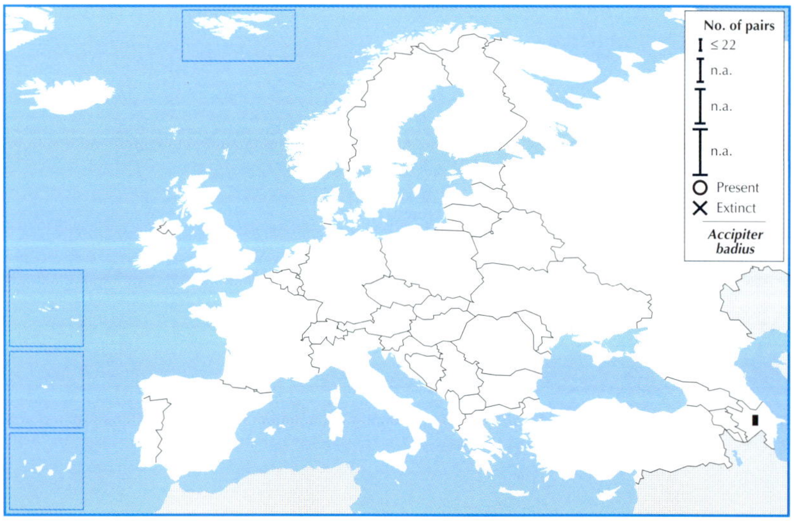

Country	Breeding pop. size (pairs)	Year(s)	Trend	Mag.%	References
Azerbaijan	(10 – 50)	96–00	(0)	(0–19)	

Total (approx.) 10 – 50 **Overall trend** Stable
Breeding range >20,000 km² **Gen. length.** 4 **% Global pop.** <5

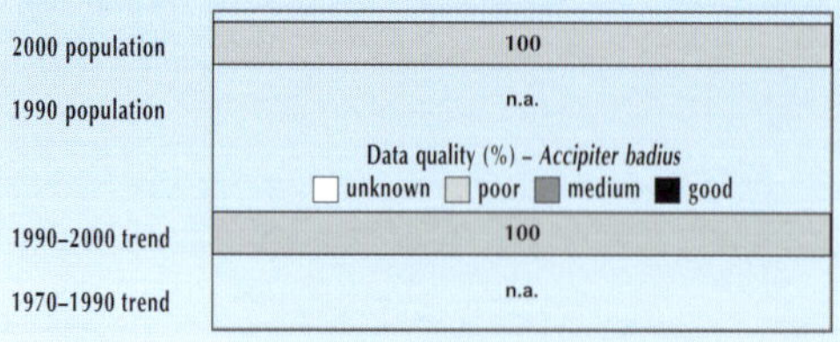

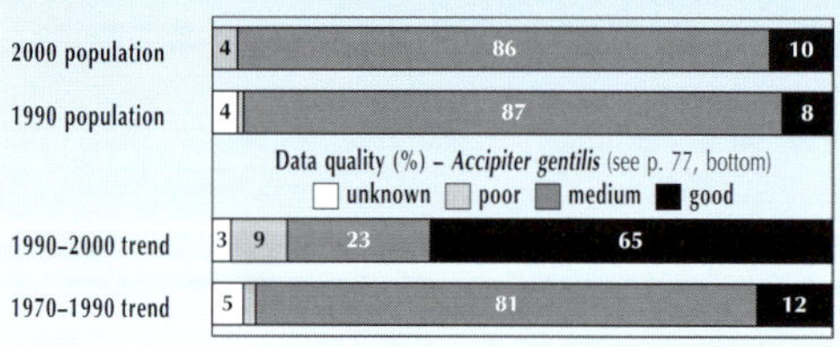

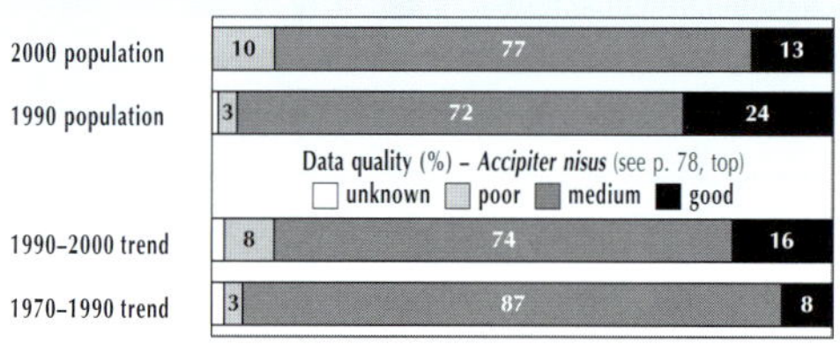

Accipiter brevipes
LEVANT SPARROWHAWK

SPEC 2 (1994: 2) **Status** (Vulnerable)
Criteria See IUCN below
European IUCN Red List Category VU
Criteria C1
Global IUCN Red List Category —
Criteria —

Country	Breeding pop. size (pairs)	Year(s)	Trend	Mag.%	References
Albania	10 – 50	02	(0)	(0–19)	
Armenia	50 – 250	01–02	+	0–19	*5*
Azerbaijan	(100 – 1,000)	96–00	(0)	(0–19)	
Bulgaria	50 – 90	96–02	(0)	(0–19)	
Croatia	(1 – 10)	02	(0)	(0–19)	*16*
Georgia	30 – 60	94–03	?	–	
Greece	(1,000 – 2,000)	95–00	(0)	(0–19)	
Hungary	0 – 5	95–02	**0**	**0–19**	*1,15*
Macedonia	40 – 100	99–00	(0)	(0–19)	*3*
Romania	(60 – 100)	99–02	(+)	(0–19)	*29,46*
Russia	1,500 – 3,000	96–02	–	30–49	*6,8,26,33,94,114*
Serbia & MN	60 – 100	97–02	0	0–19	*1,60,155,172a*
Turkey	(300 – 900)	01	?	–	
Ukraine	**30 – 45**	90–00	**0**	**0–9**	
Total (approx.)	3,200 – 7,700	**Overall trend** Moderate decline			
Breeding range	>500,000 km²	**Gen. length.** 4		**% Global pop.** 75–94	

Accipiter brevipes is a patchily distributed summer visitor to south-eastern Europe, which constitutes >75% of its global breeding range. Its European breeding population is small (as few as 3,200 pairs), but was stable between 1970–1990. Although the species remained stable or increased across the vast majority of its European range during 1990–2000, there were declines in the sizeable population in Russia, and the species underwent a moderate decline (>10%) overall. As a consequence of this decline and its small population, the species is provisionally evaluated as Vulnerable.

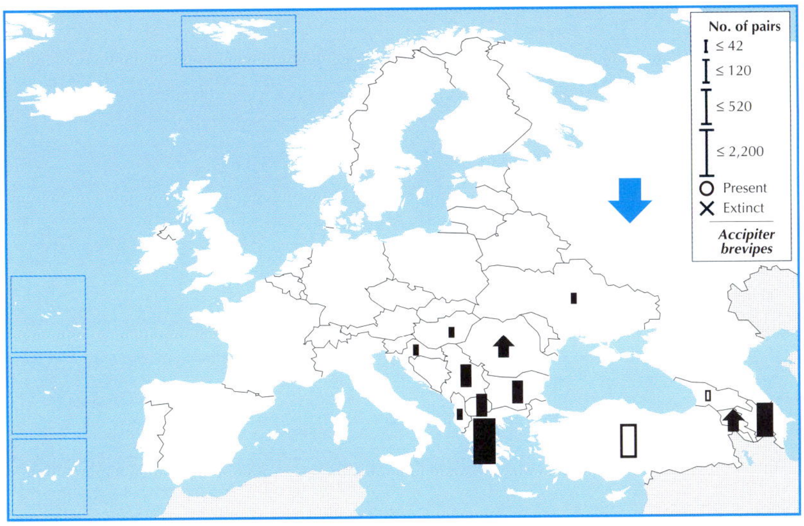

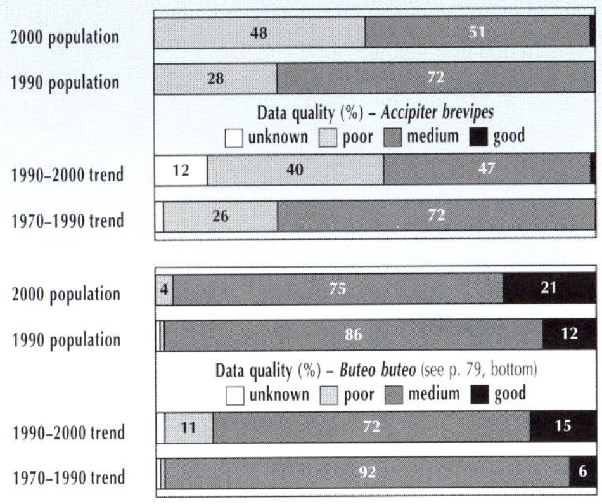

Buteo buteo
COMMON BUZZARD

Non-SPEC (1994: —) **Status** Secure
Criteria —
European IUCN Red List Category —
Criteria —
Global IUCN Red List Category —
Criteria —

Buteo buteo is a widespread breeder across most of Europe, which accounts for less than half of its global breeding range. Its European breeding population is large (>710,000 pairs), and increased between 1970–1990. Although there were declines in a few countries during 1990–2000, key populations in Russia, Germany and France were stable, and the species was stable or increased across most of the rest of Europe. Consequently, it is evaluated as Secure.

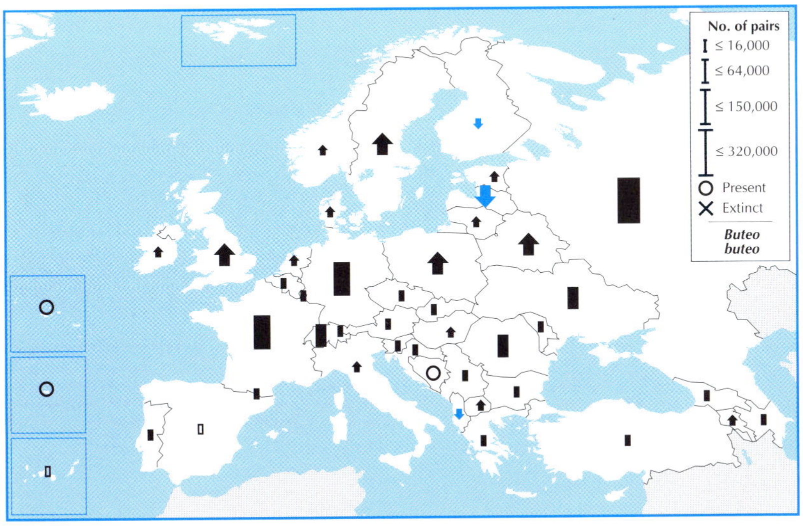

Country	Breeding pop. size (pairs)	Year(s)	Trend	Mag.%	References
Albania	300 – 1,000	96–02	(–)	(0–19)	
Andorra	20 – 30	99–02	0	0–19	*1,3*
Armenia	80 – 150	01–02	+	0–9	
Austria	8,000 – 12,000	98–02	(0)	(0–19)	
Azerbaijan	(300 – 1,000)	96–00	(0)	(0–19)	
Belarus	**18,000 – 24,000**	97–02	+	0–9	
Belgium	5,300 – 7,900	01–02	0	0–19	*1*
Bosnia & HG	Present	90–03	?	–	
Bulgaria	1,200 – 2,500	96–02	0	0–9	
Croatia	(10,000 – 15,000)	02	(0)	(0–19)	*54,26*
Czech Rep.	**10,000 – 13,000**	00	**0**	**0–9**	
Denmark	6,000 – 6,000	00	+	20	*14*
Estonia	4,500 – 6,500	98	+	20–29	*1*
Finland	5,000 – 7,000	98–02	–	**25**	
France	125,000 – 163,000	00–02	0	0–19	*8,7*
Georgia	900 – 1,000	94–03	0	0–19	
Germany	67,000 – 110,000	95–99	0	0–19	
Greece	3,000 – 5,000	95–00	0	0–19	
Hungary	10,000 – 20,000	99–02	+	**>80**	*19,1*
Rep. Ireland	138 – 200	88–91	+	30–49	
Italy	(4,000 – 8,000)	03	(+)	(0–19)	*6*
Latvia	18,000 20,000	90 00		20 29	*23,18,2*
Liechtenstein	15 – 20	98–00	0	0–19	
Lithuania	3,500 – 6,000	99–01	(+)	(20–29)	*20*
Luxembourg	900 – 1,000	00–02	**0**	**10–19**	
Macedonia	1,000 – 1,400	99–00	(+)	(20–29)	*3*
Moldova	**200 – 300**	90–00	**0**	**0–19**	
Netherlands	**8,000 – 10,000**	98–00	+	**187**	*1*
Norway	(1,000 – 2,000)	90–02	+	0–19	
Poland	**50,000 – 80,000**	00	(+)	(0–19)	*23*
Portugal	(1,500 – 3,000)	02	(0)	(0–19)	
Azores	Present	02	?	–	
Madeira	Present	02	?	–	
Romania	**28,000 – 34,000**	96–02	0	0–19	*61*
Russia	200,000 – 500,000	94–02	0	0–19	*104,134*
Serbia & MN	2,600 – 3,500	97–02	0	0–19	*1,160,155,172a, 67a*
Slovakia	5,000 – 7,000	90–99	0	0–19	
Slovenia	2,000 – 4,000	94	(0)	(0–19)	
Spain	(13,000 – 18,000)	98–02	?	–	*10,16*
Canary Is.	250 – 1,000	97–03	?	–	*28,25*
Sweden	25,000 – 35,000	99–00	+	**19**	
Switzerland	**20,000 – 25,000**	93–96	0	0–19	
Turkey	(2,500 – 3,500)	01	(0)	(0–19)	
Ukraine	22,500 – 32,500	90–00	**0**	**0–19**	
UK	**31,100 – 44,000**	00	+	**159**	*5,31*
Total (approx.)	710,000 – 1,200,000	**Overall trend** Small increase			
Breeding range	>7,000,000 km²	**Gen. length.** 8		**% Global pop.** 25–49	

(See p. 79, top, for data quality graph)

Buteo rufinus
LONG-LEGGED BUZZARD

SPEC 3 (1994: 3) **Status** (Vulnerable)
Criteria See IUCN below
European IUCN Red List Category VU
Criteria A2b
Global IUCN Red List Category —
Criteria —

Buteo rufinus is a widespread breeder in south-eastern Europe, which accounts for less than a quarter of its global breeding range. Its European breeding population is small (as few as 8,700 pairs), and declined substantially between 1970–1990. Although the species was stable or increased across the majority of its European range during 1990–2000, it declined in its Turkish stronghold, and probably declined overall at a rate that—on top of earlier declines—equates to a large decline (>30%) over three generations. Consequently, it is provisionally evaluated as Vulnerable.

Country	Breeding pop. size (pairs)	Year(s)	Trend	Mag.%	References
Albania	(0 – 10)	02	?	–	
Armenia	50 – 250	99–02	0	0–19	
Azerbaijan	(1,000 – 2,500)	96–00	(0)	(0–19)	
Bulgaria	**250 – 400**	98–02	+	50–79	
Cyprus	10 – 20	92–02	+	**N**	
Georgia	50 – 60	94–03	**0**	**0–19**	
Greece	200 – 300	95–00	0	0–19	
Hungary	**2 – 7**	98–02	+	**>80**	*1*
Macedonia	50 – 80	99–00	(+)	(20–29)	*3*
Romania	65 – 110	00–02	+	20–29	*54*
Russia	(1,000 – 2,000)	94–01	(0)	(0–19)	*6,26,33,93,74*
Serbia & MN	**22 – 28**	97–02	+	**20–29**	*1,61,200,48,36,155*
Turkey	(6,000 – 9,000)	01	(–)	(20–29)	
Ukraine	**50 – 150**	90–00	+	**30–49**	
Total (approx.)	8,700 – 15,000	**Overall trend** Large decline			
Breeding range	>500,000 km²	**Gen. length.** 8		**% Global pop.** 5–24	

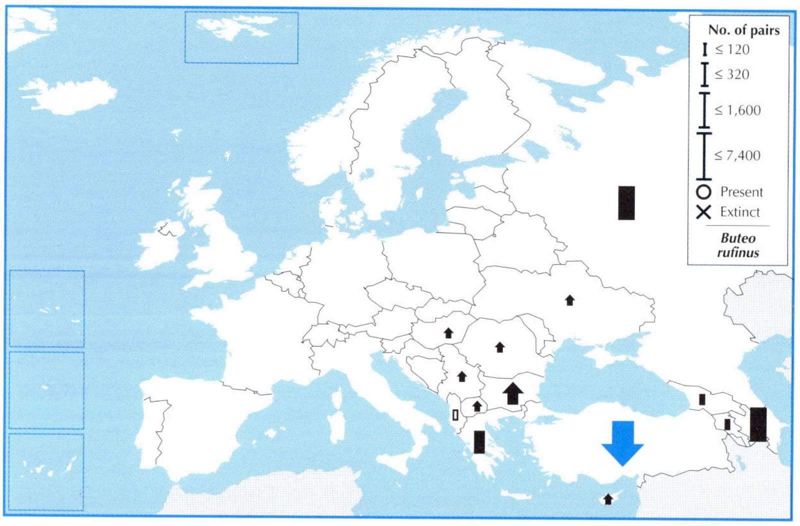

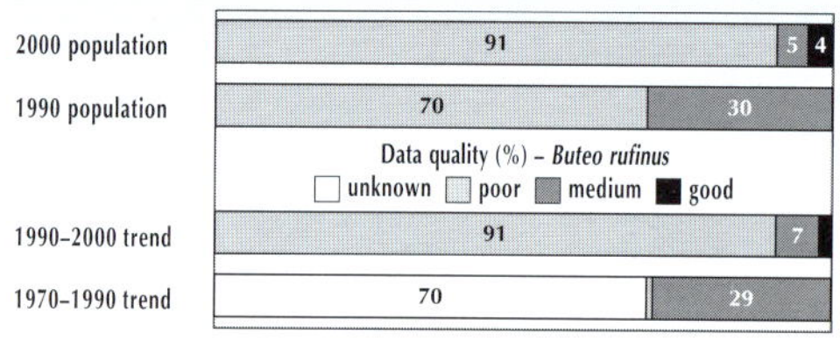

Buteo lagopus
ROUGH-LEGGED HAWK

Non-SPEC (1994: —) **Status** (Secure)
Criteria —
European IUCN Red List Category —
Criteria —
Global IUCN Red List Category —
Criteria —

Buteo lagopus is a widespread breeder in Fennoscandia and northern Russia, with Europe accounting for less than a quarter of its global breeding range. Its European breeding population is relatively small (<79,000 pairs), but was stable between 1970–1990. Although the species declined slightly in Sweden during 1990–2000, populations elsewhere—including the sizeable one in Russia—were broadly stable, and the species probably remained stable overall. Consequently, it is provisionally evaluated as Secure.

Country	Breeding pop. size (pairs)	Year(s)	Trend	Mag.%	References
Finland	500 – 4,000	98–02	(F)	(200–250)	
Norway	(5,000 – 10,000)	90–03	(F)	(50–79)	*59*
Russia	(30,000 – 60,000)	94–01	(0)	(0–19)	*138*
Sweden	2,000 – 5,000	99–00	–	10–19	
Total (approx.)	38,000 – 79,000	**Overall trend** Fluctuating			
Breeding range	>1,000,000 km²	**Gen. length.** 8		**% Global pop.** 5–24	

Country	Winter pop. size (individuals)	Year(s)	Trend	Mag.%	References
Armenia	(50 – 250)	94–02	?	–	
Austria	(10 – 100)	98–02	(F)	(>80)	
Azerbaijan	(10 – 100)	96–02	(0)	(0–19)	
Belarus	250 – 1,000	97–02	?	–	
Bulgaria	500 – 2,000	98–02	(F)	(30–49)	
Croatia	(1 – 10)	02	?	–	*26*
Czech Rep.	1,000 – 2,000	90–00	–	30–49	
Denmark	(500 – 1,500)	99–00	–	50–79	*23*
Estonia	10 – 50	98	0	0–19	1
Finland	(10 – 100)	98–02	(0)	(0–19)	
Greece	(0 – 10)	95–00	(F)	(>80)	
Hungary	1,500 – 3,000	00–02	?	–	*24,25*
Latvia	50 – 200	90–98	?	–	*28,36*
Lithuania	(500 – 5,000)	92–02	(F)	(>80)	
Moldova	150 – 250	90–00	+	20–29	
Netherlands	75 – 150	00	–	50–79	*4,5,6,7,9,10*
Poland	50,000 – 100,000	90–00	F	30–49	*114,121*
Romania	(500 – 2,000)	90–00	–	20–29	*60*
Slovakia	800 – 2,500	90–99	?	–	*4*
Slovenia	0 – 5	90–00	F	0–100	
Ukraine	13,000 – 230,000	90–00	F	20–29	
UK	**5 – 119**	90–00	(F)	(–)	*46*
Total (approx.)	>69,000	**Overall trend** Fluctuating			
% in European IBAs	Unknown	**Gen. length** 8		**% Global pop.** 5–24	

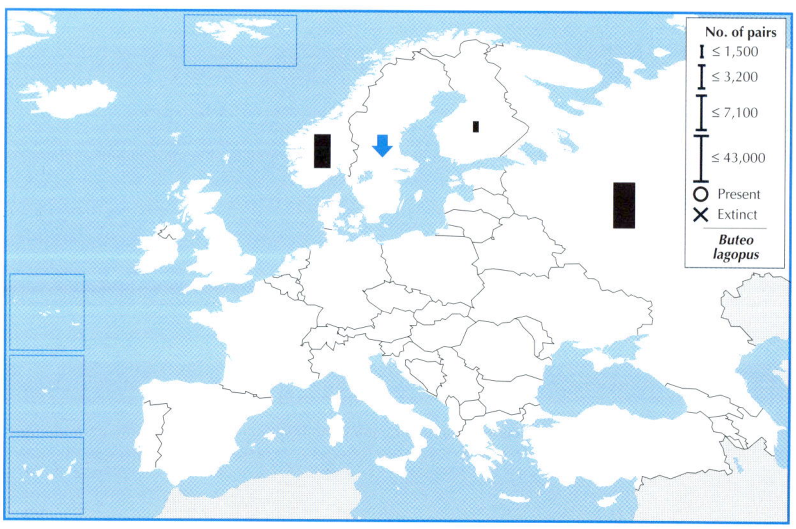

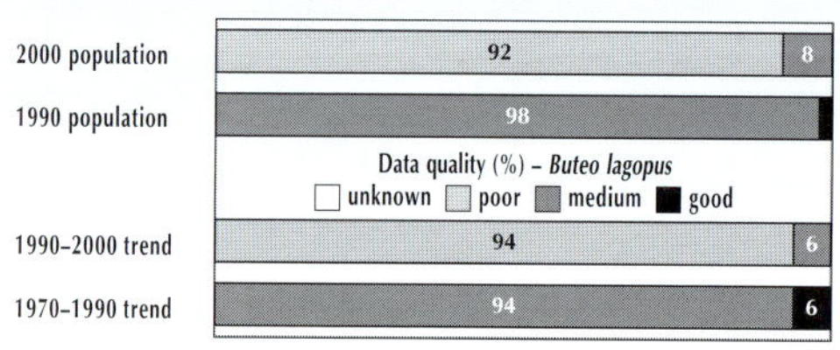

Aquila pomarina
LESSER SPOTTED EAGLE

SPEC 2 (1994: 3) **Status** (Declining)
Criteria Moderate recent decline
European IUCN Red List Category —
Criteria —
Global IUCN Red List Category —
Criteria —

Country	Breeding pop. size (pairs)	Year(s)	Trend	Mag.%	References
Albania	2 – 10	96–02	?	–	
Armenia	20 – 30	99–02	**-**	**0–19**	
Azerbaijan	(20 – 100)	96–00	(0)	(0–19)	
Belarus	**3,200 – 3,800**	97–02	**0**	**0–19**	
Bulgaria	120 – 200	96–02	+	30–49	
Croatia	60 – 70	02	(–)	(>80)	*70,16*
Czech Rep.	Present	00	?	–	
Estonia	480 – 600	98	**0**	**0–19**	*1*
Georgia	10 – 70	94–03	(+)	(–)	
Germany	**134 – 143**	95–99	**-**	**0–19**	
Greece	**67 – 90**	97	**-**	**0–19**	
Hungary	**30 – 60**	95–02	**-**	**50–79**	*1*
Latvia	**2,800 – 5,200**	90–00	–	0–19	*21*
Lithuania	900 – 1,200	99–01	+	20–29	*20*
Macedonia	3 – 10	99–00	(0)	(0–19)	*3*
Moldova	2 – 5	90–00	–	30–49	
Poland	**1,700 – 1,900**	98	**0**	**0–19**	*33*
Romania	**2,500 – 2,800**	96–02	0	0–19	*62,63*
Russia	300 – 400	96–00	+	20–29	*60,63,74,163*
Serbia & MN	**20 – 25**	97–02	**-**	**0–19**	*1,161,155,172a*
Slovakia	800 – 900	90–99	0	0–19	
Slovenia	**2 – 4**	99–00	0	0–19	
Turkey	(80 – 120)	01	–	20–29	
Ukraine	500 – 1,000	90–00	+	0–19	
Total (approx.)	14,000 – 19,000	**Overall trend** Moderate decline			
Breeding range	>1,000,000 km²	**Gen. length.** 11		**% Global pop.** >95	

Aquila pomarina is a summer visitor to east-central Europe, which constitutes more than 95% of its global breeding range. Its European breeding population is relatively small (<19,000 pairs), but was stable between 1970–1990. Although the species increased or was stable across much of its core range during 1990–2000, the key Latvian population declined—along with many smaller peripheral populations—and it probably underwent a moderate decline overall. Consequently, the species is provisionally evaluated as Declining.

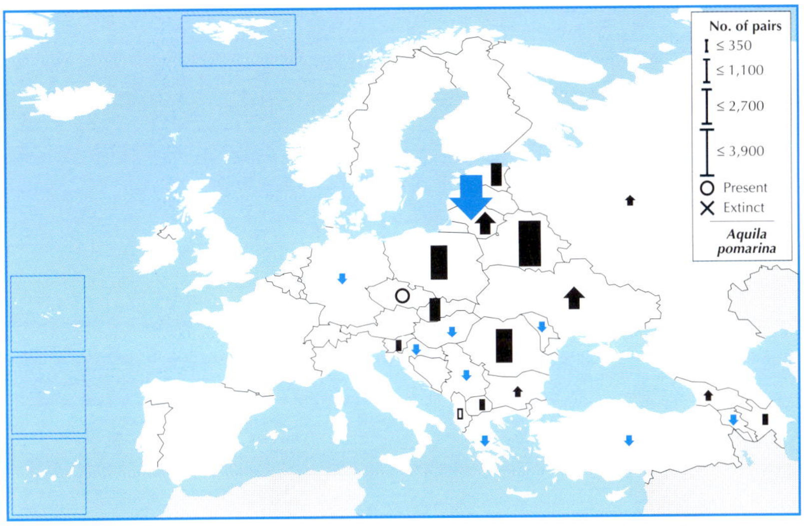

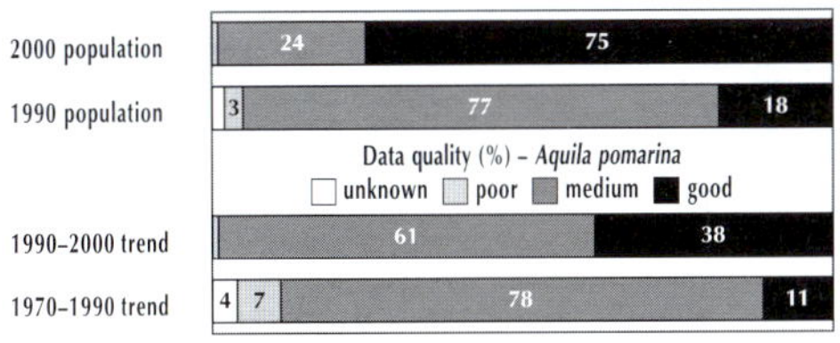

Aquila clanga
GREATER SPOTTED EAGLE

SPEC 1 (1994: 1) **Status** Endangered
Criteria See IUCN below
European IUCN Red List Category EN
Criteria C1
Global IUCN Red List Category VU
Criteria C1

Country	Breeding pop. size (pairs)	Year(s)	Trend	Mag.%	References
Belarus	**150 – 200**	97–02	**0**	**0–19**	
Estonia	**15 – 30**	98	0	0–19	*1*
Finland	(0 – 1)	98–02	0	0–19	
Latvia	(0 – 5)	90–00	?	–	*18,21*
Moldova	0 – 0	96–00	–	X	
Poland	**15 – 15**	98	**0**	**0–19**	*33*
Romania	**0 – 2**	99–02	(0)	(0–19)	*62*
Russia	600 – 800	96–00	–	20–29	*60,63,74,162,163*
Ukraine	30 – 45	90–00	**-**	**20–29**	
Total (approx.)	810 – 1,100	**Overall trend** Large decline			
Breeding range	>1,000,000 km²	**Gen. length.** 11		**% Global pop.** 25–49	

Aquila clanga is a summer visitor to north-eastern Europe, which accounts for less than half of its global breeding range. Its European breeding population is very small (as few as 810 pairs), and declined substantially between 1970–1990. It continued to decline in its Russian stronghold during 1990–2000, and declined overall at a rate that—on top of earlier declines—equates to a large decline (>20%) over two generations. As a consequence of this continuing decline and its very small population, this globally threatened species is evaluated as Endangered in Europe.

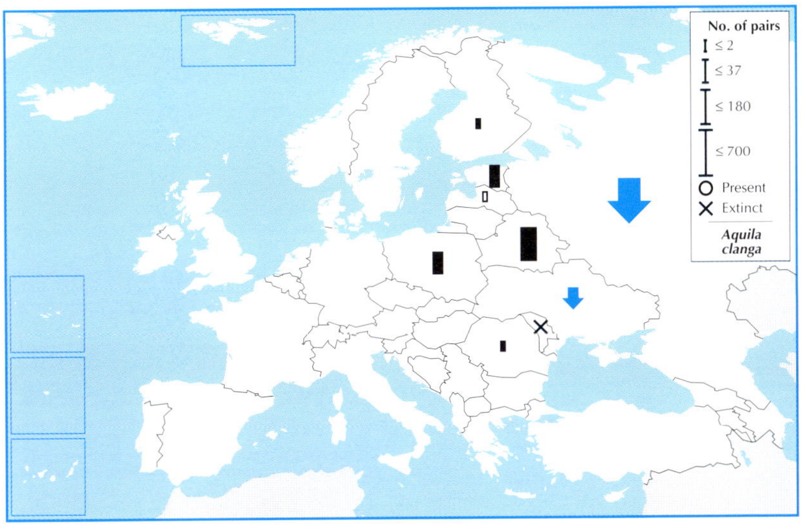

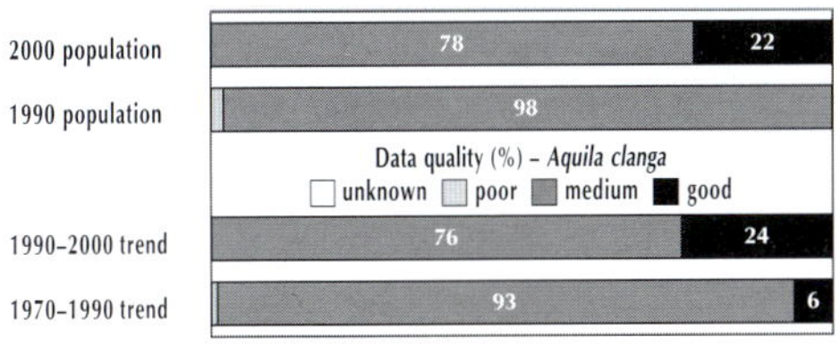

Aquila nipalensis
STEPPE EAGLE

SPEC 3 (1994: 3) **Status** (Endangered)
Criteria See IUCN below
European IUCN Red List Category EN
Criteria A2b
Global IUCN Red List Category —
Criteria —

Country	Breeding pop. size (pairs)	Year(s)	Trend	Mag.%	References
Russia	(5,000 – 20,000)	94–01	(–)	(20–29)	*6,26,33,52,74*
Turkey	3 – 10	01	(–)	(30–49)	
Ukraine	**0 – 0**	00	**-**	**X**	
Total (approx.)	5,000 – 20,000	**Overall trend** Large decline			
Breeding range	>500,000 km²	**Gen. length.** 11		**% Global pop.** <5	

Aquila nipalensis has a predominantly Asian distribution, which just extends into easternmost Europe. Its European breeding population is small (as few as 5,000 pairs), and declined substantially between 1970–1990. The species continued to decline in its Russian stronghold during 1990–2000, and declined overall at a rate that—on top of earlier declines—probably equates to a very large decline (>50%) over three generations. Consequently, the species is provisionally evaluated as Endangered.

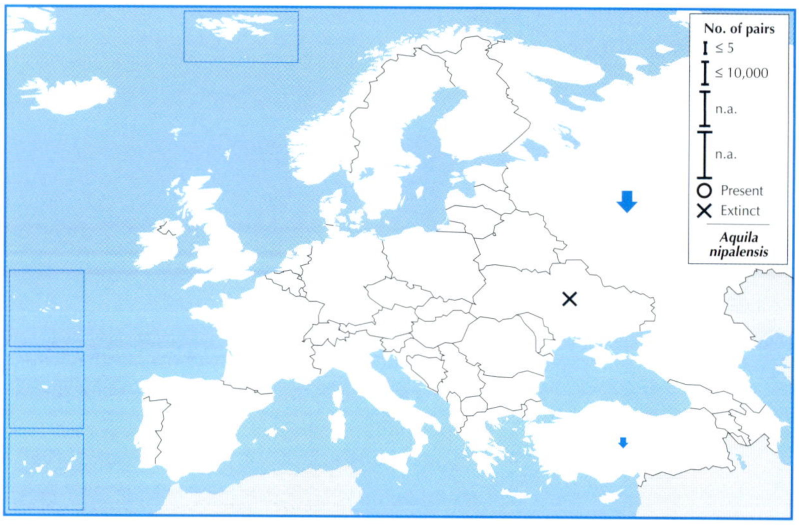

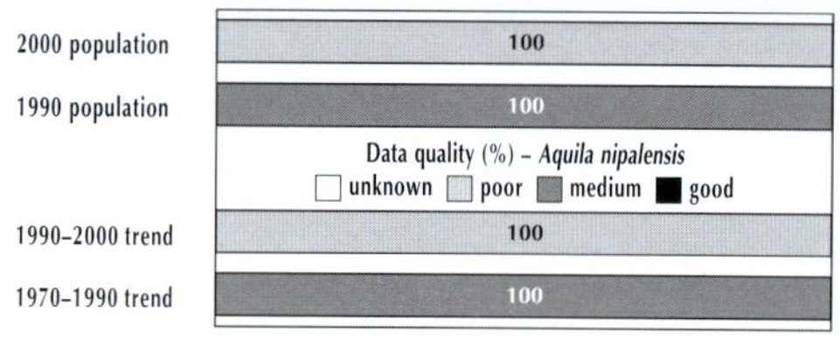

Aquila heliaca
IMPERIAL EAGLE

SPEC 1 (1994: 1) **Status** Rare
Criteria <10,000 pairs
European IUCN Red List Category —
Criteria —
Global IUCN Red List Category VU
Criteria C1

Country	Breeding pop. size (pairs)	Year(s)	Trend	Mag.%	References
Armenia	(0 – 50)	97–02	?	–	
Austria	**1 – 1**	98–02	**+**	**N**	
Azerbaijan	(5 – 25)	96–00	(0)	(0–19)	
Bulgaria	**20 – 25**	98–02	**0**	**0–19**	
Croatia	(0 – 1)	02	(0)	(0–19)	*26*
Cyprus	0 – 2	94–02	–	>80	
Czech Rep.	**1 – 2**	00	**+**	**N**	
Georgia	**10 – 15**	94–03	–	0–19	
Greece	(0 – 1)	95–00	(0)	(0–19)	
Hungary	**50 – 65**	98–02	**+**	**>80**	*2,1*
Macedonia	**35 – 45**	98–00	0	0–19	
Moldova	0 – 0	96–00	–	X	
Romania	**5 – 10**	90–02	(F)	(20–29)	*62*
Russia	**600 – 900**	96–01	**0**	**0–19**	*74,64*
Serbia & MN	**2 – 4**	00–02	**-**	**30–49**	*72,154,226a*
Slovakia	**35 – 40**	80–99	+	50–79	
Turkey	42 – 180	01	–	20–29	
Ukraine	45 – 75	90–00	0	0–19	
Total (approx.)	850 – 1,400	**Overall trend** Stable			
Breeding range	>1,000,000 km²	**Gen. length.** 11		**% Global pop.** 25–49	

Aquila heliaca is a summer visitor to eastern and south-eastern Europe, which accounts for less than half of its global breeding range. Its European breeding population is very small (as few as 850 pairs), and declined substantially between 1970–1990. Although declines continued in south-easternmost Europe during 1990–2000, the species was stable in its Russian stronghold, and was thus stable overall. Consequently, this globally threatened species, which was previously assessed as Endangered in Europe, is now evaluated as Rare.

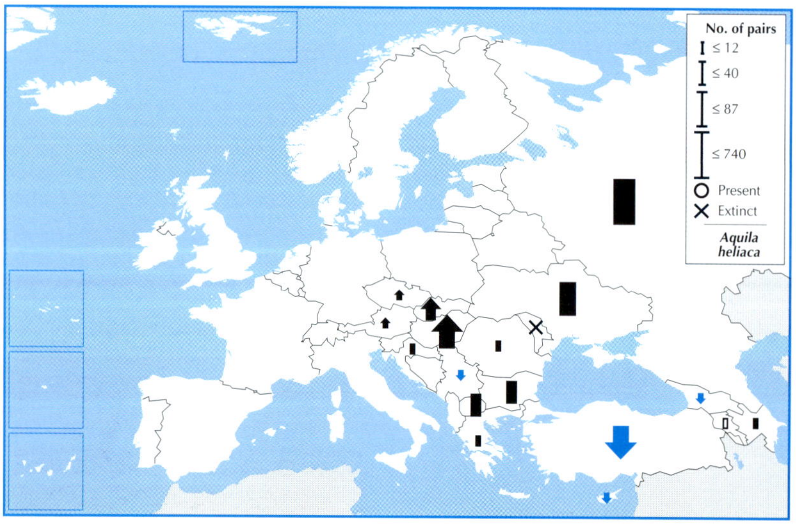

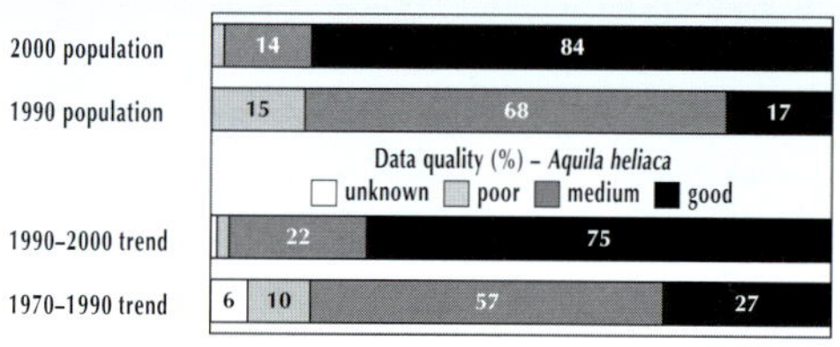

Aquila adalberti
SPANISH IMPERIAL EAGLE

SPEC 1 (1994: 1) **Status** (Endangered)
Criteria See IUCN below
European IUCN Red List Category EN
Criteria C1; C2a(i)
Global IUCN Red List Category EN
Criteria C1; C2a(i)

Country	Breeding pop. size (pairs)	Year(s)	Trend	Mag.%	References
Portugal	1 - 3	03	+	N	
Spain	175 - 175	02	+	0–19	*10,16*
Total (approx.)	175 - 180	Overall trend Small increase			
Breeding range	>100,000 km²	Gen. length. 11		% Global pop. 100	

Aquila adalberti is endemic to Europe, where its entire global range is restricted to central and south-western Iberia. Its European breeding population is very small (as few as 175 pairs), but underwent a large increase between 1970–1990. The stronghold population in Spain increased slightly overall during 1990–2000, and hence would qualify as Vulnerable. However, it apparently declined slightly towards the end of the decade. As a consequence of this decline and its very small population, this globally threatened species is provisionally evaluated as Endangered in Europe.

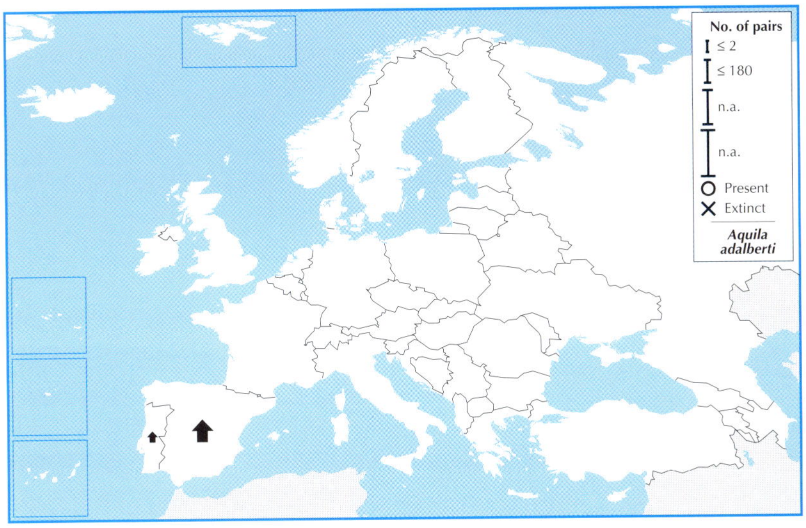

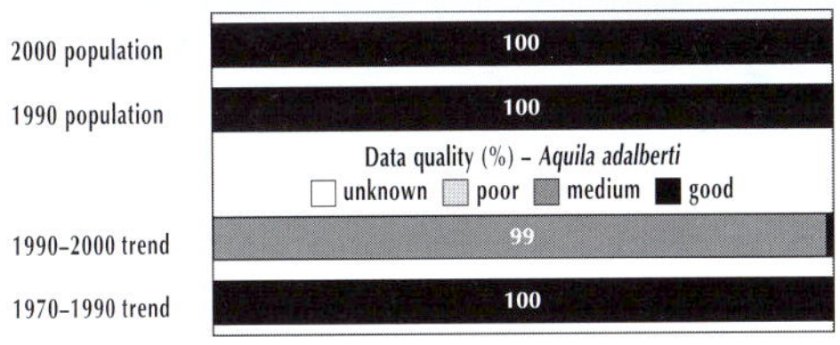

Aquila chrysaetos
GOLDEN EAGLE

SPEC 3 (1994: 3) **Status** Rare
Criteria <10,000 pairs
European IUCN Red List Category —
Criteria —
Global IUCN Red List Category —
Criteria —

Country	Breeding pop. size (pairs)	Year(s)	Trend	Mag.%	References
Albania	50 – 200	96–02	(–)	(0–19)	
Andorra	3 - 3	99–01	0	0–19	*1,3*
Armenia	20 - 80	98–02	?	–	
Austria	260 - 360	98–02	0	0–19	
Azerbaijan	(30 – 100)	96–00	(0)	(0–19)	
Belarus	25 - 35	97–02	0	0–19	
Bosnia & HG	Present	90–03	?	–	
Bulgaria	150 - 170	96–02	+	0–9	
Croatia	90 – 110	02	(–)	(20–29)	*70*
Denmark	1 - 1	99–01	+	N	*4,5,6,7*
Estonia	35 - 45	98	0	0–19	*1*
Finland	300 – 350	98–02	+	20	
France	390 - 460	00–02	+	0–19	*8,7*
Georgia	30 - 35	94–03	0	0–19	
Germany	45 - 50	95–99	0	0–19	
Greece	100 – 150	95–00	(–)	(0–9)	
Hungary	3 - 5	98–02	+	>80	*1*
Italy	476 - 541	03	0	0–19	*6*
Latvia	5 - 10	90–00	–	0–19	*21*
Liechtenstein	1 - 2	98–00	0	0–19	
Macedonia	60 – 100	99–00	(0)	(0–19)	*3*
Norway	860 - 1,040	03	0	0–19	*21*
Poland	35 - 40	98	+	30–49	*33,35,37*
Portugal	46 - 48	02	0	0–19	*16,11,12*
Romania	85 - 130	90–02	0	0–19	*62*
Russia	500 – 1,000	96–00	0	0–19	*74,26,33,60,63, 72,103,113,141*
Serbia & MN	90 - 110	97–02	0	0–19	*1,45,59,227,31,33*
Slovakia	90 - 95	90–99	0	0–19	
Slovenia	25 - 35	90–00	0	0–19	
Spain	1,300 – 1,300	98–02	0	0–19	*10,16*
Sweden	550 - 610	99–00	+	0–9	
Switzerland	300 - 310	93–96	0	0–9	
Turkey	2,000 – 3,000	01	0	0–19	
Ukraine	(3 – 5)	90–00	(0)	(0–9)	
UK	422 - 422	92	0	1	
Total (approx.)	8,400 – 11,000	Overall trend Stable			
Breeding range	>3,000,000 km²	Gen. length. 11		% Global pop. 5–24	

Aquila chrysaetos has a widespread but discontinuous distribution across much of Europe, which accounts for less than a quarter of its global breeding range. Its European breeding population is small (as few as 8,400 pairs), but was stable between 1970–1990. Although the species declined in a handful of countries during 1990–2000, populations were stable or increased across the majority of Europe, and the species remained stable overall. Nevertheless, its population size renders it susceptible to the risks affecting small populations, and consequently it is evaluated as Rare.

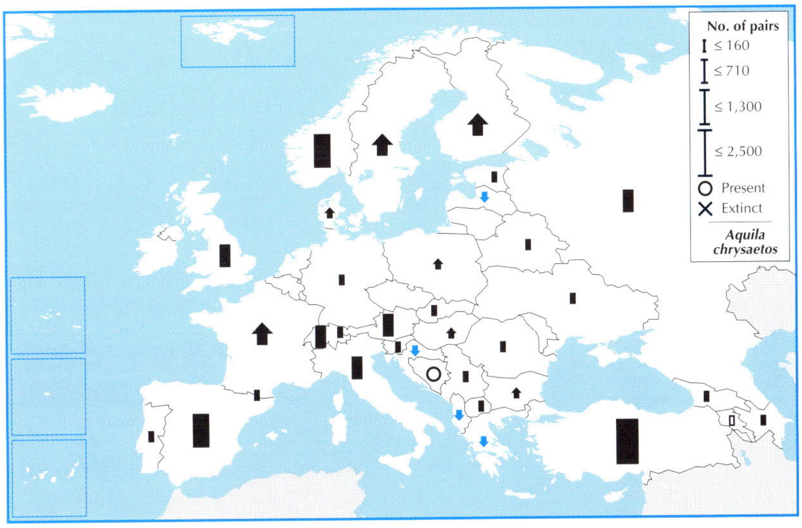

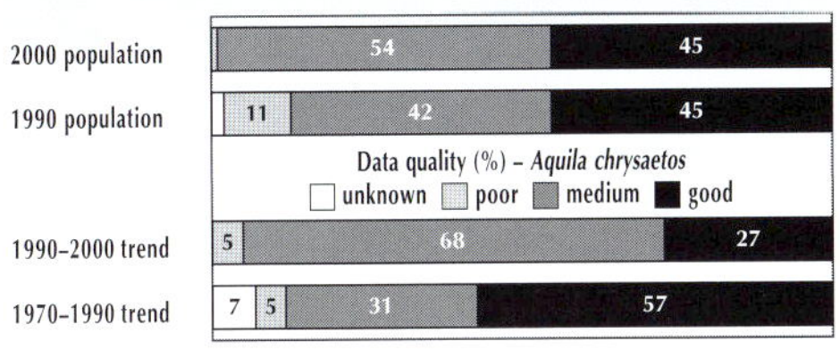

Hieraaetus pennatus
BOOTED EAGLE

SPEC 3 (1994: 3) **Status** (Rare)
Criteria <10,000 pairs
European IUCN Red List Category —
Criteria —
Global IUCN Red List Category —
Criteria —

Hieraaetus pennatus is a widespread summer visitor to much of south-west and eastern Europe, which accounts for less than half of its global breeding range. Its European breeding population is small (as few as 4,400 pairs), but was stable between 1970–1990. Trend data were not available for the key Spanish population during 1990–2000, but despite declines in much of south-east Europe, the species probably remained stable overall. Nevertheless, its population size still renders it susceptible to the risks affecting small populations, and consequently it is provisionally evaluated as Rare.

Country	Breeding pop. size (pairs)	Year(s)	Trend	Mag.%	References
Albania	5 – 20	96–02	(–)	(0–19)	
Armenia	10 – 15	98–02	0	0–19	
Austria	1 – 2	98–02	+	N	
Azerbaijan	(30 – 200)	96–00	(0)	(0–19)	
Belarus	(25 – 40)	97–02	0	0–19	
Bulgaria	60 – 120	96–02	+	50–79	
Croatia	(5 – 10)	02	?	–	*70*
France	380 – 650	00–02	(0)	(0–19)	*8,7*
Georgia	(70 – 150)	94–03	?	–	
Germany	**0 – 1**	95–99	**+**	**N**	
Greece	(50 – 100)	95–00	(–)	(20–29)	
Hungary	**0 – 4**	98–02	**–**	**>80**	*1*
Macedonia	15 – 25	99–00	(–)	(20–29)	*3*
Moldova	15 – 25	90–00	+	0–19	
Poland	1 – 5	95–00	**0**	**0–19**	*38*
Portugal	(300 – 1,000)	02	(0)	(0–19)	
Romania	(80 – 120)	90–02	(–)	(0–19)	*62*
Russia	**800 – 1,500**	94–02	**+**	**20–29**	*3,6,34,67,141*
Serbia & MN	11 – 16	90–02	–	10–29	*1,111,50,227*
Slovakia	0 – 1	80–91	–	50–79	
Spain	(2,000 – 4,000)	98–02	?	–	*10*
Turkey	300 – 500	01	–	20–29	
Ukraine	**240 – 380**	90–00	**–**	**0–19**	
Total (approx.)	4,400 – 8,900	**Overall trend** Stable			
Breeding range	>2,000,000 km²	**Gen. length.** 11		**% Global pop.** 25–49	

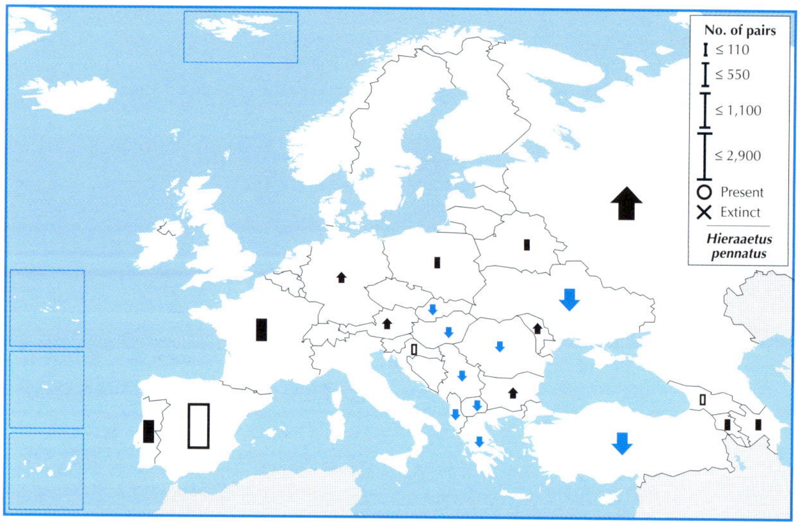

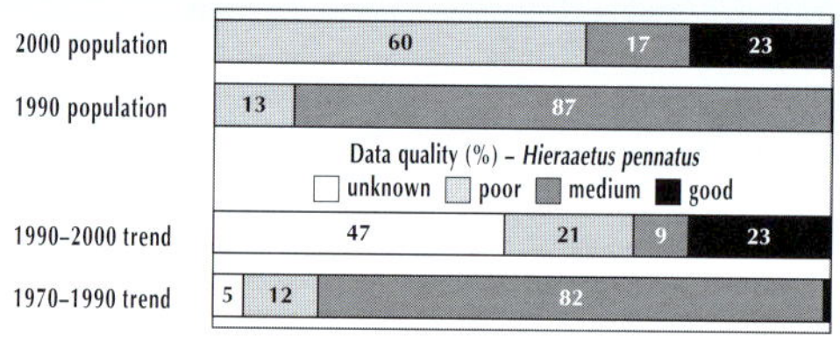

Hieraaetus fasciatus
BONELLI'S EAGLE

SPEC 3 (1994: 3) **Status** Endangered
Criteria See IUCN below
European IUCN Red List Category EN
Criteria C1
Global IUCN Red List Category —
Criteria —

Hieraaetus fasciatus is a patchily distributed resident in southern Europe, which accounts for less than a quarter of its global range. Its European breeding population is very small (as few as 920 pairs), and declined substantially between 1970–1990. Although some populations were stable during 1990–2000, the species continued to decline in its Spanish stronghold, and declined overall at a rate that—on top of earlier declines—equates to a large decline (>20%) over two generations. As a consequence of this decline and its very small population, it is evaluated as Endangered.

Country	Breeding pop. size (pairs)	Year(s)	Trend	Mag.%	References
Albania	(2 – 10)	95–02	?	–	
Azerbaijan	(0 – 5)	96–00	?	–	
Bosnia & HG	Present	85–89	?	–	
Bulgaria	1 – 3	96–02	(F)	(>80)	
Croatia	1 – 5	02	(0)	(0–19)	*70,16,54*
Cyprus	**20 – 40**	94–02	**0**	**0–9**	
France	**23 – 23**	03	**–**	**10–19**	*8,7*
Greece	**85 – 105**	97	(0)	(0–19)	
Italy	13 – 18	03	(–)	(10–19)	*6*
Macedonia	(1 – 3)	99–00	(–)	(20–29)	*3*
Portugal	**85 – 100**	02	**0**	**0–19**	
Serbia & MN	4 – 6	95–02	0	0–19	*1,50,227,156a*
Spain	650 – 713	98–02	–	20–29	*10,16*
Turkey	30 – 40	01	–	0–19	
Total (approx.)	920 – 1,100	**Overall trend** Large decline			
Breeding range	>750,000 km²	**Gen. length.** 11		**% Global pop.** 5–24	

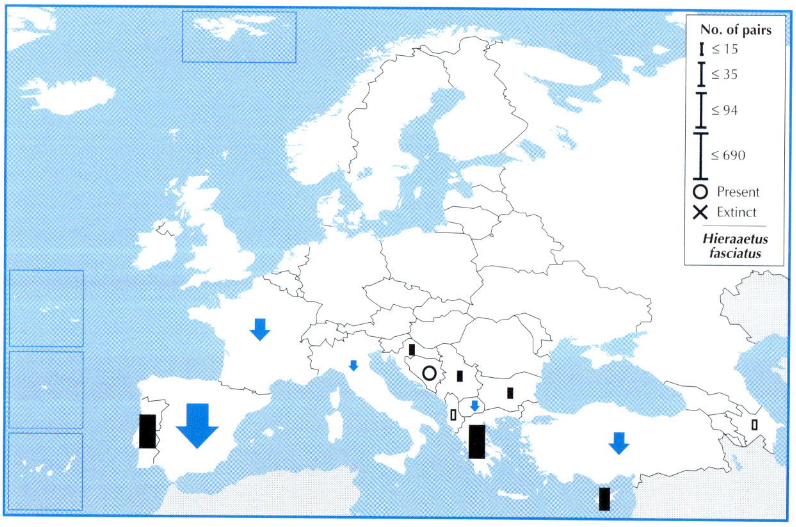

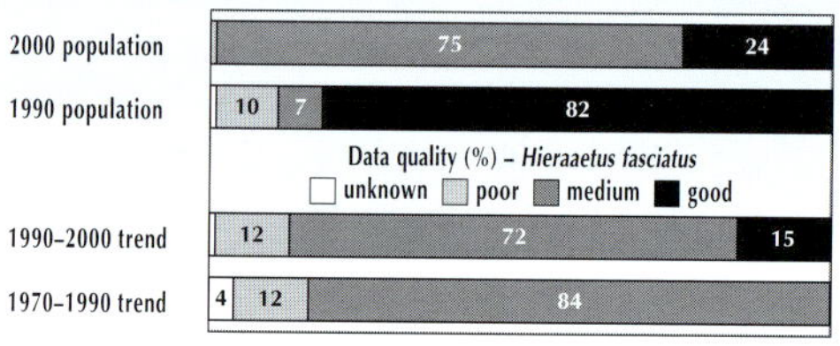

Pandion haliaetus
OSPREY

SPEC 3 (1994: 3) **Status** Rare
Criteria <10,000 pairs
European IUCN Red List Category —
Criteria —
Global IUCN Red List Category —
Criteria —

Pandion haliaetus is a widespread summer visitor to much of northern Europe (occurring only patchily farther south), which accounts for less than a quarter of its global breeding range. Its European breeding population is small (as few as 7,600 pairs), but increased markedly between 1970–1990. Most European populations increased or were stable during 1990–2000, and the species underwent a moderate increase overall. Nevertheless, its population size still renders it susceptible to the risks affecting small populations, and consequently it is evaluated as Rare.

Country	Breeding pop. size (pairs)	Year(s)	Trend	Mag.%	References
Armenia	1 – 4	99–02	(–)	(10–19)	
Azerbaijan	(0 – 5)	96–00	(0)	(0–19)	
Belarus	120 – 180	97–02	0	0–19	
Bulgaria	**1 – 3**	96–02	?	–	
Denmark	1 – 3	98–01	0	0–19	*4,5,6,7*
Estonia	**40 – 45**	98	+	**20–29**	*1*
Finland	**1,150 – 1,300**	01–02	+	**20**	
France	**40 – 45**	03	+	**50–79**	*8,7*
Germany	**350 – 380**	95–99	+	**30–49**	
Latvia	100 – 150	90–00	+	0–19	*17*
Lithuania	**30 – 40**	99–01	**0**	**0–19**	*20*
Moldova	0 – 2	90–00	0	0–19	
Norway	150 – 200	90–02	+	**0–19**	*52,63*
Poland	**70 – 75**	98	+	**20–39**	*33,35*
Portugal	**0 – 0**	02	–	**X**	
Russia	2,000 – 4,000	94–02	0	0–19	*74,63,65,86, 103,113*
Spain	**16 – 18**	00–02	+	**30–49**	*10,16*
Canary Is.	**15 – 20**	97–01	+	**0–19**	*28*
Sweden	**3,400 – 4,100**	99–00	0	0–19	
Turkey	(0 – 10)	01	(–)	(–)	
Ukraine	(0 – 2)	90–00	–	>80	
UK	**127 – 127**	96–00	+	**144**	
Total (approx.)	7,600 – 11,000	**Overall trend** Moderate increase			
Breeding range	>3,000,000 km²	**Gen. length.** 9		**% Global pop.** 5–24	

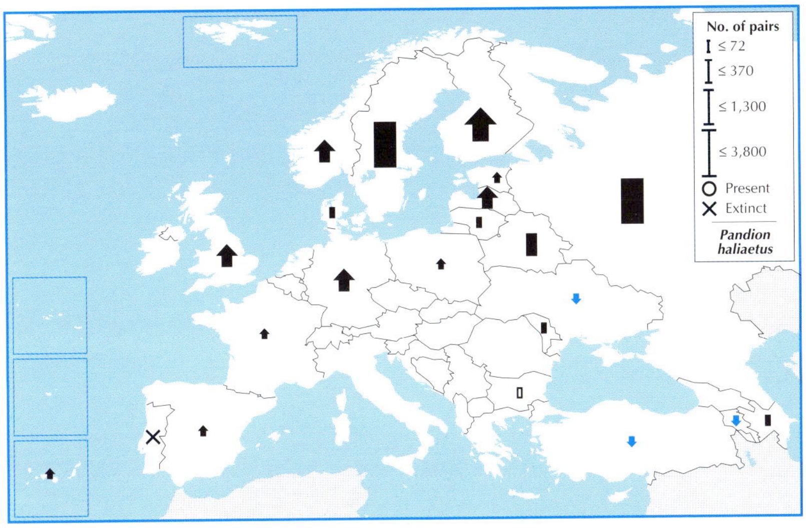

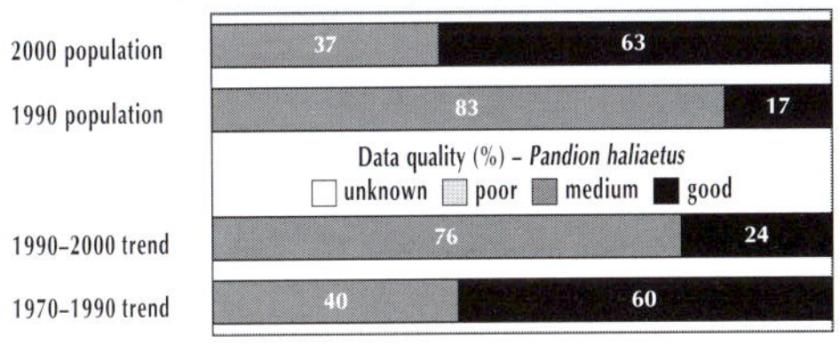

Falco naumanni
LESSER KESTREL

SPEC 1 (1994: 1) **Status** Depleted
Criteria Large historical decline
European IUCN Red List Category —
Criteria —
Global IUCN Red List Category VU
Criteria A2b,c,e; A3b,c,e

Falco naumanni is a summer visitor to southern Europe, which accounts for less than half of its global breeding range. Its European breeding population is relatively small (<42,000 pairs), and declined substantially between 1970–1990. The species was stable or increased in south-western Europe during 1990–2000, but many south-eastern populations continued to decline, and the species underwent a small decline overall. Its total population size remains far below the level that preceded its decline, and consequently this globally threatened species is evaluated as Depleted in Europe.

Country	Breeding pop. size (pairs)	Year(s)	Trend	Mag.%	References
Albania	0 – 20	98–01	?	–	
Armenia	15 – 60	00–02	?	–	*10*
Azerbaijan	(500 – 3,000)	96–00	(0)	(0–19)	
Bosnia & HG	(0 – 250)	90–00	?	–	
Bulgaria	**0 – 5**	95–00	–	**>80**	
Croatia	0 – 0	02	(–)	(X)	*70*
France	**72 – 72**	03	+	**>80**	*8,7*
Georgia	20 – 100	94–03	–	**>80**	
Greece	2,000 – 3,480	95–00	(0)	(0–19)	
Italy	**3,640 – 3,840**	01	+	20–29	*6*
Macedonia	**1,500 – 3,000**	02	(–)	(10–19)	*1*
Moldova	3 – 6	90–00	**0**	**0–19**	
Portugal	**286 – 293**	01	(+)	(–)	
Romania	(0 – 5)	90–02	(F)	(20–29)	*29*
Russia	400 – 600	98–02	+	20–29	*6,45,46,73,74,141*
Serbia & MN	0 – 6	90–02	(–)	(>80)	*1,169,104*
Slovenia	**0 – 0**	94–00	–	**X**	
Spain	12,000 – 20,000	94–02	0	0–19	*10,16*
Turkey	5,000 – 7,000	01	–	20–29	
Ukraine	(5 – 10)	90–00	–	50–79	
UK					
Gibraltar	**4 – 10**	00	**0**	**0–19**	
Total (approx.)	25,000 – 42,000	**Overall trend** Small decline			
Breeding range	>1,000,000 km²	**Gen. length.** <3.3		**% Global pop.** 25–49	

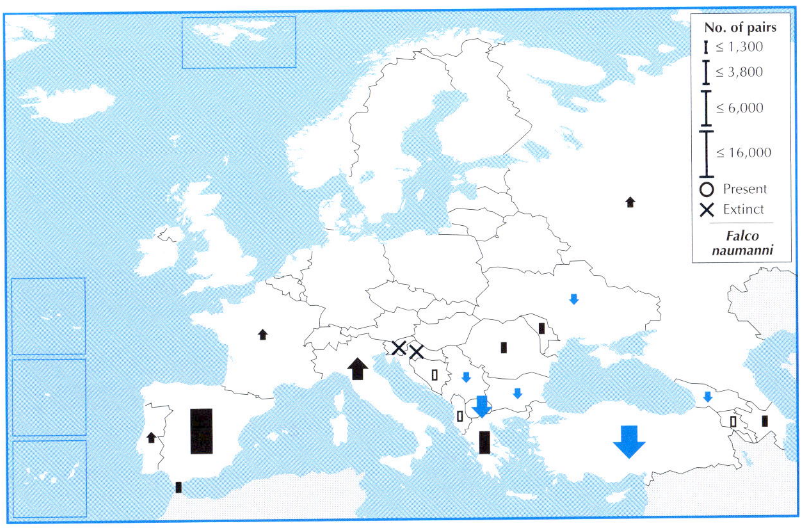

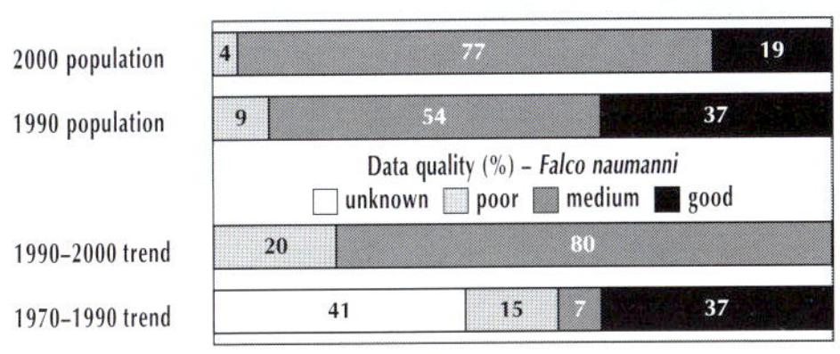

Falco tinnunculus
COMMON KESTREL

SPEC 3 (1994: 3) **Status** Declining
Criteria Moderate continuing decline
European IUCN Red List Category —
Criteria —
Global IUCN Red List Category —
Criteria —

Falco tinnunculus is a widespread breeder across most of Europe, which accounts for less than a quarter of its global breeding range. Its European breeding population is large (>330,000 pairs), but underwent a moderate decline between 1970–1990. Although trends were stable or increased in many countries during 1990–2000, key populations in the United Kingdom, France and Russia declined, and the species underwent a moderate decline (>10%) overall. Consequently, it is evaluated as Declining.

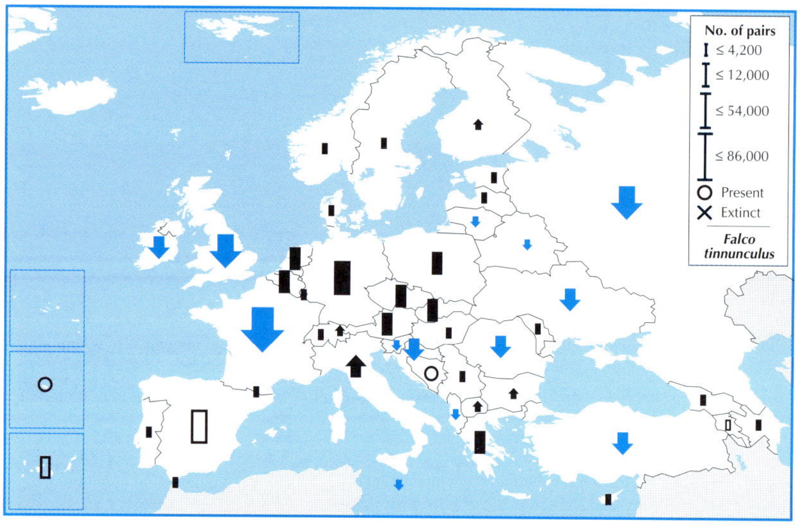

Country	Breeding pop. size (pairs)	Year(s)	Trend	Mag.%	References
Albania	500 – 2,000	96–02	(–)	(0–19)	
Andorra	35 – 90	99–01	(0)	(0–19)	*1,3*
Armenia	260 – 330	00–02	?	–	
Austria	(5,000 – 10,000)	98–02	(0)	(0–19)	
Azerbaijan	(1,000 – 5,000)	96–00	(0)	(0–19)	
Belarus	1,200 – 1,700	97–02	–	0–9	
Belgium	3,800 – 6,000	01–02	0	0–19	*1*
Bosnia & HG	Present	90–03	?	–	
Bulgaria	2,500 – 5,000	96–02	+	0–19	
Croatia	(5,000 – 8,000)	02	(–)	(50–79)	*26,54,16*
Cyprus	(2,000 – 4,000)	94–02	(0)	(0–9)	
Czech Rep.	**9,000 – 13,000**	00	**0**	**0–19**	
Denmark	2,500 – 2,500	00	F	30–49	*14*
Estonia	300 – 450	98	0	0–19	*1*
Finland	2,000 – 3,000	98–02	+	**25**	
France	72,000 – 101,000	00–02	–	22	*8,2*
Georgia	(2,000 – 3,000)	94–03	0	0–19	
Germany	41,500 – 68,000	95–99	0	0–19	
Greece	(5,000 – 10,000)	95–00	(0)	(0–19)	
Hungary	3,500 – 5,000	99–02	0	0–19	*19,1*
Rep. Ireland	2,500 – 10,000	88–91	–	0–19	
Italy	(8,000 – 12,000)	03	(+)	(0–19)	*6*
Latvia	100 – 200	90–00	(0)	(0–19)	*23,16*
Liechtenstein	20 – 30	98–00	(+)	(0–9)	
Lithuania	200 – 300	99–01	–	20–29	*20*
Luxembourg	500 – 700	00–02	**0**	**0–19**	
Macedonia	(1,000 – 2,500)	99–00	(+)	(20–29)	*3*
Malta	**1 – 2**	90–02	**–**	**>80**	*1*
Moldova	**250 – 350**	90–00	**0**	**0–19**	
Netherlands	**5,000 – 7,500**	98–00	**0**	**17**	*1*
Norway	(2,000 – 4,000)	90–03	(0)	(0–19)	
Poland	5,000 – 10,000	00–03	(0)	(0–19)	*23*
Portugal	(1,000 – 3,000)	02	(0)	(0–19)	
Madeira	Present	02	?	–	
Romania	10,000 – 14,000	90–02	(–)	(0–19)	*48*
Russia	(40,000 – 60,000)	94–02	(–)	(20–29)	*8,22,104*
Serbia & MN	3,400 – 4,500	90–02	0	0–19	*1,112,172a*
Slovakia	4,000 – 6,000	90–99	0	0–19	
Slovenia	1,500 – 2,000	94	(–)	(0–19)	
Spain	(25,000 – 30,000)	92	?	–	*13,12,16,10*
Canary Is.	2,500 – 10,000	97–03	?	–	*28,25*
Sweden	**2,500 – 3,500**	99–00	0	0–19	
Switzerland	**3,000 – 5,000**	93–96	0	0–19	
Turkey	(9,000 – 14,000)	01	(–)	(0–19)	
Ukraine	9,000 – 14,400	90–00	–	0–19	
UK	**36,800 – 36,800**	00	**–**	**28**	*5,11*
Gibraltar	**6 – 8**	00	**0**	**0–19**	
Total (approx.)	330,000 – 500,000	**Overall trend** Moderate decline			
Breeding range	>8,000,000 km²	**Gen. length.** <3.3 **% Global pop.** 5–24			

(See p. 86, bottom, for data quality graph)

Falco vespertinus
RED-FOOTED FALCON

SPEC 3 (1994: 3) **Status** (Vulnerable)
Criteria See IUCN below
European IUCN Red List Category VU
Criteria A2b
Global IUCN Red List Category —
Criteria —

Falco vespertinus is a widespread summer visitor to much of eastern Europe, which accounts for less than half of its global breeding range. Its European breeding population is relatively small (<39,000 pairs), and underwent a large decline between 1970–1990. Although the species was stable or increased in several marginal countries during 1990–2000, it continued to decline in its Russian stronghold and other sizeable populations in eastern Europe, and underwent a large decline (>30%) overall. Consequently, it is provisionally evaluated as Vulnerable.

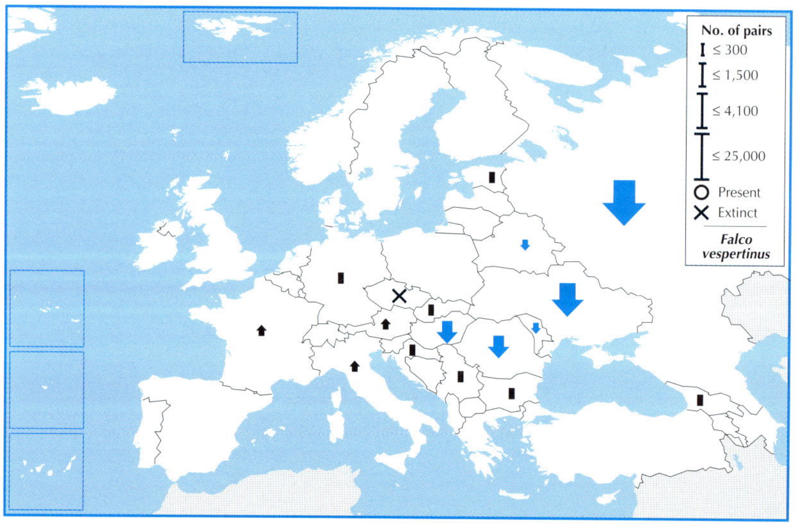

Country	Breeding pop. size (pairs)	Year(s)	Trend	Mag.%	References
Austria	1 – 5	98–02	+	N	
Belarus	(10 – 15)	97–02	–	0–9	
Bulgaria	50 – 150	95–02	F	30–49	
Croatia	(1 – 10)	02	(0)	(0–19)	*70,44,57*
Czech Rep.	**0 – 0**	00	**–**	**X**	
Estonia	(0 – 10)	98	F	20–29	*1*
France	**1 – 5**	00–02	+	**0–19**	*8,7*
Georgia	10 – 50	94–03	(F)	(>80)	
Germany	**0 – 1**	95–99	**F**	**>80**	
Hungary	800 – 1,500	95–02	**–**	**50–79**	*1*
Italy	**70 – 70**	03	+	**>80**	*6*
Moldova	**70 – 90**	90–00	**–**	**20–29**	
Romania	(1,300 – 1,600)	90–02	(–)	(0–19)	*62*
Russia	(20,000 – 30,000)	94–02	(–)	(20–29)	*8,24,33,67,73,94*
Serbia & MN	250 – 350	97–02	F	20–29	*1,73,132,246,12*
Slovakia	**20 – 100**	80–99	F	30–49	
Ukraine	3,200 – 5,100	90–00	–	30–49	
Total (approx.)	26,000 – 39,000	**Overall trend** Large decline			
Breeding range	>2,000,000 km²	**Gen. length.** <3.3 **% Global pop.** 25–49			

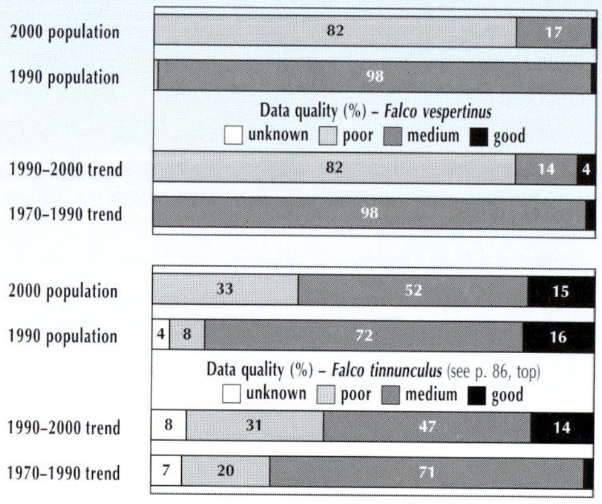

Falco columbarius
MERLIN

Non-SPEC (1994: —) **Status** (Secure)
Criteria —
European IUCN Red List Category —
Criteria —
Global IUCN Red List Category —
Criteria —

Falco columbarius is a widespread breeder in northern Europe, which accounts for less than a quarter of its global breeding range. Its European breeding population is relatively small (<49,000 pairs), but was stable between 1970–1990. Although there was a slight decline in Sweden during 1990–2000, other significant populations in Iceland, the United Kingdom, Norway, Finland and Russia were stable or increased, and the species probably remained stable overall. Consequently, it is provisionally evaluated as Secure.

Country	Breeding pop. size (pairs)	Year(s)	Trend	Mag.%	References
Belarus	300 – 350	97–02	0	0–19	
Denmark					
Faroe Is.	15 – 15	95	(0)	(0–19)	
Estonia	10 – 20	98	–	50–79	*1*
Finland	2,000 – 3,000	98–02	0	0–19	
Iceland	1,000 – 2,000	78–94	+	0–19	*30*
Rep. Ireland	110 – 130	88–91	(0)	(0–19)	
Latvia	(10 – 30)	90–00	(–)	(30–49)	*23,21*
Lithuania	(5 – 10)	99–01	(F)	(20–29)	*20*
Norway	(2,500 – 6,500)	02	(0)	(0–19)	*42*
Russia	(20,000 – 30,000)	94–00	(0)	(0–19)	*103,113,138*
Sweden	**4,200 – 5,700**	99–00	–	**0–9**	
UK	**1,300 – 1,300**	90–94	+	**117**	*28*
Total (approx.)	31,000 – 49,000	**Overall trend** Stable			
Breeding range	>3,000,000 km²	**Gen. length.** <3.3		**% Global pop.** 5–24	

Country	Winter pop. size (individuals)	Year(s)	Trend	Mag.%	References
Armenia	(250 – 1,000)	94–02	?	–	
Austria	(100 – 300)	98–02	(F)	(>80)	
Bulgaria	100 – 500	98–02	(0)	(0–19)	
Croatia	(50 – 100)	02	?	–	*70*
Czech Rep.	50 – 100	90–00	–	20–29	
Denmark	(50 – 150)	99–00	?	–	
Finland	(100 – 500)	98–02	0	0–19	
Greece	(100 – 300)	95–00	(F)	(30–49)	
Hungary	400 – 700	00–02	?	–	*24,25*
Iceland	(10 – 100)	98	?	–	*64*
Lithuania	(50 – 250)	92–02	?	–	
Netherlands	500 – 1,000	00	–	20–29	*4,5,6,7,9,10*
Poland	(250 – 1,000)	90–00	?	–	*121*
Romania	(400 – 1,500)	90–00	(–)	(0–19)	*60*
Slovakia	100 – 350	90–99	?	–	*4*
Ukraine	1,500 – 2,000	90–00	0	0–19	
Total (approx.)	>4,000	**Overall Trend** Unknown			
% in European IBAs	5–7	**Gen. length** <3.3		**% Global pop.** 5–24	

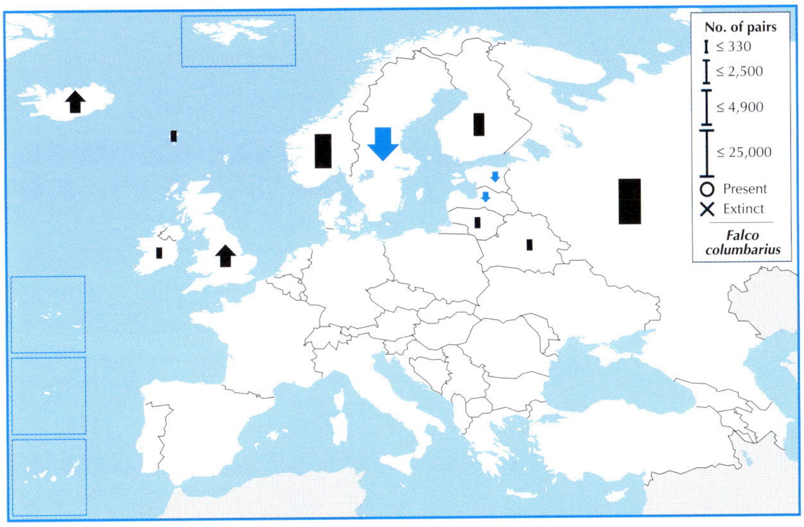

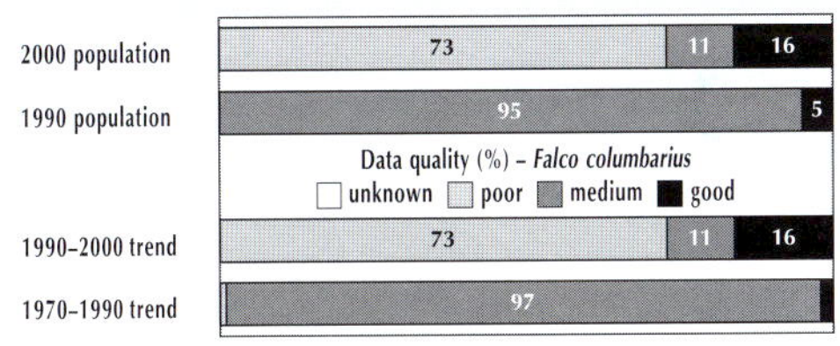

Falco subbuteo
EURASIAN HOBBY

Non-SPEC (1994: —) **Status** (Secure)
Criteria —
European IUCN Red List Category —
Criteria —
Global IUCN Red List Category —
Criteria —

Falco subbuteo is a widespread summer visitor to much of Europe, which accounts for less than a quarter of its global breeding range. Its European breeding population is relatively small (<120,000 pairs), but was stable between 1970–1990. Although there were declines in certain countries—most notably Germany and Finland—during 1990–2000, populations were stable or increased elsewhere in Europe, and the species probably remained stable overall. Consequently, it is provisionally evaluated as Secure.

Country	Breeding pop. size (pairs)	Year(s)	Trend	Mag.%	References
Albania	10 – 50	98–02	(–)	(0–19)	
Andorra	(0 – 2)	99–01	(0)	(0–19)	*1,3*
Armenia	50 – 250	97–02	?	–	
Austria	(800 – 1,200)	98–02	(0)	(0–19)	
Azerbaijan	(100 – 500)	96–00	(0)	(0–19)	
Belarus	**2,500 – 2,700**	97–02	**0**	**0–19**	
Belgium	593 – 991	01–02	+	50–79	*1*
Bosnia & HG	Present	85–89	?	–	
Bulgaria	600 – 1,200	96–02	0	0–9	
Croatia	(500 – 1,000)	02	(+)	(30–49)	*16*
Cyprus	20 – 70	98–02	+	N	
Czech Rep.	**200 – 300**	00	+	**30–49**	
Denmark	1 – 11	98–01	(0)	(0–19)	*4,5,6,7*
Estonia	500 – 1,000	98	+	20–29	*1*
Finland	**2,000 – 3,000**	98–02	–	**10**	
France	6,500 – 9,600	00–02	(+)	(0–19)	*8,7*
Georgia	100 – 300	94–03	?	–	
Germany	2,700 – 3,600	95–99	–	20–29	
Greece	(400 – 600)	95–00	(–)	(0–19)	
Hungary	900 – 1,900	99–02	+	50–79	*19,1*
Italy	(500 – 1,000)	03	(+)	(0–19)	*6*
Latvia	(400 – 1,000)	90–00	0	0–19	*23,18*
Liechtenstein	(1 – 3)	98–00	(0)	(0–19)	
Lithuania	400 – 700	99–01	(–)	(20–29)	*20*
Luxembourg	20 – 25	00–02	+	**0–19**	
Macedonia	(100 – 200)	99–00	(0)	(0–19)	*3*
Moldova	50 – 80	90–00	–	**30–49**	
Netherlands	**750 – 1,000**	98–00	–	50–79	*1*
Norway	70 – 75	90–02	**0**	**0–19**	
Poland	3,000 – 5,000	00–02	(0)	(0–19)	*2,23*
Portugal	(100 – 500)	02	(0)	(0–19)	
Romania	3,200 – 4,000	90–02	(0)	(0–19)	*7,41*
Russia	(30,000 – 60,000)	94–02	(0)	(0–19)	*104*
Serbia & MN	450 – 620	90–02	0	0–19	*1,162,172a*
Slovakia	600 – 800	90–99	0	0–19	
Slovenia	100 – 200	94	(0)	(0–19)	
Spain	(2,300 – 3,000)	98–02	?	–	*10,16*
Sweden	**1,800 – 2,200**	99–00	0	0–19	
Switzerland	**400 – 600**	93–96	+	10–19	
Turkey	(4,000 – 8,000)	01	(0)	(0–19)	
Ukraine	2,000 – 3,000	90–00	0	0–19	
UK	**2,200 – 2,200**	01	+	**61**	*4,11*
Total (approx.)	71,000 – 120,000	**Overall trend** Stable			
Breeding range	>6,000,000 km²	**Gen. length.** 5		**% Global pop.** 5–24	

(See p. 88, top, for data quality graph)

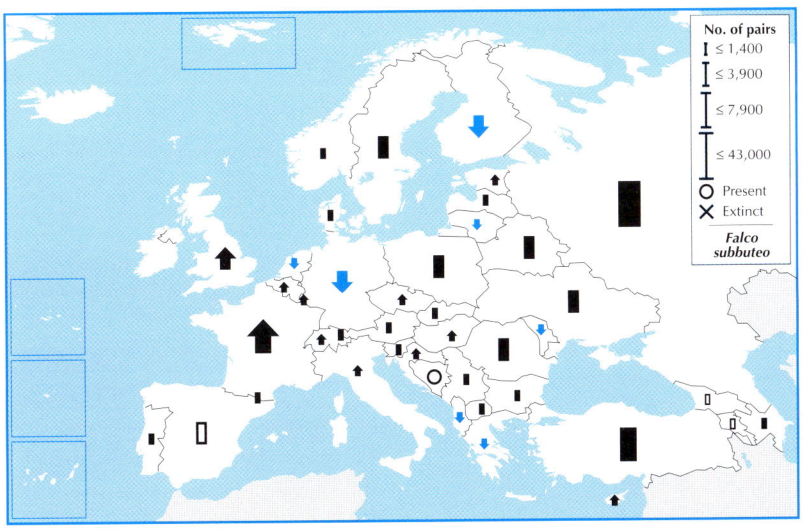

Falco eleonorae
ELEONORA'S FALCON

SPEC 2 (1994: 2) **Status** Declining
Criteria Moderate recent decline
European IUCN Red List Category —
Criteria —
Global IUCN Red List Category —
Criteria —

Country	Breeding pop. size (pairs)	Year(s)	Trend	Mag.%	References
Croatia	**60 – 70**	02	(-)	(30–49)	*70*
Cyprus	**140 – 160**	02	**0**	**0–9**	
Greece	**4,500 – 4,500**	99	-	0–19	
Italy	**500 – 600**	03	**0**	**0–19**	*6*
Serbia & MN	1 – 2	00–02	0	0–19	*1,156a*
Spain	(485 – 535)	01	?	-	*10,16*
Canary Is.	**200 – 200**	00	+	**20–29**	*30*
Turkey	(20 – 100)	01	-	30–49	
Total (approx.)	5,900 – 6,200	**Overall trend** Moderate decline			
Breeding range	>100,000 km²	**Gen. length.** 5		**% Global pop.** >95	

Falco eleonorae is a patchily distributed summer visitor to rocky coasts and islands in the Mediterranean, with Europe constituting >95% of its global breeding range. Its European breeding population is small (as few as 5,900 pairs), but was stable between 1970–1990. Although some populations were stable or increased during 1990–2000 (the trend in Spain was unknown), the species declined in its Greek stronghold, and underwent a moderate decline (>10%) overall. Consequently, this previously Rare species is now evaluated as Declining.

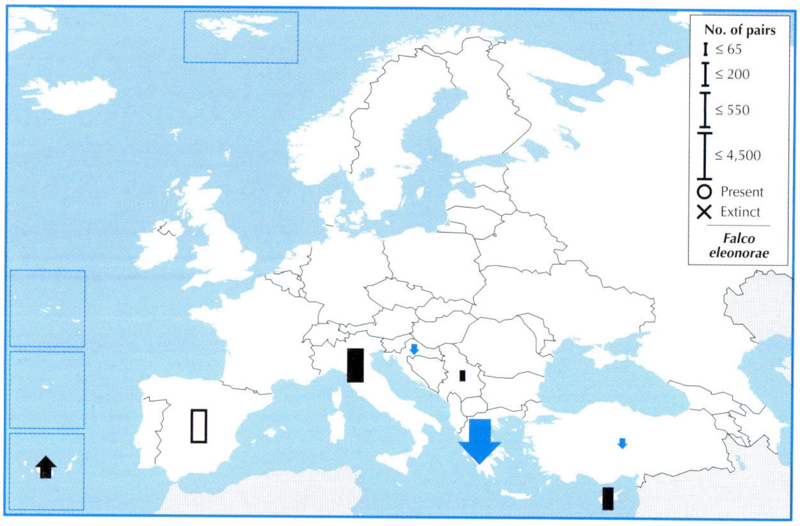

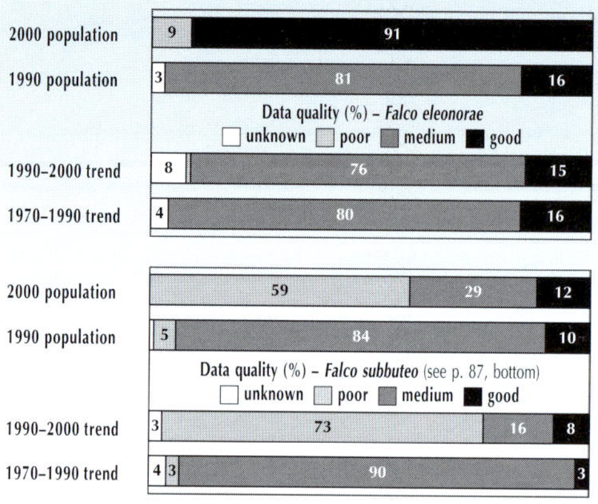

Falco biarmicus
LANNER FALCON

SPEC 3 (1994: 3) **Status** Vulnerable
Criteria See IUCN below
European IUCN Red List Category VU
Criteria C1; C2a(i); D1
Global IUCN Red List Category —
Criteria —

Country	Breeding pop. size (pairs)	Year(s)	Trend	Mag.%	References
Armenia	20 – 30	97–02	?	–	
Azerbaijan	(5 – 20)	96–00	(0)	(0–19)	
Bosnia & HG	Present	90–03	?	–	
Bulgaria	**2 – 3**	01–02	+	**N**	
Croatia	(1 – 5)	02	(-)	(50–79)	*70*
Georgia	**1 – 3**	94–03	**0**	**0–19**	
Greece	**36 – 55**	99	0	0–19	
Italy	**100 – 140**	02	-	0–19	*11,12*
Macedonia	10 – 30	99–00	(+)	(20–29)	*3*
Serbia & MN	8 – 12	97–02	-	0–19	*1,110,39*
Turkey	300 – 600	01	-	0–19	
Total (approx.)	480 – 900	**Overall trend** Moderate decline			
Breeding range	>250,000 km²	**Gen. length.** 5		**% Global pop.** <5	

Falco biarmicus is a patchily distributed resident in southern Europe, which accounts for a tiny proportion of its global range. Its European breeding population is very small (as few as 480 pairs), and underwent a large decline between 1970–1990. Although several small populations were stable or increased during 1990–2000, key populations in Italy and Turkey suffered declines, and the species underwent a moderate decline (>10%) overall. As a consequence of this continuing decline and its small population, the species is evaluated as Vulnerable.

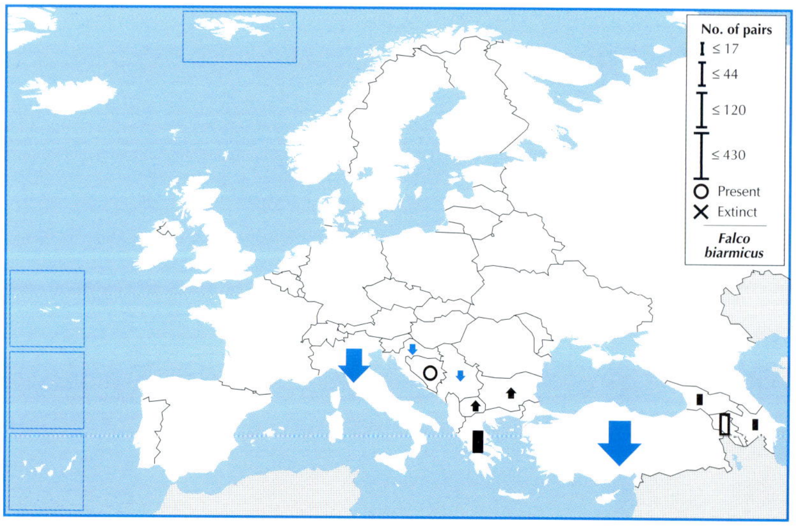

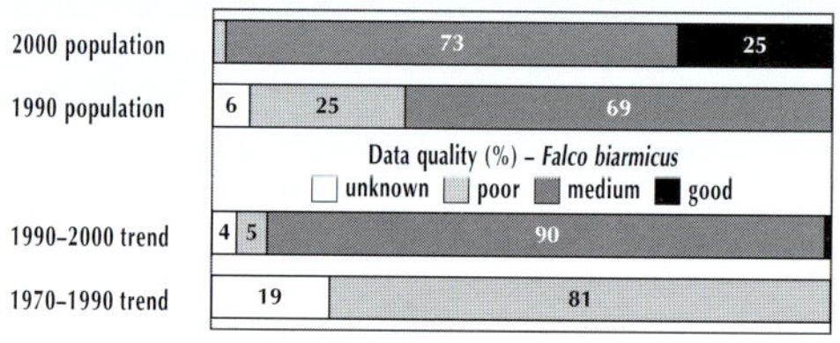

Falco cherrug
SAKER FALCON

SPEC 1 (1994: 3) **Status** Endangered
Criteria See IUCN below
European IUCN Red List Category EN
Criteria C1
Global IUCN Red List Category EN
Criteria A2b,c,d; A3b,c,d

Country	Breeding pop. size (pairs)	Year(s)	Trend	Mag.%	References
Austria	15 – 20	98–02	0	0–19	
Bulgaria	**2 – 6**	97–02	**-**	**>80**	
Croatia	5 – 10	02	(–)	(30–49)	*70*
Czech Rep.	**8 – 10**	00	**0**	**0–19**	
Georgia	**1 – 3**	00–03	**0**	**0–19**	
Germany	**1 – 1**	95–99	**+**	**N**	
Hungary	**130 – 150**	97–02	**+**	**20–29**	*1*
Macedonia	(0 – 3)	99–00	(–)	(20–29)	*5*
Moldova	4 – 7	90–00	0	0–19	
Poland	**0 – 2**	95–00	+	N	*68*
Romania	(5 – 10)	90–02	(–)	(0–19)	*62*
Russia	**30 – 60**	94–02	**-**	**>80**	*74,67,76,158*
Serbia & MN	**52 – 64**	97–02	**0**	**0–19**	*1,74,180,67a*
Slovakia	**10 – 40**	80–99	+	20–29	
Turkey	50 – 70	01	-	30–49	
Ukraine	45 – 80	90–00	-	30–49	
Total (approx.)	360 – 540	**Overall trend** Large decline			
Breeding range	>500,000 km²	**Gen. length.** 5		**% Global pop.** <5	

Falco cherrug breeds mainly in south-east Europe, which holds only a tiny proportion of its global breeding population. Its European breeding population is very small (as few as 360 pairs), and declined substantially between 1970–1990. Although several central European populations were stable or increased during 1990–2000, the species continued to decline throughout eastern Europe, and underwent a large decline overall (>20% in two generations). As a consequence of this decline and its very small population, this globally threatened species is evaluated as Endangered in Europe.

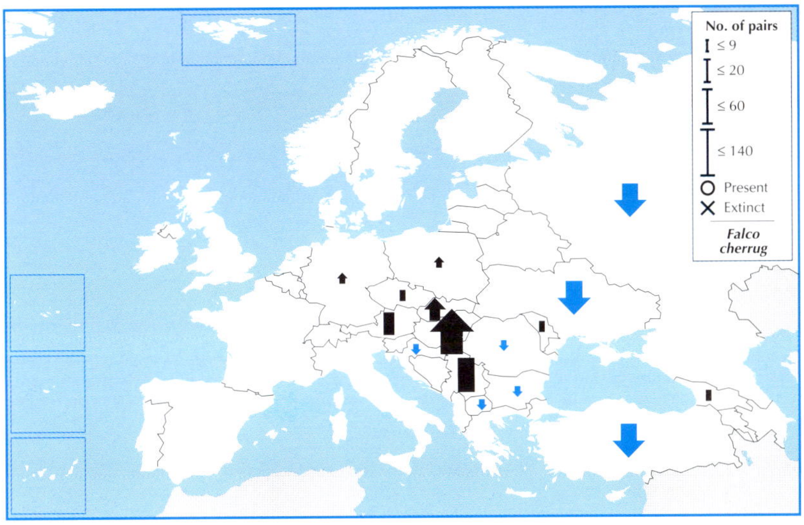

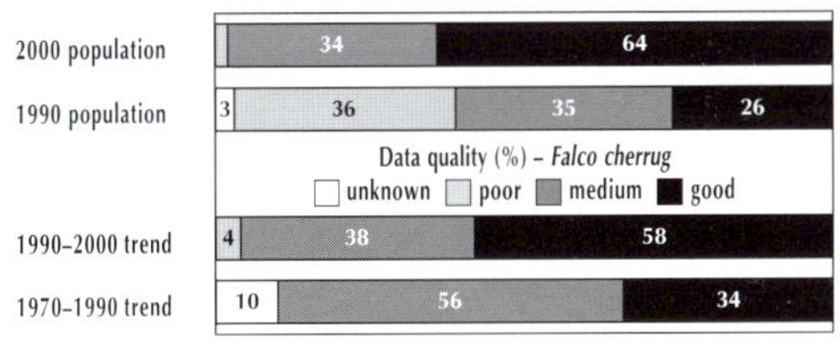

Falco rusticolus
GYRFALCON

SPEC 3 (1994: 3) **Status** (Rare)
Criteria <10,000 pairs
European IUCN Red List Category —
Criteria —
Global IUCN Red List Category —
Criteria —

Country	Breeding pop. size (pairs)	Year(s)	Trend	Mag.%	References
Denmark					
Greenland	(500 – 1,000)	90–00	(0)	(0–19)	
Finland	**25 – 35**	98–02	0	0–19	
Iceland	300 – 400	00	(F)	(–)	*21*
Norway	**300 – 481**	98	**0**	**0–19**	*61*
Russia	(100 – 200)	94–00	(–)	(20–29)	*74,106*
Sweden	80 – 135	99–00	F	20–29	
Total (approx.)	1,300 – 2,300	**Overall trend** Stable			
Breeding range	>500,000 km²	**Gen. length.** 5		**% Global pop.** <5	

Falco rusticolus breeds in Greenland, Iceland, Fennoscandia and arctic Russia, with Europe accounting for a tiny proportion of its global breeding range. Its European breeding population is small (as few as 1,300 pairs), but was stable between 1970–1990. Although the species declined in Russia during 1990–2000, populations elsewhere in Europe were stable or fluctuating, and the species probably remained stable overall. Nevertheless, its population size renders it susceptible to the risks affecting small populations, and consequently it is provisionally evaluated as Rare.

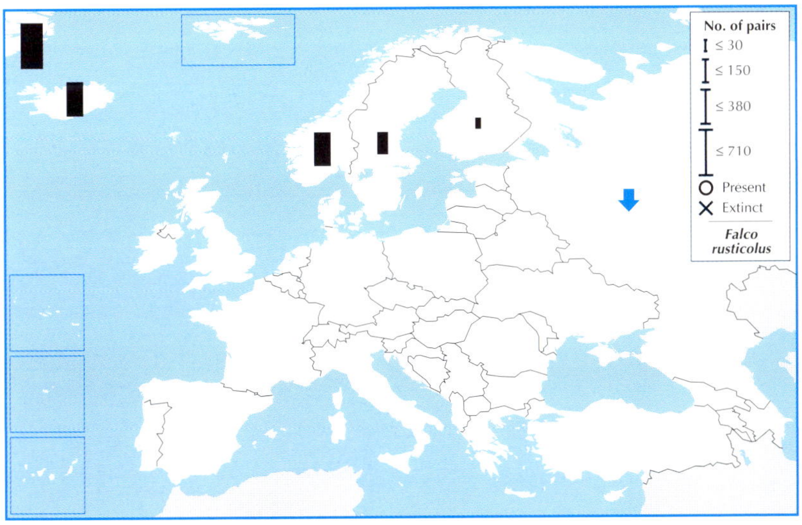

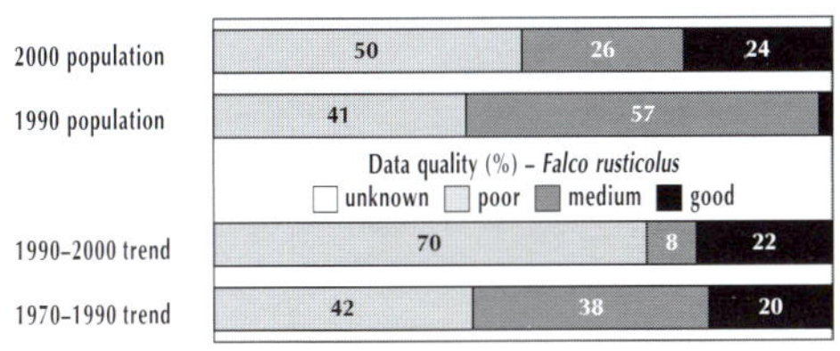

Falco peregrinus
PEREGRINE FALCON

Non-SPEC (1994: 3) **Status** Secure
Criteria —
European IUCN Red List Category —
Criteria —
Global IUCN Red List Category —
Criteria —

Falco peregrinus is a widespread but patchily distributed breeder across much of Europe, which accounts for less than a quarter of its global range. Its European breeding population is relatively small (<25,000 pairs), but increased markedly between 1970–1990. Although the sizeable Turkish population declined during 1990–2000, virtually every other population in its European range increased, and the species underwent a moderate increase overall. Consequently, this previously Rare species is now evaluated as Secure.

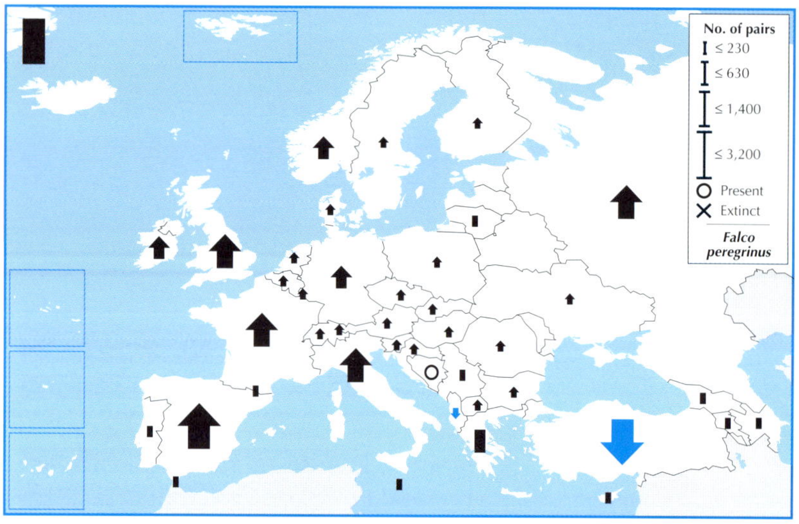

Country	Breeding pop. size (pairs)	Year(s)	Trend	Mag.%	References
Albania	50 – 200	98–02	(–)	(0–19)	
Andorra	3 – 6	98	0	0–19	1,3
Armenia	0 – 50	99–02	0	0–19	4
Austria	200 – 250	98–02	+	20–29	
Azerbaijan	(20 – 50)	96–00	(0)	(0–19)	
Belgium	24 – 26	01–02	+	>80	1
Bosnia & HG	Present	85–89	?	–	
Bulgaria	80 – 130	96–02	+	>80	
Croatia	160 – 200	02	(+)	(0–19)	70
Cyprus	20 – 40	94–02	0	0–19	
Czech Rep.	15 – 17	00	+	>80	
Denmark	1 – 1	01	+	N	7
Greenland	(1,000 – 10,000)	90–00	(0)	(0–19)	
Finland	120 – 150	98–02	+	10	
France	1,100 – 1,400	00–02	+	20–29	8,7
Georgia	60 – 70	94–03	0	0–19	
Germany	620 – 640	95–99	+	30–49	
Greece	(200 – 500)	95–00	(0)	(0–19)	
Hungary	1 – 7	97–02	+	N	1
Rep. Ireland	319 – 350	02	+	0–19	
Italy	787 – 991	01	+	50–79	1
Liechtenstein	2 – 4	98–00	(+)	(30–49)	
Lithuania	(1 – 10)	99–01	F	>80	20
Luxembourg	8 – 15	00–02	+	N	
Macedonia	50 – 150	99–00	(+)	(20–29)	3
Malta	1 – 1	90–02	0	0–19	1
Netherlands	5 – 7	98–00	+	700	1
Norway	350 – 500	98	+	30–49	60
Poland	5 – 8	00	+	200–300	2
Portugal	50 – 100	02	0	0–19	
Romania	8 – 15	90–02	(+)	(0–19)	62
Russia	1,000 – 1,200	94–02	+	20–29	74,63,106
Serbia & MN	90 – 110	95–02	0	10–19	1,113,126,227
Slovakia	1 – 24	80–99	+	20–29	
Slovenia	60 – 80	96–00	+	10–19	
Spain	2,400 – 2,700	98–02	+	0–19	10
Sweden	85 – 100	99–00	+	60–69	
Switzerland	200 – 200	93–96	+	10–19	
Turkey	1,500 – 3,000	01	(–)	(0–19)	
Ukraine	10 – 50	90–00	+	50–79	
UK	1,400 – 1,400	02	+	9	1
Gibraltar	4 – 6	00	0	0–19	
Total (approx.)	12,000 – 25,000	Overall trend Moderate increase			
Breeding range	>3,000,000 km²	Gen. length. 5		% Global pop. 5–24	

(See p. 90, bottom, for data quality graph)

Falco pelegrinoides
BARBARY FALCON

Non-SPEC (1994: —) **Status** Secure
Criteria —
European IUCN Red List Category —▼▼▼
Criteria —
Global IUCN Red List Category —
Criteria —

Falco pelegrinoides has a predominantly North African distribution, which extends into Europe in the Canary Islands and Turkey. Its European breeding population is extremely small (as few as 75 pairs), but was stable between 1970–1990. The Canary Islands population increased during 1990–2000, and the species underwent a moderate increase overall. Although the size of the European population could render it susceptible to the risks affecting small populations, it is marginal to a much larger non-European population. Consequently, the species is evaluated as Secure.

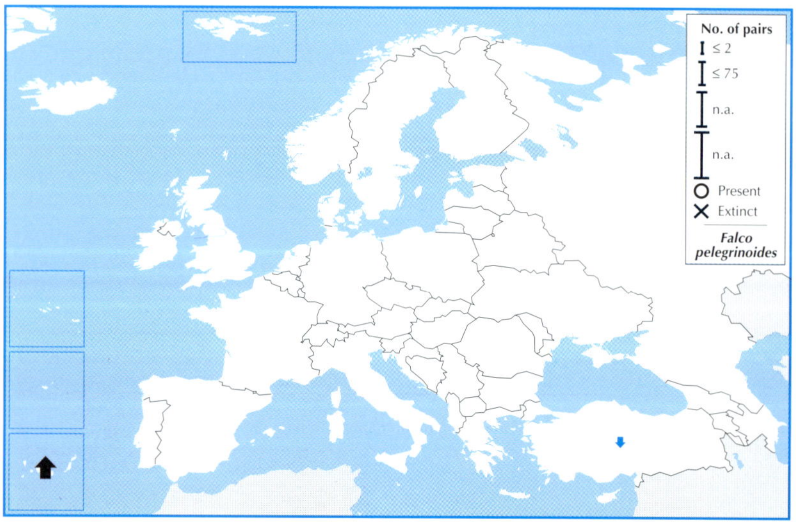

Country	Breeding pop. size (pairs)	Year(s)	Trend	Mag.%	References
Spain					
Canary Is.	75 – 75	97–02	+	20–29	28
Turkey	(0 – 5)	01	(–)	(–)	
Total (approx.)	75 – 80	Overall trend Moderate increase			
Breeding range	>20,000 km²	Gen. length. 5		% Global pop. <5	

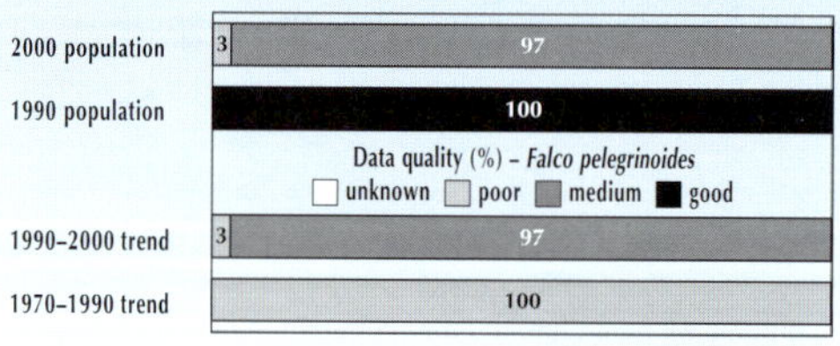

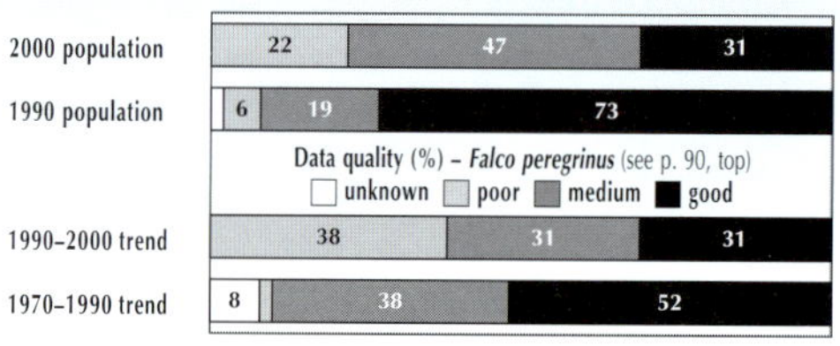

Bonasa bonasia
HAZEL GROUSE

Non-SPEC (1994: —) **Status** Secure
Criteria —
European IUCN Red List Category —
Criteria —
Global IUCN Red List Category —
Criteria —

Bonasa bonasia is a widespread resident in northern Europe, but occurs more patchily farther south, with Europe accounting for less than half of its global range. Its European breeding population is very large (>2,500,000 pairs), and was stable between 1970–1990. Although there were declines in several countries—most notably Finland—during 1990–2000, these were more than compensated for by increases in the stronghold population in Russia, and the species underwent a small increase overall. Consequently, it is evaluated as Secure.

Country	Breeding pop. size (pairs)	Year(s)	Trend	Mag.%	References
Albania	(10 – 30)	02	(–)	(30–49)	
Austria	(5,000 – 20,000)	98–02	(0)	(0–19)	
Belarus	**80,000 – 100,000**	99	0	0–19	
Belgium	50 – 250	95–02	–	0–19	*1*
Bosnia & HG	Present	90–03	?	–	
Bulgaria	500 – 2,000	95–02	(0)	(0–19)	
Croatia	(2,500 – 5,000)	02	(+)	(>80)	*16*
Czech Rep.	**900 – 1,800**	00	**0**	**10–19**	
Estonia	15,000 – 25,000	98	–	20–29	*1*
Finland	300,000 – 500,000	98–02	**-**	**10**	
France	(4,000 – 10,000)	98–02	(–)	(0–9)	*4*
Germany	1,300 – 1,800	95–99	(–)	(0–19)	
Greece	(100 – 200)	95–00	(–)	(0–19)	
Hungary	50 – 100	98–02	(0)	(0–19)	
Italy	5,000 – 6,000	97–03	0	0–19	*8*
Latvia	10,000 – 12,000	90–00	(0)	(0–19)	*23,18*
Liechtenstein	(5 – 10)	98–00	(0)	(0–19)	
Lithuania	(5,000 – 10,000)	99–01	(0)	(0–19)	*20*
Luxembourg	60 – 100	02	0	0–19	
Macedonia	(50 – 250)	90–00	(0)	(0–19)	
Norway	(10,000 – 38,000)	90–02	**-**	**0–9**	*47,8*
Poland	(35,000 – 45,000)	90–00	(0)	(0–19)	1
Romania	10,000 – 13,000	98–02	0	0–19	
Russia	**1,900,000 – 2,200,000**	00	**+**	**20–29**	*97,169*
Serbia & MN	1,000 – 1,700	90–02	+	10–19	*1,143,62,227, 225,155*
Slovakia	3,000 – 5,000	80–99	–	30–49	
Slovenia	(1,000 – 2,000)	00	(–)	(30–49)	
Sweden	**80,000 – 120,000**	99–00	(F)	(20–29)	
Switzerland	**7,500 – 9,000**	93–96	(–)	(0–9)	
Ukraine	2,600 – 5,100	90–00	–	30–49	
Total (approx.)	2,500,000 – 3,100,000	**Overall trend** Small increase			
Breeding range	>4,000,000 km²	**Gen. length.** <3.3	**% Global pop.** 25–49		

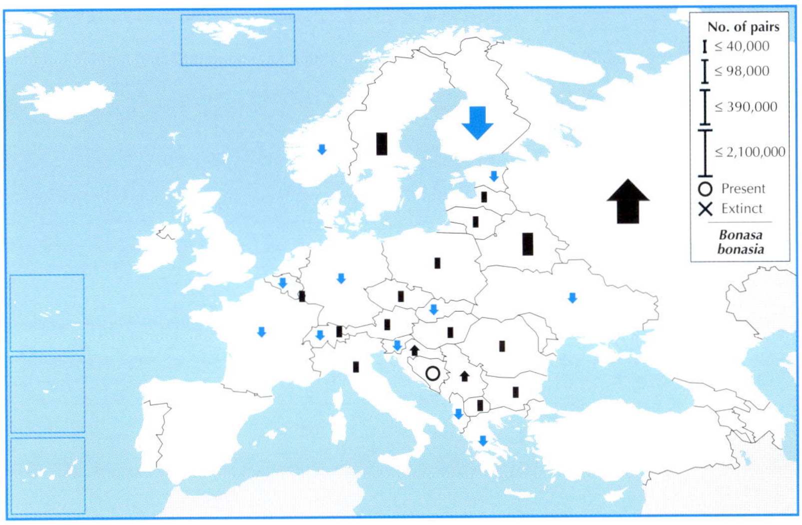

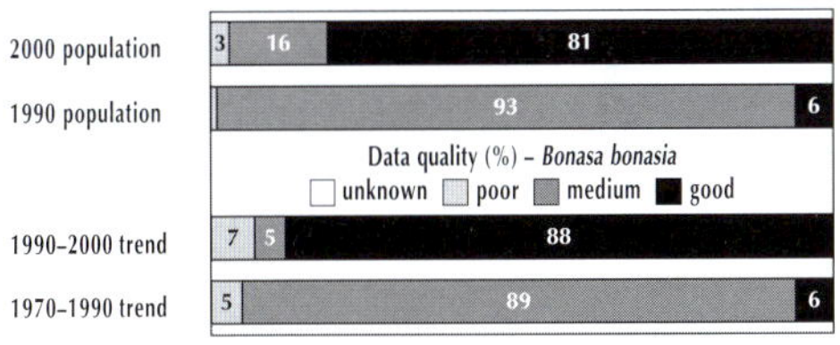

Lagopus lagopus
WILLOW PTARMIGAN

Non-SPEC (1994: —) **Status** Secure
Criteria —
European IUCN Red List Category —
Criteria —
Global IUCN Red List Category —
Criteria —

Lagopus lagopus is a widespread resident across much of northern Europe, which accounts for less than half of its global range. Its European breeding population is very large (>2,100,000 pairs), and was stable between 1970–1990. Although there were declines in the United Kingdom and Finland during 1990–2000, other significant populations in Norway, Sweden and Russia were stable or fluctuating, and the species remained broadly stable overall. Consequently, it is evaluated as Secure.

Country	Breeding pop. size (pairs)	Year(s)	Trend	Mag.%	References
Belarus	110 – 200	97–00	0	0–19	
Estonia	100 – 150	98	+	20–29	*1*
Finland	50,000 – 120,000	98–02	**-**	**10**	
Rep. Ireland	1,000 – 2,500	88–91	(–)	(20–29)	
Latvia	(5 – 20)	90–00	(0)	(0–19)	*23*
Lithuania	Present	99–01	?	–	
Norway	500,000 – 1,000,000	90–03	0	0–19	*47,8*
Russia	1,300,000 – 1,600,000	00	F	30–49	*97,169*
Sweden	100,000 – 400,000	99–00	(F)	(30–49)	
UK	155,000 – 155,000	00	**-**	**38**	
Total (approx.)	2,100,000 – 3,300,000	**Overall trend** Fluctuating			
Breeding range	>3,000,000 km²	**Gen. length.** <3.3	**% Global pop.** 25–49		

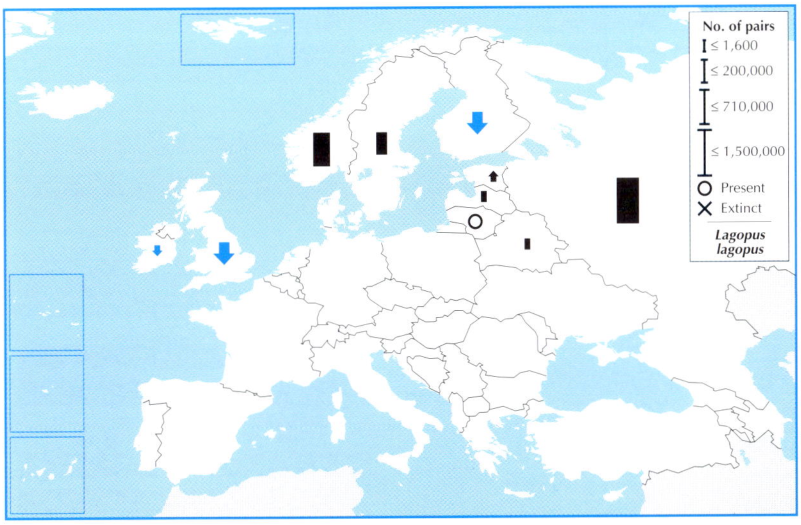

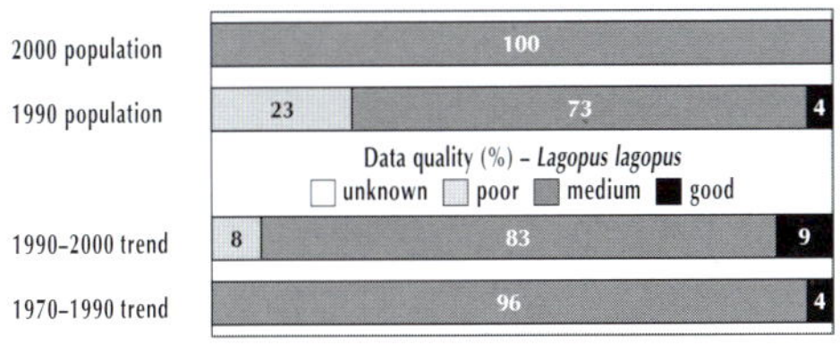

Lagopus mutus
ROCK PTARMIGAN

Non-SPEC (1994: —) **Status** Secure
Criteria —
European IUCN Red List Category —
Criteria —
Global IUCN Red List Category —
Criteria —

Country	Breeding pop. size (pairs)	Year(s)	Trend	Mag.%	References
Andorra	30 – 60	01	(0)	(0–19)	*1,3*
Austria	(8,000 – 16,000)	98–02	(0)	(0–19)	
Denmark					
Greenland	(50,000 – 500,000)	90–00	(F)	(–)	
Finland	1,000 – 3,000	98–02	(0)	(0–19)	
France	5,000 – 8,000	98–00	**0**	**0–19**	*4*
Germany	300 – 600	95–99	0	0–19	
Iceland	**50,000 – 200,000**	02	**F**	**>80**	*30*
Italy	5,000 – 8,000	97–03	–	0–19	*8*
Liechtenstein	(25 – 50)	98–00	(–)	(0–19)	
Norway	200,000 – 500,000	90–02	0	0–19	*47,8*
Svalbard	(1,000 – 10,000)	90–02	0	0–19	
Russia	45,000 – 68,000	00	F	30–49	*97,169*
Slovenia	300 – 500	00	0	0–19	
Spain	442 – 738	98–02	–	0–19	*10,16*
Sweden	40,000 – 80,000	99–00	(–)	(20–29)	
Switzerland	**12,000 – 15,000**	93–96	0	0–19	
UK	10,000 – 10,000	90	**–**	**0–19**	*27*
Total (approx.)	430,000 – 1,400,000	**Overall trend** Fluctuating			
Breeding range	>2,000,000 km²	**Gen. length.** <3.3		**% Global pop.** 25–49	

Lagopus mutus has a patchy distribution in northern Europe and in alpine areas of south-central Europe, with Europe accounting for less than half of its global range. Its European breeding population is large (>430,000 pairs), and was stable between 1970–1990. Although there were declines in a number of countries during 1990–2000, populations across most of the European range—including sizeable ones in Greenland, Iceland and Norway—were stable or fluctuating, and the species remained broadly stable overall. Consequently, it is evaluated as Secure.

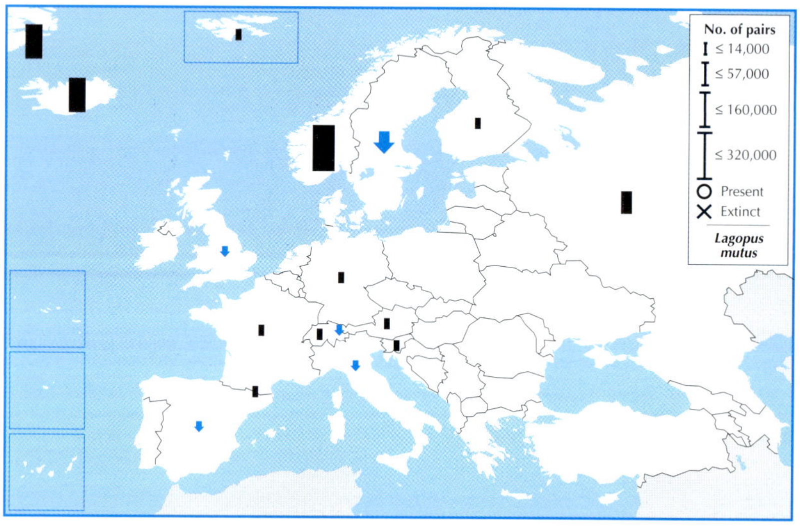

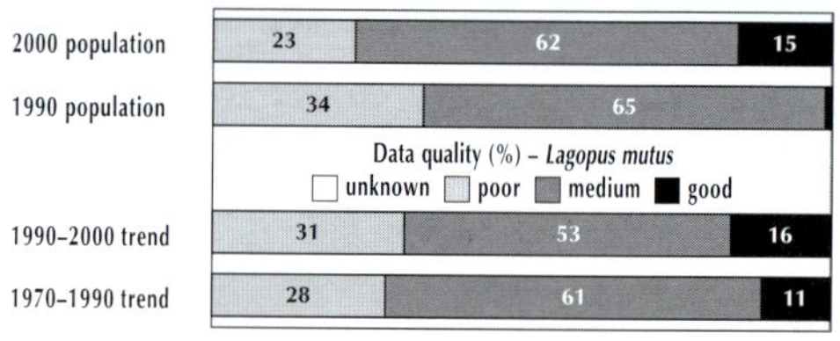

Tetrao tetrix
BLACK GROUSE

SPEC 3 (1994: 3) **Status** Depleted
Criteria Large historical decline
European IUCN Red List Category —
Criteria —
Global IUCN Red List Category —
Criteria —

Country	Breeding pop. size (pairs)	Year(s)	Trend	Mag.%	References
Albania	(0 – 50)	02	?	–	
Austria	10,000 – 15,000	98–02	0	0–19	
Belarus	**20,000 – 30,000**	97–00	0	0–19	
Belgium	**15 – 30**	01–03	**–**	**30–49**	*1*
Czech Rep.	**800 – 1,000**	00	**–**	**50–79**	
Denmark	**0 – 0**	01	**–**	**X**	*4,5,6,7*
Estonia	8,000 – 12,000	98	0	0–19	*1*
Finland	350,000 – 550,000	98–02	**–**	**15**	
France	7,000 – 11,000	98–02	–	0–9	*4*
Germany	1,100 – 1,600	95–99	0	0–19	
Italy	8,000 – 10,000	97–03	–	0–19	*8*
Latvia	5,000 – 10,000	90–00	(0)	(0–19)	*23,12*
Liechtenstein	40 – 70	98–00	–	10–19	
Lithuania	2,000 – 3,000	99–01	+	20–29	*20*
Netherlands	**15 – 23**	98–00	**–**	**46**	*1*
Norway	100,000 – 200,000	92–02	0	0–19	*8*
Poland	800 – 900	97–99	**–**	**50–79**	*40,41,42*
Romania	60 – 80	98–02	(F)	(20–29)	*4*
Russia	**1,850,000 – 2,150,000**	00	**+**	**0–19**	*97,169*
Slovakia	200 – 300	80–99	–	50–79	
Slovenia	1,000 – 2,000	94	(+)	(0–19)	
Sweden	150,000 – 200,000	99–00	**–**	**40–49**	
Switzerland	**7,500 – 10,000**	93–96	–	0–9	
Ukraine	1,600 – 3,000	90–00	–	30–49	
UK	5,000 – 8,100	95–96	**–**	**74**	*13*
Total (approx.)	2,500,000 – 3,200,000	**Overall trend** Small decline			
Breeding range	>4,000,000 km²	**Gen. length.** 4		**% Global pop.** 25–49	

Tetrao tetrix is a widespread resident in northern Europe, occurring more patchily further south, with Europe accounting for less than half of its global range. Its European breeding population is very large (>2,500,000 pairs), but declined substantially between 1970–1990. Although most populations continued to decline during 1990–2000, the species increased in its Russian stronghold, and probably only underwent a slight decline overall. Nevertheless, it total population size clearly remains far below the level that preceded its decline, and consequently it is evaluated as Depleted.

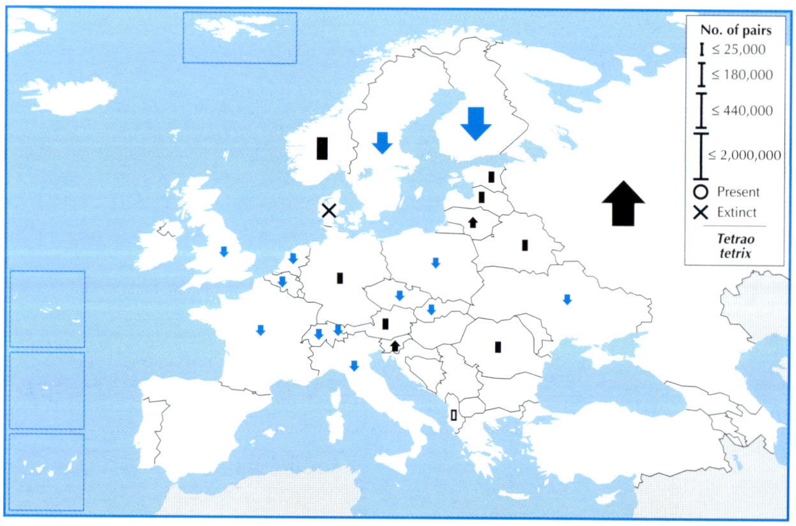

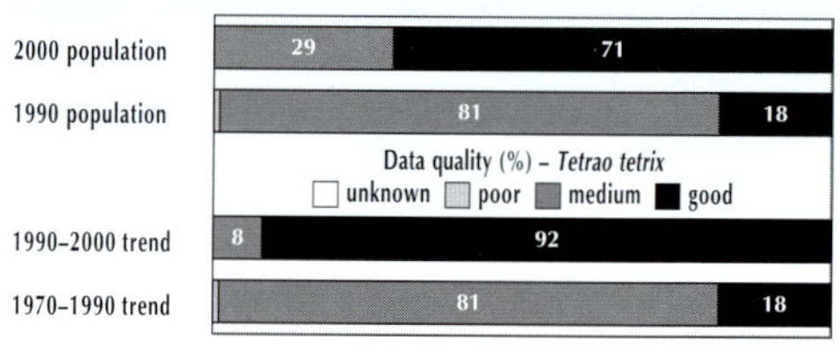

Tetrao mlokosiewiczi
CAUCASIAN GROUSE

SPEC 1 (1994: 2) **Status** Data Deficient
Criteria —
European IUCN Red List Category DD
Criteria —
Global IUCN Red List Category DD
Criteria —

Tetrao mlokosiewiczi is a poorly known resident in north-eastern Turkey and the Caucasus mountains of Russia, Georgia, Armenia and Azerbaijan, with Europe constituting >95% of its global range. Its European breeding population is small (<110,000 individuals), but its trend between 1970–1990 was unknown. Although there were declines in some areas during 1990–2000, the trend of the key Georgian population remained unknown, and insufficient data were available to make a comprehensive assessment. Consequently, the species is evaluated as Data Deficient.

Country	Breeding pop. size (individuals)	Year(s)	Trend	Mag.%	References
Armenia	400 – 800	94–02	–	10–19	*14*
Azerbaijan	700 – 3,000	96–00	(–)	(20–29)	
Georgia	**40,000 – 50,000**	94–02	?	–	
Russia	(15,000 – 50,000)	95–01	0	0–19	*26,33,60,72,74, 169*
Turkey	3,000 – 4,500	01	(–)	(0–19)	
Total (approx.)	59,000 – 110,000	**Overall trend** Unknown			
Breeding range	>100,000 km²	**Gen. length.** 4		**% Global pop.** >95	

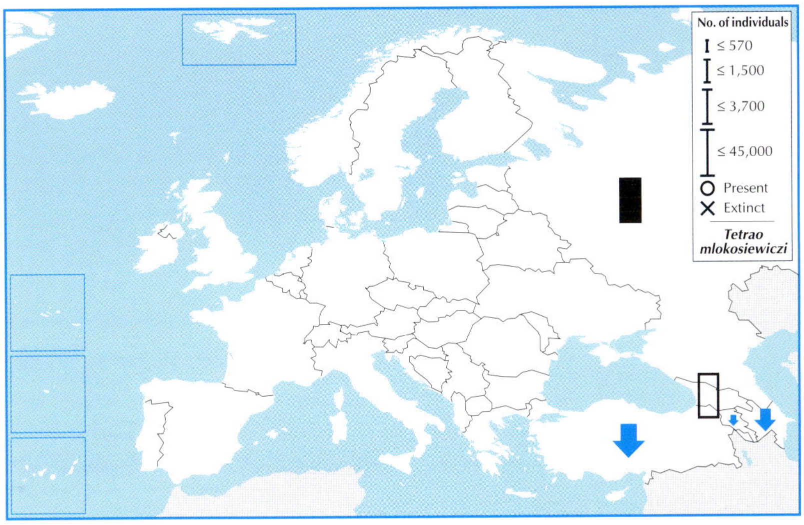

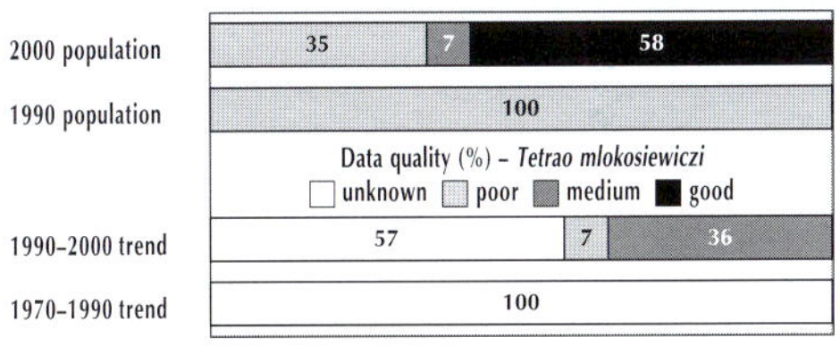

Tetrao urogallus
WESTERN CAPERCAILLIE

Non-SPEC (1994: —) **Status** (Secure)
Criteria —
European IUCN Red List Category —
Criteria —
Global IUCN Red List Category —
Criteria —

Tetrao urogallus is a widespread resident in northern Europe, but occurs more patchily farther south, with Europe accounting for less than half of its global range. Its European breeding population is large (>760,000 pairs), and was stable between 1970–1990. Although the species declined across much of its European range—notably in Finland and Sweden—during 1990–2000, the key Russian population increased and another sizeable population, in Norway, was stable. Consequently, the species probably remained stable overall, and is provisionally evaluated as Secure.

Country	Breeding pop. size (pairs)	Year(s)	Trend	Mag.%	References
Albania	10 – 30	02–03	(–)	(30–49)	*7*
Andorra	60 – 80	96	(0)	(0–19)	*1,3*
Austria	4,000 – 8,000	98–02	–	30–49	
Belarus	**2,500 – 3,000**	97–00	0	0–19	
Bosnia & HG	Present	90–03	?	–	
Bulgaria	500 – 800	98–02	–	0–19	
Croatia	(35 – 140)	02	(–)	(50–79)	*70*
Czech Rep.	**75 – 100**	00	**–**	**20–29**	
Estonia	**1,200 – 2,000**	98	–	20–29	*1*
Finland	200,000 – 300,000	98–02	**–**	**10**	
France	1,300 – 3,000	98–00	–	0–19	*4*
Germany	**650 – 970**	95–99	–	30–49	
Greece	225 – 313	98	?	–	
Italy	2,000 – 2,500	97–03	–	0–19	*8*
Latvia	**1,500 – 2,500**	90–00	(0)	(0–19)	*18,23*
Liechtenstein	(4 – 8)	98–00	–	50–79	
Lithuania	**300 – 400**	99–01	–	20–29	*20*
Macedonia	(50 – 80)	90–00	(0)	(0–19)	
Norway	(50,000 – 100,000)	90–01	(0)	(0–19)	*8*
Poland	220 – 400	95–00	–	20–29	*43*
Romania	4,500 – 5,200	98–02	0	0–19	
Russia	**400,000 – 450,000**	00	+	**20–29**	*97,169*
Serbia & MN	170 – 220	90–02	–	20–29	*1,140a,62,143, 10b,225,227,155*
Slovakia	500 – 700	80–99	–	50–79	
Slovenia	**550 – 600**	98–00	–	10–19	
Spain	1,900 – 2,000	98–02	–	30–49	*10,16*
Sweden	84,000 – 110,000	99–00	–	0–19	
Switzerland	450 – 500	98–02	–	10–19	
Ukraine	800 – 1,000	90–00	–	30–49	
UK	536 – 536	98–99	**–**	**51**	*36*
Total (approx.)	760,000 – 1,000,000	**Overall trend** Stable			
Breeding range	>4,000,000 km²	**Gen. length.** 4		**% Global pop.** 25–49	

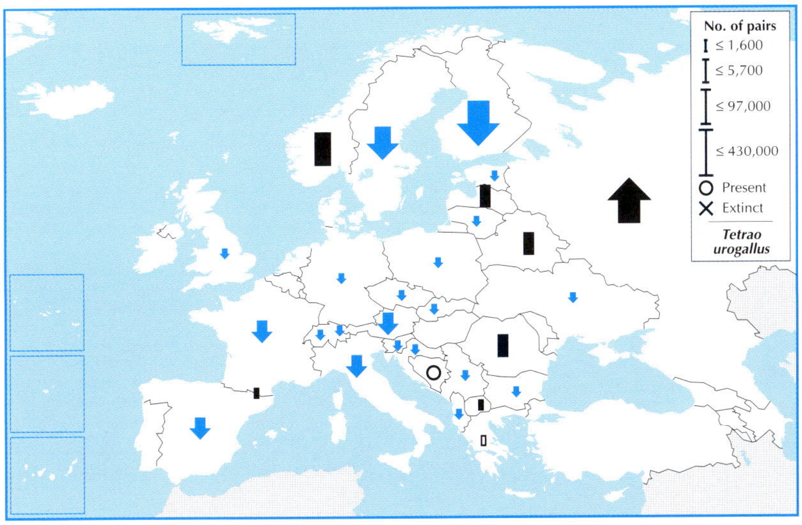

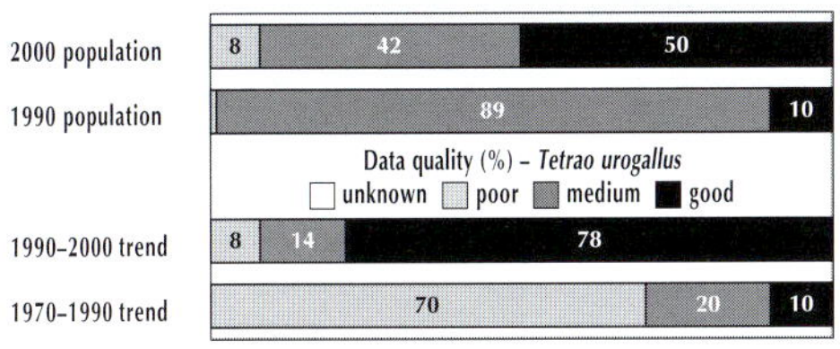

Tetraogallus caucasicus
CAUCASIAN SNOWCOCK

Non-SPEC[E] (1994: 4) **Status** Secure
Criteria —
European IUCN Red List Category —
Criteria —
Global IUCN Red List Category —
Criteria —

Country	Breeding pop. size (pairs)	Year(s)	Trend	Mag.%	References
Azerbaijan	(300 – 1,000)	96–00	(0)	(0–19)	
Georgia	25,000 – 50,000	94–02	0	0–19	
Russia	(6,700 – 16,000)	95–01	0	0–19	*33,60,72,169*
Total (approx.)	32,000 – 67,000	**Overall trend** Stable			
Breeding range	>50,000 km²	**Gen. length.** <3.3		**% Global pop.** 100	

Tetraogallus caucasicus is endemic to Europe, where it is confined to the Greater Caucasus Mountains of Russia, Georgia and Azerbaijan. Its breeding population is relatively small (<67,000 pairs), but the trend between 1970–1990 was unknown. Nevertheless, the species was thought to have been stable throughout its range during 1990–2000, and consequently it is evaluated as Secure.

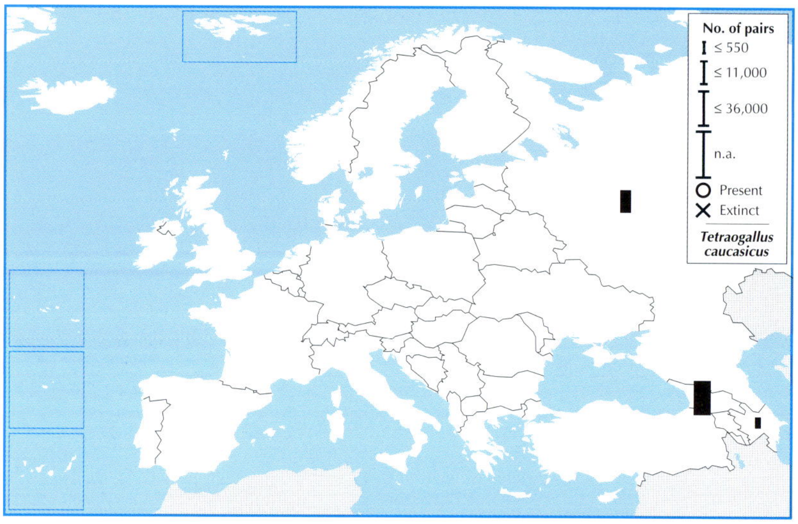

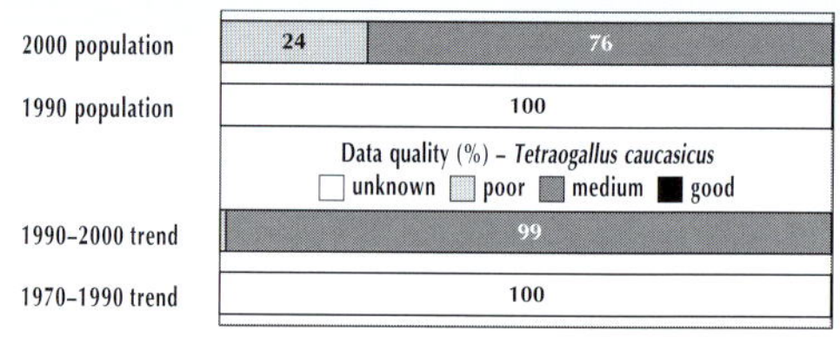

Tetraogallus caspius
CASPIAN SNOWCOCK

SPEC 3 (1994: 3) **Status** (Vulnerable)
Criteria See IUCN below
European IUCN Red List Category VU
Criteria C1
Global IUCN Red List Category —
Criteria —

Country	Breeding pop. size (pairs)	Year(s)	Trend	Mag.%	References
Armenia	200 – 270	98–02	–	0–9	
Azerbaijan	(50 – 250)	96–00	(–)	(30–49)	
Georgia	(250 – 500)	94–02	?	–	
Turkey	(2,000 – 8,000)	01	(–)	(10–19)	
Total (approx.)	2,500 – 9,000	**Overall trend** Moderate decline			
Breeding range	>50,000 km²	**Gen. length.** <3.3		**% Global pop.** 5–24	

Tetraogallus caspius is resident in parts of Turkey and the Caucasus, with Europe accounting for less than a quarter of its global range. Its European breeding population is small (as few as 2,500 pairs), but its trend between 1970–1990 was unknown. Although no trend data were available for the Georgian population during 1990–2000, the species declined elsewhere in the Caucasus and in Turkey, and probably underwent a moderate decline (>10%) overall. As a consequence of this decline and its small population, this poorly known species is provisionally evaluated as Vulnerable.

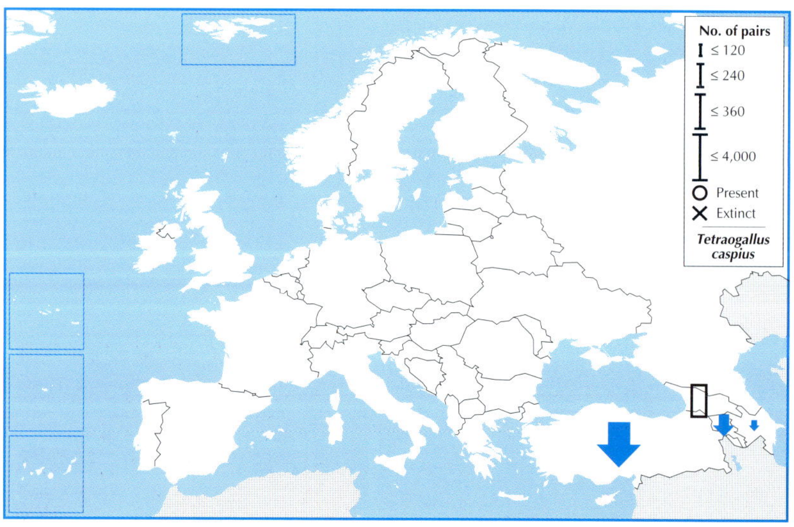

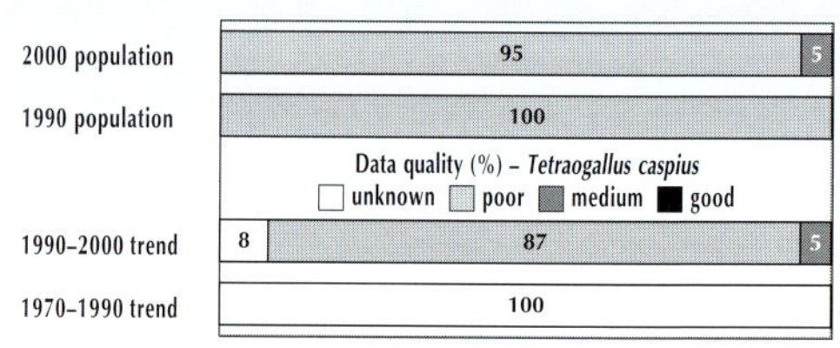

Alectoris chukar
CHUKAR

SPEC 3 (1994: 3) **Status** (Vulnerable)
Criteria See IUCN below
European IUCN Red List Category VU
Criteria A2b
Global IUCN Red List Category —
Criteria —

Country	Breeding pop. size (pairs)	Year(s)	Trend	Mag.%	References
Armenia	**6,500 – 14,000**	97–02	**0**	**0–19**	
Azerbaijan	8,000 – 25,000	96–00	(0)	(0–19)	
Bulgaria	1,500 – 3,000	95–02	0	0–19	
Cyprus	100,000 – 200,000	91–02	0	0–19	
Georgia	(30,000 – 100,000)	94–02	(F)	(20–29)	
Greece	10,000 – 10,000	01	0	0–19	
Russia	(2,000 – 7,000)	95–01	?	–	*61,169*
Turkey	(500,000 – 1,500,000)	01	(–)	(30–49)	
Ukraine	**2,000 – 2,500**	95–00	0	10–19	
Total (approx.)	660,000 – 1,900,000	**Overall trend** Large decline			
Breeding range	>1,000,000 km²	**Gen. length.** <3.3 **% Global pop.** 5–24			

Alectoris chukar is resident in south-eastern Europe, which accounts for less than a quarter of its global range. Its European breeding population is large (>660,000 pairs), but underwent a large decline between 1970–1990. Although the species was stable in much of its European range during 1990–2000, it declined substantially in its Turkish stronghold, and probably underwent a large decline (>30%) overall. Consequently, it is provisionally evaluated as Vulnerable.

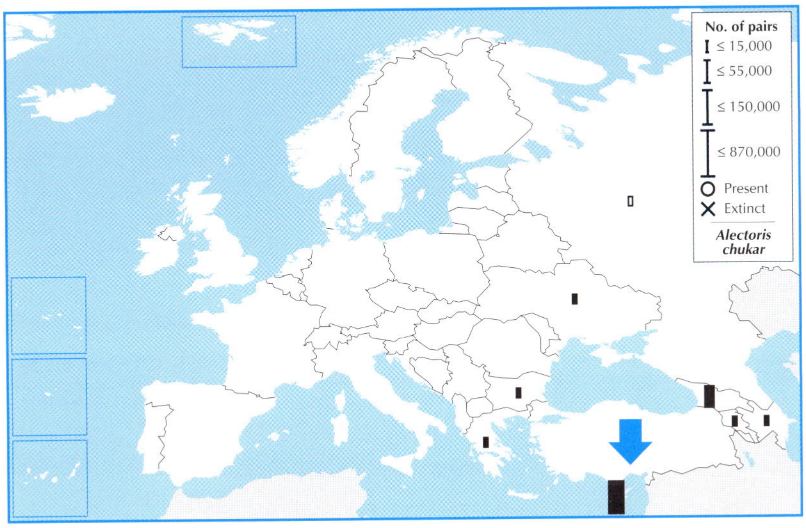

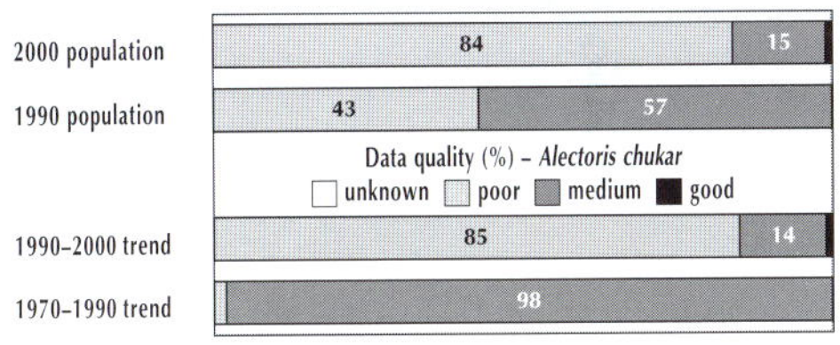

Alectoris graeca
ROCK PARTRIDGE

SPEC 2 (1994: 2) **Status** (Declining)
Criteria Moderate continuing decline
European IUCN Red List Category —
Criteria —
Global IUCN Red List Category —
Criteria —

Country	Breeding pop. size (pairs)	Year(s)	Trend	Mag.%	References
Albania	1,000 – 3,000	95–02	(–)	(20–29)	
Austria	(900 – 1,200)	98–02	(0)	(0–19)	
Bosnia & HG	Present	90–03	?	–	
Bulgaria	800 – 1,500	95–02	–	>80	
Croatia	(5,000 – 10,000)	02	(–)	(30–49)	*6,54*
France	2,000 – 3,000	00	?	–	*1*
Greece	7,000 – 13,000	01	+	20–29	
Italy	(10,000 – 20,000)	03	(–)	(0–19)	
Macedonia	5,000 – 15,000	90–00	(0)	(0–19)	
Serbia & MN	5,000 – 7,000	90–02	–	30–39	*1,126b,62,225, 227,155*
Slovenia	100 – 150	00	–	30–49	
Switzerland	**3,000 – 4,000**	93–96	(0)	(0–9)	
Total (approx.)	40,000 – 78,000	**Overall trend** Moderate decline			
Breeding range	>500,000 km²	**Gen. length.** <3.3 **% Global pop.** 100			

Alectoris graeca is endemic to Europe, occurring only in the Alps and mountainous parts of Italy and the Balkans. Its European breeding population is relatively small (<78,000 pairs), and underwent a large decline between 1970–1990. Although certain populations—notably sizeable ones in the Former Yugoslav Republic of Macedonia and Greece—were stable or increased during 1990–2000, the species continued to decline across most of its European range, and underwent a moderate decline (>10%) overall. Consequently, it is provisionally evaluated as Declining.

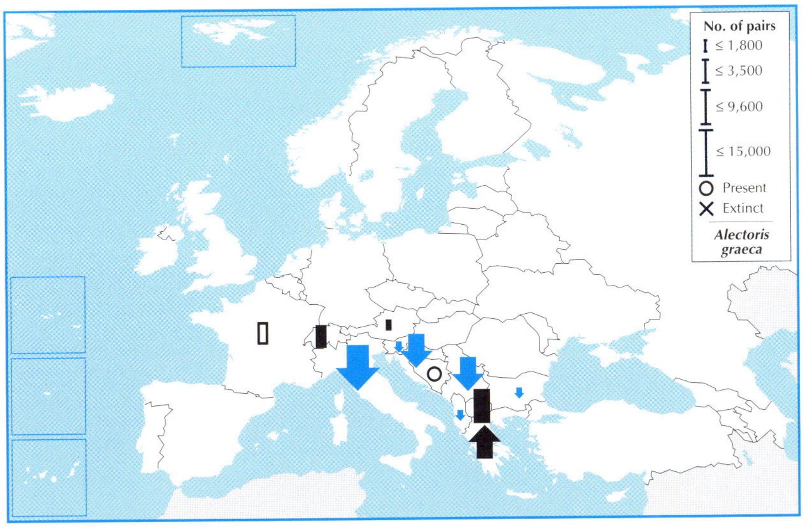

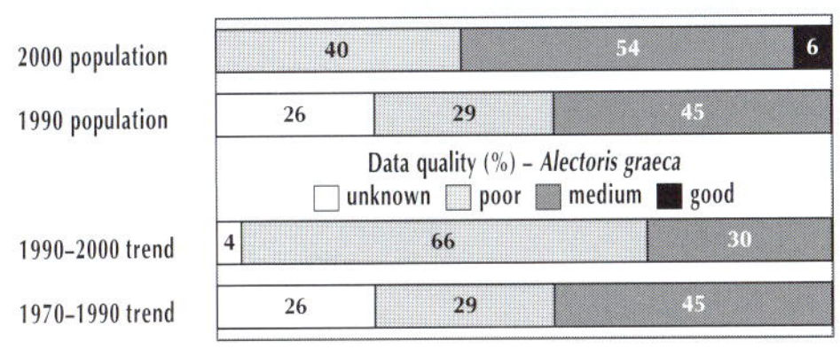

Alectoris rufa
RED-LEGGED PARTRIDGE

SPEC 2 (1994: 2) **Status** (Declining)
Criteria Moderate continuing decline
European IUCN Red List Category —
Criteria —
Global IUCN Red List Category —
Criteria —

Country	Breeding pop. size (pairs)	Year(s)	Trend	Mag.%	References
Andorra	7 – 12	99–01	-	5–19	*1,3*
France	(100,000 – 500,000)	98–02	(-)	(28)	*4,2*
Rep. Ireland	(0 – 50)	90–00	?	-	
Italy	(1,500 – 2,000)	03	(0)	(0–19)	
Portugal	(10,000 – 100,000)	02	(0)	(0–19)	
Madeira	Present	02	?	-	
Spain	(1,778,000 – 3,683,000)	92	(-)	(20–29)	*13,12,16,10*
UK	**72,000 – 200,000**	00	**-**	**20**	*5,31*
Total (approx.)	2,000,000 – 4,500,000	**Overall trend** Moderate decline			
Breeding range	>1,000,000 km²	**Gen. length.** <3.3	**% Global pop.** 100		

Alectoris rufa is endemic to Europe, where it is confined to south-western parts of the continent. Its European breeding population is very large (>2,000,000 pairs), but underwent a large decline between 1970–1990. Although populations were stable in Portugal and Italy during 1990–2000, the species continued to decline across most of its European range—including in its Spanish stronghold—and underwent a moderate decline (>10%) overall. Consequently, it is provisionally evaluated as Declining.

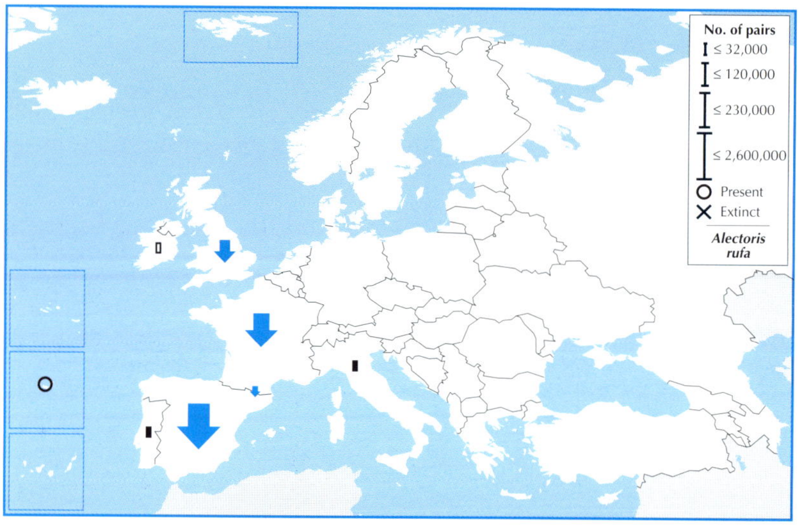

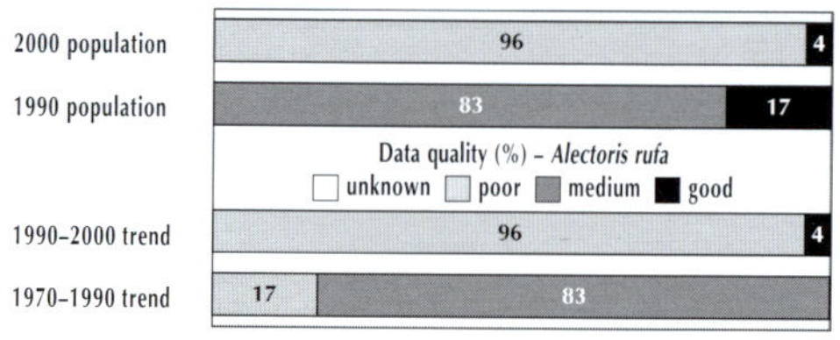

Alectoris barbara
BARBARY PARTRIDGE

SPEC 3 (1994: 3) **Status** (Rare)
Criteria <10,000 pairs
European IUCN Red List Category —
Criteria —
Global IUCN Red List Category —
Criteria —

Country	Breeding pop. size (pairs)	Year(s)	Trend	Mag.%	References
Italy	(5,000 – 10,000)	03	?	–	
Spain					
Canary Is.	(2,500 – 10,000)	97–03	(+)	(0–19)	*28,25*
UK					
Gibraltar	30 – 60	00	0	0–19	
Total (approx.)	7,500 – 20,000	**Overall trend** Unknown			
Breeding range	>50,000 km²	**Gen. length.** <3.3	**% Global pop.** <5		

Alectoris barbara has a predominantly North African distribution, but also occurs in Europe in the Canary Islands, Gibraltar and Sardinia (Italy). Its European breeding population is small (as few as 7,500 pairs), and underwent a large decline between 1970–1990. Although the species increased in the Canary Islands during 1990–2000, the trend of the other key population in Sardinia was unknown. Nevertheless, its population size probably still renders it susceptible to the risks affecting small populations, and consequently it is provisionally evaluated as Rare.

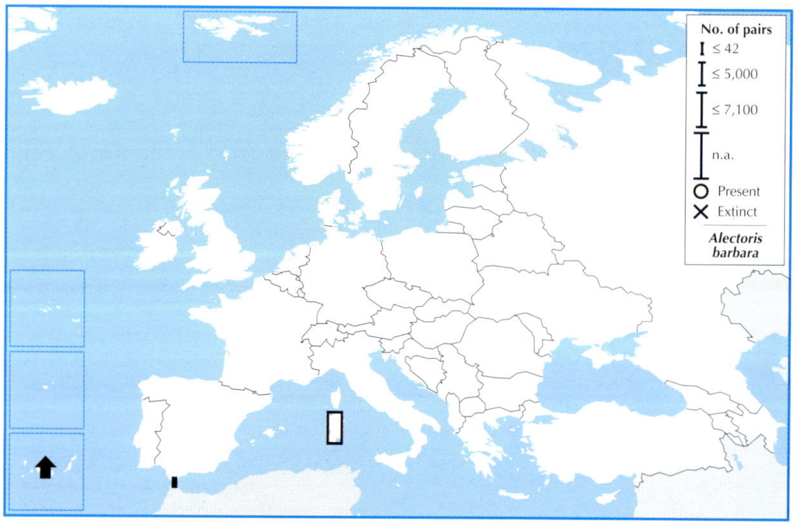

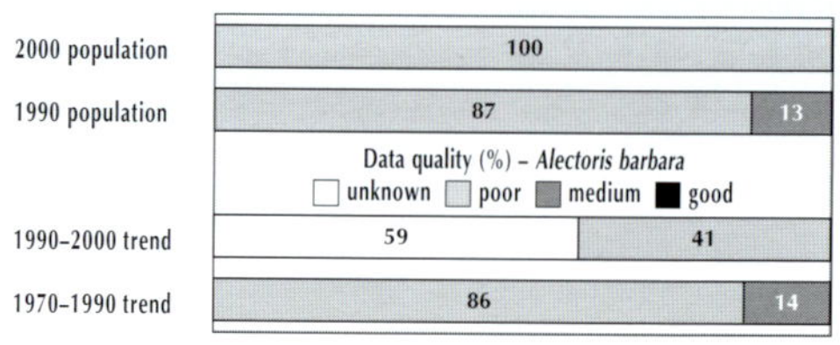

Ammoperdix griseogularis
SEE-SEE PARTRIDGE

SPEC 3 (1994: —) **Status** Vulnerable
Criteria See IUCN below
European IUCN Red List Category VU
Criteria C1
Global IUCN Red List Category —
Criteria —

Country	Breeding pop. size (pairs)	Year(s)	Trend	Mag.%	References
Azerbaijan	(5 – 50)	96–02	(0)	(0–19)	
Turkey	(1,000 – 2,000)	01	-	20–29	
Total (approx.)	1,000 – 2,100	**Overall trend** Moderate decline			
Breeding range	>20,000 km²	**Gen. length.** <3.3	**% Global pop.** <5		

Ammoperdix griseogularis is resident in Azerbaijan and southern Turkey, with Europe accounting for a tiny proportion of its global range. Its European breeding population is small (as few as 1,000 pairs), but was stable between 1970–1990. Although the tiny population in Azerbaijan was stable during 1990–2000, the species declined markedly in Turkey, and underwent a moderate decline (>10%) overall. As a consequence of this decline and its small population, this previously Secure species is now evaluated as Vulnerable.

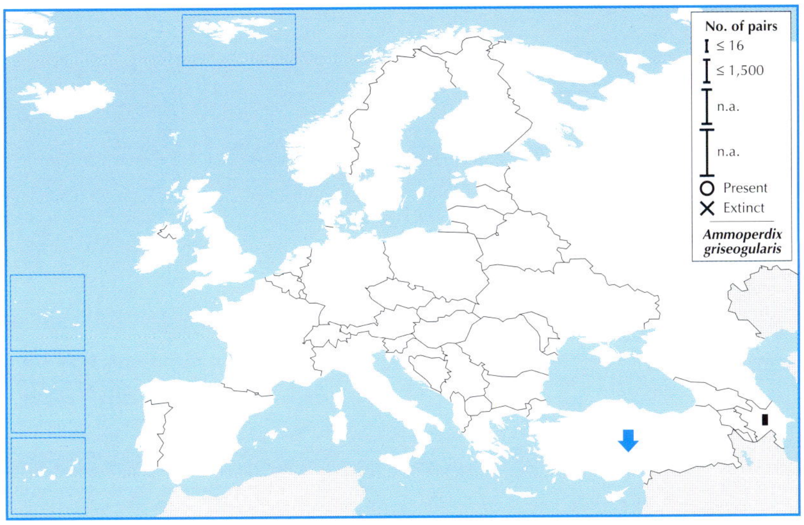

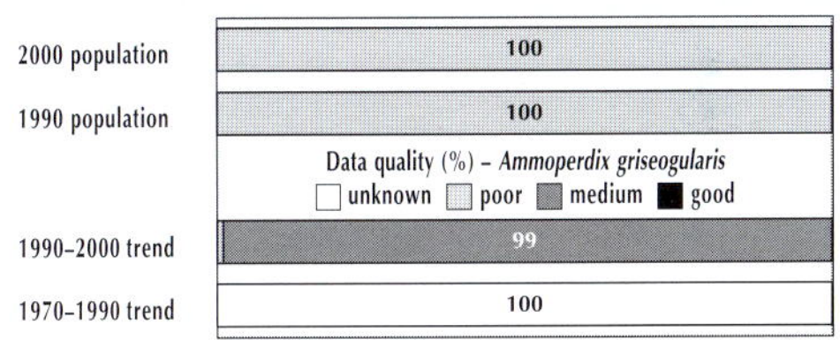

Francolinus francolinus
BLACK FRANCOLIN

SPEC 3 (1994: 3) **Status** Declining
Criteria Moderate continuing decline
European IUCN Red List Category —
Criteria —
Global IUCN Red List Category —
Criteria —

Country	Breeding pop. size (pairs)	Year(s)	Trend	Mag.%	References
Azerbaijan	(2,500 – 10,000)	96–00	-	20–29	
Cyprus	2,000 – 5,000	98–02	+	10–19	
Georgia	500 – 1,000	94–02	F	20–29	
Turkey	1,000 – 2,000	01	(+)	(0–19)	
Total (approx.)	6,000 – 18,000	**Overall trend** Moderate decline			
Breeding range	>50,000 km²	**Gen. length.** <3.3	**% Global pop.** <5		

Francolinus francolinus is resident in Azerbaijan, Georgia, Turkey and Cyprus, with Europe accounting for a tiny proportion of its global range. Its European breeding population is small (as few as 6,000 pairs), and underwent a large decline between 1970–1990. Although the Cypriot and Turkish populations increased slightly during 1990–2000, the key population in Azerbaijan declined, and the species underwent a moderate decline (>10%) overall. Consequently, it is evaluated as Declining.

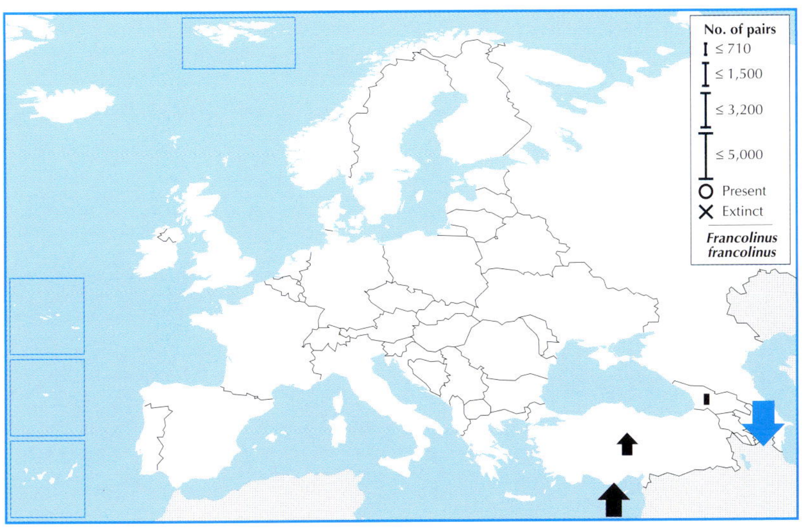

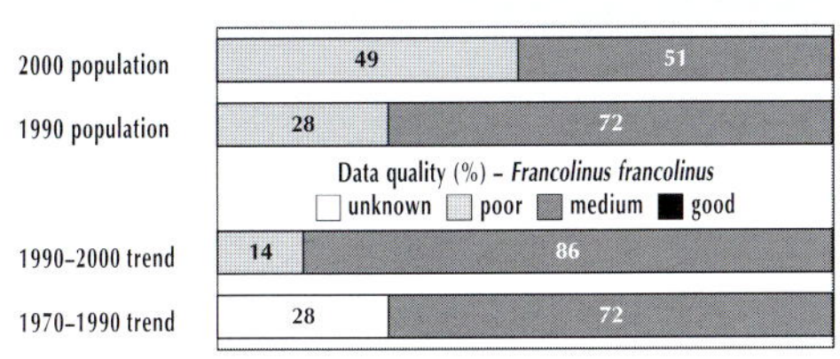

Perdix perdix
GREY PARTRIDGE

SPEC 3 (1994: 3) **Status** Vulnerable
Criteria See IUCN below
European IUCN Red List Category VU
Criteria A2b
Global IUCN Red List Category —
Criteria —

Perdix perdix is a widespread resident across much of Europe, which accounts for less than half of its global range. Its European breeding population is very large (>1,600,000 pairs), but underwent a large decline between 1970–1990. Although the species was stable or increased in many eastern European countries during 1990–2000, it continued to decline throughout most of western and central Europe—including sizeable populations in France and Poland—and underwent a large decline (>30%) overall. Consequently, it is evaluated as Vulnerable.

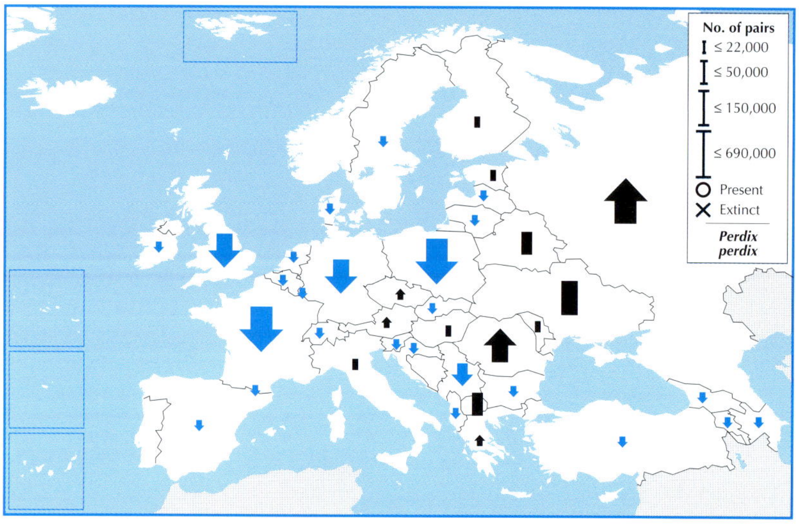

Country	Breeding pop. size (pairs)	Year(s)	Trend	Mag.%	References
Albania	100 – 300	95–02	(–)	(20–29)	
Andorra	35 – 70	99–01	–	20–29	*1,3*
Armenia	**830 – 3,400**	97–02	**–**	**10–19**	
Austria	(6,000 – 12,000)	98–02	(+)	(20–29)	
Azerbaijan	(2,000 – 5,000)	96–00	(–)	(20–29)	
Belarus	25,000 – 50,000	97–02	0	0–19	
Belgium	5,600 – 11,000	01–02	–	30–49	*1*
Bulgaria	5,000 – 10,000	95–02	–	0–19	
Croatia	(5,000 – 10,000)	02	(–)	(50–79)	*70,16,54,44*
Czech Rep.	**12,000 – 24,000**	00	**+**	**30–49**	
Denmark	(10,000 – 15,000)	00	–	50	
Estonia	4,000 – 7,000	98	F	20–29	*1*
Finland	(3,000 – 5,000)	98–02	0	0–19	
France	(200,000 – 800,000)	98–02	**–**	**49**	*4,2*
Georgia	(2,000 – 10,000)	94–02	(–)	(20–29)	
Germany	56,000 – 91,000	95–99	(–)	(50–79)	
Greece	2,500 – 4,000	01	+	20–29	
Hungary	**14,000 – 33,000**	99–02	(0)	(0–19)	*19*
Rep. Ireland	**2 – 20**	88–91	**–**	**50–79**	
Italy	(2,000 – 4,000)	03	(0)	(0–19)	
Latvia	(500 – 5,000)	90–00	(–)	(0–19)	*23,16*
Lithuania	(10,000 – 20,000)	99–01	(–)	(20–29)	*20*
Luxembourg	40 – 60	02	**–**	**50–79**	
Macedonia	(25,000 – 100,000)	90–00	(0)	(0–19)	
Moldova	1,800 – 3,200	90–00	**0**	**0–19**	
Netherlands	**9,000 – 13,000**	98–00	**–**	**50**	*1*
Poland	**300,000 – 600,000**	00–02	**–**	**60–74**	*23,109,110*
Romania	120,000 – 180,000	00–02	(+)	(0–19)	*22*
Russia	580,000 – 800,000	00	**+**	**30–49**	*97,169*
Serbia & MN	30,000 – 50,000	90–02	–	20–29	*1,172a,143a, 67a,227,155*
Slovakia	3,000 – 8,000	80–99	–	50–79	
Slovenia	(800 – 1,200)	94	–	20–29	
Spain	(2,000 – 6,000)	92	(–)	(20–29)	*13,12,16,10*
Sweden	**7,000 – 8,000**	99–00	–	10–19	
Switzerland	**1 – 2**	98–02	**–**	**>80**	
Turkey	(4,000 – 8,000)	01	(–)	(30–49)	
Ukraine	60,000 – 90,000	90–00	F	30–49	
UK	**70,000 – 75,000**	00	**–**	**50**	*5,31*
Total (approx.)	1,600,000 – 3,100,000	**Overall trend** Large decline			
Breeding range	>5,000,000 km²	**Gen. length.** <3.3		**% Global pop.** 25–49	

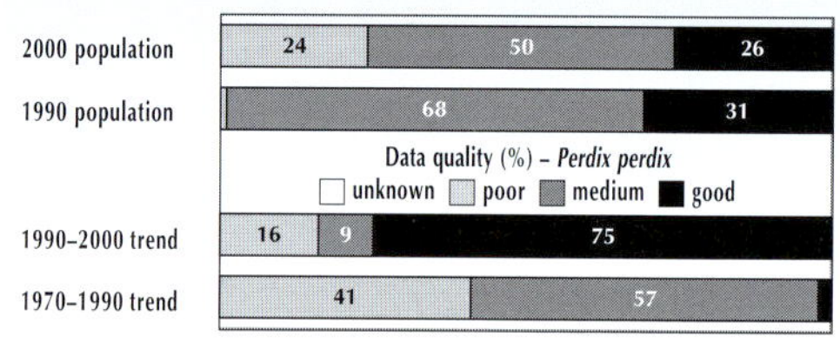

Coturnix coturnix
COMMON QUAIL

SPEC 3 (1994: 3) **Status** (Depleted)
Criteria Large historical decline
European IUCN Red List Category —
Criteria —
Global IUCN Red List Category —
Criteria —

Coturnix coturnix is a widespread summer visitor to much of Europe, which accounts for less than a quarter of its global breeding range. Its European breeding population is very large (>2,800,000 pairs) and fluctuates, but underwent a large decline during 1970–1990, especially in central and eastern Europe. Although the species increased in northern and central Europe during 1990–2000, declines continued in south-eastern Europe, and the total population size probably remains below the level that preceded its decline. Consequently, it is provisionally evaluated as Depleted.

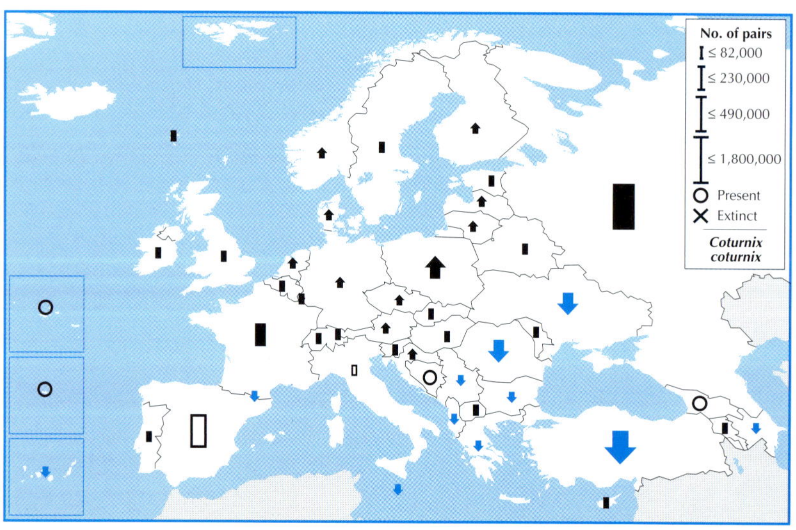

Country	Breeding pop. size (pairs)	Year(s)	Trend	Mag.%	References
Albania	800 – 1,500	96–02	(–)	(30–49)	
Andorra	6 – 12	98	–	20–29	*1,3*
Armenia	16,500 – 24,000	97–02	0	0–19	
Austria	(5,000 – 15,000)	98–02	(+)	(20–29)	
Azerbaijan	(10,000 – 50,000)	96–00	(–)	(20–29)	
Belarus	15,000 – 30,000	97–02	0	0–19	
Belgium	2,400 – 5,700	01–02	(0)	(0–19)	*1*
Bosnia & HG	Present	90–03	?	–	
Bulgaria	8,000 – 15,000	95–02	–	0–19	
Croatia	(10,000 – 15,000)	02	(+)	(50–79)	*70,16,54*
Cyprus	(1,000 – 4,000)	94–02	(0)	(0–9)	
Czech Rep.	**5,000 – 10,000**	00	**+**	**50–79**	
Denmark	200 – 600	00–01	**+**	**>80**	*4,5,6,7*
Faroe Is.	0 – 1	95	(0)	(0–19)	
Estonia	(10 – 50)	98	0	0–19	*1*
Finland	(10 – 100)	98–02	+	500	
France	(100,000 – 500,000)	00	(F)	(20–29)	*1,2*
Georgia	Present	03	?	–	
Germany	12,000 – 32,000	95–99	(+)	(0–19)	
Greece	(2,000 – 5,000)	95–00	(–)	(0–19)	
Hungary	**70,000 – 94,000**	99–02	(0)	(0–19)	*19*
Rep. Ireland	0 – 20	88–91	F	20–29	
Italy	(5,000 – 20,000)	03	?	–	
Latvia	20 – 500	90–00	+	>80	*22,2*
Liechtenstein	**5 – 20**	98–00	(F)	(–)	
Lithuania	(1,000 – 2,000)	99–01	+	30–49	*20*
Luxembourg	10 – 25	02	**0**	**0–19**	
Macedonia	2,000 – 3,000	90–00	(0)	(0–19)	
Malta	1 – 3	90–02	–	30–49	*1*
Moldova	3,500 – 4,500	90–00	**0**	**0–19**	
Netherlands	**2,000 – 6,500**	98–00	**+**	**64**	*1*
Norway	(50 – 300)	02	**+**	**0–19**	
Poland	**100,000 – 150,000**	00–02	(+)	(–)	*23,111*
Portugal	(5,000 – 50,000)	02	(0)	(0–19)	
Azores	Present	02	?	–	
Madeira	Present	02	?	–	
Romania	160,000 – 220,000	99–02	(–)	(0–19)	*48*
Russia	1,500,000 – 2,000,000	00	F	30–49	*8,35*
Serbia & MN	10,000 – 15,000	90–02	–	10–29	*1,172a,67a,143a, 227,155*
Slovakia	2,000 – 6,000	90–99	0	0–19	
Slovenia	1,000 – 2,000	99	0	0–19	
Spain	(320,000 – 435,000)	92	?	–	*13,12,16,10*
Canary Is.	(2,500 – 10,000)	97–03	(–)	(0–19)	*28,25*
Sweden	(10 – 40)	99–00	(F)	(20–29)	
Switzerland	**1,500 – 2,000**	93–96	F	30–49	
Turkey	(300,000 – 800,000)	01	(–)	(20–29)	
Ukraine	100,000 – 160,000	90–00	–	0–19	
UK	(5 – 450)	96–00	**0**	**1**	
Total (approx.)	2,800,000 – 4,700,000	**Overall trend** Fluctuating			
Breeding range	>6,000,000 km²	**Gen. length.** <3.3		**% Global pop.** 5–24	

(See p. 99, bottom, for data quality graph)

Phasianus colchicus
COMMON PHEASANT

Non-SPEC (1994: —) **Status** (Secure)
Criteria —
European IUCN Red List Category —
Criteria —
Global IUCN Red List Category —
Criteria —

Phasianus colchicus is a widespread resident across much of Europe, which accounts for less than a quarter of its global range. Its European breeding population is very large (>3,400,000 pairs), and was probably stable between 1970–1990. Although the species declined in parts of north-central Europe during 1990–2000, most populations—including the key one in the United Kingdom—were stable or increased. The status of the truly wild population of this species in Europe is obscured by confusion with introduced birds. Nevertheless, the species is provisionally evaluated as Secure.

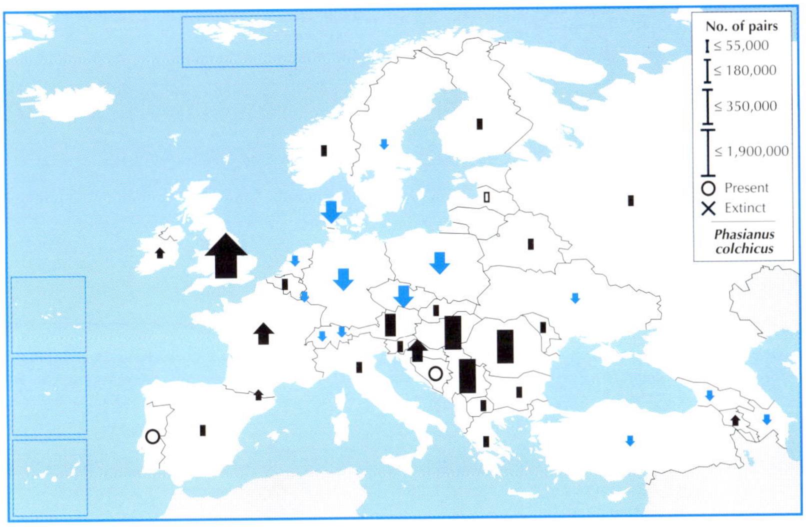

Country	Breeding pop. size (pairs)	Year(s)	Trend	Mag.%	References
Andorra	(0 – 5)	01	(+)	(N)	*1,3*
Armenia	**80 – 200**	97–02	**+**	**10–19**	
Austria	(50,000 – 150,000)	98–02	(F)	(20–29)	
Azerbaijan	(1,000 – 2,500)	96–00	(–)	(20–29)	
Belarus	30 – 50	90–97	(F)	(–)	
Belgium	20,000 – 100,000	01–02	(0)	(0–19)	*1*
Bosnia & HG	Present	90–03	?	–	
Bulgaria	4,000 – 10,000	95–02	0	0–19	
Croatia	(100,000 – 150,000)	02	(+)	(20–29)	*16*
Czech Rep.	**75,000 – 150,000**	00	**–**	**50–79**	
Denmark	(100,000 – 200,000)	00	–	30	
Finland	10,000 – 20,000	98–02	0	0–19	
France	(100,000 – 300,000)	00	(+)	(29)	*1,2*
Georgia	5,000 – 10,000	94–02	–	50–79	
Germany	120,000 – 210,000	95–99	(–)	(30–49)	
Greece	(50 – 800)	95–00	(0)	(0–19)	
Hungary	**320,000 – 380,000**	99–02	(0)	(0–19)	*19*
Rep. Ireland	20,000 – 100,000	88–91	+	30–49	
Italy	(1,000 – 100,00)	03	(0)	(0–19)	
Latvia	(0 – 20)	90–00	?	–	*16*
Liechtenstein	**2 – 5**	98–00	**–**	**50–79**	
Luxembourg	80 – 100	02	–	50–79	
Macedonia	(5,000 – 20,000)	90–00	(F)	(20–29)	
Moldova	8,500 – 10,000	90–00	**0**	**0–19**	
Netherlands	**50,000 – 60,000**	98–00	**–**	**32**	*1*
Norway	(500 – 1,000)	90–03	(0)	(0–19)	
Poland	150,000 – 200,000	00–02	**–**	**30–39**	*23,109*
Portugal	Present	02	?	–	
Romania	(200,000 – 300,000)	00–02	(0)	(0–19)	*22*
Russia	20,000 – 30,000	90–00	(F)	(20–29)	*97,169*
Serbia & MN	200,000 – 250,000	90–02	(F)	(30–49)	*1,138,67a,172a, 227,155*
Slovakia	20,000 – 40,000	90–99	0	0–19	
Slovenia	4,000 – 5,000	94	(0)	(0–19)	
Spain	(250 – 1,000)	98–02	(0)	(0–19)	*10*
Sweden	40,000 – 50,000	99–00	**–**	**26**	
Switzerland	(100 – 500)	98–02	(–)	(>80)	
Turkey	(1,000 – 3,000)	01	(–)	(–)	
Ukraine	23,000 – 30,000	90–00	–	50–79	
UK	**1,800,000 – 1,900,000**	00	**+**	**10**	*5*
Total (approx.)	3,400,000 – 4,700,000	**Overall trend** Unknown			
Breeding range	>4,000,000 km²	**Gen. length.** <3.3		**% Global pop.** 5–24	

(See p. 99, bottom, for data quality graph)

Turnix sylvatica
SMALL BUTTONQUAIL

SPEC 3 (1994: 3) **Status** Critically Endangered
Criteria See IUCN below
European IUCN Red List Category CR
Criteria D1
Global IUCN Red List Category —
Criteria —

Country	Breeding pop. size (pairs)	Year(s)	Trend	Mag.%	References
Spain	0 – 1	98–02	?	–	*10,16*
Total (approx.)	0 – 1	**Overall trend** Unknown			
Breeding range	<100 km²	**Gen. length.** <3.3		**% Global pop.** <5	

Turnix sylvatica has a predominantly African and Asian distribution, which just extends into Europe in southernmost Spain. Its European population is extremely poorly known, but probably numbers no more than a few pairs. Although its trend during 1990–2000 was unknown, the species declined substantially between 1970–1990, and has virtually disappeared from Europe. As a consequence of its tiny population, it is evaluated as Critically Endangered.

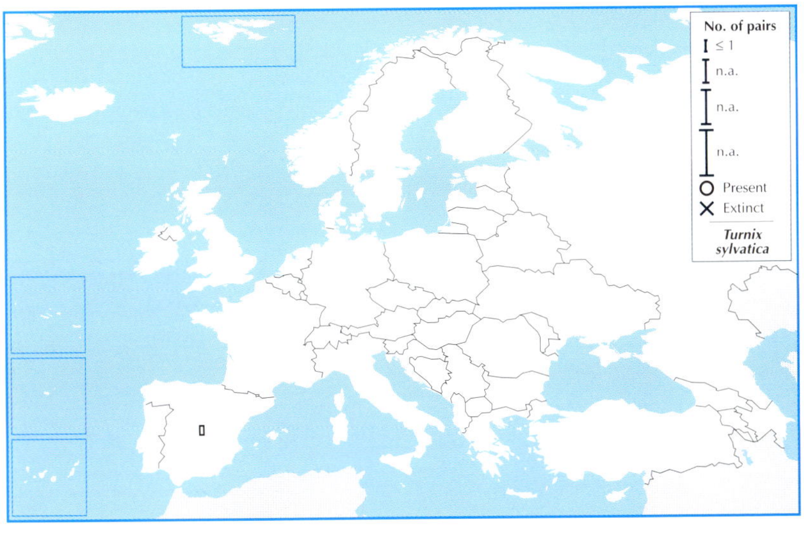

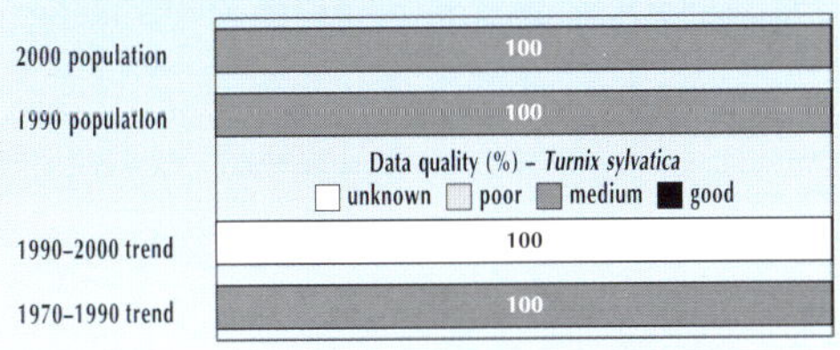

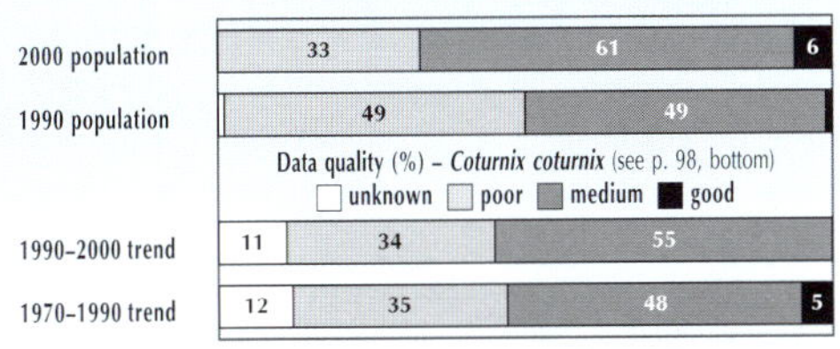

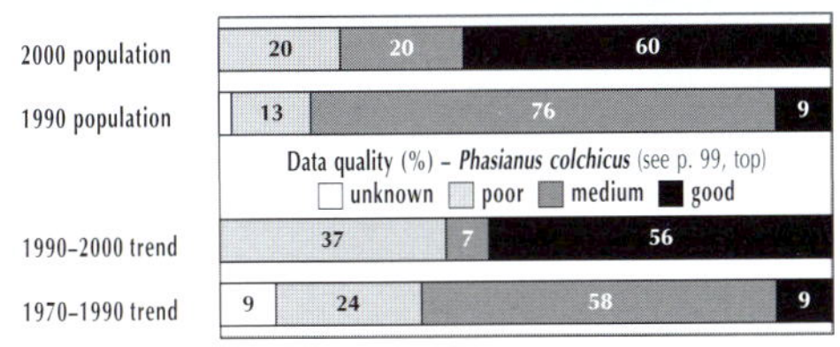

Rallus aquaticus
WATER RAIL

Non-SPEC (1994: —) **Status** (Secure)
Criteria —
European IUCN Red List Category —
Criteria —
Global IUCN Red List Category —
Criteria —

Rallus aquaticus is a widespread breeder across much of Europe, which accounts for less than half of its global breeding range. Its European breeding population is large (>140,000 pairs), and was stable between 1970–1990. Although some populations declined during 1990–2000—notably the sizeable one in Ukraine—and trends were not known for the key populations in Spain and Russia, the species was stable in most parts of Europe, and probably underwent only a small decline overall. Consequently, it is provisionally evaluated as Secure.

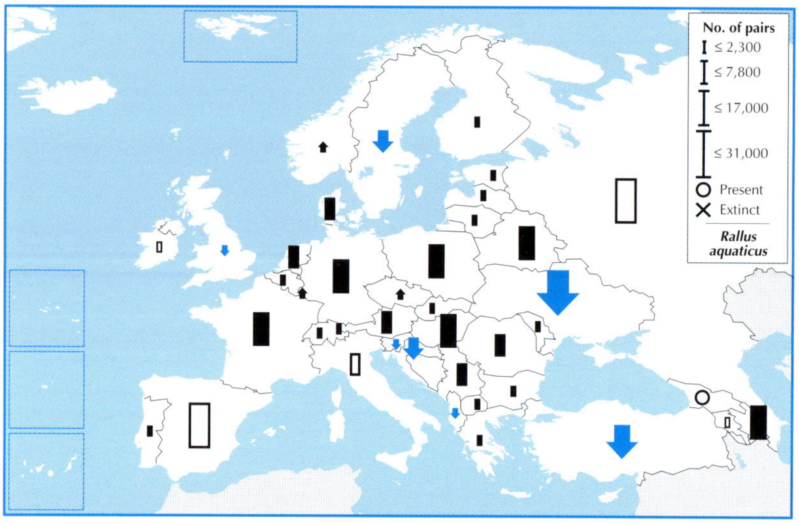

Country	Breeding pop. size (pairs)	Year(s)	Trend	Mag.%	References
Albania	(100 – 300)	93–02	(–)	(0–19)	
Armenia	330 – 830	98–02	?	–	
Austria	3,500 – 7,000	98–02	(0)	(0–19)	
Azerbaijan	(5,000 – 15,000)	96–00	(0)	(0–19)	
Belarus	8,000 – 14,000	97–02	0	0–19	
Belgium	750 – 1,200	95–02	(0)	(0–19)	*1*
Bosnia & HG	Present	90–03	?	–	
Bulgaria	700 – 1,300	96–02	0	0–19	
Croatia	(2,000 – 5,000)	02	(–)	(50–79)	*70,16,54*
Czech Rep.	**800 – 1,200**	00	**+**	**50–79**	
Denmark	(2,000 – 5,000)	93–96	(0)	(0–19)	*3*
Estonia	(300 – 1,000)	98	0	0–19	*1*
Finland	300 – 600	98–02	0	0–19	
France	(8,000 – 35,000)	98–02	(0)	(0–19)	*4*
Georgia	Present	03	?	–	
Germany	7,500 – 14,700	95–99	(0)	(0–19)	
Greece	(1,000 – 3,000)	95–00	(0)	(0–19)	
Hungary	(10,000 – 20,000)	90–93	(0)	(0–19)	*14*
Rep. Ireland	(1,000 – 2,500)	88–91	?	–	
Italy	(3,000 – 6,000)	03	?	–	
Latvia	(500 – 1,500)	90–00	(0)	(0–19)	*23,17*
Liechtenstein	1 – 2	98–00	(F)	(–)	
Lithuania	(2,000 – 2500)	99–01	(0)	(0–19)	*20*
Luxembourg	**30 – 40**	00–02	**+**	**0–19**	
Macedonia	100 – 1,000	90–00	(0)	(0–19)	
Moldova	150 – 200	90–00	0	0–19	
Netherlands	**2,500 – 3,200**	98–00	**F**	**28**	1
Norway	70 – 170	02	+	20–29	
Poland	(10,000 – 20,000)	90–00	(0)	(0–19)	*1*
Portugal	(500 – 2,000)	02	(0)	(0–19)	
Romania	(5,000 – 12,000)	90–02	(0)	(0–19)	
Russia	(15,000 – 50,000)	90–00	?	–	*8,73,108,125*
Serbia & MN	3,500 – 5,000	90–02	0	0–19	*1,29,227,155, 101,13,78*
Slovakia	300 – 600	90–99	0	0–19	
Slovenia	(100 – 200)	94	(–)	(0–19)	
Spain	(11,400 – 68,800)	92	?	–	*13,12,10*
Sweden	4,500 – 5,500	99–00	–	30–49	
Switzerland	**1,000 – 1,500**	93–96	0	0–19	
Turkey	(8,000 – 16,000)	01	(–)	(0–19)	
Ukraine	24,000 – 39,000	90–00	–	0–19	
UK	**700 – 1,400**	88–91	(–)	(30–49)	*8*

Total (approx.) 140,000 – 360,000 **Overall trend** Small decline
Breeding range >5,000,000 km² **Gen. length**. <3.3 **% Global pop.** 25–49
(See p. 101, top, for data quality graph)

Winter pop. size (individuals)

Total (approx.) >20,000 **Overall trend** Unknown
% in European IBAs Unknown **Gen. length** <3.3 **% Global pop.** 25–49

Porzana porzana
SPOTTED CRAKE

Non-SPEC[E] (1994: 4) **Status** (Secure)
Criteria —
European IUCN Red List Category —
Criteria —
Global IUCN Red List Category —
Criteria —

Porzana porzana is scattered thinly across much of Europe, which constitutes >50% of its global breeding range. The European breeding population is relatively large (>120,000 pairs), and was stable between 1970–1990. Although there were declines in certain countries—notably Ukraine—during 1990–2000, key populations in Russia, Belarus and Romania were broadly stable or increased, and the species remained stable overall. Consequently, it is provisionally evaluated as Secure.

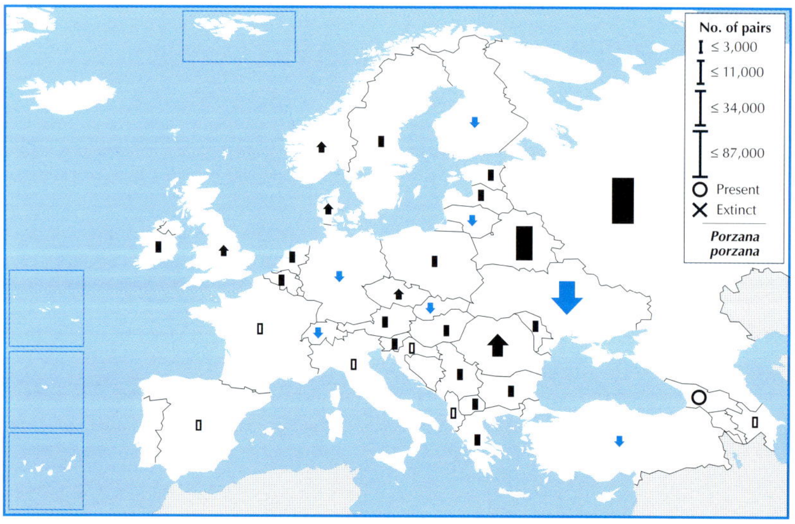

Country	Breeding pop. size (pairs)	Year(s)	Trend	Mag.%	References
Albania	(0 – 50)	92–93	?	–	
Austria	10 – 100	98–02	(F)	(>80)	
Azerbaijan	(0 – 100)	96–00	?	–	
Belarus	25,000 – 30,000	97–02	0	0–19	
Belgium	67 – 87	95–02	(0)	(0–19)	*1*
Bulgaria	50 – 150	96–02	(F)	(20–29)	
Croatia	(10 – 50)	02	?	–	*70*
Czech Rep.	**40 – 80**	00	**+**	**>80**	
Denmark	90 – 121	98–01	**+**	**50–79**	*4,5,6,7*
Estonia	2,000 – 4,000	98	F	20–29	*1*
Finland	500 – 1,500	98–02	**–**	**30**	
France	(250 – 1,000)	00	?	–	*1*
Georgia	Present	03	?	–	
Germany	540 – 1,030	95–99	–	20–29	
Greece	(0 – 10)	95–00	(0)	(0–19)	
Hungary	(500 – 600)	90–93	(0)	(0–19)	*14*
Rep. Ireland	2 – 7	88–91	(F)	(–)	
Italy	(50 – 200)	03	?	–	
Latvia	500 – 1,000	90–00	(0)	(0–19)	
Lithuania	1,000 – 1,500	99–01	(–)	(0–19)	*20*
Macedonia	(0 – 10)	90–00	(0)	(0–19)	
Moldova	200 – 250	90–00	(0)	(0–19)	
Netherlands	**150 – 300**	98–00	**F**	**22**	*1*
Norway	(30 – 100)	90–03	**+**	**0–19**	
Poland	2,500 – 3,500	90–00	(0)	(0–19)	*1,2*
Romania	(8,000 – 15,000)	90–02	(+)	(0–19)	
Russia	(50,000 – 150,000)	90–00	(F)	(30–49)	*3,18,104,108, 114,125*
Serbia & MN	(500 – 700)	90–02	(0)	(10–19)	*1,155,144,78, 227,203,13*
Slovakia	60 – 200	80–99	–	30–49	
Slovenia	(10 – 60)	90–00	F	>80	
Spain	(1 – 53)	98–02	?	–	*12,16,10*
Sweden	100 – 150	99–00	0	0–19	
Switzerland	**10 – 20**	93–96	(–)	(10–19)	
Turkey	(10 – 100)	01	(–)	(–)	
Ukraine	26,000 – 43,000	90–00	–	0–19	
UK	**73 – 73**	99	**+**	**68**	*9*

Total (approx.) 120,000 – 260,000 **Overall trend** Fluctuating
Breeding range >4,000,000 km² **Gen. length**. <3.3 **% Global pop.** 50–74

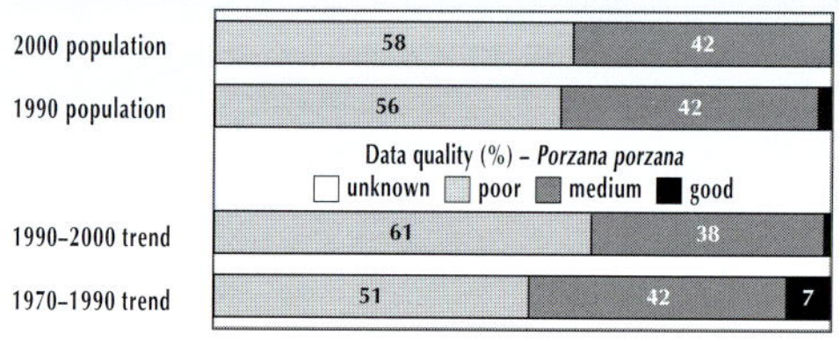

Porzana parva
LITTLE CRAKE

Non-SPEC[E] (1994: 4) **Status** (Secure)
Criteria —
European IUCN Red List Category —
Criteria —
Global IUCN Red List Category —
Criteria —

Porzana parva is a widespread but patchily distributed summer visitor to Europe, which constitutes >75% of its global breeding range. Its European breeding population is relatively small (<140,000 pairs), but increased substantially between 1970–1990. Although trends were stable across much of Europe during 1990–2000, some national populations declined—including the sizeable Ukrainian one—and the species probably declined slightly overall. Nevertheless, this recent decline is outweighed by earlier increases, and consequently the species is provisionally evaluated as Secure.

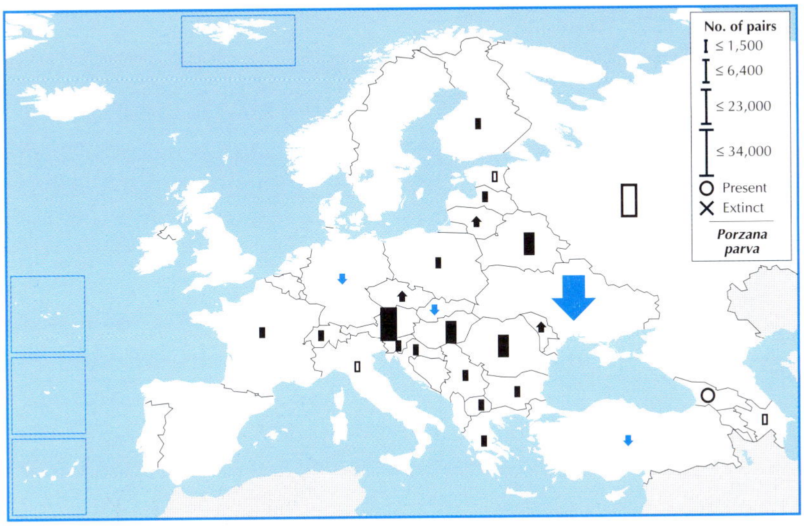

Country	Breeding pop. size (pairs)	Year(s)	Trend	Mag.%	References
Austria	12,000 – 22,000	98–02	(0)	(0–19)	
Azerbaijan	(50 – 500)	96–00	?	–	
Belarus	2,000 – 3,000	97–02	0	0–19	
Bulgaria	100 – 300	96–02	(0)	(0–19)	
Croatia	(100 – 500)	02	(0)	(0–19)	*70,16*
Czech Rep.	**15 – 30**	00	+	**>80**	
Estonia	(20 – 50)	98	?	–	*1*
Finland	1 – 10	98–02	F	>80	
France	3 – 10	98–00	(0)	(0–19)	*6*
Georgia	Present	03	?	–	
Germany	50 – 100	95–99	–	20–29	
Greece	(10 – 50)	95–00	(0)	(0–19)	
Hungary	(3,000 – 5,000)	90–93	(0)	(0–19)	*14*
Italy	(20 – 60)	03	?	–	
Latvia	(50 – 200)	90–00	(0)	(0–19)	*23*
Lithuania	250 – 400	99–01	(+)	(20–29)	*20*
Macedonia	(0 – 50)	90–00	(0)	(0–19)	
Moldova	150 – 200	90–00	(+)	(0–19)	
Poland	1,200 – 1,800	95–00	(0)	(0–19)	*45*
Romania	(5,000 – 8,000)	90–02	(0)	(0–19)	
Russia	(10,000 – 50,000)	90–00	?	–	*3,4,8,108,121,125*
Serbia & MN	(400 – 600)	90–02	(0)	(0–19)	*1,155,78,227*
Slovakia	70 – 150	80–99	–	30–49	
Slovenia	10 – 20	90–00	F	>80	
Switzerland	**0 – 10**	93–96	(0)	(0–19)	
Turkey	(10 – 100)	01	(–)	(–)	
Ukraine	26,000 – 43,000	90–00	–	0–19	
Total (approx.)	61,000 – 140,000	**Overall trend** Small decline			
Breeding range	>1,000,000 km²	**Gen. length.** <3.3	**% Global pop.** 75–94		

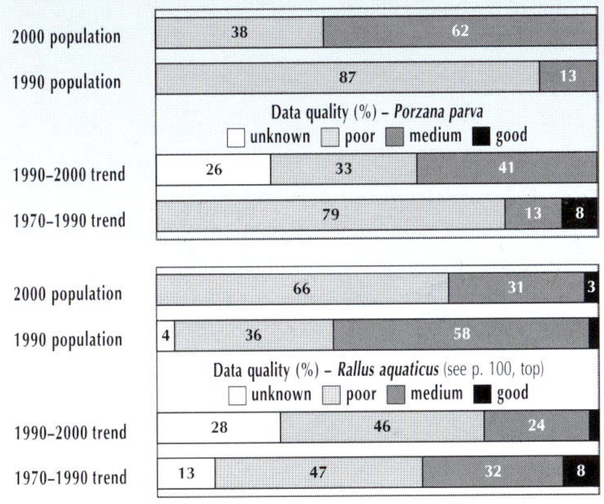

Porzana pusilla
BAILLON'S CRAKE

SPEC 3 (1994: 3) **Status** (Rare)
Criteria <10,000 pairs
European IUCN Red List Category —
Criteria —
Global IUCN Red List Category —
Criteria —

Porzana pusilla is a widespread but patchily distributed summer visitor to much of southern and eastern Europe, which accounts for less than a quarter of its global breeding range. Its European breeding population is very small (as few as 760 pairs), but was stable between 1970–1990. Although the species declined in a number of countries during 1990–2000, no trend data were available for its Russian stronghold. Nevertheless, its population size renders it susceptible to the risks affecting small populations, and consequently the species is provisionally evaluated as Rare.

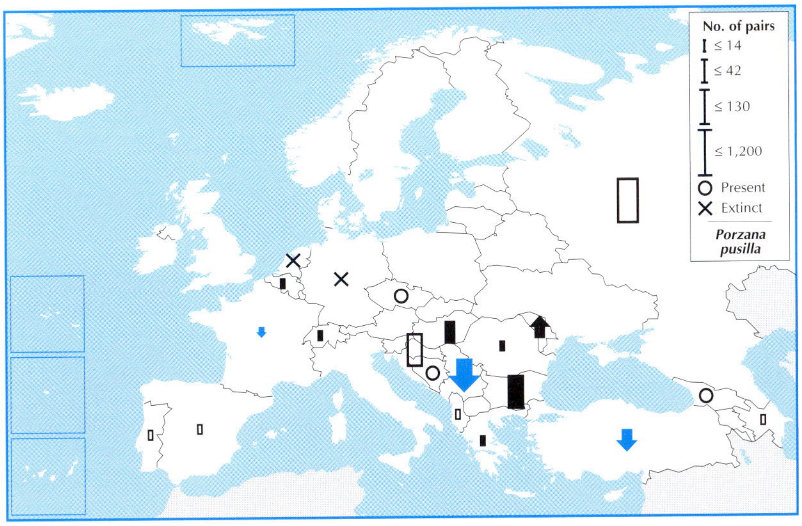

Country	Breeding pop. size (pairs)	Year(s)	Trend	Mag.%	References
Albania	(0 – 50)	92–93	?	–	
Azerbaijan	(0 – 100)	96–00	?	–	
Belgium	1 – 3	00–02	(0)	(0–19)	*1*
Bosnia & HG	Present	90–03	?	–	
Bulgaria	50 – 100	96–02	(F)	(20–29)	
Croatia	(50 – 100)	02	?	–	*70,16*
Czech Rep.	Present	00	?	–	
France	0 – 3	96–00	(–)	(0–19)	*6*
Georgia	Present	03	?	–	
Germany	**0 – 0**	00	–	**X**	
Greece	(0 – 10)	95–00	(0)	(0–19)	
Hungary	10 – 40	98–01	F	20–49	
Moldova	30 – 60	90–00	(+)	(0–19)	
Netherlands	**0 – 0**	98–00	–	**X**	*1*
Portugal	(0 – 5)	02	?	–	
Romania	(10 – 20)	90–02	(0)	(0–19)	
Russia	(500 – 2,500)	90–00	?	–	*67,144,145*
Serbia & MN	(100 – 150)	90–02	(–)	(10–29)	*1,155,227*
Spain	(0 – 50)	98–02	?	–	*10,12*
Switzerland	**0 – 5**	93–96	**0**	**0–19**	
Turkey	(10 – 50)	01	(–)	(–)	
Total (approx.)	760 – 3,200	**Overall trend** Unknown			
Breeding range	>500,000 km²	**Gen. length.** <3.3	**% Global pop.** 5–24		

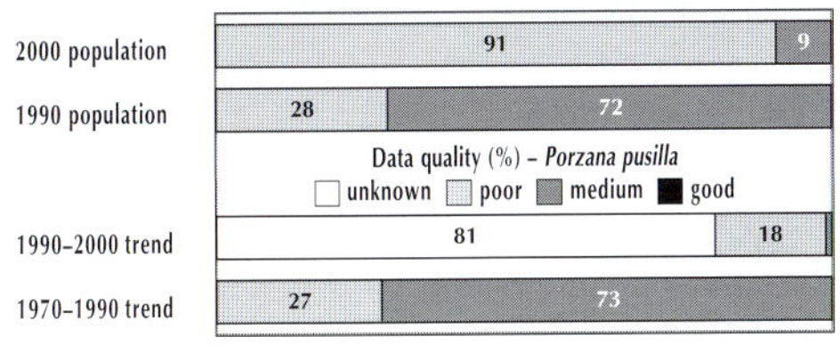

Crex crex
CORNCRAKE

SPEC 1 (1994: 1) **Status** Depleted
Criteria Large historical decline
European IUCN Red List Category —
Criteria —
Global IUCN Red List Category NT
Criteria A3c

Crex crex is a widespread summer visitor to middle latitudes of Europe, which constitutes >50% of its global breeding range. Its European breeding population is very large (>1,300,000 pairs), but declined substantially between 1970–1990. Although many populations increased during 1990–2000, the species fluctuated in its Russian stronghold, and was broadly stable overall. Nevertheless, its total population size clearly remains far below the level that preceded its decline. Consequently, this globally Near Threatened species is evaluated as Depleted in Europe.

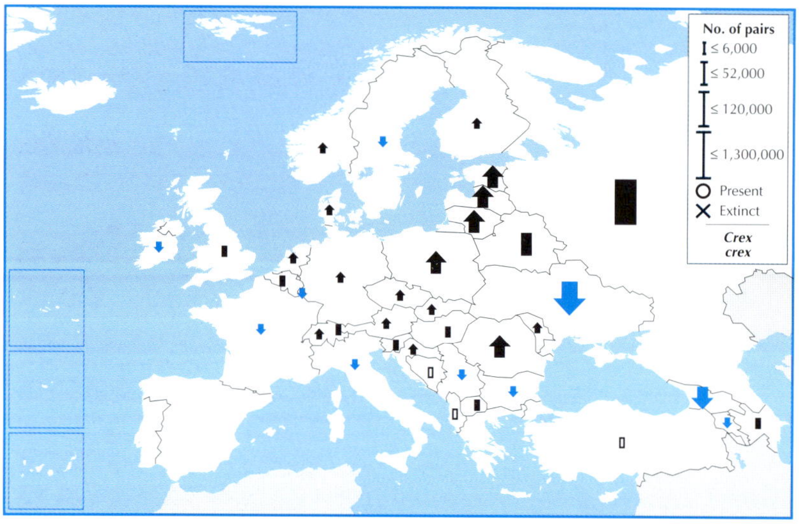

Country	Breeding pop. size (pairs)	Year(s)	Trend	Mag.%	References
Albania	(0 – 20)	96–02	?	–	
Armenia	330 – 830	98–02	–	20–29	
Austria	150 – 400	98–02	+	20–29	
Azerbaijan	(0 – 100)	96–00	(0)	(0–19)	
Belarus	25,000 – 60,000	97–02	0	0–19	
Belgium	**21 – 44**	95–02	**F**	**20–29**	*1*
Bosnia & HG	(0 – 250)	90–00	?	–	
Bulgaria	**4,000 – 8,800**	96–97	(–)	(0–19)	
Croatia	1,000 – 1,500	02	(+)	(50–79)	*70,16*
Czech Rep.	**1,500 – 1,700**	00	**+**	**>80**	
Denmark	(50 – 250)	98–01	**+**	**>80**	*4,5,6,7*
Estonia	15,000 – 25,000	98	+	20–29	*1*
Finland	2,000 – 8,000	98–02	**+**	**500**	
France	551 – 599	02	**–**	**20–49**	*6*
Georgia	(10,000 – 50,000)	94–02	(–)	(20–29)	
Germany	2,000 – 3,100	95–99	+	0–19	
Hungary	500 – 1,200	98–02	**F**	**20–49**	*4*
Rep. Ireland	**139 – 157**	98–02	**–**	**20–29**	
Italy	200 – 450	03	–	0–19	*20,19*
Latvia	**26,000 – 38,000**	95–03	**+**	**30–49**	*9,11*
Liechtenstein	**1 – 4**	98–00	(F)	(–)	
Lithuania	25,000 – 30,000	99–01	+	30–49	*20*
Luxembourg	**0 – 5**	00–02	**–**	**50–79**	
Macedonia	(50 – 150)	90–00	(F)	(20–29)	
Moldova	100 – 150	90–00	**+**	**0–19**	
Netherlands	**240 – 700**	98–00	**+**	**573**	*1*
Norway	20 – 40	95–03	**+**	**30–49**	*18*
Poland	**30,000 – 45,000**	97–98	(+)	(20–29)	*44*
Romania	44,000 – 60,000	90–02	+	0–19	*48,18*
Russia	**1,000,000 – 1,500,000**	90–00	F	30–49	*105*
Serbia & MN	800 – 1,400	95–02	–	10–29	*1,234,67a,155, 37,227,225*
Slovakia	1,400 – 1,700	80–99	+	20–29	
Slovenia	**500 – 600**	92–99	0	10–19	
Sweden	150 – 200	99–00	–	0–19	
Switzerland	**10 – 50**	98–02	**+**	**30–49**	
Turkey	(25 – 100)	01	?	–	
Ukraine	**83,400 – 154,000**	00–01	–	0–19	
UK	**589 – 589**	98	**0**	**3**	*10*

Total (approx.) 1,300,000 – 2,000,000 **Overall trend** Fluctuating
Breeding range >4,000,000 km² **Gen. length.** <3.3 **% Global pop.** 50–74

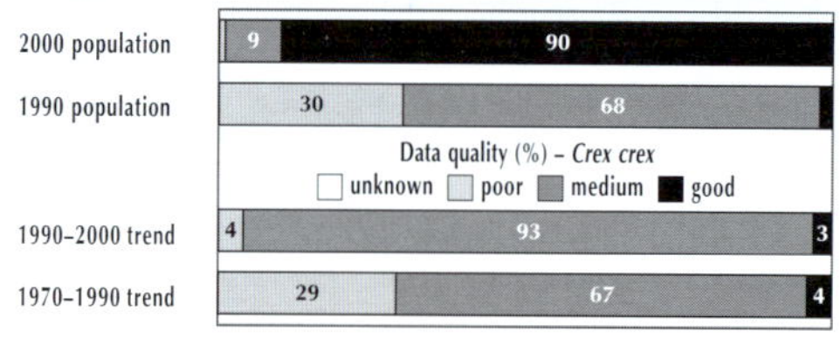

Gallinula chloropus
COMMON MOORHEN

Non-SPEC (1994: —) **Status** Secure
Criteria —
European IUCN Red List Category —
Criteria —
Global IUCN Red List Category —
Criteria —

Gallinula chloropus is a widespread breeder across much of Europe, which accounts for less than a quarter of its global breeding range. Its European breeding population is large (>900,000 pairs), and was stable between 1970–1990. Although there were declines in a few countries during 1990–2000, populations were stable across most of Europe, and the sizeable population in France increased. The species hence remained stable overall, and is consequently evaluated as Secure.

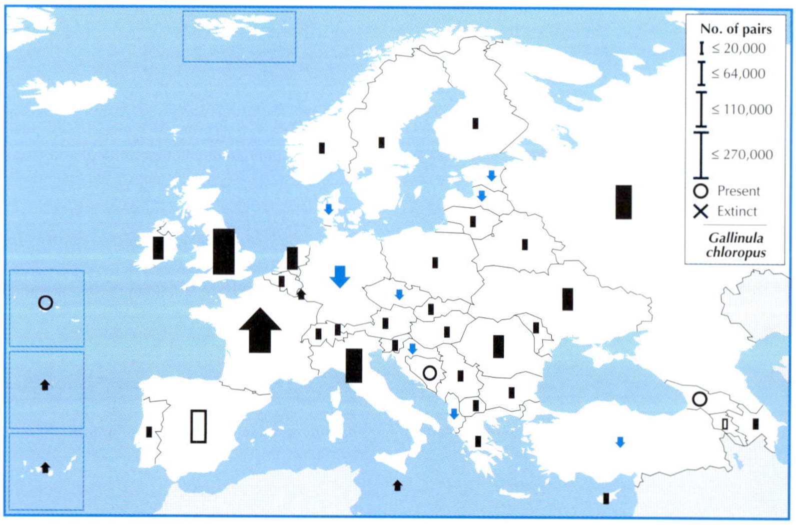

Country	Breeding pop. size (pairs)	Year(s)	Trend	Mag.%	References
Albania	200 – 700	96–02	–	0–19	*6*
Armenia	850 – 3,300	98–02	?	–	
Austria	(1,400 – 2,500)	98–02	(0)	(0–19)	
Azerbaijan	(5,000 – 15,000)	96–00	(0)	(0–19)	
Belarus	15,000 – 20,000	97–02	0	0–19	
Belgium	10,000 – 20,000	01–02	(0)	(0–19)	*1*
Bosnia & HG	Present	90–03	?	–	
Bulgaria	3,000 – 8,000	96–02	0	0–9	
Croatia	(5,000 – 10,000)	02	(–)	(30–49)	*54,26,16*
Cyprus	(300 – 600)	94–02	(0)	(0–9)	
Czech Rep.	**4,000 – 8,000**	00	**–**	**10–19**	
Denmark	(5,000 – 20,000)	00	–	30–49	
Estonia	(1,000 – 2,000)	98	–	20–29	*1*
Finland	50 – 200	98–02	0	0–19	
France	(100,000 – 400,000)	98–00	(+)	(46)	*4,2*
Georgia	Present	03	?	–	
Germany	33,000 – 50,000	95–99	(–)	(20–29)	
Greece	(5,000 – 10,000)	95–00	(0)	(0–19)	
Hungary	6,000 – 12,000	95–02	0	0–19	
Rep. Ireland	20,000 – 100,000	88–91	(0)	(0–19)	
Italy	(80,000 – 120,000)	03	(0)	(0–19)	
Latvia	(500 – 2,000)	90–00	(–)	(30–49)	*17*
Liechtenstein	**2 – 4**	98–00	**0**	**0–19**	
Lithuania	(3,000 – 4,000)	99–01	(0)	(0–19)	*20*
Luxembourg	**400 – 600**	02	+	0–19	
Macedonia	1,000 – 2,500	95–99	(0)	(0–19)	
Malta	**10 – 25**	90–02	**+**	**0–19**	*1*
Moldova	1,100 – 1,700	90–00	**0**	**0–19**	
Netherlands	**40,000 – 55,000**	98–00	**0**	**2**	*1*
Norway	(500 – 2,000)	90–02	(0)	(0–19)	*27*
Poland	10,000 – 20,000	90–00	0	0–19	*27*
Portugal	(5,000 – 50,000)	02	(0)	(0–19)	
Azores	Present	02	?	–	
Madeira	**0 – 10**	02	**+**	**N**	
Romania	(50,000 – 80,000)	90–02	(0)	(0–19)	
Russia	(70,000 – 150,000)	90–00	0	0–19	*8,51,52,104,108, 114,125*
Serbia & MN	6300 – 7,500	90–02	0	0–19	*1,155,227,144, 203,13,225*
Slovakia	1,000 – 2,500	90–99	0	0–19	
Slovenia	500 – 1,000	94	(0)	(0–19)	
Spain	(90,000 – 120,000)	92	?	–	*13,12,10*
Canary Is.	250 – 1,000	97–03	+	0–19	*28,25*
Sweden	**2,000 – 4,000**	99–00	(F)	(20–29)	
Switzerland	**1,000 – 2,500**	93–96	0	10–19	
Turkey	(15,000 – 25,000)	01	(–)	(0–19)	
Ukraine	40,000 – 66,000	90–00	0	0–19	
UK	**270,000 – 270,000**	00	**0**	**4**	*5,31*

Total (approx.) 900,000 – 1,700,000 **Overall trend** Stable
Breeding range >7,000,000 km² **Gen. length.** <3.3 **% Global pop.** 5–24
(See p. 103, top, for data quality graph)

Winter pop. size (individuals)

Total (approx.) >280,000 **Overall trend** Stable
% in European IBAs Unknown **Gen. length** <3.3 **% Global pop.** 5–24

Porphyrio porphyrio
PURPLE SWAMPHEN

SPEC 3 (1994: 3) **Status** Localised
Criteria ≥90% breed at ≤10 sites
European IUCN Red List Category —
Criteria —
Global IUCN Red List Category —
Criteria —

Country	Breeding pop. size (pairs)	Year(s)	Trend	Mag.%	References
Azerbaijan	5,000 – 25,000	96–00	(F)	(–)	
Croatia	3 – 5	02	(+)	(N)	*70,16*
France	**10 – 15**	98–00	+	**0–19**	*6*
Italy	450 – 600	99	+	0–19	*3*
Portugal	**49 – 67**	02	(+)	(–)	
Russia	(500 – 1,500)	95–02	+	50–79	*33,74,172*
Spain	6,400 – 6,700	98–02	+	>80	*10*
Turkey	**750 – 1,000**	01	+	**20–29**	
Total (approx.)	13,000 – 35,000	**Overall trend** Moderate increase			
Breeding range	>100,000 km²	**Gen. length.** <3.3 **% Global pop.** 5–24			

Porphyrio porphyrio breeds locally in southern Europe, which accounts for less than a quarter of its global range. The European population is relatively small (<35,000 pairs), but increased substantially between 1970–1990. The species continued to increase across most of its European range during 1990–2000—including key populations in Spain and Azerbaijan—and underwent a moderate increase overall. Nevertheless, more than 90% of the European breeding population is confined to just 10 sites (see Appendix 5), and consequently the species is evaluated as Localised.

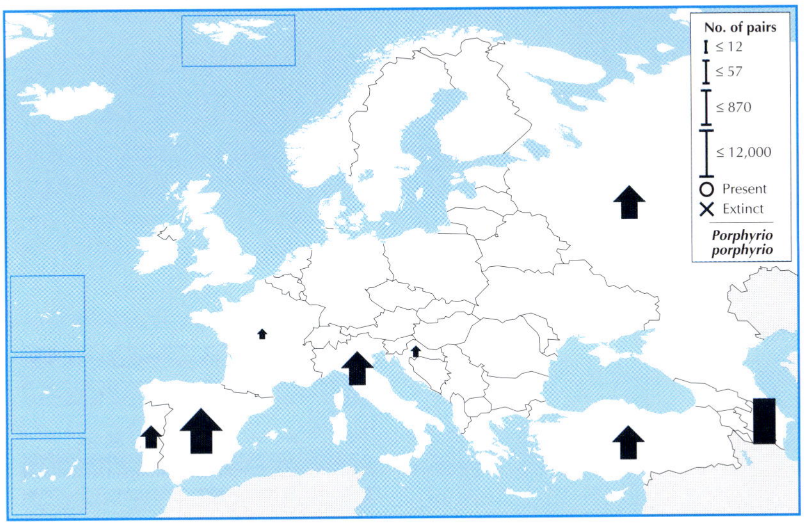

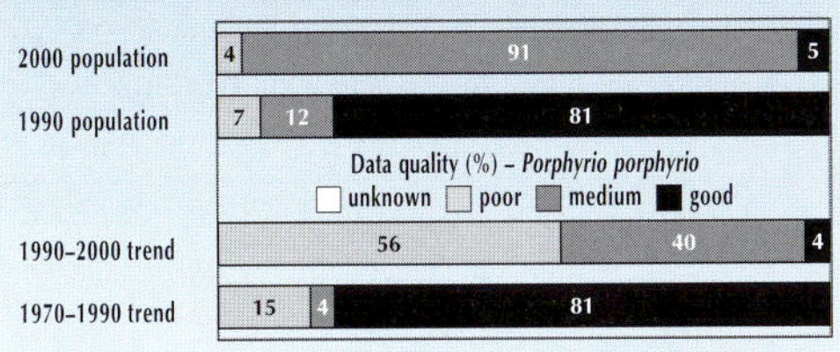

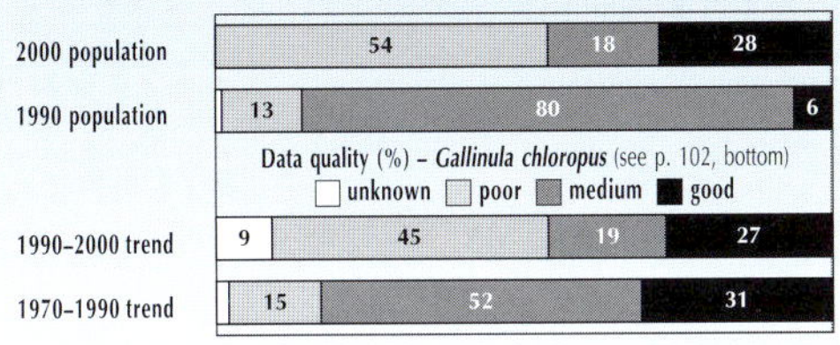

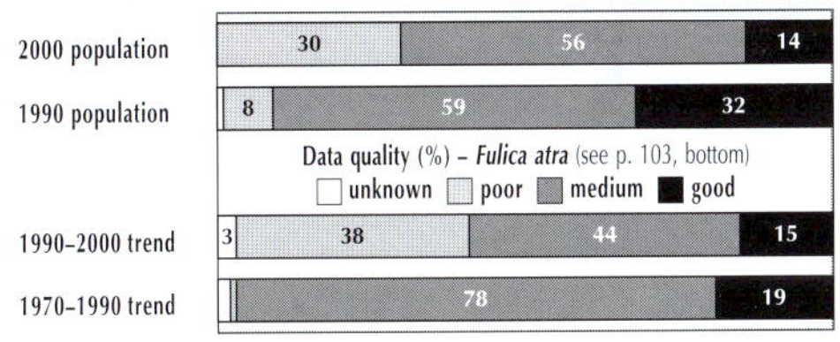

Fulica atra
COMMON COOT

Non-SPEC (1994: —) **Status** (Secure)
Criteria —
European IUCN Red List Category —
Criteria —
Global IUCN Red List Category —
Criteria —

Country	Breeding pop. size (pairs)	Year(s)	Trend	Mag.%	References
Albania	100 – 500	96–02	–	0–19	*6*
Armenia	3,300 – 6,500	99–02	–	0–9	
Austria	(2,000 – 3,500)	98–02	(0)	(0–19)	
Azerbaijan	(10,000 – 20,000)	96–00	(0)	(0–19)	
Belarus	**14,000 – 17,000**	97–02	**0**	**0–19**	
Belgium	5,500 – 8,000	01–02	+	0–19	*1*
Bosnia & HG	Present	90–03	?	–	
Bulgaria	**1,200 – 2,500**	96–02	0	0–9	
Croatia	(2,500 – 5,000)	02	(–)	(50–79)	*26*
Cyprus	(200 – 400)	94–02	(+)	(0–9)	
Czech Rep.	**20,000 – 40,000**	00	**–**	**20–29**	
Denmark	15,000 – 20,000	00	(–)	(10–19)	*3*
Estonia	3,000 – 5,000	98	–	20–29	*1*
Finland	8,000 – 15,000	98–02	**–**	**20**	
France	(50,000 – 150,000)	98–02	(0)	(0–19)	*4*
Georgia	Present	03	?	–	
Germany	61,000 – 140,000	95–99	(0)	(0–19)	
Greece	(2,000 – 5,000)	95–00	(0)	(0–19)	
Hungary	(80,000 – 120,000)	90–93	(–)	(0–19)	*14*
Rep. Ireland	(2,500 – 10,000)	88–91	?	–	
Italy	(8,000 – 12,000)	03	(0)	(0–19)	
Latvia	8,000 – 10,000	90–00	0	0–19	*23,17*
Liechtenstein	8 – 15	98–00	**+**	**0–9**	
Lithuania	**20,000 – 30,000**	99–01	(0)	(0–19)	*20*
Luxembourg	**40 – 60**	02	+	**0–19**	
Macedonia	500 – 1,000	90–00	(0)	(0–19)	
Moldova	**6,000 – 6,300**	90–00	**0**	**0–19**	
Netherlands	**130,000 – 180,000**	98–00	**0**	**2**	*1*
Norway	(500 – 2,000)	90–02	+	0–19	
Poland	(100,000 – 250,000)	90–00	(–)	(20–29)	*1*
Portugal	(200 – 2,000)	02	(0)	(0–19)	
Madeira	**0 – 10**	02	+	**N**	
Romania	(80,000 – 120,000)	90–02	(0)	(0–19)	
Russia	450,000 – 850,000	90–00	–	20–29	*85,129,114*
Serbia & MN	15,000 – 22,000	90–02	+	0–19	*1,155,227,144, 143a,203,13,78, 225*
Slovakia	4,000 – 8,000	90–99	0	0–19	
Slovenia	300 – 500	94	(0)	(0–19)	
Spain	(16,520 – 20,000)	98–02	?	–	*10*
Canary Is.	50 – 250	97–03	+	0–19	*28,25*
Sweden	30,000 – 40,000	99–00	+	**25–29**	
Switzerland	**4,500 – 7,000**	93–96	0	0–19	
Turkey	20,000 – 40,000	01	(0)	(0–19)	
Ukraine	70,000 – 90,000	90–00	F	20–29	
UK	22,600 – 28,800	00	?	–	*5,31*
Total (approx.)	1,300,000 – 2,300,000	**Overall trend** Moderate decline			
Breeding range	>7,000,000 km²	**Gen. length.** <3.3 **% Global pop.** 25–49			

(See p. 103, top, for data quality graph)

	Winter pop. size (individuals)	
Total (approx.)	>2,500,000	**Overall trend** Small increase
% in European IBAs	76–87	**Gen. length** <3.3 **% Global pop.** 25–49

Fulica atra is a widespread breeder across much of Europe, which accounts for less than half of its global breeding range. Its European breeding population is very large (>1,300,000 pairs), and increased markedly between 1970–1990. Although the species was stable (or increasing) across much of Europe during 1990–2000, key populations in Russia, Hungary and Poland suffered declines, and the species underwent a moderate decline (>10%) overall. Nevertheless, this recent decline is still outweighed by earlier increases, and consequently the species is provisionally evaluated as Secure.

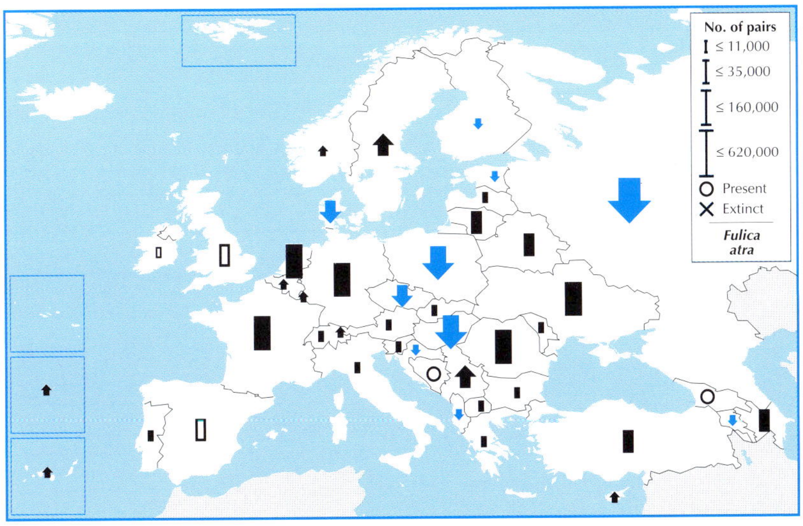

Fulica cristata
RED-KNOBBED COOT

SPEC 3 (1994: 3) **Status** Critically Endangered
Criteria See IUCN below
European IUCN Red List Category CR
Criteria C2a(ii); C2b
Global IUCN Red List Category —
Criteria —

Country	Breeding pop. size (pairs)	Year(s)	Trend	Mag.%	References
Spain	**80 – 80**	02	**F**	**>80**	*10,16*
Total (approx.)	80 – 80	**Overall trend** Fluctuating			
Breeding range	<20,000 km²	**Gen. length.** <3.3	**% Global pop.** <5		

Country	Winter pop. size (individuals)	Year(s)	Trend	Mag.%	References
Spain	**0 – 31**	90–01	-	**0–19**	*19*
Total (approx.)	0–31	**Overall trend** Small decline			
% in European IBAs	Unknown	**Gen. length** <3.3	**% Global pop.** <5		

Fulica cristata has a predominantly African distribution, which just extends into Europe in southernmost Spain. Its European breeding population is extremely small (c.80 pairs), and fluctuated widely between 1970–1990 in response to weather conditions. These extreme fluctuations continued during 1990–2000, when several droughts caused the population to decline slightly overall (see winter data). As a consequence of this decline and its extremely small and fluctuating population, the species is evaluated as Critically Endangered.

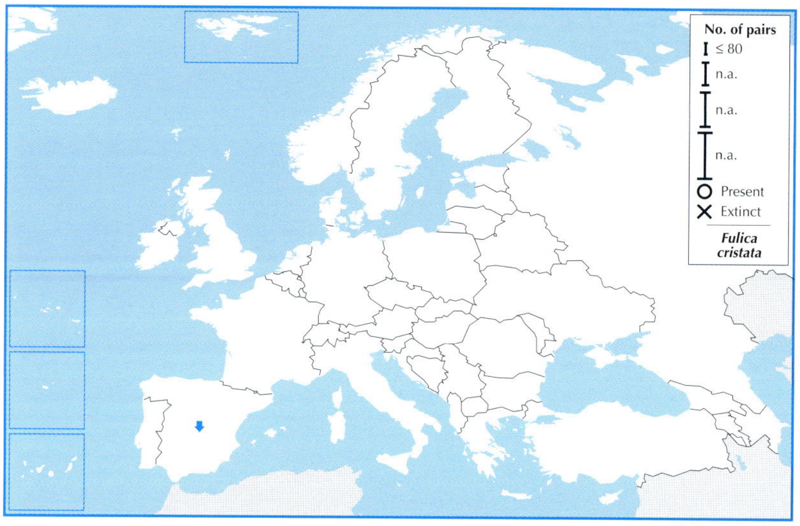

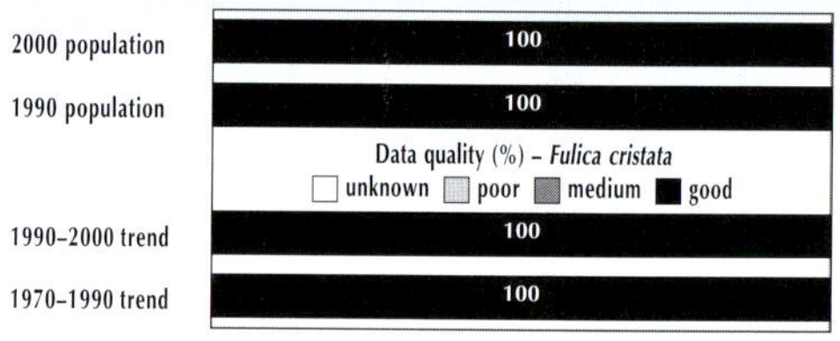

Grus grus
COMMON CRANE

SPEC 2 (1994: 3) **Status** (Depleted)
Criteria Large historical decline
European IUCN Red List Category —
Criteria —
Global IUCN Red List Category —
Criteria —

Country	Breeding pop. size (pairs)	Year(s)	Trend	Mag.%	References
Armenia	10 – 20	97–02	?	–	
Belarus	800 – 1,500	97–02	0	0–19	
Czech Rep.	**9 – 13**	00	+	**>80**	
Denmark	**16 – 28**	99–01	+	**>80**	*5,6,7*
Estonia	1,500 – 2,000	98	+	20–29	*1*
Finland	15,000 – 20,000	98–02	+	20	
France	**1 – 3**	96–00	**0**	**0–19**	*5*
Georgia	**7 – 10**	94–02	-	**20–29**	
Germany	**3,000 – 3,100**	95–99	+	**30–49**	
Latvia	1,000 – 2,500	90–00	+	>80	
Lithuania	900 – 1,200	99–01	+	50–79	*20*
Netherlands	**0 – 1**	98–00	?	–	*1*
Norway	1,000 – 3,000	02	+	**30–49**	*9*
Poland	**10,000 – 12,000**	01–03	+	**>80**	*13*
Romania	(0 – 2)	90–02	(F)	(50–79)	
Russia	25,000 – 40,000	90–00	0	0–19	*31,36,37,47,51, 52,75,104*
Sweden	15,000 – 20,000	99–00	+	10–19	
Turkey	190 – 270	01	-	30–49	
Ukraine	700 – 850	90–00	+	**20–29**	
UK	**3 – 3**	96–00	**F**	**300**	
Total (approx.)	74,000 – 110,000	**Overall trend** Large increase			
Breeding range	>4,000,000 km²	**Gen. length.** 14	**% Global pop.** 50–74		

Country	Winter pop. size (individuals)	Year(s)	Trend	Mag.%	References
Azerbaijan	(0 – 200)	96–02	(0)	(0–19)	
Croatia	(100 – 1,000)	02	(F)	(–)	*26*
France	**20,000 – 28,000**	98–02	+	**50–79**	*9*
Italy	10 – 120	02	F	20–29	*22*
Luxembourg	**10 – 100**	00–02	+	**50–79**	
Poland	50 – 100	90–00	(+)	(>80)	*114*
Portugal	**2,000 – 3,200**	90–02	**0**	**0–19**	
Serbia & MN	**0 – 1,000**	90–02	**F**	**50–100**	
Spain	75,000 – 82,000	99–00	(F)	(–)	*20*
Turkey	**900 – 1,200**	91–01	0	0–19	
Total (approx.)	>98,000	**Overall trend** Moderate increase			
% in European IBAs	>80	**Gen. length** 14	**% Global pop.** 5–24		

Grus grus is a widespread summer visitor to northern Europe (occurring more patchily farther south), which holds >50% of its global breeding population. Its European breeding population is relatively small (<110,000 pairs), and underwent a large decline between 1970–1990. Although the species increased substantially overall during 1990–2000—with increasing or stable trends across most of its European range—its population has probably not yet recovered to the level that preceded its decline. Consequently, it is provisionally evaluated as Depleted.

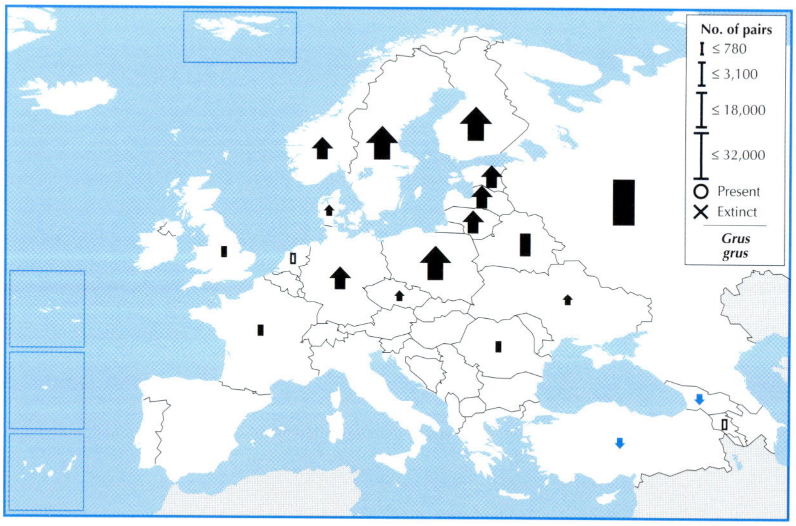

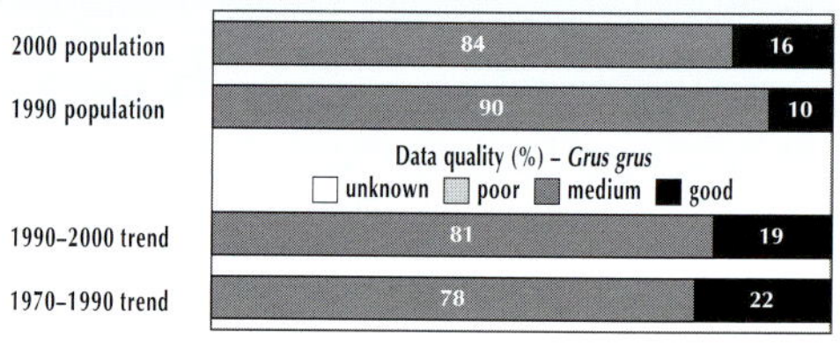

Grus virgo
DEMOISELLE CRANE

Non-SPEC (1994: —) **Status** Secure
Criteria —
European IUCN Red List Category —
Criteria —
Global IUCN Red List Category —
Criteria —

Country	Breeding pop. size (pairs)	Year(s)	Trend	Mag.%	References
Russia	20,000 – 25,000	90–00	+	20–29	*37,69,76,149*
Turkey	10 – 20	01	(-)	(30–49)	
Ukraine	**200 – 250**	90–00	**0**	**0–19**	
Total (approx.)	20,000 – 25,000	**Overall trend** Large increase			
Breeding range	>250,000 km²	**Gen. length.** 6		**% Global pop.** 25–49	

Grus virgo is a summer visitor to Turkey, Ukraine and southern Russia, with Europe accounting for less than half of its global breeding range. Its European breeding population is relatively small (<25,000 pairs), but increased between 1970–1990. Although the small population in Turkey declined during 1990–2000, the species was stable in Ukraine and increased in the Russian stronghold, and hence continued to increase overall. Consequently, it is evaluated as Secure.

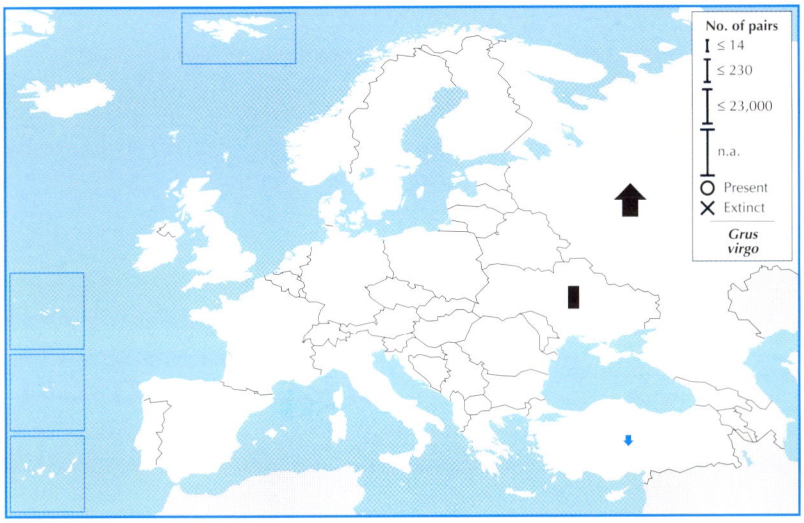

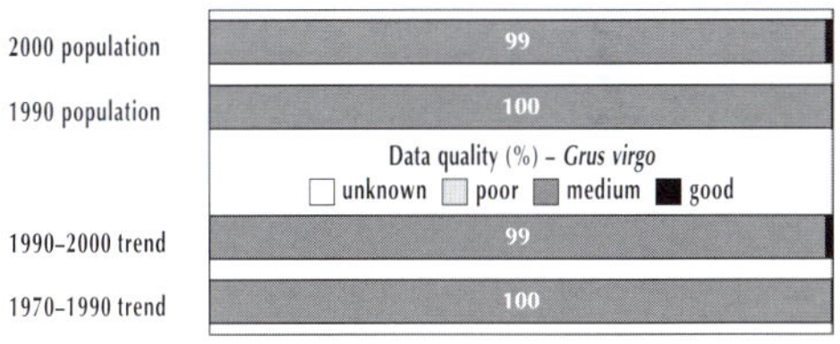

Tetrax tetrax
LITTLE BUSTARD

SPEC 1 (1994: 2) **Status** Vulnerable
Criteria See IUCN below
European IUCN Red List Category VU
Criteria A2b
Global IUCN Red List Category NT
Criteria A2c,d; A3c,d

Country	Breeding pop. size (individuals)	Year(s)	Trend	Mag.%	References
France	**2,700 – 4,000**	99–00	-	**20–49**	*6*
Italy	1,000 – 1,500	03	0	0–19	
Macedonia	(0 – 10)	90–00	(F)	(50–79)	
Portugal	(10,000 – 20,000)	02	(0)	(0–19)	
Russia	10,000 – 20,000	90–00	+	**50–79**	*6,74,76,149,178*
Spain	(100,000 – 250,000)	98–02	(-)	(20–29)	*10,16*
Turkey	(30 – 60)	01	-	**>80**	
Ukraine	**100 – 110**	90–00	+	**20–29**	*3*
Total (approx.)	120,000 – 300,000	**Overall trend** Moderate decline			
Breeding range	>750,000 km²	**Gen. length.** 5		**% Global pop.** 75–94	

Tetrax tetrax breeds mainly in south-west Europe and southern Russia, with Europe constituting >75% of its global breeding range. Its European breeding population is large (>120,000 individuals), but declined substantially between 1970–1990. Although the species was stable in Portugal and increased in Russia during 1990–2000, it declined in its Spanish stronghold, and declined overall at a rate that—on top of earlier declines—equates to a large decline (>30%) over three generations. Consequently, this globally Near Threatened species is evaluated as Vulnerable in Europe.

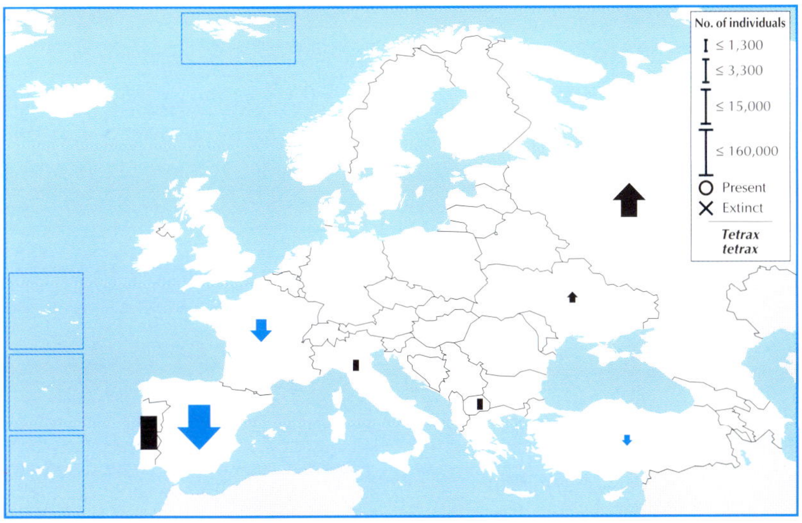

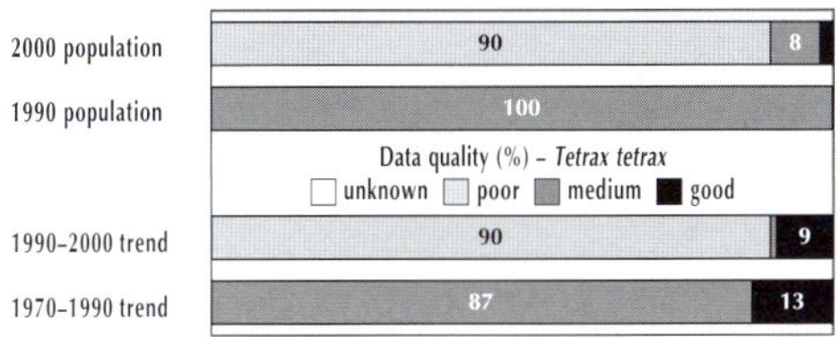

Chlamydotis undulata
HOUBARA BUSTARD

SPEC 1 (1994: 3) **Status** (Vulnerable)
Criteria See IUCN below
European IUCN Red List Category VU
Criteria D1
Global IUCN Red List Category VU
Criteria A2b,c,d; A3b,c,d

Country	Breeding pop. size (individuals)	Year(s)	Trend	Mag.%	References
Azerbaijan	(0 – 10)	96–00	(0)	(0–19)	
Russia	(1 – 20)	90–00	?	–	*2,74*
Spain					
Canary Is.	**527 – 527**	94	?	–	*31*
Total (approx.)	530 – 560	**Overall trend** Unknown			
Breeding range	<20,000 km²	**Gen. length.** 6		**% Global pop.** <5	

Chlamydotis undulata has a predominantly North African distribution, which just extends into Europe in the Canary Islands (Fuerteventura and Lanzarote), southernmost Russia and Azerbaijan. Its European breeding population is very small (as few as 530 individuals), and its trend during 1970–1990 and 1990–2000 was unknown. The resident subspecies endemic to the Canary Islands (*C. u. fuerteventurae*) is clearly susceptible to the risks affecting small populations. Consequently, this globally threatened species is provisionally evaluated as Vulnerable in Europe.

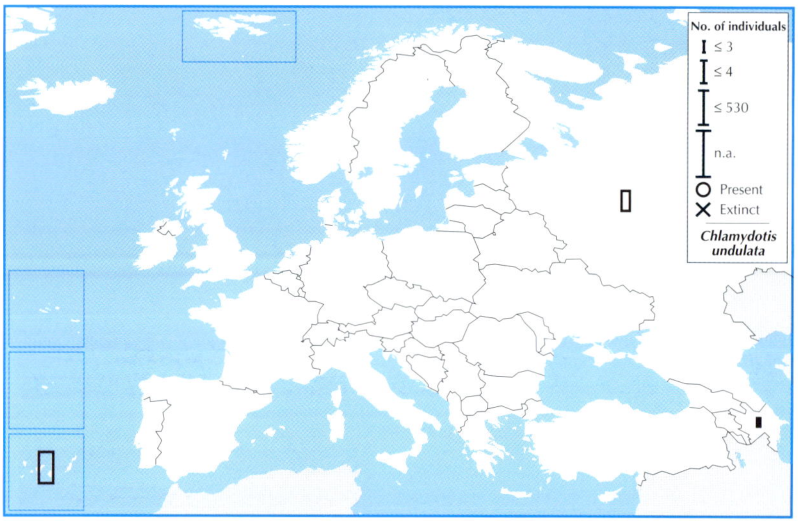

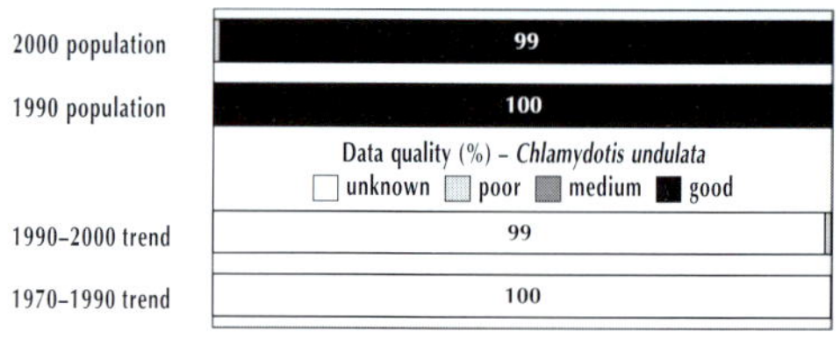

Otis tarda
GREAT BUSTARD

SPEC 1 (1994: 1) **Status** Vulnerable
Criteria See IUCN below
European IUCN Red List Category VU
Criteria A2b
Global IUCN Red List Category VU
Criteria A3c

Country	Breeding pop. size (individuals)	Year(s)	Trend	Mag.%	References
Austria	**74 – 140**	98–02	+	**30–49**	
Bulgaria	(0 – 10)	95–02	?	–	
Czech Rep.	**1 – 10**	00	–	**>80**	
Germany	**73 – 95**	95–99	–	**20–29**	
Hungary	**1,100 – 1,200**	98–02	**0**	**0–19**	
Moldova	0 – 0	96–00	–	X	
Portugal	(500 – 1,500)	02	(0)	(0–19)	
Romania	(0 – 5)	90–02	(F)	(20–29)	*8,28*
Russia	**5,500 – 8,000**	95–00	+	**20–29**	*2,74,115,149*
Serbia & MN	**30 – 36**	00–02	**0**	**0–19**	*1,19,29,155*
Slovakia	8 – 16	00–03	–	50–79	
Spain	23,000 – 23,000	98–02	0	0–19	*10,16*
Turkey	500 – 1,000	01	–	**20–29**	
Ukraine	**500 – 720**	90–00	**0**	**0–19**	*3*
Total (approx.)	31,000 – 36,000	**Overall trend** Stable			
Breeding range	>750,000 km²	**Gen. length.** 14		**% Global pop.** 50–74	

Otis tarda has a disjunct breeding distribution in south-west, central and eastern Europe, which probably holds >50% of its global population. Its European breeding population is relatively small (<36,000 individuals), and declined markedly between 1970–1990. Although the species stabilised in its Iberian stronghold during 1990–2000 (and increased in Russia), several smaller populations declined further, and it has clearly undergone a large decline (>30%) overall during the last three generations. Consequently, this globally threatened species is evaluated as Vulnerable in Europe.

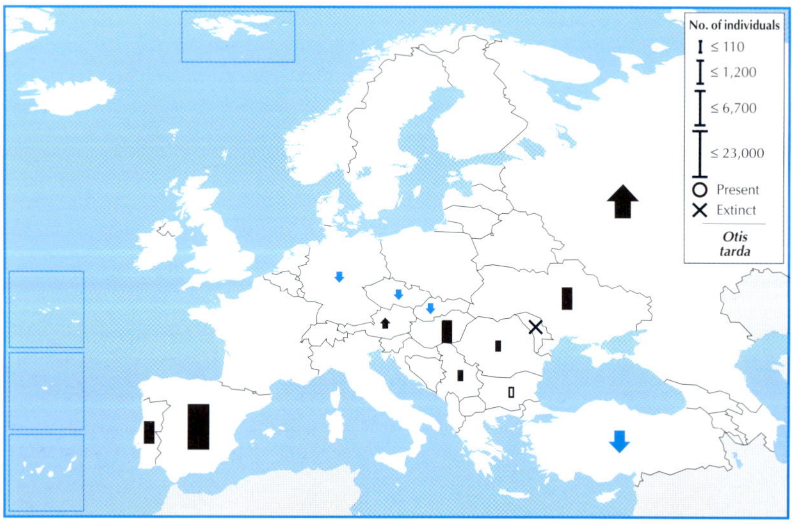

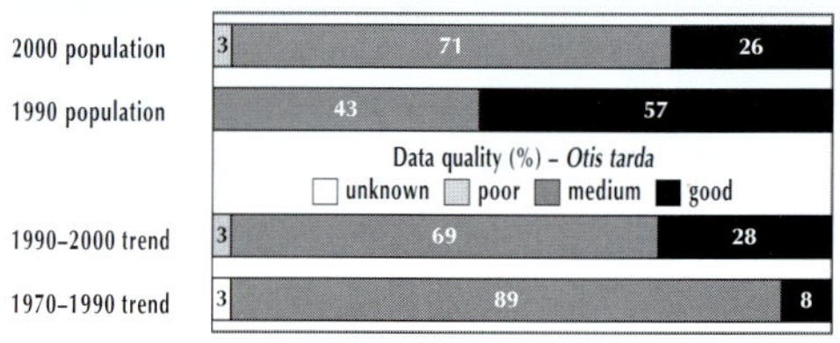

Haematopus ostralegus
EURASIAN OYSTERCATCHER

Non-SPEC[E] (1994: —) **Status** (Secure)
Criteria —
European IUCN Red List Category —
Criteria —
Global IUCN Red List Category —
Criteria —

Haematopus ostralegus is a widespread but patchily distributed breeder across much of Europe (especially the north-west), which constitutes more than half of its global breeding range. Its European breeding population is large (>300,000 pairs), and increased substantially between 1970–1990. Although most populations were stable or increased during 1990–2000, the species declined in the Netherlands, and underwent a moderate decline (>10%) overall. Nevertheless, this decline is outweighed by earlier increases, and consequently the species is provisionally evaluated as Secure.

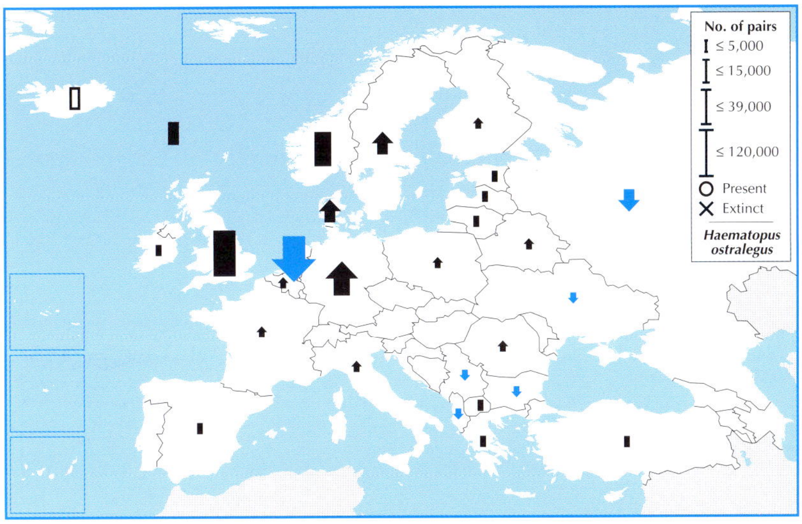

Country	Breeding pop. size (pairs)	Year(s)	Trend	Mag.%	References
Albania	10 – 15	96–02	–	0–19	6
Belarus	230 – 300	97–00	+	0–9	
Belgium	1,500 – 2,100	01–02	+	0–19	1
Bulgaria	25 – 50	96–02	–	>80	
Denmark	10,000 – 14,500	93–01	+	30–39	22
Faroe Is.	10,000 – 10,000	95	(0)	(0–19)	
Estonia	(3,000 – 4,000)	98	0	0–19	1
Finland	4,000 – 5,000	98–02	+	25	
France	1,000 – 1,200	98–02	+	0–9	4
Germany	31,000 – 36,000	95–99	+	20–29	
Greece	30 – 60	95–00	(0)	(0–19)	
Iceland	10,000 – 20,000	90–00	?	–	41
Rep. Ireland	2,500 – 10,000	88–91	(0)	(0–19)	
Italy	129 – 130	00	+	20–29	18
Latvia	50 – 80	90–00	0	0–19	23
Lithuania	20 – 40	99–01	0	0–19	20
Macedonia	(0 – 5)	99–00	(F)	(50–79)	
Netherlands	80,000 – 130,000	98–00	–	40	1,2
Norway	(30,000 – 50,000)	00–03	(0)	(0–19)	11
Poland	20 – 30	97–00	+	40–59	46,60–69
Romania	(50 – 150)	90–02	(+)	(0–19)	
Russia	7,000 – 23,000	90–00	–	20–29	28,29,60
Serbia & MN	1 – 3	97–02	–	50–79	1,148,179,156, 227,135a
Spain	46 – 58	98–02	0	0–19	10,16
Sweden	12,000 – 18,000	99–00	+	0–9	
Turkey	600 – 1,200	01	(0)	(0–19)	
Ukraine	650 – 880	90–00	–	0–19	
UK	98,500 – 127,000	00	0	3	5,23
Total (approx.)	300,000 – 450,000	Overall trend Moderate decline			
Breeding range	>3,000,000 km²	Gen. length. 11		% Global pop. 50–74	

Country	Winter pop. size (individuals)	Year(s)	Trend	Mag.%	References
Denmark	45,000 – 50,000	99–00	+	50–79	25
France	50,000 – 70,000	98–02	+	50–79	9
Germany	130,000 – 320,000	95–00	–	11–42	2
Rep. Ireland	40,000 – 44,000	94–00	0	0–19	
Netherlands	230,000 – 230,000	99–01	–	38	3,4,8,9,11,12
UK	338,700 – 338,700	94–99	–	12	44,52
Total (approx.)	>840,000	Overall trend Small decline			
% in European IBAs	55–61	Gen. length 11		% Global pop. 50–74	

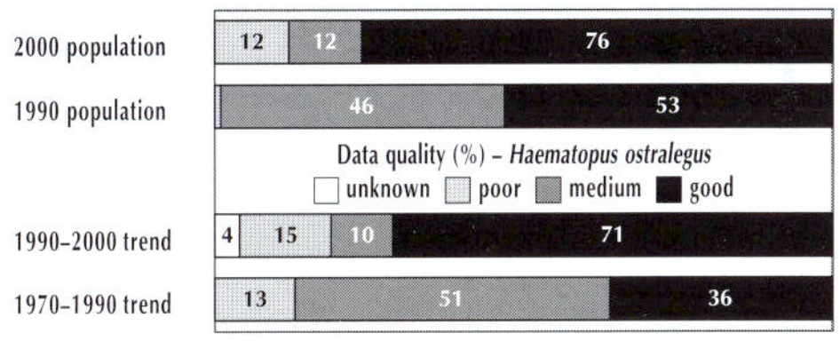

Himantopus himantopus
BLACK-WINGED STILT

Non-SPEC (1994: —) **Status** Secure
Criteria —
European IUCN Red List Category —
Criteria —
Global IUCN Red List Category —
Criteria —

Himantopus himantopus is a widespread but patchily distributed summer visitor to southern and eastern Europe, which accounts for less than a quarter of its global breeding range. Its European breeding population is relatively small (<64,000 pairs), but was stable between 1970–1990. Although there were declines in a few countries—including Turkey—during 1990–2000, populations across most of its European range were stable or (as in the case of the key Spanish population) increased, and the species remained broadly stable overall. Consequently, it is evaluated as Secure.

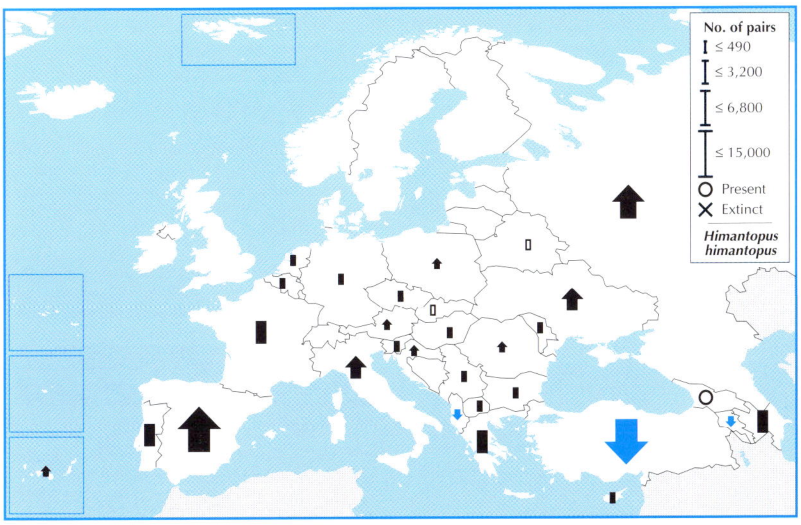

Country	Breeding pop. size (pairs)	Year(s)	Trend	Mag.%	References
Albania	50 – 250	96–02	–	0–19	
Armenia	80 – 350	00–02	–	0–9	11
Austria	10 – 25	98–02	+	N	
Azerbaijan	(1,000 – 5,000)	96–00	(0)	(0–19)	
Belarus	0 – 10	96–99	?	–	
Belgium	10 – 13	90–02	0	0–19	1
Bulgaria	130 – 340	96–02	0	0–19	
Croatia	20 – 30	02	(+)	(N)	70
Cyprus	40 – 140	94–02	(0)	(0–9)	
Czech Rep.	0 – 2	00	F	>80	
France	1,500 – 2,000	97–02	F	20–29	4
Georgia	Present	03	?	–	
Germany	0 – 3	95–99	F	>80	
Greece	1,000 – 3,000	95–00	F	>80	
Hungary	180 – 400	98–02	F	50–79	
Italy	1,700 – 4,000	03	+	20–29	
Macedonia	10 – 50	97–00	F	30–49	
Moldova	20 – 25	90–00	0	0–19	
Netherlands	1 – 32	98–00	F	116	1
Poland	0 – 5	90–00	+	N	2,58,59,60–69
Portugal	2,000 – 5,000	02	(F)	(–)	
Romania	400 – 600	90–02	(+)	(0–19)	53
Russia	3,800 – 12,000	90–00	+	20–29	28,29,60
Serbia & MN	160 – 280	98–02	F	20–39	1,155,148,14,177, 179,29,214
Slovakia	0 – 5	80–99	?	–	
Slovenia	5 – 32	90–00	F	>80	
Spain	(14,000 – 15,500)	98–00	+	0–19	10,5
Canary Is.	0 – 50	97–03	+	N	28,25
Turkey	9,000 – 12,000	01	(–)	(0–19)	
Ukraine	1,500 – 3,300	90–00	+	20–29	
Total (approx.)	37,000 – 64,000	Overall trend Fluctuating			
Breeding range	>1,000,000 km²	Gen. length. 5		% Global pop. 5–24	

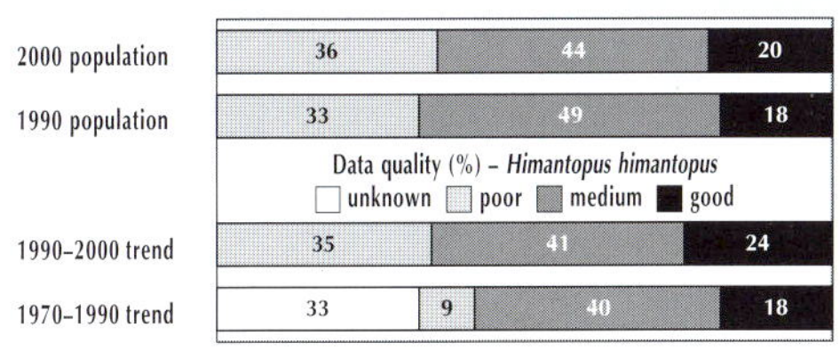

Recurvirostra avosetta
PIED AVOCET

Non-SPEC (1994: 4/3W) **Status** Secure
Criteria —
European IUCN Red List Category —
Criteria —
Global IUCN Red List Category —
Criteria —

Recurvirostra avosetta is a widespread but patchily distributed breeder across much of Europe (except the far north), which accounts for less than half of its global breeding range. Its European breeding population is relatively small (<57,000 pairs), but increased substantially between 1970–1990. Although several populations declined during 1990–2000, most key ones increased or were stable, and the species was stable overall. Improved knowledge of its winter distribution has shown that it no longer qualifies as Localised, and consequently it is now evaluated as Secure.

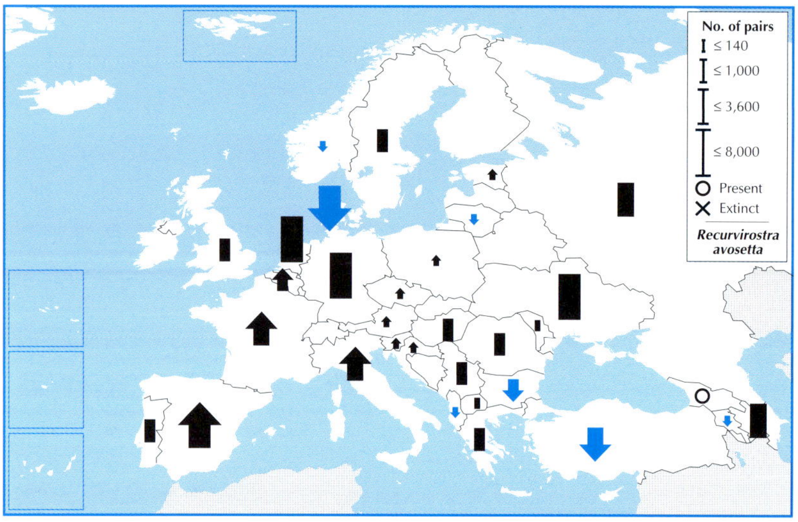

Country	Breeding pop. size (pairs)	Year(s)	Trend	Mag.%	References
Albania	**20 – 40**	96–00	–	0–19	
Armenia	20 – 80	00–02	–	0–9	
Austria	**91 – 198**	98–02	**+**	**>80**	
Azerbaijan	(500 – 5,000)	96–00	(0)	(0–19)	
Belgium	**650 – 690**	95–02	**+**	**0–19**	*1*
Bulgaria	**210 – 530**	96–02	**-**	**20–29**	
Croatia	**1 – 2**	02	**+**	**N**	*26*
Czech Rep.	**1 – 5**	00	**+**	**N**	
Denmark	**4,100 – 4,600**	93–99	–	10–19	*22*
Estonia	100 – 150	98	+	20–29	*1*
France	**2,000 – 3,000**	97–02	**+**	**0–9**	*4*
Georgia	Present	03	?	–	
Germany	**6,100 – 6,500**	95–99	**0**	**0–19**	
Greece	300 – 500	95–00	F	20–29	
Hungary	**400 – 800**	98–02	F	50–79	
Italy	**1,800 – 2,000**	03	**+**	**0–19**	
Lithuania	**10 – 15**	99–01	**-**	**30–49**	*20*
Macedonia	(0 – 10)	97–00	(F)	(30–49)	*7*
Moldova	**0 – 8**	90–00	**0**	**0–19**	
Netherlands	**7,000 – 9,000**	98–00	**0**	**18**	*1*
Norway	**0 – 2**	02	**-**	**>80**	
Poland	**0 – 4**	94–02	**+**	**N**	*60–69*
Portugal	**350 – 500**	02	(F)	(–)	*3*
Romania	300 – 500	90–02	(0)	(0–19)	*53*
Russia	**1,100 – 2,700**	90–00	**0**	**0–19**	*28,29,60*
Serbia & MN	**250 – 350**	98–02	F	10–29	*1,155,98,80,6,15, 83,148,29,214*
Slovakia	**0 – 20**	80–99	F	30–49	
Slovenia	**1 – 1**	00	**+**	**N**	
Spain	5,500 – 6,000	98–02	+	20–29	*10*
Sweden	**900 – 1,100**	99–00	0	0–19	
Turkey	2,500 – 5,000	01	(–)	(0–19)	
Ukraine	3,000 – 7,000	90–00	F	50–79	
UK	**743 – 743**	96–00	**F**	**176**	

Total (approx.) 38,000 – 57,000 **Overall trend** Stable
Breeding range >1,000,000 km² **Gen. length.** 5 **% Global pop.** 25–49
(See p. 108, bottom, for data quality graph)

Country	Winter pop. size (individuals)	Year(s)	Trend	Mag.%	References
Albania	**350 – 1,600**	95–02	**-**	**50–79**	
Armenia	1,000 – 2,499	00–02	0	0–9	
Azerbaijan	1,000 – 16,000	96–02	–	>80	
Belgium	**290 – 950**	99–01	**0**	**0–19**	*1*
France	**17,700 – 24,000**	98–02	**F**	**50–79**	*9*
Greece	**2,000 – 4,900**	95–99	**0**	**0–19**	
Italy	**1,200 – 6,500**	02	**+**	**>80**	*22*
Netherlands	**1,300 – 1,300**	99–01	**0**	**0–19**	*3,4,8,9,11,12*
Portugal	9,300 – 11,900	95–97	(F)	(–)	*19,20,21*
Spain	**5,500 – 24,000**	90–01	**+**	**30–49**	*19*
Turkey	1,500 – 2,500	91–01	**0**	**0–19**	
UK	**3,400 – 3,400**	94–99	**+**	**320**	*44,52*

Total (approx.) >45,000 **Overall trend** Small decline
% in European IBAs 72–96 **Gen. length** 5 **% Global pop.** 5–24

Burhinus oedicnemus
EURASIAN THICK-KNEE

SPEC 3 (1994: 3) **Status** (Vulnerable)
Criteria See IUCN below
European IUCN Red List Category VU
Criteria A2b
Global IUCN Red List Category —
Criteria —

Burhinus oedicnemus is a widespread but patchily distributed breeder in much of the southern half of Europe, which accounts for less than half of its global breeding range. Its European breeding population is relatively small (<78,000 pairs), and underwent a large decline between 1970–1990. Although some populations were stable or increased during 1990–2000, the species continued to decline across much of Europe—including in its Spanish stronghold—and underwent a large decline (>30%) overall. Consequently, it is provisionally evaluated as Vulnerable.

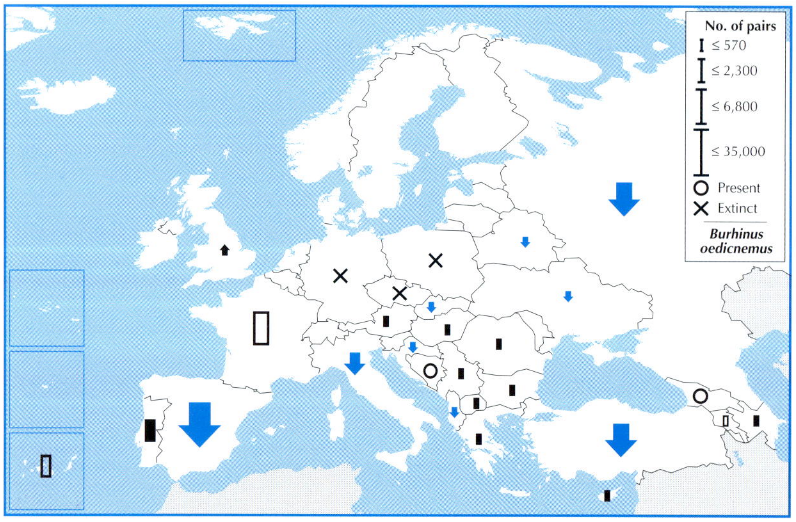

Country	Breeding pop. size (pairs)	Year(s)	Trend	Mag.%	References
Albania	30 – 100	96–02	–	0–19	
Armenia	80 – 300	99–02	?	–	
Austria	**11 – 13**	98–02	**0**	**0–19**	
Azerbaijan	(200 – 500)	96–00	(0)	(0–19)	
Belarus	**1 – 10**	97–00	**-**	**50–79**	
Bosnia & HG	Present	90–03	?	–	
Bulgaria	**130 – 250**	96–02	0	0–9	
Croatia	(30 – 50)	02	(–)	(50–79)	*70,54,26*
Cyprus	(200 – 1,000)	94–02	(0)	(0–9)	
Czech Rep.	**0 – 0**	00	**-**	**X**	
France	5,000 – 9,000	00	?	–	*1*
Georgia	Present	03	?	–	
Germany	**0 – 0**	00	**-**	**X**	
Greece	(300 – 500)	95–00	(0)	(0–19)	
Hungary	150 – 250	95–02	0	0–19	
Italy	(800 – 1,200)	03	(–)	(0–9)	
Macedonia	100 – 300	93–99	F	30–49	
Poland	**0 – 0**	00–03	**-**	**X**	*70,71*
Portugal	(1,000 – 5,000)	02	(0)	(0–19)	
Romania	(400 – 800)	90–02	(0)	(0–19)	*53*
Russia	(3,000 – 10,000)	90–00	–	20–29	*8,28,60,74*
Serbia & MN	**30 – 45**	98–02	F	10–29	*1,29,84,81,148, 227,156,135a*
Slovakia	0 – 5	80–99	–	30–49	
Spain	(30,000 – 40,000)	98–02	(–)	(0–19)	*10,16*
Canary Is.	(1,000 – 2,500)	97–03	?	–	*25*
Turkey	(3,000 – 6,000)	01	(–)	(0–19)	
Ukraine	(100 – 150)	90–00	–	30–49	
UK	**214 – 227**	96–00	**+**	**81**	

Total (approx.) 46,000 – 78,000 **Overall trend** Large decline
Breeding range >2,000,000 km² **Gen. length.** 9 **% Global pop.** 25–49

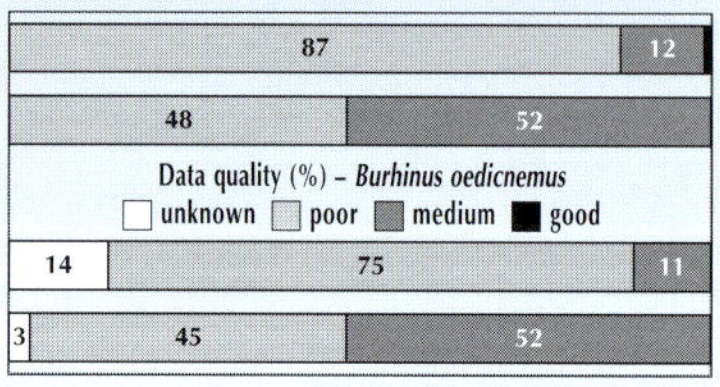

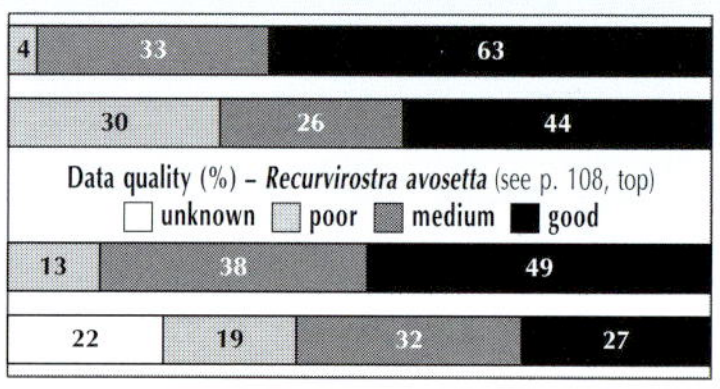

Cursorius cursor
CREAM-COLOURED COURSER

SPEC 3 (1994: 3) **Status** (Endangered)
Criteria See IUCN below
European IUCN Red List Category EN
Criteria D1
Global IUCN Red List Category —
Criteria —

Cursorius cursor has a predominantly African breeding distribution, which just extends into Europe in the Canary Islands, southern Spain and south-eastern Turkey. Its European breeding population is extremely small (as few as 100 pairs), and declined substantially between 1970–1990. Although the trend in its Canary Islands stronghold during 1990–2000 was unknown, this small, isolated population is clearly susceptible to the risks affecting small populations. Consequently, the species is provisionally evaluated as Endangered.

Country	Breeding pop. size (pairs)	Year(s)	Trend	Mag.%	References
Spain	(2 – 2)	98–02	+	N	*10*
Canary Is.	(99 – 594)	97–03	?	–	*25*
Turkey	(0 – 20)	01	(–)	(–)	
Total (approx.)	100 – 620	**Overall trend** Unknown			
Breeding range	<20,000 km²	**Gen. length.** 4		**% Global pop.** <5	

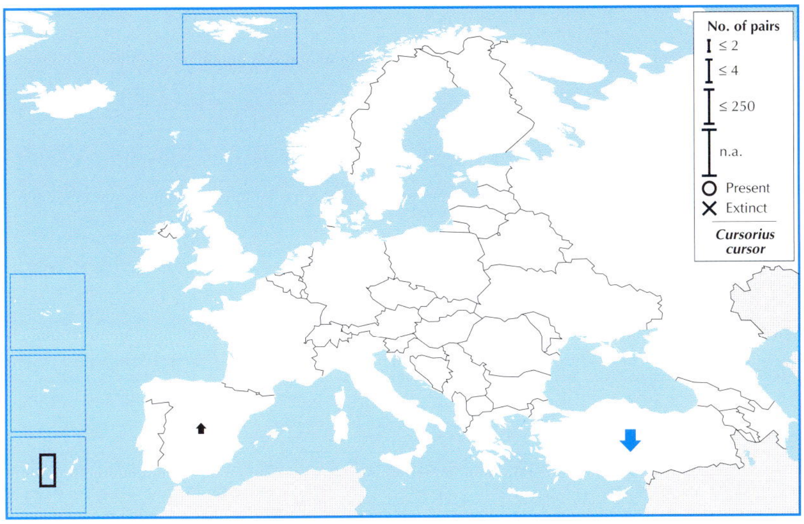

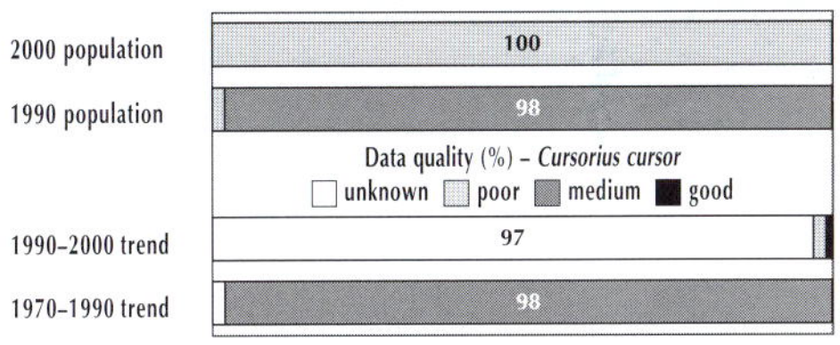

Glareola pratincola
COLLARED PRATINCOLE

SPEC 3 (1994: 3) **Status** Declining
Criteria Moderate continuing decline
European IUCN Red List Category —
Criteria —
Global IUCN Red List Category —
Criteria —

Glareola pratincola is a widespread but patchily distributed summer visitor to southern Europe, which accounts for less than a quarter of its global breeding range. Its European breeding population is relatively small (<18,000 pairs), and underwent a large decline between 1970–1990. Although trends were stable in several countries across its European range during 1990–2000, key populations in Spain and Turkey declined, and the species underwent a moderate decline (>10%) overall. Consequently, it is evaluated as Declining.

Country	Breeding pop. size (pairs)	Year(s)	Trend	Mag.%	References
Albania	100 – 300	96–02	–	0–19	
Armenia	10 – 30	97–02	–	0–9	
Azerbaijan	(500 – 1,000)	96–00	(0)	(0–19)	
Bulgaria	**25 – 55**	96–02	**0**	**0–19**	
France	**41 – 41**	01	**F**	**20–49**	*6*
Greece	(500 – 1,000)	95–00	(–)	(0–19)	
Hungary	**20 – 90**	98–02	(F)	(–)	
Italy	**103 – 130**	00	**0**	**0–19**	*18*
Macedonia	(0 – 5)	98–00	(F)	(–)	
Portugal	(250 – 1,000)	02	(F)	(–)	
Romania	450 – 800	90–02	(0)	(0–19)	*53*
Russia	320 – 1,250	90–00	–	20–29	*8,28,33*
Serbia & MN	**70 – 80**	00–02	F	30–49	*1,148,179,227, 156,135a*
Spain	(4,600 – 4,700)	86–01	(–)	(20–29)	*12,16,10,6*
Turkey	3,000 – 6,000	01	(–)	(0–19)	
Ukraine	120 – 1,200	90–00	F	50–79	
Total (approx.)	10,000 – 18,000	**Overall trend** Moderate decline			
Breeding range	>500,000 km²	**Gen. length.** <3.3		**% Global pop.** 5–24	

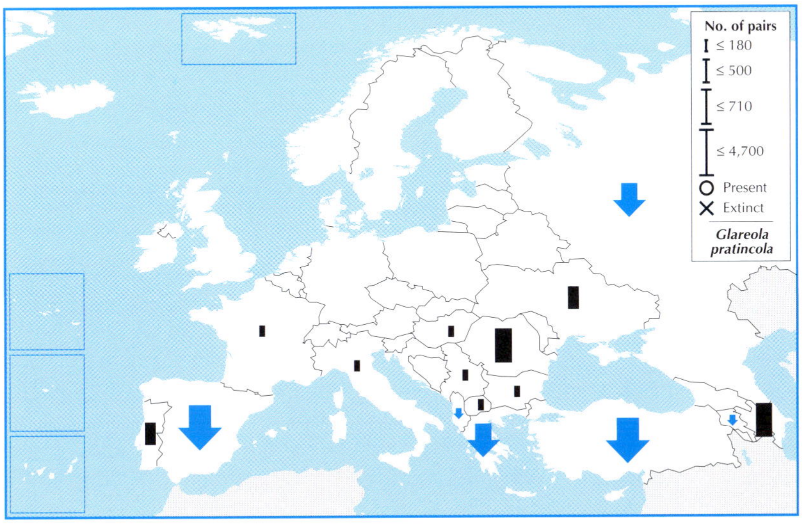

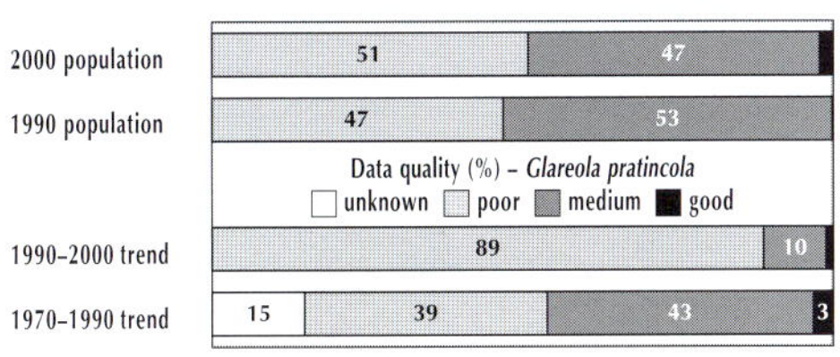

Glareola nordmanni
BLACK-WINGED PRATINCOLE

SPEC 1 (1994: 3) **Status** Endangered
Criteria See IUCN below
European IUCN Red List Category EN
Criteria A2b
Global IUCN Red List Category DD
Criteria DD

Country	Breeding pop. size (pairs)	Year(s)	Trend	Mag.%	References
Bulgaria	(0 – 1)	95–02	?	–	
Hungary	(0 – 1)	98–02	(F)	(–)	*10*
Romania	**0 – 0**	90–02	**–**	**X**	*27*
Russia	**2,500 – 5,000**	90–00	**–**	**50–79**	*156*
Turkey	(0 – 50)	01	(–)	(–)	
Ukraine	0 – 50	90–00	–	50–79	
Total (approx.)	2,500 – 5,100	**Overall trend** Large decline			
Breeding range	>250,000 km²	**Gen. length.** <3.3 **% Global pop.** 25–49			

Glareola nordmanni is a summer visitor to south-east Europe, which accounts for less than half of its global breeding range. Its European breeding population is small (as few as 2,500 pairs), but was stable between 1970–1990. However, the species declined across most of its European range during 1990–2000—most crucially in its Russian stronghold—and underwent a very large decline (>50%) overall. Consequently, this previously Rare species (categorised as Data Deficient at a global level) is now evaluated as Endangered.

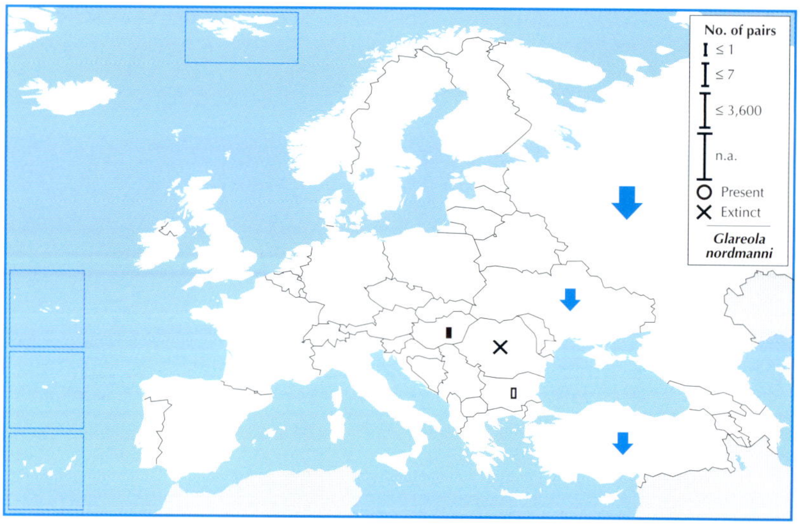

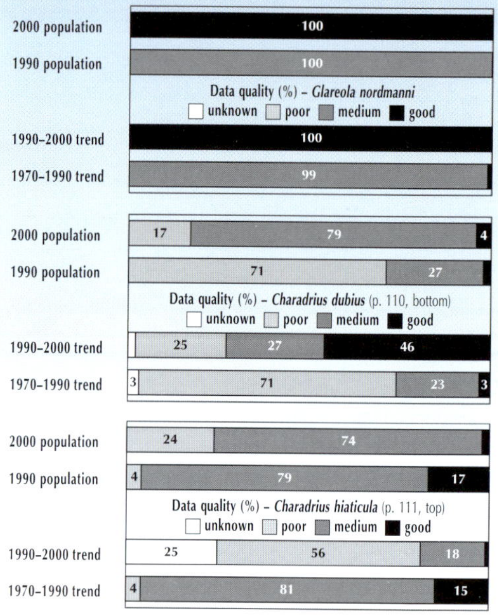

Charadrius dubius
LITTLE RINGED PLOVER

Non-SPEC (1994: —) **Status** (Secure)
Criteria —
European IUCN Red List Category —
Criteria —
Global IUCN Red List Category —
Criteria —

Country	Breeding pop. size (pairs)	Year(s)	Trend	Mag.%	References
Albania	300 – 1,000	96–02	–	0–19	*6*
Armenia	330 – 830	98–02	(0)	(0–19)	
Austria	(300 – 550)	98–02	(0)	(0–19)	
Azerbaijan	(500 – 1,000)	96–00	(0)	(0–19)	
Belarus	8,500 – 12,000	97–02	0	0–19	
Belgium	370 – 500	95–02	0	0–19	*1*
Bosnia & HG	Present	85–89	?	–	
Bulgaria	**700 – 1,200**	96–02	0	0–9	
Croatia	(1,000 – 5,000)	02	(+)	(>80)	*70,16*
Cyprus	0 – 4	94–00	**+**	**N**	
Czech Rep.	**800 – 1,400**	00	**0**	**0–9**	
Denmark	**350 – 500**	93–99	(+)	(20–29)	*22*
Estonia	400 – 800	98	–	50–79	*1*
Finland	4,000 – 5,000	98–02	–	10	
France	6,000 – 7,500	98–02	0	0–19	*4*
Georgia	Present	03	?	–	
Germany	4,300 – 6,800	95–99	(0)	(0–19)	
Greece	(3,000 – 5,000)	95–00	(0)	(0–19)	
Hungary	1,600 – 2,300	95–02	(F)	(20–49)	
Italy	3,000 – 4,000	03	–	0–19	
Latvia	1,500 – 2,000	90–00	0	0–19	*23*
Liechtenstein	**15 – 20**	98–00	**0**	**0–19**	
Lithuania	(3,000 – 5,000)	99–01	(–)	(20–29)	*20*
Luxembourg	**20 – 30**	00–02	**+**	**0–19**	
Macedonia	(200 – 500)	90–00	(0)	(0–19)	
Malta	**1 – 2**	90–02	**F**	**>80**	*1*
Moldova	15 – 30	90–00	–	30–49	
Netherlands	**750 – 1,100**	98–00	**0**	**10**	*1*
Norway	(200 – 300)	90–03	(0)	(0–19)	
Poland	3,000 – 4,000	97–00	(0)	(0–19)	*2*
Portugal	(1,000 – 5,000)	02	(0)	(0–19)	
Romania	(3,000 – 5,000)	90–02	(0)	(0–19)	*32*
Russia	40,000 – 125,000	90–00	**0**	**0–19**	*28,29,60*
Serbia & MN	1,100 – 1,500	90–02	–	0–19	*1,83,175,193, 155,227,135a*
Slovakia	2,000 – 3,000	90–99	0	0–19	
Slovenia	400 – 600	00	(–)	(0–19)	
Spain	(2,500 – 3,300)	98–02	?	–	*10,6*
Canary Is.	50 – 250	97–03	F	20–29	*28,25*
Sweden	**1,300 – 1,600**	99–00	(–)	(0–19)	
Switzerland	**100 – 120**	93–96	**+**	**20–29**	
Turkey	(3,000 – 12,000)	01	(–)	(20–29)	
Ukraine	7,000 – 10,600	90–00	0	0–19	
UK	**825 – 1,070**	88–91	(+)	(30–49)	*8*
Total (approx.)	110,000 – 240,000	**Overall trend** Small decline			
Breeding range	>7,000,000 km²	**Gen. length.** <3.3 **% Global pop.** 25–49			

(See p. 110, top, for data quality graph)

Charadrius dubius is a widespread summer visitor to much of Europe, which accounts for less than half of its global breeding range. Its European breeding population is large (>110,000 pairs), and increased between 1970–1990. Although there were declines in a number of countries—most notably Turkey – during 1990–2000, key populations in countries such as Russia, Belarus and Ukraine were stable, and the species probably declined only slightly overall. Consequently, it is provisionally evaluated as Secure.

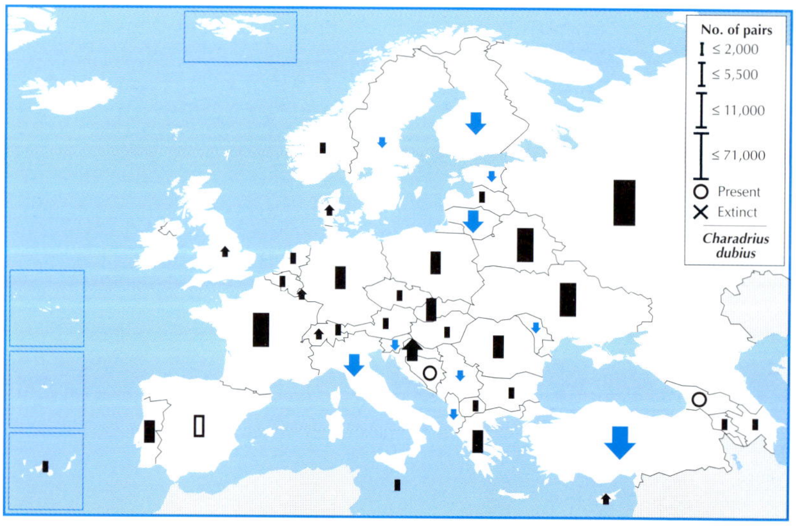

Charadrius hiaticula
COMMON RINGED PLOVER

Non-SPEC[E] (1994: —) **Status** (Secure)
Criteria —
European IUCN Red List Category —
Criteria —
Global IUCN Red List Category —
Criteria —

Charadrius hiaticula is a widespread breeder in northern Europe (occurring more patchily further south), which probably constitutes >50% of its global breeding range. Its European breeding population is large (>120,000 pairs), and was stable between 1970–1990. Although there were declines in a few countries during 1990–2000, the species was stable or increased across most of its European range (the trend in Iceland was unknown), and probably underwent only a small decline overall. Consequently, it is provisionally evaluated as Secure.

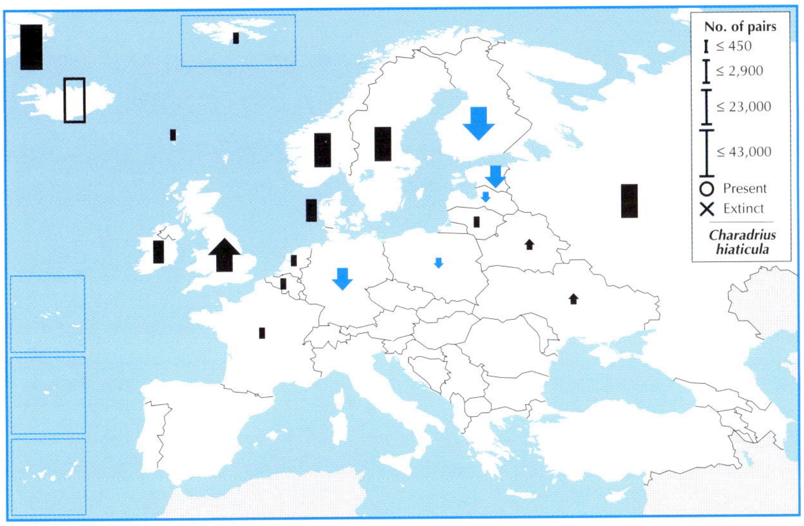

Country	Breeding pop. size (pairs)	Year(s)	Trend	Mag.%	References
Belarus	**150 – 250**	97–00	+	**10–19**	
Belgium	**19 – 21**	00–02	0	0–19	*1*
Denmark	**1,900 – 2,500**	94–01	0	0–19	*22*
Faroe Is.	15 – 15	95	(0)	(0–19)	
Greenland	30,000 – 60,000	95–00	(0)	(0–19)	*11*
Estonia	(2,000 – 4,000)	98	–	20–29	*1*
Finland	8,000 – 11,000	98–02	–	10	
France	**100 – 130**	98–02	0	0–19	*4*
Germany	1,000 – 1,700	95–99	**–**	**20–29**	
Iceland	30,000 – 50,000	99–01	?	–	*41*
Rep. Ireland	1,000 – 2,500	88–91	(0)	(0–19)	
Latvia	60 – 80	90–00	(–)	(0–19)	*20,23,36*
Lithuania	(30 – 50)	99–01	(0)	(0–19)	*20*
Netherlands	**430 – 470**	98–00	**0**	**10**	*1*
Norway	(10,000 – 15,000)	90–03	(0)	(0–19)	
Svalbard	50 – 400	90–03	(0)	(0–19)	
Poland	**350 – 400**	97–00	**–**	**20–29**	*47*
Russia	(13,000 – 40,000)	90–00	(F)	(20–29)	*28,29,60*
Sweden	10,000 – 20,000	99–00	0	0–19	
Ukraine	0 – 100	90–00	+	N	
UK	8,600 – 8,600	84	(+)	(0–19)	*8,26*

Total (approx.) 120,000 – 220,000 **Overall trend** Small decline
Breeding range >3,000,000 km² **Gen. length.** <3.3 **% Global pop.** 50–74
(See p. 110, top, for data quality graph)

Country	Winter pop. size (individuals)	Year(s)	Trend	Mag.%	References
Albania	**0 – 9**	95–02	(F)	(–)	
Azerbaijan	100 – 500	96–02	(0)	(0–19)	
Belgium	20 – 60	99–01	0	0–19	*1*
Cyprus	20 – 200	94–02	(0)	(0–19)	
Denmark	(0 – 500)	99–00	?	–	
Faroe Is.	0 – 5	92	?	–	*3*
France	**12,000 – 15,000**	98–02	**F**	**30–49**	*9*
Germany	**0 – 90**	95–00	?	–	*2*
Greece	**1 – 15**	95–99	(0)	(0–19)	
Rep. Ireland	**11,000 – 12,000**	94–00	0	0–19	
Italy	100 – 200	02	0	0–19	*22*
Netherlands	**390 – 390**	99–01	(F)	(–)	*3,4,8,9,11,12*
Portugal	1,800 – 6,000	95–97	(F)	(–)	*19,20,21*
Spain	**950 – 6,400**	90–01	+	**>80**	*19*
Turkey	100 – 500	91–01	**F**	**>80**	
UK	**35,000 – 35,000**	94–99	**–**	**28**	*44,52*

Total (approx.) >62,000 **Overall trend** Small decline
% in European IBAs 58–61 **Gen. length** <3.3 **% Global pop.** 5–24

Charadrius alexandrinus
KENTISH PLOVER

SPEC 3 (1994: 3) **Status** (Declining)
Criteria Moderate continuing decline
European IUCN Red List Category —
Criteria —
Global IUCN Red List Category —
Criteria —

Charadrius alexandrinus is a widespread breeder in coastal areas of western and southern Europe, which accounts for less than a quarter of its global breeding range. Its European breeding population is relatively small (<35,000 pairs), and underwent a moderate decline between 1970–1990. Although some populations were stable or increased during 1990–2000, the species declined across most of its European range—most notably in Spain and Turkey—and underwent a moderate decline (>10%) overall. Consequently, it is provisionally evaluated as Declining.

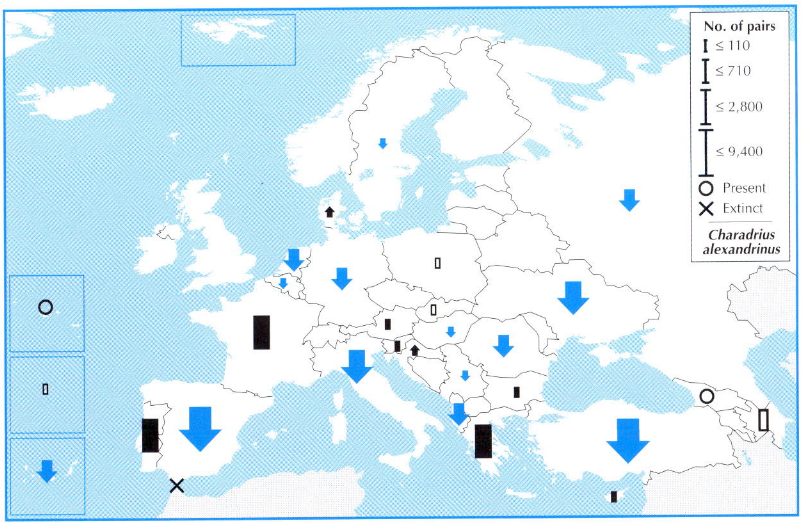

Country	Breeding pop. size (pairs)	Year(s)	Trend	Mag.%	References
Albania	150 – 500	96–02	–	0–19	*6*
Austria	**25 – 37**	98–02	**0**	**0–19**	
Azerbaijan	(500 – 1,000)	96–00	?	–	
Belgium	**80 – 85**	00–02	**–**	**0–19**	*1*
Bulgaria	**60 – 100**	95–02	**0**	**0–19**	
Croatia	27 – 34	02	(+)	(>80)	*70*
Cyprus	(60 – 120)	94–02	(0)	(0–9)	
Denmark	**100 – 115**	99–01	+	**>80**	*5,6,7*
France	**1,200 – 1,500**	98–02	0	0–19	*4*
Georgia	Present	03	?	–	
Germany	**380 – 400**	95–99	**–**	**50–79**	
Greece	(1,000 – 2,000)	95–00	(0)	(0–19)	
Hungary	15 – 30	98–02	**–**	**50–79**	
Italy	(1,400 – 2,000)	03	(–)	(0–19)	
Netherlands	**270 – 320**	98–00	**–**	**40**	*1*
Poland	**0 – 1**	90–03	?	–	*48*
Portugal	(1,500 – 5,000)	02	(0)	(0–19)	
Azores	Present	02	?	–	
Madeira	(0 – 50)	02	?	–	
Romania	(400 – 700)	90–02	(–)	(0–19)	*53*
Russia	(150 – 1,300)	90–00	(–)	(20–29)	*28,29,60*
Serbia & MN	**60 – 80**	97–02	–	0–19	*1,29,148,155,156, 101,227,135a*
Slovakia	**0 – 1**	80–99	?	–	
Slovenia	**15 – 40**	90–00	**F**	**>80**	
Spain	(5,200 – 5,700)	86–01	(–)	(0–19)	*1,16,10,6*
Canary Is.	250 – 1,000	97–03	–	0–19	*28,25*
Sweden	**1 – 1**	99–00	**–**	**50–79**	
Turkey	(8,000 – 11,000)	01	–	0–19	
Ukraine	1600 – 2,000	90–00	–	30–49	
UK					
Gibraltar	**0 – 0**	00	**–**	**X**	

Total (approx.) 22,000 – 35,000 **Overall trend** Moderate decline
Breeding range >1,000,000 km² **Gen. length.** <3.3 **% Global pop.** 5–24
(See p. 112, top, for data quality graph)

Country	Winter pop. size (individuals)	Year(s)	Trend	Mag.%	References
Greece	(1,650 – 5,000)	95–00	(0)	(0–19)	
Italy	(2,500 – 3,500)	02	0	0–19	*22*
Portugal	1,000 – 4,400	98–02	(F)	(–)	*22*
Spain	**980 – 6,900**	90–01	+	**>80**	*19*
Turkey	(1,500 – 3,000)	91–01	(–)	(20–29)	

Total (approx.) >8,100 **Overall trend** Stable
% in European IBAs >70 **Gen. length** <3.3 **% Global pop.** <5

Charadrius leschenaultii
GREATER SAND PLOVER

SPEC 3 (1994: 3) **Status** (Endangered)
Criteria See IUCN below
European IUCN Red List Category EN
Criteria C1
Global IUCN Red List Category —
Criteria —

Country	Breeding pop. size (pairs)	Year(s)	Trend	Mag.%	References
Russia	(200 – 600)	90–00	?	–	*28,29,60*
Turkey	800 – 1,200	01	(-)	(20–29)	
Total (approx.)	1,000 – 1,800	**Overall trend** Moderate decline			
Breeding range	>20,000 km²	**Gen. length.** 5		**% Global pop.** <5	

Charadrius leschenaultii has a predominantly Asian distribution, although it also occurs as a summer visitor to parts of Turkey and southern Russia. Its European breeding population is very small (as few as 1,000 pairs), but the trend between 1970–1990 was unknown. Although the trend of the Russian population during 1990–2000 was also unknown, the species underwent a moderate decline in Turkey, and probably underwent a moderate decline (>10%) overall. As a consequence of this continuing decline and its small population, the species is provisionally evaluated as Endangered.

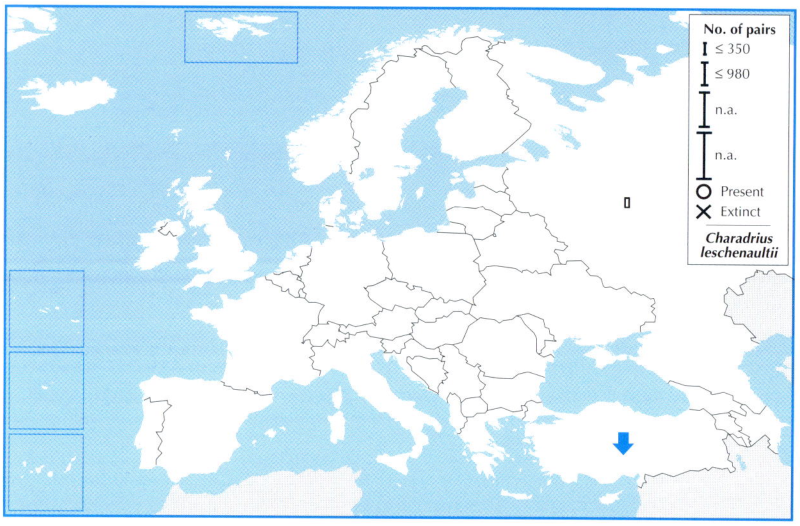

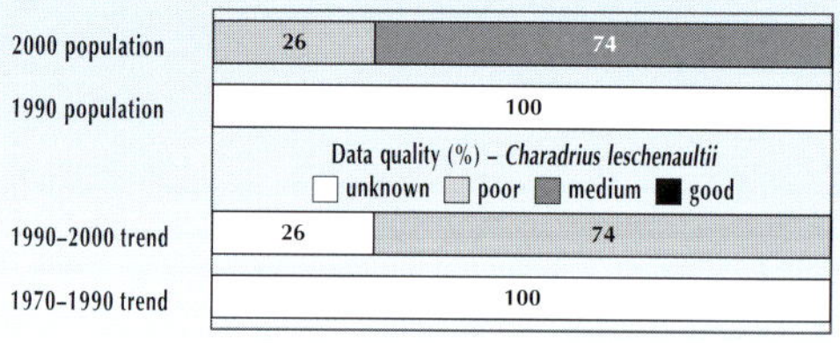

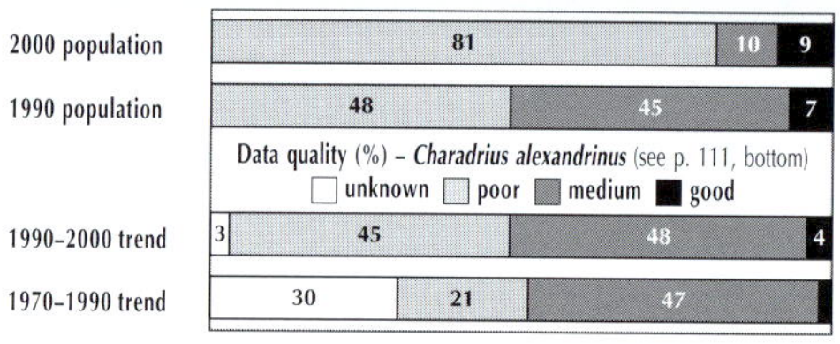

Charadrius asiaticus
CASPIAN PLOVER

SPEC 3 (1994: 3) **Status** Endangered
Criteria See IUCN below
European IUCN Red List Category EN
Criteria C2a(ii)
Global IUCN Red List Category —
Criteria —

Country	Breeding pop. size (pairs)	Year(s)	Trend	Mag.%	References
Russia	(130 – 500)	90–00	?	–	*28,29,60*
Total (approx.)	130 – 500	**Overall trend** Unknown			
Breeding range	>50,000 km²	**Gen. length.** 5		**% Global pop.** <5	

Charadrius asiaticus has a predominantly Central Asian breeding distribution, which just extends into Europe in southern Russia. Its European breeding population is very small (as few as 130 pairs), and declined substantially between 1970–1990. Although its trend in Russia during 1990–2000 was not known, there is no evidence to suggest that earlier declines ceased, and the species probably continued to decline. As a consequence of this inferred continuing decline and its small population size, the species is provisionally evaluated as Endangered.

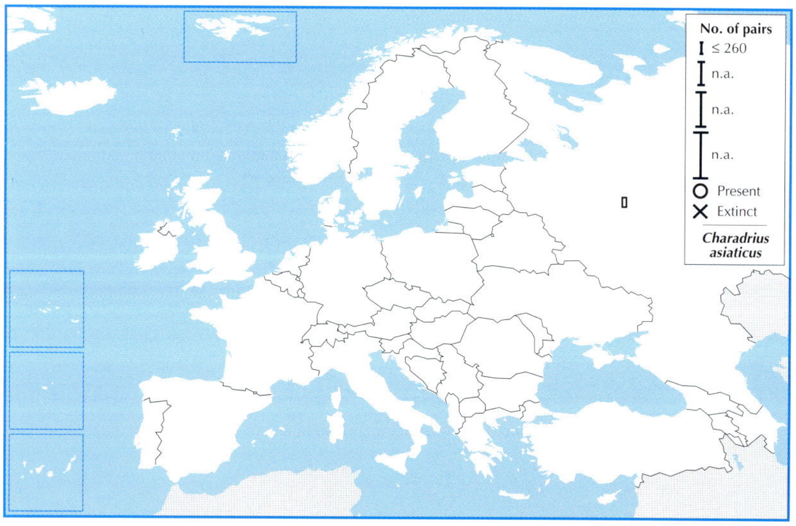

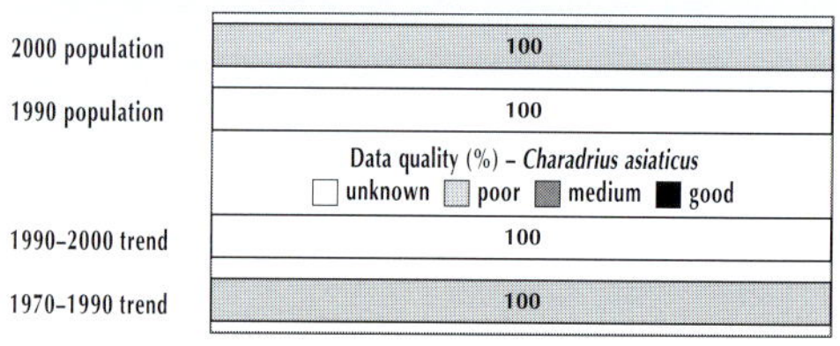

Eudromias morinellus
EURASIAN DOTTEREL

Non-SPEC (1994: —) **Status** (Secure)
Criteria —
European IUCN Red List Category —
Criteria —
Global IUCN Red List Category —
Criteria —

Eudromias morinellus is a summer visitor to northern Europe, but occurs more patchily farther south, with Europe accounting for less than half of its global breeding range. Its European breeding population is relatively small (<42,000 pairs), but was stable between 1970–1990. Although there were declines in Finland and the United Kingdom during 1990–2000, key populations in Norway and Sweden were stable (though the trend in Russia was unknown), and the species probably declined only slightly overall. Consequently, it is provisionally evaluated as Secure.

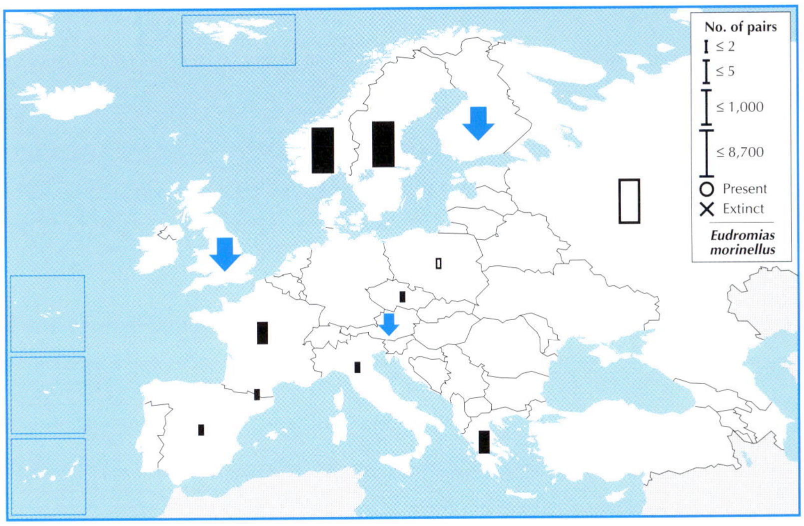

Country	Breeding pop. size (pairs)	Year(s)	Trend	Mag.%	References
Andorra	**0 – 2**	99–01	0	0–19	*1,3*
Austria	4 – 6	98–02	-	30–49	
Czech Rep.	**0 – 1**	00	**0**	**0–19**	
Finland	500 – 2,000	98–02	-	40	
France	2 – 10	98–02	0	0–19	*4*
Greece	(2 – 10)	95–00	(0)	(0–19)	
Italy	1 – 5	03	0	0–19	
Norway	(5,000 – 15,000)	90–03	(0)	(0–19)	
Poland	0 – 1	90–00	?	–	*2,49*
Russia	(2,000 – 14,000)	90–00	?	–	*28,29,60*
Spain	0 – 5	98–02	(F)	(–)	*10,16*
Sweden	3,000 – 10,000	99–00	0	0–19	
UK	510 – 750	99	**-**	**24**	35
Total (approx.)	11,000 – 42,000	**Overall trend** Small decline			
Breeding range	>500,000 km²	**Gen. length.** 5		**% Global pop.** 25–49	

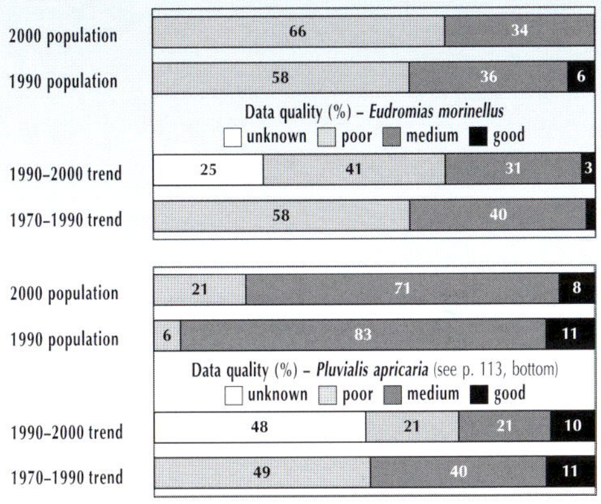

Pluvialis apricaria
EURASIAN GOLDEN-PLOVER

Non-SPEC[E] (1994: 4) **Status** (Secure)
Criteria —
European IUCN Red List Category —
Criteria —
Global IUCN Red List Category —
Criteria —

Pluvialis apricaria is a widespread breeder in northern Europe, which constitutes >50% of its global breeding range. Its European breeding population is large (>460,000 pairs), and was broadly stable between 1970–1990. Although the trend of the stronghold population in Iceland during 1990–2000 was unknown, the small declines in Sweden and the United Kingdom were at least partly compensated for by increases in Finland, and the species probably declined only slightly overall. Consequently, it is provisionally evaluated as Secure.

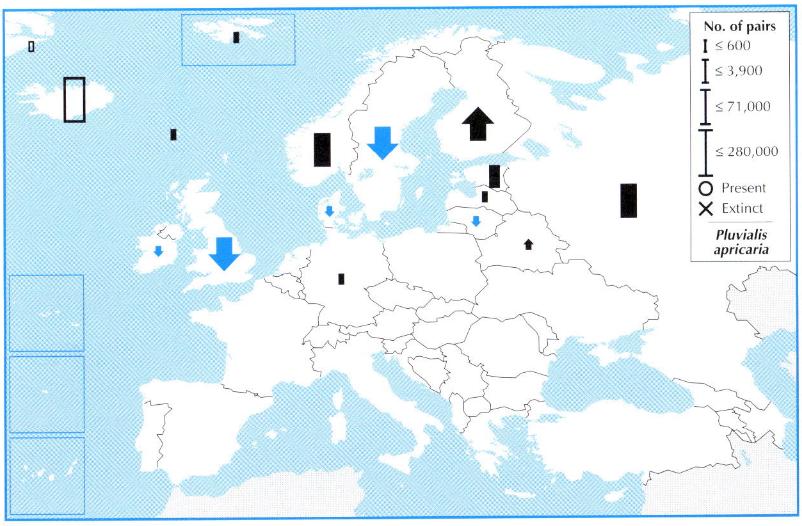

Country	Breeding pop. size (pairs)	Year(s)	Trend	Mag.%	References
Belarus	110 – 140	97–00	+	0–9	
Denmark	**4 – 5**	01	**-**	**50–79**	*7,1*
Faroe Is.	600 – 600	95	(0)	(0–19)	
Greenland	(25 – 50)	80–90	?	–	
Estonia	3,000 – 5,000	98	0	0–19	*1*
Finland	40,000 – 80,000	98–02	**+**	**10**	
Germany	**22 – 22**	95–99	**0**	**0–19**	
Iceland	250,000 – 310,000	99–01	?	–	*41*
Rep. Ireland	200 – 400	88–91	-	30–49	
Latvia	350 – 450	90–00	0	0–19	*23*
Lithuania	**40 – 50**	99–01	**-**	**0–19**	*20*
Norway	(50,000 – 100,000)	90–03	(0)	(0–19)	
Svalbard	1 – 10	90–03	(0)	(0–19)	
Russia	(24,000 – 95,000)	90–00	(0)	(0–19)	*28,29,60*
Sweden	50,000 – 90,000	99–00	-	0–9	
UK	**38,400 – 59,400**	80–00	-	12	*22,31*
Total (approx.)	460,000 – 740,000	**Overall trend** Unknown			
Breeding range	>2,000,000 km²	**Gen. length.** 4		**% Global pop.** 50–74	

(See p. 113, top, for data quality graph)

Country	Winter pop. size (individuals)	Year(s)	Trend	Mag.%	References
Albania	**240 – 2,300**	95–02	(F)	( )	
Azerbaijan	350 – 1,500	96–02	(0)	(0–19)	
Belgium	400 – 400	95–00	(0)	(0–19)	*1*
Bulgaria	0 – 50	97–01	F	>80	
Cyprus	(500 – 5,000)	94–02	(0)	(0–19)	
Denmark	**1,500 – 2,500**	99–00	(–)	(0–19)	*25*
Faroe Is.	0 – 500	92	?	–	*3*
France	(200,000 – 1,000,000)	98–02	?	–	*9*
Germany	(2,500 – 10,000)	95–00	(–)	(20–29)	*2*
Greece	**500 – 3,000**	95–99	F	>80	
Rep. Ireland	**100,000 – 120,000**	94–00	F	20–29	
Italy	(3,000 – 10,000)	02	(+)	(20–29)	*22*
Netherlands	50,000 – 60,000	99–01	**0**	**0–19**	*3,4,8,9,11,12*
Portugal	(130,000 – 300,000)	99–02	(F)	(–)	*23*
Serbia & MN	**200 – 1,000**	90–02	F	40–59	
Slovakia	0 – 20	90–99	?	–	*4*
Slovenia	0 – 2	90–00	F	0–100	
Spain	(20,000 – 100,000)	90–01	?	–	
Turkey	900 – 1,800	91–01	(0)	(0–19)	
Ukraine	(0 – 50)	90–00	(F)	(20–29)	
UK	310,000 – 310,000	81–92	+	109	*52*
Total (approx.)	>820,000	**Overall trend** Moderate increase			
% in European IBAs	17–36	**Gen. length** 4		**% Global pop.** 75–94	

Pluvialis squatarola
GREY PLOVER

Non-SPEC (1994: —) **Status** (Secure)
Criteria —
European IUCN Red List Category —▼
Criteria —
Global IUCN Red List Category —
Criteria —

Country	Breeding pop. size (pairs)	Year(s)	Trend	Mag.%	References
Russia	(2,100 – 10,500)	91–00	?	–	*28,29,60*
Total (approx.)	2,100 – 11,000	**Overall trend** Unknown			
Breeding range	>100,000 km²	**Gen. length.** 5		**% Global pop.** <5	

Country	Winter pop. size (individuals)	Year(s)	Trend	Mag.%	References
Albania	**300 – 1,400**	95–02	**+**	**30–49**	
Azerbaijan	100 – 1,500	96–02	(0)	(0–19)	
Belgium	**250 – 290**	99–01	0	0–19	*1*
Bulgaria	**14 – 171**	97–01	**F**	**>80**	
Croatia	(50 – 80)	02	?	–	*70*
Cyprus	10 – 100	94–02	(0)	(0–9)	
Denmark	**250 – 300**	99–00	(+)	(20–29)	*25*
France	**31,196 – 50,000**	98–02	**+**	**50–79**	*9*
Germany	**2,500 – 30,000**	95–00	**F**	**>80**	*2*
Greece	**500 – 1,000**	95–99	**F**	**30–49**	
Rep. Ireland	**5,500 – 6,600**	94–00	**0**	**0–19**	
Italy	(500 – 4,000)	02	(+)	(0–19)	22
Netherlands	**19,000 – 19,000**	99–01	**0**	**0–19**	*3,4,8,9,11,12*
Portugal	4,120 – 9,100	95–97	(F)	(–)	*19,20,21*
Serbia & MN	**50 – 200**	90–02	F	30–49	
Slovenia	0 – 10	90–00	F	0–100	
Spain	**2,300 – 7,200**	90–01	**+**	**>80**	*19*
Turkey	300 – 900	91–01	**F**	**>80**	
Ukraine	(0 – 50)	90–00	(F)	(20–29)	
UK	**53,300 – 53,300**	94–99	**0**	**1**	*44,52*
Total (approx.)	>120,000	**Overall trend** Moderate increase			
% in European IBAs	66–78	**Gen. length** 5		**% Global pop.** 5–24	

Pluvialis squatarola breeds in arctic Russia, with Europe accounting for a tiny proportion of its global breeding range. Its European breeding population is small (as few as 2,100 pairs), but was stable between 1970–1990. No trend data were available for 1990–2000, but there is no evidence to suggest that the species declined. Although the size of the European population could render it susceptible to the risks affecting small populations, it is marginal to a much larger non-European population. Consequently, the species is provisionally evaluated as Secure.

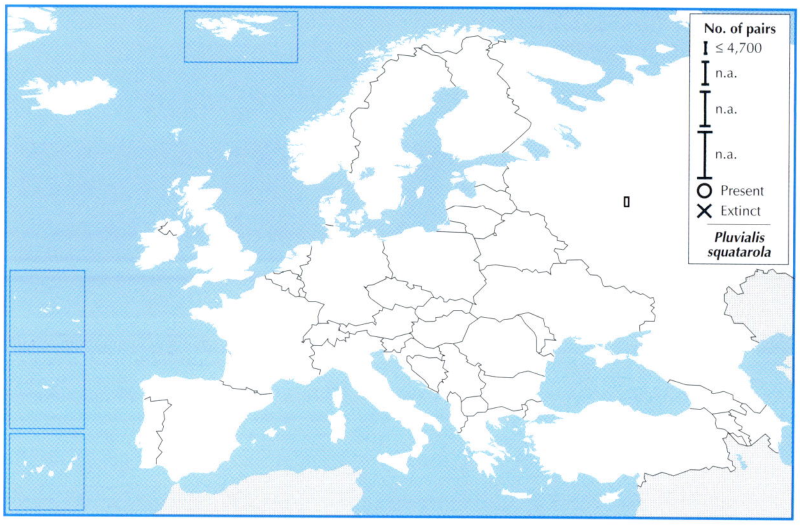

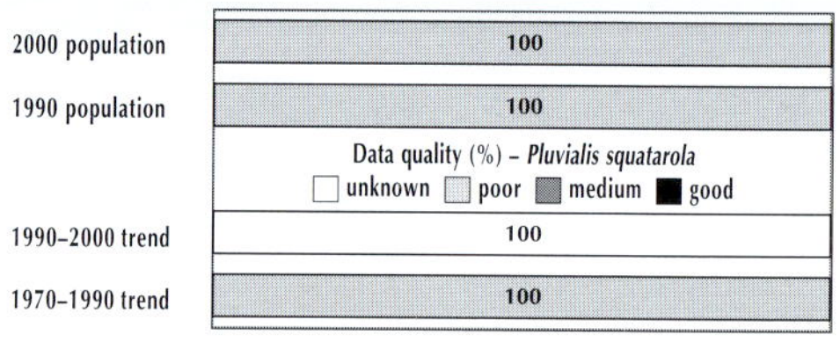

Vanellus spinosus
SPUR-WINGED LAPWING

SPEC 3 (1994: 3) **Status** Vulnerable
Criteria See IUCN below
European IUCN Red List Category VU
Criteria C1; C2a(i)
Global IUCN Red List Category —
Criteria —

Country	Breeding pop. size (pairs)	Year(s)	Trend	Mag.%	References
Cyprus	20 – 60	94–02	0	0–9	
Greece	20 – 50	95–00	(0)	(0–19)	
Turkey	1,000 – 1,500	01	**-**	**10–19**	
Total (approx.)	1,000 – 1,600	**Overall trend** Moderate decline			
Breeding range	>50,000 km²	**Gen. length.** 5		**% Global pop.** <5	

Vanellus spinosus has a predominantly African distribution, but is also a summer visitor to parts of Greece, Turkey and Cyprus. Its European breeding population is very small (as few as 1,000 pairs), and underwent a large decline between 1970–1990. Although the small populations in Cyprus and Greece were stable during 1990–2000, the species declined in its Turkish stronghold, and underwent a moderate decline (>10%) overall. As a consequence of this continuing decline and its small population, the species is evaluated as Vulnerable.

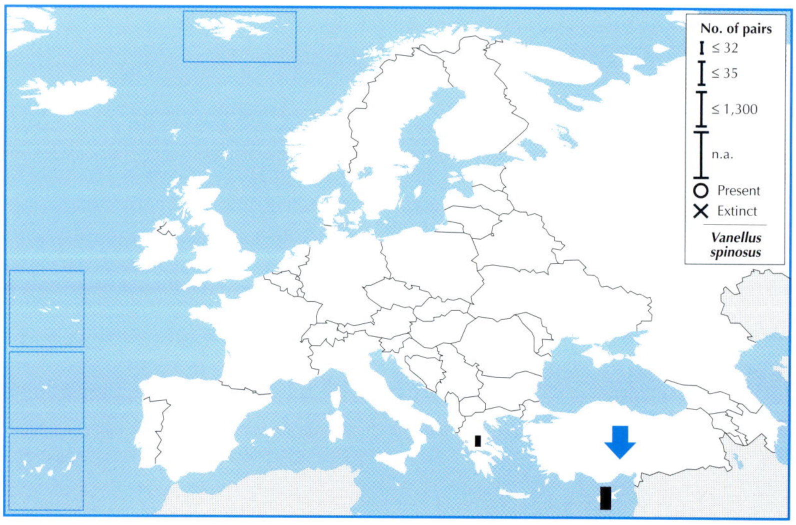

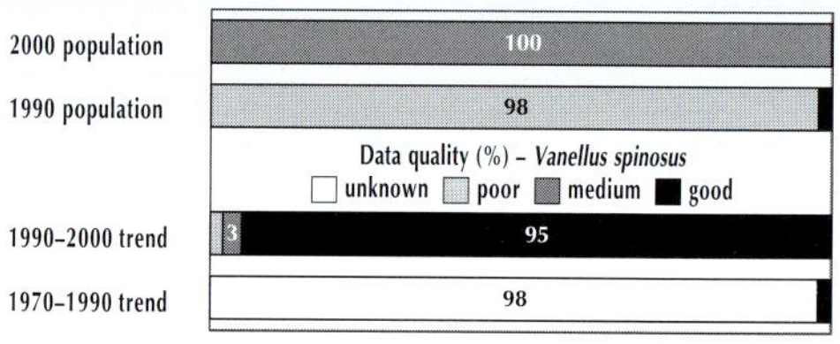

Vanellus indicus
RED-WATTLED LAPWING

SPEC 3 (1994: —) **Status** (Vulnerable)
Criteria See IUCN below
European IUCN Red List Category VU
Criteria A3c
Global IUCN Red List Category —
Criteria —

Country	Breeding pop. size (pairs)	Year(s)	Trend	Mag.%	References
Turkey	40 – 80	01	(0)	(0-19)	
Total (approx.)	40 – 80	Overall trend Stable			
Breeding range	<20,000 km²	Gen. length 5		% Global pop. <5	

Country	Winter pop. size (individuals)	Year(s)	Trend	Mag.%	References
Turkey	(0 – 50)	91-01	?	–	
Total (approx.)	<50	Overall trend Unknown			
% in European IBAs	Unknown	Gen. length 5		% Global pop. <5	

Vanellus indicus has a predominantly Asian distribution, which just extends into Europe in extreme south-eastern Turkey. Its European breeding population is extremely small (as few as 40 pairs), but increased substantially between 1970–1990. The species was stable in Turkey during 1990–2000, and hence would qualify as provisionally Secure. However, it is predicted to undergo a large future decline (>30%) owing to habitat loss resulting from dam construction. Consequently, this previously Secure species is now provisionally evaluated as Vulnerable.

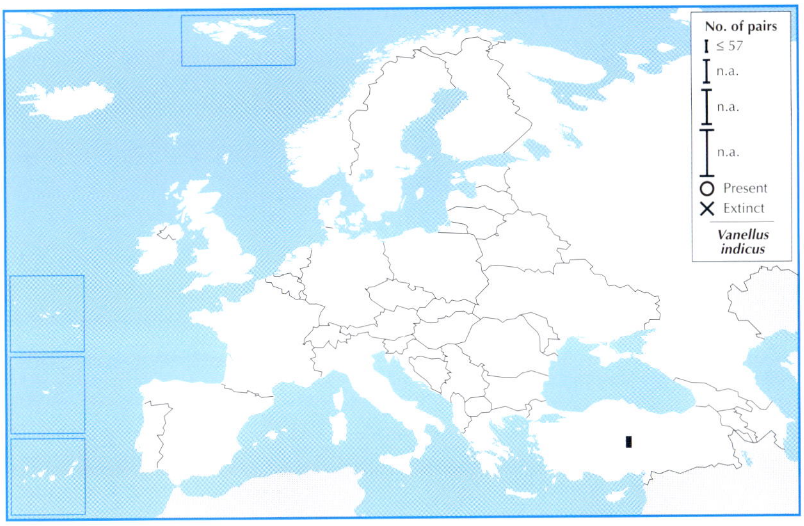

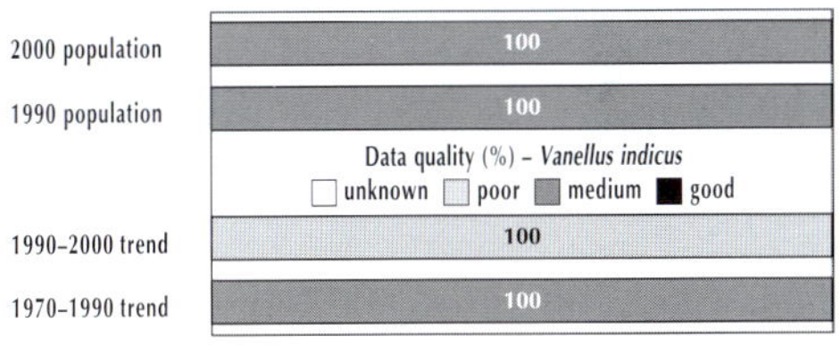

Vanellus gregarius
SOCIABLE LAPWING

SPEC 1 (1994: 1) **Status** Critically Endangered
Criteria See IUCN below
European IUCN Red List Category CR
Criteria A2b; C1
Global IUCN Red List Category CR
Criteria A3b,c

Country	Breeding pop. size (pairs)	Year(s)	Trend	Mag.%	References
Russia	**25 – 80**	90-00	**–**	**>80**	*28,29,60,168*
Total (approx.)	25 – 80	Overall trend Large decline			
Breeding range	>100,000 km²	Gen. length 5		% Global pop. 5–24	

Country	Winter pop. size (individuals)	Year(s)	Trend	Mag.%	References
Azerbaijan	0 – 10	96-02	(0)	(0-19)	
Total (approx.)	<10	Overall Trend Unknown			
% in European IBAs	Unknown	Gen. length 5		% Global pop. <5	

Vanellus gregarius breeds mainly in Kazakhstan, but its global breeding range just extends into Europe in southern Russia. Its European breeding population is extremely small (as few as 25 pairs), and underwent a large decline between 1970–1990. Subsequently, the Russian population underwent an extremely large decline (>80%) during 1990–2000. As a consequence of this continuing decline and its extremely small population, this globally threatened species, which was previously assessed as Endangered in Europe, is now evaluated as Critically Endangered.

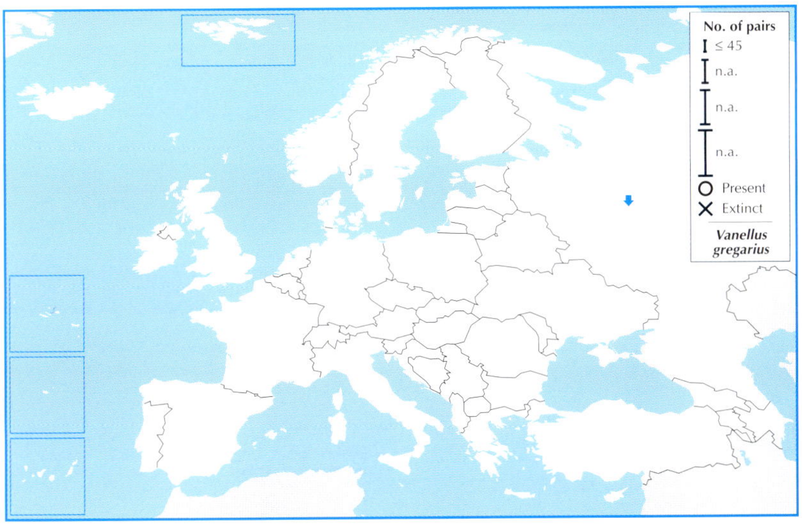

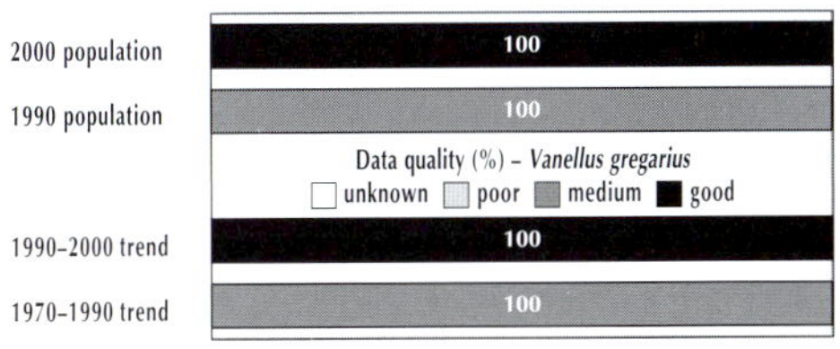

Vanellus leucurus
WHITE-TAILED LAPWING

Non-SPEC (1994: —) **Status** Secure
Criteria —
European IUCN Red List Category —▼▼▼
Criteria —
Global IUCN Red List Category —
Criteria —

Country	Breeding pop. size (pairs)	Year(s)	Trend	Mag.%	References
Armenia	**2 – 15**	99–02	**0**	**0–19**	*10*
Azerbaijan	50 – 150	96–00	(+)	(10–19)	
Romania	**0 – 23**	90–02	+	N	*29*
Russia	(30 – 120)	90–00	+	20–29	*33*
Turkey	0 – 10	01	–	30–49	
Ukraine	**0 – 2**	98–00	+	**>80**	
Total (approx.)	80 – 320	**Overall trend** Moderate increase			
Breeding range	>20,000 km²	**Gen. length.** 5		**% Global pop.** <5	

Country	Winter pop. size (individuals)	Year(s)	Trend	Mag.%	References
Azerbaijan	(10 – 100)	96–02	(0)	(0–19)	
Total (approx.)	>10	**Overall trend** Unknown			
% in European IBAs	Unknown	**Gen. length** 5		**% Global pop.** <5	

Vanellus leucurus has a predominantly Central Asian breeding distribution, which just extends into south-eastern Europe. Its European breeding population is extremely small (as few as 80 pairs), and though its trend between 1970–1990 was unknown, the species increased or was stable in most of its European range during 1990–2000, and underwent a moderate increase overall. Although the size of the European population could render it susceptible to the risks affecting small populations, it is marginal to a much a larger non-European population. Consequently, it is evaluated as Secure.

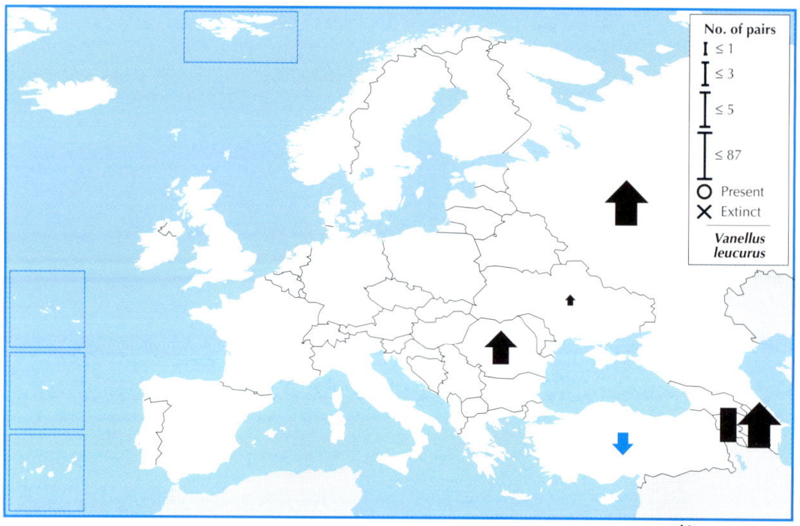

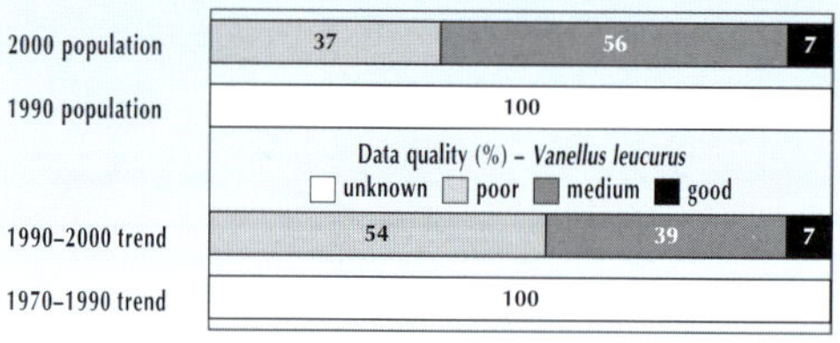

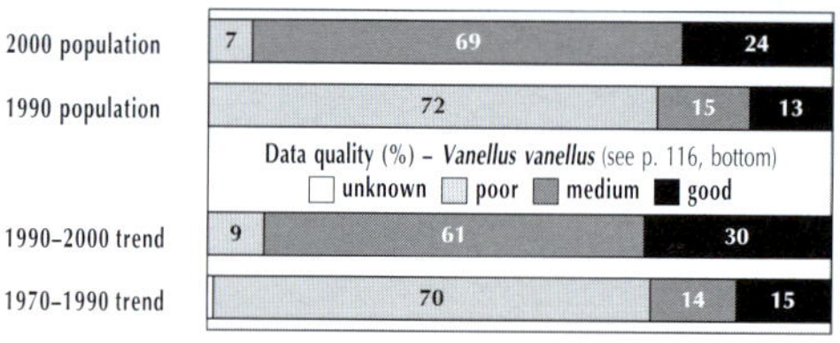

Vanellus vanellus
NORTHERN LAPWING

SPEC 2 (1994: —) **Status** Vulnerable
Criteria See IUCN below
European IUCN Red List Category VU
Criteria A2b; A3b,c
Global IUCN Red List Category —
Criteria —

Country	Breeding pop. size (pairs)	Year(s)	Trend	Mag.%	References
Albania	10 – 45	96–02	–	0–19	*2,6*
Armenia	350 – 850	98–02	(–)	(–)	
Austria	3,000 – 6,000	98–02	(0)	(0–19)	
Azerbaijan	(5,00 – 5,000)	96–00	(0)	(0–19)	
Belarus	100,000 – 160,000	97–00	**–**	**0–9**	
Belgium	17,000 – 24,000	01–02	(0)	(0–19)	*1*
Bosnia & HG	Present	85–89	?	–	
Bulgaria	600 – 1,000	96–02	+	0–19	
Croatia	(4,000 – 5,000)	02	(0)	(0–19)	*54*
Czech Rep.	**7,000 – 10,000**	00	**–**	**50–79**	
Denmark	30,000 – 45,000	95–00	**–**	**30**	*22*
Faroe Is.	10 – 10	95	(0)	(0–19)	
Estonia	(15,000 – 30,000)	98	0	0–19	*1*
Finland	50,000 – 80,000	98–02	0	0–19	
France	17,000 – 20,000	98–02	(–)	(18)	*4,2*
Georgia	Present	03	?	–	
Germany	67,000 – 104,000	95–99	–	50–79	
Greece	(50 – 100)	95–00	(–)	(0–19)	
Hungary	93,000 – 150,000	99–02	–	0–19	*19*
Iceland	**0 – 2**	90–00	(F)	(–)	*32*
Rep. Ireland	(2,500 – 10,000)	88–91	(–)	(>80)	
Italy	1,700 – 1,900	03	+	0–19	
Latvia	12,000 – 15,000	90–00	(0)	(0–19)	*23,16,2*
Liechtenstein	**0 – 1**	98–00	**–**	**30–49**	
Lithuania	(18,000 – 20,000)	99–01	(–)	(30–49)	*20*
Luxembourg	**20 – 30**	98–02	**–**	**80–89**	
Macedonia	50 – 150	90–00	F	30–49	
Moldova	350 – 470	90–00	+	20–29	
Netherlands	**200,000 – 300,000**	98–00	**–**	**8**	*1,2*
Norway	(40,000 – 80,000)	90–03	(–)	(0–19)	*37*
Poland	**100,000 – 150,000**	00–02	–	0–19	*23*
Portugal	0 – 25	02	+	**N**	
Romania	(40,000 – 60,000)	90–02	(–)	(0–19)	*48*
Russia	600,000 – 1,100,000	90–00	–	20–29	*28,29,60*
Serbia & MN	2,000 – 2,500	95–02	–	0–19	*1,135a,227,155, 100,101,78,204*
Slovakia	2,500 – 5,000	80–99	–	30–49	
Slovenia	2,000 – 3,000	94	(0)	(0–19)	
Spain	(1,000 – 2,500)	98–02	(F)	(–)	*10,16*
Sweden	50,000 – 100,000	99–00	**0**	**0**	
Switzerland	250 – 400	98–02	–	30–49	
Turkey	10,000 – 20,000	01	–	0–19	
Ukraine	65,000 – 124,000	90–00	–	0–19	
UK	**137,000 – 174,000**	00	**–**	**15**	*5,23*
Total (approx.)	1,700,000 – 2,800,000	**Overall trend** Large decline			
Breeding range	>7,000,000 km²	**Gen. length.** 5		**% Global pop.** 50–74	

(See p. 116, top, for data quality graph)

	Winter pop. size (individuals)				
Total (approx.)	>2,800,000	**Overall trend** Small increase			
% in European IBAs	19–27	**Gen. length** 5		**% Global pop.** 50–74	

Vanellus vanellus is a widespread breeder across most of Europe, which probably holds >50% of its global population. Its European breeding population is very large (>1,700,000 pairs), and was stable between 1970–1990. Although several small populations were stable or increased during 1990–2000, the species suffered declines across much of Europe—most notably sizeable populations in the United Kingdom, Netherlands and Russia—and underwent a large decline (>30%) overall. Consequently, this previously Secure species is now evaluated as Vulnerable.

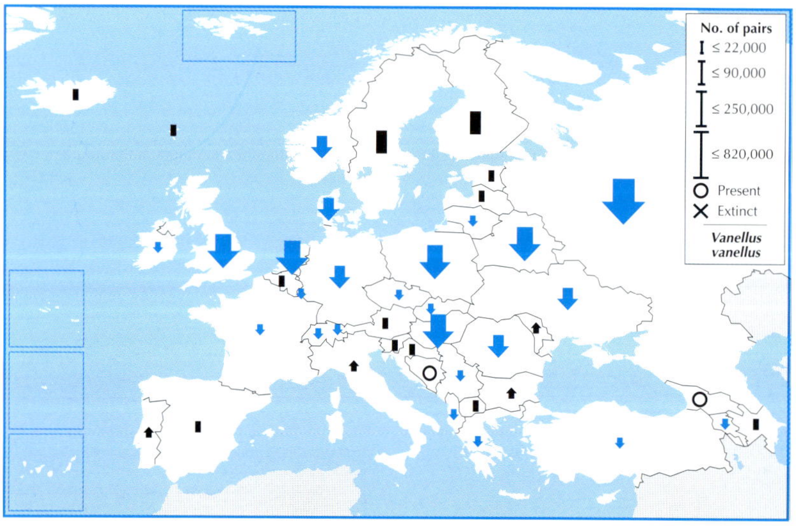

Calidris canutus
RED KNOT

SPEC 3W (1994: 3W) **Status** Declining
Criteria Moderate recent decline
European IUCN Red List Category —
Criteria —
Global IUCN Red List Category —
Criteria —

Calidris canutus breeds in Greenland, but winters mainly in coastal areas of western Europe, which accounts for less than half of its global wintering range. Its European wintering population is large (>470,000 individuals), and was stable between 1970–1990. Although the species remained broadly stable across most of its European range during 1990–2000, the key wintering population in the United Kingdom declined, and the species underwent a moderate decline (>10%) overall. Consequently, it is evaluated as Declining.

Country	Winter pop. size (individuals)	Year(s)	Trend	Mag.%	References
Belgium	20 – 20	95–00	(0)	(0–19)	*1*
Bulgaria	**0 – 1**	97–01	?	–	
Denmark	**20,000 – 25,000**	99–00	(+)	(>80)	*25*
Estonia	0 – 1	98	?	–	*1*
France	**28,000 – 38,000**	98–02	**F**	**50–79**	*9*
Germany	2,000 – 35,000	95–00	F	>80	*2*
Greece	**15 – 48**	95–99	**F**	**30–49**	
Iceland	**10 – 100**	03	?	–	*55*
Rep. Ireland	**17,000 – 27,000**	94–00	**0**	**0–19**	
Italy	(0 – 100)	02	(0)	(0–19)	*22*
Netherlands	**110,000 – 110,000**	99–01	**F**	**36**	*3,4,8,9,11,12*
Portugal	130 – 3,700	98–01	(F)	(–)	*22*
Spain	(50 – 250)	90–01	(0)	(0–19)	
Turkey	**100 – 200**	91–01	**0**	**0–19**	
UK	**295,000 – 295,000**	94–99	**-**	**23**	*44,52*
Total (approx.)	>470,000	**Overall trend** Moderate decline			
% in European IBAs	>90	**Gen. length** 5		**% Global pop.** 25–49	

Country	Breeding pop. size (pairs)	Year(s)	Trend	Mag.%	References
Denmark					
Greenland	15,000 – 30,000	90–00	?	–	*11*
Norway					
Svalbard	(2 – 20)	01–03	(0)	(0–19)	
Total (approx.)	15,000 – 30,000	**Overall trend** Unknown			
Breeding range	>500,000 km²	**Gen. length.** 5		**% Global pop.** 5–24	

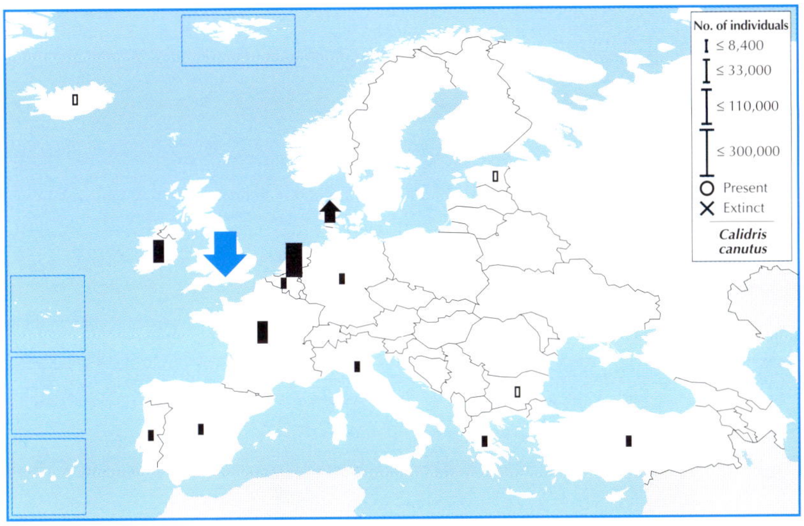

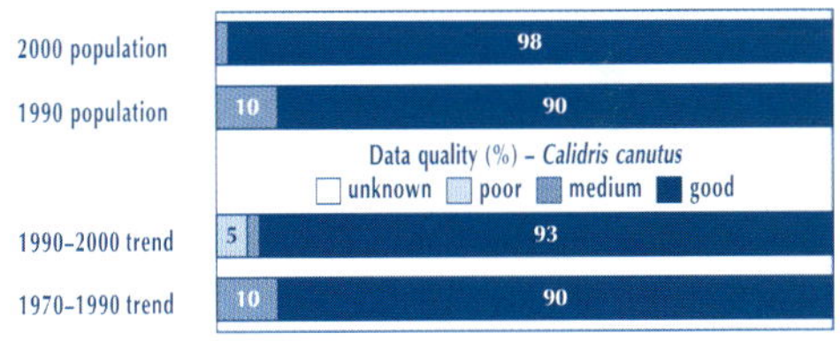

Calidris alba
SANDERLING

Non-SPEC (1994: —) **Status** (Secure)
Criteria —
European IUCN Red List Category —
Criteria —
Global IUCN Red List Category —
Criteria —

Calidris alba breeds in the high Arctic in northern Greenland and Svalbard, with Europe accounting for less than a quarter of its global breeding range. Its European breeding population is relatively small (<50,000 pairs), but was stable between 1970–1990. Although the trend of the stronghold population in Greenland during 1990–2000 was unknown, there was no evidence to suggest that the species's status deteriorated since 1990, and the European wintering population increased markedly during this period. Consequently, the species is provisionally evaluated as Secure.

Country	Breeding pop. size (pairs)	Year(s)	Trend	Mag.%	References
Denmark					
Greenland	25,000 – 50,000	90–00	?	–	*11*
Norway					
Svalbard	20 – 100	95	(0)	(0–19)	
Total (approx.)	25,000 – 50,000	**Overall trend** Unknown			
Breeding range	>500,000 km²	**Gen. length.** 5		**% Global pop.** 5–24	

Country	Winter pop. size (individuals)	Year(s)	Trend	Mag.%	References
Belgium	**500 – 500**	95–00	+	**20–29**	*1*
Denmark	500 – 1,000	99–00	(0)	(0–19)	*25*
France	11,800 – 22,000	98–02	+	**30–49**	*9*
Germany	1,000 – 3,000	95–00	F	>80	*2*
Rep. Ireland	**5,700 – 6,100**	94–00	0	0–19	
Italy	(100 – 500)	02	?	–	*22*
Netherlands	**5,900 – 5,900**	99–01	+	**83**	*3,4,8,9,11,12*
Serbia & MN	50 – 100	90–02	F	30–49	
Spain	**690 – 2,800**	90–01	+	**20–29**	*19*
Turkey	(50 – 500)	91–01	**F**	**>80**	
UK	**20,700 – 20,700**	94–99	+	**20**	*44,52*
Total (approx.)	>47,000	**Overall trend** Large increase			
% in European IBAs	33–36	**Gen. length** 5		**% Global pop.** 5–24	

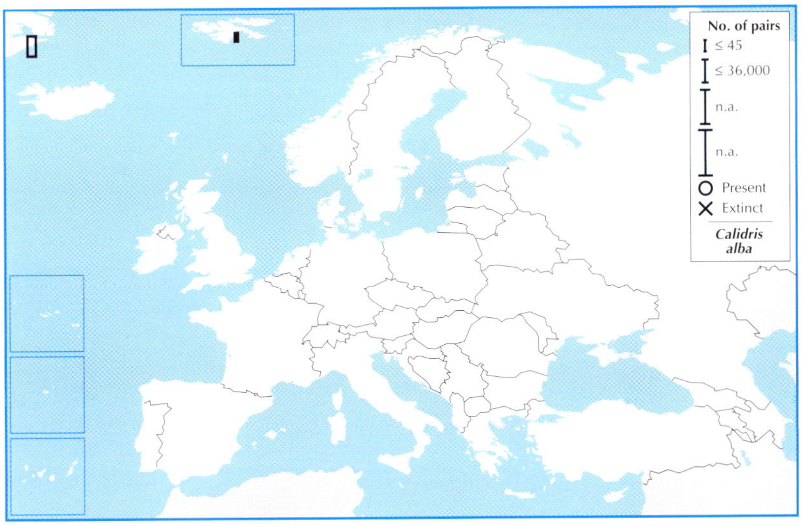

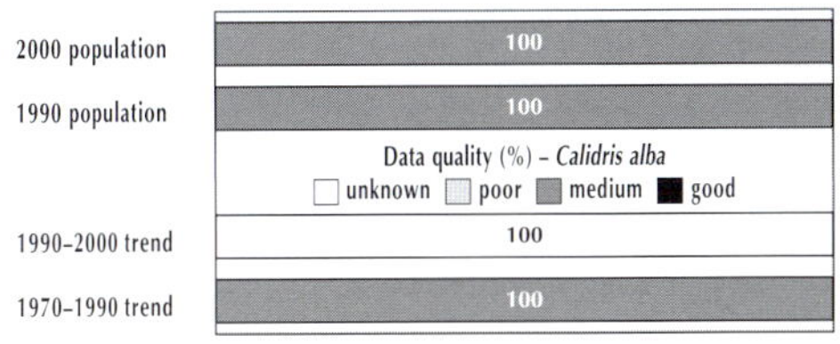

Calidris minuta
LITTLE STINT

Non-SPEC (1994: —) **Status** (Secure)
Criteria —
European IUCN Red List Category —
Criteria —
Global IUCN Red List Category —
Criteria —

Calidris minuta breeds in the arctic north of Norway and Russia, with Europe accounting for less than half of its global breeding range. Although estimates of its European breeding population vary widely, it is probably relatively large (as many as 460,000 pairs), and was stable between 1970–1990. Although there were fluctuations in Norway during 1990–2000, the stronghold population in Russia was stable, and the species probably remained stable overall. Consequently, it is provisionally evaluated as Secure.

Country	Breeding pop. size (pairs)	Year(s)	Trend	Mag.%	References
Finland	(0 – 5)	98–02	(0)	(0–19)	
Norway	500 – 5,000	00	F	30–49	64
Russia	(45,000 – 450,000)	90–00	(0)	(0–19)	28,29,60
Total (approx.)	46,000 – 460,000	**Overall trend** Stable			
Breeding range	>100,000 km²	**Gen. length.** <3.3	**% Global pop.** 25–49		

Country	Winter pop. size (individuals)	Year(s)	Trend	Mag.%	References
Albania	**120 – 2,100**	95–02	(F)	(–)	
Azerbaijan	(50 – 500)	96–02	(0)	(0–19)	
Belgium	**0 – 5**	95–00	0	0–19	*1*
Cyprus	50 – 500	94–02	(0)	(0–19)	
France	**1,350 – 3,000**	98–02	?	–	*9*
Greece	**800 – 4,600**	95–99	**F**	**30–49**	
Rep. Ireland	**103 – 431**	94–00	?	–	
Italy	1,000 – 4,000	02	–	0–19	*22*
Netherlands	**2 – 2**	99–01	?	–	*3,4,8,9,11,12*
Portugal	Present	95–97	?	–	*19,20,21*
Serbia & MN	40 – 100	90–02	F	40–59	
Slovenia	Present	90–00	?	–	
Spain	**630 – 18,000**	90–01	**+**	**>80**	*19*
Turkey	5,000 – 8,000	91–01	(–)	(0–19)	
Total (approx.)	>9,100	**Overall trend** Stable			
% in European IBAs	>40	**Gen. length** <3.3	**% Global pop.** <5		

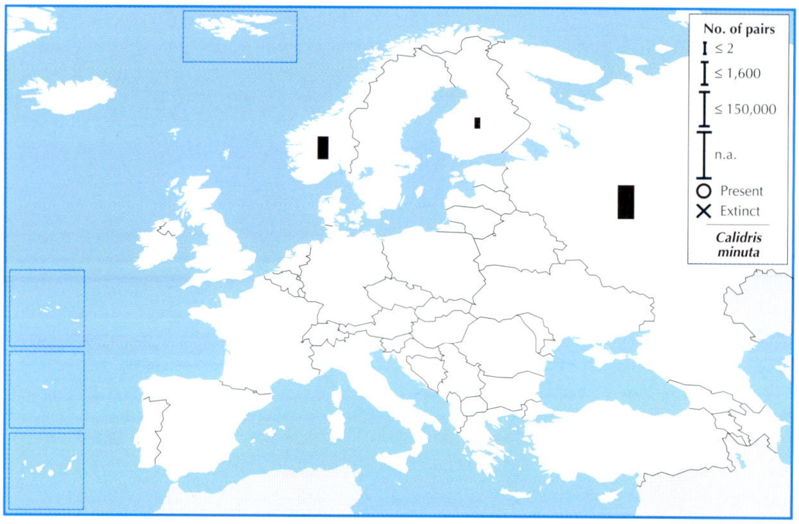

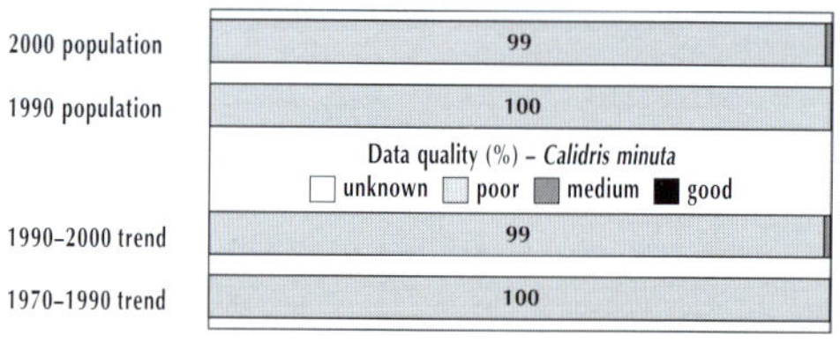

Calidris temminckii
TEMMINCK'S STINT

Non-SPEC (1994: —) **Status** (Secure)
Criteria —
European IUCN Red List Category —
Criteria —
Global IUCN Red List Category —
Criteria —

Calidris temminckii breeds mainly in Fennoscandia and arctic Russia, with Europe accounting for less than half of its global breeding range. Estimates of its European breeding population vary widely, but it is probably relatively large (as many as 420,000 pairs), and was stable between 1970–1990. Although there were declines in Sweden and Finland during 1990–2000, populations in Norway and the Russian stronghold were stable, and the species probably remained stable overall. Consequently, it is provisionally evaluated as Secure.

Country	Breeding pop. size (pairs)	Year(s)	Trend	Mag.%	References
Estonia	0 – 1	98	?	–	*1*
Finland	500 – 1500	98–02	–	50	
Norway	(3,000 – 10,000)	90–03	(0)	(0–19)	
Russia	(75,000 – 400,000)	90–00	(0)	(0–19)	*28,29,60*
Sweden	**6,000 – 7,000**	99–00	(–)	(0–9)	
UK	**2 – 4**	96–00	**–**	**18**	
Total (approx.)	85,000 – 420,000	**Overall trend** Stable			
Breeding range	>500,000 km²	**Gen. length.** <3.3	**% Global pop.** 25–49		

Country	Winter pop. size (individuals)	Year(s)	Trend	Mag.%	References
Azerbaijan	(50 – 500)	96–02	(0)	(0–19)	
Cyprus	0 – 20	94–02	?	–	
France	5 – 15	98–02	?	–	*9*
Italy	(0 – 9)	02	?	–	*22*
Spain	(0 – 50)	90–01	(0)	(0–19)	
Turkey	(20 – 200)	91–01	(F)	(>80)	
Total (approx.)	>80	**Overall trend** Fluctuating			
% in European IBAs	Unknown	**Gen. length** <3.3	**% Global pop.** <5		

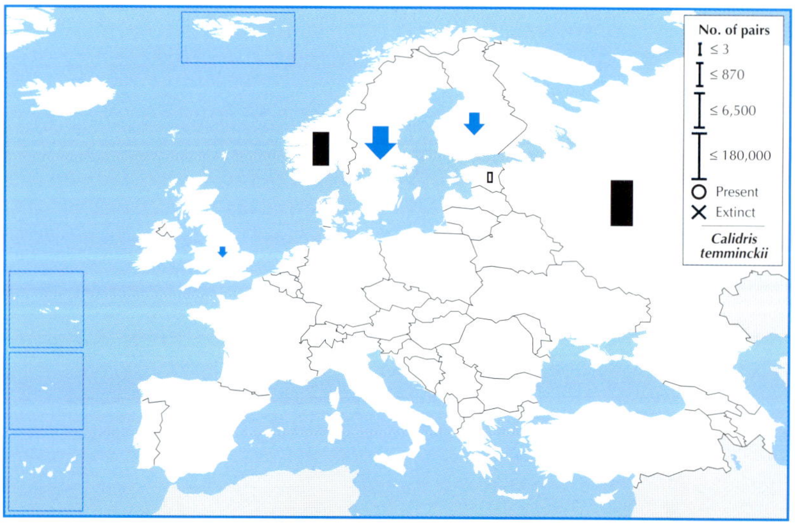

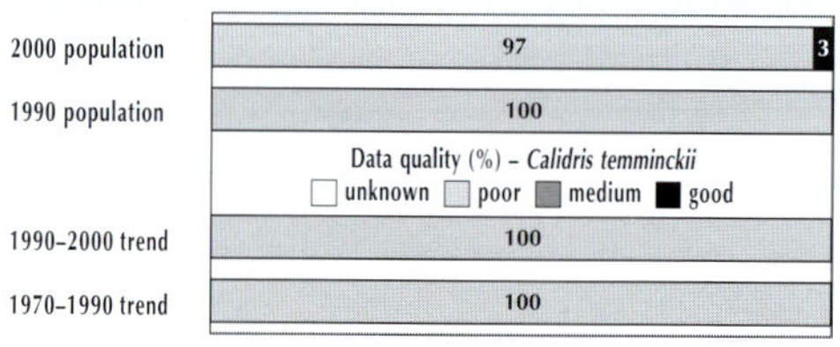

Calidris bairdii
BAIRD'S SANDPIPER

Non-SPEC (1994: NE) **Status** (Secure)
Criteria —
European IUCN Red List Category —▼
Criteria —
Global IUCN Red List Category —
Criteria —

Country	Breeding pop. size (pairs)	Year(s)	Trend	Mag.%	References
Denmark					
Greenland	(500 – 1,000)	90–00	?	-	
Total (approx.)	500 – 1,000	**Overall trend** Unknown			
Breeding range	>20,000 km²	**Gen. length** 5		**% Global pop.** <5	

Calidris bairdii has a predominantly North American breeding distribution, which just extends into Europe in north-west Greenland. Its European breeding population is very small (as few as 500 pairs), but its trend between 1970–1990 was unknown. Trend data were also unavailable during 1990–2000, but there was no evidence to suggest that it declined. Although the size of the European population could render it susceptible to the risks affecting small populations, it is marginal to a much a larger non-European population. Consequently, it is provisionally evaluated as Secure.

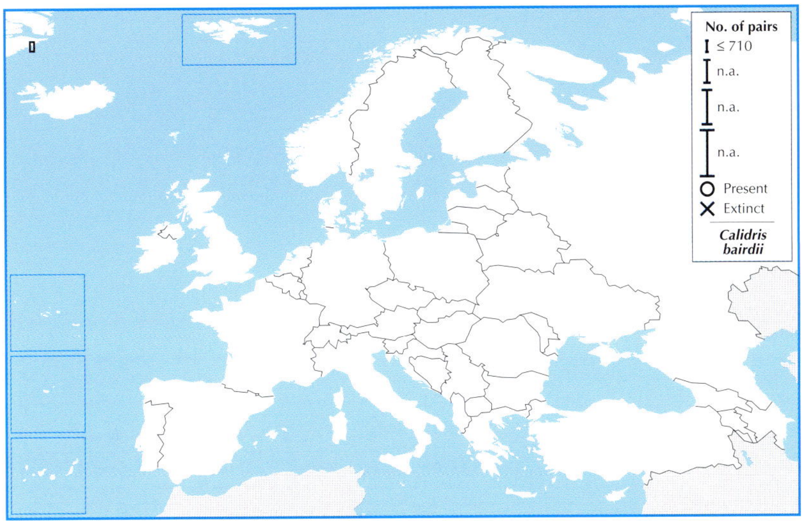

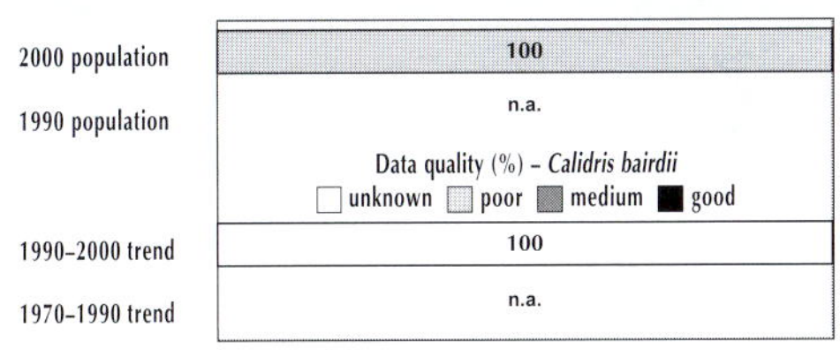

Calidris ferruginea
CURLEW SANDPIPER

NE (1994: NE) **Status** Not Evaluated
Criteria —
European IUCN Red List Category —
Criteria —
Global IUCN Red List Category —
Criteria —

Country	Winter pop. size (individuals)	Year(s)	Trend	Mag.%	References
Portugal	50 – 200	95–97	(F)	(-)	*19,20,21*
Spain	**0 – 1,800**	90–01	**-**	**50–79**	*19*
Total (approx.)	>50	**Overall trend** Unknown			
% in European IBAs	>10	**Gen. length** 5		**% Global pop.** <5	

Calidris ferruginea breeds in a narrow latitudinal range in the central Siberian Arctic, and winters mainly in western and sub-Saharan Africa. Although a tiny proportion of its global population occasionally winters in Iberia, the species is primarily a passage visitor to Europe. Consequently, its status in Europe is Not Evaluated.

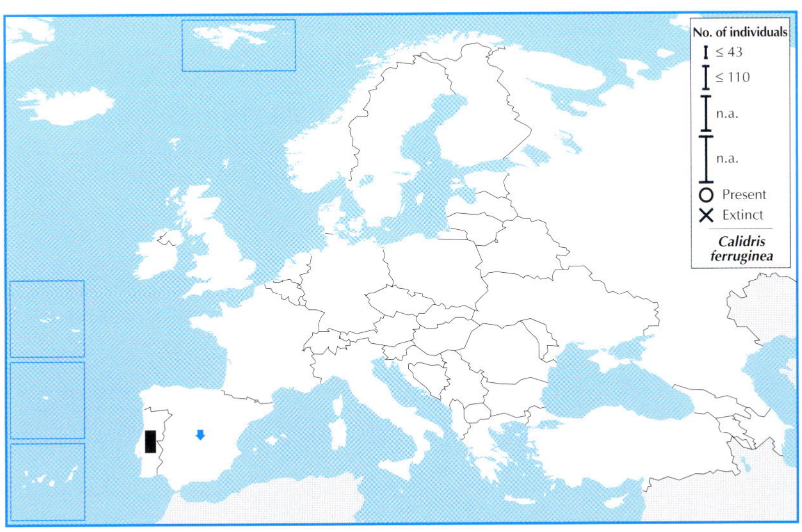

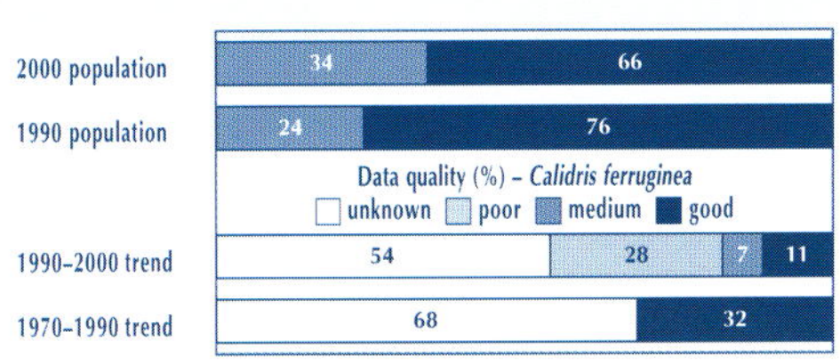

Calidris maritima
PURPLE SANDPIPER

Non-SPEC[E] (1994: 4) **Status** (Secure)
Criteria —
European IUCN Red List Category —
Criteria —
Global IUCN Red List Category —
Criteria —

Calidris maritima breeds in Greenland, Scandinavia and arctic Russia, with >75% of the global breeding range confined to Europe. The European breeding population is relatively small (<75,000 pairs), but appeared to be stable between 1970–1990. Trends were not available for the key populations in Greenland and Iceland during 1990–2000, but the species was stable elsewhere, and consequently it is provisionally evaluated as Secure.

Country	Breeding pop. size (pairs)	Year(s)	Trend	Mag.%	References
Denmark					
Faroe Is.	25 – 25	95	(0)	(0–19)	
Greenland	(10,000 – 20,000)	90–00	?	–	
Finland	(5 – 20)	98–02	(0)	(0–19)	
Iceland	10,000 – 30,000	99–01	?	–	41
Norway	(5,000 – 10,000)	95	(0)	(0–19)	
Svalbard	(2,000 – 10,000)	95	(0)	(0–19)	
Russia	(400 – 1,700)	90–00	?	–	28,29,60
Sweden	1,000 – 3,000	99–00	(0)	(0–19)	
UK	**1 – 3**	96–00	?	–	
Total (approx.)	28,000 – 75,000	**Overall trend** Unknown			
Breeding range	>1,000,000 km²	**Gen. length.** 5		**% Global pop.** 75–94	

Country	Winter pop. size (individuals)	Year(s)	Trend	Mag.%	References
Belgium	**80 – 80**	95–00	**–**	**50–79**	1
Denmark	200 – 300	99–00	0	0–19	24
Faroe Is.	500 – 1,200	92	?	–	3
Greenland	(5,000 – 25,000)	90–00	?	–	
Estonia	0 – 200	98	(+)	(N)	1
Finland	300 – 500	98–02	(+)	(30–100)	
France	600 – 1,100	98–02	+	30–49	9
Germany	**20 – 150**	95–00	?	–	
Iceland	(10,000 – 100,000)	78–94	?	–	48
Rep. Ireland	**3,000 – 3,200**	94–00	(0)	(0–19)	
Netherlands	**300 – 300**	99–01	0	0–19	3,4,8,9,11,12
Norway	40,000 – 80,000	93	(0)	(0–19)	68
Poland	0 – 10	90–00	?	–	114
Portugal	(20 – 50)	02	?	–	
Russia	62 – 62	97	?	–	179
Spain	(62 – 380)	90–01	(0)	(0–19)	
Sweden	50 – 300	98–01	0	0–19	
UK	**17,800 – 17,800**	94–99	**–**	**18**	52
Total (approx.)	>78,000	**Overall trend** Small decline			
% in European IBAs	5–14	**Gen. length** 5		**% Global pop.** 75–94	

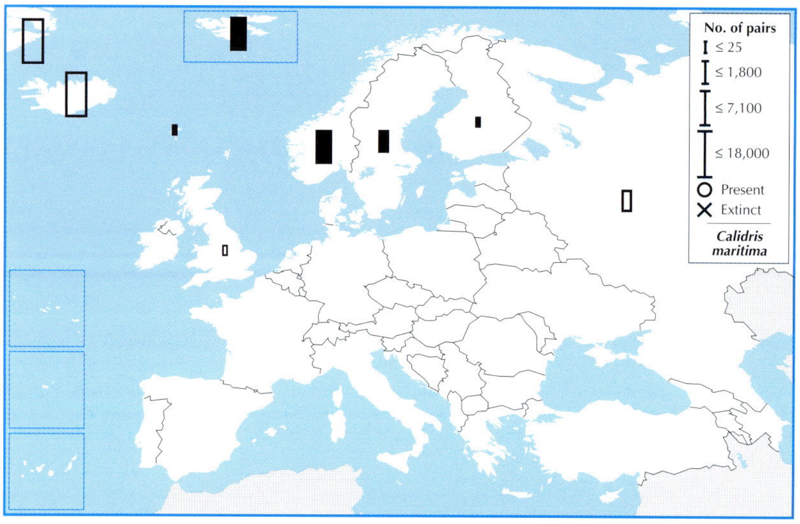

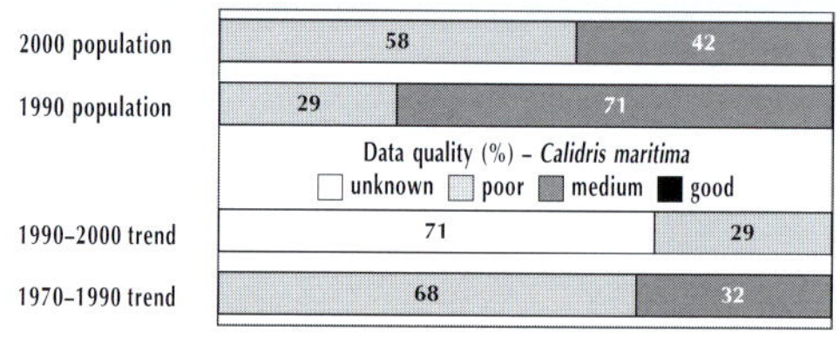

Calidris alpina
DUNLIN

SPEC 3 (1994: 3W) **Status** (Depleted)
Criteria Moderate historical decline
European IUCN Red List Category —
Criteria —
Global IUCN Red List Category —
Criteria —

Calidris alpina breeds mainly in northern Europe, which accounts for less than half of its global breeding range. Its European breeding population is large (>300,000 pairs), and was probably stable between 1970–1990 (although the European wintering population declined markedly). The species was stable in north-west Europe during 1990–2000, but declined around the Baltic and in Russia. Although the trend in Iceland was unknown, the species's population has clearly not yet recovered to the level that preceded its decline. Consequently, it is provisionally evaluated as Depleted.

Country	Breeding pop. size (pairs)	Year(s)	Trend	Mag.%	References
Belarus	0 – 10	97–02	?	–	
Denmark	**340 – 360**	02	**–**	**50**	21
Faroe Is.	10 – 10	95	(0)	(0–19)	
Greenland	7,000 – 15,000	90–00	(0)	(0–19)	11
Estonia	400 – 500	98	–	20–29	1
Finland	300 – 600	98–02	–	20	
Germany	**39 – 39**	95–99	**–**	**50–79**	
Iceland	200,000 – 300,000	99–01	?	–	41
Rep. Ireland	50 – 250	88–91	?	–	
Latvia	(1 – 5)	90–00	(–)	(0–19)	23
Lithuania	25 – 30	99–01	–	20–29	20
Norway	(30,000 – 40,000)	90–03	(0)	(0–19)	
Svalbard	(100 – 200)	90–03	(0)	(0–19)	
Poland	**10 – 20**	97–03	**–**	**70–79**	54,55
Russia	(15,000 – 130,000)	90–00	(–)	(20–29)	28,29,60
Sweden	**30,000 – 50,000**	99–00	(–)	(0–19)	
UK	**18,300 – 33,500**	00	(+)	(0–19)	8
Total (approx.)	300,000 – 570,000	**Overall trend** Unknown			
Breeding range	>1,000,000 km²	**Gen. length.** 5		**% Global pop.** 25–49	

Country	Winter pop. size (individuals)	Year(s)	Trend	Mag.%	References
Denmark	**45,000 – 50,000**	99–00	(0)	(0–19)	25
France	**322,000 – 440,000**	98–02	**+**	**30–49**	9
Germany	**20,000 – 130,000**	95–00	**F**	**50–79**	2
Rep. Ireland	**89,000 – 120,000**	94–00	**–**	**30–49**	
Italy	20,000 – 80,000	02	0	0–19	22
Netherlands	**170,000 – 170,000**	99–01	**0**	**0–19**	3,4,8,9,11,12
Portugal	28,000 – 63,000	95–97	(F)	(–)	19,20,21
Spain	**13,000 – 100,000**	90–01	**+**	**50–79**	19
UK	**577,000 – 577,000**	94–99	**–**	**15**	44,52
Total (approx.)	>1,300,000	**Overall trend** Small decline			
% in European IBAs	80–95	**Gen. length** 5		**% Global pop.** 25–49	

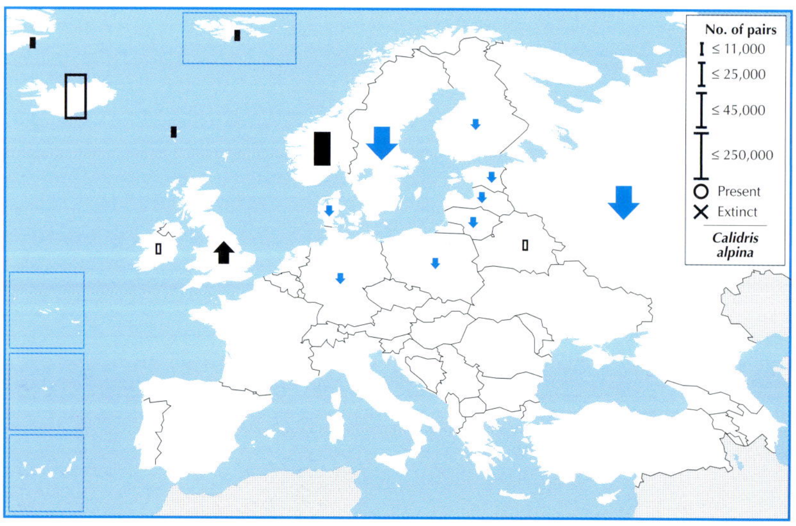

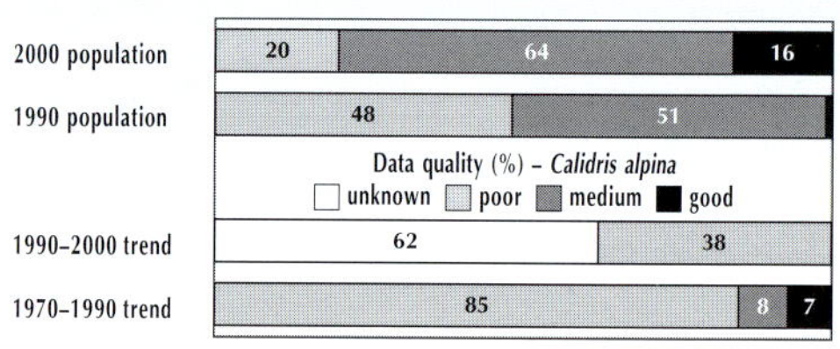

Limicola falcinellus
BROAD-BILLED SANDPIPER

SPEC 3 (1994: 3) **Status** (Declining)
Criteria Moderate continuing decline
European IUCN Red List Category —
Criteria —
Global IUCN Red List Category —
Criteria —

Country	Breeding pop. size (pairs)	Year(s)	Trend	Mag.%	References
Finland	5,000 – 15,000	98–02	-	15	
Norway	**1,000 – 1,500**	99	(0)	(0–19)	*39*
Russia	(200 – 1,200)	90–00	(0)	(0–19)	*28,29,60*
Sweden	3,000 – 4,500	99–00	?	–	
Total (approx.)	9,200 – 22,000	**Overall trend** Moderate decline			
Breeding range	>500,000 km²	**Gen. length.** 5		**% Global pop.** 25–49	

Limicola falcinellus is a widespread summer visitor to Fennoscandia and northern Russia, with Europe accounting for less than half of its global breeding range. Its European breeding population is small (as few as 9,200 pairs), and underwent a large decline between 1970–1990. Although its populations in Norway and Russia were stable during 1990–2000, the species continued to decline in its Finnish stronghold (the trend in Sweden was unknown), and it probably underwent a moderate decline (>10%) overall. Consequently, it is provisionally evaluated as Declining.

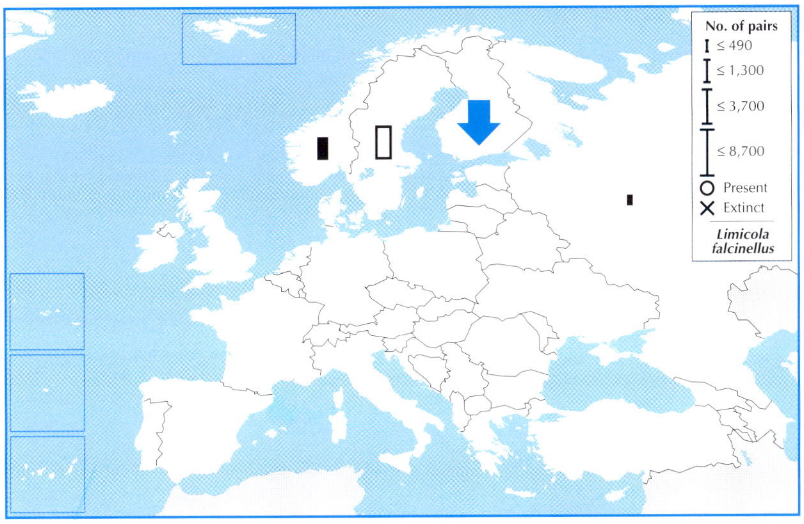

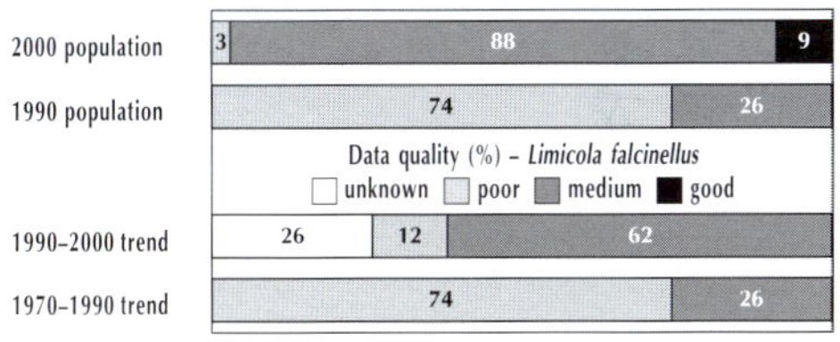

Philomachus pugnax
RUFF

SPEC 2 (1994: 4) **Status** (Declining)
Criteria Moderate recent decline
European IUCN Red List Category —
Criteria —
Global IUCN Red List Category —
Criteria —

Country	Breeding pop. size (pairs)	Year(s)	Trend	Mag.%	References
Belarus	2,000 – 2,400	97–02	0	0–19	
Croatia	(1 – 10)	02	(+)	(N)	*70,57*
Denmark	**150 – 150**	00–02	**-**	**>80**	*21*
Estonia	200 – 500	98	-	50–79	*1*
Finland	10,000 – 20,000	98–02	**-**	**60**	
France	**0 – 5**	98–01	**F**	**>80**	*4*
Germany	**109 – 124**	95–99	**-**	**50–79**	
Hungary	Present	90–92	?	–	
Latvia	(50 – 200)	90–00	(-)	(0–19)	23,16
Lithuania	(100 – 200)	99–01	(-)	(20–29)	*20*
Netherlands	**100 – 140**	98–00	**-**	**78**	*1*
Norway	(10,000 – 20,000)	90–02	(0)	(0–19)	
Poland	**5 – 50**	96–03	-	>80	*56,57*
Russia	(140,000 – 420,000)	90–00	-	20–29	*28,29,60*
Sweden	40,000 – 50,000	99–00	(-)	(0–19)	
Ukraine	100 – 150	90–00	**-**	**20–29**	
UK	3 – 4	96–00	?	–	
Total (approx.)	200,000 – 510,000	**Overall trend** Moderate decline			
Breeding range	>2,000,000 km²	**Gen. length.** <3.3		**% Global pop.** 50–74	

Country	Winter pop. size (individuals)	Year(s)	Trend	Mag.%	References
Belgium	1,200 – 2,600	99–01	0	0–19	*1*
Bulgaria	**0 – 1**	97–01	?	–	
Cyprus	50 – 500	94–02	(0)	(0–19)	
France	**221 – 525**	98–02	**F**	**50–79**	*9*
Greece	**0 – 102**	95–00	**F**	**30–49**	
Rep. Ireland	**34 – 62**	94–00	?	–	
Italy	50 – 200	02	?	–	*22*
Netherlands	**470 – 470**	99–01	0	0–19	*3,4,8,9,11,12*
Portugal	(15 – 131)	95–97	(F)	(-)	*19,20,21*
Serbia & MN	300 – 700	90–02	F	40–59	
Slovenia	Present	90–00	?	–	
Spain	**260 – 3,000**	90–01	**-**	**30–49**	*19*
Turkey	100 – 150	91–01	(0)	(0–19)	
Ukraine	(0 – 50)	90–00	(-)	(50–79)	
UK	700 – 700	89–94	+	6	*52*
Total (approx.)	>3,400	**Overall trend** Small decline			
% in European IBAs	>60	**Gen. length** <3.3		**% Global pop.** <5	

Philomachus pugnax is a widespread breeder in much of northern Europe, which constitutes >50% of its global breeding range. Its European breeding population is large (>200,000 pairs), and was stable between 1970–1990. Although the species was stable in a few countries during 1990–2000, most populations across its European range—including key ones in Russia and Sweden—suffered declines, and the species underwent a moderate decline (>10%) overall. Consequently, this previously Secure species is now provisionally evaluated as Declining.

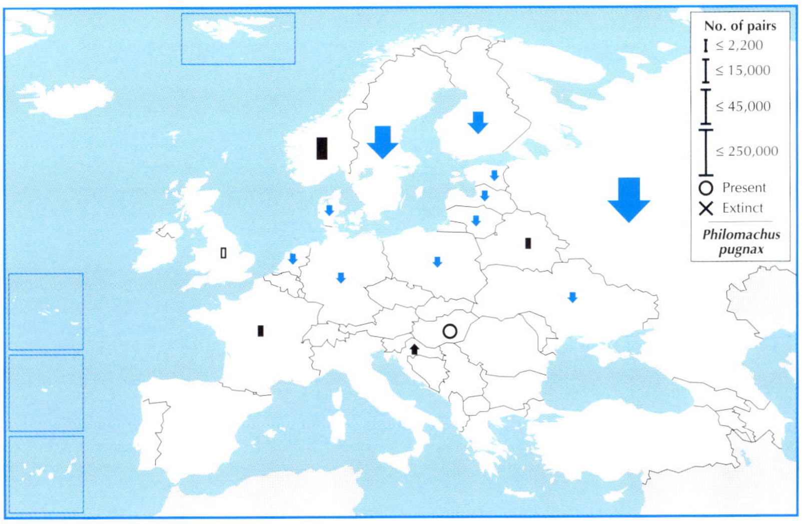

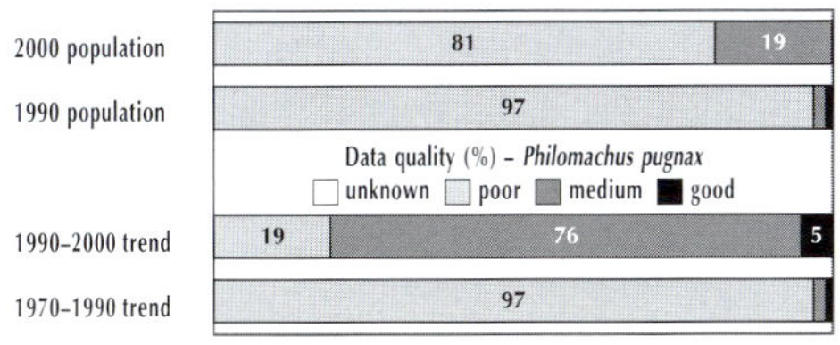

Lymnocryptes minimus
JACK SNIPE

SPEC 3 (1994: 3W) **Status** (Declining)
Criteria Moderate recent decline
European IUCN Red List Category —
Criteria —
Global IUCN Red List Category —
Criteria —

Country	Breeding pop. size (pairs)	Year(s)	Trend	Mag.%	References
Belarus	**0 – 20**	97–00	**0**	**0–19**	
Estonia	20 – 30	98	0	0–19	*1*
Finland	10,000 – 15,000	98–02	0	5	
Latvia	(0 – 10)	90–00	(0)	(0–19)	*22*
Lithuania	(0 – 5)	99–01	(0)	(0–19)	*20*
Norway	(1,000 – 5,000)	90–03	(0)	(0–19)	
Russia	(4,800 – 46,000)	91–00	(–)	(20–29)	*28,29,60*
Sweden	2,000 – 4,000	99–00	(F)	(20–29)	
Total (approx.)	18,000 – 70,000	**Overall trend** Moderate decline			
Breeding range	>1,000,000 km²	**Gen. length.** <3.3 **% Global pop.** 25–49			

Country	Winter pop. size (individuals)	Year(s)	Trend	Mag.%	References
Azerbaijan	(5 – 1,000)	96–02	(0)	(0–19)	
Belarus	**0 – 50**	91–00	?	–	
Croatia	(500 – 1,000)	02	?	–	*16*
Cyprus	(5 – 50)	94–02	(0)	(0–9)	
Denmark	(100 – 10,000)	99–00	?	–	
France	(8,000 – 45,000)	98–02	?	–	*9*
Germany	(100 – 1,000)	95–00	?	–	
Greece	**10 – 500**	95–99	?	–	
Italy	(100 – 500)	02	?	–	*22*
Lithuania	(0 – 50)	92–02	0	0–19	
Luxembourg	**30 – 50**	02	**0**	**0–19**	
Netherlands	(250 – 250)	78–83	(+)	(–)	*3,4,8,9,11,12*
Poland	1,000 – 2,500	90–00	?	–	*122,114*
Romania	(50 – 300)	90–00	(F)	(>80)	*60*
Spain	(250 – 1,000)	90–01	?	–	
Turkey	(250 – 1,000)	91–01	?	–	
Ukraine	(0 – 50)	90–00	(0)	(0–19)	
UK	1,000 – 10,000	89–94	+	11	*52*
Total (approx.)	>12,000	**Overall trend** Unknown			
% in European IBAs	Unknown	**Gen. length** <3.3 **% Global pop.** <5			

Lymnocryptes minimus breeds in Fennoscandia and north-eastern Europe, which accounts for less than a half of its global breeding range. Its European breeding population is relatively small (<70,000 pairs), but was stable between 1970–1990. Although trends were stable in most countries during 1990–2000, the sizeable population in Russia declined, and the species underwent a moderate decline (>10%) overall. Consequently, it is provisionally evaluated as Declining.

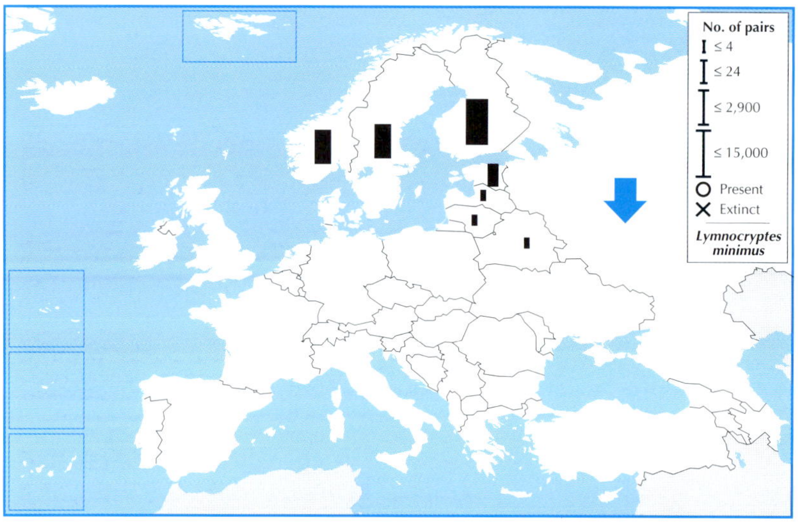

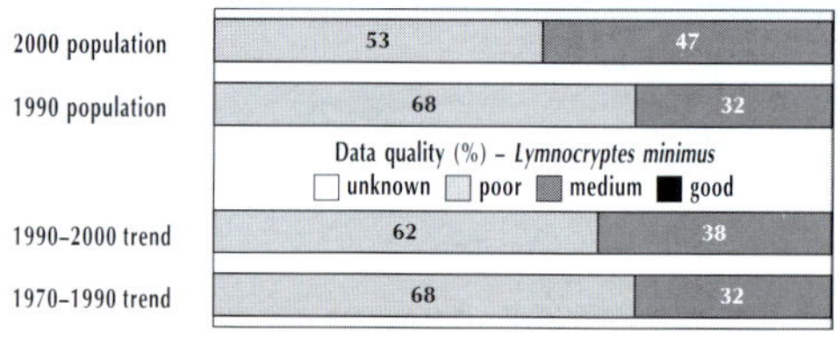

Gallinago gallinago
COMMON SNIPE

SPEC 3 (1994: —) **Status** (Declining)
Criteria Moderate recent decline
European IUCN Red List Category —
Criteria —
Global IUCN Red List Category —
Criteria —

Country	Breeding pop. size (pairs)	Year(s)	Trend	Mag.%	References
Austria	80 – 120	98–02	(–)	(0–19)	
Belarus	70,000 – 90,000	97–00	**0**	**0–19**	
Belgium	58 – 90	95–02	–	>80	1
Bulgaria	**2 – 10**	97–00	(F)	(>80)	
Croatia	(5 – 10)	02	(0)	(0–19)	*70*
Czech Rep.	**500 – 800**	00	**–**	**50–79**	
Denmark	2,500 – 3,000	93–96	(–)	(–)	3
Faroe Is.	2,500 – 2,500	95	(0)	(0–19)	
Estonia	20,000 – 30,000	98	0	0–19	*1*
Finland	80,000 – 120,000	98–02	**–**	**25**	
France	300 – 350	98–02	–	0–19	*4,11*
Germany	6,200 – 9,800	95–99	–	50–79	
Hungary	650 – 1,200	95–02	F	20–49	
Iceland	180,000 – 300,000	99–01	?	–	*41*
Rep. Ireland	2,500 – 10,000	88–91	(–)	(50–79)	
Latvia	5,000 – 8,000	90–00	0	0–19	*23,16*
Liechtenstein	**0 – 5**	98–00	**–**	**50–79**	
Lithuania	**10,000 – 20,000**	99–01	(–)	(20–29)	*20*
Luxembourg	**0 – 0**	00–02	**–**	**X**	
Macedonia	(0 – 2)	90–00	(F)	(20–29)	
Netherlands	**1,200 – 1,500**	98–00	**–**	**50**	*1*
Norway	(70,000 – 150,000)	90–03	(0)	(0–19)	
Poland	**15,000 – 30,000**	00–02	0	0–19	*23*
Portugal	2 – 20	02	(–)	(–)	
Azores	Present	02	?	–	
Romania	(150 – 350)	90–02	(0)	(0–19)	
Russia	300,000 – 850,000	90–00	(0)	(0–19)	*20,28,29*
Serbia & MN	**20 – 35**	98–02	0	10–19	*1,200,29,155, 223,127*
Slovakia	100 – 250	80–99	–	30–49	
Slovenia	15 – 40	90–00	–	50–69	
Spain	(50 – 75)	98–02	(0)	(0–19)	*10,16*
Sweden	100,000 – 150,000	99–00	**–**	**4**	
Switzerland	**0 – 3**	93–96	**–**	**30–49**	
Turkey	(5 – 20)	01	–	50–79	
Ukraine	13,000 – 15,000	90–00	–	0–19	
UK	**52,600 – 69,000**	00	**–**	**32**	*5,23*
Total (approx.)	930,000 – 1,900,000	**Overall trend** Moderate decline			
Breeding range	>6,000,000 km²	**Gen. length.** <3.3 **% Global pop.** 5–24			

(See p. 123, top, for data quality graph)

	Winter pop. size (individuals)	
Total (approx.)	>310,000	**Overall trend** Unknown
% in European IBAs	Unknown	**Gen. length** <3.3 **% Global pop.** 5–24

Gallinago gallinago is a widespread breeder across much of Europe, which accounts for less than a quarter of its global breeding range. Its European breeding population is large (>930,000 pairs), and was stable between 1970–1990. Although it remained stable in much of eastern Europe—including the key Russian population—during 1990–2000, the species suffered declines in most of the rest of Europe, and underwent a moderate decline (>10%) overall. Consequently, this previously Secure species is now provisionally evaluated as Declining.

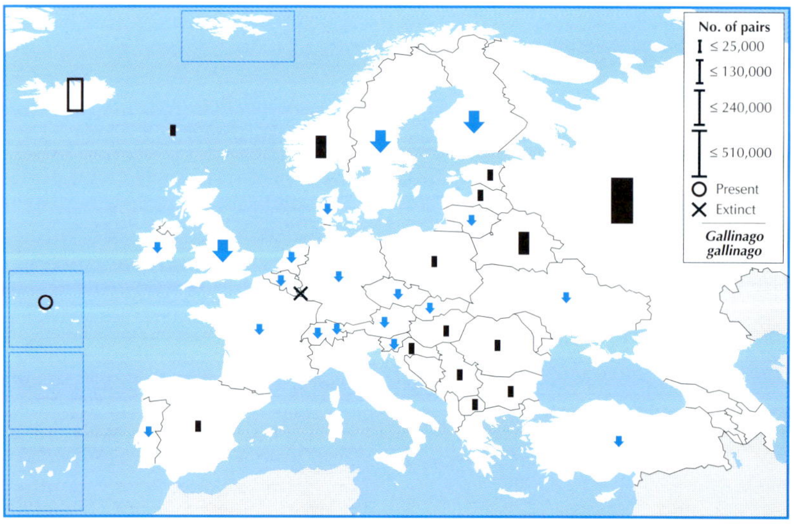

Gallinago media
GREAT SNIPE

SPEC 1 (1994: 2) **Status** Declining
Criteria Moderate continuing decline
European IUCN Red List Category —
Criteria —
Global IUCN Red List Category NT
Criteria A2c,d; A3c,d

Country	Breeding pop. size (pairs)	Year(s)	Trend	Mag.%	References
Belarus	**4,600 – 6,000**	00–01	**0**	**0–19**	
Estonia	**500 – 700**	98	-	20–29	*1*
Finland	(0 – 5)	98–02	0	0–19	
Latvia	**200 – 300**	90–00	-	20–29	*1,2*
Lithuania	200 – 300	99–01	F	20–29	*20*
Norway	5,000 – 15,000	00	(0)	(0–19)	*33*
Poland	**750 – 850**	00–02	**0**	**0–19**	*50,51*
Russia	50,000 – 140,000	90–00	-	20–29	*20,28,29,60*
Sweden	(500 – 2,000)	99–00	(F)	(20–29)	
Ukraine	**500 – 700**	90–00	**-**	**30–49**	
Total (approx.)	62,000 – 170,000	**Overall trend** Moderate decline			
Breeding range	>2,000,000 km²	**Gen. length.** <3.3 **% Global pop.** 50–74			

Gallinago media is a widespread summer visitor to Fennoscandia and north-east Europe, which constitutes >50% of its global breeding range. Its European breeding population is relatively small (<170,000 pairs), but underwent a large decline between 1970–1990. Although populations in most European countries were stable during 1990–2000, the stronghold population in Russia declined, and the species underwent a moderate decline (>10%) overall. Consequently, this globally Near Threatened species is evaluated as Declining in Europe.

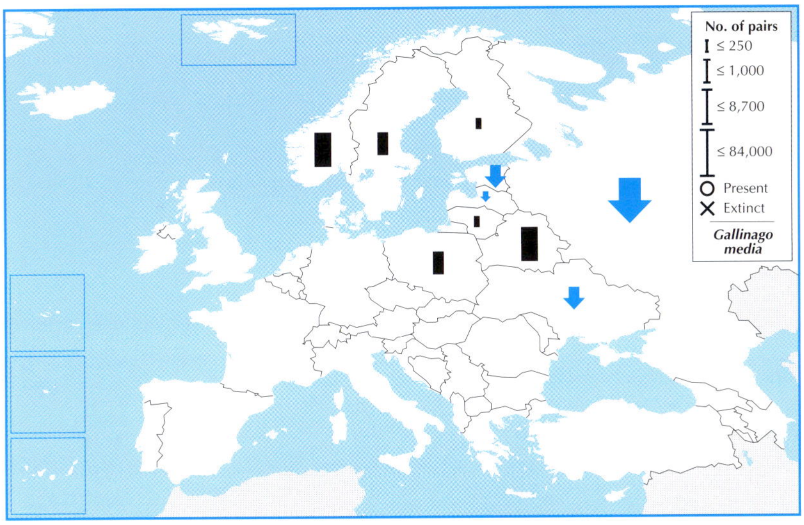

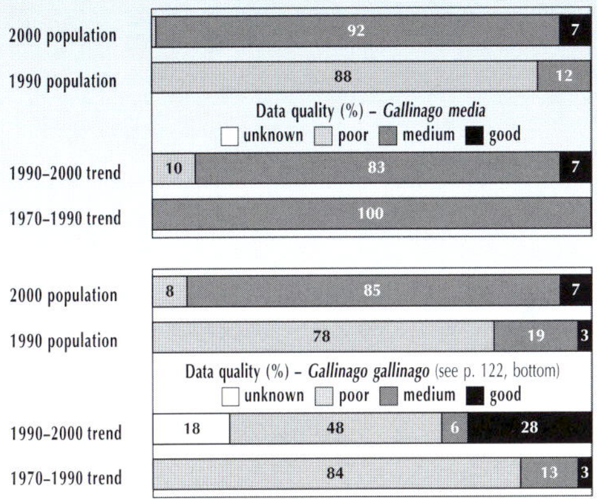

Gallinago stenura
PINTAIL SNIPE

Non-SPEC (1994: —) **Status** (Secure)
Criteria —
European IUCN Red List Category —▼
Criteria —
Global IUCN Red List Category —
Criteria —

Country	Breeding pop. size (pairs)	Year(s)	Trend	Mag.%	References
Russia	(1,000 – 2,500)	90–00	?	-	*28,29,60*
Total (approx.)	1,000 – 2,500	**Overall trend** Unknown			
Breeding range	>100,000 km²	**Gen. length.** <3.3 **% Global pop.** <5			

Gallinago stenura has a predominantly Asian breeding distribution, which just extends into Europe in northern Russia. Its European breeding population is small (as few as 1,000 pairs), but was broadly stable between 1970–1990. No trend data were available for 1990–2000, but there is no evidence to suggest that the species declined. Although the size of the European population could render it susceptible to the risks affecting small populations, it is marginal to a much larger non-European population. Consequently, the species is provisionally evaluated as Secure.

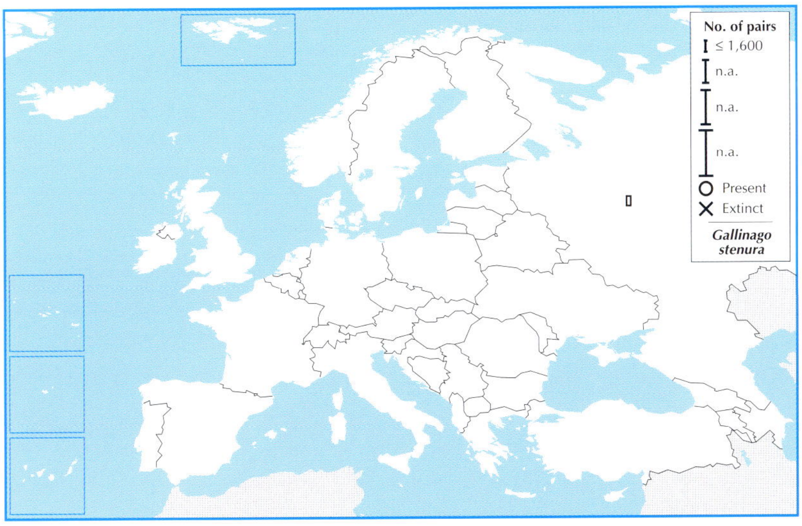

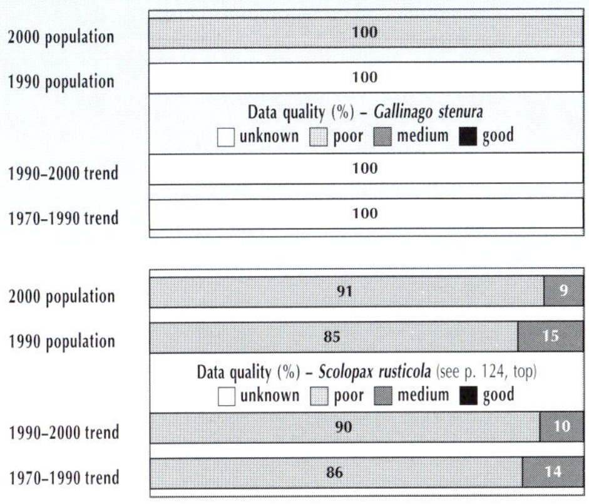

Scolopax rusticola
EURASIAN WOODCOCK

SPEC 3 (1994: 3W) **Status** (Declining)
Criteria Moderate recent decline
European IUCN Red List Category —
Criteria —
Global IUCN Red List Category —
Criteria —

Scolopax rusticola is a widespread breeder across much of Europe (particularly northern and central areas), which accounts for less than half of its global breeding range. Its European breeding population is very large (>1,800,000 pairs), and was broadly stable between 1970–1990. Although trends were stable across most of its European range during 1990–2000, the species declined in its Russian stronghold, and underwent a moderate decline (>10%) overall. Consequently, it is provisionally evaluated as Declining.

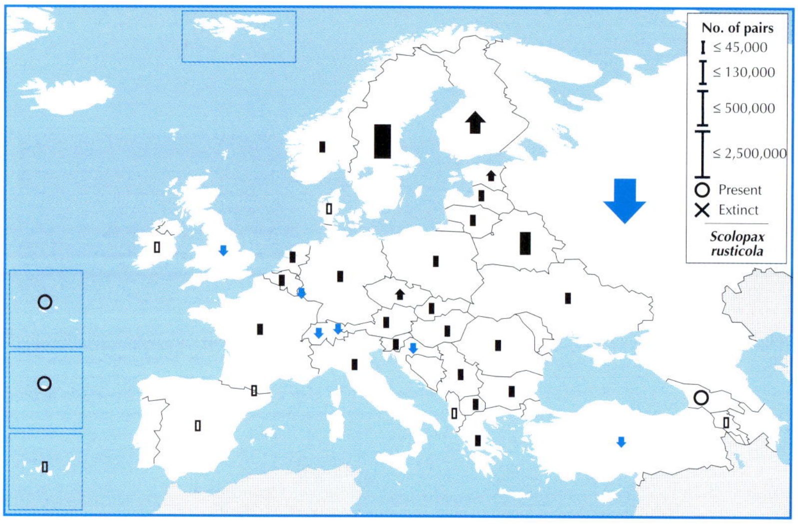

Country	Breeding pop. size (pairs)	Year(s)	Trend	Mag.%	References
Albania	50 – 100	96–02	?	–	
Andorra	(15 – 30)	99–01	?	–	*1,3*
Armenia	350 – 850	97–02	?	–	
Austria	(1,500 – 4,500)	98–02	(0)	(0–19)	
Belarus	100,000 – 120,000	97–00	0	0–19	
Belgium	1,600 – 2,800	01–02	(0)	(0–19)	*1*
Bulgaria	0 – 20	96–02	(F)	(>80)	
Croatia	(1 – 50)	02	(–)	(–)	*70,16*
Czech Rep.	**2,000 – 4,000**	00	**+**	**30–49**	
Denmark	2,000 – 3,500	89–96	?	–	*22*
Estonia	20,000 – 50,000	98	+	20–29	*1*
Finland	100,000 – 150,000	98–02	+	10	
France	(10,000 – 30,000)	98–02	0	0–19	*4*
Georgia	Present	03	?	–	
Germany	(12,000 – 24,000)	95–99	(0)	(0–19)	
Greece	(0 – 10)	95–00	(0)	(0–19)	
Hungary	(10 – 60)	95–02	F	20–49	
Rep. Ireland	2,500 – 10,000	88–91	?	–	
Italy	(80 – 150)	03	(0)	(0–19)	
Latvia	20,000 – 30,000	90–00	0	0–19	*23,18*
Liechtenstein	3 – 8	98–00	–	0–19	
Lithuania	(10,000 – 20,000)	99–01	(0)	(0–19)	*20*
Luxembourg	80 – 100	00–02	–	20–29	
Macedonia	(30 – 90)	90–00	(0)	(0–19)	
Netherlands	**2,000 – 3,000**	98–00	0	10	*1*
Norway	(30,000 – 50,000)	90–02	(0)	(0–19)	
Poland	(20,000 – 100,000)	90–00	(0)	(0–19)	*1*
Portugal					
Azores	Present	02	?	–	
Madeira	Present	02	?	–	
Romania	(6,000 – 9,000)	90–02	(0)	(0–19)	
Russia	(1,200,000 – 5,000,000)	90–00	(–)	(20–29)	*29,143,161*
Serbia & MN	**600 – 1,000**	90–02	F	20–29	*1,142a,92,235, 143a,227,155*
Slovakia	1,300 – 2,500	90–99	0	0–19	
Slovenia	(100 – 200)	99–00	(0)	(0–19)	
Spain	(2,310 – 2,590)	80–01	?	–	*13,10,6*
Canary Is.	(1,000 – 2,500)	97–03	?	–	*28,25*
Sweden	(250,000 – 1,000,000)	99–00	(F)	(20–29)	
Switzerland	**1,130 – 1,630**	93–96	(–)	(0–19)	
Turkey	(0 – 50)	01	(–)	(–)	
Ukraine	7,600 – 12,800	90–00	0	0–19	
UK	5,400 – 13,700	00	–	40	*5,31*
Total (approx.)	1,800,000 – 6,600,000	**Overall trend** Moderate decline			
Breeding range	>6,000,000 km²	**Gen. length.** <3.3 **% Global pop.** 25–49			

(See p. 123, bottom, for data quality graph)

	Winter pop. size (individuals)	
Total (approx.)	>490,000	**Overall trend** Unknown
% in European IBAs	Unknown	**Gen. length** <3.3 **% Global pop.** 25–49

Limosa limosa
BLACK-TAILED GODWIT

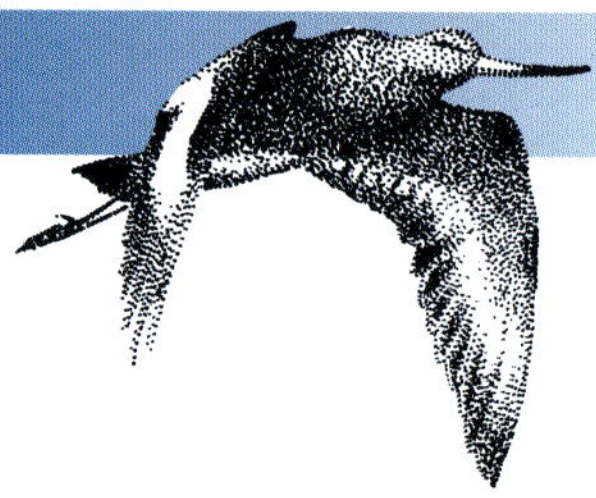

SPEC 2 (1994: 2) **Status** Vulnerable
Criteria See IUCN below
European IUCN Red List Category VU
Criteria A2b
Global IUCN Red List Category —
Criteria —

Limosa limosa is a widespread but patchily distributed breeder in eastern and parts of north-west Europe, which holds more than half of its global breeding population. Its European breeding population is relatively large (>99,000 pairs), but underwent a large decline between 1970–1990. Although the species was stable or increased in several countries—notably Iceland—during 1990–2000, key populations in the Netherlands and Russia continued to decline, and the species underwent a large decline (>30%) overall. Consequently, it is evaluated as Vulnerable.

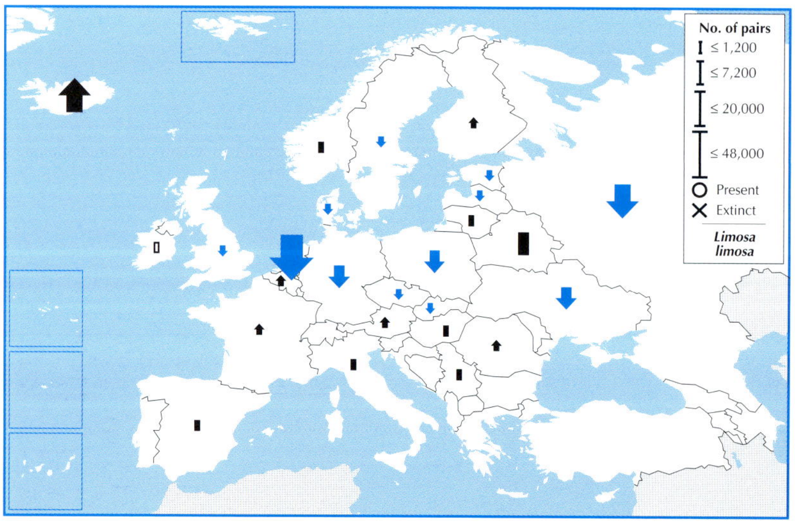

Country	Breeding pop. size (pairs)	Year(s)	Trend	Mag.%	References
Austria	**100 – 160**	98–02	**+**	**30–49**	
Belarus	**6,000 – 8,500**	97–00	**0**	**0–19**	
Belgium	**1,100 – 1,300**	00–02	**+**	**0–19**	*1*
Czech Rep.	**10 – 20**	00	**–**	**50–79**	
Denmark	**700 – 725**	00–02	**–**	**20–29**	*21*
Estonia	500 – 1,000	98	–	20–29	*1*
Finland	**40 – 60**	98–02	**+**	**60**	
France	**160 – 170**	97–00	**+**	**10–19**	*4*
Germany	6,000 – 7,300	95–99	–	50–79	
Hungary	400 – 1,500	95–02	F	20–49	
Iceland	15,000 – 25,000	99–01	**+**	**0–19**	*41*
Rep. Ireland	1 – 1	88–91	?	–	
Italy	**10 – 12**	00	**0**	**0–19**	*18*
Latvia	80 – 100	90–00	–	20–29	*23,16*
Lithuania	**300 – 450**	99–01	F	30–49	*20*
Netherlands	**45,000 – 50,000**	98–00	**–**	**27**	*1,2*
Norway	40 – 100	90–03	(0)	(0–19)	
Poland	**5,000 – 6,000**	95–00	**–**	**0–19**	*52*
Romania	**60 – 80**	90–02	+	0–19	*53*
Russia	**13,000 – 30,000**	90–00	**–**	**20–29**	*28,29,60*
Serbia & MN	20 – 40	90–02	0	0–19	*1,29,77a,155, 13,204*
Slovakia	5 – 40	80–99	–	50–79	
Spain	4 – 4	98–02	(F)	(–)	*16,10,6*
Sweden	**100 – 250**	99–00	–	10–19	
Ukraine	5,000 – 9,000	90–00	–	30–49	
UK	**40 – 48**	96–00	**–**	**19**	
Total (approx.)	99,000 – 140,000	**Overall trend** Large decline			
Breeding range	>2,000,000 km²	**Gen. length.** 5 **% Global pop.** 50–74			

(See p. 125, top, for data quality graph)

Country	Winter pop. size (individuals)	Year(s)	Trend	Mag.%	References
Azerbaijan	1,000 – 1,500	96–02	0	0–19	
Croatia	(1,500 – 2,000)	02	?	–	*16*
France	**11,000 – 17,500**	98–02	**+**	**50–79**	*9*
Rep. Ireland	**10,000 – 11,000**	94–00	**+**	**0–19**	
Netherlands	**260 – 260**	99–01	(F)	(–)	*3,4,8,9,11,12*
Portugal	13,000 – 39,000	95–00	(F)	(–)	*19,20,21*
Spain	**11,000 – 61,000**	90–01	**–**	**30–49**	*19*
Turkey	1,000 – 1,500	91–01	0	0–19	
UK	**15,400 – 15,400**	94–99	**+**	**89**	*44,52*
Total (approx.)	>64,000	**Overall trend** Small decline			
% in European IBAs	>90	**Gen. length** 5 **% Global pop.** 5–24			

Limosa lapponica
BAR-TAILED GODWIT

Non-SPEC (1994: 3W) **Status** (Secure)
Criteria —
European IUCN Red List Category —▼
Criteria —
Global IUCN Red List Category —
Criteria —

Country	Breeding pop. size (pairs)	Year(s)	Trend	Mag.%	References
Finland	(100 – 300)	98–02	–	30	
Norway	(1,000 – 3,000)	90–03	(0)	(0–19)	
Russia	(300 – 4,000)	90–00	(0)	(0–19)	*28,29,60*
Sweden	(5 – 50)	99–00	(0)	(0–19)	
Total (approx.)	1,400 – 7,400	**Overall trend** Stable			
Breeding range	>100,000 km²	**Gen. length.** 5		**% Global pop.** <5	

Country	Winter pop. size (individuals)	Year(s)	Trend	Mag.%	References
Azerbaijan	0 – 50	96–02	(0)	(0–19)	
Croatia	(50 – 100)	02	?	–	*16*
Denmark	**1,000 – 1,500**	99–00	?	–	*25*
France	**6,900 – 8,000**	98–02	F	**30–49**	*9*
Germany	1,000 – 15,000	95–00	–	60–90	*2*
Greece	**5 – 10**	95–99	F	**30–49**	
Iceland	**4 – 11**	90–99	0	**0–19**	*65,66,69,70,71, 72,73,74,58,59*
Rep. Ireland	**10,000 – 18,000**	94–00	0	**0–19**	
Italy	0 – 10	02	?	–	*22*
Netherlands	**35,000 – 35,000**	99–01	+	**44**	*3,4,8,9,11,12*
Portugal	1,800 – 6,000	95–97	(F)	(–)	*19,20,21*
Spain	**300 – 3,200**	90–01	+	**0–19**	*19*
UK	**61,600 – 61,600**	94–99	–	**9**	*44,52*
Total (approx.)	>120,000	**Overall trend** Small decline			
% in European IBAs	>90	**Gen. length** 5		**% Global pop.** 5–24	

Limosa lapponica has a predominantly Siberian breeding distribution, which extends into Europe in northern Fennoscandia and parts of adjacent Russia. Its European breeding population is small (as few as 1,400 pairs), but was stable between 1970–1990. Although the species declined in Finland during 1990–2000, it was stable across the rest of its European range, and probably remained stable overall. An improved knowledge of the species's winter distribution has shown that it no longer qualifies as Localised, and consequently it is provisionally evaluated as Secure.

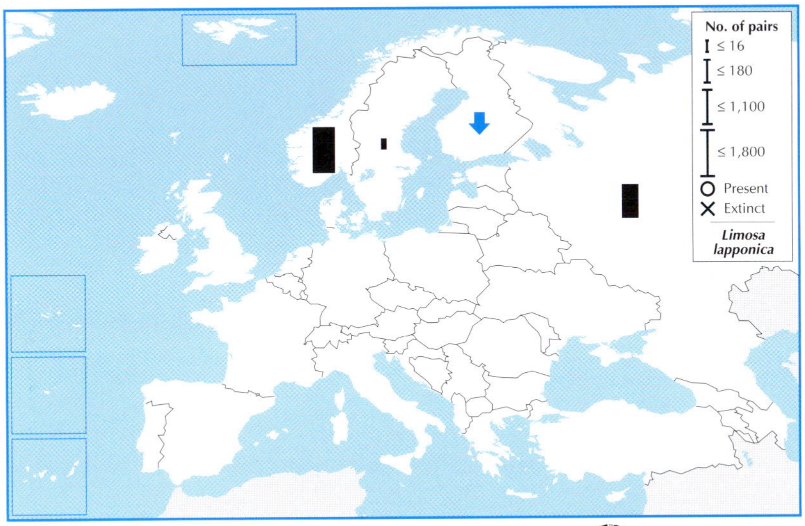

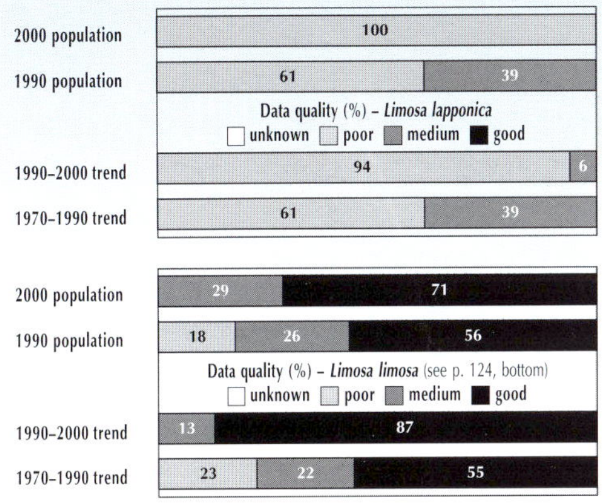

Numenius phaeopus
WHIMBREL

Non-SPECᴱ (1994: 4) **Status** (Secure)
Criteria —
European IUCN Red List Category —
Criteria —
Global IUCN Red List Category —
Criteria —

Country	Breeding pop. size (pairs)	Year(s)	Trend	Mag.%	References
Belarus	100 – 170	97–00	0	0–19	
Denmark					
Faroe Is.	2,500 – 2,500	95	(–)	(20–29)	
Greenland	(50 – 100)	80–90	?	–	
Estonia	400 – 500	98	+	20–29	*1*
Finland	30,000 – 50,000	98–02	–	10	
Iceland	100,000 – 250,000	99–01	?	–	*41*
Latvia	60 – 100	90–00	(0)	(0–19)	*23*
Norway	(10,000 – 20,000)	90–03	(0)	(0–19)	
Russia	(6,000 – 30,000)	90–00	(F)	(20–29)	*28,29,60*
Sweden	9,000 – 10,000	99–00	–	0–19	
UK	**530 – 530**	89–92	+	**8**	
Total (approx.)	160,000 – 360,000	**Overall trend** Unknown			
Breeding range	>2,000,000 km²	**Gen. length.** 5		**% Global pop.** 50–74	

Numenius phaeopus is a widespread summer visitor to northern Europe, which constitutes >50% of its global breeding range. Its European breeding population is large (>160,000 pairs), and was stable between 1970–1990. Although the species declined in a few countries—notably Finland – during 1990–2000, it was stable or increased across much of its European range. Trend data were not available for the stronghold population in Iceland, but there was no evidence to suggest that the species declined significantly overall. Consequently, it is provisionally evaluated as Secure.

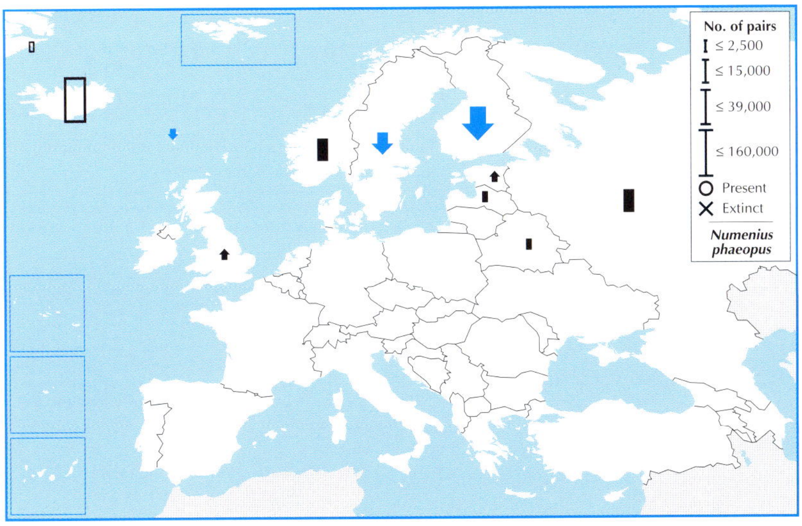

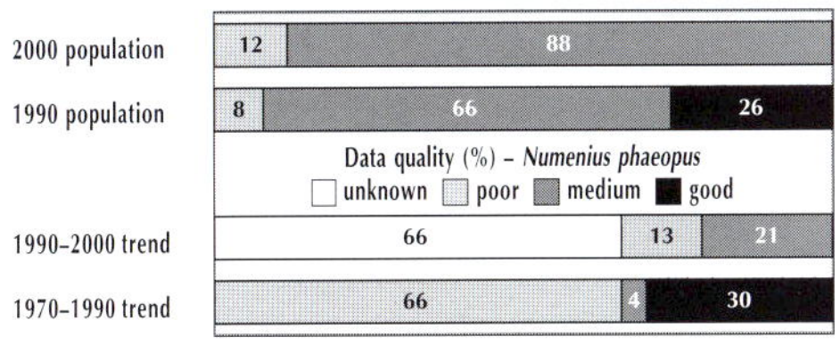

Numenius tenuirostris
SLENDER-BILLED CURLEW

SPEC 1 (1994: 1) **Status** Not Evaluated
Criteria —
European IUCN Red List Category —
Criteria —
Global IUCN Red List Category CR
Criteria C2a(ii); D1

Numenius tenuirostris occurs in Europe as an extremely rare passage migrant (and very occasionally as a winter visitor). The species is very poorly known, and its breeding area remains unknown. The number of verified records declined further during 1990–2000, in particular after 1995. The last flock recorded was of 19 birds wintering in Italy (1995–1996), while the last two verified records (United Kingdom 1998 and Greece 1999) were of single birds. Given its mostly passage occurrence, the European status of this globally threatened species is Not Evaluated.

Country	1970–1979		1980–1989		1990–2003	
	Records	No. of birds	Records	No. of birds	Records	No. of birds
Albania	–	–	–	–	2	6
Austria	3	3	1	1	–	–
Bulgaria	–	–	2	2–4	6	6–8
Cyprus	1	1	–	–	–	–
France	–	–	–	–	1	1
Greece	16	43–44	34	99–107	15	35
Hungary	8	16	–	–	8	16
Italy	5	5	6	10	3	21
Malta	1	1	–	–	–	–
Poland	2	2	–	–	–	–
Romania	10	67	3	4	1	1
Russia	1	1	1	1	–	–
Switzerland	1	1	–	–	–	–
Turkey	7	10	6	9	1	1
UK	–	–	–	–	1	1
Ukraine	6	55	4	4	14	14
(Former) Yugoslavia	1	7	1	5	–	–

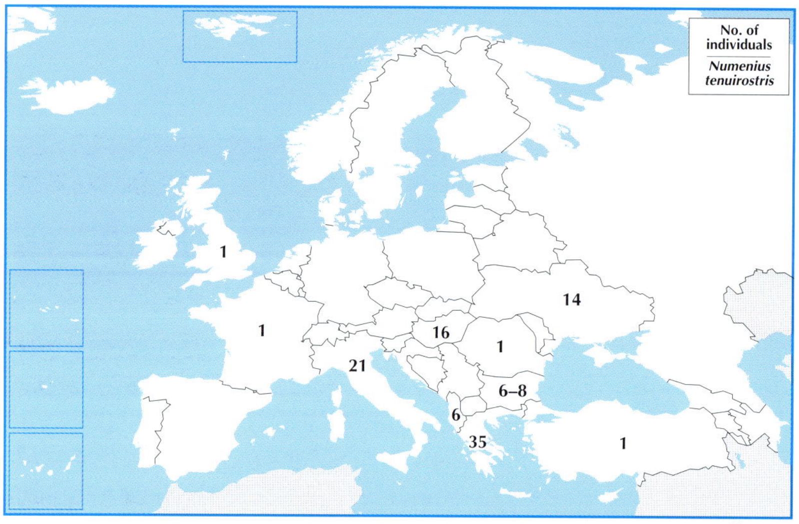

Numenius arquata
EURASIAN CURLEW

SPEC 2 (1994: 3W) **Status** Declining
Criteria Moderate continuing decline
European IUCN Red List Category —
Criteria —
Global IUCN Red List Category —
Criteria —

Numenius arquata is a widespread breeder across much of northern Europe (occurring more patchily farther south), which probably holds >75% of its global breeding population. Its European breeding population is large (>220,000 pairs), but underwent a moderate decline between 1970–1990 (based mainly on wintering data). Although some smaller populations were stable or increased during 1990–2000, key populations in the United Kingdom, Finland and Russia all declined, and the species underwent a moderate decline (>10%) overall. Consequently, it is evaluated as Declining.

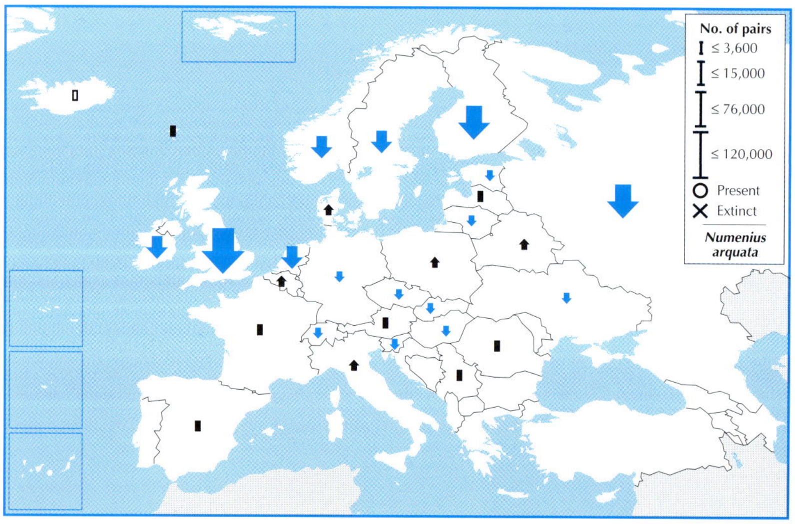

Country	Breeding pop. size (pairs)	Year(s)	Trend	Mag.%	References
Austria	**100 – 120**	98–02	**0**	**0–19**	
Belarus	**950 – 1,200**	97–00	**+**	**0–9**	
Belgium	**500 – 700**	00–02	**+**	**0–19**	*1*
Czech Rep.	**1 – 3**	00	**-**	**50–79**	
Denmark	**390 – 550**	90–97	(+)	(–)	*22*
Faroe Is.	0 – 3	90	(0)	(0–19)	
Estonia	(1,000 – 3,000)	98	–	20–29	*1*
Finland	35,000 – 50,000	98–02	–	10	
France	**1,500 – 1,800**	98–02	**0**	**0–19**	*4*
Germany	3,200 – 4,000	95–99	–	20–29	
Hungary	(20 – 60)	90–93	–	0–19	*14*
Iceland	(0 – 50)	90–00	?	–	*32,33*
Rep. Ireland	2,500 – 10,000	88–91	–	>80	
Italy	**1 – 1**	00	+	0–19	*18*
Latvia	150 – 200	90–00	(0)	(0–19)	*23*
Lithuania	(300 – 400)	99–01	(–)	(20–29)	*20*
Netherlands	**6,400 – 7,400**	98–00	**-**	**41**	*1*
Norway	(5,000 – 10,000)	97–02	(–)	(0–19)	*12,27,30*
Poland	**650 – 700**	95–00	**+**	**0–19**	*2*
Romania	(40 – 60)	90–02	(F)	(20–29)	
Russia	**48,000 – 120,000**	90–00	**-**	**20–29**	*28,29,60*
Serbia & MN	(0 – 5)	90–02	(0)	(0–19)	*1,29,125,155*
Slovakia	3 – 30	80–99	–	80–89	
Slovenia	**5 – 15**	90–00	–	30–49	
Spain	1 – 5	98–02	(F)	(–)	*16,10,6*
Sweden	10,000 – 20,000	99–00	**-**	**26**	
Switzerland	**1 – 4**	98–02	**-**	**30–49**	
Ukraine	50 – 100	90–00	–	30–49	
UK	**99,500 – 125,000**	00	**-**	**13**	*5,23*
Total (approx.)	220,000 – 360,000	**Overall trend** Moderate decline			
Breeding range	>4,000,000 km²	**Gen. length.** 5		**% Global pop.** 50–74	

Country	Winter pop. size (individuals)	Year(s)	Trend	Mag.%	References
Belgium	**3,000 – 3,000**	95–00	+	0–19	*1*
France	**18,000 – 22,000**	98–02	**F**	**50–79**	*9*
Germany	**40,000 – 150,000**	95–00	**F**	**20–29**	*2*
Rep. Ireland	**41,000 – 50,000**	94–00	**0**	**0–19**	
Netherlands	140,000 – 150,000	99–01	**0**	**0–19**	*3,4,8,9,11,12*
Spain	**2,000 – 8,700**	90–01	**+**	**50–79**	*19*
UK	**164,700 – 164,700**	94–99	**+**	**18**	*44,52*
Total (approx.)	>420,000	**Overall trend** Stable			
% in European IBAs	70–81	**Gen. length** 5		**% Global pop.** 25–49	

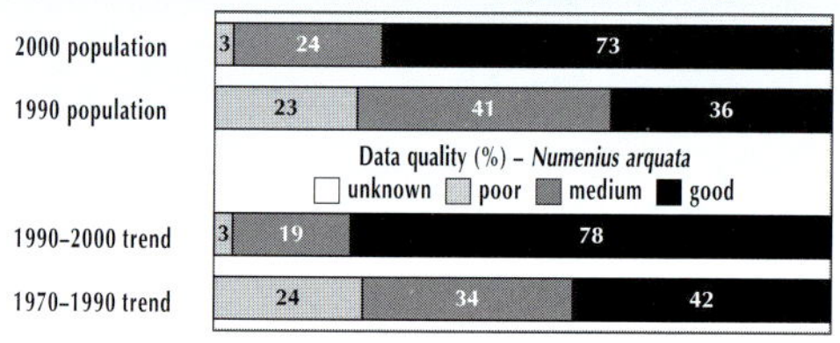

Tringa erythropus
SPOTTED REDSHANK

SPEC 3 (1994: —) **Status** (Declining)
Criteria Moderate recent decline
European IUCN Red List Category —
Criteria —
Global IUCN Red List Category —
Criteria —

Tringa erythropus breeds in Fennoscandia and northern Russia, with Europe accounting for less than a quarter of its global breeding range. Its European breeding population is relatively small (<42,000 pairs), but was stable between 1970–1990. Although the species remained broadly stable in Russia and Norway during 1990–2000, the key population in Finland declined (no trend data were available for Sweden), and the species underwent a moderate decline (>10%) overall. Consequently, this previously Secure species is now provisionally evaluated as Declining.

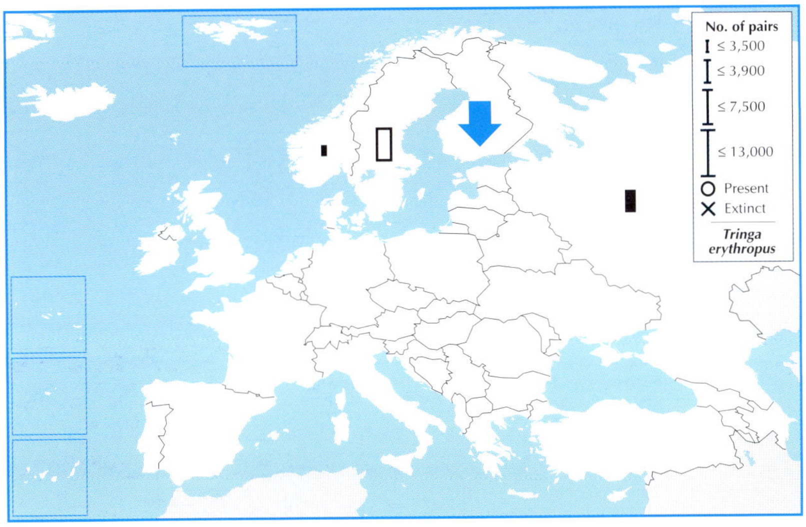

Country	Breeding pop. size (pairs)	Year(s)	Trend	Mag.%	References
Finland	10,000 – 15,000	98–02	-	**20**	
Norway	(2,000 – 6,000)	90–03	(0)	(0–19)	
Russia	(1,500 – 10,000)	90–00	(F)	(20–29)	*28,29,60*
Sweden	5,000 – 11,000	99–00	?	–	
Total (approx.)	19,000 – 42,000	**Overall trend** Moderate decline			
Breeding range	>500,000 km²	**Gen. length.** <3.3 **% Global pop.** 5–24			

Country	Winter pop. size (individuals)	Year(s)	Trend	Mag.%	References
Albania	**0 – 400**	95–02	-	**50–79**	
Belgium	**150 – 150**	95–00	(0)	(0–19)	*1*
France	316 – 550	98–02	?	–	*9*
Greece	**12 – 610**	95–99	**F**	**30–49**	
Rep. Ireland	**34 – 106**	94–00	0	0–19	
Italy	200 – 1,000	02	?	–	*22*
Netherlands	**95 – 95**	99–01	(F)	(–)	*3,4,8,9,11,12*
Portugal	12 – 68	98–02	(F)	(–)	*22*
Serbia & MN	200 – 700	90–02	F	30–59	
Slovenia	0 – 3	90–00	F	0–100	
Spain	**34 – 2,700**	90–01	-	**20–29**	*19*
Turkey	(200 – 600)	91–01	(0)	(0–19)	
UK	138 – 138	94–99	+	35	*48,52*
Total (approx.)	>1,400	**Overall trend** Small decline			
% in European IBAs	>70	**Gen. length** <3.3 **% Global pop.** <5			

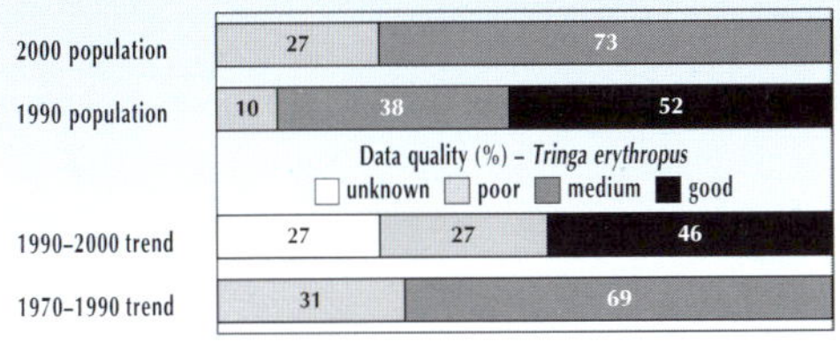

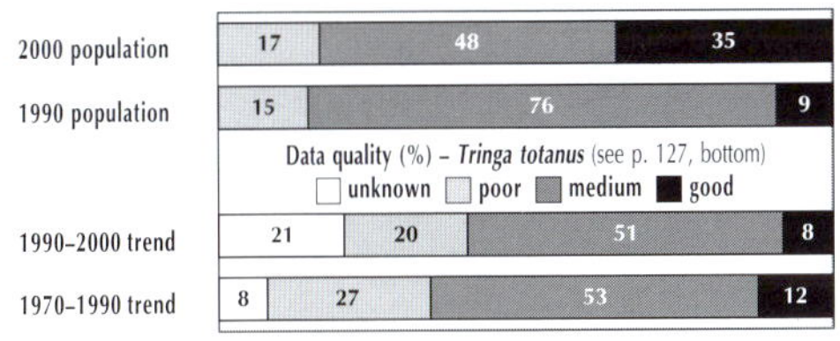

Tringa totanus
COMMON REDSHANK

SPEC 2 (1994: 2) **Status** Declining
Criteria Moderate continuing decline
European IUCN Red List Category —
Criteria —
Global IUCN Red List Category —
Criteria —

Tringa totanus is a widespread breeder across much of Europe, which constitutes >50% of its global breeding range. Its European breeding population is large (>280,000 pairs), but underwent a moderate decline between 1970–1990. Although several populations—notably sizeable ones in Russia and Norway—were stable or increased during 1990–2000, the species continued to decline across much of its European range, and underwent a moderate decline (>10%) overall. Consequently, it is evaluated as Declining.

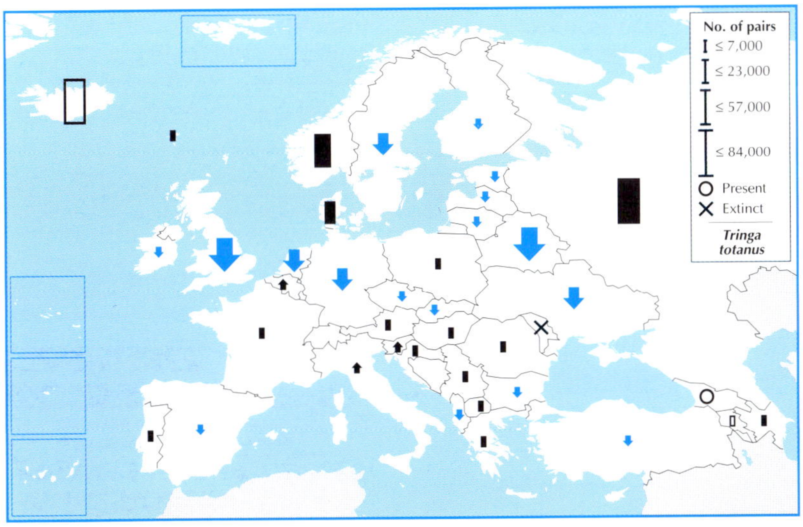

Country	Breeding pop. size (pairs)	Year(s)	Trend	Mag.%	References
Albania	30 – 90	96–02	–	0–19	*4*
Armenia	350 – 850	98–02	?	–	
Austria	**120 – 190**	98–02	**0**	**0–19**	
Azerbaijan	500 – 5,000	96–00	(0)	(0–19)	
Belarus	**40,000 – 70,000**	97–00	–	0–9	
Belgium	**413 – 456**	00–02	+	**0–19**	*1*
Bulgaria	**5 – 20**	95–02	–	>80	
Croatia	**6 – 10**	02	(0)	(0–19)	*70*
Czech Rep.	**25 – 40**	00	-	**30–49**	
Denmark	**12,000 – 15,000**	99	(F)	(–)	*22*
Faroe Is.	25 – 25	95	(0)	(0–19)	
Estonia	(5,000 – 6,000)	98	–	20–29	*1*
Finland	6,000 – 8,000	98–02	-	**20**	
France	**1,300 – 1,500**	98–02	**0**	**0–19**	*4*
Georgia	Present	03	?	–	
Germany	**9,700 – 12,000**	95–99	–	20–29	
Greece	(400 – 800)	95–00	(0)	(0–19)	
Hungary	(600 – 800)	90–93	0	0–19	*14*
Iceland	50,000 – 140,000	99–01	?	–	*41*
Rep. Ireland	250 – 1,000	88–91	–	50–79	
Italy	**800 – 1,200**	03	+	**20–29**	
Latvia	300 – 700	90–00	–	30–49	*16,24*
Lithuania	600 – 800	99–01	(–)	(20–29)	*20*
Macedonia	(10 – 20)	90–00	(0)	(0–19)	
Moldova	**0 – 0**	00	-	**X**	
Netherlands	**20,000 – 25,000**	98–00	-	**9**	*1*
Norway	(40,000 – 80,000)	90–03	(0)	(0–19)	
Poland	**2,000 – 2,500**	95–00	0	0–19	*2*
Portugal	(10 – 50)	02	(F)	(–)	
Romania	(800 – 1,200)	90–02	(0)	(0–19)	*53*
Russia	30,000 – 140,000	90–00	0	0–19	*28,29,60*
Serbia & MN	**360 – 470**	90–02	0	0–19	*1,179,148,200, 29,77a,155,204, 13,78,214*
Slovakia	35 – 70	80–99	–	30–49	
Slovenia	5 – 15	90–00	+	50–79	
Spain	(650 – 850)	98–02	(–)	(0–19)	*5,16,10*
Sweden	10,000 – 20,000	99–00	–	10–19	
Turkey	(4,000 – 8,000)	01	(–)	(0–19)	
Ukraine	14,000 – 23,000	90–00	–	0–19	
UK	**31,400 – 44,400**	00	–	48	*5,23*
Total (approx.)	280,000 – 610,000	**Overall trend** Moderate decline			
Breeding range	>4,000,000 km²	**Gen. length.** <3.3 **% Global pop.** 50–74			

(See p. 127, top, for data quality graph)

	Winter pop. size (individuals)		
Total (approx.)	>180,000	**Overall trend** Small decline	
% in European IBAs	63–74	**Gen. length** <3.3	**% Global pop.** 5–24

Tringa stagnatilis
MARSH SANDPIPER

Non-SPEC (1994: —) **Status** (Secure)
Criteria —
European IUCN Red List Category —
Criteria —
Global IUCN Red List Category —
Criteria —

Country	Breeding pop. size (pairs)	Year(s)	Trend	Mag.%	References
Belarus	**40 – 70**	96–00	+	10–19	
Finland	0 – 5	98–02	0	0–19	
Latvia	10 – 30	90–00	F	20–29	*23,17*
Lithuania	Present	99–01	?	–	*20*
Poland	**3 – 5**	95–00	+	**>80**	*2*
Romania	(20 – 50)	90–02	(0)	(0–19)	*53*
Russia	12,000 – 32,000	90–00	-	20–29	*28,29,60*
Ukraine	(50 – 100)	90–00	-	20–29	
Total (approx.)	12,000 – 32,000	**Overall trend** Moderate decline			
Breeding range	>500,000 km²	**Gen. length.** <3.3		**% Global pop.** 25–49	

Tringa stagnatilis is a summer visitor to central Russia and parts of eastern Europe, which accounts for less than half of its global breeding range. Its European breeding population is relatively small (<32,000 pairs), but increased substantially between 1970–1990. Although some marginal populations were stable or increased during 1990–2000, the species declined in its Russian stronghold, and underwent a moderate decline (>10%) overall. Nevertheless, this decline is probably outweighed by the earlier increase, and consequently it is provisionally evaluated as Secure.

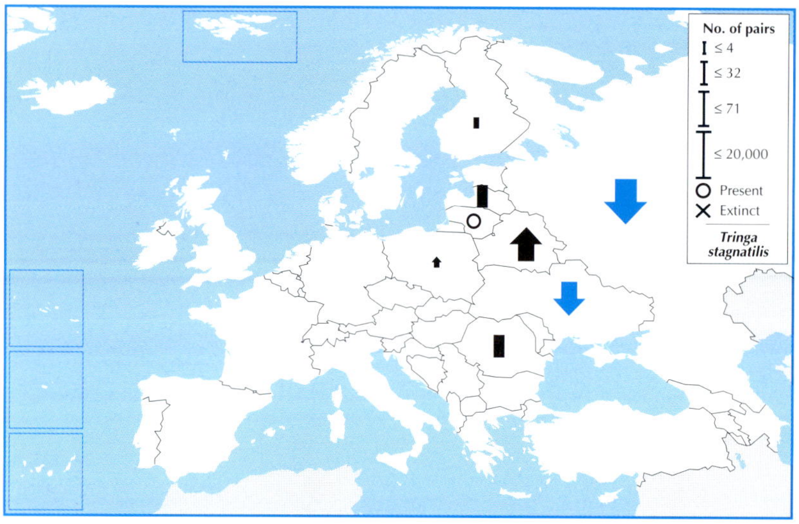

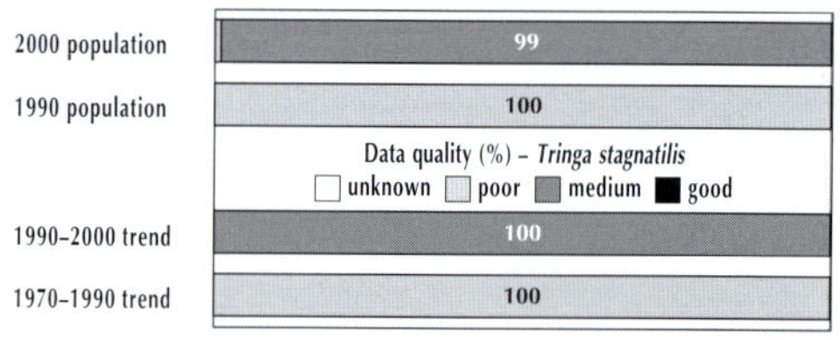

Tringa nebularia
COMMON GREENSHANK

Non-SPEC (1994: —) **Status** Secure
Criteria —
European IUCN Red List Category —
Criteria —
Global IUCN Red List Category —
Criteria —

Country	Breeding pop. size (pairs)	Year(s)	Trend	Mag.%	References
Belarus	200 – 250	97–00	+	10–19	
Estonia	150 – 200	98	+	50–79	*1*
Finland	30,000 – 40,000	98–02	+	**10**	
Germany	**0 – 1**	95–99	+	**N**	
Rep. Ireland	0 – 1	88–91	(F)	(–)	
Latvia	(0 – 5)	90–00	?	–	23,26
Lithuania	Present	02	?	–	*20*
Norway	(15,000 – 30,000)	90–02	(0)	(0–19)	
Russia	(14,000 – 65,000)	90–00	(F)	(20–29)	*28,29,60*
Sweden	15,000 – 25,000	99–00	(0)	(0–19)	
UK	**720 – 1,500**	95	(–)	(0–19)	*14*
Total (approx.)	75,000 – 160,000	**Overall trend** Stable			
Breeding range	>2,000,000 km²	**Gen. length.** 5		**% Global pop.** 5–24	

Country	Winter pop. size (individuals)	Year(s)	Trend	Mag.%	References
Albania	**1 – 22**	95–02	(F)	(–)	
Azerbaijan	(100 – 1,000)	96–02	(0)	(0–19)	
Belgium	20 – 20	95–00	(0)	(0–19)	*1*
Bulgaria	**0 – 3**	97–01	?	–	
Cyprus	10 – 100	94–02	(0)	(0–9)	
France	177 – 312	98–02	?	–	*9*
Greece	**17 – 86**	95–99	**F**	**30–49**	
Rep. Ireland	**970 – 1,100**	94–00	**0**	**0–19**	
Italy	100 – 300	02	0	0–19	*22*
Netherlands	**11 – 11**	99–01	(F)	(–)	*3,4,8,9,11,12*
Portugal	74 – 256	95–00	(F)	(–)	*19,20,21,22*
Serbia & MN	10 – 50	90–02	F	30–59	
Slovenia	0 – 4	90–00	F	0–100	
Spain	**260 – 1,300**	90–01	+	**>80**	*19*
Turkey	(150 – 300)	91–01	(0)	(0–19)	
Ukraine	(0 – 50)	90–00	(0)	(0–19)	
UK	701 – 701	94–99	+	55	*48,52*
Total (approx.)	>2,600	**Overall trend** Small increase			
% in European IBAs	8–16	**Gen. length** 5		**% Global pop.** <5	

Tringa nebularia is a widespread breeder in northern Europe, which accounts for less than a quarter of its global breeding range. Its European breeding population is relatively small (<160,000 pairs), but was stable between 1970–1990. Although the species declined in the United Kingdom during 1990–2000, populations in the rest of its European range were stable or increased, and the species remained stable overall. Consequently, it is evaluated as Secure.

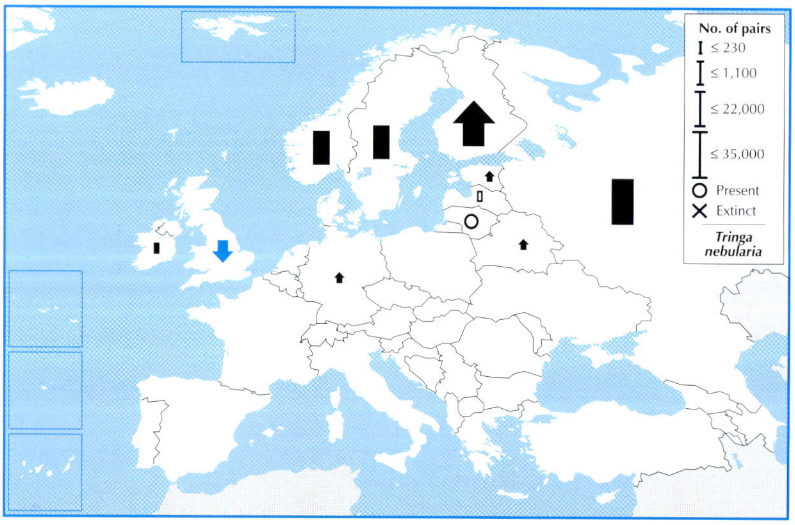

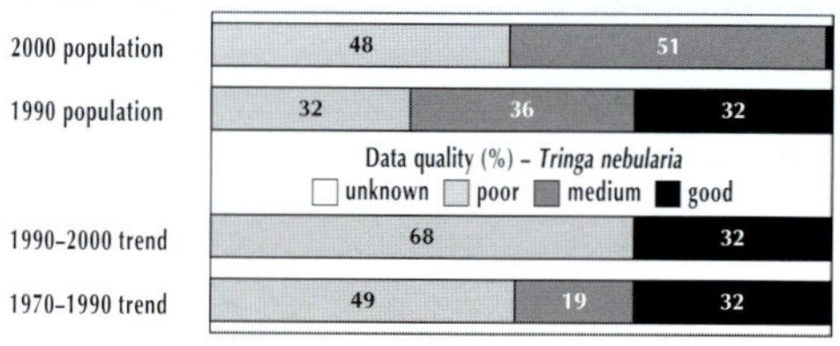

Tringa ochropus
GREEN SANDPIPER

Non-SPEC (1994: —) **Status** Secure
Criteria —
European IUCN Red List Category —
Criteria —
Global IUCN Red List Category —
Criteria —

Tringa ochropus is a widespread breeder in northern and central Europe, which accounts for less than half of its global breeding range. Its European breeding population is large (>330,000 pairs), and was stable between 1970–1990. The species remained stable overall during 1990–2000, with all national trends either stable, fluctuating (as in the Russian stronghold) or increasing. Consequently, the species is evaluated as Secure.

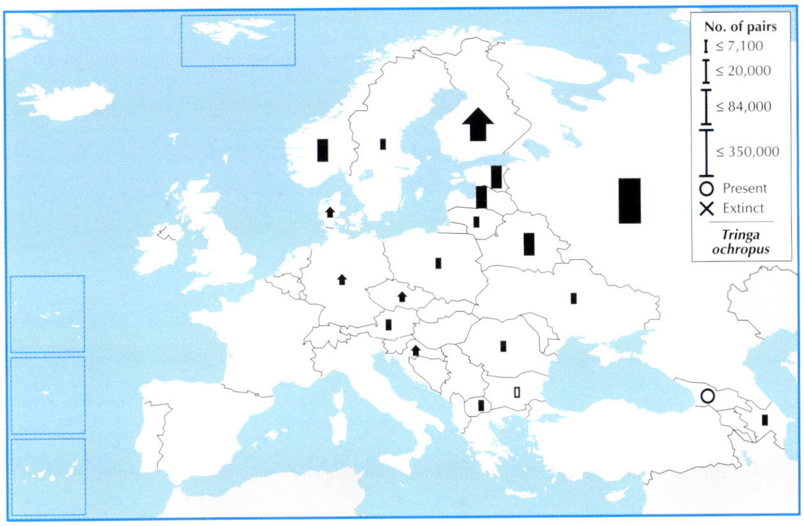

Country	Breeding pop. size (pairs)	Year(s)	Trend	Mag.%	References
Austria	(0 – 3)	98–02	(0)	(0–19)	
Azerbaijan	(10 – 100)	96–00	(0)	(0–19)	
Belarus	**10,000 – 15,000**	97–02	0	0–19	
Bulgaria	(5 – 30)	96–02	?	–	
Croatia	(1 – 10)	02	(+)	(N)	*70,16*
Czech Rep.	**40 – 70**	00	**+**	**>80**	
Denmark	11 – 20	02	(+)	(–)	*3,8*
Estonia	(10,000 – 15,000)	98	0	0–19	*1*
Finland	70,000 – 100,000	98–02	**+**	**30**	
Georgia	Present	03	?	–	
Germany	650 – 1,050	95–99	+	20–29	
Latvia	15,000 – 25,000	90–00	0	0–19	*23,18*
Lithuania	(5,000 – 7,000)	99–01	(0)	(0–19)	*20*
Macedonia	(10 – 50)	90–00	(F)	(20–29)	
Norway	(10,000 – 20,000)	90–02	(0)	(0–19)	
Poland	**5,000 – 10,000**	00–02	0	0–19	*23*
Romania	(5 – 10)	90–02	(F)	(20–29)	
Russia	200,000 – 600,000	90–00	(F)	(20–29)	*28,29,60*
Sweden	(3,000 – 10,000)	99–00	**0**	**0**	
Ukraine	850 – 1,400	90–00	0	0–19	
Total (approx.)	330,000 – 800,000	**Overall trend** Fluctuating			
Breeding range	>4,000,000 km²	**Gen. length.** <3.3		**% Global pop.** 25–49	

Country	Winter pop. size (individuals)	Year(s)	Trend	Mag.%	References
Azerbaijan	0 – 100	96–02	(0)	(0–19)	
Croatia	(500 – 1,000)	02	?	–	*16*
Cyprus	(10 – 100)	94–02	(0)	(0–9)	
Czech Rep.	**40 – 100**	90–00	0	0–19	
France	500 – 1,000	98–02	?	–	*9*
Germany	(250 – 1,000)	95–00	?	–	
Greece	**500 – 1,000**	95–99	**F**	**30–49**	
Italy	(500 – 1,000)	02	0	0–19	*22*
Netherlands	93 – 93	99–01	(+)	(–)	*3,4,8,9,11,12*
Romania	(60 – 300)	90–00	(F)	(>80)	*60*
Serbia & MN	100 – 200	90–02	(F)	(30–49)	
Spain	**25 – 770**	90–01	(+)	(>80)	**19**
Turkey	250 – 1,000	91–01	(0)	(0–19)	
UK	780 – 780	89–94	0	4	*51*
Total (approx.)	>3,700	**Overall trend** Stable			
% in European IBAs	5–7	**Gen. length** <3.3		**% Global pop.** <5	

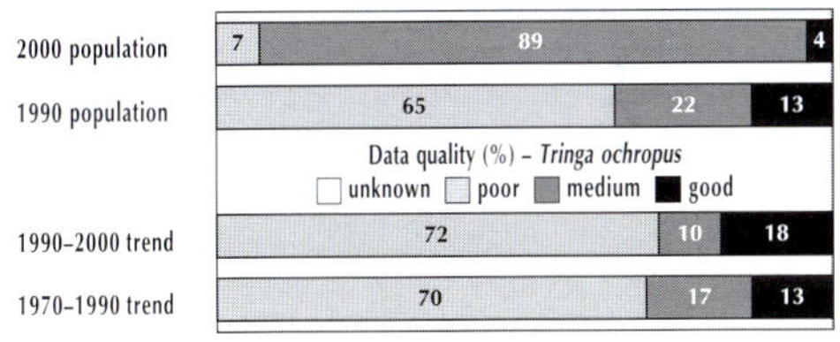

Tringa glareola
WOOD SANDPIPER

SPEC 3 (1994: 3) **Status** Depleted
Criteria Moderate historical decline
European IUCN Red List Category —
Criteria —
Global IUCN Red List Category —
Criteria —

Tringa glareola is a widespread summer visitor to northern Europe, which accounts for less than half of its global breeding range. Its European breeding population is large (>350,000 pairs), but underwent a moderate decline between 1970–1990. Although the species was stable overall during 1990–2000—with key populations in Finland and Russia stable or fluctuating—its population has clearly not yet recovered to the level that preceded its decline. Consequently, it is evaluated as Depleted.

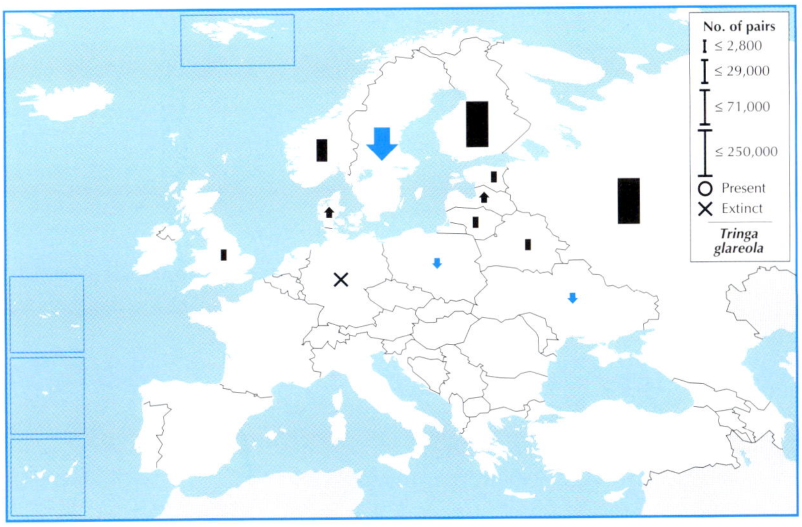

Country	Breeding pop. size (pairs)	Year(s)	Trend	Mag.%	References
Belarus	**2,500 – 3,000**	97–00	**0**	**0–19**	
Denmark	**100 – 100**	01	**+**	**30–49**	*7*
Estonia	2,000 – 3,000	98	0	0–19	*1*
Finland	200,000 – 300,000	98–02	**0**	**5**	
Germany	**0 – 0**	95–99	**-**	**X**	
Latvia	800 – 1,000	90–00	(+)	(50–79)	*23*
Lithuania	100 – 150	99–01	0	0–19	*20*
Norway	(20,000 – 40,000)	90–02	(0)	(0–19)	
Poland	**0 – 5**	98–00	**-**	**50–79**	*53*
Russia	75,000 – 750,000	90–00	(F)	(20–29)	*28,29,60*
Sweden	(50,000 – 100,000)	99–00	-	0–19	
Ukraine	(0 – 10)	90–00	(–)	(30–49)	*14*
UK	**4 – 8**	96–00	**F**	**144**	
Total (approx.)	350,000 – 1,200,000	**Overall trend** Stable			
Breeding range	>3,000,000 km²	**Gen. length.** <3.3		**% Global pop.** 25–49	

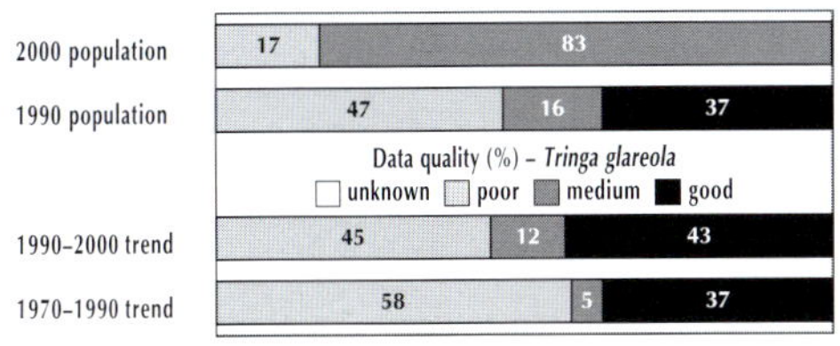

Xenus cinereus
TEREK SANDPIPER

Non-SPEC (1994: —) **Status** (Secure)
Criteria —
European IUCN Red List Category —
Criteria —
Global IUCN Red List Category —
Criteria —

Country	Breeding pop. size (pairs)	Year(s)	Trend	Mag.%	References
Belarus	**100 – 150**	97–00	+	**0–9**	
Finland	**15 – 20**	98–02	–	**25**	
Latvia	0 – 10	90–00	?	–	*17,4,5,37*
Russia	15,000 – 80,000	90–00	(F)	(20–29)	*28,29,60*
Ukraine	**300 – 500**	90–00	+	**0–9**	
Total (approx.)	15,000 – 81,000	**Overall trend** Fluctuating			
Breeding range	>500,000 km²	**Gen. length.** <3.3		**% Global pop.** 5–24	

Xenus cinereus breeds mainly in central and northern Russia, but also patchily in adjacent countries, with Europe accounting for less than a quarter of its global breeding range. Its European breeding population is relatively small (<81,000 pairs), but was stable between 1970–1990. The species remained stable overall during 1990–2000, with the stronghold population in Russia fluctuating, but staying broadly stable. Consequently, the species is provisionally evaluated as Secure.

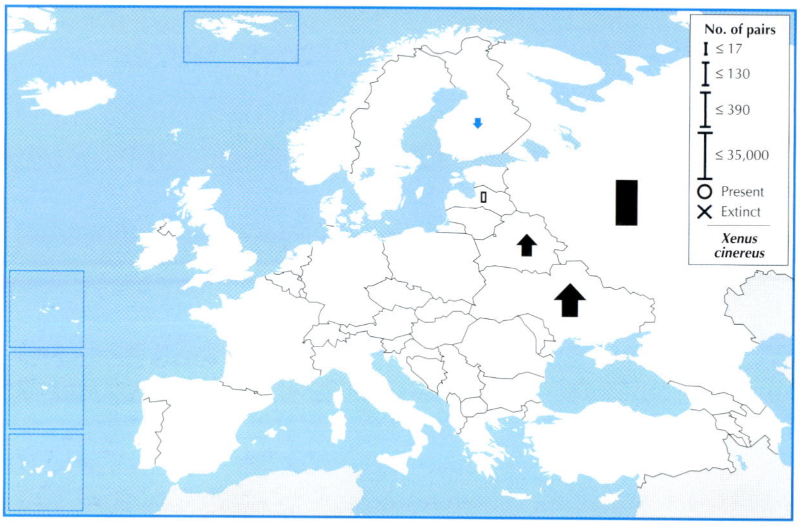

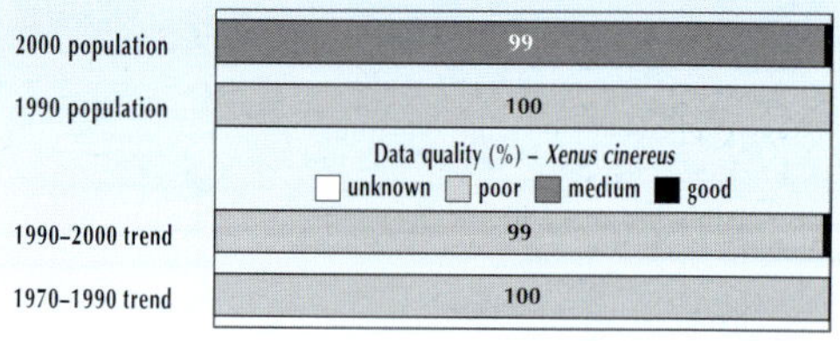

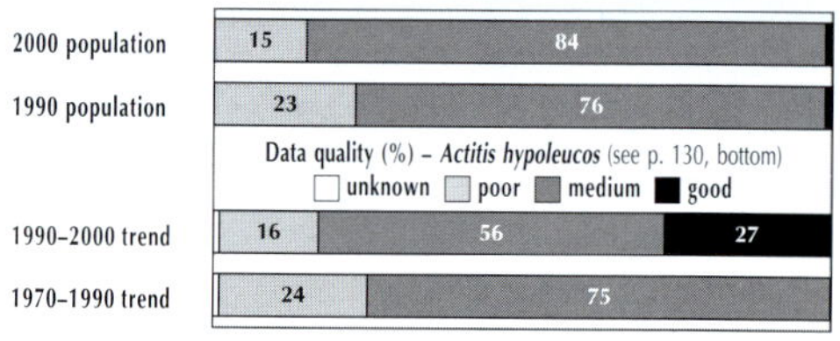

Actitis hypoleucos
COMMON SANDPIPER

SPEC 3 (1994: —) **Status** (Declining)
Criteria Moderate recent decline
European IUCN Red List Category —
Criteria —
Global IUCN Red List Category —
Criteria —

Country	Breeding pop. size (pairs)	Year(s)	Trend	Mag.%	References
Albania	(20 – 100)	96–02	–	0–19	
Armenia	350 – 850	98–02	?	–	
Austria	200 – 300	98–02	0	0–19	
Azerbaijan	(500 – 2,000)	96–00	(0)	(0–19)	
Belarus	**12,000 – 14,500**	97–00	0	0–19	
Belgium	**7 – 8**	00–03	+	0–19	*1*
Bosnia & HG	Present	90–03	?	–	
Bulgaria	100 – 250	96–02	0	0–19	
Croatia	300 – 380	02	(+)	(20–29)	*70*
Czech Rep.	**200 – 400**	00	–	**30–49**	
Denmark	0 – 1	98–01	**0**	**0–19**	*4,5,6,7*
Estonia	(5,000 – 10,000)	98	0	0–19	*1*
Finland	150,000 – 250,000	98–02	–	**20**	
France	800 – 1,000	98–02	0	0–19	*4*
Georgia	Present	03	?	–	
Germany	240 – 350	95–99	–	0–19	
Greece	(50 – 100)	95–00	(0)	(0–19)	
Hungary	150 – 180	95–02	0	0–19	
Rep. Ireland	1,000 – 2,500	88–91	?	–	
Italy	(500 – 1,000)	03	(0)	(0–19)	
Latvia	3,500 – 5,000	90–00	0	0–19	*23,17*
Lithuania	(1,500 – 2,000)	99–01	(–)	(20–29)	*20*
Macedonia	(50 – 100)	90–00	(0)	(0–19)	
Moldova	70 – 100	90–00	–	30–49	
Netherlands	**5 – 10**	98–00	+	305	*1*
Norway	(100,000 – 200,000)	90–03	(0)	(0–19)	
Poland	1,000 – 2,000	97–00	(0)	(0–19)	*1*
Portugal	(250 – 2,000)	02	(0)	(0–19)	
Romania	16,000 – 22,000	90–02	(+)	(0–19)	*32*
Russia	350,000 – 900,000	90–00	0	0–19	*28,29,60*
Serbia & MN	700 – 1,100	90–02	(–)	(0–19)	*1,155,227,62, 225,152*
Slovakia	700 – 1,300	80–99	+	20–29	
Slovenia	250 – 500	94	(–)	(0–19)	
Spain	(2,500 – 10,000)	98–02	?	–	*12,1*
Sweden	50,000 – 125,000	99–00	–	**33**	
Switzerland	**80 – 100**	93–96	–	20–29	
Turkey	(1,500 – 4,500)	01	(–)	(0–19)	
Ukraine	5,500 – 8,300	90–00	–	10–19	
UK	12,000 – 12,000	00	–	**24**	*5,31*
Total (approx.)	720,000 – 1,600,000	**Overall trend** Moderate decline			
Breeding range	>6,000,000 km²	**Gen. length.** 5		**% Global pop.** 25–49	

(See p. 130, top, for data quality graph)

	Winter pop. size (individuals)			
Total (approx.)	>1,600	**Overall trend** Moderate increase		
% in European IBAs	5–8	**Gen. length** 5		**% Global pop.** <5

Actitis hypoleucos is a widespread breeder across most of Europe, which accounts for less than half of its global breeding range. Its European breeding population is large (>720,000 pairs), and was stable between 1970–1990. Although many European populations, including key ones in Russia and Norway, were stable or increased during 1990–2000, the species suffered widespread declines—most notably in Sweden and Finland—and underwent a moderate decline (>10%) overall. Consequently, this previously Secure species is now provisionally evaluated as Declining.

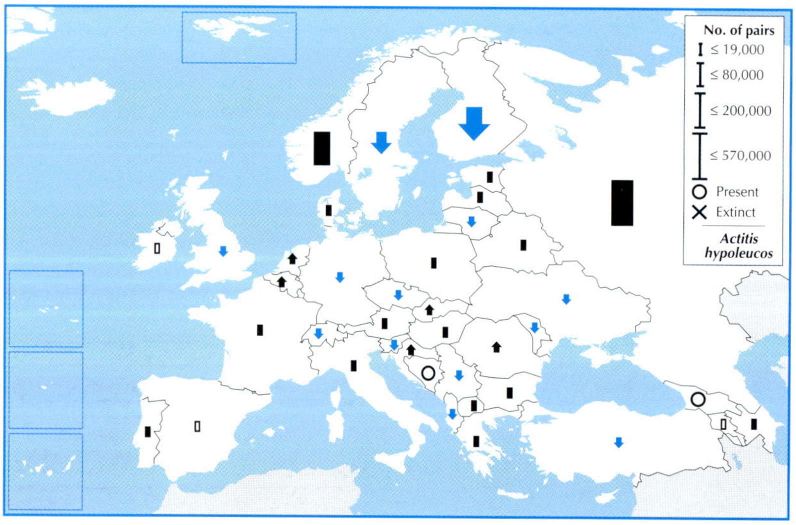

Arenaria interpres
RUDDY TURNSTONE

Non-SPEC (1994: —) **Status** (Secure)
Criteria —
European IUCN Red List Category —
Criteria —
Global IUCN Red List Category —
Criteria —

Arenaria interpres breeds in coastal areas of northern Europe, which accounts for less than a quarter of its global breeding range. Its European breeding population is relatively small (<81,000 pairs), but was stable between 1970–1990. Although there were declines in a couple of countries during 1990–2000, the species was stable or increased across most of its European range—including the key population in Greenland—and remained stable overall. Consequently, it is provisionally evaluated as Secure.

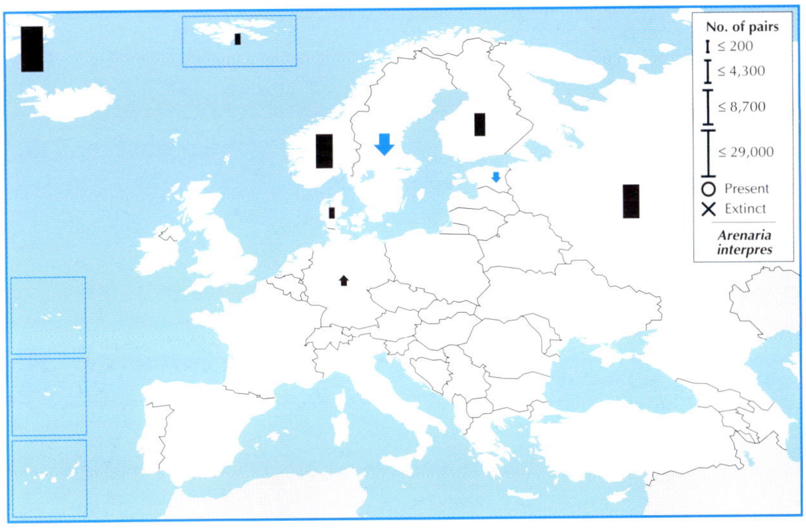

Country	Breeding pop. size (pairs)	Year(s)	Trend	Mag.%	References
Denmark	**37 – 41**	00	**0**	**0–19**	*6*
Greenland	20,000 – 40,000	90–00	(0)	(0–19)	*11*
Estonia	150 – 250	98	–	20–29	*1*
Finland	4,000 – 4,500	98–02	**0**	**0–19**	
Germany	**4 – 4**	95–99	+	**N**	
Norway	(5,000 – 15,000)	90–03	0	0–19	*4*
Svalbard	50 – 500	90–03	0	0–19	
Russia	2,700 – 17,000	90–00	(F)	(20–29)	*28,29,60*
Sweden	2,000 – 4,000	99–00	–	30–49	
Total (approx.)	34,000 – 81,000	**Overall trend** Stable			
Breeding range	>1,000,000 km²	**Gen. length.** 5		**% Global pop.** 5–24	

Country	Winter pop. size (individuals)	Year(s)	Trend	Mag.%	References
Azerbaijan	(0 – 50)	96–02	(0)	(0–19)	
Belgium	**1,300 – 1,300**	95–00	+	**>80**	*1*
Bulgaria	**0 – 2**	97–01	?	–	
Cyprus	5 – 50	94–02	(0)	(0–9)	
Denmark	10 – 50	99–00	(0)	(0–19)	*25*
Faroe Is.	500 – 2,000	92	?	–	*3*
Estonia	(0 – 1)	98	?	–	*1*
France	**9,580 – 15,000**	98–02	+	50–79	*9*
Germany	**500 – 2,500**	95–00	–	**39–80**	*2*
Greece	**0 – 4**	95–99	F	**30–49**	
Iceland	(2,000 – 5,000)	78–94	?	–	*48*
Rep. Ireland	**8,400 – 8,900**	94–00	0	0–19	
Italy	(20 – 70)	02	F	20–29	*22*
Latvia	(1 – 2)	90–00	?	–	
Netherlands	**4,400 – 4,400**	99–01	–	**45**	*3,4,8,9,11,12*
Portugal	Present	02	?	–	
Spain	**590 – 1,700**	90–01	+	**>80**	*19*
Turkey	20 – 80	91–01	F	**>80**	
UK	**52,400 – 52,400**	94–99	–	**28**	*44,52*
Gibraltar	0 – 3	00–03	0	0–19	
Total (approx.)	>80,000	**Overall trend** Small decline			
% in European IBAs	48–49	**Gen. length** 5		**% Global pop.** 5–24	

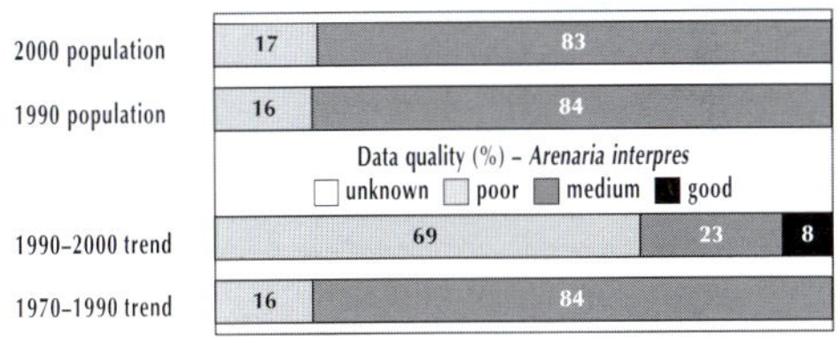

Phalaropus lobatus
RED-NECKED PHALAROPE

Non-SPEC (1994: —) **Status** (Secure)
Criteria —
European IUCN Red List Category —
Criteria —
Global IUCN Red List Category —
Criteria —

Phalaropus lobatus is a widespread summer visitor to northernmost Europe, which accounts for less than a quarter of its global breeding range. Its European breeding population is relatively large (>85,000 pairs), and was stable between 1970–1990. Although the species declined in Iceland and Finland during 1990–2000, it remained stable across much of its European range (its trend in Greenland and Sweden was unknown), and probably declined only slightly overall. Consequently, the species is provisionally evaluated as Secure.

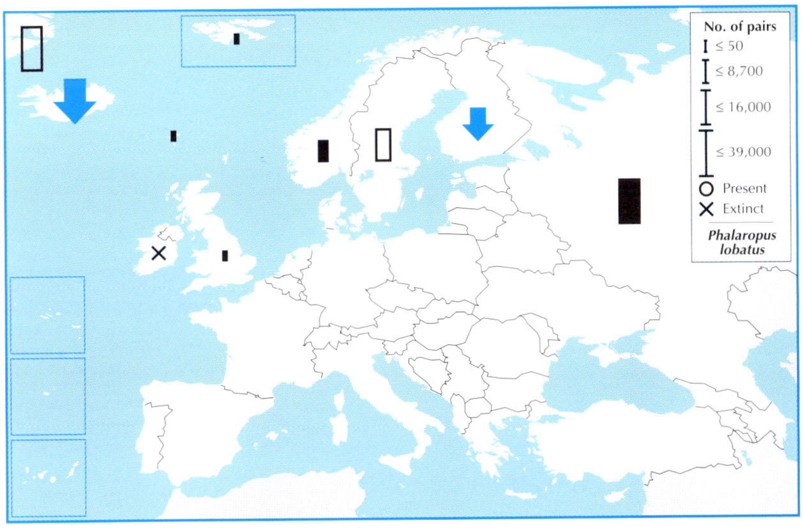

Country	Breeding pop. size (pairs)	Year(s)	Trend	Mag.%	References
Denmark					
Faroe Is.	50 – 50	95	(0)	(0–19)	
Greenland	(20,000 – 40,000)	90–00	?	–	
Finland	10,000 – 20,000	98–02	–	**10**	
Iceland	30,000 – 50,000	99–01	(–)	(–)	*41*
Rep. Ireland	0 – 0	00–02	–	X	
Norway	(5,000 – 15,000)	90–03	(0)	(0–19)	
Svalbard	(10 – 100)	90–03	(0)	(0–19)	
Russia	(10,000 – 70,000)	90–00	(0)	(0–19)	*28,29,60*
Sweden	(10,000 – 25,000)	99–00	?	–	
UK	25 – 25	96–00	F	**30**	
Total (approx.)	85,000 – 220,000	**Overall trend** Small decline			
Breeding range	>1,000,000 km²	**Gen. length.** <3.3		**% Global pop.** 5–24	

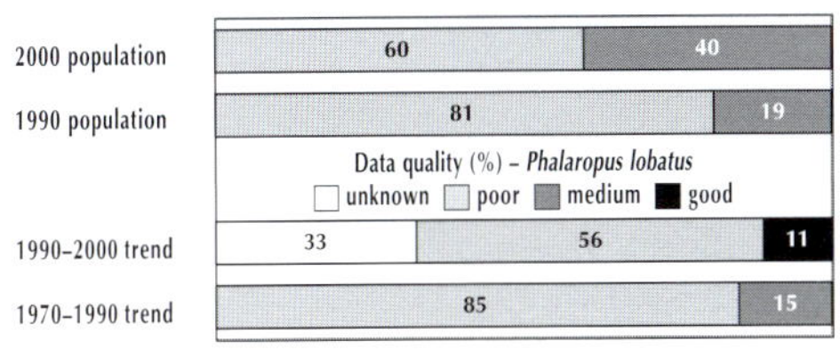

Phalaropus fulicarius
GREY PHALAROPE

Non-SPEC (1994: —) **Status** Secure
Criteria —
European IUCN Red List Category —▼▼
Criteria —
Global IUCN Red List Category —
Criteria —

Country	Breeding pop. size (pairs)	Year(s)	Trend	Mag.%	References
Denmark					
Greenland	(150 – 500)	90–00	?	–	
Iceland	**40 – 50**	87–97	(F)	(–)	*21,41*
Norway					
Svalbard	200 – 1,000	95	(0)	(0–19)	
Russia	(2 – 150)	90–02	?	–	
Total (approx.)	390 – 1,700	**Overall trend** Stable			
Breeding range	>100,000 km²	**Gen. length.** <3.3		**% Global pop.** <5	

Phalaropus fulicarius is a summer visitor to the European Arctic, which accounts for a tiny proportion of its global breeding range. Its European breeding population is very small (as few as 390 pairs), but was stable between 1970–1990. The key Svalbard population remained stable during 1990–2000, and the species was probably stable overall. Although the size of the European population could render it susceptible to the risks affecting small populations, it is marginal to a much larger non-European population. Consequently, the species is provisionally evaluated as Secure.

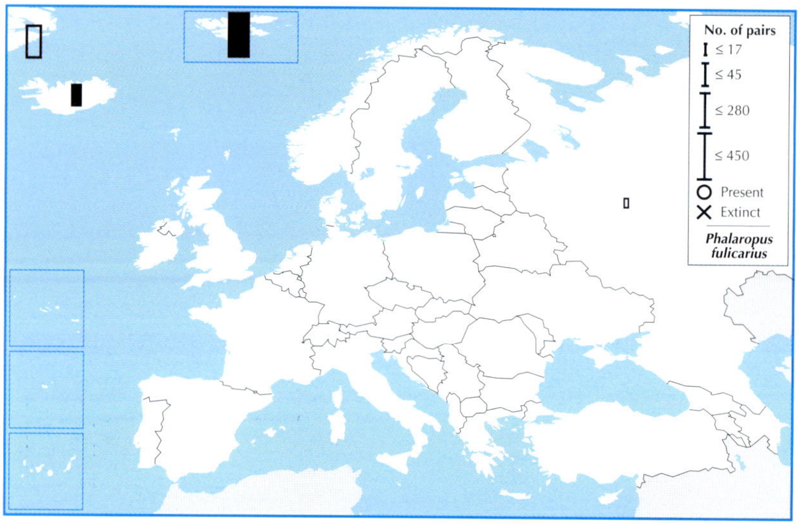

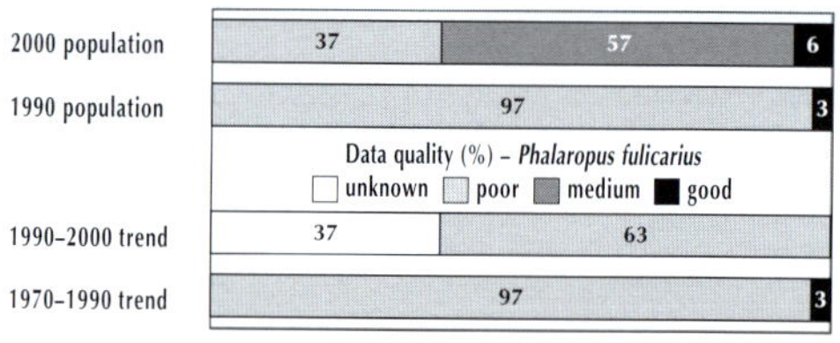

Stercorarius pomarinus
POMARINE JAEGER

Non-SPEC (1994: —) **Status** (Secure)
Criteria —
European IUCN Red List Category —
Criteria —
Global IUCN Red List Category —
Criteria —

Country	Breeding pop. size (pairs)	Year(s)	Trend	Mag.%	References
Russia	(20,000 – 50,000)	90–02	?	–	81
Total (approx.)	20,000 – 50,000	**Overall trend** Unknown			
Breeding range	>100,000 km²	**Gen. length.** 11		**% Global pop.** 5–24	

Stercorarius pomarinus breeds in arctic Russia, with Europe accounting for less than a quarter of its global breeding range. Its European breeding population is relatively small (<50,000 pairs), but was broadly stable between 1970–1990. No trend data were available for 1990–2000, but there was no evidence to suggest that the species's status had deteriorated since 1990, and it is provisionally evaluated as Secure.

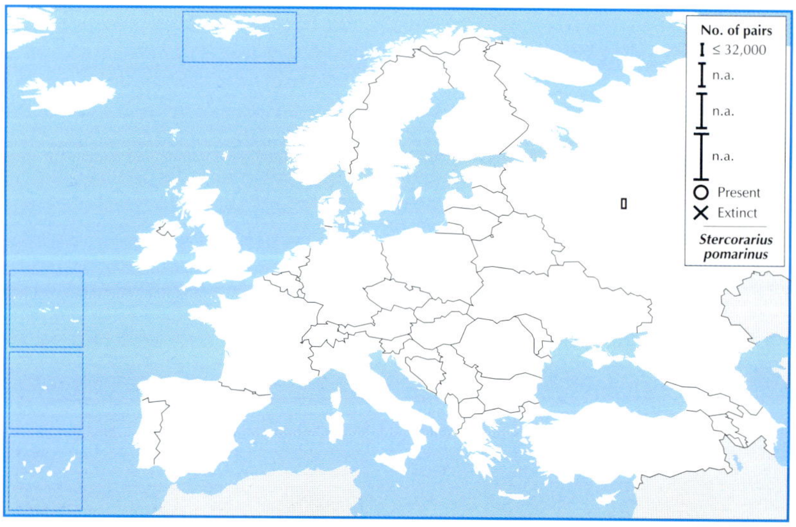

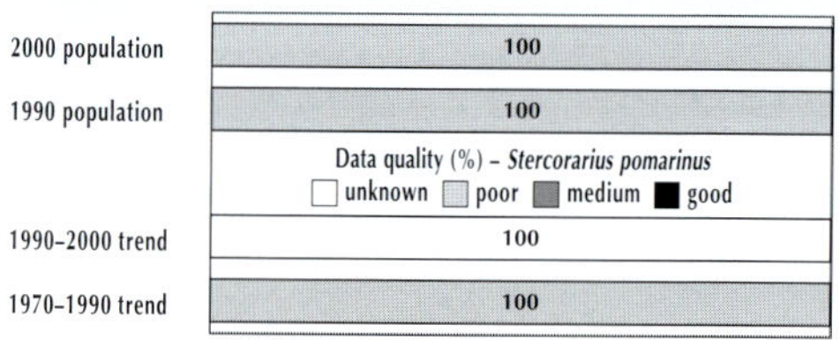

Stercorarius parasiticus
PARASITIC JAEGER

Non-SPEC (1994: —) **Status** (Secure)
Criteria —
European IUCN Red List Category —
Criteria —
Global IUCN Red List Category —
Criteria —

Country	Breeding pop. size (pairs)	Year(s)	Trend	Mag.%	References
Denmark					
Faroe Is.	900 – 900	95	(0)	(0–19)	
Greenland	(1,000 – 10,000)	90–00	?	–	
Finland	**500 – 550**	98–02	**0**	**0–19**	
Iceland	5,000 – 10,000	87–94	?	–	*1*
Norway	9,000 – 14,000	95	–	0–19	*4,27*
Svalbard	1,000 – 2,000	95–03	–	0–19	
Russia	(20,000 – 100,000)	90–02	?	–	*11,82,154*
Sweden	**450 – 600**	99–00	+	0–19	
UK	**2,100 – 2,100**	98–02	–	**37**	*20*
Total (approx.)	40,000 – 140,000	**Overall trend** Unknown			
Breeding range	>750,000 km²	**Gen. length.** 11		**% Global pop.** 5–24	

Stercorarius parasiticus is a widespread breeder in coastal areas of northern Europe, which accounts for less than a quarter of its global breeding range. Its European breeding population is relatively small (<140,000 pairs), but was stable between 1970–1990. Although the species declined in Norway, Svalbard and the United Kingdom during 1990–2000, trend data were not available for key populations in Greenland, Iceland and Russia, and there is no evidence to suggest that the species declined significantly overall. Consequently, it is provisionally evaluated as Secure.

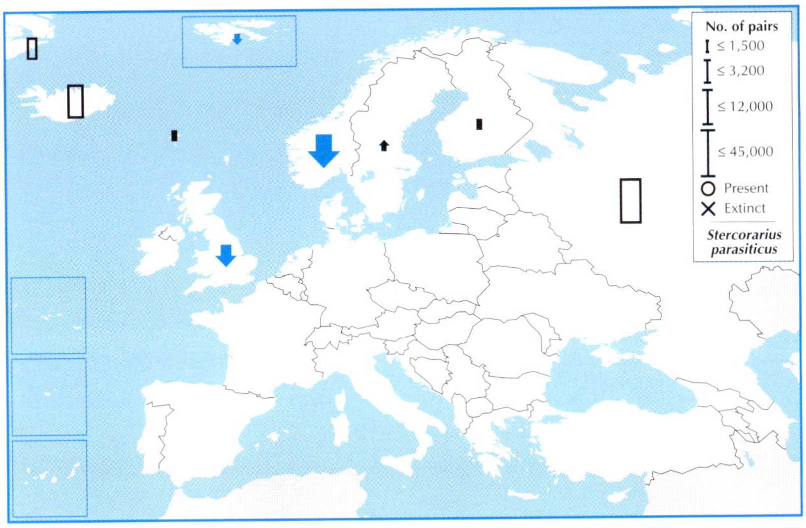

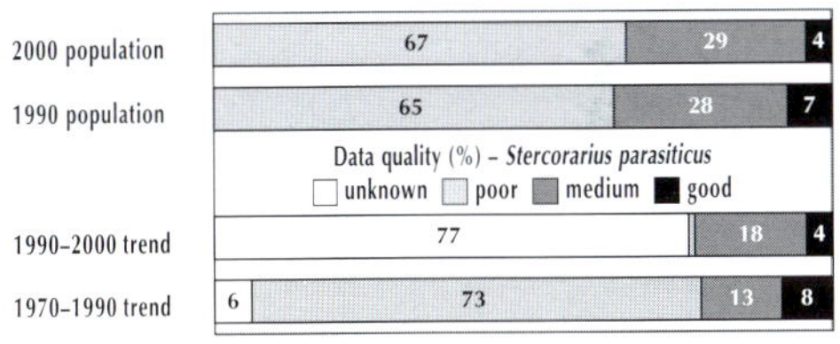

Stercorarius longicaudus
LONG-TAILED JAEGER

Non-SPEC (1994: —) **Status** (Secure)
Criteria —
European IUCN Red List Category —
Criteria —
Global IUCN Red List Category —
Criteria —

Country	Breeding pop. size (pairs)	Year(s)	Trend	Mag.%	References
Denmark					
Greenland	(1,000 – 20,000)	90–00	?	–	
Finland	100 – 2,000	98–02	F	>80	
Norway	(0 – 1,000)	01	F	30–49	*15*
Svalbard	3 – 20	01–03	F	20–29	
Russia	(10,000 – 50,000)	90–02	?	–	*81*
Sweden	500 – 5,000	99–00	F	30–49	
Total (approx.)	12,000 – 78,000	**Overall trend** Fluctuating			
Breeding range	>500,000 km²	**Gen. length.** 11		**% Global pop.** 5–24	

Stercorarius longicaudus breeds in Greenland, Svalbard, Fennoscandia and arctic Russia, with Europe accounting for less than a quarter of its global breeding range. Its European breeding population is relatively small (<78,000 pairs), but was broadly stable between 1970–1990. Trend data were not available for the key populations in Russia and Greenland during 1990–2000, but populations were broadly stable elsewhere in Europe, and consequently the species is provisionally evaluated as Secure.

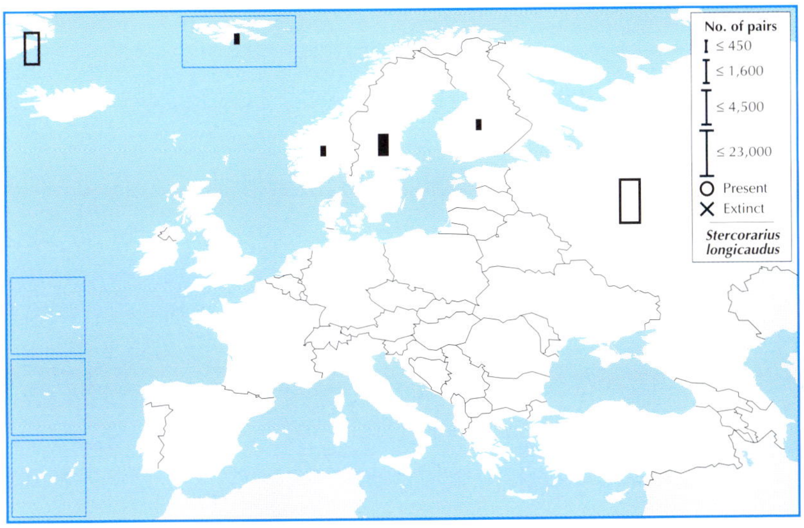

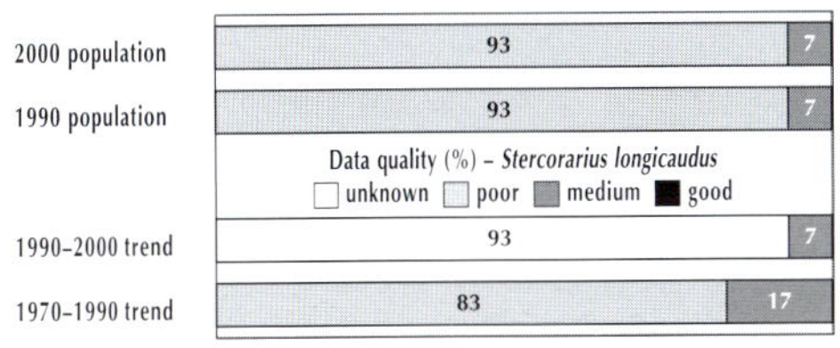

Catharacta skua
GREAT SKUA

Non-SPEC[E] (1994: 4) **Status** Secure
Criteria —
European IUCN Red List Category —
Criteria —
Global IUCN Red List Category —
Criteria —

Country	Breeding pop. size (pairs)	Year(s)	Trend	Mag.%	References
Denmark					
Faroe Is.	450 – 450	95	(0)	(0–19)	
Iceland	**5,400 – 5,400**	84–85	?	–	*27*
Rep. Ireland	**1 – 2**	00–01	+	N	
Norway	**45 – 50**	98–01	+	**0–19**	*35*
Svalbard	**200 – 500**	98–03	+	20–29	
Russia	**30 – 50**	90–02	+	**20–29**	*79,80,154*
UK	**9,600 – 9,600**	98–02	+	**26**	*20*
Total (approx.)	16,000 – 16,000	**Overall trend** Large increase			
Breeding range	>100,000 km²	**Gen. length.** 15		**% Global pop.** 100	

Catharacta skua breeds only in northernmost Europe, with the entire global breeding range confined to the region. The European breeding population is relatively small (16,000 pairs), but increased between 1970–1990. Although the trend of the sizeable population in Iceland was unknown, the species continued to increase in most of the rest of its range during 1990–2000, and consequently it is evaluated as Secure.

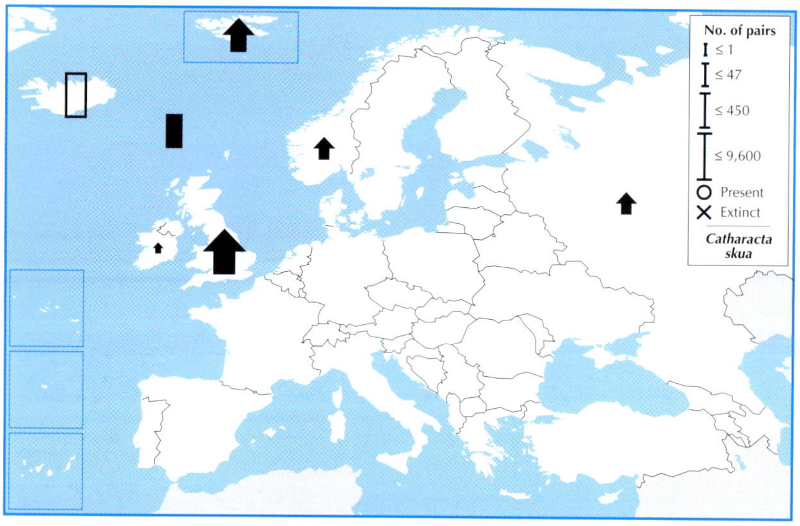

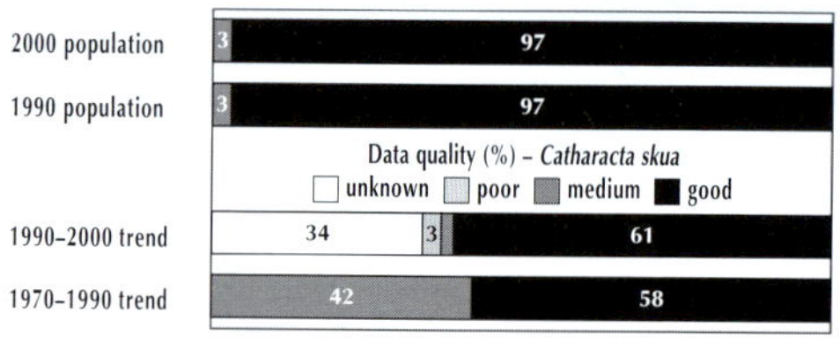

Larus ichthyaetus
GREAT BLACK-HEADED GULL

Non-SPEC (1994: —) **Status** (Secure)
Criteria —
European IUCN Red List Category —
Criteria —
Global IUCN Red List Category —
Criteria —

Country	Breeding pop. size (pairs)	Year(s)	Trend	Mag.%	References
Russia	15,000 – 25,000	90–02	–	20–29	*6,33,60,74,120*
Ukraine	1,000 – 1,500	90–00	+	10–19	
Total (approx.)	16,000 – 27,000	**Overall trend** Moderate decline			
Breeding range	>50,000 km²	**Gen. length.** 13		**% Global pop.** 5–24	

Country	Winter pop. size (individuals)	Year(s)	Trend	Mag.%	References
Armenia	**250 – 1,000**	97–02	+	**10–19**	
Azerbaijan	50 – 280	96–02	(0)	(0–19)	
Bulgaria	**0 – 3**	96–02	?	–	
Georgia	**13 – 13**	02	?	–	*1*
Italy	(0 – 7)	02	?	–	*22*
Romania	2 – 20	90–00	0	0–19	*25,60*
Slovenia	Present	90–00	?	–	
Turkey	200 – 400	91–01	+	20–29	
Ukraine	50 – 250	90–00	(+)	(20–29)	
Total (approx.)	>600	**Overall trend** Unknown			
% in European IBAs	15–35	**Gen. length** 13		**% Global pop.** <5	

Larus ichthyaetus breeds patchily in Ukraine and southern European Russia, which together account for less than a quarter of its global breeding range. Its European breeding population is relatively small (<27,000 pairs), but increased substantially between 1970–1990. Although the species increased in Ukraine during 1990–2000, it declined in its Russian stronghold, and underwent a moderate decline (>10%) overall. Nevertheless, this decline is probably outweighed by the earlier increase, and consequently the species is provisionally evaluated as Secure.

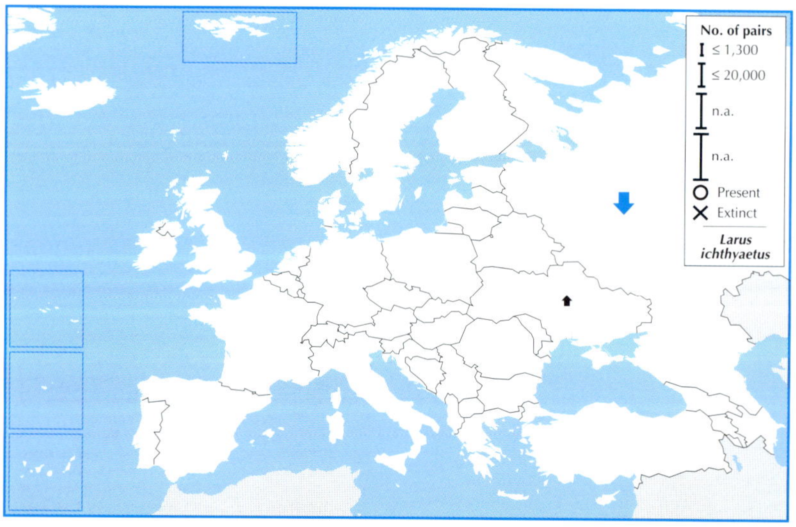

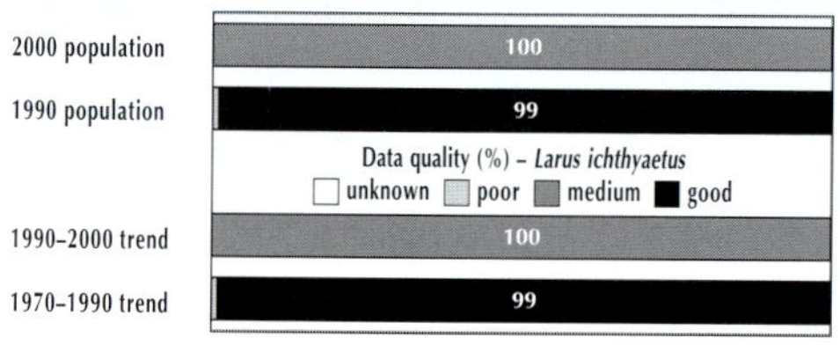

Larus melanocephalus
MEDITERRANEAN GULL

Non-SPEC[E] (1994: 4) **Status** Secure
Criteria —
European IUCN Red List Category —
Criteria —
Global IUCN Red List Category —
Criteria —

Larus melanocephalus has a widespread but very patchy breeding distribution in Europe, which contains the entire global breeding range. Its European breeding population is large (>120,000 pairs), and increased between 1970–1990. The species continued to increase across most of its range during 1990–2000, although the stronghold population in Ukraine was stable, and therefore the species was stable overall. Consequently, it is evaluated as Secure.

Country	Breeding pop. size (pairs)	Year(s)	Trend	Mag.%	References
Austria	**20 – 40**	98–02	+	**50–79**	
Azerbaijan	**250 – 250**	96–00	+	**>80**	
Belarus	0 – 20	92–00	?	–	
Belgium	**1,450 – 1,450**	00–02	+	**50–79**	*1*
Bulgaria	0 – 25	96–02	F	>80	
Czech Rep.	**30 – 40**	00	+	**>80**	
Denmark	**0 – 6**	98–01	+	50–79	*4,5,6,7*
Estonia	0 – 1	98	?	–	*1*
France	**2,228 – 2,300**	00–01	+	**>80**	*6*
Germany	**113 – 113**	95–99	+	**50–79**	
Greece	**1,000 – 1,350**	98–00	–	0–19	
Hungary	110 – 230	98–02	0	0–19	*18*
Rep. Ireland	**1 – 3**	96–02	+	N	
Italy	**1,980 – 1,980**	00	**0**	**0–19**	18
Netherlands	**416 – 850**	98–00	+	**630**	1
Poland	**30 – 50**	00–03	+	**200–300**	*74,60–69*
Romania	**120 – 250**	90–02	+	0–19	*10,39*
Russia	3,000 – 10,000	90–00	(F)	(30–49)	*8,60*
Serbia & MN	**30 – 50**	00–02	+	30–49	*1,22,155,214*
Slovakia	**0 – 70**	80–99	+	N	
Spain	**2 – 3**	98–02	**0**	**0–19**	*10*
Switzerland	**0 – 2**	93–96	**0**	**0–19**	
Turkey	**4,900 – 5,500**	01	**0**	**0–19**	
Ukraine	100,000 – 300,000	90–00	F	20–29	
UK	**110 – 110**	98–02	+	**686**	*20*
Total (approx.)	120,000 – 320,000	**Overall trend** Stable			
Breeding range	>250,000 km²	**Gen. length.** 6		**% Global pop.** 100	

Country	Winter pop. size (individuals)	Year(s)	Trend	Mag.%	References
France	3,600 – 9,800	98–02	+	30–49	*9*
Greece	**1,000 – 5,000**	95–99	F	>80	
Italy	(5,000 – 20,000)	02	+	0–19	*22*
Malta	**300 – 1,000**	95–02	**0**	**0–19**	*4*
Turkey	(2,000 – 3,000)	91–01	(0)	(0–19)	
Ukraine	250 – 1,000	90–00	0	0–19	
Total (approx.)	>12,000	**Overall trend** Unknown			
% in European IBAs	>90	**Gen. length** 6		**% Global pop.** 50–74	

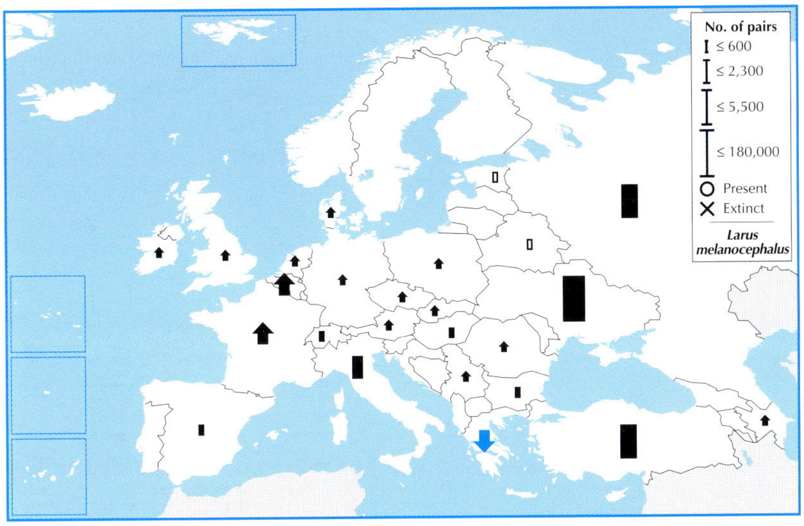

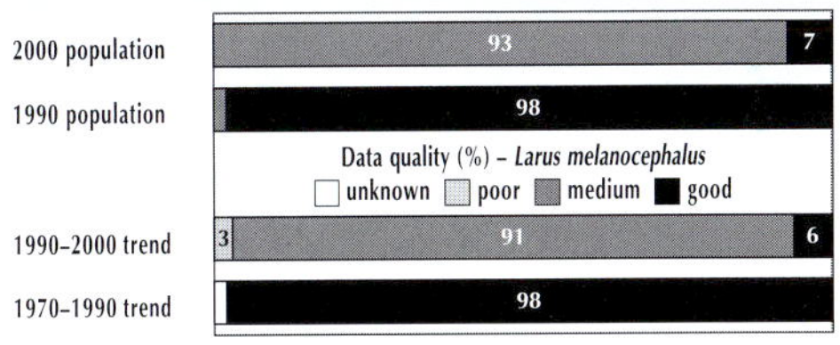

Larus minutus
LITTLE GULL

SPEC 3 (1994: 3) **Status** (Depleted)
Criteria Moderate historical decline
European IUCN Red List Category —
Criteria —
Global IUCN Red List Category —
Criteria —

Larus minutus breeds mainly in north-east Europe, which accounts for less than half of its global breeding range. Its European breeding population is relatively small (<58,000 pairs), and underwent a moderate decline between 1970–1990. Although the species increased markedly overall during 1990–2000—with stable, fluctuating or increasing trends across the vast majority of its European range—its population has probably not yet recovered to the level that preceded its decline. Consequently, it is provisionally evaluated as Depleted.

Country	Breeding pop. size (pairs)	Year(s)	Trend	Mag.%	References
Belarus	1,000 – 2,000	97–02	0	0–19	
Denmark	**1 – 1**	99–01	**0**	**0–19**	*5,6,7*
Estonia	(1,000 – 2,000)	98	0	0–19	*1*
Finland	10,000 – 20,000	98–02	+	50	
Germany	**2 – 2**	95–99	**0**	**0–19**	
Latvia	700 – 2,000	90–00	–	20–29	*17*
Lithuania	200 – 400	99–01	F	>80	*20*
Netherlands	**5 – 8**	98–00	(F)	(–)	*1*
Norway	**45 – 50**	02	+	**>80**	*23*
Poland	**0 – 20**	95–00	–	50–79	*2,60–69*
Russia	(10,000 – 30,000)	90–00	(F)	(30–49)	*4,42,60,104,135*
Sweden	**300 – 600**	99–00	+	20–29	
Ukraine	**300 – 750**	90–00	**F**	**20–29**	
Total (approx.)	24,000 – 58,000	**Overall trend** Moderate increase			
Breeding range	>2,000,000 km²	**Gen. length.** 6		**% Global pop.** 25–49	

Country	Winter pop. size (individuals)	Year(s)	Trend	Mag.%	References
Azerbaijan	(1,000 – 2,500)	96–02	(0)	(0–19)	
Cyprus	50 – 500	94–02	0	0–19	
Estonia	(0 – 500)	98	(0)	(0–19)	*1*
Finland	50 – 500	98–02	0	0–19	
France	(175 – 536)	90–02	?	–	*9*
Germany	**2,100 – 2,100**	87–03	?	–	*4*
Greece	**100 – 300**	95–99	F	>80	
Italy	(100 – 500)	02	(0)	(0–19)	*22*
Latvia	**500 – 500**	92–94	(0)	(0–19)	*32*
Netherlands	(5,000 – 5,000)	85–99	(0)	(0–19)	*3,4,5,8,9,11,12*
Poland	(50 – 250)	90–00	(+)	(>80)	*114*
Turkey	1,500 – 2,500	91–01	?	–	
Ukraine	(50 – 250)	90–00	(0)	(0–19)	
Total (approx.)	>11,000	**Overall trend** Fluctuating			
% in European IBAs	26–31	**Gen. length** 6		**% Global pop.** 5–24	

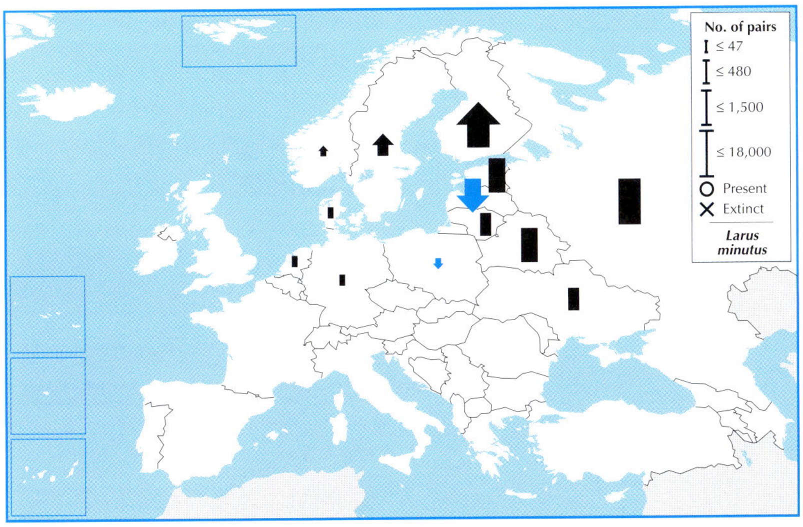

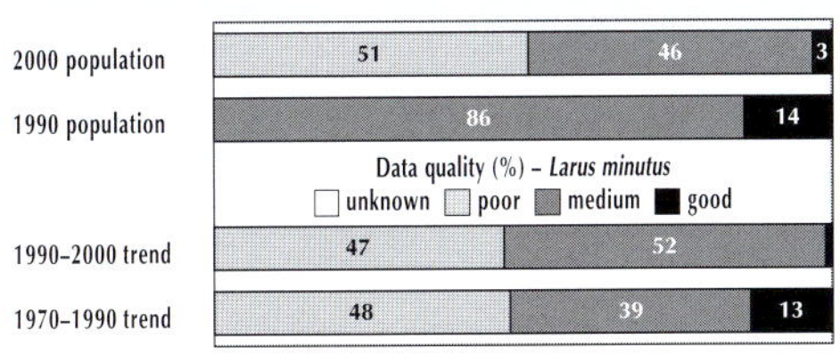

Larus ridibundus
COMMON BLACK-HEADED GULL

Non-SPEC[E] (1994: —) **Status** (Secure)
Criteria —
European IUCN Red List Category —
Criteria —
Global IUCN Red List Category —
Criteria —

Larus ridibundus is a widespread breeder across much of Europe, which is now thought to hold >50% of its global breeding population. Its European breeding population is very large (>1,500,000 pairs), and increased substantially between 1970–1990. Although the species declined in north-central Europe—particularly in countries bordering the Baltic Sea—during 1990–2000, and underwent a moderate decline (>10%) overall, this decline is probably outweighed by the earlier increase. Consequently, the species is provisionally evaluated as Secure.

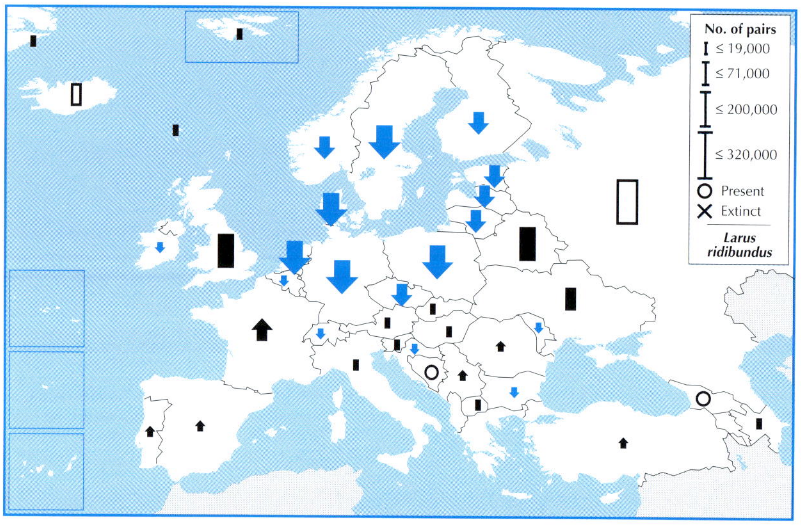

Country	Breeding pop. size (pairs)	Year(s)	Trend	Mag.%	References
Austria	6,000 – 8,000	98–02	0	0–19	
Azerbaijan	(0 – 100)	96–00	(0)	(0–19)	
Belarus	180,000 – 220,000	90–00	0	0–19	
Belgium	**18,000 – 19,000**	95–02	**–**	**0–19**	*1*
Bosnia & HG	Present	90–03	?	–	
Bulgaria	**40 – 100**	98–02	–	>80	
Croatia	(500 – 2,000)	02	(–)	(50–79)	*26,25,29*
Czech Rep.	**50,000 – 100,000**	00	**–**	**50–79**	
Denmark	**110,000 – 125,000**	98	**–**	**30–49**	*11*
Faroe Is.	250 – 250	95	(0)	(0–19)	
Greenland	(5 – 50)	90–00	(0)	(0–19)	
Estonia	(50,000 – 80,000)	98	–	20–29	*1*
Finland	50,000 – 100,000	98–02	**–**	**30**	
France	**29,100 – 32,500**	99–00	**+**	**100**	*1,2*
Georgia	Present	03	?	–	
Germany	**136,000 – 167,000**	95–99	**–**	**20–29**	
Hungary	7,000 – 15,000	95–02	0	0–19	
Iceland	(25,000 – 30,000)	78–94	?	–	*1*
Rep. Ireland	3,900 – 3,900	99–02	–	50–79	
Italy	**500 – 1,000**	03	**0**	**0–19**	
Latvia	**32,000 – 32,000**	90–00	**–**	**20–29**	*25*
Lithuania	**30,000 – 60,000**	99–01	–	20–29	*20*
Macedonia	(50 – 250)	96–00	(F)	(–)	*7*
Moldova	15 – 30	90–00	–	30–49	
Netherlands	**132,000 – 137,000**	98–00	**–**	**26**	*1*
Norway	(20,000 – 60,000)	90–02	–	0–19	*27*
Svalbard	0 – 1	95–03	0	0–19	
Poland	110,000 – 120,000	95–00	–	30–49	*2*
Portugal	**0 – 5**	02	**+**	**N**	
Romania	14,000 – 16,000	90–02	+	0–19	
Russia	200,000 – 500,000	90–00	?	–	*4,24,42,60,104, 112,113,114,135, 139*
Serbia & MN	**2,500 – 3,700**	98–02	+	10–19	*1,138,29,155, 204,78,214*
Slovakia	9,000 – 17,000	90–99	0	0–19	
Slovenia	**150 – 300**	95–00	**F**	**>80**	
Spain	2,500 – 10,000	98–02	+	0–19	*10,11*
Sweden	75,000 – 125,000	99–00	–	10–19	
Switzerland	**700 – 1,200**	98–02	**–**	**50–79**	
Turkey	4,000 – 8,000	01	(+)	(0–19)	
Ukraine	35,000 – 70,000	90–00	F	30–49	
UK	**138,000 – 138,000**	98–02	0	4	*20*
Total (approx.)	1,500,000 – 2,200,000	**Overall trend** Moderate decline			
Breeding range	>6,000,000 km²	**Gen. length.** 6		**% Global pop.** 50-74	

(See p. 136, bottom, for data quality graph)

	Winter pop. size (individuals)		
Total (approx.)	>3,200,000	**Overall trend** Small decline	
% in European IBAs	>10	**Gen. length** 6	**% Global pop.** 25–49

Larus genei
SLENDER-BILLED GULL

SPEC 3 (1994: —) **Status** Localised
Criteria ≥90% breed at ≤10 sites
European IUCN Red List Category —
Criteria —
Global IUCN Red List Category —
Criteria —

Larus genei breeds locally in coastal areas of the Mediterranean and Black Seas and in Turkey, with Europe accounting for less than half of its global breeding range. Its European breeding population is relatively small (<56,000 pairs), but increased substantially between 1970–1990. Although the species declined in Russia during 1990–2000, it was stable or increased elsewhere in Europe, and hence remained stable overall. Nevertheless, more than 90% of the European breeding population occurs at just 10 sites (see Appendix 5), and consequently the species is evaluated as Localised.

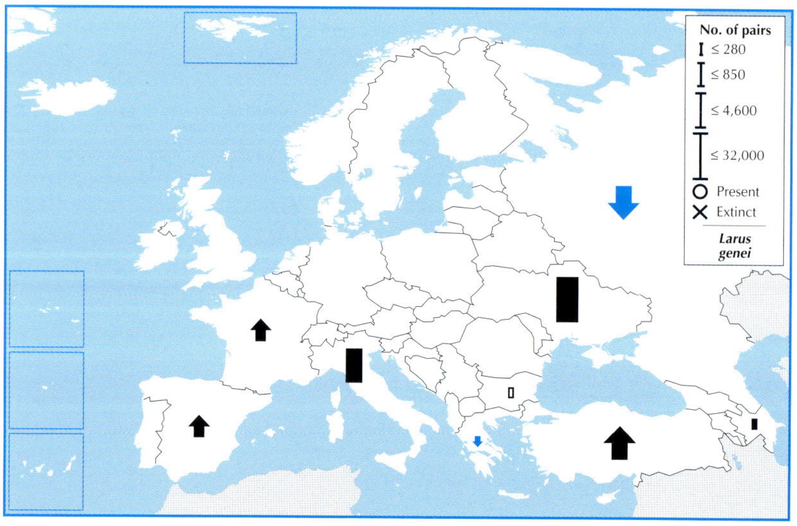

Country	Breeding pop. size (pairs)	Year(s)	Trend	Mag.%	References
Azerbaijan	**250 – 300**	96–00	(0)	(0–19)	
Bulgaria	**1 – 1**	96	?	–	
France	**818 – 877**	01	**+**	**>80**	*6*
Greece	(0 – 10)	95–00	(–)	(20–29)	
Italy	**3900 – 3,900**	00	**0**	**0–19**	*18*
Russia	(2,000 – 5,000)	90–02	–	20–29	*8,60*
Spain	800 – 900	98–02	+	>80	*10,16*
Turkey	**3,800 – 5,500**	01	**+**	**0–19**	
Ukraine	25,000 – 40,000	90–00	F	20–29	
Total (approx.)	37,000 – 56,000	**Overall trend** Stable			
Breeding range	>100,000 km²	**Gen. length.** 6		**% Global pop.** 25–49	

Country	Winter pop. size (individuals)	Year(s)	Trend	Mag.%	References
Albania	**107 – 959**	95–02	(F)	(–)	
Azerbaijan	(10 – 250)	96–02	(0)	(0–19)	
Bulgaria	**4 – 34**	97–01	**F**	**>80**	
Cyprus	10 – 50	94–02	0	0–9	
France	10 – 30	98–02	?	–	*9*
Greece	**1,670 – 4,060**	95–99	**0**	**0–19**	
Italy	(1,300 – 3,000)	02	+	0–19	*22*
Serbia & MN	(6 – 10)	90–02	(F)	(20–49)	
Turkey	**800 – 1600**	91–01	**F**	**>80**	
Ukraine	0 – 50	90–00	(0)	(0–19)	
Total (approx.)	>4,000	**Overall trend** Unknown			
% in European IBAs	>50	**Gen. length** 6		**% Global pop.** 5–24	

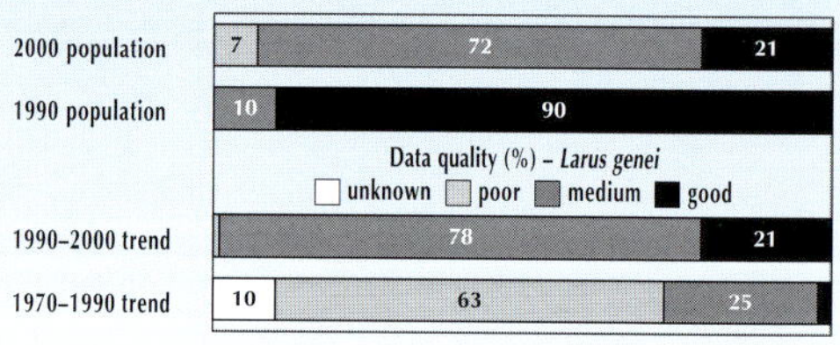

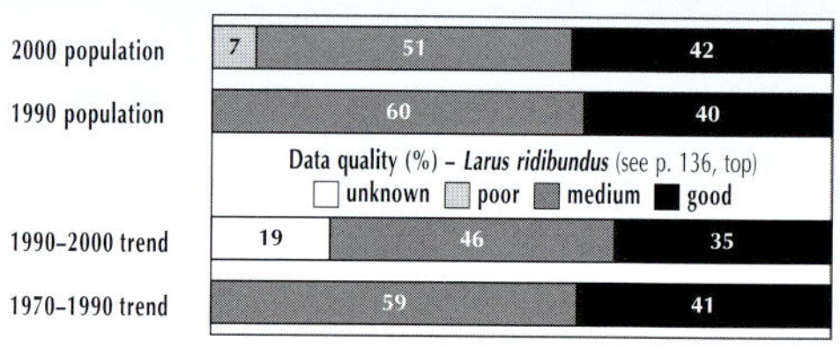

Larus audouinii
AUDOUIN'S GULL

SPEC 1 (1994: 1) **Status** Localised
Criteria ≥90% breed at ≤10 sites
European IUCN Red List Category —
Criteria —
Global IUCN Red List Category NT
Criteria A3c

Larus audouinii breeds locally in coastal areas and on islands of the Mediterranean, with Europe holding >75% of its global breeding population. Its European breeding population is relatively small (<19,000 pairs), but increased substantially between 1970–1990. The stronghold population in Spain continued to increase during 1990–2000, and the species underwent a large increase overall. Nevertheless, more than 90% of the European breeding population occurs at just 10 sites (see Appendix 5), and consequently the species is evaluated as Localised.

Country	Breeding pop. size (pairs)	Year(s)	Trend	Mag.%	References
Croatia	**65 – 70**	02	+	N	*33*
Cyprus	(15 – 30)	98–02	(0)	(0–9)	
France	**56 – 92**	98–01	F	**20–49**	*6*
Greece	**750 – 900**	95–00	0	0–19	
Italy	**510 – 982**	99–00	F	**20–29**	*17*
Portugal	**25 – 30**	02–03	+	N	
Spain	**17,000 – 17,000**	00	+	**30–49**	*10,16*
Turkey	**50 – 100**	01	+	>80	
Total (approx.)	18,000 – 19,000	**Overall trend** Large increase			
Breeding range	>100,000 km²	**Gen. length.** 13		**% Global pop.** 75–94	

Country	Winter pop. size (individuals)	Year(s)	Trend	Mag.%	References
Cyprus	(40 – 80)	94–02	0	0–9	
Greece	(200 – 1,000)	95–00	(0)	(0–19)	
Italy	(500 – 1,000)	02	(0)	(0–19)	*22*
Turkey	20 – 40	91–01	F	>80	
UK					
Gibraltar	**5 – 20**	00–03	**0**	**0–19**	
Total (approx.)	>800	**Overall trend** Unknown			
% in European IBAs	>90	**Gen. length** 13		**% Global pop.** 25–49	

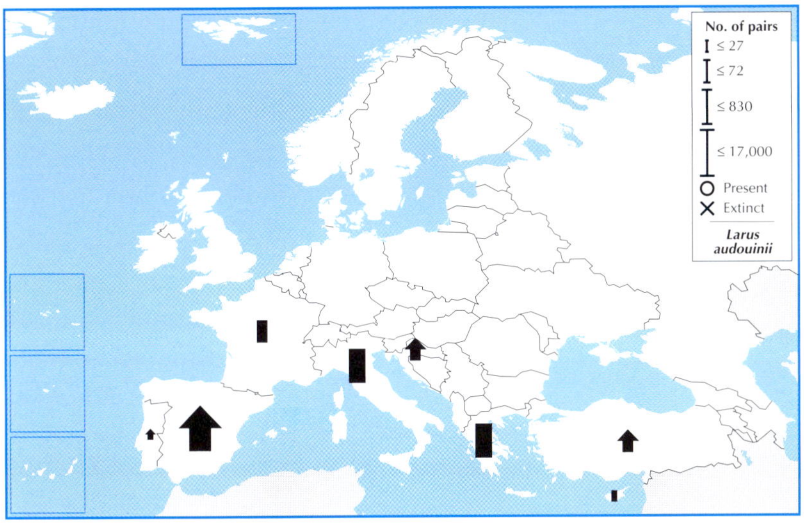

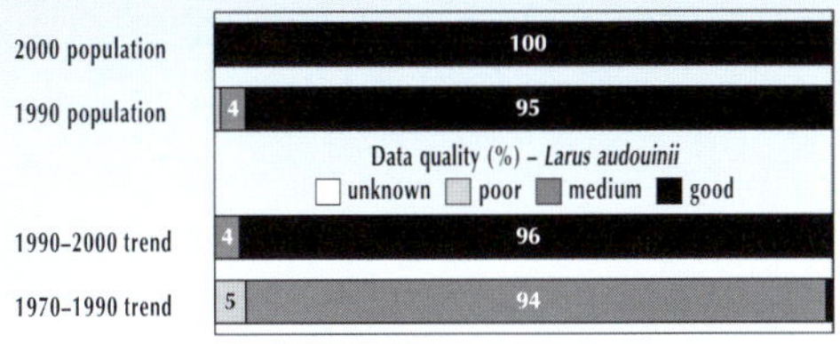

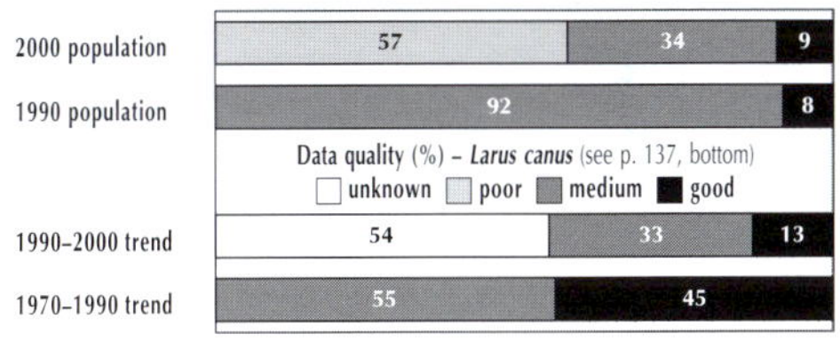

Larus canus
MEW GULL

SPEC 2 (1994: 2) **Status** (Depleted)
Criteria Moderate historical decline
European IUCN Red List Category —
Criteria —
Global IUCN Red List Category —
Criteria —

Larus canus is a widespread breeder across much of northern Europe, which constitutes >50% of its global breeding range. Its European breeding population is large (>590,000 pairs), but underwent a moderate decline between 1970–1990. Although the species was stable or increased in some countries during 1990–2000, it declined across much of north-west Europe, and the trend in its Russian stronghold was unknown. Nevertheless, its population has clearly not yet recovered to the level that preceded its decline, and consequently the species is provisionally evaluated as Depleted.

Country	Breeding pop. size (pairs)	Year(s)	Trend	Mag.%	References
Austria	**1 – 5**	98–02	**0**	**0–19**	
Belarus	500 – 1,200	97–02	**–**	**10–19**	
Belgium	**66 – 77**	01–02	+	**0–19**	*1*
Czech Rep.	**2 – 3**	00	**–**	**30–49**	
Denmark	(25,000 – 30,000)	00	0	0–19	*3*
Faroe Is.	1,000 – 1,000	95	(0)	(0–19)	
Estonia	10,000 – 20,000	98	0	0–19	*1*
Finland	60,000 – 80,000	98–02	+	15	
France	**2 – 35**	96–01	F	**20–49**	*6*
Germany	**19,000 – 25,000**	95–99	+	**0–19**	
Hungary	1 – 2	95–02	0	0–19	*14,3*
Iceland	400 – 500	00	+	0–19	*16*
Rep. Ireland	1,060 – 1,060	99–02	–	30–49	
Latvia	500 – 600	90–00	0	0–19	*23,17*
Lithuania	**120 – 150**	99–01	+	**0–19**	*20*
Netherlands	**5,600 – 6,500**	98–00	**–**	**46**	*1*
Norway	60,000 – 120,000	89–01	**–**	**30–49**	*35,27*
Svalbard	1 – 5	89–01	–	20–29	
Poland	**3,000 – 3,500**	97–00	**–**	**20–29**	*2*
Russia	(250,000 – 1,000,000)	90–00	?	–	*4,24,40,42,51, 52,104,141,145*
Slovakia	**0 – 3**	80–99	F	30–49	
Sweden	100,000 – 200,000	99–00	–	0–19	
Switzerland	**2 – 5**	98–02	**0**	**0–19**	
Ukraine	50 – 150	90–00	+	**0–19**	
UK	**48,700 – 48,700**	98–02	–	29	*18,20*
Total (approx.)	590,000 – 1,500,000	**Overall trend** Unknown			
Breeding range	>3,000,000 km²	**Gen. length.** 8		**% Global pop.** 50–74	

(See p. 137, top, for data quality graph)

Country	Winter pop. size (individuals)	Year(s)	Trend	Mag.%	References
Belgium	**112,000 – 112,000**	95–00	(0)	(0–19)	*1*
Finland	(10,000 – 100,000)	98–02	0	0–19	
France	(20,000 – 80,000)	98–02	?	–	*9*
Latvia	10,000 – 20,000	90–99	(0)	(0–19)	*29*
Netherlands	270,000 – 350,000	99–01	**0**	**0–19**	*3,4,8,9,11,12*
Poland	(20,000 – 40,000)	90–00	(0)	(0–19)	
Romania	5,000 – 120,000	90–00	(F)	(20–29)	*60*
Turkey	8,000 – 12,000	91–01	F	**30–49**	
Ukraine	10,000 – 20,000	90–00	(0)	(0–19)	
UK	430,000 – 430,000	93	–	0–19	*45*
Total (approx.)	>910,000	**Overall trend** Small decline			
% in European IBAs	13–17	**Gen. length** 8		**% Global pop.** 25–49	

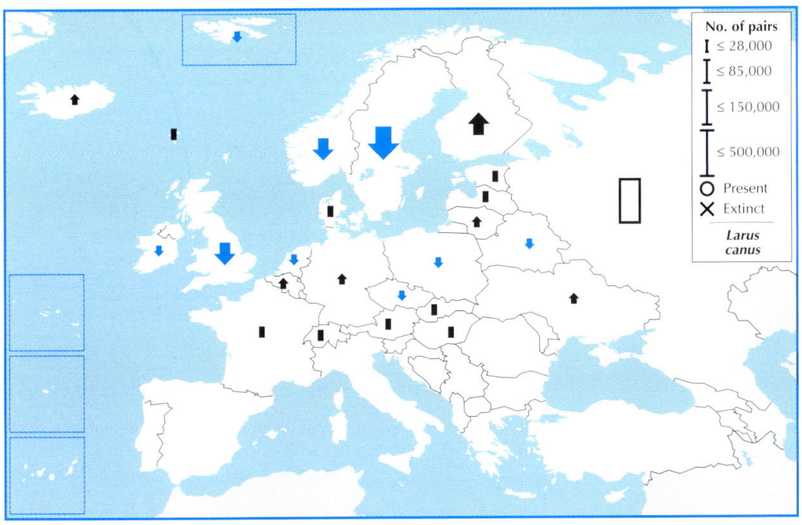

Larus fuscus
LESSER BLACK-BACKED GULL

Non-SPEC^E (1994: 4) **Status** Secure
Criteria —
European IUCN Red List Category —
Criteria —
Global IUCN Red List Category —
Criteria —

Larus fuscus is a widespread breeder in coastal areas of northern and western Europe, which constitutes >75% of its global breeding range. Its European breeding population is large (>300,000 pairs), and increased between 1970–1990. This trend continued during 1990–2000, with declines in a few countries in the north-east of its range more than compensated for by increasing or stable trends elsewhere. Consequently, it is evaluated as Secure.

Country	Breeding pop. size (pairs)	Year(s)	Trend	Mag.%	References
Belgium	**2,800 – 2,800**	00–02	+	**>80**	*1*
Denmark	4,000 – 6,000	00	(+)	(50–79)	*3*
Faroe Is.	9,000 – 9,000	95	(0)	(0–19)	
Greenland	(500 – 1,000)	90–00	(+)	(0–19)	*3,15*
Estonia	80 – 120	98	–	20–29	*1*
Finland	5,000 – 7,000	98–02	–	15	
France	22,500 – 23,030	97–00	0	0–19	*1*
Germany	**23,000 – 29,000**	95–99	+	**50–79**	
Iceland	25,000 – 35,000	97	+	0–19	*16*
Rep. Ireland	2,900 – 2,900	99–02	+	**50–79**	
Netherlands	**58,500 – 72,000**	98–00	+	**184**	*1*
Norway	**30,000 – 40,000**	90–01	+	**0–19**	*35*
Svalbard	0 – 2	95–03	0	0–19	
Poland	**0 – 3**	90–00	**0**	**0–19**	*72,73*
Portugal	**5 – 10**	02	**0**	**0–19**	
Russia	**4,000 – 5,700**	90–00	–	**20–29**	*59,65,96*
Spain	480 – 500	98–02	+	20–29	*10*
Canary Is.	15 – 15	97–03	(+)	(N)	*30*
Sweden	2,000 – 5,000	99–00	–	50–79	
UK	**114,000 – 114,000**	98–02	+	29	*20*
Total (approx.)	300,000 – 350,000	**Overall trend** Large increase			
Breeding range	>1,000,000 km²	**Gen. length.** 11		**% Global pop.** 75–94	

Country	Winter pop. size (individuals)	Year(s)	Trend	Mag.%	References
Belgium	1,000 – 1,000	95–00	(0)	(0–19)	*1*
Croatia	(50 – 100)	02	?	–	*16*
Cyprus	50 – 500	94–02	0	0–9	
France	(50,000 – 150,000)	98–02	?	–	*9*
Greece	(50 – 200)	95–00	(F)	(>80)	
Italy	(300 – 1,500)	02	(0)	(0–19)	*22*
Netherlands	600 – 700	99–01	+	**258**	*3,4,8,9,11,12*
Romania	50 – 100	90–00	(0)	(0–19)	*60*
Spain	(17,000 – 79,000)	90–01	(+)	(>80)	*19*
Turkey	**30 – 90**	91–01	F	>80	
UK	60,800 – 60,800	93	+	0–19	*45*
Total (approx.)	>130,000	**Overall trend** Large increase			
% in European IBAs	16–20	**Gen. length** 11		**% Global pop.** 25–49	

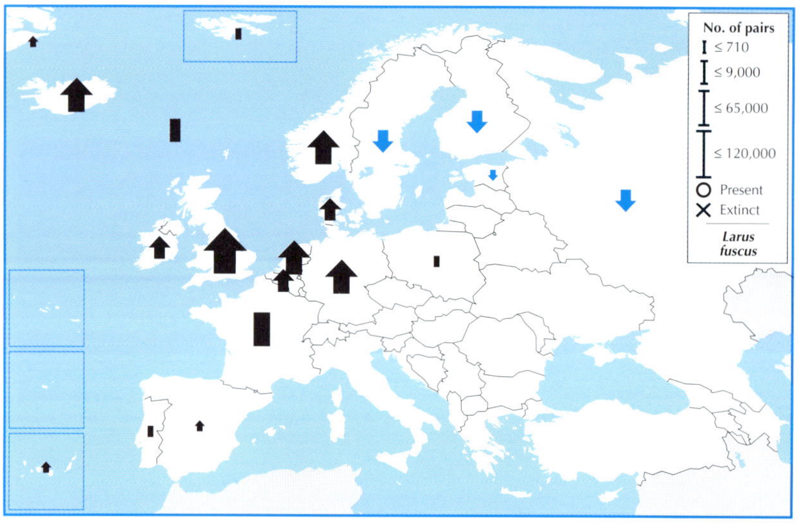

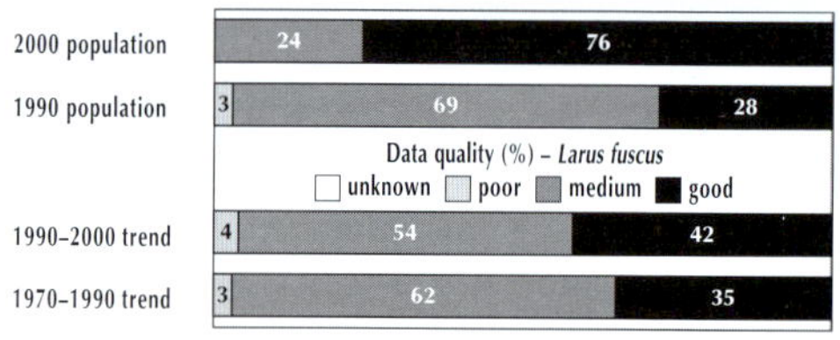

Larus argentatus
HERRING GULL

Non-SPEC^E (1994: —) **Status** Secure
Criteria —
European IUCN Red List Category —
Criteria —
Global IUCN Red List Category —
Criteria —

Larus argentatus is a widespread breeder across much of northern Europe, which probably holds >50% of its global population. Its European breeding population is large (>760,000 pairs), and increased between 1970–1990. Although populations in the Republic of Ireland, United Kingdom and Netherlands declined during 1990–2000, these losses were more than compensated for by increases in most other European populations – notably those in Norway and Russia—and the species increased overall. Consequently, it is evaluated as Secure.

Country	Breeding pop. size (pairs)	Year(s)	Trend	Mag.%	References
Belarus	**350 – 500**	97–02	+	**50–79**	
Belgium	**1,500 – 1,600**	00–02	+	**50–79**	*1*
Denmark	55,000 – 60,000	99–00	+	10–19	*3*
Faroe Is.	1,500 – 1,500	95	(0)	(0–19)	
Greenland	(0 – 10)	90–00	(+)	(0–19)	*5*
Estonia	35,000 – 50,000	98	+	20–29	*1*
Finland	30,000 – 40,000	98–02	+	**15**	
France	74,000 – 77,000	97–00	(+)	(32)	*1,2*
Germany	**39,000 – 46,000**	95–99	+	**0–19**	
Iceland	(5,000 – 10,000)	97	?	–	*16*
Rep. Ireland	5,500 – 5,500	99–02	–	**81**	
Latvia	800 – 1,000	90–00	+	30–49	*17*
Lithuania	**10 – 20**	99–01	**0**	**0–19**	*20*
Netherlands	**62,000 – 67,000**	98–00	–	**18**	*1*
Norway	(150,000 – 250,000)	90–01	+	**0–19**	*35*
Svalbard	1 – 5	95–03	0	0–19	
Poland	**1,700 – 2,200**	97–00	+	0–19	*1,2*
Russia	(100,000 – 500,000)	90–00	+	20–29	11,40,82,84,81, 96,139
Sweden	50,000 – 100,000	99–00	+	10–19	
UK	**144,000 – 144,000**	98–02	–	**6**	*20*
Total (approx.)	760,000 – 1,400,000	**Overall trend** Moderate increase			
Breeding range	>2,000,000 km²	**Gen. length.** 13		**% Global pop.** 50–74	

Country	Winter pop. size (individuals)	Year(s)	Trend	Mag.%	References
Belgium	5,000 – 27,800	95–00	(0)	(0–19)	*1*
Estonia	10,000 – 20,000	98	+	20–29	*1*
Finland	(10,000 – 10,000)	98–02	+	**15**	
France	(200,000 – 800,000)	98–02	(–)	(30–49)	*9*
Iceland	(15,000 – 25,000)	78–94	?	–	*48*
Latvia	20,000 – 50,000	90–99	(0)	(0–19)	*29*
Netherlands	150,000 – 200,000	99–01	–	**31**	*3,4,8,9,11,12*
Poland	40,000 – 80,000	90–02	+	**50–79**	*114,124*
UK	376,000 – 376,000	93	+	30–49	*45*
Total (approx.)	>800,000	**Overall trend** Small decline			
% in European IBAs	10–12	**Gen. length** 13		**% Global pop.** 50–74	

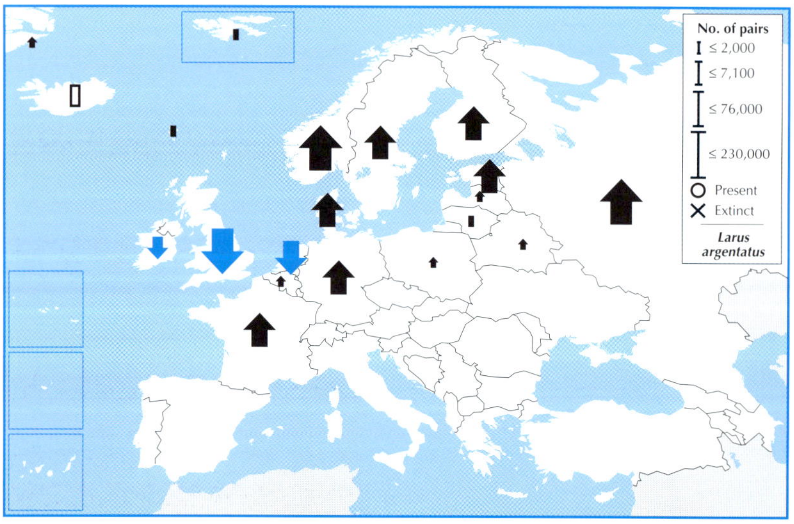

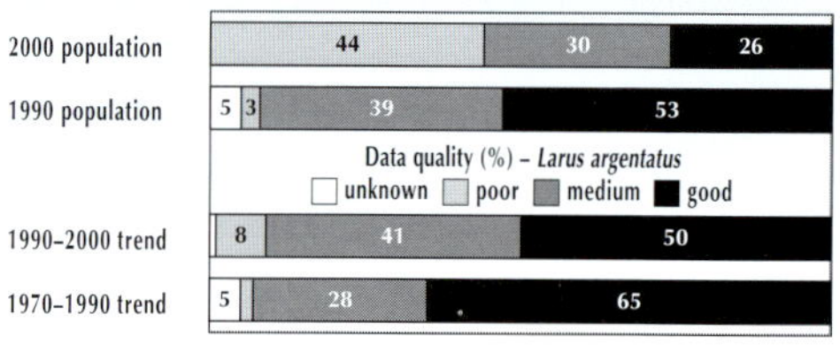

Larus cachinnans
YELLOW-LEGGED GULL

Non-SPEC[E] (1994: —) **Status** Secure
Criteria —
European IUCN Red List Category —
Criteria —
Global IUCN Red List Category —
Criteria —

Larus cachinnans is a widespread breeder in coastal areas of southern and eastern Europe, which constitutes >50% of its global breeding range. Its European breeding population is large (>310,000 pairs), and increased between 1970–1990. Although there were declines in Croatia and Georgia during 1990–2000, populations across the rest of its European range increased or were stable, and the species showed a marked increase overall. Consequently, it is evaluated as Secure.

Country	Breeding pop. size (pairs)	Year(s)	Trend	Mag.%	References
Albania	**90 – 110**	96–02	**0**	**0–19**	
Austria	10 – 25	98–02	+	**>80**	
Azerbaijan	10,000 – 15,000	96–00	(0)	(0–19)	
Belarus	**150 – 400**	97–02	+	10–19	
Belgium	**2 – 2**	00–02	+	**N**	*1*
Bosnia & HG	Present	90–03	?	–	
Bulgaria	5,000 – 7,000	96–02	+	0–19	
Croatia	(25,000 – 50,000)	02	(–)	(0–19)	*16*
Cyprus	(100 – 200)	98–02	(+)	(0–19)	
Czech Rep.	**0 – 5**	00	+	**N**	
France	40,000 – 45,000	97–00	+	50–79	*1*
Georgia	300 – 1,000	94–02	–	20–29	
Germany	**79 – 89**	95–99	+	**50–79**	
Greece	(3,000 – 5,000)	95–00	(+)	(0–19)	
Hungary	3 – 9	95–02	(F)	(–)	*14,7*
Italy	40,000 – 50,000	03	+	30–49	
Macedonia	50 – 250	90–00	(F)	(–)	*7*
Malta	**150 – 180**	90–02	+	**0–19**	*1*
Moldova	0 – 30	90–00	F	20–29	
Netherlands	**16 – 32**	98–00	?	–	*1*
Poland	**150 – 230**	97–02	+	**N**	*2*
Portugal	20,000 – 30,000	02	(+)	(–)	*2,5*
Azores	Present	02	?	–	
Madeira	(5,000 – 10,000)	02	(0)	(0–19)	
Romania	3,500 – 5,000	90–02	+	0–19	*19,30*
Russia	20,000 – 40,000	90–00	0	0–19	*8,24,60,120, 141,145*
Serbia & MN	230 – 270	98–02	+	20–49	*1,227*
Slovakia	**0 – 15**	80–99	+	N	
Slovenia	60 – 80	90–00	(0)	(0–19)	
Spain	(100,000 – 250,000)	98–02	(+)	(0–19)	*10*
Canary Is.	2,500 – 10,000	97–03	+	30–49	*28,30,25*
Switzerland	**300 – 600**	98–02	+	**>80**	
Turkey	20,000 – 30,000	01	+	20–29	
Ukraine	12,500 – 17,500	90–00	+	20–29	
UK	**1 – 3**	99–01	?	–	*20*
Gibraltar	**5,000 – 7,000**	00	**0**	**0–19**	

Total (approx.) 310,000 – 580,000 **Overall trend** Large increase
Breeding range >1,000,000 km² **Gen. length**. 13 **% Global pop.** 50–74
(See p. 139, bottom, for data quality graph)

Winter pop. size (individuals)
Total (approx.) >380,000 **Overall trend** Large increase
% in European IBAs 22–35 **Gen. length** 13 **% Global pop.** 25–49

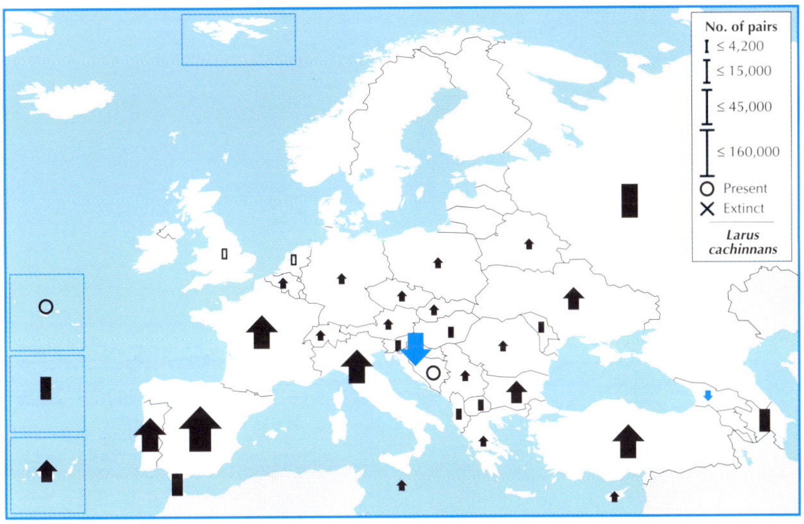

Larus armenicus
ARMENIAN GULL

SPEC 2 (1994: —) **Status** Localised
Criteria ≥90% breed at ≤10 sites
European IUCN Red List Category —
Criteria —
Global IUCN Red List Category —
Criteria —

Country	Breeding pop. size (pairs)	Year(s)	Trend	Mag.%	References
Armenia	**10,000 – 12,000**	97–02	**0**	**0–19**	
Georgia	Present	03	?	–	
Turkey	5,000 – 10,000	01	+	20–29	

Total (approx.) 15,000 – 22,000 **Overall trend** Moderate increase
Breeding range >20,000 km² **Gen. length**. 13 **% Global pop.** 50–74

Country	Winter pop. size (individuals)	Year(s)	Trend	Mag.%	References
Cyprus	10 – 100	97–02	0	0–9	
Turkey	(2,000 – 8,000)	91–01	+	20–29	

Total (approx.) >2,000 **Overall trend** Unknown
% in European IBAs 4–15 **Gen. length** 13 **% Global pop.** 25–49

Larus armenicus breeds locally in Turkey, Armenia and Georgia, with Europe now known to constitute >50% of its global breeding range. Its European breeding population is relatively small (<22,000 pairs), but the trend between 1970–1990 was not known. The trend in Georgia during 1990–2000 was also unknown, but the species was stable or increased in Armenia and Turkey, and underwent a moderate increase overall. Nevertheless, more than 90% of the European breeding population occurs at just 10 sites (see Appendix 5), and consequently it is evaluated as Localised.

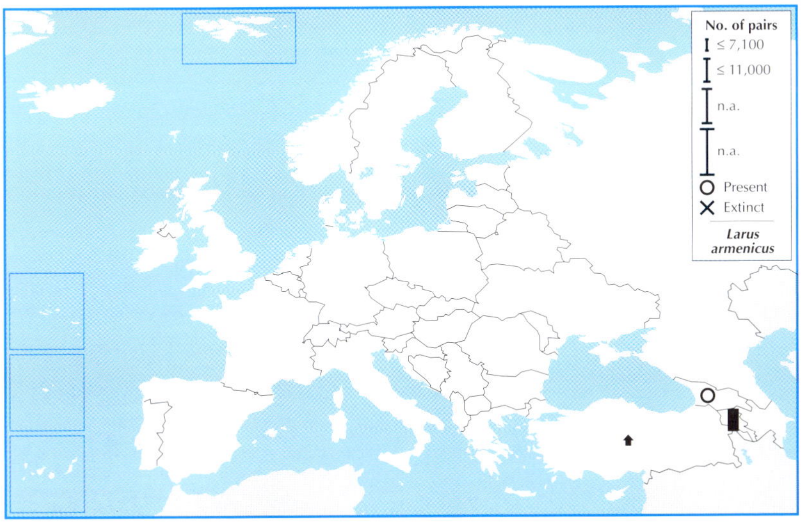

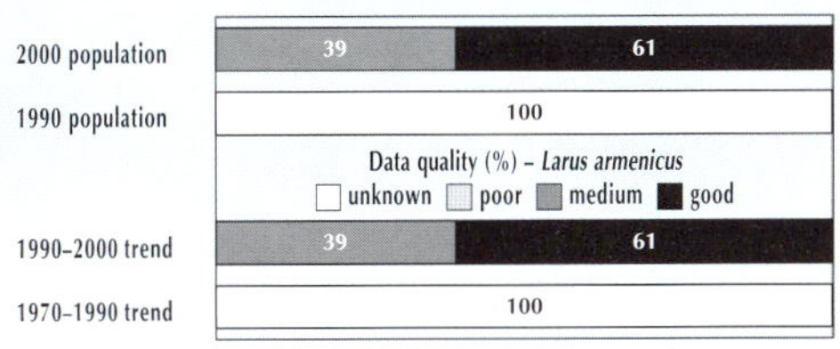

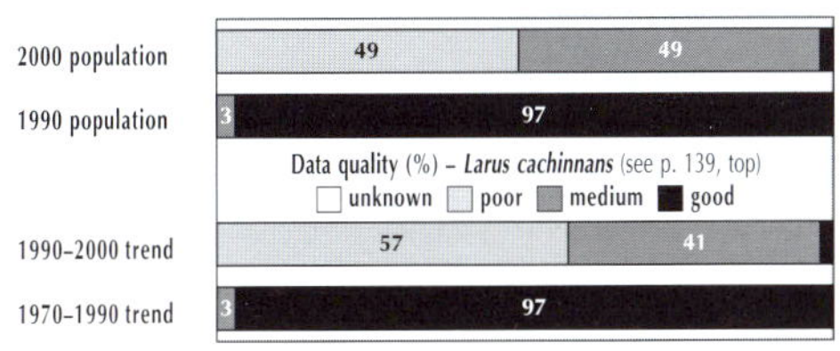

Larus glaucoides
ICELAND GULL

Non-SPEC[E] (1994: —) **Status** (Secure)
Criteria —
European IUCN Red List Category —
Criteria —
Global IUCN Red List Category —
Criteria —

Larus glaucoides breeds in southern and western Greenland, which is now thought to constitute >50% of its global breeding range. Its European breeding population is relatively small (<100,000 pairs), but probably remained stable during 1970–1990 and 1990–2000. Consequently, the species is provisionally evaluated as Secure.

Country	Breeding pop. size (pairs)	Year(s)	Trend	Mag.%	References
Denmark					
Greenland	(30,000 – 100,000)	90–00	(0)	(0–19)	*5*
Total (approx.)	30,000 – 100,000	**Overall trend** Stable			
Breeding range	>250,000 km²	**Gen. length** 11		**% Global pop.** 50–74	

Country	Winter pop. size (individuals)	Year(s)	Trend	Mag.%	References
Denmark					
Faroe Is.	10 – 100	92	?	–	*3*
Greenland	(300,000 – 500,000)	90–00	(0)	(0–19)	
Germany	**0 – 50**	95–00	?	–	
Iceland	(15,000 – 20,000)	99	?	–	*54*
Total (approx.)	>320,000	**Overall trend** Unknown			
% in European IBAs	Unknown	**Gen. length** 11		**% Global pop.** 50–74	

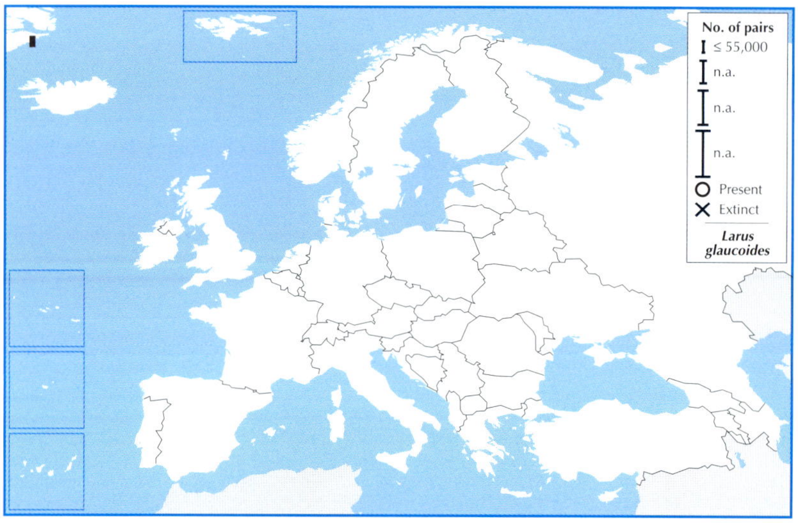

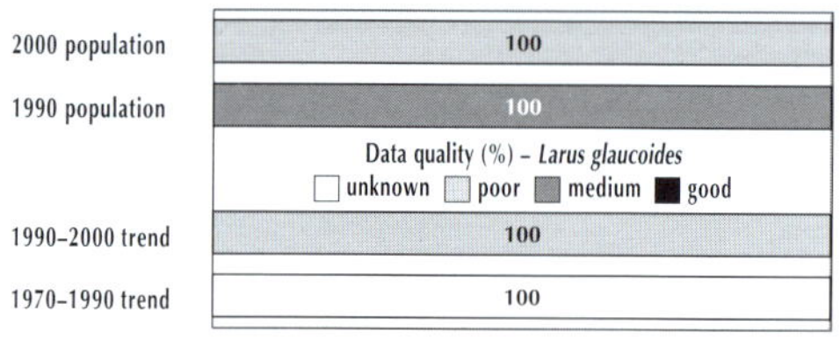

Larus hyperboreus
GLAUCOUS GULL

Non-SPEC (1994: —) **Status** (Secure)
Criteria —
European IUCN Red List Category —
Criteria —
Global IUCN Red List Category —
Criteria —

Larus hyperboreus breeds in Greenland, Svalbard, Iceland and arctic Russia, with Europe accounting for less than a quarter of its global breeding range. Its European breeding population is relatively small (<140,000 pairs), but was stable between 1970–1990. Although the trends of the populations in Iceland and Russia during 1990–2000 were unknown, the species was stable in Svalbard and the stronghold Greenland, and hence probably remained stable overall. Consequently, it is provisionally evaluated as Secure.

Country	Breeding pop. size (pairs)	Year(s)	Trend	Mag.%	References
Denmark					
Greenland	(30,000 – 100,000)	90–00	(0)	(0–19)	*5*
Iceland	(10,000 – 15,000)	97	?	–	*16*
Norway					
Svalbard	(4,000 – 10,000)	70–96	(0)	(0–19)	
Russia	(2,500 – 10,000)	90–00	?	–	*81,154,155*
Total (approx.)	47,000 – 140,000	**Overall trend** Stable			
Breeding range	>500,000 km²	**Gen. length** 13		**% Global pop.** 5–24	

Country	Winter pop. size (individuals)	Year(s)	Trend	Mag.%	References
Denmark					
Faroe Is.	30 – 200	92	?	–	*3*
Greenland	(300,000 – 500,000)	90–00	(0)	(0–19)	
Estonia	0 – 5	98	(+)	(N)	*1*
France	20 – 80	98–02	?	–	*9*
Germany	**0 – 50**	95–00	?	–	
Iceland	(30,000 – 50,000)	78–94	?	–	*48*
Latvia	**0 – 10**	90–01	(0)	(0–19)	*32*
Netherlands	**2 – 2**	99–01	?	–	*3,4,8,9,11,12*
Poland	**5 – 30**	90–00	**F**	**>80**	*123*
Romania	**0 – 1**	90–00	?	–	
Total (approx.)	>330,000	**Overall trend** Unknown			
% in European IBAs	Unknown	**Gen. length** 13		**% Global pop.** 25–49	

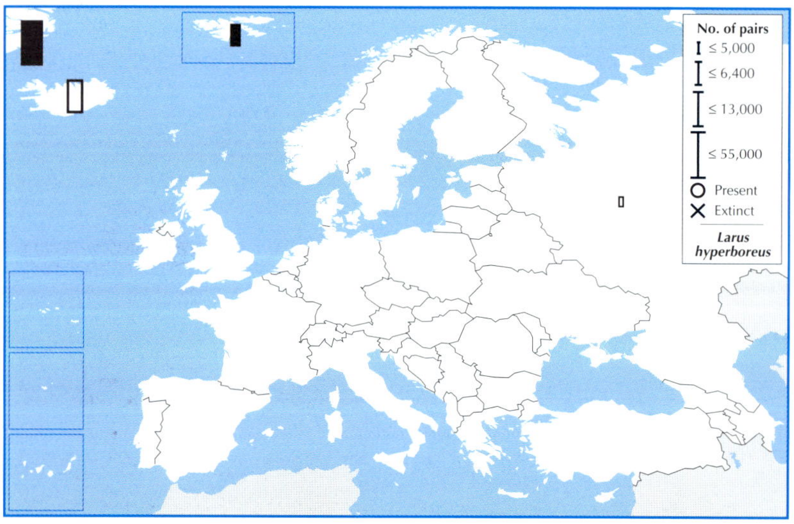

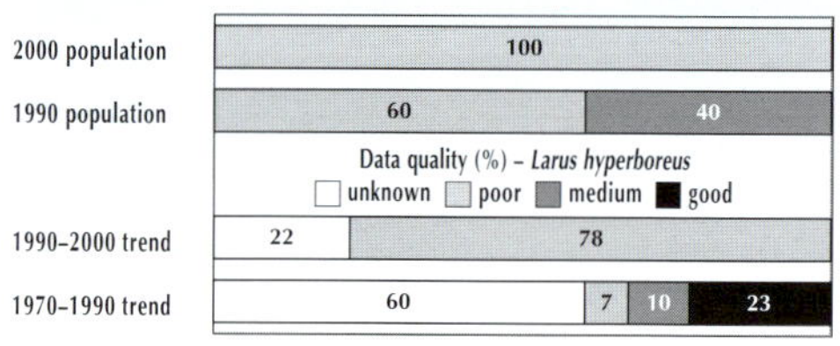

Larus marinus
GREAT BLACK-BACKED GULL

Non-SPEC[E] (1994: 4) **Status** Secure
Criteria —
European IUCN Red List Category —
Criteria —
Global IUCN Red List Category —
Criteria —

Larus marinus is a widespread breeder in coastal areas of northern and western Europe, which constitutes >50% of its global breeding range. Its European breeding population is large (>110,000 pairs), and was stable between 1970–1990. Although there were declines in Iceland and the Republic of Ireland during 1990–2000, populations in the rest of its range increased or were stable, and the species increased overall. Consequently, it is evaluated as Secure.

Country	Breeding pop. size (pairs)	Year(s)	Trend	Mag.%	References
Denmark	(2,000 – 3,000)	00	+	50–79	*3*
Faroe Is.	1,200 – 1,200	95	(0)	(0–19)	
Greenland	(3,000 – 5,000)	90–00	(+)	(0–19)	*5*
Estonia	3,000 – 5,000	98	+	20–29	*1*
Finland	3,000 – 4,000	98–02	+	15	
France	3,700 – 4,000	97–00	+	0–9	*4*
Germany	**22 – 24**	95–99	+	**50–79**	
Iceland	(15,000 – 20,000)	00	(–)	(0–19)	*21*
Rep. Ireland	2,200 – 2,200	99–02	–	**28**	
Latvia	(0 – 5)	90–00	?	(N)	*22*
Netherlands	**11 – 15**	98–00	?	–	*1*
Norway	50,000 – 100,000	90–02	+	**20–29**	*4*
Svalbard	50 – 150	90–02	+	0–19	
Russia	**3,000 – 4,000**	90–02	**0**	**0–19**	*11,82,84,154,155*
Sweden	10,000 – 15,000	99–00	+	0–19	
UK	**17,500 – 17,500**	98–02	**0**	**3**	*20*
Total (approx.)	110,000 – 180,000	**Overall trend** Large increase			
Breeding range	>1,000,000 km²	**Gen. length.** 13		**% Global pop.** 50–74	

(See p. 141, bottom, for data quality graph)

Country	Winter pop. size (individuals)	Year(s)	Trend	Mag.%	References
Albania	**0 – 1**	95–02	(F)	(–)	
Belgium	300 – 300	95–00	(0)	(0–19)	*1*
Bulgaria	**0 – 2**	97–01	?	–	
Denmark	Present	99–00	?	–	
Faroe Is.	2,500 – 5,000	92	?	–	*3*
Greenland	(20,000 – 50,000)	90–00	(+)	(0–19)	
Estonia	1,000 – 3,000	98	0	0–19	*1*
Finland	(1,000 – 10,000)	98–02	+	10–19	
France	(15,000 – 60,000)	98–02	?	–	*9*
Iceland	(50,000 – 80,000)	78–94	?	–	*48*
Latvia	1,000 – 2,000	90–01	(0)	(0–19)	*32*
Lithuania	(250 – 1,000)	92–02	?	–	
Netherlands	10,000 – 25,000	85–01	**0**	**0–19**	*3,4,8,9,11,12*
Poland	2,000 – 6,000	90–00	(0)	(0–19)	
Portugal	Present	02	?	–	
Romania	**0 – 4**	90–00	?	–	
Slovakia	0 – 2	90–99	?	–	4
Slovenia	Present	90–00	?	–	
Switzerland	**0 – 3**	98–02	**0**	**0–19**	
UK	43,000 – 43,000	93	+	0–19	*45*
Total (approx.)	>150,000	**Overall trend** Small increase			
% in European IBAs	3–6	**Gen. length** 13		**% Global pop.** 50–74	

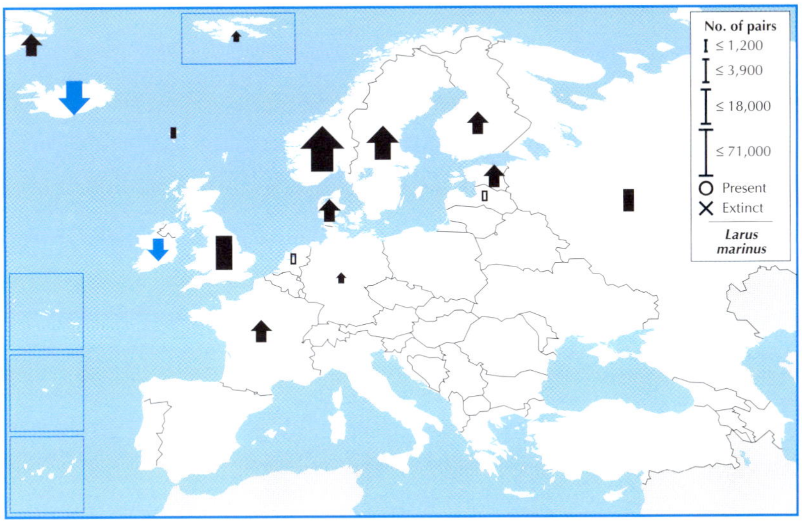

Xema sabini
SABINE'S GULL

Non-SPEC (1994: —) **Status** Secure
Criteria —
European IUCN Red List Category —▼▼▼
Criteria —
Global IUCN Red List Category —
Criteria —

Xema sabini has a predominantly Siberian and North American breeding distribution, which just extends into Europe in Greenland. Its European breeding population is very small (as few as 100 pairs), but increased substantially between 1970–1990. The species continued to increase (albeit more slowly) during 1990–2000. Although the size of the European population still renders it susceptible to the risks affecting small populations, it is marginal to a much a larger non-European population. Consequently, the species is evaluated as Secure.

Country	Breeding pop. size (pairs)	Year(s)	Trend	Mag.%	References
Denmark					
Greenland	(100 – 500)	90–00	(+)	(0–19)	
Total (approx.)	100 – 500	**Overall trend** Small increase			
Breeding range	>100,000 km²	**Gen. length.** 6		**% Global pop.** <5	

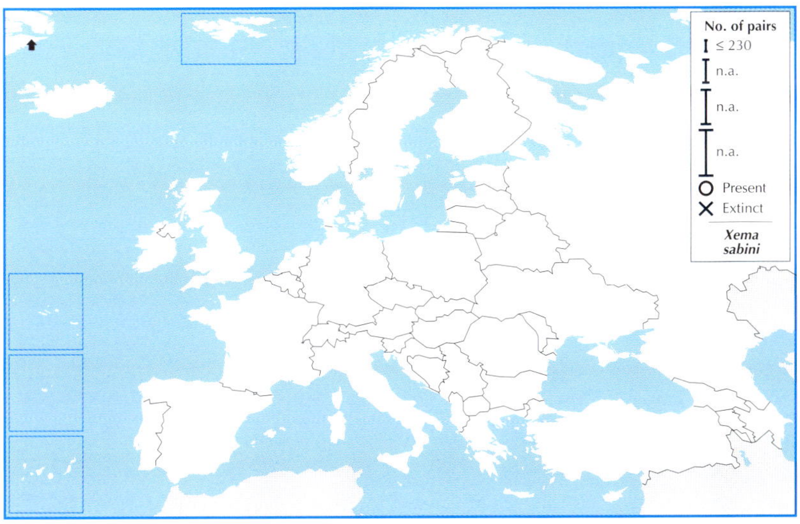

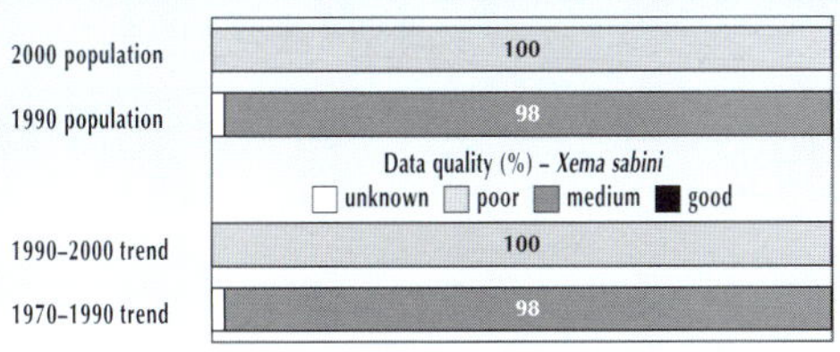

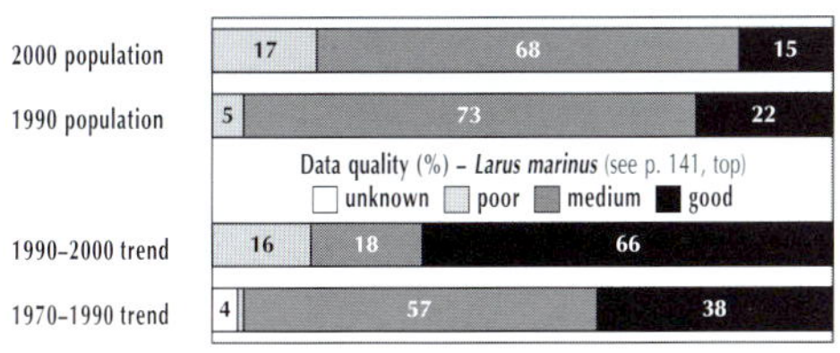

Rhodostethia rosea
ROSS'S GULL

Non-SPEC (1994: —) **Status** (Secure)
Criteria —
European IUCN Red List Category —▼▼▼▼
Criteria —
Global IUCN Red List Category —
Criteria —

Country	Breeding pop. size (pairs)	Year(s)	Trend	Mag.%	References
Denmark					
Greenland	(0 – 2)	90–00	?	–	
Total (approx.)	0 – 2	**Overall trend** Unknown			
Breeding range	<20,000 km²	**Gen. length.** 6		**% Global pop.** <5	

Rhodostethia rosea breeds mostly in the tundra of Siberia and north-eastern Canada, with one or two pairs breeding within Europe in Greenland. This tiny European breeding population was stable between 1970–1990, but its trend during 1990–2000 was unknown. Although the size of the European population renders it susceptible to the risks affecting small populations, it is marginal to a much a larger non-European population. Consequently, the species is provisionally evaluated as Secure.

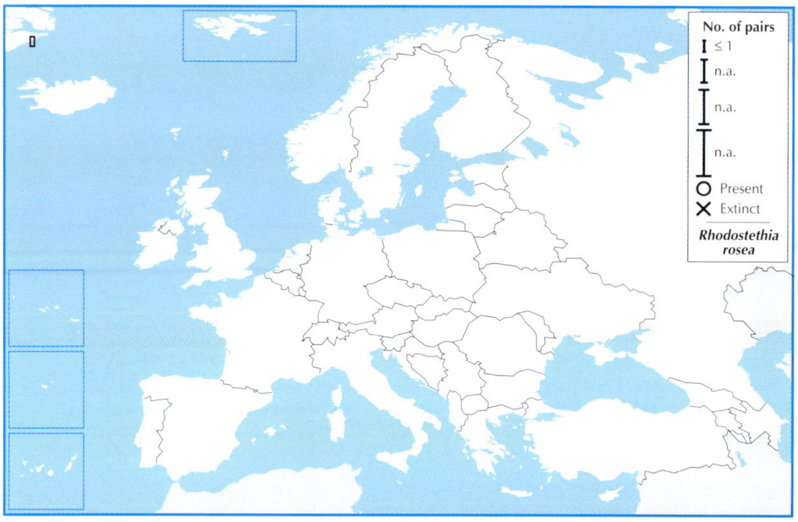

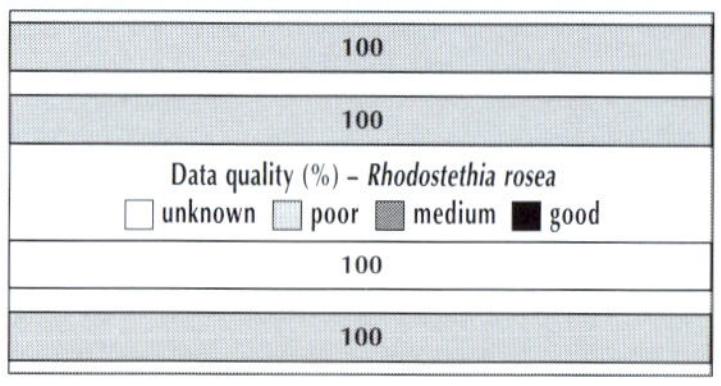

Rissa tridactyla
BLACK-LEGGED KITTIWAKE

Non-SPEC (1994: —) **Status** (Secure)
Criteria —
European IUCN Red List Category —
Criteria —
Global IUCN Red List Category —
Criteria —

Country	Breeding pop. size (pairs)	Year(s)	Trend	Mag.%	References
Denmark	**430 – 500**	99–00	–	20–29	*16*
Faroe Is.	230,000 – 230,000	95	(0)	(0–19)	
Greenland	150,000 – 300,000	90–00	(–)	(0–19)	*5,17*
France	**5,000 – 5,600**	96–00	**0**	**0–9**	*4*
Germany	**7,300 – 7,300**	95–99	**+**	**30–49**	
Iceland	**631,000 – 631,000**	83–85	**+**	**0–19**	*8*
Rep. Ireland	**33,000 – 33,000**	99–02	**0**	**3**	
Netherlands	**0 – 3**	98–00	?	–	*1*
Norway	(300,000 – 600,000)	90–01	**-**	**20–29**	*4,35*
Svalbard	**265,000 – 275,000**	80–01	F	20–29	
Russia	(100,000 – 500,000)	90–02	(0)	(0–19)	*11,79,83,155, 165,166,167*
Spain	50 – 50	98–02	–	>80	*10,16*
Sweden	**25 – 30**	99–00	**0**	**0–19**	
UK	**380,000 – 380,000**	98–02	**-**	**25**	*20*
Total (approx.)	2,100,000 – 3,000,000	**Overall trend** Moderate decline			
Breeding range	>1,000,000 km²	**Gen. length.** 10		**% Global pop.** 25–49	

Country	Winter pop. size (individuals)	Year(s)	Trend	Mag.%	References
Czech Rep.	0 – 10	90–00	0	0–19	
Denmark	Present	99–00	?	–	
Faroe Is.	10,000 – 100,000	92	?	–	*3*
Greenland	(0 – 5,000)	90–00	(F)	(–)	
France	(10,000 – 40,000)	98–02	?	–	*9*
Germany	**18,800 – 18,800**	87–03	?	–	*4*
Greece	(5 – 10)	95–00	(F)	(>80)	
Iceland	(10,000 – 1,000,000)	78–94	?	–	*48*
Lithuania	(0 – 5)	92–02	(0)	(0–19)	
Moldova	0 – 12	90–00	+	50–79	
Netherlands	(9,000 – 9,000)	85–93	(0)	(0–19)	*3,4,5,8,9,11,12*
Poland	10 – 160	90–00	F	>80	*114,128*
Portugal	Present	02	?	–	
Slovakia	0 – 30	90–99	?	–	*4*
Slovenia	Present	90–00	?	–	
Sweden	100,000 – 120,000	98–01	0	0–19	
Turkey	(30 – 60)	91–01	?	–	
UK					
Gibraltar	**1 – 3**	00–03	**0**	**0–19**	
Total (approx.)	>200,000	**Overall trend** Unknown			
% in European IBAs	>10	**Gen. length** 10		**% Global pop.** 5–24	

Rissa tridactyla is a widespread but patchily distributed breeder along the Atlantic coasts of western and northern Europe, which accounts for less than half of its global breeding range. Its European breeding population is very large (>2,100,000 pairs), and underwent a moderate increase between 1970–1990. Although the species declined in Greenland, Norway and the United Kingdom during 1990–2000, and underwent a moderate decline (>10%) overall, this decline is probably outweighed by the earlier increase. Consequently, the species is provisionally evaluated as Secure.

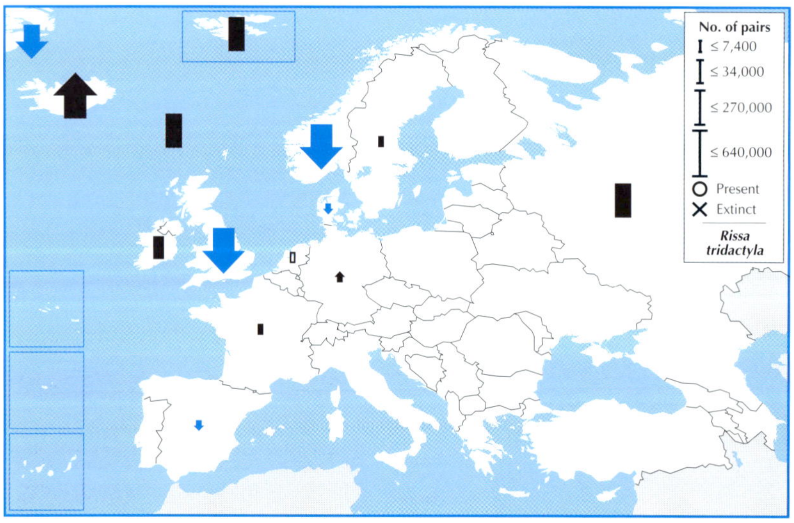

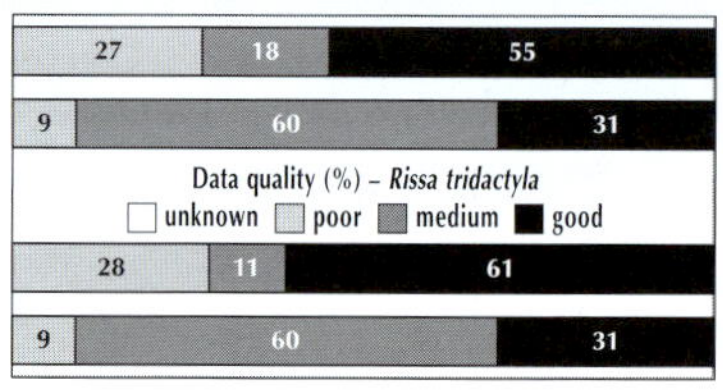

Pagophila eburnea
IVORY GULL

SPEC 3 (1994: 3) **Status** (Rare)
Criteria <10,000 pairs
European IUCN Red List Category —
Criteria —
Global IUCN Red List Category —
Criteria —

Country	Breeding pop. size (pairs)	Year(s)	Trend	Mag.%	References
Denmark					
Greenland	(500 – 1,000)	90–00	?	–	*16*
Norway					
Svalbard	(50 – 200)	80–03	(-)	(0–19)	
Russia	(2,500 – 10,000)	90–02	?	–	*154,155*
Total (approx.)	3,100 – 11,000	Overall trend Unknown			
Breeding range	>250,000 km²	Gen. length 12		% Global pop. 5–24	

Country	Winter pop. size (individuals)	Year(s)	Trend	Mag.%	References
Denmark					
Greenland	(100 – 1,000)	90–00	?	–	
Total (approx.)	>100	Overall trend Unknown			
% in European IBAs	Unknown	Gen. length 12		% Global pop. <5	

Pagophila eburnea breeds in Greenland, Svalbard and arctic Russia, with Europe accounting for less than a quarter of its global breeding range. Its European breeding population is small (as few as 3,100 pairs), and underwent a large decline between 1970–1990. Although the species declined in Svalbard during 1990–2000, trend data were not available for its key populations in Greenland and Russia, and its overall trend was unknown. Nevertheless, its population size renders it susceptible to the risks affecting small populations, and consequently it is provisionally evaluated as Rare.

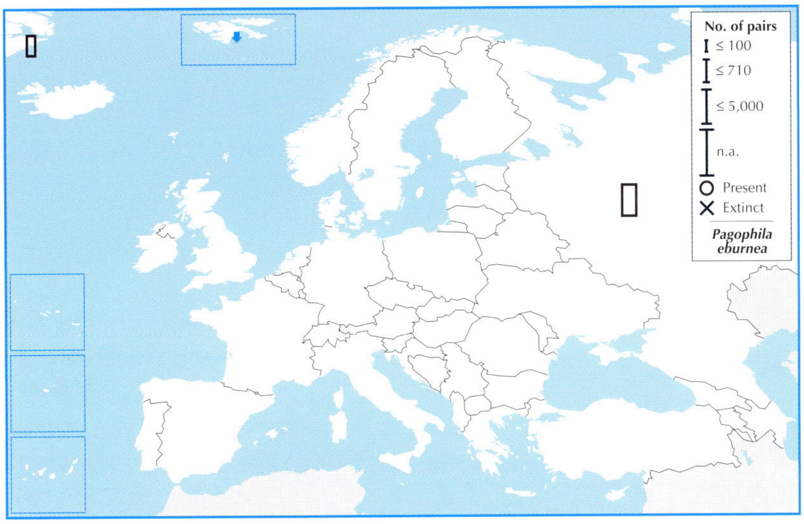

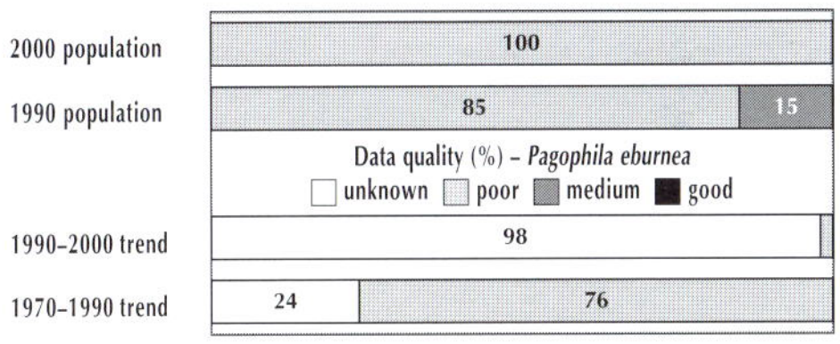

Sterna nilotica
GULL-BILLED TERN

SPEC 3 (1994: 3) **Status** (Vulnerable)
Criteria See IUCN below
European IUCN Red List Category VU
Criteria A2b
Global IUCN Red List Category —
Criteria —

Country	Breeding pop. size (pairs)	Year(s)	Trend	Mag.%	References
Albania	(0 – 5)	96–02	–	0–19	
Armenia	20 – 80	97–02	?	–	
Bulgaria	2 – 15	96–02	F	>80	
Denmark	**1 – 2**	00–01	**-**	**>80**	*6,7*
France	269 – 370	98–01	**F**	**20–29**	*6*
Germany	**60 – 60**	95–99	**0**	**0–19**	
Greece	(50 – 100)	95–00	(-)	(0–19)	
Italy	**382 – 382**	00	**0**	**0–19**	*18*
Macedonia	(1 – 5)	90–00	(F)	(20–29)	
Portugal	**40 – 55**	03	**+**	**N**	
Romania	(12 – 50)	90–02	–	20–29	
Russia	2,000 – 5,000	90–00	(F)	(30–49)	*8,33,60*
Spain	3,000 – 3,500	98–02	(F)	(-)	*10,16*
Turkey	2,000 – 6,000	01	**-**	**20–29**	
Ukraine	4,000 – 6,000	90–00	–	0–9	
Total (approx.)	12,000 – 22,000	Overall trend Moderate decline			
Breeding range	>100,000 km²	Gen. length 9		% Global pop. 5–24	

Sterna nilotica is a patchily distributed breeder in southern and eastern Europe, which accounts for less than a quarter of its global breeding range. Its European breeding population is relatively small (<22,000 pairs), and declined substantially between 1970–1990. Although the species was broadly stable in south-west Europe and Russia during 1990–2000, it declined in south-eastern Europe, and continued to decline overall at a rate that—on top of earlier declines—probably equates to a large decline (>30%) over three generations. Consequently, it is provisionally evaluated as Vulnerable.

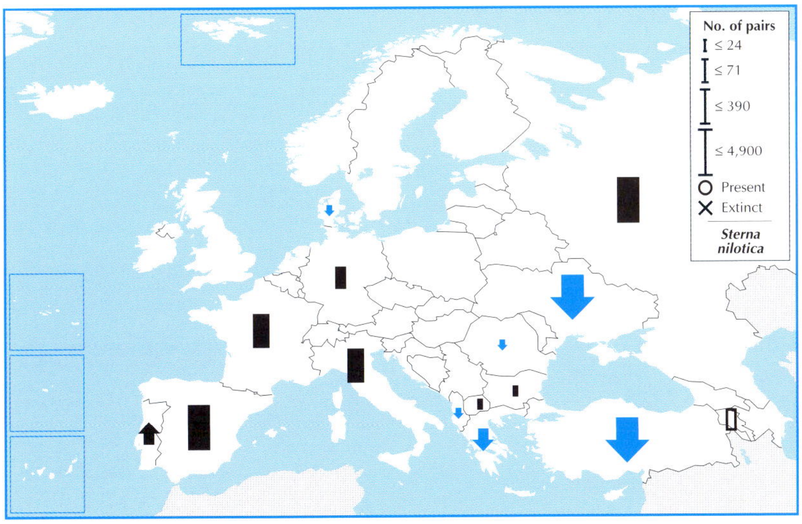

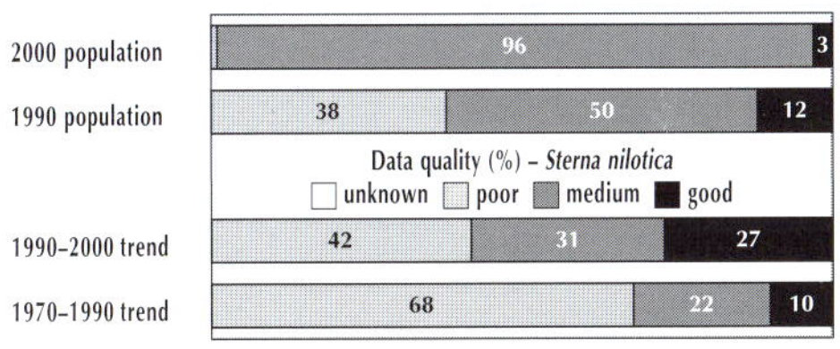

Sterna caspia
CASPIAN TERN

SPEC 3 (1994: 3) **Status** Rare
Criteria <10,000 pairs
European IUCN Red List Category —
Criteria —
Global IUCN Red List Category —
Criteria —

Country	Breeding pop. size (pairs)	Year(s)	Trend	Mag.%	References
Azerbaijan	(50 – 250)	96–00	(0)	(0–19)	
Estonia	250 – 400	98	-	20–29	*1*
Finland	**800 – 900**	00–01	**+**	**20**	
Germany	**0 – 2**	95–99	**0**	**0–19**	
Russia	2,000 – 5,500	90–02	+	20–29	*6,33,60,70,120, 178*
Sweden	**450 – 470**	99–00	**F**	**20–29**	
Turkey	**300 – 600**	01	**+**	**20–29**	
Ukraine	**800 – 1,200**	90–00	-	0–9	
Total (approx.)	4,700 – 9,300	**Overall trend** Large increase			
Breeding range	>100,000 km²	**Gen. length.** 11		**% Global pop.** 5–24	

Sterna caspia breeds patchily along the Baltic Sea coast and in south-east Europe, which accounts for less than a quarter of its global breeding range. Its European breeding population is small (as few as 4,700 pairs), and underwent a large decline between 1970–1990. Although the species increased substantially overall during 1990–2000—with increasing, stable or fluctuating trends across most of its European range—its population size still renders it susceptible to the risks affecting small populations, and consequently it is evaluated as Rare.

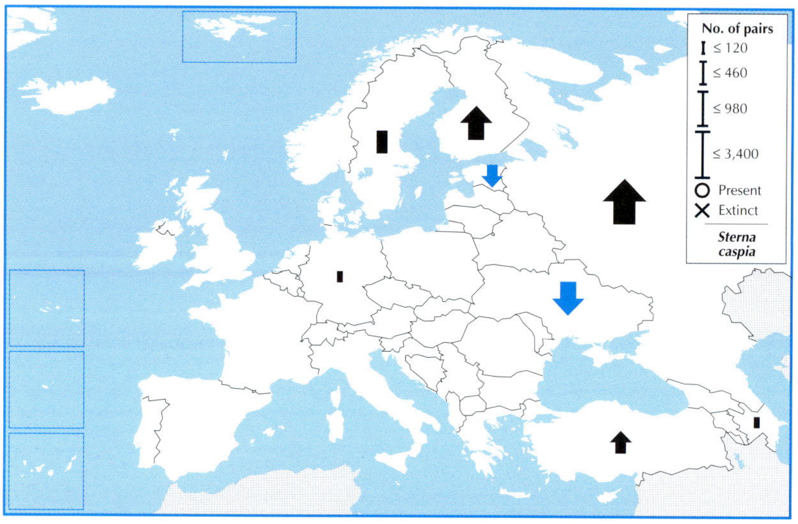

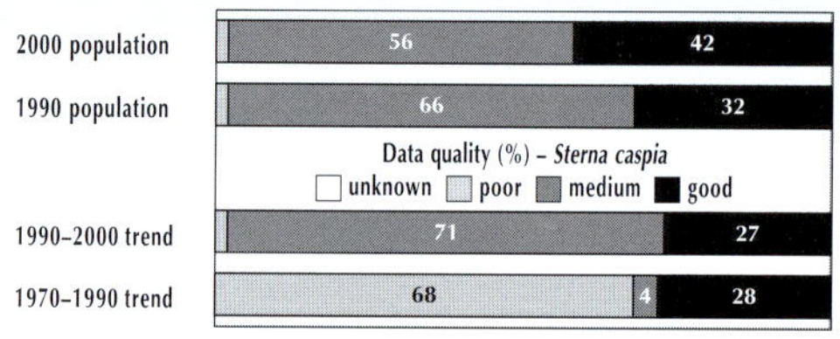

Sterna bengalensis
LESSER CRESTED-TERN

Non-SPEC (1994: NE) **Status** (Secure)
Criteria —
European IUCN Red List Category —▼▼▼▼
Criteria —
Global IUCN Red List Category —
Criteria —

Country	Breeding pop. size (pairs)	Year(s)	Trend	Mag.%	References
Italy	**1 – 1**	00	?	-	18
Spain	1 – 2	98–02	(F)	(-)	10
Total (approx.)	2 – 3	**Overall trend** Unknown			
Breeding range	<100 km²	**Gen. length.** 9		**% Global pop.** <5	

Sterna bengalensis has a predominantly tropical and subtropical breeding distribution, although two or three pairs do breed in Europe at two sites in north-eastern Spain and northern Italy. Although this tiny European population probably remained broadly stable overall during 1990–2000, its size could render it susceptible to the risks affecting small populations. Nevertheless, as it is marginal to a far larger non-European population, the species is provisionally evaluated as Secure.

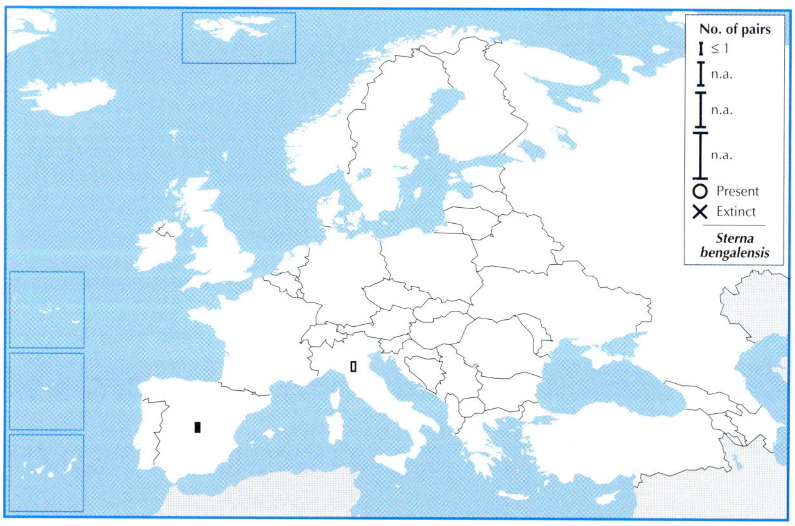

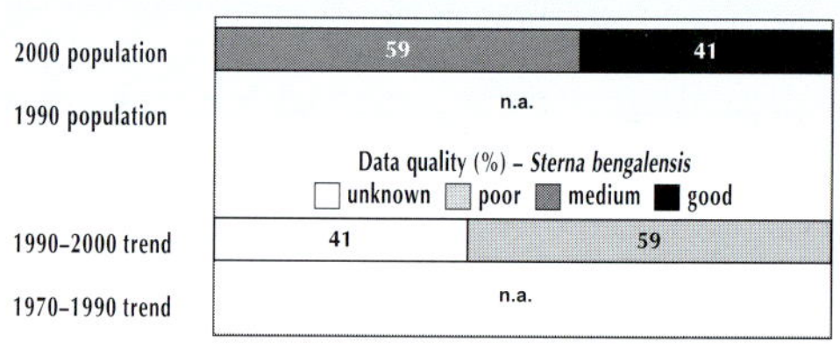

Sterna sandvicensis
SANDWICH TERN

SPEC 2 (1994: 2) **Status** Depleted
Criteria Moderate historical decline
European IUCN Red List Category —
Criteria —
Global IUCN Red List Category —
Criteria —

Sterna sandvicensis is a widespread but very dispersed breeder in coastal areas of Europe, which holds >50% of its global population. Its European breeding population is relatively small (<130,000 pairs), and underwent a moderate decline between 1970–1990. Although there were declines in a few countries during 1990–2000, key populations in the Netherlands, Ukraine and Russia fluctuated, and the species declined only slightly overall. Nevertheless, its population still remains below the level that preceded its decline, and consequently it is evaluated as Depleted.

Country	Breeding pop. size (pairs)	Year(s)	Trend	Mag.%	References
Azerbaijan	**6,500 – 10,000**	96–00	(F)	(–)	
Belgium	**1,550 – 1,550**	00–02	**F**	**>80**	*1*
Bulgaria	**150 – 500**	96–02	**+**	**>80**	
Denmark	**4,000 – 4,500**	02	**-**	**0–19**	*2*
Estonia	500 – 900	98	-	20–29	*1*
France	6,800 – 6,800	00	**0**	**0–19**	*1*
Germany	**9,700 – 10,500**	95–99	**0**	**0–19**	
Greece	(30 – 50)	95–00	(F)	(50–79)	
Rep. Ireland	**1,800 – 1,800**	99–02	**0**	**0–19**	
Italy	**610 – 612**	00	**0**	**0–19**	*18*
Netherlands	**14,500 – 14,500**	98–00	**F**	**34**	*1*
Poland	**0 – 24**	90–00	**-**	**>80**	*75*
Romania	40 – 60	90–02	(F)	(20–29)	
Russia	15,000 – 25,000	90–02	(F)	(30–49)	*60,120*
Spain	3,000 – 3,000	01	+	>80	*10,16*
Sweden	**300 – 350**	99–00	**-**	**30–49**	
Turkey	**50 – 60**	01	**0**	**0–19**	
Ukraine	5,000 – 40,000	90–00	**F**	**30–49**	
UK	**12,500 – 12,500**	00–01	**-**	**15**	*20*
Total (approx.)	82,000 – 130,000	**Overall trend** Small decline			
Breeding range	>250,000 km²	**Gen. length.** 9		**% Global pop.** 50–74	

Country	Winter pop. size (individuals)	Year(s)	Trend	Mag.%	References
Albania	**171 – 314**	95–02	(F)	(–)	
Croatia	(150 – 300)	02	?	–	*16,26*
Cyprus	20 – 60	94–02	0	0–9	
France	(500 – 1,500)	98–02	?	–	*9*
Greece	**1,000 – 2,000**	95–99	**0**	**0–19**	
Italy	(500 – 1,500)	02	0	0–19	*22*
Malta	**2 – 5**	95–02	**0**	**0–19**	*4*
Netherlands	**2 – 2**	99–01	?	–	*3,4,8,9,11,12*
Portugal	Present	02	?	–	
Slovenia	15 – 35	95–00	F	30–49	
Turkey	(800 – 1,200)	91–01	(0)	(0–19)	
UK					
Gibraltar	**10 – 20**	00–03	**0**	**0–19**	
Total (approx.)	>3,200	**Overall trend** Unknown			
% in European IBAs	46–60	**Gen. length** 9		**% Global pop.** <5	

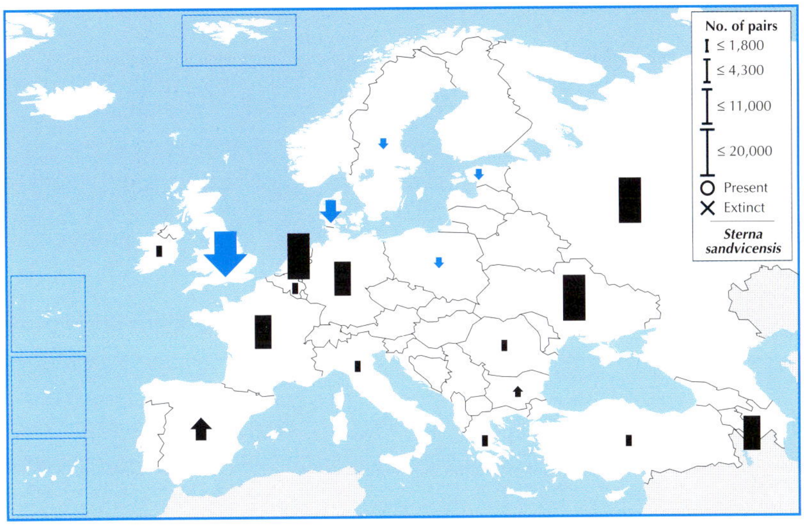

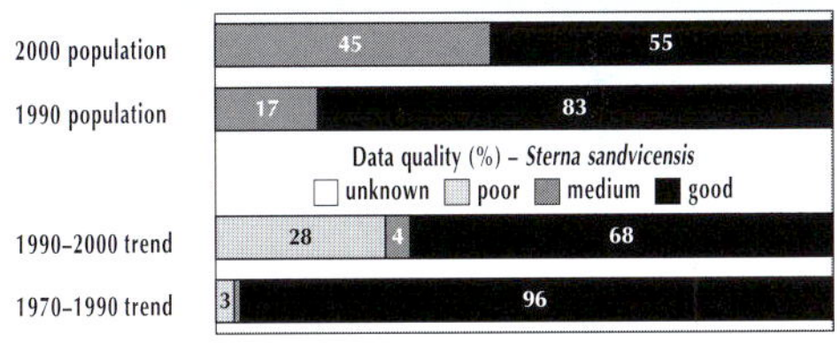

Sterna dougallii
ROSEATE TERN

SPEC 3 (1994: 3) **Status** Rare
Criteria <10,000 pairs
European IUCN Red List Category —
Criteria —
Global IUCN Red List Category —
Criteria —

Sterna dougallii breeds in the Macaronesian islands and patchily in coastal areas of north-west Europe, which accounts for a tiny proportion of its global breeding range. Its European breeding population is small (as few as 1,800 pairs), and underwent a large decline between 1970–1990. Although the species was stable overall during 1990–2000—with key populations in the Azores and the Republic of Ireland stable and increasing respectively—its population size still renders it susceptible to the risks affecting small populations, and consequently it is evaluated as Rare.

Country	Breeding pop. size (pairs)	Year(s)	Trend	Mag.%	References
France	**70 – 91**	00	**0**	**10–19**	5
Rep. Ireland	**719 – 719**	99–02	**+**	**>80**	
Portugal					
Azores	**1,000 – 1,000**	94–00	**0**	**0–19**	34,39
Madeira	(0 – 15)	93	(F)	(–)	
Spain					
Canary Is.	(1 – 2)	97–03	?	–	28,25
UK	**56 – 56**	99–00	**-**	**83**	20
Total (approx.)	1,800 – 1,900	**Overall trend** Stable			
Breeding range	>100,000 km²	**Gen. length.** 9		**% Global pop.** <5	

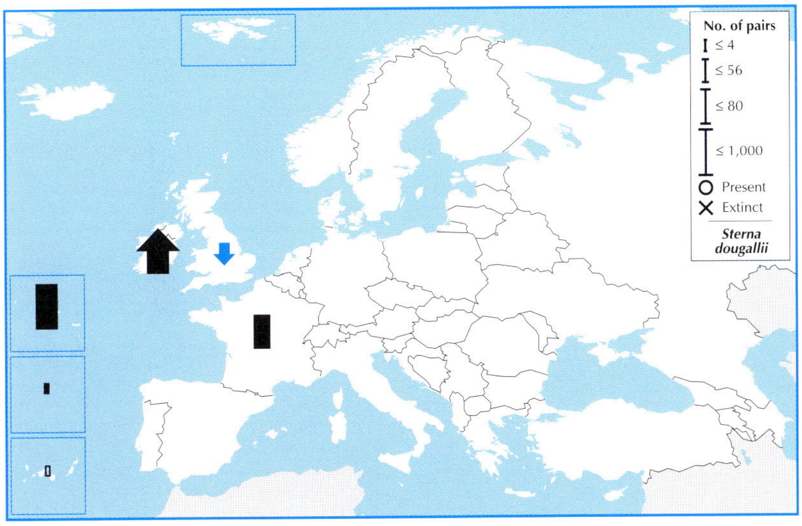

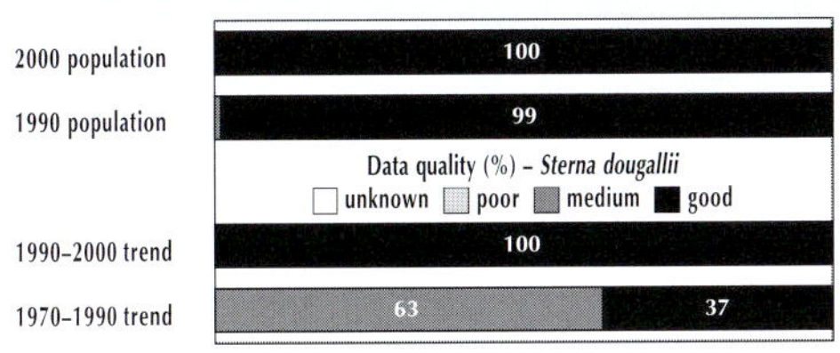

Sterna hirundo
COMMON TERN

Non-SPEC (1994: —) **Status** Secure
Criteria —
European IUCN Red List Category —
Criteria —
Global IUCN Red List Category —
Criteria —

Sterna hirundo is a widespread but patchily distributed breeder across much of Europe, which accounts for less than half of its global breeding range. Its European breeding population is large (>270,000 pairs), and was stable between 1970–1990. Although there were declines in a number of countries during 1990–2000, populations across much of its European range—including sizeable ones in Ukraine, Finland and Russia—were fluctuating, stable or increased, and the species remained stable overall. Consequently, it is evaluated as Secure.

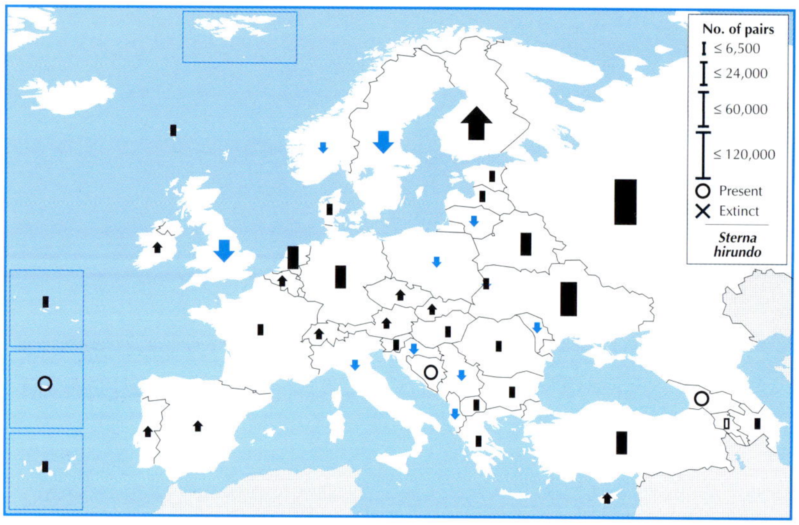

Country	Breeding pop. size (pairs)	Year(s)	Trend	Mag.%	References
Albania	**100 – 200**	96–02	**–**	**0–19**	*4*
Armenia	20 – 80	98–02	?	–	
Austria	**200 – 300**	98–02	**+**	**50–79**	
Azerbaijan	3,000 – 5,000	96–00	(0)	(0–19)	
Belarus	14,000 – 40,000	97–02	0	0–19	
Belgium	**2,500 – 2,600**	00–02	**+**	**30–49**	*1*
Bosnia & HG	Present	85–89	?	–	
Bulgaria	**180 – 540**	96–02	**0**	**0–9**	
Croatia	(500 – 1,000)	02	(–)	(50–79)	*70,16*
Cyprus	**1 – 1**	02	**+**	**N**	
Czech Rep.	**400 – 600**	00	**+**	**>80**	
Denmark	1000 – 1,000	93–96	(0)	(0–19)	*3*
Faroe Is.	(0 – 1)	81	(0)	(0–19)	
Estonia	4,500 – 6,000	98	0	0–19	*1*
Finland	50,000 – 70,000	98–02	+	10	
France	4,500 – 5,000	97–00	0	0–19	*4*
Georgia	Present	03	?	–	
Germany	**8,900 – 9,600**	95–99	**0**	**0–19**	
Greece	1,000 – 1,500	95–00	(0)	(0–19)	
Hungary	**700 – 1,200**	98–02	0	0–19	
Rep. Ireland	**2,500 – 2,500**	99–02	**+**	**50–79**	
Italy	4,000 – 6,000	03	–	0–19	
Latvia	1,500 – 2,500	90–00	0	0–19	*25*
Lithuania	**2,000 – 3,000**	99–01	**–**	**20–29**	*20*
Macedonia	30 – 100	90–00	(0)	(0–19)	
Moldova	40 – 80	90–00	–	30–49	
Netherlands	**18,000 – 19,500**	98–00	**0**	**10**	*1*
Norway	(4,000 – 10,000)	89–02	**–**	**30–49**	*27*
Poland	**4,000 – 4,500**	97–00	–	20–29	*2*
Portugal	**0 – 5**	02	**+**	**N**	
Azores	**1,500 – 2,500**	94–00	**0**	**0–19**	*34,39*
Madeira	Present	02	?	–	
Romania	5,500 – 7,500	90–02	(0)	(0–19)	
Russia	(50,000 – 250,000)	90–02	(F)	(20–29)	*8,23,24,42,104, 112,120*
Serbia & MN	**330 – 420**	00–02	–	20–39	*1,148,138,179, 155,156,78,214*
Slovakia	150 – 250	80–99	+	20–29	
Slovenia	**100 – 200**	95–00	**F**	**>80**	
Spain	2,500 – 10,000	98–02	+	50–79	*10,16*
Canary Is.	50 – 93	97–03	0	0–19	*28,25*
Sweden	**20,000 – 25,000**	99–00	–	0–9	
Switzerland	**350 – 450**	98–02	**+**	**10–19**	
Turkey	8,000 – 12,000	01	(0)	(0–19)	
Ukraine	40,000 – 55,000	90–00	**F**	**20–29**	
UK	**12,000 – 12,000**	98–02	**–**	**10**	*20*

Total (approx.) 270,000 – 570,000 **Overall trend** Stable
Breeding range >5,000,000 km² **Gen. length**. 9 **% Global pop.** 25–49
(See p. 146, bottom, for data quality graph)

Sterna paradisaea
ARCTIC TERN

Non-SPEC (1994: —) **Status** (Secure)
Criteria —
European IUCN Red List Category —
Criteria —
Global IUCN Red List Category —
Criteria —

Sterna paradisaea is a widespread summer visitor to the Atlantic seaboard of north-west and northern Europe, which accounts for less than a quarter of its global breeding range. Its European breeding population is large (>500,000 pairs), and was stable between 1970–1990. Although there were declines in Greenland, Norway and the United Kingdom during 1990–2000, trend data were not available for the stronghold population in Iceland, and most of the populations were stable or increased. Consequently, the species is provisionally evaluated as Secure.

Country	Breeding pop. size (pairs)	Year(s)	Trend	Mag.%	References
Denmark	8,000 – 9,000	93–96	0	0–19	*3*
Faroe Is.	2,000 – 2,000	95	(0)	(0–19)	
Greenland	(30,000 – 100,000)	90–00	(–)	(0–19)	*5*
Estonia	6,000 – 9,000	98	0	0–19	*1*
Finland	60,000 – 90,000	98–02	+	20	
France	**0 – 1**	96–01	**F**	**>80**	*6*
Germany	**6,100 – 6,700**	95–99	**0**	**0–19**	
Iceland	(250,000 – 500,000)	78–94	?	–	*1*
Rep. Ireland	**2,700 – 2,700**	99–02	**+**	**50–79**	
Latvia	25 – 50	90–00	–	50–79	
Netherlands	**1,900 – 2,300**	98–00	**0**	**18**	*1*
Norway	30,000 – 40,000	95	**–**	**0–19**	*4*
Svalbard	5,000 – 10,000	95	0	0–19	
Russia	20,000 – 50,000	90–02	(F)	(20–29)	*11,59,82,154,155*
Sweden	20,000 – 25,000	99–00	+	0–9	
UK	**53,400 – 53,400**	98–02	**–**	**31**	*20*

Total (approx.) 500,000 – 900,000 **Overall trend** Unknown
Breeding range >3,000,000 km² **Gen. length**. 14 **% Global pop.** 5–24

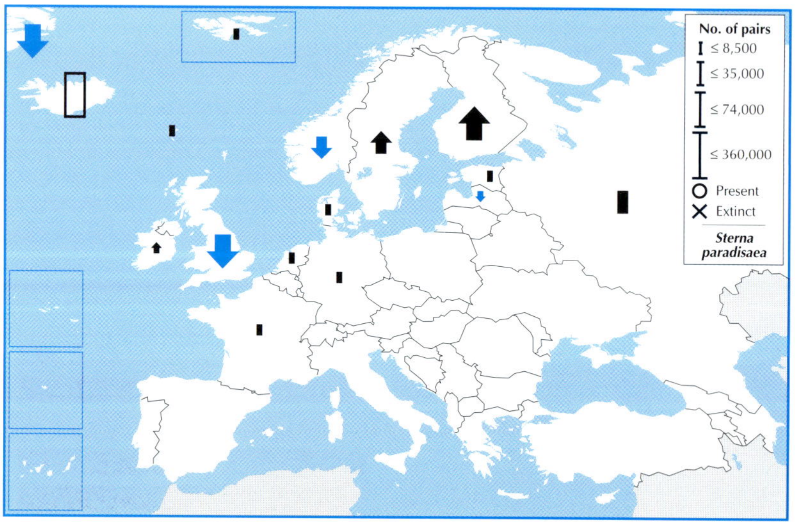

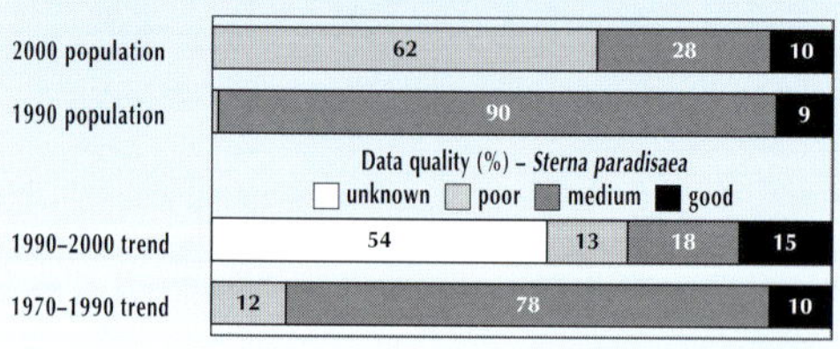

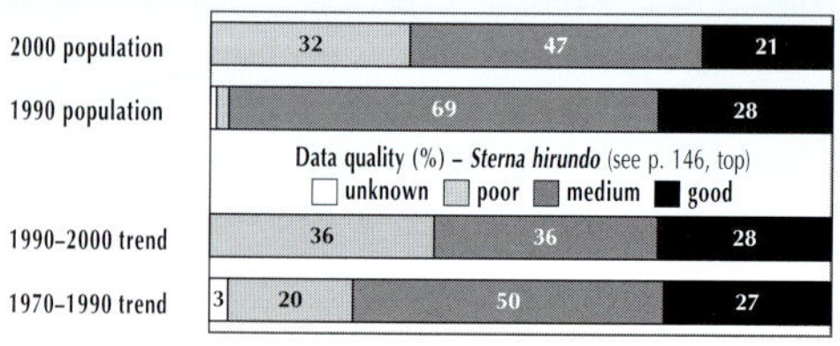

Sterna albifrons
LITTLE TERN

SPEC 3 (1994: 3) **Status** Declining
Criteria Moderate continuing decline
European IUCN Red List Category —
Criteria —
Global IUCN Red List Category —
Criteria —

Sterna albifrons is a widespread but patchily distributed summer visitor to much of Europe, which accounts for less than half of its global breeding range. Its European breeding population is relatively small (<55,000 pairs), and underwent a moderate decline between 1970–1990. Although the species was stable or increased across much of Europe during 1990–2000, sizeable populations in Russia, Turkey and Italy declined, and the species underwent a moderate decline (>10%) overall. Consequently, it is evaluated as Declining.

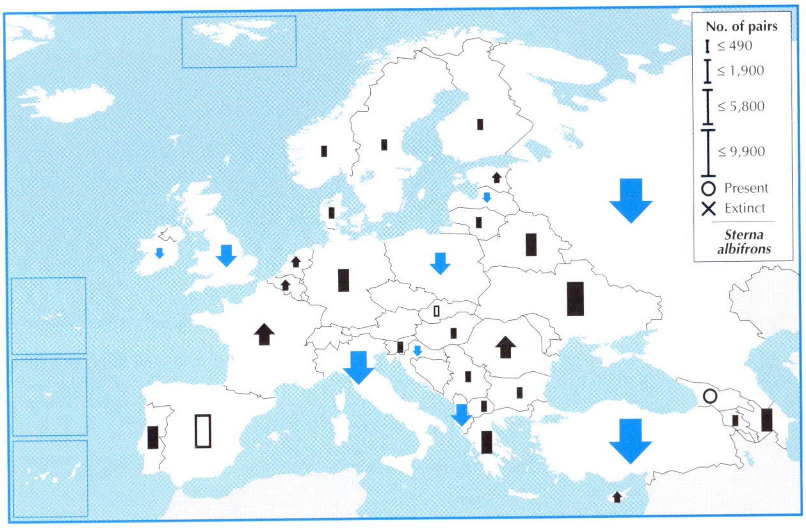

Country	Breeding pop. size (pairs)	Year(s)	Trend	Mag.%	References
Albania	**600 – 1,000**	96–02	-	**0–19**	*4*
Armenia	80 – 300	98–02	0	0–19	
Azerbaijan	500 – 1,000	96–00	(0)	(0–19)	
Belarus	900 – 1,100	97–02	0	0–19	
Belgium	**224 – 224**	00–02	+	**30–49**	*1*
Bulgaria	**60 – 150**	96–02	0	0–19	
Croatia	45 – 70	02	(–)	(50–79)	*70,16*
Cyprus	**1 – 1**	02	+	**N**	
Denmark	**450 – 470**	99–01	0	0–19	*5,6,7*
Estonia	300 – 500	98	+	20–29	*1*
Finland	**45 – 55**	98–02	0	5	
France	1,500 – 1,700	00–02	+	10–19	*4*
Georgia	Present	03	?	–	
Germany	**730 – 770**	95–99	**0**	**0–19**	
Greece	1,000 – 2,000	95–00	(F)	(30–49)	
Hungary	**5 – 10**	98–01	**0**	**0–19**	
Rep. Ireland	**206 – 206**	99–02	**-**	**30–49**	
Italy	2,500 – 6,000	03	–	0–19	
Latvia	(150 – 200)	90–00	(–)	(20–29)	*23*
Lithuania	150 – 200	99–01	**F**	**50–79**	*20*
Macedonia	(3 – 10)	90–00	(F)	(20–29)	
Netherlands	**463 – 512**	98–00	+	**20**	*1*
Norway	**0 – 2**	90–03	**F**	**>80**	*63*
Poland	**900 – 900**	97–99	**-**	**10–29**	*76*
Portugal	**700 – 733**	03	(F)	(–)	
Romania	500 – 800	96–02	+	0–19	
Russia	7,000 – 14,000	90–02	–	20–29	*51,52,60,69,73, 74,75,102,112, 117,128,141*
Serbia & MN	**70 – 80**	00–02	**F**	**30–49**	*1,148,179,156, 13,227*
Slovakia	**0 – 5**	80–99	?	–	
Slovenia	**2 – 3**	90–00	**0**	**0–19**	
Spain	5,500 – 6,000	95	?	–	*12,16,10*
Sweden	**400 – 500**	99–00	F	20–29	
Turkey	6,000 – 10,000	01	(–)	(0–19)	
Ukraine	**2,500 – 4,000**	90–00	**0**	**0–19**	
UK	**1,900 – 1,900**	99–02	**-**	**24**	*20*
Total (approx.)	35,000 – 55,000	**Overall trend** Moderate decline			
Breeding range	>2,000,000 km²	**Gen. length.** 8		**% Global pop.** 25–49	

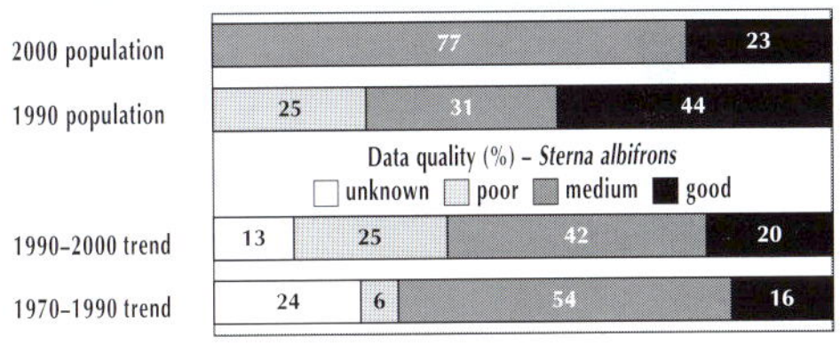

Chlidonias hybrida
WHISKERED TERN

SPEC 3 (1994: 3) **Status** Depleted
Criteria Moderate historical decline
European IUCN Red List Category —
Criteria —
Global IUCN Red List Category —
Criteria —

Chlidonias hybrida is a widespread but patchily distributed summer visitor to southern and eastern Europe, which accounts for less than half of its global breeding range. Its European breeding population is relatively small (<87,000 pairs), and underwent a moderate decline between 1970–1990. Although the species was broadly stable overall during 1990–2000—with stable, fluctuating or increasing trends across most of its European range—its population has not yet recovered to the level that preceded its decline. Consequently, it is evaluated as Depleted.

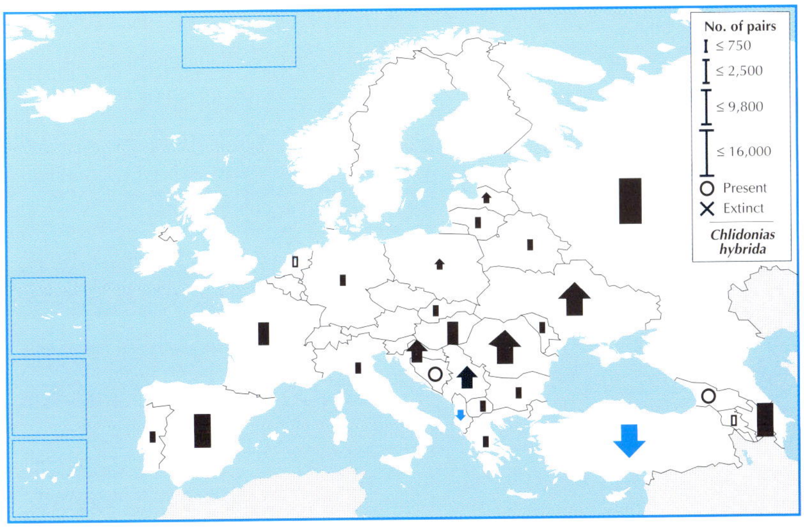

Country	Breeding pop. size (pairs)	Year(s)	Trend	Mag.%	References
Albania	(200 – 500)	96–02	(–)	(0–19)	*6*
Armenia	20 – 25	97–02	?	–	
Azerbaijan	3,000 – 10,000	96–00	(0)	(0–19)	
Belarus	**200 – 400**	95–00	**0**	**0–19**	
Bosnia & HG	Present	85–89	?	–	
Bulgaria	**180 – 460**	96–02	**F**	**20–29**	
Croatia	1,600 – 1,800	02	(+)	(>80)	*70*
France	**2,300 – 2,400**	02	**F**	**20–49**	*6*
Georgia	Present	03	?	–	
Germany	**2 – 6**	95–99	**F**	**>80**	
Greece	(250 – 400)	95–00	(0)	(0–19)	
Hungary	**1,600 – 3,800**	95–02	**F**	**50–79**	*11*
Italy	**423 – 458**	00	**F**	**20–29**	*18*
Latvia	1 – 10	92–03	+	N	*23,17,26*
Lithuania	5 – 20	99–01	F	30–49	*20*
Macedonia	(5 – 30)	90–00	(F)	(20–29)	
Moldova	**200 – 300**	90–00	0	0–19	
Netherlands	**0 – 1**	98–00	?	–	*1*
Poland	**700 – 800**	00	**+**	**400–2,000**	*2,77,60–69*
Portugal	0 – 200	02	(F)	(–)	
Romania	8,000 – 12,000	90–02	+	0–19	
Russia	(10,000 25,000)	90–02	(F)	(30–49)	*8,52,67,145*
Serbia & MN	**1,700 – 2,100**	97–02	**+**	**30–49**	*1,138,29,155, 78,227*
Slovakia	50 – 200	80–99	F	30–49	
Spain	2,500 – 10,000	98–02	(F)	(–)	*10*
Turkey	4,000 – 8,000	01	(–)	(0–19)	
Ukraine	**5,000 – 8,500**	90–00	**+**	**5–14**	
Total (approx.)	42,000 – 87,000	**Overall trend** Fluctuating			
Breeding range	>750,000 km²	**Gen. length.** 9		**% Global pop.** 25–49	

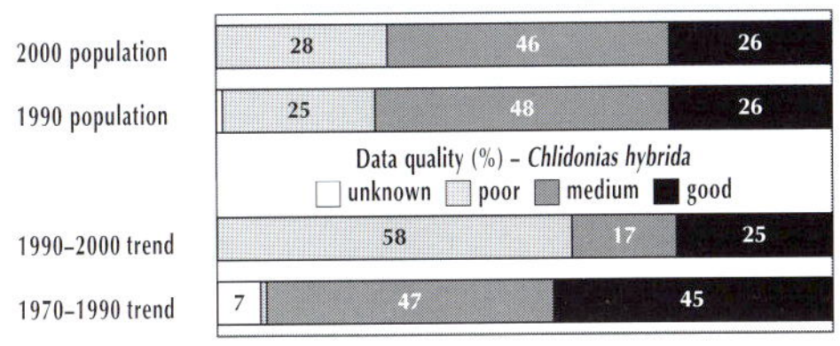

Chlidonias niger
BLACK TERN

SPEC 3 (1994: 3) **Status** (Depleted)
Criteria Moderate historical decline
European IUCN Red List Category —
Criteria —
Global IUCN Red List Category —
Criteria —

Chlidonias niger is a widespread but patchily distributed summer visitor to much of Europe, which accounts for less than half of its global breeding range. Its European breeding population is relatively large (>83,000 pairs), but underwent a moderate decline between 1970–1990. The species declined across much of Europe (particularly the east) during 1990–2000, but the trend in its Russian stronghold was unknown. Nevertheless, its population has clearly not yet recovered to the level that preceded its decline, and consequently the species is provisionally evaluated as Depleted.

Country	Breeding pop. size (pairs)	Year(s)	Trend	Mag.%	References
Albania	(0 – 10)	96–02	(–)	(0–19)	
Belarus	6,000 – 22,000	97–02	–	0–9	
Bulgaria	0 – 20	96–02	(–)	(>80)	
Croatia	(10 – 100)	02	(–)	(30–49)	70,16
Czech Rep.	10 – 20	00	–	50–79	
Denmark	58 – 60	01	–	50	7
Estonia	(2,000 – 3,000)	98	0	0–19	1
Finland	25 – 35	98–02	+	30–49	
France	179 – 190	02	F	50–79	6
Georgia	Present	03	?	–	
Germany	860 – 1,000	95–99	0	0–19	
Greece	(50 – 150)	95–00	(0)	(0–19)	
Hungary	400 – 1,000	98–02	F	50–79	11
Italy	150 – 200	00	0	0–19	18
Latvia	2,000 – 3,000	90–00	F	50–79	23,17,25
Lithuania	2,000 – 4,000	99–01	–	20–29	20
Macedonia	(5 – 20)	90–00	(F)	(20–29)	
Moldova	300 – 350	90–00	0	0–19	
Netherlands	1,000 – 1250	98–00	F	23	1
Poland	4,000 – 5,000	95–00	–	20–29	2
Romania	1,200 – 2,500	90–02	–	0–19	
Russia	50,000 – 100,000	90–02	?	–	3,8,23,24,51,60, 104,112,133
Serbia & MN	50 – 100	97–02	–	10–29	1,29,155,13,227
Slovakia	40 – 150	80–99	–	50–79	
Spain	30 – 60	98–02	–	50–79	10
Sweden	190 – 200	99–00	0	0–19	
Turkey	(200 – 600)	01	(–)	(30–49)	
Ukraine	12,000 – 26,000	90–00	F	30–49	
Total (approx.)	83,000 – 170,000	**Overall trend** Unknown			
Breeding range	>2,000,000 km²	**Gen. length.** 9		**% Global pop.** 25–49	

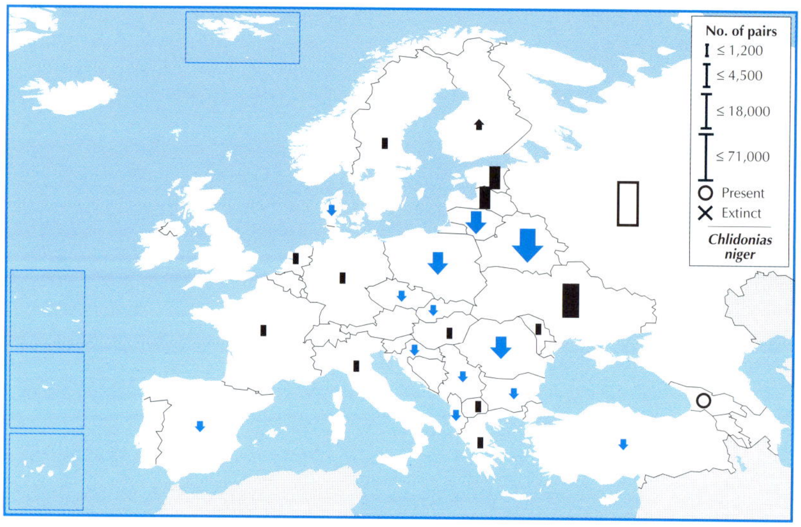

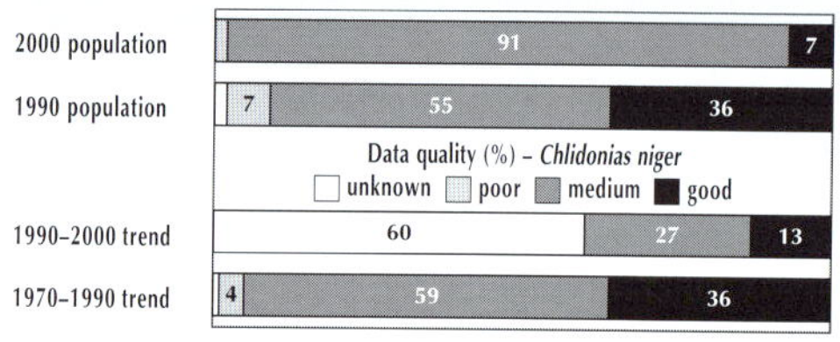

Chlidonias leucopterus
WHITE-WINGED TERN

Non-SPEC (1994: —) **Status** (Secure)
Criteria —
European IUCN Red List Category —
Criteria —
Global IUCN Red List Category —
Criteria —

Chlidonias leucopterus is a widespread summer visitor to eastern Europe, which accounts for less than a quarter of its global breeding range. Its European breeding population is relatively large (>74,000 pairs), and increased between 1970–1990. Although there were declines in Romania and Turkey during 1990–2000, populations were fluctuating or stable across the rest of its European range, and the species was probably stable overall. Consequently, it is provisionally evaluated as Secure.

Country	Breeding pop. size (pairs)	Year(s)	Trend	Mag.%	References
Azerbaijan	1,000 – 10,000	96	(0)	(0–19)	
Belarus	8,000 – 30,000	97–02	0	0–19	
Bulgaria	(0 – 5)	96–02	?	–	
Croatia	0 – 20	02	?	–	70
Estonia	0 – 10	98	0	0–19	1
Georgia	Present	03	?	–	
Germany	1 – 24	95–99	F	>80	
Hungary	50 – 2,000	95–02	F	>80	11
Italy	10 – 15	00	0	0–19	18
Latvia	50 – 300	90–00	F	50–79	17
Lithuania	10 – 100	99–01	F	>80	20
Macedonia	10 – 50	90–00	F	50–79	
Poland	50 – 4,500	90–00	F	>80	2,78,60-69
Romania	40 – 80	90–02	–	20–29	
Russia	50,000 – 120,000	90–02	(F)	(30–49)	3,24,51,112,133
Serbia & MN	0 – 10	97–02	F	50–79	1,155,13,203,227
Turkey	(40 – 120)	01	(–)	(20–29)	
Ukraine	15,000 – 45,000	90–00	F	30–49	
Total (approx.)	74,000 – 210,000	**Overall trend** Fluctuating			
Breeding range	>1,000,000 km²	**Gen. length.** 9		**% Global pop.** 5–24	

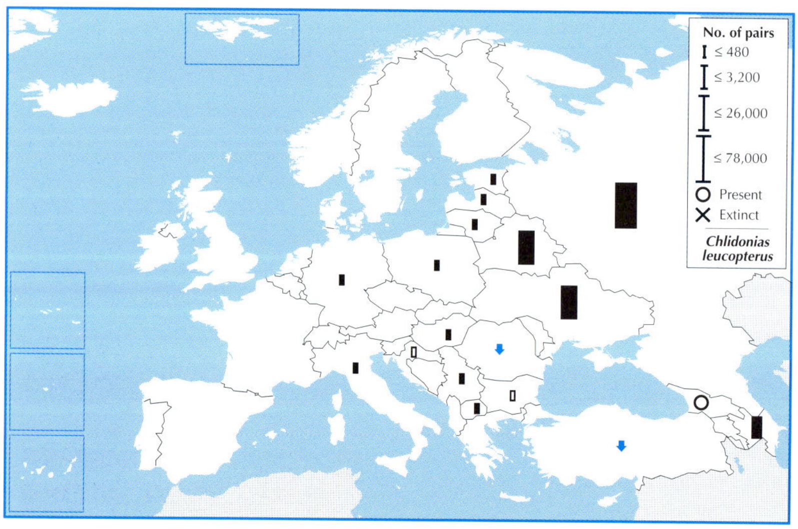

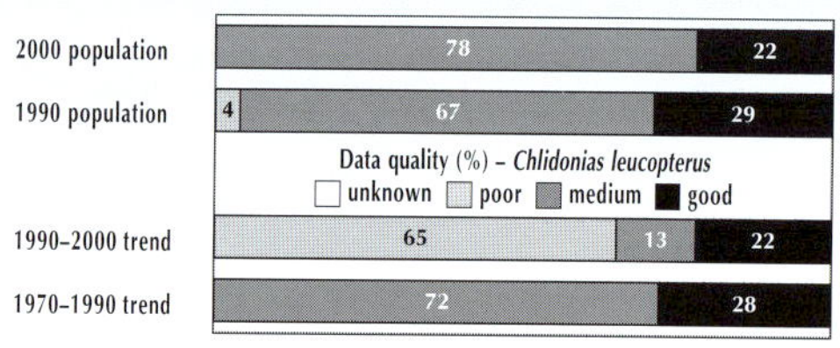

Uria aalge
COMMON MURRE

Non-SPEC (1994: —) **Status** (Secure)
Criteria —
European IUCN Red List Category —
Criteria —
Global IUCN Red List Category —
Criteria —

Uria aalge is a widespread but patchily distributed breeder in coastal areas of western and northern Europe, which accounts for less than half of its global breeding range. Its European breeding population is very large (>2,000,000 pairs), and was stable between 1970–1990. Although the trend of the large population in Iceland during 1990–2000 was unknown, declines in countries such as Svalbard and Norway were more than compensated for by significant increases in the United Kingdom and Republic of Ireland. Consequently, the species is provisionally evaluated as Secure.

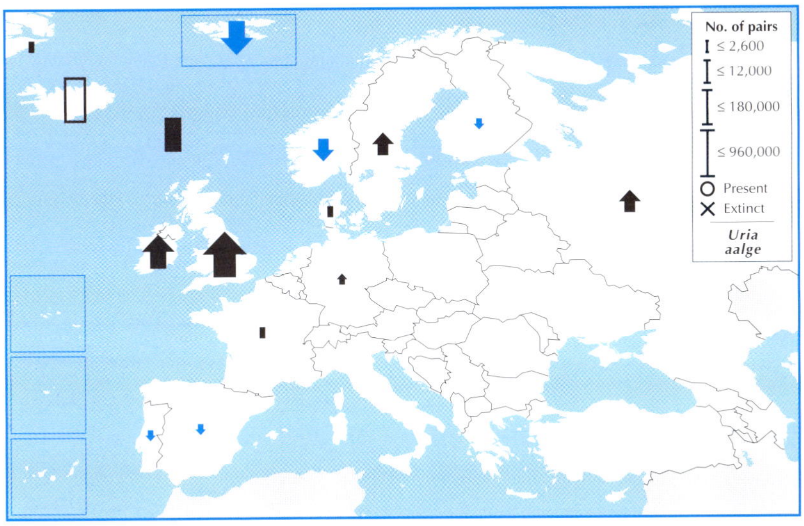

Country	Breeding pop. size (pairs)	Year(s)	Trend	Mag.%	References
Denmark	**2,500 - 2,500**	99-00	0	0-19	16
Faroe Is.	175,000 - 175,000	95	(0)	(0-19)	
Greenland	(1,000 - 2,000)	90-00	(0)	(0-19)	5
Finland	**25 - 35**	98-01	**-**	**25**	
France	**248 - 248**	00	**F**	**20-29**	6
Germany	**2,600 - 2,600**	95-99	**+**	**30-49**	
Iceland	**656,000 - 1,380,000**	75-94	?	-	6
Rep. Ireland	**88,000 - 88,000**	99-02	**+**	**50-79**	
Norway	5,000 - 11,000	74-96	**-**	**20-29**	4
Svalbard	100,000 - 100,000	74-96	**-**	**30-49**	
Portugal	**2 - 10**	02	(-)	(-)	
Russia	**5,000 - 7,500**	90-02	**+**	**20-29**	82,154,155
Spain	0 - 50	98-02	(-)	(0-19)	10,16
Sweden	**11,000 - 12,000**	99-00	**+**	**0-9**	
UK	**952,000 - 952,000**	98-02	**+**	**31**	20
Total (approx.)	2,000,000 - 2,700,000	**Overall trend** Large increase			
Breeding range	>250,000 km²	**Gen. length** 16		**% Global pop.** 5-24	

Country	Winter pop. size (individuals)	Year(s)	Trend	Mag.%	References
Denmark	(170,000 - 300,000)	87-91	?	-	27
Faroe Is.	5,000 - 50,000	92	?	-	3
Greenland	(100 - 5,000)	90-00	?	-	
Estonia	(0 - 25)	98	(0)	(0-19)	1
France	(4,000 - 20,000)	98-02	?	-	9
Germany	**33,500 - 33,500**	87-03	?	-	4
Iceland	(4,000,000 - 5,000,000)	78-94	?	-	48
Latvia	(0 - 100)	90-02	(0)	(0-19)	32
Lithuania	(0 - 50)	92-02	?	-	
Portugal	Present	02	?	-	
Sweden	100,000 - 110,000	98-01	0	0-19	
UK					
Gibraltar	**0 - 1**	00-03	**0**	**0-19**	
Total (approx.)	>4,300,000	**Overall trend** Unknown			
% in European IBAs	3-4	**Gen. length** 16		**% Global pop.** 5-24	

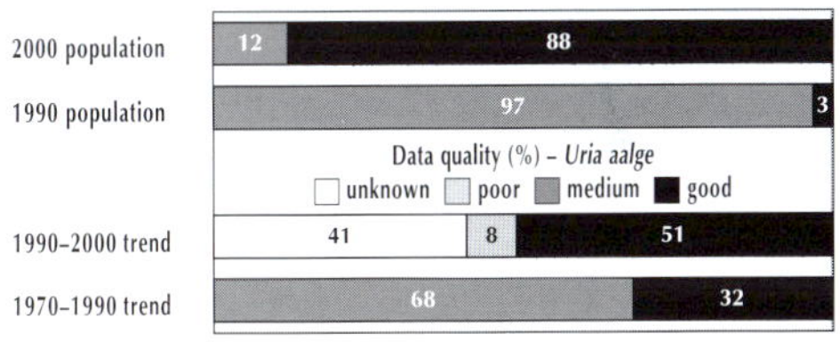

Uria lomvia
THICK-BILLED MURRE

SPEC 3 (1994: —) **Status** (Vulnerable)
Criteria See IUCN below
European IUCN Red List Category VU
Criteria A4b
Global IUCN Red List Category —
Criteria —

Uria lomvia breeds locally in coastal areas of northernmost Europe, which accounts for less than a quarter of its global breeding range. Its European breeding population is very large (>1,800,000 pairs), and was stable between 1970–1990. Although the large population in Svalbard was broadly stable during 1990–2000, the species suffered declines in Iceland and Greenland, and declined overall at a rate that—if sustained—would equate to a large decline (>30%) over three generations. Consequently, this previously Secure species is now provisionally evaluated as Vulnerable.

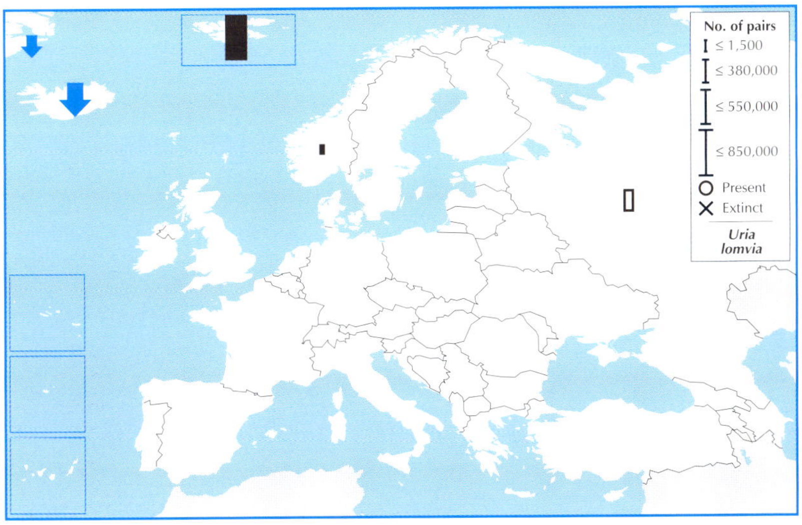

Country	Breeding pop. size (pairs)	Year(s)	Trend	Mag.%	References
Denmark					
Greenland	350,000 - 400,000	90-00	-	0-19	8
Iceland	**341,000 - 861,000**	75-94	-	30-49	6,14
Norway	(1,000 - 2,000)	94	**0**	**0-19**	4
Svalbard	(850,000 - 850,000)	94	**F**	**20-29**	
Russia	(250,000 - 500,000)	90-02	?	-	79,154,155,165,167
Total (approx.)	1,800,000 - 2,600,000	**Overall trend** Large decline			
Breeding range	>100,000 km²	**Gen. length** 16		**% Global pop.** 5-24	

Country	Winter pop. size (individuals)	Year(s)	Trend	Mag.%	References
Denmark					
Greenland	(1,000,000 - 5,000,000)	90-00	(-)	(0-19)	
Germany	**0 - 50**	95-00	?	-	
Iceland	(10,000 - 1,000,000)	78-94	?	-	48
Total (approx.)	>1,000,000	**Overall trend** Unknown			
% in European IBAs	Unknown	**Gen. length** 16		**% Global pop.** 5-24	

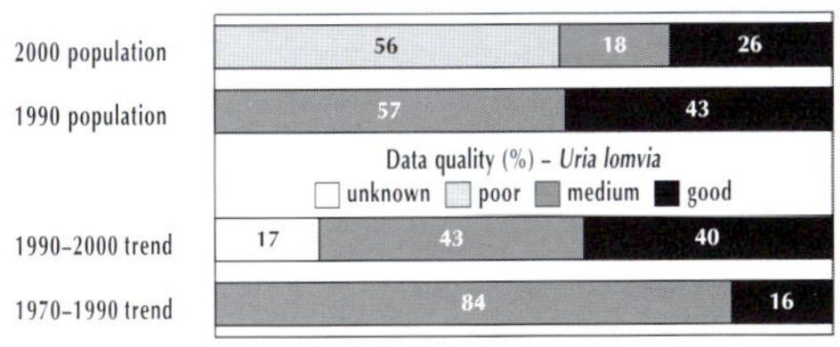

Alca torda
RAZORBILL

Non-SPEC^E (1994: 4) **Status** (Secure)
Criteria —
European IUCN Red List Category —
Criteria —
Global IUCN Red List Category —
Criteria —

Alca torda is a widespread breeder in coastal areas of north-west Europe, which constitutes >75% of its global breeding range. Its European breeding population is large (>430,000 pairs), and increased between 1970–1990. Although the trend of the stronghold population in Iceland during 1990–2000 was unknown, the species was increasing or stable in most of the rest of its range, and consequently it is provisionally evaluated as Secure.

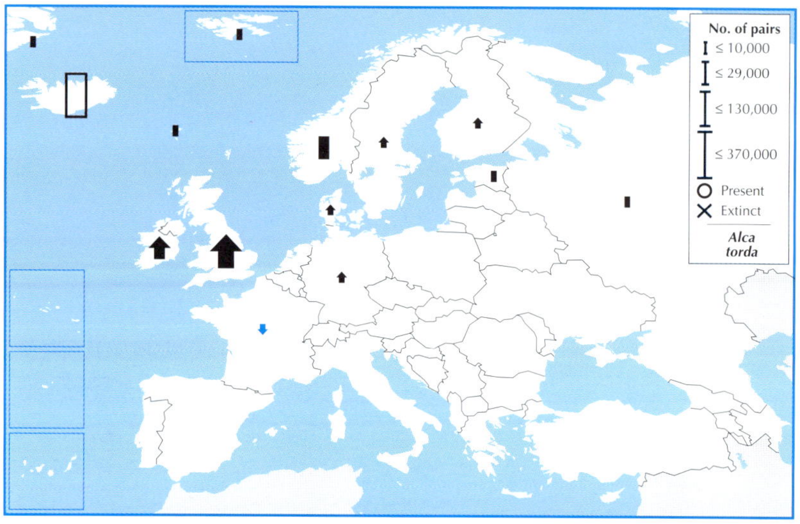

Country	Breeding pop. size (pairs)	Year(s)	Trend	Mag.%	References
Denmark	**650 – 750**	99–00	+	**50–79**	*16*
Faroe Is.	4,500 – 4,500	95	(0)	(0–19)	
Greenland	(1,500 – 5,500)	90–00	(0)	(0–19)	*5*
Estonia	(1 – 10)	98	0	0–19	*1*
Finland	6,000 – 8,500	98–02	+	30	
France	**26 – 27**	00	-	**0–9**	*5*
Germany	**11 – 11**	95–99	+	**50–79**	
Iceland	**247,000 – 548,000**	75–94	?	–	*6*
Rep. Ireland	**17,400 – 17,400**	99–02	+	**50–79**	
Norway	(20,000 – 40,000)	96–02	**0**	**0–19**	*35*
Svalbard	(100 – 100)	96–02	0	0–19	
Russia	2,500 – 10,000	90–02	0	0–19	*11,82,154*
Sweden	**9,000 – 11,000**	99–00	+	0–19	
UK	**126,000 – 126,000**	98–02	+	**22**	*20*
Total (approx.)	430,000 – 770,000	**Overall trend** Unknown			
Breeding range	>500,000 km²	**Gen. length.** 16		**% Global pop.** 75–94	

Country	Winter pop. size (individuals)	Year(s)	Trend	Mag.%	References
Denmark	(120,000 – 400,000)	87–91	?	–	*27*
Faroe Is.	1,000 – 10,000	92	?	–	*3*
Greenland	(100 – 5,000)	90–00	(0)	(0–19)	
Estonia	(300 – 1,000)	98	(0)	(0–19)	*1*
France	(2,000 – 12,000)	98–02	?	–	*9*
Germany	**8,300 – 8,300**	87–03	?	–	
Iceland	(300,000 – 600,000)	78–94	(0)	(0–19)	*48*
Latvia	**2,000 – 2,300**	92–94	(0)	(0–19)	*27*
Lithuania	(0 – 50)	92–02	?	–	
Portugal	Present	02	?	–	
Sweden	90,000 – 110,000	98–01	0	0–19	
UK					
Gibraltar	**10 – 30**	00–03	**0**	**0–19**	
Total (approx.)	>500,000	**Overall trend** Unknown			
% in European IBAs	21–45	**Gen. length** 16		**% Global pop.** 75–94	

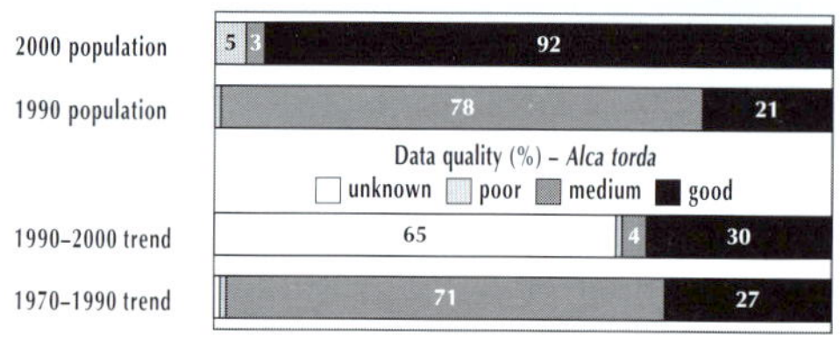

Cepphus grylle
BLACK GUILLEMOT

SPEC 2 (1994: 2) **Status** Depleted
Criteria Moderate historical decline
European IUCN Red List Category —
Criteria —
Global IUCN Red List Category —
Criteria —

Cepphus grylle is a widespread breeder in coastal areas of northern Europe, which constitutes >50% of its global breeding range. Its European breeding population is large (>130,000 pairs), but underwent a moderate decline between 1970–1990. Although the populations in Norway and Sweden continued to decline during 1990–2000, the species was stable, fluctuated or increased across most of its European range, and remained stable overall. Nevertheless, its population has not yet recovered to the level that preceded its decline, and consequently it is evaluated as Depleted.

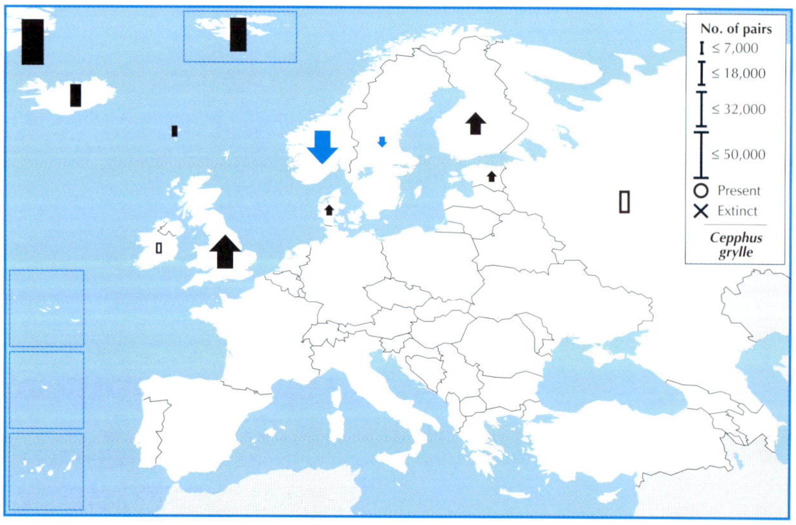

Country	Breeding pop. size (pairs)	Year(s)	Trend	Mag.%	References
Denmark	**950 – 1,150**	00–01	+	**50–79**	*16*
Faroe Is.	3,500 – 3,500	95	(0)	(0–19)	
Greenland	(25,000 – 100,000)	90–00	(0)	(0–19)	*5*
Estonia	**10 – 20**	98	+	20–29	*1*
Finland	**17,000 – 19,000**	98–01	+	**30**	
Iceland	10,000 – 20,000	78–94	0	0–19	*1*
Rep. Ireland	**3,400 – 3,400**	99–02	?	–	
Norway	(15,000 – 35,000)	90–00	**-**	**0–19**	*4*
Svalbard	(20,000 – 50,000)	90–00	(F)	(20–29)	
Russia	(10,000 – 20,000)	90–02	?	–	*11,82,154,155*
Sweden	6,000 – 8,000	99–00	-	20–29	
UK	**19,700 – 39,300**	99–03	+	**4**	*20*
Total (approx.)	130,000 – 300,000	**Overall trend** Stable			
Breeding range	1,000,000 km²	**Gen. length.** 9		**% Global pop.** 50–74	

Country	Winter pop. size (individuals)	Year(s)	Trend	Mag.%	References
Denmark	(5,000 – 7,000)	92–93	?	–	*27*
Faroe Is.	10,000 – 20,000	92	?	–	*3*
Greenland	(250,000 – 500,000)	90–00	(0)	(0–19)	
Estonia	**1,000 – 3,000**	98	(0)	(0–19)	*1*
Germany	**750 – 750**	87–03	?	–	*4*
Iceland	(50,000 – 100,000)	78–94	(0)	(0–19)	*48*
Latvia	**1,500 – 2,000**	92–94	(0)	(0–19)	*27*
Lithuania	(0 – 50)	92–02	?	–	
Norway	25,000 – 40,000	93	(0)	(0–19)	*68*
Slovenia	Present	90–00	?	–	
Total (approx.)	>340,000	**Overall trend** Unknown			
% in European IBAs	2–3	**Gen. length** 9		**% Global pop.** 50–74	

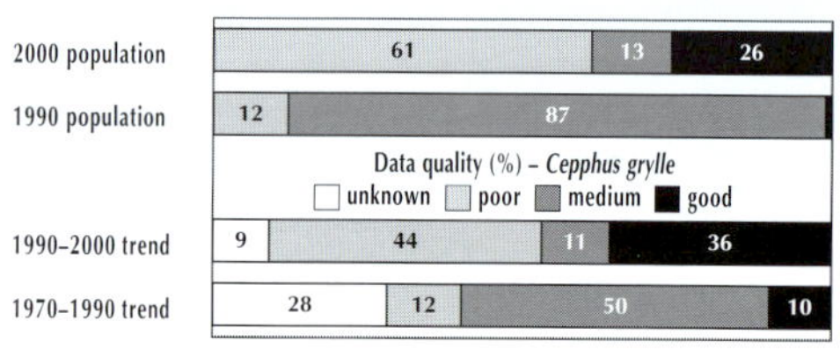

Alle alle
DOVEKIE

Non-SPEC (1994: —) **Status** (Secure)
Criteria —
European IUCN Red List Category —
Criteria —
Global IUCN Red List Category —
Criteria —

Country	Breeding pop. size (pairs)	Year(s)	Trend	Mag.%	References
Denmark					
Greenland	(10,000,000 – 40,000,000)	90–00	(0)	(0–19)	*6*
Iceland	**0 – 0**	97	-	X	*21*
Norway					
Svalbard	(1,000,000 – 3,000,000)	94	(0)	(0–19)	
Russia	(100,000 – 500,000)	90–02	?	-	*154,155*
Total (approx.)	11,000,000 – 44,000,000	**Overall trend** Stable			
Breeding range	>250,000 km²	**Gen. length.** 16		**% Global pop.** 5–24	

Alle alle breeds in Greenland, Svalbard, and on the Russian islands of the high Arctic, with Europe accounting for less than a quarter of its global breeding range. Its European breeding population is extremely large (>11,000,000 pairs), and was stable between 1970–1990. Although the tiny population off the coast of northern Iceland went extinct during 1990–2000, populations were stable in Svalbard and the Greenland stronghold, and the species probably remained stable overall. Consequently, it is provisionally evaluated as Secure.

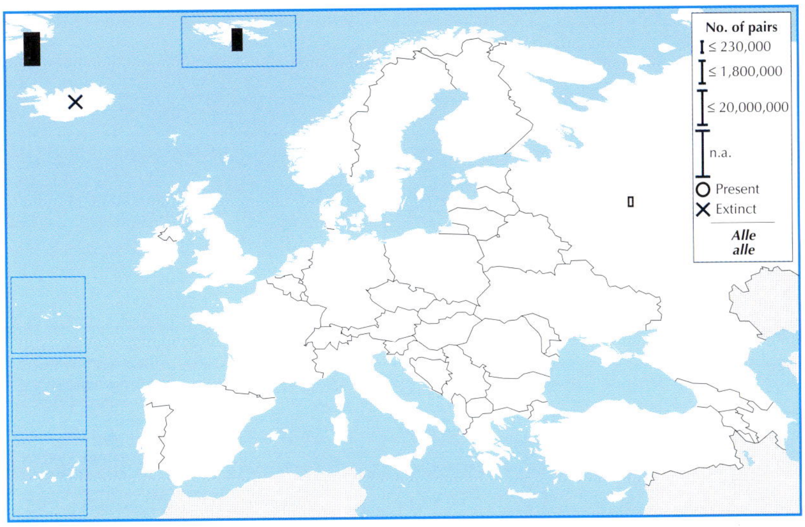

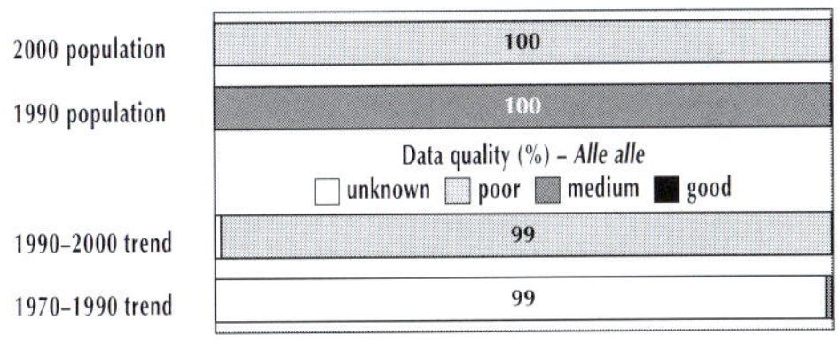

Fratercula arctica
ATLANTIC PUFFIN

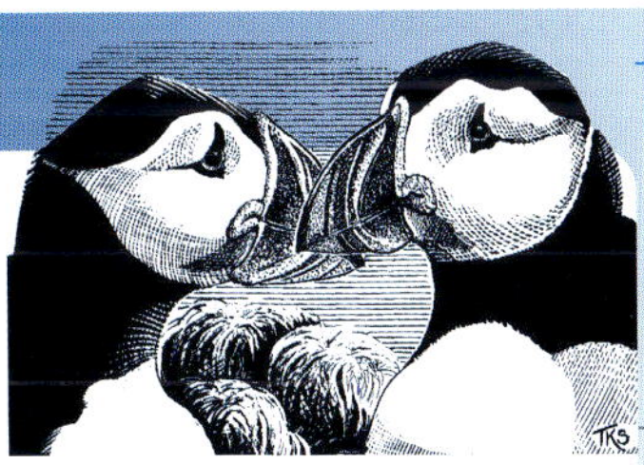

SPEC 2 (1994: 2) **Status** (Depleted)
Criteria Large historical decline
European IUCN Red List Category —
Criteria —
Global IUCN Red List Category —
Criteria —

Country	Breeding pop. size (pairs)	Year(s)	Trend	Mag.%	References
Denmark					
Faroe Is.	550,000 – 550,000	95	(0)	(0–19)	
Greenland	(1,500 – 3,000)	90–00	(+)	(0–19)	*5*
France	**229 – 229**	01	**0**	**0–19**	*6*
Iceland	3,000,000 – 4,000,000	02	?	-	*14*
Rep. Ireland	20,000 – 20,000	99–02	0	0–19	
Norway	1,500,000 – 2,000,000	90–02	-	**20**	*5*
Svalbard	(10,000 – 100,000)	90–02	**0**	**0–19**	
Russia	**5,000 – 6,000**	90–02	**0**	**0–19**	*82,154*
UK	**621,000 – 621,000**	98–03	+	**27**	*20*
Total (approx.)	5,700,000 – 7,300,000	**Overall trend** Unknown			
Breeding range	>250,000 km²	**Gen. length.** 22		**% Global pop.** 75–94	

Fratercula arctica is a widespread but patchily distributed breeder in coastal north-west Europe, which constitutes >75% of its global breeding range. Its European breeding population is very large (>5,700,000 pairs), but underwent a large decline between 1970–1990. Although the species was stable or increased in most parts of its range (except Norway) during 1990–2000, the trend in its Icelandic stronghold was unknown. Nevertheless, its total population probably remains below the level that preceded its decline. Consequently, the species is provisionally evaluated as Depleted.

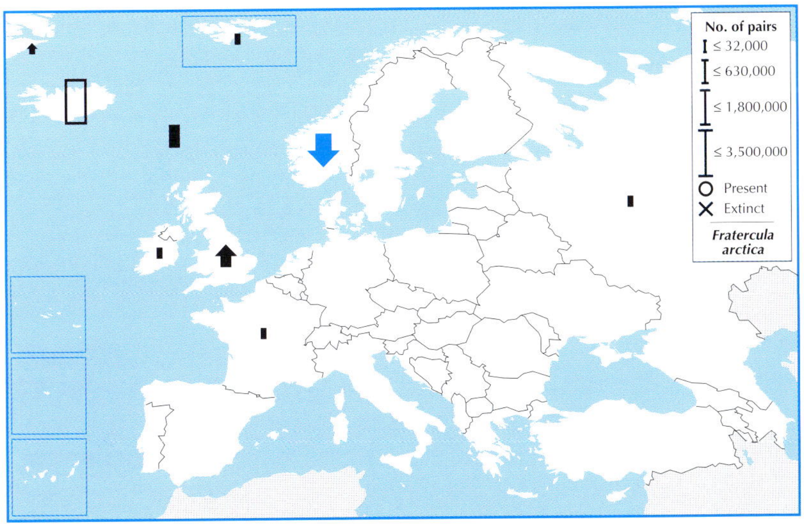

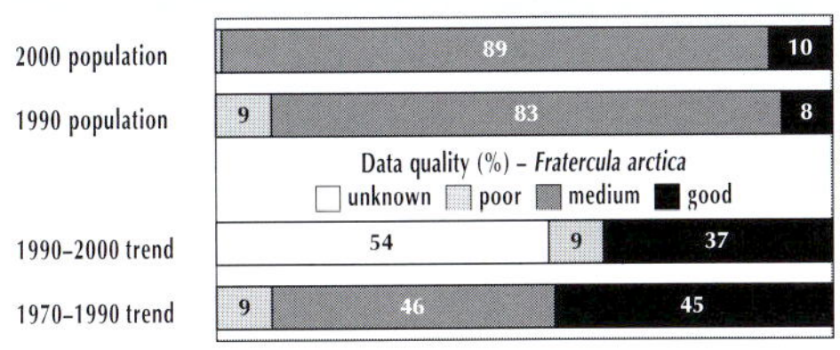

Pterocles orientalis
BLACK-BELLIED SANDGROUSE

SPEC 3 (1994: 3) **Status** (Declining)
Criteria Moderate continuing decline
European IUCN Red List Category —
Criteria —
Global IUCN Red List Category —
Criteria —

Country	Breeding pop. size (pairs)	Year(s)	Trend	Mag.%	References
Armenia	300 – 850	97–02	–	0–9	
Azerbaijan	(50 – 250)	96–00	(–)	(20–29)	
Cyprus	(0 – 20)	94–02	(–)	(>80)	
Portugal	100 – 300	02	(–)	(–)	
Russia	5 – 50	90–01	(F)	(30–49)	*15,176*
Spain	2,500 – 10,000	98–02	(–)	(0–19)	*10,16*
Canary Is.	(250 – 1,000)	97–03	?	–	*22,28,25*
Turkey	(25,000 – 50,000)	01	(–)	(0–19)	
Total (approx.)	28,000 – 62,000	**Overall trend** Moderate decline			
Breeding range	>250,000 km²	**Gen. length.** <3.3 **% Global pop.** 5–24			

Pterocles orientalis breeds in Iberia, the Canary Islands and parts of south-east Europe, which together account for less than a quarter of its global breeding range. Its European breeding population is relatively small (<62,000 pairs), and underwent a large decline between 1970–1990. The species continued to decline across most of its European range—including its Turkish stronghold—during 1990–2000, and underwent a moderate decline (>10%) overall. Consequently, it is provisionally evaluated as Declining.

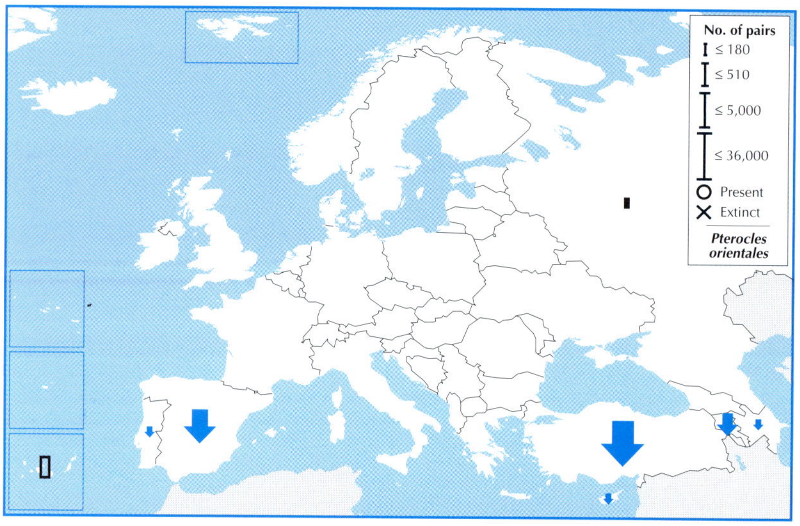

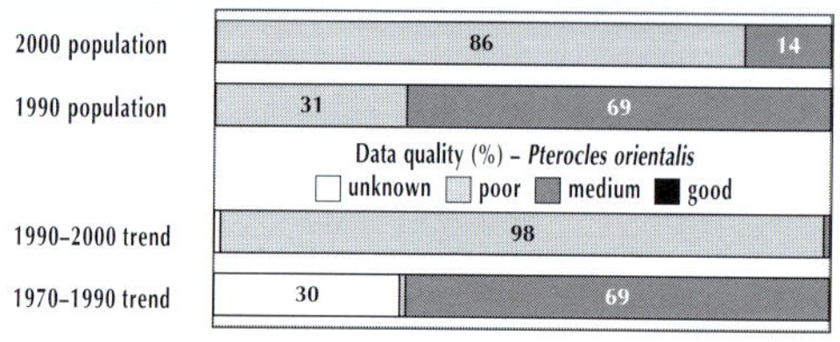

Pterocles alchata
PIN-TAILED SANDGROUSE

SPEC 3 (1994: 3) **Status** (Declining)
Criteria Moderate continuing decline
European IUCN Red List Category —
Criteria —
Global IUCN Red List Category —
Criteria —

Country	Breeding pop. size (pairs)	Year(s)	Trend	Mag.%	References
France	**115 – 180**	98–02	**F**	**30–49**	*4*
Portugal	0 – 6	02	–	0–19	
Spain	10,000 – 20,000	98–02	(–)	(0–19)	*10,16*
Turkey	50 – 500	01	–	>80	
Total (approx.)	10,000 – 21,000	**Overall trend** Moderate decline			
Breeding range	>100,000 km²	**Gen. length.** <3.3 **% Global pop.** 5–24			

Pterocles alchata is resident in Iberia, south-eastern France and southern Turkey, with Europe accounting for less than a quarter of its global range. Its European breeding population is relatively small (<21,000 pairs), and underwent a large decline between 1970–1990. Although the population in France was stable during 1990–2000, the species continued to decline in Portugal, Turkey and its Spanish stronghold, and underwent a moderate decline (>10%) overall. Consequently, it is provisionally evaluated as Declining.

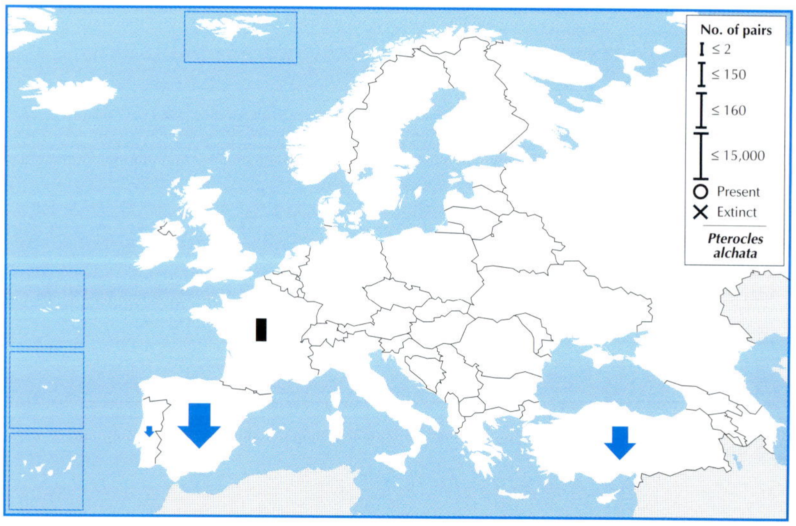

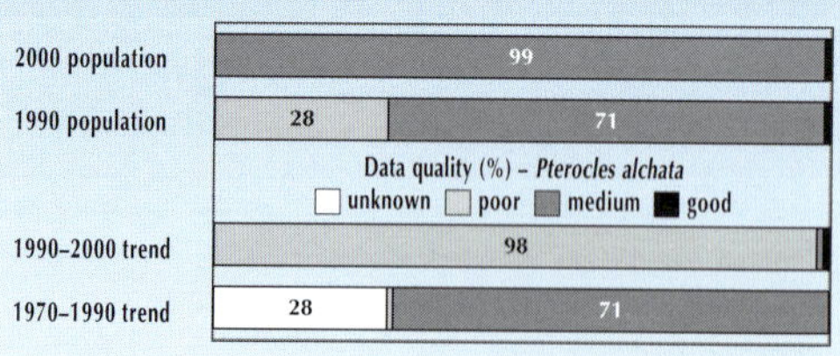

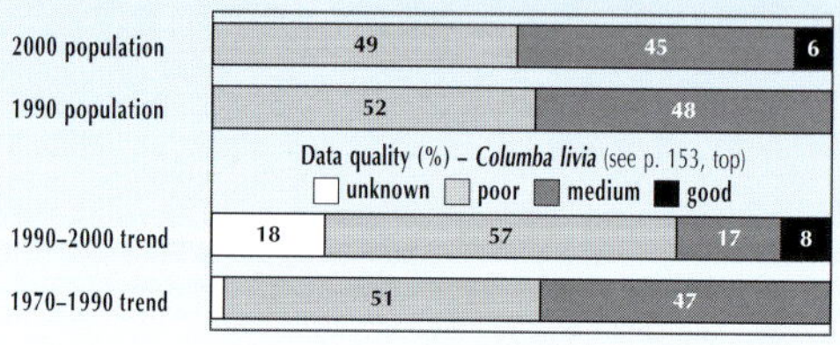

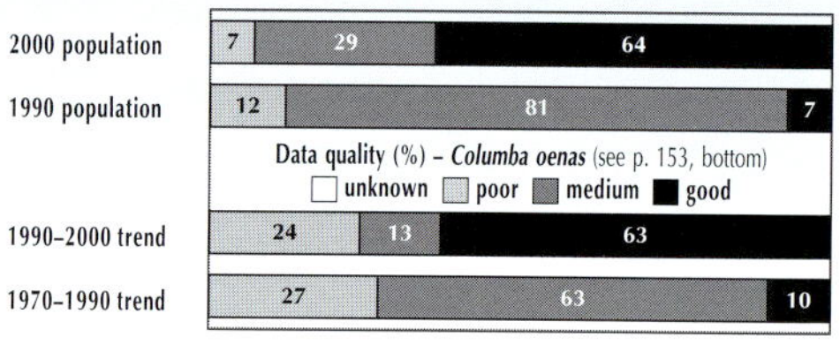

Columba livia
ROCK PIGEON

Non-SPEC (1994: —) **Status** (Secure)
Criteria —
European IUCN Red List Category —
Criteria —
Global IUCN Red List Category —
Criteria —

Columba livia is a widespread resident across much of Europe (except the far north), which accounts for less than half of its global range. Its European breeding population is very large (>9,300,000 pairs), and was probably stable between 1970–1990. Although there were declines in several countries—notably Russia—during 1990–2000, the species was stable or increased across most of Europe. The status of the truly wild population of this species in Europe is obscured by confusion with feral birds. Nevertheless, the species is provisionally evaluated as Secure.

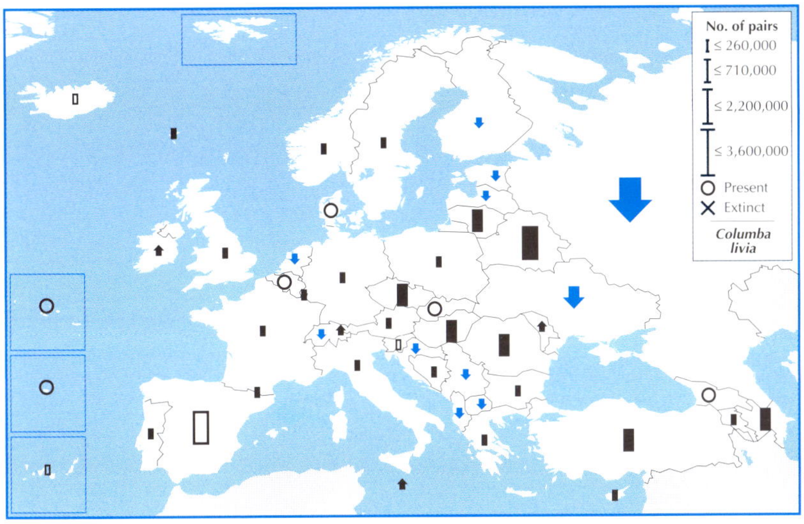

Country	Breeding pop. size (pairs)	Year(s)	Trend	Mag.%	References
Albania	1,000 – 3,000	02	(–)	(0–19)	
Andorra	(50 – 250)	01	(0)	(0–19)	*1,3*
Armenia	800 – 3,500	97–02	**0**	**0–19**	
Austria	(100,000 – 150,000)	98–02	(0)	(0–19)	
Azerbaijan	(250,000 – 500,000)	96–00	(0)	(0–19)	
Belarus	1,200,000 – 1,500,000	97–02	0	0–19	
Belgium	Present	00	?	–	
Bosnia & HG	(2,500 – 10,000)	90–00	(0)	(0–19)	
Bulgaria	400 – 800	96–02	(0)	(0–19)	
Croatia	(1,000 – 5,000)	02	(–)	(50–79)	*16*
Cyprus	(2,500 – 10,000)	94–02	(0)	(0–9)	
Czech Rep.	**500,000 – 1,000,000**	00	**0**	**0–19**	
Denmark	Present	00	?	–	
Faroe Is.	500 – 500	95	(0)	(0–19)	
Estonia	(100,000 – 150,000)	98	–	20–29	*1*
Finland	20,000 – 40,000	98–02	**–**	**20**	
France	800 – 2,000	98–00	(0)	(0–19)	*4*
Georgia	Present	03	?	–	
Germany	200,000 – 330,000	95–99	(0)	(0–19)	
Greece	(10,000 – 30,000)	95–00	(0)	(0–19)	
Hungary	(400,000 – 500,000)	90–00	0	0–19	*14*
Iceland	(50 – 250)	02	?	–	*40,26*
Rep. Ireland	(1,000 – 2,500)	88–91	+	0–19	
Italy	(3,000 – 7,000)	03	(0)	(0–19)	
Latvia	100,000 – 200,000	90–00	(–)	(50–79)	*16*
Liechtenstein	50 – 100	98–00	+	0–19	
Lithuania	(500,000 – 700,000)	99–01	(0)	(0–19)	*20*
Luxembourg	(5,000 – 10,000)	02	(0)	(0–19)	
Macedonia	150 – 250	90–00	–	20–29	
Malta	1,000 – 2,500	90–02	+	20–29	
Moldova	6,000 – 8,000	90–00	+	0–19	
Netherlands	(50,000 – 200,000)	98–00	(–)	(0–19)	
Norway	(15,000 – 30,000)	90–03	(0)	(0–19)	
Poland	(100,000 – 250,000)	90–00	(0)	(0–19)	*2*
Portugal	(100 – 1,000)	02	(0)	(0–19)	
Azores	Present	02	?	–	
Madeira	Present	02	?	–	
Romania	(400,000 – 600,000)	98–02	(0)	(0–19)	
Russia	2,500,000 – 5,000,000	90–00	(–)	(20–29)	*23,122*
Serbia & MN	1,500 – 2,500	90–02	–	20–29	*1,227,67a,155, 225*
Slovakia	Present	90–99	?	–	
Slovenia	100 – 200	90–00	?	–	
Spain	(2,000,000 – 2,280,000)	92	?	–	*13,12,10*
Canary Is.	(20,000 – 100,000)	97–03	?	–	*28,25*
Sweden	30,000 – 50,000	99–00	0	0–19	
Switzerland	**20,000 – 40,000**	93–96	(–)	(0–19)	
Turkey	(400,000 – 800,000)	01	(0)	(0–19)	
Ukraine	(300,000 – 500,000)	90–00	(–)	(0–19)	
UK	(100,000 – 250,000)	88–91	**0**	**3**	*22*
Total (approx.)	9,300,000 – 15,000,000	**Overall trend** Unknown			
Breeding range	>8,000,000 km²	**Gen. length.** <3.3 **% Global pop.** 25–49			

(See p. 152, bottom, for data quality graph)

Columba oenas
STOCK PIGEON

Non-SPEC[E] (1994: 4) **Status** Secure
Criteria —
European IUCN Red List Category —
Criteria —
Global IUCN Red List Category —
Criteria —

Columba oenas is a widespread breeder across much of Europe, which constitutes >75% of its global breeding range. Its European breeding population is large (>520,000 pairs), and increased between 1970–1990. Although there were declines across much of its range during 1990–2000—particularly in eastern Europe—key populations in France, Germany, Netherlands and the United Kingdom all increased, and the species underwent a moderate increase overall. Consequently, it is evaluated as Secure.

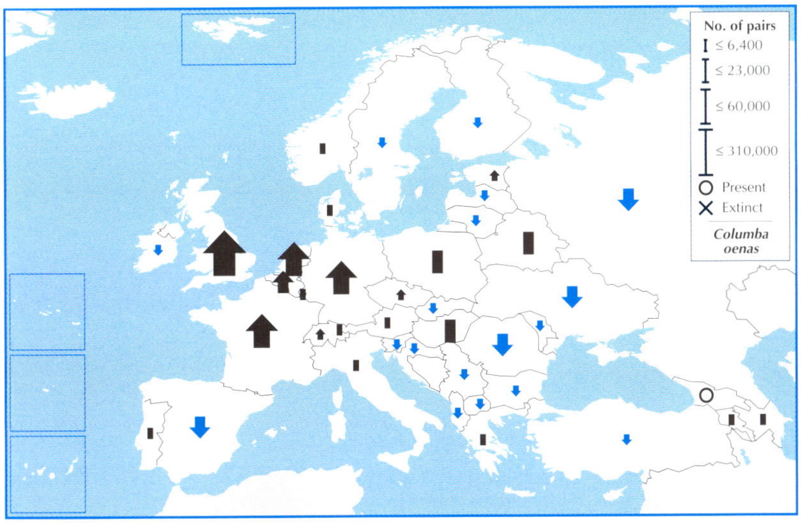

Country	Breeding pop. size (pairs)	Year(s)	Trend	Mag.%	References
Albania	(200 – 500)	02	(–)	(0–19)	
Armenia	300 – 800	99–02	0	0–19	
Austria	(2,500 – 4,500)	98–02	(0)	(0–19)	
Azerbaijan	(1,000 – 5,000)	96–00	(0)	(0–19)	
Belarus	14,000 – 25,000	97–02	0	0–19	
Belgium	10,000 – 20,000	01–02	+	0–19	*1*
Bulgaria	50 – 120	96–02	–	>80	
Croatia	(500 – 600)	02	(–)	(30–49)	*70,16*
Czech Rep.	**4,000 – 7,000**	00	**+**	**20–29**	
Denmark	800 – 1,000	00	(0)	(0–19)	*3*
Estonia	500 – 1,000	98	+	20–29	*1*
Finland	2,000 – 4,000	98–02	**–**	**60**	
France	20,000 – 80,000	98–02	(+)	(57)	*4,2*
Georgia	Present	03	?	–	
Germany	34,000 – 56,000	95–99	(+)	(20–29)	
Greece	(1,000 – 1,500)	95–00	(0)	(0–19)	
Hungary	**4,300 – 14,200**	99–02	0	0–19	*19*
Rep. Ireland	2,500 – 10,000	88–91	–	0–19	
Italy	(100 – 300)	03	(0)	(0–19)	
Latvia	(200 – 500)	90–00	(–)	(0–19)	*18*
Liechtenstein	2 – 5	98–00	0	0–19	
Lithuania	400 – 600	99–01	(–)	(0–19)	*20*
Luxembourg	250 – 400	02	**0**	**0–19**	
Macedonia	200 – 300	90–00	–	20–29	
Moldova	9 – 12	90–00	–	20–29	
Netherlands	**50,000 – 70,000**	98–00	**+**	**12**	*1*
Norway	(1,000 – 5,000)	90–03	0	0–19	*1*
Poland	10,000 – 20,000	00–02	(0)	(0–19)	*2,27*
Portugal	(100 – 500)	02	(0)	(0–19)	
Romania	(6,000 – 10,000)	90–02	(–)	(0–19)	
Russia	6,000 – 20,000	90–00	–	20–29	*23,24,67,72,103, 110,128,140*
Serbia & MN	300 – 600	90–02	–	20–39	*1,29,155,67a,62, 227,225*
Slovakia	3,500 – 5,500	80–99	–	30–49	
Slovenia	200 – 300	94	(–)	(0–19)	
Spain	(20,000 – 25,000)	98–02	(–)	(0–19)	*10,16*
Sweden	4,000 – 10,000	99–00	**–**	**56**	
Switzerland	**1,500 – 2,500**	93–96	+	0–19	
Turkey	(2.000 – 8.000)	01	(–)	(0–19)	
Ukraine	8,400 – 11,700	90–00	–	0–19	
UK	**309,000 – 309,000**	00	**+**	**29**	*31*
Total (approx.)	520,000 – 730,000	**Overall trend** Moderate increase			
Breeding range	>5,000,000 km²	**Gen. length.** <3.3 **% Global pop.** 75–94			

(See p. 152, bottom, for data quality graph)

Columba palumbus
COMMON WOOD-PIGEON

Non-SPEC[E] (1994: 4) **Status** Secure
Criteria —
European IUCN Red List Category —
Criteria —
Global IUCN Red List Category —
Criteria —

Columba palumbus is a widespread breeder throughout Europe, which constitutes >75% of its global breeding range. Its European breeding population is very large (>9,000,000 pairs), and was stable between 1970–1990. The key populations in Germany and the United Kingdom increased during 1990–2000, and the species increased slightly overall. Consequently, it is evaluated as Secure.

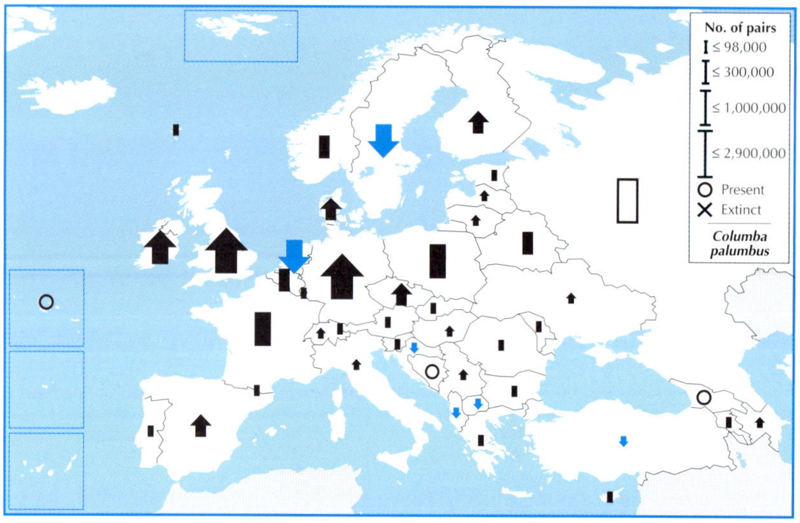

Country	Breeding pop. size (pairs)	Year(s)	Trend	Mag.%	References
Albania	2,000 – 5,000	02	(–)	(0–19)	
Andorra	(60 – 120)	99–01	(0)	(0–19)	*1,3*
Armenia	1,500 – 5,000	99–02	0	0–19	
Austria	(20,000 – 40,000)	98–02	(0)	(0–19)	
Azerbaijan	(2,000 – 10,000)	96–00	(+)	(0–19)	
Belarus	140,000 – 160,000	97–02	0	0–19	
Belgium	100,000 – 250,000	01–02	0	0–19	*1*
Bosnia & HG	Present	90–03	?	–	
Bulgaria	15,000 – 35,000	96–02	0	0–19	
Croatia	(10,000 – 20,000)	02	(–)	(50–79)	*16*
Cyprus	(15,000 – 30,000)	94–02	(0)	(0–19)	
Czech Rep.	**150,000 – 300,000**	00	**+**	**20–29**	
Denmark	250,000 – 350,000	00	+	10–19	*12*
Faroe Is.	0 – 2	95	(0)	(0–19)	
Estonia	40,000 – 80,000	98	0	0–19	*1*
Finland	150,000 – 200,000	98–02	**+**	**10**	
France	(500,000 – 2,000,000)	98–02	**0**	**0–19**	*4*
Georgia	Present	03	?	–	
Germany	1,700,000 – 2,900,000	95–99	+	20–29	
Greece	(5,000 – 8,000)	95–00	(0)	(0–19)	
Hungary	**77,000 – 110,000**	99–02	+	20–49	*19*
Rep. Ireland	500,000 – 1,000,000	88–91	+	0–19	
Italy	(20,000 – 50,000)	03	(+)	(0–19)	
Latvia	40,000 – 60,000	90–00	+	20–29	*23,18,2*
Liechtenstein	60 – 100	98–00	0	0–19	
Lithuania	80,000 – 120,000	99–01	(+)	(20–29)	*20*
Luxembourg	9,000 – 10,000	02	**0**	**0–19**	
Macedonia	30,000 – 60,000	90–00	(–)	(20–29)	
Moldova	**4,000 – 4,500**	90–00	0	0–19	
Netherlands	**400,000 – 500,000**	98–00	**–**	**13**	*1*
Norway	(100,000 – 500,000)	90–02	(0)	(0–19)	*8*
Poland	**400,000 – 600,000**	00–02	**0**	**0–19**	*23*
Portugal	(10,000 – 100,000)	02	(0)	(0–19)	
Azores	Present	02	?	–	
Romania	(12,000 – 16,000)	90–02	(0)	(0–19)	
Russia	1,000,000 – 2,500,000	90–00	?	–	*3,8,24,104,111, 114,122*
Serbia & MN	70,000 – 90,000	90–02	+	10–29	*1,67a,172a,155, 143a,78,152, 227,225*
Slovakia	60,000 – 120,000	90–99	0	0–19	
Slovenia	5,000 – 10,000	94	(0)	(0–19)	
Spain	(100,000 – 250,000)	98–02	(+)	(0–19)	*10*
Sweden	300,000 – 1,000,000	99–00	**–**	**28**	
Switzerland	**50,000 – 60,000**	93–96	+	20–29	
Turkey	(5,000 – 15,000)	01	(–)	(20–29)	
Ukraine	77,000 – 122,000	90–00	**+**	**0–19**	
UK	**2,570,000 – 3,160,000**	00	**+**	**17**	*5,31*

Total (approx.) 9,000,000 – 17,000,000 **Overall trend** Small increase
Breeding range >8,000,000 km² **Gen. length.** <3.3 **% Global pop.** 75–94
(See p. 154, bottom, for data quality graph)

Columba trocaz
MADEIRA LAUREL PIGEON

SPEC 1 (1994: 1) **Status** (Rare)
Criteria <10,000 pairs
European IUCN Red List Category —
Criteria —
Global IUCN Red List Category NT
Criteria B1a+b(i,ii,iii,iv,v); B2a+b(i,ii,iii,iv,v)

Columba trocaz is endemic to Europe, where it has a very small range (<500 km²) on Madeira. Its breeding population is small (as few as 4,100 individuals), but was stable between 1970–1990. Although the species increased slightly during 1990–2000, its population size still renders it susceptible to the risks affecting small populations, and consequently this globally Near Threatened species is provisionally evaluated as Rare in Europe.

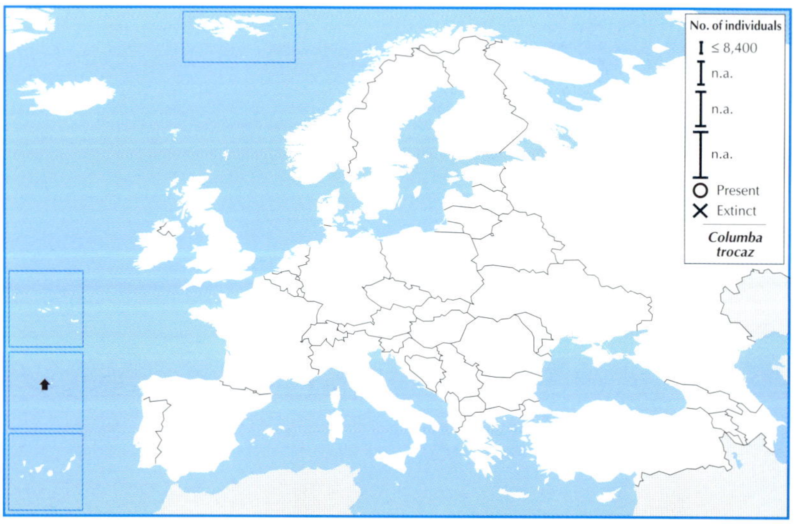

Country	Breeding pop. size (individuals)	Year(s)	Trend	Mag.%	References
Portugal					
Madeira	**4,100 – 17,000**	02	+	0–19	

Total (approx.) 4,100 – 17,000 **Overall trend** Small increase
Breeding range <500 km² **Gen. length.** 4 **% Global pop.** 100

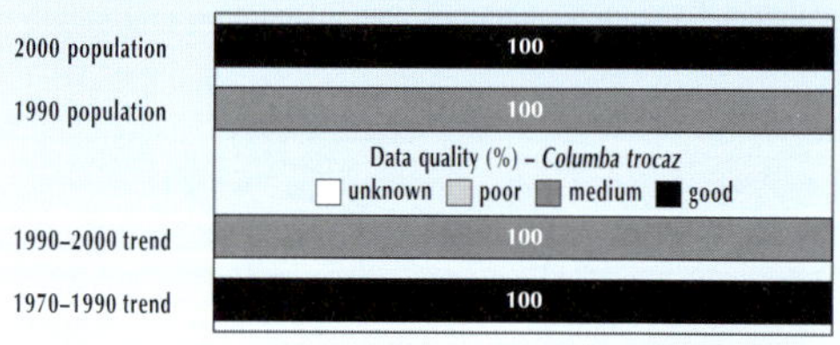

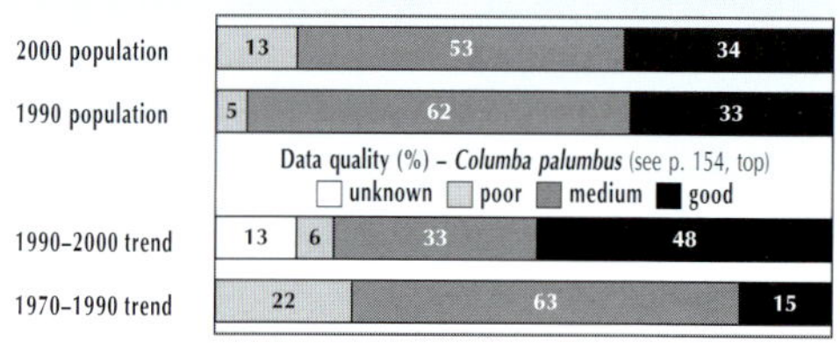

Columba bollii
DARK-TAILED LAUREL PIGEON

SPEC 1 (1994: 1) **Status** (Rare)
Criteria <10,000 pairs
European IUCN Red List Category —
Criteria —
Global IUCN Red List Category NT
Criteria C2a(i)

Country	Breeding pop. size (pairs)	Year(s)	Trend	Mag.%	References
Spain					
Canary Is.	(2,500 – 10,000)	96–03	(+)	(0–19)	*27,28*
Total (approx.)	2,500 – 10,000	**Overall trend** Small increase			
Breeding range	<2,500 km²	**Gen. length.** 4		**% Global pop.** 100	

Columba bollii is endemic to Europe, where it has a very small range (<2,500 km²) on the Canary Islands of Tenerife, La Gomera, El Hierro and La Palma. Its breeding population is small (as few as 2,500 pairs), but was stable between 1970–1990, and probably increased slightly during 1990–2000. Nevertheless, its population size renders it susceptible to the risks affecting small populations, and consequently this globally Near Threatened species is provisionally evaluated as Rare in Europe.

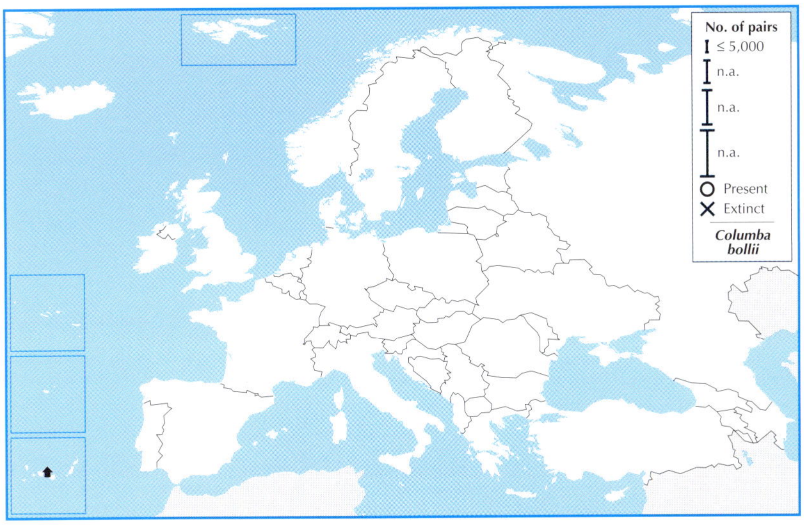

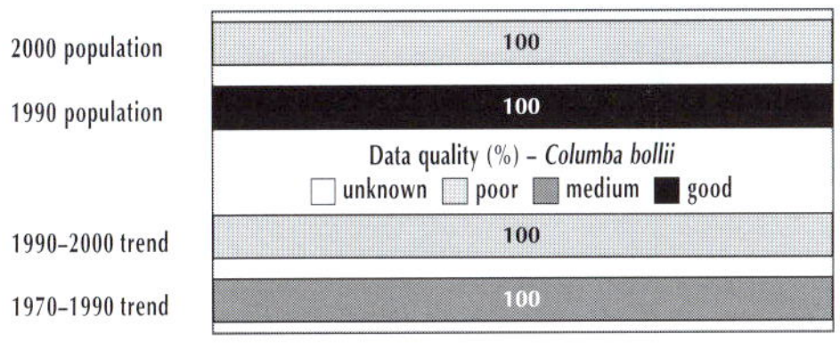

Columba junoniae
WHITE-TAILED LAUREL PIGEON

SPEC 1 (1994: 1) **Status** Endangered
Criteria See IUCN below
European IUCN Red List Category EN
Criteria B1a+b(iii)
Global IUCN Red List Category EN
Criteria B1a+b(iii)

Country	Breeding pop. size (pairs)	Year(s)	Trend	Mag.%	References
Spain					
Canary Is.	(1,000 – 2,500)	96–03	?	–	*27,28*
Total (approx.)	1,000 – 2,500	**Overall trend** Unknown			
Breeding range	<2,500 km²	**Gen. length.** 4		**% Global pop.** 100	

Columba junoniae is endemic to Europe, where it has a very small range (<2,500 km²) on the Canary Islands of La Palma, La Gomera, Tenerife and El Hierro. Its breeding population is small (as few as 1,000 pairs), and though it was stable between 1970–1990, its trend during 1990–2000 was unknown. Nevertheless, as a consequence of its very small range, within which its habitat is declining in quality, this globally threatened species is evaluated as Endangered in Europe.

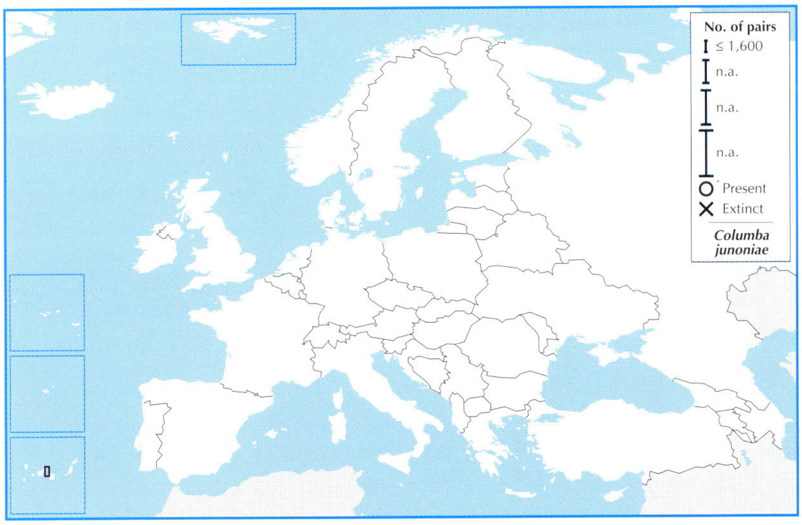

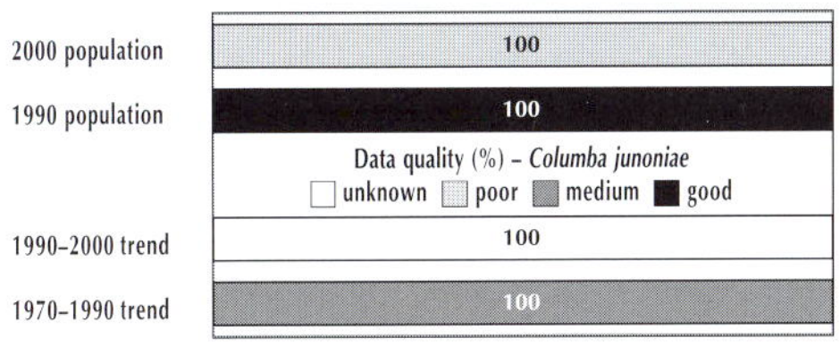

Streptopelia decaocto
EURASIAN COLLARED-DOVE

Non-SPEC (1994: —) **Status** Secure
Criteria —
European IUCN Red List Category —
Criteria —
Global IUCN Red List Category —
Criteria —

Streptopelia decaocto is a widespread resident across much of Europe, which accounts for less than half of its global range. Its European breeding population is very large (>4,700,000 pairs), and was stable between 1970–1990. Although there were declines in several countries during 1990–2000, particularly in the north of its European range, key populations in France, Romania, Russia and Turkey increased or were stable, and the species increased overall. Consequently, it is evaluated as Secure.

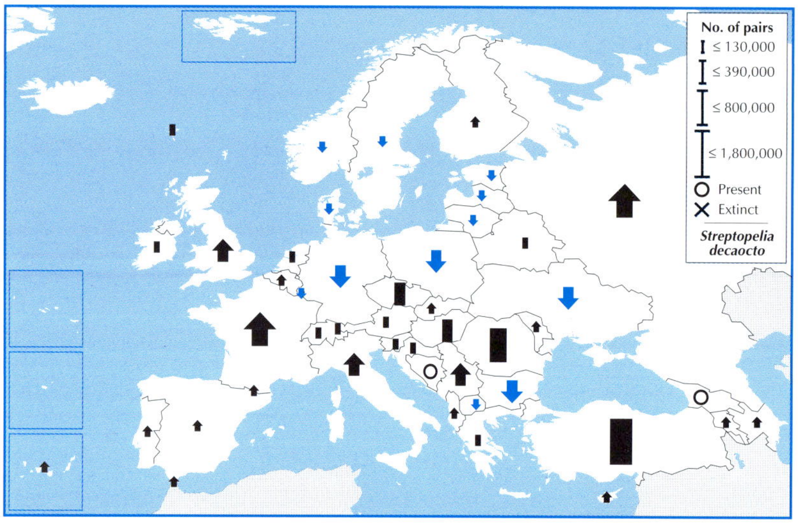

Country	Breeding pop. size (pairs)	Year(s)	Trend	Mag.%	References
Albania	5,000 – 10,000	02	(+)	(0–19)	
Andorra	0 – 6	02	+	N	*1,3*
Armenia	150 – 300	99–02	+	20–29	
Austria	(20,000 – 40,000)	98–02	(0)	(0–19)	
Azerbaijan	(10,000 – 100,000)	96–00	(+)	(>80)	
Belarus	4,500 – 7,000	97–02	0	0–19	
Belgium	70,000 – 100,000	01–02	+	0–19	*1*
Bosnia & HG	Present	90–03	?	–	
Bulgaria	100,000 – 500,000	96–02	–	0–19	
Croatia	(100,000 – 150,000)	02	(0)	(0–19)	*16*
Cyprus	6,000 – 12,000	98–02	+	30–49	
Czech Rep.	**180,000 – 360,000**	00	**0**	**0–9**	
Denmark	20,000 – 50,000	00	(–)	(10–19)	*3,2*
Faroe Is.	10 – 10	95	(0)	(0–19)	
Estonia	(400 – 800)	98	–	20–29	*1*
Finland	100 – 150	98–02	+	**20**	
France	400,000 – 1,600,000	98–02	+	**100**	*4,2*
Georgia	Present	03	?	–	
Germany	270,000 – 440,000	95–99	(–)	(20–29)	
Greece	(10,000 – 50,000)	95–00	(0)	(0–19)	
Hungary	**160,000 – 220,000**	99–02	0	0–19	*19*
Rep. Ireland	20,000 – 100,000	88–91	0	0–19	
Italy	(300,000 – 500,000)	03	(+)	(10–19)	
Latvia	200 – 500	90–00	–	0–19	*16,26,36*
Liechtenstein	5 – 10	98–00	+	0–19	
Lithuania	(10,000 – 20,000)	99–01	(–)	(20–29)	*20*
Luxembourg	200 – 300	02	–	30–49	
Macedonia	5,000 – 10,000	90–00	–	10–19	
Moldova	**40,000 – 80,000**	90–00	+	30–49	
Netherlands	**50,000 – 100,000**	98–00	**F**	**26**	*1*
Norway	500 – 3,000	90–03	–	20–29	
Poland	**200,000 – 400,000**	00–02	(–)	(0–19)	*23*
Portugal	(5,000 – 50,000)	02	(+)	(–)	
Romania	(400,000 – 800,000)	90–02	(0)	(0–19)	*48*
Russia	(500,000 – 1,000,000)	90–00	+	20–29	*8,93,122,123,124*
Serbia & MN	130,000 – 360,000	90–02	+	10–19	*1,143a,172a,155, 67a,78,227*
Slovakia	40,000 – 80,000	80–99	+	20–29	
Slovenia	40,000 – 60,000	94	(0)	(0–19)	
Spain	(20,000 – 100,000)	98–02	+	>80	*10*
Canary Is.	(1,000 – 2,500)	97–03	+	N	*28,25*
Sweden	**2,000 – 3,000**	99–00	–	20–29	
Switzerland	**15,000 – 20,000**	93–96	0	0–19	
Turkey	(1,000,000 – 3,000,000)	01	(0)	(0–19)	
Ukraine	280,000 – 435,000	90–00	–	0–19	
UK	**298,000 – 298,000**	00	+	**42**	*5,31*
Gibraltar	**80 – 100**	00	+	**N**	
Total (approx.)	4,700,000 – 11,000,000	**Overall trend** Moderate increase			
Breeding range	>6,000,000 km²	**Gen. length.** 9		**% Global pop.** 25–49	

(See p. 157, top, for data quality graph)

Streptopelia turtur
EUROPEAN TURTLE-DOVE

SPEC 3 (1994: 3) **Status** Declining
Criteria Moderate continuing decline
European IUCN Red List Category —
Criteria —
Global IUCN Red List Category —
Criteria —

Streptopelia turtur is a widespread summer visitor to much of Europe, which accounts for less than half of its global breeding range. Its European breeding population is very large (>3,500,000 pairs), but underwent a moderate decline between 1970–1990. Although the species was stable or increased in various countries, especially in central Europe, during 1990–2000, most populations—including sizeable ones in Spain, Russia and Turkey—declined, and the species underwent a moderate decline (>10%) overall. Consequently, it is evaluated as Declining.

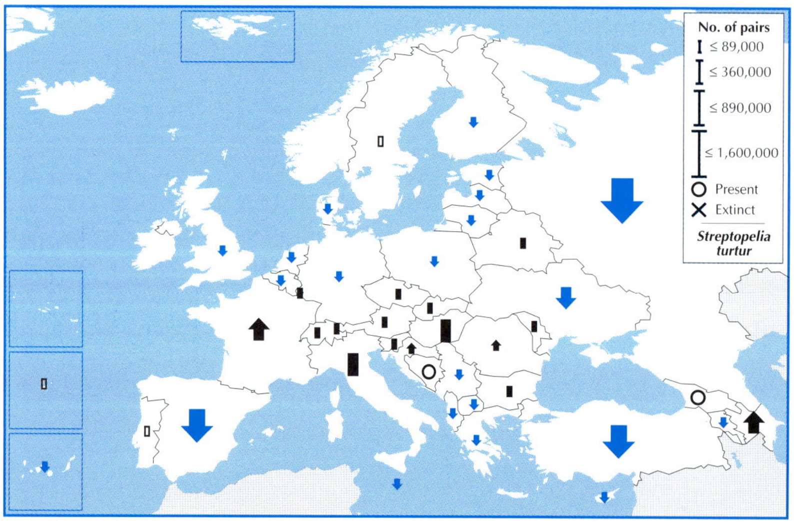

Country	Breeding pop. size (pairs)	Year(s)	Trend	Mag.%	References
Albania	1,000 – 5,000	02	(–)	(0–19)	
Armenia	650 – 1,000	98–02	–	0–19	
Austria	(8,000 – 15,000)	98–02	(0)	(0–19)	
Azerbaijan	(250,000 – 500,000)	96–00	(+)	(20–29)	
Belarus	40,000 – 60,000	97–02	0	0–19	
Belgium	5,800 – 9,600	01–02	–	50–79	*1*
Bosnia & HG	Present	90–03	?	–	
Bulgaria	20,000 – 100,000	96–02	0	0–19	
Croatia	(50,000 – 100,000)	02	(+)	(0–19)	*16*
Cyprus	5,000 – 15,000	94–02	–	0–19	
Czech Rep.	**60,000 – 120,000**	00	**0**	**0–19**	
Denmark	(25 – 75)	98–01	(–)	(50)	*4,5,6,7*
Estonia	4,000 – 8,000	98	–	20–29	*1*
Finland	5 – 30	98–02	–	>80	
France	150,000 – 450,000	98–02	(+)	(10)	*4,2*
Georgia	Present	03	?	–	
Germany	55,000 – 81,000	95–99	(–)	(20–29)	
Greece	(10,000 – 30,000)	95–00	(–)	(0–19)	
Hungary	**165,000 – 215,000**	99–02	0	0–19	*19*
Italy	(200,000 – 400,000)	03	(0)	(0–19)	
Latvia	(500 – 2,000)	90–00	–	50–79	*18*
Liechtenstein	1 – 3	98–00	0	0–19	
Lithuania	(2,000 – 5,000)	99–01	(–)	(30–49)	*20*
Luxembourg	1,800 – 2,000	02	**0**	**0–19**	
Macedonia	25,000 – 60,000	90–00	–	20–29	
Malta	2 – 5	90–02	–	0–19	*1*
Moldova	**3,000 – 3,500**	90–00	0	0–19	
Netherlands	**10,000 – 12,000**	98–00	**–**	**53**	*1*
Poland	**40,000 – 70,000**	00–02	(–)	(0–19)	*23*
Portugal	(10,000 – 100,000)	02	?	–	
Madeira	(0 – 5)	02	?	–	
Romania	(15,000 – 25,000)	90–02	(+)	(0–19)	*48*
Russia	1,000,000 – 2,500,000	90–00	–	20–29	*8,35,39,43,111, 114,122*
Serbia & MN	65,000 – 120,000	90–02	–	10–19	*1,143a,172a,155, 67a,227,225*
Slovakia	15,000 – 30,000	90–99	0	0–19	
Slovenia	2,000 – 3,000	94	(0)	(0–19)	
Spain	790,000 – 1,000,000	92	–	30–49	*13,12,16,10*
Canary Is.	(2,500 – 10,000)	97–03	(–)	(0–19)	*28,25*
Sweden	0 – 1	99–00	?	–	
Switzerland	**1,000 – 2,500**	93–96	0	0–19	
Turkey	(300,000 – 900,000)	01	(–)	(0–19)	
Ukraine	110,000 – 176,000	90–00	–	0–19	
UK	**44,000 – 44,000**	00	**–**	**42**	*5,31*
Total (approx.)	3,500,000 – 7,200,000	**Overall trend** Moderate decline			
Breeding range	>7,000,000 km²	**Gen. length.** <3.3		**% Global pop.** 25–49	

(See p. 157, top, for data quality graph)

Streptopelia senegalensis
LAUGHING DOVE

Non-SPEC (1994: —) **Status** Secure
Criteria —
European IUCN Red List Category —
Criteria —
Global IUCN Red List Category —
Criteria —

Streptopelia senegalensis has an extensive global distribution, which just extends into Europe in Turkey and parts of the Caucasus. Its European breeding population is relatively small (<210,000 pairs), but increased substantially between 1970–1990. Although the trend in Georgia during 1990–2000 was unknown, the species increased elsewhere within its European range, and underwent a large increase overall. Consequently, it is evaluated as Secure.

Country	Breeding pop. size (pairs)	Year(s)	Trend	Mag.%	References
Armenia	**800 – 3,500**	99–02	+	**0–19**	1
Azerbaijan	(10,000 – 100,000)	96–00	(+)	(30–49)	
Georgia	Present	03	?	–	
Russia	(500 – 1,500)	90–00	+	**50–79**	*172*
Turkey	(50,000 – 100,000)	01	(+)	(30–49)	
Total (approx.)	61,000 – 210,000	**Overall trend** Large increase			
Breeding range	>20,000 km²	**Gen. length.** <3.3 **% Global pop.** <5			

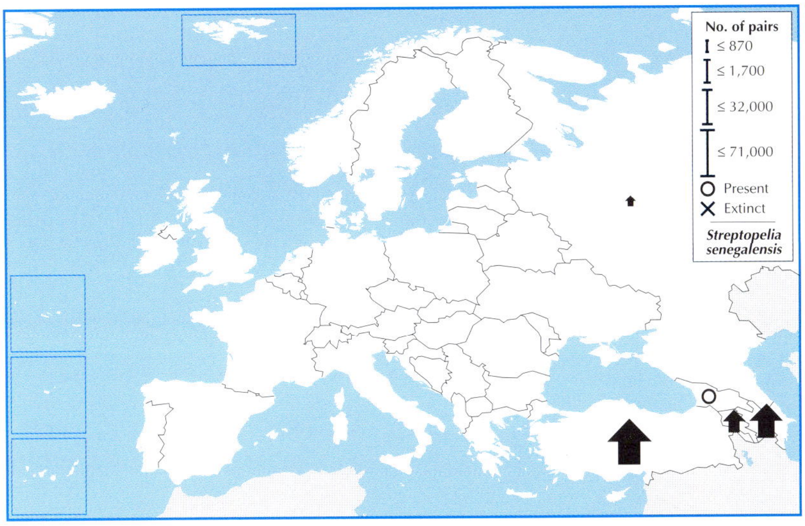

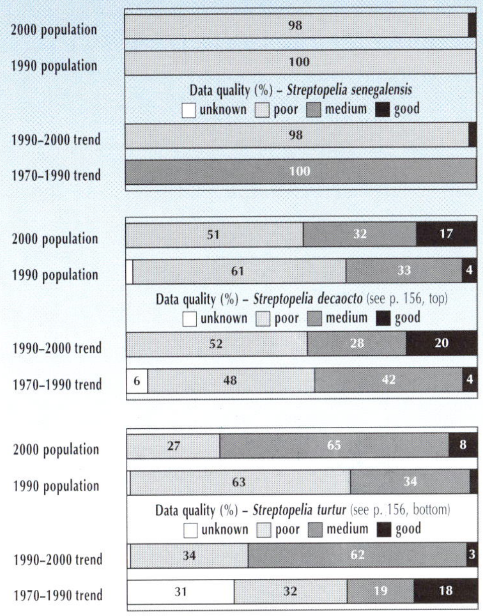

Clamator glandarius
GREAT SPOTTED CUCKOO

Non-SPEC (1994: —) **Status** (Secure)
Criteria —
European IUCN Red List Category —
Criteria —
Global IUCN Red List Category —
Criteria —

Clamator glandarius is a widespread summer visitor to southern Europe, which accounts for less than a quarter of its global breeding range. Its European breeding population is relatively small (<77,000 pairs), but increased between 1970–1990. Although the trend of the stronghold population in Spain during 1990–2000 was unknown, the species was stable or increased across the rest of its European range, and is provisionally evaluated as Secure.

Country	Breeding pop. size (pairs)	Year(s)	Trend	Mag.%	References
Bulgaria	(0 – 5)	96–02	(F)	(>80)	
Croatia	(3 – 4)	02	(+)	(0–19)	*70*
Cyprus	(400 – 4,000)	94–02	(+)	(0–19)	
France	250 – 600	00	F	50–79	*1*
Greece	(10 – 20)	95–00	(0)	(0–19)	
Italy	(10 – 15)	03	0	0–19	
Macedonia	(5 – 10)	90–00	(F)	(20–29)	
Portugal	(500 – 1,500)	02	(0)	(0–19)	
Spain	(55,000 – 64,500)	92	?	–	*13,12,10*
Turkey	(2,000 – 6,000)	01	(0)	(0–19)	
Total (approx.)	58,000 – 77,000	**Overall trend** Unknown			
Breeding range	>750,000 km²	**Gen. length.** <3.3 **% Global pop.** 5–24			

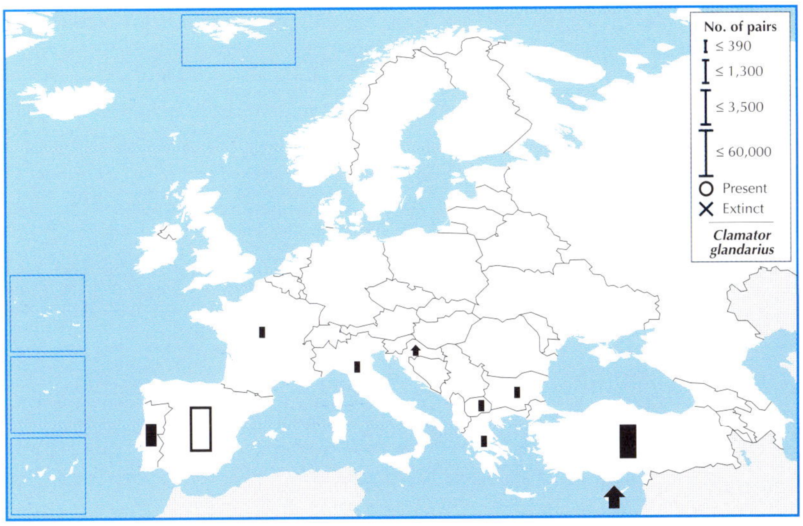

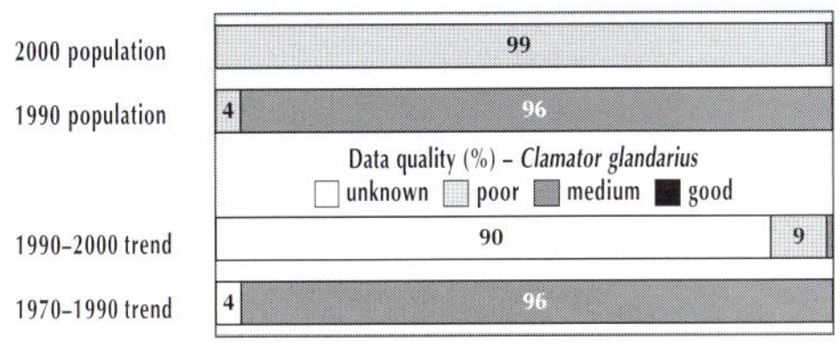

Cuculus canorus
COMMON CUCKOO

Non-SPEC (1994: —) **Status** Secure
Criteria —
European IUCN Red List Category —
Criteria —
Global IUCN Red List Category —
Criteria —

Cuculus canorus is a widespread summer visitor to Europe, which accounts for less than half of its global breeding range. Its European breeding population is very large (>4,200,000 pairs), and was stable between 1970–1990. Although there were declines in many western populations—most notably France—during 1990–2000, most populations in the east, including key ones in Russia and Romania, were stable, and the species underwent only a slight decline overall. Consequently, it is evaluated as Secure.

Country	Breeding pop. size (pairs)	Year(s)	Trend	Mag.%	References
Albania	2,000 – 5,000	02	(0)	(0–19)	
Andorra	(50 – 150)	99–01	(0)	(0–19)	*1,3*
Armenia	3,500 – 5,000	00–02	**0**	**0–19**	
Austria	(20,000 – 40,000)	98–02	(0)	(0–19)	
Azerbaijan	(10,000 – 50,000)	96–00	(0)	(0–19)	
Belarus	50,000 – 80,000	97–02	0	0–19	
Belgium	6,000 – 8,000	01–02	–	0–19	*1*
Bosnia & HG	Present	90–03	?	–	
Bulgaria	20,000 – 50,000	96–02	0	0–19	
Croatia	(5,000 – 10,000)	02	(–)	(50–79)	*16*
Czech Rep.	**35,000 – 70,000**	00	**0**	**0–19**	
Denmark	(20,000 – 40,000)	00	**0**	**0–19**	
Estonia	20,000 – 30,000	98	–	20–29	*1*
Finland	50,000 – 80,000	98–02	**0**	**5**	
France	(250,000 – 750,000)	98–02	**–**	**28**	*4*
Georgia	Present	03	?	–	
Germany	60,000 – 114,000	95–99	(–)	(20–29)	
Greece	(5,000 – 10,000)	95–00	(0)	(0–19)	
Hungary	55,000 – 65,000	99–02	0	0–19	*19*
Rep. Ireland	2,500 – 10,000	88–91	–	0–19	
Italy	(50,000 – 100,000)	03	(0)	(0–19)	
Latvia	25,000 – 35,000	90–00	0	0–19	*23,18*
Liechtenstein	20 – 30	98–00	–	0–19	
Lithuania	(15,000 – 25,000)	99–01	(0)	(0–19)	*20*
Luxembourg	60 – 80	02	**–**	**30–49**	
Macedonia	(5,000 – 12,000)	90–00	(0)	(0–19)	
Malta	**1 – 1**	90–02	**0**	**0–19**	*1*
Moldova	10,000 – 12,000	90–00	0	0–19	
Netherlands	**6,000 – 8,000**	98–00	**–**	**6**	*1*
Norway	(1,000 – 3,000)	95–02	(0)	(0–19)	*30*
Poland	**150,000 – 200,000**	00–02	(0)	(0–19)	*23*
Portugal	(10,000 – 100,000)	02	(0)	(0–19)	
Romania	400,000 – 800,000	90–02	(0)	(0–19)	*48*
Russia	2,500,000 – 5,000,000	90–00	0	0–19	*35,39,43,114,122*
Serbia & MN	18,000 – 25,000	90–02	0	0–19	*1,155,29,172a, 67a,78,227,225*
Slovakia	7,000 – 14,000	90–99	0	0–19	
Slovenia	5,000 – 10,000	94	(0)	(0–19)	
Spain	(20,000 – 100,000)	98–02	?	–	*10*
Sweden	30,000 – 70,000	99–00	**–**	**5**	
Switzerland	**20,000 – 30,000**	93–96	–	10–19	
Turkey	(200,000 – 500,000)	01	?	–	
Ukraine	98,000 – 147,000	90–00	(0)	(0–19)	
UK	**9,600 – 20,000**	00	**–**	**26**	*5,31*
Total (approx.)	4,200,000 – 8,600,000	**Overall trend** Small decline			
Breeding range	>8,000,000 km²	**Gen. length.** <3.3 **% Global pop.** 25–49			

(See p. 158, bottom, for data quality graph)

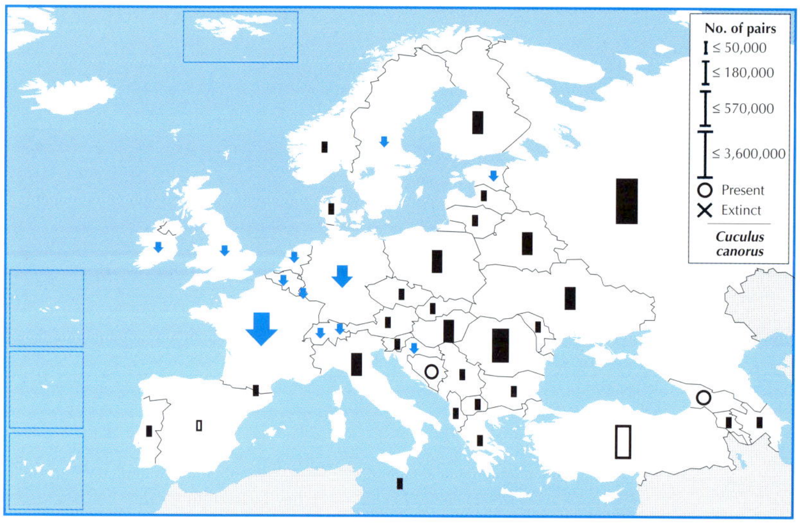

Cuculus saturatus
ORIENTAL CUCKOO

Non-SPEC (1994: —) **Status** (Secure)
Criteria —
European IUCN Red List Category —
Criteria —
Global IUCN Red List Category —
Criteria —

Cuculus saturatus has a predominantly Asian breeding distibution, which just extends into Europe in northern Russia. The European breeding population is large (>250,000 pairs), and was stable between 1970–1990. The species remained stable during 1990–2000, and consequently it is provisionally evaluated as Secure.

Country	Breeding pop. size (pairs)	Year(s)	Trend	Mag.%	References
Russia	(250,000 – 500,000)	90–00	(0)	(0–19)	*4,122*
Total (approx.)	250,000 – 500,000	**Overall trend** Stable			
Breeding range	>750,000 km²	**Gen. length.** <3.3 **% Global pop.** <5			

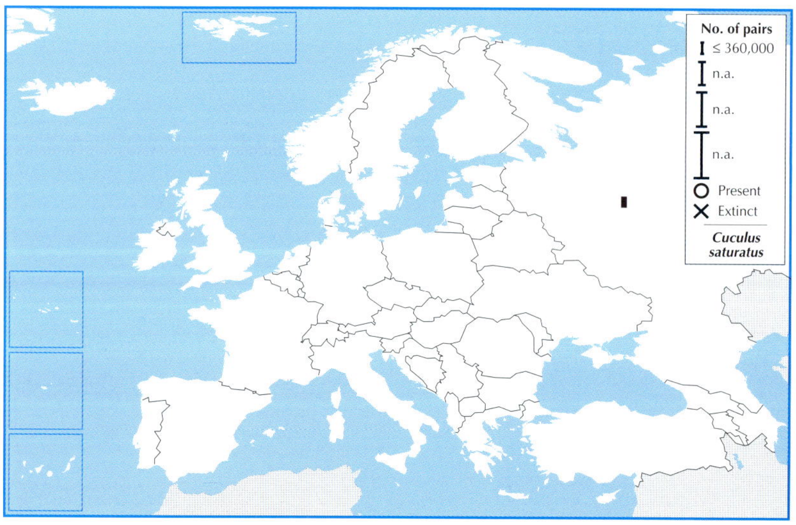

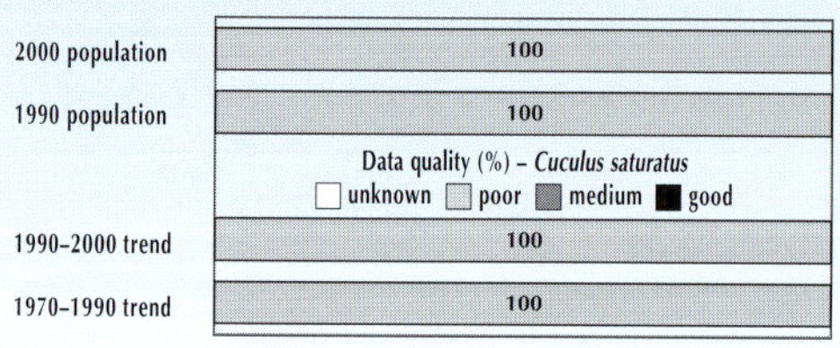

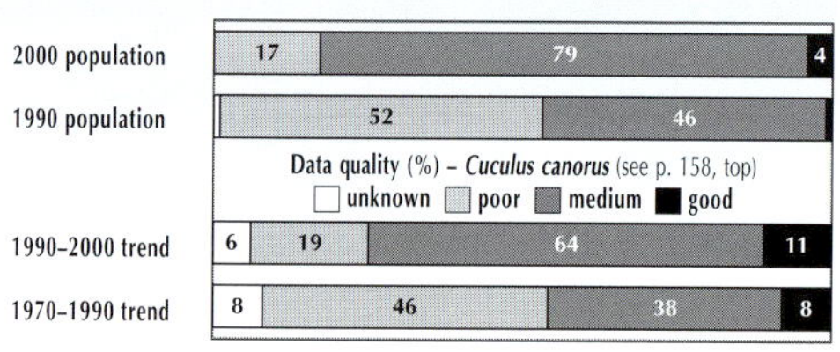

Tyto alba
BARN OWL

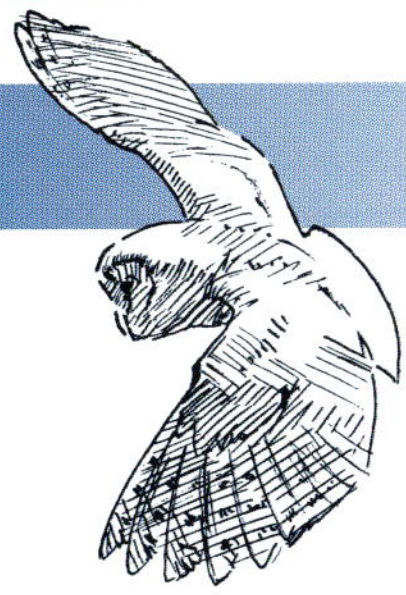

SPEC 3 (1994: 3) **Status** (Declining)
Criteria Moderate continuing decline
European IUCN Red List Category —
Criteria —
Global IUCN Red List Category —
Criteria —

Tyto alba is a widespread resident across much of Europe, which accounts for less than a quarter of its vast global range. Its European breeding population is large (>110,000 pairs), but underwent a moderate decline between 1970–1990. Although the species was stable or increased across much of Europe during 1990–2000, there were declines in a number of countries—including the key population in Spain—and the species probably underwent a moderate decline (>10%) overall. Consequently, it is provisionally evaluated as Declining.

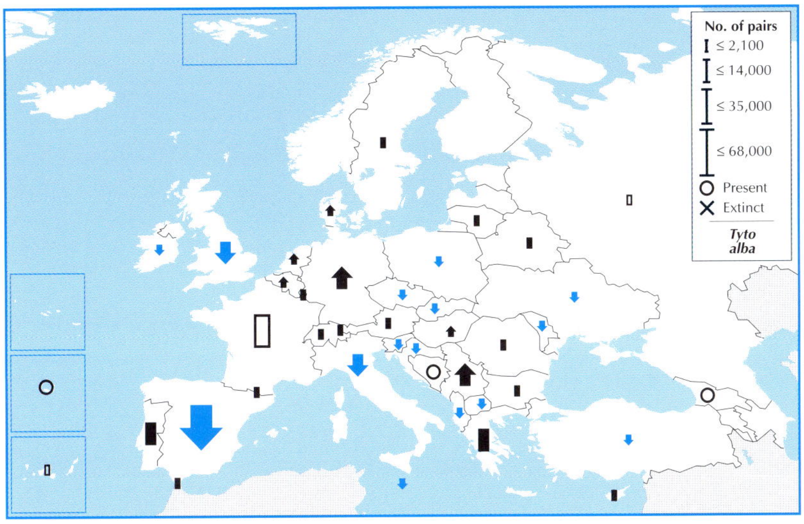

Country	Breeding pop. size (pairs)	Year(s)	Trend	Mag.%	References
Albania	300 – 800	02	(–)	(0–19)	
Andorra	0 – 2	98	0	0–19	*1,3*
Austria	(20 – 40)	98–02	(0)	(0–19)	
Belarus	5 – 15	97–02	0	0–19	
Belgium	1,600 – 2,600	95–02	+	0–19	*1*
Bosnia & HG	Present	90–03	?	–	
Bulgaria	150 – 500	96–02	(0)	(0–19)	
Croatia	(500 – 1,000)	02	(–)	(50–79)	*16*
Cyprus	(200 – 600)	94–02	(0)	(0–19)	
Czech Rep.	**200 – 350**	00	**–**	**50–79**	
Denmark	**104 – 104**	01	**+**	**>80**	*7*
France	20,000 – 60,000	98–00	?	–	*4*
Georgia	Present	03	?	–	
Germany	11,000 – 17,000	95–99	+	20–29	
Greece	(3,000 – 6,000)	95–00	(0)	(0–19)	
Hungary	1,000 – 1,500	99–02	+	20–29	
Rep. Ireland	250 – 1,000	88–91	(–)	(20–29)	
Italy	(6,000 – 12,000)	03	(–)	(0–19)	
Liechtenstein	**0 – 2**	98–00	**0**	**0–19**	
Lithuania	(30 – 70)	99–01	(F)	(30–49)	*20*
Luxembourg	200 – 300	00–02	**0**	**0–19**	
Macedonia	(180 – 300)	90–00	(–)	(20–29)	
Malta	0 – 1	90–02	–	>80	*1*
Moldova	**1 – 2**	90–00	–	50–79	
Netherlands	**1,150 – 2,000**	98–00	**+**	**137**	*1*
Poland	1,000 – 2,000	95–00	–	20–29	*27*
Portugal	(2,000 – 6,000)	02	(0)	(0–19)	
Madeira	Present	02	?	–	
Romania	800 – 1,200	90–02	0	0–19	
Russia	(20 – 70)	90–00	?	–	*171,174*
Serbia & MN	3,200 – 4,400	90–02	+	0–19	*1,78,29,172a,155*
Slovakia	400 – 600	80–99	–	50–79	
Slovenia	50 – 80	00	–	30–49	
Spain	(50,400 – 90,500)	92	(–)	(0–19)	*13,12,16,10*
Canary Is.	(250 – 1,000)	97–03	?	–	*25*
Sweden	**1 – 2**	99–00	**0**	**0–19**	
Switzerland	**1,000 – 1,500**	93–96	F	20–29	
Turkey	(800 – 1,200)	01	(–)	(0–19)	
Ukraine	(5 – 10)	90–00	(–)	(10–19)	
UK	**3,000 – 5,000**	95–97	(–)	(0–19)	*33*
Gibraltar	(0 – 1)	00	(0)	(0–19)	
Total (approx.)	110,000 – 220,000	**Overall trend** Moderate decline			
Breeding range	>3,000,000 km²	**Gen. length.** <3.3 **% Global pop.** 5–24			

(See p. 159, bottom, for data quality graph)

Otus brucei
PALLID SCOPS-OWL

SPEC 3 (1994: —) **Status** Critically Endangered
Criteria See IUCN below
European IUCN Red List Category CR
Criteria C2a(i); D1
Global IUCN Red List Category —
Criteria —

Otus brucei has a predominantly Asian and Middle Eastern distribution, which just extends into Europe in south-eastern Turkey. Its European breeding population is tiny (as few as 10 pairs), and although its trend between 1970–1990 was unknown, it underwent a large decline (>30%) overall during 1990–2000. As a consequence of this decline and its tiny population, this previously Secure species is now evaluated as Critically Endangered.

Country	Breeding pop. size (pairs)	Year(s)	Trend	Mag.%	References
Turkey	10 – 50	01	(–)	(30–49)	
Total (approx.)	10 – 50	**Overall trend** Large decline			
Breeding range	<20,000 km²	**Gen. length.** <3.3 **% Global pop.** <5			

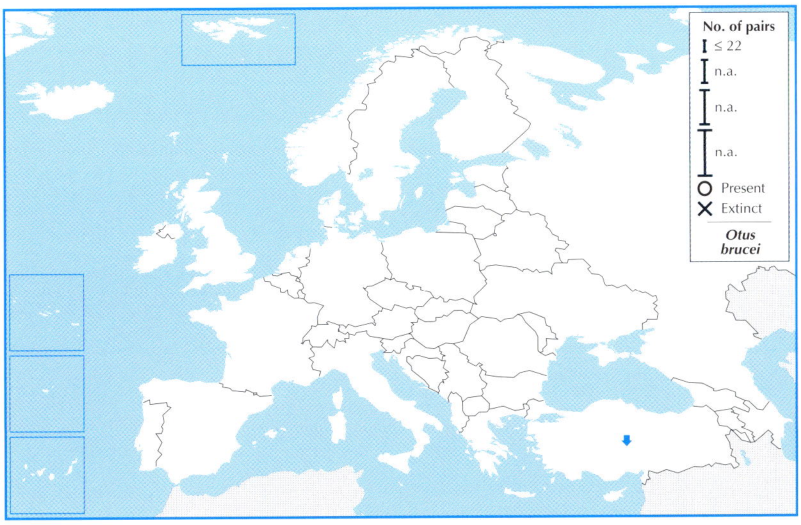

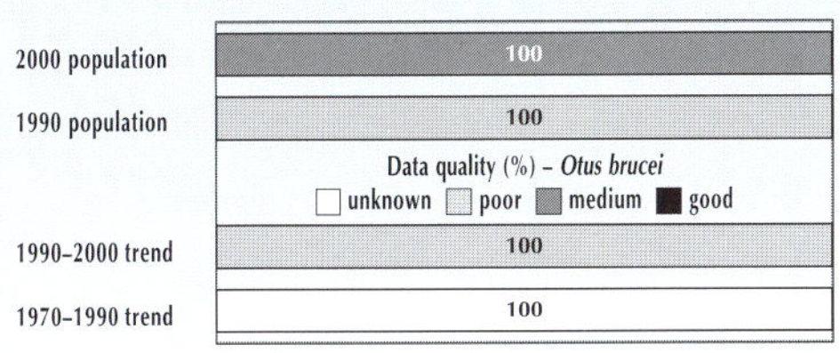

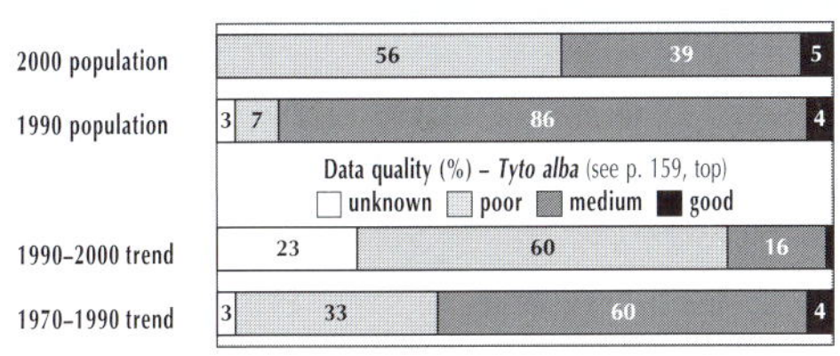

Otus scops
COMMON SCOPS-OWL

SPEC 2 (1994: 2) **Status** (Depleted)
Criteria Moderate historical decline
European IUCN Red List Category —
Criteria —
Global IUCN Red List Category —
Criteria —

Otus scops is a widespread breeder across much of southern and eastern Europe, which constitutes >50% of its global breeding range. Its European breeding population is large (>210,000 pairs), but underwent a moderate decline between 1970–1990. The species was stable or increased in some countries during 1990–2000, but it declined in many others. Although trends were not available for the key populations in Russia and Spain, the species's population has clearly not yet recovered to the level that preceded its decline. Consequently, it is provisionally evaluated as Depleted.

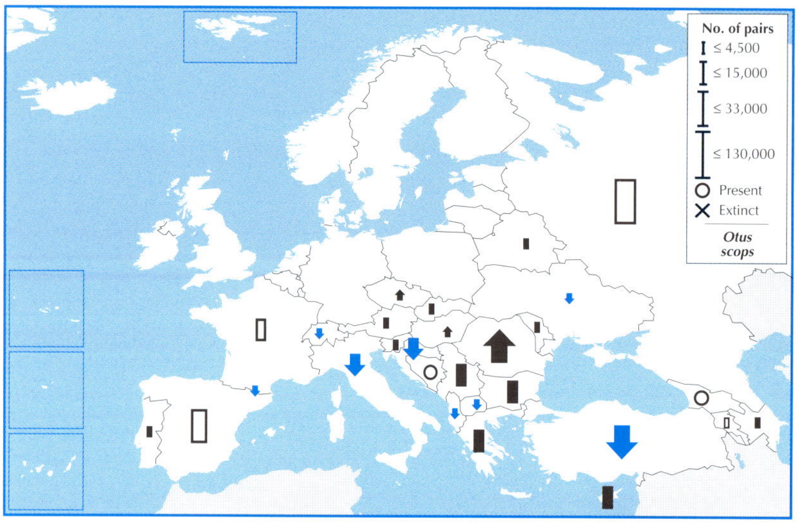

Country	Breeding pop. size (pairs)	Year(s)	Trend	Mag.%	References
Albania	1,500 – 3,000	02	(–)	(0–19)	
Andorra	2 – 3	98	–	20–29	*1,3*
Armenia	850 – 1,500	97–02	?	–	
Austria	**40 – 60**	98–02	**0**	**0–19**	
Azerbaijan	(1,000 – 10,000)	96–00	(0)	(0–19)	
Belarus	10 – 50	97–02	0	0–19	
Bosnia & HG	Present	90–03	?	–	
Bulgaria	**6,000 – 9,000**	96–02	0	0–19	
Croatia	(5,000 – 10,000)	02	(–)	(50–79)	*70,16*
Cyprus	(10,000 – 20,000)	94–02	(0)	(0–19)	
Czech Rep.	**0 – 1**	00	+	**N**	
France	3,000 – 12,000	98–00	?	–	*4*
Georgia	Present	03	?	–	
Greece	(5,000 – 20,000)	95–00	(0)	(0–19)	
Hungary	500 – 600	99–02	+	50–79	
Italy	(5,000 – 10,000)	03	(–)	(0–19)	
Macedonia	1,800 – 4,000	90–00	(–)	(20–29)	
Moldova	280 – 340	90–00	**0**	**0–19**	
Portugal	(2,000 – 6,000)	02	(0)	(0–19)	
Romania	25,000 – 40,000	90–02	+	0–19	
Russia	(80,000 – 200,000)	90–00	?	–	*67,73*
Serbia & MN	10,000 – 14,000	90–02	0	0–19	*1,78,155,62,152, 225,227,91*
Slovakia	40 – 80	80–99	F	20–29	
Slovenia	800 – 1,300	99–00	(0)	(0–19)	
Spain	(30,000 – 35,000)	92	?	–	*13,12,10*
Switzerland	**5 – 10**	98–02	**-**	**30–49**	
Turkey	(20,000 – 40,000)	01	(–)	(0–19)	
Ukraine	(4,200 – 4,700)	90–00	(–)	(0–19)	
Total (approx.)	210,000 – 440,000	**Overall trend** Unknown			
Breeding range	>3,000,000 km²	**Gen. length.** <3.3		**% Global pop.** 50–74	

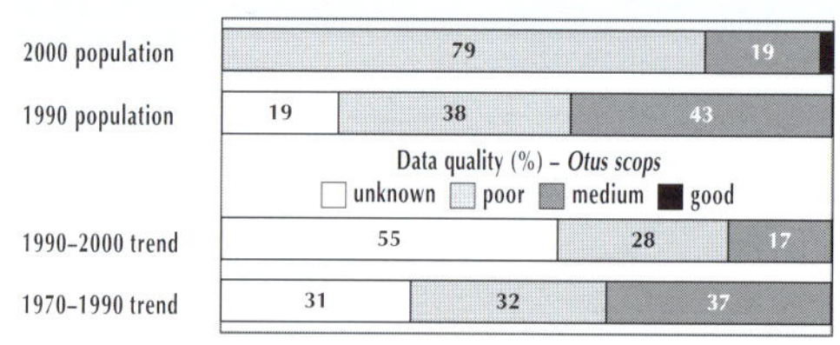

Bubo bubo
EURASIAN EAGLE-OWL

SPEC 3 (1994: 3) **Status** (Depleted)
Criteria Large historical decline
European IUCN Red List Category —
Criteria —
Global IUCN Red List Category —
Criteria —

Bubo bubo is a widespread but patchily distributed resident across much of Europe, which accounts for less than a quarter of its global range. Its European breeding population is relatively small (<38,000 pairs), and underwent a large decline during 1970–1990. Although there were declines in a few countries during 1990–2000, the species increased or was stable across most of its European range, and was stable overall. Nevertheless, its total population size probably still remains below the level that preceded its decline, and consequently it is provisionally evaluated as Depleted.

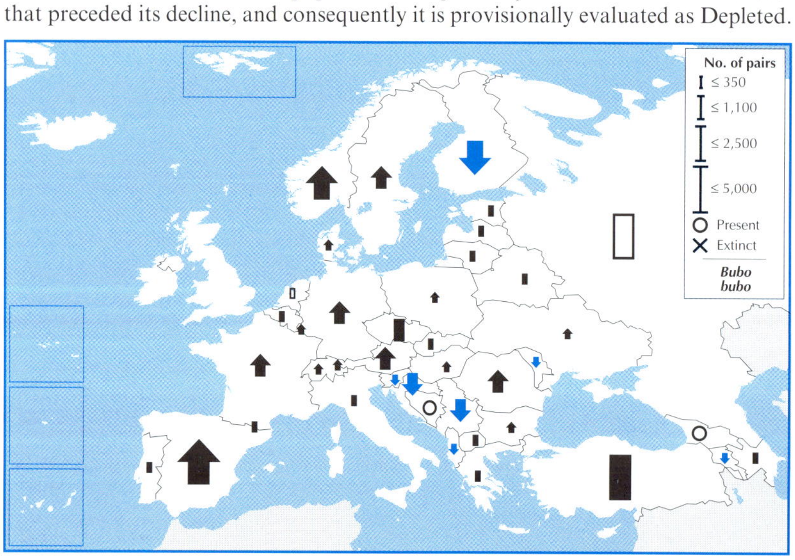

Country	Breeding pop. size (pairs)	Year(s)	Trend	Mag.%	References
Albania	50 – 100	98–02	(–)	(0–19)	
Andorra	2 – 3	98	0	0–19	*1,3*
Armenia	70 – 150	97–02	–	0–19	
Austria	400 – 600	98–02	+	20–29	
Azerbaijan	(100 – 1,000)	96–00	(0)	(0–19)	
Belarus	250 – 400	97–02	0	0–19	
Belgium	**25 – 30**	95–02	**0**	**0–19**	*1*
Bosnia & HG	Present	90–03	?	–	
Bulgaria	**180 – 260**	96–02	+	**50–79**	
Croatia	(500 – 1,000)	02	(–)	(30–49)	*70,16*
Czech Rep.	**600 – 800**	00	**0**	**0–9**	
Denmark	**22 – 22**	01	+	**>80**	*7*
Estonia	120 – 200	98	0	0–19	1
Finland	2,000 – 3,000	98–01	–	20	
France	**1,000 – 1,200**	98–00	+	0–19	*4*
Georgia	Present	03	?	–	
Germany	**660 – 780**	95–99	+	**30–49**	
Greece	(200 – 500)	95–00	(0)	(0–19)	
Hungary	10 – 30	95–02	+	>80	
Italy	250 – 300	03	0	0–19	
Latvia	30 – 50	90–00	0	0–19	*23,18*
Liechtenstein	**3 – 4**	98–00	+	**20–29**	
Lithuania	(10 – 20)	99–01	0	0–19	*20*
Luxembourg	**8 – 20**	00–02	+	**10–19**	
Macedonia	30 – 80	90–00	(0)	(0–19)	
Moldova	**5 – 10**	90–00	**-**	**>80**	
Netherlands	**1 – 2**	98–00	?	–	*1*
Norway	(1,000 – 2,000)	90–02	(+)	(0–19)	*58*
Poland	**250 – 270**	95–00	+	**30–39**	*79*
Portugal	200 – 500	02	(0)	(0–19)	
Romania	750 – 1,000	90–02	+	0–19	
Russia	(3,000 – 5,500)	90–00	?	–	*60,74,103,142*
Serbia & MN	**450 – 700**	90–02	–	10–19	*1,62,155,152, 225,227*
Slovakia	300 – 400	90–99	0	0–19	
Slovenia	**30 – 50**	96–00	–	0–9	
Spain	(2,500 – 10,000)	98–02	+	0–19	*10*
Sweden	**500 – 1,000**	99–00	+	**30–49**	
Switzerland	**120 – 120**	93–96	+	0–9	
Turkey	(3,000 – 6,000)	01	(0)	(0–19)	
Ukraine	**150 – 200**	90–00	+	**0–19**	
Total (approx.)	19,000 – 38,000	**Overall trend** Stable			
Breeding range	>5,000,000 km²	**Gen. length.** 9		**% Global pop.** 5–24	

(See p. 161, top, for data quality graph)

Ketupa zeylonensis
BROWN FISH-OWL

SPEC 3 (1994: —) **Status** Critically Endangered
Criteria See IUCN below
European IUCN Red List Category CR
Criteria B1a+b(i,ii,iii,iv,v); C1; C2a(i); D1
Global IUCN Red List Category —
Criteria —

Ketupa zeylonensis has a predominantly Asian distribution, but also an extremely small (<100 km²) and disjunct range in south-eastern Turkey. The European breeding population is tiny (fewer than 10 pairs), and although its trend between 1970–1990 was unknown, it underwent a very large decline (>50%) during 1990–2000. As a consequence of this decline, its small population size and its extremely small and declining range, this previously Secure species is now evaluated as Critically Endangered.

Country	Breeding pop. size (pairs)	Year(s)	Trend	Mag.%	References
Turkey	(1 – 10)	01	-	50–79	
Total (approx.)	1 – 10	**Overall trend** Large decline			
Breeding range	<100 km²	**Gen. length.** 5		**% Global pop.** <5	

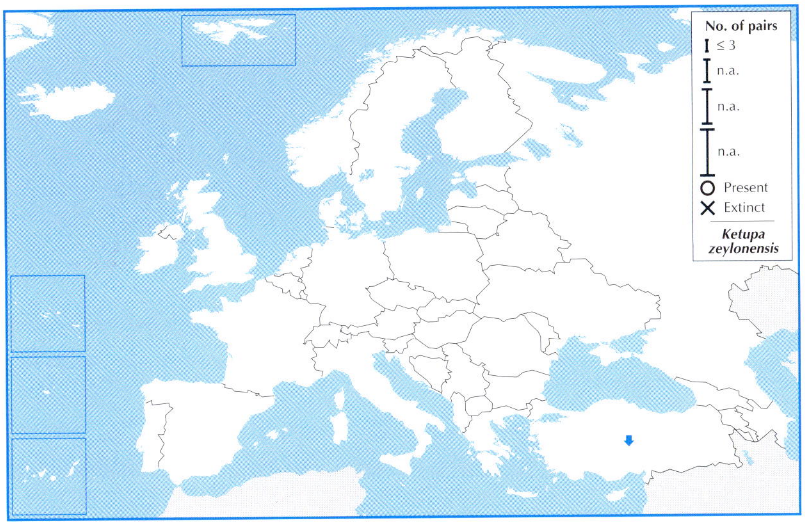

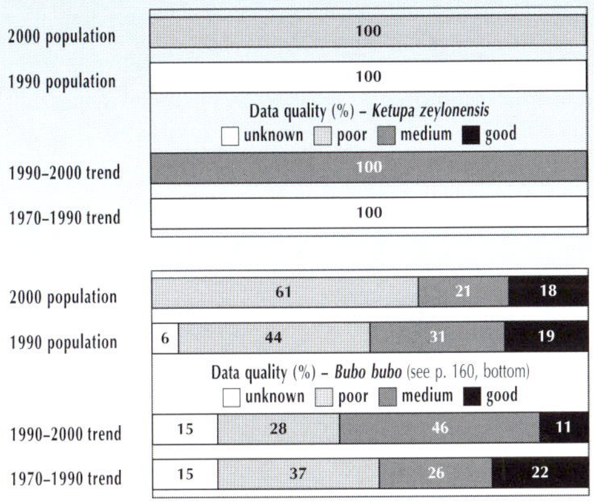

Nyctea scandiaca
SNOWY OWL

SPEC 3 (1994: 3) **Status** (Rare)
Criteria <10,000 pairs
European IUCN Red List Category —
Criteria —
Global IUCN Red List Category —
Criteria —

Nyctea scandiaca is an irregular breeder in Greenland, Iceland, Fennoscandia and arctic Russia, with Europe accounting for less than a quarter of its global breeding range. Its European breeding population is small (as few as 1,400 pairs), but fluctuated widely between 1970–1990. Although populations continued to fluctuate in most European countries during 1990–2000, the species remained broadly stable overall. Nevertheless, its population size still renders it susceptible to the risks affecting small populations, and consequently it is provisionally evaluated as Rare.

Country	Breeding pop. size (pairs)	Year(s)	Trend	Mag.%	References
Denmark					
Greenland	(50 – 1,000)	90–00	(F)	(–)	
Finland	0 – 20	98–02	**0**	**0–19**	
Iceland	**0 – 1**	90–00	**F**	**>80**	*21,23,24,25,32, 33,42,43,44,45, 46,47*
Norway	0 – 5	90–03	**F**	**>80**	*63*
Russia	(1,300 – 4,500)	90–00	(F)	(–)	*60,139*
Sweden	(0 – 2)	99–00	(–)	(>80)	
Total (approx.)	1,400 – 5,500	**Overall trend** Fluctuating			
Breeding range	>500,000 km²	**Gen. length.** 7		**% Global pop.** 5–24	

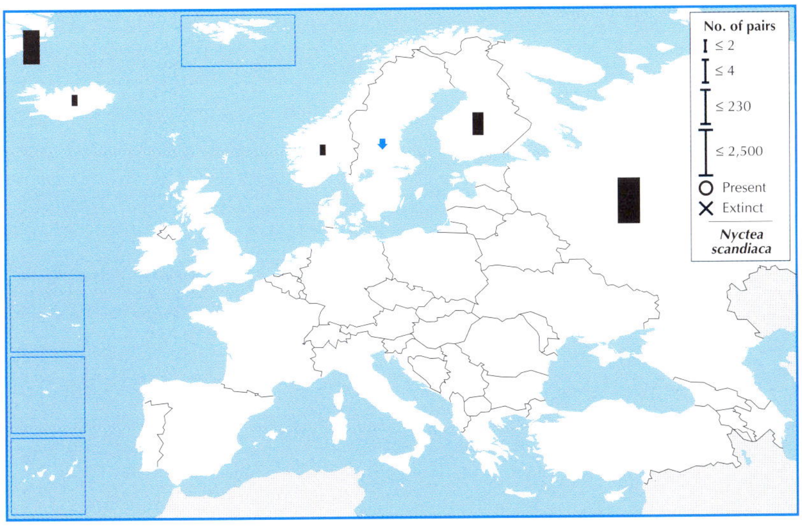

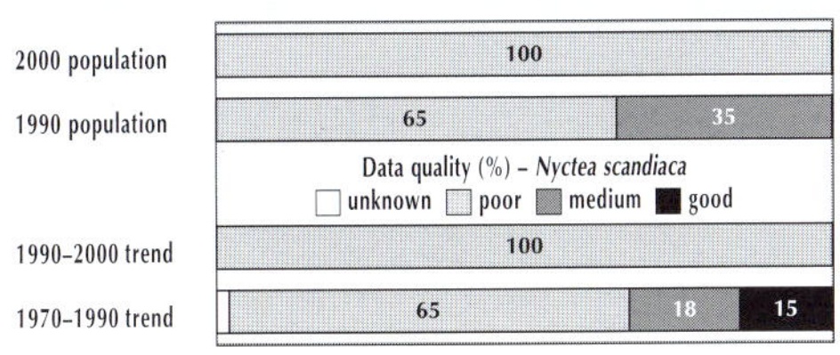

Surnia ulula
NORTHERN HAWK OWL

Non-SPEC (1994: —) **Status** (Secure)
Criteria —
European IUCN Red List Category —▼
Criteria —
Global IUCN Red List Category —
Criteria —

Surnia ulula is a widespread resident in northern Europe, which accounts for less than a quarter of its global range. Its European breeding population is small (as few as 9,200 pairs), but was broadly stable between 1970–1990. The extent of the fluctuation in Russia during 1990–2000 was unknown, but the species remained broadly stable overall. Although the size of the European population could render it susceptible to the risks affecting small populations, it is marginal to a much larger non-European population. Consequently, the species is provisionally evaluated as Secure.

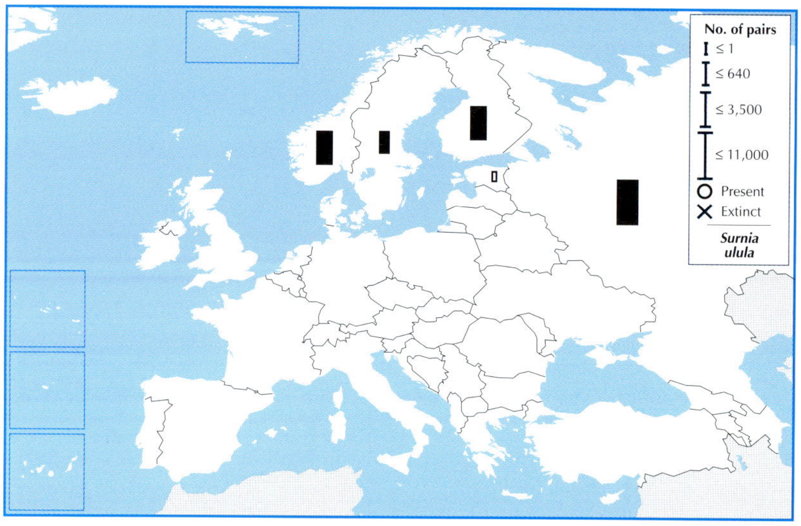

Country	Breeding pop. size (pairs)	Year(s)	Trend	Mag.%	References
Estonia	0 – 1	98	?	–	*1*
Finland	2,000 – 6,000	98–02	F	200–1,000	
Norway	(1,000 – 10,000)	90–02	F	30–49	*56*
Russia	(6,000 – 20,000)	90–00	(F)	(–)	*40,125,126,139*
Sweden	(200 – 2,000)	99–00	(F)	(>80)	
Total (approx.)	9,200 – 38,000	**Overall trend** Fluctuating			
Breeding range	>2,000,000 km²	**Gen. length.** <3.3		**% Global pop.** 5–24	

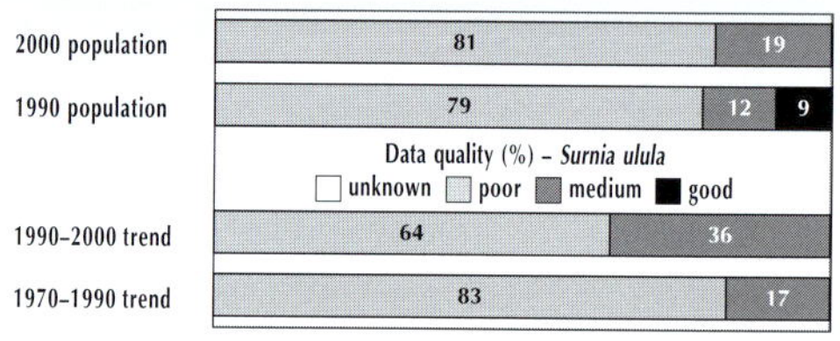

Glaucidium passerinum
EURASIAN PYGMY-OWL

Non-SPEC (1994: —) **Status** Secure
Criteria —
European IUCN Red List Category —
Criteria —
Global IUCN Red List Category —
Criteria —

Glaucidium passerinum is a widespread resident across much of northern and parts of central Europe, which accounts for less than half of its global range. Its European breeding population is relatively small (<110,000 pairs), but was stable between 1970–1990. Although the key population in Russia declined during 1990–2000, the species was stable or increased across the rest of its European range—including sizeable populations in Norway, Sweden and Finland—and probably remained stable overall. Consequently, the species is evaluated as Secure.

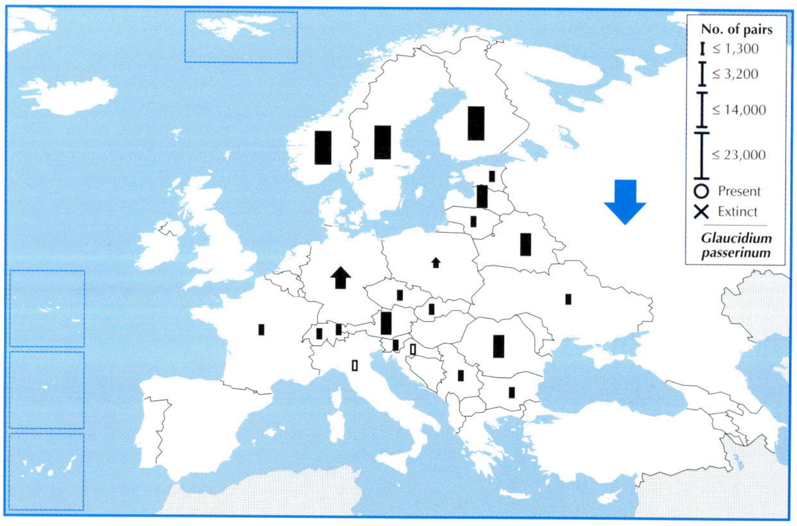

Country	Breeding pop. size (pairs)	Year(s)	Trend	Mag.%	References
Austria	(2,000 – 3,500)	98–02	(0)	(0–19)	
Belarus	1,200 – 2,000	97–02	0	0–19	
Bulgaria	20 – 100	96–02	(0)	(0–19)	
Croatia	200 – 350	02	?	–	*70*
Czech Rep.	**900 – 1,300**	00	**0**	**0–19**	
Estonia	200 – 500	98	0	0–19	*1*
Finland	8,000 – 13,000	98–01	0	0–19	
France	100 – 500	97–00	(0)	(0–19)	*4*
Germany	1,400 – 3,200	95–99	+	20–29	
Italy	(500 – 1,000)	03	?	–	
Latvia	1,000 – 2,500	90–00	0	0–19	*3*
Liechtenstein	(5 – 10)	98–00	(0)	(0–19)	
Lithuania	(50 – 150)	99–01	(0)	(0–19)	*20*
Norway	5,000 – 10,000	90–03	F	30–49	*56*
Poland	250 – 400	95–00	+	30–49	*2,81*
Romania	2,500 – 4,000	90–02	(0)	(0–19)	
Russia	(10,000 – 50,000)	90–00	(–)	(20–29)	*60,104,125,126, 159*
Serbia & MN	(10 – 15)	90–02	(0)	(10–19)	*1,117a,155*
Slovakia	1,000 – 1,500	90–99	0	0–19	
Slovenia	200 – 300	00	0	0–19	
Sweden	12,000 – 16,000	99–00	0	0–19	
Switzerland	**300 – 500**	93–96	(0)	(0–19)	
Ukraine	(150 – 350)	90–00	F	20–29	
Total (approx.)	47,000 – 110,000	**Overall trend** Stable			
Breeding range	>3,000,000 km²	**Gen. length.** <3.3		**% Global pop.** 25–49	

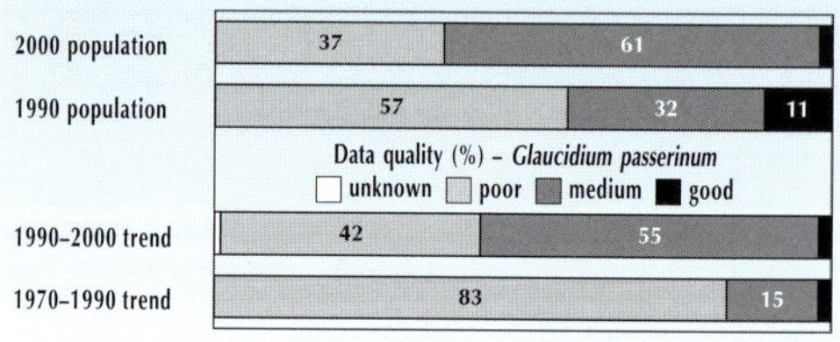

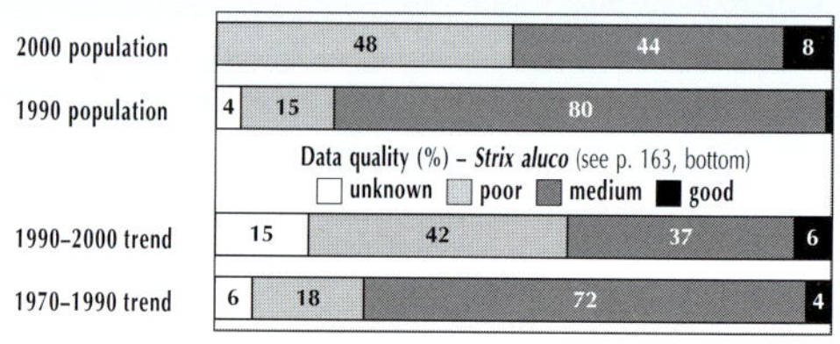

Athene noctua
LITTLE OWL

SPEC 3 (1994: 3) **Status** (Declining)
Criteria Moderate continuing decline
European IUCN Red List Category —
Criteria —
Global IUCN Red List Category —
Criteria —

Athene noctua is a widespread resident across much of Europe (except the north), which accounts for less than half of its global range. Its European breeding population is large (>560,000 pairs), but underwent a moderate decline between 1970–1990. Although the species was stable or increased across much of its European range during 1990–2000, several populations declined—notably the key one in Turkey—and the species probably underwent a moderate decline (>10%) overall. Consequently, it is provisionally evaluated as Declining.

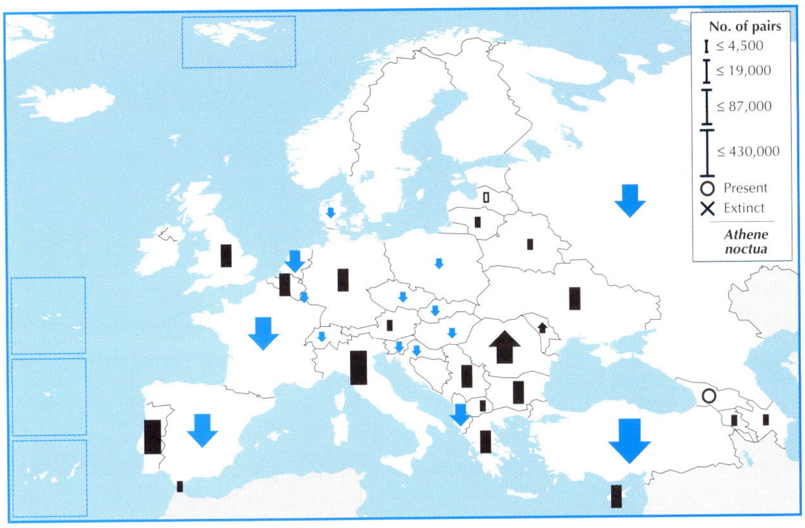

Country	Breeding pop. size (pairs)	Year(s)	Trend	Mag.%	References
Albania	4,000 – 8,000	02	(–)	(0–19)	
Armenia	800 – 1,500	97–02	0	0–19	
Austria	**70 – 100**	98–02	0	0–19	
Azerbaijan	(2,000 – 10,000)	96–00	(0)	(0–19)	
Belarus	400 – 1,000	97–02	0	0–19	
Belgium	12,500 – 14,000	01–02	0	0–19	*1*
Bulgaria	**5,000 – 8,000**	96–02	0	0–19	
Croatia	(500 – 1,000)	02	(–)	(50–79)	*70,16*
Cyprus	(5,000 – 15,000)	94–02	(0)	(0–19)	
Czech Rep.	**200 – 400**	00	**–**	**>80**	
Denmark	100 – 200	00	(–)	(10–19)	*3*
France	20,000 – 60,000	98–00	(–)	(0–19)	*4*
Georgia	Present	03	?	–	
Germany	5,800 – 6,100	95–99	0	0–19	
Greece	(5,000 – 15,000)	95–00	(0)	(0–19)	
Hungary	1,500 – 2,500	95–02	–	0–19	
Italy	(30,000 – 50,000)	03	(0)	(0–19)	
Latvia	(10 – 30)	90–00	?	–	*16*
Lithuania	(5 – 10)	99–01	(0)	(0–19)	*20*
Luxembourg	**40 – 80**	00–02	**–**	**50–79**	
Macedonia	1,400 – 2,400	90–00	(0)	(0–19)	
Moldova	**3,200 – 4,200**	90–00	**+**	**0–19**	
Netherlands	**5,500 – 6,500**	98–00	**–**	**29**	*1*
Poland	1,000 – 2,000	90–00	(–)	(20–29)	*27*
Portugal	(50,000 – 150,000)	02	(0)	(0–19)	
Romania	40,000 – 60,000	90–02	+	0–19	*7*
Russia	(10,000 – 100,000)	90–00	(–)	(20–29)	*8,73,141*
Serbia & MN	10,000 – 15,000	90–02	0	10–19	*1,190,29,172a, 155,78*
Slovakia	800 – 1,000	80–99	–	30–49	
Slovenia	150 – 200	99–00	–	30–49	
Spain	(20,000 – 100,000)	98–02	(–)	(0–19)	*10*
Switzerland	**60 – 70**	93–96	–	10–19	
Turkey	(300,000 – 600,000)	01	(–)	(0–19)	
Ukraine	15,000 – 22,000	90–00	0	0–19	
UK	**5,800 – 11,600**	00	**0**	**3**	*5,31*
Gibraltar	5 – 10	00	0	0–19	

Total (approx.) 560,000 – 1,300,000 **Overall trend** Moderate decline
Breeding range >5,000,000 km² **Gen. length.** <3.3 **% Global pop.** 25–49

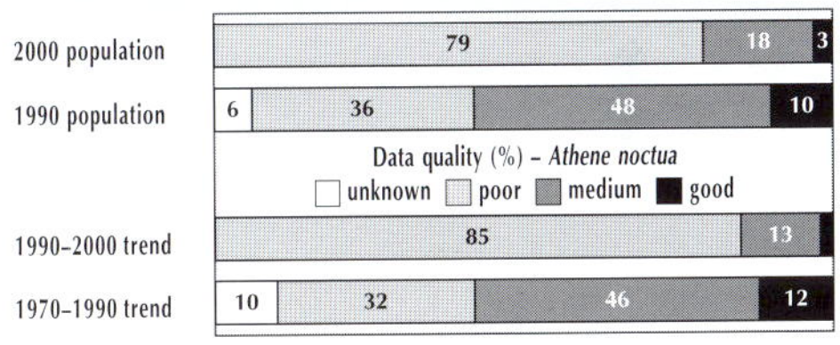

Strix aluco
TAWNY OWL

Non-SPEC[E] (1994: 4) **Status** Secure
Criteria —
European IUCN Red List Category —
Criteria —
Global IUCN Red List Category —
Criteria —

Strix aluco is a widespread resident across much of Europe, which constitutes >50% of its global range. Its European breeding population is large (>480,000 pairs), and was stable between 1970–1990. Although there were declines in countries such as Croatia and Sweden during 1990–2000, key populations in France, Germany, Poland and Romania were stable, and the species remained stable overall. Consequently, it is evaluated as Secure.

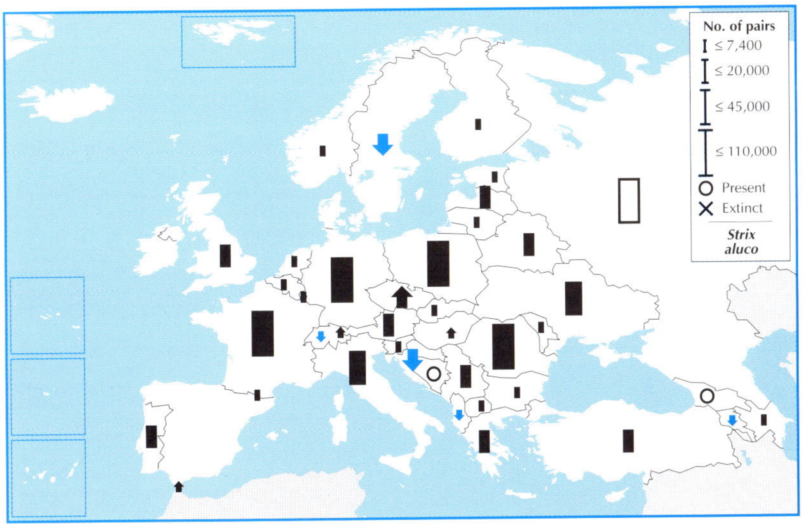

Country	Breeding pop. size (pairs)	Year(s)	Trend	Mag.%	References
Albania	1,000 – 3,000	02	(–)	(0–19)	
Andorra	50 – 80	99–01	0	0–19	*1,3*
Armenia	80 – 300	97–02	(–)	(–)	
Austria	(9,000 – 16,000)	98–02	(0)	(0–19)	
Azerbaijan	(1,000 – 10,000)	96–00	(0)	(0–19)	
Belarus	8,500 – 12,000	97–02	0	0–19	
Belgium	6,000 – 9,000	01–02	0	0–19	*1*
Bosnia & HG	Present	90–03	?	–	
Bulgaria	**4,000 – 8,000**	96–02	0	0–19	
Croatia	(10,000 – 15,000)	02	(–)	(50–79)	*16*
Czech Rep.	**10,000 – 18,000**	00	**+**	**>80**	
Denmark	(3,000 – 4,000)	00	(–)	(0–19)	*20*
Estonia	1,100 – 1,600	98	0	0–19	*1*
Finland	1,500 – 2,500	98–01	0	0–19	
France	(60,000 – 200,000)	98–00	(0)	(0–19)	*4*
Georgia	Present	03	?	–	
Germany	48,000 – 73,000	95–99	0	0–19	
Greece	(10,000 – 20,000)	95–00	(0)	(0–19)	
Hungary	5,000 – 8,000	95–02	+	20–29	
Italy	(20,000 – 40,000)	03	(0)	(0–19)	
Latvia	15,000 – 20,000	90–00	0	0–19	*17*
Liechtenstein	15 – 25	98–00	+	0–9	
Lithuania	2,000 – 4,000	99–01	0	0–19	*20*
Luxembourg	400 – 500	02	**0**	**0–19**	
Macedonia	(2,000 – 5,000)	90–00	(0)	(0–19)	
Moldova	**220 – 380**	90–00	**0**	**0–19**	
Netherlands	**4,500 – 5,500**	98–00	**0**	**10**	*1*
Norway	**2,500 – 6,000**	02–03	0	0–19	*57*
Poland	65,000 – 75,000	90–00	0	0–19	*1*
Portugal	(10,000 – 40,000)	02	(0)	(0–19)	
Romania	50,000 – 80,000	90–02	(0)	(0–19)	
Russia	(35,000 – 100,000)	90–00	?	–	*8,39,104,114*
Serbia & MN	10,000 – 14,000	90–02	0	0–19	*1,53,236,29, 172a,155,227*
Slovakia	2,500 – 3,000	90–99	0	0–19	
Slovenia	1,000 – 5,000	94	(0)	(0–19)	
Spain	(20,000 – 100,000)	98–02	?	–	*10*
Sweden	10,000 – 20,000	99–00	–	0–19	
Switzerland	**5,000 – 6,000**	93–96	–	0–9	
Turkey	(10,000 – 20,000)	01	(0)	(0–19)	
Ukraine	22,000 – 34,000	90–00	0	0–19	
UK	**19,400 – 19,400**	00	**0**	**3**	*5,31*
Gibraltar	0 – 2	00	+	N	

Total (approx.) 480,000 – 1,000,000 **Overall trend** Stable
Breeding range >6,000,000 km² **Gen. length.** 4 **% Global pop.** 50–74
(See p. 162, bottom, for data quality graph)

Strix uralensis
URAL OWL

Non-SPEC (1994: —) **Status** (Secure)
Criteria —
European IUCN Red List Category —
Criteria —
Global IUCN Red List Category —
Criteria —

Strix uralensis is a widespread resident across much of northern and parts of central Europe, which accounts for less than a quarter of its global range. Its European breeding population is relatively small (<140,000 pairs), but was stable between 1970–1990. The species remained stable overall during 1990–2000, with no significant change in the stronghold population in Russia, and stable or increasing trends elsewhere within its European range. Consequently, it is provisionally evaluated as Secure.

Country	Breeding pop. size (pairs)	Year(s)	Trend	Mag.%	References
Austria	(2 – 4)	98–02	+	N	
Belarus	1,200 – 1,800	97–02	+	0–9	
Bulgaria	**15 – 60**	96–02	**0**	**0–9**	
Croatia	(250 – 700)	02	(+)	(>80)	*16*
Czech Rep.	**5 – 10**	00	+	**>80**	
Estonia	1,600 – 2,300	98	0	0–19	*1*
Finland	2,500 – 4,000	98–00	0	0–19	
Germany	**5 – 5**	95–99	**0**	**0–19**	
Hungary	50 – 200	95–02	+	>80	
Italy	**1 – 5**	03	+	0–19	
Latvia	1,500 – 2,500	90–00	+	50–79	
Lithuania	(15 – 30)	99–01	F	30–49	*20*
Macedonia	(0 – 50)	90–00	?	–	*7*
Norway	**1 – 12**	02	**0**	**0–19**	*43*
Poland	**450 – 700**	95–00	+	0–19	*80*
Romania	12,000 – 20,000	90–02	+	0–19	
Russia	(30,000 – 100,000)	90–00	(0)	(0–19)	*104,125,126,159*
Serbia & MN	90 – 140	90–02	(0)	(0–19)	*1,172,143,117a, 155,67a,227*
Slovakia	700 – 1,000	80–99	+	20–29	
Slovenia	**500 – 600**	00	0	0–9	
Sweden	2,000 – 2,700	99–00	+	0–19	
Ukraine	(400 – 900)	90–00	(0)	(0–19)	
Total (approx.)	53,000 – 140,000	**Overall trend** Stable			
Breeding range	>2,000,000 km²	**Gen. length.** 9		**% Global pop.** 5–24	

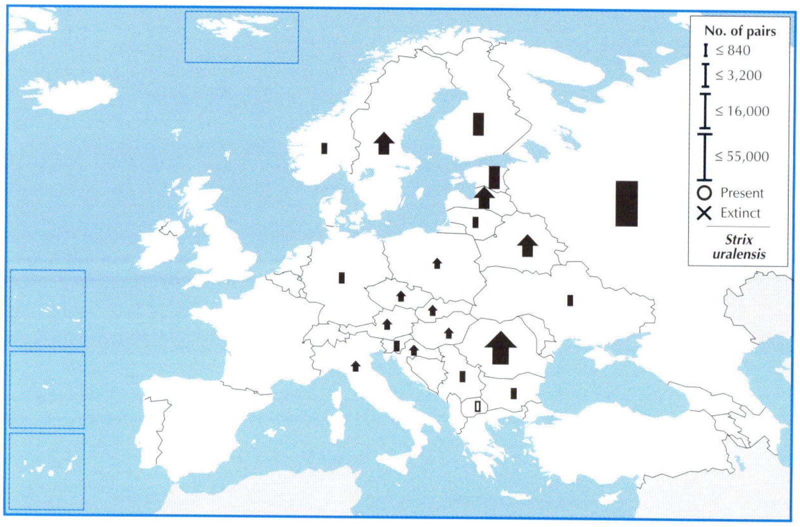

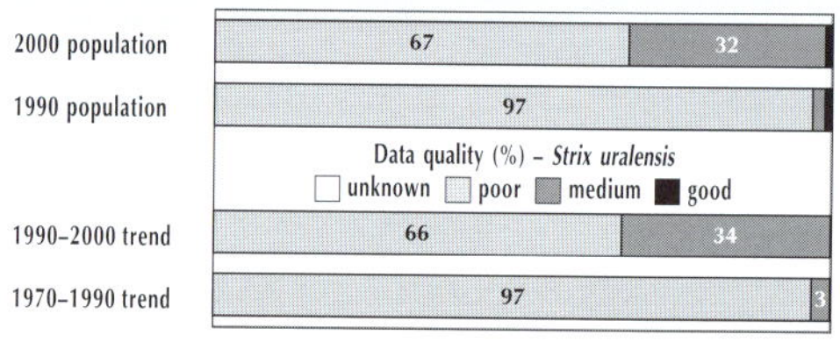

Strix nebulosa
GREAT GREY OWL

Non-SPEC (1994: —) **Status** (Secure)
Criteria —
European IUCN Red List Category —▼
Criteria —
Global IUCN Red List Category —
Criteria —

Strix nebulosa is resident in the boreal zone of Europe, which accounts for less than a quarter of its global range. Its European breeding population is small (as few as 2,100 pairs), but was broadly stable between 1970–1990. No trend was available for the key population in Russia during 1990–2000, but the species probably remained broadly stable overall. Although the size of the European population could render it susceptible to the risks affecting small populations, it is marginal to a much larger non-European population. Consequently, the species is provisionally evaluated as Secure.

Country	Breeding pop. size (pairs)	Year(s)	Trend	Mag.%	References
Belarus	50 – 100	97–02	0	0–19	
Finland	300 – 1,500	98–01	F	200–1,000	
Norway	**0 – 10**	90–03	**F**	**>80**	
Russia	(1,500 – 4,500)	90–00	?	–	*2,60,103,125, 126,128*
Sweden	**250 – 500**	99–00	0	0–19	
Ukraine	**15 – 100**	90–00	+	**50–79**	
Total (approx.)	2,100 – 6,700	**Overall trend** Unknown			
Breeding range	>2,000,000 km²	**Gen. length.** 5		**% Global pop.** 5–24	

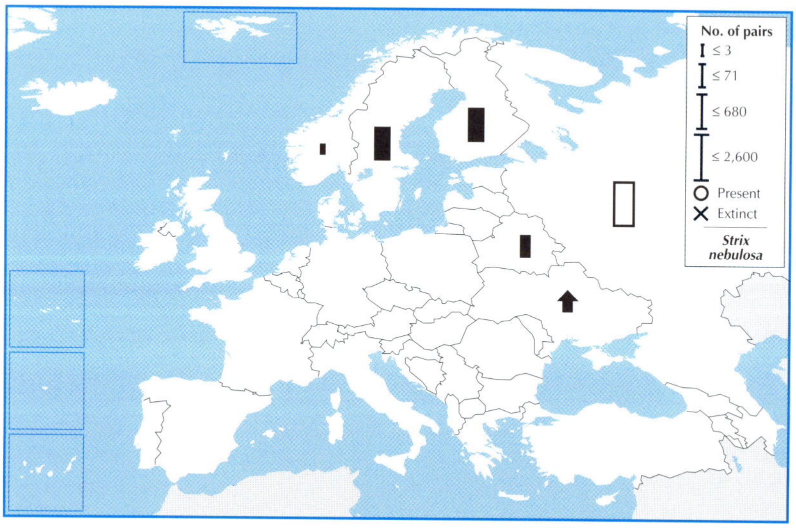

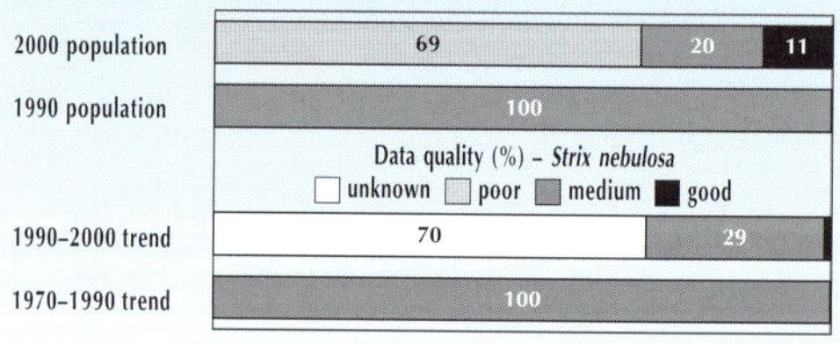

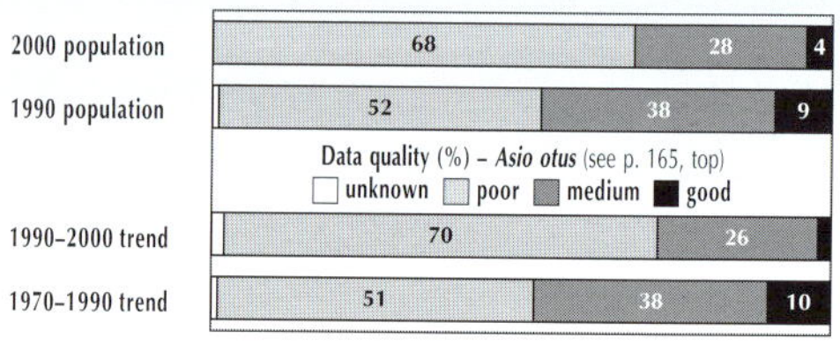

Asio otus
LONG-EARED OWL

Non-SPEC (1994: —) **Status** (Secure)
Criteria —
European IUCN Red List Category —
Criteria —
Global IUCN Red List Category —
Criteria —

Asio otus is a widespread breeder across much of Europe, which accounts for less than half of its global breeding range. Its European breeding population is large (>380,000 pairs), and was stable between 1970–1990. Although there were declines in a few countries during 1990–2000, most populations—including key ones in Russia and Romania—were stable, and the species remained stable overall. Consequently, it is provisionally evaluated as Secure.

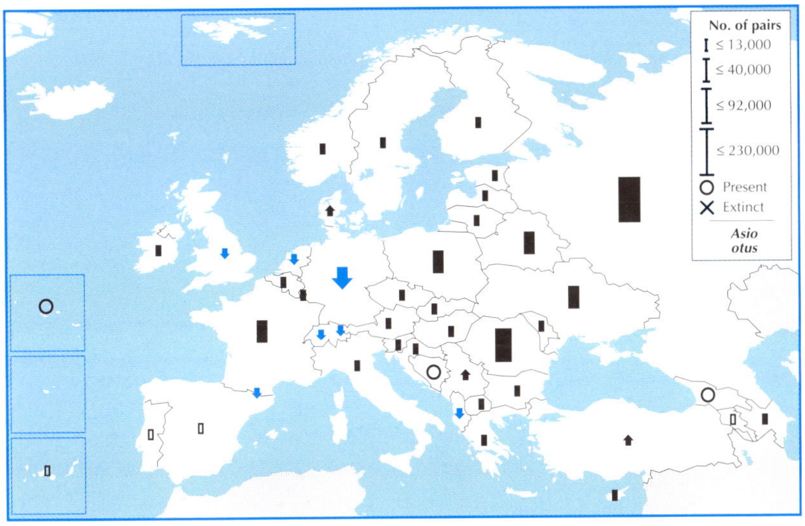

Country	Breeding pop. size (pairs)	Year(s)	Trend	Mag.%	References
Albania	(500 – 1,000)	02	(–)	(0–19)	
Andorra	3 – 10	99–01	–	0–19	*1,3*
Armenia	150 – 330	97–02	?	–	
Austria	(2,000 – 5,000)	98–02	(0)	(0–19)	
Azerbaijan	(1,000 – 10,000)	96–00	(0)	(0–19)	
Belarus	12,000 – 20,000	97–02	0	0–19	
Belgium	3,500 – 6,500	01–02	(0)	(0–19)	*1*
Bosnia & HG	Present	90–03	?	–	
Bulgaria	**3,000 – 5,000**	96–02	0	0–19	
Croatia	(3,000 – 5,000)	02	(0)	(0–19)	*16,54*
Cyprus	(20 – 40)	94–02	(0)	(0–19)	
Czech Rep.	**4,000 – 8,000**	00	**0**	**0–19**	
Denmark	(2,000 – 3,000)	00	(+)	(0–19)	*3*
Estonia	800 – 2,000	98	F	20–29	*1*
Finland	2,000 – 10,000	98–01	F	200–1,000	
France	(10,000 – 30,000)	00	(0)	(0–19)	*1*
Georgia	Present	03	?	–	
Germany	25,000 – 40,000	95–99	–	0–19	
Greece	(2,000 – 5,000)	95–00	(0)	(0–19)	
Hungary	6,500 – 12,000	95–02	0	0–19	
Rep. Ireland	1,000 – 2500	88–91	(0)	(0–19)	
Italy	(5,000 – 10,000)	03	(0)	(0–19)	
Latvia	2,000 – 4,000	90–00	(0)	(0–19)	*3*
Liechtenstein	3 – 6	98–00	–	20–29	
Lithuania	3,000 – 7,000	99–01	(0)	(0–19)	*20*
Luxembourg	**400 – 500**	02	**0**	**0–19**	
Macedonia	(200 – 1,000)	90–00	(0)	(0–19)	
Moldova	**1,300 – 2,600**	90–00	**0**	**0–19**	
Netherlands	**5,000 – 6,000**	98–00	**–**	**32**	*1*
Norway	(1,000 – 10,000)	90–03	F	30–49	*56*
Poland	8,000 – 25,000	90–00	F	20–29	*1*
Portugal	(100 – 1,000)	02	?	–	
Azores	Present	02	?	–	
Romania	(70,000 – 120,000)	90–02	(0)	(0–19)	
Russia	(150,000 – 350,000)	90–00	(0)	(0–19)	*104,150*
Serbia & MN	10,000 – 15,000	90–02	+	10–19	*1,165,78,93,25, 197,29,172a,155*
Slovakia	2,500 – 4,000	90–99	0	0–19	
Slovenia	1,800 – 2,200	94	(0)	(0–19)	
Spain	(2,500 – 10,000)	98–02	?	–	*10*
Canary Is.	(2,500 – 10,000)	97–03	?	–	*28,25*
Sweden	2,000 – 10,000	99–00	F	30–49	
Switzerland	**2,500 – 3,000**	93–96	–	0–19	
Turkey	(2,000 – 6,000)	01	(+)	(0–19)	
Ukraine	33,000 – 47,000	90–00	0	0–9	
UK	1,400 – 4,800	88–91	(–)	(20–29)	*8*
Total (approx.)	380,000 – 810,000	**Overall trend** Stable			
Breeding range	>6,000,000 km²	**Gen. length.** <3.3		**% Global pop.** 25–49	

(See p. 164, bottom, for data quality graph)

Asio flammeus
SHORT-EARED OWL

SPEC 3 (1994: 3) **Status** (Depleted)
Criteria Large historical decline
European IUCN Red List Category —
Criteria —
Global IUCN Red List Category —
Criteria —

Asio flammeus is a widespread but patchily distributed breeder across much of Europe, which accounts for less than a quarter of its global breeding range. Its European breeding population is relatively large (>58,000 pairs), but underwent a large decline between 1970–1990. Although declines continued in a few countries during 1990–2000, they abated across most of Europe, and the species was broadly stable overall. Nevertheless, its total population size remains below the level that preceded its decline, and consequently it is provisionally evaluated as Depleted.

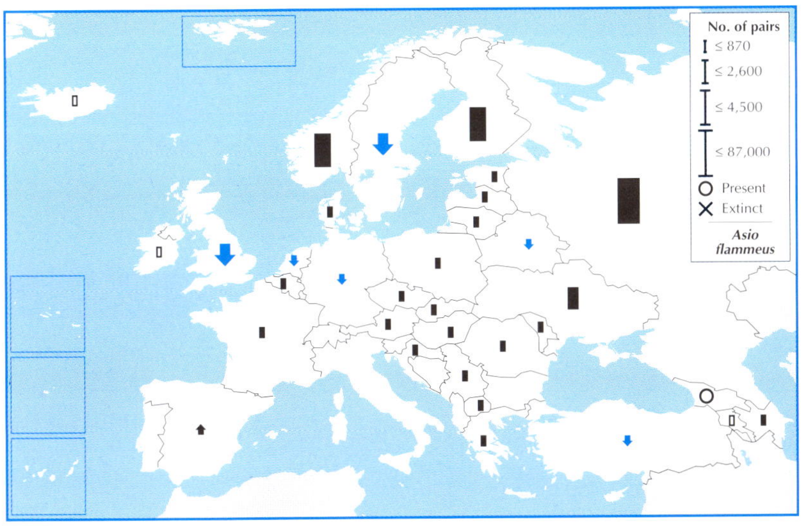

Country	Breeding pop. size (pairs)	Year(s)	Trend	Mag.%	References
Armenia	15 – 30	95–02	?	–	
Austria	2 – 15	98–02	**F**	**>80**	
Azerbaijan	(5 – 50)	96–00	(0)	(0–19)	
Belarus	500 – 1,500	97–02	–	0–9	
Belgium	0 – 2	95–03	0	0–19	*1*
Croatia	(10 – 20)	02	(0)	(0–19)	*44,57*
Czech Rep.	**0 – 10**	00	**0**	**0–19**	
Denmark	4 – 15	98–01	F	>80	*4,5,6,7*
Estonia	50 – 150	98	0	0–19	*1*
Finland	2,000 – 10,000	98–01	F	200–1,000	
France	50 – 250	98–02	F	>80	*4*
Georgia	Present	03	?	–	
Germany	**30 – 140**	95–99	–	30–49	
Greece	(0 – 10)	95–00	(0)	(0–19)	
Hungary	50 – 200	95–02	(F)	(–)	
Iceland	(100 – 200)	00	?	–	*21*
Rep. Ireland	2 – 2	88–91	?	–	
Latvia	(0 – 10)	90–00	(0)	(0–19)	*23,3*
Lithuania	50 – 200	99–01	F	>80	*20*
Macedonia	(0 – 3)	90–00	(0)	(0–19)	
Moldova	**8 – 15**	90–00	**0**	**0–19**	
Netherlands	**35 – 45**	98–00	**–**	**61**	*1*
Norway	(1,000 – 10,000)	90–03	F	30–49	*56*
Poland	20 – 100	90–00	F	30–49	*2,83*
Romania	(100 – 400)	90–02	(F)	(50–79)	
Russia	(50,000 – 150,000)	90–00	(F)	(–)	*8,104,114,159*
Serbia & MN	20 – 30	90–02	F	10–29	*1,124,200,29,155*
Slovakia	0 – 50	80–99	F	20–29	
Spain	(360 – 360)	94	(+)	(0–19)	*10,16*
Sweden	1,500 – 4,200	99–00	–	0–19	
Turkey	(20 – 80)	01	**–**	**30–49**	
Ukraine	850 – 1,700	90–00	F	20–29	
UK	1,000 – 3,500	88–91	(–)	(0–19)	*8*
Total (approx.)	58,000 – 180,000	**Overall trend** Fluctuating			
Breeding range	>4,000,000 km²	**Gen. length.** <3.3		**% Global pop.** 5–24	

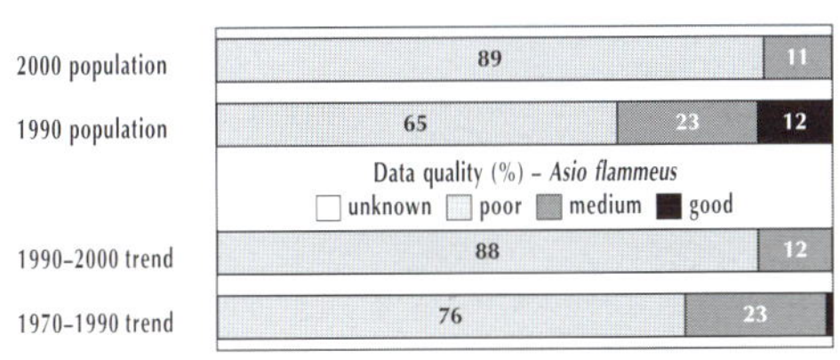

Aegolius funereus
BOREAL OWL

Non-SPEC (1994: —) **Status** (Secure)
Criteria —
European IUCN Red List Category —
Criteria —
Global IUCN Red List Category —
Criteria —

Aegolius funereus is a widespread resident in northern Europe, and occurs more patchily farther south, with Europe accounting for less than a quarter of its global range. Its European breeding population is large (>110,000 pairs), and was stable between 1970–1990. Although there were declines in a small number of countries during 1990–2000, key populations in Fennoscandia, Romania and the stronghold Russia were stable, and trends were stable or increasing in most of the rest of its European range. Consequently, the species is provisionally evaluated as Secure.

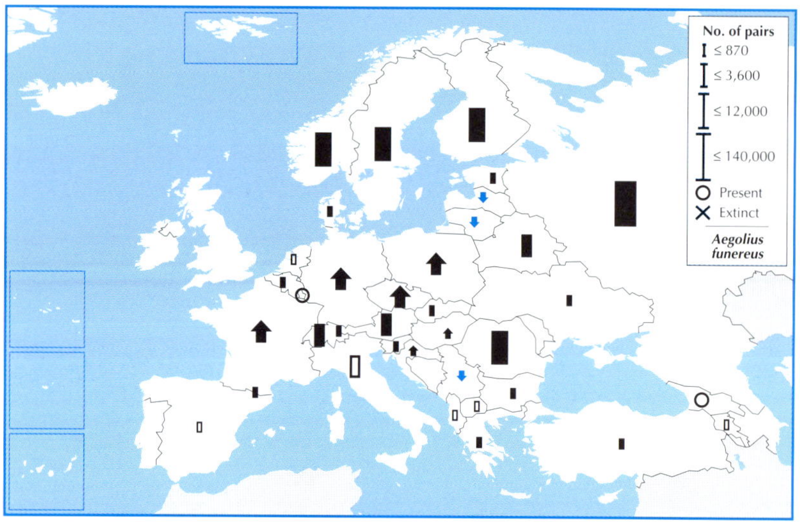

Country	Breeding pop. size (pairs)	Year(s)	Trend	Mag.%	References
Albania	(0 – 30)	96–02	?	–	
Andorra	**25 – 35**	98–01	0	0–19	*1,3*
Armenia	5 – 15	94–02	?	–	
Austria	(1,100 – 2,200)	98–02	(0)	(0–19)	
Belarus	2,500 – 5,000	97–02	0	0–19	
Belgium	**30 – 120**	95–02	**0**	**0–19**	*1*
Bulgaria	**600 – 900**	96–02	0	0–9	
Croatia	(250 – 500)	02	(+)	(>80)	*70,16*
Czech Rep.	**1,500 – 2,000**	00	+	**>80**	
Denmark	0 – 4	98–01	0	0–19	*4,5,6,7*
Estonia	200 – 500	98	0	0–19	*1*
Finland	7,000 – 20,000	98–01	F	200–250	
France	2,000 – 4,000	98–00	+	20–49	*4*
Georgia	Present	03	?	–	
Germany	1,600 – 2,900	95–99	+	0–19	
Greece	(10 – 100)	95–00	(0)	(0–19)	
Hungary	0 – 2	97–02	+	N	*17*
Italy	1,500 – 3,500	03	?	–	
Latvia	500 – 1,500	90–00	(–)	(20–29)	*23,18*
Liechtenstein	(10 – 20)	99–00	(0)	(0–19)	
Lithuania	300 – 500	99–01	(–)	(20–29)	*20*
Luxembourg	Present	02	?	–	
Macedonia	(0 – 50)	90–00	?	–	*7*
Netherlands	**0 – 1**	98–00	?	–	*1*
Norway	2,000 – 20,000	90–03	F	30–49	*56*
Poland	1,000 – 2,000	97–00	+	30–49	*82*
Romania	(6,000 – 10,000)	90–02	(0)	(0–19)	
Russia	(70,000 – 250,000)	90–00	(0)	(0–19)	*104,125,126,139, 159*
Serbia & MN	110 – 220	90–02	–	10–19	*1,155,117a,225, 227,143*
Slovakia	400 – 600	90–99	0	0–19	
Slovenia	300 – 400	00	0	0–19	
Spain	50 – 250	02	?	–	*10*
Sweden	(5,000 – 20,000)	99–00	F	20–29	
Switzerland	**1,000 – 1,500**	93–96	F	20–29	
Turkey	(50 – 500)	01	(0)	(0–19)	
Ukraine	150 – 350	90–00	F	20–29	
Total (approx.)	110,000 – 350,000	**Overall trend** Stable			
Breeding range	>3,000,000 km²	**Gen. length.** <3.3		**% Global pop.** 5–24	

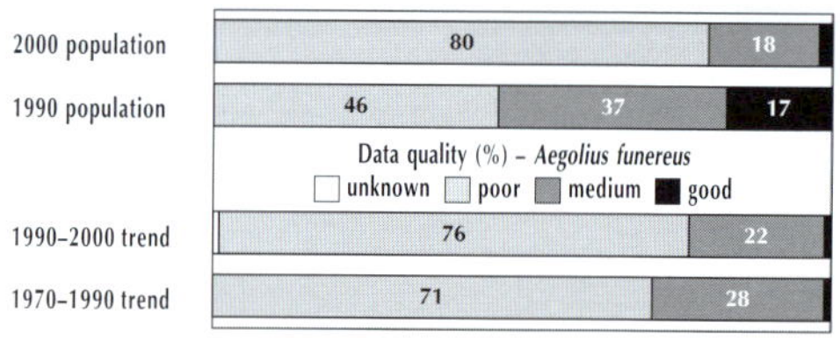

Caprimulgus europaeus
EURASIAN NIGHTJAR

SPEC 2 (1994: 2) **Status** (Depleted)
Criteria Moderate historical decline
European IUCN Red List Category —
Criteria —
Global IUCN Red List Category —
Criteria —

Caprimulgus europaeus is a widespread summer visitor to much of Europe, which constitutes >50% of its global breeding range. Its European breeding population is large (>470,000 pairs), but underwent a moderate decline between 1970–1990. Although many of these declines abated during 1990–2000, several populations—notably the sizeable one in Turkey—suffered declines, and the species declined slightly overall. Its population has clearly not yet recovered to the level that preceded the initial decline, and consequently it is provisionally evaluated as Depleted.

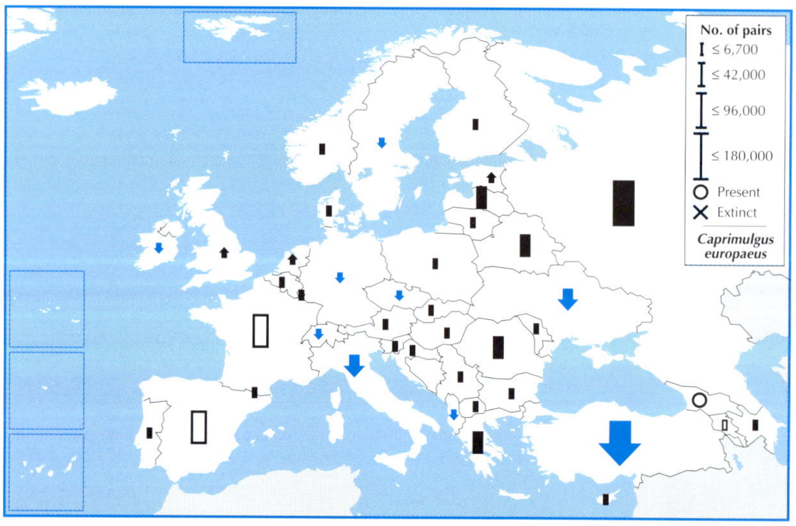

Country	Breeding pop. size (pairs)	Year(s)	Trend	Mag.%	References
Albania	3,000 – 8,000	96–02	(–)	(0–19)	
Andorra	20 – 50	99–01	(0)	(0–19)	*1,3*
Armenia	830 – 3,000	97–02	?	–	
Austria	300 – 500	98–02	0	0–19	
Azerbaijan	(1,000 – 10,000)	96–00	(0)	(0–19)	
Belarus	35,000 – 50,000	97–02	0	0–19	
Belgium	381 – 510	95–02	**F**	**30–49**	*1*
Bulgaria	2,000 – 8,000	96–02	0	0–19	
Croatia	(1,000 – 5,000)	02	(0)	(0–19)	*16*
Cyprus	(500 – 1,500)	94–02	(0)	(0–19)	
Czech Rep.	**400 – 700**	00	**–**	**30–49**	
Denmark	500 – 600	92–95	(0)	(0–19)	*13*
Estonia	3,000 – 5,000	98	+	20–29	*1*
Finland	3,000 – 5,000	98–02	0	0–19	
France	(40,000 – 160,000)	00–02	?	–	*4*
Georgia	Present	03	?	–	
Germany	3,100 – 4,400	95–99	–	20–29	
Greece	(10,000 – 30,000)	95–00	(0)	(0–19)	
Hungary	3,500 – 6,000	98–01	0	0–19	
Rep. Ireland	(5 – 30)	88–91	(–)	(30–49)	
Italy	(8,000 – 20,000)	03	(–)	(0–19)	
Latvia	15,000 – 23,000	90–00	0	0–19	*18*
Lithuania	(4,000 – 6,000)	99–01	(0)	(0–19)	*20*
Luxembourg	1 – 5	00–02	0	0–19	
Macedonia	(2,000 – 5,000)	90–00	(0)	(0–19)	
Moldova	280 – 350	90–00	0	0–19	
Netherlands	**950 – 1,150**	98–00	+	**34**	*1*
Norway	140 – 400	90–02	**0**	**0–19**	
Poland	4,000 – 6,000	90–00	(0)	(0–19)	*84*
Portugal	(1,000 – 10,000)	02	(0)	(0–19)	
Romania	(12,000 – 15,000)	00–02	(0)	(0–19)	
Russia	100,000 – 300,000	90–00	(0)	(0–19)	*122*
Serbia & MN	5,500 – 8,000	90–02	0	0–19	*1,62,172a,155, 67a,227*
Slovakia	1,000 – 2,000	90–99	0	0–19	
Slovenia	1,000 – 1,500	99–00	0	0–19	
Spain	(82,000 – 112,000)	92	?	–	*13,12,10*
Sweden	1500 – 2500	99–00	–	30–49	
Switzerland	**50 – 70**	93–96	–	10–19	
Turkey	(100,000 – 200,000)	01	(–)	(0–19)	
Ukraine	16,000 – 23,000	90–00	–	20–29	
UK	**3,400 – 3,400**	92	+	**55**	
Total (approx.)	470,000 – 1,000,000	**Overall trend** Small decline			
Breeding range	>6,000,000 km²	**Gen. length.** 4		**% Global pop.** 50–74	

(See p. 167, top, for data quality graph)

Caprimulgus ruficollis
RED-NECKED NIGHTJAR

Non-SPEC (1994: —) **Status** (Secure)
Criteria —
European IUCN Red List Category —
Criteria —
Global IUCN Red List Category —
Criteria —

Country	Breeding pop. size (pairs)	Year(s)	Trend	Mag.%	References
Portugal	(1,000 – 10,000)	02	(0)	(0–19)	
Spain	(20,000 – 100,000)	92	?	–	*13,12,10*
Total (approx.)	21,000 – 110,000	**Overall trend** Unknown			
Breeding range	>250,000 km²	**Gen. length.** 4		**% Global pop.** 25–49	

Caprimulgus ruficollis is a summer visitor to Iberia, with Europe accounting for less than half of its global breeding range. Its European breeding population is relatively small (<110,000 pairs), but was stable between 1970–1990. Although the trend of the stronghold population in Spain during 1990–2000 was unknown, the species remained stable in Portugal, and is provisionally evaluated as Secure.

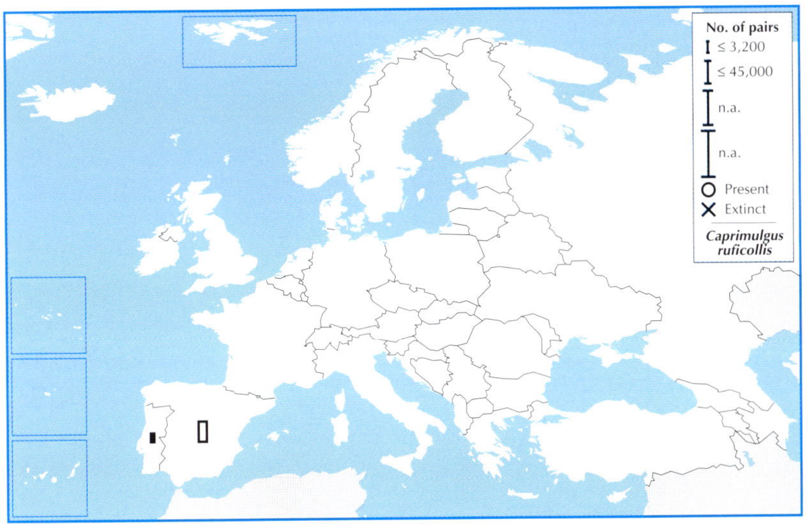

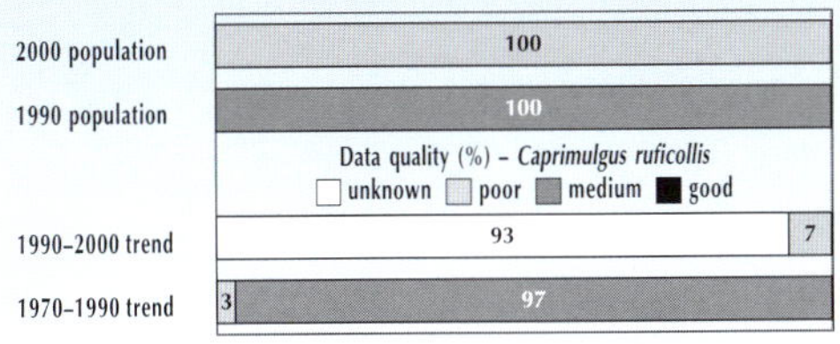

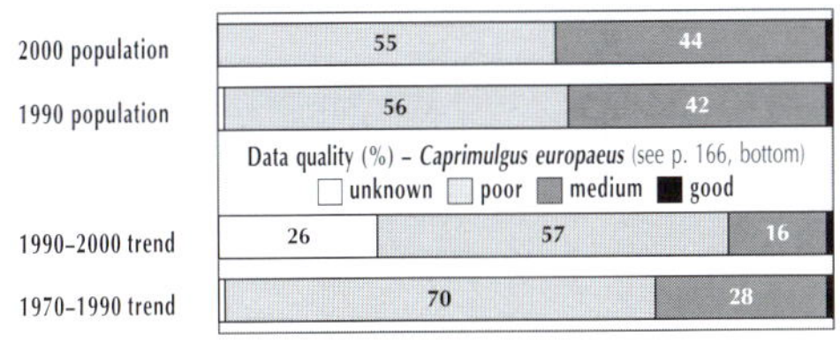

Tachymarptis melba
ALPINE SWIFT

Non-SPEC (1994: —) **Status** Secure
Criteria —
European IUCN Red List Category —
Criteria —
Global IUCN Red List Category —
Criteria —

Country	Breeding pop. size (pairs)	Year(s)	Trend	Mag.%	References
Albania	2,000 – 5,000	02	(0)	(0–19)	
Armenia	300 – 830	00–02	0	0–19	
Austria	(150 – 350)	98–02	(0)	(0–19)	
Azerbaijan	(3,000 – 10,000)	96–00	(0)	(0–19)	
Bosnia & HG	Present	90–03	?	–	
Bulgaria	2,500 – 7,000	96–02	0	0–19	
Croatia	(10,000 – 50,000)	02	(0)	(0–19)	*16*
Cyprus	(200 – 600)	94–02	(0)	(0–19)	
France	4,000 – 8,000	00	+	0–19	*1*
Georgia	Present	03	?	–	
Germany	**75 – 85**	95–99	+	**20–29**	
Greece	(1,000 – 5,000)	95–00	(0)	(0–19)	
Italy	(5,000 – 10,000)	03	(0)	(0–19)	
Macedonia	(1,000 – 2,500)	90–00	(0)	(0–19)	
Portugal	(100 – 1,000)	02	(–)	(–)	
Romania	700 – 1,100	99–02	+	0–19	
Russia	(3,000 – 10,000)	90–00	?	–	*61*
Serbia & MN	**360 – 600**	90–02	0	10–19	*1,221,238,155, 227,34*
Slovenia	**250 – 350**	99	(0)	(0–19)	
Spain	(2,500 – 10,000)	98–02	?	–	*110*
Switzerland	**1,200 – 1,300**	93–96	+	10–19	
Turkey	(100,000 – 200,000)	01	(+)	(0–19)	
Ukraine	(500 – 3,000)	90–00	(0)	(0–19)	
UK					
Gibraltar	5 – 20	00	(0)	(0–19)	
Total (approx.)	140,000 – 330,000	**Overall trend** Small increase			
Breeding range	>2,000,000 km²	**Gen. length.** 7		**% Global pop.** 25–49	

Tachymarptis melba is a widespread summer visitor to southern Europe, which accounts for less than half of its global breeding range. Its European breeding population is large (>140,000 pairs), and was stable between 1970–1990. The stronghold population in Turkey underwent a slight increase during 1990–2000, and trends were stable or increasing across most of the rest of its European range. Consequently, the species is evaluated as Secure.

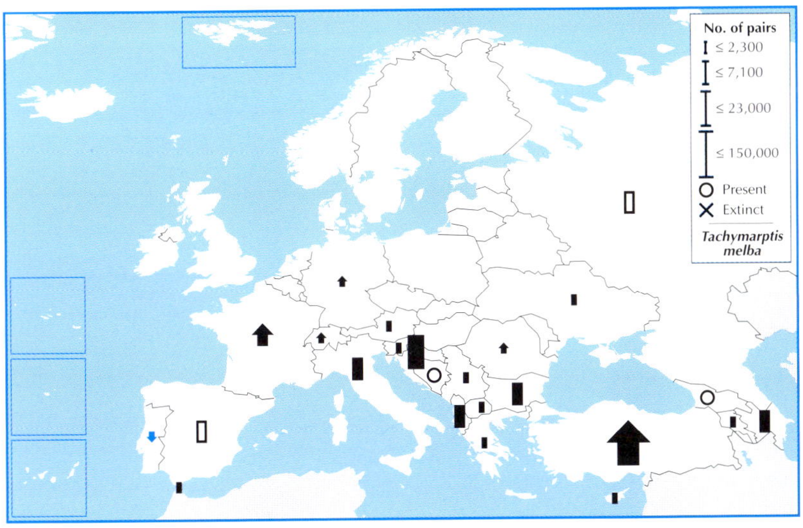

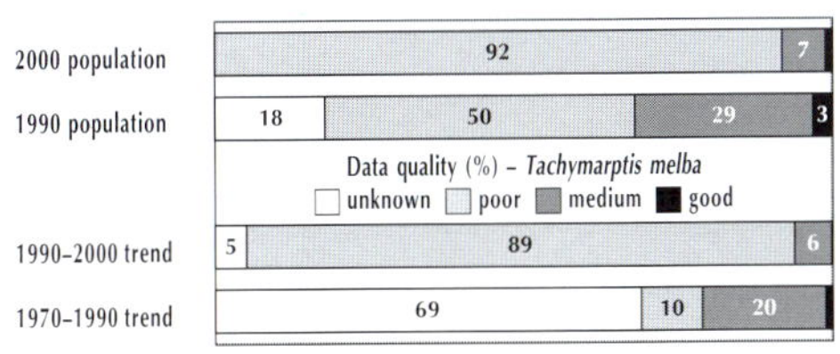

Apus unicolor
PLAIN SWIFT

SPEC 2 (1994: 4) **Status** (Rare)
Criteria <10,000 pairs
European IUCN Red List Category —
Criteria —
Global IUCN Red List Category —
Criteria —

Country	Breeding pop. size (pairs)	Year(s)	Trend	Mag.%	References
Portugal					
Madeira	Present	02	?	–	
Spain					
Canary Is.	(2,500 – 10,000)	97–03	?	–	25
Total (approx.)	2,500 – 10,000	**Overall trend** Unknown			
Breeding range	>20,000 km²	**Gen. length.** 7		**% Global pop.** 100	

Apus unicolor breeds only in Europe, with its entire global breeding range confined to Madeira and the Canary Islands. The current size of the population on Madeira is unknown, but the total breeding population is small (possibly as few as 2,500 pairs), and was probably stable between 1970–1990. Although trend data were not available during 1990–2000, there was no evidence to suggest that the species declined. Nevertheless, its population size renders it susceptible to the risks affecting small populations, and consequently it is provisionally evaluated as Rare.

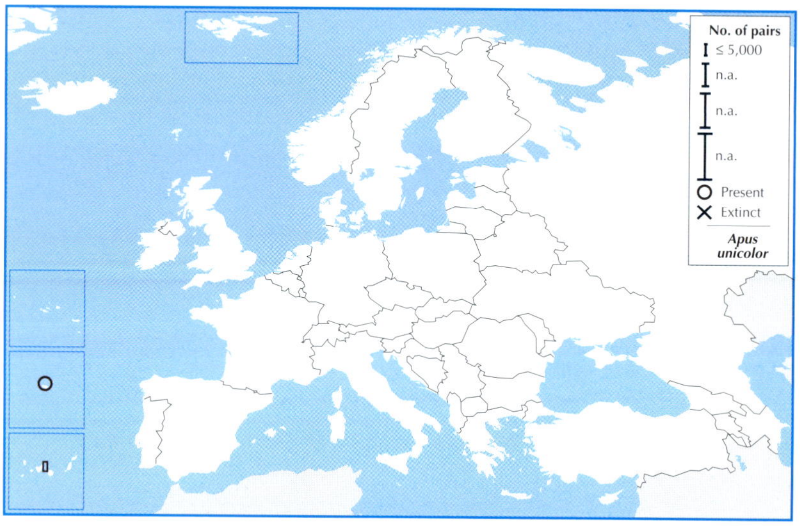

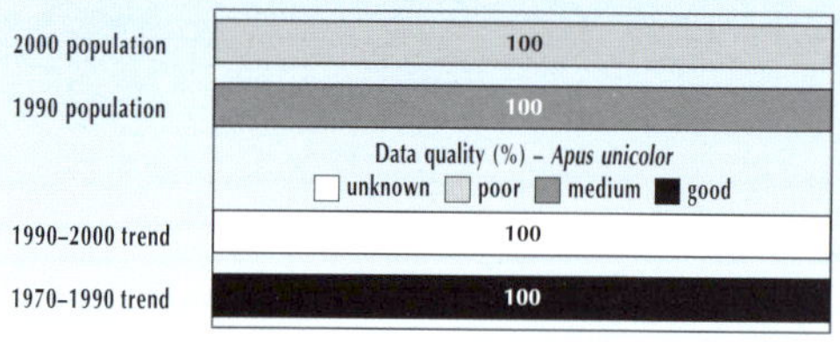

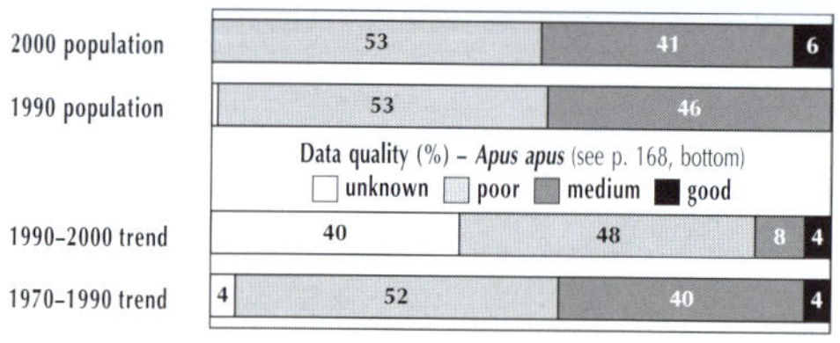

Apus apus
COMMON SWIFT

Non-SPEC (1994: —) **Status** (Secure)
Criteria —
European IUCN Red List Category —
Criteria —
Global IUCN Red List Category —
Criteria —

Apus apus is a widespread summer visitor to most of Europe, which accounts for less than half of its global breeding range. Its European breeding population is very large (>6,900,000 pairs), and was stable between 1970–1990. Although there were declines in a number of countries during 1990–2000, these were set against stable or increasing trends elsewhere in Europe – notably in the sizeable populations in France, Italy and Turkey—and the species probably remained stable overall. Consequently, it is provisionally evaluated as Secure.

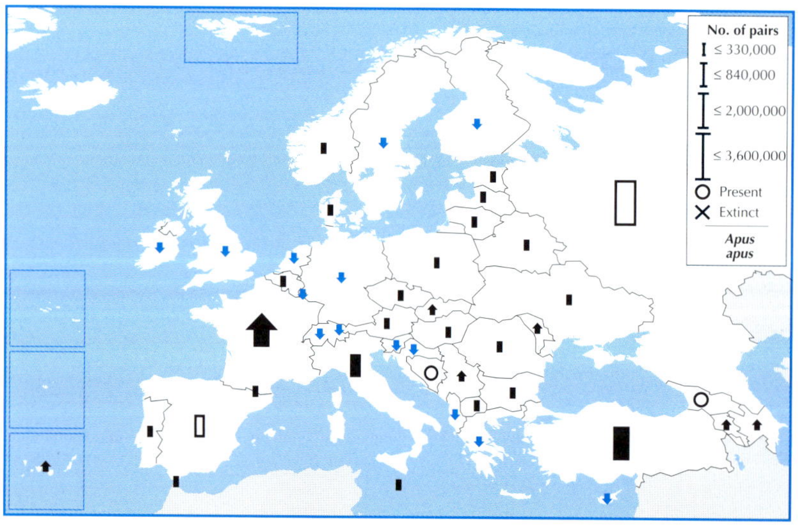

Country	Breeding pop. size (pairs)	Year(s)	Trend	Mag.%	References
Albania	5,000 – 20,000	02	(–)	(0–19)	
Andorra	(300 – 400)	99–01	(0)	(0–19)	*1,3*
Armenia	100,000 – 150,000	01–02	(+)	(–)	
Austria	(25,000 – 50,000)	98–02	(0)	(0–19)	
Azerbaijan	(10,000 – 100,000)	96–00	(+)	(20–29)	
Belarus	140,000 – 160,000	97–02	0	0–19	
Belgium	25,000 – 50,000	01–02	(0)	(0–19)	*1*
Bosnia & HG	Present	90–03	?	–	
Bulgaria	7,500 – 15,000	96–02	0	0–19	
Croatia	(5,000 – 10,000)	02	(–)	(>80)	*16*
Cyprus	(5,000 – 50,000)	94–02	(–)	(0–19)	
Czech Rep.	**60,000 – 120,000**	00	**0**	**0–19**	
Denmark	(20,000 – 80,000)	00	0	0–19	
Estonia	(30,000 – 70,000)	98	0	0–19	*1*
Finland	30,000 – 60,000	98–02	**–**	**10**	
France	(800,000 – 3,000,000)	00–02	(+)	(6)	*4,2*
Georgia	Present	03	?	–	
Germany	230,000 – 460,000	95–99	–	20–29	
Greece	(50,000 – 70,000)	95–00	(–)	(0–19)	
Hungary	3,000 – 5,000	95–02	0	0–19	*19*
Rep. Ireland	10,000 – 20,000	88–91	–	0–19	
Italy	(700,000 – 1,000,000)	03	(0)	(0–19)	
Latvia	40,000 – 100,000	90–00	0	0–19	*23,16*
Liechtenstein	80 – 200	98–00	–	0–9	
Lithuania	(50,000 – 100,000)	99–01	(0)	(0–19)	*20*
Luxembourg	2,500 – 3,000	02	**–**	**30–49**	
Macedonia	(10,000 – 20,000)	90–00	(0)	(0–19)	
Malta	**1 – 1**	90–02	**0**	**0–19**	*1*
Moldova	**800 – 1,200**	90–00	+	20–29	
Netherlands	**30,000 – 60,000**	98–00	–	0–19	*1*
Norway	(10,000 – 100,000)	90–03	(0)	(0–19)	
Poland	**100,000 – 300,000**	90–00	(0)	(0–19)	*23*
Portugal	(20,000 – 100,000)	02	(0)	(0–19)	
Romania	(10,000 – 12,000)	99–02	(0)	(0–19)	
Russia	2,500,000 – 5,000,000	90–00	?	–	*8,23,53,122*
Serbia & MN	3,000 – 4,700	90–02	+	30–49	*1,11,191,29,172a, 155,189,227*
Slovakia	30,000 – 60,000	80–99	+	20–29	
Slovenia	1,500 – 3,000	00	(–)	(0–19)	
Spain	(500,000 – 1,000,000)	98–02	?	–	*10*
Canary Is.	(0 – 50)	97–03	(+)	(N)	*28,25*
Sweden	**200,000 – 400,000**	99–00	**–**	**11**	
Switzerland	**50,000 – 75,000**	93–96	–	0–9	
Turkey	(1,000,000 – 4,000,000)	01	(0)	(0–19)	
Ukraine	(70,000 – 110,000)	90–00	(0)	(0–19)	
UK	(20,000 – 100,000)	00	–	**18**	*22,31*
Gibraltar	(1,000 – 2,000)	00	0	0–19	
Total (approx.)	6,900,000 – 17,000,000	**Overall trend** Small decline			
Breeding range	>8,000,000 km²	**Gen. length.** 7		**% Global pop.** 25–49	

(See p. 168, top, for data quality graph)

Apus pallidus
PALLID SWIFT

Non-SPEC (1994: —) **Status** (Secure)
Criteria —
European IUCN Red List Category —
Criteria —
Global IUCN Red List Category —
Criteria —

Country	Breeding pop. size (pairs)	Year(s)	Trend	Mag.%	References
Albania	500 – 1,000	02	(–)	(0–19)	
Bulgaria	**800 – 1,300**	96–02	+	>80	
Croatia	(1,000 – 5,000)	02	(–)	(30–49)	*16*
Cyprus	(100 – 1,500)	94–02	(0)	(0–19)	
France	1,500 – 2,500	00–02	(0)	(0–19)	*4*
Greece	(1,000 – 3,000)	95–00	(0)	(0–19)	
Italy	(5,000 – 10,000)	03	(0)	(0–19)	
Portugal	(5,000 – 20,000)	02	(0)	(0–19)	
Madeira	Present	02	?	–	
Serbia & MN	(200 – 400)	90–02	(0)	(10–19)	*1,155,227,91*
Spain	(20,000 – 100,000)	98–02	?	–	*10*
Canary Is.	(50 – 250)	97–03	?	–	*25*
Switzerland	**11 – 20**	93–96	**0**	**0–19**	
Turkey	(1,000 – 10,000)	01	?	–	
UK					
Gibraltar	2,500 – 3,000	00	0	0–19	
Total (approx.)	39,000 – 160,000	**Overall trend** Unknown			
Breeding range	>500,000 km²	**Gen. length** 7		**% Global pop.** 25–49	

Apus pallidus is a widespread but patchily distributed summer visitor to much of southern Europe, which accounts for less than half of its global breeding range. Its European breeding population is relatively small (<160,000 pairs), but increased between 1970–1990. Although the trend of the key Spanish population during 1990–2000 was unknown, the species was stable in most of the rest of its European range, and is provisionally evaluated as Secure.

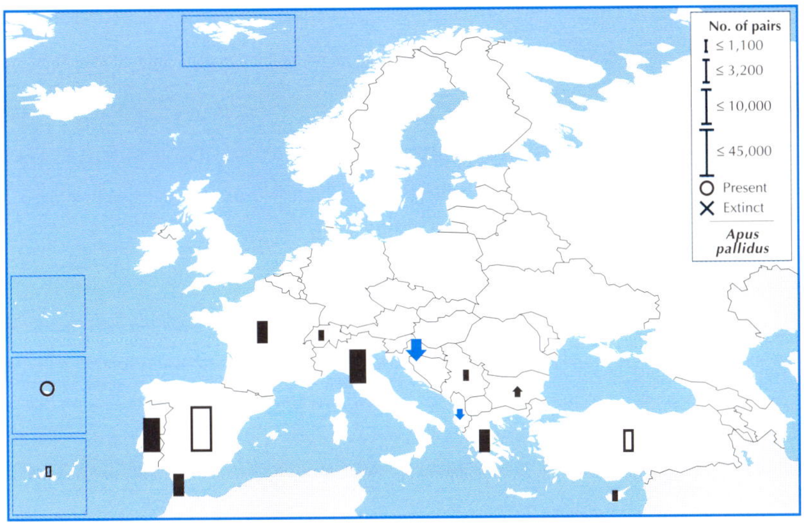

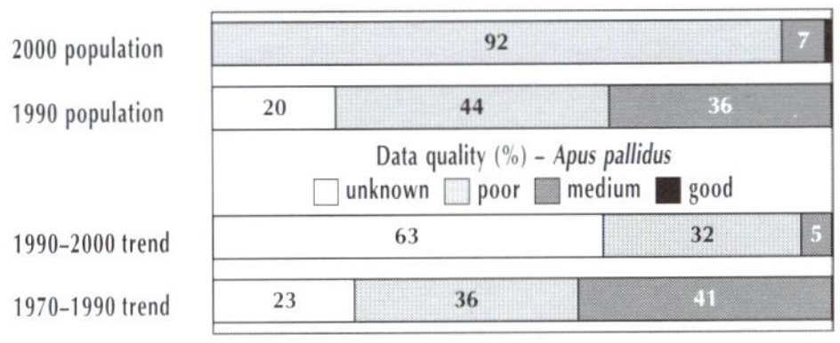

Apus caffer
WHITE-RUMPED SWIFT

Non-SPEC (1994: —) **Status** Secure
Criteria —
European IUCN Red List Category —▼▼▼
Criteria —
Global IUCN Red List Category —
Criteria —

Country	Breeding pop. size (pairs)	Year(s)	Trend	Mag.%	References
Portugal	1 – 10	02	+	N	
Spain	100 – 150	98–02	+	>80	*10,16*
Total (approx.)	100 – 160	**Overall trend** Large increase			
Breeding range	>20,000 km²	**Gen. length** 7		**% Global pop.** <5	

Apus caffer has a predominantly African breeding distribution, which just extends into Europe in southern Spain and Portugal. Its European breeding population is very small (as few as 100 pairs), but increased markedly between 1970–1990. The species continued to increase in Iberia during 1990–2000, and underwent a large increase overall. Although the size of the European population could render it susceptible to the risks affecting small populations, it is marginal to a much larger non-European population. Consequently, the species is evaluated as Secure.

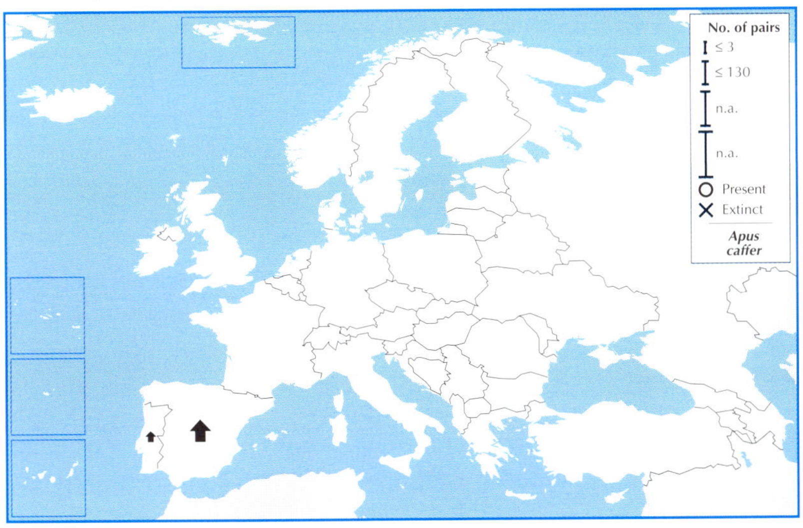

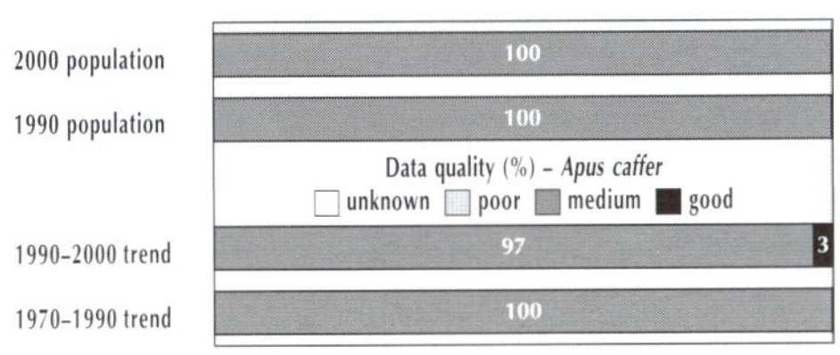

Apus affinis
LITTLE SWIFT

Country	Breeding pop. size (pairs)	Year(s)	Trend	Mag.%	References
Spain	1 - 1	98–02	+	N	10
Turkey	(900 - 2,500)	01	(-)	(0–19)	
Total (approx.)	900 - 2,500	**Overall trend** Small decline			
Breeding range	>20,000 km²	**Gen. length.** 7		**% Global pop.** <5	

SPEC 3 (1994: —) **Status** (Endangered)
Criteria See IUCN below
European IUCN Red List Category EN
Criteria A3c
Global IUCN Red List Category —
Criteria —

Apus affinis has a predominantly Asian and African distribution, which just extends into Europe in south-eastern Turkey and (recently) Spain. Its European breeding population is very small (as few as 900 pairs), but its trend between 1970–1990 was unknown. Although it bred in Spain for the first time during 1990–2000, the Turkish population declined, and the species probably declined slightly overall. However, it is predicted to undergo a large future decline (>50%) in Turkey owing to habitat loss resulting from dam construction. Consequently, it is provisionally evaluated as Endangered.

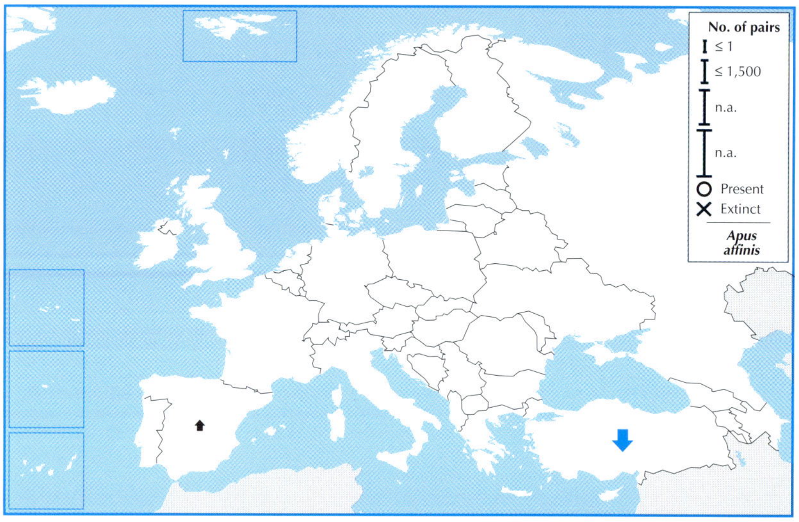

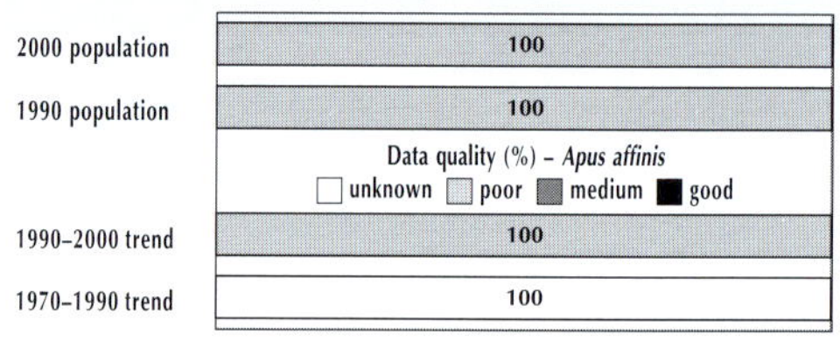

Halcyon smyrnensis
WHITE-THROATED KINGFISHER

Country	Breeding pop. size (pairs)	Year(s)	Trend	Mag.%	References
Azerbaijan	(10 - 50)	96–00	(0)	(0–19)	
Turkey	80 - 120	01	-	30–49	
Total (approx.)	90 - 170	**Overall trend** Large decline			
Breeding range	>20,000 km²	**Gen. length.** <3.3		**% Global pop.** <5	

SPEC 3 (1994: —) **Status** Endangered
Criteria See IUCN below
European IUCN Red List Category EN
Criteria C2a(i); D1
Global IUCN Red List Category —
Criteria —

Halcyon smyrnensis has a predominantly Asian distribution, which just extends into Europe in southern Turkey and Azerbaijan. Its European breeding population is extremely small (as few as 90 pairs), but its trend between 1970–1990 was unknown. Although the population in Azerbaijan was stable during 1990–2000, the species declined in Turkey, and underwent a large decline (>30%) overall. As a consequence of this decline and its small population size, this previously Secure species is now evaluated as Endangered.

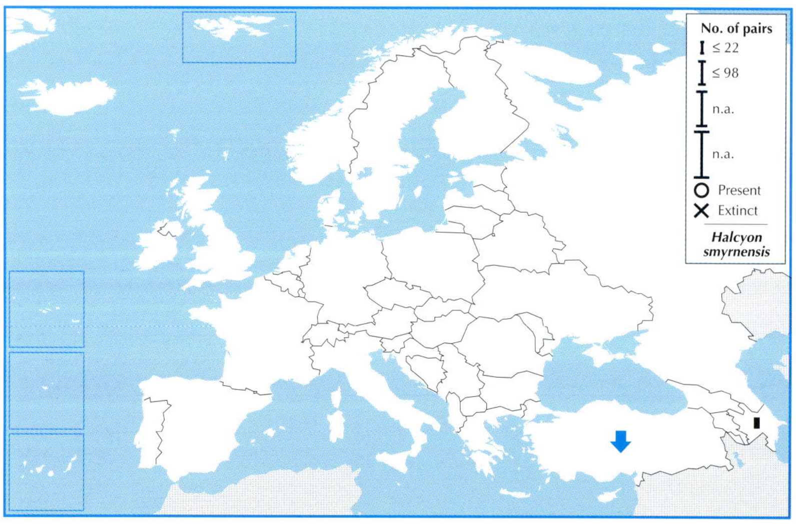

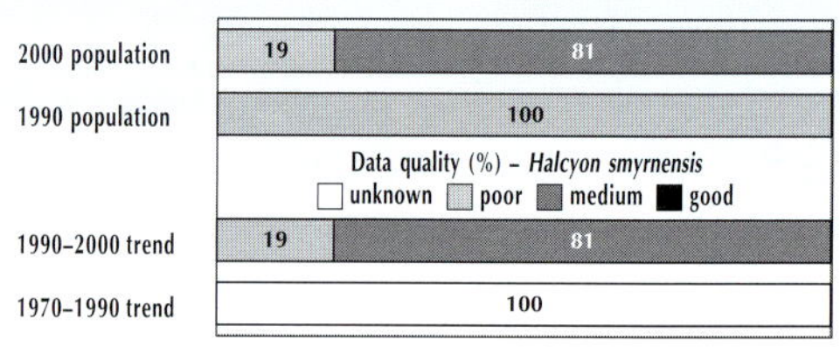

Alcedo atthis
COMMON KINGFISHER

SPEC 3 (1994: 3) **Status** Depleted
Criteria Moderate historical decline
European IUCN Red List Category —
Criteria —
Global IUCN Red List Category —
Criteria —

Country	Breeding pop. size (pairs)	Year(s)	Trend	Mag.%	References
Albania	100 – 500	02	(F)	(>80)	
Armenia	300 – 1,000	98–02	?	–	
Austria	300 – 500	98–02	0	0–19	
Azerbaijan	(500 – 5,000)	96–00	(0)	(0–19)	
Belarus	3,000 – 6,000	97–02	0	0–19	
Belgium	900 – 1,500	01–02	F	20–29	*1*
Bosnia & HG	Present	90–03	?	–	
Bulgaria	800 – 1,200	96–02	0	0–19	
Croatia	(1,500 – 2,500)	02	(0)	(0–19)	*70,54*
Czech Rep.	**600 – 1,000**	00	**F**	**>80**	
Denmark	(250 – 400)	00	+	50–79	*3*
Estonia	200 – 500	98	F	20–29	*1*
Finland	5 – 30	98–02	+	100	
France	10,000 – 30,000	00	(F)	(20–49)	*1*
Georgia	Present	03	?	–	
Germany	4,500 – 7,000	95–99	0	0–19	
Greece	(100 – 300)	95–00	(–)	(0–19)	
Hungary	900 – 1,600	95–02	0	0–19	
Rep. Ireland	(1,000 – 2,500)	88–91	?	–	
Italy	(5,000 – 10,000)	03	(0)	(0–19)	
Latvia	600 – 1,500	90–00	**F**	**50–79**	*17*
Liechtenstein	**0 – 1**	99–00	**0**	**0–19**	
Lithuania	700 – 1,000	99–01	(0)	(0–19)	*20*
Luxembourg	**50 – 80**	00–02	**0**	**0–19**	
Macedonia	(100 – 400)	90–00	(0)	(0–19)	
Moldova	310 – 390	90–00	**0**	**0–19**	
Netherlands	**70 – 300**	98–00	+	**22**	*1*
Norway	0 – 5	99	**F**	**30–49**	*63*
Poland	2,500 – 6,000	90–00	F	50–79	*94*
Portugal	(2,000 – 10,000)	02	(0)	(0–19)	
Romania	12,000 – 15,000	98–02	0	0–19	
Russia	12,000 – 25,000	90–00	0	0–19	*8,55,73,117,118*
Serbia & MN	1,500 – 2,400	90–02	0	10–19	*1,205,156,155, 78,227,225*
Slovakia	700 – 1,300	80–99	–	30–49	
Slovenia	200 – 300	99–00	–	10–19	
Spain	(3,600 – 7,000)	98–02	(–)	(0–19)	*10,16*
Sweden	**200 – 220**	99–00	0	0–19	
Switzerland	**300 – 350**	93–96	+	0–9	
Turkey	(1,000 – 2,000)	01	(–)	(0–19)	
Ukraine	**6,000 – 9,500**	90–00	0	0–19	
UK	**4,800 – 8,000**	00	+	**29**	*5*
Total (approx.)	79,000 – 160,000	**Overall trend** Fluctuating			
Breeding range	>5,000,000 km²	**Gen. length** <3.3 **% Global pop.** 25–49			

(See p. 171, bottom, for data quality graph)

Alcedo atthis is a widespread breeder across much of Europe, which accounts for less than half of its global breeding range. Its European breeding population is relatively small (<160,000 pairs), and underwent a moderate decline between 1970–1990. Although the species was broadly stable overall during 1990–2000—with stable, fluctuating or increasing trends across the vast majority of Europe—its population has not yet recovered to the level that preceded its decline. Consequently, it is evaluated as Depleted.

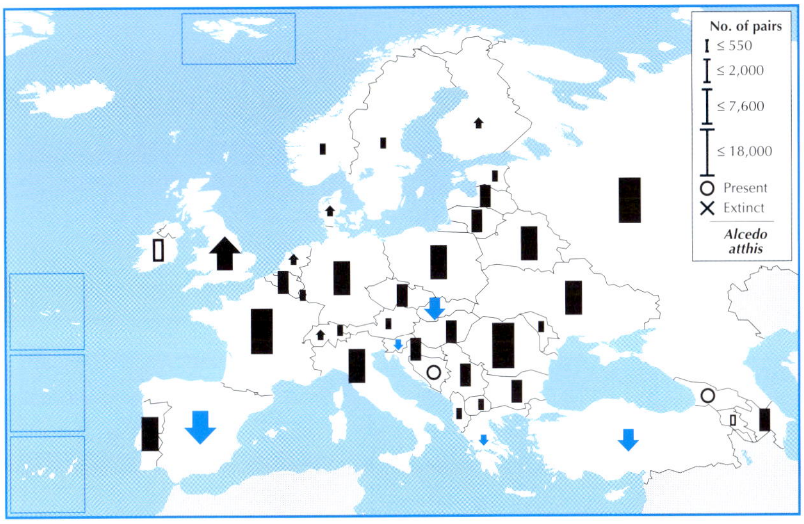

Ceryle rudis
PIED KINGFISHER

SPEC 3 (1994: —) **Status** (Critically Endangered)
Criteria See IUCN below
European IUCN Red List Category CR
Criteria A2c; C1
Global IUCN Red List Category —
Criteria —

Country	Breeding pop. size (pairs)	Year(s)	Trend	Mag.%	References
Cyprus	**1 – 1**	96	+	N	
Turkey	100 – 200	01	(–)	(>80)	
Total (approx.)	100 – 200	**Overall trend** Large decline			
Breeding range	>20,000 km²	**Gen. length** <3.3 **% Global pop.** <5			

Ceryle rudis has a predominantly African and Asian distribution, which just extends into Europe in Turkey and (recently) Cyprus. Its European breeding population is extremely small (as few as 100 pairs), but its trend between 1970–1990 was unknown. Although it bred in Cyprus for the first time during 1990–2000, the Turkish population declined substantially, and the species underwent an extremely large decline (>80%) overall. As a consequence of this decline and its small population size, this previously Secure species is now provisionally evaluated as Critically Endangered.

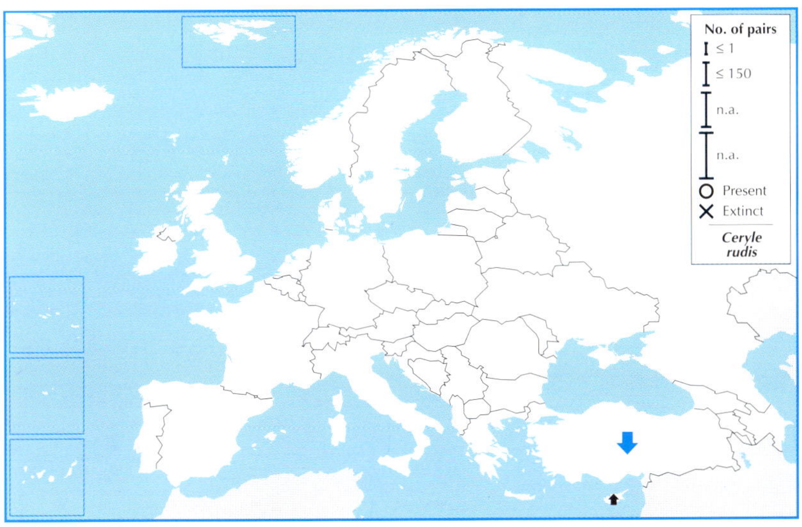

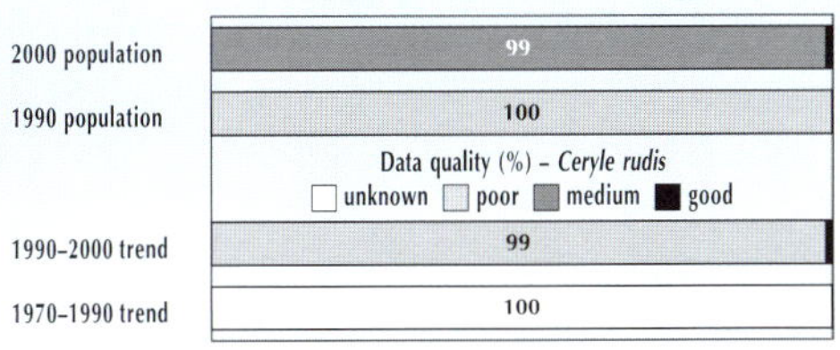

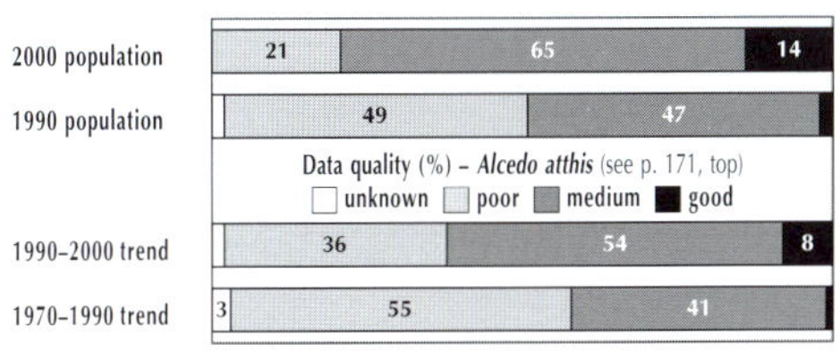

Merops persicus
BLUE-CHEEKED BEE-EATER

Non-SPEC (1994: —) **Status** (Secure)
Criteria —
European IUCN Red List Category —▼
Criteria —
Global IUCN Red List Category —
Criteria —

Merops persicus has an extensive global breeding distribution, which just extends into south-east Europe. Its European breeding population is small (as few as 3,100 pairs), but was stable between 1970–1990. Despite some declines during 1990–2000, key populations in Russia and Azerbaijan increased, and the species increased overall. Although the size of the European population could render it susceptible to the risks affecting small populations, it is marginal to a much larger non-European population. Consequently, the species is provisionally evaluated as Secure.

Country	Breeding pop. size (pairs)	Year(s)	Trend	Mag.%	References
Armenia	3 – 5	99–02	–	20–29	
Azerbaijan	(1,000 – 5,000)	96–00	(+)	(0–19)	
Russia	(2,000 – 5,000)	90–00	(+)	(20–29)	*172*
Turkey	60 – 150	01	–	**50–79**	
Total (approx.)	3,100 – 10,000	**Overall trend** Moderate increase			
Breeding range	>20,000 km²	**Gen. length.** <3.3 **% Global pop.** <5			

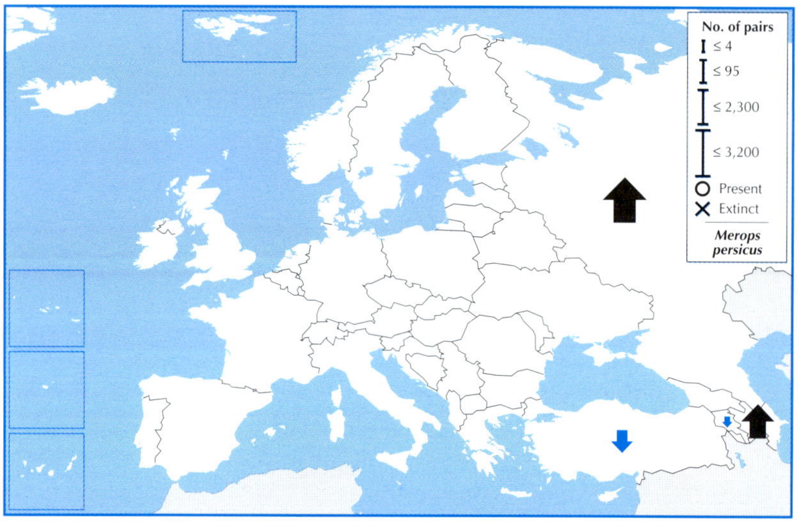

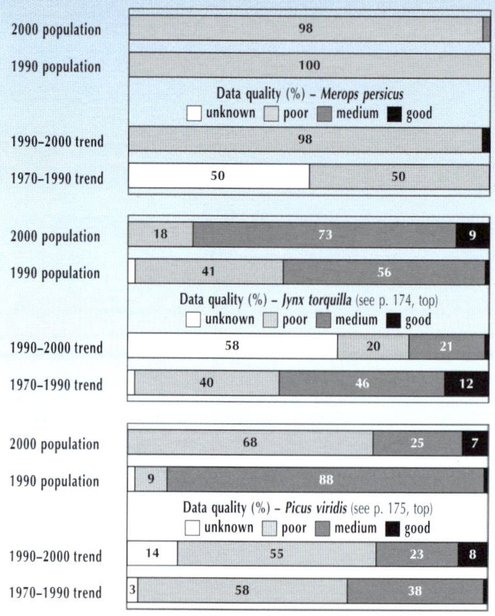

Merops apiaster
EUROPEAN BEE-EATER

SPEC 3 (1994: 3) **Status** (Depleted)
Criteria Moderate historical decline
European IUCN Red List Category —
Criteria —
Global IUCN Red List Category —
Criteria —

Merops apiaster is a widespread summer visitor to southern and eastern Europe, which accounts for less than half of its global breeding range. Its European breeding population is large (>480,000 pairs), but underwent a moderate decline between 1970–1990. Although the species increased overall during 1990–2000—with increasing or stable trends across most of its European range—its population probably has not yet recovered to the level that preceded its decline. Consequently, it is provisionally evaluated as Depleted.

Country	Breeding pop. size (pairs)	Year(s)	Trend	Mag.%	References
Albania	300 – 1,000	98–02	(–)	(20–29)	
Armenia	3,000 – 6,500	00–02	+	0–9	
Austria	**150 – 300**	98–02	F	**>80**	
Azerbaijan	(5,000 – 10,000)	96–00	(0)	(0–19)	
Belarus	30 – 60	97–02	0	0–19	
Belgium	**6 – 11**	90–02	+	**0–19**	*1*
Bosnia & HG	(0 – 70)	90–00	(+)	(0–9)	
Bulgaria	15,000 – 30,000	96–02	+	0–19	
Croatia	(1,000 – 5,000)	02	(0)	(0–19)	*16*
Cyprus	80 – 200	93–02	+	0–9	
Czech Rep.	**36 – 40**	00	+	**300**	
Denmark	**4 – 6**	98–01	+	N	*4,5,6,7*
France	4,000 – 12,000	00–02	?	–	*4,2*
Georgia	Present	03	?	–	
Germany	**120 – 190**	95–99	+	**30–49**	
Greece	(2,000 – 3,000)	95–00	(0)	(0–19)	
Hungary	**20,000 – 40,000**	97–01	+	20–29	*5,19*
Italy	5,000 – 10,000	03	+	10–19	
Latvia	**0 – 3**	03	?	–	*8,22*
Macedonia	4,000 – 20,000	90–00	(0)	(0–9)	
Moldova	8,000 – 10,000	90–00	+	30–49	
Poland	**20 – 50**	90–00	F	**30–49**	*2*
Portugal	(5,000 – 25,000)	02	(0)	(0–19)	
Romania	15,000 – 20,000	00–02	–	0–19	
Russia	(200,000 – 400,000)	90–00	(+)	(20–29)	*2,8,73*
Serbia & MN	2,800 – 3,800	90–02	–	0–19	*1,155,134,191, 29,67a,78,227*
Slovakia	700 – 1,300	80–99	+	20–29	
Slovenia	**30 – 40**	90–00	0	0–19	
Spain	(100,000 – 250,000)	98–02	?	–	*10*
Switzerland	**5 – 15**	98–02	F	**20–29**	
Turkey	(60,000 – 120,000)	01	(–)	(0–19)	
Ukraine	25,000 – 45,000	90–00	0	0–9	
Total (approx.)	480,000 – 1,000,000	**Overall trend** Moderate increase			
Breeding range	>3,000,000 km²	**Gen. length.** <3.3 **% Global pop.** 25–49			

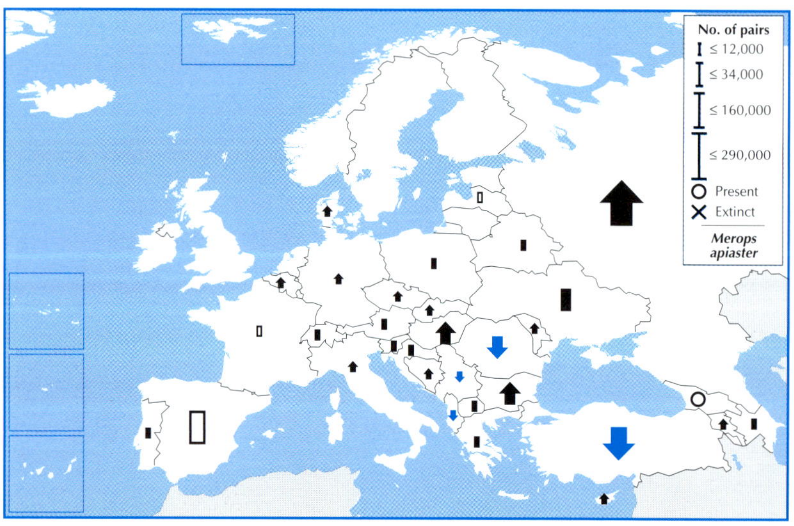

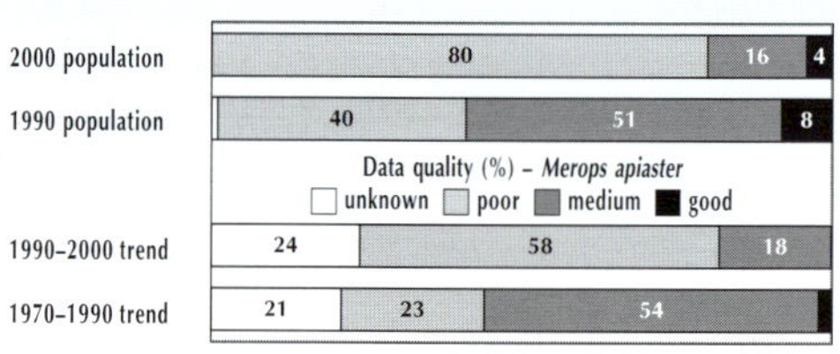

Coracias garrulus
EUROPEAN ROLLER

SPEC 2 (1994: 2) **Status** Vulnerable
Criteria See IUCN below
European IUCN Red List Category VU
Criteria A2b
Global IUCN Red List Category —
Criteria —

Coracias garrulus is a widespread summer visitor to southern and eastern Europe, which constitutes >50% of its global breeding range. Its European breeding population is relatively small (<110,000 pairs), and underwent a moderate decline between 1970–1990. With few exceptions, the species continued to decline across most of its European range during 1990–2000—including key populations in Turkey and Russia—and probably underwent a large decline (>30%) overall. Consequently, it is provisionally evaluated as Vulnerable.

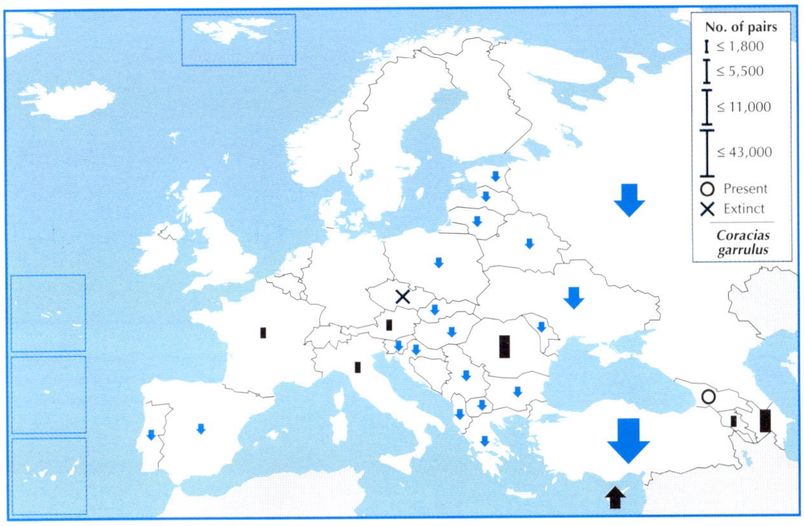

Country	Breeding pop. size (pairs)	Year(s)	Trend	Mag.%	References
Albania	10 – 50	02	(–)	(30–49)	
Armenia	300 – 650	00–02	0	0–19	
Austria	**7 – 10**	01–03	**0**	**0–19**	
Azerbaijan	(1,000 – 5,000)	96–00	(0)	(0–19)	
Belarus	600 – 900	97–02	–	50–79	
Bulgaria	1,000 – 3,000	96–02	–	20–29	
Croatia	(0 – 5)	02	(–)	(>80)	*70,54*
Cyprus	(2,000 – 4,000)	94–02	(+)	(0–9)	
Czech Rep.	**0 – 0**	00	**–**	**X**	
Estonia	50 – 100	98	–	20–29	*1*
France	520 – 620	00	(0)	(0–19)	*1*
Georgia	Present	03	?	–	
Greece	(200 – 300)	95–00	(–)	(0–19)	
Hungary	400 – 700	95–02	–	0–19	
Italy	(300 – 500)	03	(0)	(0–19)	
Latvia	**25 – 50**	98–00	–	30–49	*18,22*
Lithuania	100 – 150	99–01	–	>80	*20*
Macedonia	(300 – 1,000)	90–00	(–)	(20–29)	
Moldova	50 – 80	90–00	–	30–49	
Poland	**100 – 150**	98–00	**–**	**50–59**	*85*
Portugal	100 – 300	02	(–)	(–)	*10*
Romania	4,600 – 6,500	00–02	0	0–19	*48*
Russia	(6,000 – 20,000)	90–00	–	20–29	*8,43,73,114*
Serbia & MN	**65 – 85**	98–02	**–**	**50–69**	*1,133,116,177,89, 237,77a,155,227*
Slovakia	10 – 40	80–99	–	50–79	
Slovenia	**3 – 5**	98–00	**–**	**80–89**	
Spain	(1,000 – 2,500)	98–02	(–)	(0–19)	*10,16*
Turkey	(30,000 – 60,000)	01	(–)	(0–19)	
Ukraine	4,000 – 5,000	90–00	**–**	**30–49**	
Total (approx.)	53,000 – 110,000	**Overall trend** Large decline			
Breeding range	>3,000,000 km²	**Gen. length.** 5		**% Global pop.** 50–74	

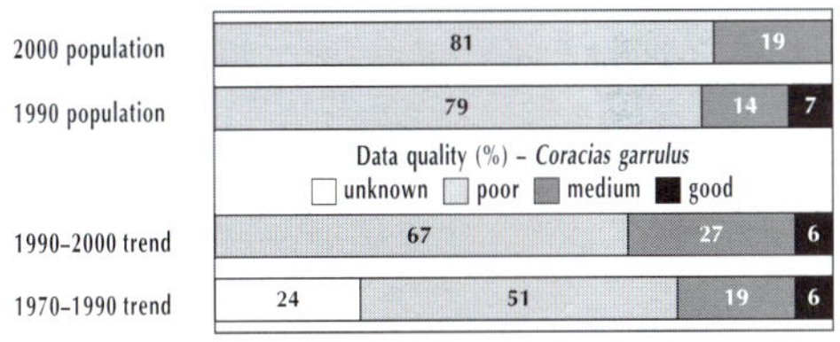

Upupa epops
EURASIAN HOOPOE

SPEC 3 (1994: —) **Status** (Declining)
Criteria Moderate recent decline
European IUCN Red List Category —
Criteria —
Global IUCN Red List Category —
Criteria —

Upupa epops is a widespread breeder across much of the southern half of Europe, which accounts for less than a quarter of its global breeding range. Its European breeding population is large (>890,000 pairs), and was stable between 1970–1990. Although the species was stable across much of Europe during 1990–2000, several populations suffered declines—most notably the sizeable one in Turkey—and the species probably underwent a moderate decline (>10%) overall. Consequently, this previously Secure species is now provisionally evaluated as Declining.

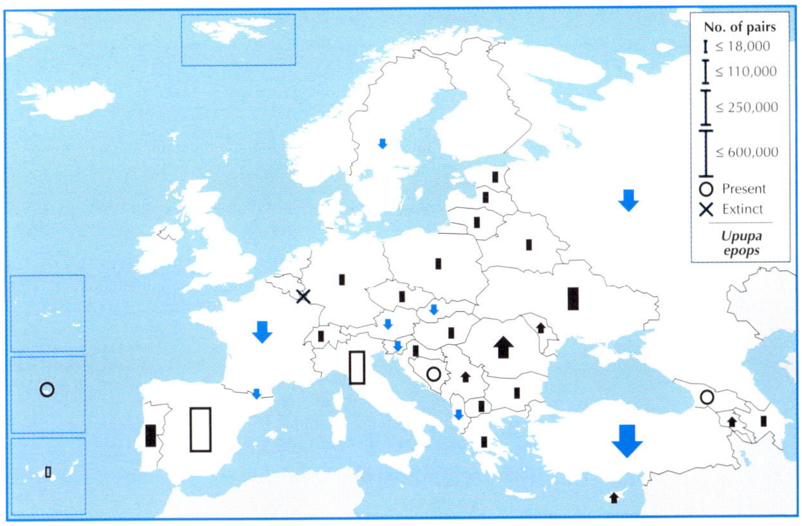

Country	Breeding pop. size (pairs)	Year(s)	Trend	Mag.%	References
Albania	1,000 – 2,000	02	(–)	(20–29)	
Andorra	(0 – 2)	99–01	(–)	(30–49)	*1,3*
Armenia	650 – 1,500	01–02	+	0–9	
Austria	400 – 600	98–02	(–)	(30–49)	
Azerbaijan	(10,000 – 20,000)	96–00	(0)	(0–19)	
Belarus	14,000 – 22,000	97–02	0	0–19	
Bosnia & HG	Present	90–03	?	–	
Bulgaria	4,000 – 8,000	96–02	0	0–9	
Croatia	(3,500 – 7,000)	02	(0)	(0–19)	*70,16,54*
Cyprus	(300 – 1,000)	91–02	(+)	(0–9)	
Czech Rep.	**70 – 140**	00	**0**	**0–19**	
Estonia	(5 – 20)	98	F	20–29	*1*
France	30,000 – 50,000	00	**–**	**51**	*4,2*
Georgia	Present	03	?	–	
Germany	**310 – 460**	95–99	**0**	**0–19**	
Greece	(5,000 – 20,000)	95–00	(0)	(0–19)	
Hungary	10,000 – 17,000	99–02	0	0–19	*19*
Italy	(20,000 – 50,000)	03	?	–	
Latvia	100 – 300	90–00	(0)	(0–19)	*23,16*
Lithuania	(200 – 300)	99–01	(0)	(0–19)	*20*
Luxembourg	**0 – 0**	02	**–**	**X**	
Macedonia	(3,000 – 10,000)	90–00	(0)	(0–19)	
Moldova	1,800 – 2,300	90–00	+	20–29	
Poland	**10,000 – 15,000**	00–02	0	0–19	*23*
Portugal	(10,000 – 100,000)	02	(0)	(0–19)	
Madeira	Present	02	?	–	
Romania	24,000 – 42,000	00–02	+	0–19	*48,7*
Russia	(60,000 – 200,000)	90–00	(–)	(20–29)	*8,24,39,43,114*
Serbia & MN	10,000 – 14,000	90–02	+	0–19	*1,29,155,67a, 78,227,225*
Slovakia	600 – 1,000	80–99	–	30–49	
Slovenia	600 – 1,000	99–00	–	20–29	
Spain	(497,000 – 710,000)	92	?	–	*13,12,10*
Canary Is.	(2,500 – 10,000)	97–03	?	–	*28,25*
Sweden	**2 – 3**	99–00	–	30–49	
Switzerland	**100 – 150**	93–96	**F**	**20–29**	
Turkey	(150,000 – 400,000)	01	(–)	(0–19)	
Ukraine	24,000 – 42,000	90–00	0	0–19	
Total (approx.)	890,000 – 1,700,000	**Overall trend** Moderate decline			
Breeding range	>5,000,000 km²	**Gen. length.** <3.3		**% Global pop.** 5–24	

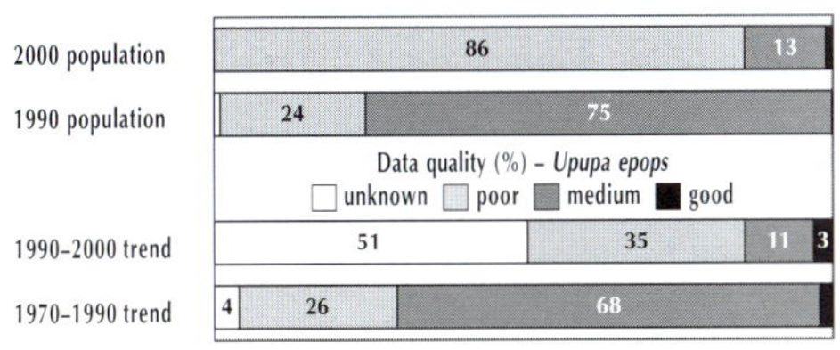

Jynx torquilla
EURASIAN WRYNECK

SPEC 3 (1994: 3) **Status** (Declining)
Criteria Moderate continuing decline
European IUCN Red List Category —
Criteria —
Global IUCN Red List Category —
Criteria —

Jynx torquilla is a widespread summer visitor to much of Europe, which accounts for less than half of its global breeding range. Its European breeding population is large (>580,000 pairs), but underwent a moderate decline between 1970–1990. Although several populations in eastern Europe were stable during 1990–2000, and the trend of the key population in Russia was unknown, the species continued to decline across most of its European range, and probably underwent a moderate decline (>10%) overall. Consequently, it is provisionally evaluated as Declining.

Country	Breeding pop. size (pairs)	Year(s)	Trend	Mag.%	References
Albania	(50 – 150)	96–02	(–)	(20–29)	
Andorra	2 – 5	99–01	(0)	(0–19)	*1,3*
Armenia	70 – 150	95–02	?	–	
Austria	(2,000 – 5,000)	98–02	(0)	(0–19)	
Azerbaijan	(1,000 – 5,000)	96–00	(0)	(0–19)	
Belarus	30,000 – 40,000	97–02	0	0–19	
Belgium	13 – 23	95–02	0	0–19	*1*
Bosnia & HG	Present	90–03	?	–	
Bulgaria	5,000 – 10,000	96–02	0	0–19	
Croatia	(5,000 – 10,000)	02	(0)	(0–19)	*16*
Czech Rep.	**2,500 – 5,000**	00	**0**	**0–19**	
Denmark	25 – 50	01–02	–	50–79	*7,3*
Estonia	4,000 – 8,000	98	–	50–79	*1*
Finland	5,000 – 15,000	98–02	–	50	
France	7,000 – 20,000	98–02	(–)	(38)	*4,2*
Georgia	Present	03	?	–	
Germany	12,000 – 21,000	95–99	–	30–49	
Greece	(100 – 200)	95–00	(0)	(0–19)	
Hungary	**48,000 – 81,000**	99–02	0	0–19	*19*
Italy	(50,000 – 100,000)	03	(–)	(0–19)	
Latvia	2,000 – 5,000	90–00	(0)	(0–19)	*23,18*
Liechtenstein	5 – 10	98–00	–	30–49	
Lithuania	(4,000 – 10,000)	99–01	(0)	(0–19)	*20*
Luxembourg	50 – 100	00–02	**–**	**30–49**	
Macedonia	(1,000 – 2,500)	90–00	(–)	(10–19)	
Moldova	1,900 – 2,500	90–00	–	0–19	
Netherlands	**50 – 65**	98–00	**–**	**10**	*1*
Norway	(2,000 – 4,000)	03	(–)	(0–19)	*38*
Poland	**8,000 – 20,000**	00–02	(–)	(0–19)	*23*
Portugal	(1,000 – 5,000)	02	(0)	(0–19)	
Romania	28,000 – 36,000	00–02	+	0–19	*48*
Russia	300,000 – 800,000	90–00	?	–	*3,39,43,114,122*
Serbia & MN	6,000 – 9,000	90–02	–	0–19	*1,29,172a,155, 67a,227*
Slovakia	2,500 – 4,000	80–99	–	30–49	
Slovenia	2,000 – 3,000	99–00	(–)	(0–19)	
Spain	(10,000 – 20,000)	98–02	?	–	*10,16*
Sweden	5,000 – 10,000	99–00	**–**	**50**	
Switzerland	**2,000 – 3,000**	93–96	–	10–19	
Turkey	(500 – 2,000)	01	(0)	(0–19)	
Ukraine	(35,000 – 55,000)	90–00	(0)	(0–19)	
UK	**0 – 3**	96–00	**–**	**65**	
Total (approx.)	580,000 – 1,300,000	**Overall trend** Moderate decline			
Breeding range	>6,000,000 km²	**Gen. length.** <3.3 **% Global pop.** 25–49			

(See p. 172, top, for data quality graph)

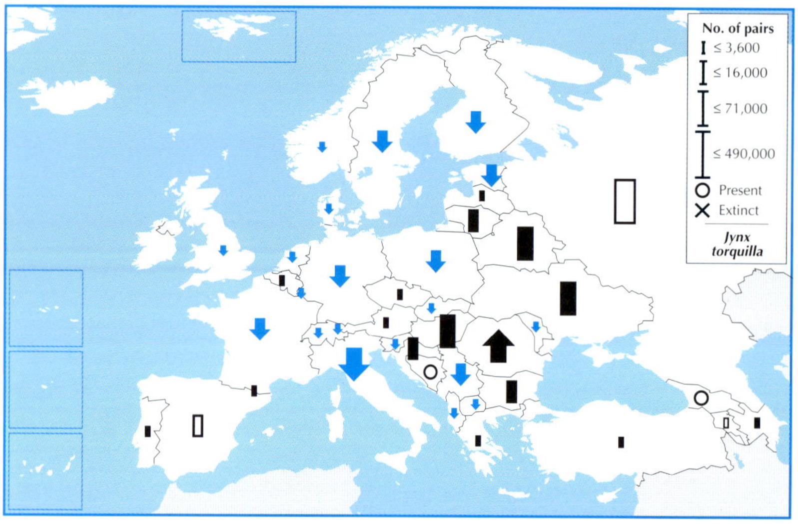

Picus canus
GREY-FACED WOODPECKER

SPEC 3 (1994: 3) **Status** (Depleted)
Criteria Moderate historical decline
European IUCN Red List Category —
Criteria —
Global IUCN Red List Category —
Criteria —

Picus canus is a widespread resident across much of continental Europe, which accounts for less than a quarter of its global range. Its European breeding population is large (>180,000 pairs), but underwent a moderate decline between 1970–1990. Although the species was stable overall during 1990–2000—with stable or increasing trends across most of its European range—its population has clearly not yet recovered to the level that preceded its decline. Consequently, it is provisionally evaluated as Depleted.

Country	Breeding pop. size (pairs)	Year(s)	Trend	Mag.%	References
Albania	10 – 50	98–02	?	–	
Austria	1,900 – 3,200	98–02	(0)	(0–19)	
Belarus	8,000 – 12,000	97–02	0	0–19	
Belgium	**10 – 30**	95–02	+	0–19	*1*
Bosnia & HG	Present	90–03	?	–	
Bulgaria	**1,500 – 3,000**	96–02	0	0–19	
Croatia	(1,000 – 5,000)	02	(+)	(0–19)	*70,16,44*
Czech Rep.	**3,000 – 5,000**	00	**0**	**0–19**	
Estonia	1,000 – 2,000	98	0	0–19	*1*
Finland	1,500 – 2,500	98–02	0	0–19	
France	5,000 – 15,000	98–02	?	–	*4*
Germany	12,500 – 18,000	95–99	–	20–29	
Greece	(50 – 200)	95–00	(0)	(0–19)	
Hungary	2,000 – 3,000	98–01	0	0–19	
Italy	(700 – 1,500)	03	(0)	(0–19)	
Latvia	2,000 – 3,000	90–00	(+)	(50–79)	*18*
Liechtenstein	**0 – 3**	98–00	**0**	**0–19**	
Lithuania	500 – 700	99–01	(0)	(0–19)	*20*
Luxembourg	30 – 50	02	**–**	**30–49**	
Macedonia	(50 – 250)	90–00	(–)	(0–9)	
Moldova	700 – 900	90–00	**0**	**0–19**	
Norway	2,000 – 4,000	03	(–)	(0–19)	*38*
Poland	2,000 – 3,000	90–00	+	0–19	*27*
Romania	45,000 – 60,000	00–02	0	0–19	
Russia	(70,000 – 150,000)	90–00	?	–	
Serbia & MN	2,600 – 3,500	90–02	0	0–19	*1,62,155,67a, 225,227*
Slovakia	1,500 – 2,000	90–99	0	0–19	
Slovenia	1,000 – 2,000	99–00	(0)	(0–19)	
Sweden	700 – 900	99–00	+	10–19	
Switzerland	**1,000 – 2,000**	93–96	–	10–19	
Turkey	(500 – 1,500)	01	?	–	
Ukraine	11,000 – 15,000	90–00	0	0–9	
Total (approx.)	180,000 – 320,000	**Overall trend** Stable			
Breeding range	>3,000,000 km²	**Gen. length.** <3.3 **% Global pop.** 5–24			

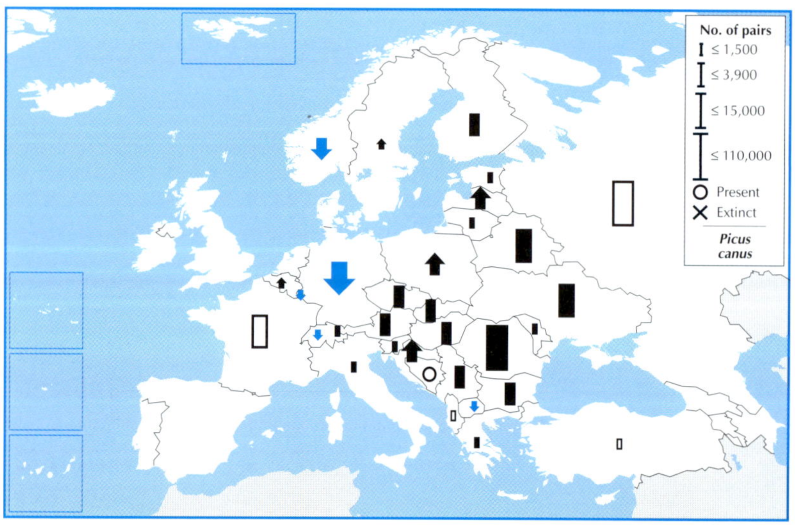

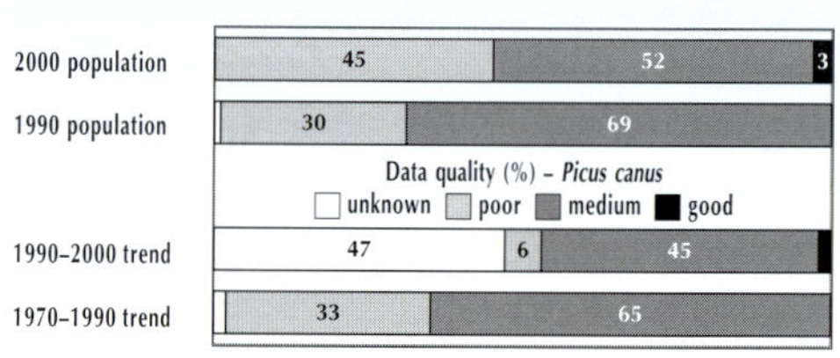

Picus viridis
EURASIAN GREEN WOODPECKER

SPEC 2 (1994: 2) **Status** (Depleted)
Criteria Moderate historical decline
European IUCN Red List Category —
Criteria —
Global IUCN Red List Category —
Criteria —

Picus viridis is a widespread resident across much of Europe, which constitutes >75% of its global range. Its European breeding population is large (>590,000 pairs), but underwent a moderate decline between 1970–1990. Although there were declines in some smaller populations during 1990–2000, many populations—including key ones in France and Romania—were stable or increased, and the species was probably stable overall. Nevertheless, its population has not yet recovered to the level that preceded the decline, and consequently the species is provisionally evaluated as Depleted.

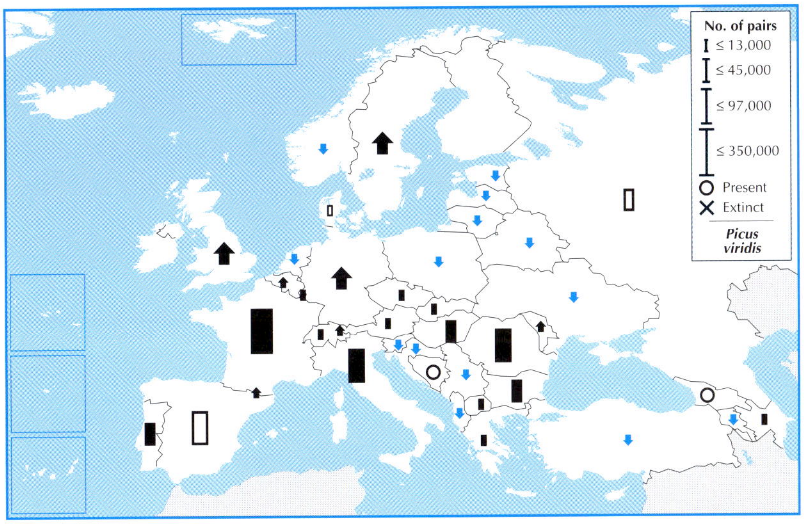

Country	Breeding pop. size (pairs)	Year(s)	Trend	Mag.%	References
Albania	1,000 – 2,000	98–02	(–)	(0–19)	
Andorra	50 – 100	99–01	+	0–19	*1,3*
Armenia	300 – 500	97–02	–	0–9	
Austria	(7,000 – 14,000)	98–02	(0)	(0–19)	
Azerbaijan	(2,000 – 8,000)	96–00	(0)	(0–19)	
Belarus	3,000 – 5,000	97–02	(–)	(0–9)	
Belgium	8,800 – 13,000	01–02	+	0–19	*1*
Bosnia & HG	Present	90–03	?	–	
Bulgaria	10,000 – 30,000	96–02	0	0–9	
Croatia	(5,000 – 10,000)	02	(–)	(50–79)	*70,16*
Czech Rep.	**9,000 – 18,000**	00	**0**	**0–19**	
Denmark	750 – 1,000	93–96	?	–	*3*
Estonia	200 – 300	98	–	20–29	*1*
France	(200,000 – 600,000)	98–02	(0)	(1)	*4,2*
Georgia	Present	03	?	–	
Germany	23,000 – 35,000	95–99	+	0–19	
Greece	(5,000 – 10,000)	95–00	(0)	(0–19)	
Hungary	**12,000 – 20,000**	99–02	0	0–19	*19*
Italy	(40,000 – 80,000)	03	(0)	(0–19)	
Latvia	10 – 100	90–00	–	50–79	*18,22*
Liechtenstein	25 – 40	98–00	+	0–9	
Lithuania	(300 – 500)	99–01	(–)	(20–29)	*20*
Luxembourg	250 – 400	02	**0**	**0–19**	
Macedonia	(2,500 – 10,000)	90–00	(0)	(0–19)	
Moldova	4 – 8	90–00	+	**20–29**	
Netherlands	**4,500 – 5,500**	98–00	–	**5**	*1*
Norway	3,000 – 6,000	03	(–)	(0–19)	*38*
Poland	4,000 – 8,000	00–02	–	0–19	*23*
Portugal	(5,000 – 50,000)	02	(0)	(0–19)	
Romania	85,000 – 110,000	00–02	0	0–19	
Russia	(20,000 – 100,000)	90–00	?	–	*67,116,122,136*
Serbia & MN	9,000 – 13,000	90–02	–	10–19	*1,29,172a,155, 67a,78,225,227*
Slovakia	1,200 – 2,000	90–99	0	0–19	
Slovenia	1,000 – 2,000	99–00	(–)	(0–19)	
Spain	(65,500 – 95,600)	92	?	–	*13,12,10*
Sweden	20,000 – 40,000	99–00	+	**18**	
Switzerland	**5,000 – 10,000**	93–96	0	0–19	
Turkey	(8,000 – 12,000)	01	(–)	(0–19)	
Ukraine	(500 – 800)	90–00	–	30–49	
UK	**24,200 – 24,200**	00	+	**61**	*31*
Total (approx.)	590,000 – 1,300,000	**Overall trend** Stable			
Breeding range	>5,000,000 km²	**Gen. length.** <3.3 **% Global pop.** 75–94			

(See p. 172, top, for data quality graph)

Dryocopus martius
BLACK WOODPECKER

Non-SPEC (1994: —) **Status** Secure
Criteria —
European IUCN Red List Category —
Criteria —
Global IUCN Red List Category —
Criteria —

Dryocopus martius is a widespread resident across much of Europe, which accounts for less than half of its global range. Its European breeding population is large (>740,000 pairs), and was stable between 1970–1990. Although there were declines in a few countries during 1990–2000, most populations—including the Russian stronghold—were stable or increased, and the species remained stable overall. Consequently, it is evaluated as Secure.

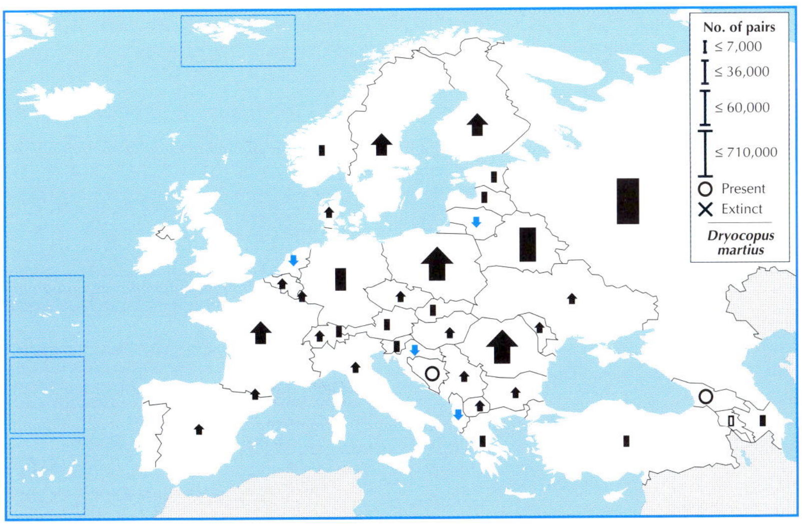

Country	Breeding pop. size (pairs)	Year(s)	Trend	Mag.%	References
Albania	300 – 1,000	98–02	(–)	(0–19)	
Andorra	**25 – 30**	99–01	+	20–29	*1,3*
Armenia	80 – 150	97–02	?	–	
Austria	4500 – 8,000	98–02	(0)	(0–19)	
Azerbaijan	(1,000 – 2,000)	96–00	(0)	(0–19)	
Belarus	45,000 – 80,000	97–02	0	0–19	
Belgium	1,200 – 2,600	95–02	+	0–19	*1*
Bosnia & HG	Present	90–03	?	–	
Bulgaria	**2,000 – 3,000**	96–02	+	**>80**	
Croatia	(1,000 – 1500)	02	(–)	(30–49)	*16,44*
Czech Rep.	**4,000 – 8,000**	00	+	**30–49**	
Denmark	200 – 300	00	(+)	(0–19)	*3*
Estonia	2,000 – 4,000	98	0	0–19	*1*
Finland	10,000 – 20,000	98–02	+	**35**	
France	8,000 – 32,000	98–02	+	0–19	*4*
Georgia	Present	03	?	–	
Germany	28,000 – 44,000	95–99	0	0–19	
Greece	(1,000 – 2,000)	95–00	(0)	(0–19)	
Hungary	**5,000 – 9,000**	99–02	+	0–19	*19*
Italy	(1,000 – 4,000)	03	(+)	(0–19)	
Latvia	6,000 – 8,000	90–00	(0)	(0–19)	*18*
Liechtenstein	30 – 50	98–00	(0)	(0–19)	
Lithuania	(3,000 – 6,000)	99–01	(–)	(0–19)	*20*
Luxembourg	100 – 150	02	+	**0–19**	
Macedonia	1,500 – 5,000	90–00	(+)	(0–19)	
Moldova	2 – 6	90–00	+	**30–49**	
Netherlands	**1,100 – 1,600**	98–00	–	**22**	*1*
Norway	2,000 – 4,000	03	(F)	(20–29)	38
Poland	35,000 – 70,000	00–02	+	0–19	*23*
Romania	40,000 – 60,000	00–02	+	0–19	
Russia	(500,000 – 1,000,000)	90–00	0	0–19	*104,122*
Serbia & MN	1,900 – 2,600	90–02	+	0–19	*1,29,155,172a, 225,227*
Slovakia	1,500 – 2,500	90–99	0	0–19	
Slovenia	1,500 – 2,500	94	(0)	(0–19)	
Spain	**1,000 – 1,400**	98–02	+	0–19	*10*
Sweden	**20,000 – 30,000**	99–00	+	**3**	
Switzerland	**3,000 – 5,000**	93–96	+	0–9	
Turkey	(500 – 1,500)	01	(0)	(0–19)	
Ukraine	5,000 – 9,400	90–00	+	0–19	
Total (approx.)	740,000 – 1,400,000	**Overall trend** Stable			
Breeding range	>5,000,000 km²	**Gen. length.** <3.3 **% Global pop.** 25–49			

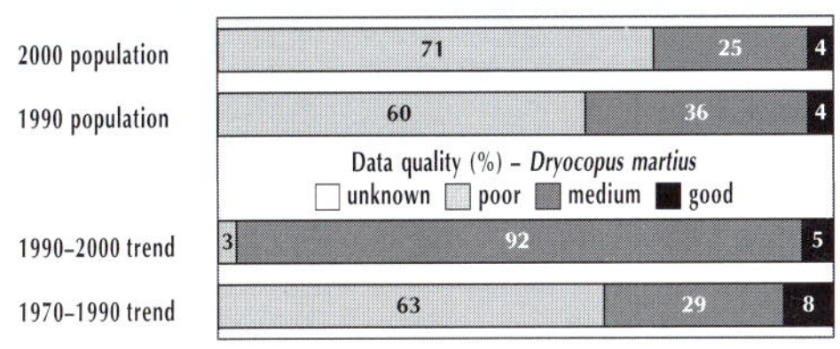

Dendrocopos major
GREAT SPOTTED WOODPECKER

Non-SPEC (1994: —) **Status** Secure
Criteria —
European IUCN Red List Category —
Criteria —
Global IUCN Red List Category —
Criteria —

Dendrocopos major is a widespread resident across most of Europe, which accounts for less than half of its global range. Its European breeding population is extremely large (>12,000,000 pairs), and was stable between 1970–1990. Although there were declines in a few countries during 1990–2000, populations were stable or increased across the majority of its European range—including sizeable ones in France, Germany, Poland, Ukraine and Russia—and the species remained stable overall. Consequently, it is evaluated as Secure.

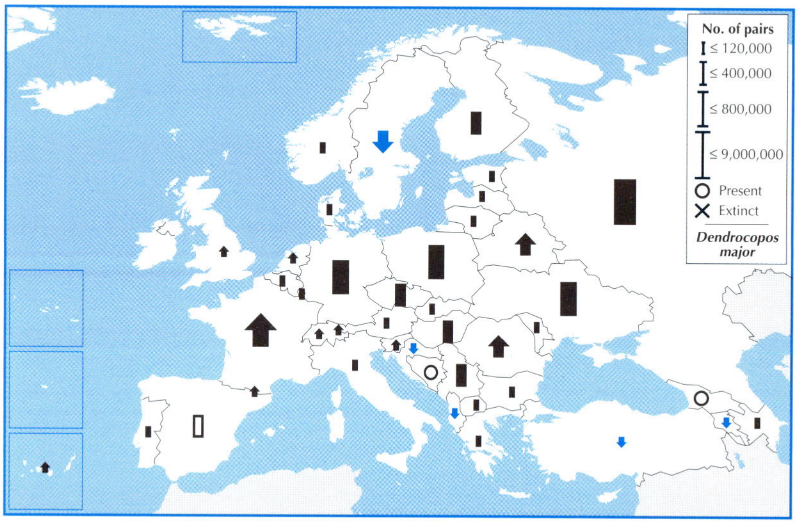

Country	Breeding pop. size (pairs)	Year(s)	Trend	Mag.%	References
Albania	500 – 1,000	98–02	(–)	(0–19)	
Andorra	200 – 300	99–01	(+)	(20–29)	*1,3*
Armenia	1,000 – 2,500	95–02	–	0–19	
Austria	(60,000 – 120,000)	98–02	(0)	(0–19)	
Azerbaijan	(5,000 – 20,000)	96–00	(0)	(0–19)	
Belarus	300,000 – 500,000	97–02	+	50–79	
Belgium	25,000 – 40,000	01–02	(0)	(0–19)	*1*
Bosnia & HG	Present	90–03	?	–	
Bulgaria	90,000 – 150,000	96–02	0	0–9	
Croatia	(10,000 – 25,000)	02	(–)	(50–79)	*16*
Czech Rep.	**200,000 – 400,000**	00	**0**	**0–19**	
Denmark	30,000 – 80,000	00	0	0–19	*3*
Estonia	20,000 – 50,000	98	0	0–19	*1*
Finland	100,000 – 350,000	98–02	**0**	**5**	
France	(400,000 – 1,600,000)	00	(+)	(10)	*1,2*
Georgia	Present	03	?	–	
Germany	450,000 – 830,000	95–99	0	0–19	
Greece	(1,000 – 2,000)	95–00	(0)	(0–19)	
Hungary	**250,000 – 350,000**	99–02	0	0–19	*19*
Italy	(50,000 – 100,000)	03	(0)	(0–19)	
Latvia	60,000 – 100,000	90–00	0	0–19	*18*
Liechtenstein	200 – 300	98–00	(+)	(0–9)	
Lithuania	(40,000 – 60,000)	99–01	(0)	(0–19)	*20*
Luxembourg	3,000 – 5,000	02	**0**	**0–19**	
Macedonia	(100 – 500)	90–00	(0)	(0–19)	
Moldova	3,100 – 3,900	90–00	0	0–19	
Netherlands	**55,000 – 65,000**	98–00	**+**	**32**	*1*
Norway	5,000 – 15,000	03	(0)	(0–19)	*38*
Poland	**400,000 – 800,000**	00–02	(0)	(0–19)	*23*
Portugal	(10,000 – 100,000)	02	(0)	(0–19)	
Romania	340,000 – 460,000	00–02	+	0–19	*48*
Russia	8,000,000 – 10,000,000	90–00	(0)	(0–19)	*39,114,122*
Serbia & MN	150,000 – 240,000	90–02	0	0–19	*1,29,172a,67a, 225,227*
Slovakia	30,000 – 60,000	90–99	0	0–19	
Slovenia	10,000 – 20,000	94	(+)	(0–19)	
Spain	(143,000 – 190,000)	92	?	–	13,12,10
Canary Is.	250 – 1,000	97–03	(+)	(0–19)	28,25
Sweden	100,000 – 250,000	99–00	**–**	**14**	
Switzerland	**35,000 – 55,000**	93–96	+	10–19	
Turkey	(50,000 – 100,000)	01	(–)	(0–19)	
Ukraine	(450,000 – 1,130,000)	90–00	(0)	(0–19)	
UK	**37,000 – 44,400**	00	**+**	**48**	*5,31*

Total (approx.) 12,000,000 – 18,000,000 **Overall trend** Stable
Breeding range >7,000,000 km² **Gen. length.** <3.3 **% Global pop.** 25–49
(See p. 176, bottom, for data quality graph)

Dendrocopos syriacus
SYRIAN WOODPECKER

Non-SPEC[E] (1994: 4) **Status** (Secure)
Criteria —
European IUCN Red List Category —
Criteria —
Global IUCN Red List Category —
Criteria —

Dendrocopos syriacus is a widespread resident across much of central and south-eastern Europe, which constitutes >50% of its global range. Its European breeding population is large (>530,000 pairs), and increased between 1970–1990. Although the species increased or was stable in most of its range during 1990–2000, some populations experienced declines—notably the Turkish stronghold—and the species probably declined slightly overall. Nevertheless, this recent decline is still outweighed by earlier increases, and consequently the species is provisionally evaluated as Secure.

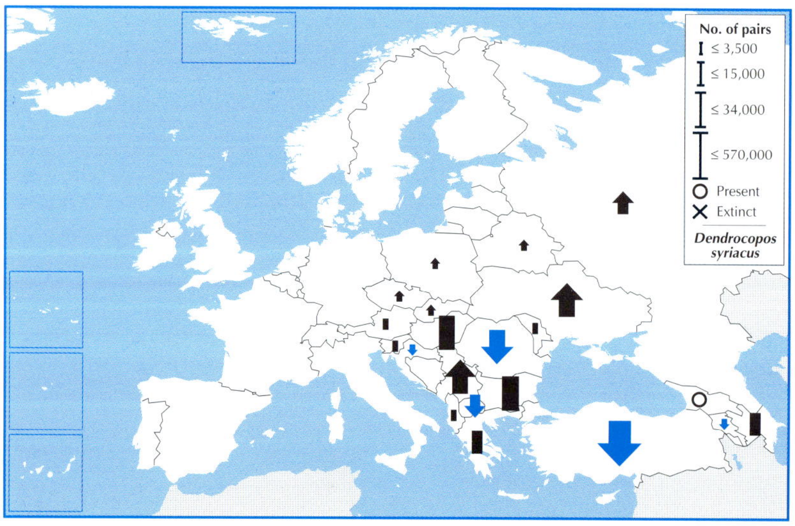

Country	Breeding pop. size (pairs)	Year(s)	Trend	Mag.%	References
Albania	1,000 – 4,000	98–02	(0)	(0–19)	
Armenia	1,000 – 2,000	90–02	–	0–19	
Austria	(2,000 – 4,000)	98–02	(0)	(0–19)	
Azerbaijan	(5,000 – 10,000)	96–00	(0)	(0–19)	
Belarus	**50 – 100**	93–94	+	0–9	
Bulgaria	10,000 – 50,000	98–02	0	0–19	
Croatia	(500 – 1,000)	02	(–)	(30–49)	*70,16*
Czech Rep.	**1,000 – 1,400**	00	**+**	**>80**	
Georgia	Present	03	?	–	
Greece	(10,000 – 20,000)	95–00	(0)	(0–19)	
Hungary	**15,000 – 35,000**	99–02	0	0–19	*19*
Macedonia	(3,000 – 15,000)	90–00	(–)	(20–29)	
Moldova	3,200 – 3,800	90–00	0	0–19	
Poland	300 – 800	97–02	+	50–79	*86*
Romania	24,000 – 32,000	00–02	–	20–29	*48*
Russia	(5,000 – 10,000)	90–00	**+**	**20–29**	*8,19*
Serbia & MN	25,000 – 33,000	90–02	+	0–19	*1,29,155,172a, 67a,78*
Slovakia	1,500 – 2,500	80–99	+	20–29	
Slovenia	(5 – 10)	97–00	(0)	(0–19)	
Turkey	(400,000 – 800,000)	01	(–)	(0–19)	
Ukraine	(25,000 – 45,000)	90–00	+	0–19	

Total (approx.) 530,000 – 1,100,000 **Overall trend** Small decline
Breeding range >2,000,000 km² **Gen. length.** <3.3 **% Global pop.** 50–74

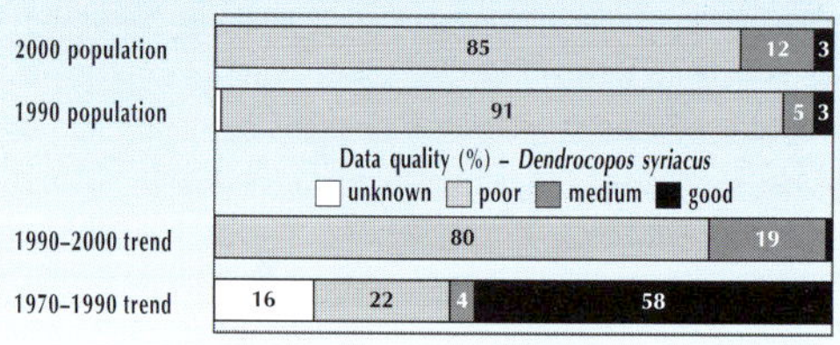

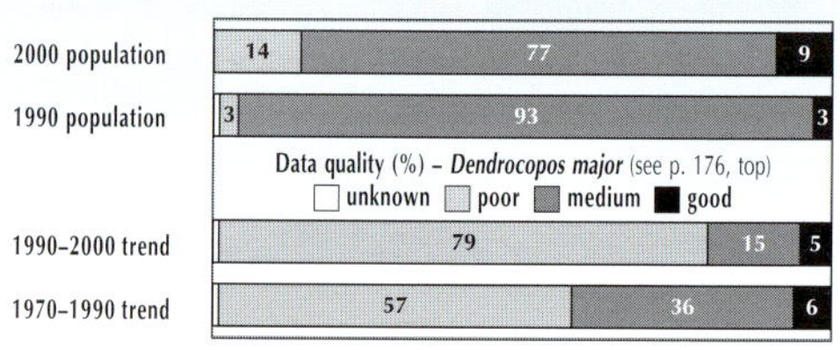

Dendrocopos medius
MIDDLE SPOTTED WOODPECKER

Non-SPEC[E] (1994: 4) **Status** (Secure)
Criteria —
European IUCN Red List Category —
Criteria —
Global IUCN Red List Category —
Criteria —

Dendrocopos medius is a widespread resident across much of central and south-eastern Europe, which constitutes >95% of its global range. Its European breeding population is large (>140,000 pairs), and was stable between 1970–1990. Although there were declines in south-eastern Europe—notably in Romania—during 1990–2000, key populations in Germany, Poland and Greece were stable (though the trend in France was unknown), and the species probably remained stable overall. Consequently, it is provisionally evaluated as Secure.

Country	Breeding pop. size (pairs)	Year(s)	Trend	Mag.%	References
Albania	500 – 1,000	98–02	(–)	(0–19)	
Armenia	1,000 – 1500	90–02	–	0–19	
Austria	2,900 – 4,300	98–02	(0)	(0–19)	
Azerbaijan	(500 – 5,000)	96–00	(0)	(0–19)	
Belarus	5,000 – 9,000	97–02	0	0–19	
Belgium	1,515 – 3,017	95–02	+	>80	*1*
Bulgaria	8,000 – 15,000	96–02	0	0–9	
Croatia	(3,000 – 5,000)	02	(–)	(30–49)	*70,54,16,57,44,57*
Czech Rep.	**3,500 – 7,000**	00	+	**>80**	
Estonia	1 – 10	00	+	N	*1*
France	(25,000 – 100,000)	98–02	?	–	*4*
Georgia	Present	03	?	–	
Germany	9,700 – 16,000	95–99	0	0–19	
Greece	(10,000 – 30,000)	95–00	(0)	(0–19)	
Hungary	7,000 – 16,000	99–02	0	0–19	*19*
Italy	(400 – 600)	03	?	–	
Latvia	1,500 – 2,000	90–00	+	>80	*18*
Lithuania	2,000 – 2,500	99–01	(0)	(0–19)	*20*
Luxembourg	200 – 300	02	**0**	**0–19**	
Macedonia	(3,000 – 10,000)	90–00	(–)	(20–29)	
Moldova	50 – 80	90–00	–	30–49	
Netherlands	**10 – 15**	98–00	?	–	*1*
Poland	**10,000 – 20,000**	00–02	(0)	(0–19)	*23*
Romania	(20,000 – 24,000)	00–02	–	0–19	*48*
Russia	2,000 – 4,000	90–00	?	–	*5,74,136,160*
Serbia & MN	5,500 – 9,000	90–02	–	0–19	*1,29,172a,67a, 155,227*
Slovakia	2,500 – 4,000	90–99	0	0–19	
Slovenia	700 – 900	99–00	(–)	(0–19)	
Spain	1,045 – 1,205	98–02	?	–	*10,16*
Switzerland	**250 – 300**	93–96	–	0–9	
Turkey	(5,000 – 10,000)	01	(0)	(0–19)	
Ukraine	6,000 – 9,500	90–00	0	0–9	
Total (approx.)	140,000 – 310,000	**Overall trend** Stable			
Breeding range	>2,000,000 km²	**Gen. length.** <3.3 **% Global pop.** >95			

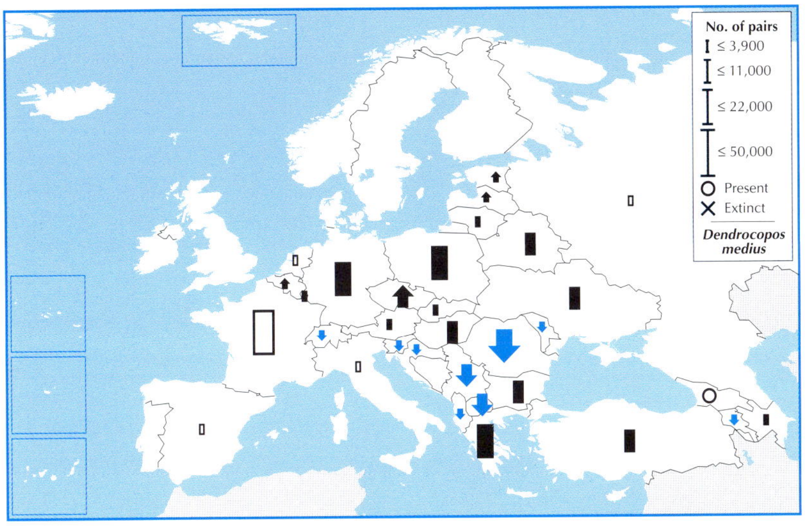

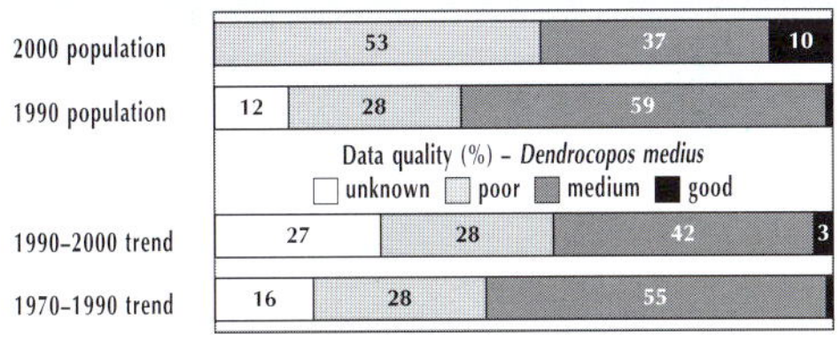

Dendrocopos leucotos
WHITE-BACKED WOODPECKER

Non-SPEC (1994: —) **Status** (Secure)
Criteria —
European IUCN Red List Category —
Criteria —
Global IUCN Red List Category —
Criteria —

Dendrocopos leucotos is a widespread resident in eastern Europe, and occurs more patchily elsewhere, with Europe accounting for less than half of its global range. Its European breeding population is large (>180,000 pairs), and was stable between 1970–1990. Although there were declines in a few countries during 1990–2000, populations were stable or increased across most of its European range, and the species probably remained stable overall. Consequently, it is provisionally evaluated as Secure.

Country	Breeding pop. size (pairs)	Year(s)	Trend	Mag.%	References
Albania	200 – 500	98–02	(–)	(0–19)	
Austria	800 – 1,500	98–02	(0)	(0–19)	
Azerbaijan	(0 – 100)	96–00	(0)	(0–19)	
Belarus	5,000 – 6,000	97–02	0	0–19	
Bulgaria	**1,200 – 1,700**	96–02	0	0–9	
Croatia	(100 – 500)	02	(–)	(50–79)	*70,16*
Czech Rep.	**150 – 250**	00	**0**	**0–19**	
Estonia	350 – 700	98	–	20–29	*1*
Finland	**30 – 50**	98–02	+	**100**	
France	200 – 450	00	0	0–19	*1*
Georgia	Present	03	?	–	
Germany	250 – 400	95–99	0	0–19	
Greece	(500 – 2,000)	95–00	(0)	(0–19)	
Hungary	**250 – 400**	98–02	0	0–19	
Italy	300 – 500	03	0	0–19	
Latvia	2,000 – 2,500	90–00	(+)	(30–49)	*18*
Liechtenstein	(4 – 8)	98–00	(0)	(0–19)	
Lithuania	900 – 1,200	99–01	(0)	(0–19)	*20*
Macedonia	(100 – 1,000)	90–00	(0)	(0–19)	
Norway	**1,700 – 1,800**	90–00	**–**	**0–19**	*62*
Poland	**400 – 600**	90–00	0	0–19	*2*
Romania	16,000 – 24,000	00–02	0	0–19	*1*
Russia	150,000 – 500,000	90–00	(0)	(0–19)	*39,122*
Serbia & MN	500 – 1,000	90–02	0	0–19	*1,172a,155,227, 225*
Slovakia	1,500 – 2,500	90–99	0	0–19	
Slovenia	20 – 30	99–00	(0)	(0–19)	
Spain	**78 – 95**	98–02	?	–	*10,16*
Sweden	**4 – 6**	99–00	**–**	**>80**	
Switzerland	5 – 10	98–02	(+)	(N)	
Turkey	(500 – 900)	01	(–)	(0–19)	
Ukraine	**570 – 930**	90–00	**–**	**30–49**	
Total (approx.)	180,000 – 550,000	**Overall trend** Stable			
Breeding range	>2,000,000 km²	**Gen. length.** <3.3 **% Global pop.** 25–49			

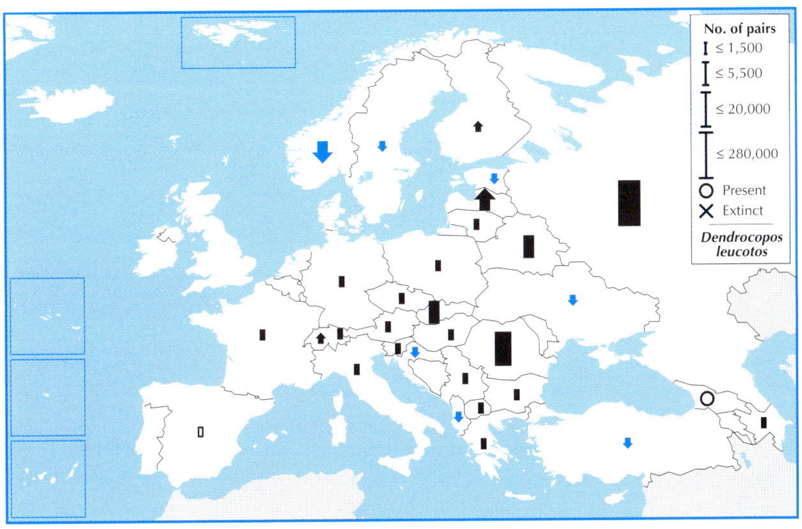

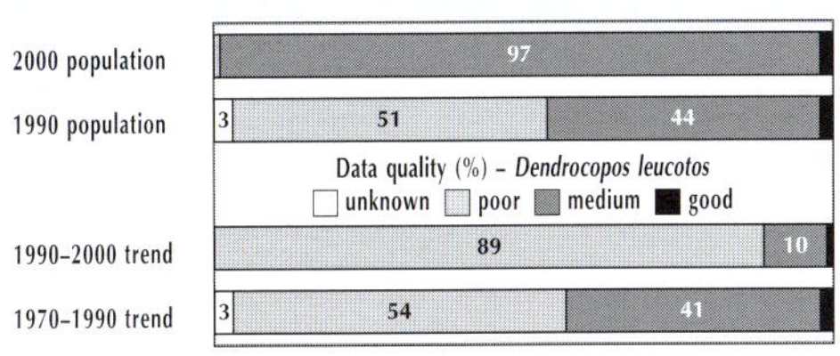

Dendrocopos minor
LESSER SPOTTED WOODPECKER

Non-SPEC (1994: —) **Status** (Secure)
Criteria —
European IUCN Red List Category —
Criteria —
Global IUCN Red List Category —
Criteria —

Dendrocopos minor is a widespread resident across most of Europe, which accounts for less than half of its global range. Its European breeding population is large (>450,000 pairs), and was stable between 1970–1990. Although the species declined in some countries during 1990–2000, and the trend of the key population in Russia was unknown, most populations across Europe—including the sizeable one in France—were stable. The species probably remained stable overall, and is provisionally evaluated as Secure.

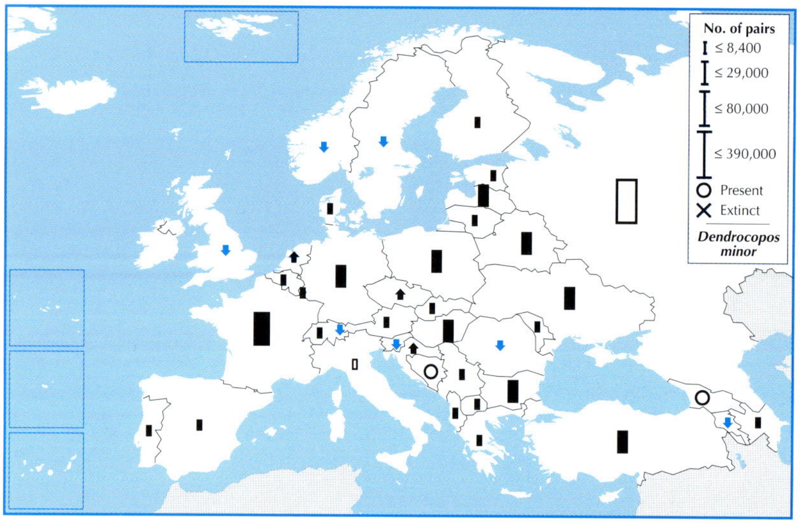

Country	Breeding pop. size (pairs)	Year(s)	Trend	Mag.%	References
Albania	100 – 500	98–02	(0)	(0–19)	
Armenia	(1,000 – 2,500)	95–02	(–)	(0–19)	
Austria	(2,200 – 4,500)	98–02	(0)	(0–19)	
Azerbaijan	(500 – 1,000)	96–00	(0)	(0–19)	
Belarus	12,000 – 20,000	97–00	0	0–19	
Belgium	1,700 – 3,800	01–02	(0)	(0–19)	*1*
Bosnia & HG	Present	90–03	?	–	
Bulgaria	10,000 – 20,000	96–02	0	0–9	
Croatia	(2,000 – 5,000)	02	(+)	(0–19)	*70,16,44,57*
Czech Rep.	**3,000 – 6,000**	00	**+**	**30–49**	
Denmark	(33 – 55)	98	(0)	(0–19)	*4*
Estonia	2,000 – 5,000	98	0	0–19	1
Finland	4,000 – 7,000	98–02	0	0–19	
France	(40,000 – 160,000)	98–02	(0)	(0–19)	*4*
Georgia	Present	03	?	–	
Germany	16,000 – 32,000	95–99	0	0–19	
Greece	(500 – 1,000)	95–00	(0)	(0–19)	
Hungary	**7,000 – 40,000**	99–02	0	0–19	*19*
Italy	(3,000 – 5,000)	03	?	–	
Latvia	10,000 – 15,000	90–00	0	0–19	*23,18*
Liechtenstein	10 – 15	98–00	–	0–9	
Lithuania	(7,000 – 10,000)	99–01	(0)	(0–19)	*20*
Luxembourg	300 – 400	02	**0**	**0–19**	
Macedonia	(2,000 – 5,000)	90–00	(0)	(0–19)	
Moldova	210 – 290	90–00	0	0–19	
Netherlands	**4,500 – 5,500**	98–00	**+**	**86**	*1*
Norway	**2,000 – 3,000**	90–00	–	0–19	*38*
Poland	**20,000 – 40,000**	00–02	(0)	(0–19)	*23*
Portugal	(1,000 – 5,000)	02	(0)	(0–19)	
Romania	(6,000 – 8,000)	00–02	–	20–29	*48*
Russia	250,000 – 600,000	90–00	?	–	*116,122*
Serbia & MN	3500 – 5,000	90–02	0	0–19	*1,29,155,172a, 67a,227,91*
Slovakia	2,000 – 4,000	90–99	0	0–19	
Slovenia	500 – 1,000	94	(–)	(0–19)	
Spain	2,971 – 4,044	98–02	0	0–19	*10*
Sweden	3,000 – 6,000	99–00	–	20–29	
Switzerland	**2,500 – 3,000**	93–96	0	0–19	
Turkey	(8,000 – 16,000)	01	(0)	(0–19)	
Ukraine	15,000 – 23,000	90–00	0	0–19	
UK	**1,400 – 2,900**	00	**–**	**51**	*31*

Total (approx.) 450,000 – 1,100,000 **Overall trend** Unknown
Breeding range >6,000,000 km² **Gen. length.** <3.3 **% Global pop.** 25–49
(See p. 178, bottom, for data quality graph)

Picoides tridactylus
THREE-TOED WOODPECKER

SPEC 3 (1994: 3) **Status** (Depleted)
Criteria Moderate historical decline
European IUCN Red List Category —
Criteria —
Global IUCN Red List Category —
Criteria —

Picoides tridactylus is a resident in northern and parts of central Europe, which accounts for less than a quarter of its global range. Its European breeding population is large (>350,000 pairs), but underwent a moderate decline between 1970–1990. Although declines continued in a few countries during 1990–2000, the species was stable across most of its European range (the trend in its Russian stronghold was unknown). Nevertheless, its total population size probably remains below the level that preceded its decline, and consequently the species is provisionally evaluated as Depleted.

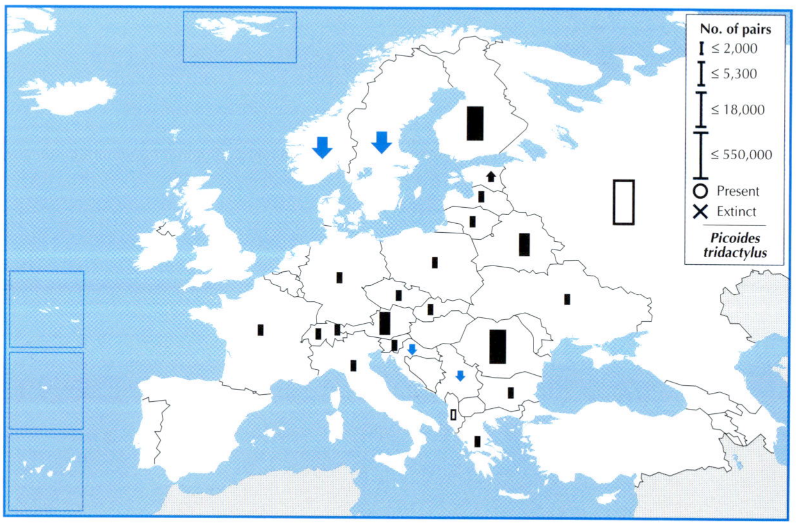

Country	Breeding pop. size (pairs)	Year(s)	Trend	Mag.%	References
Albania	(20 – 100)	98–02	?	–	
Austria	2,200 – 4,600	98–02	(0)	(0–19)	
Belarus	3,000 – 5,000	97–02	0	0–19	
Bulgaria	80 – 200	96–02	(0)	(0–19)	
Croatia	(10 – 100)	02	(–)	(>80)	*70,16*
Czech Rep.	**300 – 500**	00	**0**	**0–19**	
Estonia	500 – 1,200	98	+	20–29	*1*
Finland	15,000 – 20,000	98–02	0	0–19	
France	50 – 100	98–02	(0)	(0–19)	*1*
Germany	720 – 1,030	95–99	**0**	**0–19**	
Greece	(50 – 100)	95–00	(0)	(0–19)	
Italy	80 – 150	03	(0)	(0–19)	
Latvia	(1,500 – 2,500)	90–00	(0)	(0–19)	*18*
Liechtenstein	20 – 40	98–00	0	0–19	
Lithuania	(150 – 200)	99–01	(0)	(0–19)	*20*
Norway	3,000 – 6,000	03	(–)	(0–19)	*38*
Poland	300 – 700	95–00	0	0–19	*2,87*
Romania	15,000 – 20,000	00–02	0	0–19	*4*
Russia	(300,000 – 1,000,000)	90–00	?	–	*104,122*
Serbia & MN	**50 – 90**	95–02	–	0–19	*1,62,10b,155, 225,227,228*
Slovakia	800 – 1,200	90–99	0	0–19	
Slovenia	200 – 300	99–00	(0)	(0–19)	
Sweden	4,000 – 7,000	99–00	–	10–19	
Switzerland	**1,000 – 1,500**	93–96	(0)	(0–19)	
Ukraine	340 – 530	90–00	(0)	(0–19)	

Total (approx.) 350,000 – 1,100,000 **Overall trend** Unknown
Breeding range >2,000,000 km² **Gen. length.** <3.3 **% Global pop.** 5–24

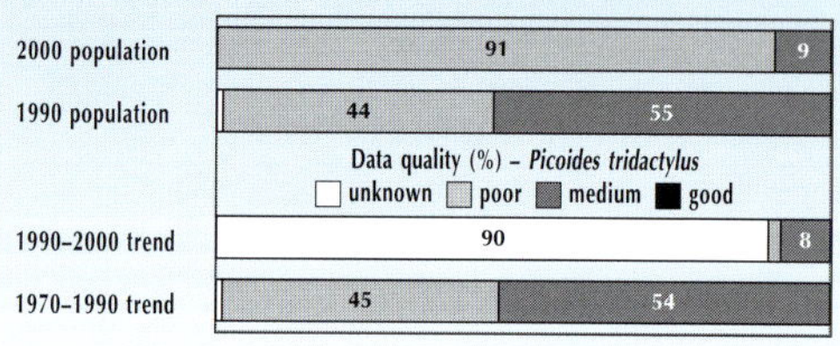

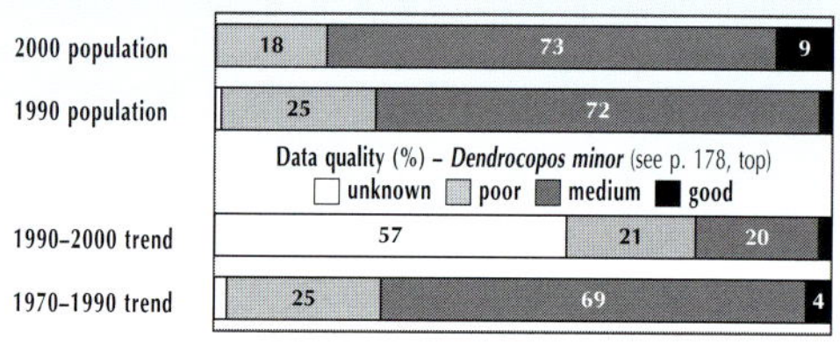

Ammomanes deserti
DESERT LARK

SPEC 3 (1994: —) **Status** (Endangered)
Criteria See IUCN below
European IUCN Red List Category EN
Criteria C1; D1
Global IUCN Red List Category —
Criteria —

Country	Breeding pop. size (pairs)	Year(s)	Trend	Mag.%	References
Turkey	(30 – 300)	01	(–)	(30–49)	
Total (approx.)	30 – 300	**Overall trend** Large decline			
Breeding range	<20,000 km²	**Gen. length** <3.3	**% Global pop.** <5		

Ammomanes deserti has a predominantly North African and Middle Eastern distribution, which just extends into Europe in south-eastern Turkey. The European breeding population is extremely small (as few as 30 pairs), and though the trend between 1970–1990 was unknown, it underwent a large decline (>30%) during 1990–2000. As a consequence of this decline and its small population size, this previously Secure species is now provisionally evaluated as Endangered.

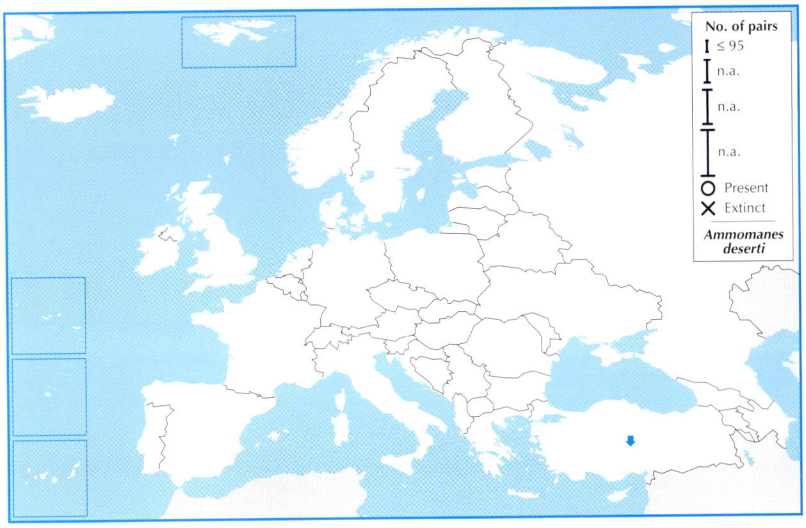

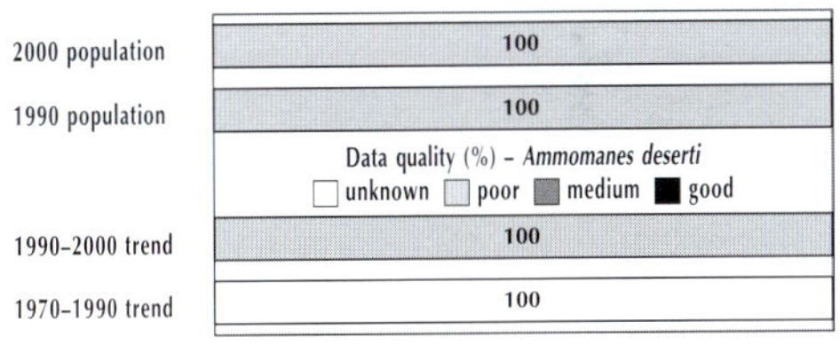

Chersophilus duponti
DUPONT'S LARK

SPEC 3 (1994: 3) **Status** (Depleted)
Criteria Large historical decline
European IUCN Red List Category —
Criteria —
Global IUCN Red List Category —
Criteria —

Country	Breeding pop. size (pairs)	Year(s)	Trend	Mag.%	References
Spain	13,000 – 15,000	88	(–)	(0–19)	*4,12,16,12*
Total (approx.)	13,000 – 15,000	**Overall trend** Small decline			
Breeding range	>50,000 km²	**Gen. length** <3.3	**% Global pop.** 25–49		

Chersophilus duponti occurs mainly in North Africa, but is also patchily distributed in parts of north-eastern and south-eastern Spain. Its European breeding population is relatively small (<15,000 pairs), and underwent a large decline between 1970–1990. The species continued to decline—albeit at a reduced rate—during 1990–2000, and the population probably underwent a small decline overall. Its population size clearly remains far below the level that preceded its decline, and consequently the species is provisionally evaluated as Depleted.

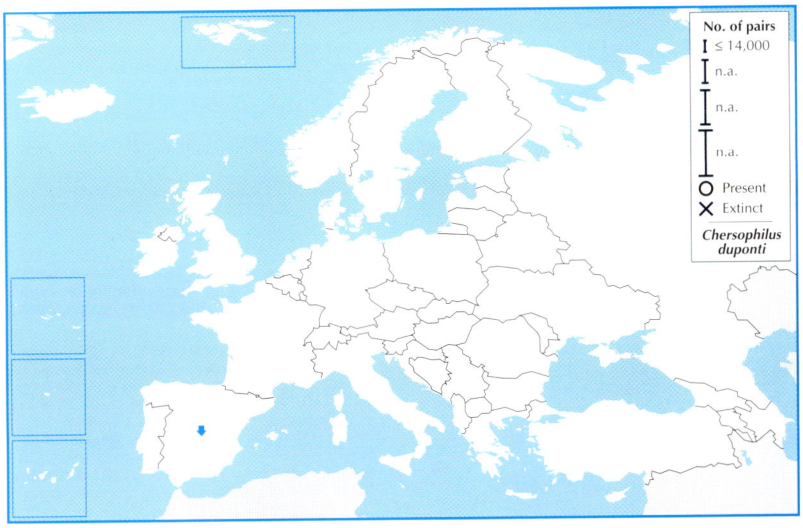

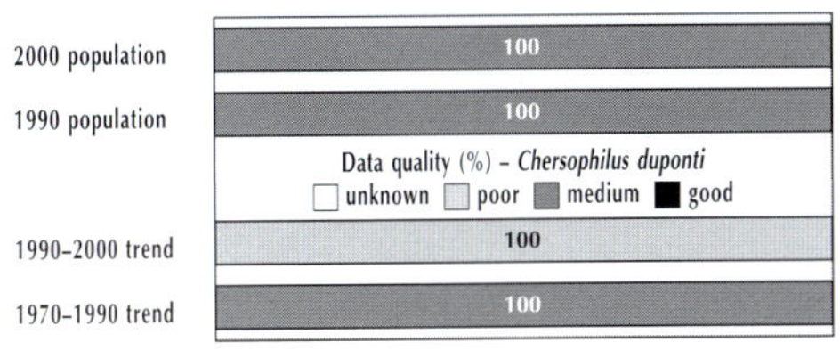

Melanocorypha calandra
CALANDRA LARK

SPEC 3 (1994: 3) **Status** (Declining)
Criteria Moderate continuing decline
European IUCN Red List Category —
Criteria —
Global IUCN Red List Category —
Criteria —

Melanocorypha calandra is a widespread breeder in southern and south-eastern Europe, which accounts for less than half of its global breeding range. Its European breeding population is extremely large (>10,000,000 pairs), but underwent a moderate decline between 1970–1990. Although certain populations—notably the sizeable Turkish one—were stable or increased during 1990–2000, the species declined across much of its European range, and probably underwent a moderate decline (>10%) overall. Consequently, it is provisionally evaluated as Declining.

Country	Breeding pop. size (pairs)	Year(s)	Trend	Mag.%	References
Albania	(500 – 2,000)	95–02	(–)	(0–19)	
Azerbaijan	(5,000 – 10,000)	96–00	(0)	(0–19)	
Bulgaria	3,000 – 5,000	96–02	+	0–19	
Croatia	(100 – 200)	02	(0)	(0–19)	*70*
Cyprus	(4,000 – 12,000)	94–02	(0)	(0–19)	
France	40 – 50	00	–	**20–49**	*1*
Georgia	Present	03	?	–	
Greece	(3,000 – 5,000)	95–00	(–)	(0–19)	
Italy	(7,000 – 15,000)	03	(–)	(0–19)	
Macedonia	(50,000 – 100,000)	90–00	(0)	(0–19)	
Moldova	**15 – 25**	90–00	–	20–29	
Portugal	(1,000 – 10,000)	02	(–)	(–)	
Romania	85,000 – 105,000	00–02	0	0–19	*48*
Russia	(5,000,000 – 10,000,000)	90–00	–	20–29	*8*
Serbia & MN	100 – 200	90–02	–	10–29	*1,156,227*
Spain	(1,030,000 – 3,400,000)	92	?	–	*13,12,10*
Turkey	(4,000,000 – 10,000,000)	01	(0)	(0–19)	
Ukraine	(40,000 – 80,000)	90–00	–	10–19	
Total (approx.)	10,000,000 – 24,000,000	**Overall trend** Moderate decline			
Breeding range	>2,000,000 km²	**Gen. length.** <3.3		**% Global pop.** 25–49	

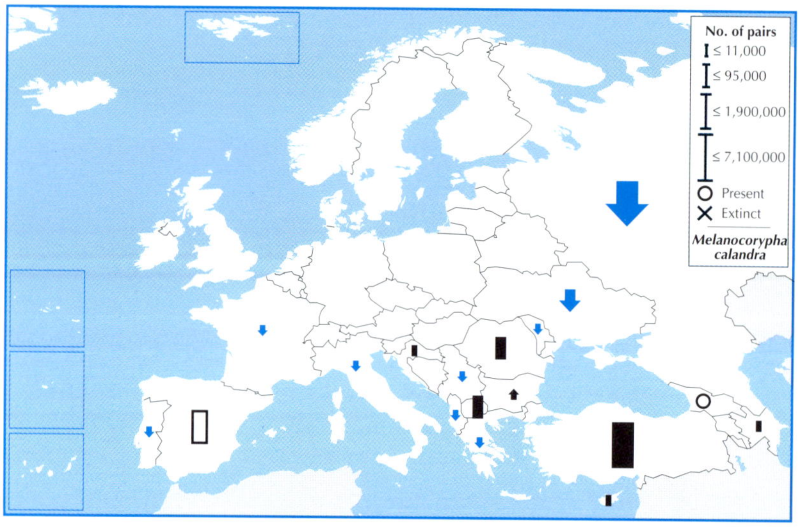

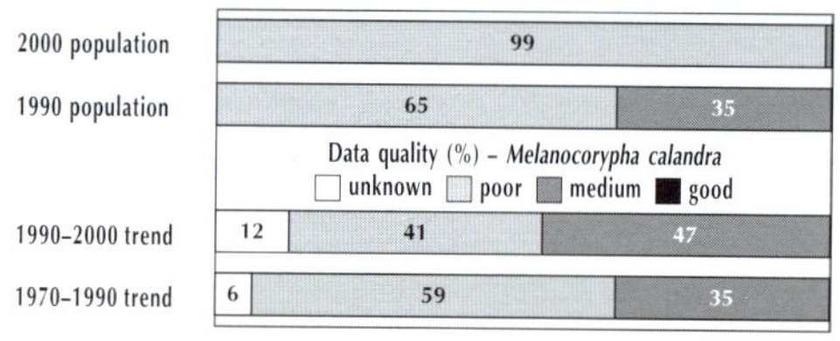

Melanocorypha bimaculata
BIMACULATED LARK

Non-SPEC (1994: —) **Status** Secure
Criteria —
European IUCN Red List Category —
Criteria —
Global IUCN Red List Category —
Criteria —

Melanocorypha bimaculata is a widespread summer visitor to Turkey and parts of the Caucasus, with Europe accounting for less than a quarter of its global breeding range. Its European breeding population is very large (>1,000,000 pairs), but its trend between 1970–1990 was unknown. The species was stable in Armenia and Azerbaijan during 1990–2000, but the stronghold population in Turkey increased, and the species probably underwent a small increase overall. Consequently, it is evaluated as Secure.

Country	Breeding pop. size (pairs)	Year(s)	Trend	Mag.%	References
Armenia	30,000 – 150,000	99–02	0	0–19	
Azerbaijan	(0 – 50)	96–00	(0)	(0–19)	
Georgia	Present	03	?	–	
Turkey	(1,000,000 – 2,000,000)	01	(+)	(0–19)	
Total (approx.)	1,000,000 – 2,200,000	**Overall trend** Small increase			
Breeding range	>500,000 km²	**Gen. length.** <3.3		**% Global pop.** 5–24	

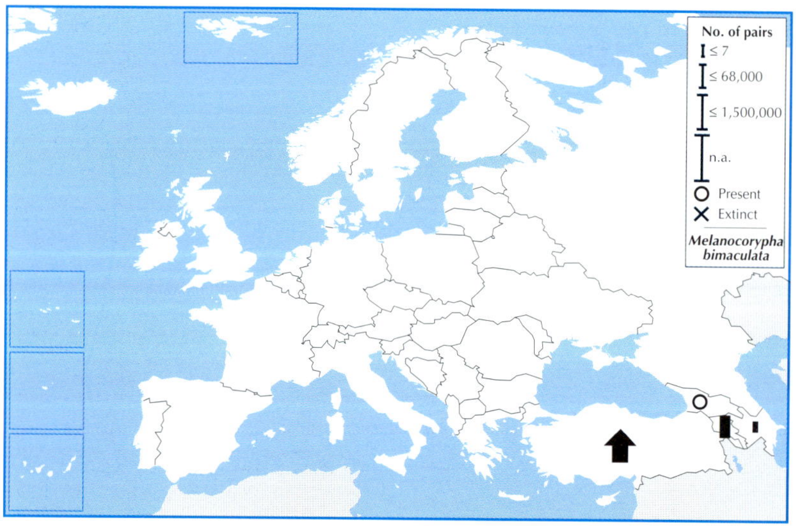

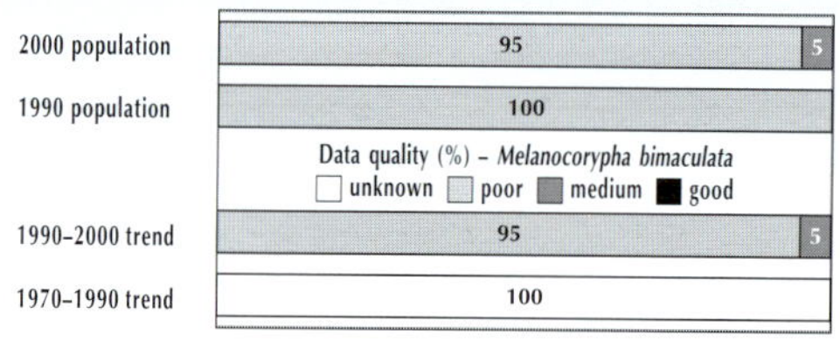

Melanocorypha leucoptera
WHITE-WINGED LARK

Non-SPEC^E W (1994: 4W) **Status** (Secure)
Criteria —
European IUCN Red List Category —
Criteria —
Global IUCN Red List Category —
Criteria —

Country	Breeding pop. size (pairs)	Year(s)	Trend	Mag.%	References
Russia	(20,000 – 65,000)	90–00	(F)	(30–49)	*8,9*
Total (approx.)	20,000 – 65,000	Overall trend Fluctuating			
Breeding range	>500,000 km²	Gen. length. <3.3	% Global pop. 25–49		

Melanocorypha leucoptera breeds in south-eastern European Russia, but winters more widely in the steppe region to the north and west of the Black Sea, with Europe constituting >50% of its global wintering range. Insufficient information was available to assess the species's status using wintering population data, but its European breeding population is relatively small (<65,000 pairs), and was stable between 1970–1990. Although it fluctuated markedly during 1990–2000, the population remained broadly stable overall, and consequently the species is provisionally evaluated as Secure.

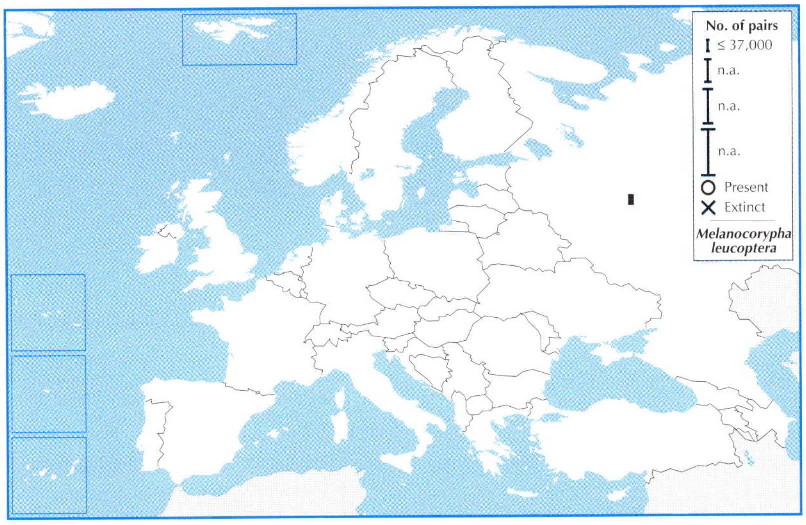

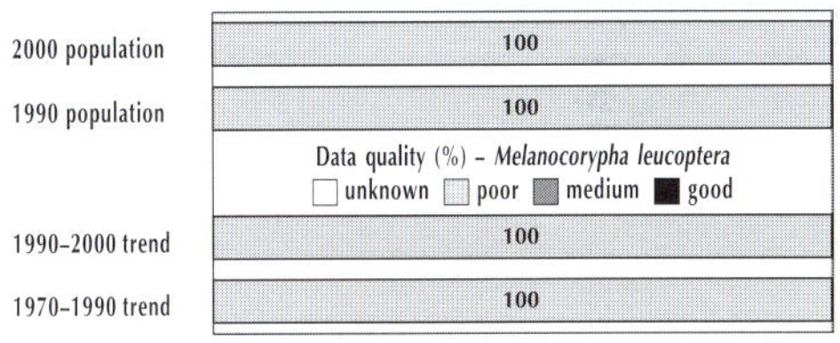

Melanocorypha yeltoniensis
BLACK LARK

SPEC 3 (1994: 3) **Status** Endangered
Criteria See IUCN below
European IUCN Red List Category EN
Criteria A2b
Global IUCN Red List Category —
Criteria —

Country	Breeding pop. size (pairs)	Year(s)	Trend	Mag.%	References
Russia	(4,000 – 7,000)	90–00	-	**50–79**	*178*
Total (approx.)	4,000 – 7,000	Overall trend Large decline			
Breeding range	>500,000 km²	Gen. length. <3.3	% Global pop. 5–24		

Melanocorypha yeltoniensis breeds in the steppe zone of south-eastern European Russia, which accounts for less than a quarter of its global breeding range. Its European breeding population is small (as few as 4,000 pairs), and underwent a large decline between 1970–1990. The species continued to decline in Europe during 1990–2000, undergoing a very large decline (>50%) overall. Consequently, it is evaluated as Endangered.

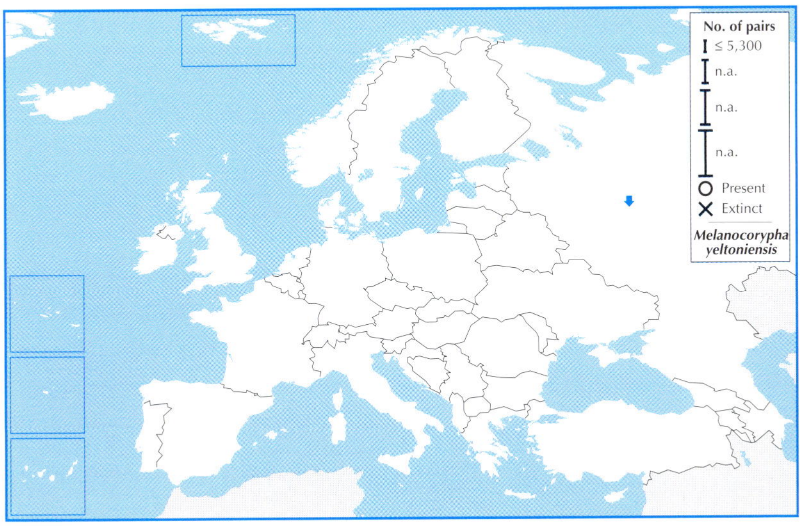

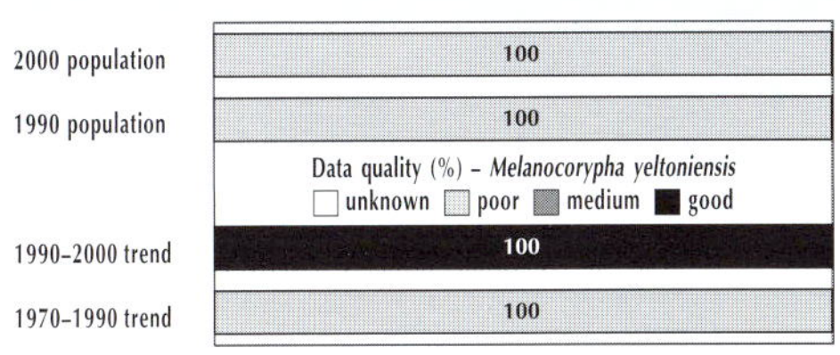

Calandrella brachydactyla
GREATER SHORT-TOED LARK

SPEC 3 (1994: 3) **Status** Declining
Criteria Moderate continuing decline
European IUCN Red List Category —
Criteria —
Global IUCN Red List Category —
Criteria —

Calandrella brachydactyla is a widespread summer visitor to southern and south-eastern Europe, which accounts for less than a quarter of its global breeding range. Its European breeding population is very large (>7,300,000 pairs), but underwent a large decline between 1970–1990. Although several populations—notably the key one in Turkey—were stable or increased during 1990–2000, other sizeable populations in Spain and Russia declined, and the species underwent a moderate decline (>10%) overall. Consequently, it is evaluated as Declining.

Country	Breeding pop. size (pairs)	Year(s)	Trend	Mag.%	References
Albania	(2,000 – 5,000)	02	?	–	
Armenia	(50,000 – 100,000)	90–02	0	0–19	
Azerbaijan	(10,000 – 100,000)	96–00	(0)	(0–19)	
Bulgaria	1,500 – 3,000	96–02	0	0–9	
Croatia	(100 – 500)	02	(–)	(50–79)	*70,16*
Cyprus	(10 – 1,000)	94–02	(0)	(0–19)	
France	3,000 – 6,000	98–02	?	–	*4*
Georgia	Present	03	?	–	
Greece	(20,000 – 30,000)	95–00	(–)	(0–19)	
Hungary	20 – 50	95–02	0	0–19	
Italy	(15,000 – 30,000)	03	(–)	(0–19)	
Macedonia	(10,000 – 30,000)	90–00	(0)	(0–9)	
Malta	**300 – 450**	90–00	**–**	**30–49**	*1*
Portugal	(2,000 – 20,000)	02	(–)	(–)	
Romania	(10,000 – 12,000)	00–02	(0)	(0–19)	
Russia	(2,000,000 – 5,000,000)	90–00	**–**	**20–29**	*8,67*
Serbia & MN	130 – 250	90–02	–	10–19	*1,150,29,155, 225,227*
Slovakia	0 – 10	80–99	?	N	
Spain	(2,200,000 – 2,600,000)	92	(–)	(30–49)	*13,12,16,10*
Turkey	(3,000,000 – 6,000,000)	01	(+)	(0–19)	
Ukraine	6,000 – 10,000	90–00	–	30–49	
Total (approx.)	7,300,000 – 14,000,000	**Overall trend** Moderate decline			
Breeding range	>2,000,000 km²	**Gen. length.** <3.3	**% Global pop.** 5–24		

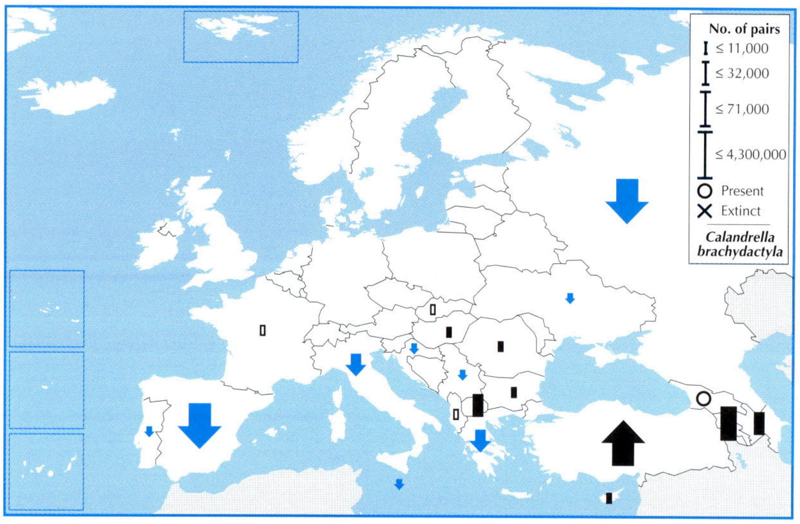

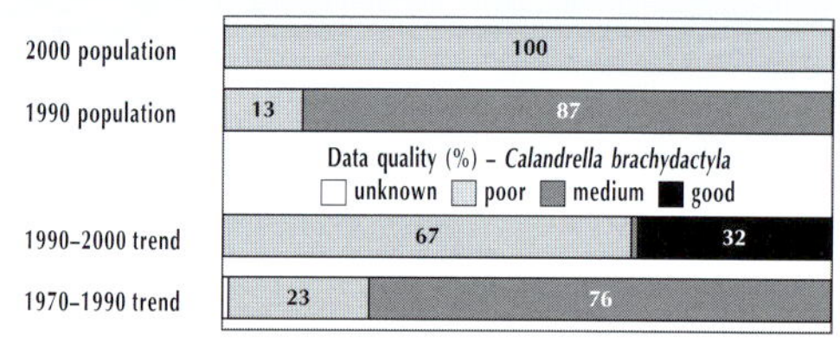

Calandrella rufescens
LESSER SHORT-TOED LARK

SPEC 3 (1994: 3) **Status** Declining
Criteria Moderate continuing decline
European IUCN Red List Category —
Criteria —
Global IUCN Red List Category —
Criteria —

Calandrella rufescens breeds in Iberia and the Canary Islands as well as parts of south-east of Europe, which accounts for less than half of its global breeding range. Its European breeding population is very large (>1,600,000 pairs), but underwent a large decline between 1970–1990. Although the comparatively small populations in Portugal and the Caucasus were stable during 1990–2000, the species continued to decline across most of its European range, and underwent a moderate decline (>10%) overall. Consequently, it is evaluated as Declining.

Country	Breeding pop. size (pairs)	Year(s)	Trend	Mag.%	References
Armenia	20,000 – 50,000	90–02	0	0–19	
Azerbaijan	(10,000 – 100,000)	96–00	(0)	(0–19)	
Georgia	Present	03	?	–	
Portugal	**70 – 76**	02	**0**	**0–19**	*6*
Russia	(1,000,000 – 3,000,000)	90–00	–	20–29	*8*
Spain	(230,000 – 260,000)	92	(–)	(0–19)	*13,12,16,10*
Canary Is.	(10,000 – 20,000)	97–03	(–)	(0–19)	*28,25*
Turkey	(300,000 – 600,000)	01	(–)	(0–19)	
Ukraine	(200 – 2,000)	90–00	(–)	(20–29)	
Total (approx.)	1,600,000 – 4,000,000	**Overall trend** Moderate decline			
Breeding range	>500,000 km²	**Gen. length.** <3.3	**% Global pop.** 25–49		

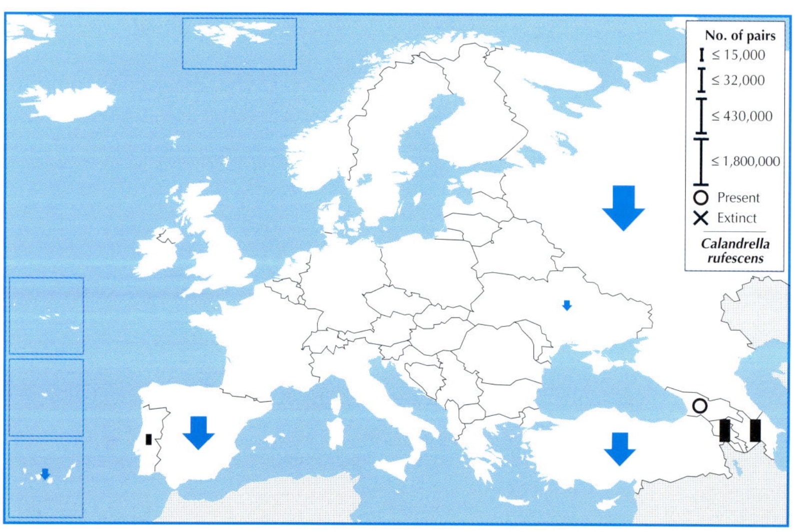

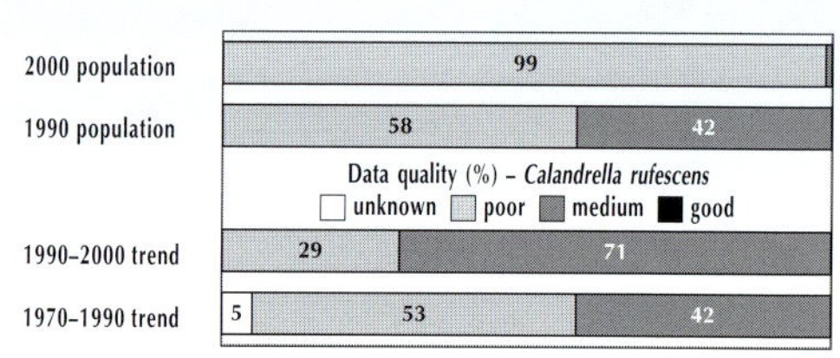

Calandrella cheleensis
ASIAN SHORT-TOED LARK

SPEC 3 (1994: NE) **Status** (Vulnerable)
Criteria See IUCN below
European IUCN Red List Category VU
Criteria A3c
Global IUCN Red List Category —
Criteria —

Country	Breeding pop. size (pairs)	Year(s)	Trend	Mag.%	References
Turkey	10,000 – 30,000	01	(+)	(0–19)	

Total (approx.) 10,000 – 30,000 **Overall trend** Small increase
Breeding range >20,000 km² **Gen. length.** <3.3 **% Global pop.** <5

Calandrella cheleensis (recently split from *C. rufescens*) has a predominantly Asian distribution, but also occurs within Europe on the central plateau in Turkey. Its European breeding population is relatively small (<30,000 pairs), but its trend between 1970–1990 was unknown. The Turkish population increased slightly during 1990–2000, and hence would qualify as provisionally Secure. However, the species is predicted to undergo a large future decline (>30%) owing to habitat loss resulting from dam construction, and consequently it is provisionally evaluated as Vulnerable.

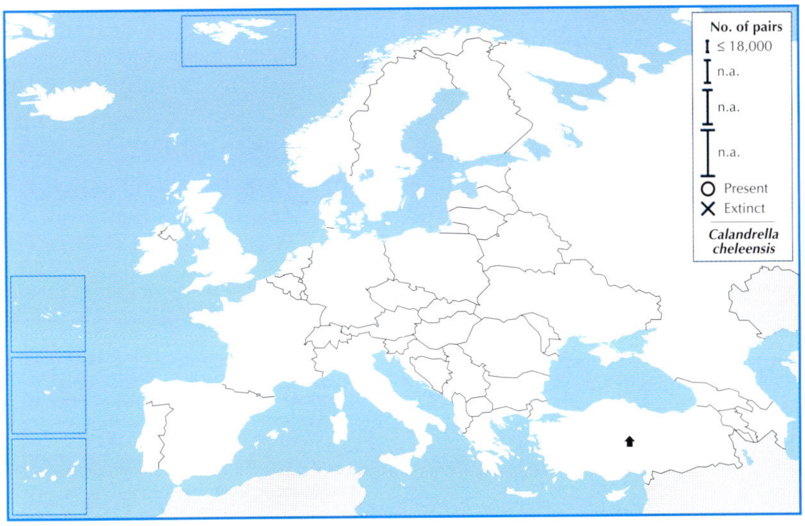

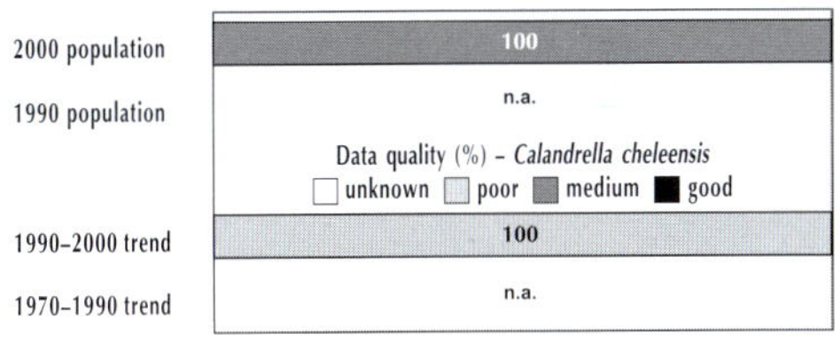

Galerida cristata
CRESTED LARK

SPEC 3 (1994: 3) **Status** (Depleted)
Criteria Moderate historical decline
European IUCN Red List Category —
Criteria —
Global IUCN Red List Category —
Criteria —

Country	Breeding pop. size (pairs)	Year(s)	Trend	Mag.%	References
Albania	10,000 – 20,000	02	?	–	
Armenia	50,000 – 100,000	90–02	0	0–19	
Austria	350 – 500	98–02	(0)	(0–19)	
Azerbaijan	(100,000 – 300,000)	96–00	(0)	(0–19)	
Belarus	1,000 – 1,500	97–02	(0)	(0–19)	
Belgium	**50 – 79**	00–02	–	0–19	*1*
Bosnia & HG	Present	90–03	?	–	
Bulgaria	40,000 – 100,000	96–02	0	0–19	
Croatia	(10,000 – 20,000)	02	(–)	(30–49)	*70,16,54*
Cyprus	(50,000 – 100,000)	94–02	(0)	(0–9)	
Czech Rep.	600 – 1,200	00	–	30–49	
Denmark	6 – 13	00–01	**–**	**>80**	*6,7*
Estonia	(20 – 50)	98	–	20–29	*1*
France	(10,000 – 40,000)	98–02	(–)	(0–19)	*4*
Georgia	Present	03	?	–	
Germany	6,600 – 14,000	95–99	–	50–79	
Greece	(50,000 – 100,000)	95–00	(0)	(0–19)	
Hungary	**190,000 – 340,000**	99–02	(+)	(20–49)	*19*
Italy	(200,000 – 400,000)	03	(0)	(0–19)	
Latvia	(0 – 10)	90–00	(–)	(30–49)	*22,26*
Lithuania	(0 – 10)	99–01	(–)	(>80)	*20*
Macedonia	(50,000 – 150,000)	90–00	(0)	(0–19)	
Moldova	17,000 – 22,000	90–00	0	0–19	
Netherlands	**60 – 80**	98–00	**–**	**84**	*1*
Poland	**7,000 – 10,000**	00–02	–	30–49	*23*
Portugal	(10,000 – 100,000)	02	(0)	(0–19)	
Romania	220,000 – 312,000	00–02	+	0–19	*48*
Russia	(70,000 – 300,000)	90–00	?	–	*8,122*
Serbia & MN	50,000 – 70,000	90–02	–	10–19	*1,29,172a,67a, 155,78,227*
Slovakia	3,000 – 6,000	80–99	–	30–49	
Slovenia	800 – 1,000	94	(–)	(0–19)	
Spain	(400,000 – 1,000,000)	92	?	–	*10*
Sweden	**0 – 1**	99–00	**–**	**>80**	
Turkey	(2,000,000 – 4,000,000)	01	(0)	(0–19)	
Ukraine	50,000 – 70,000	90–00	–	20–29	

Total (approx.) 3,600,000 – 7,600,000 **Overall trend** Stable
Breeding range >5,000,000 km² **Gen. length.** <3.3 **% Global pop.** 25–49

Galerida cristata is a widespread resident across much of Europe (except the north), which accounts for less than half of its global range. Its European breeding population is very large (>3,600,000 pairs), but underwent a moderate decline between 1970–1990. Although the species was stable overall during 1990–2000—with stable or increasing trends in south-eastern Europe compensating for declines farther north—its population has clearly not yet recovered to the level that preceded its decline. Consequently, it is provisionally evaluated as Depleted.

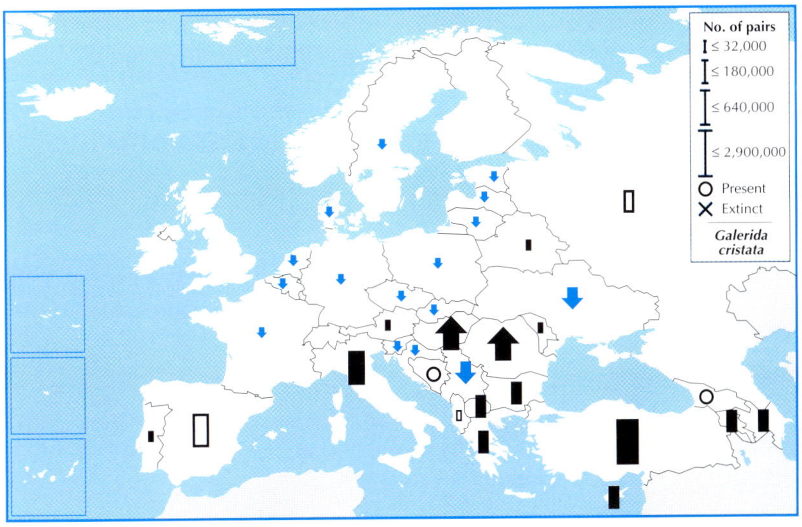

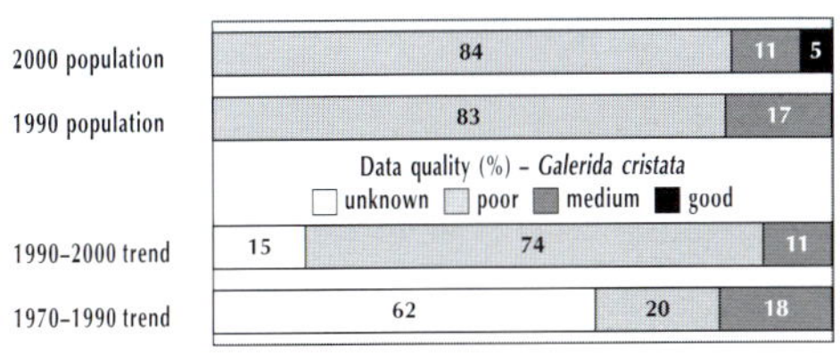

Galerida theklae
THEKLA LARK

SPEC 3 (1994: 3) **Status** (Depleted)
Criteria Large historical decline
European IUCN Red List Category —
Criteria —
Global IUCN Red List Category —
Criteria —

Country	Breeding pop. size (pairs)	Year(s)	Trend	Mag.%	References
France	300 – 350	00–03	(0)	(0–19)	*10,4*
Portugal	(50,000 – 500,000)	02	(0)	(0–19)	
Spain	(1,400,000 – 1,600,000)	92	?	–	*13,12,10*
Total (approx.)	1,500,000 – 2,100,000	**Overall trend** Unknown			
Breeding range	>250,000 km²	**Gen. length.** <3.3 **% Global pop.** 25–49			

Galerida theklae is a widespread resident in Iberia and southernmost France, with Europe accounting for less than half of its global range. Its European breeding population is very large (>1,500,000 pairs), but underwent a large decline between 1970–1990. Although the species was stable in Portugal and France during 1990–2000, the trend of its stronghold population in Spain was unknown. Nevertheless, its total population size almost certainly remains far below the level that preceded its decline, and consequently the species is provisionally evaluated as Depleted.

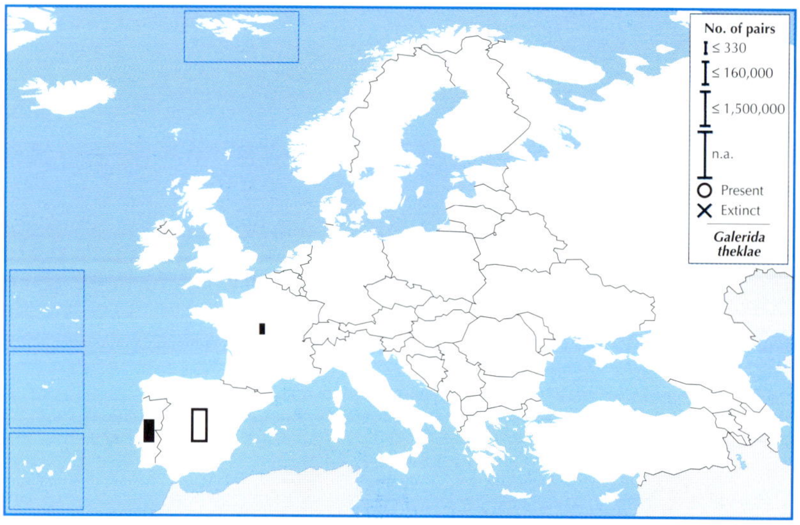

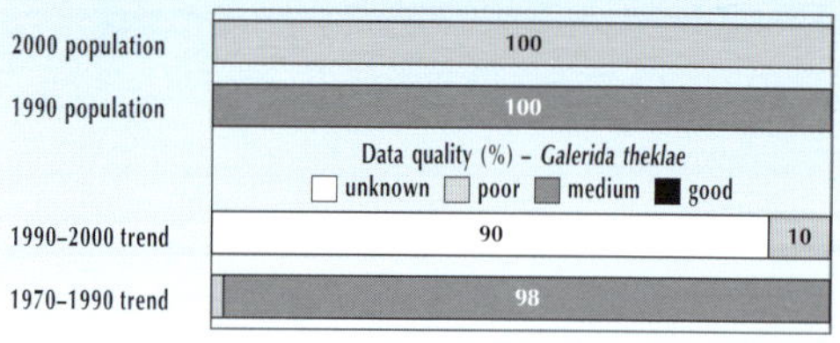

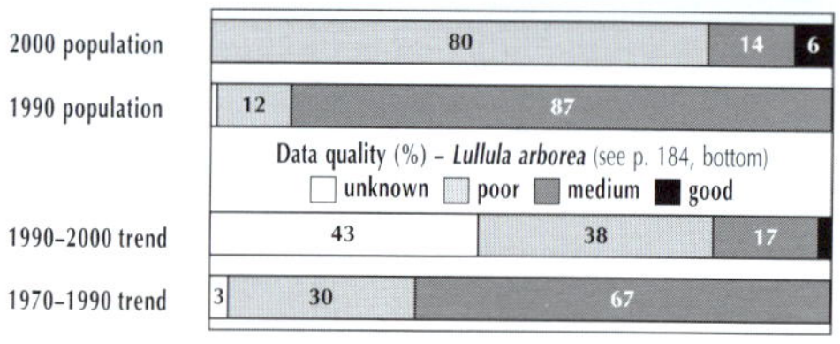

Lullula arborea
WOOD LARK

SPEC 2 (1994: 2) **Status** Depleted
Criteria Large historical decline
European IUCN Red List Category —
Criteria —
Global IUCN Red List Category —
Criteria —

Lullula arborea is a widespread breeder across much of Europe (except the far north), which constitutes >75% of its global breeding range. Its European breeding population is very large (>1,300,000 pairs), but underwent a large decline between 1970–1990. Although declines continued in a few countries during 1990–2000, they abated across most of Europe (the trend in Spain was unknown), and the species was stable overall. Nevertheless, its population size remains far below the level that preceded its decline, and consequently the species is evaluated as Depleted.

Country	Breeding pop. size (pairs)	Year(s)	Trend	Mag.%	References
Albania	3,000 – 10,000	02	?	–	
Andorra	(20 – 75)	99–01	(+)	(0–19)	*1,3*
Armenia	25,000 – 50,000	95–02	0	0–19	
Austria	700 – 900	98–02	+	**>80**	
Azerbaijan	(5,000 – 10,000)	96–00	(0)	(0–19)	
Belarus	20,000 – 35,000	97–02	**0**	**0–19**	
Belgium	600 – 900	95–02	+	0–19	*1*
Bulgaria	40,000 – 80,000	96–02	0	0–9	
Croatia	(5,000 – 10,000)	02	(–)	(30–49)	*70,16*
Cyprus	(600 – 2,000)	94–02	(0)	(0–9)	
Czech Rep.	500 – 1,000	00	0	0–19	
Denmark	300 – 300	93–96	(0)	(0–19)	*3*
Estonia	1,000 – 3,000	98	0	0–19	*1*
Finland	800 – 1,000	98–02	+	**10**	
France	(50,000 – 200,000)	98–02	(+)	(33)	*4,2*
Georgia	Present	03	?	–	
Germany	25,000 – 45,000	95–99	–	20–29	
Greece	(5,000 – 20,000)	95–00	(0)	(0–19)	
Hungary	**30,000 – 75,000**	99–02	0	0–19	*19*
Italy	(50,000 – 100,000)	03	(0)	(0–19)	
Latvia	2,000 – 6,000	90–00	(0)	(0–19)	*22*
Lithuania	(15,000 – 20,000)	99–01	(0)	(0–19)	*20*
Luxembourg	**25 – 30**	00–02	**–**	**50–79**	
Macedonia	(5,000 – 15,000)	90–00	(0)	(0–19)	
Moldova	1,600 – 2,200	90–00	0	0–19	
Netherlands	**5,000 – 6,000**	98–00	+	**141**	*1*
Norway	**150 – 260**	02–03	+	**0–19**	*54*
Poland	**50,000 – 80,000**	00–02	0	0–19	*23*
Portugal	(50,000 – 500,000)	02	(0)	(0–19)	
Romania	65,000 – 87,000	00–02	+	0–19	*48*
Russia	(100,000 – 250,000)	90–00	(0)	(0–19)	*8,24,35,39,61, 73,122*
Serbia & MN	10,000 – 15,000	90–02	–	10–19	*1,155,172a,67a, 225,227*
Slovakia	1,500 – 3,000	80–99	0	0–19	
Slovenia	3,000 – 4,000	99–00	(0)	(0–19)	
Spain	(560,000 – 1,300,000)	92	?	–	*13,12,10*
Sweden	5,000 – 10,000	99–00	+	**50–79**	
Switzerland	**250 – 500**	93–96	0	0–19	
Turkey	(150,000 – 350,000)	01	(0)	(0–19)	
Ukraine	8,000 – 12,000	90–00	–	20–29	
UK	**1,400 – 1,600**	97	+	**498**	*41*
Total (approx.)	1,300,000 – 3,300,000	**Overall trend** Stable			
Breeding range	>5,000,000 km²	**Gen. length.** <3.3 **% Global pop.** 75–94			

(See p. 184, top, for data quality graph)

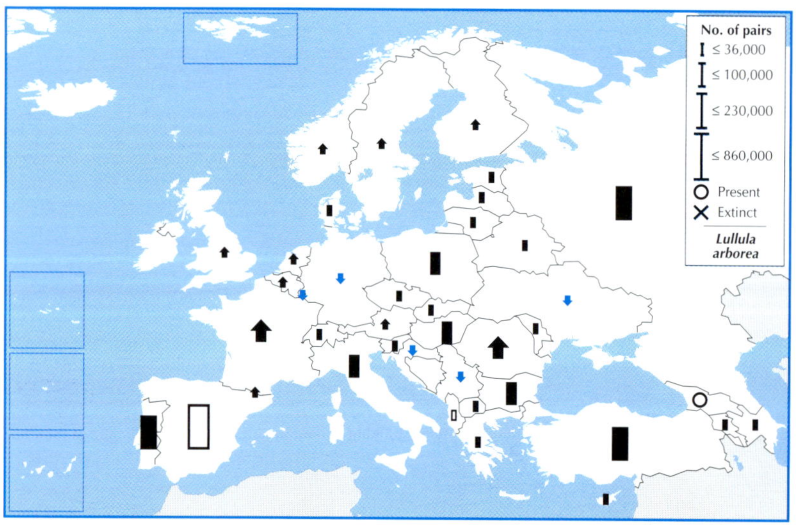

Alauda arvensis
EURASIAN SKYLARK

SPEC 3 (1994: 3) **Status** (Depleted)
Criteria Large historical decline
European IUCN Red List Category —
Criteria —
Global IUCN Red List Category —
Criteria —

Alauda arvensis is a widespread breeder across most of Europe, which accounts for less than half of its global breeding range. Its European breeding population is extremely large (>40,000,000 pairs), but underwent a large decline between 1970–1990. Although declines continued in many western European countries during 1990–2000, key eastern populations remained stable, and the species probably declined only slightly overall. Nevertheless, its population size remains far below the level that preceded its decline, and consequently the species is provisionally evaluated as Depleted.

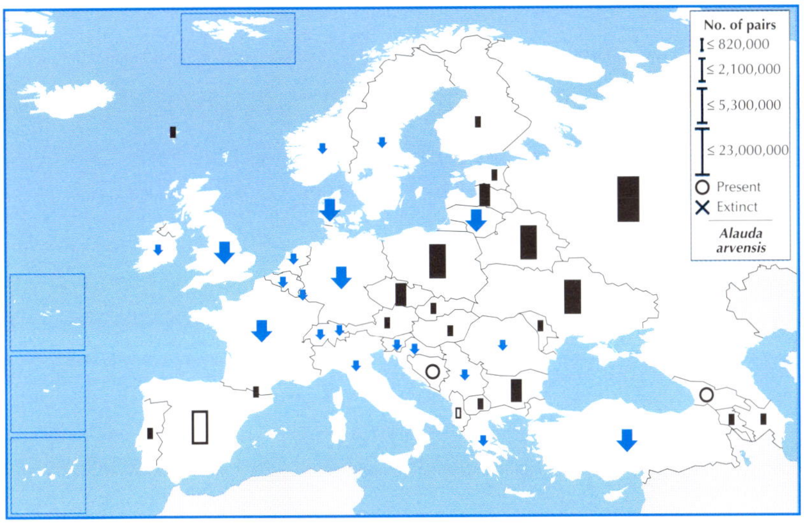

Country	Breeding pop. size (pairs)	Year(s)	Trend	Mag.%	References
Albania	500 – 1,000	02	?	–	
Andorra	(500 – 750)	99–01	(0)	(0–19)	*1,3*
Armenia	250,000 – 500,000	90–02	0	0–19	
Austria	120,000 – 240,000	98–02	(0)	(0–19)	
Azerbaijan	(10,000 – 100,000)	96–00	(0)	(0–19)	
Belarus	2,300,000 – 3,000,000	97–02	0	0–19	
Belgium	29,000 – 52,000	01–02	–	20–29	*1*
Bosnia & HG	Present	90–03	?	–	
Bulgaria	800,000 – 2,500,000	96–02	0	0–9	
Croatia	(50,000 – 100,000)	02	(–)	(50–79)	*70,16*
Czech Rep.	800,000 – 1,600,000	00	0	0–19	
Denmark	1,100,000 – 1,300,000	00	–	0–9	*12*
Faroe Is.	10 – 10	95	(0)	(0–19)	
Estonia	150,000 – 350,000	98	0	0–19	*1*
Finland	300,000 – 400,000	98–02	0	5	
France	(800,000 – 3,000,000)	98–02	**–**	**14**	*4,2*
Georgia	Present	03	?	–	
Germany	1,600,000 – 2,700,000	95–99	–	20–29	
Greece	(2,000 – 5,000)	95–00	(–)	(0–19)	
Hungary	**730,000 – 900,000**	99–02	0	0–19	*19*
Rep. Ireland	250,000 – 500,000	88–91	–	0–19	
Italy	(500,000 – 1,000,000)	03	(–)	(0–19)	
Latvia	1,100,000 – 1,800,000	90–00	0	0–19	*23,16*
Liechtenstein	**4 – 8**	98–00	**–**	**30–49**	
Lithuania	(1,100,000 – 1,500,000)	99–01	(–)	(0–19)	*20*
Luxembourg	7,000 – 8,000	02	**–**	**30–49**	
Macedonia	(70,000 – 200,000)	90–00	(0)	(0–19)	
Moldova	50,000 – 55,000	90–00	0	0–19	
Netherlands	**50,000 – 70,000**	98–00	**–**	**28**	*1*
Norway	(100,000 – 400,000)	90–01	(–)	(0–19)	*27*
Poland	**4,000,000 – 7,000,000**	00–02	0	0–19	*23*
Portugal	(1,000 – 10,000)	02	(0)	(0–19)	
Romania	460,000 – 850,000	00–02	–	0–19	*48*
Russia	15,000,000 – 35,000,000	90–00	(0)	(0–19)	*3,8,23,32,104*
Serbia & MN	350,000 – 500,000	90–02	–	10–19	*1,29,155,172a, 67a,78,225*
Slovakia	200,000 – 400,000	80–99	0	0–19	
Slovenia	8,000 – 12,000	94	(–)	(0–19)	
Spain	(2,000,000 – 6,000,000)	92	?	–	*13,12,10*
Sweden	500,000 – 1,000,000	99–00	**–**	**28**	
Switzerland	**40,000 – 50,000**	93–96	–	10–19	
Turkey	(900,000 – 1,800,000)	01	(–)	(0–19)	
Ukraine	2,200,000 – 3,100,000	90–00	0	5–19	
UK	**1,785,000 – 1,785,000**	00	**–**	**15**	*5,31*
Total (approx.)	40,000,000 – 80,000,000	**Overall trend** Small decline			
Breeding range	>8,000,000 km²	**Gen. length.** <3.3 **% Global pop.** 25–49			

(See p. 185, bottom, for data quality graph)

Eremophila alpestris
HORNED LARK

Non-SPEC (1994: —) **Status** (Secure)
Criteria —
European IUCN Red List Category —
Criteria —
Global IUCN Red List Category —
Criteria —

Eremophila alpestris has a patchy breeding distribution in parts of south-eastern Europe, as well as Fennoscandia and arctic Russia, with Europe accounting for a tiny proportion of its global breeding range. Its European breeding population is very large (>2,200,000 pairs), and was stable between 1970–1990. Although there were declines in certain countries during 1990–2000, populations were stable across most of its European range—including the stronghold in Turkey—and the species probably remained stable overall. Consequently, it is provisionally evaluated as Secure.

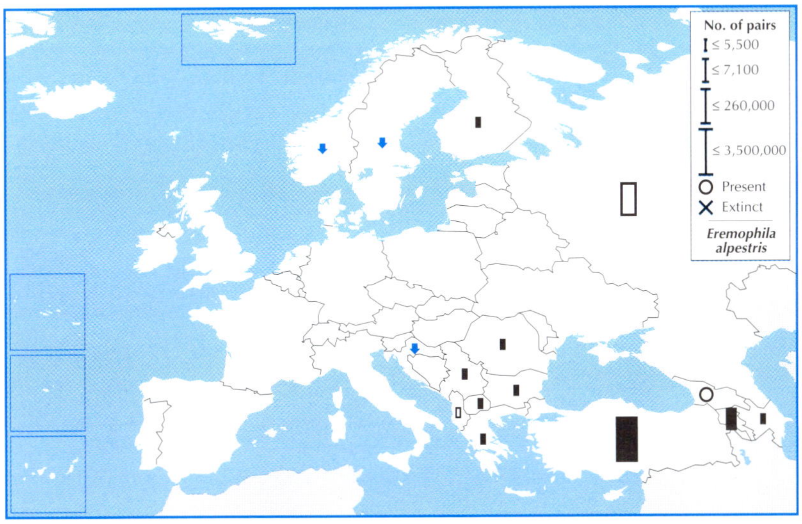

Country	Breeding pop. size (pairs)	Year(s)	Trend	Mag.%	References
Albania	(100 – 500)	02	?	–	
Armenia	50,000 – 100,000	90–02	0	0–19	
Azerbaijan	(1,000 – 10,000)	96–00	(0)	(0–19)	
Bulgaria	**4,500 – 6,500**	95–02	**0**	**0–9**	
Croatia	(50 – 100)	02	(–)	(0–19)	*70*
Finland	(0 – 10)	98–02	(0)	(0–19)	
Georgia	Present	03	?	–	
Greece	(500 – 800)	95–00	(0)	(0–19)	
Macedonia	(500 – 3,000)	90–00	(0)	(0–19)	
Norway	1,000 – 5,000	02	(–)	(0–19)	*65*
Romania	150 – 250	00–02	0	0–19	*1*
Russia	(130,000 – 500,000)	90–00	?	–	*17,116,122*
Serbia & MN	**800 – 1,000**	90–02	**0**	**0–19**	*1,135,150,152, 153,235,62, 155,225*
Sweden	150 – 250	99–00	–	30–49	
Turkey	(2,000,000 – 6,000,000)	01	(0)	(0–19)	
Total (approx.)	2,200,000 – 6,600,000	**Overall trend** Stable			
Breeding range	>500,000 km²	**Gen. length.** <3.3 **% Global pop.** <5			

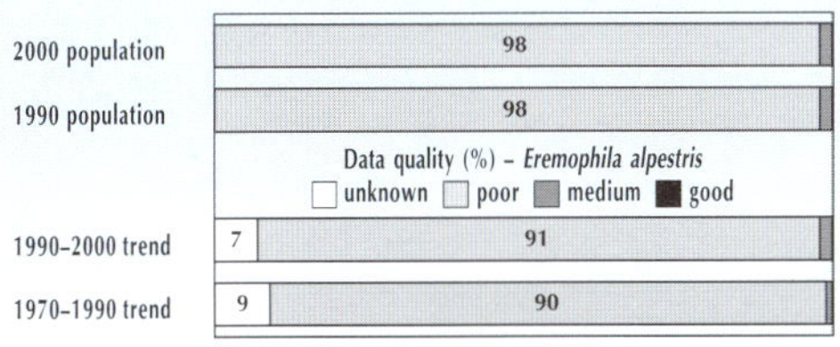

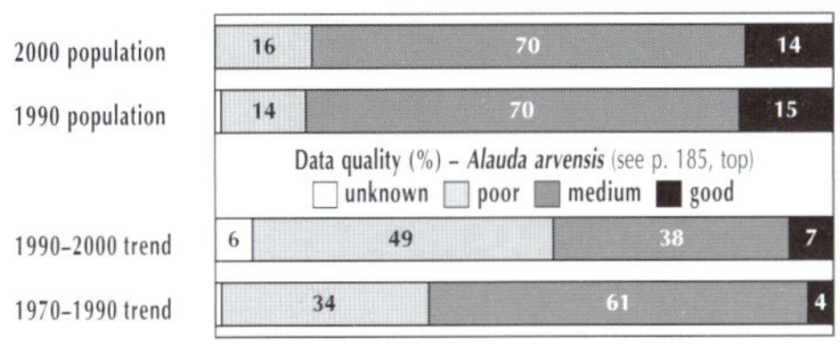

Riparia riparia
SAND MARTIN

SPEC 3 (1994: 3) **Status** (Depleted)
Criteria Moderate historical decline
European IUCN Red List Category —
Criteria —
Global IUCN Red List Category —
Criteria —

Riparia riparia is a widespread summer visitor to Europe, which accounts for less than a quarter of its global breeding range. Its European breeding population is very large (>5,400,000 pairs), but underwent a moderate decline between 1970–1990. Although declines continued in a number of countries during 1990–2000, the species was stable across most of Europe (the trend in Russia was unknown), and probably declined only slightly overall. Nevertheless, its population size remains far below the level that preceded its decline, and consequently it is provisionally evaluated as Depleted.

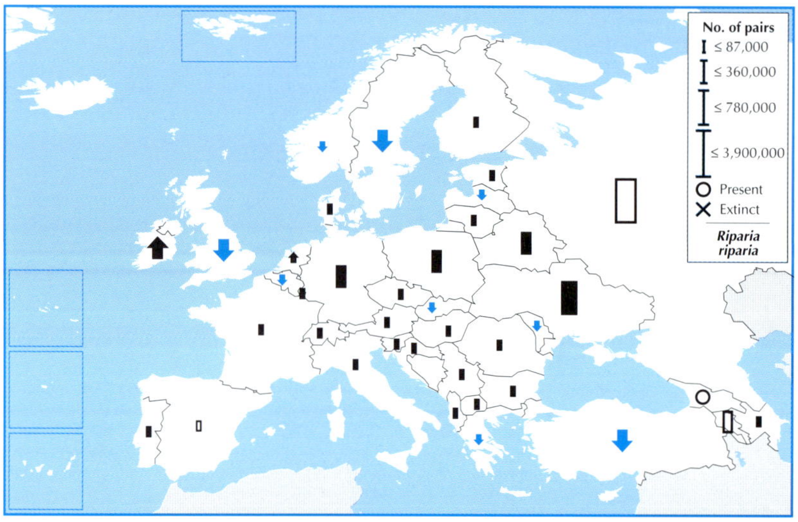

Country	Breeding pop. size (pairs)	Year(s)	Trend	Mag.%	References
Albania	1,500 – 4,000	02	(0)	(0–19)	
Armenia	250,000 – 500,000	90–02	?	–	
Austria	**8,000 – 12,000**	98–02	(0)	(0–19)	
Azerbaijan	(10,000 – 50,000)	96–00	(0)	(0–19)	
Belarus	200,000 – 300,000	97–02	(0)	(0–19)	
Belgium	**9,400 – 11,000**	95–02	–	20–29	*1*
Bulgaria	20,000 – 80,000	96–02	0	0–19	
Croatia	(25,000 – 40,000)	02	(0)	(0–19)	*70,16,44*
Czech Rep.	15,000 – 30,000	00	0	10–19	
Denmark	20,000 – 40,000	00	F	50–79	
Estonia	20,000 – 50,000	98	0	0–19	*1*
Finland	50,000 – 100,000	98–02	**0**	**5**	
France	50,000 – 150,000	98–02	(F)	(20–49)	*4*
Georgia	Present	03	?	–	
Germany	106,000 – 210,000	95–99	0	0–19	
Greece	(10,000 – 20,000)	95–00	(–)	(0–19)	
Hungary	40,000 – 165,000	99–02	0	0–19	*20*
Rep. Ireland	100,000 – 250,000	88–91	+	20–29	
Italy	8,000 – 9,000	03	F	20–29	
Latvia	(30,000 – 100,000)	90–00	(–)	(0–19)	*23,17*
Lithuania	(50,000 – 150,000)	99–01	(0)	(0–19)	*20*
Luxembourg	**100 – 200**	02	**0**	**0–19**	
Macedonia	(1,000 – 5,000)	90–00	(0)	(0–19)	
Moldova	7,200 – 7,800	90–00	–	30–49	
Netherlands	**18,500 – 32,000**	98–00	**+**	**63**	*1*
Norway	(50,000 – 100,000)	90–03	–	20–29	*27*
Poland	(150,000 – 300,000)	90–00	0	0–19	*1*
Portugal	(10,000 – 100,000)	02	(0)	(0–19)	
Romania	55,000 – 80,000	00–02	0	0–19	*32,55*
Russia	(3,000,000 – 5,000,000)	90–00	?	–	*8,104,122*
Serbia & MN	**50,000 – 65,000**	97–02	0	10–19	*1,114,174,29, 155,67a,78,214*
Slovakia	10,000 – 20,000	80–99	–	30–49	
Slovenia	**200 – 800**	99–00	**F**	**>80**	
Spain	(20,000 – 100,000)	90–00	?	–	*13,12,10,7*
Sweden	100,000 – 150,000	99–00	–	30–49	
Switzerland	**5,500 – 6,500**	93–96	**F**	**20–29**	
Turkey	(100,000 – 250,000)	01	(–)	(20–29)	
Ukraine	750,000 – 800,000	90–00	0	0–19	
UK	**66,300 – 211,000**	00	**–**	**22**	*5,31*

Total (approx.) 5,400,000 – 9,500,000 **Overall trend** Unknown
Breeding range >7,000,000 km² **Gen. length.** <3.3 **% Global pop.** 5–24
(See p. 186, bottom, for data quality graph)

Hirundo rupestris
EURASIAN CRAG-MARTIN

Non-SPEC (1994: —) **Status** Secure
Criteria —
European IUCN Red List Category —
Criteria —
Global IUCN Red List Category —
Criteria —

Hirundo rupestris is a widespread breeder across much of southern Europe, which accounts for less than half of its global breeding range. Its European breeding population is large (>120,000 pairs), and was stable between 1970–1990. Although the trend of the sizeable Spanish population during 1990–2000 was unknown, populations were stable or increasing across the vast majority of its European range, and the species remained stable overall. Consequently, it is evaluated as Secure.

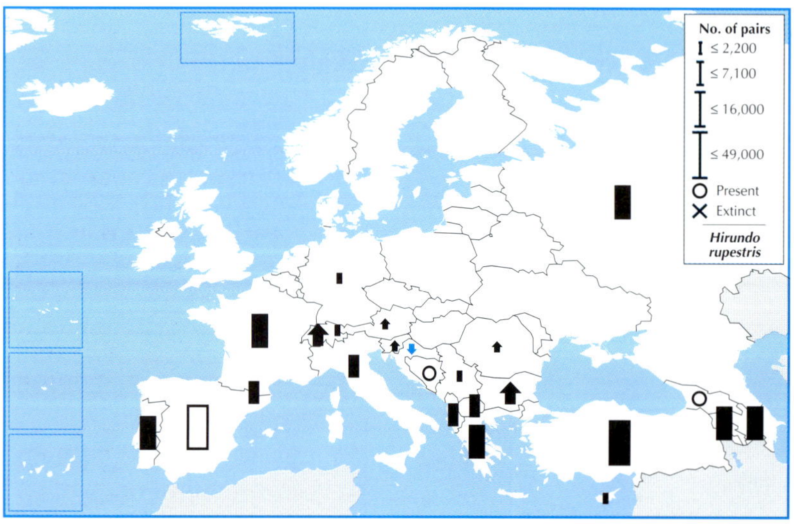

Country	Breeding pop. size (pairs)	Year(s)	Trend	Mag.%	References
Albania	2,000 – 5,000	02	0	0–19	
Andorra	3,000 – 4,000	99–01	(0)	(0–19)	*1,3*
Armenia	10,000 – 15,000	90–02	0	0–19	
Austria	(1,500 – 3,000)	98–02	(+)	(20–29)	
Azerbaijan	(10,000 – 20,000)	96–00	(0)	(0–19)	
Bosnia & HG	Present	90–03	?	–	*1*
Bulgaria	5,000 – 8,000	96–02	+	0–19	
Croatia	(500 – 1,000)	02	(–)	(30–49)	*16*
Cyprus	(200 – 1,000)	94–02	(0)	(0–9)	
France	7,500 – 15,000	98–02	(0)	(0–19)	*4*
Georgia	Present	03	?	–	
Germany	**30 – 50**	95–99	**0**	**0–19**	
Greece	(5,000 – 20,000)	95–00	(0)	(0–19)	
Italy	(5,000 – 10,000)	03	(0)	(0–19)	
Liechtenstein	**10 – 15**	98–00	0	0–19	
Macedonia	(3,000 – 4,000)	90–00	(0)	(0–19)	
Portugal	(5,000 – 50,000)	02	(0)	(0–19)	
Romania	650 – 850	00–02	+	**20–29**	
Russia	(5,000 – 20,000)	90–00	(0)	(0–19)	*50*
Serbia & MN	1,500 – 2,500	90–02	0	0–19	*1,219,155,225, 227*
Slovenia	500 – 1,000	99–00	+	>80	
Spain	(20,000 – 100,000)	98–02	?	–	*10*
Switzerland	**4,000 – 5,000**	93–96	+	20–29	
Turkey	(30,000 – 80,000)	01	(0)	(0–19)	

Total (approx.) 120,000 – 370,000 **Overall trend** Stable
Breeding range >2,000,000 km² **Gen. length.** <3.3 **% Global pop.** 25–49

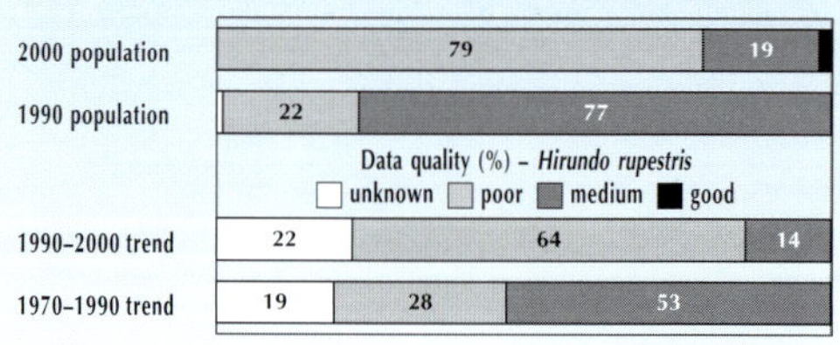

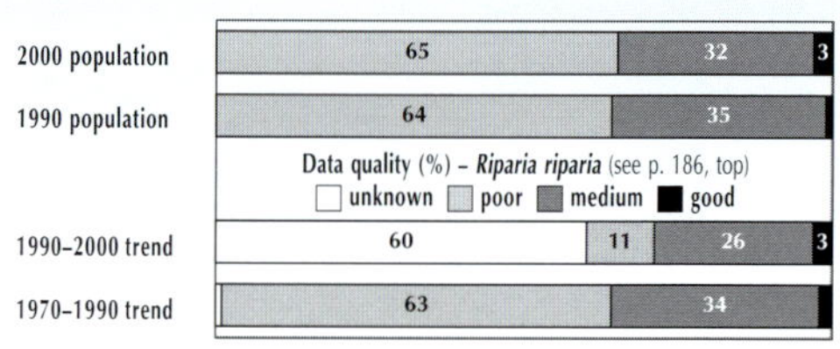

Hirundo rustica
BARN SWALLOW

SPEC 3 (1994: 3) **Status** Depleted
Criteria Moderate historical decline
European IUCN Red List Category —
Criteria —
Global IUCN Red List Category —
Criteria —

Hirundo rustica is a widespread summer visitor to most of Europe, which accounts for less than a quarter of its global breeding range. Its European breeding population is extremely large (>16,000,000 pairs), but underwent a moderate decline between 1970–1990. Although declines abated or even reversed in certain countries during 1990–2000, the species continued to decline across much of Europe, and underwent a small decline overall. Its population has clearly not yet recovered to the level that preceded its initial decline, and consequently it is evaluated as Depleted.

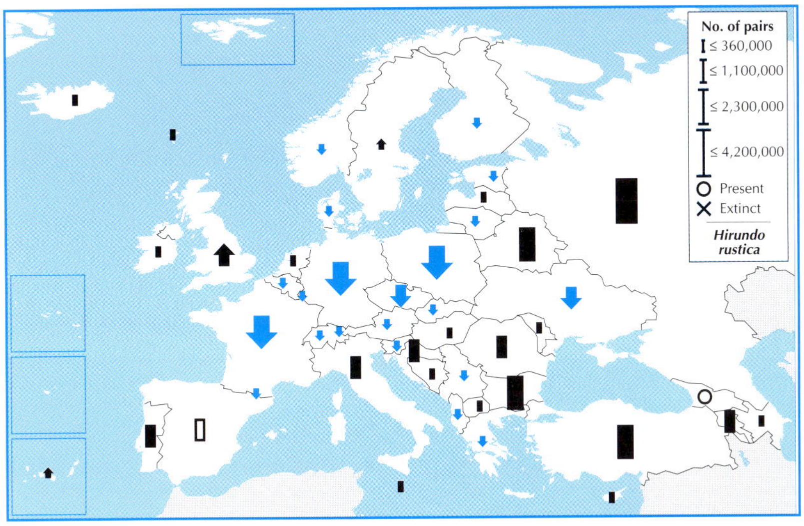

Country	Breeding pop. size (pairs)	Year(s)	Trend	Mag.%	References
Albania	20,000 – 40,000	02	(–)	(0–19)	
Andorra	**15 – 20**	99–01	–	20–29	1,3
Armenia	250,000 – 750,000	90–02	0	0–19	
Austria	(90,000 – 170,000)	98–02	(–)	(30–49)	
Azerbaijan	(250,000 – 500,000)	96–00	(0)	(0–19)	
Belarus	1,000,000 – 1,500,000	97–02	(0)	(0–19)	
Belgium	30,000 – 70,000	01–02	–	0–19	1
Bosnia & HG	(50,000 – 100,000)	90–00	(0)	(0–19)	
Bulgaria	800,000 – 3,000,000	96–02	0	0–9	
Croatia	(500,000 – 1,000,000)	02	(0)	(0–19)	70,16
Cyprus	(50,000 – 100,000)	94–02	(0)	(0–9)	
Czech Rep.	320,000 – 640,000	00	–	0–19	
Denmark	(200,000 – 250,000)	00	–	0–19	12
Faroe Is.	0 – 5	95	(0)	(0–19)	
Estonia	150,000 – 250,000	98	–	20–29	1
Finland	130,000 – 180,000	98–02	–	10	
France	(800,000 – 3,200,000)	98–02	(–)	(19)	4,2
Georgia	Present	03	?	–	
Germany	950,000 – 1,600,000	95–99	–	20–29	
Greece	(50,000 – 200,000)	95–00	(–)	(0–19)	
Hungary	220,000 – 320,000	99–02	0	0–19	19
Iceland	**0 – 6**	90–00	(F)	(–)	*23,24,25,32,33, 42,43,44,45,46,47*
Rep. Ireland	100,000 – 250,000	88–91	0	0–19	
Italy	(500,000 – 1,000,000)	03	(0)	(0–19)	
Latvia	117,000 – 475,000	90–00	(0)	(0–19)	*23,16*
Liechtenstein	50 – 70	98–00	–	0–9	
Lithuania	(150,000 – 300,000)	99–01	(–)	(0–19)	*20*
Luxembourg	6,000 – 8,000	00–02	**–**	**30–49**	
Macedonia	(20,000 – 50,000)	90–00	(0)	(0–19)	
Malta	1 – 2	90–02	0	0–19	*1*
Moldova	15,000 – 18,000	90–00	0	0–19	
Netherlands	**100,000 – 200,000**	98–00	0	4	*1*
Norway	40,000 – 50,000	95	–	20–29	
Poland	**1,800,000 – 2,700,000**	00–02	–	0–19	*23*
Portugal	(500,000 – 2,000,000)	02	(0)	(0–19)	
Romania	800,000 – 1,000,000	00–02	0	0–19	
Russia	2,500,000 – 7,000,000	90–00	0	0–19	*8,122*
Serbia & MN	250,000 – 400,000	90–02	–	10–19	*1,29,155,172a, 67a,78,225*
Slovakia	200,000 – 400,000	80–99	–	30–49	
Slovenia	100,000 – 200,000	00	(–)	(0–19)	
Spain	(500,000 – 1,000,000)	98–02	?	–	*10*
Canary Is.	2 – 3	97–03	(+)	(N)	*28,25*
Sweden	**150,000 – 300,000**	99–00	+	**18**	
Switzerland	60,000 – 90,000	98–02	–	10–19	
Turkey	(1,000,000 – 2,500,000)	01	(0)	(0–19)	
Ukraine	(870,000 – 1,300,000)	90–00	–	0–19	
UK	726,000 – 726,000	00	+	**18**	
Total (approx.)	16,000,000 – 36,000,000	**Overall trend** Small decline			
Breeding range	>8,000,000 km²	**Gen. length.** <3.3 **% Global pop.** 5–24			

(See p. 187, bottom, for data quality graph)

Hirundo daurica
RED-RUMPED SWALLOW

Non-SPEC (1994: —) **Status** (Secure)
Criteria —
European IUCN Red List Category —
Criteria —
Global IUCN Red List Category —
Criteria —

Hirundo daurica is a widespread but somewhat patchily distributed summer visitor to southern Europe, which accounts for less than a quarter of its global breeding range. Its European breeding population is large (>100,000 pairs), and increased between 1970–1990. Although the species declined slightly in Greece and Albania during 1990–2000, these losses were compensated for by increases elsewhere—notably in Spain and Portugal—and it was stable overall. Consequently, the species is evaluated as Secure.

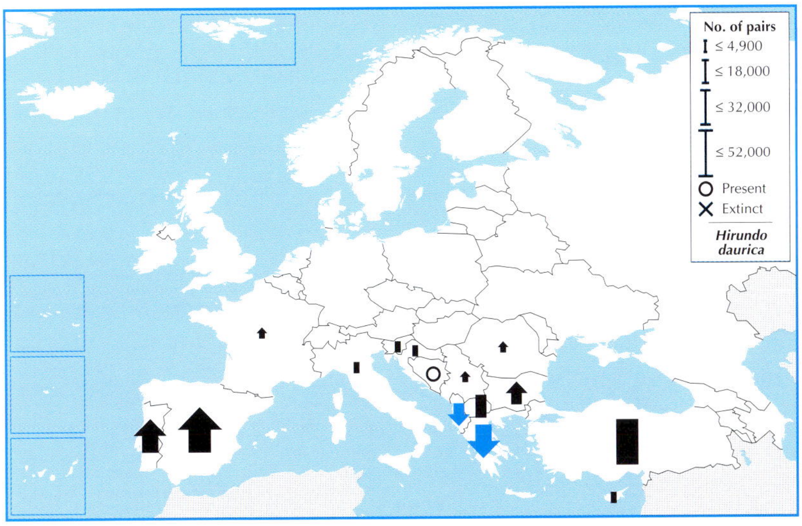

Country	Breeding pop. size (pairs)	Year(s)	Trend	Mag.%	References
Albania	10,000 – 30,000	02	(–)	(0–19)	
Bosnia & HG	Present	85–89	?	–	
Bulgaria	10,000 – 20,000	96–02	+	0–19	
Croatia	(100 – 500)	02	(0)	(0–19)	*70,16*
Cyprus	(3,000 – 8,000)	94–02	(0)	(0–19)	
France	**71 – 81**	00	+	**>80**	*5*
Greece	(10,000 – 50,000)	95–00	(–)	(0–19)	
Italy	(20 – 50)	03	(0)	(0–19)	
Macedonia	(10,000 – 30,000)	90–00	(0)	(0–19)	
Portugal	(10,000 – 100,000)	02	(+)	(–)	
Romania	160 – 240	00–02	+	20–29	*52*
Serbia & MN	**1,000 – 1,600**	95–02	+	10–19	*1,199,217,155*
Slovenia	(1 – 3)	00	(F)	(>80)	
Spain	(20,000 – 100,000)	98–02	(+)	(0–19)	*10*
Turkey	(30,000 – 90,000)	01	(0)	(0–19)	
Total (approx.)	100,000 – 430,000	**Overall trend** Stable			
Breeding range	>1,000,000 km²	**Gen. length.** <3.3 **% Global pop.** 5–24			

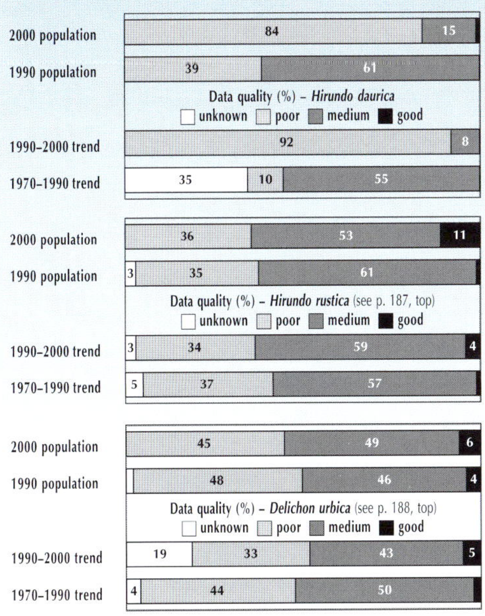

Delichon urbica
NORTHERN HOUSE-MARTIN

SPEC 3 (1994: —) **Status** (Declining)
Criteria Moderate recent decline
European IUCN Red List Category —
Criteria —
Global IUCN Red List Category —
Criteria —

Delichon urbica is a widespread summer visitor to most of Europe, which accounts for less than half of its global breeding range. Its European breeding population is very large (>9,900,000 pairs), and was stable between 1970–1990. Although the species remained stable in several countries during 1990–2000, it declined across much of its European range—including the sizeable populations in Turkey, France and Germany—and probably underwent a moderate decline (>10%) overall. Consequently, this previously Secure species is now provisionally evaluated as Declining.

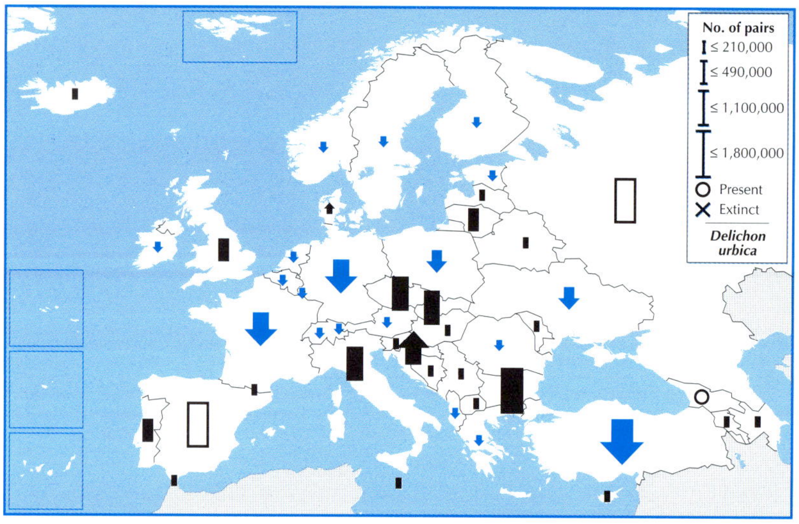

Country	Breeding pop. size (pairs)	Year(s)	Trend	Mag.%	References
Albania	40,000 – 80,000	02	(–)	(0–19)	
Andorra	3,000 – 5,000	99–01	(0)	(0–19)	*1,3*
Armenia	30,000 – 150,000	00–02	0	0–19	
Austria	(50,000 – 100,000)	98–02	(–)	(30–49)	
Azerbaijan	(100,000 – 250,000)	96–00	(0)	(0–19)	
Belarus	100,000 – 150,000	97–02	0	0–19	
Belgium	28,000 – 51,000	01–02	–	20–29	*1*
Bosnia & HG	(50,000 – 100,000)	90–00	(F)	(–)	
Bulgaria	800,000 – 4,000,000	96–02	0	0–9	
Croatia	(500,000 – 1,000,000)	02	(+)	(>80)	*16*
Cyprus	(10,000 – 50,000)	94–02	(0)	(0–19)	
Czech Rep.	600,000 – 1,200,000	00	0	0–19	
Denmark	(50,000 – 100,000)	93–96	+	30–49	
Estonia	150,000 – 250,000	98	–	20–29	*1*
Finland	80,000 – 120,000	98–02	**–**	**20**	
France	(400,000 – 1,600,000)	98–02	(–)	(41)	*4,2*
Georgia	Present	03	?	–	
Germany	820,000 – 1,400,000	95–99	–	20–29	
Greece	(50,000 – 200,000)	95–00	(–)	(0–19)	
Hungary	100,000 – 200,000	99–02	0	0–19	*19*
Iceland	**0 – 2**	90–00	(F)	(–)	*23,24,25,32,33, 42,43,44,45,46,47*
Rep. Ireland	20,000 – 100,000	88–91	–	0–19	
Italy	(500,000 – 1,000,000)	03	(0)	(0–19)	
Latvia	90,000 – 320,000	90–00	(0)	(0–19)	*23,16*
Liechtenstein	500 – 1,000	98–00	–	0–19	
Lithuania	(200,000 – 400,000)	99–01	(0)	(0–19)	*20*
Luxembourg	4,000 – 5,000	00–02	**–**	**30–49**	
Macedonia	(25,000 – 35,000)	90–00	(0)	(0–19)	
Malta	1 – 2	90–02	**0**	**0–19**	*1*
Moldova	14,000 – 16,000	90–00	0	0–19	
Netherlands	**60,000 – 125,000**	98–00	**–**	**15**	*1*
Norway	(30,000 – 80,000)	90–02	–	20–29	
Poland	**350,000 – 600,000**	00–02	–	0–19	*23*
Portugal	(100,000 – 1,000,000)	02	(0)	(0–19)	
Romania	120,000 – 220,000	00–02	–	0–19	
Russia	800,000 – 2,300,000	90–00	?	–	*53,122*
Serbia & MN	160,000 – 260,000	90–02	0	0–19	*1,29,172a,78, 225,227*
Slovakia	500,000 – 1,000,000	90–99	0	0–19	
Slovenia	50,000 – 100,000	94	(0)	(0–19)	
Spain	(1,140,000 – 2,160,000)	92	?	–	*13,12,10*
Sweden	100,000 – 200,000	99–00	**–**	**3**	
Switzerland	50,000 – 100,000	98–02	–	10–19	
Turkey	(1,000,000 – 2,000,000)	01	(–)	(20–29)	
Ukraine	(400,000 – 580,000)	90–00	–	20–29	
UK	**273,000 – 535,000**	00	**0**	**1**	*31*
Gibraltar	**5 – 10**	00	0	0–19	

Total (approx.) 9,900,000 – 24,000,000 **Overall trend** Moderate decline
Breeding range >8,000,000 km² **Gen. length**. <3.3 **% Global pop.** 25–49
(See p. 187, bottom, for data quality graph)

Anthus campestris
TAWNY PIPIT

SPEC 3 (1994: 3) **Status** (Declining)
Criteria Moderate continuing decline
European IUCN Red List Category —
Criteria —
Global IUCN Red List Category —
Criteria —

Anthus campestris is a fairly widespread summer visitor to much of Europe, which accounts for less than half of its global breeding range. Its European breeding population is very large (>1,000,000 pairs), but underwent a large decline between 1970–1990. Although trends were not available for key populations in Spain and Russia during 1990–2000, the species declined across much of its European range—including the sizeable population in Turkey—and probably underwent a moderate decline (>10%) overall. Consequently, it is provisionally evaluated as Declining.

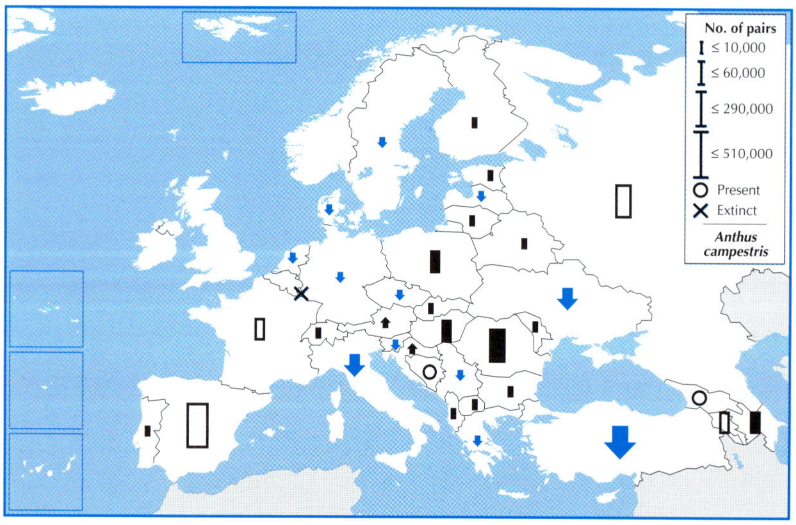

Country	Breeding pop. size (pairs)	Year(s)	Trend	Mag.%	References
Albania	(2,000 – 5,000)	96–02	(0)	(0–19)	
Armenia	35,000 – 100,000	97–02	?	–	
Austria	**40 – 60**	98–02	+	**>80**	
Azerbaijan	(10,000 – 50,000)	96–00	(0)	(0–19)	
Belarus	2,000 – 3,000	97–02	0	0–19	
Bosnia & HG	Present	90–03	?	–	
Bulgaria	500 – 1,000	96–02	0	0–9	
Croatia	(5,000 – 10,000)	02	(+)	(50–79)	*70,16*
Czech Rep.	**30 – 60**	00	**–**	**20–29**	
Denmark	10 – 12	01	**–**	**50–79**	*7*
Estonia	(30 – 80)	98	0	0–19	*1*
Finland	(0 – 3)	98–02	(0)	(0–19)	
France	10,000 – 36,000	98–02	?	–	*4*
Georgia	Present	03	?	–	
Germany	940 – 1,650	95–99	–	30–49	
Greece	(5,000 – 20,000)	95–00	(–)	(0–19)	
Hungary	**13,000 – 40,000**	99–02	0	0–19	*19*
Italy	(15,000 – 40,000)	03	(–)	(0–19)	
Latvia	(150 – 300)	90–00	(–)	(20–29)	*23*
Lithuania	400 – 600	99–01	(0)	(0–19)	*20*
Luxembourg	**0 – 0**	02	**–**	**X**	
Macedonia	(5,000 – 20,000)	90–00	(0)	(0–19)	
Moldova	4,000 – 5,000	90–00	0	0–19	
Netherlands	**25 – 30**	98–00	**–**	**57**	*1*
Poland	**15,000 – 30,000**	00–02	(0)	(0–19)	*23*
Portugal	(1,000 – 10,000)	02	(0)	(0–19)	
Romania	(150,000 – 220,000)	00–02	0	0–19	*48*
Russia	(130,000 – 250,000)	90–00	?	–	
Serbia & MN	3500 – 5,000	90–02	–	0–19	*1,150,29,155, 172a,67a*
Slovakia	70 – 150	90–99	0	0–19	
Slovenia	**30 – 50**	99–00	–	30–49	
Spain	(400,000 – 640,000)	92	?	–	*13,12,10*
Sweden	**20 – 30**	99–00	**–**	**>80**	
Switzerland	**0 – 2**	93–96	**0**	**0–19**	
Turkey	(200,000 – 400,000)	01	(–)	(20–29)	
Ukraine	27,500 – 44,500	90–00	–	5–9	

Total (approx.) 1,000,000 – 1,900,000 **Overall trend** Unknown
Breeding range >3,000,000 km² **Gen. length**. <3.3 **% Global pop.** 25–49

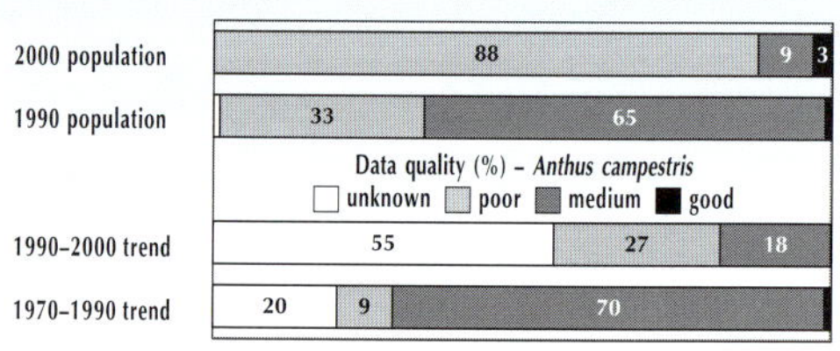

Anthus berthelotii
BERTHELOT'S PIPIT

Non-SPEC[E] (1994: 4) **Status** (Secure)
Criteria —
European IUCN Red List Category —
Criteria —
Global IUCN Red List Category —
Criteria —

Country	Breeding pop. size (pairs)	Year(s)	Trend	Mag.%	References
Portugal					
Madeira	Present	02	?	-	
Spain					
Canary Is.	(20,000 – 100,000)	97–03	?	-	*28,25*
Total (approx.)	20,000 – 100,000	**Overall trend** Unknown			
Breeding range	>20,000 km²	**Gen. length.** <3.3	**% Global pop.** 100		

Anthus berthelotii is endemic to Europe, where its entire global range is confined to the Canary Islands and Madeira. Its breeding population is relatively small (probably <100,000 pairs), although recent data on the size of the Madeira population were not available. Populations on both island groups were stable between 1970–1990, and although their trends during 1990–2000 were not known, there was no evidence to suggest that the species declined. Consequently, it is provisionally evaluated as Secure.

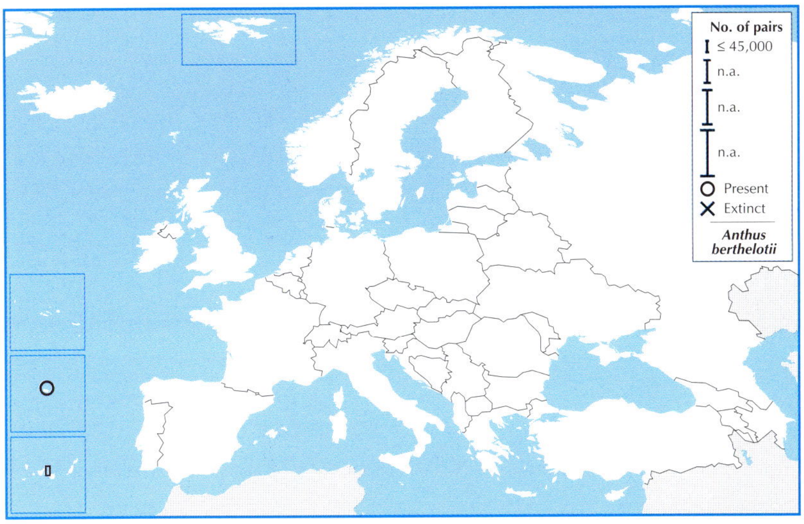

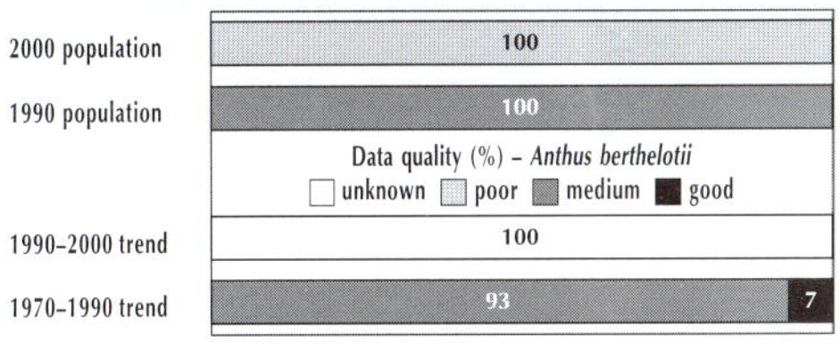

Anthus hodgsoni
OLIVE-BACKED PIPIT

Non-SPEC (1994: —) **Status** (Secure)
Criteria —
European IUCN Red List Category —
Criteria —
Global IUCN Red List Category —
Criteria —

Country	Breeding pop. size (pairs)	Year(s)	Trend	Mag.%	References
Russia	(25,000 – 50,000)	90–00	?	-	*51*
Total (approx.)	25,000 – 50,000	**Overall trend** Unknown			
Breeding range	>100,000 km²	**Gen. length.** <3.3	**% Global pop.** <5		

Anthus hodgsoni has a predominantly Asian distribution, but its breeding range extends just west of the Urals into north-east European Russia. Its European breeding population is relatively small (<50,000 pairs), but was stable between 1970–1990. No trend data were available for the Russian population during 1990–2000, but there was no evidence to suggest that its status had deteriorated significantly since 1990, and consequently the species is provisionally evaluated as Secure.

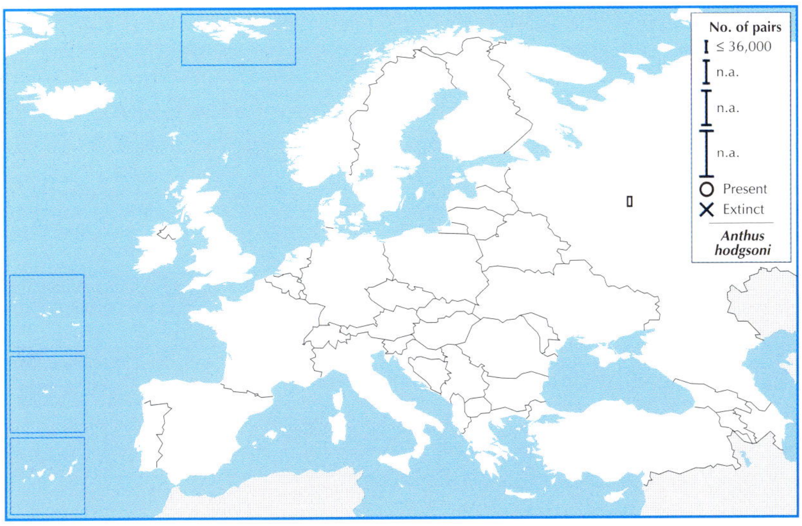

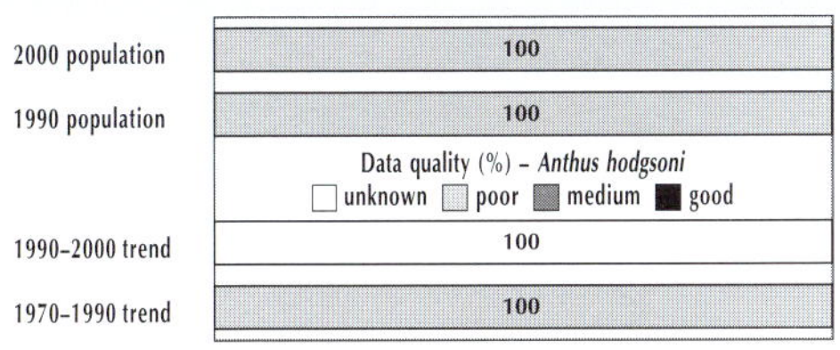

Anthus trivialis
TREE PIPIT

Non-SPEC (1994: —) **Status** Secure
Criteria —
European IUCN Red List Category —
Criteria —
Global IUCN Red List Category —
Criteria —

Anthus trivialis is a widespread summer visitor to most of Europe, which accounts for less than half of its global breeding range. Its European breeding population is extremely large (>27,000,000 pairs), and was stable between 1970–1990. Although there were declines in parts of western Europe and Fennoscandia (notably Sweden) during 1990–2000, most populations to the east—including the key one in Russia—were stable or increased, and the species underwent only a small decline overall. Consequently, it is evaluated as Secure.

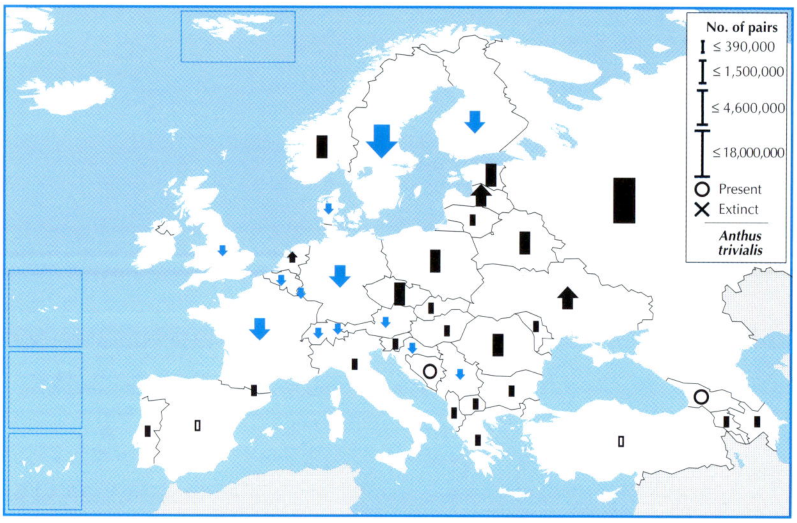

Country	Breeding pop. size (pairs)	Year(s)	Trend	Mag.%	References
Albania	(1,000 – 2,000)	02	(0)	(0–19)	
Andorra	750 – 1,000	99–01	(0)	(0–19)	*1,3*
Armenia	150,000 – 350,000	97–02	0	0–19	
Austria	(35,000 – 70,000)	98–02	(–)	(30–49)	
Azerbaijan	(10,000 – 50,000)	96–00	(0)	(0–19)	
Belarus	800,000 – 1,200,000	97–02	0	0–19	
Belgium	6,900 – 11,000	01–02	–	0–19	*1*
Bosnia & HG	Present	90–03	?	–	
Bulgaria	10,000 – 18,000	96–02	0	0–9	
Croatia	(10,000 – 50,000)	02	(–)	(20–29)	*16*
Czech Rep.	500,000 – 1,000,000	00	0	0–19	
Denmark	(10,000 – 50,000)	00	(–)	(20)	
Estonia	450,000 – 800,000	98	0	0–19	*1*
Finland	1,300,000 – 1,700,000	98–02	**–**	**30**	
France	(250,000 – 1,000,000)	98–02	**–**	**41**	*4,3*
Georgia	Present	03	?	–	
Germany	500,000 – 880,000	95–99	–	20–29	
Greece	(400 – 800)	95–00	(0)	(0–19)	
Hungary	**130,000 – 235,000**	99–02	0	0–19	*19*
Italy	(40,000 – 80,000)	03	(0)	(0–19)	
Latvia	500,000 – 900,000	90–00	+	20–29	*23,18,2*
Liechtenstein	70 – 120	98–00	–	20–29	
Lithuania	(300,000 – 500,000)	99–01	(0)	(0–19)	*20*
Luxembourg	5,000 – 6,000	02	–	0–19	
Macedonia	(2,000 – 3,000)	90–00	(0)	(0–19)	
Moldova	15,000 – 20,000	90–00	0	0–19	
Netherlands	**35,000 – 45,000**	98–00	**+**	**16**	*1*
Norway	(1,000,000 – 2,000,000)	90–03	(0)	(0–19)	*30*
Poland	**500,000 – 800,000**	00–02	(0)	(0–19)	*23*
Portugal	(50 – 100)	02	(0)	(0–19)	
Romania	600,000 – 850,000	00–02	0	0–19	*4,46,48*
Russia	15,000,000 – 20,000,000	90–00	0	0–19	*122*
Serbia & MN	35,000 – 50,000	90–02	–	0–19	*1,29,172a,67a, 225,227*
Slovakia	200,000 – 400,000	90–99	0	0–19	
Slovenia	20,000 – 30,000	94	(0)	(0–19)	
Spain	(300,000 – 400,000)	98–02	?	–	*10,13*
Sweden	**3,000,000 – 7,000,000**	99–00	**–**	**31**	
Switzerland	**50,000 – 70,000**	93–96	–	30–49	
Turkey	(10,000 – 30,000)	01	?	–	
Ukraine	850,000 – 1,240,000	90–00	+	5–19	
UK	**74,400 – 74,400**	00	**–**	**38**	*31*
Total (approx.)	27,000,000 – 42,000,000	**Overall trend** Small decline			
Breeding range	>7,000,000 km²	**Gen. length.** <3.3 **% Global pop.** 25–49			

(See p. 190, bottom, for data quality graph)

Anthus gustavi
PECHORA PIPIT

Non-SPEC (1994: —) **Status** (Secure)
Criteria —
European IUCN Red List Category —▼
Criteria —
Global IUCN Red List Category —
Criteria —

Anthus gustavi has a predominantly Siberian breeding distribution, which extends just west of the Urals into northern European Russia. Its European breeding population is small (as few as 1,000 pairs), but was stable between 1970–1990. Trend data were not available during 1990–2000, but there was no evidence to suggest that the species declined. Although the size of the European population could make it susceptible to the risks affecting small populations, it is marginal to a much larger non-European population. Consequently, it is provisionally evaluated as Secure.

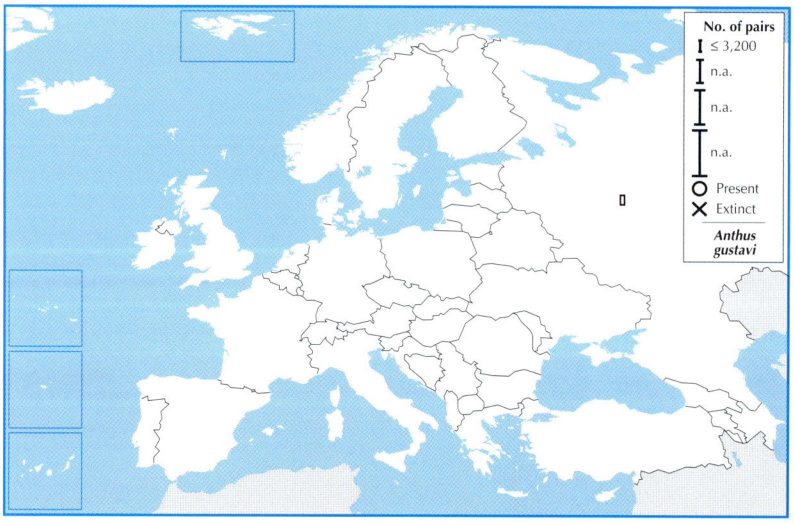

Country	Breeding pop. size (pairs)	Year(s)	Trend	Mag.%	References
Russia	(1,000 – 10,000)	90–00	?	–	
Total (approx.)	1,000 – 10,000	**Overall trend** Unknown			
Breeding range	>20,000 km²	**Gen. length.** <3.3 **% Global pop.** <5			

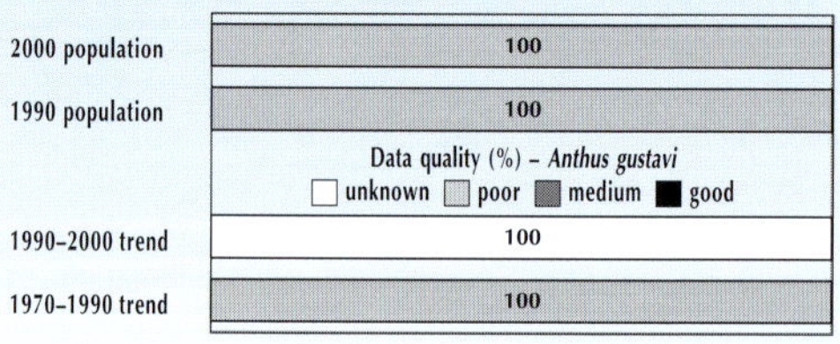

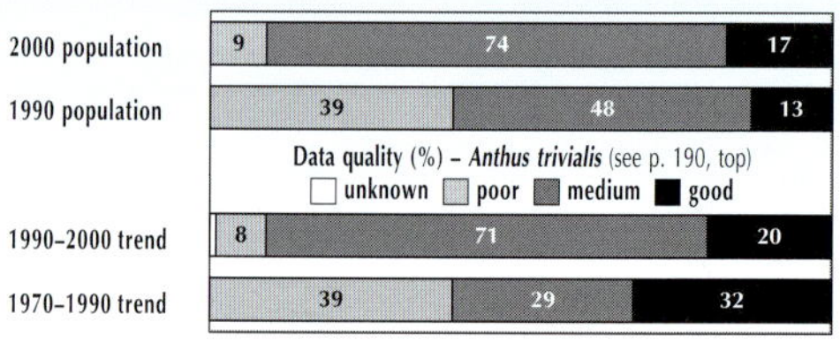

Anthus pratensis
MEADOW PIPIT

Non-SPEC[E] (1994: 4) **Status** (Secure)
Criteria —
European IUCN Red List Category —
Criteria —
Global IUCN Red List Category —
Criteria —

Anthus pratensis is a widespread breeder across much of central and northern Europe, which constitutes >75% of its global breeding range. Its European breeding population is very large (>7,000,000 pairs), and was stable between 1970–1990. Although there were declines in countries such as the United Kingdom, France and Sweden during 1990–2000, the species was stable or increased across most of its European range—including in Norway and Russia—and probably declined only slightly overall. Consequently, it is provisionally evaluated as Secure.

Country	Breeding pop. size (pairs)	Year(s)	Trend	Mag.%	References
Andorra	(0 – 10)	99–01	(0)	(0–19)	*1,3*
Armenia	15,000 – 50,000	97–02	?	–	
Austria	1,300 – 2,300	98–02	**+**	**20–29**	
Belarus	180,000 – 230,000	97–02	0	0–19	
Belgium	6,800 – 12,000	01–02	–	30–49	*1*
Czech Rep.	35,000 – 70,000	00	0	10–19	
Denmark	30,000 – 45,000	00	**–**	**20–29**	*12*
Faroe Is.	1,500 – 1,500	95	(0)	(0–19)	
Greenland	(10 – 500)	90–00	?	–	
Estonia	50,000 – 100,000	98	–	20–29	*1*
Finland	700,000 – 1,200,000	98–02	**0**	**5**	
France	(250,000 – 1,000,000)	98–02	(–)	(35)	*4,2*
Georgia	Present	03	?	–	
Germany	64,000 – 120,000	95–99	0	0–19	
Iceland	(500,000 – 1,000,000)	78–94	?	–	*1*
Rep. Ireland	500,000 – 1,000,000	88–91	0	0–19	
Latvia	50,000 – 100,000	90–00	–	30–49	*23,16,2*
Lithuania	(200,000 – 300,000)	99–01	(–)	(0–19)	*20*
Luxembourg	**250 – 350**	00–02	**–**	**30–49**	
Moldova	500 – 600	90–00	0	0–19	
Netherlands	**70,000 – 80,000**	98–00	**F**	**24**	*1*
Norway	(1,000,000 – 5,000,000)	90–03	(0)	(0–19)	
Poland	**150,000 – 300,000**	00–02	0	0–19	*23*
Russia	1,000,000 – 2,500,000	90–00	(0)	(0–19)	*122*
Slovakia	250 – 500	80–99	+	30–49	
Slovenia	**0 – 0**	94	**–**	**X**	
Spain	(50 – 100)	98–02	?	–	*10*
Sweden	500,000 – 1,000,000	99–00	–	10–19	
Switzerland	**500 – 500**	93–96	+	0–19	
Ukraine	21,000 – 26,000	90–00	+	5–9	
UK	**1,680,000 – 1,680,000**	00	**–**	**16**	*5,31*

Total (approx.) 7,000,000 – 16,000,000 **Overall trend** Small decline
Breeding range >5,000,000 km² **Gen. length.** <3.3 **% Global pop.** 75–94

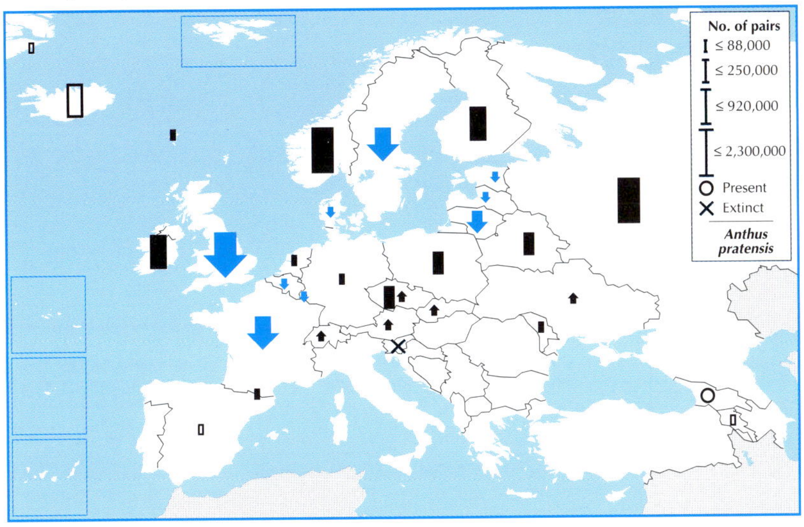

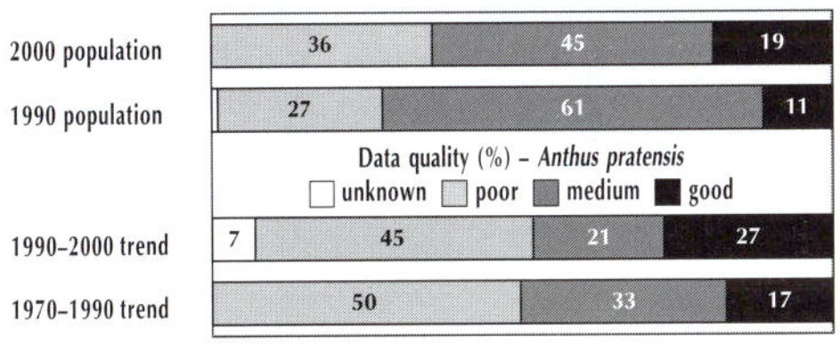

Anthus cervinus
RED-THROATED PIPIT

Non-SPEC (1994: —) **Status** (Secure)
Criteria —
European IUCN Red List Category —
Criteria —
Global IUCN Red List Category —
Criteria —

Anthus cervinus is a widespread summer visitor to northern Fennoscandia and Russia, with Europe accounting for less than a quarter of its global breeding range. Its European breeding population is very large (>1,000,000 pairs), and was stable between 1970–1990. The relatively small populations in Sweden and Finland declined during 1990–2000, whereas the trend for the Russian stronghold was unknown, but there was no evidence to suggest that its status deteriorated significantly. Consequently, the species is provisionally evaluated as Secure.

Country	Breeding pop. size (pairs)	Year(s)	Trend	Mag.%	References
Finland	500 – 3,000	98–02	–	**60**	
Norway	(5,000 – 20,000)	90–03	(0)	(0–19)	
Russia	1,000,000 – 3,000,000	90–00	?	–	*122*
Sweden	**300 – 900**	99–00	–	10–19	

Total (approx.) 1,000,000 – 3,000,000 **Overall trend** Unknown
Breeding range >250,000 km² **Gen. length.** <3.3 **% Global pop.** 5–24

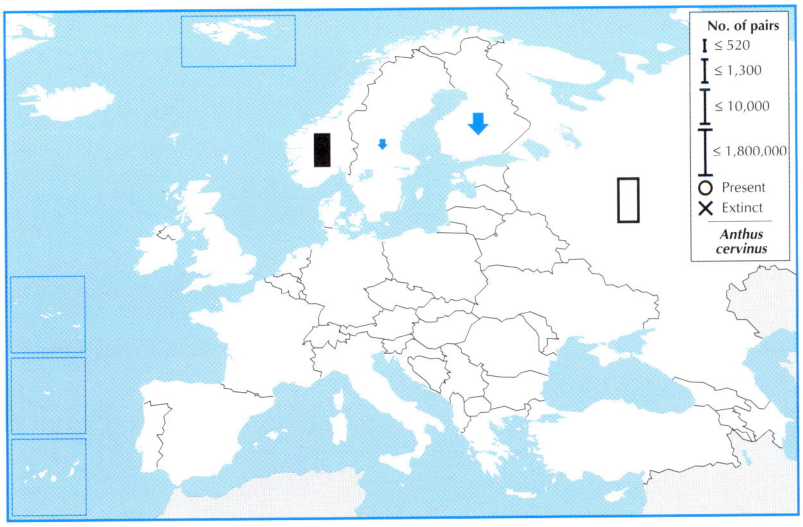

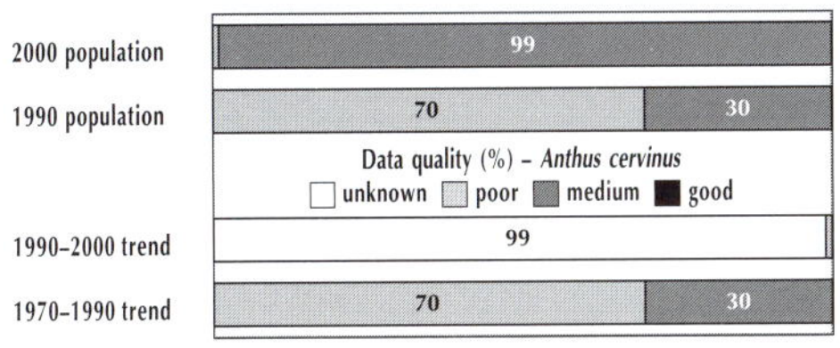

Anthus spinoletta
WATER PIPIT

Non-SPEC (1994: —) **Status** (Secure)
Criteria —
European IUCN Red List Category —
Criteria —
Global IUCN Red List Category —
Criteria —

Anthus spinoletta has a fragmented breeding distribution in mountainous areas of central and southern Europe, which accounts for less than a quarter of its global breeding range. Its European breeding population is large (>640,000 pairs), and was stable between 1970–1990. Although the species declined in certain countries (notably Russia) during 1990–2000, populations were stable across most of its range—including sizeable ones in Switzerland and Romania—and it probably declined slightly overall. Consequently, the species is provisionally evaluated as Secure.

Country	Breeding pop. size (pairs)	Year(s)	Trend	Mag.%	References
Albania	(500 – 1,000)	02	(0)	(0–19)	
Andorra	2,000 – 3,000	99-01	(0)	(0–19)	*1,3*
Armenia	50,000 – 100,000	97-02	0	0–19	
Austria	(60,000 – 120,000)	98-02	(0)	(0–19)	
Azerbaijan	(10,000 – 50,000)	96-00	(0)	(0–19)	
Bulgaria	**12,000 – 20,000**	96-02	0	0–19	
Croatia	(500 – 1,000)	02	(–)	(30–49)	*70,16*
Czech Rep.	110 – 150	00	–	50–79	
France	10,000 – 20,000	00	?	–	*1*
Georgia	Present	03	?	–	
Germany	2,100 – 6,100	95-99	(0)	(0–19)	
Greece	(200 – 500)	95-00	(0)	(0–19)	
Italy	(30,000 – 70,000)	03	(0)	(0–19)	
Liechtenstein	300 – 600	98-00	–	0–19	
Macedonia	(800 – 1,200)	90-00	(0)	(0–19)	
Poland	2,200 – 2,700	90-00	(0)	(0–19)	*2*
Portugal	0 – 5	02	(0)	(0–19)	
Romania	160,000 – 190,000	00-02	0	0–19	
Russia	(100,000 – 1,000,000)	90-00	(–)	(20–29)	*17,50,61,116,136*
Serbia & MN	5,000 – 8,000	90-02	0	0–19	*1,136,155,225, 227*
Slovakia	700 – 1,100	90-99	0	0–19	
Slovenia	1,500 – 3,000	99-00	(0)	(0–19)	
Spain	(20,000 – 100,000)	98-02	?	–	*10*
Switzerland	150,000 – 200,000	98-02	0	0–19	
Turkey	(20,000 – 60,000)	01	(0)	(0–19)	
Ukraine	1,300 – 2,400	90-00	0	0–19	
Total (approx.)	640,000 – 2,000,000	**Overall trend** Stable			
Breeding range	>1,000,000 km²	**Gen. length.** <3.3 **% Global pop.** 5–24			

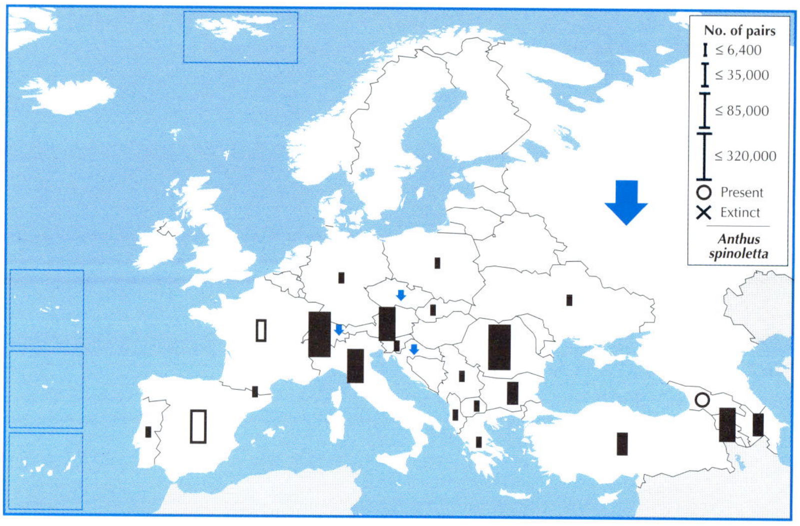

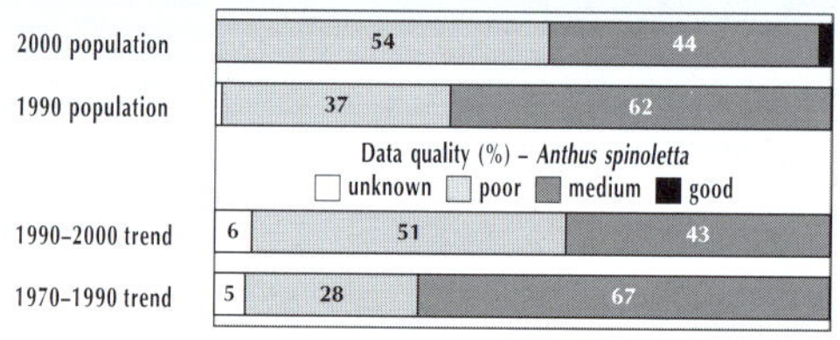

Anthus petrosus
ROCK PIPIT

Non-SPEC[E] (1994: —) **Status** (Secure)
Criteria —
European IUCN Red List Category —
Criteria —
Global IUCN Red List Category —
Criteria —

Anthus petrosus breeds only in Europe, where it occurs in coastal areas of the north-west and Fennoscandia. Its European breeding population is relatively large (>110,000 pairs), and was stable between 1970–1990. Although the species declined in the United Kingdom and Sweden during 1990–2000, populations were stable or increased across the majority of its range—including the Norwegian stronghold—and the species probably declined only slightly overall. Consequently, it is provisionally evaluated as Secure.

Country	Breeding pop. size (pairs)	Year(s)	Trend	Mag.%	References
Denmark	(100 – 200)	00	(+)	(0–19)	*3*
Faroe Is.	2,500 – 2,500	95	(0)	(0–19)	
Estonia	3 – 10	98	0	0–19	*1*
Finland	1,700 – 2,000	98-02	+	**10**	
France	10,000 – 20,000	00	(F)	(20–49)	*1*
Germany	**0 – 1**	95-99	+	**N**	
Iceland	**0 – 2**	87-99	F	>80	*21,23,24,25,32, 33,42,43,44,45, 46,47*
Rep. Ireland	10,000 – 20,000	88-91	?	–	
Norway	(50,000 – 200,000)	90-03	(0)	(0–19)	
Russia	(250 – 2,500)	90-00	?	–	
Sweden	3,000 – 5,000	99-00	–	30–49	
UK	**35,700 – 35,700**	88-91	(–)	(0–19)	*8*
Total (approx.)	110,000 – 290,000	**Overall trend** Small decline			
Breeding range	>500,000 km²	**Gen. length.** <3.3 **% Global pop.** 100			

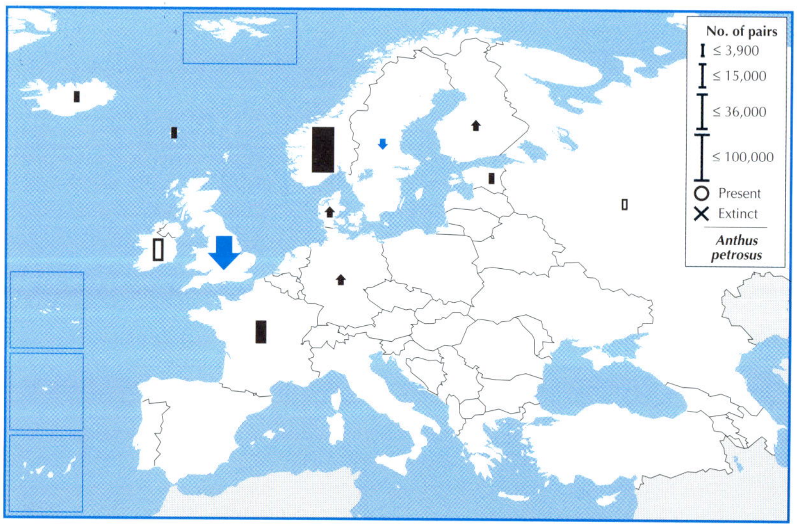

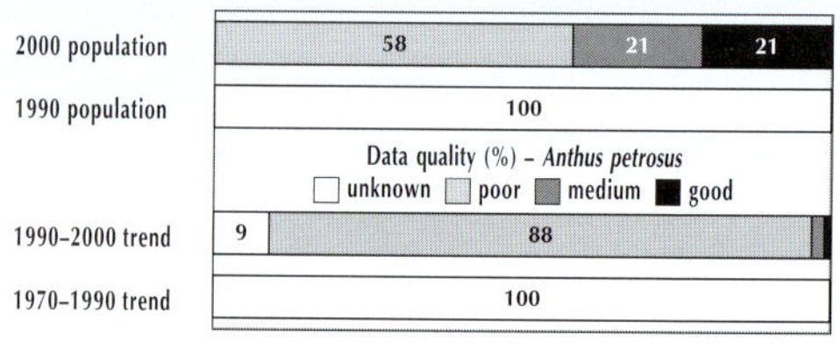

Motacilla flava
YELLOW WAGTAIL

Non-SPEC (1994: —) **Status** (Secure)
Criteria —
European IUCN Red List Category —
Criteria —
Global IUCN Red List Category —
Criteria —

Motacilla flava is a widespread summer visitor to most of Europe, which accounts for less than half of its global breeding range. Its European breeding population is very large (>7,900,000 pairs), and was stable between 1970–1990. Although there were widespread declines—most notably in Romania—during 1990–2000, other key populations (such as those in Russia, Belarus, Poland, Ukraine and Bulgaria) were stable, and the species probably underwent only a slight decline overall. Consequently, it is provisionally evaluated as Secure.

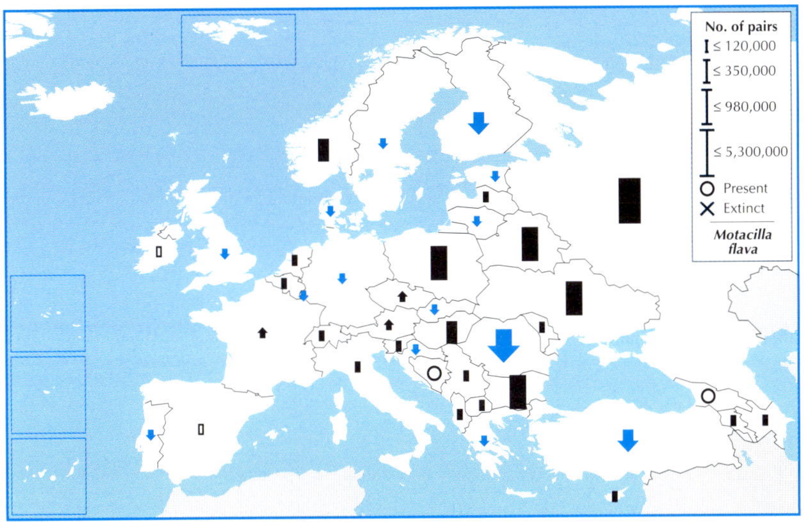

Country	Breeding pop. size (pairs)	Year(s)	Trend	Mag.%	References
Albania	2,000 – 5,000	96–02	0	0–19	
Armenia	33,000 – 150,000	99–02	0	0–19	
Austria	400 – 700	98–02	+	20–29	
Azerbaijan	(20,000 – 200,000)	96–00	(0)	(0–19)	
Belarus	470,000 – 530,000	97–02	0	0–19	
Belgium	10,000 – 20,000	01–02	0	0–19	*1*
Bosnia & HG	Present	90–03	?	–	
Bulgaria	500,000 – 1,000,000	96–02	0	0–19	
Croatia	(1,000 – 5,000)	02	(–)	(50–79)	*70,16*
Cyprus	(10 – 40)	94–02	(0)	(0–9)	
Czech Rep.	800 – 1,600	00	+	30–49	
Denmark	5,000 – 10,000	00	(–)	(30–49)	*12,3*
Estonia	10,000 – 20,000	98	–	20–29	*1*
Finland	250,000 – 400,000	98–02	**–**	**50**	
France	(60,000 – 240,000)	98–02	**+**	**68**	*4,2*
Georgia	Present	03	?	–	
Germany	73,000 – 115,000	95–99	–	20–29	
Greece	(10,000 – 20,000)	95–00	(–)	(0–19)	
Hungary	**150,000 – 225,000**	99–02	0	0–19	*19*
Rep. Ireland	2 – 2	88–91	?	–	
Italy	(20,000 – 40,000)	03	(0)	(0–19)	
Latvia	10,000 – 25,000	90–00	(0)	(0–19)	*23,16*
Lithuania	(20,000 – 30,000)	99–01	(–)	(30–49)	*20*
Luxembourg	**50 – 70**	00–02	**–**	**>80**	
Macedonia	(10,000 – 20,000)	90–00	(0)	(0–19)	
Moldova	10,000 – 12,000	90–00	0	0–19	
Netherlands	**40,000 – 50,000**	98–00	**0**	**6**	*1*
Norway	(100,000 – 500,000)	90–02	(0)	(0–19)	*46*
Poland	**400,000 – 800,000**	00–02	0	0–19	*23*
Portugal	(5,000 – 50,000)	02	(–)	(–)	
Romania	800,000 – 1,200,000	97–02	–	0–19	*48*
Russia	4,000,000 – 7,000,000	90–00	(0)	(0–19)	*114,32,57,122*
Serbia & MN	20,000 – 25,000	90–02	0	0–19	*1,29,172a,67a, 155,78,225,227*
Slovakia	2,500 – 4,000	80–99	–	30–49	
Slovenia	300 – 400	00	0	0–19	
Spain	(20,000 – 100,000)	98–02	?	–	*10*
Sweden	80,000 – 170,000	99–00	**–**	**55**	
Switzerland	**150 – 200**	93–96	F	20–29	
Turkey	(200,000 – 600,000)	01	(–)	(0–19)	
Ukraine	535,000 – 740,000	90–00	0	0–19	
UK	**11,500 – 26,500**	00	**–**	**50–79**	*5,31*
Total (approx.)	7,900,000 – 14,000,000	**Overall trend** Small decline			
Breeding range	>7,000,000 km²	**Gen. length.** <3.3 **% Global pop.** 25–49			

(See p. 193, bottom, for data quality graph)

Motacilla citreola
CITRINE WAGTAIL

Non-SPEC (1994: —) **Status** (Secure)
Criteria —
European IUCN Red List Category —
Criteria —
Global IUCN Red List Category —
Criteria —

Motacilla citreola is a summer visitor to northern Russia and parts of central and eastern Europe, which accounts for less than a quarter of its global breeding range. Its European breeding population is large (>210,000 pairs), and was stable between 1970–1990. Although the trend of the stronghold population in Russia during 1990–2000 was unknown, the species increased or was stable across the rest of its European range, and there was no evidence to suggest that its status deteriorated significantly. Consequently, it is provisionally evaluated as Secure.

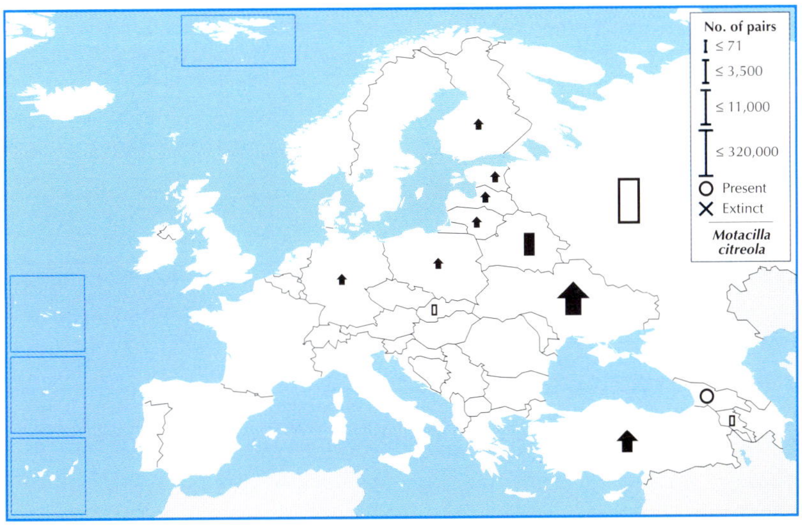

Country	Breeding pop. size (pairs)	Year(s)	Trend	Mag.%	References
Armenia	**0 – 50**	00–02	?	–	*9*
Belarus	1,000 – 2,500	97–02	0	0–19	
Estonia	**1 – 10**	98	+	N	*1*
Finland	5 – 30	98–02	(+)	(0–19)	
Georgia	Present	03	?	–	
Germany	**0 – 1**	95–99	**+**	**N**	
Latvia	5 – 20	90–00	+	>80	*16*
Lithuania	50 – 100	99–01	+	0–19	*20*
Poland	**25 – 50**	98–02	**+**	**N**	*60–69,88*
Russia	(200,000 – 500,000)	90–00	?	–	*23,32*
Slovakia	0 – 2	80–99	?	–	
Turkey	2,000 – 6,000	01	(+)	(0–19)	
Ukraine	8,300 – 13,800	90–00	**+**	**20–29**	
Total (approx.)	210,000 – 520,000	**Overall trend** Unknown			
Breeding range	>1,000,000 km²	**Gen. length.** <3.3 **% Global pop.** 5–24			

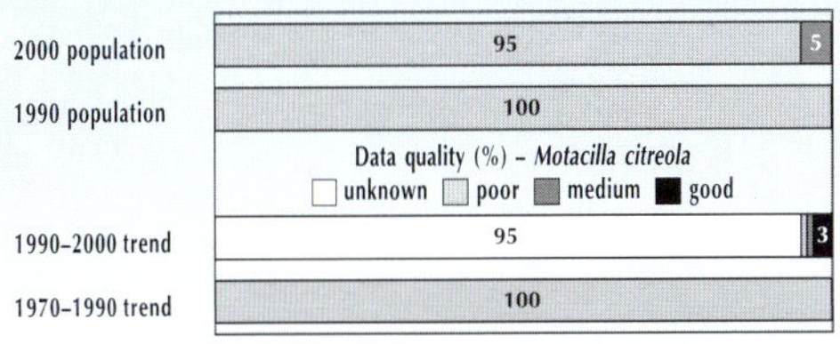

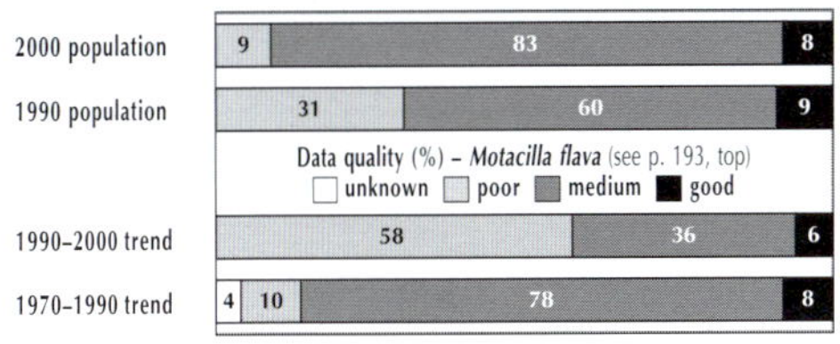

Motacilla cinerea
GREY WAGTAIL

Non-SPEC (1994: —) **Status** Secure
Criteria —
European IUCN Red List Category —
Criteria —
Global IUCN Red List Category —
Criteria —

Motacilla cinerea has a discontinuous breeding distribution in western and southern Europe, the Caucasus and the western Urals, with Europe accounting for less than a quarter of its global breeding range. Its European breeding population is large (>740,000 pairs), and was stable between 1970–1990. The species remained stable overall during 1990–2000, with the vast majority of national populations stable or increasing—including the key one in Romania. Consequently, it is evaluated as Secure.

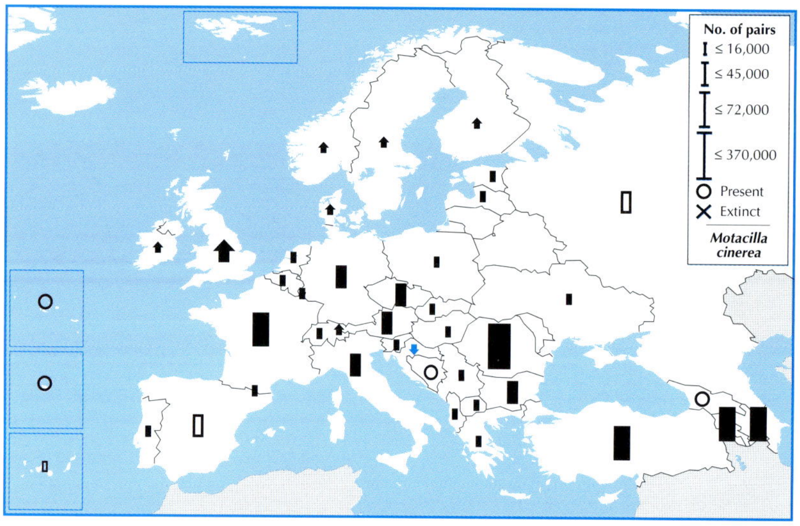

Country	Breeding pop. size (pairs)	Year(s)	Trend	Mag.%	References
Albania	800 – 1,600	96–02	(0)	(0–19)	
Andorra	70 – 100	99–01	(0)	(0–19)	*1,3*
Armenia	30,000 – 170,000	99–02	0	0–19	
Austria	25,000 – 50,000	98–02	(0)	(0–19)	
Azerbaijan	(20,000 – 200,000)	96–00	(0)	(0–19)	
Belgium	5,100 – 8,300	01–02	0	0–19	*1*
Bosnia & HG	Present	90–03	?	–	
Bulgaria	20,000 – 30,000	96–02	0	0–9	
Croatia	(5,000 – 10,000)	02	(–)	(30–49)	*16*
Czech Rep.	20,000 – 40,000	00	0	0–19	
Denmark	(450 – 500)	93–96	(+)	(30–49)	*3*
Estonia	0 – 5	98	0	0–19	*1*
Finland	10 – 30	98–02	+	**300**	
France	(40,000 – 120,000)	98–02	(0)	(0–19)	*4*
Georgia	Present	03	?	–	
Germany	27,000 – 43,000	95–99	0	0–19	
Greece	(5,000 – 10,000)	95–00	(0)	(0–19)	
Hungary	190 – 300	95–02	0	0–19	
Rep. Ireland	10,000 – 20,000	88–91	+	0–19	
Italy	(20,000 – 50,000)	03	(0)	(0–19)	
Latvia	(1 – 10)	90–00	(0)	(0–19)	*17*
Liechtenstein	50 – 80	98–00	+	0–19	
Luxembourg	300 – 400	02	**0**	**0–19**	
Macedonia	(3,000 – 5,000)	90–00	(0)	(0–19)	
Netherlands	**240 – 300**	98–00	0	0–19	*1*
Norway	600 – 1,500	90–02	+	0–19	*27*
Poland	**2,000 – 4,000**	90–00	0	0–19	*27*
Portugal	(5,000 – 50,000)	02	(0)	(0–19)	
Azores	Present	02	?	–	
Madeira	Present	02	?	–	
Romania	**340,000 – 385,000**	00–02	0	0–19	
Russia	(30,000 – 55,000)	90–00	?	–	*51,52,118*
Serbia & MN	8,000 – 12,000	90–02	0	0–19	*1,155,172a,91, 43,152*
Slovakia	10,000 – 20,000	90–99	0	0–19	
Slovenia	3,000 – 6,000	00	(0)	(0–19)	
Spain	(20,000 – 100,000)	98–02	?	–	*10*
Canary Is.	(2,500 – 10,000)	97–03	?	–	*28,25*
Sweden	500 – 2,000	99–00	+	**50–79**	
Switzerland	**8,000 – 11,000**	93–96	0	0–19	
Turkey	(30,000 – 120,000)	01	(0)	(0–19)	
Ukraine	**9,000 – 13,000**	90–00	**0**	**0–19**	
UK	**38,400 – 46,200**	00	+	**0–19**	*5,31*
Total (approx.)	740,000 – 1,600,000	**Overall trend** Stable			
Breeding range	>4,000,000 km²	**Gen. length.** <3.3 **% Global pop.** 5–24			

(See p. 195, top, for data quality graph)

Motacilla alba
WHITE WAGTAIL

Non-SPEC (1994: —) **Status** Secure
Criteria —
European IUCN Red List Category —
Criteria —
Global IUCN Red List Category —
Criteria —

Motacilla alba is a widespread breeder across most of Europe, which accounts for less than a quarter of its global breeding range. Its European breeding population is extremely large (>13,000,000 pairs), and was stable between 1970–1990. Although there were declines in Sweden, Finland and Latvia during 1990–2000, populations were stable across the vast majority of its European range, and the species remained stable overall. Consequently, it is evaluated as Secure.

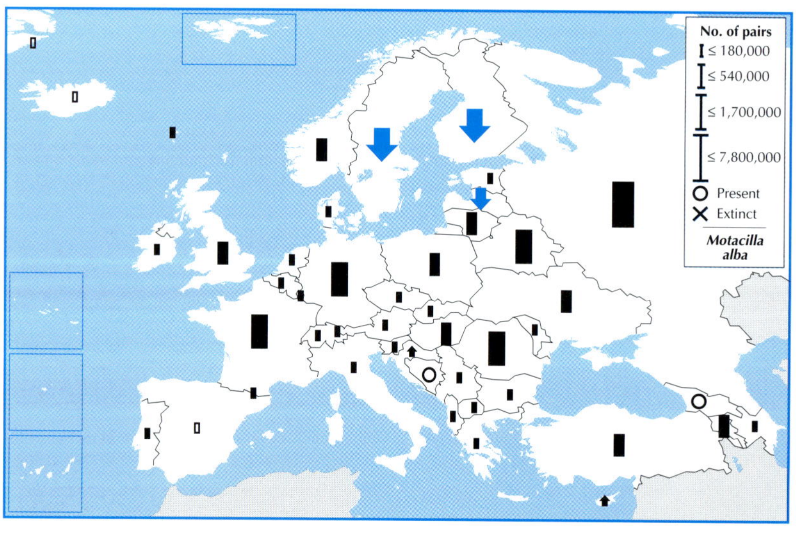

Country	Breeding pop. size (pairs)	Year(s)	Trend	Mag.%	References
Albania	2,000 – 5,000	96–02	(0)	(0–19)	
Andorra	100 – 200	99–01	(0)	(0–19)	*1,3*
Armenia	300,000 – 500,000	99–02	0	0–19	
Austria	(50,000 – 100,000)	98–02	(0)	(0–19)	
Azerbaijan	(100,000 – 250,000)	96–00	(0)	(0–19)	
Belarus	800,000 – 1,200,000	97–02	0	0–19	
Belgium	30,000 – 40,000	01–02	(0)	(0–19)	*1*
Bosnia & HG	Present	90–03	?	–	
Bulgaria	100,000 – 300,000	96–02	0	0–9	
Croatia	(50,000 – 100,000)	02	(+)	(30–49)	*16*
Cyprus	**1 – 1**	98	+	**N**	
Czech Rep.	100,000 – 200,000	00	0	0–19	
Denmark	100,000 – 150,000	00	**0**	**0–19**	*12*
Faroe Is.	(2 – 5)	81	(0)	(0–19)	
Greenland	(1 – 50)	90–00	?	–	
Estonia	100,000 – 150,000	98	0	0–19	*1*
Finland	600,000 – 900,000	98–02	**–**	**10**	
France	(400,000 – 1,600,000)	98–02	(0)	(3)	*4,2*
Georgia	Present	03	?	–	
Germany	670,000 – 1,000,000	95–99	0	0–19	
Greece	(5,000 – 10,000)	95–00	(0)	(0–19)	
Hungary	**160,000 – 265,000**	99–02	0	0–19	*19*
Iceland	(20,000 – 50,000)	00	?	–	*21*
Rep. Ireland	100,000 – 250,000	88–91	0	0–19	
Italy	(60,000 – 150,000)	03	(0)	(0–19)	
Latvia	150,000 – 300,000	90–00	–	30–49	*23,16,2*
Liechtenstein	200 – 400	98–00	0	0–19	
Lithuania	(400,000 – 500,000)	99–01	(0)	(0–19)	*20*
Luxembourg	8,000 – 10,000	02	0	0–19	
Macedonia	(5,000 – 10,000)	90–00	(0)	(0–19)	
Moldova	24,000 – 26,000	90–00	0	0–19	
Netherlands	**70,000 – 140,000**	98–00	**0**	**10**	*1*
Norway	(100,000 – 500,000)	90–03	(0)	(0–19)	*37,63,27*
Poland	**150,000 – 300,000**	00–02	(0)	(0–19)	*23*
Portugal	(10,000 – 100,000)	02	(0)	(0–19)	
Romania	1,500,000 – 1,900,000	00–02	0	0–19	*48*
Russia	5,000,000 – 12,000,000	90–00	0	0–19	*122*
Serbia & MN	30,000 – 50,000	90–02	0	0–19	*1,29,67a,78, 225,227*
Slovakia	50,000 – 100,000	90–99	0	0–19	
Slovenia	5,000 – 10,000	94	(0)	(0–19)	
Spain	(100,000 – 250,000)	98–02	?	–	*10*
Sweden	500,000 – 1,000,000	99–00	**–**	**19**	
Switzerland	**100,000 – 150,000**	93–96	0	0–19	
Turkey	(300,000 – 600,000)	01	(0)	(0–19)	
Ukraine	(465,000 – 620,000)	90–00	(0)	(0–19)	
UK	**272,000 – 352,000**	00	**0**	**0–19**	*5,31*
Total (approx.)	13,000,000 – 26,000,000	**Overall trend** Stable			
Breeding range	>8,000,000 km²	**Gen. length.** <3.3 **% Global pop.** 5–24			

(See p. 195, bottom, for data quality graph)

Pycnonotus xanthopygos
WHITE-SPECTACLED BULBUL

Non-SPEC (1994: —) **Status** Secure
Criteria —
European IUCN Red List Category —▼
Criteria —
Global IUCN Red List Category —
Criteria —

Country	Breeding pop. size (pairs)	Year(s)	Trend	Mag.%	References
Turkey	6,000 – 18,000	01	0	0–19	
Total (approx.)	6,000 – 18,000	Overall trend Stable			
Breeding range	>20,000 km²	Gen. length. <3.3	% Global pop. 5–24		

Pycnonotus xanthopygos is a resident of southernmost coastal Turkey, with Europe accounting for less than a quarter of its global range. Its European breeding population is small (as few as 6,000 pairs), and its trend between 1970–1990 was unknown, but it was stable during 1990–2000. Although the size of the European population could make it susceptible to the risks affecting small populations, it is marginal to a much larger non-European population. Consequently, the species is evaluated as Secure.

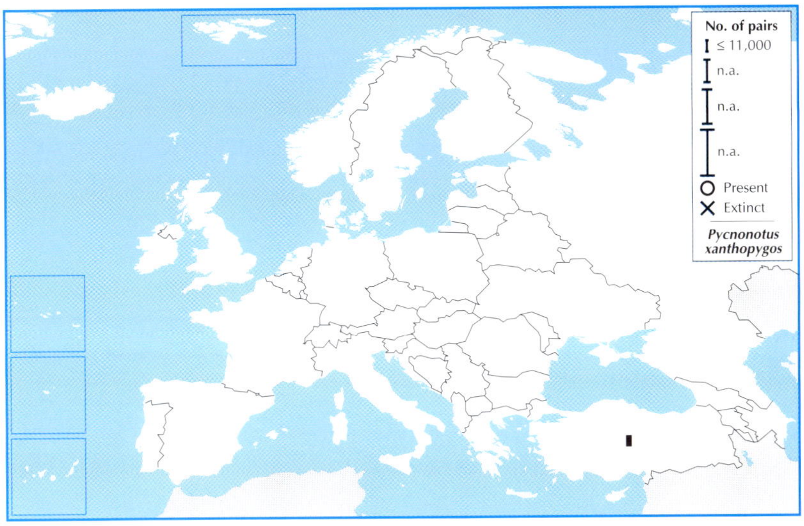

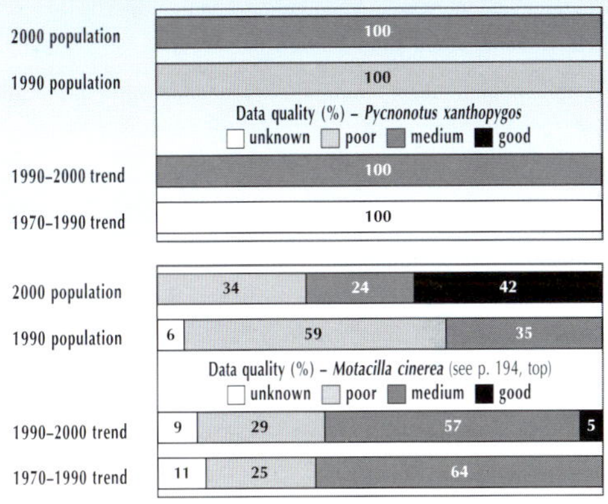

Bombycilla garrulus
BOHEMIAN WAXWING

Non-SPEC (1994: —) **Status** (Secure)
Criteria —
European IUCN Red List Category —
Criteria —
Global IUCN Red List Category —
Criteria —

Country	Breeding pop. size (pairs)	Year(s)	Trend	Mag.%	References
Estonia	(0 – 1)	98	?	–	1
Finland	30,000 – 150,000	98–02	+	200	
Norway	(0 – 2,000)	90–02	(F)	(20–29)	24
Russia	(100,000 – 500,000)	90–00	(F)	(20–29)	40
Sweden	(1,000 – 50,000)	99–00	(F)	(50–79)	
Total (approx.)	130,000 – 700,000	Overall trend Fluctuating			
Breeding range	>1,000,000 km²	Gen. length. <3.3	% Global pop. 5–24		

Bombycilla garrulus is a widespread breeder in northern Fennoscandia and Russia, with Europe accounting for less than a quarter of its global breeding range. Its European breeding population is large (>130,000 pairs), and was stable between 1970–1990. Although overall trends are complicated by sizeable fluctuations, the species increased significantly in Finland during 1990–2000, and was broadly stable in Sweden, Norway and the stronghold Russia. Consequently, it is provisionally evaluated as Secure.

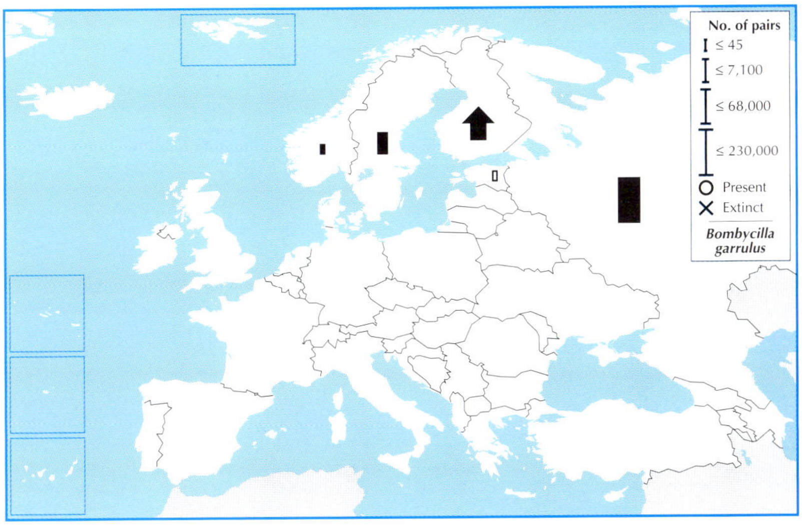

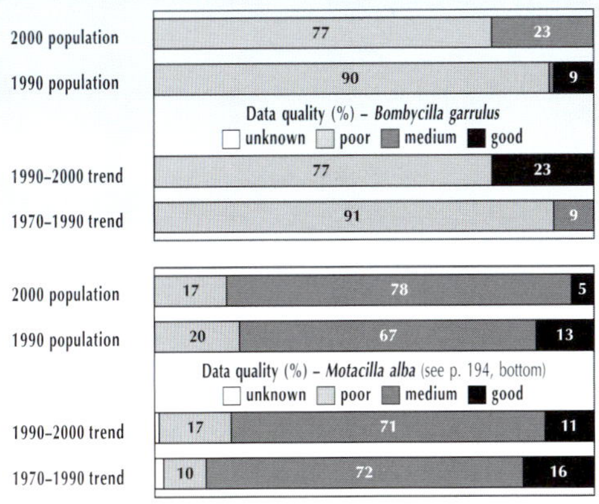

Cinclus cinclus
WHITE-THROATED DIPPER

Non-SPEC (1994: —) **Status** Secure
Criteria —
European IUCN Red List Category —
Criteria —
Global IUCN Red List Category —
Criteria —

Cinclus cinclus is a widespread but patchily distributed resident in the more mountainous areas of Europe, which accounts for less than half of its global range. Its European breeding population is large (>170,000 pairs), and was stable between 1970–1990. Although there were declines in a few countries during 1990–2000, populations were stable across the majority of its European range—including the key one in Romania—and the species remained stable overall. Consequently, it is evaluated as Secure.

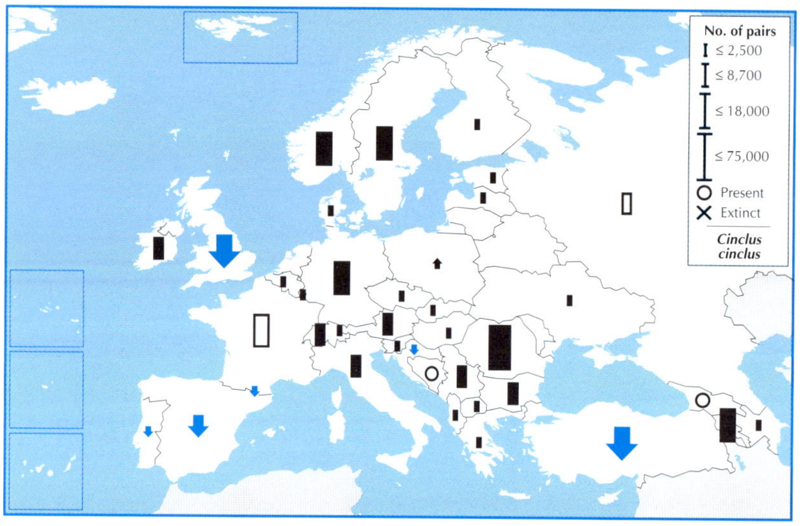

Country	Breeding pop. size (pairs)	Year(s)	Trend	Mag.%	References
Albania	200 – 1,000	02	(0)	(0–19)	
Andorra	**70 – 120**	99–01	(–)	(10–29)	*1,3*
Armenia	10,000 – 15,000	97–02	0	0–19	
Austria	6,000 – 9,000	98–02	(0)	(0–19)	
Azerbaijan	(500 – 5,000)	96–00	(0)	(0–19)	
Belgium	800 – 1,500	01–02	0	0–19	*1*
Bosnia & HG	Present	90–03	?	–	
Bulgaria	**4,000 – 6,000**	96–02	0	0–9	
Croatia	(100 – 500)	02	(–)	(50–79)	*16*
Czech Rep.	1,000 – 2,000	00	0	0–19	
Denmark	**4 – 6**	98–01	0	0–19	*15*
Estonia	**0 – 10**	98	0	0–19	*1*
Finland	250 – 350	98–02	(0)	(0–19)	
France	10,000 – 30,000	98–02	?	–	*4*
Georgia	Present	03	?	–	
Germany	8,600 – 12,000	95–99	0	0–19	
Greece	(500 – 2,000)	95–00	(0)	(0–19)	
Hungary	10 – 20	98–02	F	50–79	*8*
Rep. Ireland	2,500 – 10,000	88–91	(0)	(0–19)	
Italy	(4,000 – 8,000)	03	(0)	(0–19)	
Latvia	0 – 10	90–00	F	30–49	*17*
Liechtenstein	15 – 20	98–00	0	0–19	
Luxembourg	**150 – 200**	02	**0**	**0–19**	
Macedonia	(500 – 800)	90–00	(0)	(0–19)	
Norway	10,000 – 15,000	90–03	0	0–19	*37,40*
Poland	**1,000 – 1,500**	90–01	+	0–19	*91,92*
Portugal	(500 – 2,500)	02	(–)	(–)	
Romania	**68,000 – 81,000**	96–02	0	0–19	*21*
Russia	(5,000 – 15,000)	90–00	?	–	*51,52*
Serbia & MN	3,200 – 4,500	90–02	0	0–19	*1,155,225,227, 40,197,62,152*
Slovakia	2,000 – 3,000	90–99	0	0–19	
Slovenia	1,000 – 3,000	94	(0)	(0–19)	
Spain	(2,500 – 10,000)	98–02	(–)	(0–19)	*10*
Sweden	(5,000 – 50,000)	99–00	(F)	(20–29)	
Switzerland	**3,000 – 5,000**	93–96	0	0–19	
Turkey	(7,000 – 14,000)	01	–	0–19	
Ukraine	1,150 – 1,600	90–00	0	0–19	
UK	**6,800 – 20,000**	00	**–**	**8**	*5*
Total (approx.)	170,000 – 330,000	**Overall trend** Stable			
Breeding range	>4,000,000 km²	**Gen. length**. <3.3 **% Global pop.** 25–49			

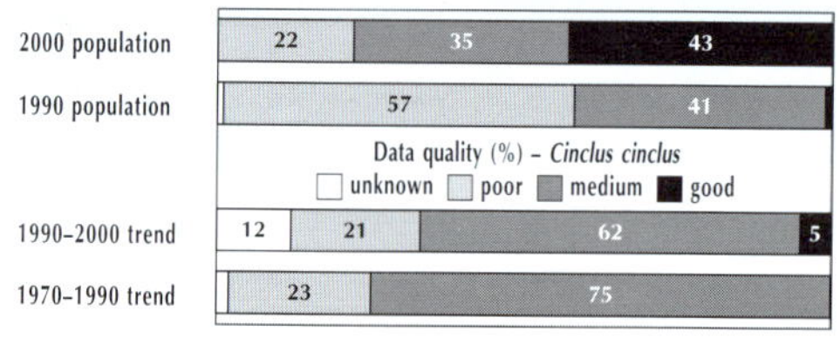

Troglodytes troglodytes
WINTER WREN

Non-SPEC (1994: —) **Status** Secure
Criteria —
European IUCN Red List Category —
Criteria —
Global IUCN Red List Category —
Criteria —

Troglodytes troglodytes is a widespread breeder across most of Europe, which accounts for less than a quarter of its global breeding range. Its European breeding population is extremely large (>23,000,000 pairs), and was stable between 1970–1990. The species underwent a slight increase overall during 1990–2000, with populations across the vast majority of its European range—including the sizeable one in the United Kingdom—stable or increasing. Consequently, it is evaluated as Secure.

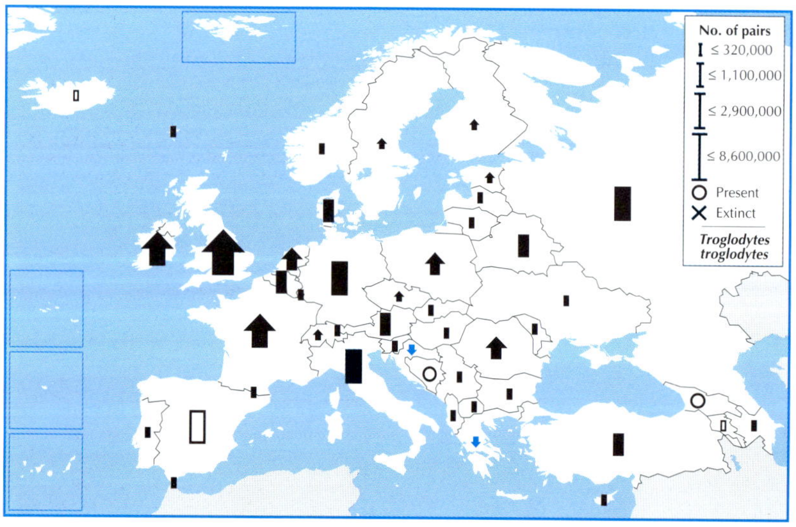

Country	Breeding pop. size (pairs)	Year(s)	Trend	Mag.%	References
Albania	1,000 – 5,000	02	(0)	(0–19)	
Andorra	500 – 700	99–01	(0)	(0–19)	*1,3*
Armenia	15,000 – 25,000	97–02	?	–	
Austria	(350,000 – 700,000)	98–02	(0)	(0–19)	
Azerbaijan	(20,000 – 100,000)	96–00	(0)	(0–19)	
Belarus	450,000 – 500,000	97–02	0	0–19	
Belgium	250,000 – 500,000	01–02	(0)	(0–19)	*1*
Bosnia & HG	Present	90–03	?	–	
Bulgaria	100,000 – 500,000	96–02	0	0–19	
Croatia	(50,000 – 100,000)	02	(–)	(50–79)	*16*
Cyprus	(1,000 – 3,000)	94–02	(0)	(0–19)	
Czech Rep.	150,000 – 300,000	00	+	50–79	
Denmark	300,000 – 500,000	00	**F**	**>80**	*12*
Faroe Is.	500 – 500	95	(0)	(0–19)	
Estonia	70,000 – 150,000	98	+	20–29	*1*
Finland	60,000 – 170,000	98–02	+	**180**	
France	(1,500,000 – 5,000,000)	98–02	(+)	(25)	*4,3*
Georgia	Present	03	?	–	
Germany	1,500,000 – 2,200,000	95–99	0	0–19	
Greece	(50,000 – 100,000)	95–00	(–)	(0–19)	
Hungary	**27,000 – 53,000**	99–02	0	0–19	*19*
Iceland	(3,000 – 5,000)	00	?	–	*21*
Rep. Ireland	1,000,000 – 2,500,000	88–91	+	0–19	
Italy	(1,000,000 – 2,500,000)	03	(0)	(0–19)	
Latvia	200,000 – 300,000	90–00	(0)	(0–19)	*23,18*
Liechtenstein	500 – 1,000	98–00	0	0–19	
Lithuania	(100,000 – 250,000)	99–01	(0)	(0–19)	*20*
Luxembourg	10,000 – 15,000	02	0	0–19	
Macedonia	(120,000 – 400,000)	90–00	(0)	(0–19)	
Moldova	8,000 – 10,000	90–00	0	0–19	
Netherlands	**500,000 – 600,000**	98–00	+	**42**	*1*
Norway	(100,000 – 500,000)	90–02	F	20–29	*27*
Poland	**400,000 – 1,000,000**	90–00	(+)	(0–19)	*23*
Portugal	(100,000 – 1,000,000)	02	(0)	(0–19)	
Romania	860,000 – 1,240,000	00–02	+	0–19	*48*
Russia	2,000,000 – 4,000,000	90–00	(0)	(0–19)	122
Serbia & MN	100,000 – 140,000	90–02	0	0–19	*1,29,155,172a, 225,227*
Slovakia	100,000 – 200,000	90–99	0	0–19	
Slovenia	50,000 – 70,000	94	(0)	(0–19)	
Spain	(2,100,000 – 4,000,000)	92	?	–	*13,12*
Sweden	(100,000 – 500,000)	99–00	+	**76**	
Switzerland	**250,000 – 350,000**	93–96	+	0–19	
Turkey	(400,000 – 1,200,000)	01	(0)	(0–19)	
Ukraine	140,000 – 160,000	90–00	0	0–19	
UK	**8,512,000 – 8,512,000**	00	+	**12**	*5,31*
Gibraltar	100 – 150	00	**0**	**0–19**	
Total (approx.)	23,000,000 – 40,000,000	**Overall trend** Small increase			
Breeding range	>7,000,000 km²	**Gen. length**. <3.3 **% Global pop.** 5–24			

(See p. 197, bottom, for data quality graph)

Prunella modularis
HEDGE ACCENTOR

Non-SPEC[E] (1994: 4) **Status** Secure
Criteria —
European IUCN Red List Category —
Criteria —
Global IUCN Red List Category —
Criteria —

Prunella modularis is a widespread breeder across much of Europe, which constitutes >95% of its global breeding range. Its European breeding population is extremely large (>12,000,000 pairs), and was stable between 1970–1990. Although there were declines in France and Sweden during 1990–2000, populations increased or were stable elsewhere in Europe, and the species remained stable overall. Consequently, it is evaluated as Secure.

Country	Breeding pop. size (pairs)	Year(s)	Trend	Mag.%	References
Albania	(500 – 1,000)	02	(0)	(0–19)	
Andorra	5,000 – 7,000	99–01	(0)	(0–19)	*1,3*
Armenia	3,300 – 6,500	97–02	?	–	
Austria	(200,000 – 400,000)	98–02	(0)	(0–19)	
Azerbaijan	(5,000 – 50,000)	96–00	(0)	(0–19)	
Belarus	120,000 – 200,000	97–02	(0)	(0–19)	
Belgium	250,000 – 500,000	01–02	(0)	(0–19)	*1*
Bulgaria	20,000 – 60,000	96–02	0	0–9	
Croatia	(5,000 – 10,000)	02	(0)	(0–19)	*16*
Czech Rep.	250,000 – 500,000	00	+	20–29	
Denmark	80,000 – 120,000	00	**F**	**50–79**	*12*
Estonia	50,000 – 100,000	98	0	0–19	*1*
Finland	300,000 – 500,000	98–02	**0**	**0–19**	
France	(1,500,000 – 7,000,000)	98–02	(–)	(15)	*4,3*
Georgia	Present	03	?	–	
Germany	1,300,000 – 2,000,000	95–99	0	0–19	
Greece	(1,000 – 5,000)	95–00	(0)	(0–19)	
Hungary	(1,200 – 1,500)	90–93	+	50–79	*14*
Rep. Ireland	500,000 – 1,000,000	88–91	+	0–19	
Italy	(100,000 – 200,000)	03	(0)	(0–19)	
Latvia	150,000 – 300,000	90–00	(0)	(0–19)	*23*
Liechtenstein	1,400 – 2,000	98–00	0	0–19	
Lithuania	(150,000 – 300,000)	99–01	(0)	(0–19)	*20*
Luxembourg	10,000 – 15,000	02	0	0–19	
Macedonia	(1,000 – 3,000)	90–00	(0)	(0–19)	
Netherlands	**200,000 – 250,000**	98–00	**0**	**1**	*1*
Norway	(500,000 – 1,500,000)	90–03	(0)	(0–19)	
Poland	**150,000 – 300,000**	90–00	(+)	(0–19)	*23*
Portugal	(5,000 – 50,000)	02	(0)	(0–19)	
Romania	640,000 – 915,000	00–02	0	0–19	*48*
Russia	1,500,000 – 3,000,000	90–00	?	–	*122*
Serbia & MN	9,000 – 14,000	90–02	+	0–19	*1,139,155,225, 227*
Slovakia	300,000 – 500,000	90–99	0	0–19	
Slovenia	20,000 – 30,000	94	(0)	(0–19)	
Spain	(400,000 – 1,000,000)	92	?	–	*13,12,10*
Sweden	(1,000,000 – 2,500,000)	99–00	**–**	**38**	
Switzerland	**110,000 – 220,000**	93–96	0	0–19	
Turkey	(9,000 – 25,000)	01	(0)	(0–19)	
Ukraine	24,500 – 34,500	90–00	0	0–19	
UK	**2,163,000 – 2,163,000**	00	**0**	**3**	*5,31*

Total (approx.) 12,000,000 – 26,000,000 **Overall trend** Stable
Breeding range >6,000,000 km² **Gen. length.** <3.3 **% Global pop.** >95
(See p. 197, bottom, for data quality graph)

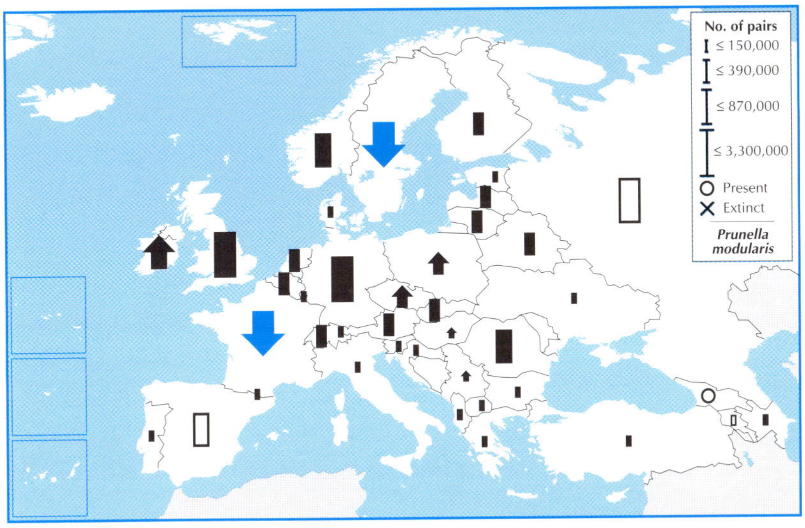

Prunella montanella
SIBERIAN ACCENTOR

Non-SPEC (1994: —) **Status** (Secure)
Criteria —
European IUCN Red List Category —▼▼▼
Criteria —
Global IUCN Red List Category —
Criteria —

Country	Breeding pop. size (pairs)	Year(s)	Trend	Mag.%	References
Russia	(100 – 500)	90–00	?	–	*51*

Total (approx.) 100 – 500 **Overall trend** Unknown
Breeding range >100,000 km² **Gen. length.** <3.3 **% Global pop.** <5

Prunella montanella has a predominantly Asian breeding distribution, which just extends into Europe in northern Russia. Its European breeding population is very small (as few as 100 pairs), but was stable between 1970–1990. No trend data were available for 1990–2000, but there was no evidence to suggest that the species declined. Although the size of the European population could render it susceptible to the risks affecting small populations, it is marginal to a much larger non-European population. Consequently, the species is provisionally evaluated as Secure.

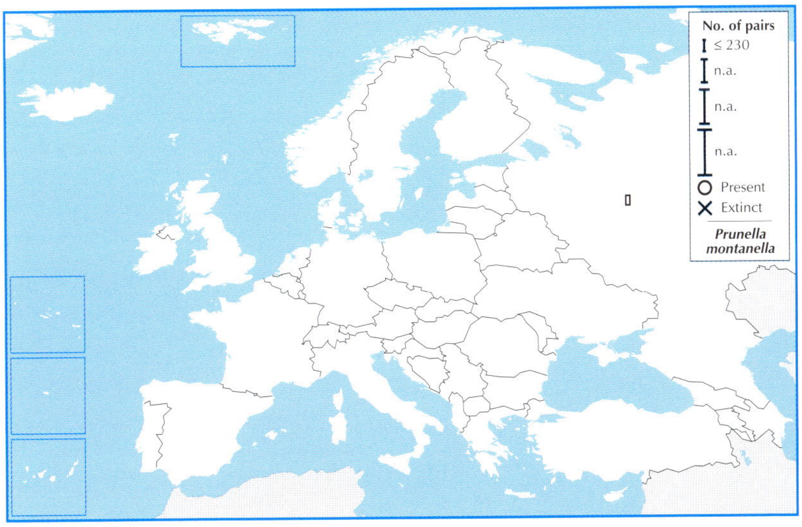

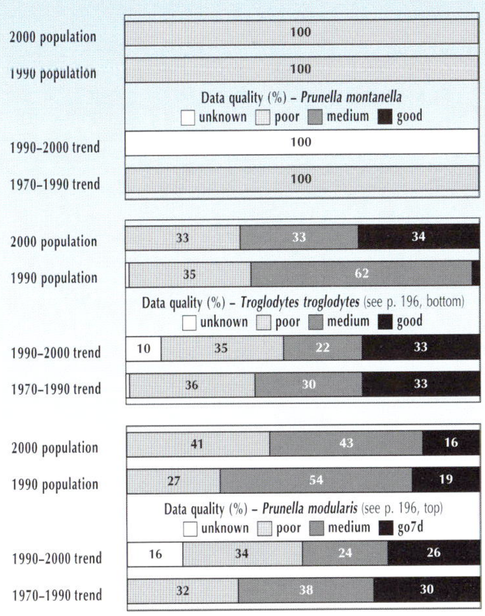

Prunella ocularis
RADDE'S ACCENTOR

Non-SPEC[E] (1994: 3) **Status** (Secure)
Criteria —
European IUCN Red List Category —
Criteria —
Global IUCN Red List Category —
Criteria —

Country	Breeding pop. size (pairs)	Year(s)	Trend	Mag.%	References
Armenia	830 – 3,000	97–02	?	–	
Azerbaijan	(50 – 250)	96–00	(0)	(0-19)	
Georgia	Present	03	?	–	
Turkey	(10,000 – 30,000)	01	?	–	
Total (approx.)	11,000 – 33,000	Overall trend Unknown			
Breeding range	>20,000 km²	Gen. length. <3.3	% Global pop. 50–74		

Prunella ocularis is a resident of the high mountains of Turkey and parts of the Caucasus, with Europe now known to constitute >50% of its global range. Its European breeding population is relatively small (<33,000 pairs), but is now known to exceed 10,000 pairs, and its trend between 1970–1990 was unknown. Trend data were also unavailable for Turkey, Armenia and Georgia during 1990–2000, but the population in Azerbaijan was stable, and there was no evidence to suggest that the species declined overall. Consequently, it is provisionally evaluated as Secure.

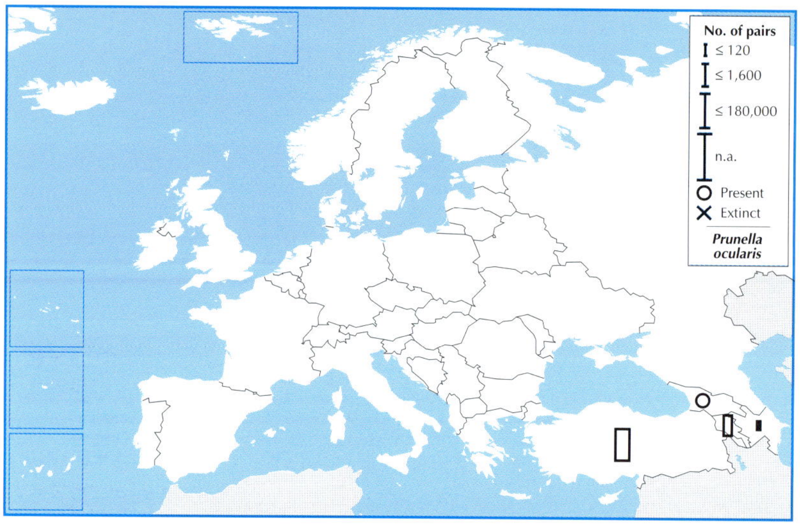

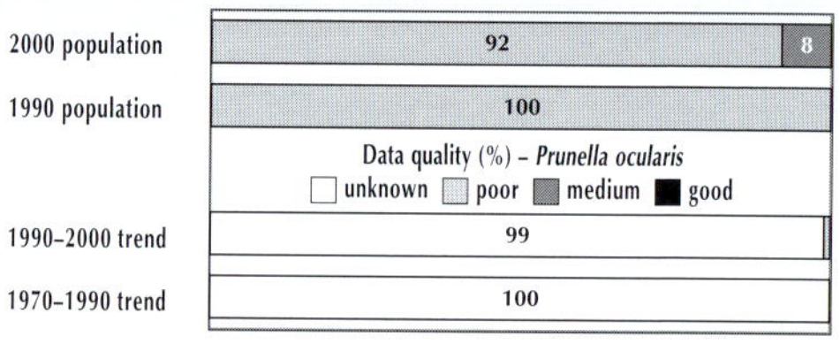

Prunella atrogularis
BLACK-THROATED ACCENTOR

SPEC 3 (1994: 3) **Status** (Rare)
Criteria <10,000 pairs
European IUCN Red List Category —
Criteria —
Global IUCN Red List Category —
Criteria —

Country	Breeding pop. size (pairs)	Year(s)	Trend	Mag.%	References
Russia	(1,500 – 2,500)	90–00	?	–	51,52
Total (approx.)	1,500 – 2,500	Overall trend Unknown			
Breeding range	>50,000 km²	Gen. length. <3.3	% Global pop. 5–24		

Prunella atrogularis is a patchily distributed summer visitor to north-east European Russia, which accounts for less than a quarter of its global breeding range. Its European breeding population is small (as few as 1,500 pairs), but its trend during 1970–1990 was unknown. Trend data were also unavailable for 1990–2000, although there is no evidence to suggest that the species is declining. Nevertheless, its population size still renders it susceptible to the risks affecting small populations, and consequently it is provisionally evaluated as Rare.

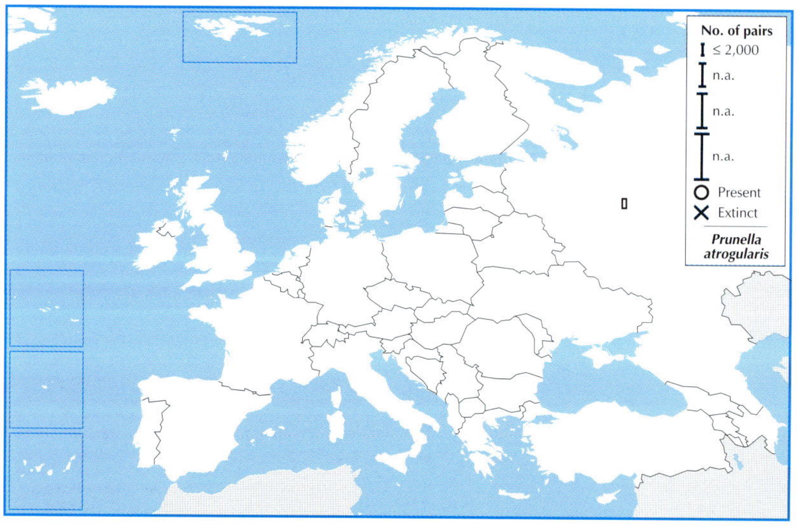

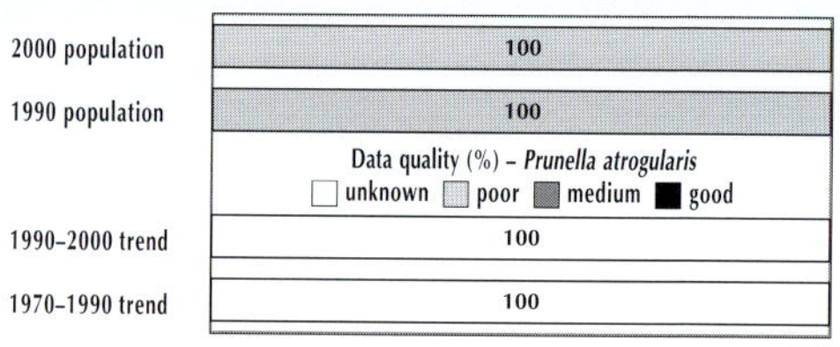

Prunella collaris
ALPINE ACCENTOR

Non-SPEC (1994: —) **Status** (Secure)
Criteria —
European IUCN Red List Category —
Criteria —
Global IUCN Red List Category —
Criteria —

Prunella collaris is a widespread but patchily distributed resident in mountainous areas of central and southern Europe, which accounts for less than half of its global range. Its European breeding population is relatively large (>100,000 pairs), and was stable between 1970–1990. Although there was no trend information available for the sizeable populations in Spain, France, Italy and Russia, trends elsewhere in Europe during 1990–2000 were stable, and the species probably remained stable overall. Consequently, it is provisionally evaluated as Secure.

Country	Breeding pop. size (pairs)	Year(s)	Trend	Mag.%	References
Albania	(500 – 1,000)	02	(0)	(0–19)	
Andorra	200 – 300	99–01	(0)	(0–19)	*1,3*
Armenia	800 – 3,300	97–02	?	–	
Austria	(8,000 – 16,000)	98–02	(0)	(0–19)	
Azerbaijan	(1,000 – 5,000)	96–00	(0)	(0–19)	
Bulgaria	**4,000 – 5,500**	95–02	0	0–9	
Croatia	(50 – 100)	02	(–)	(>80)	*70,16*
Czech Rep.	**15 – 20**	00	**0**	**0–19**	
France	(5,000 – 25,000)	98–02	?	–	*4*
Georgia	Present	03	?	–	
Germany	700 – 1,200	95–99	0	0–19	
Greece	(2,000 – 5,000)	95–00	(0)	(0–19)	
Italy	(10,000 – 20,000)	03	?	–	
Liechtenstein	100 – 150	98–00	0	0–19	
Macedonia	(250 – 1,000)	90–00	(0)	(0–19)	
Poland	**350 – 400**	90–00	0	0–19	*2,93*
Romania	**16,000 – 19,500**	00–02	0	0–19	*48*
Russia	(10,000 – 20,000)	90–00	?	–	*60*
Serbia & MN	**300 – 400**	94–02	0	0–19	*1,139,62,155, 225,227*
Slovakia	300 – 400	90–99	0	0–19	
Slovenia	1,000 – 2,000	99–00	(0)	(0–19)	
Spain	(17,000 – 20,000)	92	?	–	*13,12,10*
Switzerland	**15,000 – 20,000**	93–96	0	0–19	
Turkey	(8,000 – 16,000)	01	(0)	(0–19)	
Ukraine	60 – 70	90–00	(0)	(0–19)	
Total (approx.)	100,000 – 180,000	**Overall trend** Stable			
Breeding range	>750,000 km²	**Gen. length.** <3.3	**% Global pop.** 25–49		

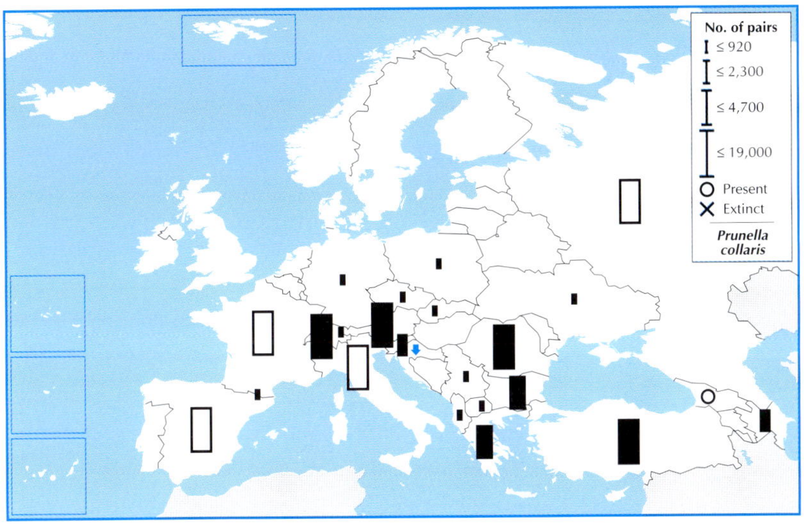

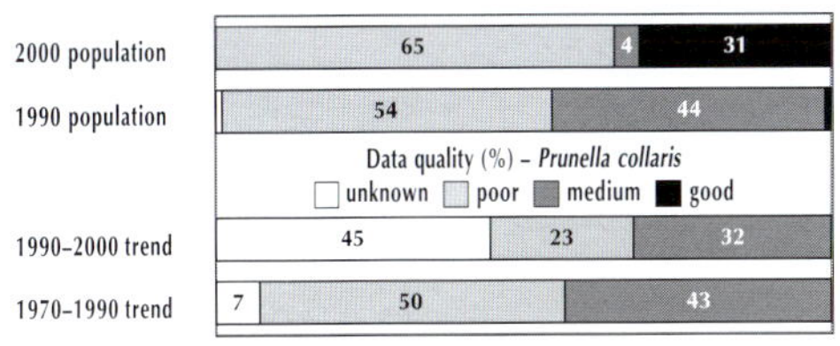

Erythropygia galactotes
RUFOUS-TAILED SCRUB-ROBIN

SPEC 3 (1994: —) **Status** (Vulnerable)
Criteria See IUCN below
European IUCN Red List Category VU
Criteria A2b
Global IUCN Red List Category —
Criteria —

Erythropygia galactotes is a widespread but patchily distributed summer visitor to southernmost Europe, which accounts for less than half of its global breeding range. Its European breeding population is relatively small (<96,000 pairs), but was stable between 1970–1990. Although the species was stable in a few countries during 1990–2000, key populations in Turkey and Spain suffered declines, and the species underwent a large decline (>30%) overall. Consequently, this previously Secure species is now provisionally evaluated as Vulnerable.

Country	Breeding pop. size (pairs)	Year(s)	Trend	Mag.%	References
Albania	(0 – 50)	02	?	–	
Armenia	830 – 3,300	98–02	0	0–19	
Azerbaijan	(1,000 – 10,000)	96–00	(0)	(0–19)	
Bulgaria	0 – 2	96–02	?	–	
Croatia	(1 – 10)	02	?	–	*70,16*
Georgia	Present	03	?	–	
Greece	(1,000 – 3,000)	95–00	(0)	(0–19)	
Portugal	(500 – 5,000)	02	(–)	(–)	
Russia	10 – 50	90–00	(–)	(50–79)	*33*
Spain	(8,400 – 14,300)	98–02	(–)	(50–79)	*10,16*
Turkey	(20,000 – 60,000)	01	(–)	(0–19)	
Total (approx.)	32,000 – 96,000	**Overall trend** Large decline			
Breeding range	>500,000 km²	**Gen. length.** <3.3	**% Global pop.** 25–49		

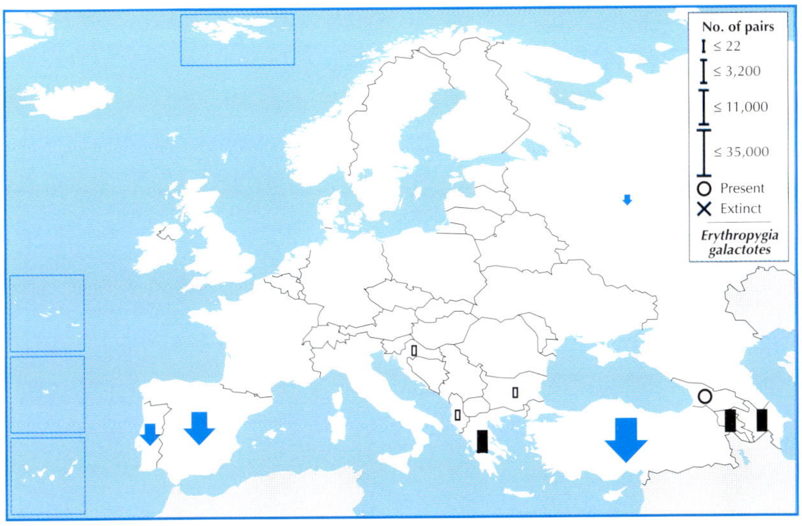

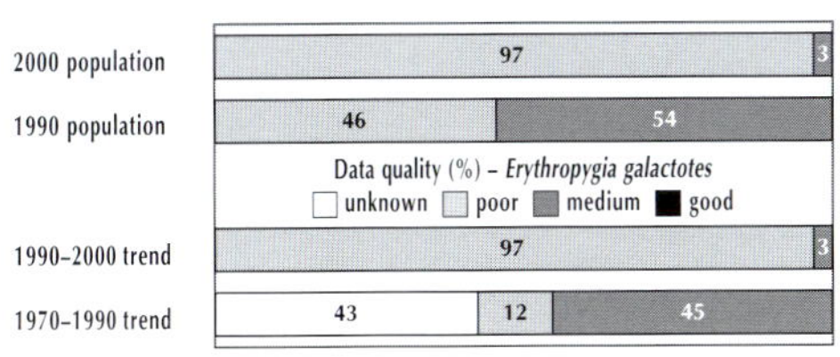

Erithacus rubecula
EUROPEAN ROBIN

Non-SPEC^E (1994: 4) **Status** Secure
Criteria —
European IUCN Red List Category —
Criteria —
Global IUCN Red List Category —
Criteria —

Erithacus rubecula is a widespread breeder across most of Europe, which constitutes >75% of its global breeding range. Its European breeding population is extremely large (>43,000,000 pairs), and was stable between 1970–1990. Although the species declined in Sweden during 1990–2000, these losses were compensated for by increases in the sizeable populations in France and the United Kingdom, and the species underwent a small increase overall. Consequently, it is evaluated as Secure.

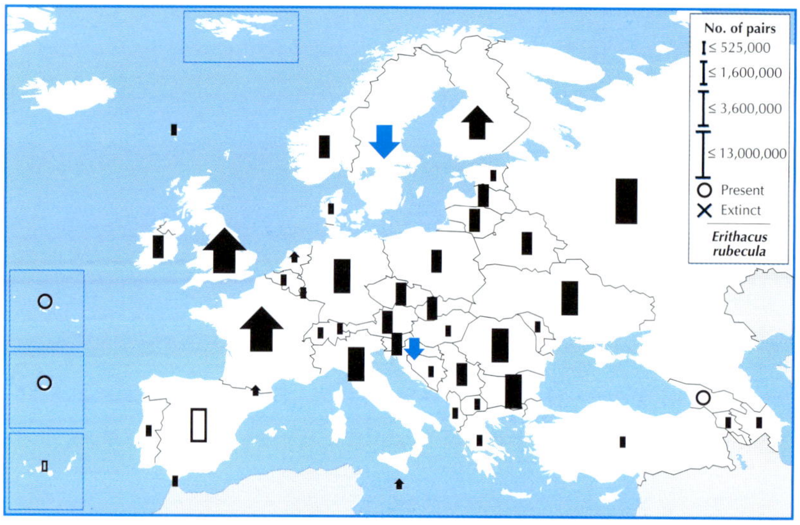

Country	Breeding pop. size (pairs)	Year(s)	Trend	Mag.%	References
Albania	10,000 – 20,000	02	(0)	(0–19)	
Andorra	(700 – 1,000)	99–01	(+)	(0–19)	*1,3*
Armenia	3300 – 5,000	99–02	0	0–19	
Austria	(700,000 – 1,400,000)	98–02	(0)	(0–19)	
Azerbaijan	(50,000 – 100,000)	96–00	(0)	(0–19)	
Belarus	1,200,000 – 1,500,000	97–02	0	0–19	
Belgium	100,000 – 250,000	01–02	(0)	(0–19)	*1*
Bosnia & HG	(50,000 – 100,000)	90–00	(0)	(0–19)	
Bulgaria	1,500,000 – 3,500,000	96–02	0	0–9	
Croatia	(500,000 – 1,000,000)	02	(–)	(20–29)	*16*
Czech Rep.	600,000 – 1,200,000	00	0	0–19	
Denmark	200,000 – 300,000	00	**F**	**50–79**	*12*
Faroe Is.	0 – 2	92–93	(0)	(0–19)	
Estonia	250,000 – 500,000	98	0	0–19	*1*
Finland	1,200,000 – 3,300,000	98–02	**+**	**65**	
France	(3,000,000 – 15,000,000)	98–02	**+**	**68**	*4,3*
Georgia	Present	03	?	–	
Germany	2,500,000 – 4,000,000	95–99	0	0–19	
Greece	(50,000 – 100,000)	95–00	(0)	(0–19)	
Hungary	**255,000 – 400,000**	99–02	0	0–19	*19*
Rep. Ireland	1,000,000 – 2,500,000	88–91	0	0–19	
Italy	(1,000,000 – 3,000,000)	03	(0)	(0–19)	
Latvia	700,000 – 1,000,000	90–00	(0)	(0–19)	*23,18*
Liechtenstein	(1,000 – 2,500)	98–00	(0)	(0–19)	
Lithuania	(800,000 – 1,500,000)	99–01	(0)	(0–19)	*20*
Luxembourg	15,000 – 20,000	02	0	0–19	
Macedonia	(300,000 – 700,000)	90–00	(0)	(0–19)	
Malta	1 – 2	90–02	+	0–19	*1*
Moldova	32,000 – 38,000	90–00	0	0–19	
Netherlands	**350,000 – 450,000**	98–00	**+**	**33**	*1*
Norway	(500,000 – 1,500,000)	90–02	(0)	(0–19)	
Poland	**1,000,000 – 1,500,000**	00–02	0	0–19	*23*
Portugal	(50,000 – 500,000)	02	(0)	(0–19)	
Azores	Present	02	?	–	
Madeira	Present	02	?	–	
Romania	1,854,000 – 2,670,000	00–02	0	0–19	*48*
Russia	10,000,000 – 15,000,000	90–00	(0)	(0–19)	*122*
Serbia & MN	600,000 – 800,000	90–02	0	0–19	*1,29,155,172a, 67a,78,225,227*
Slovakia	500,000 – 1,000,000	90–99	0	0–19	
Slovenia	700,000 – 1,000,000	00	(0)	(0–19)	
Spain	(1,200,000 – 3,000,000)	92	?	–	*10,13*
Canary Is.	(2,500 – 10,000)	97–03	?	–	*28,25*
Sweden	2,500,000 – 5,000,000	99–00	**–**	**20**	
Switzerland	**450,000 – 600,000**	93–96	0	0–19	
Turkey	(100,000 – 400,000)	01	(0)	(0–19)	
Ukraine	1,650,000 – 2,000,000	90–00	0	0–19	
UK	**5,895,000 – 5,895,000**	00	**+**	**31**	*5,31*
Gibraltar	**0 – 2**	00	**0**	**0–19**	
Total (approx.)	43,000,000 – 83,000,000	**Overall trend** Small increase			
Breeding range	>8,000,000 km²	**Gen. length.** <3.3		**% Global pop.** 75–94	

(See p. 200, bottom, for data quality graph)

Luscinia luscinia
THRUSH NIGHTINGALE

Non-SPEC^E (1994: 4) **Status** Secure
Criteria —
European IUCN Red List Category —
Criteria —
Global IUCN Red List Category —
Criteria —

Luscinia luscinia is a widespread summer visitor to north-eastern Europe, which constitutes >50% of its global breeding range. Its European breeding population is very large (>3,700,000 pairs), and was stable between 1970–1990. Although there were declines in some marginal populations during 1990–2000, key populations in the core of the breeding range—notably Russia—were stable, and the species remained stable overall. Consequently, it is evaluated as Secure.

Country	Breeding pop. size (pairs)	Year(s)	Trend	Mag.%	References
Belarus	160,000 – 200,000	97–02	0	0–19	
Bulgaria	(0 – 5)	96–02	?	–	
Czech Rep.	**0 – 5**	00	**0**	**0–19**	
Denmark	(10,000 – 50,000)	00	**–**	**30–49**	
Estonia	40,000 – 80,000	98	+	20–29	*1*
Finland	15,000 – 20,000	98–02	**–**	**10**	
Georgia	Present	03	?	–	
Germany	9,600 – 36,000	95–99	–	20–29	
Hungary	0 – 10	95–02	–	20–49	
Latvia	50,000 – 150,000	90–00	+	30–49	*23,16,2*
Lithuania	(40,000 – 70,000)	99–01	(0)	(0–19)	*20*
Moldova	40,000 – 50,000	90–00	0	0–19	
Netherlands	**0 – 2**	98–00	?	–	*1*
Norway	300 – 1,000	90–03	0	0–19	
Poland	**200,000 – 400,000**	00–02	0	0–19	*23*
Romania	86,000 – 120,000	00–02	+	0–19	*48*
Russia	2,500,000 – 5,000,000	90–00	(0)	(0–19)	*122*
Slovakia	1,000 – 1,500	90–99	0	0–19	
Sweden	20,000 – 50,000	99–00	**–**	**38**	
Ukraine	520,000 – 680,000	90–00	0	0–9	
Total (approx.)	3,700,000 – 6,900,000	**Overall trend** Stable			
Breeding range	>4,000,000 km²	**Gen. length.** <3.3		**% Global pop.** 50–74	

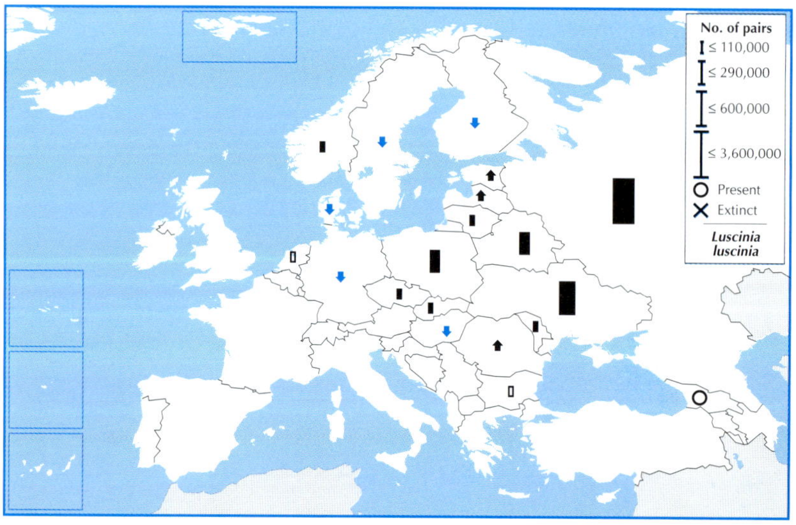

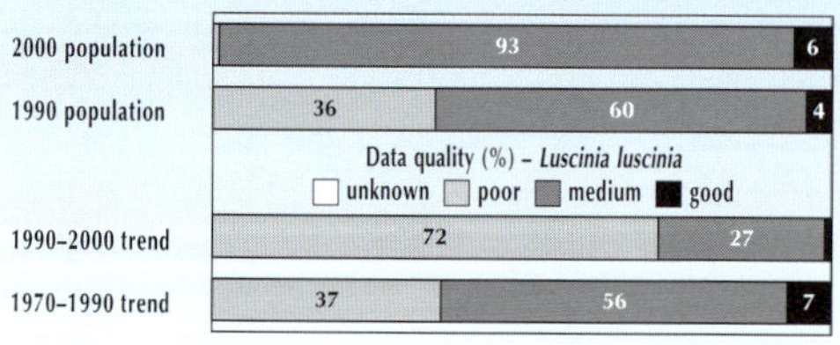

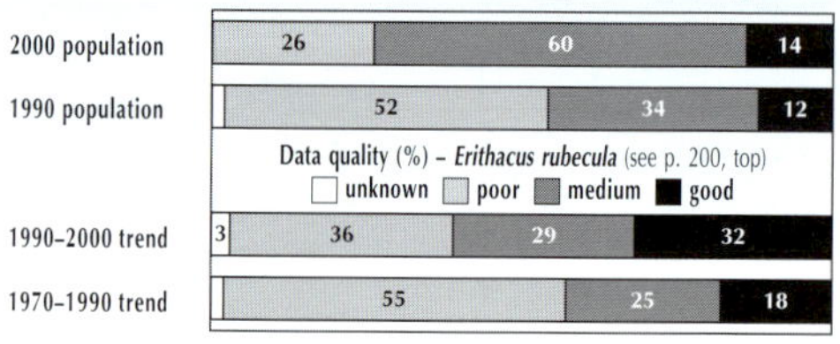

Luscinia megarhynchos
COMMON NIGHTINGALE

Non-SPEC[E] (1994: 4) **Status** (Secure)
Criteria —
European IUCN Red List Category —
Criteria —
Global IUCN Red List Category —
Criteria —

Country	Breeding pop. size (pairs)	Year(s)	Trend	Mag.%	References
Albania	20,000 – 50,000	02	(0)	(0–19)	
Andorra	(30 – 75)	99–01	(–)	(0–19)	*1,3*
Armenia	830 – 1,600	99–02	0	0–19	
Austria	(2,000 – 4,000)	98–02	(0)	(0–19)	
Azerbaijan	(100,000 – 500,000)	96–00	(0)	(0–19)	
Belgium	1,370 – 2,480	01–02	–	20–29	*1*
Bosnia & HG	Present	90–03	?	–	
Bulgaria	200,000 – 800,000	96–02	0	0–9	
Croatia	(500,000 – 1,000,000)	02	(+)	(>80)	*16*
Cyprus	(1,000 – 3,000)	94–02	(0)	(0–19)	
Czech Rep.	8,000 – 16,000	00	+	30–49	
France	(300,000 – 1,500,000)	98–02	**–**	**15**	*4,3*
Georgia	Present	03	?	–	
Germany	80,000 – 130,000	95–99	0	0–19	
Greece	(100,000 – 200,000)	95–00	(–)	(0–19)	
Hungary	**225,000 – 280,000**	99–02	0	0–19	*19*
Italy	(500,000 – 1,000,000)	03	(0)	(0–19)	
Liechtenstein	**5 – 10**	98–00	**0**	**0–19**	
Luxembourg	300 – 500	02	0	0–19	
Macedonia	(30,000 – 100,000)	90–00	(0)	(0–19)	
Netherlands	**6,500 – 7,500**	98–00	+	**23**	*1*
Poland	**100,000 – 150,000**	00–02	0	0–19	*23*
Portugal	(100,000 – 1,000,000)	02	(0)	(0–19)	
Romania	(216,000 – 340,000)	00–02	0	0–19	*48*
Russia	(100,000 – 500,000)	90–00	?	–	*8,61,136*
Serbia & MN	160,000 – 270,000	90–02	–	0–19	*1,29,155,78*
Slovakia	10,000 – 20,000	90–99	0	0–19	
Slovenia	2,000 – 4,000	94	(0)	(0–19)	
Spain	(450,000 – 1,700,000)	92	?	–	*13,12,10*
Switzerland	**2,000 – 2,500**	93–96	0	0–19	
Turkey	(1,000,000 – 2,000,000)	01	(–)	(0–19)	
Ukraine	14,700 – 15,300	90–00	(0)	(0–19)	
UK	**5,600 – 9,400**	99	+	**11**	*38*
Total (approx.)	4,200,000 – 12,000,000	**Overall trend** Stable			
Breeding range	>4,000,000 km²	**Gen. length.** <3.3		**% Global pop.** 50–74	

Luscinia megarhynchos is a widespread summer visitor to southern and western Europe, which constitutes >50% of its global breeding range. Its European breeding population is very large (>4,200,000 pairs), and was stable between 1970–1990. Although there were declines in France and Turkey during 1990–2000, these were set against stable trends or increases in other key populations such as Italy and Croatia, and the species probably remained stable overall. Consequently, it is provisionally evaluated as Secure.

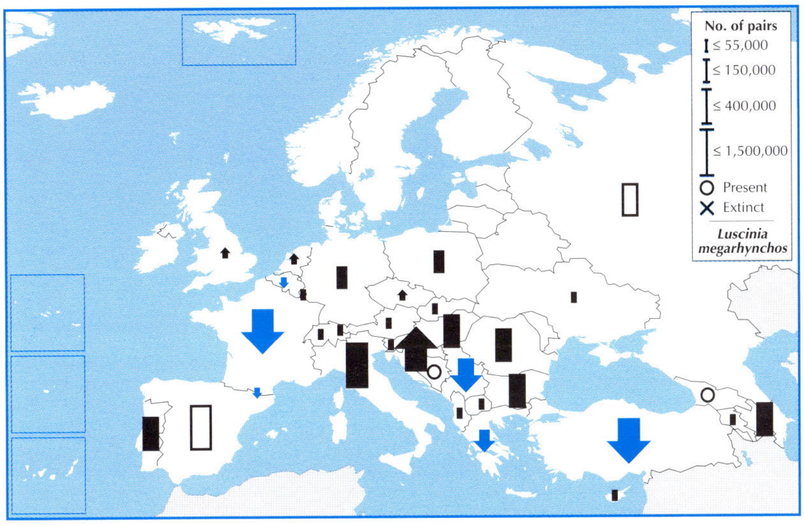

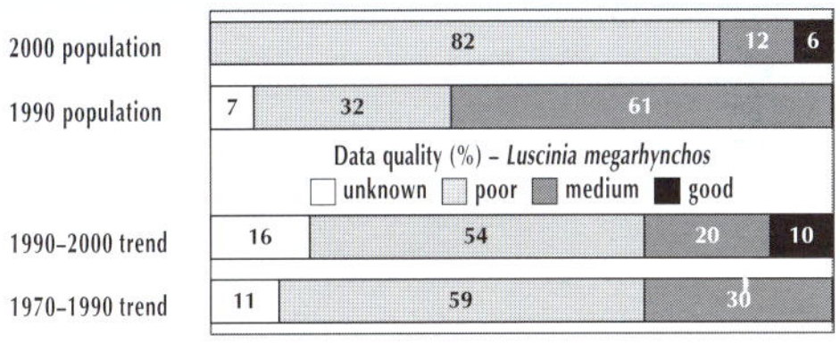

Luscinia calliope
SIBERIAN RUBYTHROAT

Non-SPEC (1994: —) **Status** (Secure)
Criteria —
European IUCN Red List Category —▼
Criteria —
Global IUCN Red List Category —
Criteria —

Country	Breeding pop. size (pairs)	Year(s)	Trend	Mag.%	References
Russia	(800 – 1,000)	90–00	?	–	*51,52*
Total (approx.)	800 – 1,000	**Overall trend** Unknown			
Breeding range	>50,000 km²	**Gen. length.** <3.3		**% Global pop.** <5	

Luscinia calliope has a predominantly Asian breeding distribution, which just extends west of the Urals into European Russia. Its European breeding population is very small (as few as 800 pairs), but was stable between 1970–1990. No trend data were available for 1990–2000, but there was no evidence to suggest that the species declined. Although the size of the European population could render it susceptible to the risks affecting small populations, it is marginal to a much larger non-European population. Consequently, the species is provisionally evaluated as Secure.

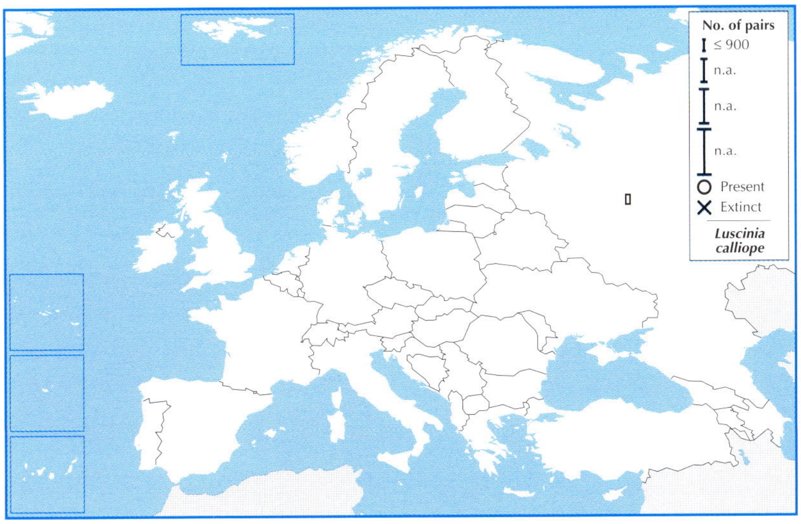

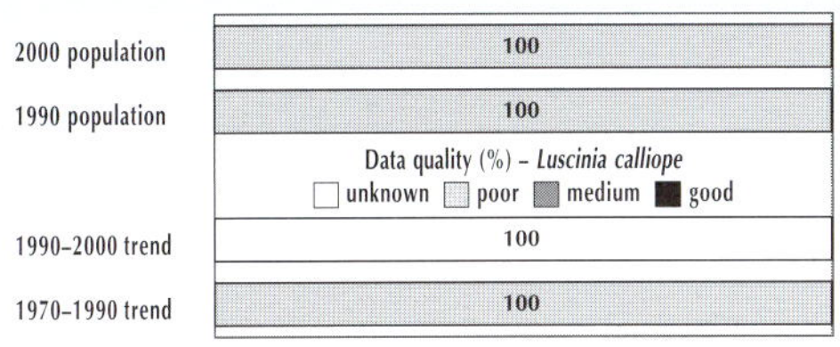

Luscinia svecica
BLUETHROAT

Non-SPEC (1994: —) **Status** (Secure)
Criteria —
European IUCN Red List Category —
Criteria —
Global IUCN Red List Category —
Criteria —

Luscinia svecica is a widespread breeder across much of northern Europe, and occurs more patchily farther south, with Europe accounting for less than a quarter of its global breeding range. Its European breeding population is very large (>4,500,000 pairs), and was stable between 1970–1990. Although there were declines in a few countries during 1990–2000, the stronghold population in Russia probably remained stable, and trends were stable or increasing across most of its European range. Consequently, the species is provisionally evaluated as Secure.

Country	Breeding pop. size (pairs)	Year(s)	Trend	Mag.%	References
Armenia	1,600 – 3,300	98–02	?	–	
Austria	250 – 350	98–02	–	20–29	
Belarus	5,000 – 10,000	97–02	0	0–19	
Belgium	**3,100 – 4,100**	95–02	**+**	**30–49**	*1*
Croatia	(50 – 75)	02	(–)	(20–29)	*70*
Czech Rep.	**400 – 600**	00	**+**	**>80**	
Denmark	22 – 22	01	**+**	**>80**	7
Estonia	(30 – 100)	98	–	20–29	*1*
Finland	100,000 – 200,000	98–02	**0**	**5**	
France	10,000 – 40,000	98–02	?	–	*4,3*
Georgia	Present	03	?	–	
Germany	3,300 – 4,600	95–99	+	30–49	
Hungary	500 – 800	98–01	0	0–19	
Latvia	(50 – 200)	90–00	(0)	(0–19)	*23,17*
Lithuania	(100 – 200)	99–01	(F)	(20–29)	*20*
Netherlands	**9,000 – 11,000**	98–00	**+**	**89**	*1*
Norway	(500,000 – 1,000,000)	90–03	0	0–19	
Poland	**1,300 – 1,800**	95–00	–	0–19	*2,95*
Portugal	0 – 5	02	**?**	**N**	
Romania	(1,500 – 3,500)	00–02	+	0–19	
Russia	3,500,000 – 6,000,000	90–00	(0)	(0–19)	*122*
Serbia & MN	**100 – 150**	98–02	+	0–9	*1,29,155,78*
Slovakia	**15 – 30**	90–99	+	20–29	
Spain	(9,000 – 128,00)	92	?	–	*13,12,10*
Sweden	**140,000 – 250,000**	99–00	(0)	(0–19)	
Switzerland	1 – 7	98–02	0	0–19	
Turkey	(1,000 – 3,000)	01	(–)	(0–19)	
Ukraine	(235,000 – 280,000)	90–00	0	0–9	
Total (approx.)	4,500,000 – 7,800,000	**Overall trend** Stable			
Breeding range	>4,000,000 km²	**Gen. length.** <3.3		**% Global pop.** 5–24	

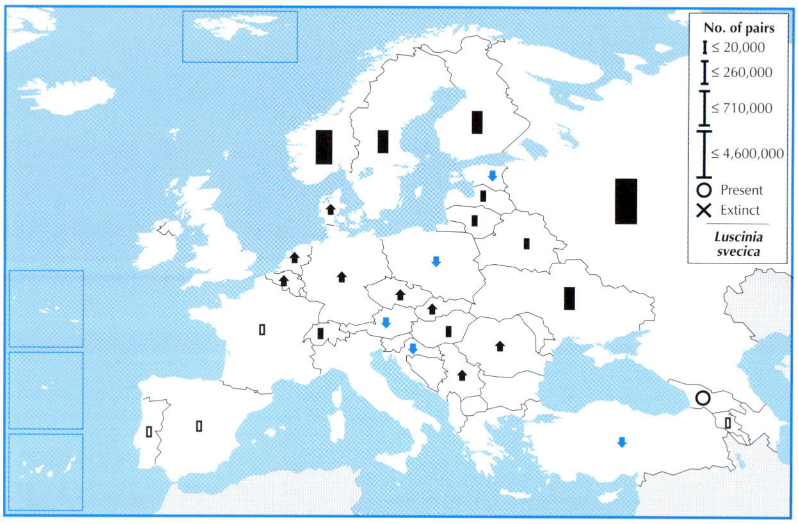

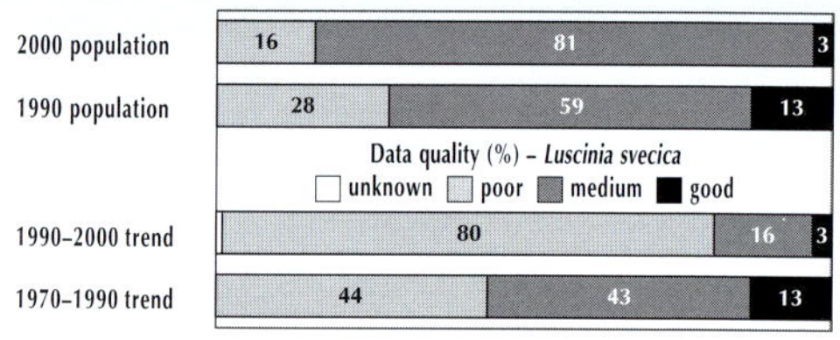

Tarsiger cyanurus
ORANGE-FLANKED BUSH-ROBIN

Non-SPEC (1994: —) **Status** (Secure)
Criteria —
European IUCN Red List Category —
Criteria —
Global IUCN Red List Category —
Criteria —

Tarsiger cyanurus has a predominantly Asian distribution, but its breeding range also extends into the boreal zone of European Russia and Finland. Its European breeding population is relatively small (<21,000 pairs), but was stable between 1970–1990. No trend data were available for the stronghold Russian population during 1990–2000, but the small population in Finland increased, and consequently the species is provisionally evaluated as Secure.

Country	Breeding pop. size (pairs)	Year(s)	Trend	Mag.%	References
Finland	50 – 500	98–02	+	100	
Russia	(10,000 – 20,000)	90–00	?	–	*51,52*
Total (approx.)	10,000 – 21,000	**Overall trend** Unknown			
Breeding range	>250,000 km²	**Gen. length.** <3.3		**% Global pop.** <5	

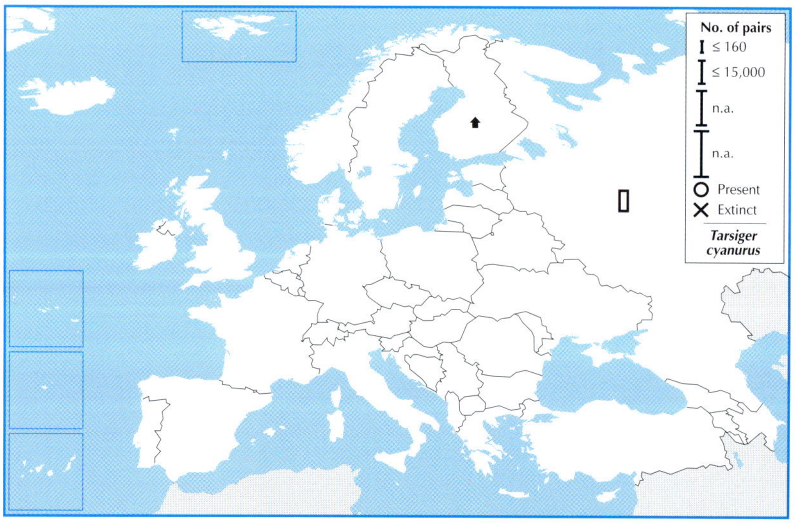

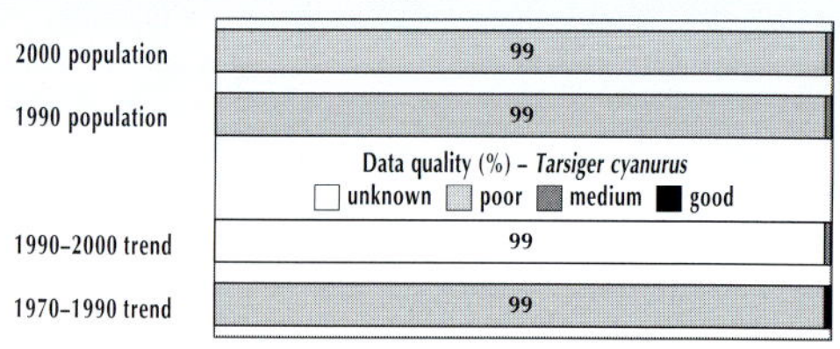

Irania gutturalis
WHITE-THROATED ROBIN

Non-SPEC (1994: —) **Status** (Secure)
Criteria —
European IUCN Red List Category —
Criteria —
Global IUCN Red List Category —
Criteria —

Country	Breeding pop. size (pairs)	Year(s)	Trend	Mag.%	References
Armenia	6,500 – 10,000	97–02	0	0–19	
Azerbaijan	(1,000 – 10,000)	96–00	(0)	(0–19)	
Greece	(0 – 5)	95–00	?	(N)	
Turkey	(400,000 – 900,000)	01	(+)	(0–19)	
Total (approx.)	410,000 – 920,000	**Overall trend** Small increase			
Breeding range	>250,000 km²	**Gen. length.** <3.3	**% Global pop.** 25–49		

Irania gutturalis is a patchily distributed summer visitor to Turkey and parts of the Caucasus (with a tiny new population in Greece), with Europe accounting for less than half of its global breeding range. Its European breeding population is large (>410,000 pairs), but its trend between 1970–1990 was unknown. Although the species was stable in Armenia and Azerbaijan during 1990–2000, the stronghold population in Turkey increased, and the species probably underwent a small increase overall. Consequently, it is provisionally evaluated as Secure.

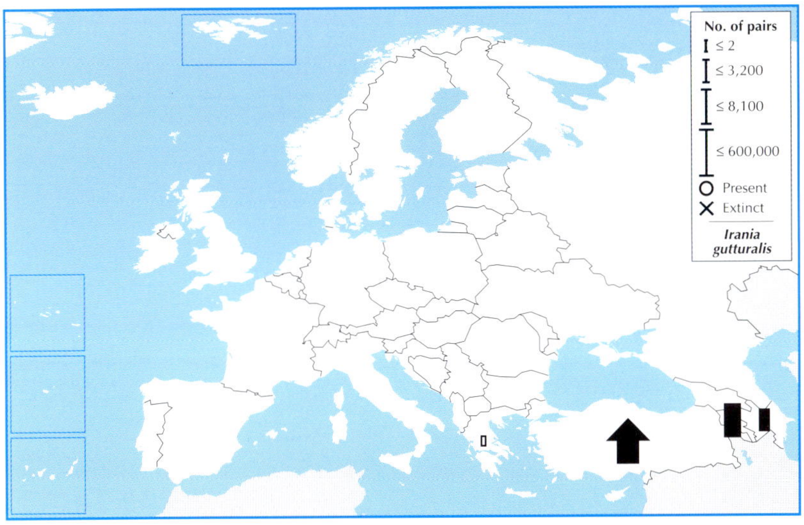

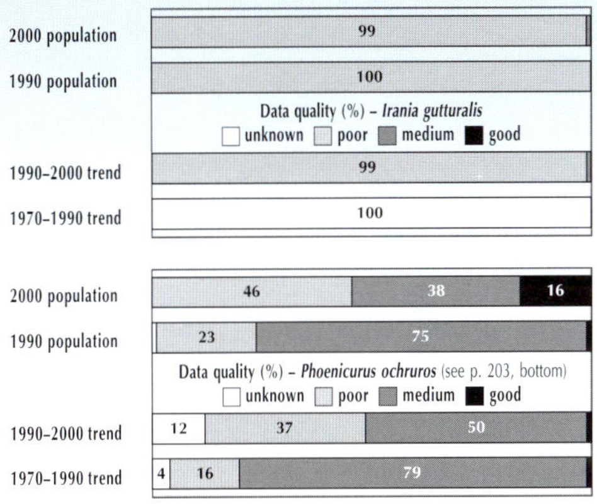

Phoenicurus ochruros
BLACK REDSTART

Non-SPEC (1994: —) **Status** Secure
Criteria —
European IUCN Red List Category —
Criteria —
Global IUCN Red List Category —
Criteria —

Phoenicurus ochruros is a widespread breeder across much of Europe, which accounts for less than half of its global breeding range. Its European breeding population is very large (>4,000,000 pairs), and was stable between 1970–1990. Although there were declines in a few populations during 1990–2000—notably the sizeable one in France—trends were stable or increasing across most of Europe, and the species underwent a small increase overall. Consequently, it is evaluated as Secure.

Country	Breeding pop. size (pairs)	Year(s)	Trend	Mag.%	References
Albania	2,000 – 7,000	02	(0)	(0–19)	
Andorra	(6,000 – 10,000)	99–01	(0)	(0–19)	*1,3*
Armenia	50,000 – 65,000	00–02	0	0–19	
Austria	(100,000 – 200,000)	98–02	(0)	(0–19)	
Azerbaijan	(10,000 – 100,000)	96–00	(0)	(0–19)	
Belarus	20,000 – 35,000	97–02	0	0–19	
Belgium	18,000 – 35,000	01–02	0	0–19	*1*
Bosnia & HG	Present	90–03	?	–	
Bulgaria	50,000 – 150,000	96–02	0	0–19	
Croatia	(25,000 – 50,000)	02	(0)	(0–19)	*16*
Czech Rep.	200,000 – 400,000	00	0	0–19	
Denmark	750 – 1,000	93–96	+	0–9	*3*
Estonia	300 – 600	98	+	20–29	*1*
Finland	20 – 100	98–02	(+)	(300)	
France	(400,000 – 1,600,000)	98–02	(–)	(12)	*4,3*
Georgia	Present	03	?	–	
Germany	600,000 – 1,000,000	95–99	+	0–19	
Greece	(10,000 – 30,000)	95–00	(–)	(0–19)	
Hungary	**170,000 – 270,000**	99–02	+	20–49	*19*
Italy	(200,000 – 400,000)	03	(0)	(0–19)	
Latvia	800 – 2,000	00–03	+	50–79	*22*
Liechtenstein	300 – 500	98–00	0	0–19	
Lithuania	(10,000 – 30,000)	99–01	(+)	(0–19)	*20*
Luxembourg	8,000 – 10,000	02	0	0–19	
Macedonia	(5,000 – 15,000)	90–00	(0)	(0–19)	
Moldova	6,000 – 8,500	90–00	+	30–49	
Netherlands	**27,000 – 37,000**	98–00	+	**14**	*1*
Norway	10 – 50	90–03	**0**	**0–19**	
Poland	**250,000 – 400,000**	00–02	0	0–19	*23*
Portugal	(20,000 – 200,000)	02	(0)	(0–19)	
Romania	240,000 – 410,000	00–02	+	0–19	*48*
Russia	(50,000 – 250,000)	90–00	?	–	*8,17,61,116*
Serbia & MN	40,000 – 65,000	95–02	+	**50–79**	*1,29,155,172a, 78,225,227*
Slovakia	100,000 – 200,000	90–99	0	0–19	
Slovenia	100,000 – 200,000	00	(+)	(0–19)	
Spain	(400,000 – 900,000)	92	?	–	*13,12,10*
Sweden	**500 – 1,000**	99–00	-	10–19	
Switzerland	**250,000 – 500,000**	93–96	+	0–9	
Turkey	(400,000 – 900,000)	01	(+)	(0–19)	
Ukraine	200,000 – 355,000	90–00	+	30–49	
UK	**26 – 74**	96–00	-	**29**	
Total (approx.)	4,000,000 – 8,800,000	**Overall trend** Small increase			
Breeding range	>5,000,000 km²	**Gen. length.** <3.3	**% Global pop.** 25–49		

(See p. 203, top, for data quality graph)

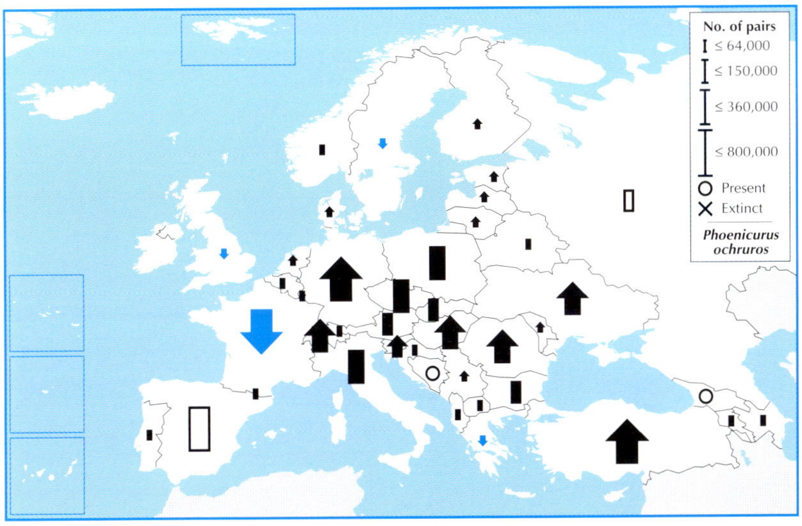

Phoenicurus phoenicurus
COMMON REDSTART

SPEC 2 (1994: 2) **Status** (Depleted)
Criteria Large historical decline
European IUCN Red List Category —
Criteria —
Global IUCN Red List Category —
Criteria —

Phoenicurus phoenicurus is a widespread summer visitor to Europe, which constitutes >50% of its global breeding range. Its European breeding population is very large (>6,800,000 pairs), but underwent a large decline between 1970–1990. Although declines continued in various countries during 1990–2000, they abated in much of Europe—most notably in Russia—and the species was probably stable overall. Nevertheless, its population size remains far below the level that preceded its decline, and consequently the species is provisionally evaluated as Depleted.

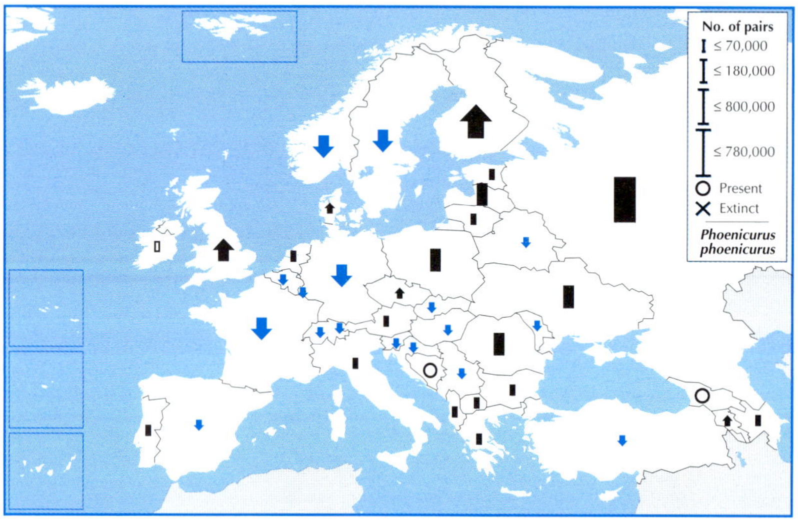

Country	Breeding pop. size (pairs)	Year(s)	Trend	Mag.%	References
Albania	(500 – 1,000)	02	(0)	(0–19)	
Armenia	10,000 – 15,000	00–02	+	0–9	
Austria	(6,000 – 12,000)	98–02	(0)	(0–19)	
Azerbaijan	(10,000 – 100,000)	96–00	(0)	(0–19)	
Belarus	40,000 – 60,000	97–02	–	0–9	
Belgium	3,600 – 6,300	01–02	–	0–19	*1*
Bosnia & HG	Present	85–89	?	–	
Bulgaria	2,000 – 5,000	96–02	(0)	(0–19)	
Croatia	(5,000 – 6,000)	02	(–)	(0–19)	*70,16*
Czech Rep.	4,000 – 8,000	00	+	30–49	
Denmark	40,000 – 60,000	00	+	**20–29**	*12*
Estonia	30,000 – 60,000	98	0	0–19	*1*
Finland	700,000 – 900,000	98–02	+	**35**	
France	(80,000 – 320,000)	98–02	(–)	(17)	*4,3*
Georgia	Present	03	?	–	
Germany	94,000 – 185,000	95–99	–	20–29	
Greece	(2,000 – 5,000)	95–00	(0)	(0–19)	
Hungary	6,000 – 19,000	99–02	–	20–49	*19*
Rep. Ireland	5 – 10	88–91	?	–	
Italy	(30,000 – 60,000)	03	(0)	(0–19)	
Latvia	60,000 – 100,000	90–00	(0)	(0–19)	*23,16*
Liechtenstein	10 – 20	98–00	–	0–19	
Lithuania	(25,000 – 50,000)	99–01	(0)	(0–19)	*20*
Luxembourg	400 – 500	02	**–**	**30–49**	
Macedonia	(2,000 – 3,000)	90–00	(0)	(0–19)	
Moldova	5,000 – 5,800	90–00	–	20–29	
Netherlands	**23,000 – 30,000**	98–00	**0**	**12**	*1*
Norway	(50,000 – 500,000)	90–02	(–)	(0–19)	*27*
Poland	**80,000 – 150,000**	00–02	0	0–19	*23*
Portugal	(1,000 – 10,000)	02	(0)	(0–19)	
Romania	160,000 – 190,000	00–02	0	0–19	*48*
Russia	5,000,000 – 12,000,000	90–00	(0)	(0–19)	*122*
Serbia & MN	9,000 – 12,000	90–02	–	0–19	*1,29,166,155, 172a,67a,225,227*
Slovakia	10,000 – 15,000	90–99	–	30–49	
Slovenia	2,000 – 4,000	99–00	(–)	(10–29)	
Spain	(10,000 – 20,000)	98–02	(–)	(30–49)	*10,16*
Sweden	100,000 – 300,000	99–00	**–**	**10**	
Switzerland	**10,000 – 15,000**	93–96	–	0–19	
Turkey	(40,000 – 120,000)	01	(–)	(0–19)	
Ukraine	(78,000 – 92,000)	90–00	0	0–19	
UK	**101,000 – 101,000**	00	+	**12**	*5,31*

Total (approx.) 6,800,000 – 16,000,000 **Overall trend** Stable
Breeding range >7,000,000 km² **Gen. length.** <3.3 **% Global pop.** 50–74
(See p. 204, bottom, for data quality graph)

Phoenicurus erythrogastrus
WHITE-WINGED REDSTART

SPEC 3 (1994: 3) **Status** (Rare)
Criteria <10,000 pairs
European IUCN Red List Category —
Criteria —
Global IUCN Red List Category —
Criteria —

Phoenicurus erythrogastrus has a predominantly Asian distribution, which just extends into Europe in the Caucasus. Its European breeding population is small (as few as 2,100 pairs), but its trend between 1970–1990 was unknown. Although the species remains poorly known, with no trend data available for Georgia during 1990–2000, it was stable in Russia and Azerbaijan, and was probably stable overall. Nevertheless, its population size renders it susceptible to the risks affecting small populations, and consequently it is provisionally evaluated as Rare.

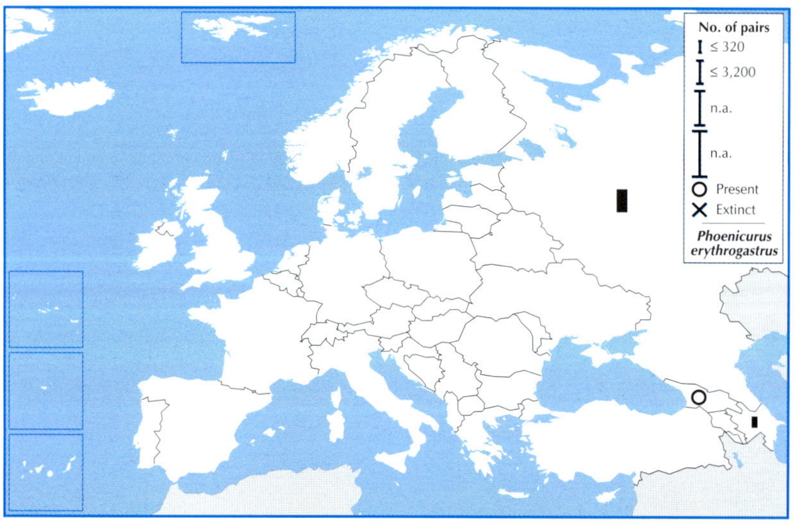

Country	Breeding pop. size (pairs)	Year(s)	Trend	Mag.%	References
Azerbaijan	(100 – 1,000)	96–00	(0)	(0–19)	
Georgia	Present	03	?	–	
Russia	(2,000 – 5,000)	90–00	(0)	(0–19)	*72*

Total (approx.) 2,100 – 6,000 **Overall trend** Stable
Breeding range >20,000 km² **Gen. length.** <3.3 **% Global pop.** <5

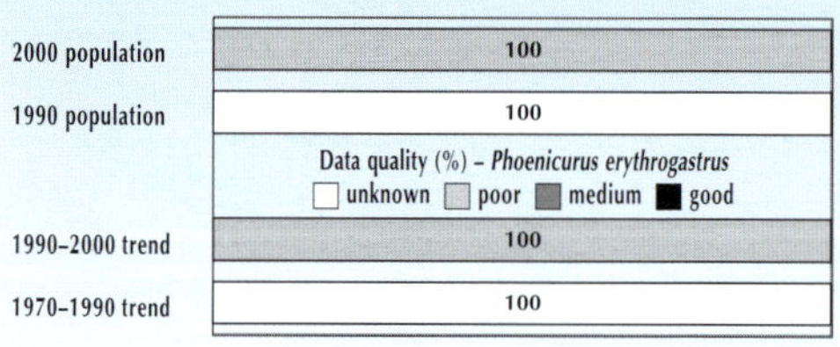

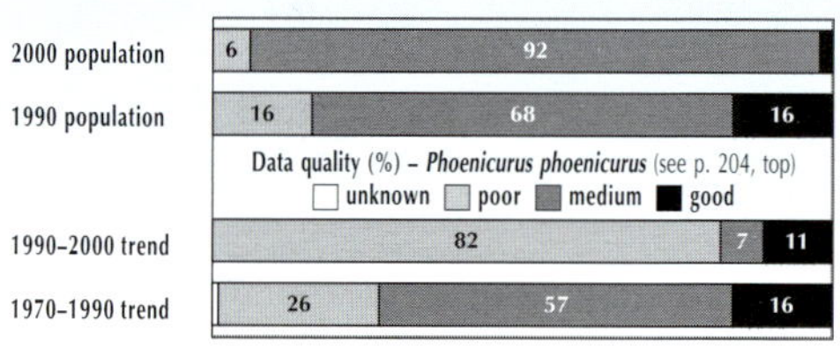

Saxicola rubetra
WHINCHAT

Non-SPEC[E] (1994: 4) **Status** (Secure)
Criteria —
European IUCN Red List Category —
Criteria —
Global IUCN Red List Category —
Criteria —

Saxicola rubetra is a widespread summer visitor to much of Europe, which constitutes >75% of its global breeding range. Its European breeding population is very large (>5,400,000 pairs), and was stable between 1970–1990. Although the species declined in many countries—notably Finland and Ukraine—during 1990–2000, the key population in Russia was stable, with other sizeable populations such as Belarus, Poland and Romania also stable or increasing, and the species probably declined only slightly overall. Consequently, it is provisionally evaluated as Secure.

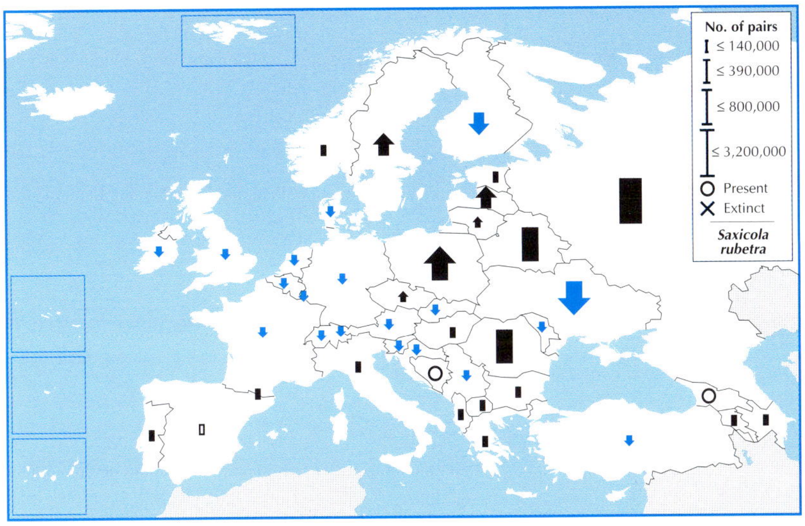

Country	Breeding pop. size (pairs)	Year(s)	Trend	Mag.%	References
Albania	(500 – 1,000)	02	(0)	(0–19)	
Andorra	75 – 200	99–01	(0)	(0–19)	*1,3*
Armenia	15,000 – 20,000	97–02	0	0–19	
Austria	3,500 – 7,000	98–02	(–)	(50–79)	
Azerbaijan	(50,000 – 100,000)	96–00	(0)	(0–19)	
Belarus	550,000 – 650,000	97–02	0	0–19	
Belgium	**168 – 325**	01–02	–	**20–29**	*1*
Bosnia & HG	Present	90–03	?	–	
Bulgaria	1,500 – 5,000	96–02	0	0–9	
Croatia	(5,000 – 10,000)	02	(–)	(30–49)	*70,16,44*
Czech Rep.	15,000 – 30,000	00	+	30–49	
Denmark	5,000 – 10,000	00	–	**50**	*12*
Estonia	(50,000 – 150,000)	98	0	0–19	*1*
Finland	300,000 – 400,000	98–02	–	**30**	
France	15,000 – 45,000	98–02	–	**54**	*4,2*
Georgia	Present	03	?	–	
Germany	37,000 – 90,000	95–99	–	30–49	
Greece	(500 – 1,000)	95–00	(0)	(0–19)	
Hungary	**95,000 – 180,000**	99–02	0	0–19	*19*
Rep. Ireland	1,000 – 2,500	88–91	(–)	(–)	
Italy	(10,000 – 20,000)	03	(0)	(0–19)	
Latvia	300,000 – 500,000	90–00	+	20–29	*23,16,2*
Liechtenstein	**25 – 35**	98–00	–	**10–19**	
Lithuania	(50,000 – 120,000)	99–01	(+)	(0–19)	*20*
Luxembourg	**20 – 40**	00–02	–	**90**	
Macedonia	(250 – 1,000)	90–00	(0)	(0–19)	
Moldova	1,500 – 2,000	90–00	–	20–29	
Netherlands	**500 – 700**	98–00	–	**52**	*1*
Norway	(50,000 – 300,000)	90–03	(0)	(0–19)	
Poland	**400,000 – 600,000**	00–02	+	0–19	*23*
Portugal	10 – 100	02	0	0–19	
Romania	**715,000 – 882,000**	00–02	0	0–19	*48*
Russia	2,000,000 – 5,000,000	90–00	0	0–19	*122*
Serbia & MN	6,000 – 8,000	90–02	–	0–19	*1,29,155,172a, 67a,78,225,227*
Slovakia	10,000 – 20,000	90–99	–	30–49	
Slovenia	3,000 – 5,000	99–00	(–)	(10–19)	
Spain	(15,000 – 20,000)	92	?	–	*13,12,10*
Sweden	200,000 – 400,000	99–00	+	**11**	
Switzerland	**10,000 – 15,000**	93–96	–	20–29	
Turkey	(2,000 – 8,000)	01	(–)	(0–19)	
Ukraine	(435,000 – 640,000)	90–00	–	20–29	
UK	**11,000 – 22,100**	00	–	**21**	*22,31*
Total (approx.)	5,400,000 – 10,000,000	**Overall trend** Small decline			
Breeding range	>7,000,000 km²	**Gen. length.** <3.3 **% Global pop.** 75–94			

(See p. 205, bottom, for data quality graph)

Saxicola dacotiae
FUERTEVENTURA CHAT

SPEC 1 (1994: 2) **Status** Endangered
Criteria See IUCN below
European IUCN Red List Category EN
Criteria B1a+b(ii,iii,iv,v); C2a(ii)
Global IUCN Red List Category EN
Criteria B1a+b(ii,iii,iv,v); C2a(ii)

Country	Breeding pop. size (pairs)	Year(s)	Trend	Mag.%	References
Spain					
Canary Is.	(<1,300)	97–03	(–)	(0–19)	*28,25*
Total (approx.)	<1,300	**Overall trend** Small decline			
Breeding range	<1,000 km²	**Gen. length.** <3.3 **% Global pop.** 100			

Saxicola dacotiae is endemic to Europe, where it has a very small range (<1,000 km²) on the Canary Island of Fuerteventura. Its breeding population is very small (probably fewer than 1,250 pairs), and was stable between 1970–1990, but underwent a small decline during 1990–2000. As a consequence of its very small and declining population, and its very small range (which is declining owing to habitat degradation and destruction) on a single island, this globally threatened species is evaluated as Endangered.

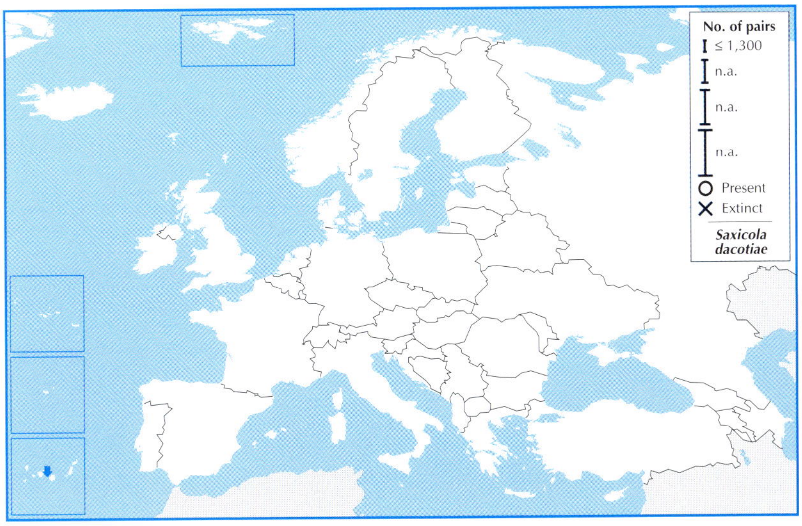

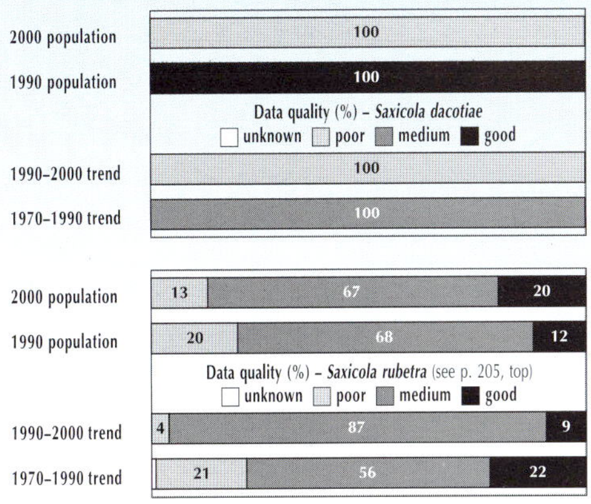

Saxicola torquata
COMMON STONECHAT

Non-SPEC (1994: 3) **Status** (Secure)
Criteria —
European IUCN Red List Category —
Criteria —
Global IUCN Red List Category —
Criteria —

Saxicola torquata is a widespread breeder across much of Europe, which accounts for less than a quarter of its global breeding range. Its European breeding population is very large (>2,000,000 pairs), but declined markedly between 1970–1990. Although the species declined in a few countries during 1990–2000, the majority of European populations increased or were stable (trends were not known for Spain or Russia), and it underwent a moderate increase overall. This increase probably outweighs the earlier decline, and consequently the species is provisionally evaluated as Secure.

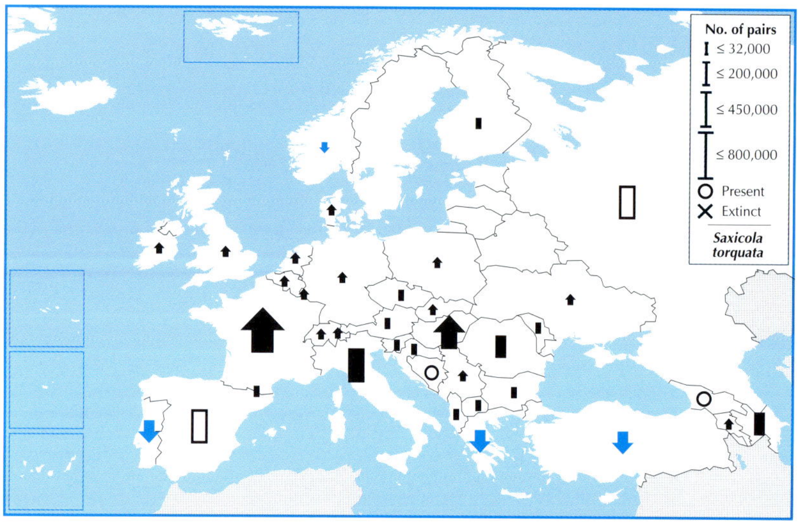

Country	Breeding pop. size (pairs)	Year(s)	Trend	Mag.%	References
Albania	3,000 – 10,000	02	(0)	(0–19)	
Andorra	50 – 80	99–01	(0)	(0–19)	*1,3*
Armenia	20,000 – 25,000	97–02	+	0–9	
Austria	4,500 – 9,000	98–02	(0)	(0–19)	
Azerbaijan	(50,000 – 100,000)	96–00	(0)	(0–19)	
Belgium	2,900 – 4,700	95–02	+	0–19	*1*
Bosnia & HG	Present	90–03	?	–	
Bulgaria	3,000 – 8,000	96–02	0	0–9	
Croatia	(10,000 – 15,000)	02	(0)	(0–19)	*70,16,44*
Czech Rep.	2,500 – 5,000	00	0	0–19	
Denmark	**25 – 31**	00	+	**>80**	*7*
Finland	(0 – 5)	98–02	(0)	(0–19)	
France	(400,000 – 1,600,000)	98–02	+	62	*4,3*
Georgia	Present	03	?	–	
Germany	3,500 – 4,900	95–99	+	20–29	
Greece	(50,000 – 100,000)	95–00	(–)	(0–19)	
Hungary	**390,000 – 515,000**	99–02	+	20–49	*19*
Rep. Ireland	10,000 – 20,000	88–91	+	0–19	
Italy	(200,000 – 300,000)	03	(0)	(0–19)	
Liechtenstein	**30 – 50**	98–00	+	**20–29**	
Luxembourg	400 – 500	02	+	**10–19**	
Macedonia	(1,000 – 2,500)	90–00	(0)	(0–19)	
Moldova	3,200 – 3,800	90–00	0	0–19	
Netherlands	**6,500 – 7,000**	98–00	+	**150**	*1*
Norway	**0 – 5**	90–03	–	**>80**	*37*
Poland	**25,000 – 35,000**	97–02	+	0–19	*23*
Portugal	(25,000 – 250,000)	02	(–)	(–)	
Romania	164,000 – 240,000	00–02	0	0–19	*48*
Russia	200,000 – 500,000	90–00	?	–	*122*
Serbia & MN	5,500 – 9,000	90–02	+	10–29	*1,29,155,172a, 67a,78,227*
Slovakia	20,000 – 40,000	90–99	+	20–29	
Slovenia	15,000 – 20,000	00	(0)	(0–19)	
Spain	(250,000 – 500,000)	98–02	?	–	*10*
Switzerland	**500 – 500**	93–96	+	30–49	
Turkey	(120,000 – 200,000)	01	(–)	(20–29)	
Ukraine	**26,000 – 38,000**	90–00	+	**20–29**	
UK	**19,300 – 49,400**	00	+	**115**	*22,31*
Total (approx.)	2,000,000 – 4,600,000	**Overall trend** Moderate increase			
Breeding range	>5,000,000 km²	**Gen. length.** <3.3	**% Global pop.** 5–24		

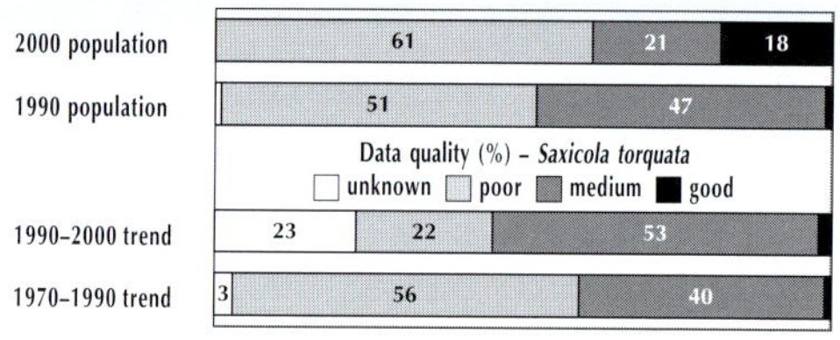

Oenanthe isabellina
ISABELLINE WHEATEAR

Non-SPEC (1994: —) **Status** (Secure)
Criteria —
European IUCN Red List Category —
Criteria —
Global IUCN Red List Category —
Criteria —

Oenanthe isabellina is a widespread summer visitor to south-eastern Europe, which accounts for less than a quarter of its global breeding range. Its European breeding population is very large (>2,100,000 pairs), and was stable between 1970–1990. Although there were declines in Russia during 1990–2000, the stronghold population in Turkey was stable, and trends were stable or increasing across the rest of its European range. The species hence remained stable overall, and consequently is provisionally evaluated as Secure.

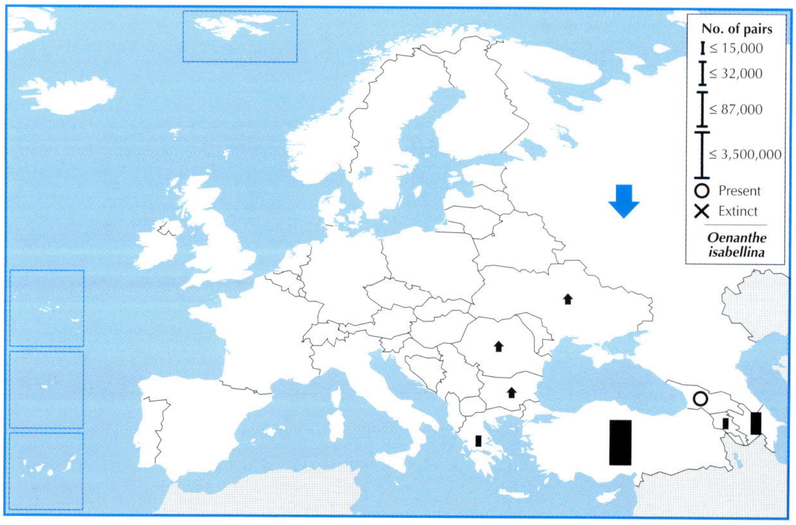

Country	Breeding pop. size (pairs)	Year(s)	Trend	Mag.%	References
Armenia	13,000 – 16,000	98–02	0	0–19	
Azerbaijan	(10,000 – 100,000)	96–00	(0)	(0–19)	
Bulgaria	**1,500 – 4,000**	96–02	+	0–19	
Georgia	Present	03	?	–	
Greece	(50 – 200)	95–00	(0)	(0–19)	
Romania	**500 – 700**	00–02	+	20–29	*17,32*
Russia	(50,000 – 150,000)	90–00	–	20–29	*8*
Turkey	(2,000,000 – 6,000,000)	01	(0)	(0–19)	
Ukraine	4,000 – 7,000	90–00	+	**5–9**	
Total (approx.)	2,100,000 – 6,300,000	**Overall trend** Stable			
Breeding range	>1,000,000 km²	**Gen. length.** <3.3	**% Global pop.** 5–24		

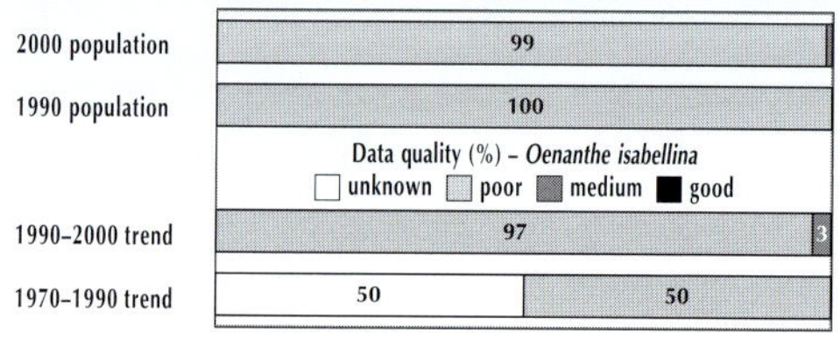

Oenanthe oenanthe
NORTHERN WHEATEAR

SPEC 3 (1994: —) **Status** (Declining)
Criteria Moderate recent decline
European IUCN Red List Category —
Criteria —
Global IUCN Red List Category —
Criteria —

Oenanthe oenanthe is a widespread summer visitor to most of Europe, which accounts for less than half of its global breeding range. Its European breeding population is very large (>4,600,000 pairs), and was stable between 1970–1990. Although it remained stable in various countries—particularly in eastern Europe—during 1990–2000, the species suffered widespread declines, including in the key Turkish population, and underwent a moderate decline (>10%) overall. Consequently, this previously Secure species is now provisionally evaluated as Declining.

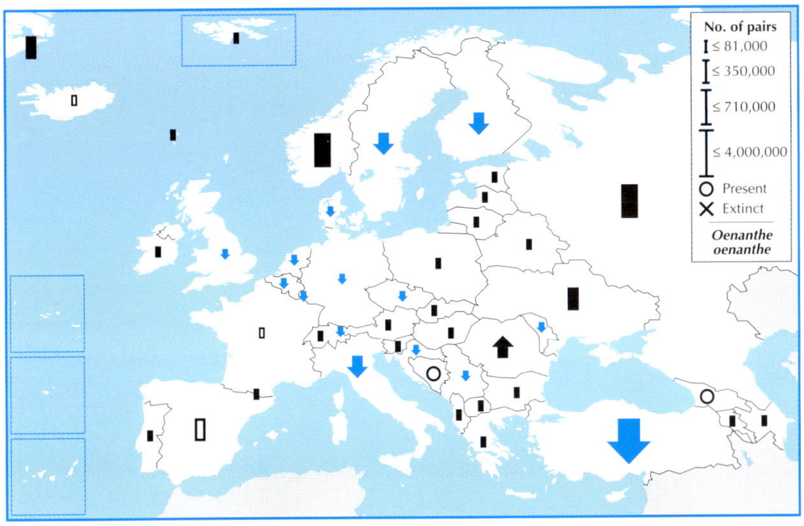

Country	Breeding pop. size (pairs)	Year(s)	Trend	Mag.%	References
Albania	5,000 – 10,000	02	(0)	(0–19)	
Andorra	2,500 – 4,000	99–01	(0)	(0–19)	*1,3*
Armenia	65,000 – 100,000	99–02	0	0–19	*2*
Austria	(4,500 – 9,000)	98–02	(0)	(0–19)	
Azerbaijan	(20,000 – 50,000)	96–00	(0)	(0–19)	
Belarus	50,000 – 70,000	97–02	0	0–19	
Belgium	**28 – 31**	00–02	**–**	**30–49**	*1*
Bosnia & HG	Present	85–89	?	–	
Bulgaria	20,000 – 60,000	96–02	0	0–19	
Croatia	(5,000 – 6,000)	02	(–)	(50–79)	*70,16*
Czech Rep.	**200 – 400**	00	**–**	**50–79**	
Denmark	(1,000 – 2,000)	00	(–)	(30–49)	*3*
Faroe Is.	3,000 – 3,000	95	(0)	(0–19)	
Greenland	(100,000 – 500,000)	90–00	(0)	(0–19)	
Estonia	(10,000 – 20,000)	98	0	0–19	*1*
Finland	150,000 – 200,000	98–02	**–**	**40**	
France	(15,000 – 45,000)	98–02	?	–	*4*
Georgia	Present	03	?	–	
Germany	7,000 – 13,000	95–99	–	30–49	
Greece	(30,000 – 100,000)	95–00	(0)	(0–19)	
Hungary	**28,000 – 57,000**	99–02	0	0–19	*19*
Iceland	(10,000 – 50,000)	78–94	?	–	*16*
Rep. Ireland	2,500 – 10,000	88–91	0	0–19	
Italy	(100,000 – 200,000)	03	(–)	(0–19)	
Latvia	(10,000 – 30,000)	90–00	(0)	(0–19)	*23,16*
Liechtenstein	3 – 5	98–00	(–)	(0–19)	
Lithuania	(5,000 – 10,000)	99–01	(0)	(0–19)	*20*
Luxembourg	**5 – 10**	00–02	**–**	**50**	
Macedonia	(10,000 – 30,000)	90–00	(0)	(0–19)	
Moldova	3,500 – 4,000	90–00	–	0–19	
Netherlands	**600 – 800**	98–00	**–**	**56**	*1*
Norway	(500,000 – 1,000,000)	90–03	(0)	(0–19)	
Svalbard	(0 – 20)	90–03	(0)	(0–19)	
Poland	**20,000 – 50,000**	00–02	(0)	(0–19)	*23*
Portugal	(500 – 5,000)	02	(0)	(0–19)	
Romania	225,000 – 343,000	00–02	+	0–19	*48*
Russia	(500,000 – 1,000,000)	90–00	(0)	(0–19)	*8,61,122*
Serbia & MN	11,000 – 16,000	90–02	–	0–19	*1,29,155,172a,67a, 78,225,227,184*
Slovakia	6,000 – 9,000	90–99	0	0–19	
Slovenia	300 – 500	00	(0)	(0–19)	
Spain	(326,000 – 361,000)	92	?	–	*13,12,10*
Sweden	(100,000 – 500,000)	99–00	**–**	**33**	
Switzerland	20,000 – 30,000	98–02	0	0–19	
Turkey	(2,000,000 – 8,000,000)	01	(–)	(0–19)	
Ukraine	140,000 – 170,000	90–00	0	10–19	
UK	**52,500 – 52,500**	00	**–**	**6**	*22,31*
Total (approx.)	4,600,000 – 13,000,000	**Overall trend** Moderate decline			
Breeding range	>8,000,000 km²	**Gen. length.** <3.3 **% Global pop.** 25–49			

(See p. 207, bottom, for data quality graph)

Oenanthe pleschanka
PIED WHEATEAR

Non-SPEC (1994: —) **Status** (Secure)
Criteria —
European IUCN Red List Category —
Criteria —
Global IUCN Red List Category —
Criteria —

Oenanthe pleschanka is a widespread summer visitor to much of south-eastern Europe, which accounts for less than a quarter of its global breeding range. Its European breeding population is relatively small (<140,000 pairs), but was stable between 1970–1990. The species remained stable overall during 1990–2000, with all populations across its European range either stable or increasing. Consequently, it is provisionally evaluated as Secure.

Country	Breeding pop. size (pairs)	Year(s)	Trend	Mag.%	References
Azerbaijan	(5,000 – 50,000)	96–00	(0)	(0–19)	
Bulgaria	**350 – 500**	96–02	**0**	**0–9**	
Croatia	(1 – 5)	02	(+)	(N)	*70,16*
Georgia	Present	03	?	–	
Romania	680 – 750	00–02	0	0–19	
Russia	(15,000 – 50,000)	90–00	(0)	(0–19)	*8*
Turkey	(8,000 – 30,000)	01	(0)	(0–19)	
Ukraine	3,000 – 5,500	90–00	**+**	**5–9**	
Total (approx.)	32,000 – 140,000	**Overall trend** Stable			
Breeding range	>500,000 km²	**Gen. length.** <3.3 **% Global pop.** 5–24			

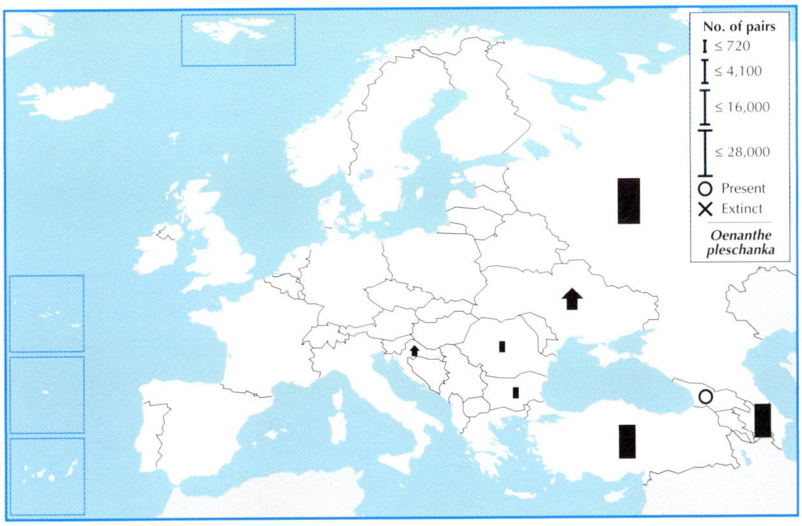

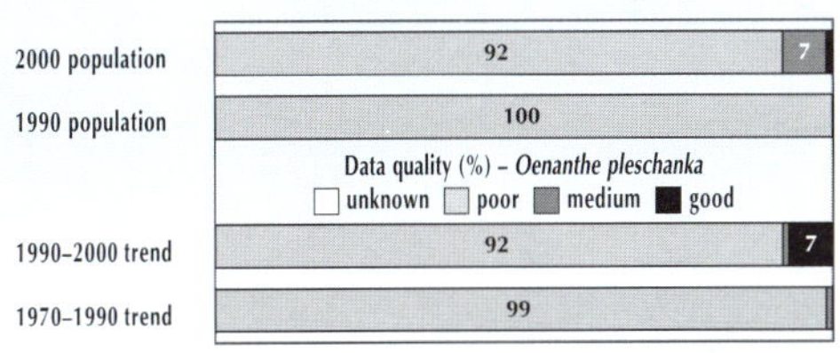

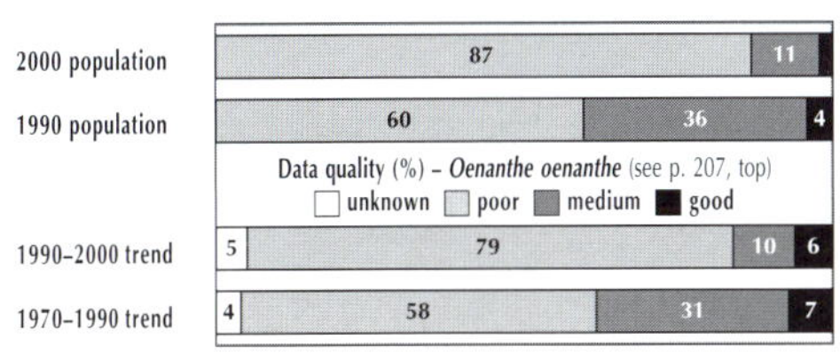

Oenanthe cypriaca
CYPRUS WHEATEAR

Non-SPEC^E (1994: 2) **Status** (Secure)
Criteria —
European IUCN Red List Category —
Criteria —
Global IUCN Red List Category —
Criteria —

Country	Breeding pop. size (pairs)	Year(s)	Trend	Mag.%	References
Cyprus	(90,000 – 180,000)	94–02	(0)	(0–19)	
Total (approx.)	90,000 – 180,000	**Overall trend** Stable			
Breeding range	>20,000 km²	**Gen. length.** <3.3	**% Global pop.** 100		

Oenanthe cypriaca breeds entirely within Europe, with its global breeding range confined to the island of Cyprus. Its breeding population was stable between 1970–1990, and remained stable during 1990–2000. Although the species was previously classified as Rare, recent surveys have revealed its breeding population to be far larger (>90,000 pairs) than was formerly believed. Consequently, it is now provisionally evaluated as Secure.

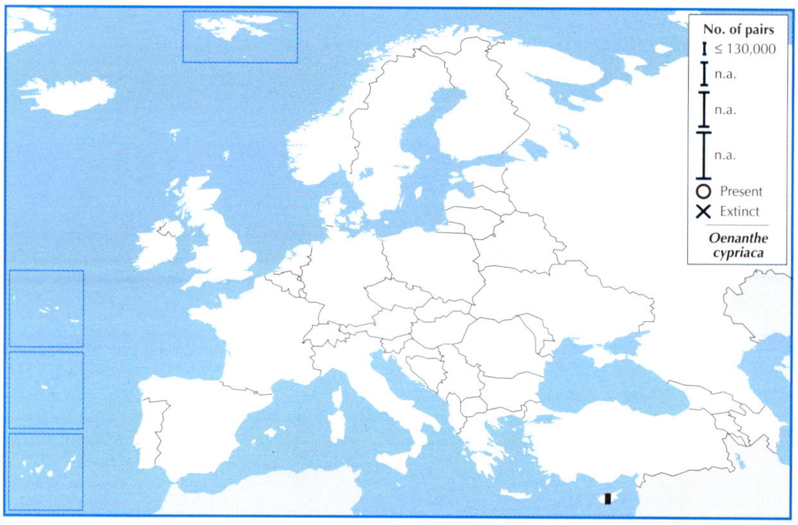

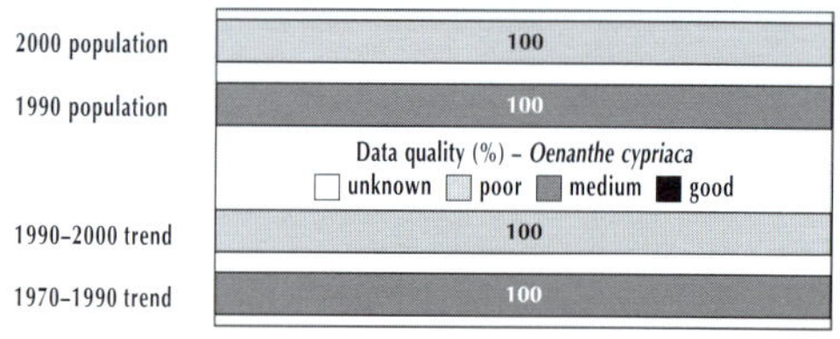

Oenanthe hispanica
BLACK-EARED WHEATEAR

SPEC 2 (1994: 2) **Status** (Depleted)
Criteria Large historical decline
European IUCN Red List Category —
Criteria —
Global IUCN Red List Category —
Criteria —

Country	Breeding pop. size (pairs)	Year(s)	Trend	Mag.%	References
Albania	5,000 – 15,000	02	(0)	(0–19)	
Armenia	26,000 – 33,000	97–02	0	0–19	*2*
Bosnia & HG	Present	90–03	?	–	
Bulgaria	2,000 – 4,000	96–02	0	0–19	
Croatia	(5,000 – 10,000)	02	(–)	(50–79)	*16,54*
France	1,000 – 4,000	98–02	(0)	(0–19)	*4*
Georgia	Present	03	?	–	
Greece	(50,000 – 150,000)	95–00	(0)	(0–19)	
Italy	(1,000 – 2,000)	03	(0)	(0–19)	
Macedonia	(1,000 – 2,500)	90–00	(0)	(0–19)	
Portugal	(2,000 – 20,000)	02	(–)	(–)	
Romania	**55 – 80**	00–02	+	0–19	
Russia	(2,500 – 5,000)	90–00	(+)	(20–29)	*172*
Serbia & MN	1,000 – 1,600	95–02	–	0–19	*1,62,247,50,155, 225,227,10,141*
Spain	(513,000 – 620,000)	92	(–)	(0–19)	*13,12,16,10*
Turkey	(800,000 – 2,400,000)	01	(0)	(0–19)	
Ukraine	(0 – 50)	90–00	(0)	(0–19)	
Total (approx.)	1,400,000 – 3,300,000	**Overall trend** Small decline			
Breeding range	>1,000,000 km²	**Gen. length.** <3.3	**% Global pop.** 50–74		

Oenanthe hispanica is a widespread summer visitor to southern Europe, which constitutes >50% of its global breeding range. Its European breeding population is very large (>1,400,000 pairs), but underwent a large decline between 1970–1990. Although the species was stable or increased in most of its range during 1990–2000, declines continued in a few countries—notably Spain—and it probably underwent a small decline overall. Its population size has clearly not yet recovered to the level that preceded its decline, and consequently the species is provisionally evaluated as Depleted.

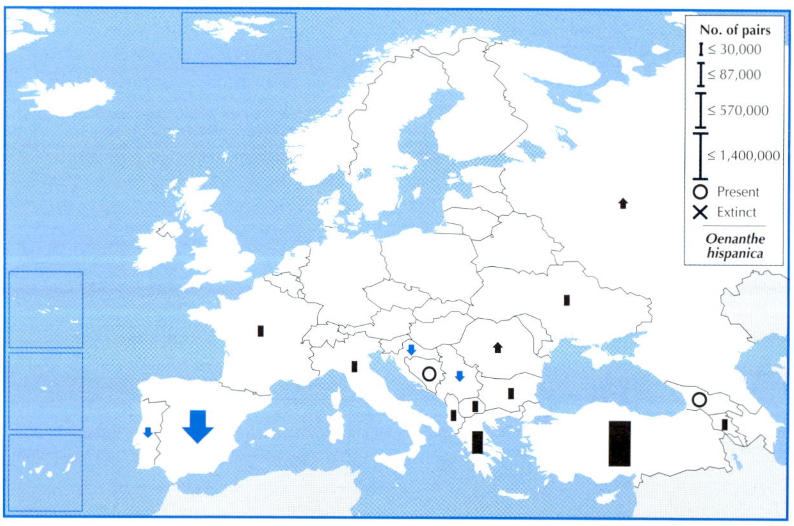

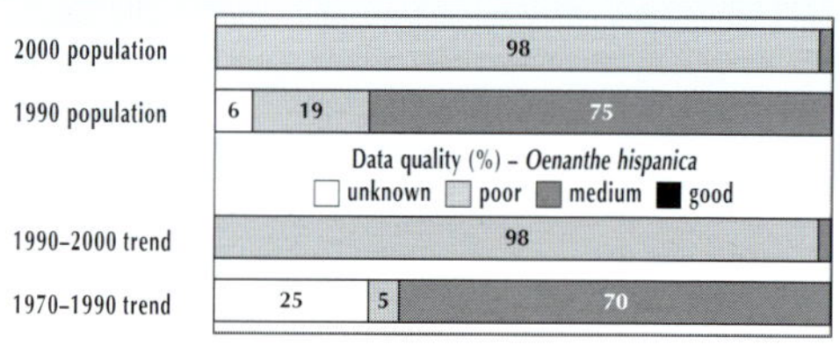

Oenanthe deserti
DESERT WHEATEAR

Non-SPEC (1994: —) **Status** (Secure)
Criteria —
European IUCN Red List Category —▼▼▼
Criteria —
Global IUCN Red List Category —
Criteria —

Country	Breeding pop. size (pairs)	Year(s)	Trend	Mag.%	References
Azerbaijan	(100 - 1,000)	96–00	(0)	(0–19)	
Turkey	(10 - 100)	01	(-)	(-)	
Total (approx.)	110 - 1,100	Overall trend Stable			
Breeding range	>20,000 km²	Gen. length <3.3	% Global pop. <5		

Oenanthe deserti breeds in Azerbaijan and Turkey, with Europe accounting for a tiny proportion of its global breeding range. Its European breeding population is very small (as few as 110 pairs), and its trend between 1970–1990 was unknown. The species declined in Turkey during 1990–2000, but was stable in Azerbaijan, and probably remained stable overall. Although the size of the European population could make it susceptible to the risks affecting small populations, it is marginal to a much larger non-European population. Consequently, it is provisionally evaluated as Secure.

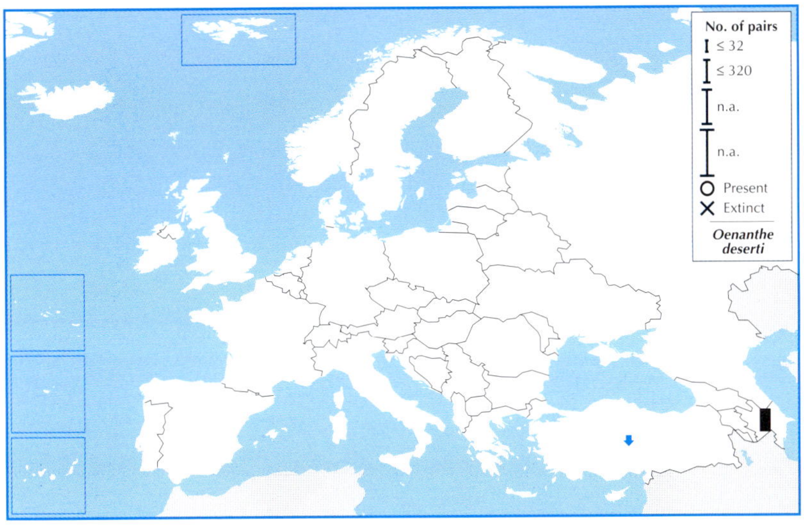

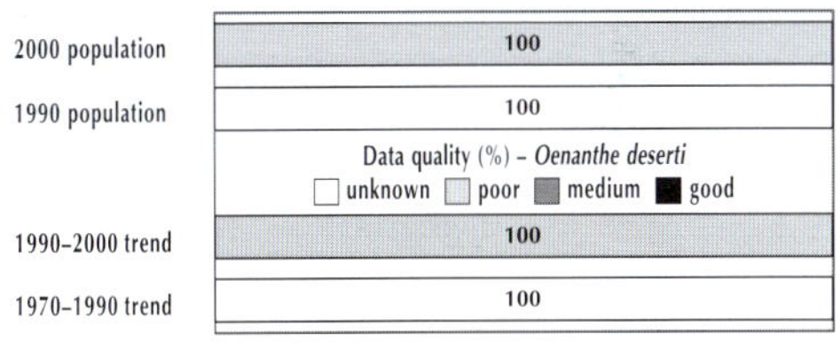

Oenanthe finschii
FINSCH'S WHEATEAR

Non-SPEC (1994: —) **Status** (Secure)
Criteria —
European IUCN Red List Category —
Criteria —
Global IUCN Red List Category —
Criteria —

Country	Breeding pop. size (pairs)	Year(s)	Trend	Mag.%	References
Armenia	1,300 - 1,600	97–02	0	0–19	
Azerbaijan	(1,000 - 10,000)	96–00	(0)	(0–19)	
Georgia	Present	03	?	-	
Turkey	(100,000 - 300,000)	01	(0)	(0–19)	
Total (approx.)	100,000 - 310,000	Overall trend Stable			
Breeding range	>250,000 km²	Gen. length <3.3	% Global pop. 5–24		

Oenanthe finschii breeds in Turkey and the Caucasus, with Europe accounting for less than a quarter of its global breeding range. Its European breeding population is large (>100,000 pairs), but its trend between 1970–1990 was unknown. Although no trend was available for Georgia during 1990–2000, the species was stable across the rest of its European range, and consequently it is provisionally evaluated as Secure.

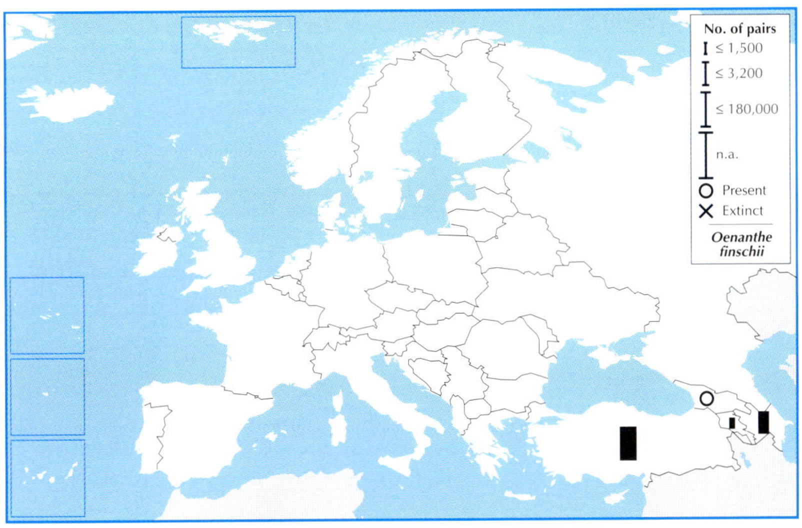

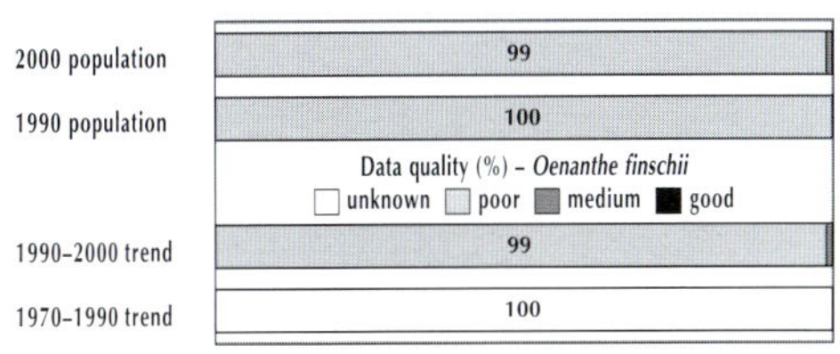

Oenanthe xanthoprymna
RUFOUS-TAILED WHEATEAR

SPEC 3 (1994: —) **Status** (Vulnerable)
Criteria See IUCN below
European IUCN Red List Category VU
Criteria C1
Global IUCN Red List Category —
Criteria —

Country	Breeding pop. size (pairs)	Year(s)	Trend	Mag.%	References
Armenia	**15 – 20**	95–00	0	0–19	*2*
Azerbaijan	(100 – 1,000)	96–00	(0)	(0–19)	
Georgia	Present	03	?	–	
Turkey	(2,000 – 6,000)	01	(–)	(10–19)	
Total (approx.)	2,100 – 7,000	**Overall trend** Moderate decline			
Breeding range	>100,000 km²	**Gen. length.** <3.3 **% Global pop.** 25–49			

Oenanthe xanthoprymna is a patchily distributed summer visitor to Turkey and the Caucasus, with Europe accounting for less than half of its global breeding range. Its European breeding population is small (as few as 2,100 pairs), but was stable between 1970–1990. Although the species was stable in Armenia and Azerbaijan during 1990–2000 (the trend in Georgia was unknown), it declined in its Turkish stronghold, and probably underwent a moderate decline (>10%) overall. As a consequence of its small population and this continuing decline, it is provisionally evaluated as Vulnerable.

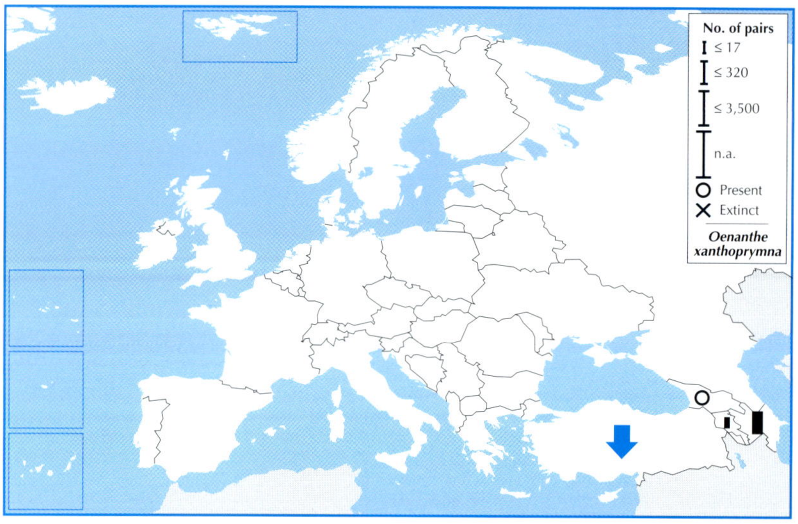

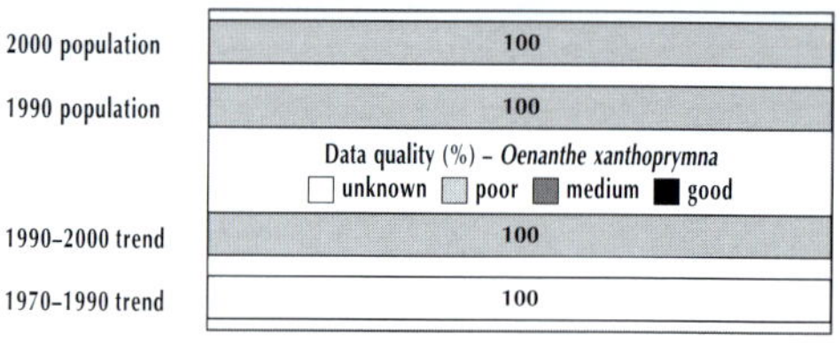

Oenanthe leucura
BLACK WHEATEAR

SPEC 3 (1994: 3) **Status** (Rare)
Criteria <10,000 pairs
European IUCN Red List Category —
Criteria —
Global IUCN Red List Category —
Criteria —

Country	Breeding pop. size (pairs)	Year(s)	Trend	Mag.%	References
France	**0 – 0**	97–02	–	X	*5*
Portugal	(50 – 500)	02	(–)	(–)	
Spain	(4,000 – 15,000)	92	?	–	*13,12,10*
Total (approx.)	4,100 – 16,000	**Overall trend** Unknown			
Breeding range	>250,000 km²	**Gen. length.** <3.3 **% Global pop.** 25–49			

Oenanthe leucura is resident in Iberia and France, with Europe accounting for less than half of its global range. Its European breeding population is small (as few as 4,100 pairs), and underwent a large decline between 1970–1990. Although the species declined in Portugal (and went extinct in France) during 1990–2000, the trend of the stronghold population in Spain was unknown. Nevertheless, its population size renders it susceptible to the risks affecting small populations, and consequently the species is provisionally evaluated as Rare.

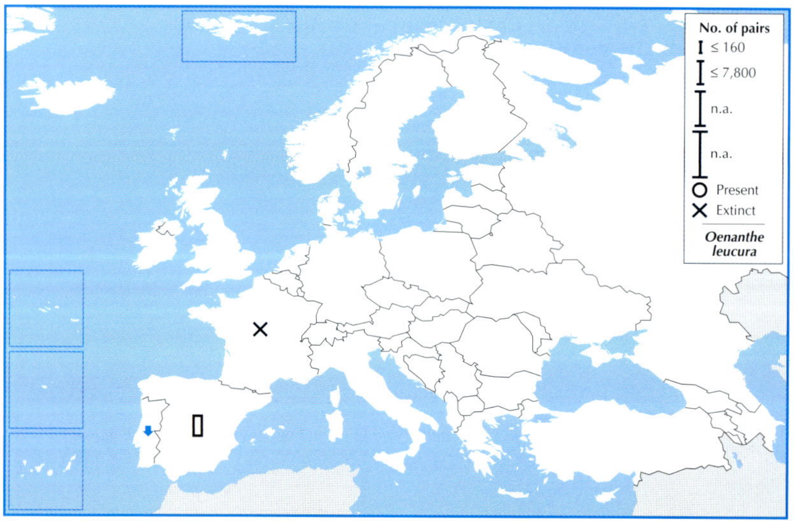

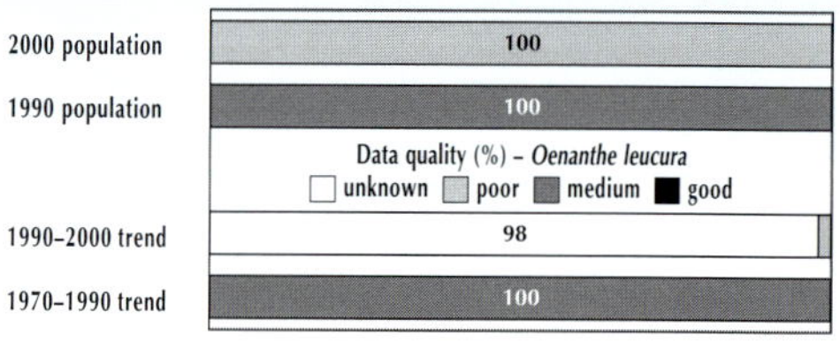

Monticola saxatilis
RUFOUS-TAILED ROCK-THRUSH

SPEC 3 (1994: 3) **Status** (Depleted)
Criteria Moderate historical decline
European IUCN Red List Category —
Criteria —
Global IUCN Red List Category —
Criteria —

Monticola saxatilis is a widespread but patchily distributed summer visitor to much of southern Europe, which accounts for less than half of its global breeding range. Its European breeding population is large (>100,000 pairs), but underwent a moderate decline between 1970–1990. Although the species was stable or increased across much of its European range during 1990–2000, it probably underwent a small decline overall, and its population has clearly not yet recovered to the level that preceded its initial decline. Consequently, it is provisionally evaluated as Depleted.

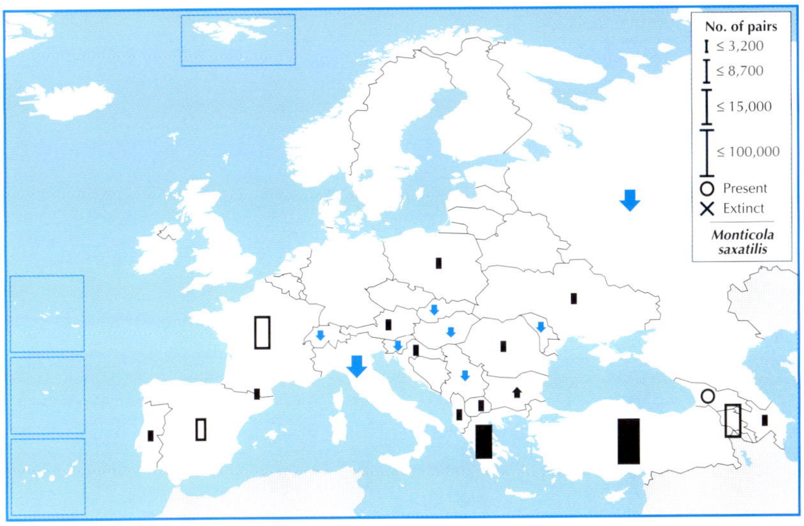

Country	Breeding pop. size (pairs)	Year(s)	Trend	Mag.%	References
Albania	300 – 1,000	02	(0)	(0–19)	
Andorra	50 – 100	99–01	(0)	(0–19)	*1,3*
Armenia	10,000 – 15,000	97–02	?	–	
Austria	(30 – 150)	98–02	(0)	(0–19)	
Azerbaijan	(1,000 – 10,000)	96–00	(0)	(0–19)	
Bulgaria	**1,200 – 2,500**	96–02	+	0–19	
Croatia	(1,000 – 5,000)	02	(0)	(0–19)	*16*
France	10,000 – 20,000	00	?	–	1
Georgia	Present	03	?	–	
Greece	(10,000 – 20,000)	95–00	(0)	(0–19)	
Hungary	20 – 30	98–02	–	50–79	
Italy	(5,000 – 10,000)	03	(–)	(0–19)	
Macedonia	(1,000 – 2,500)	90–00	(0)	(0–19)	
Moldova	5 – 25	90–00	(–)	(>80)	
Poland	0 – 10	90–00	0	0–19	*2,112*
Portugal	(50 – 500)	02	(0)	(0–19)	
Romania	(1,200 – 1,500)	00–02	0	0–19	*48*
Russia	(5,000 – 15,000)	90–00	(–)	(20–29)	*17,61,116*
Serbia & MN	1,000 – 1,400	90–02	–	0–19	*1,155,150,201,225, 227,224,62,40*
Slovakia	15 – 30	90–99	–	50–79	
Slovenia	150 – 250	99–00	(–)	(0–19)	
Spain	(2,500 – 10,000)	98–02	?	–	*10*
Switzerland	**800 – 1,200**	93–96	–	0–19	
Turkey	(50,000 – 200,000)	01	(0)	(0–19)	
Ukraine	850 – 900	90–00	(0)	(0–19)	
Total (approx.)	100,000 – 320,000	**Overall trend** Small decline			
Breeding range	>1,000,000 km²	**Gen. length.** <3.3 **% Global pop.** 25–49			

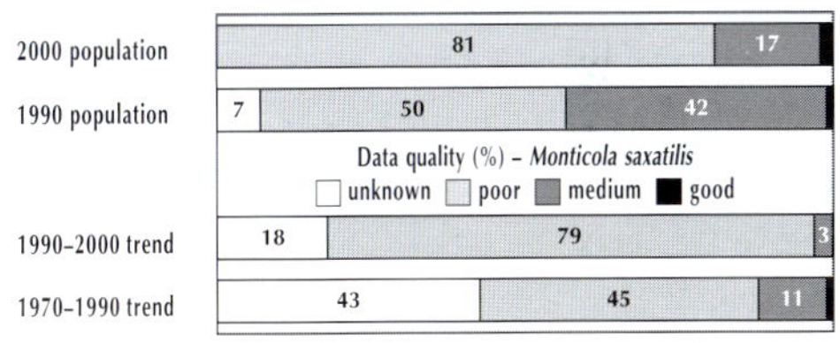

Monticola solitarius
BLUE ROCK-THRUSH

SPEC 3 (1994: 3) **Status** (Depleted)
Criteria Large historical decline
European IUCN Red List Category —
Criteria —
Global IUCN Red List Category —
Criteria —

Monticola solitarius is a widespread but patchily distributed breeder in southern Europe, which accounts for less than half of its global breeding range. Its European breeding population is large (>120,000 pairs), but underwent a large decline between 1970–1990. The species was stable across most of its European range during 1990–2000—including the key Turkish population—and thus was probably stable overall. Nevertheless, its population size has not yet recovered to the level that preceded its decline, and consequently it is provisionally evaluated as Depleted.

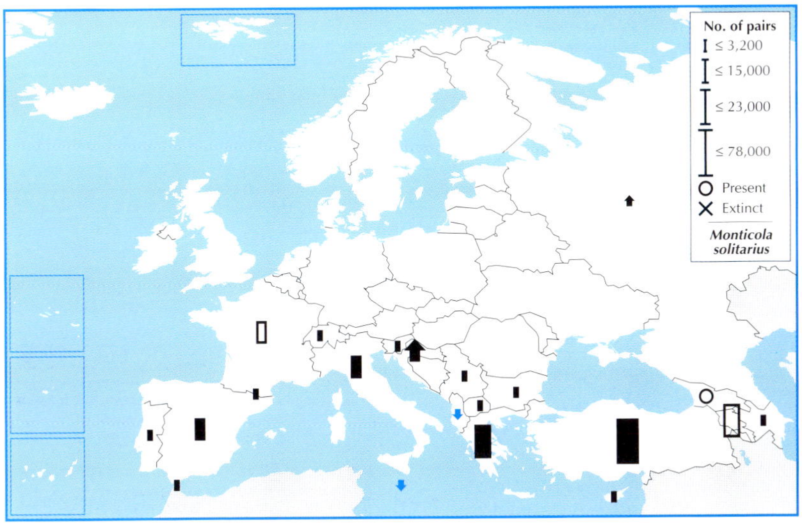

Country	Breeding pop. size (pairs)	Year(s)	Trend	Mag.%	References
Albania	(1,000 – 2,000)	02	(–)	(0–19)	
Andorra	**5 – 10**	99–01	(0)	(0–19)	*1,3*
Armenia	20,000 – 25,000	97–02	?	–	
Azerbaijan	(500 – 5,000)	96–00	(0)	(0–19)	
Bulgaria	**250 – 450**	96–02	0	0–9	
Croatia	(10,000 – 15,000)	02	(+)	(>80)	*16*
Cyprus	200 – 400	94–02	0	0–9	
France	5,000 – 10,000	00	?	–	*1*
Georgia	Present	03	?	–	
Greece	(10,000 – 30,000)	95–00	(0)	(0–19)	
Italy	(10,000 – 20,000)	03	(0)	(0–19)	
Macedonia	(400 – 1,000)	90–00	(0)	(0–19)	
Malta	**150 – 180**	90–02	–	**20–29**	*1*
Portugal	(1,000 – 10,000)	02	(0)	(0–19)	
Russia	1,000 – 3,000	90–00	(+)	(20–29)	*33*
Serbia & MN	200 – 500	95–02	(0)	(0–19)	*1,50,156,246*
Slovenia	**20 – 30**	99–00	0	0–19	
Spain	(10,000 – 20,000)	98–02	(0)	(0–19)	*10*
Switzerland	**20 – 25**	93–96	0	0–19	
Turkey	(50,000 – 120,000)	01	(0)	(0–19)	
UK					
Gibraltar	(20 – 35)	00	0	0–19	
Total (approx.)	120,000 – 260,000	**Overall trend** Stable			
Breeding range	>1,000,000 km²	**Gen. length.** <3.3 **% Global pop.** 25–49			

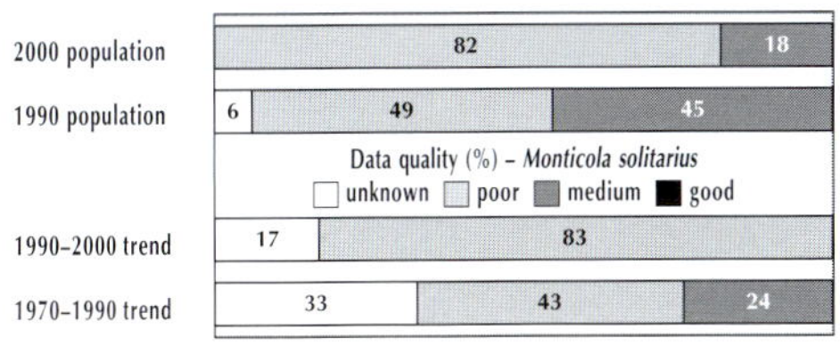

Zoothera dauma
SCALY THRUSH

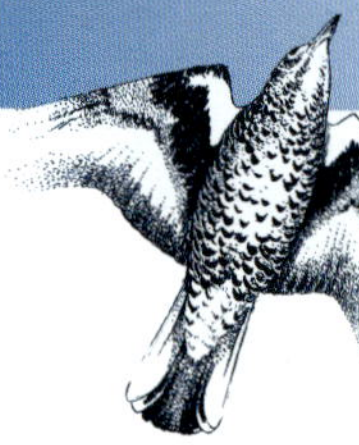

Non-SPEC (1994: —) **Status** (Secure)
Criteria —
European IUCN Red List Category —
Criteria —
Global IUCN Red List Category —
Criteria —

Zoothera dauma has an extensive global distribution, which just extends into Europe in the western foothills of the Urals. Its European breeding population is relatively small (<100,000 pairs), but was stable between 1970–1990. Trend data were not available for the Russian population during 1990–2000, but there was no evidence to suggest that its status had deteriorated significantly since 1990, and consequently the species is provisionally evaluated as Secure.

Country	Breeding pop. size (pairs)	Year(s)	Trend	Mag.%	References
Russia	(25,000 – 100,000)	90–00	?	–	*51,52,122*
Total (approx.)	25,000 – 100,000	**Overall trend** Unknown			
Breeding range	>100,000 km²	**Gen. length.** <3.3	**% Global pop.** <5		

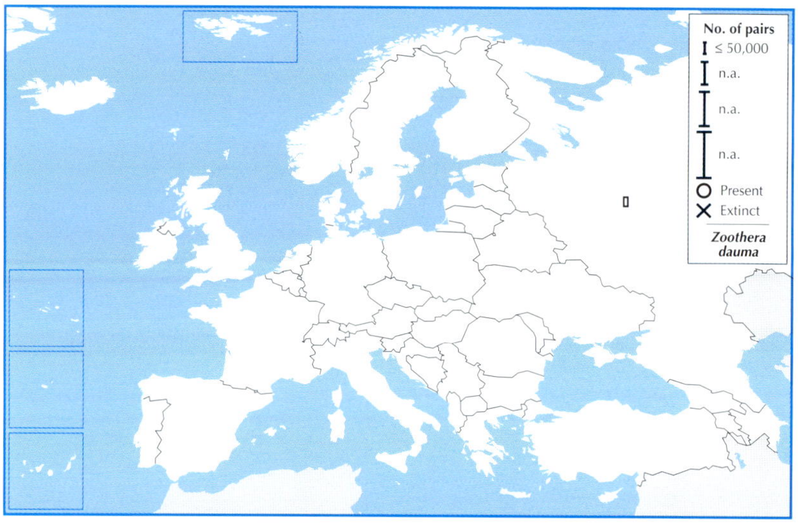

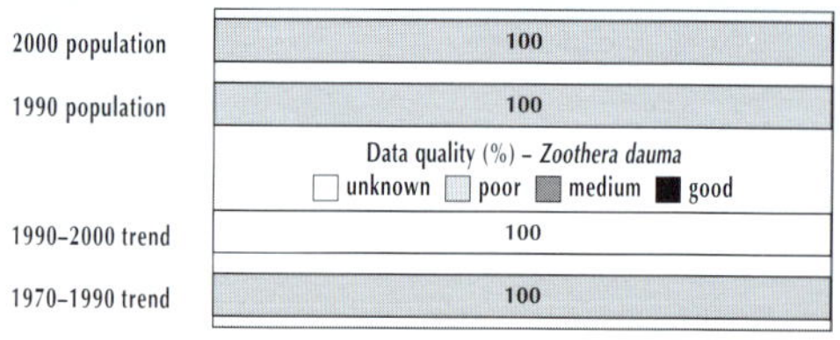

Turdus torquatus
RING OUZEL

Non-SPEC[E] (1994: 4) **Status** Secure
Criteria —
European IUCN Red List Category —
Criteria —
Global IUCN Red List Category —
Criteria —

Turdus torquatus has a patchy breeding distribution in Europe, which constitutes >95% of its global breeding range. Its European breeding population is large (>310,000 pairs), and was stable between 1970–1990. Although there were marked declines in certain countries—notably the United Kingdom and Czech Republic—during 1990–2000, key populations in Austria, Switzerland, Romania and Russia were stable, and the species remained stable overall. Consequently, it is evaluated as Secure.

Country	Breeding pop. size (pairs)	Year(s)	Trend	Mag.%	References
Albania	(20 – 50)	02	?	–	
Andorra	150 – 300	99–01	(0)	(0–19)	*1,3*
Armenia	13,000 – 16,000	97–02	0	0–19	
Austria	(50,000 – 100,000)	98–02	(0)	(0–19)	
Azerbaijan	(5,000 – 10,000)	96–00	(0)	(0–19)	
Belgium	5 – 20	95–02	0	0–19	*1*
Bulgaria	15,000 – 30,000	96–02	0	0–19	
Croatia	(100 – 500)	02	(–)	(50–79)	*16*
Czech Rep.	**700 – 1,400**	00	**–**	**50–79**	
Denmark					
Faroe Is.	(0 – 2)	82	(0)	(0–19)	
Finland	100 – 200	98–02	(0)	(0–19)	
France	4,000 – 10,000	00	?	–	*1*
Georgia	Present	03	?	–	
Germany	11,000 – 16,000	95–99	0	0–19	
Greece	(50 – 100)	95–00	(0)	(0–19)	
Rep. Ireland	50 – 250	01–03	(–)	(50–79)	
Italy	(10,000 – 20,000)	03	(0)	(0–19)	
Liechtenstein	500 – 1,000	98–00	–	0–19	
Macedonia	(1,000 – 2,500)	90–00	(0)	(0–19)	
Norway	(10,000 – 100,000)	90–03	(0)	(0–19)	
Poland	1,000 – 3,000	90–00	(0)	(0–19)	*27*
Romania	80,000 – 112,000	00–02	0	0–19	*4,46*
Russia	25,000 – 100,000	90–00	0	0–19	*61,116,136*
Serbia & MN	3,000 – 5,000	90–02	0	0–19	*1,155,117a,143, 225,227,152*
Slovakia	3,000 – 5,000	90–99	0	0–19	
Slovenia	3,000 – 5,000	99–00	(0)	(0–19)	
Spain	(6,000 – 7,000)	92	?	–	*13,12,10*
Sweden	3,000 – 10,000	99–00	0	0–19	
Switzerland	**40,000 – 60,000**	93–96	0	0–19	
Turkey	(10,000 – 40,000)	01	(0)	(0–19)	
Ukraine	4,500 – 8,000	90–00	0	0–19	
UK	**6,200 – 7,600**	99	**–**	**58**	*8,42*
Total (approx.)	310,000 – 670,000	**Overall trend** Stable			
Breeding range	>1,000,000 km²	**Gen. length.** <3.3	**% Global pop.** >95		

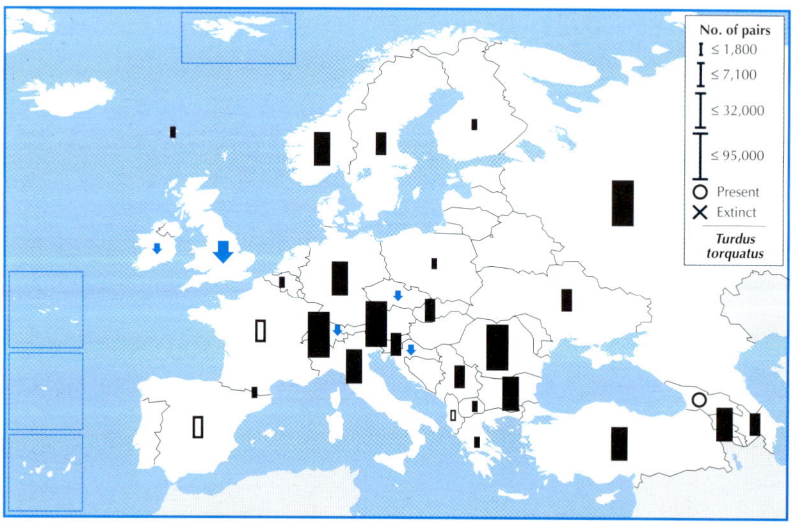

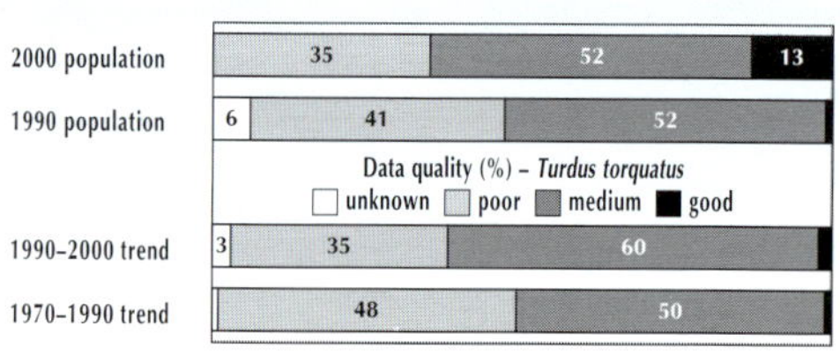

Turdus merula
EURASIAN BLACKBIRD

Non-SPEC[E] (1994: 4) **Status** Secure
Criteria —
European IUCN Red List Category —
Criteria —
Global IUCN Red List Category —
Criteria —

Turdus merula is a widespread breeder across most of Europe, which constitutes >50% of its global breeding range. Its European breeding population is extremely large (>40,000,000 pairs), and was stable between 1970–1990. The species underwent a small increase during 1990–2000, with increases in the sizeable populations in Germany, France and Italy, and stable or increasing trends across most of the rest of Europe. Consequently, the species is evaluated as Secure.

Country	Breeding pop. size (pairs)	Year(s)	Trend	Mag.%	References
Albania	200,000 – 500,000	02	(–)	(0–19)	
Andorra	(400 – 800)	99–01	(+)	(30–49)	*1,3*
Armenia	25,000 – 35,000	99–02	0	0–19	
Austria	(350,000 – 700,000)	98–02	(0)	(0–19)	
Azerbaijan	(100,000 – 500,000)	96–00	(0)	(0–19)	
Belarus	500,000 – 650,000	97–02	0	0–19	
Belgium	(500,000 – 1,000,000)	95–02	(0)	(0–19)	*1*
Bosnia & HG	(100,000 – 500,000)	90–00	(0)	(0–19)	
Bulgaria	1,000,000 – 5,000,000	96–02	0	0–9	
Croatia	(1,000,000 – 1,500,000)	02	(–)	(30–49)	*16*
Cyprus	5 – 20	94–02	+	N	
Czech Rep.	2,000,000 – 4,000,000	00	0	0–19	
Denmark	2,000,000 – 2,500,000	00	**0**	**0–19**	*12*
Faroe Is.	400 – 400	95	(0)	(0–19)	
Estonia	150,000 – 250,000	98	+	20–29	*1*
Finland	300,000 – 450,000	98–02	+	**>80**	
France	(1,200,000 – 6,000,000)	98–02	+	12	*4,3*
Georgia	Present	03	?	–	
Germany	8,000,000 – 16,000,000	95–99	+	0–19	
Greece	(800,000 – 2,000,000)	95–00	(0)	(0–19)	
Hungary	**1,100,000 – 1,450,000**	99–02	+	20–49	*19*
Iceland	(0 – 50)	90–00	(+)	(>80)	*15,33,45,46,47*
Rep. Ireland	1,000,000 – 2,500,000	88–91	0	0–19	
Italy	(2,000,000 – 5,000,000)	03	(+)	(0–19)	
Latvia	150,000 – 200,000	90–00	(0)	(0–19)	*23,18*
Liechtenstein	1,000 – 2,000	98–00	0	0–19	
Lithuania	(250,000 – 350,000)	99–01	(+)	(0–19)	*20*
Luxembourg	**50,000 – 60,000**	02	**0**	**0–19**	
Macedonia	(100,000 – 300,000)	90–00	(0)	(0–19)	
Moldova	40,000 – 45,000	90–00	0	0–19	
Netherlands	**900,000 – 1,200,000**	98–00	+	**28**	*1*
Norway	(100,000 – 1,000,000)	90–02	(0)	(0–19)	*27*
Poland	**1,000,000 – 2,000,000**	00–02	0	0–19	*23*
Portugal	(200,000 – 2,000,000)	02	(0)	(0–19)	
Azores	Present	02	?	–	
Madeira	Present	02	?	–	
Romania	890,000 – 1,780,000	00–02	+	0–19	*48*
Russia	2,000,000 – 3,000,000	90–00	(0)	(0–19)	*8,116,122*
Serbia & MN	500,000 – 800,000	90–02	0	0–19	*1,29,172a,67a, 78,225,227*
Slovakia	400,000 – 800,000	90–99	0	0–19	
Slovenia	350,000 – 550,000	00	(0)	(0–19)	
Spain	(2,300,000 – 5,900,000)	92	?	–	*13,12,10*
Canary Is.	(20,000 – 100,000)	97–03	?	–	*28,25*
Sweden	1,000,000 – 2,500,000	99–00	+	**25**	
Switzerland	400,000 – 600,000	98–02	0	0–19	
Turkey	(800,000 – 2,000,000)	01	(–)	(0–19)	
Ukraine	830,000 – 1,000,000	90–00	0	0–4	
UK	**4,935,000 – 4,935,000**	00	**0**	**5**	*5,31*
Gibraltar	150 – 250	00	**0**	**0–19**	
Total (approx.)	40,000,000 – 82,000,000	**Overall trend** Small increase			
Breeding range	>8,000,000 km²	**Gen. length** <3.3	**% Global pop.** 50–74		

(See p. 213, bottom, for data quality graph)

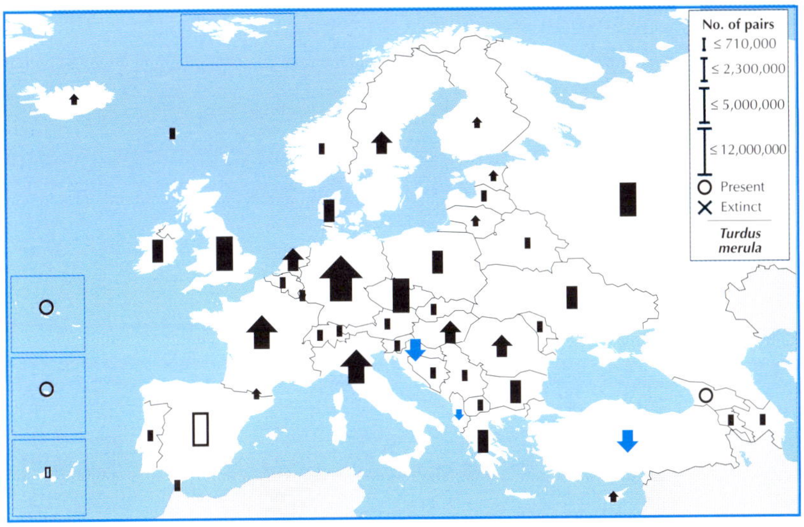

Turdus ruficollis
DARK-THROATED THRUSH

Non-SPEC (1994: —) **Status** (Secure)
Criteria —
European IUCN Red List Category —▼
Criteria —
Global IUCN Red List Category —
Criteria —

Country	Breeding pop. size (pairs)	Year(s)	Trend	Mag.%	References
Russia	5,000 – 20,000	90–00	?	–	*51,52*
Total (approx.)	5,000 – 20,000	**Overall trend** Unknown			
Breeding range	>100,000 km²	**Gen. length** <3.3	**% Global pop.** <5		

Turdus ruficollis has a predominantly Asian breeding distribution, which just extends west of the Urals into European Russia. Its European breeding population is small (as few as 5,000 pairs), but was stable between 1970–1990. No trend data were available for 1990–2000, but there was no evidence to suggest that the species declined. Although the size of the European population could render it susceptible to the risks affecting small populations, it is marginal to a much larger non-European population. Consequently, the species is provisionally evaluated as Secure.

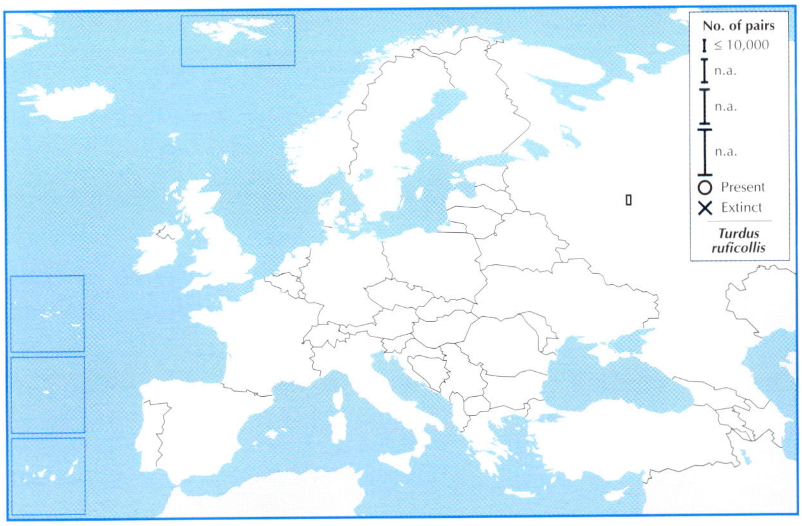

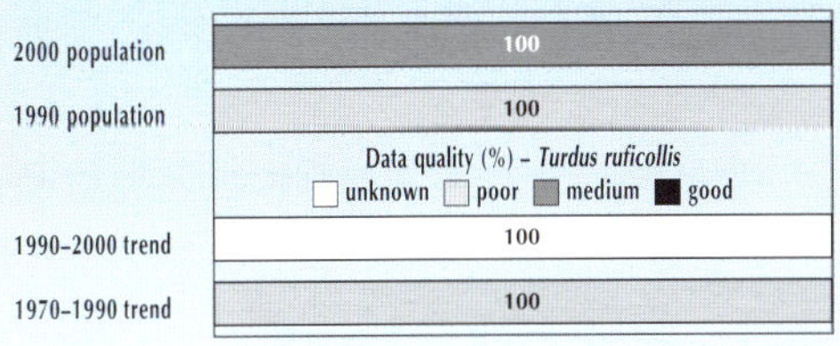

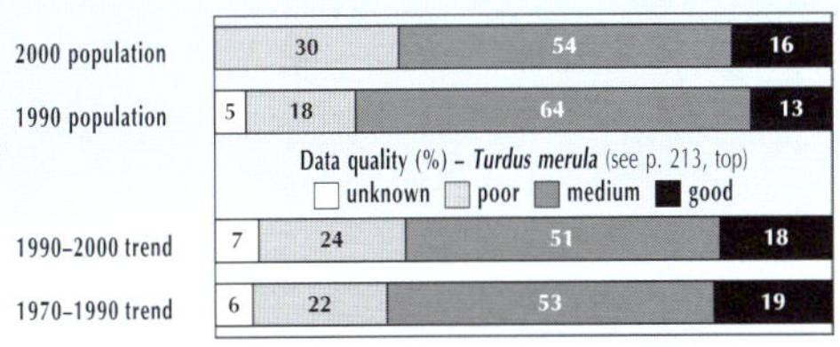

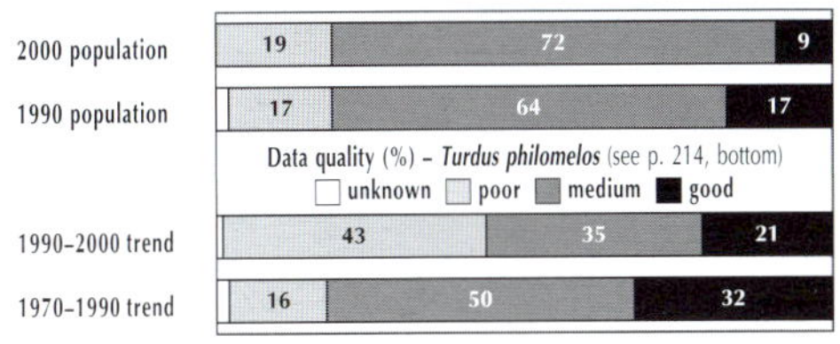

Turdus pilaris
FIELDFARE

Non-SPEC^{E}W (1994: 4W) **Status** (Secure)
Criteria —
European IUCN Red List Category —
Criteria —
Global IUCN Red List Category —
Criteria —

Turdus pilaris is a widespread breeder in central and northern Europe, but winters across much of the continent, which constitutes >75% of its global wintering range. Insufficient information was available to assess the species's status using wintering population data, but its European breeding population is extremely large (>14,000,000 pairs), and was stable between 1970–1990. Breeding populations in most countries (including Russia) were stable or increased during 1990–2000, and the species probably remained stable overall. Consequently, it is provisionally evaluated as Secure.

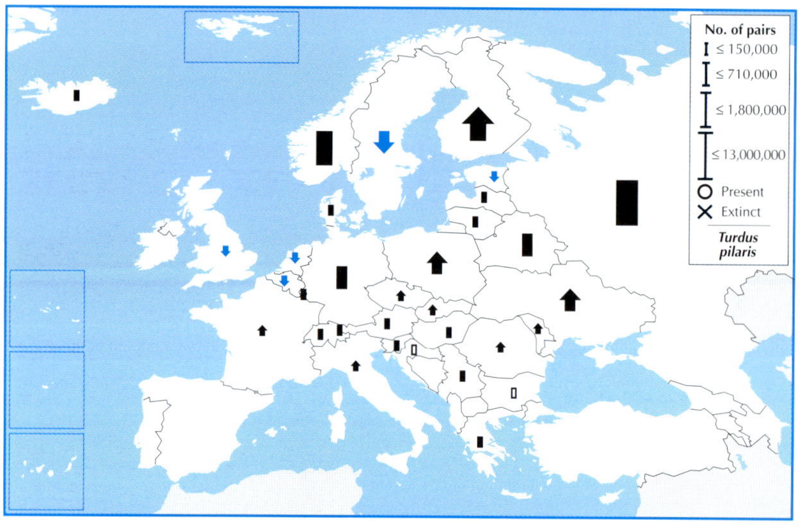

Country	Breeding pop. size (pairs)	Year(s)	Trend	Mag.%	References
Austria	(30,000 – 60,000)	98–02	(0)	(0–19)	
Belarus	300,000 – 500,000	97–02	**0**	**0–19**	
Belgium	10,500 – 20,000	01–02	–	0–19	*1*
Bulgaria	1 – 1	93	?	–	
Croatia	**3 – 5**	02	?	–	*5,62*
Czech Rep.	80,000 – 160,000	00	+	10–19	
Denmark	2,000 – 5,000	00	(F)	(30–49)	*3*
Estonia	(100,000 – 200,000)	98	–	20–29	*1*
Finland	1,000,000 – 2,000,000	98–02	**+**	**30**	
France	10,000 – 20,000	00	+	0–19	*1*
Germany	350,000 – 600,000	95–99	0	0–19	
Greece	(10 – 100)	95–00	(0)	(0–19)	
Hungary	30 – 100	98–02	0	0–19	
Iceland	**0 – 1**	90–00	(F)	(–)	*33,45,46,47*
Italy	(5,000 – 10,000)	03	(+)	(0–19)	
Latvia	40,000 – 150,000	90–00	(0)	(0–19)	*23,16*
Liechtenstein	300 – 1,000	98–00	0	0–19	
Lithuania	(80,000 – 120,000)	99–01	(0)	(0–19)	*20*
Luxembourg	2,000 – 3,000	02	**0**	**0–19**	
Moldova	**0 – 5**	90–00	+	N	
Netherlands	**150 – 200**	98–00	–	>80	*1*
Norway	(1,000,000 – 3,000,000)	95–02	F	30–49	*30*
Poland	**200,000 – 400,000**	00–02	+	20–29	*23*
Romania	64,000 – 78,000	00–02	+	0–19	*48,12*
Russia	10,000,000 – 15,000,000	90–00	(0)	(0–19)	*122*
Serbia & MN	0 – 10	95–02	F	50–79	*1*
Slovakia	10,000 – 20,000	90–99	+	20–29	
Slovenia	1,000 – 3,000	94	(0)	(0–19)	
Sweden	500,000 – 1,000,000	99–00	**–**	**6**	
Switzerland	60,000 – 100,000	98–02	0	0–19	
Ukraine	280,000 – 365,000	90–00	+	10–19	
UK	**1 – 4**	96–00	**–**	**50**	
Total (approx.)	14,000,000 – 24,000,000	**Overall trend** Stable			
Breeding range	>5,000,000 km²	**Gen. length**. <3.3	**% Global pop.** 25–49		

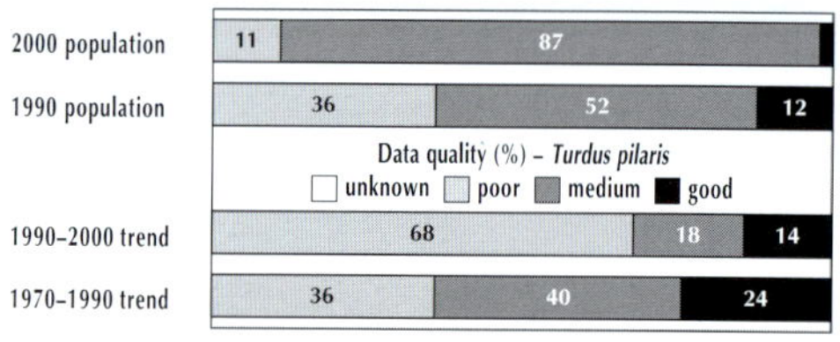

Turdus philomelos
SONG THRUSH

Non-SPEC^{E} (1994: 4) **Status** Secure
Criteria —
European IUCN Red List Category —
Criteria —
Global IUCN Red List Category —
Criteria —

Turdus philomelos is a widespread breeder across most of Europe, which constitutes >50% of its global breeding range. Its European breeding population is extremely large (>20,000,000 pairs), and was stable between 1970–1990. Although there were declines in Germany during 1990–2000, these were compensated for by increases in two other key populations, in France and Norway, and the species remained stable overall. Consequently, it is evaluated as Secure.

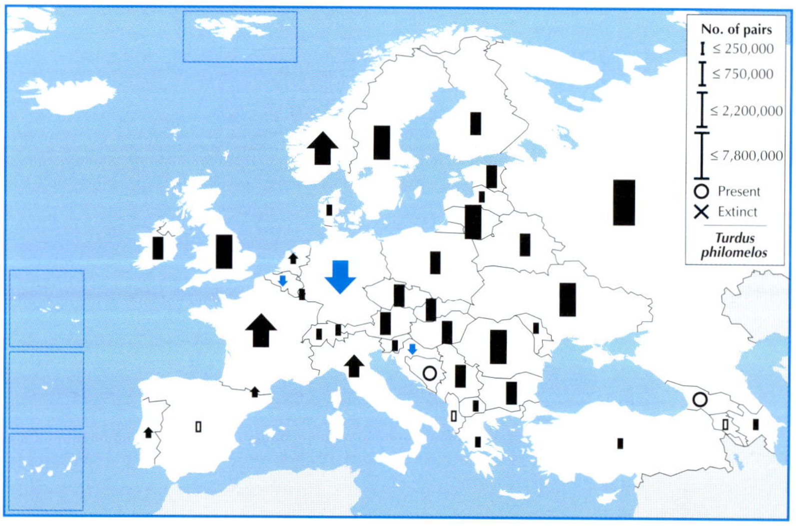

Country	Breeding pop. size (pairs)	Year(s)	Trend	Mag.%	References
Albania	(1,000 – 5,000)	02	?	–	
Andorra	200 – 300	99–01	(+)	(0–19)	*1,3*
Armenia	6,500 – 10,000	98–02	?	–	
Austria	(250,000 – 500,000)	98–02	(0)	(0–19)	
Azerbaijan	(50,000 – 100,000)	96–00	(0)	(0–19)	
Belarus	700,000 – 800,000	97–02	0	0–19	
Belgium	20,000 – 100,000	01–02	–	0–19	*1*
Bosnia & HG	Present	90–03	?	–	
Bulgaria	200,000 – 800,000	96–02	0	0–9	
Croatia	(50,000 – 100,000)	02	(–)	(50–79)	*16*
Czech Rep.	400,000 – 800,000	00	0	0–19	
Denmark	200,000 – 300,000	00	F	20–29	*12*
Estonia	200,000 – 400,000	98	0	0–19	*1*
Finland	600,000 – 900,000	98–02	**0**	**5**	
France	(600,000 – 3,000,000)	98–02	**+**	**59**	*4,3*
Georgia	Present	03	?	–	
Germany	1,200,000 – 2,500,000	95–99	–	0–19	
Greece	(1,000 – 3,000)	95–00	(0)	(0–19)	
Hungary	**270,000 – 410,000**	99–02	0	0–19	*19*
Rep. Ireland	250,000 – 500,000	88–91	0	0–19	
Italy	(200,000 – 400,000)	03	(+)	(0–19)	
Latvia	200,000 – 250,000	90–00	(0)	(0–19)	*23,18*
Liechtenstein	1,000 – 1,500	98–00	0	0–19	
Lithuania	(1,000,000 – 2,000,000)	99–01	(0)	(0–19)	*20*
Luxembourg	5,000 – 8,000	02	0	0–19	
Macedonia	(30,000 – 100,000)	90–00	(0)	(0–19)	
Moldova	35,000 – 40,000	90–00	0	0–19	
Netherlands	**120,000 – 160,000**	98–00	**+**	**45**	*1*
Norway	(1,000,000 – 1,500,000)	90–03	(+)	(0–19)	*30*
Poland	**500,000 – 800,000**	00–02	0	0–19	*23*
Portugal	(100 – 1,000)	02	(+)	(–)	
Romania	1,420,000 – 2,150,000	00–02	0	0–19	*48*
Russia	6,000,000 – 10,000,000	90–00	(0)	(0–19)	*122*
Serbia & MN	250,000 – 400,000	90–02	0	0–19	*1,29,155,172a, 67a,225,227*
Slovakia	300,000 – 600,000	90–99	0	0–19	
Slovenia	100,000 – 200,000	00	(0)	(0–19)	
Spain	(100,000 – 250,000)	98–02	?	–	*10*
Sweden	1,500,000 – 3,000,000	99–00	**0**	**0**	
Switzerland	**200,000 – 250,000**	93–96	0	0–19	
Turkey	(20,000 – 60,000)	01	(0)	(0–19)	
Ukraine	1,000,000 – 1,200,000	90–00	0	0–19	
UK	**1,144,000 – 1,144,000**	00	**0**	**4**	*5,31*
Total (approx.)	20,000,000 – 36,000,000	**Overall trend** Stable			
Breeding range	>7,000,000 km²	**Gen. length**. <3.3	**% Global pop.** 50–74		

(See p. 213, bottom, for data quality graph)

Turdus iliacus
REDWING

Non-SPEC^EW (1994: 4W) **Status** (Secure)
Criteria —
European IUCN Red List Category —
Criteria —
Global IUCN Red List Category —
Criteria —

Turdus iliacus is a widespread breeder in northern Europe, but winters across much of the continent, which constitutes >50% of its global wintering range. Insufficient information was available to assess the species's status using wintering population data, but its European breeding population is extremely large (>16,000,000 pairs), and was stable between 1970–1990. Breeding populations in most countries (including Russia) were stable or increased during 1990–2000, and the species probably remained stable overall. Consequently, it is provisionally evaluated as Secure.

Country	Breeding pop. size (pairs)	Year(s)	Trend	Mag.%	References
Belarus	70,000 – 140,000	97–02	0	0–19	
Czech Rep.	**0 – 5**	00	F	**50–79**	
Denmark					
Faroe Is.	30 – 30	95	(0)	(0–19)	
Greenland	(1 – 10)	90–00	?	–	
Estonia	(100,000 – 200,000)	98	+	20–29	*1*
Finland	1,500,000 – 2,500,000	98–02	**0**	**0–19**	
Iceland	(100,000 – 200,000)	78–94	?	–	*1*
Latvia	60,000 – 100,000	90–00	(0)	(0–19)	*23,18*
Lithuania	(30,000 – 40,000)	99–01	(0)	(0–19)	*20*
Norway	(1,000,000 – 1,500,000)	95–02	(+)	(0–19)	*30*
Svalbard	**0 – 1**	95–03	0	0–19	
Poland	500 – 1,000	90–00	(–)	(0–19)	*1*
Russia	12,000,000 – 15,000,000	90–00	(0)	(0–19)	*122*
Slovakia	0 – 5	90–99	?	–	
Sweden	750,000 – 1,500,000	99–00	-	**21**	
Ukraine	(9,500 – 12,500)	90–00	(0)	(10–19)	
UK	**2 – 22**	96–00	-	**48**	
Total (approx.)	16,000,000 – 21,000,000	**Overall trend** Stable			
Breeding range	>4,000,000 km²	**Gen. length.** <3.3		**% Global pop.** 25–49	

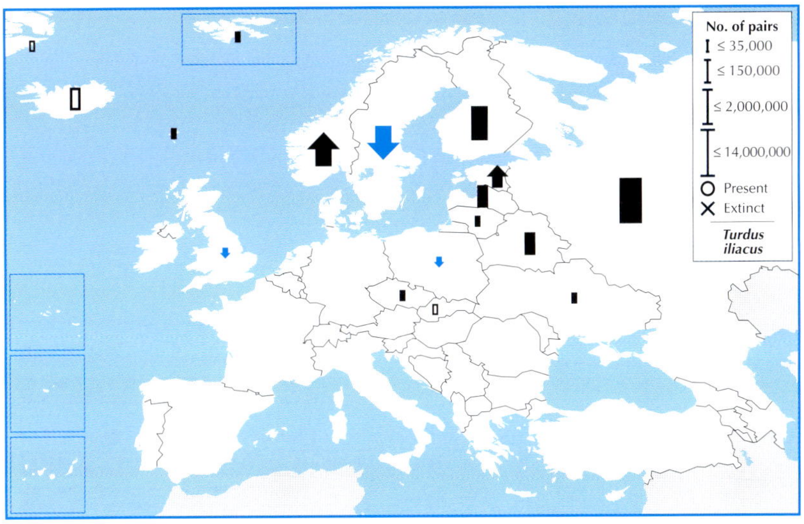

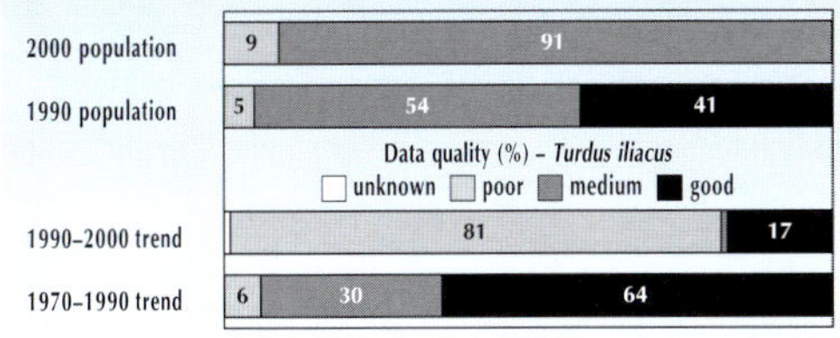

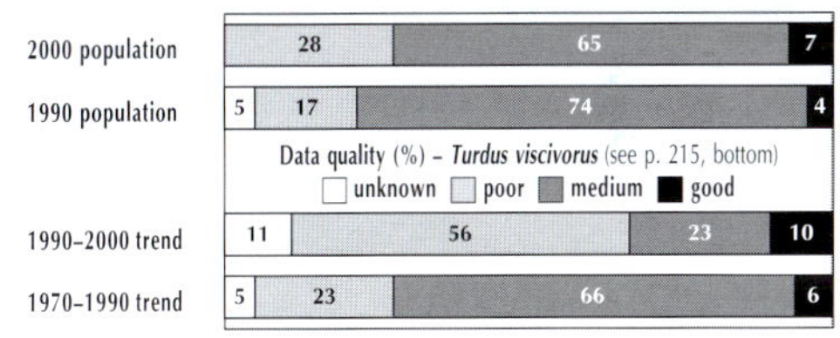

Turdus viscivorus
MISTLE THRUSH

Non-SPEC^E (1994: 4) **Status** Secure
Criteria —
European IUCN Red List Category —
Criteria —
Global IUCN Red List Category —
Criteria —

Turdus viscivorus is a widespread breeder across much of Europe, with >50% of its global breeding range occurring in the region. Its European breeding population is very large (>3,000,000 pairs), and was stable between 1970–1990. The species remained stable overall during 1990–2000, with increases in the sizeable German population at least partly compensating for declines elsewhere—notably in France and the United Kingdom. Crucially, the key population in Russia remained stable. Consequently, the species is evaluated as Secure.

Country	Breeding pop. size (pairs)	Year(s)	Trend	Mag.%	References
Albania	2,000 – 5,000	02	?	–	
Andorra	(300 – 600)	99–01	(0)	(0–19)	*1,3*
Armenia	13,000 – 16,000	99–02	?	–	
Austria	(80,000 – 120,000)	98–02	(0)	(0–19)	
Azerbaijan	(5,000 – 50,000)	96–00	(0)	(0–19)	
Belarus	20,000 – 35,000	97–02	0	0–19	
Belgium	20,000 – 100,000	01–02	0	0–19	*1*
Bosnia & HG	Present	90–03	?	–	
Bulgaria	50,000 – 150,000	96–02	0	0–9	
Croatia	(5,000 – 10,000)	02	(–)	(>80)	*16*
Czech Rep.	40,000 – 80,000	00	+	10–19	
Denmark	20,000 – 30,000	00	-	30–49	*12*
Estonia	5,000 – 10,000	98	0	0–19	*1*
Finland	60,000 – 100,000	98–02	+	**50**	
France	(100,000 – 500,000)	98–02	(–)	(19)	*4,2*
Georgia	Present	03	?	–	
Germany	300,000 – 550,000	95–99	+	0–19	
Greece	(20,000 – 50,000)	95–00	(0)	(0–19)	
Hungary	4,000 – 25,000	99–02	+	50–79	*19*
Rep. Ireland	20,000 – 100,000	88–91	-	0–19	
Italy	(50,000 – 100,000)	03	(–)	(0–19)	
Latvia	30,000 – 60,000	90–00	(0)	(0–19)	*23,18*
Liechtenstein	200 – 500	98–00	0	0–19	
Lithuania	(10,000 – 20,000)	99–01	(0)	(0–19)	*20*
Luxembourg	2,000 – 3,000	02	0	0–19	
Macedonia	(100,000 – 300,000)	90–00	(0)	(0–19)	
Netherlands	**13,000 – 17,000**	98–00	**0**	**4**	*1*
Norway	(10,000 – 50,000)	90–02	+	0–19	
Poland	**50,000 – 100,000**	00–02	0	0–19	*23*
Portugal	(10,000 – 100,000)	02	(0)	(0–19)	
Romania	28,000 – 42,000	00–02	0	0–19	*48*
Russia	1,000,000 – 3,000,000	90–00	(0)	(0–19)	*122*
Serbia & MN	35,000 – 60,000	90–02	0	0–19	*1,155,225,227, 184,185,152*
Slovakia	40,000 – 80,000	90–99	0	0–19	
Slovenia	20,000 – 25,000	94	(0)	(0–19)	
Spain	(330,000 – 790,000)	92	?	–	*13,12,10*
Sweden	75,000 – 200,000	99–00	+	**10–19**	
Switzerland	80,000 – 120,000	98–02	+	0–9	
Turkey	(80,000 – 140,000)	01	(0)	(0–19)	
Ukraine	25,000 – 28,000	90–00	0	0–19	
UK	**222,500 – 222,500**	00	-	**11**	*5,31*
Total (approx.)	3,000,000 – 7,400,000	**Overall trend** Stable			
Breeding range	>5,000,000 km²	**Gen. length.** <3.3		**% Global pop.** 50–74	

(See p. 215, top, for data quality graph)

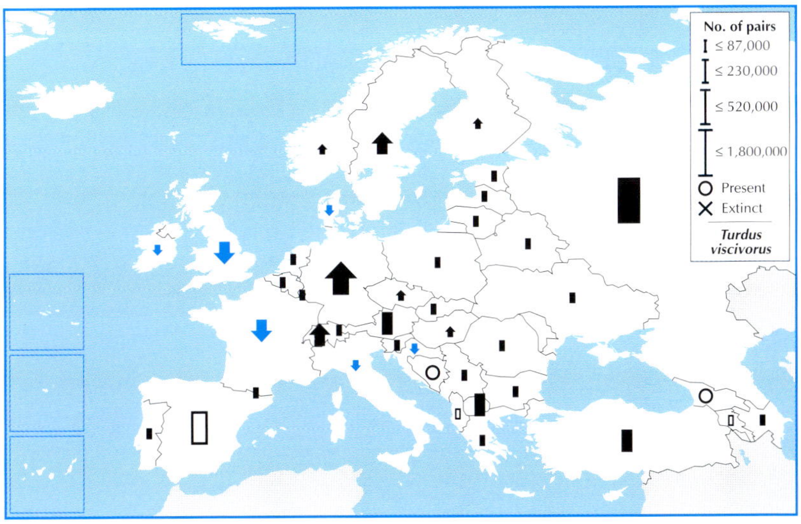

Cettia cetti
CETTI'S WARBLER

Non-SPEC (1994: —) **Status** Secure
Criteria —
European IUCN Red List Category —
Criteria —
Global IUCN Red List Category —
Criteria —

Cettia cetti is a widespread breeder across much of southern Europe, which accounts for less than half of its global breeding range. Its European breeding population is large (>600,000 pairs), and increased between 1970–1990. Although the species declined slightly in Greece during 1990–2000, key populations in Italy and Turkey were stable, and trends were increasing or stable across the rest of its European range. The species hence underwent a small increase overall, and consequently is evaluated as Secure.

Country	Breeding pop. size (pairs)	Year(s)	Trend	Mag.%	References
Albania	5,000 – 20,000	02	(0)	(0–19)	
Armenia	20,000 – 25,000	98–02	?	–	
Azerbaijan	(20,000 – 50,000)	96–00	(0)	(0–19)	
Belgium	**60 – 91**	01–02	+	**0–19**	*1*
Bosnia & HG	Present	90–03	?	–	
Bulgaria	**150 – 250**	96–02	+	0–19	
Croatia	(10,000 – 20,000)	02	(+)	(>80)	*16*
Cyprus	(10,000 – 20,000)	94–02	(0)	(0–19)	
France	(50,000 – 250,000)	98–02	+	100	*4,3*
Georgia	Present	03	?	–	
Greece	(50,000 – 200,000)	95–00	(–)	(0–19)	
Hungary	0 – 1	00–03	?	N	
Italy	(200,000 – 400,000)	03	(0)	(0–19)	
Macedonia	(3,000 – 10,000)	90–00	(0)	(0–19)	
Malta	300 – 500	95–02	+	0–19	*1*
Netherlands	**0 – 1**	98–00	?	–	*1*
Portugal	(10,000 – 100,000)	02	(0)	(0–19)	
Romania	(10 – 50)	00–02	(0)	(0–19)	*16,15*
Russia	(5,000 – 15,000)	90–00	+	20–29	*8*
Serbia & MN	100 – 200	90–02	F	20–29	*1,215,224,227,155*
Slovenia	**110 – 130**	98–00	0	0–19	
Spain	(20,000 – 100,000)	98–02	?	–	*10*
Switzerland	**0 – 3**	93–96	**0**	**0–19**	
Turkey	(200,000 – 400,000)	01	(0)	(0–19)	
UK	**534 – 534**	97–01	+	**144**	
Total (approx.)	600,000 – 1,600,000	**Overall trend** Small increase			
Breeding range	>1,000,000 km²	**Gen. length.** <3.3	**% Global pop.** 25–49		

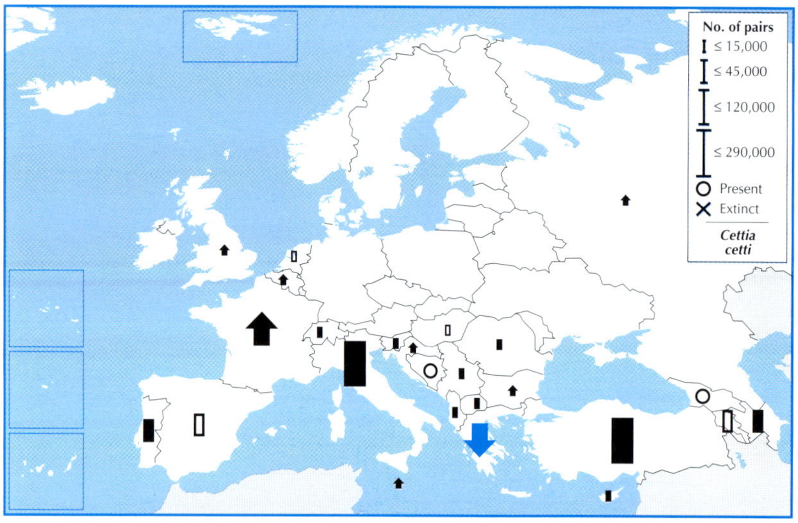

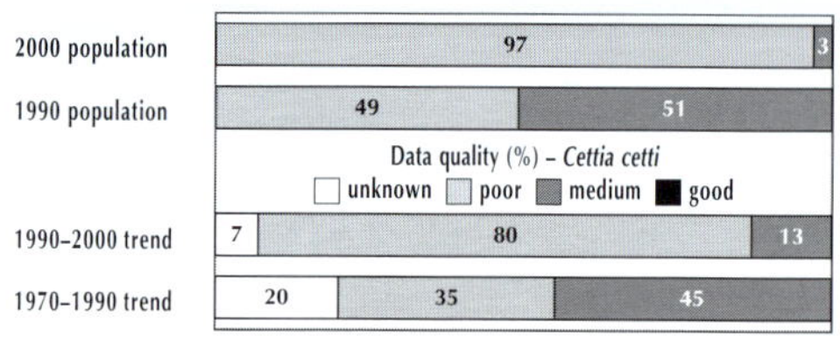

Cisticola juncidis
ZITTING CISTICOLA

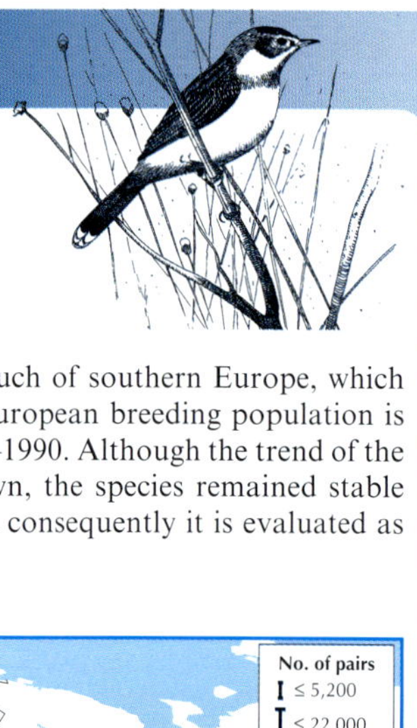

Non-SPEC (1994: —) **Status** Secure
Criteria —
European IUCN Red List Category —
Criteria —
Global IUCN Red List Category —
Criteria —

Cisticola juncidis is a widespread resident across much of southern Europe, which accounts for less than 5% of its global range. Its European breeding population is large (>230,000 pairs), and was stable between 1970–1990. Although the trend of the Spanish population during 1990–2000 was unknown, the species remained stable across the vast majority of its European range, and consequently it is evaluated as Secure.

Country	Breeding pop. size (pairs)	Year(s)	Trend	Mag.%	References
Albania	(1,000 – 3,000)	02	(0)	(0–19)	
Belgium	**14 – 22**	95–02	**0**	**0–19**	*1*
Croatia	(500 – 1,000)	02	(0)	(0–19)	*70,16*
Cyprus	(15,000 – 30,000)	94–02	(0)	(0–19)	
France	30,000 – 150,000	98–02	F	>80	*4*
Greece	(10,000 – 20,000)	95–00	(0)	(0–19)	
Italy	(100,000 – 300,000)	03	(0)	(0–19)	
Malta	1,000 – 2,000	90–02	**–**	**50–79**	*1*
Netherlands	**3 – 31**	98–00	?	–	*1*
Portugal	(50,000 – 500,000)	02	(0)	(0–19)	
Serbia & MN	(5 – 10)	98–02	(0)	(0–19)	*1,50*
Slovenia	**50 – 60**	98–00	0	0–19	
Spain	(20,000 – 100,000)	98–02	?	–	*10*
Turkey	(3,000 – 9,000)	01	(0)	(0–19)	
UK					
Gibraltar	**5 – 10**	00	+	**20–29**	
Total (approx.)	230,000 – 1,100,000	**Overall trend** Stable			
Breeding range	>1,000,000 km²	**Gen. length.** <3.3	**% Global pop.** <5		

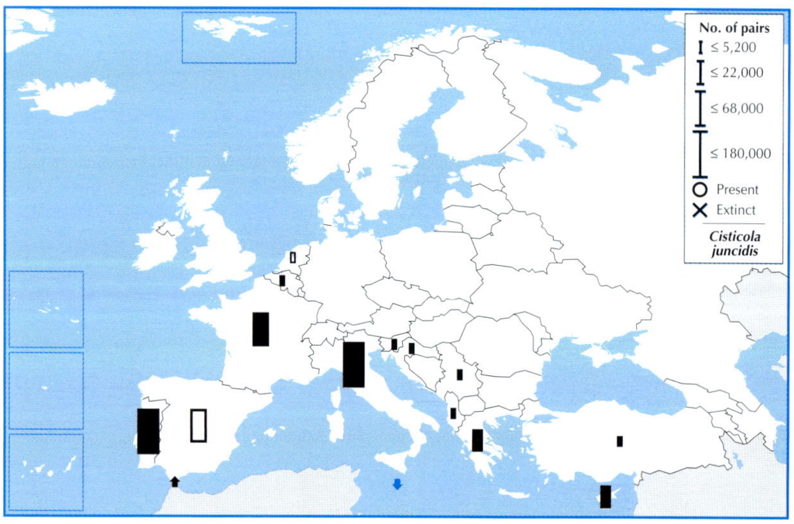

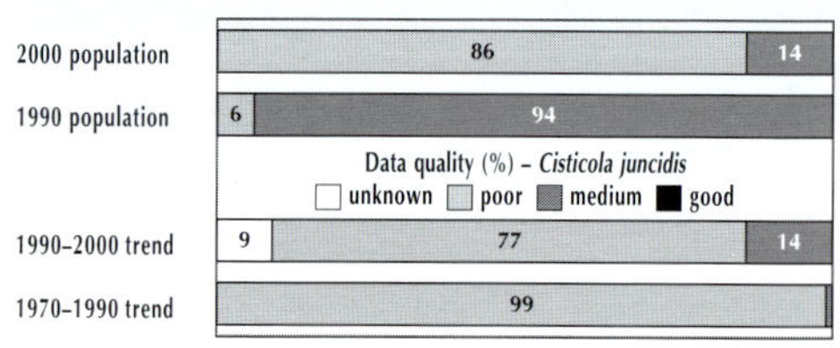

Prinia gracilis
GRACEFUL PRINIA

SPEC 3 (1994: —) **Status** (Vulnerable)
Criteria See IUCN below
European IUCN Red List Category VU
Criteria C1
Global IUCN Red List Category —
Criteria —

Country	Breeding pop. size (pairs)	Year(s)	Trend	Mag.%	References
Turkey	(1,000 – 4,000)	01	(-)	(10–19)	
Total (approx.)	1,000 – 4,000	Overall trend Moderate decline			
Breeding range	>20,000 km²	Gen. length <3.3		% Global pop. <5	

Prinia gracilis has a predominantly Middle Eastern distribution, which just extends into Europe in southern Turkey. The European breeding population is small (as few as 1,000 pairs), and was stable between 1970–1990, but underwent a moderate decline (>10%) during 1990–2000. As a consequence of its small population and this continuing decline, the species is provisionally evaluated as Vulnerable.

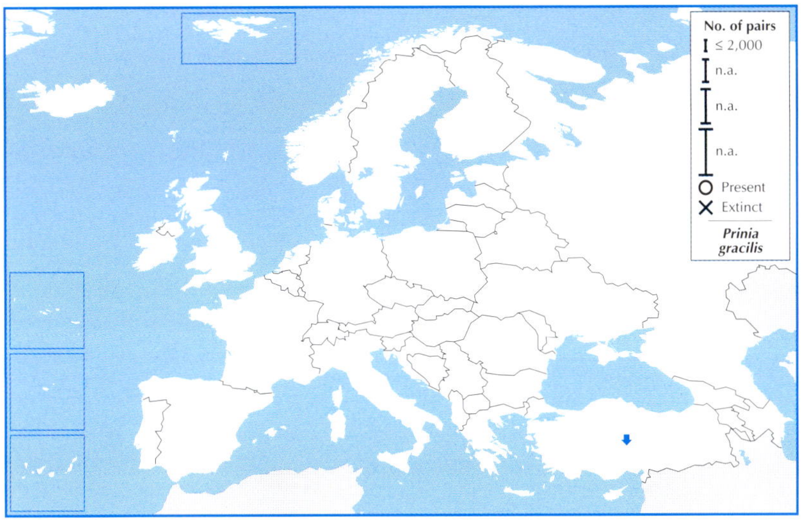

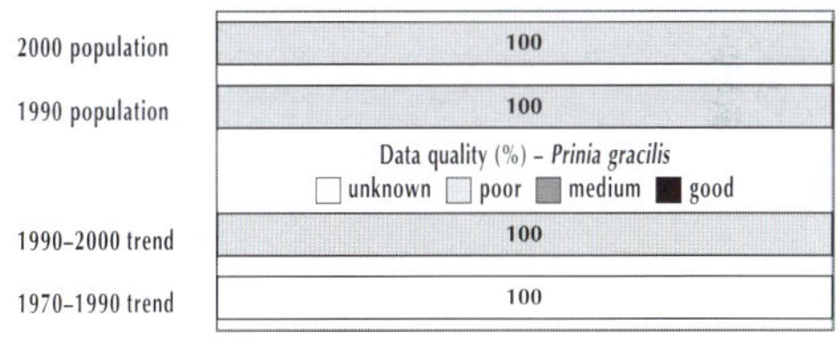

Locustella lanceolata
LANCEOLATED WARBLER

Non-SPEC (1994: —) **Status** (Secure)
Criteria —
European IUCN Red List Category —
Criteria —
Global IUCN Red List Category —
Criteria —

Country	Breeding pop. size (pairs)	Year(s)	Trend	Mag.%	References
Russia	(50,000 – 100,000)	90–00	?	-	*51,52*
Total (approx.)	50,000 – 100,000	Overall trend Unknown			
Breeding range	>100,000 km²	Gen. length <3.3		% Global pop. <5	

Locustella lanceolata has a predominantly Asian distribution, but its breeding range extends just west of the Urals into north-east European Russia. Its European breeding population is relatively small (<100,000 pairs), but was stable between 1970–1990. No trend data were available for the Russian population during 1990–2000, but there was no evidence to suggest that its status deteriorated significantly. Consequently, the species is provisionally evaluated as Secure.

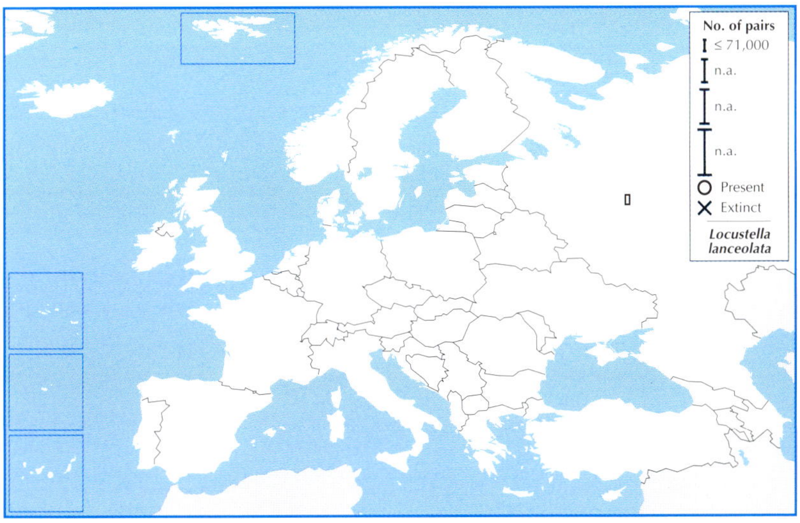

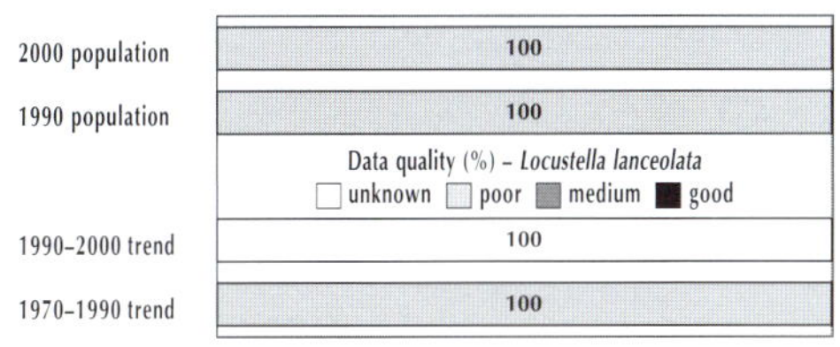

Locustella naevia
COMMON GRASSHOPPER-WARBLER

Non-SPEC[E] (1994: 4) **Status** (Secure)
Criteria —
European IUCN Red List Category —
Criteria —
Global IUCN Red List Category —
Criteria —

Locustella naevia is a fairly widespread summer visitor to much of Europe, which constitutes >50% of its global breeding range. Its European breeding population is large (>840,000 pairs), and was stable between 1970–1990. Although there were declines in a number of countries during 1990–2000, the sizeable population in Poland was stable, and the species increased or was stable elsewhere. The trend of the stronghold population in Russia was unknown, but the species probably remained stable overall, and consequently it is provisionally evaluated as Secure.

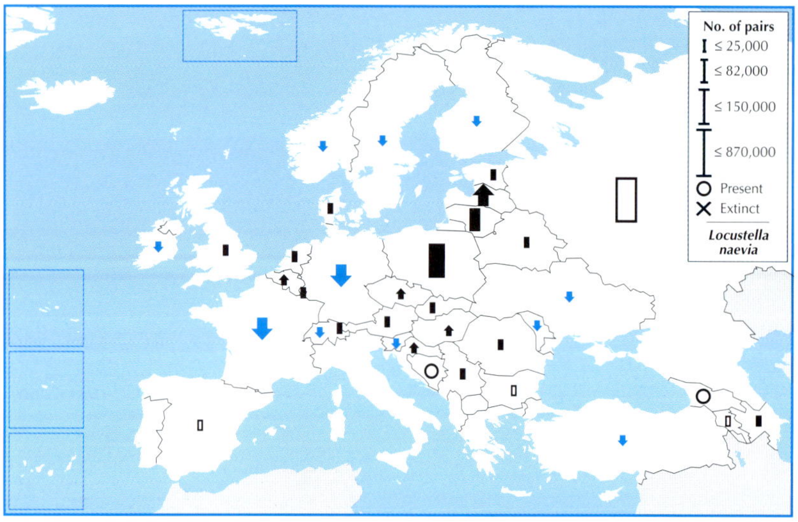

Country	Breeding pop. size (pairs)	Year(s)	Trend	Mag.%	References
Armenia	30 – 80	99–02	?	–	
Austria	(1,200 – 2,500)	98–02	(0)	(0–19)	
Azerbaijan	(1,000 – 10,000)	96–00	(0)	(0–19)	
Belarus	12,000 – 17,000	97–02	0	0–19	
Belgium	1,500 – 3,800	01–02	+	0–19	*1*
Bosnia & HG	Present	85–89	?	–	
Bulgaria	(0 – 2)	96–02	?	–	
Croatia	(1,000 – 2,000)	02	(+)	(>80)	*70,16,54*
Czech Rep.	17,000 – 34,000	00	+	10–19	
Denmark	1,000 – 1,000	93–96	0	0–19	*3*
Estonia	5,000 – 15,000	98	0	0–19	*1*
Finland	2,000 – 4,000	98–02	–	40	
France	(15,000 – 75,000)	98–02	(–)	(29)	*4,3*
Georgia	Present	03	?	–	
Germany	55,000 – 120,000	95–99	–	0–19	
Hungary	9,000 – 30,000	99–02	+	50–79	*19*
Rep. Ireland	2,500 – 10,000	88–91	–	0–19	
Latvia	30,000 – 80,000	90–00	+	50–79	*23,16,2*
Liechtenstein	**25 – 30**	98–00	**0**	**0–19**	
Lithuania	50,000 – 60,000	99–01	(0)	(0–19)	*20*
Luxembourg	350 – 500	02	0	0–19	
Moldova	500 – 800	90–00	–	20–29	
Netherlands	**4,000 – 6,000**	98–00	**F**	**24**	*1*
Norway	**30 – 100**	99	**–**	**0–19**	*63*
Poland	**100,000 – 200,000**	00–02	(0)	(0–19)	23
Romania	(3,000 – 5,000)	00–02	(0)	(0–19)	
Russia	500,000 – 1,500,000	90–00	?	–	*122*
Serbia & MN	30 – 40	98–02	F	20–29	*1,29,78,155*
Slovakia	300 – 500	90–99	0	0–19	
Slovenia	150 – 300	93–00	(–)	(0–19)	
Spain	(5,000 – 10,000)	98–02	?	–	*10,16*
Sweden	3,000 – 6,000	99–00	–	30–49	
Switzerland	**200 – 250**	93–96	–	0–9	
Turkey	(50 – 500)	01	(–)	(–)	
Ukraine	5,000 – 8,000	90–00	–	0–19	
UK	**12,300 – 12,300**	00	**0**	**5**	*22*
Total (approx.)	840,000 – 2,200,000	**Overall trend** Unknown			
Breeding range	>5,000,000 km²	**Gen. length.** <3.3		**% Global pop.** 50–74	

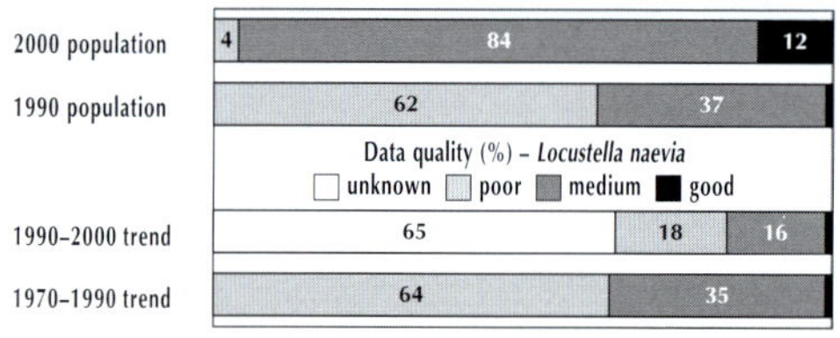

Locustella fluviatilis
EURASIAN RIVER WARBLER

Non-SPEC[E] (1994: 4) **Status** (Secure)
Criteria —
European IUCN Red List Category —
Criteria —
Global IUCN Red List Category —
Criteria —

Locustella fluviatilis is a widespread summer visitor to central and eastern Europe, which constitutes >75% of its global breeding range. Its European breeding population is very large (>1,900,000 pairs), and was stable between 1970–1990. The species remained stable overall during 1990–2000, with the stronghold population in Russia probably stable, and stable or increasing trends across much of the rest of Europe. Consequently, the species is provisionally evaluated as Secure.

Country	Breeding pop. size (pairs)	Year(s)	Trend	Mag.%	References
Austria	(4,000 – 8,000)	98–02	(0)	(0–19)	
Belarus	100,000 – 140,000	97–02	0	0–19	
Bosnia & HG	Present	85–89	?	–	
Bulgaria	(0 – 10)	96–02	?	–	
Croatia	(3,000 – 4,000)	02	(–)	(30–49)	*57*
Czech Rep.	10,000 – 20,000	00	0	0–19	
Denmark	0 – 34	95	(F)	(>80)	*3*
Estonia	4,000 – 10,000	98	+	20–29	*1*
Finland	500 – 1,000	98–00	+	100	
Germany	3,500 – 10,000	95–99	+	0–19	
Hungary	40,000 – 75,000	99–02	0	0–19	*19*
Latvia	50,000 – 100,000	90–00	0	0–19	*23,16,2*
Lithuania	30,000 – 50,000	99–01	(–)	(0–19)	*20*
Moldova	600 – 800	90–00	–	20–29	
Poland	**50,000 – 80,000**	00–02	(0)	(0–19)	*23*
Romania	(23,000 – 35,000)	00–02	(–)	(0–19)	
Russia	1,500,000 – 4,000,000	90–00	(0)	(0–19)	*122*
Serbia & MN	400 – 500	95–02	–	10–29	*1,29,155*
Slovakia	10,000 – 15,000	90–99	0	0–19	
Slovenia	1,500 – 2,500	99–00	(0)	(0–19)	
Sweden	**40 – 60**	99–00	+	10–19	
Ukraine	45,000 – 60,000	90–00	0	0–9	
Total (approx.)	1,900,000 – 4,600,000	**Overall trend** Stable			
Breeding range	>4,000,000 km²	**Gen. length.** <3.3		**% Global pop.** 75–94	

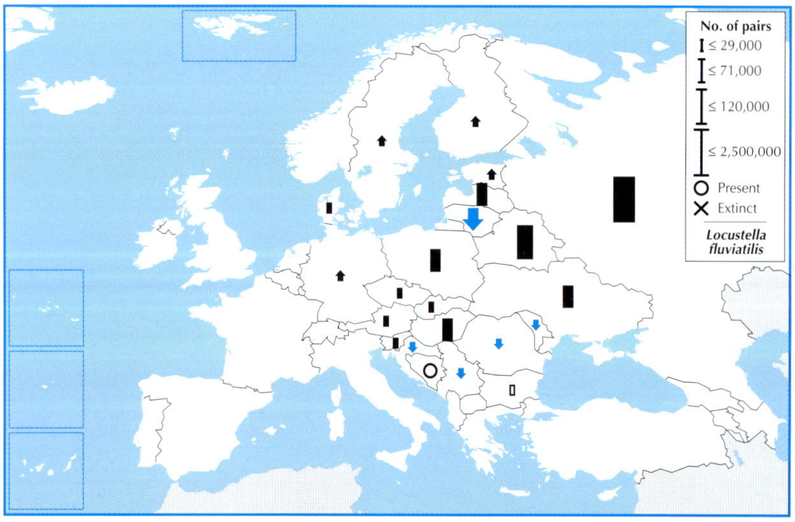

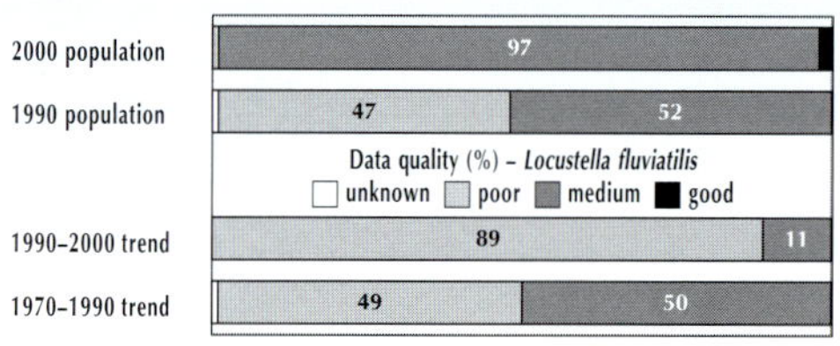

Locustella luscinioides
SAVI'S WARBLER

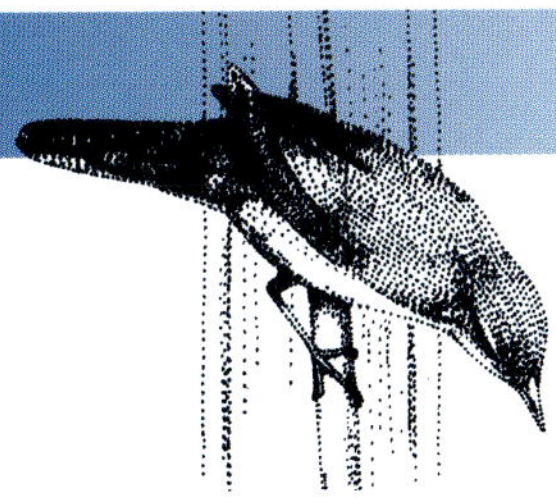

Non-SPEC[E] (1994: 4) **Status** (Secure)
Criteria —
European IUCN Red List Category —
Criteria —
Global IUCN Red List Category —
Criteria —

Locustella luscinioides has a widespread but patchy breeding distribution in Europe, which constitutes >50% of its global breeding range. Its European breeding population is large (>530,000 pairs), and was stable between 1970–1990. Although there were declines in some marginal populations during 1990–2000, key populations in the core of the breeding range—notably in Romania—were stable, and the species probably remained stable overall. Consequently, the species is provisionally evaluated as Secure.

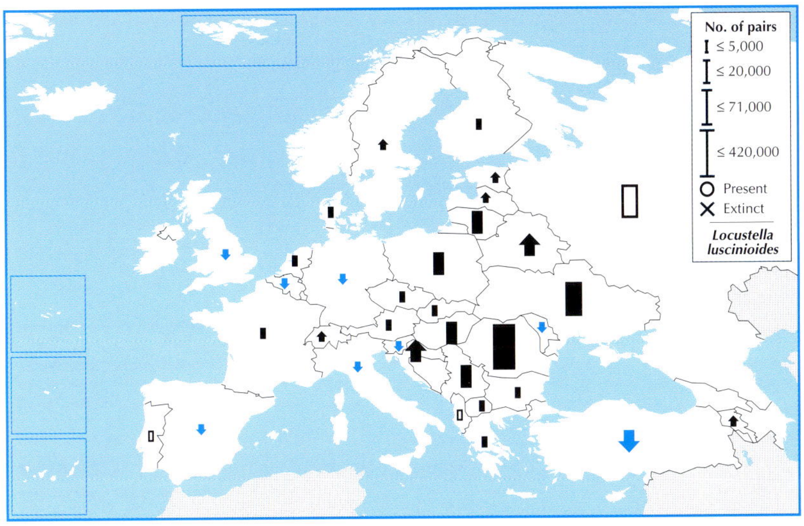

Country	Breeding pop. size (pairs)	Year(s)	Trend	Mag.%	References
Albania	(50 – 200)	02	?	–	
Armenia	0 – 50	00–02	+	30–49	*9*
Austria	1,700 – 2,400	98–02	(0)	(0–19)	
Belarus	6,000 – 10,000	97–02	(+)	(0–9)	
Belgium	**30 – 37**	95–02	**–**	**0–19**	*1*
Bulgaria	**300 – 500**	96–02	0	0–9	
Croatia	(5,000 – 10,000)	02	(+)	(>80)	*70,16,54*
Czech Rep.	400 – 800	00	0	0–9	
Denmark	(5 – 28)	98–01	(F)	(–)	*4,5,6,7*
Estonia	150 – 300	98	+	50–79	*1*
Finland	0 – 3	98–02	**0**	**0–19**	
France	(2,000 – 10,000)	98–02	(0)	(0–19)	*4*
Germany	3,300 – 7,500	95–99	–	0–19	
Greece	(500 – 2,000)	95–00	(0)	(0–19)	
Hungary	13,000 – 30,000	99–02	0	0–19	*19*
Italy	(1,000 – 3,000)	03	(–)	(0–19)	
Latvia	400 – 800	90–00	+	0–19	*23,17*
Lithuania	(6,000 – 10,000)	99–01	(0)	(0–19)	*20*
Macedonia	(50 – 250)	98–00	(0)	(0–19)	
Moldova	600 – 800	90–00	–	20–29	
Netherlands	**1,700 – 2,100**	98–00	**0**	**7**	*1*
Poland	10,000 – 30,000	00–02	(0)	(0–19)	*23*
Portugal	(50 – 500)	02	?	–	
Romania	(360,000 – 480,000)	00–02	(0)	(0–19)	
Russia	(50,000 – 100,000)	90–00	?	–	*3,8,32,114*
Serbia & MN	6,300 – 7,500	95–02	0	0–19	*1,200,29,155,78, 203,243,144,227*
Slovakia	1,000 – 2,000	90–99	0	0–19	
Slovenia	100 – 200	99–00	(–)	(0–19)	
Spain	(1,000 – 2,500)	98–02	(–)	(0–19)	*10,16*
Sweden	**5 – 10**	99–00	**+**	**30–49**	
Switzerland	**250 – 300**	93–96	+	30–49	
Turkey	(4,000 – 9,000)	01	(–)	(0–19)	
Ukraine	55,000 – 75,000	90–00	0	5–14	
UK	**0 – 4**	96–00	**–**	**71**	
Total (approx.)	530,000 – 800,000	**Overall trend** Stable			
Breeding range	>3,000,000 km²	**Gen. length.** <3.3		**% Global pop.** 50–74	

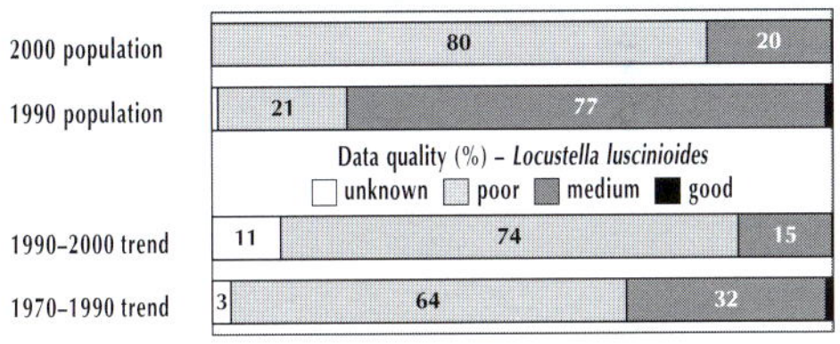

Acrocephalus melanopogon
MOUSTACHED WARBLER

Non-SPEC (1994: —) **Status** (Secure)
Criteria —
European IUCN Red List Category —
Criteria —
Global IUCN Red List Category —
Criteria —

Acrocephalus melanopogon is a widespread but patchily distributed breeder in southern and eastern Europe, which accounts for less than half of its global breeding range. Its European breeding population is large (>150,000 pairs), and was stable between 1970–1990. Although the trend in its Russian stronghold during 1990–2000 was unknown, populations were stable across most of the rest of its European range—including the sizeable one in Romania—and the species remained stable overall. Consequently, it is provisionally evaluated as Secure.

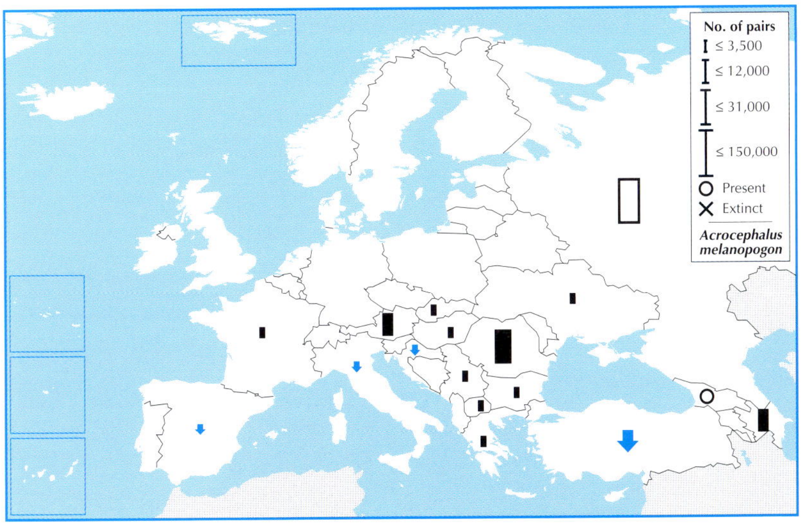

Country	Breeding pop. size (pairs)	Year(s)	Trend	Mag.%	References
Austria	9,000 – 16,000	98–02	(0)	(0–19)	
Azerbaijan	(5,000 – 15,000)	96–00	(0)	(0–19)	
Bulgaria	0 – 5	96–02	(F)	(>80)	
Croatia	10 – 20	02	(–)	(>80)	*16,70,57*
France	2,000 – 6,000	98–02	(0)	(0–19)	*4*
Georgia	Present	03	?	–	
Greece	(50 – 200)	95–00	(0)	(0–19)	
Hungary	600 – 1,000	98–01	0	0–19	
Italy	600 – 1,000	03	–	0–19	
Macedonia	10 – 100	90–00	(0)	(0–19)	
Romania	(24,000 – 40,000)	00–02	(0)	(0–19)	
Russia	(100,000 – 200,000)	90–00	?	–	*8,92*
Serbia & MN	**70 – 100**	97–02	0	0–19	*1,21,29,155,78*
Slovakia	**10 – 20**	90–99	0	0–19	
Spain	(1,000 – 2,500)	98–02	(–)	(0–19)	*10,16*
Turkey	(8,000 – 12,000)	01	(–)	(0–19)	
Ukraine	**1,000 – 2,000**	99–00	(0)	(0–19)	
Total (approx.)	150,000 – 300,000	**Overall trend** Stable			
Breeding range	>250,000 km²	**Gen. length.** <3.3		**% Global pop.** 25–49	

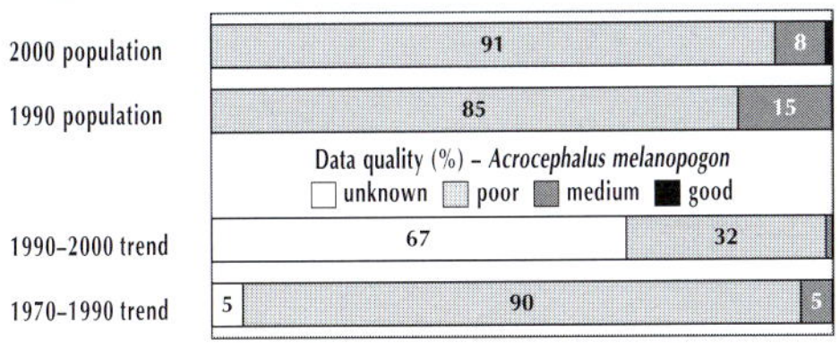

Acrocephalus paludicola
AQUATIC WARBLER

SPEC 1 (1994: 1) **Status** (Vulnerable)
Criteria See IUCN below
European IUCN Red List Category VU
Criteria A3c
Global IUCN Red List Category VU
Criteria A2c; A3c

Country	Breeding pop. size (pairs)	Year(s)	Trend	Mag.%	References
Belarus	**6,500 – 12,500**	97–02	**F**	**20–49**	
Germany	12 – 15	95–99	**–**	**50–79**	
Hungary	**350 – 700**	98–01	**+**	**>80**	
Latvia	0 – 10	90–00	(0)	(0–19)	*17,10,26*
Lithuania	**250 – 300**	99–01	–	50–79	*20*
Poland	**2,700 – 2,750**	03	**–**	**20–29**	*96,99,100*
Russia	**5 – 50**	90–02	F	20–29	*59,128,173*
Ukraine	**2,600 – 3,400**	99–00	**0**	**0–9**	
Total (approx.)	12,000 – 20,000	**Overall trend** Moderate decline			
Breeding range	>50,000 km²	**Gen. length.** <3.3	**% Global pop.** >95		

Acrocephalus paludicola is a widespread summer visitor to north-central and eastern Europe, which constitutes >95% of its global breeding range. Its European breeding population is relatively small (<20,000 pairs), and underwent a large decline between 1970–1990. Although key populations in Belarus and Ukraine fluctuated during 1990–2000, the species continued to decline in Poland, Lithuania and Germany, and is predicted to decline further (>30%) owing to ongoing habitat loss. Consequently, this globally threatened species is provisionally evaluated as Vulnerable in Europe.

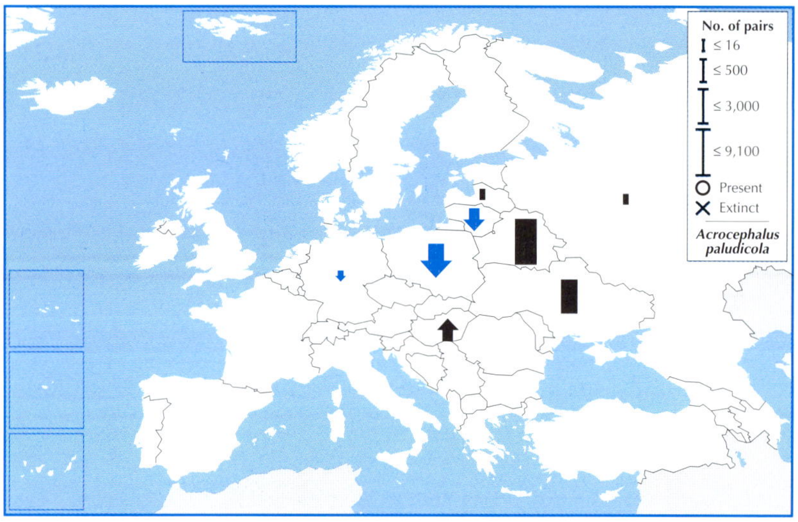

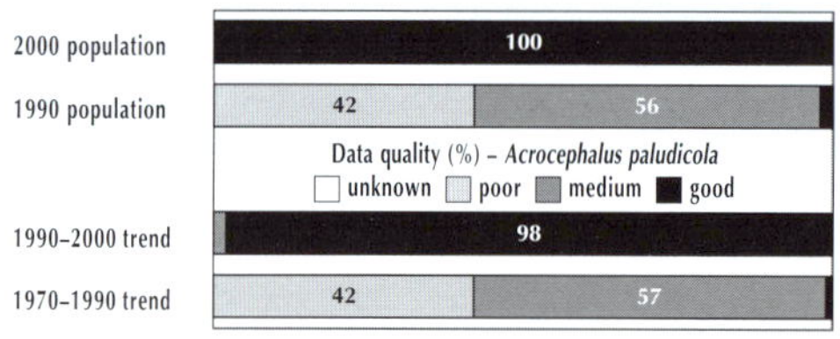

Acrocephalus schoenobaenus
SEDGE WARBLER

Non-SPEC[E] (1994: 4) **Status** Secure
Criteria —
European IUCN Red List Category —
Criteria —
Global IUCN Red List Category —
Criteria —

Country	Breeding pop. size (pairs)	Year(s)	Trend	Mag.%	References
Albania	(50 – 200)	02	(0)	(0–19)	
Armenia	6500 – 16,000	98–02	?	–	
Austria	(7,000 – 13,000)	98–02	(0)	(0–19)	
Azerbaijan	(5,000 – 50,000)	96–00	(0)	(0–19)	
Belarus	600,000 – 900,000	97–02	0	0–19	
Belgium	**2,200 – 2,800**	01–02	**–**	**0–19**	*1*
Bosnia & HG	Present	85–89	?	–	
Bulgaria	100 – 500	96–02	(F)	(20–29)	
Croatia	(10,000 – 15,000)	02	(–)	(20–29)	*16*
Czech Rep.	40,000 – 80,000	00	0	0–19	
Denmark	3,000 – 4,000	00	**–**	**>80**	*12*
Estonia	70,000 – 120,000	98	0	0–19	*1*
Finland	200,000 – 400,000	98–02	–	25	
France	30,000 – 120,000	98–02	+	100	*4,3*
Georgia	Present	03	?	–	
Germany	6,000 – 12,000	95–99	–	30–49	
Greece	(500 – 1,000)	95–00	(–)	(0–19)	
Hungary	170,000 – 325,000	99–02	0	0–19	*19*
Rep. Ireland	20,000 – 100,000	88–91	+	0–19	
Italy	(30 – 100)	03	(–)	(0–9)	
Latvia	80,000 – 200,000	90–00	0	0–19	*23,17,2*
Lithuania	300,000 – 400,000	99–01	(0)	(0–19)	*20*
Luxembourg	**0 – 0**	00	**–**	**X**	
Macedonia	(250 – 1,000)	00	(0)	(0–19)	
Moldova	500 – 700	90–00	–	20–29	
Netherlands	**20,000 – 25,000**	98–00	**+**	**108**	*1*
Norway	(20,000 – 70,000)	93–03	+	0–19	
Poland	**60,000 – 120,000**	00–02	0	0–19	*23*
Romania	(850,000 – 1,100,000)	00–02	(0)	(0–19)	
Russia	1,300,000 – 2,500,000	90–00	0	0–19	*122*
Serbia & MN	12,000 – 15,000	95–02	+	0–19	*1,200,29,78, 155,243,227*
Slovakia	10,000 – 16,000	90–99	0	0–19	
Slovenia	500 – 1,000	99–00	(–)	(0–19)	
Sweden	**50,000 – 200,000**	99–00	**+**	**42**	
Turkey	(5,000 – 10,000)	01	(–)	(20–29)	
Ukraine	200,000 – 300,000	99–00	0	0–19	
UK	**321,000 – 321,000**	00	**+**	**19**	*5,31*
Total (approx.)	4,400,000 – 7,400,000	**Overall trend** Stable			
Breeding range	>6,000,000 km²	**Gen. length.** <3.3	**% Global pop.** 50–74		

Acrocephalus schoenobaenus is a widespread summer visitor to much of Europe, which constitutes >50% of its global breeding range. Its European breeding population is very large (>4,400,000 pairs), and was stable between 1970–1990. Although there were declines in certain countries—most notably Finland—during 1990–2000, key populations to the east in Russia, Romania and Belarus were stable, and the species remained stable overall. Consequently, it is evaluated as Secure.

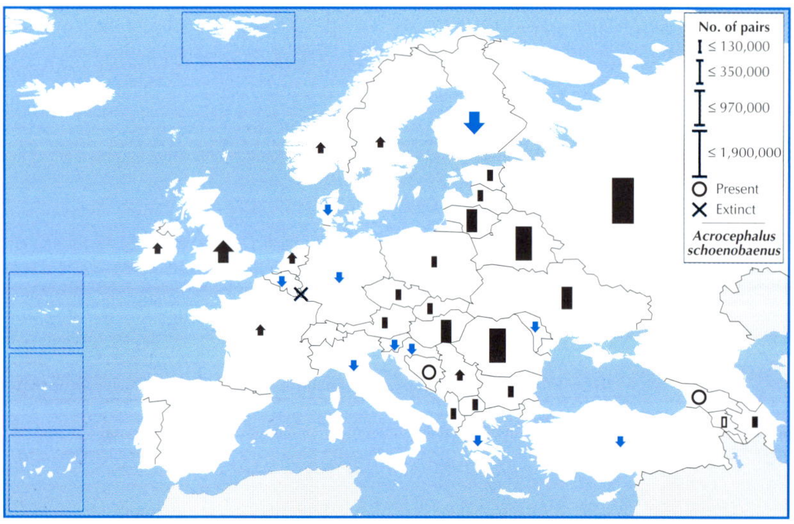

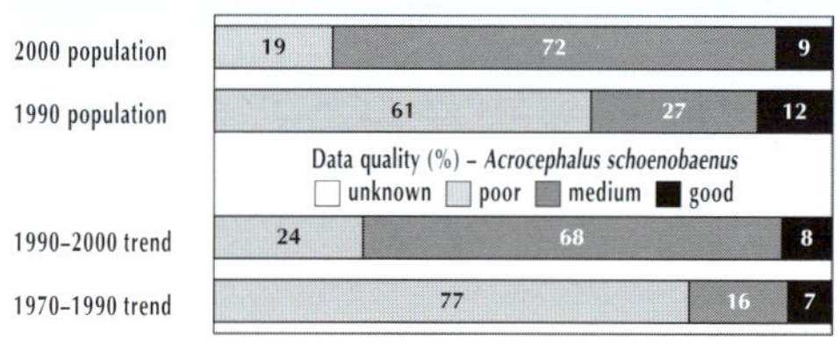

Acrocephalus agricola
PADDYFIELD WARBLER

Non-SPEC (1994: —) **Status** (Secure)
Criteria —
European IUCN Red List Category —
Criteria —
Global IUCN Red List Category —
Criteria —

Country	Breeding pop. size (pairs)	Year(s)	Trend	Mag.%	References
Armenia	50 – 250	99–02	+	20–29	*10*
Bulgaria	**45 – 70**	96–00	0	0–9	
Georgia	Present	03	?	–	
Romania	(350,000 – 650,000)	00–02	(0)	(0–19)	
Russia	(100,000 – 200,000)	90–00	(+)	(20–29)	*8,32,92*
Turkey	(50 – 500)	01	(–)	(20–29)	
Ukraine	(17,000 – 25,000)	99–00	+	10–19	
Total (approx.)	470,000 – 880,000	**Overall trend** Small increase			
Breeding range	>250,000 km²	**Gen. length.** <3.3		**% Global pop.** <5	

Acrocephalus agricola is a summer visitor to the eastern European countries adjoining the Black Sea, with Europe accounting for a tiny proportion of its global breeding range. Its European breeding population is large (>470,000 pairs), and increased between 1970–1990. Although the small population in Turkey declined during 1990–2000, key populations in Romania and Russia were stable or increased respectively, and the species probably increased slightly overall. Consequently, it is provisionally evaluated as Secure.

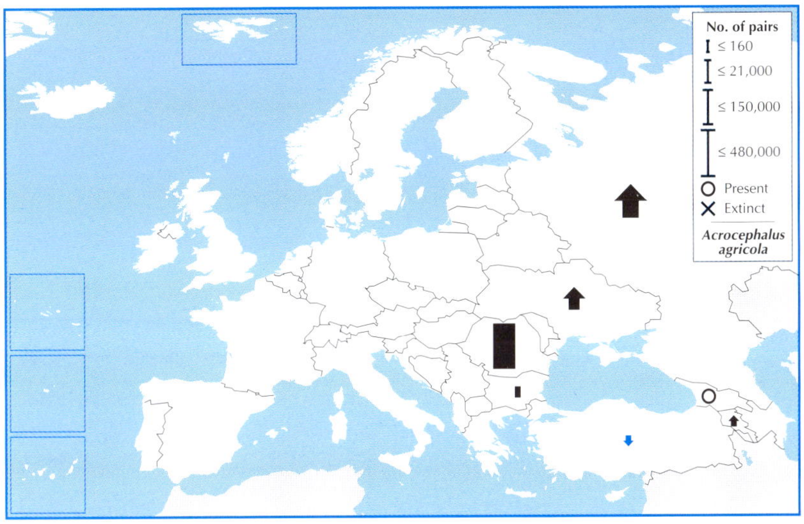

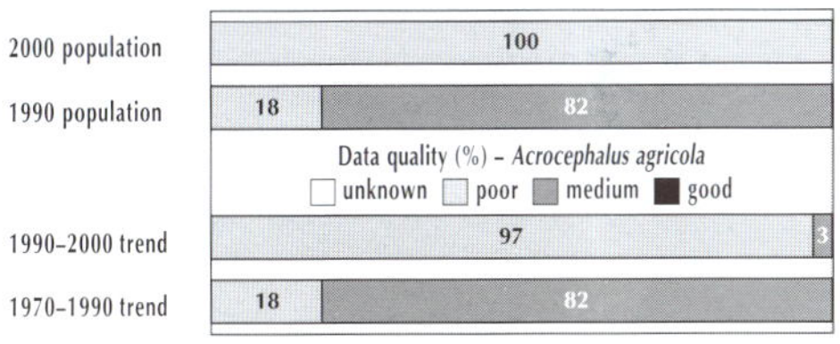

Acrocephalus dumetorum
BLYTH'S REED-WARBLER

Non-SPEC (1994: —) **Status** Secure
Criteria —
European IUCN Red List Category —
Criteria —
Global IUCN Red List Category —
Criteria —

Country	Breeding pop. size (pairs)	Year(s)	Trend	Mag.%	References
Belarus	1,000 – 2,000	97–02	0	0–19	
Estonia	2,000 – 4,000	98	0	0–19	*1*
Finland	5,000 – 8,000	98–02	0	0–19	
Latvia	(3,000 – 6,000)	90–00	(0)	(0–19)	*34*
Lithuania	(200 – 300)	99–01	(0)	(0–19)	*20*
Poland	0 – 50	95–00	+	N	
Russia	2,000,000 – 5,000,000	90–00	0	0–19	*122*
Sweden	**5 – 15**	99–00	+	**20–29**	
Ukraine	(10 – 50)	99–00	(+)	(50–79)	
Total (approx.)	2,000,000 – 5,000,000	**Overall trend** Stable			
Breeding range	>1,000,000 km²	**Gen. length.** <3.3		**% Global pop.** 25–49	

Acrocephalus dumetorum is a widespread summer visitor to central Russia and adjoining areas of northern and central Europe, which accounts for less than half of its global breeding range. Its European breeding population is very large (>2,000,000 pairs), and was stable between 1970–1990. The stronghold population in Russia remained stable during 1990–2000, with trends also stable or increasing elsewhere in Europe, and consequently the species is evaluated as Secure.

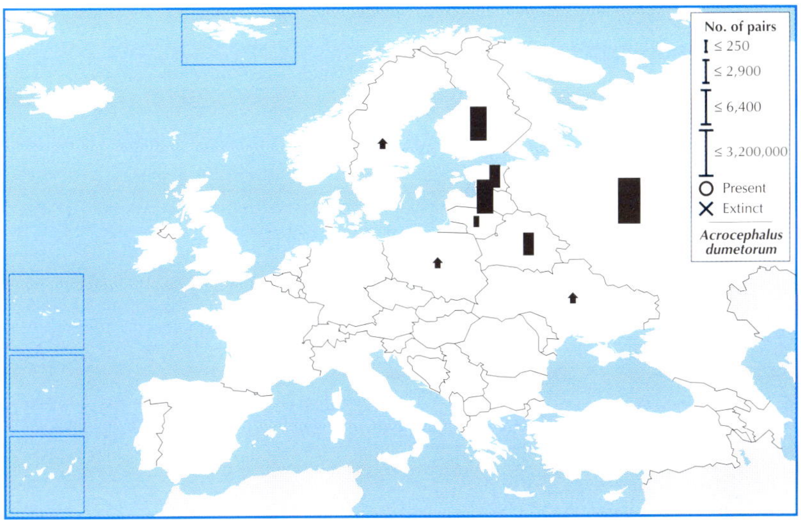

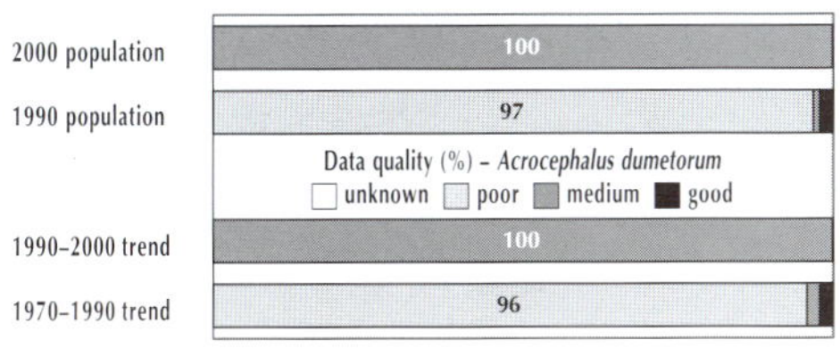

Acrocephalus palustris
MARSH WARBLER

Non-SPEC[E] (1994: 4) **Status** (Secure)
Criteria —
European IUCN Red List Category —
Criteria —
Global IUCN Red List Category —
Criteria —

Acrocephalus palustris is a widespread summer visitor to central and eastern Europe, which constitutes >75% of its global breeding range. Its European breeding population is very large (>3,200,000 pairs), and was stable between 1970–1990. Although there were declines in Germany during 1990–2000, other populations increased or remained stable—including the key populations in Poland, Romania and Russia—and the species probably remained stable overall. Consequently, it is provisionally evaluated as Secure.

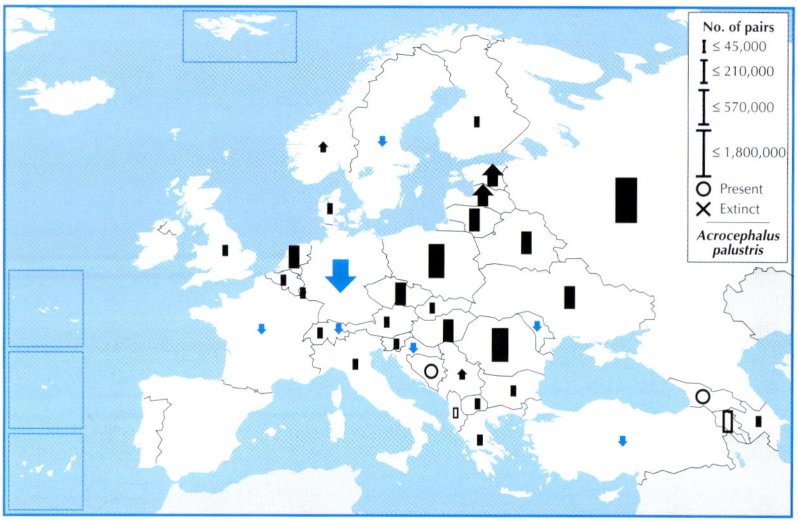

Country	Breeding pop. size (pairs)	Year(s)	Trend	Mag.%	References
Albania	(0 – 50)	02	?	–	
Armenia	30,000 – 150,000	99–02	?	–	
Austria	(20,000 – 40,000)	98–02	(0)	(0–19)	
Azerbaijan	(1,000 – 10,000)	96–00	(0)	(0–19)	
Belarus	80,000 – 130,000	97–02	0	0–19	
Belgium	8,000 – 20,000	01–02	(0)	(0–19)	*1*
Bosnia & HG	Present	85–89	?	–	
Bulgaria	3,000 – 7,000	96–02	0	0–19	
Croatia	(25,000 – 50,000)	02	(–)	(50–79)	*16,54,57*
Czech Rep.	80,000 – 160,000	00	0	0–19	
Denmark	30,000 – 40,000	00	0	0–19	*12*
Estonia	(50,000 – 100,000)	98	+	20–29	*1*
Finland	4,000 – 6,000	98–02	0	0–19	
France	(20,000 – 80,000)	98–02	–	34	*4,3*
Georgia	Present	03	?	–	
Germany	400,000 – 800,000	95–99	–	0–19	
Greece	(200 – 2,000)	95–00	(0)	(0–19)	
Hungary	65,000 – 130,000	99–02	0	0–19	*19*
Italy	(10,000 – 30,000)	03	(0)	(0–19)	
Latvia	70,000 – 120,000	90–00	+	30–49	*23,16,2*
Liechtenstein	80 – 100	00–02	–	0–9	
Lithuania	100,000 – 150,000	99–01	(0)	(0–19)	*20*
Luxembourg	800 – 1,000	02	0	0–19	
Macedonia	(50 – 250)	90–00	(0)	(0–19)	
Moldova	2,000 – 2,500	90–00	–	0–19	
Netherlands	**70,000 – 110,000**	98–00	**0**	**6**	*1*
Norway	400 – 900	02	**+**	**0–19**	
Poland	**400,000 – 600,000**	00–02	(0)	(0–19)	*23*
Romania	(450,000 – 650,000)	00–02	(0)	(0–19)	
Russia	1,000,000 – 3,000,000	90–00	(0)	(0–19)	*122*
Serbia & MN	19,000 – 24,000	95–02	+	0–9	*1,200,29,78,155, 203,101,152*
Slovakia	40,000 – 50,000	90–99	0	0–19	
Slovenia	5,000 – 10,000	94	(0)	(0–19)	
Sweden	**15,000 – 20,000**	99–00	(–)	(0–19)	
Switzerland	**3,000 – 6,000**	93–96	0	0–19	
Turkey	(8,000 – 25,000)	01	(–)	(0–19)	
Ukraine	180,000 – 230,000	90–00	(0)	(0–19)	
UK	**4 – 28**	96–00	**0**	**4**	
Total (approx.)	3,200,000 – 6,800,000	**Overall trend** Stable			
Breeding range	>5,000,000 km²	**Gen. length.** <3.3		**% Global pop.** 75–94	

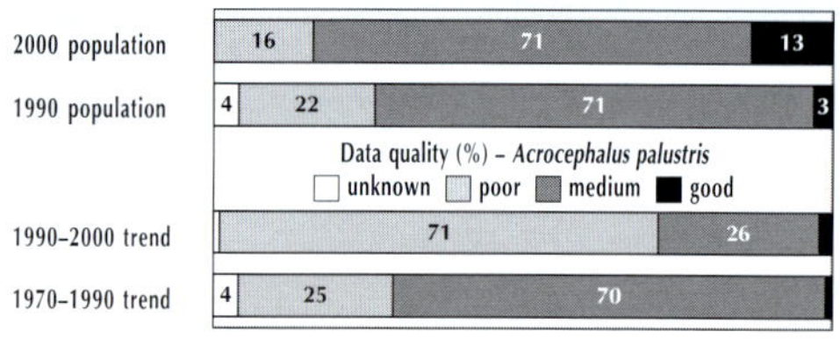

Acrocephalus scirpaceus
COMMON REED-WARBLER

Non-SPEC[E] (1994: 4) **Status** Secure
Criteria —
European IUCN Red List Category —
Criteria —
Global IUCN Red List Category —
Criteria —

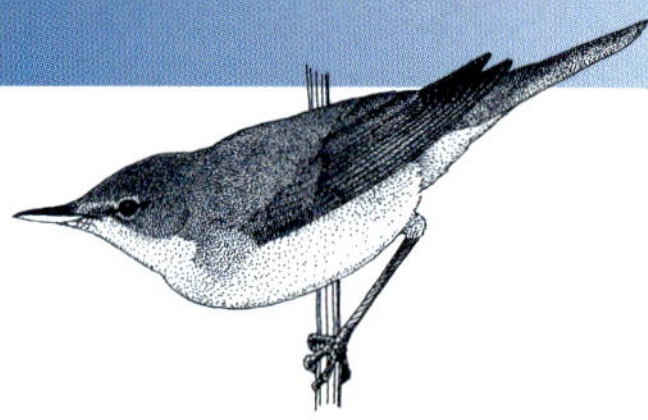

Acrocephalus scirpaceus is a widespread summer visitor to much of Europe, which constitutes >50% of its global breeding range. Its European breeding population is very large (>2,700,000 pairs), and was stable between 1970–1990. The species remained stable overall during 1990–2000, with the majority of national trends stable or increasing—including sizeable populations in Romania and Sweden. Consequently, the species is evaluated as Secure.

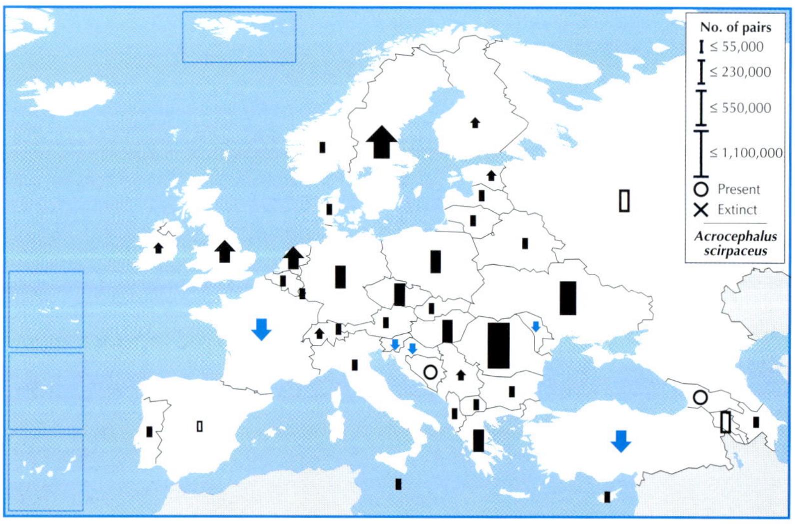

Country	Breeding pop. size (pairs)	Year(s)	Trend	Mag.%	References
Albania	1,000 – 3,000	96–02	(0)	(0–19)	
Armenia	30,000 – 150,000	99–02	?	–	
Austria	(20,000 – 40,000)	98–02	(0)	(0–19)	
Azerbaijan	(1,000 – 10,000)	96–00	(0)	(0–19)	
Belarus	15,000 – 25,000	97–02	0	0–19	
Belgium	3,500 – 7,000	01–02	0	0–19	*1*
Bosnia & HG	Present	90–03	?	–	
Bulgaria	**1,200 – 2,500**	96–02	0	0–9	
Croatia	(5,000 – 10,000)	02	(–)	(0–19)	*16,57*
Cyprus	(100 – 300)	94–02	(0)	(0–19)	
Czech Rep.	50,000 – 100,000	00	0	0–19	
Denmark	50,000 – 60,000	00	(0)	(0–19)	*12*
Estonia	20,000 – 40,000	98	+	20–29	*1*
Finland	20,000 – 30,000	98–02	+	60	
France	(40,000 – 200,000)	98–02	–	19	*4,3*
Georgia	Present	03	?	–	
Germany	120,000 – 250,000	95–99	0	0–19	
Greece	(50,000 – 100,000)	95–00	(0)	(0–19)	
Hungary	75,000 – 160,000	99–02	0	0–19	*19*
Rep. Ireland	55 – 97	95–96	+	0–19	
Italy	(30,000 – 60,000)	03	(0)	(0–19)	
Latvia	20,000 – 40,000	90–00	(0)	(0–19)	*23,17*
Liechtenstein	8 – 10	00–01	0	0–19	
Lithuania	30,000 – 50,000	99–01	(0)	(0–19)	*20*
Luxembourg	**200 – 300**	02	**0**	**0–19**	
Macedonia	(2,000 – 5,000)	90–00	(0)	(0–19)	
Malta	**4 – 7**	90–02	**0**	**0–19**	*1*
Moldova	1,000 – 1,200	90–00	–	20–29	
Netherlands	**150,000 – 250,000**	98–00	**+**	**20**	*1*
Norway	**1,600 – 2,400**	00–03	**0**	**0–9**	*25*
Poland	(40,000 – 200,000)	90–00	0	0–19	*1,23*
Portugal	(1,000 – 10,000)	02	(0)	(0–19)	
Romania	(800,000 – 1,300,000)	00–02	(0)	(0–19)	
Russia	(100,000 – 500,000)	90–00	?	–	*8,92,114*
Serbia & MN	15,000 – 20,000	95–02	+	0–19	*1,29,155,78, 227,243*
Slovakia	10,000 – 20,000	90–99	0	0–19	
Slovenia	150 – 250	99–00	(–)	(0–19)	
Spain	(20,000 – 100,000)	98–02	?	–	*10*
Sweden	**500,000 – 600,000**	99–00	**+**	**3**	
Switzerland	**7,000 – 9,000**	93–96	**+**	**10–19**	
Turkey	(80,000 – 160,000)	01	(–)	(0–19)	
Ukraine	(280,000 – 400,000)	99–00	(0)	(0–9)	
UK	**60,800 – 122,000**	00	**+**	**52**	*5,31*
Total (approx.)	2,700,000 – 5,000,000	**Overall trend** Stable			
Breeding range	>5,000,000 km²	**Gen. length.** <3.3		**% Global pop.** 50–74	

(See p. 223, bottom, for data quality graph)

Acrocephalus arundinaceus
GREAT REED-WARBLER

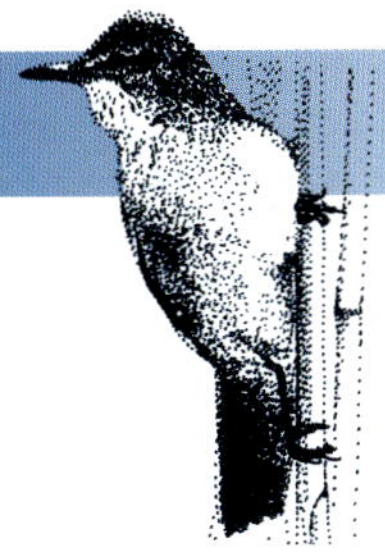

Non-SPEC (1994: —) **Status** (Secure)
Criteria —
European IUCN Red List Category —
Criteria —
Global IUCN Red List Category —
Criteria —

Acrocephalus arundinaceus is a widespread summer visitor to much of Europe, which accounts for less than half of its global breeding range. Its European breeding population is very large (>1,500,000 pairs), and was stable between 1970–1990. Although the species declined in various countries during 1990–2000, trends in the east of its European range—including key populations in Russia, Ukraine and Romania—were stable, and the species probably declined only slightly overall. Consequently, it is provisionally evaluated as Secure.

Country	Breeding pop. size (pairs)	Year(s)	Trend	Mag.%	References
Albania	5,000 – 10,000	96–02	(0)	(0–19)	
Armenia	30,000 – 160,000	98–02	0	0–19	
Austria	1,200 – 2,300	98–02	(0)	(0–19)	
Azerbaijan	(15,000 – 50,000)	96–00	(0)	(0–19)	
Belarus	60,000 – 90,000	97–02	0	0–19	
Belgium	**2 – 5**	95–02	**–**	**>80**	*1*
Bosnia & HG	Present	90–03	?	–	
Bulgaria	**20,000 – 40,000**	96–02	+	0–19	
Croatia	(20,000 – 30,000)	02	(0)	(0–19)	*57*
Czech Rep.	1,200 – 2,400	00	–	10–19	
Denmark	5 – 14	01	0	0–19	*7*
Estonia	4,000 – 8,000	98	0	0–19	*1*
Finland	30 – 100	98–02	–	40	
France	3,000 – 8,000	98–02	–	20–49	*4*
Georgia	Present	03	?	–	
Germany	4,500 – 7,000	95–99	–	20–29	
Greece	(50,000 – 100,000)	95–00	(–)	(0–19)	
Hungary	70,000 – 110,000	99–02	0	0–19	*19*
Italy	(20,000 – 40,000)	03	(–)	(0–19)	
Latvia	10,000 – 20,000	90–00	(0)	(0–19)	*17,23*
Lithuania	40,000 – 80,000	99–01	(0)	(0–19)	*20*
Luxembourg	**1 – 3**	02	**0**	**0–19**	
Macedonia	(3,000 – 10,000)	90–00	(0)	(0–19)	
Moldova	30,000 – 35,000	90–00	0	0–19	
Netherlands	**250 – 300**	98–00	**–**	**46**	*1*
Poland	20,000 – 50,000	00–02	0	0–19	*23*
Portugal	(1,000 – 10,000)	02	(0)	(0–19)	
Romania	(145,000 – 260,000)	00–02	(0)	(0–19)	
Russia	(500,000 – 1,000,000)	90–00	(0)	(0–19)	*3,8,23,92,114*
Serbia & MN	24,000 – 30,000	95–02	+	0–19	*1,29,155,144,78, 225,152,243,227*
Slovakia	1,000 – 2,000	90–99	–	30–49	
Slovenia	250 – 350	99–00	(–)	(0–19)	
Spain	(10,000 – 20,000)	98–02	?	–	*10*
Sweden	**400 – 500**	99–00	**0**	**0–19**	
Switzerland	**200 – 250**	93–96	–	0–9	
Turkey	(100,000 – 300,000)	01	(–)	(0–19)	
Ukraine	(275,000 – 380,000)	90–00	0	0–19	
Total (approx.)	1,500,000 – 2,900,000	**Overall trend** Small decline			
Breeding range	>5,000,000 km²	**Gen. length.** <3.3	**% Global pop.** 25–49		

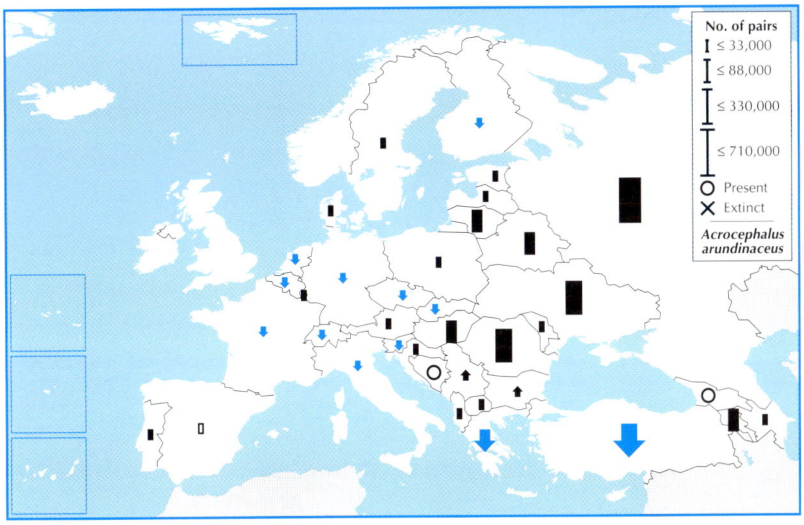

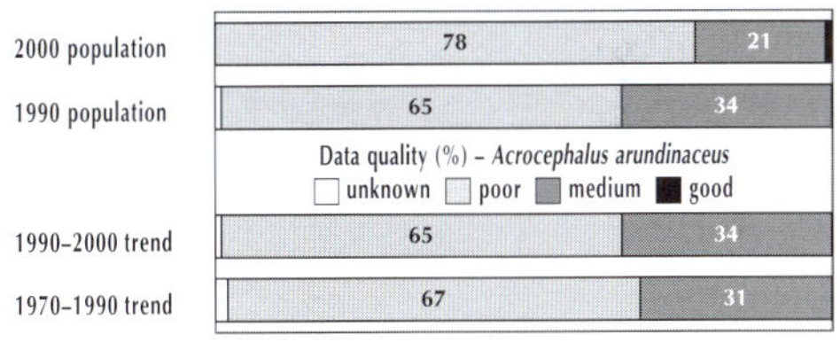

Hippolais pallida
OLIVACEOUS WARBLER

SPEC 3 (1994: 3) **Status** (Depleted)
Criteria Large historical decline
European IUCN Red List Category —
Criteria —
Global IUCN Red List Category —
Criteria —

Hippolais pallida is a summer visitor to south-eastern Europe and Iberia, with Europe accounting for less than half of its global breeding range. Its European breeding population is very large (>3,300,000 pairs), but underwent a large decline between 1970–1990. Although it declined in a few countries—notably Greece—during 1990–2000, the stronghold population in Turkey increased, and the species was stable overall. However, its population size has probably not yet recovered to the level that preceded its decline, and consequently it is provisionally evaluated as Depleted.

Country	Breeding pop. size (pairs)	Year(s)	Trend	Mag.%	References
Albania	10,000 – 20,000	02	(0)	(0–19)	
Armenia	15,000 – 30,000	98–02	?	–	
Azerbaijan	(100,000 – 250,000)	96–00	(0)	(0–19)	
Bosnia & HG	Present	90–03	?	–	
Bulgaria	5,000 – 10,000	96–02	+	0–19	
Croatia	(5,000 – 10,000)	02	(0)	(0–19)	*16*
Cyprus	(40,000 – 80,000)	94–02	(0)	(0–19)	
Georgia	Present	03	?	–	
Greece	(50,000 – 200,000)	95–00	(–)	(0–19)	
Hungary	(200 – 500)	90–93	+	50–79	*14*
Macedonia	(5,000 – 20,000)	90–00	(0)	(0–19)	
Portugal	(5 – 50)	02	(–)	(–)	
Romania	(6,000 – 8,000)	00–02	(0)	(0–19)	
Russia	(20,000 – 50,000)	90–00	(–)	(20–29)	*88,172*
Serbia & MN	1,700 – 3,300	90–02	–	10–19	*1,29,155,215, 145,227,141*
Slovenia	**3 – 5**	97	(0)	(0–19)	
Spain	(2,500 – 10,000)	98–02	(–)	(0–19)	*10,16*
Turkey	(3,000,000 – 6,000,000)	01	(+)	(0–19)	
Ukraine	**500 – 1,000**	90–00	(0)	(0–19)	
Total (approx.)	3,300,000 – 6,700,000	**Overall trend** Stable			
Breeding range	>1,000,000 km²	**Gen. length.** <3.3	**% Global pop.** 25–49		

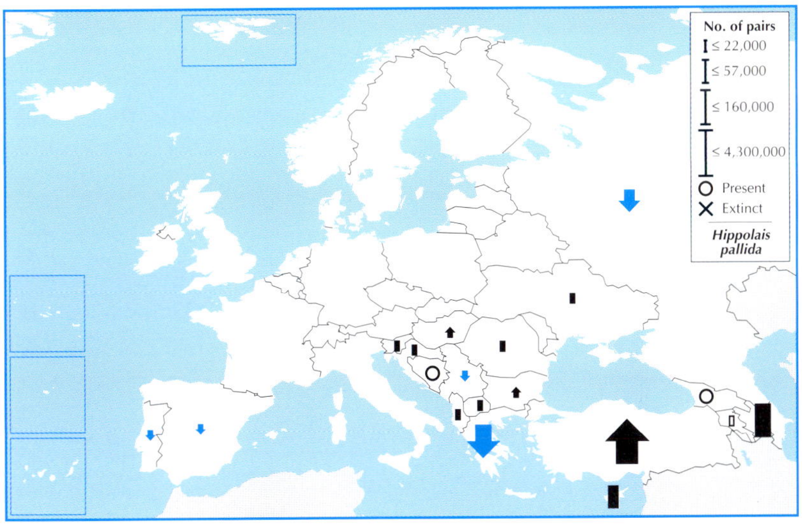

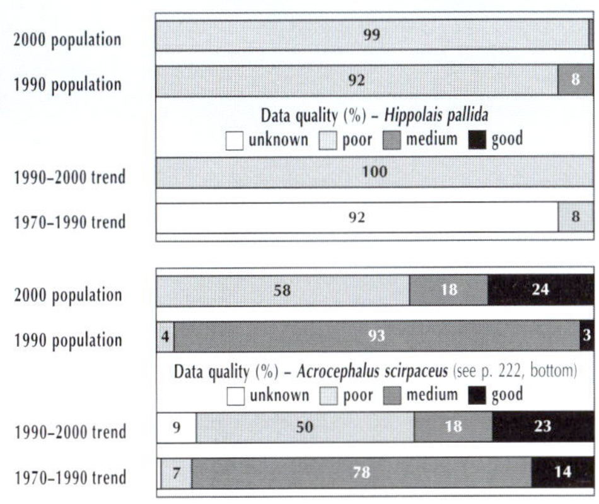

Hippolais caligata
BOOTED WARBLER

Non-SPEC (1994: —) **Status** (Secure)
Criteria —
European IUCN Red List Category —
Criteria —
Global IUCN Red List Category —
Criteria —

Country	Breeding pop. size (pairs)	Year(s)	Trend	Mag.%	References
Finland	0 – 30	00–02	+	N	
Russia	3,000 – 80,000	90–00	?	–	*122*
Total (approx.)	30,000 – 80,000	**Overall trend** Unknown			
Breeding range	>500,000 km²	**Gen. length.** <3.3	**% Global pop.** 5–24		

Hippolais caligata is a summer visitor to central and eastern Russia, with Europe accounting for less than a quarter of its global breeding range. Its European breeding population is relatively small (<80,000 pairs), and was stable between 1970–1990. Trend data were not available for the Russian population during 1990–2000 (although the population in Finland increased), but there was no evidence to suggest that its status deteriorated significantly. Consequently, the species is provisionally evaluated as Secure.

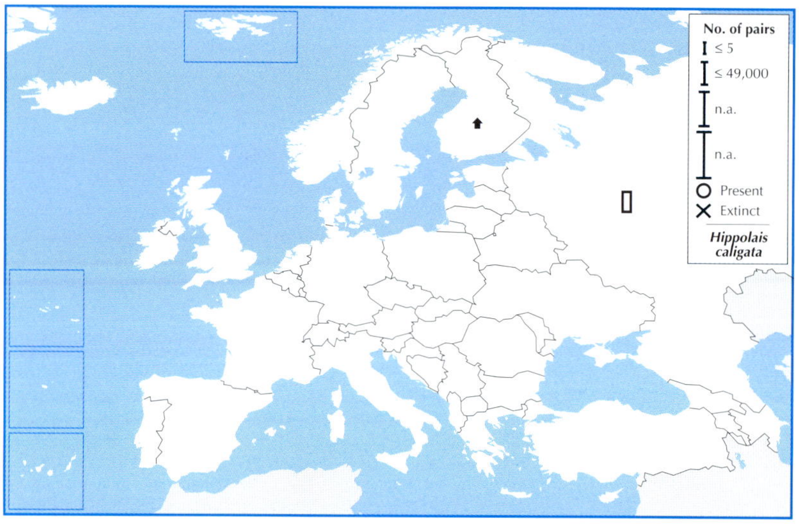

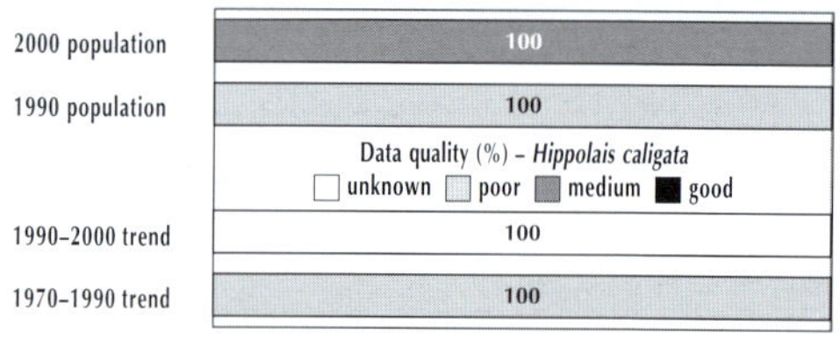

Hippolais languida
UPCHER'S WARBLER

Non-SPEC (1994: —) **Status** Secure
Criteria —
European IUCN Red List Category —
Criteria —
Global IUCN Red List Category —
Criteria —

Country	Breeding pop. size (pairs)	Year(s)	Trend	Mag.%	References
Armenia	3,500 – 6,500	98–02	?	–	
Azerbaijan	(0 – 100)	96–00	(0)	(0–19)	
Georgia	Present	03	?	–	
Turkey	(10,000 – 30,000)	01	(+)	(20–29)	
Total (approx.)	14,000 – 37,000	**Overall trend** Moderate increase			
Breeding range	>100,000 km²	**Gen. length.** <3.3	**% Global pop.** 5–24		

Hippolais languida is a summer visitor to parts of southern Turkey and the Caucasus, which together account for less than a quarter of its global breeding range. Its European breeding population is relatively small (<37,000 pairs), but its trend between 1970–1990 was unknown. Although trend data were not available for Armenia and Georgia during 1990–2000, the stronghold population in Turkey increased, and the species probably underwent a moderate increase overall. Consequently, it is evaluated as Secure.

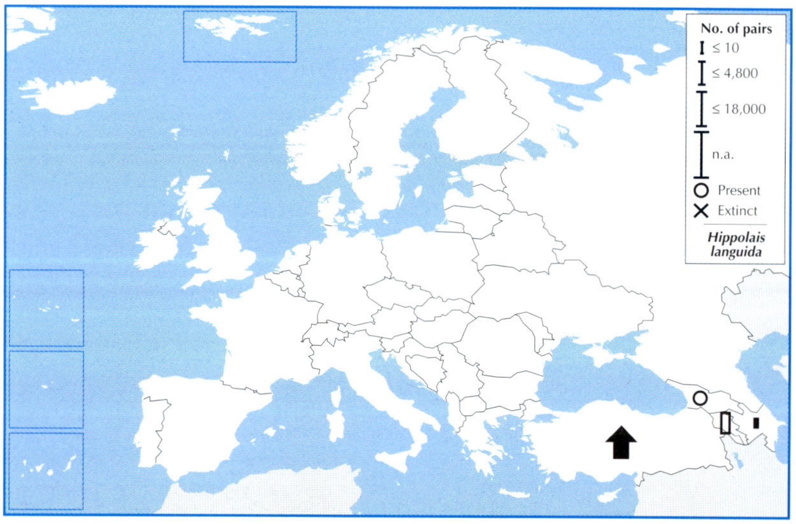

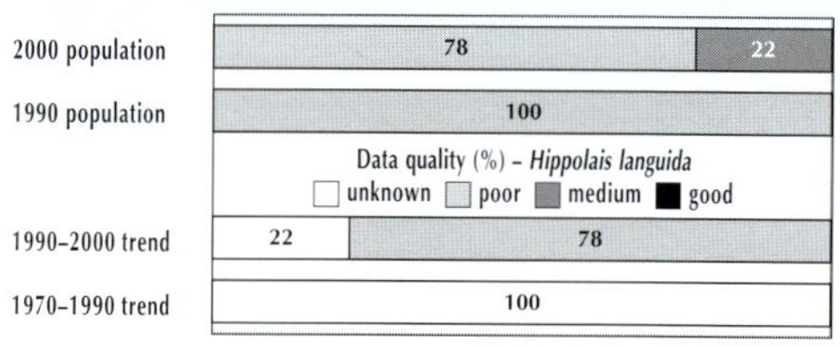

Hippolais olivetorum
OLIVE-TREE WARBLER

Non-SPEC[E] (1994: 2) **Status** (Secure)
Criteria —
European IUCN Red List Category —
Criteria —
Global IUCN Red List Category —
Criteria —

Country	Breeding pop. size (pairs)	Year(s)	Trend	Mag.%	References
Albania	(1,000 – 3,000)	02	(0)	(0–19)	
Bulgaria	**500 – 900**	96–02	+	0–9	
Croatia	(500 – 750)	02	(–)	(0–19)	*70*
Greece	(3,000 – 5,000)	95–00	(–)	(0–19)	
Macedonia	(500 – 3,000)	90–00	(0)	(0–19)	
Serbia & MN	5 – 10	95–02	+	N	*1,141,155,227*
Turkey	(5,000 – 10,000)	01	(0)	(0–19)	
Total (approx.)	11,000 – 23,000	**Overall trend** Stable			
Breeding range	>250,000 km²	**Gen. length.** <3.3	**% Global pop.** >95		

Hippolais olivetorum is a summer visitor to south-eastern Europe, which constitutes >95% of its global breeding range. Its European breeding population is relatively small (<23,000 pairs), but was stable between 1970–1990. Despite declines in Greece and Croatia during 1990–2000, the species was stable or increased elsewhere within its European range, and probably remained stable overall. Although it was previously classified as Rare, the species's European breeding population is now known to exceed 10,000 pairs, and consequently it is provisionally evaluated as Secure.

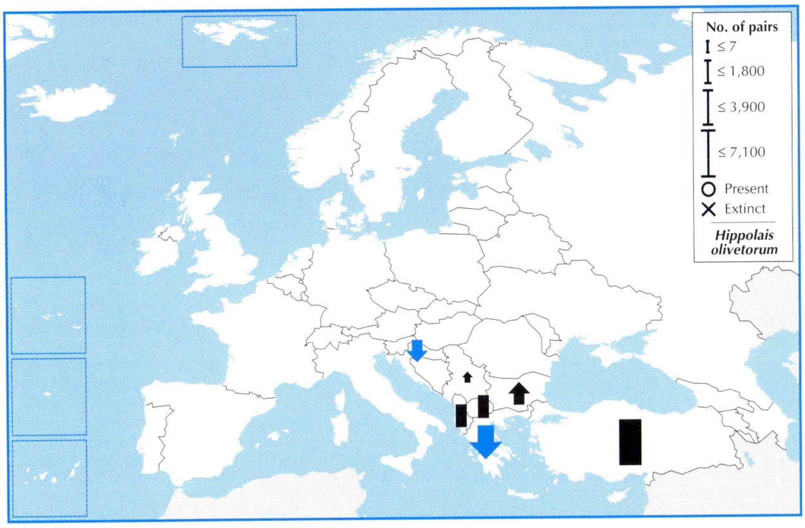

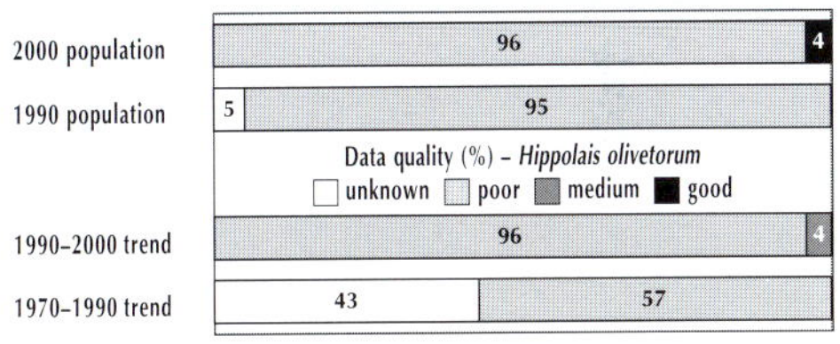

Hippolais icterina
ICTERINE WARBLER

Non-SPEC[E] (1994: 4) **Status** (Secure)
Criteria —
European IUCN Red List Category —
Criteria —
Global IUCN Red List Category —
Criteria —

Country	Breeding pop. size (pairs)	Year(s)	Trend	Mag.%	References
Austria	(10,000 – 20,000)	98–02	(0)	(0–19)	
Azerbaijan	(1,000 – 10,000)	96–00	(0)	(0–19)	
Belarus	100,000 – 180,000	97–02	0	0–19	
Belgium	3,500 – 7,000	01–02	–	0–19	*1*
Bosnia & HG	Present	85–89	?	–	
Bulgaria	150 – 300	96–02	(F)	(20–29)	
Croatia	(50 – 75)	02	(–)	(>80)	*70*
Czech Rep.	50,000 – 100,000	00	0	0–19	
Denmark	(10,000 – 50,000)	00	–	**20–29**	
Estonia	(20,000 – 50,000)	98	+	20–29	*1*
Finland	10,000 – 15,000	98–02	–	15	
France	(15,000 – 60,000)	98–02	?	–	*4*
Georgia	Present	03	?	–	
Germany	200,000 – 400,000	95–99	–	0–19	
Hungary	3,000 – 12,000	99–02	0	0–19	
Latvia	50,000 – 110,000	90–00	(0)	(0–19)	*23,18*
Liechtenstein	**8 – 12**	98–00	0	0–19	
Lithuania	60,000 – 90,000	99–01	(0)	(0–19)	*20*
Luxembourg	1 – 5	02	–	**50–79**	
Macedonia	(5 – 10)	98–00	(0)	(0–19)	
Moldova	1,000 – 1,200	90–00	0	0–19	
Netherlands	**17,000 – 25,000**	98–00	–	**21**	*1*
Norway	(50,000 – 300,000)	90–03	(0)	(0–19)	
Poland	**200,000 – 400,000**	00–02	0	0–19	*23*
Romania	(60,000 – 75,000)	00–02	(0)	(0–19)	*48*
Russia	2,500,000 – 5,000,000	90–00	(0)	(0–19)	*89,122*
Serbia & MN	1,000 – 1,500	90–02	–	10–29	*1,29,67a,155, 215,87,224*
Slovakia	10,000 – 20,000	90–99	0	0–19	
Sweden	**40,000 – 100,000**	99–00	+	**23**	
Switzerland	**200 – 500**	93–96	–	20–29	
Turkey	(50 – 500)	01	?	–	
Ukraine	57,000 – 78,000	90–00	0	0–19	
Total (approx.)	3,500,000 – 7,100,000	**Overall trend** Small decline			
Breeding range	>4,000,000 km²	**Gen. length.** <3.3	**% Global pop.** 75–94		

Hippolais icterina is a widespread summer visitor to much of Europe, except for the far west and south, with the region constituting >75% of its global breeding range. Its European breeding population is very large (>3,500,000 pairs), and was stable between 1970–1990. Although there were declines in some western populations—most notably in Germany—during 1990–2000, key populations to the east, including the stronghold in Russia, were stable, and the species probably remained stable overall. Consequently, it is provisionally evaluated as Secure.

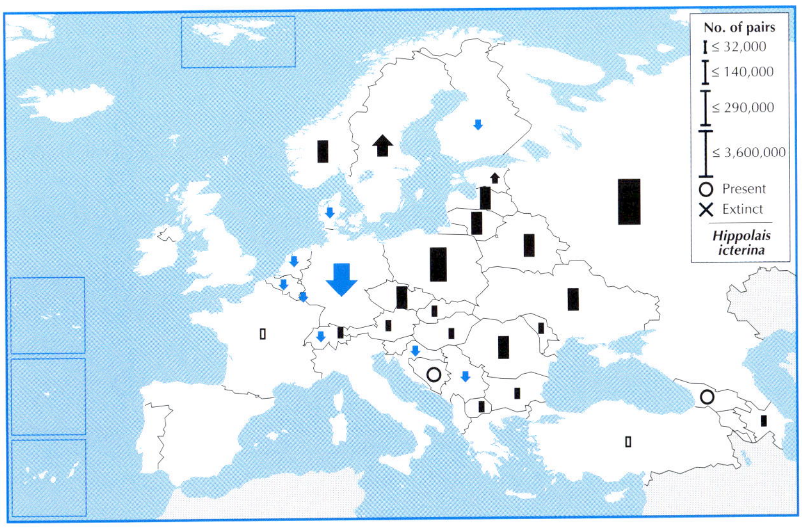

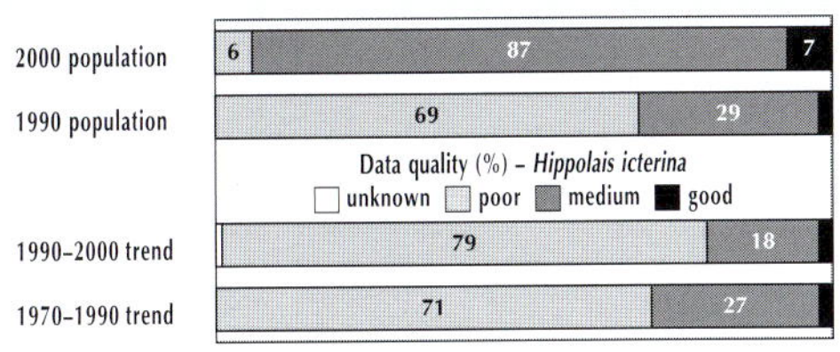

Hippolais polyglotta
MELODIOUS WARBLER

Non-SPEC[E] (1994: 4) **Status** (Secure)
Criteria —
European IUCN Red List Category —
Criteria —
Global IUCN Red List Category —
Criteria —

Country	Breeding pop. size (pairs)	Year(s)	Trend	Mag.%	References
Andorra	(0 – 1)	99–01	(0)	(0–19)	*1,3*
Belgium	1,000 – 2,000	01–02	+	0–19	*1*
Croatia	(5,000 – 10,000)	02	(+)	(50–79)	*16*
France	(200,000 – 800,000)	98–02	-	17	*4,3*
Germany	**480 – 690**	95–99	+	50–79	
Italy	(50,000 – 150,000)	03	(0)	(0–19)	
Luxembourg	30 – 40	02	+	10–19	
Netherlands	**0 – 2**	98–00	?	–	*1*
Portugal	(50,000 – 500,000)	02	(0)	(0–19)	
Slovenia	1,000 – 2,000	94	(0)	(0–19)	
Spain	(700,000 – 1,500,000)	92	?	–	*13,12,10*
Switzerland	**300 – 500**	93–96	+	0–19	
Total (approx.)	1,000,000 – 3,000,000	**Overall trend** Unknown			
Breeding range	>1,000,000 km²	**Gen. length.** <3.3	**% Global pop.** 75–94		

Hippolais polyglotta is a widespread summer visitor to south-western Europe, which constitutes >75% of its global breeding range. Its European breeding population is very large (>1,000,000 pairs), and was stable between 1970–1990. Although most European populations increased or were stable during 1990–2000, the sizeable population in France declined, and no trend data were available for the Spanish stronghold. Consequently, the species is provisionally evaluated as Secure.

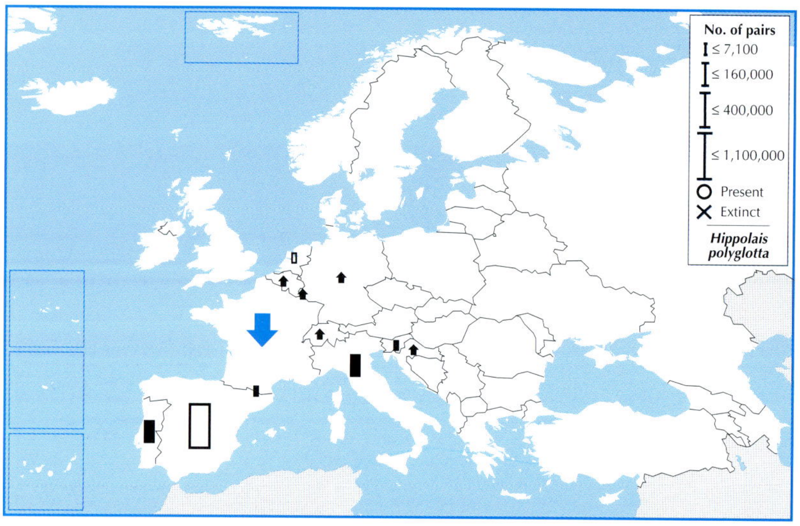

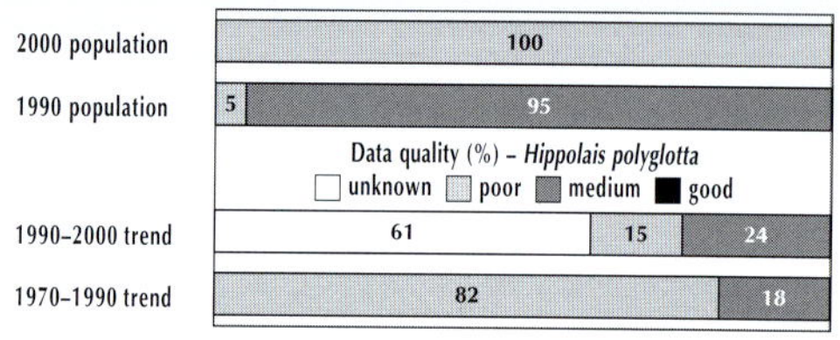

Sylvia sarda
MARMORA'S WARBLER

Non-SPEC[E] (1994: 4) **Status** (Secure)
Criteria —
European IUCN Red List Category —
Criteria —
Global IUCN Red List Category —
Criteria —

Country	Breeding pop. size (pairs)	Year(s)	Trend	Mag.%	References
France	(10,000 – 40,000)	98–02	?	–	*4*
Italy	(5,000 – 10,000)	03	(0)	(0–19)	
Spain	(14,000 – 25,000)	92	?	–	*13,12,10*
Total (approx.)	29,000 – 75,000	**Overall trend** Unknown			
Breeding range	>50,000 km²	**Gen. length.** <3.3	**% Global pop.** 100		

Sylvia sarda breeds only in Europe, where it is confined to the islands and islets of the western Mediterranean—notably Corsica, Sardinia and the Balearic Islands. Its breeding population is relatively small (<75,000 pairs), but increased between 1970–1990. Trends were not available for the Spanish and French populations during 1990–2000, but the Italian population was stable, and the species is provisionally evaluated as Secure.

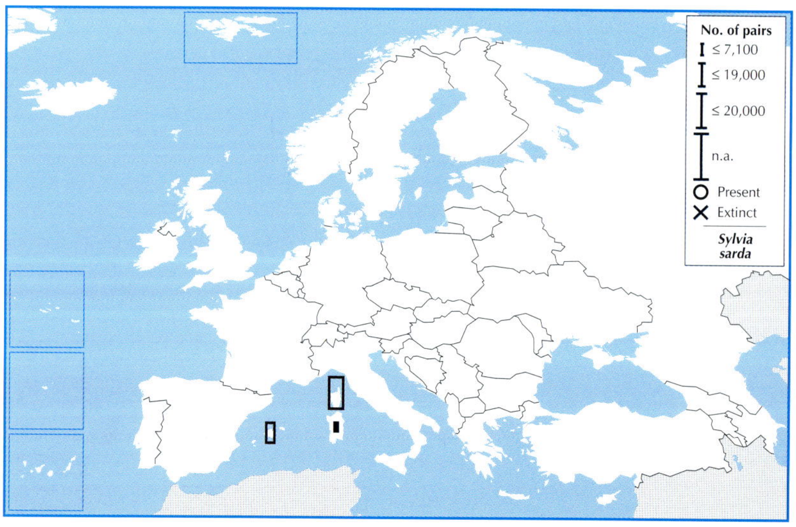

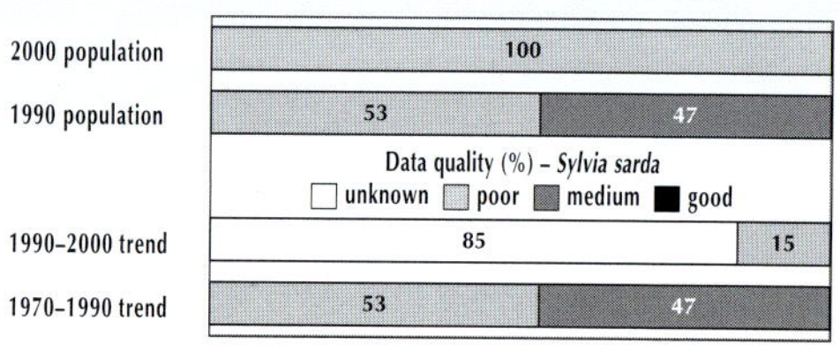

Sylvia undata
DARTFORD WARBLER

SPEC 2 (1994: 2) **Status** (Depleted)
Criteria Large historical decline
European IUCN Red List Category —
Criteria —
Global IUCN Red List Category —
Criteria —

Country	Breeding pop. size (pairs)	Year(s)	Trend	Mag.%	References
Andorra	20 – 30	99–01	(0)	(0–19)	*1,3*
France	(150,000 – 600,000)	98–02	(F)	(20–49)	*4*
Italy	(10,000 – 30,000)	03	(0)	(0–19)	
Portugal	(10,000 – 100,000)	02	(0)	(0–19)	
Spain	(1,700,000 – 3,000,000)	92	?	–	*13,12,10*
UK	**1,600 – 1,900**	94	+	**187**	
Total (approx.)	1,900,000 – 3,700,000	**Overall trend** Unknown			
Breeding range	>1,000,000 km²	**Gen. length.** <3.3	**% Global pop.** >95		

Sylvia undata is a patchily distributed resident across parts of south-western Europe, which constitutes >95% of its global range. Its European breeding population is very large (>1,900,000 pairs), but underwent a large decline between 1970–1990. Although the trend of the stronghold population in Spain during 1990–2000 was unknown, the species was stable or increased elsewhere in its European range. Nevertheless, its total population size has probably not recovered to the level that preceded its decline, and consequently the species is provisionally evaluated as Depleted.

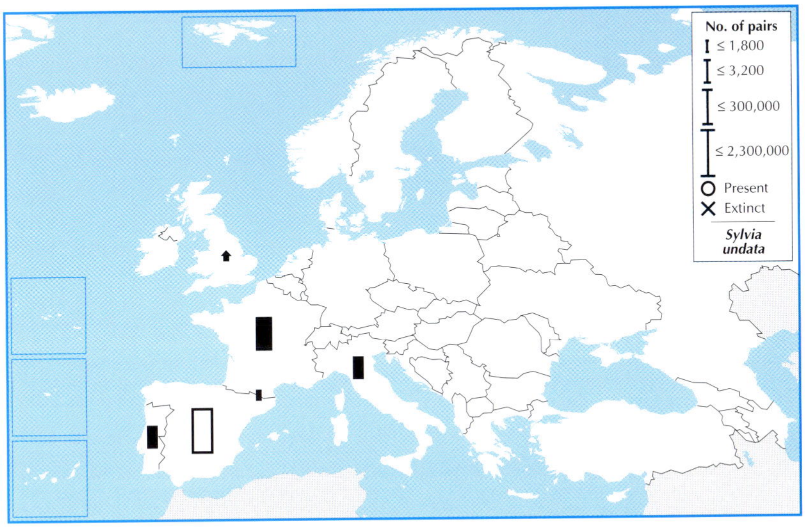

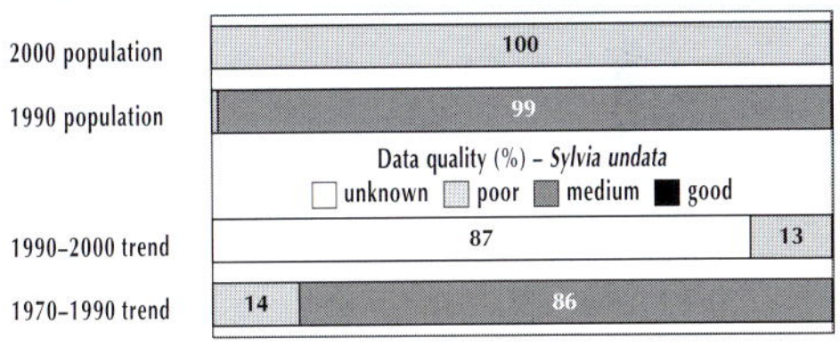

Sylvia conspicillata
SPECTACLED WARBLER

Non-SPEC (1994: —) **Status** (Secure)
Criteria —
European IUCN Red List Category —
Criteria —
Global IUCN Red List Category —
Criteria —

Country	Breeding pop. size (pairs)	Year(s)	Trend	Mag.%	References
Cyprus	(4,000 – 8,000)	94–02	(0)	(0–9)	
France	(2,000 – 10,000)	98–02	?	–	*4*
Italy	(10,000 – 20,000)	03	?	–	
Malta	**200 – 250**	90–02	-	**20–29**	*1*
Portugal	(250 – 2,500)	02	(0)	(0–19)	
Madeira	Present	02	?	–	
Spain	(140,000 – 300,000)	92	?	–	*13,12,10*
Canary Is.	(20,000 – 100,000)	97–03	?	–	*28,25*
Turkey	20 – 200	01	?	–	
Total (approx.)	180,000 – 440,000	**Overall trend** Unknown			
Breeding range	>500,000 km²	**Gen. length.** <3.3	**% Global pop.** 25–49		

Sylvia conspicillata is a widespread summer visitor to southern Europe, which accounts for less than half of its global breeding range. Its European breeding population is large (>180,000 pairs), and was stable between 1970–1990. Trend data were not available for several countries during 1990–2000—including the stronghold population in Spain—but there was no evidence to suggest that the species's status deteriorated significantly. Consequently, it is provisionally evaluated as Secure.

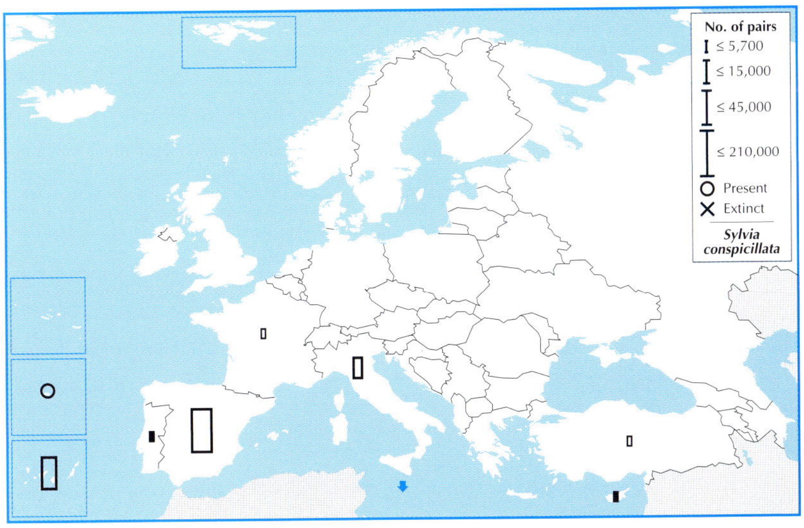

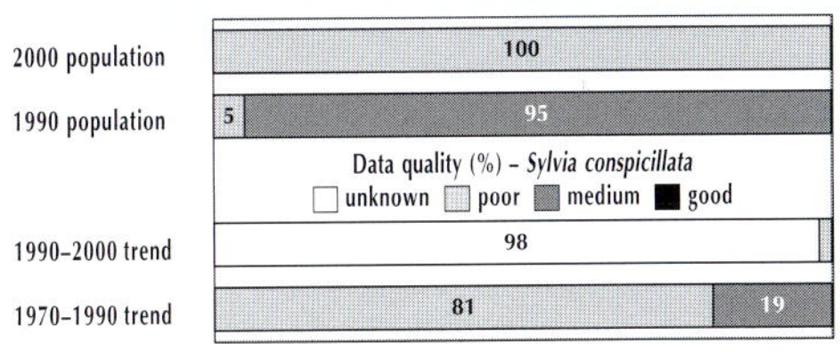

Sylvia cantillans
SUBALPINE WARBLER

Non-SPEC[E] (1994: 4) **Status** (Secure)
Criteria —
European IUCN Red List Category —
Criteria —
Global IUCN Red List Category —
Criteria —

Country	Breeding pop. size (pairs)	Year(s)	Trend	Mag.%	References
Albania	10,000 – 30,000	02	?	–	
Andorra	60 – 100	99–01	(+)	(0–19)	*1,3*
Bosnia & HG	Present	90–03	?	–	
Bulgaria	3,000 – 8,000	96–02	0	0–9	
Croatia	(50,000 – 100,000)	02	(0)	(0–19)	*16*
France	(30,000 – 120,000)	98–02	–	16	*4,2*
Greece	(200,000 – 500,000)	95–00	(–)	(0–19)	
Italy	(10,000 – 40,000)	03	(0)	(0–19)	
Macedonia	(2,000 – 5,000)	90–00	(0)	(0–19)	
Portugal	(10,000 – 100,000)	02	(0)	(0–19)	
Serbia & MN	(5,000 – 10,000)	90–02	(0)	(0–19)	*1,155,216,215, 224,220*
Slovenia	100 – 300	98	(0)	(0–19)	
Spain	(1,100,000 – 2,300,000)	92	?	–	*13,12,10*
Turkey	(6,000 – 18,000)	01	(0)	(0–19)	
Total (approx.)	1,400,000 – 3,200,000	**Overall trend** Unknown			
Breeding range	>1,000,000 km²	**Gen. length.** <3.3		**% Global pop.** 75–94	

Sylvia cantillans is a widespread summer visitor to southern Europe, which constitutes >75% of its global breeding range. Its European breeding population is very large (>1,400,000 pairs), and was stable between 1970–1990. Although the sizeable populations in Greece and France declined to some extent during 1990–2000, and the trend of the stronghold population in Spain was unknown, the species remained stable in most other countries in its European range. Consequently, it is provisionally evaluated as Secure.

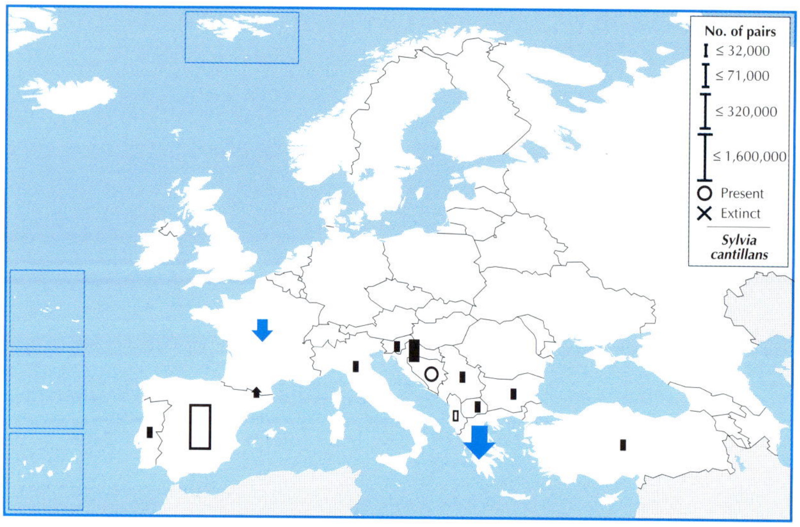

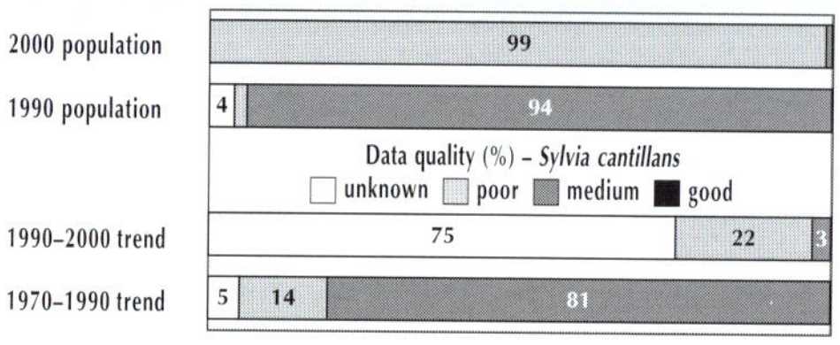

Sylvia mystacea
MÉNÉTRIES'S WARBLER

Non-SPEC (1994: —) **Status** (Secure)
Criteria —
European IUCN Red List Category —
Criteria —
Global IUCN Red List Category —
Criteria —

Country	Breeding pop. size (pairs)	Year(s)	Trend	Mag.%	References
Armenia	3,000 – 6,000	97–02	?	–	
Azerbaijan	(5,000 – 50,000)	96–00	(0)	(0–19)	
Georgia	Present	03	?	–	
Russia	(20,000 – 50,000)	90–00	(+)	(20–29)	*172*
Turkey	(15,000 – 45,000)	01	-	**0–19**	
Total (approx.)	43,000 – 150,000	**Overall trend** Stable			
Breeding range	>250,000 km²	**Gen. length.** <3.3		**% Global pop.** <5	

Sylvia mystacea has a predominantly Middle Eastern breeding distribution, which extends into south-eastern Europe in southern Russia, the Caucasus and parts of Turkey. Its European breeding population is relatively small (<150,000 pairs), but was probably stable between 1970–1990. Although there were declines in Turkey during 1990–2000 (trends were not available for Armenia and Georgia), the sizeable population in Russia increased, and the species probably remained stable overall. Consequently, it is provisionally evaluated as Secure.

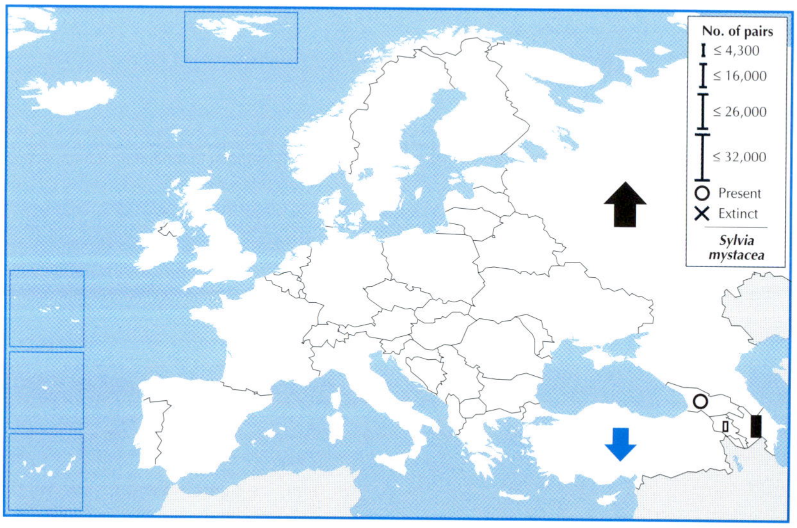

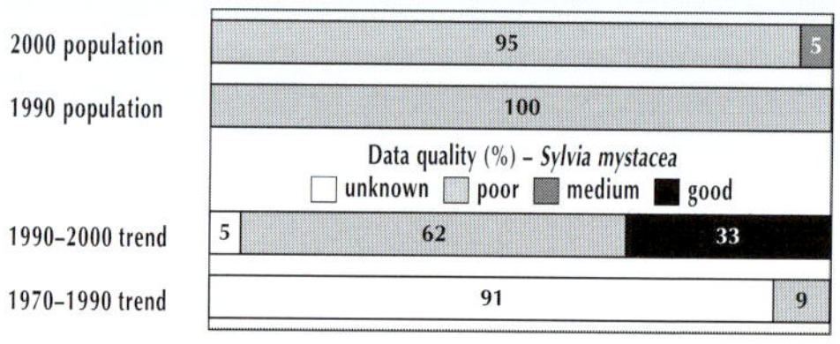

Sylvia melanocephala
SARDINIAN WARBLER

Non-SPEC[E] (1994: 4) **Status** (Secure)
Criteria —
European IUCN Red List Category —
Criteria —
Global IUCN Red List Category —
Criteria —

Country	Breeding pop. size (pairs)	Year(s)	Trend	Mag.%	References
Albania	10,000 – 20,000	02	?	–	
Bosnia & HG	Present	90–03	?	–	
Bulgaria	800 – 1,200	96–02	(0)	(0–19)	
Croatia	(50,000 – 100,000)	02	(+)	(50–79)	*16*
Cyprus	6,000 – 12,000	94–02	+	20–29	
France	(150,000 – 600,000)	98–02	(+)	(44)	*4,2*
Greece	(500,000 – 1,000,000)	95–00	(0)	(0–19)	
Italy	(500,000 – 1,000,000)	03	(0)	(0–19)	
Macedonia	(250 – 1,000)	90–00	(0)	(0–19)	
Malta	3,000 – 5,000	90–02	**0**	**0–19**	*1*
Portugal	(100,000 – 1,000,000)	02	(0)	(0–19)	
Serbia & MN	2,000 – 3,000	90–02	(0)	(0–19)	*1,155,141,216,227*
Slovenia	100 – 300	98	(0)	(0–19)	
Spain	(990,000 – 1,900,000)	92	?	–	*13,12,10*
Canary Is.	(20,000 – 100,000)	97–03	?	–	*28,25*
Turkey	(800,000 – 2,400,000)	01	(0)	(0–19)	
UK					
Gibraltar	100 – 200	00	**0**	**0–19**	
Total (approx.)	3,100,000 – 8,100,000	**Overall trend** Stable			
Breeding range	>1,000,000 km²	**Gen. length.** <3.3		**% Global pop.** 50–74	

Sylvia melanocephala is a widespread breeder in southern Europe, which constitutes >50% of its global breeding range. Its European breeding population is very large (>3,100,000 pairs), and increased between 1970–1990. Populations were stable or increased across most of its range during 1990–2000, although the trend of the sizeable Spanish population was unknown—but there was no evidence to suggest that the species's status deteriorated significantly. Consequently, it is provisionally evaluated as Secure.

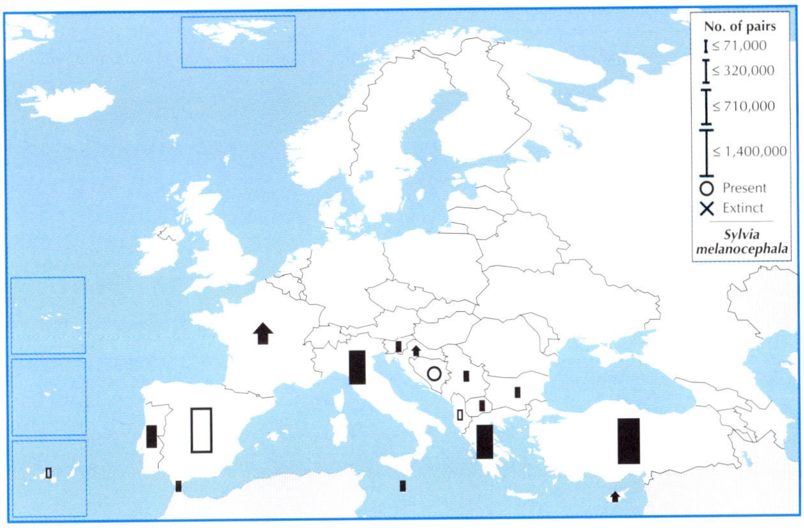

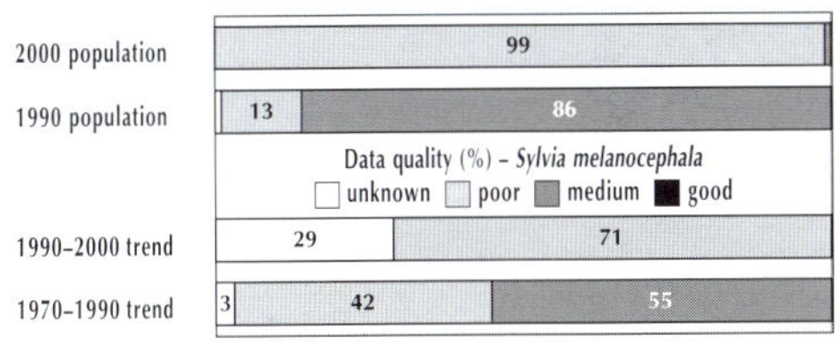

Sylvia melanothorax
CYPRUS WARBLER

Non-SPEC[E] (1994: 2) **Status** (Secure)
Criteria —
European IUCN Red List Category —
Criteria —
Global IUCN Red List Category —
Criteria —

Country	Breeding pop. size (pairs)	Year(s)	Trend	Mag.%	References
Cyprus	(70,000 – 140,000)	94–02	(0)	(0–9)	
Total (approx.)	70,000 – 140,000	**Overall trend** Stable			
Breeding range	>20,000 km²	**Gen. length.** <3.3		**% Global pop.** 100	

Sylvia melanothorax breeds entirely within Europe, with its global breeding range confined to the island of Cyprus. Its breeding population underwent a large increase between 1970–1990, but was thought to have been stable during 1990–2000. Although the species was previously classified as Rare, recent surveys have revealed its breeding population to be far larger (>70,000 pairs) than was formerly believed. Consequently, it is now provisionally evaluated as Secure.

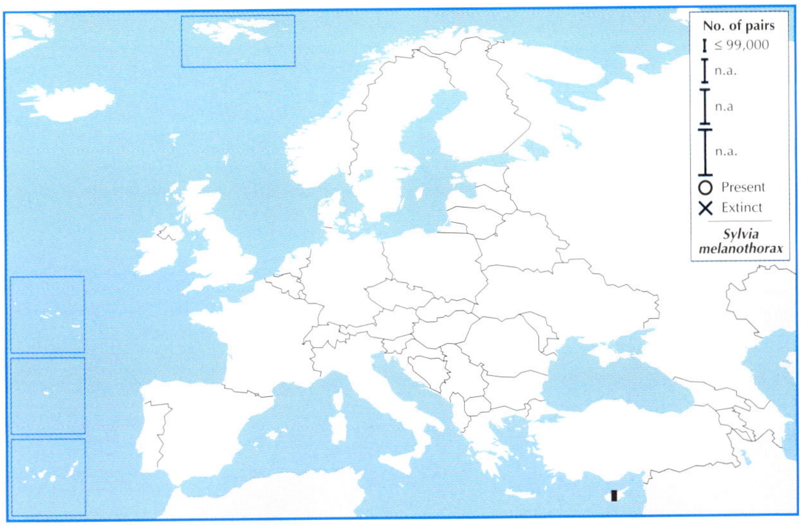

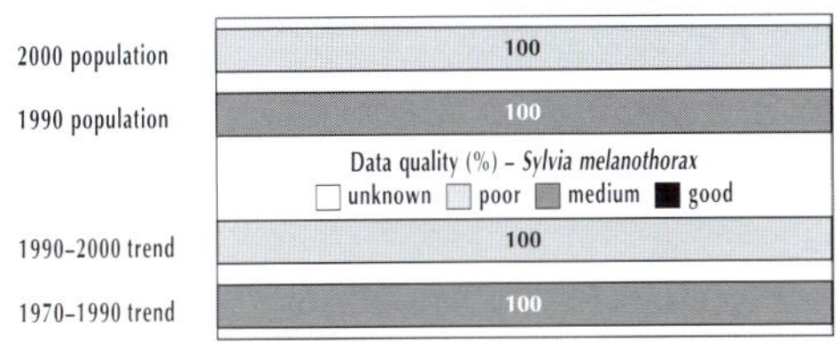

Sylvia rueppelli
RÜPPELL'S WARBLER

Non-SPEC[E] (1994: 4) **Status** (Secure)
Criteria —
European IUCN Red List Category —
Criteria —
Global IUCN Red List Category —
Criteria —

Country	Breeding pop. size (pairs)	Year(s)	Trend	Mag.%	References
Greece	(3,000 – 10,000)	95–00	(0)	(0–19)	
Turkey	(200,000 – 600,000)	01	(–)	(0–19)	
Total (approx.)	200,000 – 610,000	**Overall trend** Small decline			
Breeding range	>100,000 km²	**Gen. length.** <3.3	**% Global pop.** >95		

Sylvia rueppelli breeds almost entirely in Europe, where it is a summer visitor to Greece and Turkey. Its European breeding population is large (>200,000 pairs), and was stable between 1970–1990. Although the species declined to some extent in its Turkish stronghold 1990–2000, it remained stable in Greece, and probably underwent only a slight decline overall. Consequently, it is provisionally evaluated as Secure.

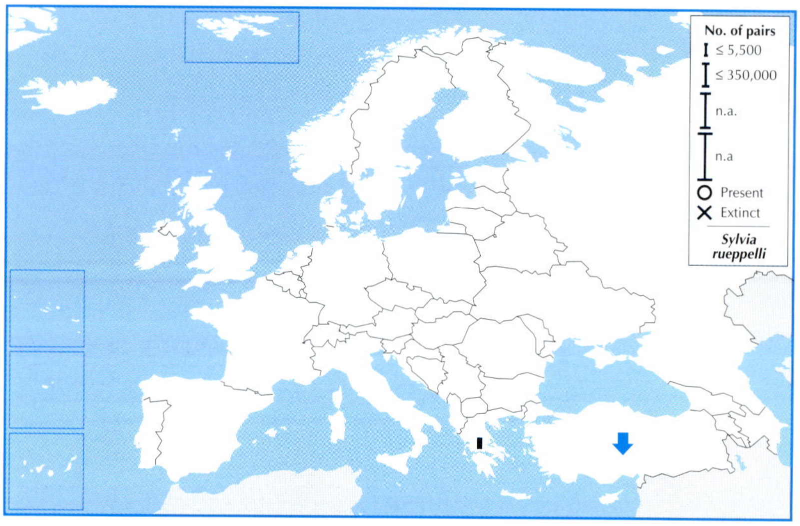

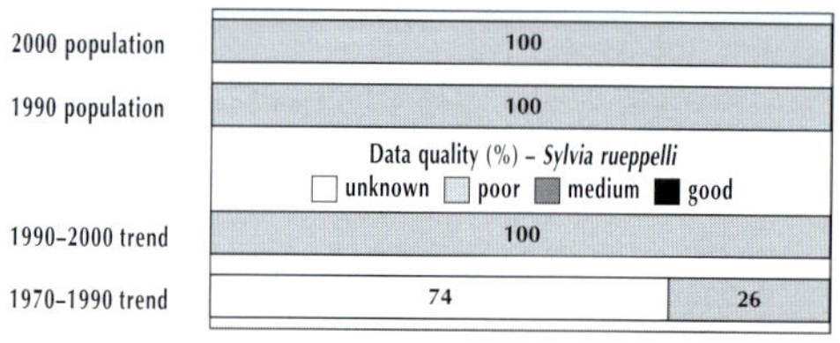

Sylvia nana
DESERT WARBLER

Non-SPEC (1994: —) **Status** (Secure)
Criteria —
European IUCN Red List Category —▼
Criteria —
Global IUCN Red List Category —
Criteria —

Country	Breeding pop. size (pairs)	Year(s)	Trend	Mag.%	References
Russia	(1,000 – 5,000)	90–00	(0)	(0–19)	
Total (approx.)	1,000 – 5,000	**Overall trend** Stable			
Breeding range	>50,000 km²	**Gen. length.** <3.3	**% Global pop.** <5		

Sylvia nana has a predominantly North African and Central Asian breeding distribution, which just extends into Europe in south-western Russia. Its European breeding population is very small (as few as 1,000 pairs), but probably remained stable during both 1970–1990 and 1990–2000. Although the size of the European population could render it susceptible to the risks affecting small populations, it is marginal to a much larger non-European population. Consequently, the species is provisionally evaluated as Secure.

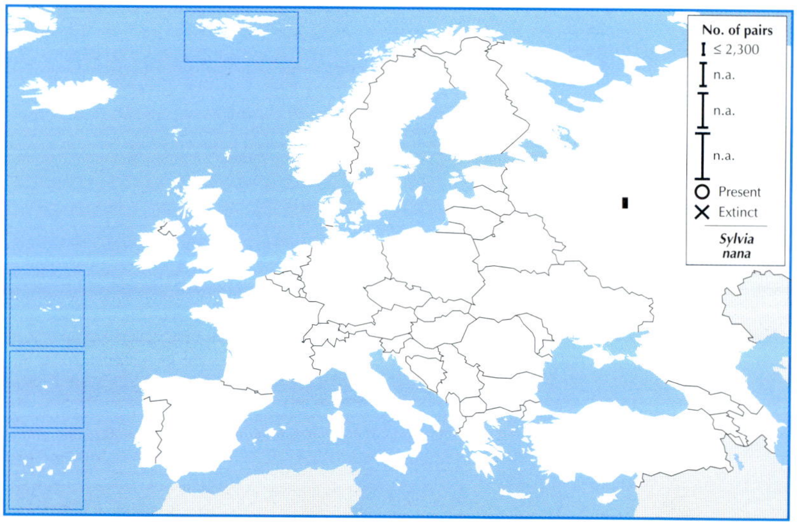

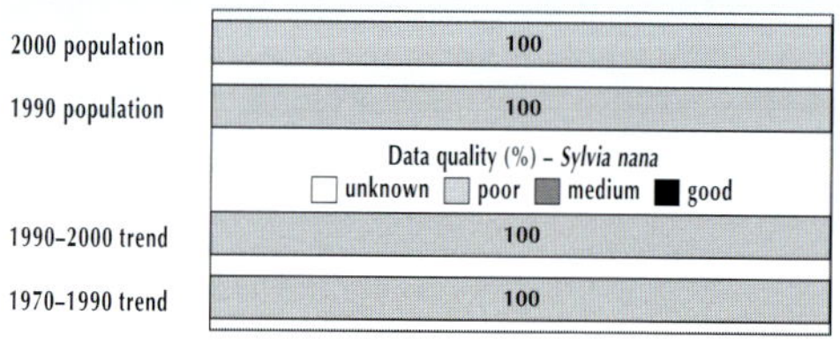

Sylvia hortensis
ORPHEAN WARBLER

SPEC 3 (1994: 3) **Status** (Depleted)
Criteria Large historical decline
European IUCN Red List Category —
Criteria —
Global IUCN Red List Category —
Criteria —

Sylvia hortensis is a widespread but patchily distributed summer visitor to southern Europe, which accounts for less than half of its global breeding range. Its European breeding population is large (>170,000 pairs), but underwent a large decline between 1970–1990. Although the trend of the key Spanish population during 1990–2000 was unknown, the species declined in Turkey, and probably underwent a small decline overall. Its total population size has almost certainly not recovered to the level that preceded its decline, and consequently the species is provisionally evaluated as Depleted.

Country	Breeding pop. size (pairs)	Year(s)	Trend	Mag.%	References
Albania	(1,000 – 2,000)	02	?	–	
Andorra	(0 – 3)	99–01	(F)	(20–29)	*1,3*
Armenia	300 – 800	97–02	?	–	
Azerbaijan	(5,000 – 10,000)	96–00	(0)	(0–19)	
Bosnia & HG	Present	90–03	?	–	
Bulgaria	400 – 600	96–02	+	0–19	
Croatia	(10,000 – 50,000)	02	(0)	(0–19)	*16*
France	(5,000 – 25,000)	98–02	+	100	*4,2*
Georgia	Present	03	?	–	
Greece	(5,000 – 10,000)	95–00	(0)	(0–19)	
Italy	(1,000 – 2,000)	03	(–)	(0–19)	
Macedonia	(2,000 – 5,000)	90–00	(0)	(0–19)	
Portugal	(250 – 2,500)	02	(0)	(0–19)	
Russia	(10 – 100)	90–00	?	–	*109*
Serbia & MN	300 – 600	90–02	(0)	(0–19)	*1,76,227*
Spain	(100,000 – 250,000)	98–02	?	–	*10*
Switzerland	**1 – 5**	93–96	**–**	**0–19**	
Turkey	(40,000 – 120,000)	01	(–)	(0–19)	
Total (approx.)	170,000 – 480,000	**Overall trend** Small decline			
Breeding range	>1,000,000 km²	**Gen. length.** <3.3	**% Global pop.** 25–49		

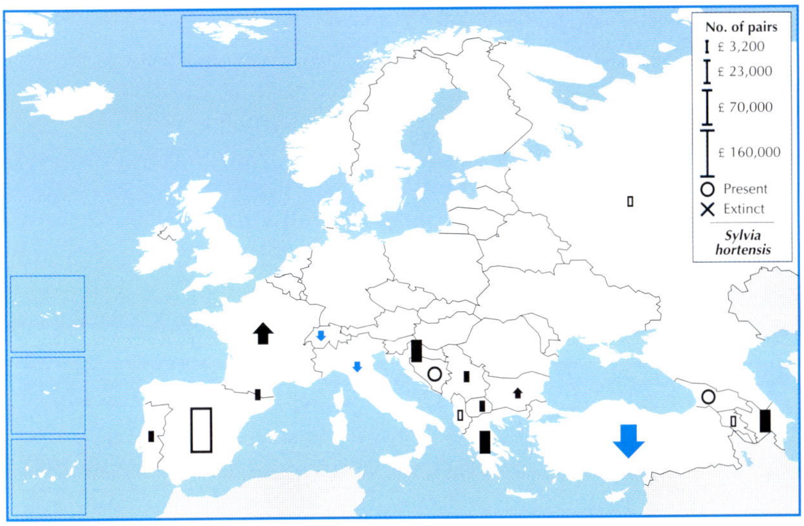

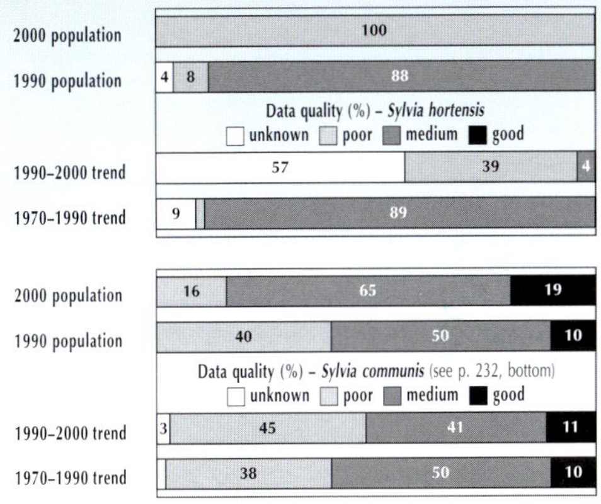

Sylvia nisoria
BARRED WARBLER

Non-SPEC[E] (1994: 4) **Status** (Secure)
Criteria —
European IUCN Red List Category —
Criteria —
Global IUCN Red List Category —
Criteria —

Sylvia nisoria is a widespread summer visitor to central and eastern Europe, which constitutes >50% of its global breeding range. Its European breeding population is large (>460,000 pairs), and was stable between 1970–1990. Although there were declines in some marginal populations during 1990–2000, most populations—including the sizeable one in Ukraine—were stable or increasing. Nevertheless, the trend of the Russian stronghold population was unknown, and consequently the species is provisionally evaluated as Secure.

Country	Breeding pop. size (pairs)	Year(s)	Trend	Mag.%	References
Albania	(100 – 500)	02	?	–	
Armenia	300 – 800	97–02	?	–	
Austria	(1,100 – 2,000)	98–02	(0)	(0–19)	
Azerbaijan	(1,000 – 5,000)	96–00	(0)	(0–19)	
Belarus	8,000 – 15,000	97–02	(0)	(0–19)	
Bulgaria	4,000 – 10,000	96–02	0	0–9	
Croatia	(5,000 – 10,000)	02	(+)	(>80)	*16,57*
Czech Rep.	3,000 – 6,000	00	+	>80	
Denmark	**0 – 0**	99–01	**–**	**X**	*5,6,7*
Estonia	(10,000 – 20,000)	98	0	0–19	*1*
Finland	500 – 1,500	98–02	–	60	
Georgia	Present	03	?	–	
Germany	7,000 – 16,000	95–99	–	0–19	
Greece	(100 – 1,000)	95–00	(0)	(0–19)	
Hungary	**32,000 – 65,000**	99–02	+	20–49	*19*
Italy	(1,000 – 2,000)	03	(–)	(0–19)	
Latvia	(1,000 – 6,000)	90–00	(0)	(0–19)	*23,16*
Lithuania	(2,000 – 5,000)	99–01	(0)	(0–19)	*20*
Macedonia	(1,000 – 2,500)	90–00	(0)	(0–19)	
Moldova	7,000 – 9,000	90–00	–	20–29	
Norway	10 – 20	99	–	30–49	*63*
Poland	**20,000 – 50,000**	00–02	0	0–19	*23*
Romania	(25,000 – 40,000)	00–02	(0)	(0–19)	*48*
Russia	250,000 – 650,000	90–00	?	–	*8,39,114*
Serbia & MN	4,500 – 6,000	95–02	(+)	(0–19)	*1,29,155,67a, 225,227,215,224*
Slovakia	3,000 – 6,000	90–99	0	0–19	
Slovenia	600 – 1,000	99–00	(–)	(0–19)	
Sweden	**250 – 350**	99–00	**–**	**20–29**	
Switzerland	1 – 10	98–02	F	20–29	
Turkey	(2,000 – 8,000)	01	(–)	(–)	
Ukraine	67,000 – 102,000	90–00	0	5–14	
Total (approx.)	460,000 – 1,000,000	**Overall trend** Unknown			
Breeding range	>3,000,000 km²	**Gen. length.** <3.3	**% Global pop.** 50–74		

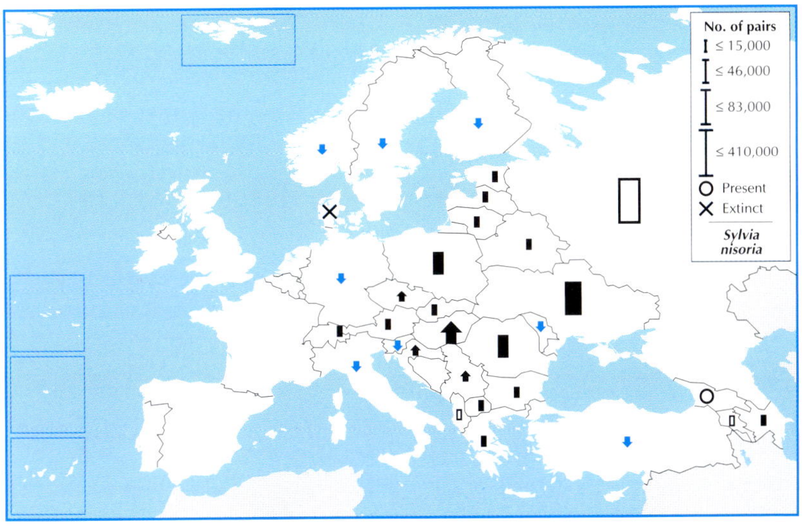

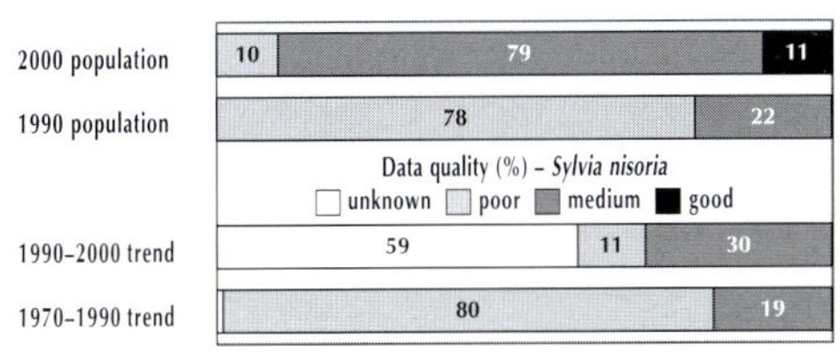

Sylvia curruca
LESSER WHITETHROAT

Non-SPEC (1994: —) **Status** Secure
Criteria —
European IUCN Red List Category —
Criteria —
Global IUCN Red List Category —
Criteria —

Sylvia curruca is a widespread summer visitor to much of Europe, which accounts for less than half of its global breeding range. Its European breeding population is very large (>4,800,000 pairs), and was stable between 1970–1990. Although the species declined in a number of countries during 1990–2000, key populations in Russia, Romania and Poland were stable, and trends were stable or increasing across the majority of its European range. The species hence remained stable overall, and consequently is evaluated as Secure.

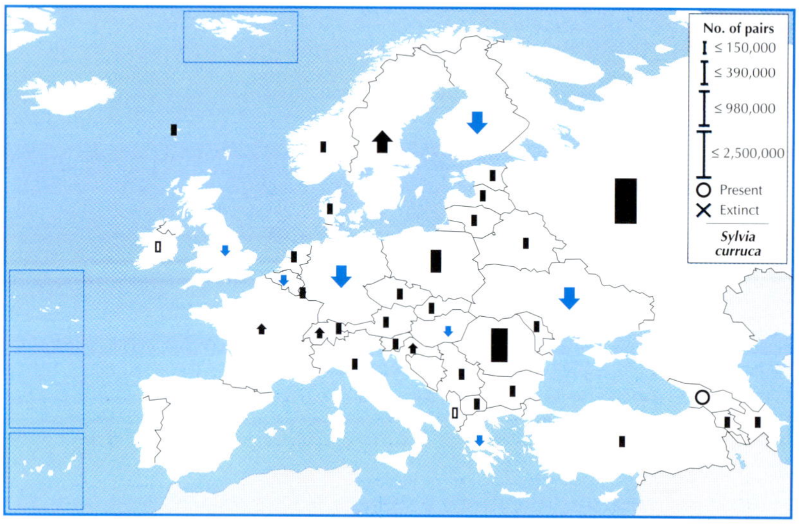

Country	Breeding pop. size (pairs)	Year(s)	Trend	Mag.%	References
Albania	(1,000 – 2,000)	02	?	–	
Armenia	66,000 – 120,000	99–02	0	0–19	
Austria	(35,000 – 70,000)	98–02	(0)	(0–19)	
Azerbaijan	(1,000 – 10,000)	96–00	(0)	(0–19)	
Belarus	120,000 – 150,000	97–02	0	0–19	
Belgium	2,500 – 10,000	01–02	–	20–29	*1*
Bulgaria	10,000 – 30,000	96–02	0	0–9	
Croatia	(5,000 – 10,000)	02	(+)	(50–79)	*16*
Czech Rep.	60,000 – 120,000	00	0	0–19	
Denmark	100,000 – 200,000	00	F	20–29	*12*
Faroe Is.	(0 – 1)	81	(0)	(0–19)	
Estonia	40,000 – 70,000	98	0	0–19	*1*
Finland	200,000 – 300,000	98–02	–	10	
France	20,000 – 80,000	98–02	+	11	*4,2*
Georgia	Present	03	?	–	
Germany	250,000 – 500,000	95–99	–	0–19	
Greece	(5,000 – 20,000)	95–00	(–)	(0–19)	
Hungary	**61,000 – 115,000**	99–02	–	20–49	*19*
Rep. Ireland	0 – 3	88–91	?	–	
Italy	(10,000 – 40,000)	03	(0)	(0–19)	
Latvia	40,000 – 80,000	90–00	(0)	(0–19)	*23,16*
Liechtenstein	100 – 150	98–00	0	0–19	
Lithuania	50,000 – 100,000	99–01	(0)	(0–19)	*20*
Luxembourg	1,000 – 2,000	02	0	0–19	
Macedonia	(2,500 – 10,000)	90–00	(0)	(0–19)	
Moldova	3,000 – 5,000	90–00	0	0–19	
Netherlands	**13,000 – 20,000**	98–00	**F**	**31**	*1*
Norway	(10,000 – 100,000)	90–03	(0)	(0–19)	
Poland	**300,000 – 500,000**	00–02	0	0–19	*23*
Romania	860,000 – 1,100,000	00–02	0	0–19	*48*
Russia	2,000,000 – 3,000,000	90–00	(0)	(0–19)	*122*
Serbia & MN	18,000 – 26,000	90–02	0	0–19	*1,29,155,172a, 67a,78,225,227*
Slovakia	40,000 – 80,000	90–99	0	0–19	
Slovenia	3,000 – 5,000	94	(0)	(0–19)	
Sweden	150,000 – 400,000	99–00	+	**32**	
Switzerland	15,000 – 20,000	98–02	+	0–9	
Turkey	(50,000 – 200,000)	01	(0)	(0–19)	
Ukraine	155,000 – 230,000	90–00	–	5–14	
UK	**64,000 – 64,000**	00	–	**20**	*5,31*

Total (approx.) 4,800,000 – 7,800,000 **Overall trend** Stable
Breeding range >6,000,000 km² **Gen. length.** <3.3 **% Global pop.** 25–49

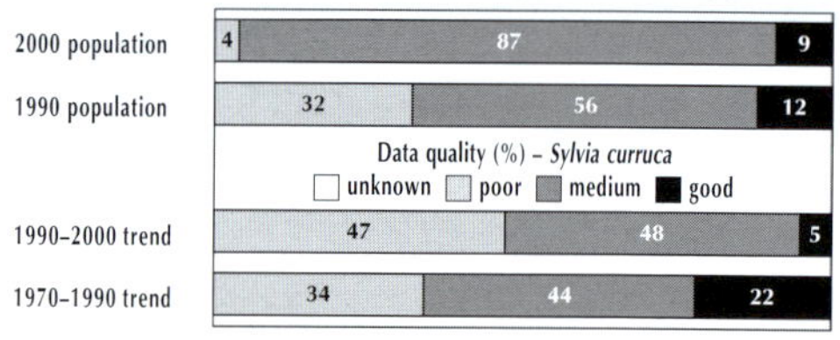

Sylvia communis
COMMON WHITETHROAT

Non-SPEC[E] (1994: 4) **Status** Secure
Criteria —
European IUCN Red List Category —
Criteria —
Global IUCN Red List Category —
Criteria —

Sylvia communis is a widespread summer visitor to most of Europe, which constitutes >50% of its global breeding range. Its European breeding population is extremely large (>14,000,000 pairs), and was stable between 1970–1990. The species underwent a small increase overall during 1990–2000, with the majority of national trends increasing or—as in the case of the key population in Russia—stable. Consequently, the species is evaluated as Secure.

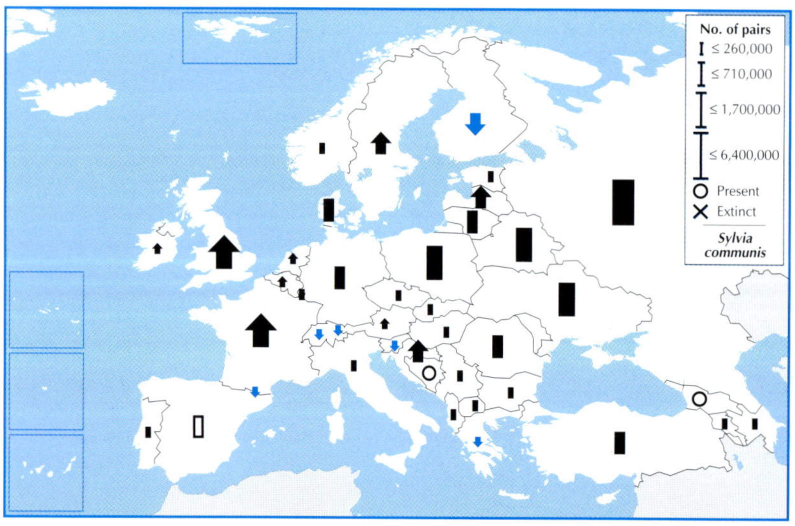

Country	Breeding pop. size (pairs)	Year(s)	Trend	Mag.%	References
Albania	5,000 – 15,000	02	(0)	(0–19)	
Andorra	(0 – 2)	99–01	(–)	(>80)	*1,3*
Armenia	130,000 – 230,000	99–02	0	0–19	
Austria	(20,000 – 40,000)	98–02	(+)	(20–29)	
Azerbaijan	(10,000 – 100,000)	96–00	(0)	(0–19)	
Belarus	800,000 – 900,000	97–02	0	0–19	
Belgium	20,000 – 100,000	01–02	+	20–29	*1*
Bosnia & HG	Present	90–03	?	–	
Bulgaria	75,000 – 200,000	96–02	0	0–9	
Croatia	(500,000 – 700,000)	02	(+)	(>80)	*16*
Czech Rep.	100,000 – 200,000	00	0	10–19	
Denmark	300,000 – 450,000	00	F	20–29	*12*
Estonia	(100,000 – 200,000)	98	0	0–19	*1*
Finland	250,000 – 400,000	98–02	–	**10**	
France	(400,000 – 2,000,000)	98–02	+	9	*4,2*
Georgia	Present	03	?	–	
Germany	250,000 – 500,000	95–99	0	0–19	
Greece	(50,000 – 100,000)	95–00	(–)	(20–29)	
Hungary	**210,000 – 320,000**	99–02	0	0–19	*19*
Rep. Ireland	20,000 – 100,000	88–91	+	0–19	
Italy	(50,000 – 200,000)	03	(0)	(0–19)	
Latvia	300,000 – 500,000	90–00	+	50–79	*23,16,2*
Liechtenstein	**0 – 2**	98–00	–	**30–49**	
Lithuania	400,000 – 500,000	99–01	(0)	(0–19)	*20*
Luxembourg	2,000 – 3,000	02	0	0–19	
Macedonia	(10,000 – 20,000)	90–00	(0)	(0–19)	
Moldova	30,000 – 35,000	90–00	+	0–19	
Netherlands	**130,000 – 150,000**	98–00	+	**100**	*1*
Norway	(50,000 – 300,000)	90–03	(0)	(0–19)	
Poland	**1,000,000 – 2,000,000**	00–02	0	0–19	*23*
Portugal	(2,500 – 25,000)	02	(0)	(0–19)	
Romania	500,000 – 800,000	00–02	0	0–19	*48*
Russia	5,000,000 – 8,000,000	90–00	(0)	(0–19)	*122*
Serbia & MN	40,000 – 60,000	95–02	0	0–19	*1,29,155,172a, 67a,78,225,227*
Slovakia	60,000 – 120,000	90–99	0	0–19	
Slovenia	8,000 – 10,000	99–00	(–)	(0–19)	
Spain	(450,000 – 600,000)	92	?	–	*13,12,10*
Sweden	**500,000 – 1,000,000**	99–00	+	**31**	
Switzerland	**1,000 – 2,000**	93–96	–	0–19	
Turkey	(300,000 – 900,000)	01	(0)	(0–19)	
Ukraine	1,460,000 – 1,900,000	90–00	F	20–29	
UK	**945,000 – 945,000**	00	+	**41**	*5,31*

Total (approx.) 14,000,000 – 25,000,000 **Overall trend** Small increase
Breeding range >8,000,000 km² **Gen. length.** <3.3 **% Global pop.** 50–74

(See p. 231, top, for data quality graph)

Sylvia borin
GARDEN WARBLER

Non-SPEC[E] (1994: 4) **Status** (Secure)
Criteria —
European IUCN Red List Category —
Criteria —
Global IUCN Red List Category —
Criteria —

Sylvia borin is a widespread summer visitor to much of Europe, which constitutes >75% of its global breeding range. Its European breeding population is extremely large (>17,000,000 pairs), and was stable between 1970–1990. Although there were declines in France during 1990–2000, other key populations in Sweden, Finland, Germany, Romania and Russia were stable, and the species probably remained stable overall. Consequently, it is provisionally evaluated as Secure.

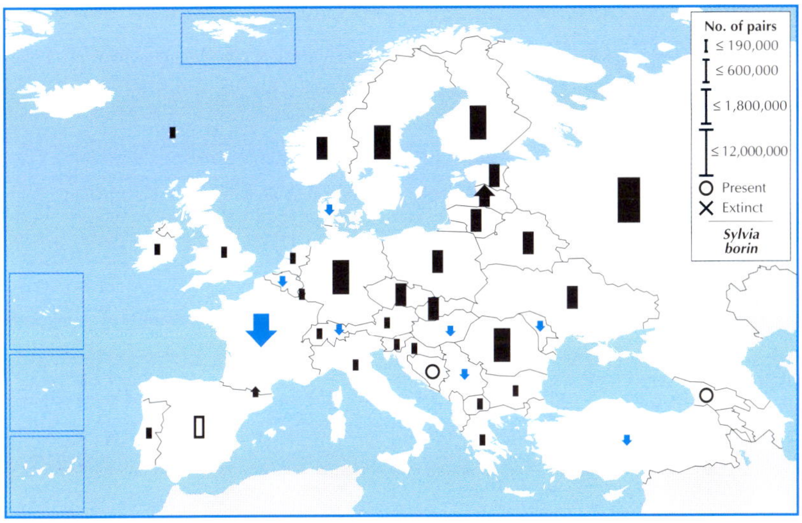

Country	Breeding pop. size (pairs)	Year(s)	Trend	Mag.%	References
Andorra	80 – 150	99–01	(+)	(0–19)	*1,3*
Armenia	3,300 – 6,600	97–02	?	–	
Austria	(10,000 – 20,000)	98–02	(0)	(0–19)	
Azerbaijan	(1,000 – 10,000)	96–00	(0)	(0–19)	
Belarus	500,000 – 600,000	97–02	0	0–19	
Belgium	20,000 – 100,000	01–02	(–)	(0–19)	*1*
Bosnia & HG	Present	85–89	?	–	
Bulgaria	50 – 150	96–02	(F)	(20–29)	
Croatia	(1,000 – 3,000)	02	(0)	(0–19)	*16*
Czech Rep.	200,000 – 400,000	00	0	0–19	
Denmark	150,000 – 200,000	00	–	30–49	*12*
Faroe Is.	(0 – 1)	92–93	(0)	(0–19)	
Estonia	200,000 – 400,000	98	0	0–19	*1*
Finland	800,000 – 1,200,000	98–02	**0**	**0–19**	
France	(800,000 – 3,200,000)	98–02	–	13	*4,3*
Georgia	Present	03	?	–	
Germany	800,000 – 1,400,000	95–99	0	0–19	
Greece	(100 – 1,000)	95–00	(0)	(0–19)	
Hungary	**18,000 – 40,000**	99–02	–	20–49	*19*
Rep. Ireland	330 – 400	95–97	(0)	(0–19)	
Italy	(10,000 – 50,000)	03	(0)	(0–19)	
Latvia	500,000 – 700,000	90–00	+	0–19	*23,18,2*
Liechtenstein	30 – 80	98–00	–	0–9	
Lithuania	350,000 – 450,000	99–01	(0)	(0–19)	*20*
Luxembourg	2,000 – 3,000	02	0	0–19	
Macedonia	(250 – 1,000)	90–00	(0)	(0–19)	
Moldova	5,000 – 5,500	90–00	–	0–19	
Netherlands	**120,000 – 150,000**	98–00	**0**	**7**	*1*
Norway	(200,000 – 700,000)	95–02	0	0–19	*30*
Poland	**300,000 – 500,000**	00–02	0	0–19	*23*
Portugal	(100 – 1,000)	02	(0)	(0–19)	
Romania	800,000 – 1,100,000	00–02	0	0–19	*48*
Russia	8,500,000 – 15,000,000	90–00	(0)	(0–19)	*122*
Serbia & MN	100 – 600	95–02	(–)	(10–29)	*1,29,172a,155, 62,224,227*
Slovakia	200,000 – 400,000	90–99	0	0–19	
Slovenia	(2,000 – 3,000)	94	(0)	(0–19)	
Spain	(400,000 – 700,000)	92	?	–	*13,12,10*
Sweden	1,000,000 – 3,000,000	99–00	**0**	**0**	
Switzerland	**100,000 – 150,000**	93–96	0	0–9	
Turkey	(10,000 – 25,000)	01	(–)	(0–19)	
Ukraine	460,000 – 670,000	90–00	0	0–9	
UK	**190,000 – 190,000**	00	**0**	**5**	*5,31*

Total (approx.) 17,000,000 – 31,000,000 **Overall trend** Stable
Breeding range >7,000,000 km² **Gen. length.** <3.3 **% Global pop.** 75–94
(See p. 234, top, for data quality graph)

Sylvia atricapilla
BLACKCAP

Non-SPEC[E] (1994: 4) **Status** Secure
Criteria —
European IUCN Red List Category —
Criteria —
Global IUCN Red List Category —
Criteria —

Sylvia atricapilla is a widespread breeder across much of Europe, which constitutes >75% of its global breeding range. Its European breeding population is extremely large (>25,000,000 pairs), and was stable between 1970–1990. The species underwent a slight increase overall during 1990–2000, with almost all national trends either stable or increasing, including those of key populations in France, Germany, Italy and Russia. Consequently, the species is evaluated as Secure.

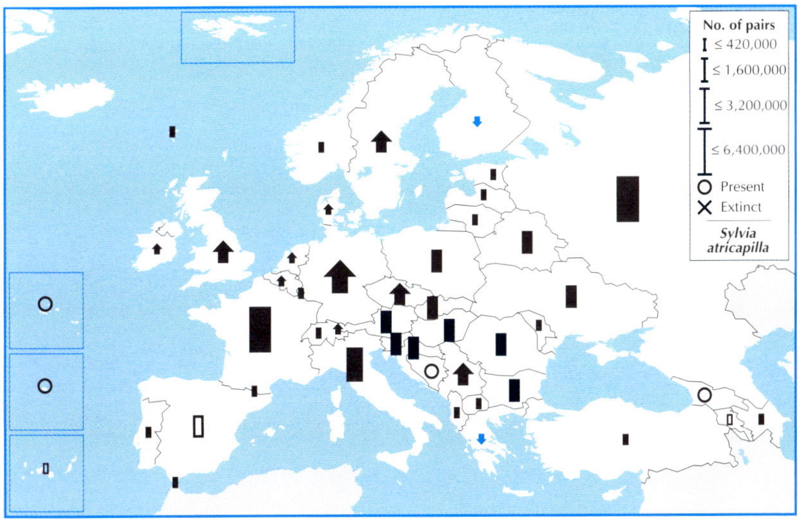

Country	Breeding pop. size (pairs)	Year(s)	Trend	Mag.%	References
Albania	20,000 – 50,000	02	(0)	(0–19)	
Andorra	200 – 300	99–01	(0)	(0–19)	*1,3*
Armenia	6,500 – 16,000	98–02	?	–	
Austria	(700,000 – 1,400,000)	98–02	(0)	(0–19)	
Azerbaijan	(10,000 – 50,000)	96–00	(0)	(0–19)	
Belarus	600,000 – 750,000	97–02	0	0–19	
Belgium	100,000 – 250,000	01–02	(+)	(0–19)	*1*
Bosnia & HG	Present	90–03	?	–	
Bulgaria	800,000 – 2,000,000	96–02	0	0–9	
Croatia	(500,000 – 1,000,000)	02	(0)	(0–19)	*16*
Czech Rep.	800,000 – 1,600,000	00	+	30–49	
Denmark	300,000 – 450,000	00	+	**30**	*12*
Faroe Is.	(0 – 1)	85	(0)	(0–19)	
Estonia	150,000 – 250,000	98	0	0–19	*1*
Finland	30,000 – 50,000	98–02	**–**	**30**	
France	(2,500,000 – 10,000,000)	98–02	0	0–19	*4,3*
Georgia	Present	03	?	–	
Germany	2,000,000 – 3,500,000	95–99	+	0–19	
Greece	(5,000 – 20,000)	95–00	(–)	(0–19)	
Hungary	**790,000 – 1,050,000**	99–02	0	0–19	*19*
Rep. Ireland	10,000 – 20,000	88–91	+	20–29	
Italy	(2,000,000 – 5,000,000)	03	(0)	(0–19)	
Latvia	350,000 – 500,000	90–00	(0)	(0–19)	*23,18*
Liechtenstein	400 – 600	98–00	+	0–9	
Lithuania	300,000 – 400,000	99–01	(0)	(0–19)	*20*
Luxembourg	15,000 – 18,000	02	0	0–19	
Macedonia	(60,000 – 150,000)	90–00	(0)	(0–19)	
Moldova	65,000 – 70,000	90–00	0	0–19	
Netherlands	**270,000 – 320,000**	98–00	+	**61**	*1*
Norway	(200,000 – 700,000)	90–03	(0)	(0–19)	
Poland	**1,200,000 – 2,000,000**	00–02	0	0–19	*23*
Portugal	(50,000 – 500,000)	02	(0)	(0–19)	
Azores	Present	02	?	–	
Madeira	Present	02	?	–	
Romania	650,000 – 900,000	00–02	0	0–19	*48*
Russia	5,000,000 – 8,000,000	90–00	(0)	(0–19)	*122*
Serbia & MN	1,000,000 – 1,500,000	90–02	+	0–19	*1,29,172a,67a, 78,225,227,152*
Slovakia	800,000 – 1,000,000	90–99	0	0–19	
Slovenia	600,000 – 900,000	00	(0)	(0–19)	
Spain	(850,000 – 1,500,000)	98–02	?	–	*13,12*
Canary Is.	(10,000 – 20,000)	97–03	?	–	*28,25*
Sweden	**400,000 – 1,000,000**	99–00	+	**32**	
Switzerland	**200,000 – 300,000**	93–96	0	0–19	
Turkey	(80,000 – 160,000)	01	(0)	(0–19)	
Ukraine	930,000 – 1,150,000	90–00	0	0–19	
UK	**932,000 – 932,000**	00	+	**58**	*5,31*
Gibraltar	100 – 200	00	**0**	**0–19**	

Total (approx.) 25,000,000 – 49,000,000 **Overall trend** Small increase
Breeding range >7,000,000 km² **Gen. length.** <3.3 **% Global pop.** 75–94
(See p. 234, bottom, for data quality graph)

Phylloscopus trochiloides
GREENISH WARBLER

Non-SPEC (1994: —) **Status** (Secure)
Criteria —
European IUCN Red List Category —
Criteria —
Global IUCN Red List Category —
Criteria —

Phylloscopus trochiloides (which here includes 'Green Warbler' *P. t. nitidus*) is a widespread summer visitor to the Caucasus and north-eastern Europe, which accounts for less than a quarter of its global breeding range. Its European breeding population is extremely large (>12,000,000 pairs), and was stable between 1970–1990. The species probably remained stable overall during 1990–2000, with all populations either stable—as in the Russian stronghold—or increasing. Consequently, it is provisionally evaluated as Secure.

Country	Breeding pop. size (pairs)	Year(s)	Trend	Mag.%	References
Armenia	16,000 – 33,000	98–02	0	0–19	
Azerbaijan	(100,000 – 250,000)	96–00	(0)	(0–19)	
Belarus	2,000 – 3,000	97–02	0	0–19	
Czech Rep.	**5 – 10**	00	+	**>80**	
Denmark	0 – 1	99–00	**0**	**0–19**	*16*
Estonia	(1,000 – 5,000)	98	F	20–29	*1*
Finland	3,000 – 10,000	98–02	+	30	
Georgia	Present	03	?	–	
Germany	**6 – 6**	95–99	+	**30–49**	
Latvia	500 – 2,000	90–00	(0)	(0–19)	*23,18*
Lithuania	(1,500 – 2,500)	99–01	(0)	(0–19)	*20*
Poland	**0 – 30**	95–00	**F**	**50–79**	*2,60–69*
Russia	12,000,000 – 15,000,000	90–00	(0)	(0–19)	*122*
Slovakia	10 – 100	90–99	+	N	
Sweden	40 – 100	99–00	F	20–29	
Turkey	(3,000 – 12,000)	01	(0)	(0–19)	
Ukraine	40 – 70	90–00	+	**>80**	
Total (approx.)	12,000,000 – 15,000,000	**Overall trend** Stable			
Breeding range	>2,000,000 km²	**Gen. length.** <3.3		**% Global pop.** 5–24	

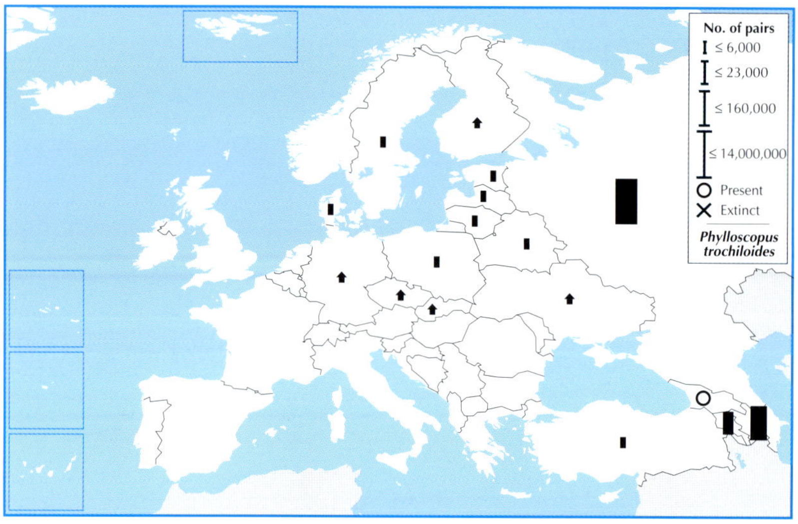

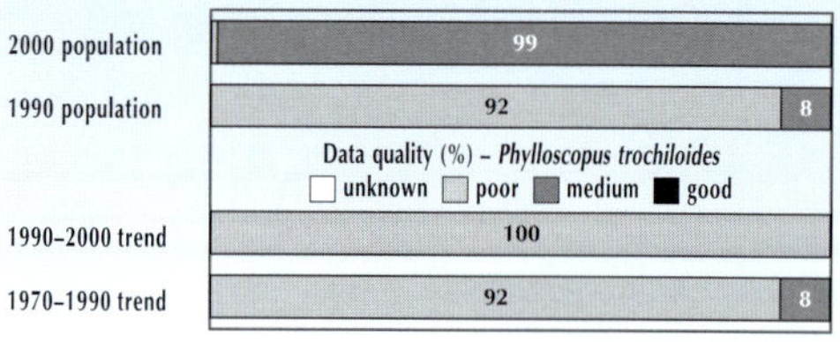

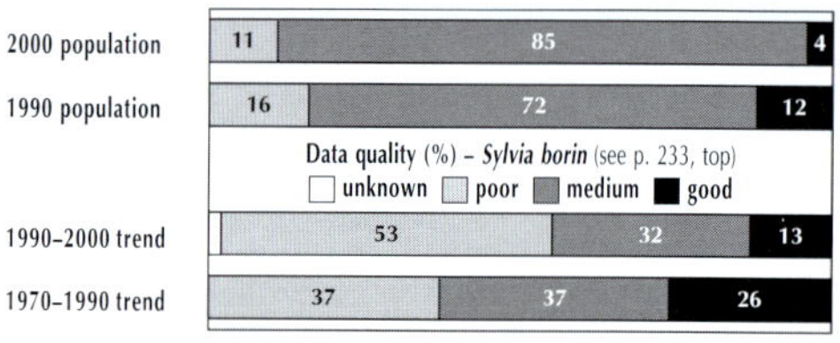

Phylloscopus borealis
ARCTIC WARBLER

Non-SPEC (1994: —) **Status** (Secure)
Criteria —
European IUCN Red List Category —
Criteria —
Global IUCN Red List Category —
Criteria —

Phylloscopus borealis is a widespread summer visitor to the boreal zone of Fennoscandia and Russia, with Europe accounting for less than a quarter of its global breeding range. Its European breeding population is very large (>4,500,000 pairs), and was stable between 1970–1990. No trend data were available for the stronghold population in Russia during 1990–2000, but the small populations in Norway, Sweden and Finland were all stable, and the species is provisionally evaluated as Secure.

Country	Breeding pop. size (pairs)	Year(s)	Trend	Mag.%	References
Finland	2,000 – 5,000	98–02	(0)	(0–19)	
Norway	(10 – 100)	90–02	0	0–19	*26*
Russia	4,500,000 – 8,000,000	90–00	?	–	*122*
Sweden	**10 – 100**	99–00	(F)	(20–29)	
Total (approx.)	4,500,000 – 8,000,000	**Overall trend** Unknown			
Breeding range	>1,000,000 km²	**Gen. length.** <3.3		**% Global pop.** 5–24	

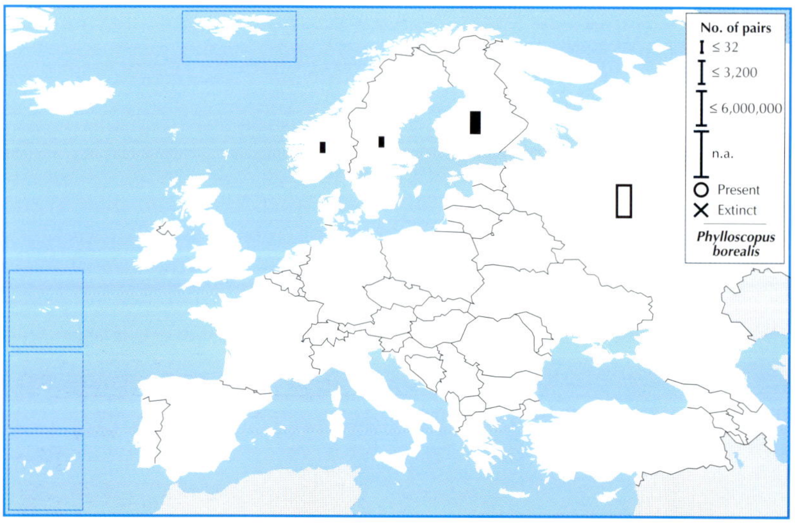

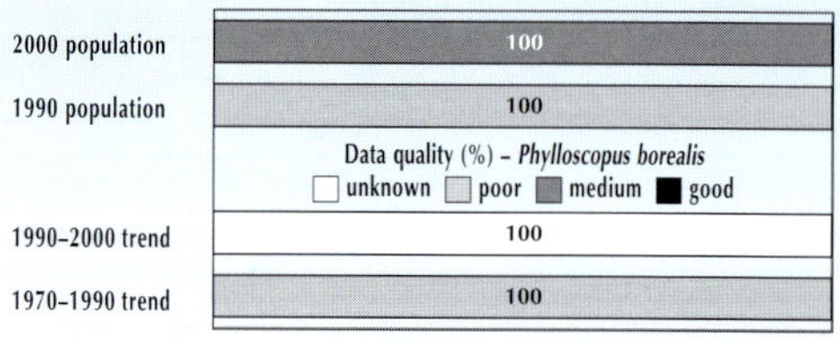

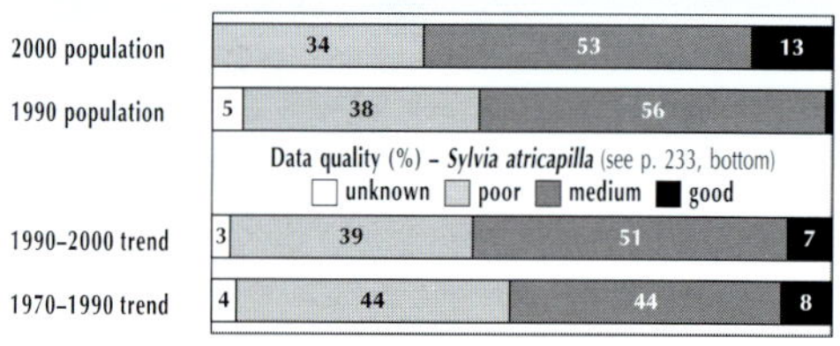

Phylloscopus inornatus
INORNATE WARBLER

Non-SPEC (1994: —) **Status** (Secure)
Criteria —
European IUCN Red List Category —▼
Criteria —
Global IUCN Red List Category —
Criteria —

Country	Breeding pop. size (pairs)	Year(s)	Trend	Mag.%	References
Russia	(5,000 – 35,000)	90–00	?	–	*122*
Total (approx.)	5,000 – 35,000	**Overall trend** Unknown			
Breeding range	>100,000 km²	**Gen. length.** <3.3		**% Global pop.** <5	

Phylloscopus inornatus has a predominantly Asian breeding distribution, which just extends into Europe in northern Russia. Its European breeding population is small (as few as 5,000 pairs), but the trend between 1970–1990 was unknown. No trend data were available for 1990–2000, but there was no evidence to suggest that the species declined. Although the size of the European population could render it susceptible to the risks affecting small populations, it is marginal to a much larger non-European population. Consequently, the species is provisionally evaluated as Secure.

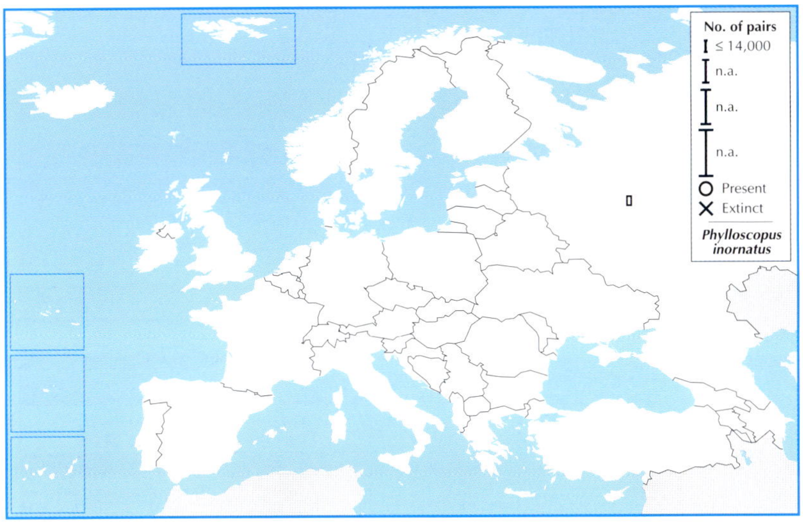

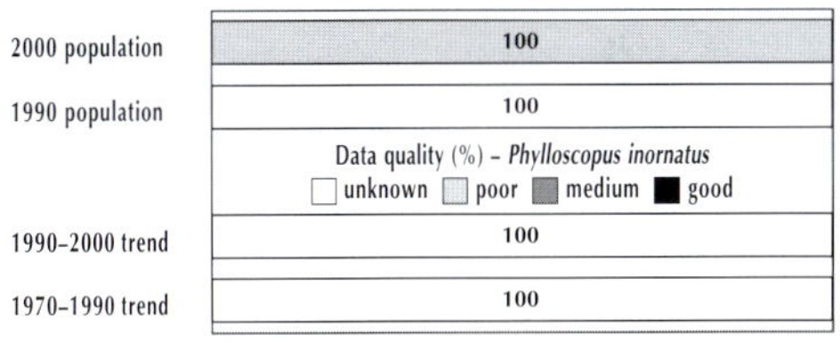

Phylloscopus bonelli
BONELLI'S WARBLER

SPEC 2 (1994: 4) **Status** (Declining)
Criteria Moderate recent decline
European IUCN Red List Category —
Criteria —
Global IUCN Red List Category —
Criteria —

Country	Breeding pop. size (pairs)	Year(s)	Trend	Mag.%	References
Albania	(100 – 500)	02	?	–	
Andorra	250 – 400	99–01	(0)	(0–19)	*1,3*
Austria	(35,000 – 65,000)	98–02	(0)	(0–19)	
Bulgaria	5,000 – 20,000	96–02	+	30–49	
Croatia	(0 – 20)	02	(–)	(–)	*70*
France	(100,000 – 400,000)	98–02	–	51	*4,3*
Germany	22,000 – 40,000	95–99	–	0–19	
Greece	(10,000 – 30,000)	95–00	(0)	(0–19)	
Italy	(50,000 – 100,000)	03	?	–	
Liechtenstein	50 – 100	98–00	0	0–19	
Macedonia	(3,000 – 8,000)	90–00	(0)	(0–19)	
Netherlands	**0 – 3**	98–00	?	–	*1*
Portugal	(10,000 – 100,000)	02	(0)	(0–19)	
Romania	(10 – 25)	00–02	F	50–79	*59*
Serbia & MN	(50 – 100)	90–02	(–)	(0–19)	*1,117a*
Slovenia	300 – 500	99–00	(0)	(0–19)	
Spain	(1,100,000 – 2,700,000)	92	?	–	*13,12,10*
Switzerland	15,000 – 30,000	98–02	0	0–19	
Turkey	(10,000 – 50,000)	01	(–)	(0–19)	
Total (approx.)	1,400,000 – 3,500,000	**Overall trend** Moderate decline			
Breeding range	>1,000,000 km²	**Gen. length.** <3.3		**% Global pop.** 75–94	

Phylloscopus bonelli is a widespread summer visitor to southern Europe, which constitutes >75% of its global breeding range. Its European breeding population is very large (>1,400,000 pairs), and was stable between 1970–1990. Although trend data were not available for the large Spanish population during 1990–2000, the species declined in a number of countries, and the large French population halved in size. Hence, this previously Secure species probably underwent a moderate decline (>10%) overall, and consequently it is now provisionally evaluated as Declining.

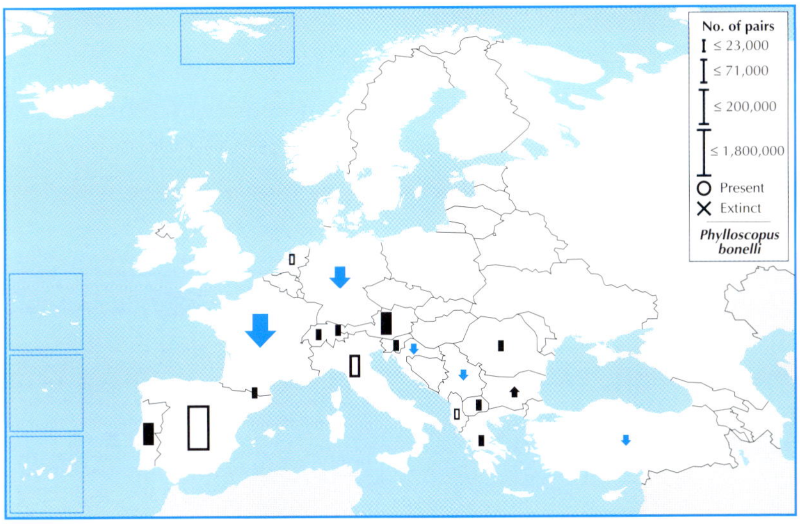

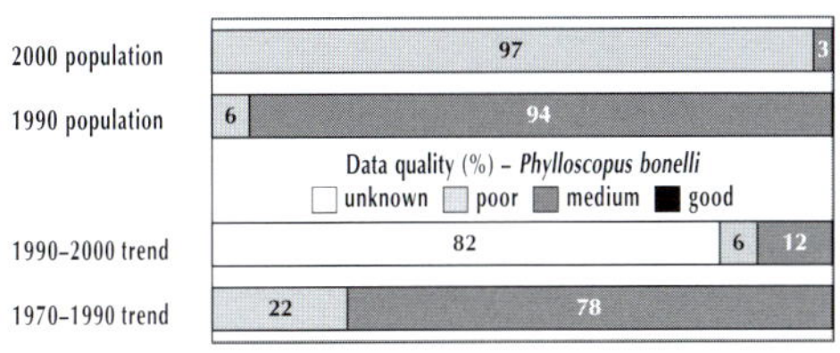

Phylloscopus sibilatrix
WOOD WARBLER

SPEC 2 (1994: 4) **Status** (Declining)
Criteria Moderate recent decline
European IUCN Red List Category —
Criteria —
Global IUCN Red List Category —
Criteria —

Phylloscopus sibilatrix is a widespread summer visitor to much of Europe, which constitutes >75% of its global breeding range. Its European breeding population is extremely large (>14,000,000 pairs), and was stable between 1970–1990. Although most populations in eastern Europe remained stable during 1990–2000 (the trend of the key population in Russia was unknown), the species declined in the north and west, and underwent a moderate decline (>10%) overall. Consequently, this previously Secure species is now provisionally evaluated as Declining.

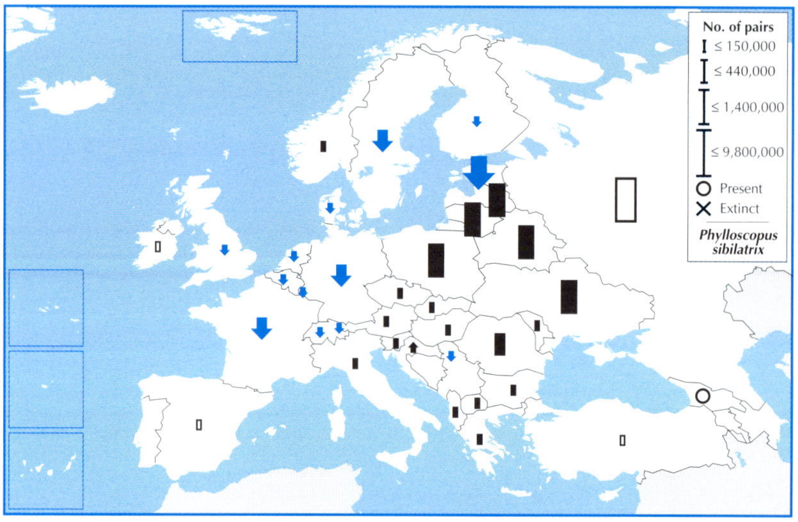

Country	Breeding pop. size (pairs)	Year(s)	Trend	Mag.%	References
Albania	(0 – 100)	02	(0)	(0–19)	
Austria	(50,000 – 90,000)	98–02	(0)	(0–19)	
Belarus	1,200,000 – 1,500,000	97–02	0	0–19	
Belgium	5,300 – 15,000	01–02	–	0–19	*1*
Bulgaria	10,000 – 25,000	96–02	0	0–9	
Croatia	(1,500 – 2,000)	02	(+)	(20–29)	*70,16*
Czech Rep.	80,000 – 160,000	00	0	0–19	
Denmark	5,000 – 15,000	00	**–**	**50–79**	*12*
Estonia	500,000 – 1,000,000	98	–	20–29	*1*
Finland	100,000 – 200,000	98–02	**–**	**50**	
France	(100,000 – 400,000)	98–02	**–**	**72**	*4,3*
Georgia	Present	03	?	–	
Germany	320,000 – 600,000	95–99	–	0–19	
Greece	(500 – 2,000)	95–00	(0)	(0–19)	
Hungary	**59,000 – 135,000**	99–02	0	0–19	*19*
Rep. Ireland	10 – 20	88–91	?	–	
Italy	(10,000 – 50,000)	03	(0)	(0–19)	
Latvia	1,000,000 – 1,300,000	90–00	(0)	(0–19)	*23,18*
Liechtenstein	30 – 60	98–00	–	20–29	
Lithuania	500,000 – 1,000,000	99–01	(0)	(0–19)	*20*
Luxembourg	3,000 – 5,000	02	–	20–29	
Macedonia	(100 – 300)	90–00	(0)	(0–19)	
Moldova	15,000 – 25,000	90–00	0	0–19	
Netherlands	**2,000 – 3,500**	98–00	**–**	**29**	*1*
Norway	(1,000 – 10,000)	90–02	0	0–19	
Poland	**700,000 – 1,000,000**	00–02	0	0–19	*23*
Romania	(260,000 – 460,000)	00–02	0	0–19	*48*
Russia	8,000,000 – 12,000,000	90–00	?	–	*122*
Serbia & MN	3,500 – 6,000	90–02	(–)	(10–19)	*1,29,155,172a,67a, 227,117a,215,224*
Slovakia	100,000 – 200,000	90–99	0	0–19	
Slovenia	3,000 – 5,000	94	(0)	(0–19)	
Spain	(1 – 3)	98–02	?	–	*10*
Sweden	**200,000 – 250,000**	99–00	**–**	**13**	
Switzerland	10,000 – 15,000	98–02	–	50–79	
Turkey	(50 – 500)	01	?	–	
Ukraine	(950,000 – 1,250,000)	90–00	0	5–9	
UK	9,000 – 10,500	00	**–**	**43**	*22,31*
Total (approx.)	14,000,000 – 22,000,000	**Overall trend** Moderate decline			
Breeding range	>6,000,000 km²	**Gen. length.** <3.3 **% Global pop.** 75–94			

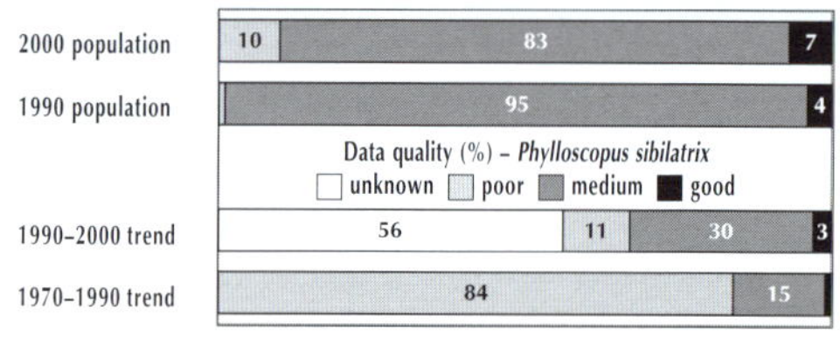

Phylloscopus sindianus
MOUNTAIN CHIFFCHAFF

SPEC 3 (1994: —) **Status** Declining
Criteria Moderate recent decline
European IUCN Red List Category —
Criteria —
Global IUCN Red List Category —
Criteria —

Phylloscopus sindianus breeds in the Caucasus and parts of Turkey, with Europe accounting for less than a quarter of its global breeding range. Its European breeding population is relatively large (>71,000 pairs), and was stable between 1970–1990. Although the species was stable in Azerbaijan during 1990–2000, and trends for Armenia, Georgia and Turkey were unknown, the Russian stronghold declined, and the species underwent a moderate decline (>10%) overall. Consequently, this previously Secure species is now evaluated as Declining.

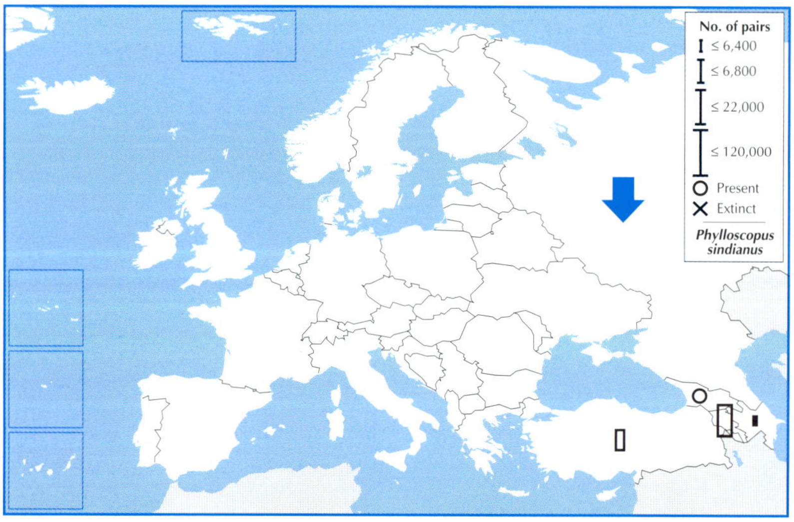

Country	Breeding pop. size (pairs)	Year(s)	Trend	Mag.%	References
Armenia	16,000 – 30,000	98–02	?	–	
Azerbaijan	(2,000 – 20,000)	96–00	(0)	(0–19)	
Georgia	Present	03	?	–	
Russia	(50,000 – 250,000)	90–00	–	20–29	*61,116,136*
Turkey	(3,000 – 15,000)	01	?	–	
Total (approx.)	71,000 – 320,000	**Overall trend** Moderate decline			
Breeding range	>100,000 km²	**Gen. length.** <3.3 **% Global pop.** 5–24			

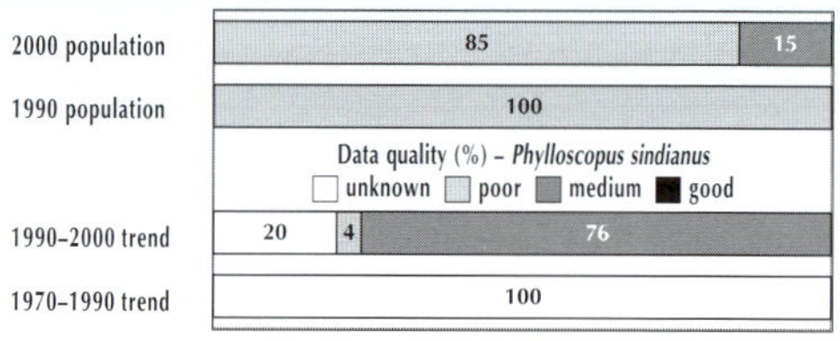

Phylloscopus collybita
COMMON CHIFFCHAFF

Non-SPEC (1994: —) **Status** Secure
Criteria —
European IUCN Red List Category —
Criteria —
Global IUCN Red List Category —
Criteria —

Phylloscopus collybita is a widespread breeder across much of Europe, which accounts for less than half of its global breeding range. Its European breeding population is extremely large (>30,000,000 pairs), and was stable between 1970–1990. Although there were declines in a few countries—notably France—during 1990–2000, populations were stable or increased across the majority of its European range, and the species remained stable overall. Consequently, it is evaluated as Secure.

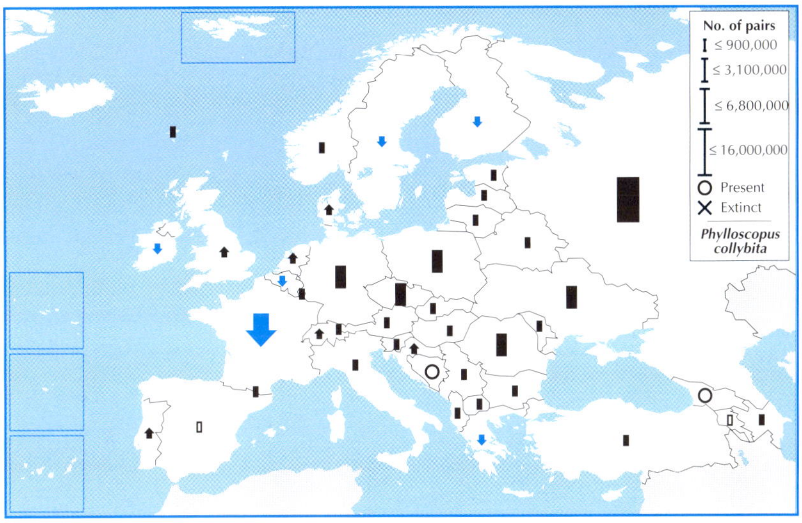

Country	Breeding pop. size (pairs)	Year(s)	Trend	Mag.%	References
Albania	2,000 – 5,000	02	(0)	(0–19)	
Andorra	300 – 500	99–01	(0)	(0–19)	*1,3*
Armenia	830 – 3,300	98–02	?	–	
Austria	(600,000 – 1,000,000)	98–02	(0)	(0–19)	
Azerbaijan	(10,000 – 100,000)	96–00	(0)	(0–19)	
Belarus	850,000 – 950,000	97–02	0	0–19	
Belgium	100,000 – 250,000	01–02	–	0–19	*1*
Bosnia & HG	Present	85–89	?	–	
Bulgaria	300,000 – 1,000,000	96–02	0	0–9	
Croatia	(500,000 – 800,000)	02	(+)	(>80)	*16,57*
Czech Rep.	800,000 – 1,600,000	00	0	0–19	
Denmark	200,000 – 300,000	00	+	**10–19**	*12*
Faroe Is.	(0 – 1)	82	(0)	(0–19)	
Estonia	(200,000 – 500,000)	98	0	0–19	*1*
Finland	100,000 – 150,000	98–02	–	**40**	
France	(3,000,000 – 15,000,000)	98–02	–	**15**	*4*
Georgia	Present	03	?	–	
Germany	2,400,000 – 4,000,000	95–99	0	0–19	
Greece	(20,000 – 50,000)	95–00	(–)	(0–19)	
Hungary	**520,000 – 720,000**	99–02	0	0–19	*19*
Rep. Ireland	20,000 – 100,000	88–91	–	0–19	
Italy	(300,000 – 800,000)	03	(0)	(0–19)	
Latvia	500,000 – 600,000	90–00	(0)	(0–19)	*23,18*
Liechtenstein	400 – 600	98–00	0	0–19	
Lithuania	400,000 – 800,000	99–01	(0)	(0–19)	*20*
Luxembourg	25,000 – 30,000	02	0	0–19	
Macedonia	(50,000 – 300,000)	90–00	(0)	(0–19)	
Moldova	65,000 – 70,000	90–00	0	0–19	
Netherlands	**550,000 – 600,000**	98–00	+	**88**	*1*
Norway	(100,000 – 500,000)	90–02	0	0–19	
Poland	**1,000,000 – 1,500,000**	00–02	0	0–19	*23*
Portugal	(10 – 100)	02	+	N	
Romania	980,000 – 1,600,000	00–02	(0)	(0–19)	*48*
Russia	1,200,0000 – 20,000,000	90–00	0	0–19	*122*
Serbia & MN	450,000 – 750,000	90–02	0	0–19	*1,29,172a,67a, 78,225,227,37, 62,144,117a*
Slovakia	600,000 – 1,000,000	90–99	0	0–19	
Slovenia	400,000 – 600,000	00	(0)	(0–19)	
Spain	(40,000 – 60,000)	98–02	?	–	*10*
Sweden	**100,000 – 400,000**	99–00	–	**29**	
Switzerland	200,000 – 300,000	98–02	+	10–19	
Turkey	(100,000 – 400,000)	01	(0)	(0–19)	
Ukraine	1,600,000 – 2,200,000	90–00	0	0–19	
UK	**807,000 – 807,000**	00	+	**17**	*5,31*

Total (approx.) 30,000,000 – 60,000,000 **Overall trend** Stable
Breeding range >7,000,000 km² **Gen. length.** <3.3 **% Global pop.** 25–49
(See p. 237, bottom, for data quality graph)

Phylloscopus ibericus
IBERIAN CHIFFCHAFF

Non-SPEC[E] (1994: NE) **Status** (Secure)
Criteria —
European IUCN Red List Category —
Criteria —
Global IUCN Red List Category —
Criteria —

Phylloscopus ibericus breeds in Portugal, Spain and south-west France, which together constitute >90% of its global breeding range. Its European breeding population is large (>360,000 pairs), but the trend between 1970–1990 was unknown (due to confusion with *P. collybita*, from which it has only recently been separated). Although trend data were again unavailable for the key populations in Spain and France during 1990–2000, the species was stable in Portugal, and there was no evidence to suggest that it declined overall. Consequently, it is provisionally evaluated as Secure.

Country	Breeding pop. size (pairs)	Year(s)	Trend	Mag.%	References
France	(10,000 – 30,000)	98–02	?	–	*4*
Portugal	(10,000 – 100,000)	02	(0)	(0–19)	
Spain	(340,000 – 400,000)	98–02	?	–	*10*

Total (approx.) 360,000 – 530,000 **Overall trend** Unknown
Breeding range >500,000 km² **Gen. length.** <3.3 **% Global pop.** >90

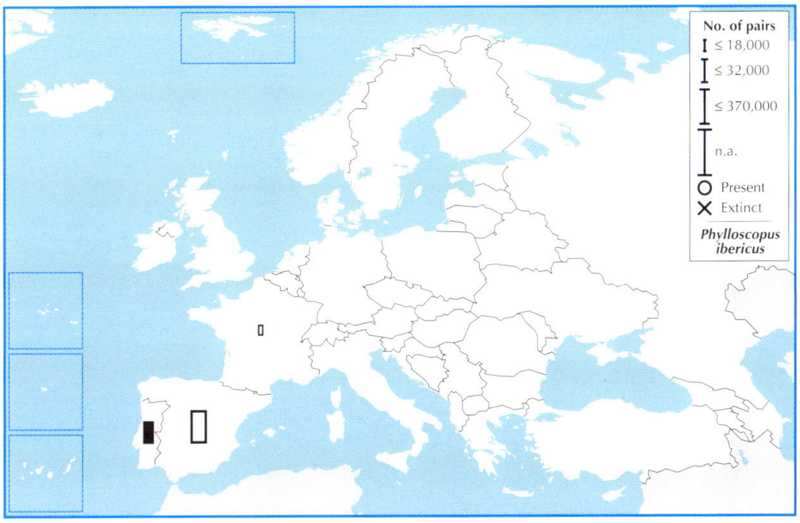

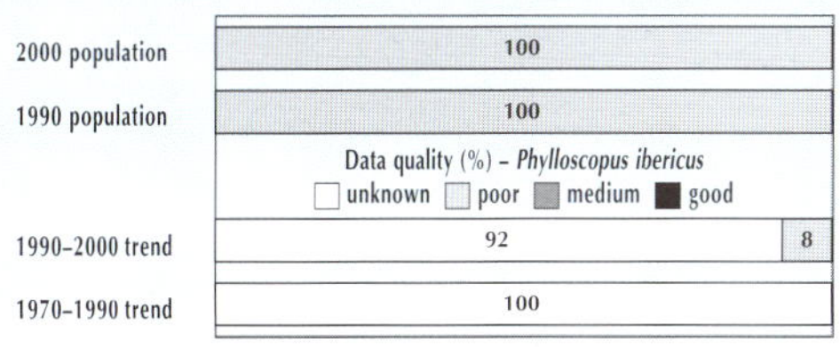

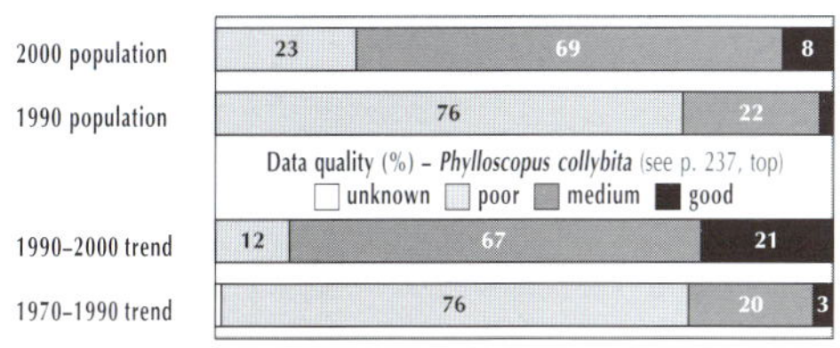

Phylloscopus canariensis
CANARY ISLANDS CHIFFCHAFF

Non-SPEC[E] (1994: NE) **Status** (Secure)
Criteria —
European IUCN Red List Category —
Criteria —
Global IUCN Red List Category —
Criteria —

Country	Breeding pop. size (pairs)	Year(s)	Trend	Mag.%	References
Spain					
Canary Is.	(20,000 – 100,000)	97–03	?	–	*28,25*
Total (approx.)	20,000 – 100,000	**Overall trend** Unknown			
Breeding range	>20,000 km²	**Gen. length.** <3.3	**% Global pop.** 100		

Phylloscopus canariensis (which has only recently been split as a separate species from *P. collybita*) is endemic to Europe, where it is confined to the Canary Islands. Its breeding population is relatively small (<100,000 pairs), but its trend between 1970–1990 was unknown. Trend data were also unavailable during 1990–2000, but there was no evidence to suggest that the species declined, and consequently it is provisionally evaluated as Secure.

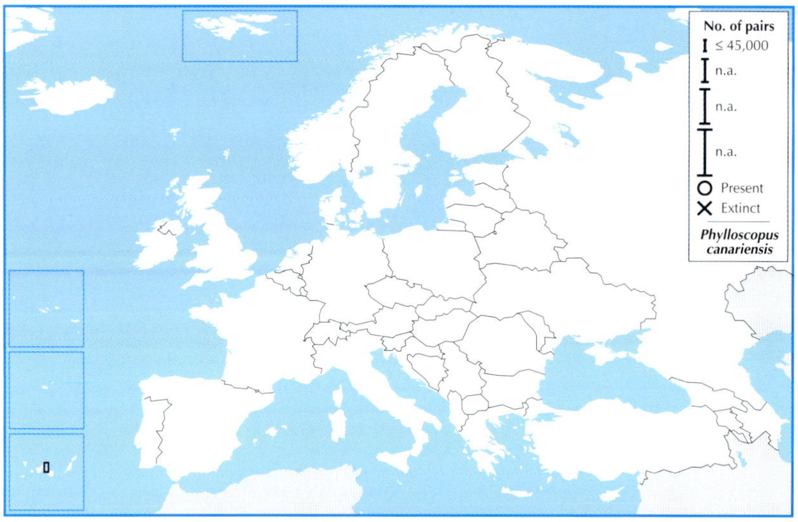

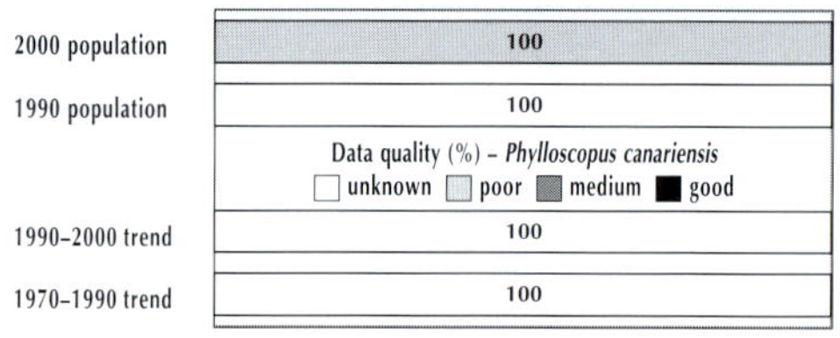

Phylloscopus trochilus
WILLOW WARBLER

Non-SPEC (1994: —) **Status** Secure
Criteria —
European IUCN Red List Category —
Criteria —
Global IUCN Red List Category —
Criteria —

Country	Breeding pop. size (pairs)	Year(s)	Trend	Mag.%	References
Albania	(0 – 50)	02	?	–	
Austria	(20,000 – 40,000)	98–02	(0)	(0–19)	
Belarus	950,000 – 1,100,000	97–02	0	0–19	
Belgium	20,000 – 100,000	01–02	–	0–19	*1*
Bulgaria	(0 – 5)	96–02	?	–	
Croatia	50 – 100	02	(+)	(–)	*16,54*
Czech Rep.	500,000 – 1,000,000	00	0	0–19	
Denmark	400,000 – 600,000	00	**–**	**10–19**	*12*
Faroe Is.	(0 – 1)	85	(0)	(0–19)	
Estonia	(800,000 – 2,000,000)	98	+	20–29	*1*
Finland	7,000,000 – 11,000,000	98–02	**–**	**15**	
France	(1,500,000 – 7,500,000)	98–02	**–**	**51**	*4,3*
Georgia	Present	03	?	–	
Germany	1,700,000 – 2,800,000	95–99	–	0–19	
Greece	(10 – 100)	95–00	(0)	(0–19)	
Hungary	54,000 – 115,000	99–02	+	50–79	*19*
Rep. Ireland	500,000 – 1,000,000	88–91	0	0–19	
Latvia	500,000 – 600,000	90–00	(0)	(0–19)	*23,18*
Liechtenstein	30 – 60	98–00	+	0–19	
Lithuania	(400,000 – 600,000)	99–01	(0)	(0–19)	*20*
Luxembourg	8,000 – 12,000	02	0	0–19	
Moldova	1,000 – 1,500	90–00	–	20–29	
Netherlands	**450,000 – 550,000**	98–00	**0**	**5**	*1*
Norway	(2,000,000 – 10,000,000)	95–02	0	0–19	*30*
Poland	**1,000,000 – 2,000,000**	00–02	0	0–19	*23*
Romania	60,000 – 85,000	00–02	(0)	(0–19)	*48*
Russia	25,000,000 – 40,000,000	90–00	0	0–19	*122*
Slovakia	400,000 – 600,000	90–99	0	0–19	
Slovenia	200 – 300	99–00	–	10–19	
Spain	(50 – 250)	98–02	?	–	*10*
Sweden	**10,000,000 – 16,000,000**	99–00	**–**	**14**	
Switzerland	4,000 – 8,000	98–02	–	30–49	
Ukraine	420,000 – 760,000	90–00	0	0–19	
UK	**2,125,000 – 2,125,000**	00	**–**	**15**	*31*
Total (approx.)	56,000,000 – 100,000,000	**Overall trend** Small decline			
Breeding range	>6,000,000 km²	**Gen. length.** <3.3	**% Global pop.** 25–49		

Phylloscopus trochilus is a widespread summer visitor to much of Europe (except the south), which accounts for less than half of its global breeding range. Its European breeding population is extremely large (>56,000,000 pairs), and was stable between 1970–1990. Although there were declines in a number of countries—notably Sweden and Finland—during 1990–2000, populations were stable in Russia and across much of central and eastern Europe, and the species underwent only a small decline overall. Consequently, it is evaluated as Secure.

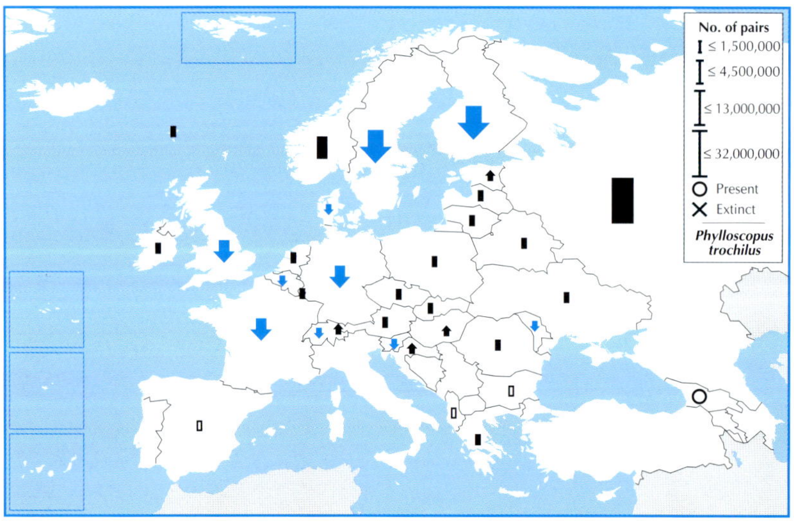

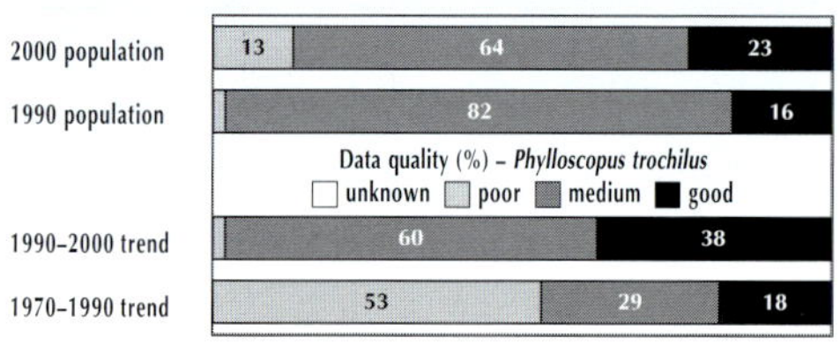

Regulus regulus
GOLDCREST

Non-SPEC[E] (1994: 4) **Status** Secure
Criteria —
European IUCN Red List Category —
Criteria —
Global IUCN Red List Category —
Criteria —

Regulus regulus is a widespread breeder across much of Europe, which constitutes >50% of its global range. Its European breeding population is extremely large (>19,000,000 pairs), and was stable between 1970–1990. Although there were some declines in Sweden and Germany during 1990–2000, populations were stable or increased across most of the rest of Europe—including the key one in Russia—and the species remained stable overall. Consequently, it is evaluated as Secure.

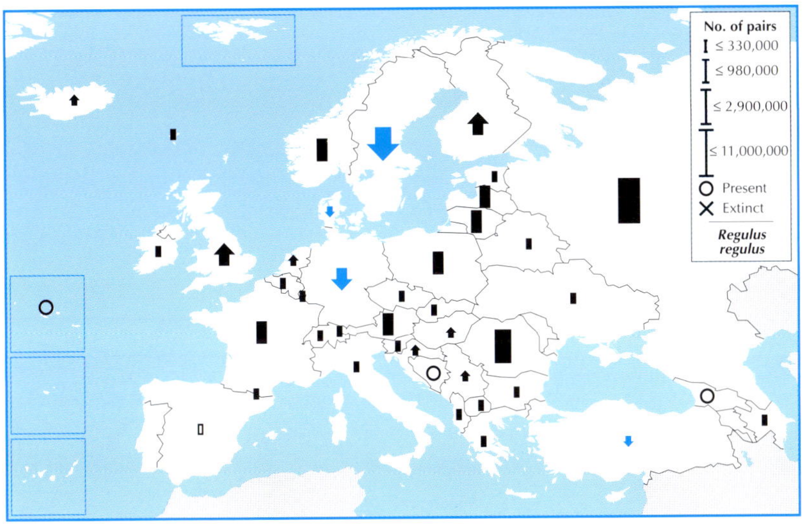

Country	Breeding pop. size (pairs)	Year(s)	Trend	Mag.%	References
Albania	(1,000 – 4,000)	02	(0)	(0–19)	
Andorra	(500 – 1,000)	99–01	(0)	(0–19)	*1,3*
Austria	(600,000 – 1,200,000)	98–02	(0)	(0–19)	
Azerbaijan	(500 – 5,000)	96–00	(0)	(0–19)	
Belarus	300,000 – 350,000	97–02	0	0–19	
Belgium	20,000 – 100,000	01–02	(0)	(0–19)	*1*
Bosnia & HG	Present	90–03	?	–	
Bulgaria	150,000 – 300,000	96–02	0	0–9	
Croatia	(5,000 – 10,000)	02	(+)	(>80)	*16*
Czech Rep.	200,000 – 400,000	00	0	0–19	
Denmark	50,000 – 150,000	00	–	30–49	*12*
Faroe Is.	10 – 10	95	(0)	(0–19)	
Estonia	100,000 – 300,000	98	F	20–29	*1*
Finland	600,000 – 1,600,000	98–02	+	**25**	
France	(250,000 – 1,000,000)	98–02	0	1	*4,2*
Georgia	Present	03	?	–	
Germany	740,000 – 1,200,000	95–99	–	0–19	
Greece	(1,000 – 5,000)	95–00	(0)	(0–19)	
Hungary	(1,000 – 5,000)	90–93	+	50–79	*14*
Iceland	(50 – 250)	00	(+)	(N)	*3,24,42,44*
Rep. Ireland	100,000 – 250,000	88–91	0	0–19	
Italy	(200,000 – 400,000)	03	(0)	(0–19)	
Latvia	500,000 – 700,000	90–00	(0)	(0–19)	*23,18*
Liechtenstein	600 – 1,000	98–00	0	0–19	
Lithuania	(400,000 – 600,000)	99–01	(0)	(0–19)	*20*
Luxembourg	12,000 – 15,000	02	0	0–19	
Macedonia	(2,000 – 10,000)	90–00	(0)	(0–19)	
Netherlands	**40,000 – 50,000**	98–00	+	**45**	*1*
Norway	(500,000 – 1,000,000)	90–02	(F)	(30–49)	
Poland	300,000 – 600,000	00–02	(0)	(0–19)	*23*
Portugal					
Azores	Present	02	?	–	
Romania	1,640,000 – 2,450,000	00–02	0	0–19	*48*
Russia	8,000,000 – 15,000,000	90–00	(0)	(0–19)	*122*
Serbia & MN	11,000 – 17,000	90–02	+	0–19	*1,225,117a,155, 62,152,228*
Slovakia	200,000 – 400,000	90–99	0	0–19	
Slovenia	200,000 – 300,000	00	(0)	(0–19)	
Spain	(170,000 – 580,000)	92	?	–	*13,12,10*
Sweden	2,000,000 – 4,000,000	99–00	–	**30**	
Switzerland	180,000 – 300,000	98–02	0	0–19	
Turkey	(150,000 – 600,000)	01	(–)	(0–19)	
Ukraine	40,000 – 65,000	90–00	0	0–19	
UK	**842,000 – 842,000**	00	+	**38**	*5,31*
Total (approx.)	19,000,000 – 35,000,000	**Overall trend** Stable			
Breeding range	>6,000,000 km²	**Gen. length.** <3.3 **% Global pop.** 50–74			

(See p. 239, bottom, for data quality graph)

Regulus teneriffae
CANARY ISLANDS KINGLET

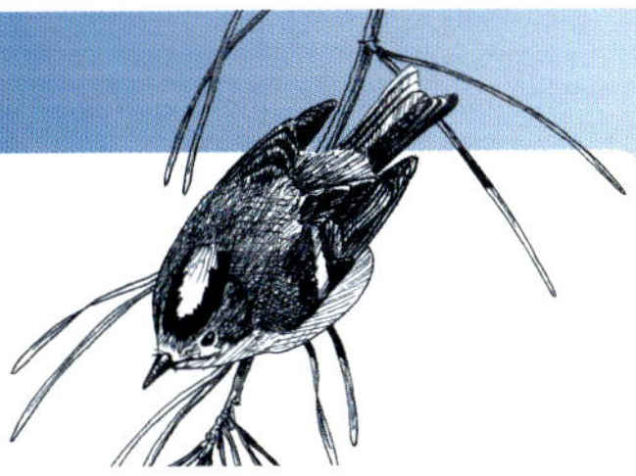

Non-SPEC[E] (1994: 4) **Status** (Secure)
Criteria —
European IUCN Red List Category —
Criteria —
Global IUCN Red List Category —
Criteria —

Regulus teneriffae is endemic to Europe, where it is confined to the Canary Islands of Tenerife, La Gomera, La Palma and El Hierro (but absent from Gran Canaria, Lanzarote and Fuerteventura). Its breeding population is relatively small (<20,000 pairs), but was probably stable between 1970–1990. Although trend data were not available during 1990–2000, there was no evidence to suggest that the species declined, and consequently it is provisionally evaluated as Secure.

Country	Breeding pop. size (pairs)	Year(s)	Trend	Mag.%	References
Spain					
Canary Is.	(10,000 – 20,000)	97–03	?	–	*28,25*
Total (approx.)	10,000 – 20,000	**Overall trend** Unknown			
Breeding range	>20,000 km²	**Gen. length.** <3.3 **% Global pop.** 100			

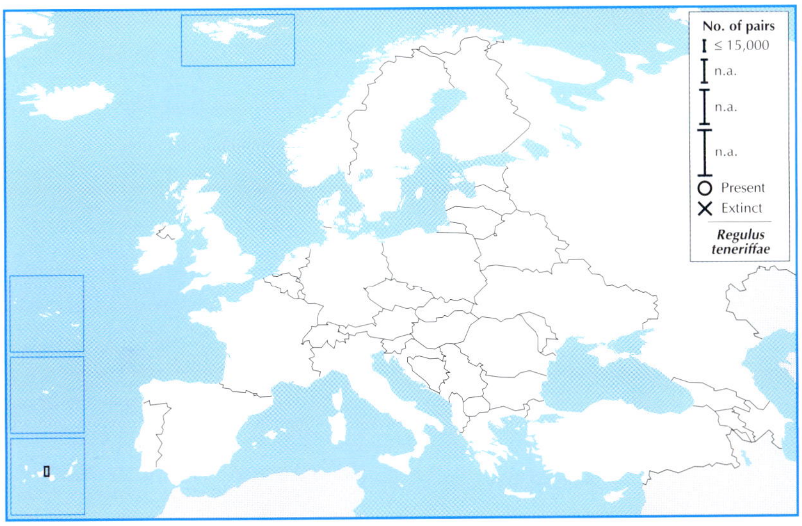

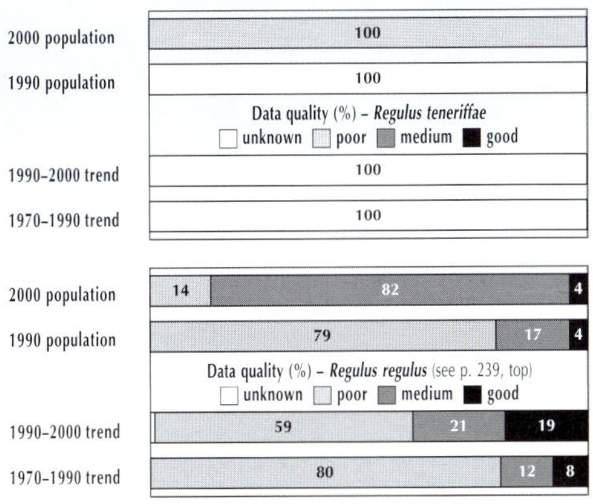

Regulus ignicapilla
FIRECREST

Non-SPEC[E] (1994: 4) **Status** (Secure)
Criteria —
European IUCN Red List Category —
Criteria —
Global IUCN Red List Category —
Criteria —

Regulus ignicapilla is a widespread breeder in west-central and southern Europe, which constitutes >75% of its global breeding range. Its European breeding population is very large (>3,300,000 pairs), and was stable between 1970–1990. Although the trend of the key population in Spain during 1990–2000 was unknown, most other populations—including those in Germany, Italy and Romania—were stable, and the species probably remained stable overall. Consequently, it is provisionally evaluated as Secure.

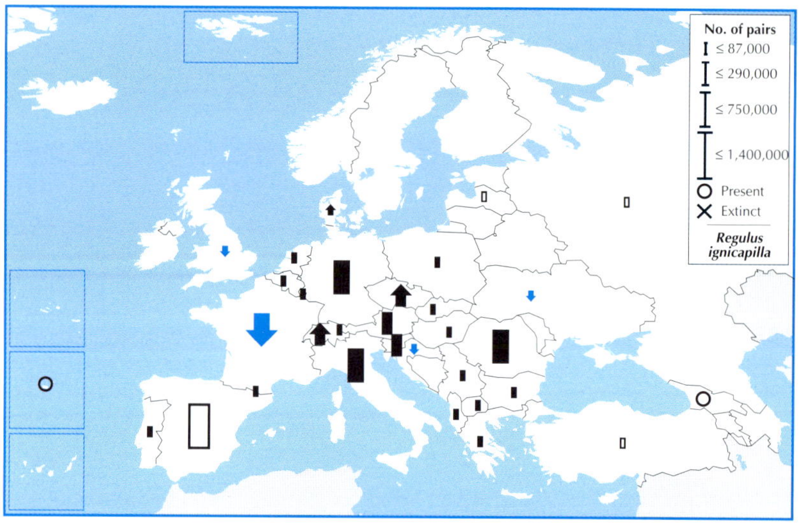

Country	Breeding pop. size (pairs)	Year(s)	Trend	Mag.%	References
Albania	(500 – 2,000)	02	(0)	(0–19)	
Andorra	(500 – 1,000)	99–01	(0)	(0–19)	*1,3*
Austria	(200,000 – 400,000)	98–02	(0)	(0–19)	
Belgium	20,000 – 100,000	01–02	(0)	(0–19)	*1*
Bulgaria	50,000 – 150,000	96–02	0	0–9	
Croatia	(5,000 – 10,000)	02	(–)	(30–49)	*16*
Czech Rep.	70,000 – 140,000	00	+	30–49	
Denmark	(15 – 50)	98–01	(+)	(50–79)	*4,5,6,7*
France	(200,000 – 800,000)	98–02	–	23	*4,2*
Georgia	Present	03	?	–	
Germany	520,000 – 830,000	95–99	0	0–19	
Greece	(20,000 – 100,000)	95–00	(0)	(0–19)	
Hungary	(400 – 500)	90–93	0	0–19	*14*
Italy	(300,000 – 600,000)	03	(0)	(0–19)	
Latvia	(0 – 10)	90–00	?	–	*23,18,26*
Liechtenstein	400 – 600	98–00	0	0–19	
Luxembourg	10,000 – 12,000	02	0	0–19	
Macedonia	(10,000 – 20,000)	90–00	(0)	(0–19)	
Netherlands	**5,000 – 7,000**	98–00	**F**	**22**	*1*
Poland	60,000 – 100,000	00–02	(0)	(0–19)	*23*
Portugal	(10,000 – 100,000)	02	(0)	(0–19)	
Madeira	Present	02	?	–	
Romania	(650,000 – 850,000)	00–02	(0)	(0–19)	*48*
Russia	(50 – 250)	90–02	?	–	*171,177*
Serbia & MN	2,200 – 4,200	90–02	0	0–19	*1,225,117a,155, 62,152,228*
Slovakia	30,000 – 60,000	90–99	0	0–19	
Slovenia	100,000 – 200,000	00	(0)	(0–19)	
Spain	(910,000 – 2,000,000)	92	?	–	*13,12,10*
Switzerland	150,000 – 250,000	98–02	+	0–9	
Turkey	(1,000 – 10,000)	01	?	–	
Ukraine	(0 – 400)	90–00	(–)	(>80)	
UK	2 – 75	96–00	**–**	**11**	
Total (approx.)	3,300,000 – 6,700,000	**Overall trend** Stable			
Breeding range	>2,000,000 km²	**Gen. length.** <3.3		**% Global pop.** 75–94	

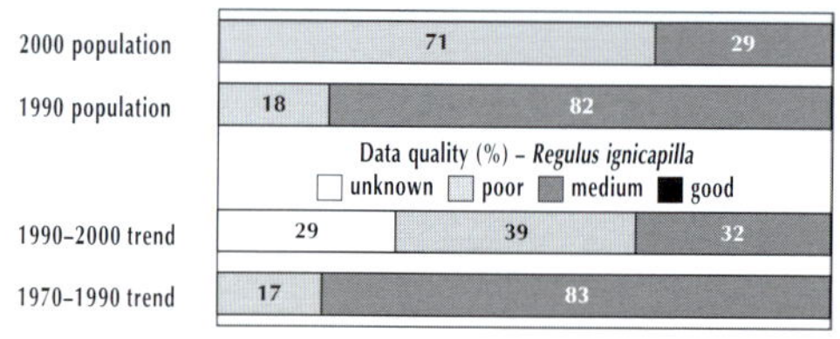

Muscicapa striata
SPOTTED FLYCATCHER

SPEC 3 (1994: 3) **Status** Depleted
Criteria Moderate historical decline
European IUCN Red List Category —
Criteria —
Global IUCN Red List Category —
Criteria —

Muscicapa striata is a widespread summer visitor to Europe, which accounts for less than half of its global breeding range. Its European breeding population is extremely large (>14,000,000 pairs), but underwent a moderate decline between 1970–1990. Although there were declines in a small number of countries during 1990–2000, the species was stable or increased across most of its European range, and declined only slightly overall. Nevertheless, its population has clearly not yet recovered to the level that preceded its initial decline. Consequently, it is evaluated as Depleted.

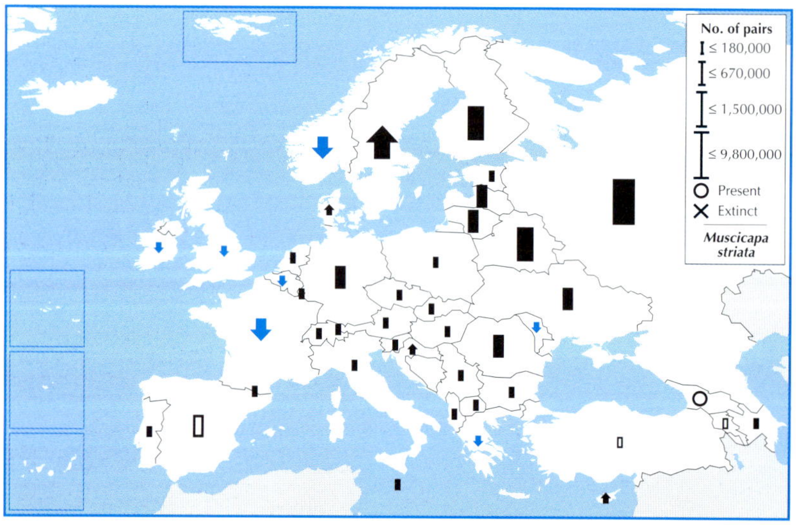

Country	Breeding pop. size (pairs)	Year(s)	Trend	Mag.%	References
Albania	(2,000 – 5,000)	02	(0)	(0–19)	
Andorra	(20 – 50)	99–01	(0)	(0–19)	*1,3*
Armenia	830 – 3,000	98–02	?	–	
Austria	(20,000 – 50,000)	98–02	(0)	(0–19)	
Azerbaijan	(10,000 – 100,000)	96–00	(0)	(0–19)	
Belarus	700,000 – 1,000,000	97–02	0	0–19	
Belgium	4,500 – 15,000	01–02	–	0–19	*1*
Bulgaria	10,000 – 25,000	96–02	0	0–9	
Croatia	(25,000 – 50,000)	02	(+)	(30–49)	*70,16*
Cyprus	(1,000 – 3,000)	94–02	(+)	(10–19)	
Czech Rep.	30,000 – 60,000	00	0	0–19	
Denmark	15,000 – 25,000	00	(+)	(30–49)	*12*
Estonia	(100,000 – 200,000)	98	0	0–19	*1*
Finland	1,300,000 – 1,700,000	98–02	0	5	
France	(200,000 – 1,000,000)	98–02	–	51	*4*
Georgia	Present	03	?	–	
Germany	200,000 – 440,000	95–99	0	0–19	
Greece	(10,000 – 20,000)	95–00	(–)	(0–19)	
Hungary	**65,000 – 115,000**	99–02	0	0–19	*19*
Rep. Ireland	20,000 – 100,000	88–91	–	20–29	
Italy	(100,000 – 300,000)	03	(0)	(0–19)	
Latvia	200,000 – 400,000	90–00	(0)	(0–19)	*23,16*
Liechtenstein	150 – 250	98–00	0	0–19	
Lithuania	(200,000 – 500,000)	99–01	(0)	(0–19)	*20*
Luxembourg	600 – 800	02	0	0–19	
Macedonia	(1,000 – 3,000)	90–00	(0)	(0–19)	
Malta	**6 – 7**	90–02	**0**	**0–19**	*1*
Moldova	6,000 – 7,500	90–00	–	20–29	
Netherlands	**20,000 – 30,000**	98–00	**F**	**23**	*1*
Norway	(100,000 – 500,000)	90–03	–	0–19	*30*
Poland	**100,000 – 300,000**	00–02	0	0–19	*23*
Portugal	(500 – 5,000)	02	(0)	(0–19)	
Romania	(350,000 – 450,000)	00–02	(0)	(0–19)	*48*
Russia	8,000,000 – 12,000,000	90–00	0	0–19	*122*
Serbia & MN	7,000 – 10,000	90–02	(0)	(0–19)	*1,29,155,172a, 67a,78,227*
Slovakia	65,000 – 150,000	90–99	0	0–19	
Slovenia	15,000 – 20,000	94	(0)	(0–19)	
Spain	(640,000 – 690,000)	92	?	–	*13,12,10*
Sweden	**500,000 – 1,200,000**	99–00	**+**	**6**	
Switzerland	**30,000 – 60,000**	93–96	0	0–19	
Turkey	(30,000 – 90,000)	01	?	–	
Ukraine	555,000 – 625,000	90–00	F	20–29	
UK	**63,700 – 63,700**	00	**–**	**51**	*5,31*
Total (approx.)	14,000,000 – 22,000,000	**Overall trend** Small decline			
Breeding range	>8,000,000 km²	**Gen. length.** <3.3		**% Global pop.** 25–49	

(See p. 241, bottom, for data quality graph)

Ficedula parva
RED-BREASTED FLYCATCHER

Non-SPEC (1994: —) **Status** (Secure)
Criteria —
European IUCN Red List Category —
Criteria —
Global IUCN Red List Category —
Criteria —

Ficedula parva is a widespread summer visitor to north-eastern and central Europe, which accounts for less than half of its global breeding range. Its European breeding population is very large (>3,200,000 pairs), and was stable between 1970–1990. Although there were declines in a few countries during 1990–2000, populations were stable across the majority of its European range—including in the Russian stronghold—and the species probably remained stable overall. Consequently, it is provisionally evaluated as Secure.

Country	Breeding pop. size (pairs)	Year(s)	Trend	Mag.%	References
Albania	(0 – 50)	02	?	–	
Armenia	800 – 3,300	98–02	?	–	
Austria	1,500 – 3,000	98–02	(0)	(0–19)	
Azerbaijan	(10,000 – 50,000)	96–00	(0)	(0–19)	
Belarus	60,000 – 100,000	97–02	0	0–19	
Bulgaria	3,500 – 5,500	95–02	0	0–19	
Croatia	(10,000 – 15,000)	02	(+)	(>80)	*70,57*
Czech Rep.	1,200 – 2,400	00	+	50–79	
Estonia	20,000 – 50,000	98	0	0–19	*1*
Finland	1,000 – 2,000	98–02	0	0–19	
Georgia	Present	03	?	–	
Germany	2,000 – 4,500	95–99	0	0–19	
Greece	(5 – 10)	95–00	(0)	(0–19)	
Hungary	(300 – 500)	90–93	0	0–19	*14*
Latvia	50,000 – 80,000	90–00	(0)	(0–19)	*23,18*
Lithuania	(15,000 – 25,000)	99–01	(–)	(0–19)	*20*
Moldova	3,500 – 3,900	90–00	0	0–19	
Norway	0 – 10	90–03	**F**	**30–49**	*63*
Poland	**20,000 – 40,000**	00–02	0	0–19	*23*
Romania	360,000 – 512,000	00–02	0	0–19	
Russia	2,500,000 – 3,500,000	90–00	(0)	(0–19)	*122*
Serbia & MN	500 – 800	90–02	(–)	(0–19)	*1,54,225,227, 152,155,149,185*
Slovakia	5,000 – 10,000	90–99	0	0–19	
Slovenia	70 – 180	99–00	(0)	(0–19)	
Sweden	**500 – 1,000**	99–00	(0)	(0–19)	
Turkey	(1,000 – 5,000)	01	(0)	(0–19)	
Ukraine	(150,000 – 200,000)	90–00	(0)	(10–19)	
Total (approx.)	3,200,000 – 4,600,000	**Overall trend** Stable			
Breeding range	>4,000,000 km²	**Gen. length.** <3.3		**% Global pop.** 25–49	

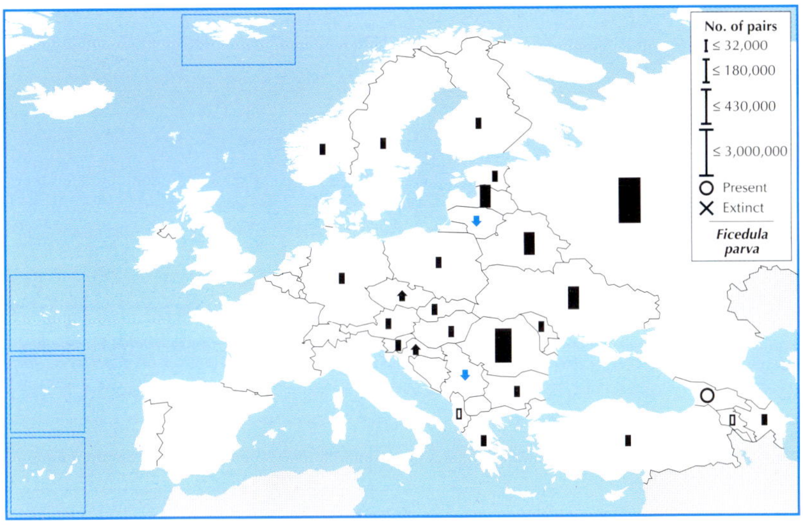

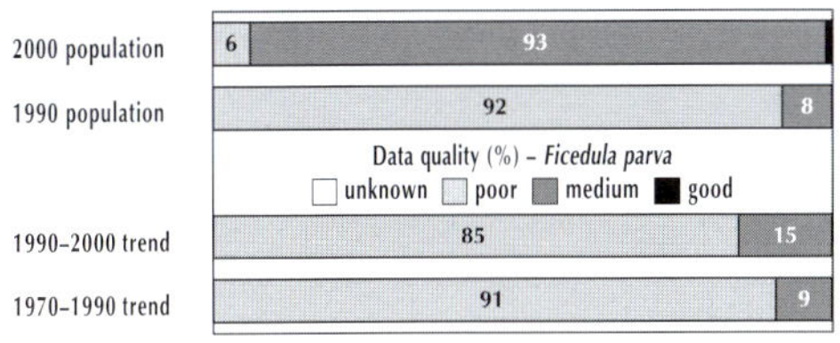

Ficedula semitorquata
SEMICOLLARED FLYCATCHER

SPEC 2 (1994: 2) **Status** Declining
Criteria Moderate continuing decline
European IUCN Red List Category —
Criteria —
Global IUCN Red List Category —
Criteria —

Ficedula semitorquata is a summer visitor to parts of south-eastern Europe, which constitutes >50% of its global breeding range. Its European breeding population is relatively small (<53,000 pairs), and underwent a large decline between 1970–1990. Although it was stable in Bulgaria and the Former Yugoslav Republic of Macedonia during 1990–2000, there were declines elsewhere in its European range—including the key Turkish and Russian populations—and the species underwent a moderate decline (>10%) overall. Consequently, it is evaluated as Declining.

Country	Breeding pop. size (pairs)	Year(s)	Trend	Mag.%	References
Albania	(0 – 100)	02	?	–	
Armenia	300 – 800	98–02	–	10–19	
Bulgaria	1,500 – 3,500	95–02	0	0–9	
Georgia	Present	03	?	–	
Greece	(1,000 – 3,000)	95–00	(–)	(0–19)	
Macedonia	(100 – 1,000)	90–00	(0)	(0–19)	
Russia	(10,000 – 20,000)	90–00	(–)	(20–29)	*50,61,116,136*
Turkey	(2,500 – 25,000)	01	(–)	(0–19)	
Total (approx.)	15,000 – 53,000	**Overall trend** Moderate decline			
Breeding range	>500,000 km²	**Gen. length.** <3.3		**% Global pop.** 50–74	

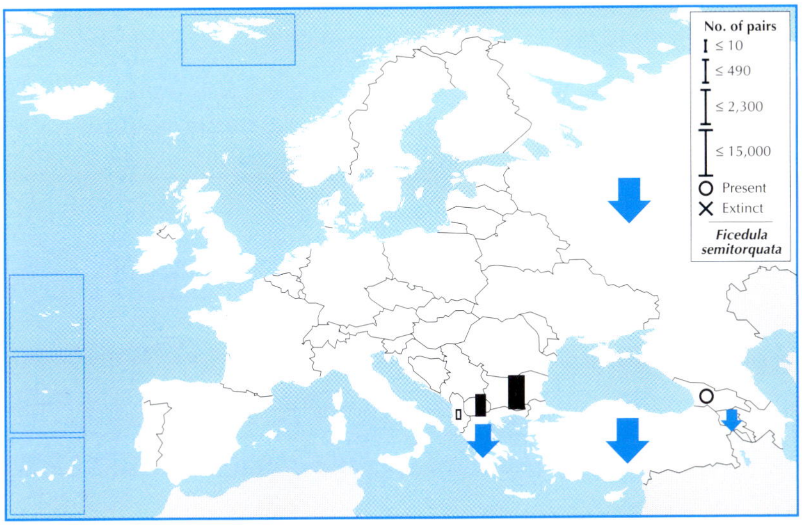

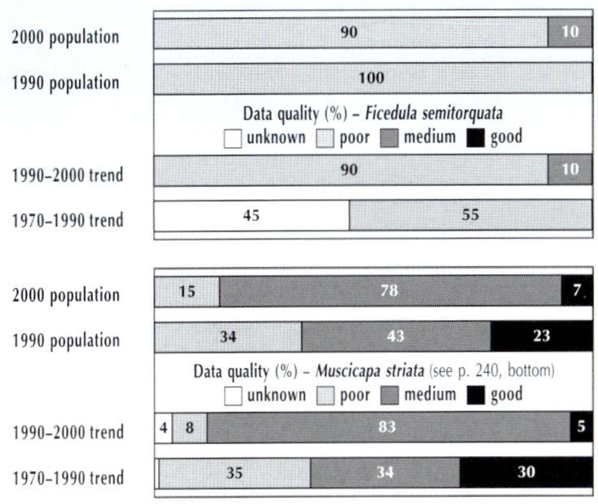

Ficedula albicollis
COLLARED FLYCATCHER

Non-SPEC[E] (1994: 4) **Status** Secure
Criteria —
European IUCN Red List Category —
Criteria —
Global IUCN Red List Category —
Criteria —

Ficedula albicollis is a summer visitor, mainly to eastern and central Europe, with its entire global breeding range confined to the region. Its European breeding population is very large (>1,400,000 pairs), and was stable between 1970–1990. Although there were declines in a few marginal populations during 1990–2000, key populations in the core of the range—notably Romania, Russia and Ukraine – were stable or increased, and the species underwent a small increase overall. Consequently, it is evaluated as Secure.

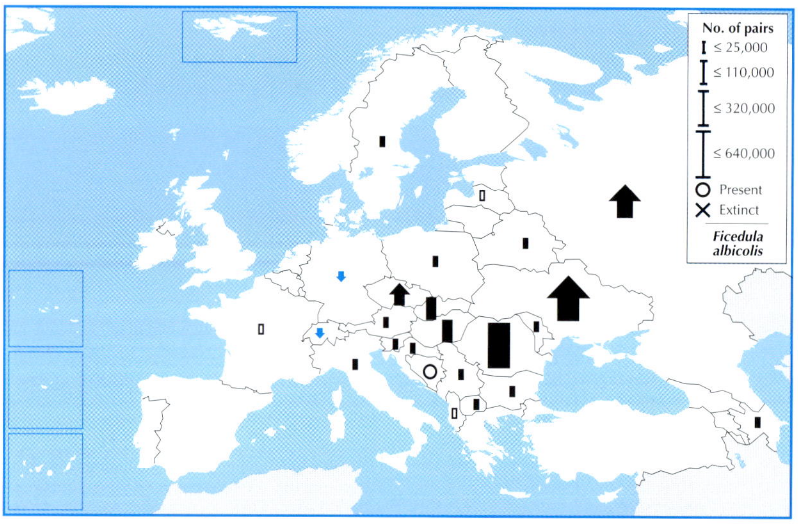

Country	Breeding pop. size (pairs)	Year(s)	Trend	Mag.%	References
Albania	(10 – 100)	02	?	–	
Austria	9,000 – 18,000	98–02	(0)	(0–19)	
Azerbaijan	(10,000 – 60,000)	96–00	(0)	(0–19)	
Belarus	3,000 – 6,000	97–02	0	0–19	
Bosnia & HG	Present	90–03	?	–	
Bulgaria	(0 – 10)	96–02	(F)	(>80)	
Croatia	(10,000 – 50,000)	02	(0)	(0–19)	*70,16,54*
Czech Rep.	30,000 – 60,000	00	+	10–19	
France	4,000 – 12,000	98–02	?	–	*4*
Germany	2,800 – 3,900	95–99	–	20–29	
Hungary	**18,000 – 92,000**	99–02	0	0–19	*19,1*
Italy	2,000 – 4,000	03	(0)	(0–19)	
Latvia	(0 – 10)	00–03	?	–	*26,18,22*
Macedonia	(2,000 – 8,000)	90–00	(0)	(0–19)	
Moldova	20,000 – 25,000	90–00	0	0–19	
Poland	**2,500 – 10,000**	90–00	0	0–19	*2*
Romania	460,000 – 712,000	00–02	0	0–19	
Russia	200,000 – 500,000	90–00	+	20–29	*8,89,114,122*
Serbia & MN	8,000 – 12,000	95–02	0	10–19	*1,29,155,172a, 225,227,152*
Slovakia	70,000 – 150,000	90–99	0	0–19	
Slovenia	2,500 – 3,500	99–00	(0)	(0–19)	
Sweden	**4,600 – 5,700**	99–00	**0**	**0–19**	
Switzerland	**20 – 25**	93–96	–	0–9	
Ukraine	(580,000 – 700,000)	90–00	(+)	(10–19)	
Total (approx.)	1,400,000 – 2,400,000	**Overall trend** Small increase			
Breeding range	>1,000,000 km²	**Gen. length.** <3.3 **% Global pop.** 100			

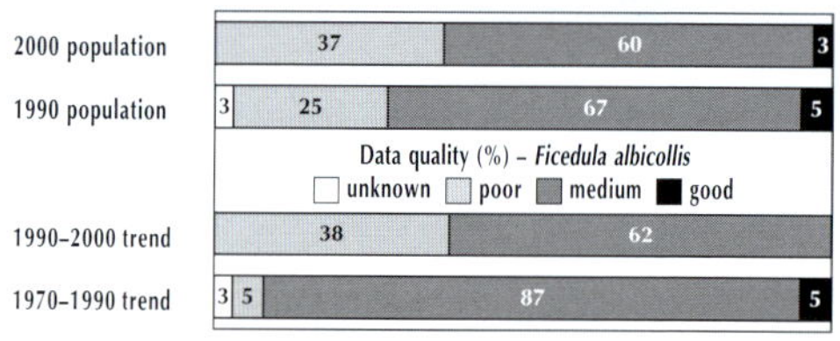

Ficedula hypoleuca
EUROPEAN PIED FLYCATCHER

Non-SPEC[E] (1994: 4) **Status** Secure
Criteria —
European IUCN Red List Category —
Criteria —
Global IUCN Red List Category —
Criteria —

Ficedula hypoleuca is a widespread summer visitor to much of Europe (occurring more patchily in the south and west), with the region constituting >75% of its global breeding range. Its European breeding population is extremely large (>12,000,000 pairs), and was stable between 1970–1990. Although the species declined in several countries during 1990–2000—most notably in Fennoscandia—the stronghold population in Russia and other eastern populations were stable, and the species declined only slightly overall. Consequently, it is evaluated as Secure.

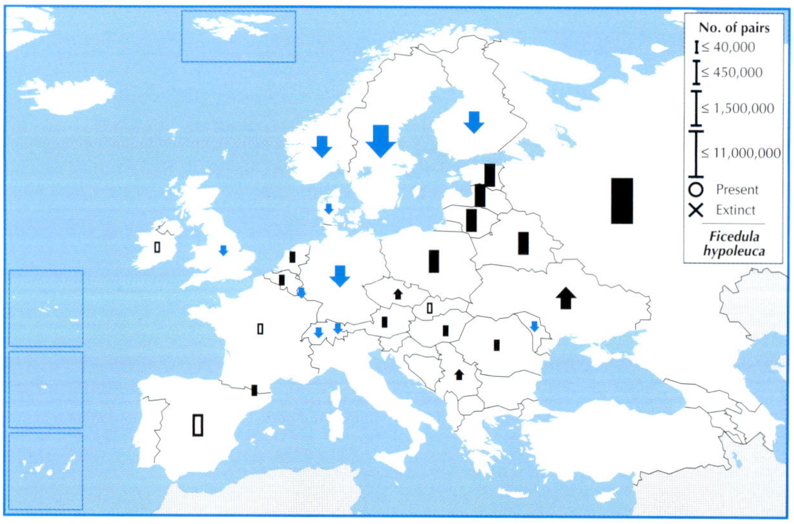

Country	Breeding pop. size (pairs)	Year(s)	Trend	Mag.%	References
Andorra	(0 – 5)	99–01	(0)	(0–19)	*1,3*
Austria	(250 – 600)	98–02	(0)	(0–19)	
Belarus	250,000 – 400,000	97–02	0	0–19	
Belgium	1,100 – 3,000	95–02	0	0–19	*1*
Czech Rep.	12,000 – 24,000	00	+	10–19	
Denmark	10,000 – 20,000	00	–	20–29	*12,3*
Estonia	200,000 – 500,000	98	0	0–19	*1*
Finland	250,000 – 700,000	98–02	–	20	
France	(20,000 – 80,000)	98–02	?	–	*4*
Germany	170,000 – 300,000	95–99	–	0–19	
Hungary	30 – 60	95–02	0	0–19	
Rep. Ireland	1 – 10	88–91	?	–	
Latvia	300,000 – 400,000	90–00	(0)	(0–19)	*23,18*
Liechtenstein	(0 – 3)	98–00	(–)	(30–49)	
Lithuania	150,000 – 250,000	99–01	(0)	(0–19)	*20*
Luxembourg	2,000 – 2,500	02	–	0–19	
Moldova	350 – 500	90–00	–	30–49	
Netherlands	**14,000 – 18,000**	98–00	**0**	**2**	*1*
Norway	(200,000 – 1,000,000)	90–03	–	0–19	*30*
Poland	**100,000 – 300,000**	00–02	0	0–19	*23*
Romania	(200 – 400)	00–02	0	0–19	
Russia	9,000,000 – 13,000,000	90–00	0	0–19	*122*
Serbia & MN	**1 – 1**	03	+	N	*1*
Slovakia	100 – 200	90–99	?	–	
Spain	(130,000 – 350,000)	92	?	–	*13,12,10*
Sweden	**1,000,000 – 2,000,000**	99–00	**–**	**24**	
Switzerland	**10,000 – 20,000**	93–96	–	10–19	
Ukraine	(265,000 – 335,000)	90–00	(+)	(10–19)	
UK	**29,500 – 33,800**	00	**–**	**16**	*22,31*
Total (approx.)	12,000,000 – 20,000,000	**Overall trend** Small decline			
Breeding range	>5,000,000 km²	**Gen. length.** <3.3 **% Global pop.** 75–94			

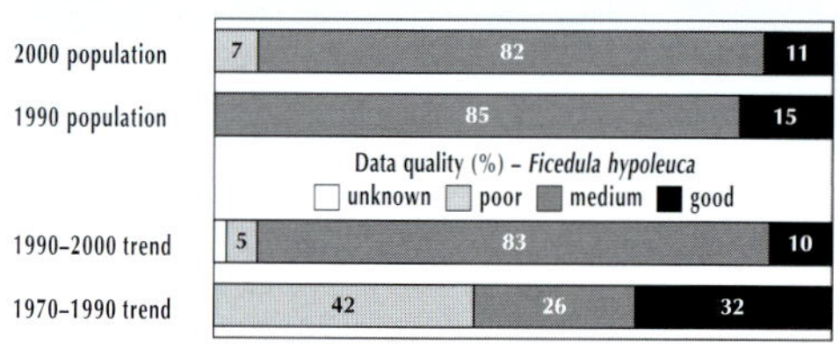

Panurus biarmicus
BEARDED PARROTBILL

Non-SPEC (1994: —) **Status** (Secure)
Criteria —
European IUCN Red List Category —
Criteria —
Global IUCN Red List Category —
Criteria —

Panurus biarmicus is a widespread but very patchily distributed resident across much of Europe, which accounts for less than half of its global range. Its European breeding population is large (>240,000 pairs), and was stable between 1970–1990. Although there were declines in a number of countries during 1990–2000, the majority of European populations—including key ones in Russia, Ukraine and Romania—increased or were stable, and the species probably remained stable overall. Consequently, it is provisionally evaluated as Secure.

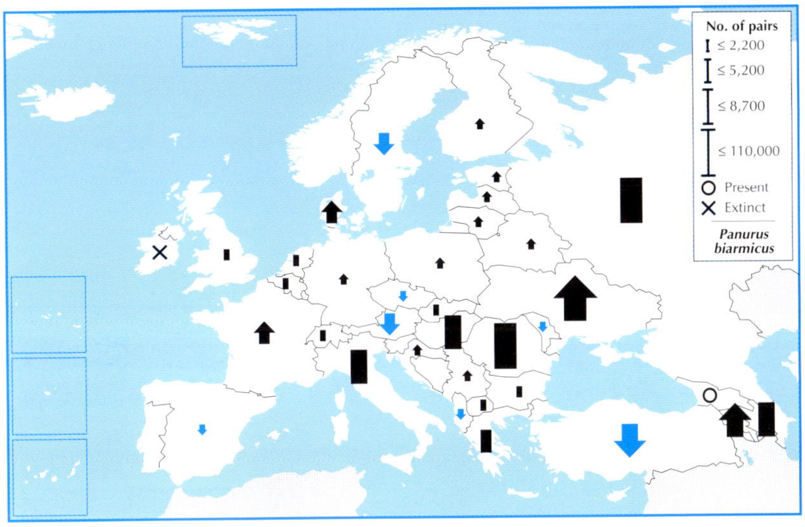

Country	Breeding pop. size (pairs)	Year(s)	Trend	Mag.%	References
Albania	300 – 1,000	02	(–)	(0–19)	
Armenia	6,500 – 10,000	99–02	+	0–9	
Austria	(3,000 – 6,000)	98–02	(–)	(20–29)	
Azerbaijan	(5,000 – 15,000)	96–00	(0)	(0–19)	
Belarus	**20 – 100**	93–99	+	10–19	
Belgium	**30 – 33**	00–02	0	0–19	*1*
Bulgaria	**80 – 200**	96–02	0	0–9	
Croatia	30 – 60	02	(+)	(>80)	*70*
Czech Rep.	**80 – 160**	00	–	**20–29**	
Denmark	(2,000 – 10,000)	00	(+)	(>80)	*3,12,16*
Estonia	500 – 1,500	98	+	50–79	*1*
Finland	400 – 800	98–02	+	200	
France	3,000 – 9,000	00	(+)	(0–19)	*1*
Georgia	Present	03	?	–	
Germany	1,400 – 2,700	95–99	+	30–49	
Greece	(2,000 – 5,000)	95–00	(0)	(0–19)	
Hungary	6,000 – 9,000	95–02	0	0–19	
Rep. Ireland	0 – 0	00–02	–	X	
Italy	(4,000 – 10,000)	03	(0)	(0–19)	
Latvia	600 – 1,000	90–00	+	20–29	*17*
Lithuania	(300 – 600)	99–01	(+)	(0–19)	*20*
Macedonia	(50 – 250)	90–00	(F)	(20–29)	
Moldova	300 – 500	90–00	–	20–29	
Netherlands	**1,200 – 2,000**	98–00	**0**	**10**	*1*
Poland	1,800 – 2,500	95–00	+	**50–59**	*2,103*
Romania	(85,000 – 125,000)	00–02	(0)	(0–19)	
Russia	(50,000 – 150,000)	90–00	(0)	(0–19)	*8*
Serbia & MN	700 – 900	97–02	+	0–19	*1,78,29,101,203, 243,20,155*
Slovakia	150 – 400	90–99	F	30–49	
Spain	(650 – 1,100)	98–02	(–)	(0–19)	*10,16*
Sweden	3,000 – 8,000	99–00	–	30–49	
Switzerland	**30 – 40**	93–96	**F**	**20–29**	
Turkey	(6,000 – 12,000)	01	(–)	(0–19)	
Ukraine	(60,000 – 95,000)	90–00	+	0–19	
UK	**348 – 362**	97–01	**0**	**3**	
Total (approx.)	240,000 – 480,000	**Overall trend** Stable			
Breeding range	>1,000,000 km²	**Gen. length.** <3.3		**% Global pop.** 25–49	

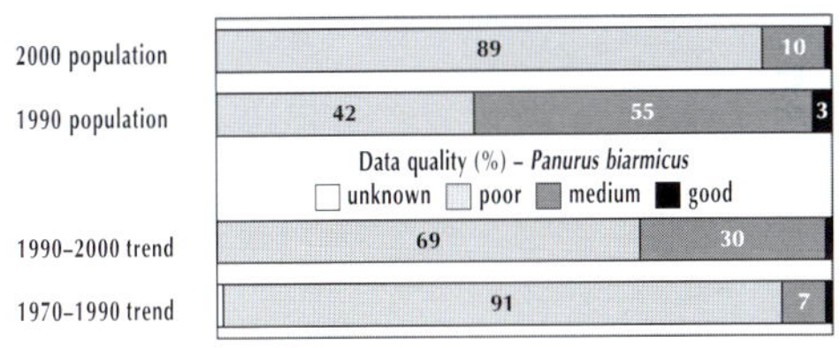

Aegithalos caudatus
LONG-TAILED TIT

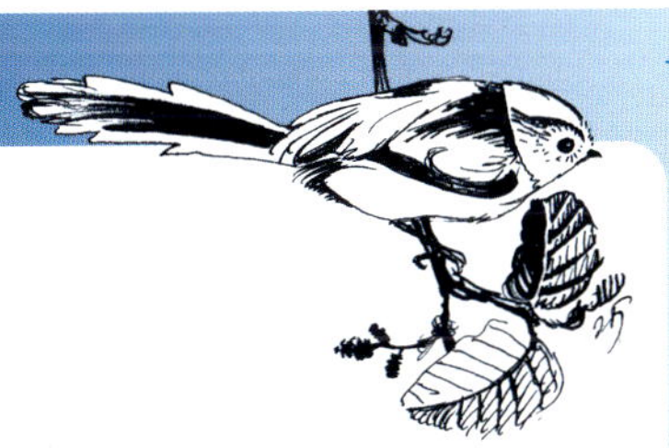

Non-SPEC (1994: —) **Status** Secure
Criteria —
European IUCN Red List Category —
Criteria —
Global IUCN Red List Category —
Criteria —

Aegithalos caudatus is a widespread resident across most of Europe, which accounts for less than half of its global range. Its European breeding population is very large (>5,000,000 pairs), and was stable between 1970–1990. Although there were slight declines in the sizeable French and Turkish populations during 1990–2000, trends were stable or increasing across most of Europe, and the species remained stable overall. Consequently, it is evaluated as Secure.

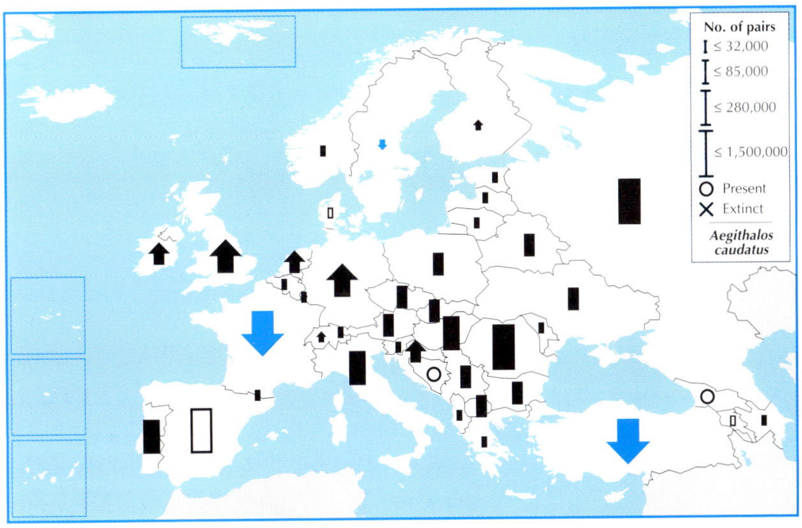

Country	Breeding pop. size (pairs)	Year(s)	Trend	Mag.%	References
Albania	(1,000 – 3,000)	02	(0)	(0–19)	
Andorra	100 – 200	99–01	(0)	(0–19)	*1,3*
Armenia	800 – 3,000	98–02	?	–	
Austria	(25,000 – 50000)	98–02	(0)	(0–19)	
Azerbaijan	(10,000 – 100,000)	96–00	(0)	(0–19)	
Belarus	35,000 – 50,000	97–02	0	0–19	
Belgium	17,000 – 35,000	01–02	0	0–19	*1*
Bosnia & HG	Present	90–03	?	–	
Bulgaria	20,000 – 80,000	96–02	0	0–9	
Croatia	(50,000 – 100,000)	02	(+)	(>80)	*16*
Czech Rep.	50,000 – 100,000	00	0	0–9	
Denmark	8,000 – 12,000	00	?	–	*12*
Estonia	(5,000 – 15,000)	98	0	0–19	*1*
Finland	20,000 – 40,000	98–02	+	**300**	
France	(500,000 – 2,000,000)	98–02	–	19	*4,3*
Georgia	Present	03	?	–	
Germany	120,000 – 360,000	95–99	+	0–19	
Greece	(20,000 – 50,000)	95–00	(0)	(0–19)	
Hungary	**105,000 – 180,000**	99–02	0	0–19	*19*
Rep. Ireland	20,000 – 100,000	88–91	+	0–19	
Italy	(100,000 – 500,000)	03	(0)	(0–19)	
Latvia	20,000 – 40,000	90–00	(0)	(0–19)	*23,18*
Liechtenstein	(20 – 30)	98–00	(0)	(0–19)	
Lithuania	(10,000 – 20,000)	99–01	(0)	(0–19)	*20*
Luxembourg	3,000 – 4,000	02	0	0–19	
Macedonia	(20,000 – 70,000)	90–00	(0)	(0–19)	
Moldova	450 – 550	90–00	0	0–19	
Netherlands	**30,000 – 40,000**	98–00	+	**32**	*1*
Norway	(5,000 – 15,000)	90–03	(0)	(0–19)	
Poland	**30,000 – 80,000**	00–02	0	0–19	*23*
Portugal	(50,000 – 500,000)	02	(0)	(0–19)	
Romania	(740,000 – 910,000)	00–02	(0)	(0–19)	*48*
Russia	1,000,000 – 2,000,000	90–00	0	0–19	*122*
Serbia & MN	45,000 – 60,000	95–02	0	0–19	*1,29,155,172a, 67a,225,227*
Slovakia	60,000 – 120,000	90–99	0	0–19	
Slovenia	15,000 – 20,000	94	(0)	(0–19)	
Spain	(700,000 – 1,650,000)	92	?	–	*13,12,10*
Sweden	**10,000 – 50,000**	99–00	(–)	(10–19)	
Switzerland	**10,000 – 15,000**	93–96	+	30–49	
Turkey	(800,000 – 2,500,000)	01	(–)	(0–19)	
Ukraine	(68,000 – 87,000)	90–00	0	0–19	
UK	**273,000 – 273,000**	00	+	**24**	*5,31*
Total (approx.)	5,000,000 – 12,000,000	**Overall trend** Stable			
Breeding range	>7,000,000 km²	**Gen. length.** <3.3		**% Global pop.** 25–49	

(See p. 244, bottom, for data quality graph)

Parus palustris
MARSH TIT

SPEC 3 (1994: —) **Status** Declining
Criteria Moderate recent decline
European IUCN Red List Category —
Criteria —
Global IUCN Red List Category —
Criteria —

Parus palustris is a widespread resident across much of Europe, which accounts for less than half of its global range. Its European breeding population is very large (>3,000,000 pairs), and was stable between 1970–1990. Although the species remained stable across the majority of Europe during 1990–2000, the sizeable population in France suffered a substantial decline and there were also declines in other countries in north-west Europe, and the species underwent a moderate decline (>10%) overall. Consequently, this previously Secure species is now evaluated as Declining.

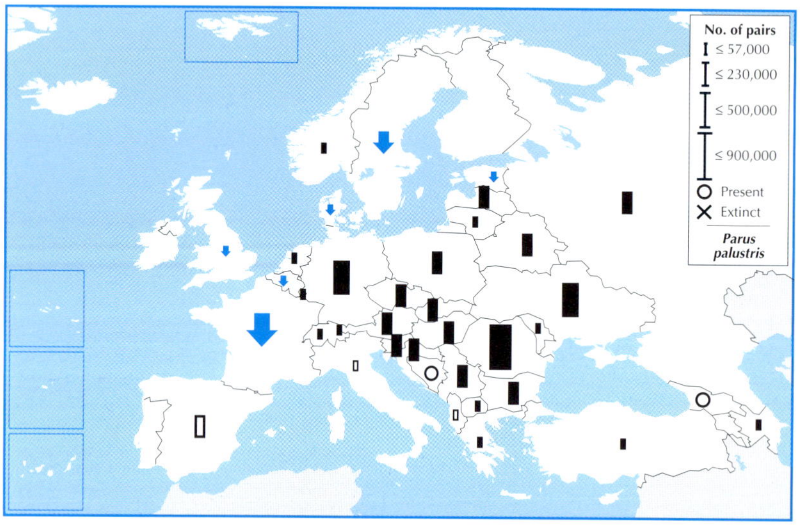

Country	Breeding pop. size (pairs)	Year(s)	Trend	Mag.%	References
Albania	(100 – 300)	02	?	–	
Austria	(50,000 – 100,000)	98–02	(0)	(0–19)	
Azerbaijan	(0 – 50)	96–00	(0)	(0–19)	
Belarus	100,000 – 120,000	97–02	0	0–19	
Belgium	10,000 – 20,000	01–02	–	0–19	*1*
Bosnia & HG	Present	90–03	?	–	
Bulgaria	100,000 – 200,000	96–02	0	0–9	
Croatia	(100,000 – 500,000)	02	(0)	(0–19)	*16*
Czech Rep.	60,000 – 120,000	00	0	0–19	
Denmark	10,000 – 20,000	00	**–**	**30–49**	*12*
Estonia	40,000 – 70,000	98	–	20–29	*1*
France	(250,000 – 1,000,000)	98–02	**–**	**59**	*4,3*
Georgia	Present	03	?	–	
Germany	250,000 – 640,000	95–99	0	0–19	
Greece	(2,000 – 10,000)	95–00	(0)	(0–19)	
Hungary	**58,000 – 105,000**	99–02	0	0–19	*19*
Italy	(30,000 – 100,000)	03	?	–	
Latvia	100,000 – 200,000	90–00	(0)	(0–19)	*23,18*
Liechtenstein	200 – 400	98–00	0	0–19	
Lithuania	(40,000 – 60,000)	99–01	(0)	(0–19)	*20*
Luxembourg	5,000 – 6,000	02	0	0–19	
Macedonia	(20,000 – 60,000)	90–00	(0)	(0–19)	
Moldova	5,500 – 6,200	90–00	0	0–19	
Netherlands	**12,000 – 15,000**	98–00	**0**	**10**	*1*
Norway	(20,000 – 80,000)	90–03	(0)	(0–19)	
Poland	**100,000 – 200,000**	00–02	0	0–19	*23*
Romania	870,000 – 920,000	00–02	0	0–19	*48*
Russia	100,000 – 350,000	90–00	(0)	(0–19)	*122*
Serbia & MN	70,000 – 100,000	90–02	0	0–19	*1,155,172a,67a, 225,227,117a*
Slovakia	80,000 – 160,000	90–99	0	0–19	
Slovenia	50,000 – 100,000	00	(0)	(0–19)	
Spain	(82,000 – 96,000)	92	?	–	*13,12,10*
Sweden	75,000 – 150,000	99–00	**–**	**23**	
Switzerland	**40,000 – 80,000**	93–96	0	0–19	
Turkey	(5,000 – 20,000)	01	(0)	(0–19)	
Ukraine	255,000 – 380,000	90–00	0	0–19	
UK	**52,800 – 52,800**	00	**–**	**12**	*5,31*

Total (approx.) 3,000,000 – 6,000,000 **Overall trend** Moderate decline
Breeding range >4,000,000 km² **Gen. length.** <3.3 **% Global pop.** 25–49

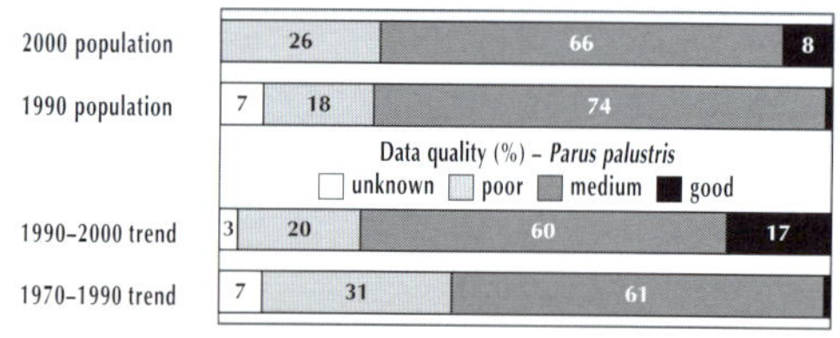

Parus lugubris
SOMBRE TIT

Non-SPEC[E] (1994: 4) **Status** (Secure)
Criteria —
European IUCN Red List Category —
Criteria —
Global IUCN Red List Category —
Criteria —

Parus lugubris is a resident across much of south-eastern Europe, which constitutes >75% of its global range. Its European breeding population is large (>450,000 pairs), and was stable between 1970–1990. Although there was a marked decline in the small Croatian population during 1990–2000, populations were stable (or increasing) across the rest of its European range—including in the Turkish stronghold—and the species probably remained stable overall. Consequently, it is provisionally evaluated as Secure.

Country	Breeding pop. size (pairs)	Year(s)	Trend	Mag.%	References
Albania	(2,000 – 5,000)	02	(0)	(0–19)	
Armenia	**400 – 600**	00–02	0	0–19	*7*
Azerbaijan	(1,000 – 5,000)	96–00	(0)	(0–19)	
Bosnia & HG	Present	90–03	?	–	
Bulgaria	10,000 – 30,000	96–02	(+)	(0–19)	
Croatia	(500 – 1,000)	02	(–)	(50–79)	*16*
Georgia	Present	03	?	–	
Greece	(10,000 – 30,000)	95–00	(0)	(0–19)	
Macedonia	(2,500 – 10,000)	90–00	(0)	(0–19)	
Romania	(15,000 – 25,000)	00–02	0	0–19	*57*
Serbia & MN	9,000 – 17,000	90–02	0	0–19	*1,155,67a,225, 227,117a*
Slovenia	(5 – 10)	00	(0)	(0–19)	
Turkey	(400,000 – 1,200,000)	01	(0)	(0–19)	

Total (approx.) 450,000 – 1,300,000 **Overall trend** Stable
Breeding range >750,000 km² **Gen. length.** <3.3 **% Global pop.** 75–94

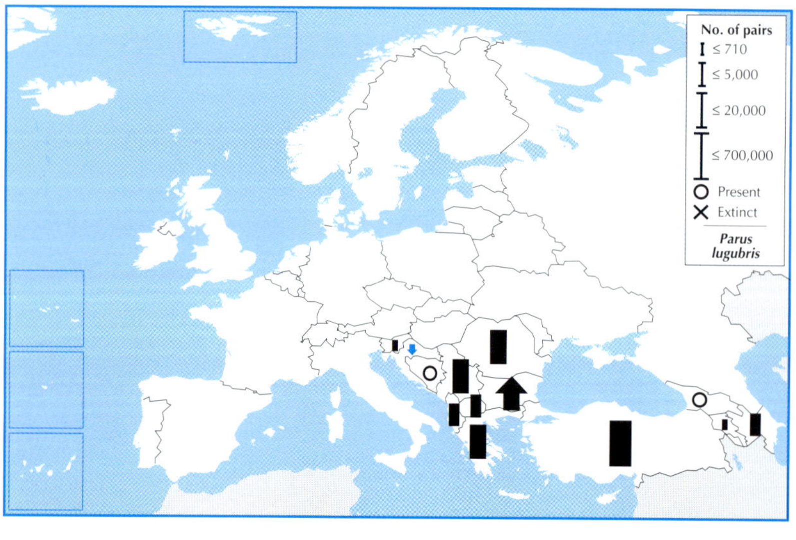

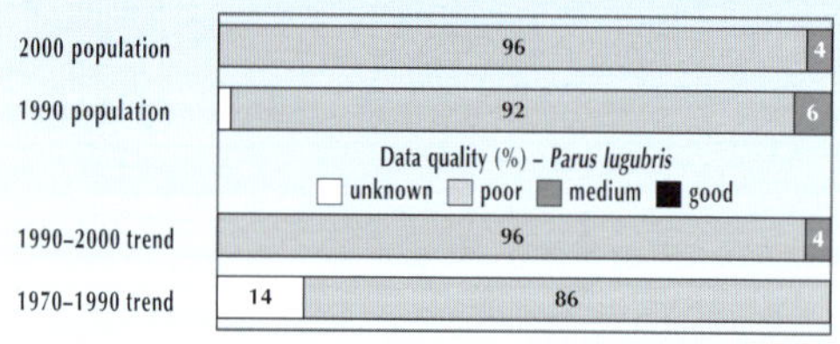

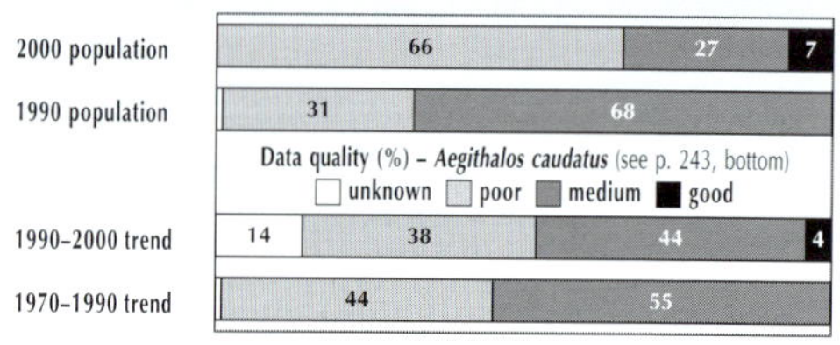

Parus montanus
WILLOW TIT

Non-SPEC (1994: —) **Status** Secure
Criteria —
European IUCN Red List Category —
Criteria —
Global IUCN Red List Category —
Criteria —

Parus montanus is a widespread resident across most of central and northern Europe, which accounts for less than half of its global range. Its European breeding population is extremely large (>24,000,000 pairs), and was stable between 1970–1990. Although there were declines in Fennoscandia and certain western populations during 1990–2000, the species was stable in its Russian stronghold and elsewhere across eastern Europe, and hence underwent only a small decline overall. Consequently, it is evaluated as Secure.

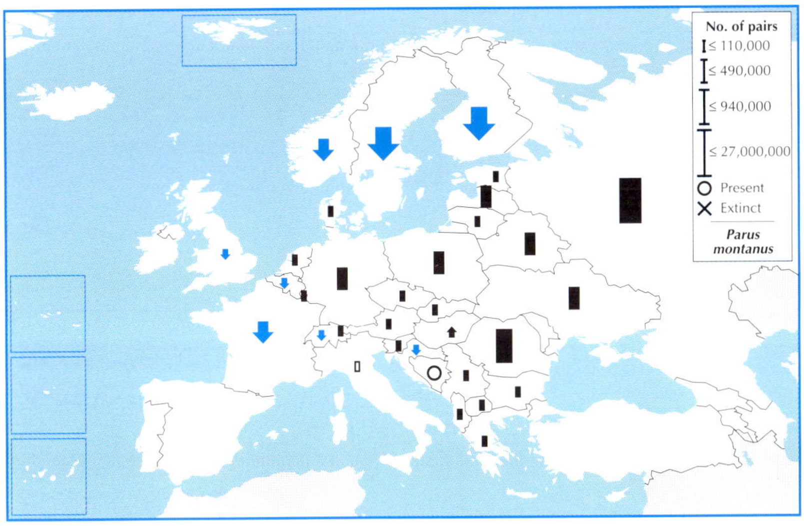

Country	Breeding pop. size (pairs)	Year(s)	Trend	Mag.%	References
Albania	(50 – 250)	02	(0)	(0–19)	
Austria	(80,000 – 140,000)	98–02	(0)	(0–19)	
Belarus	400,000 – 600,000	97–02	0	0–19	
Belgium	10,000 – 20,000	01–02	–	0–19	*1*
Bosnia & HG	Present	90–03	?	–	
Bulgaria	40,000 – 80,000	96–02	0	0–9	
Croatia	(1,000 – 5,000)	02	(–)	(50–79)	*16*
Czech Rep.	40,000 – 80,000	00	0	0–19	
Denmark	100 – 300	99	(0)	(0–19)	*5*
Estonia	50,000 – 100,000	98	0	0–19	*1*
Finland	400,000 – 900,000	98–02	**–**	**40**	
France	(150,000 – 750,000)	98–02	**–**	**50**	*4,3*
Germany	150,000 – 380,000	95–99	0	0–19	
Greece	(200 – 1,000)	95–00	(0)	(0–19)	
Hungary	(300 – 500)	90–93	+	50–79	*14*
Italy	(30,000 – 50,000)	03	?	–	
Latvia	180,000 – 250,000	90–00	(0)	(0–19)	*23,18*
Liechtenstein	400 – 600	98–00	(0)	(0–19)	
Lithuania	(60,000 – 80,000)	99–01	(0)	(0–19)	*20*
Luxembourg	5,000 – 6,000	02	0	0–19	
Macedonia	(250 – 1,000)	90–00	(0)	(0–19)	
Netherlands	**20,000 – 30,000**	98–00	**0**	**3**	*1*
Norway	(200,000 – 800,000)	95–02	–	0–19	*30,28*
Poland	**200,000 – 300,000**	00–02	0	0–19	*23*
Romania	710,000 – 1,240,000	00–02	0	0–19	*48*
Russia	20,000,000 – 35,000,000	90–00	0	0–19	*122*
Serbia & MN	7,000 – 10,000	90–02	0	0–19	*1,155,225,227, 117a,224,185*
Slovakia	60,000 – 150,000	90–99	0	0–19	
Slovenia	10,000 – 20,000	94	(0)	(0–19)	
Sweden	500,000 – 1,000,000	99–00	**–**	**53**	
Switzerland	**30,000 – 70,000**	93–96	–	0–9	
Ukraine	170,000 – 210,000	90–00	0	0–19	
UK	**8,500 – 8,500**	00	**–**	**66**	*5,31*
Total (approx.)	24,000,000 – 42,000,000	**Overall trend** Small decline			
Breeding range	>5,000,000 km²	**Gen. length.** <3.3 **% Global pop.** 25–49			

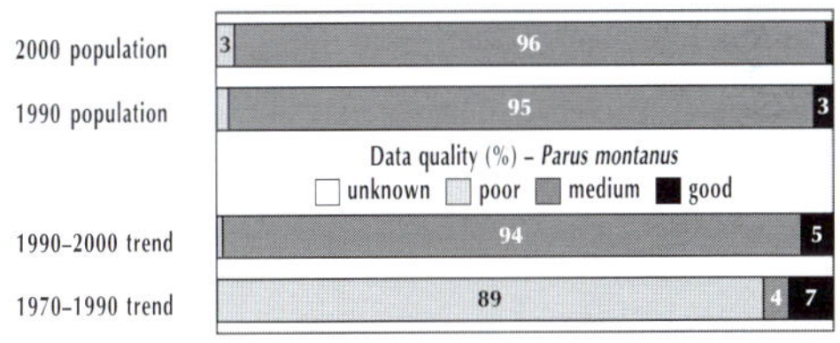

Parus cinctus
SIBERIAN TIT

Non-SPEC (1994: —) **Status** (Secure)
Criteria —
European IUCN Red List Category —
Criteria —
Global IUCN Red List Category —
Criteria —

Parus cinctus is a widespread resident in the boreal zone of Fennoscandia and Russia, with Europe accounting for less than a quarter of its global range. Its European breeding population is large (>870,000 pairs), and was stable between 1970–1990. Although the Swedish population declined during 1990–2000 and the trend of the stronghold population in Russia was unknown, the species was stable in Norway and Finland, and there is no evidence to suggest that it declined overall. Consequently, it is provisionally evaluated as Secure.

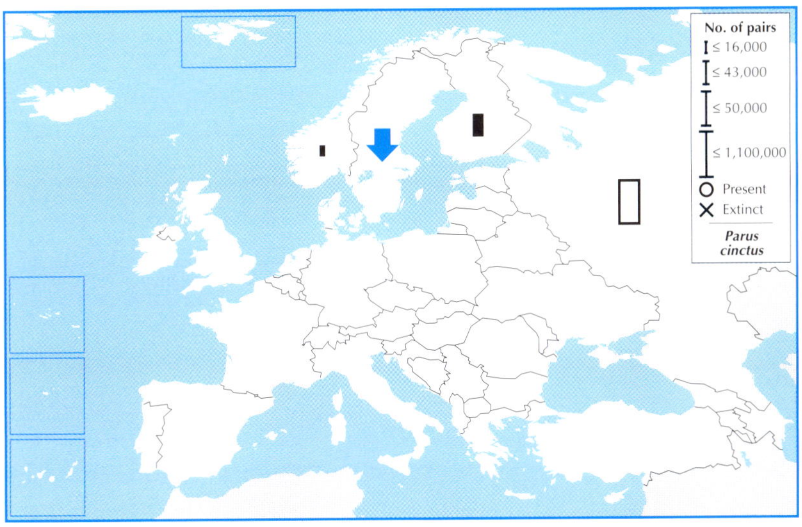

Country	Breeding pop. size (pairs)	Year(s)	Trend	Mag.%	References
Finland	30,000 – 60,000	98–02	0	0–19	
Norway	(10,000 – 25,000)	90–03	(0)	(0–19)	
Russia	800,000 – 1,500,000	90–00	?	–	*122*
Sweden	25,000 – 100,000	99–00	(–)	(10–19)	
Total (approx.)	870,000 – 1,700,000	**Overall trend** Unknown			
Breeding range	>750,000 km²	**Gen. length.** <3.3 **% Global pop.** 5–24			

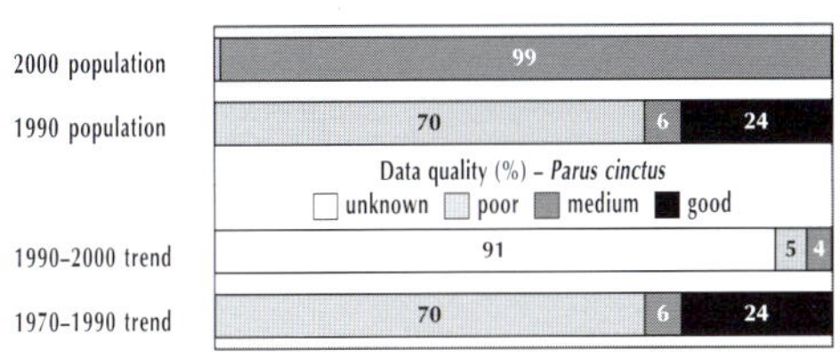

Parus cristatus
CRESTED TIT

SPEC 2 (1994: 4) **Status** (Declining)
Criteria Moderate recent decline
European IUCN Red List Category —
Criteria —
Global IUCN Red List Category —
Criteria —

Parus cristatus is a widespread resident across much of Europe, which constitutes >95% of its global range. Its European breeding population is very large (>6,100,000 pairs), and was stable between 1970–1990. Although most European populations remained stable or increased during 1990–2000, the sizeable population in Russia declined, and the species underwent a moderate decline (>10%) overall. Consequently, this previously Secure species is now provisionally evaluated as Declining.

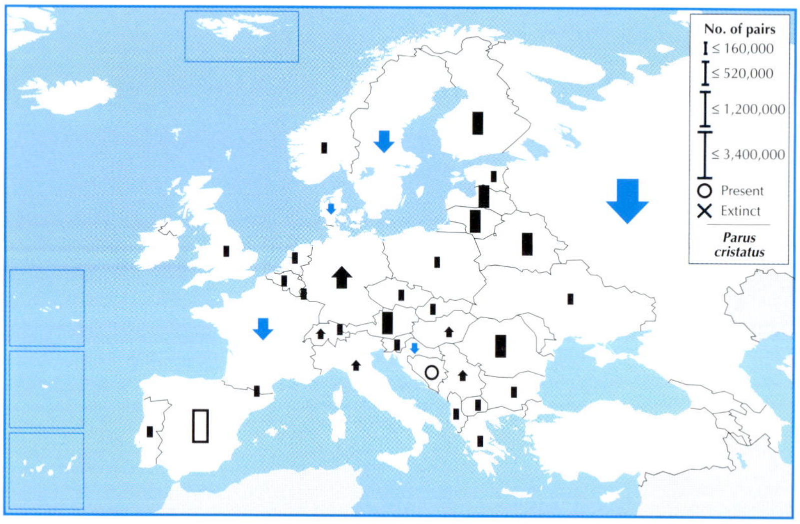

Country	Breeding pop. size (pairs)	Year(s)	Trend	Mag.%	References
Albania	(1,000 – 3,000)	02	(0)	(0–19)	
Andorra	(1,000 – 3,000)	99–01	(0)	(0–19)	*1,3*
Austria	(150,000 – 300,000)	98–02	(0)	(0–19)	
Belarus	200,000 – 350,000	97–02	0	0–19	
Belgium	10,000 – 20,000	01–02	(0)	(0–19)	*1*
Bosnia & HG	Present	90–03	?	–	
Bulgaria	7,500 – 15,000	96–02	0	0–9	
Croatia	(500 – 1,000)	02	(–)	(30–49)	*16*
Czech Rep.	80,000 – 160,000	00	0	0–19	
Denmark	20,000 – 30,000	00	**–**	**20–29**	*12*
Estonia	(20,000 – 50,000)	98	0	0–19	*1*
Finland	200,000 – 400,000	98–02	**0**	**0–19**	
France	(200,000 – 1,000,000)	98–02	–	27	*4,3*
Germany	200,000 – 450,000	95–99	+	0–19	
Greece	(2,000 – 5,000)	95–00	(0)	(0–19)	
Hungary	(3,500 – 5,000)	90–93	+	50–79	*14*
Italy	(20,000 – 40,000)	03	(+)	(0–19)	
Latvia	300,000 – 370,000	90–00	(0)	(0–19)	*23,18*
Liechtenstein	500 – 700	98–00	0	0–19	
Lithuania	(200,000 – 400,000)	99–01	(0)	(0–19)	*20*
Luxembourg	2,500 – 3,000	02	0	0–19	
Macedonia	(1,000 – 2,500)	90–00	(0)	(0–19)	
Netherlands	**20,000 – 30,000**	98–00	**F**	**23**	*1*
Norway	(50,000 – 200,000)	90–03	(0)	(0–19)	
Poland	**100,000 – 200,000**	00–02	0	0–19	*23*
Portugal	(50,000 – 500,000)	02	(0)	(0–19)	
Romania	450,000 – 600,000	00–02	0	0–19	*48*
Russia	2,500,000 – 4,500,000	90–00	(–)	(20–29)	*122*
Serbia & MN	3,500 – 6,500	90–02	(+)	(0–19)	*1,155,225,227, 117a,215*
Slovakia	25,000 – 50,000	90–99	0	0–19	
Slovenia	50,000 – 100,000	00	(0)	(0–19)	
Spain	(860,000 – 1,500,000)	92	?	–	*13,12,10*
Sweden	150,000 – 500,000	99–00	**–**	**16**	
Switzerland	**120,000 – 160,000**	93–96	+	0–9	
Ukraine	75,000 – 95,000	90–00	0	0–19	
UK	2,400 – 2,400	95	(0)	(0–19)	*32*
Total (approx.)	6,100,000 – 12,000,000	**Overall trend** Moderate decline			
Breeding range	>5,000,000 km²	**Gen. length.** <3.3 **% Global pop.** >95			

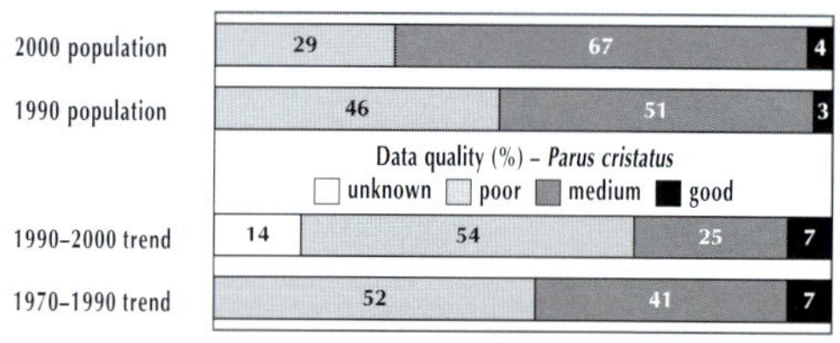

Parus ater
COAL TIT

Non-SPEC (1994: —) **Status** (Secure)
Criteria —
European IUCN Red List Category —
Criteria —
Global IUCN Red List Category —
Criteria —

Parus ater is a widespread resident across much of Europe, which accounts for less than half of its global range. Its European breeding population is extremely large (>12,000,000 pairs), and increased between 1970–1990. Although trends were not available for key populations in Spain, Italy and Russia during 1990–2000, the species was stable or increased across most of the rest of Europe—including sizeable populations in Germany and Romania—and was probably stable overall. Consequently, it is provisionally evaluated as Secure.

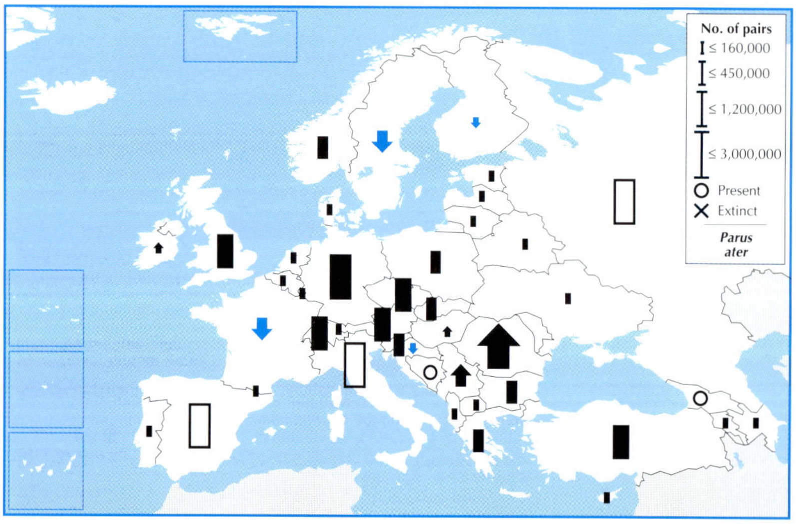

Country	Breeding pop. size (pairs)	Year(s)	Trend	Mag.%	References
Albania	(3,000 – 6,000)	02	(0)	(0–19)	
Andorra	(10,000 – 20,000)	99–01	(0)	(0–19)	*1,3*
Armenia	3,000 – 6,500	98–02	0	0–19	
Austria	(800,000 – 1,600,000)	98–02	(0)	(0–19)	
Azerbaijan	(20,000 – 100,000)	96–00	(0)	(0–19)	
Belarus	50,000 – 100,000	97–02	0	0–19	
Belgium	20,000 – 100,000	01–02	(0)	(0–19)	*1*
Bosnia & HG	Present	90–03	?	–	
Bulgaria	300,000 – 600,000	96–02	0	0–9	
Croatia	(50,000 – 100,000)	02	(–)	(50–79)	*16*
Cyprus	(5,000 – 15,000)	94–02	(0)	(0–19)	
Czech Rep.	450,000 – 900,000	00	0	0–19	
Denmark	100,000 – 200,000	00	**F**	**30–39**	*12*
Estonia	5,000 – 20,000	98	F	20–29	*1*
Finland	20,000 – 60,000	98–02	**–**	**40**	
France	(200,000 – 1,000,000)	98–02	**–**	**37**	*4,2*
Georgia	Present	03	?	–	
Germany	1,000,000 – 2,400,000	95–99	0	0–19	
Greece	(100,000 – 500,000)	95–00	(0)	(0–19)	
Hungary	4,000 – 20,000	99–02	+	50–79	*19*
Rep. Ireland	100,000 – 250,000	88–91	+	0–19	
Italy	(1,000,000 – 3,000,000)	03	?	–	
Latvia	60,000 – 100,000	90–00	(0)	(0–19)	*23,18*
Liechtenstein	1,000 – 1,500	98–00	0	0–19	
Lithuania	(50,000 – 100,000)	99–01	(0)	(0–19)	*20*
Luxembourg	10,000 – 12,000	02	0	0–19	
Macedonia	(25,000 – 100,000)	90–00	(0)	(0–19)	
Netherlands	**30,000 – 40,000**	98–00	**0**	**1**	*1*
Norway	(100,000 – 500,000)	90–02	(0)	(0–19)	
Poland	**200,000 – 400,000**	00–02	0	0–19	*23*
Portugal	(50,000 – 500,000)	02	(0)	(0–19)	
Romania	1,200,000 – 2,220,000	00–02	+	0–19	*48*
Russia	1,000,000 – 3,500,000	90–00	?	–	*122*
Serbia & MN	200,000 – 400,000	90–02	+	10–19	*1,155,75,67a, 117a,152,185*
Slovakia	250,000 – 500,000	90–99	0	0–19	
Slovenia	300,000 – 500,000	00	(0)	(0–19)	
Spain	(1,660,000 – 5,200,000)	92	?	–	*13,12,10*
Sweden	300,000 – 600,000	99–00	**–**	**49**	
Switzerland	**500,000 – 800,000**	93–96	0	0–19	
Turkey	(750,000 – 1,800,000)	01	(0)	(0–19)	
Ukraine	(85,000 – 115,000)	90–00	(0)	(5–14)	
UK	**653,000 – 653,000**	00	**0**	**1**	*5,31*
Total (approx.)	12,000,000 – 29,000,000	**Overall trend** Stable			
Breeding range	>6,000,000 km²	**Gen. length.** <3.3 **% Global pop.** 25–49			

(See p. 247, bottom, for data quality graph)

Parus caeruleus
BLUE TIT

Non-SPEC[E] (1994: 4) **Status** Secure
Criteria —
European IUCN Red List Category —
Criteria —
Global IUCN Red List Category —
Criteria —

Parus caeruleus is a widespread resident across most of Europe, which constitutes >75% of its global range. Its European breeding population is extremely large (>20,000,000 pairs), and was stable between 1970–1990. Although there were slight declines in France and Sweden during 1990–2000, populations were stable or increasing across the rest of its European range, and the species remained stable overall. Consequently, it is evaluated as Secure.

Country	Breeding pop. size (pairs)	Year(s)	Trend	Mag.%	References
Albania	10,000 – 30,000	02	(0)	(0–19)	
Andorra	(200 – 300)	99–01	(0)	(0–19)	*1,3*
Armenia	33,000 – 150,000	00–02	0	0–19	
Austria	(200,000 – 400,000)	98–02	(0)	(0–19)	
Azerbaijan	(100,000 – 300,000)	96–00	(0)	(0–19)	
Belarus	350,000 – 400,000	97–02	0	0–19	
Belgium	20,000 – 100,000	01–02	(0)	(0–19)	*1*
Bosnia & HG	Present	90–03	?	–	
Bulgaria	300,000 – 800,000	96–02	0	0–9	
Croatia	(500,000 – 1,000,000)	02	(+)	(>80)	*16*
Czech Rep.	800,000 – 1,600,000	00	0	0–19	
Denmark	200,000 – 250,000	00	**F**	**>80**	*12*
Estonia	50,000 – 100,000	98	+	20–29	*1*
Finland	400,000 – 650,000	98–02	+	**220**	
France	(2,000,000 – 10,000,000)	98–02	–	8	*4,3*
Georgia	Present	03	?	–	
Germany	2,000,000 – 4,200,000	95–99	0	0–19	
Greece	(500,000 – 1,000,000)	95–00	(0)	(0–19)	
Hungary	**170,000 – 270,000**	99–02	0	0–19	*19*
Rep. Ireland	500,000 – 1,000,000	88–91	0	0–19	
Italy	(500,000 – 1,000,000)	03	(0)	(0–19)	
Latvia	100,000 – 140,000	90–00	(0)	(0–19)	*23,18*
Liechtenstein	140 – 200	98–00	+	0–9	
Lithuania	(70,000 – 110,000)	99–01	(0)	(0–19)	*20*
Luxembourg	20,000 – 30,000	02	0	0–19	
Macedonia	(120,000 – 200,000)	90–00	(0)	(0–19)	
Moldova	38,000 – 46,000	90–00	0	0–19	
Netherlands	**275,000 – 325,000**	98–00	+	**27**	*1*
Norway	(100,000 – 200,000)	90–02	+	0–19	*27*
Poland	**500,000 – 1,000,000**	00–02	0	0–19	*23*
Portugal	(500,000 – 2,500,000)	02	(0)	(0–19)	
Romania	840,000 – 1,100,000	00–02	0	0–19	*48*
Russia	1,000,000 – 2,000,000	90–00	0	0–19	*122*
Serbia & MN	300,000 – 500,000	90–02	(0)	(0–19)	*1,29,172a,67a,78, 225,227,185,152*
Slovakia	700,000 – 1,400,000	90–99	0	0–19	
Slovenia	50,000 – 100,000	00	(0)	(0–19)	
Spain	(930,000 – 3,600,000)	92	?	–	*13,12,10*
Canary Is.	(20,000 – 100,000)	97–03	?	–	*28,25*
Sweden	750,000 – 1,250,000	99–00	–	**3**	
Switzerland	**150,000 – 250,000**	93–96	+	0–9	
Turkey	(700,000 – 1,800,000)	01	(0)	(0–19)	
Ukraine	(390,000 – 515,000)	90–00	0	0–19	
UK	**3,535,000 – 3,535,000**	00	**0**	**1**	*5,31*
Gibraltar	30 – 80	00	**0**	**0–19**	
Total (approx.)	20,000,000 – 44,000,000	**Overall trend** Stable			
Breeding range	>7,000,000 km²	**Gen. length.** <3.3 **% Global pop.** 75–94			

(See p. 247, bottom, for data quality graph)

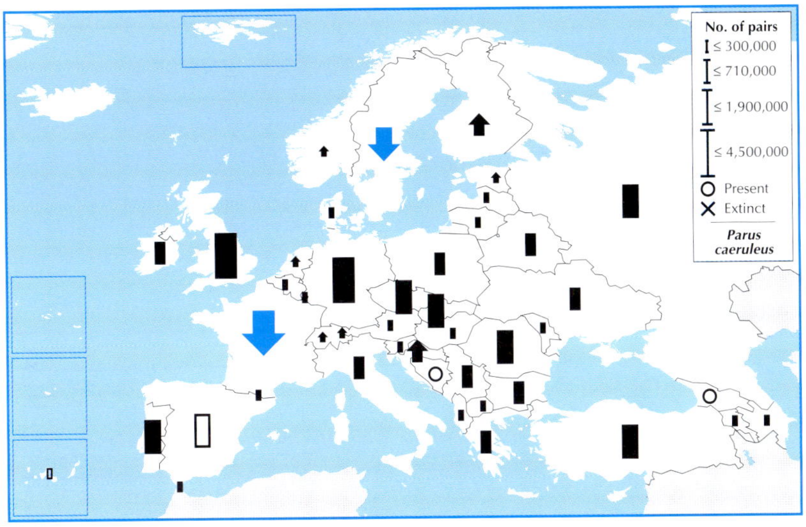

Parus cyanus
AZURE TIT

Non-SPEC (1994: —) **Status** (Secure)
Criteria —
European IUCN Red List Category —▼
Criteria —
Global IUCN Red List Category —
Criteria —

Country	Breeding pop. size (pairs)	Year(s)	Trend	Mag.%	References
Belarus	400 – 800	97–02	(0)	(0–19)	
Russia	(2,500 – 10,000)	90–00	?	–	
Ukraine	**0 – 30**	90–00	+	N	
Total (approx.)	2,900 – 11,000	**Overall trend** Unknown			
Breeding range	>500,000 km²	**Gen. length.** <3.3 **% Global pop.** 5–24			

Parus cyanus is resident in Russia, Belarus and Ukraine, with Europe accounting for less than a quarter of its global range. Its European breeding population is small (as few as 2,900 pairs), but was stable between 1970–1990. No trend data were available for Russia during 1990–2000, but there was no evidence to suggest the species declined. Although the size of the European population could render it susceptible to the risks affecting small populations, it is marginal to a much larger non-European population. Consequently, the species is provisionally evaluated as Secure.

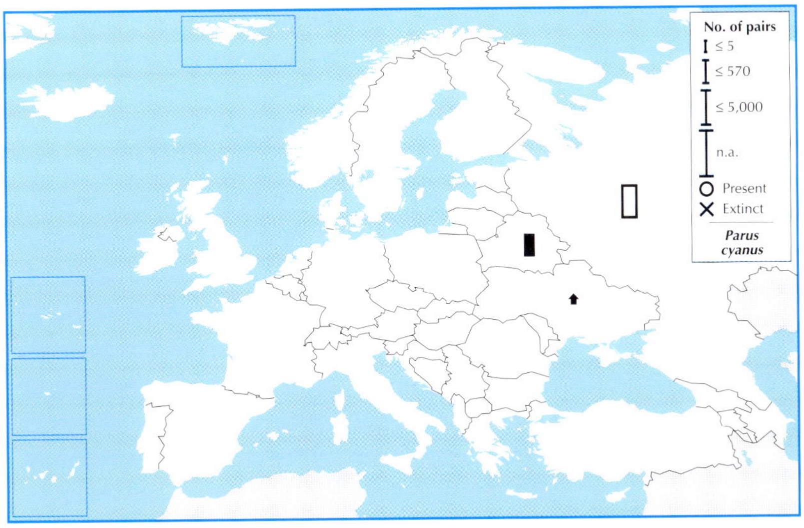

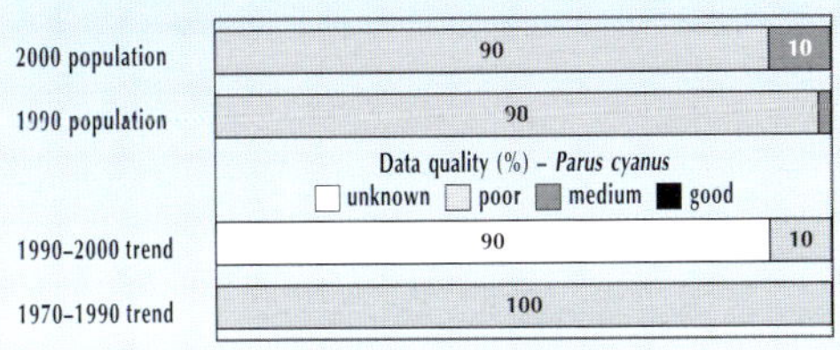

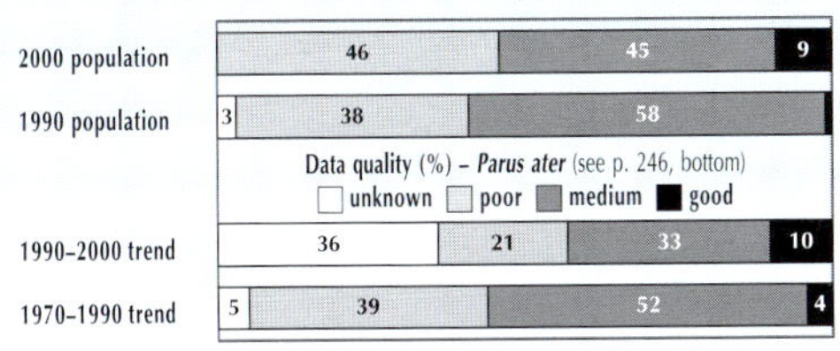

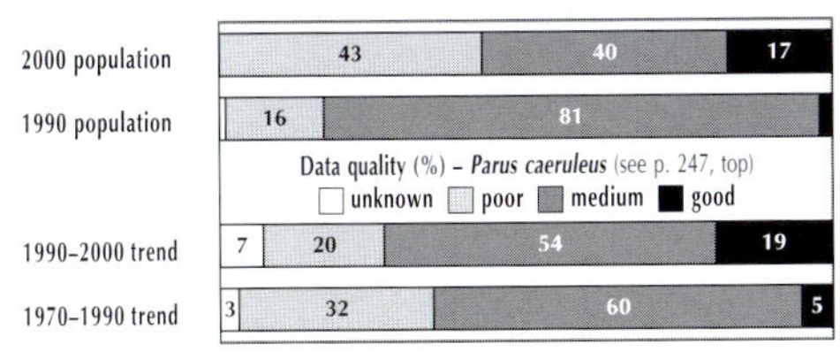

Parus major
GREAT TIT

Non-SPEC (1994: —) **Status** Secure
Criteria —
European IUCN Red List Category —
Criteria —
Global IUCN Red List Category —
Criteria —

Parus major is a widespread resident across most of Europe, which accounts for less than half of its global range. Its European breeding population is extremely large (>46,000,000 pairs), and was stable between 1970–1990. Although there were declines in a handful of countries during 1990–2000, populations were stable across the vast majority of Europe, and the species remained stable overall. Consequently, it is evaluated as Secure.

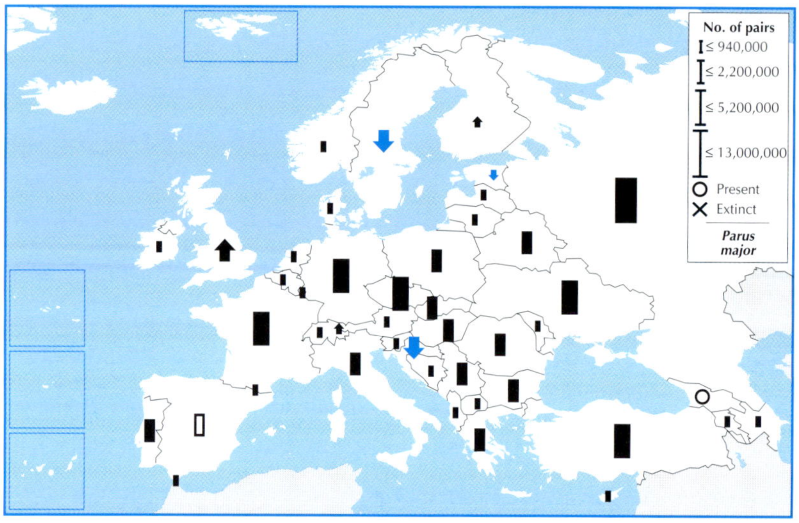

Country	Breeding pop. size (pairs)	Year(s)	Trend	Mag.%	References
Albania	30,000 – 50,000	02	(0)	(0–19)	
Andorra	(500 – 1,000)	99–01	(0)	(0–19)	*1,3*
Armenia	30,000 – 150,000	00–02	0	0–19	
Austria	(400,000 – 800,000)	98–02	(0)	(0–19)	
Azerbaijan	(200,000 – 400,000)	96–00	(0)	(0–19)	
Belarus	1,500,000 – 1,700,000	97–02	0	0–19	
Belgium	100,000 – 250,000	01–02	(0)	(0–19)	*1*
Bosnia & HG	(100,000 – 500,000)	90–00	(0)	(0–19)	
Bulgaria	800,000 – 2,500,000	96–02	0	0–9	
Croatia	(1,000,000 – 1,500,000)	02	(–)	(50–79)	*16*
Cyprus	(100,000 – 200,000)	94–02	(0)	(0–19)	
Czech Rep.	3,000,000 – 6,000,000	00	0	0–19	
Denmark	700,000 – 1,000,000	00	**0**	**0–19**	*12*
Estonia	150,000 – 250,000	98	–	20–29	*1*
Finland	**800,000 – 1,100,000**	98–02	**+**	**25**	
France	(2,000,000 – 10,000,000)	98–02	0	5	*4,3*
Georgia	Present	03	?	–	
Germany	3,500,000 – 7,600,000	95–99	0	0–19	
Greece	(1,000,000 – 2,000,000)	95–00	(0)	(0–19)	
Hungary	**1,050,000 – 1,350,000**	99–02	0	0–19	*19*
Rep. Ireland	250,000 – 500,000	88–91	0	0–19	
Italy	(1,000,000 – 2,000,000)	03	(0)	(0–19)	
Latvia	400,000 – 500,000	90–00	(0)	(0–19)	*23,18*
Liechtenstein	300 – 500	98–00	+	0–9	
Lithuania	(500,000 – 1,000,000)	99–01	(0)	(0–19)	*20*
Luxembourg	40,000 – 50,000	02	0	0–19	
Macedonia	(120,000 – 200,000)	90–00	(0)	(0–19)	
Moldova	80,000 – 100,000	90–00	0	0–19	
Netherlands	**500,000 – 600,000**	98–00	**0**	**6**	*1*
Norway	(500,000 – 1,000,000)	90–03	(0)	(0–19)	
Poland	**1,000,000 – 3,000,000**	00–02	0	0–19	*23*
Portugal	(500,000 – 2,500,000)	02	(0)	(0–19)	
Romania	1,850,000 – 2,600,000	00–02	0	0–19	*48*
Russia	10,000,000 – 15,000,000	90–00	0	0–19	*122*
Serbia & MN	900,000 – 1,400,000	90–02	0	0–19	*1,29,172a,67a, 225,227,117a*
Slovakia	1,500,000 – 3,000,000	90–99	0	0–19	
Slovenia	400,000 – 700,000	00	(0)	(0–19)	
Spain	(930,000 – 3,600,000)	92	?	–	*13,12,10*
Sweden	**1,500,000 – 3,000,000**	99–00	**–**	**12**	
Switzerland	350,000 – 500,000	98–02	0	0–19	
Turkey	(2,000,000 – 6,000,000)	01	(0)	(0–19)	
Ukraine	(3,000,000 – 4,000,000)	90–00	(0)	(0–19)	
UK	**2,074,000 – 2,074,000**	00	+	**22**	*5,31*
Gibraltar	**5 – 15**	00	**0**	**0–19**	
Total (approx.)	46,000,000 – 91,000,000	**Overall trend** Stable			
Breeding range	>8,000,000 km²	**Gen. length.** <3.3 **% Global pop.** 25–49			

(See p. 248, bottom, for data quality graph)

Sitta krueperi
KRÜPER'S NUTHATCH

SPEC 2 (1994: 4) **Status** (Declining)
Criteria Moderate recent decline
European IUCN Red List Category —
Criteria —
Global IUCN Red List Category —
Criteria —

Sitta krueperi is endemic to Europe, occurring only on the Greek island of Lesbos, and in parts of Turkey and the Caucasus. Its European breeding population is relatively large (>80,000 pairs), and was stable between 1970–1990. Although the species remained stable in Greece during 1990–2000, the key populations in Turkey and Russia both declined, and the species underwent a moderate decline (>10%) overall. Consequently, this previously Secure species is now provisionally evaluated as Declining.

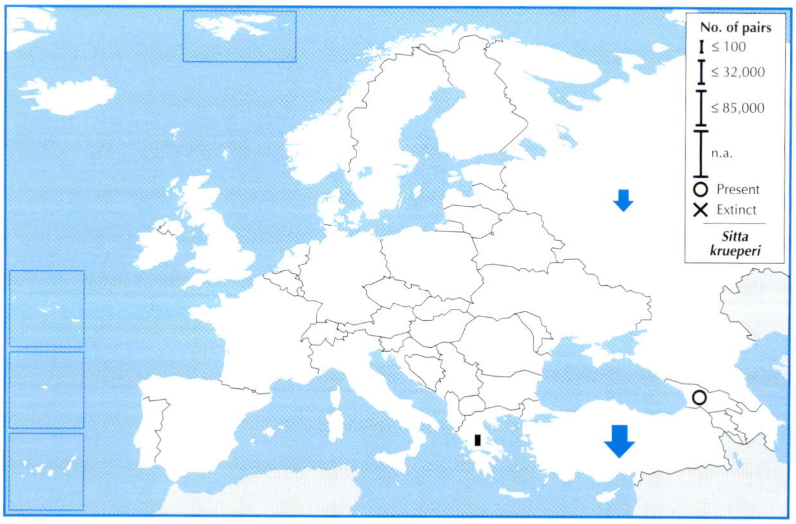

Country	Breeding pop. size (pairs)	Year(s)	Trend	Mag.%	References
Georgia	Present	03	?	–	
Greece	(50 – 200)	95–00	(0)	(0–19)	
Russia	20,000 – 50,000	90–00	(–)	(20–29)	*60,116,136*
Turkey	(60,000 – 120,000)	01	(–)	(0–19)	
Total (approx.)	80,000 – 170,000	**Overall trend** Moderate decline			
Breeding range	>250,000 km²	**Gen. length.** <3.3 **% Global pop.** 100			

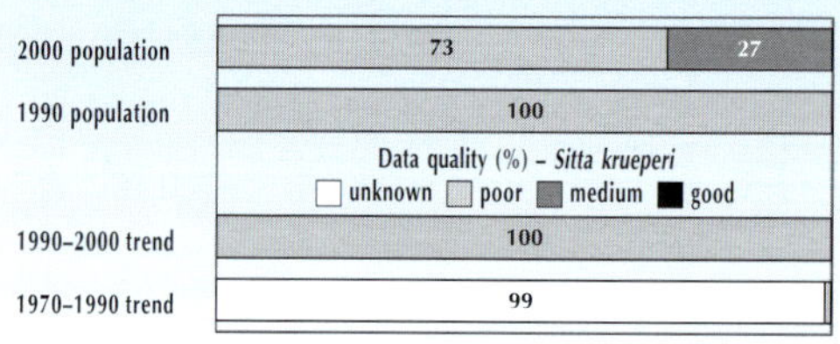

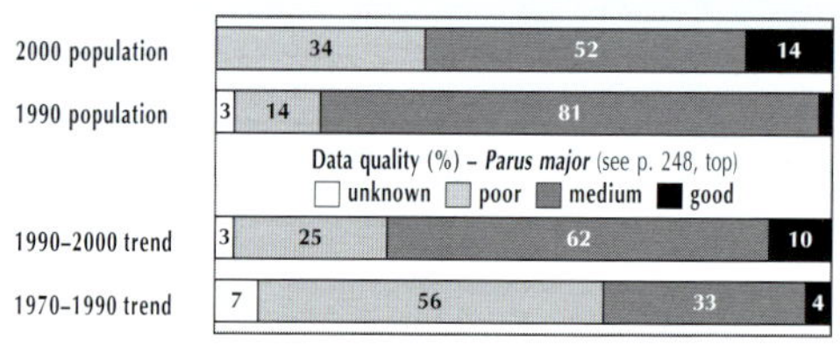

Sitta whiteheadi
CORSICAN NUTHATCH

SPEC 2 (1994: 2) **Status** Rare
Criteria <10,000 pairs
European IUCN Red List Category —
Criteria —
Global IUCN Red List Category —
Criteria —

Country	Breeding pop. size (pairs)	Year(s)	Trend	Mag.%	References
France	1,500 – 4,500	98–02	0	0–19	*4*
Total (approx.)	1,500 – 4,500	**Overall trend** Stable			
Breeding range	<500 km²	**Gen. length.** <3.3	**% Global pop.** 100		

Sitta whiteheadi is endemic to Europe, where it is confined to the island of Corsica (France). Its European breeding population is small (as few as 1,500 pairs), but was stable between 1970–1990. Although the species remained stable during 1990–2000, its population size still renders it susceptible to the risks affecting small populations, and consequently it is evaluated as Rare.

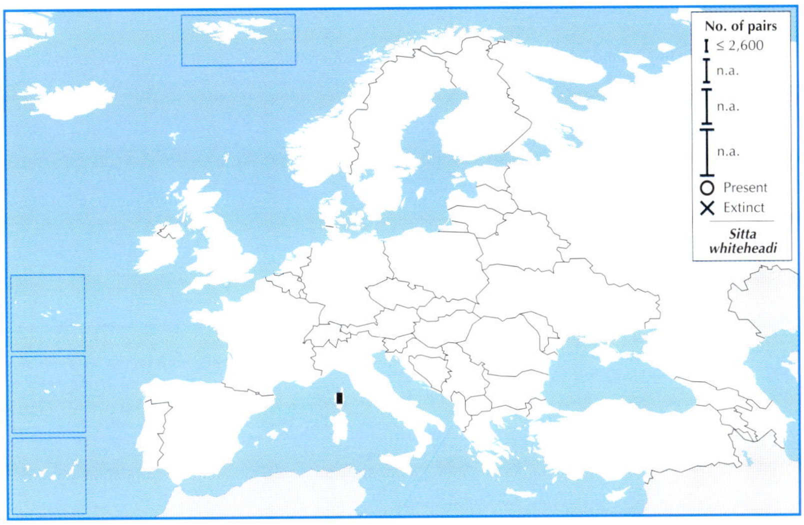

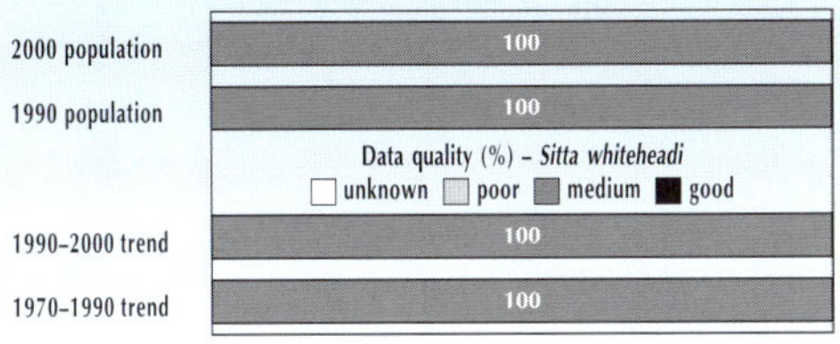

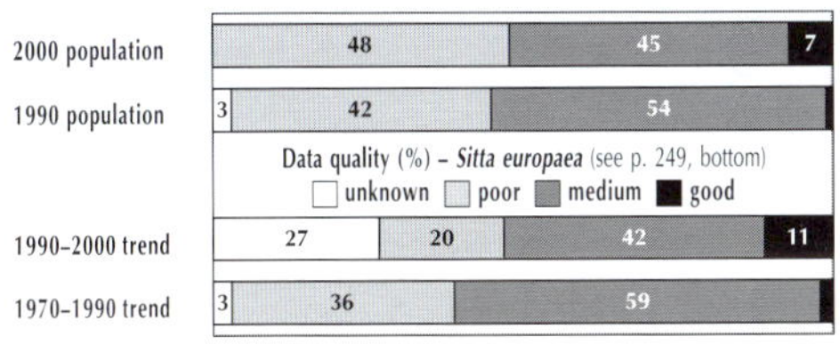

Sitta europaea
WOOD NUTHATCH

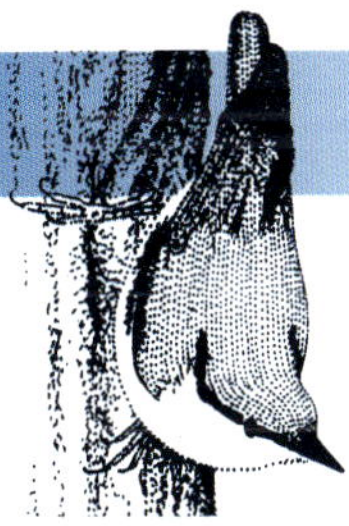

Non-SPEC (1994: —) **Status** Secure
Criteria —
European IUCN Red List Category —
Criteria —
Global IUCN Red List Category —
Criteria —

Sitta europaea is a widespread resident across much of Europe, which accounts for less than half of its global range. Its European breeding population is very large (>7,500,000 pairs), and was stable between 1970–1990. Although there were declines in a handful of countries—most notably France—during 1990–2000, populations were stable or increased across most of Europe, and the species remained stable overall. Consequently, it is evaluated as Secure.

Country	Breeding pop. size (pairs)	Year(s)	Trend	Mag.%	References
Albania	2,000 – 5,000	98–02	?	–	
Andorra	(150 – 300)	99–01	(0)	(0–19)	*1,3*
Armenia	15,000 – 30,000	99–02	0	0–19	
Austria	(200,000 – 400,000)	98–02	(0)	(0–19)	
Azerbaijan	(10,000 – 100,000)	96–00	(0)	(0–19)	
Belarus	200,000 – 280,000	97–02	0	0–19	
Belgium	23,000 – 45,000	01–02	(0)	(0–19)	*1*
Bosnia & HG	(50,000 – 100,000)	90–00	(0)	(0–19)	
Bulgaria	300,000 – 800,000	96–02	0	0–9	
Croatia	(500,000 – 1,000,000)	02	(+)	(>80)	*16*
Czech Rep.	600,000 – 1,200,000	00	0	0–19	
Denmark	20,000 – 30,000	00	-	**30–49**	*12*
Estonia	(20,000 – 50,000)	98	+	20–29	*1*
Finland	(0 – 20)	98–02	0	0–19	
France	(400,000 – 2,000,000)	98–02	-	**55**	*4,3*
Georgia	Present	03	?	–	
Germany	600,000 – 1,400,000	95–99	+	20–29	
Greece	(10,000 – 50,000)	95–00	(0)	(0–19)	
Hungary	**140,000 – 235,000**	99–02	0	0–19	*19*
Italy	(50,000 – 200,000)	03	(0)	(0–19)	
Latvia	40,000 – 100,000	90–00	(0)	(0–19)	*23,18*
Liechtenstein	150 – 250	98–00	0	0–19	
Lithuania	(30,000 – 50,000)	99–01	(0)	(0–19)	*20*
Luxembourg	15,000 – 20,000	02	0	0–19	
Macedonia	(50,000 – 150,000)	90–00	(0)	(0–19)	
Moldova	18,000 – 20,000	90–00	0	0–19	
Netherlands	**16,000 – 20,000**	98–00	+	**23**	*1*
Norway	(10,000 – 50,000)	90–02	(0)	(0–19)	
Poland	**300,000 – 400,000**	00–02	0	0–19	*23*
Portugal	(10,000 – 100,000)	02	(0)	(0–19)	
Romania	480,000 – 660,000	00–02	(0)	(0–19)	*48*
Russia	(1,000,000 – 5,000,000)	90–00	?	–	*122*
Serbia & MN	350,000 – 500,000	90–02	0	0–19	*1,29,172a,155,67a, 225,227,185,152*
Slovakia	700,000 – 1,000,000	90–99	0	0–19	
Slovenia	100,000 – 200,000	00	(0)	(0–19)	
Spain	(550,000 – 1,200,000)	92	?	–	*13,12,10*
Sweden	150,000 – 300,000	99–00	-	**19**	
Switzerland	**70,000 – 120,000**	93–96	0	0–19	
Turkey	(50,000 – 150,000)	01	(0)	(0–19)	
Ukraine	300,000 – 400,000	90–00	0	0–19	
UK	**144,000 – 144,000**	00	+	**11**	*5,31*
Total (approx.)	7,500,000 – 19,000,000	**Overall trend** Stable			
Breeding range	>6,000,000 km²	**Gen. length.** <3.3	**% Global pop.** 25–49		

(See p. 249, top, for data quality graph)

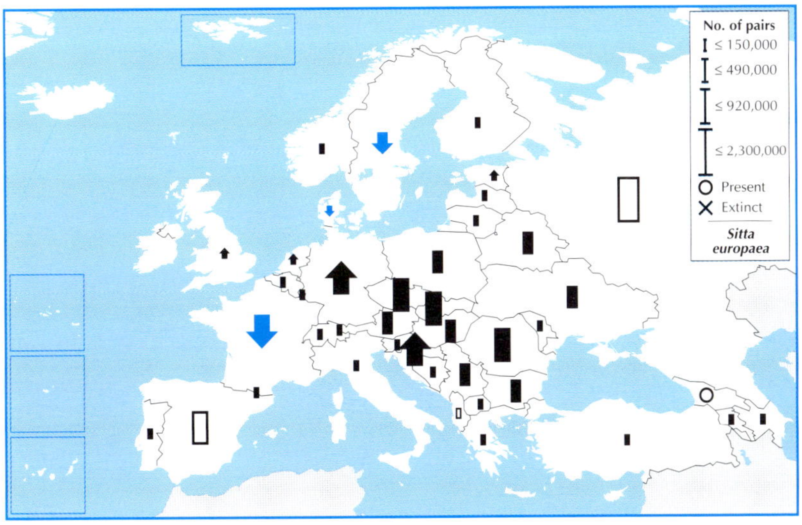

Sitta tephronota
EASTERN ROCK-NUTHATCH

Non-SPEC (1994: —) **Status** (Secure)
Criteria —
European IUCN Red List Category —
Criteria —
Global IUCN Red List Category —
Criteria —

Country	Breeding pop. size (pairs)	Year(s)	Trend	Mag.%	References
Armenia	1,500 – 2,500	99–02	0	0–19	
Azerbaijan	(100 – 1,000)	96–00	(0)	(0–19)	
Turkey	(20,000 – 100,000)	01	(–)	(0–19)	
Total (approx.)	22,000 – 100,000	**Overall trend** Small decline			
Breeding range	>100,000 km²	**Gen. length.** <3.3		**% Global pop.** <5	

Sitta tephronota has a predominantly Asian distribution, but its range also extends into Europe in parts of south-eastern Turkey and the Caucasus. Its European breeding population is relatively small (<100,000 pairs), but was stable between 1970–1990. The species was stable in Armenia and Azerbaijan during 1990–2000, but the Turkish stronghold declined slightly, and the species probably underwent a small decline. Consequently, it is provisionally evaluated as Secure.

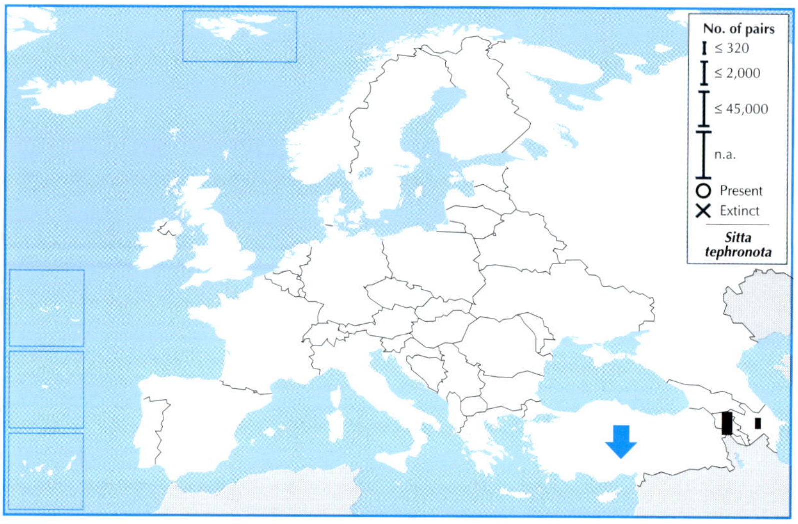

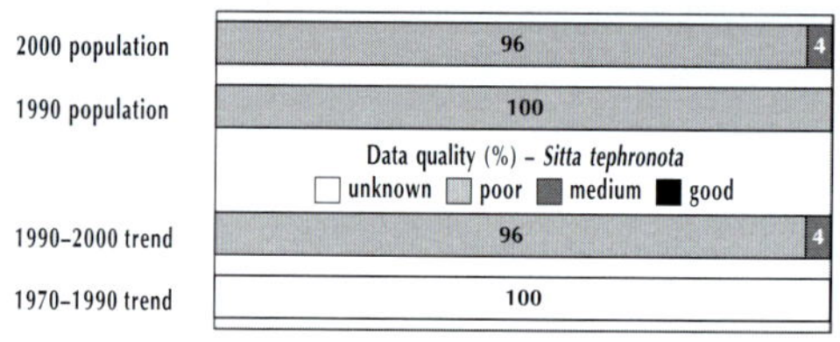

Sitta neumayer
WESTERN ROCK-NUTHATCH

Non-SPEC[E] (1994: 4) **Status** (Secure)
Criteria —
European IUCN Red List Category —
Criteria —
Global IUCN Red List Category —
Criteria —

Country	Breeding pop. size (pairs)	Year(s)	Trend	Mag.%	References
Albania	(2,000 – 5,000)	02	(0)	(0–19)	
Armenia	6,000 – 10,000	98–00	0	0–19	
Azerbaijan	(10,000 – 30,000)	96–00	(0)	(0–19)	
Bosnia & HG	Present	85–89	?	–	
Bulgaria	200 – 500	96–02	0	0–9	
Croatia	(5,000 – 10,000)	02	(+)	(50–79)	*16*
Georgia	Present	03	?	–	
Greece	(10,000 – 30,000)	95–00	(0)	(0–19)	
Macedonia	500 – 2,000	90–00	(0)	(0–19)	
Serbia & MN	400 – 600	97–02	0	0–19	*1,50,225,227, 156,216,91*
Turkey	(2,000,000 – 6,000,000)	01	(0)	(0–19)	
Total (approx.)	2,000,000 – 6,100,000	**Overall trend** Stable			
Breeding range	>1,000,000 km²	**Gen. length.** <3.3		**% Global pop.** 75–94	

Sitta neumayer is distributed patchily across much of south-eastern Europe, which constitutes >75% of its global range. Its European breeding population is very large (>2,000,000 pairs), and was stable between 1970–1990. The stronghold population in Turkey was probably stable during 1990–2000, and populations were also stable (or increased) across the rest of its European range. Consequently, it is provisionally evaluated as Secure.

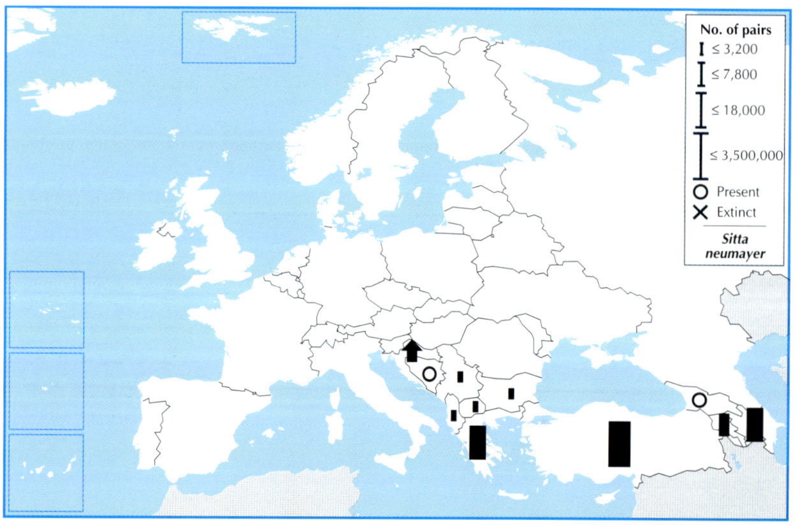

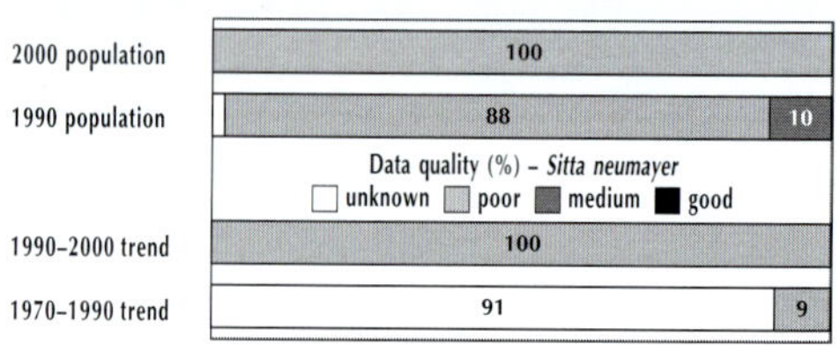

Tichodroma muraria
WALLCREEPER

Non-SPEC (1994: —) **Status** (Secure)
Criteria —
European IUCN Red List Category —
Criteria —
Global IUCN Red List Category —
Criteria —

Tichodroma muraria has a fragmented breeding distribution in mountainous areas of central and southern Europe, which accounts for less than half of its global breeding range. Its European breeding population is relatively small (<100,000 pairs), but was stable between 1970–1990. Although trends were not available for key populations in Spain and France during 1990–2000, the species was probably stable throughout the rest of its European range. Consequently it is provisionally evaluated as Secure.

Country	Breeding pop. size (pairs)	Year(s)	Trend	Mag.%	References
Albania	(10 – 100)	02	?	–	
Andorra	(4 – 8)	99–01	(0)	(0–19)	*1,3*
Armenia	150 – 200	97–02	0	0–19	
Austria	(700 – 1,400)	98–02	(0)	(0–19)	
Azerbaijan	(100 – 1,000)	96–00	(0)	(0–19)	
Bulgaria	250 – 400	96–02	0	0–9	
Croatia	(1 – 10)	02	(0)	(0–19)	*70,16*
France	4,000 – 20,000	98–02	?	–	*4*
Georgia	Present	03	?	–	
Germany	**100 – 200**	95–99	0	0–19	
Greece	(100 – 300)	95–00	(0)	(0–19)	
Italy	(2,000 – 6,000)	03	(0)	(0–19)	
Liechtenstein	(20 – 40)	98–00	0	0–19	
Macedonia	(50 – 100)	90–00	(0)	(0–19)	
Poland	**15 – 20**	95–00	(0)	(0–19)	*2,101*
Romania	(1,200 – 1,500)	00–02	(0)	(0–19)	
Russia	(15,000 – 30,000)	90–00	(0)	(0–19)	*33,60,61*
Serbia & MN	**45 – 60**	95–02	0	0–19	*1,55,62,235,225, 227,155,40,197*
Slovakia	30 – 50	90–99	0	0–19	
Slovenia	200 – 300	00	(0)	(0–19)	
Spain	(9,000 – 12,000)	92	?	–	*13,12,10*
Switzerland	**500 – 1,000**	93–96	(0)	(0–19)	
Turkey	(5,000 – 25,000)	01	(0)	(0–19)	
Total (approx.)	38,000 – 100,000	**Overall trend** Stable			
Breeding range	>750,000 km²	**Gen. length.** <3.3	**% Global pop.** 25–49		

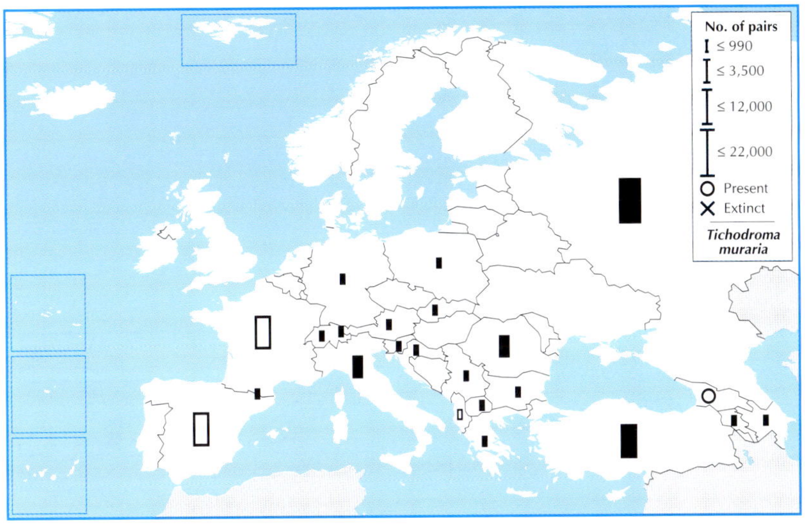

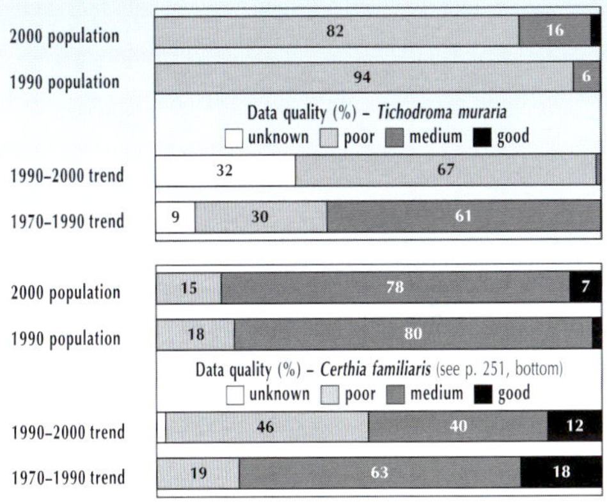

Certhia familiaris
EURASIAN TREE-CREEPER

Non-SPEC (1994: —) **Status** Secure
Criteria —
European IUCN Red List Category —
Criteria —
Global IUCN Red List Category —
Criteria —

Certhia familiaris is a widespread resident across much of Europe, which accounts for less than half of its global range. Its European breeding population is very large (>5,700,000 pairs), and was stable between 1970–1990. Although there were declines in a couple of countries during 1990–2000, populations across most of Europe were stable or increased, and the species remained stable overall. Consequently, it is evaluated as Secure.

Country	Breeding pop. size (pairs)	Year(s)	Trend	Mag.%	References
Albania	(500 – 2,000)	02	?	–	
Andorra	(50 – 100)	99–01	(0)	(0–19)	*1,3*
Armenia	1,000 – 15,000	99–02	?	–	
Austria	(150,000 – 300,000)	98–02	(0)	(0–19)	
Azerbaijan	(10,000 – 50,000)	96–00	(0)	(0–19)	
Belarus	350,000 – 400,000	97–02	0	0–19	
Belgium	2,500 – 10,000	95–02	0	0–19	*1*
Bosnia & HG	Present	85–89	?	–	
Bulgaria	75,000 – 200,000	96–02	0	0–9	
Croatia	(5,000 – 7,000)	02	(–)	(50–79)	*70,16*
Czech Rep.	300,000 – 600,000	00	0	0–19	
Denmark	15,000 – 25,000	00	F	50–79	*12*
Estonia	40,000 – 80,000	98	+	20–29	*1*
Finland	150,000 – 250,000	98–02	+	**>80**	
France	(50,000 – 250,000)	00–03	?	–	*4*
Georgia	Present	03	?	–	
Germany	160,000 – 460,000	95–99	0	0–19	
Greece	(2,000 – 5,000)	95–00	(0)	(0–19)	
Hungary	5,000 – 13,000	99–02	+	50–79	*19*
Rep. Ireland	20,000 – 100,000	88–91	(0)	(0–19)	
Italy	(30,000 – 100,000)	03	(0)	(0–19)	
Latvia	110,000 – 160,000	90–00	(0)	(0–19)	*23,18*
Liechtenstein	400 – 600	98–00	0	0–19	
Lithuania	(70,000 – 100,000)	99–01	(0)	(0–19)	*20*
Luxembourg	2,000 – 3,000	02	0	0–19	
Macedonia	(15,000 – 50,000)	90–00	(0)	(0–19)	
Moldova	3,500 – 4,500	90–00	0	0–19	
Netherlands	**75 – 100**	98–00	?	–	*1*
Norway	(20,000 – 100,000)	90–03	(0)	(0–19)	
Poland	**200,000 – 300,000**	00–02	0	0–19	*23*
Romania	950,000 – 1,280,000	00–02	0	0–19	
Russia	2,000,000 – 4,000,000	90–00	(0)	(0–19)	*122*
Serbia & MN	40,000 – 60,000	90–02	0	0–19	*1,172a,225,227, 185,152,117a*
Slovakia	100,000 – 150,000	90–99	0	0–19	
Slovenia	50,000 – 100,000	00	(0)	(0–19)	
Spain	(32,000 – 38,000)	92	?	–	*13,12,10*
Sweden	(350,000 – 800,000)	99–00	**–**	**6**	
Switzerland	**60,000 – 80,000**	93–96	0	0–19	
Turkey	(2,000 – 8,000)	01	(0)	(0–19)	
Ukraine	150,000 – 195,000	90–00	0	0–9	
UK	**214,000 – 214,000**	00	**0**	**2**	*5,31*
Total (approx.)	5,700,000 – 11,000,000	**Overall trend** Stable			
Breeding range	>5,000,000 km²	**Gen. length.** <3.3	**% Global pop.** 25–49		

(See p. 251, top, for data quality graph)

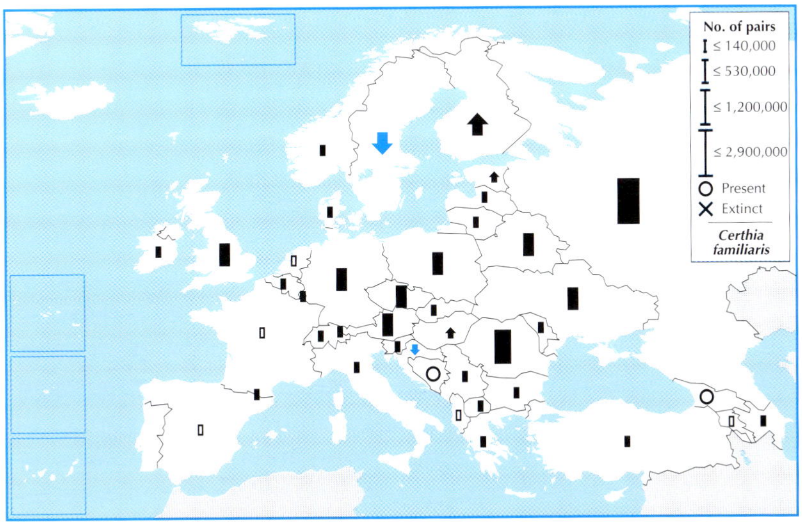

Certhia brachydactyla
SHORT-TOED TREE-CREEPER

Non-SPEC[E] (1994: 4) **Status** (Secure)
Criteria —
European IUCN Red List Category —
Criteria —
Global IUCN Red List Category —
Criteria —

Certhia brachydactyla is a widespread resident across much of central and southern Europe, which constitutes >75% of its global range. Its European breeding population is very large (>2,700,000 pairs), and was stable between 1970–1990. Although the trend of the Spanish population during 1990–2000 was unknown, the other key population—in France—increased, and trends were stable or increasing across most of the rest of its European range. Consequently, the species is provisionally evaluated as Secure.

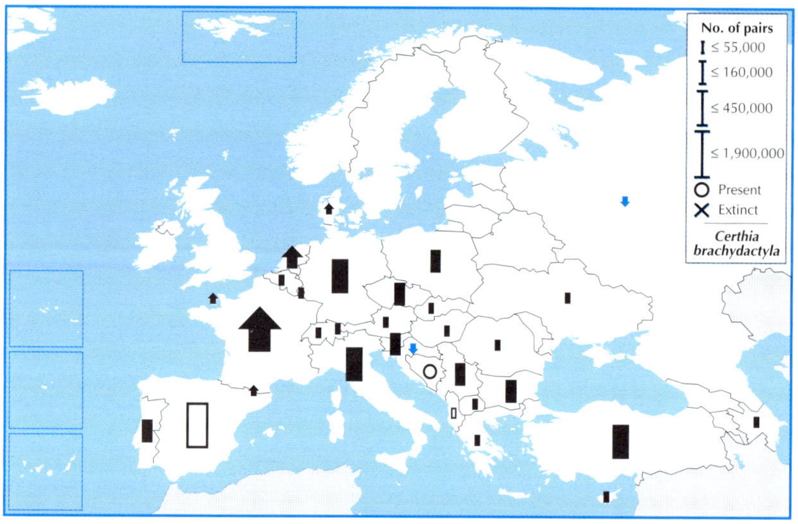

Country	Breeding pop. size (pairs)	Year(s)	Trend	Mag.%	References
Albania	(2,000 – 5,000)	02	?	–	
Andorra	(300 – 400)	99–01	(+)	(20–29)	*1,3*
Austria	(9,000 – 18,000)	98–02	(0)	(0–19)	
Azerbaijan	(0 – 200)	96–00	(0)	(0–19)	
Belgium	20,000 – 100,000	01–02	(0)	(0–19)	*1*
Bosnia & HG	Present	90–03	?	–	
Bulgaria	75,000 – 200,000	96–02	0	0–9	
Croatia	(7,000 – 12,000)	02	(–)	(50–79)	*70,16*
Cyprus	(2,000 – 4,000)	94–02	(0)	(0–19)	
Czech Rep.	75,000 – 150,000	00	0	0–19	
Denmark	(1,500 – 3,000)	00	+	30–49	*3*
France	(600,000 – 3,000,000)	98–02	+	19	*4,3*
Germany	270,000 – 740,000	95–99	0	0–19	
Greece	(30,000 – 100,000)	95–00	(0)	(0–19)	
Hungary	**19,000 – 54,000**	99–02	0	0–19	*19*
Italy	(100,000 – 500,000)	03	(0)	(0–19)	
Liechtenstein	30 – 50	98–00	0	0–19	
Luxembourg	5,000 – 8,000	02	0	0–19	
Macedonia	(15,000 – 30,000)	90–00	(0)	(0–19)	
Netherlands	**80,000 – 120,000**	98–00	+	**41**	*1*
Poland	**50,000 – 150,000**	00–02	0	0–19	*23*
Portugal	(50,000 – 500,000)	02	(0)	(0–19)	
Romania	4,000 – 7,000	00–02	0	0–19	
Russia	1,000 – 1,500	90–00	(–)	(20–29)	*177*
Serbia & MN	60,000 – 85,000	90–02	0	0–19	*1,29,225,227, 155,152,117a*
Slovakia	1,000 – 3,000	90–99	0	0–19	
Slovenia	50,000 – 100,000	00	(0)	(0–19)	
Spain	(1,000,000 – 3,300,000)	92	?	–	*13,12,10*
Switzerland	**30,000 – 60,000**	93–96	0	0–19	
Turkey	(150,000 – 450,000)	01	(0)	(0–19)	
Ukraine	700 – 1,700	90–00	0	0–19	
UK	475 – 475	00	(+)	(–)	

Total (approx.) 2,700,000 – 9,700,000 **Overall trend** Small increase
Breeding range >3,000,000 km² **Gen. length.** <3.3 **% Global pop.** 75–94

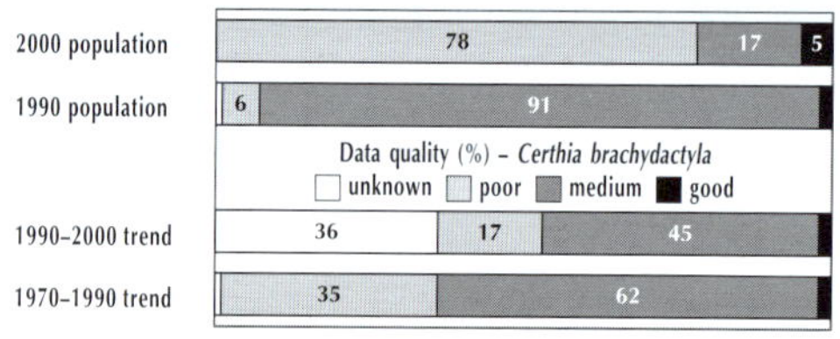

Remiz pendulinus
EURASIAN PENDULINE-TIT

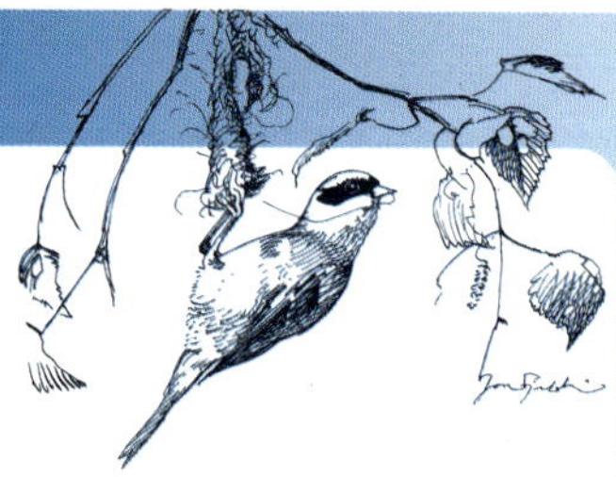

Non-SPEC (1994: —) **Status** (Secure)
Criteria —
European IUCN Red List Category —
Criteria —
Global IUCN Red List Category —
Criteria —

Remiz pendulinus has a widespread but patchy breeding distribution across much of central and southern Europe, which accounts for less than half of its global breeding range. Its European breeding population is large (>210,000 pairs), and was stable between 1970–1990. Although there were declines in a number of countries, notably Turkey, during 1990–2000, populations across most of Europe—including sizeable ones in Romania and Italy—were stable or increased, and the species probably remained stable overall. Consequently, it is provisionally evaluated as Secure.

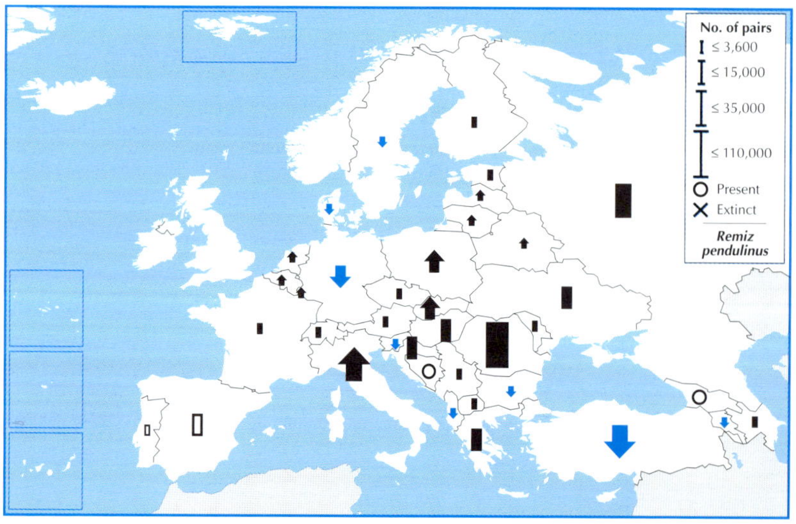

Country	Breeding pop. size (pairs)	Year(s)	Trend	Mag.%	References
Albania	(500 – 1,000)	95–02	(–)	(0–19)	
Armenia	800 – 3,000	99–02	–	0–9	
Austria	(1,000 – 1,500)	98–02	(0)	(0–19)	
Azerbaijan	(1,000 – 10,000)	96–00	(0)	(0–19)	
Belarus	2,500 – 4,000	97–00	+	**10–19**	
Belgium	13 – 20	95–02	+	**N**	*1*
Bosnia & HG	Present	90–03	?	–	
Bulgaria	400 – 800	96–02	–	0–19	
Croatia	(5,000 – 7,000)	02	(0)	(0–19)	*26,16*
Czech Rep.	2,500 – 5,000	00	0	0–19	
Denmark	25 – 50	99–00	**–**	**50–79**	*16*
Estonia	(100 – 300)	98	0	0–19	*1*
Finland	0 – 3	98–02	0	0–19	
France	100 – 300	98–02	(0)	(0–19)	*4*
Georgia	Present	03	?	–	
Germany	4,400 – 8,800	95–99	–	0–19	
Greece	(5,000 – 30,000)	95–00	(0)	(0–19)	
Hungary	4,500 – 13,000	99–02	0	0–19	*19*
Italy	(20,000 – 30,000)	03	(+)	(0–19)	
Latvia	300 – 800	90–00	(+)	(0–19)	*23,17*
Lithuania	(1,000 – 1,500)	99–01	(+)	(20–29)	*20*
Luxembourg	**1 – 5**	02	+	**50–79**	
Macedonia	(200 – 500)	90–00	(0)	(0–19)	
Moldova	1,100 – 1,300	90–00	0	0–19	
Netherlands	**140 – 210**	98–00	+	30–49	*1*
Poland	**10,000 – 20,000**	00–02	+	0–19	*23*
Portugal	0 – 5	02	?	N	
Romania	(90,000 – 125,000)	00–02	(0)	(0–19)	*48*
Russia	(10,000 – 50,000)	90–00	(0)	(0–19)	*8,54,141*
Serbia & MN	1,900 – 2,400	95–02	0	0–19	*1,194,155,191, 29,78,203,227,13*
Slovakia	5,000 – 10,000	90–99	+	20–29	
Slovenia	100 – 200	99–00	–	10–19	
Spain	(12,400 – 14,600)	92	?	–	*13,12,10*
Sweden	50 – 100	99–00	**–**	**20–29**	
Switzerland	**0 – 5**	93–96	**0**	**0–19**	
Turkey	(20,000 – 60,000)	01	(–)	(0–19)	
Ukraine	11,000 – 17,000	90–00	0	10–19	

Total (approx.) 210,000 – 420,000 **Overall trend** Stable
Breeding range >3,000,000 km² **Gen. length.** <3.3 **% Global pop.** 25–49

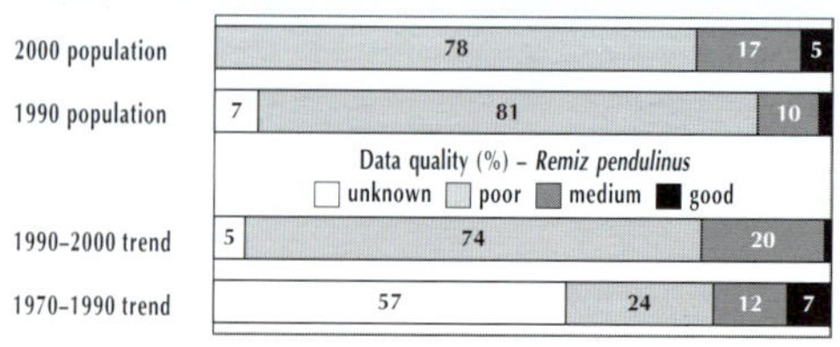

Oriolus oriolus
EURASIAN GOLDEN-ORIOLE

Non-SPEC (1994: —) **Status** Secure
Criteria —
European IUCN Red List Category —
Criteria —
Global IUCN Red List Category —
Criteria —

Oriolus oriolus is a widespread summer visitor to much of Europe, which accounts for less than half of its global breeding range. Its European breeding population is very large (>3,400,000 pairs), and was stable between 1970–1990. Although there were declines in a number of countries—notably France and Turkey—during 1990–2000, the vast majority of populations in the east of its European range, including key ones in Russia and Romania, were stable, and the species declined only slightly overall. Consequently, it is evaluated as Secure.

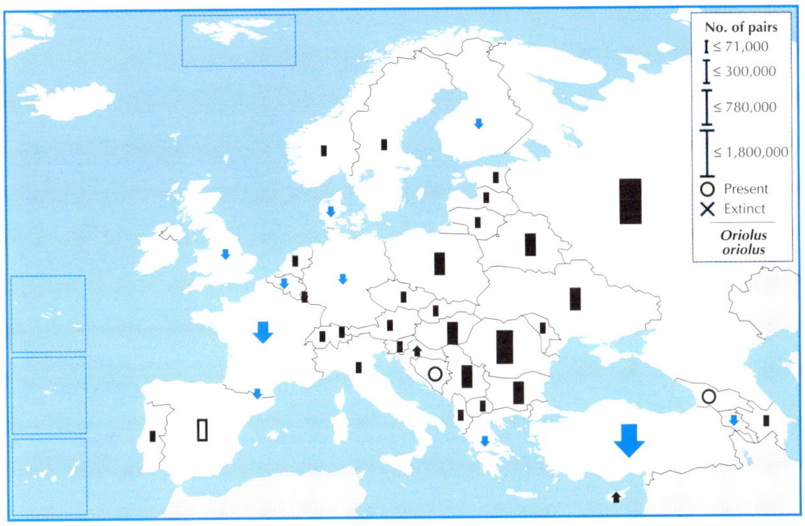

Country	Breeding pop. size (pairs)	Year(s)	Trend	Mag.%	References
Albania	5,000 – 10,000	02	(0)	(0–19)	
Andorra	(0 – 2)	99–01	(–)	(20–29)	*1,3*
Armenia	1,000 – 1,500	98–02	–	0–9	
Austria	(6,000 – 12,000)	98–02	(0)	(0–19)	
Azerbaijan	(50,000 – 100,000)	96–00	(0)	(0–19)	
Belarus	120,000 – 160,000	97–02	(0)	(0–19)	
Belgium	1,600 – 3,800	95–02	–	0–19	*1*
Bosnia & HG	Present	90–03	?	–	
Bulgaria	100,000 – 300,000	96–02	0	0–9	
Croatia	(50,000 – 100,000)	02	(+)	(>80)	*26,54,44*
Cyprus	**1 – 2**	94–96	**+**	**N**	
Czech Rep.	8,000 – 16,000	00	0	0–19	
Denmark	13 – 19	99–00	**-**	**50–79**	*5,6*
Estonia	(7,000 – 15,000)	98	0	0–19	*1*
Finland	2,000 – 4,000	98–02	**-**	**40**	
France	(150,000 – 600,000)	98–02	–	21	*4,2*
Georgia	Present	03	?	–	
Germany	40,000 – 90,000	95–99	–	20–29	
Greece	(20,000 – 30,000)	95–00	(–)	(0–19)	
Hungary	**190,000 – 240,000**	99–02	0	0–19	*19*
Italy	(20,000 – 50,000)	03	(0)	(0–19)	
Latvia	10,000 – 30,000	90–00	(0)	(0–19)	*18*
Liechtenstein	**8 – 12**	98–00	**0**	**0–19**	
Lithuania	(15,000 – 25,000)	99–01	(0)	(0–19)	*20*
Luxembourg	100 – 150	02	0	0–19	
Macedonia	(20,000 – 35,000)	90–00	(0)	(0–19)	
Moldova	15,000 – 18,000	90–00	0	0–19	
Netherlands	**4,000 – 5,000**	98–00	**0**	**10**	*1*
Norway	0 – 5	99	**0**	**0–19**	
Poland	**80,000 – 150,000**	00–02	0	0–19	*23*
Portugal	(10,000 – 100,000)	02	(0)	(0–19)	
Romania	680,000 – 890,000	00–02	0	0–19	*48*
Russia	1,000,000 – 3,000,000	90–00	(0)	(0–19)	*122*
Serbia & MN	100,000 – 140,000	90–02	0	0–19	*1,29,155,172a, 67a,78,225,227*
Slovakia	7,000 – 15,000	90–99	0	0–19	
Slovenia	4,000 – 8,000	94	(0)	(0–19)	
Spain	(150,000 – 200,000)	92	?	–	*13,12,10*
Sweden	**50 – 100**	99–00	0	0–19	
Switzerland	**1,000 – 2,000**	93–96	0	0–19	
Turkey	(350,000 – 600,000)	01	(–)	(0–19)	
Ukraine	140,000 – 190,000	90–00	0	0–19	
UK	**7 – 22**	96–00	**-**	**10**	
Total (approx.)	3,400,000 – 7,100,000	**Overall trend** Small decline			
Breeding range	>7,000,000 km²	**Gen. length.** <3.3 **% Global pop.** 25–49			

(See p. 254, top, for data quality graph)

Lanius collurio
RED-BACKED SHRIKE

SPEC 3 (1994: 3) **Status** (Depleted)
Criteria Moderate historical decline
European IUCN Red List Category —
Criteria —
Global IUCN Red List Category —
Criteria —

Lanius collurio is a widespread summer visitor to much of Europe, which accounts for less than half of its global breeding range. Its European breeding population is very large (>6,300,000 pairs), but underwent a moderate decline between 1970–1990. Although declines continued in several countries during 1990–2000, most eastern populations remained stable, and trend data were not available for the key populations in Russia and Spain. Nevertheless, the species probably declined only slightly overall, and consequently it is provisionally evaluated as Depleted.

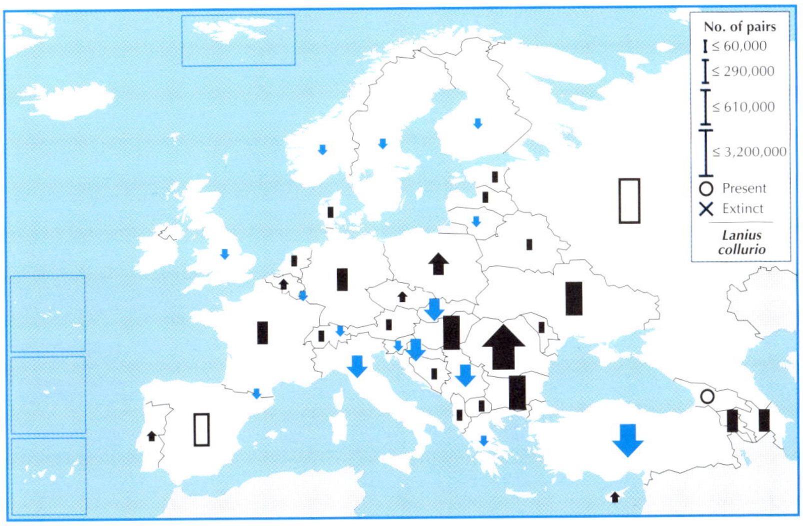

Country	Breeding pop. size (pairs)	Year(s)	Trend	Mag.%	References
Albania	5,000 – 20,000	02	(0)	(0–19)	
Andorra	20 – 50	99–01	(–)	(10–29)	*1,3*
Armenia	65,000 – 150,000	98–02	0	0–19	
Austria	(20,000 – 40,000)	98–02	(0)	(0–19)	
Azerbaijan	(50,000 – 200,000)	96–00	(0)	(0–19)	
Belarus	50,000 – 70,000	97–02	0	0–19	
Belgium	**1,500 – 2,200**	95–02	**+**	**50–79**	*1*
Bosnia & HG	(20,000 – 50,000)	90–00	(F)	(–)	
Bulgaria	300,000 – 1,000,000	96–02	0	0–9	
Croatia	(70,000 – 150,000)	02	(–)	(50–79)	*16*
Cyprus	**1 – 2**	99–02	**+**	**N**	
Czech Rep.	30,000 – 60,000	00	+	0–19	
Denmark	1,500 – 3,000	93–96	(F)	(–)	*3*
Estonia	(10,000 – 20,000)	98	0	0–19	*1*
Finland	30,000 – 60,000	98–02	**-**	**30**	
France	120,000 – 360,000	98–02	**0**	**0–19**	*4*
Georgia	Present	03	?	–	
Germany	90,000 – 190,000	95–99	0	0–19	
Greece	(10,000 – 30,000)	95–00	(–)	(20–29)	
Hungary	**540,000 – 670,000**	99–02	0	0–19	*19*
Italy	(50,000 – 120,000)	03	(–)	(0–19)	
Latvia	20,000 – 40,000	90–00	0	0–19	*23,16,2*
Liechtenstein	10 – 20	98–00	–	50–79	
Lithuania	(30,000 – 50,000)	99–01	(–)	(0–19)	*20*
Luxembourg	**1,500 – 2,000**	02	**-**	**10–19**	
Macedonia	(15,000 – 50,000)	90–00	(0)	(0–19)	
Moldova	40,000 – 50,000	90–00	0	0–19	
Netherlands	**160 – 200**	98–00	**0**	**18**	*1*
Norway	(1,000 – 5,000)	90–03	–	0–19	*3*
Poland	**200,000 – 400,000**	00–03	+	0–19	*23,102*
Portugal	600 – 1,100	02	(+)	(–)	
Romania	**1,380,000 – 2,600,000**	00–02	+	0–19	*48*
Russia	2,000,000 – 5,000,000	90–00	?	–	*122*
Serbia & MN	70,000 – 100,000	95–02	–	0–19	*1,29,155,172a, 67a,78,225,227,91*
Slovakia	65,000 – 130,000	90–99	–	30–49	
Slovenia	20,000 – 30,000	94	(–)	(0–19)	
Spain	(240,000 – 500,000)	95	?	–	*10*
Sweden	**26,000 – 34,000**	99–00	**-**	**21**	
Switzerland	**20,000 – 25,000**	93–96	0	0–19	
Turkey	(400,000 – 800,000)	01	(–)	(0–19)	
Ukraine	350,000 – 460,000	90–00	0	0–19	
UK	**0 – 5**	96–00	**-**	**45**	
Total (approx.)	6,300,000 – 13,000,000	**Overall trend** Small decline			
Breeding range	>7,000,000 km²	**Gen. length.** <3.3 **% Global pop.** 25–49			

(See p. 254, bottom, for data quality graph)

Lanius minor
LESSER GREY SHRIKE

SPEC 2 (1994: 2) **Status** (Declining)
Criteria Moderate continuing decline
European IUCN Red List Category —
Criteria —
Global IUCN Red List Category —
Criteria —

Lanius minor is a widespread but patchily distributed summer visitor to much of southern and eastern Europe, which constitutes >50% of its global breeding range. Its European breeding population is large (>620,000 pairs), but underwent a moderate decline between 1970–1990. Although certain populations were stable or increased during 1990–2000, there were widespread declines across most of Europe—including in the Romanian stronghold—and the species underwent a moderate decline (>10%) overall. Consequently, it is provisionally evaluated as Declining.

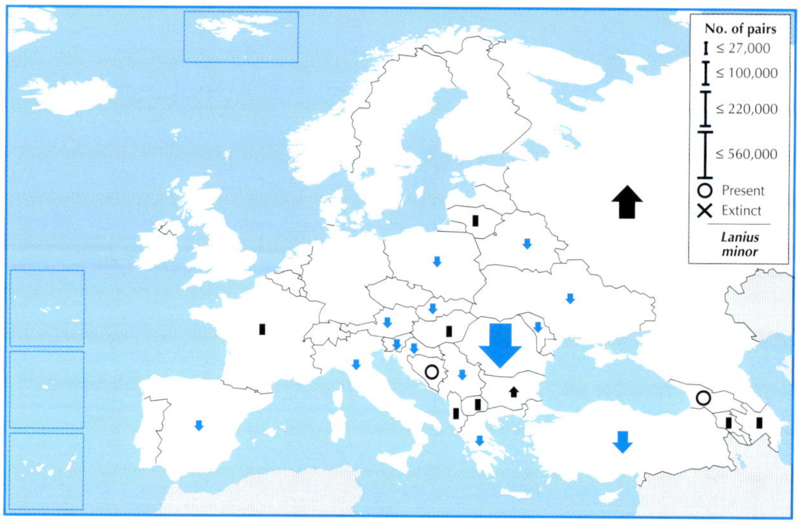

Country	Breeding pop. size (pairs)	Year(s)	Trend	Mag.%	References
Albania	(200 – 500)	02	(0)	(0–19)	
Armenia	15,000 – 30,000	97–02	0	0–19	
Austria	0 – 2	98–02	**-**	**>80**	
Azerbaijan	(5,000 – 50,000)	96–00	(0)	(0–19)	
Belarus	**50 – 200**	97–02	-	0–9	
Bosnia & HG	Present	90–03	?	–	
Bulgaria	5,000 – 15,000	96–02	+	0–19	
Croatia	(1,000 – 1,500)	02	(–)	(50–79)	*70,16,54*
France	**45 – 46**	00	**F**	**20–29**	*5*
Georgia	Present	03	?	–	
Greece	(2,000 – 3,000)	95–00	(–)	(0–19)	
Hungary	2,800 – 3,700	98–02	0	0–19	*13*
Italy	(1,000 – 2500)	03	(–)	(0–19)	
Lithuania	(1 – 10)	99–01	(F)	(>80)	*20*
Macedonia	(2,000 – 5,000)	90–00	(0)	(0–19)	
Moldova	5,300 – 5,800	90–00	-	20–29	
Poland	**5 – 10**	97–00	**-**	**70–79**	*89*
Romania	**364,000 – 857,000**	00–02	-	0–19	*48*
Russia	(150,000 – 300,000)	90–00	+	20–29	*8,24*
Serbia & MN	**400 – 500**	95–02	-	20–29	*1,245,155,178, 62,197,191,29, 77a,172a,78,225*
Slovakia	**600 – 800**	90–99	-	50–79	
Slovenia	**10 – 20**	99–00	**-**	**50–79**	
Spain	**0 – 50**	98–01	**-**	**30–49**	*10,16*
Turkey	(50,000 – 200,000)	01	(–)	(0–19)	
Ukraine	20,000 – 35,000	90–00	**-**	**0–19**	

Total (approx.) 620,000 – 1,500,000 **Overall trend** Moderate decline
Breeding range >3,000,000 km² **Gen. length.** <3.3 **% Global pop.** 50–74

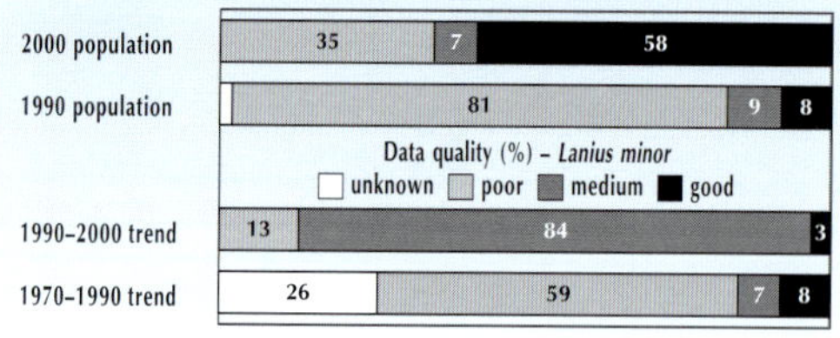

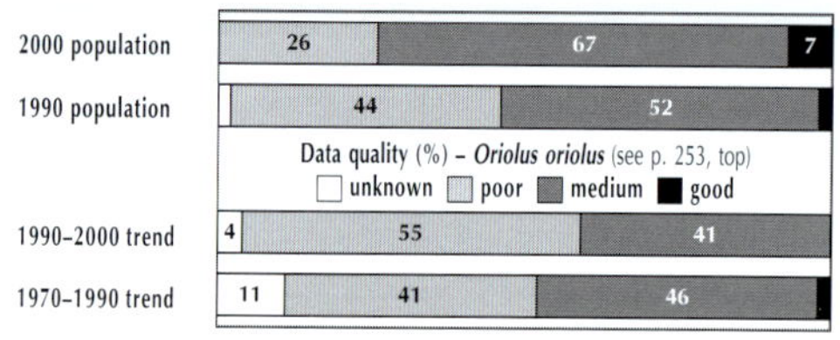

Lanius excubitor
GREAT GREY SHRIKE

SPEC 3 (1994: 3) **Status** (Depleted)
Criteria Moderate historical decline
European IUCN Red List Category —
Criteria —
Global IUCN Red List Category —
Criteria —

Lanius excubitor is a widespread breeder across much of Europe, which accounts for less than a quarter of its global breeding range. Its European breeding population is large (>250,000 pairs), but underwent a moderate decline between 1970–1990. Although the species was stable or increased across much of its European range during 1990–2000, many western and north-western populations—including its Spanish stronghold—continued to decline, and the species probably underwent a slight decline overall. Consequently, it is provisionally evaluated as Depleted.

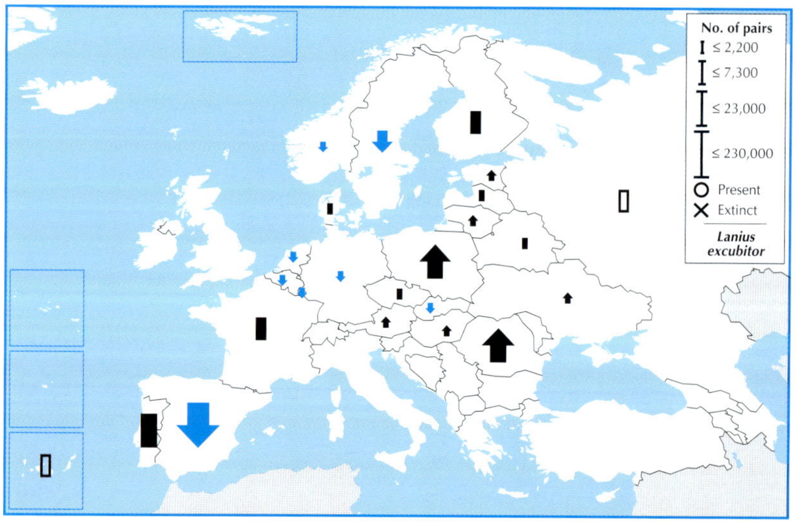

Country	Breeding pop. size (pairs)	Year(s)	Trend	Mag.%	References
Austria	**33 – 50**	98–02	+	**>80**	
Belarus	600 – 1,200	97–02	0	0–19	
Belgium	**101 – 251**	95–02	**-**	**30–49**	*1*
Czech Rep.	1,000 – 2,000	00	0	0–19	
Denmark	**20 – 30**	98–01	(0)	(0–9)	*4,5,6,7*
Estonia	400 – 500	98	+	20–29	*1*
Finland	5,000 – 8,000	98–02	0	0–19	
France	**2,800 – 7,500**	96–00	F	>80	*1*
Germany	1,800 – 2,500	95–99	-	30–49	
Hungary	1 – 6	00–03	+	N	
Latvia	(100 – 150)	90–00	(0)	(0–19)	*23*
Lithuania	(100 – 150)	99–01	(+)	(30–49)	*20*
Luxembourg	**50 – 80**	00–02	**-**	**10–19**	
Netherlands	**1 – 4**	98–00	**-**	**88**	*1*
Norway	400 – 800	03	-	0–19	*38*
Poland	**10,000 – 20,000**	00–02	+	20–29	*23*
Portugal	(10,000 – 50,000)	02	(0)	(0–19)	
Romania	14,000 – 17,000	00–02	+	0–19	*48,13*
Russia	(3,500 – 15,000)	90–00	?	–	*74*
Slovakia	500 – 1,000	90–99	-	30–49	
Spain	(200,000 – 250,000)	92	(–)	(0–19)	*13,12,16,10*
Canary Is.	(2,500 – 10,000)	97–03	?	–	*25*
Sweden	(1,000 – 10,000)	99–00	(–)	(0–19)	
Ukraine	**600 – 900**	90–00	+	**5–19**	

Total (approx.) 250,000 – 400,000 **Overall trend** Moderate decline
Breeding range >4,000,000 km² **Gen. length.** <3.3 **% Global pop.** 5–24

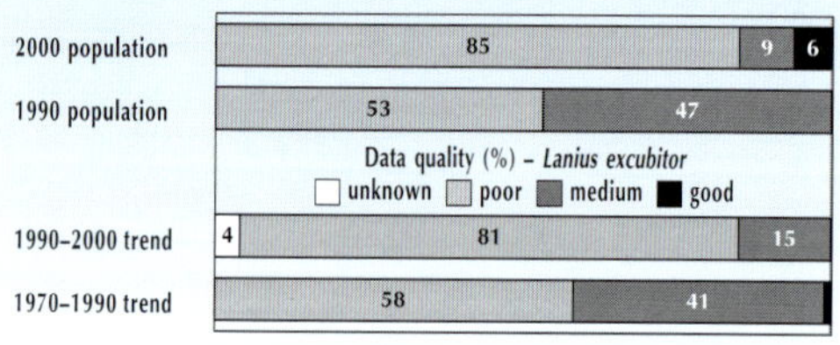

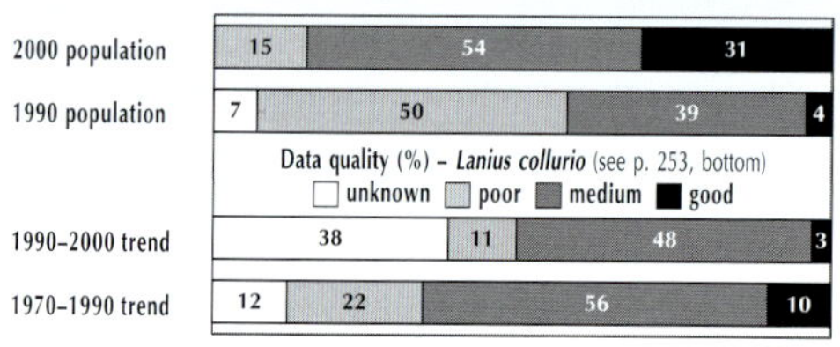

Lanius senator
WOODCHAT SHRIKE

SPEC 2 (1994: 2) **Status** (Declining)
Criteria Moderate continuing decline
European IUCN Red List Category —
Criteria —
Global IUCN Red List Category —
Criteria —

Lanius senator is a widespread summer visitor to much of southern and central Europe, which constitutes >50% of its global breeding range. Its European breeding population is large (>480,000 pairs), but underwent a large decline between 1970–1990. Although some populations—notably in south-east Europe—were stable or increased during 1990–2000, the species continued to decline across most of its European range, including in its Spanish stronghold, and underwent a moderate decline (>10%) overall. Consequently, it is provisionally evaluated as Declining.

Country	Breeding pop. size (pairs)	Year(s)	Trend	Mag.%	References
Albania	(2,000 – 4,000)	02	(0)	(0–19)	
Andorra	**0 – 2**	97–98	(0)	(0–19)	*1,3*
Armenia	50 – 100	98–02	-	50–79	
Azerbaijan	(5,000 – 20,000)	96–00	(0)	(0–19)	
Bosnia & HG	Present	90–03	?	–	
Bulgaria	5,000 – 10,000	96–02	+	50–79	
Croatia	(2,500 – 5,000)	02	(-)	(50–79)	*16,44*
Cyprus	**25 – 50**	94–02	+	**30–49**	
Czech Rep.	**0 – 0**	00	**-**	**X**	
France	10,000 – 30,000	00–02	-	0–19	*4*
Georgia	Present	03	?	–	
Germany	**11 – 19**	95–99	**-**	**>80**	
Greece	(10,000 – 30,000)	95–00	(-)	(0–19)	
Italy	(10,000 – 20,000)	03	(-)	(0–19)	
Luxembourg	**0 – 0**	00	**-**	**X**	
Macedonia	(5,000 – 15,000)	90–00	(0)	(0–19)	
Malta	1 – 3	90–02	-	0–19	*1*
Poland	**5 – 10**	90–00	**-**	**70–79**	*90*
Portugal	(10,000 – 100,000)	02	(-)	(-)	
Romania	40 – 80	00–02	+	0–19	*48*
Russia	20 – 50	90–02	+	20–29	*33*
Serbia & MN	210 – 310	95–02	-	0–19	*1,62,197,155, 225,227,152*
Slovakia	0 – 0	90–99	-	X	
Spain	(390,000 – 860,000)	92	(-)	(0–19)	*13,12,16,10*
Switzerland	**3 – 10**	98–02	**-**	**>80**	
Turkey	(30,000 – 90,000)	01	(0)	(0–19)	
Total (approx.)	480,000 – 1,200,000	**Overall trend** Moderate decline			
Breeding range	>2,000,000 km²	**Gen. length.** <3.3 **% Global pop.** 50–74			

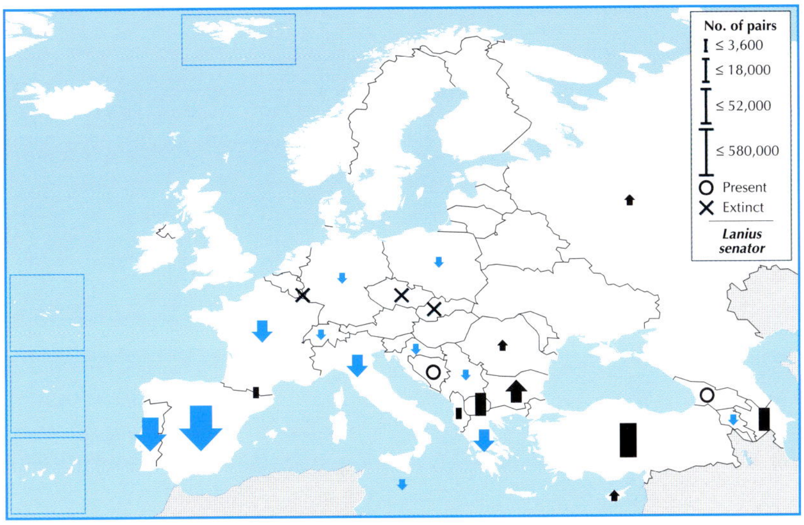

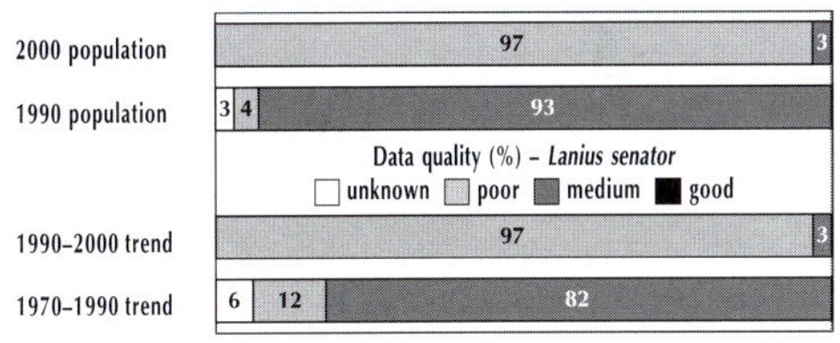

Lanius nubicus
MASKED SHRIKE

SPEC 2 (1994: 2) **Status** (Declining)
Criteria Moderate continuing decline
European IUCN Red List Category —
Criteria —
Global IUCN Red List Category —
Criteria —

Lanius nubicus is a summer visitor to south-eastern Europe, which constitutes >50% of its global breeding range. Its European breeding population is relatively small (<100,000 pairs), and underwent a large decline between 1970–1990. Although populations in Cyprus, Bulgaria and the Former Yugoslav Republic of Macedonia were stable or increased during 1990–2000, the species declined in Greece and its Turkish stronghold, and underwent a moderate decline (>10%) overall. Consequently, it is provisionally evaluated as Declining.

Country	Breeding pop. size (pairs)	Year(s)	Trend	Mag.%	References
Bulgaria	50 – 100	96–02	(F)	(20–29)	
Cyprus	(4,000 – 10,000)	94–02	(0)	(0–19)	
Greece	(500 – 2,000)	95–00	(-)	(0–19)	
Macedonia	100 – 150	90–00	(+)	(0–19)	
Turkey	(30,000 – 90,000)	01	(-)	(0–19)	
Total (approx.)	35,000 – 100,000	**Overall trend** Moderate decline			
Breeding range	>500,000 km²	**Gen. length.** <3.3 **% Global pop.** 50–74			

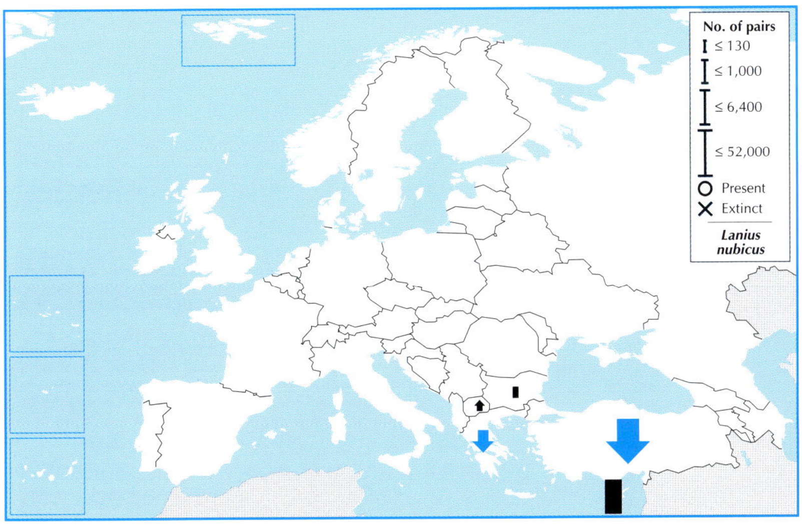

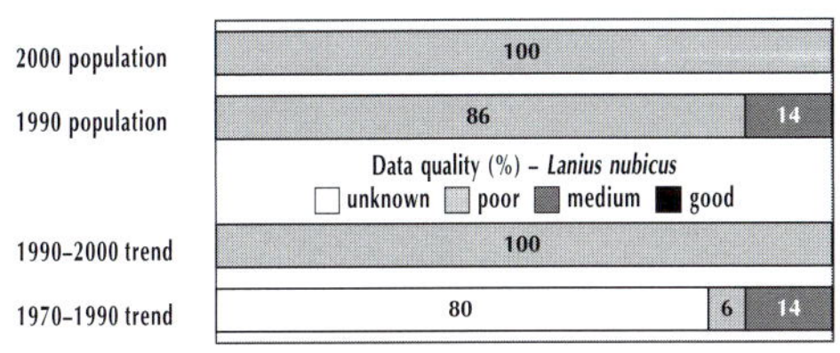

Garrulus glandarius
EURASIAN JAY

Non-SPEC (1994: —) **Status** Secure
Criteria —
European IUCN Red List Category —
Criteria —
Global IUCN Red List Category —
Criteria —

Garrulus glandarius is a widespread resident across most of Europe, which accounts for less than half of its global range. Its European breeding population is very large (>6,000,000 pairs), and was stable between 1970–1990. Although there were declines in a few countries during 1990–2000, populations across the vast majority of Europe—including key ones in France, Russia and Turkey—were stable or increasing, and the species remained stable overall. Consequently, it is evaluated as Secure.

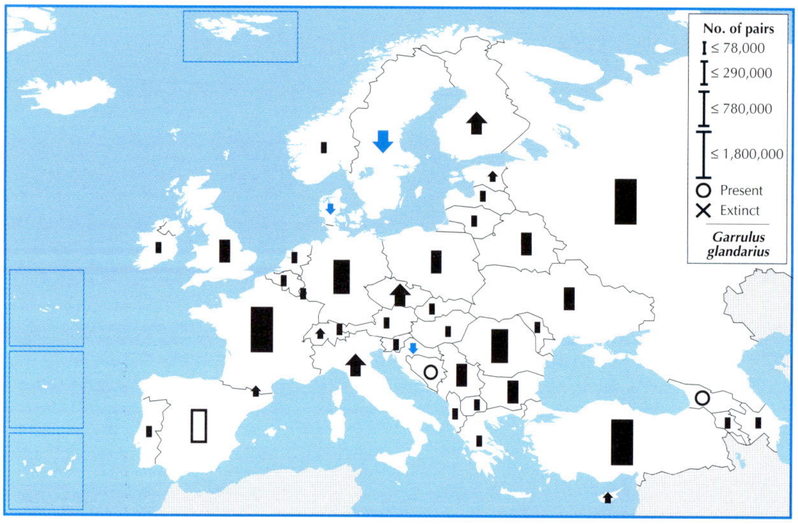

Country	Breeding pop. size (pairs)	Year(s)	Trend	Mag.%	References
Albania	5,000 – 10,000	02	(0)	(0–19)	
Andorra	(300 – 400)	99–01	(+)	(20–29)	*1,3*
Armenia	3,000 – 5,000	00–02	0	0–19	
Austria	(20,000 – 40,000)	98–02	(0)	(0–19)	
Azerbaijan	(10,000 – 50,000)	96–00	(0)	(0–19)	
Belarus	220,000 – 250,000	97–02	(0)	(0–19)	
Belgium	30,000 – 40,000	01–02	0	0–19	*1*
Bosnia & HG	Present	90–03	?	–	
Bulgaria	100,000 – 400,000	96–02	0	0–9	
Croatia	(50,000 – 100,000)	02	(–)	(30–49)	*16*
Cyprus	(5,000 – 10,000)	94–02	(+)	(0–9)	
Czech Rep.	170,000 – 340,000	00	**+**	**0–19**	
Denmark	30,000 – 50,000	00	**-**	**30**	*12*
Estonia	(20,000 – 40,000)	98	+	20–29	*1*
Finland	150,000 – 200,000	98–02	**+**	**20**	
France	(500,000 – 2,000,000)	98–02	(F)	(20–29)	*4,2*
Georgia	Present	03	?	–	
Germany	300,000 – 760,000	95–99	0	0–19	
Greece	(20,000 – 50,000)	95–00	(0)	(0–19)	
Hungary	**58,000 – 81,000**	99–02	0	0–19	*19*
Rep. Ireland	2,500 – 10,000	88–91	0	0–19	
Italy	(200,000 – 400,000)	03	(+)	(0–19)	
Latvia	20,000 – 40,000	90–00	(0)	(0–19)	*23,18*
Liechtenstein	150 – 200	98–00	(0)	(0–19)	
Lithuania	(60,000 – 100,000)	99–01	(0)	(0–19)	*20*
Luxembourg	3,000 – 4,000	02	0	0–19	
Macedonia	(40,000 – 80,000)	90–00	(0)	(0–19)	
Moldova	15,000 – 20,000	90–00	0	0–19	
Netherlands	**40,000 – 60,000**	98–00	**0**	**10**	*1*
Norway	(10,000 – 100,000)	90–03	(0)	(0–19)	
Poland	**200,000 – 400,000**	00–02	0	0–19	*23*
Portugal	(10,000 – 100,000)	02	(0)	(0–19)	
Romania	(460,000 – 580,000)	00–02	(0)	(0–19)	*48*
Russia	1,200,000 – 2,500,000	90–00	(0)	(0–19)	*2,89,114,122,152*
Serbia & MN	200,000 – 300,000	90–02	0	0–19	*1,29,172a,67a, 227,78,152*
Slovakia	15,000 – 30,000	90–99	0	0–19	
Slovenia	10,000 – 20,000	00	(0)	(0–19)	
Spain	(540,000 – 1,100,000)	92	?	–	*13,12,10*
Sweden	**200,000 – 400,000**	99–00	**-**	**10**	
Switzerland	**50,000 – 70,000**	93–96	+	0–9	
Turkey	(6,00000 – 1,800,000)	01	(0)	(0–19)	
Ukraine	225,000 – 320,000	90–00	0	0–19	
UK	**160,000 – 160,000**	00	**0**	**0**	*5,31*
Total (approx.)	6,000,000 – 13,000,000	**Overall trend** Stable			
Breeding range	>8,000,000 km²	**Gen. length.** 4		**% Global pop.** 25–49	

(See p. 256, bottom, for data quality graph)

Perisoreus infaustus
SIBERIAN JAY

SPEC 3 (1994: 3) **Status** (Depleted)
Criteria Moderate historical decline
European IUCN Red List Category —
Criteria —
Global IUCN Red List Category —
Criteria —

Perisoreus infaustus is a widespread resident in Fennoscandia and Russia, with Europe accounting for less than half of its global range. Its European breeding population is large (>340,000 pairs), but underwent a moderate decline between 1970–1990. Although there were declines in Sweden during 1990–2000, and the trend in Russia was unknown, there was no evidence to suggest that it declined significantly overall. Nevertheless, its total population size probably remains below the level that preceded its decline, and consequently the species is provisionally evaluated as Depleted.

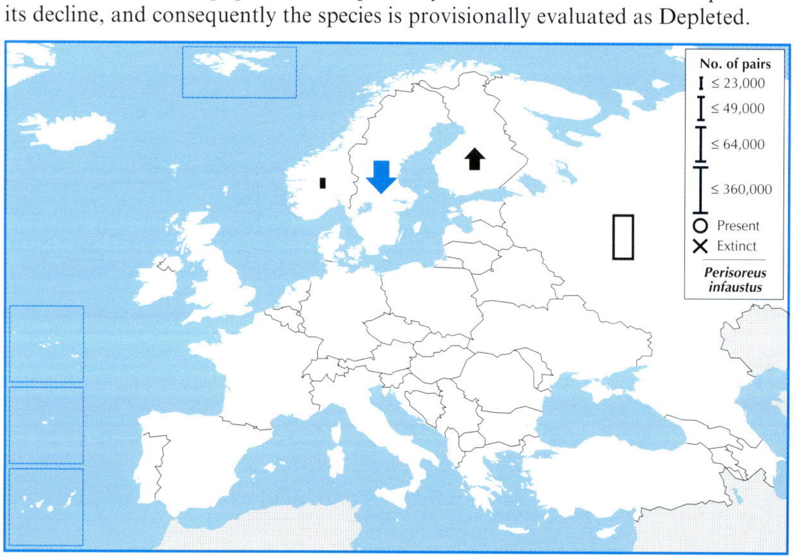

Country	Breeding pop. size (pairs)	Year(s)	Trend	Mag.%	References
Finland	40,000 – 60,000	98–02	**+**	**30**	
Norway	(10,000 – 50,000)	90–03	(0)	(0–19)	
Russia	(250,000 – 500,000)	90–00	?	–	
Sweden	**40,000 – 100,000**	99–00	(–)	(0–19)	
Total (approx.)	340,000 – 710,000	**Overall trend** Unknown			
Breeding range	>1,000,000 km²	**Gen. length.** 4		**% Global pop.** 25–49	

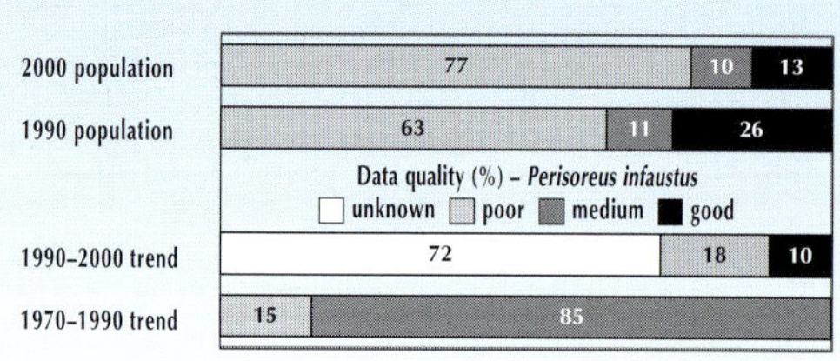

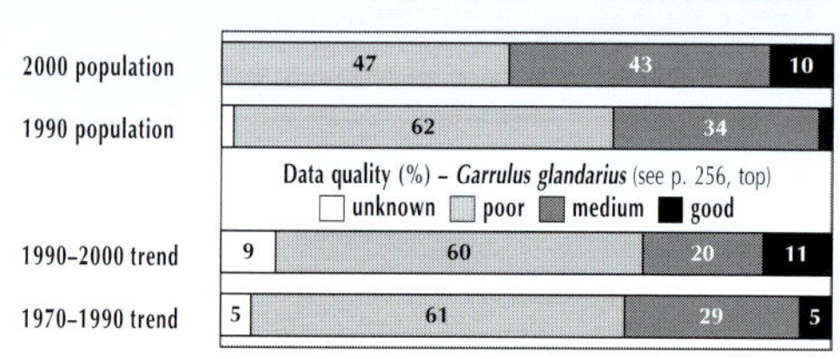

Cyanopica cyanus
AZURE-WINGED MAGPIE

Non-SPEC (1994: —) **Status** (Secure)
Criteria —
European IUCN Red List Category —
Criteria —
Global IUCN Red List Category —
Criteria —

Cyanopica cyanus is resident in Iberia, with Europe accounting for less than a quarter of its disjunct global range (the rest—well separated—occurring in eastern Asia). Its European breeding population is large (>260,000 pairs), and was stable between 1970–1990. Although the trend of the key Spanish population during 1990–2000 was unknown, the population in Portugal was stable, and there was no evidence to suggest that the species's status deteriorated overall. Consequently, it is provisionally evaluated as Secure.

Country	Breeding pop. size (pairs)	Year(s)	Trend	Mag.%	References
Portugal	(20,000 – 200,000)	02	(0)	(0–19)	
Spain	(240,000 – 260,000)	92	?	–	*13,12,10*
Total (approx.)	260,000 – 460,000	**Overall trend** Unknown			
Breeding range	>250,000 km²	**Gen. length.** <3.3		**% Global pop.** 5–24	

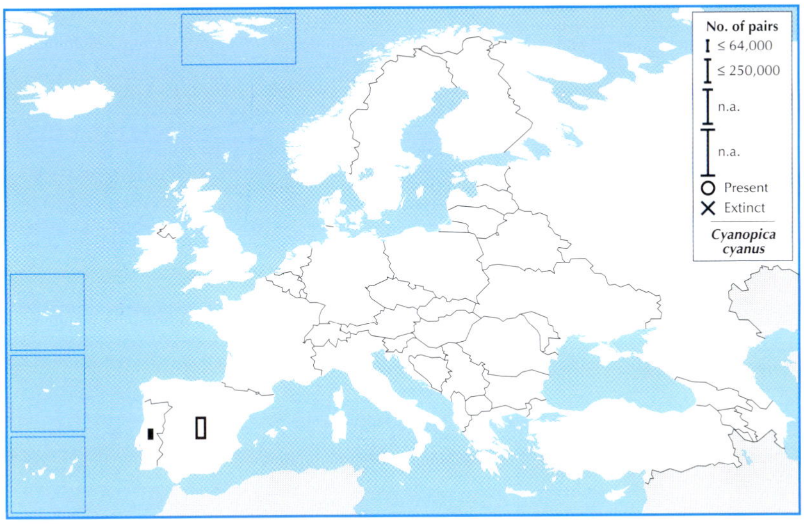

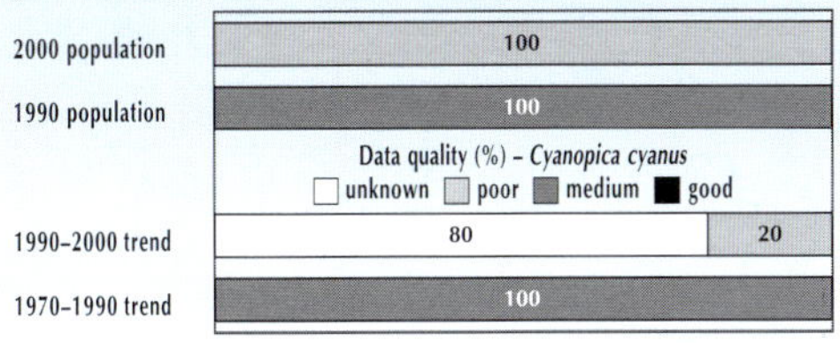

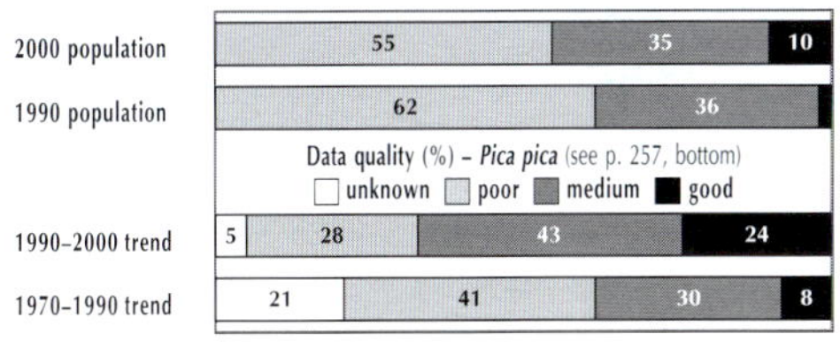

Pica pica
BLACK-BILLED MAGPIE

Non-SPEC (1994: —) **Status** Secure
Criteria —
European IUCN Red List Category —
Criteria —
Global IUCN Red List Category —
Criteria —

Pica pica is a widespread resident across most of Europe, which accounts for less than half of its global range. Its European breeding population is very large (>7,500,000 pairs), and increased between 1970–1990. Although most European populations—including the sizeable one in Turkey—were stable or increased during 1990–2000, key populations in France and Russia declined sharply, and the species probably declined overall. Nevertheless, this recent decline is still outweighed by earlier increases, and consequently the species is provisionally evaluated as Secure.

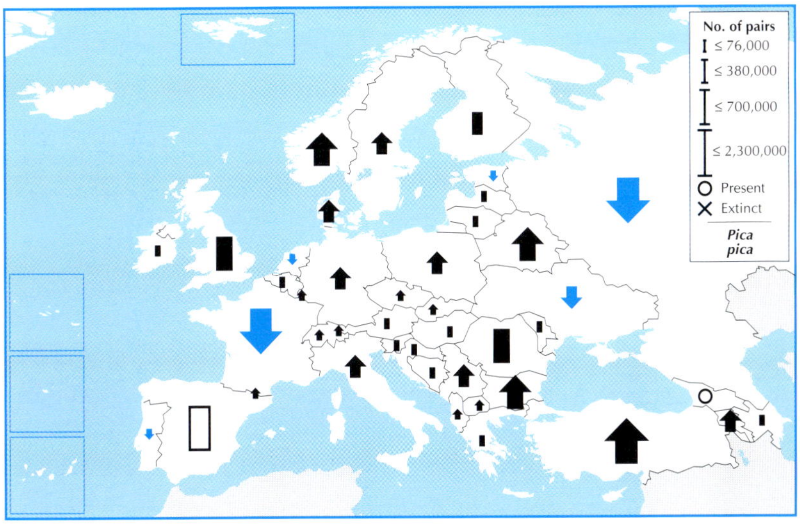

Country	Breeding pop. size (pairs)	Year(s)	Trend	Mag.%	References
Albania	5,000 – 10,000	02	(+)	(0–19)	
Andorra	20 – 35	99–01	(+)	(30–49)	*1,3*
Armenia	250,000 – 300,000	01–02	+	0–9	
Austria	(8,000 – 16,000)	98–02	(0)	(0–19)	
Azerbaijan	(50,000 – 100,000)	96–00	(0)	(0–19)	
Belarus	480,000 – 500,000	97–02	(+)	(0–9)	
Belgium	20,000 – 100,000	01–02	(0)	(0–19)	*1*
Bosnia & HG	(20,000 – 50,000)	90–00	(0)	(0–19)	
Bulgaria	300,000 – 800,000	96–02	+	0–19	
Croatia	(50,000 – 60,000)	02	(0)	(0–19)	*26*
Cyprus	40,000 – 120,000	94–02	(+)	(0–19)	
Czech Rep.	50,000 – 100,000	00	+	20–29	
Denmark	200,000 – 300,000	00	+	**10–19**	*12*
Estonia	(30,000 – 60,000)	98	–	20–29	*1*
Finland	150,000 – 200,000	98–02	**0**	**5**	
France	(600,000 – 2,400,000)	98–02	–	**59**	*4,2*
Georgia	Present	03	?	–	
Germany	180,000 – 500,000	95–99	+	0–19	
Greece	(10,000 – 50,000)	95–00	(0)	(0–19)	
Hungary	**64,000 – 89,000**	99–02	0	0–19	*19*
Rep. Ireland	20,000 – 100,000	88–91	0	0–19	
Italy	(200,000 – 500,000)	03	(+)	(0–9)	
Latvia	10,000 – 20,000	90–00	(0)	(0–19)	*23,16*
Liechtenstein	30 – 50	98–00	+	0–19	
Lithuania	(15,000 – 25,000)	99–01	(0)	(0–19)	*20*
Luxembourg	8,000 – 10,000	02	+	0–19	
Macedonia	(10,000 – 30,000)	90–00	(+)	(20–29)	
Moldova	65,000 – 70,000	90–00	0	0–19	
Netherlands	**40,000 – 60,000**	98–00	**–**	**20**	*1*
Norway	(300,000 – 600,000)	90–03	+	0–19	
Poland	**200,000 – 500,000**	00–02	+	0–19	*23*
Portugal	(5,000 – 50,000)	02	(–)	(–)	
Romania	624,000 – 780,000	00–02	(0)	(0–19)	*48*
Russia	(1,000,000 – 5,000,000)	90–00	–	30–49	*8,23,39,122, 151,152*
Serbia & MN	150,000 – 200,000	90–02	+	10–29	*1,29,172a,67a, 78,227,225,152*
Slovakia	30,000 – 60,000	80–99	+	20–29	
Slovenia	8,000 – 12,000	94	(0)	(0–19)	
Spain	(220,000 – 1,200,000)	92	?	–	*13,12,10*
Sweden	200,000 – 700,000	99–00	+	**19**	
Switzerland	**20,000 – 40,000**	93–96	+	0–19	
Turkey	(900,000 – 1,800,000)	01	(+)	(0–19)	
Ukraine	285,000 – 360,000	90–00	–	5–9	
UK	**650,000 – 650,000**	00	**0**	**0**	*5,31*
Total (approx.)	7,500,000 – 19,000,000	**Overall trend** Moderate decline			
Breeding range	>8,000,000 km²	**Gen. length.** 4		**% Global pop.** 25–49	

(See p. 257, top, for data quality graph)

Nucifraga caryocatactes
SPOTTED NUTCRACKER

Non-SPEC (1994: —) **Status** Secure
Criteria —
European IUCN Red List Category —
Criteria —
Global IUCN Red List Category —
Criteria —

Nucifraga caryocatactes is a widespread resident in the boreal zone, and also occurs more patchily in the mountains of central and south-eastern Europe, which accounts for less than half of its global range. Its European breeding population is large (>400,000 pairs), and was stable between 1970–1990. Although no trend was available for the sizeable Russian population during 1990–2000, the key population in Romania remained stable, and the species was stable or increased in the vast majority of its European range. Consequently, it is evaluated as Secure.

Country	Breeding pop. size (pairs)	Year(s)	Trend	Mag.%	References
Albania	(100 – 500)	02	?	–	
Austria	(20,000 – 40,000)	98–02	(0)	(0–19)	
Belarus	1,500 – 4,000	97–02	0	0–19	
Belgium	200 – 500	95–02	0	0–19	*1*
Bosnia & HG	Present	90–03	?	–	
Bulgaria	10,000 – 30,000	96–02	0	0–19	
Croatia	(500 – 800)	02	(–)	(>80)	*16*
Czech Rep.	2,500 – 5,000	00	0	0–19	
Denmark	(1 – 7)	98–01	(F)	(–)	*4,5,6,7*
Estonia	5,000 – 10,000	98	+	20–29	*1*
Finland	1,500 – 2,500	98–02	+	10	
France	2,500 – 10,000	98–02	F	20–49	*4*
Germany	8,000 – 17,000	95–99	+	0–19	
Greece	(50 – 100)	95–00	(0)	(0–19)	
Italy	(10,000 – 30,000)	03	(0)	(0–19)	
Latvia	5,000 – 20,000	90–00	(0)	(0–19)	*23,18*
Liechtenstein	40 – 50	98–00	0	0–19	
Lithuania	(5,000 – 10,000)	99–01	(0)	(0–19)	*20*
Macedonia	(500 – 2,000)	90–00	(0)	(0–19)	
Norway	(100 – 1,000)	90–03	+	0–19	*63*
Poland	(2,000 – 5,000)	95–00	(0)	(0–19)	*1,27*
Romania	(240,000 – 360,000)	00–02	(0)	(0–19)	*48*
Russia	(50,000 – 250,000)	90–00	?	–	
Serbia & MN	**1,600 – 2,200**	95–02	+	0–19	*1,149,155,227, 225,150*
Slovakia	4,000 – 8,000	90–99	0	0–19	
Slovenia	2,000 – 3,000	94	(0)	(0–19)	
Sweden	5,000 – 15,000	99–00	F	30–49	
Switzerland	**20,000 – 30,000**	93–96	0	0–19	
Ukraine	**3,300 – 3,600**	90–00	**0**	**0–19**	
Total (approx.)	400,000 – 860,000	**Overall trend** Stable			
Breeding range	>2,000,000 km²	**Gen. length.** 5		**% Global pop.** 25–49	

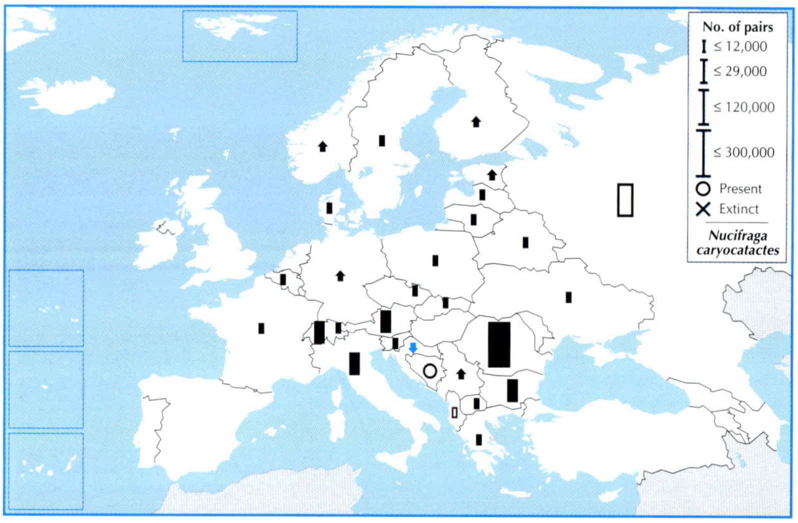

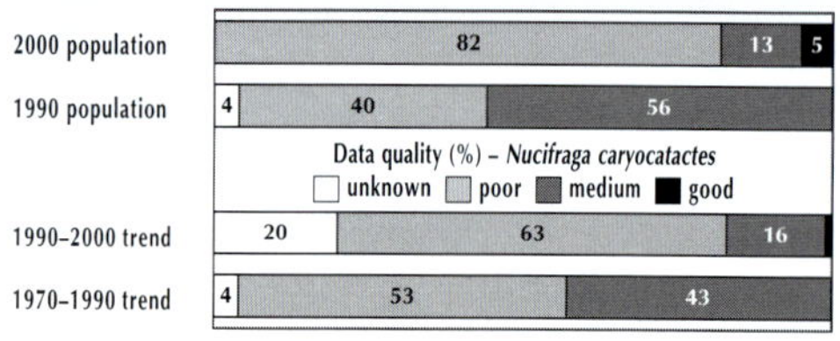

Pyrrhocorax graculus
YELLOW-BILLED CHOUGH

Non-SPEC (1994: —) **Status** (Secure)
Criteria —
European IUCN Red List Category —
Criteria —
Global IUCN Red List Category —
Criteria —

Pyrrhocorax graculus has a fragmented distribution in the mountains of central and southern Europe, which accounts for less than half of its global range. Its European breeding population is large (>130,000 pairs), and was stable between 1970–1990. This trend continued during 1990–2000, with the vast majority of European populations—including key ones in France, Russia and Turkey—probably remaining stable overall. Consequently, the species is provisionally evaluated as Secure.

Country	Breeding pop. size (pairs)	Year(s)	Trend	Mag.%	References
Albania	(2,000 – 5,000)	02	?	–	
Andorra	(200 – 300)	99–01	(0)	(0–19)	*1,3*
Armenia	500 – 600	01–02	?	–	
Austria	(8,000 – 16,000)	98–02	(0)	(0–19)	
Azerbaijan	(1,000 – 10,000)	96–00	(0)	(0–19)	
Bosnia & HG	Present	90–03	?	–	
Bulgaria	1,500 – 3,000	96–02	(0)	(0–19)	
Croatia	(1,000 – 5,000)	02	(0)	(0–19)	*70,16*
France	15,000 – 45,000	00–02	(0)	(0–19)	*4*
Georgia	Present	03	?	–	
Germany	2,000 – 4,000	95–99	0	0–19	
Greece	(2,000 – 10,000)	95–00	(–)	(0–19)	
Italy	(5,000 – 10,000)	03	(0)	(0–19)	
Liechtenstein	40 – 50	98–00	0	0–19	
Macedonia	(500 – 1,000)	90–00	(0)	(0–19)	
Russia	(30,000 – 50,000)	90–00	(0)	(0–19)	*60,61*
Serbia & MN	**1,300 – 1,900**	95–02	**0**	**10–19**	*1,62,50,235,227, 40,225,97,155,91*
Slovenia	800 – 1,200	94	(0)	(0–19)	
Spain	(10,000 – 11,000)	92	(0)	(0–19)	*13,12,10*
Switzerland	**10,000 – 15,000**	93–96	0	0–19	
Turkey	(40,000 – 120,000)	01	(0)	(0–19)	
Total (approx.)	130,000 – 310,000	**Overall trend** Stable			
Breeding range	>750,000 km²	**Gen. length.** 7		**% Global pop.** 25–49	

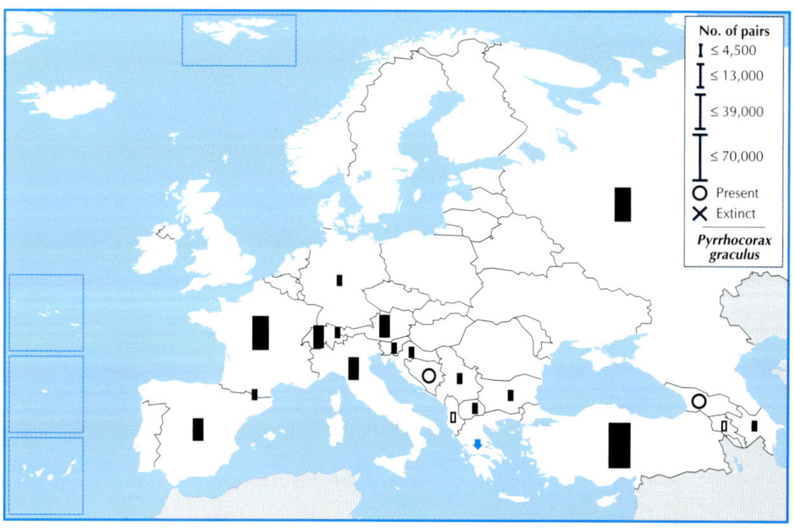

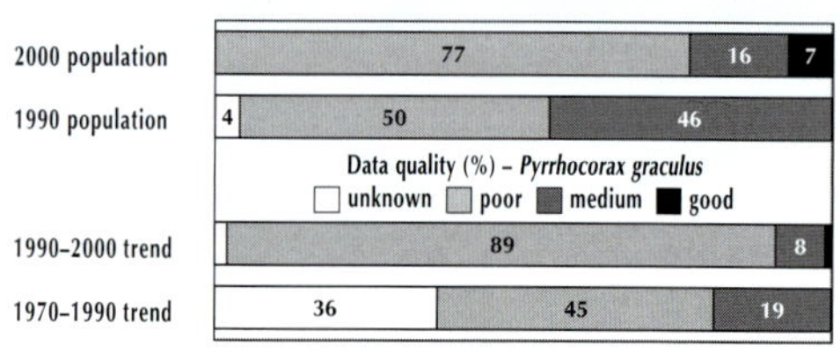

Pyrrhocorax pyrrhocorax
RED-BILLED CHOUGH

SPEC 3 (1994: 3) **Status** Declining
Criteria Moderate continuing decline
European IUCN Red List Category —
Criteria —
Global IUCN Red List Category —
Criteria —

Pyrrhocorax pyrrhocorax is resident in parts of Britain and Ireland and across much of southern Europe, which accounts for less than half of its global range. Its European breeding population is relatively small (<110,000 pairs), and underwent a large decline between 1970–1990. Although populations were stable across much of its European range during 1990–2000, key populations in Spain and Turkey declined, and the species underwent a moderate decline (>10%) overall. Consequently, it is evaluated as Declining.

Country	Breeding pop. size (pairs)	Year(s)	Trend	Mag.%	References
Albania	(0 – 50)	02	?	–	
Andorra	(200 – 300)	99–01	(–)	(0–19)	*1,3*
Armenia	1,000 – 1,500	00–02	?	–	
Azerbaijan	(2,000 – 20,000)	96–00	(0)	(0–19)	
France	1,000 – 2,000	00	–	0–19	*1*
Georgia	Present	03	?	–	
Greece	(1,100 – 1,800)	95–00	(–)	(0–19)	
Rep. Ireland	**445 – 838**	02–03	**0**	**0–19**	
Italy	1,500 – 2,000	03	0	0–9	
Macedonia	(40 – 120)	90–00	(0)	(0–19)	
Portugal	(150 – 300)	02	(–)	(–)	
Russia	(15,000 – 25,000)	90–00	(0)	(0–19)	*50,61*
Serbia & MN	**30 – 40**	97–98	0	0–19	*1,62,227,225, 40,155,152*
Spain	(10,000 – 20,000)	98–02	(–)	(0–19)	*12,16,10*
Canary Is.	(250 – 1,000)	97–02	?	–	*28,25*
Switzerland	**40 – 60**	93–96	0	0–19	
Turkey	(10,000 – 30,000)	01	(–)	(0–19)	
UK	**450 – 500**	02	+	**45**	
Total (approx.)	43,000 – 110,000	**Overall trend** Moderate decline			
Breeding range	>1,000,000 km²	**Gen. length.** 7		**% Global pop.** 25–49	

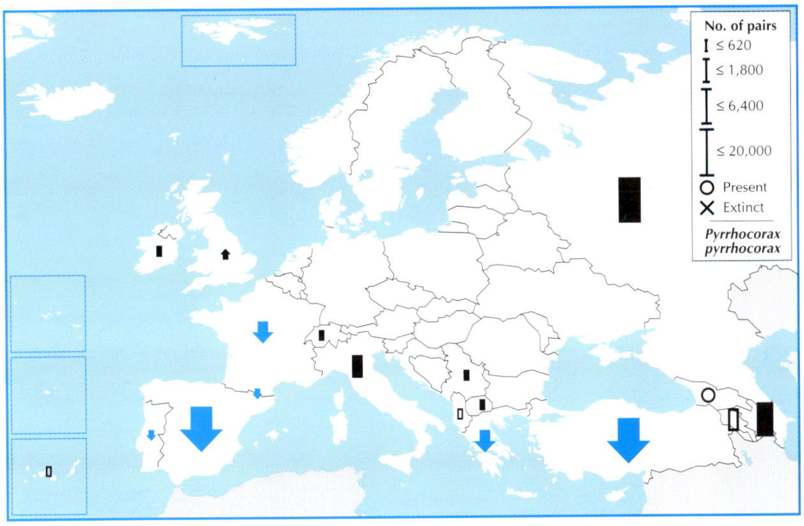

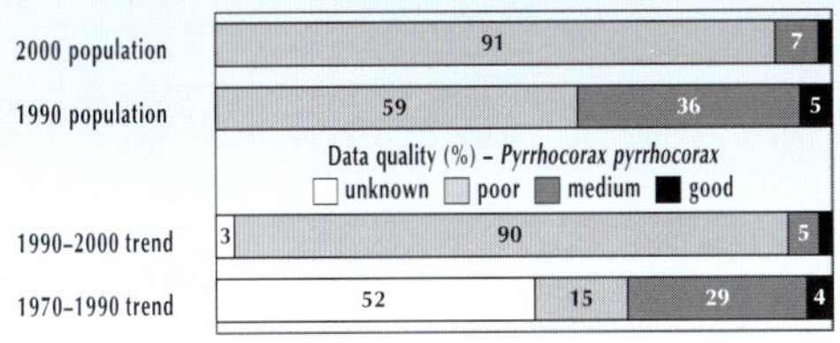

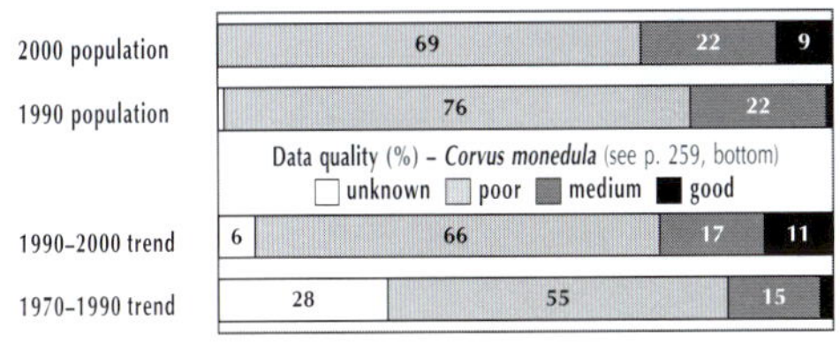

Corvus monedula
EURASIAN JACKDAW

Non-SPEC[E] (1994: 4) **Status** (Secure)
Criteria —
European IUCN Red List Category —
Criteria —
Global IUCN Red List Category —
Criteria —

Corvus monedula is a widespread breeder across most of Europe, which constitutes >50% of its global range. Its European breeding population is very large (>5,200,000 pairs), and was stable between 1970–1990. Although the species declined in several countries during 1990–2000—notably Turkey and France—the Russian population was stable, and other key populations (such as those in Belarus, Romania and the United Kingdom) were also stable or increasing. The species probably remained stable overall, and consequently it is provisionally evaluated as Secure.

Country	Breeding pop. size (pairs)	Year(s)	Trend	Mag.%	References
Albania	10,000 – 30,000	02	(–)	(0–19)	
Armenia	6,000 – 10,000	00–02	0	0–19	
Austria	3,600 – 4,600	98–02	(0)	(0–19)	
Azerbaijan	(1,000 – 10,000)	96–00	(0)	(0–19)	
Belarus	350,000 – 400,000	97–02	(0)	(0–19)	
Belgium	20,000 – 100,000	01–02	(0)	(0–19)	*1*
Bosnia & HG	(85 – 100)	90–00	(0)	(0–19)	
Bulgaria	75,000 – 250,000	96–02	0	0–19	
Croatia	(2,000 – 3,000)	02	(–)	(>80)	*16,57,26*
Cyprus	(10,000 – 20,000)	94–02	(0)	(0–9)	
Czech Rep.	10,000 – 20,000	00	0	0–19	
Denmark	(100,000 – 200,000)	00	0	0–19	
Estonia	(30,000 – 40,000)	98	0	0–19	*1*
Finland	80,000 – 130,000	98–02	+	**110**	
France	(150,000 – 600,000)	98–02	–	35	*4,2*
Georgia	Present	03	?	–	
Germany	50,000 – 110,000	95–99	–	0–19	
Greece	(100,000 – 200,000)	95–00	(0)	(0–19)	
Hungary	1,000 – 14,000	99–02	0	0–19	*19*
Rep. Ireland	100,000 – 250,000	88–91	0	0–19	
Italy	50,000 – 100,000	03	(0)	(0–19)	
Latvia	10,000 – 50,000	90–00	(0)	(0–19)	*23,16*
Liechtenstein	**10 – 15**	98–00	**0**	**0–19**	
Lithuania	(50,000 – 60,000)	99–01	(0)	(0–19)	*20*
Luxembourg	2,500 – 3,000	02	0	0–19	
Macedonia	(40,000 – 100,000)	90–00	(+)	(20–29)	
Moldova	250 – 350	90–00	+	**20–29**	
Netherlands	**180,000 – 220,000**	98–00	0	5	*1*
Norway	(3,500 – 20,000)	90–03	+	0–19	
Poland	100,000 – 300,000	00–02	(0)	(0–19)	*23*
Portugal	(1,000 – 10,000)	02	(–)	(–)	
Romania	(250,000 – 400,000)	00–02	(+)	(0–19)	*48*
Russia	(1,000,000 – 5,000,000)	90–00	(0)	(0–19)	*151,152*
Serbia & MN	150,000 – 250,000	90–02	–	0–19	*1,29,155,172, 67a,78,227,225*
Slovakia	3,000 – 5,000	80–99	–	30–49	
Slovenia	2,000 – 3,000	00	(–)	(0–19)	
Spain	(423,600 – 533,000)	92	?	–	*13,12,10*
Sweden	150,000 – 400,000	99–00	+	**10**	
Switzerland	**1,100 – 1,200**	93–96	–	**0–9**	
Turkey	(1,000,000 – 4,000,000)	01	(–)	(0–19)	
Ukraine	135,000 – 175,000	90–00	0	0–4	
UK	**555,000 – 555,000**	00	+	**29**	*5,31*
Total (approx.)	5,200,000 – 15,000,000	**Overall trend** Stable			
Breeding range	>7,000,000 km²	**Gen. length.** 5		**% Global pop.** 50–74	

(See p. 259, top, for data quality graph)

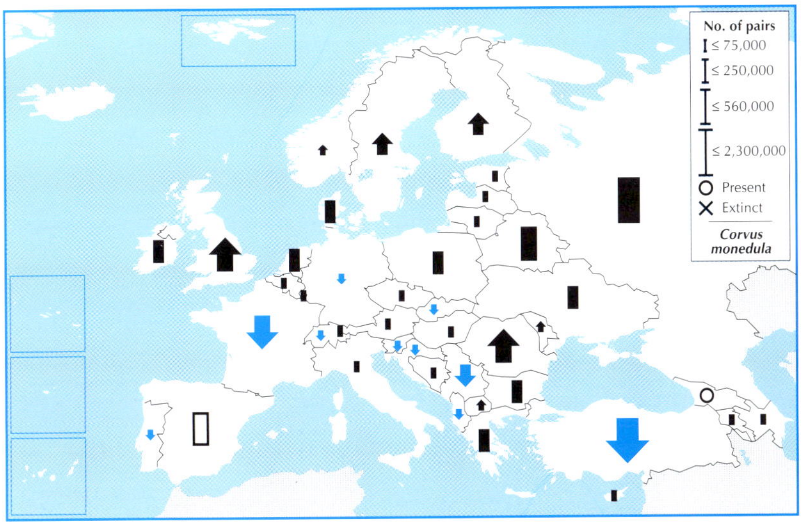

Corvus frugilegus
ROOK

Non-SPEC (1994: —) **Status** (Secure)
Criteria —
European IUCN Red List Category —
Criteria —
Global IUCN Red List Category —
Criteria —

Corvus frugilegus is a widespread resident across much of Europe, which accounts for less than half of its global range. Its European breeding population is extremely large (>10,000,000 pairs), and increased between 1970–1990. Although the species declined in a few countries during 1990–2000, most European populations were stable or increased. The trend of the key population in Russia was unknown, but the species was probably stable overall. Consequently, it is provisionally evaluated as Secure.

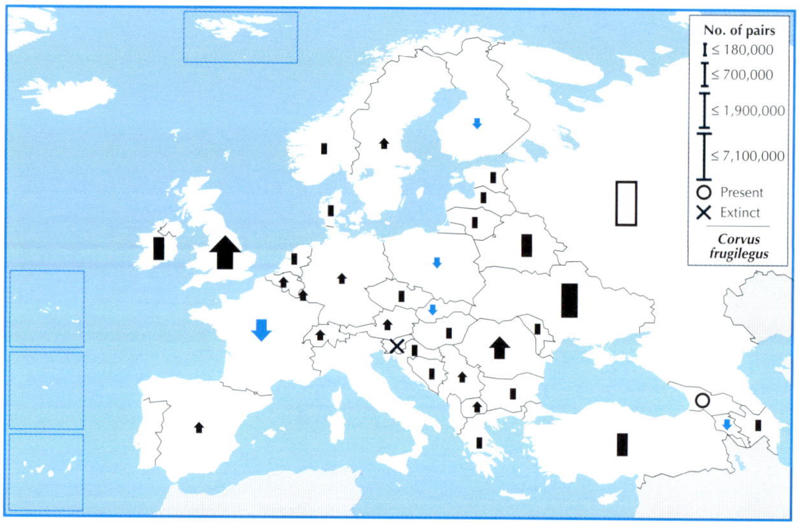

Country	Breeding pop. size (pairs)	Year(s)	Trend	Mag.%	References
Armenia	6,500 – 15,000	00–02	–	10–19	
Austria	800 – 1,000	98–02	+	50–79	
Azerbaijan	(5,000 – 50,000)	96–00	(0)	(0–19)	
Belarus	600,000 – 800,000	97–02	0	0–19	
Belgium	**15,000 – 20,000**	95–02	+	**20–29**	*1*
Bosnia & HG	(2,500 – 10,000)	90–00	(F)	(–)	
Bulgaria	**5,000 – 8,000**	96–02	F	20–29	
Croatia	(5,000 – 6,000)	02	(0)	(0–19)	
Czech Rep.	3,000 – 4,000	00	0	0–19	
Denmark	40,000 – 50,000	00	**0**	**0–19**	*3*
Estonia	5,000 – 10,000	98	0	0–19	*1*
Finland	1,000 – 1200	98–02	**–**	**10**	
France	(200,000 – 800,000)	98–02	–	11	*4,2*
Georgia	Present	03	?	–	
Germany	54,000 – 64,000	95–99	+	**30–49**	
Greece	(500 – 2,000)	95–00	(0)	(0–19)	
Hungary	(30,000 – 35,000)	90–93	0	0–19	*14*
Rep. Ireland	250,000 – 500,000	88–91	0	0–19	
Latvia	7,000 – 9,000	90–00	(0)	(0–19)	*23,16*
Lithuania	(80,000 – 100,000)	99–01	(0)	(0–19)	*20*
Luxembourg	**3,500 – 4,000**	00–02	+	**50–79**	
Macedonia	(2,000 – 5,000)	90–00	(+)	(20–29)	
Moldova	120,000 – 150,000	90–00	**0**	**0–19**	
Netherlands	**60,000 – 64,000**	98–00	**0**	**10**	*1*
Norway	(500 – 1,000)	90–03	0	0–19	
Poland	**150,000 – 200,000**	01–02	(–)	(0–19)	*13*
Romania	350,000 – 420,000	90–02	+	0–19	*48,5*
Russia	(5,000,000 – 10,000,000)	90–00	?	–	*2,8,58,130,151,152*
Serbia & MN	40,000 – 80,000	90–02	+	0–19	*1,29,172,78,101,13*
Slovakia	10,000 – 17,000	80–99	–	30–49	
Slovenia	**0 – 0**	00	–	**X**	
Spain	**2,100 – 2,100**	98–02	+	**30–49**	*10*
Sweden	**35,000 – 40,000**	99–00	+	**0–19**	
Switzerland	**1,000 – 1,600**	98–02	+	**>80**	
Turkey	300,000 – 600,000	01	(0)	(0–19)	
Ukraine	(1,500,000 – 2,200,000)	90–00	(0)	(0–9)	
UK	**1,120,000 – 1,430,000**	96	+	**6**	*19*
Total (approx.)	10,000,000 – 18,000,000	**Overall trend** Unknown			
Breeding range	>5,000,000 km²	**Gen. length.** 5		**% Global pop.** 25–49	

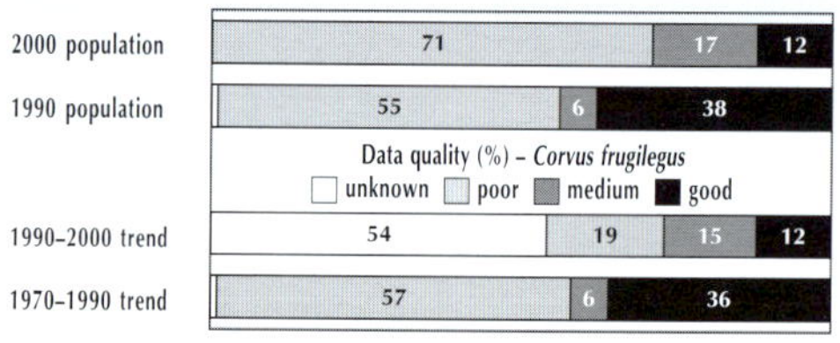

Corvus corone
CARRION CROW

Non-SPEC (1994: —) **Status** Secure
Criteria —
European IUCN Red List Category —
Criteria —
Global IUCN Red List Category —
Criteria —

Corvus corone is a widespread breeder across most of Europe, which accounts for less than half of its global breeding range. Its European breeding population is very large (>7,000,000 pairs), and underwent a large increase between 1970–1990. Although the species declined in a number of countries during 1990–2000, these declines were set against stable or increasing trends in other key populations, such as those in the United Kingdom and Turkey. Despite the lack of recent trend data from Russia, the species was stable overall, and consequently it is evaluated as Secure.

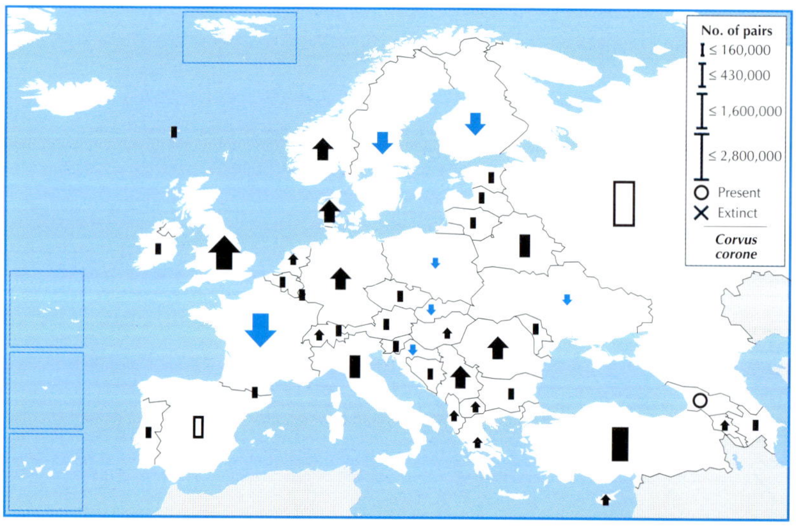

Country	Breeding pop. size (pairs)	Year(s)	Trend	Mag.%	References
Albania	10,000 – 30,000	02	(+)	(0–19)	
Andorra	(100 – 200)	99–01	(0)	(0–19)	*1,3*
Armenia	10,000 – 15,000	01–02	+	0–9	
Austria	(40,000 – 80,000)	98–02	(0)	(0–19)	
Azerbaijan	(2,000 – 20,000)	96–00	(0)	(0–19)	
Belarus	280,000 – 320,000	97–02	(0)	(0–19)	
Belgium	20,000 – 100,000	01–02	(0)	(0–19)	*1*
Bosnia & HG	(20,000 – 50,000)	90–00	(0)	(0–19)	
Bulgaria	50,000 – 120,000	96–02	0	0–9	
Croatia	(40,000 – 50,000)	02	(–)	(>80)	
Cyprus	(20,000 – 60,000)	94–02	(+)	(10–19)	
Czech Rep.	12,000 – 24,000	00	0	0–19	
Denmark	(150,000 – 300,000)	00	+	**10–19**	*3*
Faroe Is.	1,500 – 1,500	95	(0)	(0–19)	
Estonia	(50,000 – 100,000)	98	0	0–19	*1*
Finland	160,000 – 230,000	98–02	–	**20**	
France	(800,000 – 3,200,000)	98–02	–	9	*4,2*
Georgia	Present	03	?	–	
Germany	300,000 – 600,000	95–99	+	0–19	
Greece	(50,000 – 100,000)	95–00	(+)	(0–19)	
Hungary	**51,000 – 84,000**	99–02	+	20–49	*19*
Rep. Ireland	100,000 – 250,000	88–91	0	0–19	
Italy	(110,000 – 520,000)	03	(0)	(0–19)	
Latvia	20,000 – 60,000	90–00	(0)	(0–19)	*23,16*
Liechtenstein	40 – 80	98–00	0	0–19	
Lithuania	(50,000 – 70,000)	99–01	(0)	(0–19)	*20*
Luxembourg	3,000 – 4,000	02	0	0–19	
Macedonia	(20,000 – 50,000)	90–00	(+)	(20–29)	
Moldova	8,000 – 8,500	90–00	**0**	**0–19**	
Netherlands	**70,000 – 100,000**	98–00	+	**33**	*1*
Norway	(200,000 – 700,000)	95–02	+	0–19	*30*
Poland	50,000 – 150,000	00–02	(–)	(0–19)	*23*
Portugal	(5,000 – 50,000)	02	(0)	(0–19)	
Romania	240,000 – 320,000	98–02	+	0–19	*48*
Russia	(1,500,000 – 5,000,000)	90–00	?	–	*151,152*
Serbia & MN	160,000 – 250,000	90–02	+	0–9	*1,29,172a,67a,78,227,225*
Slovakia	8,000 – 15,000	80–99	–	30–49	
Slovenia	10 – 50	94	(0)	(0–19)	
Spain	(310,000 – 530,000)	92	?	–	*13,12,10*
Sweden	240,000 – 500,000	99–00	–	**7**	
Switzerland	**80,000 – 150,000**	93–96	+	10–19	
Turkey	(500,000 – 1,500,000)	01	(0)	(0–19)	
Ukraine	(95,000 – 130,000)	90–00	–	20–29	
UK	**1,202,000 – 1,202,000**	00	+	**25**	*5,31*
Total (approx.)	7,000,000 – 17,000,000	**Overall trend** Stable			
Breeding range	>8,000,000 km²	**Gen. length.** 5		**% Global pop.** 25–49	

(See p. 262, top, for data quality graph)

Corvus corax
COMMON RAVEN

Non-SPEC (1994: —) **Status** Secure
Criteria —
European IUCN Red List Category —
Criteria —
Global IUCN Red List Category —
Criteria —

Corvus corax is a widespread resident across most of Europe, which accounts for less than half of its global range. Its European breeding population is large (>450,000 pairs), and underwent a large increase between 1970–1990. The vast majority of European populations were stable or increased during 1990–2000, and the species underwent a slight increase overall. Consequently, it is evaluated as Secure.

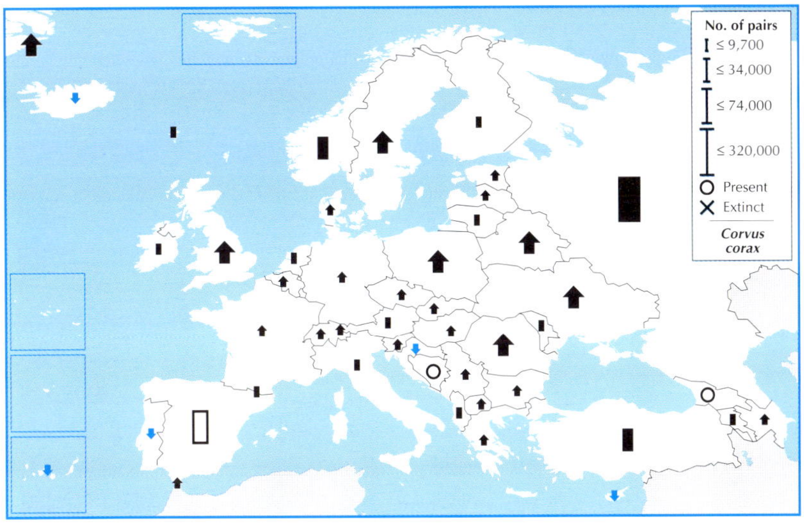

Country	Breeding pop. size (pairs)	Year(s)	Trend	Mag.%	References
Albania	2,000 – 5,000	02	(0)	(0–19)	
Andorra	50 – 150	99–01	(0)	(0–19)	*1,3*
Armenia	600 – 1,100	99–02	0	0–19	
Austria	2,000 – 4,000	98–02	(0)	(0–19)	
Azerbaijan	(1,000 – 10,000)	96–00	(+)	(10–19)	
Belarus	14,000 – 22,000	97–02	+	10–19	
Belgium	**20 – 40**	95–02	**+**	**20–29**	*1*
Bosnia & HG	Present	90–03	?	–	
Bulgaria	500 – 1,000	96–02	+	0–19	
Croatia	(1,000 – 3,000)	02	(–)	(50–79)	*16*
Cyprus	8 – 20	98–02	–	>80	
Czech Rep.	**800 – 1,200**	00	**+**	**>80**	
Denmark	(500 – 700)	00	+	10–19	*3*
Faroe Is.	500 – 500	95	(0)	(0–19)	
Greenland	(5,000 – 50,000)	90–00	(+)	(0–19)	
Estonia	4,000 – 6,000	98	+	20–29	*1*
Finland	5,000 – 7,000	98–02	**0**	**0–19**	
France	4,000 – 5,000	00	+	0–19	*1*
Georgia	Present	03	?	–	
Germany	6,700 – 14,000	95–99	+	20–29	
Greece	(5,000 – 10,000)	95–00	(+)	(0–19)	
Hungary	**2,000 – 3,100**	99–02	+	>80	
Iceland	2,000 – 2,000	81–91	–	0–19	*17,21,37,17*
Rep. Ireland	2,500 – 10,000	88–91	0	0–19	
Italy	(3,000 – 5,000)	03	(0)	(0–19)	
Latvia	3,000 – 5,000	90–00	+	20–29	*18*
Liechtenstein	10 – 15	98–00	+	0–19	
Lithuania	(4,000 – 7,000)	99–01	(0)	(0–19)	*20*
Macedonia	1,000 – 3,000	90–00	(+)	(20–29)	
Moldova	800 – 1200	90–00	**0**	**0–19**	
Netherlands	**90 – 100**	98–00	**0**	**10**	*1*
Norway	(20,000 – 50,000)	90–02	0	0–19	
Poland	**10,000 – 20,000**	95–00	+	>80	*2,23*
Portugal	(300 – 1,000)	02	(–)	(–)	
Romania	24,000 – 35,000	00–02	+	0–19	*48*
Russia	(200,000 – 500,000)	90–00	(0)	(0–19)	*122,151,152*
Serbia & MN	1,200 – 1,500	95–02	+	**0–19**	*1,29,172a,67a, 227,225,155,91*
Slovakia	2,000 – 4,000	80–99	+	20–29	
Slovenia	1,000 – 1,500	00	(+)	(0–19)	
Spain	(60,000 – 90,000)	92	?	–	*13,12,16,10*
Canary Is.	400 – 400	97–03	–	20–29	*28,25*
Sweden	**10,000 – 20,000**	99–00	+	**55**	
Switzerland	**1,500 – 2,500**	93–96	+	10–19	
Turkey	(10,000 – 20,000)	01	(0)	(0–19)	
Ukraine	**29,000 – 38,000**	90–00	+	**0–19**	
UK	**12,900 – 12,900**	00	+	**64**	*22,31*
Gibraltar	**1 – 1**	00	+	**>80**	

Total (approx.) 450,000 – 970,000 **Overall trend** Small increase
Breeding range >7,000,000 km² **Gen. length.** 7 **% Global pop.** 25–49
(See p. 262, top, for data quality graph)

Sturnus vulgaris
COMMON STARLING

SPEC 3 (1994: —) **Status** Declining
Criteria Moderate recent decline
European IUCN Red List Category —
Criteria —
Global IUCN Red List Category —
Criteria —

Sturnus vulgaris is a widespread breeder across most of Europe, which accounts for less than half of its global breeding range. Its European breeding population is extremely large (>23,000,000 pairs), and was stable between 1970–1990. Although the species was stable or increased in much of southern and central Europe during 1990–2000, it declined in Turkey, Russia and most countries in the north and north-west, and underwent a moderate decline (>10%) overall. Consequently, this previously Secure species is now evaluated as Declining.

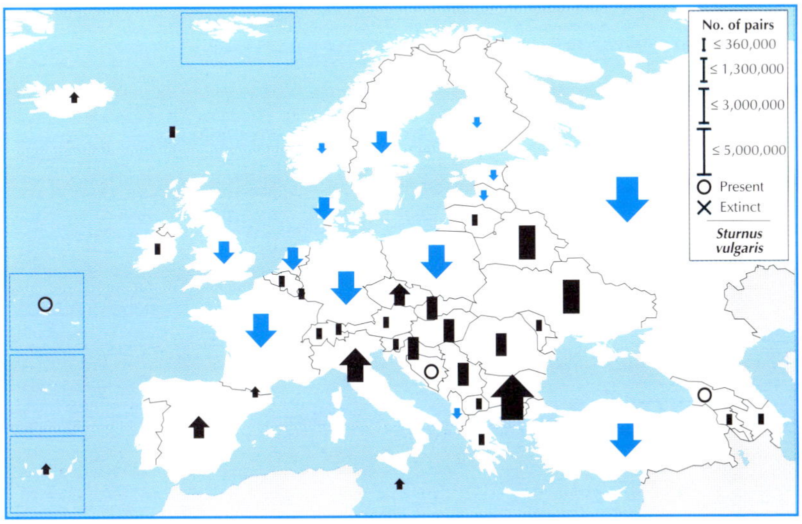

Country	Breeding pop. size (pairs)	Year(s)	Trend	Mag.%	References
Albania	500 – 3,000	02	(–)	(0–19)	
Andorra	(5 – 10)	99–01	(+)	(N)	*1,2,3*
Armenia	100,000 – 150,000	00–02	0	0–19	
Austria	(100,000 – 200,000)	98–02	(0)	(0–19)	
Azerbaijan	(100,000 – 500,000)	96–00	(0)	(0–19)	
Belarus	1,500,000 – 1,700,000	97–02	(0)	(0–19)	
Belgium	20,000 – 100,000	01–02	(0)	(0–19)	*1*
Bosnia & HG	Present	90–03	?	–	
Bulgaria	2,000,000 – 6,000,000	96–02	+	0–19	
Croatia	(1,000,000 – 1,500,000)	02	(0)	(0–19)	*16*
Czech Rep.	900,000 – 1,800,000	00	+	0–19	
Denmark	400,000 – 600,000	00	**–**	**20–29**	*12*
Faroe Is.	25,000 – 25,000	95	(0)	(0–19)	
Estonia	(20,000 – 50,000)	98	–	20–29	*1*
Finland	30,000 – 60,000	98–02	**–**	**30**	
France	(1,500,000 – 6,000,000)	98–02	–	14	*4,2*
Georgia	Present	03	?	–	
Germany	1,700,000 – 4,300,000	95–99	–	0–19	
Greece	(10,000 – 20,000)	95–00	(0)	(0–19)	
Hungary	710,000 – 990,000	99–02	0	0–19	*19*
Iceland	(3,000 – 4,000)	00	(+)	(0–19)	*21*
Rep. Ireland	250,000 – 500,000	88 91	0	0 19	
Italy	(1,000,000 – 3,000,000)	03	(+)	(0–19)	
Latvia	50,000 – 250,000	90–00	(–)	(0–19)	*23,16,2*
Liechtenstein	300 – 500	98–00	0	0–19	
Lithuania	(250,000 – 300,000)	99–01	(0)	(0–19)	*20*
Luxembourg	30,000 – 40,000	02	0	0–19	
Macedonia	(100,000 – 500,000)	90–00	(0)	(0–19)	
Malta	**5 – 10**	95–02	**+**	**0–19**	*1*
Moldova	150,000 – 200,000	90–00	0	0–19	
Netherlands	**500,000 – 900,000**	98–00	–	24	*1*
Norway	(200,000 – 500,000)	90–03	–	20–29	
Poland	**1,500,000 – 3,000,000**	00–02	–	20–29	23
Portugal					
Azores	Present	02	?	–	
Romania	840,000 – 1,224,000	00–02	(0)	(0–19)	*48*
Russia	(2,500,000 – 10,000,000)	90–00	(–)	(20–29)	*8,23,39,122*
Serbia & MN	400,000 – 700,000	90–02	F	10–29	*1,29,172a,67a, 78,227,152*
Slovakia	400,000 – 800,000	90–99	0	0–19	
Slovenia	80,000 – 100,000	94	(0)	(0–19)	
Spain	(400,000 – 1,200,000)	98–02	(+)	(0–19)	*10*
Canary Is.	(50 – 250)	97–03	(+)	(0–19)	*28,25*
Sweden	**750,000 – 1,500,000**	99–00	**–**	**12**	
Switzerland	**150,000 – 220,000**	93–96	0	0–19	
Turkey	(1,000,000 – 4,000,000)	01	(–)	(0–19)	
Ukraine	1,400,000 – 1,900,000	90–00	F	20–29	
UK	**804,000 – 804,000**	00	**–**	**33**	5,31

Total (approx.) 23,000,000 – 56,000,000 **Overall trend** Moderate decline
Breeding range >7,000,000 km² **Gen. length.** <3.3 **% Global pop.** 25–49
(See p. 262, bottom, for data quality graph)

Sturnus unicolor
SPOTLESS STARLING

Non-SPEC[E] (1994: 4) **Status** Secure
Criteria —
European IUCN Red List Category —
Criteria —
Global IUCN Red List Category —
Criteria —

Country	Breeding pop. size (pairs)	Year(s)	Trend	Mag.%	References
Andorra	(0 – 5)	99–01	(+)	(N)	*1,3*
France	2,500 – 12,500	98–02	+	0–19	*4*
Italy	(50,000 – 100,000)	03	(0)	(0–19)	
Portugal	(50,000 – 500,000)	02	(0)	(0–19)	
Spain	(2,000,000 – 2,500,000)	92	(+)	(0–19)	*13,12,10*
UK					
Gibraltar	50 – 80	00	**0**	**0–19**	
Total (approx.)	2,100,000 – 3,100,000	**Overall trend** Small increase			
Breeding range	>500,000 km²	**Gen. length.** <3.3 **% Global pop.** 50–74			

Sturnus unicolor is resident in south-west Europe and on the islands of Corsica, Sardinia and Sicily, with Europe constituting >50% of its global range. Its European breeding population is very large (>2,100,000 pairs), and increased between 1970–1990. The stronghold population in Spain continued to increase during 1990–2000, and populations increased or remained stable across the rest of its European range. Consequently, the species is evaluated as Secure.

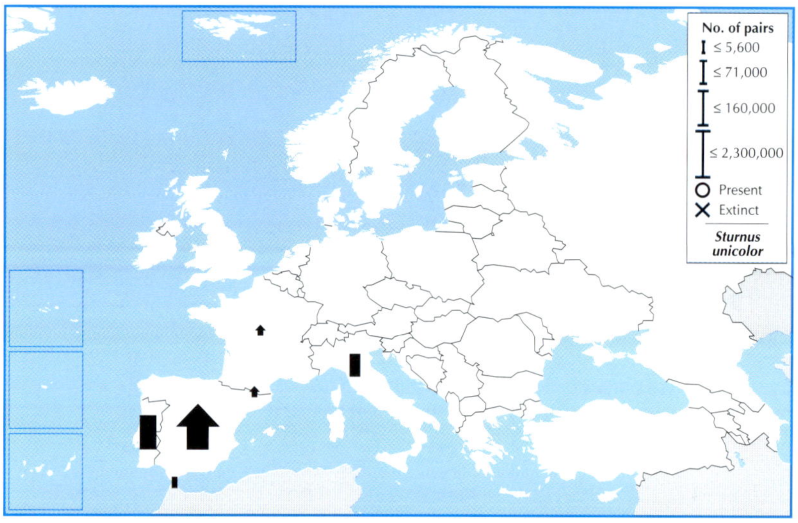

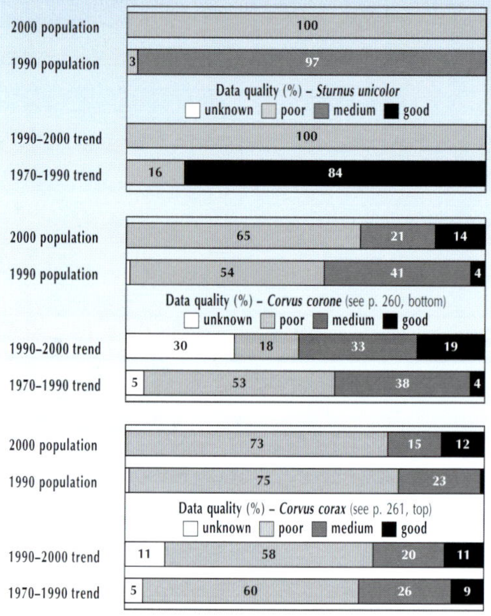

Sturnus roseus
ROSY STARLING

Non-SPEC (1994: —) **Status** Secure
Criteria —
European IUCN Red List Category —
Criteria —
Global IUCN Red List Category —
Criteria —

Country	Breeding pop. size (pairs)	Year(s)	Trend	Mag.%	References
Albania	(0 – 500)	02	?	–	
Armenia	35,000 – 70,000	00–02	0	0–19	
Azerbaijan	(1,000 – 20,000)	96–00	(F)	(–)	
Bulgaria	**500 – 3,000**	96–02	F	>80	
Georgia	Present	03	?	–	
Greece	(0 – 1,000)	95–00	(F)	(>80)	
Macedonia	(0 – 5,000)	90–00	(F)	(50–79)	
Romania	250 – 24,000	00–02	F	50–79	*3*
Russia	(15,000 – 50,000)	90–00	F	50–79	*8*
Serbia & MN	0 – 200	90–02	(F)	(50–79)	*1,130,94,155,117*
Turkey	(3,000 – 30,000)	01	(0)	(0–19)	
Ukraine	**3,000 – 7,000**	90–00	**F**	**30–49**	
Total (approx.)	58,000 – 210,000	**Overall trend** Fluctuating			
Breeding range	>1,000,000 km²	**Gen. length.** <3.3 **% Global pop.** 25–49			

Sturnus roseus is a widespread summer visitor to south-eastern Europe, which accounts for less than half of its global breeding range. Its European breeding population is relatively small (<210,000 pairs), and was broadly stable between 1970–1990. Although most European populations experienced sizeable fluctuations during 1990–2000, the species remained broadly stable overall, and consequently is evaluated as Secure.

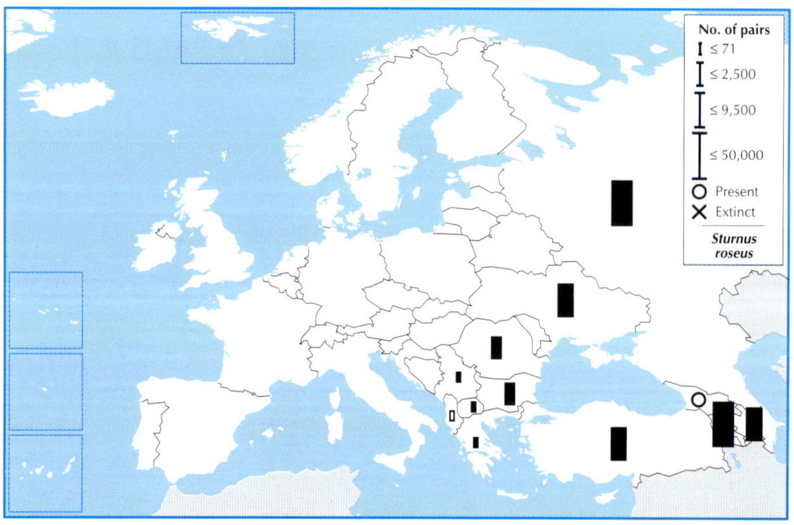

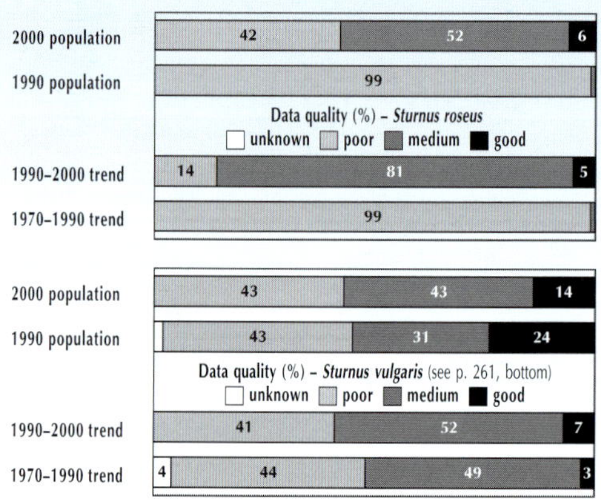

Passer domesticus
HOUSE SPARROW

SPEC 3 (1994: —) **Status** Declining
Criteria Moderate recent decline
European IUCN Red List Category —
Criteria —
Global IUCN Red List Category —
Criteria —

Passer domesticus is a widespread resident across most of Europe, which accounts for less than half of its global range. Its European breeding population is extremely large (>63,000,000 pairs), and was stable between 1970–1990. Although the species remained stable or increased in numerous countries during 1990–2000, there were declines across much of Europe—most notably in the sizeable Turkish population—and the species underwent a moderate decline (>10%) overall. Consequently, this previously Secure species is now evaluated as Declining.

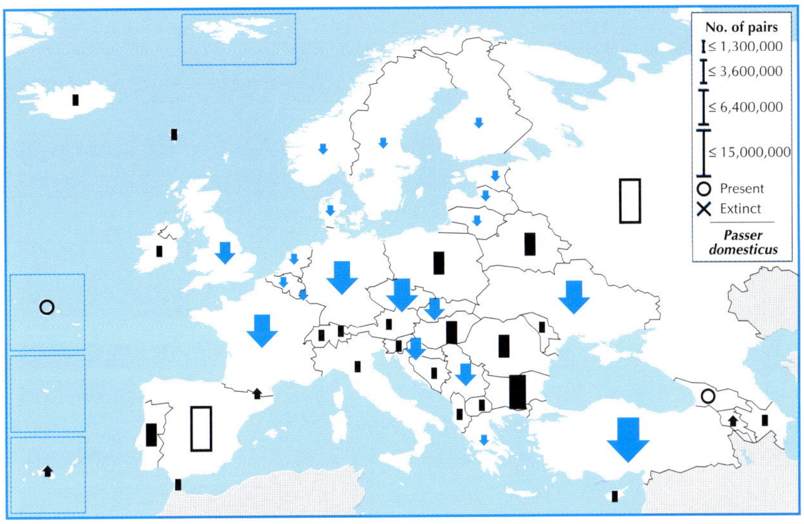

Country	Breeding pop. size (pairs)	Year(s)	Trend	Mag.%	References
Albania	100,000 – 500,000	02	(0)	(0–19)	
Andorra	(500 – 1,000)	99–01	(+)	(0–19)	*1,3*
Armenia	**350,000 – 500,000**	95–02	**+**	**0–9**	
Austria	(350,000 – 700,000)	98–02	(0)	(0–19)	
Azerbaijan	(500,000 – 1,000,000)	96–00	(0)	(0–19)	
Belarus	2,100,000 – 2,300,000	97–02	(0)	(0–19)	
Belgium	250,000 – 500,000	01–02	–	0–19	*1*
Bosnia & HG	(500,000 – 1,000,000)	90–00	(0)	(0–19)	
Bulgaria	4,000,000 – 8,000,000	96–02	0	0–19	
Croatia	(1,000,000 – 2,000,000)	02	(–)	(30–49)	*16*
Cyprus	(400,000 – 800,000)	94–02	(0)	(0–19)	
Czech Rep.	2,800,000 – 5,600,000	00	–	0–9	
Denmark	500,000 – 1,000,000	00	**–**	**20–29**	*12*
Faroe Is.	**2,500 – 2,700**	01–02	**0**	**0–19**	*1*
Estonia	(100,000 – 200,000)	98	–	20–29	*1*
Finland	200,000 – 400,000	98–02	**–**	**25**	
France	(2,000,000 – 10,000,000)	98–02	–	16	*4,3*
Georgia	Present	03	?	–	
Germany	4,000,000 – 10,000,000	95–99	–	20–29	
Greece	(200,000 – 1,000,000)	95–00	(–)	(0–19)	
Hungary	1,900,000 – 2,750,000	99–02	0	0–19	*19*
Iceland	**3 – 8**	90–00	**0**	**0–19**	*21,23,24,25,32,33, 42,43,44,45,46,47*
Rep. Ireland	500,000 – 1,000,000	88–91	0	0–19	
Italy	(50,000 – 100,000)	03	(0)	(0–19)	
Latvia	(750,000 – 2,000,000)	90–00	(–)	(0–19)	*23,16*
Liechtenstein	1,000 – 2,500	98–00	0	0–19	
Lithuania	(200,000 – 300,000)	99–01	(–)	(0–19)	*20*
Luxembourg	35,000 – 40,000	02	–	20–29	
Macedonia	(500,000 – 1,000,000)	90–00	(0)	(0–19)	
Moldova	280,000 – 350,000	90–00	0	0–19	
Netherlands	**500,000 – 1,000,000**	98–00	**–**	**28**	*1*
Norway	(200,000 – 700,000)	90–03	(–)	(20–29)	
Poland	(2,500,000 – 5,000,000)	00–02	(0)	(0–19)	*23*
Portugal	(1,000,000 – 10,000,000)	02	(0)	(0–19)	
Azores	Present	02	?	–	
Romania	1,200,000 – 3,400,000	00–02	(0)	(0–19)	*48*
Russia	10,000,000 – 20,000,000	90–00	?	–	*8,23,122*
Serbia & MN	1,500,000 – 2,500,000	90–02	(–)	(0–19)	*1,29,172a,67a, 225,227,117*
Slovakia	1,200,000 – 1,800,000	90–99	–	20–29	
Slovenia	500,000 – 800,000	94	(0)	(0–19)	
Spain	(9,307,000 – 10,000,000)	92	?	–	*13,12,10*
Canary Is.	(0 – 50)	97–03	(+)	(N)	*28,25*
Sweden	200,000 – 500,000	99–00	**–**	**46**	
Switzerland	400,000 – 500,000	98–02	0	0–19	
Turkey	(5,000,000 – 15,000,000)	01	(–)	(0–19)	
Ukraine	(4,000,000 – 5,300,000)	90–00	(–)	(5–19)	
UK	**2,100,000 – 3,675,000**	00	**–**	**25**	*5,31*
Gibraltar	200 – 400	00	**0**	**0–19**	
Total (approx.)	63,000,000 – 130,000,000	**Overall trend** Moderate decline			
Breeding range	>8,000,000 km²	**Gen. length.** <3.3		**% Global pop.** 25–49	

(See p. 263, bottom, for data quality graph)

Passer hispaniolensis
SPANISH SPARROW

Non-SPEC (1994: —) **Status** (Secure)
Criteria —
European IUCN Red List Category —
Criteria —
Global IUCN Red List Category —
Criteria —

Passer hispaniolensis is a widespread breeder in parts of southern Europe, which accounts for less than half of its global range. Its European breeding population is very large (>2,800,000 pairs), and increased between 1970–1990. Although there were slight declines in Greece and Turkey during 1990–2000, populations across most of Europe—including sizeable ones in Italy and Bulgaria—were stable or increased, and the species was probably stable overall. Consequently, it is provisionally evaluated as Secure.

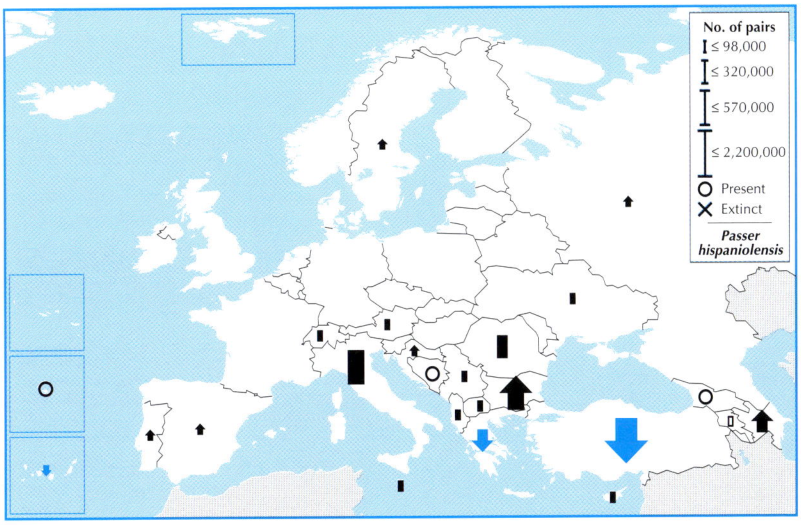

Country	Breeding pop. size (pairs)	Year(s)	Trend	Mag.%	References
Albania	5,000 – 20,000	02	(0)	(0–19)	
Armenia	80 – 300	95–02	?	–	
Austria	(10 – 30)	98–02	(0)	(0–19)	
Azerbaijan	(100,000 – 500,000)	96–00	(+)	(30–49)	
Bosnia & HG	Present	90–03	?	–	
Bulgaria	400,000 – 800,000	96–02	+	0–19	
Croatia	(10,000 – 50,000)	02	(+)	(>80)	*16*
Cyprus	(60,000 – 160,000)	94–02	(0)	(0–19)	
Georgia	Present	03	?	–	
Greece	(200,000 – 500,000)	95–00	(–)	(0–19)	
Italy	(300,000 – 500,000)	03	(0)	(0–19)	
Macedonia	(10,000 – 20,000)	90–00	(0)	(0–19)	
Malta	(60,000 – 100,000)	90–02	**0**	**0–19**	*1*
Portugal	(10,000 – 100,000)	02	(+)	(–)	
Madeira	Present	02	?	–	
Romania	(125,000 – 254,000)	00–02	(0)	(0–19)	*48*
Russia	(2,000 – 5,000)	90–00	(+)	(20–29)	*8*
Serbia & MN	2,000 – 3,000	95–02	F	20–49	*1,155,141,227,117*
Spain	(20,000 – 100,000)	92	(+)	(0–19)	*13,12,10*
Canary Is.	(20,000 – 100,000)	97–03	(–)	(0–19)	*28,25*
Switzerland	(10,000 – 25,000)	98–02	(0)	(0–19)	
Turkey	(1,500,000 – 3,000,000)	01	(–)	(0–19)	
Ukraine	50 – 100	90–00	F	>80	
Total (approx.)	2,800,000 – 6,200,000	**Overall trend** Stable			
Breeding range	>1,000,000 km²	**Gen. length.** <3.3		**% Global pop.** 25–49	

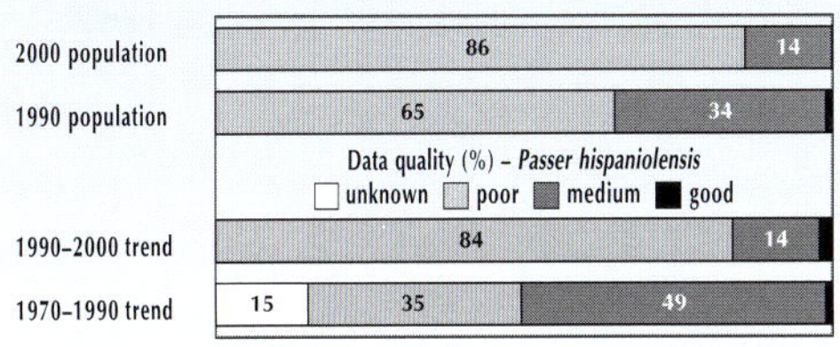

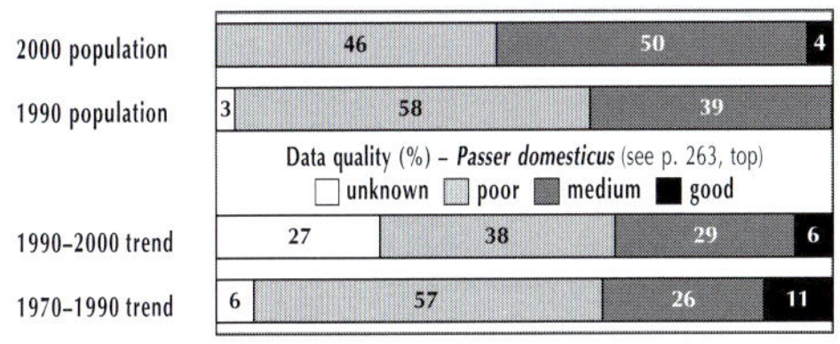

Passer moabiticus
DEAD SEA SPARROW

Non-SPEC (1994: —) **Status** (Secure)
Criteria —
European IUCN Red List Category —▼
Criteria —
Global IUCN Red List Category —
Criteria —

Country	Breeding pop. size (pairs)	Year(s)	Trend	Mag.%	References
Cyprus	(0 – 2)	94–02	?	–	
Turkey	(900 – 3,000)	01	(–)	(20–29)	
Total (approx.)	900 – 3,000	**Overall trend** Moderate decline			
Breeding range	>20,000 km²	**Gen. length.** <3.3 **% Global pop.** 5–24			

Passer moabiticus breeds locally in Cyprus and southern Turkey, with Europe accounting for less than a quarter of its global breeding range. Its European breeding population is very small (as few as 900 pairs), but underwent a large increase between 1970–1990, which outweighed the moderate decline (>10%) it suffered during 1990–2000. Although the size of the European population could render it susceptible to the risks affecting small populations, it is marginal to a much larger non-European population. Consequently, the species is provisionally evaluated as Secure.

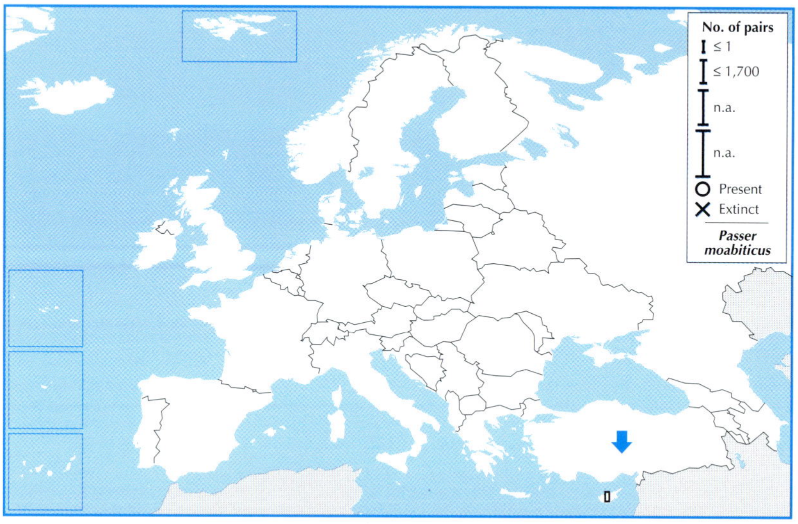

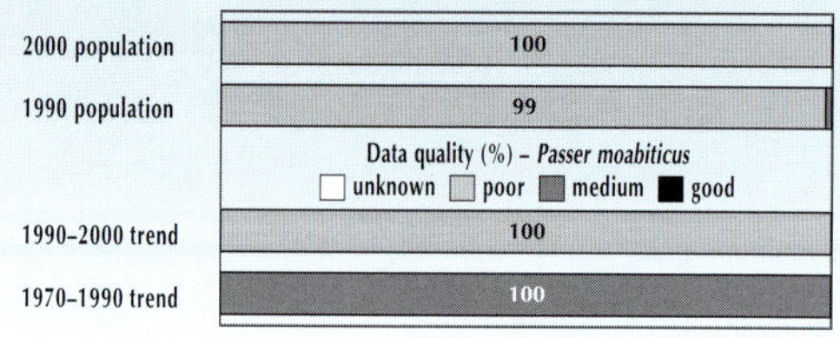

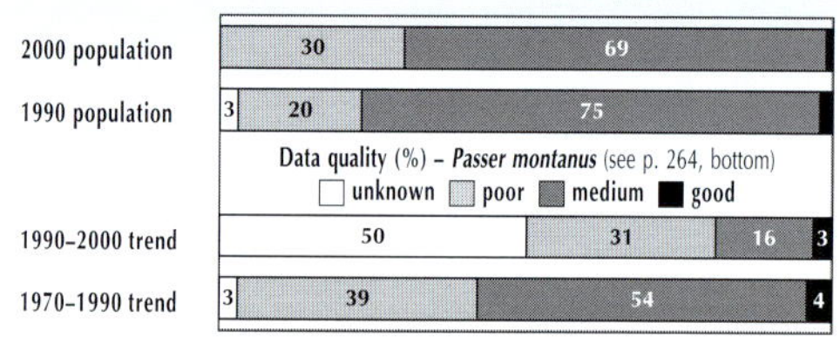

Passer montanus
EURASIAN TREE SPARROW

SPEC 3 (1994: —) **Status** (Declining)
Criteria Moderate recent decline
European IUCN Red List Category —
Criteria —
Global IUCN Red List Category —
Criteria —

Country	Breeding pop. size (pairs)	Year(s)	Trend	Mag.%	References
Albania	5,000 – 20,000	02	(0)	(0–19)	
Andorra	20 – 50	99–01	(–)	(20–29)	*1,3*
Armenia	3,500 – 5,000	98–02	0	0–19	
Austria	(80,000 – 160,000)	98–02	(0)	(0–19)	
Azerbaijan	(200,000 – 500,000)	96–00	(0)	(0–19)	
Belarus	900,000 – 950,000	97–02	(0)	(0–19)	
Belgium	15,000 – 30,000	01–02	–	0–19	*1*
Bosnia & HG	(50,000 – 100,000)	90–00	(0)	(0–19)	
Bulgaria	300,000 – 600,000	96–02	(0)	(0–19)	
Croatia	(500,000 – 1,000,000)	02	(0)	(0–19)	*16*
Czech Rep.	400,000 – 800,000	00	**–**	**0–19**	
Denmark	400,000 – 600,000	00	F	20–29	*12*
Estonia	(50,000 – 100,000)	98	0	0–19	*1*
Finland	20,000 – 40,000	98–02	**+**	**300**	
France	(180,000 – 900,000)	98–02	–	29	*4,2*
Georgia	Present	03	?	–	
Germany	900,000 – 2,100,000	95–99	–	30–49	
Greece	(10,000 – 30,000)	95–00	(–)	(0–19)	
Hungary	2,000,000 – 2,800,000	99–02	0	0–19	*19*
Rep. Ireland	1,000 – 2,500	88–91	0	0–19	
Italy	(500,000 – 1,000,000)	03	(–)	(0–19)	
Latvia	(150,000 – 300,000)	90–00	(0)	(0–19)	*23,16*
Liechtenstein	200 – 300	98–00	+	0–19	
Lithuania	(300,000 – 500,000)	99–01	(–)	(0–19)	*20*
Luxembourg	6,000 – 15,000	02	–	0–19	
Macedonia	(15,000 – 40,000)	90–00	(0)	(0–19)	
Malta	200 – 300	90–02	0	0–19	*1*
Moldova	250,000 – 310,000	90–00	0	0–19	
Netherlands	**50,000 – 150,000**	98–00	**–**	**6**	*1*
Norway	(50,000 – 150,000)	90–03	+	20–29	
Poland	500,000 – 1,000,000	00–02	(–)	(20–29)	*23*
Portugal	(50,000 – 500,000)	02	(–)	(–)	
Romania	1,354,000 – 2,873,000	00–02	(0)	(0–19)	*48*
Russia	10,000,000 – 20,000,000	90–00	?	–	*8,23,122*
Serbia & MN	(350,000 – 550,000)	90–02	(0)	(0–19)	*1,29,172a,67a, 225,227,117*
Slovakia	300,000 – 600,000	90–99	0	0–19	
Slovenia	100,000 – 200,000	94	(0)	(0–19)	
Spain	(2,500,000 – 4,100,000)	92	?	–	*13,12,10*
Canary Is.	(250 – 1,000)	97–03	(+)	(20–29)	*28,25*
Sweden	300,000 – 700,000	99–00	**–**	**32**	
Switzerland	**70,000 – 100,000**	93–96	+	0–19	
Turkey	(20,000 – 100,000)	01	?	–	
Ukraine	(3,000,000 – 4,000,000)	90–00	(0)	(0–19)	
UK	**68,000 – 68,000**	00	**–**	**38**	*31*
Total (approx.)	26,000,000 – 48,000,000	**Overall trend** Moderate decline			
Breeding range	>7,000,000 km²	**Gen. length.** <3.3 **% Global pop.** 25–49			

(See p. 264, top, for data quality graph)

Passer montanus is a widespread resident across much of Europe, which accounts for less than half of its global range. Its European breeding population is extremely large (>26,000,000 pairs), and was stable between 1970–1990. Although populations were stable or increased across parts of Europe—particularly eastern areas—during 1990–2000, the species suffered widespread declines in western and north-west Europe, and underwent a moderate decline (>10%) overall. Consequently, this previously Secure species is now provisionally evaluated as Declining.

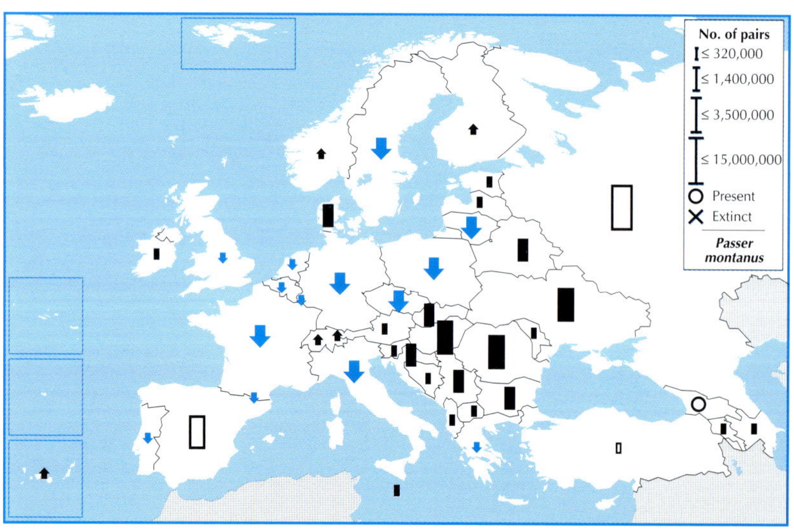

Petronia brachydactyla
PALE ROCK-FINCH

Non-SPEC (1994: —) **Status** Secure
Criteria —
European IUCN Red List Category —▼
Criteria —
Global IUCN Red List Category —
Criteria —

Country	Breeding pop. size (pairs)	Year(s)	Trend	Mag.%	References
Armenia	6,000 – 15,000	00–02	+	50–79	*10*
Azerbaijan	(50 – 250)	96–00	(0)	(0–19)	
Russia	(5 – 50)	90–00	(F)	(30–49)	*15,170*
Turkey	(900 – 3,000)	01	?	–	
Total (approx.)	7,000 – 18,000	**Overall trend** Large increase			
Breeding range	>20,000 km²	**Gen. length.** <3.3	**% Global pop.** 5–24		

Petronia brachydactyla is a summer visitor to south-eastern Turkey and parts of the Caucasus, with Europe accounting for less than a quarter of its global breeding range. Its European breeding population is small (as few as 7,000 pairs), but was probably broadly stable between 1970–1990, and underwent a large increase overall during 1990–2000. Although the size of the European population could render it susceptible to the risks affecting small populations, it is marginal to a much larger non-European population. Consequently, the species is evaluated as Secure.

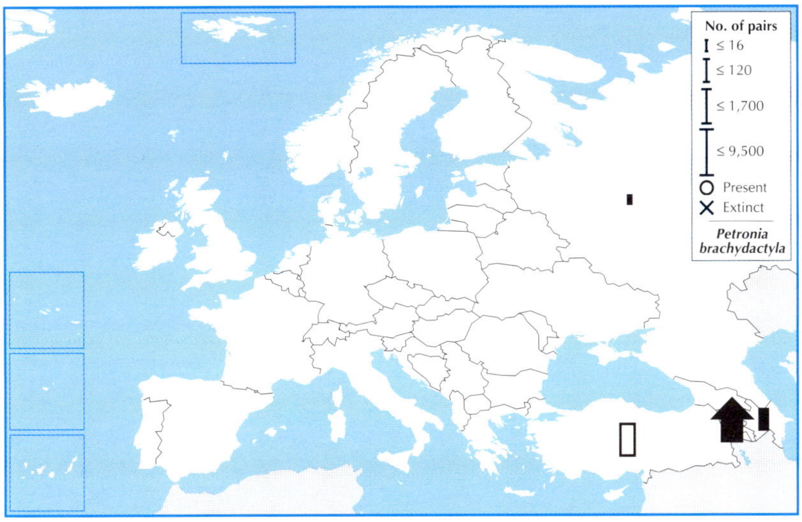

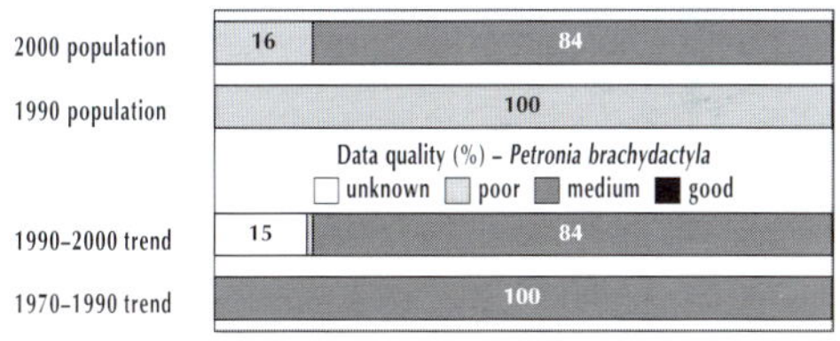

Petronia xanthocollis
CHESTNUT-SHOULDERED PETRONIA

Non-SPEC (1994: —) **Status** (Secure)
Criteria —
European IUCN Red List Category —▼
Criteria —
Global IUCN Red List Category —
Criteria —

Country	Breeding pop. size (pairs)	Year(s)	Trend	Mag.%	References
Turkey	(500 – 2,000)	01	(+)	(–)	
Total (approx.)	500 – 2,000	**Overall trend** Increase			
Breeding range	>20,000 km²	**Gen. length.** <3.3	**% Global pop.** <5		

Petronia xanthocollis has a predominantly Middle Eastern and Asian breeding distribution, which just extends into Europe in south-eastern Turkey. Its European breeding population is small (as few as 500 pairs), and its trend between 1970–1990 was unknown, but it probably increased during 1990–2000. Although the size of the European population could render it susceptible to the risks affecting small populations, it is marginal to a much larger non-European population. Consequently, the species is provisionally evaluated as Secure.

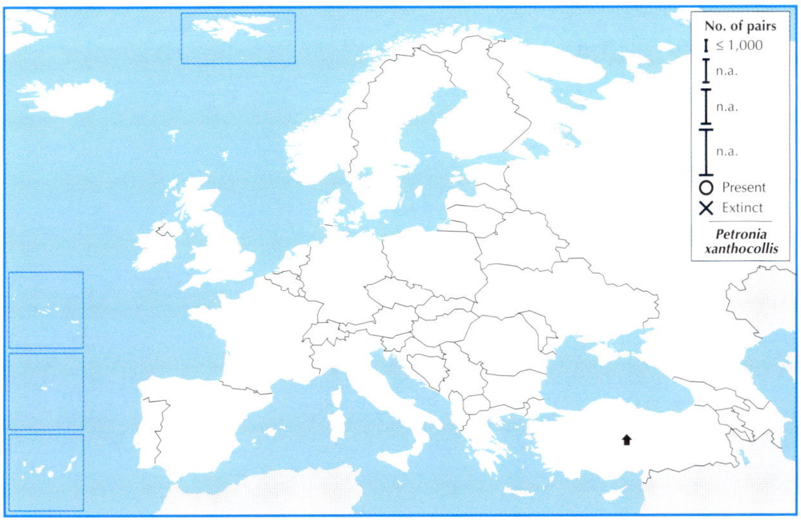

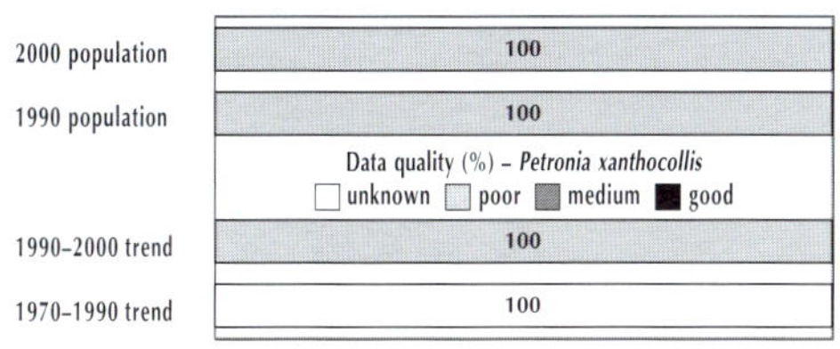

Petronia petronia
ROCK SPARROW

Non-SPEC (1994: —) **Status** (Secure)
Criteria —
European IUCN Red List Category —
Criteria —
Global IUCN Red List Category —
Criteria —

Petronia petronia is a widespread resident across much of southern Europe, which accounts for less than a quarter of its global range. Its European breeding population is very large (>1,700,000 pairs), and was stable between 1970–1990. Although there were declines in a few countries during 1990–2000, the key population in Turkey was stable (the trend of the other sizeable population, in Spain, was unknown), and the species probably remained stable overall. Consequently, it is provisionally evaluated as Secure.

Country	Breeding pop. size (pairs)	Year(s)	Trend	Mag.%	References
Albania	(300 – 1,000)	02	?	–	
Andorra	15 – 30	99–01	(0)	(0–19)	*1,3*
Armenia	10,000 – 25,000	99–02	+	20–29	
Azerbaijan	(2,000 – 15,000)	96–00	(0)	(0–19)	
Bulgaria	15 – 100	96–02	(F)	(>80)	
France	15,000 – 75,000	98–02	(–)	(0–19)	*4*
Georgia	Present	03	?	–	
Greece	(2,000 – 5,000)	95–00	(–)	(0–19)	
Italy	(10,000 – 20,000)	03	(0)	(0–19)	
Macedonia	(2,000 – 5,000)	90–00	(0)	(0–19)	
Portugal	(10,000 – 100,000)	02	(0)	(0–19)	
Madeira	Present	02	?	–	
Russia	(5,000 – 15,000)	90–00	(–)	(20–29)	*50*
Spain	(825,000 – 1,150,000)	92	?	–	*13,12,10*
Canary Is.	(2,500 – 10,000)	97–03	(–)	(0–19)	*28,25*
Turkey	(800,000 – 2,400,000)	01	(0)	(0–19)	
Total (approx.)	1,700,000 – 3,800,000	**Overall trend** Stable			
Breeding range	>1,000,000 km²	**Gen. length.** <3.3	**% Global pop.** 5–24		

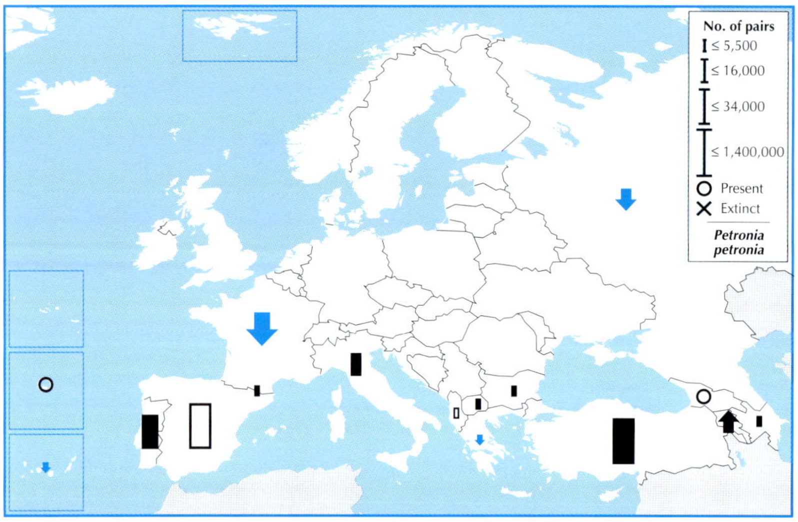

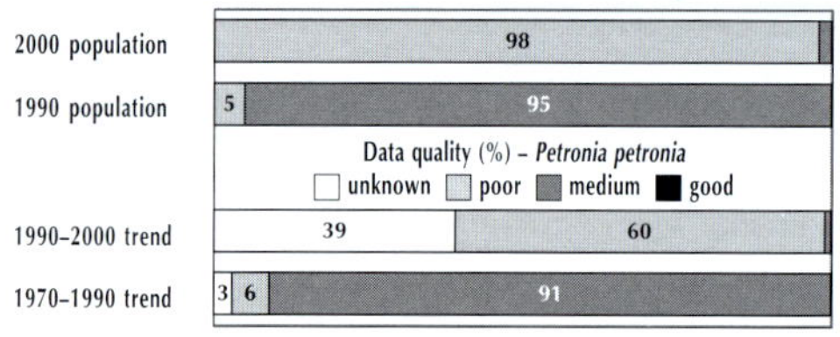

Montifringilla nivalis
WHITE-WINGED SNOWFINCH

Non-SPEC (1994: —) **Status** (Secure)
Criteria —
European IUCN Red List Category —
Criteria —
Global IUCN Red List Category —
Criteria —

Montifringilla nivalis has a fragmented distribution in the mountains of central and southern Europe, which accounts for less than half of its global range. Its European breeding population is large (>520,000 pairs), and was stable between 1970–1990. This trend continued during 1990–2000, with populations across the entire European range—including the Turkish stronghold – remaining stable. Consequently, the species is provisionally evaluated as Secure.

Country	Breeding pop. size (pairs)	Year(s)	Trend	Mag.%	References
Albania	250 – 1,000	81–02	?	–	
Andorra	(0 – 10)	99–01	(0)	(0–19)	*1,3*
Armenia	800 – 3,000	98–02	0	0–19	
Austria	(2,000 – 5,000)	98–02	(0)	(0–19)	
Azerbaijan	(5,000 – 20,000)	96–00	(0)	(0–19)	
France	2,500 – 12,500	00–02	(0)	(0–19)	*4*
Georgia	Present	03	?	–	
Germany	**200 – 400**	95–99	0	0–19	
Greece	**200 – 300**	96	(0)	(0–19)	
Italy	(3,000 – 6,000)	03	(0)	(0–19)	
Liechtenstein	4 – 8	98–00	0	0–19	
Macedonia	(80 – 150)	90–00	(0)	(0–19)	
Russia	(1,000 – 5,000)	90–00	(0)	(0–19)	*17,60*
Serbia & MN	**60 – 100**	97–00	**0**	**0–19**	*1,62,40,155, 225,227*
Slovenia	100 – 300	99–00	(0)	(0–19)	
Spain	(4,500 – 6,000)	98–02	(0)	(0–19)	*10*
Switzerland	**4,000 – 8,000**	93–96	0	0–19	
Turkey	(500,000 – 1,500,000)	01	(0)	(0–19)	
Total (approx.)	520,000 – 1,600,000	**Overall trend** Stable			
Breeding range	>500,000 km²	**Gen. length.** <3.3	**% Global pop.** 25–49		

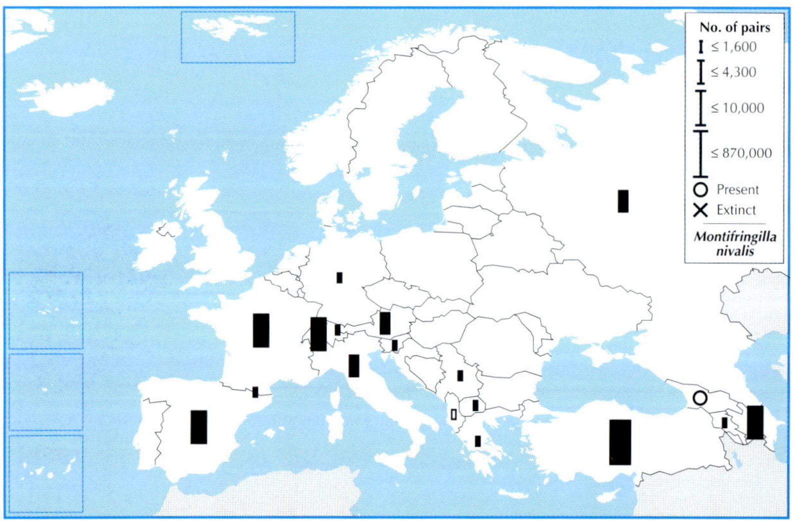

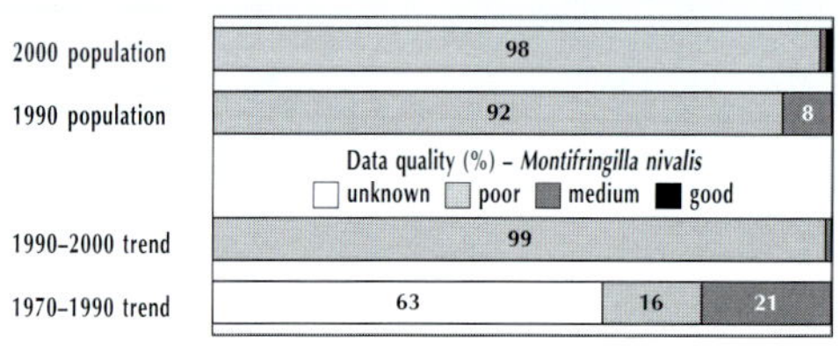

Fringilla coelebs
CHAFFINCH

Non-SPEC[E] (1994: 4) **Status** Secure
Criteria —
European IUCN Red List Category —
Criteria —
Global IUCN Red List Category —
Criteria —

Fringilla coelebs is a widespread breeder across most of Europe, which constitutes >50% of its global range. Its European breeding population is extremely large (>130,000,000 pairs), and was stable between 1970–1990. Although there were slight declines in France and Sweden during 1990–2000, populations were stable across most of the rest of Europe—including the key one in Russia—and the species remained stable overall. Consequently, it is evaluated as Secure.

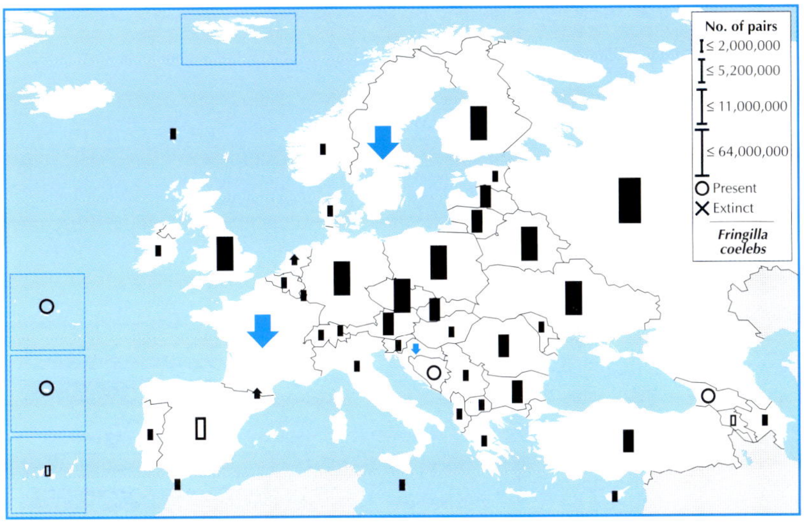

Country	Breeding pop. size (pairs)	Year(s)	Trend	Mag.%	References
Albania	50,000 – 90,000	02	(0)	(0–19)	
Andorra	(5,000 – 8,000)	99–01	(+)	(0–19)	*1,3*
Armenia	160,000 – 260,000	00–02	?	–	
Austria	(1,600,000 – 3,200,000)	98–02	(0)	(0–19)	
Azerbaijan	(100,000 – 500,000)	96–00	(0)	(0–19)	
Belarus	7,500,000 – 8,500,000	97–02	0	0–19	
Belgium	100,000 – 250,000	01–02	(0)	(0–19)	*1*
Bosnia & HG	Present	90–03	?	–	
Bulgaria	2,000,000 – 6,000,000	96–02	0	0–19	
Croatia	(1,000,000 – 1,500,000)	02	(–)	(50–79)	*16*
Cyprus	(20,000 – 40,000)	94–02	(0)	(0–19)	
Czech Rep.	4,000,000 – 8,000,000	00	0	0–19	
Denmark	1,500,000 – 2,000,000	00	**0**	**0–19**	*12*
Faroe Is.	(0 – 1)	81	(0)	(0–19)	
Estonia	1,500,000 – 2,500,000	98	0	0–19	*1*
Finland	5,000,000 – 7,000,000	98–02	**0**	**0–19**	
France	(4,000,000 – 15,000,000)	98–02	–	12	*4,2*
Georgia	Present	03	?	–	
Germany	5,500,000 – 12,500,000	95–99	0	0–19	
Greece	(1,000,000 – 3,000,000)	95–00	(0)	(0–19)	
Hungary	**940,000 – 1,230,000**	99–02	0	0–19	*19,1*
Rep. Ireland	1,000,000 – 2,500,000	88–91	0	0–19	
Italy	(1,000,000 – 2,000,000)	03	(0)	(0–19)	
Latvia	2,600,000 – 3,200,000	90–00	(0)	(0–19)	*23,18*
Liechtenstein	2,000 – 3,000	98–00	(0)	(0–19)	
Lithuania	(2,500,000 – 3,500,000)	99–01	(0)	(0–19)	*20*
Luxembourg	50,000 – 70,000	02	0	0–19	
Macedonia	(300,000 – 500,000)	90–00	(0)	(0–19)	
Malta	2 – 3	90–02	**0**	**0–19**	*1*
Moldova	200,000 – 250,000	90–00	0	0–19	
Netherlands	**600,000 – 700,000**	98–00	+	**55**	*1*
Norway	(1,000,000 – 1,500,000)	90–02	(0)	(0–19)	
Poland	**5,000,000 – 10,000,000**	00–02	0	0–19	*23*
Portugal	(500,000 – 2,500,000)	02	(0)	(0–19)	
Azores	Present	02	?	–	
Madeira	Present	02	?	–	
Romania	2,450,000 – 6,300,000	99–02	0	0–19	*48*
Russia	50,000,000 – 80,000,000	90–00	0	0–19	*122*
Serbia & MN	1,500,000 – 2,000,000	90–02	0	0–19	*1,29,172a,67a,78, 227,225,185,117a*
Slovakia	3,000,000 – 5,000,000	90–99	0	0–19	
Slovenia	1,000,000 – 1,500,000	00	(0)	(0–19)	
Spain	(2,600,000 – 6,400,000)	92	?	–	*13,12,10*
Canary Is.	(2,500 – 10,000)	97–03	?	–	*28,25*
Sweden	7,500,000 – 15,000,000	99–00	**–**	**9**	
Switzerland	900,000 – 1,200,000	98–02	0	0–19	
Turkey	(3,000,000 – 9,000,000)	01	(0)	(0–19)	
Ukraine	5,300,000 – 6,700,000	90–00	0	0–19	
UK	**5,974,000 – 5,974,000**	00	**0**	**3**	*5,31*
Gibraltar	**0 – 4**	00	**0**	**0–19**	

Total (approx.) 130,000,000 – 240,000,000 **Overall trend** Stable
Breeding range >8,000,000 km² **Gen. length.** <3.3 **% Global pop.** 50–74
(See p. 267, bottom, for data quality graph)

Fringilla teydea
BLUE CHAFFINCH

SPEC 1 (1994: 1) **Status** Rare
Criteria <10,000 pairs
European IUCN Red List Category —
Criteria —
Global IUCN Red List Category NT
Criteria B1a+b(ii,iii,v); B2a+b(ii,iii,v); C2a(ii)

Fringilla teydea is endemic to Europe, where it has a very small range (<500 km²) on the Canary Islands of Tenerife and Gran Canaria. Its European breeding population is small (as few as 1,000 pairs), but was stable between 1970–1990. The species remained stable overall during 1990–2000, with the decline of the small Gran Canaria population outweighed by the stable trend of the larger Tenerife population. Nevertheless, its population size renders it susceptible to the risks affecting small populations, and consequently this globally Near Threatened species is evaluated as Rare.

Country	Breeding pop. size (pairs)	Year(s)	Trend	Mag.%	References
Spain					
Canary Is.	(1,000 – 2,500)	97–03	0	0–19	*28,25*

Total (approx.) 1,000 – 2,500 **Overall trend** Stable
Breeding range <500 km² **Gen. length.** 4 **% Global pop.** 100

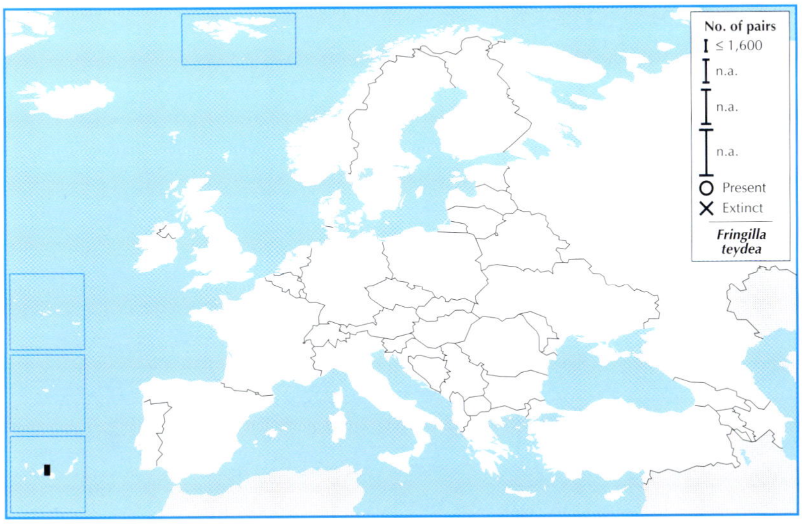

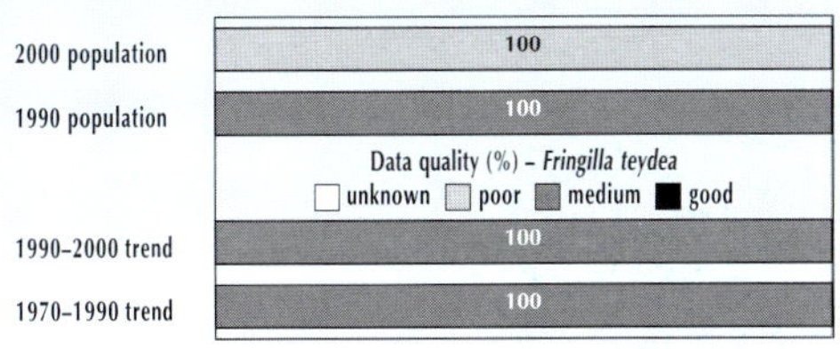

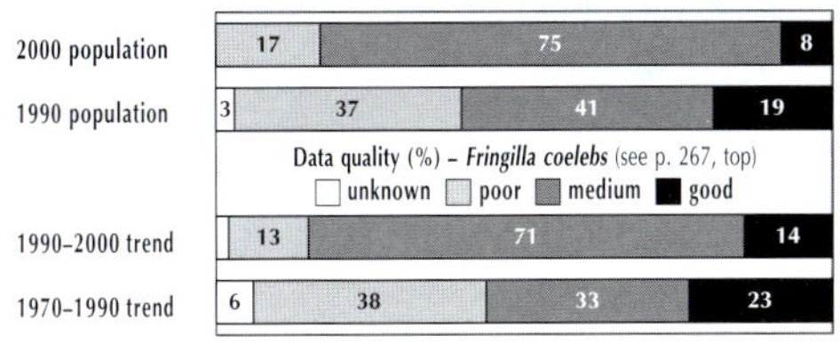

Fringilla montifringilla
BRAMBLING

Non-SPEC (1994: —) **Status** Secure
Criteria —
European IUCN Red List Category —
Criteria —
Global IUCN Red List Category —
Criteria —

Fringilla montifringilla is a widespread breeder in Fennoscandia and northern Russia, but also occurs very patchily farther south, with Europe accounting for less than half of its global range. Its European breeding population is extremely large (>13,000,000 pairs), and was stable between 1970–1990. This trend continued during 1990–2000, with all European populations—including the key northern ones – remaining stable. Consequently, the species is evaluated as Secure.

Country	Breeding pop. size (pairs)	Year(s)	Trend	Mag.%	References
Belarus	0 – 10	97–00	?	–	
Denmark	0 – 3	99–00	**0**	**0–19**	*16*
Faroe Is.	(0 – 2)	81	(0)	(0–19)	
Estonia	50 – 500	98	F	20–29	*1*
Finland	1,000,000 – 2,500,000	98–02	**0**	**5**	
Germany	**2 – 2**	95–99	F	>80	
Iceland	**0 – 4**	90–00	(F)	(–)	*24,25,32,33,45, 46,47,42,43,44*
Latvia	(10 – 100)	90–00	(0)	(0–19)	*23,18*
Lithuania	(0 – 5)	99–01	(0)	(0–19)	*20*
Netherlands	**20 – 45**	98–00	(F)	(–)	*1*
Norway	(1,000,000 – 2,000,000)	90–02	(0)	(0–19)	
Russia	10,000,000 – 15,000,000	90–00	(0)	(0–19)	*122*
Slovenia	(0 – 10)	00	(F)	(>80)	
Sweden	500,000 – 2,000,000	99–00	(0)	(0–19)	
Total (approx.)	13,000,000 – 22,000,000	**Overall trend** Stable			
Breeding range	>2,000,000 km²	**Gen. length.** <3.3		**% Global pop.** 25–49	

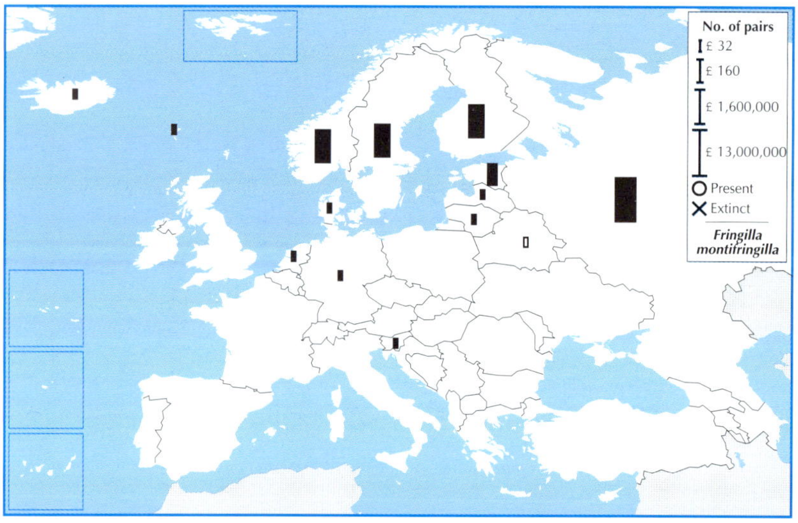

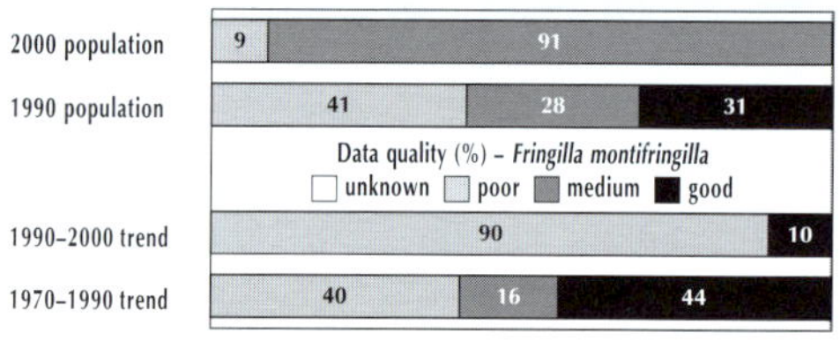

Serinus pusillus
FIRE-FRONTED SERIN

Non-SPEC (1994: —) **Status** (Secure)
Criteria —
European IUCN Red List Category —
Criteria —
Global IUCN Red List Category —
Criteria —

Serinus pusillus is a widespread resident in the Caucasus and adjacent parts of Russia and Turkey, with Europe accounting for less than a quarter of its global range. Its European breeding population is large (>280,000 pairs), and was stable between 1970–1990. Although the species declined slightly in Armenia during 1990–2000, these losses were set against increases in Russia, and stable trends in Azerbaijan and the Turkish stronghold, and the species probably remained stable overall. Consequently, it is provisionally evaluated as Secure.

Country	Breeding pop. size (pairs)	Year(s)	Trend	Mag.%	References
Armenia	15,000 – 27,000	00–02	–	0–9	
Azerbaijan	(5,000 – 20,000)	96–00	(0)	(0–19)	
Georgia	Present	03	?	–	
Russia	(10,000 – 25,000)	90–00	(+)	(20–29)	*61,116*
Turkey	(250,000 – 750,000)	01	(0)	(0–19)	
Total (approx.)	280,000 – 820,000	**Overall trend** Stable			
Breeding range	>250,000 km²	**Gen. length.** <3.3		**% Global pop.** 5–24	

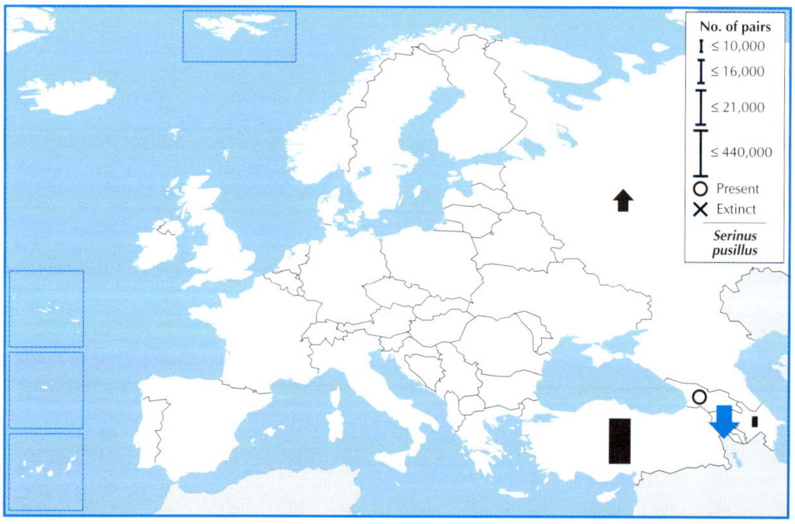

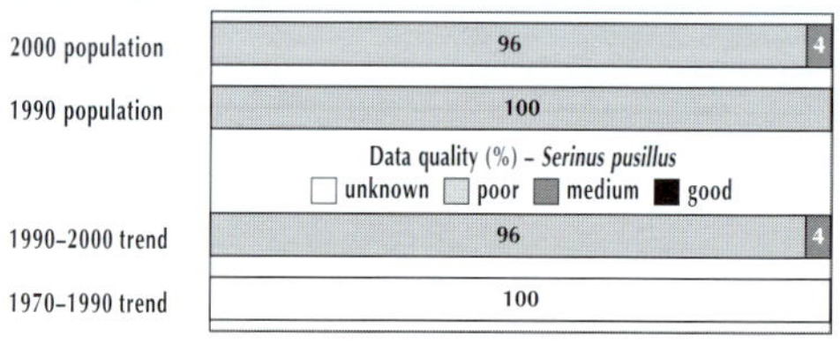

Serinus serinus
EUROPEAN SERIN

Non-SPEC[E] (1994: 4) **Status** Secure
Criteria —
European IUCN Red List Category —
Criteria —
Global IUCN Red List Category —
Criteria —

Serinus serinus is a widespread breeder across most of Europe (except for the north and east), which constitutes >75% of its global range. Its European breeding population is very large (>8,300,000 pairs), and increased between 1970–1990. Although there were declines in France and Malta during 1990–2000, populations increased or were stable elsewhere in Europe—including the key Spanish population—and the species was stable overall. Consequently, it is evaluated as Secure.

Country	Breeding pop. size (pairs)	Year(s)	Trend	Mag.%	References
Albania	5,000 – 20,000	02	(0)	(0–19)	
Andorra	(200 – 700)	99–01	(+)	(0–19)	*1,3*
Austria	(45,000 – 90,000)	98–02	(0)	(0–19)	
Belarus	3,000 – 8,000	97–02	0	0–19	
Belgium	1,200 – 3,200	01–02	0	0–19	*1*
Bosnia & HG	Present	90–03	?	–	
Bulgaria	50,000 – 150,000	96–02	+	0–19	
Croatia	(150,000 – 200,000)	02	(+)	(0–19)	57
Cyprus	(2,500 – 10,000)	94–02	(0)	(0–9)	
Czech Rep.	450,000 – 900,000	00	0	0–19	
Denmark	1 – 11	99–01	(+)	(30–49)	*5,6,7*
Estonia	50 – 100	98	0	0–19	*1*
Finland	(0 – 3)	98–02	0	0–19	
France	(400,000 – 2,000,000)	98–02	–	37	*4,2*
Germany	200,000 – 420,000	95–99	0	0–19	
Greece	(10,000 – 30,000)	95–00	(0)	(0–19)	
Hungary	**160,000 – 245,000**	99–02	+	20–49	*19*
Italy	(500,000 – 1,000,000)	03	(+)	(0–19)	
Latvia	50 – 300	90–00	+	0–19	*23,36*
Liechtenstein	80 – 120	98–00	+	0–19	
Lithuania	(2,000 – 4,000)	99–01	(0)	(0–19)	*20*
Luxembourg	2,000 – 2,500	02	0	0–19	
Macedonia	(3,000 – 10,000)	90–00	(0)	(0–19)	
Malta	2 – 5	90–02	**–**	**0–19**	*1*
Moldova	100 – 200	90–00	+	50–79	
Netherlands	**400 – 450**	98–00	+	50–79	*1*
Poland	**150,000 – 250,000**	00–02	0	0–19	*23*
Portugal	(1,000,000 – 5,000,000)	02	(0)	(0–19)	
Romania	(245,000 – 650,000)	00–02	(0)	(0–19)	*48*
Russia	(100 – 1,000)	90–00	?	–	*171*
Serbia & MN	60,000 – 90,000	95–02	+	0–19	*1,29,172a,155, 67a,78,227,225, 117a,164*
Slovakia	50,000 – 100,000	90–99	0	0–19	
Slovenia	30,000 – 40,000	94	(0)	(0–19)	
Spain	(4,100,000 – 6,600,000)	92	(0)	(0–19)	*13,12,10*
Canary Is.	(1,000 – 2,500)	97–03	?	–	*28,25*
Sweden	**1 – 15**	99–00	+	**10–19**	
Switzerland	**20,000 – 40,000**	93–96	+	0–19	
Turkey	(500,000 – 2,000,000)	01	(0)	(0–19)	
Ukraine	155,000 – 215,000	90–00	+	0–19	
UK					
Gibraltar	**2 – 10**	00	**0**	**0–19**	
Total (approx.)	8,300,000 – 20,000,000	**Overall trend** Stable			
Breeding range	>4,000,000 km²	**Gen. length.** <3.3 **% Global pop.** 75–94			

(See p. 269, bottom, for data quality graph)

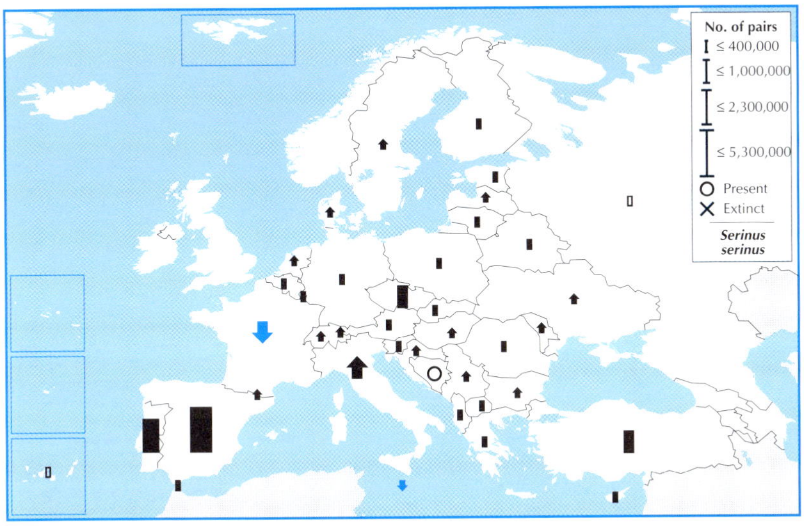

Serinus canaria
ISLAND CANARY

Non-SPEC[E] (1994: 4) **Status** (Secure)
Criteria —
European IUCN Red List Category —
Criteria —
Global IUCN Red List Category —
Criteria —

Serinus canaria is endemic to Europe, where it is confined to the Canary Islands, Madeira and the Azores. Its population is relatively small (<100,000 pairs), but was apparently stable between 1970–1990. Trends were not available for any of the species's populations during 1990–2000, but there was no evidence to suggest that its status deteriorated. Consequently, it is provisionally evaluated as Secure.

Country	Breeding pop. size (pairs)	Year(s)	Trend	Mag.%	References
Portugal					
Azores	Present	02	?	–	
Madeira	Present	02	?	–	
Spain					
Canary Is.	(20,000 – 100,000)	97–03	?	–	*28,25*
Total (approx.)	20,000 – 100,000	**Overall trend** Unknown			
Breeding range	>20,000 km²	**Gen. length.** <3.3 **% Global pop.** 100			

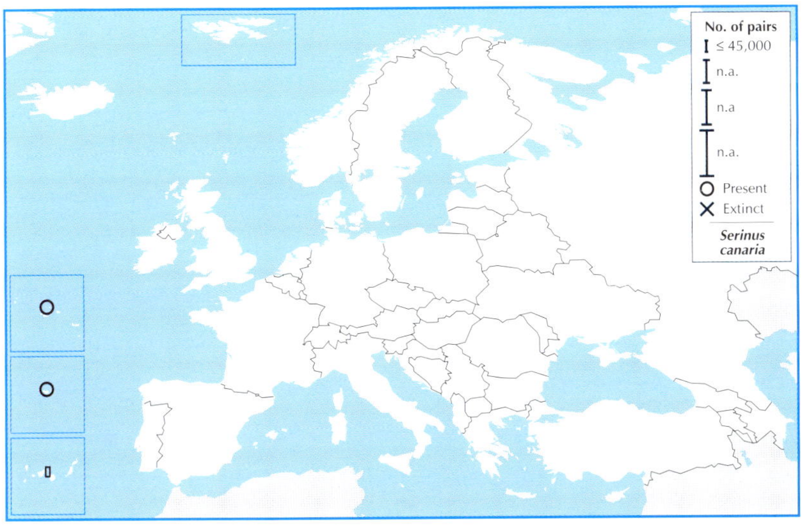

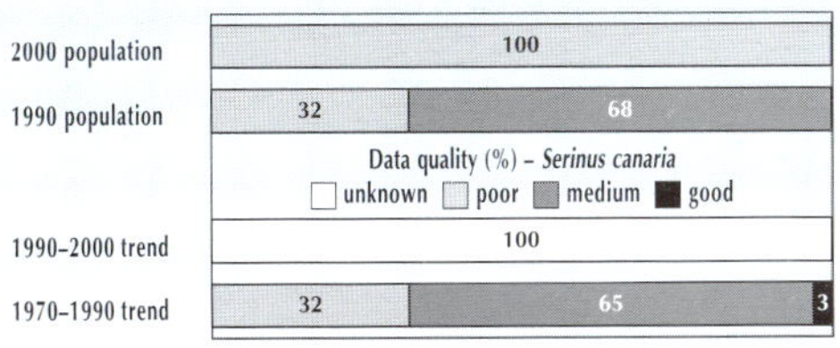

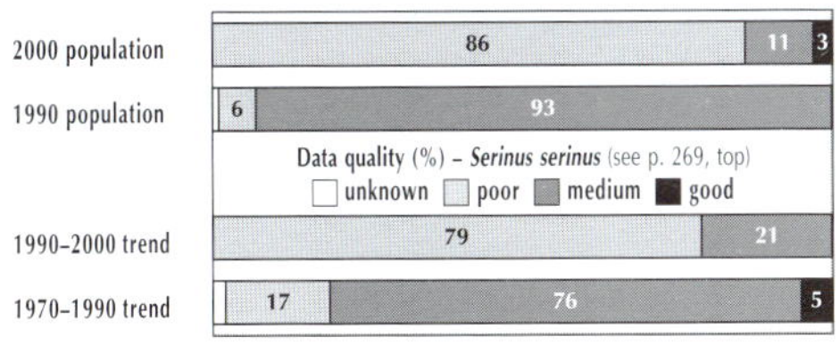

Serinus citrinella
CITRIL FINCH

Non-SPEC^E (1994: 4) **Status** (Secure)
Criteria —
European IUCN Red List Category —
Criteria —
Global IUCN Red List Category —
Criteria —

Country	Breeding pop. size (pairs)	Year(s)	Trend	Mag.%	References
Andorra	(500 – 1,000)	99–01	(0)	(0–19)	*1,3*
Austria	(1,500 – 5,000)	98–02	(0)	(0–19)	
France	8,000 – 40,000	98–02	?	–	*4*
Germany	3,600 – 5,700	95–99	–	0–19	
Italy	(2,500 – 5,000)	03	(0)	(0–19)	*4*
Liechtenstein	150 – 200	98–00	–	0–19	
Spain	(225,000 – 230,000)	92	?	–	*13,12,10*
Switzerland	**5,000 – 30,000**	93–96	0	0–19	
Total (approx.)	250,000 – 320,000	**Overall trend** Unknown			
Breeding range	>250,000 km²	**Gen. length** <3.3 **% Global pop.** 100			

Serinus citrinella is endemic to Europe, with its fragmented distribution confined to central and south-west regions of the continent. Its population is large (>250,000 pairs), and increased between 1970–1990. Although trends were not available for the populations in France or the Spanish stronghold during 1990–2000, there was no evidence to suggest that the species's status deteriorated. Consequently, it is provisionally evaluated as Secure.

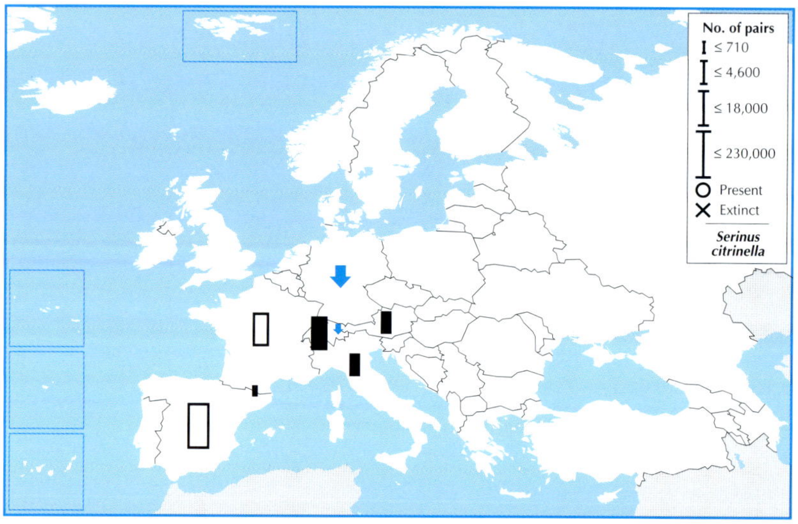

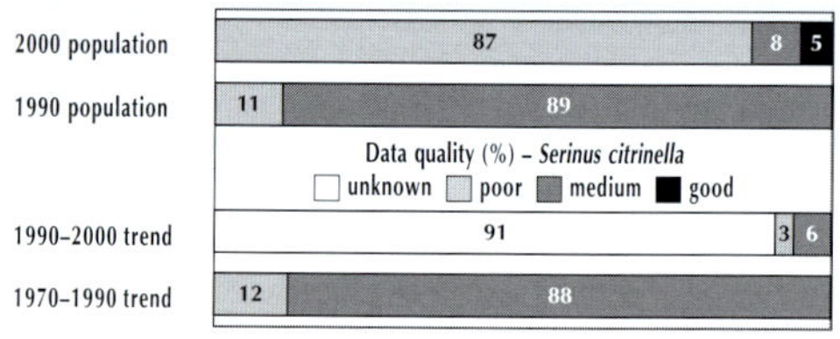

Serinus corsicana
CORSICAN FINCH

Non-SPEC^E (1994: NE) **Status** (Secure)
Criteria —
European IUCN Red List Category —
Criteria —
Global IUCN Red List Category —
Criteria —

Country	Breeding pop. size (pairs)	Year(s)	Trend	Mag.%	References
France	15,000 – 75,000	98–02	(0)	(0–19)	*4*
Italy	(4,000 – 10,000)	03	(0)	(0–19)	
Total (approx.)	19,000 – 85,000	**Overall trend** Stable			
Breeding range	>50,000 km²	**Gen. length** <3.3 **% Global pop.** 100			

Serinus corsicana is endemic to Europe, where it is confined to the Mediterranean islands of Sardinia, Corsica, Capraia and Elba. It was formerly treated as a subspecies of *S. citrinella*, but is now considered to be a species in its own right. Its breeding population is relatively small (<85,000 pairs), but was probably stable during 1990–2000. Consequently, the species is provisionally evaluated as Secure.

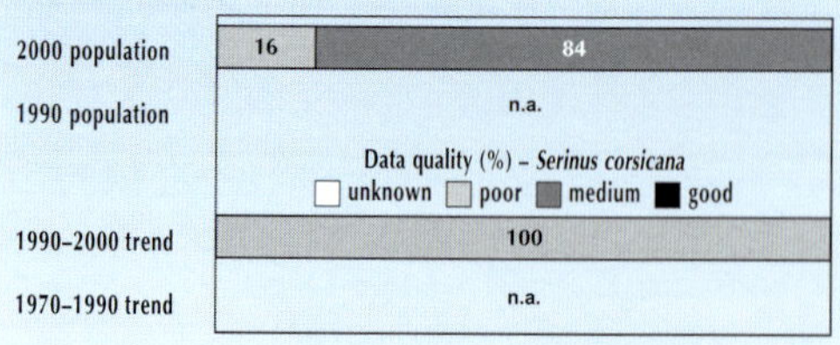

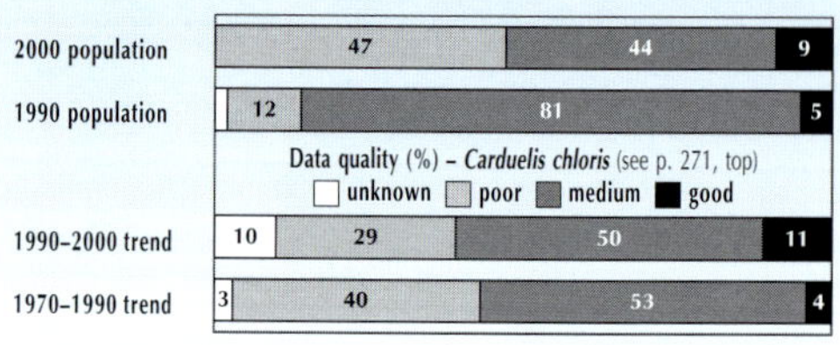

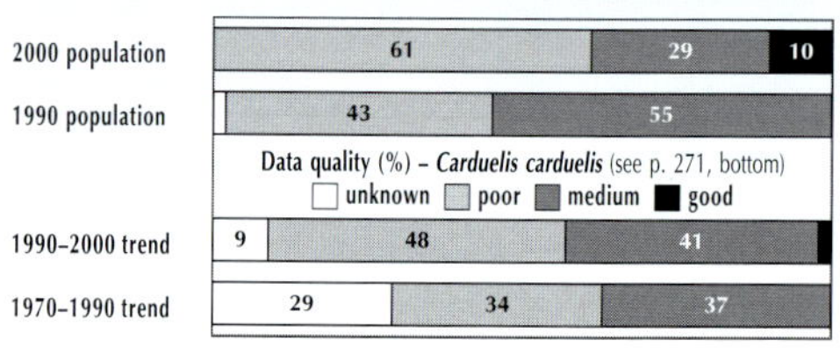

Carduelis chloris
EUROPEAN GREENFINCH

Non-SPEC[E] (1994: 4) **Status** Secure
Criteria —
European IUCN Red List Category —
Criteria —
Global IUCN Red List Category —
Criteria —

Carduelis chloris is a widespread breeder across most of Europe, which constitutes >75% of its global range. Its European breeding population is extremely large (>14,000,000 pairs), and was stable between 1970–1990. Although there were declines in a few countries—most notably France—during 1990–2000, key populations in Germany and Russia were stable, and trends were stable or increasing across most of the rest of Europe. The species hence remained stable overall, and consequently is evaluated as Secure.

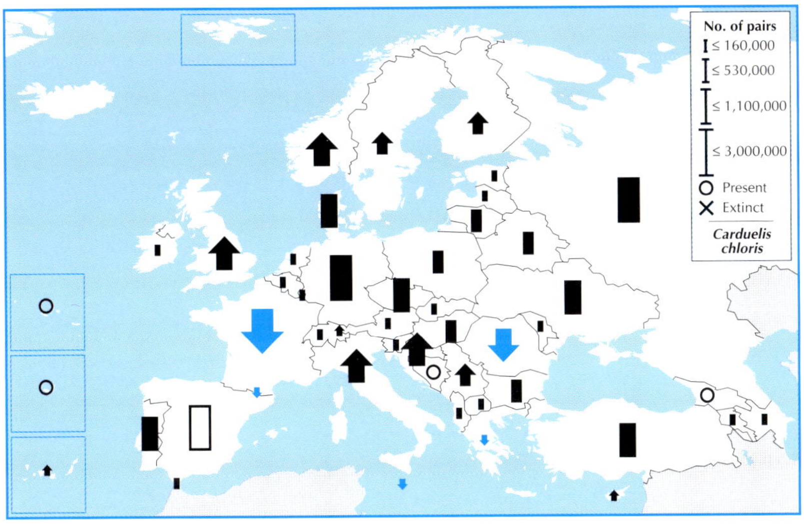

Country	Breeding pop. size (pairs)	Year(s)	Trend	Mag.%	References
Albania	20,000 – 50,000	02	(0)	(0–19)	
Andorra	50 – 100	99–01	(–)	(0–19)	*1,3*
Armenia	30,000 – 150,000	00–02	0	0–19	
Austria	(110,000 – 220,000)	98–02	(0)	(0–19)	
Azerbaijan	(100,000 – 200,000)	96–00	(0)	(0–19)	
Belarus	200,000 – 250,000	97–02	0	0–19	
Belgium	40,000 – 80,000	01–02	0	0–19	*1*
Bosnia & HG	Present	90–03	?	–	
Bulgaria	200,000 – 600,000	96–02	0	0–9	
Croatia	(500,000 – 1,000,000)	02	(+)	(>80)	*16*
Cyprus	(80,000 – 200,000)	94–02	(+)	(0–19)	
Czech Rep.	500,000 – 1,000,000	00	**0**	**0–19**	
Denmark	500,000 – 700,000	00	0	0–19	*12*
Estonia	(20,000 – 50,000)	98	0	0–19	*1*
Finland	300,000 – 400,000	98–02	**+**	**180**	
France	(1,500,000 – 6,000,000)	98–02	–	33	*4,2*
Georgia	Present	03	?	–	
Germany	1,500,000 – 3,000,000	95–99	0	0–19	
Greece	(50,000 – 200,000)	95–00	(–)	(0–19)	
Hungary	**445,000 – 585,000**	99–02	0	0–19	*19*
Rep. Ireland	100,000 – 250,000	88–91	0	0–19	
Italy	(400,000 – 800,000)	03	(+)	(0–19)	
Latvia	10,000 – 20,000	90–00	(0)	(0–19)	*23,16*
Liechtenstein	200 – 400	98–00	(+)	(0–9)	
Lithuania	(200,000 – 400,000)	99–01	(0)	(0–19)	*20*
Luxembourg	15,000 – 20,000	02	0	0–19	
Macedonia	(30,000 – 50,000)	90–00	(0)	(0–19)	
Malta	2 – 3	90–02	–	0–19	*1*
Moldova	35,000 – 40,000	90–00	0	0–19	
Netherlands	**50,000 – 100,000**	98–00	0	4	*1*
Norway	(500,000 – 1,000,000)	90–02	+	30–49	
Poland	**400,000 – 700,000**	00–02	0	0–19	*23*
Portugal	(200,000 – 2,000,000)	02	(0)	(0–19)	
Azores	Present	02	?	–	
Madeira	Present	02	?	–	
Romania	850,000 – 910,000	99–02	–	0–19	*48*
Russia	1,000,000 – 2,500,000	90–00	(0)	(0–19)	*122*
Serbia & MN	200,000 – 300,000	90–02	(+)	(0–9)	*1,29,172a,67a,78, 227,225,117a,164*
Slovakia	100,000 – 130,000	90–99	0	0–19	
Slovenia	50,000 – 80,000	94	(0)	(0–19)	
Spain	(1,060,000 – 3,600,000)	92	?	–	*13,12,15,10*
Canary Is.	(2,500 – 10,000)	97–03	(+)	(0–19)	*28,25*
Sweden	2500,00 – 650,000	99–00	**+**	**3**	
Switzerland	80,000 – 150,000	98–02	0	0–19	
Turkey	(600,000 – 2,000,000)	01	(0)	(0–19)	
Ukraine	640,000 – 820,000	90–00	0	5–9	
UK	**734,000 – 734,000**	00	**+**	**31**	*5,31*
Gibraltar	10 – 30	00	**0**	**0–19**	
Total (approx.)	14,000,000 – 32,000,000	**Overall trend** Stable			
Breeding range	>7,000,000 km²	**Gen. length.** <3.3		**% Global pop.** 75–94	

(See p. 270, bottom, for data quality graph)

Carduelis carduelis
EUROPEAN GOLDFINCH

Non-SPEC (1994: —) **Status** Secure
Criteria —
European IUCN Red List Category —
Criteria —
Global IUCN Red List Category —
Criteria —

Carduelis carduelis is a widespread resident across most of Europe, which accounts for less than half of its global range. Its European breeding population is extremely large (>12,000,000 pairs), and was stable between 1970–1990. Although there were declines in a few countries—notably Turkey—during 1990–2000, populations were stable or increased across the vast majority of Europe, and the species underwent a slight increase overall. Consequently, it is evaluated as Secure.

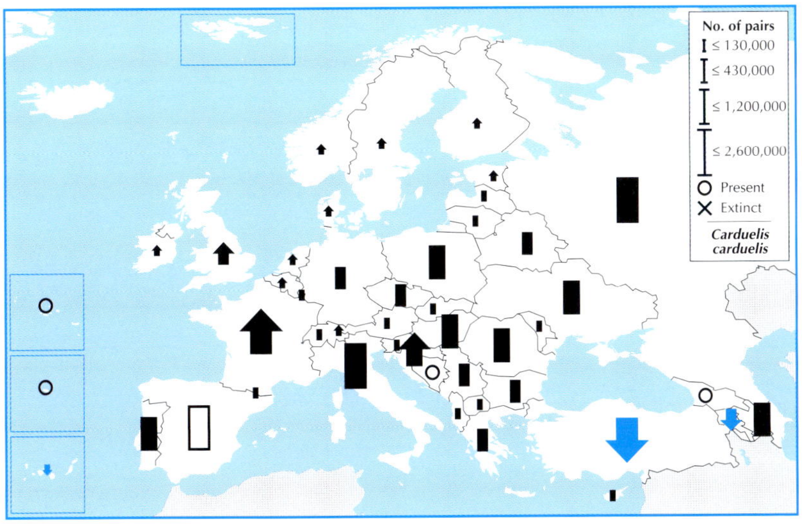

Country	Breeding pop. size (pairs)	Year(s)	Trend	Mag.%	References
Albania	20,000 – 50,000	02	(0)	(0–19)	
Andorra	(400 – 1,000)	99–01	(0)	(0–19)	*1,3*
Armenia	150,000 – 300,000	00–02	–	0–9	
Austria	(25,000 – 50,000)	98–02	(0)	(0–19)	
Azerbaijan	(500,000 – 1,000,000)	96–00	(0)	(0–19)	
Belarus	180,000 – 200,000	97–02	0	0–19	
Belgium	2,500 – 10,000	01–02	+	0–19	*1*
Bosnia & HG	Present	90–03	?	–	
Bulgaria	200,000 – 600,000	96–02	0	0–9	
Croatia	(500,000 – 1,000,000)	02	(+)	(>80)	*16*
Cyprus	(50,000 – 150,000)	94–02	(0)	(0–9)	
Czech Rep.	200,000 – 400,000	00	0	0–19	
Denmark	30,000 – 50,000	00	**+**	**>80**	*12*
Estonia	(20,000 – 30,000)	98	+	20–29	*1*
Finland	10,000 – 20,000	98–02	**+**	**150**	
France	(1,000,000 – 5,000,000)	00	+	26	*1,2*
Georgia	Present	03	?	–	
Germany	300,000 – 600,000	95–99	0	0–19	
Greece	(100,000 – 500,000)	95–00	(0)	(0–19)	
Hungary	**690,000 – 910,000**	99–02	0	0–19	*19*
Rep. Ireland	20,000 – 100,000	88–91	+	20–29	
Italy	(1,000,000 – 2,000,000)	03	(0)	(0–19)	
Latvia	15,000 – 50,000	90–00	0	0–19	*23,16,2*
Liechtenstein	30 – 60	98–00	+	0–19	
Lithuania	(50,000 – 100,000)	99–01	(0)	(0–19)	*20*
Luxembourg	6,000 – 8,000	02	0	0–19	
Macedonia	(10,000 – 50,000)	90–00	(0)	(0–19)	
Moldova	40,000 – 45,000	90–00	0	0–19	
Netherlands	**15,000 – 20,000**	98–00	**+**	**116**	*1*
Norway	(1,000 – 5,000)	90–02	+	0–19	*49*
Poland	**400,000 – 800,000**	00–02	0	0–19	*23*
Portugal	(500,000 – 2,500,000)	02	(0)	(0–19)	
Azores	Present	02	?	–	
Madeira	Present	02	?	–	
Romania	887,000 – 964,000	99–02	0	0–19	*48*
Russia	1,000,000 – 2,000,000	90–00	(0)	(0–19)	*122*
Serbia & MN	250,000 – 350,000	90–02	0	0–19	*1,29,172a,67a,78, 227,225,117a,164*
Slovakia	100,000 – 150,000	90–99	0	0–19	
Slovenia	50,000 – 60,000	94	(0)	(0–19)	
Spain	(800,000 – 2,900,000)	92	?	–	*13,12,15,10*
Canary Is.	(2,500 – 10,000)	97–03	(–)	(0–19)	*28,25*
Sweden	5,000 – 10,000	99–00	+	10–19	
Switzerland	**20,000 – 50,000**	93–96	0	0–19	
Turkey	(1,500,000 – 4,500,000)	01	(–)	(0–19)	
Ukraine	560,000 – 720,000	90–00	0	5–9	
UK	**313,000 – 313,000**	00	**+**	**36**	*5,31*
Total (approx.)	12,000,000 – 29,000,000	**Overall trend** Small increase			
Breeding range	>7,000,000 km²	**Gen. length.** <3.3		**% Global pop.** 25–49	

(See p. 270, bottom, for data quality graph)

Carduelis spinus
EURASIAN SISKIN

Non-SPEC[E] (1994: 4) **Status** Secure
Criteria —
European IUCN Red List Category —
Criteria —
Global IUCN Red List Category —
Criteria —

Carduelis spinus is a widespread breeder across most of Europe, which constitutes >75% of its global range. Its European breeding population is extremely large (>10,000,000 pairs), and was stable between 1970–1990. The stronghold population in Russia fluctuated during 1990–2000, and most other European populations—including sizeable ones in Finland and Sweden—either increased or were stable. The species probably remained broadly stable overall, and consequently is provisionally evaluated as Secure.

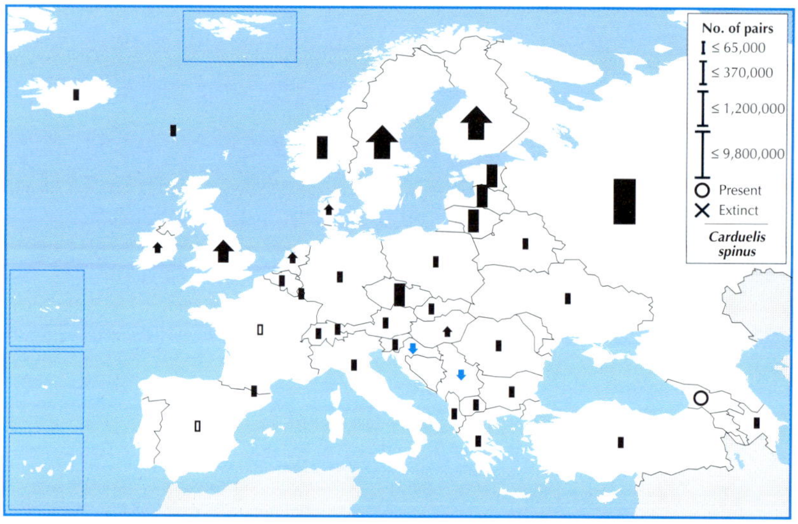

Country	Breeding pop. size (pairs)	Year(s)	Trend	Mag.%	References
Albania	(0 – 50)	02	(0)	(0–19)	
Andorra	0 – 20	99–01	(0)	(0–19)	*1,3*
Austria	(30,000 – 140,000)	98–02	(F)	(>80)	
Azerbaijan	(1,000 – 10,000)	96–00	(0)	(0–19)	
Belarus	10,000 – 40,000	97–02	0	0–19	
Belgium	160 – 1,100	95–02	0	0–19	*1*
Bulgaria	2,000 – 4,000	96–02	0	0–19	
Croatia	(5,000 – 10,000)	02	(–)	(30–49)	*70,16*
Czech Rep.	90,000 – 180,000	00	0	0–19	
Denmark	(200 – 2,000)	00	(+)	(–)	*12,3*
Faroe Is.	(0 – 4)	85	(0)	(0–19)	
Estonia	100,000 – 250,000	98	F	20–29	*1*
Finland	700,000 – 2,000,000	98–02	+	**35**	
France	500 – 2,500	98–02	?	–	*4*
Georgia	Present	03	?	–	
Germany	25,000 – 100,000	95–99	F	50–79	
Greece	(500 – 2,000)	95–00	(F)	(>80)	
Hungary	(200 – 300)	90–93	+	20–49	*14*
Iceland	**0 – 4**	90–00	(F)	(–)	*24,25,45,46,47, 42,43,44*
Rep. Ireland	20,000 – 100,000	88–91	+	20–29	
Italy	(4,000 – 15,000)	03	(0)	(0–19)	
Latvia	100,000 – 200,000	90–00	(0)	(0–19)	*23,18*
Liechtenstein	(10 – 20)	98–00	(F)	(–)	
Lithuania	(100,000 – 300,000)	99–01	(0)	(0–19)	*20*
Luxembourg	1 – 10	02	0	0–19	
Macedonia	(250 – 1,000)	90–00	(0)	(0–19)	
Netherlands	**500 – 1,200**	98–00	+	**236**	*1*
Norway	(100,000 – 1,000,000)	90–03	0	0–19	*27*
Poland	(10,000 – 20,000)	00–02	0	0–19	*23*
Romania	24,000 – 65,000	99–02	(0)	(0–19)	*48*
Russia	8,000,000 – 12,000,000	90–00	(F)	(20–29)	*122*
Serbia & MN	60 – 100	90–02	(–)	(0–9)	*1,241,155,225, 227,62,117a*
Slovakia	20,000 – 40,000	90–99	0	0–19	
Slovenia	2,000 – 3,000	94	(0)	(0–19)	
Spain	(500 – 1,100)	92	?	–	*13,12,10*
Sweden	500,000 – 1,000,000	99–00	+	**10–19**	
Switzerland	**5,000 – 10,000**	93–96	(F)	(30–49)	
Turkey	(5,000 – 20,000)	01	(0)	(0–19)	
Ukraine	(4,000 – 20,000)	90–00	(F)	(30–49)	
UK	**369,000 – 369,000**	00	+	**19**	*22,31*

Total (approx.) 10,000,000 – 18,000,000 **Overall trend** Fluctuating
Breeding range >5,000,000 km² **Gen. length.** <3.3 **% Global pop.** 75–94
(See p. 273, top, for data quality graph)

Carduelis cannabina
EURASIAN LINNET

SPEC 2 (1994: 4) **Status** Declining
Criteria Moderate recent decline
European IUCN Red List Category —
Criteria —
Global IUCN Red List Category —
Criteria —

Carduelis cannabina is a widespread breeder across much of Europe, which constitutes >50% of its global breeding range. Its European breeding population is extremely large (>10,000,000 pairs), and was stable between 1970–1990. Although it was stable or increased across much of Europe during 1990–2000, there were declines in several north-western European populations—most notably the sizeable one in France—and the species underwent a moderate decline (>10%) overall. Consequently, this previously Secure species is now evaluated as Declining.

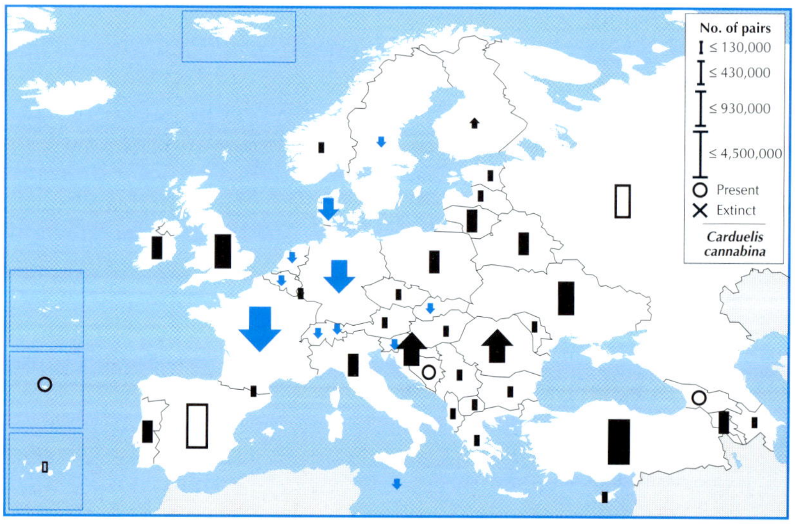

Country	Breeding pop. size (pairs)	Year(s)	Trend	Mag.%	References
Albania	3,000 – 10,000	02	(0)	(0–19)	
Andorra	(300 – 1,000)	99–01	(0)	(0–19)	*1,3*
Armenia	150,000 – 300,000	00–02	0	0–19	
Austria	(12,000 – 24,000)	98–02	(0)	(0–19)	
Azerbaijan	(50,000 – 200,000)	96–00	(0)	(0–19)	
Belarus	130,000 – 180,000	97–02	0	0–19	
Belgium	30,000 – 60,000	01–02	–	0–19	*1*
Bosnia & HG	Present	90–03	?	–	
Bulgaria	40,000 – 100,000	96–02	0	0–9	
Croatia	(500,000 – 800,000)	02	(+)	(>80)	*16*
Cyprus	(10,000 – 30,000)	94–02	(0)	(0–9)	
Czech Rep.	60,000 – 120,000	00	0	0–19	
Denmark	150,000 – 300,000	00	–	20–29	*12*
Estonia	(20,000 – 40,000)	98	0	0–19	*1*
Finland	20,000 – 30,000	98–02	+	**100**	
France	(1,000,000 – 5,000,000)	98–02	**–**	**49**	*4,2*
Georgia	Present	03	?	–	
Germany	380,000 – 830,000	95–99	–	20–29	
Greece	(50,000 – 100,000)	95–00	(0)	(0–19)	
Hungary	**85,000 – 150,000**	99–02	0	0–19	*19*
Rep. Ireland	100,000 – 250,000	88–91	0	0–19	
Italy	(100,000 – 400,000)	03	(0)	(0–19)	
Latvia	10,000 – 25,000	90–00	0	0–19	*23,16,2*
Liechtenstein	50 – 80	98–00	–	0–19	
Lithuania	(150,000 – 300,000)	99–01	(0)	(0–19)	*20*
Luxembourg	14,000 – 18,000	02	0	0–19	
Macedonia	(5,000 – 20,000)	90–00	(0)	(0–19)	
Malta	5 – 10	90–02	**–**	**20–29**	*1*
Moldova	25,000 – 30,000	90–00	0	0–19	
Netherlands	**40,000 – 50,000**	98–00	**–**	**17**	*1*
Norway	(10,000 – 15,000)	90–02	(0)	(0–19)	
Poland	**300,000 – 600,000**	00–02	0	0–19	*23*
Portugal	(100,000 – 1,000,000)	02	(0)	(0–19)	
Madeira	Present	02	?	–	
Romania	780,000 – 1,100,000	99–02	+	0–19	*48*
Russia	500,000 – 1,000,000	90–00	?	–	*122*
Serbia & MN	35,000 – 50,000	95–02	(0)	(0–19)	*1,29,155,225,227, 40,62,152,184, 185,117a*
Slovakia	40,000 – 60,000	90–99	–	20–29	
Slovenia	5,000 – 10,000	00	(–)	(0–19)	
Spain	(1,700,000 – 3,300,000)	92	?	–	*13,12,15,10*
Canary Is.	(10,000 – 20,000)	97–03	?	–	*28,25*
Sweden	100,000 – 150,000	99–00	**–**	**37**	
Switzerland	**30,000 – 60,000**	93–96	–	10–19	
Turkey	(2,000,000 – 10,000,000)	01	(0)	(0–19)	
Ukraine	800,000 – 1,000,000	90–00	0	0–19	
UK	**556,000 – 556,000**	00	**0**	**3**	*5,31*

Total (approx.) 10,000,000 – 28,000,000 **Overall trend** Moderate decline
Breeding range >7,000,000 km² **Gen. length.** <3.3 **% Global pop.** 50–74
(See p. 273, top, for data quality graph)

Carduelis flavirostris
TWITE

Non-SPEC (1994: —) **Status** Secure
Criteria —
European IUCN Red List Category —
Criteria —
Global IUCN Red List Category —
Criteria —

Country	Breeding pop. size (pairs)	Year(s)	Trend	Mag.%	References
Armenia	20,000 – 30,000	99–02	0	0–19	
Azerbaijan	(1,000 – 10,000)	96–00	(0)	(0–19)	
Finland	(0 – 10)	98–02	(0)	(0–19)	
Georgia	Present	03	?	–	
Rep. Ireland	250 – 1,000	88–91	(–)	(50–79)	
Norway	(100,000 – 500,000)	90–03	0	0–19	
Russia	(2,000 – 5,000)	90–00	?	–	*17*
Sweden	(10 – 100)	99–00	?	–	
Turkey	(40,000 – 200,000)	01	(0)	(0–19)	
UK	**7,600 – 16,700**	99	(–)	(0–19)	*17*
Total (approx.)	170,000 – 760,000	**Overall trend** Stable			
Breeding range	>750,000 km²	**Gen. length.** <3.3 **% Global pop.** 25–49			

Carduelis flavirostris has a disjunct breeding distribution in Europe, occupying the north-west, the Caucasus and adjacent parts of Russia and Turkey, with Europe accounting for less than half of its global breeding range. Its European breeding population is large (>170,000 pairs), and was stable between 1970–1990. Although there were declines in the Republic of Ireland and the United Kingdom during 1990–2000, key populations in Norway and Turkey were stable, and the species remained stable overall. Consequently, it is evaluated as Secure.

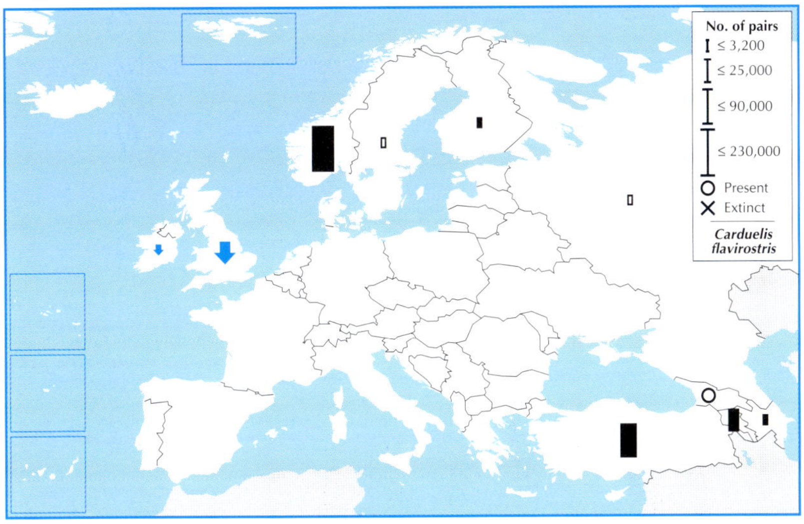

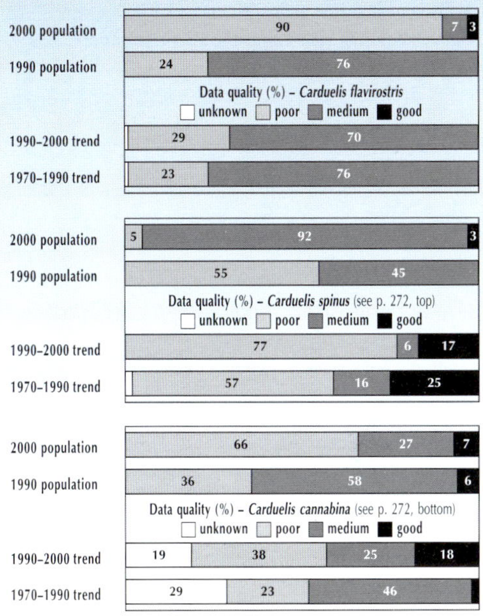

Carduelis flammea
COMMON REDPOLL

Non-SPEC (1994: —) **Status** (Secure)
Criteria —
European IUCN Red List Category —
Criteria —
Global IUCN Red List Category —
Criteria —

Country	Breeding pop. size (pairs)	Year(s)	Trend	Mag.%	References
Austria	(15,000 – 30,000)	98–02	(0)	(0–19)	
Belgium	50 – 250	95–02	–	0–19	*1*
Czech Rep.	6,000 – 12,000	00	0	0–19	
Denmark	(10,000 – 20,000)	93–96	(+)	(50–79)	*3*
Greenland	(100,000 – 500,000)	90–00	(0)	(0–19)	
Finland	200,000 – 500,000	98–02	**–**	**15**	
France	(5,000 – 10,000)	00	?	–	*1*
Germany	7,000 – 17,000	95–99	+	30–49	
Iceland	**2,000 – 20,000**	78–94	?	–	
Rep. Ireland	20,000 – 100,000	88–91	+	20–29	
Italy	(20,000 – 50,000)	03	(0)	(0–19)	
Liechtenstein	100 – 150	98–00	–	0–19	
Netherlands	**200 – 300**	98–00	–	50–79	*1*
Norway	(100,000 – 2,000,000)	90–02	F	30–49	
Svalbard	0 – 2	95–03	0	0–19	
Poland	**120 – 180**	95–00	+	0–19	*2,104,105*
Russia	(8,000,000 – 15,000,000)	90–00	?	–	*122*
Slovakia	300 – 600	90–99	0	0–19	
Slovenia	500 – 1,000	94	(0)	(0–19)	
Sweden	250,000 – 1,000,000	99–00	(F)	(20–29)	
Switzerland	10,000 – 15,000	93–96	0	0–19	
UK	26,900 – 26,900	00	–	84	*31*
Total (approx.)	8,800,000 – 19,000,000	**Overall trend** Unknown			
Breeding range	>2,000,000 km²	**Gen. length.** <3.3 **% Global pop.** 25–49			

Carduelis flammea is a widespread resident across much of northern Europe (occurring more patchily farther south), which accounts for less than half of its global range. Its European breeding population is very large (>8,800,000 pairs), and was broadly stable between 1970–1990. Although the trend of the stronghold population in Russia during 1990–2000 was unknown, the species was broadly stable in most other key populations in its European range, and there was no evidence to suggest that it declined significantly overall. Consequently, it is provisionally evaluated as Secure.

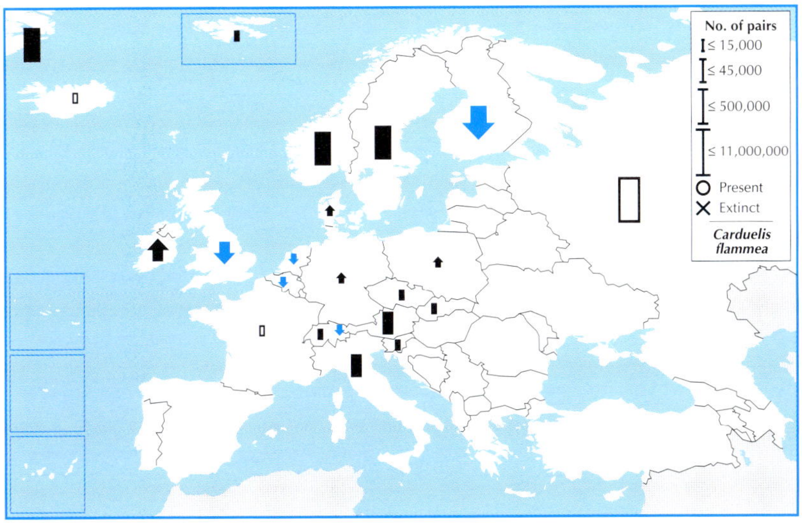

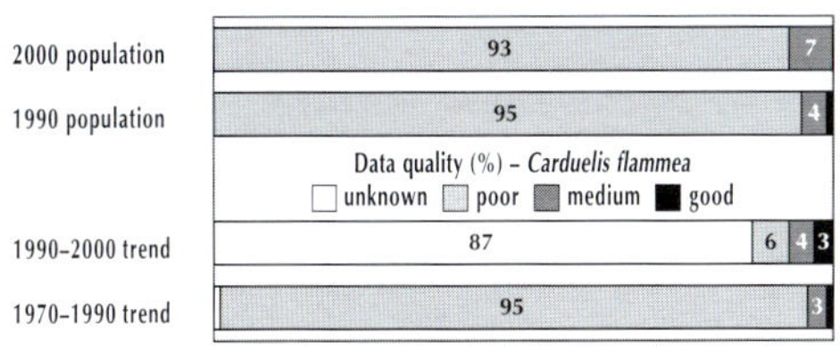

Carduelis hornemanni
HOARY REDPOLL

Non-SPEC (1994: —) **Status** (Secure)
Criteria —
European IUCN Red List Category —
Criteria —
Global IUCN Red List Category —
Criteria —

Country	Breeding pop. size (pairs)	Year(s)	Trend	Mag.%	References
Denmark					
Greenland	(1,000 – 10,000)	90–00	(0)	(0–19)	
Finland	(1,000 – 5,000)	98–02	(0)	(0–19)	
Norway	(1,000 – 10,000)	02	F	30–49	*66*
Russia	(80,000 – 150,000)	90–00	?	–	*122*
Sweden	1,000 – 5,000	99–00	?	–	
Total (approx.)	84,000 – 180,000	**Overall trend** Unknown			
Breeding range	>250,000 km²	**Gen. length.** <3.3 **% Global pop.** 25–49			

Carduelis hornemanni is resident in Greenland, northern Fennoscandia and arctic Russia, with Europe accounting for less than half of its global range. Its European breeding population is relatively small (<180,000 pairs), but was stable between 1970–1990. Although the trend of the stronghold population in Russia during 1990–2000 was unknown, the species was stable in Greenland, Sweden and Finland, and there was no evidence to suggest that it declined overall. Consequently, it is provisionally evaluated as Secure.

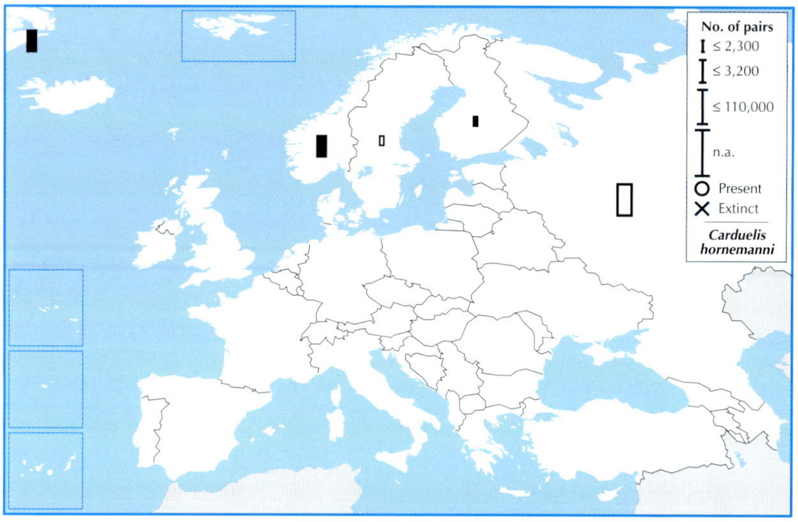

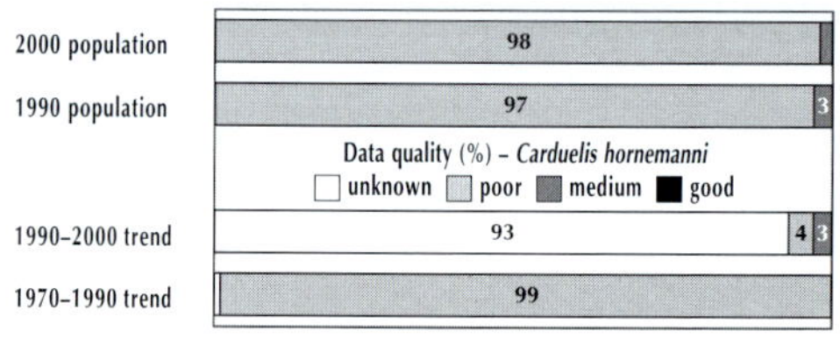

Loxia leucoptera
TWO-BARRED CROSSBILL

Non-SPEC (1994: —) **Status** (Secure)
Criteria —
European IUCN Red List Category —
Criteria —
Global IUCN Red List Category —
Criteria —

Country	Breeding pop. size (pairs)	Year(s)	Trend	Mag.%	References
Finland	1,000 – 20,000	98–02	**+**	**50**	
Norway	(0 – 500)	90–03	**F**	**>80**	
Russia	(1,000,000 – 2,850,000)	90–00	(F)	(–)	*122*
Sweden	(500 – 5,000)	99–00	(F)	(>80)	
Total (approx.)	1,000,000 – 2,900,000	**Overall trend** Fluctuating			
Breeding range	>750,000 km²	**Gen. length.** <3.3 **% Global pop.** 5–24			

Loxia leucoptera is a widespread resident in the boreal zone of Russia and Fennoscandia, with Europe accounting for less than a quarter of its global range. Its European breeding population is very large (>1,000,000 pairs), and was stable between 1970–1990. The species remained broadly stable overall during 1990–2000, with fluctuations in Russia, Sweden and Norway, and increases in Finland. Consequently, the species is provisionally evaluated as Secure.

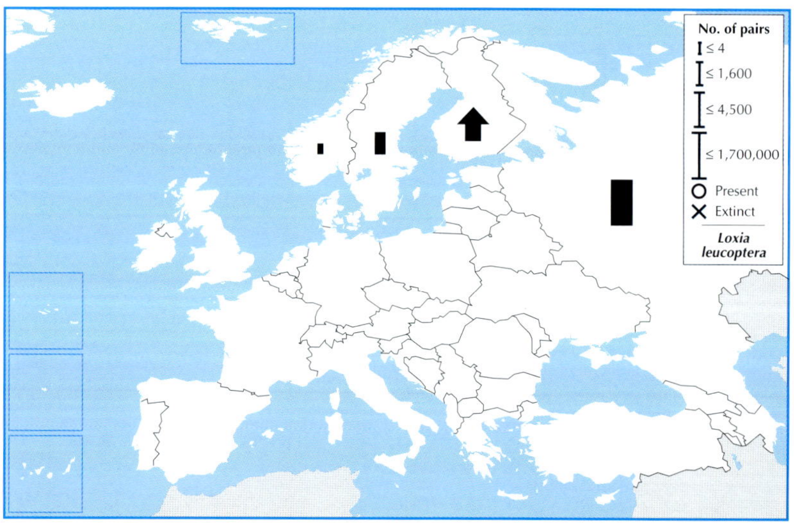

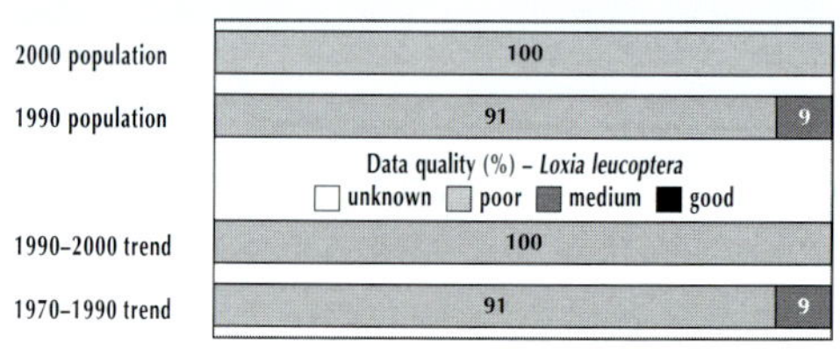

Loxia curvirostra
RED CROSSBILL

Non-SPEC (1994: —) **Status** (Secure)
Criteria —
European IUCN Red List Category —
Criteria —
Global IUCN Red List Category —
Criteria —

Loxia curvirostra is a widespread resident across much of northern Europe, and also occurs more patchily farther south, with Europe accounting for less than a quarter of its global range. Its European breeding population is very large (>5,800,000 pairs), and was broadly stable between 1970–1990. Although there were widespread fluctuations during 1990–2000, the vast majority of national populations—including key ones in Russia and Fennoscandia—were broadly stable, and consequently the species is provisionally evaluated as Secure.

Country	Breeding pop. size (pairs)	Year(s)	Trend	Mag.%	References
Albania	(500 – 2,000)	02	(0)	(0–19)	
Andorra	(300 – 1,000)	99–01	(0)	(0–19)	*1,3*
Armenia	50 – 100	95–02	?	–	
Austria	(60,000 – 150,000)	98–02	(F)	(>80)	
Azerbaijan	(0 – 250)	96–00	(0)	(0–19)	
Belarus	2,000 – 20,000	97–02	0	0–19	
Belgium	2,500 – 10,000	01–02	(0)	(0–19)	*1*
Bulgaria	3,000 – 6,000	96–02	F	30–49	
Croatia	(2,500 – 5,000)	02	(–)	(30–49)	*16*
Cyprus	(1,000 – 3,000)	94–02	(0)	(0–9)	
Czech Rep.	30,000 – 100,000	00	0	0–19	
Denmark	(1,000 – 3,000)	00	(F)	(–)	*12,3*
Estonia	(1,000 – 50,000)	98	F	20–29	*1*
Finland	50,000 – 400,000	98–02	**F**	**500–1,000**	
France	15,000 – 75,000	98–02	F	20–49	*4*
Georgia	Present	03	?	–	
Germany	27,000 – 100,000	95–99	F	50–79	
Greece	(5,000 – 10,000)	95–00	(0)	(0–19)	
Hungary	200 – 400	95–02	F	50–79	
Rep. Ireland	(1,000 – 2,500)	88–91	(F)	(50–79)	
Italy	(30,000 – 60,000)	03	(0)	(0–19)	
Latvia	1,000 – 15,000	90–00	(F)	(–)	*18*
Liechtenstein	(50 – 100)	98–00	(0)	(0–19)	
Lithuania	(3,000 – 30,000)	99–01	(0)	(0–19)	*20*
Luxembourg	(0 – 50)	02	F	>80	
Macedonia	(1,000 – 5,000)	90–00	(0)	(0–19)	
Netherlands	**100 – 3,500**	98–00	**0**	**4**	*1*
Norway	(100,000 – 500,000)	90–02	**F**	**>80**	
Poland	(2,500 – 10,000)	90–00	F	>80	*2*
Portugal	(0 – 100)	02	(F)	(–)	
Romania	(125,000 – 680,000)	00–02	(0)	(0–19)	
Russia	(5,000,000 – 10,000,000)	90–00	(F)	(20–29)	*122*
Serbia & MN	4,000 – 6,700	95–02	(F)	(10–39)	*1,172a,67a,155, 225,227,117a,185, 62*
Slovakia	25,000 – 50,000	90–99	F	20–29	
Slovenia	5,000 – 10,000	00	(0)	(0–19)	
Spain	(140,000 – 190,000)	92	?	–	*13,12,10*
Sweden	100,000 – 500,000	99–00	(F)	(>80)	
Switzerland	30,000 – 60,000	98–02	F	30–49	
Turkey	(10,000 – 30,000)	01	(0)	(0–19)	
Ukraine	(3,600 – 8,000)	90–00	F	30–49	
UK	**1,000 – 20,000**	68–90	(0)	(0–19)	

Total (approx.) 5,800,000 – 13,000,000 **Overall trend** Fluctuating
Breeding range >5,000,000 km² **Gen. length.** <3.3 **% Global pop.** 5–24
(See p. 275, bottom, for data quality graph)

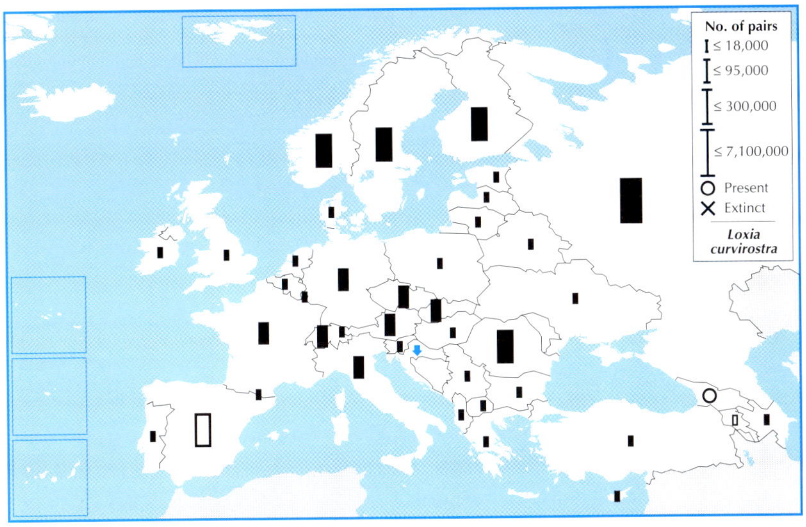

Loxia scotica
SCOTTISH CROSSBILL

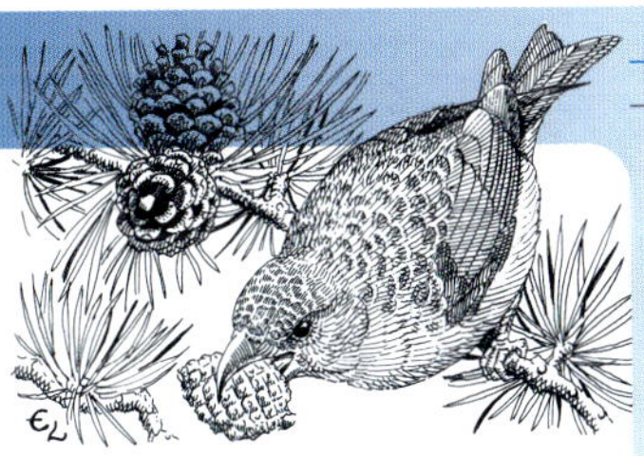

SPEC 1 (1994: 1) **Status** Data Deficient
Criteria —
European IUCN Red List Category DD
Criteria —
Global IUCN Red List Category DD
Criteria —

Country	Breeding pop. size (pairs)	Year(s)	Trend	Mag.%	References
UK	300 – 1,250	88	?	–	

Total (approx.) 300 – 1,250 **Overall trend** Unknown
Breeding range >20,000 km² **Gen. length.** <3.3 **% Global pop.** 100

Loxia scotica is endemic to Europe, where it is resident in northern Scotland (United Kingdom). Its breeding population is probably very small (as few as 300 pairs), but trend data were not available for either 1970–1990 or 1990–2000. Consequently, this poorly known species is categorised as Data Deficient.

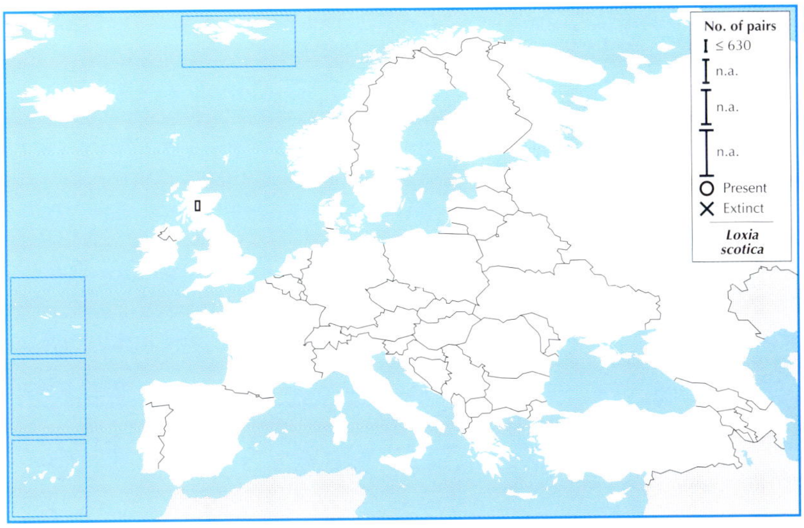

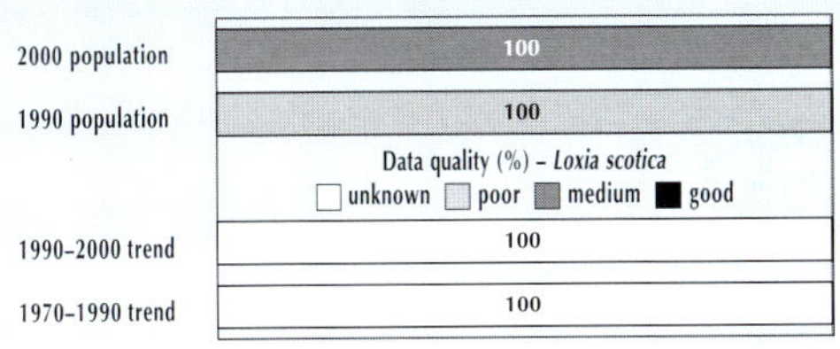

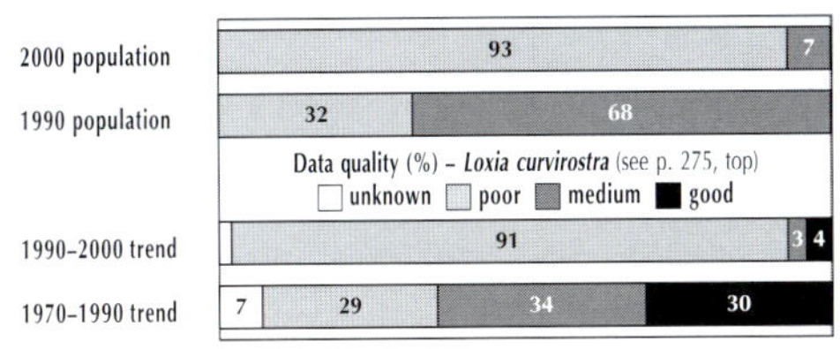

Loxia pytyopsittacus
PARROT CROSSBILL

Non-SPEC[E] (1994: 4) **Status** (Secure)
Criteria —
European IUCN Red List Category —
Criteria —
Global IUCN Red List Category —
Criteria —

Country	Breeding pop. size (pairs)	Year(s)	Trend	Mag.%	References
Denmark	(0 – 1)	99	**0**	**0–19**	*5*
Estonia	(2,000 – 5,000)	98	F	20–29	*1*
Finland	10,000 – 100,000	98–02	**0**	**0–19**	
Germany	**1 – 1**	95–99	F	>80	
Latvia	(400 – 1,000)	90–00	(F)	(–)	*23,18*
Netherlands	**0 – 10**	98–00	?	–	*1*
Norway	(10,000 – 100,000)	90–03	**F**	**>80**	
Russia	(230,000 – 850,000)	90–00	(F)	(20–29)	*122*
Sweden	10,000 – 50,000	99–00	F	>80	
UK	(0 – 50)	89–01	?	–	*43*
Total (approx.)	260,000 – 1,100,000	**Overall trend** Fluctuating			
Breeding range	>2,000,000 km²	**Gen. length.** <3.3	**% Global pop.** >95		

Loxia pytyopsittacus is a widespread resident across much of northern Europe, which constitutes >95% of its global range. Its European breeding population is large (>260,000 pairs), and was broadly stable between 1970–1990. Despite fluctuations, the stronghold population in Russia was broadly stable during 1990–2000, as were populations in Fennoscandia and elsewhere. Consequently, the species is provisionally evaluated as Secure.

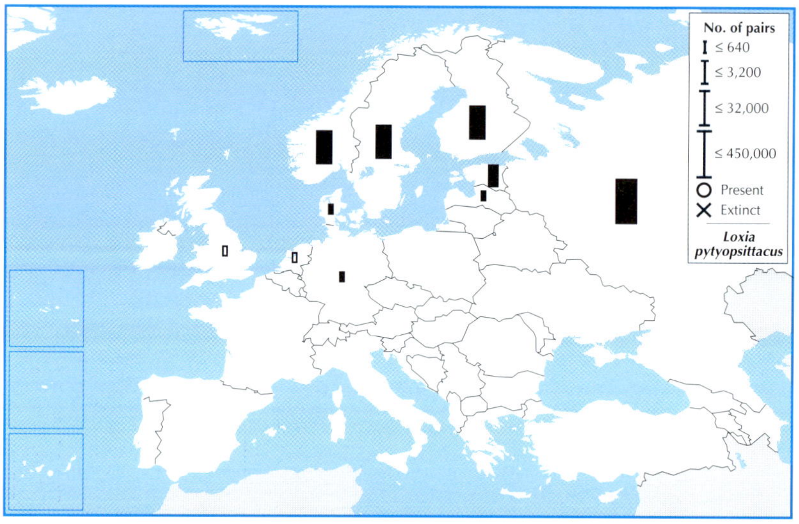

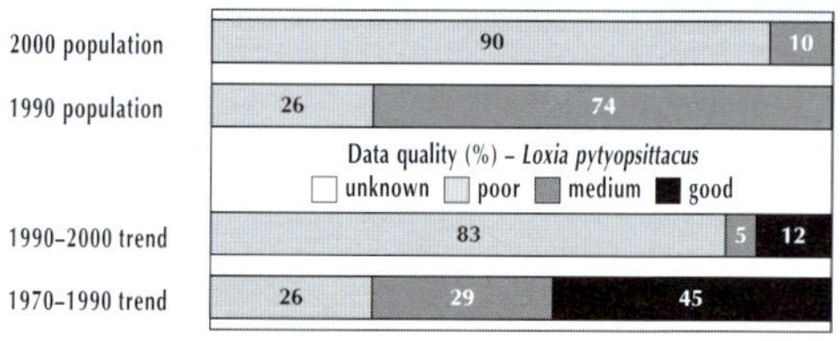

Rhodopechys sanguinea
CRIMSON-WINGED FINCH

Non-SPEC (1994: —) **Status** Secure
Criteria —
European IUCN Red List Category —
Criteria —
Global IUCN Red List Category —
Criteria —

Country	Breeding pop. size (pairs)	Year(s)	Trend	Mag.%	References
Armenia	1,000 – 1,500	97–02	+	0–9	
Azerbaijan	(1,000 – 10,000)	96–00	(0)	(0–19)	
Georgia	Present	03	?	–	
Turkey	(300,000 – 900,000)	01	+	0–19	
Total (approx.)	300,000 – 910,000	**Overall trend** Small increase			
Breeding range	>250,000 km²	**Gen. length.** <3.3	**% Global pop.** 25–49		

Rhodopechys sanguinea is a patchily distributed resident in Turkey and the Caucasus, with Europe accounting for less than half of its global range. Its European breeding population is large (>300,000 pairs), but its trend between 1970–1990 was unknown. Although trend data were not available for Georgia during 1990–2000, the species increased in Armenia and Turkey, and probably underwent a small increase overall. Consequently, it is evaluated as Secure.

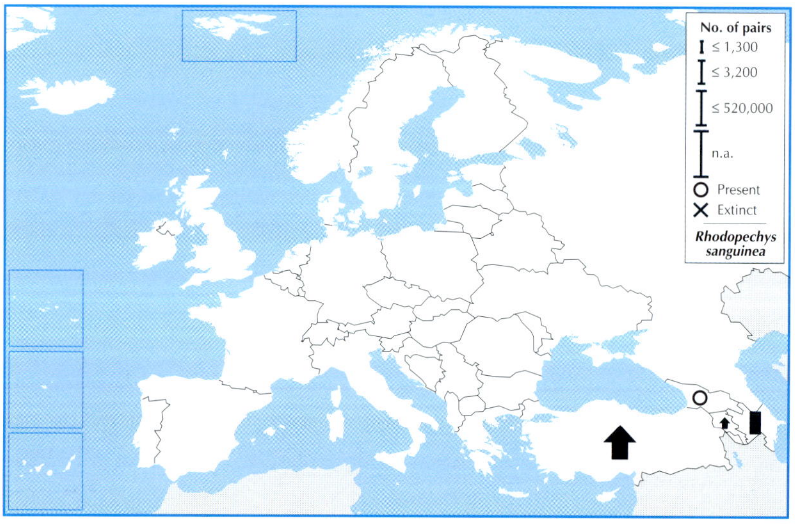

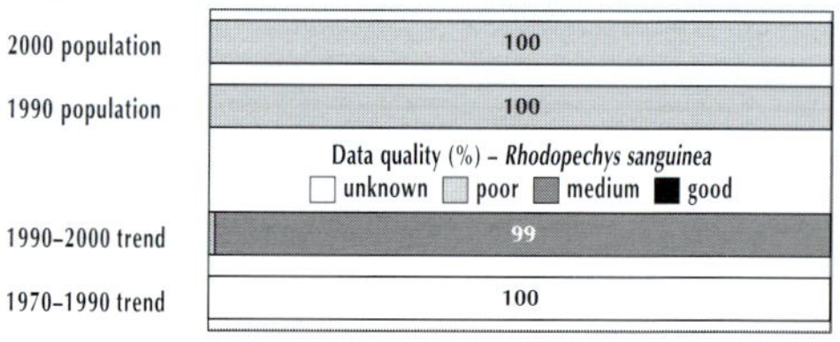

Rhodopechys obsoleta
DESERT FINCH

Non-SPEC (1994: —) **Status** (Secure)
Criteria —
European IUCN Red List Category —▼
Criteria —
Global IUCN Red List Category —
Criteria —

Country	Breeding pop. size (pairs)	Year(s)	Trend	Mag.%	References
Turkey	(3,000 – 6,000)	01	(+)	(0–19)	
Total (approx.)	3,000 – 6,000	**Overall trend** Small increase			
Breeding range	>20,000 km²	**Gen. length** <3.3	**% Global pop.** <5		

Rhodopechys obsoleta has a predominantly Asian distribution, which just extends into Europe in southern Turkey. Its European breeding population is small (as few as 3,000 pairs), and although its trend between 1970–1990 was unknown, it probably increased slightly during 1990–2000. Although the size of the Turkish population could render it susceptible to the risks affecting small populations, it is marginal to a much larger non-European population. Consequently, the species is provisionally evaluated as Secure.

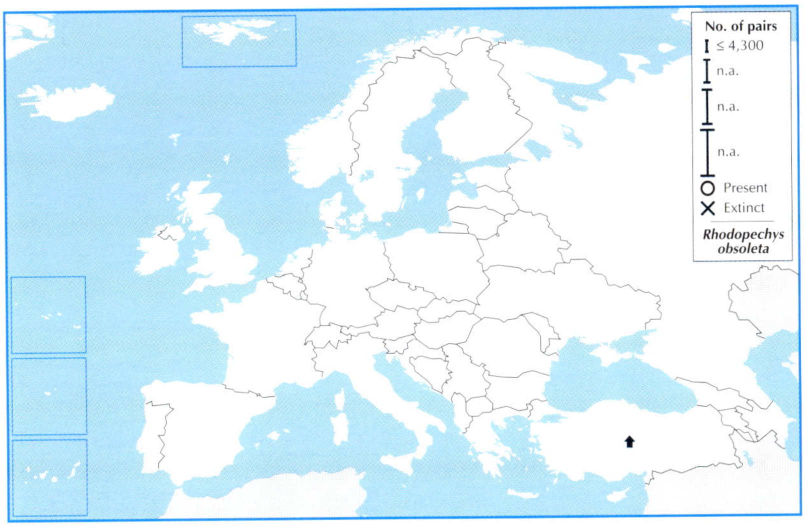

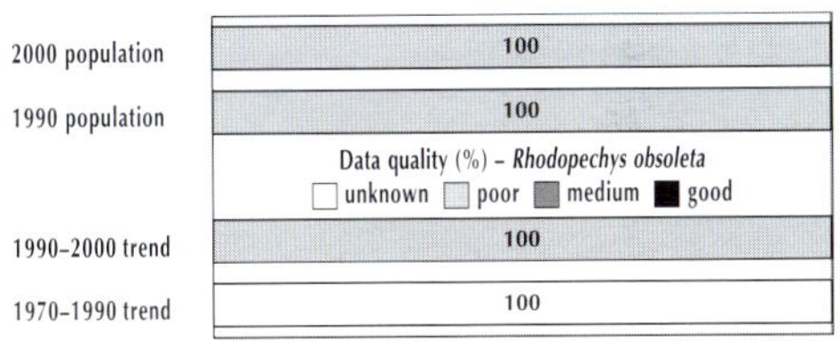

Rhodopechys mongolica
MONGOLIAN FINCH

Non-SPEC (1994: —) **Status** (Secure)
Criteria —
European IUCN Red List Category —▼
Criteria —
Global IUCN Red List Category —
Criteria —

Country	Breeding pop. size (pairs)	Year(s)	Trend	Mag.%	References
Armenia	**20 – 25**	00–02	?	-	*13*
Turkey	(500 – 5,000)	01	(+)	(0–19)	
Total (approx.)	520 – 5,000	**Overall trend** Small increase			
Breeding range	>20,000 km²	**Gen. length** <3.3	**% Global pop.** <5		

Rhodopechys mongolica has a predominantly Asian distribution, which just extends into Europe in Armenia and eastern Turkey. Its European breeding population is small (as few as 520 pairs), but its trend between 1970–1990 was unknown. The species increased in its Turkish stronghold during 1990–2000, and probably increased slightly overall. Although the size of the European population could render it susceptible to the risks affecting small populations, it is marginal to a much larger non-European population. Consequently, the species is provisionally evaluated as Secure.

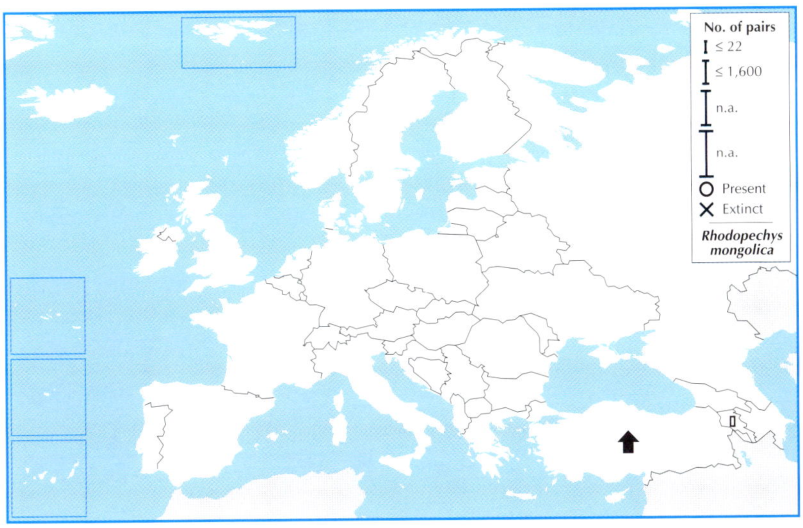

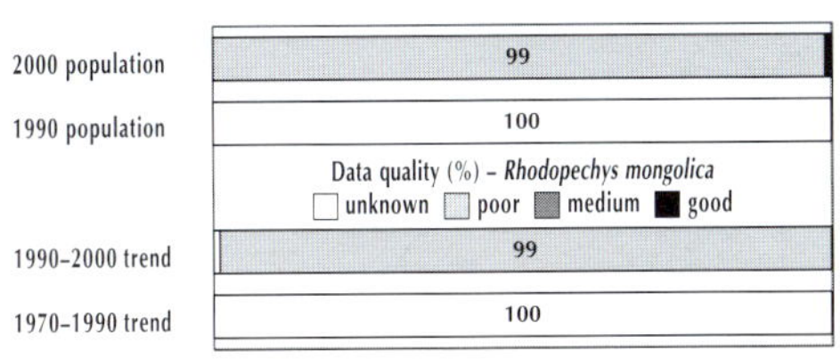

Bucanetes githagineus
TRUMPETER FINCH

Non-SPEC (1994: 3) **Status** (Secure)
Criteria —
European IUCN Red List Category —
Criteria —
Global IUCN Red List Category —
Criteria —

Country	Breeding pop. size (pairs)	Year(s)	Trend	Mag.%	References
Armenia	50 – 100	95–02	0	0–19	
Azerbaijan	(10 – 100)	96–00	(0)	(0–19)	
Spain	(300 – 500)	92	(+)	(0–19)	*8,12,10*
Canary Is.	(10,000 – 20,000)	97–03	(–)	(0–19)	*28,25*
Turkey	(200 – 800)	01	?	–	
Total (approx.)	11,000 – 21,000	**Overall trend** Small decline			
Breeding range	>20,000 km²	**Gen. length.** <3.3	**% Global pop.** <5		

Bucanetes githagineus breeds patchily in the Canary Islands, Spain, Turkey, Azerbaijan and Armenia, with Europe accounting for a tiny proportion of its global breeding range. Its European breeding population is relatively small, but is now known to exceed 10,000 pairs and was stable between 1970–1990. Although it may have declined in its Canary Islands stronghold during 1990–2000, most European populations were stable or increased, and the species probably declined only slightly overall. Consequently, this previously Rare species is now provisionally evaluated as Secure.

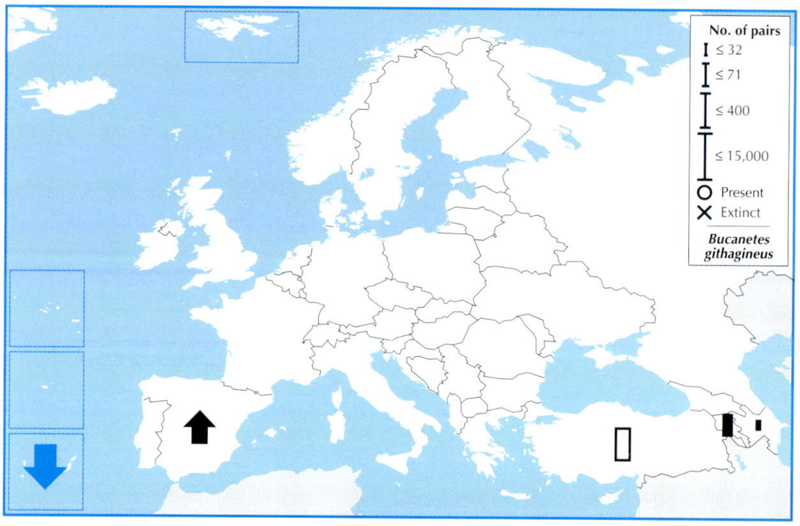

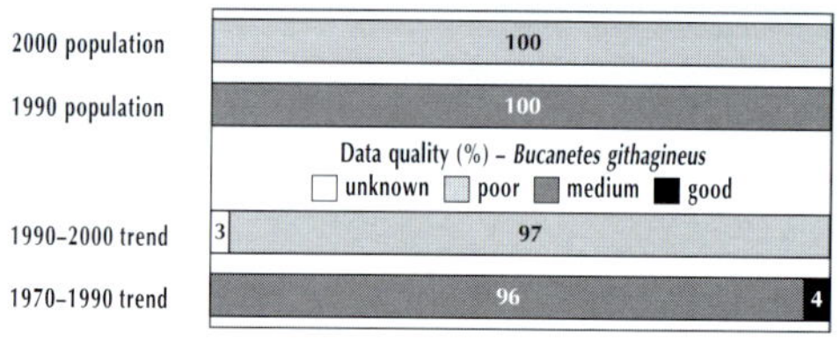

Carpodacus erythrinus
COMMON ROSEFINCH

Non-SPEC (1994: —) **Status** (Secure)
Criteria —
European IUCN Red List Category —
Criteria —
Global IUCN Red List Category —
Criteria —

Country	Breeding pop. size (pairs)	Year(s)	Trend	Mag.%	References
Armenia	30,000 – 160,000	99–02	–	0–19	
Austria	170 – 300	98–02	0	0–19	
Azerbaijan	(5,000 – 50,000)	96–00	(0)	(0–19)	
Belarus	70,000 – 120,000	97–02	(0)	(0–19)	
Belgium	**1 – 1**	01–02	**0**	**0–19**	*1*
Bulgaria	**1 – 3**	00–02	(0)	(0–19)	
Czech Rep.	**1,300 – 2,000**	00	+	**>80**	
Denmark	50 – 250	99–00	(–)	(10–19)	*16*
Estonia	(50,000 – 80,000)	98	+	20–29	*1*
Finland	250,000 – 350,000	98–02	**–**	**25**	
France	5 – 15	98–01	+	20–49	*6*
Georgia	Present	03	?	–	
Germany	800 – 970	95–99	0	0–19	
Rep. Ireland	(1 – 3)	98–01	(+)	(N)	
Latvia	20,000 – 50,000	90–00	0	0–19	*23,16*
Lithuania	(50,000 – 120,000)	99–01	(0)	(0–19)	20
Netherlands	**10 – 15**	98–00	(F)	(–)	*1*
Norway	500 – 2,000	90–03	**–**	**20–29**	
Poland	**10,000 – 30,000**	00–02	+	0–19	*23*
Romania	**740 – 1,200**	96–02	**+**	**20–29**	*31*
Russia	(2,500,000 – 5,000,000)	90–00	(0)	(0–19)	*8,39,43,89,122*
Slovakia	500 – 1,000	90–99	+	20–29	
Slovenia	**10 – 15**	99–00	–	30–49	
Sweden	8,000 – 25,000	99–00	+	0–19	
Switzerland	**10 – 20**	93–96	F	20–29	
Turkey	(20,000 – 60,000)	01	(0)	(0–19)	
Ukraine	20,000 – 30,000	90–00	0	5–9	
UK	**0 – 7**	96–00	**F**	**N**	
Total (approx.)	3,000,000 – 6,100,000	**Overall trend** Stable			
Breeding range	>4,000,000 km²	**Gen. length.** <3.3	**% Global pop.** 5–24		

Carpodacus erythrinus is a widespread summer visitor to most of northern Europe, occurring more patchily farther south, with Europe accounting for less than a quarter of its global breeding range. Its European breeding population is very large (>3,000,000 pairs), and was stable between 1970–1990. Although there were declines in a few countries—notably Finland—during 1990–2000, the species was stable in its Russian stronghold, and was stable or increased across much of Europe. It probably remained stable overall, and consequently is provisionally evaluated as Secure.

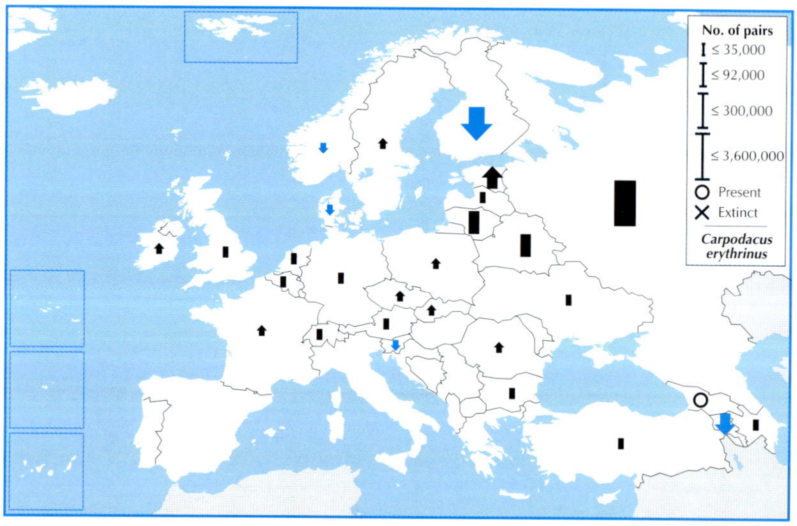

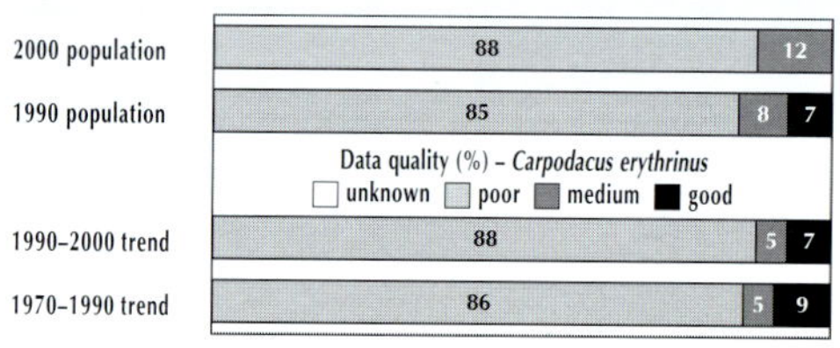

Carpodacus rubicilla
GREAT ROSEFINCH

SPEC 3 (1994: 3) **Status** (Rare)
Criteria <10,000 pairs
European IUCN Red List Category —
Criteria —
Global IUCN Red List Category —
Criteria —

Country	Breeding pop. size (pairs)	Year(s)	Trend	Mag.%	References
Azerbaijan	(50 – 250)	96–00	(0)	(0–19)	
Georgia	Present	03	?	–	
Russia	5,000 – 10,000	90–00	?	–	*17,61,72,60*
Total (approx.)	5,100 – 10,000	**Overall trend** Unknown			
Breeding range	>20,000 km²	**Gen. length.** <3.3	**% Global pop.** <5		

Carpodacus rubicilla has a predominantly Asian distribution, which just extends into Europe in the Caucasus. Its European breeding population is small (as few as 5,100 pairs), and underwent a large decline between 1970–1990. Although the species was stable in Azerbaijan during 1990–2000, no trend data were available for Russia or Georgia, and its overall trend was unknown. Nevertheless, its population size renders it susceptible to the risks affecting small populations, and consequently it is provisionally evaluated as Rare.

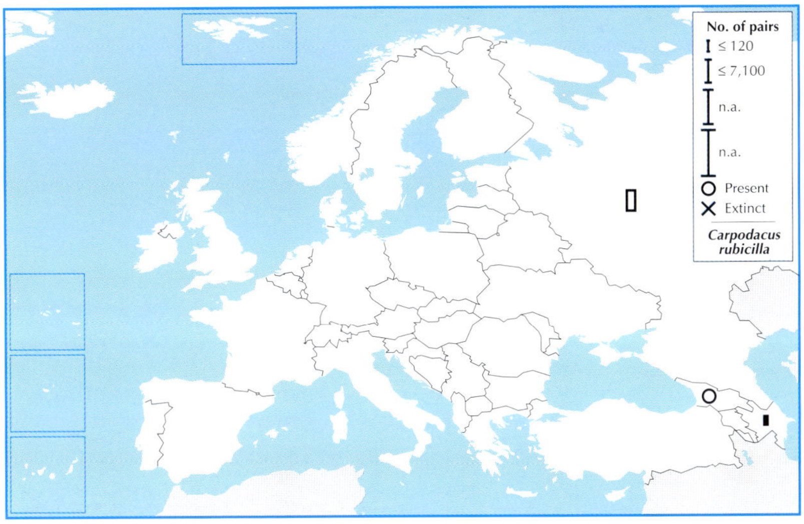

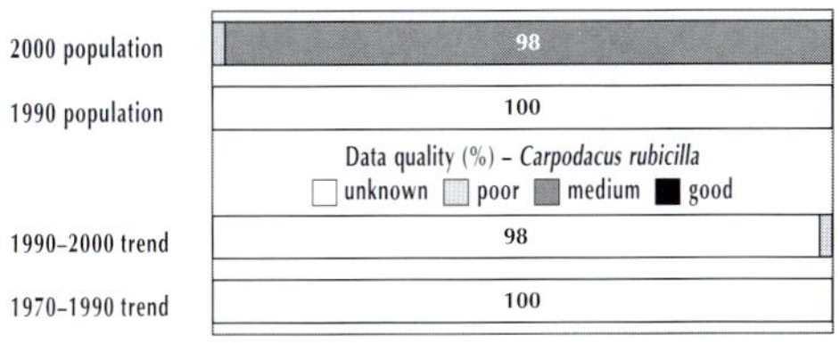

Pinicola enucleator
PINE GROSBEAK

Non-SPEC (1994: —) **Status** (Secure)
Criteria —
European IUCN Red List Category —
Criteria —
Global IUCN Red List Category —
Criteria —

Country	Breeding pop. size (pairs)	Year(s)	Trend	Mag.%	References
Estonia	0 – 1	98	?	–	1
Finland	5,000 – 20,000	98–02	**-**	**70**	
Norway	(500 – 1,000)	90–03	**F**	**>80**	
Russia	(100,000 – 250,000)	90–00	?	–	*122*
Sweden	3,000 – 15,000	99–00	(F)	(>80)	
Total (approx.)	110,000 – 290,000	**Overall trend** Unknown			
Breeding range	>750,000 km²	**Gen. length.** <3.3	**% Global pop.** 5–24		

Pinicola enucleator is a widespread resident in the boreal zone of Fennoscandia and Russia, with Europe accounting for less than a quarter of its global range. Its European breeding population is relatively large (>110,000 pairs), and was stable between 1970–1990. Although the species underwent a substantial decline in Finland during 1990–2000, and the trend of the stronghold population in Russia was unknown, there was no evidence to suggest that the species declined overall. Consequently, it is provisionally evaluated as Secure.

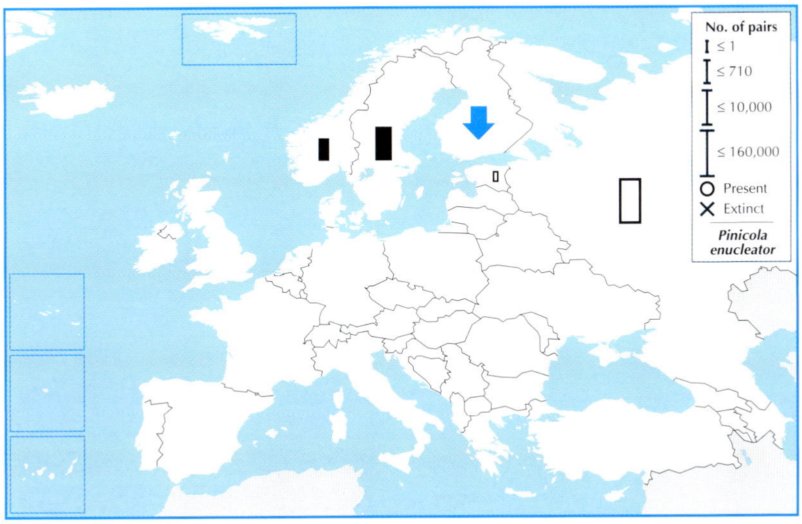

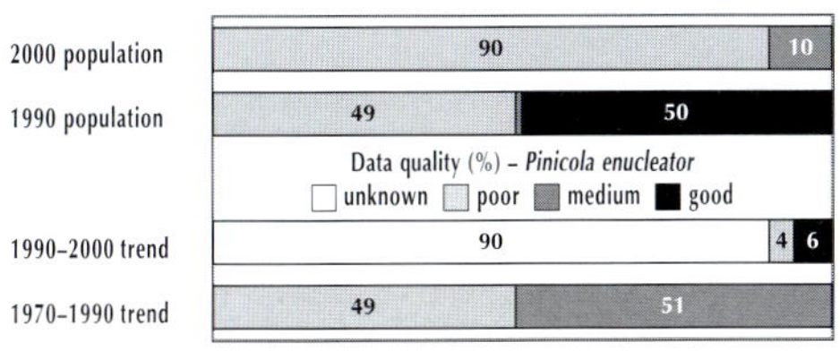

Pyrrhula pyrrhula
EURASIAN BULLFINCH

Non-SPEC (1994: —) **Status** (Secure)
Criteria —
European IUCN Red List Category —
Criteria —
Global IUCN Red List Category —
Criteria —

Pyrrhula pyrrhula is a widespread breeder across much of Europe, which accounts for less than half of its global breeding range. Its European breeding population is very large (>7,300,000 pairs), and was stable between 1970–1990. Although there were declines in a few countries—notably France—during 1990–2000, the key population in Russia was stable, and trends were stable or increasing across most of Europe. The species probably remained stable overall, and consequently it is provisionally evaluated as Secure.

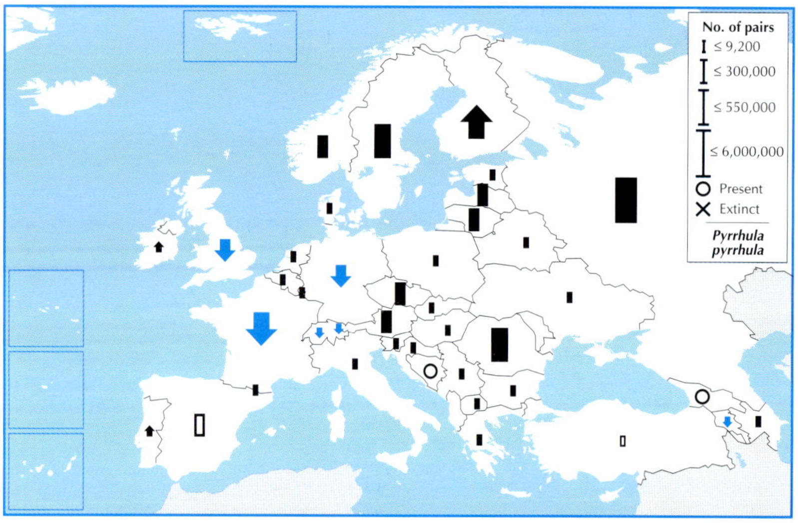

Country	Breeding pop. size (pairs)	Year(s)	Trend	Mag.%	References
Andorra	200 – 400	99–01	(0)	(0–19)	*1,3*
Armenia	800 – 3,000	99–02	–	0–9	
Austria	(150,000 – 250,000)	98–02	(0)	(0–19)	
Azerbaijan	(1,000 – 10,000)	96–00	(0)	(0–19)	
Belarus	40,000 – 60,000	97–02	0	0–19	
Belgium	5,200 – 15,000	01–02	0	0–19	*1*
Bosnia & HG	Present	90–03	?	–	
Bulgaria	15,000 – 25,000	96–02	0	0–9	
Croatia	(10,000 – 50,000)	02	(0)	(0–19)	*16*
Czech Rep.	170,000 – 340,000	00	0	0–19	
Denmark	25,000 – 50,000	00	**F**	**50–79**	*12*
Estonia	30,000 – 80,000	98	0	0–19	*1*
Finland	300,000 – 500,000	98–02	**+**	**70**	
France	(200,000 – 1,000,000)	98–02	–	42	*4,2*
Georgia	Present	03	?	–	
Germany	200,000 – 450,000	95–99	–	0–19	
Greece	(500 – 2,000)	95–00	(0)	(0–19)	
Hungary	100 – 250	95–02	F	50–79	
Rep. Ireland	20,000 – 100,000	88–91	+	0–19	
Italy	(30,000 – 60,000)	03	(0)	(0–19)	
Latvia	160,000 – 210,000	90–00	(0)	(0–19)	*23,18*
Liechtenstein	(120 – 180)	98–00	(–)	(0–19)	
Lithuania	(100,000 – 300,000)	99–01	(0)	(0–19)	*20*
Luxembourg	3,000 – 4,000	02	0	0–19	
Macedonia	(10,000 – 30,000)	90–00	(0)	(0–19)	
Netherlands	**7,000 – 9,000**	98–00	**0**	**17**	*1*
Norway	(100,000 – 500,000)	90–02	(0)	(0–19)	
Poland	**50,000 – 100,000**	00–02	(0)	(0–19)	*23*
Portugal	(200 – 2,000)	02	(+)	(–)	
Romania	450,000 – 650,000	00–02	(0)	(0–19)	*48*
Russia	(4,500,000 – 8,000,000)	90–00	(0)	(0–19)	*122*
Serbia & MN	4,000 – 6,000	95–02	0	0–19	*1,155,225,227, 62,40,152,223*
Slovakia	70,000 – 120,000	90–99	0	0–19	
Slovenia	10,000 – 20,000	94	(0)	(0–19)	
Spain	(118,000 – 170,000)	92	?	–	*10,13*
Sweden	250,000 – 500,000	99–00	**0**	**0–19**	
Switzerland	**50,000 – 80,000**	93–96	–	0–19	
Turkey	(6,000 – 12,000)	01	?	–	
Ukraine	(7,000 – 14,000)	90–00	F	20–29	
UK	**166,000 – 166,000**	00	**–**	**17**	*5,31*
Total (approx.)	7,300,000 – 14,000,000	**Overall trend** Stable			
Breeding range	>6,000,000 km²	**Gen. length.** <3.3	**% Global pop.** 25–49		

(See p. 280, bottom, for data quality graph)

Pyrrhula murina
AZORES BULLFINCH

SPEC 1 (1994: NE) **Status** (Endangered)
Criteria See IUCN below
European IUCN Red List Category EN
Criteria D1
Global IUCN Red List Category EN
Criteria D1

Pyrrhula murina is endemic to Europe, where it has an extremely small range (<50 km²) on the island of São Miguel in the Azores. Its European breeding population is very small (c.240 birds), but its trend between 1970–1990 was unknown. Although the species probably declined during 1990–2000, no quantitative trend data were available for this period. If this trend were confirmed, the species could qualify as Critically Endangered. In the meantime, as a consequence of its very small population, this globally threatened species is provisionally evaluated as Endangered.

Country	Breeding pop. size (individuals)	Year(s)	Trend	Mag.%	References
Portugal					
Azores	(238 – 238)	02	(–)	(–)	
Total (approx.)	240 – 240	**Overall trend** Decline			
Breeding range	<50 km²	**Gen. length.** <3.3	**% Global pop.** 100		

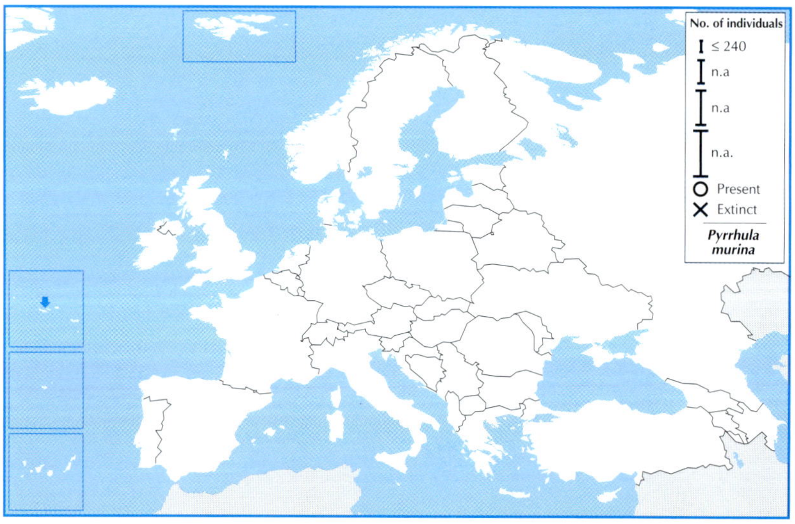

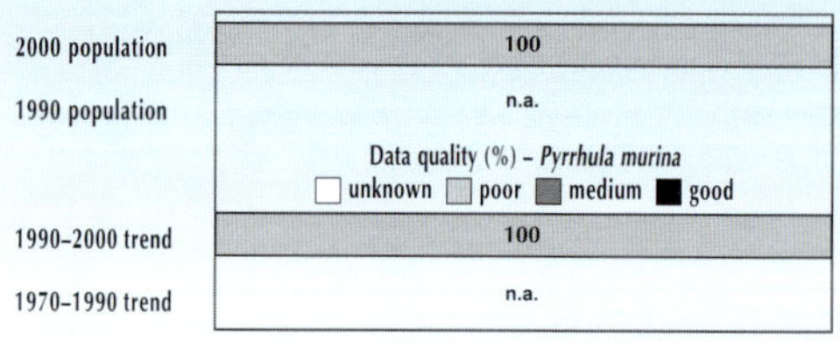

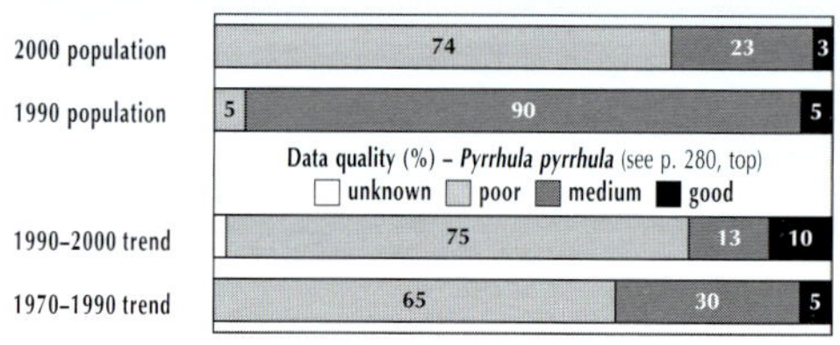

Coccothraustes coccothraustes
HAWFINCH

Non-SPEC (1994: —) **Status** Secure
Criteria —
European IUCN Red List Category —
Criteria —
Global IUCN Red List Category —
Criteria —

Coccothraustes coccothraustes is a widespread breeder across much of Europe, which accounts for less than half of its global breeding range. Its European breeding population is very large (>2,400,000 pairs), and was stable between 1970–1990. Although there were declines in a few countries during 1990–2000, populations were stable or increased across the vast majority of Europe, and the species remained stable overall. Consequently, it is evaluated as Secure.

Country	Breeding pop. size (pairs)	Year(s)	Trend	Mag.%	References
Armenia	830 – 3,300	01–02	–	0–9	
Austria	(25,000 – 50,000)	98–02	(0)	(0–19)	
Azerbaijan	(5,000 – 50,000)	96–00	(0)	(0–19)	
Belarus	30,000 – 70,000	97–02	(0)	(0–19)	
Belgium	5,200 – 15,000	01–02	–	0–19	*1*
Bosnia & HG	Present	90–03	?	–	
Bulgaria	50,000 – 150,000	96–02	0	0–9	
Croatia	(200,000 – 300,000)	02	(+)	(>80)	*57*
Czech Rep.	140,000 – 280,000	00	0	0–9	
Denmark	15,000 – 30,000	00	+	**50–79**	*12*
Estonia	5,000 – 10,000	98	+	20–29	*1*
Finland	400 – 800	98–02	+	100	
France	(50,000 – 250,000)	98–02	(+)	(62)	*4,2*
Georgia	Present	03	?	–	
Germany	160,000 – 350,000	95–99	0	0–19	
Greece	(5,000 – 20,000)	95–00	(0)	(0–19)	
Hungary	**82,000 – 145,000**	99–02	0	0–19	*19*
Italy	(5,000 – 15,000)	03	(0)	(0–19)	
Latvia	3,000 – 10,000	90–00	(0)	(0–19)	*23,16*
Liechtenstein	30 – 50	98–00	+	0–19	
Lithuania	(40,000 – 60,000)	99–01	(0)	(0–19)	*20*
Luxembourg	3,000 – 4,000	02	0	0–19	
Macedonia	(2,000 – 8,000)	90–00	(0)	(0–19)	
Moldova	13,000 – 15,000	90–00	0	0–19	
Netherlands	**8,000 – 10,000**	98–00	+	**122**	*1*
Norway	(1,000 – 5,000)	90–02	+	**20–29**	*3,27*
Poland	**200,000 – 400,000**	00–02	(0)	(0–19)	*23*
Portugal	(1,000 – 10,000)	02	(0)	(0–19)	
Romania	760,000 – 940,000	00–02	(0)	(0–19)	*48*
Russia	(100,000 – 250,000)	90–00	(0)	(0–19)	*8,23,39,89,114,122*
Serbia & MN	12,000 – 18,000	90–02	0	0–19	*1,209,29,172a,155, 67a,225,227,152*
Slovakia	110,000 – 220,000	90–99	0	0–19	
Slovenia	10,000 – 20,000	00	(0)	(0–19)	
Spain	(2,500 – 10,000)	98–02	?	–	*10*
Sweden	**5,000 – 15,000**	99–00	+	10–19	
Switzerland	**4,000 – 8,000**	93–96	0	0–19	
Turkey	(5,000 – 15,000)	01	(0)	(0–19)	
Ukraine	345,000 – 460,000	90–00	0	0–9	
UK	**3,000 – 6,500**	88–91	–	37	*8,16*
Total (approx.)	2,400,000 – 4,200,000	**Overall trend** Stable			
Breeding range	>5,000,000 km²	**Gen. length.** <3.3 **% Global pop.** 25–49			

(See p. 281, bottom, for data quality graph)

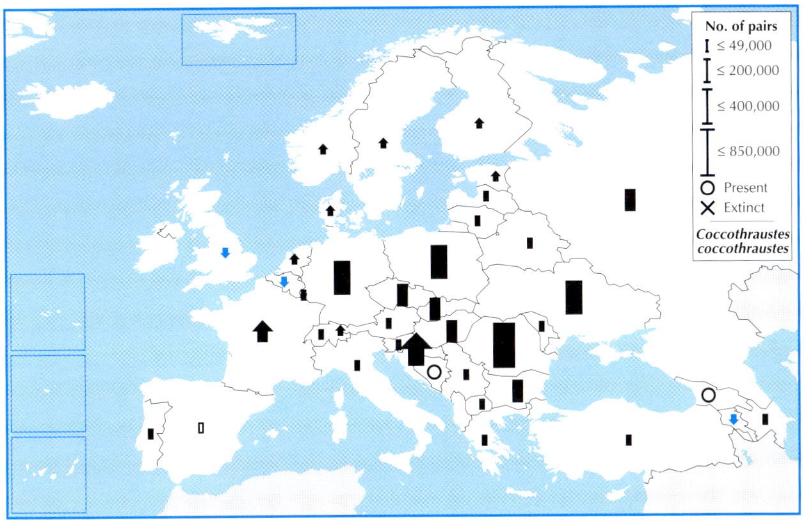

Calcarius lapponicus
LAPLAND LONGSPUR

Non-SPEC (1994: —) **Status** (Secure)
Criteria —
European IUCN Red List Category —
Criteria —
Global IUCN Red List Category —
Criteria —

Calcarius lapponicus is a widespread summer visitor to Greenland, Fennoscandia and arctic Russia, with Europe accounting for less than a quarter of its global breeding range. Its European breeding population is very large (>5,800,000 pairs), and was stable between 1970–1990. Although the trend of the stronghold population in Russia was unknown, the species remained stable in Greenland, Norway and Sweden during 1990–2000, and there was no evidence to suggest that it declined significantly overall. Consequently, the species is provisionally evaluated as Secure.

Country	Breeding pop. size (pairs)	Year(s)	Trend	Mag.%	References
Denmark					
Greenland	(500,000 – 1,000,000)	90–00	(0)	(0–19)	
Finland	20,000 – 50,000	98–02	–	15	
Norway	(200,000 – 500,000)	90–02	(0)	(0–19)	
Russia	(5,000,000 – 9,000,000)	90–00	?	–	*122*
Sweden	100,000 – 400,000	99–00	(0)	(0–19)	
Total (approx.)	5,800,000 – 11,000,000	**Overall trend** Unknown			
Breeding range	>1,000,000 km²	**Gen. length.** <3.3 **% Global pop.** 5–24			

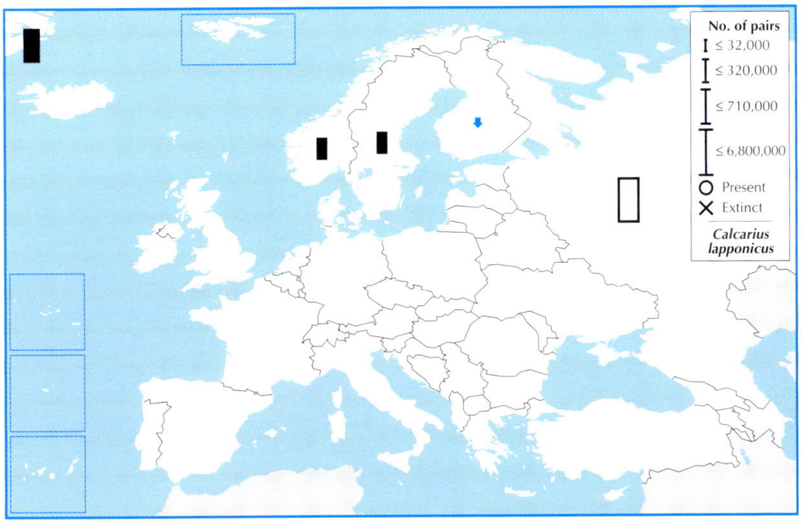

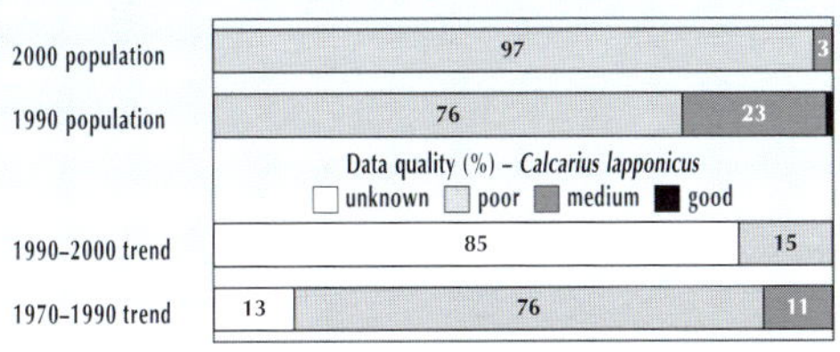

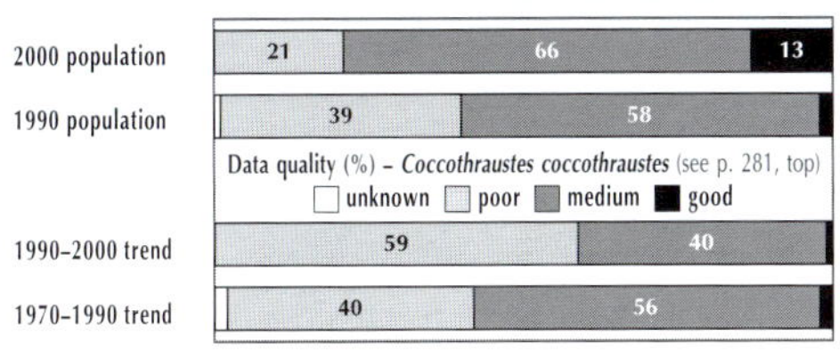

Plectrophenax nivalis
SNOW BUNTING

Non-SPEC (1994: —) **Status** (Secure)
Criteria —
European IUCN Red List Category —
Criteria —
Global IUCN Red List Category —
Criteria —

Country	Breeding pop. size (pairs)	Year(s)	Trend	Mag.%	References
Denmark					
Faroe Is.	(0 – 10)	81	(0)	(0–19)	
Greenland	(500,000 – 1,000,000)	90–00	(0)	(0–19)	
Finland	2,000 – 4,000	98–02	–	40	
Iceland	**50,000 – 100,000**	78–94	?	–	*1*
Norway	(100,000 – 500,000)	90–03	(0)	(0–19)	
Svalbard	(10,000 – 50,000)	90–03	(0)	(0–19)	
Russia	(1,500 – 15,000)	90–00	(F)	(20–29)	*122*
Sweden	20,000 – 50,000	99–00	(–)	(10–19)	
UK	**70 – 100**	88–91	+	**20**	*8*
Total (approx.)	680,000 – 1,700,000	**Overall trend** Stable			
Breeding range	>1,000,000 km²	**Gen. length.** <3.3	**% Global pop.** 5–24		

Plectrophenax nivalis is a widespread breeder in northernmost Europe, which accounts for less than a quarter of its global breeding range. Its European breeding population is large (>680,000 pairs), and was stable between 1970–1990. Although there were declines in Sweden and Finland during 1990–2000, key populations in Greenland and Norway were stable, and the species probably remained stable overall. Consequently, it is provisionally evaluated as Secure.

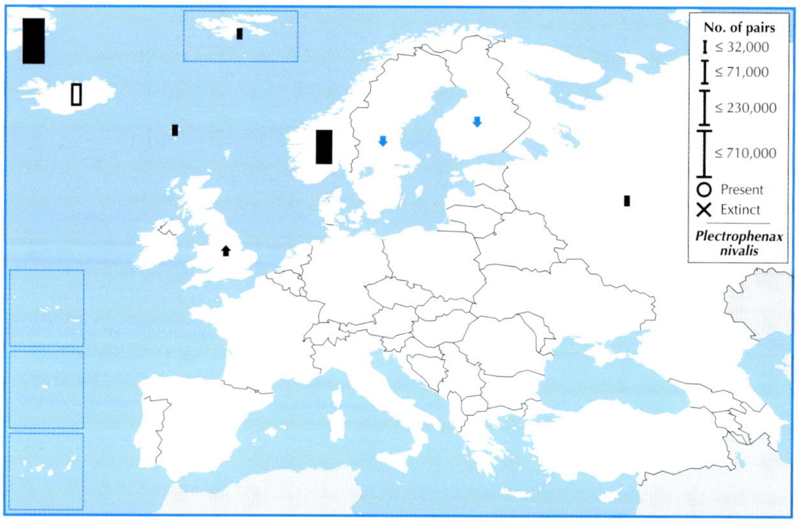

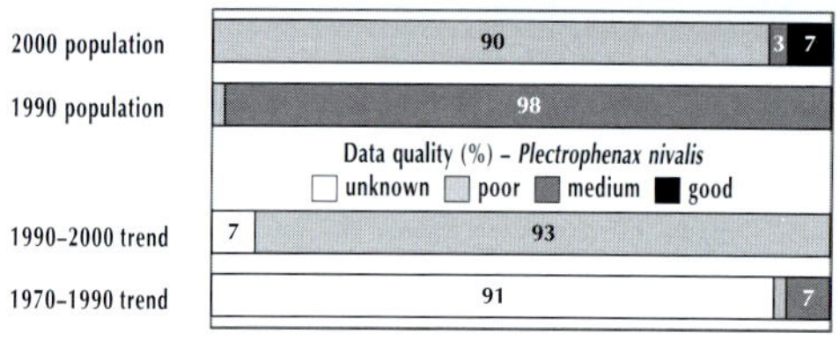

Emberiza leucocephalos
PINE BUNTING

Non-SPEC (1994: NE) **Status** (Secure)
Criteria —
European IUCN Red List Category —▼▼▼
Criteria —
Global IUCN Red List Category —
Criteria —

Country	Breeding pop. size (pairs)	Year(s)	Trend	Mag.%	References
Russia	(50 – 100)	90–00	?	–	*51,52*
Total (approx.)	50 – 100	**Overall trend** Unknown			
Breeding range	>20,000 km²	**Gen. length.** <3.3	**% Global pop.** <5		

Emberiza leucocephalos has a predominantly Asian breeding distribution, which just extends into Europe in the foothills of the Urals. Its European breeding population is extremely small (as few as 50 pairs), but no trend was available for 1970–1990. Its trend during 1990–2000 was also unknown, but there is no evidence to suggest that the species declined. Although the size of the European population could render it susceptible to the risks affecting small populations, it is marginal to a much larger non-European population. Consequently, it is provisionally evaluated as Secure.

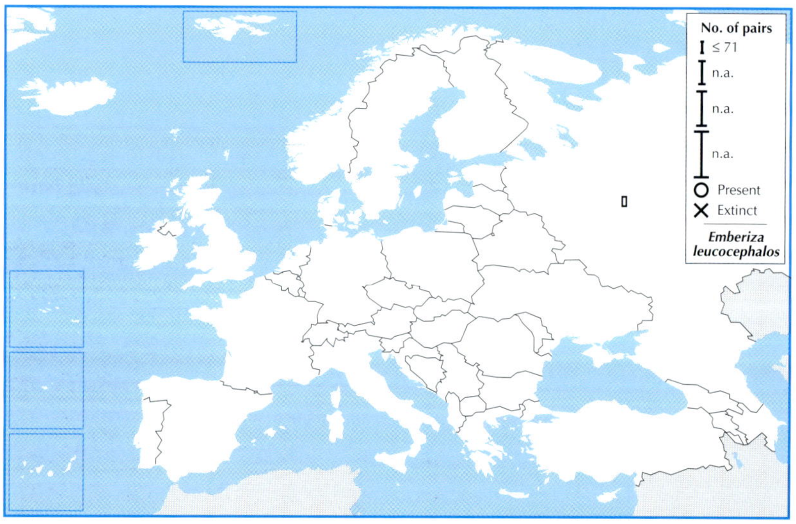

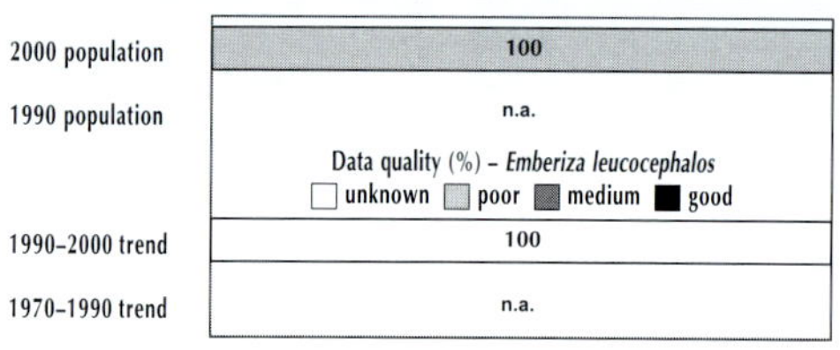

Emberiza citrinella
YELLOWHAMMER

Non-SPEC[E] (1994: 4) **Status** (Secure)
Criteria —
European IUCN Red List Category —
Criteria —
Global IUCN Red List Category —
Criteria —

Emberiza citrinella is a widespread breeder across much of Europe, which constitutes >50% of its global breeding range. Its European breeding population is extremely large (>18,000,000 pairs), and was stable between 1970–1990. Although there were declines in Fennoscandia and western Europe during 1990–2000, most central and eastern European populations—including sizeable ones in Germany, Czech Republic, Poland and Ukraine—were stable (the Russian trend was unknown). The species probably declined only slightly overall, and is hence provisionally evaluated as Secure.

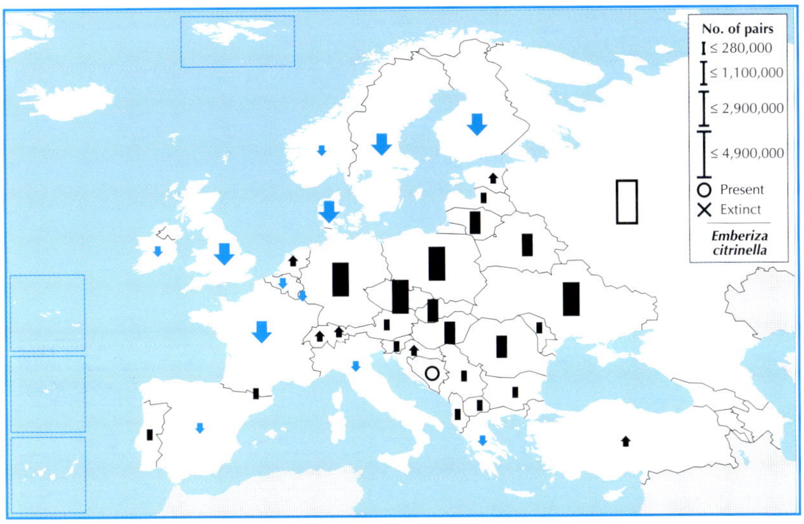

Country	Breeding pop. size (pairs)	Year(s)	Trend	Mag.%	References
Albania	(2,000 – 4,000)	02	(0)	(0–19)	
Andorra	50 – 150	99–01	(0)	(0–19)	*1,3*
Austria	60,000 – 120,000	98–02	(0)	(0–19)	
Belarus	600,000 – 800,000	97–02	0	0–19	
Belgium	18,000 – 35,000	01–02	–	0–19	*1*
Bosnia & HG	Present	90–03	?	–	
Bulgaria	50,000 – 80,000	96–02	0	0–9	
Croatia	(100,000 – 150,000)	02	(+)	(30–49)	*16*
Czech Rep.	2,000,000 – 4,000,000	00	0	0–19	
Denmark	400,000 – 600,000	00	**–**	**10–19**	*12*
Estonia	(100,000 – 200,000)	98	+	20–29	*1*
Finland	700,000 – 1,100,000	98–02	**–**	**10**	
France	(400,000 – 1,600,000)	98–02	**–**	**29**	*4,2*
Germany	1,000,000 – 2,800,000	95–99	0	0–19	
Greece	(2,000 – 5,000)	95–00	(–)	(0–19)	
Hungary	**630,000 – 855,000**	99–02	0	0–19	*19*
Rep. Ireland	20,000 – 100,000	88–91	–	50–79	
Italy	(20,000 – 50,000)	03	(–)	(0–19)	
Latvia	80,000 – 160,000	90–00	0	0–19	*23,16,2*
Liechtenstein	40 – 60	98–00	+	0–19	
Lithuania	(600,000 – 750,000)	99–01	(0)	(0–19)	*20*
Luxembourg	12,000 – 15,000	02	–	0–19	
Macedonia	(30,000 – 100,000)	90–00	(0)	(0–19)	
Moldova	40,000 – 50,000	90–00	0	0–19	
Netherlands	**22,000 – 28,000**	98–00	**+**	**66**	*1*
Norway	(150,000 – 500,000)	95–02	–	0–19	*30*
Poland	**2,000,000 – 4,000,000**	00–02	0	0–19	*23*
Portugal	(100 – 1,000)	02	(0)	(0–19)	
Romania	760,000 – 890,000	00–02	0	0–19	*48*
Russia	4,000,000 – 6,000,000	90–00	?	–	*122*
Serbia & MN	50,000 – 70,000	90–02	0	0–19	*1,29,172a,67a,225, 227,185,152,56,43*
Slovakia	800,000 – 1,500,000	90–99	0	0–19	
Slovenia	30,000 – 50,000	94	(0)	(0–19)	
Spain	(140,000 – 170,000)	92	(–)	(0–19)	*13,12,10*
Sweden	500,000 – 1,500,000	99–00	**–**	**23**	
Switzerland	**50,000 – 80,000**	93–96	+	10–19	
Turkey	(1,500 – 2,500)	01	(+)	(0–19)	
Ukraine	1,500,000 – 2,000,000	90–00	0	0–19	
UK	**792,000 – 792,000**	00	**–**	**34**	*5,31*
Total (approx.)	18,000,000 – 31,000,000	**Overall trend** Small decline			
Breeding range	>7,000,000 km²	**Gen. length.** <3.3 **% Global pop.** 50–74			

(See p. 283, bottom, for data quality graph)

Emberiza cirlus
CIRL BUNTING

Non-SPEC[E] (1994: 4) **Status** Secure
Criteria —
European IUCN Red List Category —
Criteria —
Global IUCN Red List Category —
Criteria —

Emberiza cirlus is a widespread resident across much of southern Europe, which constitutes >95% of its global range. Its European breeding population is very large (>2,000,000 pairs), and was stable between 1970–1990. Although the trend of the Spanish population during 1990–2000 was unknown, other key populations in France and Croatia increased, and the Italian population remained stable. The species hence increased moderately overall, and is consequently evaluated as Secure.

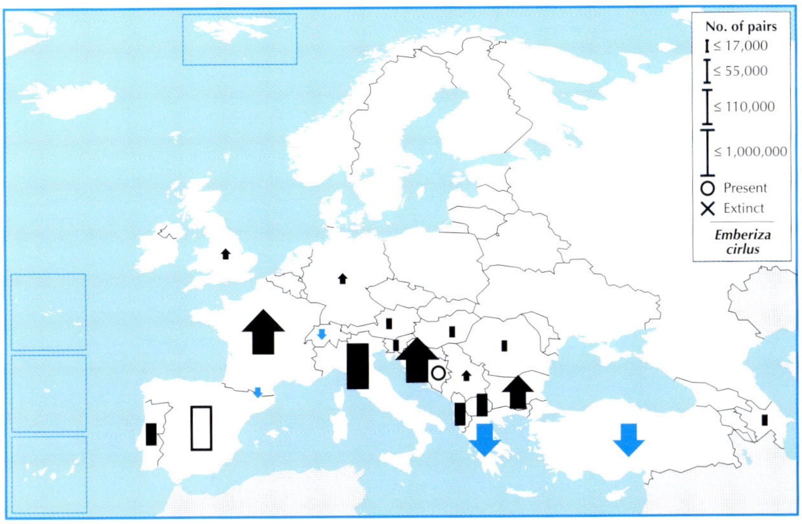

Country	Breeding pop. size (pairs)	Year(s)	Trend	Mag.%	References
Albania	20,000 – 50,000	02	(0)	(0–19)	
Andorra	20 – 40	99–01	(–)	(0–19)	*1,3*
Austria	5 – 10	98–02	0	0–19	
Azerbaijan	(0 – 250)	96–00	(0)	(0–19)	
Bosnia & HG	Present	90–03	?	–	
Bulgaria	50,000 – 100,000	96–02	+	0–19	
Croatia	(500,000 – 800,000)	02	(+)	(>80)	*16*
France	(500,000 – 2,000,000)	98–02	**+**	**42**	*4,2*
Germany	**100 – 140**	95–99	**+**	**20–29**	
Greece	(50,000 – 200,000)	95–00	(–)	(0–19)	
Hungary	(0 – 10)	95–02	0	0–19	
Italy	(300,000 – 800,000)	03	(0)	(0–19)	
Macedonia	(30,000 – 100,000)	90–00	(0)	(0–19)	
Portugal	(10,000 – 100,000)	02	(0)	(0–19)	
Romania	(9,000 – 11,000)	00–02	(0)	(0–19)	
Serbia & MN	13,000 – 20,000	95–02	+	0–19	*1,181,186,155, 67a,225,227,141*
Slovenia	1,000 – 1,500	99–00	(0)	(0–19)	
Spain	(500,000 – 800,000)	92	?	–	*13,12,10*
Switzerland	**800 – 1,200**	93–96	–	0–9	
Turkey	(60,000 – 180,000)	01	(–)	(0–19)	
UK	**415 – 504**	98	**+**	**448**	*40*
Total (approx.)	2,000,000 – 5,200,000	**Overall trend** Moderate increase			
Breeding range	>2,000,000 km²	**Gen. length.** <3.3 **% Global pop.** >95			

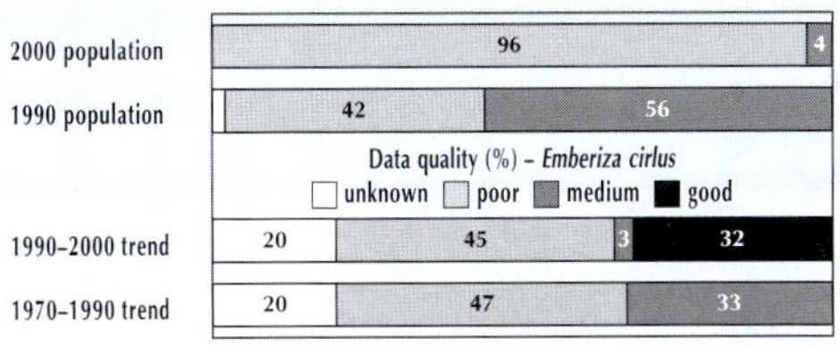

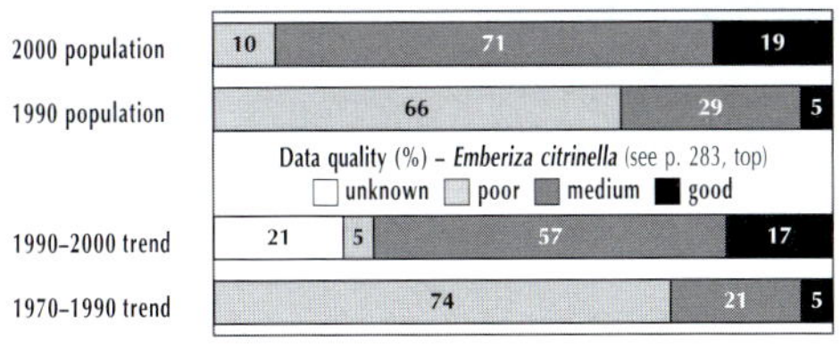

Emberiza cia
ROCK BUNTING

SPEC 3 (1994: 3) **Status** (Depleted)
Criteria Large historical decline
European IUCN Red List Category —
Criteria —
Global IUCN Red List Category —
Criteria —

Emberiza cia is a widespread resident across much of southern Europe, which accounts for less than half of its global range. Its European breeding population is very large (>1,300,000 pairs), but underwent a large decline between 1970–1990. Although the trend of the key population in Spain was unknown during 1990–2000, the species was stable across most of its European range, and was probably stable overall. Nevertheless, its total population size probably remains below the level that preceded its decline, and consequently the species is provisionally evaluated as Depleted.

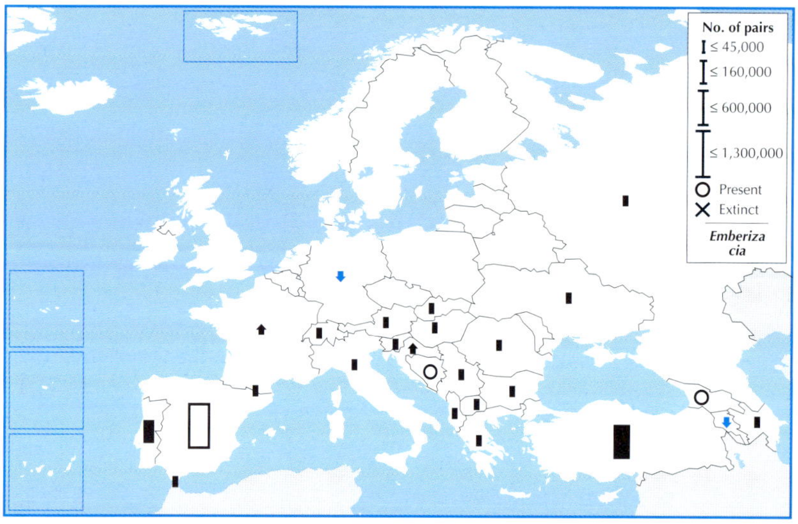

Country	Breeding pop. size (pairs)	Year(s)	Trend	Mag.%	References
Albania	(5,000 – 10,000)	02	(0)	(0–19)	
Andorra	(200 – 1,000)	99–01	(0)	(0–19)	*1,3*
Armenia	15,000 – 20,000	99–02	–	0–9	
Austria	(500 – 1,000)	98–02	(0)	(0–19)	
Azerbaijan	(10,000 – 50,000)	96–00	(0)	(0–19)	
Bosnia & HG	Present	90–03	?	–	
Bulgaria	3,000 – 6,000	96–02	0	0–9	
Croatia	(1,000 – 5,000)	02	(+)	(0–19)	*16*
France	(20,000 – 100,000)	98–02	+	0–19	*4*
Georgia	Present	03	?	–	
Germany	**290 – 360**	95–99	–	**30–49**	
Greece	(10,000 – 20,000)	95–00	(0)	(0–19)	
Hungary	340 – 620	97–02	0	0–19	
Italy	(22,000 – 90,000)	03	(0)	(0–19)	
Macedonia	(2,000 – 5,000)	90–00	(0)	(0–19)	
Portugal	(50,000 – 500,000)	02	(0)	(0–19)	
Romania	(12,000 – 14,500)	00–02	(0)	(0–19)	
Russia	(10,000 – 50,000)	90–00	(0)	(0–19)	*26,50*
Serbia & MN	3,500 – 5,500	90–02	0	0–19	*1,233,62,155,225, 227,117a,43,40, 143,152,34*
Slovakia	150 – 250	90–99	0	0–19	
Slovenia	5,000 – 10,000	00–03	(0)	(0–19)	
Spain	(820,000 – 2,000,000)	92	?	–	*13,12,10*
Switzerland	**4,000 – 8,000**	93–96	0	0–19	
Turkey	(300,000 – 1,200,000)	01	(0)	(0–19)	
Ukraine	**2,000 – 3,000**	90–00	**0**	**0–19**	
UK					
Gibraltar	**0 – 4**	00	**0**	**0–19**	
Total (approx.)	1,300,000 – 4,100,000	**Overall trend** Stable			
Breeding range	>2,000,000 km²	**Gen. length** <3.3		**% Global pop.** 25–49	

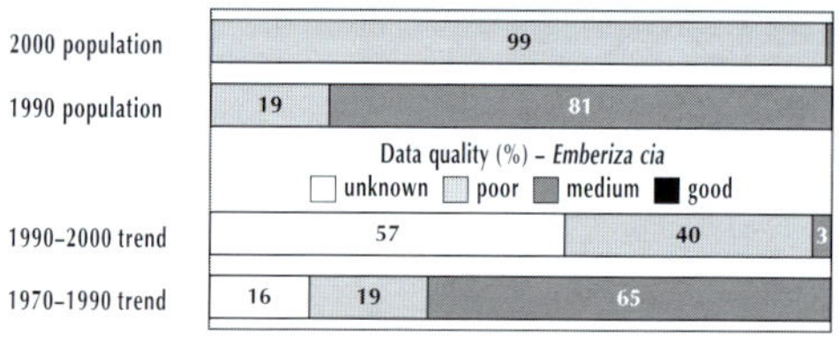

Emberiza cineracea
CINEREOUS BUNTING

SPEC 1 (1994: 2) **Status** (Rare)
Criteria <10,000 pairs
European IUCN Red List Category —
Criteria —
Global IUCN Red List Category NT
Criteria D1

Emberiza cineracea is a summer visitor to Turkey and Greece, with Europe constituting >90% of its global breeding range. Its European breeding population is small (as few as 2,600 pairs), but was stable between 1970–1990. Although the species may have declined in its Turkish stronghold during 1990–2000, it was stable in Greece and probably underwent only a small decline overall. Nevertheless, its population size still renders it susceptible to the risks affecting small populations. Consequently, this globally Near Threatened species is provisionally evaluated as Rare in Europe.

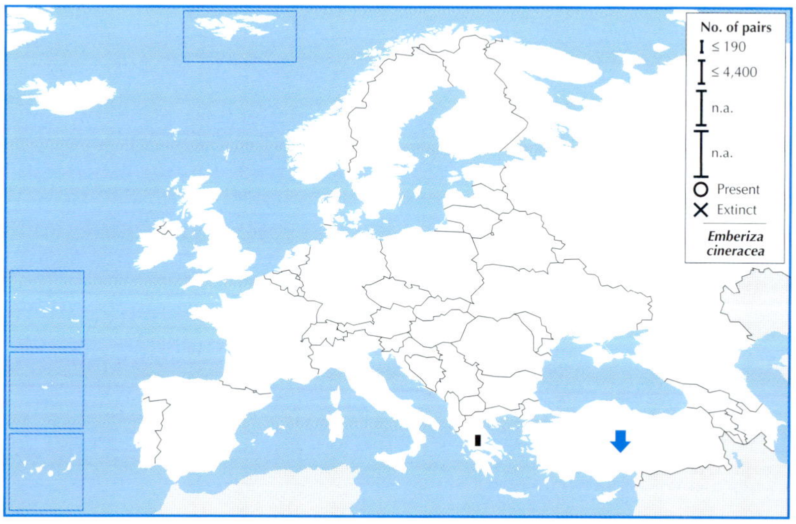

Country	Breeding pop. size (pairs)	Year(s)	Trend	Mag.%	References
Greece	**115 – 305**	95–00	(0)	(0–19)	
Turkey	(2,500 – 7,500)	01	(-)	(0–19)	
Total (approx.)	2,600 – 7,800	**Overall trend** Small decline			
Breeding range	>20,000 km²	**Gen. length** <3.3		**% Global pop.** >90	

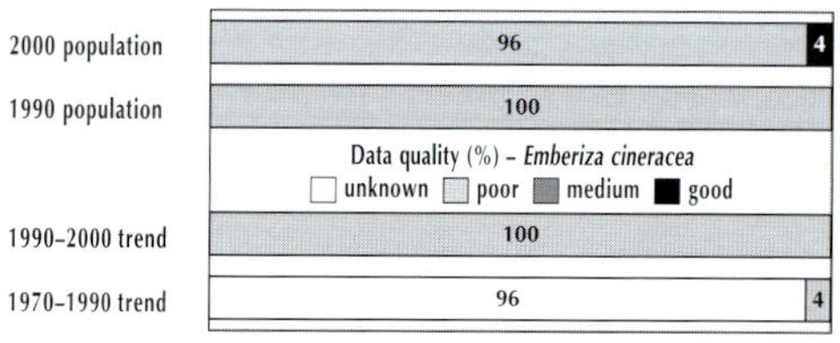

Emberiza hortulana
ORTOLAN BUNTING

SPEC 2 (1994: 2) **Status** (Depleted)
Criteria Large historical decline
European IUCN Red List Category —
Criteria —
Global IUCN Red List Category —
Criteria —

Emberiza hortulana is a widespread summer visitor to much of Europe, which constitutes >50% of its global breeding range. Its European breeding population is very large (>5,200,000 pairs), but underwent a large decline between 1970–1990. Although the species was stable in some countries—most notably its Turkish stronghold—during 1990–2000, it continued to decline across much of Europe, and underwent a small decline overall. Its population has clearly not yet recovered to the level that preceded its decline, and consequently the species is provisionally evaluated as Depleted.

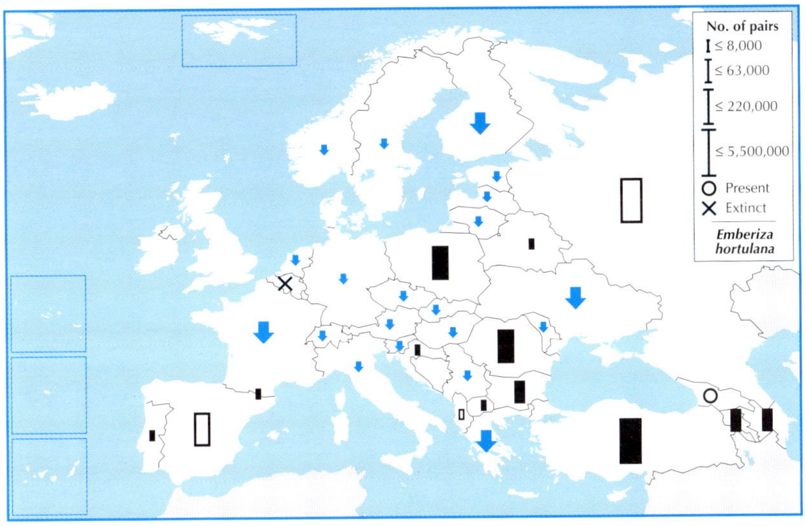

Country	Breeding pop. size (pairs)	Year(s)	Trend	Mag.%	References
Albania	(1,000 – 2,000)	02	?	–	
Andorra	**4 – 10**	99–01	(0)	(0–19)	*1,3*
Armenia	15,000 – 30,000	99–02	0	0–19	
Austria	**15 – 25**	98–02	**-**	**30–49**	
Azerbaijan	(20,000 – 100,000)	96–00	(0)	(0–19)	
Belarus	2,500 – 4,000	98–02	0	0–19	
Belgium	**0 – 0**	00	**-**	**X**	*1*
Bulgaria	25,000 – 50,000	96–02	0	0–9	
Croatia	(1,000 – 5,000)	02	(0)	(0–19)	*70,16*
Czech Rep.	**100 – 200**	00	**-**	**30–49**	
Estonia	(2,000 – 4,000)	98	-	20–29	*1*
Finland	30,000 – 50,000	98–02	**-**	**75**	
France	10,000 – 40,000	98–02	-	0–19	*4*
Georgia	Present	03	?	–	
Germany	5,600 – 7,000	95–99	-	30–49	
Greece	(20,000 – 50,000)	95–00	(-)	(0–19)	
Hungary	10 – 15	98–02	-	20–49	
Italy	(4,000 – 16,000)	03	(-)	(0–19)	
Latvia	500 – 2,000	90–00	(-)	(0–19)	*23,16*
Lithuania	(200 – 800)	99–01	(-)	(0–19)	*20*
Macedonia	(3,000 – 10,000)	90–00	(0)	(0–19)	
Moldova	4,500 – 5,000	90–00	-	20–29	
Netherlands	**0 – 5**	98–00	-	72	*1*
Norway	**150 – 155**	02	-	30–49	*14*
Poland	**150,000 – 300,000**	00–02	0	0–19	*23,106*
Portugal	(500 – 2,500)	02	(0)	(0–19)	
Romania	(125,000 – 255,000)	00–02	(0)	(0–19)	*48*
Russia	1,500,000 – 5,000,000	90–00	?	–	*122*
Serbia & MN	4,500 – 6,500	97–02	-	0–19	*1,155,150,147, 247,172a,67a, 227,201*
Slovakia	0 – 5	90–99	-	30–49	
Slovenia	200 – 300	00	-	30–49	
Spain	(200,000 – 225,000)	92	?	–	*13,12,10*
Sweden	**2,000 – 7,000**	99–00	-	>80	
Switzerland	100 – 150	98–02	-	20–29	
Turkey	(3,000,000 – 10,000,000)	01	(0)	(0–19)	
Ukraine	58,000 – 67,000	90–00	-	0–19	

Total (approx.) 5,200,000 – 16,000,000 **Overall trend** Small decline
Breeding range >5,000,000 km² **Gen. length.** <3.3 **% Global pop.** 50–74

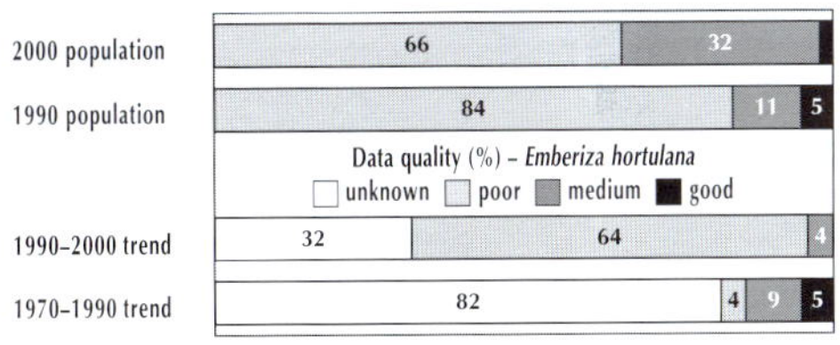

Emberiza buchanani
GREY-NECKED BUNTING

Non-SPEC (1994: —) **Status** (Secure)
Criteria —
European IUCN Red List Category —▼
Criteria —
Global IUCN Red List Category —
Criteria —

Country	Breeding pop. size (pairs)	Year(s)	Trend	Mag.%	References
Armenia	300 – 600	98–02	-	10–19	
Azerbaijan	(100 – 1,000)	96–00	(0)	(0–19)	
Turkey	(6,000 – 18,000)	01	(0)	(0–19)	

Total (approx.) 6,400 – 20,000 **Overall trend** Stable
Breeding range >20,000 km² **Gen. length.** <3.3 **% Global pop.** 5–24

Emberiza buchanani is a summer visitor to Turkey and the Caucasus, with Europe accounting for less than a quarter of its global breeding range. Its European breeding population is small (as few as 6,400 pairs), and its trend between 1970–1990 was unknown. However, the species was probably stable overall during 1990–2000. Although the size of the European population could render it susceptible to the risks affecting small populations, it is marginal to a much larger non-European population. Consequently, it is provisionally evaluated as Secure.

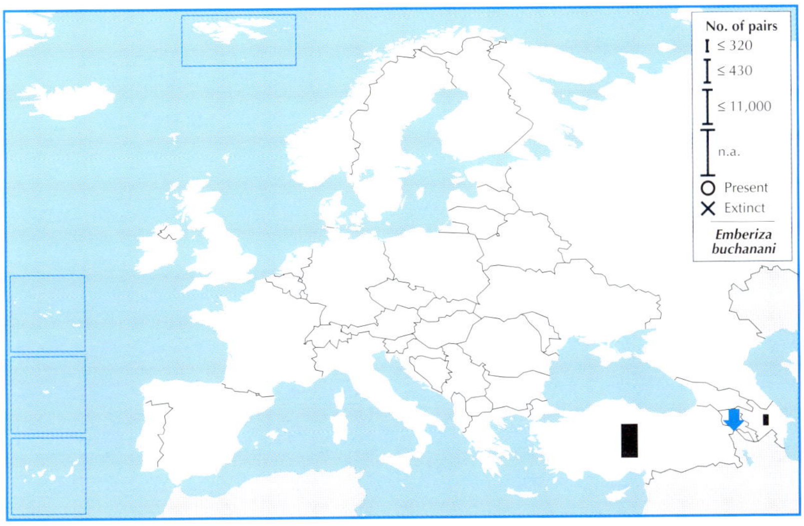

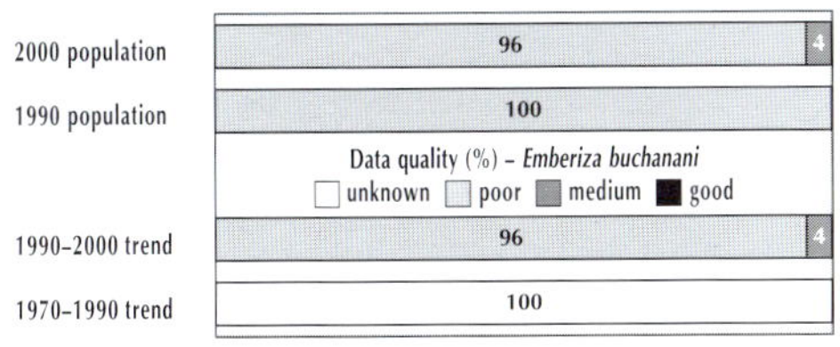

Emberiza caesia
CRETZSCHMAR'S BUNTING

Non-SPEC[E] (1994: 4) **Status** (Secure)
Criteria —
European IUCN Red List Category —
Criteria —
Global IUCN Red List Category —
Criteria —

Country	Breeding pop. size (pairs)	Year(s)	Trend	Mag.%	References
Albania	(50 – 200)	02	?	–	
Azerbaijan	(0 – 250)	96–00	(0)	(0–19)	
Cyprus	(8,000 – 24,000)	94–02	(0)	(0–9)	
Greece	(5,000 – 20,000)	95–00	(0)	(0–19)	
Turkey	(125,000 – 185,000)	01	(–)	(0–19)	
Total (approx.)	140,000 – 230,000	**Overall trend** Small decline			
Breeding range	>500,000 km²	**Gen. length** <3.3	**% Global pop.** 75–94		

Emberiza caesia is a summer visitor to south-east Europe, which constitutes >75% of its global breeding range. Its European breeding population is large (>140,000 pairs), and was stable between 1970–1990. Although the species underwent a small decline in its Turkish stronghold during 1990–2000, its populations in Greece and Cyprus remained stable, and it probably declined only slightly overall. Consequently, it is provisionally evaluated as Secure.

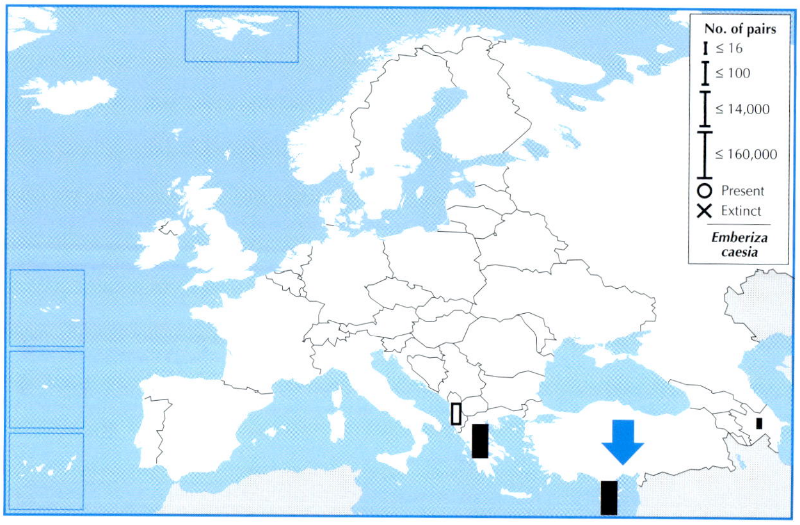

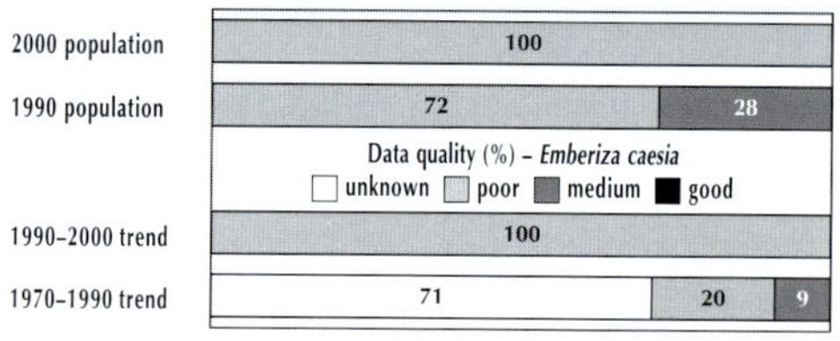

Emberiza rustica
RUSTIC BUNTING

Non-SPEC (1994: —) **Status** (Secure)
Criteria —
European IUCN Red List Category —
Criteria —
Global IUCN Red List Category —
Criteria —

Country	Breeding pop. size (pairs)	Year(s)	Trend	Mag.%	References
Estonia	0 – 10	98	–	20–29	*1*
Finland	100,000 – 200,000	98–02	**–**	**35**	
Latvia	(0 – 10)	90–00	?	–	*23,26*
Norway	100 – 500	90–02	**0**	**0–19**	*10*
Russia	6,000,000 – 10,000,000	90–00	?	–	*122*
Sweden	30,000 – 100,000	99–00	(–)	(0–9)	
Total (approx.)	6,100,000 – 10,000,000	**Overall trend** Unknown			
Breeding range	>2,000,000 km²	**Gen. length** <3.3	**% Global pop.** 25–49		

Emberiza rustica is a widespread summer visitor to Fennoscandia and northern Russia, with Europe accounting for less than half of its global breeding range. Its European breeding population is very large (>6,100,000 pairs), and was stable between 1970–1990. Although the species declined in Finland and Sweden during 1990–2000, the trend of the stronghold population in Russia was unknown, and there was no evidence to suggest that the species declined overall. Consequently, it is provisionally evaluated as Secure.

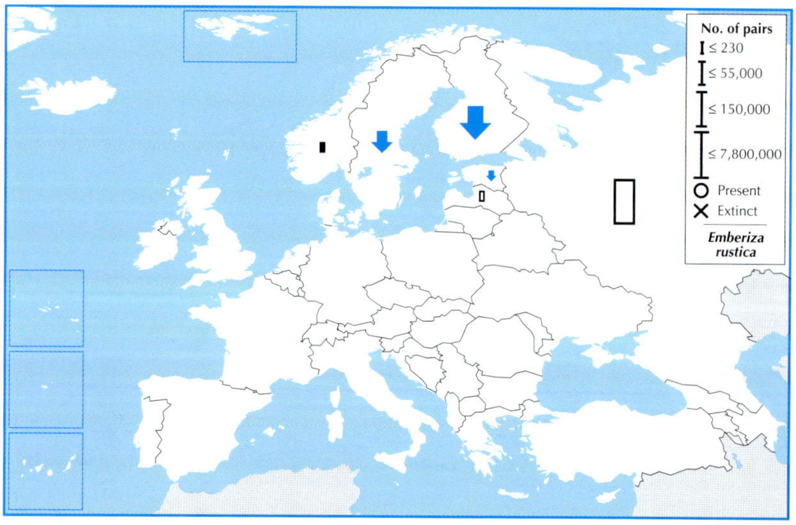

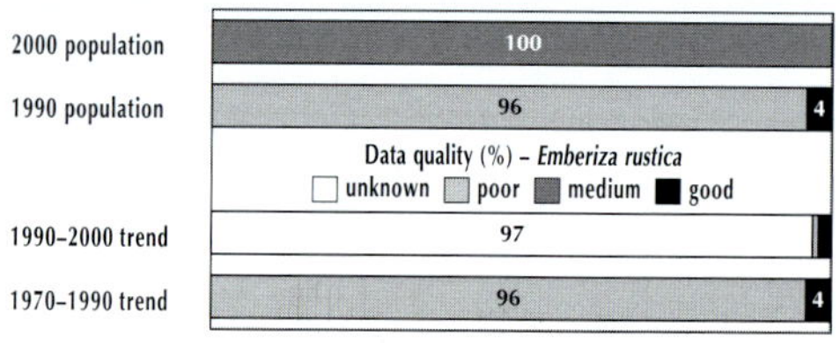

Emberiza pusilla
LITTLE BUNTING

Country	Breeding pop. size (pairs)	Year(s)	Trend	Mag.%	References
Finland	1,000 – 5,000	98–02	–	50	
Norway	0 – 30	02	F	>80	26
Russia	5,000,000 – 8,000,000	90–00	?	–	122
Sweden	(10 – 150)	99–00	(F)	(20–29)	
Total (approx.)	5,000,000 – 8,000,000	Overall trend Unknown			
Breeding range	>1,000,000 km²	Gen. length <3.3	% Global pop. 25–49		

Non-SPEC (1994: —) **Status** (Secure)
Criteria —
European IUCN Red List Category —
Criteria —
Global IUCN Red List Category —
Criteria —

Emberiza pusilla is a widespread summer visitor to northern Fennoscandia and Russia, with Europe accounting for less than half of its global breeding range. Its European breeding population is very large (>5,000,000 pairs), and was stable between 1970–1990. Although the species underwent a large decline in Finland during 1990–2000, the trend of the stronghold population in Russia was unknown, and there was no evidence to suggest that the species declined overall. Consequently, it is provisionally evaluated as Secure.

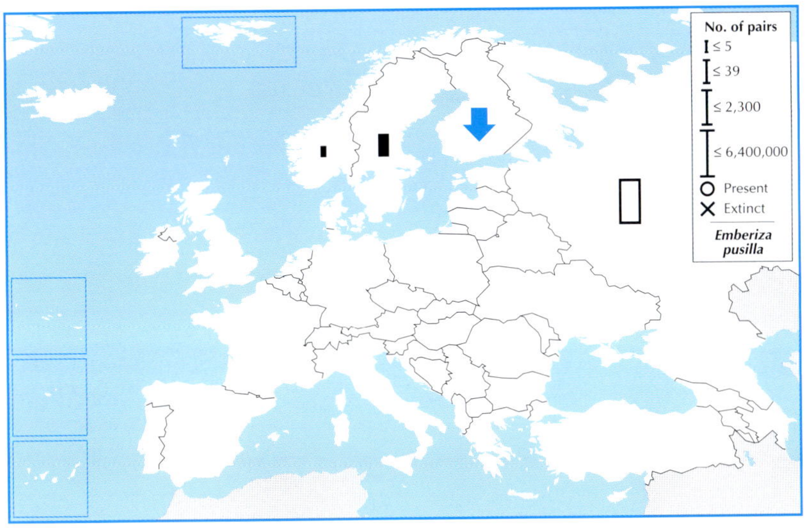

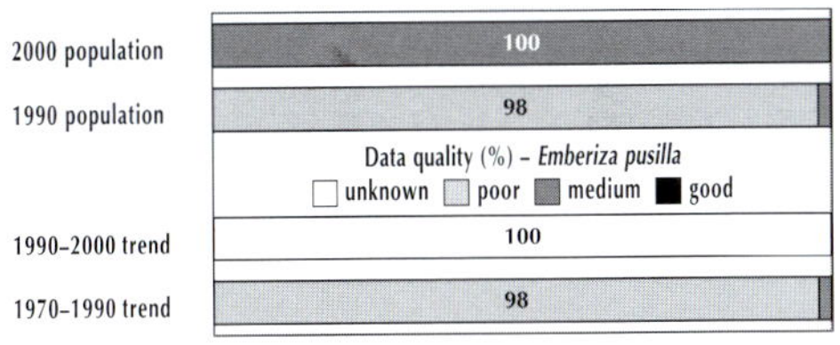

Emberiza aureola
YELLOW-BREASTED BUNTING

Country	Breeding pop. size (pairs)	Year(s)	Trend	Mag.%	References
Finland	**0 – 10**	98–02	–	**95**	
Russia	(20,000 – 100,000)	90–00	–	20–29	
Ukraine	**10 – 50**	90–00	**F**	**>80**	
Total (approx.)	20,000 – 100,000	Overall trend Moderate decline			
Breeding range	>1,000,000 km²	Gen. length <3.3	% Global pop. 25–49		

SPEC 1 (1994: —) **Status** Declining
Criteria Moderate recent decline
European IUCN Red List Category —
Criteria —
Global IUCN Red List Category NT
Criteria A2d; A3d

Emberiza aureola is a widespread summer visitor to north-eastern Europe (particularly Russia), which accounts for less than half of its global breeding range. Its European breeding population is relatively small (<100,000 pairs), but was stable between 1970–1990. However, the stronghold population in Russia declined markedly during 1990–2000, and the species underwent a moderate decline (>10%) overall. Consequently, this globally Near Threatened species, which was previously assessed as Secure in Europe, is now evaluated as Declining.

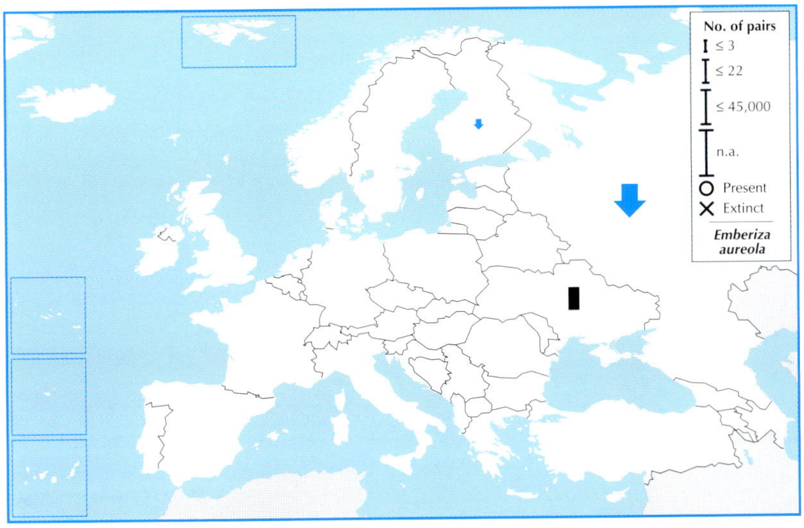

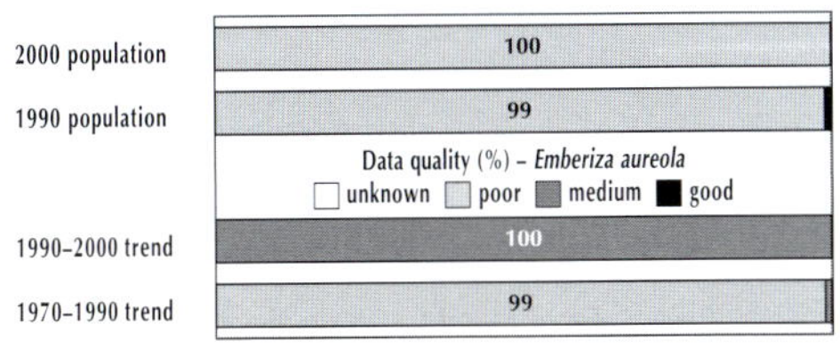

Emberiza schoeniclus
REED BUNTING

Non-SPEC (1994: —) **Status** Secure
Criteria —
European IUCN Red List Category —
Criteria —
Global IUCN Red List Category —
Criteria —

Emberiza schoeniclus is a widespread breeder across much of Europe, which accounts for less than half of its global breeding range. Its European breeding population is very large (>4,800,000 pairs), and was stable between 1970–1990. Although the species declined in several countries—notably Norway and Sweden—during 1990–2000, populations were stable across most of Europe, including key ones in Russia and Romania, and the species declined only slightly overall. Consequently, it is evaluated as Secure.

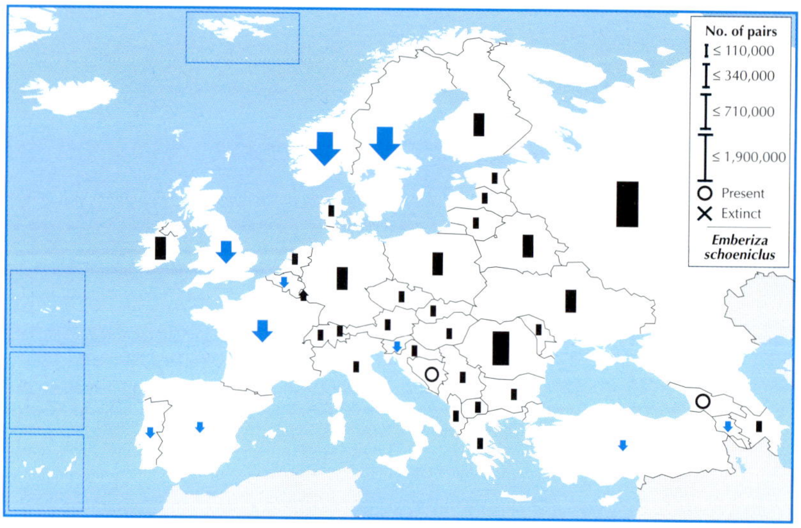

Country	Breeding pop. size (pairs)	Year(s)	Trend	Mag.%	References
Albania	(1,000 – 2,000)	02	(0)	(0–19)	
Armenia	200 – 500	97–02	–	20–29	
Austria	7,000 – 12,000	98–02	(0)	(0–19)	
Azerbaijan	(1,000 – 10,000)	96–00	(0)	(0–19)	
Belarus	200,000 – 350,000	97–02	0	0–19	
Belgium	3,100 – 6,200	01–02	–	0–19	*1*
Bosnia & HG	Present	85–89	?	–	
Bulgaria	**200 – 400**	96–02	0	0–9	
Croatia	(10,000 – 12,000)	02	(0)	(0–19)	*70,16*
Czech Rep.	40,000 – 80,000	00	0	0–19	
Denmark	40,000 – 60,000	00	**0**	**0–19**	*12*
Estonia	(50,000 – 80,000)	98	0	0–19	*1*
Finland	200,000 – 400,000	98–02	**0**	**5**	
France	(80,000 – 400,000)	98–02	**–**	**37**	*4,2*
Georgia	Present	03	?	–	
Germany	160,000 – 400,000	95–99	0	0–19	
Greece	(300 – 500)	95–00	(0)	(0–19)	
Hungary	**43,000 – 80,000**	99–02	0	0–19	*19*
Rep. Ireland	100,000 – 250,000	88–91	0	0–19	
Italy	(50,000 – 100,000)	03	(0)	(0–19)	
Latvia	50,000 – 70,000	90–00	(0)	(0–19)	*23,17,2*
Liechtenstein	25 – 35	98–00	0	0–19	
Lithuania	(80,000 – 150,000)	99–01	(0)	(0–19)	*20*
Luxembourg	800 – 1,000	02	+	0–19	
Macedonia	(250 – 1,000)	90–00	(0)	(0–19)	
Moldova	2,000 – 2,500	90–00	0	0–19	
Netherlands	**70,000 – 100,000**	98–00	**0**	**6**	*1*
Norway	(500,000 – 1,000,000)	90–00	–	0–19	*27*
Poland	**200,000 – 500,000**	00–02	0	0–19	*23*
Portugal	(100 – 1,000)	02	(–)	(–)	
Romania	580,000 – 745,000	00–02	(0)	(0–19)	*48*
Russia	1,400,000 – 2,500,000	90–00	0	0–19	*122*
Serbia & MN	5,000 – 7,000	95–02	(0)	(0–19)	*1,29,78,227, 203,101,99,144*
Slovakia	20,000 – 40,000	90–99	0	0–19	
Slovenia	100 – 200	99–00	(–)	(0–19)	
Spain	239 – 399	98–02	–	>80	*10,16*
Sweden	400,000 – 800,000	99–00	**–**	**18**	
Switzerland	**3,000 – 5,000**	93–96	0	0–19	
Turkey	(5,000 – 10,000)	01	(–)	(0–19)	
Ukraine	300,000 – 380,000	90–00	0	0–19	
UK	**192,000 – 211,000**	00	**–**	**12–20**	*5,31*
Total (approx.)	4,800,000 – 8,800,000	**Overall trend** Small decline			
Breeding range	>7,000,000 km²	**Gen. length.** <3.3		**% Global pop.** 25–49	

(See p. 288, bottom, for data quality graph)

Emberiza pallasi
PALLAS'S BUNTING

Non-SPEC (1994: NE) **Status** (Secure)
Criteria —
European IUCN Red List Category —
Criteria —
Global IUCN Red List Category —
Criteria —

Emberiza pallasi has a predominantly Asian breeding distribution, which just extends into Europe in northernmost Russia. Its European breeding population is relatively small (<150,000 pairs), but its trend between 1970–1990 was unknown. Trend data were also unavailable for 1990–2000, but there is no evidence to suggest that the species declined. Consequently, it is provisionally evaluated as Secure.

Country	Breeding pop. size (pairs)	Year(s)	Trend	Mag.%	References
Russia	(80,000 – 150,000)	90–00	?	–	*122*
Total (approx.)	80,000 – 150,000	**Overall trend** Unknown			
Breeding range	>20,000 km²	**Gen. length.** <3.3		**% Global pop.** <5	

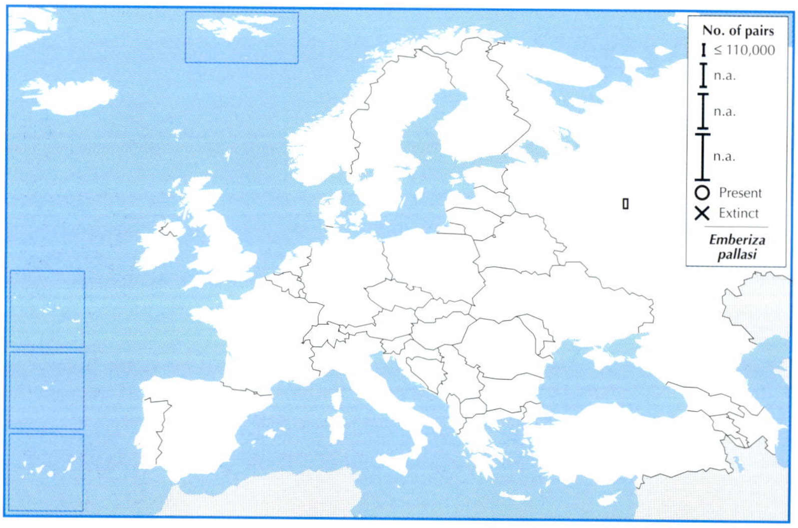

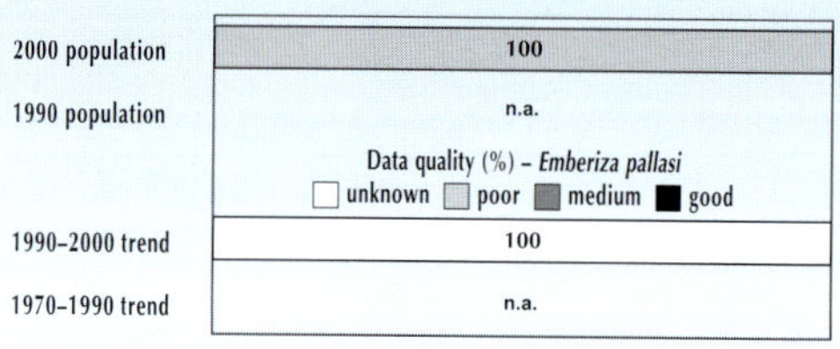

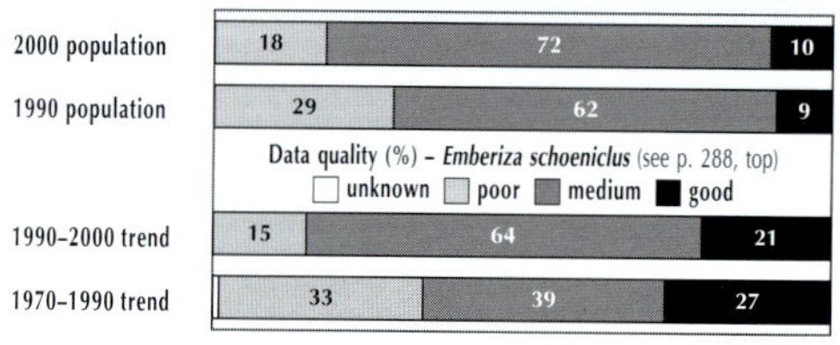

Emberiza bruniceps
RED-HEADED BUNTING

Non-SPEC (1994: —) **Status** (Secure)
Criteria —
European IUCN Red List Category —▼▼
Criteria —
Global IUCN Red List Category —
Criteria —

Country	Breeding pop. size (pairs)	Year(s)	Trend	Mag.%	References
Russia	(250 – 1,000)	90–00	?	–	*15,170*
Total (approx.)	250 – 1,000	**Overall trend** Unknown			
Breeding range	>20,000 km²	**Gen. length.** <3.3	**% Global pop.** <5		

Emberiza bruniceps has a predominantly Asian breeding distribution, which just extends into Europe in southern Russia. Its European breeding population is very small (as few as 250 pairs), but underwent a large increase between 1970–1990. No trend data were available for 1990–2000, but there is no evidence to suggest that it declined. Although the size of the European population could render it susceptible to the risks affecting small populations, it is marginal to a much larger non-European population. Consequently, the species is provisionally evaluated as Secure.

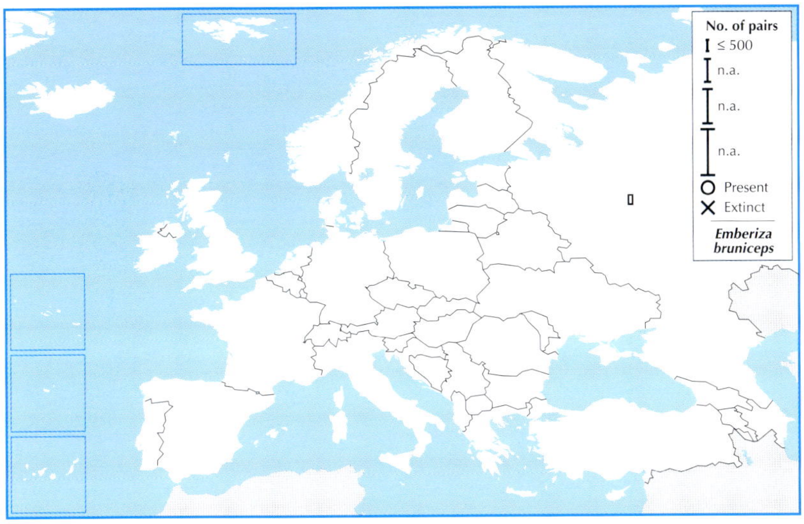

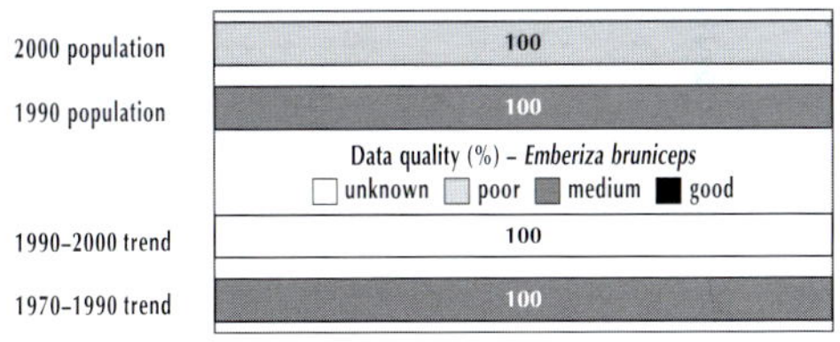

Emberiza melanocephala
BLACK-HEADED BUNTING

SPEC 2 (1994: 2) **Status** (Depleted)
Criteria Large historical decline
European IUCN Red List Category —
Criteria —
Global IUCN Red List Category —
Criteria —

Country	Breeding pop. size (pairs)	Year(s)	Trend	Mag.%	References
Albania	10,000 – 20,000	02	(0)	(0–19)	
Armenia	30,000 – 150,000	99–02	?	–	
Azerbaijan	(100,000 – 200,000)	96–00	(0)	(0–19)	
Bosnia & HG	(1,000 – 2,500)	90–00	(F)	(–)	
Bulgaria	12,000 – 25,000	98–02	+	30–49	
Croatia	(5,000 – 8,000)	02	(–)	(50–79)	*16*
Cyprus	(6,000 – 20,000)	94–02	(0)	(0–9)	
Georgia	Present	03	?	–	
Greece	(30,000 – 100,000)	95–00	(–)	(20–29)	
Italy	4,000 – 16,000	03	(–)	(0–19)	
Macedonia	(10,000 – 30,000)	90–00	(0)	(0–19)	
Romania	29 – 45	00–02	+	0–19	
Russia	(100,000 – 200,000)	90–00	?	–	*8*
Serbia & MN	550 – 850	95–03	+	0–19	*1,86,62,197, 155,141,91*
Turkey	(2,500,000 – 8,500,000)	01	(+)	(0–19)	
Ukraine	100 – 200	90–00	+	0–9	
Total (approx.)	2,800,000 – 9,300,000	**Overall trend** Small increase			
Breeding range	>1,000,000 km²	**Gen. length.** <3.3	**% Global pop.** 50–74		

Emberiza melanocephala is a widespread summer visitor to south-east Europe, which constitutes >50% of its global breeding range. Its European breeding population is very large (>2,800,000 pairs), but underwent a large decline between 1970–1990. Although the species increased slightly overall during 1990–2000—mainly due to the increase of the stronghold population in Turkey—its population has probably not yet recovered to the level that preceded its decline. Consequently, it is provisionally evaluated as Depleted.

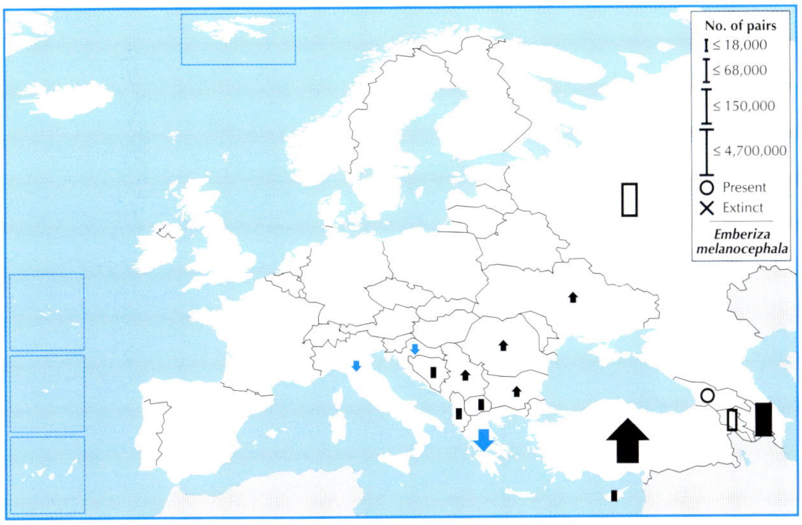

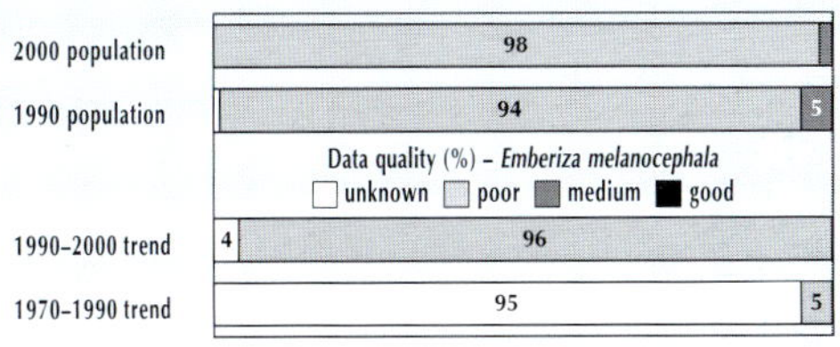

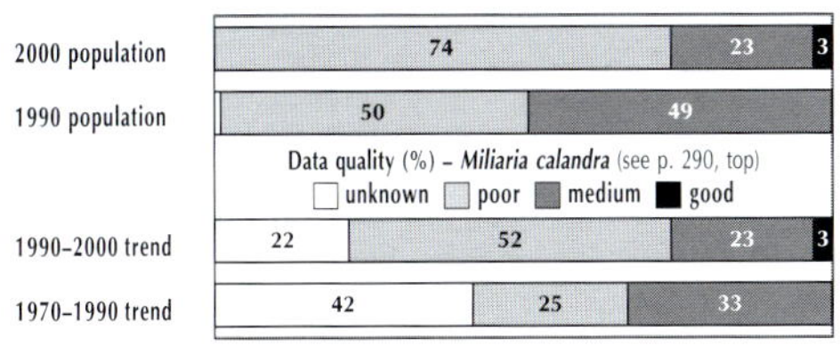

Miliaria calandra
CORN BUNTING

SPEC 2 (1994: 4) **Status** (Declining)
Criteria Moderate recent decline
European IUCN Red List Category —
Criteria —
Global IUCN Red List Category —
Criteria —

Miliaria calandra is a widespread breeder across much of Europe (except the north), which constitutes >50% of its global breeding range. Its European breeding population is very large (>7,900,000 pairs), and was stable between 1970–1990. Although certain populations—notably in Bulgaria and Romania—remained stable or increased during 1990–2000, the species declined across much of Europe, including the key population in Turkey, and underwent a moderate decline (>10%) overall. Consequently, this previously Secure species is now provisionally evaluated as Declining.

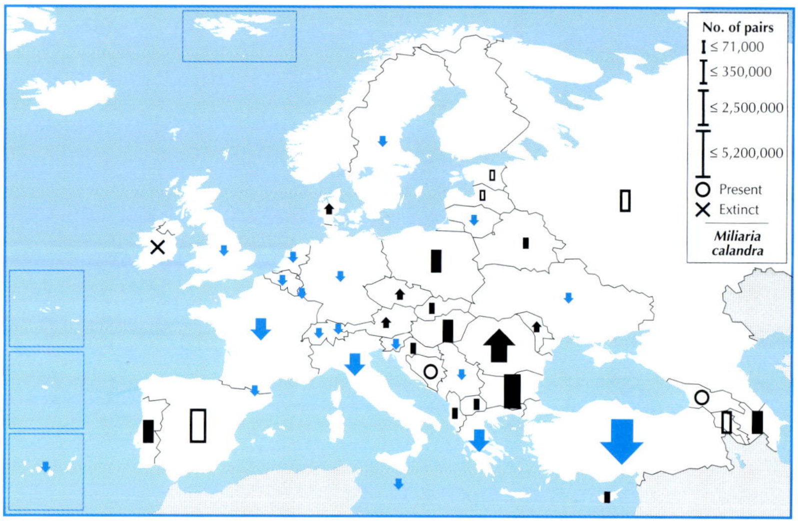

Country	Breeding pop. size (pairs)	Year(s)	Trend	Mag.%	References
Albania	20,000 – 40,000	02	(0)	(0–19)	
Andorra	0 – 5	99–01	(-)	(20–29)	*1,3*
Armenia	170,000 – 300,000	00–02	?	–	
Austria	(3,500 – 7,000)	98–02	+	**>80**	
Azerbaijan	(200,000 – 500,000)	96–00	(0)	(0–19)	
Belarus	600 – 1,000	97–02	0	0–19	
Belgium	1,800 – 4,200	01–02	–	30–49	*1*
Bosnia & HG	Present	90–03	?	–	
Bulgaria	800,000 – 2,500,000	96–02	0	0–19	
Croatia	(50,000 – 100,000)	02	(0)	(0–19)	*70,16,54*
Cyprus	(25,000 – 100,000)	94–02	(0)	(0–19)	
Czech Rep.	**4,000 – 8,000**	00	+	**>80**	
Denmark	40,000 – 60,000	00	+	**30–49**	*12*
Estonia	0 – 5	98	?	–	*1*
France	(150,000 – 600,000)	98–02	–	**24**	*4,2*
Georgia	Present	03	?	–	
Germany	13,000 – 320,00	95–99	–	0–19	
Greece	(200,000 – 500,000)	95–00	(-)	(0–19)	
Hungary	**165,000 – 225,000**	99–02	0	0–19	*19*
Rep. Ireland	0 – 0	00–02	–	X	
Italy	(200,000 – 600,000)	03	(-)	(0–9)	
Latvia	(0 – 10)	90–00	?	–	*23,16*
Liechtenstein	**2 – 5**	98–00	–	**50–79**	
Lithuania	(10 – 30)	99–01	(-)	(0–19)	*20*
Luxembourg	**20 – 40**	02	–	**30–49**	
Macedonia	(30,000 – 100,000)	90–00	(0)	(0–19)	
Malta	50 – 100	90–02	–	**50–79**	*1*
Moldova	1,500 – 2,000	90–00	+	0–19	
Netherlands	**50 – 100**	98–00	–	**69**	*1*
Poland	**150,000 – 400,000**	00–02	0	0–19	*23,107*
Portugal	(100,000 – 1,000,000)	02	(0)	(0–19)	
Romania	940,000 – 1,200,000	00–02	+	0–19	*48*
Russia	100,000 – 300,000	90–00	?	–	*8,122*
Serbia & MN	25,000 – 40,000	90–02	–	0–19	*1,29,172a,67a,155, 78,227,225,198*
Slovakia	4,000 – 8,000	90–99	0	0–19	
Slovenia	2,500 – 3,500	99–00	(-)	(0–19)	
Spain	(1,440,000 – 4,300,000)	92	?	–	*13,12,10*
Canary Is.	(2,500 – 10,000)	97–03	(-)	(0–19)	*28,25*
Sweden	**3 – 5**	99–00	–	**20–29**	
Switzerland	**400 – 600**	93–96	–	10–19	
Turkey	(3,000,000 – 9,000,000)	01	(-)	(0–19)	
Ukraine	(30,000 – 50,000)	90–00	(-)	(0–19)	
UK	**8,500 – 12,200**	00	–	**47**	*5,31*
Total (approx.)	7,900,000 – 22,000,000	**Overall trend** Moderate decline			
Breeding range	>5,000,000 km²	**Gen. length**. <3.3 **% Global pop.** 50–74			

(See p. 289, bottom, for data quality graph)

APPENDICES

Appendix 1. Conservation status, population estimates and trends of all European species.

Species	SPEC Category	European Threat Status (ETS)	ETS Criteria	European population size	1970–1990 European population trend	1990–2000 European population trend	2004 Global IUCN Red List Category	% Global population in Europe
Gavia stellata Red-throated Loon	SPEC 3	(H)	Large historical decline	32,000 – 92,000	Large decline	Stable	—	5–24
Gavia arctica Arctic Loon	SPEC 3	(VU)	A2b	51,000 – 92,000	Large decline	Large decline	—	5–24
Gavia immer Common Loon	Non-SPEC	(S)	—[▼]	700 – 2,300	Stable	Unknown	—	<5
Gavia adamsii Yellow-billed Loon	Non-SPEC	(S)	—[▼▼]	>500 **ind** [W]	Unknown	Stable	—	5–24 [W]
Tachybaptus ruficollis Little Grebe	Non-SPEC	S	—	99,000 – 170,000	Stable	Stable	—	5–24
Podiceps cristatus Great Crested Grebe	Non-SPEC	S	—	300,000 – 450,000	Large increase	Moderate decline	—	25–49
Podiceps grisegena Red-necked Grebe	Non-SPEC	S	—	32,000 – 56,000	Stable	Small decline	—	5–24
Podiceps auritus Horned Grebe	SPEC 3	D	Moderate recent decline	6,300 – 11,000	Stable	Moderate decline	—	5–24
Podiceps nigricollis Black-necked Grebe	Non-SPEC	S	—	53,000 – 96,000	Large increase	Small decline	—	5–24
Fulmarus glacialis Northern Fulmar	Non-SPEC	S	—	2,800,000 – 4,400,000	Large increase	Large increase	—	25–49
Pterodroma madeira Zino's Petrel	SPEC 1	(CR)	D1	30 – 40	Stable	Stable	CR: D1	100
Pterodroma feae Fea's Petrel	SPEC 1	VU	D1; D2	170 – 260	Stable	Stable	NT: D1; D2	25–49
Bulweria bulwerii Bulwer's Petrel	SPEC 3	(R)	<10,000 pairs	7,000 – 9,000	Moderate decline	Stable	—	5–24
Calonectris diomedea Cory's Shearwater	SPEC 2	(VU)	A4b	270,000 – 290,000	Large decline	Small decline	—	75–94
Puffinus gravis Great Shearwater	—	NE	—	—	—	—	—	—
Puffinus griseus Sooty Shearwater	SPEC 1	NE	—	—	—	—	NT: A2d,e; A3d,e	—
Puffinus puffinus Manx Shearwater	SPEC 2	(L)	≥90% breed at ≤10 sites	350,000 – 390,000	Stable	Unknown	—	>95
Puffinus mauretanicus Balearic Shearwater	SPEC 1	CR	A4b,c,e; B2a+b(ii,iii,iv,v)	1,700 – 2,000	Moderate decline	Large decline	CR: A4b,c,e; B2a+b(ii,iii,iv,v)	100
Puffinus yelkouan Yelkouan Shearwater	Non-SPEC[E]	S	—	13,000 – 33,000	Stable	Stable	—	>95
Puffinus assimilis Little Shearwater	SPEC 3	(R)	<10,000 pairs	5,200 – 6,900	Moderate decline	Stable	—	5–24
Pelagodroma marina White-faced Storm-petrel	SPEC 3	VU	D2	61,000 – 61,000	Stable	Stable	—	5–24
Hydrobates pelagicus European Storm-petrel	Non-SPEC[E]	(S)	—	430,000 – 510,000	Stable	Stable	—	>95
Oceanodroma leucorhoa Leach's Storm-petrel	SPEC 3	(L)	≥90% breed at ≤10 sites	120,000 – 220,000	Stable	Unknown	—	5–24
Oceanodroma castro Band-rumped Storm-petrel	SPEC 3	(R)	<10,000 pairs	3,700 – 4,800	Moderate decline	Stable	—	25–49
Morus bassanus Northern Gannet	Non-SPEC[E]	S	—	300,000 – 310,000	Large increase	Large increase	—	75–94
Phalacrocorax carbo Great Cormorant	Non-SPEC	S	—	310,000 – 370,000	Large increase	Large increase	—	25–49
Phalacrocorax aristotelis European Shag	Non-SPEC[E]	(S)	—	75,000 – 81,000	Large increase	Moderate decline	—	75–94
Phalacrocorax pygmeus Pygmy Cormorant	SPEC 1	S	—	28,000 – 39,000	Moderate decline	Moderate increase	NT: A2c; A3c	75–94
Pelecanus onocrotalus Great White Pelican	SPEC 3	R	<10,000 pairs	4,100 – 5,100	Large increase	Stable	—	5–24
Pelecanus crispus Dalmatian Pelican	SPEC 1	R	<10,000 pairs	1,600 – 2,000	Large increase	Moderate increase	VU: A2c; A3c	25–49
Botaurus stellaris Great Bittern	SPEC 3	H	Large historical decline	34,000 – 54,000	Large decline	Stable	—	25–49
Ixobrychus minutus Little Bittern	SPEC 3	(H)	Large historical decline	60,000 – 120,000	Large decline	Stable	—	5–24
Nycticorax nycticorax Black-crowned Night-heron	SPEC 3	H	Moderate historical decline	63,000 – 87,000	Moderate decline	Stable	—	5–24
Ardeola ralloides Squacco Heron	SPEC 3	(D)	Moderate continuing decline	18,000 – 27,000	Large decline	Moderate decline	—	5–24
Bubulcus ibis Cattle Egret	Non-SPEC	S	—	54,000 – 150,000	Large increase	Small increase	—	5–24
Egretta garzetta Little Egret	Non-SPEC	S	—	68,000 – 94,000	Large increase	Small increase	—	5–24
Casmerodius albus Great Egret	Non-SPEC	S	—	11,000 – 24,000	Large increase	Large increase	—	<5
Ardea cinerea Grey Heron	Non-SPEC	S	—	210,000 – 290,000	Large increase	Moderate increase	—	5–24
Ardea purpurea Purple Heron	SPEC 3	(D)	Moderate continuing decline	29,000 – 42,000	Large decline	Moderate decline	—	5–24
Ciconia nigra Black Stork	SPEC 2	R	<10,000 pairs	7,800 – 12,000	Stable	Stable	—	50–74
Ciconia ciconia White Stork	SPEC 2	H	Large historical decline	180,000 – 220,000	Large decline	Moderate increase	—	75–94
Plegadis falcinellus Glossy Ibis	SPEC 3	(D)	Moderate continuing decline	16,000 – 22,000	Moderate decline	Moderate decline	—	<5
Geronticus eremita Northern Bald Ibis	SPEC 1	CR	D1	15 – 15	Large decline	Stable	CR: C2a(ii)	5–24
Platalea leucorodia Eurasian Spoonbill	SPEC 2	R	<10,000 pairs	8,900 – 15,000	Large decline	Stable	—	50–74
Phoenicopterus roseus Greater Flamingo	SPEC 3	L	≥90% breed at ≤10 sites	56,000 – 58,000	Large increase	Large increase	—	25–49
Cygnus olor Mute Swan	Non-SPEC[E]	S	—	86,000 – 120,000	Moderate increase	Large increase	—	50–74
Cygnus columbianus Tundra Swan	SPEC 3W	VU	A2b	>23,000 **ind** [W]	Stable	Large decline	—	5–24 [W]
Cygnus cygnus Whooper Swan	Non-SPEC[E]W	S	—	>65,000 **ind** [W]	Large increase	Large increase	—	50–74 [W]
Anser fabalis Bean Goose	Non-SPEC[E]W	S	—	>390,000 **ind** [W]	Stable	Stable	—	50–74 [W]
Anser brachyrhynchus Pink-footed Goose	Non-SPEC[E]	S	—	50,000 – 69,000	Large increase	Moderate increase	—	100
Anser albifrons Greater White-fronted Goose	Non-SPEC	S	—	62,000 – 72,000	Stable	Large increase	—	5–24
Anser erythropus Lesser White-fronted Goose	SPEC 1	EN	C1	240 – 460	Stable	Moderate decline	VU: A2b,c,d; A3b,c,d	5–24
Anser anser Greylag Goose	Non-SPEC	S	—	120,000 – 190,000	Large increase	Large increase	—	25–49
Chen caerulescens Snow Goose	Non-SPEC	(S)	—[▼▼]	200 – 2,000	Unknown	Small increase	—	<5
Branta canadensis Canada Goose	Non-SPEC	(S)	—[▼]	2,500 – 10,000	Unknown	Small increase	—	<5
Branta leucopsis Barnacle Goose	Non-SPEC[E]	S	—	41,000 – 54,000	Large increase	Large increase	—	100
Branta bernicla Brent Goose	SPEC 3W	VU	A2b	>240,000 **ind** [W]	Large increase	Large decline	—	25–49 [W]
Branta ruficollis Red-breasted Goose	SPEC 1W	VU	B2a+b(iii)	>27,000 **ind** [W]	Large increase	Fluctuating	VU: B2a+b(iii)	>95 [W]
Tadorna ferruginea Ruddy Shelduck	SPEC 3	(VU)	A2b	19,000 – 33,000	Large decline	Large decline	—	5–24
Tadorna tadorna Common Shelduck	Non-SPEC	S	—	42,000 – 65,000	Moderate increase	Stable	—	25–49
Anas penelope Eurasian Wigeon	Non-SPEC[E]W	S	—	>1,700,000 **ind** [W]	Moderate increase	Stable	—	50–74 [W]

Appendix 1 ...continued. Conservation status, population estimates and trends of all European species.

Species	SPEC Category	European Threat Status (ETS)	ETS Criteria	European population size	1970–1990 European population trend	1990–2000 European population trend	2004 Global IUCN Red List Category	% Global population in Europe
Anas strepera Gadwall	SPEC 3	(H)	Large historical decline	60,000 – 96,000	Large decline	Unknown	—	5–24
Anas crecca Common Teal	Non-SPEC	(S)	—	920,000 – 1,200,000	Stable	Unknown	—	5–24
Anas platyrhynchos Mallard	Non-SPEC	(S)	—	3,300,000 – 5,100,000	Stable	Small decline	—	5–24
Anas acuta Northern Pintail	SPEC 3	(D)	Moderate continuing decline	320,000 – 360,000	Large decline	Moderate decline	—	5–24
Anas querquedula Garganey	SPEC 3	(D)	Moderate continuing decline	390,000 – 590,000	Large decline	Moderate decline	—	25–49
Anas clypeata Northern Shoveler	SPEC 3	(D)	Moderate recent decline	170,000 – 210,000	Stable	Moderate decline	—	5–24
Marmaronetta angustirostris Marbled Teal	SPEC 1	(VU)	C1; C2a(i); D1	390 – 1,000	Large decline	Moderate decline	VU: A2c,d; A3c,d	5–24
Netta rufina Red-crested Pochard	Non-SPEC	(S)	—	27,000 – 59,000	Moderate decline	Moderate increase	—	25–49
Aythya ferina Common Pochard	SPEC 2	(D)	Moderate recent decline	210,000 – 440,000	Stable	Moderate decline	—	50–74
Aythya nyroca Ferruginous Duck	SPEC 1	(VU)	A2b	12,000 – 18,000	Large decline	Large decline	NT: A2c,d; A3c,d	25–49
Aythya fuligula Tufted Duck	SPEC 3	(D)	Moderate recent decline	730,000 – 880,000	Stable	Moderate decline	—	25–49
Aythya marila Greater Scaup	SPEC 3W	EN	A2b L	>120,000 **ind** [W]	Stable	Large decline	—	25–49 [W]
Somateria mollissima Common Eider	Non-SPEC[E]	S	—	840,000 – 1,200,000	Large increase	Small increase	—	50–74
Somateria spectabilis King Eider	Non-SPEC	(S)	—	37,000 – 46,000	Stable	Stable	—	<5
Polysticta stelleri Steller's Eider	SPEC 3W	L	≥90% winter at ≤10 sites	>7,700 **ind** [W]	Moderate increase	Fluctuating	—	<5 [W]
Histrionicus histrionicus Harlequin Duck	SPEC 3	(R)	<10,000 pairs	4,000 – 10,000	Large increase	Unknown	—	<5
Clangula hyemalis Long-tailed Duck	Non-SPEC	(S)	—	690,000 – 750,000	Stable	Stable	—	25–49
Melanitta nigra Black Scoter	Non-SPEC	(S)	—	100,000 – 130,000	Stable	Stable	—	5–24
Melanitta fusca White-winged Scoter	SPEC 3	(D)	Moderate recent decline L	85,000 – 100,000	Stable	Moderate decline	—	5–24
Bucephala islandica Barrow's Goldeneye	SPEC 3	VU	D2	500 – 600	Stable	Fluctuating	—	<5
Bucephala clangula Common Goldeneye	Non-SPEC	(S)	—	490,000 – 590,000	Moderate increase	Small increase	—	25–49
Mergellus albellus Smew	SPEC 3	(D)	Moderate continuing decline	5,300 – 8,400	Large decline	Moderate decline	—	5–24
Mergus serrator Red-breasted Merganser	Non-SPEC	(S)	—	73,000 – 120,000	Stable	Small decline	—	5–24
Mergus merganser Common Merganser	Non-SPEC	(S)	—	47,000 – 74,000	Moderate increase	Small decline	—	5–24
Oxyura leucocephala White-headed Duck	SPEC 1	VU	A3e	550 – 1,400	Moderate increase	Moderate increase	EN: A2b,c,d,e	5–24
Pernis apivorus European Honey-buzzard	Non-SPEC[E]	(S)	—	110,000 – 160,000	Stable	Stable	—	75–94
Elanus caeruleus Black-winged Kite	SPEC 3	R	<10,000 pairs	810 – 2,000	Large increase	Small increase	—	<5
Milvus migrans Black Kite	SPEC 3	(VU)	A2b	64,000 – 100,000	Large decline	Large decline	—	5–24
Milvus milvus Red Kite	SPEC 2	D	Moderate recent decline	19,000 – 25,000	Stable	Moderate decline	—	>95
Haliaeetus albicilla White-tailed Eagle	SPEC 1	R	<10,000 pairs	5,000 – 6,600	Large increase	Large increase	NT: C2a(i)	50–74
Gypaetus barbatus Lammergeier	SPEC 3	(VU)	C1; C2a(i)	610 – 1,000	Stable	Moderate decline	—	5–24
Neophron percnopterus Egyptian Vulture	SPEC 3	EN	A2b	3,500 – 5,600	Large decline	Large decline	—	25–49
Gyps fulvus Eurasian Griffon	Non-SPEC	S	—	19,000 – 21,000	Large increase	Large increase	—	25–49
Aegypius monachus Cinereous Vulture	SPEC 1	R	<10,000 pairs	1,800 – 1,900	Moderate increase	Large increase	NT: C1	25–49
Circaetus gallicus Short-toed Snake-eagle	SPEC 3	(R)	<10,000 pairs	8,400 – 13,000	Stable	Small decline	—	25–49
Circus aeruginosus Western Marsh-harrier	Non-SPEC	S	—	93,000 – 140,000	Large increase	Moderate increase	—	25–49
Circus cyaneus Northern Harrier	SPEC 3	H	Large historical decline	32,000 – 59,000	Large decline	Small decline	—	5–24
Circus macrourus Pallid Harrier	SPEC 1	(EN)	C1	310 – 1,200	Large decline	Large decline	NT: A2c,d,e; A3c,d,e	25–49
Circus pygargus Montagu's Harrier	Non-SPEC[E]	S	—	35,000 – 65,000	Large increase	Moderate increase	—	50–74
Accipiter gentilis Northern Goshawk	Non-SPEC	S	—	160,000 – 210,000	Large increase	Moderate increase	—	25–49
Accipiter nisus Eurasian Sparrowhawk	Non-SPEC	S	—	340,000 – 450,000	Large increase	Small increase	—	25–49
Accipiter badius Shikra	Non-SPEC	(S)	—▼▼▼▼	10 – 50	Unknown	Stable	—	<5
Accipiter brevipes Levant Sparrowhawk	SPEC 2	(VU)	C1	3,200 – 7,700	Stable	Moderate decline	—	75–94
Buteo buteo Common Buzzard	Non-SPEC	S	—	710,000 – 1,200,000	Moderate increase	Small increase	—	25–49
Buteo rufinus Long-legged Buzzard	SPEC 3	(VU)	A2b	8,700 – 15,000	Large decline	Large decline	—	5–24
Buteo lagopus Rough-legged Hawk	Non-SPEC	(S)	—	38,000 – 79,000	Stable	Fluctuating	—	5–24
Aquila pomarina Lesser Spotted Eagle	SPEC 2	(D)	Moderate recent decline	14,000 – 19,000	Stable	Moderate decline	—	>95
Aquila clanga Greater Spotted Eagle	SPEC 1	EN	C1	810 – 1,100	Large decline	Large decline	VU: C1	25–49
Aquila nipalensis Steppe Eagle	SPEC 3	(EN)	A2b	5,000 – 20,000	Large decline	Large decline	—	<5
Aquila heliaca Imperial Eagle	SPEC 1	R	<10,000 pairs	850 – 1,400	Large decline	Stable	VU: C1	25–49
Aquila adalberti Spanish Imperial Eagle	SPEC 1	(EN)	C1; C2a(i)	180 – 180	Large increase	Small increase	EN: C1; C2a(i)	100
Aquila chrysaetos Golden Eagle	SPEC 3	R	<10,000 pairs	8,400 – 11,000	Stable	Stable	—	5–24
Hieraaetus pennatus Booted Eagle	SPEC 3	(R)	<10,000 pairs	4,400 – 8,900	Stable	Stable	—	25–49
Hieraaetus fasciatus Bonelli's Eagle	SPEC 3	EN	C1	920 – 1,100	Large decline	Large decline	—	5–24
Pandion haliaetus Osprey	SPEC 3	R	<10,000 pairs	7,600 – 11,000	Moderate increase	Moderate increase	—	5–24
Falco naumanni Lesser Kestrel	SPEC 1	H	Large historical decline	25,000 – 42,000	Large decline	Small decline	VU: A2b,c,e; A3b,c,e	25–49
Falco tinnunculus Common Kestrel	SPEC 3	D	Moderate continuing decline	330,000 – 500,000	Moderate decline	Moderate decline	—	5–24
Falco vespertinus Red-footed Falcon	SPEC 3	(VU)	A2b	26,000 – 39,000	Large decline	Large decline	—	25–49
Falco columbarius Merlin	Non-SPEC	(S)	—	31,000 – 49,000	Stable	Stable	—	5–24
Falco subbuteo Eurasian Hobby	Non-SPEC	(S)	—	71,000 – 120,000	Stable	Stable	—	5–24
Falco eleonorae Eleonora's Falcon	SPEC 2	D	Moderate recent decline	5,900 – 6,200	Stable	Moderate decline	—	>95
Falco biarmicus Lanner Falcon	SPEC 3	VU	C1; C2a(i); D1	480 – 900	Large decline	Moderate decline	—	<5
Falco cherrug Saker Falcon	SPEC 1	EN	C1	360 – 540	Large decline	Large decline	EN: A2b,c,d; A3b,c,d	<5
Falco rusticolus Gyrfalcon	SPEC 3	(R)	<10,000 pairs	1,300 – 2,300	Stable	Stable	—	<5
Falco peregrinus Peregrine Falcon	Non-SPEC	S	—	12,000 – 25,000	Moderate increase	Moderate increase	—	5–24
Falco pelegrinoides Barbary Falcon	Non-SPEC	S	—▼▼▼	75 – 80	Stable	Moderate increase	—	<5

Appendix 1 ...continued. Conservation status, population estimates and trends of all European species.

Species	SPEC Category	European Threat Status (ETS)	ETS Criteria	European population size	1970–1990 European population trend	1990–2000 European population trend	2004 Global IUCN Red List Category	% Global population in Europe
Bonasa bonasia Hazel Grouse	Non-SPEC	S	—	2,500,000 – 3,100,000	Stable	Small increase	—	25–49
Lagopus lagopus Willow Ptarmigan	Non-SPEC	S	—	2,100,000 – 3,300,000	Stable	Fluctuating	—	25–49
Lagopus mutus Rock Ptarmigan	Non-SPEC	S	—	430,000 – 1,400,000	Stable	Fluctuating	—	25–49
Tetrao tetrix Black Grouse	SPEC 3	H	Large historical decline	2,500,000 – 3,200,000	Large decline	Small decline	—	25–49
Tetrao mlokosiewiczi Caucasian Grouse	SPEC 1	DD	—	59,000 – 110,000 **ind.**	Unknown	Unknown	DD	>95
Tetrao urogallus Western Capercaillie	Non-SPEC	(S)	—	760,000 – 1,000,000	Stable	Stable	—	25–49
Tetraogallus caucasicus Caucasian Snowcock	Non-SPEC[E]	S	—	32,000 – 67,000	Stable	Stable	—	100
Tetraogallus caspius Caspian Snowcock	SPEC 3	(VU)	C1	2,500 – 9,000	Unknown	Moderate decline	—	5–24
Alectoris chukar Chukar	SPEC 3	(VU)	A2b	660,000 – 1,900,000	Large decline	Large decline	—	5–24
Alectoris graeca Rock Partridge	SPEC 2	(D)	Moderate continuing decline	40,000 – 78,000	Large decline	Moderate decline	—	100
Alectoris rufa Red-legged Partridge	SPEC 2	(D)	Moderate continuing decline	2,000,000 – 4,500,000	Large decline	Moderate decline	—	100
Alectoris barbara Barbary Partridge	SPEC 3	(R)	<10,000 pairs	7,500 – 20,000	Large decline	Unknown	—	<5
Ammoperdix griseogularis See-see Partridge	SPEC 3	VU	C1	1,000 – 2,100	Unknown	Moderate decline	—	<5
Francolinus francolinus Black Francolin	SPEC 3	D	Moderate continuing decline	6,000 – 18,000	Large decline	Moderate decline	—	<5
Perdix perdix Grey Partridge	SPEC 3	VU	A2b	1,600,000 – 3,100,000	Large decline	Large decline	—	25–49
Coturnix coturnix Common Quail	SPEC 3	(H)	Large historical decline	2,800,000 – 4,700,000	Large decline	Fluctuating	—	5–24
Phasianus colchicus Common Pheasant	Non-SPEC	(S)	—	3,400,000 – 4,700,000	Stable	Unknown	—	5–24
Turnix sylvatica Small Buttonquail	SPEC 3	CR	D1	0 – 1	Large decline	Unknown	—	<5
Rallus aquaticus Water Rail	Non-SPEC	(S)	—	140,000 – 360,000	Stable	Small decline	—	25–49
Porzana porzana Spotted Crake	Non-SPEC[E]	(S)	—	120,000 – 260,000	Stable	Fluctuating	—	50–74
Porzana parva Little Crake	Non-SPEC[E]	(S)	—	61,000 – 140,000	Large increase	Small decline	—	75–94
Porzana pusilla Baillon's Crake	SPEC 3	(R)	<10,000 pairs	760 – 3,200	Stable	Unknown	—	5–24
Crex crex Corncrake	SPEC 1	H	Large historical decline	1,300,000 – 2,000,000	Large decline	Fluctuating	NT: A3c	50–74
Gallinula chloropus Common Moorhen	Non-SPEC	S	—	900,000 – 1,700,000	Stable	Stable	—	5–24
Porphyrio porphyrio Purple Swamphen	SPEC 3	L	≥90% breed at ≤10 sites [L]	13,000 – 35,000	Large increase	Moderate increase	—	5–24
Fulica atra Common Coot	Non-SPEC	(S)	—	1,300,000 – 2,300,000	Moderate increase	Moderate decline	—	25–49
Fulica cristata Red-knobbed Coot	SPEC 3	CR	C2a(ii); C2b	80 – 80	Stable	Fluctuating	—	<5
Grus grus Common Crane	SPEC 2	(H)	Large historical decline	74,000 – 110,000	Large decline	Large increase	—	50–74
Grus virgo Demoiselle Crane	Non-SPEC	S	—	20,000 – 25,000	Large increase	Large increase	—	25–49
Tetrax tetrax Little Bustard	SPEC 1	VU	A2b	120,000 – 300,000 **ind.**	Large decline	Moderate decline	NT: A2c,d; A3c,d	75–94
Chlamydotis undulata Houbara Bustard	SPEC 1	(VU)	D1	530 – 560 **ind.**	Unknown	Unknown	VU: A2b,c,d; A3b,c,d	<5
Otis tarda Great Bustard	SPEC 1	VU	A2b	31,000 – 36,000 **ind.**	Moderate decline	Stable	VU: A3c	50–74
Haematopus ostralegus Eurasian Oystercatcher	Non-SPEC[E]	(S)	—	300,000 – 450,000	Large increase	Moderate decline	—	50–74
Himantopus himantopus Black-winged Stilt	Non-SPEC	S	—	37,000 – 64,000	Stable	Fluctuating	—	5–24
Recurvirostra avosetta Pied Avocet	Non-SPEC	S	—	38,000 – 57,000	Large increase	Stable	—	25–49
Burhinus oedicnemus Eurasian Thick-knee	SPEC 3	(VU)	A2b	46,000 – 78,000	Large decline	Large decline	—	25–49
Cursorius cursor Cream-coloured Courser	SPEC 3	(EN)	D1	100 – 620	Large decline	Unknown	—	<5
Glareola pratincola Collared Pratincole	SPEC 3	D	Moderate continuing decline	10,000 – 18,000	Large decline	Moderate decline	—	5–24
Glareola nordmanni Black-winged Pratincole	SPEC 1	EN	A2b	2,500 – 5,100	Stable	Large decline	DD	25–49
Charadrius dubius Little Ringed Plover	Non-SPEC	(S)	—	110,000 – 240,000	Moderate increase	Small decline	—	25–49
Charadrius hiaticula Common Ringed Plover	Non-SPEC[E]	(S)	—	120,000 – 220,000	Stable	Small decline	—	50–74
Charadrius alexandrinus Kentish Plover	SPEC 3	(D)	Moderate continuing decline	22,000 – 35,000	Moderate decline	Moderate decline	—	5–24
Charadrius leschenaultii Greater Sand Plover	SPEC 3	(EN)	C1	1,000 – 1,800	Unknown	Moderate decline	—	<5
Charadrius asiaticus Caspian Plover	SPEC 3	EN	C2a(ii)	130 – 500	Large decline	Unknown	—	<5
Eudromias morinellus Eurasian Dotterel	Non-SPEC	(S)	—	11,000 – 42,000	Stable	Small decline	—	25–49
Pluvialis apricaria Eurasian Golden-plover	Non-SPEC[E]	(S)	—	460,000 – 740,000	Stable	Unknown	—	50–74
Pluvialis squatarola Grey Plover	Non-SPEC	(S)	—▼	2,100 – 11,000	Stable	Unknown	—	<5
Vanellus spinosus Spur-winged Lapwing	SPEC 3	VU	C1; C2a(i)	1,000 – 1,600	Large decline	Moderate decline	—	<5
Vanellus indicus Red-wattled Lapwing	SPEC 3	(VU)	A3c	40 – 80	Large increase	Stable	—	<5
Vanellus gregarius Sociable Lapwing	SPEC 1	CR	A2b; C1	25 – 80	Large decline	Large decline	CR: A3b,c	5–24
Vanellus leucurus White-tailed Lapwing	Non-SPEC	S	—▼▼▼	80 – 320	Unknown	Moderate increase	—	<5
Vanellus vanellus Northern Lapwing	SPEC 2	VU	A2b; A3b,c	1,700,000 – 2,800,000	Stable	Large decline	—	50–74
Calidris canutus Red Knot	SPEC 3W	D	Moderate recent decline	>470,000 **ind** [W]	Stable	Moderate decline	—	25–49 [W]
Calidris alba Sanderling	Non-SPEC	(S)	—	25,000 – 50,000	Stable	Unknown	—	5–24
Calidris minuta Little Stint	Non-SPEC	(S)	—	46,000 – 460,000	Stable	Stable	—	25–49
Calidris temminckii Temminck's Stint	Non-SPEC	(S)	—	85,000 – 420,000	Stable	Stable	—	25–49
Calidris bairdii Baird's Sandpiper	Non-SPEC	(S)	—▼	500 – 1,000	Unknown	Unknown	—	<5
Calidris ferruginea Curlew Sandpiper	—	NE	—	—	—	—	—	—
Calidris maritima Purple Sandpiper	Non-SPEC[E]	(S)	—	28,000 – 75,000	Stable	Unknown	—	75–94
Calidris alpina Dunlin	SPEC 3	(H)	Large historical decline	300,000 – 570,000	Large decline [W]	Unknown	—	25–49
Limicola falcinellus Broad-billed Sandpiper	SPEC 3	(D)	Moderate continuing decline	9,200 – 22,000	Large decline	Moderate decline	—	25–49
Philomachus pugnax Ruff	SPEC 2	(D)	Moderate recent decline	200,000 – 510,000	Stable	Moderate decline	—	50–74
Lymnocryptes minimus Jack Snipe	SPEC 3	(D)	Moderate recent decline	18,000 – 70,000	Large decline [W]	Moderate decline	—	25–49
Gallinago gallinago Common Snipe	SPEC 3	(D)	Moderate recent decline	930,000 – 1,900,000	Stable	Moderate decline	—	5–24
Gallinago media Great Snipe	SPEC 1	D	Moderate continuing decline	62,000 – 170,000	Large decline	Moderate decline	NT: A2c,d; A3c,d	50–74
Gallinago stenura Pintail Snipe	Non-SPEC	(S)	—▼	1,000 – 2,500	Unknown	Unknown	—	<5

Appendix 1 ...continued. Conservation status, population estimates and trends of all European species.

Species	SPEC Category	European Threat Status (ETS)	ETS Criteria	European population size	1970–1990 European population trend	1990–2000 European population trend	2004 Global IUCN Red List Category	% Global population in Europe
Scolopax rusticola Eurasian Woodcock	SPEC 3	(D)	Moderate recent decline	1,800,000 – 6,600,000	Large decline [W]	Moderate decline	—	25–49
Limosa limosa Black-tailed Godwit	SPEC 2	VU	A2b	99,000 – 140,000	Large decline	Large decline	—	50–74
Limosa lapponica Bar-tailed Godwit	Non-SPEC	(S)	—[▼]	1,400 – 7,400	Stable	Stable	—	<5
Numenius phaeopus Whimbrel	Non-SPEC[E]	(S)	—	160,000 – 360,000	Stable	Unknown	—	50–74
Numenius tenuirostris Slender-billed Curlew	SPEC 1	NE	—	—	—	—	CR: C2a(ii); D1	—
Numenius arquata Eurasian Curlew	SPEC 2	D	Moderate continuing decline	220,000 – 360,000	Moderate decline [W]	Moderate decline	—	50–74
Tringa erythropus Spotted Redshank	SPEC 3	(D)	Moderate recent decline	19,000 – 42,000	Stable	Moderate decline	—	5–24
Tringa totanus Common Redshank	SPEC 2	D	Moderate continuing decline	280,000 – 610,000	Moderate decline	Moderate decline	—	50–74
Tringa stagnatilis Marsh Sandpiper	Non-SPEC	(S)	—	12,000 – 32,000	Large increase	Moderate decline	—	25–49
Tringa nebularia Common Greenshank	Non-SPEC	S	—	75,000 – 160,000	Stable	Stable	—	5–24
Tringa ochropus Green Sandpiper	Non-SPEC	S	—	330,000 – 800,000	Stable	Fluctuating	—	25–49
Tringa glareola Wood Sandpiper	SPEC 3	H	Moderate historical decline	350,000 – 1,200,000	Moderate decline	Stable	—	25–49
Xenus cinereus Terek Sandpiper	Non-SPEC	(S)	—	15,000 – 81,000	Stable	Fluctuating	—	5–24
Actitis hypoleucos Common Sandpiper	SPEC 3	(D)	Moderate recent decline	720,000 – 1,600,000	Stable	Moderate decline	—	25–49
Arenaria interpres Ruddy Turnstone	Non-SPEC	(S)	—	34,000 – 81,000	Stable	Stable	—	5–24
Phalaropus lobatus Red-necked Phalarope	Non-SPEC	(S)	—	85,000 – 220,000	Stable	Small decline	—	5–24
Phalaropus fulicarius Grey Phalarope	Non-SPEC	S	—[▼▼]	390 – 1,700	Stable	Stable	—	<5
Stercorarius pomarinus Pomarine Jaeger	Non-SPEC	(S)	—	20,000 – 50,000	Stable	Unknown	—	5–24
Stercorarius parasiticus Parasitic Jaeger	Non-SPEC	(S)	—	40,000 – 140,000	Stable	Unknown	—	5–24
Stercorarius longicaudus Long-tailed Jaeger	Non-SPEC	(S)	—	12,000 – 78,000	Stable	Fluctuating	—	5–24
Catharacta skua Great Skua	Non-SPEC[E]	S	—	16,000 – 16,000	Moderate increase	Large increase	—	100
Larus ichthyaetus Great Black-headed Gull	Non-SPEC	(S)	—	16,000 – 27,000	Large increase	Moderate decline	—	5–24
Larus melanocephalus Mediterranean Gull	Non-SPEC[E]	S	—	120,000 – 320,000	Large increase	Stable	—	100
Larus minutus Little Gull	SPEC 3	(H)	Moderate historical decline	24,000 – 58,000	Moderate decline	Moderate increase	—	25–49
Larus ridibundus Common Black-headed Gull	Non-SPEC[E]	(S)	—	1,500,000 – 2,200,000	Large increase	Moderate decline	—	50–74
Larus genei Slender-billed Gull	SPEC 3	L	≥90% breed at ≤10 sites	37,000 – 56,000	Large increase	Stable	—	25–49
Larus audouinii Audouin's Gull	SPEC 1	L	≥90% breed at ≤10 sites	18,000 – 19,000	Large increase	Large increase	NT: A3c	75–94
Larus canus Mew Gull	SPEC 2	(H)	Moderate historical decline	590,000 – 1,500,000	Moderate decline	Unknown	—	50–74
Larus fuscus Lesser Black-backed Gull	Non-SPEC[E]	S	—	300,000 – 350,000	Large increase	Large increase	—	75–94
Larus argentatus Herring Gull	Non-SPEC[E]	S	—	760,000 – 1,400,000	Large increase	Moderate increase	—	50–74
Larus cachinnans Yellow-legged Gull	Non-SPEC[E]	S	—	310,000 – 580,000	Large increase	Large increase	—	50–74
Larus armenicus Armenian Gull	SPEC 2	L	≥90% breed at ≤10 sites	15,000 – 22,000	Unknown	Moderate increase	—	50–74
Larus glaucoides Iceland Gull	Non-SPEC[E]	(S)	—	30,000 – 100,000	Stable	Stable	—	50–74
Larus hyperboreus Glaucous Gull	Non-SPEC	(S)	—	47,000 – 140,000	Stable	Stable	—	5–24
Larus marinus Great Black-backed Gull	Non-SPEC[E]	S	—	110,000 – 180,000	Stable	Large increase	—	50–74
Xema sabini Sabine's Gull	Non-SPEC	S	—[▼▼▼]	100 – 500	Large increase	Small increase	—	<5
Rhodostethia rosea Ross's Gull	Non-SPEC	(S)	—[▼▼▼▼]	0 – 2	Stable	Unknown	—	<5
Rissa tridactyla Black-legged Kittiwake	Non-SPEC	(S)	—	2,100,000 – 3,000,000	Moderate increase	Moderate decline	—	25–49
Pagophila eburnea Ivory Gull	SPEC 3	(R)	<10,000 pairs	3,100 – 11,000	Large decline	Unknown	—	5–24
Sterna nilotica Gull-billed Tern	SPEC 3	(VU)	A2b	12,000 – 22,000	Large decline	Moderate decline	—	25–49
Sterna caspia Caspian Tern	SPEC 3	R	<10,000 pairs	4,700 – 9,300	Large decline	Large increase	—	5–24
Sterna bengalensis Lesser Crested-tern	Non-SPEC	(S)	—[▼▼▼▼]	2 – 3	Unknown	Unknown	—	<5
Sterna sandvicensis Sandwich Tern	SPEC 2	H	Moderate historical decline	82,000 – 130,000	Moderate decline	Small decline	—	50–74
Sterna dougallii Roseate Tern	SPEC 3	R	<10,000 pairs	1,800 – 1,900	Large decline	Stable	—	5–24
Sterna hirundo Common Tern	Non-SPEC	S	—	270,000 – 570,000	Stable	Stable	—	25–49
Sterna paradisaea Arctic Tern	Non-SPEC	(S)	—	500,000 – 900,000	Stable	Unknown	—	5–24
Sterna albifrons Little Tern	SPEC 3	D	Moderate continuing decline	35,000 – 55,000	Moderate decline	Moderate decline	—	25–49
Chlidonias hybrida Whiskered Tern	SPEC 3	H	Moderate historical decline	42,000 – 87,000	Moderate decline	Fluctuating	—	25–49
Chlidonias niger Black Tern	SPEC 3	(H)	Moderate historical decline	83,000 – 170,000	Moderate decline	Unknown	—	25–49
Chlidonias leucopterus White-winged Tern	Non-SPEC	(S)	—	74,000 – 210,000	Moderate increase	Fluctuating	—	25–49
Uria aalge Common Murre	Non-SPEC	(S)	—	2,000,000 – 2,700,000	Stable	Large increase	—	5–24
Uria lomvia Thick-billed Murre	SPEC 3	(VU)	A4b	1,800,000 – 2,600,000	Stable	Large decline	—	5–24
Alca torda Razorbill	Non-SPEC[E]	(S)	—	430,000 – 770,000	Large increase	Unknown	—	75–94
Cepphus grylle Black Guillemot	SPEC 2	H	Moderate historical decline	130,000 – 300,000	Moderate decline	Stable	—	50–74
Alle alle Dovekie	Non-SPEC	(S)	—	11,000,000 – 44,000,000	Stable	Stable	—	5–24
Fratercula arctica Atlantic Puffin	SPEC 2	(H)	Large historical decline	5,700,000 – 7,300,000	Large decline	Unknown	—	75–94
Pterocles orientalis Black-bellied Sandgrouse	SPEC 3	(D)	Moderate continuing decline	28,000 – 62,000	Large decline	Moderate decline	—	5–24
Pterocles alchata Pin-tailed Sandgrouse	SPEC 3	(D)	Moderate continuing decline	10,000 – 21,000	Large decline	Moderate decline	—	5–24
Columba livia Rock Pigeon	Non-SPEC	(S)	—	9,300,000 – 15,000,000	Stable	Unknown	—	25–49
Columba oenas Stock Pigeon	Non-SPEC[E]	S	—	520,000 – 730,000	Large increase	Moderate increase	—	75–94
Columba palumbus Common Wood-pigeon	Non-SPEC[E]	S	—	9,000,000 – 17,000,000	Stable	Small increase	—	75–94
Columba trocaz Madeira Laurel Pigeon	SPEC 1	(R)	<10,000 pairs	4,100 – 17,000 **ind.**	Stable	Small increase	NT: B1a+b(i,ii,iii,iv,v); B2a+b(i,ii,iii,iv,v)	100
Columba bollii Dark-tailed Laurel Pigeon	SPEC 1	(R)	<10,000 pairs	2,500 – 10,000	Stable	Small increase	NT: C2a(i)	100
Columba junoniae White-tailed Laurel Pigeon	SPEC 1	EN	B1a+b(iii)	1,000 – 2,500	Stable	Unknown	EN: B1a+b(iii)	100
Streptopelia decaocto Eurasian Collared-dove	Non-SPEC	S	—	4,700,000 – 11,000,000	Stable	Moderate increase	—	25–49

Appendix 1 ...continued. Conservation status, population estimates and trends of all European species.

Species	SPEC Category	European Threat Status (ETS)	ETS Criteria	European population size	1970–1990 European population trend	1990–2000 European population trend	2004 Global IUCN Red List Category	% Global population in Europe
Streptopelia turtur European Turtle-dove	SPEC 3	D	Moderate continuing decline	3,500,000 – 7,200,000	Moderate decline	Moderate decline	—	25–49
Streptopelia senegalensis Laughing Dove	Non-SPEC	S	—	61,000 – 210,000	Large increase	Large increase	—	<5
Clamator glandarius Great Spotted Cuckoo	Non-SPEC	(S)	—	58,000 – 77,000	Large increase	Unknown	—	5–24
Cuculus canorus Common Cuckoo	Non-SPEC	S	—	4,200,000 – 8,600,000	Stable	Small decline	—	25–49
Cuculus saturatus Oriental Cuckoo	Non-SPEC	(S)	—	250,000 – 500,000	Stable	Stable	—	<5
Tyto alba Barn Owl	SPEC 3	(D)	Moderate continuing decline	110,000 – 220,000	Moderate decline	Moderate decline	—	5–24
Otus brucei Pallid Scops-owl	SPEC 3	CR	C2a(i); D1	10 – 50	Unknown	Large decline	—	<5
Otus scops Common Scops-owl	SPEC 2	(H)	Moderate historical decline	210,000 – 440,000	Moderate decline	Unknown	—	50–74
Bubo bubo Eurasian Eagle-owl	SPEC 3	(H)	Large historical decline	19,000 – 38,000	Large decline	Stable	—	5–24
Ketupa zeylonensis Brown Fish-owl	SPEC 3	CR	B1a+b(i,ii,iii,iv,v); C1; C2a(i); D1	1 – 10	Unknown	Large decline	—	<5
Nyctea scandiaca Snowy Owl	SPEC 3	(R)	<10,000 pairs	1,400 – 5,500	Stable	Fluctuating	—	5–24
Surnia ulula Northern Hawk Owl	Non-SPEC	(S)	—▼	9,200 – 38,000	Stable	Fluctuating	—	5–24
Glaucidium passerinum Eurasian Pygmy-owl	Non-SPEC	S	—	47,000 – 110,000	Stable	Stable	—	25–49
Athene noctua Little Owl	SPEC 3	(D)	Moderate continuing decline	560,000 – 1,300,000	Moderate decline	Moderate decline	—	25–49
Strix aluco Tawny Owl	Non-SPEC[E]	S	—	480,000 – 1,000,000	Stable	Stable	—	50–74
Strix uralensis Ural Owl	Non-SPEC	(S)	—	53,000 – 140,000	Stable	Stable	—	5–24
Strix nebulosa Great Grey Owl	Non-SPEC	(S)	—▼	2,100 – 6,700	Stable	Unknown	—	5–24
Asio otus Long-eared Owl	Non-SPEC	(S)	—	380,000 – 810,000	Stable	Stable	—	25–49
Asio flammeus Short-eared Owl	SPEC 3	(H)	Large historical decline	58,000 – 180,000	Large decline	Fluctuating	—	5–24
Aegolius funereus Boreal Owl	Non-SPEC	(S)	—	110,000 – 350,000	Stable	Stable	—	5–24
Caprimulgus europaeus Eurasian Nightjar	SPEC 2	(H)	Moderate historical decline	470,000 – 1,000,000	Moderate decline	Small decline	—	50–74
Caprimulgus ruficollis Red-necked Nightjar	Non-SPEC	(S)	—	21,000 – 110,000	Stable	Unknown	—	25–49
Tachymarptis melba Alpine Swift	Non-SPEC	S	—	140,000 – 330,000	Stable	Small increase	—	25–49
Apus unicolor Plain Swift	SPEC 2	(R)	<10,000 pairs	2,500 – 10,000	Stable	Unknown	—	100
Apus apus Common Swift	Non-SPEC	(S)	—	6,900,000 – 17,000,000	Stable	Small decline	—	25–49
Apus pallidus Pallid Swift	Non-SPEC	(S)	—	39,000 – 160,000	Moderate increase	Unknown	—	25–49
Apus caffer White-rumped Swift	Non-SPEC	S	—▼▼▼	100 – 160	Large increase	Large increase	—	<5
Apus affinis Little Swift	SPEC 3	(EN)	A3c	900 – 2,500	Unknown	Small decline	—	<5
Halcyon smyrnensis White-throated Kingfisher	SPEC 3	EN	C2a(i); D1	90 – 170	Unknown	Large decline	—	<5
Alcedo atthis Common Kingfisher	SPEC 3	H	Moderate historical decline	79,000 – 160,000	Moderate decline	Fluctuating	—	25–49
Ceryle rudis Pied Kingfisher	SPEC 3	(CR)	A2c; C1	100 – 200	Unknown	Large decline	—	<5
Merops persicus Blue-cheeked Bee-eater	Non-SPEC	(S)	—▼	3,100 – 10,000	Stable	Moderate increase	—	<5
Merops apiaster European Bee-eater	SPEC 3	(H)	Moderate historical decline	480,000 – 1,000,000	Moderate decline	Moderate increase	—	25–49
Coracias garrulus European Roller	SPEC 2	VU	A2b	53,000 – 110,000	Moderate decline	Large decline	—	50–74
Upupa epops Eurasian Hoopoe	SPEC 3	(D)	Moderate recent decline	890,000 – 1,700,000	Stable	Moderate decline	—	5–24
Jynx torquilla Eurasian Wryneck	SPEC 3	(D)	Moderate continuing decline	580,000 – 1,300,000	Moderate decline	Moderate decline	—	25–49
Picus canus Grey-faced Woodpecker	SPEC 3	(H)	Moderate historical decline	180,000 – 320,000	Moderate decline	Stable	—	5–24
Picus viridis Eurasian Green Woodpecker	SPEC 2	(H)	Moderate historical decline	590,000 – 1,300,000	Moderate decline	Stable	—	75–94
Dryocopus martius Black Woodpecker	Non-SPEC	S	—	740,000 – 1,400,000	Stable	Stable	—	25–49
Dendrocopos major Great Spotted Woodpecker	Non-SPEC	S	—	12,000,000 – 18,000,000	Stable	Stable	—	25–49
Dendrocopos syriacus Syrian Woodpecker	Non-SPEC[E]	(S)	—	530,000 – 1,100,000	Large increase	Small decline	—	50–74
Dendrocopos medius Middle Spotted Woodpecker	Non-SPEC[E]	(S)	—	140,000 – 310,000	Stable	Stable	—	>95
Dendrocopos leucotos White-backed Woodpecker	Non-SPEC	(S)	—	180,000 – 550,000	Stable	Stable	—	25–49
Dendrocopos minor Lesser Spotted Woodpecker	Non-SPEC	(S)	—	450,000 – 1,100,000	Stable	Unknown	—	25–49
Picoides tridactylus Three-toed Woodpecker	SPEC 3	(H)	Moderate historical decline	350,000 – 1,100,000	Moderate decline	Unknown	—	5–24
Ammomanes deserti Desert Lark	SPEC 3	(EN)	C1; D1	30 – 300	Unknown	Large decline	—	<5
Chersophilus duponti Dupont's Lark	SPEC 3	(H)	Large historical decline	13,000 – 15,000	Large decline	Small decline	—	25–49
Melanocorypha calandra Calandra Lark	SPEC 3	(D)	Moderate continuing decline	10,000,000 – 24,000,000	Moderate decline	Moderate decline	—	25–49
Melanocorypha bimaculata Bimaculated Lark	Non-SPEC	S	—	1,000,000 – 2,200,000	Stable	Small increase	—	5–24
Melanocorypha leucoptera White-winged Lark	Non-SPEC[E]W	(S)	—	20,000 – 65,000	Stable	Fluctuating	—	50–74[W]
Melanocorypha yeltoniensis Black Lark	SPEC 3	EN	A2b	4,000 – 7,000	Large decline	Large decline	—	5–24
Calandrella brachydactyla Greater Short-toed Lark	SPEC 3	D	Moderate continuing decline	7,300,000 – 14,000,000	Large decline	Moderate decline	—	5–24
Calandrella rufescens Lesser Short-toed Lark	SPEC 3	D	Moderate continuing decline	1,600,000 – 4,000,000	Large decline	Moderate decline	—	25–49
Calandrella cheleensis Asian Short-toed Lark	SPEC 3	(VU)	A3c	10,000 – 30,000	Unknown	Small increase	—	<5
Galerida cristata Crested Lark	SPEC 3	(H)	Moderate historical decline	3,600,000 – 7,600,000	Moderate decline	Stable	—	25–49
Galerida theklae Thekla Lark	SPEC 3	(H)	Large historical decline	1,500,000 – 2,100,000	Large decline	Unknown	—	25–49
Lullula arborea Wood Lark	SPEC 2	H	Large historical decline	1,300,000 – 3,300,000	Large decline	Stable	—	75–94
Alauda arvensis Eurasian Skylark	SPEC 3	(H)	Large historical decline	40,000,000 – 80,000,000	Large decline	Small decline	—	25–49
Eremophila alpestris Horned Lark	Non-SPEC	(S)	—	2,200,000 – 6,600,000	Stable	Stable	—	<5
Riparia riparia Sand Martin	SPEC 3	(H)	Moderate historical decline	5,400,000 – 9,500,000	Moderate decline	Unknown	—	5–24
Hirundo rupestris Eurasian Crag-martin	Non-SPEC	S	—	120,000 – 370,000	Stable	Stable	—	25–49
Hirundo rustica Barn Swallow	SPEC 3	H	Moderate historical decline	16,000,000 – 36,000,000	Moderate decline	Small decline	—	5–24
Hirundo daurica Red-rumped Swallow	Non-SPEC	(S)	—	100,000 – 430,000	Moderate increase	Stable	—	5–24
Delichon urbica Northern House-martin	SPEC 3	(D)	Moderate recent decline	9,900,000 – 24,000,000	Stable	Moderate decline	—	25–49
Anthus campestris Tawny Pipit	SPEC 3	(D)	Moderate continuing decline	1,000,000 – 1,900,000	Large decline	Moderate decline	—	25–49
Anthus berthelotii Berthelot's Pipit	Non-SPEC[E]	(S)	—	20,000 – 100,000	Stable	Unknown	—	100

Appendix 1 ...continued. Conservation status, population estimates and trends of all European species.

Species	SPEC Category	European Threat Status (ETS)	ETS Criteria	European population size	1970–1990 European population trend	1990–2000 European population trend	2004 Global IUCN Red List Category	% Global population in Europe
Anthus hodgsoni Olive-backed Pipit	Non-SPEC	(S)	—	25,000 - 50,000	Stable	Unknown	—	<5
Anthus trivialis Tree Pipit	Non-SPEC	S	—	27,000,000 - 42,000,000	Stable	Small decline	—	25–49
Anthus gustavi Pechora Pipit	Non-SPEC	(S)	—▼	1,000 - 10,000	Stable	Unknown	—	<5
Anthus pratensis Meadow Pipit	Non-SPEC^E	(S)	—	7,000,000 - 16,000,000	Stable	Small decline	—	75–94
Anthus cervinus Red-throated Pipit	Non-SPEC	(S)	—	1,000,000 - 3,000,000	Stable	Unknown	—	5–24
Anthus spinoletta Water Pipit	Non-SPEC	(S)	—	640,000 - 2,000,000	Stable	Stable	—	5–24
Anthus petrosus Rock Pipit	Non-SPEC^E	(S)	—	110,000 - 290,000	Stable	Small decline	—	100
Motacilla flava Yellow Wagtail	Non-SPEC	(S)	—	7,900,000 - 14,000,000	Stable	Small decline	—	25–49
Motacilla citreola Citrine Wagtail	Non-SPEC	(S)	—	210,000 - 520,000	Stable	Unknown	—	5–24
Motacilla cinerea Grey Wagtail	Non-SPEC	S	—	740,000 - 1,600,000	Stable	Stable	—	5–24
Motacilla alba White Wagtail	Non-SPEC	S	—	13,000,000 - 26,000,000	Stable	Stable	—	5–24
Pycnonotus xanthopygos White-spectacled Bulbul	Non-SPEC	S	—▼	6,000 - 18,000	Unknown	Stable	—	5–24
Bombycilla garrulus Bohemian Waxwing	Non-SPEC	(S)	—	130,000 - 700,000	Stable	Fluctuating	—	5–24
Cinclus cinclus White-throated Dipper	Non-SPEC	S	—	170,000 - 330,000	Stable	Stable	—	25–49
Troglodytes troglodytes Winter Wren	Non-SPEC	S	—	23,000,000 - 40,000,000	Stable	Small increase	—	5–24
Prunella modularis Hedge Accentor	Non-SPEC^E	S	—	12,000,000 - 26,000,000	Stable	Stable	—	>95
Prunella montanella Siberian Accentor	Non-SPEC	(S)	—▼▼▼	100 - 500	Stable	Unknown	—	<5
Prunella ocularis Radde's Accentor	Non-SPEC^E	(S)	—	11,000 - 33,000	Unknown	Unknown	—	50–74
Prunella atrogularis Black-throated Accentor	SPEC 3	R	<10,000 pairs	1,500 - 2,500	Unknown	Unknown	—	5–24
Prunella collaris Alpine Accentor	Non-SPEC	(S)	—	100,000 - 180,000	Stable	Stable	—	25–49
Erythropygia galactotes Rufous-tailed Scrub-robin	SPEC 3	VU	A2b	32,000 - 96,000	Stable	Large decline	—	25–49
Erithacus rubecula European Robin	Non-SPEC^E	S	—	43,000,000 - 83,000,000	Stable	Small increase	—	75–94
Luscinia luscinia Thrush Nightingale	Non-SPEC^E	S	—	3,700,000 - 6,900,000	Stable	Stable	—	50–74
Luscinia megarhynchos Common Nightingale	Non-SPEC^E	(S)	—	4,200,000 - 12,000,000	Stable	Stable	—	50–74
Luscinia calliope Siberian Rubythroat	Non-SPEC	(S)	—▼	800 - 1,000	Stable	Unknown	—	<5
Luscinia svecica Bluethroat	Non-SPEC	S	—	4,500,000 - 7,800,000	Stable	Stable	—	5–24
Tarsiger cyanurus Orange-flanked Bush-robin	Non-SPEC	(S)	—	10,000 - 21,000	Stable	Unknown	—	<5
Irania gutturalis White-throated Robin	Non-SPEC	(S)	—	410,000 - 920,000	Unknown	Small increase	—	25–49
Phoenicurus ochruros Black Redstart	Non-SPEC	S	—	4,000,000 - 8,800,000	Stable	Small increase	—	25–49
Phoenicurus phoenicurus Common Redstart	SPEC 2	(H)	Large historical decline	6,800,000 - 16,000,000	Large decline	Stable	—	50–74
Phoenicurus erythrogastrus White-winged Redstart	SPEC 3	(R)	<10,000 pairs	2,100 - 6,000	Unknown	Stable	—	<5
Saxicola rubetra Whinchat	Non-SPEC^E	(S)	—	5,400,000 - 10,000,000	Stable	Small decline	—	75–94
Saxicola dacotiae Fuerteventura Chat	SPEC 1	EN	B1a+b(ii,iii,iv,v); C2a(ii)	<1,300	Stable	Small decline	EN: B1a+b(ii,iii,iv,v); C2a(ii)	100
Saxicola torquata Common Stonechat	Non-SPEC	(S)	—	2,000,000 - 4,600,000	Moderate decline	Moderate increase	—	5–24
Oenanthe isabellina Isabelline Wheatear	Non-SPEC	(S)	—	2,100,000 - 6,300,000	Stable	Stable	—	5–24
Oenanthe oenanthe Northern Wheatear	SPEC 3	(D)	Moderate recent decline	4,600,000 - 13,000,000	Stable	Moderate decline	—	25–49
Oenanthe pleschanka Pied Wheatear	Non-SPEC	(S)	—	32,000 - 140,000	Stable	Stable	—	5–24
Oenanthe cypriaca Cyprus Wheatear	Non-SPEC^E	(S)	—	90,000 - 180,000	Stable	Stable	—	100
Oenanthe hispanica Black-eared Wheatear	SPEC 2	(H)	Large historical decline	1,400,000 - 3,300,000	Large decline	Small decline	—	50–74
Oenanthe deserti Desert Wheatear	Non-SPEC	(S)	—▼▼▼	110 - 1,100	Unknown	Stable	—	<5
Oenanthe finschii Finsch's Wheatear	Non-SPEC	(S)	—	100,000 - 310,000	Unknown	Stable	—	5–24
Oenanthe xanthoprymna Rufous-tailed Wheatear	SPEC 3	(VU)	C1	2,100 - 7,000	Unknown	Moderate decline	—	25–49
Oenanthe leucura Black Wheatear	SPEC 3	(R)	<10,000 pairs	4,100 - 16,000	Large decline	Unknown	—	25–49
Monticola saxatilis Rufous-tailed Rock-thrush	SPEC 3	(H)	Moderate historical decline	100,000 - 320,000	Moderate decline	Small decline	—	25–49
Monticola solitarius Blue Rock-thrush	SPEC 3	(H)	Large historical decline	120,000 - 260,000	Large decline	Stable	—	25–49
Zoothera dauma Scaly Thrush	Non-SPEC	(S)	—	25,000 - 100,000	Stable	Unknown	—	<5
Turdus torquatus Ring Ouzel	Non-SPEC^E	S	—	310,000 - 670,000	Stable	Stable	—	>95
Turdus merula Eurasian Blackbird	Non-SPEC^E	S	—	40,000,000 - 82,000,000	Stable	Small increase	—	50–74
Turdus ruficollis Dark-throated Thrush	Non-SPEC	(S)	—▼	5,000 - 20,000	Stable	Unknown	—	<5
Turdus pilaris Fieldfare	Non-SPEC^{E}W	(S)	—	14,000,000 - 24,000,000	Stable	Stable	—	75–94 ^{W}
Turdus philomelos Song Thrush	Non-SPEC^E	S	—	20,000,000 - 36,000,000	Stable	Stable	—	50–74
Turdus iliacus Redwing	Non-SPEC^{E}W	(S)	—	16,000,000 - 21,000,000	Stable	Stable	—	50–74 ^{W}
Turdus viscivorus Mistle Thrush	Non-SPEC^E	S	—	3,000,000 - 7,400,000	Stable	Stable	—	50–74
Cettia cetti Cetti's Warbler	Non-SPEC	S	—	600,000 - 1,600,000	Moderate increase	Small increase	—	25–49
Cisticola juncidis Zitting Cisticola	Non-SPEC	S	—	230,000 - 1,100,000	Stable	Stable	—	<5
Prinia gracilis Graceful Prinia	SPEC 3	(VU)	C1	1,000 - 4,000	Unknown	Moderate decline	—	<5
Locustella lanceolata Lanceolated Warbler	Non-SPEC	(S)	—	50,000 - 100,000	Stable	Unknown	—	<5
Locustella naevia Common Grasshopper-warbler	Non-SPEC^E	(S)	—	840,000 - 2,200,000	Stable	Small decline	—	50–74
Locustella fluviatilis Eurasian River Warbler	Non-SPEC^E	(S)	—	1,900,000 - 4,600,000	Stable	Stable	—	75–94
Locustella luscinioides Savi's Warbler	Non-SPEC^E	(S)	—	530,000 - 800,000	Stable	Stable	—	50–74
Acrocephalus melanopogon Moustached Warbler	Non-SPEC	(S)	—	150,000 - 300,000	Stable	Stable	—	25–49
Acrocephalus paludicola Aquatic Warbler	SPEC 1	(VU)	A3c	12,000 - 20,000	Large decline	Moderate decline	VU: A2c; A3c	>95
Acrocephalus schoenobaenus Sedge Warbler	Non-SPEC^E	S	—	4,400,000 - 7,400,000	Stable	Stable	—	50–74
Acrocephalus agricola Paddyfield Warbler	Non-SPEC	(S)	—	470,000 - 880,000	Moderate increase	Small increase	—	<5
Acrocephalus dumetorum Blyth's Reed-warbler	Non-SPEC	S	—	2,000,000 - 5,000,000	Stable	Stable	—	25–49
Acrocephalus palustris Marsh Warbler	Non-SPEC^E	(S)	—	3,200,000 - 6,800,000	Stable	Stable	—	75–94

Appendix 1 ...continued. Conservation status, population estimates and trends of all European species.

Species	SPEC Category	European Threat Status (ETS)	ETS Criteria	European population size	1970–1990 European population trend	1990–2000 European population trend	2004 Global IUCN Red List Category	% Global population in Europe
Acrocephalus scirpaceus Common Reed-warbler	Non-SPEC^{E}	S	—	2,700,000 – 5,000,000	Stable	Stable	—	50–74
Acrocephalus arundinaceus Great Reed-warbler	Non-SPEC	(S)	—	1,500,000 – 2,900,000	Stable	Small decline	—	25–49
Hippolais pallida Olivaceous Warbler	SPEC 3	(H)	Large historical decline	3,300,000 – 6,700,000	Large decline	Stable	—	25–49
Hippolais caligata Booted Warbler	Non-SPEC	(S)	—	30,000 – 80,000	Stable	Unknown	—	5–24
Hippolais languida Upcher's Warbler	Non-SPEC	S	—	14,000 – 37,000	Unknown	Moderate increase	—	5–24
Hippolais olivetorum Olive-tree Warbler	Non-SPEC^{E}	(S)	—	11,000 – 23,000	Stable	Stable	—	>95
Hippolais icterina Icterine Warbler	Non-SPEC^{E}	(S)	—	3,500,000 – 7,100,000	Stable	Small decline	—	75–94
Hippolais polyglotta Melodious Warbler	Non-SPEC^{E}	(S)	—	1,000,000 – 3,000,000	Stable	Unknown	—	75–94
Sylvia sarda Marmora's Warbler	Non-SPEC^{E}	(S)	—	29,000 – 75,000	Moderate increase	Unknown	—	100
Sylvia undata Dartford Warbler	SPEC 2	H	Large historical decline	1,900,000 – 3,700,000	Large decline	Unknown	—	>95
Sylvia conspicillata Spectacled Warbler	Non-SPEC	(S)	—	180,000 – 440,000	Stable	Unknown	—	25–49
Sylvia cantillans Subalpine Warbler	Non-SPEC^{E}	(S)	—	1,400,000 – 3,200,000	Stable	Unknown	—	75–94
Sylvia mystacea Ménétries's Warbler	Non-SPEC	(S)	—	43,000 – 150,000	Stable	Stable	—	<5
Sylvia melanocephala Sardinian Warbler	Non-SPEC^{E}	(S)	—	3,100,000 – 8,100,000	Moderate increase	Stable	—	50–74
Sylvia melanothorax Cyprus Warbler	Non-SPEC^{E}	(S)	—	70,000 – 140,000	Large increase	Stable	—	100
Sylvia rueppelli Rüppell's Warbler	Non-SPEC^{E}	(S)	—	200,000 – 610,000	Stable	Small decline	—	>95
Sylvia nana Desert Warbler	Non-SPEC	(S)	—▼	1,000 – 5,000	Stable	Stable	—	<5
Sylvia hortensis Orphean Warbler	SPEC 3	H	Large historical decline	170,000 – 480,000	Large decline	Small decline	—	25–49
Sylvia nisoria Barred Warbler	Non-SPEC^{E}	S	—	460,000 – 1,000,000	Stable	Unknown	—	50–74
Sylvia curruca Lesser Whitethroat	Non-SPEC	S	—	4,800,000 – 7,800,000	Stable	Stable	—	25–49
Sylvia communis Common Whitethroat	Non-SPEC^{E}	S	—	14,000,000 – 25,000,000	Stable	Small increase	—	50–74
Sylvia borin Garden Warbler	Non-SPEC^{E}	S	—	17,000,000 – 31,000,000	Stable	Stable	—	75–94
Sylvia atricapilla Blackcap	Non-SPEC^{E}	S	—	25,000,000 – 49,000,000	Stable	Small increase	—	75–94
Phylloscopus trochiloides Greenish Warbler	Non-SPEC	S	—	12,000,000 – 15,000,000	Stable	Stable	—	5–24
Phylloscopus borealis Arctic Warbler	Non-SPEC	S	—	4,500,000 – 8,000,000	Stable	Unknown	—	5–24
Phylloscopus inornatus Inornate Warbler	Non-SPEC	(S)	—▼	5,000 – 35,000	Unknown	Unknown	—	<5
Phylloscopus bonelli Bonelli's Warbler	SPEC 2	D	Moderate recent decline	1,400,000 – 3,500,000	Stable	Moderate decline	—	75–94
Phylloscopus sibilatrix Wood Warbler	SPEC 2	D	Moderate recent decline	14,000,000 – 22,000,000	Stable	Moderate decline	—	75–94
Phylloscopus sindianus Mountain Chiffchaff	SPEC 3	D	Moderate recent decline	71,000 – 320,000	Unknown	Moderate decline	—	5–24
Phylloscopus collybita Common Chiffchaff	Non-SPEC	S	—	30,000,000 – 60,000,000	Stable	Stable	—	25–49
Phylloscopus ibericus Iberian Chiffchaff	Non-SPEC^{E}	(S)	—	360,000 – 530,000	Unknown	Unknown	—	>95
Phylloscopus canariensis Canary Islands Chiffchaff	Non-SPEC^{E}	(S)	—	20,000 – 100,000	Unknown	Unknown	—	100
Phylloscopus trochilus Willow Warbler	Non-SPEC	S	—	56,000,000 – 100,000,000	Stable	Small decline	—	25–49
Regulus regulus Goldcrest	Non-SPEC^{E}	S	—	19,000,000 – 35,000,000	Stable	Stable	—	50–74
Regulus teneriffae Canary Islands Kinglet	Non-SPEC^{E}	(S)	—	10,000 – 20,000	Unknown	Unknown	—	100
Regulus ignicapilla Firecrest	Non-SPEC^{E}	(S)	—	3,300,000 – 6,700,000	Stable	Stable	—	75–94
Muscicapa striata Spotted Flycatcher	SPEC 3	H	Moderate historical decline	14,000,000 – 22,000,000	Moderate decline	Small decline	—	25–49
Ficedula parva Red-breasted Flycatcher	Non-SPEC	(S)	—	3,200,000 – 4,600,000	Stable	Stable	—	25–49
Ficedula semitorquata Semicollared Flycatcher	SPEC 2	D	Moderate continuing decline	15,000 – 53,000	Large decline	Moderate decline	—	50–74
Ficedula albicollis Collared Flycatcher	Non-SPEC^{E}	S	—	1,400,000 – 2,400,000	Stable	Small increase	—	100
Ficedula hypoleuca European Pied Flycatcher	Non-SPEC^{E}	S	—	12,000,000 – 20,000,000	Stable	Small decline	—	75–94
Panurus biarmicus Bearded Parrotbill	Non-SPEC	(S)	—	240,000 – 480,000	Stable	Stable	—	25–49
Aegithalos caudatus Long-tailed Tit	Non-SPEC	S	—	5,000,000 – 12,000,000	Stable	Stable	—	25–49
Parus palustris Marsh Tit	SPEC 3	D	Moderate recent decline	3,000,000 – 6,000,000	Stable	Moderate decline	—	25–49
Parus lugubris Sombre Tit	Non-SPEC^{E}	(S)	—	450,000 – 1,300,000	Stable	Stable	—	75–94
Parus montanus Willow Tit	Non-SPEC	S	—	24,000,000 – 42,000,000	Stable	Small decline	—	25–49
Parus cinctus Siberian Tit	Non-SPEC	(S)	—	870,000 – 1,700,000	Stable	Unknown	—	5–24
Parus cristatus Crested Tit	SPEC 2	(D)	Moderate recent decline	6,100,000 – 12,000,000	Stable	Moderate decline	—	>95
Parus ater Coal Tit	Non-SPEC	(S)	—	12,000,000 – 29,000,000	Moderate increase	Stable	—	25–49
Parus caeruleus Blue Tit	Non-SPEC^{E}	S	—	20,000,000 – 44,000,000	Stable	Stable	—	75–94
Parus cyanus Azure Tit	Non-SPEC	(S)	—▼	2,900 – 11,000	Stable	Unknown	—	5–24
Parus major Great Tit	Non-SPEC	S	—	46,000,000 – 91,000,000	Stable	Stable	—	25–49
Sitta krueperi Krüper's Nuthatch	SPEC 2	(D)	Moderate recent decline	80,000 – 170,000	Stable	Moderate decline	—	100
Sitta whiteheadi Corsican Nuthatch	SPEC 2	R	<10,000 pairs	1,500 – 4,500	Stable	Stable	—	100
Sitta europaea Wood Nuthatch	Non-SPEC	S	—	7,500,000 – 19,000,000	Stable	Stable	—	25–49
Sitta tephronota Eastern Rock-nuthatch	Non-SPEC	(S)	—	22,000 – 100,000	Unknown	Small decline	—	<5
Sitta neumayer Western Rock-nuthatch	Non-SPEC^{E}	(S)	—	2,000,000 – 6,100,000	Stable	Stable	—	75–94
Tichodroma muraria Wallcreeper	Non-SPEC	(S)	—	38,000 – 100,000	Stable	Stable	—	25–49
Certhia familiaris Eurasian Tree-creeper	Non-SPEC	S	—	5,700,000 – 11,000,000	Stable	Stable	—	25–49
Certhia brachydactyla Short-toed Tree-creeper	Non-SPEC^{E}	(S)	—	2,700,000 – 9,700,000	Stable	Small increase	—	75–94
Remiz pendulinus Eurasian Penduline-tit	Non-SPEC	(S)	—	210,000 – 420,000	Stable	Stable	—	25–49
Oriolus oriolus Eurasian Golden-oriole	Non-SPEC	S	—	3,400,000 – 7,100,000	Stable	Small decline	—	25–49
Lanius collurio Red-backed Shrike	SPEC 3	(H)	Moderate historical decline	6,300,000 – 13,000,000	Moderate decline	Small decline	—	25–49
Lanius minor Lesser Grey Shrike	SPEC 2	(D)	Moderate continuing decline	620,000 – 1,500,000	Moderate decline	Moderate decline	—	50–74
Lanius excubitor Great Grey Shrike	SPEC 3	(H)	Moderate historical decline	250,000 – 400,000	Moderate decline	Small decline	—	5–24
Lanius senator Woodchat Shrike	SPEC 2	(D)	Moderate continuing decline	480,000 – 1,200,000	Large decline	Moderate decline	—	50–74

Appendix 1 ...continued. Conservation status, population estimates and trends of all European species.

Species	SPEC Category	European Threat Status (ETS)	ETS Criteria	European population size	1970–1990 European population trend	1990–2000 European population trend	2004 Global IUCN Red List Category	% Global population in Europe
Lanius nubicus Masked Shrike	SPEC 2	(D)	Moderate continuing decline	35,000 – 100,000	Large decline	Moderate decline	—	50–74
Garrulus glandarius Eurasian Jay	Non-SPEC	S	—	6,000,000 – 13,000,000	Stable	Stable	—	25–49
Perisoreus infaustus Siberian Jay	SPEC 3	(H)	Moderate historical decline	340,000 – 710,000	Moderate decline	Unknown	—	25–49
Cyanopica cyanus Azure-winged Magpie	Non-SPEC	(S)	—	260,000 – 460,000	Stable	Unknown	—	5–24
Pica pica Black-billed Magpie	Non-SPEC	S	—	7,500,000 – 19,000,000	Large increase	Moderate decline	—	25–49
Nucifraga caryocatactes Spotted Nutcracker	Non-SPEC	S	—	400,000 – 860,000	Stable	Stable	—	25–49
Pyrrhocorax graculus Yellow-billed Chough	Non-SPEC	(S)	—	130,000 – 310,000	Stable	Stable	—	25–49
Pyrrhocorax pyrrhocorax Red-billed Chough	SPEC 3	D	Moderate continuing decline	43,000 – 110,000	Large decline	Moderate decline	—	25–49
Corvus monedula Eurasian Jackdaw	Non-SPEC^E	(S)	—	5,200,000 – 15,000,000	Stable	Stable	—	50–74
Corvus frugilegus Rook	Non-SPEC	(S)	—	10,000,000 – 18,000,000	Large increase	Unknown	—	25–49
Corvus corone Carrion Crow	Non-SPEC	S	—	7,000,000 – 17,000,000	Large increase	Stable	—	25–49
Corvus corax Common Raven	Non-SPEC	S	—	450,000 – 970,000	Large increase	Small increase	—	25–49
Sturnus vulgaris Common Starling	SPEC 3	D	Moderate recent decline	23,000,000 – 56,000,000	Stable	Moderate decline	—	25–49
Sturnus unicolor Spotless Starling	Non-SPEC^E	S	—	2,100,000 – 3,100,000	Large increase	Small increase	—	50–74
Sturnus roseus Rosy Starling	Non-SPEC	S	—	58,000 – 210,000	Stable	Fluctuating	—	25–49
Passer domesticus House Sparrow	SPEC 3	D	Moderate recent decline	63,000,000 – 130,000,000	Stable	Moderate decline	—	25–49
Passer hispaniolensis Spanish Sparrow	Non-SPEC	(S)	—	2,800,000 – 6,200,000	Moderate increase	Stable	—	25–49
Passer moabiticus Dead Sea Sparrow	Non-SPEC	(S)	—▼	900 – 3,000	Large increase	Moderate decline	—	5–24
Passer montanus Eurasian Tree Sparrow	SPEC 3	(D)	Moderate recent decline	26,000,000 – 48,000,000	Stable	Moderate decline	—	25–49
Petronia brachydactyla Pale Rock-finch	Non-SPEC	S	—▼	7,000 – 18,000	Stable	Large increase	—	5–24
Petronia xanthocollis Chestnut-shouldered Petronia	Non-SPEC	(S)	—▼	500 – 2,000	Unknown	Increase	—	<5
Petronia petronia Rock Sparrow	Non-SPEC	(S)	—	1,700,000 – 3,800,000	Stable	Stable	—	5–24
Montifringilla nivalis White-winged Snowfinch	Non-SPEC	(S)	—	520,000 – 1,600,000	Stable	Stable	—	25–49
Fringilla coelebs Chaffinch	Non-SPEC^E	S	—	130,000,000 – 240,000,000	Stable	Stable	—	50–74
Fringilla teydea Blue Chaffinch	SPEC 1	R	<10,000 pairs	1,000 – 2,500	Stable	Stable	NT: B1a+b(ii,iii,v); B2a+b(ii,iii,v); C2a(ii)	100
Fringilla montifringilla Brambling	Non-SPEC	S	—	13,000,000 – 22,000,000	Stable	Stable	—	25–49
Serinus pusillus Fire-fronted Serin	Non-SPEC	(S)	—	280,000 – 820,000	Unknown	Stable	—	5–24
Serinus serinus European Serin	Non-SPEC^E	S	—	8,300,000 – 20,000,000	Moderate increase	Stable	—	75–94
Serinus canaria Island Canary	Non-SPEC^E	(S)	—	20,000 – 100,000	Stable	Unknown	—	100
Serinus citrinella Citril Finch	Non-SPEC^E	(S)	—	250,000 – 320,000	Large increase	Unknown	—	100
Serinus corsicana Corsican Finch	Non-SPEC^E	(S)	—	19,000 – 85,000	Unknown	Stable	—	100
Carduelis chloris European Greenfinch	Non-SPEC^E	S	—	14,000,000 – 32,000,000	Stable	Stable	—	75–94
Carduelis carduelis European Goldfinch	Non-SPEC	S	—	12,000,000 – 29,000,000	Stable	Small increase	—	25–49
Carduelis spinus Eurasian Siskin	Non-SPEC^E	S	—	10,000,000 – 18,000,000	Stable	Fluctuating	—	75–94
Carduelis cannabina Eurasian Linnet	SPEC 2	D	Moderate recent decline	10,000,000 – 28,000,000	Stable	Moderate decline	—	50–74
Carduelis flavirostris Twite	Non-SPEC	S	—	170,000 – 760,000	Stable	Stable	—	25–49
Carduelis flammea Common Redpoll	Non-SPEC	(S)	—	8,800,000 – 19,000,000	Stable	Unknown	—	25–49
Carduelis hornemanni Hoary Redpoll	Non-SPEC	(S)	—	84,000 – 180,000	Stable	Unknown	—	25–49
Loxia leucoptera Two-barred Crossbill	Non-SPEC	(S)	—	1,000,000 – 2,900,000	Stable	Fluctuating	—	5–24
Loxia curvirostra Red Crossbill	Non-SPEC	(S)	—	5,800,000 – 13,000,000	Stable	Fluctuating	—	5–24
Loxia scotica Scottish Crossbill	SPEC 1	DD	—	300 – 1,250	Unknown	Unknown	DD	100
Loxia pytyopsittacus Parrot Crossbill	Non-SPEC^E	(S)	—	260,000 – 1,100,000	Stable	Fluctuating	—	>95
Rhodopechys sanguinea Crimson-winged Finch	Non-SPEC	S	—	300,000 – 910,000	Unknown	Small increase	—	25–49
Rhodopechys obsoleta Desert Finch	Non-SPEC	(S)	—▼	3,000 – 6,000	Unknown	Small increase	—	<5
Rhodopechys mongolica Mongolian Finch	Non-SPEC	(S)	—▼	520 – 5,000	Unknown	Small increase	—	<5
Bucanetes githagineus Trumpeter Finch	Non-SPEC	(S)	—	11,000 – 21,000	Stable	Small decline	—	<5
Carpodacus erythrinus Common Rosefinch	Non-SPEC	(S)	—	3,000,000 – 6,100,000	Stable	Stable	—	5–24
Carpodacus rubicilla Great Rosefinch	SPEC 3	(R)	<10,000 pairs	5,100 – 10,000	Large decline	Unknown	—	<5
Pinicola enucleator Pine Grosbeak	Non-SPEC	(S)	—	110,000 – 290,000	Stable	Unknown	—	5–24
Pyrrhula pyrrhula Eurasian Bullfinch	Non-SPEC	(S)	—	7,300,000 – 14,000,000	Stable	Stable	—	25–49
Pyrrhula murina Azores Bullfinch	SPEC 1	(EN)	D1	240 – 240 **ind.**	Unknown	Decline	EN: D1	100
Coccothraustes coccothraustes Hawfinch	Non-SPEC	S	—	2,400,000 – 4,200,000	Stable	Stable	—	25–49
Calcarius lapponicus Lapland Longspur	Non-SPEC	(S)	—	5,800,000 – 11,000,000	Stable	Unknown	—	5–24
Plectrophenax nivalis Snow Bunting	Non-SPEC	(S)	—	680,000 – 1,700,000	Stable	Stable	—	5–24
Emberiza leucocephalos Pine Bunting	Non-SPEC	(S)	—▼▼▼	50 – 100	Unknown	Unknown	—	<5
Emberiza citrinella Yellowhammer	Non-SPEC^E	(S)	—	18,000,000 – 31,000,000	Stable	Small decline	—	50–74
Emberiza cirlus Cirl Bunting	Non-SPEC^E	S	—	2,000,000 – 5,200,000	Stable	Moderate increase	—	>95
Emberiza cia Rock Bunting	SPEC 3	(H)	Large historical decline	1,300,000 – 4,100,000	Large decline	Stable	—	25–49
Emberiza cineracea Cinereous Bunting	SPEC 1	(R)	<10,000 pairs	2,600 – 7,800	Stable	Small decline	NT: D1	>95
Emberiza hortulana Ortolan Bunting	SPEC 2	(H)	Large historical decline	5,200,000 – 16,000,000	Large decline	Small decline	—	50–74
Emberiza buchanani Grey-necked Bunting	Non-SPEC	(S)	—▼	6,400 – 20,000	Unknown	Stable	—	5–24
Emberiza caesia Cretzschmar's Bunting	Non-SPEC^E	(S)	—	140,000 – 230,000	Stable	Small decline	—	75–94
Emberiza rustica Rustic Bunting	Non-SPEC	(S)	—	6,100,000 – 10,000,000	Stable	Unknown	—	25–49
Emberiza pusilla Little Bunting	Non-SPEC	(S)	—	5,000,000 – 8,000,000	Stable	Unknown	—	25–49

Appendix 1 ...continued. Conservation status, population estimates and trends of all European species.

Species	SPEC Category	European Threat Status (ETS)	ETS Criteria	European population size	1970–1990 European population trend	1990–2000 European population trend	2004 Global IUCN Red List Category	% Global population in Europe
Emberiza aureola Yellow-breasted Bunting	SPEC 1	D	Moderate recent decline	20,000 - 100,000	Stable	Moderate decline	NT: A2d; A3d	25–49
Emberiza schoeniclus Reed Bunting	Non-SPEC	S	—	4,800,000 - 8,800,000	Stable	Small decline	—	25–49
Emberiza pallasi Pallas's Bunting	Non-SPEC	(S)	—	80,000 - 150,000	Unknown	Unknown	—	<5
Emberiza bruniceps Red-headed Bunting	Non-SPEC	(S)	—▼▼	250 - 1,000	Large increase	Unknown	—	<5
Emberiza melanocephala Black-headed Bunting	SPEC 2	(H)	Large historical decline	2,800,000 - 9,300,000	Large decline	Small increase	—	50–74
Miliaria calandra Corn Bunting	SPEC 2	(D)	Moderate recent decline	7,900,000 - 22,000,000	Stable	Moderate decline	—	50–74

Notes.

SPEC Category

SPEC 1: Species of global conservation concern, i.e. classified as Threatened, Near Threatened or Data Deficient under the IUCN Red List Criteria at global level (BirdLife International 2004; IUCN 2004).

SPEC 2: Species concentrated in Europe and with an Unfavourable Conservation Status in Europe.

SPEC 3: Species not concentrated in Europe but with an Unfavourable Conservation Status in Europe.

Non-SPEC[E]: Species concentrated in Europe but with a Favourable Conservation Status in Europe.

Non-SPEC: Species not concentrated in Europe and with a Favourable Conservation Status in Europe.

W: Category relates to winter populations.

European Threat Status

CR Critically Endangered
EN Endangered
VU Vulnerable
D Declining
R Rare
H Depleted
L Localised
DD Data Deficient
S Secure
NE Not Evaluated (occurs in the region on passage only)
( ) Status provisional

Criteria

Summary of criteria upon which species qualify as having an Unfavourable Conservation Status in Europe. For details of IUCN Red List Criteria codes, see Appendix 7.

▼ Small population size criteria not met because the species's European population is marginal to a large non-European population, which could 'rescue' it if it declined. The number of ▼ symbols denotes how many steps the species's status has been downgraded by, e.g. ▼▼▼▼denotes a 4-step downgrade, from Critically Endangered to Secure. Species were downgraded following IUCN (2003) *Guidelines for application of IUCN Red List Criteria at regional levels: Version 3.0.* IUCN Species Survival Commission. Gland, Switzerland and Cambridge, U.K.: IUCN.

European population size

Unless stated otherwise, all figures refer to breeding pairs and are rounded to two significant figures.

ind Figures refer to breeding individuals, not pairs.

ind [W] Figures refer to wintering individuals.

1970–1990 European population trend

Calculated using the method developed by Tucker and Heath (1994). For details, see 'Status assessment' and Appendix 6.

1990–2000 European population trend

Calculated using the method described under 'Status assessment'.

2004 Global IUCN Red List Category and Criteria

As published by BirdLife International (2004) and IUCN (2004).

% Global population in Europe

Based on the proportion of each species's global range (or population, where known) occurring in Europe.

[W] Figures refer to wintering populations.

Appendix 2. Occurrence of all European species by country.

SPEC Category	Species	Albania	Andorra	Armenia	Austria	Azerbaijan	Belarus	Belgium	Bosnia and Herzegovina	Bulgaria	Croatia	Cyprus	Czech Republic	Denmark	*Faroe Islands*	*Greenland*	Estonia	Finland	France	Georgia	Germany	Greece	Hungary	Iceland	Republic of Ireland	Italy	Latvia
3	***Gavia stellata***	✓			✓	✓		✓		✓	✓		✓	✓	✓	✓	✓	✓	✓		✓	✓	✓	✓	✓	✓	✓
3	***Gavia arctica***	✓			✓	✓	✓	✓		✓	✓		✓	✓			✓	✓	✓	✓	✓	✓	✓		✓	✓	✓
—	*Gavia immer*							✓							✓	✓			✓		✓			✓	✓	✓	
—	*Gavia adamsii*																		✓		✓						
—	*Tachybaptus ruficollis*	✓		✓	✓	✓	✓	✓	✓	✓	✓	✓	✓	✓			✓	✓	✓	✓	✓	✓	✓		✓	✓	✓
—	*Podiceps cristatus*	✓		✓	✓	✓	✓	✓	✓	✓	✓	✓	✓	✓			✓	✓	✓	✓	✓	✓	✓		✓	✓	✓
—	*Podiceps grisegena*	✓		✓	✓	✓	✓	✓		✓	✓		✓	✓			✓	✓	✓	✓	✓	✓	✓		✓	✓	✓
3	***Podiceps auritus***	✓		✓	✓	✓		✓		✓	✓		✓	✓			✓	✓	✓		✓	✓	✓	✓	✓	✓	✓
—	*Podiceps nigricollis*	✓		✓	✓	✓	✓	✓	✓	✓	✓	✓	✓	✓			✓		✓	✓	✓	✓	✓		✓	✓	✓
—	*Fulmarus glacialis*													✓	✓	✓			✓		✓			✓	✓		
1	***Pterodroma madeira***																										
1	***Pterodroma feae***																										
3	***Bulweria bulwerii***																										
2	***Calonectris diomedea***										✓								✓			✓				✓	
NE	*Puffinus gravis*															✓											
1	***Puffinus griseus***																										
2	***Puffinus puffinus***														✓				✓					✓	✓		
1	***Puffinus mauretanicus***																										
—[E]	*Puffinus yelkouan*	✓								✓	✓								✓			✓				✓	
3	***Puffinus assimilis***																										
3	***Pelagodroma marina***																										
—[E]	*Hydrobates pelagicus*										✓				✓				✓			✓		✓	✓	✓	
3	***Oceanodroma leucorhoa***														✓									✓	✓		
3	***Oceanodroma castro***																										
—[E]	*Morus bassanus*													✓	✓				✓		✓	✓		✓	✓		
—	*Phalacrocorax carbo*	✓	✓	✓	✓	✓	✓	✓	✓	✓	✓	✓	✓	✓	✓	✓	✓	✓	✓	✓	✓	✓	✓	✓	✓	✓	✓
—[E]	*Phalacrocorax aristotelis*	✓								✓	✓	✓			✓				✓			✓		✓	✓	✓	
1	***Phalacrocorax pygmeus***	✓		✓		✓			✓	✓	✓									✓		✓	✓			✓	
3	***Pelecanus onocrotalus***					✓				✓		✓								✓		✓					
1	***Pelecanus crispus***	✓		✓		✓				✓										✓		✓					
3	***Botaurus stellaris***	✓		✓	✓	✓	✓	✓	✓	✓	✓		✓	✓			✓	✓	✓		✓	✓	✓		✓	✓	✓
3	***Ixobrychus minutus***	✓		✓	✓	✓	✓	✓	✓	✓	✓		✓						✓	✓	✓	✓	✓			✓	✓
3	***Nycticorax nycticorax***	✓		✓	✓	✓	✓	✓	✓	✓	✓		✓						✓	✓	✓	✓	✓			✓	
3	***Ardeola ralloides***	✓		✓		✓			✓	✓	✓								✓	✓		✓	✓			✓	
—	*Bubulcus ibis*			✓		✓													✓	✓		✓				✓	
—	*Egretta garzetta*	✓		✓	✓	✓		✓	✓	✓	✓	✓	✓						✓	✓	✓	✓	✓		✓	✓	
—	*Casmerodius albus*	✓			✓	✓	✓	✓		✓	✓	✓	✓						✓	✓		✓	✓			✓	✓
—	*Ardea cinerea*	✓	✓	✓	✓	✓	✓	✓	✓	✓	✓	✓	✓	✓	✓		✓	✓	✓	✓	✓	✓	✓	✓	✓	✓	✓
3	***Ardea purpurea***	✓		✓	✓	✓		✓	✓	✓	✓		✓						✓	✓	✓	✓	✓			✓	
2	***Ciconia nigra***	✓		✓	✓	✓	✓	✓	✓	✓	✓		✓	✓			✓		✓	✓	✓	✓	✓			✓	✓
2	***Ciconia ciconia***	✓		✓	✓	✓	✓	✓	✓	✓	✓		✓	✓			✓		✓	✓	✓	✓	✓			✓	✓
3	***Plegadis falcinellus***	✓		✓		✓			✓	✓	✓								✓			✓	✓		✓	✓	
1	***Geronticus eremita***																										
2	***Platalea leucorodia***	✓			✓	✓		✓	✓	✓	✓	✓	✓	✓					✓		✓	✓	✓		✓	✓	
3	***Phoenicopterus roseus***	✓				✓						✓							✓			✓				✓	
—[E]	*Cygnus olor*	✓		✓	✓	✓	✓	✓		✓	✓	✓	✓	✓			✓	✓	✓	✓	✓	✓	✓		✓	✓	✓
3W	***Cygnus columbianus***	✓		✓		✓		✓		✓		✓		✓		✓	✓		✓		✓	✓	✓		✓		✓
—[E]W	*Cygnus cygnus*	✓		✓	✓	✓		✓		✓	✓	✓	✓	✓	✓		✓	✓	✓	✓	✓	✓	✓	✓	✓		✓
—[E]W	*Anser fabalis*	✓					✓			✓	✓		✓	✓			✓	✓	✓		✓		✓		✓	✓	✓
—[E]	*Anser brachyrhynchus*							✓						✓		✓			✓		✓		✓	✓	✓		
—	*Anser albifrons*	✓		✓		✓	✓	✓		✓	✓	✓	✓	✓		✓			✓		✓	✓	✓		✓	✓	✓
1	***Anser erythropus***					✓		✓		✓	✓							✓			✓	✓	✓				
—	*Anser anser*	✓		✓	✓	✓	✓	✓		✓	✓	✓	✓	✓	✓		✓	✓	✓	✓	✓	✓	✓	✓	✓	✓	✓
—	*Chen caerulescens*															✓											
—	*Branta canadensis*															✓											
—[E]	*Branta leucopsis*							✓		✓	✓			✓	✓	✓	✓	✓	✓		✓		✓	✓	✓		✓
3W	***Branta bernicla***									✓				✓		✓			✓		✓				✓		
1W	***Branta ruficollis***	✓				✓				✓	✓	✓							✓		✓	✓	✓		✓		
3	***Tadorna ferruginea***	✓		✓		✓				✓		✓								✓		✓					
—	*Tadorna tadorna*	✓		✓	✓	✓		✓		✓	✓	✓	✓	✓			✓	✓	✓		✓	✓	✓	✓	✓	✓	✓
—[E]W	*Anas penelope*	✓		✓	✓	✓	✓	✓	✓	✓	✓	✓	✓	✓			✓	✓	✓		✓	✓	✓	✓	✓	✓	✓
3	***Anas strepera***	✓		✓	✓	✓	✓	✓	✓	✓	✓	✓	✓	✓			✓	✓	✓	✓	✓	✓	✓	✓	✓	✓	✓
—	*Anas crecca*	✓		✓	✓	✓	✓	✓	✓	✓	✓	✓	✓	✓	✓	✓	✓	✓	✓	✓	✓	✓	✓	✓	✓	✓	✓

Appendix 2 ...continued. Occurrence of all European species by country.

SPEC Category	Species	Liechtenstein	Lithuania	Luxembourg	FYR of Macedonia	Malta	Moldova	Netherlands	Norway	*Svalbard*	Poland	Portugal	*Azores*	*Madeira*	Romania	Russia	Serbia and Montenegro	Slovakia	Slovenia	Spain	*Canary Islands*	Sweden	Switzerland	Turkey	Ukraine	United Kingdom	*Gibraltar*	TOTAL
3	***Gavia stellata***		✓				✓	✓	✓	✓	✓	✓			✓	✓	✓	✓	✓			✓	✓	✓	✓	✓		37
3	***Gavia arctica***		✓		✓		✓	✓	✓		✓				✓	✓	✓	✓	✓			✓	✓	✓	✓	✓		35
—	*Gavia immer*							✓	✓	✓		✓			✓			✓					✓			✓		16
—	*Gavia adamsii*		✓						✓									✓										5
—	*Tachybaptus ruficollis*	✓	✓	✓	✓		✓	✓	✓		✓	✓			✓	✓	✓	✓	✓	✓		✓	✓	✓	✓	✓		42
—	*Podiceps cristatus*		✓	✓	✓		✓	✓	✓		✓	✓			✓	✓	✓	✓	✓	✓		✓	✓	✓	✓	✓		41
—	*Podiceps grisegena*		✓		✓			✓	✓		✓				✓	✓	✓	✓	✓			✓	✓	✓	✓	✓		35
3	***Podiceps auritus***		✓		✓			✓	✓		✓				✓	✓		✓	✓			✓	✓	✓	✓	✓		33
—	*Podiceps nigricollis*		✓		✓	✓	✓	✓			✓	✓			✓	✓	✓	✓	✓	✓		✓	✓	✓	✓	✓		39
—	*Fulmarus glacialis*								✓	✓						✓										✓		11
1	***Pterodroma madeira***													✓														1
1	***Pterodroma feae***												✓	✓														2
3	***Bulweria bulwerii***												✓	✓							✓							3
2	***Calonectris diomedea***					✓						✓	✓	✓			✓			✓	✓			✓			✓	13
NE	*Puffinus gravis*																											1
1	***Puffinus griseus***																											0
2	***Puffinus puffinus***											✓	✓	✓							✓					✓		9
1	***Puffinus mauretanicus***											✓								✓							✓	3
—[E]	*Puffinus yelkouan*					✓														✓				✓	✓			10
3	***Puffinus assimilis***												✓	✓							✓							3
3	***Pelagodroma marina***													✓							✓							2
—[E]	*Hydrobates pelagicus*					✓			✓			✓								✓	✓					✓		13
3	***Oceanodroma leucorhoa***								✓			✓														✓		6
3	***Oceanodroma castro***											✓	✓	✓							✓							4
—[E]	*Morus bassanus*					✓			✓			✓				✓								✓		✓	✓	14
—	*Phalacrocorax carbo*		✓	✓	✓	✓	✓	✓	✓		✓	✓			✓	✓	✓	✓	✓	✓		✓	✓	✓	✓	✓		46
—[E]	*Phalacrocorax aristotelis*							✓	✓			✓				✓	✓	✓	✓	✓				✓	✓	✓	✓	22
1	***Phalacrocorax pygmeus***				✓		✓								✓	✓	✓	✓	✓					✓	✓			19
3	***Pelecanus onocrotalus***				✓										✓	✓	✓							✓	✓			11
1	***Pelecanus crispus***				✓										✓	✓	✓							✓	✓			12
3	***Botaurus stellaris***		✓	✓	✓		✓	✓			✓	✓			✓	✓	✓	✓	✓	✓		✓	✓	✓	✓	✓		38
3	***Ixobrychus minutus***		✓	✓	✓	✓	✓	✓			✓	✓			✓	✓	✓	✓	✓	✓	✓		✓	✓	✓			35
3	***Nycticorax nycticorax***				✓		✓	✓			✓	✓			✓	✓	✓	✓	✓	✓				✓	✓			29
3	***Ardeola ralloides***				✓		✓					✓			✓	✓	✓	✓		✓				✓	✓			21
—	*Bubulcus ibis*			✓				✓				✓			✓	✓				✓	✓			✓			✓	15
—	*Egretta garzetta*				✓		✓	✓			✓	✓			✓	✓	✓	✓	✓	✓	✓			✓	✓	✓	✓	33
—	*Casmerodius albus*		✓				✓	✓			✓				✓	✓	✓	✓	✓	✓			✓	✓	✓			28
—	*Ardea cinerea*	✓	✓	✓	✓		✓	✓	✓		✓	✓			✓	✓	✓	✓	✓	✓		✓	✓	✓	✓	✓	✓	46
3	***Ardea purpurea***				✓		✓	✓			✓	✓			✓	✓	✓	✓		✓				✓	✓			27
2	***Ciconia nigra***		✓	✓	✓		✓				✓	✓			✓	✓	✓	✓	✓	✓		✓		✓	✓			34
2	***Ciconia ciconia***		✓		✓		✓	✓			✓	✓			✓	✓	✓	✓	✓	✓		✓	✓	✓	✓			35
3	***Plegadis falcinellus***				✓		✓								✓	✓	✓			✓				✓	✓			19
1	***Geronticus eremita***																							✓				1
2	***Platalea leucorodia***				✓		✓	✓				✓			✓	✓	✓	✓		✓				✓	✓	✓		28
3	***Phoenicopterus roseus***											✓							✓	✓				✓				10
—[E]	*Cygnus olor*	✓	✓	✓	✓		✓	✓	✓		✓				✓	✓	✓	✓	✓	✓		✓	✓	✓	✓	✓		40
3W	***Cygnus columbianus***		✓					✓	✓		✓				✓	✓		✓				✓	✓	✓	✓	✓		27
—[E]W	*Cygnus cygnus*		✓		✓		✓	✓	✓		✓				✓	✓	✓	✓	✓			✓	✓	✓	✓	✓		37
—[E]W	*Anser fabalis*		✓	✓				✓	✓		✓				✓	✓	✓	✓	✓	✓		✓	✓	✓	✓	✓		30
—[E]	*Anser brachyrhynchus*							✓		✓					✓				✓			✓				✓		14
—	*Anser albifrons*		✓	✓	✓		✓	✓			✓				✓	✓	✓	✓	✓	✓		✓	✓	✓	✓	✓		35
1	***Anser erythropus***				✓			✓	✓						✓	✓		✓				✓		✓	✓			17
—	*Anser anser*		✓		✓		✓	✓	✓	✓	✓	✓			✓	✓	✓	✓	✓	✓		✓	✓	✓	✓	✓		42
—	*Chen caerulescens*																											1
—	*Branta canadensis*																											1
—[E]	*Branta leucopsis*							✓	✓	✓	✓					✓				✓		✓				✓		22
3W	***Branta bernicla***							✓		✓	✓				✓	✓										✓		12
1W	***Branta ruficollis***				✓		✓								✓		✓	✓						✓	✓			17
3	***Tadorna ferruginea***				✓		✓								✓	✓					✓			✓	✓			14
—	*Tadorna tadorna*		✓					✓	✓		✓	✓			✓	✓	✓	✓	✓	✓		✓	✓	✓	✓	✓		36
—[E]W	*Anas penelope*			✓	✓		✓	✓	✓		✓	✓			✓	✓	✓	✓	✓	✓		✓	✓	✓	✓	✓		40
3	***Anas strepera***		✓	✓	✓		✓	✓	✓		✓	✓			✓	✓	✓	✓	✓	✓		✓	✓	✓	✓	✓		42
—	*Anas crecca*		✓	✓	✓		✓	✓	✓	✓	✓	✓			✓	✓	✓	✓	✓	✓		✓	✓	✓	✓	✓		45

Appendix 2 ...continued. Occurrence of all European species by country.

SPEC Category	Species	Albania	Andorra	Armenia	Austria	Azerbaijan	Belarus	Belgium	Bosnia and Herzegovina	Bulgaria	Croatia	Cyprus	Czech Republic	Denmark	*Faroe Islands*	*Greenland*	Estonia	Finland	France	Georgia	Germany	Greece	Hungary	Iceland	Republic of Ireland	Italy	Latvia
—	Anas platyrhynchos	✓		✓	✓	✓	✓	✓	✓	✓	✓	✓	✓	✓	✓	✓	✓	✓	✓	✓	✓	✓	✓	✓	✓	✓	✓
3	**Anas acuta**	✓		✓	✓	✓	✓	✓	✓	✓	✓	✓	✓	✓	✓	✓	✓	✓	✓		✓	✓	✓	✓	✓	✓	✓
3	**Anas querquedula**	✓		✓	✓	✓	✓	✓	✓	✓	✓	✓	✓	✓			✓	✓	✓	✓	✓	✓	✓		✓	✓	✓
3	**Anas clypeata**	✓		✓	✓	✓	✓	✓	✓	✓	✓	✓	✓	✓			✓	✓	✓	✓	✓	✓	✓	✓	✓	✓	✓
1	**Marmaronetta angustirostris**			✓		✓																				✓	
—	Netta rufina	✓		✓	✓	✓		✓		✓	✓	✓	✓	✓					✓	✓	✓	✓	✓			✓	✓
2	**Aythya ferina**	✓		✓	✓	✓	✓	✓	✓	✓	✓	✓	✓	✓			✓	✓	✓	✓	✓	✓	✓		✓	✓	✓
1	**Aythya nyroca**	✓		✓	✓	✓	✓	✓	✓	✓	✓	✓	✓						✓	✓	✓	✓	✓			✓	✓
3	**Aythya fuligula**	✓		✓	✓	✓	✓	✓		✓	✓	✓	✓	✓	✓		✓	✓	✓	✓	✓	✓	✓	✓	✓	✓	✓
3W	**Aythya marila**	✓			✓	✓		✓		✓	✓		✓	✓	✓		✓	✓	✓		✓	✓	✓	✓	✓	✓	✓
—[E]	Somateria mollissima	✓			✓		✓	✓		✓			✓	✓	✓	✓	✓	✓	✓		✓	✓	✓	✓	✓	✓	✓
—	Somateria spectabilis															✓	✓							✓			
3W	**Polysticta stelleri**																✓	✓									✓
3	**Histrionicus histrionicus**															✓								✓			
—	Clangula hyemalis	✓			✓	✓	✓			✓	✓		✓	✓	✓	✓	✓	✓	✓		✓	✓	✓	✓	✓	✓	✓
—	Melanitta nigra	✓			✓			✓		✓	✓			✓			✓	✓	✓		✓	✓	✓	✓	✓	✓	✓
3	**Melanitta fusca**	✓			✓		✓	✓		✓	✓		✓	✓			✓	✓	✓	✓	✓	✓	✓		✓	✓	✓
3	**Bucephala islandica**																							✓			
—	Bucephala clangula	✓			✓	✓	✓	✓		✓	✓	✓	✓	✓	✓		✓	✓	✓		✓	✓	✓	✓	✓	✓	✓
3	**Mergellus albellus**	✓			✓	✓	✓	✓		✓	✓		✓	✓			✓	✓	✓	✓	✓	✓	✓		✓	✓	✓
—	Mergus serrator	✓			✓	✓	✓	✓		✓	✓	✓	✓	✓	✓	✓	✓	✓	✓		✓	✓	✓	✓	✓	✓	✓
—	Mergus merganser	✓			✓	✓	✓	✓		✓	✓		✓	✓			✓	✓	✓		✓	✓	✓	✓	✓	✓	✓
1	**Oxyura leucocephala**	✓		✓		✓				✓		✓							✓	✓		✓					
—[E]	Pernis apivorus	✓		✓	✓	✓	✓	✓		✓	✓		✓	✓			✓	✓	✓	✓	✓	✓	✓			✓	✓
3	**Elanus caeruleus**																		✓								
3	**Milvus migrans**	✓		✓	✓	✓	✓	✓	✓	✓	✓		✓				✓	✓	✓	✓	✓	✓	✓			✓	✓
2	**Milvus milvus**				✓	✓	✓	✓			✓		✓	✓					✓		✓	✓	✓			✓	✓
1	**Haliaeetus albicilla**	✓		✓	✓	✓	✓			✓	✓		✓	✓		✓	✓	✓	✓	✓	✓	✓	✓	✓		✓	✓
3	**Gypaetus barbatus**	✓	✓	✓		✓													✓	✓		✓				✓	
3	**Neophron percnopterus**	✓		✓		✓			✓	✓									✓	✓		✓				✓	
—	Gyps fulvus	✓		✓		✓				✓	✓	✓							✓	✓		✓				✓	
1	**Aegypius monachus**			✓		✓				✓									✓	✓		✓					
3	**Circaetus gallicus**	✓	✓	✓		✓	✓		✓	✓	✓						✓		✓	✓		✓	✓			✓	✓
—	Circus aeruginosus	✓		✓	✓	✓	✓	✓	✓	✓	✓	✓	✓	✓			✓	✓	✓	✓	✓	✓	✓			✓	✓
3	**Circus cyaneus**	✓		✓	✓	✓	✓	✓		✓	✓	✓	✓	✓			✓	✓	✓		✓	✓	✓		✓	✓	✓
1	**Circus macrourus**					✓																					
—[E]	Circus pygargus			✓	✓	✓	✓	✓		✓	✓		✓	✓			✓	✓	✓	✓	✓	✓	✓			✓	✓
—	Accipiter gentilis	✓	✓	✓	✓	✓	✓	✓	✓	✓	✓	✓	✓	✓			✓	✓	✓	✓	✓	✓	✓		✓	✓	✓
—	Accipiter nisus	✓	✓	✓	✓	✓	✓	✓	✓	✓	✓	✓	✓	✓			✓	✓	✓	✓	✓	✓	✓		✓	✓	✓
—	Accipiter badius					✓																					
2	**Accipiter brevipes**	✓		✓		✓				✓	✓									✓		✓	✓				
—	Buteo buteo	✓	✓	✓	✓	✓	✓	✓	✓	✓	✓	✓	✓	✓			✓	✓	✓	✓	✓	✓	✓		✓	✓	✓
3	**Buteo rufinus**	✓		✓		✓				✓		✓								✓		✓	✓				
—	Buteo lagopus			✓	✓	✓	✓			✓	✓		✓	✓			✓	✓				✓	✓				✓
2	**Aquila pomarina**	✓		✓		✓	✓			✓	✓		✓				✓			✓	✓	✓	✓				✓
1	**Aquila clanga**	✓				✓	✓			✓	✓						✓	✓	✓			✓	✓			✓	✓
3	**Aquila nipalensis**																						✓				
1	**Aquila heliaca**	✓		✓	✓	✓				✓	✓	✓	✓							✓		✓	✓				
1	**Aquila adalberti**																										
3	**Aquila chrysaetos**	✓	✓	✓	✓	✓	✓		✓	✓	✓		✓	✓			✓	✓	✓	✓	✓	✓	✓			✓	✓
3	**Hieraaetus pennatus**	✓		✓	✓	✓	✓			✓	✓								✓	✓	✓	✓	✓				
3	**Hieraaetus fasciatus**	✓				✓			✓	✓	✓	✓							✓			✓				✓	
3	**Pandion haliaetus**	✓		✓		✓	✓			✓				✓			✓	✓	✓		✓	✓				✓	✓
1	**Falco naumanni**	✓		✓		✓			✓	✓	✓								✓	✓		✓				✓	
3	**Falco tinnunculus**	✓	✓	✓	✓	✓	✓	✓	✓	✓	✓	✓	✓	✓			✓	✓	✓	✓	✓	✓	✓		✓	✓	✓
3	**Falco vespertinus**				✓		✓			✓	✓		✓				✓		✓	✓	✓		✓			✓	
—	Falco columbarius			✓	✓		✓			✓	✓	✓	✓	✓	✓		✓	✓				✓	✓	✓	✓		✓
—	Falco subbuteo	✓	✓	✓	✓	✓	✓	✓	✓	✓	✓	✓	✓	✓			✓	✓	✓	✓	✓	✓	✓			✓	✓
2	**Falco eleonorae**										✓	✓										✓				✓	
3	**Falco biarmicus**			✓		✓			✓	✓	✓									✓		✓				✓	
1	**Falco cherrug**				✓	✓				✓	✓		✓							✓	✓	✓	✓				
3	**Falco rusticolus**															✓	✓	✓						✓			
—	Falco peregrinus	✓	✓	✓	✓	✓	✓	✓	✓	✓	✓	✓	✓	✓		✓	✓	✓	✓	✓	✓	✓	✓		✓	✓	

Appendix 2 ...continued. Occurrence of all European species by country.

SPEC Category	Species	Liechtenstein	Lithuania	Luxembourg	FYR of Macedonia	Malta	Moldova	Netherlands	Norway	*Svalbard*	Poland	Portugal	*Azores*	*Madeira*	Romania	Russia	Serbia and Montenegro	Slovakia	Slovenia	Spain	*Canary Islands*	Sweden	Switzerland	Turkey	Ukraine	United Kingdom	*Gibraltar*	TOTAL
—	*Anas platyrhynchos*	✓	✓	✓	✓		✓	✓	✓		✓	✓	✓		✓	✓	✓	✓	✓	✓		✓	✓	✓	✓	✓		46
3	***Anas acuta***		✓					✓	✓	✓	✓	✓			✓	✓	✓	✓	✓	✓		✓	✓	✓	✓	✓		**41**
3	***Anas querquedula***		✓	✓	✓		✓	✓	✓		✓	✓			✓	✓	✓	✓	✓	✓		✓	✓	✓	✓	✓		**41**
3	***Anas clypeata***		✓	✓	✓			✓	✓		✓	✓			✓	✓	✓	✓	✓	✓		✓	✓	✓	✓	✓		**41**
1	***Marmaronetta angustirostris***															✓				✓	✓			✓				**7**
—	*Netta rufina*				✓			✓			✓	✓			✓	✓	✓	✓	✓	✓			✓	✓	✓			30
2	***Aythya ferina***		✓	✓	✓		✓	✓	✓		✓	✓			✓	✓	✓	✓	✓	✓		✓	✓	✓	✓	✓		**41**
1	***Aythya nyroca***		✓	✓	✓		✓	✓			✓	✓			✓	✓	✓	✓	✓	✓			✓	✓	✓			**34**
3	***Aythya fuligula***	✓	✓	✓			✓	✓	✓		✓	✓			✓	✓	✓	✓	✓	✓		✓	✓	✓	✓	✓		**42**
3W	***Aythya marila***		✓				✓	✓	✓		✓	✓			✓	✓	✓	✓	✓	✓		✓	✓	✓	✓	✓		**36**
—[E]	*Somateria mollissima*		✓					✓	✓	✓	✓				✓	✓	✓	✓	✓			✓	✓	✓	✓	✓		34
—	*Somateria spectabilis*								✓	✓						✓												6
3W	***Polysticta stelleri***		✓						✓							✓						✓						**7**
3	***Histrionicus histrionicus***																											**2**
—	*Clangula hyemalis*		✓					✓	✓	✓	✓				✓	✓	✓	✓	✓			✓	✓	✓	✓	✓		35
—	*Melanitta nigra*		✓					✓	✓	✓	✓	✓			✓	✓		✓	✓	✓		✓		✓	✓	✓		31
3	***Melanitta fusca***		✓				✓	✓	✓		✓				✓	✓	✓	✓	✓	✓		✓	✓	✓	✓	✓		**34**
3	***Bucephala islandica***																											**1**
—	*Bucephala clangula*		✓	✓	✓		✓	✓	✓		✓				✓	✓	✓	✓	✓			✓	✓	✓	✓	✓		38
3	***Mergellus albellus***		✓	✓	✓		✓	✓	✓		✓				✓	✓	✓	✓	✓			✓	✓	✓	✓	✓		**36**
—	*Mergus serrator*		✓				✓	✓	✓		✓	✓			✓	✓	✓	✓	✓	✓		✓	✓	✓	✓	✓		39
—	*Mergus merganser*		✓	✓	✓		✓	✓	✓		✓				✓	✓	✓	✓	✓			✓	✓	✓	✓	✓		36
1	***Oxyura leucocephala***														✓	✓				✓				✓	✓			**13**
—	*Pernis apivorus*	✓	✓	✓	✓		✓	✓	✓		✓	✓			✓	✓	✓	✓	✓	✓		✓	✓	✓	✓	✓		39
3	***Elanus caeruleus***											✓								✓								**3**
3	***Milvus migrans***	✓	✓	✓	✓		✓	✓			✓	✓			✓	✓	✓	✓	✓	✓		✓	✓	✓	✓			**37**
2	***Milvus milvus***		✓	✓			✓	✓			✓	✓			✓	✓	✓	✓		✓		✓	✓	✓	✓	✓		**29**
1	***Haliaeetus albicilla***		✓				✓	✓	✓		✓				✓	✓	✓	✓	✓			✓		✓	✓	✓		**34**
3	***Gypaetus barbatus***				✓											✓				✓			✓	✓				**13**
3	***Neophron percnopterus***				✓		✓					✓				✓	✓			✓	✓			✓				**17**
—	*Gyps fulvus*				✓							✓				✓	✓			✓				✓	✓			17
1	***Aegypius monachus***				✓							✓				✓				✓				✓	✓			**12**
3	***Circaetus gallicus***				✓		✓				✓	✓			✓	✓	✓	✓	✓	✓				✓	✓		✓	**28**
—	*Circus aeruginosus*		✓		✓		✓	✓	✓		✓	✓			✓	✓	✓	✓	✓	✓		✓		✓	✓	✓		38
3	***Circus cyaneus***		✓	✓			✓	✓	✓		✓	✓			✓	✓	✓	✓		✓		✓	✓	✓	✓	✓		**37**
1	***Circus macrourus***						✓								✓	✓								✓	✓			**6**
—[E]	*Circus pygargus*		✓	✓	✓		✓	✓			✓	✓			✓	✓	✓	✓		✓		✓		✓	✓	✓		34
—	*Accipiter gentilis*	✓	✓	✓	✓		✓	✓	✓		✓	✓			✓	✓	✓	✓	✓	✓		✓	✓	✓	✓	✓		43
—	*Accipiter nisus*	✓	✓	✓	✓		✓	✓	✓		✓	✓		✓	✓	✓	✓	✓	✓	✓	✓	✓	✓	✓	✓	✓		45
—	*Accipiter badius*																											1
2	***Accipiter brevipes***				✓		✓								✓	✓	✓							✓	✓			**15**
—	*Buteo buteo*	✓	✓	✓	✓		✓	✓	✓		✓	✓	✓	✓	✓	✓	✓	✓	✓	✓	✓	✓	✓	✓	✓	✓	✓	47
3	***Buteo rufinus***				✓										✓	✓	✓							✓	✓			**14**
—	*Buteo lagopus*		✓				✓	✓	✓		✓				✓	✓		✓	✓			✓			✓	✓		25
2	***Aquila pomarina***		✓		✓		✓				✓				✓	✓	✓	✓	✓					✓	✓			**24**
1	***Aquila clanga***				✓		✓				✓				✓	✓	✓		✓				✓	✓	✓			**22**
3	***Aquila nipalensis***															✓								✓	✓			**4**
1	***Aquila heliaca***				✓		✓								✓	✓	✓	✓						✓	✓			**19**
1	***Aquila adalberti***											✓								✓								**2**
3	***Aquila chrysaetos***	✓	✓		✓		✓		✓		✓	✓			✓	✓	✓	✓	✓	✓		✓	✓	✓	✓	✓		**38**
3	***Hieraaetus pennatus***				✓		✓				✓	✓			✓	✓	✓	✓		✓				✓	✓		✓	**24**
3	***Hieraaetus fasciatus***				✓							✓					✓			✓				✓				**14**
3	***Pandion haliaetus***		✓				✓	✓	✓		✓	✓				✓			✓	✓	✓	✓		✓	✓	✓		**27**
1	***Falco naumanni***				✓		✓					✓			✓	✓	✓		✓	✓				✓	✓		✓	**21**
3	***Falco tinnunculus***	✓	✓	✓	✓	✓	✓	✓	✓		✓	✓		✓	✓	✓	✓	✓	✓	✓	✓	✓	✓	✓	✓	✓	✓	**47**
3	***Falco vespertinus***						✓								✓	✓	✓	✓							✓			**17**
—	*Falco columbarius*		✓	✓	✓		✓	✓	✓		✓	✓			✓	✓		✓	✓			✓	✓		✓	✓		32
—	*Falco subbuteo*	✓	✓	✓	✓		✓	✓	✓		✓	✓			✓	✓	✓	✓	✓	✓		✓	✓	✓	✓	✓		42
2	***Falco eleonorae***																✓			✓	✓			✓				**8**
3	***Falco biarmicus***				✓												✓							✓				**11**
1	***Falco cherrug***				✓		✓				✓				✓	✓	✓	✓						✓	✓			**18**
3	***Falco rusticolus***								✓							✓						✓						**7**
—	*Falco peregrinus*	✓	✓	✓	✓	✓	✓	✓	✓		✓	✓			✓	✓	✓	✓	✓	✓		✓	✓	✓	✓	✓	✓	45

Appendix 2 ...continued. Occurrence of all European species by country.

SPEC Category	Species	Albania	Andorra	Armenia	Austria	Azerbaijan	Belarus	Belgium	Bosnia and Herzegovina	Bulgaria	Croatia	Cyprus	Czech Republic	Denmark	*Faroe Islands*	*Greenland*	Estonia	Finland	France	Georgia	Germany	Greece	Hungary	Iceland	Republic of Ireland	Italy	Latvia
—	*Falco pelegrinoides*																										
—	*Bonasa bonasia*	✔			✔		✔	✔	✔	✔	✔		✔				✔	✔	✔		✔	✔	✔			✔	✔
—	*Lagopus lagopus*		✔				✔										✔	✔							✔		✔
—	*Lagopus mutus*		✔		✔											✔		✔	✔		✔			✔		✔	
3	***Tetrao tetrix***	✔			✔		✔	✔					✔	✔			✔	✔	✔		✔					✔	✔
1	***Tetrao mlokosiewiczi***			✔		✔														✔							
—	*Tetrao urogallus*	✔	✔		✔		✔		✔	✔	✔		✔				✔	✔	✔		✔	✔	✔			✔	✔
—[E]	*Tetraogallus caucasicus*					✔														✔							
3	***Tetraogallus caspius***			✔		✔														✔							
3	***Alectoris chukar***			✔		✔				✔	✔	✔								✔		✔					
2	***Alectoris graeca***	✔			✔				✔	✔	✔								✔			✔				✔	
2	***Alectoris rufa***		✔																✔						✔	✔	
3	***Alectoris barbara***																									✔	
3	***Ammoperdix griseogularis***					✔																					
3	***Francolinus francolinus***					✔						✔								✔							
3	***Perdix perdix***	✔	✔	✔	✔	✔	✔	✔		✔	✔		✔	✔			✔	✔	✔	✔	✔	✔	✔		✔	✔	✔
3	***Coturnix coturnix***	✔	✔	✔	✔	✔	✔	✔	✔	✔	✔	✔	✔	✔	✔		✔	✔	✔	✔	✔	✔	✔		✔	✔	✔
—	*Phasianus colchicus*		✔	✔	✔	✔	✔	✔	✔	✔	✔		✔	✔			✔	✔	✔	✔	✔	✔	✔		✔	✔	✔
3	***Turnix sylvatica***																										
—	*Rallus aquaticus*	✔		✔	✔	✔	✔	✔	✔	✔	✔	✔	✔	✔	✔		✔	✔	✔	✔	✔	✔	✔	✔	✔	✔	✔
—[E]	*Porzana porzana*	✔			✔	✔	✔	✔		✔	✔		✔	✔			✔	✔	✔	✔	✔	✔	✔		✔	✔	✔
—[E]	*Porzana parva*				✔	✔	✔			✔	✔		✔				✔	✔	✔	✔	✔	✔	✔			✔	✔
3	***Porzana pusilla***	✔				✔		✔	✔	✔	✔		✔						✔	✔	✔	✔	✔				
1	***Crex crex***	✔		✔	✔	✔	✔	✔	✔	✔	✔		✔	✔			✔	✔	✔	✔	✔		✔		✔	✔	✔
—	*Gallinula chloropus*	✔		✔	✔	✔	✔	✔	✔	✔	✔	✔	✔	✔			✔	✔	✔	✔	✔	✔	✔		✔	✔	✔
3	***Porphyrio porphyrio***					✔					✔								✔							✔	
—	*Fulica atra*	✔		✔	✔	✔	✔	✔	✔	✔	✔	✔	✔	✔	✔		✔	✔	✔	✔	✔	✔	✔		✔	✔	✔
3	***Fulica cristata***																										
2	***Grus grus***			✔		✔	✔			✔	✔		✔	✔			✔	✔	✔	✔	✔	✔	✔			✔	✔
—	*Grus virgo*																										
1	***Tetrax tetrax***					✔													✔							✔	
1	***Chlamydotis undulata***					✔																					
1	***Otis tarda***				✔	✔				✔			✔								✔	✔	✔				
—[E]	*Haematopus ostralegus*	✔					✔	✔		✔				✔	✔		✔	✔	✔		✔	✔		✔	✔	✔	✔
—	*Himantopus himantopus*	✔		✔	✔	✔	✔	✔		✔	✔	✔	✔						✔	✔	✔	✔	✔			✔	
—	*Recurvirostra avosetta*	✔		✔	✔	✔		✔		✔	✔	✔	✔	✔			✔		✔	✔	✔	✔	✔			✔	
3	***Burhinus oedicnemus***	✔		✔	✔	✔	✔		✔	✔	✔	✔	✔						✔	✔	✔	✔	✔			✔	
3	***Cursorius cursor***																										
3	***Glareola pratincola***	✔		✔		✔				✔									✔			✔	✔			✔	
1	***Glareola nordmanni***									✔													✔				
—	*Charadrius dubius*	✔		✔	✔	✔	✔	✔	✔	✔	✔	✔	✔	✔			✔	✔	✔	✔	✔	✔	✔			✔	✔
—[E]	*Charadrius hiaticula*	✔				✔	✔	✔				✔		✔	✔	✔	✔	✔	✔		✔	✔		✔	✔	✔	✔
3	***Charadrius alexandrinus***	✔			✔	✔		✔		✔	✔	✔		✔					✔	✔	✔	✔	✔			✔	
3	***Charadrius leschenaultii***											✔															
3	***Charadrius asiaticus***																										
—	*Eudromias morinellus*		✔		✔	✔							✔					✔	✔			✔				✔	
—[E]	*Pluvialis apricaria*	✔				✔	✔	✔		✔		✔		✔	✔	✔	✔	✔	✔		✔	✔		✔	✔	✔	✔
—	*Pluvialis squatarola*	✔				✔		✔		✔	✔	✔		✔					✔		✔	✔			✔	✔	
3	***Vanellus spinosus***											✔										✔					
3	***Vanellus indicus***																										
1	***Vanellus gregarius***					✔																					
—	*Vanellus leucurus*			✔		✔																					
2	***Vanellus vanellus***	✔		✔	✔	✔	✔	✔	✔	✔	✔	✔	✔	✔	✔		✔	✔	✔	✔	✔	✔	✔	✔	✔	✔	✔
3W	***Calidris canutus***							✔		✔				✔		✔	✔		✔		✔	✔		✔	✔	✔	
—	*Calidris alba*							✔		✔		✔		✔	✔	✔			✔		✔	✔			✔	✔	
—	*Calidris minuta*	✔				✔		✔				✔						✔	✔			✔			✔	✔	
—	*Calidris temminckii*					✔						✔					✔	✔	✔							✔	
—	*Calidris bairdii*															✔											
NE	*Calidris ferruginea*																										
—[E]	*Calidris maritima*							✔						✔	✔	✔	✔	✔	✔		✔			✔	✔		
3	***Calidris alpina***	✔				✔	✔	✔		✔	✔	✔		✔	✔	✔	✔	✔	✔		✔	✔		✔	✔	✔	✔
3	***Limicola falcinellus***																	✔									
2	***Philomachus pugnax***						✔	✔		✔	✔	✔		✔			✔	✔	✔		✔	✔	✔		✔	✔	✔

Appendix 2 ...continued. Occurrence of all European species by country.

SPEC Category	Species	Liechtenstein	Lithuania	Luxembourg	FYR of Macedonia	Malta	Moldova	Netherlands	Norway	*Svalbard*	Poland	Portugal	*Azores*	*Madeira*	Romania	Russia	Serbia and Montenegro	Slovakia	Slovenia	Spain	*Canary Islands*	Sweden	Switzerland	Turkey	Ukraine	United Kingdom	*Gibraltar*	TOTAL
—	*Falco pelegrinoides*																				✓			✓				2
—	*Bonasa bonasia*	✓	✓	✓	✓				✓		✓				✓	✓	✓	✓	✓			✓	✓		✓			30
—	*Lagopus lagopus*		✓						✓							✓						✓				✓		11
—	*Lagopus mutus*	✓							✓	✓						✓			✓	✓		✓	✓			✓		17
3	***Tetrao tetrix***	✓	✓					✓	✓		✓				✓	✓		✓	✓			✓	✓		✓	✓		25
1	***Tetrao mlokosiewiczi***															✓								✓				5
—	*Tetrao urogallus*	✓	✓		✓				✓		✓				✓	✓	✓	✓	✓	✓		✓	✓		✓	✓		31
—[E]	*Tetraogallus caucasicus*															✓												3
3	***Tetraogallus caspius***																							✓				4
3	***Alectoris chukar***															✓								✓	✓			10
2	***Alectoris graeca***				✓												✓		✓				✓					12
2	***Alectoris rufa***											✓		✓						✓						✓		8
3	***Alectoris barbara***																				✓						✓	3
3	***Ammoperdix griseogularis***																							✓				2
3	***Francolinus francolinus***																							✓				4
3	***Perdix perdix***		✓	✓	✓		✓	✓			✓				✓	✓	✓	✓	✓	✓		✓	✓	✓	✓	✓		38
3	***Coturnix coturnix***	✓	✓	✓	✓	✓	✓	✓	✓		✓	✓	✓	✓	✓	✓	✓	✓	✓	✓	✓	✓	✓	✓	✓	✓		48
—	*Phasianus colchicus*	✓		✓	✓		✓	✓	✓		✓	✓			✓	✓	✓	✓	✓	✓		✓	✓	✓	✓	✓		40
3	***Turnix sylvatica***																			✓								1
—	*Rallus aquaticus*	✓	✓	✓	✓	✓	✓	✓	✓		✓	✓			✓	✓	✓	✓	✓	✓		✓	✓	✓	✓	✓		45
—[E]	*Porzana porzana*		✓		✓	✓	✓	✓	✓		✓	✓			✓	✓	✓	✓	✓	✓		✓	✓	✓	✓	✓		38
—[E]	*Porzana parva*		✓		✓		✓				✓				✓	✓	✓	✓	✓				✓	✓	✓			27
3	***Porzana pusilla***						✓	✓				✓			✓	✓	✓			✓			✓	✓				21
1	***Crex crex***	✓	✓	✓	✓		✓	✓	✓		✓				✓	✓	✓	✓	✓			✓	✓	✓	✓	✓		38
—	*Gallinula chloropus*	✓	✓	✓	✓	✓	✓	✓	✓		✓	✓	✓	✓	✓	✓	✓	✓	✓	✓	✓	✓	✓	✓	✓	✓		46
3	***Porphyrio porphyrio***											✓				✓				✓				✓				8
—	*Fulica atra*	✓	✓	✓	✓	✓	✓	✓	✓		✓	✓		✓	✓	✓	✓	✓	✓	✓	✓	✓	✓	✓	✓	✓		46
3	***Fulica cristata***																			✓								1
2	***Grus grus***		✓	✓				✓	✓		✓	✓			✓	✓	✓			✓		✓		✓	✓	✓		30
—	*Grus virgo*															✓								✓	✓			3
1	***Tetrax tetrax***				✓							✓				✓				✓				✓	✓			9
1	***Chlamydotis undulata***															✓					✓							3
1	***Otis tarda***						✓					✓			✓	✓	✓	✓	✓	✓				✓	✓			17
—[E]	*Haematopus ostralegus*		✓		✓			✓	✓		✓	✓			✓	✓	✓			✓		✓		✓	✓	✓		29
—	*Himantopus himantopus*				✓		✓	✓			✓	✓			✓	✓	✓	✓	✓	✓	✓			✓	✓			30
—	*Recurvirostra avosetta*		✓		✓		✓	✓	✓		✓	✓			✓	✓	✓	✓	✓	✓		✓		✓	✓	✓		34
3	***Burhinus oedicnemus***				✓						✓	✓			✓	✓	✓	✓		✓	✓			✓	✓	✓		28
3	***Cursorius cursor***																			✓	✓			✓				3
3	***Glareola pratincola***				✓							✓			✓	✓	✓			✓				✓	✓			16
1	***Glareola nordmanni***														✓	✓								✓	✓			6
—	*Charadrius dubius*	✓	✓	✓	✓	✓	✓	✓	✓		✓	✓			✓	✓	✓	✓	✓	✓	✓	✓	✓	✓	✓	✓		43
—[E]	*Charadrius hiaticula*		✓					✓	✓	✓	✓	✓				✓				✓		✓		✓	✓	✓		29
3	***Charadrius alexandrinus***							✓			✓	✓	✓	✓	✓	✓	✓	✓	✓	✓	✓	✓		✓	✓		✓	30
3	***Charadrius leschenaultii***															✓								✓				3
3	***Charadrius asiaticus***															✓												1
—	*Eudromias morinellus*								✓		✓					✓				✓		✓				✓		14
—[E]	*Pluvialis apricaria*		✓					✓	✓	✓		✓				✓	✓	✓	✓	✓		✓		✓	✓	✓		32
—	*Pluvialis squatarola*							✓				✓				✓	✓		✓	✓				✓	✓	✓		21
3	***Vanellus spinosus***																							✓				3
3	***Vanellus indicus***																							✓				1
1	***Vanellus gregarius***															✓												2
—	*Vanellus leucurus*														✓	✓								✓	✓			6
2	***Vanellus vanellus***	✓	✓	✓	✓	✓	✓	✓	✓		✓	✓			✓	✓	✓	✓	✓	✓		✓	✓	✓	✓	✓		45
3W	***Calidris canutus***							✓		✓		✓								✓				✓		✓		17
—	*Calidris alba*							✓		✓	✓	✓					✓			✓				✓		✓		19
—	*Calidris minuta*							✓	✓			✓				✓	✓		✓	✓				✓				17
—	*Calidris temminckii*								✓							✓				✓		✓		✓		✓		12
—	*Calidris bairdii*																											1
NE	*Calidris ferruginea*											✓								✓								2
—[E]	*Calidris maritima*							✓	✓	✓	✓	✓				✓				✓		✓				✓		19
3	***Calidris alpina***		✓					✓	✓	✓	✓	✓				✓	✓		✓	✓		✓		✓	✓	✓		33
3	***Limicola falcinellus***								✓							✓						✓						4
2	***Philomachus pugnax***		✓					✓	✓		✓	✓				✓	✓		✓	✓		✓		✓	✓	✓		28

Appendix 2 ...continued. Occurrence of all European species by country.

SPEC Category	Species	Albania	Andorra	Armenia	Austria	Azerbaijan	Belarus	Belgium	Bosnia and Herzegovina	Bulgaria	Croatia	Cyprus	Czech Republic	Denmark	*Faroe Islands*	*Greenland*	Estonia	Finland	France	Georgia	Germany	Greece	Hungary	Iceland	Republic of Ireland	Italy	Latvia
3	***Lymnocryptes minimus***	✓				✓	✓	✓		✓	✓	✓	✓	✓	✓		✓	✓	✓		✓	✓	✓		✓	✓	✓
3	***Gallinago gallinago***	✓			✓	✓	✓	✓	✓	✓	✓	✓	✓	✓	✓		✓	✓	✓		✓	✓	✓	✓	✓	✓	✓
1	***Gallinago media***						✓										✓	✓									✓
—	*Gallinago stenura*																										
3	***Scolopax rusticola***	✓	✓	✓	✓	✓	✓	✓		✓	✓	✓	✓	✓	✓		✓	✓	✓	✓	✓	✓	✓		✓	✓	✓
2	***Limosa limosa***	✓			✓	✓	✓	✓		✓	✓	✓	✓	✓			✓	✓	✓		✓	✓	✓	✓	✓	✓	✓
—	*Limosa lapponica*					✓					✓			✓				✓	✓		✓	✓		✓	✓	✓	
—[E]	*Numenius phaeopus*						✓								✓	✓	✓	✓	✓					✓	✓	✓	✓
1	***Numenius tenuirostris***																					✓				✓	
2	***Numenius arquata***	✓			✓	✓	✓	✓		✓	✓	✓	✓	✓	✓		✓	✓	✓		✓	✓	✓	✓	✓	✓	✓
3	***Tringa erythropus***	✓						✓										✓	✓			✓			✓	✓	
2	***Tringa totanus***	✓		✓	✓	✓	✓	✓		✓	✓	✓	✓	✓	✓		✓	✓	✓	✓	✓	✓	✓	✓	✓	✓	✓
—	*Tringa stagnatilis*					✓	✓											✓	✓								✓
—	*Tringa nebularia*	✓				✓	✓	✓		✓		✓					✓	✓	✓		✓	✓			✓	✓	✓
—	*Tringa ochropus*	✓			✓	✓	✓	✓		✓	✓	✓	✓	✓			✓	✓	✓	✓	✓	✓	✓		✓	✓	✓
3	***Tringa glareola***						✓							✓			✓	✓	✓		✓				✓		✓
—	*Xenus cinereus*						✓											✓									✓
3	***Actitis hypoleucos***	✓		✓	✓	✓	✓	✓	✓	✓	✓	✓	✓	✓			✓	✓	✓	✓	✓	✓	✓		✓	✓	✓
—	*Arenaria interpres*					✓		✓		✓		✓		✓	✓	✓	✓	✓	✓		✓	✓		✓	✓	✓	✓
—	*Phalaropus lobatus*														✓	✓		✓	✓					✓	✓		
—	*Phalaropus fulicarius*															✓			✓					✓			
—	*Stercorarius pomarinus*																		✓								
—	*Stercorarius parasiticus*														✓	✓		✓						✓			
—	*Stercorarius longicaudus*															✓		✓									
—[E]	*Catharacta skua*														✓				✓					✓	✓		
—	*Larus ichthyaetus*			✓		✓				✓										✓						✓	
—[E]	*Larus melanocephalus*	✓			✓	✓	✓	✓		✓	✓	✓	✓	✓			✓		✓		✓	✓	✓		✓	✓	
3	***Larus minutus***	✓				✓	✓			✓	✓	✓		✓			✓	✓	✓		✓	✓	✓			✓	✓
—[E]	*Larus ridibundus*	✓			✓	✓	✓	✓	✓	✓	✓	✓	✓	✓	✓	✓	✓	✓	✓	✓	✓	✓	✓	✓	✓	✓	✓
3	***Larus genei***	✓				✓				✓		✓							✓			✓				✓	
1	***Larus audouinii***										✓	✓							✓			✓				✓	
2	***Larus canus***	✓			✓	✓	✓	✓		✓	✓	✓	✓	✓	✓		✓	✓	✓		✓	✓	✓	✓	✓	✓	✓
—[E]	*Larus fuscus*	✓						✓		✓	✓	✓		✓	✓	✓	✓	✓	✓		✓	✓	✓	✓	✓	✓	✓
—[E]	*Larus argentatus*						✓	✓					✓	✓	✓	✓	✓	✓	✓		✓		✓	✓	✓	✓	✓
—[E]	*Larus cachinnans*	✓			✓	✓	✓	✓	✓	✓	✓	✓	✓						✓	✓	✓	✓	✓			✓	
2	***Larus armenicus***			✓								✓								✓							
—[E]	*Larus glaucoides*														✓	✓					✓			✓			
—	*Larus hyperboreus*														✓	✓	✓		✓		✓			✓			✓
—[E]	*Larus marinus*	✓						✓		✓				✓	✓	✓	✓	✓	✓		✓			✓	✓		✓
—	*Xema sabini*															✓			✓								
—	*Rhodostethia rosea*															✓											
—	*Rissa tridactyla*												✓	✓	✓	✓			✓		✓	✓		✓	✓		
3	***Pagophila eburnea***															✓											
3	***Sterna nilotica***	✓		✓						✓				✓					✓		✓	✓				✓	
3	***Sterna caspia***	✓				✓											✓	✓			✓	✓				✓	
—	*Sterna bengalensis*																									✓	
2	***Sterna sandvicensis***	✓				✓		✓		✓	✓	✓		✓			✓		✓		✓	✓			✓	✓	
3	***Sterna dougallii***																		✓						✓		
—	*Sterna hirundo*	✓		✓	✓	✓	✓	✓	✓	✓	✓	✓	✓	✓	✓		✓	✓	✓	✓	✓	✓	✓		✓	✓	✓
—	*Sterna paradisaea*													✓	✓	✓	✓	✓	✓		✓			✓	✓		✓
3	***Sterna albifrons***	✓		✓		✓	✓	✓		✓	✓	✓		✓			✓	✓	✓	✓	✓	✓	✓		✓	✓	✓
3	***Chlidonias hybrida***	✓		✓		✓	✓		✓	✓	✓								✓	✓	✓	✓	✓			✓	✓
3	***Chlidonias niger***	✓					✓			✓	✓		✓	✓			✓	✓	✓	✓	✓	✓	✓			✓	✓
—	*Chlidonias leucopterus*					✓	✓			✓	✓						✓			✓	✓		✓			✓	✓
—	*Uria aalge*													✓	✓	✓	✓	✓	✓		✓			✓	✓		✓
3	***Uria lomvia***															✓					✓			✓			
—[E]	*Alca torda*													✓	✓	✓	✓	✓	✓		✓			✓	✓		✓
2	***Cepphus grylle***													✓	✓	✓	✓	✓	✓		✓			✓	✓		✓
—	*Alle alle*													✓	✓	✓			✓					✓			
2	***Fratercula arctica***													✓	✓	✓			✓		✓			✓	✓		
3	***Pterocles orientalis***			✓		✓						✓															
3	***Pterocles alchata***																		✓								
—	*Columba livia*	✓	✓	✓	✓	✓	✓	✓	✓	✓	✓	✓	✓	✓	✓		✓	✓	✓	✓	✓	✓	✓	✓	✓	✓	✓

Appendix 2 ...continued. Occurrence of all European species by country.

SPEC Category	Species	Liechtenstein	Lithuania	Luxembourg	FYR of Macedonia	Malta	Moldova	Netherlands	Norway	*Svalbard*	Poland	Portugal	*Azores*	*Madeira*	Romania	Russia	Serbia and Montenegro	Slovakia	Slovenia	Spain	*Canary Islands*	Sweden	Switzerland	Turkey	Ukraine	United Kingdom	*Gibraltar*	TOTAL
3	**Lymnocryptes minimus**		✓	✓		✓		✓	✓		✓	✓			✓	✓		✓	✓	✓		✓		✓	✓	✓		35
3	**Gallinago gallinago**	✓	✓	✓	✓	✓	✓	✓	✓		✓	✓	✓		✓	✓	✓	✓	✓	✓		✓	✓	✓	✓	✓		44
1	**Gallinago media**		✓						✓		✓					✓						✓			✓			10
—	Gallinago stenura															✓												1
3	**Scolopax rusticola**	✓	✓	✓	✓		✓	✓	✓		✓	✓	✓	✓	✓	✓	✓	✓	✓	✓	✓	✓	✓	✓	✓	✓	✓	47
2	**Limosa limosa**		✓					✓	✓		✓	✓			✓	✓	✓	✓		✓		✓		✓	✓	✓		34
—	Limosa lapponica							✓	✓			✓				✓				✓		✓				✓		17
—[E]	Numenius phaeopus								✓			✓				✓				✓		✓		✓		✓	✓	18
1	**Numenius tenuirostris**																											2
2	**Numenius arquata**		✓					✓	✓		✓	✓			✓	✓	✓	✓	✓	✓		✓	✓	✓	✓	✓		37
3	**Tringa erythropus**							✓	✓			✓				✓	✓		✓	✓		✓		✓		✓		17
2	**Tringa totanus**		✓		✓		✓	✓	✓		✓	✓			✓	✓	✓	✓	✓	✓		✓		✓	✓	✓		40
—	Tringa stagnatilis		✓								✓				✓	✓	✓								✓			11
—	Tringa nebularia		✓					✓	✓			✓				✓	✓		✓	✓		✓		✓	✓	✓		26
—	Tringa ochropus		✓	✓	✓		✓	✓	✓		✓	✓			✓	✓	✓	✓	✓	✓		✓		✓	✓	✓		38
3	**Tringa glareola**		✓						✓		✓	✓				✓						✓		✓	✓	✓		17
—	Xenus cinereus															✓									✓			5
3	**Actitis hypoleucos**		✓		✓		✓	✓	✓		✓	✓			✓	✓	✓	✓	✓	✓		✓	✓	✓	✓	✓	✓	41
—	Arenaria interpres							✓	✓	✓		✓				✓				✓		✓		✓		✓	✓	26
—	Phalaropus lobatus								✓	✓						✓						✓				✓		11
—	Phalaropus fulicarius									✓						✓												5
—	Stercorarius pomarinus															✓												2
—	Stercorarius parasiticus								✓	✓		✓				✓	✓					✓		✓	✓	✓	✓	14
—	Stercorarius longicaudus								✓	✓						✓						✓						6
—[E]	Catharacta skua								✓	✓		✓				✓							✓			✓	✓	11
—	Larus ichthyaetus														✓	✓			✓					✓	✓			10
—[E]	Larus melanocephalus				✓	✓		✓			✓	✓			✓	✓	✓	✓	✓	✓			✓	✓	✓	✓	✓	33
3	**Larus minutus**		✓			✓		✓	✓		✓	✓			✓	✓	✓	✓	✓			✓	✓	✓	✓			30
—[E]	Larus ridibundus		✓	✓	✓	✓	✓	✓	✓	✓	✓	✓			✓	✓	✓	✓	✓	✓		✓	✓	✓	✓	✓		45
3	**Larus genei**															✓	✓			✓				✓	✓			12
1	**Larus audouinii**											✓								✓				✓			✓	9
2	**Larus canus**		✓	✓			✓	✓	✓	✓	✓	✓			✓	✓	✓	✓	✓			✓	✓	✓	✓	✓		39
—[E]	Larus fuscus		✓			✓	✓	✓	✓	✓	✓	✓			✓	✓		✓	✓	✓	✓	✓	✓	✓	✓	✓	✓	38
—[E]	Larus argentatus		✓				✓	✓	✓	✓	✓				✓	✓		✓	✓			✓	✓			✓		28
—[E]	Larus cachinnans				✓	✓	✓	✓			✓	✓	✓	✓	✓	✓	✓	✓	✓	✓	✓		✓	✓	✓	✓	✓	36
2	**Larus armenicus**																							✓				4
—[E]	Larus glaucoides																											4
—	Larus hyperboreus							✓		✓	✓				✓	✓												12
—[E]	Larus marinus		✓					✓	✓	✓	✓	✓			✓	✓		✓	✓			✓	✓			✓		26
—	Xema sabini																											2
—	Rhodostethia rosea																											1
—	Rissa tridactyla		✓				✓	✓	✓	✓	✓	✓				✓		✓	✓	✓		✓		✓		✓	✓	24
3	**Pagophila eburnea**									✓						✓												3
3	**Sterna nilotica**				✓							✓			✓	✓				✓				✓	✓			15
3	**Sterna caspia**											✓				✓						✓		✓	✓			12
—	Sterna bengalensis																			✓								2
2	**Sterna sandvicensis**					✓		✓			✓	✓			✓	✓			✓	✓		✓		✓	✓	✓	✓	26
3	**Sterna dougallii**												✓	✓							✓					✓		6
—	Sterna hirundo		✓		✓		✓	✓	✓		✓	✓	✓	✓	✓	✓	✓	✓	✓	✓	✓	✓	✓	✓	✓	✓		44
—	Sterna paradisaea							✓	✓	✓						✓						✓				✓		16
3	**Sterna albifrons**		✓		✓			✓	✓		✓	✓			✓	✓	✓	✓	✓	✓		✓		✓	✓	✓		3
3	**Chlidonias hybrida**		✓		✓		✓	✓			✓	✓			✓	✓	✓	✓		✓				✓	✓			27
3	**Chlidonias niger**		✓		✓		✓	✓			✓				✓	✓	✓	✓		✓		✓		✓	✓			28
—	Chlidonias leucopterus		✓		✓						✓				✓	✓	✓							✓	✓			18
—	Uria aalge		✓						✓	✓		✓				✓				✓		✓				✓	✓	19
3	**Uria lomvia**								✓	✓						✓												6
—[E]	Alca torda		✓						✓	✓		✓				✓						✓				✓	✓	18
2	**Cepphus grylle**		✓						✓	✓						✓			✓			✓				✓		17
—	Alle alle									✓						✓												7
2	**Fratercula arctica**								✓	✓		✓				✓										✓		12
3	**Pterocles orientalis**											✓				✓				✓	✓			✓				8
3	**Pterocles alchata**											✓								✓				✓				4
—	Columba livia	✓	✓	✓	✓	✓	✓	✓	✓		✓	✓	✓	✓	✓	✓	✓	✓	✓	✓	✓	✓	✓	✓	✓	✓		49

Appendix 2 ...continued. Occurrence of all European species by country.

SPEC Category	Species	Albania	Andorra	Armenia	Austria	Azerbaijan	Belarus	Belgium	Bosnia and Herzegovina	Bulgaria	Croatia	Cyprus	Czech Republic	Denmark	*Faroe Islands*	*Greenland*	Estonia	Finland	France	Georgia	Germany	Greece	Hungary	Iceland	Republic of Ireland	Italy	Latvia
—[E]	*Columba oenas*	✓		✓	✓	✓	✓	✓		✓	✓		✓	✓			✓	✓	✓	✓	✓	✓	✓		✓	✓	✓
—[E]	*Columba palumbus*	✓	✓	✓	✓	✓	✓	✓	✓	✓	✓	✓	✓	✓	✓		✓	✓	✓	✓	✓	✓	✓		✓	✓	✓
1	***Columba trocaz***																										
1	***Columba bollii***																										
1	***Columba junoniae***																										
—	*Streptopelia decaocto*	✓	✓	✓	✓	✓	✓	✓	✓	✓	✓	✓	✓	✓	✓		✓	✓	✓	✓	✓	✓	✓		✓	✓	✓
3	***Streptopelia turtur***	✓		✓	✓	✓	✓	✓	✓	✓	✓	✓	✓	✓			✓	✓	✓	✓	✓	✓	✓			✓	✓
—	*Streptopelia senegalensis*			✓		✓														✓							
—	*Clamator glandarius*									✓	✓	✓							✓			✓				✓	
—	*Cuculus canorus*	✓	✓	✓	✓	✓	✓	✓	✓	✓	✓		✓	✓			✓	✓	✓	✓	✓	✓	✓		✓	✓	✓
—	*Cuculus saturatus*																										
3	***Tyto alba***	✓	✓		✓		✓	✓	✓	✓	✓	✓	✓	✓					✓	✓	✓	✓	✓		✓	✓	
3	***Otus brucei***																										
2	***Otus scops***	✓	✓	✓	✓	✓	✓		✓	✓	✓	✓	✓						✓	✓		✓	✓			✓	
3	***Bubo bubo***	✓	✓	✓	✓	✓	✓	✓	✓	✓	✓		✓	✓			✓	✓	✓	✓	✓	✓	✓			✓	✓
3	***Ketupa zeylonensis***																										
3	***Nyctea scandiaca***															✓	✓	✓						✓			
—	*Surnia ulula*																✓	✓									
—	*Glaucidium passerinum*				✓		✓			✓	✓		✓				✓	✓	✓		✓					✓	✓
3	***Athene noctua***	✓		✓	✓	✓	✓	✓		✓	✓	✓	✓	✓					✓	✓	✓	✓	✓			✓	✓
—[E]	*Strix aluco*	✓	✓	✓	✓	✓	✓	✓	✓	✓	✓		✓	✓			✓	✓	✓	✓	✓	✓	✓			✓	✓
—	*Strix uralensis*				✓		✓			✓	✓		✓				✓	✓			✓		✓			✓	✓
—	*Strix nebulosa*						✓										✓	✓									
—	*Asio otus*	✓	✓	✓	✓	✓	✓	✓	✓	✓	✓	✓	✓	✓	✓		✓	✓	✓	✓	✓	✓	✓		✓	✓	✓
3	***Asio flammeus***			✓	✓	✓	✓	✓			✓		✓	✓			✓	✓	✓	✓	✓	✓	✓	✓	✓		✓
—	*Aegolius funereus*	✓	✓	✓	✓		✓	✓		✓	✓		✓	✓			✓	✓	✓	✓	✓	✓	✓			✓	✓
2	***Caprimulgus europaeus***	✓	✓	✓	✓	✓	✓	✓		✓	✓	✓	✓	✓			✓	✓	✓	✓	✓	✓	✓		✓	✓	✓
—	*Caprimulgus ruficollis*																										
—	*Tachymarptis melba*	✓		✓	✓	✓			✓	✓	✓	✓							✓	✓	✓	✓				✓	
2	***Apus unicolor***																										
—	*Apus apus*	✓	✓	✓	✓	✓	✓	✓	✓	✓	✓	✓	✓	✓			✓	✓	✓	✓	✓	✓	✓		✓	✓	✓
—	*Apus pallidus*	✓								✓	✓	✓							✓			✓				✓	
—	*Apus caffer*																										
3	***Apus affinis***																										
3	***Halcyon smyrnensis***					✓																					
3	***Alcedo atthis***	✓		✓	✓	✓	✓	✓	✓	✓	✓	✓	✓	✓			✓	✓	✓	✓	✓	✓	✓		✓	✓	✓
3	***Ceryle rudis***											✓															
—	*Merops persicus*			✓		✓																					
3	***Merops apiaster***	✓		✓	✓	✓	✓	✓	✓	✓	✓	✓	✓	✓					✓	✓	✓	✓	✓			✓	✓
2	***Coracias garrulus***	✓		✓	✓	✓	✓			✓	✓	✓	✓				✓		✓	✓		✓	✓			✓	✓
3	***Upupa epops***	✓	✓	✓	✓	✓	✓		✓	✓	✓	✓	✓				✓		✓	✓	✓	✓	✓			✓	✓
3	***Jynx torquilla***	✓	✓	✓	✓	✓	✓	✓	✓	✓	✓		✓	✓			✓	✓	✓	✓	✓	✓	✓			✓	✓
3	***Picus canus***	✓			✓		✓	✓	✓	✓	✓		✓				✓	✓	✓		✓	✓	✓			✓	✓
2	***Picus viridis***	✓	✓	✓	✓	✓	✓	✓	✓	✓	✓		✓	✓			✓		✓	✓	✓	✓	✓			✓	✓
—	*Dryocopus martius*	✓	✓	✓	✓	✓	✓	✓	✓	✓	✓		✓	✓			✓	✓	✓	✓	✓	✓	✓			✓	✓
—	*Dendrocopos major*	✓	✓	✓	✓	✓	✓	✓	✓	✓	✓		✓	✓			✓	✓	✓	✓	✓	✓	✓			✓	✓
—[E]	*Dendrocopos syriacus*	✓		✓	✓	✓	✓			✓	✓		✓							✓		✓	✓				
—[E]	*Dendrocopos medius*	✓		✓	✓	✓	✓	✓		✓	✓		✓				✓		✓	✓	✓	✓	✓			✓	✓
—	*Dendrocopos leucotos*	✓			✓	✓	✓			✓	✓		✓				✓	✓	✓	✓	✓	✓	✓			✓	✓
—	*Dendrocopos minor*	✓		✓	✓	✓	✓	✓	✓	✓	✓		✓	✓			✓	✓	✓	✓	✓	✓	✓			✓	✓
3	***Picoides tridactylus***	✓			✓		✓			✓	✓		✓				✓	✓	✓		✓	✓				✓	✓
3	***Ammomanes deserti***																										
3	***Chersophilus duponti***																										
3	***Melanocorypha calandra***	✓				✓				✓	✓	✓							✓	✓		✓				✓	
—	*Melanocorypha bimaculata*			✓		✓														✓							
—[E]W	*Melanocorypha leucoptera*					✓																					
3	***Melanocorypha yeltoniensis***																										
3	***Calandrella brachydactyla***	✓		✓		✓				✓	✓	✓							✓	✓		✓	✓			✓	
3	***Calandrella rufescens***			✓		✓														✓							
3	***Calandrella cheleensis***																										
3	***Galerida cristata***	✓		✓	✓	✓	✓	✓	✓	✓	✓	✓	✓	✓			✓		✓	✓	✓	✓	✓			✓	✓
3	***Galerida theklae***																		✓								
2	***Lullula arborea***	✓	✓	✓	✓	✓	✓	✓	✓	✓	✓	✓	✓	✓			✓	✓	✓	✓	✓	✓	✓			✓	✓

Appendix 2 ...continued. Occurrence of all European species by country.

SPEC Category	Species	Liechtenstein	Lithuania	Luxembourg	FYR of Macedonia	Malta	Moldova	Netherlands	Norway	*Svalbard*	Poland	Portugal	*Azores*	*Madeira*	Romania	Russia	Serbia and Montenegro	Slovakia	Slovenia	Spain	*Canary Islands*	Sweden	Switzerland	Turkey	Ukraine	United Kingdom	*Gibraltar*	TOTAL
—[E]	Columba oenas	✓	✓	✓	✓		✓	✓	✓		✓	✓			✓	✓	✓	✓	✓	✓		✓	✓	✓	✓	✓		40
—[E]	Columba palumbus	✓	✓	✓	✓		✓	✓	✓		✓	✓	✓		✓	✓	✓	✓	✓	✓		✓	✓	✓	✓	✓		45
1	**Columba trocaz**													✓														**1**
1	**Columba bollii**																				✓							**1**
1	**Columba junoniae**																				✓							**1**
—	Streptopelia decaocto	✓	✓	✓	✓		✓	✓	✓		✓	✓			✓	✓	✓	✓	✓	✓	✓	✓	✓	✓	✓	✓	✓	46
3	**Streptopelia turtur**	✓	✓	✓	✓	✓	✓	✓			✓	✓		✓	✓	✓	✓	✓	✓	✓	✓	✓	✓	✓	✓	✓		**43**
—	Streptopelia senegalensis															✓								✓				5
—	Clamator glandarius				✓							✓								✓				✓				10
—	Cuculus canorus	✓	✓	✓	✓	✓	✓	✓	✓		✓	✓			✓	✓	✓	✓	✓	✓		✓	✓	✓	✓	✓		43
—	Cuculus saturatus															✓												1
3	**Tyto alba**	✓	✓	✓	✓	✓	✓	✓			✓	✓		✓	✓	✓	✓	✓	✓	✓	✓	✓	✓	✓	✓	✓	✓	**41**
3	**Otus brucei**																							✓				**1**
2	**Otus scops**				✓		✓					✓			✓	✓	✓	✓	✓	✓			✓	✓	✓			**28**
3	**Bubo bubo**	✓	✓	✓	✓		✓	✓	✓		✓	✓			✓	✓	✓	✓	✓	✓		✓	✓	✓	✓			**40**
3	**Ketupa zeylonensis**																							✓				**1**
3	**Nyctea scandiaca**		✓						✓							✓		✓				✓			✓			**10**
—	Surnia ulula								✓							✓		✓				✓			✓			7
—	Glaucidium passerinum	✓	✓						✓		✓				✓	✓	✓	✓	✓			✓	✓		✓			23
3	**Athene noctua**		✓	✓	✓		✓	✓			✓	✓			✓	✓	✓	✓	✓	✓			✓	✓	✓	✓	✓	**36**
—[E]	Strix aluco	✓	✓	✓	✓		✓	✓	✓		✓	✓			✓	✓	✓	✓	✓	✓		✓	✓	✓	✓	✓	✓	42
—	Strix uralensis		✓		✓				✓		✓				✓	✓	✓	✓	✓			✓			✓			22
—	Strix nebulosa								✓							✓						✓			✓			7
—	Asio otus	✓	✓	✓	✓		✓	✓	✓		✓	✓	✓		✓	✓	✓	✓	✓	✓	✓	✓	✓	✓	✓	✓		46
3	**Asio flammeus**		✓		✓		✓	✓	✓		✓	✓			✓	✓	✓	✓		✓		✓		✓	✓	✓		**34**
—	Aegolius funereus	✓	✓	✓	✓			✓	✓		✓				✓	✓	✓	✓	✓	✓		✓	✓	✓	✓			36
2	**Caprimulgus europaeus**		✓	✓	✓		✓	✓	✓		✓	✓			✓	✓	✓	✓	✓	✓		✓	✓	✓	✓	✓		**41**
—	Caprimulgus ruficollis											✓								✓								2
—	Tachymarptis melba				✓							✓			✓	✓	✓		✓	✓			✓	✓	✓		✓	24
2	**Apus unicolor**													✓							✓							**2**
—	Apus apus	✓	✓	✓	✓	✓	✓	✓	✓		✓	✓			✓	✓	✓	✓	✓	✓	✓	✓	✓	✓	✓	✓	✓	46
—	Apus pallidus											✓		✓			✓			✓	✓		✓	✓			✓	15
—	Apus caffer											✓								✓								2
3	**Apus affinis**																			✓				✓				**2**
3	**Halcyon smyrnensis**																							✓				**2**
3	**Alcedo atthis**	✓	✓	✓	✓	✓	✓	✓	✓		✓	✓			✓	✓	✓	✓	✓	✓		✓	✓	✓	✓	✓	✓	**44**
3	**Ceryle rudis**																							✓				**2**
—	Merops persicus															✓								✓				4
3	**Merops apiaster**				✓		✓				✓	✓			✓	✓	✓	✓	✓	✓			✓	✓	✓			**32**
2	**Coracias garrulus**		✓		✓		✓				✓	✓			✓	✓	✓	✓	✓	✓				✓	✓			**29**
3	**Upupa epops**		✓	✓	✓		✓				✓	✓		✓	✓	✓	✓	✓	✓	✓	✓	✓	✓	✓	✓			**37**
3	**Jynx torquilla**	✓	✓	✓	✓	✓	✓	✓	✓		✓	✓			✓	✓	✓	✓	✓	✓		✓	✓	✓	✓	✓		**42**
3	**Picus canus**	✓	✓	✓	✓		✓		✓		✓				✓	✓	✓	✓	✓			✓	✓	✓	✓			**32**
2	**Picus viridis**	✓	✓	✓	✓		✓	✓	✓		✓	✓			✓	✓	✓	✓	✓	✓		✓	✓	✓	✓	✓		**40**
—	Dryocopus martius	✓	✓	✓	✓		✓	✓	✓		✓				✓	✓	✓	✓	✓	✓		✓	✓	✓	✓			39
—	Dendrocopos major	✓	✓	✓	✓		✓	✓	✓		✓	✓			✓	✓	✓	✓	✓	✓	✓	✓	✓	✓	✓	✓		42
—[E]	Dendrocopos syriacus				✓		✓				✓				✓	✓	✓	✓	✓					✓	✓			21
—[E]	Dendrocopos medius		✓	✓	✓		✓	✓			✓				✓	✓	✓	✓	✓	✓			✓	✓	✓			32
—	Dendrocopos leucotos	✓	✓		✓				✓		✓				✓	✓	✓	✓	✓	✓		✓	✓	✓	✓			31
—	Dendrocopos minor	✓	✓	✓	✓		✓	✓	✓		✓	✓			✓	✓	✓	✓	✓	✓		✓	✓	✓	✓	✓		40
3	**Picoides tridactylus**	✓	✓						✓		✓				✓	✓	✓	✓	✓			✓	✓		✓			**25**
3	**Ammomanes deserti**																							✓				**1**
3	**Chersophilus duponti**																			✓								**1**
3	**Melanocorypha calandra**				✓		✓					✓			✓	✓	✓			✓				✓	✓			**18**
—	Melanocorypha bimaculata																							✓				4
—[E]W	Melanocorypha leucoptera															✓									✓			3
3	**Melanocorypha yeltoniensis**															✓									✓			**2**
3	**Calandrella brachydactyla**				✓	✓						✓			✓	✓	✓	✓		✓				✓	✓			**21**
3	**Calandrella rufescens**											✓				✓				✓	✓			✓	✓			**9**
3	**Calandrella cheleensis**																							✓				**1**
3	**Galerida cristata**		✓		✓		✓	✓			✓	✓			✓	✓	✓	✓	✓	✓		✓		✓	✓			**35**
3	**Galerida theklae**											✓								✓							✓	**4**
2	**Lullula arborea**		✓	✓	✓		✓	✓	✓		✓	✓			✓	✓	✓	✓	✓	✓		✓	✓	✓	✓	✓		**41**

Appendix 2 ...continued. Occurrence of all European species by country.

SPEC Category	Species	Albania	Andorra	Armenia	Austria	Azerbaijan	Belarus	Belgium	Bosnia and Herzegovina	Bulgaria	Croatia	Cyprus	Czech Republic	Denmark	*Faroe Islands*	*Greenland*	Estonia	Finland	France	Georgia	Germany	Greece	Hungary	Iceland	Republic of Ireland	Italy	Latvia
3	**Alauda arvensis**	✔	✔	✔	✔	✔	✔	✔	✔	✔	✔	✔	✔	✔	✔		✔	✔	✔	✔	✔	✔	✔		✔	✔	✔
—	*Eremophila alpestris*	✔		✔		✔				✔	✔							✔		✔		✔					
3	**Riparia riparia**	✔		✔	✔	✔	✔	✔		✔	✔		✔	✔			✔	✔	✔	✔	✔	✔	✔		✔	✔	✔
—	*Hirundo rupestris*	✔	✔	✔	✔	✔			✔	✔	✔	✔							✔	✔	✔	✔				✔	
3	**Hirundo rustica**	✔	✔	✔	✔	✔	✔	✔	✔	✔	✔	✔	✔	✔	✔		✔	✔	✔	✔	✔	✔	✔	✔	✔	✔	✔
—	*Hirundo daurica*	✔							✔	✔	✔	✔							✔			✔				✔	
3	**Delichon urbica**	✔	✔	✔	✔	✔	✔	✔	✔	✔	✔	✔	✔	✔			✔	✔	✔	✔	✔	✔	✔	✔	✔	✔	✔
3	**Anthus campestris**	✔		✔	✔	✔	✔		✔	✔	✔		✔	✔			✔	✔	✔	✔	✔	✔	✔			✔	✔
—[E]	*Anthus berthelotii*																										
—	*Anthus hodgsoni*																										
—	*Anthus trivialis*	✔	✔	✔	✔	✔	✔	✔	✔	✔	✔		✔	✔			✔	✔	✔	✔	✔	✔	✔			✔	✔
—	*Anthus gustavi*																										
—[E]	*Anthus pratensis*		✔	✔	✔	✔	✔	✔					✔	✔	✔	✔	✔	✔	✔	✔	✔	✔		✔	✔		✔
—	*Anthus cervinus*																	✔									
—	*Anthus spinoletta*	✔	✔	✔	✔	✔				✔	✔		✔						✔	✔	✔	✔				✔	
—[E]	*Anthus petrosus*													✔	✔		✔	✔	✔		✔			✔	✔		
—	*Motacilla flava*	✔		✔	✔	✔	✔	✔	✔	✔	✔	✔	✔	✔			✔	✔	✔	✔	✔	✔	✔		✔	✔	✔
—	*Motacilla citreola*			✔			✔										✔	✔		✔	✔						✔
—	*Motacilla cinerea*	✔	✔	✔	✔	✔		✔	✔	✔	✔		✔	✔			✔	✔	✔	✔	✔	✔	✔		✔	✔	✔
—	*Motacilla alba*	✔	✔	✔	✔	✔	✔	✔	✔	✔	✔	✔	✔	✔	✔	✔	✔	✔	✔	✔	✔	✔	✔	✔	✔	✔	✔
—	*Pycnonotus xanthopygos*																										
—	*Bombycilla garrulus*					✔	✔										✔	✔									
—	*Cinclus cinclus*	✔	✔	✔	✔	✔	✔	✔	✔	✔	✔		✔	✔			✔	✔	✔	✔	✔	✔	✔		✔	✔	✔
—	*Troglodytes troglodytes*	✔	✔	✔	✔	✔	✔	✔	✔	✔	✔	✔	✔	✔	✔		✔	✔	✔	✔	✔	✔	✔	✔	✔	✔	✔
—[E]	*Prunella modularis*	✔	✔	✔	✔	✔	✔	✔		✔	✔		✔	✔			✔	✔	✔	✔	✔	✔	✔		✔	✔	✔
—	*Prunella montanella*																										
—[E]	*Prunella ocularis*			✔		✔														✔							
3	**Prunella atrogularis**																										
—	*Prunella collaris*	✔	✔	✔	✔	✔				✔	✔		✔						✔	✔	✔	✔				✔	
3	**Erythropygia galactotes**	✔		✔		✔				✔	✔									✔		✔					
—[E]	*Erithacus rubecula*	✔	✔	✔	✔	✔	✔	✔	✔	✔	✔		✔	✔	✔		✔	✔	✔	✔	✔	✔	✔		✔	✔	✔
—[E]	*Luscinia luscinia*						✔			✔			✔	✔			✔	✔		✔	✔		✔				✔
—[E]	*Luscinia megarhynchos*	✔	✔	✔	✔	✔		✔	✔	✔	✔	✔	✔						✔	✔	✔	✔	✔			✔	
—	*Luscinia calliope*																										
—	*Luscinia svecica*			✔	✔	✔	✔	✔			✔		✔	✔			✔	✔	✔	✔	✔		✔				✔
—	*Tarsiger cyanurus*																	✔									
—	*Irania gutturalis*			✔		✔																✔					
—	*Phoenicurus ochruros*	✔	✔	✔	✔	✔	✔	✔	✔	✔	✔		✔	✔			✔	✔	✔	✔	✔	✔	✔			✔	✔
2	**Phoenicurus phoenicurus**	✔		✔	✔	✔	✔	✔	✔	✔	✔		✔	✔			✔	✔	✔	✔	✔	✔	✔		✔	✔	✔
3	**Phoenicurus erythrogastrus**					✔														✔							
—[E]	*Saxicola rubetra*	✔	✔	✔	✔	✔	✔	✔	✔	✔	✔		✔	✔			✔	✔	✔	✔	✔	✔	✔		✔	✔	✔
1	**Saxicola dacotiae**																										
—	*Saxicola torquata*	✔	✔	✔	✔	✔		✔	✔	✔	✔		✔	✔				✔	✔	✔	✔	✔	✔		✔	✔	
—	*Oenanthe isabellina*			✔		✔				✔										✔		✔					
3	**Oenanthe oenanthe**	✔	✔	✔	✔	✔	✔	✔	✔	✔	✔		✔	✔	✔	✔	✔	✔	✔	✔	✔	✔	✔	✔	✔	✔	✔
—	*Oenanthe pleschanka*					✔				✔	✔									✔							
—[E]	*Oenanthe cypriaca*											✔															
2	**Oenanthe hispanica**	✔		✔					✔	✔	✔								✔	✔		✔				✔	
—	*Oenanthe deserti*					✔																					
—	*Oenanthe finschii*			✔		✔														✔							
3	**Oenanthe xanthoprymna**			✔		✔														✔							
3	**Oenanthe leucura**																		✔								
3	**Monticola saxatilis**	✔	✔	✔	✔	✔				✔	✔								✔	✔		✔	✔			✔	
3	**Monticola solitarius**	✔	✔	✔		✔				✔	✔	✔							✔	✔		✔				✔	
—	*Zoothera dauma*																										
—[E]	*Turdus torquatus*	✔	✔	✔	✔	✔		✔		✔	✔		✔		✔			✔	✔	✔	✔	✔			✔	✔	
—[E]	*Turdus merula*	✔	✔	✔	✔	✔	✔	✔	✔	✔	✔	✔	✔	✔	✔		✔	✔	✔	✔	✔	✔	✔	✔	✔	✔	✔
—	*Turdus ruficollis*					✔																					
—[E]W	*Turdus pilaris*				✔	✔	✔	✔		✔	✔		✔	✔	✔	✔	✔	✔	✔		✔	✔	✔	✔		✔	✔
—[E]	*Turdus philomelos*	✔	✔	✔	✔	✔	✔	✔	✔	✔	✔		✔	✔			✔	✔	✔	✔	✔	✔	✔		✔	✔	✔
—[E]W	*Turdus iliacus*					✔	✔				✔		✔		✔	✔	✔	✔				✔		✔			✔
—[E]	*Turdus viscivorus*	✔	✔	✔	✔	✔	✔	✔	✔	✔	✔		✔	✔			✔	✔	✔	✔	✔	✔	✔		✔	✔	✔
—	*Cettia cetti*	✔		✔		✔		✔	✔	✔	✔	✔							✔	✔		✔	✔			✔	

Appendix 2 ...continued. Occurrence of all European species by country.

SPEC Category	Species	Liechtenstein	Lithuania	Luxembourg	FYR of Macedonia	Malta	Moldova	Netherlands	Norway	*Svalbard*	Poland	Portugal	*Azores*	*Madeira*	Romania	Russia	Serbia and Montenegro	Slovakia	Slovenia	Spain	*Canary Islands*	Sweden	Switzerland	Turkey	Ukraine	United Kingdom	*Gibraltar*	TOTAL
3	***Alauda arvensis***	✓	✓	✓	✓	✓	✓	✓	✓		✓	✓			✓	✓	✓	✓	✓	✓		✓	✓	✓	✓	✓	✓	**46**
—	*Eremophila alpestris*		✓		✓		✓		✓						✓	✓	✓	✓				✓		✓	✓	✓		20
3	***Riparia riparia***		✓	✓	✓		✓	✓	✓		✓	✓			✓	✓	✓	✓	✓	✓		✓	✓	✓	✓	✓		**39**
—	*Hirundo rupestris*	✓			✓							✓			✓	✓	✓		✓	✓			✓	✓			✓	25
3	***Hirundo rustica***	✓	✓	✓	✓	✓	✓	✓	✓		✓	✓			✓	✓	✓	✓	✓	✓	✓	✓	✓	✓	✓	✓		**47**
—	*Hirundo daurica*				✓							✓			✓		✓		✓	✓				✓				15
3	***Delichon urbica***	✓	✓	✓	✓	✓	✓	✓	✓		✓	✓			✓	✓	✓	✓	✓	✓		✓	✓	✓	✓	✓	✓	**46**
3	***Anthus campestris***		✓	✓	✓		✓	✓			✓	✓			✓	✓	✓	✓	✓	✓		✓	✓	✓	✓			**36**
—[E]	*Anthus berthelotii*													✓							✓							2
—	*Anthus hodgsoni*															✓												1
—	*Anthus trivialis*	✓	✓	✓	✓		✓	✓	✓		✓	✓			✓	✓	✓	✓	✓	✓		✓	✓	✓	✓	✓		41
—	*Anthus gustavi*															✓												1
—[E]	*Anthus pratensis*		✓	✓		✓	✓	✓	✓		✓	✓			✓	✓		✓	✓	✓		✓	✓		✓	✓	✓	37
—	*Anthus cervinus*					✓			✓							✓						✓			✓			6
—	*Anthus spinoletta*	✓		✓	✓	✓					✓	✓			✓	✓	✓	✓	✓	✓			✓	✓	✓	✓		29
—[E]	*Anthus petrosus*								✓			✓				✓						✓				✓		13
—	*Motacilla flava*		✓	✓	✓		✓	✓	✓		✓	✓			✓	✓	✓	✓	✓	✓		✓	✓	✓	✓	✓		41
—	*Motacilla citreola*		✓								✓					✓		✓						✓	✓			13
—	*Motacilla cinerea*	✓		✓	✓	✓	✓	✓	✓		✓	✓	✓	✓	✓	✓	✓	✓	✓	✓	✓	✓	✓	✓	✓	✓	✓	45
—	*Motacilla alba*	✓	✓	✓	✓	✓	✓	✓	✓		✓	✓			✓	✓	✓	✓	✓	✓		✓	✓	✓	✓	✓	✓	48
—	*Pycnonotus xanthopygos*																							✓				1
—	*Bombycilla garrulus*			✓			✓		✓						✓	✓		✓				✓			✓	✓		13
—	*Cinclus cinclus*	✓	✓	✓	✓				✓		✓	✓			✓	✓	✓	✓	✓	✓		✓	✓	✓	✓	✓		40
—	*Troglodytes troglodytes*	✓	✓	✓	✓		✓	✓	✓		✓	✓			✓	✓	✓	✓	✓	✓		✓	✓	✓	✓	✓	✓	46
—[E]	*Prunella modularis*	✓	✓	✓	✓	✓	✓	✓	✓		✓	✓			✓	✓	✓	✓	✓	✓		✓	✓	✓	✓	✓	✓	43
—	*Prunella montanella*															✓												1
—[E]	*Prunella ocularis*																							✓				4
3	***Prunella atrogularis***															✓												**1**
—	*Prunella collaris*	✓			✓						✓	✓			✓	✓	✓	✓	✓	✓			✓	✓	✓		✓	27
3	***Erythropygia galactotes***											✓				✓				✓				✓				**11**
—[E]	*Erithacus rubecula*	✓	✓	✓	✓	✓	✓	✓	✓		✓	✓	✓	✓	✓	✓	✓	✓	✓	✓	✓	✓	✓	✓	✓	✓	✓	48
—[E]	*Luscinia luscinia*		✓				✓	✓	✓		✓				✓	✓		✓				✓			✓			20
—[E]	*Luscinia megarhynchos*	✓		✓	✓			✓			✓	✓			✓	✓	✓	✓	✓	✓			✓	✓	✓	✓		33
—	*Luscinia calliope*															✓												1
—	*Luscinia svecica*		✓					✓	✓		✓	✓			✓	✓	✓	✓		✓		✓	✓	✓	✓			29
—	*Tarsiger cyanurus*															✓												2
—	*Irania gutturalis*																							✓				4
—	*Phoenicurus ochruros*	✓	✓	✓	✓	✓	✓	✓	✓		✓	✓			✓	✓	✓	✓	✓	✓		✓	✓	✓	✓	✓	✓	43
2	***Phoenicurus phoenicurus***	✓	✓	✓	✓		✓	✓	✓		✓	✓			✓	✓	✓	✓	✓	✓		✓	✓	✓	✓	✓		**41**
3	***Phoenicurus erythrogastrus***															✓												**3**
—[E]	*Saxicola rubetra*	✓	✓	✓	✓		✓	✓	✓		✓	✓			✓	✓	✓	✓	✓	✓		✓	✓	✓	✓	✓		42
1	***Saxicola dacotiae***																				✓							**1**
—	*Saxicola torquata*	✓		✓	✓	✓	✓	✓	✓		✓	✓			✓	✓	✓	✓	✓	✓			✓	✓	✓	✓	✓	39
—	*Oenanthe isabellina*														✓	✓								✓	✓			9
3	***Oenanthe oenanthe***	✓	✓	✓	✓		✓	✓	✓	✓	✓	✓			✓	✓	✓	✓	✓	✓		✓	✓	✓	✓	✓		**46**
—	*Oenanthe pleschanka*														✓	✓								✓	✓			8
—[E]	*Oenanthe cypriaca*																											1
2	***Oenanthe hispanica***				✓							✓			✓	✓	✓			✓				✓	✓			**17**
—	*Oenanthe deserti*																							✓				2
—	*Oenanthe finschii*																							✓				4
3	***Oenanthe xanthoprymna***																							✓				**4**
3	***Oenanthe leucura***											✓								✓								**3**
3	***Monticola saxatilis***				✓		✓				✓	✓			✓	✓	✓	✓	✓	✓			✓	✓	✓			**25**
3	***Monticola solitarius***				✓	✓						✓				✓	✓		✓	✓			✓	✓			✓	**21**
—	*Zoothera dauma*															✓												1
—[E]	*Turdus torquatus*	✓			✓				✓		✓	✓			✓	✓	✓	✓	✓	✓		✓	✓	✓	✓	✓		33
—[E]	*Turdus merula*	✓	✓	✓	✓	✓	✓	✓	✓		✓	✓	✓	✓	✓	✓	✓	✓	✓	✓	✓	✓	✓	✓	✓	✓	✓	50
—	*Turdus ruficollis*															✓												2
—[E]W	*Turdus pilaris*	✓	✓	✓			✓	✓	✓		✓	✓			✓	✓	✓	✓	✓			✓	✓		✓	✓		36
—[E]	*Turdus philomelos*	✓	✓	✓	✓	✓	✓	✓	✓		✓	✓			✓	✓	✓	✓	✓	✓		✓	✓	✓	✓	✓	✓	44
—[E]W	*Turdus iliacus*		✓	✓			✓		✓	✓	✓	✓			✓	✓		✓				✓			✓	✓	✓	25
—[E]	*Turdus viscivorus*	✓	✓	✓	✓		✓	✓	✓		✓	✓			✓	✓	✓	✓	✓	✓		✓	✓	✓	✓	✓		42
—	*Cettia cetti*				✓	✓		✓				✓			✓	✓	✓		✓	✓			✓	✓		✓		25

Appendix 2 ...continued. Occurrence of all European species by country.

SPEC Category	Species	Albania	Andorra	Armenia	Austria	Azerbaijan	Belarus	Belgium	Bosnia and Herzegovina	Bulgaria	Croatia	Cyprus	Czech Republic	Denmark	*Faroe Islands*	*Greenland*	Estonia	Finland	France	Georgia	Germany	Greece	Hungary	Iceland	Republic of Ireland	Italy	Latvia
—	*Cisticola juncidis*	✓						✓			✓	✓							✓			✓				✓	
3	***Prinia gracilis***																										
—	*Locustella lanceolata*																										
—[E]	*Locustella naevia*			✓	✓	✓	✓	✓	✓	✓	✓		✓	✓			✓	✓	✓	✓	✓		✓		✓		✓
—[E]	*Locustella fluviatilis*				✓		✓		✓	✓	✓		✓	✓			✓	✓			✓		✓				✓
—[E]	*Locustella luscinioides*	✓		✓	✓		✓	✓		✓	✓		✓	✓			✓	✓	✓		✓	✓	✓			✓	✓
—	*Acrocephalus melanopogon*				✓	✓				✓	✓								✓	✓		✓	✓			✓	
1	***Acrocephalus paludicola***						✓														✓		✓				✓
—[E]	*Acrocephalus schoenobaenus*	✓		✓	✓	✓	✓	✓	✓	✓	✓		✓	✓			✓	✓	✓	✓	✓	✓	✓		✓	✓	✓
—	*Acrocephalus agricola*			✓						✓										✓							
—	*Acrocephalus dumetorum*						✓										✓	✓									✓
—[E]	*Acrocephalus palustris*	✓		✓	✓	✓	✓	✓	✓	✓	✓		✓	✓			✓	✓	✓	✓	✓	✓	✓			✓	✓
—[E]	*Acrocephalus scirpaceus*	✓		✓	✓	✓	✓	✓	✓	✓	✓	✓	✓	✓			✓	✓	✓	✓	✓	✓	✓		✓	✓	✓
—	*Acrocephalus arundinaceus*	✓		✓	✓	✓	✓	✓	✓	✓	✓		✓	✓			✓	✓	✓	✓	✓	✓	✓			✓	✓
3	***Hippolais pallida***	✓		✓		✓			✓	✓	✓	✓								✓		✓	✓				
—	*Hippolais caligata*																	✓									
—	*Hippolais languida*			✓		✓														✓							
—[E]	*Hippolais olivetorum*	✓								✓	✓											✓					
—[E]	*Hippolais icterina*				✓	✓	✓	✓	✓	✓	✓		✓	✓			✓	✓	✓	✓	✓		✓				✓
—[E]	*Hippolais polyglotta*		✓					✓			✓								✓		✓					✓	
—[E]	*Sylvia sarda*																		✓							✓	
2	***Sylvia undata***		✓																✓							✓	
—	*Sylvia conspicillata*											✓							✓							✓	
—[E]	*Sylvia cantillans*	✓	✓						✓	✓	✓								✓			✓				✓	
—	*Sylvia mystacea*			✓		✓														✓							
—[E]	*Sylvia melanocephala*	✓							✓	✓	✓	✓							✓			✓				✓	
—[E]	*Sylvia melanothorax*											✓															
—[E]	*Sylvia rueppelli*																					✓					
—	*Sylvia nana*																										
3	***Sylvia hortensis***	✓	✓	✓		✓			✓	✓	✓								✓	✓		✓				✓	
—[E]	*Sylvia nisoria*	✓		✓	✓	✓	✓			✓	✓		✓	✓			✓	✓		✓	✓	✓	✓			✓	✓
—	*Sylvia curruca*	✓		✓	✓	✓	✓	✓		✓	✓		✓	✓	✓		✓	✓	✓	✓	✓	✓	✓		✓	✓	✓
—[E]	*Sylvia communis*	✓	✓	✓	✓	✓	✓	✓	✓	✓	✓		✓	✓			✓	✓	✓	✓	✓	✓	✓		✓	✓	✓
—[E]	*Sylvia borin*		✓	✓	✓	✓	✓	✓	✓	✓	✓		✓	✓	✓		✓	✓	✓	✓	✓	✓	✓		✓	✓	✓
—[E]	*Sylvia atricapilla*	✓	✓	✓	✓	✓	✓	✓	✓	✓	✓		✓	✓	✓		✓	✓	✓	✓	✓	✓	✓		✓	✓	✓
—	*Phylloscopus trochiloides*			✓		✓	✓						✓	✓			✓	✓		✓	✓						✓
—	*Phylloscopus borealis*																	✓									
—	*Phylloscopus inornatus*																										
2	***Phylloscopus bonelli***	✓	✓		✓					✓	✓								✓		✓	✓				✓	
2	***Phylloscopus sibilatrix***	✓			✓		✓	✓		✓	✓		✓	✓			✓	✓	✓	✓	✓	✓	✓		✓	✓	✓
3	***Phylloscopus sindianus***			✓		✓														✓							
—	*Phylloscopus collybita*	✓	✓	✓	✓	✓	✓	✓	✓	✓	✓		✓	✓	✓		✓	✓	✓	✓	✓	✓	✓		✓	✓	✓
—[E]	*Phylloscopus ibericus*																		✓								
—[E]	*Phylloscopus canariensis*																										
—	*Phylloscopus trochilus*	✓			✓		✓	✓		✓	✓		✓	✓	✓		✓	✓	✓	✓	✓	✓	✓		✓		✓
—[E]	*Regulus regulus*	✓	✓		✓	✓	✓	✓	✓	✓	✓		✓	✓	✓		✓	✓	✓	✓	✓	✓	✓	✓	✓	✓	✓
—[E]	*Regulus teneriffae*																										
—[E]	*Regulus ignicapilla*	✓	✓		✓			✓		✓	✓		✓	✓					✓	✓	✓	✓	✓			✓	✓
3	***Muscicapa striata***	✓	✓	✓	✓	✓	✓	✓		✓	✓	✓	✓	✓			✓	✓	✓	✓	✓	✓	✓		✓	✓	✓
—	*Ficedula parva*	✓		✓	✓	✓	✓			✓	✓		✓				✓	✓		✓	✓	✓	✓				✓
2	***Ficedula semitorquata***	✓		✓						✓										✓		✓					
—[E]	*Ficedula albicollis*	✓			✓	✓	✓		✓	✓	✓		✓						✓		✓		✓			✓	✓
—[E]	*Ficedula hypoleuca*		✓		✓		✓	✓					✓	✓			✓	✓	✓		✓		✓		✓		✓
—	*Panurus biarmicus*	✓		✓	✓	✓	✓	✓		✓	✓		✓	✓			✓	✓	✓	✓	✓	✓	✓		✓	✓	✓
—	*Aegithalos caudatus*	✓	✓	✓	✓	✓	✓	✓	✓	✓	✓		✓	✓			✓	✓	✓	✓	✓	✓	✓		✓	✓	✓
3	***Parus palustris***	✓			✓	✓	✓	✓	✓	✓	✓		✓	✓			✓		✓	✓	✓	✓	✓			✓	✓
—[E]	*Parus lugubris*	✓		✓		✓			✓	✓	✓									✓		✓					
—	*Parus montanus*	✓			✓		✓	✓	✓	✓	✓		✓	✓			✓	✓	✓		✓	✓	✓			✓	✓
—	*Parus cinctus*																	✓									
2	***Parus cristatus***	✓	✓		✓		✓	✓	✓	✓	✓		✓	✓			✓	✓	✓		✓	✓	✓			✓	✓
—	*Parus ater*	✓	✓	✓	✓	✓	✓	✓	✓	✓	✓	✓	✓	✓			✓	✓	✓	✓	✓	✓	✓		✓	✓	✓
—[E]	*Parus caeruleus*	✓	✓	✓	✓	✓	✓	✓	✓	✓	✓		✓	✓			✓	✓	✓	✓	✓	✓	✓		✓	✓	✓
—	*Parus cyanus*						✓																				

Appendix 2 ...continued. Occurrence of all European species by country.

SPEC Category	Species	Liechtenstein	Lithuania	Luxembourg	FYR of Macedonia	Malta	Moldova	Netherlands	Norway	*Svalbard*	Poland	Portugal	*Azores*	*Madeira*	Romania	Russia	Serbia and Montenegro	Slovakia	Slovenia	Spain	*Canary Islands*	Sweden	Switzerland	Turkey	Ukraine	United Kingdom	*Gibraltar*	TOTAL
—	*Cisticola juncidis*					✔		✔				✔					✔		✔	✔				✔			✔	15
3	***Prinia gracilis***																							✔				**1**
—	*Locustella lanceolata*															✔												1
—[E]	*Locustella naevia*	✔	✔	✔			✔	✔	✔		✔				✔	✔	✔	✔	✔	✔		✔	✔	✔	✔	✔		36
—[E]	*Locustella fluviatilis*		✔				✔				✔				✔	✔	✔	✔	✔			✔			✔			22
—[E]	*Locustella luscinioides*		✔		✔		✔	✔			✔	✔			✔	✔	✔	✔	✔	✔		✔	✔	✔	✔	✔		34
—	*Acrocephalus melanopogon*				✔	✔									✔	✔	✔	✔		✔				✔	✔			18
1	***Acrocephalus paludicola***		✔								✔					✔									✔			**8**
—[E]	*Acrocephalus schoenobaenus*		✔	✔	✔		✔	✔	✔		✔				✔	✔	✔	✔	✔			✔		✔	✔	✔		37
—	*Acrocephalus agricola*														✔	✔								✔	✔			7
—	*Acrocephalus dumetorum*		✔								✔					✔						✔			✔			9
—[E]	*Acrocephalus palustris*	✔	✔	✔	✔		✔	✔	✔		✔				✔	✔	✔	✔	✔			✔	✔	✔	✔	✔		38
—[E]	*Acrocephalus scirpaceus*	✔	✔	✔	✔	✔	✔	✔	✔		✔	✔			✔	✔	✔	✔	✔	✔		✔	✔	✔	✔	✔		43
—	*Acrocephalus arundinaceus*		✔	✔	✔		✔	✔			✔	✔			✔	✔	✔	✔	✔	✔		✔	✔	✔	✔			37
3	***Hippolais pallida***				✔							✔			✔	✔	✔		✔	✔				✔	✔			**19**
—	*Hippolais caligata*															✔												2
—	*Hippolais languida*																							✔				4
—[E]	*Hippolais olivetorum*				✔												✔							✔				7
—[E]	*Hippolais icterina*	✔	✔	✔	✔		✔	✔	✔		✔				✔	✔	✔	✔				✔	✔	✔	✔			32
—[E]	*Hippolais polyglotta*			✔				✔				✔							✔	✔			✔					12
—[E]	*Sylvia sarda*																			✔								3
2	***Sylvia undata***					✔						✔								✔						✔		**7**
—	*Sylvia conspicillata*					✔						✔		✔						✔	✔			✔				9
—[E]	*Sylvia cantillans*				✔							✔					✔		✔	✔				✔				14
—	*Sylvia mystacea*															✔								✔				5
—[E]	*Sylvia melanocephala*				✔	✔						✔					✔		✔	✔	✔			✔			✔	17
—[E]	*Sylvia melanothorax*																											1
—[E]	*Sylvia rueppelli*																							✔				2
—	*Sylvia nana*															✔												1
3	***Sylvia hortensis***				✔							✔				✔	✔			✔			✔	✔				**18**
—[E]	*Sylvia nisoria*		✔		✔		✔		✔		✔				✔	✔	✔	✔	✔			✔	✔	✔	✔			31
—	*Sylvia curruca*	✔	✔	✔	✔		✔	✔	✔		✔				✔	✔	✔	✔	✔			✔	✔	✔	✔	✔		39
—[E]	*Sylvia communis*	✔	✔	✔	✔		✔	✔	✔		✔	✔			✔	✔	✔	✔	✔	✔		✔	✔	✔	✔	✔		42
—[E]	*Sylvia borin*	✔	✔	✔	✔		✔	✔	✔		✔	✔			✔	✔	✔	✔	✔	✔		✔	✔	✔	✔	✔		42
—[E]	*Sylvia atricapilla*	✔	✔	✔	✔	✔	✔	✔	✔		✔	✔	✔	✔	✔	✔	✔	✔	✔	✔	✔	✔	✔	✔	✔	✔	✔	48
—	*Phylloscopus trochiloides*		✔								✔					✔		✔				✔		✔	✔			17
—	*Phylloscopus borealis*								✔							✔						✔						4
—	*Phylloscopus inornatus*															✔												1
2	***Phylloscopus bonelli***	✔			✔			✔				✔			✔		✔		✔	✔			✔	✔				**19**
2	***Phylloscopus sibilatrix***	✔	✔	✔	✔		✔	✔	✔		✔				✔	✔	✔	✔	✔	✔		✔	✔	✔	✔	✔		**37**
3	***Phylloscopus sindianus***															✔								✔				**5**
—	*Phylloscopus collybita*	✔	✔	✔	✔	✔	✔	✔	✔		✔	✔			✔	✔	✔	✔	✔	✔		✔	✔	✔	✔	✔	✔	45
—[E]	*Phylloscopus ibericus*											✔								✔								3
—[E]	*Phylloscopus canariensis*																				✔							1
—	*Phylloscopus trochilus*	✔	✔	✔			✔	✔	✔		✔				✔	✔		✔	✔	✔		✔	✔		✔	✔		34
—[E]	*Regulus regulus*	✔	✔	✔	✔	✔	✔	✔	✔		✔	✔	✔		✔	✔	✔	✔	✔	✔		✔	✔	✔	✔	✔		45
—[E]	*Regulus teneriffae*																				✔							1
—[E]	*Regulus ignicapilla*	✔		✔	✔	✔	✔	✔			✔	✔		✔	✔	✔	✔	✔	✔	✔			✔	✔	✔	✔	✔	35
3	***Muscicapa striata***	✔	✔	✔	✔	✔	✔	✔	✔		✔	✔			✔	✔	✔	✔	✔	✔		✔	✔	✔	✔	✔		**43**
—	*Ficedula parva*		✔				✔		✔		✔				✔	✔	✔	✔	✔			✔		✔	✔			27
2	***Ficedula semitorquata***				✔											✔								✔				**8**
—[E]	*Ficedula albicollis*				✔		✔				✔				✔	✔	✔	✔	✔			✔	✔		✔			24
—[E]	*Ficedula hypoleuca*	✔	✔	✔			✔	✔	✔		✔				✔	✔	✔	✔		✔		✔	✔		✔	✔		29
—	*Panurus biarmicus*		✔		✔		✔	✔			✔				✔	✔	✔	✔		✔		✔	✔	✔	✔	✔		35
—	*Aegithalos caudatus*	✔	✔	✔	✔		✔	✔	✔		✔	✔			✔	✔	✔	✔	✔	✔		✔	✔	✔	✔	✔		42
3	***Parus palustris***	✔	✔	✔	✔		✔	✔	✔		✔				✔	✔	✔	✔	✔	✔		✔	✔	✔	✔	✔		**37**
—[E]	*Parus lugubris*				✔										✔		✔		✔					✔				13
—	*Parus montanus*	✔	✔	✔	✔			✔	✔		✔				✔	✔	✔	✔	✔			✔	✔		✔	✔		33
—	*Parus cinctus*								✔							✔						✔						4
2	***Parus cristatus***	✔	✔	✔	✔			✔	✔		✔	✔			✔	✔	✔	✔	✔	✔		✔	✔		✔	✔		**36**
—	*Parus ater*	✔	✔	✔	✔		✔	✔	✔		✔	✔			✔	✔	✔	✔	✔	✔		✔	✔	✔	✔	✔		43
—[E]	*Parus caeruleus*	✔	✔	✔	✔		✔	✔	✔		✔	✔			✔	✔	✔	✔	✔	✔	✔	✔	✔	✔	✔	✔	✔	44
—	*Parus cyanus*															✔									✔			3

Appendix 2 ...continued. Occurrence of all European species by country.

SPEC Category	Species	Albania	Andorra	Armenia	Austria	Azerbaijan	Belarus	Belgium	Bosnia and Herzegovina	Bulgaria	Croatia	Cyprus	Czech Republic	Denmark	*Faroe Islands*	*Greenland*	Estonia	Finland	France	Georgia	Germany	Greece	Hungary	Iceland	Republic of Ireland	Italy	Latvia
—	*Parus major*	✔	✔	✔	✔	✔	✔	✔	✔	✔	✔	✔	✔	✔			✔	✔	✔	✔	✔	✔	✔		✔	✔	✔
2	***Sitta krueperi***																			✔		✔					
2	***Sitta whiteheadi***																		✔								
—	*Sitta europaea*	✔	✔	✔	✔	✔	✔	✔	✔	✔	✔		✔	✔			✔	✔	✔	✔	✔	✔	✔			✔	✔
—	*Sitta tephronota*			✔		✔																					
—[E]	*Sitta neumayer*	✔		✔		✔			✔	✔	✔									✔		✔					
—	*Tichodroma muraria*	✔	✔	✔	✔	✔				✔	✔								✔	✔	✔	✔				✔	
—	*Certhia familiaris*	✔	✔	✔	✔	✔	✔	✔	✔	✔	✔		✔	✔			✔	✔	✔	✔	✔	✔	✔		✔	✔	✔
—[E]	*Certhia brachydactyla*	✔	✔		✔	✔		✔	✔	✔	✔	✔	✔	✔					✔		✔	✔	✔			✔	
—	*Remiz pendulinus*	✔		✔	✔	✔	✔	✔	✔	✔	✔		✔	✔			✔	✔	✔	✔	✔	✔	✔			✔	✔
—	*Oriolus oriolus*	✔	✔	✔	✔	✔	✔	✔	✔	✔	✔	✔	✔	✔			✔	✔	✔	✔	✔	✔	✔			✔	✔
3	***Lanius collurio***	✔	✔	✔	✔	✔	✔	✔	✔	✔	✔	✔	✔	✔			✔	✔	✔	✔	✔	✔	✔			✔	✔
2	***Lanius minor***	✔		✔	✔	✔	✔		✔	✔	✔								✔	✔		✔	✔			✔	
3	***Lanius excubitor***				✔	✔	✔	✔					✔	✔			✔	✔	✔		✔	✔	✔				✔
2	***Lanius senator***	✔	✔	✔		✔			✔	✔	✔	✔	✔						✔	✔	✔	✔				✔	
2	***Lanius nubicus***									✔		✔										✔					
—	*Garrulus glandarius*	✔	✔	✔	✔	✔	✔	✔	✔	✔	✔	✔	✔	✔			✔	✔	✔	✔	✔	✔	✔		✔	✔	✔
3	***Perisoreus infaustus***																	✔									
—	*Cyanopica cyanus*																										
—	*Pica pica*	✔	✔	✔	✔	✔	✔	✔	✔	✔	✔	✔	✔	✔			✔	✔	✔	✔	✔	✔	✔		✔	✔	✔
—	*Nucifraga caryocatactes*	✔			✔		✔	✔	✔	✔	✔		✔	✔			✔	✔	✔		✔	✔				✔	✔
—	*Pyrrhocorax graculus*	✔	✔	✔	✔	✔			✔	✔	✔								✔	✔	✔	✔				✔	
3	***Pyrrhocorax pyrrhocorax***	✔	✔	✔		✔													✔	✔		✔			✔	✔	
—[E]	*Corvus monedula*	✔		✔	✔	✔	✔	✔	✔	✔	✔	✔	✔	✔			✔	✔	✔	✔	✔	✔	✔		✔	✔	✔
—	*Corvus frugilegus*			✔	✔	✔	✔	✔	✔	✔	✔		✔	✔	✔		✔	✔	✔	✔	✔	✔	✔		✔		✔
—	*Corvus corone*	✔	✔	✔	✔	✔	✔	✔	✔	✔	✔	✔	✔	✔	✔		✔	✔	✔	✔	✔	✔	✔		✔	✔	✔
—	*Corvus corax*	✔	✔	✔	✔	✔	✔	✔	✔	✔	✔	✔	✔	✔	✔	✔	✔	✔	✔	✔	✔	✔	✔	✔	✔	✔	✔
3	***Sturnus vulgaris***	✔	✔	✔	✔	✔	✔	✔	✔	✔	✔	✔	✔	✔	✔		✔	✔	✔	✔	✔	✔	✔	✔	✔	✔	✔
—[E]	*Sturnus unicolor*		✔																✔							✔	
—	*Sturnus roseus*	✔		✔		✔				✔										✔		✔					
3	***Passer domesticus***	✔	✔	✔	✔	✔	✔	✔	✔	✔	✔	✔	✔	✔	✔		✔	✔	✔	✔	✔	✔	✔	✔	✔	✔	✔
—	*Passer hispaniolensis*	✔		✔	✔	✔			✔	✔	✔	✔								✔		✔				✔	
—	*Passer moabiticus*											✔															
3	***Passer montanus***	✔	✔	✔	✔	✔	✔	✔	✔	✔	✔		✔	✔			✔	✔	✔	✔	✔	✔	✔		✔	✔	✔
—	*Petronia brachydactyla*			✔		✔																					
—	*Petronia xanthocollis*																										
—	*Petronia petronia*	✔	✔	✔		✔				✔									✔	✔		✔				✔	
—	*Montifringilla nivalis*	✔	✔	✔	✔	✔													✔	✔	✔	✔				✔	
—[E]	*Fringilla coelebs*	✔	✔	✔	✔	✔	✔	✔	✔	✔	✔	✔	✔	✔	✔		✔	✔	✔	✔	✔	✔	✔		✔	✔	✔
1	***Fringilla teydea***																										
—	*Fringilla montifringilla*					✔	✔							✔	✔		✔	✔			✔	✔		✔			✔
—	*Serinus pusillus*			✔		✔														✔							
—[E]	*Serinus serinus*	✔	✔		✔		✔	✔	✔	✔	✔	✔	✔	✔			✔	✔	✔		✔	✔	✔			✔	✔
—[E]	*Serinus canaria*																										
—[E]	*Serinus citrinella*		✔		✔														✔		✔					✔	
—[E]	*Serinus corsicana*																		✔							✔	
—[E]	*Carduelis chloris*	✔	✔	✔	✔	✔	✔	✔	✔	✔	✔	✔	✔	✔			✔	✔	✔	✔	✔	✔	✔		✔	✔	✔
—	*Carduelis carduelis*	✔	✔	✔	✔	✔	✔	✔	✔	✔	✔	✔	✔	✔			✔	✔	✔	✔	✔	✔	✔		✔	✔	✔
—[E]	*Carduelis spinus*	✔	✔	✔	✔	✔	✔	✔		✔	✔		✔	✔	✔		✔	✔	✔	✔	✔	✔	✔	✔	✔	✔	✔
2	***Carduelis cannabina***	✔	✔	✔	✔	✔	✔	✔	✔	✔	✔	✔	✔	✔			✔	✔	✔	✔	✔	✔	✔		✔	✔	✔
—	*Carduelis flavirostris*			✔		✔											✔	✔		✔					✔		
—	*Carduelis flammea*				✔		✔	✔					✔	✔	✔	✔	✔	✔	✔		✔			✔	✔	✔	
—	*Carduelis hornemanni*															✔	✔	✔									
—	*Loxia leucoptera*																✔	✔									
—	*Loxia curvirostra*	✔	✔	✔	✔	✔	✔	✔		✔	✔	✔	✔	✔			✔	✔	✔	✔	✔	✔	✔		✔	✔	✔
1	***Loxia scotica***																										
—[E]	*Loxia pytyopsittacus*													✔			✔	✔			✔						✔
—	*Rhodopechys sanguinea*			✔		✔														✔							
—	*Rhodopechys obsoleta*																										
—	*Rhodopechys mongolica*			✔																							
—	*Bucanetes githagineus*			✔		✔																					
—	*Carpodacus erythrinus*			✔	✔	✔	✔	✔		✔			✔	✔			✔	✔	✔	✔	✔				✔		✔
3	***Carpodacus rubicilla***					✔														✔							

Appendix 2 ...continued. Occurrence of all European species by country.

SPEC Category	Species	Liechtenstein	Lithuania	Luxembourg	FYR of Macedonia	Malta	Moldova	Netherlands	Norway	*Svalbard*	Poland	Portugal	*Azores*	*Madeira*	Romania	Russia	Serbia and Montenegro	Slovakia	Slovenia	Spain	*Canary Islands*	Sweden	Switzerland	Turkey	Ukraine	United Kingdom	*Gibraltar*	TOTAL
—	*Parus major*	✔	✔	✔	✔		✔	✔	✔		✔	✔			✔	✔	✔	✔	✔	✔		✔	✔	✔	✔	✔	✔	44
2	***Sitta krueperi***															✔								✔				**4**
2	***Sitta whiteheadi***																											**1**
—	*Sitta europaea*	✔	✔	✔	✔		✔	✔	✔		✔	✔			✔	✔	✔	✔	✔	✔		✔	✔	✔	✔	✔		41
—	*Sitta tephronota*																							✔				3
—[E]	*Sitta neumayer*				✔												✔							✔				11
—	*Tichodroma muraria*	✔			✔						✔				✔	✔	✔	✔	✔	✔			✔	✔				23
—	*Certhia familiaris*	✔	✔	✔	✔		✔	✔	✔		✔				✔	✔	✔	✔	✔	✔		✔	✔	✔	✔	✔		41
—[E]	*Certhia brachydactyla*	✔		✔	✔			✔			✔	✔			✔	✔	✔	✔	✔	✔			✔	✔	✔	✔		32
—	*Remiz pendulinus*		✔	✔	✔		✔	✔			✔	✔			✔	✔	✔	✔	✔	✔		✔	✔	✔	✔			37
—	*Oriolus oriolus*	✔	✔	✔	✔		✔	✔	✔		✔	✔			✔	✔	✔	✔	✔	✔		✔	✔	✔	✔	✔		42
3	***Lanius collurio***	✔	✔	✔	✔		✔	✔	✔		✔	✔			✔	✔	✔	✔	✔	✔		✔	✔	✔	✔	✔		**42**
2	***Lanius minor***		✔		✔		✔				✔				✔	✔	✔	✔	✔	✔				✔	✔			**25**
3	***Lanius excubitor***		✔	✔			✔	✔	✔		✔	✔			✔	✔		✔		✔	✔	✔			✔	✔		**28**
2	***Lanius senator***			✔	✔	✔					✔	✔			✔	✔	✔	✔		✔			✔	✔				**26**
2	***Lanius nubicus***				✔																			✔				**5**
—	*Garrulus glandarius*	✔	✔	✔	✔		✔	✔	✔		✔	✔			✔	✔	✔	✔	✔	✔		✔	✔	✔	✔	✔		43
3	***Perisoreus infaustus***								✔							✔						✔						**4**
—	*Cyanopica cyanus*											✔								✔								2
—	*Pica pica*	✔	✔	✔	✔		✔	✔	✔		✔	✔			✔	✔	✔	✔	✔	✔		✔	✔	✔	✔	✔		43
—	*Nucifraga caryocatactes*	✔	✔	✔	✔				✔		✔				✔	✔	✔	✔	✔			✔	✔		✔			30
—	*Pyrrhocorax graculus*	✔			✔											✔	✔		✔	✔			✔	✔				21
3	***Pyrrhocorax pyrrhocorax***				✔							✔				✔	✔			✔	✔		✔	✔		✔		**18**
—[E]	*Corvus monedula*	✔	✔	✔	✔		✔	✔	✔		✔	✔			✔	✔	✔	✔	✔	✔		✔	✔	✔	✔	✔		42
—	*Corvus frugilegus*		✔	✔	✔		✔	✔	✔		✔				✔	✔	✔	✔	✔	✔		✔	✔	✔	✔	✔		38
—	*Corvus corone*	✔	✔	✔	✔		✔	✔	✔		✔	✔			✔	✔	✔	✔	✔	✔		✔	✔	✔	✔	✔		44
—	*Corvus corax*	✔	✔		✔		✔	✔	✔		✔	✔			✔	✔	✔	✔	✔	✔	✔	✔	✔	✔	✔	✔	✔	47
3	***Sturnus vulgaris***	✔	✔	✔	✔	✔	✔	✔	✔		✔	✔	✔		✔	✔	✔	✔	✔	✔	✔	✔	✔	✔	✔	✔	✔	**48**
—[E]	*Sturnus unicolor*											✔								✔							✔	6
—	*Sturnus roseus*				✔										✔	✔	✔							✔	✔			12
3	***Passer domesticus***	✔	✔	✔	✔		✔	✔	✔		✔	✔	✔		✔	✔	✔	✔	✔	✔	✔	✔	✔	✔	✔	✔	✔	**48**
—	*Passer hispaniolensis*				✔	✔						✔		✔	✔	✔	✔			✔	✔		✔	✔	✔			23
—	*Passer moabiticus*																							✔				2
3	***Passer montanus***	✔	✔	✔	✔	✔	✔	✔	✔		✔	✔			✔	✔	✔	✔	✔	✔	✔	✔	✔	✔	✔	✔		**44**
—	*Petronia brachydactyla*															✔								✔				4
—	*Petronia xanthocollis*																							✔				1
—	*Petronia petronia*				✔							✔		✔		✔				✔	✔			✔				16
—	*Montifringilla nivalis*	✔			✔										✔	✔	✔		✔	✔			✔	✔				19
—[E]	*Fringilla coelebs*	✔	✔	✔	✔	✔	✔	✔	✔		✔	✔	✔	✔	✔	✔	✔	✔	✔	✔	✔	✔	✔	✔	✔	✔	✔	49
1	***Fringilla teydea***																				✔							**1**
—	*Fringilla montifringilla*		✔	✔			✔	✔	✔			✔			✔	✔		✔	✔			✔			✔	✔		23
—	*Serinus pusillus*															✔								✔				5
—[E]	*Serinus serinus*	✔	✔	✔	✔	✔	✔	✔			✔	✔			✔	✔	✔	✔	✔	✔	✔	✔	✔	✔	✔		✔	40
—[E]	*Serinus canaria*												✔	✔							✔							3
—[E]	*Serinus citrinella*	✔																		✔			✔					8
—[E]	*Serinus corsicana*																											2
—[E]	*Carduelis chloris*	✔	✔	✔	✔	✔	✔	✔	✔		✔	✔	✔	✔	✔	✔	✔	✔	✔	✔	✔	✔	✔	✔	✔	✔	✔	48
—	*Carduelis carduelis*	✔	✔	✔	✔		✔	✔	✔		✔	✔	✔	✔	✔	✔	✔	✔	✔	✔	✔	✔	✔	✔	✔	✔	✔	47
—[E]	*Carduelis spinus*	✔	✔	✔	✔		✔	✔	✔		✔	✔			✔	✔	✔	✔	✔	✔		✔	✔	✔	✔	✔	✔	44
2	***Carduelis cannabina***	✔	✔	✔	✔	✔	✔	✔	✔		✔	✔		✔	✔	✔	✔	✔	✔	✔	✔	✔	✔	✔	✔	✔	✔	**47**
—	*Carduelis flavirostris*								✔						✔	✔		✔				✔		✔		✔		13
—	*Carduelis flammea*	✔		✔			✔	✔	✔	✔	✔				✔	✔		✔	✔			✔	✔		✔	✔		29
—	*Carduelis hornemanni*								✔							✔		✔				✔						7
—	*Loxia leucoptera*								✔							✔		✔				✔						6
—	*Loxia curvirostra*	✔	✔	✔	✔		✔	✔	✔		✔	✔			✔	✔	✔	✔	✔	✔		✔	✔	✔	✔	✔		42
1	***Loxia scotica***																									✔		**1**
—	*Loxia pytyopsittacus*		✔					✔	✔							✔						✔				✔		11
—	*Rhodopechys sanguinea*																							✔				4
—	*Rhodopechys obsoleta*																							✔				1
—	*Rhodopechys mongolica*																							✔				2
—	*Bucanetes githagineus*																			✔	✔			✔				5
—	*Carpodacus erythrinus*		✔					✔	✔		✔				✔	✔		✔	✔			✔	✔	✔	✔	✔		28
3	***Carpodacus rubicilla***															✔												**3**

Appendix 2 ...continued. Occurrence of all European species by country.

SPEC Category	Species	Albania	Andorra	Armenia	Austria	Azerbaijan	Belarus	Belgium	Bosnia and Herzegovina	Bulgaria	Croatia	Cyprus	Czech Republic	Denmark	*Faroe Islands*	*Greenland*	Estonia	Finland	France	Georgia	Germany	Greece	Hungary	Iceland	Republic of Ireland	Italy	Latvia
–	*Pinicola enucleator*																✔	✔									
–	*Pyrrhula pyrrhula*		✔	✔	✔	✔	✔	✔	✔	✔	✔		✔	✔	✔		✔	✔	✔	✔	✔	✔	✔		✔	✔	✔
1	***Pyrrhula murina***																										
–	*Coccothraustes coccothraustes*			✔	✔	✔	✔	✔	✔	✔	✔		✔	✔			✔	✔	✔	✔	✔	✔	✔			✔	✔
–	*Calcarius lapponicus*															✔		✔									
–	*Plectrophenax nivalis*						✔								✔	✔	✔	✔						✔			
–	*Emberiza leucocephalos*																										
–[E]	*Emberiza citrinella*	✔	✔		✔	✔	✔	✔	✔	✔	✔		✔	✔			✔	✔	✔		✔	✔	✔		✔	✔	✔
–[E]	*Emberiza cirlus*	✔	✔		✔	✔			✔	✔	✔								✔		✔	✔	✔			✔	
3	***Emberiza cia***	✔	✔	✔	✔	✔			✔	✔	✔								✔	✔	✔	✔	✔			✔	
1	***Emberiza cineracea***																					✔					
2	***Emberiza hortulana***	✔	✔	✔	✔	✔	✔	✔		✔	✔		✔				✔	✔	✔	✔	✔	✔	✔			✔	✔
–	*Emberiza buchanani*			✔		✔																					
–[E]	*Emberiza caesia*	✔				✔						✔										✔					
–	*Emberiza rustica*																✔	✔									✔
–	*Emberiza pusilla*																	✔									
1	***Emberiza aureola***																	✔									
–	*Emberiza schoeniclus*	✔		✔	✔	✔	✔	✔	✔	✔	✔		✔	✔			✔	✔	✔	✔	✔	✔	✔		✔	✔	✔
–	*Emberiza pallasi*																										
–	*Emberiza bruniceps*																										
2	***Emberiza melanocephala***	✔		✔		✔			✔	✔	✔	✔								✔		✔				✔	
2	***Miliaria calandra***	✔	✔	✔	✔	✔	✔	✔	✔	✔	✔	✔	✔	✔			✔		✔	✔	✔	✔	✔		✔	✔	✔
	TOTAL	272	114	243	233	313	232	219	164	302	282	158	228	226	91	65	246	248	332	251	280	314	247	92	295	238	128
	TOTAL SPEC 1	10	0	11	6	20	6	3	4	15	11	5	6	2	0	1	4	6	10	11	8	16	12	1	10	6	1
	TOTAL SPEC 2	30	12	23	24	26	24	21	17	33	34	20	24	23	7	2	23	16	36	25	27	37	26	8	34	23	8
	TOTAL SPEC 3	83	25	68	56	91	58	54	45	83	74	47	55	55	15	12	59	55	90	72	70	89	69	22	81	58	25
	TOTAL Non-SPEC[E]	55	29	36	51	53	52	54	34	62	60	23	53	57	31	16	57	56	73	43	67	61	54	27	60	56	33
	TOTAL Non-SPEC	94	48	105	96	123	92	87	64	109	103	63	90	89	38	33	103	115	123	100	108	111	86	34	110	95	61

Notes.

This appendix includes all species known to breed or winter regularly in the geopolitical units covered by this review. All Species of European Conservation Concern (i.e. SPECs 1, 2 and 3) are highlighted in blue to indicate particular national responsibilities. Abbreviated SPEC categories are given for each species as follows:

- 1 Species of global conservation concern, i.e. classified as Threatened, Near Threatened or Data Deficient under the IUCN Red List Criteria at global level (BirdLife International 2004; IUCN 2004).
- 2 Species concentrated in Europe and with an Unfavourable Conservation Status in Europe.
- 3 Species not concentrated in Europe but with an Unfavourable Conservation Status in Europe.
- –[E] Species concentrated in Europe but with a Favourable Conservation Status in Europe.
- – Species not concentrated in Europe and with a Favourable Conservation Status in Europe.
- NE Not Evaluated.
- W Category relates to winter populations.

Appendix 2 ...continued. Occurrence of all European species by country.

SPEC Category	Species	Liechtenstein	Lithuania	Luxembourg	FYR of Macedonia	Malta	Moldova	Netherlands	Norway	*Svalbard*	Poland	Portugal	*Azores*	*Madeira*	Romania	Russia	Serbia and Montenegro	Slovakia	Slovenia	Spain	*Canary Islands*	Sweden	Switzerland	Turkey	Ukraine	United Kingdom	*Gibraltar*	TOTAL
–	*Pinicola enucleator*								✓							✓		✓				✓						6
–	*Pyrrhula pyrrhula*	✓	✓	✓	✓		✓	✓	✓		✓	✓			✓	✓	✓	✓	✓	✓		✓	✓	✓	✓	✓		42
1	***Pyrrhula murina***												✓															**1**
–	*Coccothraustes coccothraustes*	✓	✓	✓	✓		✓	✓	✓		✓	✓			✓	✓	✓	✓	✓	✓		✓	✓	✓	✓	✓		39
–	*Calcarius lapponicus*		✓						✓						✓	✓		✓				✓			✓	✓		10
–	*Plectrophenax nivalis*						✓		✓	✓					✓	✓		✓				✓			✓	✓		15
–	*Emberiza leucocephalos*															✓												1
–^{E}	*Emberiza citrinella*	✓	✓	✓	✓		✓	✓	✓		✓	✓			✓	✓	✓	✓	✓	✓		✓	✓	✓	✓	✓		40
–^{E}	*Emberiza cirlus*				✓							✓			✓		✓		✓	✓			✓	✓		✓		21
3	***Emberiza cia***				✓							✓			✓	✓	✓	✓	✓	✓			✓	✓	✓		✓	**26**
1	***Emberiza cineracea***																							✓				**2**
2	***Emberiza hortulana***		✓		✓		✓	✓	✓		✓	✓			✓	✓	✓	✓	✓	✓		✓	✓	✓	✓			**36**
–	*Emberiza buchanani*																							✓				3
–^{E}	*Emberiza caesia*																							✓				5
–	*Emberiza rustica*								✓							✓						✓						6
–	*Emberiza pusilla*								✓							✓						✓						4
1	***Emberiza aureola***															✓									✓			**3**
–	*Emberiza schoeniclus*	✓	✓	✓	✓	✓	✓	✓	✓		✓	✓			✓	✓	✓	✓	✓	✓		✓	✓	✓	✓	✓		42
–	*Emberiza pallasi*															✓												1
–	*Emberiza bruniceps*															✓												1
2	***Emberiza melanocephala***				✓										✓	✓	✓							✓	✓			**16**
2	***Miliaria calandra***	✓	✓	✓	✓	✓	✓	✓			✓	✓			✓	✓	✓	✓	✓	✓	✓	✓	✓	✓	✓	✓	✓	**44**
	TOTAL	231	160	246	69	213	235	241	45	255	265	31	41	185	292	403	281	261	250	290	77	257	218	363	308	237	73	
	TOTAL SPEC 1	5	2	12	0	11	4	4	0	7	8	2	3	2	15	23	11	9	7	10	6	4	3	22	20	3	3	
	TOTAL SPEC 2	24	15	27	7	21	22	18	3	25	31	2	5	18	30	34	33	24	25	31	6	22	19	36	29	24	4	
	TOTAL SPEC 3	58	36	66	18	58	58	52	8	65	80	10	12	43	75	106	76	66	59	86	27	61	54	109	83	52	17	
	TOTAL Non-SPEC^E	51	40	50	21	48	57	55	13	59	55	9	9	48	59	69	58	57	58	57	15	57	53	62	59	61	24	
	TOTAL Non-SPEC	93	67	91	23	75	94	112	21	99	90	8	12	74	113	171	103	105	101	105	23	113	89	134	117	97	25	

Appendix 3. Protection status of all European species under international legislation, conventions and agreements.

Species	SPEC Category	European Threat Status (ETS)	Global IUCN Red List Category	Birds Directive	Bern Convention	Emerald Network	CMS	AEWA	CITES	Species Action Plan	Notes (e.g. Annex restrictions)
Gavia stellata	3	(H)		I	II	✔	II	✔			
Gavia arctica	3	(VU)		I	II	✔	II*	✔			* *G. a. arctica* only
Gavia immer	—	(S)		I	II	✔	II*	✔			* *G. a. immer* only
Gavia adamsii	—	(S)			II	✔	II	✔			
Tachybaptus ruficollis	—	S			II						
Podiceps cristatus	—	S			III						
Podiceps grisegena	—	S			II		II*	✔			* *P. g. grisegena* only
Podiceps auritus	3	D		I	II	✔	II	✔			
Podiceps nigricollis	—	S			II*						* *P. n. caspicus* only
Fulmarus glacialis	—	S			III						
Pterodroma madeira	1	(CR)	CR	I	II	✔				Ornis; Bern	
Pterodroma feae	1	VU	NT	I	II	✔				Ornis; Bern	
Bulweria bulwerii	3	(R)		I	II	✔					
Calonectris diomedea	2	(VU)		I	II	✔					
Puffinus gravis	NE	NE			III						
Puffinus griseus	1	NE	NT		III						
Puffinus puffinus	2	(L)			II						
Puffinus mauretanicus	1	CR	CR	I	II	✔				Ornis; Bern	
Puffinus yelkouan	—[E]	S		I	II	✔					
Puffinus assimilis	3	(R)		I	II*	✔					* *P. a. baroli* only
Pelagodroma marina	3	VU		I	II	✔					
Hydrobates pelagicus	—[E]	(S)		I	II	✔					
Oceanodroma leucorhoa	3	(L)		I	II	✔					
Oceanodroma castro	3	(R)		I	II	✔					
Morus bassanus	—[E]	S			III						
Phalacrocorax carbo	—	S			III						
Phalacrocorax aristotelis	—[E]	(S)		I*	III	✔*				Ornis*; Bern*	* *P. a. desmarestii* only
Phalacrocorax pygmeus	1	S	NT	I	III	✔	II	✔		Ornis; Bern; Bonn	
Pelecanus onocrotalus	3	R		I	III	✔	I; II*	✔			* Western Palearctic population only
Pelecanus crispus	1	R	VU	I	III	✔	I; II	✔	I	Ornis; Bern; Bonn	
Botaurus stellaris	3	H		I	III	✔	II*	✔		Ornis	* *B. s. stellaris* only
Ixobrychus minutus	3	(H)		I	III	✔	II*	✔			* *I. m. minutus* only
Nycticorax nycticorax	3	H		I	III	✔					
Ardeola ralloides	3	(D)		I	III	✔					
Bubulcus ibis	—	S			III						
Egretta garzetta	—	S		I	III	✔					
Casmerodius albus	—	S		I	III	✔	II*	✔			* *C. a. albus* only
Ardea cinerea	—	S			III						
Ardea purpurea	3	(D)		I	III	✔	II*	✔			* *A. p. purpurea* only
Ciconia nigra	2	R		I	III	✔	II	✔	II		
Ciconia ciconia	2	H		I	III	✔	II	✔			
Plegadis falcinellus	3	(D)		I	III	✔	II	✔			
Geronticus eremita	1	CR	CR		III		I; II	✔	I		
Platalea leucorodia	2	R		I	III	✔	II	✔	II		
Phoenicopterus roseus	3	L		I	III	✔	II	✔	II		
Cygnus olor	—[E]	S		II/2	III		II	✔			
Cygnus columbianus	3W	VU		I	III	✔	II	✔			
Cygnus cygnus	—[E]W	S		I	III	✔	II	✔			
Anser fabalis	—[E]W	S		II/1	III		II	✔			
Anser brachyrhynchus	—[E]	S		II/2	III		II	✔			
Anser albifrons	—	S		I*; II/2; III/2**	III	✔*	II	✔			* *A. a. flavirostris* only; ** *A. a. albifrons* only
Anser erythropus	1	EN	VU	I	III	✔	I; II	✔		Ornis; Bern; Bonn	
Anser anser	—	S		II/1; III/2	III		II	✔			
Chen caerulescens	—	(S)			III		II*				* wild European populations only
Branta canadensis	—	(S)		II/1	III		II*				* wild European populations only
Branta leucopsis	—[E]	S		I	III	✔	II	✔			
Branta bernicla	3W	VU		II/2	III		II	✔			
Branta ruficollis	1W	VU	VU	I	III	✔	I; II	✔	II	Ornis; Bern; Bonn	
Tadorna ferruginea	3	(VU)		I	III	✔	II	✔			
Tadorna tadorna	—	S			III		II	✔			
Anas penelope	—[E]W	S		II/1; III/2	III		II	✔			
Anas strepera	3	(H)		II/1	III		II	✔			
Anas crecca	—	(S)		II/1; III/2	III		II	✔			
Anas platyrhynchos	—	(S)		II/1; III/1	III		II	✔			

Appendix 3 ...continued. Protection status of all European species under international legislation, conventions and agreements.

Species	SPEC Category	European Threat Status (ETS)	Global IUCN Red List Category	Birds Directive	Bern Convention	Emerald Network	CMS	AEWA	CITES	Species Action Plan	Notes (e.g. Annex restrictions)
Anas acuta	3	(D)		II/1; III/2	III		II	✔			
Anas querquedula	3	(D)		II/1	III		II	✔			
Anas clypeata	3	(D)		II/1; III/2	III		II	✔			
Marmaronetta angustirostris	1	(VU)	VU	I	III	✔	I; II	✔		Ornis; Bern; Bonn	
Netta rufina	—	(S)		II/2	III		II	✔			
Aythya ferina	2	(D)		II/1; III/2	III		II	✔			
Aythya nyroca	1	(VU)	NT	I	III	✔	I; II	✔		Ornis; Bern; Bonn	
Aythya fuligula	3	(D)		II/1; III/2	III		II	✔			
Aythya marila	3W	EN		II/2; III/2	III		II	✔			
Somateria mollissima	—[E]	S		II/2; III/2	III		II	✔			
Somateria spectabilis	—	(S)			III		II	✔			
Polysticta stelleri	3W	L		I	III		I; II	✔		Ornis; Bern; Bonn	
Histrionicus histrionicus	3	(R)			III	✔	II				
Clangula hyemalis	—	(S)		II/2	III		II	✔			
Melanitta nigra	—	(S)		II/2; III/2	III		II	✔			
Melanitta fusca	3	(D)		II/2	III		II	✔			
Bucephala islandica	3	VU			III	✔	II				
Bucephala clangula	—	(S)		II/2	III		II	✔			
Mergellus albellus	3	(D)		I	III	✔	II	✔			
Mergus serrator	—	(S)		II/2	III		II	✔			
Mergus merganser	—	(S)		II/2	III		II	✔			
Oxyura leucocephala	1	VU	EN	I	III	✔	I; II	✔	II	Ornis; Bern; Bonn	
Pernis apivorus	—[E]	(S)		I	III	✔	II		II		
Elanus caeruleus	3	R		I	III	✔	II		II		
Milvus migrans	3	(VU)		I	III	✔	II		II		
Milvus milvus	2	D		I	III	✔	II		II		
Haliaeetus albicilla	1	R	NT	I	III	✔	I; II		I	Bern	
Gypaetus barbatus	3	(VU)		I	III	✔	II		II	Ornis; Bern	
Neophron percnopterus	3	EN		I	III	✔	II		II		
Gyps fulvus	—	S		I	III	✔	II		II		
Aegypius monachus	1	R	NT	I	III	✔	II		II	Ornis; Bern; Bonn	
Circaetus gallicus	3	(R)		I	III	✔	II		II		
Circus aeruginosus	—	S		I	III	✔	II		II		
Circus cyaneus	3	H		I	III	✔	II		II		
Circus macrourus	1	(EN)	NT	I	III	✔	II		II	Bern	
Circus pygargus	—[E]	S		I	III	✔	II		II		
Accipiter gentilis	—	S		I*	III	✔*	II		II	Ornis*; Bern*	* *A. g. arrigonii* only
Accipiter nisus	—	S		I*	III	✔*	II		II	Ornis*; Bern*	* *A. n. granti* only
Accipiter badius	—	(S)			III		II		II		
Accipiter brevipes	2	(VU)		I	III	✔	II		II		
Buteo buteo	—	S			III		II		II		
Buteo rufinus	3	(VU)		I	III	✔	II		II		
Buteo lagopus	—	(S)			III		II		II		
Aquila pomarina	2	(D)		I	III	✔	II		II	Ornis; Bern	
Aquila clanga	1	EN	VU	I	III	✔	I; II		II	Ornis; Bern; Bonn	
Aquila nipalensis	3	(EN)			III	✔	II		II		
Aquila heliaca	1	R	VU	I	III	✔	I; II		I	Ornis; Bern; Bonn	
Aquila adalberti	1	(EN)	EN	I	III	✔	I; II		I	Ornis; Bern; Bonn	
Aquila chrysaetos	3	R		I	III	✔	II		II		
Hieraaetus pennatus	3	(R)		I	III	✔	II		II		
Hieraaetus fasciatus	3	EN		I	III	✔	II		II	Ornis; Bern	
Pandion haliaetus	3	R		I	III	✔	II		II		
Falco naumanni	1	H	VU	I	III	✔	I; II		II	Ornis; Bern; Bonn	
Falco tinnunculus	3	D			III		II		II		
Falco vespertinus	3	(VU)		I	III	✔	II		II		
Falco columbarius	—	(S)		I	III	✔	II		II		
Falco subbuteo	—	(S)			III		II		II		
Falco eleonorae	2	D		I	III	✔	II		II	Ornis; Bern	
Falco biarmicus	3	VU		I	III	✔	II		II	Ornis; Bern	
Falco cherrug	1	EN	EN	I	III		II		II		
Falco rusticolus	3	(R)		I	III	✔	II		I	Ornis; Bern	
Falco peregrinus	—	S		I	III	✔	II		I		
Falco pelegrinoides	—	S			III		II		I		
Bonasa bonasia	—	S		I; II/2	III	✔					
Lagopus lagopus	—	S		II/1*; II/2**; III/1***	III						* *L. l. scoticus* + *L. l. hibernicus* only; ** *L. l. lagopus* only;

Appendix 3 ...continued. Protection status of all European species under international legislation, conventions and agreements.

Species	SPEC Category	European Threat Status (ETS)	Global IUCN Red List Category	Birds Directive	Bern Convention	Emerald Network	CMS	AEWA	CITES	Species Action Plan	Notes (e.g. Annex restrictions)
Lagopus lagopus continued											*** *L. l. lagopus, scoticus + hibernicus* only
Lagopus mutus	—	S		I*; II/1; III/2	III	✔*					* *L. m. pyrenaicus + L. m. helveticus* only
Tetrao tetrix	3	H		I*; II/2**; III/2***	III	✔*					* *T. t. tetrix* only; ** all others; *** *T. t. britannicus* only
Tetrao mlokosiewiczi	1	DD	DD		III						
Tetrao urogallus	—	(S)		I; II/2; III/2	II*; III**	✔					* *T. u. cantabricus* only; ** all others
Tetraogallus caucasicus	—^{E}	S			III						
Tetraogallus caspius	3	(VU)			III				I		
Alectoris chukar	3	(VU)		II/2	III						
Alectoris graeca	2	(D)		I*; II/1**	III	✔*				Ornis***; Bern***	* *A. g. saxatilis + A. g. whitakeri* only; ** all others; *** *A. g. whitakeri* only
Alectoris rufa	2	(D)		II/1; III/1	III						
Alectoris barbara	3	(R)		I; II/2; III/1	III	✔					
Ammoperdix griseogularis	3	VU			III						
Francolinus francolinus	3	D		II/2	III						
Perdix perdix	3	VU		I*; II/1**; III/1	III	✔*				Ornis***; Bern***	* *P. p. italica + P. p. hispaniensis* only; ** all others; *** *P. p. italica* only
Coturnix coturnix	3	(H)		II/2	III		II*				* *C. c. coturnix* only
Phasianus colchicus	—	(S)		II/1; III/1	III						
Turnix sylvatica	3	CR		I	III	✔					
Rallus aquaticus	—	(S)		II/2	III						
Porzana porzana	—^{E}	(S)		I	III	✔	II	✔			
Porzana parva	—^{E}	(S)		I	III	✔	II*	✔			* *P. p. parva* only
Porzana pusilla	3	(R)		I	III	✔	II*	✔			* *P. p. intermedia* only
Crex crex	1	H	NT	I	III	✔	II			Ornis; Bern; Bonn	
Gallinula chloropus	—	S		II/2	III						
Porphyrio porphyrio	3	L		I	III	✔				Ornis; Bern	
Fulica atra	—	(S)		II/1; III/2	III		II*	✔			* Mediterranean and Black Sea populations of *F. a. atra* only
Fulica cristata	3	CR		I	III	✔				Ornis; Bern	
Grus grus	2	(H)		I	III	✔	II	✔	II		
Grus virgo	—	S			III		II	✔	II		
Tetrax tetrax	1	VU	NT	I	III	✔			II	Ornis; Bern	
Chlamydotis undulata	1	(VU)	VU	I	III	✔	I*; II**		I	Ornis; Bern	* north-west African populations only ** Asian populations only
Otis tarda	1	VU	VU	I	III	✔	I*; II		II	Ornis; Bern; Bonn	* Middle European populations only
Haematopus ostralegus	—^{E}	(S)		II/2	III		II				
Himantopus himantopus	—	S		I	III	✔	II	✔			
Recurvirostra avosetta	—	S		I	III	✔	II	✔			
Burhinus oedicnemus	3	(VU)		I	III	✔	II				
Cursorius cursor	3	(EN)		I	III	✔				Ornis; Bern	
Glareola pratincola	3	D		I	III	✔	II	✔			
Glareola nordmanni	1	EN	DD		III	✔	II	✔		Bern; AEWA	
Charadrius dubius	—	(S)			III		II	✔			
Charadrius hiaticula	—^{E}	(S)			III		II	✔			
Charadrius alexandrinus	3	(D)		I	III		II	✔			
Charadrius leschenaultii	3	(EN)			III	✔	II	✔			
Charadrius asiaticus	3	EN			III	✔	II	✔			
Eudromias morinellus	—	(S)		I	III	✔	II	✔			
Pluvialis apricaria	—^{E}	(S)		I; II/2; III/2	III	✔	II	✔			
Pluvialis squatarola	—	(S)		II/2	III		II	✔			
Vanellus spinosus	3	VU		I	III	✔	II	✔			
Vanellus indicus	3	(VU)			III		II				
Vanellus gregarius	1	CR	CR		III	✔	I; II	✔		Bern; AEWA	
Vanellus leucurus	—	S			III		II	✔			
Vanellus vanellus	2	VU		II/2	III		II	✔			
Calidris canutus	3W	D		II/2	III		II	✔			
Calidris alba	—	(S)			III		II	✔			
Calidris minuta	—	(S)			III		II	✔			
Calidris temminckii	—	(S)			III		II	✔			
Calidris bairdii	—	(S)			III		II				
Calidris ferruginea	NE	NE			III		II	✔			
Calidris maritima	—^{E}	(S)			III		II	✔			
Calidris alpina	3	(H)		I*	III		II	✔			* *C. a. schinzii* only

Appendix 3 ...continued. Protection status of all European species under international legislation, conventions and agreements.

Species	SPEC Category	European Threat Status (ETS)	Global IUCN Red List Category	Birds Directive	Bern Convention	Emerald Network	CMS	AEWA	CITES	Species Action Plan	Notes (e.g. Annex restrictions)
Limicola falcinellus	3	(D)			III		II	✔			
Philomachus pugnax	2	(D)		I; II/2	III	✔	II	✔			
Lymnocryptes minimus	3	(D)		II/1; III/2	III		II	✔			
Gallinago gallinago	3	(D)		II/1; III/2	III		II	✔			
Gallinago media	1	D	NT	I	III	✔	II	✔		Bern; AEWA	
Gallinago stenura	—	(S)			III		II				
Scolopax rusticola	3	(D)		II/1; III/2	III		II				
Limosa limosa	2	VU		II/2	III		II	✔			
Limosa lapponica	—	(S)		I; II/2	III	✔	II	✔			
Numenius phaeopus	—^{E}	(S)		II/2	III		II	✔			
Numenius tenuirostris	1	NE	CR	I	III	✔	I; II	✔	I	Ornis; Bern; Bonn	
Numenius arquata	2	D		II/2	III		II	✔			
Tringa erythropus	3	(D)		II/2	III		II	✔			
Tringa totanus	2	D		II/2	III		II	✔			
Tringa stagnatilis	—	(S)			III		II	✔			
Tringa nebularia	—	S		II/2	III		II	✔			
Tringa ochropus	—	S			III		II	✔			
Tringa glareola	3	H		I	III	✔	II	✔			
Xenus cinereus	—	(S)		I	III	✔	II	✔			
Actitis hypoleucos	3	(D)			III		II	✔			
Arenaria interpres	—	(S)			III		II	✔			
Phalaropus lobatus	—	(S)		I	III	✔	II	✔			
Phalaropus fulicarius	—	S			III	✔	II	✔			
Stercorarius pomarinus	—	(S)			III						
Stercorarius parasiticus	—	(S)			III						
Stercorarius longicaudus	—	(S)			III						
Catharacta skua	—^{E}	S			III						
Larus ichthyaetus	—	(S)			III		II	✔			
Larus melanocephalus	—^{E}	S		I	III	✔	II	✔			
Larus minutus	3	(H)		I	III						
Larus ridibundus	—^{E}	(S)		II/2	III						
Larus genei	3	L		I	III	✔	II	✔			
Larus audouinii	1	L	NT	I	III	✔	I; II	✔		Ornis; Bern; Bonn	
Larus canus	2	(H)		II/2	III						
Larus fuscus	—^{E}	S		II/2							
Larus argentatus	—^{E}	S		II/2							
Larus cachinnans	—^{E}	S		II/2	III						
Larus armenicus	2	L			III		II	✔			
Larus glaucoides	—^{E}	(S)			III						
Larus hyperboreus	—	(S)			III						
Larus marinus	—^{E}	S		II/2							
Xema sabini	—	S			III						
Rhodostethia rosea	—	(S)			III						
Rissa tridactyla	—	(S)			III						
Pagophila eburnea	3	(R)			III	✔					
Sterna nilotica	3	(VU)		I	III	✔	II*	✔			* *S. n. nilotica* only
Sterna caspia	3	R		I	III	✔	II	✔			
Sterna bengalensis	—	(S)			III		II*	✔			* African and south-west Asian populations only
Sterna sandvicensis	2	H		I	III	✔	II*	✔			* *S. s. sandvicensis* only
Sterna dougallii	3	R		I	III	✔	II	✔		Ornis; Bern	
Sterna hirundo	—	S		I	III	✔	II*	✔			* *S. h. hirundo* only
Sterna paradisaea	—	(S)		I	III	✔	II	✔			
Sterna albifrons	3	D		I	III	✔	II	✔			
Chlidonias hybrida	3	H		I	III	✔					
Chlidonias niger	3	(H)		I	III	✔	II*	✔			* *C. n. niger* only
Chlidonias leucopterus	—	(S)			III	✔	II	✔			
Uria aalge	—	(S)		I*	III	✔*					* *U. a. ibericus* only
Uria lomvia	3	(VU)			III						
Alca torda	—^{E}	(S)			III						
Cepphus grylle	2	H			III						
Alle alle	—	(S)			III						
Fratercula arctica	2	(H)			III						
Pterocles orientalis	3	(D)		I	III	✔					
Pterocles alchata	3	(D)		I	III	✔					
Columba livia	—	(S)		II/1	III						

Appendix 3 ...continued. Protection status of all European species under international legislation, conventions and agreements.

Species	SPEC Category	European Threat Status (ETS)	Global IUCN Red List Category	Birds Directive	Bern Convention	Emerald Network	CMS	AEWA	CITES	Species Action Plan	Notes (e.g. Annex restrictions)
Columba oenas	—[E]	S		II/2	III						
Columba palumbus	—[E]	S		I*; II/I**; III/I		✔*					* *C. p. azorica* only; ** all others
Columba trocaz	1	(R)	NT	I	III	✔				Ornis; Bern	
Columba bollii	1	(R)	NT	I	III	✔				Ornis; Bern	
Columba junoniae	1	EN	EN	I	III	✔				Ornis; Bern	
Streptopelia decaocto	—	S		II/2	III						
Streptopelia turtur	3	D		II/2	III		II*				* *S. t. turtur* only
Streptopelia senegalensis	—	S			III						
Clamator glandarius	—	(S)			III						
Cuculus canorus	—	S			III						
Cuculus saturatus	—	(S)			III						
Tyto alba	3	(D)			III				I		
Otus brucei	3	CR			III				II		
Otus scops	2	(H)			III				II		
Bubo bubo	3	(H)		I	III	✔			II		
Ketupa zeylonensis	3	CR			III	✔			II		
Nyctea scandiaca	3	(R)		I	III	✔			II		
Surnia ulula	—	(S)		I	III	✔			II		
Glaucidium passerinum	—	S		I	III	✔			II		
Athene noctua	3	(D)			III				II		
Strix aluco	—[E]	S			III				II		
Strix uralensis	—	(S)		I	III	✔			II		
Strix nebulosa	—	(S)		I	III	✔			II		
Asio otus	—	(S)			III				II		
Asio flammeus	3	(H)		I	III	✔			II		
Aegolius funereus	—	(S)		I	III	✔			II		
Caprimulgus europaeus	2	(H)		I	III	✔					
Caprimulgus ruficollis	—	(S)			III						
Tachymarptis melba	—	S			III						
Apus unicolor	2	(R)			III						
Apus apus	—	(S)			III						
Apus pallidus	—	(S)			III						
Apus caffer	—	S		I	III	✔					
Apus affinis	3	(EN)			III						
Halcyon smyrnensis	3	EN			III	✔					
Alcedo atthis	3	H		I	III	✔					
Ceryle rudis	3	(CR)			III						
Merops persicus	—	(S)			III						
Merops apiaster	3	(H)			III		II				
Coracias garrulus	2	VU		I	III	✔	II				
Upupa epops	3	(D)			III						
Jynx torquilla	3	(D)			III						
Picus canus	3	(H)		I	III	✔					
Picus viridis	2	(H)			III						
Dryocopus martius	—	S		I	III	✔					
Dendrocopos major	—	S		I*	III	✔*				Ornis*; Bern*	* *D. m. canariensis* + *D. m. thanneri* only
Dendrocopos syriacus	—[E]	(S)		I	III	✔					
Dendrocopos medius	—[E]	(S)		I	III	✔					
Dendrocopos leucotos	—	(S)		I	III	✔					
Dendrocopos minor	—	(S)			III						
Picoides tridactylus	3	(H)		I	III	✔					
Ammomanes deserti	3	(EN)			III						
Chersophilus duponti	3	(H)		I	III	✔					
Melanocorypha calandra	3	(D)		I	III	✔					
Melanocorypha bimaculata	—	S			III						
Melanocorypha leucoptera	—[E]W	(S)			III						
Melanocorypha yeltoniensis	3	EN			III	✔					
Calandrella brachydactyla	3	D		I	III	✔					
Calandrella rufescens	3	D			III						
Calandrella cheleensis	3	(VU)			III						
Galerida cristata	3	(H)			III						
Galerida theklae	3	(H)		I	III	✔					
Lullula arborea	2	H		I	III	✔					
Alauda arvensis	3	(H)		II/2	III						
Eremophila alpestris	—	(S)			III						

Appendix 3 ...continued. Protection status of all European species under international legislation, conventions and agreements.

Species	SPEC Category	European Threat Status (ETS)	Global IUCN Red List Category	Birds Directive	Bern Convention	Emerald Network	CMS	AEWA	CITES	Species Action Plan	Notes (e.g. Annex restrictions)
Riparia riparia	3	(H)			III						
Hirundo rupestris	—	S			III						
Hirundo rustica	3	H			III						
Hirundo daurica	—	(S)			III						
Delichon urbica	3	(D)			III						
Anthus campestris	3	(D)		I	III	✔					
Anthus berthelotii	—^{E}	(S)			III						
Anthus hodgsoni	—	(S)			III						
Anthus trivialis	—	S			III						
Anthus gustavi	—	(S)			III						
Anthus pratensis	—^{E}	(S)			III						
Anthus cervinus	—	(S)			III						
Anthus spinoletta	—	(S)			III						
Anthus petrosus	—^{E}	(S)			III						
Motacilla flava	—	(S)			III						
Motacilla citreola	—	(S)			III						
Motacilla cinerea	—	S			III						
Motacilla alba	—	S			III						
Pycnonotus xanthopygos	—	S			III						
Bombycilla garrulus	—	(S)			III						
Cinclus cinclus	—	S			III						
Troglodytes troglodytes	—	S		I*	III	✔*					* *T. t. fridariensis* only
Prunella modularis	—^{E}	S			III						
Prunella montanella	—	(S)			III						
Prunella ocularis	—^{E}	(S)			III						
Prunella atrogularis	3	R			III						
Prunella collaris	—	(S)			III						
Erythropygia galactotes	3	VU			III		II				
Erithacus rubecula	—^{E}	S			III		II				
Luscinia luscinia	—^{E}	S			III		II				
Luscinia megarhynchos	—^{E}	(S)			III		II				
Luscinia calliope	—	(S)			III		II				
Luscinia svecica	—	S		I	III	✔	II				
Tarsiger cyanurus	—	(S)			III		II				
Irania gutturalis	—	(S)			III		II				
Phoenicurus ochruros	—	S			III		II				
Phoenicurus phoenicurus	2	(H)			III		II				
Phoenicurus erythrogastrus	3	(R)			III		II				
Saxicola rubetra	—^{E}	(S)			III		II				
Saxicola dacotiae	1	EN	EN	I	III	✔	II			Bern	
Saxicola torquata	—	(S)			III		II				
Oenanthe isabellina	—	(S)			III		II				
Oenanthe oenanthe	3	(D)			III		II				
Oenanthe pleschanka	—	(S)		I	III		II				
Oenanthe cypriaca	—^{E}	(S)		I	III	✔	II				
Oenanthe hispanica	2	(H)			III		II				
Oenanthe deserti	—	(S)			III		II				
Oenanthe finschii	—	(S)			III		II				
Oenanthe xanthoprymna	3	(VU)			III		II				
Oenanthe leucura	3	(R)		I	III	✔	II				
Monticola saxatilis	3	(H)			III		II				
Monticola solitarius	3	(H)			III		II				
Zoothera dauma	—	(S)			III		II				
Turdus torquatus	—^{E}	S			III		II				
Turdus merula	—^{E}	S		II/2	III		II				
Turdus ruficollis	—	(S)			III		II				
Turdus pilaris	—^{E}W	(S)		II/2	III		II				
Turdus philomelos	—^{E}	S		II/2	III		II				
Turdus iliacus	—^{E}W	(S)		II/2	III		II				
Turdus viscivorus	—^{E}	S		II/2	III		II				
Cettia cetti	—	S			III		II				
Cisticola juncidis	—	S			III		II				
Prinia gracilis	3	(VU)			III		II				
Locustella lanceolata	—	(S)			III		II				
Locustella naevia	—^{E}	(S)			III		II				
Locustella fluviatilis	—^{E}	(S)			III		II				

Appendix 3 ...continued. Protection status of all European species under international legislation, conventions and agreements.

Species	SPEC Category	European Threat Status (ETS)	Global IUCN Red List Category	Birds Directive	Bern Convention	Emerald Network	CMS	AEWA	CITES	Species Action Plan	Notes (e.g. Annex restrictions)
Locustella luscinioides	—[E]	(S)			III		II				
Acrocephalus melanopogon	—	(S)		I	III	✔	II				
Acrocephalus paludicola	1	(VU)	VU	I	III	✔	I; II			Ornis; Bern; Bonn	
Acrocephalus schoenobaenus	—[E]	S			III		II				
Acrocephalus agricola	—	(S)			III		II				
Acrocephalus dumetorum	—	S			III		II				
Acrocephalus palustris	—[E]	(S)			III		II				
Acrocephalus scirpaceus	—[E]	S			III		II				
Acrocephalus arundinaceus	—	(S)			III		II				
Hippolais pallida	3	(H)			III		II				
Hippolais caligata	—	(S)			III		II				
Hippolais languida	—	S			III		II				
Hippolais olivetorum	—[E]	(S)		I	III	✔	II				
Hippolais icterina	—[E]	(S)			III		II				
Hippolais polyglotta	—[E]	(S)			III		II				
Sylvia sarda	—[E]	(S)		I	III	✔	II				
Sylvia undata	2	H		I	III	✔	II				
Sylvia conspicillata	—	(S)			III		II				
Sylvia cantillans	—[E]	(S)			III		II				
Sylvia mystacea	—	(S)			III		II				
Sylvia melanocephala	—[E]	(S)			III		II				
Sylvia melanothorax	—[E]	(S)		I	III		II				
Sylvia rueppelli	—[E]	(S)		I	III	✔	II				
Sylvia nana	—	(S)			III		II				
Sylvia hortensis	3	H			III		II				
Sylvia nisoria	—[E]	S		I	III	✔	II				
Sylvia curruca	—	S			III		II				
Sylvia communis	—[E]	S			III		II				
Sylvia borin	—[E]	S			III		II				
Sylvia atricapilla	—[E]	S			III		II				
Phylloscopus trochiloides	—	S			III		II				
Phylloscopus borealis	—	S			III		II				
Phylloscopus inornatus	—	(S)			III		II				
Phylloscopus bonelli	2	D			III		II				
Phylloscopus sibilatrix	2	D			III		II				
Phylloscopus sindianus	3	D			III		II				
Phylloscopus collybita	—	S			III		II				
Phylloscopus ibericus	—[E]	(S)			III		II				
Phylloscopus canariensis	—[E]	(S)			III		II				
Phylloscopus trochilus	—	S			III		II				
Regulus regulus	—[E]	S			III		II				
Regulus teneriffae	—[E]	(S)			III		II				
Regulus ignicapilla	—[E]	(S)			III		II				
Muscicapa striata	3	H			III		II				
Ficedula parva	—	(S)		I	III	✔	II				
Ficedula semitorquata	2	D		I	III	✔	II				
Ficedula albicollis	—[E]	S		I	III	✔	II				
Ficedula hypoleuca	—[E]	S			III		II				
Panurus biarmicus	—	(S)			III		II				
Aegithalos caudatus	—	S			III						
Parus palustris	3	D			III						
Parus lugubris	—[E]	(S)			III						
Parus montanus	—	S			III						
Parus cinctus	—	(S)			III						
Parus cristatus	2	(D)			III						
Parus ater	—	(S)		I*	III						* *P. a. cypriotes* only
Parus caeruleus	—[E]	S			III						
Parus cyanus	—	(S)			III						
Parus major	—	S			III						
Sitta krueperi	2	(D)		I	III	✔					
Sitta whiteheadi	2	R		I	III	✔					
Sitta europaea	—	S			III						
Sitta tephronota	—	(S)			III						
Sitta neumayer	—[E]	(S)			III						
Tichodroma muraria	—	(S)			III						
Certhia familiaris	—	S			III						

Appendix 3...continued. Protection status of all European species under international legislation, conventions and agreements.

Species	SPEC Category	European Threat Status (ETS)	Global IUCN Red List Category	Birds Directive	Bern Convention	Emerald Network	CMS	AEWA	CITES	Species Action Plan	Notes (e.g. Annex restrictions)
Certhia brachydactyla	—[E]	(S)		I*	III						* *C. b. dorotheae* only
Remiz pendulinus	—	(S)			III						
Oriolus oriolus	—	S			III						
Lanius collurio	3	(H)		I	III	✔					
Lanius minor	2	(D)		I	III	✔					
Lanius excubitor	3	(H)			III						
Lanius senator	2	(D)			III						
Lanius nubicus	2	(D)		I	III						
Garrulus glandarius	—	S		II/2							
Perisoreus infaustus	3	(H)			III						
Cyanopica cyanus	—	(S)			III						
Pica pica	—	S		II/2							
Nucifraga caryocatactes	—	S			III						
Pyrrhocorax graculus	—	(S)			III						
Pyrrhocorax pyrrhocorax	3	D		I	III	✔					
Corvus monedula	—[E]	(S)		II/2							
Corvus frugilegus	—	(S)		II/2							
Corvus corone	—	S		II/2							
Corvus corax	—	S			III						
Sturnus vulgaris	3	D		II/2							
Sturnus unicolor	—[E]	S			III						
Sturnus roseus	—	S			III						
Passer domesticus	3	D									
Passer hispaniolensis	—	(S)			III						
Passer moabiticus	—	(S)			III						
Passer montanus	3	(D)			III						
Petronia brachydactyla	—	S			III						
Petronia xanthocollis	—	(S)			III						
Petronia petronia	—	(S)			III						
Montifringilla nivalis	—	(S)			III						
Fringilla coelebs	—[E]	S		I*	III	✔*					* *F. c. ombriosa* only
Fringilla teydea	1	R	NT	I	III	✔				Ornis; Bern	
Fringilla montifringilla	—	S			III						
Serinus pusillus	—	(S)			III						
Serinus serinus	—[E]	S			III						
Serinus canaria	—[E]	(S)			III						
Serinus citrinella	—[E]	(S)			III						
Serinus corsicana	—[E]	(S)			III						
Carduelis chloris	—[E]	S			III						
Carduelis carduelis	—	S			III						
Carduelis spinus	—[E]	S			III						
Carduelis cannabina	2	D			III						
Carduelis flavirostris	—	S			III						
Carduelis flammea	—	(S)			III						
Carduelis hornemanni	—	(S)			III						
Loxia leucoptera	—	(S)			III						
Loxia curvirostra	—	(S)			III						
Loxia scotica	1	DD	DD	I	III	✔				Ornis; Bern	
Loxia pytyopsittacus	—[E]	(S)			III						
Rhodopechys sanguinea	—	S			III						
Rhodopechys obsoleta	—	(S)			III						
Rhodopechys mongolica	—	(S)			III						
Bucanetes githagineus	—	(S)		I	III	✔					
Carpodacus erythrinus	—	(S)			III						
Carpodacus rubicilla	3	(R)			III						
Pinicola enucleator	—	(S)			III						
Pyrrhula pyrrhula	—	(S)			III						
Pyrrhula murina	1	(EN)	EN	I	III	✔				Ornis; Bern	
Coccothraustes coccothraustes	—	S			III						
Calcarius lapponicus	—	(S)			III						
Plectrophenax nivalis	—	(S)			III						
Emberiza leucocephalos	—	(S)			III						
Emberiza citrinella	—[E]	(S)			III						
Emberiza cirlus	—[E]	S			III						
Emberiza cia	3	(H)			III						
Emberiza cineracea	1	(R)	NT	I	III	✔				Bern	

Appendix 3 ...continued. Protection status of all European species under international legislation, conventions and agreements.

Species	SPEC Category	European Threat Status (ETS)	Global IUCN Red List Category	Birds Directive	Bern Convention	Emerald Network	CMS	AEWA	CITES	Species Action Plan	Notes (e.g. Annex restrictions)
Emberiza hortulana	2	(H)		I	III	✔					
Emberiza buchanani	—	(S)			III						
Emberiza caesia	—[E]	(S)		I	III	✔					
Emberiza rustica	—	(S)			III						
Emberiza pusilla	—	(S)			III						
Emberiza aureola	1	D	NT		III						
Emberiza schoeniclus	—	S			III						
Emberiza pallasi	—	(S)			III						
Emberiza bruniceps	—	(S)			III						
Emberiza melanocephala	2	(H)			III						
Miliaria calandra	2	(D)			III						

Notes.

This appendix details the protection status of all European species according to the annexes and appendices of the Birds Directive, Bern Convention, Emerald Network, Bonn Convention, AEWA and CITES. It also identifies those species covered by Species Action Plans. For definitions of SPEC categories, European Threat Status and global IUCN Red List categories, see Appendix 1 (p. 291).

EU Council Directive on the Conservation of Wild Birds (79/409/EEC, 'Birds Directive')

Aim: to protect wild birds and their habitats, e.g. through the designation of Special Protection Areas (SPAs). Details: http://europa.eu.int/comm/environment/nature/nature_conservation/eu_nature_legislation/birds_directive/index_en.htm

This appendix incorporates all amendments to the Annexes of the Council Directive up to 23 September 2003.

Annex I

The directive requires that species listed in Annex I 'shall be subject of special conservation measures concerning their habitat in order to ensure their survival and reproduction in their area of distribution' and that 'Member States shall classify in particular the most suitable territories in number and size as special protection areas for the conservation of these species, taking into account their protection requirements in the geographical sea and land area where this Directive applies'.

Annex II

'The species referred to in Annex II/1may be hunted in the geographical sea and land area where the Directive applies'. 'Species referred to in Annex II/2 may be hunted only in Member States in respect of which they are indicated'.

Annex III

For species listed in Annex III/1, Member States shall not prohibit 'trade activities'. Member States may, for the species listed in Annex III/2, allow 'trade activities'. These activities are prohibited for all other species of naturally occurring wild birds in the European territory of EU Member States.

Convention on the Conservation of European Wildlife and Natural Habitats (Bern Convention)

Aim: to maintain populations of wild flora and fauna with particular emphasis on endangered and vulnerable species, including migratory species. Details: http://conventions.coe.int/Treaty/Commun/QueVoulezVous.asp?NT=104&CM=1&CL=ENG

This appendix incorporates all revisions made by the Standing Committee up to 1 March 2002.

The Convention lists protected species on four Appendices: Appendix II lists strictly protected fauna species, and Appendix III lists protected fauna species.

Emerald Network (under Bern Convention)

Aim: to develop an ecological network of 'areas of special conservation interest'. The Emerald Network falls under the Bern Convention, and involves all European Union Member States, several non-Community states and a number of African states. It is based on the same principles as the Natura 2000 network (as established by the EU), and represents its *de facto* extension to non-Community countries. Details: http://www.coe.int/t/e/Cultural_Co-operation/Environment/Nature_and_biological_diversity/Ecological_networks/The_Emerald_Network/

This appendix incorporates all revisions made by the Standing Committee up to 4 December 1998

The species indicated in this appendix have been identified as requiring specific habitat conservation measures.

Convention on the Conservation of Migratory Species of Wild Animals (CMS, or Bonn Convention)

Aim: to provide a framework for the conservation of migratory species and their habitats by means of, as appropriate, strict protection and the conclusion of international agreements. Details: http://www.cms.int/about/intro.htm

This appendix incorporates all amendments by the Conference of the Parties up to 23 December 2002.

Appendix I: *'Species in danger of extinction throughout all or major parts of their range'*

Parties to the Convention undertake to provide immediate protection to species included in Appendix I, and Range States should conserve and, where feasible and appropriate, restore those habitats of the species which are of importance in removing it from danger of extinction.

Appendix II: *'Species which would benefit from international cooperation in their conservation and management'*

Parties to the Convention shall 'conclude Agreements covering the conservation and management of migratory species included in Appendix III. Each Agreement should, where appropriate, provide for the 'maintenance of a network of suitable habitats appropriately disposed in relation to migratory routes'.

Agreement on the Conservation of African-Eurasian Migratory Waterbirds (AEWA, under CMS)

Aim: the conservation of African-Eurasian migratory waterbirds through coordinated measures to restore species to a favourable conservation status or to maintain them in such a status. The Agreement falls under the Bonn Convention. Details: http://www.unep-aewa.org/index2.html

This appendix incorporates all revisions made by the Meeting of the Parties up to 27 September 2002.

The species indicated in this appendix are those to which the agreement applies.

Convention on International Trade in Endangered Species of Wild Fauna and Flora (CITES)

Aim: to ensure that international trade in specimens of wild animals and plants does not threaten their survival. Details: http://www.cites.org/

The list below incorporates all amendments by the Conference of the Parties up to 16 October 2003.

Appendix I lists species that are the most endangered among CITES-listed animals and plants. These are threatened with extinction and CITES generally prohibits commercial international trade in specimens of these species.

Appendix II lists species that are not necessarily now threatened with extinction but that may become so unless trade is closely controlled. It also includes so-called "look-alike species", i.e. species of which the specimens in trade look like those of species listed for conservation reasons.

Species Action Plan

Aim: to identify the priorities for conservation action to halt the decline and restore the populations of Europe's most endangered bird species. SAPs are conservation documents that are based on the most up-to-date information available at the time of compilation about limiting factors, threats and actions needed. SAPs are 'endorsed' by various international treaties: the ORNIS Committee of the European Union, the Standing Committee of the Bern Convention, the Convention on Migratory Species (Bonn), and the African-Eurasian Migratory Waterbird Agreement (AEWA). Details: http://www.birdlife.org/action/science/species/esap/esap_intro.html

This appendix incorporates all species for which SAPs have been officially approved up to 2 February 2004.

Appendix 4. Additional attributes of all European species, including key cross-references.

										Page number in key publications					
Species name in this book (BiE2)	2004 SPEC	2004 ETS	1994 SPEC	1994 ETS	Habitat association	Migratory status	Generation length (years)	EURING code	European breeding range (km²)	BiE2 (2004)	BiE1 (1994)	EBPET (2000)	EBCC (1997)	BWPC (1998)	Species name in BiE1 (Tucker and Heath 1994)
Gavia stellata Red-throated Loon	3	(H)	3	V	4	3	7	00020	>2,000,000	28	56	12	1	3	***Gavia stellata* Red-throated Diver**
Gavia arctica Arctic Loon	3	(VU)	3	V	3	3	7	00030	>3,000,000	28	58	12	2	6	***Gavia arctica* Black-throated Diver**
Gavia immer Common Loon	—	(S)	—	(S)	4	3	7	00040	>100,000	29	—	12	4	8	***Gavia immer* Great Northern Diver**
Gavia adamsii Yellow-billed Loon	—	(S)	—	(S)	10	3	7	00050	—	29	—	12	5	10	***Gavia adamsii* White-billed Diver**
Tachybaptus ruficollis Little Grebe	—	S	—	S	3	2	4	00070	>4,000,000	30	—	12	6	14	*Tachybaptus ruficollis* Little Grebe
Podiceps cristatus Great Crested Grebe	—	S	—	S	3	3	5	00090	>5,000,000	30	—	13	8	17	*Podiceps cristatus* Great Crested Grebe
Podiceps grisegena Red-necked Grebe	—	S	—	S	3	3	5	00100	>3,000,000	31	—	13	10	20	*Podiceps grisegena* Red-necked Grebe
Podiceps auritus Horned Grebe	3	D	—	(S)	3	3	5	00110	>2,000,000	31	—	13	12	22	***Podiceps auritus* Slavonian Grebe**
Podiceps nigricollis Black-necked Grebe	—	S	—	S	3	3	5	00120	>2,000,000	32	—	13	14	24	*Podiceps nigricollis* Black-necked Grebe
Fulmarus glacialis Northern Fulmar	—	S	—	S	1	2	31	00220	>500,000	32	—	14	16	34	***Fulmarus glacialis* Fulmar**
Pterodroma madeira Zino's Petrel	1	(CR)	1	E	1	4	16	00263	<20	33	62	14	18	37	***Pterodroma madeira* Freira**
Pterodroma feae Fea's Petrel	1	VU	1	E	1	2	16	00262	<5,000	33	60	14	18	39	***Pterodroma feae* Gon-gon**
Bulweria bulwerii Bulwer's Petrel	3	(R)	3	V	1	4	24	00340	>20,000	34	64	14	19	41	*Bulweria bulwerii* Bulwer's Petrel
Calonectris diomedea Cory's Shearwater	2	(VU)	2	(V)	1	4	18	00360	>100,000	34	66	14	20	42	*Calonectris diomedea* Cory's Shearwater
Puffinus gravis Great Shearwater	NE	NE	—	—	1	3	18	00400	—	35	—	14	—	47	*Puffinus gravis* Great Shearwater
Puffinus griseus Sooty Shearwater	1	NE	—	—	1	3	18	00430	—	35	—	14	—	49	*Puffinus griseus* Sooty Shearwater
Puffinus puffinus Manx Shearwater	2	(L)	2	(L)	1	5	18	00461	>50,000	35	68	15	22	51	*Puffinus puffinus* Manx Shearwater
Puffinus mauretanicus Balearic Shearwater	1	CR	—	—	1	5	18	00463	<100	36	439	15	22	53	**Not included (recent split)**
Puffinus yelkouan Yelkouan Shearwater	—[E]	S	4	S	1	2	18	00462	>100,000	36	439	15	22	54	***Puffinus yelkouan* Mediterranean Shearwater**
Puffinus assimilis Little Shearwater	3	(R)	3	V	1	1	18	00480	>20,000	37	70	15	21	55	*Puffinus assimilis* Little Shearwater
Pelagodroma marina White-faced Storm-petrel	3	VU	3	L	1	2	16	00510	<5,000	37	72	15	26	60	***Pelagodroma marina* White-faced Petrel**
Hydrobates pelagicus European Storm-petrel	—[E]	(S)	2	(L)	1	3	14	00520	>100,000	38	74	15	24	62	***Hydrobates pelagicus* Storm Petrel**
Oceanodroma leucorhoa Leach's Storm-petrel	3	(L)	3	(L)	1	5	9	00550	>20,000	38	76	15	27	64	***Oceanodroma leucorhoa* Leach's Petrel**
Oceanodroma castro Band-rumped Storm-petrel	3	(R)	3	V	1	2	16	00580	>20,000	39	78	16	28	67	***Oceanodroma castro* Madeiran Petrel**
Morus bassanus Northern Gannet	—[E]	S	2	L	1	3	21	00710	>50,000	39	80	16	30	77	***Sula bassana* Gannet**
Phalacrocorax carbo Great Cormorant	—	S	—	S	9	2	11	00720	>1,000,000	40	—	16	34	80	***Phalacrocorax carbo* Cormorant**
Phalacrocorax aristotelis European Shag	—[E]	(S)	4	S	1	1	10	00800	>750,000	40	439	16	36	83	***Phalacrocorax aristotelis* Shag**
Phalacrocorax pygmeus Pygmy Cormorant	1	S	2	V	9	2	5	00820	>100,000	41	82	16	38	87	*Phalacrocorax pygmeus* Pygmy Cormorant
Pelecanus onocrotalus Great White Pelican	3	R	3	R	2	4	11	00880	>50,000	41	84	17	32	93	***Pelecanus onocrotalus* White Pelican**
Pelecanus crispus Dalmatian Pelican	1	R	1	V	9	2	11	00890	>50,000	42	86	17	33	96	*Pelecanus crispus* Dalmatian Pelican
Botaurus stellaris Great Bittern	3	H	3	(V)	3	2	4	00950	>3,000,000	42	88	17	40	101	***Botaurus stellaris* Bittern**
Ixobrychus minutus Little Bittern	3	(H)	3	(V)	3	5	<3.3	00980	>3,000,000	43	90	17	42	105	*Ixobrychus minutus* Little Bittern
Nycticorax nycticorax Black-crowned Night-heron	3	H	3	D	7	5	5	01040	>1,000,000	43	92	17	44	109	***Nycticorax nycticorax* Night Heron**
Ardeola ralloides Squacco Heron	3	(D)	3	V	3	5	5	01080	>500,000	44	94	18	39	113	*Ardeola ralloides* Squacco Heron
Bubulcus ibis Cattle Egret	—	S	—	S	7	2	5	01110	>100,000	44	—	18	54	116	*Bubulcus ibis* Cattle Egret
Egretta garzetta Little Egret	—	S	—	S	3	2	5	01190	>1,000,000	45	—	18	46	122	*Egretta garzetta* Little Egret
Casmerodius albus Great Egret	—	S	—	S	3	2	5	01210	>250,000	45	—	18	49	126	***Egretta alba* Great White Egret**
Ardea cinerea Grey Heron	—	S	—	S	3	2	5	01220	>5,000,000	46	—	19	50	128	*Ardea cinerea* Grey Heron
Ardea purpurea Purple Heron	3	(D)	3	V	3	5	5	01240	>1,000,000	46	96	19	52	132	*Ardea purpurea* Purple Heron
Ciconia nigra Black Stork	2	R	3	R	9	5	8	01310	>3,000,000	47	98	19	56	138	*Ciconia nigra* Black Stork
Ciconia ciconia White Stork	2	H	2	V	7	5	8	01340	>4,000,000	47	100	19	58	140	*Ciconia ciconia* White Stork
Plegadis falcinellus Glossy Ibis	3	(D)	3	D	9	5	5	01360	>100,000	48	102	20	55	144	*Plegadis falcinellus* Glossy Ibis
Geronticus eremita Northern Bald Ibis	1	CR	—	—	7	1	8	01400	<100	48	—	—	—	146	**Not included (no data)**
Platalea leucorodia Eurasian Spoonbill	2	R	2	E	2	4	9	01440	>250,000	49	104	20	60	149	***Platalea leucorodia* Spoonbill**
Phoenicopterus roseus Greater Flamingo	3	L	3	L	2	2	16	01470	>20,000	49	106	20	62	153	***Phoenicopterus ruber* Greater Flamingo**
Cygnus olor Mute Swan	—[E]	S	—	S	3	2	7	01520	>3,000,000	50	—	20	64	162	*Cygnus olor* Mute Swan
Cygnus columbianus Tundra Swan	3W	VU	3W	LW	4	3	9	01530	>100,000	50	108	21	63	166	***Cygnus columbianus* Bewick's Swan**
Cygnus cygnus Whooper Swan	—[E]W	S	4W	S	9	3	9	01540	>2,000,000	51	440	21	66	168	*Cygnus cygnus* Whooper Swan
Anser fabalis Bean Goose	—[E]W	S	—	S	4	3	7	01570	>1,000,000	51	—	21	68	170	*Anser fabalis* Bean Goose
Anser brachyrhynchus Pink-footed Goose	—[E]	S	4	S	9	3	7	01580	>100,000	52	440	21	70	173	*Anser brachyrhynchus* Pink-footed Goose
Anser albifrons Greater White-fronted Goose	—	S	—	S	4	3	7	01590	>20,000	52	—	22	71	175	***Anser albifrons* White-fronted Goose**
Anser erythropus Lesser White-fronted Goose	1	EN	1	V	4	3	7	01600	>100,000	53	110	22	74	178	*Anser erythropus* Lesser White-fronted Goose
Anser anser Greylag Goose	—	S	—	S	9	2	7	01610	>2,000,000	53	—	22	72	179	*Anser anser* Greylag Goose
Chen caerulescens Snow Goose	—	(S)	—	—	4	5	7	01630	>20,000	54	—	—	—	182	**Not included (no data)**
Branta canadensis Canada Goose	—	(S)	—	—	3	2	7	01660	>20,000	54	—	—	75	185	**Not included (no data)**
Branta leucopsis Barnacle Goose	—[E]	S	4/2	LW	4	3	7	01670	>100,000	55	112	22	76	187	*Branta leucopsis* Barnacle Goose
Branta bernicla Brent Goose	3W	VU	3	V	4	3	7	01680	>50,000	55	114	22	78	189	*Branta bernicla* Brent Goose
Branta ruficollis Red-breasted Goose	1W	VU	1	LW	10	3	7	01690	—	56	116	23	—	192	*Branta ruficollis* Red-breasted Goose
Tadorna ferruginea Ruddy Shelduck	3	(VU)	3	V	3	2	6	01710	>250,000	56	118	23	82	197	*Tadorna ferruginea* Ruddy Shelduck
Tadorna tadorna Common Shelduck	—	S	—	S	2	2	6	01730	>1,000,000	57	—	23	80	198	***Tadorna tadorna* Shelduck**
Anas penelope Eurasian Wigeon	—[E]W	S	—	S	2	2	<3.3	01790	>3,000,000	57	—	23	86	207	***Anas penelope* Wigeon**
Anas strepera Gadwall	3	(H)	3	V	9	2	<3.3	01820	>3,000,000	58	120	23	88	212	*Anas strepera* Gadwall
Anas crecca Common Teal	—	(S)	—	S	3	2	<3.3	01840	>6,000,000	58	—	24	90	215	***Anas crecca* Teal**

Appendix 4 ...continued. Additional attributes of all European species, including key cross-references.

Species name in this book (BiE2)	2004 SPEC	2004 ETS	1994 SPEC	1994 ETS	Habitat association	Migratory status	Generation length (years)	EURING code	European breeding range (km²)	Page number in key publications					Species name in BiE1 (Tucker and Heath 1994)
										BiE2 (2004)	BiE1 (1994)	EBPET (2000)	EBCC (1997)	BWPC (1998)	
Anas platyrhynchos Mallard	—	(S)	—	S	3	2	<3.3	01860	>8,000,000	59	—	24	92	218	*Anas platyrhynchos* Mallard
Anas acuta Northern Pintail	3	(D)	3	V	9	3	<3.3	01890	>1,000,000	59	122	24	94	223	***Anas acuta* Pintail**
Anas querquedula Garganey	3	(D)	3	V	9	5	<3.3	01910	>5,000,000	60	124	25	96	226	*Anas querquedula* Garganey
Anas clypeata Northern Shoveler	3	(D)	—	S	3	2	<3.3	01940	>5,000,000	60	—	25	98	230	***Anas clypeata* Shoveler**
Marmaronetta angustirostris Marbled Teal	1	(VU)	1	E	9	2	<3.3	01950	>50,000	61	126	25	83	232	***Marmaronetta angustirostris* Marbled Duck**
Netta rufina Red-crested Pochard	—	(S)	3	D	9	2	<3.3	01960	>750,000	61	128	26	100	236	*Netta rufina* Red-crested Pochard
Aythya ferina Common Pochard	2	(D)	4	S	3	2	<3.3	01980	>5,000,000	62	440	26	102	238	***Aythya ferina* Pochard**
Aythya nyroca Ferruginous Duck	1	(VU)	1	V	3	4	<3.3	02020	>1,000,000	62	130	26	104	242	*Aythya nyroca* Ferruginous Duck
Aythya fuligula Tufted Duck	3	(D)	—	S	3	2	<3.3	02030	>5,000,000	63	—	26	106	244	*Aythya fuligula* Tufted Duck
Aythya marila Greater Scaup	3W	EN	3W	LW	4	3	<3.3	02040	>1,000,000	63	132	26	108	247	***Aythya marila* Scaup**
Somateria mollissima Common Eider	—[E]	S	—	S	1	2	5	02060	>1,000,000	64	—	27	110	251	***Somateria mollissima* Eider**
Somateria spectabilis King Eider	—	(S)	—	S	1	3	5	02070	>100,000	64	—	27	112	253	*Somateria spectabilis* King Eider
Polysticta stelleri Steller's Eider	3W	L	1	LW	1	4	4	02090	<20,000	65	134	27	113	256	*Polysticta stelleri* Steller's Eider
Histrionicus histrionicus Harlequin Duck	3	(R)	3	V	1	2	4	02110	>100,000	65	136	28	114	259	*Histrionicus histrionicus* Harlequin Duck
Clangula hyemalis Long-tailed Duck	—	(S)	—	S	4	2	4	02120	>750,000	66	—	28	115	261	*Clangula hyemalis* Long-tailed Duck
Melanitta nigra Black Scoter	—	(S)	—	S	4	3	4	02130	>1,000,000	66	—	28	116	263	***Melanitta nigra* Common Scoter**
Melanitta fusca White-winged Scoter	3	(D)	3W	LW	4	3	4	02150	>1,000,000	67	138	28	117	266	***Melanitta fusca* Velvet Scoter**
Bucephala islandica Barrow's Goldeneye	3	VU	3	E	4	1	4	02170	<20,000	67	140	28	119	269	*Bucephala islandica* Barrow's Goldeneye
Bucephala clangula Common Goldeneye	—	(S)	—	S	3	3	4	02180	>3,000,000	68	—	29	120	272	***Bucephala clangula* Goldeneye**
Mergellus albellus Smew	3	(D)	3	V	5	3	4	02200	>1,000,000	68	142	29	122	274	***Mergus albellus* Smew**
Mergus serrator Red-breasted Merganser	—	(S)	—	S	1	2	4	02210	>2,000,000	69	—	29	124	277	*Mergus serrator* Red-breasted Merganser
Mergus merganser Common Merganser	—	(S)	—	S	3	2	4	02230	>3,000,000	69	—	29	126	279	***Mergus merganser* Goosander**
Oxyura leucocephala White-headed Duck	1	VU	1	E	3	2	<3.3	02260	>100,000	70	144	29	129	285	*Oxyura leucocephala* White-headed Duck
Pernis apivorus European Honey-buzzard	—[E]	(S)	4	S	5	5	9	02310	>5,000,000	70	441	30	130	290	***Pernis apivorus* Honey Buzzard**
Elanus caeruleus Black-winged Kite	3	R	3	V	7	1	5	02350	>100,000	71	146	30	153	294	***Elanus caeruleus* Black-shouldered Kite**
Milvus migrans Black Kite	3	(VU)	3	V	9	5	6	02380	>5,000,000	71	148	30	132	295	*Milvus migrans* Black Kite
Milvus milvus Red Kite	2	D	4	S	5	2	6	02390	>1,000,000	72	441	30	134	298	*Milvus milvus* Red Kite
Haliaeetus albicilla White-tailed Eagle	1	R	3	R	9	2	16	02430	>2,000,000	72	150	31	136	303	*Haliaeetus albicilla* White-tailed Eagle
Gypaetus barbatus Lammergeier	3	(VU)	3	E	8	1	15	02460	>250,000	73	152	31	138	308	*Gypaetus barbatus* Lammergeier
Neophron percnopterus Egyptian Vulture	3	EN	3	E	9	5	14	02470	>1,000,000	73	154	31	140	311	*Neophron percnopterus* Egyptian Vulture
Gyps fulvus Eurasian Griffon	—	S	3	R	9	2	16	02510	>1,000,000	74	156	31	141	315	***Gyps fulvus* Griffon Vulture**
Aegypius monachus Cinereous Vulture	1	R	3	V	6	1	16	02550	>500,000	74	158	31	142	321	***Aegypius monachus* Black Vulture**
Circaetus gallicus Short-toed Snake-eagle	3	(R)	3	R	6	5	13	02560	>3,000,000	75	160	32	144	324	***Circaetus gallicus* Short-toed Eagle**
Circus aeruginosus Western Marsh-harrier	—	S	—	S	3	2	6	02600	>5,000,000	75	—	32	146	328	***Circus aeruginosus* Marsh Harrier**
Circus cyaneus Northern Harrier	3	H	3	V	9	2	6	02610	>4,000,000	76	162	32	148	331	***Circus cyaneus* Hen Harrier**
Circus macrourus Pallid Harrier	1	(EN)	3	E	7	5	6	02620	>1,000,000	76	164	32	152	335	*Circus macrourus* Pallid Harrier
Circus pygargus Montagu's Harrier	—[E]	S	4	S	7	5	6	02630	>3,000,000	77	441	33	150	337	*Circus pygargus* Montagu's Harrier
Accipiter gentilis Northern Goshawk	—	S	—	S	5	2	6	02670	>7,000,000	77	—	33	154	342	***Accipiter gentilis* Goshawk**
Accipiter nisus Eurasian Sparrowhawk	—	S	—	S	9	2	4	02690	>7,000,000	78	—	33	156	345	***Accipiter nisus* Sparrowhawk**
Accipiter badius Shikra	—	(S)	—	—	9	4	4	02720	>20,000	78	—	—	—	348	**Not included (no data)**
Accipiter brevipes Levant Sparrowhawk	2	(VU)	2	R	9	5	4	02730	>500,000	79	166	34	158	349	*Accipiter brevipes* Levant Sparrowhawk
Buteo buteo Common Buzzard	—	S	—	S	9	2	8	02870	>7,000,000	79	168	34	160	353	***Buteo buteo* Buzzard**
Buteo rufinus Long-legged Buzzard	3	(VU)	3	(E)	7	2	8	02880	>500,000	80	168	34	159	359	*Buteo rufinus* Long-legged Buzzard
Buteo lagopus Rough-legged Hawk	—	(S)	—	S	4	3	8	02900	>1,000,000	80	—	34	162	360	***Buteo lagopus* Rough-legged Buzzard**
Aquila pomarina Lesser Spotted Eagle	2	(D)	3	R	5	5	11	02920	>1,000,000	81	170	34	164	363	*Aquila pomarina* Lesser Spotted Eagle
Aquila clanga Greater Spotted Eagle	1	EN	1	E	9	4	11	02930	>1,000,000	81	172	35	166	365	***Aquila clanga* Spotted Eagle**
Aquila nipalensis Steppe Eagle	3	(EN)	3	V	7	5	11	02942	>500,000	82	174	35	163	370	*Aquila nipalensis* Steppe Eagle
Aquila heliaca Imperial Eagle	1	R	1	E	7	2	11	02951	>1,000,000	82	176	35	168	371	*Aquila heliaca* Imperial Eagle
Aquila adalberti Spanish Imperial Eagle	1	(EN)	1	E	6	1	11	02952	>100,000	83	178	35	196	375	*Aquila adalberti* Spanish Imperial Eagle
Aquila chrysaetos Golden Eagle	3	R	3	R	9	2	11	02960	>3,000,000	83	180	35	170	376	*Aquila chrysaetos* Golden Eagle
Hieraaetus pennatus Booted Eagle	3	(R)	3	R	6	5	11	02980	>2,000,000	84	182	36	172	381	*Hieraaetus pennatus* Booted Eagle
Hieraaetus fasciatus Bonelli's Eagle	3	EN	3	E	6	1	11	02990	>750,000	84	184	36	174	383	***Hieraaetus fasciatus* Bonelli's Eagle**
Pandion haliaetus Osprey	3	R	3	R	3	5	9	03010	>3,000,000	85	186	36	176	386	*Pandion haliaetus* Osprey
Falco naumanni Lesser Kestrel	1	H	1	(V)	7	5	<3.3	03030	>1,000,000	85	188	36	178	391	*Falco naumanni* Lesser Kestrel
Falco tinnunculus Common Kestrel	3	D	3	D	7	2	<3.3	03040	>8,000,000	86	190	36	180	393	***Falco tinnunculus* Kestrel**
Falco vespertinus Red-footed Falcon	3	(VU)	3	V	7	5	<3.3	03070	>2,000,000	86	192	37	182	397	*Falco vespertinus* Red-footed Falcon
Falco columbarius Merlin	—	(S)	—	S	4	2	<3.3	03090	>3,000,000	87	—	37	184	400	*Falco columbarius* Merlin
Falco subbuteo Eurasian Hobby	—	(S)	—	S	9	5	5	03100	>6,000,000	87	—	37	186	404	***Falco subbuteo* Hobby**
Falco eleonorae Eleonora's Falcon	2	D	2	R	2	5	5	03110	>100,000	88	194	37	188	407	*Falco eleonorae* Eleonora's Falcon
Falco biarmicus Lanner Falcon	3	VU	3	(E)	7	1	5	03140	>250,000	88	196	38	189	411	***Falco biarmicus* Lanner**
Falco cherrug Saker Falcon	1	EN	3	E	7	2	5	03160	>500,000	89	198	38	190	414	***Falco cherrug* Saker**
Falco rusticolus Gyrfalcon	3	(R)	3	V	4	2	5	03180	>250,000	89	200	38	191	417	*Falco rusticolus* Gyrfalcon
Falco peregrinus Peregrine Falcon	—	S	3	R	9	2	5	03200	>3,000,000	90	202	38	192	419	***Falco peregrinus* Peregrine**

Appendix 4 ...continued. Additional attributes of all European species, including key cross-references.

Species name in this book (BiE2)	2004 SPEC	2004 ETS	1994 SPEC	1994 ETS	Habitat association	Migratory status	Generation length (years)	EURING code	European breeding range (km^2)	Page number in key publications: BiE2 (2004)	BiE1 (1994)	EBPET (2000)	EBCC (1997)	BWPC (1998)	Species name in BiE1 (Tucker and Heath 1994)
Falco pelegrinoides Barbary Falcon	—	S	—	S	10	2	5	03210	>20,000	90	—	38	—	424	*Falco pelegrinoides* Barbary Falcon
Bonasa bonasia Hazel Grouse	—	S	—	S	5	1	<3.3	03260	>4,000,000	91	—	39	194	428	*Bonasa bonasia* Hazel Grouse
Lagopus lagopus Willow Ptarmigan	—	S	—	S	4	1	<3.3	03290	>3,000,000	91	—	39	196	430	***Lagopus lagopus* Red/Willow Grouse**
Lagopus mutus Rock Ptarmigan	—	S	—	S	4	1	<3.3	03300	>1,000,000	92	—	39	198	434	***Lagopus mutus* Ptarmigan**
Tetrao tetrix Black Grouse	3	H	3	V	5	1	4	03320	>4,000,000	92	204	39	200	437	*Tetrao tetrix* Black Grouse
Tetrao mlokosiewiczi Caucasian Grouse	1	DD	2	Ins	8	1	4	03330	>100,000	93	206	40	202	440	***Tetrao mlokosiewiczi* Caucasian Black Grouse**
Tetrao urogallus Western Capercaillie	—	(S)	—	(S)	5	1	4	03350	>4,000,000	93	—	40	204	442	***Tetrao urogallus* Capercaillie**
Tetraogallus caucasicus Caucasian Snowcock	—^{E}	S	4	S	8	1	<3.3	03500	>50,000	94	442	40	202	449	*Tetraogallus caucasicus* Caucasian Snowcock
Tetraogallus caspius Caspian Snowcock	3	(VU)	3	Ins	8	1	<3.3	03510	>50,000	94	208	40	203	451	*Tetraogallus caspius* Caspian Snowcock
Alectoris chukar Chukar	3	(VU)	3	V	6	1	<3.3	03550	>1,000,000	95	210	40	206	452	*Alectoris chukar* Chukar
Alectoris graeca Rock Partridge	2	(D)	2	(V)	9	1	<3.3	03570	>500,000	95	212	40	207	453	*Alectoris graeca* Rock Partridge
Alectoris rufa Red-legged Partridge	2	(D)	2	V	7	1	<3.3	03580	>1,000,000	96	214	40	208	455	*Alectoris rufa* Red-legged Partridge
Alectoris barbara Barbary Partridge	3	(R)	3	(E)	6	1	<3.3	03590	>50,000	96	216	41	210	457	*Alectoris barbara* Barbary Partridge
Ammoperdix griseogularis See-see Partridge	3	VU	—	(S)	10	1	<3.3	03620	>20,000	97	—	41	—	459	*Ammoperdix griseogularis* See-see Partridge
Francolinus francolinus Black Francolin	3	D	3	V	6	1	<3.3	03640	>50,000	97	218	41	211	461	*Francolinus francolinus* Black Francolin
Perdix perdix Grey Partridge	3	VU	3	V	7	1	<3.3	03670	>5,000,000	98	220	41	212	464	*Perdix perdix* Grey Partridge
Coturnix coturnix Common Quail	3	(H)	3	V	7	5	<3.3	03700	>6,000,000	98	222	41	214	467	***Coturnix coturnix* Quail**
Phasianus colchicus Common Pheasant	—	(S)	—	S	9	1	<3.3	03940	>4,000,000	99	—	42	218	472	***Phasianus colchicus* Pheasant**
Turnix sylvatica Small Buttonquail	3	CR	3	E	6	1	<3.3	04000	<100	99	224	42	221	482	***Turnix sylvatica* Andalusian Hemipode**
Rallus aquaticus Water Rail	—	(S)	—	(S)	3	2	<3.3	04070	>5,000,000	100	—	42	222	484	*Rallus aquaticus* Water Rail
Porzana porzana Spotted Crake	—^{E}	(S)	4	S	3	5	<3.3	04080	>4,000,000	100	442	42	224	488	*Porzana porzana* Spotted Crake
Porzana parva Little Crake	—^{E}	(S)	4	(S)	3	5	<3.3	04100	>1,000,000	101	442	43	226	490	*Porzana parva* Little Crake
Porzana pusilla Baillon's Crake	3	(R)	3	R	3	5	<3.3	04110	>500,000	101	226	43	228	494	*Porzana pusilla* Baillon's Crake
Crex crex Corncrake	1	H	1	V	7	5	<3.3	04210	>4,000,000	102	228	43	230	496	*Crex crex* Corncrake
Gallinula chloropus Common Moorhen	—	S	—	S	3	2	<3.3	04240	>7,000,000	102	—	43	232	499	***Gallinula chloropus* Moorhen**
Porphyrio porphyrio Purple Swamphen	3	L	3	R	2	1	<3.3	04270	>100,000	103	230	44	234	504	***Porphyrio porphyrio* Purple Gallinule**
Fulica atra Common Coot	—	(S)	—	S	3	2	<3.3	04290	>7,000,000	103	—	44	236	506	***Fulica atra* Coot**
Fulica cristata Red-knobbed Coot	3	CR	3	E	3	1	<3.3	04310	<20,000	104	232	44	238	509	***Fulica cristata* Crested Coot**
Grus grus Common Crane	2	(H)	3	V	9	4	14	04330	>4,000,000	104	234	44	240	511	***Grus grus* Crane**
Grus virgo Demoiselle Crane	—	S	—	S	7	5	6	04410	>250,000	105	—	45	239	516	***Anthropoides virgo* Demoiselle Crane**
Tetrax tetrax Little Bustard	1	VU	2	V	7	2	5	04420	>750,000	105	236	45	242	519	*Tetrax tetrax* Little Bustard
Chlamydotis undulata Houbara Bustard	1	(VU)	3	(E)	7	1	6	04440	<20,000	106	238	45	—	525	*Chlamydotis undulata* Houbara Bustard
Otis tarda Great Bustard	1	VU	1	D	7	2	14	04460	>750,000	106	240	45	244	529	*Otis tarda* Great Bustard
Haematopus ostralegus Eurasian Oystercatcher	—^{E}	(S)	—	S	2	2	11	04500	>3,000,000	107	—	45	246	535	***Haematopus ostralegus* Oystercatcher**
Himantopus himantopus Black-winged Stilt	—	S	—	S	3	5	5	04550	>1,000,000	107	—	46	248	484	*Himantopus himantopus* Black-winged Stilt
Recurvirostra avosetta Pied Avocet	—	S	4/3W	LW	2	2	5	04560	>1,000,000	108	242	46	250	542	***Recurvirostra avosetta* Avocet**
Burhinus oedicnemus Eurasian Thick-knee	3	(VU)	3	V	7	2	9	04590	>2,000,000	108	244	46	252	546	***Burhinus oedicnemus* Stone-curlew**
Cursorius cursor Cream-coloured Courser	3	(EN)	3	V	7	2	4	04640	<20,000	109	246	46	—	552	*Cursorius cursor* Cream-coloured Courser
Glareola pratincola Collared Pratincole	3	D	3	E	9	5	<3.3	04650	>500,000	109	248	47	254	555	*Glareola pratincola* Collared Pratincole
Glareola nordmanni Black-winged Pratincole	1	EN	3	R	7	5	<3.3	04670	>250,000	110	250	47	255	557	*Glareola nordmanni* Black-winged Pratincole
Charadrius dubius Little Ringed Plover	—	(S)	—	(S)	3	5	<3.3	04690	>7,000,000	110	—	47	256	561	*Charadrius dubius* Little Ringed Plover
Charadrius hiaticula Common Ringed Plover	—^{E}	(S)	—	S	2	2	<3.3	04700	>2,000,000	111	—	47	258	564	***Charadrius hiaticula* Ringed Plover**
Charadrius alexandrinus Kentish Plover	3	(D)	3	D	2	2	<3.3	04770	>1,000,000	111	252	47	260	569	*Charadrius alexandrinus* Kentish Plover
Charadrius leschenaultii Greater Sand Plover	3	(EN)	3	(E)	7	5	5	04790	>20,000	112	254	48	262	574	*Charadrius leschenaultii* Greater Sand Plover
Charadrius asiaticus Caspian Plover	3	EN	3	(V)	7	5	5	04800	>50,000	112	255	48	263	575	*Charadrius asiaticus* Caspian Plover
Eudromias morinellus Eurasian Dotterel	—	(S)	—	(S)	4	4	5	04820	>500,000	113	—	48	264	577	***Charadrius morinellus* Dotterel**
Pluvialis apricaria Eurasian Golden-plover	—^{E}	(S)	4	S	4	3	4	04850	>2,000,000	113	443	48	266	581	***Pluvialis apricaria* Golden Plover**
Pluvialis squatarola Grey Plover	—	(S)	—	(S)	4	4	5	04860	>100,000	114	—	48	268	584	*Pluvialis squatarola* Grey Plover
Vanellus spinosus Spur-winged Lapwing	3	VU	3	(E)	2	2	5	04870	>50,000	114	256	48	269	587	***Hoplopterus spinosus* Spur-winged Plover**
Vanellus indicus Red-wattled Lapwing	3	(VU)	—	S	10	1	5	04900	<20,000	115	—	49	—	589	***Hoplopterus indicus* Red-wattled Plover**
Vanellus gregarius Sociable Lapwing	1	CR	1	E	7	5	5	04910	>100,000	115	258	49	270	590	***Chettusia gregaria* Sociable Plover**
Vanellus leucurus White-tailed Lapwing	—	S	—	(S)	3	2	5	04920	>20,000	116	—	49	271	592	***Chettusia leucura* White-tailed Plover**
Vanellus vanellus Northern Lapwing	2	VU	—	(S)	7	2	5	04930	>7,000,000	116	—	49	272	593	***Vanellus vanellus* Lapwing**
Calidris canutus Red Knot	3W	D	3W	LW	4	5	5	04960	>500,000	117	260	49	274	599	***Calidris canutus* Knot**
Calidris alba Sanderling	—	(S)	—	S	4	4	5	04970	>500,000	117	—	49	275	602	*Calidris alba* Sanderling
Calidris minuta Little Stint	—	(S)	—	(S)	4	5	<3.3	05010	>100,000	118	—	50	276	607	*Calidris minuta* Little Stint
Calidris temminckii Temminck's Stint	—	(S)	—	(S)	4	5	<3.3	05020	>500,000	118	—	50	277	610	*Calidris temminckii* Temminck's Stint
Calidris bairdii Baird's Sandpiper	—	(S)	—	—	4	5	5	05060	>20,000	119	—	—	—	614	**Not included (no data)**
Calidris ferruginea Curlew Sandpiper	NE	NE	—	—	10	5	5	05090	—	119	—	50	—	617	*Calidris ferruginea* Curlew Sandpiper
Calidris maritima Purple Sandpiper	—^{E}	(S)	4	(S)	4	3	5	05100	>500,000	120	443	50	279	618	*Calidris maritima* Purple Sandpiper
Calidris alpina Dunlin	3	(H)	3W	VW	4	3	5	05120	>1,000,000	120	262	50	280	620	*Calidris alpina* Dunlin
Limicola falcinellus Broad-billed Sandpiper	3	(D)	3	(V)	4	5	5	05140	>500,000	121	264	50	282	624	*Limicola falcinellus* Broad-billed Sandpiper
Philomachus pugnax Ruff	2	(D)	4	(S)	9	5	<3.3	05170	>2,000,000	121	443	51	284	628	*Philomachus pugnax* Ruff

Appendix 4 ...continued. Additional attributes of all European species, including key cross-references.

Species name in this book (BiE2)	2004 SPEC	2004 ETS	1994 SPEC	1994 ETS	Habitat association	Migratory status	Generation length (years)	EURING code	European breeding range (km²)	Page number in key publications: BiE2 (2004)	BiE1 (1994)	EBPET (2000)	EBCC (1997)	BWPC (1998)	Species name in BiE1 (Tucker and Heath 1994)
Lymnocryptes minimus Jack Snipe	3	(D)	3W	(V)W	4	3	<3.3	05180	>1,000,000	122	266	51	286	633	*Lymnocryptes minimus* Jack Snipe
Gallinago gallinago Common Snipe	3	(D)	—	(S)	9	2	<3.3	05190	>6,000,000	122	—	51	288	635	***Gallinago gallinago* Snipe**
Gallinago media Great Snipe	1	D	2	(V)	7	5	<3.3	05200	>2,000,000	123	268	51	290	638	*Gallinago media* Great Snipe
Gallinago stenura Pintail Snipe	—	(S)	—	S	10	5	<3.3	05210	>100,000	123	—	52	291	640	*Gallinago stenura* Pintail Snipe
Scolopax rusticola Eurasian Woodcock	3	(D)	3W	VW	5	2	<3.3	05290	>6,000,000	124	270	52	292	644	***Scolopax rusticola* Woodcock**
Limosa limosa Black-tailed Godwit	2	VU	2	V	7	4	5	05320	>2,000,000	124	272	52	294	647	*Limosa limosa* Black-tailed Godwit
Limosa lapponica Bar-tailed Godwit	—	(S)	3W	LW	4	3	5	05340	>100,000	125	274	53	296	650	*Limosa lapponica* Bar-tailed Godwit
Numenius phaeopus Whimbrel	—[E]	(S)	4	(S)	4	5	5	05380	>2,000,000	125	444	53	298	655	*Numenius phaeopus* Whimbrel
Numenius tenuirostris Slender-billed Curlew	1	NE	1	—	10	4	5	05400	—	126	276	53	765	657	*Numenius tenuirostris* Slender-billed Curlew
Numenius arquata Eurasian Curlew	2	D	3W	DW	2	4	5	05410	>4,000,000	126	278	53	300	658	***Numenius arquata* Curlew**
Tringa erythropus Spotted Redshank	3	(D)	—	S	4	4	<3.3	05450	>500,000	127	—	54	304	662	*Tringa erythropus* Spotted Redshank
Tringa totanus Common Redshank	2	D	2	D	2	2	<3.3	05460	>4,000,000	127	280	54	302	665	***Tringa totanus* Redshank**
Tringa stagnatilis Marsh Sandpiper	—	(S)	—	(S)	3	5	<3.3	05470	>500,000	128	—	54	305	668	*Tringa stagnatilis* Marsh Sandpiper
Tringa nebularia Common Greenshank	—	S	—	S	2	5	5	05480	>2,000,000	128	—	54	306	670	***Tringa nebularia* Greenshank**
Tringa ochropus Green Sandpiper	—	S	—	(S)	3	3	<3.3	05530	>4,000,000	129	—	55	308	674	*Tringa ochropus* Green Sandpiper
Tringa glareola Wood Sandpiper	3	H	3	D	4	5	<3.3	05540	>3,000,000	129	282	55	310	676	*Tringa glareola* Wood Sandpiper
Xenus cinereus Terek Sandpiper	—	(S)	—	(S)	4	5	<3.3	05550	>500,000	130	—	55	316	678	*Xenus cinereus* Terek Sandpiper
Actitis hypoleucos Common Sandpiper	3	(D)	—	S	3	5	5	05560	>6,000,000	130	—	55	312	680	*Actitis hypoleucos* Common Sandpiper
Arenaria interpres Ruddy Turnstone	—	(S)	—	S	2	4	5	05610	>500,000	131	—	55	314	686	***Arenaria interpres* Turnstone**
Phalaropus lobatus Red-necked Phalarope	—	(S)	—	(S)	4	5	<3.3	05640	>750,000	131	—	56	317	692	*Phalaropus lobatus* Red-necked Phalarope
Phalaropus fulicarius Grey Phalarope	—	S	—	(S)	4	4	<3.3	05650	>50,000	132	—	56	318	691	*Phalaropus fulicarius* Grey Phalarope
Stercorarius pomarinus Pomarine Jaeger	—	(S)	—	(S)	1	5	11	05660	>100,000	132	—	56	319	696	***Stercorarius pomarinus* Pomarine Skua**
Stercorarius parasiticus Parasitic Jaeger	—	(S)	—	(S)	1	4	11	05670	>750,000	133	—	56	320	699	***Stercorarius parasiticus* Arctic Skua**
Stercorarius longicaudus Long-tailed Jaeger	—	(S)	—	(S)	1	4	11	05680	>500,000	133	—	56	321	701	***Stercorarius longicaudus* Long-tailed Skua**
Catharacta skua Great Skua	—[E]	S	4	S	1	4	15	05690	>100,000	134	444	56	322	704	***Stercorarius skua* Great Skua**
Larus ichthyaetus Great Black-headed Gull	—	(S)	—	S	2	4	13	05730	>50,000	134	—	56	323	712	*Larus ichthyaetus* Great Black-headed Gull
Larus melanocephalus Mediterranean Gull	—[E]	S	4	S	2	3	6	05750	>250,000	135	444	57	324	714	*Larus melanocephalus* Mediterranean Gull
Larus minutus Little Gull	3	(H)	3	D	3	3	6	05780	>2,000,000	135	284	57	326	719	*Larus minutus* Little Gull
Larus ridibundus Common Black-headed Gull	—[E]	(S)	—	S	3	2	6	05820	>6,000,000	136	—	57	328	724	***Larus ridibundus* Black-headed Gull**
Larus genei Slender-billed Gull	3	L	—	(S)	1	2	6	05850	>100,000	137	—	57	331	729	*Larus genei* Slender-billed Gull
Larus audouinii Audouin's Gull	1	L	1	L	1	2	13	05880	>100,000	137	286	58	334	732	*Larus audouinii* Audouin's Gull
Larus canus Mew Gull	2	(H)	2	D	9	3	8	05900	>3,000,000	138	288	58	332	735	***Larus canus* Common Gull**
Larus fuscus Lesser Black-backed Gull	—[E]	S	4	S	1	3	11	05910	>1,000,000	138	445	58	336	737	*Larus fuscus* Lesser Black-backed Gull
Larus argentatus Herring Gull	—[E]	S	—	S	1	2	13	05920	>2,000,000	139	—	58	338	741	*Larus argentatus* Herring Gull
Larus cachinnans Yellow-legged Gull	—[E]	S	—	(S)	1	2	13	05927	>1,000,000	139	—	58	340	746	*Larus cachinnans* Yellow-legged Gull
Larus armenicus Armenian Gull	2	L	—	(S)	3	2	13	05929	>20,000	140	—	59	335	748	*Larus armenicus* Armenian Gull
Larus glaucoides Iceland Gull	—[E]	(S)	—	(S)	1	2	11	05980	>250,000	140	—	59	342	749	*Larus glaucoides* Iceland Gull
Larus hyperboreus Glaucous Gull	—	(S)	—	S	1	2	13	05990	>250,000	141	—	59	343	752	*Larus hyperboreus* Glaucous Gull
Larus marinus Great Black-backed Gull	—[E]	S	4	S	1	2	13	06000	>1,000,000	141	445	59	344	754	*Larus marinus* Great Black-backed Gull
Xema sabini Sabine's Gull	—	S	—	S	4	5	6	05790	>100,000	136	—	57	330	721	***Larus sabini* Sabine's Gull**
Rhodostethia rosea Ross's Gull	—	(S)	—	S	4	4	6	06010	<20,000	142	—	59	—	757	*Rhodostethia rosea* Ross's Gull
Rissa tridactyla Black-legged Kittiwake	—	(S)	—	S	1	3	10	06020	>500,000	142	—	59	346	758	***Rissa tridactyla* Kittiwake**
Pagophila eburnea Ivory Gull	3	(R)	3	(E)	1	1	12	06040	>50,000	143	290	60	348	761	*Pagophila eburnea* Ivory Gull
Sterna nilotica Gull-billed Tern	3	(VU)	3	(E)	9	5	9	06050	>100,000	143	292	60	349	764	***Gelochelidon nilotica* Gull-billed Tern**
Sterna caspia Caspian Tern	3	R	3	(E)	2	5	11	06060	>100,000	144	294	60	350	766	*Sterna caspia* Caspian Tern
Sterna bengalensis Lesser Crested-tern	—	(S)	—	—	1	2	9	06090	<20,000	144	—	—	354	772	**Not included (new breeder)**
Sterna sandvicensis Sandwich Tern	2	H	2	D	2	5	9	06110	>250,000	145	296	60	352	773	*Sterna sandvicensis* Sandwich Tern
Sterna dougallii Roseate Tern	3	R	3	E	2	4	9	06140	>100,000	145	298	60	354	777	*Sterna dougallii* Roseate Tern
Sterna hirundo Common Tern	—	S	—	S	9	5	9	06150	>5,000,000	146	—	60	356	779	*Sterna hirundo* Common Tern
Sterna paradisaea Arctic Tern	—	(S)	—	S	2	5	14	06160	>2,000,000	146	—	61	358	782	*Sterna paradisaea* Arctic Tern
Sterna albifrons Little Tern	3	D	3	D	2	5	8	06240	>2,000,000	147	300	61	360	790	*Sterna albifrons* Little Tern
Chlidonias hybrida Whiskered Tern	3	H	3	D	3	5	9	06260	>750,000	147	302	61	362	794	***Chlidonias hybridus* Whiskered Tern**
Chlidonias niger Black Tern	3	(H)	3	D	3	5	9	06270	>2,000,000	148	304	62	364	799	*Chlidonias niger* Black Tern
Chlidonias leucopterus White-winged Tern	—	(S)	—	S	3	5	9	06280	>1,000,000	148	—	62	366	796	***Chlidonias leucopterus* White-winged Black Tern**
Uria aalge Common Murre	—	(S)	—	S	1	3	16	06340	>250,000	149	—	62	368	806	***Uria aalge* Guillemot**
Uria lomvia Thick-billed Murre	3	(VU)	—	S	1	3	16	06350	>100,000	149	—	62	370	810	***Uria lomvia* Brunnich's Guillemot**
Alca torda Razorbill	—[E]	(S)	4	S	1	3	16	06360	>500,000	150	445	62	372	812	*Alca torda* Razorbill
Cepphus grylle Black Guillemot	2	H	2	D	1	3	9	06380	>750,000	150	306	63	374	815	*Cepphus grylle* Black Guillemot
Alle alle Dovekie	—	(S)	—	(S)	1	3	16	06470	>100,000	151	—	63	371	817	***Alle alle* Little Auk**
Fratercula arctica Atlantic Puffin	2	(H)	2	V	1	3	22	06540	>250,000	151	308	63	376	821	***Fratercula arctica* Puffin**
Pterocles orientalis Black-bellied Sandgrouse	3	(D)	3	V	7	1	<3.3	06610	>250,000	152	310	63	378	832	*Pterocles orientalis* Black-bellied Sandgrouse
Pterocles alchata Pin-tailed Sandgrouse	3	(D)	3	E	7	1	<3.3	06620	>100,000	152	312	63	379	834	*Pterocles alchata* Pin-tailed Sandgrouse

Appendix 4 ...continued. Additional attributes of all European species, including key cross-references.

Species name in this book (BiE2)	2004 SPEC	2004 ETS	1994 SPEC	1994 ETS	Habitat association	Migratory status	Generation length (years)	EURING code	European breeding range (km²)	Page number in key publications					Species name in BiE1 (Tucker and Heath 1994)
										BiE2 (2004)	BiE1 (1994)	EBPET (2000)	EBCC (1997)	BWPC (1998)	
Columba livia Rock Pigeon	—	(S)	—	S	9	1	<3.3	06650	>8,000,000	153	—	63	380	839	***Columba livia* Rock Dove**
Columba oenas Stock Pigeon	—[E]	S	4	S	9	2	<3.3	06680	>5,000,000	153	446	64	382	842	***Columba oenas* Stock Dove**
Columba palumbus Common Wood-pigeon	—[E]	S	4	S	7	2	<3.3	06700	>8,000,000	154	446	64	384	846	***Columba palumbus* Woodpigeon**
Columba trocaz Madeira Laurel Pigeon	1	(R)	1	V	6	1	4	06710	<500	154	314	64	386	848	***Columba trocaz* Long-toed Pigeon**
Columba bollii Dark-tailed Laurel Pigeon	1	(R)	1	V	6	1	4	06720	<2,500	155	316	65	—	848	***Columba bollii* Bolle's Laurel Pigeon**
Columba junoniae White-tailed Laurel Pigeon	1	EN	1	V	6	1	4	06730	<2,500	155	318	65	—	850	***Columba junoniae* Laurel Pigeon**
Streptopelia decaocto Eurasian Collared-dove	—	S	—	(S)	9	1	9	06840	>6,000,000	156	—	65	388	853	***Streptopelia decaocto* Collared Dove**
Streptopelia turtur European Turtle-dove	3	D	3	D	7	5	<3.3	06870	>7,000,000	156	320	65	390	856	***Streptopelia turtur* Turtle Dove**
Streptopelia senegalensis Laughing Dove	—	S	—	(S)	6	1	<3.3	06900	>20,000	157	—	66	387	860	*Streptopelia senegalensis* Laughing Dove
Clamator glandarius Great Spotted Cuckoo	—	(S)	—	S	6	5	<3.3	07160	>750,000	157	—	66	394	872	***Clamator glandarius* Great-spotted Cuckoo**
Cuculus canorus Common Cuckoo	—	S	—	S	9	5	<3.3	07240	>8,000,000	158	—	66	396	875	***Cuculus canorus* Cuckoo**
Cuculus saturatus Oriental Cuckoo	—	(S)	—	(S)	10	5	<3.3	07250	>750,000	158	—	66	395	878	*Cuculus saturatus* Oriental Cuckoo
Tyto alba Barn Owl	3	(D)	3	D	7	1	<3.3	07350	>3,000,000	159	322	66	398	886	*Tyto alba* Barn Owl
Otus brucei Pallid Scops-owl	3	CR	—	(S)	10	4	<3.3	07380	<20,000	159	—	67	—	890	***Otus brucei* Striated Scops Owl**
Otus scops Common Scops-owl	2	(H)	2	(D)	9	2	<3.3	07390	>3,000,000	160	324	67	400	891	***Otus scops* Scops Owl**
Bubo bubo Eurasian Eagle-owl	3	(H)	3	V	9	1	9	07440	>5,000,000	160	326	67	402	893	***Bubo bubo* Eagle Owl**
Ketupa zeylonensis Brown Fish-owl	3	CR	—	(S)	10	1	5	07470	<100	161	—	67	—	896	***Ketupa zeylonensis* Brown Fish Owl**
Nyctea scandiaca Snowy Owl	3	(R)	3	V	4	2	7	07490	>250,000	161	328	67	404	897	*Nyctea scandiaca* Snowy Owl
Surnia ulula Northern Hawk Owl	—	(S)	—	(S)	5	2	<3.3	07500	>2,000,000	162	—	67	405	899	***Surnia ulula* Hawk Owl**
Glaucidium passerinum Eurasian Pygmy-owl	—	S	—	(S)	5	1	<3.3	07510	>3,000,000	162	—	68	406	901	***Glaucidium passerinum* Pygmy Owl**
Athene noctua Little Owl	3	(D)	3	D	7	1	<3.3	07570	>5,000,000	163	330	68	408	903	*Athene noctua* Little Owl
Strix aluco Tawny Owl	—[E]	S	4	S	5	1	4	07610	>6,000,000	163	447	68	410	907	*Strix aluco* Tawny Owl
Strix uralensis Ural Owl	—	(S)	—	(S)	5	1	9	07650	>2,000,000	164	—	68	412	911	*Strix uralensis* Ural Owl
Strix nebulosa Great Grey Owl	—	(S)	—	S	5	1	5	07660	>2,000,000	164	—	69	414	913	*Strix nebulosa* Great Grey Owl
Asio otus Long-eared Owl	—	(S)	—	S	9	2	<3.3	07670	>6,000,000	165	—	69	416	915	*Asio otus* Long-eared Owl
Asio flammeus Short-eared Owl	3	(H)	3	(V)	7	2	<3.3	07680	>4,000,000	165	332	69	418	918	*Asio flammeus* Short-eared Owl
Aegolius funereus Boreal Owl	—	(S)	—	(S)	5	1	<3.3	07700	>3,000,000	166	—	69	420	923	***Aegolius funereus* Tengmalm's Owl**
Caprimulgus europaeus Eurasian Nightjar	2	(H)	2	(D)	9	5	4	07780	>6,000,000	166	334	70	422	929	***Caprimulgus europaeus* Nightjar**
Caprimulgus ruficollis Red-necked Nightjar	—	(S)	—	S	6	5	4	07790	>250,000	167	—	70	424	932	*Caprimulgus ruficollis* Red-necked Nightjar
Tachymarptis melba Alpine Swift	—	S	—	(S)	9	4	7	07980	>2,000,000	167	—	71	429	940	***Apus melba* Alpine Swift**
Apus unicolor Plain Swift	2	(R)	4	S	9	4	7	07940	>20,000	168	447	70	425	942	*Apus unicolor* Plain Swift
Apus apus Common Swift	—	(S)	—	S	9	5	7	07950	>8,000,000	168	—	70	426	943	***Apus apus* Swift**
Apus pallidus Pallid Swift	—	(S)	—	(S)	9	5	7	07960	>500,000	169	—	71	428	945	*Apus pallidus* Pallid Swift
Apus caffer White-rumped Swift	—	S	—	S	6	4	7	07990	>20,000	169	—	71	430	947	*Apus caffer* White-rumped Swift
Apus affinis Little Swift	3	(EN)	—	(S)	6	2	7	08000	>20,000	170	—	71	—	949	*Apus affinis* Little Swift
Halcyon smyrnensis White-throated Kingfisher	3	EN	—	(S)	3	1	<3.3	08270	>20,000	170	—	71	—	953	***Halcyon smyrnensis* White-breasted Kingfisher**
Alcedo atthis Common Kingfisher	3	H	3	D	3	2	<3.3	08310	>5,000,000	171	336	71	434	956	***Alcedo atthis* Kingfisher**
Ceryle rudis Pied Kingfisher	3	(CR)	—	(S)	3	1	<3.3	08330	>20,000	171	—	72	—	959	*Ceryle rudis* Pied Kingfisher
Merops persicus Blue-cheeked Bee-eater	—	(S)	—	(S)	10	5	<3.3	08390	>20,000	172	—	72	431	964	***Merops superciliosus* Blue-cheeked Bee-eater**
Merops apiaster European Bee-eater	3	(H)	3	D	9	5	<3.3	08400	>3,000,000	172	338	72	432	966	***Merops apiaster* Bee-eater**
Coracias garrulus European Roller	2	VU	2	(D)	9	5	5	08410	>3,000,000	173	340	72	436	970	***Coracias garrulus* Roller**
Upupa epops Eurasian Hoopoe	3	(D)	—	S	9	5	<3.3	08460	>5,000,000	173	—	72	438	976	***Upupa epops* Hoopoe**
Jynx torquilla Eurasian Wryneck	3	(D)	3	D	5	5	<3.3	08480	>6,000,000	174	342	73	440	980	***Jynx torquilla* Wryneck**
Picus canus Grey-faced Woodpecker	3	(H)	3	D	5	1	<3.3	08550	>3,000,000	174	344	73	442	983	***Picus canus* Grey-headed Woodpecker**
Picus viridis Eurasian Green Woodpecker	2	(H)	2	D	5	1	<3.3	08560	>5,000,000	175	346	73	444	986	***Picus viridis* Green Woodpecker**
Dryocopus martius Black Woodpecker	—	S	—	S	5	1	<3.3	08630	>5,000,000	175	—	74	446	989	*Dryocopus martius* Black Woodpecker
Dendrocopos major Great Spotted Woodpecker	—	S	—	S	9	1	<3.3	08760	>7,000,000	176	448	74	448	993	*Dendrocopos major* Great Spotted Woodpecker
Dendrocopos syriacus Syrian Woodpecker	—[E]	(S)	4	(S)	9	1	<3.3	08780	>2,000,000	176	447	74	450	996	*Dendrocopos syriacus* Syrian Woodpecker
Dendrocopos medius Middle Spotted Woodpecker	—[E]	(S)	4	S	5	1	<3.3	08830	>2,000,000	177	448	74	452	998	*Dendrocopos medius* Middle Spotted Woodpecker
Dendrocopos leucotos White-backed Woodpecker	—	(S)	—	S	5	1	<3.3	08840	>2,000,000	177	—	75	454	1000	*Dendrocopos leucotos* White-backed Woodpecker
Dendrocopos minor Lesser Spotted Woodpecker	—	(S)	—	S	5	1	<3.3	08870	>6,000,000	178	—	75	456	1002	*Dendrocopos minor* Lesser Spotted Woodpecker
Picoides tridactylus Three-toed Woodpecker	3	(H)	3	D	5	1	<3.3	08980	>2,000,000	178	348	75	458	1005	*Picoides tridactylus* Three-toed Woodpecker
Ammomanes deserti Desert Lark	3	(EN)	—	(S)	10	1	<3.3	09570	<20,000	179	—	76	—	1017	*Ammomanes deserti* Desert Lark
Chersophilus duponti Dupont's Lark	3	(H)	3	V	7	1	<3.3	09590	>50,000	179	350	76	460	1020	*Chersophilus duponti* Dupont's Lark
Melanocorypha calandra Calandra Lark	3	(D)	3	(D)	7	2	<3.3	09610	>2,000,000	180	352	76	461	1024	*Melanocorypha calandra* Calandra Lark
Melanocorypha bimaculata Bimaculated Lark	—	S	—	(S)	10	4	<3.3	09620	>500,000	180	—	76	766	1026	*Melanocorypha bimaculata* Bimaculated Lark
Melanocorypha leucoptera White-winged Lark	—[E]W	(S)	4W	(S)	7	4	<3.3	09650	>500,000	181	448	76	462	1027	*Melanocorypha leucoptera* White-winged Lark
Melanocorypha yeltoniensis Black Lark	3	EN	3	(V)	7	2	<3.3	09660	>500,000	181	354	76	766	1029	*Melanocorypha yeltoniensis* Black Lark
Calandrella brachydactyla Greater Short-toed Lark	3	D	3	V	7	5	<3.3	09680	>2,000,000	182	356	76	466	1032	***Calandrella brachydactyla* Short-toed Lark**
Calandrella rufescens Lesser Short-toed Lark	3	D	3	V	7	2	<3.3	09701	>500,000	182	358	76	467	1034	*Calandrella rufescens* Lesser Short-toed Lark

Appendix 4 ...continued. Additional attributes of all European species, including key cross-references.

Species name in this book (BiE2)	2004 SPEC	2004 ETS	1994 SPEC	1994 ETS	Habitat association	Migratory status	Generation length (years)	EURING code	European breeding range (km²)	Page number in key publications: BiE2 (2004)	BiE1 (1994)	EBPET (2000)	EBCC (1997)	BWPC (1998)	Species name in BiE1 (Tucker and Heath 1994)
Calandrella cheleensis Asian Short-toed Lark	3	(VU)	—	—	7	2	<3.3	09702	>20,000	183	—	—	—	1036	**Not included (recent split)**
Galerida cristata Crested Lark	3	(H)	3	(D)	7	1	<3.3	09720	>5,000,000	183	360	77	464	1037	*Galerida cristata* Crested Lark
Galerida theklae Thekla Lark	3	(H)	3	V	6	1	<3.3	09730	>250,000	184	362	77	463	1040	*Galerida theklae* Thekla Lark
Lullula arborea Wood Lark	2	H	2	V	5	2	<3.3	09740	>5,000,000	184	364	77	468	1041	***Lullula arborea*** **Woodlark**
Alauda arvensis Eurasian Skylark	3	(H)	3	V	7	2	<3.3	09760	>8,000,000	185	366	77	470	1043	***Alauda arvensis*** **Skylark**
Eremophila alpestris Horned Lark	—	(S)	—	(S)	4	2	<3.3	09780	>500,000	185	—	78	472	1047	***Eremophila alpestris*** **Shore Lark**
Riparia riparia Sand Martin	3	(H)	3	D	3	5	<3.3	09810	>7,000,000	186	368	78	474	1055	*Riparia riparia* Sand Martin
Hirundo rupestris Eurasian Crag-martin	—	S	—	S	6	2	<3.3	09910	>2,000,000	186	—	78	476	1059	***Ptyonoprogne rupestris*** **Crag Martin**
Hirundo rustica Barn Swallow	3	H	3	D	7	5	<3.3	09920	>8,000,000	187	370	78	478	1061	***Hirundo rustica*** **Swallow**
Hirundo daurica Red-rumped Swallow	—	(S)	—	S	6	5	<3.3	09950	>1,000,000	187	—	79	477	1064	*Hirundo daurica* Red-rumped Swallow
Delichon urbica Northern House-martin	3	(D)	—	S	9	5	<3.3	10010	>8,000,000	188	—	79	480	1066	***Delichon urbica*** **House Martin**
Anthus campestris Tawny Pipit	3	(D)	3	V	9	5	<3.3	10050	>3,000,000	188	372	79	482	1072	*Anthus campestris* Tawny Pipit
Anthus berthelotii Berthelot's Pipit	—[E]	(S)	4	S	9	1	<3.3	10060	>20,000	189	448	80	483	1075	*Anthus berthelotii* Berthelot's Pipit
Anthus hodgsoni Olive-backed Pipit	—	(S)	—	(S)	5	5	<3.3	10080	>100,000	189	—	80	484	1077	*Anthus hodgsoni* Olive-backed Pipit
Anthus trivialis Tree Pipit	—	S	—	S	5	5	<3.3	10090	>7,000,000	190	—	80	486	1079	*Anthus trivialis* Tree Pipit
Anthus gustavi Pechora Pipit	—	(S)	—	(S)	4	5	<3.3	10100	>20,000	190	—	80	—	1081	*Anthus gustavi* Pechora Pipit
Anthus pratensis Meadow Pipit	—[E]	(S)	4	S	9	2	<3.3	10110	>5,000,000	191	449	80	488	1082	*Anthus pratensis* Meadow Pipit
Anthus cervinus Red-throated Pipit	—	(S)	—	(S)	4	5	<3.3	10120	>250,000	191	—	80	485	1084	*Anthus cervinus* Red-throated Pipit
Anthus spinoletta Water Pipit	—	(S)	—	S	8	2	<3.3	10140	>1,000,000	192	—	80	490	1086	*Anthus spinoletta* Water Pipit
Anthus petrosus Rock Pipit	—[E]	(S)	—	S	2	2	<3.3	10142	>500,000	192	—	80	492	1089	*Anthus petrosus* Rock Pipit
Motacilla flava Yellow Wagtail	—	(S)	—	S	7	5	<3.3	10170	>7,000,000	193	—	81	494	1094	*Motacilla flava* Yellow Wagtail
Motacilla citreola Citrine Wagtail	—	(S)	—	(S)	4	5	<3.3	10180	>1,000,000	193	—	81	500	1098	*Motacilla citreola* Citrine Wagtail
Motacilla cinerea Grey Wagtail	—	S	—	(S)	3	2	<3.3	10190	>4,000,000	194	—	82	496	1100	*Motacilla cinerea* Grey Wagtail
Motacilla alba White Wagtail	—	S	—	S	9	2	<3.3	10200	>8,000,000	194	—	82	498	1103	***Motacilla alba*** **Pied Wagtail**
Pycnonotus xanthopygos White-spectacled Bulbul	—	S	—	(S)	10	1	<3.3	10360	>20,000	195	—	82	—	1110	***Pycnonotus xanthopygos*** **Yellow-vented Bulbul**
Bombycilla garrulus Bohemian Waxwing	—	(S)	—	(S)	5	2	<3.3	10480	>1,000,000	195	—	82	501	1113	***Bombycilla garrulus*** **Waxwing**
Cinclus cinclus White-throated Dipper	—	S	—	(S)	3	2	<3.3	10500	>4,000,000	196	—	82	502	1118	***Cinclus cinclus*** **Dipper**
Troglodytes troglodytes Winter Wren	—	S	—	S	9	2	<3.3	10660	>7,000,000	196	—	82	504	1122	***Troglodytes troglodytes*** **Wren**
Prunella modularis Hedge Accentor	—[E]	S	4	S	5	2	<3.3	10840	>6,000,000	197	449	83	506	1128	***Prunella modularis*** **Dunnock**
Prunella montanella Siberian Accentor	—	(S)	—	(S)	10	5	<3.3	10860	>100,000	197	—	83	508	1131	*Prunella montanella* Siberian Accentor
Prunella ocularis Radde's Accentor	—[E]	(S)	3	(V)	8	1	<3.3	10880	>20,000	198	374	84	767	1132	*Prunella ocularis* Radde's Accentor
Prunella atrogularis Black-throated Accentor	3	R	3	(V)	5	4	<3.3	10900	>50,000	198	376	84	509	1133	*Prunella atrogularis* Black-throated Accentor
Prunella collaris Alpine Accentor	—	(S)	—	S	8	2	<3.3	10940	>750,000	199	—	84	510	1134	*Prunella collaris* Alpine Accentor
Erythropygia galactotes Rufous-tailed Scrub-robin	3	VU	—	S	6	5	<3.3	10950	>500,000	199	—	84	511	1137	***Cercotrichas galactotes*** **Rufous Bush Robin**
Erithacus rubecula European Robin	—[E]	S	4	S	9	2	<3.3	10990	>8,000,000	200	450	84	512	1140	***Erithacus rubecula*** **Robin**
Luscinia luscinia Thrush Nightingale	—[E]	S	4	S	5	5	<3.3	11030	>4,000,000	200	450	84	514	1143	*Luscinia luscinia* Thrush Nightingale
Luscinia megarhynchos Common Nightingale	—[E]	(S)	4	(S)	5	5	<3.3	11040	>4,000,000	201	450	85	516	1145	***Luscinia megarhynchos*** **Nightingale**
Luscinia calliope Siberian Rubythroat	—	(S)	—	(S)	5	5	<3.3	11050	>50,000	201	—	85	518	1147	*Luscinia calliope* Siberian Rubythroat
Luscinia svecica Bluethroat	—	S	—	S	4	4	<3.3	11060	>4,000,000	202	—	85	520	1149	*Luscinia svecica* Bluethroat
Tarsiger cyanurus Orange-flanked Bush-robin	—	(S)	—	(S)	5	5	<3.3	11130	>250,000	202	—	85	519	1153	***Tarsiger cyanurus*** **Red-flanked Bluetail**
Irania gutturalis White-throated Robin	—	(S)	—	(S)	6	5	<3.3	11170	>250,000	203	—	85	767	1154	*Irania gutturalis* White-throated Robin
Phoenicurus ochruros Black Redstart	—	S	—	S	9	2	<3.3	11210	>5,000,000	203	—	86	522	1157	*Phoenicurus ochruros* Black Redstart
Phoenicurus phoenicurus Common Redstart	2	(H)	2	V	5	5	<3.3	11220	>7,000,000	204	378	86	524	1161	***Phoenicurus phoenicurus*** **Redstart**
Phoenicurus erythrogastrus White-winged Redstart	3	(R)	3	Ins	10	2	<3.3	11280	>20,000	204	389	86	768	1164	***Phoenicurus erythrogaster*** **Guldenstadt's Redstart**
Saxicola rubetra Whinchat	—[E]	(S)	4	S	7	5	<3.3	11370	>7,000,000	205	451	86	526	1167	*Saxicola rubetra* Whinchat
Saxicola dacotiae Fuerteventura Chat	1	EN	2	V	10	1	<3.3	11380	<1,000	205	380	87	—	1169	***Saxicola dacotiae*** **Canary Islands Stonechat**
Saxicola torquata Common Stonechat	—	(S)	3	(D)	9	2	<3.3	11390	>5,000,000	206	382	87	528	1170	***Saxicola torquata*** **Stonechat**
Oenanthe isabellina Isabelline Wheatear	—	(S)	—	(S)	7	4	<3.3	11440	>1,000,000	206	—	87	532	1175	*Oenanthe isabellina* Isabelline Wheatear
Oenanthe oenanthe Northern Wheatear	3	(D)	—	S	9	5	<3.3	11460	>8,000,000	207	—	87	530	1178	***Oenanthe oenanthe*** **Wheatear**
Oenanthe pleschanka Pied Wheatear	—	(S)	—	(S)	7	5	<3.3	11470	>500,000	207	—	88	533	1180	*Oenanthe pleschanka* Pied Wheatear
Oenanthe cypriaca Cyprus Wheatear	—[E]	(S)	2	R	6	5	<3.3	11475	>20,000	208	384	88	—	1183	***Oenanthe cypriaca*** **Cyprus Pied Wheatear**
Oenanthe hispanica Black-eared Wheatear	2	(H)	2	V	6	5	<3.3	11480	>1,000,000	208	386	88	534	1183	*Oenanthe hispanica* Black-eared Wheatear
Oenanthe deserti Desert Wheatear	—	(S)	—	(S)	7	2	<3.3	11490	>20,000	209	—	88	535	1185	*Oenanthe deserti* Desert Wheatear
Oenanthe finschii Finsch's Wheatear	—	(S)	—	(S)	10	2	<3.3	11500	>250,000	209	—	88	536	1187	*Oenanthe finschii* Finsch's Wheatear
Oenanthe xanthoprymna Rufous-tailed Wheatear	3	(VU)	—	(S)	10	5	<3.3	11520	>100,000	210	—	88	537	1190	***Oenanthe xanthoprymna*** **Red-tailed Wheatear**
Oenanthe leucura Black Wheatear	3	(R)	3	E	6	1	<3.3	11580	>250,000	210	388	88	538	1198	*Oenanthe leucura* Black Wheatear
Monticola saxatilis Rufous-tailed Rock-thrush	3	(H)	3	(D)	9	5	<3.3	11620	>1,000,000	211	390	88	540	1201	***Monticola saxatilis*** **Rock Thrush**
Monticola solitarius Blue Rock-thrush	3	(H)	3	(V)	6	2	<3.3	11660	>1,000,000	211	392	89	541	1204	***Monticola solitarius*** **Blue Rock Thrush**
Zoothera dauma Scaly Thrush	—	(S)	—	(S)	5	5	<3.3	11700	>100,000	212	—	89	539	1206	***Zoothera dauma*** **White's Thrush**
Turdus torquatus Ring Ouzel	—[E]	S	4	S	9	2	<3.3	11860	>1,000,000	212	451	89	542	1212	*Turdus torquatus* Ring Ouzel
Turdus merula Eurasian Blackbird	—[E]	S	4	S	9	2	<3.3	11870	>8,000,000	213	452	89	544	1215	***Turdus merula*** **Blackbird**

Appendix 4 ...continued. Additional attributes of all European species, including key cross-references.

Species name in this book (BiE2)	2004 SPEC	2004 ETS	1994 SPEC	1994 ETS	Habitat association	Migratory status	Generation length (years)	EURING code	European breeding range (km^2)	Page number in key publications: BiE2 (2004)	BiE1 (1994)	EBPET (2000)	EBCC (1997)	BWPC (1998)	Species name in BiE1 (Tucker and Heath 1994)
Turdus ruficollis Dark-throated Thrush	—	(S)	—	(S)	10	5	<3.3	11970	>100,000	213	—	90	554	1220	***Turdus ruficollis*** **Black-throated Thrush**
Turdus pilaris Fieldfare	—[E]W	(S)	4W	S	7	3	<3.3	11980	>5,000,000	214	452	90	546	1222	*Turdus pilaris* Fieldfare
Turdus philomelos Song Thrush	—[E]	S	4	S	9	2	<3.3	12000	>7,000,000	214	452	90	548	1225	*Turdus philomelos* Song Thrush
Turdus iliacus Redwing	—[E]W	(S)	4W	S	5	3	<3.3	12010	>4,000,000	215	453	90	550	1228	*Turdus iliacus* Redwing
Turdus viscivorus Mistle Thrush	—[E]	S	4	S	9	2	<3.3	12020	>5,000,000	215	453	91	552	1230	*Turdus viscivorus* Mistle Thrush
Cettia cetti Cetti's Warbler	—	S	—	S	3	1	<3.3	12200	>1,000,000	216	—	91	555	1235	*Cettia cetti* Cetti's Warbler
Cisticola juncidis Zitting Cisticola	—	S	—	(S)	3	1	<3.3	12260	>1,000,000	216	—	91	556	1237	***Cisticola juncidis*** **Fan-tailed Warbler**
Prinia gracilis Graceful Prinia	3	(VU)	—	(S)	10	1	<3.3	12270	>20,000	217	—	91	—	1239	***Prinia gracilis*** **Graceful Warbler**
Locustella lanceolata Lanceolated Warbler	—	(S)	—	(S)	10	5	<3.3	12350	>100,000	217	—	92	557	1244	*Locustella lanceolata* Lanceolated Warbler
Locustella naevia Common Grasshopper-warbler	—[E]	(S)	4	S	7	5	<3.3	12360	>5,000,000	218	453	92	558	1245	***Locustella naevia*** **Grasshopper Warbler**
Locustella fluviatilis Eurasian River Warbler	—[E]	(S)	4	S	5	5	<3.3	12370	>4,000,000	218	454	92	560	1247	***Locustella fluviatilis*** **River Warbler**
Locustella luscinioides Savi's Warbler	—[E]	(S)	4	(S)	3	5	<3.3	12380	>3,000,000	219	454	92	562	1249	*Locustella luscinioides* Savi's Warbler
Acrocephalus melanopogon Moustached Warbler	—	(S)	—	(S)	3	2	<3.3	12410	>250,000	219	—	92	568	1251	*Acrocephalus melanopogon* Moustached Warbler
Acrocephalus paludicola Aquatic Warbler	1	(VU)	1	E	7	5	<3.3	12420	>50,000	220	394	93	564	1254	*Acrocephalus paludicola* Aquatic Warbler
Acrocephalus schoenobaenus Sedge Warbler	—[E]	S	4	(S)	3	5	<3.3	12430	>6,000,000	220	455	93	566	1255	*Acrocephalus schoenobaenus* Sedge Warbler
Acrocephalus agricola Paddyfield Warbler	—	(S)	—	S	3	5	<3.3	12470	>250,000	221	—	93	569	1258	*Acrocephalus agricola* Paddyfield Warbler
Acrocephalus dumetorum Blyth's Reed-warbler	—	S	—	(S)	5	5	<3.3	12480	>1,000,000	221	—	93	576	1260	***Acrocephalus dumetorum*** **Blyth's Reed Warbler**
Acrocephalus palustris Marsh Warbler	—[E]	(S)	4	S	9	5	<3.3	12500	>5,000,000	222	455	93	570	1263	*Acrocephalus palustris* Marsh Warbler
Acrocephalus scirpaceus Common Reed-warbler	—[E]	S	4	S	3	5	<3.3	12510	>5,000,000	222	456	94	572	1265	***Acrocephalus scirpaceus*** **Reed Warbler**
Acrocephalus arundinaceus Great Reed-warbler	—	(S)	—	(S)	3	5	<3.3	12530	>5,000,000	223	—	94	574	1269	***Acrocephalus arundinaceus*** **Great Reed Warbler**
Hippolais pallida Olivaceous Warbler	3	(H)	3	(V)	9	5	<3.3	12550	>1,000,000	223	396	94	577	1273	*Hippolais pallida* Olivaceous Warbler
Hippolais caligata Booted Warbler	—	(S)	—	(S)	10	5	<3.3	12560	>500,000	224	—	95	578	1277	*Hippolais caligata* Booted Warbler
Hippolais languida Upcher's Warbler	—	S	—	(S)	10	5	<3.3	12570	>100,000	224	—	95	768	1278	*Hippolais languida* Upcher's Warbler
Hippolais olivetorum Olive-tree Warbler	—[E]	(S)	2	(R)	6	5	<3.3	12580	>250,000	225	398	95	579	1280	*Hippolais olivetorum* Olive-tree Warbler
Hippolais icterina Icterine Warbler	—[E]	(S)	4	S	5	5	<3.3	12590	>4,000,000	225	456	95	580	1282	*Hippolais icterina* Icterine Warbler
Hippolais polyglotta Melodious Warbler	—[E]	(S)	4	(S)	9	5	<3.3	12600	>1,000,000	226	456	95	582	1284	*Hippolais polyglotta* Melodious Warbler
Sylvia sarda Marmora's Warbler	—[E]	(S)	4	(S)	6	2	<3.3	12610	>50,000	226	457	95	584	1286	*Sylvia sarda* Marmora's Warbler
Sylvia undata Dartford Warbler	2	H	2	V	6	1	<3.3	12620	>1,000,000	227	400	96	585	1288	*Sylvia undata* Dartford Warbler
Sylvia conspicillata Spectacled Warbler	—	(S)	—	(S)	6	2	<3.3	12640	>500,000	227	—	96	586	1291	*Sylvia conspicillata* Spectacled Warbler
Sylvia cantillans Subalpine Warbler	—[E]	(S)	4	S	6	5	<3.3	12650	>1,000,000	228	457	96	587	1293	*Sylvia cantillans* Subalpine Warbler
Sylvia mystacea Ménétries's Warbler	—	(S)	—	(S)	6	4	<3.3	12660	>250,000	228	—	96	588	1295	*Sylvia mystacea* Ménétries's Warbler
Sylvia melanocephala Sardinian Warbler	—[E]	(S)	4	S	6	2	<3.3	12670	>1,000,000	229	457	96	589	1296	*Sylvia melanocephala* Sardinian Warbler
Sylvia melanothorax Cyprus Warbler	—[E]	(S)	2	R	6	2	<3.3	12680	>20,000	229	437	96	—	1299	*Sylvia melanothorax* Cyprus Warbler
Sylvia rueppelli Rüppell's Warbler	—[E]	(S)	4	(S)	6	5	<3.3	12690	>100,000	230	457	97	590	1300	***Sylvia rueppelli*** **Rüppell's Warbler**
Sylvia nana Desert Warbler	—	(S)	—	(S)	10	2	<3.3	12700	>50,000	230	—	97	769	1301	*Sylvia nana* Desert Warbler
Sylvia hortensis Orphean Warbler	3	H	3	V	6	5	<3.3	12720	>1,000,000	231	402	97	591	1305	*Sylvia hortensis* Orphean Warbler
Sylvia nisoria Barred Warbler	—[E]	S	4	(S)	7	5	<3.3	12730	>3,000,000	231	458	97	592	1306	*Sylvia nisoria* Barred Warbler
Sylvia curruca Lesser Whitethroat	—	S	—	S	9	5	<3.3	12740	>6,000,000	232	—	97	594	1308	*Sylvia curruca* Lesser Whitethroat
Sylvia communis Common Whitethroat	—[E]	S	4	S	7	5	<3.3	12750	>8,000,000	232	458	98	596	1310	***Sylvia communis*** **Whitethroat**
Sylvia borin Garden Warbler	—[E]	S	4	S	5	5	<3.3	12760	>7,000,000	233	459	98	598	1314	*Sylvia borin* Garden Warbler
Sylvia atricapilla Blackcap	—[E]	S	4	S	9	2	<3.3	12770	>7,000,000	233	459	98	600	1316	*Sylvia atricapilla* Blackcap
Phylloscopus trochiloides Greenish Warbler	—	S	—	(S)	5	5	<3.3	12930	>2,000,000	234	—	99	602	1320	*Phylloscopus trochiloides* Greenish Warbler
Phylloscopus borealis Arctic Warbler	—	S	—	(S)	5	5	<3.3	12950	>1,000,000	234	—	99	604	1322	*Phylloscopus borealis* Arctic Warbler
Phylloscopus inornatus Inornate Warbler	—	(S)	—	S	5	5	<3.3	13000	>100,000	235	—	99	605	1325	***Phylloscopus inornatus*** **Yellow-browed Warbler**
Phylloscopus bonelli Bonelli's Warbler	2	D	4	S	9	5	<3.3	13070	>1,000,000	235	460	99	606	1329	*Phylloscopus bonelli* Bonelli's Warbler
Phylloscopus sibilatrix Wood Warbler	2	D	4	(S)	5	5	<3.3	13080	>6,000,000	236	460	100	608	1333	*Phylloscopus sibilatrix* Wood Warbler
Phylloscopus sindianus Mountain Chiffchaff	3	D	—	(S)	10	2	<3.3	13100	>100,000	236	—	100	614	1336	*Phylloscopus sindianus* Mountain Chiffchaff
Phylloscopus collybita Common Chiffchaff	—	S	—	(S)	5	2	<3.3	13110	>7,000,000	237	—	100	610	1337	***Phylloscopus collybita*** **Chiffchaff**
Phylloscopus ibericus Iberian Chiffchaff	—[E]	(S)	—	—	6	2	<3.3	13115	>500,000	237	—	100	610	1337	**Not included (recent split)**
Phylloscopus canariensis Canary Islands Chiffchaff	—[E]	(S)	—	—	10	1	<3.3	13117	>20,000	238	—	100	610	1337	**Not included (recent split)**
Phylloscopus trochilus Willow Warbler	—	S	—	S	9	5	<3.3	13120	>6,000,000	238	—	100	612	1340	*Phylloscopus trochilus* Willow Warbler
Regulus regulus Goldcrest	—[E]	S	4	(S)	5	2	<3.3	13140	>6,000,000	239	460	101	624	1342	*Regulus regulus* Goldcrest
Regulus teneriffae Canary Islands Kinglet	—[E]	(S)	4	S	10	1	<3.3	13142	>20,000	239	461	101	—	1345	***Regulus teneriffae*** **Tenerife Goldcrest**
Regulus ignicapilla Firecrest	—[E]	(S)	4	S	9	2	<3.3	13150	>2,000,000	240	461	101	626	1346	***Regulus ignicapillus*** **Firecrest**
Muscicapa striata Spotted Flycatcher	3	H	3	D	5	5	<3.3	13350	>8,000,000	240	404	101	622	1349	*Muscicapa striata* Spotted Flycatcher
Ficedula parva Red-breasted Flycatcher	—	(S)	—	(S)	5	5	<3.3	13430	>4,000,000	241	—	102	616	1353	*Ficedula parva* Red-breasted Flycatcher
Ficedula semitorquata Semicollared Flycatcher	2	D	2	(E)	9	5	<3.3	13470	>500,000	241	406	102	615	1355	***Ficedula semitorquata*** **Semi-collared Flycatcher**
Ficedula albicollis Collared Flycatcher	—[E]	S	4	S	5	5	<3.3	13480	>1,000,000	242	461	102	618	1357	*Ficedula albicollis* Collared Flycatcher
Ficedula hypoleuca European Pied Flycatcher	—[E]	S	4	S	5	5	<3.3	13490	>5,000,000	242	462	102	620	1358	***Ficedula hypoleuca*** **Pied Flycatcher**
Panurus biarmicus Bearded Parrotbill	—	(S)	—	(S)	3	1	<3.3	13640	>1,000,000	243	—	103	628	1362	***Panurus biarmicus*** **Bearded Tit**

Appendix 4 ...continued. Additional attributes of all European species, including key cross-references.

Species name in this book (BiE2)	2004 SPEC	2004 ETS	1994 SPEC	1994 ETS	Habitat association	Migratory status	Generation length (years)	EURING code	European breeding range (km^2)	Page number in key publications: BiE2 (2004)	BiE1 (1994)	EBPET (2000)	EBCC (1997)	BWPC (1998)	Species name in BiE1 (Tucker and Heath 1994)
Aegithalos caudatus Long-tailed Tit	—	S	—	S	9	1	<3.3	14370	>7,000,000	243	—	103	630	1372	*Aegithalos caudatus* Long-tailed Tit
Parus palustris Marsh Tit	3	D	—	S	5	1	<3.3	14400	>4,000,000	244	—	103	632	1375	*Parus palustris* Marsh Tit
Parus lugubris Sombre Tit	—[E]	(S)	4	(S)	6	1	<3.3	14410	>750,000	244	462	103	636	1378	*Parus lugubris* Sombre Tit
Parus montanus Willow Tit	—	S	—	(S)	5	1	<3.3	14420	>5,000,000	245	—	103	634	1379	*Parus montanus* Willow Tit
Parus cinctus Siberian Tit	—	(S)	—	(S)	5	1	<3.3	14480	>750,000	245	—	104	637	1382	*Parus cinctus* Siberian Tit
Parus cristatus Crested Tit	2	(D)	4	S	9	1	<3.3	14540	>5,000,000	246	462	104	638	1383	*Parus cristatus* Crested Tit
Parus ater Coal Tit	—	(S)	—	S	5	1	<3.3	14610	>6,000,000	246	—	105	640	1385	*Parus ater* Coal Tit
Parus caeruleus Blue Tit	—[E]	S	4	S	5	1	<3.3	14620	>7,000,000	247	463	105	642	1388	*Parus caeruleus* Blue Tit
Parus cyanus Azure Tit	—	(S)	—	(S)	5	1	<3.3	14630	>500,000	247	—	105	646	1392	*Parus cyanus* Azure Tit
Parus major Great Tit	—	S	—	S	9	1	<3.3	14640	>8,000,000	248	—	105	644	1393	*Parus major* Great Tit
Sitta krueperi Krüper's Nuthatch	2	(D)	4	(S)	6	1	<3.3	14690	>250,000	248	463	106	650	1398	***Sitta krueperi* Krüper's Nuthatch**
Sitta whiteheadi Corsican Nuthatch	2	R	2	V	6	1	<3.3	14700	<500	249	408	106	650	1399	*Sitta whiteheadi* Corsican Nuthatch
Sitta europaea Wood Nuthatch	—	S	—	S	5	1	<3.3	14790	>6,000,000	249	—	106	648	1402	***Sitta europaea* Nuthatch**
Sitta tephronota Eastern Rock-nuthatch	—	(S)	—	(S)	10	1	<3.3	14800	>100,000	250	—	106	770	1404	***Sitta tephronota* Great Rock Nuthatch**
Sitta neumayer Western Rock-nuthatch	—[E]	(S)	4	(S)	6	1	<3.3	14810	>1,000,000	250	463	106	647	1406	***Sitta neumayer* Rock Nuthatch**
Tichodroma muraria Wallcreeper	—	(S)	—	(S)	8	2	<3.3	14820	>750,000	251	—	107	651	1408	*Tichodroma muraria* Wallcreeper
Certhia familiaris Eurasian Tree-creeper	—	S	—	S	5	1	<3.3	14860	>5,000,000	251	—	107	652	1411	***Certhia familiaris* Treecreeper**
Certhia brachydactyla Short-toed Tree-creeper	—[E]	(S)	4	S	5	1	<3.3	14870	>3,000,000	252	464	107	654	1414	***Certhia brachydactyla* Short-toed Treecreeper**
Remiz pendulinus Eurasian Penduline-tit	—	(S)	—	(S)	3	2	<3.3	14900	>3,000,000	252	—	107	656	1416	***Remiz pendulinus* Penduline Tit**
Oriolus oriolus Eurasian Golden-oriole	—	S	—	S	5	5	<3.3	15080	>7,000,000	253	—	108	658	1424	***Oriolus oriolus* Golden Oriole**
Lanius collurio Red-backed Shrike	3	(H)	3	(D)	7	5	<3.3	15150	>7,000,000	253	410	108	660	1433	*Lanius collurio* Red-backed Shrike
Lanius minor Lesser Grey Shrike	2	(D)	2	(D)	7	5	<3.3	15190	>3,000,000	254	412	108	662	1436	*Lanius minor* Lesser Grey Shrike
Lanius excubitor Great Grey Shrike	3	(H)	3	D	9	2	<3.3	15200	>4,000,000	254	414	108	664	1440	*Lanius excubitor* Great Grey Shrike
Lanius senator Woodchat Shrike	2	(D)	2	V	7	5	<3.3	15230	>2,000,000	255	416	109	666	1445	*Lanius senator* Woodchat Shrike
Lanius nubicus Masked Shrike	2	(D)	2	(V)	6	5	<3.3	15240	>500,000	255	418	109	668	1447	*Lanius nubicus* Masked Shrike
Garrulus glandarius Eurasian Jay	—	S	—	(S)	5	1	4	15390	>8,000,000	256	—	109	670	1450	***Garrulus glandarius* Jay**
Perisoreus infaustus Siberian Jay	3	(H)	3	(D)	5	1	4	15430	>1,000,000	256	436	109	669	1454	*Perisoreus infaustus* Siberian Jay
Cyanopica cyanus Azure-winged Magpie	—	(S)	—	S	6	1	<3.3	15470	>250,000	257	—	110	674	1456	*Cyanopica cyanus* Azure-winged Magpie
Pica pica Black-billed Magpie	—	S	—	S	9	1	4	15490	>8,000,000	257	—	110	672	1457	***Pica pica* Magpie**
Nucifraga caryocatactes Spotted Nutcracker	—	S	—	(S)	5	1	5	15570	>2,000,000	258	—	110	678	1460	***Nucifraga caryocatactes* Nutcracker**
Pyrrhocorax graculus Yellow-billed Chough	—	(S)	—	(S)	8	1	7	15580	>750,000	258	—	110	675	1464	***Pyrrhocorax graculus* Alpine Chough**
Pyrrhocorax pyrrhocorax Red-billed Chough	3	D	3	V	9	1	7	15590	>1,000,000	259	420	111	676	1466	***Pyrrhocorax pyrrhocorax* Chough**
Corvus monedula Eurasian Jackdaw	—[E]	(S)	4	(S)	9	2	5	15600	>7,000,000	259	464	111	680	1468	***Corvus monedula* Jackdaw**
Corvus frugilegus Rook	—	(S)	—	S	7	2	5	15630	>5,000,000	260	—	111	682	1475	*Corvus frugilegus* Rook
Corvus corone Carrion Crow	—	S	—	S	9	1	5	15670	>8,000,000	260	—	111	684	1478	***Corvus corone* Carion Crow**
Corvus corax Common Raven	—	S	—	(S)	9	1	7	15720	>7,000,000	261	—	112	686	1483	***Corvus corax* Raven**
Sturnus vulgaris Common Starling	3	D	—	S	7	2	<3.3	15820	>7,000,000	261	—	112	688	1492	***Sturnus vulgaris* Starling**
Sturnus unicolor Spotless Starling	—[E]	S	4	S	9	1	<3.3	15830	>500,000	262	465	112	690	1496	*Sturnus unicolor* Spotless Starling
Sturnus roseus Rosy Starling	—	S	—	(S)	7	5	<3.3	15840	>1,000,000	262	—	113	691	1498	***Sturnus roseus* Rose-coloured Starling**
Passer domesticus House Sparrow	3	D	—	S	9	1	<3.3	15910	>8,000,000	263	—	113	694	1503	*Passer domesticus* House Sparrow
Passer hispaniolensis Spanish Sparrow	—	(S)	—	(S)	9	2	<3.3	15920	>1,000,000	263	—	113	696	1506	*Passer hispaniolensis* Spanish Sparrow
Passer moabiticus Dead Sea Sparrow	—	(S)	—	(S)	10	1	<3.3	15950	>20,000	264	—	113	—	1509	*Passer moabiticus* Dead Sea Sparrow
Passer montanus Eurasian Tree Sparrow	3	(D)	—	S	7	2	<3.3	15980	>7,000,000	264	—	113	692	1513	***Passer montanus* Tree Sparrow**
Petronia brachydactyla Pale Rock-finch	—	S	—	(S)	10	4	<3.3	16010	>20,000	265	—	114	—	1517	***Carpospiza brachydactyla* Pale Rock Sparrow**
Petronia xanthocollis Chestnut-shouldered Petronia	—	(S)	—	(S)	10	5	<3.3	16020	>20,000	265	—	114	—	1518	***Petronia xanthocollis* Yellow-throated Sparrow**
Petronia petronia Rock Sparrow	—	(S)	—	S	6	1	<3.3	16040	>1,000,000	266	—	114	699	1519	*Petronia petronia* Rock Sparrow
Montifringilla nivalis White-winged Snowfinch	—	(S)	—	(S)	8	1	<3.3	16110	>500,000	266	—	114	700	1522	***Montifringilla nivalis* Snowfinch**
Fringilla coelebs Chaffinch	—[E]	S	4	S	9	2	<3.3	16360	>8,000,000	267	465	114	702	1539	*Fringilla coelebs* Chaffinch
Fringilla teydea Blue Chaffinch	1	R	1	V	10	1	4	16370	<500	267	422	115	—	1544	***Fringilla teydea* Canary Islands Chaffinch**
Fringilla montifringilla Brambling	—	S	—	S	5	3	<3.3	16380	>2,000,000	268	—	115	704	1545	*Fringilla montifringilla* Brambling
Serinus pusillus Fire-fronted Serin	—	(S)	—	(S)	8	2	<3.3	16390	>250,000	268	—	115	707	1548	***Serinus pusillus* Red-fronted Serin**
Serinus serinus European Serin	—[E]	S	4	S	9	2	<3.3	16400	>4,000,000	269	465	115	708	1550	***Serinus serinus* Serin**
Serinus canaria Island Canary	—[E]	(S)	4	S	7	1	<3.3	16420	>20,000	269	466	115	710	1554	***Serinus canaria* Canary**
Serinus citrinella Citril Finch	—[E]	(S)	4	S	5	2	<3.3	16441	>250,000	270	466	116	711	1556	*Serinus citrinella* Citril Finch
Serinus corsicana Corsican Finch	—[E]	(S)	—	—	6	1	<3.3	16442	>50,000	270	466	116	711	1556	**Not included (recent split)**
Carduelis chloris European Greenfinch	—[E]	S	4	S	9	2	<3.3	16490	>7,000,000	271	466	116	712	1557	***Carduelis chloris* Greenfinch**
Carduelis carduelis European Goldfinch	—	S	—	(S)	7	2	<3.3	16530	>7,000,000	271	—	116	714	1561	***Carduelis carduelis* Goldfinch**
Carduelis spinus Eurasian Siskin	—[E]	S	4	S	5	2	<3.3	16540	>5,000,000	272	467	116	716	1564	***Carduelis spinus* Siskin**
Carduelis cannabina Eurasian Linnet	2	D	4	S	9	2	<3.3	16600	>7,000,000	272	467	117	718	1568	***Carduelis cannabina* Linnet**
Carduelis flavirostris Twite	—	S	—	S	4	2	<3.3	16620	>750,000	273	—	117	720	1571	*Carduelis flavirostris* Twite
Carduelis flammea Common Redpoll	—	(S)	—	(S)	5	2	<3.3	16630	>2,000,000	273	—	117	722	1574	***Carduelis flammea* Redpoll**
Carduelis hornemanni Hoary Redpoll	—	(S)	—	(S)	4	2	<3.3	16640	>250,000	274	—	118	724	1578	***Carduelis hornemanni* Arctic Redpoll**

Appendix 4 ...continued. Additional attributes of all European species, including key cross-references.

										Page number in key publications					
Species name in this book (BiE2)	2004 SPEC	2004 ETS	1994 SPEC	1994 ETS	Habitat association	Migratory status	Generation length (years)	EURING code	European breeding range (km^2)	BiE2 (2004)	BiE1 (1994)	EBPET (2000)	EBCC (1997)	BWPC (1998)	Species name in BiE1 (Tucker and Heath 1994)
Loxia leucoptera Two-barred Crossbill	—	(S)	—	(S)	5	1	<3.3	16650	>750,000	274	—	118	725	1580	*Loxia leucoptera* Two-barred Crossbill
Loxia curvirostra Red Crossbill	—	(S)	—	S	5	1	<3.3	16660	>5,000,000	275	—	118	726	1582	***Loxia curvirostra* Crossbill**
Loxia scotica Scottish Crossbill	1	DD	1	Ins	5	1	<3.3	16670	>20,000	275	424	118	730	1586	*Loxia scotica* Scottish Crossbill
Loxia pytyopsittacus Parrot Crossbill	—[E]	(S)	4	S	5	1	<3.3	16680	>2,000,000	276	468	118	728	1587	*Loxia pytyopsittacus* Parrot Crossbill
Rhodopechys sanguinea Crimson-winged Finch	—	S	—	(S)	8	2	<3.3	16730	>250,000	276	—	118	772	1589	*Rhodopechys sanguinea* Crimson-winged Finch
Rhodopechys obsoleta Desert Finch	—	(S)	—	(S)	10	1	<3.3	16740	>20,000	277	—	119	—	1591	***Rhodospiza obsoleta* Desert Finch**
Rhodopechys mongolica Mongolian Finch	—	(S)	—	(S)	10	2	<3.3	16750	>20,000	277	—	119	734	1592	***Bucanetes mongolicus* Mongolian Trumpeter Finch**
Bucanetes githagineus Trumpeter Finch	—	(S)	3	R	7	1	<3.3	16760	>20,000	278	426	119	735	1594	*Bucanetes githagineus* Trumpeter Finch
Carpodacus erythrinus Common Rosefinch	—	(S)	—	(S)	9	5	<3.3	16790	>4,000,000	278	—	119	732	1596	***Carpodacus erythrinus* Scarlet Rosefinch**
Carpodacus rubicilla Great Rosefinch	3	(R)	3	(E)	8	2	<3.3	16960	>20,000	279	438	119	731	1600	*Carpodacus rubicilla* Great Rosefinch
Pinicola enucleator Pine Grosbeak	—	(S)	—	S	5	2	<3.3	16990	>750,000	279	—	119	736	1602	*Pinicola enucleator* Pine Grosbeak
Pyrrhula pyrrhula Eurasian Bullfinch	—	(S)	—	S	5	2	<3.3	17100	>6,000,000	280	—	119	738	1606	***Pyrrhula pyrrhula* Bullfinch**
Pyrrhula murina Azores Bullfinch	1	(EN)	—	—	10	1	<3.3	17105	<50	280	—	120	737	1609	**Not included (recent split)**
Coccothraustes coccothraustes Hawfinch	—	S	—	S	5	2	<3.3	17170	>5,000,000	281	—	120	740	1610	*Coccothraustes coccothraustes* Hawfinch
Calcarius lapponicus Lapland Longspur	—	(S)	—	(S)	4	3	<3.3	18470	>500,000	281	—	120	744	1640	***Calcarius lapponicus* Lapland Bunting**
Plectrophenax nivalis Snow Bunting	—	(S)	—	(S)	4	3	<3.3	18500	>750,000	282	—	120	742	1642	*Plectrophenax nivalis* Snow Bunting
Emberiza leucocephalos Pine Bunting	—	(S)	—	—	5	5	<3.3	18560	>20,000	282	—	—	745	1646	**Not included (no data)**
Emberiza citrinella Yellowhammer	—[E]	(S)	4	(S)	7	2	<3.3	18570	>7,000,000	283	468	120	746	1648	*Emberiza citrinella* Yellowhammer
Emberiza cirlus Cirl Bunting	—[E]	S	4	(S)	9	1	<3.3	18580	>2,000,000	283	468	121	748	1651	*Emberiza cirlus* Cirl Bunting
Emberiza cia Rock Bunting	3	(H)	3	V	9	2	<3.3	18600	>2,000,000	284	428	121	749	1653	*Emberiza cia* Rock Bunting
Emberiza cineracea Cinereous Bunting	1	(R)	2	(V)	6	4	<3.3	18650	>20,000	284	430	121	750	1658	*Emberiza cineracea* Cinereous Bunting
Emberiza hortulana Ortolan Bunting	2	(H)	2	(V)	7	5	<3.3	18660	>5,000,000	285	432	121	752	1659	*Emberiza hortulana* Ortolan Bunting
Emberiza buchanani Grey-necked Bunting	—	(S)	—	(S)	10	5	<3.3	18670	>20,000	285	—	121	772	1662	*Emberiza buchanani* Grey-necked Bunting
Emberiza caesia Cretzschmar's Bunting	—[E]	(S)	4	(S)	6	4	<3.3	18680	>500,000	286	469	121	751	1663	*Emberiza caesia* Cretzschmar's Bunting
Emberiza rustica Rustic Bunting	—	(S)	—	(S)	5	5	<3.3	18730	>2,000,000	286	—	121	754	1665	*Emberiza rustica* Rustic Bunting
Emberiza pusilla Little Bunting	—	(S)	—	(S)	4	5	<3.3	18740	>1,000,000	287	—	121	755	1667	*Emberiza pusilla* Little Bunting
Emberiza aureola Yellow-breasted Bunting	1	D	—	(S)	7	5	<3.3	18760	>1,000,000	287	—	121	756	1670	*Emberiza aureola* Yellow-breasted Bunting
Emberiza schoeniclus Reed Bunting	—	S	—	S	3	2	<3.3	18770	>7,000,000	288	—	121	758	1672	*Emberiza schoeniclus* Reed Bunting
Emberiza pallasi Pallas's Bunting	—	(S)	—	—	9	5	<3.3	18780	>20,000	288	—	—	757	1676	**Not included (no data)**
Emberiza bruniceps Red-headed Bunting	—	(S)	—	S	7	5	<3.3	18800	>20,000	289	—	121	760	1677	*Emberiza bruniceps* Red-headed Bunting
Emberiza melanocephala Black-headed Bunting	2	(H)	2	(V)	9	5	<3.3	18810	>1,000,000	289	434	121	761	1679	*Emberiza melanocephala* Black-headed Bunting
Miliaria calandra Corn Bunting	2	(D)	4	(S)	7	2	<3.3	18820	>5,000,000	290	469	121	762	1681	*Miliaria calandra* Corn Bunting

Notes.

2004 SPEC and **2004 ETS (European Threat Status)** – see Appendix 1 for details.

1994 SPEC and **1994 ETS (European Threat Status)** – see Appendix 6 for details.

Habitat association

Species were classified to habitats using the assessment of Tucker, G. M. and Evans. M. I. (1997) [*Habitats for birds in Europe: a conservation strategy for the wider environment.* Cambridge, UK: BirdLife International (BirdLife Conservation Series no. 6)]. All species with more than 75% of their population occurring in one of the following eight habitats were classified as specialists of that habitat: marine; coastal; inland wetland; tundra, mires and moorland; boreal and temperate forests; Mediterranean forest, shrubland and rocky habitats; agricultural and grassland habitats; and montane grassland. The last of these was treated as a sub-class of agricultural and grassland habitats by Tucker and Evans (1997), but is separated here because it supports an avifauna distinct from that occurring in other agricultural habitats; the trends of these species are therefore not representative of agricultural habitats in general. In addition, species with 10–75% of their population using only one of the above types were classed as specialists in that habitat, following Tucker and Evans (1997) for SPECs and descriptions in Snow, D. W. and Perrins, C. M. (1998) [*The birds of the western Palearctic: concise edition.* 2 volumes. Oxford, U.K.: Oxford University Press] for Non-SPECs. Remaining species with 10–75% of their population occurring in more than one habitat were classed as non-specialists. Any species that did not meet the criteria above (due to insufficient data) remained unclassified. Tucker and Evans (1997) included a further habitat of lowland Atlantic heathland; however, no species met the criteria to be classed as a specialist of this habitat.

1. Marine habitats
2. Coastal habitats
3. Inland wetlands
4. Tundra, mires and moorland
5. Boreal and temperate forests
6. Mediterranean habitats
7. Agricultural and grassland habitats
8. Montane grassland
9. Associated with more than one habitat
10. Unclassified

Migratory status

1. Resident
2. Partial migrant within Europe
3. Full migrant within Europe
4. Short-distance migrant (winters just outside Europe, e.g. North Africa)
5. Long-distance migrant (winters in sub-Saharan Africa or Asia)

Generation length (years) – see data analysis for details

EURING code

EURING is the coordinating organisation for European bird-ringing schemes, through which each species has been allocated a unique 5-digit code. Details: http://www.euring.org

European breeding range (km^2) – see data analysis for details (derived from Hagemeijer and Blair 1997)

Page number in key references

BiE2 (2004) – the page number on which each species appears in this book.

BiE1 (1994) – the page number on which each species appears in Tucker, G. M. and Heath, M. F. (1994) *Birds in Europe: their conservation status.* Cambridge, U.K.: BirdLife International (BirdLife Conservation Series No. 3).

EBPET (2000) – the page number on which each species appears in BirdLife International/European Bird Census Council (2000) *European bird populations: estimates and trends.* Cambridge, U.K.: BirdLife International (BirdLife Conservation Series No. 10).

EBCC (1997) – the page number on which each species appears in Hagemeijer, E. J. M. and Blair, M., eds. (1997) *The EBCC Atlas of European breeding birds: their distribution and abundance.* London: T. and A. D. Poyser.

BWPC (1998) – the page number on which each species appears in Snow, D. W. and Perrins, C. M. (1998) *The Birds of the Western Palearctic: Concise Edition.* Oxford and New York: Oxford University Press.

Species name in BiE1 (Tucker and Heath 1994)

The scientific and common names of many European species have changed since *BiE1* was published. The index to this book includes only those names in current use, i.e. those appearing at the top of each species account on pages 27–290. To allow easy cross-referencing, this column lists all the names used in *BiE1* (Tucker and Heath 1994), highlighting those that have changed in the interim or were not covered in that book.

Appendix 5. Important Birds Areas in Europe for species classified as Localised.

Puffinus puffinus **(breeding; see p. 35)**

Country	IBA code	International IBA name
Faroe Islands (DK)	001	Mykines and Mykineshólmur
Faroe Islands (DK)	014	Sandoy
Faroe Islands (DK)	016	Skúvoy
Iceland	028	Vestmannaeyjar
Ireland	073	Blasket Islands
Ireland	075	Puffin Island
Ireland	076	The Skelligs: Great Skellig and Little Skellig
United Kingdom	088	Glannau Aberdaron and Ynys Enlli
United Kingdom	094	Skokholm and Skomer
United Kingdom	236	Rum

Oceanodroma leucorhoa **(breeding; see p. 38)**

Country	IBA code	International IBA name
Faroe Islands (DK)	001	Mykines and Mykineshólmur
Iceland	028	Vestmannaeyjar
Iceland	031	Skeidarársandur
Ireland	025	Rathlin O'Birne
Ireland	040	Stags of Broadhaven
Norway	025	Røst
Norway	031	Sklinna
United Kingdom	145	Flannan Isles
United Kingdom	228	Ramna Stacks and Gruney
United Kingdom	238	St Kilda

Phoenicopterus roseus **(breeding; see p. 49)**

Country	IBA code	International IBA name
France	239	Camargue
Italy	188	Cagliari wetlands
Spain	148	Ebro delta
Spain	165	El Hondo wetland
Spain	240	Fuente de Piedra, Gosque and Campillos lakes
Spain	259	Guadalquivir marshes
Turkey	016	Gediz delta
Turkey	053	Seyfe lake
Turkey	064	Eregli marshes
Turkey	072	Tuz lake

Branta ruficollis **(winter; see p. 56)**

Country	IBA code	International IBA name
Bulgaria	049	Shabla lake complex
Bulgaria	050	Durankulak lake
Romania	001	Danube Delta and Razelm-Sinoe complex
Romania	002	Parches-Somova wetland
Romania	003	Lake Beibugeac
Russia	140	Veselovskoye reservoir
Ukraine	067	Karkinits'ka and Dzharylgats'ka bays
Ukraine	081	Kugurluj and Kartal lakes
Ukraine	083	Kytaj lake
Ukraine	087	Shagany-Alibej-Burnas lake-system

Aythya marila **(winter; see p. 63)**

Country	IBA code	International IBA name
Germany	022	Coastline: Niendorf-Travemunde
Germany	023	Traveförde and Dassower See
Netherlands	001	Wadden Sea
Netherlands	034	Lake IJsselmeer
Netherlands	035	Makkumer-and Kooiwaard
Netherlands	037	Steile Bank and Mokkebank
Netherlands	038	Lake Markermeer
Poland	002	Szczecin lagoon
Ukraine	065	Yagorlyts'ka and Tendrivs'ka Bays
Ukraine	071	Molochnyj Liman

Polysticta stelleri **(winter; see p. 65)**

Country	IBA code	International IBA name
Estonia	011	Küdema bay
Estonia	012	Tagamõisa peninsula
Estonia	013	Vilsandi archipelago
Finland	081	Hanko western archipelago
Finland	083	Pori archipelago and wetlands
Finland	093	Föglö southern archipelago
Finland	096	Lågskär-Nyhamn
Lithuania	001	Seashore at Palanga
Norway	003	Varangerfjord
Russia	010	Gavrilovski archipelago

Melanitta fusca **(winter; see p. 67)**

Country	IBA code	International IBA name
Denmark	094	Sejrø Bay and Nekselø
Estonia	014	Irbe strait
France	040	Littoral Augeron
France	045	Estuaire et embouchure de la Seine
Germany	040	Pomeranian bay
Lithuania	002	Seashore at Nida
Netherlands	018	North Sea north of the Wadden Sea
Norway	033	Ørland wetland system
Norway	039	Smøla archipelago
Poland	080	Central Polish coastal waters

Porphyrio porphyrio **(breeding; see p. 103)**

Country	IBA code	International IBA name
Azerbaijan	030	Lake Aggel
Azerbaijan	032	Lake Sarysu
Azerbaijan	043	Shorgel lakes/Shirvan reserve
Italy	182	Oristano wetlands and Cape San Marco
Italy	188	Cagliari wetlands
Italy	190	Palmas Gulf wetlands
Spain	137	Ampurdán marshes
Spain	228	Middle section of river Guadalquivir reservoirs
Spain	259	Guadalquivir marshes
Turkey	073	Göksu delta

Larus genei **(breeding; see p. 136)**

Country	IBA code	International IBA name
France	239	Camargue
Italy	188	Cagliari wetlands
Russia	152	Kiziltash limans
Russia	156	Lake Khanskoye
Russia	165	Lake Manych-Gudilo
Turkey	070	Bolluk lake
Ukraine	063	Kinburns'kyj peninsula
Ukraine	065	Yagorlyts'ka and Tendrivs'ka Bays
Ukraine	069	Syvash Bay
Ukraine	100	Kazantip Cape, Aktash lake and Astanins'ka marsh

Larus audouinii **(breeding; see p. 137)**

Country	IBA code	International IBA name
Italy	171	Asinara island and Stintino peninsula cliffs
Spain	025	Isle of Conejera and islets of Bledes and Espartar (Ibiza)
Spain	148	Ebro delta
Spain	153	Columbretes islands
Spain	170	Littoral islets of Murcia and Almería
Spain	220	Chafarinas Islands
Spain	221	Alborán island
Spain	312	Salt-pans of Ibiza and Formentera and Freus isles
Spain	315	Dragonera Island-La Trapa
Spain	324	Archipelago of Cabrera

Larus armenicus **(breeding; see p. 139)**

Country	IBA code	International IBA name
Armenia	001	Lake Arpi
Armenia	005	Lake Sevan
Turkey	072	Tuz lake
Turkey	087	Aktas lake
Turkey	088	Çildir lake
Turkey	094	Sodaligöl
Turkey	096	Lake Van
Turkey	124	Agri plain
Turkey	127	Keban Reservoir
Turkey	128	Karakaya Reservoir

Notes.

Localised species are those whose populations are highly concentrated at a few sites at some stage during their life cycle. They have an Unfavourable Conservation Status because their small ranges render them more susceptible to accelerated declines via large-scale population fluctuations and catastrophic chance events, or through existing or potential exploitation, persecution, disturbance or interference by man. A species is considered Localised when more than 90% of its European population occurs at ten or fewer sites in Europe (provided it does not meet any IUCN Red List Criteria at European level, is neither Declining, Rare nor Depleted, and its total European population is more than 10,000 breeding pairs or 40,000 wintering individuals).

This appendix lists the 'top ten' sites for each Localised species. All sites are Important Bird Areas (IBAs), as listed in Heath, M. F. and Evans, M. I., eds. (2000) *Important Bird Areas in Europe: priority sites for conservation.* Cambridge, U.K.: BirdLife International (BirdLife Conservation Series No. 8). Sites are not listed in order of importance, but in alphabetical order by country and then by ascending order and by ascending national IBA code number.

Tables are included for three species (*Branta ruficollis, Aythya marila* and *Melanitta fusca*) that are no longer classified as Localised because they meet other criteria. Nevertheless, all three remain highly concentrated in winter, and the conservation of their key wintering sites should remain a priority.

Appendix 6. Summary of conservation status assessment in Tucker and Heath (1994).

This appendix summarises the methods used by Tucker and Heath (1994) in the first assessment of the conservation status of European birds. For details of the methods used in the current review, see the section on 'Status assessment'.

The conservation status of Europe's breeding birds was first assessed in Tucker and Heath (1994). The aim of this assessment was to identify species of conservation concern on a European scale. These birds were termed Species of European Conservation Concern (SPECs) and were divided into four categories (Box 1) depending on their global conservation status, their European Threat Status and the proportion of their world population in Europe (Figure 1). Species were considered to have an Unfavourable Conservation Status if their European Threat Status was Localised, Declining, Rare, Vulnerable, Endangered or Insufficiently Known in Europe, according to the criteria summarised in Figure 1 and described in full in Box 2 (see Box 3 for definition and explanation of terms).

Box 1. Species of European Conservation Concern (SPECs), as in Tucker and Heath (1994).

Category 1 Species of global conservation concern because they were classified as Globally Threatened, Conservation Dependent or Data Deficient in Collar *et al.* (1994).

Category 2 Species whose global populations were concentrated in Europe (i.e. more than 50% of their global population or range in Europe), and which had an Unfavourable Conservation Status in Europe.

Category 3 Species whose global populations were not concentrated in Europe, but which had an Unfavourable Conservation Status in Europe.

Category 4 Species whose global populations were concentrated in Europe (i.e. species with more than 50% of their global population or range in Europe), but which had a Favourable Conservation Status in Europe.

CALCULATION OF POPULATION CHANGES

The criteria for identifying the SPEC category and the European Threat Status of each species took into consideration the quality and availability of information on European bird populations. Calculations were based on categories of population trend, rather than on precise figures. To take account of the broad ranges involved in most estimates of population size, analyses were performed using both population minima and maxima. As a precautionary measure, the higher proportion was used in the assessment of a species' European Threat Status.

The assessment of European Threat Status was based on the percentage of the total European population occurring in countries where populations had declined. Due to data limitations, it was not possible to use an exact measure of population decline in each country to calculate the rate of a species's decline in Europe as a whole. Assessments of the decline in range were based on the percentage of the European population (rather than range) occurring in countries where range had declined. Inevitably therefore, range declines in large populations were given greater weight than those in marginal parts of the range. The assessments of European

Figure 1. Defining Species of European Conservation Concern (SPECs), as in Tucker and Heath (1994).

Species of conservation concern on a European scale were identified according to their European Threat Status and the proportion of their world population in Europe.

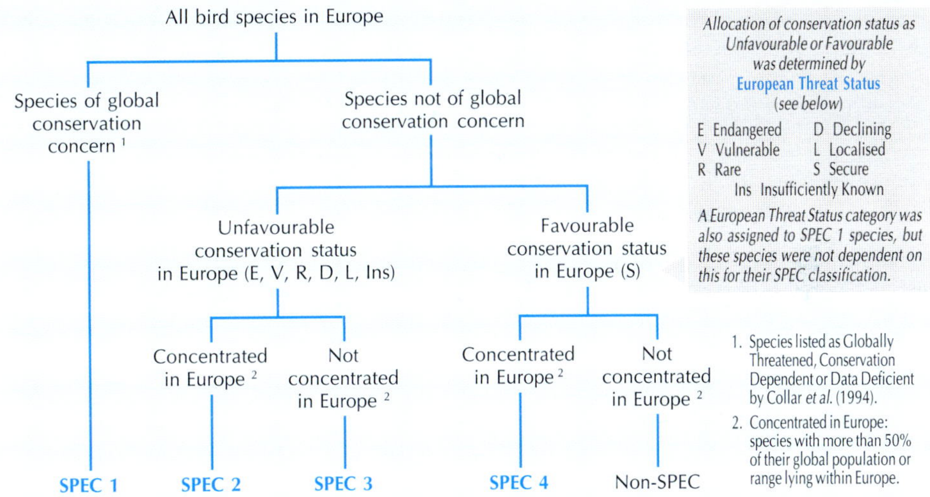

European Threat Status: summary of criteria and categories

Criteria: European population size/trend	<250 pairs	<2,500 pairs	<10,000 pairs	>10,000 pairs
Large decline [3]	ENDANGERED	ENDANGERED	ENDANGERED	VULNERABLE
Moderate decline [4]	ENDANGERED	ENDANGERED	VULNERABLE	DECLINING
No decline	ENDANGERED	VULNERABLE	RARE	SECURE

In addition, species that had more than 10,000 pairs in Europe were categorised as LOCALISED if more than 90% of the European population occurred at 10 sites or fewer in Europe. See Tucker and Heath (1994) for full details of criteria.

3. Large decline: applied to a breeding or wintering population that declined in size or range by at least 20% in at least 66% of the population, or by at least 50% in at least 25% of the population, between 1970 and 1990, and where the total size of populations that declined was greater than the total size of populations that increased.
4. Moderate decline: applied to a breeding or wintering population which has declined in size or range by at least 20% in 33–65% of the population or by at least 50% in 12–24% of the population between 1970 and 1990, and where the total size of populations that declined is greater than the total size of populations that increased.

Note that:

- Winter population criteria used flyway population levels of less than 1,000, 10,000 and 40,000 individuals as respective equivalents to the figures of 250, 2,500 and 10,000 pairs used above for breeding populations.
- Due to inadequate data for most species, declines in winter populations were only considered for Anatidae, Haematopodidae, Charadriidae and Scolopacidae.

Appendix 6 ...continued. Summary of conservation status assessment in Tucker and Heath (1994).

Box 2. Classification of European Threat Status, as in Tucker and Heath (1994).

All population size thresholds refer to minimum population estimates.

Insufficiently Known Suspected to be Localised, Declining, Rare, Vulnerable or Endangered (as below), but insufficient information was available to attribute a European Threat Status, even provisionally.

Remaining categories are ranked here in ascending order of threat.

Secure Population more than 10,000 breeding pairs or 40,000 wintering birds, and neither in moderate or large decline nor Localised. Secure species had a Favourable Conservation Status.

Localised Population more than 10,000 breeding pairs or 40,000 wintering birds, and neither in moderate nor in large decline, but with more than 90% of the population occurring at 10 or fewer sites (Important Bird Areas), as listed in Grimmett and Jones (1989).

Declining Population in moderate decline and population more than 10,000 breeding pairs or 40,000 wintering individuals.

Rare Population neither in moderate nor in large decline but fewer than 10,000 breeding pairs and not marginal to a larger non-European population; or European wintering population and entire flyway population less than 40,000 birds and therefore at risk due to the susceptibility of small populations to:
- break-up of social structure;
- loss of genetic diversity;
- large-scale population fluctuations and chance events;
- existing or potential exploitation, persecution, disturbance and interference by man.

Vulnerable Any of the following.
- Population in large decline and of more than 10,000 breeding pairs or 40,000 wintering individuals.
- Population in moderate decline *and* population fewer than 10,000 breeding pairs and not marginal to a larger non-European population or European wintering and entire flyway population fewer than 40,000 birds.
- Population neither in moderate nor in large decline but fewer than 2,500 breeding pairs and not marginal to a larger non-European population; or European wintering and entire flyway population fewer than 10,000 birds and therefore at risk due to the susceptibility of small populations to the factors described under 'Rare' above.

Endangered Any of the following.
- Population in large decline *and* population fewer than 10,000 breeding pairs and not marginal to a larger non-European population or European wintering and entire flyway population fewer than 40,000 birds.
- Population in moderate decline *and* population fewer than 2,500 breeding pairs and not marginal to a larger non-European population or European wintering and entire flyway population fewer than 10,000 birds.
- Population neither in moderate nor in large decline but fewer than 250 breeding pairs and not marginal to a larger non-European population; or European wintering and entire flyway population fewer than 1,000 birds and therefore at risk due to the susceptibility of small populations to factors described above.

Conservation Status were made on the basis of breeding season data unless a species qualified on winter data.

European Threat Status categories were only allocated provisionally when more than 50% of the European population size data were of poor or unknown quality, or were missing. Assessment of overall population trends, solely on the basis of the available quantitative data would have restricted analyses to those countries with such data. This would have introduced regional biases and therefore been invalid. Bird populations in east and south-east Europe, for example, were particularly poorly known, largely due to the lower number of ornithologists in these regions. Clearly, the population trends in these areas may often have been different from those in other parts of Europe.

References

Collar, N. J., Crosby, M. J. and Stattersfield, A. J. (1994) *Birds to watch 2: the world list of threatened birds*. Cambridge, U.K.: BirdLife International (BirdLife Conservation Series No. 4).

Cramp, S. *et al.* (1977–1994) *The birds of the western Palearctic*. Volumes 1–9. Oxford, U.K.: Oxford University Press.

Grimmett, R. F. A. and Jones, T. A. (1989) *Important Bird Areas in Europe*. Cambridge, U.K.: International Council for Bird Preservation (Techn. Publ. 9).

Harrison, C. (1982) *An atlas of the birds of the western Palearctic*. London: Collins.

Tucker, G. M. and Heath, M. F. (1994) *Birds in Europe: their conservation status*. Cambridge, U.K.: BirdLife International (BirdLife Conservation Series No. 3).

Box 3. Definitions of terms used in the criteria, as in Tucker and Heath (1994).

Concentrated in Europe More than 50% of the species' breeding or wintering population or range occurred in Europe, according to range maps in Cramp *et al.* (1977–1993) or Harrison (1982) or to global population estimates where available.

Conservation Dependent A species which did not qualify as Globally Threatened but was the focus of a continuing conservation programme, the cessation of which would have resulted in the species qualifying as Globally Threatened (Collar *et al.* 1994).

Data Deficient A species for which there was inadequate information to make a direct or indirect assessment of its risk of global extinction (Collar *et al.* 1994).

Favourable Conservation Status The European Threat Status was classed as Secure.

Large decline Applied to a breeding or wintering population that declined in size or range by at least 20% in at least 66% of the population, or by at least 50% in at least 25% of the population between 1970 and 1990, and where the total size of populations that declined was greater than the total size of populations that increased. Only wintering populations of waterbirds of the families Anatidae, Haematopodidae, Charadriidae and Scolopacidae were considered because these were typically the only species that had well monitored winter populations.

Marginal population European populations that were considered to have adequate potential for repopulation from large non-European populations (the combined total of which was 10,000 pairs or more) and which were therefore not at risk from small population size.

Moderate decline Applied to a breeding or wintering population that declined in size or range by at least 20% in 33–65% of the population, or by at least 50% in 12–24% of the population, between 1970 and 1990, and where the total size of populations that declined was greater than the total size of populations that increased. See also 'Large decline' (above) for restrictions on species covered.

Unfavourable Conservation Status The European Threat Status is classed as Endangered, Vulnerable, Rare, Declining, Localised or Insufficiently Known.

Appendix 7. IUCN Red List Categories and Criteria (2001).

Type of criteria	Main criteria	Sub-criteria	Qualifiers	Codes
A. REDUCTION IN POPULATION SIZE	Reduction ≥90% in 10 years or 3 generations (**CR**) Reduction ≥70% in 10 years or 3 generations (**EN**) Reduction ≥50% in 10 years or 3 generations (**VU**)	**1.** Reduction in the past (observed, estimated, inferred or suspected), where the causes are clearly reversible AND understood AND ceased, *based on a–e opposite*:	**a.** Direct observation	A1a
			b. Index of abundance	A1b
			c. Decline in area of occupancy, extent of occurrence, and/or quality of habitat	A1c
			d. Actual or potential levels of exploitation	A1d
			e. Effects of introduced taxa, hybridisation, pathogens, pollutants, competitors or parasites	A1e
	Reduction ≥80% in 10 years or 3 generations (**CR**) Reduction ≥50% in 10 years or 3 generations (**EN**) Reduction ≥30% in 10 years or 3 generations (**VU**)	**2.** Reduction in the past (observed, estimated, inferred or suspected), where the reduction or its causes may not be reversible OR understood OR have ceased, *based on a–e opposite*:	**a.** As a above	A2a
			b. As b above	A2b
			c. As c above	A2c
			d. As d above	A2d
			e. As e above	A2e
	Reduction ≥80% in 10 years or 3 generations (**CR**) *to 100 years max* Reduction ≥50% in 10 years or 3 generations (**EN**) *to 100 years max* Reduction ≥30% in 10 years or 3 generations (**VU**) *to 100 years max*	**3.** Reduction in the future (projected or suspected), *based on b–e opposite:*	**b.** As b above	A3b
			c. As c above	A3c
			d. As d above	A3d
			e. As e above	A3e
	Reduction ≥80% in 10 years or 3 generations (**CR**) *to 100 years max* Reduction ≥50% in 10 years or 3 generations (**EN**) *to 100 years max* Reduction ≥30% in 10 years or 3 generations (**VU**) *to 100 years max*	**4.** Reduction includes the past and the future (observed, estimated, inferred, projected or suspected) where the reduction or its causes may not be reversible OR understood OR have ceased, *based on a–e opposite*:	**a.** As a above	A4a
			b. As b above	A4b
			c. As c above	A4c
			d. As d above	A4d
			e. As e above	A4e
B. SMALL RANGE fragmented, declining or fluctuating	**1.** Extent of occurrence estimated <100 km² (**CR**) *with at least two of a,b or c:* Extent of occurrence estimated <5,000 km² (**EN**) *with at least two of a, b or c:* Extent of occurrence estimated <20,000 km² (**VU**) *with at least two of a, b or c*:	**a.** Severe fragmented *or* At 1 location (**CR**) At ≥5 locations (**EN**) At ≥10 locations (**VU**)	None	B1a
		b. Continuing decline (observed, inferred or projected) *in any of i–v opposite*:	**i.** Extent of occurrence	B1bi
			ii. Area of occupancy	B1bii
			iii. Area, extent and/or quality of habitat	B1biii
			iv. Number of locations or subpopulations	B1biv
			v. Number of mature individuals	B1bv
		c. Extreme fluctuations *in any of i–iv opposite:*	**i.** Extent of occurrence	B1ci
			ii. Area of occupancy	B1cii
			iii. Number of locations or subpopulations	B1ciii
			iv. Number of mature individuals	B1civ
	2. Area of occupancy estimated <10 km² (**CR**) *with at least two of a, b or c:* Area of occupancy estimated <500 km² (**EN**) *with at least two of a, b or c:* Area of occupancy estimated <2000 km² (**VU**) *with at least two of a, b or c*:	**a.** As a above	None	B2a
		b. As b above *in any of i–v opposite:*	**i.** Extent of occurrence	B2bi
			ii. Area of occupancy	B2bii
			iii. Area, extent and/or quality of habitat	B2biii
			iv. Number of locations or subpopulations	B2biv
			v. Number of mature individuals	B2bv
		c. As c above *in any of i–iv opposite:*	**i.** Extent of occurrence	B2ci
			ii. Area of occupancy	B2cii
			iii. Number of locations or subpopulations	B2ciii
			iv. Number of mature individuals	B2civ
C. SMALL POPULATION declining or fluctuating	Population <250 mature individuals (**CR**) *and either 1 or 2* Population <2,500 mature individuals (**EN**) *and either 1 or 2* Population <10,000 mature individuals (**VU**) *and either 1 or 2*	**1.** Continuing decline ≥25% in 3 years or 1 generation (**CR**) *to 100 years max* Continuing decline ≥20% in 5 years or 2 generations (**EN**) *to 100 years max* Continuing decline ≥10% in 10 years or 3 generations (**VU**) *to 100 years max*	None	C1
		2. Continuing decline (observed, projected or inferred) and *a and/or b opposite:*	**ai.** all sub-pops ≥50 (**CR**) all sub-pops ≥250 (**EN**) all sub-pops ≥1,000 (**VU**)	C2ai
			aii. ≥90% mature individuals in 1 sub-pop (**CR**) ≥95% mature individuals in 1 sub-pop (**EN**) all individuals in 1 sub-pop (**VU**)	C2aii
			b. Extreme fluctuations in number of mature individuals	C2b
D1. V. SMALL POPULATION	Population <50 mature individuals (**CR**) Population <250 mature individuals (**EN**) Population <1,000 mature individuals (**VU**)	None	None	D1

Appendix 7...continued IUCN Red List Categories and Criteria (2001).

Type of criteria	Main criteria	Sub-criteria	Qualifiers	Codes
D2. V. SMALL RANGE	Area of occupancy typically <20 km² or typically <6 locations **(VU** only – capable of becoming CR or EX in very short time)	None	None	D2
E. QUANTITATIVE ANALYSIS	Probability of extinction in wild >20% in 20 years or 5 gens **(EN)** *to 100 years max* Probability of extinction in wild is 10% in 100 years **(VU)**	None	None	E

The IUCN Red List Categories and Criteria

The IUCN Red List Categories and Criteria are an objective and widely accepted system for classifying species at high risk of extinction. The criteria (summarised in the table above) are based on biological factors related to extinction risk, and include rate of decline, population size, area of geographic distribution, and degree of population and distribution fragmentation. BirdLife coordinates the assessment of the status of the world's birds using these categories and criteria and, as the official Red Listing Authority for birds for the IUCN Red List, submits this information to be included on the IUCN Red List (along with that of other animals and plants). The IUCN Red List Categories and Criteria (in English, French or Spanish) can be viewed and downloaded at: http://iucn.org/themes/ssc/redlists/RLcats2001booklet.html. Detailed guidelines on their use can also be found at: http://iucn.org/themes/ssc/redlists/Redlistguidelines.pdf.

Ranked from the highest relative extinction risk to the lowest, the three threat categories are:

CRITICALLY ENDANGERED (CR)

A species is Critically Endangered when the best available evidence indicates that it meets any of the criteria A to E for Critically Endangered, and is therefore considered to be facing an extremely high risk of extinction in the wild.

ENDANGERED (EN)

A species is Endangered when the best available evidence indicates that it meets any of the criteria A to E for Endangered, and is therefore considered to be facing a very high risk of extinction in the wild.

VULNERABLE (VU)

A species is Vulnerable when the best available evidence indicates that it meets any of the criteria A to E for Vulnerable, and is therefore considered to be facing a high risk of extinction in the wild.

Definitions of terms used by the criteria

Population and mature individuals

Population is defined as the total number of mature individuals, i.e. the number of individuals known, estimated or inferred to be capable of reproduction. But note the following:

- Where the population is characterised by natural fluctuations, the minimum number is used.
- Individuals that are environmentally, behaviourally or otherwise reproductively suppressed in the wild are excluded.
- Re-introduced individuals must have produced viable offspring before they are counted as mature individuals.
- In the case of populations with biased adult or breeding sex ratios, lower estimates for the number of mature individuals are used which take this into account (e.g. the estimated effective population size).

Subpopulations

Subpopulations are defined as geographically or otherwise distinct groups in the population between which there is little exchange (typically one successful migrant individual or gamete per year or less).

Generation length

Generation is defined as the average age of parents in the population. Generation length is used in preference to the specified time period if the time span of the specified number of generations is longer.

Reduction

A reduction is a decline in the number of mature individuals of at least the amount (%) stated over the time period (years) specified, although the decline need not still be continuing.

Continuing decline

A continuing decline is a recent, current or projected future decline whose causes are not known or not adequately controlled, and so is liable to continue unless remedial measures are taken.

Extreme fluctuations

Extreme fluctuations occur where population size or distribution area varies widely, rapidly and frequently, typically with a variation greater than one order of magnitude (i.e. a tenfold increase or decrease).

Severely fragmented

Severely fragmented refers to the situation where increased extinction risks to the species result from the fact that most individuals within a species are found in small and relatively isolated subpopulations. These small subpopulations may go extinct, with a reduced probability of recolonisation.

Extent of occurrence

Extent of Occurrence (EOO) is defined as the area contained within the shortest continuous imaginary boundary which can be drawn to encompass all the known, inferred or projected sites of present occurrence of a species, excluding cases of vagrancy. This measure may exclude discontinuities or disjunctions within the overall distributions of species (e.g. large areas of obviously unsuitable habitat) (but see "Area of Occupancy" below).

Area of occupancy

Area of Occupancy (AOO) is defined as the area within its "Extent of Occurrence" (EOO, see definition above) which is occupied by a species, excluding cases of vagrancy. The measure reflects the fact that a species will not usually occur throughout the area of its EOO, which may, for example, contain unsuitable habitats. The AOO is the smallest area essential at any stage to the survival of existing populations of a species (e.g. colonial nesting sites, feeding sites for migratory species).

Location

Location defines a geographically or ecologically distinct area in which a single threatening event (e.g. pollution) can rapidly affect all individuals of the species present. The size of the location depends on the area covered by the threatening event and may include part of one or many subpopulations.

DATA REFERENCES BY COUNTRY

This list comprises details of the literature sources used by national coordinators to compile the population tables for individual species. The code numbers given cross-refer to those in the final column of each data table.

Albania

Breeding

1. Bego, F., Bino, T. and Jorgo, G. (1998) Faunistical values of Drin delta. Pp.30–49 in ECAT, eds. *Conservation and wise use of wetlands in the Mediterranean basin. Focus on the Kune-Vaini lagoon, Lezha, Albania.* Tirana.
2. Bino, T. (2000) Monitoring of waterbirds in wetlands of Kune-Vain, Patoku Rrushkull and Karavasta. Report from the MNS – Tirana.
3. Bino, T. (2001) Monitoring of waterbirds in wetlands of Kune-Vain, Patoku, Karavasta and Butrint. Report from the MNS – Tirana.
4. Bino, T. (2002) Monitoring of waterbirds in wetlands of Velipoja, Kune-Vain, Patoku, Karavasta and Butrint. Report from the MNS – Tirana.
5. Bino, T. (2003) Monitoring of waterbirds in wetlands of Velipoja, Patoku and Karavasta. Report from the MNS – Tirana.
6. Zekhuist, M. J. and Tempelman, D. (1998) *Breeding waterbirds of the Albanian wetlands, spring 1996*. WIWO-report Nr. 64, Zeist.
7. Zoto, H and Bino, T. (2003) Action plan for the conservation of Capercaille *Tetrao urogallus* in Albania. Dajti 2000. Tirana.

Winter

8. Bego, F., Bino, T. and Jorgo, G. (1998) Faunistical values of Drin delta. Pp.30–49 in ECAT, eds. *Conservation and wise use of wetlands in the Mediterranean basin. Focus on the Kune-Vaini lagoon, Lezha, Albania.* Tirana.
9. Bino, T. and Jorgo, G. (2001) *Conservation status and threats over Albanian wetlands*. ASPBM and BirdLife International report.
10. Bino, T. (1998) [L'avifaune aquatique du systeme lagunaire méditerranéen de Karavasta (Albanie) et sa conservation.] Université de Montpellier II Sciences et Techniques du Languedoc (Thèse de Doctorat). (In French.)
11. Bino, T. (2000) Monitoring of waterbirds in wetlands of Kune-Vain, Patoku Rrushkull and Karavasta. Report from the MNS – Tirana.
12. Bino, T. (2001) Monitoring of waterbirds in wetlands of Kune-Vain, Patoku, Karavasta and Butrint. Report from the MNS – Tirana.
13. Bino, T. (2002) Monitoring of waterbirds in wetlands of Velipoja, Kune-Vain, Patoku, Karavasta and Butrint. Report from the MNS – Tirana.
14. Bino, T. (2003) Monitoring of waterbirds in wetlands of Velipoja, Patoku and Karavasta. Report from the MNS – Tirana.
15. Bino, T., Tourenq, C., Kayser, Y., Bussutil, S., Crozier, J., Dore, B. J. and Bego, F. (1996) [Recensement des oiseaux d'eau hivernants en Albanie (14–31 janvier 1996).] Rapport de la Station Biologique de la Tour du Valat, Museum des Sciences de la Nature de Tirana, ASPBM et LIPU. (In French.)
16. Kayser, Y., Bino, T. and Gauthier-Clerc, M. (1995) [Recensement des oiseaux d'eau hivernants en Albanie. 17 janvier–7 février 1995.] Rapport de la Station Biologique de la Tour du Valat. (In French.)
17. Kayser, Y., Bino, T., Bego, F., Fremuth, W. and Jorgo. G. (1997) [Recensement des oiseaux d'eau hivernants en Albanie (3–19 Janvier 1997).] Rapport de la Station Biologique de la Tour du Valat (France) et Museum des Sciences de la Nature (Tirana; Albanie). (In French.)

Andorra

Breeding

1. Associació per a la Defensa de la Natura (ADN) (2002a) *Atles dels ocells nidificants d'Andorra – atlas of the breeding Bbirds of Andorra.* Andorra la Vella: ADN, Govern d'Andorra and Andbanc.
2. Associació per a la Defensa de la Natura (ADN) (2002b) [Noticiari naturalista del Principat d'Andorra. Informe any 2000.] *Aigüerola* 11: 13–16. (In Catalan.)
3. Dalmau, J. and Mariné, R. (1998) [Distribution, status and conservation of boreal owl in the Principality of Andorra.] Ministeri d'Agricultura i Medi Ambient, Govern d'Andorra. Unpublished report. (In Catalan.)

Winter

4. Associació per a la Defensa de la Natura (ADN) (2002b) [Noticiari naturalista del Principat d'Andorra. Informe any 2000.] *Aigüerola* 11: 13–16. (In Catalan.)

Armenia

Breeding

1. Adamian, M. S. and Klem, D., Jr. (1999) *Handbook of the birds of Armenia.* Oakland: American University of Armenia. California, USA.
2. Aghababian, K. E. (1999) Fauna and some peculiarities of ecology of the representatives of Oenanthe genus (Turdidae, Passeriformes) of the South of Syunik region (former Meghry region). Collection of articles of the post graduates and young researchers of Yerevan State University (YSU).
3. Aghababian, K. E. (1999) *The ecology of the Lammergeier* (*Gypaetus barbatus aureus*, Hablizl) in the Meghry region of Armenia. Vestnik *MANEB* 5(17): 43–44. St. Petersburg.
4. Aghababian, K. E. (1999) *The ecology of the Peregrine Falcon* (*Falco peregrinus brookei*, Sharpe) in the Meghry region of Armenia. Vestnik *MANEB* 5(17): 44–45. St. Petersburg.
5. Aghababian, K. E. (2001) *The ecology of the Levant Sparrowhawk* (*Accipiter brevipes*, Severtzov, 1850) in Meghry region of Armenia. Vestnik *MANEB* 6(42): 40–41. St. Petersburg.
6. Aghababyan K., Ghasabian, M., Gavashelishvili, A., Bildstein, K. and McGrady, M. (2003) A pilot survey of Griffon Vultures in Armenia. 6th world conference on birds of prey and owls, Budapest, Hungary, 18 – 23 May 2003.
7. Aghababyan, K. (2000) Some aspects of vertical landscape distribution and ecology of birds in Meghry region of Armenia. (Thesis for the degree of candidate of biological sciences.) Yerevan.
8. Ananian, V. and Busuttil, S. (2003) Further observations and probable breeding of Mongolian Trumpeter Finch *Bucanetes mongolicus* in Armenia. *Sandgrouse* 25: 148–150.
9. Ananian, V. and Busuttil, S. (2002) The first breeding records of Citrine Wagtail *Motacilla citreola* and Savi's Warbler *Locustella luscinioides* in Armenia. *Sandgrouse* 24: 52–53.
10. Ananian V., Busuttil S. and Finn M. (2002) Recent observations of some rare breeding birds in Armenia. *Sandgrouse* 24: 46–48.
11. Balian L. V., Ghasabian M. G., Adamian M. S. and Klem, D., Jr. (2002) Changes in the waterbird community of the Lake Sevan – Lake Gilli area, Republic of Armenia: a case for restoration. *Biological Conservation* 106: 157–163.
12. Balyan, L. V. (2000) Changes in the waterfowl population of Armenia as a result of lowering of the water level in Lake Sevan. Central European University, Environmental Sciences and Policy Department (Master thesis). Budapest, Hungary.
13. Beddard R., Ananian V. and Finn M. (2002) The first Mongolian Trumpeter Finch *Bucanetes mongolicus* in Armenia. *Sandgrouse* 24: 144–147.
14. Solokha, A. V. (2001) Study and conservation of Caucasian Grouse in Armenia: actual problems of study and conservation of birds of eastern Europe and North Asia. *Proceedings of XI International Ornithological conference. Kazan. 29 January–3 February 2001.* 'Matbugat Yorti' Digest.

Winter

15. Adamian, M. S. and Klem, D., Jr. (1999) *Handbook of the birds of Armenia.* Oakland: American University of Armenia. California, USA.
16. Balian L. V., Ghasabian M. G., Adamian M. S. and Klem, D., Jr. (2002) Changes in the waterbird community of the Lake Sevan – Lake Gilli area, Republic of Armenia: a case for restoration. *Biological Conservation* 106: 157–163.
17. Balyan, L. V. (2000) Changes in the waterfowl population of Armenia as a result of lowering of the water level in Lake Sevan. Central European University, Environmental Sciences and Policy Department (Master thesis). Budapest, Hungary.

Austria

Breeding

1. Aubrecht, G. and Winkler, H. (1997) [*Analyse der Internationalen Wasservogelzählungen (IWC) in Österreich 1970–1995 – Trends und Bestände.*] Österr. Akademie d. Wiss. Biosystematics and Ecology Series No. 13. Wien. (In German.)
2. Brader, M. and Aubrecht, G. (2003) [*Atlas der Brutvögel Oberösterreichs.*] Denisia 7, zugleich Kataloge der OÖ Landesmuseen, N. F. Nr. 194. (In German.)
3. Dvorak, M. (2003) [Die Brutbestände der Schwimmvögel an den Lacken des Seewinkels im Jahr 2002.] Pp.3–12 in [*Ornithologisches monitoring im Nationalpark Neusiedler See – Seewinkel.*] Bericht über das Jahr 2002. Bericht im Auftrag des Nationalparks Neusiedler See – Seewinkel. BirdLife Österreich, Wien. (In German.)
4. Dvorak, M. and Tebbich, S. (1998) [*Bestand, bestandsentwicklung und schutzsituation der moorente (*Aythya nyroca*) im Neusiedler See-Gebiet. Ergebnisse von Bestandsaufnahmen in den Jahren 1996 und 1997.*] Bericht an die Biol. Station Illmitz und die Abt. IV (Naturschutz) des Amtes der Burgenländ. Landesregierung. BirdLife Österreich, Wien. (In German.)

5. Dvorak, M., Nemeth, E., Tebbich, S., Rössler, M. and Busse, K. (1997) [Verbreitung, bestand und habitatwahl schilfbewohnender vogelarten in der naturzone des Nationalparks Neusiedler See – Seewinkel]. *Biol. Forschungsinstitut Burgenland – Bericht* 86: 1–69. (In German.)
6. Kilzer, R., Amann, G. and Kilzer, G. (2002) [*Rote Liste gefährdeter Brutvögel Vorarlbergs*.] Vorarlberger Naturschau, Dornbirn. (In German.)
7. Kohler, B. (2003) [Der Brutbestand des Säbelschnäblers (*Recurvirostra avosetta*) 2002 im Seewinkel.] Pp.17–23 in [*Ornithologisches monitoring im Nationalpark Neusiedler See – Seewinkel. Bericht über das Jahr 2002*.] Bericht im Auftrag des Nationalparks Neusiedler See – Seewinkel. BirdLife Österreich, Wien. (In German.)
8. Kohler, B. and Rauer, G. (2003) [Ergebnisse des wiesenlimikolen-monitorings im Seewinkel im Jahr 2002.] Pp.24–32 in [*Ornithologisches monitoring im Nationalpark Neusiedler See – Seewinkel. Bericht über das Jahr 2002*.] Bericht im Auftrag des Nationalparks Neusiedler See – Seewinkel. BirdLife Österreich, Wien. (In German.)
9. Laber, J. (2000) [Die brutbestandsentwicklung der Schwarzkopfmöwe (*Larus melanocephalus*) im Seewinkel.] *Egretta* 43: 112–118. (In German.)
10. Landmann, A. and Lentner, R. (2001) [*Die Brutvögel Tirols. Bestand, Gefährdung, Schutz und Rote Liste*.] Ber. nat.-med. Verein Innsbruck. Suppl. 14. Universitätsverlag Wagner, Innsbruck. (In German.)
11. Nemeth, E., Grubbauer, P., Rössler, M. and Schuster, A. (In print.) [Ökologie der Reiher und Löffler des Neusiedler See-Gebietes. Habitatwahl, Nahrungsökologie, Bruterfolg, Populationsentwicklung und Schutz der in Kolonien brütenden Schreitvögel.] *Biol. Forschungsinstitut Burgenland – Bericht* ##. (In German.)
12. Sabathy, E. (2001) Verbreitung und bestand der Brutvögel an den Gewässern in Wien 1995–1999. *Egretta* 44: 89–138. (In German.)
13. Sackl, P. and Samwald, O. (1997) [*Atlas der Brutvögel der Steiermark. Ergebnisse der Steirischen Brutvogelkartierung*.] BirdLife Österreich & Steiermärkisches Landesmuseum Joanneum, Graz. (In German.)
14. Schuster, A. (2001) [*Brutvogelkartierung Reichersberger Au (Innstausee Schärding-Neuhaus, Oberösterreich)*.] Unveröff. Bericht im Auftrag des Landes Oberösterreich, LIFE-Projekt B4-3200/98/480. Wien. (In German.)
15. Wendelin, B. (2003) [Der Brutbestand der Flussseeschwalbe (Sterna hirundo) im Seewinkel im Jahr 2002]. Pp.50–59 in [*Ornithologisches monitoring im Nationalpark Neusiedler See – Seewinkel. Bericht über das Jahr 2002*.] Bericht im Auftrag des Nationalparks Neusiedler See – Seewinkel. BirdLife Österreich, Wien. (In German.)
16. Zuna-Kratky, T., Kalivodova, E., Kürthy, A., Horal, D. and Horák, P. (2000) [*Die vögel der March-Thaya-Auen im österreichisch-slowakisch-tschechischen Grenzraum*.] Distelverein, Deutsch-Wagram. (In German.)

Winter

17. Probst, R. (2003) [Verbreitung und Häufigkeit des Seeadlers (*Haliaeetus albicilla*) im Mitwinter 2001/02 in Österreich]. *Egretta* 46: 92–97. (In German.)

Azerbaijan

Winter

1. Sultanov, E. H. (1998) [The role of seed-eating and insectivorous birds of Azerbaijan for biocoenosis.] *Caucasian Ornithological Society, Stavropol* 10: 126–129. (In Russian.)
2. Sultanov, E. H., Karabanova, N. I., Humbatova, S. E. and Karabeyli, F.A. [Materials about predators of Caspian coast of Azerbaijan.] Pp.141–143 in [*Proc. Of 3rd conf. of young scientists in Ukrain. Chernovtsi, 2998*.] (In Russian.)
3. Sultanov, E. H., Muhtarov, H. Kerimov, T. and Huseynov, R. (1998) The main places of concentration for waterbirds in northern-east of Azerbaijan. XIV Ulusal bioloji kongresi, Samsun.
4. Sultanov, E. H. and Agayeva, N. (1998) The birds of place localization of Sangachal oil terminal. XIV Ulusal bioloji kongresi, Samsun.
5. Sultanov, E. H. (1998) Mass mortality of birds from oil near the Caspian coast of Azerbaijan in March 1998. Strategies for conserving migratory waterbirds. Pp.69–70 in *Proceedings of Workshop 2 of the 2nd International Conference of Wetlands and Development held in Dakar, Senegal, 1998*. Wetlands International Publication No. 55, Wageningen, The Netherlands.
6. Agayeva, H. and Sultanov, E.H. (1998) [Dates about wintering of birds around Sangachal Gulf.] Pp.43–45 in [*Countrylogy and invironmental protection in Azerbaijan*.] Abstracts of a conference. Baku. (in Russian)
7. Sultanov, E. H., Agayeva, N. and Kerimov, T. (1998) Condition of population of ferruginous duck and other rare species in Azerbaijan. Pp.60–61 in [Countryside and environmental protection in Azerbaijan.] Abstracts of a conference. Baku. (In Russian.)
8. Sultanov, E. H. and Kerimov, T. (1998) Pp.61–63 in [Countryside and environmental protection in Azerbaijan.] Abstracts of a conference. Baku. (In Russian.)
9. Sultanov, E. H. and Kerimov, T. (1998) Modern conditions of nesting fauna of waterbirds at the islands of Baku and Absheron archipelago. Abstracts of Workshop: Ecological problems of Caspian sea and ecological education in Caspian countries, Baku.
10. Sultanov, E. H., Mosley, P., Paynter, D. and Aarvak, T. (1998) Results of geese and swans count during wintering in Azerbaijan in 1993 and 1996. *Berkut* 7(1/2): 30–31.
11. Musayev, M. and Sultanov, E.H. (1999) *Ornithological resources of the Azerbaijan sector of the Caspian Sea*. Volga. Official publication of MIOGE'99. The 5th International Oil and Gas Exhibition.
12. Agayeva, N. and Sultanov, E. H. (1999) Ornithofauna around sangachal oil terminal in changed ecological conditions during nesting period. Pp.17–22 in *Ecological monitoring of great baku and Sumgait*. Baku. (In Russian.)
13. Sultanov, E. H. (1999) The places of mass concentration of birds on Absheron archipelago and their annual dynamics. Pp.130–131 in *Ecological monitoring of great baku and Sumgait*. Baku. (In Azeri.)
14. Sultanov, E. H. and Muxtarov, H. (1999) [Waterbirds of Azerbaijan and their meaning.] *X National Symposium for water animals*. Adana. (In Turkish.)
15. Sultanov E. H. (1999) [Distribution of waterbirds along western oil pipeline.] *X National Symposium for water animals*. Adana. (In Turkish.)
16. Sultanov, E. H., Gavashelishvili, A., Kerimov, T., Javakhi, Z., Agayeva, N. and Humbatova, S. (1999) [*Ecological monitoring with the public's participation at sites noted for high biodiversity and vulnerability of ecosystems along Baku-Supsa oil pipeline*.] Baku-Tbilisi. (In Azeri, Georgian, English and Russian.)
17. Sultanov, E. H. and Musayev, A. M. (2000) [Anseriformes.] Pp.290–301 in [*Animal world of Azerbaijan. Vol.3. Vertebrates*.] Baku, Elm. (In Russian.)
18. Sultanov, E. H. (2000) [Charadriiformes.] Pp.334–351 in [*Animal world of Azerbaijan. Vol.3. Vertebrates*.] Baku, Elm. (In Russian.)
19. Sultanov, E. H. [Hirundinidae.] Pp.366–370 in [*Animal world of Azerbaijan. Vol.3. Vertebrates*.] Baku, Elm. (In Russian.)
20. Sultanov, E. H. [Alaudidae.] Pp.370–372 in [*Animal world of Azerbaijan. Vol.3. Vertebrates*.] Baku, Elm. (In Russian.)
21. Sultanov, E. H. [Oriolidae.] Pp.378–379 in [*Animal world of Azerbaijan. Vol.3. Vertebrates*.] Baku, Elm. (In Russian.)
22. Sultanov, E. H. [Trogloditidae.] Pp.385–386 in [*Animal world of Azerbaijan. Vol.3. Vertebrates*.] Baku, Elm. (In Russian.)
23. Sultanov, E. H. [Prunellidae.] Pp.386–387 in [*Animal world of Azerbaijan. Vol.3. Vertebrates*.] Baku, Elm. (In Russian.)
24. Sultanov, E. H. [Silviidae.] Pp.387–396 in [*Animal world of Azerbaijan. Vol.3. Vertebrates*.] Baku, Elm. (In Russian.)
25. Sultanov, E. H. [Muscicapidae.] Pp.396–398 in [*Animal world of Azerbaijan. Vol.3. Vertebrates*.] Baku, Elm. (In Russian.)
26. Sultanov, E. H. [Turdidae.] Pp.398–408 in [*Animal world of Azerbaijan. Vol.3. Vertebrates*.] Baku, Elm. (In Russian.)
27. Sultanov, E. H. [Paridae.] Pp.408–412 in [*Animal world of Azerbaijan. Vol.3. Vertebrates*.] Baku, Elm. (In Russian.)
28. Mustafayev, G. T. and Sultanov, E. H. [Emberizidae. Pp.426–430 in [*Animal world of Azerbaijan. Vol.3. Vertebrates*.] Baku, Elm. (In Russian.)
29. Sultanov, E. H., Kerimov, T. Aliyev, S. and Humbatova, S., eds. (2000) [*Potential Ramsar sites of Azerbaijan*]. Wetlands International – AEME publ., Baku. (In Azeri and Russian.)
30. Sultanov E. H. (2000) *Methods of monitoring in Caspian Sea*. Gapp-Poligraph, Baku.
31. Sultanov, E. H. (2001) Status of the Lesser White-fronted Geese (Anser erythropus) in Azerbaijan. Pp.8–12 in *Focus on the Caspian Sea. A regional action plan for protection of the Lesser White-fronted Goose (*Anser erythropus). Bulletin of the Regional Environmental Agency of Hame, Finland.
32. Sultanov, E. H. (2001) Oil platforms. A birds eye view. *Azerbaijan International*, 2: 32–34.
33. Sultanov, E. H. (2001) Wetlands ecosystems and problems of biodiversity in Azerbaijan. *Asian Wetlands Symposium 2001: Bringing Partnerships Into Wetlands Practies. 27–30*, Penang, Malaysia.
34. Shelton, N. and Sultanov, E. H. (2001) *Where to watch birds in Azerbaijan*. "Azerbaijan" Publishing House, Baku.
35. Suktanov E. H., Kerimov T. A., Agayeva, N. Ch. and Talibov, Sh. T. [*Let's protect rare waterbirds of Azerbaijan!*] Baku, Sada. (In Azeri and Russian.)
36. Sultanov, E. H. (2002) Oil and problems of biodiversity of Caspian region. *Abstracts of 1st International Symposium: "The European Union and the Caspian sea region in dialogue – sustainability, environment policy and Caspian region development". 11–13 March 2002*, Berlin.
37. Sultanov, E. H. (2001) Status of White-Headed Ducks in Azerbaijan. *TWSG News* 13: 44–45.
38. Sultanov E.H. (2001) Problems of biodiversity and bird study in Azerbaijan. Pp.51 in Abstract of the International Conference on ecology and environmental management in Caucasus, Tbilisi, Georgia, 6–7 October 2001.
39. Sultanov E. H. (2001) Problems of biodiversity and bird study in Azerbaijan. Pp.88–92 in International Conference on ecology and environmental management in Caucasus, Tbilisi, Georgia, 6–7 October 2001.

40. Sultanov E. H. and Mukhtarov Kh. Sh. (2001) [Concentration especialities of birds along North and Western Oil Pipelines.] Pp.350–352 in *Proc. of conference in honor of 80-years academition M.A. Musayev.* Baku, Elm. (In Azeri.)
41. Sultanov, E. H. (2002) [Dynamics of waterbirds along Caspian coast of Azerbaijan.] Pp.103–105 in *Proceedings of the all-Russian scientific conference celebrating the 175th anniversary of the birth of P.A. Semyonov-Tyanshanski.* Lipetsk. (In Russian.)
42. Sultanov, E. H. and Agayeva, N. (2003) The current breeding status of Ferruginous Duck *Aythya nyroca* in Azerbaijan. *Sandgrouse* 25(1): 41–49.
43. Sultanov, E. H. (2003) [Oil pollution and the fate of waterfowl on the Caspian Sea.] *Caucasus Environment* 1(2): 36–40. (In English and Russian.)
44. Sultanov, E. H. (2003) Current Status and Coservation of the Ferruginous Duck in Azerbaijan. Pp.80–83 in N. Petkov, B. Huges and U. Gallo-Orsi, eds. *Ferruginous duck. From Research to Conservation.* Conservation Series No. 6. BirdLife International – BSPB – TWSG, Sofia.
45. Sultanov E. H., Kerimov T. A. and Agayeva N. Ch. (2003) [Modern summer and winter ornithocomplexes of lake Sarisu.] Pp.440–445 in Proc.of 1-st Congress of Azerbaijan Zoological Society. Baku, 2003. (In Russian.)
46. Kerimov T. A., Sultanov E. H. (2003) [Migration, dynamics and territorial distribution of swans along Azerbaijan coast of Caspian Sea. Pp.483–489 in Proc. of 1-st Congress of Azerbaijan Zoological Society. Baku, 2003. (In Azeri.)
47. Sultanov, E. H. Kerimov, T., Klaus, S. and Etzold, Y. (2003) [*Caucasian Black Grouse.*] Baku, Asada. (In Azeri.)
48. Klaus, S., Bergmann, H-H., Wiesner, J., Vitovich, O. A., Etzold, J. und Sultanov, E. H. 2003. [Verhaltenund Okolgie des Kaukasusbirkhuhns *Tetrao mlokosiewiczi* – stumme Balz am steilen Hang.] *Limicola* 7(5): 225–268. (In German.)

Belarus

Breeding

1. Nikiforov, M. E. and Kozulin, A. V., eds. (1997) *Belarussian birds at the beginning of XXI century: status, numbers, distribution.*
2. Samusenko, I. (2003) Great Cormorant Breeding Population in Belarus. *Vogelwelt* 124 (suppl.): 87–91.

Winter

3. Kozulin, A. V. (2001) Changes in species composition, numbers and habitat selection of wintering waterfowl in Belarus in 1967–2000. In S. Svazas, Meissner, W., Serebryakov, W., Kozulin, A. and Grishanov, G. eds. *Changes of wintering sites of waterfowl in central and Eastern Europe.* OMPO Special Publication. OMPO Vilnius and Lithuanian Institute of Ecology. Vilnius, Lithuania.
4. Nikiforov, M. E. and Kozulin, A. V. eds. (1997) *Belarussian birds at the beginning of XXI century: status, numbers, distribution.*

Belgium

General

1. Anselin, A. and Jacob, J.-P. (2003) Belgian contribution to the project "Birds in Europe 2" of Birdlife International. Data compilation of bird monitoring and survey programmes in Flanders (Vermeersch, G., Devos, K. and Anselin, A., coord.), Wallonia and the Brussels Region (Jacob, J.-P., coord.). Unpublished data. Instituut voor Natuurbehoud, Brussels and Aves (convention Région wallonne), Liège.

Bosnia and Herzegovina

Breeding

1. Gašić, B. (2001) [Rezultati novih istraživanja faune ptica Republike Srpske.] *Ciconia* 10:108–128. (In Serbian)
2. Kotrošan, D. (2003) [Ornitofauna hercegovačkog vinograda "željuša".] Unpublished data (Master thesis). (In Bosnian)
3. Marinković, S and Orlandić, Lj. (1989) [Prvi nalaz planinskog orla (*Hieraaetus fasciatus* V.) na gniježđenju u Jugoslaviji.] *Larus*, 40: 179–183. (In Serbian)
4. Mulaomerović, J., Kotrošan, D., and Bajraktarević, A. (2002) [Dodatak poznavanju gnježđenja žutokljune galice, *Pyrrhocorax graculus* (Linnaeus, 1758), na području planine Vranice.] *Naš Krš*, vol 22, 35:65–73. (In Bosnian)
5. Obratil, S. (1987) [Naselja ptica (*Aves*) u životnim zajednicama na trajnim plohama nacionalnog parka "Sutjeska".] *Godišnjak biološkog instituta*, 40: 73–87. (In Croatian)
6. Obratil, S. (1988) [Fauna ptica (*Aves*) Vran planine i međugorske depresije Dugo polje.] *Glasnik Zemaljskog Muzeja BiH* (PN) NS 27: 161–188. (In Croatian)
7. Obratil, S. (1992) [Prilog poznavanju avifaune Buškog jezera (Bosna) i užeg priobalnog područja.] *Glasnik Zemaljskog Muzeja BiH* (PN) NS 30: 155–168. (In Croatian)
8. Obratil, S. (1996) [Prva istraživanja ornitofaune Hutova blata poslije izgradnje akumulacionog jezera PHE "Čapljina".] *Glasnik Zemaljskog Muzeja BiH* (PN) NS 31: 403–429. (In Croatian)
9. Obratil, S. (1999) The Corncrake (*Crex crex*) in Bosnia-Herzegovina. Pp.19–21 in N. Schäffer and U. Mammen, eds. *Proceedings International Corncrake Workshop 1998.* Hilpoltstein, Germany.
10. Obratil, S. (2001) Research of the ornithofauna in Hutovo blato in the period January–December 2000. Report for project "Development of new management policy for the hutovo blato marshland". Unpublished data.

Winter

11. Gašić, B. (2001) [Rezultati novih istraživanja faune ptica Republike Srpske.] *Ciconia* 10:108–128. (In Serbian)
12. Kotrošan, D. (2003) [Ornitofauna hercegovačkog vinograda "željuša".] Unpublished data (Master thesis). (In Bosnian)
13. Obratil, S. (1987) [Naselja ptica (*Aves*) u životnim zajednicama na trajnim plohama nacionalnog parka "Sutjeska".] *Godišnjak biološkog instituta* 40: 73–87. (In Croatian)
14. Obratil, S. (1988) [Fauna ptica (*Aves*) Vran planine i međugorske depresije Dugo polje.] *Glasnik Zemaljskog Muzeja BiH* (PN) NS 27: 161–188. (In Croatian)
15. Obratil, S. (1992) [Prilog poznavanju avifaune Buškog jezera (Bosna) i užeg priobalnog područja.] *Glasnik Zemaljskog Muzeja BiH* (PN) NS 30: 155–168. (In Croatian)
16. Obratil, S. (1996) [Prva istraživanja ornitofaune Hutova blata poslije izgradnje akumulacionog jezera PHE "Èapljina".] *Glasnik Zemaljskog Muzeja BiH* (PN) NS 31: 403–429. (In Croatian)
17. Obratil, S. (2001) Research of the ornithofauna in Hutovo blato in the period January–December 2000. Report for project "Development of new management policy for the Hutovo blato marshland". Unpublished data.

Bulgaria

General

1. Antonov, A. and Atanasova, D. (2002) Cohabitation and nest-site selection of Common Swift (*Apus apus*) and Pallid Swift (*Apus pallidus*). *Vogelwarte* 41: 231–239.
2. Barov, B. (2002) [National Action Plan for the conservation of the Lesser Kestrel (*Falco naumanni*) in Bulgaria, 2002–2006.] Pp.161–182 in P. Iankov, ed. [*Globally threatened species of birds in Bulgaria. National action plans for their conservation.*] Part 1. BSBP – MoEW, Conservation Series, Book 4, BSPB, Sofia. (In Bulgarian.)
3. Delov, V. and Petkov, N. (2002) [National Action Plan for the Conservation of the Corncrake (*Crex crex*) in Bulgaria, 2002–2006.] Pp.183–203 in P. Iankov, ed. [*Globally threatened species of birds in Bulgaria. National action plans for their conservation.*] Part 1. BSBP – MoEW, Conservation Series, Book 4, BSPB, Sofia. (In Bulgarian.)
4. Dereliev, S. (2000a) [Dynamics of the numbers and distribution of the Red-breasted Goose *Branta ruficollis* (Pallas, 1769) in its main wintering area in the region of the lakes Shabla and Durankulak.] MSc thesis. Faculty of Biology, Sofia University "St. Kliment Ohridski", Bulgaria. (In Bulgarian.)
5. Dereliev, S., Hulea, D., Ivanov, B., Sutherland, W. and Summers, R. (2000b) The number and distribution of Red-breasted Geese (*Branta ruficollis*) at winter roosts in Romania and Bulgaria. *Acta Ornitologica* 35: 63–66.
6. Georgiev D., ed. (2002) [Management Plan of Durankulak Lake Protected Site.] Bulgarian Biodiversity Foundation – BSBCP and MoEW. Sofia. (In Bulgarian.)
7. Georgiev D., ed. (2003) [Management Plan of Shabla Lake Protected Site.] Bulgarian Biodiversity Foundation – BSBCP and MoEW. Sofia. (In Bulgarian.)
8. Iankov P., ed. (In print.) *The Atlas of the Breeding Birds in Bulgaria.* BSBP Conservation Series. BSPB, Sofia.
9. Iankov P., ed. (2003) [Management plan of the Rilski Monastery Nature Park.] ARD and MoEW. Sofia. (In Bulgarian.)
10. Ivanov, B., Karaivanov, N. and Nonev, S. (1998) Breeding Birds communities in the steppe habitats in Dobrudja, Bulgaria. *Acta Zool. Bulg.* 50(2/3): 67–77.
11. Ivanov, B. and Nonev, S. (1997) [Breeding birds in the region of Kaliakra, Bulgaria.] Pp.99–107 in Dobrudja and Kaliakra, Scientific Reports, Agriculture University, Plovdiv. (In Bulgarian.)
12. Ivanov, B. and Nonev, S. (1997) [Breeding birds in the steppe areas along the Bulgarian coast between the town of Balchik and Durankulak Lake.] Pp.108–125 in Dobrudja and Kaliakra, Scientific Reports, Agriculture University, Plovdiv. (In Bulgarian.)
13. Ivanov, B. and Muraveev, J. (2002) [National Action Plan for the Conservation of the Pygmy Cormorant (*Phalacrocorax pygmeus*) in Bulgaria, 2002–2006.] Pp.13–37 in P. Iankov, ed. [*Globally threatened species of birds in Bulgaria. National action plans for their conservation.*] Part 1. BSBP – MoEW, Conservation Series, Book 4, BSPB, Sofia. (In Bulgarian.)
14. Kostadinova, I., comp. (1997) *Important Bird Areas in Bulgaria.* BSBP Conservation Series. Book 1. BSPB, Sofia. (In Bulgarian with English summary.)

15. Kostadonova, I. and Dereliev, S. (2001) *Results from the mid-winter counts of waterbirds in Bulgaria for the period 1997–2001*. BSBP Conservation Series. Book 3, BSPB, Sofia.
16. Miltschev, B., Georgiev, V. and Kovachev, A. (2002) [Breeding status and the nest site selection of the Barn Owl (*Tyto alba*) in South-eastern Bulgaria.] *Egretta* 45: 114–120. (In German with English summary.)
17. Mitchev, T. and Profirov, L. (2003) *Mid-winter numbers of waterbirds in Bulgaria (1977–2001)*. Pensoft Publishers. Sofia.
18. Mitchev, T. and Crivelli, A. (1998) [Les effectifs reroducteurrs du pelican Frise (Pelecanus crispus Bruch, 1832) dans la reserve de Biosphere de "Srebarna, 1955–1998.] *Branta No 3/1998*. Le Balkan Bulgaria, Sofia. (In French.)
19. Nankinov, D. (2002) [Contemporary state of the owls populations in Bulgaria.] *Berkut* 11(1): 48–60. (In Russian with English summary.)
20. Nankinov, D., Shurulinkov, P., Nikolov, B., Nikolov, I., Dalakchieva, S., Hristov, I., Stanchev, R., Rogev, A., Dutsov, A. and Sarov, M. (1998) Studies of the Waders (*Charadriiformes*) on the wetlands around Sofia (Bulgaria). *Riv. Ital. Orn.* 68(1): 63–83.
21. Nankinov, D. and Nikolov, B. (2003) On the subspecies, breeding, migration and wintering of the Great Grey Shrike (*Lanius excubitor*) in Bulgaria. *Biota* 4(1).
22. Peev, D., ed. (2002) [*Management plan of the Rila National Park.*] ARD and MoEW. Sofia. (In Bulgarian.)
23. Petkov, N. (2002) [National Action Plan for the conservation of the Ferruginous Duck (*Aythya nyroca*) in Bulgaria, 2002–2006.] Pp.61–84 in P. Iankov, ed. [*Globally threatened species of birds in Bulgaria. National action plans for their conservation.*] Part 1. BSBP – MoEW, Conservation Series, Book 4, BSPB, Sofia. (In Bulgarian.)
24. Petkov, N. (2003) Status and distribution of breeding Ferruginous Duck in Bulgaria. Pp.22–27 in N. Petkov, B. Hughes. and U. Gallo-Orsi, eds. *Ferruginous Duck: from research to conservation*. Conservation Series No 6, BirdLife International – BSPB – TWSG, Sofia.
25. Petrov, Tz., ed. (1997) *The White Stork (*Ciconia ciconia*) in Bulgaria*. BSBP Conservation Series, Book 2, BSPB, Plovdiv. (In Bulgarian with English summary.)
26. Petrov, Tz. and Stoychev, S. (2002) [National Action Plan for the Conservation of the Imperial Eagle (*Aquila heliaca*) in Bulgaria, 2002–2006.] Pp.132–160 in P. Iankov, ed. [*Globally threatened species of birds in Bulgaria. National action plans for their conservation.*] Part 1. BSBP – MoEW, Conservation Series, Book 4, BSPB, Sofia. (In Bulgarian.)
27. Simeonov, P., Mitchev, T., Dobler, G. and Crivelli, A. J. (2000) White Pelicans Breeding in Bulgaria. *British Birds* 94: 390.
28. Shurulinkov, P., Nikolov, B. and Tsonev, R. (2001) On the Distribution of the Black-headed Bunting (*E. melanocephala*) in Bulgaria. Pp.81–87 in P. Tryjanowski , T. Osiejuk , M. Kupczyk, eds. *Bunting Studies in Europe*. Bogucki Wyd. Nauk., Poznan.
29. Shurulinkov, P. and Hristov, I. (2001) [*The birds of the Vitosha Nature Park.*] "Vitosha" Library. Vitosha Nature Park Directorate. Sofia. (In Bulgarian.)

Croatia

General

1. Belić, M. and Šafarek, G. (2002) Finding of the Dotterel *Charadrius morinellus* on the Velebit Mountain. *Larus* 48: 92–94.
2. Crnković, R., Grlica, I. D., Šetina, M. and Barać, D. (2002) Breeding of the black-vinged Stilt *Himantopus himantopus* (L.) in Croatia. *Larus* 48: 85–88.
3. Dolenec, Z. (2002) Intraclutch egg dimensions variation in the tree sparrow *Passer montanus*. *Larus* 48: 47–54.
4. Gec, D. (1993) Occurence of Summer Chafer and Lesser Castrel in Osijek during 1984. *Troglodytes* 6: 105–108.
5. Grošelj, P. (Unknown) Breeding of the Fieldfare pilars in Croatia. *Larus* 48: 90–91.
6. Hutuzović, B. H. (1995) Breeding of the bee-eater in Slavonia and Baranja. *Troglodytes* 8: 111–112.
7. Jurčević, I. (1995) Wintering of Long-eared owls in Osijek during 1992–3 and 1993–4. *Troglodytes* 8: 89–95.
8. Kralj, J., Leskovar, K., Radović, D., Tutiš, V., Pongrac, Z. and Sušić, G. (1993) Breeding of the tufted duck (*Aythia fuligula*) in the Pokupsko Basin (Croatia). *Larus* 44/45: 33–40.
9. Kralj, J. and Radović, D. (2002) Rare birds in Croatia (1st report of the Croatian Rarities Committe). *Larus* 48: 73–83.
10. Leskovar, K. (2002) [Najzapadnijegniježđenje bjelobrade čigre *Chilodonias hybridus* u Hrvatskoj.] *Larus* 48: 88–89. (In Croatian.)
11. Leskovar, K. (1999) [Gniježđenje močvarne sove (*Asio flammeus* (Pont.)) u sjeverozapadnoj Hrvatskoj.] *Larus* 47: 125–127. (In Croatian.)
12. Lukač, G., Kuèinić, M. and Vuković, V. (1993) Northern Gannet in Croatia – Third record. *Troglodytes* 6: 61–53.
13. Lukač, G. and Vujčić, S. (Unknown) [Karlo: Drugi nalaz snježne zebe (*Montifringila nivalis(L)*) u Hrvatskoj.] *Larus* 46: 126–127. (In Croatian.)
14. Lukač, G., Stipčević, M., Crnković, R. and Bem, D. (1992) Characteristics of habitat and distribution of *Sitta neumayer* Mich (Aves) in Croatia and neigbouring areas. *Natura Croatica* 1: 81–91.
15. Lukač, G., Karadžić, R. and Gottstein, S. (1993) [Uber die Farbungsvariabilitat von einigen Arten aus der Gattung Ficedula (Aves).] *Nat. Croat.* 2. (In German.)
16. Lukač, G. (2003) pers. comm. to A. Radoviæ.
17. Lukač, G., Stipčević, M. and Tvrtković, N. (1997) The status and distribution of Eleonora's falcon (*Falco eleonorae* 1834) in Croatia. *Nat. Croat.* 6(3).
18. Merić, S. and Merdić, E. (1995) Comparision of the diet of the Barn owl and the Long-eared owl in continental Croatia. *Troglodytes* 8.
19. Mikuska, J. and Mikuska, T. (1994) [Ptice Dunava na podruèju Hrvatske.] *Anali Zavoda za znanstveni rad u Osijeku, svezak* 10: 109–175. (In Croatian.)
20. Mikuska, J. and Mikuska, T. (1995) Results of mid-winter waterfowl counts in Croatia during 1990. *Troglodytes* 8: 13–50.
21. Mikuska, J. (1995) Breeding of the Bee-eater in East Croatia. *Troglodytes* 8: 113–122.
22. Mikuska, T. and Mikuska, J. (2002) From the ornithological notebook – Black-necked Grebe, *Podiceps nigricollis*. *Acrocephalus* 22(109): 239.
23. Mikuska, T., Mikuska, J. and Šetina, M. (1994) *Status of herons in Croatia during 1990–1993*. Abstract book. Colonial Waterbird Society 18th Stated Annual meeting, Wrightsville Beach, North Carolina.
24. Mikuska, T. (Unknown) The population of nesting herons in Croatia. *IWRB Spec. Publ. no. 20*: 68–69.
25. Mikuska, J. (2003) Pers. comm. to A. Radoviæ.
26. Mikuska, T. (2003) Pers. comm. to A. Radoviæ.
27. Mužinić, J., Delić, A. and Jandl, F. (1995) New localities of breeding colonies of the Grey Heron, *Ardea cinerea*, in Croatia. *Acrocephalus* 8: 9–12.
28. Mužinić, J., Lesinger, A. and Čonda, J. (1993) Breeding of the bee-eater in Međimurje (Northern Croatia). *Troglodytes* 6: 109–112.
29. Mužinić, J. (2003) pers. comm. to A. Radoviæ.
30. Purger, J. (1995) Breeding of the Bee-eater, *Merops apiaster*, near Nin. *Troglodytes* 8: 123.
31. Radović, D. and Dumbović, V. (1998) The Corncrake in Croatia. *Proceedings of the International Corncrake Workshop*. Hilpoltstein, Germany.
32. Radović, D. and Sušić, G. (1997) [Pripombe na neresnice navedene v èlanku g. Petra Trontlja "Ignoranca birokratskega aparata (IBA).] *Acrocephalus* 18: 80–81. (In Croatian.)
33. Rubinić, B. and Vrezec, A. (2000) Audoin's Gull *Larus audouinii*, a new breeding species in the Adriatic; Arocephalus 21 (102–103): 219–222.
34. Rubinić, B., Božić, L. and Sackl, P. (2002) The actual status of Slender-billed Gull on the Southern Dalmatian Coast. *Acrocephalus* 23(112): 87–90.
35. Rubinić, B. (2001) Expansion of the northwestern frontier of the Spanish Sparrow's *Passer hispaniolensis* breeding range along the Adriatic coast: behaviour of the new population in Croatian Istra. *Acrocephalus* 22(109): 207–211.
36. Rubinić, B. (2003) pers. comm. to A. Radoviæ.
37. Schneider-Jacoby, M. (2001) Lastovo – a new bottelneck site for the migratory Honey Buzzards *Pernis apivorus*? *Acrocephalus* 22(108): 163–165.
38. Schneider-Jacoby, M. (2000) [Freizeit und Entenschutz am Wasser – Sicherung der Brut- und Rastgebiete von Kolbenenten und Moorenten in Deutschland.] Bundesamt für Naturschutz, Bericht über die Tagung, Konzepte und Maßnahmen zur Erhaltung gefährdeter Wasservogelarten in Mitteleuropa: Beiträge zum Aktionsplan des afrikanische-eurasischen Wasservogelabkommens /AEWA, Schriftenreihe Naturschutz Landschaftspflege. (In German.)
39. Schneider-Jacoby, M. (1993) [Vögel als Indikatoren für das ökologische Potential der Saveauen und Möglichkeiten für deren Erhaltung.] Dissertation Resch, Überlingen. (In German.)
40. Schneider-Jacoby, M. (1994) Sava and Drava – Ecological value and future of the two main rivers in Croatia. *Period. Biol.* 96(4): 348–356.
41. Schneider-Jacoby, M. (1996) [Drau und Mur – Leben durch Flußdynamik.] Naturerbe Verlag Jürgen Resch, Überlingen. (In German.)
42. Schneider-Jacoby, M. (Unknown) Breeding distribution and ecology of the Black Stork *Ciconia nigra* in the Sava alluwial wetlands, Croatia.
43. Schneider-Jacoby, M. (2001) [Auswirkung der Jagd auf Wasservögel und die Bedeutung von Ruhezonen.] Fachtagung der ANL in Starnberg, Bayrische Beamtenhochschule. (In German.)
44. Schneider-Jacoby, M. (2003) Pers. comm. to A. Radoviæ.
45. Schneider-Jacoby, M. (1991) [Verbreitung und Bestand des Wachtelkönigs in Jugoslawien.] *Vogelwelt* 112:48–57. (In German.)
46. Schneider-Jacoby, M. (2000) Breeding distribution and ecology of the Black Stork *Ciconia nigra* in the Sava alluvial wetlands, Croatia. *Acrocephalus* 20(97): 167–176.

46a. Schneider-Jacoby, M. (1996) [Brutbestand des Seeadlers (*Haliaeetus albicilla*) und des Schreiadlers (*Aquila pomarina*) in den Save-Auen (Kroatien).] Pp.149–163 in *Eagles Studies*. World Working Group of Birds of Prey, Berlin, London & Paris. (In German.)
47. Špoljarić, N. (2000) [Gniježđenje orla štekavca, *Haliaeetus albicilla*, u Kopačkom ritu.] Diplomski rad, Pedagoški fakultet Sveučilišta u Osijeku. Unpublished report. (In Croatian.)
48. Stipčević, M. and Lukač, G. (Unknown) Status of tubenose seabirds Procellariiformes breeding in the Eastern Adriatic.
49. Stipčević, M., Lukač, G. and Radović, D. (1998) Status and the occurrence of *Sterna sandvicensis* in Croatia. *Nat. Croat*. 7(3).
50. Stipčević, M. and Lukač, G. (Unknown) Breeding-range changes of Red Swallow in Yugoslavia. *Brit. Birds* 85(1): 37–40.
51. Stipčević, M. and Lukač, G. (Unknown) Survey of spreading and breeding-range changes of Red Swallow *Hirundo daurica* on Yugoslav Territory. *Larus* 43: 37–49.
52. Stipčević, M. (1992) Šesto opazovanje afričke kukavice *Clamator glandarius* na Hrvaškem. *Acrocephalus* 13(55): 189. (In Croatian.)
53. Stipčević, M. (Unknown) Solitary breeding of the Griffon Vulture *Gyps fulvus* on the Island of Pag (Croatia) in 1997
54. Štumberger, B. (2003). Pers. comm. to A. Radoviæ.
55. Sušić, G. (Unknown) [Ptice cresko-lošinjske otočne skupine: Biologija Cresa i Lošinja.] *Otočki ljetopis Cres-Lošinj* (8): 145–157. (In Croatian.)
56. Sušić, G. (1991) [Stanje i perspektive populacije bjeloglavih supova na otoku Cresu u: Zbornik radova 4.] *Simpozijuma "Divljač i priroda"*, Ljubljana. (In Croatian.)
57. Tadić, Z. (2003). Pers. comm. to A. Radoviæ.
58. Tarbuk, Z. (1993) The birds of Jarun. *Troglodytes* 6: 95–103.
59. Tutiš, V., Bartowsky, V., Radović, D. and Sušić, G. (1990) [*Les repaces nocturnes nicheurs en Yougoslavie*.] Repaces Nocturnes. Actes du 30 Colloque interregional d'ornitologie. (In French.)
60. Vrazec, A. (2001) The breeding density of Eurasian Scops Owl *Otus scops* in urban areas of Pelješac Peninsula in Southern Dalmatia. *Acrocephalus* 22(108): 149–154.
61. Vrazec, A. (2003). Pers. comm. to A. Radoviæ.
62. Žganec, K. (2002) The Fieldfaire *Turdus pilaris*, a new breeding species in Croatia. *Larus* 48: 91–92.
63. Muzinic, J. (1999) The population of the White Stork, *Ciconia ciconia*, in the Croatia. Pp.213–217 in H. Shultz, ed. [*Weisstorch im Aufwing? – ' White Stork on up?'*] *Proceedings Internat. Symp. on the White Stork, Hamburg 1996*. NABU (Naturschutzbund Deutchland e. v.), Bonn. (In German.)
64. Mužinić, J. (1999) The population of the White Stork, *Ciconia ciconia*, in Croatia. In: Shultz, H. ed [*Weisstorch im Aufwing? – White Stork on up?*] *Proceedings Internat. Symp. On the White Stork, Hamburg 1996*. NABU.
65. Radovic, D., Kralj, J., Tutis, V. and Cikovic, D. (2003) [*Crvena knjiga ugrozenih ptica Hrvatske. – 'Red data book of birds of Croatia.'*] Ministarstvo zastite okolisa prostornog uredenja, Zagreb. (In Croatian.)

Cyprus

Breeding

1. Cyprus Ornithological Society (1957) Annual Bird Reports (1970–2003).
2. Cyprus Ornithological Society (1957) Monthly Newsletters.
3. Cyprus Ornithological Society (1970) Annual Bird Reports.
4. Flint, P. R. and Stewart, P. F. (1992) *The Birds of Cyprus*. British Ornithologists Union,Check List No 6, 11.5. Tring, UK.
5. Iezekiel, S. Forestry Department records.
6. Kasinis, N. Game Fund records.
7. Kourtellarides, L. (1998) *Breeding Birds of Cyprus*. Bank of Cyprus, Nicosia.
8. Mullarney, K., Svensson, L., Zetterstrom, D. and Grant, P. J. (2000) *Collins Bird Guide*. Harper Collins,London, UK.
9. Pomeroy, D. (1998) Counting whilst atlasing in Cyprus. *Bird Census News* vol.10(i).

Winter

10. Cyprus Ornithological Society (1957) Annual Bird Reports (1970–2003).
11. Cyprus Ornithological Society (1957) Monthly Newsletters.
12. Cyprus Ornithological Society (1970) Annual Bird Reports (1970–1990).
13. Flint, P. R. and Stewart, P. F. (1992) *The Birds of Cyprus*. British Ornithologists Union, Check List No 6, Tring, UK.
14. Kourtellarides, L. (1998) *Breeding Birds of Cyprus*. Bank of Cyprus, Nicosia.
15. Mullarney, K., Svensson, L., Zetterstrom, D. and Grant, P. J. (2000) *Collins Bird Guide*. Harper Collins,London, UK.

Czech Republic

Breeding

1. Hudec, K., ed. (1994) *Fauna ČR a SR – Ptáci 1*. [*Fauna of the Czech Republic and Slovak Republic – Birds 1*.] Academia Praha. (In Czech.)
2. Hudec, K., Chytil, J., Stastny, K. and Bejcek, V. (1995) Ptaci Ceske republiky. [The birds of the Czech Republic.] *Sylvia* 31: 97–149. (In Czech.)
3. Hudec, K. and Šťastný, K., eds. (in press). *Fauna ÈR – Ptáci 2/I, 2/II*. [*Fauna of the Czech Republic – Birds 2/I, 2/II*.] Academia Praha. (In Czech.)
4. Šťastný, K. and Bejček, V. (1993) Početnost hnízdních populací ptáků v České republice. [Breeding bird populations sizes in the Czech Republic.] *Sylvia* 29: 72–81. (In Czech.)
5. Šťastný, K. and Bejček, V. (2003) Cerveny seznam ptaku České republiky. In Plesnik, J., Hanzal, V. and Brejskova, L., eds. *Cerveny seznam ohrozenych druhu České republiky*. [*Red list of birds of the Czech Republic*.] Obratlovci. AOPK CR. (In Czech.)
6. Stastny, K., Bejcek, V. and Hudec, K. (1996) *Atlas hnizdniho rozsireni ptaku v Ceske republice 1985–1989*. [*The Atlas of breeding birds in the Czech Republic 1985–1989*.] HandH Praha. (In Czech.)
7. Stastny, K., Randik, A. and Hudec, K. (1987) *Atlas hnizdniho rozsireni ptaku v CSSR 1973/77*. [*The Atlas of breeding birds in Czechoslovakia 1973/77*.] Academia Praha. (In Czech.)

Winter

8. Bejcek, V., Stastny, K. and Hudec, K. (1995) *Atlas zimniho rozsireni ptaku v Ceske republice 1982–1985*. [*The Atlas of wintering birds in the Czech Republic 1982–1985*.] HandH Praha, MZP CR. (In Czech.)
9. Hudec, K., ed. (1994) *Fauna ÈR a SR – Ptáci 1*. [*Fauna of the Czech Republic and Slovak Republic – Birds 1*.] Academia Praha. (In Czech.)
10. Hudec, K., Chytil, J., Stastny, K. and Bejcek, V. (1995) Ptaci Ceske republiky. [The birds of the Czech Republic.] *Sylvia* 31: 97–149. (In Czech.)
11. Hudec, K. and Šťastný, K., eds. (in press) *Fauna ÈR – Ptáci 2/I, 2/II*. [*Fauna of the Czech Republic – Birds 2/I, 2/II*.] Academia Praha. (In Czech.)

Denmark

Breeding

1. Eskildsen, J. (2002) [Skarver 2002. Naturovervågning.] Danmarks Miljøundersøgelser. – *Arbejdsrapport fra DMU* 172: 47. (In Danish.)
2. Flensted, K. (2003) Pers. comm. to T. Nyegaard.
3. Grell, M. B. (1998) [Fuglenes Danmark.] Gads Forlag and Dansk Ornitologisk Forening. (In Danish.)
4. Grell, M. B. (1999) [Truede og sjældne ynglefugle i Danmark 1998.] *Dansk Ornitologisk Forenings tidsskrift* 93: 105–126. (In Danish.)
5. Grell, M. B. (2000) [Truede og sjældne ynglefugle i Danmark 1999.] *Dansk Ornitologisk Forenings tidsskrift* 94: 55–72. (In Danish.)
6. Grell, M. B. (2001) [Truede og sjældne ynglefugle i Danmark 2000.] *Dansk Ornitologisk Forenings tidsskrift* 95: 51–68. (In Danish.)
7. Grell, M. B. (2002) [Truede og sjældne ynglefugle i Danmark 2001.] *Dansk Ornitologisk Forenings tidsskrift* 96: 43–66. (In Danish.)
8. Grell, M. B. (2003) [Forslag til en forvaltningsplan for bevarelsen af Rød Glente Milvus milvus i Danmark.] Dansk Ornitologisk Forening. (In Danish.)
9. Grell, M. B. (in prep.) [Truede og sjældne ynglefugle i Danmark 2002.] *Dansk Ornitologisk Forenings tidsskrift*. (In Danish.)
10. Heldbjerg, H. and Grell, M. B. (2001) [Forslag til forvaltningsplan for den danske ynglebestand af Hjejle Pluvialis apricaria.] *Dansk Ornitologisk Forening*. (In Danish.)
11. Heldbjerg, H. (2001) The recent decline in the population of Black-headed Gulls *Larus ridibundus* in Denmark and its plausible causes. *Dansk Ornitologisk Forenings tidsskrift* 95: 19–27.
12. Jacobsen, E. M. (1997) [Hvor mange fugle yngler i Danmark?] *Dansk Ornitologisk Forenings tidsskrift* 91: 93–100. (In Danish.)
13. Jensen, N. O. and Jacobsen, L. B. (1996) [Ynglebestanden af Natravn Caprimulgus europaeus i Danmark, 1992–95.] *Dansk Ornitologisk Forenings tidsskrift* 90: 93–98. (In Danish.)
14. Jørgensen, H.E. (2003) Pers. comm. to T. Nyegaard.
15. Lange, P. (2003) Pers. comm. to T. Nyegaard.
16. Lange, P., Christensen, R. and Lindballe, P. (2002) [Fugle i Danmark 1999 og 2000.] *Dansk Ornitologisk Forenings tidsskrift* 96. (In Danish.)
17. Lyngs, P. (2003) Pers. comm. to T. Nyegaard.
18. Novrup, L. and Østergaard, E. (2003) Pers. comm. to T. Nyegaard.
19. Rasmussen, L. M. (2002) [Hedehøg i Sydvestjylland ynglesæsonen 2002.] *Dansk Ornitologisk Forening and Dyrenes Beskyttelse*. (In Danish).
20. Sunde, P. (2003) pers. comm. to T. Nyegaard.
21. Thorup, O. (2003) [Truede engfugle. Status for bestande og forvaltning i Danmark.] *Dansk Ornitologisk Forening*. (In Danish).
22. Thorup, O. (In prep.) [Breeding waders in Europe.]

Winter

23. Jørgensen, H. E. (2003) Pers. comm. to T. Nyegaard.
24. Lange, P., Christensen, R. and Lindballe, P. (2002) [Fugle i Danmark 1999 og 2000.] *Dansk Ornitologisk Forenings tidsskrift* 96. (In Danish.)
25. Pihl, S., Petersen, I. K., Hounisen, J. P. and Laubek, B. (2001) [Landsdækkende optælling af vandfugle, vinteren 1999/2000.] DMU. Faglig rapport fra DMU, nr. 356. (In Danish.)

26. Pihl, S., Laursen, K., Hounisen, J. P. and Frikke, J. (1992) [Landsdækkende optælling af vandfugle fra flyvemaskine, januar/februar 1991 og januar/marts 1992.] Danmarks Miljøundersøgelser. Faglig rapport fra DMU, nr. 44. (In Danish.)
27. World Bird Database (WBDB). Internet site: http://www.birdlife.info/wbdbeurope/. BirdLife International.

Faroe Islands

Breeding

1. Bengtson, S. -A., Eliasen, K., Mary Jacobsen, L. M. and Magnussen, E. (2003) A history of colonization and current status of the house sparrow (*Passer domesticus*) in the Faroe Islands. *Fróðskaparrit* 51: 237–251.
2. Bloch, D., Jensen, J. -K. and Olsen, B. (1996) List of birds seen in the Faroe Islands. Føroya Náttúrugripasavn, Føroya Fuglafrøðifelag and Føroya Skúlabókagrunnur, Tórshavn.

Winter

3. Bloch, D., Jensen, J. -K. and Olsen, B. (1996) List of birds seen in the Faroe Islands. Føroya Náttúrugripasavn, Føroya Fuglafrøðifelag and Føroya Skúlabókagrunnur, Tórshavn.

Greenland

Breeding

1. Boertmann, D. and Egevang, C. (2002) Canada geese *Branta canadensis* in West Greenland: In conflict with Greenland white-fronted geese *Anser albifrons flavrostris*? *Ardea* 90: 335–336.
2. Boertmann, D. and Glahder, C. (1999) [Grønlandske gåsebestande – en oversigt.] National Environmental Research Institute, Technical report no. 276. (In Danish.)
3. Boertmann, D. and Mosbech, A. (1999) [Bemærkelsesværdige fugleobservationer fra Vest- og Nordgrønland, 1992–1998.] – *Dansk Ornitologisk Forenings Tidsskrift* 93: 145–152. (In Danish.)
4. Boertmann, D. and Mosbech, A. (1997) Breeding distribution and abundance of the great cormorant *Phalacrocorax carbo carbo* in Greenland. *Polar Research* 16: 93–100.
5. Boertmann, D., Mosbech, A., Falk, K. and Kampp, K. (1996) Seabird colonies in western Greenland. – National Environmental Research Institute, technical report 170, Roskilde.
6. Boertmann, D. and Mosbech, A. (1998) Distribution of little auk (*Alle alle*) breeding colonies in Thule District, northwest Greenland. *Polar Biology* 19: 206–210.
7. Egevang, C., Boertmann, D., Mosbech, A. and Tamstorf, M. (2002) Estimating colony area and population size of little auks *Alle alle*, at Northumberland Island using aerial images. *Polar Biology* 26: 8–13.
8. Falk, K . and Kampp, K. (2001) [Lomvien i Grønland: mulige effekter af forskellige bestandspåvirkende faktorer, og praktiske grænser for ressourceudnyttelse.] Pinngortitaleriffik, Grønlands Naturinstitut, teknisk rapport nr.38. (In Danish.)
9. Fox, A. D., Glahder, C., Mitchell, C. R., Stroud, D. A., Boyd, H. and Frikke, J. (1996) North American Canada Geese (*Branta canadensis*) in West Greenland. *Auk* 113: 231–233.
10. Kampp, K. and Wille, F. (1990) The White-tailed Eagle population in Greenland. *Dansk Ornithologisk Forenings Tidsskrift* 84: 37–44.
11. Melfofte, H. (2001) Wader population censuses in the Arctic: getting the timing right. *Arctic* 54: 367–376.
12. Merkel, F. R. (2002) [Ederfugleoptællinger i Ilulissat, Ummannaq og Upernavik Kommuner, 1998–2001.] Pinngortitaleriffik, Grønlands Naturinstitut, teknisk rapport nr. 43. (In Danish.)
13. Boertmann, D. and Mosbech, A. (2002) Moulting Harlequin ducks in Greenland. *Waterbirds* 25: 326–332.
14. Boertmann, D. (2003) Distribution and conservation of Harlequin Duck, *Histrionicus histrionicus*, in Greenland. *Canadian Field-Naturalist* 117: 249–256.
15. Boertmann, D. (2003) Seabird colonies and moulting harlequin ducks in South Greenland. National Environmental Research Institute. Research Notes from NERI no. 191.
16. Gilg, O., Sabard, B., Sittler, B., Mariaux, F., Leguesdron, P. and Gilg, V. (2004) Ecopolaris 2003. Ecological expedition to North Greenland & Dove Bugt: North-East Greenland National Park. Field report, Groupe de Recherches en Ecologie Arctique).
17. Nyeland, J. (In print.) Apparent trends in the Black-legged Kittiwake in Greenland. *Waterbirds*.
18. Boertmann, D., Lyngs, P., Merkel, F. and Mosbech A. (2004) The significance of Southwest Greenland as winter quarters for seabirds. *Bird Conservation International* 14: 87–112.

Winter

19. Merkel, F. R. M., Mosbech, A., Boertmann, D. and Groendahl, L. (2002) Winter seabird distribution and abundance off south-western Greenland, 1999. *Polar Research* 21: 17–36.
20. Mosbech, A. and Johnson, S. (1999) Late winter distribution and abundance of sea-associated birds in Southwest Greenland, Davis Strait and Baffin Bay. *Polar Research* 18: 1–17.

Estonia

General

1. Elts, J., Kuresoo, A., Leibak, E., Leito, A., Lilleleht, V., Luigujõe, L., Lõhmus, A., Mägi, E. and Ots, M. (2003) Status and numbers of Estonian birds, 1998–2002. *Hirundo* 2: 58–83.
2. Lõhmus, A., Kuresoo, A., Leibak, E., Leito, A., Lilleleht, V., Kose, M., Leivits, A., Luigujõe, L and Sellis, U. (1998) Status and numbers of Estonian birds. *Hirundo* 2: 63– 83.

Finland

General

1. Hannula, H., Haapala, J. and Saurola, P. (2002) [Petolintuvuosi 2001 – myyrät harvojen herkkua.] *Linnut-vuosikirja* 2001: 15–25. (In Finnish.)
2. Hario, M. and Rintala, J. (2004) [Kyhmyjoutsenen, haahkan ja hanhien kannankehitys rannikolla 1986–2003.] *Linnut vuosikirja* 2003: 49–57. (In Finnish.)
3. Hario, M. and Rintala, J. (2002) [Haahkan ja lokkien kannankehitys rannikoillamme vuosina 1986–2001.] *Linnut-vuosikirja* 2001: 26–36. (In Finnish.)
4. Laine, T. (2002) [Valkoselkätikan seuranta 1999–2001: lievä nousukausi alkanut.] *Linnut-vuosikirja* 2001: 10–14. (In Finnish.)
5. Ollila, T. (2003) [Suomen maakotkat 2000–2002.] *Linnut-vuosikirja* 2002: 24–25. (In Finnish.)
6. Ollila, T. (2003) [Suomen muuttohaukkakanta kasvussa.] *Linnut-vuosikirja* 2002: 26–27. (In Finnish.)
7. Rusanen, P., Mikkola-Roos, M. and Asanti, T. (2002) [Merimetso – matkalla maineeseen.] *Linnut-vuosikirja* 2001: 37–41. (In Finnish.)
8. Saurola, P. (2003) [Suomen sääkset – tarkastustyö jatkui tehokkaana.] *Linnut-vuosikirja* 2002: 20–23. (In Finnish.)
9. Väisänen, R. A. (2003) [Yleisten talvilintujen kannanmuutokset 27 talvena Suomen eri osissa.] *Linnut-vuosikirja* 2002: 41–62. (In Finnish.)
10. Väisänen, R. A., Lammi, E. and Koskimies, P. (1998) [*Muuttuva pesimälinnusto*.] Otava, Keuruu. (in Finnish.)

France

Breeding

1. Dubois, P. J., Le Maréchal, P., Olioso, G. and Yésou, P. (2000) [*L'Inventaire des oiseaux de France*.] Avifaune de la France métropolitaine, Nathan, Paris. (In French.)
2. Jiguet, F., Julliard, R. and Weltz, M. (2003) [*État de santé des populations d'Oiseaux nicheurs en France de 1989 à 2001*.] Internet publication: http://www.mnhn.fr/mnhn/meo/crbpo/. (In French.)
3. Julliard, R. (2002) [Programme STOC-Capture Bilan 2001 pour la France.] Suivi temporel des oiseaux communs par échantillonnage par filets japonais. *Ornithos* 9: 129–137. (In French.)
4. LPO BirdLife France. (2003) Bird inventory of France and overseas territories. Database. Rochefort.
5. Sériot, J. (2002) [Suivi quantitatif annuel des espèces nicheuses rares et menacées en France – Année 2000.] Rapport LPO / MEDD. Rochefort. (In French.)
6. Sériot, J. (2003) [Suivi quantitatif annuel des espèces nicheuses rares et menacées en France – Années 2001 et 2002.] Rapport LPO / MEDD. Rochefort. (In French.)
7. Thiollay, J. -M., Bretagnolle, V. and Sériot, J. (2003) [Inventaire des rapaces de France 2000–2003.] Rapport LPO/MEDD Rochefort. (In French.)
8. Thiollay, J. -M., Bretagnolle, V. and Sériot, J. (in press) [*Atlas des rapaces nicheurs de France 2000–2003*.] (In French.)
9. Gonin, J. (2003) [Effectif et repartition du Cochevis de Thékla (*Galerida theklae*) en France. Poster 28ème Colloque Francophone d'Ornithologie. Aves Namur 28–30 novembre 2003. (In French.)
10. LPO-BirdLife France (2004) [Base de données, Zones importantes pour la conservation des oiseaux en France. 'Database, Important Bird Areas in France.'] LPO-BirdLife France, Rochefort. (In French.)

Winter

11. LPO BirdLife France. (2003) Bird inventory of France and overseas territories. Database. Rochefort

Georgia

General

1. GCCW database (2003) GCCW (Georgian Center for the Conservation of Wildlife).

Germany

Breeding

1. Bauer, H. -G., Berthold, P., Boye, P., Knief, W. Südbeck, P. and Witt, K. (2003) [Rote Liste der Brutvögel (Aves), 3. Fassung, 31.12.2001. – 'The 2002 Red list of breeding birds of Germany'] *Ber. Vogelschutz* 39: 13–60. (In German with English summary.)

Winter
2. Blew, J., Günther, K. Laursen, K., van Roomen, M. W. J. and Südbeck, P. (In press.) Overview on trends and numbers of migratory waterbirds in the Wadden Sea 1992–2000. In J. Blew and P. Südbeck, eds. *Monitoring of migratory waterbirds in the Wadden Sea 1992–2000*. Waddensea Ecosystems, Common Waddensea Secretariat, Wilhelmshaven, Germany.
3. Garthe, S. (2003) [Verteilungsmuster und bestände von seevögeln in der Ausschließlichen Wirtschaftszone (AWZ) der deutschen Nord- und Ostsee und Fachvorschläge für EU-Vogelschutzgebiete – 'Distribution patterns and numbers of seabirds in the German Exclusive Economic Zone (EEZ) of North Sea and Baltic Sea and proposals for EU Special Protection Areas'.] *Ber. Vogelschutz* 40: 15–56. (In German with English summary.)
4. Garthe, S., Ullrich, N., Weichler, T., Dierschke, V., Kubetzki, U., Kotzerka, J., Krüger, T., Sonntag, N. and Helbig, A. J. (2003) [*See- und Wasservögel der deutschen Ostsee – Verbreitung, Gefährdung, Schutz. – ' Seabirds and waterbirds of the German Baltic Sea – Distribution, threats and conservation*.'] Bundesamt für Naturschutz (Federal Agency for Nature Conservation), Bonn-Bad Godesberg, Germany. (In German.)
5. Wahl, J., Blew, J., Garthe, S., Günther, K., Mooij, J. H. and Sudfeldt, C. (2003) [Überwinternde Wasser- und Watvögel in Deutschland: Bestandsgrößen und Trends ausgewählter Vogelarten für den Zeitraum 1990–2000. – ' Waterbirds wintering in Germany – trends and population estimates for 1990–2000'] *Ber. Vogelschutz* 40: 91–103. (In German with English summary.)

Greece

Breeding
1. BirdLife International (2002) Action Plan for the conservation of White-tailed Sea Eagle (*Haliaeetus albicilla*). B. Helander and T. Stjernberg, compilers. Unpublished data.
2. BirdLife International (1999) International Species Action Plan: Eleonora's Falcon (*Falco eleonorae*). D. Ristow, compiler. Unpublished data.
3. BirdLife International (1999) International Species Action Plan: Lanner Falcon *Falco biarmicus*. M. Gustin, G. Palumbo and A. Corso, compilers. Unpublished data.
4. BirdLife International (2002) International Species Action Plan: Cinereous Bunting *Emberiza cineracea*. T. Albayrak, A. Gursoy and G.M. Kirwan, compilers. Unpublished data.
5. Bourdakis, S. and Vareltzidou, S., eds. (2000) Greece. Pp.261–333 in M. F. Heath and M. I. Evans, eds. *Important Bird Areas in Europe: Priority sites for conservation. 2: Southern Europe*. Cambridge, UK: Birdlife International Birdlife Conservation Series No. 8.
6. Bousbouras, D. and Bourdakis, S. (1999) Vol 1. Pp.324 in D. Bousbouras, ed. (1999). *Special Environmental Study for Grammos and NW Voio. LIFE-NATURE project 'Arctos'*. European Commission, DG ENV, Ministry of Environment, Physical Planning and Public Works, Ministry of Agriculture, Arctouros.
7. Bousbouras, D. (2002) Bird monitoring report in the Special Protection Area of lake Chimaditis and lake Zazari. Conservation and management of lake Chimaditis and lake Zazaris, LIFE00NAT/GR/7242.
8. Crivelli, A. J., Catsadorakis, G., Hatzilacou, D., Hulea, D., Malakou, M., Marinov, M., Naziribes, T., Peja, N., Sarigul, G. and Siki, M. (2000) Status and population development of Great white pelican *Pelecanus onocrotalus* and Dalmatian pelican *P. crispus* breeding in the Palearctic. Pp.38–45 in P. Yésou and J. Sultana, eds. *Monitoring and conservation of birds, mammals and sea turtles of the Mediterranean and Black Seas*. La Vallette, Malta.
9. Dretakis, M. (2000) Ferruginous Ducks in Crete. *TWSG News* 12: 11.
10. Gatzelia, A., ed. (1999) [Final Technical Report LIFE-Nature Project "Actions for the conservation of the Audouin's Gull, *Larus audouinii*, in Greece".] (Hellenic Ornithological Society, European Commission, DG ENV, Min. of Environment, Physical Planning and Public Works.) (In Greek.)
11. Hallmann, B. (1995) Lesser Kestrel Survey Thessaly 1995, final report. Hellenic Ornithological Society. Unpublished data.
12. Hallmann, B. (1996) *Greece's endangered birds of prey, eleventh hour for 10 species*. WWF Greece.
13. Handrinos, G. I. (1992) Birds Pp.123–243 in M. Karandeinos, and T. Legakis, eds. [*The Red data book of threatened Vertebrates in Greece*.] Hellenic Zoological Society and Hellenic Ornithological Society, Athens. (In Greek.)
14. Handrinos, G. and Akriotis, T. (1997) *The birds of Greece*. London, UK, Helm Publishers.
15. Hellenic Ornithological Society (HOS) (1994) [*Important Bird Areas in Greece*.] Hellenic Ornithological Society Special Publication, Athens. (In Greek.)
16. Heath, M., Borggreve, C. and Peet, N. (2000) *European bird populations: estimates and trends*. Birdlife Conservation Series No 10. Birdlife International, Cambridge.
17. Heredia, B., Rose, L. and Painter, M., eds. (1996) *Globally threatened birds in Europe. Action plans*. Strasbourg: Council of Europe.
18. Holzinger, J. (1993) [Schneefink (*Montifringilla nivalis*) Brutvogel in Griechenland]. *J. Orn.* 134: 405–411. (In German.)
19. Kazantzidis, S. (1998) The breeding ecology of the Little Egret (*Egretta g. garzetta*) in the Axios Delta, Macedonia, Greece. Aristotelian University of Thessaloniki (Doctorate Thesis). Unpublished data.
20. Kazantzidis, S. and Naziribes, T. (1999) [National Action Plan for the Pygmy Cormorant *Phalacrocorax pygmaeus*.] WWF-HOS-Society for the protection of Prespa. (In Greek.)
21. Kazantzidis, S. and Naziridis, T. (2003) [The effect of Cormorant *Phalacrocorax carbo sinensis* (Linnaeus, 1758) on fishery at Lake Kerkini, Greece.] Pp.235–238 in Proceedings of the 11th Pan Hellenic Ichthyologists Conference, Preveza, Greece, 10–13 April 2003. (In Greek with summary in English).
22. Goutner, V. and Isenmann, P. (1993) Breeding status of the Mediterranean Gull (*Larus melanocephalus*) in the Mediterranean basin. Pp.59–63 in J. S. Aguilar, X. Monbailliu and A. Paterson, eds. *Status and conservation of seabirds-ecogeography and Mediterranean Action Plan. Proceed. 2nd Mediterr. Seabird Symp. Calvia, 21–26 March 1989*. SEO and MEDMARAVIS.
23. Goutner V., Jerrentrup, H., Kazantzidis, S. and Poirazidis, K. (1999) Population trends, distribution, ring recoveries and conservation of Mediterranean Gull *Larus melanocephalus* in Greece. Pp.31–37 in P. L. Meininger, W. Hoogendoorn, R. Flamant and P. Raevel, eds. *Proceedings of the 1st Internationa Mediterranean Gull Meeting, Le Portel, Pas-de-Calais, France, 4–7 September 1998*. EcoNum, Bailleul.
24. Magioris, S. (1994) The avifauna of the Cyclades (Aegean Sea). *Hellen. Zool. Arch* 2: 1–16.
25. Oro, D., Baccetti, N., Boukhalfa, D., Eken, G., El Hili, A., Goutner, V., Karauz, S., Papakonstantinou, K., Recorbetr, B. and Ruiz, X. (1998) (In press.) *Current breeding distribution and status of Audouin's Gulls* Larus audouinii *in the Mediterranean*. MEDMARAVIS Conference, Malta, Oct. 1998.
26. Papakonstantinou, C., Economidou, E., Bousbouras, D., Kardakari, N., Zogaris, S., Tsiakiris, R. and Roussopoulos, Y. (1995) [*Ecological survey in northern Dodecanese*.] Hellenic Ornithological Society, Athens. (In Greek.)
27. Pergantis, F. and Akriotis, T. (1994) The breeding status of terns Sternidae in the Amvrakikos area. 5e Congres Intern. de Zoogeographie et Ecologie de la Grece et des Regions Avoisinantes. Iraklion, 16–20 Avril 1990. *Biol. Gallo-hellen.* 22: 179–184.
28. Poirazidis, K., Goutner, V., Skartsi, Th. and Stamou G. (2004) Modelling nesting habitat as a conservation tool for the Eurasian black vulture (*Aegypius monachus*) in Dadia Nature Reserve, northeastern Greece. *Biological Conservation* 118(2): 235–248.
29. Ristow, D., Feldmann, F. and Scharlau, W. (1991) Population dynamics of Cory's Shearwater (*Calonectris diomedea*) and Eleonora's falcon (*Falco eleonorae*) in Eastern Mediterranean. Pp.199–212 in A. Seitz and V. Loeschcke, eds. *Species conservation: A Population – Biological Approach*. Birkhauser Verlag, Basel.
30. Sakoulis, A., Bourdakis, S., Hallmann, B. and Alivizatos, H. (1997) The status of the Imperial Eagle (*Aquila heliaca*) in Greece. Hellenic Ornithological Society, Athens. Unpublished data.
31. Schaffer, N. and Gallo-Orsi, U., eds. (2001) *European Union action plans for eight priority species*. European Commission.
32. Skartsi, Th. and Poirazidis, K. (2002) Management plan for the Eurasian black vulture (*Aegypius monachus*) in Dadia Nature Reserve. WWF Greece, Athens. Unpublished data.
33. Skartsi Th., Vasilakis, D. and Elorriaga J. (2003) Black Vulture and Griffon Vulture monitoring in National Park of Dadia-Lefkimi-Soufli Forest. Annual Technical Report. WWF Greece. Athens. Unpublished report.
34. Thibault, J. C. (1993) Breeding distribution and numbers of Cory's Shearwater (*Calonectris diomedea*) in the Mediterranean.] Pp.25–35 in J. S. Aguilar, X. Monbailliu and A. Paterson, eds. *Status and conservation of seabirds-ecogeography and Mediterranean Action Plan. Proceed. 2nd Mediterr. Seabird Symp. Calvia, 21–26 March 1989*. SEO and MEDMARAVIS.
35. Tsounis, G. and Dimitropoulos, A. (1992a) The avifauna of the North-Eastern Aegean islands Lesvos and Chios: differences and similarities. *Biol. Gallo-hellen.* 19: 73–102.
36. Tucker, G. M. and Heath, M. F. (1994) *Birds in Europe: their conservation status*. Birdlife Conservation Series No 3. Birdlife International, Cambridge.
37. Vangeluwe, D., Kazantzidis, S. and Handrinos, G. I., eds. (1999) *Conservation action for the Slender-billed Curlew*. Final Report, LIFE NATURE Contract No. b4 3200/95/853. Grec Ministre de Agriculture, Institut Royal des Sciences Naturelles de Belgique, Federiationes Associations de Chasseurs de l' UE, Budesamt fur Naturschutz, Sea Mammal Research unit. Athens.
38. Xirouchakis, S. (2001) The Golden eagle *Aquila chrysaetos* in Crete. Distribution, population status and conservation problems. *Avocetta* 25(2): 275–281.
39. Xirouchakis, S., Sakoulis, A. and Andreou, G. (2001) The decline of the Bearded Vulture *Gypaetus barbatus* in Greece. *Ardeola* 48(2): 183–190.
40. Xirouchakis, S. (2003) The ecology of the Eurasian Griffon (*Gyps fulvus*) on the island of Crete. University of Crete (Ph.D. thesis). Unpublished data.

41. Zogaris, S. (1999) *Breeding birds as indicators for habitat conservation in Southern Evvoia, Greece, Contributions to the zoogeography and ecology of the Eastern Mediterranean region.* Vol.1. Athens University / Hellenic Zoological Society.
42. Zogaris, S. (2001) Wetland Birds at Amvrakikos (Greece): Habitat use Aassessment for monitoring threatened species. University of Wales (MSc. thesis), Bangor, UK. Unpublished data.
43. Zogaris, S. and Handrinos, G.I. (2002) The breeding status of the Ferruginous Duck in Greece and a trial habitat use assessment at its national stronghold. Ferruginous Duck Working Group, BirdLife Meeting of BSPB.
44. Zogaris, S., Papandropoulos, D., Alivizatos, H., Rigas, Y., Hatzirvassanis V. and Kardakari, N. (2003) [*Threatened birds at Amvrakikos.*] Athens, Koan. (In Greek.)

Wintering

45. Handrinos, G. I. (1990) The status of geese in Greece. *Ardea* 79: 175–178.
46. Handrinos, G. I. (1993) Midwinter numbers and distribution of Cormorants and Pygmy Cormorants in Greece. Pp.135–147 in J.S. Aguilar, X. Monbailliu and A. Paterson, eds. *Status and conservation of seabirds-ecogeography and Mediterranean Action Plan. Proceed. 2nd Mediterr. Seabird Symp. Calvia, 21–26 March 1989.* SEO and MEDMARAVIS.
47. Handrinos, G. I. (1995) The White-headed Duck *Oxyura leucocephala* in Greece. *IWRB Threatened Waterfowl Res. Group Newsletter* 7: 6–7.
48. Handrinos, G. I. (1996) Numbers and distribution of swans wintering in Greece. *Gibier fauna sauvagem, Game Wild.* 13: 463–476.
49. Hellenic Ornithological Society (HOS) Midwinter Waterfowl Census, 1995–1999. Unpublished data.
50. Kazantzidis, S. and Naziridеs, T., eds. (1999) [*National Action Plan for the Lesser White-fronted Goose (*Ánser erythropus L., *1758).*] WWF-HOS-Society for the protection of Prespa. (In Greek.)

Hungary

Breeding

1. Bagyura, J., Horváth, Z. and Kalocsa, B. (2001) [A Magyar Madártani és Természetvédelmi Egyesület Ragadozómadár-védelmi Szakosztálya 2000. évben elért eredményeinek összefoglalója.] *Túzok* 6(1): 25–29. (In Hungarian with English Summary.)
2. Bagyura, J., Szitta, T., Haraszthy, L., Firmánszky, G., Viszló, L., Kovács, A., Demeter, I. and Horváth, M. (2002) Population increase of Imperial Eagle (*Aquila heliaca*) in Hungary between 1980 and 2000. *Aquila* 107–108: 133–144.
3. Bankovics, A. (1989) [Viharsirály (*Larus canus*) fészkelése Magyarországon.] *Madártani Tájékoztató* 1989 (jul–dec): 22–23. (In Hungarian.)
4. Boldogh, S. (1999) The Corncrake (*Crex crex*) in Hungary. Pp.53–59 in N. Schaffer and U. Mammen, eds. *Proceedings International Corncrake Workshop 1998.* (In Hilpoltstein/Germany.)
5. Gyurácz, J., Nagy, K., Bagdi, A., Hadarics, T. and Ragats, Zs. (In press.) [A gyurgyalag (*Merops apiaster*) monitorozása és védelmének helyzete Magyarországon, 1997–2001.] *Természetvédelmi Közlemények* 11. (In Hungarian with English Summary.)
6. Hadarics, T. (1996) [Bütykös ásólúd (*Tadorna tadorna*) fészkelése Magyarországon.] *Túzok* 1(3): 127–127. (In Hungarian with English Summary.)
7. Hadarics, T., Mogyorósi, S. and Pellinger, A. (1993) [A sárgalábú sirály (*Larus cachinnans*) újabb költése a Fertõ-tónál.] *Madártani Tájékoztató* 1993 (jan–jun): 26–27. (In Hungarian.)
8. Horváth, R. and Szép, T. (1998) [A magyarországi vízirigók állománycsökkenése.] *Ornis Hungarica* 8(1): 9–16. (In Hungarian with English Summary.)
9. Kalocsa, B. and Tamás, E. (2002) Status of the Black Stork (*Ciconia nigra*) in Hungary in 2000. *Aquila* 107–108: 207–213.
10. Kovács, G. and Magyar, G. (1998) [A feketeszárnyú székicsér (*Glareola nordmanni*) határozása és elõfordulása Magyarországon.] *Túzok* 3(1): 1–10. (In Hungarian with English Summary.)
11. Kovács, G., Kapocsi, I. and Végvári, Zs. (2000) [Szerkõfajok rekordmennyiségû fészkelése a Hortobágyon 1999-ben.] *Túzok* 5(3–4): 97–101. (In Hungarian with English Summary.)
12. Lovászi, P. (1998) [A fehér gólya (*Ciconia ciconia*) helyzete Magyarországon az 1941–1994 közötti országos állományfelmérések eredményeinek tükrében.] *Ornis Hungarica* 8(1): 1–8. (In Hungarian with English Summary.)
13. Lovászi, P., Bártol, I. and Moskát, Cs. (2000) Nest-site selection and breeding success of Lesser Grey Shrike (*Lanius minor*) in Hungary. *Ring* 22(1): 157–164.
14. Magyar, G., Hadarics, T., Waliczky, Z., Schmidt, A., Nagy, T. and Bankovics, A. (1998) [*Nomenclator avium Hungariae. Magyarország madarainak névjegyzéke.*] Madártani Intézet – MME, Budapest. (In Hungarian and English.)
15. Marik, P., Forgách B., Tóth, I. and Tõgye, J. (1997) [Megfigyelések a kis héja (*Acccipiter brevipes*) újabb magyarországi fészkelésével kapcsolatban.] *Túzok* 2(2): 49–59. (In Hungarian with English Summary.)
16. MME / BirdLife Hungary. Databases of MME / BirdLife Hungary – Monitoring Centre. Unpublished data.
17. Petrovics, Z. (1997) [A gatyáskuvik (*Aegolius funereus*) elsõ fészkelése Magyarországon.] *Túzok* 2(3): 106–108. (In Hungarian with English Summary.)
18. Széll, A. and Bakacsi, G. (1996) [A szerecsensirály (*Larus melanocephalus*) fészkelési viszonyai Magyarországon.] *Túzok* 1(3): 105–115. (In Hungarian with English Summary).
19. Szép, T. and Nagy, K. (2002) [*Mindennapi Madaraink Monitoringja (MMM) 1999–2000.*] MME BirdLife Hungary, Budapest (In Hungarian, with English Summary.)
20. Szép, T. (1999) Monitoring of the sand martin (*Riparia riparia*), chances of forecasting a severe population decline. *Vogelwelt* 120: 349–353.

Wintering

21. Faragó, S. (1999) [A vadlúd monitoring eredményei az 1997–1998-as idényben Magyarországon.] *Magyar Vízivad Közlemények* 5: 3–62. (In Hungarian.)
22. Faragó, S. (1999) [A vízivad monitoring eredményei az 1997–1998-as idényben Magyarországon.] *Magyar Vízivad Közlemények* 5: 63–327. (In Hungarian.)
23. Magyar, G., Hadarics, T., Waliczky, Z., Schmidt, A., Nagy, T. and Bankovics, A. (1998) [*Nomenclator avium Hungariae. Magyarország madarainak névjegyzéke.*] Madártani Intézet – MME, Budapest (In Hungarian and English.)
24. MME / BirdLife Hungary. Databases of MME / BirdLife Hungary – Monitoring Centre. Unpublished data.
25. Szép, T. and Nagy, K. (2002) [*Mindennapi Madaraink Monitoringja (MMM) 1999–2000.*] MME BirdLife Hungary, Budapest (In Hungarian, with English Summary.)

Iceland

Breeding

1. Asbirk, S., Berg, L., Hardeng, G., Koskimies, P. and Petersen, Æ. (1997) *Population sizes and trends of birds in the Nordic Countries 1978–1994.* TemaNord 1997. Environment. Nordic Council of Ministers. Copenhagen.
2. Einarsson, Á., Garðarsson, A., Gíslason, G. M. and Guðbergsson, G. (In press.) *Population of ducks and trout of the River Laxá, Iceland, in relation to variation in food resources.* Unpublished.
3. Einarsson, Ó. (2000) [Músarindill ekki lengur minnsti varpfuglinn. – 'Wren no longer the smallest breeding bird.'] *Fréttabréf Fuglaverndarfélags Íslands* 13(29): 14–15. (In Icelandic.)
4. Einarsson, Ó., Kristinsson, H. Skarphéðinsson, K. H. and Ottósson, J. G. (2002) [*Verndun tegunda og svæða. Tillögur Náttúrufræðistofnunar Íslands vegna Náttúruverndaráætlunar 2002.*] [*'Conservation of species and areas. Proposals of the Icelandic Institute of Natural History for the nature conservation plan 2002.'*] Icelandic Institute of Natural History, Reykjavík. NÍ-02016. (In Icelandic.)
5. Fredriksen, M., Guðmundsson, G. A. and Skarphéðinsson, K. H. (2002) *Iceland/Greenland Pink-footed Goose population – Preliminary risk analyses. With emphasis on the impact of the Norðlingaalda-reservoir and other proposed hydropower schemes in Iceland.* Icelandic Institute of Natural History, Reykjavík.
6. Garðarsson, A. (1995a) [Svartfugl í íslenskum fuglabjörgum. – 'Numbers and distribution of Common Murre *Uria aalge*, Thick-billed Murre *U. lomvia* and Razorbill *Alca torda* in Iceland.'] *Bliki* 16: 47–65. (In Icelandic with English summary.)
7. Garðarsson, A. (1995b). [Fjöldi súlu við Ísland 1989–1994. – 'Numbers of Gannets Sula bassana in Iceland 1984–94.'] *Náttúrufræðingurinn* 64: 203–208. (In Icelandic with English summary.)
8. Garðarsson, A. (1996a) [Ritubyggðir. – 'Numbers and distribution of breeding Kittiwake *Rissa tridactyla* in Iceland.'] *Bliki* 17: 1–16. (In Icelandic with English summary.)
9. Garðarsson, A. and Einarsson, Á. (1997) [Viðkoma og fjöldi nokkurra Mývatnsanda. – 'Production and numbers of some diving ducks at Mývatn, Iceland.'] *Bliki* 18: 1–13. (In Icelandic with English summary.)
10. Garðarsson, A. and Þórarinsson, Þ. L. (2003) [Útbreiðsla og fjöldi straumandar á Íslandi að vetrarlagi. – 'Distribution and numbers of Harlequin Ducks wintering in Iceland.'] *Bliki* 23: 5–20. (In Icelandic with English summary.)
11. Garðarsson, A. (1979) [Skarfatal 1975. – 'A census of breeding Cormorants (*Phalacrocorax carbo*) and Shags (*Phalacrocorax aristotelis*) in Iceland 1975.'] *Náttúrufræðingurinn* 49: 126–154. (In Icelandic.)
12. Garðarsson, A. (1991) [Fuglalíf vid Mývatn og Laxá. – 'The birdlife of Mývatn and Laxá']. Pp.278–319 in A. Garðarsson and Á. Einarsson, eds. [*Náttúra Mývatns.*] The Icelandic Natural History Society, Reykjavík. (In Icelandic.)
13. Garðarsson, A. (1996b) [Dílaskarfsbyggðir 1975–1994. –'Numbers of breeding Cormorants *Phalacrocorax carbo* in Iceland 1975–1994.'] *Bliki* 17: 35–42. (In Icelandic.)

14. Garðarsson, A. Unpublished data.
15. Gunnarsson, H. pers. comm.
16. Gunnarsson, T. pers. comm.
17. Harðardóttir, M. and Nielsen, Ó. K. (1999) [Hröfnum fækkar í Þingeyjarsýslum. – 'Decline of the Raven population in the Thingeyjarsýslur area.'] *Náttúrufræðingurinn* 68: 147–154. (In Icelandic with English summary.)
18. Hearn, R. D. (2003) *The 2001 Icelandic-breeding Goose Census*. WWT report, Slimbridge.
19. Hersteinsson, P., Sigfússon, A. Þ. and Björnsson, Þ. Þ. (1999) [Varpstofn sílamáfs og tilraunaveiðar á máfum á suðvesturlandi árið 1990. – 'The population size and experimental culling of Lesser Black-backed Gull in Southwest Iceland in 1990.'] (In Icelandic.) Institute of Wildlife Management, Reykjavík.
20. Hilmarsson, J. Ó. and Hansen, E. S. (Census done in 1991.) A census of the nocturnal tubenoses in Iceland.
21. Icelandic Institute of Natural History (2000) [Válisti 2. Fuglar. – 'Red list 2. Birds.'] Á. Ingadóttir, ed. Icelandic Institute of Natural History, Reykjavík. (In Icelandic with English summary.)
22. Ingvarsdóttir, B. S. pers. comm.
23. Kolbeinsson, Y., Pétursson, G. and Þráinsson, G. (2001) [Sjaldgæfir fuglar á Íslandi 1998. – 'Rare birds in Iceland 1999.'] *Bliki* 22: 21–46. (In Icelandic with English summary.)
24. Kolbeinsson, Y., Pétursson, G. and Þráinsson, G. (2003a) [Sjaldgæfir fuglar á Íslandi 1999. – 'Rare birds in Iceland 1999.'] *Bliki* 23: 21–49. (In Icelandic with English summary.)
25. Kolbeinsson, Y., Pétursson, G. and Þráinsson, G. (2003b) [Sjaldgæfir fuglar á Íslandi 2000. – 'Rare birds in Iceland 2000.'] *Bliki* 24: 25–52. (In Icelandic with English summary.)
26. Kristjánsson, P. G. (2000) [Bjargdúfan *Columba livia* – villtur varpfugl á Íslandi. – 'Rock Pigeon – breeding bird in Iceland.'] *Náttúrufræðingurinn* 70: 21–36. (In Icelandic with English summary.)
27. Lund-Hansen, L. C. and Lange, P. (1991) The numbers and distribution of the Great Skua (*Stercorarius skua*) breeding in Iceland 1984–1985. *Acta Naturalia Islandica* 34: 1–16.
28. Mývatn Research Station (2003) Observed on the world-wide-web 30.6.2003. http://www.hi.is/HI/Stofn/Myvatn/engframe.htm.
29. Nielsen, Ó. K. pers. comm.
30. Nielsen, Ó. K. Unpublished data.
31. Pétursson, G. pers. comm.
32. Pétursson, G., Þráinsson, G. and Ólafsson, E. (1992) [Sjaldgæfur fuglar á Íslandi 1990. – 'Rare birds in Iceland 1990.'] *Bliki* 12: 15–54. (In Icelandic with English summary.)
33. Pétursson, Þráinsson, G., and Ólafsson, E. (1993) [Sjaldgæfir fuglar á Íslandi 1991. – 'Rare birds in Iceland 1991.'] *Bliki* 13: 11–44. (In Icelandic with English summary.)
34. Skarphéðinsson, K. H. (1994) [*Tjón af völdum arna í æðarvörpum. – 'Damage caused by White-tailed Eagles in Eider colonies.'*] Skýrsla unnin af Náttúrufræðistofnun Íslands fyrir Umhverfisráðuneytið. Umhverfisráðuneyti, Reykjavík.
35. Skarphéðinsson, K. H. (In print.) Sea Eagles in Iceland: population trends and reproduction. In B. Hellander *Proceedings of Sea Eagle 2000 conference*. Bjørkø, Sweden.
36. Skarphéðinsson, K. H. pers. comm.
37. Skarphéðinsson, K. H., Nielsen, Ó. K., Þórisson, S. and Petersen, I. K. (1992) [Varpútbreiðsla og fjöldi hrafna á Íslandi. – 'The breeding distribution and numbers of Raven in Iceland.'] *Bliki* 11: 1–26. (In Icelandic with English summary.)
38. Snorrason, Þ. (1990) [Brandendur í Eyjafirði 1990. – 'First breeding record for Shelduck in Iceland.'] *Bliki* 12: 9–10. (In Icelandic with English summary.)
39. Snow, D. W. and Perrins, C. M., eds. (1998) *Birds of the Western Palearctic*. Concised edition. Vol. 1. Oxford University Press, Oxford.
40. Þórisson, S. G. (1996) [Er bjargdúfa varpfugl á Íslandi. – 'Is Rock Pigeon a Icelandic breeding bird?'] *Fréttabréf Fuglaverndarfélag Íslands* 9 (17): 6–7. (In Icelandic.)
41. Thorup (In press.) *Breeding waders in Europe 2000*. Wader Study Group Supplement.
42. Þráinsson, G. and Pétursson, G. (1997) [Sjaldgæfir fuglar á Íslandi 1995. – 'Rare birds in Iceland 1995.'] *Bliki* 18: 23–50. (In Icelandic.)
43. Þráinsson, G. and Pétursson, G. (1998) [Sjaldgæfir fuglar á Íslandi 1996. – 'Rare birds in Iceland 1996.'] *Bliki* 19: 13–42. (In Icelandic.)
44. Þráinsson, G. and Pétursson, G. (2000) [Sjaldgæfir fuglar á Íslandi 1997. – 'Rare birds in Iceland 1997.'] *Bliki* 21: 31–58. (In Icelandic.)
45. Þráinsson, G., Pétursson, G. and Ólafsson, E. (1994) [Sjaldgæfir fuglar á Íslandi 1992. – 'Rare birds in Iceland 1992.'] *Bliki* 14: 17–48. (In Icelandic.)
46. Þráinsson, G., Pétursson, G. and Ólafsson, E. (1995a) [Sjaldgæfir fuglar á Íslandi 1993. – 'Rare birds in Iceland 1993.'] *Bliki* 15: 21–51. (In Icelandic.)
47. Þráinsson, G., Pétursson, G. and Ólafsson, E. (1995b) [Sjaldgæfir fuglar á Íslandi 1994. – 'Rare birds in Iceland 1994.'] *Bliki* 16: 11–45. (In Icelandic.)

Wintering

48. Asbirk, S., Berg, L., Hardeng, G., Koskimies, P. and Petersen, Æ. (1997) *Population sizes and trends of birds in the Nordic Countries 1978–1994*. TemaNord 1997. Environment. Nordic Council of Ministers. Copenhagen.
49. Cranswick, P., Bowler, J., Delany, S. N. Einarsson, Ó., Gardarsson, A., McElwaine, J. G., Merne, O. J., Rees, E. C. and Wells, J. H. (1997) Numbers of Whooper Swans *Cygnus cygnus* in Iceland, Ireland and Britain in January 1995: results of the international Whooper Swan census. *Wildfowl* 47: 17–30.
50. Cranswick, P., Colhoun, K., Einarsson, Ó., McElwaine, J. G., Gardarsson, A., Pollitt, M. S. and Rees, E. C. (2001) The status and distribution of the Icelandic Whooper Swans population: results of the International Whooper Swan census 2000. Í: Rees, E. C., Earnst, S. L., Coulson, J. C. Special publication 1. *Waterbirds* 25: 37–48.
51. Einarsson, Á., Garðarsson, A., Gíslason, G. M. and Guðbergsson, G. (In press.) *Population of ducks and trout of the River Laxá, Iceland, in relation to variation in food resources*. Unpublished.
52. Garðarsson, A. and Þórarinsson, Þ. L. (2003) [*Útbreiðsla og fjöldi straumandar á Íslandi að vetrarlagi. – 'Distribution and numbers of Harlequin Ducks wintering in Iceland.'*] *Bliki* 23: 5–20. (In Icelandic with English summary.)
53. Harðardóttir, M. and Nielsen, Ó. K. (1999) [Hröfnum fækkar í Þingeyjarsýslum. – 'Decline of the Raven population in the Thingeyjarsýslur area.'] *Náttúrufræðingurinn* 68: 147–154. (In Icelandic with English summary.)
54. Hilmarsson, J. Ó. (1999) [*Íslenskur fuglavísir. – 'Icelandic bird guide.'*] Iðunn, Reykjavík. (In Icelandic.)
55. Hilmarsson, J. Ó. pers. comm.
56. Icelandic Institute of Natural History (2000) [*Válisti 2. Fuglar. – 'Red list 2. Birds.'*] Á. Ingadóttir, ed. Icelandic Institute of Natural History, Reykjavík. (In Icelandic with English summary.)
57. Kirby, J. S., Rees, E. C., Merne, O. J. and Gardarsson, A. (1992) International census of Whooper Swans *C. cygnus* in Britain, Ireland and Iceland: January 1991. *Wildfowl* 43: 20–26.
58. Kolbeinsson, Y., Þráinsson, G. and Pétursson, G. (2003b) [Sjaldgæfir fuglar á Íslandi 2000. – 'Rare birds in Iceland 2000.'] *Bliki* 24: 25–52. (In Icelandic with English summary.)
59. Kolbeinsson, Y., Þráinsson, G. and Pétursson, G. (2003a) [Sjaldgæfir fuglar á Íslandi 1999. – 'Rare birds in Iceland 1999.'] *Bliki* 23: 21–49. (In Icelandic with English summary.)
60. Mývatn Research Station (2003) Observed on the world-wide-web 30.6.2003. http://www.hi.is/HI/Stofn/Myvatn/engframe.htm.
61. Nielsen, Ó. K. (1992) [Tjarnarfuglar.] Pp.93–171 in Ó.K. Nielsen, ed. [*Tjörnin. Saga og lífríki. – 'The lake Tjörnin. History and biology.'*] (In Icelandic with English summary.)
62. Nielsen, Ó. K. pers. comm.
63. Nielsen, Ó. K. Unpublished data.
64. Petersen, Æ. (1998) [*Íslenskir fuglar. – 'Icelancic birds.'*] Vaka-Helgafell, Reykjavík.
65. Pétursson, G., Þráinsson, G. and Ólafsson, E. (1992) [Sjaldgæfur fuglar á Íslandi 1990. – 'Rare birds in Iceland 1990.'] *Bliki* 12: 15–54. (In Icelandic with English summary.)
66. Pétursson, Þráinsson, G., and Ólafsson, E. (1993) [Sjaldgæfir fuglar á Íslandi 1991. – 'Rare birds in Iceland 1991.'] *Bliki* 13: 11–44. (In Icelandic with English summary.)
67. Skarphéðinsson, K. H. pers. comm.
68. Skarphéðinsson, K. H., Nielsen, Ó. K., Þórisson, S. and Petersen, I. K. (1992) [Varpútbreiðsla og fjöldi hrafna á Íslandi. – 'The breeding distribution and numbers of Raven in Iceland.'] *Bliki* 11: 1–26. (In Icelandic with English summary.)
69. Þráinsson, G. and Pétursson, G. (1997) [Sjaldgæfir fuglar á Íslandi 1995. – 'Rare birds in Iceland 1995.'] *Bliki* 18: 23–50.
70. Þráinsson, G. and Pétursson, G. (1998) [Sjaldgæfir fuglar á Íslandi 1996. – 'Rare birds in Iceland 1996.'] *Bliki* 19: 13–42.
71. Þráinsson, G. and Pétursson, G. (2000) [Sjaldgæfir fuglar á Íslandi 1997. – 'Rare birds in Iceland 1997.'] *Bliki* 21: 31–58.
72. Þráinsson, G., Pétursson, G. and Ólafsson, E. (1994) [Sjaldgæfir fuglar á Íslandi 1992. – 'Rare birds in Iceland 1992.'] *Bliki* 14: 17–48.
73. Þráinsson, G., Pétursson, G. and Ólafsson, E. (1995a) [Sjaldgæfir fuglar á Íslandi 1993. – 'Rare birds in Iceland 1993.'] *Bliki* 15: 21–51.
74. Þráinsson, G., Pétursson, G. and Ólafsson, E. (1995b) [Sjaldgæfir fuglar á Íslandi 1994. – 'Rare birds in Iceland 1994.'] *Bliki* 16: 11–45.

Ireland

Breeding

1. Coombes, R. H., Crowe, O., Lysaght, L., Noble, D., O'Halloran, J., O'Sullivan, O. and Wilson, H. J. (2002) *Countryside Bird Survey report, 1998–2000*. Dublin: BirdWatch Ireland.
2. Cromie, J. (2002) Breeding status of Red-throated Diver *Gavia stellata* in Ireland. *Irish Birds* 7: 13–20.

3. Hillis, J. P. (2003) Rare Irish breeding birds, 1992–2001. *Irish Birds* 7: 157–172.
4. Mitchell, P. I., Newton, S. F., Ratcliffe, N. and Dunne, T. E. (2004) *Seabird populations of Britain and Ireland – results of the seabird 2000 census (1998–2002)*. T and AD Poyser, London.
5. Newton, S., Donaghy, A., Allen, D. and Gibbons, D. (1999) Birds of conservation concern in Ireland. *Irish Birds* 6: 333–344.
6. Smiddy, P. (2002) Breeding of the Little Egret *Egretta garzetta* in Ireland, 1997–2001. *Irish Birds* 7: 57–60.
7. Tierney, T. D., Hudson, J. and Casey, C. (2002) Survey of breeding waders on the River Shannon Callows, 2002. *Irish Birds* 7: 21–32.

Wintering

8. Colhoun, K. and Newton, S. F. (2000) Winter waterbird populations on non-estuarine coasts in the Republic of Ireland: results of the 1997/98 Non-Estuarine Coastal Waterfowl Survey (NEWS). *Irish Birds* 6: 527–542.
9. Colhoun, K. (2002) Waterbird monitoring in Ireland 2000/01: Results of the seventh year of the Irish Wetland Bird Survey. *Irish Birds* 7: 43–52.
10. Merne, O. J. and Walsh, A. (2002) Barnacle Geese *Branta leucopsis* in Ireland, spring 1999. *Irish Birds* 7: 53–56.

Italy

Breeding

1. Allavena, S. and Brunelli, M. (2003) [Revisione delle conoscenze sulla distribuzione e la consistenza del Pellegrino Falco peregrinus in Italia.] *Avocetta* 27: 20–23. (In Italian.)
2. Allavena, S., Angelini, J. and Pellegrni, M. (2001) *Abstracts 4th Eurasian Congress On Raptors Seville-Spain, 25–29 September 2001*. Estación Biológica Doñana – Raptor Research Foundation.
3. Andreotti, P. (2001) [*Piano d'azione nazionale per il Pollo sultano (*Porphyrio porphyrio*)*.] Quad. Cons. Natura, 8, Min. Ambiente – Ist. Naz. Fauna Selvartica. (In Italian.)
4. Boano, G. and Pulcher, C. (2003) pers. comm. to M. Gustin and P. Brichetti.
5. Borgo, F., Boschetti, E., Panzarin, L., Verza, E. and Volponi, S. (2003) [Incremento del Marangone minore *Phalacrocorax pygmeus* nelle aree costiere dell'Adriatico settentrionale.] *Avocetta* 27: 133. (In Italian)
6. Brichetti, P. and Fracasso, G. (2003) [*Ornitologia italiana. 1 Gaviidae-Falconidae. Identificazione, distribuzione, consistenza e movimenti degli uccelli italiani*.] Alberto Perdisa Editore. (In Italian.)
7. Canova, L., Volponi, S. and Fasola, M. (2003) [Demografia della popolazione di Spatola Platalea leucorodia nidificante in Italia (1989–2002).] *Avocetta* 27: 130. (In Italian.)
8. De Franceschi, P. (2002) [Gestione dei Tetraonidi.] Pp.91–125 in P. Brichetti and A. L. Gariboldi. [*Manuale di Ornitologia. Vol. 3*.] Edagricole. (In Italian)
9. Fasce, P. (2003) [L'Aquila reale Aquila chrysaetos in Italia: un aggiornamento sullo status della popolazione.] *Avocetta* 27: 10–13. (In Italian)
11. Gustin, M., Palumbo, G. and Corso, A. (1999) *International Species Action Plan. Lanner Falcon* Falco biarmicus. BirdLife International, The Netherlands.
12. Leonardi, G. (2001) Falco biarmicus *Lanner Falcon*. Pp157–174. in *the birds of the Western Palearctic Update vol. 3 No. 3*. Oxford University Press.
13. Martelli, D. and Rigacci, L. (2003) [Interpretazione sintetica della demografia del Lanario *Falco biarmicus* in Italia dal 1971 al 2000.] *Avocetta* 27: 14–16. (In Italian.)
14. Melega, L. (2003) Population status and trends of Ferruginous Duck in Italy. In N. Petkov , B. Hughes and U. Gallo-Orsi, eds. *BSPB Conservation Series* No. 6. Sofia, Bulgaria.
15. Melega, L. (2003) [Recente incremento della Moretta tabaccata *Aythya nyroca* in Italia.] *Avocetta* 27: 136. (In Italian.)
16. Palumbo, G. (2003) pers. comm. to M. Gustin and P. Brichetti.
17. Serra, G., Melega, L. and Baccetti, N. (2001) [*Piano d'Azione nazionale per il Gabbiano corso (*Larus audouinii*)*.] Quad. Cons. Natura, 6, Min. Ambiente – Ist. Naz. Fauna Selvatica. (In Italian.)
18. Serra, L. and Brichetti, P. (2002) [Uccelli acquatici nidificanti: (2000).] *Avocetta* 26(2): 123–129. (In Italian.)
19. Tout, P. (1995) The status and distribution of Corncrakes (*Crex crex*) during 1995 in Fiuuli-Venezia Giulia (North-east Italy). Unpublished report.
20. Zorzenon, T. (2003) pers. comm. to M. Gustin and P. Brichetti.

Wintering

21. Azzolini, M., Kravos, K., Panzarin, L., Passarella, M. and Sgorlon, G. (2003) [Risultati del censimento nazionale delle Oche (generi *Anser* e *Branta*), febbraio 2003.] *Avocetta* 27: 29. (In Italian.)
22. Baccetti, N., Dell'Antonia, P., Magagnoli, P., Melega, L., Serra, L., Soldatini, C. and Zenatello, M. (2002) [Risultati dei censimenti degli uccelli acquatici svernanti in Italia: distribuzione, stima e trend delle popolazioni nel 1991–2000.] *Biol. Cons. Fauna*, 111: 1–234. (In Italian.)
23. Borgo, F., Boschetti, E., Panzarin, L., Verza, E. and Volponi, S. (2003) [Incremento del Marangone minore *Phalacrocorax pygmeus* nelle aree costiere dell'Adriatico settentrionale.] *Avocetta* 27: 133. (In Italian.)
24. Gustin, M., Arcamone, E., Corso, A., Passarella, M. and Pizzari, T. (2000) Recent increase in wintering and breeding of Cattle egret, *Bubulcus ibis* population in Italy. *Alauda*, 69(2): 311–318.
25. Melega, L. (2003) Population status and trends of Ferruginous Duck in Italy. In N. Petkov, B. Hughes and U. Gallo-Orsi, eds. *BSPB Conservation Series* No. 6. Sofia, Bulgaria.
26. Melega, L. (2003) [Recente incremento della Moretta tabaccata *Aythya nyroca* in Italia.] *Avocetta* 27: 136. (In Italian.)

Latvia

Breeding

1. Auniòð, A. (2001) [Íikuta populâcijas teritoriâlais izvietojums, skaits un biotopa izvçle Latvijâ: patreizçjâ situâcija (1999–2001) un vçsturiskâ informâcija.] *Putni dabâ I pielikums*: 4–12. (In Latvian.)
2. Auniòð, A. Farmland bird counts in 4 standard census plots (1995–2003, co-ordinated by A. Auniòð) Latvian Fund for Nature. Unpublished report.
3. Avotiòð, A., Graubics, G., Íemlers, A., Krçsliòð, V., Íuze, J. and Ïoïâns, U. (1999) Number and breeding densities of owls in Latvia: studies in sample plots. *Vogelwelt* 120: 333–337.
4. Baumanis, J. and Celmiòð, A. (1995) [Latvijas Ornitofaunistikas komisijas pârskats par 1993–1994.g.] *Putni dabâ* 5(2): 71–81. (In Latvian.)
5. Celmiòð, A. and Baumanis, J. (1997) [Pârskats par retiem putniem Latvijâ 1995 gadâ.] *Putni dabâ* 7(1): 27–32. (In Latvian.)
6. Celmiòð, A. and Baumanis, J. (1998) [Pârskats par retiem putniem Latvijâ 1996–97 gadâ.] *Putni dabâ* 8(2): 1–10. (In Latvian.)
7. Celmiòð, A., Baumanis, J. and Mednis, A. (1993) *List of Latvian Bird Species 1993*. Riga.
8. Celmiòð, A., Millers, K. and Roze, V. (2003) [Pirmais biðu dzeòa *Merops apiaster* ligzdoðanas gadîjums Latvijâ.] *Putni dabâ* 13(3): 6–13. (In Latvian.)
9. Keišs O. 1997. Results of a randomised Corncrake *Crex crex* survey in Latvia 1996: population estimate and habitat selection. *Vogelwelt* 118: 231–235.
10. Keišs, O. (2003) [Brief history of the Aquatic Warbler *Acrocephalus paludicola* in Latvia.] *Putni dabâ* 13(2): 9–11. (In Latvian, with English summary.)
11. Keišs, O. and Íemlers, A. (2000) Grieþu (*Crex crex*) skaita palielinâðanâs Latvijâ 1990.gados – vai varam lepoties ar sekmîgu sugas aizsardzîbu? *Putni dabâ* 10(3): 22–30. (In Latvian.)
12. Liepa, V., Raèinskis, E., Kalvâns, A. and Hofmanis, H. (2003) *Species Action Plan for the Black Grouse* Tetrao tetrix. Rîga, Latvian Ornithological Society.
13. Lipsbergs, J., Opermanis, O., Stîpniece, A. and Stîpnieks, A. (2001) [Pirmais pierâdîtais lielâ baltâ garòa *Egretta alba* ligzdoðanas gadîjums Latvijâ.] *Putni dabâ* 11(1): 22–24. (In Latvian.)
14. Lipsbergs, J. and Roze, V. (2001) [Lielâ baltâ gâròa *Egretta alba* un zivju gâròa *Ardea cinerea* hibridizâcijas otrais zinâmais gadîjums Latvijâ.] *Putni dabâ* 1: 51–53. (In Latvian.)
15. LOB (2004) The new Latvian Breeding Bird Atlas (2000–2004, co-ordinated by Latvian Ornithological Society). Unpublished data of an on-going work).
16. LOB (1998) [*Latvijas lauku putni. Rîga, Latvijas Ornitoloìijas biedrîba.*] (In Latvian.)
17. LOB (1999) [*Latvijas ûdeòu putni. Rîga, Latvijas Ornitoloìijas biedrîba.*] (In Latvian.)
18. LOB (2002) [*Latvijas meþa putni. 2. izdevums. Rîga, Latvijas Ornitoloìijas biedrîba.*] (In Latvian.)
19. Opermanis, O. (2003) Ferruginous duck in Latvia: a historical overview. Pp.77 in N. Petkov, B. Hughes and U. Gallo-Orsi, eds. *Ferruginous duck: from research to conservation*. Conservation Series No. 6. BirdLife International – BSPB – TWSG, Sofia.
20. Opermanis, O., Strazds, M., Pçterhofs, E. and Strazds, A. (1996) [Putni Latvijas piekrastç: skaita izmaiòas, izplatîba un paðreizçjais stâvoklis.] *Putni dabâ* 6(1): 2–14. (In Latvian.)
21. Petriòð, A., Strazds, M. and Bergmanis, U. (1997) [Vidçjais çrglis *Aquila clanga* Latvijâ laika ritumâ.] *Putni dabâ* 6(3): 7–14. (In Latvian.)
22. Raèinskis, E. (2000) Zaïo vârnu *Coracias garrulus* skaita izmaiòas, ligzdoðanas bioloìija un aizsardzîbas jautâjumi Latvijâ. Maìistra darbs. Latvijas Universitâte, Rîga. Unpublished report. (In Latvian.)
23. Strazds, M., Priednieks, J. and Vâveriòð, G. (1994) [Bird numbers in Latvia] *Putni dabâ* 4: 3–18. (In Latvian, with English summary.)
24. Vîksne, J. (1997) *The bird lake Engure*. Jâòa sçta, Rîga.
25. Vîksne, J. and Janaus, M. (2000) [Melnâ zìriòa *Chlidonias niger* aizsardzîbas plâns.] LU Bioloìijas institûts, Salaspils. Unpublished report. (In Latvian.)
26. Celmiòð, A. and Matrozis, R. (1998–2004) [Latvijas putni.] Internet site: http://www.putni.lv. (In Latvian.)
27. Koskimies, P. and Priednieks, J. (1997) Blyth's Reed Warbler *Acrocephalus dumetorum*. Pp.576 in W. J. M. Hagemeijer and M.J. Blair, eds. *The EBCC atlas of European breeding birds: their distribution and abundance*. T & AD Poyser, London.
28. Jaunas, M. (2003) pers. comm. to E. Raèinskis and A. Stîpniece.
29. Matrozis, R. (2003) pers. comm. to E. Raèinskis and A. Stîpniece.
30. Baumanis, J. (2003) pers. comm. to E. Raèinskis and A. Stîpniece.

Wintering

31. Durinck, J., Skov, H., Jensen, F. P. and Pihl, S. (1994) Important marine areas for wintering birds in the Baltic Sea. EU DG XI research contract no. 22421/90-09/01. Ornis Consult report 1994.Unpublished report.
32. LOB (1998) [*Latvijas lauku putni. Rîga, Latvijas Ornitoloìijas biedrîba.*] (In Latvian.)
33. LOB (1999) [*Latvijas ûdeòu putni. Rîga, Latvijas Ornitoloìijas biedrîba.*] (In Latvian.)
34. LOB (2002) [*Latvijas meþa putni. 2. izdevums. Rîga, Latvijas Ornitoloìijas biedrîba.*] (In Latvian.)
35. Raèinskis, E. (1996) [An observation of a wintering Jack Snipe.] *Putni dabâ* 6(2): 23. (In Latvian, with English summary.)
36. Stipniece, A., ed. (2004) International Waterfowl Census data collected and submitted to Wetlands International annually by Antra Stipniece (Institute of Biology, University of Latvia). Unpublished data.
37. Strazds, M., Priednieks, J. and Vâveriòð, G. (1994) [Bird numbers in Latvia.] *Putni dabâ* 4: 3–18. (In Latvian, with English summary.)
38. Jaunas, M. (2003) pers. comm. to E. Raèinskis and A. Stîpniece.
39. Matrozis, R. (2003) pers. comm. to E. Raèinskis and A. Stîpniece.
40. Baumanis, J. (2003) pers. comm. to E. Raèinskis and A. Stîpniece.

Liechtenstein

Breeding

1. Willi, G. (In prep.) [Die Vogelwelt des Fürstentums Liechtenstein.] *Ber. Bot.-Zool. Ges. Liechtenstein-Sargans-Werdenberg* 31. (In German.)

Wintering

2. Willi, G (2002) [Bemerkenswerte Beobachtungen aus der regionalen Avifauna 2001, inkl. [Wasservogelzählungen am Rhein in den Wintern 2000/2001 und 2001/2002.] *Ber. Bot.-Zool. Ges. Liechtenstein-Sargans-Werdenberg* 29. (In German.)

Lithuania

General

1. Adomaitis, E., Barauskas, R., Kirstukas, M. and Preikša, Ž. (1998) How many Corncrakes are in Lithuania? *Ciconia* 6: 5–6. (In Lithuanian.)
2. Castren, K. (2001) The expansion of White Egret (*Egretta alba*) in Lithuania. *Ciconia* 9: 41–42. (In Lithuanian.)
3. Drobelis, E., Matiukas, G. and Vaitkus, G. (1996). Results of the Black Stork census in 1995. *Ciconia* 4: 16–21. (In Lithuanian.)
4. Jusys, V., Mačiulis, M., Mečionis, R., Pođkus, A., Graţulevičius, G. and Petraitis, A. (1999) *The breeding bird atlas of the Klaipeda region*. Vilnius.
5. Kurlavičius, P. and Raudonikis, L. (1999) Assessment of the Lithuanian breeding bird abundance, 1999. *Ciconia* 7: 52–57. (In Lithuanian.)
6. Kurlavičius, P. and Raudonikis, L. (2001) Assesment of the Lithuanian breeding bird abundance, 1999–2001. *Ciconia* 9: 92–97. (In Lithuanian.)
7. Lithuanian Bird Rarities Commission report (2001) *Ciconia* 9: 98–109. (In Lithuanian.)
8. Malinauskas, V. and Dementavičius, D. (1996) Final results of the White Stork census in 1994–1995. *Ciconia* 4: 1–9. (In Lithuanian.)
9. Lithuanian Ornithofaunistic Commission (1995) *Ciconia* 3: 78–85. (In Lithuanian.)
10. Lithuanian Ornithofaunistic Commission (1996) *Ciconia* 4: 65–74. (In Lithuanian.)
11. Lithuanian Ornithofaunistic Commission (1998) *Ciconia* 6: 64–78. (In Lithuanian.)
12. Lithuanian Ornithofaunistic Commission (1999) *Ciconia* 7: 58–65. (In Lithuanian.)
13. Matiukas, G. and Vaitkus, G. (1996) Lapwing – the bird of 1995. *Ciconia* 4: 22–24. (In Lithuanian.)
14. Matiukas, G., ed. (1997) Report of the Lithuanian Ornithofaunistic Commission. *Ciconia* 5: 72–86. (In Lithuanian.)
15. Raudonikis, L., Švažas, S., Pareigis, V. and Mierauskas, P. (1989) Wintering waterfowl in Lithuania 1988. Vilnius: Institute of Zoology and parasitology. Unpublished report.
16. Raudonikis, L. (1999) Black-tailed Godwit: bird of 1998 in Lithuania. *Acta Zoologica Lituanica* 9 (1): 215–218.
17. Raudonikis, L. and Švažas, S. (2000) Population status of Great Snipe (*Gallinago media*) in Lithuania. *OMPO Newsletter* 21: 47–51.
18. Raudonikis, L. (2001) Species composition changes and their reasons in the terrestrial birds wintering in Lithuania. *Acta Zoologica Lituanica* 11(3): 309–318.
19. Raudonikis, L. and Kurlavièius, P. (2000) *Important bird areas in Lithuania*. Lithuanian Ornithological Society & Institute of Ecology. Lututë, Vilnius.
20. Raudonikis, L. (2003) Unpublished data based on the results of Lithuanian IBA inventory scheme (2000–2003). Unpublished data.
21. Raudonikis, L. (2003) Unpublished data based on the results of Lithuanian mid-winter waterfowls monitoring scheme (2003). Unpublished data.
22. Skov, H., Vaitkus., Flensted, K. N., Grishanov, G., Kalamees, A., Kondratyev, A., Leivo, M., Luigujõe, L., Mayr, C., Rasmussen, J. F., Raudonikis, L., Scheller, W., Sidlo, P. O., Stipniece, A., Struve-Juhl, B. and Welander, B. (2000) Inventory of coastal and marine important bird areas in the Baltic Sea. BirdLife International, Cambridge.
23. Stanevičius, V., Đvaţas, S., Raudonikis, L. and Dagys, M. (2002) Waterbirds of large fishponds of Lithuania. Pp.54–111 in S. Švažas and A. Kozulin *Waterbirds of large fishponds of Belarus and Lithuania*. "OMPO Vilnius" & Institute of Ecology, Vilnius.
24. Stanevičius, V., Švažas, S. and Raudonikis, L. (2003) Population of Garganey and Shoveler in Lithuania. The Garganey *Anas querquedula* and Shoveler *Anas clypeata* in the Baltic States and Belarus. "OMPO Vilnius", Vilnius.
25. Stratford, J. (1998) Tundra Swan: a new breeding species. *Ciconia* 6: 3–4.
26. Stratford, J. (1998) Marsh Sandpiper: a new breeding species. *Ciconia* 6: 4–5.
27. Švažas, S., Raudonikis, L., Vaikus, G. and Pareigis, V. (1989) Wintering waterfowl in Lithuania 1989, 32 p.
28. Švažas, S. and Raudonikis, L. (1990) Wintering waterfowl in Lithuania and at Kaliningrad region coast. *IWRB Seaduck Newsletter*, No1, 1:20–22.
29. Đvaţas, S., Drobelis, E., Balčiauskas, L. and Raudonikis, L. (1999) Important wetlands in Lithuania. Vilnius.
30. Švažas, S., Dagys, M., Žydelis, R. and Raudonikis, L. (2001) Long-term changes in numbers and distribution of wintering waterfowl in Lithuania and their reasons. Pp.38–53 in S. Švažas, W. Meissner, V. Serebryakov, A. Kozulin and G. Grishanov *Changes of wintering sites of waterfowl in Central and Eastern Europe*. OMPO Special Publication.
31. Švažas, S., Dagys, M., Žydelis, R. and Raudonikis, L. (2001) Changes in numbers and distribution of wintering waterfowl populations in Lithuania in the 20th century. *Acta Zoologica Lituanica* 11(3): 243–254.
32. Švažas, S., Jusys, V., Raudonikis, L. and Žydelis, R. (2001) *Snipes in Lithuania*. Vilnius: OMPO & Institute of Ecology.
33. Švažas, S., Raudonikis, L., Jusys, V. and Žydelis, R. (2002) Snipes *Gallinago gallinago*, *G. media* and *Lymnocryptes minimus* in Lithuania. Pp.63–81 in S. Švažas *et al.* Snipes of the Eastern Baltic region and Belarus. "OMPO Vilnius", Vilnius.
34. Švažas, S. and Raudonikis, L., eds. (2003) New wintering grounds of waders and terretrial birds in the Eastern Baltic Region. A report. OMPO Vilnius, OMPO, Lithuanian Ornithological Society. Vilnius.
35. Đvaţas, S., Kozulin, A., Grishanov, G., Maximenkov, M., Rađomavičius, V., Raudonikis, L., Baichorov, V., Skuratovich, A., Balčiauskas, L., Drobelis, E. and Grishanov, D. (2003) Important transboundary Belarusian-Lithuanian and Lithuanian-Russian wetlands. Vilnius: Institute of Ecology of Vilnius University, OMPO Vilnius.
36. Vaitkus, G. (1997) Results of national inventory of Common Rook colonies. *Ciconia* 5: 30–31. (In Lithuanian.)
37. Ţalakevičius, M., Đvaţas, S., Stanevičius, V. and Vaitkus, G. (1995) Bird migration and wintering in Lithuania. *Acta Zoologica Lituanica*, 2. Vilnius.

Luxembourg

Breeding

1. Lorgé, P. *et al.* (In prep.) Vogelfauna Luxemburgs (Birds in Luxembourg).
2. Heidt, C., Konter, A. Lorgé, P. and Weiss, J. (2002) [Tätigkeitsbericht 1985 – 1997 der AG Feldornithologie]. *Regulus – Wissenschaftliche Berichte* No. 19. (In German.)
3. Melchior, E. *et al.* (1987) Atlas of the breeding birds in Luxembourg, LNVL.
4. Ornithological database of LNVL.

Wintering

5. Lorgé, P. *et al.* (In prep.) Vogelfauna Luxemburgs (Birds in Luxembourg).
6. Heidt, A. C., Konter, P. Lorgé and J. Weiss (2002) [Tätigkeitsbericht 1985 – 1997 der AG Feldornithologie.] *Regulus – Wissenschaftliche Berichte* No. 19. (In German.)
7. Ornithological database of LNVL.

The Former Yugoslav Republic of Macedonia

Breeding

1. Anon. (2002) National Lesser Kestrel census in Macedonia. FWFF-Macedonia's Report. Unpublished report.
2. Grubach, B. (1997) The present status of vultures Aegypiinae in Central Balkans. *Actas Del II Congreso International Sobre Aves Carroneras, 1–4 May 1997.*
3. Stoynov, E. (2001) Recent status of birds of prey in Macedonia. Report of FWFF to Whitley Award Foundation. Unpublished report.
4. Stoynov, E. (2001) Monitoring of the bearded vulture in the Balkans. Report of FWFF to Friends of Vienna Zoo and FCBV. Unpublished report.
5. Stoynov, E. (2002) Conservation actions for birds of prey in Macedonia. Report of FWFF to Whitley Award Foundation. Unpublished report.
6. Vasich, V. (1987) The status of birds of prey in Yugoslavia, with particular reference to Macedonia. *Newsletter of Institut for Biological Research.*

7. Velevski, M. (2003) Some recent observations of birds around Vardar River in Macedonia. Unpublished report.

Wintering

8. Gilissen, N., Haanstra, L., Delany, S., Boere, G. and Hagemeijer, W. (2002) *Numbers and distribution of wintering waterbirds in the Western Palearctic and Southwest Asia in 1997, 1998 and 1999*. Results from the International Waterbird Census. Wetlands International Global Series No. 11, Wageningen, The Netherlands.
9. Stoynov, E. (2001) Recent status of birds of prey in Macedonia. Report of FWFF to Whitley Award Foundation. Unpublished report.

Malta

Breeding

1. BirdLife Malta. (2002) BirdLife Malta daily log of bird sightings (1990–2002). Unpublished data.
2. Borg, J. J. and Sultana, J. (2002) Status and distribution of the breeding Procellariiformes in Malta. *Il-Merill* 30: 10–15.
3. Sultana, J. (2001) [*L-Ghasafar ta' Malta*.] PIN Pub. (In Maltese.)

Wintering

4. BirdLife Malta. (2002) BirdLife Malta daily log of bird sightings (1990–2002). Unpublished data.

Moldova

General

1. Heath, M. F. and Evans, M. I., eds. (2000) *Important bird areas in Europe: Priority sites for conservation*. Vol. 2: Southern Europe. Cambridge, UK, BirdLife International (BirdLife Conservation Series No. 8).
2. Heath, M., Borggreve, C. and Peet, N. (2001) *European bird populations: Estimates and trends*. Cambridge, UK, BirdLife International (BirdLife Conservation Series No. 10).
3. Republica Moldova (2000) [*Primul Raport national cu privire la Diversitatea Biologica*.] Chisinau, Stiinta. (In Moldovian.)
4. Snow, D. W. and Perrins, C. M. (1998) *The Birds of the Western Palearctic*. Oxford University Press, Oxford.
5 Friends of the Earth and EEB (1998) *The European Union and Biodiversity*. Brussels.
6. Tucker, G. M. and Heath, M. F. (1994). *Birds in Europe: Their conservation status*. Cambridge, UK, BirdLife International (BirdLife Conservation Series No. 3).
7. Tucker, G. M. and Evans M. I. (1997). *Habitats for birds in Europe: A conservation strategy for the wider enviroment*. Cambridge, UK, BirdLife International (BirdLife Conservation Series No. 6).
8. Zubcov, N., Buciuceanu, L., Stirbu, V., Jurminschii, S. and Glavan, T. (1996) On the status of the White Stork (*Ciconia ciconia*) population in Moldova: White Storks on the up? Pp.237–239 in *Proceedings, International Symposium on the White Stork, Hamburg*. NABU, Bonn.
9. Zubcov, N., Stirbu, V. and Jurminschii, S. (1997) [Unele aspecte al migratiilor pasarilor in Republica Moldova.] Academia Romano-Americana de stiinte si arte. Universitatea de stat „Valahia", Targovaste. (In Russian.)
10. Zubkov, N. I. (2001) [Variety and change ornitofaune of the central part of Moldova.] Pp.87–89 in [*The Jubilee symposium dedicated to 30 years of the reserve "Ñodry", Vol. II, 27–28 septembrie 2001, Lozova*.] (In Russian.)
11. Ganya, I. M. and Zubkov, N. I. (1989) [*Rare and disappearing species of birds of Moldova*.] Chisinau. (In Russian.)
12. Zubkov, N. I. (1983) [Bird of prey and owls of Moldova] Autoref. dissertation kand. biol. sciences. Chisinau. (In Russian.)
13. Averin, Yu. V., Ganya, I. M., Zubkov, N. I., Muntyanu, A. I., and Uspenskii, G. A. (1981) [*The birds and animals of Moldova*.] Chisinau, Stiinta. (In Russian.)

Netherlands

Breeding

1. SOVON Vogelonderzoek Nederland (2002) [*Atlas of Dutch breeding birds 1998–2000. Dutch Fauna 5*.] Nationaal Natuurhistorisch Museum Naturalis, KNNV Uitgeverij and European Invertebrate Survey – Nederland, Leiden. (In Dutch.)
2. Teunissen, W. A., Soldaat, L, van Veller, M., Willems, F. and van Strien, A. J. (2002) [Calculation of indexes in the meadow bird monitoring scheme.] SOVON-onderzoeksrapport 02/09. SOVON, Beek-Ubbergen. (In Dutch.)

Wintering

3. Berg, A. B. van den and Bosman, C. A. W. (1999) [*Rare birds of the Netherlands (Avifauna van Nederland 1)*]. GMB Uitgeverij/KNNV Uitgeverij, Haarlem/Utrecht. (In Dutch.)
4. Bijlsma, R. G., Hustings, F. and Camphuysen, C. J. (2001) [*Common and scarce birds of the Netherlands (Avifauna van Nederland 2)*.] GMB Uitgeverij/ KNNV Uitgeverij, Haarlem/Utrecht. (In Dutch.)
5. Camphuysen, C. J. and Leopold, M. F. (1994) *Atlas of seabirds in the southern North Sea*. IBN Reserarch Report 94/6, NIOZ report 1994–8, Institute for Forestry and Nature Research, Dutch Seabird Group and Netherlands Institute for Sea Research, Texel.
6. Hustings, F. and Winden, E. van. (1998) [Peregrine falcons recovering after nadir.] *SOVON-Nieuws* 11 (4): 14–16. (In Dutch.)
7. Kleefstra, R. (2003) [Raptors in the midwinter count.] *SOVON-Nieuws* 16 (1): 13–14. (In Dutch.)
8. Roomen, M. W. J. van, Winden, E. A. J. van, Koffijberg, K., Voslamber, B., Kleefstra, R., Ottens, G. and SOVON Ganzen en Zwanenwerkgroep (2002) [Waterbirds in the Netherlands in 2000/2001.] SOVON monitoringrapport 2002/04, RIZA-rapport BM02.15. SOVON Vogelonderzoek Nederland. Beek-Ubbergen. (In Dutch.)
9. SOVON (1987) [*Atlas of the birds of the Netherlands*.] SOVON, Arnhem. (In Dutch.)
10. SOVON Vogelonderzoek Nederland (2002) [*Atlas of Dutch breeding birds 1998–2000. Dutch Fauna 5*.] Nationaal Natuurhistorisch Museum Naturalis, KNNV Uitgeverij and European invertebrate survey. Nederland, Leiden. (In Dutch.)
11. Voslamber, B., Winden, E. A. J. van and Roomen, M. W. J. van (2000) [*Midwinter count of waterbirds in the Netherlands, January 1999*.] SOVON-monitoringrapport 2000/02, RIZA-rapport BM99.14, IKC Natuurbeheer coproductie C-31. SOVON Vogelonderzoek Nederland, Beek-Ubbergen. (In Dutch.)
12. Voslamber, B., Winden, E. A. J. van and Roomen, M. W. J. van (2001) [*Midwinter count of waterbirds in the Netherlands, January 2000*.] SOVON-monitoringrapport 2001/05, RIZA-rapport BM01.11. SOVON Vogelonderzoek Nederland, Beek-Ubbergen. (In Dutch.)

Norway

Breeding

1. Andersen, G. S. and Bergan, M. (2003) pers. comm. to I. J. Øien and T. Berge.
2. Andersen, G. S. and Folvik, A. (2003) pers. comm. to I. J. Øien and T. Berge.
3. Andersen, G. S. (2003) pers. comm. to I. J. Øien and T. Berge.
4. Anker-Nilssen, T., Bakken, V, Strøm, H., Golovkin, A. N., Bianki, V. V. and Tatarinkova, I. P. (2000) *The status of marine birds breeding in the Barents Sea region*. Tromsø, Norsk Polarinstitutt.
5. Anker-Nilssen, T. and Aarvak, T. (2004) [Lundens populasjonsøkologi på Røst. Status etter hekkesesongen 2003.] NINA Oppdragsmelding 809. (In Norwegian.)
6. Anker-Nilsen, T. (2003) pers. comm. to I. J. Øien and T. Berge.
7. Anon. (1999) [Statusrapport for trua arter i Telemark.] Fagrapport 08/1999. (In Norwegian.)
8. Statistics Norway (2003). *Statistical yearbook of Norway 2003*. Oslo.
9. Bangjord, G. (2000) [Landsomfattende telling av traner høsten 2000.] *Vår Fuglefauna* 23: 85. (In Norwegian.)
10. Bekken, J. (2003) pers. comm. to I. J. Øien and T. Berge.
11. Bergan, M. (2000). Sjøfuglene i Indre Oslofjord. *Toppdykker'n* 23: 154–162.
12. Byrkjedal, I. Kyllingstad, K., Efteland, S. and Grøsfjell, S. (2003). [Bestandsutviklingen hos vipe, storspove og tjeld på Jæren, 1997–2002.] *Falco* 32: 15–21. (In Norwegian).
13. Clausen, P., Madsen, J., Percivial, S. M., Anderson, G. Q. A., Koffijberg, K., Mehlum, F. and Wangeluwe, D. (1999) In Madsen, Cracknell and Fox (eds.) *Goose populations of the Western Palearctic – a review of status and distribution*. Wetlands International publ. No. 48, Wetlands International, Wageningen, The Netherlands. National Research Institute, Rønde, Denmark.
14. Dale, S. (1997) [Hortulan – spurv i fritt fall.] Pp.176–187 in K. H. Brox, ed. [*Natur 97/89*.] Tapir forlag, Trondheim. (In Norwegian.)
15. Falkenberg, F. and Högstedt, G. (2003) pers. comm. to I. J. Øien and T. Berge.
16. Folkestad, A. O. (2003) pers. comm. to I. J. Øien and T. Berge.
17. Follestad, A. (2003) pers. comm. to I. J. Øien and T. Berge.
18. Folvik, A. (2003) pers. comm. to I. J. Øien and T. Berge.
19. Frafjord, K. (2000) [Sædgås i Finnmark.] *Lappmeisen* 24: 28–31. (In Norwegian.)
20. Fredriksen, Å. S. and Johansen, P.-A. (1999) [Storskarv *Phalacrocorax carbo* ny hekkefugl i Østfold.] *Natur i Østfold* 18: 21–24. (In Norwegian.)
21. Gjershaug, J. O. and Nygård, T. (2003) [Utredninger i forbindelse med ny rovviltmelding. Kongeørn i Norge: Bestand, predatorrolle og forvaltning.] *NINA Fagrapport 58*. (In Norwegian.)
22. Grønlien, H., ed. (2004) Hønsehauken i Norge – Bestandens status og utvikling de siste 150 år. *NOF Rapportserie 5-2004*. (In Norwegian.)
23. Günther, M. (2002) [Første hekkefunn av dvergmåke Larus minutus i Finnmark.] *Vår Fuglefauna* 23: 82–84. (In Norwegian.)
24. Günther, M. (2003) pers. comm. to I. J. Øien and T. Berge.
25. Hafstad, I. (2003) [Rørsangeren i Norge. Bestandens utbredelse, størrelse og utvikling.] *Vår Fuglefauna* 26: 155–161. (In Norwegian.)
26. Heggland, H. and Günther, M. (2003) pers. comm. to I. J. Øien and T. Berge.

27. Hogstad, O. and Øien, I. J. (2001) [Endringer i norsk fuglefauna – hva har skjedd i løpet av de siste hundre år? – 'Changes in Norwegian avifauna – What has happened during the latest 100 years?'] *Vår Fuglefauna* 24: 54–66. (In Norwegian.)
28. Hogstad, O. (2003) pers. comm. to I. J. Øien and T. Berge.
29. Holtan, D. (2001). Dvergdykkerens status i Norge (Status of the Little Grebe in Norway). *Vår Fuglefauna* 24: 100–108.
30. Husby, M., Stueflotten, S. and Husby, A. (2003) [Norsk hekkefugltaksering. Årsrapport for 2003.] *NOF Rapportserie nr. 4-2004.* (In Norwegian).
31. Isaksen, K. and Bakken, V., eds. (1995) Seabird populations in the northern Barents Sea, NP/AKUP.
32. Johansen, P. A. (2003) pers. comm. to I. J. Øien and T. Berge.
33. Kålås (2003) International Action Plan for the Great Snipe. AEWA publication. Unpublished report.
34. Lislevand, T. (1995) [Bestandsvurderingene i Norsk Hekkefuglatlas: På tide med en revidering!] *Vår Fuglefauna* 18: 236–237. (In Norwegian).
35. Lorentsen, S.-H. (2003) [Det nasjonale overvåkingsprogrammet for sjøfugl. Resultater til og med hekkesesongen 2003. – 'The national monitoring programme for seabirds. Results up to and including the breeding season 2003.'] *NINA Oppdragsmelding 803*. (In Norwegian.)
36. Madsen, J. and Tombre, I. (2002). Kortnebbgjess i Vesterålen; problemer for norsk gåseforvaltning. *Ottar* 1-2002: 22–29. (In Norwegian).
37. Mork Soot, K. (2003) pers. comm. to I. J. Øien and T. Berge.
38. Myklebust, M. (2003) pers. comm. to I. J. Øien and T. Berge.
39. Nilsen, S. Ø. and Strann, K. B. (2000) [Fjellmyrløper – en truet fugleart i Barentsregionen.] *Vår Fuglefauna* 23: 5–9. (In Norwegian.)
40. Nybø, S. (1997) *Impacts of long-range transported air pollution on birds with particular reference to the Dipper Cinclus cinclus in southern Norway*. Dr.Scient.-avhandling, NTNU, Trondheim.
41. Nygård, T. (1994) [Det nasjonale overvåkingsprogrammet for overvintrende vannfugl i Norge 1980–93. – 'The national monitoring programme for wintering waterfowl in Norway 1980–93.']. – *NINA Oppdragsmelding* 313: 1–83. (In Norwegian.)
42. Nygård, T. (2003) pers. comm. to I. J. Øien and T. Berge.
43. Nyhus, G. C. and Mæhlen, A. (2003) [Hvorfor er slagugla så sjelden i Norge? – en sammeligning av norsk og svensk naturgeografi, skogbrannregime og skoghistorie.] *Vår Fuglefauna 26*: 33–41. (In Norwegian.)
44. Øien, I. J. and Aarvak, T. (2003) [Fjällgås – Finns det hopp för Skandinaviens "sjungande gäss".] *Vår Fågelvärld* 62: 6–12. (In Norwegian.)
45. Paulsen, B. E. and Olsen, K. (2003) pers. comm. to I. J. Øien and T. Berge.
46. Paulsen, B. E., Olsen, K., Stueflotten, S. and Johansen, P. A. (2003) pers. comm. to I.J. Øien and T. Berge.
47. Pedersen, H. C. and Swensson, J. (2003) pers. comm. to I. J. Øien and T. Berge.
48. Pihl, S. (1996) *Status and trends of swans, geese and ducks in continental north-west Europe*. DMU/Nordisk Jägersamvirke.
49. Ree, V. and Andersen, G. S. (2003) pers. comm. to I. J. Øien and T. Berge.
50. Reitan, O. (2003) pers. comm. to I. J. Øien and T. Berge.
51. Selås, V. and Steen, O. F. (2003) pers. comm. to I. J. Øien and T. Berge.
52. Selås, V. (2003) pers. comm. to I. J. Øien and T. Berge.
53. Shimmings, P. (2002) [Hvitkinngås – Bestandsutvikling, trekkmønster og rasteplasser hos Svalbard-bestanden.] *Ottar* 1-2002: 17–21. (In Norwegian.)
54. Solvang, R. (2000) [Trelerke – nye funn og status i Telemark.] *Fugler i Telemark* 26: 7–15. (In Norwegian.)
55. Solvang, R. (2003) pers. comm. to I. J. Øien and T. Berge.
56. Sonerud, G. (2003) pers. comm. to I. J. Øien and T. Berge.
57. Steel, C., Øien, I. J. and Sonerud, G. (2003) pers. comm. to I. J. Øien and T. Berge.
58. Steen, O. F. and Sonerud, G. (2003) pers. comm. to I. J. Øien and T. Berge.
59. Steen, O. F. (2003) pers. comm. to I. J. Øien and T. Berge.
60. Steen, O. F. (2002) [*Prosjekt Vandrefalk Sørøst-Norge*. Årsrapport 2001. – 'Peregrine Project Southeast Norway']. Annual Report.
61. Steen, O. F. (1999) [Jaktfalk i Norge – Fylkesvis oversikt over hekkeplasser og anslag på hekkende par.] *Vandrefalken* 4: 48–51. (In Norwegian.)
62. Stenberg, I. (2001). Kvitryggspetten i Noreg. Status fram til 2001. *NOF Rapportserie* 6-2001.
63. Størkersen, Ø., Husby, M., Rudolfsen, G., Stokke, B. G. and Værnesbranden, P. I. (2002) [Fugler i Norge 1999 – rapport fra Norsk Faunakomite for fugl (NFKF).] *Ornis Norvegica* 25: 93–123. (In Norwegian.)
64. Strann, K. B.(2003) pers. comm. to I. J. Øien and T. Berge.
65. Stueflotten, S. (2003) pers. comm. to I. J. Øien and T. Berge.
66. Systad, G. H. (2003) pers. comm. to I. J. Øien and T. Berge.

Wintering

67. Anker-Nilssen, T, Bakken, V, Strøm, H., Golovkin, A. N., Bianki, V. V. and Tatarinkova, I. P. (2000) *The status of marine birds breeding in the Barents Sea region*. Tromsø, Norsk Polarinstitutt.
68. Nygård, T. (1994) [Det nasjonale overvåkingsprogrammet for overvintrende vannfugl i Norge 1980–93. – 'The national monitoring programme for wintering waterfowl in Norway 1980–93.']. *NINA Oppdragsmelding 313*: 1–83. (In Norwegian.)
69. Øien., I. J (In print) [Tjeld.] In [*Atlas of wintering birds in Norway*.] (In Norwegian.)
70. Pihl, S. (1996) *Status and trends of swans, geese and ducks in continental north-west Europe*. DMU/Nordisk Jägersamvirke.
71. Størkersen, Ø., Husby, M., Rudolfsen, G., Stokke, B. G. and Værnesbranden, P. I. (2002) [Fugler i Norge 1999 – rapport fra Norsk Faunakomite for fugl (NFKF).] *Ornis Norvegica* 25: 93–123. (In Norwegian.)
72. Stueflotten, S. (2003) pers. comm. to I. J. Øien and T. Berge.

Svalbard

General

Anker-Nilssen, T., Bakken, V., Strøm, H., Golovkin, A. N., Bianki, V. V. and Tatarinkova, I. P. (2000) *The status of marine birds breeding in the Barents Sea region*. Tromsø, Norsk Polarinstitutt.

Anon. (2003) *Statistical yearbook of Norway 2003*.

Bangjord, G. (2003) pers comm to I. J. Øien.

Isaksen, K. and Bakken,V., ed. (1995) *Seabird populations in the northern Barents Sea*. NP/AKUP.

Poland

Breeding

1. BirdLife International/European Bird Census Council. (2000) *European bird populations: Estimates and trends*. Cambridge, UK: BirdLife International (BirdLife Conservation Series No. 10). Cambridge.
2. Tomialojc, L. and Stawarczyk, T. (2003) [*The Avifauna of Poland: Distribution, numbers and trends*.] PTPP pro Natura, Wroclaw. (In Polish.)
3. Betleja, J. and Bochenski, Z. (2001) [Purple Heron (*Ardea purpurea*).] Pp.117–119 in Z. Glowacinski, ed. [*Polish red data book of animals. Vertebrates*.] PWRiL, Warszawa. (In Polish.)
4. Pugacewicz, E. and Kowalski, J. (1997) [The first 20th-century breeding of the Great White Egret *Egretta alba* in Poland.] *Not. Orn.* 38: 323–325. (In Polish.)
5. Wypychowski, K. (2003) pers. comm. to P. Chylarecki and A. Sikora.
6. Janiszewski, T. and Glubowski, M. (2002) [Breeding of the Great White Egret *Egretta alba* and Night Heron *Nycticorax nycticorax* on the Jeziorsko Reservoir in 2002.] *Not. Orn.* 43: 259–262. (In Polish.)
7. Betleja, J. and Bochenski, Z. (2001) [Night Heron (*Nycticorax nycticorax*).] Pp.115–117 in Z. Glowacinski, ed. [*Polish red data book of animals. Vertebrates*.] PWRiL, Warszawa. (In Polish.)
8. Betleja, J. (2001) [Breeding of the Night Heron *Nycticorax nycticorax* in the upper Vistula River valley.] *Not. Orn.* 42: 147–158. (In Polish.)
9. Cempulik, P and Kupczyk, M. A. (2001) [Little Bittern (*Ixobrychus minutus*).] Pp.112–114. in Z. Glowacinski, ed. [*Polish red data book of animals. Vertebrates*.] PWRiL, Warszawa. (In Polish.)
10. Sikora, A. and Chylarecki, P. (2003) Unpublished data from survey of flagship bird species in 2003. Institute of Ornithology, Polish Academy of Sciences. Unpublished data.
11. Dombrowski, A. (2001) [Bittern (*Botaurus stellaris*).] Pp.110–112 in Z. Glowacinski, ed. [*Polish red data book of animals. Vertebrates*.] PWRiL, Warszawa. (In Polish.)
12. Jakubiec, Z. and Guziak, R. (1998) [White Stork *Ciconia ciconia* in Poland in 1995, distribution, numbers, conservation problems.] *Not. Orn.* 39: 195–209. (In Polish.)
13. Sikora, A., Chylarecki, P., Gromadzki, M. and Rohde, Z. (2001) [Flagship bird species survey. Annual report 2001.] Ornithological Station, Inst. Ecology PAS, Gdańsk. Unpublished report (in Polish.)
14. Wieloch, M. and Chylarecki, P. Unpublished data from a national survey made in 1998. Unpublished data.
15. Profus, P. (1998) [Numbers and territorial expansion of the breeding population of the Whooper Swan *Cygnus cygnus* in Central Europe.] *Chr. Przyr. Ojcz.* 54, 3: 7–22. (In Polish.)
16. Sikora, A. (1994) [The Whooper Swan (*Cygnus cygnus*) – a new breeding species of the Pomerania.] *Not. Orn.* 35: 179–180. (In Polish.)
17. Czapulak, A. and Witkowski, J. (1996) [Breeding of the Whooper Swan *Cygnus cygnus* in the Barycz Valley.] *Ptaki Śląska* 11: 153–155. (In Polish.)
18. Czeraszkiewicz, R. and Oleksiak, A., eds. (2003) [*Waterfowl of Pomorze Zachodnie: Results of 2002/2003 survey, ecology and conservation*.] ZTO/PZŁ, Szczecin. (In Polish.)
19. Jakubiec, Z. and Stawarczyk, T. (2001) [Shelduck (*Tadorna tadorna*).] Pp.119–121 in Z. Glowacinski, ed. [*Polish red data book of animals. Vertebrates*.] PWRiL, Warszawa. (In Polish.)
20. Kaliciuk, J. and Staszewski, A., eds. (1997) [*Birds refuges in the Polish part of Szczecin Lagoon*.] Computer Oryginals, Szczecin. (In Polish.)
21. Bartoszewicz, M. and Wypychowski, K. and Engel, J. (2000) Numbers of some birds species in the Słońsk Nature Reserve in 1994–1997. *Biological Bulletin of Poznan* 37: 235–255.

22. Jakubiec, Z. and Stawarczyk, T. (2001) [Wigeon (*Anas penelope*).] Pp.121–123 in Z. Glowacinski, ed. [*Polish red data book of animals. Vertebrates.*] PWRiL, Warszawa. (In Polish.)
23. Chylarecki, P. Unpublished data based on results of Polish common breeding bird monitoring scheme (2000–2002). Unpublished data.
25. Wieloch, M. (In print.) *The status of the Ferruginous Duck*. BirdLife International.
26. Wieloch, M. (2003) pers. comm. to P. Chylarecki and A. Sikora.
27. Snow, D. W. and Perrins, C. M. (1998) *The birds of the Western Palearctic.* Concise edition. Vol. 1. Oxford, Oxford University Press.
28. Czyzak, M. (2001) [Common Eider *Somateria mollissima* – a new breeding species in Poland.] *Not. Orn.* 42: 63–64. (In Polish.)
29. Sikora, A. (2001) [Red-breasted Merganser (*Mergus serrator*).] Pp.129–131 in Z. Glowacinski, ed. [*Polish red data book of animals. Vertebrates.*] PWRiL, Warszawa. (In Polish.)
30. Sikora, A. (1996) [Breeding ecology of the Red-breasted Merganser (*Mergus serrator*) in the Pojezierze Kaszubskie lake district.] *Not. Orn.* 37: 5–24. (In Polish.)
31. Mizera, T., Uhlig, R., Kalisinski, M., Mundt, J. and Czeraszkiewicz, R. (1994) [Brutverbreitung, Mauser, Nichtbruter- und Winterbestand des Gansesagers *Mergus merganser* im Einzugsgebiet der Oder.] *Vogelwelt* 115: 155–162. (In German.)
33. Adamski, A., Lontkowski, J., Maciorowski, G., Mizera, T., Rodziewicz, M., Stawarczyk, T. and Waclawek, K. (1999) [Distribution and numbers of rare birds of prey in Poland at the end of the 20th century.] *Not. Orn.* 40: 1–22. (In Polish.)
34. Dyrcz, A. (2001) [Short-toed Eagle (*Circaetus gallicus*).] Pp.141–143 in Z. Glowacinski, ed. [*Polish red data book of animals. Vertebrates.*] PWRiL, Warszawa. (In Polish.)
35. Mizera, T. (2003) pers. comm. to P. Chylarecki and A. Sikora.
36. Witkowski, J. (2001) [Hen Harrier (*Circus cyaneus*).] Pp.143–145 in Z. Glowacinski, ed. [*Polish red data book of animals. Vertebrates.*] PWRiL, Warszawa. (In Polish.)
37. Stoj, M., Cwikowski, C. and Waclawek, K. (1997) [Occurence of the Golden Eagle *Aquila chrysaetos* in the Carpatians in 1993–1996.] *Not. Orn.* 38: 255–272. (In Polish.)
38. Dyrcz, A. (2001) [Booted eagle (*Hieraaetus penntaus*).] Pp.156–157 in Z. Glowacinski, ed. [*Polish red data book of animals. Vertebrates.*] PWRiL, Warszawa. (In Polish.)
40. Kamieniarz, R. and Szymkiewicz, M. (2001) [Black Grouse (*Tetrao tetrix*).] Pp.169–173 in Z. Glowacinski, ed. [*Polish red data book of animals. Vertebrates.*] PWRiL, Warszawa. (In Polish.)
41. Pugacewicz, E. (1998) [Numbers of the Black Grouse *Tetrao tetrix* in the Biebrza River valley.] *Not. Orn.* 39: 77–90. (In Polish.)
42. Chmielewski, S. and Fijewski, Z. (2000) [Numbers and distribution of the Black Grouse *Tetrao tetrix* in the Kielce region in 1997.] *Not. Orn.* 41: 246–249. (In Polish.)
43. Glowacinski, Z. and Profus, P. (2001) [Capercaillie (*Tetrao urogallus*).] Pp.173–177 in Z. Glowacinski, ed. [*Polish red data book of animals. Vertebrates.*] PWRiL, Warszawa. (In Polish.)
44. Chylarecki, P., Gromadzka, J., Gromadzki, M. and Zielinski, P. (1999) *Corncrake survey in Poland.* Part 1. Final report of the survey in 1998. RSPB, GOS IE PAN; Gdansk.
45. Dombrowski, A. (2001) [Little Crake (*Porzana parva*).] Pp.178–180. in Z. Glowacinski, ed. [*Polish red data book of animals. Vertebrates.*] PWRiL, Warszawa. (In Polish.)
46. Sikora, A. (2001) [Oystercatcher (*Haematopus ostralegus*).] Pp.187–189 in Z. Glowacinski, ed. [*Polish red data book of animals. Vertebrates.*] PWRiL, Warszawa. (In Polish.)
47. Chylarecki, P. (2001) [Ringed plover (*Charadrius hiaticula*).] Pp.191–194 in Z. Glowacinski, ed. [*Polish red data book of animals. Vertebrates.*] PWRiL, Warszawa. (In Polish.)
48. Baszanowski, P., Sikora, A. and Chylarecki, P. (1993) [First breeding record of the Kentish Plover (*Charadrius alexandrinus*) in Poland.] *Not. Orn.* 34: 376–378. (In Polish.)
49. Dyrcz, A. (2001) [Dotterel (*Charadrius morinellus*).] Pp.194–196 in Z. Glowacinski, ed. [*Polish red data book of animals. Vertebrates*. PWRiL, Warszawa. (In Polish.)
50. Dyrcz, A. (2001) [Great Snipe (*Gallinago media*).] Pp.207–208 in Z. Glowacinski, ed. [*Polish red data book of animals. Vertebrates.*] PWRiL, Warszawa. (In Polish.)
51. Pugacewicz, E. (2002) [Numbers and distribution of the Great Snipe *Gallinago media* in the Nizina Północnopodlaska lowland in 1976–2000.] *Not. Orn.* 43: 1–7. (In Polish.)
52. Chylarecki, P. Unpublished data.
53. Sikora, A. (2001) [Wood sandpiper (*Tringa glareola*).] Pp.212–214 in Z. Glowacinski, ed. [*Polish red data book of animals. Vertebrates.*] PWRiL, Warszawa. (In Polish.)
54. Wlodarczak, A. (1999) [Distribution and numbers of the Polish population of the Dunlin *Calidris alpina schinzii*.] *Not. Orn.* 40: 45–49. (In Polish.)
55. Gromadzka, J. (2001) [Dunlin (*Calidris alpina*).] Pp.198–201 in Z. Glowacinski, ed. [*Polish red data book of animals. Vertebrates.*] PWRiL, Warszawa. (In Polish.)
57. Chylarecki, P. and Winiecki, A. (2001) [Ruff (*Philomachus pugnax*).] Pp.201–205 in Z. Glowacinski, ed. [*Polish red data book of animals. Vertebrates.*] PWRiL, Warszawa. (In Polish.)
58. Staszewski, A., Giergielewicz, J. and Niedzwiecki, S. (1995) [First breeding record of the Black-winged Stilt (*Himantopus himantopus*) in Poland.] *Not. Orn.* 36: 367–368. (In Polish.)
59. Kuzniak, S., Lorek, G. and Lewandowski, M. (1997) [Occurence of the Black-winged Stilt *Himantopus himantopus* in Poland.] *Not. Orn.* 38: 131–139. (In Polish.)
60. Komisja Faunistyczna Sekcji Ornitologicznej PTZool. (1993) [Rare birds recorded in Poland in 1991. Avifaunistic Commission report No 8.] *Not. Orn.* 34: 347–358. (In Polish.)
61. Komisja Faunistyczna Sekcji Ornitologicznej PTZool. (1994) [Rare birds recorded in Poland in 1992. Avifaunistic Commission report No 9.] *Not. Orn.* 35: 157–170. (In Polish.)
62. Komisja Faunistyczna Sekcji Ornitologicznej PTZool. (1994) [Rare birds recorded in Poland in 1993. Avifaunistic Commission report No 10.] *Not. Orn.* 35: 331–346. (In Polish.)
63. Komisja Faunistyczna Sekcji Ornitologicznej PTZool. (1995) [Rare birds recorded in Poland in 1994. Avifaunistic Commission report No 11.] *Not. Orn.* 36: 343–358. (In Polish.)
64. Komisja Faunistyczna Sekcji Ornitologicznej PTZool. (1996) [Rare birds recorded in Poland in 1995. Avifaunistic Commission report No 12.] *Not. Orn.* 37: 301–317. (In Polish.)
65. Komisja Faunistyczna Sekcji Ornitologicznej PTZool. (1997) [Rare birds recorded in Poland in 1996. Avifaunistic Commission report No 13.] *Not. Orn.* 38: 291–311. (In Polish.)
66. Komisja Faunistyczna Sekcji Ornitologicznej PTZool. (1995) [Rare birds recorded in Poland in 1997. Avifaunistic Commission report No 14.] *Not. Orn.* 39: 151–174. (In Polish.)
67. Komisja Faunistyczna Sekcji Ornitologicznej PTZool. (2000) [Rare birds recorded in Poland in 1998. Avifaunistic Commission report No 15.] *Not. Orn.* 41: 29–53. (In Polish.)
68. Komisja Faunistyczna Sekcji Ornitologicznej PTZool. (2000) [Rare birds recorded in Poland in 1999. Avifaunistic Commission report No 16.] *Not. Orn.* 293–316. (In Polish.)
69. Komisja Faunistyczna Sekcji Ornitologicznej PTZool. (2001) [Rare birds recorded in Poland in 2000. Avifaunistic Commission report No 17.] *Not. Orn* 42: 193–214. (In Polish.)
70. Dombrowski, A. and Chmielewski, S. (1996) [Stone curlew (*Burhinus oedicnemus*): the most endangered bird species of Poland and Mazowsze region.] *Kulon* 1, 1–2: 3–9. (In Polish.)
71. Wesolowski, T. (2001) [Stone Curlew (*Burhinus oedicnemus*).] Pp.189–191 in Z. Glowacinski, ed. [*Polish red data book of animals. Vertebrates.*] PWRiL, Warszawa. (In Polish.)
72. Sikora, A., Cenian, Z., Mokwa, T. and Poltorak, W. (1992) [Successive nesting of the Lesser Black-backed Gull (*Larus fuscus*) in Poland.] *Not. Orn.* 33: 167–168. (In Polish.)
73. Pajkert, Z., Pajkert, W. and Gorski, W. (1992) [The Lesser Black-backed Gull (*Larus fuscus*) – a new breeding species to the avifauna of Poland.] *Not. Orn.* 33: 165–166. (In Polish.)
74. Zielinska, M. and Zielinski, P. (2003) pers. comm. to P. Chylarecki and A. Sikora.
75. Wieloch, M. (2001) [Sandwich Tern (*Sterna sanvicensis*).] Pp.216–218 in Z. Glowacinski, ed. [*Polish red data book of animals. Vertebrates.*] PWRiL, Warszawa. (In Polish.)
76. Winiecki, A. and Chylarecki, P. (2001) [Little Tern (*Sterna albifrons*).] Pp.219–223 in Z. Glowacinski, ed. [*Polish red data book of animals. Vertebrates.*] PWRiL, Warszawa. (In Polish.)
77. Dyrcz, A. (2001) [Whiskered Tern (*Chlidonias hybridus*).] Pp.223–225 in Z. Glowacinski, ed. [*Polish red data book of animals. Vertebrates.*] PWRiL, Warszawa. (In Polish.)
78. Witkowski, J. (2001) [White-winged Tern (*Chlidonias leucopterus*).] Pp.225–228 in Z. Glowacinski, ed. [*Polish red data book of animals. Vertebrates.*] PWRiL, Warszawa. (In Polish.)
79. Profus, P. (2001) [Eagle Owl (*Bubo bubo*).] Pp.228–231 in Z. Glowacinski, ed. [*Polish red data book of animals. Vertebrates.*] PWRiL, Warszawa. (In Polish.)
80. Glowacinski, Z. (2001) [Ural Owl (*Strix uralensis*).] Pp.234–236 in Z. Glowacinski, ed. [*Polish red data book of animals. Vertebrates.*] PWRiL, Warszawa. (In Polish.)
81. Mikusek, R. (2001) [Pygmy Owl (*Glaucidium passerinum*).] Pp.232–234 in Z. Glowacinski, ed. [*Polish red data book of animals. Vertebrates.*] PWRiL, Warszawa. (In Polish.)
82. Sikora, A., Blaszczyk, K. and Domaszewicz, A. (2001) [Tengmalm's owl (*Aegolius funereus*).] Pp.241–243 in Z. Glowacinski, ed. [*Polish red data book of animals. Vertebrates.*] PWRiL, Warszawa. (In Polish.)

83. Profus, P. (2001) [Short-eared Owl (*Asio flammeus*).] Pp.238–241 in Z. Glowacinski, ed. [*Polish red data book of animals. Vertebrates.*] PWRiL, Warszawa. (In Polish.)

84. Dombrowski, A. (2003) pers. comm. to P. Chylarecki and A. Sikora.

85. Gorski, A. and Szymkiewicz, M. (2001) [Roller (*Coracias garrulus*).] Pp.246–248 in Z. Glowacinski, ed. [*Polish red data book of animals. Vertebrates.*] PWRiL, Warszawa. (In Polish.)

87. Wasilewski, J. (2001) [Three-toed Woodpecker (*Picoides tridactylus*).] Pp.250–252 in Z. Glowacinski, ed. [*Polish red data book of animals. Vertebrates.*] PWRiL, Warszawa. (In Polish.)

88. Meissner, W. and Skakuj, M. (1997) [First broods of the Citrine Wagtail *Motacilla citreola* in Poland and changes in the species's breeding range in Europe.] *Not. Orn.* 38: 51–60 (In Polish.)

89. Tomialojc, L. (2001) [Lesser Grey Shrike (*Lanius minor*). Pp.266–268 in Z. Glowacinski, ed. [*Polish red data book of animals. Vertebrates.*] PWRiL, Warszawa. (In Polish.)

90. Gorski, A. (2001) [Woodchat Shrike (*Lanius senator*). Pp.269–270 in Z. Glowacinski (ed.). *Polish red data book of animals. Vertebrates.*] PWRiL, Warszawa. (In Polish.)

91. Walasz, K. and Mielczarek, P., eds. (1992) [The Atlas of Breeding Birds in Malopolska 1985–1991 (south-eastern Poland).] *Biologica Silesiae*; Wroclaw (In Polish.)

92. Czapulak, A., Fura, M., Szelag, D., Witan, K. and Gramsz, B. (2001) [Numbers and distribution of the Dipper *Cinclus cinclus* in the Polish part of the Sudety Mountains.] *Not. Orn.* 42: 159–175. (In Polish.)

93. Bochenski, Z. (2001) [Alpine Accentor (*Prunella collaris*). Pp.253–254 in Z. Glowacinski, ed. [*Polish red data book of animals. Vertebrates.*] PWRiL, Warszawa. (In Polish.)

94. Kucharski, R. (2003) pers. comm. to P. Chylarecki and A. Sikora.

95. Bednorz, J. (2001) [Bluethroat (*Luscinia svecica*). Pp.254–256 in Z. Glowacinski, ed. [*Polish red data book of animals. Vertebrates.*] PWRiL, Warszawa. (In Polish.)

96. Maniakowski, M. (2003) pers. comm. to P. Chylarecki and A. Sikora, based on results of OTOP's Aquatic Warbler 2003 survey.

97. Profus, P. (2001) [Rock Thrush (*Monticola saxatilis*). Pp.257–259 in Z. Glowacinski, ed. [*Polish red data book of animals. Vertebrates.*] PWRiL, Warszawa. (In Polish.)

98. Dyrcz, A. and Czeraszkiewicz, R. (1993) [The abundance, threats and means of protection of the breeding population of the Aquatic Warbler (*Acrocephalus paludicola*) in Poland.] *Not. Orn.* 34: 231–246. (In Polish.)

99. Aquatic Warbler Conservation Team. (1999) World population, trends and conservation status of the Aquatic warbler *Acrocephalus paludicola*. *Vogelwelt* 120: 65–85.

100. Witkowski, J. (2001) [Aquatic warbler (*Acrocephalus paludicola*).] Pp.259–261 in Z. Glowacinski, ed. [*Polish red data book of animals. Vertebrates.*] PWRiL, Warszawa. (In Polish.)

101. Glowacinski, Z. (2001) [Wallcreeper (*Tichodroma muraria*).] Pp.265–266 in Z. Glowacinski, ed. [*Polish red data book of animals. Vertebrates.*] PWRiL, Warszawa. (In Polish.)

102. Dombrowski, A., Golawski, A., Kuzniak, S. and Tryjanowski, P. (2000) [Status and threats of the Red-backed Shrike *Lanius colurio* population in Poland.] *Not. Orn.* 41: 139–148. (In Polish.)

103. Kuzniak, S. (2001) [Bearded Tit (*Panurus biarmicus*).] Pp.261–264 in: Z. Glowacinski, ed. [*Polish red data book of animals. Vertebrates.*] PWRiL, Warszawa. (In Polish.)

104. Jakubiec, Z. (2001) [Redpoll (*Carduelis flammea*).] Pp.271–272 in Z. Glowacinski ed. *Polish red data book of animals. Vertebrates.*] PWRiL, Warszawa. (In Polish.)

105. Sikora, A. (2001) [Breeding of the Redpoll *Carduelis flammea* on the Polish Baltic coast.] *Not. Orn.* 42: 73–88. (In Polish.)

106. Kuzniak, S., Dombrowski, A., Golawski, A. and Tryjanowski, P. (1997) [Present condition and threats to the Polish population of the Ortolan Bunting *Emberiza hortulana* against the species situation in Europe.] *Not. Orn.* 38: 141–150. (In Polish.)

107. Tryjanowski, P., Kuzniak, S., Kujawa, K. and Dombrowski, A. (1995) [The condition and threats to the Polish population of the Corn Bunting (*Miliaria calandra*).] *Not. Orn.* 36: 331–341. (In Polish.)

108. Betleja, J. (2003) pers. comm. to P. Chylarecki and A. Sikora.

109. Kamieniarz, R. and Panek, M. (2000) [*Monitoring of selected species of small wildlife in 1990–2000.* In: *Forestry – information and statistics.*] Main bureau of statistics, Warszawa. (In Polish.)

110. Panek, M. (2000) [Situation of Grey Partridge population in Poland in the years 1998–2000 (monitoring results).] Pp.145–154 in S. Kubiak, ed. [*Small wildlife as element of biodiversity.*] Włocławskie Towarzystwo Naukowe, Włocławek. (In Polish.)

111. Sachanowicz, K. (2003) Report on the programme modelling farmland birds distribution in Poland – activity and results of the year 2002/2003. OTOP. Gdansk. Unpublished report.

112. Profus, P. (2001) [Rock Thrush (*Monticola saxatilis*).] Pp.257–259 in Z. Glowacinski, ed. [*Polish red data book of animals. Vertebrates.*] PWRiL, Warszawa. (In Polish.)

Wintering

113. Skov, H., Vaitkus, G., Flensted, K. N., Grishanov, G., Kalamees, A., Kondratyev, A., Leivo, M., Luigujoe, L., Mayr, C., Rasmussen, J.F., Raudonikis, L., Scheller, W., Sidlo, P. O., Stipniece, A., Struwe-Juhl, B. and Welander, B. (2000) *Inventory of coastal and marine important bird areas in the Baltic Sea.* BirdLife International, Cambridge.

114. Tomialojc, L. and Stawarczyk, T. (2003) [*The avifauna of Poland: Distribution, numbers and trends.*] PTPP pro Natura, Wroclaw. (In Polish.)

115. Meissner, W., Antczak , J., Czapulak, A., Dombrowski, A., Walasz, K. and Ziółkowski, M. (2001) Long-term changes in numbers of some waterfowl species wintering in Poland. Pp.67–77 in S. Svazas, W. Meissner, V. Serebryakov, A. Kozulin and G. Grishanov, eds. *Changes of wintering sites of waterfowl in central and eastern Europe.* OMPO Special Publication. Vilnius

116. Meissner, W. (2003) pers. comm. to P. Chylarecki and A. Sikora.

117. Wieloch, M. (2002) [Wintering of swans *Cygnus* spp. in Poland – results of January censuses in 1995 and 2000.] *Swan Specialist Group Newsletter* 4–5: 55–59. (In Polish.)

118. Staszewski, A. and Czeraszkiewicz, R. (2001) [Distribution and abundance of geese during autumn migration and wintering in 1991–1997.] *Not. Orn.* 42: 15–35. (In Polish.)

119. Mokwa, T. and Sikora, A. (1999) [Wintering of the Canada Goose *Branta canadensis* in the Żulawy Wislane region.] *Not. Orn.* 40: 51–60. (In Polish.)

120. Kuczynski, L., Tryjanowski, P., Golawski, A. and Rzepala, M. (2003) Modeling winter habitat of Great Grey Shrike using generalized additive models. *Vogelwarte* 42: 35.

121. Kasprzykowski, Z. and Rzepala, M. (2002) [Numbers and habitat preferences of raptors Falconiformes wintering in central-eastern Poland.] *Not. Orn.* 43: 73–82. (In Polish.)

122. Sikora, A. and Maniakowski, M. (2000) [Wintering of the Jack Snipe *Lymnocryptes minimus* in northern and central Poland.] *Not. Orn.* 41: 225–238. (In Polish.)

123. Sikora, A. (2001) [Occurrence of the Glaucous Gull *Larus hyperboreus* in Poland.] *Not. Orn.* 42: 37–49. (In Polish.)

124. <http://www.czaplon.most.org.pl/p/zimowanie-mewy/mewy-zimowanie2002.htm>. Internet site, accessed 26 Nov 2003.

125. Faber, M. and Neubauer, G. (2001) [Wintering of Yellow-legged Gulls *Larus cachinnans* in the inland of Poland and problems with their identification.] *Not. Orn.* 42: 233–256 (In Polish.)

126. Sikora, A. (1996) [Frequency of the Lesser Black-backed Gull (*Larus fuscus*) wintering in Poland.] *Not. Orn.* 37: 338–342. (In Polish.)

127. Neubauer, G. and Maniakowski, M. (2002) [Occurrence of the Lesser Black-backed Gull *Larus fuscus* in Toruñ in 1991–2001. Taxonomy and identification of subspecies.] *Not. Orn.* 43: 61–71. (In Polish.)

128. Sikora, A., Półtorak, W. and Ziółkowski, M. (1994) [Kittiwake (*Rissa tridactyla*) invasion in Poland in the winter of 1993.] *Not. Orn.* 35: 127–138. (In Polish.)

129. Niedzwiecki, S., Kaliciuk, J., Kalisinski, M., Kozlowska, D., Staszewski, A. and Wysocki, D. (2000) [Mortality of waterfowl in the Szczecin section of the Baltic coast in 1991/1992 and 1992/1993.] *Not. Orn.* 41: 250–254 (In Polish.)

Portugal

Breeding

1. Berliner, A., Pacheco, C. and Monteiro, A. (2001) [Censo de Grifo em Portugal]. In Del Moral, J. and Martí, R., eds. [*El Buitre Leonado en la Peninsula Iberica. III Censo Nacional e I Censo Iberico coordinado.*] Monografía nº 7. SEO/BirdLife. (In Portuguese.)

2. Catry, I. (2002) [*Distribuição e efectivo das populações reprodutoras de Gaivota-de-patas-amarelas* (Larus cachinnans) *e Corvo-marinho-de-crista* (Phalacrocorax aristotelis) *na costa rochosa continental.*] Relatório não publicado para o Instituto de Conservação da Natureza, Lisboa. Unpublished data. (In Portuguese.)

3. Catry, T. (2001) [*Monitorização das populações reprodutoras de Andorinha-do-mar-anã (*Sterna albifrons*), Alfaiate (*Recurvirostra avosetta*) e Perna-longa (*Himantopus himantopus*) na Reserva Natural do Sapal de Castro Marim e Vila Real de Santo António: caracterização do sucesso reprodutor, identificação das principais ameaças e apresentação de propostas de gestão.*] SPEA. Lisboa. (In Portuguese.)

4. Granadeiro, J., Morais, L., Silva, M., Grade, N., Lecoq, M. and Santos, C. (in press) *Breeding Storm Petrels* Oceanodroma castro *in the Farilhões islands, Portugal.* Proceedings of the IV Mediterranean Seabird Symposium: Seabird Ecology and Coastal Zone Management in the Mediterranean.

5. Lecoq, M. (2002) [*Censo de Aves Marinhas no Arquipélago da Berlenga:* Calonectris diomedea, Phalacrocorax aristotelis *e* Uria aalge.] SPEA. Lisboa. Unpublished data. (In Portuguese.)

6. Leitão, D. (2002) [*Estudo das aves dos sapais secundários da RNSCMVRSA: 2º relatório de progresso – período reprodutor e pós-reprodutor.*] SPEA. Lisboa. (In Portuguese.)
7. Ministro, J. and Fernandes, J. (2002) [Avifauna nidificante na Lagoa dos Salgados (Algarve, Portugal): valor de conservação e medidas de gestão.] *Airo* 12: 17–24. (In Portuguese.)
8. Monteiro, A., Pacheco, C., Fernandes, E., Dias, C., Jambas, J., Silva, R., Miguel, L., Berliner, A. and Silva, L. (2001) [*Situação demográfica e conservação do Milhafre-real em Portugal*]. Livro de resumos do II Congresso de Conservação da Natureza. Instituto da Conservação da Natureza. (In Portuguese.)
9. Monteiro, A., Pacheco, C. and Berliner, A. (2002) [Portugal.] In Del Moral, J and Martí, R. [*El Alimoche Común en España y Portugal (I censo coordinado). Año 2000.*] Monografia nº 8. SEO/BirdLife. (In Portuguese.)
10. Moreira, M. I., Catry, I., Henriques, I., Marques, A. and Reis, S. (In press.) [A situação do Rolieiro (*Coracias garrulus*) em Portugal e dados preliminares sobre a sua biologia reprodutora na ZPE de Castro Verde.] *Airo* 14. (In Portuguese.)
11. Pacheco, C., Rosa, G., Blanco, H. and Monteiro, A. (1999) Population and distribution of some cliff nesting birds in the hydrographic basin of the Tejo river in 1997. *Airo* 10: 15–24.
12. Pacheco, C. and Monteiro, P. (1999) [Efectivo populacional de algumas espécies de aves rupícolas na área proposta para Parque Natural do Tejo Internacional.] In Beja, P. Catry, P. and Moreira, F. eds. [*II Congresso de Ornitologia da SPEA: 90–91.*] (In Portuguese.)
13. Catry, I., Reis, S., Cordeiro, A., Rocha, P. and Franco, A. (In press.) [*Será o aumento da disponibilidade de locais de nidificação uma medida de gestão eficaz para a recuperação do Peneireiro-das-torres em Portugal?*] *Airo* 14. (In Portuguese.)
14. Rosa, G., Leitão, D., Mendes, C., Courinha, F., Fernandes, C., Costa, H., Pacheco, C., and Pereira, J. (In press) [*Situação da Águia-sapeira Circus aeruginosus em Portugal: recenseamento da população nidificante (1998).*] Actas da V Reunião Ibérica sobre Tartaranhões. FAPAS and CIDN. (In Portuguese.)
15. Rosa, G., Pacheco, C., Monteiro, A., Carvalho, A. and Araújo, A. (2001) [Situação da Cegonha-preta *Ciconia nigra* em Portugal: recenseamento da população nidificante (1995–1997).] *Airo* 11: 15–22. (In Portuguese.)
16. Rosa, G., Pacheco, C., Pombal, E. and Monteiro, A. (2001) [Evolução e situação actual da Águia-real *Aquila chrysäetos* em Portugal: recenseamento da população nidificante (1997).] *Airo* 11:29–36. (In Portuguese.)

Wintering

17. Costa, H., Rosa, G. and Costa, L. (1999) [A Ocorrência da Águia-pesqueira *Pandion haliaetus* em Portugal Continental durante o Inverno (1987/88 a 1996/97).] In Beja, P. Catry, P. and Moreira, F., eds. [*II Congresso de Ornitologia da SPEA*: 28–29.] (In Portuguese.)
19. Costa, L. and Rufino, R. (1996) [Contagens de aves aquáticas em Portugal – Janeiro de 1995.] *Airo* 7 (1): 36–43. (In Portuguese.)
20. Costa, L. and Rufino, R. (1996a) [Contagens de aves aquáticas em Portugal – Janeiro de 1996.] *Airo* 7 (2): 69–76. (In Portuguese.)
21. Costa, L. and Rufino, R. (1997) [Contagens de aves aquáticas em Portugal – Janeiro de 1997.] *Airo* 8 (1/2): 25–32. (In Portuguese.)
22. Encarnação, V. (2003) pers. comm.
23. Leitão, D. and Peris, S. (2003) Distribuição e abundância do Abibe *Vanellus vanellus* e da Tarambola-dourada *Pluvialis apricaria* em Portugal. *Airo* 13: 3–16. (In Portuguese.)
24. Rosa, G., Leitão, D., Mendes, C., Courinha, F., Costa, H., Pacheco, C. and Pereira, J. (2001) [Situação da Águia-sapeira *Circus aeruginosus* em Portugal: recenseamento da população invernante (1998/99).] *Airo* 11: 23–27. (In Portuguese.)
25. Rosa, G., Encarnação, V. and Pacheco, C. (1998) [Recenseamento dos efectivos invernantes de Cegonha-branca *Ciconia ciconia* em Portugal (1995–97).] In Costa, L. T., Costa, H., Araújo, M. and Silva, M., eds. *Simpósio Sobre Aves Migradoras na Península Ibérica*: 81–85. (In Portuguese.)
26. Rosa, G., Araújo, A. and Martins, J. (In press) *The present situation of the White Stork* Ciconia ciconia *in Portugal.*
27. Rufino, R., Neves, R. and Pina, J. (1996) [Recenseamentos aéreos na costa portuguesa.] In Farinha, J.C., Almeida, J. and Costa, H., eds. *I Congresso de Ornitologia da SPEA*: 103. (In Portuguese.)
28. Sarasa, G., Bartolome, J., Fernandez-Cruz, M. and Farinha, J. (1993) [Segundo censo de Ardeidas invernantes en la Península Ibérica y Baleares (1992–93).] *Airo* 4 (2): 41–50. (In Portuguese.)

Azores

Breeding

34. Del Nevo, A., Dunn, E., Medeiros, F., Le Grand, G., Akers, P., Avery, M. and Monteiro, L. (1990) *Status, distribution and conservation of Garajau-rosado (*Sterna dougallii*) and Garajau-comum (*Sterna hirundo*) in the Açores.* RSPB and Universidade dos Açores. Unpublished data.
35. Furness, R. and Monteiro, L. (1995) Red-billed Tropicbird *Phaeton aethereus* in the Azores: first breeding record for Europe. *Bull. B.O.C.* 115 (1): 6–8.
36. Monteiro, L. and Furness, R. (1998) Speciation through temporal segregation of Madeiran Storm Petrel (*Oceanodroma castro*) populations in the Azores. *Philosophical Transactions Royal Society London* 353: 945–953.
37. Monteiro L., Ramos, J., Pereira, J., Monteiro, P., Feio, R., Thompson, D., Bearhop, S., Furness, R., Laranjo, M., Hilton, G., Neves, V., Groz, M. and Thompson, K. (1999) Status and distribution of Fea's Petrel, Bulwer's Petrel, Manx Shearwater, Little Shearwater and Band Rumped Storm-petrel in the Azores Archipelago. *Waterbirds* 22: 358–366.
38. Monteiro, L., Feio, R., Pereira, J., Monteiro, P., Ramos, J., Tavares, A., Wilson, L., Hewitson, L., Neves, V. and Furness, R. Population census of Cory's shearwater *Calonectris diomedea borealis* in the Azores archipelago. Unpublished data.
39. Pereira, J., Groz, M., Bolton, M., Silva, A., Neves, V. and Fraga, H. (2000) [Censos das populações de Garajau-rosado Sterna dougallii e de Garajau-comum *Sterna hirundo* dos Açores.] Final technical report to the European Commission (Programme LIFE, contract BA-3200/98/509). Universidade dos Açores, Horta. Unpublished data. (In Portuguese.)

Madeira

Breeding

28. Campos, A. (1998) [*Biologia de reprodução do Calcamar* Pelagodroma marina *na Selvagem Grande.*] Estágio de licenciatura de Biologia Aplicada aos Recursos Animais da Faculdade de Ciências da Universidade de Lisboa. (In Portuguese.)
29. Geraldes, P. (2002) [*Plano de acção para a Freira do Bugio* Pterodroma feae *– revisão e actualização.*] Estágio de Licenciatura de Biologia Aplicada aos Recursos Animais da Faculdade de Ciências da Universidade de Lisboa. (In Portuguese.)
30. Mougin, J.-L. and Mougin, M.-C. (2000) [L'évolution des effectifs reproducteurs des Puffins cendrés *Calonectris diomedea borealis* de Selvagem Grande (30º09'N, 15º52'W).] *Boletim do Museu Municipal do Funchal* 52: 45–50. (In French.)
31. Oliveira, P. and Moniz, P. (1995). Breeding chronology of the Little Shearwater. *Puffinus assimilis*, in Selvagem Grande. Pp.35–36. in *Proceedings of the 5th. International Seabird Group Conference.* Glasgow.
32. Zino, F., Oliveira, P., King, S., Buckle, A., Biscoito, M., Neves, H. and Vasconcelos, A. (2001) Conservation of Zino's Petrel *Pterodroma madeira* in the archipelago of Madeira. *Oryx* 35: 128–136.

Romania

General

1. Adrian, A. and Annamaria, A. (1998) Nesting birds in the Tarcu Mountains. *Analele Banatului* 4: 367–370
2. Anon. (1997a) [Nesting of Tufted Duck in Romania.] *Buletin AIA* No. 6. (In Romanian.)
3. Anon. (1997b) [The Rose-coloured Starling (*Sturnus roseus*).] *Buletin AIA* No. 6. (In Romanian.)
4. Ardelean, G. and Bereş, I. (2000) *The vertebrate fauna of Maramureº*. Ed. Dacia
5. Ballon, E. (1993) [The Buff-necked Heron (*Bubulcus ibis*) on Sachalin Island, Danube Delta.] *Buletinul de Informare S.O.R.* 3/1993 (In Romanian.)
6. Bănică, G. (1997) Contributions to the study of the dynamics of the family *Lariidae* (Aves) in the litoral ecosystem of southern Dobrogea (1985–1994). *Analele Banatului Ştiinþele Naturii* 3: 71–81
7. Brehme, S., Müller, T. and Redlich, J. (1992) Bird observations in the Danube Delta and in the Dobrodgea (Romania). *WIWO Report* No. 43. WIWO, Zeist, the Netherlands.
8. Castiov, F. (1991) [The Great Bustard in the western part of the country.] *Buletinul de Informare S.O.R.* 2/1991. (In Romanian.)
9. Ceico, T. (1993) [More on the Cattle Egret in the Danube Delta.] *Buletinul de Informare S.O.R.* 3/1993 (In Romanian.)
10. Ceico, T. and Tănase C. (1994) [Mediterranean Gull nests again at Murighiol.] *Buletinul de Informare S.O.R.* 1–2/1994. (In Romanian.)
11. Daróczi, Sz. (1996) Redpoll (*Acanthis flammea*) and Twite (*Acanthis flavirostris*) as wintertime visitors. *Milvus* 3: 28–29.
12. Daróczi, Sz. and Kerekes, I. (1993) [Breeding expansion of Fieldfare in C Transylvania.] *Buletinul de Informare S.O.R.* 4: 7. (In Romanian.)
13. Daróczi, Sz. and Moldovan I. (1994) [A few observations about the breeding of Great-Grey Shrike.] *Buletinul de Informare S.O.R.* 1–2: 8. (In Romanian.)
14. Dijksen, A. J., Lebret T., Ouweneel, G. L. and Philippona, J. (1973) Ornithological observations on the lagoons of the Dobrodgea, Romania, in autumn and winter of 1969, 1970 and 1971. *Ardea* 61: 159–178.
15. Fischer, S. (1991) [Cetti's Warbler (*Cettia cetti*) in Romania.] *Orn. Mitt.* 43: 196–197. (In German.)
16. Fischer, S. 1990. [*Cettia cetti* observed in June in the Danube Delta.] *Buletinul de Informare S.O.R.* 3/1990. (In Romanian.)
17. Fischer, W. (1974) [Isabelline Wheatear south of the Danube Delta.] *Der Falke* 21: 66. (In German.)
18. Gache, C. (1999) [National count of the Corncrake.] *Migrans* 3(1). (in Romanian.)

19. Ionescu, D. T. (1998) [Yellow-legged Gulls nest in Transylvania.] *Migrans* 2(3). (In Romanian.)

20. Ionescu, D. T. (1999) Contribution to the study of the waterfowl from Dumbravita lake and ponds (Barsa Depression). *Transylanian rewiew of systematic and ecological research* 1: 191–195

21. Kelemen, A. M. and Papp, T. (1996) Census of Dipper (*Cinclus cinclus*) on the River Mures (Romania) in the winter of 1995–96. *Milvus* 4: 24–25.

22. Kis, A. (1997) [The Order *Galliformes* in the Banat.] *Analele Banatului* 3: 185–195. (In Romanian.)

23. Kis, A. (1999) [A species on the brink of extinction: the Great Bustard in the Banat.] *Erd. Nimród* 1999/4 (In Hungarian.)

24. Kiss, J. B. and L. Szabó (2000) First breeding record of White-tailed Lapwing in Romania (and Europe outside Russia). *British Birds* 93: 400–401.

25. Kiss, J. B. (1980) [*Charadrius asiaticus Pall* and *Larus ichthyaetus Pall* in Romania.] *Trav. Mus. Hist. Nat. "Grigore Antipa"* 22: 541–545. (In French.)

26. Kiss, J. B. (2002) [The situation of Great Cormorant (*Phalacrocorax carbo*) in the Danube Delta.] *Anal INCDD* 10: 154–160. (In Romanian.)

27. Kiss, J. B. 1981. [Ornithological observations in Nord Dobrogea.] *Der Falke* 51: 236–239. (In German.)

28. Kiss, J. B. (1999) [The Cattle Egret nests in the Danube Delta.] *Erdélyi Nimród* 1(6): 18. (in Hungarian.)

29. Kiss, J. B., Szabó L. (2000) First breeding record of certain bird species in Romania, data about the nesting of rare species. *Studii şi Cercetări* 5: 119–125.

30. Klein, R. and A. Buchheim (1997) The western shore of the Black Sea as a contact zone for the large gulls of the *Larus cachinnans*-Group. *Vogelwelt* 118: 61–70.

31. Kohl, I., Sasvári, L. (1980) [The Scarlet Rosefinch (*Carpodacus erythrinus Pall*) in Romania.] *Marmaþia* 5–6: 115–119. (In Romanian.)

32. Kohl, I., Szombath, Z., Könya, I., Lörinc, I., Libus, A. and Szombath, I. (1995) The birds of the Mures river. Pp.235–244 in *The Maros/ Mures River Valley*.

33. Komáromi, I. and Miholcsa, T. (2003) [The breeding of Goosander in Romania.]. *Migrans* 7: 3. (In Romanian.)

34. Königsted, D. and Robel, D. (1992) [The Isabelline Wheatear *Oenanthe isabellina* in the Balkans.] *Limicola* 6: 22–30. (In German.)

35. Kósa, F., Munteanu, D., Pap, P. L., Sándor, D. A. and Szabó, D. Z. (1998) [Results of White Stork (*Ciconia ciconia* L) Census in Cluj county in 1996.] Stud. Univ. Babeş-Bolyai, Biol. 43: 65–70. (In Romanian.)

36. Kósa, F. and Demeter, L. (2003) [White Stork census in eastern Transylvania.] *Migrans* 4: 3. (In Romanian.)

37. Kováts, L. (1962) [New data on the breeding of Pintail (*Anas acuta acuta*) în R. P. R.] *Natura* 14: 80–81. (In Romanian.)

38. Marinov, M. (1993) [Bird Fauna in the zone of Sf. Gheorghe in wintertime (Romania).] *Buletinul de Informare S.O.R.* 1993/2. (In Romanian.)

39. Meita, L. and Ceico, T., (1997) *Larus melanocephalus* hatches again in the Danube Delta. *Analele Banatului* 3: 253–254.

40. Mitruly, A. and Weber, P. (1990) [Histria zone –waterfowl census.] *Buletinul de Informare S.O.R.* 3/1990. (In Romanian.)

41. Müller, T. and Rohde C. (1991) [Situation of Hobby (*Falco subbuteo*) in the Danube Delta (Romania).] *Buletin WWGBP* 4: 87–95. (In German.)

42. Munteanu, D. (1996) Changes and trends in Anatidae populations in Romania after the second world war. *Gibier Faune Sauvage, Game Wildl.*, 13:573–582.

43. Pap, P. L. and Szabó, Z. (1996) Results of White Stork census (*Ciconia ciconia*) in Mureş county in 1995 in comparison with years 1974, 1990. *Milvus* 4: 15–18.

44. Pap, P.L. (1994) [*Cygnus bewickii* observations in Romania.] *Buletinul de Informare S.O.R.* 1–2/1994. (In Romanian.)

45. Papp, T. (1996) Counts of Dipper (*Cinclus cinclus*) on the river Mureş in 1995. *Milvus* 3: 9–10.

46. Petrescu, A. and Bereş, I. (1997) Ornithological remarks in the basins of Iza and Săpânţa and the biosphere reserve pietrosul Rodnei, Maramureş (Romania), in 1995–1996. *Trav. Mus. Hist. Nat. „Grigore Antipa"*, 39: 379–400.

47. Platteuw, M., Kiss, J. B. and Sadoul, N. (2002) Survey of colonial breeding birds in Romanian Danube Delta. *Ann. INDDD – Tulcea*. 10: 151–153.

48. PMSC – Programul de Monitoring al Speciilor Cuibaritoare (2003) Database analysis, common breeding bird monitoring. Unpublished report Romanian Ornithological Society.

49. Romanian Ornithological Society (ROS) (1999) Project Report: Dynamics, biology and Conservation of the White-headed Duck *Oxyura leucocephala* in Romania, Plopu.

50. Sándor, D. A. (1996) Census of rookeries – preliminary results. *Milvus* 3: 23–24.

51. Sándor, D. A. (1998) Black-crowned Night herons breeding in a pine forest. *Milvus* 4: 39–42.

52. Sándor, A., Petrovici, M. (1996) Red-rumped Swallow in the valley of River Cerna. *Milvus* 3: 32.

53. Schmitz, M., Sudfeldt, C., Legge, H., Mantel, K., Weber, P. and Marinov, M. (1999) Spring migration of waders in the Razim-Sinoie loagoon system south of the Danube Delta, Romania. *Wader Study Group Bull.* 90: 59–64.

54. Schmitz, M. (1999) [Status of the Long-legged Buzzard *Buteo rufinus* in Romania.] *Limicola* 13: 24–30. (In German.)

55. Szabó, J. and Sándor, A. D. (2003) The status of Ferruginous Duck in Romania. Pp.36–39 in Petkov, N., Hughes, B. and Gallo-Orsi, U., eds. *Ferruginous Duck: From research to conservation*. Conservation Series No. 6. BirdLife International – BSPB – TWSG, Sofia.

56. Talpeş, V. H. (1999) [*The Eider in Romania*.] VPR 7/1999 p.8. (In Romanian.)

57. Weber, M. and Just, B. (1993) [*Parus lugubris* in the Trascău Mountains.] *Buletinul de Informare S.O.R.* 2/1993. (In Romanian.)

58. Weber, P. (1997) Results of White Stork (*Ciconia ciconia*) breeding in Romania. *Analele Banatului* 3: 53–58

59. Weber, P. and Mihăileanu A. (1989) [The presence of Bonelli's Warbler *Phylloscopus bonelli* in the National Reserve of "Pădurea Hagieni", Constanta county.] *Rev. Muz.* 20: 64–65. (In Romanian.)

60. Wetlands International Midwinter Waterfowl Census. Internal reports to WI. Unpublished report Romanian Ornithological Society.

61. Zeitz, R., Daróczi, S. and Sándor, D. A. (1999) Distribution and densities of breeding Common Buzzards (*Buteo buteo*) and Eurasian Goshawk (*Accipiter gentilis*) in central Transsylvania, Romania. Raptor research conference, Trebon, Czech Republic, 1999, September 23–26. *Buteo* 10: 22.

62. Zeitz, R., Daróczi, S., Sándor, D. A. and Domahidi, Z. (1999) Preliminary results of the Transsylvanian raptor census program – pilot project for long time conservation and management. Raptor Research Conference, Trebon, Czech Republic, 1999, September 23–26. *Buteo* 10: 21.

63. Zeitz, R. and Daróczi, J. Sz. (2003) *Habitat preference of the Lesser Spotted Eagle (*Aquila pomarina*) in Central Transylvania Romania.* Pp.38 in 6th world conference on birds of prey and owls. Budapest, Hungary, 18–23 May 2003.

Russia

Breeding

1. Avdanin, V. O., Vinogradov, V. G. and Krivenko, V. G. (2000) Torna-Shoina interfluvial area, Kanin peninsula. Pp.33–38 in *Wetlands in Russia. Wetlands on the Ramsar shadow list. Vol. 3.* Wetlands International Global Series No.6.

2. Anon. (2001) [Actual problems of studying and protection of birds of eastern Europe and northern Asia.] *Materials of the international conference (XI Ornithological conference)*. Kazan (In Russian.)

3. Askeev, I. V. and Askeev, O. V. (1999) [*Avifauna of Tatarstan Republic.*] Kazan. (In Russian.)

4. Bakka, S. V. and Kiseleva, N. J. (2001) [Key ornithological territories of the Nizhniy Novgorod area.] Pp.98–110 in S.A. Bukreev, ed. [*Inventory, monitoring and protection of important bird areas of Russia.*] Issue 3. Moscow. (In Russian.)

5. Belik, V. P. (1990) [Woodpeckers of the Rostov Region.] Pp.6–29 in [*Insufficiently studied birds of Northern Caucasus.*] Stavropol. (In Russian.)

6. Belik, V. P. (1996) [Birds – Aves.] Pp.272–391 in [*Rare, disappearing and requiring protection animals of the Rostov Region.*] Rostov-on-Don. (In Russian.)

7. Belik, V. P. (1999) [Some results of field researches of 1999 on the IBA program in the south of the European Russia.] *Important Bird Areas territories of Russia. Newsletter* 10: 24–26. (In Russian.)

8. Belik, V. P. (2000) [*Birds of Don steppes: Formation of fauna, its anthropogenous transformation and questions of protection.*] Rostov-on-Don. (In Russian.)

9. Belik, V. P. and Muzaev, V. M. (1995) [Recent status of White-winged Lark in the European south-east of Russia.] *Caucasian Ornithological Bulletin* 7: 7–9. (In Russian.)

10. Preobrazhensky, A. B., ed. (2000) [*White Stork in Russia: Moving Eastward.*] Kaluga. (In Russian.)

11. Bianki, V. V., Kohanov, V. D., Koryakin, A. S., Krasnov, J. V., Paneva, T. D., Tatarinkova, I. P., Chemjakin, R. G., Shkljarevich, F. N. and Shutova, E. V. (1993) [The birds of the Kola Peninsula and White Sea.] *Russian Journal of Ornithology* 2(4): 491–586. (In Russian.)

12. Bolshakov, N. M. and Kostin, A. B. (1994) [*Predatory birds of Dobrovsky Reserve and adjacent territories. Recent conditions of flora and fauna of Lipetsk region and problems of their protection.*] Part 2. Fauna and problems of its protection. Lipetsk. (In Russian.)

13. Borodin, O. V. (1994) [Birds of the Ulyanovsk region.] Series *The nature of the Ulyanovsk region.* Issue 1. Ulyanovsk. (In Russian.)

14. Borodin, O. V., Barabashin, T. O. and Kiryashin, V. V. (2001) [Ruddy Shelduck: back to the Middle Volga Region.] *Casarca. Bulletin of the goose, swan and duck of northern Eurasia* 7: 179–188. (In Russian.)

15. Bukreev, S. A. (2001) [Monitoring of Bogdinsko-Baskunchakskt IBA.] *Important bird areas of Russia. Newsletter* 13: 17. (In Russian.)

16. Bukreev, S. A. (2002) [Monitoring of IBA: Sarpinskie lakes.] *Important bird areas of Russia. Newsletter* 15: 7. (In Russian.)

17. Butjev, V. T. and Lebedeva, E. A. (1992) [Ornithological observations around Aul Kurush (the south of Dagestan).] *Caucasian Ornithological Bulletin* 4(1): 73–83. (In Russian.)

18. Butjev, V. T., Redkin, J. A. and Shitikov, D. A. (1999) [Distribution of some bird species in the Arkhangelsk region.] *Russian journal of ornithology. Express issue* 82: 18–23. (In Russian.)

19. Vengerov, P. D. and Smirnov, S. V. (2002) [Breeding the Syrian Woodpecker in the city of Voronezh and the Voronezh region.] Pp.25–26 in [*The role of especially protected natural territories of the Central Chernozem Region in preservation and studying forest-steppe biodiversity.*] Voronezh. (In Russian.)

20. Volov, A. N. (2002) [On the numbers of Great Snipe and Common Snipe in Kirov region.] Pp.27–28 in [Studies of waders in eastern Europe and northern Asia at the turn of centuries.] *Proceedings of the IV and V meetings of RWSG.* Moscow. (In Russian.)

21. Gabuzov, O. S., Kormilitsin, A. A. and Syroechkovski, Å. Å. Jr. (1997) [Canada Goose in Russia.] *Casarca. Bulletin of the Goose, Swan and Duck of Northern Eurasia* 3: 64–83. (In Russian.)

22. Galushin, V. M., Kubareva, N. J. and Romanov, M. S. (2000) [The population of birds of wood fragments on the Upper Don.] *Ornithological researches in Russia* 2: 5–25. (In Russian.)

23. Garanin, V. I., Egorov, I. J. and Rjabova, G. A. (2000) [*Fauna of the East Zakamie.*] (In Russian.)

24. Garanin, V. I., Shahtarin, G. J., Ivliev, V. G., *et al.* (1993) [Vertebrate animals of terrestrial ecosystems.] Pp.327–342 in [*The green book of Tatarstan Republic.*] Kazan. (In Russian.)

25. Gerasimov, J. N., Salnikov, G. M. and Buslaev, S. M. (2000) [*Birds of the Ivanovo Region.*] Moscow. (In Russian.)

26. Gizatulin, I. I., Hohlov, A. N. and Iljuh, M. P. (2001) [*Birds of the Chechen and Ingushetia Republics.*] Stavropol. (In Russian.)

27. Gineev, A. M. (1989) [Modern condition and changes in ecological situation in North Caucasian region, Kalmyk Republic and Astrakhan region.] Pp.312–319 in [*Environmental problems of Stavropol territory and adjacent territories. Abstracts of regional scientific – practical conference.*] Stavropol. (In Russian.)

28. Tomkovich, P. S. and Lebedeva, E. A., eds. (1998) [*Breeding waders in eastern Europe – 2000.*] Vol. 1. Russian Bird Conservation Union, Moscow. (In Russian.)

29. Tomkovich, P. S. and Lebedeva, E. A., eds. (1999) [*Breeding waders in eastern Europe – 2000.*] Vol. 2. Russian Bird Conservation Union, Moscow. (In Russian.)

30. Gorshkov, V. I. (2000) [Nesting of the Great White Heron in the Kaluga Region.] Kaluga bulletin of nature. *Ornithological Kaluga* 1: 29–30. (In Russian.)

31. Grishanov, G. V. (1994) [Breeding birds of Kaliningrad region: territorial distribution and dynamics in XIX–XX Centuries.] 1. *Non-Passeriformes.* 83–116. (In Russian.)

32. Gudina, A. N., Brigadova, O. V., Sklejmina, A. V. and Baev, A. V. (2001) [Birds of the low Karaj river (inflow of Vorona).] Pp.51–60 in [*Plot method to estimate bird abundance in Russia. Materials of the All-Russian meeting "Counts of birds on plots: perfection and unification of methods, results of their application".*] Tambov. (In Russian.)

33. Dzhamirzoev, G. S., Hohlov, A. N. and Iljuh, M. P. (2000) [*Rare and disappearing birds of Dagestan and their protection.*] Stavropol. (In Russian.)

34. Dudin, P. I., Sarychev, V. S. and Turchin, V. G. (1989) [Booted Eagle in Upper Don.] Pp.48–52 in [*Rare and in need of protection animals (Materials to the Red Book).*] Moscow. (In Russian.)

35. Ezhova, S. A. and Butjev, V. T. (1997) [The bird population of meadow communities of Lipetsk region.] Pp.27–29 in [*P.P.Semenov-Tjan-Shansky scientific heritage and its role in the development of modern science.*] Lipetsk. (In Russian.)

36. Nejfeldt, I. A., ed. (1982) [*Cranes in the USSR.*] Zoological Institute, Leningrad. (In Russian.)

37. Anon. (2002) [Cranes of Eurasia (distribution, number, biology).] Collection of papers. Moscow. (In Russian.)

38. Zavjalov, E. V., Tabachishin, V. G., Shlyakhtin, G. V., Jakushev, N. N. and Kochetova, I. B. (2000) [Cranes, rails and crakes of Saratov region.] *Berkut* 10(1): 67–83. (In Russian.)

39. Zemljanuhin, A. I. (1994) [Fauna and population of birds in the forest-park zone of Lipetsk city.] Pp.29–38 in [*Recent condition of flora and fauna of Lipetsk region and problems of their protection.*] Part 2. Fauna and problems with its protection. Lipetsk. (In Russian.)

40. Zimin, V. B., Sazonov, S. V., Lapshin, N. V., Hohlova, T. J., Artemjev, A. V., Annenkov, V. G. and Jakovleva, M. V. (1993) [*Avifauna of Karelia.*] Petrozavodsk. (In Russian.)

41. Zozulin, G. M. (1992) [*Woods of the Low Don. Rostov-on-Don.*] (In Russian.)

42. Zubakin, V. A. (1998) [Numbers and distribution of gulls and terns in the Moscow Region.] *Ornithologia* 28: 66–75. (In Russian.)

43. Ivliev, V. G. (1998) [Fauna and population of birds in woods of Tatarstan Republic.] Pp.149–155 in [*Problems of bio- and medical ecology of Tatarstan Republic.*] Kazan. (In Russian.)

44. Iljinskij, I. V., Pchelintsev, V. G. and Fetisov, S. A. (1997) [Observations on breeding of Bittern *Botaurus stellaris* in Sebezh Lakes system (Pskov Region).] *Russian journal of ornithology. Express issue* 17: 16–21. (In Russian.)

45. Iljuh, M. P. (1997) [Lesser Kestrel in central Ciscaucasian.] Pp.57–76 in [*Scientific heritage of N. J. Dinnik and his role in development of the recent natural sciences.*] Stavropol. (In Russian.)

46. Iljuh, M. P. (2001) [Unique settlement of Lesser Kestrel in Neftekumsk.] *News in the world of birds* 3: 20. (In Russian.)

47. Crane Working Group of Eurasia (2001) *Newsletter No 2 of Crane Working Group of Eurasia.* (In Russian.)

48. Kazakov, B. A. and Sidelnikov, V. V. (1999) [Anseriformes the Rostov region: species structure, numbers, territorial distribution, migrations.] Pp.111–119 in [*Inventory, monitoring and protection of important bird areas of Russia.*] Moscow. (In Russian.)

49. Kaljakin, V. N. (1993) [Fauna of birds and mammals Novozemelski region and estimation of its condition.] Pp.23–90 in [*Novaya Zemlya. Vol. 2. Proceedings of the Sea Arctic Complex Expedition.*] Moscow. (In Russian.)

50. Karavaev, A. A. (2000) [Summer population of birds of the Skalisty Ridge area of Baranaha and Bol'she mountains.] *Caucasian Ornithological Bulletin* 12: 119–133. (In Russian.)

51. Karjakin, I. V. (1998) [*The bird fauna of the Perm region.*] Alfa, Perm. (In Russian.)

52. Karjakin, I. V. (1998) [*The bird fauna of the Bashkortostan Republic.*] Field research center of the Urals Union for Animal Protection, Perm. (In Russian.)

53. Kirina, M. V. (2002) [Features of the spatial distribution of colonies of Common Swift and House-Martin in Voronezh city.] Pp.48–49 in [*The role of especially protected natural areas of the Central Chernozem region in preservation and study of biodiversity of forest-steppe.*] Voronezh. (In Russian.)

54. Klimov, S. M. (1988) [Breeding biology of Penduline tit in the central Chernozem Region.] Pp.81–85 in [*Seasonal movements and structure of populations of terrestrial vertebrate animals.*] Interuniversity collection of papers. Moscow. (In Russian.)

55. Klimov, S. M. (1991) [Distribution and numbers of Common Kingfisher in the Upper Don basin.] Pp.132–133 in [*Studying rare animals in RSFSR (Materials of the Red Data Book).*] Moscow. (In Russian.)

56. Klimov, S. M. and Alexandrov, V. N. (1988) [Seasonal dynamics of Corvidae birds numbers in the city of Lipetsk.] *Bulletin MOIP. Division of Biology* 93(3): 68–69. (In Russian.)

57. Klimov, S. M. and Ovchinnikova, N. A. (1989) [Summer fauna and population of birds of Dobrovsky Landscape Reserve.] Pp.106–108 in [*All-Union meeting on the problem of cadastre and count of fauna.*] Part 3. Ufa. (In Russian.)

58. Klimov, S. M., Sarychev, V. S. and Nedosekin, V. J . (1997) [*Distribution and biology of the Rook in Lipetsk region. Corvidae birds in an anthropogenous landscape.*] Issue 3. Lipetsk. (In Russian.)

59. Kondratyev, A. V., ed. (2000) [*Important bird areas of the Baltic region of Russia (Kaliningrad and Leningrad regions).*] (In Russian.)

60. Sviridova, T. V., ed. (2000) [*Important bird areas of Russia.*] Vol.1. Important bird areas of international importance in European Russia. Moscow. (In Russian.)

61. Komarov, J. E. and Lipkovich, A. D. (2000) [Class birds – aves.] Pp.62–198 in [*Natural resources of Northern Ossetia Republic – Alania: Fauna of Northern Ossetia Republic – Alania.*] Vladikavkaz. (In Russian.)

62. Anon. (1998) [Conference on predatory birds of eastern Europe and northern Asia, III.] *Materials of conference. Part 1.* Stavropol. (In Russian.)

63. Anon. (1998) [Conference on predatory birds of the East Europe and Northern Asia III., 1998.] *Materials of Conference. Part 2.* Stavropol. (In Russian.)

64. Belik, V. P., ed. (1999) [*Imperial Eagle. Distribution, population status and conservation perspectives within Russia.*] (In Russian.)

65. Ivanter, E. V. and Kuznetsov, O. L., eds. (1995) [*Red data book of Karelia.*] Petrozavodsk. (In Russian.)

66. Vlasov, A. A., ed. (2001) [*Red Book of Kursk region.*] Vol. 1. Rare and disappearing species of animals. Tula. (In Russian.)

67. Anon. (1997) [*Red data book of Lipetsk region.*] Rare and threatened species of animals. Lipetsk. (In Russian.)

68. Zubakin, V. A. Tihomirov, V. N., eds. (1998) [*Red data book of Moscow region.*] Moscow. (In Russian.)

69. Anon. (1998) [*Red data book of Orenburg region.*] Animals and plants. Orenburg. (In Russian.)

70. Noskov, G. A., ed. (2002) [*Red Data Book of the nature of Leningrad Region.*] Vol. 3. Animals. (In Russian.)

71. Anon. (1998) [*Red data book of Komi republic.*] Moscow – Syktyvkar. (In Russian.)
72. Anon. (1999) [*Red Data Book of Northern Ossetia republic – Alania: kinds of plants. Rare and threatened species of plants and animals.*] Vladikavkaz. (In Russian.)
73. Schepovskikh, A. I., ed. (1995) [*Red data book of Tatarstan republic.*] Kazan. (In Russian.)
74. Pavlov, D. S., ed. (2001) [*Red data book of the Russian Federation. Animals.*] Moscow. (In Russian.)
75. Ivanchev, V. P., ed. (2001) [*Red data book of Ryazan region. Rare and threatened species of animals.*] Ryazan. (In Russian.)
76. Belov, V. S., ed. (1996) [*Red data book of Saratov region.*] Saratov. (In Russian.)
77. Bolshakov, V. N. and Gorchakovsky, P. L., eds. (1996) [*Red data book of the Middle Ural (Sverdlovsk and Perm regions): rare and threatened species of animals and plants.*] Ekaterinburg. (In Russian.)
78. Anon. (2002) [*Red data book of the Udmurt republic. Animals.*] Izhevsk. (In Russian.)
79. Krasnov, Yu. V. (1995) [Seabirds (retrospective analysis of population development).] Pp.138–147 in [*Environment and ecosystems of Novaya Zemlya. Archipelago and shelf.*] Apatity. (In Russian.)
80. Krasnov, J. V., Barrett, R. T. and Nikolaeva, N. G. (2000) [Current trends on development of Barents Sea's populations of Northern Gannet *Sula bassana* and Great Skua *Catharacta skua* and determining factors.] Pp.49–51 in [Immigrant species in the European seas of Russia.] *Abstracts of the scientific seminar (Murmansk, January, 27–28, 2000).* Murmansk. (In Russian.)
81. Krasnov, J. V., Gorjaev, J. I., Shavykin, A. A., Nikolaev, Í. Ã., Gavrilo, M. V. and Chernook, V. I. (2002) [*Atlas of birds of the Pechora sea: distribution, number, dynamics, problems of protection.*] Apatity. (In Russian.)
82. Krasnov, Yu. V., Matishov, G. G., Galaktionov, K. V. and Savinova, T. N. (1995) [*The colonial seabirds of Murman.*] St. Petersburg. (In Russian.)
83. Krasnov, Yu. V. and Nikolaeva, N. G. (1998) [Some results of the multidisciplinary study of the Kittiwake in the Barents Sea.] Pp.180–260 in [*Biology and oceanography of the Kara and Barents Seas (along the Northern marina route).*] *Apatity.* (In Russian.)
84. Krasnov, Yu. V. and Nikolaeva, N. G. (1998) [Ecology and morphology of the Black-backed and Herring Gulls in the Barents Sea.] Pp.260–325 in [*Biology and oceanography of the Kara and Barents Seas (along the northern marina route).*] Apatity. (In Russian.)
85. Database on Waterfowl Resources of Russia. (2001) [First results of implementation of the interregional program "Development of monitoring system and protection of waterfowl in Russia".] Head of the Project Prof. Krivenko, V.G. Wetlands International and Centre for Studying Migratory Animals of Eurasia. Moscow. (In Russian.)
86. Kuznetsov, A. V. and Nemtsev, V. V. (2000) [History and populations of Osprey and White-tailed Eagle on Rybinskoye and Sheksninskoye reservoirs.] Pp.18–26 in [*Rare species of predatory birds north of the forest zone of European Russia: prospects for studying and methods of protection.*] Cherepovets. (In Russian.)
87. Lapshin, A. S. and Lysenkov, E. V. (2001) [*Rare birds of Mordovia.*] A.E. Lugovskoy, ed. Saransk. (In Russian.)
88. Lebedeva, E. A. (1993) [Olivaceous Warbler *Hippolais pallida* (Aves, Sylviadae) in the south of Dagestan: new data on biology of the species.] *Zoological Magazine* 72(1): 98–111. (In Russian.)
89. Lihatskij, J. P. (1983) [Biocenology characteristic of passerine birds in upland oakeries of southern forest-steppe. Ecology and protection of a wood.] Pp.59–63 in [*The interuniversity collection papers.*] Leningrad. (In Russian.)
90. Margolin, V. A. (2000) [*Birds of the Kaluga region. Part 1. Non-passeriformes.*] Kaluga. (In Russian.)
91. Margolin, V. A. (2002) [*Birds of the Kaluga region. Part 2. Passeriformes.*] Kaluga. (In Russian.)
92. Markitan, L. V. (2002) Studying population trends of reed-warblers from the genus *Acrocephalus* in East Priazovye. Pp.334–343 in [*Ecosystem researches of Sea of Azov.* Vol. 4.] Apatity. (In Russian.)
93. Anon. (1991) [*Materials of 10th all-union ornithological conference. Part 2. Posters.*] Minsk. (In Russian.)
94. Galushin, V. M., ed. (2003) [*Materials of the 4th conference on raptors of northern Eurasia. Penza, 1–3 February 2003.*] Penza. (In Russian.)
95. Medvedev, V. N. (1992) [A case of nesting of Canada Goose (*Branta canadensis*) in the Valaam Archipelago of Ladoga Lake.] *Russian journal of ornithology* 1: 113–114. (In Russian.)
96. Medvedev, N. V. and Sazonov, S. V. (1994) [Waterbirds of Valaam and Zapadny archipelagoes of Ladoga Lake.] *Russian journal of ornithology* 3(1): 71–81. (In Russian.)
97. Mezhnev, A. P. (2001) [Condition of resources of wintering Galliformes of Russia.] Pp.286–294 in [*Achievements and problems of ornithology of northern Eurasia on a boundary of centuries: The works of the international conference "Urgent problems of bird study and protection in eastern Europe and Northern Asia".*] Kazan. (In Russian.)
98. Menshikova, S. V. (1999) [Horned Grebe *Podiceps auritus* on southern coast of the Gulf of Finland (Leningrad region).] *Russian journal of ornithology. Express issue* 80: 18–20. (In Russian.)
99. Mineev, O. J. (2000) [Waterfowl of Malozemelskaya tundra (distribution, numbers, questions of protection).] Syktyvkar (Synopsis of PhD thesis). (In Russian.)
100. Mineev, J. N. (1994) [*Waterfowl of Yugor Peninsula.*] Syktyvkar. (In Russian.)
101. Mineev, J. N. (1999) [*Anseriformes* of eastern European tundras (distribution, population trends, protection).] Moscow (Synopsis of Doctor of Science thesis). (In Russian.)
102. Mironov, V. I. (1999) [Rare bird species of Kursk region.] Pp.101–111 in [*Rare bird species and valuable ornithological territories of the Central Chernozem region.*] Lipetsk. (In Russian.)
103. Mischenko, A. L. and Suhanova, O. V. (1996) [Checklist of terrestrial vertebrate animals of Novgorod Region. Stage 1.] Unpublished report. Moscow. (In Russian.)
104. Mischenko, A. L. and Suhanova, O. V. (1997) [Checklist of terrestrial vertebrate animals of Novgorod Region. Stage 2.] Unpublished report. Moscow. (In Russian.)
105. Mischenko, A. L. and Suhanova, O. V. (2000) [Results of the two year project "Corncrake survey in European Russia".] Pp.147–175 in [*Corncrake in European Russia: numbers and distribution.*] Moscow. (In Russian.)
106. Morozov, V. V. (1991) [Peregrine and Gyrfalcon at the extreme north-east of Europe.] *Bulletin MOIP. Division of Biology* 96(1): 57–65. (In Russian.)
107. Morozov, V. V. and Syroechkovsky E. E. (2002) [Lesser White-fronted Goose on the verge of the millenium] *Casarca. Bulletin of the goose, swan and duck of northern Eurasia* 8: 233–276. (In Russian.)
108. Mosalov, A. A. (1993) [Features of biology and distribution of Rallidae in Lipetsk Region.] Pp.142–145 in [*Researches on flora and fauna of northern forest-steppe of the European center of Russia.*] Lipetsk. (In Russian.)
109. Nasrulaev, N. I. (1990) [Birds of eastern high mountains of the Bogossky Ridge.] *Ornithology* 24: 154–156. (In Russian.)
110. Nedosekin, V. J. (1994) [About recent distribution and numbers of Stock Pigeon in Lipetsk region.] Pp.62–66 in [*Recent conditions of flora and fauna of Lipetsk region and problems with their protection. Materials of 3rd regional scientific-practical conferences.*] Part 2. Lipetsk. (In Russian.)
111. Nedosekin, V. J. (1996) [Distribution of pigeons in Lipetsk region.] Pp.73–75 in [*Fauna of the Central Chernozem region and formation of ecological culture. Materials of 1st regional conference.*] Part 1. Lipetsk. (In Russian.)
112. Nedosekin, V. J., Klimov, S. M. and Sarychev, V. S. (1994) [Distribution and numbers of water colonial birds in Lipetsk region.] Pp.75–79 in [*Nature of the Upper Don.*] Issue 1. Lipetsk. (In Russian.)
113. Nikolaev, Â. È. (1998) [*Birds of the marsh landscapes of national park "Zavidovo" and Upper Volga.*] Tver. (In Russian.)
114. Numerov, A. D. (1996) [Class birds aves.] Pp.48–159 in [*Natural resources of Voronezh Region.*] Vertebrate animals. Cadastre. Voronezh. (In Russian.)
115. Oparina, O. S., Hrustov, A. V., Oparin, M. L., Zemljanoj, V. L., Borovskaya, L. F. and Borovsky, A. N. (1998) [Revision of the Great Bustard numbers in Saratov region according to the spring count in 1997.] Pp.142–144 in [*Problems of biodiversity preservation of the arid regions of Russia.*] Volgograd. (In Russian.)
116. Polivanov, V. M. (2000) [The summer population of birds in Teberdinsky State Reserve.] Pp.130–181 in [*Birds of various landscapes of Russia, their ecology and protection. Proceedings of Teberdinsky State Reserve.*] Issue 18. Stavropol. (In Russian.)
117. Anon. (2000) [*Nature of the Simbirsk Volga region.*] The collection of scientific papers. Issue 1. Ulyanovsk. (In Russian.)
118. Anon. (2001) [*Nature of the Simbirsk Volga region.*] The collection of scientific papers. Issue 2. Ulyanovsk. (In Russian.)
119. Anon. (2001) [*Problems of studying and protection of Anseriformes in eastern Europe and northern Asia.*] Mosow. (In Russian.)
120. Anon. (2002) [*Birds southern Russia.*] Materials of International Ornitological Conference. Rostov on Don. (In Russian.)
121. Pukinskij, J. B. and Sagitov, R. A. (1985) [Distribution and biology of Little Crake (*Porzana parva Scop.*) in northwest Russia.] Pp.23–29 in [*Ecology of birds during the reproductive period.*] Leningrad. (In Russian.)
122. Ravkin, E. S., Byshnev, I. I., Kochanov, S. K., Ravkin, J. S., Anufriev, V. M., Guzij, A. I., Preobrazhenskaja, E. S., Sarychev, V. S., and Ivliev, V. G. (2001) [Spatial variety of the summer population of birds of the eastern European and western Siberian plains.] Pp.212–236 in [*Achievements and problems of ornithology of northern Eurasia on a boundary of centuries: The works of the international conference "Urgent problems of birds' study and protection in eastern Europe and northern Asia".*] Kazan. (In Russian.)
123. Rakhimov, I. I. (1994) [Movements of Eurasian Collared Dove in Tatarstan.] Pp.121–123 in [*Ecological and morphological variability of animals under the influence of anthropogenous factors.*] Volgograd. (In Russian.)
124. Rahimov, I. I. (1998) [Distribution of some bird species in Tatarstan.] Pp.138–141 in [*Distribution of birds in Ural and western Siberia.*] Ekaterinburg. (In Russian.)

125. ANON. (1998) [Rare bird species of the non-Chernozem center of Russia.] From the meeting "Rare birds of the center of European Russia" (Moscow, January, 25–26, 1995). Moscow. (In Russian.)

126. ANON. (1990) [Rare bird species of the central non-Chernozem region.] From the meeting "Recent status of populations of rare breeding birds in the non-Chernozem center of the USSR " (Puschino, November, 27–28, 1989). Moscow. (In Russian.)

127. ZUBAKIN, V. A. AND ANDRIENKO, T. L., eds. (1997) [*Rare and vulnerable species of plants and animals of Nerussko-Desnyansky of physical-geographical area.*] Bryansk. (In Russian.)

128. PRIKLONSKY, S. G. AND ZUBAKIN, V. A., eds. (2000) [*Rare, endangered and insufficiently studied birds of Russia.*] The collection of scientific papers. Moscow. (In Russian.)

129. SAMIGULLIN, G. M. (1998) [*Wildfowl of the Orenburg region.*] Orenburg. (In Russian.)

130. SARYCHEV, V. S. (1992) [Population dynamics of the Rook in Lipetsk region from 1980 to 1991.] Pp.78–80 in [*Ecological problems of Corvidae.*] Stavropol. (In Russian.)

131. SARYCHEV, V. S., VOROBJEV, G. P., KLIMOV, S. M. AND NEDOSEKIN, V. J. (1990) [Recent numbers of swans in Upper Don area.] Pp.116–117 in [Ecology and protection of swans in the USSR.] *Materials of II all-union meeting on swans.* Melitopol. (In Russian.)

132. SARYCHEV, V. S. AND KLIMOV, S. M. (1995) [Distribution and biology of harriers in the Top Don.] Questions concerning the natural sciences. *The collection of proceedings of young scientists.* Issue 2. Lipetsk. (In Russian.)

133. ANON. (2000) [Modern status of ecosystems in the Oka State Nature Reserve and some areas in European Russia.] *Proceedings of the Oka State Biosphere Reserve.* Vol. 20. Ryazan. (In Russian.)

134. SOTNIKOV, V. N. (1999) [*Birds of the Kirov region and adjacent territories.*] Vol. 1. Non-passeriformes. Part 1. Kirov. (In Russian.)

135. SOTNIKOV, V. N. (1999) [*Birds of the Kirov region and adjacent territories.*] Vol. 1. Non-passeriformes. Part 2. Kirov. (In Russian.)

136. TILBA, P. A. AND KAZAKOV, Á. À. (1985) [Structure of the summer birds population of the central part of the Western Caucasus.] Pp.34–53 in [Birds of north-west Caucasus.] *Collection of scientific papers.* Moscow. (In Russian.)

137. USPENSKY, S. M. AND TOMKOVICH, P. S. (1986) [Birds of Franz Josef and and their protection.] Pp.63–76 in [*Natural complexes of the Arctic region and their protection.*] Leningrad. (In Russian.)

138. POTAPOV, R. L., ed. (1995) [*Avifauna. Nonpasseriformes.*] (Fauna of European north-eastern Russia. Avifauna; Vol. 1, pt. 1). St. Petersburg, Nauka. (In Russian.)

139. POTAPOV, R. L., ed. (1999) [*Avifauna. Nonpasseriformes.*] (Fauna of European north-eastern Russia. Avifauna; Vol. 1, pt. 2). St. Petersburg, Nauka. (In Russian.)

140. ANON. (1996) [*Fauna of the Saratov region. The problem of preservation of rare and endangered species.*] Vol. 1, issue 1. Saratov. (In Russian.)

141. ANON. (1997) [Fauna, ecology and protection of rare birds of the Middle Volga region.] *Collection of papers on materials of the All-Russia scientific-practical conference "Rare birds of the Middle Volga region".* Saransk. (In Russian.)

142. VORONETSKY, V. I., ed. (1994) [Eagle Owl in Russia, Belarus and Ukraine.] *Collection of scientific papers.* Moscow. (In Russian.)

143. FOKIN, S. J., BLOHIN, J. J. AND ZVEREV, P. A. (2000) [Some provisional results from the first comprehensive survey of Woodcock in European Russia.] *Information materials of working group on waders* 13: 27–30. Moscow. (In Russian.)

144. FROLOV, V. V. (1996) [Additions to the avifauna of Penza region.] Pp.75–76 in [Area studies and the problem of ecological education.] *Abstracts of the anniversary scientific-practical conference, May, 16–17, 1996.* Penza. (In Russian.)

145. FROLOV, V. V., KORKINA, S. A., FROLOV, A. V., LYSENKOV, E. V., LAPSHIN, A. S. AND BORODIN, O. V. (2001) [The analysis of status on fauna of non-passeriformes in south of forest-steppe zone of the right-bank Volga Region in XX century.] *Berkut* 10 (2). (In Russian.)

146. HOHLOV, A. N. AND ILJUH, M. P. (1997) [Lesser Spotted Eagle in northern Caucasus.] Pp.60–61 in [*Scientific heritage of P. P. Semenov-Tyan-Shansky and his role in the development of modern science.*] Part 2. Zoology, ecology, botany. Lipetsk. (In Russian.)

147. CHERNOBAJ, V. F., SOHINA, E. N. AND SVETLICHNY, S. V. (2002) [Monitoring of IBA: Akhtuba lakes system.] *Important Bird Areas of Russia. Newsletter* 15: 8. (In Russian.)

148. CHERNOBAJ, V. F., SOHINA, E. N. AND TSABYBIN, S. A. (2001) [Monitoring of IBA: Akhtuba lakes system.] *Important Bird Areas of Russia. Newsletter* 13: 14. (In Russian.)

149. CHIBILEV, A. A. (1995) [Birds of Orenburg region and their protection.] *Materials for the red data book of Orenburg area.* Ekaterinburg. (In Russian.)

150. SHARIKOV, A. V., KONSTANTINOV, V. M., KLIMOV, S. M., LYSENKOV, E. V. AND MALOVICHKO, L. V. (2002) [Distribution and numbers of Long-eared Owl *Asio otus* in anthropogenous landscapes of I102 European Russia.] *Russian Journal of Ornithology. Express issue* 176: 135–142. (In Russian.)

151. ANON. (1996) [*Ecology and numbers of Corvidae in Russia and the adjacent countries.*] Kazan. (In Russian.)

152. ANON. (1997) [*Ecology and numbers of Corvidae in Russia and the adjacent countries.*] Kazan. (In Russian.)

153. ABULADZE, A. V. (1997) Status and conservation of raptors in Caucasia. *Newsletter, world working group on birds of prey and owls* 25–26: 15–19.

154. ANKER-NILSSEN, T., BAKKEN, V., STRØM, H., GOLOVKIN, A. N., BIANKI, V. V. AND TATARINKOVA, I. P., eds. (2000) *The status of marine birds breeding in the Barents Sea region.* Norsk Polarinstitutt, Rapport nr. 113.

155. BAKKEN, V., ed. (2000) *Seabird colony databases of the Barents Sea region and the Kara Sea.* Norsk Polarinstitutt Rapportserie Nr 115.

156. BELIK, V. P. AND LEBEDEVA, E. A. (2002) *Draft international action plan for the Black-Winged Pratincole* Glareola nordmanni. Manuscript published in Moscow by RBCU, final SAP will be published by BirdLife International.

157. GALUSHIN, V., CLARKE, R. AND DAVYGORA, A. (2002) *International action plan for the Pallid Harrier (*Circus macrourus*).* BirdLife International.

158. GALUSHIN, V., MOSEIKIN, V. AND SANIN, N. (2001) Saker Falcon breeding range and populations in European Russia. Pp.34–43 in *Saker Falcon in Mongolia: research and conservation.* Ulaanbaatar.

159. GREENCHENKO, O. S., VOLKOV, S. V. AND SVIRIDOVA, T. V. (1997) Owls in the Moscow region: the results of a 10-year study. Pp.548–550 in *Biology and conservation of owls of the Northern Hemisphere: 2nd int. symposium.* February 5–9; Winnipeg, MB. Gen. Tech. Rep. NC–190.

160. KOSSENKO, S. M. AND KAYGORODOVA, E. YU. (1999) Density and numbers of Middle Spotted Woodpecker *Dendrocopos medius* in Nerussa woodland, Bryansk region, Russia. *Vogelwelt* 120: 229–230.

161. KUZYAKIN, V. A. (2000) Recent research activities on Woodcock (*Scolopax rusticola*) in Russia. Pp.5–9 in H. Kalchreuter, ed. Fifth European Woodcock and Snipe workshop. *Proceedings of an international symposium of Wetlands International. Woodcock and Snipe Specialist Group 3–5 May 1998.* – Wetlands International Global Series No. 4, International Wader Studies 11.

162. MISCHENKO, A. L. AND GALUSHIN, V. M. (2001) Greater Spotted Eagle *Aquila clanga* in European Russia: Distribution and status. *Acta Ornithoecologica* 4(2–4): 385–391.

163. MISCHENKO, A. L., OCHAGOV, D. M., KOSTIN, A. B., GALUSHIN, V. M., NIKOLAEV, V. I., SUKHANOVA, O. V. AND EREMKIN, G. S. (2001) Spotted Eagles in the central part of European Russia: preliminary Assessment of their ranges and population status. *Acta Ornithoecologica* 4: 331–336.

164. SCOTT, D. A. AND ROSE, P. M. (1996) *Atlas of Anatidae Populations in Africa and Western Eurasia.* Wetlands International Publication 41. Oxford.

165. STRØM, H., ØIEN, I. J., OPHEIM, J., KHAKHIN, G. V., CHELTSOV, S. N. AND KUKLIN, V. (1995) Seabird Censuses on Novaya Zemlya 1995. Norwegian Ornithological Society Report No.1–1995.

166. STRØM, H., ØIEN, I. J., OPHEIM, J., KHAKHIN, G. V., CHELTSOV, S. N. AND KUKLIN, V. (1997) Seabird censuses on Novaya Zemlya 1996. Norwegian Ornithological Society Report No.1–1997.

167. STRØM, H., ØIEN, I. J., OPHEIM, J., KUZNETSOV, E. A. AND KHAKHIN, G. V. (1994) Seabird censuses on Novaya Zemlya 1994. Norwegian Ornithological Society Report No.2–1994.

168. TOMKOVICH, P. S. AND LEBEDEVA E. A. (2002) Draft International Action Plan for the Sociable Plover *Chettusia gregaria.* Manuscript published in Moscow by RBCU, final SAP will be published by BirdLife International.

169. ANON. (1990–2000) [Data of winter transect count.] Hunting Department of the Agricultural Ministry of Russian Federation. (In Russian.)

170. BUKREEV, S. A. (2003) pers. comm. to A. Mischenko, V. Belik, O. Borodin, S. Bakka, V. Sarychev, V. Galushin, E. Ravkin, J. Krasnov, E. Lebedeva, O. Sukhanova, S. Volkov and A. Mezhnev.

171. GRISHANOV, G. V. (2003) pers. comm. to A. Mischenko, V. Belik, O. Borodin, S. Bakka, V. Sarychev, V. Galushin, E. Ravkin, J. Krasnov, E. Lebedeva, O. Sukhanova, S. Volkov and A. Mezhnev.

172. DZHAMIRZOEV, G. S. (2003) pers. comm. to A. Mischenko, V. Belik, O. Borodin, S. Bakka, V. Sarychev, V. Galushin, E. Ravkin, J. Krasnov, E. Lebedeva, O. Sukhanova, S. Volkov and A. Mezhnev.

173. KALJAKIN, M. V. (2003) pers. comm. to A. Mischenko, V. Belik, O. Borodin, S. Bakka, V. Sarychev, V. Galushin, E. Ravkin, J. Krasnov, E. Lebedeva, O. Sukhanova, S. Volkov and A. Mezhnev.

174. LIPKOVITCH, A. D. (2003) pers. comm. to A. Mischenko, V. Belik, O. Borodin, S. Bakka, V. Sarychev, V. Galushin, E. Ravkin, J. Krasnov, E. Lebedeva, O. Sukhanova, S. Volkov and A. Mezhnev.

175. MOROZOV, V. V. (2003) pers. comm. to A. Mischenko, V. Belik, O. Borodin, S. Bakka, V. Sarychev, V. Galushin, E. Ravkin, J. Krasnov, E. Lebedeva, O. Sukhanova, S. Volkov and A. Mezhnev.

176. MUZAEV, V. M. (2003) pers. comm. to A. Mischenko, V. Belik, O. Borodin, S. Bakka, V. Sarychev, V. Galushin, E. Ravkin, J. Krasnov, E. Lebedeva, O. Sukhanova, S. Volkov and A. Mezhnev.

177. TILBA, P. A. (2003) pers. comm. to A. Mischenko, V. Belik, O. Borodin, S. Bakka, V. Sarychev, V. Galushin, E. Ravkin, J. Krasnov, E. Lebedeva, O. Sukhanova, S. Volkov and A. Mezhnev.

178. CHERNOBAJ, V. F. (2003) pers. comm. to A. Mischenko, V. Belik, O. Borodin, S. Bakka, V. Sarychev, V. Galushin, E. Ravkin, J. Krasnov, E. Lebedeva, O. Sukhanova, S. Volkov and A. Mezhnev.

Wintering

179. Gilissen, N., Haanstra, L., Delany, S., Boere, G. and Hagemeijer, W. (2002) *Numbers and distribution of wintering waterbirds in the Western Palearctic and Southwest Asia in 1997, 1998 and 1999*. Results from the International Waterbird Census. Wetlands International Global Series No. 11, Wageningen, The Netherlands.

Serbia and Montenegro

Breeding

1. Anon. [nepublikovani podaci-informacije. 'unpublished data-information'.] Unpublished data by project contributors or from archives of institutes and museums. (In Serbian.)
2. Antal, L., Fernbach, J., Mikuska, J., Pelle, I. and Szlivka, L. (1971) [Namen-verzeichnis der Vogel der Autonomen Provinz Vojvodina. Mit einer historischen Ubersicht von DR Andras keve.] *Larus* 23: 73–127. (In German.)
3. Balog, I. (1997) [Gnežđenje malog kormorana (*Phalacrocorax pygmaeus*) na ribnjaku, 'Bečej' – 'Breeding of Pygmy Cormorant on the Becej fishpond'.] *Ciconia* 6. (In Serbian.)
4. Balog, I. (1997) [Nalaz istočnog trstenjaka (*Acrocephalus palucicola*) u okolini Temerina – 'Aquatic Warbler near Temerin'.] *Ciconia* 6. (In Serbian.)
5. Balog, I. (1997) [Rezultati prstenovanja čaplji na ribnjaku 'Bečej' – 'Results of ringing herons at the Becej fish-pond'.] *Ciconia* 6: 17. (In Serbian.)
6. Balog, I. (2001) [Gnežđenje sabljarke (*Recurvirostra avosetta*) u okolini Temerina – 'Nesting of Avocet near Temerin'.] *Ciconia* 10. (In Serbian.)
7. Balog, I. (2002) [Gnežđenje velike bele čaplje (*Egretta alba*) u okolini Temerina – 'Nesting of Great White Egret vicinity of Temerin'.] *Ciconia* 11. (In Serbian.)
8. Barjaktarov, D. (2000) [Velika bela čaplja (*Egretta alba*) – gnezdarica Carske bare – 'Great White Heron – breeding species of Carska bara'.] *Ciconia* 9. (In Serbian.)
9. Božić, J. (1995) [Ptice sliva Jablanice i Obnice.] SNIRS, zbornik radova (Proceedings of a conference), *Beograd*, X: 37–49. (In Serbian.)
10. Božić, J. (1995) [Izveštaj o ornitofauni Durmitora od 15. do 30. jula 1994. godine.] *MI* "V.M.Manda", zbornik radova (Proceedings of a conference), *Valjevo* 12: 10–19. (In Serbian.)
11. Butorac, B. and Hullo, I. (1992) [Florističke, faunističke i ornitološke vrednosti područja Selevenjska pustara, kao podloga za zaštitu.] *Zaštita prirode* 45: 65–76. (In Serbian.)
12. Čović, S. (1992) [Novi podaci o troprstom detliæu (*Picoides trydactylus*) i velikom tetrebu (*Tetrao urogallus*) na podruèju Srbije.] *Ciconia* 4: 57–58. (In Serbian.)
13. Dajović, M. (1998) [Rezultati posmatranja ptica šireg područja Beograda u periodu 1990 – 1997 – 'Results of bird monitoring in Belgrade area in 1990 – 1997 period'.] *Ciconia* 7. (In Serbian.)
14. Đapić, D. (2002) [Gnežđenje sive vetruške *Falco vespertinus* u okolini Stanišiæa – 'Breeding of Red-footed Falcon in vicinity of Stanišić'.] *Ciconia* 11. (In Serbian.)
15. Dević, M. (1995) [Ornitofauna ribnjaka "Uzdin". 'Ornithofauna of the Fishpond "Uzdin"'), *Ciconia* 5: 32–44. (In Serbian.)
16. Dević, M. (1998) [Kretanje populacije dugonogog prudnika (*Himantopus himantopus*) na Uzdinskom ribnjaku – 'Population fluctuations of Black-winged Stilt on Uzdin fishpond'.] *Ciconia* 7. (In Serbian.)
17. Dević, M. (1998) [Prilog prouèavanju gnežđenja sabljarke (*Recurvirostra avosetta*) na ribnjaku "Uzdin" u 1992. godini – 'Notes on breeding of Avocet on Uzdin fishpond in 1992'.] *Ciconia* 7. (In Serbian.)
18. Erg, B. (2002) [Uticaj trostepenog zoniranja zaštite Specijalnog rezervata prirode "Gornje Podunavlje" na oèuvanje gnezda crne rode *Ciconia nigra* i orla belorepana *Haliaeetus albicilla* u reviru Kanlija Pesak – 'Significance of three-leveled zonation of protection of Special Nature Reserve "Gornje Podunavlje" for protection of Black Stork and White-tailed Eagle nests within area Kanlija Pesak'.] *Ciconia* 11. (In Serbian.)
19. Frank, Z. (2001) [Sastav faune ptica veštaèki stvorenih staništa: primer deponije otpadnih voda stoène farme kod Ade – 'Composition of bird fauna in artificial habitats on example of sewage water disposal at livestock farm near Ada'.] *Ciconia* 10. (In Serbian.)
20. Frank, Z. (2002) [Interesantna zapažanja u okolini ribnjaka Jazovo – 'Interesting observation on the fishpond Jazovo'.] *Ciconia* 11. (In Serbian.)
21. Garovnikov, B. (1998) [Velika droplja (*Otis tarda*) u Vojvodini – 'Great Bustard in Vojvodina'.] *Ciconia* 7. (In Serbian.)
22. Garovnikov, B. (1998) [Ptice Titelskog platoa i njegove okoline.] *Verlag Oetker/Voges*, Kiel. (In Serbian.)
23. Garovnikov, B. and Popović, E. (1986) [Ornitofauna Carske bare.] *Priroda Vojvodine* IX-XI: 39–54. (In Serbian.)
24. Gergelj, J. and Hulo, I. (1998) [Teritorijalno ponašanje i migratorni putevi trstenjaka ševariæa (*Acrocephalus melanopogon*) – 'Territorial behaviour and migration ruotes of Moustached Warbler'.] *Ciconia* 7. (In Serbian.)
25. Gergelj, J. (1995) [Gnežđenje crnoglavog galeba (*Larus melanocephalus*) u Vojvodini od 1986 do 1995 – 'The Nesting of the Mediterranean Gull in Vojvodina from 1986 to 1995'.] *Ciconia* 5. (In Serbian.)
26. Gergelj, J. (1997) [Oološka zbirka Lasla Tota iz Ade – 'Laslo Toth's zoological collection in Ada'.] *Ciconia* 6. (In Serbian.)
27. Gergelj, J. (2002) [Bela roda i kolonije èaplji – kampanja za popularizaciju zaštite. 'White Stork and Heron colonies – public information campaign'.] *Ciconia* 11. (In Serbian.)
28. Gergelj, J. (2002) [Gnežđenje sove utine *Asio otus* na tlu – 'Nesting of Long-eared Owl on the ground'.] *Ciconia* 11. (In Serbian.)
29. Gergelj, J. and Šoti, J. (1990) [Ornitofauna ribnjaka "Kapetanski rit".] *Ciconia* 2: 22–49. (In Serbian.)
30. Gergelj, J., Hulo, I. and Petrović, S. (1997) [Istorijski pregled gnežđenja i seobe plavokljune patke (*Oxyura leucocephala*) u Vojvodini – 'An historical review of breeding and migration of White-headed Duck in Voivodina'.] *Ciconia* 6. (In Serbian.)
31. Gergelj, J., Puzoviæ, S., Rašajski, J., Balog, I., Lukač, Š., Žuljeviæ, A., Tucakov, M., Matoviæ, È., Stojnić, N. and Kovačević, I. (2000) [Bela roda (*Ciconia ciconia*) u Vojvodini 2000. godine – populacija i distribucija (Preliminarni Izveštaj) – 'White Stork in Vojvodina in 2000 – Population and Distribution (Preliminary Report)']. *Ciconia* 9. (In Serbian.)
32. Gergelj, J., Tot, L. and Frank, Z. (2000) [Ptice Potisja od Kanjiže do Novog Bečeja – 'Birds of Tisza River from Kanjiza to Novi Becej'.] *Ciconia* 9. (In Serbian.)
33. Grimmett, R. and Jones, C. (1989) Important bird areas in Europe – Yugoslavia. *ICBP Technical Publication* 9: 859–880.
34. Grubač, B. (1988) The Golden Eagle (*Aquila chrysaetos*) in South-Eastern Yugoslavia. *Larus* 38(9): 95–135.
35. Grubač, B. (1989) [Prilozi za ekologiju i etologiju jastreba kokošara (*Accipiter gentilis*) u jugoistoènoj Jugoslaviji.] *Larus* 40: 97–106. (In Serbian.)
36. Grubač, B. (1991) [Istraživanja surog orla *Aquila chrysaetos* u Srbiji i Makedoniji – Nestaje kralj ptica.] *Lovac* 4: 42–47. (In Serbian.)
37. Grubač, B. (1993) [*Ptice (Aves) i ostala fauna Sićevačke klisure*.] Zavod za zaštitu prirode Srbije, Beograd. (In Serbian.)
38. Grubač, B. (1993) [*Stanje, problemi i mere zaštite ptica (Aves) i drugih životinja na području Jelašničke klisure*.] Zavod za zaštitu prirode Srbije, Beograd. (In Serbian.)
39. Grubač, B. (1995) [Prilozi o stanju i gnežđjenju ridjeg mišara (*Buteo rufinus*) Cretszchmar, 1827 u Srbiji.] *Zaštita prirode* 46–47: 157–161. (In Serbian.)
40. Grubač, B. (1995) [*Ptice (Aves) Nacionalnog parka Đerdap i njihova zaštita. Rukopis za studiju o zaštiti NP "Derdap"*.] Zavod za zaštitu prirode Srbije, Beograd. (In Serbian.)
41. Grubač, B. (1995) [Fauna ptica (Aves) i ostalih kičmenjaka kanjona Moravice.] Zavod za zaštitu prirode Srbije, Beograd. Unpublished report. (In Serbian.)
42. Grubač, B. (1996) [La Situation du Falcon lanier *Falco biarmicus* en Mediteterrančе et specialement dans le centre des Balkans.] Pp.127–134 in J. Muntaner and J. Mayol, eds. [*Biologia y Conservacion de las Rapaces Mediterraneas, 1994*.] Monografias, No. 4. SEO, Madrid. (In French.)
43. Grubač, B. (1998) [*Ornitofauna Šar planine*.] Rukopis, pp. 1–36, Zavod za zaštitu prirode Srbije, Beograd. (In Serbian.)
44. Grubač, B. (2000) [Prilog za faunu ptica južnih delova Crne Gore – 'Contribution for bird fauna of southern Montenegro'.] *Ciconia* 9. (In Serbian.)
45. Grubač, B. (1998) Population status and conservation problems of the Black vulture in Yugoslavia. Pp.69–72 in *International Symposium on the Black Vulture in South Eastern Europe and adjacent regions* (E. Tewes, J. J. Sanchez and B. Heredia, eds), Dadia, Greece, 15–16. September 1993. Black Vulture Conservation Foundation – Frankfurt Zoological Society.
46. Grubač, B. (1998) [Pregled faune ptica klisure Jerme – (The survey of ornithofauna of Jerma River Gorge).] *Ciconia* 7: 25–38. (In Serbian.)
47. Grubač, B. (1998) [Stanje, problemi i mere zaštite lešinara (*Aegypiinae*) Srbije.] *Zaštita prirode* 50: 199–205. (In Serbian, with English summary.)
48. Grubač, B. (1998) [Suri orao (*Aquila chrysaetos*) (Golden Eagle).] Zavod za zaštitu prirode Srbije, Beograd. (In Serbian, with English summary.)
49. Grubač, B. (1998) The status of the Bearded Vulture in the Balkans. *LITE-NATURE 1998:* "Conservation of the Bearded Vulture in Greece" – Bearded Vulture Workshop, 21–24th November 1998. Iraklio, Crete.
50. Grubač, B. (1999) [Bela kanja *Neophron percnopterus* (Linnaeus, 1758) u Srbiji.] *Zaštita prirode* 51(2): 123–131. (In Serbian.)
51. Grubač, B. (1999) Distrtibution and Ecology of the Long-legged Buzzard (*Buteo rufinus* Cretzschnar, 1827) in the Central Balkans. Pp.125–130 in *Contributions to the Zoogeography and Ecology of the Eastern Mediterranean Region*. Vol. I.
52. Grubač, B. (1999) [Ubijena mala droplja (*Tetrax tetrax*) u dolini Velike Morave kod sela Strižе – 'Little Bustard shot in the Velika Morava Valley near Striža Village'.] *Ciconia* 8: 127–128. (In Serbian.)
53. Grubač, B. (2000) [Prilozi za faunu ptica južnih delova Crne Gore. 'Contribution to the birds fauna of the southern parts of Montenegro'.] *Ciconia* 9: 114–120. (In Serbian, with English summary.)
54. Grubač, B. (2000) [Ptice Radana.] Zavod za zaštitu prirode Srbije, Beograd. Unpublished report. (In Serbian.)
55. Grubač, B. (2000) The Present Status of Vultures *Aegypiinae* in Central Balkans. Pp.93–103 in *Actas del II Congreso International sobre aves carroneras*. Kanizares – Solan de Cabras, Cuenca.

56. GRUBAČ, B. (2001) [Prilozi o šumskoj sovi *Strix aluco* (Linnaeus, 1758) u Srbiji i Makedoniji – 'Contributions to the study of the Tawny Owl in the Serbia and Macedonia']. *Zaštita prirode* 53(1): 55–79. (In Serbian, with English summary.)

57. GRUBAČ, B. (2001) [Prvi nalaz gnezda male muharice (*Ficedula parva*) u Srbiji – 'First recorded of the nest of Red-Brested Flycathcer in Serbia'.] *Ciconia* 10: 150–152. (In Serbian, with English summary.)

58. GRUBAČ, B. (2001) [Puzgavac *Tichodroma muraria* (Linnaeus, 1866) u Srbiji i Makedonij – 'The Wallcreeper *Tichodroma muraria* in Serbia and Macedonia'] *Zaštita prirode* 52/2: 65–78. (In Serbian, with English summary.)

59. GRUBAČ, B. (2003) [Ptice područja Ovčarsko-kablarske klisure.] Beležnik Ovčarsko-kablarske klisure, Turistička organizacija. *Èaèak* 2: 81–95. (In Serbian.)

60. GRUBAČ, B. AND GRUBAČ, S. (2001) [Bogatstvo diverziteta faune ptica centralnog Pomoravlja na primeru jezera i ribnjaka kod Paraæina – 'Diversity richness of bird fauna of central Pomoravlje area on example of lakes and fishponds near Paraæin.'] *Ciconia* 10: 77–92. (In Serbian.)

61. GRUBAČ, B. AND MARINKOVIĆ, S. (2000) [Bela kanja *Neophron percnopterus* – 'Egyptian Vulture'.] Pp.57–62 in S. Puzoviæ, ed. [*Atlas ptica grabljivica Srbije – 'Atlas of Birds of Prey of Serbia'*.] Zavod za zaštitu prirode Srbije, Beograd/Belgrade. (In Serbian, with English summary.)

62. GRUBAČ, B. AND MARINKOVIĆ, S. (2000) [Suri orao *Aquila chrysaetos* – "Golden Eagle"] Pp.129–134 in S. Puzoviæ, ed. [*Atlas ptica grabljivica Srbije – 'Atlas of Birds of Prey of Serbia'*.] Zavod za zaštitu prirode Srbije, Beograd/Belgrade. (In Serbian, with English summary.)

63. GRUBAČ, B. AND RAŠAJSKI, J. (2000) [Kratkoprsti kobac *Accipiter brevipes*. 'Lewant Sparrowhawk'.] Pp.99–104 in S. Puzović, ed. [*Atlas ptica grabljivica Srbije – 'Atlas of Birds of Prey of Serbia'*.] Zavod za zaštitu prirode Srbije, Beograd/Belgrade. (In Serbian, with English summary.)

64. GRUBAČ, B. AND RAŠAJSKI, J. (2000) [Riđi mišar *Buteo rufinus*. 'Long-legged Buzzard'.] Pp.111–116 in S. Puzoviæ, ed. [*Atlas ptica grabljivica Srbije – 'Atlas of Birds of Prey of Serbia'*.] Zavod za zaštitu prirode Srbije, Beograd/Belgrade. (In Serbian, with English summary.)

65. GRUBAČ, B. AND PUZOVIĆ, S. (1998) [*Ptice Prokletija*.] Zavod za zaštitu prirode Srbije, studija za proglašenje nacionalnog parka, Beograd. (In Serbian.)

66. GRUBAČ, B. AND PUZOVIĆ, S. (2003) [Ptice Metohijskih Prokletija.] Pp.178–197 in L. Amidžić, P. Jakšić and M. Janković, eds. [*Metohijske Prokletije – prirodna i kulturna baština*.] Zavod za zaštitu prirode Srbije, Beograd. (In Serbian.)

67. HAM, I. (1980) [Prilozi fauni ptica (*Aves*) istočne Srbije.] SANU, Zbornik radova o fauni ptica Srbije (Proceedings of a conference). *Beograd* 1: 129–145. (In Serbian.)

68. HAM, I. (1983) [O metodama istraživanja faune ptica na primeru Deliblatske peščare.] *Drugi simpozijum o fauni SR Srbije (Symposium)*, Beograd, 2: 147–150. (In Serbian.)

69. HAM, I. (1989) Heron and Cormorant colonies on the Yugoslav sector of the Danube (588 km). Arhiv biol. nauka, *Beograd* 41(3–4): 17–18.

70. HAM, I. (1986) [Vrednovanje predela Deliblatske peščare procenom ornitološke vrednosti.] *Zbornik radova "Deliblatski pesak" (Proceedings of a conference)*, Pančevo, V: 87–104. (In Serbian.)

71. HAM, I. (1989a) Changes in Heron populations on wider area of the Carska bara. *Arhiv biol.* nauka 41(3–4): 13–14. (In Serbian.)

72. HAM, I. (1986–revision 2003) [Vrednovanje predela Deliblatske peščare procenom ornitološke vrednosti.] Deliblatski pesak, zbornik radova (Proceedings of a conference). *Pančevo* 5: 87–104. (In Serbian.)

73. HAM, I. AND PUZOVIĆ, S. (2000) [Orao belorepan *Haliaeetus albicilla*.] Pp.51–56 in S. Puzović, ed. [*Atlas ptica grabljivica Srbije – 'Atlas of Birds of Prey of Serbia'*.] Zavod za zaštitu prirode Srbije, Beograd/Belgrade. (In Serbian.)

74. HAM, I. AND RAŠAJSKI, J. (2000) [Eja močvarica *Circus aeruginosus*.] Pp.75–80 in S. Puzović, ed. [*Atlas ptica grabljivica Srbije – 'Atlas of Birds of Prey of Serbia'*.] Zavod za zaštitu prirode Srbije, Beograd/Belgrade. (In Serbian.)

75. HAM, I. AND MARINKOVIĆ, S. (2000) [Eja livadarka *Circus pygargus*.] Pp.81–86 in S. Puzović, ed. [*Atlas ptica grabljivica Srbije – 'Atlas of Birds of Prey of Serbia'*.] Zavod za zaštitu prirode Srbije, Beograd/Belgrade. (In Serbian.)

76. HAM, I. AND PUZOVIĆ, S. (2000) [Jastreb *Accipiter gentilis*.] Pp.87–92 in S. Puzović, ed. [*Atlas ptica grabljivica Srbije – 'Atlas of Birds of Prey of Serbia'*.] Zavod za zaštitu prirode Srbije, Beograd/Belgrade. (In Serbian.)

77. HAM, I. AND PUZOVIĆ, S. (2000) [Orao krstaš *Aquila heliaca*.] Pp.123–128 in S. Puzović, ed. [*Atlas ptica grabljivica Srbije – 'Atlas of Birds of Prey of Serbia'*.] Zavod za zaštitu prirode Srbije, Beograd/Belgrade. (In Serbian.)

78. HAM, I. AND RAŠAJSKI, J. (2000) [Siva vetruška *Falco vespertinus*.] Pp.153–158 in S. Puzoviæ, ed. [*Atlas ptica grabljivica Srbije – 'Atlas of Birds of Prey of Serbia'*.] Zavod za zaštitu prirode Srbije, Beograd/Belgrade. (In Serbian.)

79. HAM, I. AND PUZOVIĆ, S. (2000) [Stepski soko *Falco cherrug*.] Pp.171–176 in S. Puzoviæ, ed. [*Atlas ptica grabljivica Srbije – 'Atlas of Birds of Prey of Serbia'*.] Zavod za zaštitu prirode Srbije, Beograd/Belgrade. (In Serbian.)

80. HARDI, B. AND SEKEREŠ, O. (1998) [Gnežđenje jelove senice (*Parus ater*) u Subotičkoj šumi – 'The Breeding of Coal Tit (Parus ater) in Subotička forest.'] *Ciconia* 7. (In Serbian.)

81. HILL, D. A. (1985) Some contributions to the Avifauna of southern parts of Serbia. *Glasnik Prirodnjačkog muzeja u Beogradu* 40: 191–206.

82. HORVAT, F. AND MEGADJA, B. (1997) [Prvo gnežđenje ćuka (*Otus scops*) u veštačkim jezerima na Subotičko-Horgoškoj peščari – 'Scops Owl (*Otus scops*): a new breeding species in artificial nest boxes in the Subotica-Horgos sandy area'.] *Ciconia* 6. (In Serbian.)

83. HORVAT, F. (1999) [Procena letnje ornitofaune prirodnog dobra "Selevenjske pustare".] *Ludaški zapisi* 2(3): 73–76. (In Serbian.)

84. HULO, I. AND GERGELJ, J. (2001) [Spisak ptica Specijalnog rezervata prirode "Ludaško jezero" – 'Checklist of birds of Special Nature Rezerve of Ludas lake'.] *Ciconia* 10. (In Serbian.)

85. HULO, I. AND HORVAT, F. (2002) [Interesantna ornitološka posmatranja iz okoline Subotice – 'Interesting observation of birds vicinity of Subotica.'] *Ciconia* 11. (In Serbian.)

86. HULO, I. (1995) [Gnežđenje sabljarke (*Recurvirostra avosetta*) na Palićkom jezeru – 'The Nesting of the Avocet on the Palic Lake'.] *Ciconia* 5. (In Serbian.)

87. HULO, I. (1995) [Jesenja seoba noćnog potrka (*Burhinus oedicnemus*) u severnoj Vojvodini – 'The Migration of the Stone-curlew in Northern Vojvodina during Autumn'.] *Ciconia* 5. (In Serbian.)

88. HULO, I. (1997) [Migracija predstavnika porodica *Gaviidae*, *Podicipedidae* i *Anatidae* na Palićkom jezeru u periodu 1981–1996 – 'Migration of *Gaviidae*, *Podicipedidae* and *Anatidae* at the Palic lake'.] *Ciconia* 6. (In Serbian.)

89. HULO, I. (1997) [Neuspešno gnežđenje sabljarke (*Recurvirostra avosetta*) i zujavca slepića (*Charadrius dubius*) na ribnjaku 'Kapetanski rit' – 'Fail breeding the Avocet and Little Ringed Plover on the Kapetanski rit'.] *Ciconia* 6. (In Serbian.)

90. HULO, I. (1997) [Noæni potrk (*Burhinus oedicnemus*) na Kapetanskom ritu – 'Stone-curlew on the Kapetanski rit'.] *Ciconia* 6. (In Serbian.)

91. HULLO, I. (1996) [*Ptice Paliækog jezera*.] Manuskript, Subotica. Unpublished report. (In Serbian.)

92. ILIĆ, Z. (1998) [Posmatranje ptica u Sokobanjskoj kotlini (Istoèna Srbija) – 'Bird monitoring in Sokobanjska kotlina (Sokobanja Valley) in eastern Serbia'.] *Ciconia* 7. (In Serbian.)

93. ILIĆ, Z. (2001) [Vrabac kamenjar (*Petronia petronia*) i crnoglava strnadica (*Emberiza melanocephala*) u okolini Niša – 'Rock Sparrow and Black-Headed Bunting near Niš'.] *Ciconia* 10. (In Serbian.)

94. IRŠAI, A. (1997) [Istraživačka akcija: 'Koviljski-Petrovaradinski rit' – 'Ornithological camp at the Kovilj-Petrovarad meadow'.] *Ciconia* 6. (In Serbian.)

95. KANJO, B. (1997) [Prilog za faunu ptica područja Ribareva u gornjem Podunavlju – 'Notes on ornithofauna of Ribarevo and surroundings'.] *Ciconia* 6. (In Serbian.)

96. KANJO, B. (2002) [Poslednji slučajevi gnežđenja zlatovrane *Corracias garrulus* u zapadnoj Bačkoj – 'Last cases of breeding of Rollers in western Backa'.] *Ciconia* 11. (In Serbian.)

97. KIŠ, Č., LONČAR, L. AND MIĆEVIĆ, D. (1995) [Posmatranje ptica na Maljenu u julu 1995'. i junu 1996. godine.] *Petnièke sveske* 42: 18–27.

98. KOVAČ, J. (1998) [Prilog za poznavanje ptica Herceg Novog i okoline – 'Birds of Herceg Novi and surroundings'.] *Ciconia* 7. (In Serbian.)

99. KRSTOJEVIĆ, D. AND NOVČIĆ, R. (1999) [Nove potvrde o gnežđenju šumske šljuke (*Scolopax rusticola*) na Kopaoniku – 'New proofs on Woodcock breeding on Kopaonik'.] *Ciconia* 8. (In Serbian.)

100. KULIĆ, S. (2002) [Zimovalište utine *Asio otus* u Leskovcu – 'Wintering of Long-eared Owl in Leskovac'.] *Ciconia* 11. (In Serbian.)

101. KULIĆ, S., KULIĆ, M., GRUJIĆ, LJ. AND GRUJIĆ, S. (2002) [Pojave ružičastog čvorka *Sturnus roseus* u Leskovačkoj kotlini u poslednje dve decenije – 'Appearance of Rose-colored roseus in Leskovaèka valley during last two decades'.] Ciconia 11: 171–172. (In Serbian.)

102. LAKATOŠ, J. (1979) [*Ptice Apatina*.] NIIU "Glas Komune" Apatin. (In Serbian.)

103. LAKATOŠ, J. (1992) [Ornitofauna ribnjaka u Svilojevu.] *Ciconia* 4: 28–42. (In Serbian.)

104. LJUCOVIĆ, V. (1995) [O pticama kanjona Cijevne i Kaženik planine u Istočnoj Crnoj Gori – 'The birds of Cijevna canyon and Kazenik Mountain in Eastern Montenegro'.] *Ciconia* 5. (In Serbian.)

105. LUKAČ, Š. (1998) [*Sabljarka Recurvirostra avosetta u Vojvodini*.] Trag, Novi Sad. (In Serbian.)

106. LUKAČ, Š., GUBIK, D. AND KOVAČ, S. (1995) [Građa za ornitofaunu ribnjaka "Futog".] *Ciconia* 5: 59–66. (In Serbian.)

107. LUKAČ, Š. AND LUKAČ, A. (1992) [Ornitofauna ribnjaka "Bečej".] *Ciconia* 4: 4–27. (In Serbian.)

108. LUKAČ, Š. AND TERNOVAC, T. (1990) [Beleške o ornitofauni Slanog Kopova u 1987–1989.] *Ciconia* 2: 50–63. (In Serbian.)

109. LUKAČ, Š. AND TERNOVAC, T. (1997) [Crvenovrati gnjurac (*Podiceps griseigena*) gnezdarica Carske bare – 'Breeding of Red-necked Grebe on the Carska bara'.] *Ciconia* 6. (In Serbian.)

110. LUKAČ, Š. AND TERNOVAC, T. (1997) [Kolonije čaplji na Carskoj bari – 'Heron Colony on the Carska bara'.] *Ciconia* 6. (In Serbian.)

111. Lukač, Š. (1995) [Podaci o boravku i gnežđenju stepske vetruše (*Falco naumanni*) na ostacima zidina crkve Arača (Aracs) – 'Data about staying and nesting of Lesser Kestrel in Old Church of Aracs'.] *Ciconia* 5. (In Serbian.)

112. Lukač, Š. (2000) [Brojno gnežđenje æubastog gnjurca (*Podiceps cristatus*) na Carskoj bari – 'Great Crested Grebe breeding in large numbers on Carska bara'.] *Ciconia* 9. (In Serbian.)

113. Marinković, Đ. (1997) [Pregled ptica iz reda *Anseriformes* na širem podruèju Kragujevca u Šumadiji – 'Anseriformes fauna of Kragujevac area, Sumadija'.] *Ciconia* 6. (In Serbian.)

114. Marinković, S., Mandić, R. and Đelić, P. (1982) [Ptice žagubice i njene okoline.] Rezultati bioloških istraživanja, Homolje 81 (Proceedings of a conference), *Beograd* 17–48. (In Serbian.)

115. Marinković, S. and Orlandić, Lj. (1989) [Veliki ronac *Mergus merganser* nova gnezdarica Srbije.] [Treæi simpozijum o fauni SR Srbije, uvodni referati i rezimei.] *Beograd* III: *63*. (In Serbian.)

116. Marinković, S. and Grubač, B. (2000) [Beloglavi sup *Gyps fulvus*.] Pp.63–68 in S. Puzović, ed. [*Atlas ptica grabljivica Srbije – 'Atlas of Birds of Prey of Serbia'*.] Zavod za zaštitu prirode Srbije, Beograd/Belgrade. (In Serbian, with English summary.)

117. Marinković, S. and Grubač, B. (2000) [Kraški soko *Falco biarmicus*.] Pp.165–170 in S. Puzović, ed. [*Atlas ptica grabljivica Srbije – 'Atlas of Birds of Prey of Serbia'*.] Zavod za zaštitu prirode Srbije, Beograd/Belgrade. (In Serbian, with English summary.)

118. Marinković, S. and Grubač, B. (2000) [Patuljasti orao *Hieraaetus pennatus*.] Pp.135–140 in S. Puzović, ed. [*Atlas ptica grabljivica Srbije – 'Atlas of Birds of Prey of Serbia'*.] Zavod za zaštitu prirode Srbije, Beograd/Belgrade. (In Serbian, with English summary.)

119. Marinković, S. and Puzović, S. (2000) [Obična vetruška *Falco tinnunculus*.] Pp.147–152 in S. Puzoviæ, ed. [*Atlas ptica grabljivica Srbije – 'Atlas of Birds of Prey of Serbia'*.] Zavod za zaštitu prirode Srbije, Beograd/Belgrade. (In Serbian, with English summary.)

120. Marinković, S. and Grubač, B. (2000) [Sivi soko *Falco peregrinus*.] Pp.177–182 in S. Puzović, ed. [*Atlas ptica grabljivica Srbije – 'Atlas of Birds of Prey of Serbia'*.] Zavod za zaštitu prirode Srbije, Beograd/Belgrade. (In Serbian, with English summary.)

121. Matović, Č. and Barna, K. (2001) [Akcija prstenovanja bregunica (*Riparia riparia*) na obali Tise kod Sente – 'Ringing action on Sand Martins on Tisza river bank near Senta'.] *Ciconia* 10. (In Serbian.)

122. Matović, Č. and Barna, K. (2001) [Rezultati prstenovanja čaplji u nekim gnezdećim kolonijama u Vojvodini tokom 1998 – 2000 – 'Results of ringing of herons in some breeding colonies in Vojvodina during 1998 – 2000'.] *Ciconia* 10. (In Serbian.)

123. Matović, Č. and Sekereš, O. (2002) [Popis gnezdećih parova zlatovrane *Corracias garrulus* u severoistočnoj Bačkoj i severnom Banatu – 'Census of breeding pairs of Roller in northeastern Backa and northern Banat'.] *Ciconia* 11. (In Serbian.)

124. Matvejev, S.D. (1950) [Rasprostranjenje i život ptica u Srbiji.] *SAN, posebno izdanje* 3. (In Serbian.)

125. Matvejev, S.D. (1997) [Ptice Kopaonika – sezonski pregled.] *Zavod za zaštitu prirode Srbije, posebno izdanje* 15. (In Serbian.)

126. Mikalački, V. (1997) [Interesantno gnežđenje riđovratog gnjurca (*Podiceps griseigena*) kod Bečeja 1995 godine – 'Unusual nesting of Red-necked Grebe near Becse'.] *Ciconia* 6. (In Serbian.)

127. Mikalački, V. (1997) [Ptice grabljivice na podruèju Beèeja 1994–1996 – 'Birds of prey of Becej and surroundings between 1994–1996'.] *Ciconia* 6. (In Serbian.)

128. Nikolić, N. (1999) [Gnežđenje crvenokljunog labuda (*Cygnus olor*) u Smederevu – 'Mute Swan breeding in Smederevo'.] *Ciconia* 8. (In Serbian.)

129. Novčić, I. and Ivović, M. (2000) [Rezultati prstenovanja ptica u periodu 1993–1999; Izveštaj Centra za markiranje životinja – 'Bird ringing results between 1993 – 1999; Report of Centre for Animal Marking'.] *Ciconia* 9. (In Serbian.)

130. Obradović, R. (1994) [*Katalog ptica Baèke*.] EDG "Nikola tesla", Apatin. (In Serbian.)

131. Obradović, R. and Bela, K. (1998) [Prva korekcija i dopuna "Kataloga ptica Bačke" – 'The first correction and additional data for "Katalog ptica Bačke" (Catalogue of birds of Bačka)'.] *Ciconia* 7. (In Serbian.)

132. Obradović, R. (1997) [Gnežđenje ritske sove (*Asio flammeus*) kod Svilojeva – 'Breeding of Short-eared Owl near Svilojevo'.] *Ciconia* 6. (In Serbian.)

133. Obradović, R. (1997) [Posmatranje velike zlovremenice (*Numenius arquata*) u periodu gnežđenja kod Apatina – 'Curlew in breeding season near Apatin'.] *Ciconia* 6. (In Serbian.)

134. Obradović, R. (1997) [Sivi soko (*Falco peregrinus*) posmatran u doba gnežđenja – 'Observation of Peregrines in breeding season'.] *Ciconia* 6. (In Serbian.)

135. Obradović, R. and Mirić, M. (1996) [*Siva guska*.] EKO društvo "Nikola Tesla", Apatin. (In Serbian.)

136. Pantelić, A., Ćeranić, A., Hadži-Cenić, R. and Kučančanin, S. (1996) [*Jarebica kamenjarka (Alectoris graeca) u Srbiji – rasprostranjenost, brojnost i lovni turizam*.] Lovački zavez Jugoslavije, zbornik radova, Beograd. (In Serbian.)

137. Paunović, M. (2002) [Gnežđenje bele rode *Ciconia ciconia* u selu Krivi Vir kod Boljevca – 'White Stork nests in village Krivi Vir near Boljevac.] *Ciconia* 11. (In Serbian.)

138. Pelle, I., Ham, I., Rašajski, J. and Gavrilov, T. (1977) [Pregled gnezdarica Vojvodine.] *Larus* 29–30: 171–197. (In Serbian.)

139. Pelle, Z. (1999) Status and biology of the White Stork in Yugoslavia. Pp.219–221 in H. Schulz, ed. *White Stork on the up? Proceedings, Hamburg*.

140. Popović, Z. and Cecić, ž. (1995) [Spisak faune ptica reda *Anseriformes* u okolini Vlasotinca i Leskovca (Jugoistočna Srbija) – ' List of Birds from the Order of *Anseriformes* near Vlasotince and Leskovac (Southeastern Serbia)'.] *Ciconia* 5. (In Serbian.)

141. Popović, Z. (2001) [Ružičasti čvorak (*Sturnus roseus*) u dolini Vlasine i Južne Morave – 'Rose-colored Starling in valleys of Vlasina and Južna Morava'.] *Ciconia* 10. (In Serbian.)

142. Purger, J. J. (1989) First data on ornithofauna of the fish pond near Srpski Miletić and its surroundings (western Backa). *Larus* 40: 155–161.

143. Purger, J. J. (1986) Numbers and distribution of Red-footed Falcon *Falco vespertinus* nests in Voivodina. *J. Raptor Res.* 30(3): 165–168.

144. Purger, J. (1998) [Posmatranje zlatovrane (*Coracias garrulus*) u periodu gnežđenja u Banatu – 'Observation of Roller during breeding period in Banat'.] *Ciconia* 7. (In Serbian.)

145. Purger, J. J. (2001) Numbers and breeding distribution of the Bee-eater *Merops apiaster* in province Voivodina (northern Serbia) between 1987 and 1990. *Vogelwelt* 122: 279–282.

146. Puzović, S. (1996) Balkan Shore lark (*Eremophila alpestris balcanica Reichenow* 1895) in Serbia and former Yugoslavia – distribution, habitat and population. *VII International Congress on the zoogeography and ecology of Greece and adjacent regions, Book of abstracts*. Athens.

147. Puzović, S. (1994) [Prirodne vrednosti Ulcinjskog primorja (Tu sleæu utva i flamingo.)] *Trag* 12: 28–29. (In Serbian.)

148. Puzović, S. (1996) [Prostorna struktura i brojnost populacije planinske treptaljke (*Anthus spinoletta L.*) u Srbiji.] *V Kongres Ekologa Jugoslavije, zbornik abstrakata*. Beograd. (In Serbian.)

149. Puzović, S. (1996) *Breeding Bird Species Diversity in Serbia with Special reference to National Parks*. International Scientific Conference "Forest ecosystems of the National parks", Ministry of Environment of the Republic of Serbia. Belgrade. (In Serbian.)

150. Puzović, S. (1997) [Usklađivanje intezivnog gajenja riba i očuvanja raznovrsne faune ptica na šaranskim ribnjacima u Vojvodini.] Pp.164–179 in S. Šimić and A. Ivanc, eds. [*Nauèni skup "Intezivno gazdovanje i zaštita prirode na šaranskim ribnjacima Vojvodine"*.] Institut za Biologiju, Novi Sad. (In Serbian.)

151. Puzović, S. (1997) [Planinski popiæ (*Prunella collaris*) i obièan popić (*Prunella modularis*) u Srbiji – 'Alpine Accentor (*Prunella collaris*) and Dunnock (*Prunella modularis*) in Serbia'.] *Ciconia* 6. (In Serbian.)

152. Puzović, S. (1998) [Ramsarska područja u Srbiji u funkciji oèuvanja diverziteta faune ptica vodenih staništa.] *Zaštita prirode* 50: 283–290. (In Serbian.)

153. Puzović, S. (1998) [Veliki tetreb *Tetrao urogallus* u Srbiji.] *Veterinarski informator* 12: 57–58. (In Serbian.)

154. Puzović, S. (1998) [Letnje posmatranje voljića maslinara (*Hippolais olivetorum*, Stickland, 1837) u dolini reke Pčinje – 'Summer observations of Olive-tree Warbler (*Hippolais olivetorum*, Stickland, 1837) in the Pčinja River Valley'.] *Ciconia* 7: 116–117. (In Serbian.)

155. Puzović, S. (2000) [Pojavljivanje planinskog žalara (*Charadrius morinellus*) na Staroj planini i Goču – 'Presence of Dotterel (*Charadrius morinellus*) on Stara Planina Mt. and Goč Mt'.] *Ciconia* 9. (In Serbian.)

156. Puzović, S. (2000) [*Šumska šljuka* Scolopax rusticola *L. – populacije i lovni pritisak*.] Zadužbina Andrejeviæ, biblioteca Academia, Beograd. (In Serbian.)

157. Puzović, S. (2001) [Biološka raznovrsnost i osobenosti faune kičmenjaka Nacionalnog parka "Tara".] Pp.15–35 in V. Orlović and D. Ostojić, eds [*Otkrivanje Tare*.] Mala ekološka biblioteka, Društvo ekologa Srbije i Zavod za zaštitu prirode Srbije, 4. (In Serbian.)

158. Puzović, S. (2001) [Odstrel predatora; Očuvanje i unapređenje trajno zaštićenih vrsta divljači i divlje faune – sisari i ptice; Dinamika brojnosti i odstrela vrsta ptica selica; Redukcija brojnosti nezaštićenih predatora.] Pp.45–46; 87–91; 119–126; 128–130 in V. Šelmić, ed. [*Program razvoja lovstva Srbije 2001–2010*.] Lovački savez Srbije, Beograd. (In Serbian.)

159. Puzović, S. (2001) [*Ptice Zasavice – vrednosti, istorijske promene i moguænosti unapredenja stanja*.] Monografija "Zasavica", PMF Institut za biologiju Novi Sad i Goransko-ekološki pokret Sremska Mitrovica, Sremska Mitrovica, zbornik radova. (In Serbian.)

160. Puzović, S. (2001) [Dve nove kolonije čaplji i kormorana kod Kladova – 'Two new colonies of herons and cormorants near Kladovo'.] *Ciconia* 10. (In Serbian.)

161. Puzović, S. (2002) [Nova posmatranja crvene lunje *Milvus milvus* u Srbiji tokom reproduktivnog perioda – 'New observations of Red Kite in Serbia during reproductive period'.] *Ciconia* 11. (In Serbian.)

162. Puzović, S. (2003) [*Ptice Suvobora i Rajca. Prilog za studiju zaštite područja Valjevskih planina.*] Zavod za zaštitu prirode Srbije, Beograd. (In Serbian.)

163. Puzović, S., Vasić, V. and Ham, I. (1992) [Progradacioni procesi u ornitofauni Ulcinjskih solana.] *Glasnik Republičkog zavoda za zaštitu prirode-Prirodnjačkog muzeja Crne Gore* 25: 63–75. (In Serbian.)

164. Puzović, S. and Paunović, M. (1994) [Valorizacija ornitoloških vrednosti Borskog područja.] II Naučno-stručni skup "Naša ekološka istina", zbornik saopštenja (*Proceedings of a conference*). *Bor* II-10: 139–140. (In Serbian.)

165. Puzović, S. and Paunović, M. (1996) [Sastav, brojnost i prostorni raspored gnezdilišne ornitofaune južnih padina planine Rtanj (istočna Srbija).] IV Naučno-stručni skup "Naša ekološka istina", zbornik saopštenja. *Bor* II-10: 223–227. (In Serbian.)

166. Puzović, S. and Rašajski, J. (1996) The Black Stork (*Ciconia nigra*) in Serbia and former Yugoslavia – distribution and population. *II International conference of the Black Stork, IWRB-IUCN*. Abstracts, 38, Extremadura, Spain.

167. Puzović, S. and Grubač, B. (1999) [*Fauna ptica Stare planine i Vidliča.*] Predlog za stavljanje pod zaštitu područja, Zavod za zaštitu prirode Srbije, Beograd. (In Serbian.)

168. Puzović, S. and Ilić, Z. (1999) [Zimovanje balkanske ušate ševe (*Eremophila alpestris balcanica*) na planini Devici kod Soko Banje – 'Shore Lark wintering on the Devica Mountain near Soko Banja'.] *Ciconia* 8. (In Serbian.)

169. Puzović, S. and Stefanović, T. (2002) [Status orla krstaša *Aquila heliaca* na Fruškoj gori sa posebnim osvrtom na teritorijalnost i sezonski ritam života para kod Čortanovaca – 'Status of Imperial Eagle on Fruska gora Mt. with special reference to territoriality and seasonal living cycle of breeding pair near Čortanovci village'.] *Ciconia* 11. (In Serbian.)

170. Puzović, S. and Grubač, B. (1997) *Important bird areas in Serbia*. Zavod za zaštitu prirode Srbije i – BirdLife International, Elaborat, Novi Sad – Cambridge. (In Serbian.)

171. Puzović, S. and Grubač, B. (2000) Federal Republic of Yugoslavia. Pp.725–745 in M. F. Heath and M. I. Evans, eds. *Important birds areas in Europe: Priority sites for conservation 2: Southern Europe*. Cambridge, UK: BirdLife International (BirdLife Conservation Series No 8). (In Serbian.)

172. Puzović, S., ed. (2000) [*Atlas ptica grabljivica Srbije – mape rasprostranjenosti i procene brojnosti 1977–1996.*] Zavod za zaštitu prirode Srbije, posebno izdanje, Beograd. (In Serbian.)

173. Puzović, S. and Grubač, B. (2000) [Kobac *Accipiter nisus*.] Pp.93–98 in S. Puzović, ed. [*Atlas ptica grabljivica Srbije – 'Atlas of Birds of Prey of Serbia'*.] Zavod za zaštitu prirode Srbije, Beograd/Belgrade. (In Serbian, with English summary.)

174. Puzović, S. and Rašajski, J. (2000) [Osièar *Pernis apivorus*.] Pp.33–38 in S. Puzović, ed. [*Atlas ptica grabljivica Srbije – 'Atlas of Birds of Prey of Serbia'*.] Zavod za zaštitu prirode Srbije, Beograd/Belgrade. (In Serbian, with English summary.)

175. Puzoviæ, S. and Marinkoviæ, S. (2000) [Orao zmijar *Circaetus gallicus*.] Pp.69–74 in S. Puzović, ed. [*Atlas ptica grabljivica Srbije – 'Atlas of Birds of Prey of Serbia'*.] Zavod za zaštitu prirode Srbije, Beograd/Belgrade. (In Serbian.)

176. Puzović, S. and Ham, I. (2000) [Mišar *Buteo buteo*.] Pp.105–110 in S. Puzoviæ, ed. [*Atlas ptica grabljivica Srbije – 'Atlas of Birds of Prey of Serbia'*.] Zavod za zaštitu prirode Srbije, Beograd/Belgrade. (In Serbian.)

177. Puzović, S. and Ham, I. (2000) [Orao kliktavac *Aquila pomarina*.] Pp.117–122 in S. Puzović, ed. [*Atlas ptica grabljivica Srbije – 'Atlas of Birds of Prey of Serbia'*.] Zavod za zaštitu prirode Srbije, Beograd/Belgrade. (In Serbian.)

178. Puzović, S. and Marinković, S. (2000) [Soko lastavièar *Falco subbuteo*.] Pp.159–164 in S. Puzović, ed. [*Atlas ptica grabljivica Srbije – 'Atlas of Birds of Prey of Serbia'*.] Zavod za zaštitu prirode Srbije, Beograd/Belgrade. (In Serbian.)

179. Puzović, S., Gergelj, J. and Lukač, Š. (1999) [Kolonije čaplji i kormorana u Srbiji 1998 – 'Heron and Cormorant Colonies in Serbia 1998'.] *Ciconia* 8. (In Serbian.)

180. Puzović, S. and Stojnić, N. (2003) [*Fauna ptica Fruške gore.*] Prilog za studiju izrade Prostornog plana posebne namene, Zavod za zaštitu prirode Srbije, Novi Sad. (In Serbian.)

181. Puzović, S. and Tucakov, M. (2003) Overview of the Ferruginous Duck *Aythya nyroca* in Serbia. Pp.56–61 in N. Petrov, B. Hughes and U. Gallo-Orsi, eds. *Ferruginous Duck research to conservation*. BirdLife International and BSPB, Sofia, Conservation Series 6.

182. Radišić, D. and Stojšin, A. (2001) [Sova utina *Asio otus*: životni uslovi u urbanoj sredini Novog Sada – Long-eared Owl: life conditions in urban area of Novi Sad'.] *Ciconia* 10. (In Serbian.)

183. Radišić, D. (2002) [Posmatranja vrtne crvenorepke *Phoenicurus phoenicurus* na gnežđenju u Sremskoj Kamenici – 'Observation of Redstart while breeding in region of Sremska Kamenica'.] *Ciconia* 11. (In Serbian.)

184. Raković, M. and Despotović, J. (1999) [*Ptice šljukarice Charadriformes Valjevskog kraja.*] Regionalni centar za talente, Valjevo. (In Serbian.)

185. Rašajski J. and Vučanović, M. (1999) [Orao zmijar (*Circaetus gallicus*) nova gnezdarica Vršačkih planina – 'Short-toed Eagle a new breeder of Vrsacke Mountains'.] *Ciconia* 8. (In Serbian.)

186. Rašajski, J. and Grubač, B. (2000) [Belonokta vetruška *Falco naumanni*.] Pp.141–146 in S. Puzović, ed. [*Atlas ptica grabljivica Srbije – 'Atlas of Birds of Prey of Serbia'*.] Zavod za zaštitu prirode Srbije, Beograd/Belgrade. (In Serbian, with English summary.)

187. Rašajski, J. and Ham, I. (2000) [Crna lunja *Milvus migrans*.] Pp.39–44 in S. Puzović, ed. [*Atlas ptica grabljivica Srbije – 'Atlas of Birds of Prey of Serbia'*.] Zavod za zaštitu prirode Srbije, Beograd/Belgrade. (In Serbian, with English summary.)

188. Rašajski, J. and Marinković, S. (2000) [Crvena lunja *Milvus milvus*.] Pp.45–50 in S. Puzović, ed. [*Atlas ptica grabljivica Srbije – 'Atlas of Birds of Prey of Serbia'*.] Zavod za zaštitu prirode Srbije, Beograd/Belgrade. (In Serbian.)

189. Rašajski, J. and Vučanović, M. (1998) [Uralska sova (*Strix uralensis macroura*, Wolf 1810) nova gnezdarica Vojvodine – 'Ural Owl new nester of Vojvodina'.] *Ciconia* 7. (In Serbian.)

190. Rašajski, J. and Pelle, I. (1993) [*Ptice Vršačkih planina.*] Matica Srpska, odeljenje za prirodne nauke, Novo Sad. (In Serbian.)

191. Ružić, M. (2003) [Prilog proučavanju ornitofaune Ovčarsko-kablarske klisure.] *Beležnik Ovèarsko-kablarske klisure, Turistièka organizacija* 2: 75–80. (In Serbian.)

192. Ružić, M. and Šćiban, M. (2002) [Retke gnezdarice doline Zapadne Morave kod Èaèaka – 'Rare nesting birds in the Zapadna Morava valley in vicinity of Čačak'.] *Ciconia* 11. (In Serbian.)

193. Šćiban, M. and Ružić, M. (2002) [Nalazi žalara slepića *Charadrius dubius* na gnežđenju u dolini Zapadne Morave kod Čačka – 'Findings of Little Ringed Plover in breeding period in valley of Zapadna Morava'.] *Ciconia* 11. (In Serbian.)

194. Šćiban, M. (2000) [Podaci o nekim zanimljivim vrstama ptica uz Veliki Bački kanal od Kule do Čuruga – 'Data on some interesting bird species along the Veliki Bački Channel from Kula to Čurug'.]*Ciconia* 9. (In Serbian.)

195. Šćiban, M. (2001) [Novo gnezdilište vlastelice (*Himantopus himantopus*) u Bačkoj kod Kule – 'New breeding area of Black-winged Stilt in Baèka at Kula'.] *Ciconia* 10. (In Serbian.)

196. Šćiban, M. (2002) [Dinamika brojnosti gnezdećih parova crnoglavog svračka *Lanius minor* u okolini Bačkog Gradišta – 'Population numbers of breeding pairs of Lesser Grey Shrike in vicinity of Bačko Gradište'.] *Ciconia* 11. (In Serbian.)

197. Saveljić, D. (2002) Changes in population size of some shorebirds breeding at Ulcinj salt-pans in Montenegro. *Acrocephalus* 110–111: 39–42.

198. Sekulić, G. (1995) [Prilog poznavanju ornitofaune Stare planine – 'Enclosure to the Knowledge of Ornithofauna of Stara Planina'.] *Ciconia* 5. (In Serbian.)

199. Sekulić, G., Žuljević, A. and Stefanović, T. (1998) [Crnogrla strnadica (*Emberiza cirlus*) i krstokljun (*Loxia curvirostra*) – nove vrste u fauni ptica Obedske bare – 'Cirl Bunting and Crossbill are new species in Obedska Bara ornithofauna'.] *Ciconia* 7. (In Serbian.)

200. Schneider Jacoby, M. and Vasic, V. F. (1989) The Red-crested Pochard (*Netta rufina*) breeding and wintering in Yugoslavia. *Wildfowl* 49:39–44.

201. Simić, D. and Puzović, S. (1998) [Orao belorepan (*Haliaeetus albicilla*) u okolini Beograda – 'The White-tailed Eagle in the Belgrade area'.] *Ciconia* 7. (In Serbian.)

202. Simonov, N. (1995) [*Prilog za faunu ptica Pešterske visoravni.*] Predlog za stavljanje pod zaštitu područja, Zavod za zaštitu prirode Srbije, Beograd. (In Serbian.)

203. Simonov, N. (2000) [*Prilog za faunu ptica Golije.*] Predlog za stavljanje pod zaštitu područja, Zavod za zaštitu prirode Srbije, Beograd. (In Serbian.)

204. Simonov, N. and Popović, Z. (2001) [Kolonija velikog kormorana (*Phalacrocorax carbo*) i sive čaplje (*Ardea cinerea*) na crnim borovima na Vlasinskom jezeru – 'Colony of Cormorant and Grey Heron on Black Pines on Vlasinsko lake'.] *Ciconia* 10. (In Serbian.)

205. Stanimirović, ž. (2000) [Faunistički spisak ptica Banjičke šume i naselja Banjica (predgrađe Beograda) – 'Faunistical list of birds of Banjièka forest and Banjica (suburbia of Belgrade)'.] *Ciconia* 9. (In Serbian.)

206. Stanimirović, ž. (1998) [Crnogrla strnadica (*Emberiza cirlus*) u Beogradu – 'Cirl Bunting (Emberiza cirlus) in Belgrade'.] *Ciconia* 7. (In Serbian.)

207. Stanković, B. (1988) [*Prilog poznavanju ornitofaune opštine Svetozarevo (Jagodina)*.] Maturski rad, srednja škola Svetozarevo. (In Serbian.)

208. Stanković, B. (1997) [Kukumavka (*Athene noctua*) u urbanoj sredini Jagodine – 'Little Owl (*Athene noctua*) living in urban area of Jagodina town'.] *Ciconia* 6. (In Serbian.)

209. Stanković, B. (2000) [Pregeld faune ptica šire okoline Jagodine u srednjem Pomoravju – 'The survey of bird fauna of broader Jagodina area in middle Morava valley'.] *Ciconia* 9. (In Serbian.)

210. Stanković, B. (2001) [Gnežđenje bele rode (*Ciconia ciconia*) u okolini Jagodine – 'Breeding of White Stork near Jagodina'.] *Ciconia* 10. (In Serbian.)

211. Stanković, B. (2002) [Gnežđenje žalara slepića *Charadrius dubius* na Velikoj Moravi i faktori koji utiču na njegovu uspešnost – ' Breeding of Little Ringed Plover on Velika Morava and the factors influencing its success'.] *Ciconia* 11. (In Serbian.)

212. Stanković, B. (2002) Praćenje gnežđenja bele senice Remiz pendulinus u srednjem Pomoravlju – 'Proof and follow-up of Penduline Tit breeding in mid-Pomoravlje'.] *Ciconia* 11. (In Serbian.)

213. Stefanović, T. and Puzović, S. (1997) [Gnežđenje osičara (*Pernis apivorus*) u Bestrementu kod Apatina – 'Breeding of Honey Buzzard in Bestrement, near Apatin'.] *Ciconia* 6. (In Serbian.)

214. Stefanović, T. (1998) [Gnežđenje crne rode (*Ciconia nigra*) na padinama Fruške gore kod Èortanovaca – 'The breeding of Black Stork on the Fruška gora mountain near Čortanovci'.] *Ciconia* 7. (In Serbian.)

215. Stefanović, T. (2000) [Prvi pregled faune ptica severnih ogranaka planine Crna Gora i doline kod Preševa– 'Preliminary survey of bird fauna on northern slopes of Crna Gora Mt. and the Presevo valley'.] *Ciconia* 9. (In Serbian.)

216. Stojnić, N. and Sofronić, M. (1997) [Posmatranja strnadica (*Emberizidae*) u okolini Beočina (Fruška gora) – 'Observed the Buntings near Beoèin (Fruška gora)'.] *Ciconia* 6. (In Serbian.)

217. Stojnić, N. (2000) [Daurska lasta (*Hirundo daurica*) nova gnezdarica Vojvodine – 'Red-rumped Swallow – new breeding species of Vojvodina'.] *Ciconia* 9. (In Serbian.)

218. Stojnić, N. (2000) [Fauna ptica Karajukiæa bunara na Pešteru u avgustu 2000 – 'Bird fauna of Karajukiæa bunari on Pester in August 2000'.] *Ciconia* 9. (In Serbian.)

219. Stojnić, N. (2001) [Ornitofauna površinskih kopova Fruške gore – 'Ornitofauna of open-pit mines at Fruška gora mt'.] *Ciconia* 10. (In Serbian.)

220. Stojnić, N. (2002) [Veličina i prostorni raspored populacije jastreba (*Accipiter gentilis*) i kopca (*Accipiter nisus*) na Fruškoj gori – 'Size and spatial distribution of populations of Goshawk and Sparrowhawk on Fruška gora'.] *Ciconia* 11. (In Serbian.)

221. Ternovac, T. (1991) [Podaci o ornitofauni Jegričke iz 1989 i 1990.] *Ciconia* 3: 14–24. (In Serbian.)

222. Tot, L. and Frank, Z. (1995) [Ornitofauna Karike kod Padeja (Vojvodina, Jugoslavija) u periodu od 1986 do 1991 – 'Ornithofauna of Karika near Padej (Vojvodina, Yugoslavia) during 1986–1990 Period'.] *Ciconia* 5:52–58. (In Serbian.),

223. Tot, L. (2002) [Redovno gnežđenje vodomara *Alcedo atthis* u Adi – 'Regular breeding of Kingfisher naer Ada'.] *Ciconia* 11. (In Serbian.)

224. Tot. L. (2001) [Zapažanja o fauni ptica šire okoline Prizrena i Kosovske Mitrovice u periodu 1960–1961 – 'Observations on bird fauna in broader vicinity of Prizren and Kosovska Mitrovica in period 1960–1961'.] *Ciconia* 10. (In Serbian.)

225. Tucakov, M. and Šipoš, Š. (2001) [Nova mešovita kolonija čaplji kod Kikinde – 'Record on mixed heron colony at Kikinda'.] *Ciconia* 10. (In Serbian.)

226. Tucakov, M. (1997) [Gnežđenje crne rode (*Ciconia nigra*) kod Bačkog Brega u gornjem Podunavlju – 'Breeding of Black Stork in the Breski forest, near Backi Breg'.] *Ciconia* 6. (In Serbian.)

227. Tucakov, M. (1998) [Prilog poznavanju gnežđenja i letnje distribucije batokljuna (*Coccothraustes coccothraustes*) na području Bačkog Brega – 'On Hawfinch post-breeding distribution and breeding in Bački Breg area'.] *Ciconia* 7. (In Serbian.)

228. Tucakov, M. (2001) [Nova kolonija čaplji u Pančevačkom ritu – 'New heron colony in Panèevaèki rit marsh'.] *Ciconia* 10. (In Serbian.)

229. Tucakov, M. (2002) [Gnezdilišna fauna ptica ciglanskih kopova kod Ratkova – 'Breeding fauna of birds of clay pits near Ratkovo'.] *Ciconia* 11. (In Serbian.)

230. Tucakov, M. (2002) [Nova kolonija velikog kormorana *Phalacrocorax carbo* i sive čaplje *Ardea cinerea* na dunavskoj adi žilava – 'New colony of Cormorant and Grey Heron on Danube'.] *Ciconia* 11. (In Serbian.)

231. Tucakov, M., Lukač, Š., Gergelj, J., Barna, K., Žuljević, A. and Đapić, D. (2001) [Izveštaj o realizaciji i rezultatima projekta "Edukacijom i dijalogom do usklađivanja potreba zaštite ptica i uzgoja ribe" – 'Report on realization and results of project "Through education and dialogue to fitting the needs of bird protection and fish breeding"'.] *Ciconia* 10. (In Serbian.)

232. Tucakov, M. and Žuljević, A. (2002) [Nastanak i gnezdilišna fauna ptica dvaju kolektora industrijskih otpadnih voda u Bačkoj: značaj veštačkih vlažnih staništa – 'Origin and breeding fauna of birds in two collecting facilities of industrial waste waters in Bačka: importance of artificial wetland habitats'.] *Ciconia* 11. (In Serbian.)

233. Vasić, V. F. (1980) [*Prilozi za faunu ptica južnih delova SR Srbije (prvi deo)*.] *SANU, Zbornik radova o fauni ptica Srbije.* (Proceedings of a conference). Beograd, knjiga 1: 101–128. (In Serbian.)

234. Vasić, V. F. (1980) The list of Birds of Skadar lake (Montenegro, Yugoslavia). *Larus* 31–32: 185–208.

235. Vasić, V. F. (1980) [Savremeni areal daurske laste *Hirundo daurica* Rufula Temm. u SR Srbiji i susednim zemljama. Prilozi za faunu ptica južnih delova SR Srbije.] *SANU, Zbornik radova o fauni ptica Srbije* (Proceedings of a conference), Beograd, knjiga 1:85–100. (In Serbian.)

236. Vasić, V. (1984) [Biogeografske karakteristike ptica vodenih staništa Balkanskog poluostrva.] (Doktorski rad). Prirodno-matematički fakultet, Beograd. (In Serbian.)

237. Vasić, V. F. (1985) [Geografsko rasprostranjenje gorske laste *Hirundo rupestris* Scopoli 1769 u Jugoslaviji.] Prilozi za faunu ptica južnih delova SR Srbije. *SANU, Zbornik radova o fauni ptica Srbije (Proceedings of a conference)*, Beograd, knjiga 3: 193–208. (In Serbian.)

238. Vasić, V. (1995) [Biodiverzitet u osetljivim ekosistemima i područjima od međunarodnog znaèaja.] Pp.37–42 in V. Stevanović and V. Vasić, eds. [*Biodiverzitet Jugoslavije sa pregledom vrsta od međunarodnog znaèaja.*] Biološki fakultet i Ekolibri, Beograd. (In Serbian.)

239. Vasić, V. (1986) [Osobenosti ornitofaune planinskih staništa Gornje Morače.] [*VII Kongres biologa YU, izvodi saopštenja, Budva.*] (In Serbian.)

240. Vasić, V. (1995) *Raptors in Montenegro – estimations of population sizes (number of pairs)*. Natural History Museum, Belgrade.

241. Vasić, V., Šoti, J. (1980) [Pregled faune ptica Vlasinskog jezera i okoline.] *Biosistematika* 6(1): 81–107. (In Serbian.)

242. Vasić, V. and Grubač, B. (1983) [Prilozi za faunu ptica južnih delova SR Srbije (drugi deo).] *Zbornik radova o fauni ptica Srbije, SANU* (Proceedings of a conference), Beograd, knjiga 2: 197–245. (In Serbian.)

243. Vasić, V., Marinković, S. and Vizi, O. (1990) [Ptice Durmitora i kanjona Tare.] *Fauna Durmitora* 23 (3): 9–70. (In Serbian.)

244. Vasić, V., Džukić, G., Janković, D., Simonov, N., Petrov, B. and Savić, I. (1991) [Preliminarni spisak vrsta za crnevu listu kičmenjaka Srbije.] *Zaštita prirode* 43–44: 121–132. (In Serbian.)

245. Vasić, V. F. and Misirlić, R. (2002) The Eastern Imperial Eagle *Aquila heliaca* in Yugoslavia, with reference to F.Y.R. Macedonia. *Aquila* 107–108: 145–168.

246. Vizi, O. (1984) [Zaštićene životinjske vrste u Crnoj Gori.] *Glasnik Republičkog zavoda za zaštitu prirode – Prirodnjaèkog muzeja Crne Gore* 17: 69–108. (In Serbian.)

247. Vizi, O. (1991) [Ornitološke odlike Biogradske gore i moguænosti osnivanja ornitološke osmatračke stanice.] *CANU, naučni skup "Prirodne i društvene vrijednosti Nacionalnog parka Biogradska gora"* (Proceedings of a conference), Titograd, 23: 327–337. (In Serbian.)

248. Vizi, O. (1993) [Zaštićene životinjske vrste Prokletija.] *CANU, nauèni skup "Priroda Prokletija"* (Proceedings of a conference), Titograd, 103–113. (In Serbian.)

249. Vizi, O. (1995) [Crni žar – nova kolonija močvarnih ptica na Skadarskom jezeru.] *CANU, nauèni skup "Prirodne vrijednosti i zaštita Skadarskog jezera"* (Proceedings of a conference). Podgorica, 44: 309–320. (In Serbian.)

250. Vizi, O. (1995) [Uticaj povećanog uznemiravanja na neke ugrožene vrste ptica na Skadarskom jezeru.] *CANU, naučni skup "Prirodne vrijednosti i zaštita Skadarskog jezera"* (Proceedings of a conference). Podgorica 44: 321–330. (In Serbian.)

251. Vizi, O. (1998) [Flora, vegetacija, fauna i pejzažne vrijednosti kopnenog dijela Morskog dobra. Bazna studija, Prostorni plan područja posebne namjene za Morsko dobro Crne Gore.] Unpublished report. (In Serbian.)

252. Vučanović, M. (1999) [Planinska strnadica (*Emberiza cia*) nova gnezdarica Vojvodine – 'Rock Bunting – a new breeder of Vojvodina'.] *Ciconia* 8. (In Serbian.)

253. Vučanović, M. (2001) [Gnežđenje prdavca (*Crex crex*) u Malom ritu kod Vršca – 'Breeding of Corncrake at Mali rit marsh near Vršac' .] *Ciconia* 10. (In Serbian.)

254. Vučanović, M. (2001) [Prilog za faunu ptica planine Paštrik u Metohiji – ' Supplement to bird fauna of Paštrik mt. in Metohija'.] *Ciconia* 10. (In Serbian.)

255. Vučanović, M. (2002) [Gnežđenje šumske sove *Strix aluco* u postavljenim kuæicama na Vršaèkim planinama – 'Breeding of Tawny Owl in artificial houses in Vrsacke Mts' .] *Ciconia* 11. (In Serbian.)

256. Vučanović, M. (2002) [Gnezdilišni lokaliteti i posmatranja na seobi zlatovrane *Corracias garrulus* u okolini Vršca – 'Breeding localities and observations during migration of Roller in vicinity of Vršac'.] *Ciconia* 11. (In Serbian.)

257. Vučanović, M. (2002) [Posmatranje bele čiope *Apus melba* kod Vršačke kule u reproduktivnom periodu – 'Observation of Alpine Swift near Vrsacka Tower in the reproductive period'.] *Ciconia* 11. (In Serbian.)

258. Vučanović. M. (2001) [Osičar (*Pernis apivorus*) nova gnezdarica Vršačkih Planina – 'Honey Buzzard new breeder of Vršaèke mts'.] *Ciconia* 10. (In Serbian.)

259. Vuković, N. (1999) [Distribucija ptica u modifikovanim staništima na Glediækim planinama.] (Diplomski rad, Biološki fakultet) Beograd. (In Serbian.)

260. žujević, A. (1997) [Podaci o čižku (*Carduelis spinus*) u letnjem periodu u Somboru – 'Notes on Siskin in breeding season in Sombor area'.] *Ciconia* 6. (In Serbian.)

261. žujević, A. (1997) [Podaci o nekim ptičjim vrstama u okolini Sombora – 'Data on some bird species in Sombor area'.] *Ciconia* 6. (In Serbian.)

262. Žuljević, A. and Đapić, D. (2002) [Prvi podaci o fauni ptica bare "Jezero" kod Stanišića u severozapadnoj Bačkoj – 'First data on bird fauna of pond Jezero near Stanišić in northwestern Bačka' .] *Ciconia* 11. (In Serbian.)

263. žuljević, A. (1997) [Podaci o velikom slavuju (*Luscinia luscinia*) u severozapadnoj Bačkoj – 'Notes on Thrush Nightingale in Northwest Backa'.] *Ciconia* 6. (In Serbian.)

264. Žuljević, A. (1998) [Crnoglavi svračak (*Lanius minor*) na području severozapadne Bačke – 'Lesser Grey Shrike in northwest Bačka'.] *Ciconia* 7. (In Serbian.)

265. Žuljević, A. (1998) [Gnežđenje sive vetruške (*Falco vespertinus*) u severozapadnoj Bačkoj – 'The breeding of the Red-footed Falcon in nordwest Bačka'.] *Ciconia* 7. (In Serbian.)

266. Žuljević, A. (2000) [Ornitofaunistički podaci šire okoline Niša u 1989. godini – 'Ornithofaunistical data collected in 1989 in the broader Nis area'.] *Ciconia* 9. (In Serbian.)

Slovakia

Breeding

1. Danko, Š., Darolová, A. and Krištín, A., eds. (2002) [*The distribution of birds in Slovakia.*] VEDA, SAV Publishers. Bratislava. (In Slovak, with English summaries.)
2. Trnka, A., Krištín, A., Danko, S., Harvanèík, S. Kocian, L., Karaska, D. and Murín, B. [Zoznam vtákov Slovenska.] *Tichodroma* 8: 7–21 (In Slovak.)
3. Trnka, A. (1997) [*Aktuálny prehlad vtákov Slovenska.*] Trnavská Univerzita, Trnava (In Slovak.)

Wintering

4. Danko, Š., Darolová, A. and Krištín, A., eds. (2002) [*The distribution of birds in Slovakia.*] VEDA, SAV Publishers. Bratislava. (In Slovak, with English summaries.)
5. Delany, S., Reyes, C., Hubert, E., Pihl, S., Rees, E., Haanstra, L. and van Strien, A., eds. (1999) *Results from the International Waterbird Census in the Western Palearctic and Southwest Asia 1995 and 1996.* Wetlands International Publication 54. Wageningen, The Netherlands.
6. Gilissen, N., Haanstra, L., Delany, S., Boere, G. and Hagemeijer, W. (2002) *Numbers and distribution of wintering waterbirds in the Western Palearctic and Southwest Asia in 1997, 1998 and 1999.* Results from the International Waterbird Census. Wetlands International Global Series No. 11, Wageningen, The Netherlands.

Slovenia

Breeding

1. Bozic, L. (2003) *Important Bird Areas (IBA) in Slovenia 2. Proposed Special Protected Areas (SPA) in Slovenia.* DOPPS, Monografija DOPPS st. 2, Ljubljana.
2. Bozic, L. (2002) Comparison of communities and several population parameters of birds in selected lowland riverine forests (graduation thesis). Unpublished data.
3. Cas, M. (2000) [Pregled rastisc divjega petelina *Tetrao urogallus* v Sloveniji v letih 1999 in 2000 ter analiza ogrozenosti rastisc.] Unpublished report. (In Slovenian.)
4. Denac, D. (2001) Breeding biology, phenology and distribution of White Stork *Ciconia ciconia* in Slovenia. *Acrocephalus* 22: 89–103.
5. Geister, I. (1998) Distribution of the Reed Warbler *Acrocephalus scirpaceus* and the Great Reed Warbler *Acrocephalus arundinaceus* in the Slovene littoral. *Annales* 13: 95–100.
6. Jancar, T. (2000) The Birds of Kozjansko Regional Park. *Acrocephalus* 21: 107–134.
7. Makovec, T. (1994) [Status, razsirjenost in gnezditvene navade belocelega dezevnika *Charadrius alexandrinus* na Slovenski obali.] *Annales* 4: 63–70. (In Slovenian.)
8. Mihelic, T., Vrezec, A., Perusek, M. and Svetlicic, J. (2000) Ural Owl *Strix uralensis* in Slovenia. *Acrocephalus* 21: 9–22.
9. Perusek, M. (1991) Birds of Rajhenavski Rog, the remains of Slovene virgin forest. *Acrocephalus* 12: 124–136.
10. Polak, S., ed. (2000) *Mednarodno pomembna obmocja za ptice v Sloveniji; Important Bird Areas (IBAs) in Slovenia.* DOPPS, Monografija DOPPS St. 1, Ljubljana.
11. Surina, B. (1999) Orhithofauna of the upper Reka valley and its neighbourhood. *Annales* 9: 303–314.
12. Tome, D. (2001) Some characteristics of the avifauna of Pivka intermittent lakes. *Annales* 2: 271–276.
13. Stumberger, B. (2000) Eurasian Scops Owl *Otus scops* at Goricko (NE Slovenia). *Acrocephalus* 21: 23–26.
14. Trontelj, P. (2001) The 1999 Slovenian Corncrake *Crex crex* census indicates short-term stable population. *Acrocephalus* 22: 139–147.
15. Trontelj, P. and Vogrin, M. (1993) Birds of Jovsi and proposals as to its protection. *Acrocephalus* 14: 200–209.
16. Vogrin, M. (2000) Breeding Bird Community of a *Abieti-Fagetum* prealpinum forest near Sentjoski vrh on Karst plateau Dobrovlje in Savinja Alps. *Sodobno kmetijstvo* 33: 55–61.
17. Vogrin, M. and Svetlicic, J. (2001) Pygmy Owl *Glaucidium passerinum* in Savinja Alps, northern Slovenia. *Biota* 2: 199–210.

Wintering

18. Bozic, L. (2002) Winter Census of the Water Rail *Rallus aquaticus* in Slovenia. *Acrocephalus* 23: 27–33.
19. Stumberger, B. (1997) Results of the mid-winter waterfowl census in January 1997 in Slovenia. *Acrocephalus* 18: 29–39.
20. Stumberger, B. (1998) Results of the mid-winter waterfowl census in January 1998 in Slovenia. *Acrocephalus* 19: 36–48.
21. Stumberger, B. (1999) Results of the mid-winter waterfowl census in January 1999 in Slovenia. *Acrocephalus* 20: 6–22.
22. Stumberger, B. (2000) Results of the mid-winter waterfowl census in January 2000 in Slovenia. *Acrocephalus* 21: 271–274.

Spain

Breeding

1. Amat, J. A. (1993) Status of the Kentish Plover in Spain. *Kentish Plover Project Newsletter* 2: 2–4.
2. Del Moral, J. C. and Martí, R., eds. (2001) [*The Griffon Vulture in the Iberian Peninsula. III Spanish National Census and I Iberian Coordinated Census, 1999.*] Monografía, nº 7. SEO/BirdLife. Madrid. (In Spanish.)
3. Del Moral, J. C. and Martí, R., eds. (2002) [*The Egyptian Vulture in Spain and Portugal: I Coordinated Census, 2000.*] Monografía, nº 8. SEO/BirdLife. Madrid. (In Spanish.)
4. Garza, V. and Suárez, F. (1990) [*Distribution, population and habitat selection by Dupont's Lark* (Chersophilus duponti) *in the Iberian Peninsula.*] *Ardeola* 37: 3–12. (In Spanish.)
5. Hortas, F., Arroyo, G. M. and Pérez-Hurtado, A., coords. (2000) Breeding waders in Spain. Wader Study Group Project: Breeding Waders in Europe 2000. Unpublished interim report.
6. Hortas, F., Arroyo, G. M. and Pérez-Hurtado, A. (In press.) Breeding Waders in Spain. In O. Thorup, coord.: *Breeding Waders in Europe 2000.* Wader Study Group. JNCC. Peterborough, UK.
7. López-Redondo, J., López-Septiem, J. A. and Martínez-Olivas, F., coords. (2001) [*Census of breeding Sand Martin* Riparia riparia *in the Comunidad Autonoma of Madrid.*] *Boletín S. C. V.* 8–9: 30–34. (In Spanish.)
8. Martín, A. and Lorenzo, J. A. (2001) [*Birds of the Canary Islands.*] Francisco Lemus, Editor. La Laguna. (In Spanish.)
9. Martí, R. (1999) Results of the 5th International White Stork Census (1994) in Spain. Pp.61–68 in H. Schulz, ed. *White Stork on the up?* NABU (Naturschutzbund Deustschland). Bonn.
10. Martí, R. and Del Moral, J. C., eds. (2003) [*Atlas of breeding birds in Spain.*] Dirección General de Conservación de la Naturaleza-Sociedad Española de Ornitología. Madrid. (In Spanish.)
11. Paterson, A. M. (2002) [*Seabirds from the Iberian Peninsula, Balearic and Canary Islands.*] Edilesa. León. (In Spanish.)
12. Purroy, F. J., coord. (1997) [*Atlas of the birds of Spain (1975–1995).*] SEO/BirdLife. Lynx Edicions. Barcelona. (In Spanish.)
13. SEO/BirdLife (F. J. Purroy, coord.) (1992) Dispersed Species Project. Unpublished report for Internatinal Council for Bird Preservation. Madrid.
14. SEO/BirdLife (A. Ruiz, coord.) (2001) [Project Life B-4/3200/97/246, for the conservation of *Puffinus mauretanicus* in the Balearic Islands SPAs, Spain.] Unpublished final report to Conselleria de Medi Ambient, Govern de les Illes Balears. (In Spanish.)
15. SEO/BirdLife (J. E. Gutiérrez, A. Higueras, J. C. Rodríguez and F. Muela) (2002a) [Monitoring of the populations of some Fringillidae and European Turtle Dove in Andalucía, 2002 breeding season.] Unpublished report to Consejería de Medio Ambiente, Junta de Andalucía. (In Spanish.)
16. Madroño, A., González, C. and Atienza, J. C., eds. (in press.) [*Red data book of the Birds of Spain*]. Dirección General para la Biodiversidad-Sociedad Española de Ornitología. Madrid. (In Spanish.)

Wintering

17. Bea, A. and Fernández, J. M. (2001) [Census and distribution of Wood Pigeon *Columba palumbus* wintering in the Iberian Peninsula.] Pp.103–115 in A. Bea and J. M. Fernández, eds. [*Biology and Management of wild Columbidae. II International Symposium.*] Cuadernos de Ciencias Naturales V. Sociedad de Estudios Vascos. Donostia. (In Spanish.)
18. Martí, R. and Del Moral, J. C. (1996) [I Census of wintering White Stork in Spain.] *La Garcilla* 96: 42–43. (In Spanish.)
19. Martí, R. and Del Moral, J. C., eds. (2002) [*The wintering of waterbirds in Spain.*] Dirección General de Conservación de la Naturaleza-SEO/BirdLife. Ed. Organismo Autónomo Parques Nacionales, Ministerio de Medio Ambiente. Madrid. (In Spanish.)
20. Prieta, J. and Valiente, J. (2002) [Crane Project 2001–2002 and other wintering data.] *Boletín El Primilla* 2. ADENEX. (In Spanish.)

Canary Islands

Breeding

21. Donázar, J. A., Palacios, C. J., Gangoso, L., Ceballos, O., González, M. J. and Hiraldo, F. (2002) Conservation status and limiting factors in the endangered population of Egyptian vulture (*Neophron percnopterus*) in the Canary Islands. *Biological Conservation* 107: 89–97.

22. EMMERSON, K. W. (1999) [Distribution and abundance in the Canary Islands.] Pp.109–116 in J. Herranz and F. Suárez, eds. [*Sandgrouse*, Pterocles alchata *and* P. orientalis *in Spain. Distribution, abundance, biology and conservation.*] Colección Técnica. Organismo Autónomo Parques Nacionales. Ministerio de Medio Ambiente. Madrid. (In Spanish.)
23. HERNÁNDEZ, E., MARTÍN, A., NOGALES, M., QUILIS, V., DELGADO, G. AND TRUJILLO, O. (1990) Distribution and status of Bulwer's Petrel (*Bulweria bulwerii* Jardine and Selby, 1828) in the Canary Islands. *Bol. Mus. Mun. Funchal*. 42: 5–16.
24. MADROÑO, A., GONZÁLEZ, C. AND ATIENZA, J. C. eds. (in press). [*Red data book of the Birds of Spain*]. Dirección General para la Biodiversidad-Sociedad Española de Ornitología. Madrid. (In Spanish.)
25. SEO/BIRDLIFE (J. A. LORENZO, J. R. ALONSO, R. BARONE AND C. GONZÁLEZ) (2002–2003) [Atlases of the breeding birds of the Canary Islands.] Unpublished reports to Cabildos Insulares and TRAGSA (Lanzarote). (In Spanish.)
26. MARTÍN, A., DELGADO, G., NOGALES, M., QUILIS, V., TRUJILLO, O., HERNÁNDEZ, E. AND SANTANA, F. (1989) [Premiéres données sur la nidification du Puffin des Anglais (*Puffinus puffinus*), du Pétrel-frégate (*Pelagodroma marina*) et de la Sterne de Dougall (*Sterna dougallii*) aux îles Canaries.] *L'Oiseau et R. F. O.* 59: 73–83. (In French.)
27. MARTÍN, A., HERNÁNDEZ, M. A., LORENZO, J. A., NOGALES, M. AND GONZÁLEZ, C. (2000) [*The endemic laurel pigeons of the Canary Islands.*] Viceconsejería de Medio Ambiente del Gobierno de Canarias and SEO/BirdLife. (In Spanish.)
28. MARTÍN, A. AND LORENZO, J. A. (2001) [*Birds of the Canary Islands.*]. La Laguna. (In Spanish.)
29. MARTÍN, A., LORENZO, J. A., HERNÁNDEZ, M. A., NOGALES, M., MEDINA, F. M., DELGADO, J. D., NARANJO, J. J., QUILIS, V. AND DELGADO, G. (1997) Distribution, status and conservation of the Houbara Bustard (*Chlamydotis undulata fuertaventurae*) Rothschild and Hartert, 1894, in the Canary Islands, November–December 1994. *Ardeola* 44: 61–69.
30. MARTÍN, A., NOGALES, M., ALONSO, J., RODRÍGUEZ, B., DE LEÓN, L., IZQUIERDO, C., MARTÍN, M. C., MARRERO, P., PUERTA, N., CAZORLA, J., RODRÍGUEZ, B., LÓPEZ, M., MARTÍNEZ, J. M., PÉREZ, D., GINOVÉS, J. AND GONZÁLEZ, E. (2002) [Restoration of the islets and Risco de Famara, Lanzarote.] Unpublished report to Departamento de Biología Animal (Zoología), Universidad de La Laguna. (In Spanish.)
31. MARTÍN, A., NOGALES, M., QUILIS, V., DELGADO, G., HERNÁNDEZ, E., TRUJILLO, O. AND SANTANA, F. (1987) [Distribution and status of breeding seabirds in the Canary Islands for their conservation.] Universidad de La Laguna. Unpublished report to Dirección General de Medio Ambiente y Conservación de la Naturaleza, Gobierno de Canarias. (In Spanish.)

Sweden

Breeding

1. ANON. (1995) National inventory in 1995.
2. OTTOSSON, U. (2001) [Hur många fåglar 2001 manuscript]. Unpublished report. (In Swedish.)
3. SOF (1995) [Fågelåret 1994.[Stockholm. (In Swedish.)
4. SOF (1996) [Fågelåret 1995.] Stockholm. (In Swedish.)
5. SOF (1997) [Fågelåret 1996.] Stockholm. (In Swedish.)
6. SOF (1998) [Fågelåret 1997.] Stockholm. (In Swedish.)
7. SOF (1999) [Fågelåret 1998.] Stockholm. (In Swedish.)
8. SOF (2000) [Fågelåret 1999.] Stockholm. (In Swedish.)
9. SOF (2001) [Fågelåret 2000.] Stockholm. (In Swedish.)
10. SOF (2002) [Fågelåret 2001.] Stockholm. (In Swedish.)
11. SOF (2002) [Sveriges fåglar.] 3e uppl. Stockholm. (In Swedish.)
12. SVENSSON, S. AND LINDSTRÖM Å. (2001) Övervakning av fåglarnas populationsutveckling 2000/2001. (In Swedish.)
13. SVENSSON, S. (1991) [Övervakning av fåglarnas populationsutveckling 1990/1991.] (In Swedish.)
14. SVENSSON, S., SVENSSON, M. AND TJERNBERG, M. (1999) [*Svensk fågelatlas.* Vår fågelvärld, supplement 31.] Stockholm. (In Swedish.)

Wintering

15. NILSSSON, L. (1991) International waterfowl counts in Sweden 1990/1991.
16. NILSSSON, L. (2002) International waterfowl counts in Sweden 2001/2002.

Switzerland

General

1. KELLER, V. AND BURKHARDT, M. (2003) [*Monitoring überwinternde Wasservögel: Ergebnisse der Wasservogelzählungen 2001/02 in der Schweiz.*] Schweizerische Vogelwarte. Sempach. (In German and French.)
2. SCHMID, H., LUDER, R., NAEF-DAENZER, B., GRAF, R. AND ZBINDEN, N. (1998) [*Schweizer Brutvogelatlas. Verbreitung der Brutvögel in der Schweiz und im Fürstentum Liechtenstein 1993–1996.*] Schweizerische Vogelwarte. Sempach. (In German and French.)
3. SCHMID, H., BURKHARDT, M., KELLER, V., KNAUS, P., VOLET, B. AND ZBINDEN, N. (2001) [*Die Entwicklung der Vogelwelt in der Schweiz.*] Avifauna Report Sempach 1. (In German and French.)
4. SCHMID, H. AND VOLET, B. (submitted) [Der Bestand des Rotmilans *Milvus milvus* im Winter 2002/03 der Schweiz.] *Ornithol. Beob.* 101. (In German.)
5. VOLET, B., SCHMID, H. AND THOMA, M. (2004) [Situation actuelle du Héron cendré *Ardea cinerea* en Suisse et évolution des effectifs depuis 1980.] *Nos Oiseaux* 51 : 85–93. (In French.)

Turkey

General

1. BAŞKAYA, Ţ. (2003) Distribution and principal threats to Caucasian black grouse *Tetrao mlokosiewiczi* in the Eastern Karadeniz Mountains in Turkey. *Wildlife Biology* 9: 377–383.
2. BIRICIK, M. AND KARAKAŞ, R. (2004) [Yeşil Arıkuşu'nun (*Merops persicus*) Türkiye'de Bilinen Tek Güncel Üreme Alanı.] (Poster) Pp.52 in Kiziroğlu, İ., A. Erdoğan, L. Turan and T. Albayrak, ed. [*I. Uluslararası Avrasya Ornitoloji Kongresi, Sayfa, Antalya.*] (In Turkish.)
3. DHKD (1996) International Waterfowl Census Turkey 1996. Bird Section Report No. 9, Doğal Hayatı Koruma Derneği, İstanbul.
4. DHKD (1999) International Waterfowl Census, Turkey 1999. Biodiversity Programme Report 10. Doğal Hayatı Koruma Derneği, İstanbul.
5. EKEN, G. AND MAGNIN, G. (2000) A Preliminary Biodiversity Atlas of the Konya Basin, Central Turkey, Doğal Hayatı Koruma Derneği, İstanbul.
6. ERTAN, A., KILIÇ, A. AND KASPAREK, M. (1989) [Türkiye'nin Önemli Kuş Alanları, Doğal Hayatı Koruma Derneği ve.] International Council for Bird Preservation, İstanbul. (In Turkish.)
7. KILIÇ, D. T. AND EKEN, G. E. (2004) [Türkiye'nin Önemli Kuş Alanları 2004 Güncellemesi.] Doğa Derneği, Ankara. (In Turkish.)
8. GREEN, I. A. (1997) Corncrake *Crex crex* breeding in north-east Turkey. *Sandgrouse* 19(2): 145–146.
9. KARAUZ, S. (2000) The Results of Ornithological Surveys on 10 Central Anatolian Lakes in 1998–99. Hacettepe University, Ankara.
10. KARAUZ, S., KIRAÇ, C. O. AND EKEN, G. (2000) Mediterranean Gull *Larus melanocephalus*, Slender-billed Gull *Larus genei*, Gull-billed Tern *Gelochelidon nilotica* in Turkey. *Proceedings of the 5th Medmaravis Symposium.* MEDMARAVIS – BirdLife Malta, Malta.
11. KUŞ ARAŞTIRMALARI DERNEĞI (2004) [Kara Akbaba (*Aegypius monachus*) Türkiye Ulusal Koruma Eylem Planı.] Ankara. (In Turkish.)
12. KILIÇ, A., BIRICIK, M. AND KARAKAŞ, R. (2004) [Güneydoğu Anadolu'da Leylek (*Ciconia ciconia*) Üreme Popülasyonu (poster).] Pp.45–46 in Kiziroğlu, İ., A. Erdoğan, L. Turan and T. Albayrak, ed. [*I. Uluslararası Avrasya Ornitoloji Kongresi, Antalya, Sayfa, 2004, Antalya.*] (In Turkish.)
13. KILIÇ, A., KARAKAŞ, R. AND BIRICIK, M. (2003) Observations on a newly detected breeding site of Bonelli's Eagle, *Hieraaetus fasciatus*, in south-eastern Anatolia. *Zoology in the Middle East* 30: 37–41.
14. KUŞBANK, DOĞA DERNEĞI, ERCIYES ÜNIVERSITESI, ROYAL SOCIETY FOR THE PROTECTION OF BIRDS AND BIRDLIFE INTERNATIONAL (2004) www.kusbank.org, internetsite, May 2004.
15. PARR, S. J, COLLIN, P., SILK, S., WILBRAHAM, J., WILLIAMS, N. P. AND YARAR, M. (1996) A baseline survey of White Storks in central Turkey. *Sandgrouse* 18(2): 46–51.
16. PLEASANCE, B. (1997) Pallid Harrier *Circus macrourus* breeding in Central Turkey. *Sandgrouse* 19(2): 144–145.
17. ROBINSON, K., HOLT, L., BUCKLEY, Y., PULLEN, T., BOYLA, K. A. AND CAN, O. (1998) Breeding survey of White-headed Duck *Oxyura leucocephala* on Central Plateau, Turkey. *Sandgrouse* 20(1): 27–48.
18. ROSELAAR, C. S. (1995) *Taxonomy, morphology, and distribution of the songbirds of Turkey: an atlas of biodiversity of Turkish passerine birds.* Harlem, GMB
19. WELCH, G. AND WELCH, H. (2000) Important Forest Areas of Turkey: Bird Surveys. Doğal Hayatı Koruma Derneği and Royal Society for the Protection of Birds. Unpublished report.
20. WELCH, H. (2003) South East Anatolia Biodiversity Research Project (draft). Unpublished outline Report with full recomendations. DHKD (Doğal Hayatı koruma Derneği), İstanbul.
21. YARAR, M. AND MAGNIN, G. (1997) [*Türkiye'nin Önemli Kuş Alanları.*] Doğal Hayatı Koruma Derneği, İstanbul. (In Turkish.)
22. WIRTH, H. (1996) Breeding record of the Imperial Eagle, *Aquila heliaca*, in Turkey. *Zoology in the Middle East* 12: 47–53.
23. YEŞIL ATLAS (2002) *DBR Yayıncılık, Sayı* 5, Kırmızı Boşluk. 2002. (In Turkish.)

Ukraine

Breeding

1. AFANASIEV, V. T. (1998) [*Birds of Sumy region.*] Kyiv (In Russian.)
2. AFANASIEV, V. T., GAVRIS, G. G. AND KLESTOV, N. L. (1992) [*Avifauna of the Desna River valley and its protection.*] Kiev. (In Russian.)
3. ANDRYUSCHENKO, Y. A. AND STADNICHENKO, I. S. (1999) [Great Bustard, Little Bustard and Stone Curlew in the south of the Left-Bank Ukraine: current population state.] *Branta* 2 : 135–152. (In Russian.)

4. Anon. (1991) [*The catalogue of the avifauna in the Western regions of Ukraine. Ornithological observations for 1989–1990.*] Lutsk. (In Ukrainian.)

5. Anon. (1993) [*The catalogue of the avifauna in the Western regions of Ukraine. Ornithological observations for 1991–1992.*] Lutsk. (In Ukrainian.)

6. Bulakhov, V. L. and Gubkin, A. A. (1996) [State of modern avifauna in the Dnipropetrovsk Region.] *Proceedings of the Ukrainian Ornithological Society* 1: 3–19 (In Russian.)

7. Bushta, T.-A. V. (1996) [Avifauna of the Lviv Region and its changes during 20th century.] *Proceedings of the Ukrainian Ornithological Society* 1: 19–30 (In Ukrainian.)

8. Chernichko, I. I. ed. (1999) [*Distribution of the waterfowl in the Sivash in summer–autumn period.*] Melitopol-Simpheropol. (In Russian.)

9. Chernikov, V. F. (1994) [Grebes in the Kharkov region.] Pp.38–40 in [*The birds of the Severskij Donets River basin. Materials of the 2nd Conference "Study and conservation of the birds of the Severskij Donets River basin 4–6 May 1994".*] Kharkov. (In Russian.)

10. Fesenko, G. V. and Bokotey A. A. (2002) [*Birds of Ukraine: field guide.*] Kyiv (In Ukrainian.)

11. Gawris, G. G. (1995) [Go West: Ausbreitung von Vogelarten in der Ukraine.] *Der Falke* 2: 36–41 (In German.)

12. Gavris, G. G. (1995) [Changes to the avifauna of the Desna River valley (Ukrainian part) during the last 100 years] Pp.19–23 in [*Nature conservation and biodiversity protection in the Middle Desna River.* Materials of the Russian-Ukrainian conference.] Nerussa. (In Russian.)

13. Gavris, G. G. (2002) [Spatial structure, conservation and restoration of the avifauna of the Left -bank Polissia (Forest Zone) and Lisostep (Forest-Steppe Zone) of Ukraine.] Pp.182–190 in [*Ecological study of the river basins of the Left-bank of the Ukraine.*] Sumy. (In Ukrainian.)

14. Gavris, G. G. ed. (2003) [*Birds of the Ukraine protected by the Bern Convention.*] Kyiv. (In Ukrainian.).

Gulay, V. I., Matveev, M. D. and Novak, V. O. (1996) [*Birds of the Khmelnitsky region (faunistic characteristic).*] Kamyanets-Podilsky. (In Ukrainian.)

15. Guzij, A. I. (1996) [Avifauna of the Ukrainian Roztochie and its protection problems.] *Proceedings of the Ukrainian Ornithological Society* 1: 171–192 (In Russian.)

16. Khlebeshko, V. N. and Tsitsyura, V. K. (1993) [*Phenology of breeding birds in the North-East of the Zhitomir Polissia.*] Zhitomir (In Russian.)

17. Klestov, N. L. and Fesenko, G. V. (1990) [*Gulls and Terns of the Dnipro River reservoirs.*] Kiev. (In Russian.)

18. Lebid', E. O., Knysh, M. P. and Khomenko, S. V. (1992) [Fauna and ecology of waders in the Sumy region] Pp.76–94 in [*Problems of protection and rational use of the Sumshchyna natural recourses. Collection of scientific proceedings.*] Sumy. (In Ukrainian.)

19. Krivitsky, I. A. (1993) [The modern character of avifauna of the Seversky Donets River valley in Kharkov region.] Pp.12–13 in [*The birds of the Severskij Donets River basin. Materials of the 1st Conference "Study and conservation of the birds of the Severskij Donets River basin 26–28 January 1993".*] Donetsk. (In Russian.)

20. Lugovoy, A. E., Potish, L. A., Kuzma, V. Y. and Herevich, A. V. (2001) [Changes in the avifauna of the Uzh River valley (Transcarpathians) in the second half of the 20th century.] *Berkut* 10(1): 26–31. (In Russian.)

21. Lysenko, V. I. (1991) [*Anseriformes. Fauna of Ukraine. Volume 5.*] Kiev. (In Russian.)

22. Mykytyuk, O. ed. (1999) [*IBA territory of Ukraine: important bird areas.*] Kyiv (In Ukrainian.)

23. Mykytyuk, O. ed. (2000) [*National Action Plans for globally threatened birds conservation.*] Kyiv (In Ukrainian.)

24. Novak, V. O. (1999) [Waders of Khmelnitsky region.] *Berkut* 8(2) : 137–141 (In Ukrainian.)

25. Novak, V. O. and Savchuk, O. V. (1991) [*Avifauna of the Rivne region.*] Rivne (In Ukrainian.)

26. Shcherbak, M. M. ed. (1994) [*Red Data book of Ukraine. Fauna.*] Kyiv (In Ukrainian.)

27. Shupova, T. V. (1999) [Comparative biology of the Coraciiformes and Upupiformes in the conditions of the Steppe Zone of Ukraine. *Vestnik Zoologii* 33(3): 73–80. (In Russian.)

28. Siokhin, V. D. ed. (2000) [*Number and distribution of the breeding waterfowl in the Azov-Black Sea coastal wetlands of Ukraine.*] Melitopol-Kiev (in Russian.)

29. Talposh, V. S. and Pilyavsky, B. R. (1998) [*Fauna of the Vertebrates of the Ternopil region (handbook).*] Ternopil. (In Ukrainian.)

30. Taranenko, L. I., Pisarev, S. N., Komarov, I. F., Martynov, V. V., Ivanov, K. G., Sikorsky, I. A. and Katranzhy, G. N. (1996) [Characteristics of the Anserinae in the Donetsk region. Pp.44–52 in [*The birds of the Severskij Donets River basin. Materials of the 3rd Conference "Study and conservation of the birds of the Severskij Donets River basin 13–15 September 1995".*] Kharkov. (In Russian.)

31. Vetrov, V. V. (1993) [Structure and distribution of the birds of prey in the Seversky Donets basin.] Pp.33–38 in [*The birds of the Severskij Donets River basin. Materials of the 1st Conference "Study and conservation of the birds of the Severskij Donets River basin 26–28 January 1993".*] Donetsk. (In Russian.)

Wintering

32. Anon. (1998) [Winter counts of birds in the Azov-Black Sea Coast of Ukraine.] *(Materials of the XVIII Workshop of the Azov-Black Sea Ornithological Workgroup, 4–6 February 1998.)* Alushta-Kiev (In Russian.)

33. Anon. (1999) [Winter counts of birds in the Azov-Black Sea Coast of Ukraine. Issue 2. *(Materials of the XIX Workshop of the Azov-Black Sea Ornithological Workgroup, 18–21 February 1999.)* Melitopol-Odessa-Kiev. (In Russian.)

34. Banik, M. V., Vergeles, Y. I., Kravchouk, I. N., Odukha, A .R., Romensky, M. V., Sova, M. V. and Shaparenko, S. A. (1994) [The number and distribution of wintering waterfowl in the city of Kharkov in 1990/1991.] Pp.18–19 in [*The birds of the Severskij Donets River basin. Materials of the 2nd Conference "Study and conservation of the birds of the Severskij Donets River basin 4–6 May 1994".*] Kharkov. (In Russian.)

35. Beskaravaynyj, M. M. (1996) [Wintering conditions and structure of the winter bird community in associations of the relict dendroflora in the Crimea.] *Berkut* 5(2):125–130. (In Russian.)

36. Gavrilyuk, M. N. and Grischenko, V. N. (2001) [Present winter avifauna of the east part of Cherkasy region] *Berkut* 10(2): 184–196. (In Russian.)

37. Gladunko, I., Gorban, I. and Grenjukh, P. (1995) [Wintering avifauna of Ukrainian Roztochchja.] *Scientific News. UkrDLTU* 4: 170–175. (In Ukrainian.)

38. Gorban, I. M. and Bokotej, A. A. (1989) [Mapping the distribution of breeding and wintering birds in Western Ukraine.] Pp.24 in [*The Abstracts of XI International Conference on bird census and Atlas Work.*] Praha.

39. Grytsak, V. (1994) [Wintering birds on the lower bank of the Uzh River basin.] Pp.12–13 in [*The Thesis of 48th Scientific Conference. Biological series.*] Uzhgorod. (In Ukrainian.)

40. Kabakov, A. N., Ardamatskaya, T. B., Rudenko, A. G. and Pirogov, N. G. (1992) [Results of mid-winter counts of waterfowl in the Black Sea Nature Reserve, Ukraine, 1984–1991.] *IWRB. Seaduck Bulletin* 1: 36–37.

41. Khymyn, M. V. (1993) [*Atlas of the wintering birds of the Lutsk region (1988–1992)*] Lutsk. (In Ukrainian.)

42. Kotsyuruba, V. V., Syumar, O. A., Strigunov, V. I. and Gubenko, N. A. (1997) [Results of VII and VIII over Ukrainian counts of wintering waterfowl in Krivorozhje.] Pp.106–108 in [*Bird accounting: approaches, methods, results. Materials of school for unification of methods of bird accounting in reservations of Ukraine, Ivano-Frankove settlement, April 26–28, 1995.*] Lviv-Kiev. (In Russian.)

43. May, A. I. and Gudina, A. N. (1996) [The wintering birds of forest belts on the Upper of the Kakhovskoe reservoir] *Proceedings of the Ukrainian Ornithological Society* 1: 122–124. (In Russian.)

44. Novak, V. O. (2003) [Winter avifauna of east districts of Podolia]. *Berkut* 12 (1–2): 14–21. (In Ukrainian.)

45. Prokopenko, S. P. and Grinchenko, A. B. (1999) [Winter counts of bird of prey 1998 – 1999 in Crimea.] *Branta* 2: 204–207. (In Russian.)

46. Rudenko, A. G., Yaremchenko, O. A. and Rybachjuk, K. I. (2000) [Particularity of the waterfowl wintering in the Black Sea Biosphere Reserve] *Kazarka* 6: 302–314. (In Russian.)

47. Rusev, I. T., ed. (2002) [*Monitoring of wintering birds in the Azov-Black Sea region of Ukraine.*] Odessa-Kiev. (In Russian.)

48. Rusev, I. T., Gerzhik, I. P., Vasilkov, I. A., Pavlov, A. V., Potapov, O. V. and Korzyukov, A. I. (1996) [Results of the wintering birds census in the North-West of the Black Sea region.] Pp.290–296 in [*Important bird areas of Ukraine – IBA program. Materials of the Conference.*] Kiev. (In Russian.)

49. Rusev, I. T., Zhmud, M. E., Korzyukov, A. I., Gerzhik, I. P., Pavlov, A. V. and Potapov, O. V. (1996) [Birds wintering in the North-West of the Black Sea region (20 December 1995–10 February 1996).] *Ecosystems of wild nature: conservation, use of nature, monitoring* 3: 43. (In Russian).

50. Rusev, I. T., Zhmud , M. E., Korzyukov, A. I., Gerzhik, I. P., Satsyk, S. F., Potapov, O. V. and Roman, E. G. (1997) [Results of the Christmas counts of the wintering birds in the North-West of the Bleak Sea region in 1997.] *Ecosystems of the wild nature: conservation, use of nature, monitoring* 6: 19–50. (In Russian.).

51. Skilskiy, I. V. (1996) [Peculiarities of the winter bird population of streets and parks of Chernivtsi.] Pp.161–166 in [*Materials of the 2nd Conference of young ornithologists of Ukraine.*] Chernivtsi. (In Ukrainian.)

52. Skilskiy, I. V., Godovanets, B. Y., Shkolny, I. S. and Kuchinik, L. V. (1997) [Particularity of the winter habitats of the birds in the Prut River valley (in the Chernivtsi District limit.] Pp.111–114 in [*Bird accounting: approaches, methods, results. Materials of school for unification of methods of bird accounting in reservations of Ukraine, Ivano-Frankove settlement, April 26–28, 1995.*] Lviv-Kiev. (In Russian.)

53. Vergeles, Yu. I. (1994) [The winter bird populations of the environs of the biological research station of the Kharkov University.] Pp.33–35 in [*The birds of the Severskij Donets River basin. Materials of the 2nd Conference "Study and conservation of the birds of the Severskij Donets River basin 4–6 May 1994".*] Kharkov. (In Russian.)

United Kingdom

Breeding

1. Banks, A. N., Coombes, R. H., and Crick, H. Q. P. (2003) *The Peregrine Falcon breeding population of the UK & Isle of Man in 2002.* BTO Research Report no 330. Thetford.
2. Batten, L. A. (2001) European Honey-buzzard Survey 2000 and 2001: preliminary results and request for further surveys. *British Birds* 94:143–144.
3. Brindley, E., Norriss, K., Cook, T., Babbs, S., Brown, C. F., Massey, P., Thompson, R., and Yaxley, R. (1998) The abundance and conservation status of Redshank *Tringa totanus* nesting on saltmarshes in Great Britain. *Biol. Conserv.* 86: 289–297.
4. Clements, R. (2001) The Hobby in Britain: a new population estimate. *British Birds* 94: 402–408.
5. Crick, H. Q. P., Marchant, J. H., Noble, D. G., Baillie, S. R., Balmer, D. E., Beaven, L. P., Coombes, R. H.,Downie, I. S., Freeman, S. N., Joys, A. C., Leech, D. I., Raven, M. J., Robinson, R. A. and Thewlis, R. M. (2004) *Breeding Birds in the Wider Countryside: their conservation status 2003*. BTO Research Report No. 353. BTO, Thetford. (http://www.bto.org/birdtrends).
6. Fox, A. D. (1988) Breeding status of the Gadwall in Britain and Ireland. *British Birds* 81: 51–66.
7. Gibbons, D. W., Bainbridge, I. P., Mudge, G. P., Tharme, A. P. and Ellis, P. M. (1997) The status and distribution of the Red-throated Diver *Gavia stellata* in Britain in 1994. *Bird Study* 44: 194–205.
8. Gibbons, D. W., Reid, J. B., and Chapman, R. A. (1993) *The new atlas of breeding birds in Britain and Ireland: 1988–1991.* London.
9. Gilbert, G. (2002) The status and habitat of Spotted Crakes *Porzana porzana* in Britain in 1999. *Bird Study* 49: 79–86.
10. Green, R. E. and Gibbons, D. W. (2000) The status of the Corncrake *Crex crex* in Britain in 1998. *Bird Study* 47: 129–137.
11. Greenwood, J. J. D., Crick, H. Q. P. and Bainbridge, I. P. (2003) Numbers and international importance of raptors and owls in Britain and Ireland. Pp.25–49 in Thompson, D.B.A, Redpath, S.M., Fielding, A.H., Marquiss, M. & Galbraith, C.A., eds. *Birds of prey in a changing Environment.* The Stationery Office, Edinburgh.
12. Gregory, R. D., Eaton, M. A., Noble, D. G., Robinson, J. A., Parsons, M., Baker, H., Austin, G. and Hilton, G. M. (2003) *The state of the UK's birds 2002.* The RSPB, BTO, WWT and JNCC, Sandy.
13. Hancock, M., Baines, D., Gibbons, D. W. Etheridge, B., and Shepherd, M. (1999) Status of male Black Grouse *Tetrao tetrix* in Britain in 1995–96. *Bird Study* 46: 1–15.
14. Hancock, M. H., Gibbons, D. W., and Thompson, P. S. (1997) The status of breeding Greenshank *Tringa nebularia* in the United Kingdom in 1995. *Bird Study* 44: 290–302.
15. Kirby, J. S., and Mitchell, C. (1993) Distribution and status of wintering Shovelers *Anas clypeata* in Great Britain. *Bird Study* 40: 170–180.
16. Langston, R. H., Gregory, R. D. and Adams, R. (2002) The status of the Hawfinch in the UK 1975–1999. *British Birds* 95: 166–173.
17. Langston, R. H. W., Smith, T., Brown, A. F. and Gregory, R. D. (in prep) The status of Twite *Carduelis flavirostris* in the UK in 1999.
18. Lloyd, C., Tasker, M. L. and Partridge, K. (1991) *The status of Seabirds in Britain and Ireland.* London.
19. Marchant, J. H. and Gregory, R. D. (1999) Numbers of nesting Rooks *Corvus frugilegus* in the United Kingdom in 1996. *Bird Study* 46: 258–273.
20. Mitchell, P. I., Newton, S. F., Ratcliffe, N. and Dunn, T. E. (2004) *Seabird Populations of Britain and Ireland. Results of the Seabird 2000 Census (1998–2002).* T. & A.D. Poyser, London.
21. Mitchell, C., Patterson, D., Bowyer, P., Cunningham, P., McDonald, R., Meek, E., Okill, J. D. and Symonds, F. (2000) The summer status and distribution of Greylag Geese in north and west Scotland. *Scottish Birds* 21: 69–77.
22. Noble, D. G., Raven, M. J. and Baillie, S. R. (2001) *The breeding bird survey 2000.* BTO Research Report 265. BTO.
23. O'Brien, M. (in press) *Estimating the number of farmland waders breeding in the United Kingdom.* International Wader Studies.
24. Ogilvie, M. A. and the Rare Breeding Birds Panel (2002). Rare breeding birds in the United Kingdom in 2000. *British Birds* 95: 542–582.
25. Owen, M., Atkinson-Willes, G. L. and Salmon, D. G. (1986) *Wildfowl in Great Britain.* Second edition. Cambridge University Press, Cambridge.
26. Prater, A. J. (1989) Ringed Plover *Charadrius hiaticula* breeding population of the United Kingdom in 1984. *Bird Study* 36:154–159.
27. Ratcliffe, D. (1990) *Bird Life of Mountain and Moorland.* Cambridge University Press, Cambridge.
28. Rebecca, G. W. and Bainbridge, I. P. (1998) The breeding status of the Merlin *Falco columbarius* in Britain in 1993–94. *Bird Study* 45: 172–187.
29. Rehfisch, M. M., Austin, G. E., Holloway, S. J., Allan, J. R. and O'Connell, M. (2002) An approach to the assessment of change in the numbers of Canada Geese *Branta canadensis* and Greylag Geese *Anser anser* in southern Britain. *Bird Study* 49: 50–59.
30. Sim, I. M. W., Gibbons, D. W., Bainbridge, I. P. and Mattingley, W. A. (2001) Status of the Hen Harrier *Circus cyaneus* in the UK and the Isle of Man in 1998. *Bird Study* 48: 341–353.
31. Stone, B. H., Sears, J., Cranswick, P. A., Gregory, R. D., Gibbons, D. W., Rehfisch, M. M., Aebischer, N. J. and Reid, J. B. (2001) Population estimates of birds in Britain and in the United Kingdom. *British Birds* 90: 1–22.
32. Summers, R. W., Mavor, R. A., Buckland, S. T. and MacLennan, A. M. (1999) Winter population size and habitat selection of Crested Tits *Parus cristatus* in Scotland. *Bird Study* 46: 230–234.
33. Toms, M. P., Crick, H. Q. P. and Shawyer, C. R. (2001) The status of breeding Barn Owls *Tyta alba* in the United Kingdom 1995–97. *Bird Study* 48: 23–37.
34. Underhill, M. C., Gittings, T., Callaghan, D. A., Hughes, B., Kirby, J. S. and Delany, S. (1998) Status and distribution of breeding common scoters *Melanitta nigra nigra* in Britain and Ireland in 1995. *Bird Study* 45: 146–156.
35. Whitfield, P. (2002) Status of breeding Dotterel *Charadrius morinellus* in Britain in 1999. *Bird Study* 49: 237–249.
36. Wilkinson, N. I., Langston, R. H. W., Gregory, R. D., Gibbons, D. W. and Marquiss, M. (2002) Capercaillie Tetrao urogallus abundance and habitat use in Scotland in winter 1998–99. *Bird Study* 49: 177–185.
37. Wilson, A. M. and Browne, S. J. (1999) Breeding population estimates for Lapwing, Oystercatcher and Eurasian Curlew in Scotland: results of the 1998 BTO Lapwing survey. *Scottish Birds* 20: 73–80.
38. Wilson, A. M., Henderson, A. C. B. and Fuller, R. J. (2002) Status of the Nightingale *Luscinia megarhynchos* in Britain at the end of the 20th Century with particular reference to climate change. *Bird Study* 49: 193–204.
39. Wotton, S. R., Carter, I., Cross, A. V., Etheridge, B., Snell, N., Duffy, K., Thorpe, R. and Gregory, R. D. (2002) Breeding status of the Red Kite *Milvus milvus* in Britain in 2000. *Bird Study* 49: 278–286.
40. Wotton, S. R., Gibbons, D. W. and Pierce, A. J. (2000) The status of the Cirl Bunting *Emberiza cirlus* in the UK and the Channel Islands in 1998. *Bird Study* 47: 138–146.
41. Wotton, S. R. and Gillings, S. (2000) The status of breeding Woodlarks *Lullula arborea* in Britain in 1997. *Bird Study* 47: 212–224.
42. Wotton, S. R., Langston, R. H. W. and Gregory, R. D. (2002) The breeding status of the Ring Ouzel *Turdus torquatus* in the UK in 1999. *Bird Study* 49: 26–34.
43. Summers, R. W. (2002). Parrot Crossbills breeding in Abernathy Forest, Highland. *British Birds* 95: 4–11.

Wintering

44. Austin, G. E., Armitage, M. J. S., Atkinson, P. W., Burton, N. H. K., Leech, D. I., Marshall, A. R., Mellan, H. J., Musgrove, A. J. and Rehfisch, M. M. (2003) *WeBs Alerts 1999/2000: Changes in numbers of wintering waterbirds in the UK, its constituent countries, Special Protection Areas (SPAs) and Sites of Special Scientific Interest (SSSIs).* BTO Research Report No. 306. BTO, Thetford
45. Burton, N. H. K., Musgrove, A. J., Rehfisch, M. M., Sutcliffe, A., and Waters, R. (2003) Numbers of wintering gulls in the United Kingdom, Channel Islands and Isle of Man: a review of the 1993 and previous winter gull roost surveys. *British Birds* 96: 376–401.
46. Fraser, P. A. and Rogers, M. J. (2002) Report on scarce migrant birds in Britain in 2000. *British Birds* 95: 606–630.
47. Fraser, P. A. and Ryan, J. F. (1995) Status of the Great Grey Strike in Britain and Ireland. *British Birds* 88: 478–484.
48. Gregory, R. D., Eaton, M. A., Noble, D. G., Robinson, J. A., Parsons, M., Baker, H., Austin, G. and Hilton, G. M. (2003) *The state of the UK's birds 2002.* The RSPB, BTO, WWT and JNCC, Sandy.
49. Kershaw, M. and Cranswick, P. A. (2003). Numbers of wintering waterbirds in Great Britain, 1994/1995–1998/1999: I. Wildfowl and selected waterbirds. *Biological Conservation* 111: 91–104.
50. Lack, P. (1986) *The atlas of wintering birds in Britain and Ireland.* Calton.
51. Musgrove, A. J. (2002) The non-breeding status of the Little Egret in Britain. *British Birds* 95: 62–80.
52. Rehfisch, M. M., Austin, G. E., Armitage, M. J. S., Atkinson, P. W., Holloway, S. J., Musgrove, A. J. and Pollitt, M. S. (2003). Numbers of wintering waterbirds in Great Britain and the Isle of Man (1994/1995–1998/1999): II Coastal waders (Charadrii). *Biological Conservation* 112: 329–341.
53. Rehfisch, M. M., Austin, G. E., Holloway, S. J., Allan, J. R., and O'Connell, M. (2002) An approach to the assessment of change in the numbers of Canada Geese *Branta canadensis* and Greylag Geese *Anser anser* in southern Britain. *Bird Study* 49: 50–59.
54. Stone, B. H., Sears, J., Cranswick, P. A., Gregory, R. D., Gibbons, D. W., Rehfisch, M. M., Aebischer, N. J. and Reid, J. B. (2001) Population estimates of birds in Britain and in the United Kingdom. *British Birds* 90: 1–22.

Gibraltar

General

1. Gibraltar Ornithological database (2004) Gibraltar Ornithological & Natural History Society, 1967–2004. Unpublished data.

SPECIES INDEX

R

S

T

U

V

W